U.S. Customary Units and Their SI Equivalents

Quantity	U.S. Customary Unit	SI Equivalent
Acceleration	ft/s^2	$0.3048\ m/s^2$
	$in./s^2$	$0.0254\ m/s^2$
Area	ft^2	$0.0929\ m^2$
	in^2	$645.2\ mm^2$
Energy	ft·lb	1.356 J
Force	kip	4.448 kN
	lb	4.448 N
	oz	0.2780 N
Impulse	lb·s	4.448 N·s
Length	ft	0.3048 m
	in.	25.40 mm
	mi	1.609 km
Mass	oz mass	28.35 g
	lb mass	0.4536 kg
	slug	14.59 kg
	ton	907.2 kg
Moment of a force	lb·ft	1.356 N·m
	lb·in.	0.1130 N·m
Moment of inertia		
Of an area	in^4	$0.4162 \times 10^6\ mm^4$
Of a mass	$lb \cdot ft \cdot s^2$	$1.356\ kg \cdot m^2$
Momentum	lb·s	4.448 kg·m/s
Power	ft·lb/s	1.356 W
	hp	745.7 W
Pressure or stress	lb/ft^2	47.88 Pa
	lb/in^2 (psi)	6.895 kPa
Velocity	ft/s	0.3048 m/s
	in./s	0.0254 m/s
	mi/h (mph)	0.4470 m/s
	mi/h (mph)	1.609 km/h
Volume	ft^3	$0.02832\ m^3$
	in^3	$16.39\ cm^3$
Liquids	gal	3.785 L
	qt	0.9464 L
Work	ft·lb	1.356 J

Vector Mechanics for Engineers

STATICS AND DYNAMICS

Vector Mechanics for Engineers

STATICS AND DYNAMICS

Fourth Edition

Ferdinand P. Beer
University Distinguished Professor of
Mechanical Engineering and Mechanics
Lehigh University

E. Russell Johnston, Jr.
Professor of Civil Engineering
University of Connecticut

McGraw-Hill Book Company
New York St. Louis San Francisco Auckland Bogotá Hamburg
Johannesburg London Madrid Mexico Montreal New Delhi
Panama Paris São Paulo Singapore Sydney Tokyo Toronto

Cover Photograph

One of the many inventions sketched in Leonardo da Vinci's notebooks is a mechanical, weight-driven clock. IBM maintains and exhibits a collection of working models made from the sketches, and our cover is derived from a photograph of the clock model. The clock has two separate escapements, one for minutes and one for hours. We thank IBM for their permission to photograph the model, and especially appreciate the cooperation of Mr. Herbert E. Pedersen, director of the IBM Touring Exhibition Program. (Photograph by Felix Cooper.)

VECTOR MECHANICS FOR ENGINEERS: Statics and Dynamics

1 2 3 4 5 6 7 8 9 0 KGPKGP 8 9 8 7 6 5 4

ISBN 0-07-004438-4

This book was set in Laurel by York Graphic Services, Inc.
The editors were T. Michael Slaughter and David A. Damstra;
the production supervisor was Phil Galea.
The drawings were done by Felix Cooper.
The cover was designed by Merrill Haber.
Kingsport Press, Inc., was printer and binder.

Library of Congress Cataloging in Publication Data

Beer, Ferdinand Pierre, date
Vector mechanics for engineers.

Also issued as: Vector mechanics for engineers, 4th ed., in 2 v.
Includes index.
1. Mechanics, Applied. 2. Vector analysis. 3. Statics. 4. Dynamics. I. Johnston, Elwood Russell, date II. Title.
TA350.B3552 1984b 620.1'05 84-3886
ISBN 0-07-004438-4

Contents

Preface

The main objective of a first course in mechanics should be to develop in the engineering student the ability to analyze any problem in a simple and logical manner and to apply to its solution a few, well-understood, basic principles. This text is designed for the first courses in statics and dynamics offered in the sophomore or junior year, and it is hoped that it will help the instructor achieve this goal.†

Vector analysis is introduced early in the text and used throughout the presentation of statics and dynamics. This approach leads to more concise derivations of the fundamental principles of mechanics. It also results in simpler solutions of three-dimensional problems in statics, and makes it possible to analyze many advanced problems in kinematics and kinetics, which could not be solved by scalar methods. The emphasis in this text, however, remains on the correct understanding of the principles of mechanics and on their application to the solution of engineering problems, and vector analysis is presented chiefly as a convenient tool.‡

One of the characteristics of the approach used in this book is that the mechanics of *particles* has been clearly separated from the mechanics of *rigid bodies*. This approach makes it possible to consider simple practical applications at an early stage and to postpone the introduction of the more difficult concepts. In statics, for example, the statics of particles is treated first (Chap. 2); after the rules of addition and subtraction of vectors have been introduced, the principle of equilibrium of a particle is immediately applied to practical situations involving only concurrent forces. The statics of rigid bodies is considered in Chaps. 3 and 4. In Chap. 3, the vector and scalar products of two vectors are introduced and used to define the moment of a force about a point and about an axis. The presentation of these new concepts is followed by a thorough and rigorous discussion of equivalent systems of forces leading, in Chap. 4, to many practical applications involving the equilibrium of rigid bodies under general force systems. In dynamics, the same division is observed. The basic concepts of force, mass, and acceleration, of work and energy, and of impulse and momentum are introduced and first applied to problems involving only particles. Thus, students may familiarize themselves with the three basic methods used in

† This text is also available in two separate volumes, *Vector Mechanics for Engineers: Statics,* fourth edition, and *Vector Mechanics for Engineers: Dynamics,* fourth edition.

‡ In a parallel text, *Mechanics for Engineers,* third edition, the use of vector algebra is limited to the addition and subtraction of vectors, and vector differentiation is omitted.

dynamics and learn their respective advantages before facing the difficulties associated with the motion of rigid bodies.

Since this text is designed for first courses in statics and dynamics, new concepts have been presented in simple terms and every step explained in detail. On the other hand, by discussing the broader aspects of the problems considered, and by stressing methods of general applicability, a definite maturity of approach has been achieved. For example, the concepts of partial constraints and of statical indeterminacy are introduced early in the text and used throughout statics. In dynamics, the concept of potential energy is discussed in the general case of a conservative force. Also, the study of the plane motion of rigid bodies has been designed to lead naturally to the study of their general motion in space. This is true in kinematics as well as in kinetics, where the principle of equivalence of external and effective forces is applied directly to the analysis of plane motion, thus facilitating the transition to the study of three-dimensional motion.

The fact that mechanics is essentially a *deductive* science based on a few fundamental principles has been stressed. Derivations have been presented in their logical sequence and with all the rigor warranted at this level. However, the learning process being largely *inductive,* simple applications have been considered first. Thus the statics of particles precedes the statics of rigid bodies, and problems involving internal forces are postponed until Chap. 6. Also, in Chap. 4, equilibrium problems involving only coplanar forces are considered first and solved by ordinary algebra, while problems involving three-dimensional forces and requiring the full use of vector algebra are discussed in the second part of the chapter. Again, the dynamics of particles precedes the dynamics of rigid bodies; and, in the latter, the fundamental principles of kinetics are first applied to the solution of two-dimensional problems, which can be more easily visualized by the student (Chaps. 16 and 17), while three-dimensional problems are postponed until Chap. 18.

The fourth edition of *Vector Mechanics for Engineers* retains the unified presentation of the principles of kinetics which characterized the previous two editions. The concepts of linear and angular momentum are introduced in Chap. 12 so that Newton's second law of motion may be presented not only in its conventional form $\mathbf{F} = m\mathbf{a}$, but also as a law relating, respectively, the sum of the forces acting on a particle and the sum of their moments to the rates of change of the linear and angular momentum of the particle. This makes possible an earlier introduction of the principle of conservation of angular momentum and a more meaningful discussion of the motion of a particle under a central force (Sec. 12.9). More importantly, this approach may be readily extended to the study of the motion of a system of particles (Chap. 14) and leads to a more concise and unified treatment of the kinetics of rigid bodies in two and three dimensions (Chaps. 16 through 18).

Free-body diagrams are introduced early, and their importance is emphasized throughout the text. Color has been used to distinguish forces from other elements of the free-body diagrams. This makes it easier for the students to identify the forces acting on a given particle or rigid body and to follow the discussion of sample problems and other examples given in the text.

Free-body diagrams are used not only to solve equilibrium problems but also to express the equivalence of two systems of forces or, more generally, of two systems of vectors. The advantage of this approach becomes apparent in the study of the dynamics of rigid bodies, where it is used to solve three-dimensional as well as two-dimensional problems. By placing the emphasis on "free-body-diagram equations" rather than on the standard algebraic equations of motion, a more intuitive and more complete understanding of the fundamental principles of dynamics may be achieved. This approach, which was first introduced in 1962 in the first edition of *Vector Mechanics for Engineers*, has now gained wide acceptance among mechanics teachers in this country. It is, therefore, used in preference to the method of dynamic equilibrium and to the equations of motion in the solution of all sample problems in this edition.

Because of the current trend among American engineers to adopt the international system of units (SI metric units), the SI units most frequently used in mechanics have been introduced in Chap. 1 and are used throughout the text. Approximately half the sample problems and 60 percent of the problems to be assigned have been stated in these units, while the remainder retain U.S. customary units. The authors believe that this approach will best serve the needs of the students, who will be entering the engineering profession during the period of transition from one system of units to the other. It also should be recognized that the passage from one system to the other entails more than the use of conversion factors. Since the SI system of units is an absolute system based on the units of time, length, and mass, whereas the U.S. customary system is a gravitational system based on the units of time, length, and force, different approaches are required for the solution of many problems. For example, when SI units are used, a body is generally specified by its mass expressed in kilograms; in most problems of statics it will be necessary to determine the weight of the body in newtons, and an additional calculation will be required for this purpose. On the other hand, when U.S. customary units are used, a body is specified by its weight in pounds and, in dynamics problems, an additional calculation will be required to determine its mass in slugs (or $lb \cdot s^2/ft$). The authors, therefore, believe that problem assignments should include both systems of units. The actual distribution of assigned problems between the two systems of units, however, has been left to the instructor, and a sufficient number of problems of each type have been provided so that four complete lists of assignments may be selected with the proportion of problems stated in SI units set anywhere between 50 and 75 percent. If so desired, two complete lists of assignments may also be selected from problems stated in SI units only and two others from problems stated in U.S. customary units.

A large number of optional sections have been included. These sections are indicated by asterisks and may thus easily be distinguished from those which form the core of the basic mechanics course. They may be omitted without prejudice to the understanding of the rest of the text.

Among the topics covered in these additional sections are the reduction of a system of forces to a wrench, applications to hydrostatics, shear and bending-moment diagrams for beams, equilibrium of cables, products of inertia and Mohr's circle, mass products of inertia and principal axes of inertia for three-dimensional bodies, the method of virtual work, graphical methods for the solution of rectilinear-motion problems, the trajectory of a particle under a central force, the deflection of fluid streams, problems involving jet and rocket propulsion, the kinematics and kinetics of rigid bodies in three dimensions, damped mechanical vibrations, and electrical analogues. The sections on beams are especially useful when the course in statics is immediately followed by a course in mechanics of materials, while the additional topics included in dynamics will be found of interest when dynamics is taught in the junior year.

The material presented in the text and most of the problems require no previous mathematical knowledge beyond algebra, trigonometry, and elementary calculus, and all the elements of vector algebra necessary to the understanding of the text have been carefully presented in Chaps. 2 and 3. However, special problems have been included, which make use of a more advanced knowledge of calculus, and certain sections, such as Secs. 19.8 and 19.9 on damped vibrations, should be assigned only if the students possess the proper mathematical background. In the portions of the text using elementary calculus, a greater emphasis has been placed on the correct understanding and application of the concepts of differentiation and integration than on the nimble manipulation of mathematical formulas. In this connection, it should be mentioned that the determination of the centroids of composite areas precedes the calculation of centroids by integration, thus making it possible to establish the concept of moment of area firmly before introducing the use of integration. The presentation of numerical solutions takes into account the universal use of calculators by engineering students and instructions on the proper use of calculators for the solution of typical statics problems have been included in Chap. 2.

Each chapter begins with an introductory section setting the purpose and goals of the chapter and describing in simple terms the material to be covered and its application to the solution of engineering problems. The body of the text has been divided into units, each consisting of one or several theory sections, one or several sample problems, and a large number of problems to be assigned. Each unit corresponds to a well-defined topic and generally may be covered in one lesson. In a number of cases, however, the instructor will find it desirable to devote more than one lesson to a given topic. The sample problems have been set up in much the same form that students will use in solving the assigned problems. They thus serve the double purpose of amplifying the text and demonstrating the type of neat and orderly work that students should cultivate in their own solutions. Most of the problems to be assigned are of a practical nature and should appeal to engineering students. They are primarily designed, however, to illustrate the material presented in the text and to help students understand the basic principles of mechanics. The problems have been grouped according to the portions of material they illustrate and have been arranged in order of increasing difficulty. Problems requiring special attention have been indicated by asterisks. Answers to all even-numbered problems are given at the end of the book.

The authors wish to acknowledge gratefully the many helpful comments and suggestions offered by the users of the previous editions of *Mechanics for Engineers* and of *Vector Mechanics for Engineers*.

Ferdinand P. Beer
E. Russell Johnston, Jr.

List of Symbols

$\mathbf{a}$, a	Acceleration
a	Constant; radius; distance; semimajor axis of ellipse
$\bar{\mathbf{a}}$, $\bar{a}$	Acceleration of mass center
$\mathbf{a}_{B/A}$	Acceleration of B relative to frame in translation with A
$\mathbf{a}_{P/\mathfrak{F}}$	Acceleration of P relative to rotating frame $\mathfrak{F}$
$\mathbf{a}_c$	Coriolis acceleration
$\mathbf{A}, \mathbf{B}, \mathbf{C}, \ldots$	Reactions at supports and connections
$A, B, C, \ldots$	Points
A	Area
b	Width; distance; semiminor axis of ellipse
c	Constant; coefficient of viscous damping
C	Centroid; instantaneous center of rotation; capacitance
d	Distance
$\mathbf{e}_n, \mathbf{e}_t$	Unit vectors along normal and tangent
$\mathbf{e}_r, \mathbf{e}_\theta$	Unit vectors in radial and transverse directions
e	Coefficient of restitution; base of natural logarithms
E	Total mechanical energy; voltage
f	Frequency; scalar function
$\mathbf{F}$	Force; friction force
g	Acceleration of gravity
G	Center of gravity; mass center; constant of gravitation
h	Height; sag of cable; angular momentum per unit mass
$\mathbf{H}_O$	Angular momentum about point O
$\dot{\mathbf{H}}_G$	Rate of change of angular momentum $\mathbf{H}_G$ with respect to frame of fixed orientation
$(\dot{\mathbf{H}}_G)_{Gxyz}$	Rate of change of angular momentum $\mathbf{H}_G$ with respect to rotating frame $Gxyz$
$\mathbf{i}, \mathbf{j}, \mathbf{k}$	Unit vectors along coordinate axes
i	Current
$I, I_x, \ldots$	Moment of inertia
$\bar{I}$	Centroidal moment of inertia
$I_{xy}, \ldots$	Product of inertia
J	Polar moment of inertia
k	Spring constant
k_x, k_y, k_O	Radius of gyration
$\bar{k}$	Centroidal radius of gyration
l	Length
$\mathbf{L}$	Linear momentum

L	Length; span; inductance
m	Mass
m'	Mass per unit length
$\mathbf{M}$	Couple; moment
$\mathbf{M}_O$	Moment about point O
$\mathbf{M}_O^R$	Moment resultant about point O
M	Magnitude of couple or moment; mass of earth
M_{OL}	Moment about axis OL
n	Normal direction
$\mathbf{N}$	Normal component of reaction
O	Origin of coordinates
p	Pressure; circular frequency
$\mathbf{P}$	Force; vector
$\dot{\mathbf{P}}$	Rate of change of vector $\mathbf{P}$ with respect to frame of fixed orientation
q	Mass rate of flow; electric charge
$\mathbf{Q}$	Force; vector
$\dot{\mathbf{Q}}$	Rate of change of vector $\mathbf{Q}$ with respect to frame of fixed orientation
$(\dot{\mathbf{Q}})_{Oxyz}$	Rate of change of vector $\mathbf{Q}$ with respect to frame $Oxyz$
$\mathbf{r}$	Position vector
$\mathbf{r}_{B/A}$	Position vector of B relative to A
r	Radius; distance; polar coordinate
$\mathbf{R}$	Resultant force; resultant vector; reaction
R	Radius of earth; resistance
$\mathbf{s}$	Position vector
s	Length of arc; length of cable
$\mathbf{S}$	Force; vector
t	Time; thickness; tangential direction
$\mathbf{T}$	Force
T	Tension; kinetic energy
$\mathbf{u}$	Velocity
u	Variable
U	Work
$\mathbf{v}$, v	Velocity
v	Speed
$\overline{\mathbf{v}}$, $\overline{v}$	Velocity of mass center
$\mathbf{v}_{B/A}$	Velocity of B relative to frame in translation with A
$\mathbf{v}_{P/\mathfrak{F}}$	Velocity of P relative to rotating frame $\mathfrak{F}$
$\mathbf{V}$	Vector product; shearing force
V	Volume; potential energy; shear
w	Load per unit length
$\mathbf{W}$, W	Weight; load
x, y, z	Rectangular coordinates; distances
$\dot{x}$, $\dot{y}$, $\dot{z}$	Time derivatives of coordinates x, y, z
$\overline{x}$, $\overline{y}$, $\overline{z}$	Rectangular coordinates of centroid, center of gravity, or mass center
$\boldsymbol{\alpha}$, α	Angular acceleration

α, β, γ	Angles
γ	Specific weight
δ	Elongation
$\delta\mathbf{r}$	Virtual displacement
δU	Virtual work
ε	Eccentricity of conic section or of orbit
$\boldsymbol{\lambda}$	Unit vector along a line
η	Efficiency
θ	Angular coordinate; Eulerian angle; angle; polar coordinate
μ	Coefficient of friction
ρ	Density; radius of curvature
τ	Period; periodic time
ϕ	Angle of friction; Eulerian angle; phase angle; angle
φ	Phase difference
ψ	Eulerian angle
$\boldsymbol{\omega}, \omega$	Angular velocity
ω	Circular frequency of forced vibration
$\boldsymbol{\Omega}$	Angular velocity of frame of reference

Vector Mechanics for Engineers

STATICS AND DYNAMICS

CHAPTER ONE
Introduction

1.1. What Is Mechanics? Mechanics may be defined as that science which describes and predicts the conditions of rest or motion of bodies under the action of forces. It is divided into three parts: mechanics of *rigid bodies,* mechanics of *deformable bodies,* and mechanics of *fluids.*

The mechanics of rigid bodies is subdivided into *statics* and *dynamics,* the former dealing with bodies at rest, the latter with bodies in motion. In this part of the study of mechanics, bodies are assumed to be perfectly rigid. Actual structures and machines, however, are never absolutely rigid and deform under the loads to which they are subjected. But these deformations are usually small and do not appreciably affect the conditions of equilibrium or motion of the structure under consideration. They are important, though, as far as the resistance of the structure to failure is concerned and are studied in mechanics of materials, which is a part of the mechanics of deformable bodies. The third division of mechanics, the mechanics of fluids, is subdivided into the study of *incompressible fluids* and of *compressible fluids*. An important subdivision of the study of incompressible fluids is *hydraulics,* which deals with problems involving liquids.

Mechanics is a physical science, since it deals with the study of physical phenomena. However, some associate mechanics with mathematics, while many consider it as an engineering subject. Both these views are justified in part. Mechanics is the foundation of most engineering sciences and is an indispensable prerequisite to their study. However, it does not have the *empiricism* found in some engineering sciences, i.e., it does not rely on experience or observation alone; by its rigor and the emphasis it places on deductive reasoning it resembles mathematics. But, again, it is not an *abstract* or even a *pure* science; mechanics is an *applied* science. The purpose of mechanics is to explain and predict physical phenomena and thus to lay the foundations for engineering applications.

1.2. Fundamental Concepts and Principles. Although the study of mechanics goes back to the time of Aristotle (384–322 B.C.) and Archimedes (287–212 B.C.), one has to wait until Newton (1642–1727) to find a satisfactory formulation of its fundamental principles. These principles were later expressed in a modified form by D'Alembert, Lagrange, and Hamilton. Their validity remained unchallenged, however, until Einstein formulated his *theory of relativity* (1905). While its limitations have now been recognized, *newtonian mechanics* still remains the basis of today's engineering sciences.

The basic concepts used in mechanics are *space, time, mass,* and *force.* These concepts cannot be truly defined; they should be accepted on the basis of our intuition and experience and used as a mental frame of reference for our study of mechanics.

The concept of *space* is associated with the notion of the position of a point P. The position of P may be defined by three lengths measured from a certain reference point, or *origin,* in three given directions. These lengths are known as the *coordinates* of P.

To define an event, it is not sufficient to indicate its position in space. The *time* of the event should also be given.

The concept of *mass* is used to characterize and compare bodies on the basis of certain fundamental mechanical experiments. Two bodies of the same mass, for example, will be attracted by the earth in the same manner; they will also offer the same resistance to a change in translational motion.

A *force* represents the action of one body on another. It may be exerted by actual contact or at a distance, as in the case of gravitational forces and magnetic forces. A force is characterized by its *point of application,* its *magnitude,* and its *direction;* a force is represented by a *vector* (Sec. 2.3).

In newtonian mechanics, space, time, and mass are absolute concepts, independent of each other. (This is not true in *relativistic mechanics,* where the time of an event depends upon its position, and where the mass of a body varies with its velocity.) On the other hand, the concept of force is not independent of the other three. Indeed, one of the fundamental principles of newtonian mechanics listed below indicates that the resultant force acting on a body is related to the mass of the body and to the manner in which its velocity varies with time.

We shall study the conditions of rest or motion of particles and rigid bodies in terms of the four basic concepts we have introduced. By *particle* we mean a very small amount of matter which may be assumed to occupy a single point in space. A *rigid body* is a combination of a large number of particles occupying fixed positions with respect to each other. The study of the mechanics of particles is obviously a prerequisite to that of rigid bodies. Besides, the results obtained for a particle may be used directly in a large number of problems dealing with the conditions of rest or motion of actual bodies.

The study of elementary mechanics rests on six fundamental principles based on experimental evidence.

The Parallelogram Law for the Addition of Forces. This states that two forces acting on a particle may be replaced by a single force, called their *resultant,* obtained by drawing the diagonal of the parallelogram which has sides equal to the given forces (Sec. 2.2).

The Principle of Transmissibility. This states that the conditions of equilibrium or of motion of a rigid body will remain unchanged if a force acting at a given point of the rigid body is replaced by a force of the same magnitude and same direction, but acting at a different point, provided that the two forces have the same line of action (Sec. 3.3).

Newton's Three Fundamental Laws. Formulated by Sir Isaac Newton in the latter part of the seventeenth century, these laws may be stated as follows:

FIRST LAW. If the resultant force acting on a particle is zero, the particle will remain at rest (if originally at rest) or will move with constant speed in a straight line (if originally in motion) (Sec. 2.10).

SECOND LAW. If the resultant force acting on a particle is not zero, the particle will have an acceleration proportional to the magnitude of the resultant and in the direction of this resultant force.

As we shall see in Sec. 12.2, this law may be stated as

$$\mathbf{F} = m\mathbf{a} \tag{1.1}$$

where **F**, m, and **a** represent, respectively, the resultant force acting on the particle, the mass of the particle, and the acceleration of the particle, expressed in a consistent system of units.

THIRD LAW. The forces of action and reaction between bodies in contact have the same magnitude, same line of action, and opposite sense (Sec. 6.1).

Newton's Law of Gravitation. This states that two particles of mass M and m are mutually attracted with equal and opposite forces **F** and $-$**F** (Fig. 1.1) of magnitude F given by the formula

$$F = G\frac{Mm}{r^2} \tag{1.2}$$

where r = distance between the two particles
G = universal constant called the *constant of gravitation*

Newton's law of gravitation introduces the idea of an action exerted at a distance and extends the range of application of Newton's third law: the action **F** and the reaction $-$**F** in Fig. 1.1 are equal and opposite, and they have the same line of action.

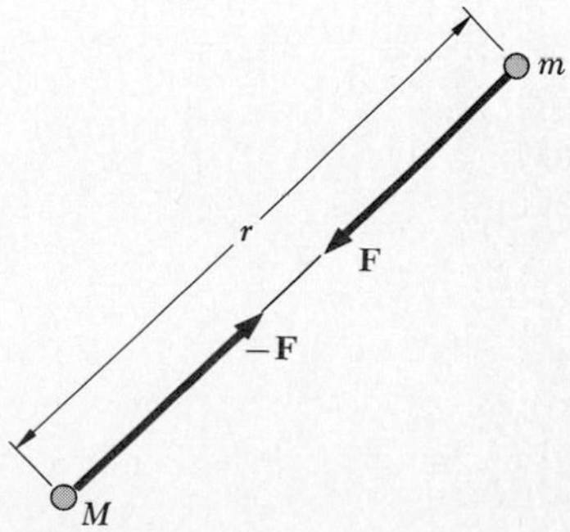

Fig. 1.1

A particular case of great importance is that of the attraction of the earth on a particle located on its surface. The force **F** exerted by the earth on the particle is then defined as the *weight* **W** of the particle. Taking M equal to the mass of the earth, m equal to the mass of the particle, and r equal to the radius R of the earth, and introducing the constant

$$g = \frac{GM}{R^2} \tag{1.3}$$

the magnitude W of the weight of a particle of mass m may be expressed as†

$$W = mg \tag{1.4}$$

The value of R in formula (1.3) depends upon the elevation of the point considered; it also depends upon its latitude, since the earth is not truly spherical. The value of g therefore varies with the position of the point considered. As long as the point actually remains on the surface of the earth, it is sufficiently accurate in most engineering computations to assume that g equals 9.81 m/s^2 or 32.2 ft/s^2.

The principles we have just listed will be introduced in the course of our study of mechanics as they are needed. The study of the statics of particles, carried out in Chap. 2, will be based on the parallelogram law of addition and on Newton's first law alone. The principle of transmissibility will be introduced in Chap. 3 as we begin the study of the statics of rigid bodies, and Newton's third law in Chap. 6 as we analyze the forces exerted on each other by the various members forming a structure. It should be noted that the four aforementioned principles underlie the entire study of the statics of particles, rigid bodies, and systems of rigid bodies. In the study of dynamics, Newton's second law and Newton's law of gravitation will be introduced. It will then be shown that Newton's first law is a particular case of Newton's second law (Sec. 12.2) and that the principle of transmissibility could be derived from the other principles and thus eliminated (Sec. 16.5). In the meantime, however, Newton's first and third laws, the parallelogram law of addition, and the principle of transmissibility will provide us with the necessary and sufficient foundation for the study of the entire field of statics.

As noted earlier, the six fundamental principles listed above are based on experimental evidence. Except for Newton's first law and the principle of transmissibility, they are independent principles which cannot be derived mathematically from each other or from any other elementary physical principle. On these principles rests most of the intricate structure of newtonian mechanics. For more than two centuries a tremendous number of problems dealing with the conditions of rest and motion of rigid bodies, deformable bodies, and fluids have been solved by applying these fundamental principles. Many of the solutions obtained could be checked exper-

† A more accurate definition of the weight **W** should take into account the rotation of the earth.

imentally, thus providing a further verification of the principles from which they were derived. It is only recently that Newton's mechanics was found at fault, in the study of the motion of atoms and in the study of the motion of certain planets, where it must be supplemented by the theory of relativity. But on the human or engineering scale, where velocities are small compared with the velocity of light, Newton's mechanics has yet to be disproved.

1.3. Systems of Units. With the four fundamental concepts introduced in the preceding section are associated the so-called *kinetic units*, i.e., the units of *length, time, mass,* and *force*. These units cannot be chosen independently if Eq. (1.1) is to be satisfied. Three of the units may be defined arbitrarily; they are then referred to as *base units*. The fourth unit, however, must be chosen in accordance with Eq. (1.1) and is referred to as a *derived unit*. Kinetic units selected in that way are said to form a *consistent system of units*.

International System of Units (SI Units†). In this system, which will be in universal use after the United States has completed its current conversion, the base units are the units of length, mass, and time, and are called, respectively, the *meter* (m), the *kilogram* (kg), and the *second* (s). All three are arbitrarily defined. The second, which is supposed to represent the 1/86,400 part of the mean solar day, is actually defined as the duration of 9,192,631,770 cycles of the radiation associated with a specified transition of the cesium atom. The meter, originally intended to represent one-ten-millionth of the distance from the equator to the pole, is now defined as 1,650,763.73 wavelengths of the orange-red line of krypton 86. The kilogram, which is approximately equal to the mass of 0.001 m^3 of water, is actually defined as the mass of a platinum standard kept at the International Bureau of Weights and Measures at Sèvres, near Paris, France. The unit of force is a derived unit. It is called the *newton* (N) and is defined as the force which gives an acceleration of 1 m/s^2 to a mass of 1 kg (Fig. 1.2). From Eq. (1.1) we write

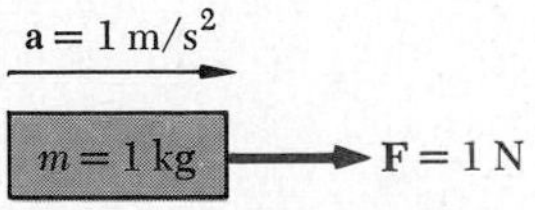

Fig. 1.2

$$1 \text{ N} = (1 \text{ kg})(1 \text{ m/s}^2) = 1 \text{ kg}\cdot\text{m/s}^2 \tag{1.5}$$

The SI units are said to form an *absolute* system of units. This means that the three base units chosen are independent of the location where measurements are made. The meter, the kilogram, and the second may be used anywhere on the earth; they may even be used on another planet. They will always have the same significance.

The *weight* of a body, or *force of gravity* exerted on that body, should, like any other force, be expressed in newtons. From Eq. (1.4) it follows that the weight of a body of mass 1 kg (Fig. 1.3) is

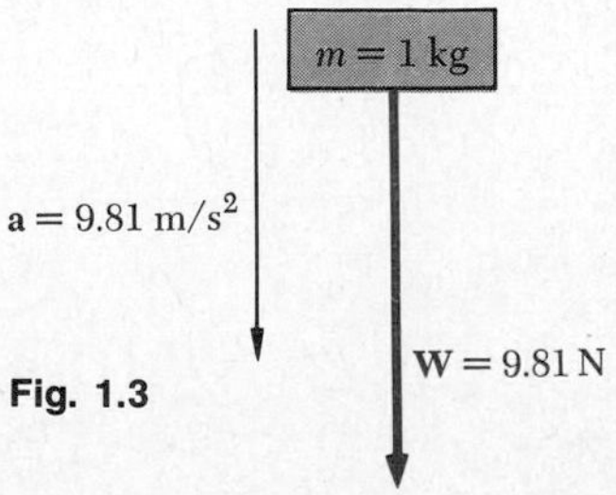

Fig. 1.3

$$\begin{aligned} W &= mg \\ &= (1 \text{ kg})(9.81 \text{ m/s}^2) \\ &= 9.81 \text{ N} \end{aligned}$$

† SI stands for *Système International d'Unités* (French).

Table 1.1 SI Prefixes

Multiplication Factor	Prefix†	Symbol
$1\ 000\ 000\ 000\ 000 = 10^{12}$	tera	T
$1\ 000\ 000\ 000 = 10^{9}$	giga	G
$1\ 000\ 000 = 10^{6}$	mega	M
$1\ 000 = 10^{3}$	kilo	k
$100 = 10^{2}$	hecto‡	h
$10 = 10^{1}$	deka‡	da
$0.1 = 10^{-1}$	deci‡	d
$0.01 = 10^{-2}$	centi‡	c
$0.001 = 10^{-3}$	milli	m
$0.000\ 001 = 10^{-6}$	micro	μ
$0.000\ 000\ 001 = 10^{-9}$	nano	n
$0.000\ 000\ 000\ 001 = 10^{-12}$	pico	p
$0.000\ 000\ 000\ 000\ 001 = 10^{-15}$	femto	f
$0.000\ 000\ 000\ 000\ 000\ 001 = 10^{-18}$	atto	a

† The first syllable of every prefix is accented so that the prefix will retain its identity. Thus, the preferred pronunciation of kilometer places the accent on the first syllable, not the second.

‡ The use of these prefixes should be avoided, except for the measurement of areas and volumes and for the nontechnical use of centimeter, as for body and clothing measurements.

Multiples and submultiples of the fundamental SI units may be obtained through the use of the prefixes defined in Table 1.1. The multiples and submultiples of the units of length, mass, and force most frequently used in engineering are, respectively, the *kilometer* (km) and the *millimeter* (mm); the *megagram*† (Mg) and the *gram* (g); and the *kilonewton* (kN). According to Table 1.1, we have

$$1\text{ km} = 1000\text{ m} \qquad 1\text{ mm} = 0.001\text{ m}$$
$$1\text{ Mg} = 1000\text{ kg} \qquad 1\text{ g} = 0.001\text{ kg}$$
$$1\text{ kN} = 1000\text{ N}$$

The conversion of these units into meters, kilograms, and newtons, respectively, can be effected by simply moving the decimal point three places to the right or to the left. For example, to convert 3.82 km into meters, one moves the decimal point three places to the right:

$$3.82\text{ km} = 3820\text{ m}$$

Similarly, 47.2 mm is converted into meters by moving the decimal point three places to the left:

$$47.2\text{ mm} = 0.0472\text{ m}$$

Using the scientific notation, one may also write

$$3.82\text{ km} = 3.82 \times 10^{3}\text{ m}$$
$$47.2\text{ mm} = 47.2 \times 10^{-3}\text{ m}$$

† Also known as a *metric ton.*

The multiples of the unit of time are the *minute* (min) and the *hour* (h). Since

$$1 \text{ min} = 60 \text{ s} \qquad \text{and} \qquad 1 \text{ h} = 60 \text{ min} = 3600 \text{ s}$$

these multiples cannot be converted as readily as the others.

By using the appropriate multiple or submultiple of a given unit, one may avoid writing very large or very small numbers. For example, one usually writes 427.2 km rather than 427 200 m, and 2.16 mm rather than 0.002 16 m.†

Units of Area and Volume. The unit of area is the *square meter* (m^2), which represents the area of a square of side 1 m; the unit of volume is the *cubic meter* (m^3), equal to the volume of a cube of side 1 m. In order to avoid exceedingly small or large numerical values in the computation of areas and volumes, one uses systems of subunits obtained by respectively squaring and cubing not only the millimeter but also two intermediate submultiples of the meter, namely, the *decimeter* (dm) and the *centimeter* (cm). Since, by definition,

$$\begin{aligned} 1 \text{ dm} &= 0.1 \text{ m} = 10^{-1} \text{ m} \\ 1 \text{ cm} &= 0.01 \text{ m} = 10^{-2} \text{ m} \\ 1 \text{ mm} &= 0.001 \text{ m} = 10^{-3} \text{ m} \end{aligned}$$

the submultiples of the unit of area are

$$\begin{aligned} 1 \text{ dm}^2 &= (1 \text{ dm})^2 = (10^{-1} \text{ m})^2 = 10^{-2} \text{ m}^2 \\ 1 \text{ cm}^2 &= (1 \text{ cm})^2 = (10^{-2} \text{ m})^2 = 10^{-4} \text{ m}^2 \\ 1 \text{ mm}^2 &= (1 \text{ mm})^2 = (10^{-3} \text{ m})^2 = 10^{-6} \text{ m}^2 \end{aligned}$$

and the submultiples of the unit of volume are

$$\begin{aligned} 1 \text{ dm}^3 &= (1 \text{ dm})^3 = (10^{-1} \text{ m})^3 = 10^{-3} \text{ m}^3 \\ 1 \text{ cm}^3 &= (1 \text{ cm})^3 = (10^{-2} \text{ m})^3 = 10^{-6} \text{ m}^3 \\ 1 \text{ mm}^3 &= (1 \text{ mm})^3 = (10^{-3} \text{ m})^3 = 10^{-9} \text{ m}^3 \end{aligned}$$

It should be noted that when the volume of a liquid is being measured, the cubic decimeter (dm^3) is usually referred to as a *liter* (L).

Other derived SI units used to measure the moment of a force, the work of a force, etc., are shown in Table 1.2. While these units will be introduced in later chapters as they are needed, we should note an important rule at this time: When a derived unit is obtained by dividing a base unit by another base unit, a prefix may be used in the numerator of the derived unit but not in its denominator. For example, the constant k of a spring which stretches 20 mm under a load of 100 N will be expressed as

$$k = \frac{100 \text{ N}}{20 \text{ mm}} = \frac{100 \text{ N}}{0.020 \text{ m}} = 5000 \text{ N/m} \qquad \text{or} \qquad k = 5 \text{ kN/m}$$

but never as $k = 5$ N/mm.

† It should be noted that when more than four digits are used on either side of the decimal point to express a quantity in SI units—as in 427 200 m or 0.002 16 m—spaces, never commas, should be used to separate the digits into groups of three. This is to avoid confusion with the comma, which is used in many countries in place of a decimal point.

Table 1.2 Principal SI Units Used in Mechanics

Quantity	Unit	Symbol	Formula
Acceleration	Meter per second squared	. . .	m/s^2
Angle	Radian	rad	†
Angular acceleration	Radian per second squared	. . .	rad/s^2
Angular velocity	Radian per second	. . .	rad/s
Area	Square meter	. . .	m^2
Density	Kilogram per cubic meter	. . .	kg/m^3
Energy	Joule	J	N·m
Force	Newton	N	$kg \cdot m/s^2$
Frequency	Hertz	Hz	s^{-1}
Impulse	Newton-second	. . .	kg·m/s
Length	Meter	m	‡
Mass	Kilogram	kg	‡
Moment of a force	Newton-meter	. . .	N·m
Power	Watt	W	J/s
Pressure	Pascal	Pa	N/m^2
Stress	Pascal	Pa	N/m^2
Time	Second	s	‡
Velocity	Meter per second	. . .	m/s
Volume, solids	Cubic meter	. . .	m^3
Liquids	Liter	L	$10^{-3}\ m^3$
Work	Joule	J	N·m

† Supplementary unit (1 revolution = 2π rad = 360°).

‡ Base unit.

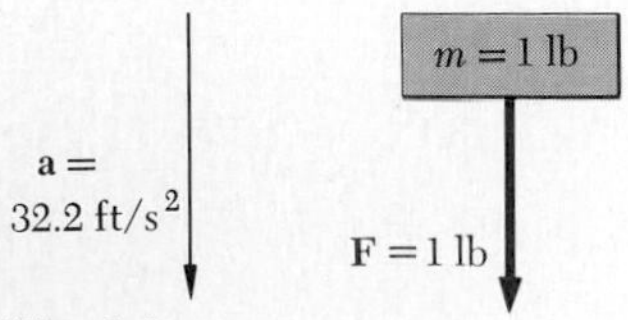

Fig. 1.4

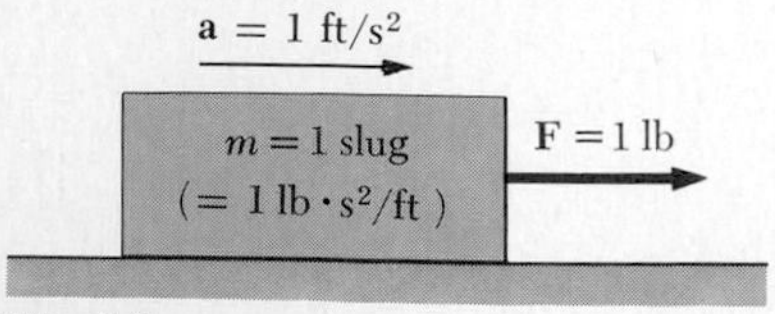

Fig. 1.5

U.S. Customary Units. Most practicing American engineers still commonly use a system in which the base units are the units of length, force, and time. These units are, respectively, the *foot* (ft), the *pound* (lb), and the *second* (s). The second is the same as the corresponding SI unit. The foot is defined as 0.3048 m. The pound is defined as the *weight* of a platinum standard, called the *standard pound* and kept at the National Bureau of Standards in Washington, the mass of which is 0.453 592 43 kg. Since the weight of a body depends upon the gravitational attraction of the earth, which varies with location, it is specified that the standard pound should be placed at sea level and at the latitude of 45° to properly define a force of 1 lb. Clearly the U.S. customary units do not form an absolute system of units. Because of their dependence upon the gravitational attraction of the earth, they form a *gravitational* system of units.

While the standard pound also serves as the unit of mass in commercial transactions in the United States, it cannot be so used in engineering computations since such a unit would not be consistent with the base units defined in the preceding paragraph. Indeed, when acted upon by a force of 1 lb, that is, when subjected to the force of gravity, the standard pound receives the acceleration of gravity, $g = 32.2\ ft/s^2$ (Fig. 1.4), not the unit acceleration required by Eq. (1.1). The unit of mass consistent with the foot, the pound, and the second is the mass which receives an acceleration of $1\ ft/s^2$ when a force of 1 lb is applied to it (Fig. 1.5). This unit, sometimes called a *slug*, can be derived from the equation $F = ma$ after substituting 1 lb and 1 ft/s² for F and a, respectively. We write

$$F = ma \qquad 1\text{ lb} = (1\text{ slug})(1\text{ ft/s}^2)$$

and obtain

$$1\text{ slug} = \frac{1\text{ lb}}{1\text{ ft/s}^2} = 1\text{ lb}\cdot\text{s}^2/\text{ft} \tag{1.6}$$

Comparing Figs. 1.4 and 1.5, we conclude that the slug is a mass 32.2 times larger than the mass of the standard pound.

The fact that in the U.S. customary system of units bodies are characterized by their weight in pounds rather than by their mass in slugs will be a convenience in the study of statics, where we shall constantly deal with weights and other forces and only seldom with masses. However, in the study of dynamics, where forces, masses, and accelerations are involved, we shall have to express the mass m in slugs of a body of which the weight W has been given in pounds. Recalling Eq. (1.4), we shall write

$$m = \frac{W}{g} \tag{1.7}$$

where g is the acceleration of gravity ($g = 32.2\text{ ft/s}^2$).

Other U.S. customary units frequently encountered in engineering problems are the *mile* (mi), equal to 5280 ft; the *inch* (in.), equal to $\frac{1}{12}$ ft; and the *kilopound* (kip), equal to a force of 1000 lb. The *ton* is often used to represent a mass of 2000 lb but, like the pound, must be converted into slugs in engineering computations.

The conversion into feet, pounds, and seconds of quantities expressed in other U.S. customary units is generally more involved and requires greater attention than the corresponding operation in SI units. If, for example, the magnitude of a velocity is given as $v = 30$ mi/h, we shall proceed as follows to convert it to ft/s. First we write

$$v = 30\,\frac{\text{mi}}{\text{h}}$$

Since we want to get rid of the unit miles and introduce instead the unit feet, we should multiply the right-hand member of the equation by an expression containing miles in the denominator and feet in the numerator. But, since we do not want to change the value of the right-hand member, the expression used should have a value equal to unity. The quotient (5280 ft)/(1 mi) is such an expression. Operating in a similar way to transform the unit hour into seconds, we write

$$v = \left(30\,\frac{\text{mi}}{\text{h}}\right)\left(\frac{5280\text{ ft}}{1\text{ mi}}\right)\left(\frac{1\text{ h}}{3600\text{ s}}\right)$$

Carrying out the numerical computations and canceling out units which appear both in the numerator and the denominator, we obtain

$$v = 44\,\frac{\text{ft}}{\text{s}} = 44\text{ ft/s}$$

1.4. Conversion from One System of Units to Another. There are many instances when an engineer wishes to convert into SI units a numerical result obtained in U.S. customary units or vice versa. Because the unit of time is the same in both systems, only two kinetic base units need be converted. Thus, since all other kinetic units can be derived from these base units, only two conversion factors need be remembered.

Units of Length. By definition the U.S. customary unit of length is

$$1 \text{ ft} = 0.3048 \text{ m} \tag{1.8}$$

It follows that

$$1 \text{ mi} = 5280 \text{ ft} = 5280(0.3048 \text{ m}) = 1609 \text{ m}$$

or

$$1 \text{ mi} = 1.609 \text{ km} \tag{1.9}$$

Also

$$1 \text{ in.} = \tfrac{1}{12} \text{ ft} = \tfrac{1}{12}(0.3048 \text{ m}) = 0.0254 \text{ m}$$

or

$$1 \text{ in.} = 25.4 \text{ mm} \tag{1.10}$$

Units of Force. Recalling that the U.S. customary unit of force (pound) is defined as the weight of the standard pound (of mass 0.4536 kg) at sea level and at the latitude of 45° (where $g = 9.807 \text{ m/s}^2$), and using Eq. (1.4), we write

$$W = mg$$
$$1 \text{ lb} = (0.4536 \text{ kg})(9.807 \text{ m/s}^2) = 4.448 \text{ kg}\cdot\text{m/s}^2$$

or, recalling (1.5),

$$1 \text{ lb} = 4.448 \text{ N} \tag{1.11}$$

Units of Mass The U.S. customary unit of mass (slug) is a derived unit. Thus, using (1.6), (1.8), and (1.11), we write

$$1 \text{ slug} = 1 \text{ lb}\cdot\text{s}^2/\text{ft} = \frac{1 \text{ lb}}{1 \text{ ft/s}^2} = \frac{4.448 \text{ N}}{0.3048 \text{ m/s}^2} = 14.59 \text{ N}\cdot\text{s}^2/\text{m}$$

and, recalling (1.5),

$$1 \text{ slug} = 1 \text{ lb}\cdot\text{s}^2/\text{ft} = 14.59 \text{ kg} \tag{1.12}$$

Although it cannot be used as a consistent unit of mass, we recall that the mass of the standard pound is, by definition,

$$1 \text{ pound mass} = 0.4536 \text{ kg} \tag{1.13}$$

This constant may be used to determine the *mass* in SI units (kilograms) of a body which has been characterized by its *weight* in U.S. customary units (pounds).

To convert into SI units a derived U.S. customary unit, one simply multiplies or divides the appropriate conversion factors. For example, to convert into SI units the moment of a force which was found to be $M = 47 \text{ lb}\cdot\text{in.}$, we use formulas (1.10) and (1.11) and write

$$\begin{aligned} M = 47 \text{ lb}\cdot\text{in.} &= 47(4.448 \text{ N})(25.4 \text{ mm}) \\ &= 5310 \text{ N}\cdot\text{mm} = 5.31 \text{ N}\cdot\text{m} \end{aligned}$$

The conversion factors given in this section may also be used to convert into U.S. customary units a numerical result obtained in SI units. For example, if the moment of a force was found to be $M = 40 \text{ N}\cdot\text{m}$, we write, following the procedure used in the last paragraph of Sec. 1.3,

$$M = 40 \text{ N}\cdot\text{m} = (40 \text{ N}\cdot\text{m})\left(\frac{1 \text{ lb}}{4.448 \text{ N}}\right)\left(\frac{1 \text{ ft}}{0.3048 \text{ m}}\right)$$

Carrying out the numerical computations and canceling out units which appear both in the numerator and the denominator, we obtain

$$M = 29.5 \text{ lb}\cdot\text{ft}$$

The U.S. customary units most frequently used in mechanics are listed in Table 1.3 with their SI equivalents.

1.5. Method of Problem Solution. The student should approach a problem in mechanics as he would approach an actual engineering situation. By drawing on his own experience and on his intuition, he will find it easier to understand and formulate the problem. Once the problem has been clearly stated, however, there is no place in its solution for the student's particular fancy. *The solution must be based on the six fundamental principles stated in Sec. 1.2 or on theorems derived from them.* Every step taken must be justified on that basis. Strict rules must be followed, which lead to the solution in an almost automatic fashion, leaving no room for the student's intuition or "feeling." After an answer has been obtained, it should be checked. Here again, the student may call upon his common sense and personal experience. If not completely satisfied with the result obtained, he should carefully check his formulation of the problem, the validity of the methods used for its solution, and the accuracy of his computations.

The *statement* of a problem should be clear and precise. It should contain the given data and indicate what information is required. A neat drawing showing all quantities involved should be included. Separate diagrams should be drawn for all bodies involved, indicating clearly the forces acting on each body. These diagrams are known as *free-body diagrams* and are described in detail in Secs. 2.11 and 4.2.

Table 1.3 U.S. Customary Units and Their SI Equivalents

Quantity	U.S. Customary Unit	SI Equivalent
Acceleration	ft/s^2	$0.3048\ m/s^2$
	$in./s^2$	$0.0254\ m/s^2$
Area	ft^2	$0.0929\ m^2$
	in^2	$645.2\ mm^2$
Energy	ft·lb	1.356 J
Force	kip	4.448 kN
	lb	4.448 N
	oz	0.2780 N
Impulse	lb·s	4.448 N·s
Length	ft	0.3048 m
	in.	25.40 mm
	mi	1.609 km
Mass	oz mass	28.35 g
	lb mass	0.4536 kg
	slug	14.59 kg
	ton	907.2 kg
Moment of a force	lb·ft	1.356 N·m
	lb·in.	0.1130 N·m
Moment of inertia		
Of an area	in^4	$0.4162 \times 10^6\ mm^4$
Of a mass	$lb\cdot ft\cdot s^2$	$1.356\ kg\cdot m^2$
Momentum	lb·s	4.448 kg·m/s
Power	ft·lb/s	1.356 W
	hp	745.7 W
Pressure or stress	lb/ft^2	47.88 Pa
	lb/in^2 (psi)	6.895 kPa
Velocity	ft/s	0.3048 m/s
	in./s	0.0254 m/s
	mi/h (mph)	0.4470 m/s
	mi/h (mph)	1.609 km/h
Volume	ft^3	$0.02832\ m^3$
	in^3	$16.39\ cm^3$
Liquids	gal	3.785 L
	qt	0.9464 L
Work	ft·lb	1.356 J

The *fundamental principles* of mechanics listed in Sec. 1.2 *will be used to write equations* expressing the conditions of rest or motion of the bodies considered. Each equation should be clearly related to one of the free-body diagrams. The student will then proceed to solve the problem, observing strictly the usual rules of algebra and recording neatly the various steps taken.

After the answer has been obtained, it should be *carefully checked*. Mistakes in *reasoning* may often be detected by checking the units. For example, to determine the moment of a force of 50 N about a point 0.60 m from its line of action, we would have written (Sec. 3.12)

$$M = Fd = (50\text{ N})(0.60\text{ m}) = 30\text{ N}\cdot\text{m}$$

The unit N·m obtained by multiplying newtons by meters is the correct unit for the moment of a force; if another unit had been obtained, we would have known that some mistake had been made.

Errors in *computation* will usually be found by substituting the numerical values obtained into an equation which has not yet been used and verifying that the equation is satisfied. The importance of correct computations in engineering cannot be overemphasized.

1.6. Numerical Accuracy. The accuracy of the solution of a problem depends upon two items: (1) the accuracy of the given data; (2) the accuracy of the computations performed.

The solution cannot be more accurate than the less accurate of these two items. For example, if the loading of a bridge is known to be 75,000 lb with a possible error of 100 lb either way, the relative error which measures the degree of accuracy of the data is

$$\frac{100\text{ lb}}{75{,}000\text{ lb}} = 0.0013 = 0.13\text{ percent}$$

In computing the reaction at one of the bridge supports, it would then be meaningless to record it as 14,322 lb. The accuracy of the solution cannot be greater than 0.13 percent, no matter how accurate the computations are, and the possible error in the answer may be as large as $(0.13/100)(14{,}322\text{ lb}) \approx 20$ lb. The answer should be properly recorded as $14{,}320 \pm 20$ lb.

In engineering problems, the data are seldom known with an accuracy greater than 0.2 percent. It is therefore seldom justified to write the answers to such problems with an accuracy greater than 0.2 percent. A practical rule is to use 4 figures to record numbers beginning with a "1" and 3 figures in all other cases. Unless otherwise indicated, the data given in a problem should be assumed known with a comparable degree of accuracy. A force of 40 lb, for example, should be read 40.0 lb, and a force of 15 lb should be read 15.00 lb.

Pocket electronic calculators are widely used by practicing engineers and engineering students. The speed and accuracy of these calculators facilitates the numerical computations in the solution of many problems. However, the student should not record more significant figures than can be justified, merely because they are easily obtained. As noted above, an accuracy greater than 0.2 percent is seldom necessary or meaningful in the solution of practical engineering problems.

CHAPTER TWO
Statics of Particles

2.1. Introduction. In this chapter we shall study the effect of forces acting on particles. First we shall learn how to replace two or more forces acting on a given particle by a single force having the same effect as the original forces. This single equivalent force is the *resultant* of the original forces acting on the particle. Later we shall derive the relations which exist among the various forces acting on a particle in a state of *equilibrium* and use these relations to determine some of the forces acting on the particle.

The use of the word "particle" does not imply that we shall restrict our study to that of small corpuscles. What it means is that the size and shape of the bodies under consideration will not significantly affect the solution of the problems treated in this chapter, and that all the forces acting on a given body will be assumed applied at the same point. Since such an assumption is verified in many practical applications, we shall be able to solve a number of engineering problems in this chapter.

The first part of the chapter is devoted to the study of forces contained in a single plane, and the second part to the analysis of forces in three-dimensional space.

FORCES IN A PLANE

2.2. Force on a Particle. Resultant of Two Forces. A force represents the action of one body on another and is generally characterized by its *point of application*, its *magnitude*, and its *direction*. Forces acting on a given particle, however, have the same point of application. Each force considered in this chapter will thus be completely defined by its magnitude and direction.

The magnitude of a force is characterized by a certain number of units. As indicated in Chap. 1, the SI units used by engineers to measure the magnitude of a force are the newton (N) and its multiple the kilonewton (kN), equal to 1000 N, while the U.S. customary units used for the same purpose are the pound (lb) and its multiple the kilopound (kip, or k), equal to 1000 lb. The direction of a force is defined by the *line of action* and the *sense* of the force. The line of action is the infinite straight line along which the force acts; it is characterized by the angle it forms with some fixed axis (Fig. 2.1). The force itself is represented by a segment of that line; through the use of an appropriate scale, the length of this segment may be chosen to represent the magnitude of the force. Finally, the sense of the force should be indicated by an arrowhead. It is important in defining a force to indicate its sense. Two forces, such as those shown in Fig. 2.1*a* and *b*, having the same magnitude and the same line of action but different sense, will have directly opposite effects on a particle.

Experimental evidence shows that two forces **P** and **Q** acting on a particle *A* (Fig. 2.2*a*) may be replaced by a single force **R** which has the same effect on the particle (Fig. 2.2*c*). This force is called the *resultant* of the forces **P** and **Q** and may be obtained, as shown in Fig. 2.2*b*, by constructing a parallelogram, using **P** and **Q** as two sides of the parallelogram. *The diagonal that passes through A represents the resultant.* This is known as the *parallelogram law* for the addition of two forces. This law is based on experimental evidence; it cannot be proved or derived mathematically.

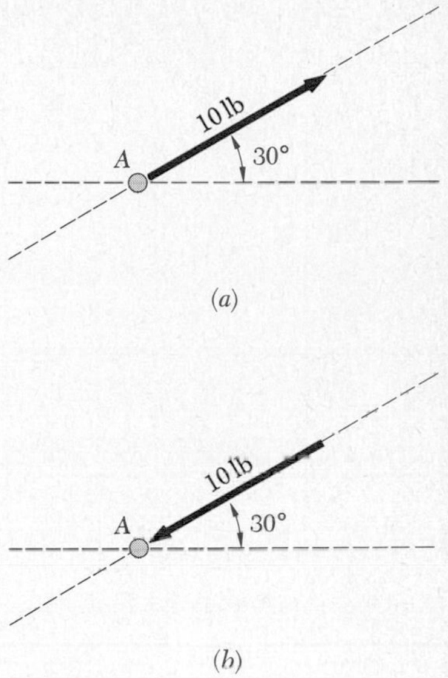

Fig. 2.1

2.3. Vectors. It appears from the above that forces do not obey the rules of addition defined in ordinary arithmetic or algebra. For example, two forces acting at a right angle to each other, one of 4 lb and the other of 3 lb, add up to a force of 5 lb, *not* to a force of 7 lb. Forces are not the only expressions which follow the parallelogram law of addition. As we shall see later, *displacements, velocities, accelerations, momenta* are other examples of physical quantities possessing magnitude and direction and which are added according to the parallelogram law. All these quantities may be represented mathematically by *vectors*, while those physical quantities which do not have direction, such as *volume, mass*, or *energy*, are represented by ordinary numbers or *scalars*.

Vectors are defined as *mathematical expressions possessing magnitude and direction, which add according to the parallelogram law*. Vectors are represented by arrows in the illustrations and will be distinguished from scalar quantities in this text through the use of boldface type (**P**). In longhand writing, a vector may be characterized by drawing a short arrow above the letter used to represent it ($\vec{P}$), or by underlining the letter ($\underline{P}$). The last method may be preferred since it can also be used on a typewriter. The magnitude of a vector defines the length of the arrow used to represent the vector. In this text, italic type will be used to denote the magnitude of a vector. Thus, the magnitude of the vector **P** will be referred to as *P*.

A vector used to represent a force acting on a given particle has a well-defined point of application, namely, the particle itself. Such a vector is said to be a *fixed*, or *bound*, vector and cannot be moved without modifying the conditions of the problem. Other physical quantities, however, such as couples (see Chap. 3), are represented by vectors which may be freely moved in space; these vectors are called *free* vectors. Still other

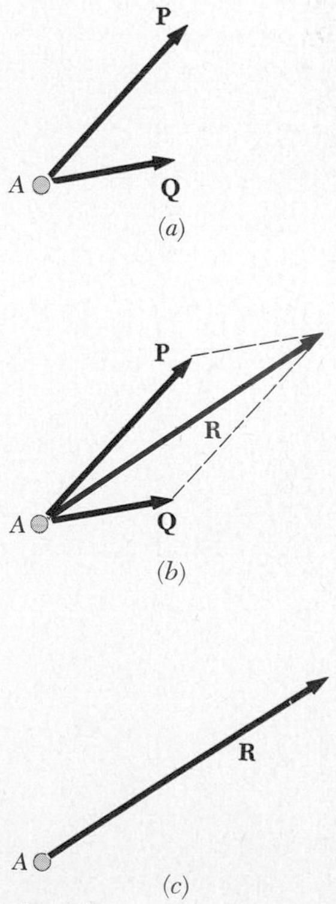

Fig. 2.2

physical quantities, such as forces acting on a rigid body (see Chap. 3), are represented by vectors which may be moved, or slid, along their line of action; they are known as *sliding* vectors.†

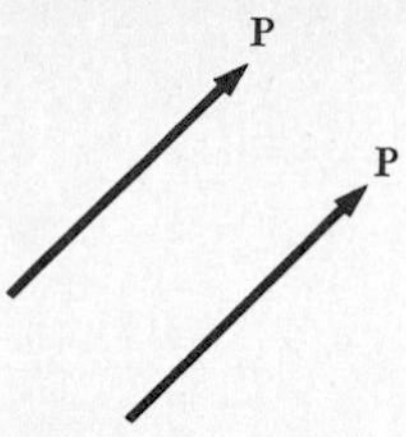

Fig. 2.4

Two vectors which have the same magnitude and the same direction are said to be *equal*, whether or not they also have the same point of application (Fig. 2.4); equal vectors may be denoted by the same letter.

The *negative vector* of a given vector **P** is defined as a vector having the same magnitude as **P** and a direction opposite to that of **P** (Fig. 2.5); the negative of the vector **P** is denoted by −**P**. The vectors **P** and −**P** are commonly referred to as *equal and opposite* vectors. Clearly, we have

$$\mathbf{P} + (-\mathbf{P}) = 0$$

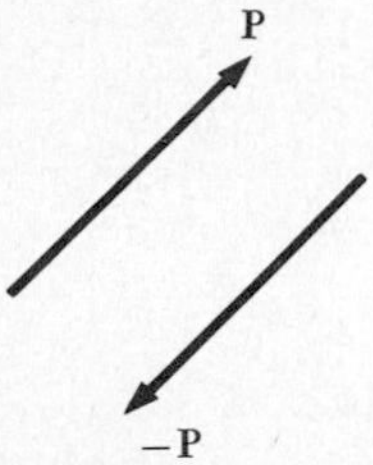

Fig. 2.5

2.4. Addition of Vectors. We saw in the preceding section that, by definition, vectors add according to the parallelogram law. Thus the sum of two vectors **P** and **Q** is obtained by attaching the two vectors to the same point A and constructing a parallelogram, using **P** and **Q** as two sides of the parallelogram (Fig. 2.6). The diagonal that passes through A represents the sum of the vectors **P** and **Q**, and this sum is denoted by **P** + **Q**. The fact that the sign + is used to denote both vector and scalar addition should not cause any confusion if vector and scalar quantities are always carefully distinguished. Thus, we should note that the magnitude of the vector **P** + **Q** is *not*, in general, equal to the sum $P + Q$ of the magnitudes of the vectors **P** and **Q**.

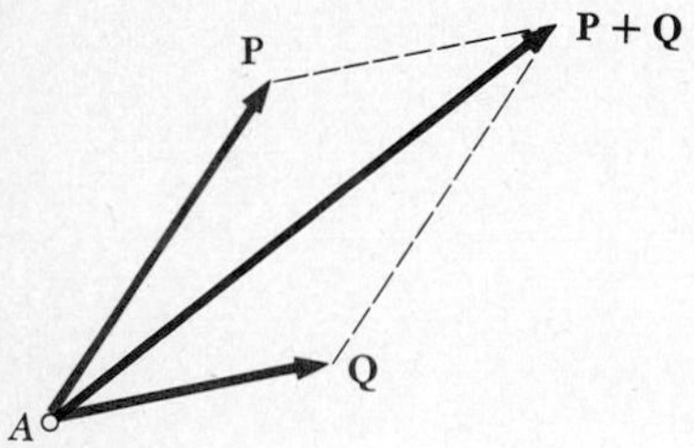

Fig. 2.6

Since the parallelogram constructed on the vectors **P** and **Q** does not depend upon the order in which **P** and **Q** are selected, we conclude that the addition of two vectors is *commutative*, and we write

$$\mathbf{P} + \mathbf{Q} = \mathbf{Q} + \mathbf{P} \tag{2.1}$$

† Some expressions have magnitude and direction, but do not add according to the parallelogram law. While these expressions may be represented by arrows, they *cannot* be considered as vectors.

A group of such expressions are the finite rotations of a rigid body. Place a closed book on a table in front of you, so that it lies in the usual fashion, with its front cover up and its binding to the left. Now rotate it through 180° about an axis parallel to the binding (Fig. 2.3*a*); this rotation may be represented by an arrow of length equal to 180 units and oriented as shown. Picking up the book as it lies in its new position, rotate it now through 180° about a horizontal

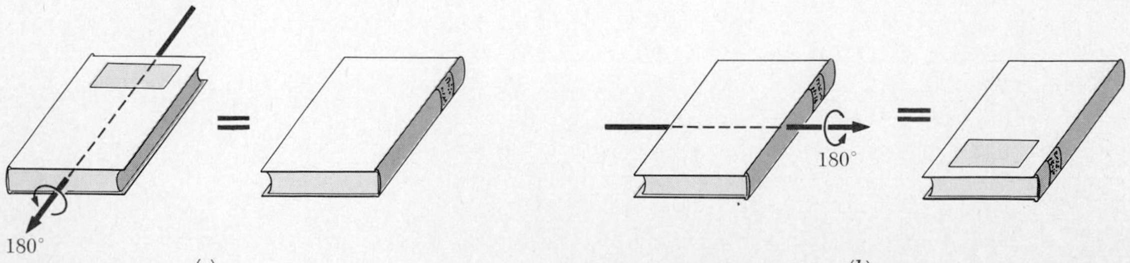

(*a*) (*b*)
Fig. 2.3 Finite rotations of a rigid body.

From the parallelogram law, we can derive an alternate method for determining the sum of two vectors. This method, known as the *triangle rule,* is derived as follows: Consider Fig. 2.6, where the sum of the vectors **P** and **Q** has been determined by the parallelogram law. Since the side of the parallelogram opposite **Q** is equal to **Q** in magnitude and direction, we could draw only half of the parallelogram (Fig. 2.7*a*). The sum of the two vectors may thus be found by *arranging* **P** *and* **Q** *in tip-to-tail fashion and then connecting the tail of* **P** *with the tip of* **Q**. In Fig. 2.7*b*, the other half of the parallelogram is considered, and the same result is obtained. This confirms the fact that vector addition is commutative.

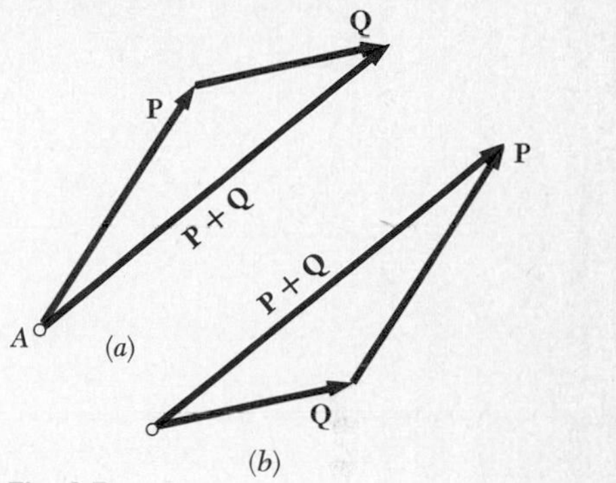

Fig. 2.7

The *subtraction* of a vector is defined as the addition of the corresponding negative vector. Thus, the vector **P** − **Q** representing the difference between the vectors **P** and **Q** is obtained by adding to **P** the negative vector −**Q** (Fig. 2.8). We write

$$\mathbf{P} - \mathbf{Q} = \mathbf{P} + (-\mathbf{Q}) \tag{2.2}$$

Here again we should observe that, while the same sign is used to denote both vector and scalar subtraction, confusion will be avoided if care is taken to distinguish between vector and scalar quantities.

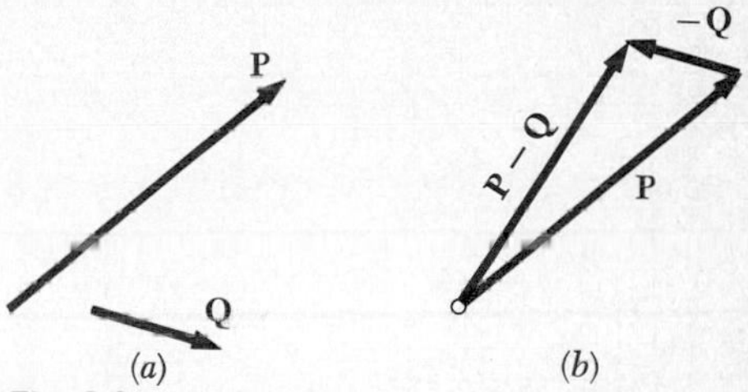

Fig. 2.8

We shall now consider the *sum of three or more vectors.* The sum of three vectors **P**, **Q**, and **S** will, *by definition,* be obtained by first adding the vectors **P** and **Q**, and then adding the vector **S** to the vector **P** + **Q**. We thus write

$$\mathbf{P} + \mathbf{Q} + \mathbf{S} = (\mathbf{P} + \mathbf{Q}) + \mathbf{S} \tag{2.3}$$

Similarly, the sum of four vectors will be obtained by adding the fourth vector to the sum of the first three. It follows that the sum of any number of vectors may be obtained by applying repeatedly the parallelogram law to successive pairs of vectors until all the given vectors are replaced by a single vector.

---

axis perpendicular to the binding (Fig. 2.3*b*); this second rotation may be represented by an arrow 180 units long and oriented as shown. But the book could have been placed in this final position through a single 180° rotation about a vertical axis (Fig. 2.3*c*). We conclude that the sum of the two 180° rotations represented by arrows directed respectively along the *z* and *x* axes is a 180° rotation represented by an arrow directed along the *y* axis (Fig. 2.3*d*). Clearly, the finite rotations of a rigid body *do not* obey the parallelogram law of addition; therefore they *cannot* be represented by vectors.

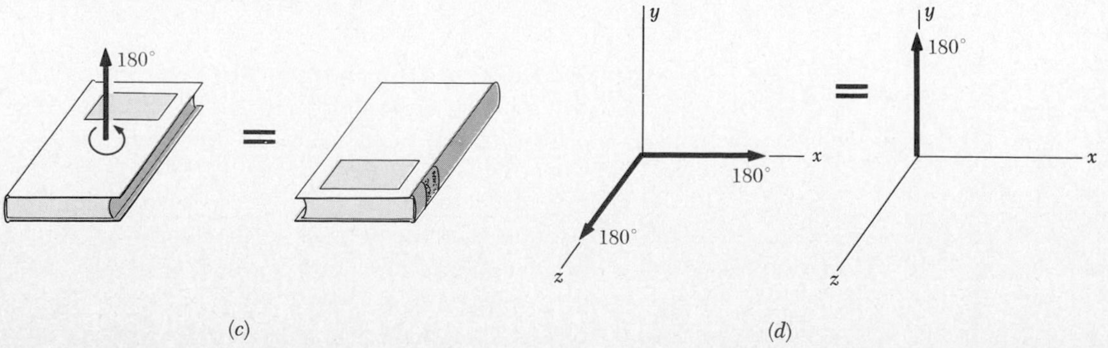

If the given vectors are *coplanar,* i.e., if they are contained in the same plane, their sum may be easily obtained graphically. In that case, the repeated application of the triangle rule will be preferred to the application of the parallelogram law. In Fig. 2.9 the sum of three vectors **P**, **Q**, and **S** was obtained in that manner. The triangle rule was first applied to obtain the sum **P** + **Q** of the vectors **P** and **Q**; it was applied again to obtain the sum of the vectors **P** + **Q** and **S**. The determination of the vector **P** + **Q**, however, could have been omitted and the sum of the three vectors could have been obtained directly, as shown in Fig. 2.10, by *arranging the given vectors in tip-to-tail fashion and connecting the tail of the first vector with the tip of the last one*. This is known as the *polygon rule* for the addition of vectors.

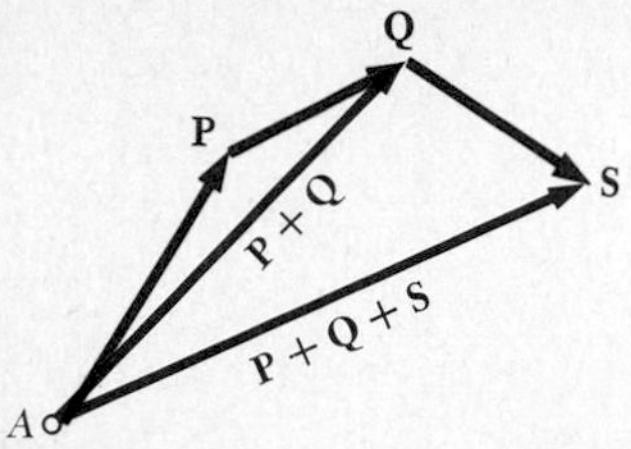

Fig. 2.9

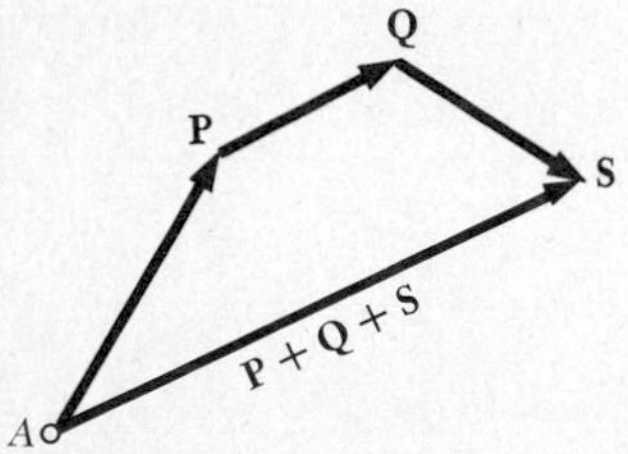

Fig. 2.10

We observe that the result obtained would have been unchanged if, as shown in Fig. 2.11, the vectors **Q** and **S** had been replaced by their sum **Q** + **S**. We may thus write

$$\mathbf{P} + \mathbf{Q} + \mathbf{S} = (\mathbf{P} + \mathbf{Q}) + \mathbf{S} = \mathbf{P} + (\mathbf{Q} + \mathbf{S}) \tag{2.4}$$

which expresses the fact that vector addition is *associative*. Recalling that vector addition has also been shown, in the case of two vectors, to be commutative, we write

$$\begin{aligned} \mathbf{P} + \mathbf{Q} + \mathbf{S} &= (\mathbf{P} + \mathbf{Q}) + \mathbf{S} = \mathbf{S} + (\mathbf{P} + \mathbf{Q}) \\ &= \mathbf{S} + (\mathbf{Q} + \mathbf{P}) = \mathbf{S} + \mathbf{Q} + \mathbf{P} \end{aligned} \tag{2.5}$$

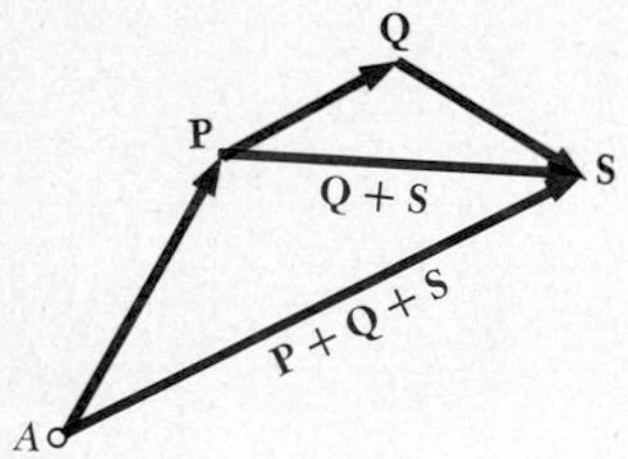

Fig. 2.11

This expression, as well as others which may be obtained in the same way, shows that the order in which several vectors are added together is immaterial (Fig. 2.12).

Product of a Scalar and a Vector. Since it is convenient to denote the sum **P** + **P** by 2**P**, the sum **P** + **P** + **P** by 3**P**, and, in general, the sum of n equal vectors **P** by the product n**P**, we shall define the product n**P** of a positive integer n and a vector **P** as a vector having the same direction as **P** and the magnitude nP. Extending this definition to include all scalars, and recalling the definition of a negative vector given in Sec. 2.3, we define the product k**P** of a scalar k and a vector **P** as a vector having the same direction as **P** (if k is positive), or a direction opposite to that of **P** (if k is negative), and a magnitude equal to the product of P and of the absolute value of k (Fig. 2.13).

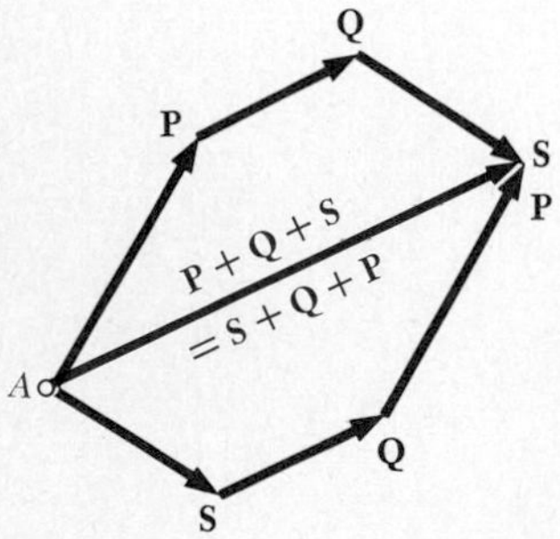

Fig. 2.12

2.5. Resultant of Several Concurrent Forces. Consider a particle A acted upon by several coplanar forces, i.e., by several forces contained in the same plane (Fig. 2.14a). Since the forces considered here all pass through A, they are also said to be *concurrent*. The vectors representing the forces acting on A may be added by the polygon rule (Fig. 2.14b). Since the use of the polygon rule is equivalent to the repeated application of the parallelogram law, the vector **R** thus obtained represents the resultant of the given concurrent forces, i.e., the single force which has

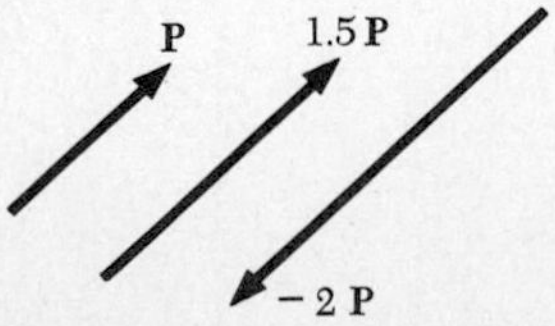

Fig. 2.13

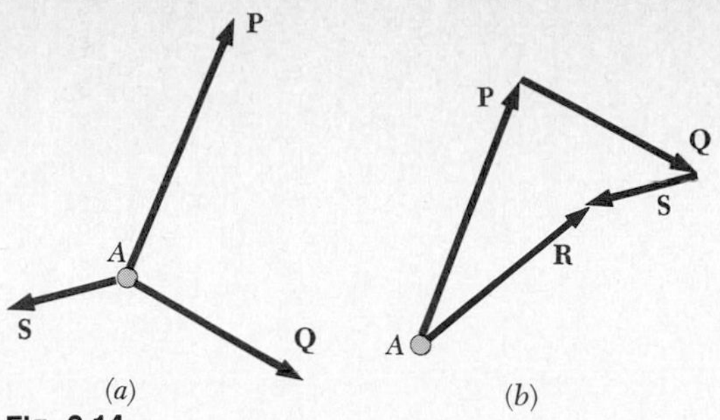

Fig. 2.14

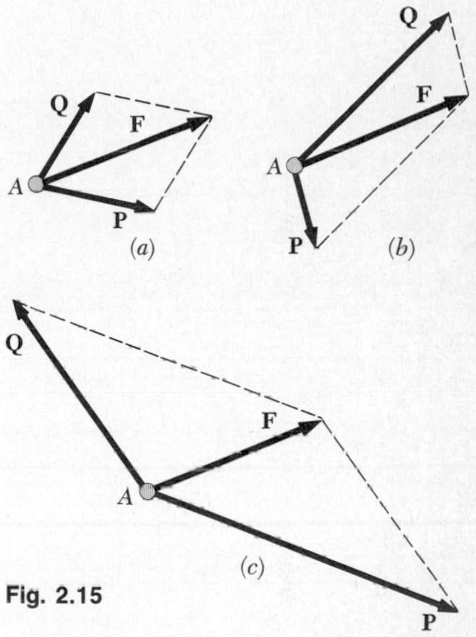

Fig. 2.15

the same effect on the particle A as the given forces. As indicated above, the order in which the vectors **P**, **Q**, and **S** representing the given forces are added together is immaterial.

2.6. Resolution of a Force Into Components. We have seen that two or more forces acting on a particle may be replaced by a single force which has the same effect on the particle. Conversely, a single force **F** acting on a particle may be replaced by two or more forces which, together, have the same effect on the particle. These forces are called the *components* of the original force **F**, and the process of substituting them for **F** is called *resolving the force* **F** *into components.*

Clearly, for each force **F** there exist an infinite number of possible sets of components. Sets of *two components* **P** *and* **Q** are the most important as far as practical applications are concerned. But, even then, the number of ways in which a given force **F** may be resolved into two components is unlimited (Fig. 2.15). Two cases are of particular interest:

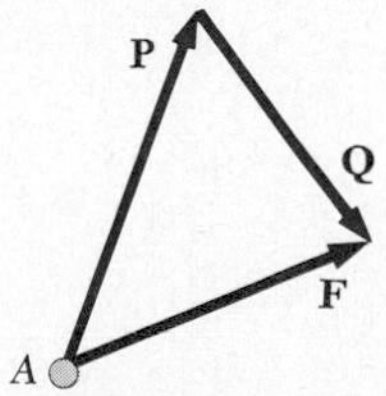

Fig. 2.16

1. *One of the Two Components,* **P**, *Is Known.* The second component, **Q**, is obtained by applying the triangle rule and joining the tip of **P** to the tip of **F** (Fig. 2.16); the magnitude and direction of **Q** are determined graphically or by trigonometry. Once **Q** has been determined, both components **P** and **Q** should be applied at A.
2. *The Line of Action of Each Component Is Known.* The magnitude and sense of the components are obtained by applying the parallelogram law and drawing lines, through the tip of **F**, parallel to the given lines of action (Fig. 2.17). This process leads to two well-defined components, **P** and **Q**, which may be determined graphically, or trigonometrically, by applying the law of sines.

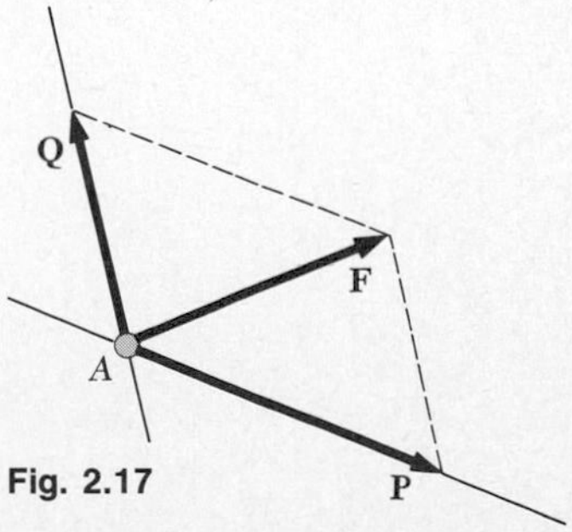

Fig. 2.17

Many other cases may be encountered; for example, the direction of one component may be known while the magnitude of the other component is to be as small as possible (see Sample Prob. 2.2). In all cases the appropriate triangle or parallelogram is drawn, which satisfies the given conditions.

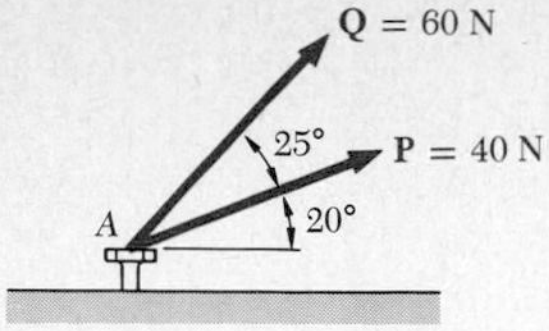

SAMPLE PROBLEM 2.1

The two forces **P** and **Q** act on a bolt A. Determine their resultant.

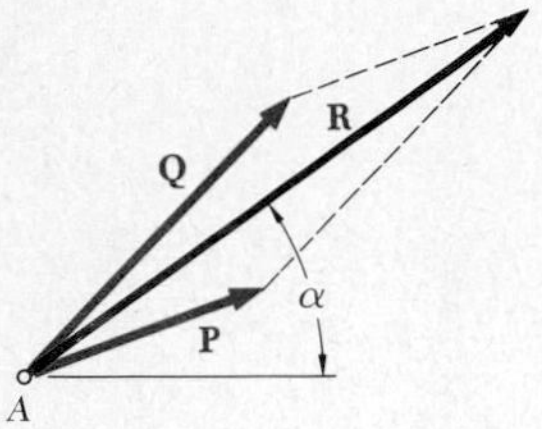

Graphical Solution. A parallelogram with sides equal to **P** and **Q** is drawn to scale. The magnitude and direction of the resultant are measured and found to be

$$R = 98 \text{ N} \qquad \alpha = 35^\circ \qquad \mathbf{R} = 98 \text{ N} \measuredangle 35^\circ$$ ◄

The triangle rule may also be used. Forces **P** and **Q** are drawn in tip-to-tail fashion. Again the magnitude and direction of the resultant are measured.

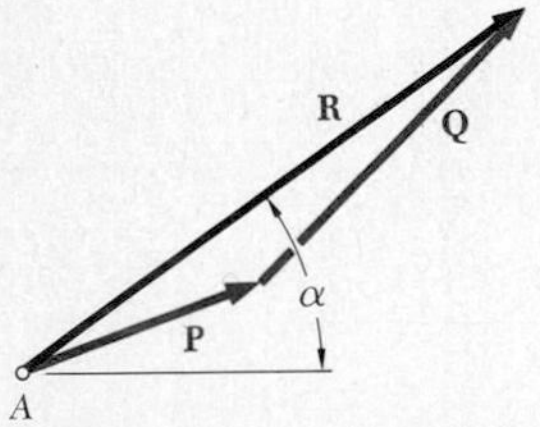

$$R = 98 \text{ N} \qquad \alpha = 35^\circ \qquad \mathbf{R} = 98 \text{ N} \measuredangle 35^\circ$$ ◄

Trigonometric Solution. The triangle rule is again used; two sides and the included angle are known. We apply the law of cosines.

$$R^2 = P^2 + Q^2 - 2PQ \cos B$$
$$R^2 = (40 \text{ N})^2 + (60 \text{ N})^2 - 2(40 \text{ N})(60 \text{ N}) \cos 155^\circ$$
$$R = 97.73 \text{ N}$$

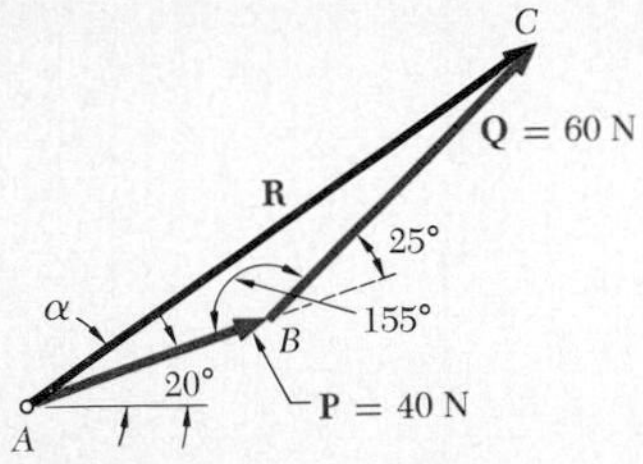

Now, applying the law of sines, we write

$$\frac{\sin A}{Q} = \frac{\sin B}{R} \qquad \frac{\sin A}{60 \text{ N}} = \frac{\sin 155^\circ}{97.73 \text{ N}} \tag{1}$$

Solving Eq. (1) for sin A, we have

$$\sin A = \frac{(60 \text{ N}) \sin 155^\circ}{97.73 \text{ N}}$$

Using a calculator, we first compute the quotient, then its arc sine, and obtain

$$A = 15.04^\circ \qquad \alpha = 20^\circ + A = 35.04^\circ$$

We use 3 significant figures to record the answer (cf. Sec. 1.6):

$$\mathbf{R} = 97.7 \text{ N} \measuredangle 35.0^\circ$$ ◄

Alternate Trigonometric Solution. We construct the right triangle BCD and compute

$$CD = (60 \text{ N}) \sin 25^\circ = 25.36 \text{ N}$$
$$BD = (60 \text{ N}) \cos 25^\circ = 54.38 \text{ N}$$

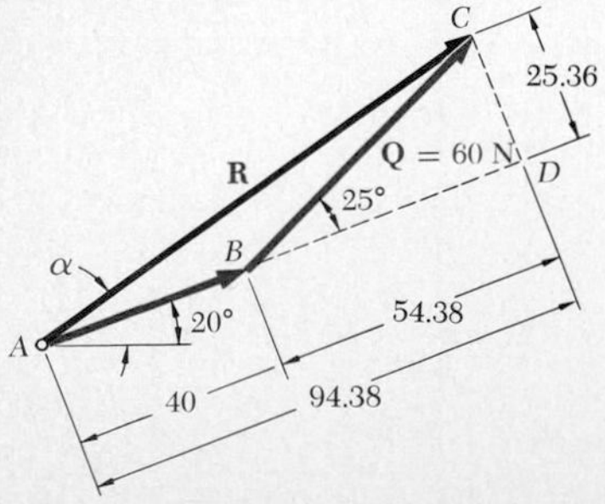

Then, using triangle ACD, we obtain

$$\tan A = \frac{25.36 \text{ N}}{94.38 \text{ N}} \qquad A = 15.04^\circ$$
$$R = \frac{25.36}{\sin A} \qquad R = 97.73 \text{ N}$$

Again, $\alpha = 20^\circ + A = 35.04^\circ \qquad \mathbf{R} = 97.7 \text{ N} \measuredangle 35.0^\circ$ ◄

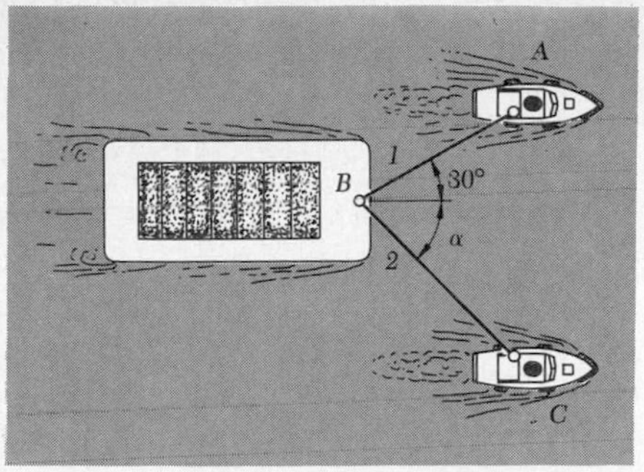

SAMPLE PROBLEM 2.2

A barge is pulled by two tugboats. If the resultant of the forces exerted by the tugboats is a 5000-lb force directed along the axis of the barge, determine (*a*) the tension in each of the ropes, knowing that $\alpha = 45°$, (*b*) the value of α such that the tension in rope *2* is minimum.

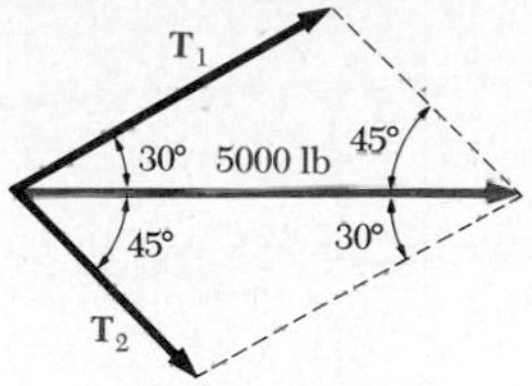

a.* Tension for $\alpha = 45°$.** ***Graphical Solution. The parallelogram law is used; the diagonal (resultant) is known to be equal to 5000 lb and to be directed to the right. The sides are drawn parallel to the ropes. If the drawing is done to scale, we measure

$$T_1 = 3700 \text{ lb} \qquad T_2 = 2600 \text{ lb} \blacktriangleleft$$

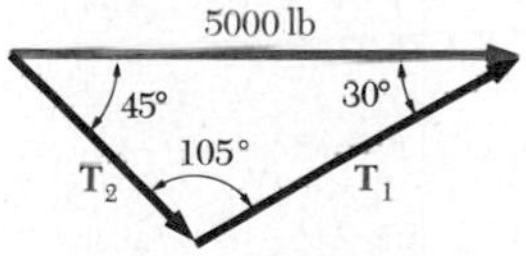

Trigonometric Solution. The triangle rule may be used. We note that the triangle shown represents half of the parallelogram shown above. Using the law of sines, we write

$$\frac{T_1}{\sin 45°} = \frac{T_2}{\sin 30°} = \frac{5000 \text{ lb}}{\sin 105°}$$

With a calculator, we first compute and store the value of the last quotient. Multiplying this value successively by sin 45° and sin 30°, we obtain

$$T_1 = 3660 \text{ lb} \qquad T_2 = 2590 \text{ lb} \blacktriangleleft$$

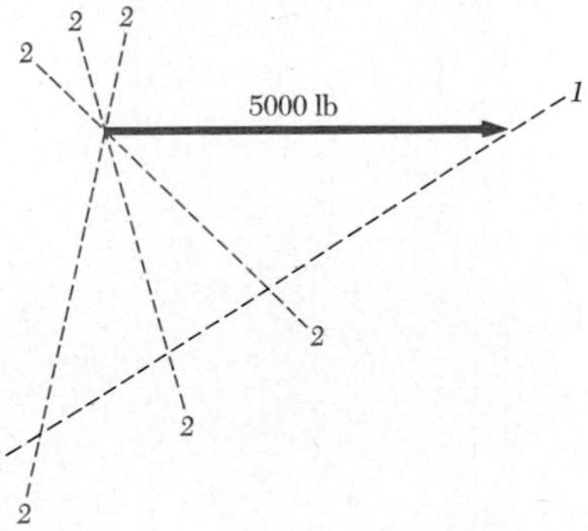

***b.* Value of α for Minimum T_2.** To determine the value of α such that the tension in rope *2* is minimum, the triangle rule is again used. In the sketch shown, line *1-1* is the known direction of $\mathbf{T}_1$. Several possible directions of $\mathbf{T}_2$ are shown by the lines 2-2. We note that the minimum value of T_2 occurs when $\mathbf{T}_1$ and $\mathbf{T}_2$ are perpendicular. The minimum value of T_2 is

$$T_2 = (5000 \text{ lb}) \sin 30° = 2500 \text{ lb}$$

Corresponding values of T_1 and α are

$$T_1 = (5000 \text{ lb}) \cos 30° = 4330 \text{ lb}$$
$$\alpha = 90° - 30° \qquad \alpha = 60° \blacktriangleleft$$

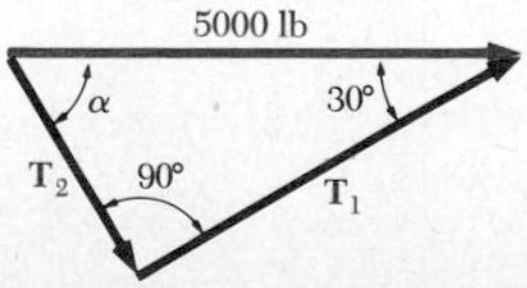

Problems†

2.1 and 2.2 Determine graphically the magnitude and direction of the resultant of the two forces shown, using (*a*) the parallelogram law, (*b*) the triangle rule.

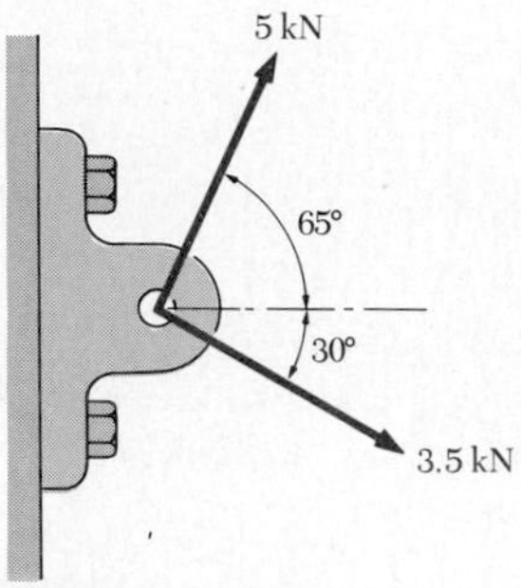

Fig. P2.1

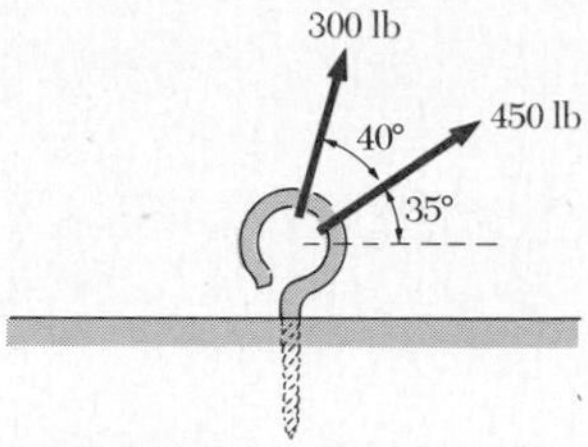

Fig. P2.2

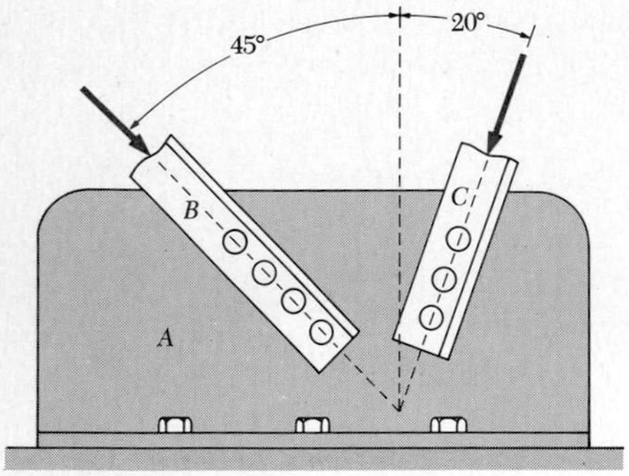

Fig. P2.3 and P2.4

2.3 Two structural members *B* and *C* are riveted to bracket *A*. Knowing that both members are in compression and that the force is 1200 lb in member *B* and 1600 lb in member *C*, determine graphically the magnitude and direction of the resultant force exerted on the bracket.

2.4 Two structural members *B* and *C* are riveted to bracket *A*. Knowing that both members are in compression and that the force is 8 kN in member *B* and 12 kN in member *C*, determine graphically the magnitude and direction of the resultant force exerted on the bracket.

2.5 The force **F** of magnitude 500 lb is to be resolved into two components along lines *a-a* and *b-b*. Determine by trigonometry the angle α, knowing that the component of **F** along line *a-a* is to be 400 lb.

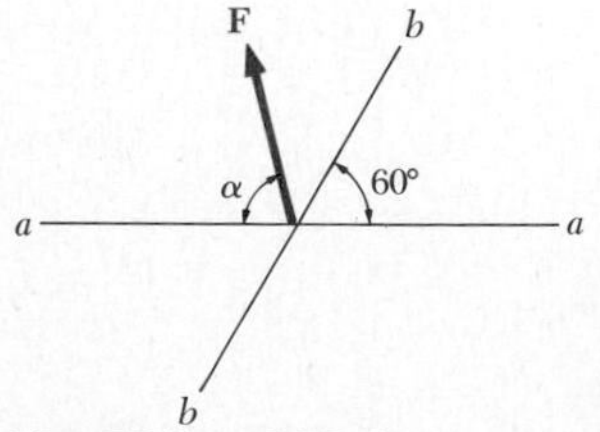

Fig. P2.5 and P2.6

2.6 The force **F** of magnitude 400 N is to be resolved into two components along lines *a-a* and *b-b*. Determine by trigonometry the angle α, knowing that the component of **F** along line *b-b* is to be 150 N.

† Answers to all even-numbered problems are given at the end of the book.

2.7 A stake is pulled out of the ground by means of two ropes as shown. (*a*) Knowing that $\alpha = 30°$, determine by trigonometry the magnitude of the force **P** so that the resultant force exerted on the stake is vertical. (*b*) What is the corresponding magnitude of the resultant?

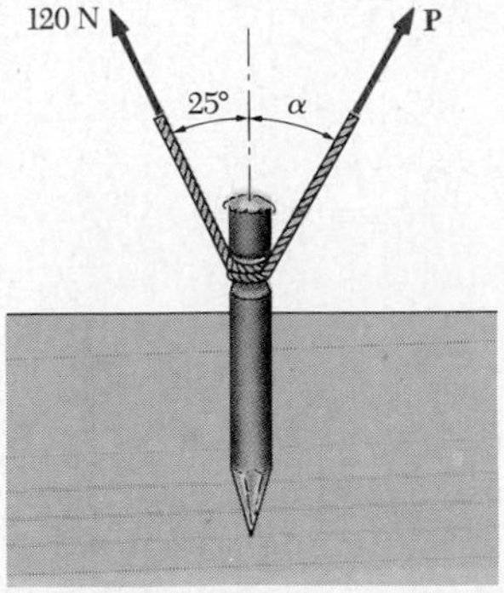

Fig. P2.7 and P2.12

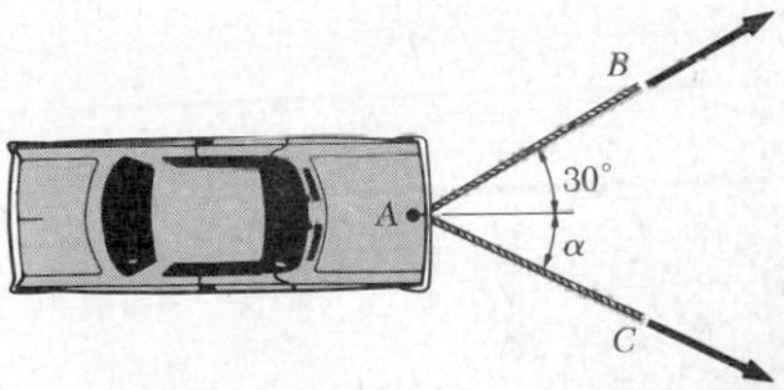

Fig. P2.8 and P2.11

2.8 A disabled automobile is pulled by means of two ropes as shown. The tension in *AB* is 400 lb and the angle α is 20°. Knowing that the resultant of the two forces applied at *A* is directed along the axis of the automobile, determine by trigonometry (*a*) the tension in rope *AC*, (*b*) the magnitude of the resultant of the two forces applied at *A*.

2.9 Solve Prob. 2.8, assuming that the tension in rope *AB* is 2.4 kN and that $\alpha = 25°$.

2.10 Solve Prob. 2.7, assuming that $\alpha = 40°$.

2.11 A disabled automobile is pulled by means of two ropes as shown. Knowing that the tension in rope *AB* is 500 lb, determine by trigonometry the tension in rope *AC* and the value of α so that the resultant force exerted at *A* is an 800-lb force directed along the axis of the automobile.

2.12 A stake is pulled out of the ground by means of two ropes as shown. Knowing that the tension in one rope is 120 N, determine by trigonometry the magnitude and direction of the force **P** so that the resultant is a vertical force of 160 N.

2.13 Solve Prob. 2.1 by trigonometry.

2.14 Determine by trigonometry the magnitude and direction of the resultant of the two forces shown.

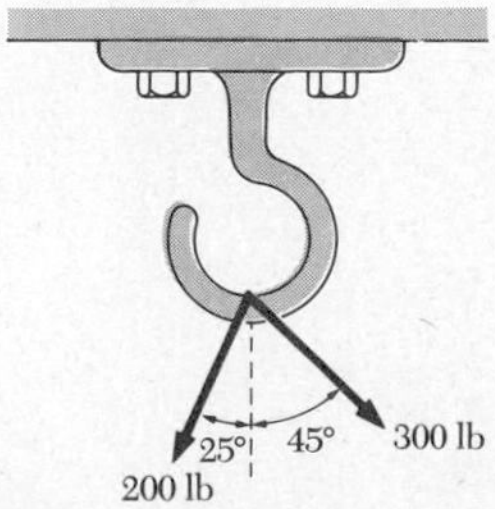

Fig. P2.14

2.15 If the resultant of the two forces exerted on the stake of Prob. 2.7 is to be vertical, find (*a*) the value of α for which the magnitude of **P** is minimum, (*b*) the corresponding magnitude of **P**.

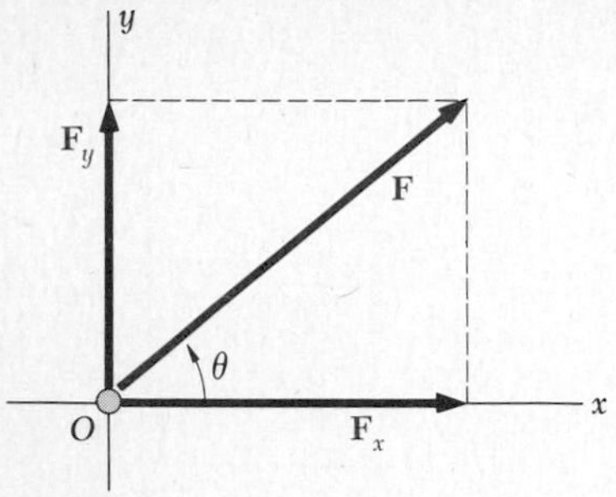

Fig. 2.18

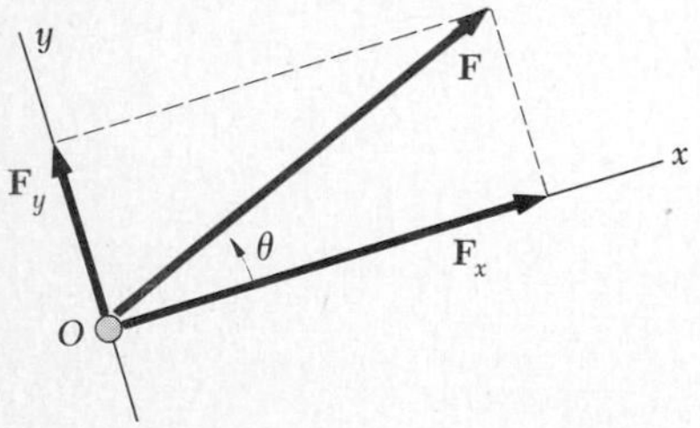

Fig. 2.19

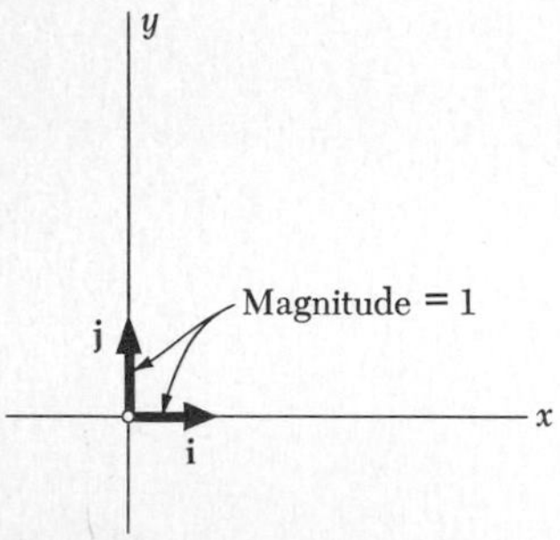

Fig. 2.20

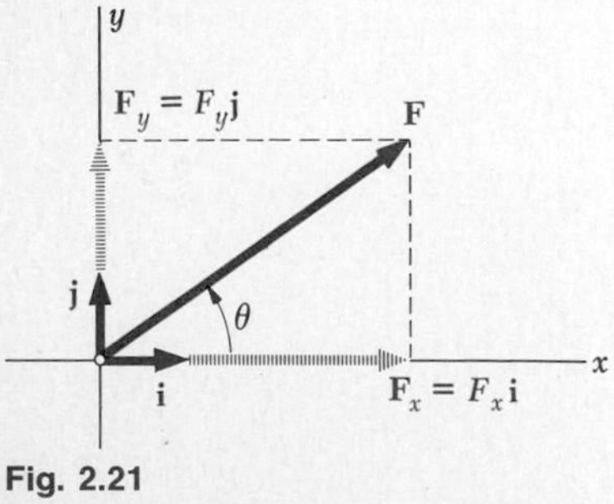

Fig. 2.21

2.7. Rectangular Components of a Force. Unit Vectors.† In many problems it will be found desirable to resolve a force into two components which are perpendicular to each other. In Fig. 2.18, the force **F** has been resolved into a component $\mathbf{F}_x$ along the x axis and a component $\mathbf{F}_y$ along the y axis. The parallelogram drawn to obtain the two components is a *rectangle,* and $\mathbf{F}_x$ and $\mathbf{F}_y$ are called *rectangular components.*

The x and y axes are usually chosen horizontal and vertical, respectively, as in Fig. 2.18; they may, however, be chosen in any two perpendicular directions, as shown in Fig. 2.19. In determining the rectangular components of a force, the student should think of the construction lines shown in Figs. 2.18 and 2.19 as being *parallel* to the x and y axes, rather than *perpendicular* to these axes. This practice will help avoid mistakes in determining *oblique* components as in Sec. 2.6.

We shall, at this point, introduce two vectors of magnitude 1, directed respectively along the positive x and y axes. These vectors are called *unit vectors* and are denoted by **i** and **j**, respectively (Fig. 2.20). Recalling the definition of the product of a scalar and a vector given in Sec. 2.4, we note that the rectangular components $\mathbf{F}_x$ and $\mathbf{F}_y$ of a force **F** may be obtained by multiplying respectively the unit vectors **i** and **j** by appropriate scalars (Fig. 2.21). We write

$$\mathbf{F}_x = F_x\mathbf{i} \qquad \mathbf{F}_y = F_y\mathbf{j} \tag{2.6}$$

and

$$\mathbf{F} = F_x\mathbf{i} + F_y\mathbf{j} \tag{2.7}$$

While the scalars F_x and F_y may be positive or negative, depending upon the sense of $\mathbf{F}_x$ and of $\mathbf{F}_y$, their absolute values are respectively equal to the magnitudes of the component forces $\mathbf{F}_x$ and $\mathbf{F}_y$. The scalars F_x and F_y are called the *scalar components* of the force **F**, while the actual component forces $\mathbf{F}_x$ and $\mathbf{F}_y$ should be referred to as the *vector components* of **F**. However, when there exists no possibility of confusion, the vector as well as the scalar components of **F** may be referred to simply as the *components* of **F**. We note that the scalar component F_x is positive when the vector component $\mathbf{F}_x$ has the same sense as the unit vector **i** (i.e., the same sense as the positive x axis) and negative when $\mathbf{F}_x$ has the opposite sense. A similar conclusion may be drawn regarding the sign of the scalar component F_y.

Denoting by F the magnitude of the force **F** and by θ the angle between **F** and the x axis, measured counterclockwise from the positive x axis (Fig. 2.21), we may express the scalar component of **F** as follows:

$$F_x = F\cos\theta \qquad F_y = F\sin\theta \tag{2.8}$$

†The properties established in Secs. 2.7 and 2.8 may be readily extended to the rectangular components of any vector quantity.

We note that the relations obtained hold for any value of the angle θ from $0°$ to $360°$, and that they define the signs as well as the absolute values of the scalar components F_x and F_y.

Example 1. A force of 800 N is exerted on a bolt A as shown in Fig. 2.22*a*. Determine the horizontal and vertical components of the force.

In order to obtain the correct sign for the scalar components F_x and F_y, the value $\theta = 180° - 35° = 145°$ should be substituted for θ in the relations (2.8). However, it will be found more practical to determine by inspection the signs of F_x and of F_y (Fig. 2.22*b*) and to use the trigonometric functions of the angle $\alpha = 35°$. We write therefore

$$F_x = -F\cos\alpha = -(800\text{ N})\cos 35° = -655\text{ N}$$
$$F_y = +F\sin\alpha = +(800\text{ N})\sin 35° = +459\text{ N}$$

The vector components of $\mathbf{F}$ are thus

$$\mathbf{F}_x = -(655\text{ N})\mathbf{i} \qquad \mathbf{F}_y = +(459\text{ N})\mathbf{j}$$

and we may write $\mathbf{F}$ in the form

$$\mathbf{F} = -(655\text{ N})\mathbf{i} + (459\text{ N})\mathbf{j}$$

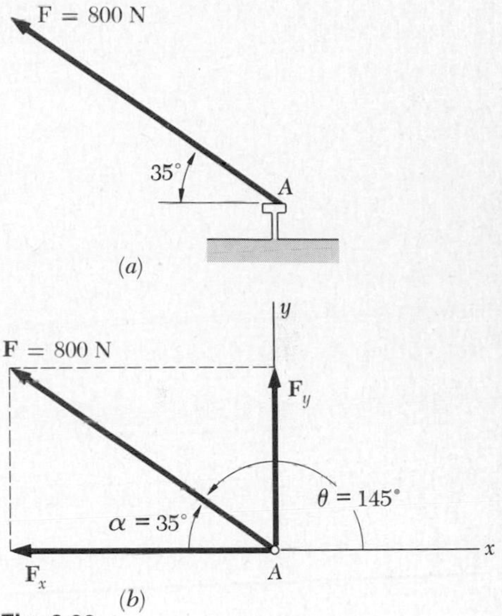

Fig. 2.22

Example 2. A man pulls with a force of 300 N on a rope attached to a building, as shown in Fig. 2.23*a*. What are the horizontal and vertical components of the force exerted by the rope at point A?

It is seen from Fig. 2.23*b* that

$$F_x = +(300\text{ N})\cos\alpha \qquad F_y = -(300\text{ N})\sin\alpha$$

Observing that $AB = 10$ m, we find from Fig. 2.23*a*

$$\cos\alpha = \frac{8\text{ m}}{AB} = \frac{8\text{ m}}{10\text{ m}} = \frac{4}{5} \qquad \sin\alpha = \frac{6\text{ m}}{AB} = \frac{6\text{ m}}{10\text{ m}} = \frac{3}{5}$$

We thus obtain

$$F_x = +(300\text{ N})\tfrac{4}{5} = +240\text{ N} \qquad F_y = -(300\text{ N})\tfrac{3}{5} = -180\text{ N}$$

and write

$$\mathbf{F} = (240\text{ N})\mathbf{i} - (180\text{ N})\mathbf{j}$$

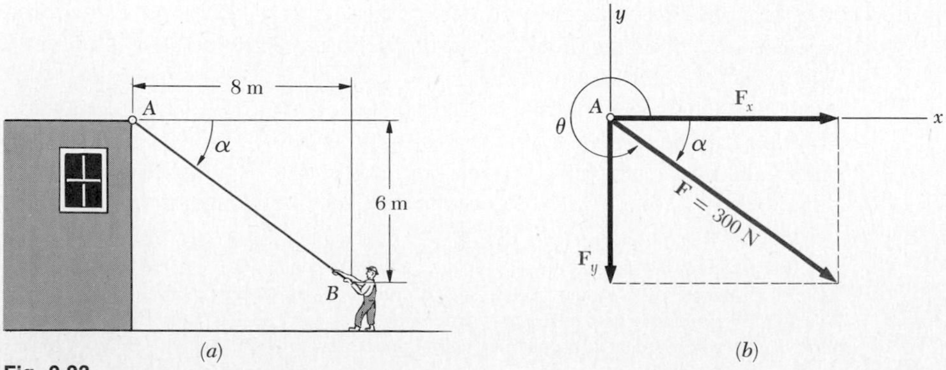

Fig. 2.23

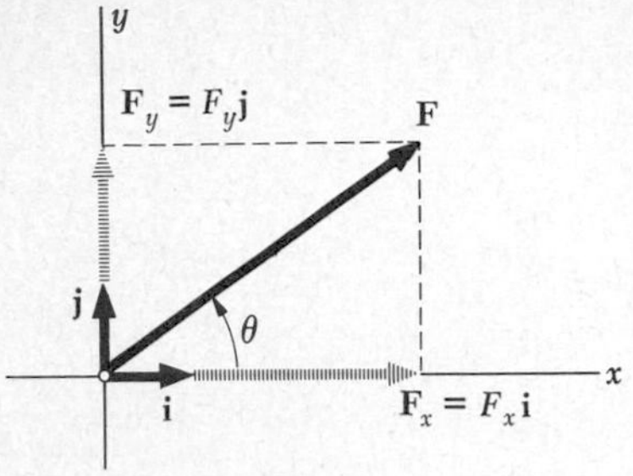

Fig. 2.21 ***(repeated)***

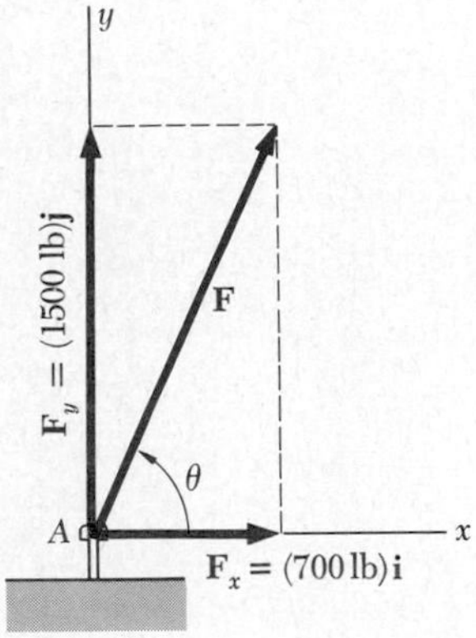

Fig. 2.24

When a force **F** is defined by its rectangular components F_x and F_y (see Fig. 2.21), the angle θ defining its direction can be obtained by writing

$$\tan \theta = \frac{F_y}{F_x} \tag{2.9}$$

The magnitude F of the force may be obtained by applying the Pythagorean theorem and writing

$$F = \sqrt{F_x^2 + F_y^2} \tag{2.10}$$

or by solving one of the formulas (2.8) for F.

Example 3. A force $\mathbf{F} = (700 \text{ lb})\mathbf{i} + (1500 \text{ lb})\mathbf{j}$ is applied to a bolt A. Determine the magnitude of the force and the angle θ it forms with the horizontal.

First we draw a diagram showing the two rectangular components of the force and the angle θ (Fig. 2.24). From Eq. (2.9), we write

$$\tan \theta = \frac{F_y}{F_x} = \frac{1500 \text{ lb}}{700 \text{ lb}}$$

Using a calculator,† we enter 1500 lb and divide by 700 lb; computing the arc tangent of the quotient, we obtain $\theta = 65.0°$. Solving the second of Eqs. (2.8) for F, we have

$$F = \frac{F_y}{\sin \theta} = \frac{1500 \text{ lb}}{\sin 65.0°} = 1655 \text{ lb}$$

The last calculation is facilitated if the value of F_y is stored when originally entered; it may then be recalled to be divided by $\sin \theta$.

2.8. Addition of Forces by Summing *x* and *y* Components. It was seen in Sec. 2.2 that forces should be added according to the parallelogram law. From this law, two other methods, more readily applicable to the *graphical* solution of problems, were derived in Secs. 2.4 and 2.5: the triangle rule for the addition of two forces and the polygon rule for the addition of three or more forces. It was also seen that the force triangle used to define the resultant of two forces could be used to obtain a *trigonometric* solution.

When three or more forces are to be added, no practical trigonometric solution may be obtained from the force polygon which defines the resultant of the forces. In this case, an *analytic* solution of the problem may be obtained by resolving each force into two rectangular components. Con-

† It is assumed that the calculator used has keys for the computation of trigonometric and inverse trigonometric functions. Some calculators also have keys for the direct conversion of rectangular coordinates into polar coordinates, and vice versa. Such calculators eliminate the need for the computation of trigonometric functions in Examples 1, 2, and 3 and in problems of the same type.

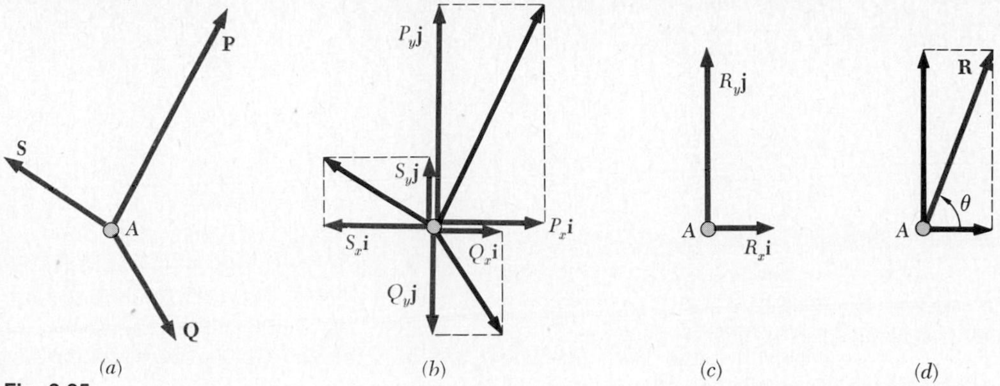

Fig. 2.25

sider, for instance, three forces **P**, **Q**, and **S** acting on a particle A (Fig. 2.25a). Their resultant **R** is defined by the relation

$$\mathbf{R} = \mathbf{P} + \mathbf{Q} + \mathbf{S} \tag{2.11}$$

Resolving each force into its rectangular components, we write

$$\begin{aligned} R_x\mathbf{i} + R_y\mathbf{j} &= P_x\mathbf{i} + P_y\mathbf{j} + Q_x\mathbf{i} + Q_y\mathbf{j} + S_x\mathbf{i} + S_y\mathbf{j} \\ &= (P_x + Q_x + S_x)\mathbf{i} + (P_y + Q_y + S_y)\mathbf{j} \end{aligned}$$

from which it follows that

$$R_x = P_x + Q_x + S_x \qquad R_y = P_y + Q_y + S_y \tag{2.12}$$

or, for short,

$$R_x = \Sigma F_x \qquad R_y = \Sigma F_y \tag{2.13}$$

We thus conclude that *the scalar components R_x and R_y of the resultant* **R** *of several forces acting on a particle are obtained by adding algebraically the corresponding scalar components of the given forces.*†

In practice, the determination of the resultant **R** is carried out in three steps as illustrated in Fig. 2.25. First, the given forces shown in Fig. 2.25a are resolved into their x and y components (Fig. 2.25b). Adding these components, we obtain the x and y components of **R** (Fig. 2.25c). Finally, the resultant $\mathbf{R} = R_x\mathbf{i} + R_y\mathbf{j}$ is determined by applying the parallelogram law (Fig. 2.25d). The procedure just described will be carried out most efficiently if the computations are arranged in a table. While it is the only practical analytic method for adding three or more forces, it is also often preferred to the trigonometric solution in the case of the addition of two forces.

†Clearly, this result also applies to the addition of other vector quantities, such as velocities, accelerations, or momenta.

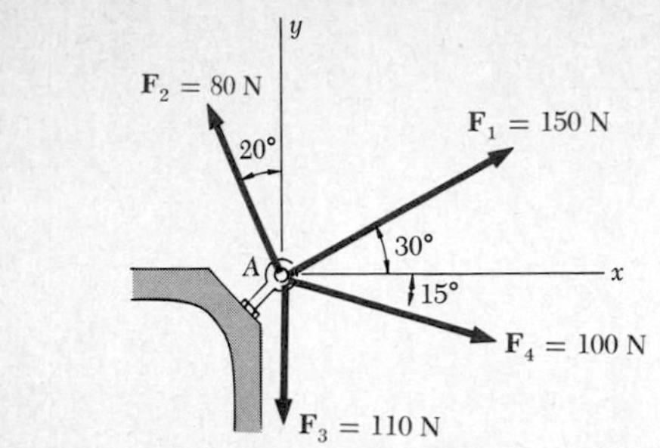

SAMPLE PROBLEM 2.3

Four forces act on bolt A as shown. Determine the resultant of the forces on the bolt.

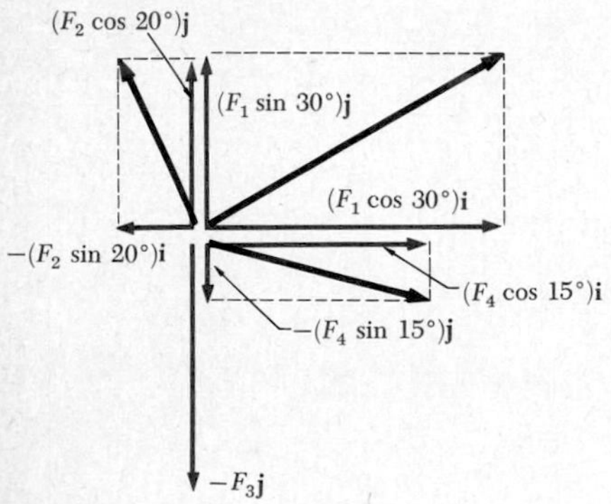

Solution. The x and y components of each force are determined by trigonometry as shown and are entered in the table below. According to the convention adopted in Sec. 2.7, the scalar number representing a force component is positive if the force component has the same sense as the corresponding coordinate axis. Thus, x components acting to the right and y components acting upward are represented by positive numbers.

Force	Magnitude, N	x Component, N	y Component, N
$\mathbf{F}_1$	150	+129.9	+75.0
$\mathbf{F}_2$	80	−27.4	+75.2
$\mathbf{F}_3$	110	0	−110.0
$\mathbf{F}_4$	100	+96.6	−25.9
		$R_x = +199.1$	$R_y = +14.3$

Thus, the resultant $\mathbf{R}$ of the four forces is

$$\mathbf{R} = R_x\mathbf{i} + R_y\mathbf{j} \qquad \mathbf{R} = (199.1\text{ N})\mathbf{i} + (14.3\text{ N})\mathbf{j} \blacktriangleleft$$

The magnitude and direction of the resultant may now be determined. From the triangle shown, we have

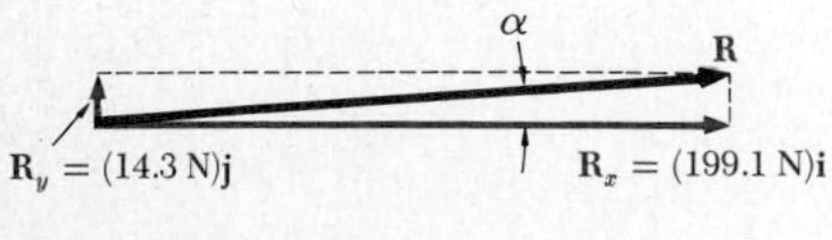

$$\tan\alpha = \frac{R_y}{R_x} = \frac{14.3\text{ N}}{199.1\text{ N}} \qquad \alpha = 4.1°$$

$$R = \frac{14.3\text{ N}}{\sin\alpha} = 199.6\text{ N} \qquad \mathbf{R} = 199.6\text{ N} \measuredangle 4.1° \blacktriangleleft$$

With a calculator, the last computation may be facilitated if the value of R_y is stored when originally entered; it may then be recalled to be divided by $\sin\alpha$. (Also see the footnote on p. 26.)

Problems

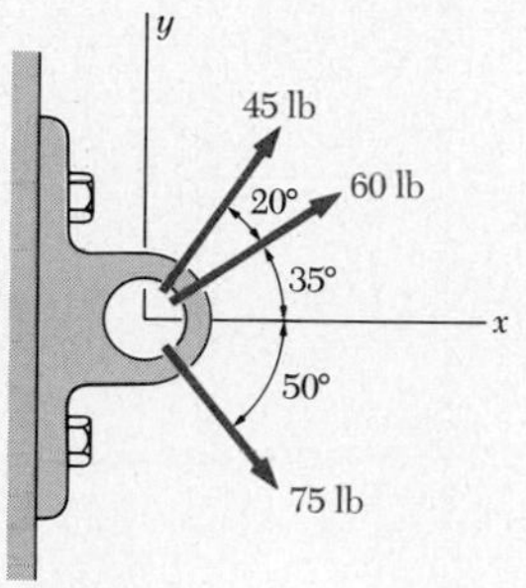

Fig. P2.16

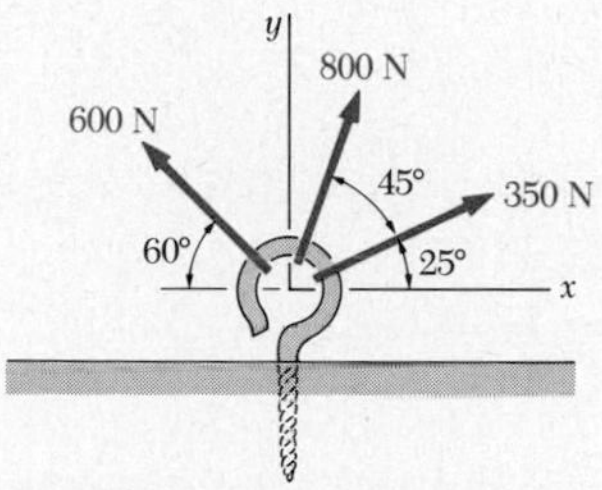

Fig. P2.17

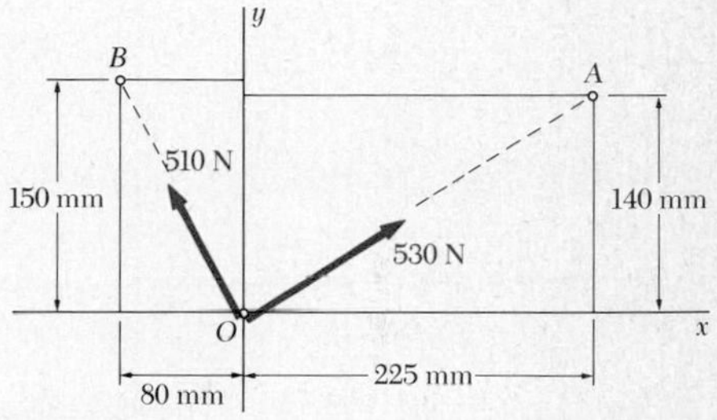

Fig. P2.18

2.16 through 2.19 Determine the x and y components of each of the forces shown.

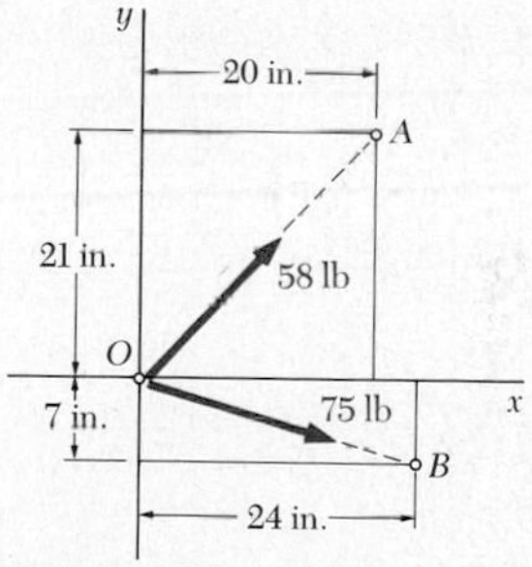

Fig. P2.19

2.20 Member CB of the vise shown exerts on block B a force $\mathbf{P}$ directed along line CB. Knowing that $\mathbf{P}$ must have a 200-lb horizontal component, determine (*a*) the magnitude of the force $\mathbf{P}$, (*b*) its vertical component.

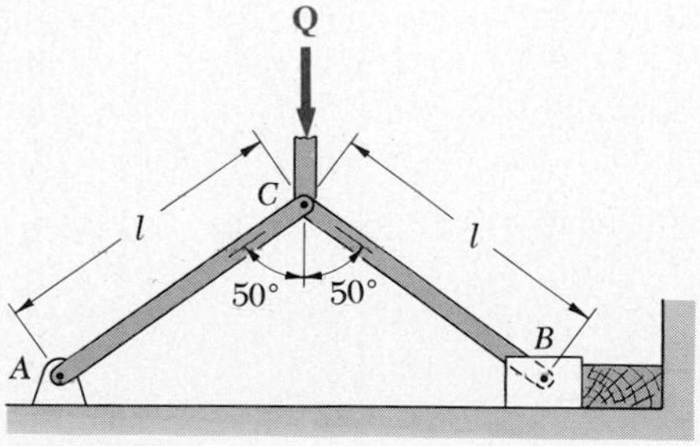

Fig. P2.20

2.21 The hydraulic cylinder GE exerts on member DF a force $\mathbf{P}$ directed along line GE. Knowing that $\mathbf{P}$ must have a 600-N component perpendicular to member DF, determine (*a*) the magnitude of the force $\mathbf{P}$, (*b*) its component parallel to DF.

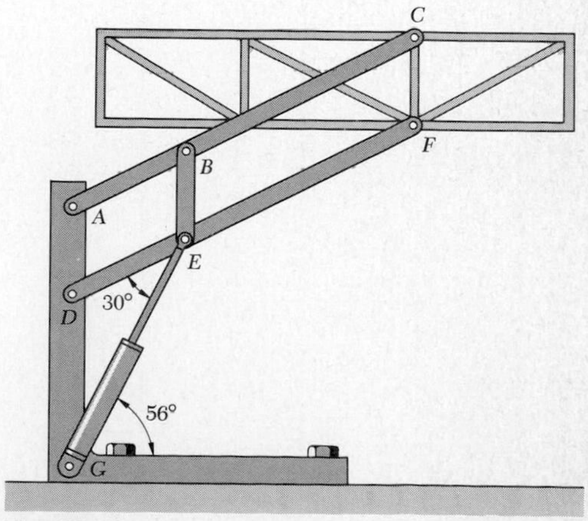

Fig. P2.21

2.22 The tension in the telephone-pole guy wire is 370 lb. Determine the horizontal and vertical components of the force exerted on the anchor at C.

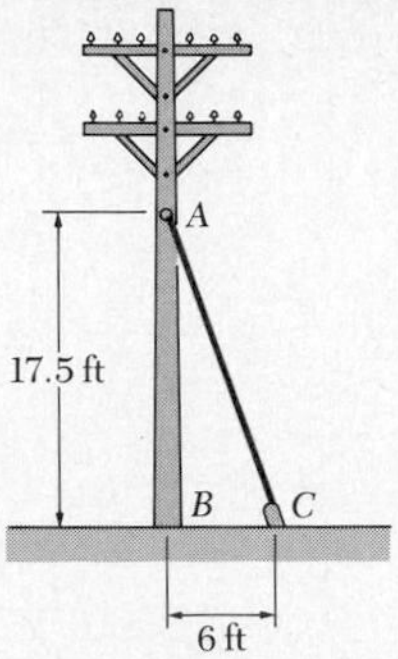

Fig. P2.22

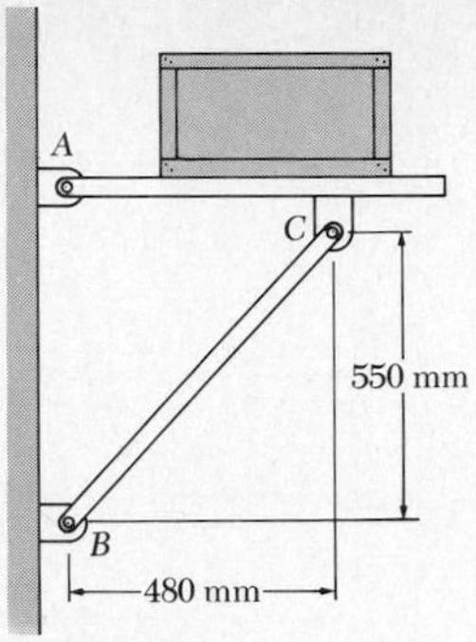

Fig. P2.23

2.23 The compression member BC exerts on the pin at C a 365-N force directed along line BC. Determine the horizontal and vertical components of that force.

2.24 Using x and y components, solve Prob. 2.2.

2.25 Using x and y components, solve Prob. 2.1.

2.26 Determine the resultant of the three forces of Prob. 2.17.

2.27 Determine the resultant of the three forces of Prob. 2.16.

2.28 Determine the resultant of the two forces of Prob. 2.18.

2.29 Determine the resultant of the two forces of Prob. 2.19.

2.30 Two cables which have known tensions are attached to the top of pylon AB. A third cable AC is used as a guy wire. Determine the tension in AC, knowing that the resultant of the forces exerted at A by the three cables must be vertical.

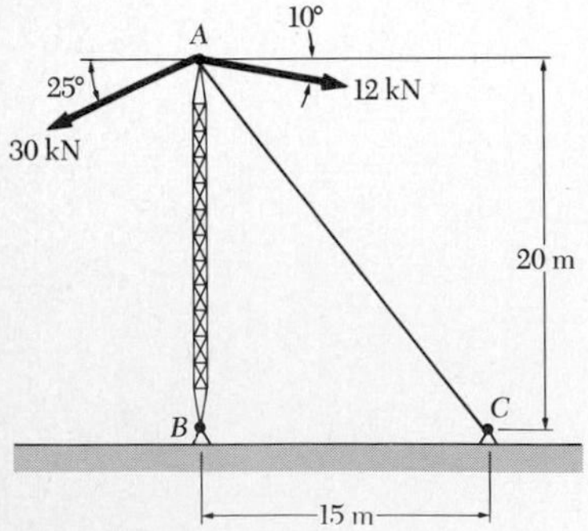

Fig. P2.30

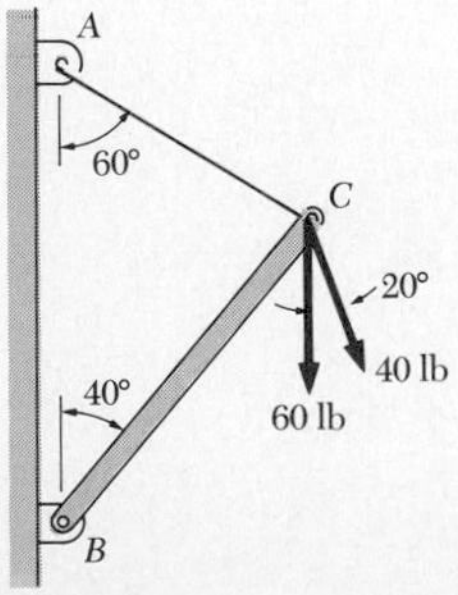

Fig. P2.31

2.31 Two loads are applied as shown to the end C of boom BC. Determine the tension in cable AC, knowing that the resultant of the three forces exerted at C must be directed along BC.

2.32 A hoist trolley is subjected to the three forces shown. Determine (*a*) the value of the angle α for which the resultant of the three forces is vertical, (*b*) the corresponding magnitude of the resultant.

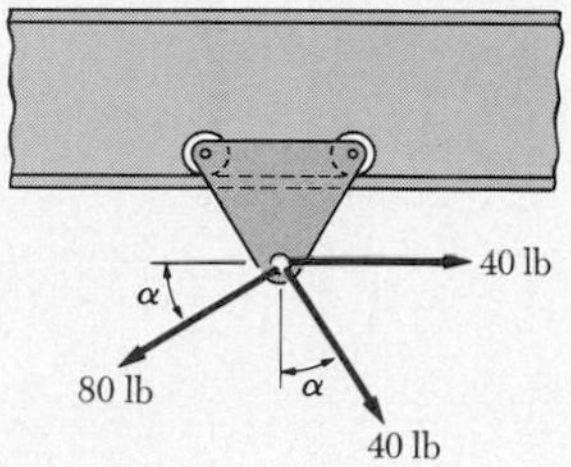

Fig. P2.32

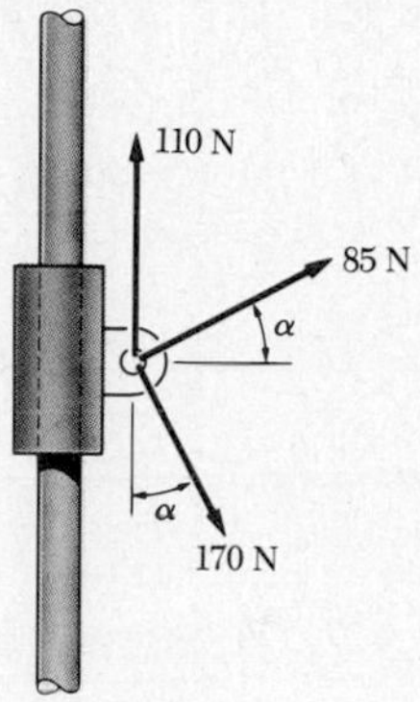

Fig. P2.33

2.33 A collar which may slide on a vertical rod is subjected to the three forces shown. Determine (*a*) the value of the angle α for which the resultant of the three forces is horizontal, (*b*) the corresponding magnitude of the resultant.

2.9. Equilibrium of a Particle. In the preceding sections, we discussed the methods for determining the resultant of several forces acting on a particle. Although it has not occurred in any of the problems considered so far, it is quite possible for the resultant to be zero. In such a case, the net effect of the given forces is zero, and the particle is said to be in equilibrium. We thus have the following definition: *When the resultant of all the forces acting on a particle is zero, the particle is in equilibrium.*

A particle which is acted upon by two forces will be in equilibrium if the two forces have the same magnitude, same line of action, and opposite sense. The resultant of the two forces is then zero. Such a case is shown in Fig. 2.26.

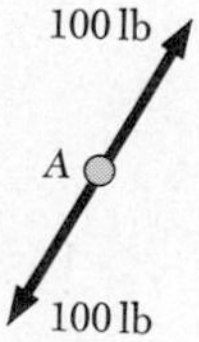

Fig. 2.26

Another case of equilibrium of a particle is represented in Fig. 2.27, where four forces are shown acting on A. In Fig. 2.28, the resultant of the given forces is determined by the polygon rule. Starting from point O with $\mathbf{F}_1$ and arranging the forces in tip-to-tail fashion, we find that the tip of $\mathbf{F}_4$ coincides with the starting point O. Thus the resultant $\mathbf{R}$ of the given system of forces is zero, and the particle is in equilibrium.

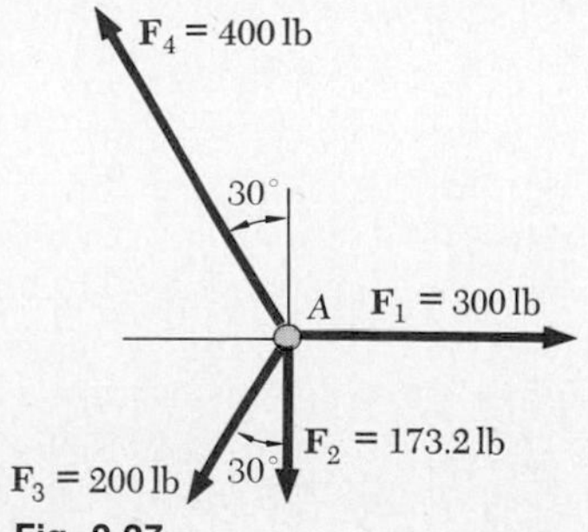

Fig. 2.27

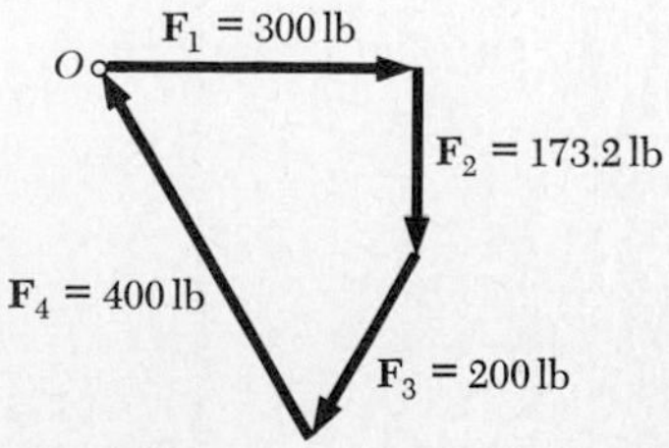

Fig. 2.28

The closed polygon drawn in Fig. 2.28 provides a *graphical* expression of the equilibrium of A. To express *algebraically* the conditions for the equilibrium of a particle, we write

$$\mathbf{R} = \Sigma \mathbf{F} = 0 \tag{2.14}$$

Resolving each force **F** into rectangular components, we have

$$\Sigma(F_x \mathbf{i} + F_y \mathbf{j}) = 0 \qquad \text{or} \qquad (\Sigma F_x)\mathbf{i} + (\Sigma F_y)\mathbf{j} = 0$$

We conclude that the necessary and sufficient conditions for the equilibrium of a particle are

$$\Sigma F_x = 0 \qquad \Sigma F_y = 0 \tag{2.15}$$

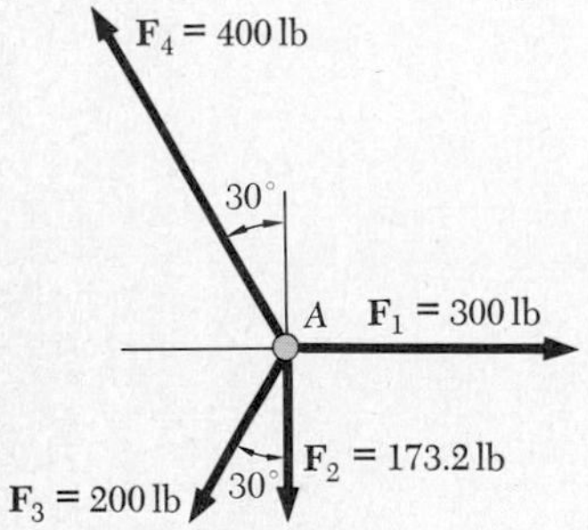

Fig. 2.27 ***(repeated)***

Returning to the particle shown in Fig. 2.27 we check that the equilibrium conditions are satisfied. We write

$$\begin{aligned}\Sigma F_x &= 300 \text{ lb} - (200 \text{ lb}) \sin 30° - (400 \text{ lb}) \sin 30° \\ &= 300 \text{ lb} - 100 \text{ lb} - 200 \text{ lb} = 0\end{aligned}$$

$$\begin{aligned}\Sigma F_y &= -173.2 \text{ lb} - (200 \text{ lb}) \cos 30° + (400 \text{ lb}) \cos 30° \\ &= -173.2 \text{ lb} - 173.2 \text{ lb} + 346.4 \text{ lb} = 0\end{aligned}$$

2.10. Newton's First Law of Motion. In the latter part of the seventeenth century, Sir Isaac Newton formulated three fundamental laws upon which the science of mechanics is based. The first of these laws can be stated as follows:

If the resultant force acting on a particle is zero, the particle will remain at rest (if originally at rest) or will move with constant speed in a straight line (if originally in motion).

From this law and from the definition of equilibrium given in Sec. 2.9, it is seen that a particle in equilibrium either is at rest or is moving in a straight line with constant speed. In the following section, various problems concerning the equilibrium of a particle will be considered.

2.11. Problems Involving the Equilibrium of a Particle. Free-Body Diagram. In practice, a problem in engineering mechanics is derived from an actual physical situation. A sketch showing the physical conditions of the problem is known as a *space diagram.*

The methods of analysis discussed in the preceding sections apply to a system of forces acting on a particle. A large number of problems involving actual structures, however, may be reduced to problems concerning the equilibrium of a particle. This is done by choosing a significant particle and drawing a separate diagram showing this particle and all the forces acting on it. Such a diagram is called a *free-body diagram.*

As an example, consider the 75-kg crate shown in the space diagram of Fig. 2.29*a*. This crate was lying between two buildings, and it is now being lifted onto a truck, which will remove it. The crate is supported by a vertical cable, which is joined at A to two ropes which pass over pulleys attached to the buildings at B and C. It is desired to determine the tension in each of the ropes AB and AC.

In order to solve this problem, a free-body diagram must be drawn, showing a particle in equilibrium. Since we are interested in the rope tensions, the free-body diagram should include at least one of these tensions and, if possible, both tensions. Point A is seen to be a good free body for this problem. The free-body diagram of point A is shown in Fig. 2.29*b*. It shows point A and the forces exerted on A by the vertical cable and the two ropes. The force exerted by the cable is directed downward and is equal to the weight W of the crate. Recalling Eq. (1.4), we write

$$W = mg = (75 \text{ kg})(9.81 \text{ m/s}^2) = 736 \text{ N}$$

and indicate this value in the free-body diagram. The forces exerted by the two ropes are not known. Since they are respectively equal in magnitude to the tension in rope AB and rope AC, we denote them by $\mathbf{T}_{AB}$ and $\mathbf{T}_{AC}$ and draw them away from A in the directions shown in the space diagram. No other detail is included in the free-body diagram.

Since point A is in equilibrium, the three forces acting on it must form a closed triangle when drawn in tip-to-tail fashion. This *force triangle* has been drawn in Fig. 2.29*c*. The values T_{AB} and T_{AC} of the tension in the ropes may be found graphically if the triangle is drawn to scale, or they may be found by trigonometry. If the latter method of solution is chosen, we use the law of sines and write

$$\frac{T_{AB}}{\sin 60°} = \frac{T_{AC}}{\sin 40°} = \frac{736 \text{ N}}{\sin 80°}$$

$$T_{AB} = 647 \text{ N} \qquad T_{AC} = 480 \text{ N}$$

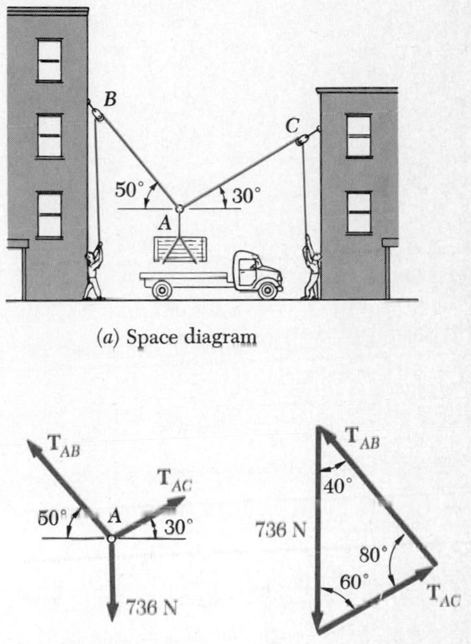

(*a*) Space diagram

(*b*) Free-body diagram (*c*) Force triangle

Fig. 2.29

When a particle is in *equilibrium under three forces,* the problem may always be solved by drawing a force triangle. When a particle is in *equilibrium under more than three forces,* the problem may be solved graphically by drawing a force polygon. If an analytic solution is desired, the *equations of equilibrium* given in Sec. 2.9 should be solved:

$$\Sigma F_x = 0 \qquad \Sigma F_y = 0 \tag{2.15}$$

These equations may be solved for no more than *two unknowns;* similarly, the force triangle used in the case of equilibrium under three forces may be solved for two unknowns.

The more common types of problems are those where the two unknowns represent (1) the two components (or the magnitude and direction) of a single force, (2) the magnitude of two forces each of known direction. Problems involving the determination of the maximum or minimum value of the magnitude of a force are also encountered (see Probs. 2.44 and 2.45).

SAMPLE PROBLEM 2.4

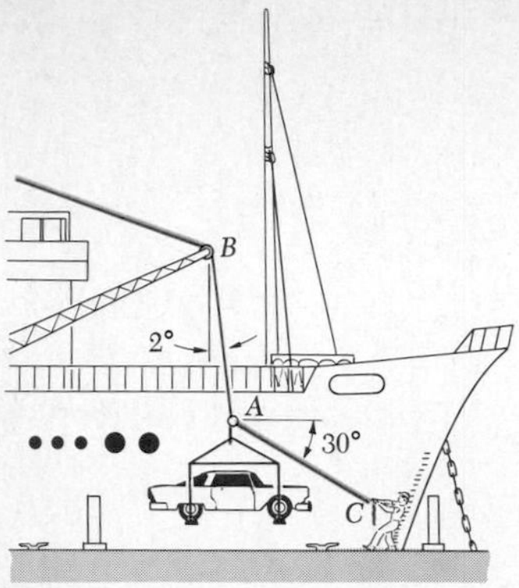

In a ship-unloading operation, a 3500-lb automobile is supported by a cable. A rope is tied to the cable at A and pulled in order to center the automobile over its intended position. The angle between the cable and the vertical is 2°, while the angle between the rope and the horizontal is 30°. What is the tension in the rope?

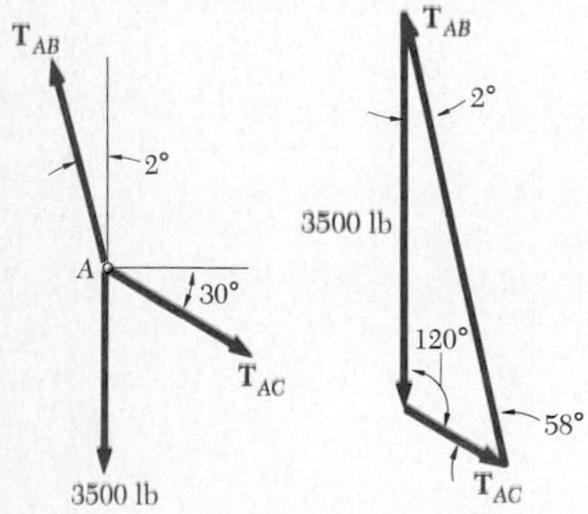

Solution. Point A is chosen as a free body, and the complete free-body diagram is drawn. T_{AB} is the tension in the cable AB, and T_{AC} is the tension in the rope.

Equilibrium Condition. Since only three forces act on the free body, we draw a force triangle to express that it is in equilibrium. Using the law of sines, we write

$$\frac{T_{AB}}{\sin 120°} = \frac{T_{AC}}{\sin 2°} = \frac{3500 \text{ lb}}{\sin 58°}$$

With a calculator, we first compute and store the value of the last quotient. Multiplying this value successively by sin 120° and sin 2°, we obtain

$$T_{AB} = 3570 \text{ lb} \qquad T_{AC} = 144 \text{ lb} \quad ◄$$

SAMPLE PROBLEM 2.5

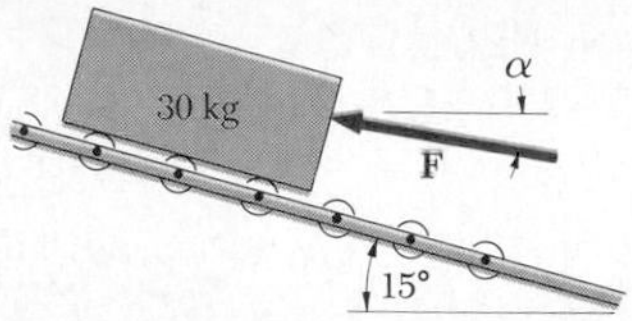

Determine the magnitude and direction of the smallest force **F** which will maintain the package shown in equilibrium. Note that the force exerted by the rollers on the package is perpendicular to the incline.

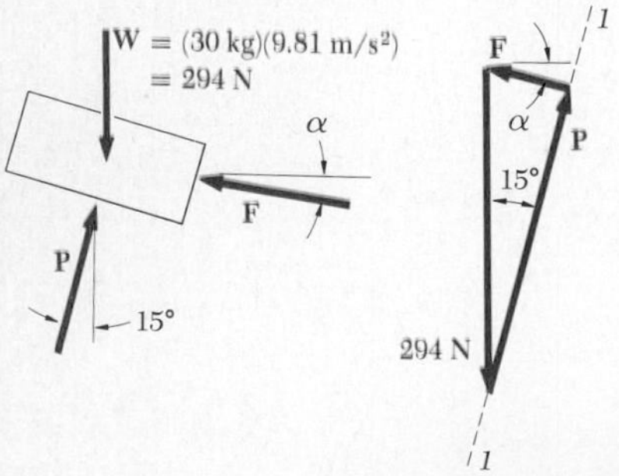

Solution. We choose the package as a free body, assuming that it may be treated as a particle. We draw the corresponding free-body diagram.

Equilibrium Condition. Since only three forces act on the free body, we draw a force triangle to express that it is in equilibrium. Line *1-1* represents the known direction of **P**. In order to obtain the minimum value of the force **F**, we choose the direction of **F** perpendicular to that of **P**. From the geometry of the triangle obtained, we find

$$F = (294 \text{ N}) \sin 15° = 76.1 \text{ N} \qquad \alpha = 15°$$

$$\mathbf{F} = 76.1 \text{ N} \measuredangle 15° \quad ◄$$

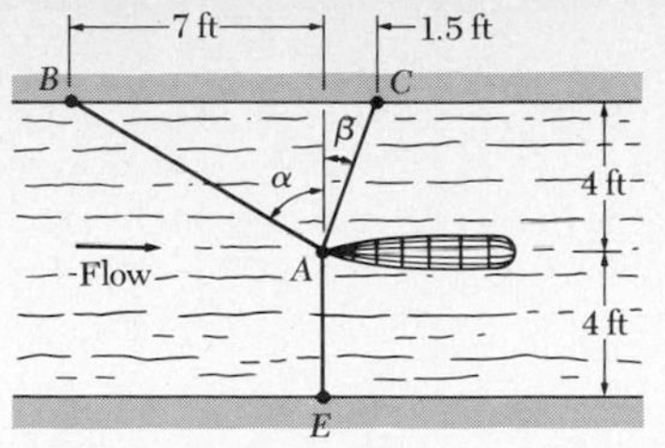

SAMPLE PROBLEM 2.6

As part of the design of a new sailboat, it is desired to determine the drag force which may be expected at a given speed. To do so, a model of the proposed hull is placed in a test channel and three cables are used to keep its bow on the centerline of the channel. Dynamometer readings indicate that for a given speed, the tension is 40 lb in cable AB and 60 lb in cable AE. Determine the drag force exerted on the hull and the tension in cable AC.

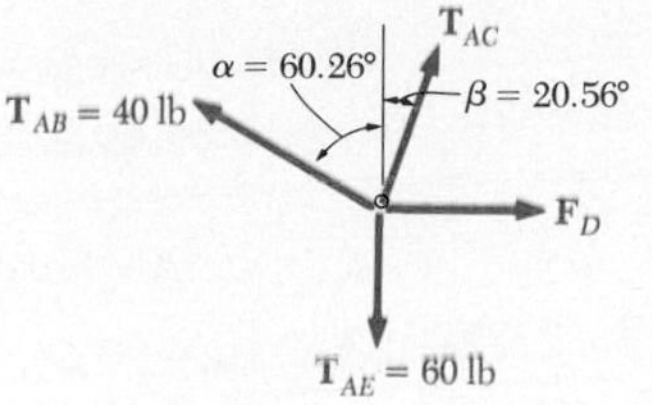

Solution. First, the angles α and β defining the direction of cables AB and AC are determined. We write

$$\tan\alpha = \frac{7\text{ ft}}{4\text{ ft}} = 1.75 \qquad \tan\beta = \frac{1.5\text{ ft}}{4\text{ ft}} = 0.375$$

$$\alpha = 60.26^\circ \qquad \beta = 20.56^\circ$$

Choosing the hull as a free body, we draw the free-body diagram shown. It includes the forces exerted by the three cables on the hull, as well as the drag force $\mathbf{F}_D$ exerted by the flow.

Equilibrium Condition. We express that the hull is in equilibrium by writing that the resultant of all forces is zero:

$$\mathbf{R} = \mathbf{T}_{AB} + \mathbf{T}_{AC} + \mathbf{T}_{AE} + \mathbf{F}_D = 0 \tag{1}$$

Since more than three forces are involved, we shall resolve the forces into x and y components:

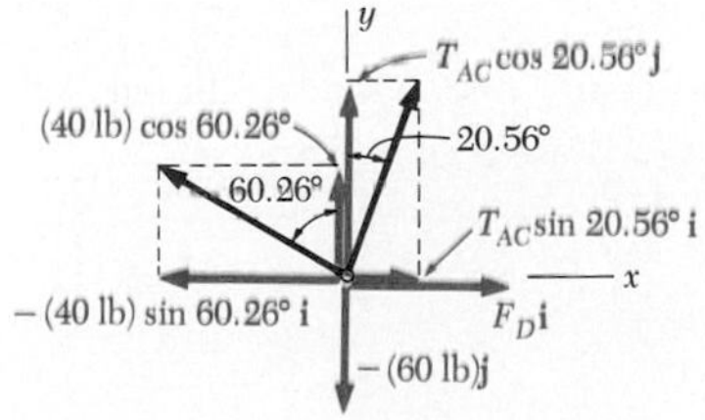

$$\begin{aligned}
\mathbf{T}_{AB} &= -(40\text{ lb})\sin 60.26^\circ\mathbf{i} + (40\text{ lb})\cos 60.26^\circ\mathbf{j} \\
&= -(34.73\text{ lb})\mathbf{i} + (19.84\text{ lb})\mathbf{j} \\
\mathbf{T}_{AC} &= T_{AC}\sin 20.56^\circ\mathbf{i} + T_{AC}\cos 20.56^\circ\mathbf{j} \\
&= 0.3512T_{AC}\mathbf{i} + 0.9363T_{AC}\mathbf{j} \\
\mathbf{T}_{AE} &= -(60\text{ lb})\mathbf{j} \\
\mathbf{F}_D &= F_D\mathbf{i}
\end{aligned}$$

Substituting the expressions obtained into Eq. (1) and factoring the unit vectors **i** and **j**, we have

$$(-34.73\text{ lb} + 0.3512T_{AC} + F_D)\mathbf{i} + (19.84\text{ lb} + 0.9363T_{AC} - 60\text{ lb})\mathbf{j} = 0$$

This equation will be satisfied if, and only if, the coefficients of **i** and **j** are equal to zero. We thus obtain the following two equilibrium equations, which express, respectively, that the sum of the x components and the sum of the y components of the given forces must be zero.

$$(\Sigma F_x = 0:) \qquad -34.73\text{ lb} + 0.3512T_{AC} + F_D = 0 \tag{2}$$

$$(\Sigma F_y = 0:) \qquad 19.84\text{ lb} + 0.9363T_{AC} - 60\text{ lb} = 0 \tag{3}$$

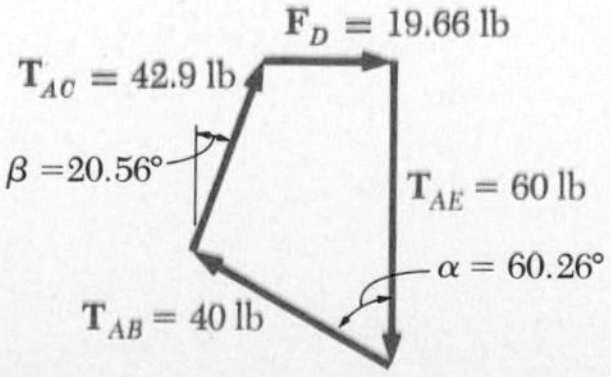

From Eq. (3) we find $\qquad T_{AC} = +42.9\text{ lb}$ ◄

and, substituting this value into Eq. (2), $\qquad F_D = +19.66\text{ lb}$ ◄

In drawing the free-body diagram, we assumed a sense for each unknown force. A positive sign in the answer indicates that the assumed sense is correct. The complete force polygon may be drawn to check the results.

Problems

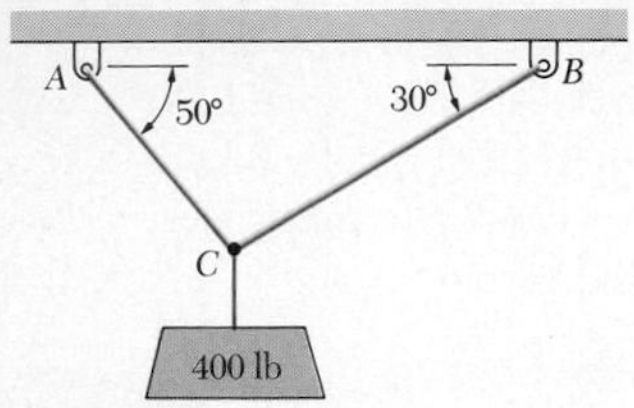

Fig. P2.34

2.34 through 2.37 Two cables are tied together at C and loaded as shown. Determine the tension in AC and BC.

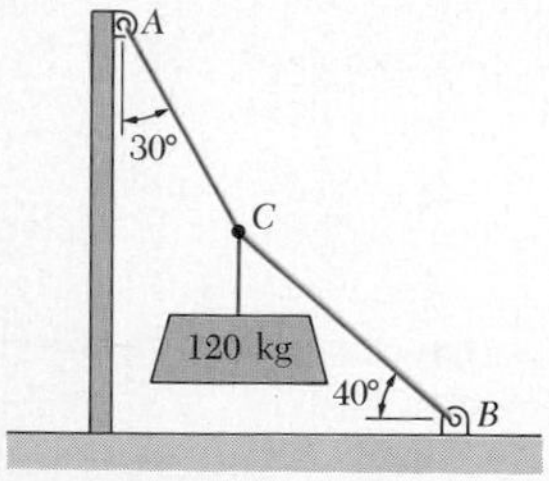

Fig. P2.35

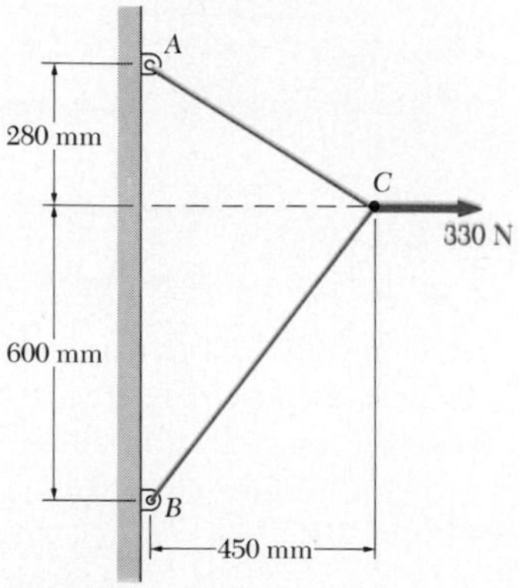

Fig. P2.36

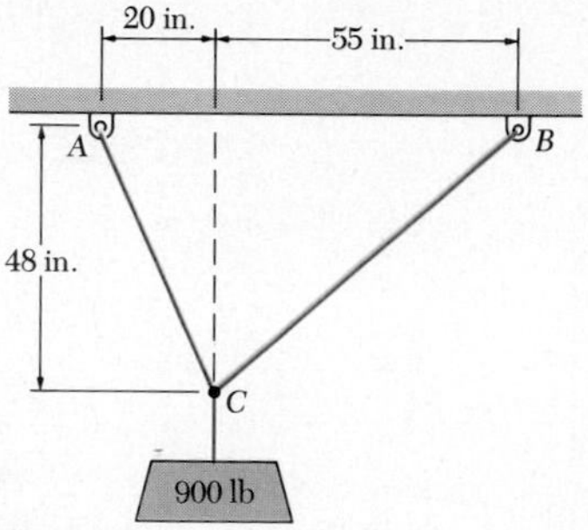

Fig. P2.37

2.38 Two cables are tied together at C and loaded as shown. Knowing that $P = 400$ N and $\alpha = 75°$, determine the tension in AC and BC.

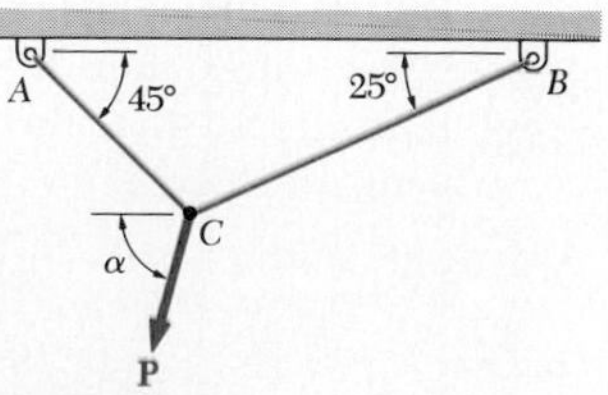

Fig. P2.38

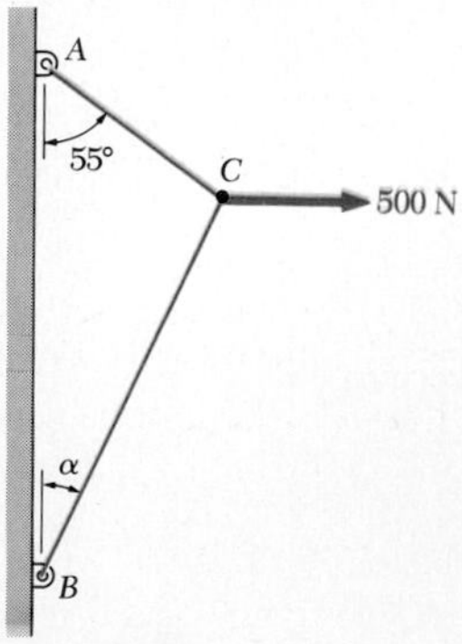

Fig. P2.39

2.39 Two cables are tied together at C and loaded as shown. Knowing that $\alpha = 25°$, determine the tension in AC and BC.

2.40 Two forces **P** and **Q** of magnitude $P = 600$ lb and $Q = 800$ lb are applied as shown to an aircraft connection. Knowing that the connection is in equilibrium, determine the tension in rods A and B.

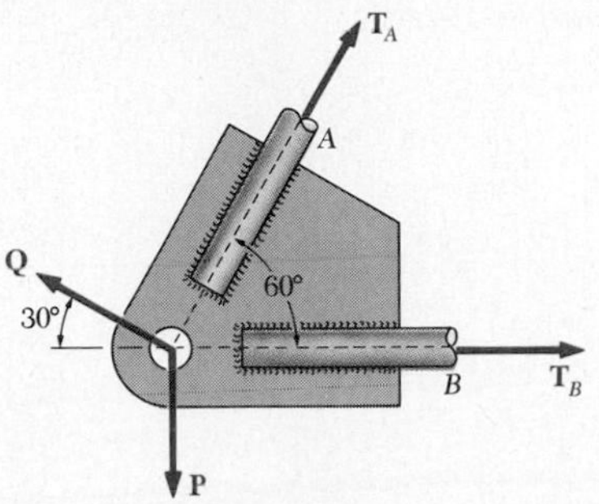

Fig. P2.40 and P2.41

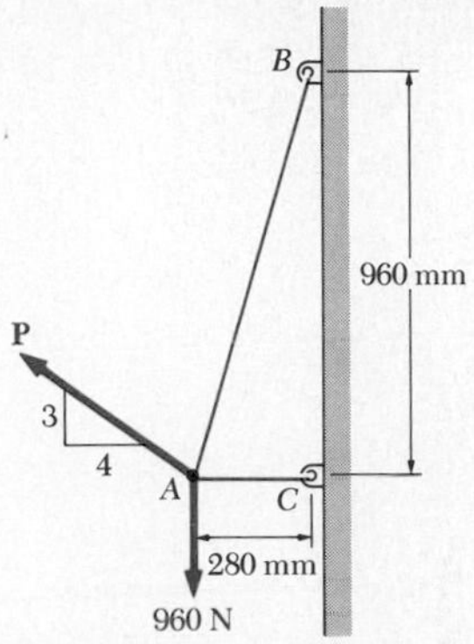

Fig. P2.42 and P2.43

2.41 Two forces **P** and **Q** are applied as shown to an aircraft connection. Determine the magnitude of each force, knowing that the connection is in equilibrium and that the tensions in rods A and B are $T_A = 240$ lb and $T_B = 500$ lb.

2.42 Two cables are tied together at A and loaded as shown. Knowing that $P = 640$ N, determine the tension in each cable.

2.43 Two cables are tied together at A and loaded as shown. Determine the range of values of P for which both cables remain taut.

2.44 For the cables of Prob. 2.39, find the value of α for which the tension is as small as possible (*a*) in cable BC, (*b*) in both cables simultaneously. In each case determine the tension in both cables.

2.45 For the cables of Prob. 2.38, it is known that the maximum allowable tension is 300 N in cable AC and 400 N in cable BC. Determine (*a*) the maximum force **P** that may be applied at C, (*b*) the corresponding value of α.

2.46 The 15-lb collar A may slide on a frictionless vertical rod and is connected as shown to a 17-lb counterweight C. Determine the value of h for which the system is in equilibrium.

2.47 A movable bin and its contents weigh 960 lb. Determine the shortest chain sling ACB which may be used to lift the loaded bin if the tension in the chain is not to exceed 730 lb.

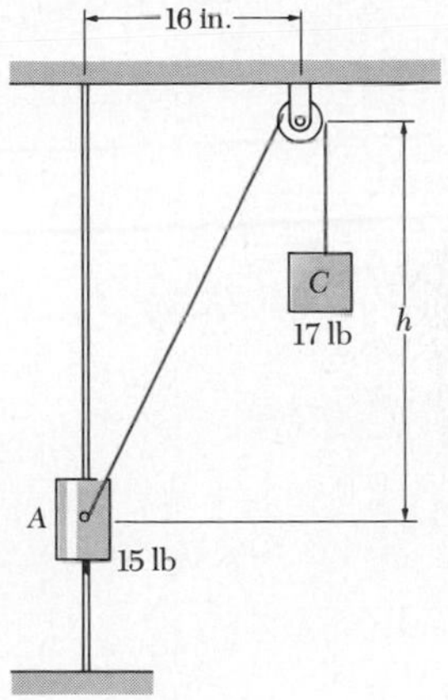

Fig. P2.46

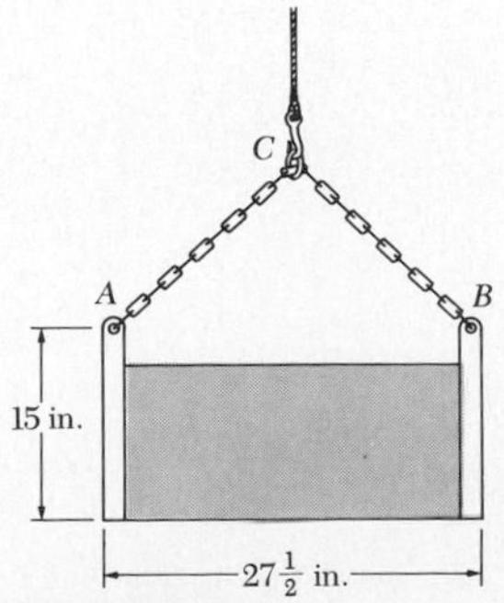

Fig. P2.47

2.48 A 300-kg crate is supported by several rope-and-pulley arrangements as shown. Determine for each arrangement the tension in the rope. (The tension in the rope is the same on each side of a simple pulley. This can be proved by the methods of Chap. 4.)

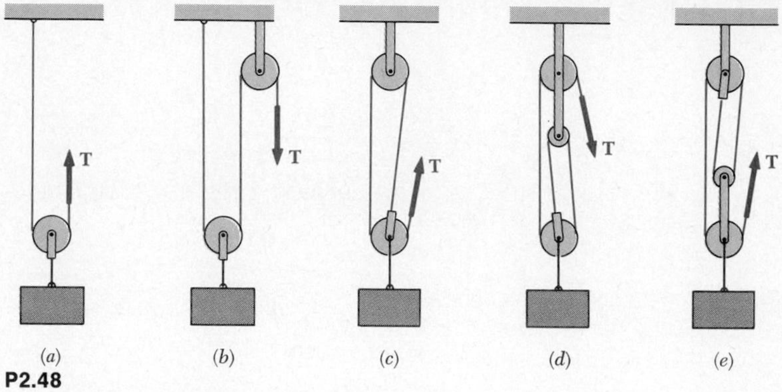

Fig. P2.48

2.49 Solve parts *b* and *d* of Prob. 2.48 assuming that the free end of the rope is attached to the crate.

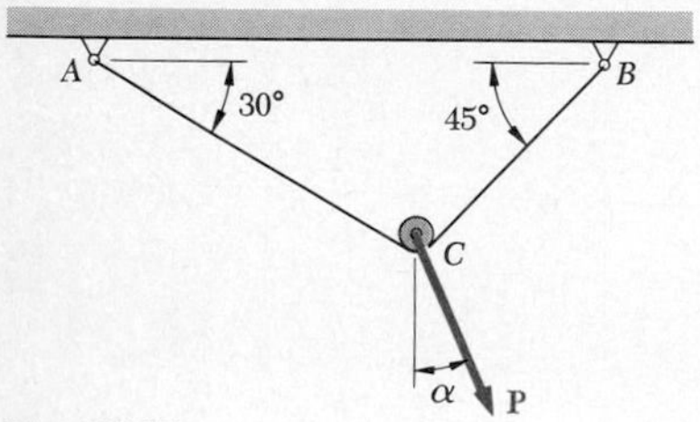

Fig. P2.50

2.50 The force **P** is applied to a small wheel which rolls on the cable *ACB*. Knowing that the tension in both parts of the cable is 750 N, determine the magnitude and direction of **P**.

2.51 A 600-lb crate is to be supported by the rope-and-pulley arrangement shown. Determine the magnitude and direction of the force **F** which should be exerted on the free end of the rope.

*** 2.52** The collar *A* may slide freely on the horizontal frictionless rod. The spring attached to the collar has a constant of 10 lb/in. and is undeformed when the collar is directly below support *B*. Determine the magnitude of the force **P** required to maintain equilibrium when (*a*) $c = 9$ in., (*b*) $c = 16$ in.

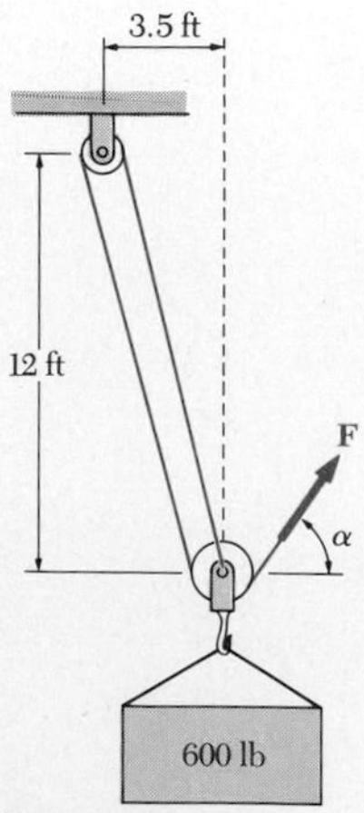

Fig. P2.51

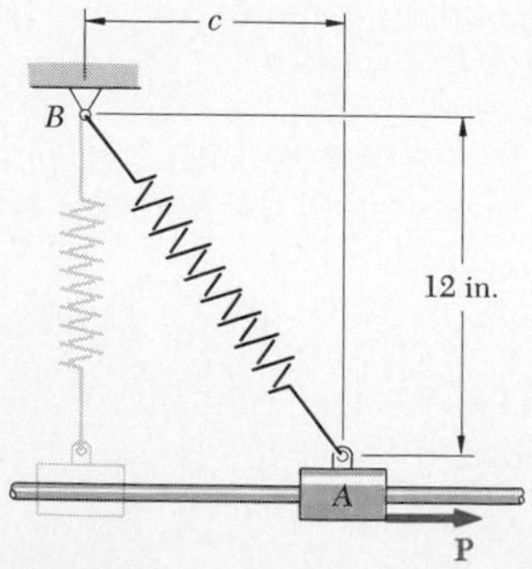

Fig. P2.52

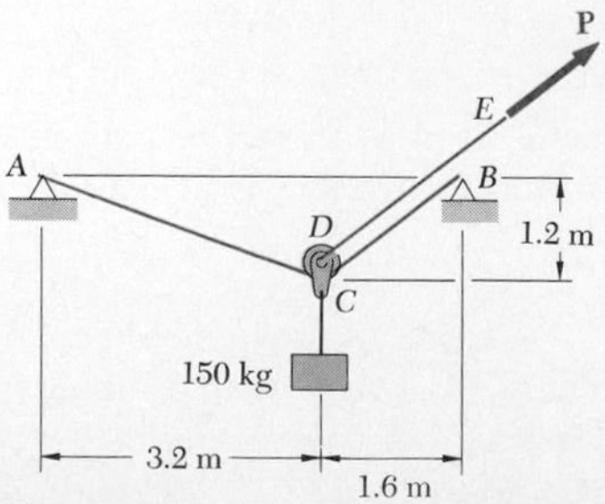

Fig. P2.53

*** 2.53** A 150-kg block is attached to a small pulley which may roll on the cable *ACB*. The pulley and load are held in the position shown by a second cable *DE* which is parallel to the portion *CB* of the main cable. Determine (*a*) the tension in cable *ACB*, (*b*) the tension in cable *DE*. Neglect the radius of the pulleys and the weight of the cables.

FORCES IN SPACE

2.12. Rectangular Components of a Force in Space. The problems considered in the first part of this chapter involved only two dimensions; they could be formulated and solved in a single plane. In this section and in the remaining sections of the chapter, we shall discuss problems involving the three dimensions of space.

Consider a force **F** acting at the origin *O* of the system of rectangular coordinates *x*, *y*, *z*. To define the direction of **F**, we may draw the vertical plane *OBAC* containing **F** and shown in Fig. 2.30*a*. This plane passes through the vertical *y* axis; its orientation is defined by the angle ϕ it forms with the *xy* plane, while the direction of **F** within the plane is defined by the angle θ_y that **F** forms with the *y* axis. The force **F** may be resolved into a vertical component $\mathbf{F}_y$ and a horizontal component $\mathbf{F}_h$; this operation, shown in Fig. 2.30*b*, is carried out in plane *OBAC* according to the rules developed in the first part of the chapter. The corresponding scalar components are

$$F_y = F\cos\theta_y \qquad F_h = F\sin\theta_y \tag{2.16}$$

But $\mathbf{F}_h$ may be resolved into two rectangular components $\mathbf{F}_x$ and $\mathbf{F}_z$ along the *x* and *z* axes, respectively. This operation, shown in Fig. 2.30*c*, is carried out in the *xz* plane. We obtain the following expressions for the corresponding scalar components:

$$\begin{aligned} F_x &= F_h\cos\phi = F\sin\theta_y\cos\phi \\ F_z &= F_h\sin\phi = F\sin\theta_y\sin\phi \end{aligned} \tag{2.17}$$

The given force **F** has thus been resolved into three rectangular vector components $\mathbf{F}_x$, $\mathbf{F}_y$, $\mathbf{F}_z$, directed along the three coordinate axes.

Applying the Pythagorean theorem to the triangles *OAB* and *OCD* of Fig. 2.30, we write

$$\begin{aligned} F^2 &= (OA)^2 = (OB)^2 + (BA)^2 = F_y^2 + F_h^2 \\ F_h^2 &= (OC)^2 = (OD)^2 + (DC)^2 = F_x^2 + F_z^2 \end{aligned}$$

Eliminating F_h^2 from these two equations and solving for *F*, we obtain the following relation between the magnitude of **F** and its rectangular scalar components:

$$F = \sqrt{F_x^2 + F_y^2 + F_z^2} \tag{2.18}$$

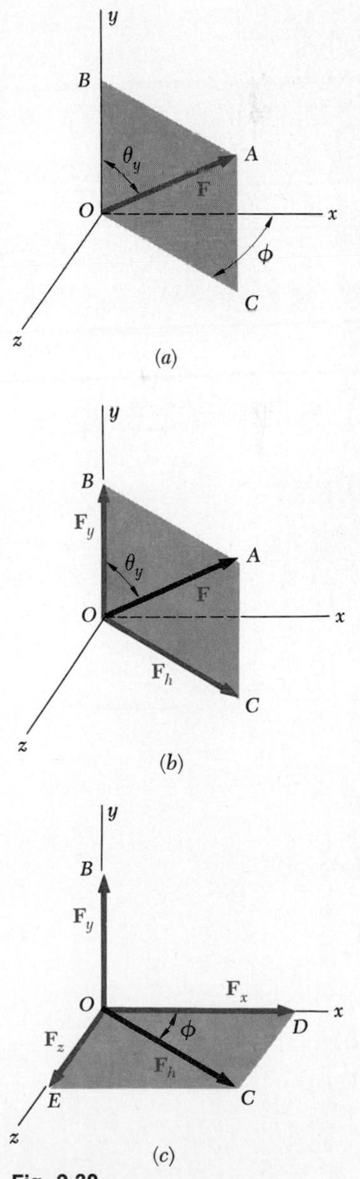

Fig. 2.30

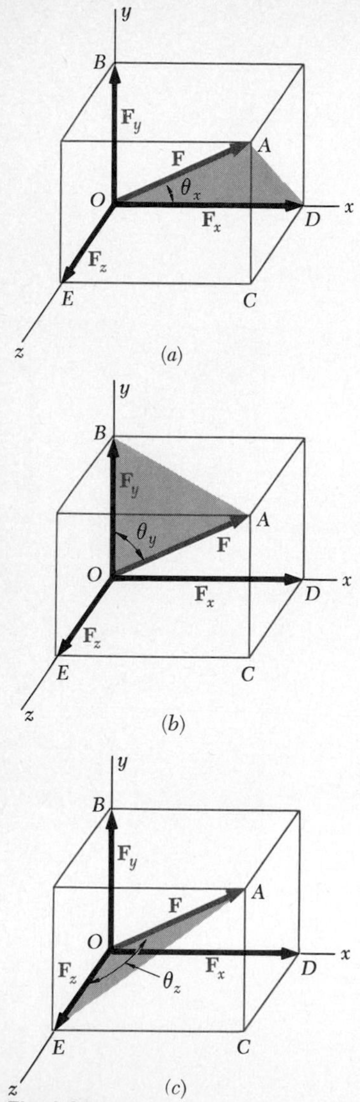

Fig. 2.31

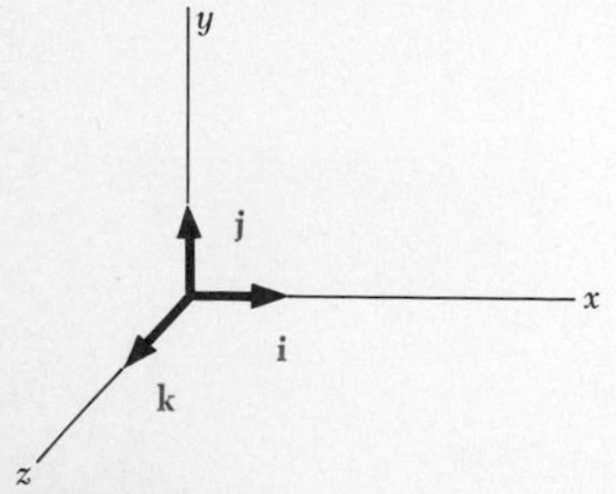

Fig. 2.32

The relationship existing between the force $\mathbf{F}$ and its three components $\mathbf{F}_x$, $\mathbf{F}_y$, $\mathbf{F}_z$ is more easily visualized if a "box" having $\mathbf{F}_x$, $\mathbf{F}_y$, $\mathbf{F}_z$ for edges is drawn as shown in Fig. 2.31. The force $\mathbf{F}$ is then represented by the diagonal OA of this box. Figure 2.31*b* shows the right triangle OAB used to derive the first of the formulas (2.16): $F_y = F\cos\theta_y$. In Fig. 2.31*a* and *c*, two other right triangles have also been drawn: OAD and OAE. These triangles are seen to occupy in the box positions comparable with that of triangle OAB. Denoting by θ_x and θ_z, respectively, the angles that $\mathbf{F}$ forms with the x and z axes, we may derive two formulas similar to $F_y = F\cos\theta_y$. We thus write

$$F_x = F\cos\theta_x \qquad F_y = F\cos\theta_y \qquad F_z = F\cos\theta_z \tag{2.19}$$

The three angles $\theta_x, \theta_y, \theta_z$ define the direction of the force $\mathbf{F}$; they are more commonly used for this purpose than the angles θ_y and ϕ introduced at the beginning of this section. The cosines of $\theta_x, \theta_y, \theta_z$ are known as the direction cosines of the force $\mathbf{F}$.

Introducing the unit vectors $\mathbf{i}$, $\mathbf{j}$, and $\mathbf{k}$, directed respectively along the x, y, and z axes (Fig. 2.32), we may express $\mathbf{F}$ in the form

$$\mathbf{F} = F_x\mathbf{i} + F_y\mathbf{j} + F_z\mathbf{k} \tag{2.20}$$

where the scalar components F_x, F_y, F_z are defined by the relations (2.19).

Example 1. A force of 500 N forms angles of 60, 45, and 120°, respectively, with the x, y, and z axes. Find the components F_x, F_y, and F_z of the force.

Substituting $F = 500$ N, $\theta_x = 60°$, $\theta_y = 45°$, $\theta_z = 120°$ into formulas (2.19), we write

$$\begin{aligned} F_x &= (500\text{ N})\cos 60° = +250\text{ N} \\ F_y &= (500\text{ N})\cos 45° = +354\text{ N} \\ F_z &= (500\text{ N})\cos 120° = -250\text{ N} \end{aligned}$$

Carrying into Eq. (2.20) the values obtained for the scalar components of $\mathbf{F}$, we have

$$\mathbf{F} = (250\text{ N})\mathbf{i} + (354\text{ N})\mathbf{j} - (250\text{ N})\mathbf{k}$$

As in the case of two-dimensional problems, the plus sign indicates that the component has the same sense as the corresponding axis, and the minus sign that it has the opposite sense.

The angle a force $\mathbf{F}$ forms with an axis should be measured from the positive side of the axis and will always be comprised between 0 and 180°. An angle θ_x smaller than 90° (acute) indicates that $\mathbf{F}$ (assumed attached to O) is on the same side of the yz plane as the positive x axis; $\cos\theta_x$ and F_x will then be positive. An angle θ_x larger than 90° (obtuse) would indicate that $\mathbf{F}$ is on the other side of the yz plane; $\cos\theta_x$ and F_x would then be negative. In Example 1 the angles θ_x and θ_y are acute, while θ_z is obtuse: consequently, F_x and F_y are positive, while F_z is negative.

Substituting into (2.20) the expressions obtained for F_x, F_y, F_z in (2.19), we write

$$\mathbf{F} = F(\cos\theta_x\mathbf{i} + \cos\theta_y\mathbf{j} + \cos\theta_z\mathbf{k}) \tag{2.21}$$

which shows that the force **F** may be expressed as the product of the scalar F and of the vector

$$\boldsymbol{\lambda} = \cos\theta_x\mathbf{i} + \cos\theta_y\mathbf{j} + \cos\theta_z\mathbf{k} \tag{2.22}$$

Clearly, the vector $\boldsymbol{\lambda}$ is a vector of magnitude equal to 1 and of the same direction as **F** (Fig. 2.33). We shall refer to $\boldsymbol{\lambda}$ as the *unit vector* along the line of action of **F**. It follows from (2.22) that the components of the unit vector $\boldsymbol{\lambda}$ are respectively equal to the direction cosines of the line of action of **F**:

$$\lambda_x = \cos\theta_x \qquad \lambda_y = \cos\theta_y \qquad \lambda_z = \cos\theta_z \tag{2.23}$$

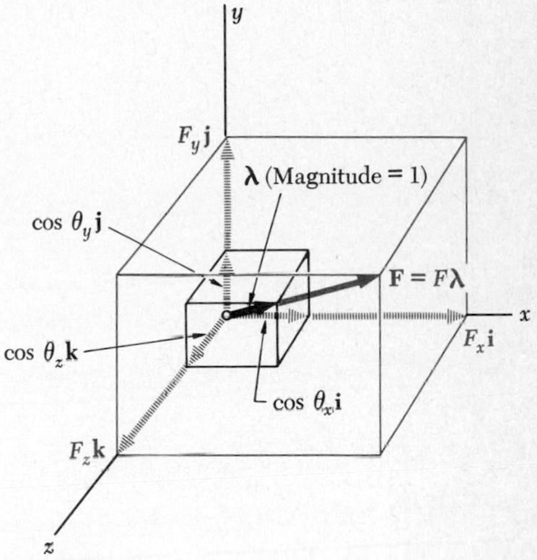

Fig. 2.33

We should observe that the values of the three angles θ_x, θ_y, θ_z are not independent. Expressing that the sum of the squares of the components of $\boldsymbol{\lambda}$ is equal to the square of its magnitude, we write

$$\lambda_x^2 + \lambda_y^2 + \lambda_z^2 = 1$$

or, substituting for λ_x, λ_y, λ_z from (2.23),

$$\cos^2\theta_x + \cos^2\theta_y + \cos^2\theta_z = 1 \tag{2.24}$$

In Example 1, for instance, once the values $\theta_x = 60°$ and $\theta_y = 45°$ have been selected, the value of θ_z *must* be equal to 60° or 120° in order to satisfy identity (2.24).

When the components F_x, F_y, F_z of a force **F** are given, the magnitude F of the force is obtained from (2.18).† The relations (2.19) may then be solved for the direction cosines,

$$\cos\theta_x = \frac{F_x}{F} \qquad \cos\theta_y = \frac{F_y}{F} \qquad \cos\theta_z = \frac{F_z}{F} \tag{2.25}$$

and the angles θ_x, θ_y, θ_z characterizing the direction of **F** may be found.

Example 2. A force **F** has the components $F_x = 20$ lb, $F_y = -30$ lb, $F_z = 60$ lb. Determine its magnitude F and the angles θ_x, θ_y, θ_z it forms with the axes of coordinates.

From formula (2.18) we obtain†

$$F = \sqrt{F_x^2 + F_y^2 + F_z^2}$$
$$= \sqrt{(20\text{ lb})^2 + (-30\text{ lb})^2 + (60\text{ lb})^2}$$
$$= \sqrt{4900}\text{ lb} = 70\text{ lb}$$

† With a calculator programmed to convert rectangular coordinates into polar coordinates, the following procedure will be found more expeditious for computing F: First determine F_h from its two rectangular components F_x and F_z (Fig. 2.30*c*), then determine F from its two rectangular components F_h and F_y (Fig. 2.30*b*). The actual order in which the three components F_x, F_y, F_z are entered is immaterial.

Substituting the values of the components and magnitude of **F** into Eqs. (2.25), we write

$$\cos\theta_x = \frac{F_x}{F} = \frac{20 \text{ lb}}{70 \text{ lb}} \qquad \cos\theta_y = \frac{F_y}{F} = \frac{-30 \text{ lb}}{70 \text{ lb}} \qquad \cos\theta_z = \frac{F_z}{F} = \frac{60 \text{ lb}}{70 \text{ lb}}$$

Calculating successively each quotient and its arc cosine, we obtain

$$\theta_x = 73.4° \qquad \theta_y = 115.4° \qquad \theta_z = 31.0°$$

The computations indicated may easily be carried out with a calculator.

2.13. Force Defined by Its Magnitude and Two Points on Its Line of Action. In many applications, the direction of a force **F** is defined by the coordinates of two points, $M(x_1, y_1, z_1)$ and $N(x_2, y_2, z_2)$, located on its line of action (Fig. 2.34). Consider the vector $\overrightarrow{MN}$ joining M

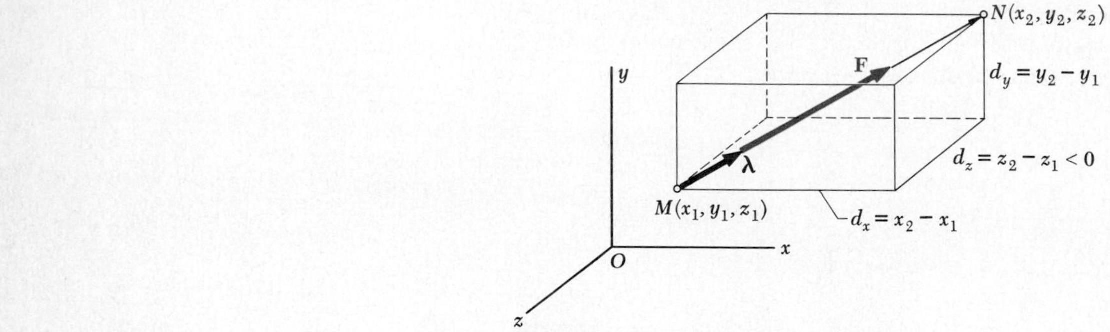

Fig. 2.34

and N and of the same sense as **F**. Denoting its scalar components by d_x, d_y, d_z, respectively, we write

$$\overrightarrow{MN} = d_x\mathbf{i} + d_y\mathbf{j} + d_z\mathbf{k} \tag{2.26}$$

The unit vector $\boldsymbol{\lambda}$ along the line of action of **F** (i.e., along the line MN) may be obtained by dividing the vector $\overrightarrow{MN}$ by its magnitude MN. Substituting for $\overrightarrow{MN}$ from (2.26) and observing that MN is equal to the distance d from M to N, we write

$$\boldsymbol{\lambda} = \frac{\overrightarrow{MN}}{MN} = \frac{1}{d}(d_x\mathbf{i} + d_y\mathbf{j} + d_z\mathbf{k}) \tag{2.27}$$

Recalling that **F** is equal to the product of F and $\boldsymbol{\lambda}$, we have

$$\mathbf{F} = F\boldsymbol{\lambda} = \frac{F}{d}(d_x\mathbf{i} + d_y\mathbf{j} + d_z\mathbf{k}) \tag{2.28}$$

from which it follows that the scalar components of **F** are, respectively,

$$F_x = \frac{Fd_x}{d} \qquad F_y = \frac{Fd_y}{d} \qquad F_z = \frac{Fd_z}{d} \tag{2.29}$$

The relations (2.29) considerably simplify the determination of the components of a force **F** of given magnitude F when the line of action of **F** is defined by two points M and N. Subtracting the coordinates of M from those of N, we first determine the components of the vector $\overrightarrow{MN}$ and the distance d from M to N:

$$d_x = x_2 - x_1 \qquad d_y = y_2 - y_1 \qquad d_z = z_2 - z_1$$
$$d = \sqrt{d_x^2 + d_y^2 + d_z^2}$$

Substituting for F and for d_x, d_y, d_z, and d into the relations (2.29), we obtain the components F_x, F_y, F_z of the force.

The angles θ_x, θ_y, θ_z that **F** forms with the coordinate axes may then be obtained from Eqs. (2.25). Comparing Eqs. (2.22) and (2.27), we may also write

$$\cos\theta_x = \frac{d_x}{d} \qquad \cos\theta_y = \frac{d_y}{d} \qquad \cos\theta_z = \frac{d_z}{d} \tag{2.30}$$

and determine the angles θ_x, θ_y, θ_z directly from the components and magnitude of the vector MN.

2.14. Addition of Concurrent Forces in Space. We shall determine the resultant **R** of two or more forces in space by summing their rectangular components. Graphical or trigonometric methods are generally not practical in the case of forces in space.

The method followed here is similar to that used in Sec. 2.8 with coplanar forces. Setting

$$\mathbf{R} = \Sigma\mathbf{F}$$

we resolve each force into its rectangular components and write

$$\begin{aligned} R_x\mathbf{i} + R_y\mathbf{j} + R_z\mathbf{k} &= \Sigma(F_x\mathbf{i} + F_y\mathbf{j} + F_z\mathbf{k}) \\ &= (\Sigma F_x)\mathbf{i} + (\Sigma F_y)\mathbf{j} + (\Sigma F_z)\mathbf{k} \end{aligned}$$

from which it follows that

$$R_x = \Sigma F_x \qquad R_y = \Sigma F_y \qquad R_z = \Sigma F_z \tag{2.31}$$

The magnitude of the resultant and the angles θ_x, θ_y, θ_z it forms with the axes of coordinates are obained by the method of Sec. 2.12. We write

$$R = \sqrt{R_x^2 + R_y^2 + R_z^2} \tag{2.32}$$

$$\cos\theta_x = \frac{R_x}{R} \qquad \cos\theta_y = \frac{R_y}{R} \qquad \cos\theta_z = \frac{R_z}{R} \tag{2.33}$$

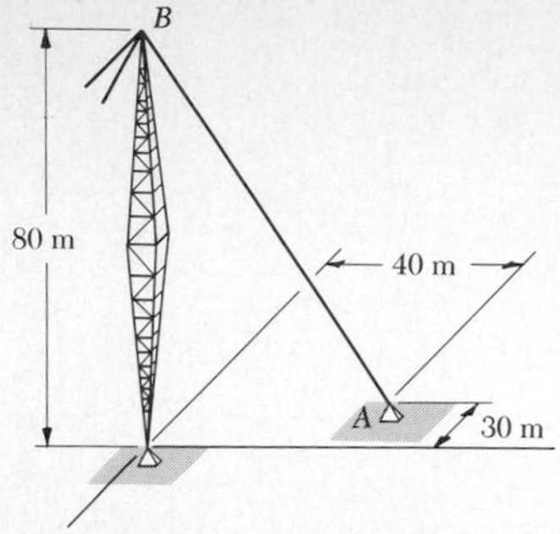

SAMPLE PROBLEM 2.7

A tower guy wire is anchored by means of a bolt at A. The tension in the wire is 2500 N. Determine (*a*) the components F_x, F_y, F_z of the force acting on the bolt, (*b*) the angles θ_x, θ_y, θ_z defining the direction of the force.

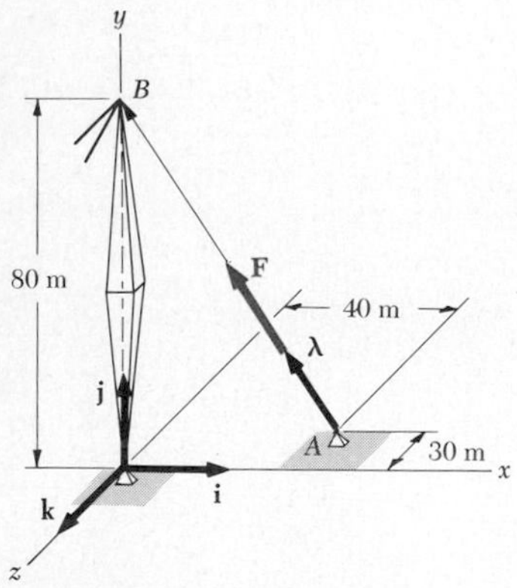

***a.* Components of the Force.** The line of action of the force acting on the bolt passes through A and B, and the force is directed from A to B. The components of the vector $\overrightarrow{AB}$, which has the same direction as the force, are

$$d_x = -40 \text{ m} \qquad d_y = +80 \text{ m} \qquad d_z = +30 \text{ m}$$

The total distance from A to B is

$$AB = d = \sqrt{d_x^2 + d_y^2 + d_z^2} = 94.3 \text{ m}$$

Denoting by **i**, **j**, **k** the unit vectors along the coordinate axes, we have

$$\overrightarrow{AB} = -(40 \text{ m})\mathbf{i} + (80 \text{ m})\mathbf{j} + (30 \text{ m})\mathbf{k}$$

Introducing the unit vector $\boldsymbol{\lambda} = \overrightarrow{AB}/AB$, we write

$$\mathbf{F} = F\boldsymbol{\lambda} = F\frac{\overrightarrow{AB}}{AB} = \frac{2500 \text{ N}}{94.3 \text{ m}}\overrightarrow{AB}$$

Substituting the expression found for $\overrightarrow{AB}$, we obtain

$$\mathbf{F} = \frac{2500 \text{ N}}{94.3 \text{ m}}[-(40 \text{ m})\mathbf{i} + (80 \text{ m})\mathbf{j} + (30 \text{ m})\mathbf{k}]$$

$$\mathbf{F} = -(1060 \text{ N})\mathbf{i} + (2120 \text{ N})\mathbf{j} + (795 \text{ N})\mathbf{k}$$

The components of **F**, therefore, are

$$F_x = -1060 \text{ N} \qquad F_y = +2120 \text{ N} \qquad F_z = +795 \text{ N} \quad ◄$$

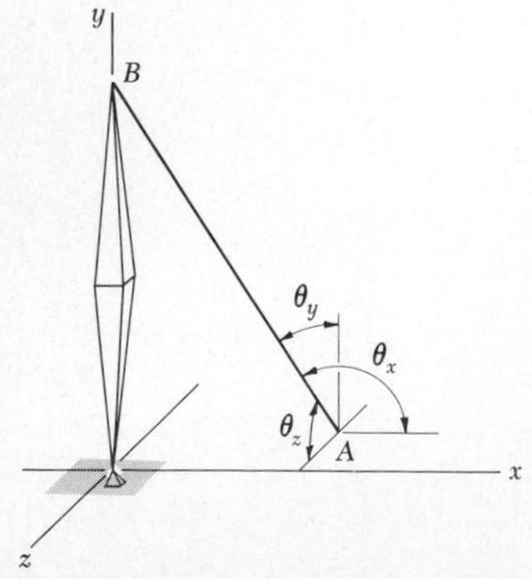

***b.* Direction of the Force.** Using Eqs. (2.25), we write

$$\cos\theta_x = \frac{F_x}{F} = \frac{-1060 \text{ N}}{2500 \text{ N}} \qquad \cos\theta_y = \frac{F_y}{F} = \frac{+2120 \text{ N}}{2500 \text{ N}}$$

$$\cos\theta_z = \frac{F_z}{F} = \frac{+795 \text{ N}}{2500 \text{ N}}$$

Calculating successively each quotient and its arc cosine, we obtain

$$\theta_x = 115.1° \qquad \theta_y = 32.0° \qquad \theta_z = 71.5° \quad ◄$$

(*Note.* This result could have been obtained by using the components and magnitude of the vector $\overrightarrow{AB}$ rather than those of the force **F**.)

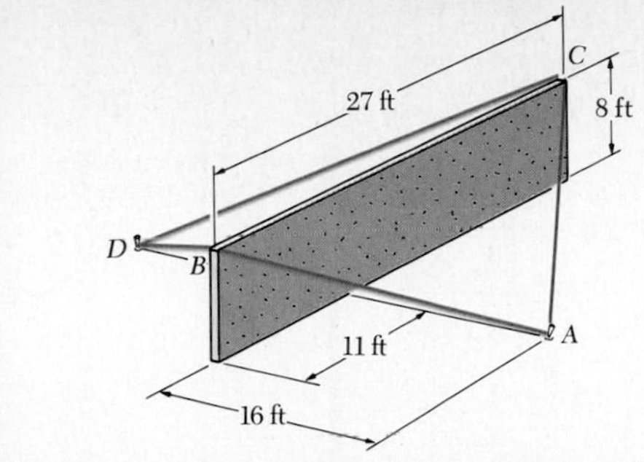

SAMPLE PROBLEM 2.8

A precast-concrete wall section is temporarily held by the cables shown. Knowing that the tension is 840 lb in cable AB and 1200 lb in cable AC, determine the magnitude and direction of the resultant of the forces exerted by cables AB and AC on stake A.

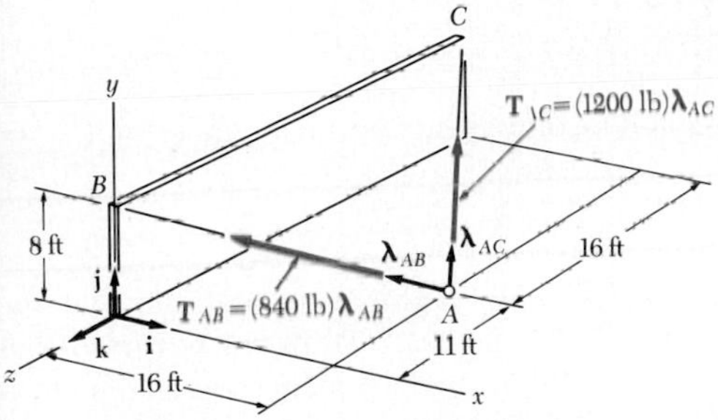

Solution. The force exerted by each cable on stake A will be resolved into x, y, and z components. We first determine the components and magnitude of the vectors $\overrightarrow{AB}$ and $\overrightarrow{AC}$, measuring them from A toward the wall section. Denoting by $\mathbf{i}$, $\mathbf{j}$, $\mathbf{k}$ the unit vectors along the coordinate axes, we write

$$\overrightarrow{AB} = -(16\text{ ft})\mathbf{i} + (8\text{ ft})\mathbf{j} + (11\text{ ft})\mathbf{k} \qquad AB = 21\text{ ft}$$
$$\overrightarrow{AC} = -(16\text{ ft})\mathbf{i} + (8\text{ ft})\mathbf{j} - (16\text{ ft})\mathbf{k} \qquad AC = 24\text{ ft}$$

Denoting by $\boldsymbol{\lambda}_{AB}$ the unit vector along AB, we have

$$\mathbf{T}_{AB} = T_{AB}\boldsymbol{\lambda}_{AB} = T_{AB}\frac{\overrightarrow{AB}}{AB} = \frac{840\text{ lb}}{21\text{ ft}}\overrightarrow{AB}$$

Substituting the expression found for $\overrightarrow{AB}$, we obtain

$$\mathbf{T}_{AB} = \frac{840\text{ lb}}{21\text{ ft}}[-(16\text{ ft})\mathbf{i} + (8\text{ ft})\mathbf{j} + (11\text{ ft})\mathbf{k}]$$

$$\mathbf{T}_{AB} = -(640\text{ lb})\mathbf{i} + (320\text{ lb})\mathbf{j} + (440\text{ lb})\mathbf{k}$$

Denoting by $\boldsymbol{\lambda}_{AC}$ the unit vector along AC, we obtain in a similar way

$$\mathbf{T}_{AC} = T_{AC}\boldsymbol{\lambda}_{AC} = T_{AC}\frac{\overrightarrow{AC}}{AC} = \frac{1200\text{ lb}}{24\text{ ft}}\overrightarrow{AC}$$

$$\mathbf{T}_{AC} = -(800\text{ lb})\mathbf{i} + (400\text{ lb})\mathbf{j} - (800\text{ lb})\mathbf{k}$$

The resultant $\mathbf{R}$ of the forces exerted by the two cables is

$$\mathbf{R} = \mathbf{T}_{AB} + \mathbf{T}_{AC} = -(1440\text{ lb})\mathbf{i} + (720\text{ lb})\mathbf{j} - (360\text{ lb})\mathbf{k}$$

The magnitude and direction of the resultant are now determined:

$$R = \sqrt{R_x^2 + R_y^2 + R_z^2} = \sqrt{(-1440)^2 + (720)^2 + (-360)^2}$$

$$R = 1650\text{ lb} \quad ◄$$

From Eqs. (2.33) we obtain

$$\cos\theta_x = \frac{R_x}{R} = \frac{-1440\text{ lb}}{1650\text{ lb}} \qquad \cos\theta_y = \frac{R_y}{R} = \frac{+720\text{ lb}}{1650\text{ lb}}$$

$$\cos\theta_z = \frac{R_z}{R} = \frac{-360\text{ lb}}{1650\text{ lb}}$$

Calculating successively each quotient and its arc cosine, we have

$$\theta_x = 150.8° \qquad \theta_y = 64.1° \qquad \theta_z = 102.6° \quad ◄$$

Problems

2.54 Determine (*a*) the x, y, and z components of the 500-N force, (*b*) the angles θ_x, θ_y, and θ_z that the force forms with the coordinate axes.

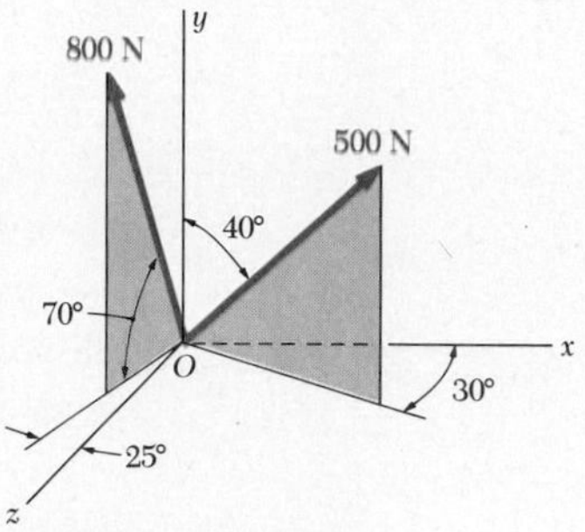

Fig. P2.54 and P2.55

2.55 Determine (*a*) the x, y, and z components of the 800-N force, (*b*) the angles θ_x, θ_y, and θ_z that the force forms with the coordinate axes.

2.56 A gun is aimed at a point A located 35° east of north. Knowing that the barrel of the gun forms an angle of 40° with the horizontal and that the maximum recoil force is 400 N, determine (*a*) the x, y, and z components of that force, (*b*) the values of the angles θ_x, θ_y, and θ_z defining the direction of the recoil force. (Assume that the x, y, and z axes are directed, respectively, east, up, and south.)

2.57 Solve Prob. 2.56, assuming that point A is located 15° north of west and that the barrel of the gun forms an angle of 25° with the horizontal.

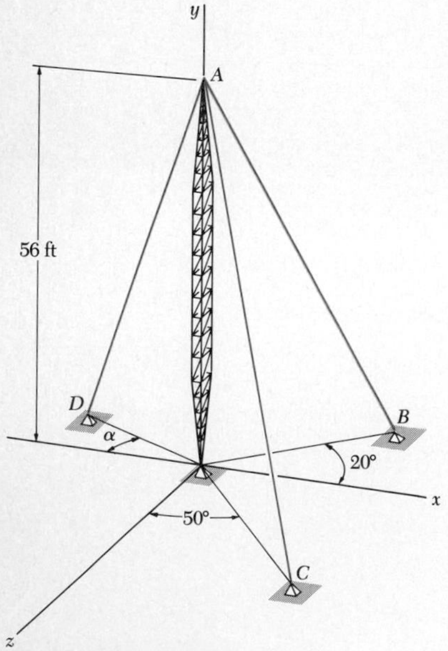

Fig. P2.58 and P2.59

2.58 Cable AB is 65 ft long and the tension in that cable is 3900 lb. Determine (*a*) the x, y, and z components of the force exerted by the cable on the anchor B, (*b*) the angles θ_x, θ_y, and θ_z defining the direction of the force exerted at B.

2.59 Cable AC is 70 ft long and the tension in that cable is 5250 lb. Determine (*a*) the x, y, and z components of the force exerted by the cable on the anchor C, (*b*) the angles θ_x, θ_y, and θ_z defining the direction of the force exerted at C.

2.60 Determine the magnitude and direction of the force $\mathbf{F} = (650\text{ N})\mathbf{i} - (320\text{ N})\mathbf{j} + (760\text{ N})\mathbf{k}$.

2.61 Determine the magnitude and direction of the force $\mathbf{F} = (580\text{ lb})\mathbf{i} + (690\text{ lb})\mathbf{j} - (300\text{ lb})\mathbf{k}$.

2.62 A force acts at the origin in a direction defined by the angles $\theta_x = 75°$ and $\theta_z = 130°$. Knowing that the y component of the force is $+300$ lb, determine (*a*) the other components and the magnitude of the force, (*b*) the value of θ_y.

2.63 A force acts at the origin in a direction defined by the angles $\theta_y = 55°$ and $\theta_z = 45°$. Knowing that the x component of the force is -500 N, determine (*a*) the other components and the magnitude of the force, (*b*) the value of θ_x.

2.64 In order to move a wrecked truck, two cables are attached at A and pulled by winches B and C as shown. Knowing that the tension in cable AB is 10 kN, determine the components of the force exerted by cable AB on the truck.

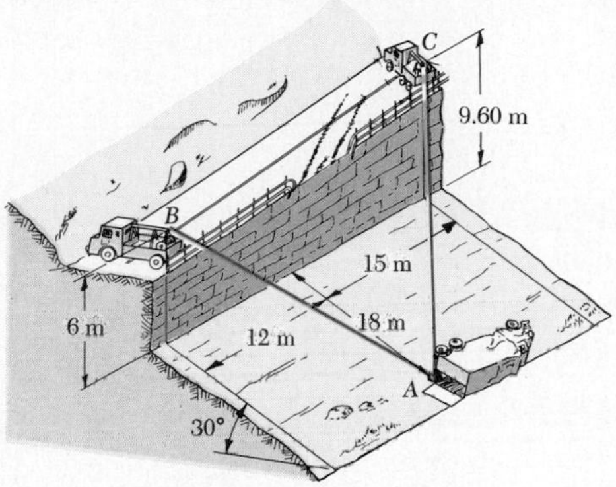

Fig. P2.64

2.65 Referring to the truck of Prob. 2.64 and knowing that the tension in cable AC is 7.5 kN, determine the components of the force exerted by cable AC on the truck.

2.66 Knowing that the tension in cable AB is 285 lb, determine the components of the force exerted on the plate at B.

2.67 Knowing that the tension in cable AC is 426 lb, determine the components of the force exerted on the plate at C.

2.68 Knowing that the tension is 285 lb in cable AB and 426 lb in cable AC, determine the magnitude and direction of the resultant of the forces exerted at A by the two cables.

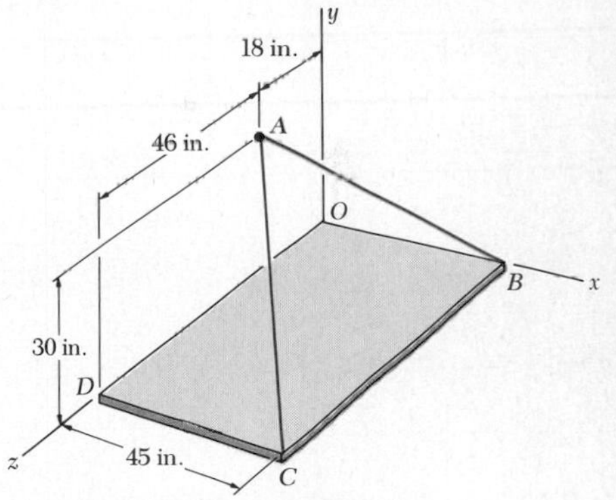

Fig. P2.66, P2.67, and P2.68

2.69 Referring to the truck of Prob. 2.64 and knowing that the tension is 10 kN in cable AB and 7.5 kN in cable AC, determine the magnitude and direction of the resultant of the forces exerted by the cables on the truck.

2.70 Determine the resultant of the two forces shown.

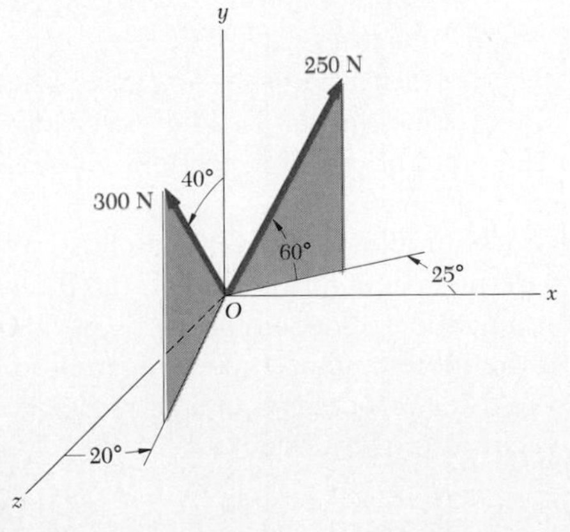

Fig. P2.70

2.71 Referring to the pylon of Probs. 2.58 and 2.59, it is recalled that the tension is 3900 lb in cable AB and 5250 lb in cable AC. (*a*) Determine the magnitude and direction of the resultant of the forces exerted at A by cables AB and AC. (*b*) Knowing that the resultant of the forces exerted at A by the three cables must be vertical, find the angle α that line OD forms with the negative x axis.

2.72 The boom OA carries a load **P** and is supported by two cables as shown. Knowing that the tension in cable AB is 850 N and that the resultant of the load **P** and of the forces exerted at A by the two cables must be directed along OA, determine the tension in cable AC.

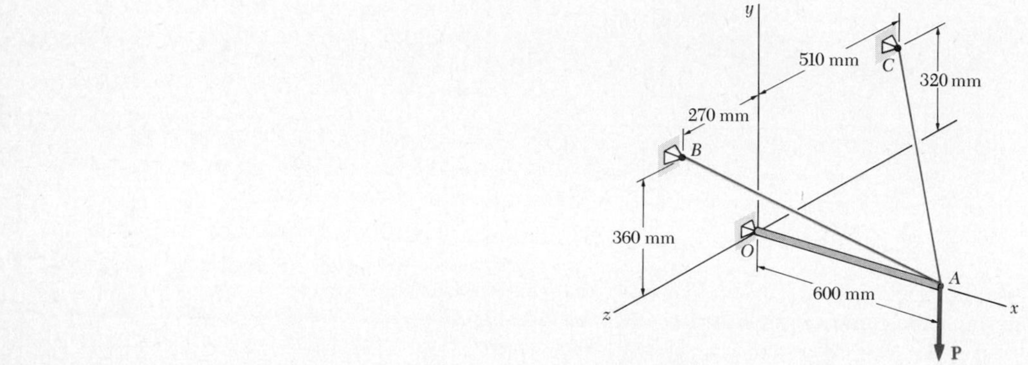

Fig. P2.72

2.73 For the boom and loading of Prob. 2.72, determine the magnitude of the load **P**.

2.15. Equilibrium of a Particle in Space.

According to the definition given in Sec. 2.9, a particle A is in equilibrium if the resultant of all the forces acting on A is zero. The components R_x, R_y, R_z of the resultant are given by the relations (2.31); expressing that the components of the resultant are zero, we write

$$\Sigma F_x = 0 \qquad \Sigma F_y = 0 \qquad \Sigma F_z = 0 \tag{2.34}$$

Equations (2.34) represent the necessary and sufficient conditions for the equilibrium of a particle in space. They may be used to solve problems dealing with the equilibrium of a particle involving no more than three unknowns.

To solve such problems, we first should draw a free-body diagram showing the particle in equilibrium and *all* the forces acting on it. We may then write the equations of equilibrium (2.34) and solve them for three unknowns. In the more common types of problems, these unknowns will represent (1) the three components of a single force or (2) the magnitude of three forces each of known direction.

SAMPLE PROBLEM 2.9

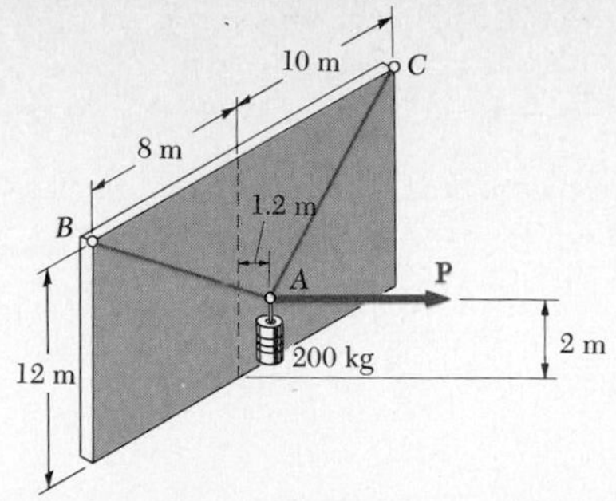

A 200-kg cylinder is hung by means of two cables *AB* and *AC*, which are attached to the top of a vertical wall. A horizontal force **P** perpendicular to the wall holds the cylinder in the position shown. Determine the magnitude of **P** and the tension in each cable.

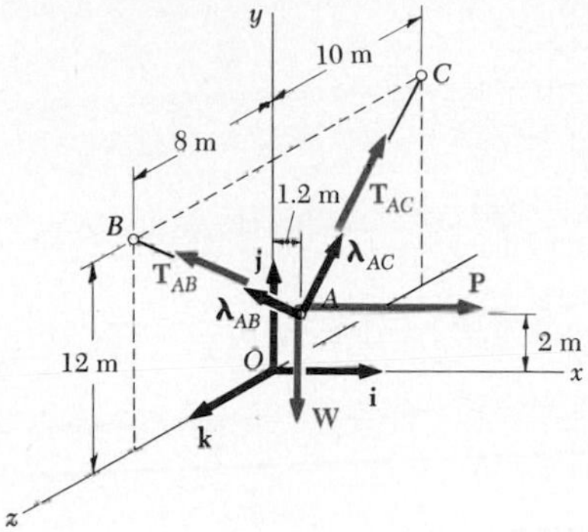

Solution. Point *A* is chosen as a free body; this point is subjected to four forces, three of which are of unknown magnitude.

Introducing the unit vectors **i**, **j**, **k**, we resolve each force into rectangular components.

$$\mathbf{P} = P\mathbf{i}$$
$$\mathbf{W} = -mg\mathbf{j} = -(200\text{ kg})(9.81\text{ m/s}^2)\mathbf{j} = -(1962\text{ N})\mathbf{j} \qquad (1)$$

In the case of $\mathbf{T}_{AB}$ and $\mathbf{T}_{AC}$, it is necessary first to determine the components and magnitudes of the vectors $\overrightarrow{AB}$ and $\overrightarrow{AC}$. Denoting by $\boldsymbol{\lambda}_{AB}$ the unit vector along *AB*, we write

$$\overrightarrow{AB} = -(1.2\text{ m})\mathbf{i} + (10\text{ m})\mathbf{j} + (8\text{ m})\mathbf{k} \qquad AB = 12.86\text{ m}$$

$$\boldsymbol{\lambda}_{AB} = \frac{\overrightarrow{AB}}{12.86\text{ m}} = -0.0933\mathbf{i} + 0.778\mathbf{j} + 0.622\mathbf{k}$$

$$\mathbf{T}_{AB} = T_{AB}\boldsymbol{\lambda}_{AB} = -0.0933T_{AB}\mathbf{i} + 0.778T_{AB}\mathbf{j} + 0.622T_{AB}\mathbf{k} \qquad (2)$$

Denoting by $\boldsymbol{\lambda}_{AC}$ the unit vector along *AC*, we write in a similar way

$$\overrightarrow{AC} = -(1.2\text{ m})\mathbf{i} + (10\text{ m})\mathbf{j} - (10\text{ m})\mathbf{k} \qquad AC = 14.19\text{ m}$$

$$\boldsymbol{\lambda}_{AC} = \frac{\overrightarrow{AC}}{14.19\text{ m}} = -0.0846\mathbf{i} + 0.705\mathbf{j} - 0.705\mathbf{k}$$

$$\mathbf{T}_{AC} = T_{AC}\boldsymbol{\lambda}_{AC} = -0.0846T_{AC}\mathbf{i} + 0.705T_{AC}\mathbf{j} - 0.705T_{AC}\mathbf{k} \qquad (3)$$

Equilibrium Condition. Since *A* is in equilibrium, we must have

$$\Sigma\mathbf{F} = 0: \qquad \mathbf{T}_{AB} + \mathbf{T}_{AC} + \mathbf{P} + \mathbf{W} = 0$$

or, substituting from (1), (2), (3) for the forces and factoring **i**, **j**, **k**,

$$(-0.0933T_{AB} - 0.0846T_{AC} + P)\mathbf{i}$$
$$+ (0.778T_{AB} + 0.705T_{AC} - 1962\text{ N})\mathbf{j} + (0.622T_{AB} - 0.705T_{AC})\mathbf{k} = 0$$

Setting the coefficients of **i**, **j**, **k** equal to zero, we write three scalar equations, which express that the sums of the *x*, *y*, and *z* components of the forces are respectively equal to zero.

$$(\Sigma F_x = 0:) \qquad -0.0933T_{AB} - 0.0846T_{AC} + P = 0$$
$$(\Sigma F_y = 0:) \qquad +0.778T_{AB} + 0.705T_{AC} - 1962\text{ N} = 0$$
$$(\Sigma F_z = 0:) \qquad +0.622T_{AB} - 0.705T_{AC} = 0$$

Solving these equations, we obtain

$$P = 235\text{ N} \qquad T_{AB} = 1401\text{ N} \qquad T_{AC} = 1236\text{ N} \blacktriangleleft$$

Problems

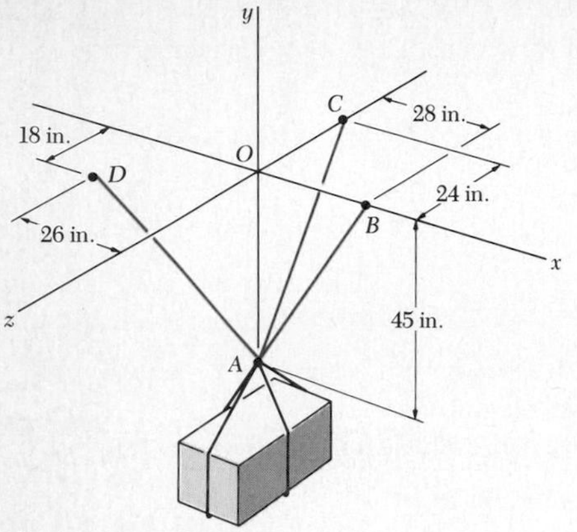

Fig. P2.74 and P2.75

2.74 A crate is supported by three cables as shown. Determine the weight W of the crate, knowing that the tension in cable AD is 924 lb.

2.75 A crate is supported by three cables as shown. Determine the weight W of the crate, knowing that the tension in cable AB is 1378 lb.

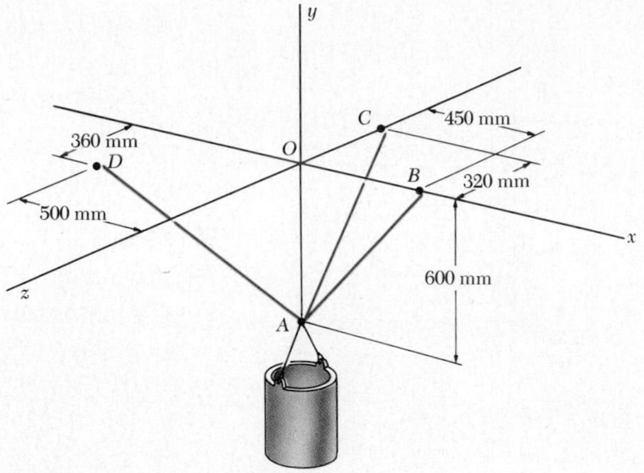

Fig. P2.76, P2.77, and P2.78

2.76 A container is supported by three cables as shown. Determine the weight W of the container, knowing that the tension in cable AB is 4 kN.

2.77 A container is supported by three cables as shown. Determine the weight W of the container, knowing that the tension in cable AD is 3.87 kN.

2.78 A container of weight $W = 1165$ N is supported by three cables as shown. Determine the tension in each cable.

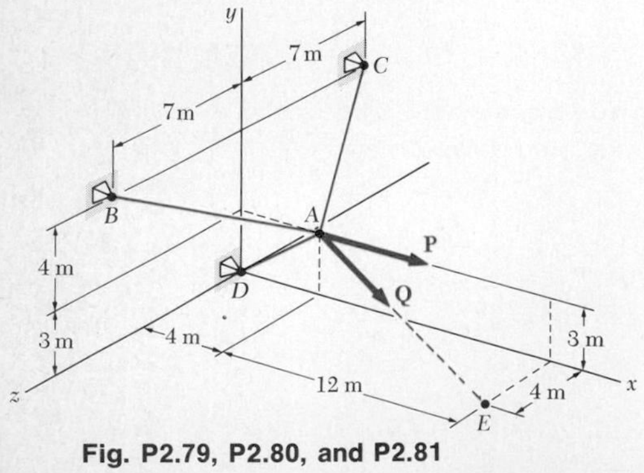

Fig. P2.79, P2.80, and P2.81

2.79 Three cables are connected at A, where the forces **P** and **Q** are applied as shown. Determine the tension in each cable when $P = 5.60$ kN and $Q = 0$.

2.80 Three cables are connected at A, where the forces **P** and **Q** are applied as shown. Determine the tension in each cable when $P = 0$ and $Q = 7.28$ kN.

2.81 Three cables are connected at A, where the forces **P** and **Q** are applied as shown. Knowing that $Q = 7.28$ kN and that the tension in cable AD is zero, determine (*a*) the magnitude and sense of **P**, (*b*) the tension in cables AB and AC.

2.82 A 12-lb circular plate of 7-in. radius is supported as shown by three wires, each of 25-in. length. Determine the tension in each wire, knowing that $\alpha = 30°$.

2.83 Solve Prob. 2.82, assuming that $\alpha = 45°$.

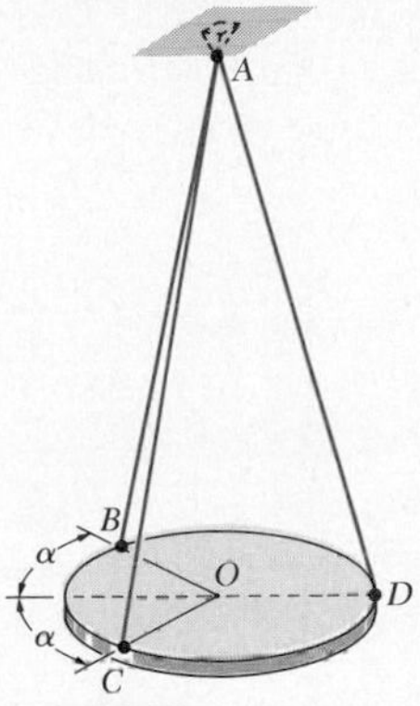

Fig. P2.82

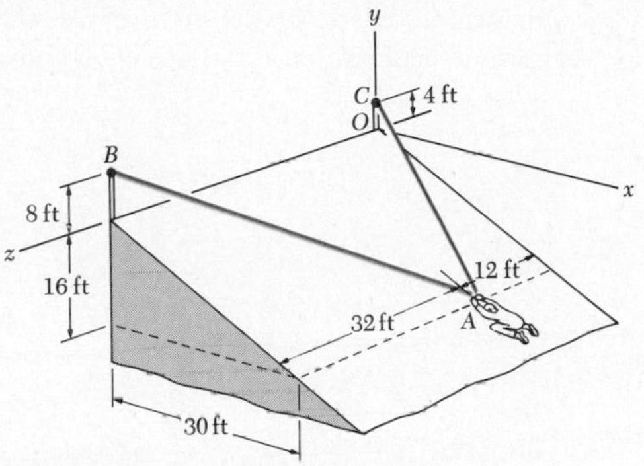

Fig. P2.84

2.84 In trying to move across a slippery icy surface, a 180-lb man uses two ropes *AB* and *AC*. Knowing that the force exerted on the man by the icy surface is perpendicular to that surface, determine the tension in each rope.

2.85 Solve Prob. 2.84, assuming that a friend is helping the man at *A* by pulling on him with a force $\mathbf{P} = -(60 \text{ lb})\mathbf{k}$.

2.86 A container of weight $W = 400$ N is supported by cables *AB* and *AC*, which are tied to ring *A*. Knowing that $\mathbf{Q} = 0$, determine (*a*) the magnitude of the force **P** which must be applied to the ring to maintain the container in the position shown, (*b*) the corresponding values of the tension in cables *AB* and *AC*.

2.87 Solve Prob. 2.86, knowing that $\mathbf{Q} = (80 \text{ N})\mathbf{k}$.

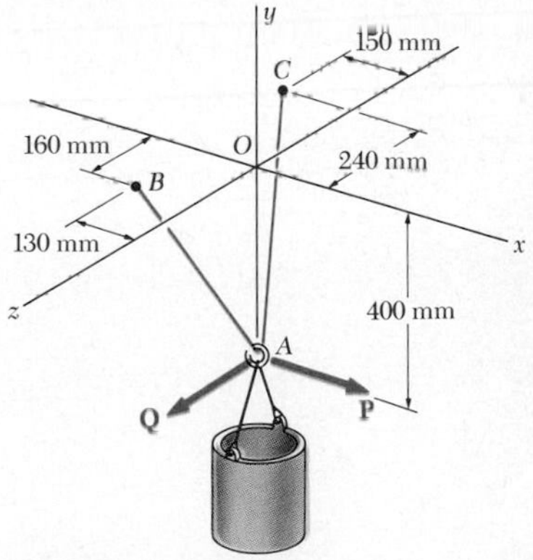

Fig. P2.86 and P2.88

2.88 A container is supported by a single cable which passes through a frictionless ring *A* and is attached to fixed points *B* and *C*. Two forces $\mathbf{P} = P\mathbf{i}$ and $\mathbf{Q} = Q\mathbf{k}$ are applied to the ring to maintain the container in the position shown. Knowing that the weight of the container is $W = 376$ N, determine the magnitudes of **P** and **Q**. (*Hint.* The tension must be the same in portions *AB* and *AC* of the cable.)

2.89 Determine the weight *W* of the container of Prob. 2.88, knowing that $P = 164$ N.

2.90 Cable *BAC* passes through a frictionless ring *A* and is attached to fixed supports at *B* and *C*, while cables *AD* and *AE* are both tied to the ring and are attached, respectively, to supports at *D* and *E*. Knowing that a 150-lb vertical load **P** is applied to ring *A*, determine the tension in each of the three cables.

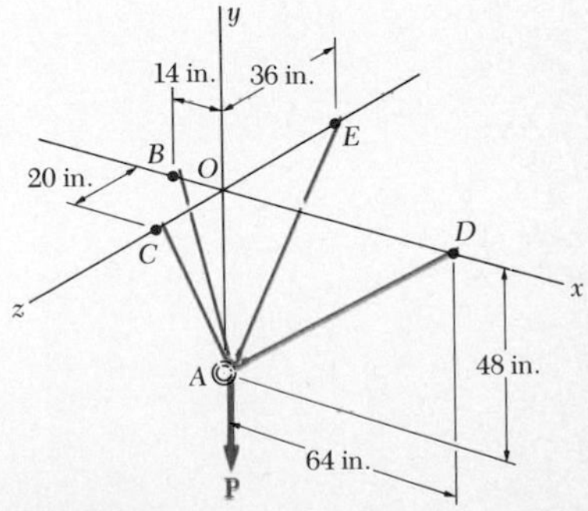

Fig. P2.90

2.91 Knowing that the tension in cable *AE* of Prob. 2.90 is 50 lb, determine (*a*) the magnitude of the load **P**, (*b*) the tension in cables *BAC* and *AD*.

2.92 A 10-kg circular plate of 250-mm radius is supported as shown by three wires of equal length L. Knowing that $\alpha = 30°$, determine the smallest permissible value of the length L if the tension is not to exceed 50 N in any of the wires.

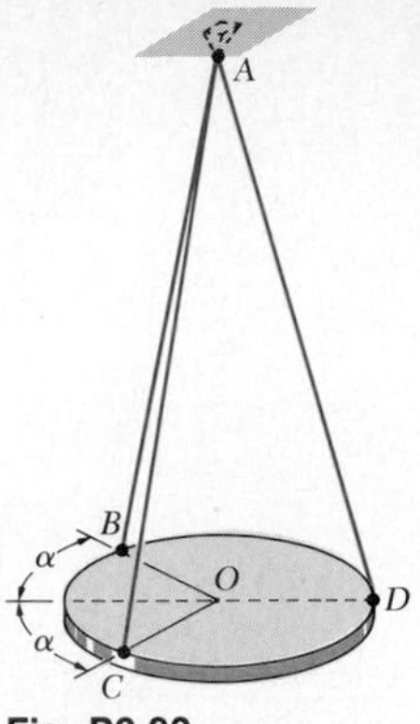

Fig. P2.92

2.93 Collars A and B are connected by a 250-mm wire and may slide freely on frictionless rods. If a force **Q** of magnitude 100 N is applied to collar B as shown, determine (*a*) the tension in the wire when $x = 90$ mm, (*b*) the corresponding magnitude of the force **P** required to maintain the equilibrium of the system.

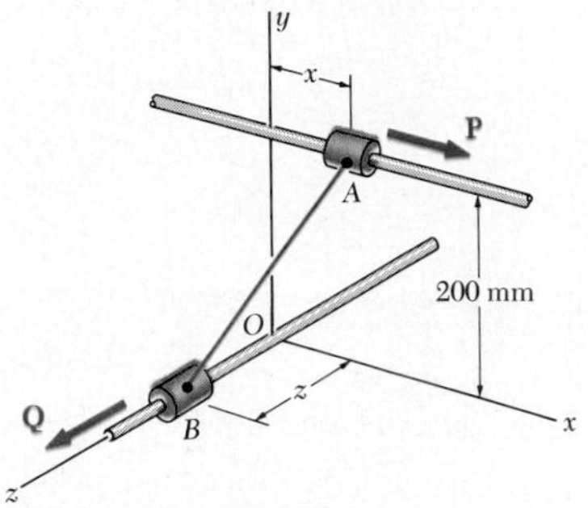

Fig. P2.93 and P2.94

* **2.94** Collars A and B are connected by a 250-mm wire and may slide freely on frictionless rods. Determine the distances x and z for which the equilibrium of the system is maintained when $P = 200$ N and $Q = 100$ N.

Review Problems

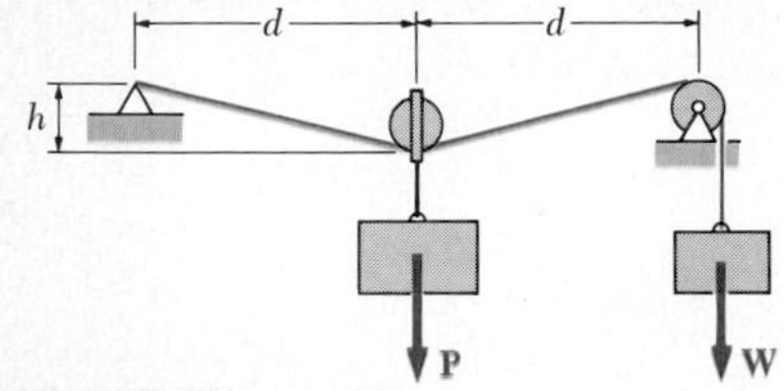

Fig. P2.95

2.95 (*a*) Express the weight W required to maintain equilibrium in terms of P, d, and h. (*b*) For $W = 500$ N, $P = 150$ N, and $d = 400$ mm, determine the value of h consistent with equilibrium.

2.96 A chain loop of length 1.25 m is placed around a 250 × 250 mm piece of timber. Knowing that the mass of the timber is 175 kg, determine the tension in the chain for each of the arrangements shown.

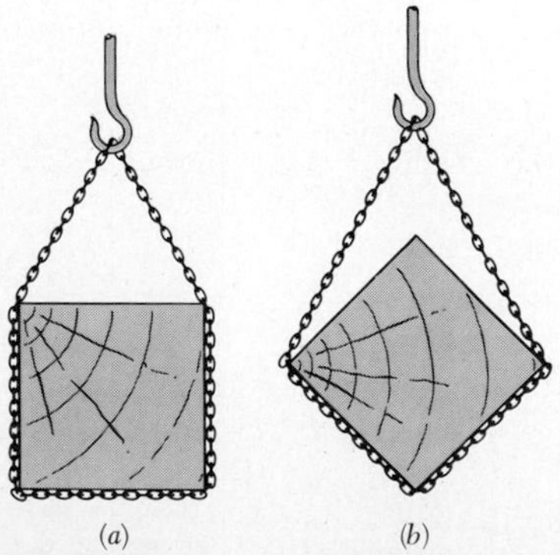

Fig. P2.96

2.97 Knowing that the magnitude of the force **P** is 100 lb, determine the resultant of the three forces applied at A.

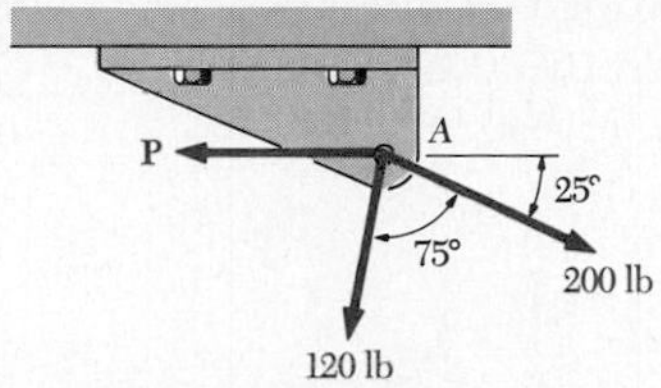

Fig. P2.97 and P2.98

2.98 Determine the range of values of P for which the resultant of the three forces applied at A does not exceed 225 lb.

2.99 A cylinder of weight $W = 650$ N is supported by two cables AC and BC, which are attached to the top of vertical posts. A horizontal force **P**, perpendicular to the plane containing the posts, holds the cylinder in the position shown. Determine (*a*) the magnitude of **P**, (*b*) the tension in each cable.

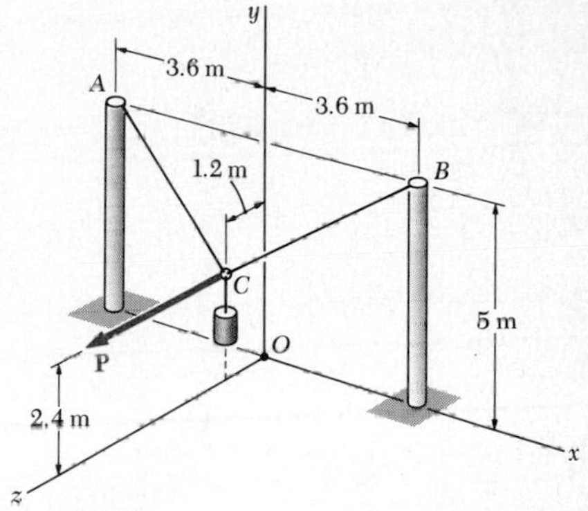

Fig. P2.99

2.100 In Prob. 2.99, determine the angles θ_x, θ_y, and θ_z for the force exerted at B by cable BC.

2.101 Knowing that $P = 400$ N, determine the tension in cables AC and BC.

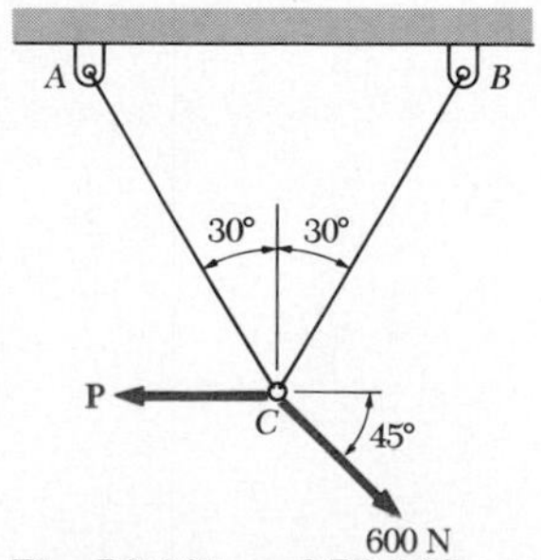

Fig. P2.101 and P2.102

2.102 Determine the range of values of **P** for which both cables remain taut.

2.103 The directions of the 75-lb forces may vary, but the angle between the forces is always 50°. Determine the value of α for which the resultant of the forces acting at A is directed horizontally to the left.

2.104 Determine the resultant of the three forces acting at A when (*a*) $\alpha = 0$, (*b*) $\alpha = 25°$.

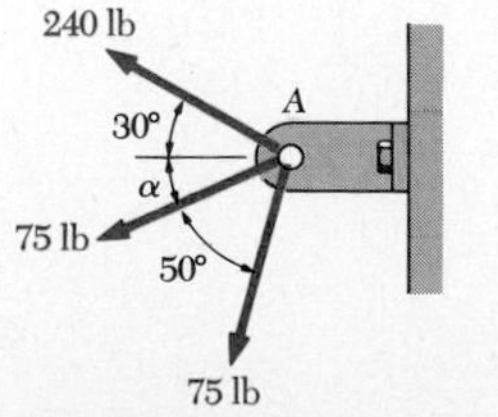

Fig. P2.103 and P2.104

2.105 An 8-kN force acts at the origin in a direction defined by the angles $\theta_y = 35°$ and $\theta_z = 65°$. It is also known that the x component of the force is positive. Determine the value of θ_x and the components of the force.

2.106 Three wires are connected at point D, which is located 9 in. below the T-shaped pipe support ABC. Determine the tension in each wire when a 60-lb cylinder is suspended from point D as shown.

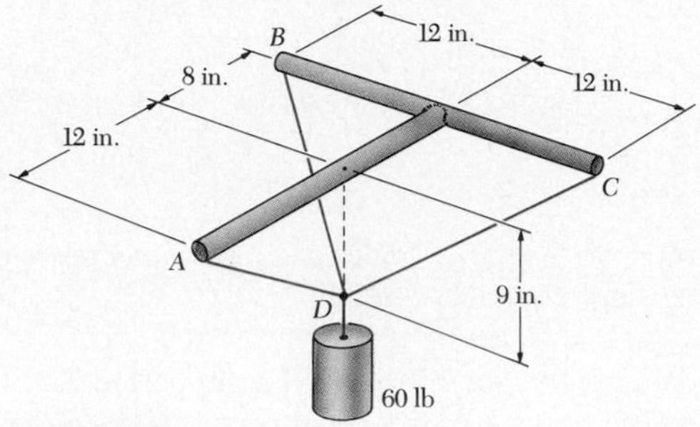

Fig. P2.106

CHAPTER THREE

Rigid Bodies: Equivalent Systems of Forces

3.1. Introduction. In the preceding chapter it was assumed that each of the bodies considered could be treated as a single particle. Such a view, however, is not always possible, and a body, in general, should be treated as a combination of a large number of particles. The size of the body will have to be taken into consideration, as well as the fact that forces will act on different particles and thus will have different points of application.

Most of the bodies considered in elementary mechanics are assumed to be *rigid*, a *rigid body* being defined as one which does not deform. Actual structures and machines, however, are never absolutely rigid and deform under the loads to which they are subjected. But these deformations are usually small and do not appreciably affect the conditions of equilibrium or motion of the structure under consideration. They are important, though, as far as the resistance of the structure to failure is concerned, and are considered in the study of mechanics of materials.

In this chapter we shall study the effect of forces exerted on a rigid body and we shall learn how to replace a given system of forces by a simpler equivalent system. Our analysis will rest on the fundamental assumption that the effect of a given force on a rigid body remains unchanged if that force is moved along its line of action (*principle of transmissibility*). It follows that forces acting on a rigid body may be represented by *sliding vectors*, as indicated earlier in Sec. 2.3.

Two important concepts associated with the effect of a force on a rigid body are the *moment of a force about a point* (Sec. 3.6) and the *moment of a force about an axis* (Sec. 3.11). Since the determination of these quantities involves the computation of vector products and scalar products of two vectors, the fundamentals of vector algebra will be introduced in this chapter and applied to the solution of problems involving forces acting on rigid bodies.

Another concept introduced in this chapter is that of a *couple*, i.e., the combination of two forces which have the same magnitude, parallel lines of action, and opposite sense (Sec. 3.12). As we shall see, any system of forces acting on a rigid body may be replaced by an equivalent system consisting of one force attached at a given point and one couple. This basic system is called a *force-couple system*. In the case of concurrent, coplanar, or parallel forces, the equivalent force-couple system may be further reduced to a single force, called the *resultant* of the system, or to a single couple, called the *resultant couple* of the system.

3.2. External and Internal Forces. Forces acting on rigid bodies may be separated into two groups: (1) *external forces,* (2) *internal forces*.

1. The *external forces* represent the action of other bodies on the rigid body under consideration. They are entirely responsible for the external behavior of the rigid body. They will either cause it to move or assure that it remains at rest. We shall be concerned only with external forces in this chapter and in Chaps. 4 and 5.
2. The *internal forces* are the forces which hold together the particles forming the rigid body. If the rigid body is structurally composed of several parts, the forces holding the component parts together are also defined as internal forces. Internal forces will be considered in Chaps. 6 and 7.

Fig. 3.1

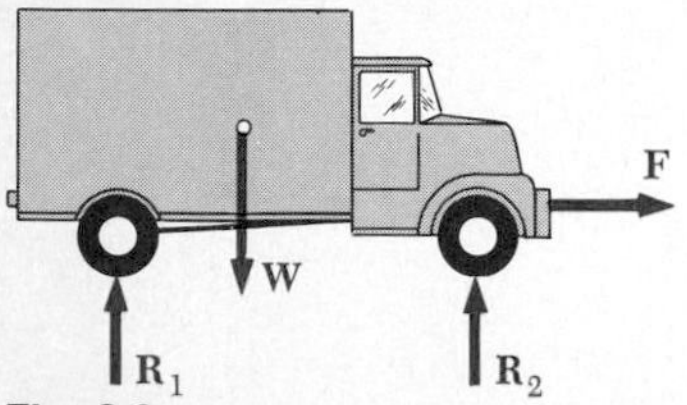

Fig. 3.2

As an example of external forces, we shall consider the forces acting on a disabled truck that men are pulling forward by means of a rope attached to the front bumper (Fig. 3.1). The external forces acting on the truck are shown in a *free-body diagram* (Fig. 3.2). Let us first consider the *weight* of the truck. Although it embodies the effect of the earth's pull on each of the particles forming the truck, the weight may be represented by the single force **W**. The *point of application* of this force, i.e., the point at which the force acts, is defined as the *center of gravity* of the truck. It will be seen in Chap. 5 how centers of gravity may be determined. The weight **W** tends to make the truck move vertically downward. In fact, it would actually cause the truck to move downward, i.e., to fall, if it were not for the presence of the ground. The ground opposes the downward motion of the truck by means of the reactions $\mathbf{R}_1$ and $\mathbf{R}_2$. These forces are exerted *by* the ground *on* the truck and must therefore be included among the external forces acting on the truck.

The men pulling on the rope exert the force **F**. The point of application of **F** is on the front bumper. The force **F** tends to make the truck move forward in a straight line and does actually make it move, since no external force opposes this motion. (Rolling resistance has been neglected here for simplicity.) This forward motion of the truck, during which all straight lines remain parallel to themselves (the floor of the truck remains horizontal, and its walls remain vertical), is known as a *translation*. Other forces might cause the truck to move differently. For example, the force exerted by a jack placed under the front axle would cause the truck to pivot about its rear axle. Such a motion is a *rotation*. It may be concluded, therefore, that each of the *external forces* acting on a *rigid body* can, if unopposed, impart to the rigid body a motion of translation or rotation, or both.

3.3. Principle of Transmissibility. Equivalent Forces. The *principle of transmissibility* states that the conditions of equilibrium or of motion of a rigid body will remain unchanged if a force **F** acting at a given point of the rigid body is replaced by a force **F′** of the same magnitude and same direction, but acting at a different point, *provided that the two forces have the same line of action* (Fig. 3.3). The two forces **F** and **F′** have the same effect on the rigid body and are said to be *equivalent*. This principle, which states in fact that the action of a force may be *transmitted* along its line of action, is based on experimental evidence. It *cannot* be derived from the properties established so far in this text and must therefore be accepted as an experimental law. However, as we shall see in Sec. 16.5, the principle of transmissibility may be derived from the study of the dynamics of rigid bodies, but this study requires the introduction of all Newton's three laws and of a number of other concepts as well. We shall therefore base our study of the statics of rigid bodies on the three principles introduced so far, i.e., the parallelogram law of addition, Newton's first law, and the principle of transmissibility.

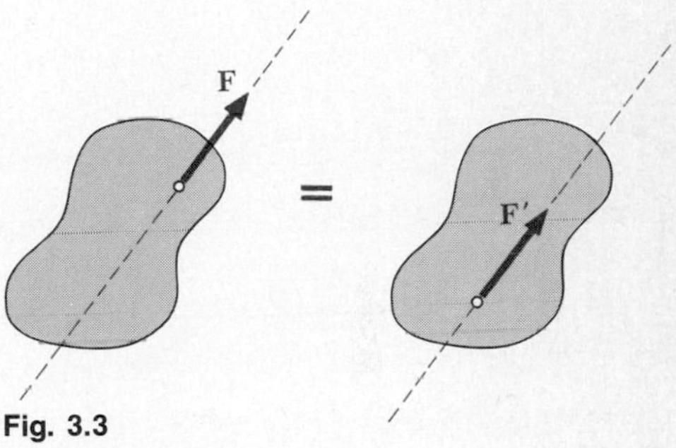

Fig. 3.3

It was indicated in Chap. 2 that the forces acting on a particle could be represented by vectors. These vectors had a well-defined point of application, namely, the particle itself, and were therefore fixed, or bound, vectors. In the case of forces acting on a rigid body, however, the point of application of the force does not matter, as long as the line of action remains unchanged. Thus forces acting on a rigid body must be represented by different kinds of vectors, known as *sliding vectors*, since these vectors may be allowed to slide along their line of action. We should note that all the properties which will be derived in the following sections for the forces acting on a rigid body will be valid more generally for any system of sliding vectors. In order to keep our presentation more intuitive, however, we shall carry it out in terms of physical forces rather than in terms of mathematical sliding vectors.

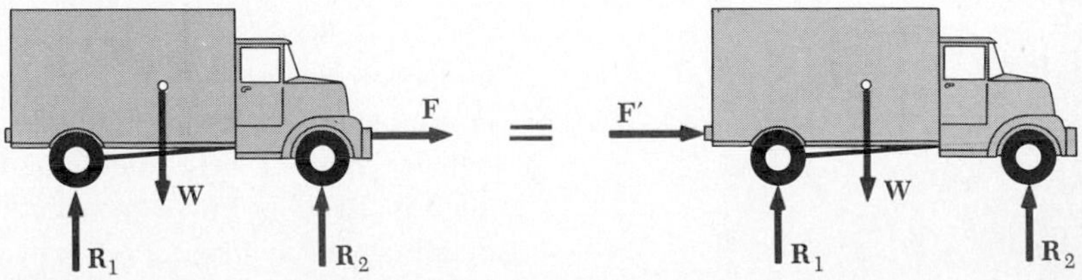

Fig. 3.4

Returning to the example of the truck, we first observe that the line of action of the force **F** is a horizontal line passing through both the front and the rear bumpers of the truck (Fig. 3.4). Using the principle of transmissibility, we may therefore replace **F** by an *equivalent force* **F′** acting on the rear bumper. In other words, the conditions of motion are unaffected, and all the other external forces acting on the truck ($\mathbf{W}, \mathbf{R}_1, \mathbf{R}_2$) remain unchanged if the men push on the rear bumper instead of pulling on the front bumper.

The principle of transmissibility and the concept of equivalent forces have limitations, however. Consider, for example, a short bar *AB* acted upon by equal and opposite axial forces $\mathbf{P}_1$ and $\mathbf{P}_2$, as shown in Fig. 3.5*a*. According to the principle of transmissibility, the force $\mathbf{P}_2$ may be replaced by a force $\mathbf{P}_2'$ having the same magnitude, same direction, and same line of action, but acting at A instead of B (Fig. 3.5*b*). The forces $\mathbf{P}_1$ and $\mathbf{P}_2'$ acting on the same particle may be added according to the rules of Chap. 2, and, being equal and opposite, their sum is found equal to zero. The original system of forces shown in Fig. 3.5*a* is thus equivalent to no force at all (Fig. 3.5*c*) from the point of view of the external behavior of the bar.

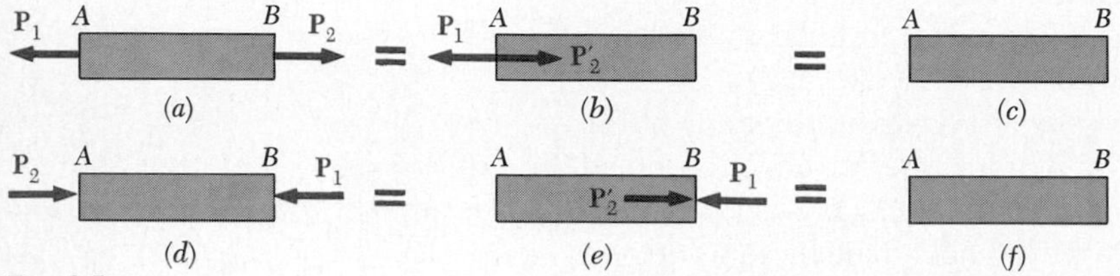

Fig. 3.5

Consider now the two equal and opposite forces $\mathbf{P}_1$ and $\mathbf{P}_2$ acting on the bar AB as shown in Fig. 3.5*d*. The force $\mathbf{P}_2$ may be replaced by a force $\mathbf{P}_2'$ having the same magnitude, same direction, and same line of action, but acting at B instead of A (Fig. 3.5*e*). The forces $\mathbf{P}_1$ and $\mathbf{P}_2'$ may then be added, and their sum is found again to be zero (Fig. 3.5*f*). From the point of view of the mechanics of rigid bodies, the systems shown in Fig. 3.5*a* and *d* are thus equivalent. But the *internal forces* and *deformations* produced by the two systems are clearly different. The bar of Fig. 3.5*a* is in *tension* and, if not absolutely rigid, will increase in length slightly; the bar of Fig. 3.5*d* is in *compression* and, if not absolutely rigid, will decrease in length slightly. Thus, while the principle of transmissibility may be used freely to determine the conditions of motion or equilibrium of rigid bodies and to compute the external forces acting on these bodies, it should be avoided, or at least used with care, in determining internal forces and deformations.

3.4. Vector Product of Two Vectors. In order to gain a better understanding of the effect of a force on a rigid body, we shall introduce a new concept, the concept of *moment of a force about a point*. This concept will be more clearly understood, and we shall be able to apply it more effectively, if we first add to the mathematical tools at our disposal by defining the *vector product* of two vectors.

The vector product of two vectors $\mathbf{P}$ and $\mathbf{Q}$ is defined as the vector $\mathbf{V}$ which satisfies the following conditions:

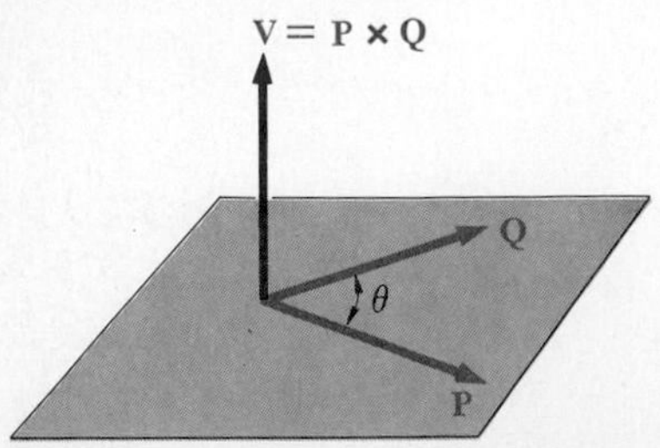

Fig. 3.6

1. The line of action of $\mathbf{V}$ is perpendicular to the plane containing $\mathbf{P}$ and $\mathbf{Q}$ (Fig. 3.6).
2. The magnitude of $\mathbf{V}$ is the product of the magnitudes of $\mathbf{P}$ and $\mathbf{Q}$ and of the sine of the angle θ formed by $\mathbf{P}$ and $\mathbf{Q}$ (the measure of which will always be 180° or less); we thus have

$$V = PQ \sin \theta \tag{3.1}$$

3. The sense of **V** is such that a person located at the tip of **V** will observe as counterclockwise the rotation through θ which brings the vector **P** in line with the vector **Q**; note that if **P** and **Q** do not have a common point of application, they should first be redrawn from the same point. The three vectors **P**, **Q**, and **V**—taken in that order—are said to form a *right-handed triad.*†

As stated above, the vector **V** satisfying these three conditions (which define it uniquely) is referred to as the vector product of **P** and **Q**; it is represented by the mathematical expression

$$\mathbf{V} = \mathbf{P} \times \mathbf{Q} \tag{3.2}$$

Because of the notation used, the vector product of two vectors **P** and **Q** is also referred to as the *cross product* of **P** and **Q**.

It follows from Eq. (3.1) that, when two vectors **P** and **Q** have either the same direction or opposite directions, their vector product is zero. In the general case when the angle θ formed by the two vectors is neither 0° nor 180°, Eq. (3.1) may be given a simple geometric interpretation: The magnitude V of the vector product of **P** and **Q** measures the area of the parallelogram which has **P** and **Q** for sides (Fig. 3.7). The vector product **P** × **Q** will therefore remain unchanged if we replace **Q** by a vector **Q′** coplanar with **P** and **Q** and such that the line joining the tips of **Q** and **Q′** is parallel to **P**. We write

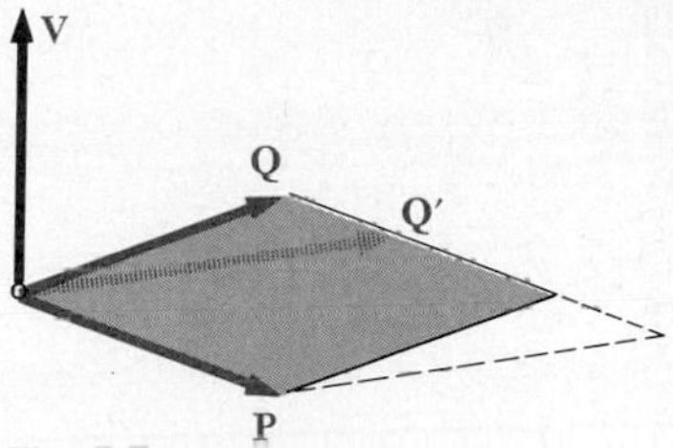

Fig. 3.7

$$\mathbf{V} = \mathbf{P} \times \mathbf{Q} = \mathbf{P} \times \mathbf{Q}' \tag{3.3}$$

From the third condition used to define the vector product **V** of **P** and **Q**, namely, the condition stating that **P**, **Q**, and **V** must form a right-handed triad, it follows that vector products *are not commutative,* i.e., **Q** × **P** is not equal to **P** × **Q**. Indeed, we may easily check that **Q** × **P** is represented by the vector −**V** equal and opposite to **V**. We thus write

$$\mathbf{Q} \times \mathbf{P} = -(\mathbf{P} \times \mathbf{Q}) \tag{3.4}$$

Example. Let us compute the vector product **V** = **P** × **Q** of the vector **P** of magnitude 6 lying in the zx plane at an angle of 30° with the x axis, and of the vector **Q** of magnitude 4 lying along the x axis (Fig. 3.8).

It follows immediately from the definition of the vector product that the vector **V** must lie along the y axis, have the magnitude

$$V = PQ \sin\theta = (6)(4) \sin 30° = 12$$

and be directed upward.

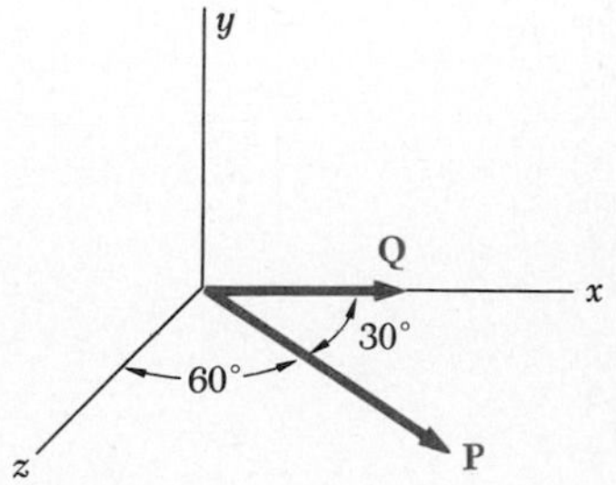

Fig. 3.8

†We should note that the x, y, and z axes used in Chap. 2 form a right-handed system of orthogonal axes and that the unit vectors **i**, **j**, **k** defined in Sec. 2.12 form a right-handed orthogonal triad.

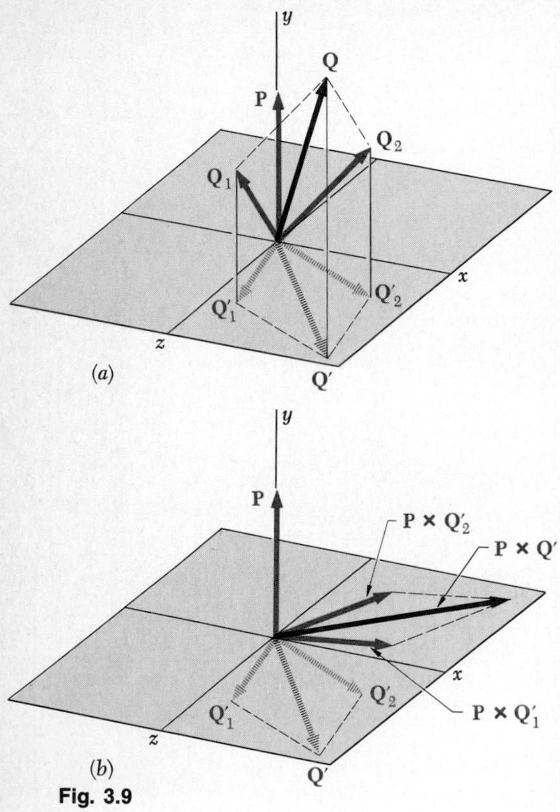

Fig. 3.9

We saw that the commutative property does not apply to vector products. We may wonder whether the *distributive* property holds, i.e., whether the relation

$$\mathbf{P} \times (\mathbf{Q}_1 + \mathbf{Q}_2) = \mathbf{P} \times \mathbf{Q}_1 + \mathbf{P} \times \mathbf{Q}_2 \tag{3.5}$$

is valid. The answer is *yes*. Many readers are probably willing to accept without formal proof an answer which they intuitively feel is correct. However, since the entire structure of vector algebra and of statics depends upon the relation (3.5), we shall take time out to derive it.

We may, without any loss of generality, assume that **P** is directed along the y axis (Fig. 3.9*a*). Denoting by **Q** the sum of $\mathbf{Q}_1$ and $\mathbf{Q}_2$, we drop perpendiculars from the tips of **Q**, $\mathbf{Q}_1$, and $\mathbf{Q}_2$ onto the zx plane, defining in this way the vectors $\mathbf{Q}'$, $\mathbf{Q}'_1$, and $\mathbf{Q}'_2$. These vectors will be referred to, respectively, as the *projections* of **Q**, $\mathbf{Q}_1$, and $\mathbf{Q}_2$ on the zx plane. Recalling the property expressed by Eq. (3.3), we note that the left-hand member of Eq. (3.5) may be replaced by $\mathbf{P} \times \mathbf{Q}'$ and that, similarly, the vector products $\mathbf{P} \times \mathbf{Q}_1$ and $\mathbf{P} \times \mathbf{Q}_2$ may respectively be replaced by $\mathbf{P} \times \mathbf{Q}'_1$ and $\mathbf{P} \times \mathbf{Q}'_2$. Thus, the relation to be proved may be written in the form

$$\mathbf{P} \times \mathbf{Q}' = \mathbf{P} \times \mathbf{Q}'_1 + \mathbf{P} \times \mathbf{Q}'_2 \tag{3.5$'$}$$

We now observe that $\mathbf{P} \times \mathbf{Q}'$ may be obtained from $\mathbf{Q}'$ by multiplying this vector by the scalar P and rotating it counterclockwise through 90° in the zx plane (Fig. 3.9*b*); the other two vector products in (3.5′) may be obtained in the same manner from $\mathbf{Q}'_1$ and $\mathbf{Q}'_2$, respectively. Now, since the projection of a parallelogram onto an arbitrary plane is a parallelogram, the projection $\mathbf{Q}'$ of the sum **Q** of $\mathbf{Q}_1$ and $\mathbf{Q}_2$ must be the sum of the projections $\mathbf{Q}'_1$ and $\mathbf{Q}'_2$ of $\mathbf{Q}_1$ and $\mathbf{Q}_2$ on the same plane (Fig. 3.9*a*). This relation between the vectors $\mathbf{Q}'$, $\mathbf{Q}'_1$, and $\mathbf{Q}'_2$ will still hold after the three vectors have been multiplied by the scalar P and rotated through 90° (Fig. 3.9*b*). Thus the relation (3.5′) has been proved and we can now be sure that the distributive property holds for vector products.

The third property, the associative property, does not apply to vector products; we have in general

$$(\mathbf{P} \times \mathbf{Q}) \times \mathbf{S} \neq \mathbf{P} \times (\mathbf{Q} \times \mathbf{S}) \tag{3.6}$$

3.5. Vector Products Expressed in Terms of Rectangular Components. We shall now determine the vector product of any two of the unit vectors, **i**, **j**, **k**, defined in Chap. 2. Consider first the product $\mathbf{i} \times \mathbf{j}$ (Fig. 3.10*a*). Since both vectors have a magnitude equal to 1 and since they are at a right angle to each other, their vector product will also be a unit vector. This unit vector must be **k**, since the vectors **i**, **j**, **k** are mutually perpendicular and form a right-handed triad. On the other hand, the product $\mathbf{j} \times \mathbf{i}$ will be equal to $-\mathbf{k}$ since the 90° rotation which brings **j** into **i** is observed as counterclockwise by a person located at the tip of $-\mathbf{k}$ (Fig. 3.10*b*). Finally it should be observed that the vector product of a unit vector by itself, such as $\mathbf{i} \times \mathbf{i}$, is equal to zero, since both vectors have the same direction. The vector products of the various possible pairs of unit vectors are

$$\begin{array}{lll} \mathbf{i} \times \mathbf{i} = 0 & \mathbf{j} \times \mathbf{i} = -\mathbf{k} & \mathbf{k} \times \mathbf{i} = \mathbf{j} \\ \mathbf{i} \times \mathbf{j} = \mathbf{k} & \mathbf{j} \times \mathbf{j} = 0 & \mathbf{k} \times \mathbf{j} = -\mathbf{i} \\ \mathbf{i} \times \mathbf{k} = -\mathbf{j} & \mathbf{j} \times \mathbf{k} = \mathbf{i} & \mathbf{k} \times \mathbf{k} = 0 \end{array} \tag{3.7}$$

By arranging in a circle and in counterclockwise order the three letters representing the unit vectors (Fig. 3.11), we may simplify the determination of the sign of the vector product of two unit vectors: The product of two unit vectors will be positive if they follow each other in counterclockwise order, and negative otherwise.

We may now easily express the vector product **V** of two given vectors **P** and **Q** in terms of the rectangular components of these vectors. Resolving **P** and **Q** into components, we first write

$$\mathbf{V} = \mathbf{P} \times \mathbf{Q} = (P_x\mathbf{i} + P_y\mathbf{j} + P_z\mathbf{k}) \times (Q_x\mathbf{i} + Q_y\mathbf{j} + Q_z\mathbf{k})$$

Making use of the distributive property, we express **V** as the sum of vector products such as $P_x\mathbf{i} \times Q_y\mathbf{j}$. Observing that each of the expressions obtained is equal to the vector product of two unit vectors, such as $\mathbf{i} \times \mathbf{j}$, multiplied by the product of two scalars, such as P_xQ_y, and recalling the identities (3.7), we obtain, after factoring **i**, **j**, and **k**,

$$\mathbf{V} = (P_yQ_z - P_zQ_y)\mathbf{i} + (P_zQ_x - P_xQ_z)\mathbf{j} + (P_xQ_y - P_yQ_x)\mathbf{k} \tag{3.8}$$

The rectangular components of the vector product **V** are thus found to be

$$\begin{aligned} V_x &= P_yQ_z - P_zQ_y \\ V_y &= P_zQ_x - P_xQ_z \\ V_z &= P_xQ_y - P_yQ_x \end{aligned} \tag{3.9}$$

Returning to Eq. (3.8), we observe that its right-hand member represents the expansion of a determinant. The vector product **V** may thus be expressed in the following form, more easily memorized:†

$$\mathbf{V} = \begin{vmatrix} \mathbf{i} & \mathbf{j} & \mathbf{k} \\ P_x & P_y & P_z \\ Q_x & Q_y & Q_z \end{vmatrix} \tag{3.10}$$

† Any determinant consisting of three rows and three columns may be evaluated by repeating the first and second columns and forming products along each diagonal line. The sum of the products obtained along the colored lines is then subtracted from the sum of the products obtained along the black lines.

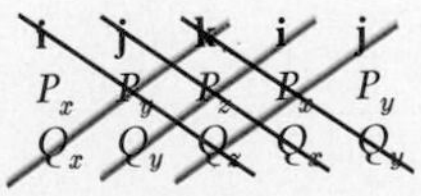

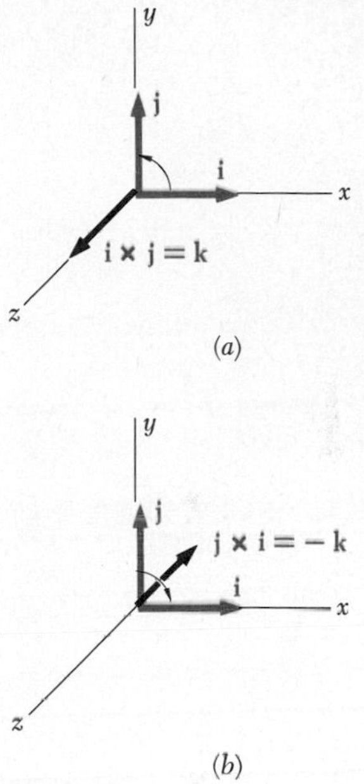

Fig. 3.10

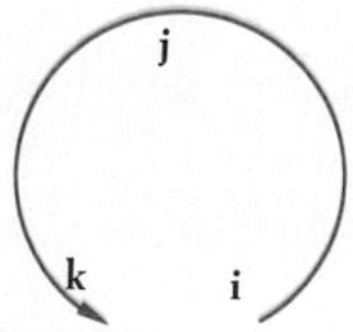

Fig. 3.11

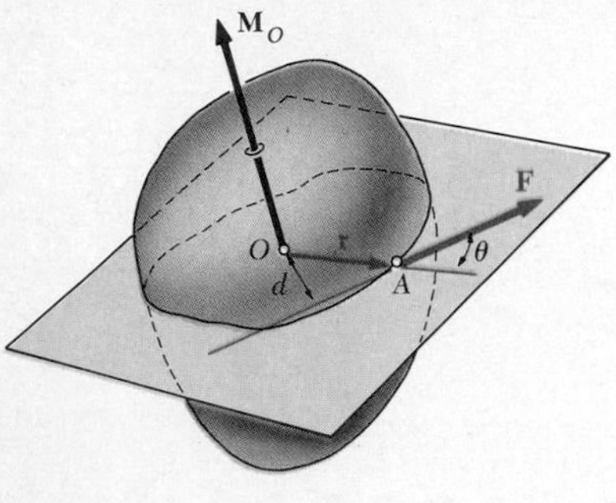

(a)

(b)

Fig. 3.12

3.6. Moment of a Force about a Point. Let us now consider a force **F** acting on a rigid body (Fig. 3.12). As we know, the force **F** is represented by a vector which defines its magnitude and direction. However, the effect of the force on the rigid body depends also upon its point of application A. The position of A may be conveniently defined by the vector **r** which joins the fixed reference point O with A; this vector is known as the *position vector* of A.† The position vector **r** and the force **F** define the plane shown in Fig. 3.12*a*.

We shall define the *moment of* **F** *about* O as the vector product of **r** and **F**:

$$\mathbf{M}_O = \mathbf{r} \times \mathbf{F} \tag{3.11}$$

According to the definition of the vector product given in Sec. 3.4, the moment $\mathbf{M}_O$ must be perpendicular to the plane containing O and the force **F**. The sense of $\mathbf{M}_O$ is defined by the sense of the rotation which would bring the vector **r** in line with the vector **F**; this rotation should be observed as *counterclockwise* by an observer located at the tip of $\mathbf{M}_O$. Another way of defining the sense of $\mathbf{M}_O$ is furnished by the *right-hand rule:* Close your right hand and hold it so that your fingers are curled in the sense of the rotation that **F** would impart to the rigid body about a fixed axis directed along the line of action of $\mathbf{M}_O$; your thumb will indicate the sense of the moment $\mathbf{M}_O$ (Fig. 3.12*b*).

Finally, denoting by θ the angle between the lines of action of the position vector **r** and the force **F**, we find that the magnitude of the moment of **F** about O is

$$M_O = rF \sin\theta = Fd \tag{3.12}$$

where d represents the perpendicular distance from O to the line of action of **F**. Since the tendency of a force **F** to make a rigid body rotate about a fixed axis perpendicular to the force depends upon the distance of **F** from that axis, as well as upon the magnitude of **F**, we note that *the magnitude of* $\mathbf{M}_O$ *measures the tendency of the force* **F** *to make the rigid body rotate about a fixed axis directed along* $\mathbf{M}_O$.

In the SI system of units, where a force is expressed in newtons (N) and a distance in meters (m), the moment of a force will be expressed in newton-meters (N·m). In the U.S. customary system of units, where a force is expressed in pounds and a distance in feet or inches, the moment of a force will be expressed in lb·ft or lb·in.

† We may easily verify that position vectors obey the law of vector addition and, thus, are truly vectors. Consider, for example, the position vectors **r** and **r**′ of A with respect to two reference points O and O', and the position vector **s** of O with respect to O' (Fig. 3.40*a*, Sec. 3.16). We check that the position vector $\mathbf{r}' = \overrightarrow{O'A}$ may be obtained from the position vectors $\mathbf{s} = \overrightarrow{O'O}$ and $\mathbf{r} = \overrightarrow{OA}$ by applying the triangle rule for the addition of vectors.

We may observe that the moment $\mathbf{M}_O$ of a force about a point, while it depends upon the magnitude, the line of action, and the sense of the force, does *not* depend upon the actual position of the point of application of the force along its line of action. Conversely, the moment $\mathbf{M}_O$ of a force $\mathbf{F}$ does not characterize the position of the point of application of $\mathbf{F}$.

However, as we shall see presently, the moment $\mathbf{M}_O$ of a force $\mathbf{F}$ of given magnitude and direction *completely defines the line of action of* $\mathbf{F}$. Indeed, the line of action of $\mathbf{F}$ must lie in a plane through O perpendicular to the moment $\mathbf{M}_O$; its distance d from O must be equal to the quotient M_O/F of the magnitudes of $\mathbf{M}_O$ and $\mathbf{F}$; and the sense of $\mathbf{M}_O$ determines whether the line of action of $\mathbf{F}$ is to be drawn on one side or the other of the point O.

We recall from Sec. 3.3 that the principle of transmissibility states that two forces $\mathbf{F}$ and $\mathbf{F}'$ are equivalent (i.e., have the same effect on a rigid body) if they have the same magnitude, same direction, and same line of action. This principle may now be restated as follows: *Two forces* $\mathbf{F}$ *and* $\mathbf{F}'$ *are equivalent if, and only if, they are equal* (i.e., have the same magnitude and same direction) *and have equal moments about a given point O.* The necessary and sufficient condition for two forces $\mathbf{F}$ and $\mathbf{F}'$ to be equivalent is thus

$$\mathbf{F} = \mathbf{F}' \qquad \text{and} \qquad \mathbf{M}_O = \mathbf{M}'_O \tag{3.13}$$

We should observe that it follows from this statement that if the relations (3.13) hold for a given point O, they will hold for any other point.

Problems Involving Only Two Dimensions. Many applications deal with two-dimensional structures, i.e., structures which have length and breadth but only negligible depth, and which are subjected to forces contained in the plane of the structure. Two-dimensional structures and the forces acting on them may be readily represented on a sheet of paper or on a blackboard. Their analysis is therefore considerably simpler than that of three-dimensional structures and forces.

Consider, for example, a rigid slab acted upon by a force $\mathbf{F}$ (Fig. 3.13). The moment of $\mathbf{F}$ about a point O chosen in the plane of the figure is represented by a vector $\mathbf{M}_O$ perpendicular to that plane and of magnitude Fd. In the case of Fig. 3.13*a* the vector $\mathbf{M}_O$ points *out* of the paper, while in the case of Fig. 3.13*b* it points *into* the paper. As we look at the figure, we observe the action of $\mathbf{F}$ in the first case as counterclockwise, and in the second case as clockwise. Therefore, it is natural to refer to the sense of the moment of $\mathbf{F}$ about O in Fig. 3.13*a* as counterclockwise ↺, and in Fig. 3.13*b* as clockwise ↻.

Since the moment of a force $\mathbf{F}$ acting in the plane of the figure must be perpendicular to that plane, we need only specify the *magnitude* and the *sense* of the moment of $\mathbf{F}$ about O. This may be done by assigning to the magnitude M_O of the moment a positive or negative sign, according to whether the vector $\mathbf{M}_O$ points out of or into the paper.

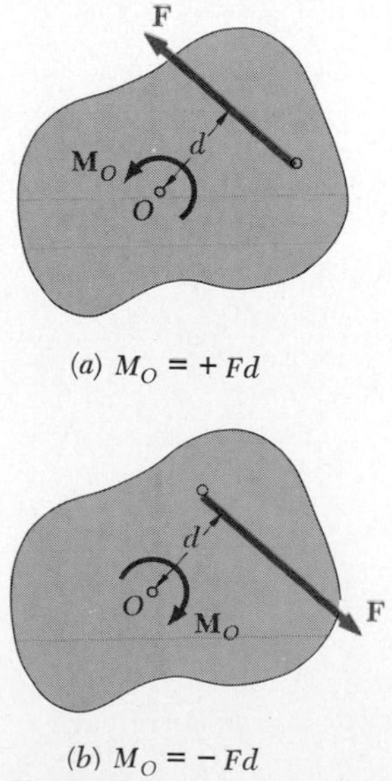

(*a*) $M_O = +Fd$

(*b*) $M_O = -Fd$

Fig. 3.13

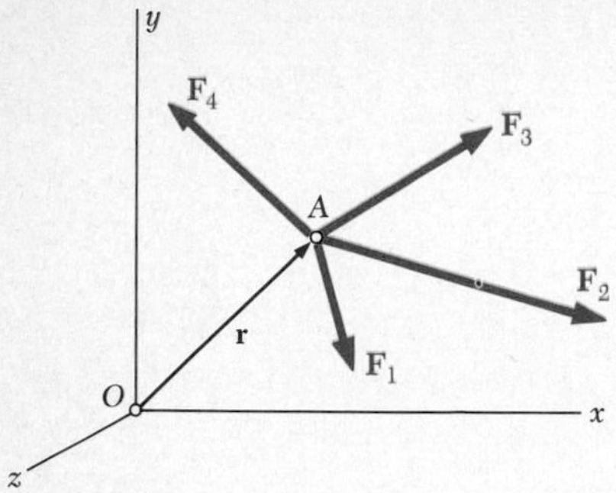

Fig. 3.14

3.7. Varignon's Theorem. The distributive property of vector products may be used to determine the moment of the resultant of several *concurrent forces.* If several forces $\mathbf{F}_1$, $\mathbf{F}_2$, . . . are applied at the same point A (Fig. 3.14), and if we denote by $\mathbf{r}$ the position vector of A, it follows immediately from formula (3.5) of Sec. 3.4 that

$$\mathbf{r} \times (\mathbf{F}_1 + \mathbf{F}_2 + \cdots) = \mathbf{r} \times \mathbf{F}_1 + \mathbf{r} \times \mathbf{F}_2 + \cdots \tag{3.14}$$

In words, *the moment about a given point O of the resultant of several concurrent forces is equal to the sum of the moments of the various forces about the same point O.* This property was originally established by the French mathematician Varignon (1654–1722), long before the introduction of vector algebra, and is known as *Varignon's theorem.*

The relation (3.14) makes it possible to replace the direct determination of the moment of a force $\mathbf{F}$ by the determination of the moments of two or more component forces. As we shall see in the next section, $\mathbf{F}$ will generally be resolved into components parallel to the coordinate axes. However, it may be found more expeditious in some instances to resolve $\mathbf{F}$ into components which are not parallel to the coordinate axes (see Sample Prob. 3.3).

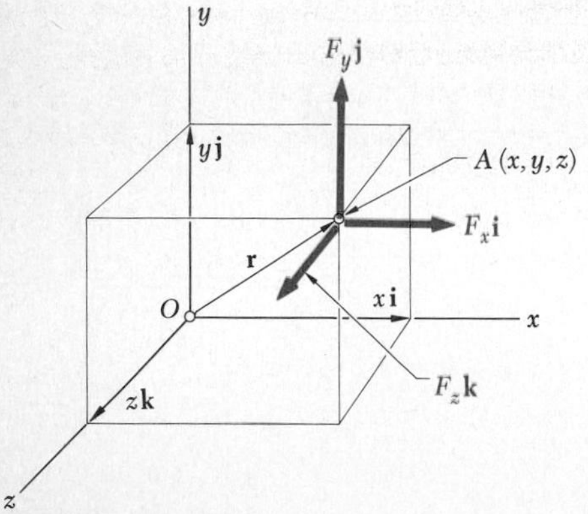

Fig. 3.15

3.8. Rectangular Components of the Moment of a Force. In general, the determination of the moment of a force in space will be considerably simplified if the force and the position vector of its point of application are resolved into rectangular x, y, and z components. Consider, for example, the moment $\mathbf{M}_O$ about O of a force $\mathbf{F}$ of components F_x, F_y, and F_z, applied at a point A of coordinates x, y, and z (Fig. 3.15). Observing that the components of the position vector $\mathbf{r}$ are respectively equal to the coordinates x, y, and z of the point A, we write

$$\mathbf{r} = x\mathbf{i} + y\mathbf{j} + z\mathbf{k} \tag{3.15}$$
$$\mathbf{F} = F_x\mathbf{i} + F_y\mathbf{j} + F_z\mathbf{k} \tag{3.16}$$

Substituting for $\mathbf{r}$ and $\mathbf{F}$ from (3.15) and (3.16) into

$$\mathbf{M}_O = \mathbf{r} \times \mathbf{F} \tag{3.11}$$

and recalling the results obtained in Sec. 3.5, we write the moment $\mathbf{M}_O$ of $\mathbf{F}$ about O in the form

$$\mathbf{M}_O = M_x\mathbf{i} + M_y\mathbf{j} + M_z\mathbf{k} \tag{3.17}$$

where the components M_x, M_y, and M_z are defined by the relations

$$\begin{aligned} M_x &= yF_z - zF_y \\ M_y &= zF_x - xF_z \\ M_z &= xF_y - yF_x \end{aligned} \tag{3.18}$$

As we shall see in Sec. 3.11, the scalar components M_x, M_y, and M_z of the moment $\mathbf{M}_O$ measure the tendency of the force $\mathbf{F}$ to impart to a rigid body a motion of rotation about the x, y, and z axes, respectively. Substituting from (3.18) into (3.17), we may also write $\mathbf{M}_O$ in the form of the determinant

$$\mathbf{M}_O = \begin{vmatrix} \mathbf{i} & \mathbf{j} & \mathbf{k} \\ x & y & z \\ F_x & F_y & F_z \end{vmatrix} \tag{3.19}$$

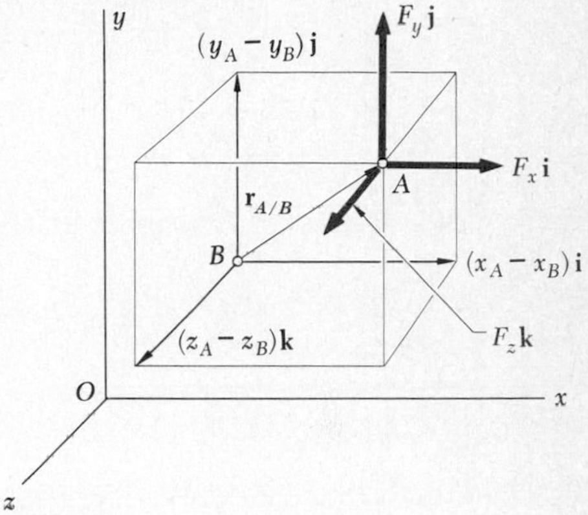

Fig. 3.16

To compute the moment $\mathbf{M}_B$ about an arbitrary point B of a force $\mathbf{F}$ applied at A (Fig. 3.16), we must replace the position vector $\mathbf{r}$ in Eq. (3.11) by a vector drawn from B to A. This vector is the *position vector of A relative to B* and will be denoted by $\mathbf{r}_{A/B}$. Observing that $\mathbf{r}_{A/B}$ may be obtained by subtracting $\mathbf{r}_B$ from $\mathbf{r}_A$, we write

$$\mathbf{M}_B = \mathbf{r}_{A/B} \times \mathbf{F} = (\mathbf{r}_A - \mathbf{r}_B) \times \mathbf{F} \tag{3.20}$$

or, using the determinant form,

$$\mathbf{M}_B = \begin{vmatrix} \mathbf{i} & \mathbf{j} & \mathbf{k} \\ x_{A/B} & y_{A/B} & z_{A/B} \\ F_x & F_y & F_z \end{vmatrix} \tag{3.21}$$

where $x_{A/B}$, $y_{A/B}$, and $z_{A/B}$ denote the components of the vector $\mathbf{r}_{A/B}$:

$$x_{A/B} = x_A - x_B \qquad y_{A/B} = y_A - y_B \qquad z_{A/B} = z_A - z_B$$

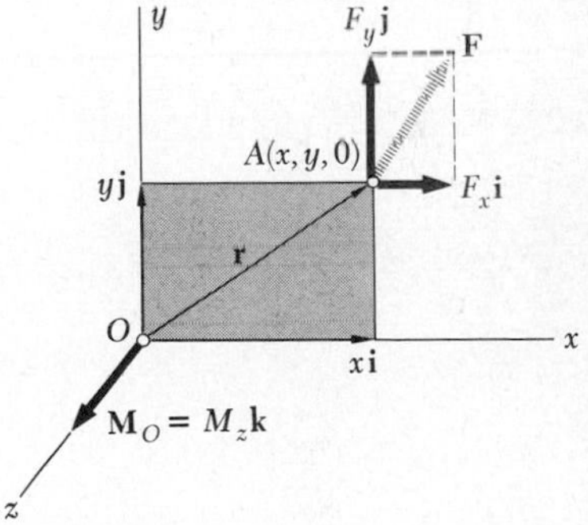

Fig. 3.17

In the case of *problems involving only two dimensions,* the force $\mathbf{F}$ may be assumed to lie in the xy plane (Fig. 3.17). Carrying $z = 0$ and $F_z = 0$ into the relations (3.19), we obtain

$$\mathbf{M}_O = (xF_y - yF_x)\mathbf{k}$$

We check that the moment of $\mathbf{F}$ about O is perpendicular to the plane of the figure and that it is completely defined by the scalar

$$M_O = M_z = xF_y - yF_x \tag{3.22}$$

As noted earlier, a positive value for M_O indicates that the vector $\mathbf{M}_O$ points out of the paper (the force $\mathbf{F}$ tends to rotate the body counterclockwise about O), and a negative value that the vector $\mathbf{M}_O$ points into the paper (the force $\mathbf{F}$ tends to rotate the body clockwise about O).

To compute the moment about $B(x_B, y_B)$ of a force lying in the xy plane and applied at $A(x_A, y_A)$ (Fig. 3.18), we carry $z_{A/B} = 0$ and $F_z = 0$ in the relations (3.21) and check that the vector $\mathbf{M}_B$ is perpendicular to the xy plane and is defined in magnitude and sense by the scalar

$$M_B = (x_A - x_B)F_y - (y_A - y_B)F_x \tag{3.23}$$

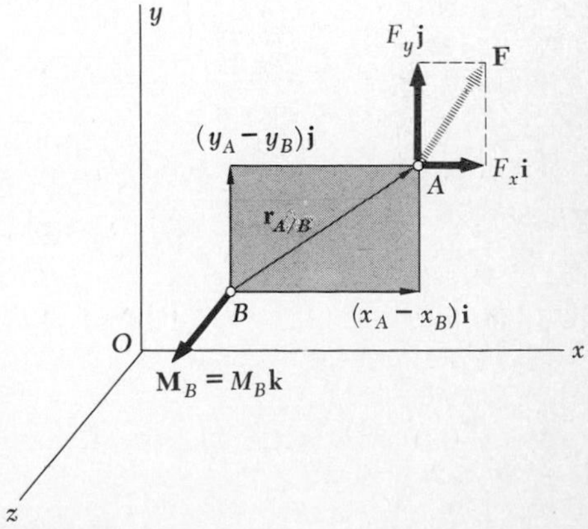

Fig. 3.18

SAMPLE PROBLEM 3.1

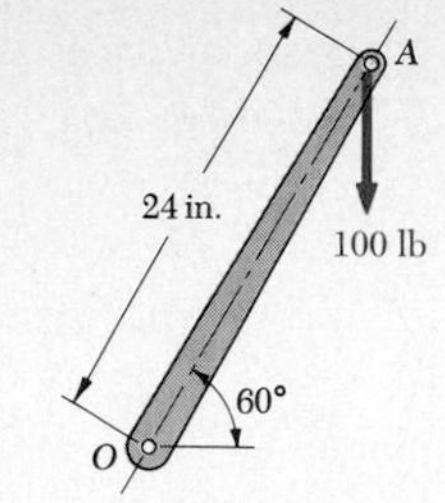

A 100-lb vertical force is applied to the end of a lever which is attached to a shaft at O. Determine (*a*) the moment of the 100-lb force about O; (*b*) the magnitude of the horizontal force applied at A which creates the same moment about O; (*c*) the smallest force applied at A which creates the same moment about O; (*d*) how far from the shaft a 240-lb vertical force must act to create the same moment about O; (*e*) whether any one of the forces obtained in parts *b*, *c*, and *d* is equivalent to the original force.

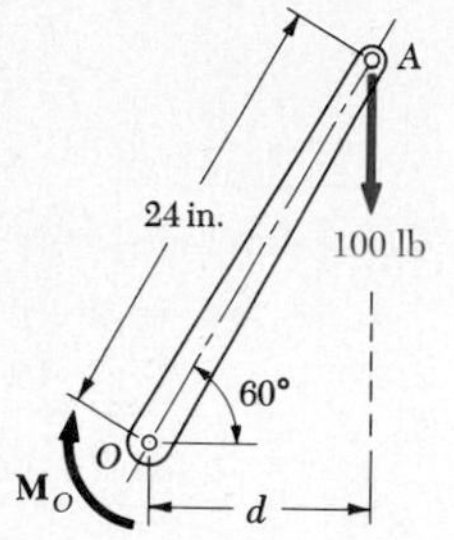

***a.* Moment about *O*.** The perpendicular distance from O to the line of action of the 100-lb force is

$$d = (24\text{ in.}) \cos 60° = 12\text{ in.}$$

The magnitude of the moment about O of the 100-lb force is

$$M_O = Fd = (100\text{ lb})(12\text{ in.}) = 1200\text{ lb}\cdot\text{in.}$$

Since the force tends to rotate the lever clockwise about O, the moment will be represented by a vector $\mathbf{M}_O$ perpendicular to the plane of the figure and pointing *into* the paper. We express this fact by writing

$$\mathbf{M}_O = 1200\text{ lb}\cdot\text{in.} \circlearrowright \quad ◄$$

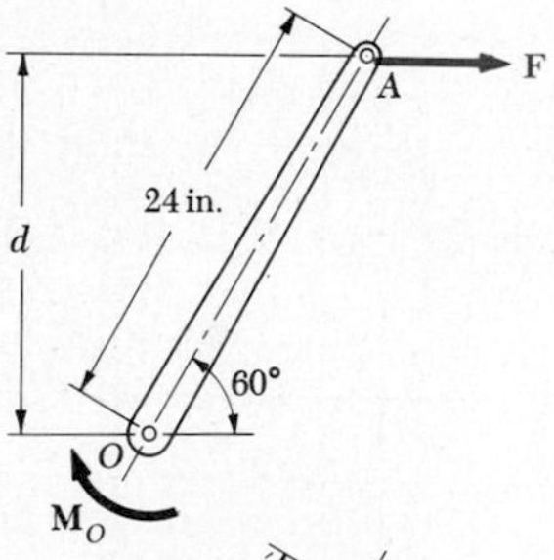

***b.* Horizontal Force.** In this case, we have

$$d = (24\text{ in.}) \sin 60° = 20.8\text{ in.}$$

Since the moment about O must be 1200 lb·in., we write

$$M_O = Fd$$
$$1200\text{ lb}\cdot\text{in.} = F(20.8\text{ in.})$$
$$F = 57.7\text{ lb} \qquad \mathbf{F} = 57.7\text{ lb} \rightarrow \quad ◄$$

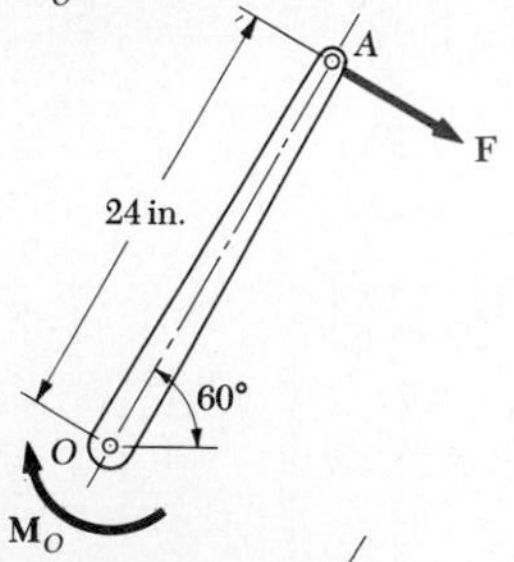

***c.* Smallest Force.** Since $M_O = Fd$, the smallest value of F occurs when d is maximum. We choose the force perpendicular to OA and find $d = 24$ in.; thus

$$M_O = Fd$$
$$1200\text{ lb}\cdot\text{in.} = F(24\text{ in.})$$
$$F = 50\text{ lb} \qquad \mathbf{F} = 50\text{ lb} \measuredangle 30° \quad ◄$$

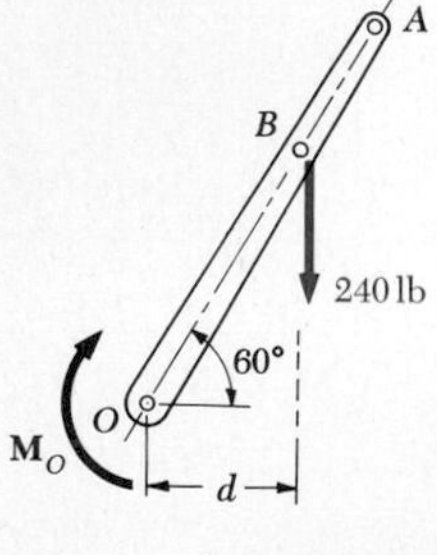

***d.* 240-lb Vertical Force.** In this case $M_O = Fd$ yields

$$1200\text{ lb}\cdot\text{in.} = (240\text{ lb})d \qquad d = 5\text{ in.}$$

but $\quad OB \cos 60° = d \qquad OB = 10\text{ in.} \quad ◄$

e. None of the forces considered in parts *b*, *c*, and *d* is equivalent to the original 100-lb force. Although they have the same moment about O, they have different x and y components. In other words, although each force tends to rotate the shaft in the same manner, each causes the lever to pull on the shaft in a different way.

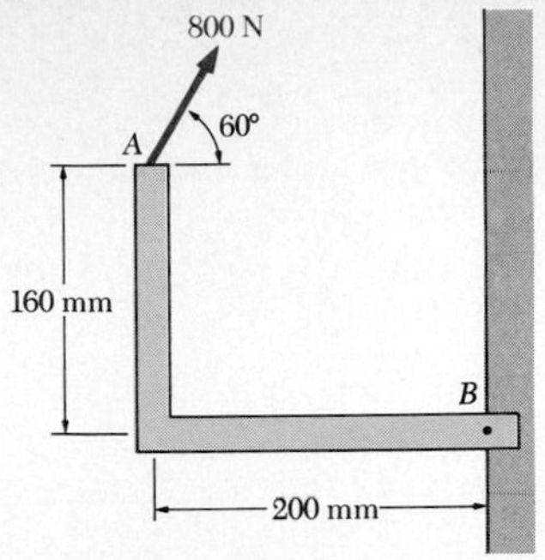

SAMPLE PROBLEM 3.2

A force of 800 N acts on a bracket as shown. Determine the moment of the force about B.

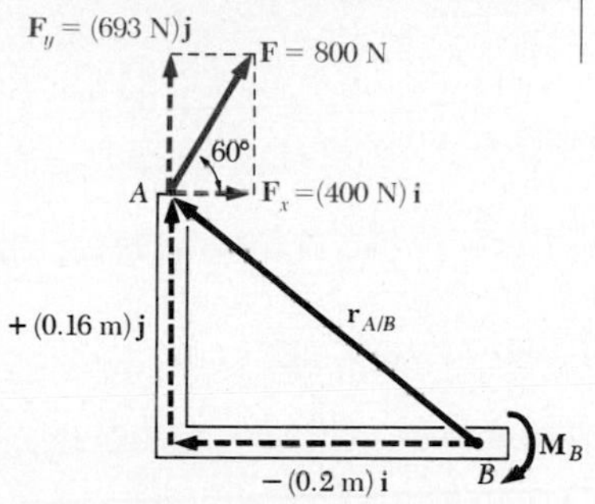

Solution. The moment $\mathbf{M}_B$ of the force $\mathbf{F}$ about B is obtained by forming the vector product

$$\mathbf{M}_B = \mathbf{r}_{A/B} \times \mathbf{F}$$

where $\mathbf{r}_{A/B}$ is the vector drawn from B to A. Resolving $\mathbf{r}_{A/B}$ and $\mathbf{F}$ into rectangular components, we have

$$\begin{aligned} \mathbf{r}_{A/B} &= -(0.2\text{ m})\mathbf{i} + (0.16\text{ m})\mathbf{j} \\ \mathbf{F} &= (800\text{ N})\cos 60°\mathbf{i} + (800\text{ N})\sin 60°\mathbf{j} \\ &= (400\text{ N})\mathbf{i} + (693\text{ N})\mathbf{j} \end{aligned}$$

Recalling the relations (3.7) for the cross products of unit vectors (Sec. 3.5), we obtain

$$\begin{aligned} \mathbf{M}_B = \mathbf{r}_{A/B} \times \mathbf{F} &= [-(0.2\text{ m})\mathbf{i} + (0.16\text{ m})\mathbf{j}] \times [(400\text{ N})\mathbf{i} + (693\text{ N})\mathbf{j}] \\ &= -(138.6\text{ N}\cdot\text{m})\mathbf{k} - (64.0\text{ N}\cdot\text{m})\mathbf{k} \\ &= -(202.6\text{ N}\cdot\text{m})\mathbf{k} \end{aligned}$$

$$M_B = 203\text{ N}\cdot\text{m} \curvearrowright \blacktriangleleft$$

The moment $\mathbf{M}_B$ is a vector perpendicular to the plane of the figure and pointing *into* the paper.

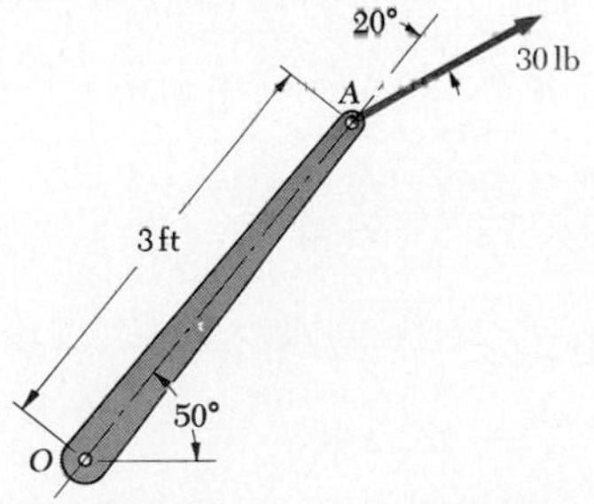

SAMPLE PROBLEM 3.3

A 30-lb force acts on the end of the 3-ft lever as shown. Determine the moment of the force about O.

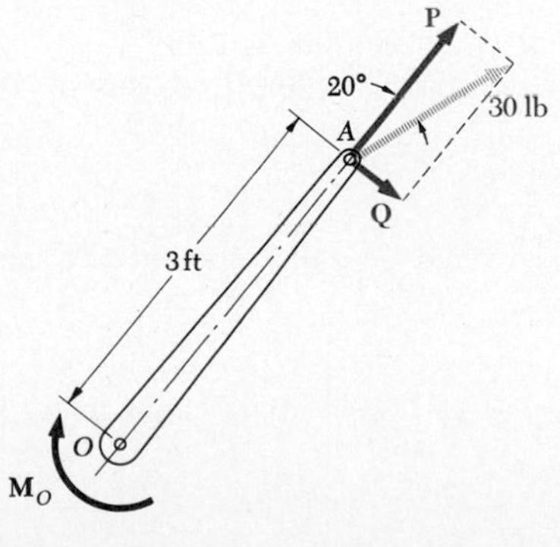

Solution. The force is replaced by two components, one component $\mathbf{P}$ in the direction of OA and one component $\mathbf{Q}$ perpendicular to OA. Since O is on the line of action of $\mathbf{P}$, the moment of $\mathbf{P}$ about O is zero and the moment of the 30-lb force reduces to the moment of $\mathbf{Q}$, which is clockwise and, thus, represented by a negative scalar.

$$\begin{aligned} Q &= (30\text{ lb})\sin 20° = 10.26\text{ lb} \\ M_O &= -Q(3\text{ ft}) = -(10.26\text{ lb})(3\text{ ft}) = -30.8\text{ lb}\cdot\text{ft} \end{aligned}$$

Since the value obtained for the scalar M_O is negative, the moment $\mathbf{M}_O$ points *into* the paper. We write

$$\mathbf{M}_O = 30.8\text{ lb}\cdot\text{ft} \curvearrowright \blacktriangleleft$$

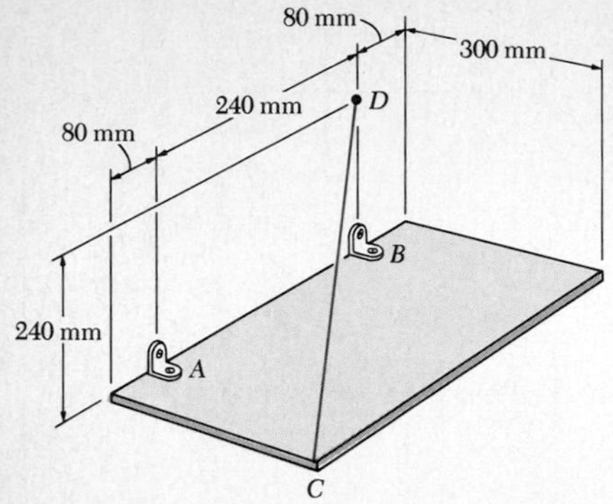

SAMPLE PROBLEM 3.4

A rectanglar plate is supported by brackets at A and B and by a wire CD. Knowing that the tension in the wire is 200 N, determine the moment about A of the force exerted by the wire on point C.

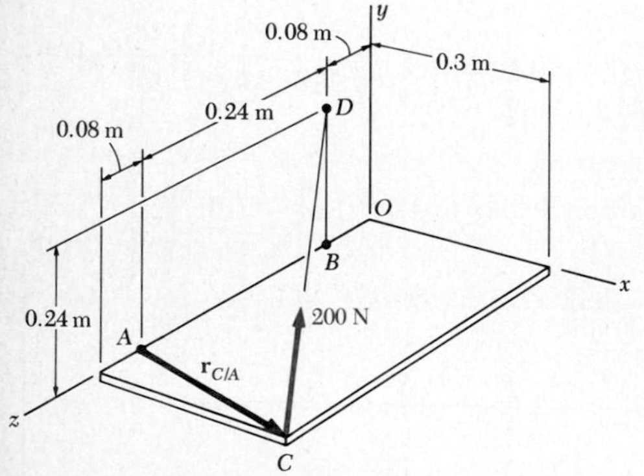

Solution. The moment $\mathbf{M}_A$ about A of the force $\mathbf{F}$ exerted by the wire on point C is obtained by forming the vector product

$$\mathbf{M}_A = \mathbf{r}_{C/A} \times \mathbf{F} \tag{1}$$

where $\mathbf{r}_{C/A}$ is the vector drawn from A to C,

$$\mathbf{r}_{C/A} = \overrightarrow{AC} = (0.3 \text{ m})\mathbf{i} + (0.08 \text{ m})\mathbf{k} \tag{2}$$

and where $\mathbf{F}$ is the 200-N force directed along CD. Introducing the unit vector $\boldsymbol{\lambda} = \overrightarrow{CD}/CD$, we write

$$\mathbf{F} = F\boldsymbol{\lambda} = (200 \text{ N})\frac{\overrightarrow{CD}}{CD} \tag{3}$$

Resolving the vector $\overrightarrow{CD}$ into rectangular components, we have

$$\overrightarrow{CD} = -(0.3 \text{ m})\mathbf{i} + (0.24 \text{ m})\mathbf{j} - (0.32 \text{ m})\mathbf{k} \qquad CD = 0.50 \text{ m}$$

Substituting into (3), we obtain

$$\begin{aligned} \mathbf{F} &= \frac{200 \text{ N}}{0.50 \text{ m}}[-(0.3 \text{ m})\mathbf{i} + (0.24 \text{ m})\mathbf{j} - (0.32 \text{ m})\mathbf{k}] \\ &= -(120 \text{ N})\mathbf{i} + (96 \text{ N})\mathbf{j} - (128 \text{ N})\mathbf{k} \end{aligned} \tag{4}$$

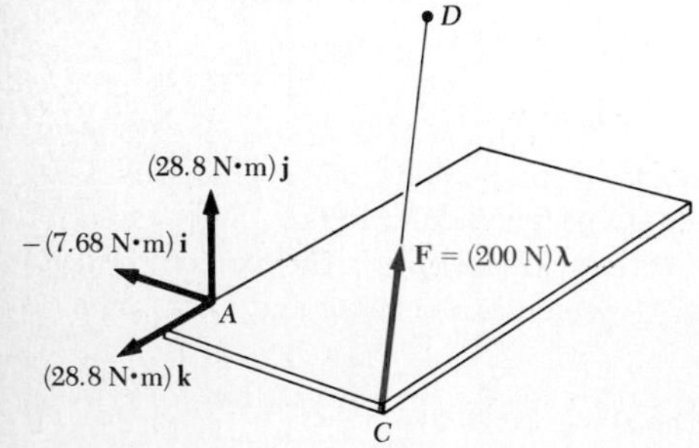

Substituting for $\mathbf{r}_{C/A}$ and $\mathbf{F}$ from (2) and (4) into (1), and recalling the relations (3.7) of Sec. 3.5, we obtain

$$\begin{aligned} \mathbf{M}_A = \mathbf{r}_{C/A} \times \mathbf{F} &= (0.3\mathbf{i} + 0.08\mathbf{k}) \times (-120\mathbf{i} + 96\mathbf{j} - 128\mathbf{k}) \\ &= (0.3)(96)\mathbf{k} + (0.3)(-128)(-\mathbf{j}) + (0.08)(-120)\mathbf{j} + (0.08)(96)(-\mathbf{i}) \end{aligned}$$

$$\mathbf{M}_A = -(7.68 \text{ N}\cdot\text{m})\mathbf{i} + (28.8 \text{ N}\cdot\text{m})\mathbf{j} + (28.8 \text{ N}\cdot\text{m})\mathbf{k} \quad ◄$$

Alternate Solution. As indicated in Sec. 3.8, the moment $\mathbf{M}_A$ may be expressed in the form of a determinant:

$$\mathbf{M}_A = \begin{vmatrix} \mathbf{i} & \mathbf{j} & \mathbf{k} \\ x_C - x_A & y_C - y_A & z_C - z_A \\ F_x & F_y & F_z \end{vmatrix} = \begin{vmatrix} \mathbf{i} & \mathbf{j} & \mathbf{k} \\ 0.3 & 0 & 0.08 \\ -120 & 96 & -128 \end{vmatrix}$$

$$\mathbf{M}_A = -(7.68 \text{ N}\cdot\text{m})\mathbf{i} + (28.8 \text{ N}\cdot\text{m})\mathbf{j} + (28.8 \text{ N}\cdot\text{m})\mathbf{k} \quad ◄$$

Problems

3.1 A 30-lb force is applied to the control rod AB as shown. Knowing that the length of the rod is 8 in. and that $\alpha = 30°$, determine the moment of the force about B by resolving the force (*a*) into horizontal and vertical components, (*b*) into components along AB and in a direction perpendicular to AB.

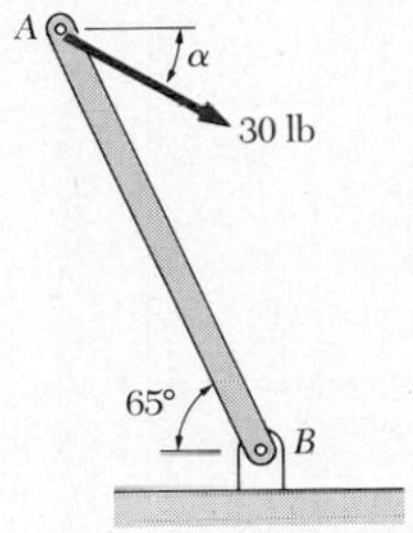

Fig. P3.1 and P3.2

3.2 A 30-lb force is applied to the control rod AB as shown. Knowing that the length of the rod is 8 in. and that the moment of the force about B is 180 lb·in., determine the value of α.

3.3 For the brake pedal shown, determine the magnitude and direction of the smallest force **P** which has a 130-N·m moment about B.

3.4 A force **P** is applied to the brake pedal at A. Knowing that $P = 450$ N and $\alpha = 30°$, determine the moment of **P** about B.

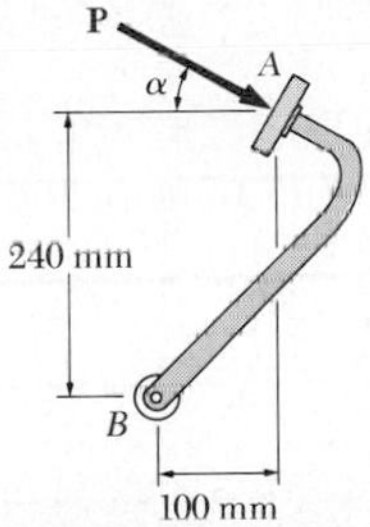

Fig. P3.3 and P3.4

3.5 A 300-N force **P** is applied at point A of the bell crank shown. (*a*) Compute the moment of the force **P** about O by resolving it into horizontal and vertical components. (*b*) Using the result of part *a*, determine the perpendicular distance from O to the line of action of **P**.

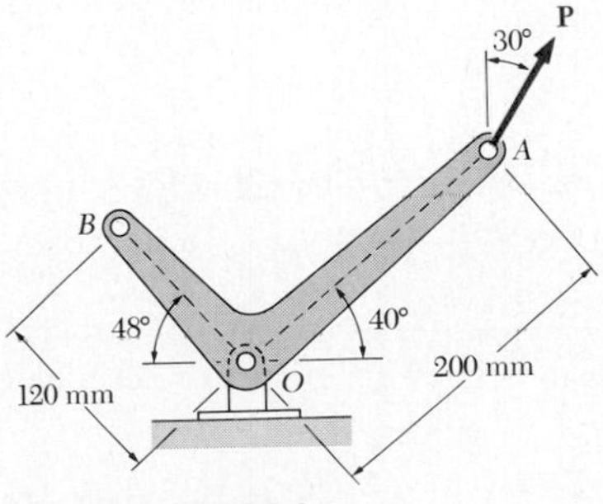

Fig. P3.5 and P3.6

3.6 A 400-N force **P** is applied at point A of the bell crank shown. (*a*) Compute the moment of the force **P** about O by resolving it into components along line OA and in a direction perpendicular to that line. (*b*) Determine the magnitude and direction of the smallest force **Q** applied at B which has the same moment as **P** about O.

3.7 and 3.8 It is known that the connecting rod AB exerts on the crank BC a 1.5-kN force directed down and to the left along the centerline of AB. Determine the moment of that force about C.

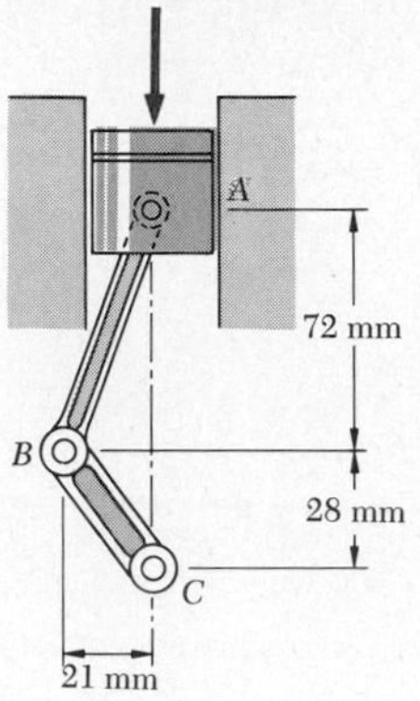

Fig. P3.7

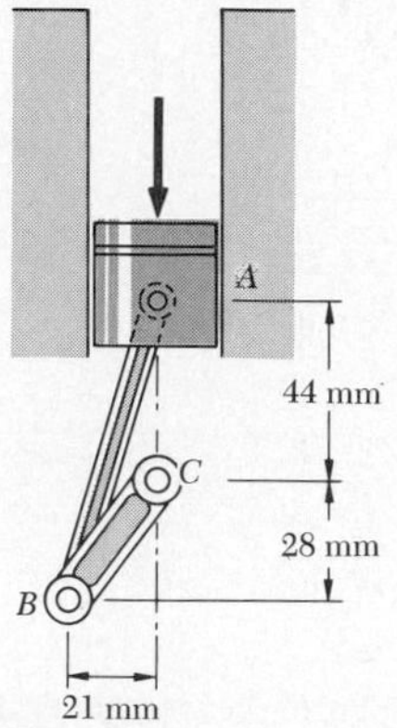

Fig. P3.8

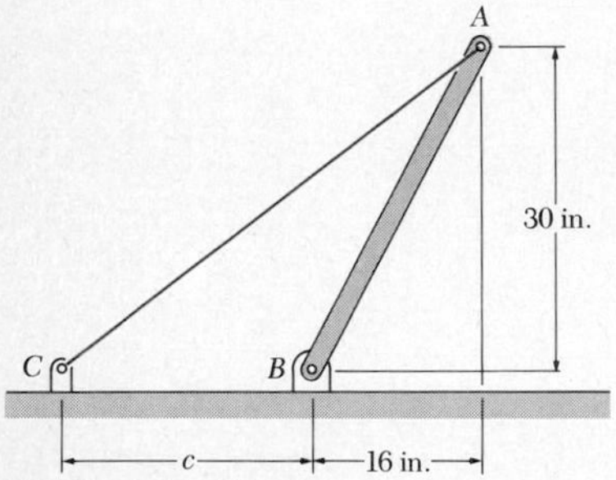

Fig. P3.9 and P3.10

3.9 Rod AB is held in place by the cord AC. Knowing that the tension in the cord is 250 lb and that $c = 24$ in., determine the moment about B of the force exerted by the cord at point A by resolving that force into horizontal and vertical components applied (*a*) at point A, (*b*) at point C.

3.10 Rod AB is held in place by the cord AC. Knowing that $c = 56$ in. and that the moment about B of the force exerted by the cord at point A is 280 lb·ft, determine the tension in the cord.

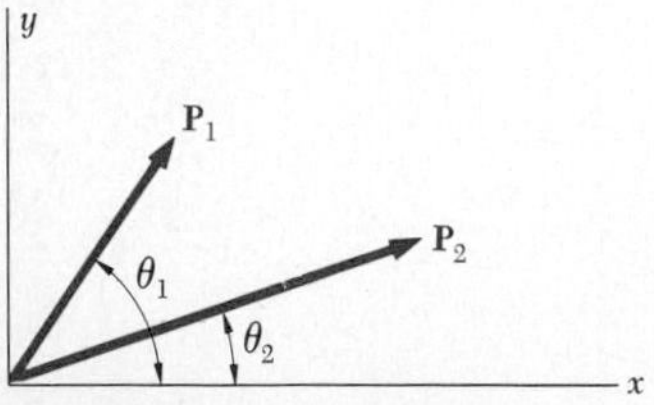

Fig. P3.11

3.11 Form the vector product $\mathbf{P}_1 \times \mathbf{P}_2$ and use the result obtained to prove the identity $\sin(\theta_1 - \theta_2) = \sin\theta_1 \cos\theta_2 - \cos\theta_1 \sin\theta_2$.

3.12 A force $\mathbf{F} = F_x\mathbf{i} + F_y\mathbf{j}$ acts at a point of coordinates x and y. Derive an expression for the perpendicular distance d from the line of action of $\mathbf{F}$ to the origin O of the system of coordinates.

3.13 The line of action of a force $\mathbf{P}$ passes through two points $A(x_1, y_1)$ and $B(x_2, y_2)$. If the force is directed from A to B, determine the moment of the force about the origin.

3.14 Determine the moment about the origin O of the force $\mathbf{F} = 4\mathbf{i} - 3\mathbf{j} + 5\mathbf{k}$ which acts at a point A. Assume that the position vector of A is (*a*) $\mathbf{r} = 2\mathbf{i} + 3\mathbf{j} - 4\mathbf{k}$, (*b*) $\mathbf{r} = -8\mathbf{i} + 6\mathbf{j} - 10\mathbf{k}$, (*c*) $\mathbf{r} = 8\mathbf{i} - 6\mathbf{j} + 5\mathbf{k}$.

3.15 Determine the moment about the origin O of the force $\mathbf{F} = 2\mathbf{i} + 3\mathbf{j} - 4\mathbf{k}$ which acts at a point A. Assume that the position vector of A is (*a*) $\mathbf{r} = 3\mathbf{i} - 6\mathbf{j} + 5\mathbf{k}$, (*b*) $\mathbf{r} = \mathbf{i} - 4\mathbf{j} - 2\mathbf{k}$, (*c*) $\mathbf{r} = 4\mathbf{i} + 6\mathbf{j} - 8\mathbf{k}$.

3.16 A 200-N force is applied as shown to the bracket ABC. Determine the moment of the force about A.

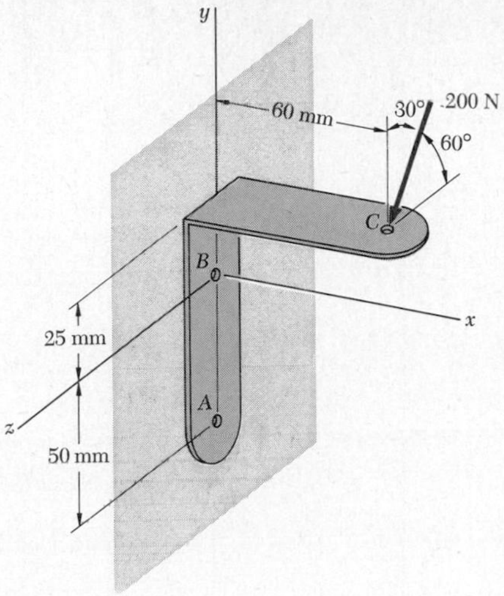

Fig. P3.16

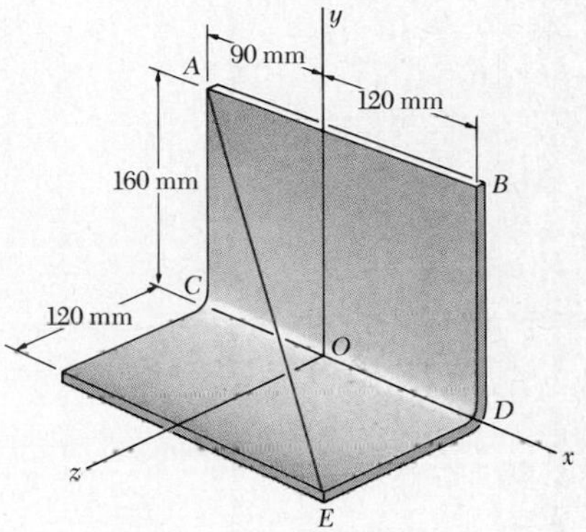

Fig. P3.17

3.17 The wire AE is stretched between the corners A and E of a bent plate. Knowing that the tension in the wire is 435 N, determine the moment about O of the force exerted by the wire (*a*) on corner A, (*b*) on corner E.

3.18 A small boat hangs from two davits, one of which is shown in the figure. The tension in line $ABAD$ is 41 lb. Determine the moment about C of the resultant force $\mathbf{R}_A$ exerted on the davit at A.

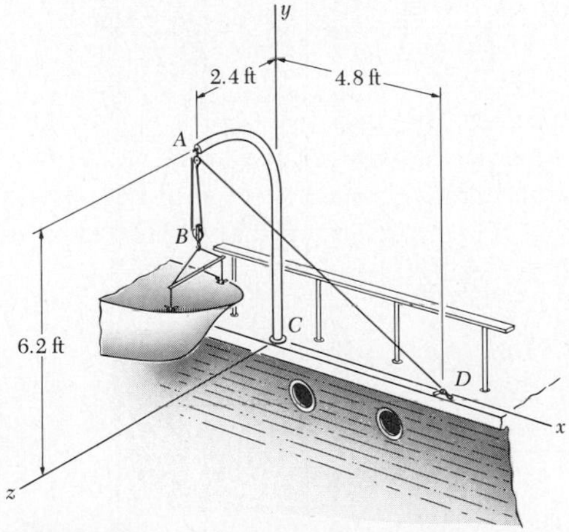

Fig. P3.18

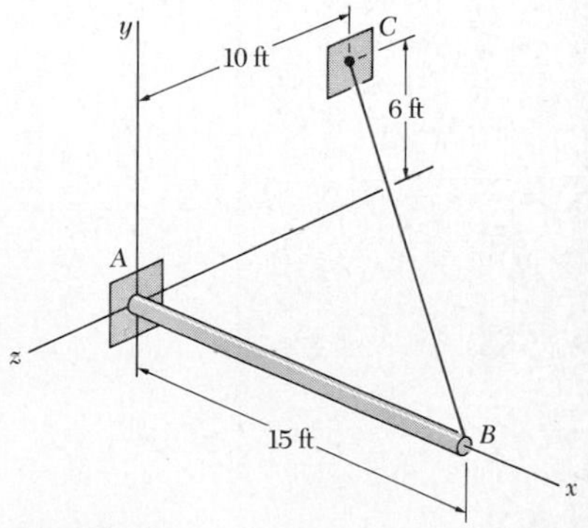

Fig. P3.19

3.19 The 15-ft boom AB has a fixed end A. A steel cable is stretched from the free end B of the boom to a point C located on the vertical wall. If the tension in the cable is 570 lb, determine the moment about A of the force exerted by the cable at B.

3.20 In Prob. 3.17, determine the perpendicular distance from point O to wire AE.

3.21 In Prob. 3.17, determine the perpendicular distance from point B to wire AE.

3.22 In Prob. 3.19, determine the perpendicular distance from point A to cable BC.

3.23 In Prob. 3.18, determine the perpendicular distance from point C to portion AD of the line $ABAD$.

3.24 In Sample Prob. 3.4, determine the perpendicular distance from point A to wire CD.

3.25 A single force $\mathbf{F}$ acts at a point A of coordinates $x = y = z = a$. Show that $M_x + M_y + M_z = 0$; that is, show that the *algebraic* sum of the rectangular components of the moment of $\mathbf{F}$ about O is zero.

3.9. Scalar Product of Two Vectors.

We shall now expand our knowledge of vector algebra and introduce the *scalar product* of two vectors.

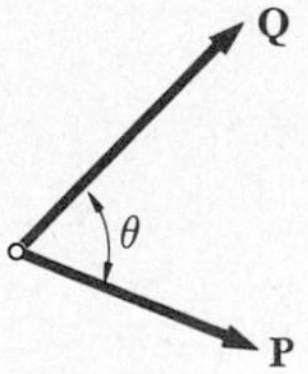

Fig. 3.19

The scalar product of two vectors $\mathbf{P}$ and $\mathbf{Q}$ is defined as the product of the magnitudes of $\mathbf{P}$ and $\mathbf{Q}$ and of the cosine of the angle θ formed by $\mathbf{P}$ and $\mathbf{Q}$ (Fig. 3.19). The scalar product of $\mathbf{P}$ and $\mathbf{Q}$ is denoted by $\mathbf{P} \cdot \mathbf{Q}$. We write therefore

$$\mathbf{P} \cdot \mathbf{Q} = PQ \cos \theta \tag{3.24}$$

Note that the expression just defined is not a vector but a *scalar*, which explains the name *scalar product;* because of the notation used, $\mathbf{P} \cdot \mathbf{Q}$ is also referred to as the *dot product* of the vectors $\mathbf{P}$ and $\mathbf{Q}$.

It follows from its very definition that the scalar product of two vectors is *commutative*, i.e., that

$$\mathbf{P} \cdot \mathbf{Q} = \mathbf{Q} \cdot \mathbf{P} \tag{3.25}$$

To prove that the scalar product is also *distributive*, we must prove the relation

$$\mathbf{P} \cdot (\mathbf{Q}_1 + \mathbf{Q}_2) = \mathbf{P} \cdot \mathbf{Q}_1 + \mathbf{P} \cdot \mathbf{Q}_2 \tag{3.26}$$

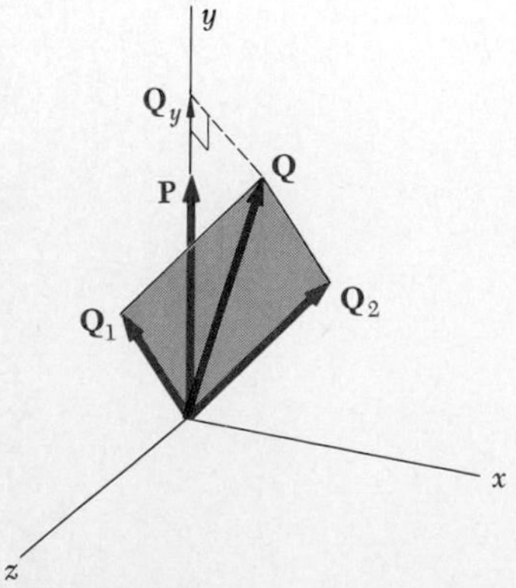

Fig. 3.20

We may, without any loss of generality, assume that $\mathbf{P}$ is directed along the y axis (Fig. 3.20). Denoting by $\mathbf{Q}$ the sum of $\mathbf{Q}_1$ and $\mathbf{Q}_2$, and by θ_y the angle $\mathbf{Q}$ forms with the y axis, we express the left-hand member of (3.26) as follows:

$$\mathbf{P} \cdot (\mathbf{Q}_1 + \mathbf{Q}_2) = \mathbf{P} \cdot \mathbf{Q} = PQ \cos \theta_y = PQ_y \tag{3.27}$$

where Q_y is the y component of $\mathbf{Q}$. We may, in a similar way, express the right-hand member of (3.26) as

$$\mathbf{P} \cdot \mathbf{Q}_1 + \mathbf{P} \cdot \mathbf{Q}_2 = P(Q_1)_y + P(Q_2)_y \tag{3.28}$$

Since $\mathbf{Q}$ is the sum of $\mathbf{Q}_1$ and $\mathbf{Q}_2$, its y component must be equal to the sum of the y components of $\mathbf{Q}_1$ and $\mathbf{Q}_2$. Thus, the expressions obtained in (3.27) and (3.28) are equal and the relation (3.26) has been proved.

As far as the third property—the associative property—is concerned, we note that this property cannot apply to scalar products. Indeed, $(\mathbf{P} \cdot \mathbf{Q}) \cdot \mathbf{S}$ has no meaning, since $\mathbf{P} \cdot \mathbf{Q}$ is not a vector but a scalar.

We shall now express the scalar product of two vectors $\mathbf{P}$ and $\mathbf{Q}$ in terms of their rectangular components. Resolving $\mathbf{P}$ and $\mathbf{Q}$ into components, we first write

$$\mathbf{P} \cdot \mathbf{Q} = (P_x\mathbf{i} + P_y\mathbf{j} + P_z\mathbf{k}) \cdot (Q_x\mathbf{i} + Q_y\mathbf{j} + Q_z\mathbf{k})$$

Making use of the distributive property, we express $\mathbf{P} \cdot \mathbf{Q}$ as the sum of scalar products such as $P_x\mathbf{i} \cdot Q_x\mathbf{i}$ and $P_x\mathbf{i} \cdot Q_y\mathbf{j}$. But we may easily check from the definition of the scalar product that the scalar products of the unit vectors are either zero or one.

$$\begin{array}{lll} \mathbf{i} \cdot \mathbf{i} = 1 & \mathbf{j} \cdot \mathbf{j} = 1 & \mathbf{k} \cdot \mathbf{k} = 1 \\ \mathbf{i} \cdot \mathbf{j} = 0 & \mathbf{j} \cdot \mathbf{k} = 0 & \mathbf{k} \cdot \mathbf{i} = 0 \end{array} \tag{3.29}$$

Thus, the expression obtained for $\mathbf{P} \cdot \mathbf{Q}$ reduces to

$$\mathbf{P} \cdot \mathbf{Q} = P_xQ_x + P_yQ_y + P_zQ_z \tag{3.30}$$

In the particular case when $\mathbf{P}$ and $\mathbf{Q}$ are equal, we check that

$$\mathbf{P} \cdot \mathbf{P} = P_x^2 + P_y^2 + P_z^2 = P^2 \tag{3.31}$$

Applications

1. *Angle formed by two given vectors.* Let two vectors be given in terms of their components:

$$\mathbf{P} = P_x\mathbf{i} + P_y\mathbf{j} + P_z\mathbf{k}$$
$$\mathbf{Q} = Q_x\mathbf{i} + Q_y\mathbf{j} + Q_z\mathbf{k}$$

To determine the angle formed by the two vectors we shall equate the expressions obtained in (3.24) and (3.30) for their scalar product and write:

$$PQ \cos \theta = P_xQ_x + P_yQ_y + P_zQ_z$$

Solving for $\cos \theta$, we have

$$\cos \theta = \frac{P_xQ_x + P_yQ_y + P_zQ_z}{PQ} \tag{3.32}$$

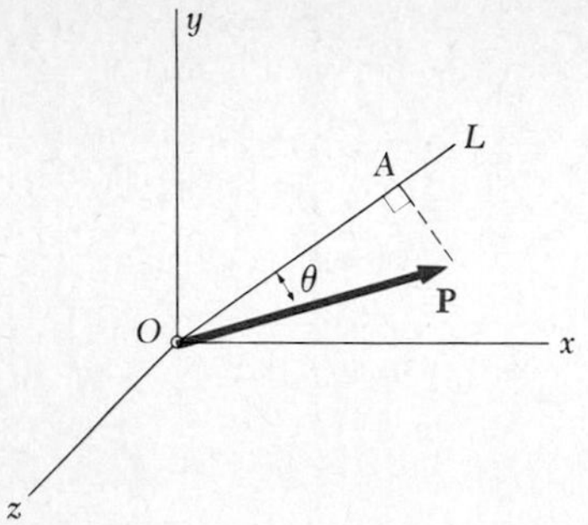

Fig. 3.21

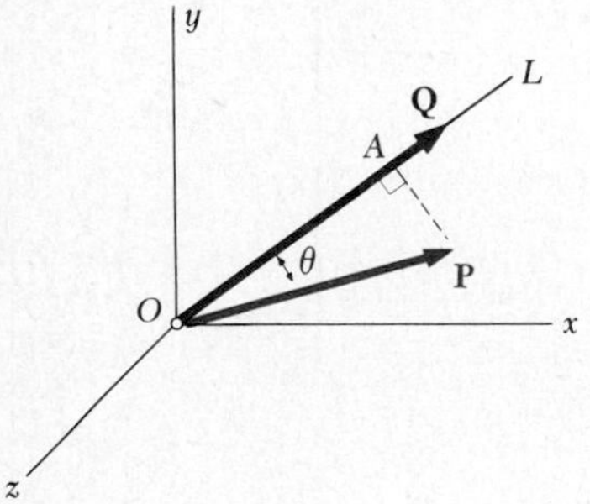

Fig. 3.22

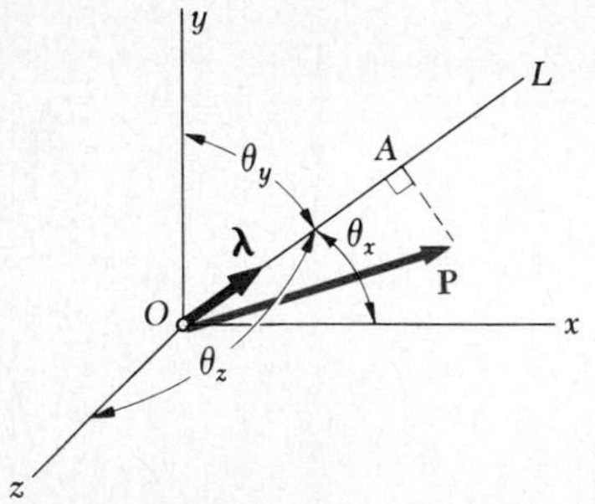

Fig. 3.23

2. *Projection of a vector on a given axis.* Consider a vector **P** forming an angle θ with an axis, or directed line, OL (Fig. 3.21). The *projection of* **P** *on the axis OL* is defined as the scalar

$$P_{OL} = P \cos \theta \tag{3.33}$$

We note that the projection P_{OL} is equal in absolute value to the length of the segment OA; it will be positive if OA has the same sense as the axis OL, that is, if θ is acute, and negative otherwise. If **P** and OL are at a right angle, the projection of **P** on OL is zero.

Consider now a vector **Q** directed along OL and of the same sense as OL (Fig. 3.22). The scalar product of **P** and **Q** may be expressed as

$$\mathbf{P} \cdot \mathbf{Q} = PQ \cos \theta = P_{OL} Q \tag{3.34}$$

from which it follows that

$$P_{OL} = \frac{\mathbf{P} \cdot \mathbf{Q}}{Q} = \frac{P_x Q_x + P_y Q_y + P_z Q_z}{Q} \tag{3.35}$$

In the particular case when the vector selected along OL is the unit vector $\boldsymbol{\lambda}$ (Fig. 3.23), we write

$$P_{OL} = \mathbf{P} \cdot \boldsymbol{\lambda} \tag{3.36}$$

Resolving **P** and $\boldsymbol{\lambda}$ into rectangular components, and recalling from Sec. 2.12 that the components of $\boldsymbol{\lambda}$ along the coordinate axes are respectively equal to the direction cosines of OL, we express the projection of **P** on OL as

$$P_{OL} = P_x \cos \theta_x + P_y \cos \theta_y + P_z \cos \theta_z \tag{3.37}$$

where θ_x, θ_y, and θ_z denote the angles that the axis OL forms with the coordinate axes.

3.10. Mixed Triple Product of Three Vectors.

We define the *mixed triple product* of the three vectors **S**, **P**, and **Q** as the scalar expression

$$\mathbf{S} \cdot (\mathbf{P} \times \mathbf{Q}) \tag{3.38}$$

obtained by forming the scalar product of **S** with the vector product of **P** and **Q**.†

A simple geometrical interpretation may be given for the mixed triple product of **S**, **P**, and **Q** (Fig. 3.24). We first recall from Sec. 3.4 that the vector $\mathbf{P} \times \mathbf{Q}$ is perpendicular to the plane containing **P** and **Q**, and that its

†Another kind of triple product will be introduced later (Chap. 15): the *vector triple product* $\mathbf{S} \times (\mathbf{P} \times \mathbf{Q})$.

magnitude is equal to the area of the parallelogram constructed on **P** and **Q**. On the other hand, Eq. (3.34) indicates that the scalar product of **S** and **P** × **Q** may be obtained by multiplying the magnitude of **P** × **Q** (i.e., the area of the parallelogram built on **P** and **Q**) by the projection of **S** on the vector **P** × **Q** (i.e., by the projection of **S** on the normal to the plane containing the parallelogram). The mixed triple product is thus equal, in absolute value, to the volume of the parallelepiped having the vectors **S**, **P**, and **Q** for sides (Fig. 3.25). We may check that the sign of the mixed triple product will be positive if **S**, **P**, and **Q** form a right-handed triad, and negative if they form a left-handed triad [that is, **S** · (**P** × **Q**) will be negative if the rotation which brings **P** into line with **Q** is observed as clockwise from the tip of **S**]. The mixed triple product will be zero if **S**, **P**, and **Q** are coplanar.

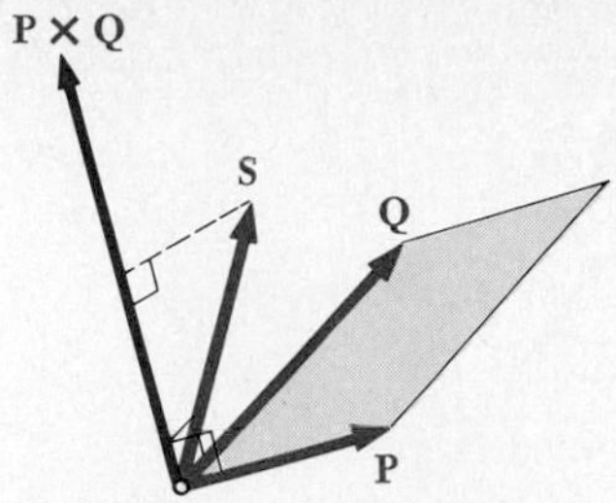

Fig. 3.24

Since the parallelepiped defined in the preceding paragraph is independent of the order in which the three vectors are taken, the six mixed triple products which may be formed with **S**, **P**, and **Q** will all have the same absolute value, although not the same sign. We check that

$$\mathbf{S}\cdot(\mathbf{P}\times\mathbf{Q}) = \mathbf{P}\cdot(\mathbf{Q}\times\mathbf{S}) = \mathbf{Q}\cdot(\mathbf{S}\times\mathbf{P})$$
$$= -\mathbf{S}\cdot(\mathbf{Q}\times\mathbf{P}) = -\mathbf{P}\cdot(\mathbf{S}\times\mathbf{Q}) = -\mathbf{Q}\cdot(\mathbf{P}\times\mathbf{S}) \qquad (3.39)$$

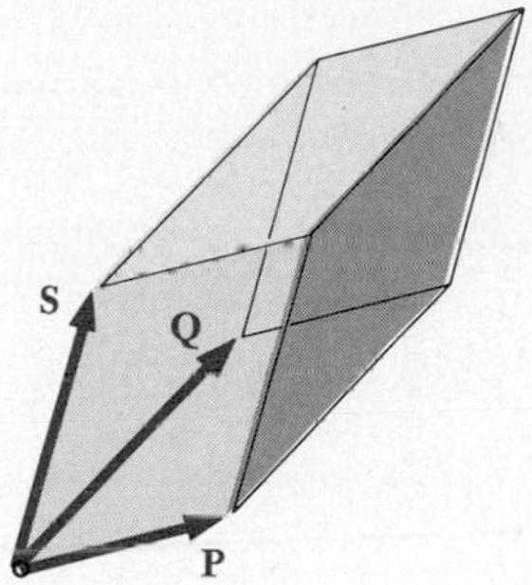

Fig. 3.25

Arranging in a circle and in counterclockwise order the letters representing the three vectors (Fig. 3.26), we note that the sign of the mixed triple product is conserved if the vectors are permuted in such a way that they are still read in counterclockwise order. Such a permutation is said to be a *circular permutation*. It also follows from (3.39) that the mixed triple product of **S**, **P**, and **Q** may be defined equally well as **S** · (**P** × **Q**) or (**S** × **P**) · **Q**.

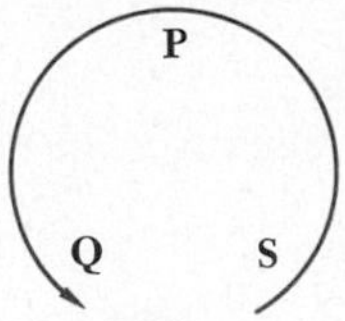

Fig. 3.26

We shall now express the mixed triple product of the vectors **S**, **P**, and **Q** in terms of the rectangular components of these vectors. Denoting **P** × **Q** by **V**, and using formula (3.30) to express the scalar product of **S** and **V**, we write

$$\mathbf{S}\cdot(\mathbf{P}\times\mathbf{Q}) = \mathbf{S}\cdot\mathbf{V} = S_xV_x + S_yV_y + S_zV_z$$

Substituting from the relations (3.9) for the components of **V**, we obtain

$$\mathbf{S}\cdot(\mathbf{P}\times\mathbf{Q}) = S_x(P_yQ_z - P_zQ_y) + S_y(P_zQ_x - P_xQ_z)$$
$$+ S_z(P_xQ_y - P_yQ_x) \qquad (3.40)$$

This expression may be written in a more compact form if we observe that it represents the expansion of a determinant:

$$\mathbf{S}\cdot(\mathbf{P}\times\mathbf{Q}) = \begin{vmatrix} S_x & S_y & S_z \\ P_x & P_y & P_z \\ Q_x & Q_y & Q_z \end{vmatrix} \qquad (3.41)$$

By applying the rules governing the permutation of rows in a determinant, we could easily verify the relations (3.39) which were derived earlier from geometrical considerations.

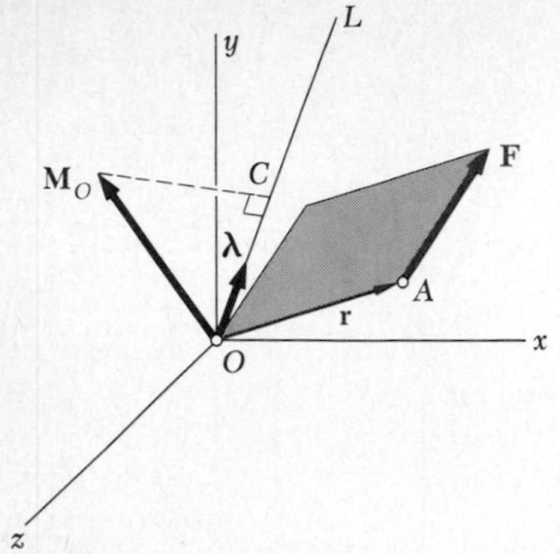

Fig. 3.27

3.11. Moment of a Force about a Given Axis. Now that we have further increased our knowledge of vector algebra, we shall introduce a new concept, the concept of *moment of a force about an axis*. Consider again a force **F** acting on a rigid body and the moment $\mathbf{M}_O$ of that force about O (Fig. 3.27). Let OL be an axis through O; *we define the moment* M_{OL} *of* **F** *about OL as the projection OC of the moment* $\mathbf{M}_O$ *on the axis OL.* Denoting by $\boldsymbol{\lambda}$ the unit vector along OL, and recalling from Secs. 3.9 and 3.6, respectively, the expressions (3.36) and (3.11) obtained for the projection of a vector on a given axis and for the moment $\mathbf{M}_O$ of a force **F**, we write

$$M_{OL} = \boldsymbol{\lambda} \cdot \mathbf{M}_O = \boldsymbol{\lambda} \cdot (\mathbf{r} \times \mathbf{F}) \tag{3.42}$$

which shows that the moment M_{OL} of **F** about the axis OL is the scalar obtained by forming the mixed triple product of $\boldsymbol{\lambda}$, **r**, and **F**. Expressing M_{OL} in the form of a determinant, we write

$$M_{OL} = \begin{vmatrix} \lambda_x & \lambda_y & \lambda_z \\ x & y & z \\ F_x & F_y & F_z \end{vmatrix} \tag{3.43}$$

where $\lambda_x, \lambda_y, \lambda_z$ = direction cosines of axis OL
x, y, z = coordinates of point of application of **F**
F_x, F_y, F_z = components of force **F**

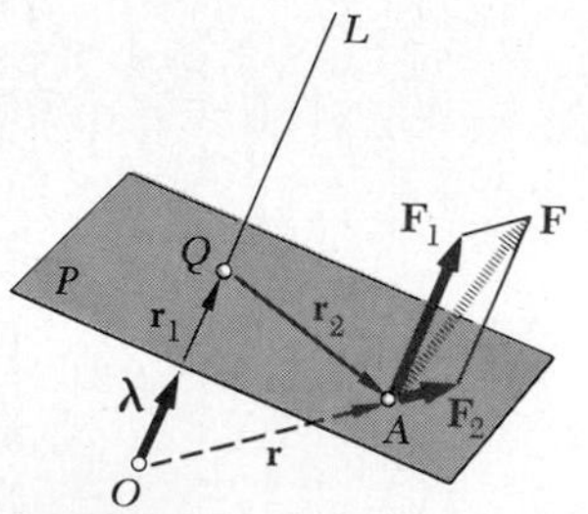

Fig. 3.28

The physical significance of the moment M_{OL} of a force **F** about a fixed axis OL becomes more apparent if we resolve **F** into two rectangular components $\mathbf{F}_1$ and $\mathbf{F}_2$, with $\mathbf{F}_1$ parallel to OL and $\mathbf{F}_2$ lying in a plane P perpendicular to OL (Fig. 3.28). Resolving **r** similarly into two components $\mathbf{r}_1$ and $\mathbf{r}_2$, and substituting for **F** and **r** into (3.42), we write

$$\begin{aligned} M_{OL} &= \boldsymbol{\lambda} \cdot [(\mathbf{r}_1 + \mathbf{r}_2) \times (\mathbf{F}_1 + \mathbf{F}_2)] \\ &= \boldsymbol{\lambda} \cdot (\mathbf{r}_1 \times \mathbf{F}_1) + \boldsymbol{\lambda} \cdot (\mathbf{r}_1 \times \mathbf{F}_2) + \boldsymbol{\lambda} \cdot (\mathbf{r}_2 \times \mathbf{F}_1) + \boldsymbol{\lambda} \cdot (\mathbf{r}_2 \times \mathbf{F}_2) \end{aligned}$$

Noting that all mixed triple products, except the last one, are equal to zero, since they involve vectors which are coplanar when drawn from a common origin (Sec. 3.10), we have

$$M_{OL} = \boldsymbol{\lambda} \cdot (\mathbf{r}_2 \times \mathbf{F}_2) \tag{3.44}$$

The vector product $\mathbf{r}_2 \times \mathbf{F}_2$ is perpendicular to the plane P and represents the moment of the component $\mathbf{F}_2$ of **F** about the point Q where OL intersects P. Therefore, the scalar M_{OL}, which will be positive if $\mathbf{r}_2 \times \mathbf{F}_2$ and OL have the same sense, and negative otherwise, measures the tendency of $\mathbf{F}_2$ to make the rigid body rotate about the fixed axis OL. Since the other component $\mathbf{F}_1$ of **F** does not tend to make the body rotate about OL, we conclude that *the moment* M_{OL} *of* **F** *about OL measures the tendency of the force* **F** *to impart to the rigid body a motion of rotation about the fixed axis OL.*

It follows from the definition of the moment of a force about an axis that the moment of **F** about a coordinate axis is equal to the component of $\mathbf{M}_O$ along that axis. Substituting successively each of the unit vectors **i**, **j**, and **k** for $\boldsymbol{\lambda}$ in (3.42), we check that the expressions thus obtained for the *moments of* **F** *about the coordinate axes* are respectively equal to the expressions obtained in Sec. 3.8 for the components of the moment $\mathbf{M}_O$ of **F** about O:

$$\begin{aligned} M_x &= yF_z - zF_y \\ M_y &= zF_x - xF_z \\ M_z &= xF_y - yF_x \end{aligned} \tag{3.18}$$

We observe that just as the components F_x, F_y, and F_z of a force **F** acting on a rigid body measure, respectively, the tendency of **F** to move the rigid body in the x, y, and z directions, the moments M_x, M_y, and M_z of **F** about the coordinate axes measure the tendency of **F** to impart to the rigid body a motion of rotation about the x, y, and z axes, respectively.

More generally, the moment of a force **F** applied at A about an axis which does not pass through the origin is obtained by choosing an arbitrary point B on the axis (Fig. 3.29) and determining the projection on the axis BL of the moment $\mathbf{M}_B$ of **F** about B. We write

$$M_{BL} = \boldsymbol{\lambda} \cdot \mathbf{M}_B = \boldsymbol{\lambda} \cdot (\mathbf{r}_{A/B} \times \mathbf{F}) \tag{3.45}$$

where $\mathbf{r}_{A/B} = \mathbf{r}_A - \mathbf{r}_B$ represents the vector drawn from B to A. Expressing M_{BL} in the form of a determinant, we have

$$M_{BL} = \begin{vmatrix} \lambda_x & \lambda_y & \lambda_z \\ x_{A/B} & y_{A/B} & z_{A/B} \\ F_x & F_y & F_z \end{vmatrix} \tag{3.46}$$

where $\lambda_x, \lambda_y, \lambda_z$ = direction cosines of axis BL

$x_{A/B} = x_A - x_B \qquad y_{A/B} = y_A - y_B \qquad z_{A/B} = z_A - z_B$

F_x, F_y, F_z = components of force **F**

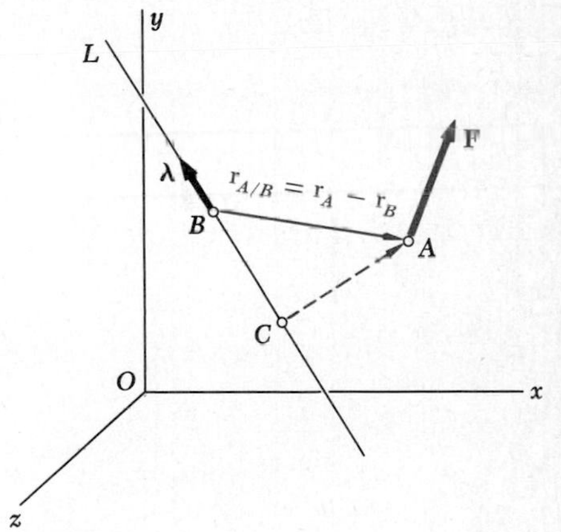

Fig. 3.29

It should be noted that the result obtained is independent of the choice of the point B on the given axis. Indeed, denoting by M_{CL} the result obtained with a different point C, we have

$$\begin{aligned} M_{CL} &= \boldsymbol{\lambda} \cdot [(\mathbf{r}_A - \mathbf{r}_C) \times \mathbf{F}] \\ &= \boldsymbol{\lambda} \cdot [(\mathbf{r}_A - \mathbf{r}_B) \times \mathbf{F}] + \boldsymbol{\lambda} \cdot [(\mathbf{r}_B - \mathbf{r}_C) \times \mathbf{F}] \end{aligned}$$

But, since the vectors $\boldsymbol{\lambda}$ and $\mathbf{r}_B - \mathbf{r}_C$ lie in the same line, the volume of the parallelepiped having the vectors $\boldsymbol{\lambda}$, $\mathbf{r}_B - \mathbf{r}_C$, and **F** for sides is zero, as is the mixed triple product of these three vectors (Sec. 3.10). The expression obtained for M_{CL} thus reduces to its first term, which is the expression used earlier to define M_{BL}.

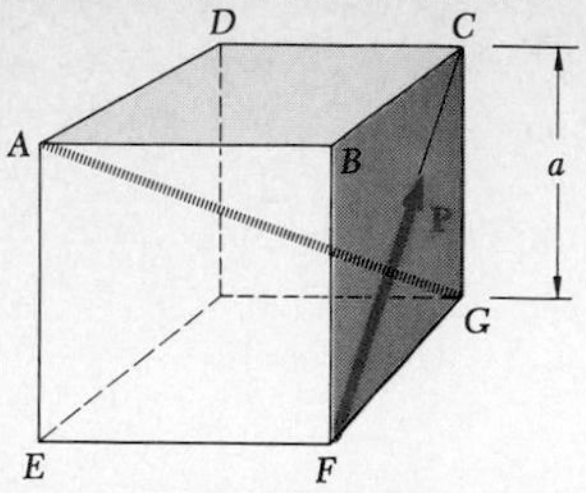

SAMPLE PROBLEM 3.5

A cube of side a is acted upon by a force **P** as shown. Determine the moment of **P** (a) about A, (b) about the edge AB, (c) about the diagonal AG of the cube. (d) Using the result of part c, determine the perpendicular distance from AG to FC.

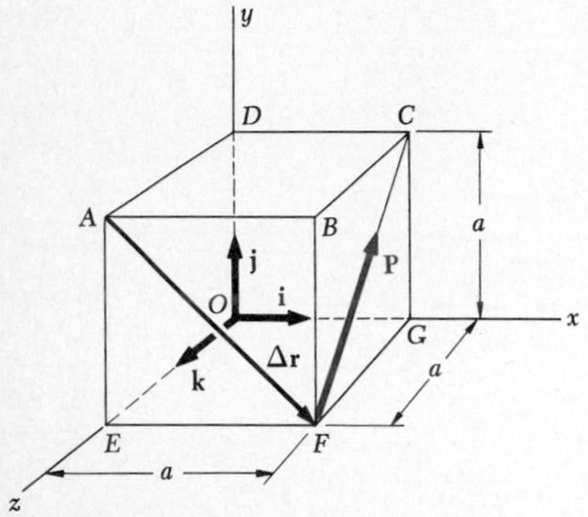

***a.* Moment about *A*.** Choosing x, y, and z axes as shown, we resolve into rectangular components the force **P** and the vector $\mathbf{r}_{F/A} = \overrightarrow{AF}$ drawn from A to the point of application F of **P**.

$$\mathbf{r}_{F/A} = a\mathbf{i} - a\mathbf{j} = a(\mathbf{i} - \mathbf{j})$$
$$\mathbf{P} = (P/\sqrt{2})\mathbf{j} - (P/\sqrt{2})\mathbf{k} = (P/\sqrt{2})(\mathbf{j} - \mathbf{k})$$

The moment of **P** about A is

$$\mathbf{M}_A = \mathbf{r}_{F/A} \times \mathbf{P} = a(\mathbf{i} - \mathbf{j}) \times (P/\sqrt{2})(\mathbf{j} - \mathbf{k})$$
$$\mathbf{M}_A = (aP/\sqrt{2})(\mathbf{i} + \mathbf{j} + \mathbf{k}) \quad ◄$$

***b.* Moment about *AB*.** Projecting $\mathbf{M}_A$ on AB, we write

$$M_{AB} = \mathbf{i} \cdot \mathbf{M}_A = \mathbf{i} \cdot (aP/\sqrt{2})(\mathbf{i} + \mathbf{j} + \mathbf{k}) \qquad M_{AB} = aP/\sqrt{2} \quad ◄$$

We verify that, since AB is parallel to the x axis, M_{AB} is also the x component of the moment $\mathbf{M}_A$.

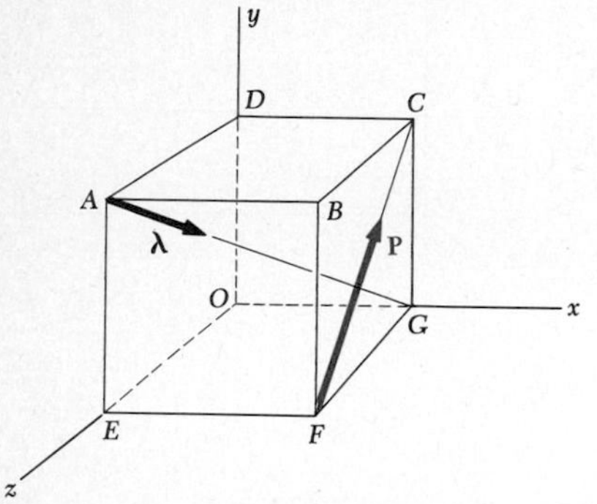

***c.* Moment about Diagonal *AG*.** The moment of **P** about AG is obtained by projecting $\mathbf{M}_A$ on AG. Denoting by $\boldsymbol{\lambda}$ the unit vector along AG, we have

$$\boldsymbol{\lambda} = \frac{\overrightarrow{AG}}{AG} = \frac{a\mathbf{i} - a\mathbf{j} - a\mathbf{k}}{a\sqrt{3}} = (1/\sqrt{3})(\mathbf{i} - \mathbf{j} - \mathbf{k})$$

$$M_{AG} = \boldsymbol{\lambda} \cdot \mathbf{M}_A = (1/\sqrt{3})(\mathbf{i} - \mathbf{j} - \mathbf{k}) \cdot (aP/\sqrt{2})(\mathbf{i} + \mathbf{j} + \mathbf{k})$$
$$M_{AG} = (aP/\sqrt{6})(1 - 1 - 1) \qquad M_{AG} = -aP/\sqrt{6} \quad ◄$$

Alternate Method. The moment of **P** about AG may also be expressed in the form of a determinant:

$$M_{AG} = \begin{vmatrix} \lambda_x & \lambda_y & \lambda_z \\ x_{F/A} & y_{F/A} & z_{F/A} \\ F_x & F_y & F_z \end{vmatrix} = \begin{vmatrix} 1/\sqrt{3} & -1/\sqrt{3} & -1/\sqrt{3} \\ a & -a & 0 \\ 0 & P/\sqrt{2} & -P/\sqrt{2} \end{vmatrix} = -aP/\sqrt{6}$$

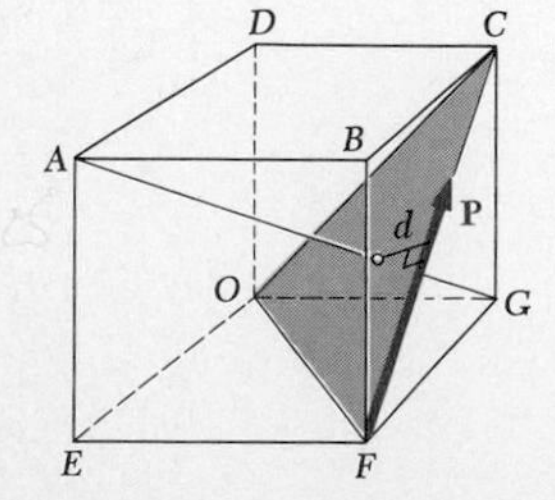

***d.* Perpendicular Distance from *AG* to *FC*.** We first observe that **P** is perpendicular to the diagonal AG. This may be checked by forming the scalar product $\mathbf{P} \cdot \boldsymbol{\lambda}$ and verifying that it is zero:

$$\mathbf{P} \cdot \boldsymbol{\lambda} = (P/\sqrt{2})(\mathbf{j} - \mathbf{k}) \cdot (1/\sqrt{3})(\mathbf{i} - \mathbf{j} - \mathbf{k}) = (P/\sqrt{6})(0 - 1 + 1) = 0$$

The moment M_{AG} may then be expressed as $-Pd$, where d is the perpendicular distance from AG to FC. (The negative sign is used since the rotation imparted to the cube by **P** appears as clockwise to an observer at G.) Recalling the value found for M_{AG} in part c,

$$M_{AG} = -Pd = -aP/\sqrt{6} \qquad d = a/\sqrt{6} \quad ◄$$

Problems

3.26 Given the vectors $\mathbf{P} = 3\mathbf{i} - \mathbf{j} + 2\mathbf{k}$, $\mathbf{Q} = 4\mathbf{i} + 5\mathbf{j} - 3\mathbf{k}$, and $\mathbf{S} = -2\mathbf{i} + 3\mathbf{j} - \mathbf{k}$, compute the scalar products $\mathbf{P} \cdot \mathbf{Q}$, $\mathbf{P} \cdot \mathbf{S}$, and $\mathbf{Q} \cdot \mathbf{S}$.

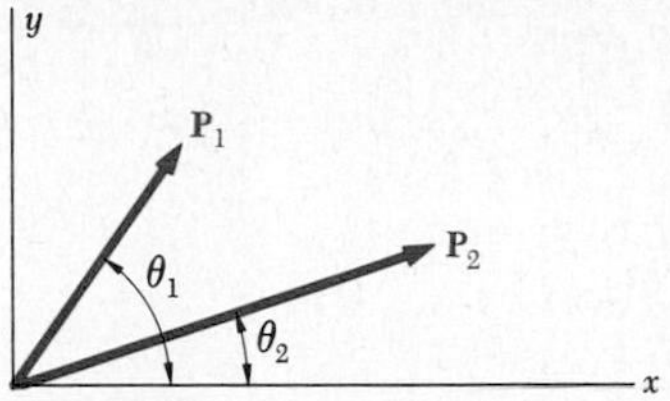

Fig. P3.27

3.27 Form the scalar product $\mathbf{P}_1 \cdot \mathbf{P}_2$ and use the result obtained to prove the identity $\cos(\theta_1 - \theta_2) = \cos\theta_1 \cos\theta_2 + \sin\theta_1 \sin\theta_2$.

3.28 Three cables are used to support a container as shown. Determine the angle formed by cables *AB* and *AD*.

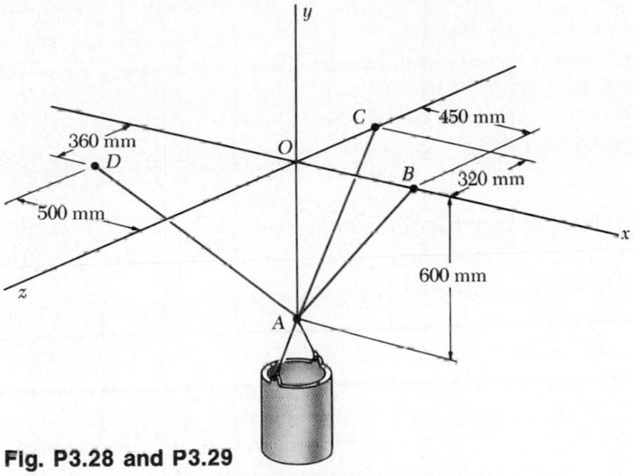

Fig. P3.28 and P3.29

3.29 Three cables are used to support a container as shown. Determine the angle formed by cables *AC* and *AD*.

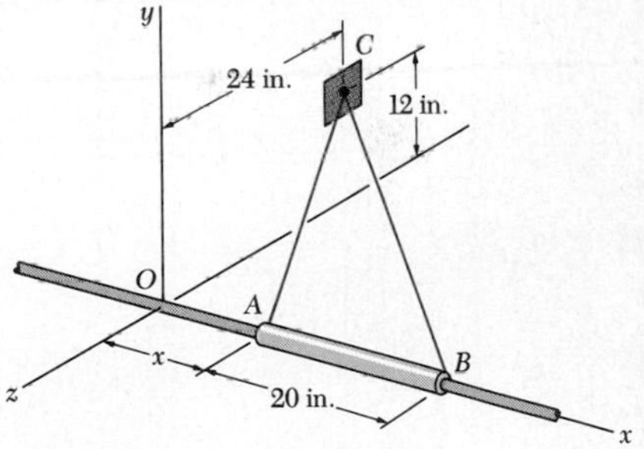

Fig. P3.30

3.30 The 20-in. tube *AB* may slide along a horizontal rod. The ends *A* and *B* of the tube are connected by elastic cords to the fixed point *C*. For the position corresponding to $x = 11$ in., determine the angle formed by the two cords, (*a*) using Eq. (3.32), (*b*) applying the law of cosines to triangle *ABC*.

3.31 Solve Prob. 3.30 for the position corresponding to $x = 4$ in.

3.32 Knowing that the tension in cable *AC* is 1260 N, determine (*a*) the angle between cable *AC* and the boom *AB*, (*b*) the projection on *AB* of the force exerted by cable *AC* at point *A*.

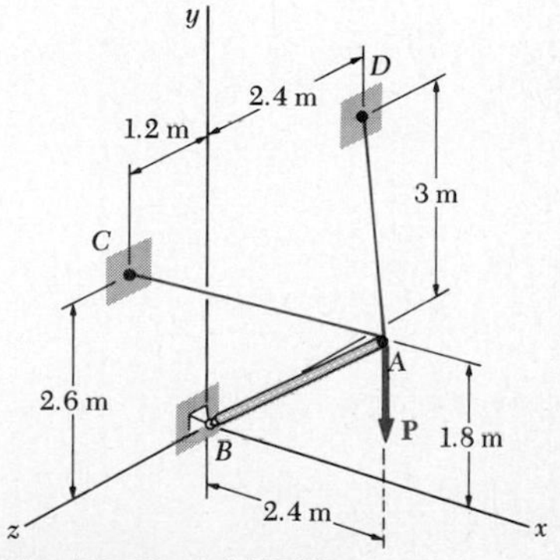

Fig. P3.32 and P3.33

3.33 Knowing that the tension in cable *AD* is 810 N, determine (*a*) the angle between cable *AD* and the boom *AB*, (*b*) the projection on *AB* of the force exerted by cable *AD* at point *A*.

3.34 Given the vectors $\mathbf{P} = 3\mathbf{i} - \mathbf{j} + 2\mathbf{k}$, $\mathbf{Q} = 4\mathbf{i} + 5\mathbf{j} - 3\mathbf{k}$, and $\mathbf{S} = -2\mathbf{i} + 3\mathbf{j} - \mathbf{k}$, compute $\mathbf{P} \cdot (\mathbf{Q} \times \mathbf{S})$, $(\mathbf{P} \times \mathbf{Q}) \cdot \mathbf{S}$, and $(\mathbf{S} \times \mathbf{Q}) \cdot \mathbf{P}$.

3.35 Given the vectors $\mathbf{P} = 4\mathbf{i} - 2\mathbf{j} + 3\mathbf{k}$, $\mathbf{Q} = 2\mathbf{i} + 4\mathbf{j} - 5\mathbf{k}$, and $\mathbf{S} = S_x\mathbf{i} - \mathbf{j} + 2\mathbf{k}$, determine the value of S_x for which the three vectors are coplanar.

3.36 Knowing that the tension in cable AB is 570 N, determine the moment about each of the coordinate axes of the force exerted on the plate at B.

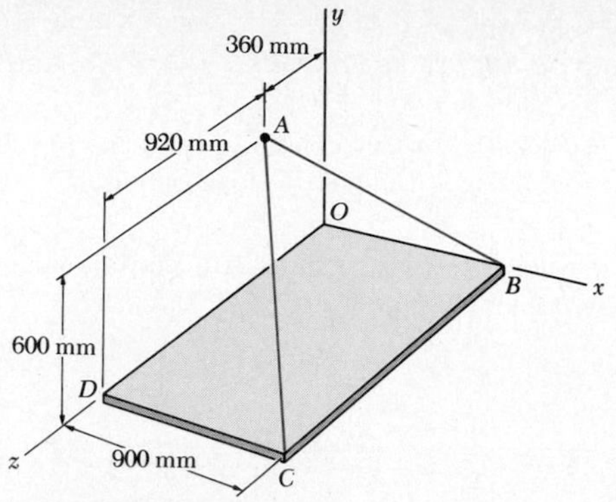

Fig. P3.36 and P3.37

3.37 Knowing that the tension in cable AC is 1065 N, determine the moment about each of the coordinate axes of the force exerted on the plate at C.

3.38 A small boat hangs from two davits, one of which is shown in the figure. It is known that the moment about the z axis of the resultant force $\mathbf{R}_A$ exerted on the davit at A must not exceed 160 lb·ft in absolute value. Determine the largest allowable tension in line $ABAD$ when $x = 4.8$ ft.

3.39 For the davit of Prob. 3.38, determine the largest allowable distance x when the tension in line $ABAD$ is 48 lb.

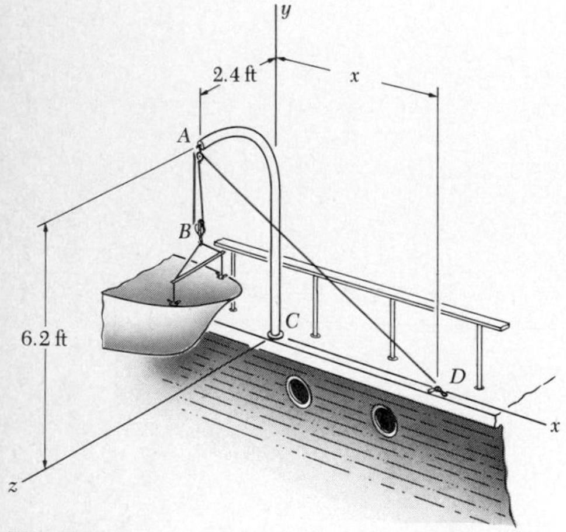

Fig. P3.38

3.40 A single force $\mathbf{P}$ acts at C in a direction perpendicular to the handle BC of the crank shown. Knowing that $M_x = +20$ N·m, $M_y = -8.75$ N·m, and $M_z = -30$ N·m, determine the magnitude of $\mathbf{P}$ and the values of ϕ and θ.

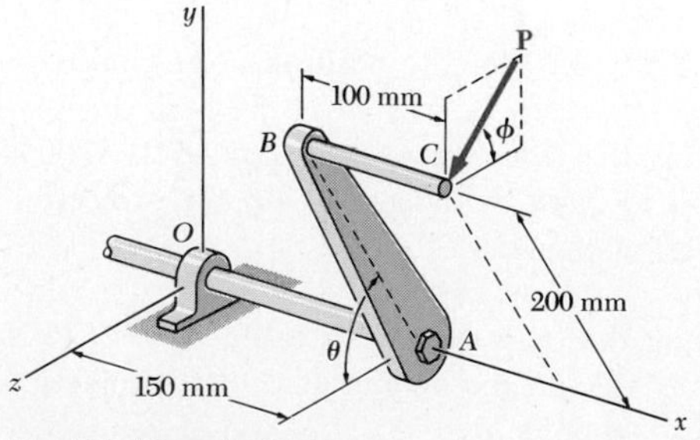

Fig. P3.40 and P3.41

3.41 A single force $\mathbf{P}$ acts at C in a direction perpendicular to the handle BC of the crank shown. Determine the moment M_x of $\mathbf{P}$ about the x axis when $\theta = 70°$, knowing that $M_y = -20$ N·m and $M_z = -37.5$ N·m.

3.42 The frame ACD is hinged at A and D and supported by a cable which passes through a ring at B and is attached to hooks at G and H. Knowing that the tension in the cable is 450 N, determine the moment about the diagonal AD of the force exerted on the frame by portion BH of the cable.

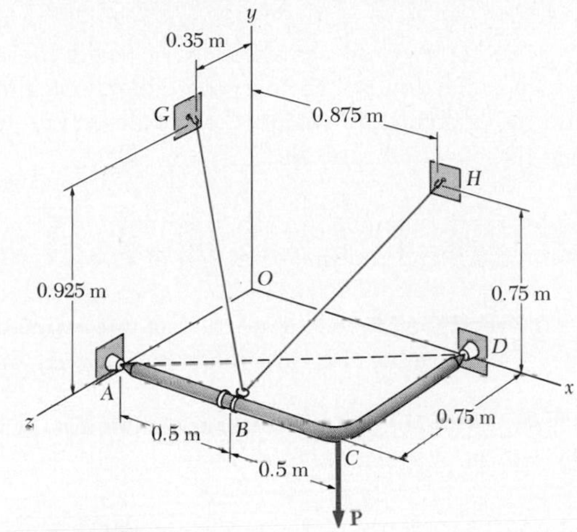

Fig. P3.42

3.43 In Prob. 3.42, determine the moment about the diagonal AD of the force exerted on the frame by portion BG of the cable.

3.44 The 23-in. vertical rod CD is welded to the midpoint C of the 50-in. rod AB. Determine the moment about AB of the 235-lb force **P**.

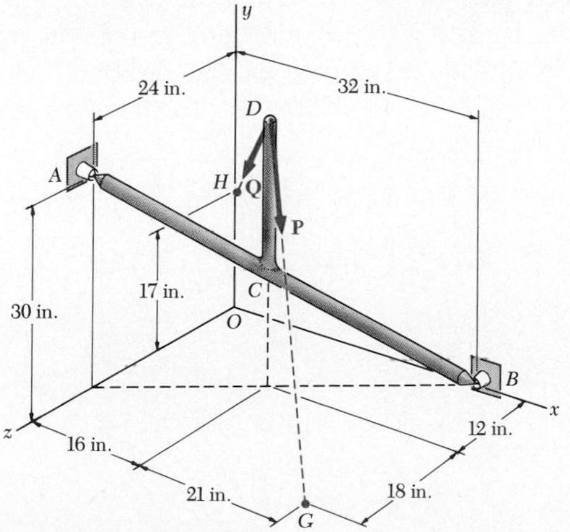

Fig. P3.44 and P3.45

3.45 The 23-in. vertical rod CD is welded to the midpoint C of the 50-in. rod AB. Determine the moment about AB of the 145-lb force **Q**.

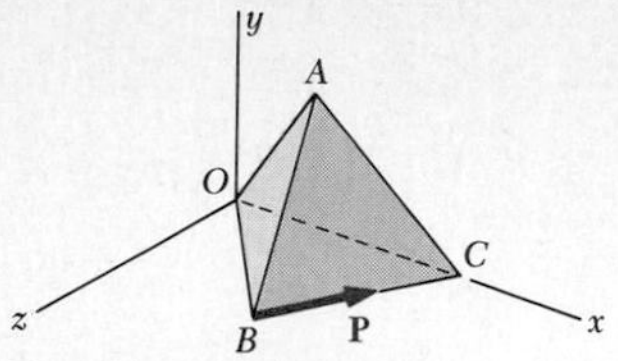

Fig. P3.46 and P3.48

3.46 A regular tetrahedron has six edges of length a. A force **P** is directed as shown along edge BC. Determine the moment of **P** about edge OA.

3.47 Two forces $\mathbf{F}_1$ and $\mathbf{F}_2$ in space have the same magnitude F. Prove that the moment of $\mathbf{F}_1$ about the line of action of $\mathbf{F}_2$ is equal to the moment of $\mathbf{F}_2$ about the line of action of $\mathbf{F}_1$.

3.48 A regular tetrahedron has six edges of length a. (*a*) Show that two opposite edges, such as OA and BC, are perpendicular to each other. (*b*) Use this property and the result obtained in Prob. 3.46 to determine the perpendicular distance between the edges OA and BC.

***3.49** Use the result obtained in Prob. 3.44 to determine the perpendicular distance between rod AB and the line of action of **P**.

***3.50** Use the result obtained in Prob. 3.42 to determine the perpendicular distance between portion BH of the cable and the diagonal AD.

***3.51** In Prob. 3.42, determine the perpendicular distance between portion BG of the cable and the diagonal AD.

***3.52** In Prob. 3.45, determine the perpendicular distance between rod AB and the line of action of **Q**.

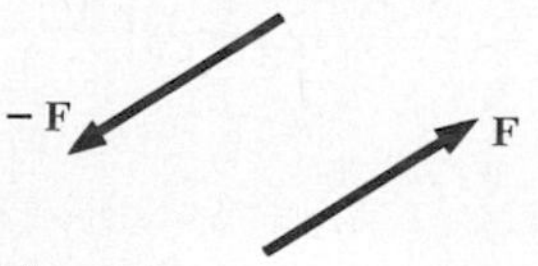

Fig. 3.30

3.12. Moment of a Couple. *Two forces* **F** *and* −**F** *having the same magnitude, parallel lines of action, and opposite sense are said to form a couple* (Fig. 3.30). Clearly, the sum of the components of the two forces in any direction is zero. The sum of the moments of the two forces about a given point, however, is not zero. While the two forces will not translate the body on which they act, they will tend to make it rotate.

Denoting by $\mathbf{r}_A$ and $\mathbf{r}_B$, respectively, the position vectors of the points of application of **F** and −**F** (Fig. 3.31), we find that the sum of the moments of the two forces about O is

$$\mathbf{r}_A \times \mathbf{F} + \mathbf{r}_B \times (-\mathbf{F}) = (\mathbf{r}_A - \mathbf{r}_B) \times \mathbf{F}$$

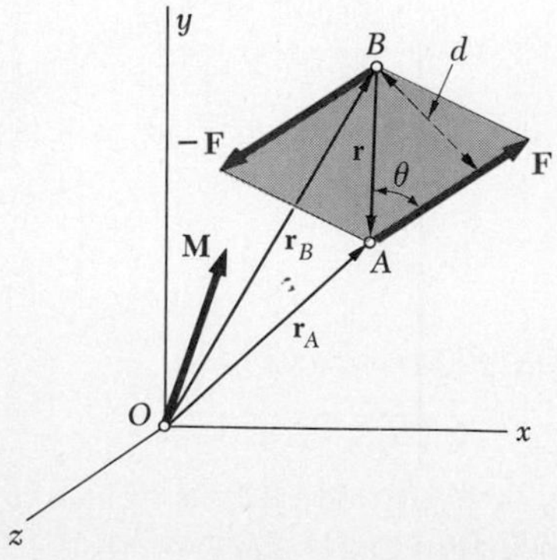

Fig. 3.31

Setting $\mathbf{r}_A - \mathbf{r}_B = \mathbf{r}$, where $\mathbf{r}$ is the vector joining the points of application of the two forces, we conclude that the sum of the moments of $\mathbf{F}$ and $-\mathbf{F}$ about O is represented by the vector

$$\mathbf{M} = \mathbf{r} \times \mathbf{F} \tag{3.47}$$

The vector $\mathbf{M}$ is called the *moment of the couple;* it is a vector perpendicular to the plane containing the two forces and its magnitude is

$$M = rF \sin\theta = Fd \tag{3.48}$$

where d is the perpendicular distance between the lines of action of $\mathbf{F}$ and $-\mathbf{F}$. The sense of $\mathbf{M}$ is defined by the right-hand rule.

Since the vector $\mathbf{r}$ in (3.47) is independent of the choice of the origin O of the coordinate axes, we note that the same result would have been obtained if the moments of $\mathbf{F}$ and $-\mathbf{F}$ had been computed about a different point O'. Thus, the moment $\mathbf{M}$ of a couple is a *free vector* (Sec. 2.3) which may be applied at any point (Fig. 3.32).

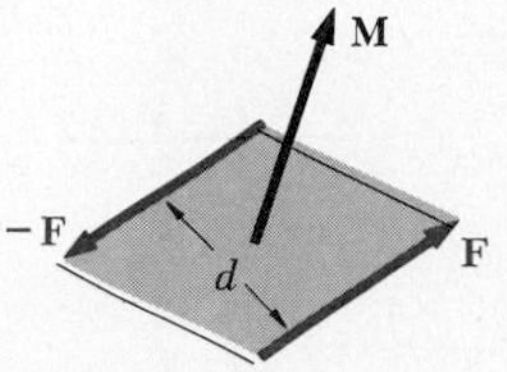

Fig. 3.32

From the definition of the moment of a couple, it also follows that two couples, one consisting of the forces $\mathbf{F}_1$ and $-\mathbf{F}_1$, the other of the forces $\mathbf{F}_2$ and $-\mathbf{F}_2$ (Fig. 3.33), will have equal moments if

$$F_1 d_1 = F_2 d_2 \tag{3.49}$$

and if the two couples lie in parallel planes (or in the same plane) and have the same sense.

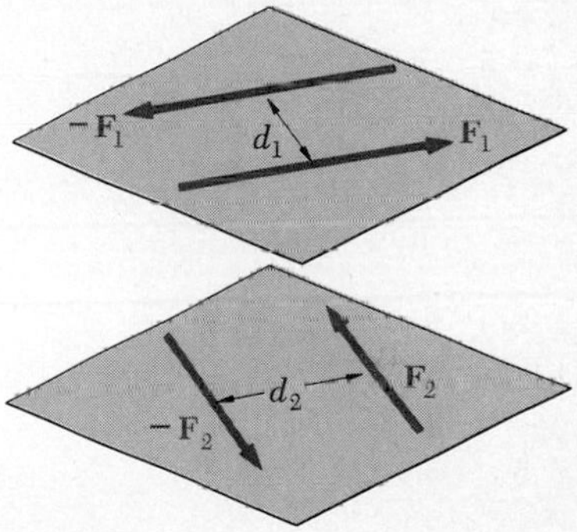

Fig. 3.33

3.13. Equivalent Couples. Consider the three couples shown in Fig. 3.34, which are made to act successively on the same rectangular box. As seen in the preceding section, the only motion a couple may impart to a rigid body is a rotation. Since each of the three couples shown has the same moment $\mathbf{M}$ (same direction and same magnitude $M = 120$ lb·in.), we may expect the three couples to have the same effect on the box.

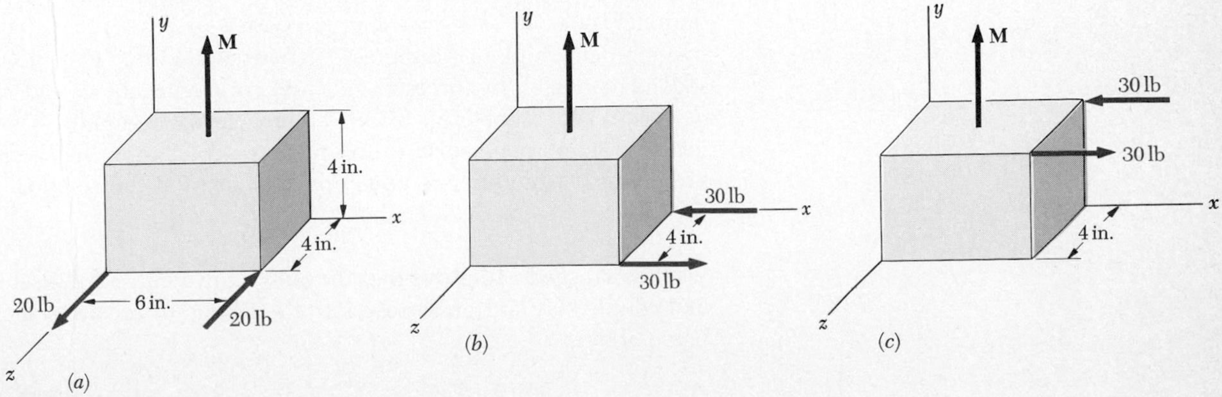

Fig. 3.34

As reasonable as this conclusion may appear, we should not accept it hastily. While intuitive feeling is of great help in the study of mechanics, it should not be accepted as a substitute for logical reasoning. Before stating that two systems (or groups) of forces have the same effect on a rigid body, we should prove that fact on the basis of the experimental evidence introduced so far. This evidence consists of the parallelogram law for the addition of two forces (Sec. 2.2) and of the principle of transmissibility (Sec. 3.3). Therefore, we shall state that *two systems of forces are equivalent* (i.e., they have the same effect on a rigid body) *if we can transform one of them into the other by means of one or several of the following operations:* (1) replacing two forces acting on the same particle by their resultant; (2) resolving a force into two components; (3) canceling two equal and opposite forces acting on the same particle; (4) attaching to the same particle two equal and opposite forces; (5) moving a force along its line of action. Each of these operations is easily justified on the basis of the parallelogram law or the principle of transmissibility.

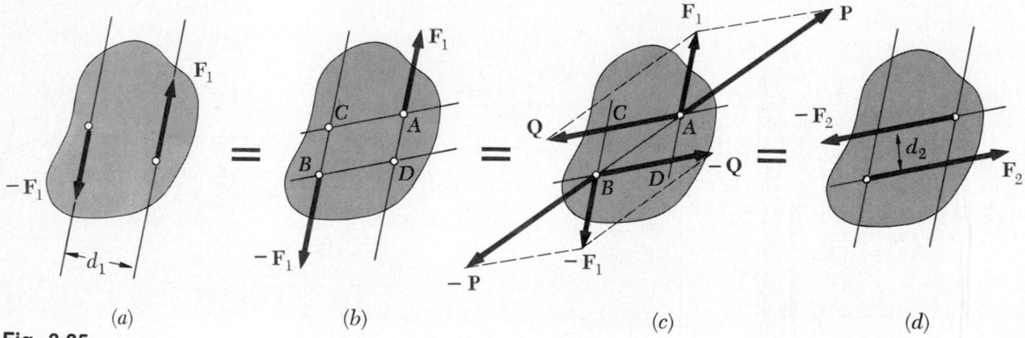

Fig. 3.35

Let us now prove that *two couples having the same moment* **M** *are equivalent.* First, we shall consider two couples contained in the same plane, and we shall assume that this plane coincides with the plane of the figure (Fig. 3.35). The first couple consists of the forces $\mathbf{F}_1$ and $-\mathbf{F}_1$, of magnitude F_1 and at a distance d_1 from each other (Fig. 3.35*a*), and the second couple of the forces $\mathbf{F}_2$ and $-\mathbf{F}_2$, of magnitude F_2 and at a distance d_2 from each other (Fig. 3.35*d*). Since the two couples have the same moment **M** perpendicular to the plane of the figure, they must have the same sense (assumed here counterclockwise) and the relation

$$F_1 d_1 = F_2 d_2 \tag{3.49}$$

must be satisfied. To prove that they are equivalent, we shall show that the first couple may be transformed into the second by means of the operations listed above.

Denoting by A, B, C, D the points of intersection of the lines of action of the two couples, we first slide the forces $\mathbf{F}_1$ and $-\mathbf{F}_1$ until they are attached, respectively, at A and B, as shown in Fig. 3.35*b*. The force $\mathbf{F}_1$ is then resolved into a component $\mathbf{P}$ along line AB and a component $\mathbf{Q}$ along AC (Fig. 3.35*c*); similarly, the force $-\mathbf{F}_1$ is resolved into $-\mathbf{P}$ along AB and $-\mathbf{Q}$ along BD. The forces $\mathbf{P}$ and $-\mathbf{P}$ have the same magnitude, same line of action, and opposite sense; they may be moved along their common line of action until they are applied at the same point and then canceled. Thus the couple formed by $\mathbf{F}_1$ and $-\mathbf{F}_1$ reduces to a couple consisting of $\mathbf{Q}$ and $-\mathbf{Q}$.

We shall now show that the forces $\mathbf{Q}$ and $-\mathbf{Q}$ are respectively equal to the forces $-\mathbf{F}_2$ and $\mathbf{F}_2$. The moment of the couple formed by $\mathbf{Q}$ and $-\mathbf{Q}$ may be obtained by computing the moment of $\mathbf{Q}$ about B; similarly, the moment of the couple formed by $\mathbf{F}_1$ and $-\mathbf{F}_1$ is the moment of $\mathbf{F}_1$ about B. But, by Varignon's theorem, the moment of $\mathbf{F}_1$ is equal to the sum of the moments of its components $\mathbf{P}$ and $\mathbf{Q}$. Since the moment of $\mathbf{P}$ about B is zero, the moment of the couple formed by $\mathbf{Q}$ and $-\mathbf{Q}$ must be equal to the moment of the couple formed by $\mathbf{F}_1$ and $-\mathbf{F}_1$. Recalling (3.49), we write

$$Qd_2 = F_1 d_1 = F_2 d_2 \qquad \text{and} \qquad Q = F_2$$

Thus the forces $\mathbf{Q}$ and $-\mathbf{Q}$ are respectively equal to the forces $-\mathbf{F}_2$ and $\mathbf{F}_2$, and the couple of Fig. 3.35*a* is equivalent to the couple of Fig. 3.35*d*.

Next we shall consider two couples contained in parallel planes P_1 and P_2 and prove that they are equivalent if they have the same moment. In view of the foregoing we may assume that the couples consist of forces of the same magnitude F acting along parallel lines (Fig. 3.36*a* and *d*). We propose to show that the couple contained in plane P_1 may be transformed into the couple contained in plane P_2 by means of the standard operations listed above.

Let us consider the two planes defined respectively by the lines of action of $\mathbf{F}_1$ and $-\mathbf{F}_2$, and of $-\mathbf{F}_1$ and $\mathbf{F}_2$ (Fig. 3.36*b*). At a point on their line of intersection we attach two forces $\mathbf{F}_3$ and $-\mathbf{F}_3$, respectively equal to $\mathbf{F}_1$ and $-\mathbf{F}_1$. The couple formed by $\mathbf{F}_1$ and $-\mathbf{F}_3$ may be replaced by a couple consisting of $\mathbf{F}_3$ and $-\mathbf{F}_2$ (Fig. 3.36*c*), since both couples have clearly the same moment and are contained in the same plane. Similarly, the couple formed by $-\mathbf{F}_1$ and $\mathbf{F}_3$ may be replaced by a couple consisting of $-\mathbf{F}_3$ and $\mathbf{F}_2$. Canceling the two equal and opposite forces $\mathbf{F}_3$ and $-\mathbf{F}_3$, we obtain the desired couple in plane P_2 (Fig. 3.36*d*). Thus, we conclude that two couples having the same moment $\mathbf{M}$ are equivalent, whether they are contained in the same plane or in parallel planes.

The property we have just established is very important for the correct understanding of the mechanics of rigid bodies. It indicates that when a couple acts on a rigid body, it does not matter where the two forces forming the couple act, or what magnitude and direction they have. The only thing which counts is the *moment* of the couple (magnitude and direction). Couples with the same moment will have the same effect on the rigid body.

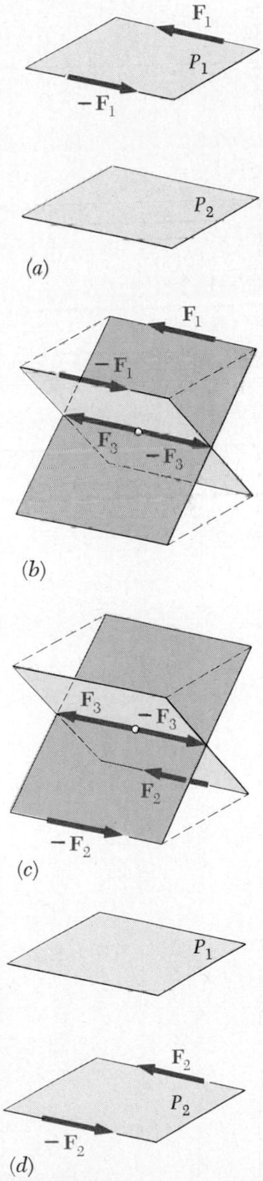

Fig. 3.36

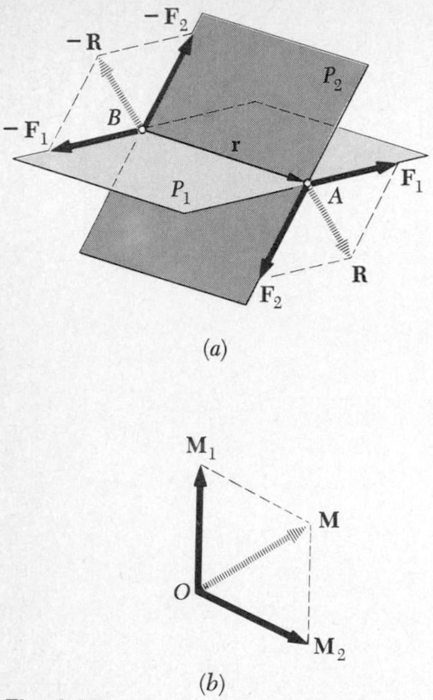

Fig. 3.37

3.14. Addition of Couples.

Consider two intersecting planes P_1 and P_2 and two couples acting respectively in P_1 and P_2. We may, without any loss of generality, assume that the couple in P_1 consists of two forces $\mathbf{F}_1$ and $-\mathbf{F}_1$ perpendicular to the line of intersection of the two planes and acting respectively at A and B (Fig. 3.37*a*). Similarly, we assume that the couple in P_2 consists of two forces $\mathbf{F}_2$ and $-\mathbf{F}_2$ perpendicular to AB and acting respectively at A and B. It is clear that the resultant $\mathbf{R}$ of $\mathbf{F}_1$ and $\mathbf{F}_2$ and the resultant $-\mathbf{R}$ of $-\mathbf{F}_1$ and $-\mathbf{F}_2$ form a couple. Denoting by $\mathbf{r}$ the vector joining B to A, and recalling the definition of the moment of a couple (Sec. 3.12), we express the moment $\mathbf{M}$ of the resulting couple as follows:

$$\mathbf{M} = \mathbf{r} \times \mathbf{R} = \mathbf{r} \times (\mathbf{F}_1 + \mathbf{F}_2)$$

and, by Varignon's theorem,

$$\mathbf{M} = \mathbf{r} \times \mathbf{F}_1 + \mathbf{r} \times \mathbf{F}_2$$

But the first term in the expression obtained represents the moment $\mathbf{M}_1$ of the couple in P_1, and the second term the moment $\mathbf{M}_2$ of the couple in P_2. We have

$$\mathbf{M} = \mathbf{M}_1 + \mathbf{M}_2 \tag{3.50}$$

and we conclude that the sum of two couples of moments $\mathbf{M}_1$ and $\mathbf{M}_2$ is a couple of moment $\mathbf{M}$ equal to the vector sum of $\mathbf{M}_1$ and $\mathbf{M}_2$ (Fig. 3.37*b*).

3.15. Couples May Be Represented by Vectors.

As we saw in Sec. 3.13, couples which have the same moment, whether they act in the same plane or in parallel planes, are equivalent. There is therefore no need to draw the actual forces forming a given couple in order to define its effect on a rigid body (Fig. 3.38*a*). It is sufficient to draw an arrow equal in magnitude and direction to the moment $\mathbf{M}$ of the couple (Fig. 3.38*b*). On the other hand, we saw in Sec. 3.14 that the sum of two couples is itself a couple and that the moment $\mathbf{M}$ of the resultant couple may be obtained by forming the vector sum of the moments $\mathbf{M}_1$ and $\mathbf{M}_2$ of the given couples.

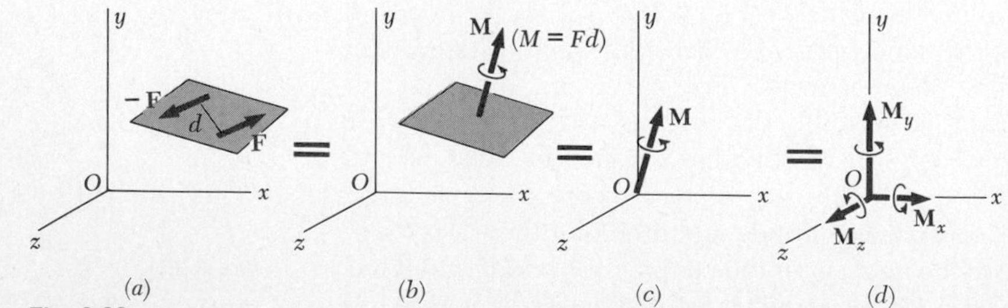

Fig. 3.38

Thus, couples obey the law of addition of vectors, and the arrow used in Fig. 3.38*b* to represent the couple defined in Fig. 3.38*a* may truly be considered a vector.

The vector representing a couple is called a *couple vector.* Note that, in Fig. 3.38, a colored arrow is used to distinguish the couple vector, *which represents the couple itself,* from the vector representing the *moment* of the couple, and that the symbol ↺ is added to avoid any confusion with vectors representing forces. A couple vector, like the moment of a couple, is a free vector. Its point of application, therefore, may be chosen at the origin of the system of coordinates, if so desired (Fig. 3.38*c*). Furthermore, the couple vector **M** may be resolved into component vectors $\mathbf{M}_x$, $\mathbf{M}_y$, and $\mathbf{M}_z$, directed along the axes of coordinates (Fig. 3.38*d*) and representing couples acting, respectively, in the *yz*, *zx*, and *xy* planes.

3.16. Resolution of a Given Force into a Force at *O* and a Couple. Consider a force **F** acting on a rigid body at a point *A* defined by the position vector **r** (Fig. 3.39*a*). Suppose that for some reason we would rather have the force act at point *O*. We know that we can move **F** along its line of action (principle of transmissibility); but we cannot move it to a point *O* away from the original line of action without modifying the action of **F** on the rigid body.

We may, however, attach two forces at point *O*, one equal to **F** and the other equal to $-\mathbf{F}$, without modifying the action of the original force on the rigid body (Fig. 3.39*b*). As a result of this tranformation, a force **F** is now applied at *O*; the other two forces form a couple of moment $\mathbf{M}_O = \mathbf{r} \times \mathbf{F}$. Thus, *any force* **F** *acting on a rigid body may be moved to an arbitrary point O, provided that a couple is added, of moment equal to the moment of* **F** *about O.* The couple tends to impart to the rigid body the same motion of rotation about *O* that the force **F** tended to produce before it was transferred to *O*. The couple is represented by a couple vector $\mathbf{M}_O$ perpendicular to the plane containing **r** and **F**. Since $\mathbf{M}_O$ is a free vector, it may be applied anywhere; for convenience, however, the couple vector is usually attached at *O*, together with **F**, and the combination obtained is referred to as a *force-couple system* (Fig. 3.39*c*).

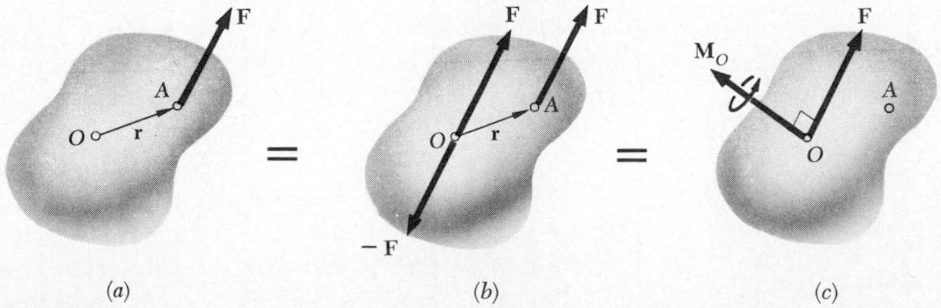

Fig. 3.39

If the force **F** had been moved from A to a different point O' (Fig. 3.40a and c), the moment $\mathbf{M}_{O'} = \mathbf{r}' \times \mathbf{F}$ of **F** about O' should have been computed, and a new force-couple system, consisting of **F** and of the couple vector $\mathbf{M}_{O'}$, would have been attached at O'. The relation existing between the moments of **F** about O and O' is obtained by writing

$$\mathbf{M}_{O'} = \mathbf{r}' \times \mathbf{F} = (\mathbf{r} + \mathbf{s}) \times \mathbf{F} = \mathbf{r} \times \mathbf{F} + \mathbf{s} \times \mathbf{F}$$

$$\mathbf{M}_{O'} = \mathbf{M}_O + \mathbf{s} \times \mathbf{F} \qquad (3.51)$$

where **s** is the vector joining O' to O. Thus, the moment $\mathbf{M}_{O'}$ of **F** about O' is obtained by adding to the moment $\mathbf{M}_O$ of **F** about O the vector product $\mathbf{s} \times \mathbf{F}$ representing the moment about O' of the force **F** applied at O.

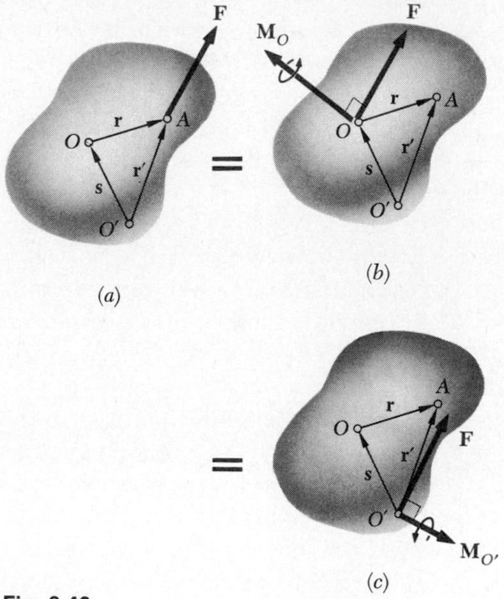

Fig. 3.40

This result could also have been established by observing that, in order to transfer to O' the force-couple system attached at O (Fig. 3.40b and c), the couple vector $\mathbf{M}_O$ may be freely moved to O'; to move the force **F** from O to O', however, it is necessary to add to **F** a couple vector $\mathbf{s} \times \mathbf{F}$ representing the moment about O' of the force **F** applied at O. Thus, the couple vector $\mathbf{M}_{O'}$ must be the sum of $\mathbf{M}_O$ and $\mathbf{s} \times \mathbf{F}$.

As noted above, the force-couple system obtained by transferring a force **F** from a point A to a point O consists of **F** and of a couple vector $\mathbf{M}_O = \mathbf{r} \times \mathbf{F}$ perpendicular to **F**. Conversely, any force-couple system consisting of a force **F** and of a couple vector $\mathbf{M}_O$ which are *mutually perpendicular* may be replaced by a single equivalent force. This is done by moving the force **F** in the plane perpendicular to $\mathbf{M}_O$ until its moment about O becomes equal to the couple vector $\mathbf{M}_O$ to be eliminated.

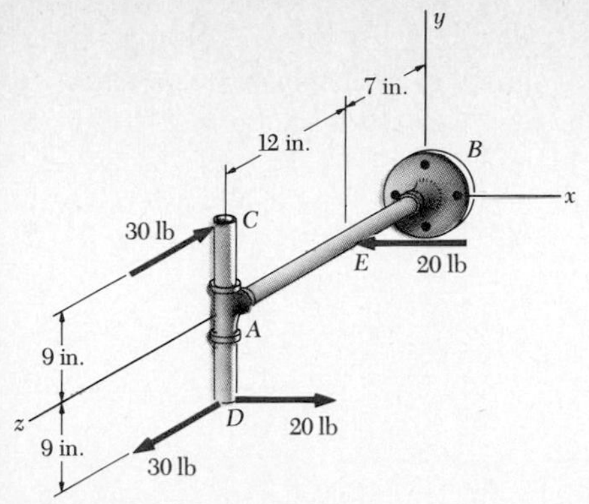

SAMPLE PROBLEM 3.6

Determine the components of the single couple equivalent to the two couples shown.

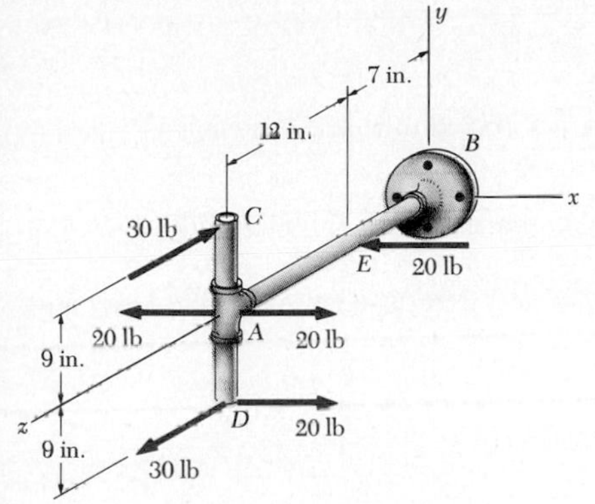

Solution. Our computations will be simplified if we attach two equal and opposite 20-lb forces at A. This enables us to replace the original 20-lb-force couple by two new 20-lb-force couples, one of which lies in the xz plane and the other in a plane parallel to the xy plane. The three couples shown in the adjoining sketch may be represented by three couple vectors $\mathbf{M}_x$, $\mathbf{M}_y$, and $\mathbf{M}_z$ directed along the coordinate axes. The corresponding moments are

$$
\begin{aligned}
M_x &= -(30 \text{ lb})(18 \text{ in.}) = -540 \text{ lb}\cdot\text{in.} \\
M_y &= +(20 \text{ lb})(12 \text{ in.}) = +240 \text{ lb}\cdot\text{in.} \\
M_z &= +(20 \text{ lb})(9 \text{ in.}) = +180 \text{ lb}\cdot\text{in.}
\end{aligned}
$$

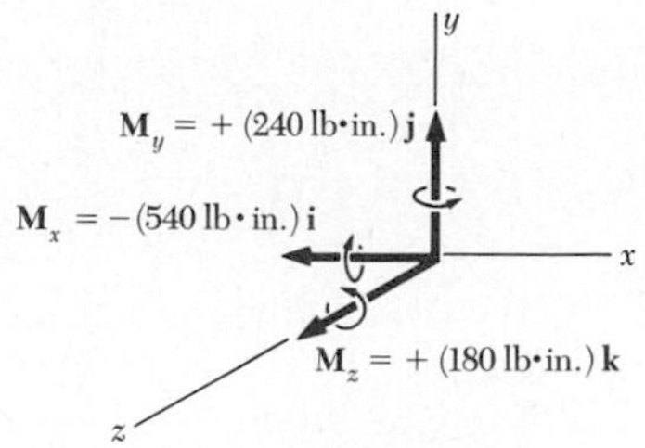

These three moments represent the components of the single couple $\mathbf{M}$ equivalent to the two given couples. We write

$$\mathbf{M} = -(540 \text{ lb}\cdot\text{in.})\mathbf{i} + (240 \text{ lb}\cdot\text{in.})\mathbf{j} + (180 \text{ lb}\cdot\text{in.})\mathbf{k} \quad ◄$$

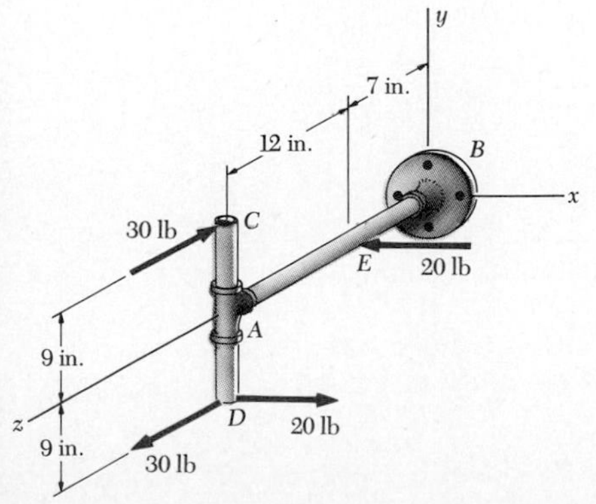

Alternate Solution. The components of the equivalent single couple $\mathbf{M}$ may also be obtained by computing the sum of the moments of the four given forces about an arbitrary point. Selecting point D, we write

$$\mathbf{M} = \mathbf{M}_D = (18 \text{ in.})\mathbf{j} \times (-30 \text{ lb})\mathbf{k} + [(9 \text{ in.})\mathbf{j} - (12 \text{ in.})\mathbf{k}] \times (-20 \text{ lb})\mathbf{i}$$

and, after computing the various cross products,

$$\mathbf{M} = -(540 \text{ lb}\cdot\text{in.})\mathbf{i} + (240 \text{ lb}\cdot\text{in.})\mathbf{j} + (180 \text{ lb}\cdot\text{in.})\mathbf{k} \quad ◄$$

SAMPLE PROBLEM 3.7

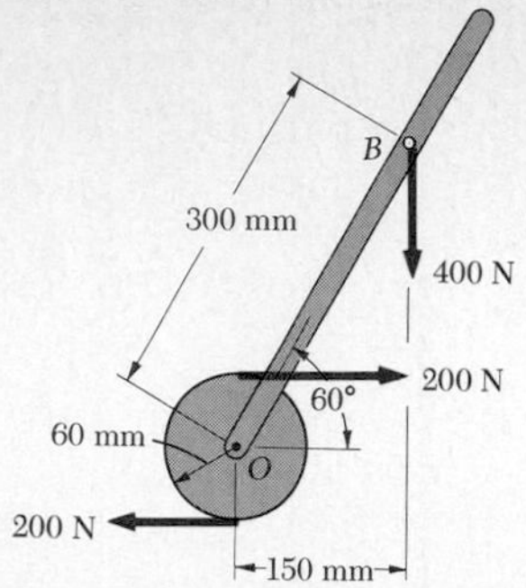

Replace the couple and force shown by an equivalent single force applied to the lever. Determine the distance from the shaft to the point of application of this equivalent force.

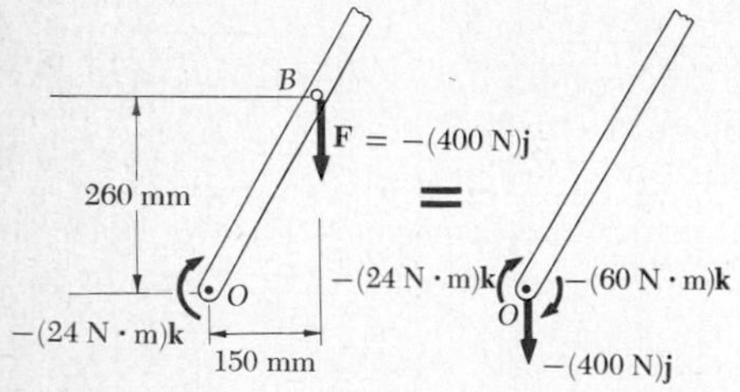

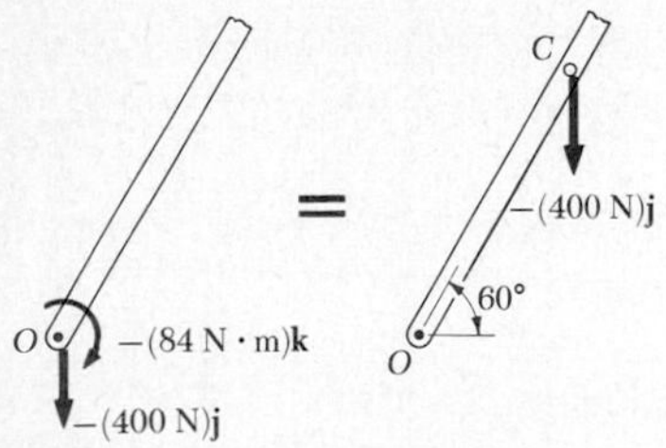

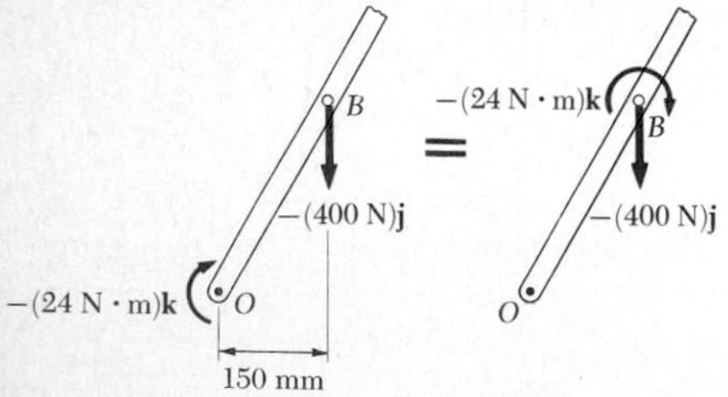

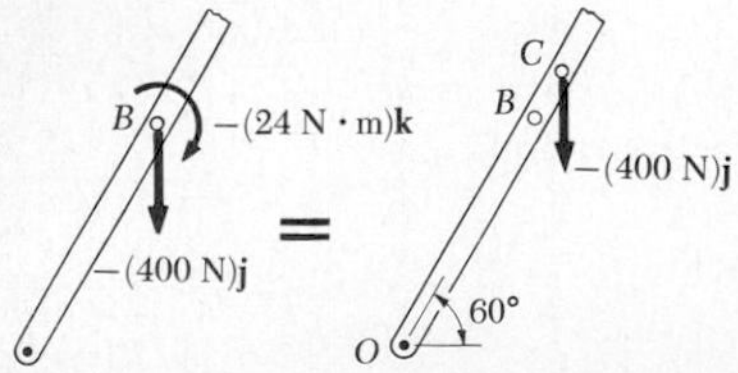

Solution. First the given force and couple are replaced by an equivalent force-couple system at O. We move the force $\mathbf{F} = -(400 \text{ N})\mathbf{j}$ to O and at the same time add a couple of moment $\mathbf{M}_O$ equal to the moment about O of the force in its original position.

$$\mathbf{M}_O = \overrightarrow{OB} \times \mathbf{F} = [(0.150 \text{ m})\mathbf{i} + (0.260 \text{ m})\mathbf{j}] \times (-400 \text{ N})\mathbf{j}$$
$$= -(60 \text{ N}\cdot\text{m})\mathbf{k}$$

This couple is added to the couple of moment $-(24 \text{ N}\cdot\text{m})\mathbf{k}$ formed by the two 200-N forces, and a couple of moment $-(84 \text{ N}\cdot\text{m})\mathbf{k}$ is obtained. This last couple may be eliminated by applying $\mathbf{F}$ at a point C chosen in such a way that

$$-(84 \text{ N}\cdot\text{m})\mathbf{k} = \overrightarrow{OC} \times \mathbf{F}$$
$$= [(OC) \cos 60°\mathbf{i} + (OC) \sin 60°\mathbf{j}] \times (-400 \text{ N})\mathbf{j}$$
$$= -(OC) \cos 60°(400 \text{ N})\mathbf{k}$$

We conclude that

$$(OC) \cos 60° = 0.210 \text{ m} = 210 \text{ mm} \qquad OC = 420 \text{ mm} \blacktriangleleft$$

Alternate Solution. Since the effect of a couple does not depend on its location, the couple of moment $-(24 \text{ N}\cdot\text{m})\mathbf{k}$ may be moved to B; we thus obtain a force-couple system at B. The couple may now be eliminated by applying $\mathbf{F}$ at a point C chosen in such a way that

$$-(24 \text{ N}\cdot\text{m})\mathbf{k} = \overrightarrow{BC} \times \mathbf{F}$$
$$= -(BC) \cos 60°(400 \text{ N})\mathbf{k}$$

We conclude that

$$(BC) \cos 60° = 0.060 \text{ m} = 60 \text{ mm} \qquad BC = 120 \text{ mm}$$
$$OC = OB + BC = 300 \text{ mm} + 120 \text{ mm} \qquad OC = 420 \text{ mm} \blacktriangleleft$$

Problems

3.53 Two 60-N forces are applied as shown to the corners A and C of a 200-mm square plate. Determine the moment of the couple formed by the two forces (*a*) by multiplying their magnitude by their perpendicular distance, (*b*) by resolving each force into horizontal and vertical components and adding the moments of the two resulting couples.

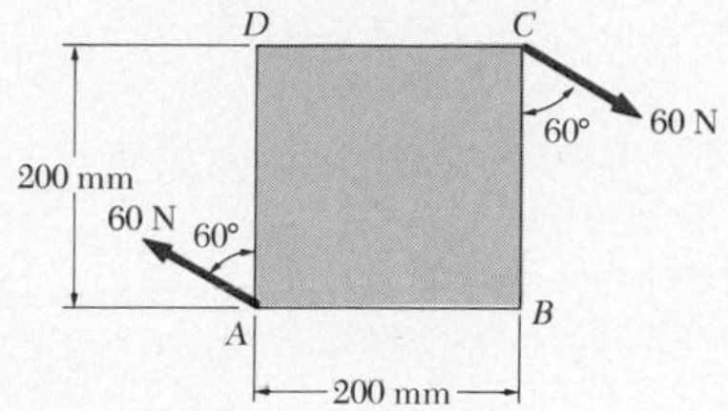

Fig. P3.53

3.54 A piece of plywood in which several holes are being drilled successively has been secured to a workbench by means of two nails. Knowing that the drill exerts a 5-N·m couple on the piece of plywood, determine the magnitude of the resulting forces applied to the nails if they are located (*a*) at A and B, (*b*) at B and C, (*c*) at A and C.

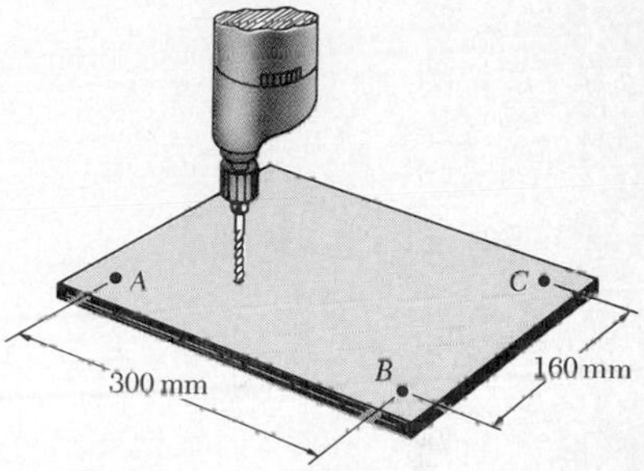

Fig. P3.54

3.55 Four 1-in.-diameter pegs are attached to a board as shown. Two strings are passed around the pegs and pulled with the forces indicated. (*a*) Determine the resultant couple acting on the board. (*b*) If only one string is used, around which pegs should it pass and in what directions should it be pulled to create the same couple with the minimum tension in the string? (*c*)What is the value of that minimum tension?

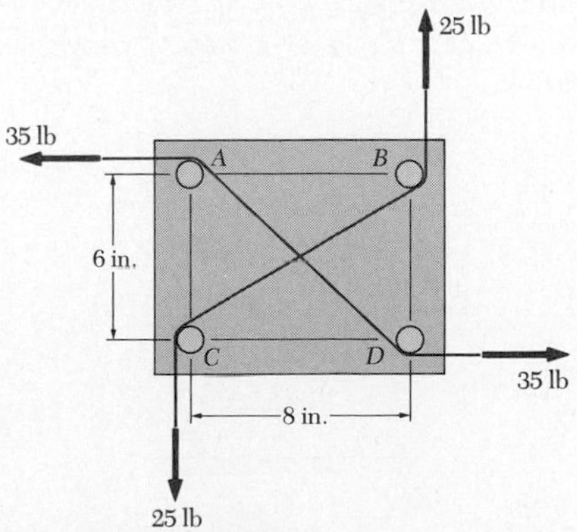

Fig. P3.55 and P3.56

3.56 Four pegs of the same diameter are attached to a board as shown. Two strings are passed around the pegs and pulled with the forces indicated. Determine the diameter of the pegs, knowing that the resultant couple applied to the board is 485 lb·in. counterclockwise.

3.57 The two shafts of a speed-reducer unit are subjected to couples of magnitude $M_1 = 15$ lb·ft and $M_2 = 3$ lb·ft, respectively. Replace the two couples by a single equivalent couple, specifying its magnitude and the direction of its axis.

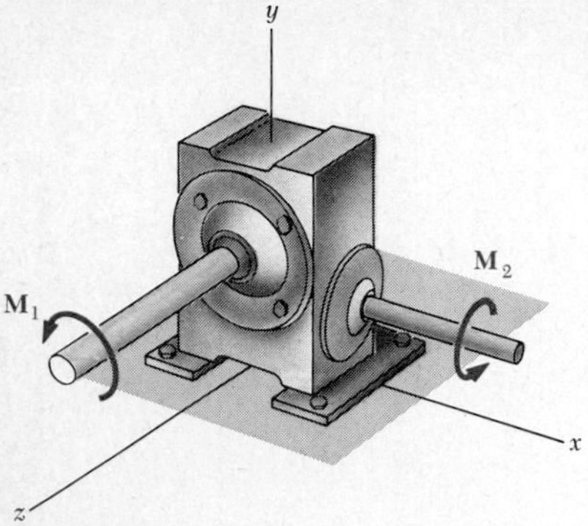

Fig. P3.57

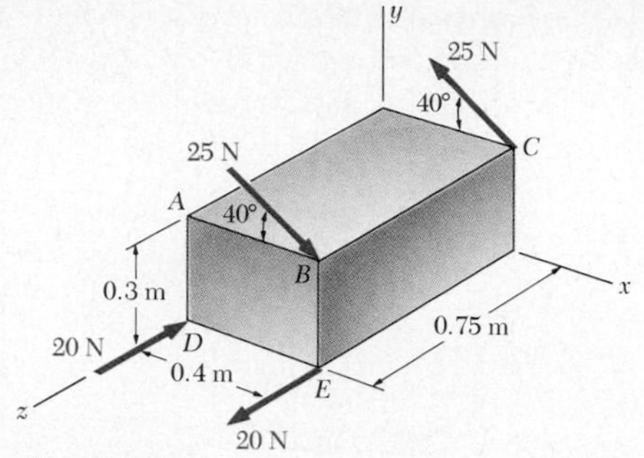

Fig. P3.58

3.58 The two couples shown are to be replaced by a single equivalent couple. Determine (*a*) the couple vector representing the equivalent couple, (*b*) the two forces acting at *B* and *C* which may be used to form that couple.

3.59 Solve part *a* of Prob. 3.58, assuming that two 15-N vertical forces have been added, one acting upward at *A* and the other downward at *C*.

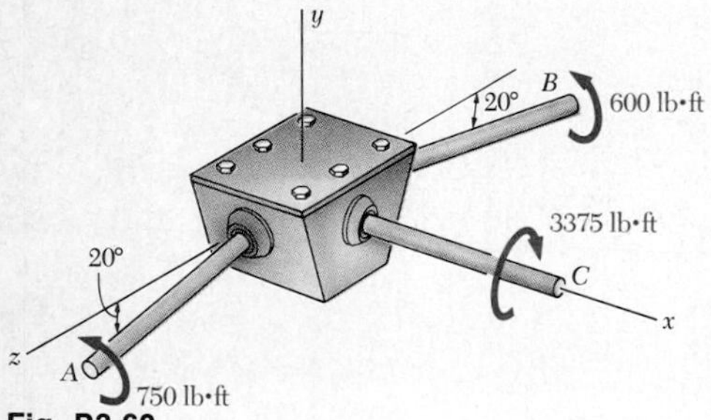

Fig. P3.60

3.60 Shafts *A* and *B* connect the gear box to the wheel assemblies of a tractor, and shaft *C* connects it to the engine. Shafts *A* and *B* lie in the vertical *yz* plane, while shaft *C* is directed along the *x* axis. Replace the couples applied to the shafts by a single equivalent couple, specifying its magnitude and the direction of its axis.

3.61 A 30-lb vertical force **P** is applied at *A* to the bracket shown, which is held by screws at *B* and *C*. (*a*) Replace **P** by an equivalent force-couple system at *B*. (*b*) Find the two horizontal forces at *B* and *C* which are equivalent to the couple obtained in part *a*.

3.62 A 260-lb force is applied at *A* to the rolled-steel section shown. Replace that force by an equivalent force-couple system at the center *C* of the section.

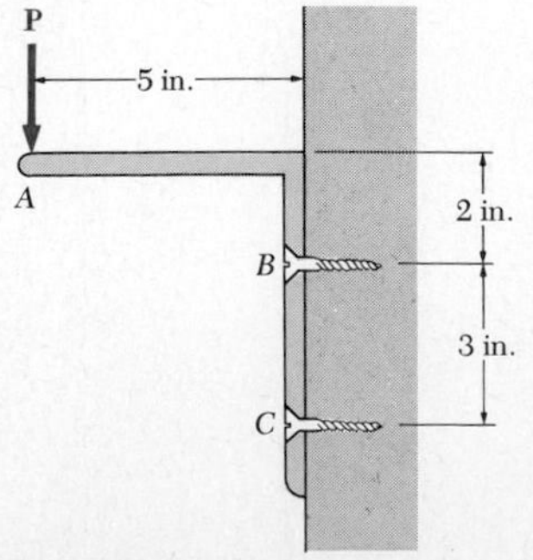

Fig. P3.61

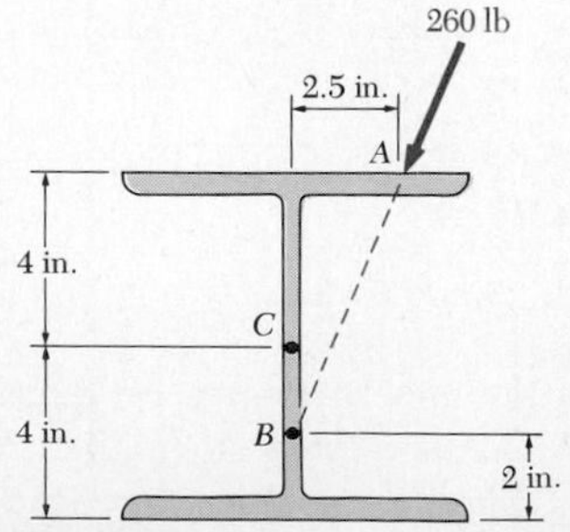

Fig. P3.62

3.63 The force **P** has a magnitude of 250 N and is applied at the end C of a 500-mm rod AC attached to a bracket at A and B. Assuming $\alpha = 30°$ and $\beta = 60°$, replace **P** by (*a*) an equivalent force-couple system at B, (*b*) an equivalent system formed by two parallel forces applied at A and B.

3.64 Solve Prob. 3.63, assuming $\alpha = \beta = 25°$.

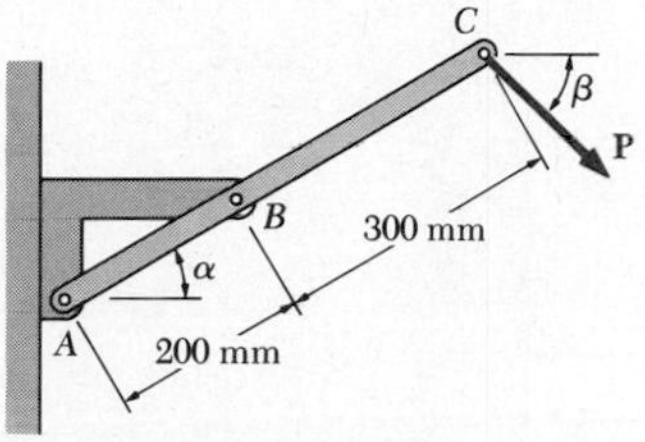

Fig. P3.63 and P3.65

3.65 The force **P** is applied at the end C of a 500-mm rod AC attached to a bracket at A and B. Replace **P** by an equivalent system formed by two parallel forces applied at A and B and show that (*a*) these forces are parallel to **P**, (*b*) their magnitudes are independent of both α and β.

3.66 The shearing forces exerted on the cross section of a steel channel may be represented by a 750-N vertical force and two 200-N horizontal forces as shown. Replace this force and couple by a single force **F** applied at point C and determine the distance x from C to line BD. (Point C is defined as the *shear center* of the section.)

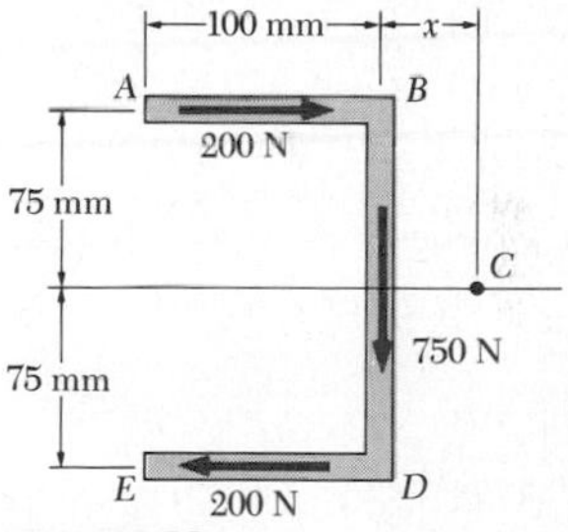

Fig. P3.66

3.67 A force and couple act as shown on a square plate of side $a = 20$ in. Knowing that $P = 50$ lb, $Q = 30$ lb, and $\alpha = 50°$, replace the given force and couple by a single force applied at a point located (*a*) on line AB, (*b*) on line AC. In each case determine the distance from A to the point of application of the force.

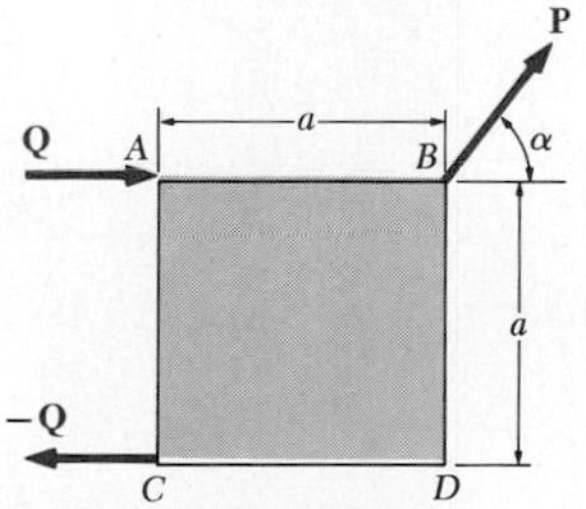

Fig. P3.67 and P3.68

3.68 The force and couple shown are to be replaced by an equivalent single force. Knowing that $P = 2Q$, determine the required value of α if the line of action of the single equivalent force is to pass through (*a*) point A, (*b*) point C.

3.69 Replace the 200-kN force **P** by an equivalent force-couple system at G.

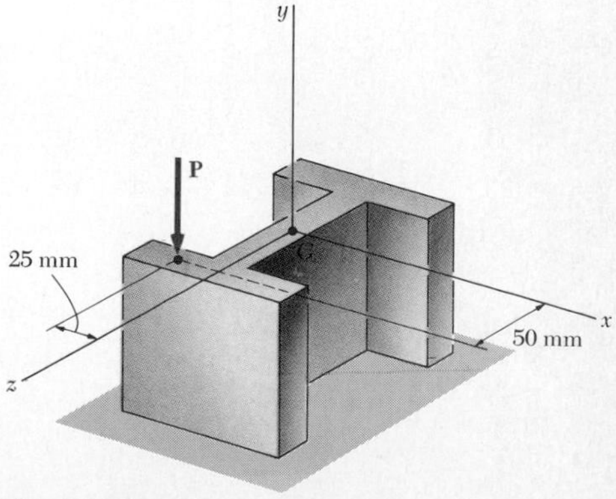

Fig. P3.69

3.70 The 15-ft boom AB has a fixed end A and the tension in cable BC is 380 lb. Replace the force that the cable exerts at B by an equivalent force-couple system at A.

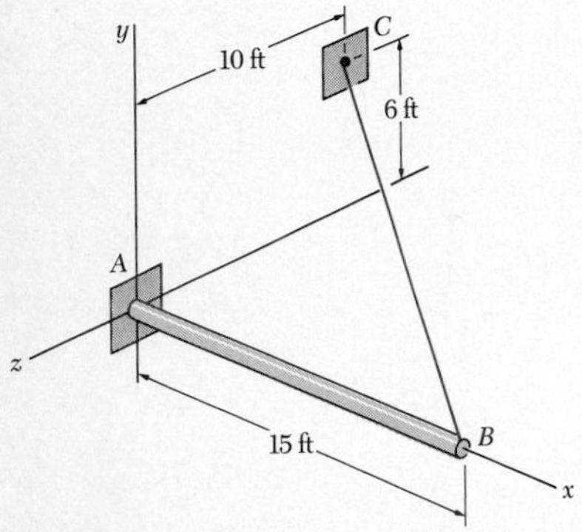

Fig. P3.70

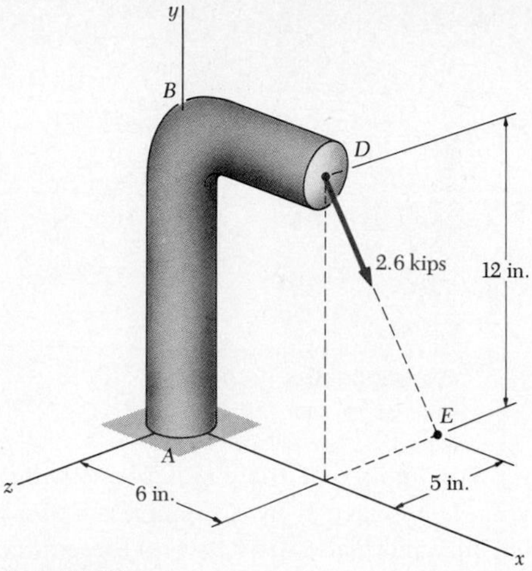

Fig. P3.71

3.71 A 2.6-kip force is applied at point D of the cast-iron post shown. Replace that force by an equivalent force-couple system at the center A of the base section.

3.72 Replace the 150-N force by an equivalent force-couple system at A.

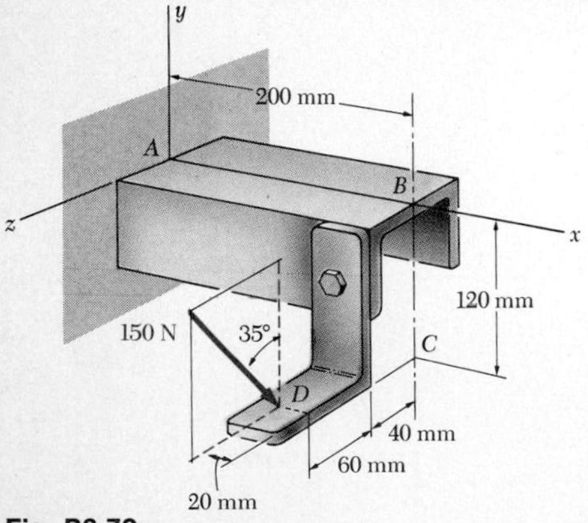

Fig. P3.72

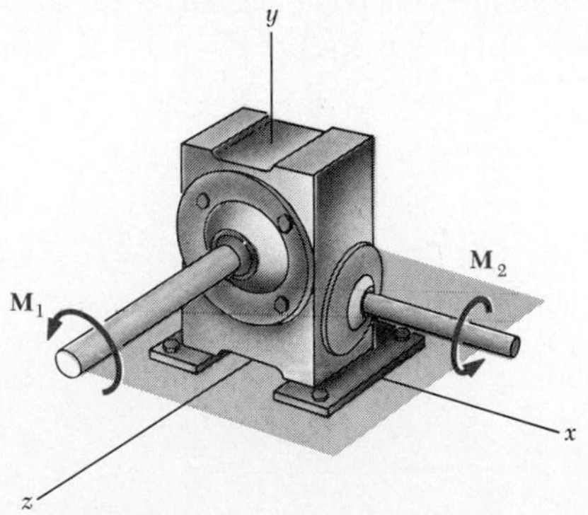

Fig. P3.73

3.73 The speed-reducer unit shown weighs 60 lb and its center of gravity is located on the y axis. Show that the weight of the unit and the two couples acting on it, of magnitude $M_1 = 15$ lb·ft and $M_2 = 3$ lb·ft, respectively, may be replaced by a single equivalent force and determine (*a*) the magnitude and direction of that force, (*b*) the point where its line of action intersects the floor.

3.74 Five separate force-couple systems act at the corners of a bent plate as shown. Find two force-couple systems which are equivalent.

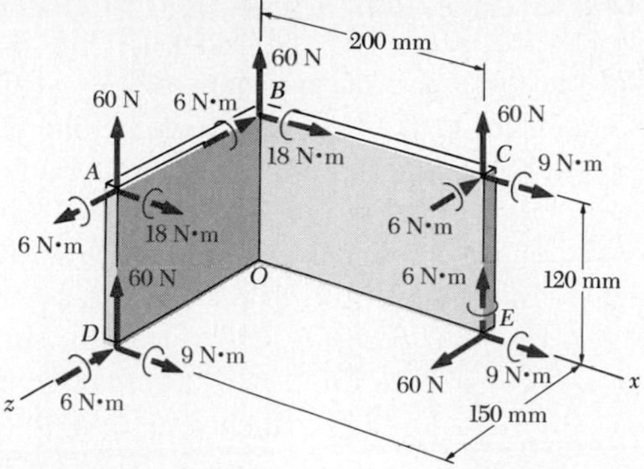

Fig. P3.74 and P3.75

3.75 Five separate force-couple systems act at the corners of a bent plate as shown. Determine which of these systems is equivalent to a force $\mathbf{F} = (60\text{ N})\mathbf{j}$ and a couple $\mathbf{M} = -(6\text{ N}\cdot\text{m})\mathbf{k}$ located at the origin O.

3.17. Reduction of a System of Forces to One Force and One Couple. Consider a system of forces $\mathbf{F}_1$, $\mathbf{F}_2$, $\mathbf{F}_3$, etc., acting on a rigid body at the points A_1, A_2, A_3, etc., defined by the position vectors $\mathbf{r}_1$, $\mathbf{r}_2$, $\mathbf{r}_3$, etc. (Fig. 3.41*a*). As seen in the preceding section, $\mathbf{F}_1$ may be moved from A_1 to a given point O if a couple of moment $\mathbf{M}_1$ equal to the moment $\mathbf{r}_1 \times \mathbf{F}_1$ of $\mathbf{F}_1$ about O is added to the original system of forces. Repeating this procedure with $\mathbf{F}_2$, $\mathbf{F}_3$, etc., we obtain the system shown in Fig. 3.41*b*, consisting of forces acting at O and of couples. Since the forces are now concurrent, they may be added vectorially and replaced by their resultant $\mathbf{R}$. Similarly, the couple vectors $\mathbf{M}_1$, $\mathbf{M}_2$, $\mathbf{M}_3$, etc., may be added vectorially and replaced by a single couple vector $\mathbf{M}_O^R$. Any system of forces, however complex, may thus be reduced to an *equivalent force-couple system acting at a given point O* (Fig. 3.41*c*). We should note that while each of the couple vectors $\mathbf{M}_1$, $\mathbf{M}_2$, $\mathbf{M}_3$, etc., in Fig. 3.41*b* is perpendicular to the corresponding force, the resultant force $\mathbf{R}$ and the resultant couple vector $\mathbf{M}_O^R$ in Fig. 3.41*c* will not, in general, be perpendicular to each other.

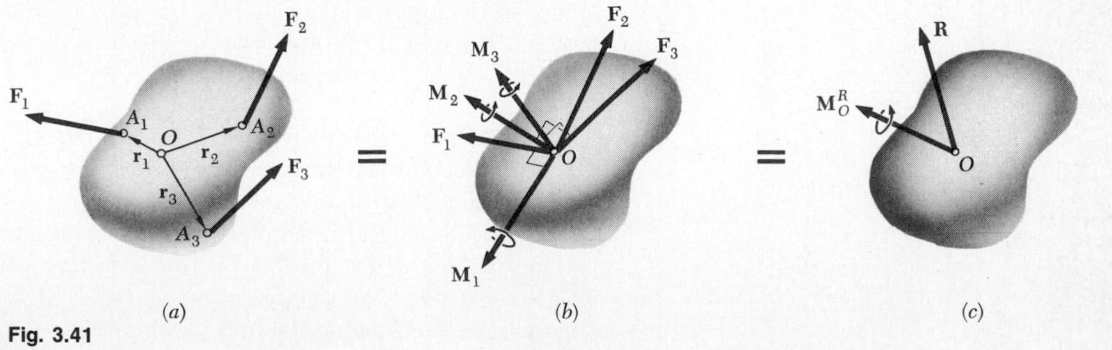

Fig. 3.41

The equivalent force-couple system is defined by the equations

$$\mathbf{R} = \Sigma\mathbf{F} \qquad \mathbf{M}_O^R = \Sigma\mathbf{M}_O = \Sigma(\mathbf{r} \times \mathbf{F}) \tag{3.52}$$

which express that the force **R** is obtained by adding all the forces of the system, while the moment $\mathbf{M}_O^R$ of the couple, called *moment resultant* of the system, is obtained by adding the moments about O of all the forces of the system.

Once a given system of forces has been reduced to a force and a couple at a point O, it may easily be reduced to a force and a couple at another point O'. While the resultant force **R** will remain unchanged, the new couple vector $\mathbf{M}_{O'}^R$ will be equal to the sum of the couple vector $\mathbf{M}_O^R$ and of the moment about O' of the force **R** attached at O (Fig. 3.42). We have

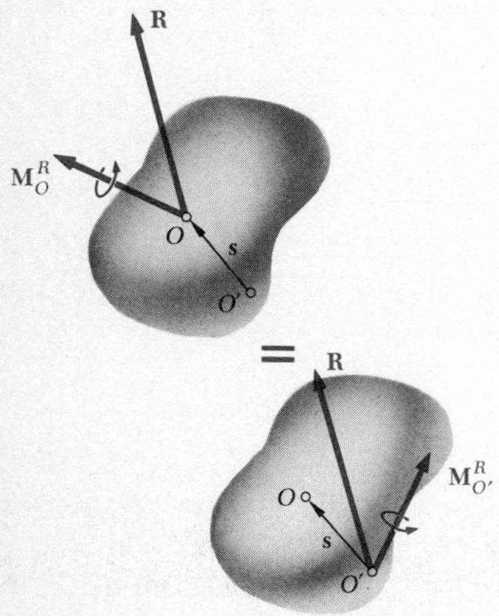

Fig. 3.42

$$\mathbf{M}_{O'}^R = \mathbf{M}_O^R + \mathbf{s} \times \mathbf{R} \tag{3.53}$$

In practice, the reduction of a given system of forces to a single force **R** at O and a couple $\mathbf{M}_O^R$ will be carried out in terms of components. Resolving each position vector **r** and each force **F** of the system into rectangular components, we write

$$\mathbf{r} = x\mathbf{i} + y\mathbf{j} + z\mathbf{k} \tag{3.54}$$
$$\mathbf{F} = F_x\mathbf{i} + F_y\mathbf{j} + F_z\mathbf{k} \tag{3.55}$$

Substituting for **r** and **F** into (3.52) and factoring the unit vectors **i**, **j**, **k**, we obtain **R** and $\mathbf{M}_O^R$ in the form

$$\mathbf{R} = R_x\mathbf{i} + R_y\mathbf{j} + R_z\mathbf{k} \qquad \mathbf{M}_O^R = M_x^R\mathbf{i} + M_y^R\mathbf{j} + M_z^R\mathbf{k} \tag{3.56}$$

The components R_x, R_y, R_z represent, respectively, the sums of the x, y, and z components of the given forces and measure the tendency of the system to impart to the rigid body a motion of translation in the x, y, or z direction. Similarly, the components M_x^R, M_y^R, M_z^R represent, respectively, the sums of the moments of the given forces about the x, y, and z axes and measure the tendency of the system to impart to the rigid body a motion of rotation about the x, y, or z axis.

If the magnitude and direction of the force **R** are desired, they may be obtained from the components R_x, R_y, R_z by means of the relations (2.18) and (2.19) of Sec. 2.12; similar computations will yield the magnitude and direction of the couple vector $\mathbf{M}_O^R$.

3.18. Equivalent Systems of Forces. We have seen in the preceding section that any system of forces acting on a rigid body may be reduced to a force-couple system at a given point O. This equivalent force-couple system characterizes completely the effect of the given system on the rigid body. *Two systems of forces are equivalent, therefore, if they may be reduced to the same force-couple system at a given point O.* Recalling that the force-couple system at O is defined by the relations (3.52), we state: *Two systems of forces* $\mathbf{F}_1$, $\mathbf{F}_2$, $\mathbf{F}_3$, *etc., and* $\mathbf{F}'_1$, $\mathbf{F}'_2$, $\mathbf{F}'_3$, *etc., are equivalent if, and only if, the sums of the forces and the sums of the moments about a given point O of the forces of the two systems are, respectively, equal.* Expressed mathematically, the necessary and sufficient conditions for the two systems of forces to be equivalent are

$$\Sigma\mathbf{F} = \Sigma\mathbf{F}' \quad \text{and} \quad \Sigma\mathbf{M}_O = \Sigma\mathbf{M}'_O \tag{3.57}$$

Note that to prove that two systems of forces are equivalent, the second of the relations (3.57) must be established with respect to *only one point O.* It will hold, however, with respect to *any point* if the two systems are equivalent.

Resolving the forces and moments in (3.57) into their rectangular components, we may express the necessary and sufficient conditions for the equivalence of two systems of forces acting on a rigid body as follows:

$$\begin{array}{lll} \Sigma F_x = \Sigma F'_x & \Sigma F_y = \Sigma F'_y & \Sigma F_z = \Sigma F'_z \\ \Sigma M_x = \Sigma M'_x & \Sigma M_y = \Sigma M'_y & \Sigma M_z = \Sigma M'_z \end{array} \tag{3.58}$$

These equations have a simple physical significance. They express that two systems of forces are equivalent if they tend to impart to the rigid body (1) the same translation in the x, y, and z directions, respectively, and (2) the same rotation about the x, y, and z axes, respectively.

3.19. Equipollent Systems of Vectors. When two systems of vectors satisfy Eqs. (3.57) or (3.58), i.e., when their resultants and their moment resultants about an arbitrary point O are respectively equal, the two systems are said to be *equipollent.* The result established in the preceding section may thus be restated as follows: *If two systems of forces acting on a rigid body are equipollent, they are also equivalent.*

It is important to note that this statement does not apply to *any* system of vectors. Consider for example a system of forces acting on a set of independent particles which do *not* form a rigid body. A different system of forces acting on the same particles may happen to be equipollent to the first one, i.e., it may have the same resultant and the same moment resultant. Yet, since different forces will now act on the various particles, their effects on these particles will be different; the two systems of forces, while equipollent, are *not equivalent.*

3.20. Further Reduction of a System of Forces. We saw in Sec. 3.17 that any given system of forces acting on a rigid body may be reduced to an equivalent force-couple system at O, consisting of a force $\mathbf{R}$ equal to the sum of the forces of the system, and of a couple vector $\mathbf{M}_O^R$ equal to the moment resultant of the system.

When $\mathbf{R} = 0$, the force-couple system reduces to the couple vector $\mathbf{M}_O^R$. The given system of forces may then be reduced to a single couple, called the *resultant couple* of the system.

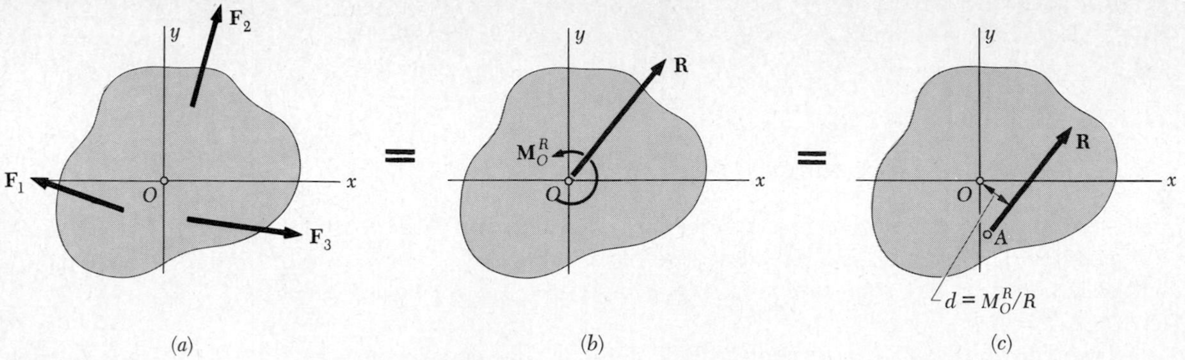

Fig. 3.43

We shall now investigate the conditions under which a given system of forces may be reduced to a single force. It follows from Sec. 3.16 that the force-couple system at O may be replaced by a single force $\mathbf{R}$ acting along a new line of action if $\mathbf{R}$ and $\mathbf{M}_O^R$ are mutually perpendicular. The systems of forces which may be reduced to a single force, or *resultant*, are therefore the systems for which the force $\mathbf{R}$ and the couple vector $\mathbf{M}_O^R$ are mutually perpendicular. While this condition *is generally not satisfied* by systems of forces in space, it *will be satisfied* by systems consisting of (1) concurrent forces, (2) coplanar forces, or (3) parallel forces. We shall discuss these cases separately.

1. *Concurrent forces* are applied at the same point and may therefore be added directly into their resultant $\mathbf{R}$. Thus, they always reduce to a single force. Concurrent forces have been discussed in detail in Chap. 2.
2. *Coplanar forces* act in the same plane, which we shall assume here to be the plane of the figure (Fig. 3.43*a*). The sum $\mathbf{R}$ of the forces of the system will also lie in the plane of the figure, while the moment of each force about O, and thus the moment resultant $\mathbf{M}_O^R$, will be perpendicular to that plane. The force-couple system at O consists therefore of a force $\mathbf{R}$ and a couple vector $\mathbf{M}_O^R$ which are mutually perpendicular (Fig. 3.43*b*).† They may be reduced to a single force $\mathbf{R}$ by moving $\mathbf{R}$ in the plane of the figure until its moment about O becomes equal to $\mathbf{M}_O^R$. The distance from O to the line of action of $\mathbf{R}$ is $d = M_O^R/R$ (Fig. 3.43*c*).

   As noted in Sec. 3.17, the reduction of a system of forces is considerably simplified if the forces are resolved into rectangular components. The force-couple system at O is then characterized by the components (Fig. 3.44*a*)

$$R_x = \Sigma F_x \qquad R_y = \Sigma F_y \qquad M_z^R = M_O^R = \Sigma M_O \qquad (3.59)$$

† Since the couple vector $\mathbf{M}_O^R$ is perpendicular to the plane of the figure, it has been represented by the symbol ↺. A counterclockwise couple ↺ corresponds to a vector pointing out of the paper, and a clockwise couple ↻ to a vector pointing into the paper.

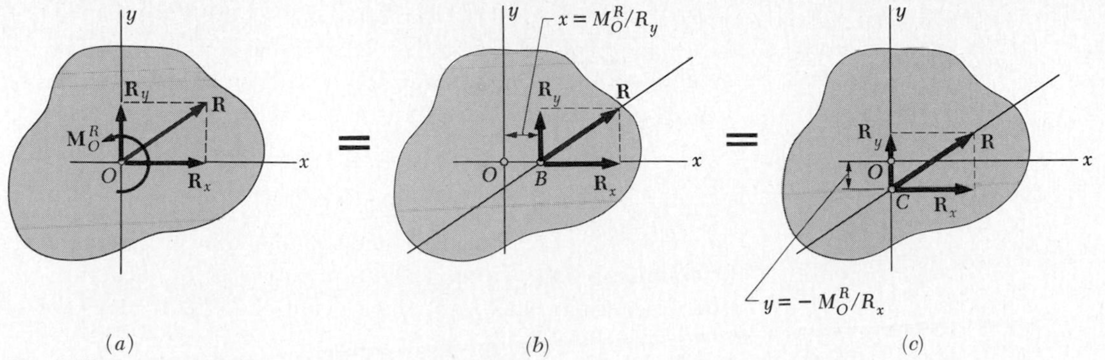

Fig. 3.44

To reduce the system to a single force **R** we shall express that the moment of **R** about O must be equal to $\mathbf{M}_O^R$. Denoting by x and y the coordinates of the point of application A of the resultant, and recalling formula (3.22) of Sec. 3.8, we write

$$xR_y - yR_x = M_O^R$$

which represents the equation of the line of action of **R**. We may also determine directly the x and y intercepts of the line of action of the resultant by noting that $\mathbf{M}_O^R$ must be equal to the moment about O of the y component of **R** when **R** is attached at B (Fig. 3.44*b*), and to the moment of its x component when **R** is attached at C (Fig. 3.44*c*).

3. *Parallel forces* have parallel lines of action and may or may not have the same sense. Assuming here that the forces are parallel to the y axis (Fig. 3.45*a*), we note that their sum **R** will also be parallel to the y axis. On the other hand, since the moment of a given force must be perpendicular to that force, the moment about O of each force of the system, and thus the moment resultant $\mathbf{M}_O^R$, will lie in the zx plane. The force-couple system at O consists therefore of a force **R** and a couple vector $\mathbf{M}_O^R$ which are mutually perpendicular (Fig. 3.45*b*). They may be reduced to a single force **R** (Fig. 3.45*c*) or, if $\mathbf{R} = 0$, to a single couple of moment $\mathbf{M}_O^R$.

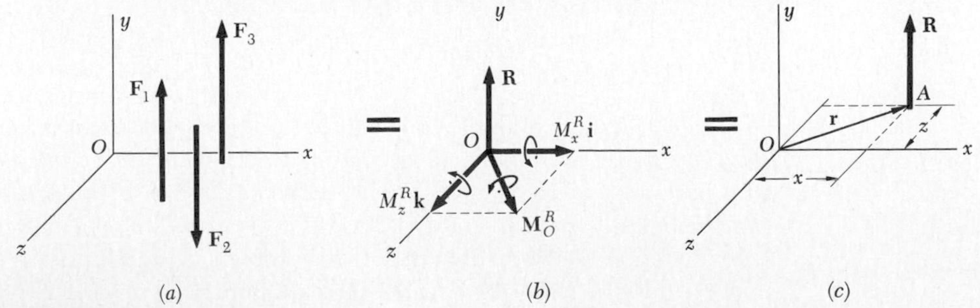

Fig. 3.45

In practice, the force-couple system at O will be characterized by the components

$$R_y = \Sigma F_y \qquad M_x^R = \Sigma M_x \qquad M_z^R = \Sigma M_z \tag{3.60}$$

The reduction of the system to a single force may be carried out by moving $\mathbf{R}$ to a new point of application $A(x, 0, z)$ chosen so that the moment of $\mathbf{R}$ about O is equal to $\mathbf{M}_O^R$. We write

$$\mathbf{r} \times \mathbf{R} = \mathbf{M}_O^R$$
$$(x\mathbf{i} + z\mathbf{k}) \times R_y\mathbf{j} = M_x^R\mathbf{i} + M_z^R\mathbf{k}$$

Computing the vector products and equating the coefficients of the corresponding unit vectors in both members of the equation, we obtain two scalar equations which define the coordinates of A:

$$-zR_y = M_x^R \qquad xR_y = M_z^R$$

These equations express that the moments of $\mathbf{R}$ about the x and z axes must, respectively, be equal to M_x^R and M_z^R.

*** 3.21. Reduction of a System of Forces to a Wrench.** In the general case of a system of forces in space, the equivalent force-couple system at O consists of a force $\mathbf{R}$ and a couple vector $\mathbf{M}_O^R$ which are not perpendicular, and neither of which is zero (Fig. 3.46*a*). Thus, the system of forces *cannot* be reduced to a single force or a single couple. The couple vector, however, may be replaced by two other couple vectors obtained by resolving $\mathbf{M}_O^R$ into a component $\mathbf{M}_1$ along $\mathbf{R}$ and a component $\mathbf{M}_2$ in a plane perpendicular to $\mathbf{R}$ (Fig. 3.46*b*). The couple vector $\mathbf{M}_2$ and the force $\mathbf{R}$ may then be replaced by a single force $\mathbf{R}$ acting along a new line of action.

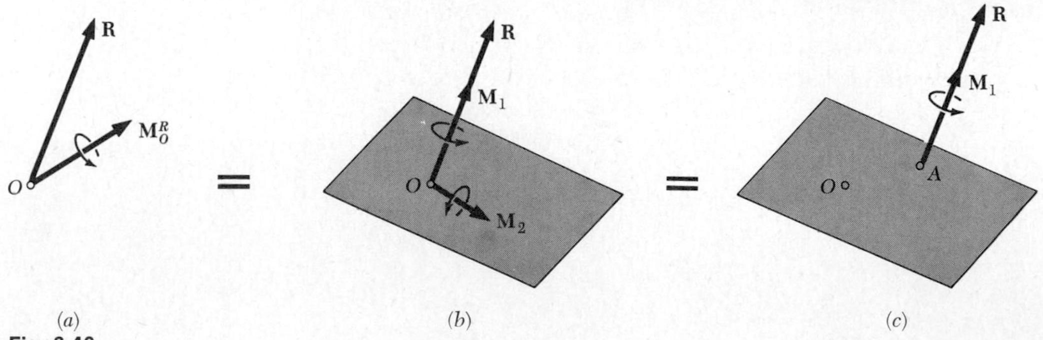

Fig. 3.46

The original system of forces thus reduces to $\mathbf{R}$ and to the couple vector $\mathbf{M}_1$ (Fig. 3.46*c*), i.e., to $\mathbf{R}$ and a couple acting in the plane perpendicular to $\mathbf{R}$. This particular force-couple system is called a *wrench* because the resulting combination of push and twist is the same that would be caused by an actual wrench. The line of action of $\mathbf{R}$ is known as the *axis of the wrench*, and the ratio $p = M_1/R$ is called the *pitch* of the wrench. A wrench, therefore, consists of two collinear vectors, namely, a force **R** and a couple vector

$$\mathbf{M}_1 = p\mathbf{R} \tag{3.61}$$

Recalling the expression (3.35) obtained in Sec. 3.9 for the projection of a vector on the line of action of another vector, we note that the projection of $\mathbf{M}_O^R$ on the line of action of **R** is

$$M_1 = \frac{\mathbf{R} \cdot \mathbf{M}_O^R}{R}$$

Thus, the pitch of the wrench may be expressed as†

$$p = \frac{M_1}{R} = \frac{\mathbf{R} \cdot \mathbf{M}_O^R}{R^2} \tag{3.62}$$

To define the axis of the wrench we may write a relation involving the position vector **r** of an arbitrary point P located on that axis. Attaching the resultant force **R** and couple vector $\mathbf{M}_1$ at P (Fig. 3.47), and expressing that the moment about O of this force-couple system is equal to the moment resultant $\mathbf{M}_O^R$ of the original force system, we write

$$\mathbf{M}_1 + \mathbf{r} \times \mathbf{R} = \mathbf{M}_O^R \tag{3.63}$$

or, recalling Eq. (3.61),

$$p\mathbf{R} + \mathbf{r} \times \mathbf{R} = \mathbf{M}_O^R \tag{3.64}$$

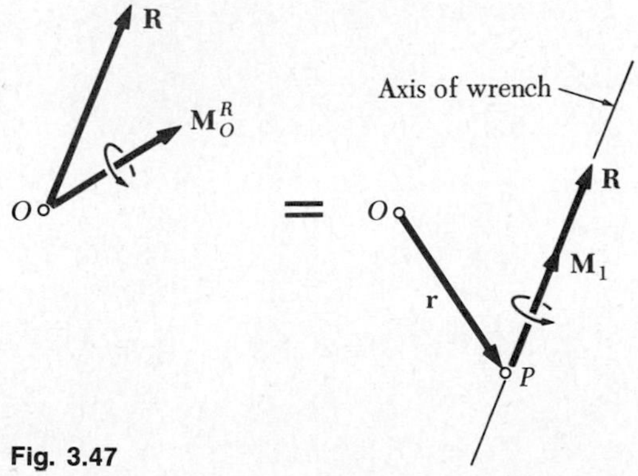

Fig. 3.47

† The expressions obtained for the projection of the couple vector on the line of action of **R** and for the pitch of the wrench are independent of the choice of point O. Using the relation (3.53) of Sec. 3.17, we check that if a different point O' had been used, the numerator in (3.62) would be

$$\mathbf{R} \cdot \mathbf{M}_{O'}^R = \mathbf{R} \cdot (\mathbf{M}_O^R + \mathbf{s} \times \mathbf{R}) = \mathbf{R} \cdot \mathbf{M}_O^R + \mathbf{R} \cdot (\mathbf{s} \times \mathbf{R})$$

Since the mixed triple product $\mathbf{R} \cdot (\mathbf{s} \times \mathbf{R})$ is identically equal to zero, we have

$$\mathbf{R} \cdot \mathbf{M}_{O'}^R = \mathbf{R} \cdot \mathbf{M}_O^R$$

Thus, the scalar product $\mathbf{R} \cdot \mathbf{M}_O^R$ is independent of the choice of point O.

SAMPLE PROBLEM 3.8

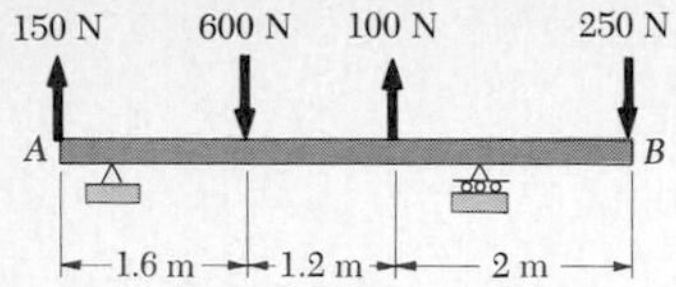

A 4.80-m beam is subjected to the forces shown. Reduce the given system of forces to (*a*) an equivalent force-couple system at *A*, (*b*) an equivalent force-couple system at *B*, (*c*) a single force or resultant.

Note. Since the reactions at the supports are not included in the given system of forces, the given system will not maintain the beam in equilibrium.

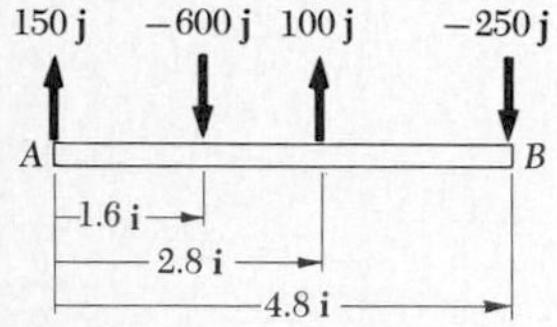

***a.* Force-Couple System at *A*.** The force-couple system at *A* equivalent to the given system of forces consists of a force **R** and a couple $\mathbf{M}_A^R$ defined as follows:

$$\mathbf{R} = \Sigma \mathbf{F}$$
$$= (150\text{ N})\mathbf{j} - (600\text{ N})\mathbf{j} + (100\text{ N})\mathbf{j} - (250\text{ N})\mathbf{j} = -(600\text{ N})\mathbf{j}$$
$$\mathbf{M}_A^R = \Sigma(\mathbf{r} \times \mathbf{F})$$
$$= (1.6\mathbf{i}) \times (-600\mathbf{j}) + (2.8\mathbf{i}) \times (100\mathbf{j}) + (4.8\mathbf{i}) \times (-250\mathbf{j})$$
$$= -(1880\text{ N}\cdot\text{m})\mathbf{k}$$

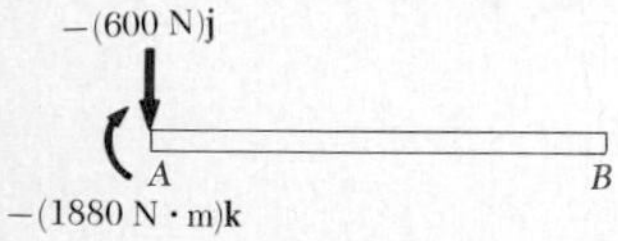

The equivalent force-couple system at *A* is thus

$$\mathbf{R} = 600\text{ N} \downarrow \qquad \mathbf{M}_A^R = 1880\text{ N}\cdot\text{m} \circlearrowright \quad ◄$$

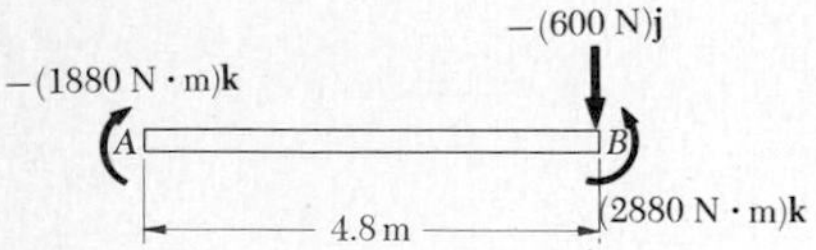

***b.* Force-Couple System at *B*.** We shall find a force-couple system at *B* equivalent to the force-couple system at *A* determined in part *a*. The force **R** is unchanged, but a new couple $\mathbf{M}_B^R$ must be determined, the moment of which is equal to the moment about *B* of the force-couple system determined in part *a*. Thus, we have

$$\mathbf{M}_B^R = \mathbf{M}_A^R + \overrightarrow{BA} \times \mathbf{R}$$
$$= -(1880\text{ N}\cdot\text{m})\mathbf{k} + (-4.8\text{ m})\mathbf{i} \times (-600\text{ N})\mathbf{j}$$
$$= -(1880\text{ N}\cdot\text{m})\mathbf{k} + (2880\text{ N}\cdot\text{m})\mathbf{k} = +(1000\text{ N}\cdot\text{m})\mathbf{k}$$

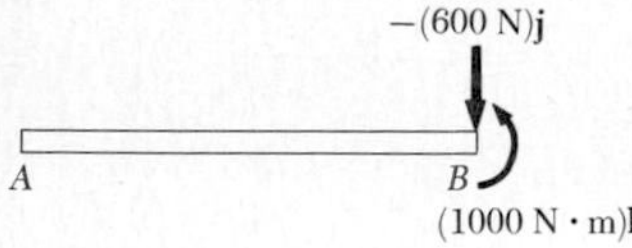

The equivalent force-couple system at *B* is thus

$$\mathbf{R} = 600\text{ N} \downarrow \qquad \mathbf{M}_B^R = 1000\text{ N}\cdot\text{m} \circlearrowleft \quad ◄$$

***c.* Single Force or Resultant.** The resultant of the given system of forces is equal to **R** and its point of application must be such that the moment of **R** about *A* is equal to $\mathbf{M}_A^R$. We write

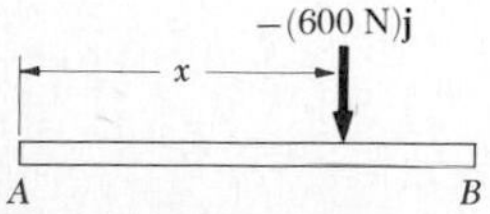

$$\mathbf{r} \times \mathbf{R} = \mathbf{M}_A^R$$
$$x\mathbf{i} \times (-600\text{ N})\mathbf{j} = -(1880\text{ N}\cdot\text{m})\mathbf{k}$$
$$-x(600\text{ N})\mathbf{k} = -(1880\text{ N}\cdot\text{m})\mathbf{k}$$

and conclude that $x = 3.13$ m. Thus, the single force equivalent to the given system is defined as

$$\mathbf{R} = 600\text{ N} \downarrow \qquad x = 3.13\text{ m} \quad ◄$$

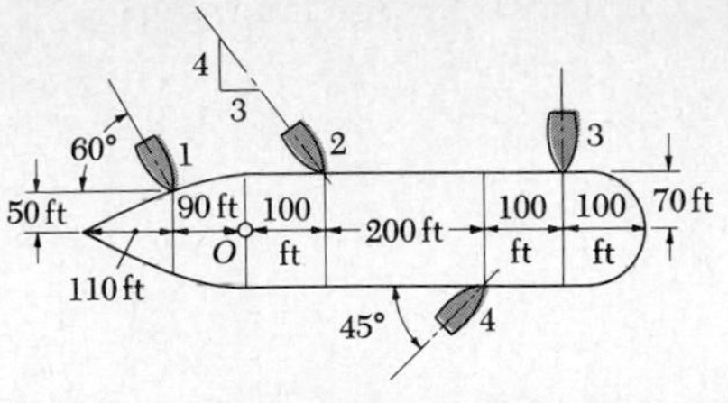

SAMPLE PROBLEM 3.9

Four tugboats are used to bring an ocean liner to its pier. Each tugboat exerts a 5000-lb force in the direction shown. Determine (*a*) the equivalent force-couple system at the foremast *O*, (*b*) the point on the hull where a single, more powerful tugboat should push to produce the same effect as the original four tugboats.

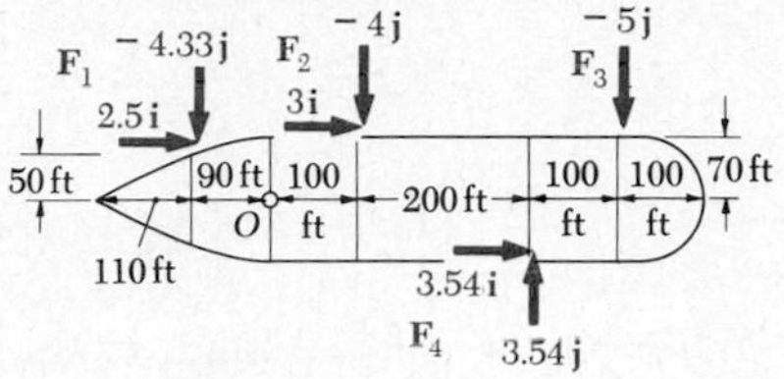

***a.* Force-Couple System at *O*.** Each of the given forces is resolved into components in the diagram shown (kip units are used). The force-couple system at *O* equivalent to the given system of forces consists of a force **R** and a couple $\mathbf{M}_O^R$ defined as follows:

$$
\begin{aligned}
\mathbf{R} &= \Sigma\mathbf{F} \\
&= (2.50\mathbf{i} - 4.33\mathbf{j}) + (3.00\mathbf{i} - 4.00\mathbf{j}) + (-5.00\mathbf{j}) + (3.54\mathbf{i} + 3.54\mathbf{j}) \\
&= 9.04\mathbf{i} - 9.79\mathbf{j} \\
\mathbf{M}_O^R &= \Sigma(\mathbf{r} \times \mathbf{F}) \\
&= (-90\mathbf{i} + 50\mathbf{j}) \times (2.50\mathbf{i} - 4.33\mathbf{j}) \\
&\quad + (100\mathbf{i} + 70\mathbf{j}) \times (3.00\mathbf{i} - 4.00\mathbf{j}) \\
&\quad + (400\mathbf{i} + 70\mathbf{j}) \times (-5.00\mathbf{j}) \\
&\quad + (300\mathbf{i} - 70\mathbf{j}) \times (3.54\mathbf{i} + 3.54\mathbf{j}) \\
&= (390 - 125 - 400 - 210 - 2000 + 1062 + 248)\mathbf{k} \\
&= -1035\mathbf{k}
\end{aligned}
$$

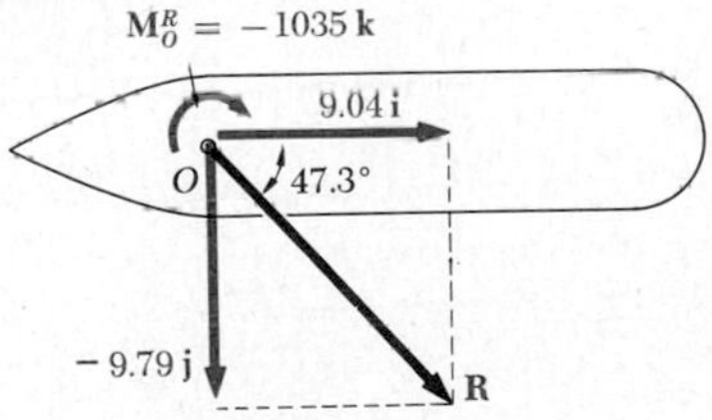

The equivalent force-couple system at *O* is thus

$$
\mathbf{R} = (9.04 \text{ kips})\mathbf{i} - (9.79 \text{ kips})\mathbf{j} \qquad \mathbf{M}_O^R = -(1035 \text{ kip}\cdot\text{ft})\mathbf{k}
$$

or

$$
\mathbf{R} = 13.33 \text{ kips} \measuredangle 47.3^\circ \qquad \mathbf{M}_O^R = 1035 \text{ kip}\cdot\text{ft} \circlearrowright \quad ◄
$$

Remark. Since all the forces are contained in the plane of the figure, we could have expected the sum of their moments to be perpendicular to that plane. Note that the moment of each force component could have been obtained directly from the diagram by forming the product of its magnitude and its perpendicular distance to *O*, and assigning to this product a positive or a negative sign, depending upon the sense of the moment.

***b.* Single Tugboat.** The force exerted by a single tugboat must be equal to **R** and its point of application *A* must be such that the moment of **R** about *O* is equal to $\mathbf{M}_O^R$. Observing that the position vector of *A* is

$$
\mathbf{r} = x\mathbf{i} + 70\mathbf{j}
$$

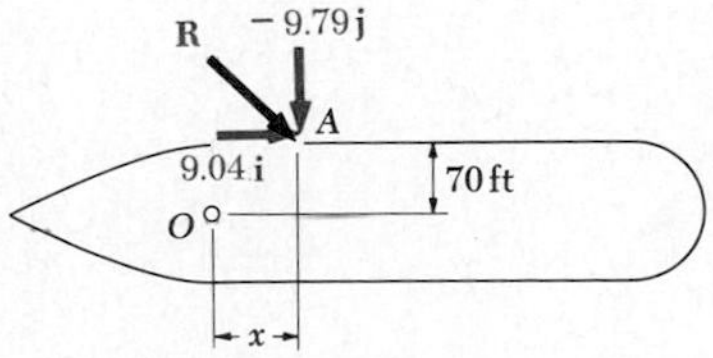

we write

$$
\begin{aligned}
\mathbf{r} \times \mathbf{R} &= \mathbf{M}_O^R \\
(x\mathbf{i} + 70\mathbf{j}) \times (9.04\mathbf{i} - 9.79\mathbf{j}) &= -1035\mathbf{k} \\
-x(9.79)\mathbf{k} - 633\mathbf{k} &= -1035\mathbf{k}
\end{aligned}
$$

$$
x = 41.1 \text{ ft} \quad ◄
$$

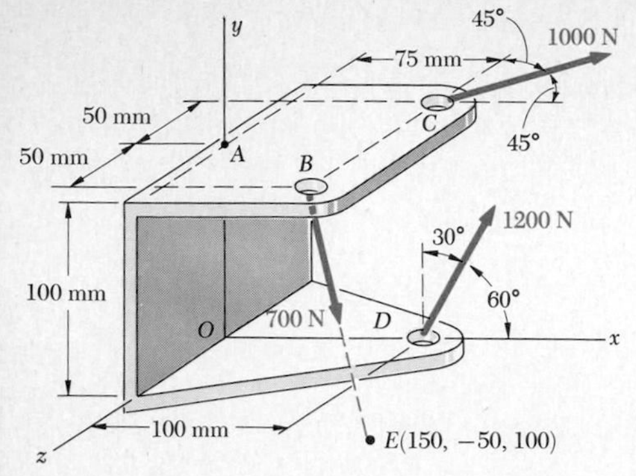

SAMPLE PROBLEM 3.10

Three cables are attached to a bracket as shown. Replace the forces exerted by the cables by an equivalent force-couple system at A.

Solution. We first determine the relative position vectors drawn from point A to the points of application of the various forces and resolve the forces into rectangular components. Observing that $\mathbf{F}_B = (700\text{ N})\boldsymbol{\lambda}_{BE}$ where

$$\boldsymbol{\lambda}_{BE} = \frac{\overrightarrow{BE}}{BE} = \frac{75\mathbf{i} - 150\mathbf{j} + 50\mathbf{k}}{175}$$

we have, using meters and newtons,

$$\begin{aligned} \mathbf{r}_{B/A} &= \overrightarrow{AB} = 0.075\mathbf{i} + 0.050\mathbf{k} & \mathbf{F}_B &= 300\mathbf{i} - 600\mathbf{j} + 200\mathbf{k} \\ \mathbf{r}_{C/A} &= \overrightarrow{AC} = 0.075\mathbf{i} - 0.050\mathbf{k} & \mathbf{F}_C &= 707\mathbf{i} - 707\mathbf{k} \\ \mathbf{r}_{D/A} &= \overrightarrow{AD} = 0.100\mathbf{i} - 0.100\mathbf{j} & \mathbf{F}_D &= 600\mathbf{i} + 1039\mathbf{j} \end{aligned}$$

The force-couple system at A equivalent to the given forces consists of a force $\mathbf{R} = \Sigma\mathbf{F}$ and a couple $\mathbf{M}_A^R = \Sigma(\mathbf{r} \times \mathbf{F})$. The force $\mathbf{R}$ is readily obtained by adding respectively the x, y, and z components of the forces:

$$\mathbf{R} = \Sigma\mathbf{F} = (1607\text{ N})\mathbf{i} + (439\text{ N})\mathbf{j} - (507\text{ N})\mathbf{k} \quad ◄$$

The computation of $\mathbf{M}_A^R$ will be facilitated if we express the moments of the forces in the form of determinants (Sec. 3.8):

$$\mathbf{r}_{B/A} \times \mathbf{F}_B = \begin{vmatrix} \mathbf{i} & \mathbf{j} & \mathbf{k} \\ 0.075 & 0 & 0.050 \\ 300 & -600 & 200 \end{vmatrix} = 30\mathbf{i} \qquad -45\mathbf{k}$$

$$\mathbf{r}_{C/A} \times \mathbf{F}_C = \begin{vmatrix} \mathbf{i} & \mathbf{j} & \mathbf{k} \\ 0.075 & 0 & -0.050 \\ 707 & 0 & -707 \end{vmatrix} = \qquad 17.68\mathbf{j}$$

$$\mathbf{r}_{D/A} \times \mathbf{F}_D = \begin{vmatrix} \mathbf{i} & \mathbf{j} & \mathbf{k} \\ 0.100 & -0.100 & 0 \\ 600 & 1039 & 0 \end{vmatrix} = \qquad 163.9\mathbf{k}$$

Adding the expressions obtained, we have

$$\mathbf{M}_A^R = \Sigma(\mathbf{r} \times \mathbf{F}) = (30\text{ N}\cdot\text{m})\mathbf{i} + (17.68\text{ N}\cdot\text{m})\mathbf{j} + (118.9\text{ N}\cdot\text{m})\mathbf{k} \quad ◄$$

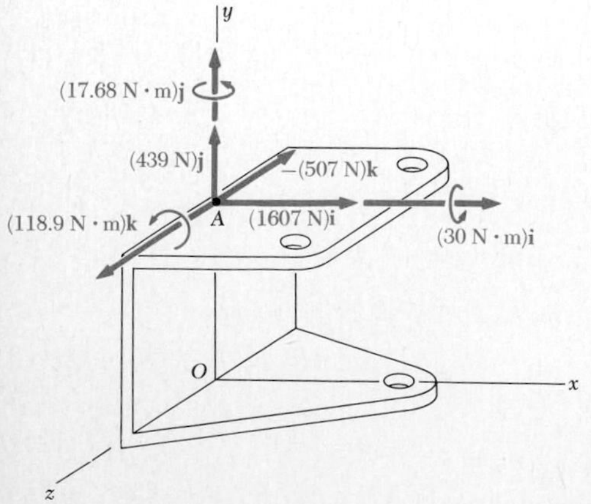

The rectangular components of the force $\mathbf{R}$ and the couple $\mathbf{M}_A^R$ are shown in the adjoining sketch.

SAMPLE PROBLEM 3.11

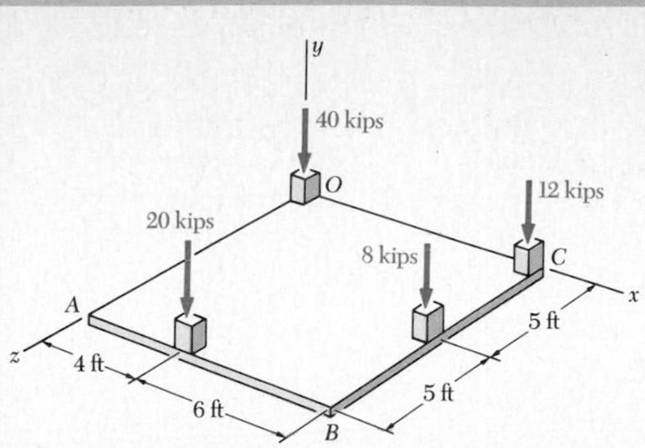

A square foundation mat supports the four columns shown. Determine the magnitude and point of application of the resultant of the four loads.

Solution. We shall first reduce the given system of forces to a force-couple system at the origin O of the coordinates. This force-couple system consists of a force $\mathbf{R}$ and a couple $\mathbf{M}_O^R$ defined as follows:

$$\mathbf{R} = \Sigma\mathbf{F} \qquad \mathbf{M}_O^R = \Sigma(\mathbf{r} \times \mathbf{F})$$

The position vectors of the points of application of the various forces are determined and the computations are arranged in tabular form.

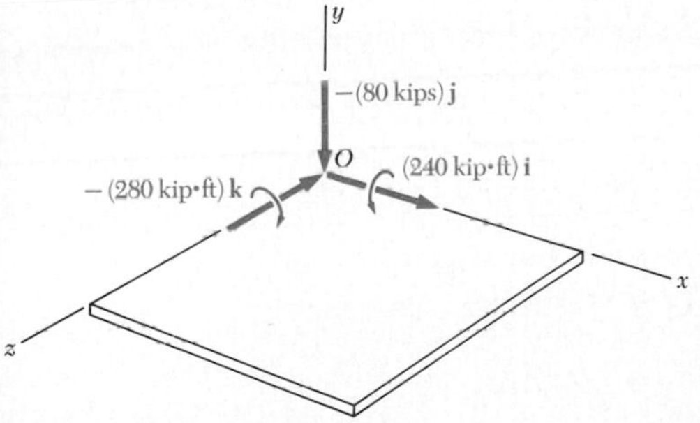

r, ft	F, kips	r × F, kip · ft
0	$-40\mathbf{j}$	0
$10\mathbf{i}$	$-12\mathbf{j}$	$-120\mathbf{k}$
$10\mathbf{i} + 5\mathbf{k}$	$-8\mathbf{j}$	$40\mathbf{i} - 80\mathbf{k}$
$4\mathbf{i} + 10\mathbf{k}$	$-20\mathbf{j}$	$200\mathbf{i} - 80\mathbf{k}$
	$\mathbf{R} = -80\mathbf{j}$	$\mathbf{M}_O^R = 240\mathbf{i} - 280\mathbf{k}$

Since the force $\mathbf{R}$ and the couple vector $\mathbf{M}_O^R$ are mutually perpendicular, the force-couple system obtained may be reduced further to a single force $\mathbf{R}$. The new point of application of $\mathbf{R}$ will be selected in the plane of the mat and in such a way that the moment of $\mathbf{R}$ about O will be equal to $\mathbf{M}_O^R$. Denoting by $\mathbf{r}$ the position vector of the desired point of application, and by x and z its coordinates, we write

$$\mathbf{r} \times \mathbf{R} = \mathbf{M}_O^R$$
$$(x\mathbf{i} + z\mathbf{k}) \times (-80\mathbf{j}) = 240\mathbf{i} - 280\mathbf{k}$$
$$-80x\mathbf{k} + 80z\mathbf{i} = 240\mathbf{i} - 280\mathbf{k}$$

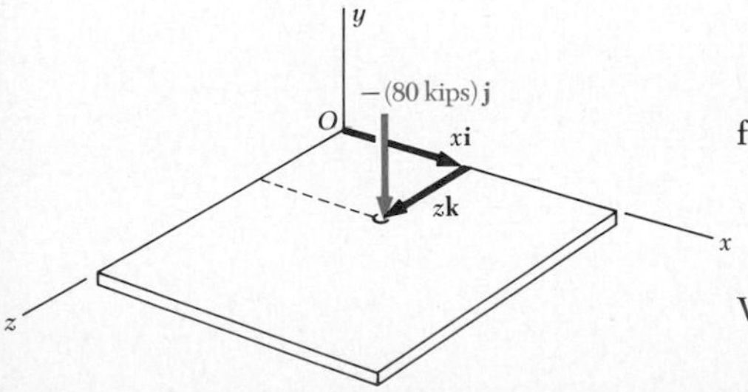

from which it follows that

$$-80x = -280 \qquad 80z = 240$$
$$x = 3.50 \text{ ft} \qquad z = 3.00 \text{ ft}$$

We conclude that the resultant of the given system of forces is

$$\mathbf{R} = 80 \text{ kips} \downarrow \quad \text{at } x = 3.50 \text{ ft}, z = 3.00 \text{ ft}$$

SAMPLE PROBLEM 3.12

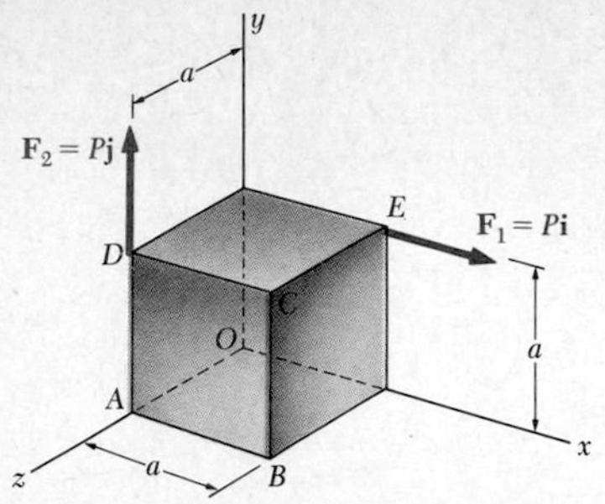

Two forces of the same magnitude P act on a cube of side a as shown. Replace the two forces by an equivalent wrench and determine (a) the magnitude and direction of the resultant force $\mathbf{R}$, (b) the pitch of the wrench, (c) the point where the axis of the wrench intersects the yz plane.

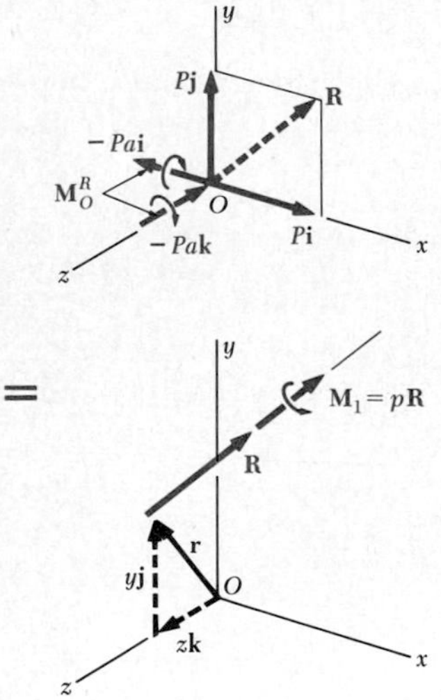

Solution. We first determine the equivalent force-couple system at the origin O. We observe that the position vectors of the points of application E and D of the two given forces are $\mathbf{r}_E = a\mathbf{i} + a\mathbf{j}$ and $\mathbf{r}_D = a\mathbf{j} + a\mathbf{k}$. The resultant $\mathbf{R}$ of the two forces and their moment resultant $\mathbf{M}_O^R$ about O are

$$\mathbf{R} = \mathbf{F}_1 + \mathbf{F}_2 = P\mathbf{i} + P\mathbf{j} = P(\mathbf{i} + \mathbf{j}) \qquad (1)$$

$$\mathbf{M}_O^R = \mathbf{r}_E \times \mathbf{F}_1 + \mathbf{r}_D \times \mathbf{F}_2 = (a\mathbf{i} + a\mathbf{j}) \times P\mathbf{i} + (a\mathbf{j} + a\mathbf{k}) \times P\mathbf{j}$$
$$= -Pa\mathbf{k} - Pa\mathbf{i} = -Pa(\mathbf{i} + \mathbf{k}) \qquad (2)$$

(a) ***Resultant Force* R.** It follows from Eq. (1) and the adjoining sketch that the resultant force $\mathbf{R}$ has the magnitude $R = P\sqrt{2}$, lies in the xy plane, and forms angles of 45° with the x and y axes. Thus

$$R = P\sqrt{2} \qquad \theta_x = \theta_y = 45° \qquad \theta_z = 90° \blacktriangleleft$$

(b) ***Pitch of Wrench.*** Recalling formula (3.62) of Sec. 3.21 and Eqs. (1) and (2) above, we write

$$p = \frac{\mathbf{R} \cdot \mathbf{M}_O^R}{R^2} = \frac{P(\mathbf{i} + \mathbf{j}) \cdot (-Pa)(\mathbf{i} + \mathbf{k})}{(P\sqrt{2})^2} = \frac{-P^2a(1 + 0 + 0)}{2P^2} \qquad p = -\frac{a}{2} \blacktriangleleft$$

(c) ***Axis of Wrench.*** It follows from the above and from Eq. (3.61) of Sec. 3.21 that the wrench consists of the force $\mathbf{R}$ found in (1) and of the couple vector

$$\mathbf{M}_1 = p\mathbf{R} = -\frac{a}{2}P(\mathbf{i} + \mathbf{j}) = -\frac{Pa}{2}(\mathbf{i} + \mathbf{j}) \qquad (3)$$

To find the point where the axis of the wrench intersects the yz plane, we shall express that the moment of the wrench about O is equal to the moment resultant $\mathbf{M}_O^R$ of the original system:

$$\mathbf{M}_1 + \mathbf{r} \times \mathbf{R} = \mathbf{M}_O^R$$

or, noting that $\mathbf{r} = y\mathbf{j} + z\mathbf{k}$ and substituting for $\mathbf{R}$, $\mathbf{M}_O^R$, and $\mathbf{M}_1$ from Eqs. (1), (2), and (3),

$$-\frac{Pa}{2}(\mathbf{i} + \mathbf{j}) + (y\mathbf{j} + z\mathbf{k}) \times P(\mathbf{i} + \mathbf{j}) = -Pa(\mathbf{i} + \mathbf{k})$$

$$-\frac{Pa}{2}\mathbf{i} - \frac{Pa}{2}\mathbf{j} - Py\mathbf{k} + Pz\mathbf{j} - Pz\mathbf{i} = -Pa\mathbf{i} - Pa\mathbf{k}$$

Equating the coefficients of $\mathbf{k}$, and then the coefficients of $\mathbf{j}$, we find

$$y = a \qquad z = a/2 \blacktriangleleft$$

Problems

3.76 A 3-m beam is loaded in the various ways represented in the figure. Find two loadings which are equivalent.

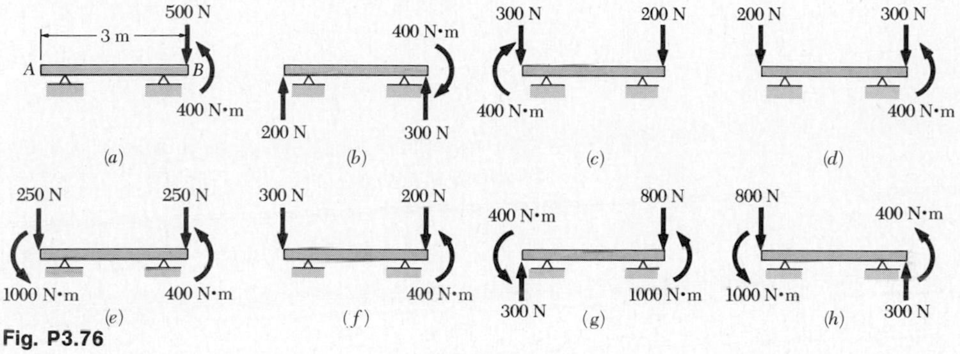

Fig. P3.76

3.77 A 3-m beam is loaded as shown. Determine the loading of Prob. 3.76 which is equivalent to this loading.

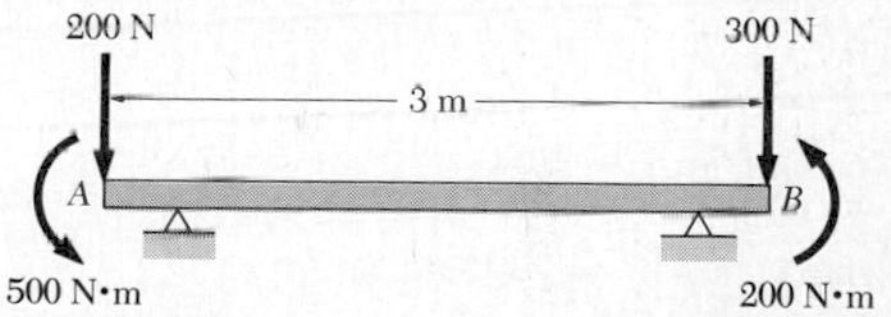

Fig. P3.77

3.78 Determine the resultant of the loads and the distance from point A to its line of action for the beam and loading of (*a*) Prob. 3.76*a*, (*b*) Prob. 3.76*b*, (*c*) Prob. 3.77.

3.79 Three horizontal forces are applied as shown to a vertical cast-iron arm. Determine the resultant of the forces and the distance from the ground to its line of action when (*a*) $P = 200$ N, (*b*) $P = 2400$ N, (*c*) $P = 1000$ N.

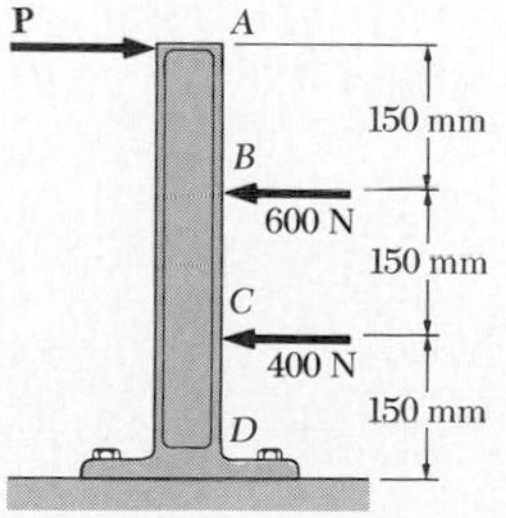

Fig. P3.79

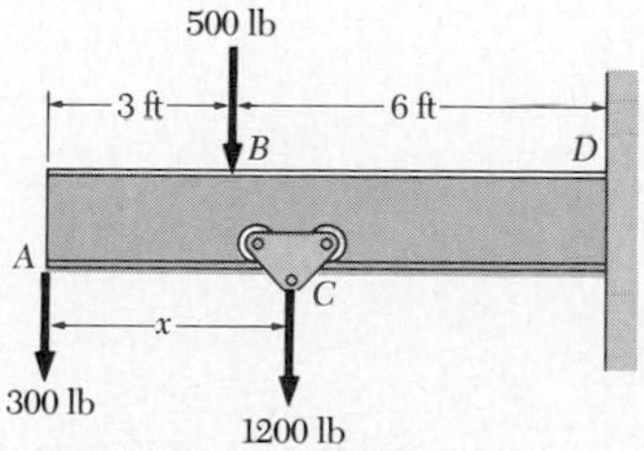

Fig. P3.80 and P3.81

3.80 Determine the distance from point A to the line of action of the resultant of the three loads shown when (*a*) $x = 1.25$ ft, (*b*) $x = 4$ ft, (*c*) $x = 8$ ft.

3.81 Express as a function of x the distance d from point A to the line of action of the resultant of the three loads shown.

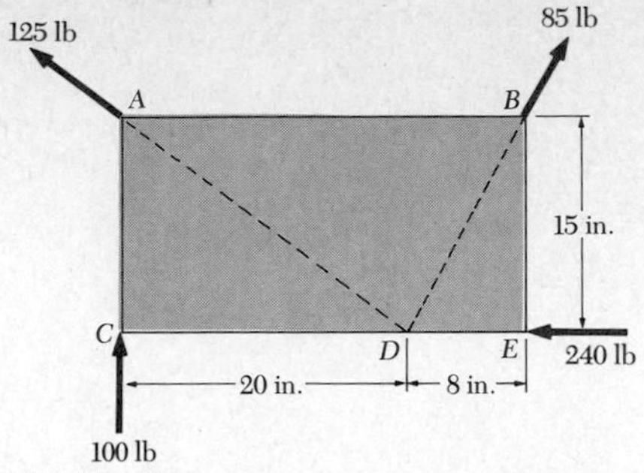

Fig. P3.82

3.82 Four forces act on a 28×15 in. plate as shown. (*a*) Find the resultant of these forces. (*b*) Locate the two points where the line of action of the resultant intersects the edge of the plate.

3.83 Solve Prob. 3.82, assuming that the 240-lb force is directed to the right.

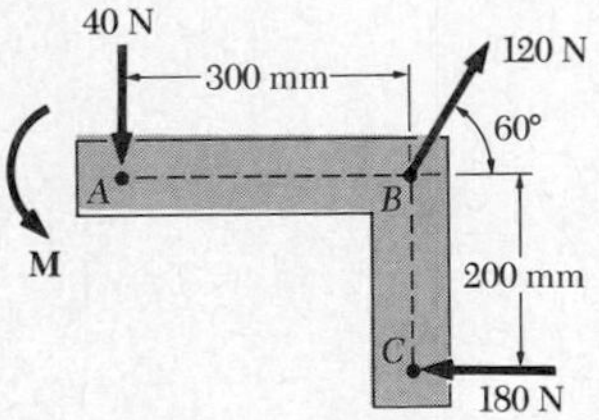

Fig. P3.84 and P3.85

3.84 The three forces shown and a couple of magnitude $M = 6\ \text{N}\cdot\text{m}$ are applied to an angle bracket. (*a*) Find the resultant of this system of forces. (*b*) Locate the points where the line of action of the resultant intersects line *AB* and line *BC*.

3.85 The three forces shown and a couple **M** are applied to an angle bracket. Find the moment of the couple if the line of action of the resultant of the force system is to pass through (*a*) point *A*, (*b*) point *B*, (*c*) point *C*.

3.86 The roof of a building frame is subjected to the wind loading shown. Determine (*a*) the equivalent force-couple system at *D*, (*b*) the resultant of the loading and its line of action.

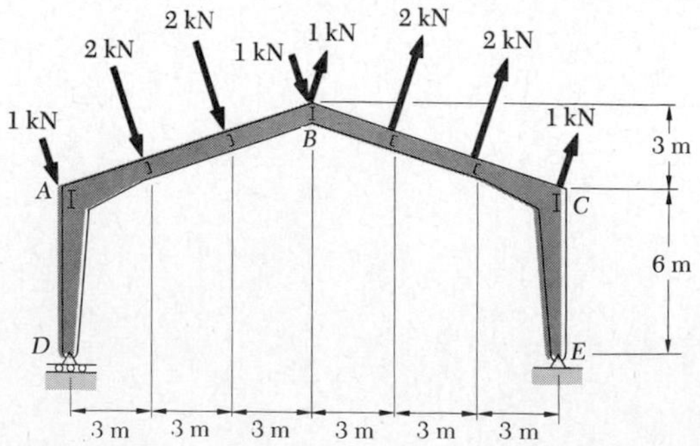

Fig. P3.86

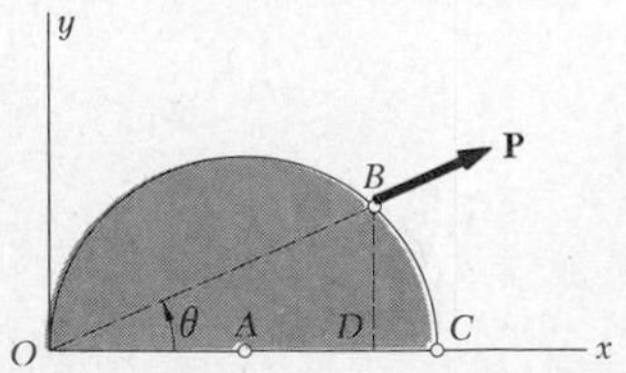

Fig. P3.87

3.87 A force **P** of given magnitude *P* is applied to the edge of a semicircular plate of radius *a* as shown. (*a*) Replace **P** by an equivalent force-couple system at the point *D* obtained by drawing the perpendicular from *B* to the *x* axis. (*b*) Determine the value of θ for which the moment of the equivalent force-couple system at *D* is maximum.

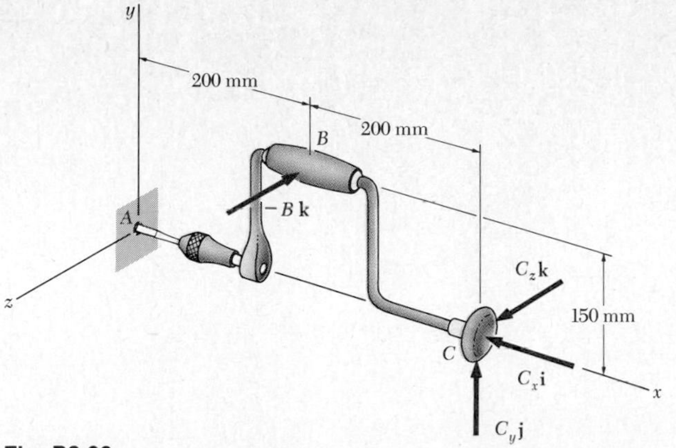

Fig. P3.88

3.88 A blade held in a brace is used to tighten a screw at A. (a) Determine the forces exerted at B and C, knowing that these forces are equivalent to a force-couple system at A consisting of $\mathbf{R} = -(30\text{ N})\mathbf{i} + R_y\mathbf{j} + R_z\mathbf{k}$ and $\mathbf{M}_A^R = -(12\text{ N}\cdot\text{m})\mathbf{i}$. ($b$) Find the corresponding values of R_y and R_z. (c) What is the orientation of the slot in the head of the screw for which the blade is least likely to slip when the brace is in the position shown?

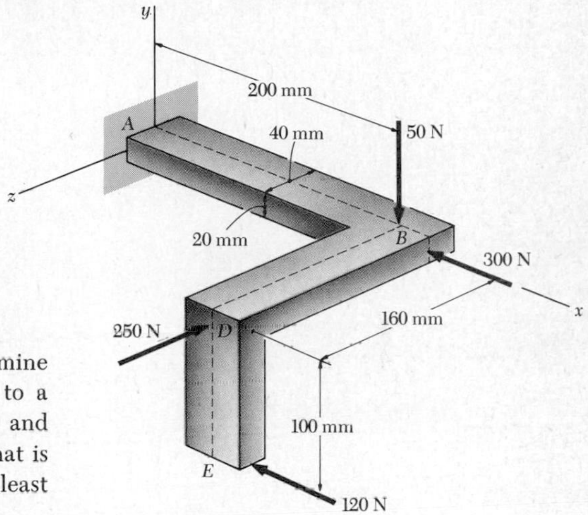

Fig. P3.89

3.89 Four forces are applied to the machine component $ABDE$ as shown. Replace these forces by an equivalent force-couple system at A.

3.90 In order to unscrew the tapped faucet A, a plumber uses two pipe wrenches as shown. By exerting a 30-lb force on each wrench, at a distance of 8 in. from the axis of the pipe and in a direction perpendicular to the pipe and to the wrench, he prevents the pipe from rotating, and thus avoids loosening or further tightening the joint between the pipe and the tapped elbow C. Determine (a) the angle θ that the wrench at A should form with the vertical if elbow C is not to rotate about the vertical, (b) the force-couple system at C equivalent to the two 30-lb forces when this condition is satisfied.

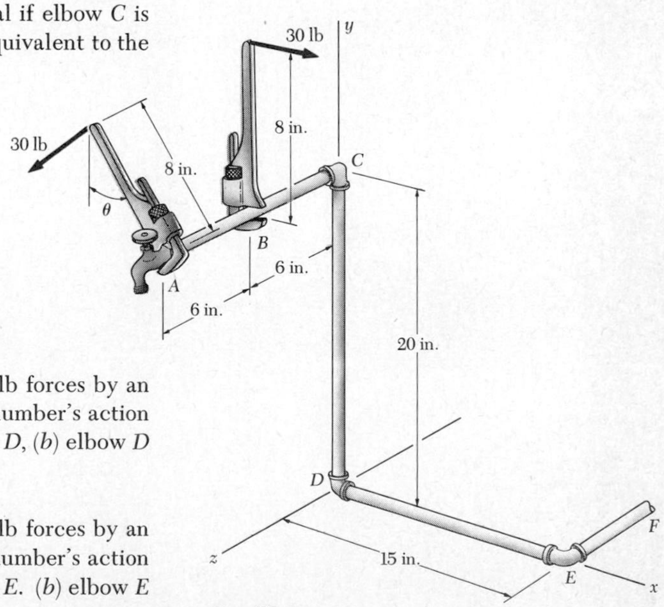

Fig. P3.90

3.91 Assuming $\theta = 60°$ in Prob. 3.90, replace the two 30-lb forces by an equivalent force-couple system at D and determine whether the plumber's action tends to tighten or loosen the joint between (a) pipe CD and elbow D, (b) elbow D and pipe DE. Assume all threads to be right-handed.

3.92 Assuming $\theta = 60°$ in Prob. 3.90, replace the two 30-lb forces by an equivalent force-couple system at E and determine whether the plumber's action tends to tighten or loosen the joint between (a) pipe DE and elbow E. (b) elbow E and pipe EF. Assume all threads to be right-handed.

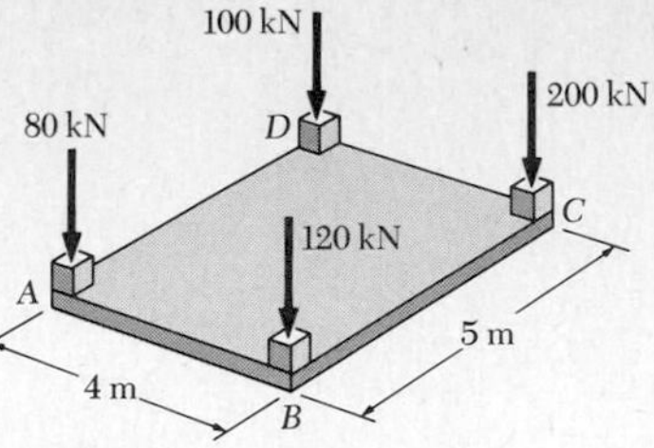

Fig. P3.93

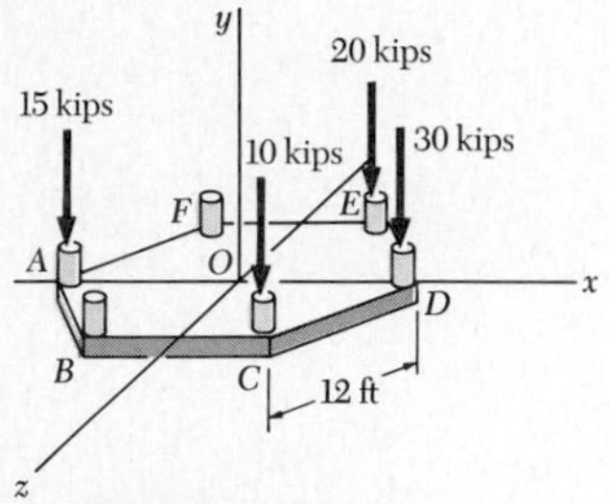

Fig. P3.94 and P3.95

3.93 A rectangular concrete foundation mat supports four column loads as shown. Determine the magnitude and point of application of the resultant of the four loads.

3.94 A concrete foundation mat in the shape of a regular hexagon of side 12 ft supports four column loads as shown. Determine the magnitude and point of application of the resultant of the four loads.

3.95 Determine the magnitudes of the additional loads which must be applied at B and F if the resultant of all six loads is to pass through the center of the mat.

3.96 In Prob. 3.93, determine the magnitude and point of application of the smallest additional load which must be applied to the foundation mat if the resultant of the five loads is to pass through the center of the mat.

* **3.97** A rectangular block is acted upon by the four forces shown, which are directed along its edges. Reduce this system of forces to (*a*) a force-couple system at the origin, (*b*) a wrench. (Specify the pitch and axis of the wrench.)

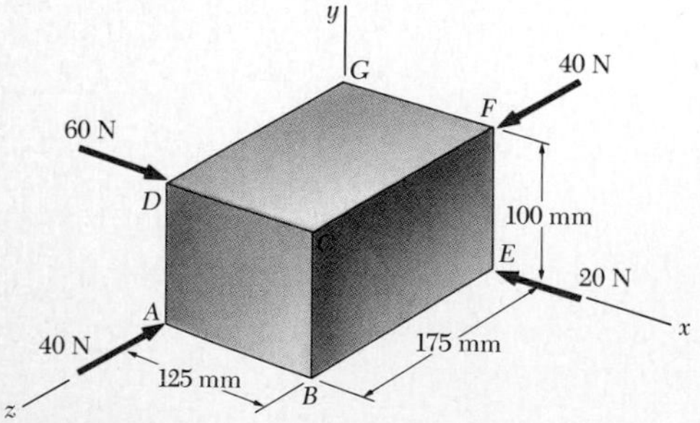

Fig. P3.97

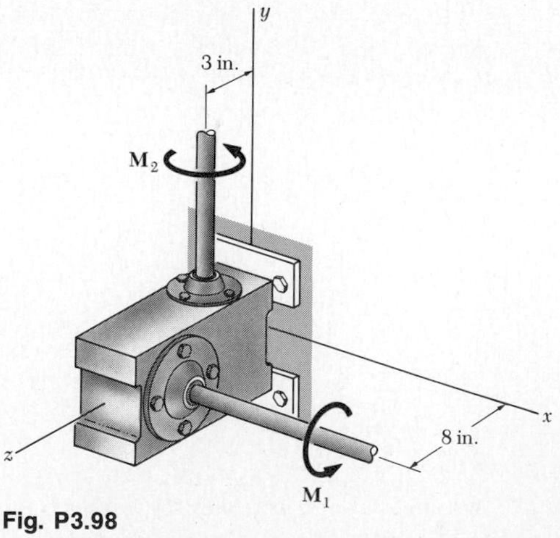

Fig. P3.98

* **3.98** The two shafts of a speed-reducer unit are subjected to couples of magnitude $M_1 = 15$ lb·ft and $M_2 = 3$ lb·ft, respectively. The unit weighs 60 lb and its center of gravity is located on the z axis at $z = 6$ in. Replace the weight and the two couples by an equivalent wrench and determine (*a*) the resultant force **R**, (*b*) the pitch of the wrench, (*c*) the point where the axis of the wrench intersects the xz plane.

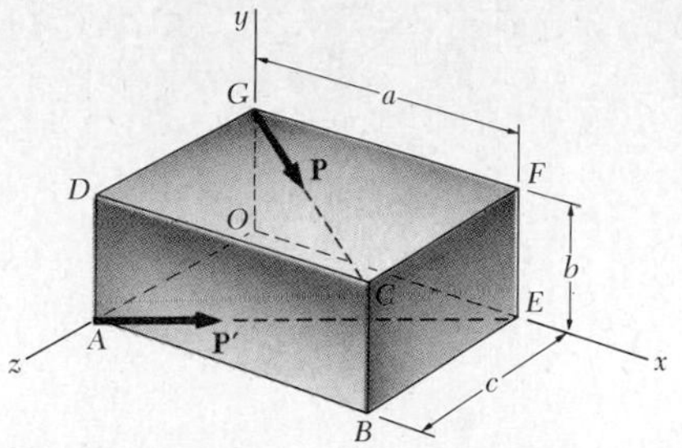

Fig. P3.99

***3.99** Two forces of magnitude P act as shown along the diagonals GC and AE of the horizontal faces of a rectangular parallelepiped. Replace the two forces by an equivalent system consisting of (*a*) a force and couple at O, (*b*) a wrench. (Specify the pitch and axis of the wrench.)

***3.100** In an automated manufacturing process three holes are drilled simultaneously as shown in an aluminum block. Each drill exerts a 40-N force and a 0.100-N·m couple on the block. Knowing that drill A rotates counterclockwise, and drills B and C clockwise (as observed from each drill), reduce the forces and couples exerted by the drills on the block to an equivalent wrench and determine (*a*) the resultant force **R**, (*b*) the pitch of the wrench, (*c*) the point where the axis of the wrench intersects the xz plane.

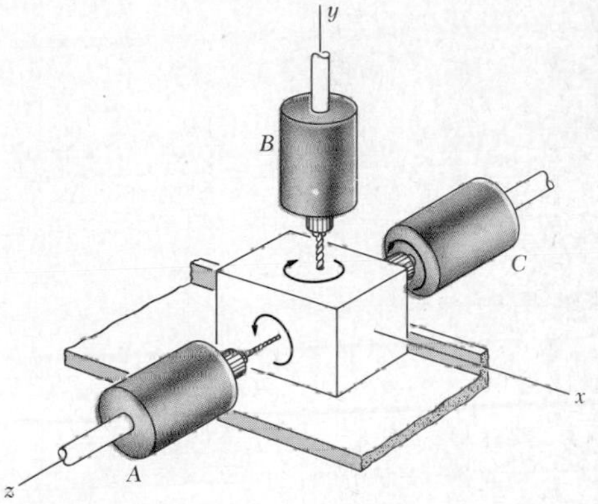

Fig. P3.100

***3.101** If only drills A and B are used in the manufacturing process of Prob. 3.100, show that the forces and couples exerted by the two drills on the aluminum block may be reduced to a single force **R** and determine (*a*) the magnitude and direction of **R**, (*b*) the line of action of **R**.

***3.102** In Sample Prob. 3.12 an additional force $\mathbf{Q} = Q\mathbf{k}$ is applied at the corner B of the cube. Knowing that the three forces acting on the cube may be reduced to a single force **R**, determine (*a*) the force **Q**, (*b*) the magnitude and direction of **R**, (*c*) the point where the line of action of **R** intersects the yz plane.

***3.103** In Prob. 3.97, assume that the force acting at A has been removed and replace the remaining three forces by an equivalent wrench. Determine (*a*) the magnitude and direction of the resultant force **R**, (*b*) the pitch of the wrench, (*c*) the point where the axis of the wrench intersects the yz plane.

***3.104** In Prob. 3.89, replace the four forces applied to the machine component by an equivalent wrench. Determine (*a*) the magnitude and direction of the resultant force **R**, (*b*) the pitch of the wrench, (*c*) the point where the axis of the wrench intersects the xz plane.

***3.105** Assuming $\theta = 60°$ in Prob. 3.90, replace the two 30-lb forces by an equivalent wrench. Determine (*a*) the magnitude and direction of the resultant force **R**, (*b*) the pitch of the wrench, (*c*) the point where the axis of the wrench intersects the yz plane.

***3.106** Replace the two wrenches shown by a single equivalent wrench and determine (*a*) the resultant force **R**, (*b*) the pitch of the single equivalent wrench, (*c*) the point where its axis intersects the xz plane.

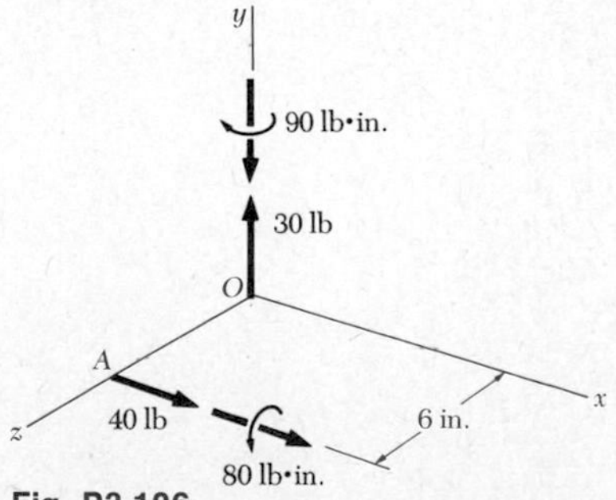

Fig. P3.106

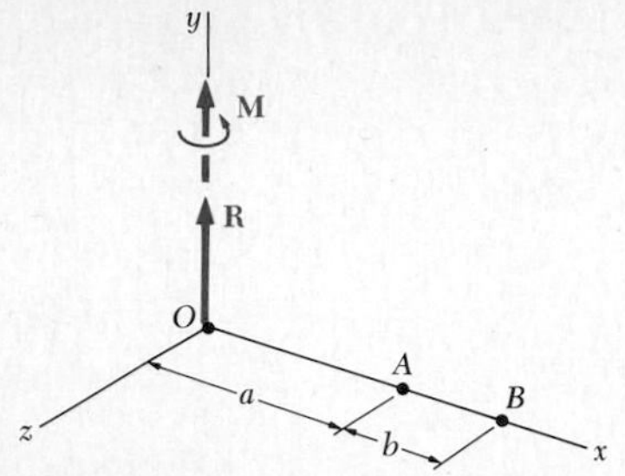

Fig. P3.107

*** 3.107** (*a*) Replace the wrench shown by an equivalent system consisting of two forces perpendicular to the x axis and applied respectively at A and B. (*b*) Solve part a assuming R = 120 N, M = 24 N·m, a = 100 mm, and b = 50 mm.

*** 3.108** Show that, in general, a wrench may be replaced by two forces chosen in such a way that one force passes through a given point while the other force lies in a given plane.

*** 3.109** Show that a wrench may be replaced by two perpendicular forces, one of which is applied at a given point.

*** 3.110** Show that a wrench may be replaced by two forces, one of which has a prescribed line of action.

Review Problems

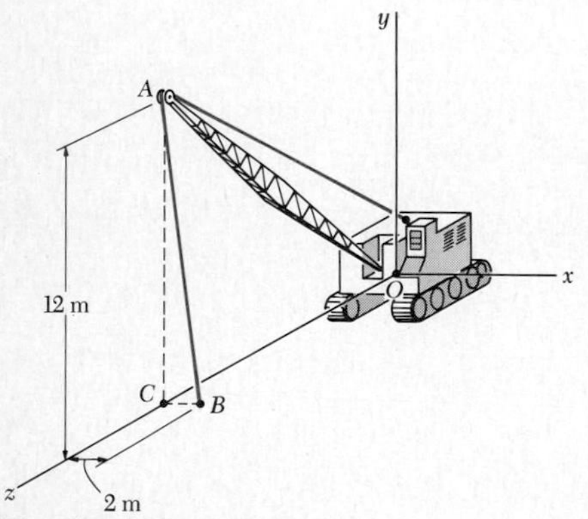

Fig. P3.111 and P3.112

3.111 A crane is oriented so that the end of the 15-m boom AO lies in the yz plane. At the instant shown the tension in cable AB is 3.5 kN. Determine the moment about each of the coordinate axes of the force exerted on A by the cable AB.

3.112 The 15-m crane boom AO lies in the yz plane. Determine the maximum permissible tension in the cable AB if the absolute value of the moments about the coordinate axes of the force exerted on A must be as follows: $|M_x| \leq 40$ kN·m, $|M_y| \leq 7.5$ kN·m, $|M_z| \leq 8$ kN·m.

3.113 A force **P** of magnitude 455 lb acts on the frame shown at point E. Determine the moment of **P** (*a*) about point D, (*b*) about a line joining points O and D.

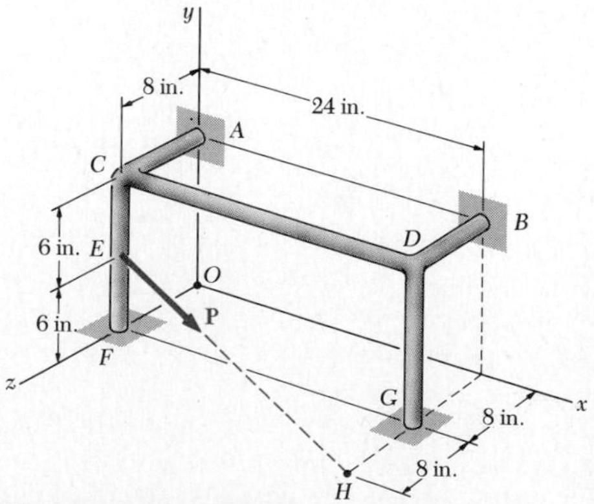

Fig. P3.113 and P3.114

3.114 A force **P** acts on the frame shown at point E. Knowing that the absolute value of the moment of **P** about a line joining points A and G is 1440 lb·in., determine the magnitude of the force **P**.

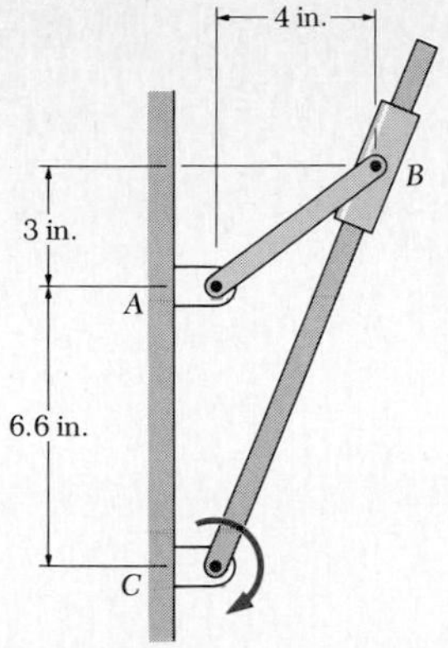

Fig. P3.115

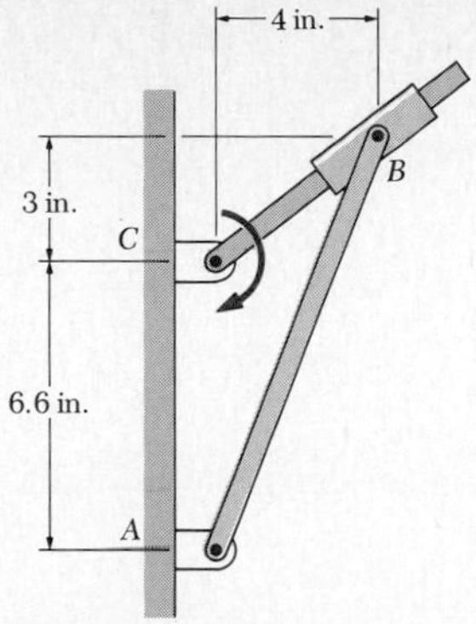

Fig. P3.116

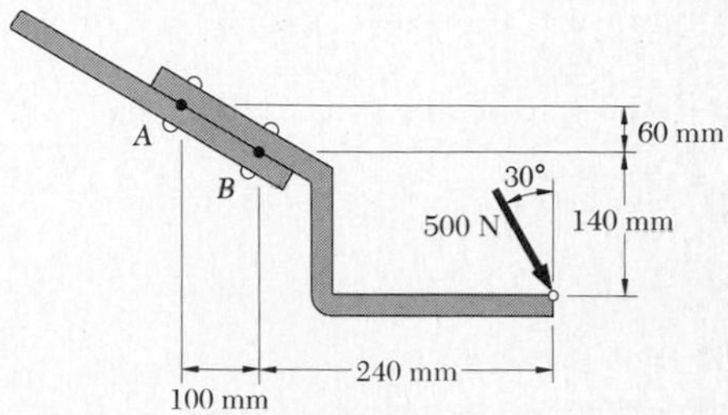

Fig. P3.117 and P3.118

3.115 and 3.116 It is known that the slider B exerts on rod AB a 130-lb force perpendicular to rod BC and directed down and to the right. Determine the moment of that force about A.

3.117 A 500-N force is applied to a bent plate as shown. Determine (*a*) an equivalent force-couple system at A, (*b*) an equivalent system formed by parallel forces acting at A and B.

3.118 A 500-N force is applied to a bent plate as shown. Determine (*a*) an equivalent force-couple system at B, (*b*) an equivalent system formed by a vertical force at A and a force at B.

3.119 A concrete foundation mat of 6-m radius supports four equally spaced columns, each of which is located 5 m from the center of the mat. Determine the magnitude and the point of application of the resultant of the four loads.

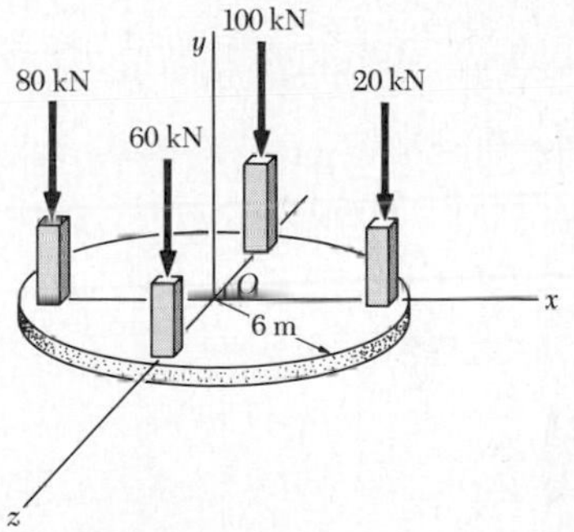

Fig. P3.119

3.120 Determine the magnitude and point of application of the smallest additional load which must be applied to the foundation mat of Prob. 3.119 if the resultant of the five loads is to pass through the center of the mat.

3.121 Replace the three forces shown by (*a*) a force-couple system at the origin, (*b*) a wrench. (Specify the pitch and axis of the wrench.)

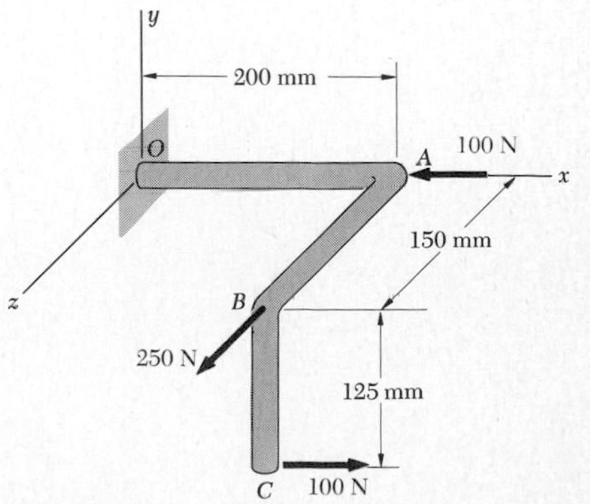

Fig. P3.121

3.122 A couple formed by two 90-lb forces is applied to the pulley assembly shown. Determine an equivalent couple which is formed by (*a*) vertical forces acting at A and C, (*b*) the smallest possible forces acting at B and D, (*c*) the smallest possible forces which can be attached to the assembly.

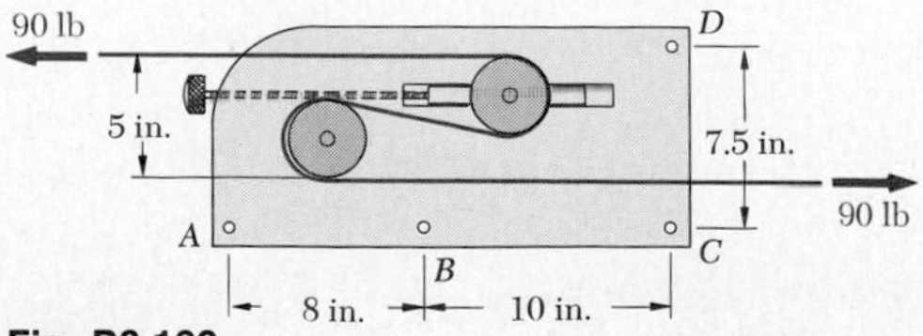

Fig. P3.122

CHAPTER FOUR

Equilibrium of Rigid Bodies

4.1. Introduction. We saw in the preceding chapter that the external forces acting on a rigid body may be reduced to a force-couple system at some arbitrary point O. When the force and the couple are both equal to zero, the external forces form a system equivalent to zero and the rigid body is said to be in *equilibrium*.

The necessary and sufficient conditions for the equilibrium of a rigid body, therefore, may be obtained by setting $\mathbf{R}$ and $\mathbf{M}_O^R$ equal to zero in the relations (3.52) of Sec. 3.17:

$$\Sigma \mathbf{F} = 0 \qquad \Sigma \mathbf{M}_O = \Sigma(\mathbf{r} \times \mathbf{F}) = 0 \tag{4.1}$$

Resolving each force and each moment into its rectangular components, we may express the necessary and sufficient conditions for the equilibrium of a rigid body by the following six scalar equations:

$$\Sigma F_x = 0 \qquad \Sigma F_y = 0 \qquad \Sigma F_z = 0 \tag{4.2}$$
$$\Sigma M_x = 0 \qquad \Sigma M_y = 0 \qquad \Sigma M_z = 0 \tag{4.3}$$

The equations obtained may be used to determine unknown forces applied to the rigid body or unknown reactions exerted on it by its supports. We note that Eqs. (4.2) express the fact that the components of the external forces in the x, y, and z directions are balanced; Eqs. (4.3) express the fact that the moments of the external forces about the x, y, and z axes are balanced. The system of the external forces, therefore, will impart no motion of translation or rotation to the rigid body considered.

In order to write the equations of equilibrium for a rigid body, it is essential to first identify correctly all the forces acting on that body and draw the corresponding *free-body diagram*. In this chapter we shall first consider the equilibrium of *two-dimensional structures* subjected to forces contained in their plane and learn to draw their free-body diagram. In addition to the forces *applied* to the structure, we shall consider the *reactions* exerted on the structure by its supports. We shall learn to associate a specific reaction with each type of support and determine whether the structure is properly supported, so that we may know in advance whether the equations of equilibrium may actually be solved for the unknown forces and reactions. Later in the chapter, we shall consider the equilibrium of three-dimensional structures and subject them and their supports to the same kind of analysis.

4.2. Free-Body Diagram. In solving a problem concerning the equilibrium of a rigid body, it is essential to consider *all* the forces acting on the body; it is equally important to exclude any force which is not directly applied on the body. Omitting a force or adding an extraneous one would destroy the conditions of equilibrium. Therefore, the first step in the solution of the problem should consist in drawing a *free-body diagram* of the rigid body under consideration. Free-body diagrams have already been used on many occasions in Chap. 2. However, in view of their importance to the solution of equilibrium problems, we shall summarize here the various steps which must be followed in drawing a free-body diagram.

First, a clear decision is made regarding the choice of the free body to be used. This body is then detached from the ground and separated from any other body. The contour of the body thus isolated is sketched.

All external forces are then indicated. These forces represent the action exerted *on* the free body *by* the ground and the bodies which have been detached; they should be applied at the various points where the free body was supported by the ground or connected to the other bodies. The *weight* of the free body should also be included among the external forces, since it represents the attraction exerted by the earth on the various particles forming the free body. As will be seen in Chap. 5, the weight should be applied at the center of gravity of the body. When the free body is made of several parts, the forces the various parts exert on each other should *not* be included among the external forces. These forces are internal forces as far as the free body is concerned.

The magnitude and direction of the *known external forces* should be clearly marked on the free-body diagram. Care should be taken to indicate the sense of the force exerted *on* the free body, not that of the force exerted *by* the free body. Known external forces generally include the *weight* of the free body and *forces applied* for a given purpose.

Unknown external forces usually consist of the *reactions*—sometimes called *constraining forces*—through which the ground and other bodies oppose a possible motion of the free body and thus constrain it to remain in the same position. Reactions are exerted at the points where the free body is *supported* or *connected* to other bodies. They will be discussed in detail in Secs. 4.3 and 4.8.

The free-body diagram should also include dimensions, since these may be needed in the computation of moments of forces. Any other detail, however, should be omitted.

4.3. Reactions at Supports and Connections for a Two-dimensional Structure. In the first part of this chapter we shall consider the equilibrium of a two-dimensional structure; i.e., we shall assume that the structure considered and the forces applied to it are contained in the plane of the figure. Clearly, the reactions needed to maintain the structure in the same position will also be contained in the plane of the figure.

The reactions exerted on a two-dimensional structure may be divided into three groups, corresponding to three types of *supports*, or *connections:*

1. *Reactions Equivalent to a Force with Known Line of Action.* Supports and connections causing reactions of this group include *rollers, rockers, frictionless surfaces, short links and cables, collars on frictionless rods,* and *frictionless pins in slots.* Each of these supports and connections can prevent motion in one direction only. They are shown in Fig. 4.1, together with the reaction they produce. Reactions of this group involve *one unknown,* namely, the magnitude of the reaction; this magnitude should be denoted by an appropriate letter. The line of action of the reaction is known and should be indicated clearly in the free-body diagram. The sense of the reaction must be as shown in Fig. 4.1 in the case of a frictionless surface (away from the surface) or of a cable (tension in the direction of the cable). The reaction may be directed either way in the case of double-track rollers, links, collars on rods, and pins in slots. Single-track rollers and rockers are generally assumed to be reversible, and thus the corresponding reactions may also be directed either way.

2. *Reactions Equivalent to a Force of Unknown Direction.* Supports and connections causing reactions of this group include *frictionless pins in fitted holes, hinges,* and *rough surfaces.* They can prevent translation of the free body in all directions, but they cannot prevent the body from rotating about the connection. Reactions of this group involve *two unknowns* and are usually represented by their x and y components. In the case of a rough surface, the component normal to the surface must be directed away from the surface.

3. *Reactions Equivalent to a Force and a Couple.* These reactions are caused by *fixed supports* which oppose any motion of the free body and thus constrain it completely. Fixed supports actually produce forces over the entire surface of contact; these forces, however, form a system which may be reduced to a force and a couple. Reactions of this group involve *three unknowns,* consisting usually of the two components of the force and the moment of the couple.

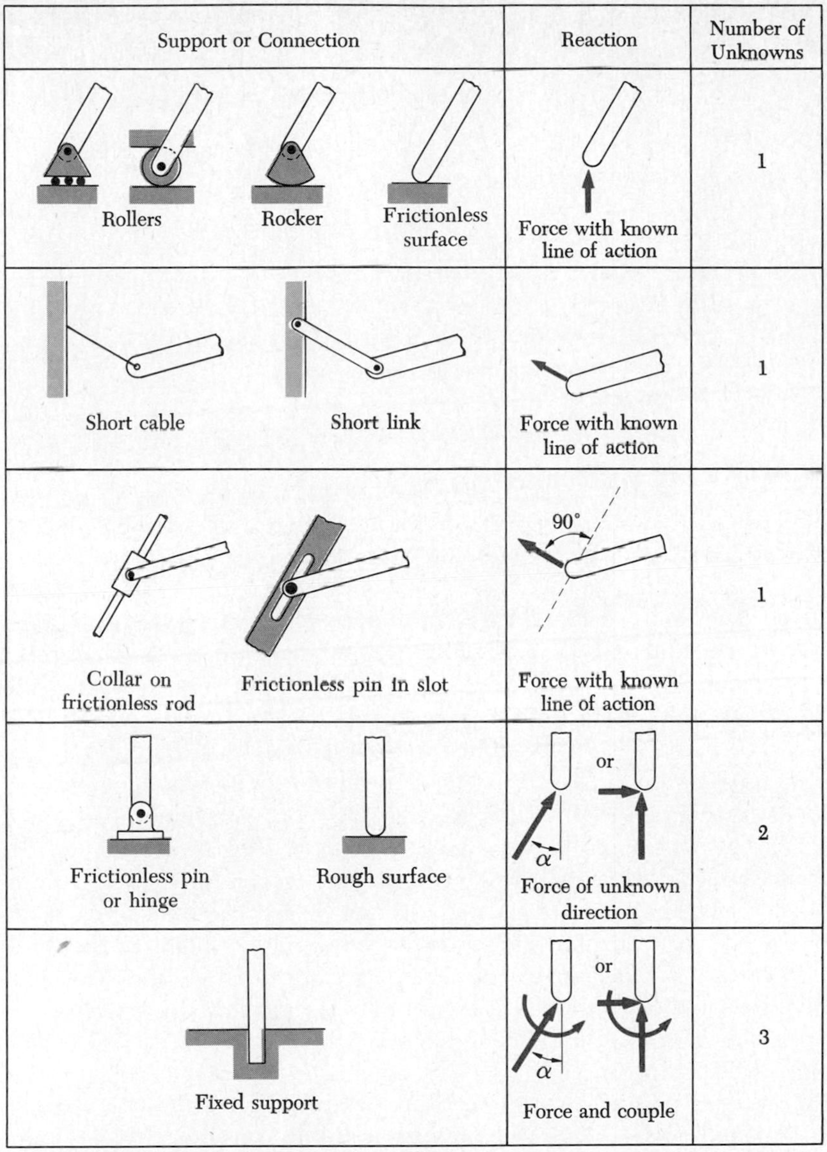

Fig. 4.1 Reactions at supports and connections.

When the sense of an unknown force or couple is not clearly apparent, no attempt should be made to determine it. Instead, the sense of the force or couple should be arbitrarily assumed; the sign of the answer obtained will indicate whether the assumption is correct or not.

4.4 Equilibrium of a Rigid Body in Two Dimensions. The conditions stated in Sec. 4.1 for the equilibrium of a rigid body become considerably simpler in the case of a two-dimensional structure. Choosing the x and y axes in the plane of the structure, we have

$$F_z = 0 \qquad M_x = M_y = 0 \qquad M_z = M_O$$

for each of the forces applied to the structure. Thus, the six equations of equilibrium derived in Sec. 4.1 reduce to

$$\Sigma F_x = 0 \qquad \Sigma F_y = 0 \qquad \Sigma M_O = 0 \tag{4.4}$$

and to three trivial identities $0 = 0$. Since the third of the equations (4.4) must be satisfied regardless of the choice of the origin O, we may write the equations of equilibrium for a two-dimensional structure in the more general form

$$\Sigma F_x = 0 \qquad \Sigma F_y = 0 \qquad \Sigma M_A = 0 \tag{4.5}$$

where A is any point in the plane of the structure. The three equations obtained may be solved for no more than *three unknowns*.

We saw in the preceding section that unknown forces usually consist of reactions, and that the number of unknowns corresponding to a given reaction depends upon the type of support or connection causing that reaction. Referring to Sec. 4.3, we check that the equilibrium equations (4.5) may be used to determine the reactions of two rollers and one cable, or of one fixed support, or of one roller and one pin in a fitted hole, etc.

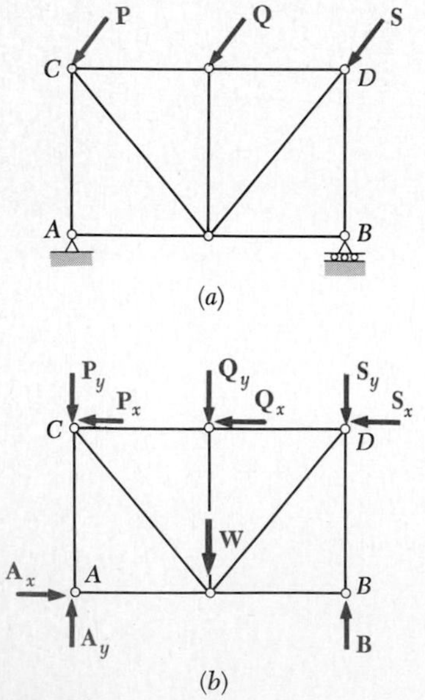

Fig. 4.2

Consider, for instance, the truss shown in Fig. 4.2*a*, which is subjected to the given forces **P**, **Q**, and **S**. The truss is held in place by a pin at A and a roller at B. The pin prevents point A from moving by exerting on the truss a force which may be resolved into the components $\mathbf{A}_x$ and $\mathbf{A}_y$; the roller keeps the truss from rotating about A by exerting the vertical force **B**. The free-body diagram of the truss is shown in Fig. 4.2*b*; it includes the reactions $\mathbf{A}_x$, $\mathbf{A}_y$, and **B** as well as the applied forces **P**, **Q**, **S**, and the weight **W** of the truss. Expressing that the sum of the moments about A of all the forces shown in Fig. 4.2*b* is zero, we write the equation $\Sigma M_A = 0$, which may be solved for the magnitude B since it does not contain A_x or A_y. Expressing, then, that the sum of the x components and the sum of the y components of the forces are zero, we write the equations $\Sigma F_x = 0$ and $\Sigma F_y = 0$, which may be solved for the components A_x and A_y, respectively.

Additional equations could be obtained by expressing that the sum of the moments of the external forces about points other than A is zero. We could write, for instance, $\Sigma M_B = 0$. Such a statement, however, does not contain any new information, since it has already been established that the system of the forces shown in Fig. 4.2*b* is equivalent to zero. The additional equation *is not independent* and cannot be used to determine a fourth unknown. It will be useful, however, for checking the solution obtained from the original three equations of equilibrium.

While the three equations of equilibrium cannot be *augmented* by additional equations, any of them may be *replaced* by another equation. Therefore, an alternate system of equations of equilibrium is

$$\Sigma F_x = 0 \qquad \Sigma M_A = 0 \qquad \Sigma M_B = 0 \tag{4.6}$$

where the line AB is chosen in a direction different from the y direction (Fig. 4.2b). These equations are sufficient conditions for the equilibrium of the truss. The first two equations indicate that the external forces must reduce to a single vertical force at A. Since the third equation requires that the moment of this force be zero about a point B which is not on its line of action, the force must be zero and the rigid body is in equilibrium.

A third possible set of equations of equilibrium is

$$\Sigma M_A = 0 \qquad \Sigma M_B = 0 \qquad \Sigma M_C = 0 \tag{4.7}$$

where the points A, B, and C are not in a straight line (Fig. 4.2b). The first equation requires that the external forces reduce to a single force at A; the second equation requires that this force pass through B; the third, that it pass through C. Since the points A, B, C are not in a straight line, the force must be zero, and the rigid body is in equilibrium.

The equation $\Sigma M_A = 0$, which expresses that the sum of the moments of the forces about pin A is zero, possesses a more definite physical meaning than either of the other two equations (4.7). These two equations express a similar idea of balance, but with respect to points about which the rigid body is not actually hinged. They are, however, as useful as the first equation, and our choice of equilibrium equations should not be unduly influenced by the physical meaning of these equations. Indeed, it will be desirable in practice to choose equations of equilibrium containing only one unknown, since this eliminates the necessity of solving simultaneous equations. Equations containing only one unknown may be obtained by summing moments about the point of intersection of the lines of action of two unknown forces or, if these forces are parallel, by summing components in a direction perpendicular to their common direction. In the case of the truss of Fig. 4.3, for example, which is held by rollers at A and B and a short link at D, the reactions at A and B may be eliminated by summing x components. The reactions at A and D will be eliminated by summing moments about C and the reactions at B and D by summing moments about D. The equations obtained are

$$\Sigma F_x = 0 \qquad \Sigma M_C = 0 \qquad \Sigma M_D = 0$$

Each of these equations contains only one unknown.

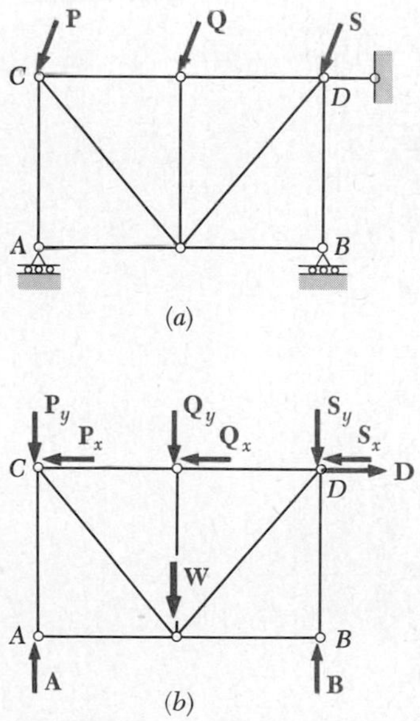

Fig. 4.3

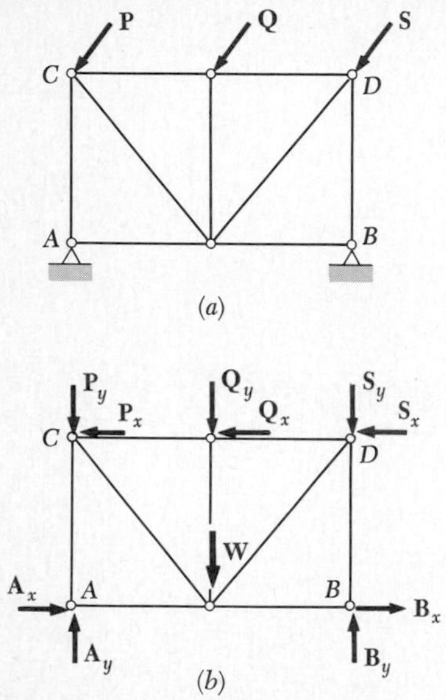

Fig. 4.4 Statically indeterminate reactions.

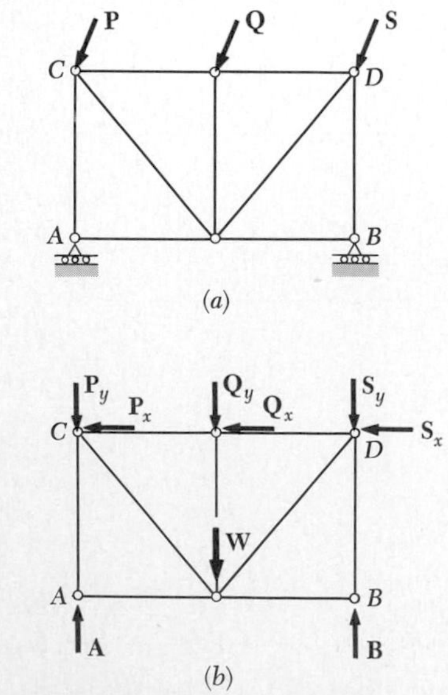

Fig. 4.5 Partial constraints.

4.5. Statically Indeterminate Reactions. Partial Constraints. In each of the two examples considered in the preceding section (Figs. 4.2 and 4.3), the types of supports used were such that the rigid body could not possibly move under the given loads or under any other loading conditions. In such cases, the rigid body is said to be *completely constrained.* We also recall that the reactions corresponding to these supports involved *three unknowns* and could be determined by solving the three equations of equilibrium. When such a situation exists, the reactions are said to be *statically determinate.*

Consider now the truss shown in Fig. 4.4*a*, which is held by pins at A and B. These supports provide more constraints than are necessary to keep the truss from moving under the given loads or under any other loading conditions. We also note from the free-body diagram of Fig. 4.4*b* that the corresponding reactions involve *four unknowns.* Since, as was pointed out in Sec. 4.4, only three independent equilibrium equations are available, there are *more unknowns than equations*, and all the unknowns cannot be determined. While the equations $\Sigma M_A = 0$ and $\Sigma M_B = 0$ yield the vertical components B_y and A_y, respectively, the equation $\Sigma F_x = 0$ gives only the sum $A_x + B_x$ of the horizontal components of the reactions at A and B. The components A_x and B_x are said to be *statically indeterminate.* They could be determined by considering the deformations produced in the truss by the given loading, but this method is beyond the scope of statics and belongs to the study of mechanics of materials.

The supports used to hold the truss shown in Fig. 4.5*a* consist of rollers at A and B. Clearly, the constraints provided by these supports are not sufficient to keep the truss from moving. While any vertical motion is prevented, the truss is free to move horizontally. The truss is said to be *partially constrained.*† Turning our attention to Fig. 4.5*b*, we note that the reactions at A and B involve only *two unknowns.* Since three equations of equilibrium must still be satisfied, there are *fewer unknowns than equations*, and one of the equilibrium equations will not be satisfied. While the equations $\Sigma M_A = 0$ and $\Sigma M_B = 0$ can be satisfied by a proper choice of reactions at A and B, the equation $\Sigma F_x = 0$ will not be satisfied unless the sum of the horizontal components of the applied forces happens to be zero. We thus check that the equilibrium of the truss of Fig. 4.5 cannot be maintained under general loading conditions.

It appears from the above that if a rigid body is to be completely constrained and if the reactions at its supports are to be statically determinate, *there must be as many unknowns as there are equations of equilibrium.* When this condition is *not* satisfied, we may be sure that the rigid body is not completely constrained, or that the reactions at its supports are not statically determinate, or both.

We should note, however, that, while *necessary,* the above condition is *not sufficient.* In other words, the fact that the number of unknowns is equal to the number of equations is no guarantee that the body is completely constrained or that the reactions at its supports are statically deter-

† Partially constrained bodies are often referred to as *unstable.* However, to avoid confusion between this type of instability, due to insufficient constraints, and the type of instability considered in Chap. 10, which relates to the behavior of a rigid body when its equilibrium is disturbed, we shall restrict the use of the words *stable* and *unstable* to the latter case.

minate. Consider the truss shown in Fig. 4.6*a*, which is held by rollers at *A*, *B*, and *E*. While there are three unknown reactions, **A**, **B**, and **E** (Fig. 4.6*b*), the equation $\Sigma F_x = 0$ will not be satisfied unless the sum of the horizontal components of the applied forces happens to be zero. There is a sufficient number of constraints, but these constraints are not properly arranged, and the truss is free to move horizontally. We say that the truss is *improperly constrained.* Since only two equilibrium equations are left for determining three unknowns, the reactions will be statically indeterminate. Thus, improper constraints also produce static indeterminacy.

Another example of improper constraints—and of static indeterminacy—is provided by the truss shown in Fig. 4.7. This truss is held by a pin at *A* and by rollers at *B* and *C*, which altogether involve four unknowns. Since only three independent equilibrium equations are available, the reactions at the supports are statically indeterminate. On the other hand, we note that the equation $\Sigma M_A = 0$ cannot be satisfied under general loading conditions, since the lines of action of the reactions **B** and **C** are required to pass through *A*. We conclude that the truss can rotate about *A* and that it is improperly constrained.†

The examples of Figs. 4.6 and 4.7 lead us to conclude that *a rigid body is improperly constrained whenever the supports,* even though they may provide a sufficient number of reactions, *are arranged in such a way that the reactions must be either concurrent or parallel.*‡

In summary, to be sure that a two-dimensional rigid body is completely constrained and that the reactions at its supports are statically determinate, we should check that the reactions involve three—and only three—unknowns, and that the supports are arranged in such a way that they do not require the reactions to be either concurrent or parallel.

Supports involving statically indeterminate reactions should be used with care in the *design* of structures, and only with a full knowledge of the problems they may cause. On the other hand, the *analysis* of structures possessing statically indeterminate reactions often may be partially carried out by the methods of statics. In the case of the truss of Fig. 4.4, for example, the vertical components of the reactions at *A* and *B* were obtained from the equilibrium equations.

For obvious reasons, supports producing partial or improper constraints should be avoided in the design of stationary structures. However, a partially or improperly constrained structure will not necessarily collapse; under particular loading conditions, equilibrium may be maintained. For example, the trusses of Figs. 4.5 and 4.6 will be in equilibrium if the applied forces **P**, **Q**, and **S** are vertical. Besides, structures which are designed to move *should* be only partially constrained. A railroad car, for instance, would be of little use if it were completely constrained by having its brakes applied permanently.

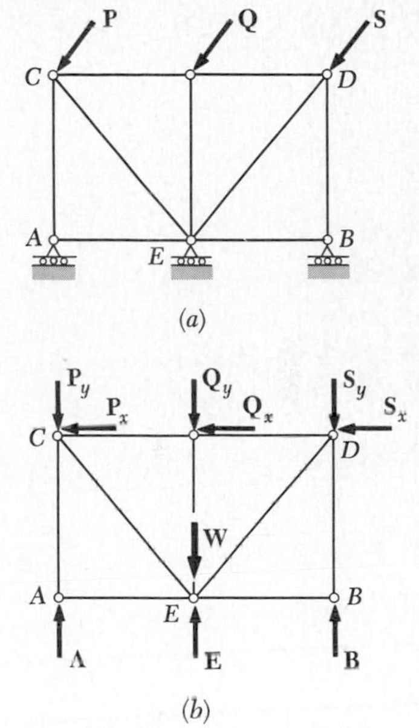

Fig. 4.6 Improper constraints.

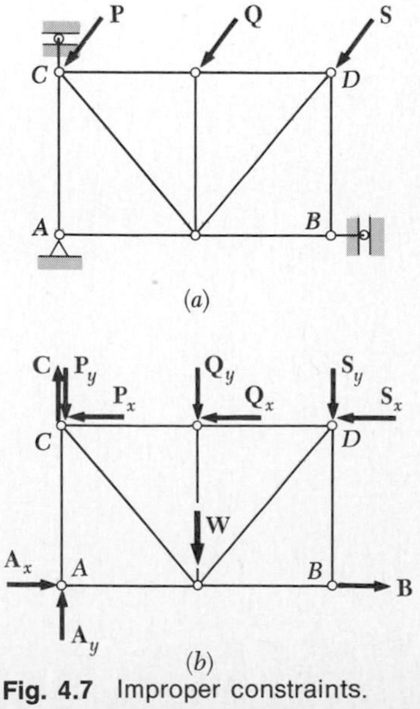

Fig. 4.7 Improper constraints.

† Rotation of the truss about *A* requires some "play" in the supports at *B* and *C*. In practice such play will always exist. Besides, we may note that if the play is kept small, the displacements of the rollers *B* and *C* and, thus, the distances from *A* to the lines of action of the reactions **B** and **C**, will also be small. The equation $\Sigma M_A = 0$ will then require that **B** and **C** be very large, a situation which may well result in the failure of the supports at *B* and *C*.

‡ Because this situation arises from an inadequate arrangement or *geometry* of the supports, it is often referred to as *geometric instability*.

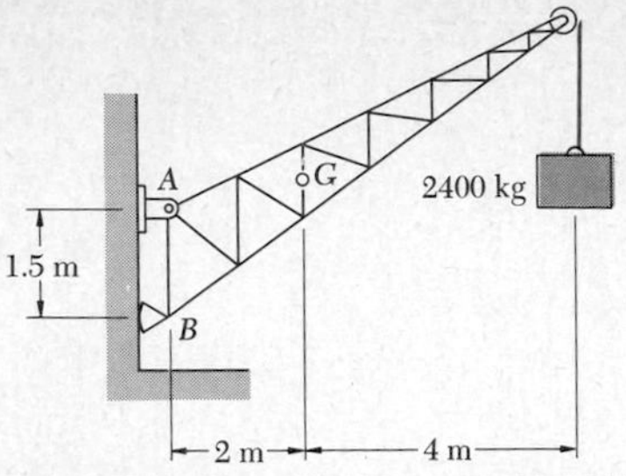

SAMPLE PROBLEM 4.1

A fixed crane has a mass of 1000 kg and is used to lift a 2400-kg crate. It is held in place by a pin at A and a rocker at B. The center of gravity of the crane is located at G. Determine the components of the reactions at A and B.

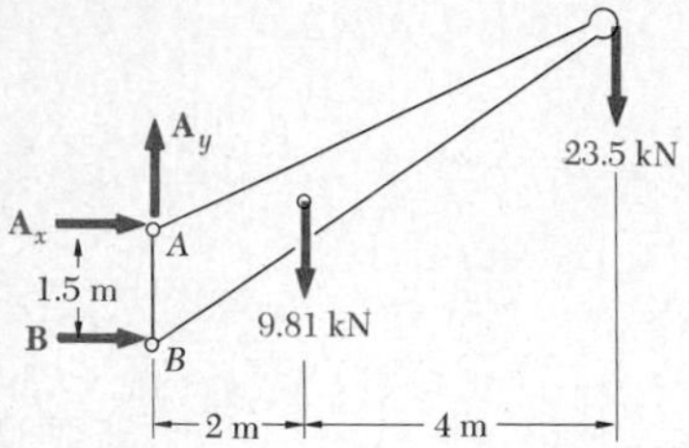

Solution. A free-body diagram of the crane is drawn. Multiplying the masses of the crane and of the crate by $g = 9.81\text{ m/s}^2$, we obtain the corresponding weights, that is, 9810 N or 9.81 kN, and 23 500 N or 23.5 kN. The reaction at pin A is a force of unknown direction, represented by its components $\mathbf{A}_x$ and $\mathbf{A}_y$. The reaction at the rocker B is perpendicular to the rocker surface; thus it is horizontal. We assume that $\mathbf{A}_x$, $\mathbf{A}_y$, and $\mathbf{B}$ act in the directions shown.

***Determination of* B.** We express that the sum of the moments of all external forces about point A is zero. The equation obtained will contain neither A_x nor A_y since the moments of $\mathbf{A}_x$ and $\mathbf{A}_y$ about A are zero. Multiplying the magnitude of each force by its perpendicular distance from B, we write

$$+\circlearrowleft \Sigma M_A = 0: \quad +B(1.5\text{ m}) - (9.81\text{ kN})(2\text{ m}) - (23.5\text{ kN})(6\text{ m}) = 0$$
$$B = +107.1\text{ kN} \qquad \mathbf{B} = 107.1\text{ kN} \rightarrow \blacktriangleleft$$

Since the result is positive, the reaction is directed as assumed.

***Determination of* $\mathbf{A}_x$.** The magnitude A_x is determined by expressing that the sum of the horizontal components of all external forces is zero.

$$\xrightarrow{+} \Sigma F_x = 0: \quad A_x + B = 0$$
$$A_x + 107.1\text{ kN} = 0$$
$$A_x = -107.1\text{ kN} \qquad \mathbf{A}_x = 107.1\text{ kN} \leftarrow \blacktriangleleft$$

Since the result is negative, the sense of $\mathbf{A}_x$ is opposite to that assumed originally.

***Determination of* $\mathbf{A}_y$.** The sum of the vertical components must also equal zero.

$$+\uparrow \Sigma F_y = 0: \quad A_y - 9.81\text{ kN} - 23.5\text{ kN} = 0$$
$$A_y = +33.3\text{ kN} \qquad \mathbf{A}_y = 33.3\text{ kN} \uparrow \blacktriangleleft$$

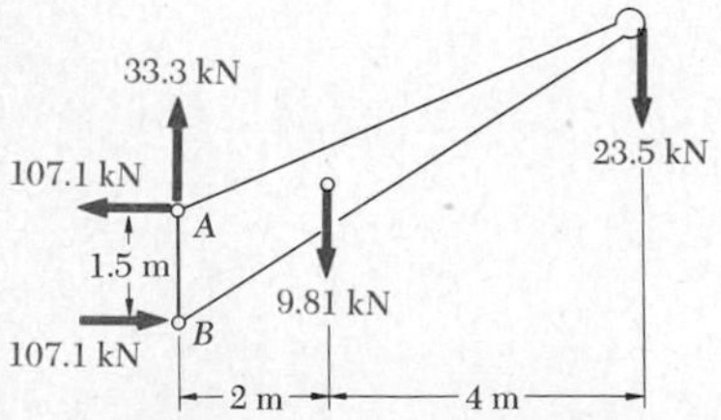

Adding vectorially the components $\mathbf{A}_x$ and $\mathbf{A}_y$, we find that the reaction at A is $112.2\text{ kN} \measuredangle 17.3°$.

Check. The values obtained for the reactions may be checked by recalling that the sum of the moments of all external forces about any point must be zero. For example, considering point B, we write

$$+\circlearrowleft \Sigma M_B = -(9.81\text{ kN})(2\text{ m}) - (23.5\text{ kN})(6\text{ m}) + (107.1\text{ kN})(1.5\text{ m}) = 0$$

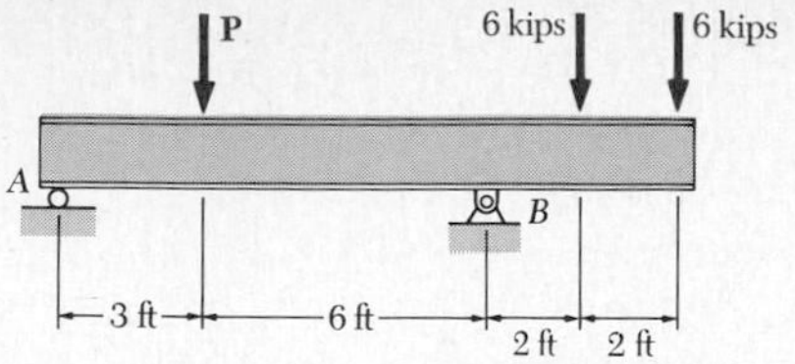

SAMPLE PROBLEM 4.2

Three loads are applied to a beam as shown. The beam is supported by a roller at A and by a pin at B. Neglecting the weight of the beam, determine the reactions at A and B when $P = 15$ kips.

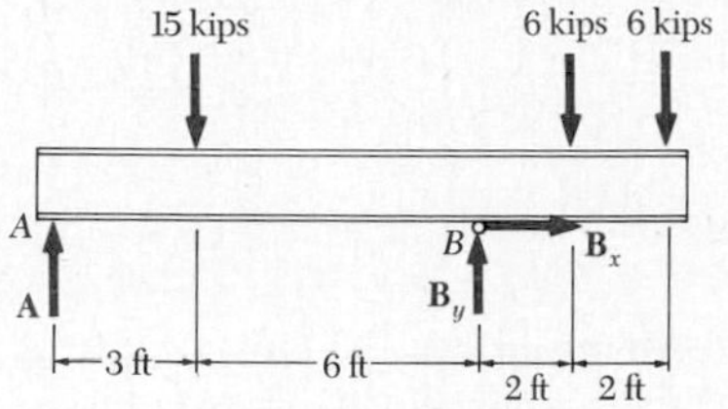

Solution. A free-body diagram of the beam is drawn. The reaction at A is vertical and is denoted by **A**. The reaction at B is represented by components $\mathbf{B}_x$ and $\mathbf{B}_y$. Each component is assumed to act in the direction shown.

Equilibrium Equations. We write the following three equilibrium equations and solve for the reactions indicated:

$$\overset{+}{\rightarrow}\Sigma F_x = 0: \qquad B_x = 0 \qquad \mathbf{B}_x = 0 \blacktriangleleft$$

$$+\circlearrowleft \Sigma M_A = 0:$$
$$-(15\text{ kips})(3\text{ ft}) + B_y(9\text{ ft}) - (6\text{ kips})(11\text{ ft}) - (6\text{ kips})(13\text{ ft}) = 0$$
$$B_y = +21.0\text{ kips} \qquad \mathbf{B}_y = 21.0\text{ kips}\uparrow \blacktriangleleft$$

$$+\circlearrowleft \Sigma M_B = 0:$$
$$-A(9\text{ ft}) + (15\text{ kips})(6\text{ ft}) - (6\text{ kips})(2\text{ ft}) - (6\text{ kips})(4\text{ ft}) = 0$$
$$A = +6.00\text{ kips} \qquad \mathbf{A} = 6.00\text{ kips}\uparrow \blacktriangleleft$$

Check. The results are checked by adding the vertical components of all the external forces:

$$+\uparrow\Sigma F_y = +6.00\text{ kips} - 15\text{ kips} + 21.0\text{ kips} - 6\text{ kips} - 6\text{ kips} = 0$$

Remark. In this problem the reactions at both A and B are vertical; however, these reactions are vertical for different reasons. At A, the beam is supported by a roller; hence the reaction cannot have any horizontal component. At B, the horizontal component of the reaction is zero because it must satisfy the equilibrium equation $\Sigma F_x = 0$ and because none of the other forces acting on the beam has a horizontal component.

We could have noticed at first glance that the reaction at B was vertical and dispensed with the horizontal component $\mathbf{B}_x$. This, however, is a bad practice. In following it, we would run the risk of forgetting the component $\mathbf{B}_x$ when the loading conditions require such a component (i.e., when a horizontal load is included). Also, the component $\mathbf{B}_x$ was found to be zero by using and solving an equilibrium equation, $\Sigma F_x = 0$. By setting $\mathbf{B}_x$ equal to zero immediately, we might not realize that we actually make use of this equation and thus might lose track of the number of equations available for solving the problem.

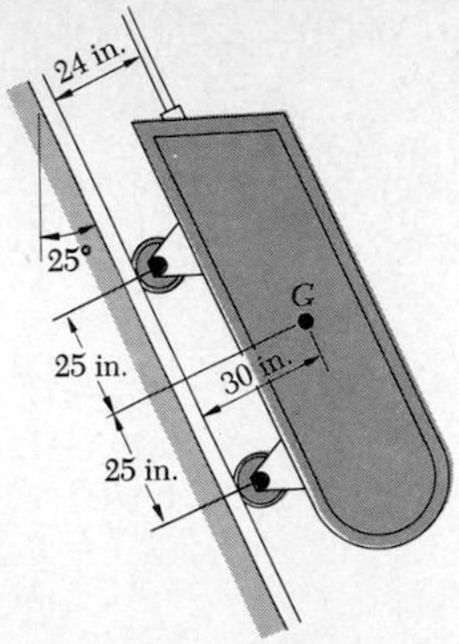

SAMPLE PROBLEM 4.3

A loading car is at rest on a track forming an angle of 25° with the vertical. The gross weight of the car and its load is 5500 lb, and it is applied at a point 30 in. from the track, halfway between the two axles. The car is held by a cable attached 24 in. from the track. Determine the tension in the cable and the reaction at each pair of wheels.

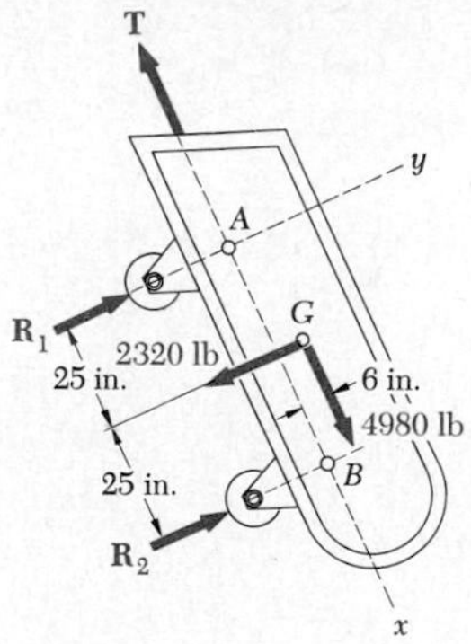

Solution. A free-body diagram of the car is drawn. The reaction at each wheel is perpendicular to the track, and the tension force **T** is parallel to the track. For convenience, we choose the x axis parallel to the track and the y axis perpendicular to the track. The 5500-lb weight is then resolved into x and y components.

$$W_x = +(5500 \text{ lb}) \cos 25° = +4980 \text{ lb}$$
$$W_y = -(5500 \text{ lb}) \sin 25° = -2320 \text{ lb}$$

Equilibrium Equations. We take moments about A to eliminate **T** and $\mathbf{R}_1$ from the computation.

$$+\circlearrowleft \Sigma M_A = 0: \quad -(2320 \text{ lb})(25 \text{ in.}) - (4980 \text{ lb})(6 \text{ in.}) + R_2(50 \text{ in.}) = 0$$
$$R_2 = +1758 \text{ lb} \qquad \mathbf{R}_2 = 1758 \text{ lb} \nearrow \blacktriangleleft$$

Now, taking moments about B to eliminate **T** and $\mathbf{R}_2$ from the computation, we write

$$+\circlearrowleft \Sigma M_B = 0: \quad (2320 \text{ lb})(25 \text{ in.}) - (4980 \text{ lb})(6 \text{ in.}) - R_1(50 \text{ in.}) = 0$$
$$R_1 = +562 \text{ lb} \qquad \mathbf{R}_1 = 562 \text{ lb} \nearrow \blacktriangleleft$$

The value of T is found by writing

$$\searrow +\Sigma F_x = 0: \quad +4980 \text{ lb} - T = 0$$
$$T = +4980 \text{ lb} \qquad \mathbf{T} = 4980 \text{ lb} \nwarrow \blacktriangleleft$$

The computed values of the reactions are shown in the adjacent sketch.

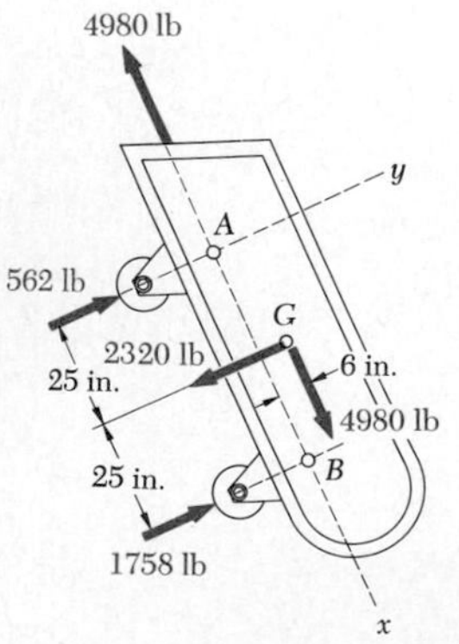

Check. The computations are verified by writing

$$\nearrow +\Sigma F_y = +562 \text{ lb} + 1758 \text{ lb} - 2320 \text{ lb} = 0$$

A check could also have been obtained by computing moments about any point except A or B.

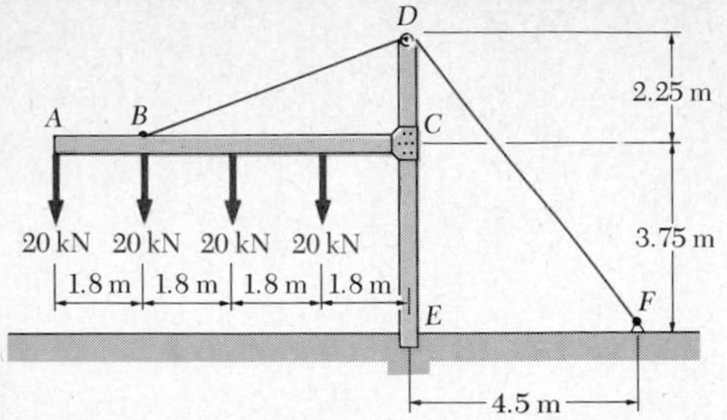

SAMPLE PROBLEM 4.4

The frame shown supports part of the roof of a small building. Knowing that the tension in the cable is 150 kN, determine the reaction at the fixed end E.

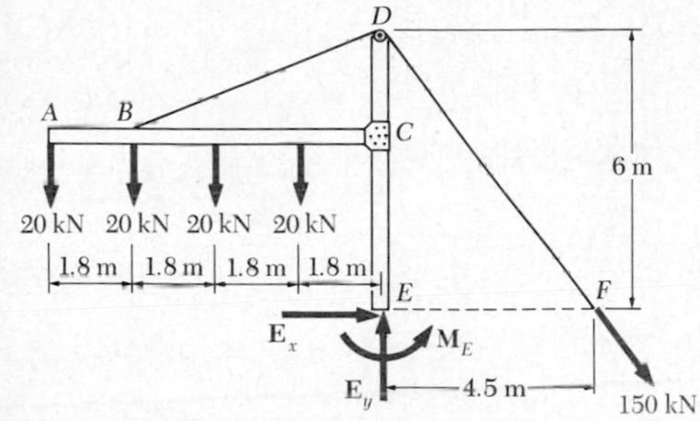

Solution. A free-body diagram of the frame and of the cable BDF is drawn. The reaction at the fixed end E is represented by the force components $\mathbf{E}_x$ and $\mathbf{E}_y$ and the couple $\mathbf{M}_E$. The other forces acting on the free body are the four 20-kN loads and the 150-kN force exerted at end F of the cable.

Equilibrium Equations. Noting that $DF = \sqrt{(4.5\text{ m})^2 + (6\text{ m})^2} = 7.5$ m, we write

$$\overset{+}{\rightarrow}\Sigma F_x = 0: \qquad E_x + \frac{4.5}{7.5}(150\text{ kN}) = 0$$

$$E_x = -90.0\text{ kN} \qquad \mathbf{E}_x = 90.0\text{ kN} \leftarrow \blacktriangleleft$$

$$+\uparrow\Sigma F_y = 0: \qquad E_y - 4(20\text{ kN}) - \frac{6}{7.5}(150\text{ kN}) = 0$$

$$E_y = +200\text{ kN} \qquad \mathbf{E}_y = 200\text{ kN}\uparrow \blacktriangleleft$$

$$+\circlearrowleft\Sigma M_E = 0: \qquad (20\text{ kN})(7.2\text{ m}) + (20\text{ kN})(5.4\text{ m}) + (20\text{ kN})(3.6\text{ m})$$

$$+ (20\text{ kN})(1.8\text{ m}) - \frac{6}{7.5}(150\text{ kN})(4.5\text{ m}) + M_E = 0$$

$$M_E = +180.0\text{ kN}\cdot\text{m} \qquad \mathbf{M}_E = 180.0\text{ kN}\cdot\text{m}\ \circlearrowleft \blacktriangleleft$$

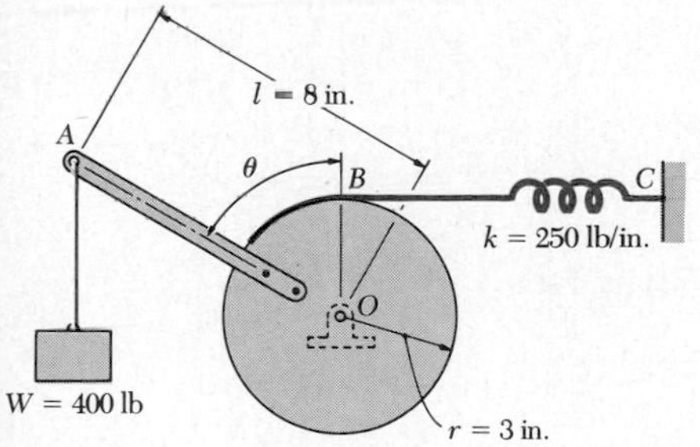

SAMPLE PROBLEM 4.5

A 400-lb weight is attached to the lever AO as shown. The constant of the spring BC is $k = 250$ lb/in., and the spring is unstretched when $\theta = 0$. Determine the position of equilibrium.

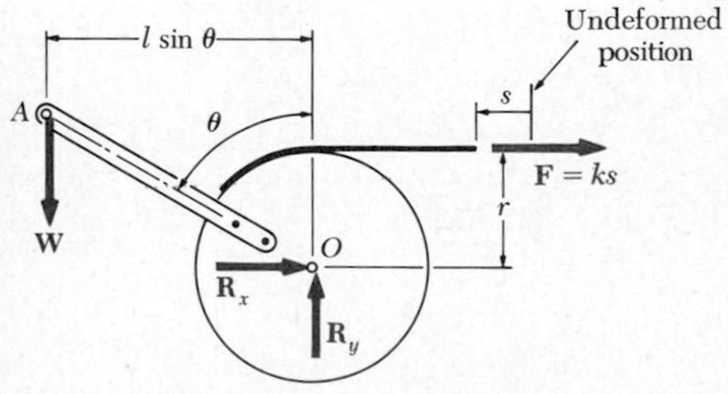

Solution. ***Force Exerted by Spring.*** Denoting by s the deflection of the spring from its undeformed position, and noting that $s = r\theta$, we write

$$F = ks = kr\theta$$

Equilibrium Equation. Summing the moments of $\mathbf{W}$ and $\mathbf{F}$ about O,

$$+\circlearrowleft\Sigma M_O = 0: \qquad Wl\sin\theta - r(kr\theta) = 0 \qquad \sin\theta = \frac{kr^2}{Wl}\theta$$

Substituting the given data, we obtain

$$\sin\theta = \frac{(250\text{ lb/in.})(3\text{ in.})^2}{(400\text{ lb})(8\text{ in.})}\theta \qquad \sin\theta = 0.703\theta$$

Solving by trial, we find $\qquad \theta = 0 \qquad \theta = 80.3° \blacktriangleleft$

Problems

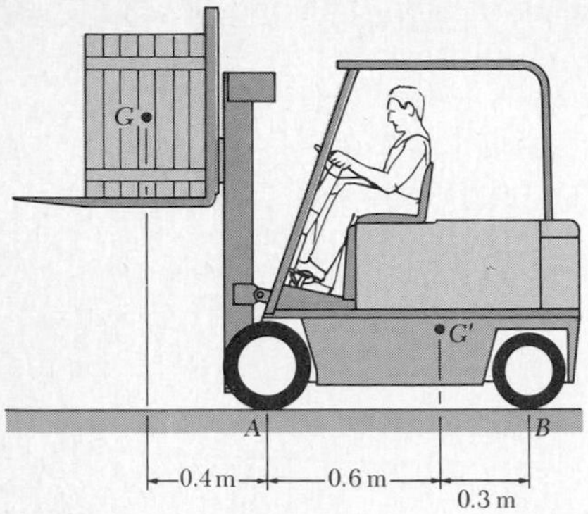

Fig. P4.1

4.1 A 2500-kg forklift truck is used to lift a 1200-kg crate. Determine the reaction at each of the two (*a*) front wheels *A*, (*b*) rear wheels *B*.

4.2 A hand truck is used to move a compressed-air cylinder. Knowing that the combined weight of the truck and cylinder is 180 lb, determine (*a*) the vertical force **P** which should be applied to the handle to maintain the cylinder in the position shown, (*b*) the corresponding reaction at each of the two wheels.

4.3 A load of lumber of weight $W = 20$ kN is being raised by a mobile crane. The weight of the boom *ABC* and the combined weight of the truck and driver are as shown. Determine the reaction at each of the two (*a*) front wheels *H*, (*b*) rear wheels *K*.

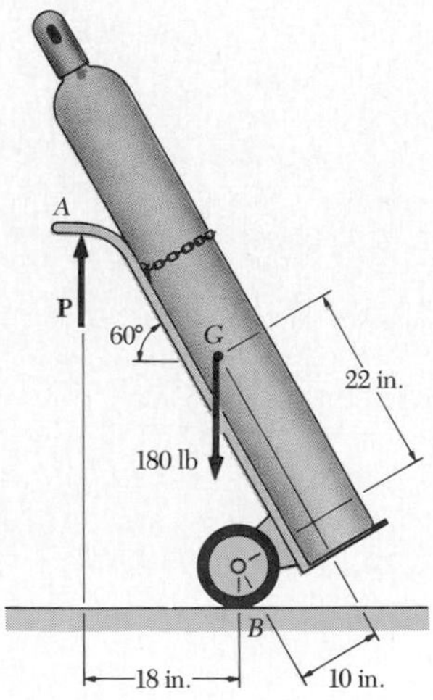

Fig. P4.2

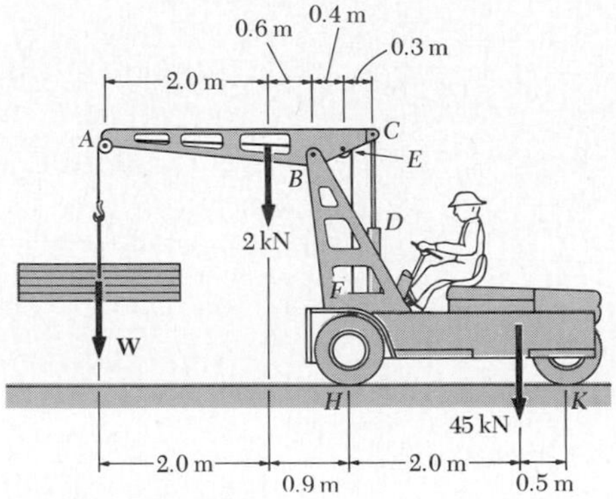

Fig. P4.3 and P4.4

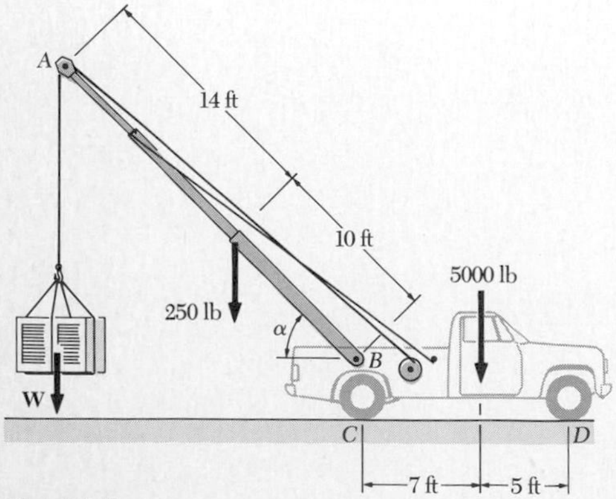

Fig. P4.5

4.4 A load of lumber of weight $W = 20$ kN is being raised as shown by a mobile crane. Knowing that the tension is 20 kN in all portions of cable *AEF* and that the weight of boom *ABC* is 2 kN, determine (*a*) the tension in rod *CD*, (*b*) the reaction at pin *B*.

4.5 A truck-mounted crane is used to lift a 600-lb compressor. The weights of the boom *AB* and of the truck are as shown, and the angle the boom forms with the horizontal is $\alpha = 45°$. Determine the reaction at each of the two (*a*) rear wheels *C*, (*b*) front wheels *D*.

4.6 For the truck-mounted crane of Prob. 4.5, determine the smallest allowable value of α if the truck is not to tip over when a 3000-lb load is lifted.

4.7 For the beam of Sample Prob. 4.2, determine the range of values of **P** for which the beam will be safe, knowing that the maximum allowable value for each of the reactions is 30 kips and that the reaction at A must be directed upward.

4.8 Three loads are applied as shown to a light beam supported by cables attached at B and D. Neglecting the weight of the beam, determine the range of values of Q for which neither cable becomes slack when $P = 0$.

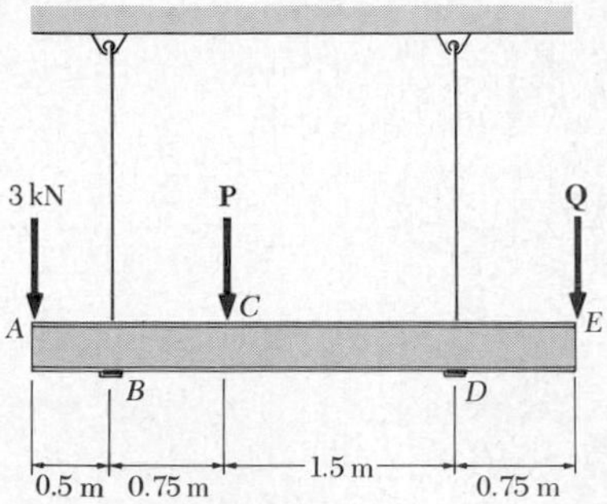

Fig. P4.8 and P4.9

4.9 Three loads are applied as shown to a light beam supported by cables attached at B and D. Knowing that the maximum allowable tension in each cable is 4 kN and neglecting the weight of the beam, determine the range of values of Q for which the loading is safe when $P = 0$.

4.10 For the beam of Prob. 4.9, determine the range of values of Q for which the loading is safe when $P = 1$ kN.

4.11 Determine the maximum load which may be raised by the mobile crane of Prob. 4.3 without tipping over, knowing that the largest force which can be exerted by the hydraulic cylinder D is 96 kN, and that the maximum allowable tension in cable AEF is 40 kN.

4.12 A T-shaped bracket supports a 200-N load as shown. Determine the reactions at A and C.

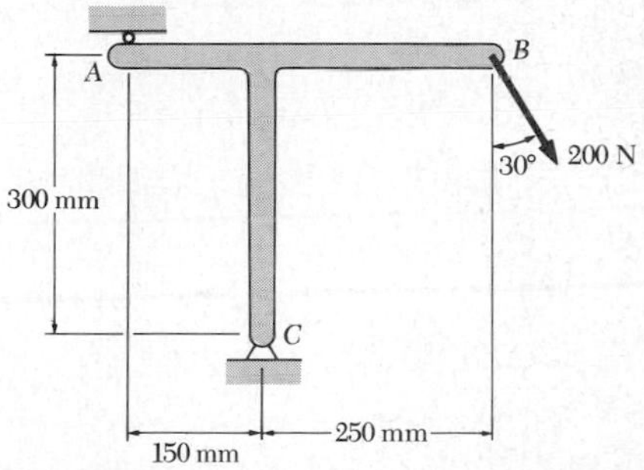

Fig. P4.12

4.13 Two links AB and DE are connected by a bell crank as shown. Knowing that the tension in link AB is 150 lb, determine (a) the tension in link DE, (b) the reaction at C.

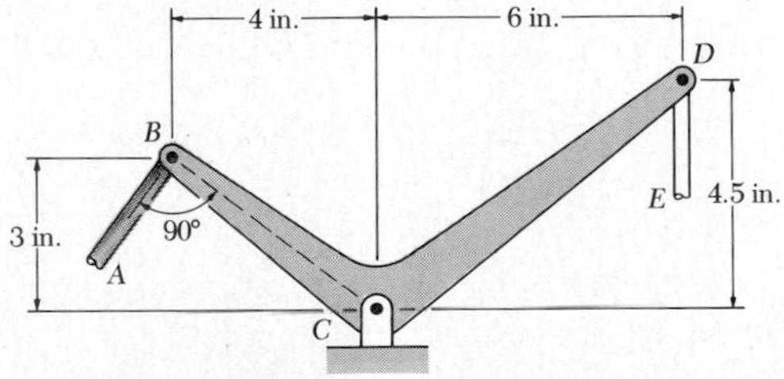

Fig. P4.13 and P4.14

4.14 Two links AB and DE are connected by a bell crank as shown. Determine the maximum force which may be safely exerted by link AB on the bell crank if the maximum allowable value for the reaction at C is 400 lb.

4.15 The lever AB is hinged at C and attached to a control cable at A. If the lever is subjected at B to a 400-N horizontal force, determine (a) the tension in the cable, (b) the reaction at C.

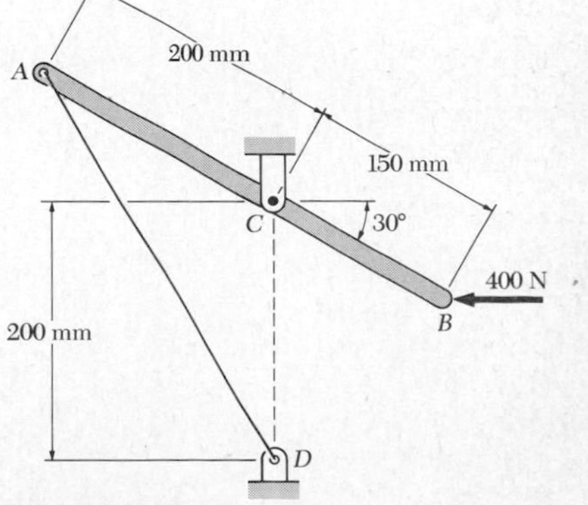

Fig. P4.15

4.16 A truss may be supported in three different ways as shown. In each case, determine the reactions at the supports.

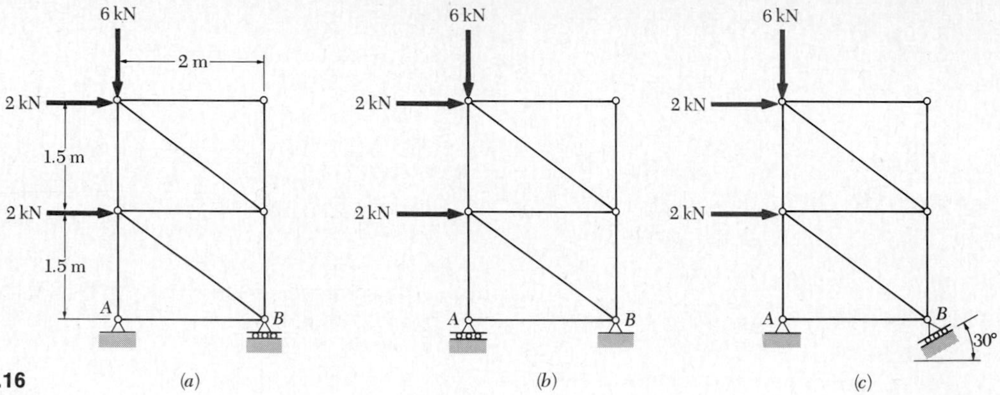

Fig. P4.16

4.17 Determine the reactions at A and B when (*a*) $\alpha = 0$, (*b*) $\alpha = 90°$, (*c*) $\alpha = 30°$.

4.18 Solve Prob. 4.17, assuming that the 50-lb force has been replaced by a 300-lb·in. clockwise couple.

4.19 Knowing that the tension in all portions of the belt is 800 N, determine the reactions at the supports A and B of the plate when (*a*) $\alpha = 0$, (*b*) $\alpha = 90°$, (*c*) $\alpha = 30°$.

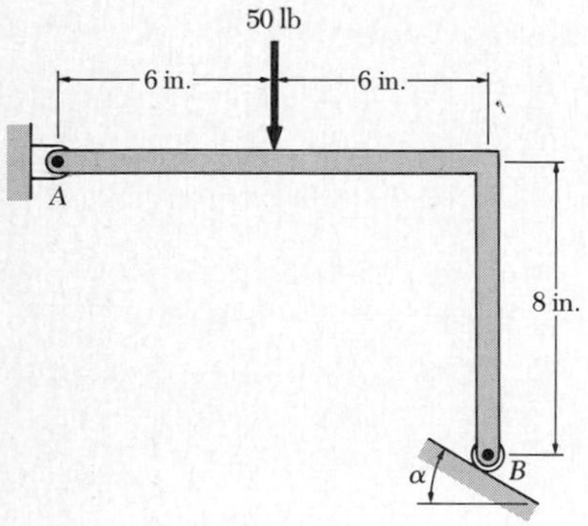

Fig. P4.17

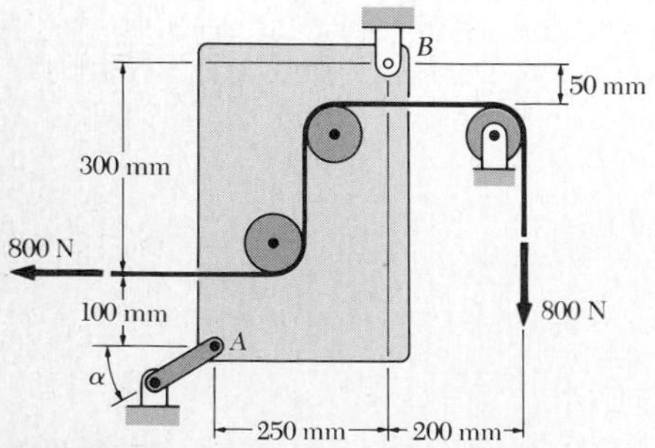

Fig. P4.19

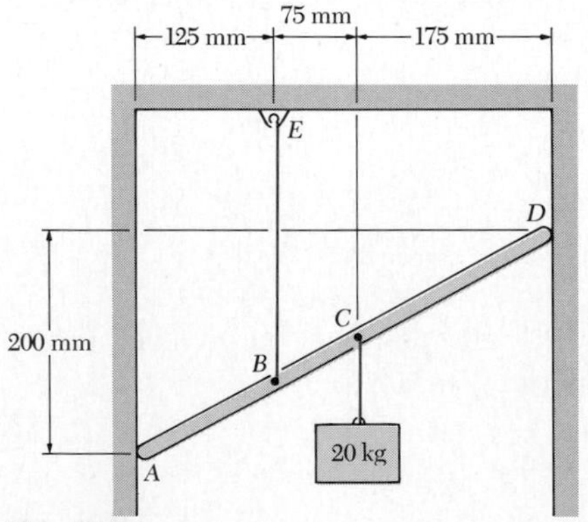

Fig. P4.20

4.20 A light bar AD is suspended from a cable BE and supports a 20-kg block at C. The extremities A and D of the bar are in contact with frictionless, vertical walls. Determine the tension in cable BE and the reactions at A and D.

4.21 A light bar AB supports a 15-kg block at its midpoint C. Rollers at A and B rest against frictionless surfaces, and a horizontal cable AD is attached at A. Determine the tension in cable AD and the reactions at A and B.

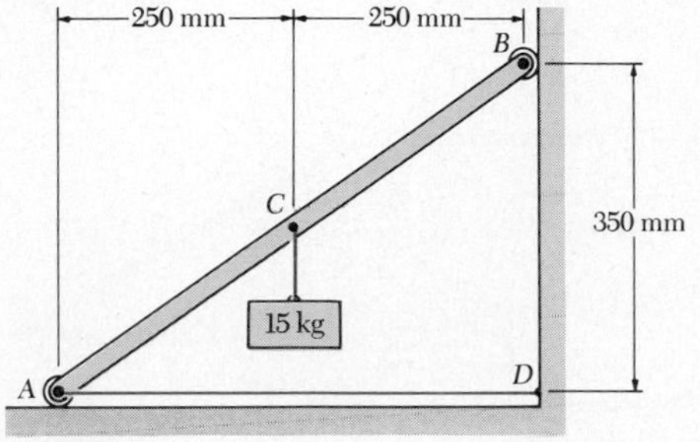

Fig. P4.21

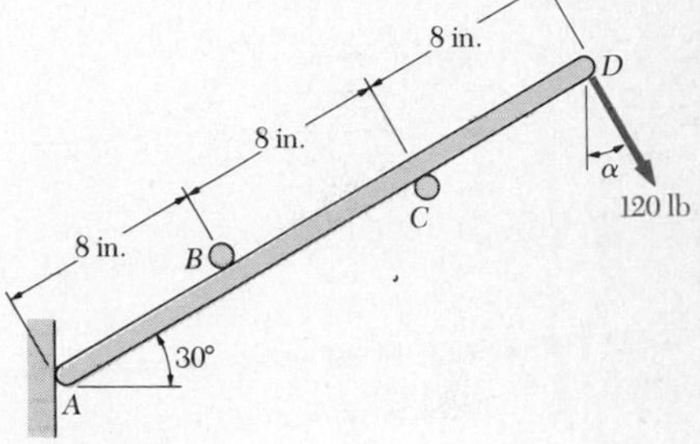

Fig. P4.22

4.22 A light rod AD is supported by frictionless pegs at B and C and rests against a frictionless wall at A. A vertical 120-lb force ($\alpha = 0$) is applied at D. Determine the reactions at A, B, and C.

4.23 Solve Prob. 4.22, assuming that the 120-lb force is applied in a direction perpendicular to the rod ($\alpha = 30°$).

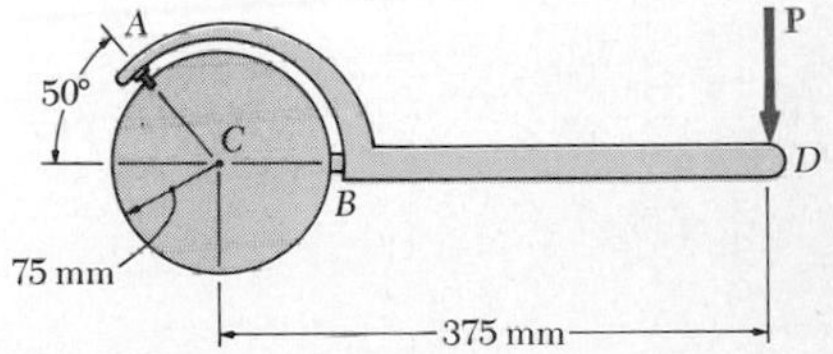

Fig. P4.24 and P4.25

4.24 The spanner shown is used to rotate a shaft. A pin fits in a hole at A, while a flat, frictionless surface rests against the shaft at B. If a 250-N force $\mathbf{P}$ is exerted on the spanner at D, find (a) the reaction at B, (b) the component of the reaction at A in a direction perpendicular to AC.

4.25 The spanner shown is used to rotate a shaft. A pin fits in a hole at A, while a flat, frictionless surface rests against the shaft at B. If the moment about C of the force exerted on the shaft at A is to be 75 N·m, find (a) the force $\mathbf{P}$ which should be exerted on the spanner at D, (b) the corresponding value of the force exerted on the spanner at B.

4.26 A 160-lb overhead garage door consists of a uniform rectangular panel AC, 84 in. high, supported by the cable AE attached at the middle of the upper edge of the door and by two sets of frictionless rollers at A and B. Each set consists of two rollers located on either side of the door. The rollers A are free to move in horizontal channels, while the rollers B are guided by vertical channels. If the door is held in the position for which $BD = 42$ in., determine (a) the tension in cable AE, (b) the reaction at each of the four rollers.

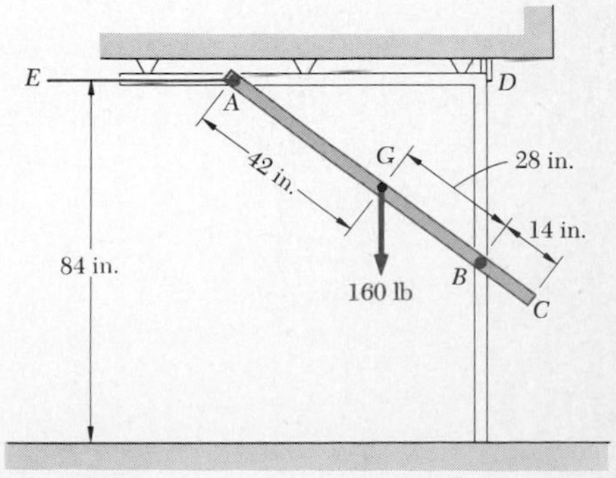

Fig. P4.26

4.27 In Prob. 4.26, determine the distance BD for which the tension in cable AE is equal to 900 lb.

4.28 The tensioner shown is used to maintain a given tension in the transmission belt *CDE*. The triangular receptacle contains a helical spring which exerts a couple **M** on arm *AB*, causing the 60-mm-radius idler pulley at *B* to press against the belt. Knowing that, in the position shown, the tension in the belt is $T = 300$ N, determine (*a*) the couple **M** exerted on arm *AB*, (*b*) the horizontal and vertical components of the force exerted at *A* on arm *AB*.

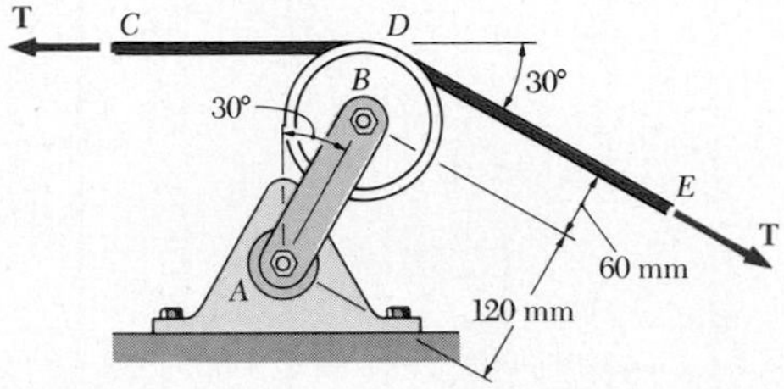

Fig. P4.28

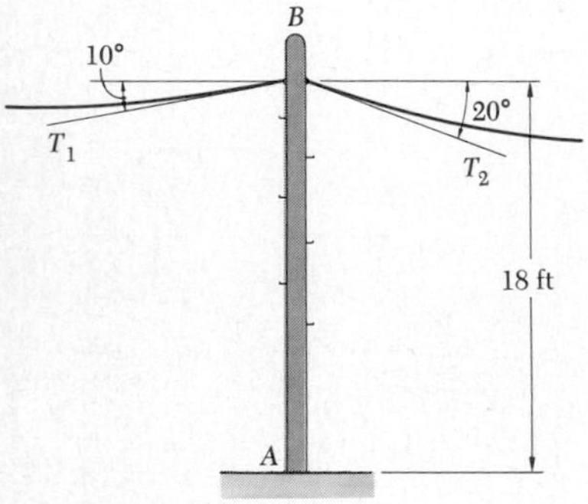

Fig. P4.29

4.29 An 18-ft telephone pole weighing 320 lb is used to support the ends of two wires. The wires form the angles shown with the horizontal and the tensions in the wires are, respectively, $T_1 = 120$ lb and $T_2 = 75$ lb. Determine the reaction at the fixed end *A*.

4.30 For the telephone pole of Prob. 4.29, determine the range of allowable values of T_2 if the magnitude of the couple at *A* is not to exceed 900 lb·ft when $T_1 = 100$ lb.

4.31 For the frame of Sample Prob. 4.4, determine the range of allowable values of the tension in cable *BDF* if the magnitude of the couple at the fixed end *E* is not to exceed 100 kN·m.

4.32 Rod *AD* is acted upon by a vertical force **P** at end *A* and by two equal and opposite horizontal forces of magnitude *Q* at points *B* and *C*. Neglecting the weight of the rod, express the angle θ corresponding to the equilibrium position in terms of *P* and *Q*.

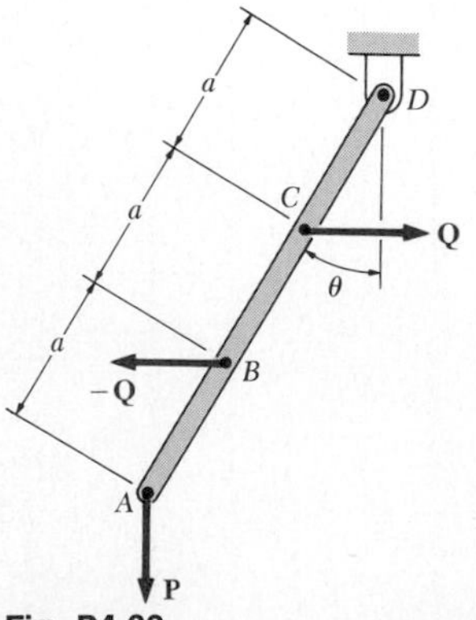

Fig. P4.32

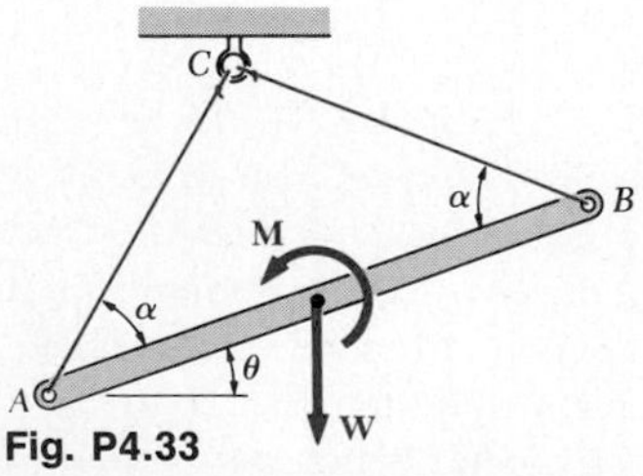

Fig. P4.33

4.33 A uniform rod *AB* of length *l* and weight *W* is suspended from two cords *AC* and *BC* of equal length. Determine the angle θ corresponding to the equilibrium position when a couple **M** is applied to the rod.

4.34 and 4.35 A vertical load **P** is applied at end B of rod BC. (a) Neglecting the weight of the rod, express the angle θ corresponding to the equilibrium position in terms of P, l, and the counterweight W. (b) Determine the value of θ corresponding to equilibrium if $P = 2W$.

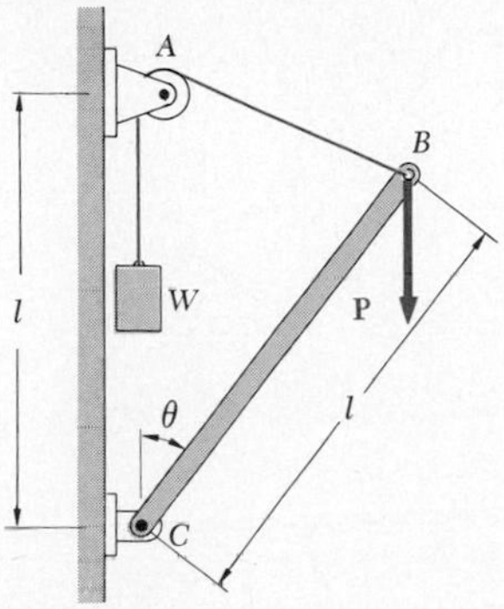

Fig. P4.34

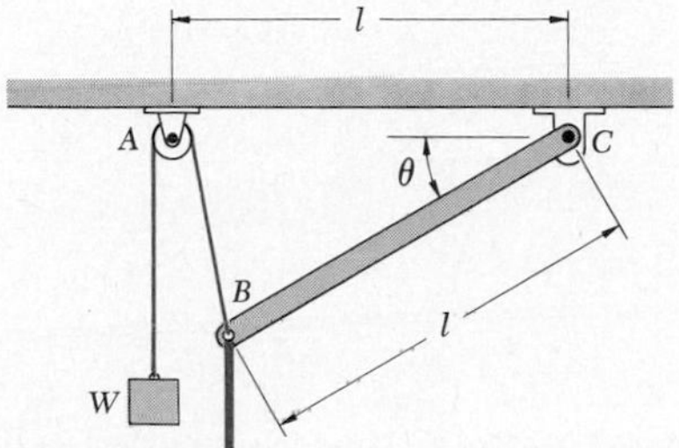

Fig. P4.35

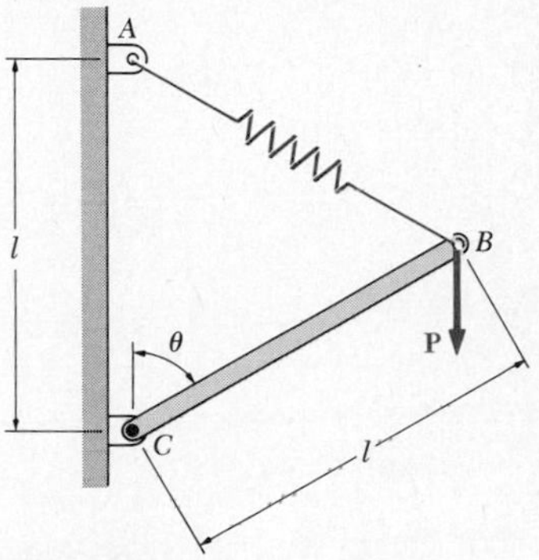

Fig. P4.36

4.36 A vertical load **P** is applied at end B of rod BC. The constant of the spring is k and the spring is unstretched when $\theta = 60°$. (a) Neglecting the weight of the rod, express the angle θ corresponding to the equilibrium position in terms of P, k, and l. (b) Determine the value of θ corresponding to equilibrium if $P = \frac{1}{4}kl$.

4.37 A vertical load **P** is applied at end B of rod BC. The constant of the spring is k and the spring is unstretched when $\theta = 0$. (a) Neglecting the weight of the rod, express the angle θ corresponding to the equilibrium position in terms of P, k, and l. (b) Determine the value of θ corresponding to equilibrium if $P = 2kl$.

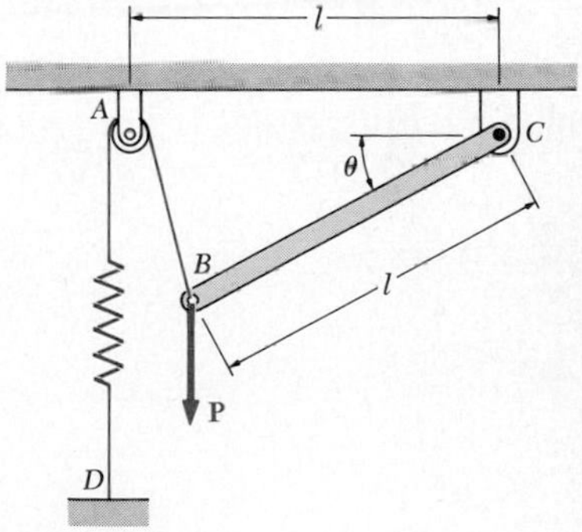

Fig. P4.37

4.38 A collar B, of weight W, may move freely along the vertical rod shown. The constant of the spring is k and the spring is unstretched when $y = 0$. (a) Derive an equation in y, W, a, and k which must be satisfied when the collar is in equilibrium. (b) Find the value of k, knowing that $W = 60$ N, $a = 400$ mm, and that the collar is in equilibrium when $y = 300$ mm.

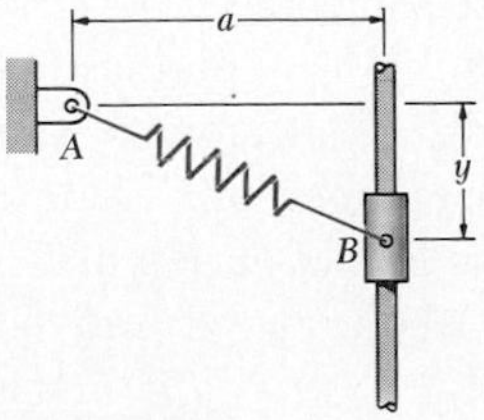

Fig. P4.38

4.39 A slender rod AB, of weight W, is attached to blocks A and B which may move freely in the guides shown. The constant of the spring is k and the spring is unstretched when AB is horizontal. (a) Neglecting the weight of the blocks, derive an equation in θ, W, l, and k which must be satisfied when the rod is in equilibrium. (b) Find the value of k, knowing that $W = 30$ N, $l = 0.5$ m, and that the rod is in equilibrium when $\theta = 30°$.

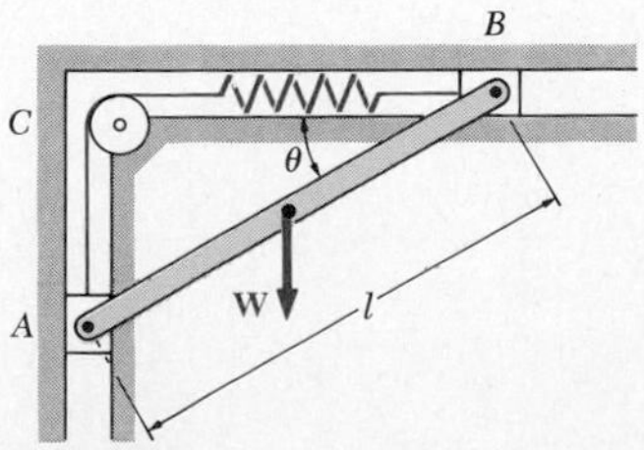

Fig. P4.39

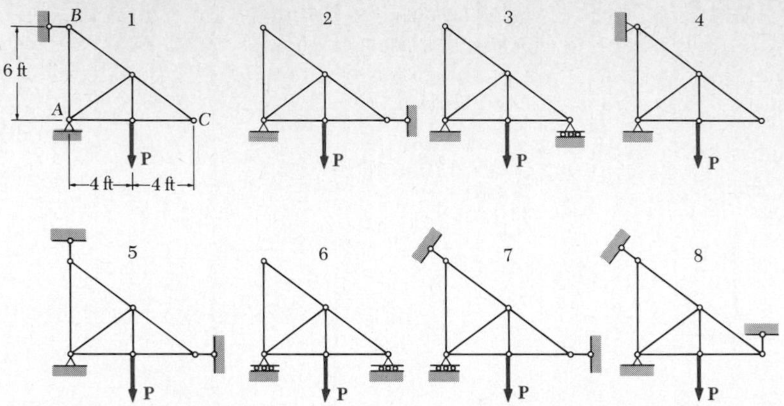

Fig. P4.40

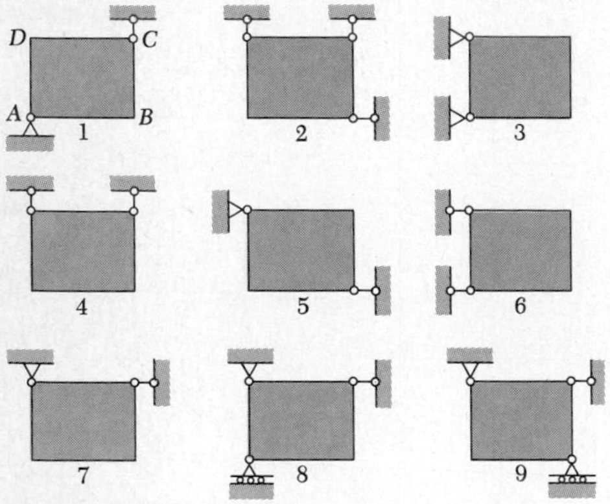

Fig. P4.41

4.40 A small truss is supported in eight different ways as shown. All connections consist of frictionless pins, rollers, and short links. In each case, determine whether (*a*) the truss is completely, partially, or improperly constrained, (*b*) the reactions are statically determinate or indeterminate, (*c*) the equilibrium of the truss is maintained in the position shown. Also, wherever possible, compute the reactions, assuming that the magnitude of the force **P** is 10 kips.

4.41 Nine identical rectangular plates, 600 × 900 mm, and each of mass $m = 50$ kg, are held in a vertical plane as shown. All connections consist of frictionless pins, rollers, or short links. For each case, answer the questions listed in Prob. 4.40, and, wherever possible, compute the reactions.

4.6. Equilibrium of a Two-Force Body. A particular case of equilibrium which is of considerable interest is that of a rigid body subjected to two forces. Such a body is commonly called a *two-force body*. We shall show that *if a two-force body is in equilibrium, the two forces must have the same magnitude, same line of action, and opposite sense.*

Consider a corner plate subjected to two forces $\mathbf{F}_1$ and $\mathbf{F}_2$ acting at A and B, respectively (Fig. 4.8*a*). If the plate is to be in equilibrium, the sum of the moments of $\mathbf{F}_1$ and $\mathbf{F}_2$ about any axis must be zero. First, we sum moments about A: Since the moment of $\mathbf{F}_1$ is obviously zero, the moment of $\mathbf{F}_2$ must also be zero and the line of action of $\mathbf{F}_2$ must pass through A (Fig. 4.8*b*). Summing moments about B, we prove similarly that the line of

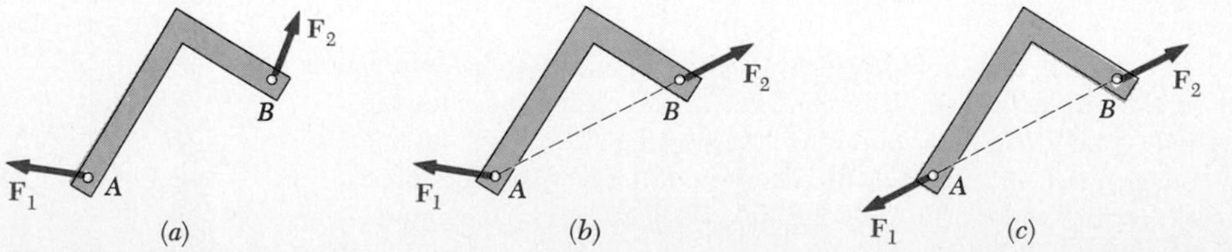

Fig. 4.8

action of $\mathbf{F}_1$ must pass through B (Fig. 4.8c). Both forces have the same line of action (line AB). From the equation $\Sigma F_x = 0$ or $\Sigma F_y = 0$, it is seen that they must have also the same magnitude but opposite sense.

If several forces act at two points A and B, the forces acting at A may be replaced by their resultant $\mathbf{F}_1$ and those acting at B by their resultant $\mathbf{F}_2$. Thus a two-force body may be more generally defined as *a rigid body subjected to forces acting at only two points.* The resultants $\mathbf{F}_1$ and $\mathbf{F}_2$ then must have the same line of action, same magnitude, and opposite sense.

In our study of structures, frames, and machines, we shall see that the recognition of two-force bodies will enable us to simplify the solution of certain problems.

4.7. Equilibrium of a Three-Force Body. Another case of equilibrium that is of great interest is that of a *three-force body,* i.e., a rigid body subjected to three forces or, more generally, *a rigid body subjected to forces acting at only three points.* Consider a rigid body subjected to a system of forces which may be reduced to three forces $\mathbf{F}_1$, $\mathbf{F}_2$, and $\mathbf{F}_3$ acting at A, B, and C, respectively (Fig. 4.9a). We shall show that if the body is in equilibrium, *the lines of action of the three forces must be either concurrent or parallel.*

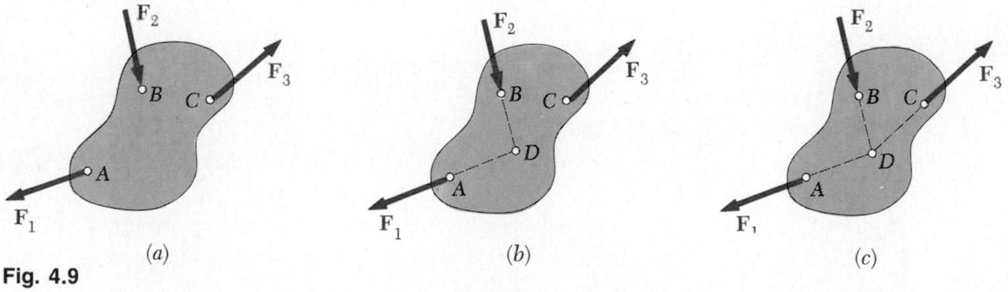

Fig. 4.9

Since the rigid body is in equilibrium, the sum of the moments of $\mathbf{F}_1$, $\mathbf{F}_2$, and $\mathbf{F}_3$ about any axis must be zero. Assuming that the lines of action of $\mathbf{F}_1$ and $\mathbf{F}_2$ intersect, and denoting their point of intersection by D, we sum moments about D (Fig. 4.9b); since the moments of $\mathbf{F}_1$ and $\mathbf{F}_2$ about D are zero, the moment of $\mathbf{F}_3$ about D must also be zero and the line of action of $\mathbf{F}_3$ must pass through D (Fig. 4.9c). The three lines of action are concurrent. The only exception occurs when none of the lines intersect; the lines of action are then parallel.

Although problems concerning three-force bodies may be solved by the general methods of Secs. 4.3 to 4.5, the property just established may be used to solve them either graphically, or mathematically from simple trigonometric or geometric relations.

SAMPLE PROBLEM 4.6

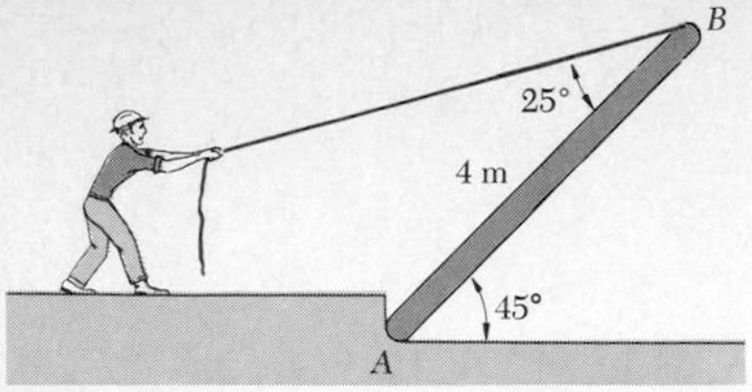

A man raises a 10-kg joist, of length 4 m, by pulling on a rope. Find the tension T in the rope and the reaction at A.

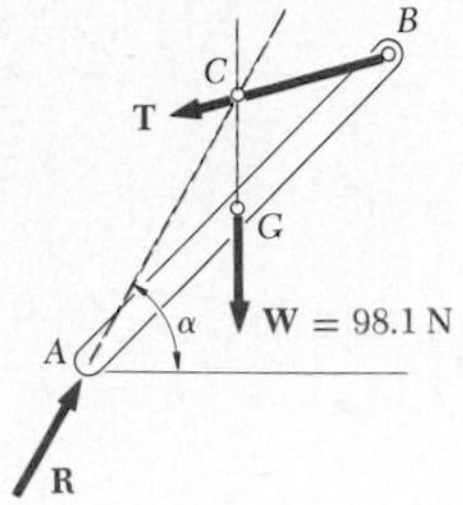

Solution. The joist is a three-force body since it is acted upon by three forces: its weight **W**, the force **T** exerted by the rope, and the reaction **R** of the ground at A. We note that

$$W = mg = (10\text{ kg})(9.81\text{ m/s}^2) = 98.1\text{ N}$$

Three-Force Body. Since the joist is a three-force body, the forces acting on it must be concurrent. The reaction **R**, therefore, will pass through the point of intersection C of the lines of action of the weight **W** and of the tension force **T**. This fact will be used to determine the angle α that **R** forms with the horizontal.

Drawing the vertical BF through B and the horizontal CD through C, we note that

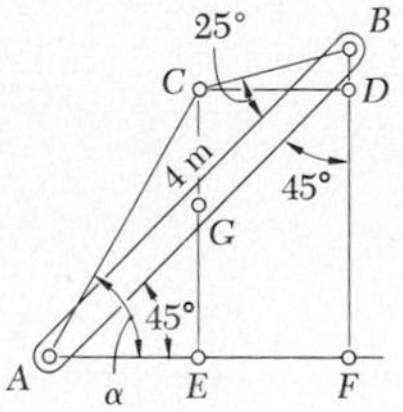

$$\begin{aligned}
AF &= BF = (AB)\cos 45^\circ = (4\text{ m})\cos 45^\circ = 2.828\text{ m}\\
CD &= EF = AE = \tfrac{1}{2}(AF) = 1.414\text{ m}\\
BD &= (CD)\cot(45^\circ + 25^\circ) = (1.414\text{ m})\tan 20^\circ = 0.515\text{ m}\\
CE &= DF = BF - BD = 2.828\text{ m} - 0.515\text{ m} = 2.313\text{ m}
\end{aligned}$$

We write

$$\tan\alpha = \frac{CE}{AE} = \frac{2.313\text{ m}}{1.414\text{ m}} = 1.636$$

$$\alpha = 58.6^\circ$$

We now know the direction of all the forces acting on the joist.

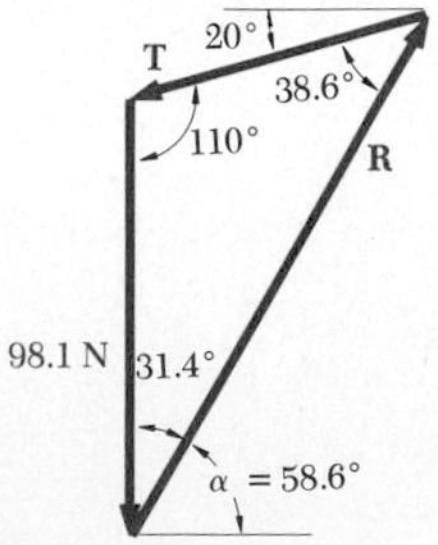

Force Triangle. A force triangle is drawn as shown, and its interior angles are computed from the known directions of the forces. Using the law of sines, we write

$$\frac{T}{\sin 31.4^\circ} = \frac{R}{\sin 110^\circ} = \frac{98.1\text{ N}}{\sin 38.6^\circ}$$

$$T = 81.9\text{ N}$$
$$\mathbf{R} = 147.8\text{ N} \measuredangle 58.6^\circ$$

Problems

4.42 The spanner shown is used to rotate a shaft. A pin fits in a hole at A, while a flat, frictionless surface rests against the shaft at B. If a 300-N force $\mathbf{P}$ is exerted on the spanner at D, find the reactions at A and B.

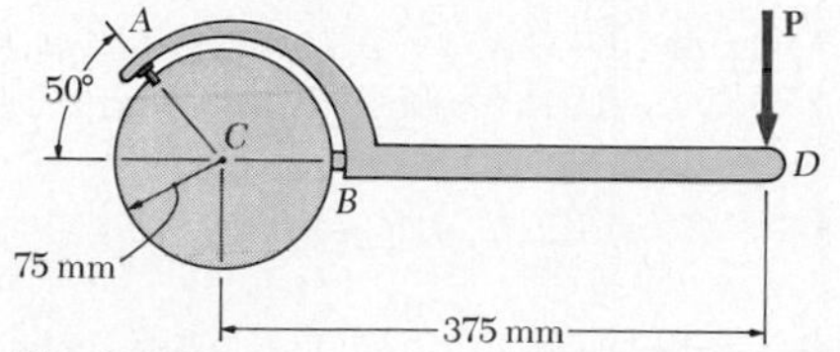

Fig. P4.42

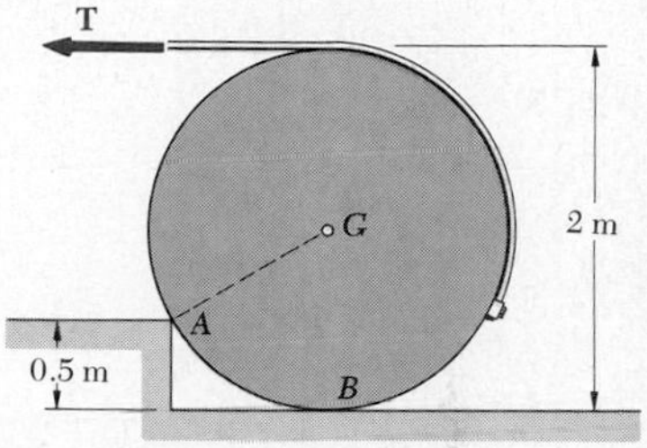

Fig. P4.43

4.43 A 250-kg cylindrical tank, 2 m in diameter, is to be raised over a 0.5-m obstruction. A cable is wrapped around the tank and pulled horizontally as shown. Knowing that the corner of the obstruction at A is rough, find the required tension in the cable and the reaction at A.

4.44 Using the method of Sec. 4.7, solve Prob. 4.17*b*.

4.45 Using the method of Sec. 4.7, solve Prob. 4.17*c*.

4.46 A load $\mathbf{P}$ is applied at joint C of the truss shown. Determine the reactions at A and B when (*a*) $\alpha = 0$, (*b*) $\alpha = 45°$.

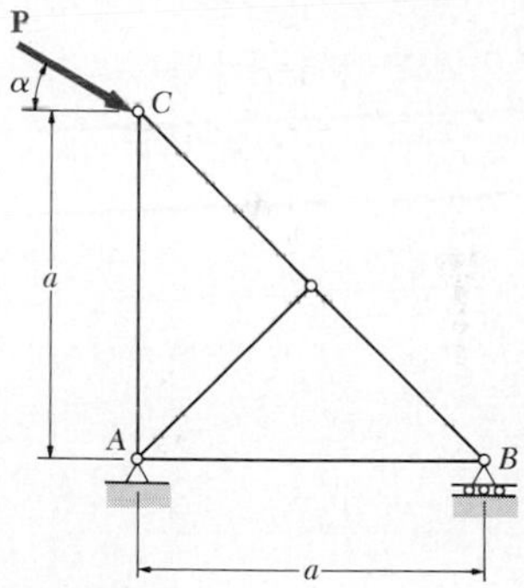

Fig. P4.46

4.47 One end of rod AB rests in the corner A and the other is attached to the cord BD. If the rod supports a 150-N load at its midpoint C, find the reaction at A and the tension in the cord.

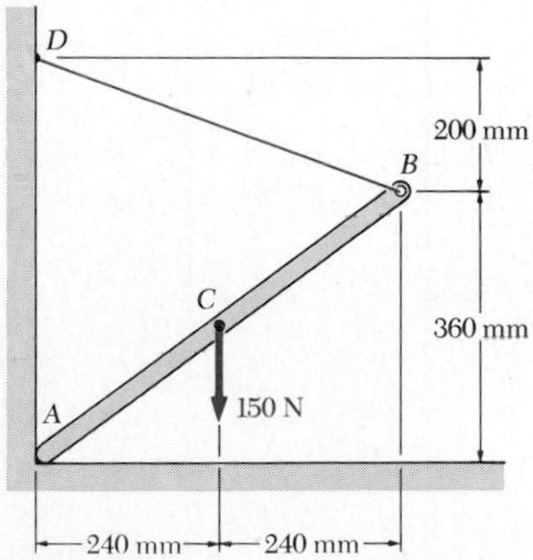

Fig. P4.47

Fig. P4.48

4.48 A 12-ft wooden beam weighing 80 lb is supported by a pin and bracket at A and by the cable BC. Find the reaction at A and the tension in the cable.

4.49 Using the method of Sec. 4.7, solve Prob. 4.13.

4.50 Using the method of Sec. 4.7, solve Prob. 4.12.

4.51 Using the method of Sec. 4.7, solve Prob. 4.15.

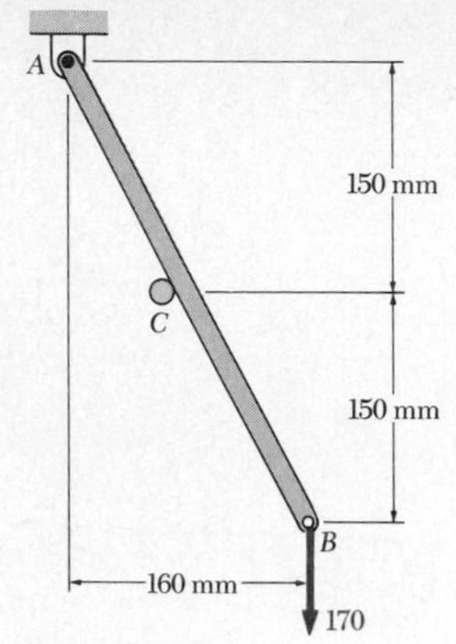

Fig. P4.52

4.52 Rod AB is supported by a pin and bracket at A and rests against a frictionless peg at C. Determine the reactions at A and C when a 170-N vertical force is applied at B.

4.53 Solve Prob. 4.52, assuming that the 170-N force applied at B is horizontal and directed to the left.

4.54 The L-shaped member ACB is supported by a pin and bracket at C and by an inextensible cord attached at A and B and passing over a frictionless pulley at D. The tension may be assumed to be the same in portions AD and BD of the cord. If the magnitudes of the forces applied at A and B are, respectively, $P = 30$ lb and $Q = 0$, determine (*a*) the tension in the cord, (*b*) the reaction at C.

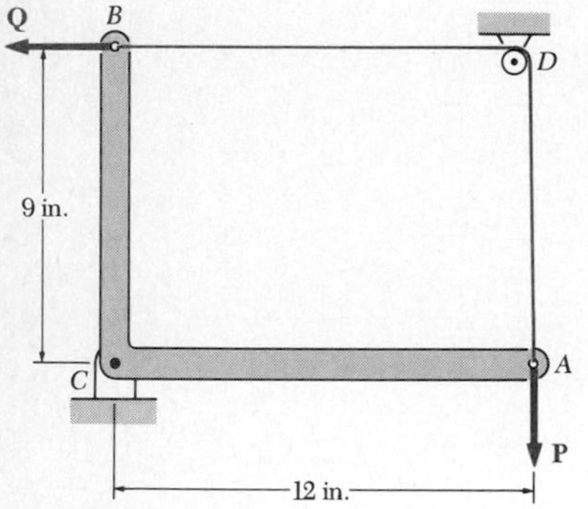

Fig. P4.54

4.55 For the L-shaped member of Prob. 4.54, (*a*) express the tension T in the cord in terms of the magnitudes P and Q of the forces applied at A and B, (*b*) assuming $P = 30$ lb, find the largest allowable value of Q if the equilibrium is to be maintained.

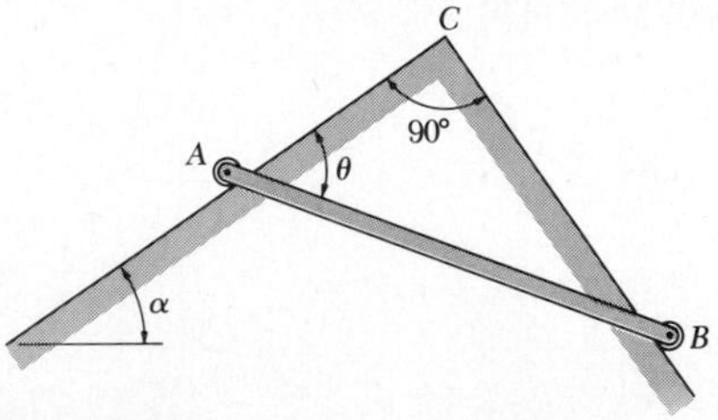

Fig. P4.56

4.56 The uniform rod AB lies in a vertical plane. Its ends are connected to rollers which rest against frictionless surfaces. Determine the relation between the angles θ and α when the rod is in equilibrium.

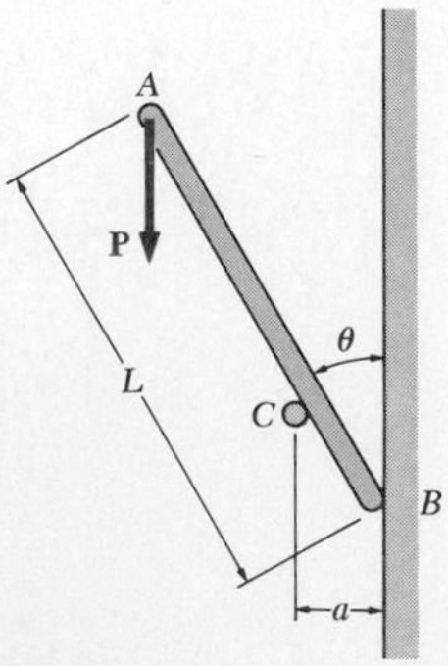

Fig. P4.57

4.57 A slender rod of length L is lodged between peg C and the vertical wall. It supports a load **P** at end A. Neglecting friction and the weight of the rod, determine the angle θ corresponding to equilibrium.

4.58 A thin ring of mass 2 kg and radius $r = 140$ mm is held against a frictionless wall by a 125-mm string AB. Determine (*a*) the distance d, (*b*) the tension in the string, (*c*) the reaction at C.

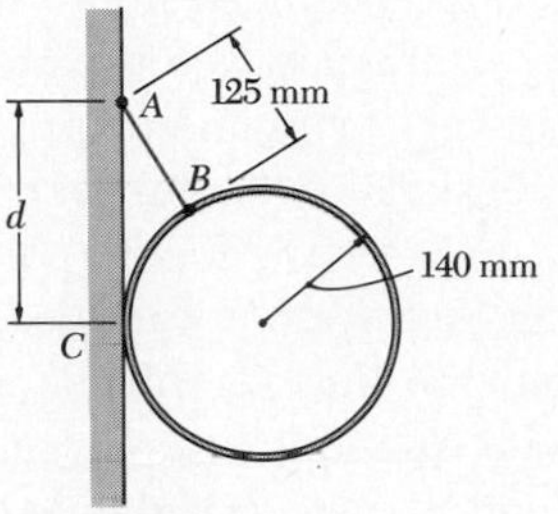

Fig. P4.58

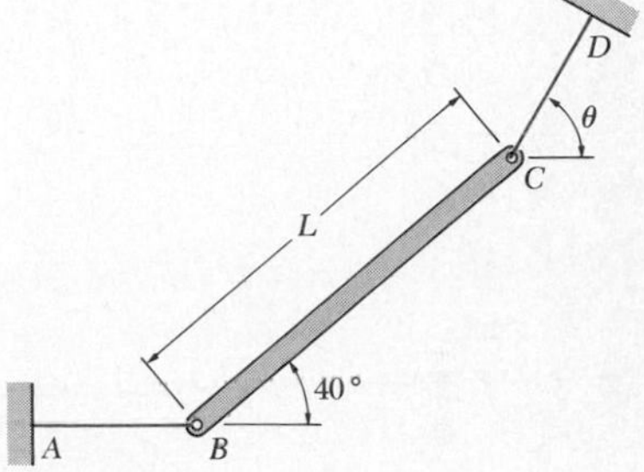

Fig. P4.59

4.59 A slender rod BC of length L and weight W is held by two cables as shown. Knowing that cable AB is horizontal and that the rod forms an angle of 40° with the horizontal, determine (*a*) the angle θ that cable CD forms with the horizontal, (*b*) the tension in each cable.

***4.60** A slender rod of length $2r$ and weight W is attached to a collar at B and rests on a circular cylinder of radius r. Knowing that the collar may slide freely along a vertical guide and neglecting friction, determine the value of θ corresponding to equilibrium.

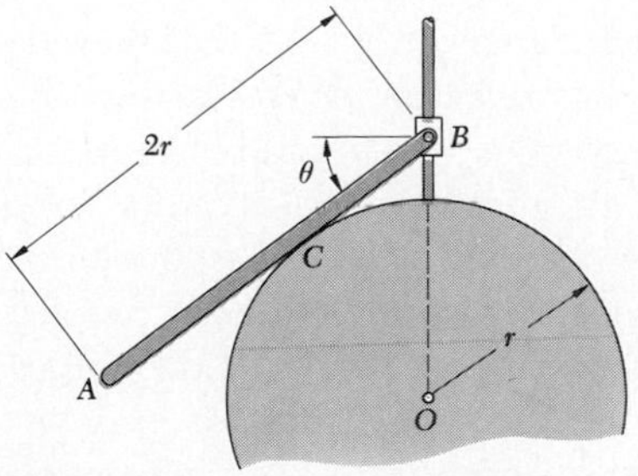

Fig. P4.60

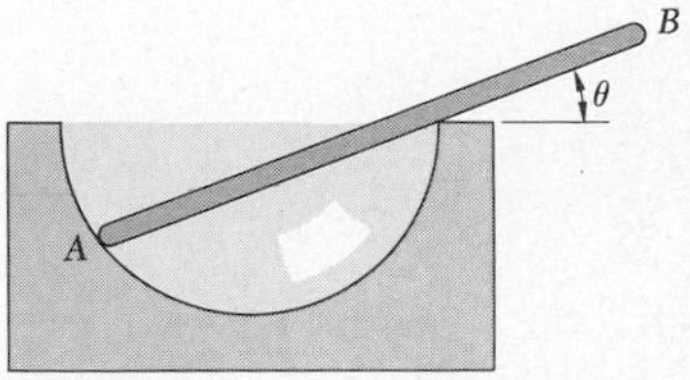

Fig. P4.61

***4.61** A uniform rod AB of length $3R$ and weight W rests inside a hemispherical bowl of radius R as shown. Neglecting friction, determine the angle θ corresponding to equilibrium.

EQUILIBRIUM IN THREE DIMENSIONS

4.8. Reactions at Supports and Connections for a Three-dimensional Structure. The reactions on a three-dimensional structure range from the single force of known direction exerted by a frictionless surface to the force-couple system exerted by a fixed support. Consequently, the number of unknowns associated with the reaction at a support or connection may vary from one to six in problems involving the equilibrium of a three-dimensional structure. Various types of supports and connections are shown in Fig. 4.10 with the corresponding reactions. A simple way of determining the type of reaction corresponding to a given support or connection and the number of unknowns involved is to find which of the six fundamental motions (translation in x, y, and z directions, rotation about the x, y, and z axes) are allowed and which motions are prevented.

Ball supports, frictionless surfaces, and cables, for example, prevent translation in one direction only and thus exert a single force of known line of action; they each involve one unknown, namely, the magnitude of the reaction. Rollers on rough surfaces and wheels on rails prevent translation in two directions; the corresponding reactions consist of two unknown force components. Rough surfaces in direct contact and ball-and-socket supports prevent translation in three directions; these supports involve three unknown force components.

Some supports and connections may prevent rotation as well as translation; the corresponding reactions include, then, couples as well as forces. The reaction at a fixed support, for example, which prevents any motion (rotation as well as translation), consists of three unknown forces and three unknown couples. A universal joint, which is designed to allow rotation about two axes, will exert a reaction consisting of three unknown force components and one unknown couple.

Other supports and connections are primarily intended to prevent translation; their design, however, is such that they also prevent some rotations. The corresponding reactions consist essentially of force components but *may* also include couples. One group of supports of this type includes hinges and bearings designed to support radial loads only (for example, journal bearings, roller bearings). The corresponding reactions consist of two force components but may also include two couples. Another group includes pin-and-bracket supports, hinges, and bearings designed to support an axial thrust as well as a radial load (for example, ball bearings). The corresponding reactions consist of three force components but may include two couples. However, these supports will not exert any appreciable couples under normal conditions of use. Therefore, *only* force components should be included in their analysis, *unless* it is found that couples are necessary to maintain the equilibrium of the rigid body, or unless the support is known to have been specifically designed to exert a couple (see Probs. 4.86 through 4.89).

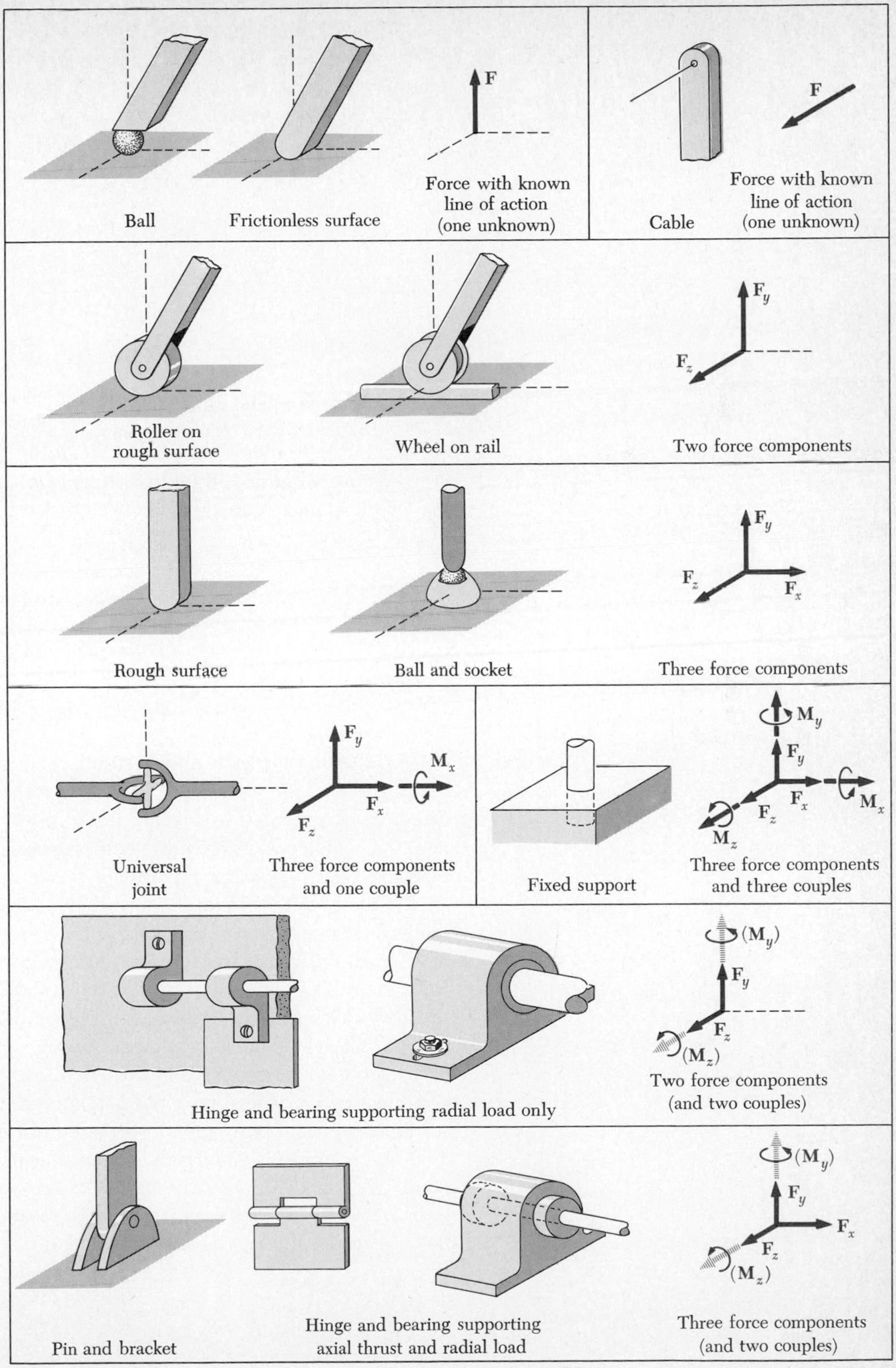

Fig. 4.10 Reactions at supports and connections.

4.9. Equilibrium of a Rigid Body in Three Dimensions.

We saw in Sec. 4.1 that six scalar equations are required to express the conditions for the equilibrium of a rigid body in the general three-dimensional case:

$$\Sigma F_x = 0 \qquad \Sigma F_y = 0 \qquad \Sigma F_z = 0 \tag{4.2}$$
$$\Sigma M_x = 0 \qquad \Sigma M_y = 0 \qquad \Sigma M_z = 0 \tag{4.3}$$

These equations may be solved for no more than *six unknowns,* which generally will represent reactions at supports or connections.

In most problems the scalar equations (4.2) and (4.3) will be more conveniently obtained if we first express in vector form the conditions for the equilibrium of the rigid body considered. We write

$$\Sigma \mathbf{F} = 0 \qquad \Sigma \mathbf{M}_O = \Sigma(\mathbf{r} \times \mathbf{F}) = 0 \tag{4.1}$$

and express the forces **F** and position vectors **r** in terms of scalar components and unit vectors. Next we compute all vector products, either by direct computation or by means of determinants (see Sec. 3.8). Equating to zero the coefficients of the unit vectors in each of the two relations (4.1), we obtain the desired scalar equations.

If the reactions involve more than six unknowns, there are more unknowns than equations, and some of the reactions are *statically indeterminate.* If the reactions involve less than six unknowns, there are more equations than unknowns, and some of the equations of equilibrium cannot be satisfied under general loading conditions; the rigid body is only *partially constrained.* Under the particular loading conditions corresponding to a given problem, however, the extra equations often reduce to trivial identities such as $0 = 0$ and may be disregarded; although only partially constrained, the rigid body remains in equilibrium (see Sample Probs. 4.7 and 4.8). Even with six or more unknowns, it is possible that some equations of equilibrium will not be satisfied. This may occur when the supports are such that the reactions are forces which must either be parallel or intersect the same line; the rigid body is then *improperly constrained.*

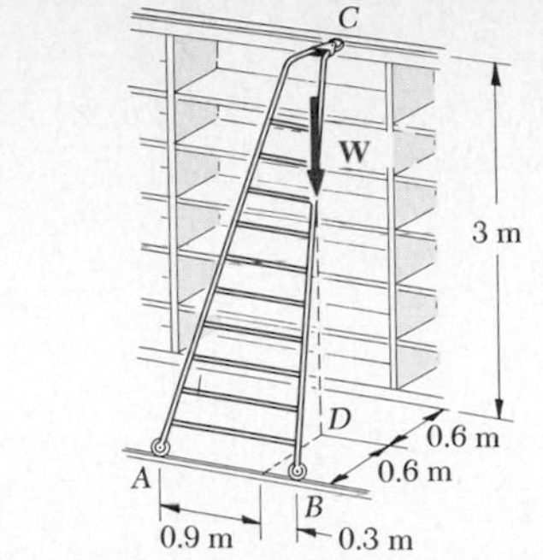

SAMPLE PROBLEM 4.7

A 20-kg ladder used to reach high shelves in a storeroom is supported by two flanged wheels A and B mounted on a rail and by an unflanged wheel C resting against a rail fixed to the wall. An 80-kg man stands on the ladder and leans to the right. The line of action of the combined weight **W** of the man and ladder intersects the floor at point D. Determine the reactions at A, B, and C.

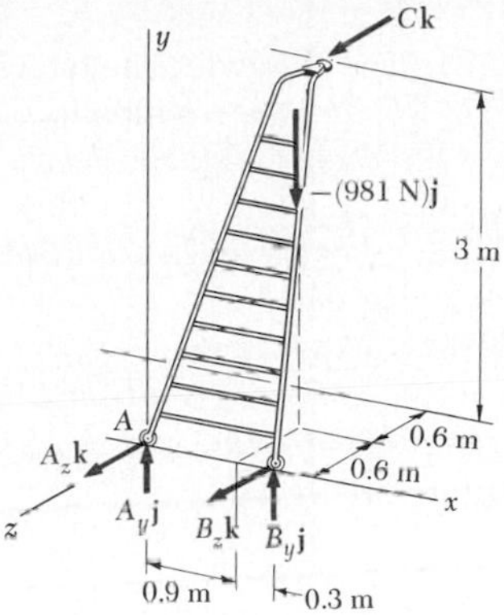

Solution. A free-body diagram of the ladder is drawn. The forces involved are the combined weight of the man and ladder,

$$\mathbf{W} = -mg\mathbf{j} = -(80\text{ kg} + 20\text{ kg})(9.81\text{ m/s}^2)\mathbf{j} = -(981\text{ N})\mathbf{j}$$

and five unknown reaction components, two at each flanged wheel and one at the unflanged wheel. The ladder is thus only partially constrained; it is free to roll along the rails. It is, however, in equilibrium under the given load since the equation $\Sigma F_x = 0$ is satisfied.

Equilibrium Equations. We express that the forces acting on the ladder form a system equivalent to zero:

$$\Sigma\mathbf{F} = 0: \qquad A_y\mathbf{j} + A_z\mathbf{k} + B_y\mathbf{j} + B_z\mathbf{k} - (981\text{ N})\mathbf{j} + C\mathbf{k} = 0$$

$$(A_y + B_y - 981\text{ N})\mathbf{j} + (A_z + B_z + C)\mathbf{k} = 0 \qquad (1)$$

$$\Sigma\mathbf{M}_A = \Sigma(\mathbf{r} \times \mathbf{F}) = 0: \qquad 1.2\mathbf{i} \times (B_y\mathbf{j} + B_z\mathbf{k}) + (0.9\mathbf{i} - 0.6\mathbf{k}) \times (-981\mathbf{j}) + (0.6\mathbf{i} + 3\mathbf{j} - 1.2\mathbf{k}) \times C\mathbf{k} = 0$$

Computing the vector products, we have†

$$1.2B_y\mathbf{k} - 1.2B_z\mathbf{j} - 882.9\mathbf{k} - 588.6\mathbf{i} - 0.6C\mathbf{j} + 3C\mathbf{i} = 0$$

$$(3C - 588.6)\mathbf{i} - (1.2B_z + 0.6C)\mathbf{j} + (1.2B_y - 882.9)\mathbf{k} = 0 \qquad (2)$$

Setting the coefficients of **i**, **j**, **k** equal to zero in Eq. (2), we obtain the following three scalar equations, which express that the sum of the moments about each coordinate axis must be zero:

$$3C - 588.6 = 0 \qquad C = +196.2\text{ N}$$
$$1.2B_z + 0.6C = 0 \qquad B_z = -98.1\text{ N}$$
$$1.2B_y - 882.9 = 0 \qquad B_y = +736\text{ N}$$

The reactions at B and C are therefore

$$\mathbf{B} = +(736\text{ N})\mathbf{j} - (98.1\text{ N})\mathbf{k} \qquad \mathbf{C} = +(196.2\text{ N})\mathbf{k} \blacktriangleleft$$

Setting the coefficients of **j** and **k** equal to zero in Eq. (1), we obtain two scalar equations expressing that the sums of the components in the y and z directions are zero. Substituting for B_y, B_z, and C the values obtained above, we write

$$A_y + B_y - 981 = 0 \qquad A_y + 736 - 981 = 0 \qquad A_y = +245\text{ N}$$
$$A_z + B_z + C = 0 \qquad A_z - 98.1 + 196.2 = 0 \qquad A_z = -98.1\text{ N}$$

We conclude that the reaction at A is $\quad \mathbf{A} = +(245\text{ N})\mathbf{j} - (98.1\text{ N})\mathbf{k} \blacktriangleleft$

† The moments in this Sample Problem and in Sample Probs. 4.8 and 4.9 could also be expressed in the form of determinants (see Sample Prob. 3.10).

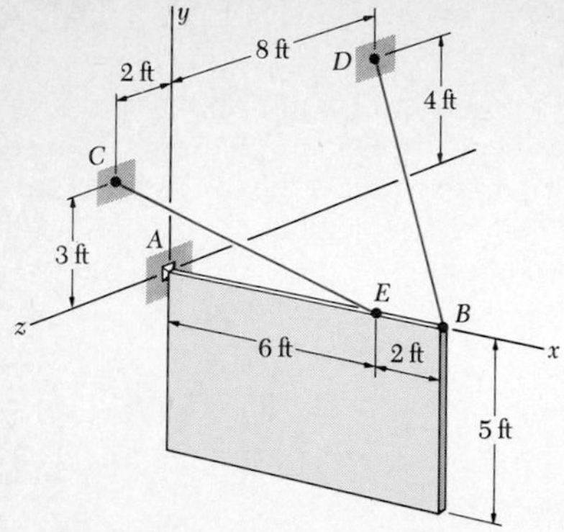

SAMPLE PROBLEM 4.8

A 5 × 8 ft sign of uniform density weighs 270 lb and is supported by a ball and socket at A and by two cables. Determine the tension in each cable and the reaction at A.

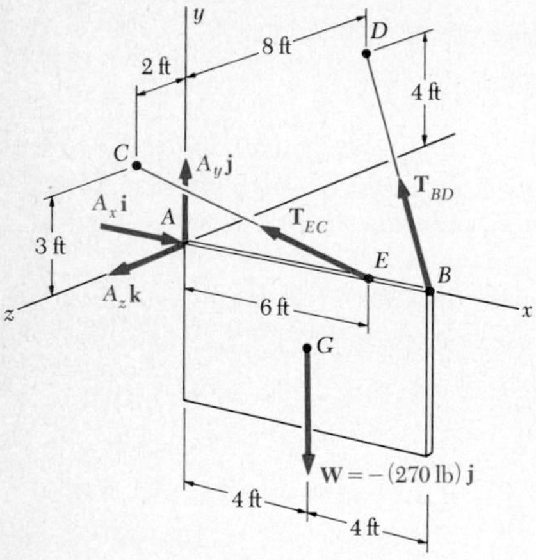

Solution. A free-body diagram of the sign is drawn. The forces acting on the free body are the weight $\mathbf{W} = -(270\text{ lb})\mathbf{j}$ and the reactions at A, B, and E. The reaction at A is a force of unknown direction and is represented by three unknown components. Since the directions of the forces exerted by the cables are known, these forces involve only one unknown each, namely the magnitudes T_{BD} and T_{EC}. Since there are only five unknowns, the sign is partially constrained. It may rotate freely about the x axis; it is, however, in equilibrium under the given loading since the equation $\Sigma M_x = 0$ is satisfied.

The components of the forces $\mathbf{T}_{BD}$ and $\mathbf{T}_{EC}$ may be expressed in terms of the unknown magnitudes T_{BD} and T_{EC} by writing

$$\overrightarrow{BD} = -(8\text{ ft})\mathbf{i} + (4\text{ ft})\mathbf{j} - (8\text{ ft})\mathbf{k} \qquad BD = 12\text{ ft}$$
$$\overrightarrow{EC} = -(6\text{ ft})\mathbf{i} + (3\text{ ft})\mathbf{j} + (2\text{ ft})\mathbf{k} \qquad EC = 7\text{ ft}$$
$$\mathbf{T}_{BD} = T_{BD}\left(\frac{\overrightarrow{BD}}{BD}\right) = T_{BD}(-\tfrac{2}{3}\mathbf{i} + \tfrac{1}{3}\mathbf{j} - \tfrac{2}{3}\mathbf{k})$$
$$\mathbf{T}_{EC} = T_{EC}\left(\frac{\overrightarrow{EC}}{EC}\right) = T_{EC}(-\tfrac{6}{7}\mathbf{i} + \tfrac{3}{7}\mathbf{j} + \tfrac{2}{7}\mathbf{k})$$

Equilibrium Equations. We express that the forces acting on the sign form a system equivalent to zero:

$$\Sigma\mathbf{F} = 0: \qquad A_x\mathbf{i} + A_y\mathbf{j} + A_z\mathbf{k} + \mathbf{T}_{BD} + \mathbf{T}_{EC} - (270\text{ lb})\mathbf{j} = 0$$
$$(A_x - \tfrac{2}{3}T_{BD} - \tfrac{6}{7}T_{EC})\mathbf{i} + (A_y + \tfrac{1}{3}T_{BD} + \tfrac{3}{7}T_{EC} - 270\text{ lb})\mathbf{j} + (A_z - \tfrac{2}{3}T_{BD} + \tfrac{2}{7}T_{EC})\mathbf{k} = 0 \qquad (1)$$

$$\Sigma\mathbf{M}_A = \Sigma(\mathbf{r}\times\mathbf{F}) = 0:$$
$$(8\text{ ft})\mathbf{i}\times T_{BD}(-\tfrac{2}{3}\mathbf{i} + \tfrac{1}{3}\mathbf{j} - \tfrac{2}{3}\mathbf{k}) + (6\text{ ft})\mathbf{i}\times T_{EC}(-\tfrac{6}{7}\mathbf{i} + \tfrac{3}{7}\mathbf{j} + \tfrac{2}{7}\mathbf{k}) + (4\text{ ft})\mathbf{i}\times(-270\text{ lb})\mathbf{j} = 0$$
$$(2.667T_{BD} + 2.571T_{EC} - 1080)\mathbf{k} + (5.333T_{BD} - 1.714T_{EC})\mathbf{j} = 0 \qquad (2)$$

Setting the coefficients of **j** and **k** equal to zero in Eq. (2), we obtain two scalar equations which may be solved for T_{BD} and T_{EC}:

$$T_{BD} = 101.3\text{ lb} \qquad T_{EC} = 315\text{ lb} \quad ◄$$

Setting the coefficients of **i**, **j**, and **k** equal to zero in Eq. (1), we obtain three more equations which yield the components of **A**. We have

$$\mathbf{A} = +(338\text{ lb})\mathbf{i} + (101.2\text{ lb})\mathbf{j} - (22.5\text{ lb})\mathbf{k} \quad ◄$$

SAMPLE PROBLEM 4.9

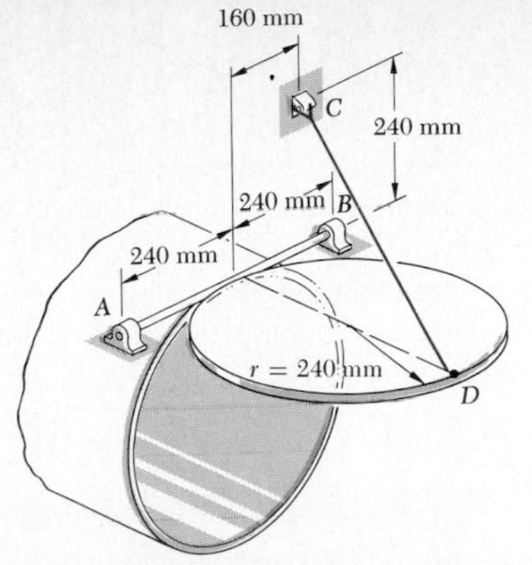

A uniform pipe cover of radius $r = 240$ mm and mass 30 kg is held in a horizontal position by the cable CD. Assuming that the bearing at B does not exert any axial thrust, determine the tension in the cable and the reactions at A and B.

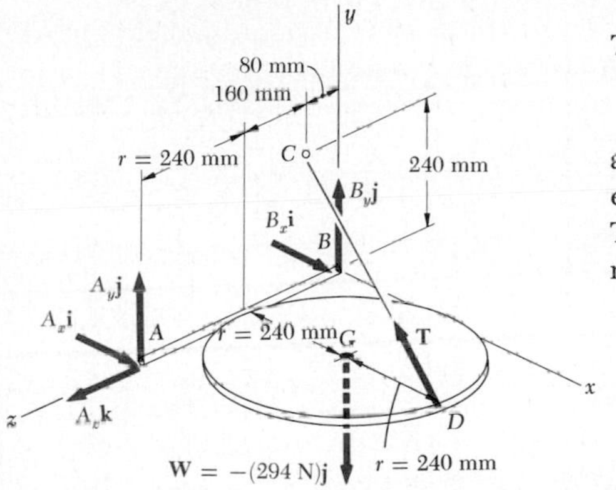

Solution. A free-body diagram is drawn with the coordinate axes shown. The forces acting on the free body are the weight of the cover,

$$\mathbf{W} = -mg\mathbf{j} = -(30 \text{ kg})(9.81 \text{ m/s}^2)\mathbf{j} = -(294 \text{ N})\mathbf{j}$$

and reactions involving six unknowns, namely, the magnitude of the force $\mathbf{T}$ exerted by the cable, three force components at hinge A, and two at hinge B. The components of $\mathbf{T}$ are expressed in terms of the unknown magnitude T by resolving the vector $\overrightarrow{DC}$ into rectangular components and writing

$$\overrightarrow{DC} = -(480 \text{ mm})\mathbf{i} + (240 \text{ mm})\mathbf{j} - (160 \text{ mm})\mathbf{k} \qquad DC = 560 \text{ mm}$$

$$\mathbf{T} = T\frac{\overrightarrow{DC}}{DC} = -\tfrac{6}{7}T\mathbf{i} + \tfrac{3}{7}T\mathbf{j} - \tfrac{2}{7}T\mathbf{k}$$

Equilibrium Equations. We express that the forces acting on the pipe cover form a system equivalent to zero:

$$\Sigma\mathbf{F} = 0: \qquad A_x\mathbf{i} + A_y\mathbf{j} + A_z\mathbf{k} + B_x\mathbf{i} + B_y\mathbf{j} + \mathbf{T} - (294 \text{ N})\mathbf{j} = 0$$

$$(A_x + B_x - \tfrac{6}{7}T)\mathbf{i} + (A_y + B_y + \tfrac{3}{7}T - 294 \text{ N})\mathbf{j} + (A_z - \tfrac{2}{7}T)\mathbf{k} = 0 \qquad (1)$$

$$\Sigma\mathbf{M}_B = \Sigma(\mathbf{r} \times \mathbf{F}) = 0:$$

$$2r\mathbf{k} \times (A_x\mathbf{i} + A_y\mathbf{j} + A_z\mathbf{k})$$
$$+ (2r\mathbf{i} + r\mathbf{k}) \times (-\tfrac{6}{7}T\mathbf{i} + \tfrac{3}{7}T\mathbf{j} - \tfrac{2}{7}T\mathbf{k})$$
$$+ (r\mathbf{i} + r\mathbf{k}) \times (-294 \text{ N})\mathbf{j} = 0$$

$$(-2A_y - \tfrac{3}{7}T + 294 \text{ N})r\mathbf{i} + (2A_x - \tfrac{2}{7}T)r\mathbf{j} + (\tfrac{6}{7}T - 294 \text{ N})r\mathbf{k} = 0 \qquad (2)$$

Setting the coefficients of the unit vectors equal to zero in Eq. (2), we write three scalar equations, which yield

$$A_x = +49.0 \text{ N} \qquad A_y = +73.5 \text{ N} \qquad T = 343 \text{ N} \blacktriangleleft$$

Setting the coefficients of the unit vectors equal to zero in Eq. (1), we obtain three more scalar equations. After substituting the values of T, A_x, and A_y into these equations, we obtain

$$A_z = +98.0 \text{ N} \qquad B_x = +245 \text{ N} \qquad B_y = +73.5 \text{ N}$$

The reactions at A and B are therefore

$$\mathbf{A} = +(49.0 \text{ N})\mathbf{i} + (73.5 \text{ N})\mathbf{j} + (98.0 \text{ N})\mathbf{k} \qquad \mathbf{B} = +(245 \text{ N})\mathbf{i} + (73.5 \text{ N})\mathbf{j} \blacktriangleleft$$

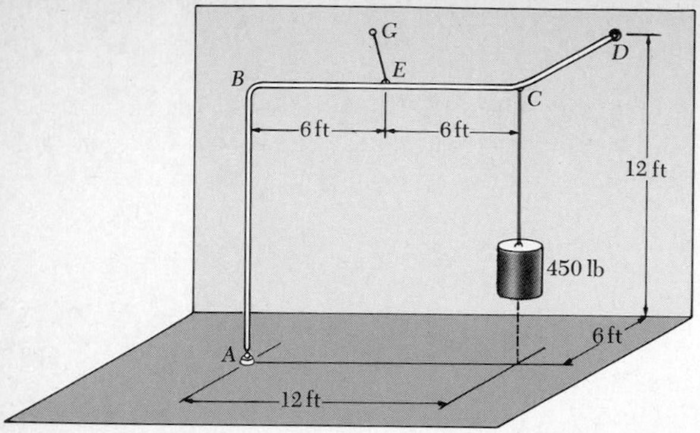

SAMPLE PROBLEM 4.10

A 450-lb load hangs from the corner C of a rigid piece of pipe $ABCD$ which has been bent as shown. The pipe is supported by the ball-and-socket joints A and D fastened, respectively, to the floor and to a vertical wall, and by a cable attached at the midpoint E of the portion BC of the pipe and at a point G on the wall. Determine (*a*) where G should be located if the tension in the cable is to be minimum, (*b*) the corresponding minimum value of the tension.

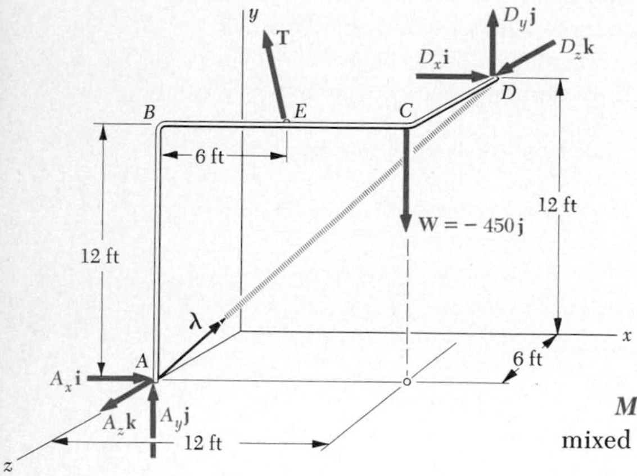

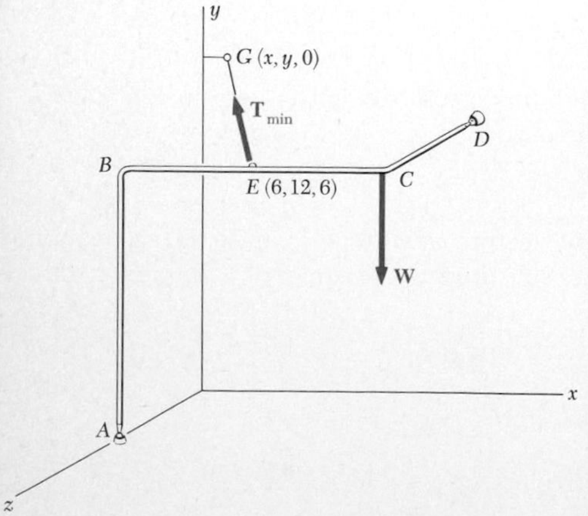

Solution. The free-body diagram of the pipe includes the load $\mathbf{W} = -450\mathbf{j}$, the reactions at A and D, and the force $\mathbf{T}$ exerted by the cable. To eliminate the reactions at A and D from the computations, we express that the sum of the moments of the forces about AD is zero. Denoting by $\boldsymbol{\lambda}$ the unit vector along AD, we write

$$\Sigma M_{AD} = 0: \qquad \boldsymbol{\lambda} \cdot (\overrightarrow{AE} \times \mathbf{T}) + \boldsymbol{\lambda} \cdot (\overrightarrow{AC} \times \mathbf{W}) = 0 \tag{1}$$

The second term in Eq. (1) may be computed as follows:

$$\overrightarrow{AC} \times \mathbf{W} = (12\mathbf{i} + 12\mathbf{j}) \times (-450\mathbf{j}) = -5400\mathbf{k}$$

$$\boldsymbol{\lambda} = \frac{\overrightarrow{AD}}{AD} = \frac{12\mathbf{i} + 12\mathbf{j} - 6\mathbf{k}}{18} = \tfrac{2}{3}\mathbf{i} + \tfrac{2}{3}\mathbf{j} - \tfrac{1}{3}\mathbf{k}$$

$$\boldsymbol{\lambda} \cdot (\overrightarrow{AC} \times \mathbf{W}) = (\tfrac{2}{3}\mathbf{i} + \tfrac{2}{3}\mathbf{j} - \tfrac{1}{3}\mathbf{k}) \cdot (-5400\mathbf{k}) = +1800$$

Substituting the value obtained into Eq. (1), we write

$$\boldsymbol{\lambda} \cdot (\overrightarrow{AE} \times \mathbf{T}) = -1800 \text{ lb} \cdot \text{ft} \tag{2}$$

Minimum Value of Tension. Recalling the commutative property for mixed triple products, we rewrite Eq. (2) in the form

$$\mathbf{T} \cdot (\boldsymbol{\lambda} \times \overrightarrow{AE}) = -1800 \text{ lb} \cdot \text{ft} \tag{3}$$

which shows that the projection of $\mathbf{T}$ on the vector $\boldsymbol{\lambda} \times \overrightarrow{AE}$ is a constant. It follows that $\mathbf{T}$ is minimum when parallel to the vector

$$\boldsymbol{\lambda} \times \overrightarrow{AE} = (\tfrac{2}{3}\mathbf{i} + \tfrac{2}{3}\mathbf{j} - \tfrac{1}{3}\mathbf{k}) \times (6\mathbf{i} + 12\mathbf{j}) = 4\mathbf{i} - 2\mathbf{j} + 4\mathbf{k}$$

The corresponding unit vector being $\tfrac{2}{3}\mathbf{i} - \tfrac{1}{3}\mathbf{j} + \tfrac{2}{3}\mathbf{k}$, we write

$$\mathbf{T}_{\min} = T(\tfrac{2}{3}\mathbf{i} - \tfrac{1}{3}\mathbf{j} + \tfrac{2}{3}\mathbf{k}) \tag{4}$$

Substituting for $\mathbf{T}$ and $\boldsymbol{\lambda} \times \overrightarrow{AE}$ in Eq. (3), we find $T = -300$. Carrying this value into (4), we obtain

$$\mathbf{T}_{\min} = -200\mathbf{i} + 100\mathbf{j} - 200\mathbf{k} \qquad T_{\min} = 300 \text{ lb} \blacktriangleleft$$

Location of G. Since the vector $\overrightarrow{EG}$ and the force $\mathbf{T}_{\min}$ have the same direction, their components must be proportional. Denoting by x, y, 0 the coordinates of G, we write

$$\frac{x-6}{-200} = \frac{y-12}{+100} = \frac{0-6}{-200} \qquad x = 0 \qquad y = 15 \text{ ft} \blacktriangleleft$$

Problems

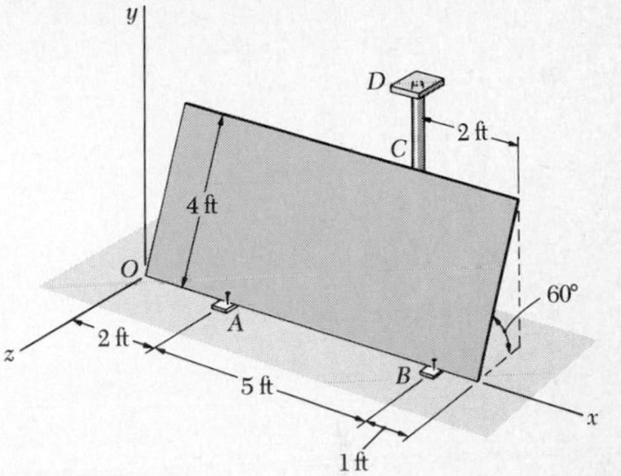

Fig. P4.62

4.62 A 4 × 8 ft sheet of plywood weighing 50 lb has been temporarily propped against column *CD*. It rests at *A* and *B* on small wooden blocks and against protruding nails. Neglecting friction at all surfaces of contact, determine the reactions at *A*, *B*, and *C*.

4.63 Two transmission belts pass over a double-sheaved pulley which is attached to an axle supported by bearings at *A* and *D*. The radius of the inner sheave is 150 mm and the radius of the outer sheave is 300 mm. Knowing that when the system is at rest, the tension is 75 N in both portions of belt *B* and 135 N in both portions of belt *C*, determine the reactions at *A* and *D*. Assume that the bearing at *D* does not exert any axial thrust.

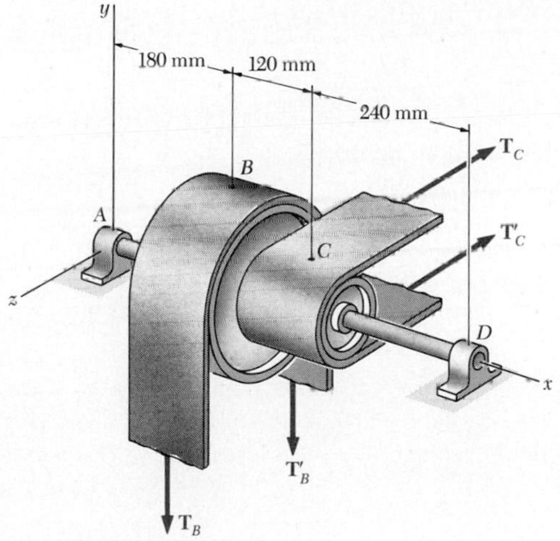

Fig. P4.63

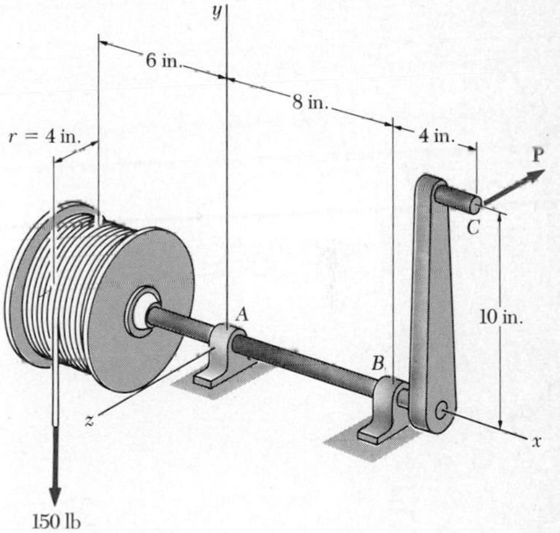

Fig. P4.65

4.64 Solve Prob. 4.63, assuming that the pulley rotates at a constant rate and that $T_B = 87$ N, $T'_B = 69$ N, and $T_C = 162$ N.

4.65 A small winch is used to raise a 150-lb load as shown. Find (*a*) the magnitude of the horizontal force **P** which should be applied at *C* to maintain equilibrium, (*b*) the reactions at *A* and *B*, assuming that the bearing at *B* does not exert any axial thrust.

4.66 A 250-mm lever and a 300-mm-diameter pulley are welded to the axle *BE* which is supported by bearings at *C* and *D*. If a 450-N vertical load is applied at *A* when the lever is horizontal, determine (*a*) the tension in the cord, (*b*) the reactions at *C* and *D*. Assume that the bearing at *D* does not exert any axial thrust.

4.67 Solve Prob. 4.66, assuming that the axle has been rotated clockwise in its bearings by 60°.

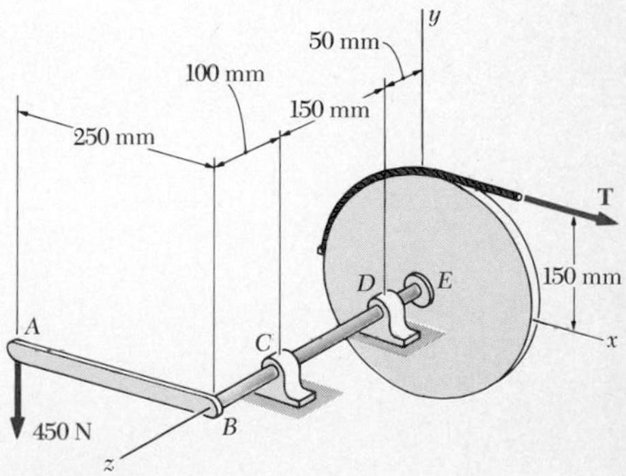

Fig. P4.66

4.68 The 20-kg square plate shown is supported by three wires. Determine the tension in each wire.

4.69 The rectangular plate shown weighs 60 lb and is supported by three wires. Determine the tension in each wire.

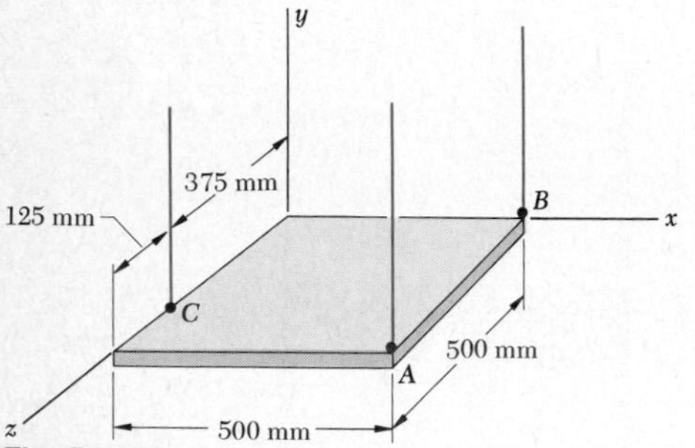

Fig. P4.68

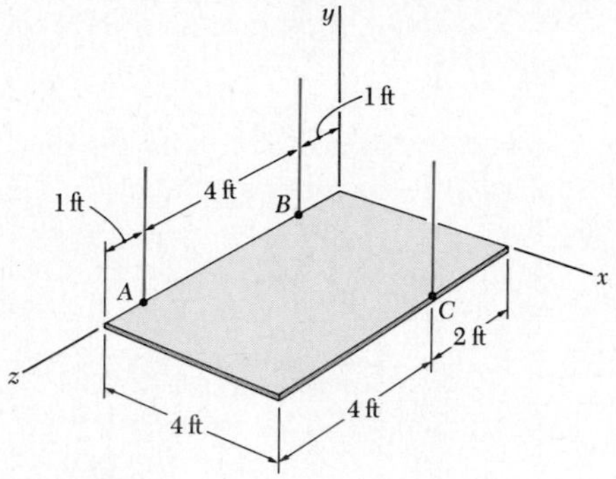

Fig. P4.69

4.70 A load W is to be placed on the 60-lb plate of Prob. 4.69. Determine the magnitude of W and the point where it should be placed if the tension is to be 50 lb in each of the three wires.

4.71 Determine the mass and location of the smallest block which should be placed on the 20-kg plate of Prob. 4.68 if the tensions in the three wires are to be equal.

4.72 The derrick shown supports a 2000-kg crate. It is held by a ball and socket at A and by two cables attached at D and E. Knowing that the derrick stands in the vertical xy plane ($\phi = 0$), determine (*a*) the tension in each cable, (*b*) the reaction at A.

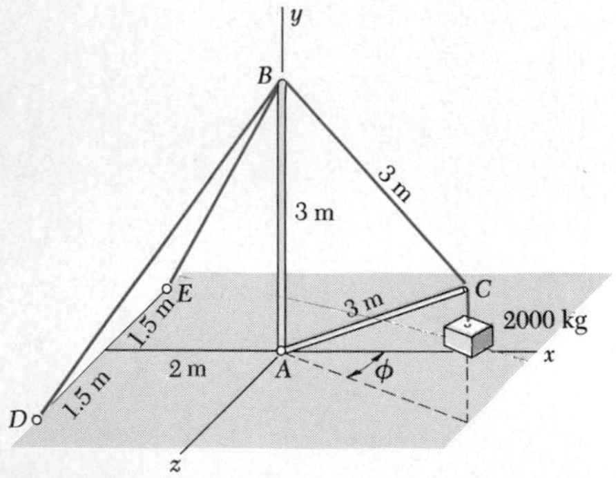

Fig. P4.72

4.73 Solve Prob. 4.72, assuming that the derrick stands in a vertical plane forming an angle $\phi = 20°$ with the xy plane.

4.74 The 12-ft boom AB is acted upon by the 850-lb force shown. Determine (*a*) the tension in each cable, (*b*) the reaction at the ball and socket at A.

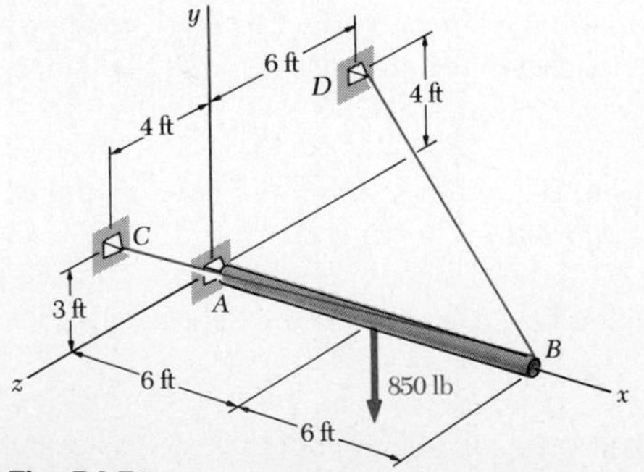

Fig. P4.74

4.75 Solve Prob. 4.74, assuming that the 850-lb load is applied at point *B*.

4.76 A 200-kg crate hangs from a cable which passes over a pulley *B* and is attached to a support at *H*. The 100-kg boom *AB* is supported by a ball and socket at *A* and by two cables *DE* and *DF*. The center of gravity of the boom is located at *G*. Determine (*a*) the tension in cables *DE* and *DF*, (*b*) the reaction at *A*.

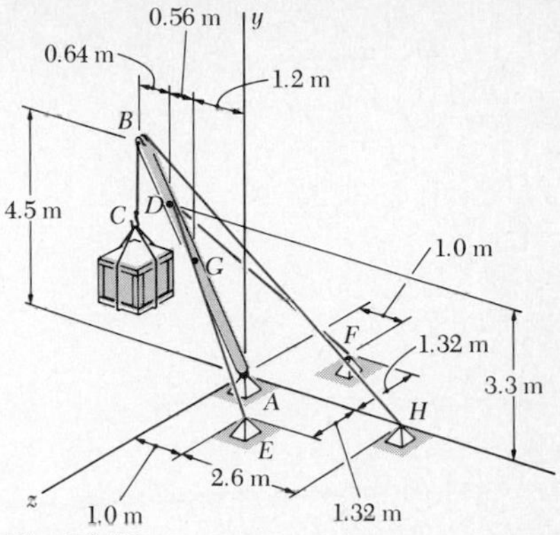

Fig. P4.76

4.77 In Prob. 4.72, determine the angle ϕ for which the tension in *BE* is maximum (*a*) if neither cable may become slack, (*b*) if cables *BE* and *BD* are replaced by rods which can withstand either a tension or a compression force.

4.78 The lid of a roof scuttle weighs 90 lb. It is hinged at the corners *A* and *B* and maintained in the desired position by a rod *CD* pivoted at *C*; a pin at end *D* of the rod fits into one of several holes drilled in the edge of the lid. For $\alpha = 45°$, determine (*a*) the magnitude of the force exerted by rod *CD*, (*b*) the reactions at the hinges. Assume that the hinge at *B* does not exert any axial thrust.

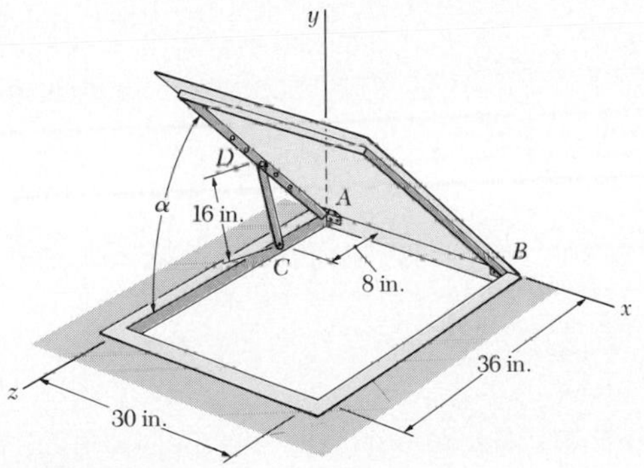

Fig. P4.78

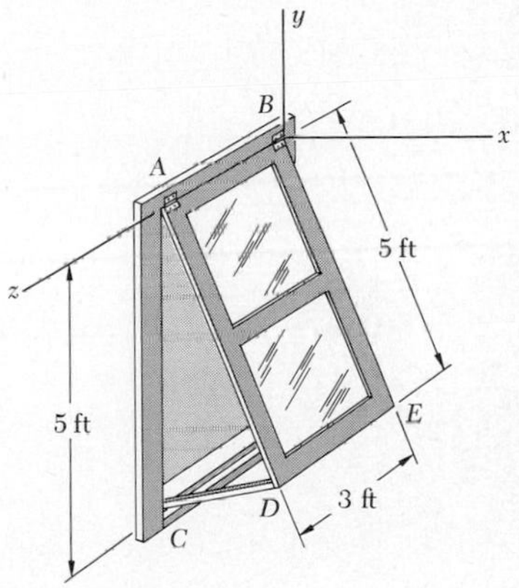

Fig. P4.80

4.79 Solve Prob. 4.78, assuming $\alpha = 60°$.

4.80 A 3×5 ft storm window weighing 15 lb is held by hinges at *A* and *B*. In the position shown, it is held away from the side of the house by a 2-ft stick *CD*. Assuming that the hinge at *A* does not exert any axial thrust, determine (*a*) the magnitude of the force exerted by the stick, (*b*) the reactions at *A* and *B*.

4.81 A 1.2×2.4 m sheet of plywood is temporarily held by nails at *D* and *E* and by two wooden braces nailed at *A*, *B*, and *C*. Wind is blowing on the hidden face of the plywood sheet and it is assumed that its effect may be represented by a force $P\mathbf{k}$ applied at the center of the sheet. Knowing that each brace becomes unsafe with respect to buckling when subjected to a 1.5-kN axial force, determine (*a*) the maximum allowable value of the magnitude *P* of the wind force, (*b*) the corresponding value of the *z* component of the reaction at *E*. Assume that the nails are loose and do not exert any couple.

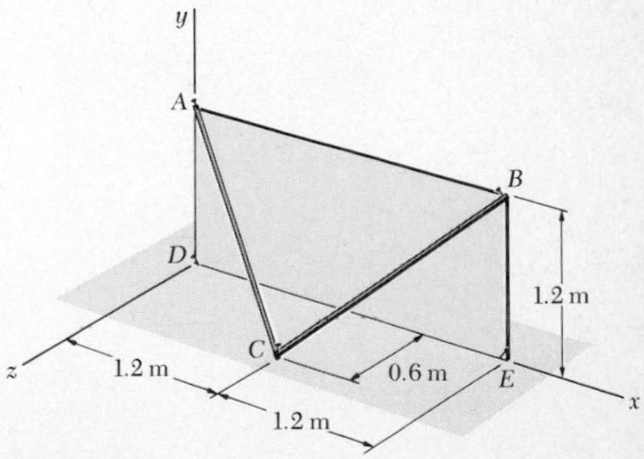

Fig. P4.81

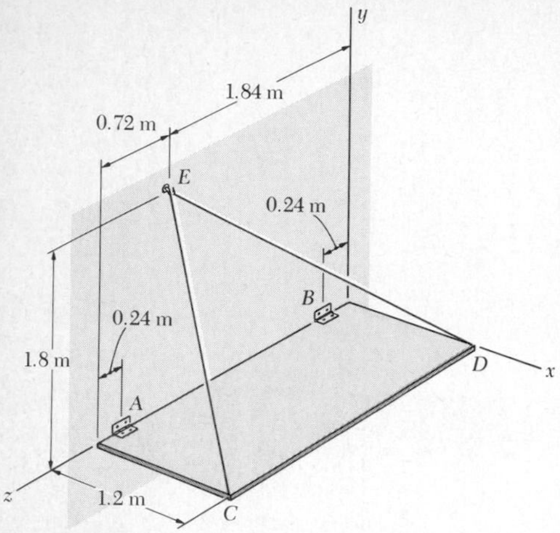

Fig. P4.82

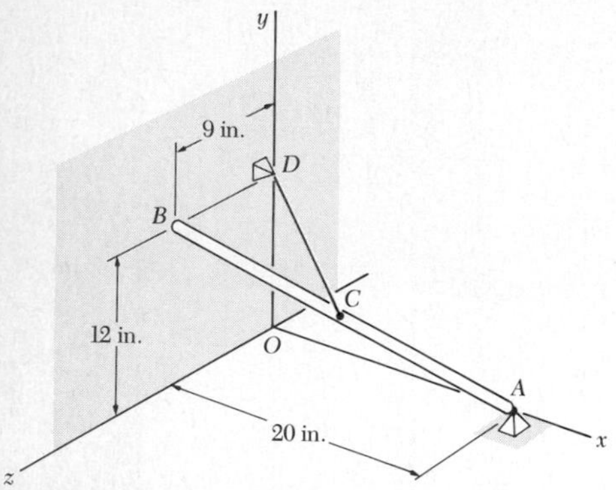

Fig. P4.90

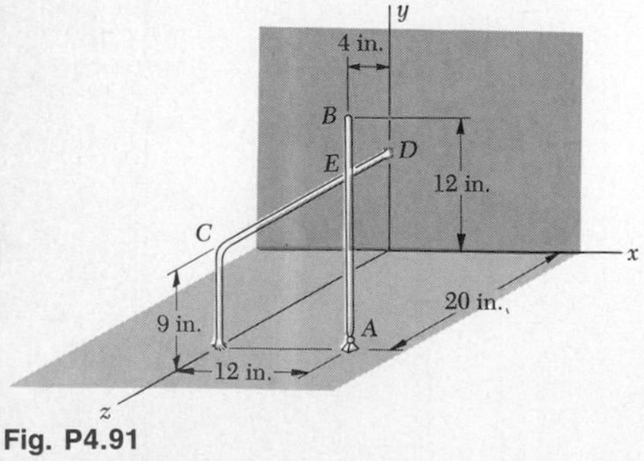

Fig. P4.91

4.82 A 200-kg uniform rectangular platform, 2.56 m long and 1.20 m wide, is supported by hinges at A and B and by a cable attached to the corners C and D of the platform and passing over a frictionless hook E. Assuming that the tension is the same in both parts of the cable and that the hinge at A does not exert any axial thrust, determine (*a*) the tension in the cable, (*b*) the reactions at A and B.

4.83 Solve Prob. 4.82, assuming that cable CED is replaced by a cable attached to corner C and to hook E.

4.84 Solve Prob. 4.82, assuming that cable CED is replaced by a cable attached to corner D and to hook E.

4.85 Solve Prob. 4.82, assuming that cable CED is replaced by two cables, CE and DE, attached, respectively, to corners C and D and tied to hook E, and that the hinge at A is replaced by a roller exerting on the platform a force parallel to the x axis.

4.86 Solve Prob. 4.80, assuming that the hinge at A has been removed and that the hinge at B can exert couples about axes parallel to the x and y axes, respectively.

4.87 Solve Prob. 4.80, assuming that the hinge at B has been removed and that the hinge at A can exert couples about axes parallel to the x and y axes, respectively.

4.88 Solve Prob. 4.82, assuming that the hinge at A has been removed and that the hinge at B can exert couples about axes parallel to the x and y axes, respectively.

4.89 Solve Prob. 4.82, assuming that the hinge at B has been removed and that the hinge at A can exert an axial thrust, as well as couples about axes parallel to the x and y axes, respectively.

4.90 The uniform rod AB weighs 12 lb. It is supported by a ball and socket at A and by the cord CD which is attached to the midpoint C of the rod. Knowing that the rod leans against a frictionless vertical wall at B, determine (*a*) the tension in the cord, (*b*) the reactions at A and B.

4.91 The uniform rod AB weighs 22.5 lb. It is supported by a ball and socket at A and leans against both the rod CD and the vertical wall. Assuming the wall and the rods to be frictionless, determine (*a*) the force which rod CD exerts on AB, (*b*) the reactions at A and B. (*Hint.* The force exerted by CD on AB must be perpendicular to both rods.)

4.92 A 200-kg catamaran is sailing in a straight course and at a constant speed. The effect of the wind on the sail may be represented by the force $\mathbf{F} = (450\text{ N})\mathbf{i} - (600\text{ N})\mathbf{k}$ applied at the point of coordinates $x = -0.48$ m, $y = 3$ m, $z = -0.36$ m. Due to the efforts of the crew, the mast is maintained in a vertical position and the twin hulls are subjected to equal forces, consisting of buoyant forces $\mathbf{B} = B\mathbf{j}$ along the lines of action $x = -0.6$ m, $z = \pm 1.2$ m, drag forces $\mathbf{D} = -D\mathbf{i}$ along the lines of action $y = -0.6$ m, $z = \pm 1.2$ m, and side forces $\mathbf{S} = S\mathbf{k}$ along the common line of action $x = -d$, $y = -0.6$ m. Determine (*a*) the magnitude D of the drag on each of the hulls, (*b*) the magnitude S of each of the side forces, and the distance d defining their line of action.

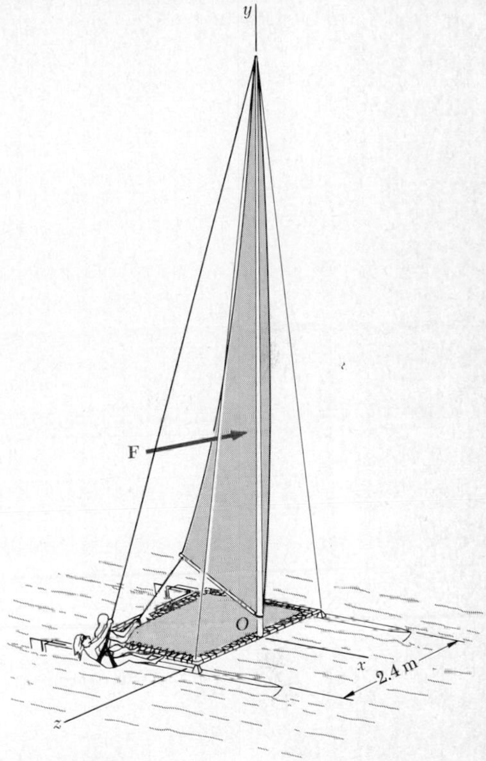

Fig. P4.92

4.93 The catamaran of Prob. 4.92 is manned by two sailors, one of mass 80 kg and the other 65 kg. The center of gravity of the boat is located at $x = -0.9$ m, $z = 0$, and the 80-kg sailor is sitting at $x = -1.5$ m, $z = +1.2$ m, while the 65-kg sailor leans overboard to maintain the mast in a vertical position. Determine the x and z coordinates of the center of gravity of the second sailor.

4.94 In order to clean the clogged drainpipe AE, a plumber has disconnected both ends of the pipe and introduced a power snake through the opening at A. The cutting head of the snake is connected by a heavy cable to an electric motor which keeps it rotating at a constant speed while the plumber forces the cable into the pipe. The forces exerted by the plumber and the motor on the end of the cable may be represented by the wrench $\mathbf{F} = -(6.5\text{ lb})\mathbf{j}$, $\mathbf{M} = -(97.5\text{ lb}\cdot\text{ft})\mathbf{j}$. Determine the additional reactions at B, C, and D caused by the cleaning operation. Assume that the reaction at each support consists of two force components perpendicular to the pipe.

4.95 Solve Prob. 4.94, assuming that the plumber exerts no force on the end of the cable and that the motor exerts a couple $\mathbf{M} = -(97.5\text{ lb}\cdot\text{ft})\mathbf{j}$.

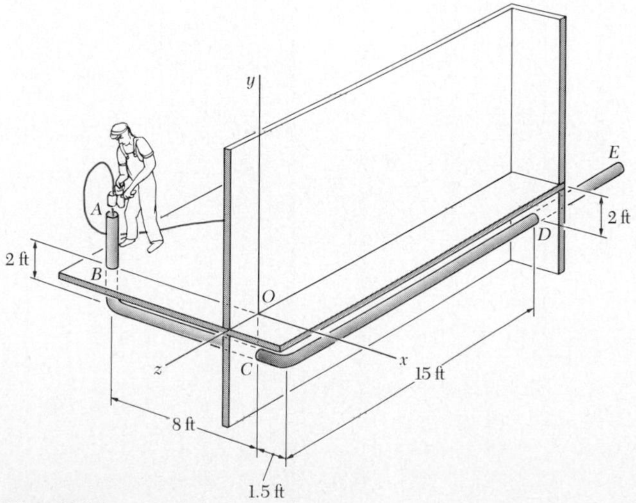

Fig. P4.94

4.96 The rigid L-shaped member ABC is supported by a ball and socket at A and by three cables. If a 2.4-kN load is applied at F, determine the tension in each cable.

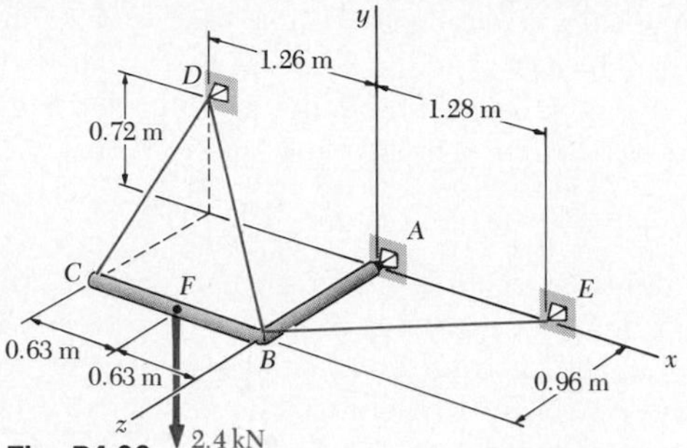

Fig. P4.96

4.97 Solve Prob. 4.96, assuming that the 2.4-kN load is applied at C.

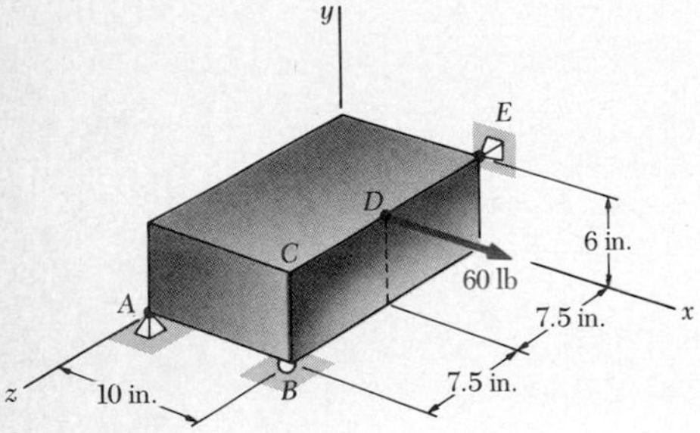

Fig. P4.98

4.98 A rectangular box is supported by balls and sockets at A and E and by a roller on a horizontal surface at B. Determine the reaction at B when a 60-lb horizontal force is applied at the midpoint D of edge CE.

4.99 Solve Prob. 4.98, assuming that the 60-lb force applied at D is directed vertically downward.

4.100 The bent rod ABC is hinged to a vertical wall by means of two brackets and bears at C against another vertical wall. Neglecting friction, determine the reaction at C when a 150-N load is applied at D as shown.

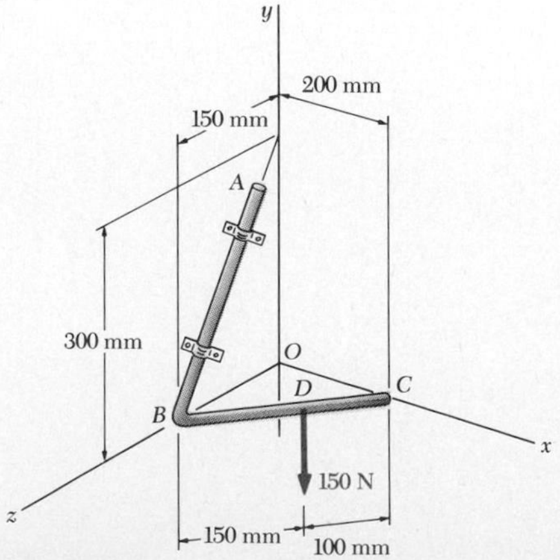

Fig. P4.100

4.101 The frame ACD is supported by ball-and-socket joints at A and D and by a cable which passes through a ring at B and is attached to hooks at G and H. Knowing that the frame supports at point C a load of magnitude $P = 268$ N, determine the tension in the cable.

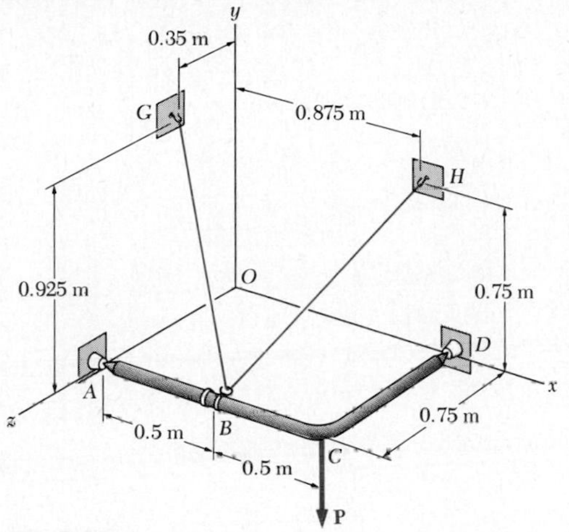

Fig. P4.101

4.102 Solve Prob. 4.101, assuming that cable GBH is replaced by a cable GB attached at G and B.

4.103 Solve Prob. 4.101, assuming that cable GBH is replaced by a cable BH attached at B and H.

Review Problems

4.104 A movable bracket is held at rest by a cable attached at C and by frictionless rollers at A and B. For the loading shown, determine (*a*) the tension in the cable, (*b*) the reactions at A and B.

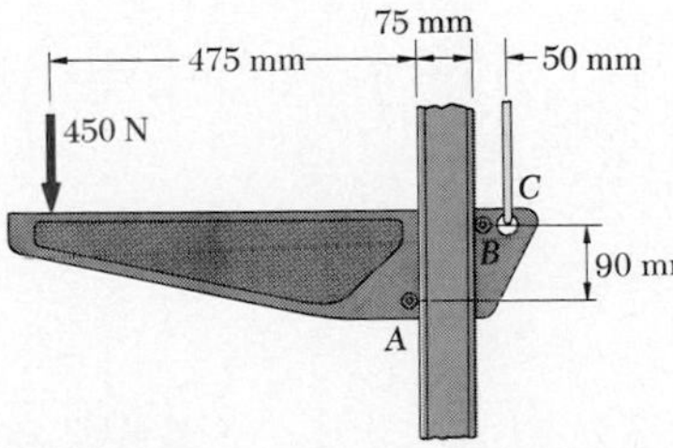

Fig. P4.104

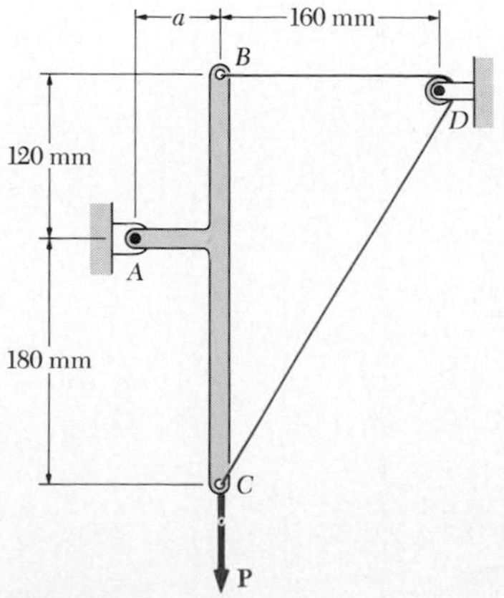

Fig. P4.105

4.105 A force **P** of magnitude 450 N is applied to member ABC which is supported by a frictionless pin at A and by the cable BDC. Since the cable passes over a pulley at D, the tension may be assumed to be the same in the portions BD and CD of the cable. Determine (*a*) the tension in the cable, (*b*) the reaction at A when $a = 60$ mm.

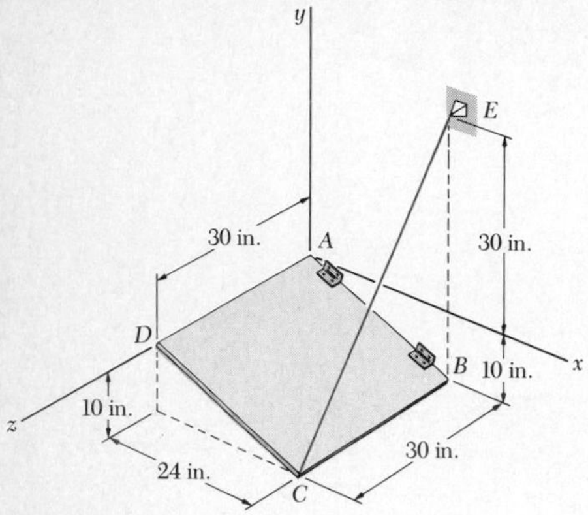

Fig. P4.106

4.106 The 26 × 30 in. plate *ABCD* is supported by hinges along edge *AB* and by the wire *CE*. Knowing that the plate is uniform and weighs 100 lb, determine the tension in the wire.

4.107 Solve Prob. 4.106, assuming that wire *CE* is replaced by a wire connecting *E* and *D*.

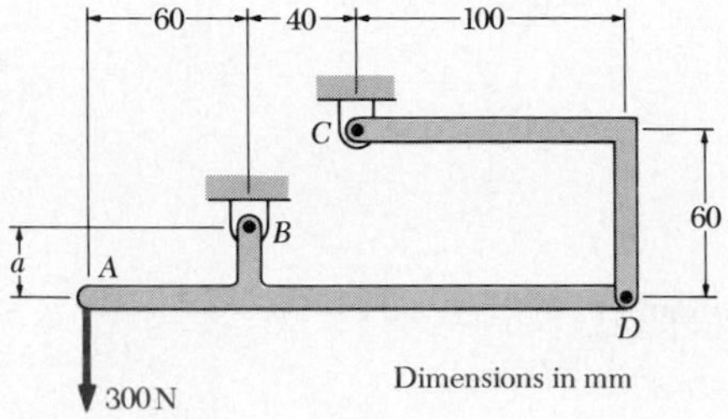

Fig. P4.108

4.108 Determine the reactions at *B* and *C* when $a = 25$ mm.

4.109 A light rod *AD* supports a vertical load **P** and is attached to collars *B* and *C* which may slide freely on the rods shown. Knowing that the wire attached at *A* forms an angle $\alpha = 30°$ with the horizontal, determine (*a*) the tension in the wire, (*b*) the reactions at *B* and *C*.

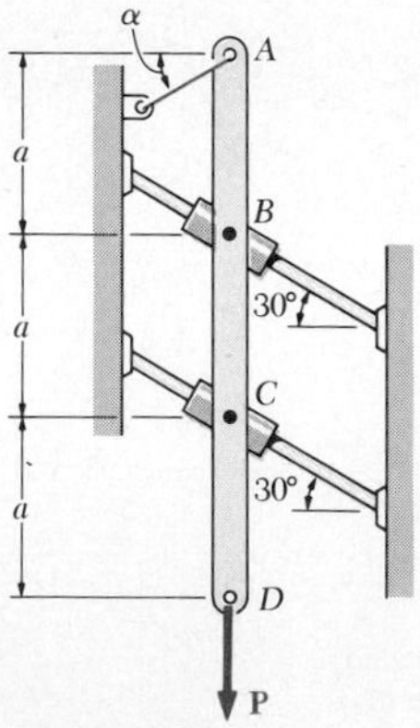

Fig. P4.109

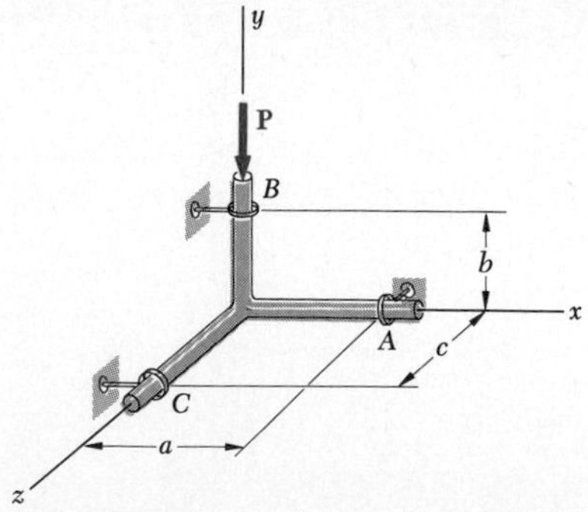

Fig. P4.110

4.110 Three rods are welded together to form the "corner" shown. The corner is supported by three frictionless eyebolts. Determine the reactions at *A*, *B*, and *C* when $P = 750$ N, $a = 250$ mm, $b = 150$ mm, and $c = 200$ mm.

4.111 Solve Prob. 4.110, assuming that the force **P** is removed and is replaced by the couple $\mathbf{M} = +(240\ \text{N}\cdot\text{m})\mathbf{j}$ acting at *B*.

4.112 A force **P** of magnitude 60 lb is applied to end E of cable CDE which passes under pulley D and is attached to the mechanism at C. Neglecting the weight of the mechanism and the radius of the pulley, determine the value of θ corresponding to equilibrium. The constant of the spring is $k = 20$ lb/in., and the spring is unstretched when $\theta = 90°$.

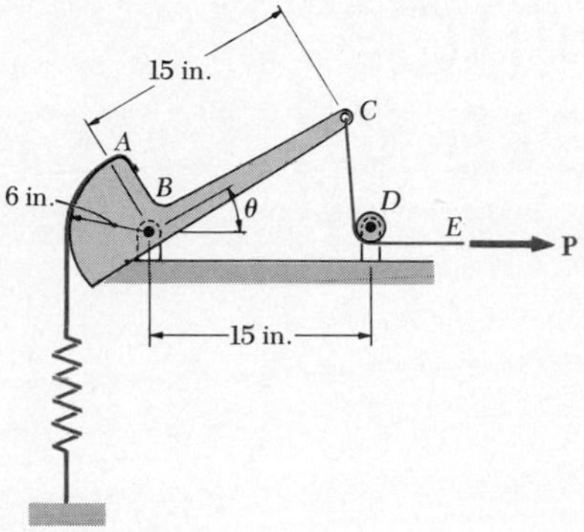

Fig. P4.112

4.113 A force **P** is applied to a bent rod AD which may be supported in four different ways as shown. In each case determine the reactions at the supports.

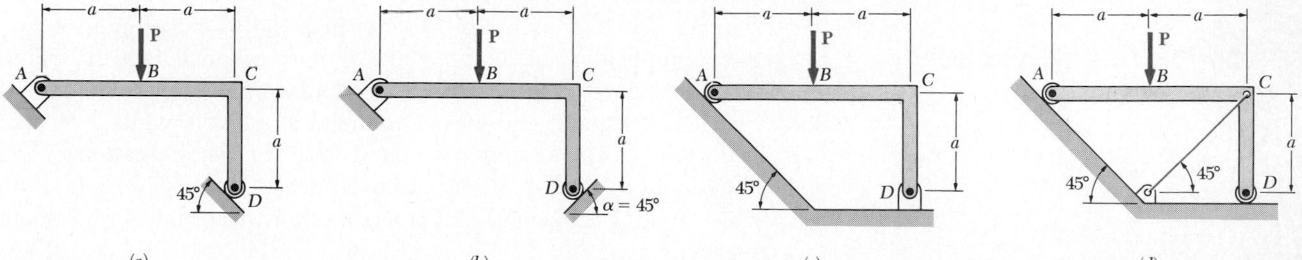

Fig. P4.113

4.114 Determine the reactions at A and B for the truss and loading shown.

***4.115** In the problems listed below, the rigid bodies considered were completely constrained, and the reactions were statically determinate. For each of these rigid bodies it is possible to create an improper set of constraints by changing either a dimension of the body or the direction of a reaction. In each problem determine the value of α or of a which results in improper constraints. (a) Prob. 4.105, (b) Prob. 4.108, (c) Prob. 4.109, (d) Prob. 4.113b, (e) Prob. 4.114.

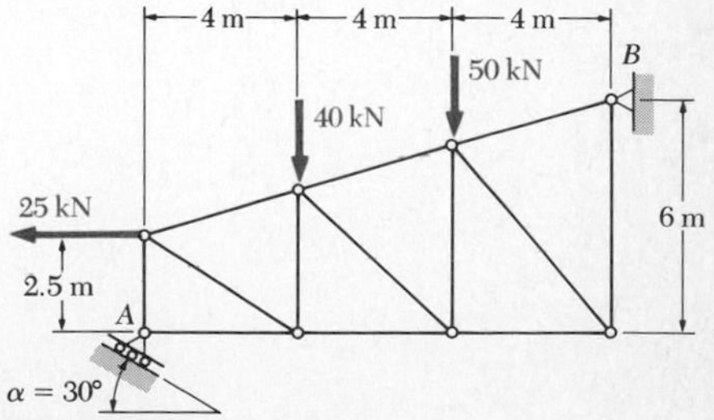

Fig. P4.114

CHAPTER FIVE

Distributed Forces: Centroids and Centers of Gravity

5.1. Introduction. We have assumed so far that the attraction exerted by the earth on a rigid body could be represented by a single force **W**. This force, called the force of gravity, or the weight of the body, was to be applied at the *center of gravity* of the body (Sec. 3.2). Actually, the earth exerts a force on each of the particles forming the body. The action of the earth on a rigid body should thus be represented by a large number of small forces distributed over the entire body. We shall see in this chapter, however, that all these small forces may be replaced by a single equivalent force **W**. We shall also learn to determine the center of gravity, i.e., the point of application of the resultant **W**, for various shapes of bodies.

In the first part of the chapter, we shall consider two-dimensional bodies, such as flat plates and wires contained in a given plane. We shall introduce two concepts closely associated with the determination of the center of gravity of a plate or wire, namely, the concept of *centroid* of an area or line, and the concept of *first moment* of an area or line with respect to a given axis.

We shall also find that the computation of the area of a surface of revolution or of the volume of a body of revolution is directly related to the determination of the centroid of the line or area used to generate that surface or body of revolution (Theorems of Pappus-Guldinus). And, as we shall see in Secs. 5.8 and 5.9, the determination of the centroid of an area simplifies the analysis of beams subjected to distributed loads and the computation of the forces exerted on submerged rectangular surfaces such as hydraulic gates and portions of dams.

In the last part of the chapter, we shall learn to determine the center of gravity of a three-dimensional body, as well as the centroid of a volume and the first moments of that volume with respect to the coordinate planes.

5.2. Center of Gravity of a Two-dimensional Body. Let us first consider a flat horizontal plate (Fig. 5.1). We may divide the plate into n small elements. The coordinates of the first element are denoted by x_1 and y_1, those of the second element by x_2 and y_2, etc. The forces exerted by the earth on the elements of plate will be denoted, respectively, by $\Delta\mathbf{W}_1$, $\Delta\mathbf{W}_2, \ldots, \Delta\mathbf{W}_n$. These forces or weights are directed toward the center of the earth; however, for all practical purposes they may be assumed parallel. Their resultant is therefore a single force in the same direction. The magnitude W of this force is obtained by adding the magnitudes of the elementary weights,

$$\Sigma F_z: \qquad W = \Delta W_1 + \Delta W_2 + \cdots + \Delta W_n$$

To obtain the coordinates $\bar{x}$ and $\bar{y}$ of the point G where the resultant $\mathbf{W}$ should be applied, we write that the moments of $\mathbf{W}$ about the y and x axes are equal to the sum of the corresponding moments of the elementary weights,

$$\begin{aligned} \Sigma M_y: &\qquad \bar{x}W = x_1\,\Delta W_1 + x_2\,\Delta W_2 + \cdots + x_n\,\Delta W_n \\ \Sigma M_x: &\qquad \bar{y}W = y_1\,\Delta W_1 + y_2\,\Delta W_2 + \cdots + y_n\,\Delta W_n \end{aligned} \qquad (5.1)$$

If we now increase the number of elements into which the plate is divided and simultaneously decrease the size of each element, we obtain at the limit the following expressions:

$$W = \int dW \qquad \bar{x}W = \int x\,dW \qquad \bar{y}W = \int y\,dW \qquad (5.2)$$

These equations define the weight $\mathbf{W}$ and the coordinates $\bar{x}$ and $\bar{y}$ of the center of gravity G of a flat plate. The same equations may be derived for a wire lying in the xy plane (Fig. 5.2). We shall observe, in the latter case, that the center of gravity G will generally not be located on the wire.

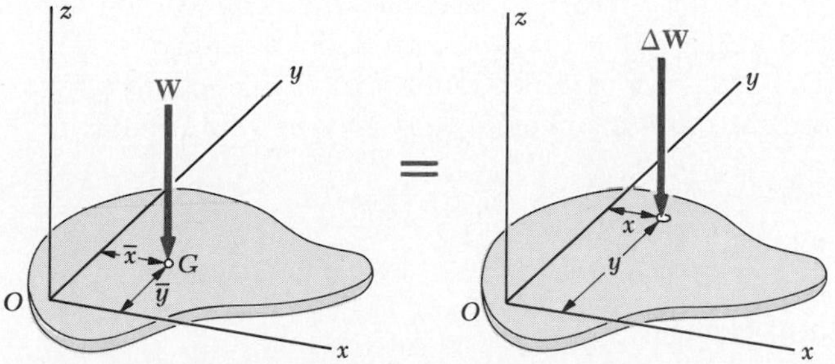

ΣM_y: $\bar{x}W = \Sigma x\,\Delta W$
ΣM_x: $\bar{y}W = \Sigma y\,\Delta W$

Fig. 5.1 Center of gravity of a plate.

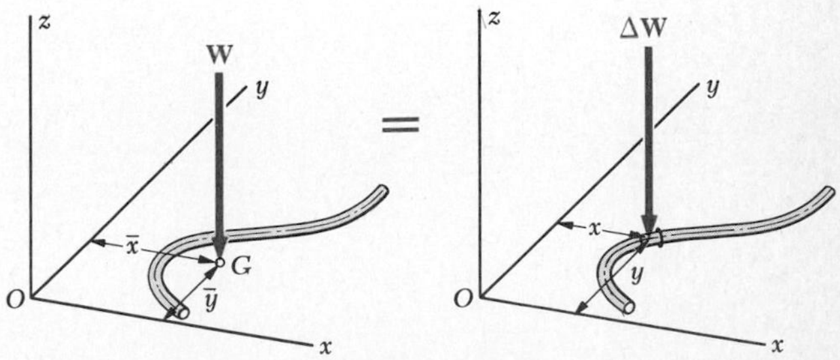

ΣM_y: $\bar{x}W = \Sigma x\,\Delta W$
ΣM_x: $\bar{y}W = \Sigma y\,\Delta W$

Fig. 5.2 Center of gravity of a wire.

5.3. Centroids of Areas and Lines. In the case of a homogeneous plate of uniform thickness, the magnitude ΔW of the weight of an element of plate may be expressed as

$$\Delta W = \gamma t\, \Delta A$$

where γ = specific weight (weight per unit volume) of material
t = thickness of plate
ΔA = area of element

Similarly, we may express the magnitude W of the weight of the entire plate in the form

$$W = \gamma t A$$

where A is the total area of the plate.

If U.S. customary units are used, the specific weight γ should be expressed in lb/ft^3, the thickness t in feet, and the areas ΔA and A in square feet. We check that ΔW and W will then be expressed in pounds. If SI units are used, γ should be expressed in N/m^3, t in meters, and the areas ΔA and A in square meters; the weights ΔW and W will then be expressed in newtons.†

Substituting for ΔW and W in the moment equations (5.1) and dividing throughout by γt, we write

$$\Sigma M_y: \quad \bar{x}A = x_1\,\Delta A_1 + x_2\,\Delta A_2 + \cdots + x_n\,\Delta A_n$$
$$\Sigma M_x: \quad \bar{y}A = y_1\,\Delta A_1 + y_2\,\Delta A_2 + \cdots + y_n\,\Delta A_n$$

If we increase the number of elements into which the area A is divided and simultaneously decrease the size of each element, we obtain at the limit

$$\bar{x}A = \int x\,dA \qquad \bar{y}A = \int y\,dA \tag{5.3}$$

These equations define the coordinates $\bar{x}$ and $\bar{y}$ of the center of gravity of a homogeneous plate. The point of coordinates $\bar{x}$ and $\bar{y}$ is also known as the *centroid C of the area A* of the plate (Fig. 5.3). If the plate is not homogeneous, the equations cannot be used to determine the center of gravity of the plate; they still define, however, the centroid of the area.

In the case of a homogeneous wire of uniform cross section, the magnitude ΔW of the weight of an element of wire may be expressed as

$$\Delta W = \gamma a\, \Delta L$$

where γ = specific weight of material
a = cross-sectional area of wire
ΔL = length of element

† It should be noted that in the SI system of units a given material is generally characterized by its density ρ (mass per unit volume) rather than by its specific weight γ. The specific weight of the material can then be obtained by writing

$$\gamma = \rho g$$

where $g = 9.81\ \text{m/s}^2$. Since ρ is expressed in kg/m^3, we check that γ will be expressed in $(\text{kg/m}^3)(\text{m/s}^2)$, that is, in N/m^3.

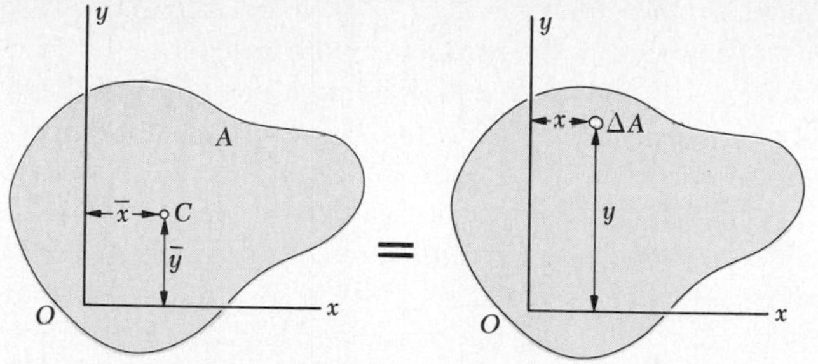

ΣM_y: $\bar{x}A = \Sigma x\,\Delta A$
ΣM_x: $\bar{y}A = \Sigma y\,\Delta A$

Fig. 5.3 Centroid of an area.

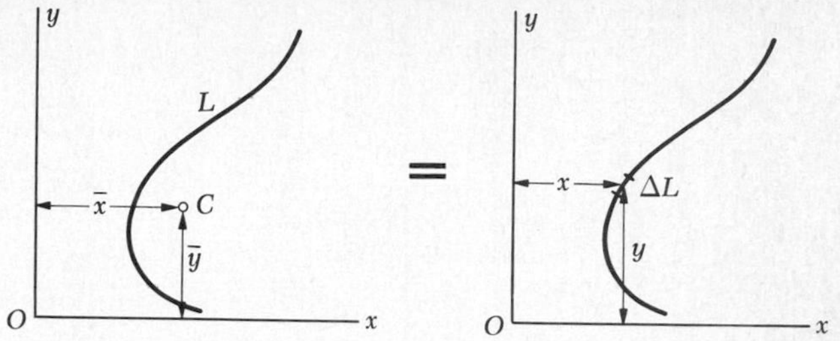

ΣM_y: $\bar{x}L = \Sigma x\,\Delta L$
ΣM_x: $\bar{y}L = \Sigma y\,\Delta L$

Fig. 5.4 Centroid of a line.

The center of gravity of the wire then coincides with the *centroid C of the line L* defining the shape of the wire (Fig. 5.4). The coordinates $\bar{x}$ and $\bar{y}$ of the centroid of the line L are obtained from the equations

$$\bar{x}L = \int x\,dL \qquad \bar{y}L = \int y\,dL \tag{5.4}$$

5.4. First Moments of Areas and Lines. The integral $\int x\,dA$ in Eqs. (5.3) of the preceding section is known as the *first moment of the area A with respect to the y axis* and is denoted by Q_y. Similarly, the integral $\int y\,dA$ defines the *first moment of A with respect to the x axis* and is denoted by Q_x. We write

$$Q_y = \int x\,dA \qquad Q_x = \int y\,dA \tag{5.5}$$

Comparing Eqs. (5.3) with Eqs. (5.5), we note that the first moments of the area A may be expressed as the products of the area and of the coordinates of its centroid:

$$Q_y = \bar{x}A \qquad Q_x = \bar{y}A \tag{5.6}$$

It follows from Eqs. (5.6) that the coordinates of the centroid of an area may be obtained by dividing the first moments of that area by the area itself. The computation of the first moments of an area is also useful in the determination of shearing stresses under transverse loadings in mechanics of materials. Finally, we observe from Eqs. (5.6) that if the centroid of an area is located on a coordinate axis, the first moment of the area with respect to that axis is zero, and conversely.

Relations similar to Eqs. (5.5) and (5.6) may be used to define the first moments of a line with respect to the coordinate axes and to express these moments as the products of the length L of the line and of the coordinates $\bar{x}$ and $\bar{y}$ of its centroid.

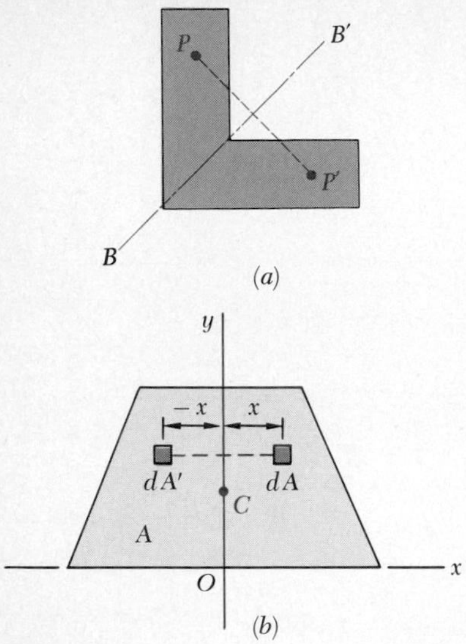

Fig. 5.5

An area A is said to be *symmetric with respect to an axis BB′* if to every point P of the area corresponds a point P' of the same area such that the line PP' is perpendicular to BB' and is divided into two equal parts by that axis (Fig. 5.5*a*). A line L is said to be symmetric with respect to BB' if it satisfies similar conditions. When an area A or a line L possesses an axis of symmetry BB', its first moment with respect to BB' is zero and its centroid is located on that axis. Indeed, considering the area A of Fig. 5.5*b*, which is symmetric with respect to the y axis, we observe that to every element of area dA of abscissa x corresponds an element of area dA' of abscissa $-x$. It follows that the integral in the first of Eqs. (5.5) is zero and, thus, that $Q_y = 0$. It also follows from the first of the relations (5.3) that $\bar{x} = 0$. Thus, if an area A or a line L possesses an axis of symmetry, its centroid C is located on that axis.

We further note that if an area or line possesses two axes of symmetry, its centroid C must be located at the intersection of the two axes of symmetry (Fig. 5.6). This property enables us to determine immediately the centroid of areas such as circles, ellipses, squares, rectangles, equilateral triangles, or other symmetric figures, as well as the centroid of lines in the shape of the circumference of a circle, the perimeter of a square, etc.

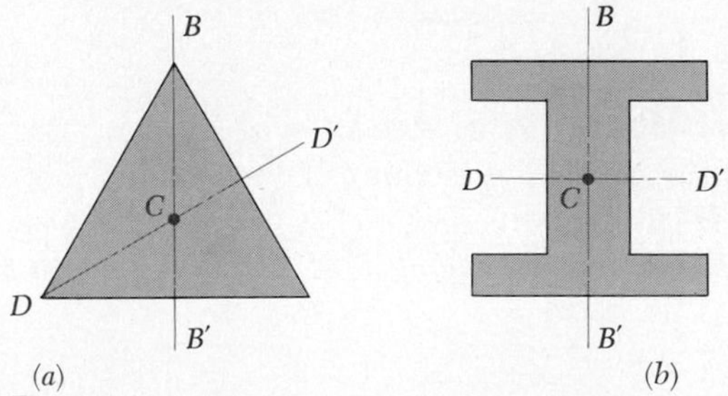

Fig. 5.6

An area A is said to be *symmetric with respect to a center O* if to every element of area dA of coordinates x and y corresponds an element of area dA' of coordinates $-x$ and $-y$ (Fig. 5.7). It then follows that the integrals in Eqs. (5.5) are both zero, and that $Q_x = Q_y = 0$. It also follows from Eqs. (5.3) that $\bar{x} = \bar{y} = 0$, that is, the centroid of the area coincides with its center of symmetry O. It may be shown in a similar way that the centroid of a line symmetric with respect to a center O coincides with O.

It should be noted that a figure possessing a center of symmetry does not necessarily possess an axis of symmetry (Fig. 5.7), while a figure possessing two axes of symmetry does not necessarily possess a center of symmetry (Fig. 5.6*a*). However, if a figure possesses two axes of symmetry at a right angle to each other, the point of intersection of these axes will be a center of symmetry (Fig. 5.6*b*).

Centroids of unsymmetrical areas and lines and of areas and lines possessing only one axis of symmetry will be determined by the methods of Secs. 5.6 and 5.7. Centroids of common shapes of areas and lines are shown in Fig. 5.8*A* and *B*. The formulas defining these centroids will be derived in the Sample Problems and Problems following Secs. 5.6 and 5.7.

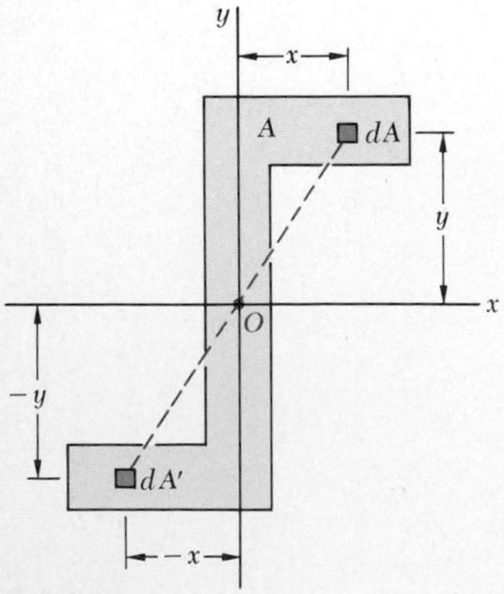

Fig. 5.7

Shape		$\bar{x}$	$\bar{y}$	Area
Triangular area	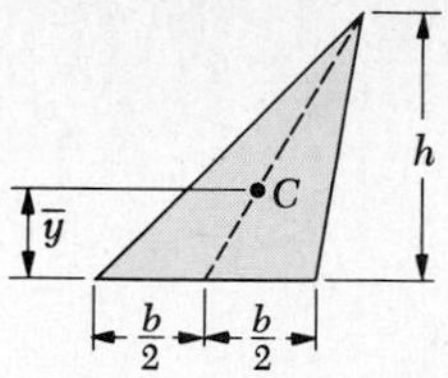		$\dfrac{h}{3}$	$\dfrac{bh}{2}$
Quarter-circular area	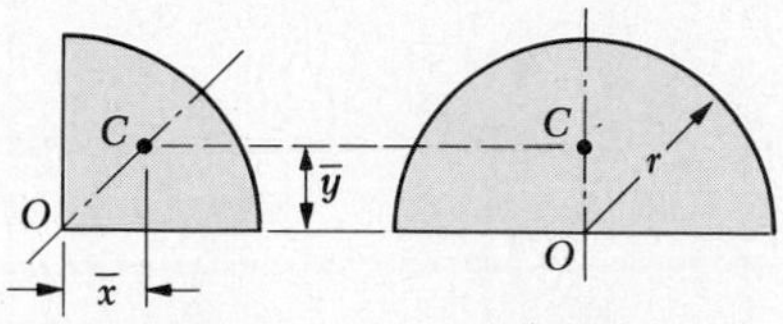	$\dfrac{4r}{3\pi}$	$\dfrac{4r}{3\pi}$	$\dfrac{\pi r^2}{4}$
Semicircular area		0	$\dfrac{4r}{3\pi}$	$\dfrac{\pi r^2}{2}$
Quarter-elliptical area	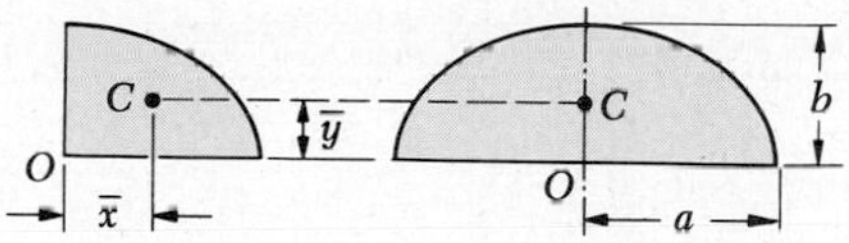	$\dfrac{4a}{3\pi}$	$\dfrac{4b}{3\pi}$	$\dfrac{\pi ab}{4}$
Semielliptical area		0	$\dfrac{4b}{3\pi}$	$\dfrac{\pi ab}{2}$
Semiparabolic area	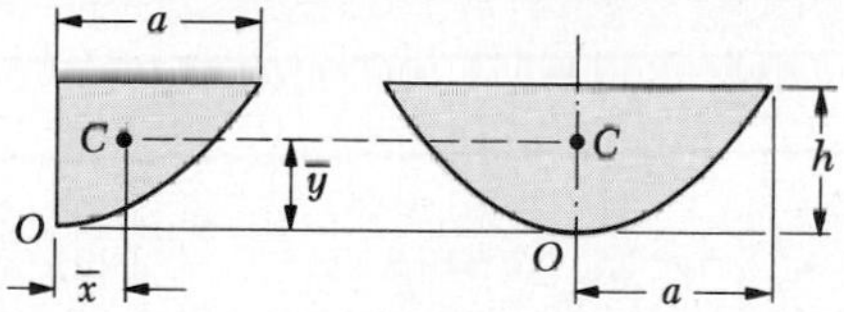	$\dfrac{3a}{8}$	$\dfrac{3h}{5}$	$\dfrac{2ah}{3}$
Parabolic area		0	$\dfrac{3h}{5}$	$\dfrac{4ah}{3}$
Parabolic spandrel	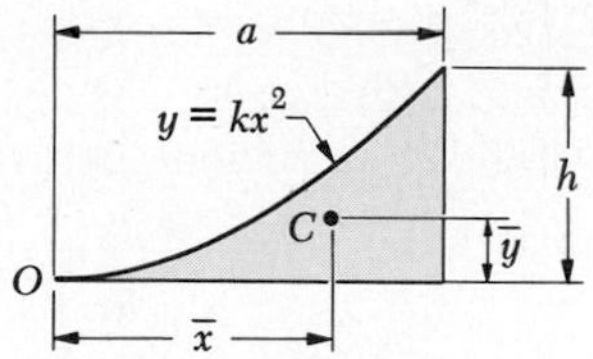	$\dfrac{3a}{4}$	$\dfrac{3h}{10}$	$\dfrac{ah}{3}$
General spandrel	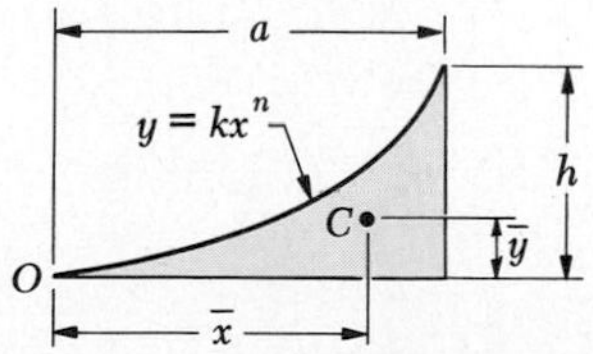	$\dfrac{n+1}{n+2}a$	$\dfrac{n+1}{4n+2}h$	$\dfrac{ah}{n+1}$
Circular sector	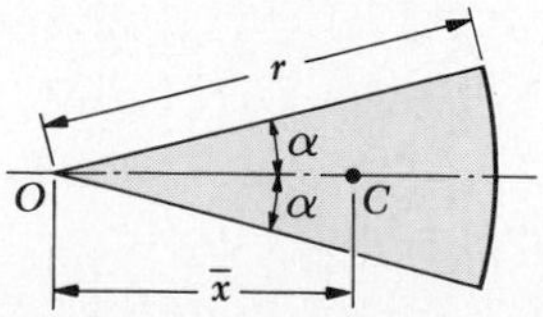	$\dfrac{2r \sin \alpha}{3\alpha}$	0	αr^2

Fig. 5.8A Centroids of common shapes of areas.

Shape		$\bar{x}$	$\bar{y}$	Length
Quarter-circular arc	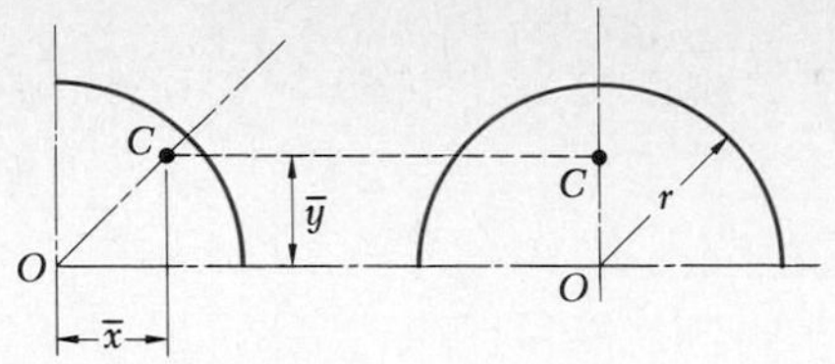	$\frac{2r}{\pi}$	$\frac{2r}{\pi}$	$\frac{\pi r}{2}$
Semicircular arc		0	$\frac{2r}{\pi}$	πr
Arc of circle	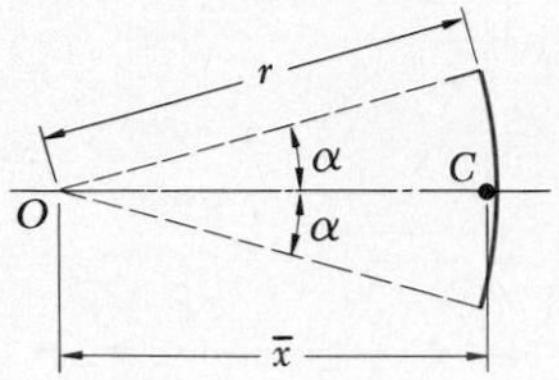	$\frac{r \sin \alpha}{\alpha}$	0	$2\alpha r$

Fig. 5.8B Centroids of common shapes of lines.

5.5. Composite Plates and Wires. In many instances, a flat plate may be divided into rectangles, triangles, or other common shapes shown in Fig. 5.8*A*. The abscissa $\bar{X}$ of its center of gravity G may be determined from the abscissas $\bar{x}_1, \bar{x}_2, \ldots$ of the centers of gravity of the various parts by expressing that the moment of the weight of the whole plate about the y axis is equal to the sum of the moments of the weights of the various parts about the same axis (Fig. 5.9). The ordinate $\bar{Y}$ of the center of gravity of the plate is found in a similar way by equating moments about the x axis. We write

$$\Sigma M_y: \quad \bar{X}(W_1 + W_2 + \cdots + W_n) = \bar{x}_1 W_1 + \bar{x}_2 W_2 + \cdots + \bar{x}_n W_n$$
$$\Sigma M_x: \quad \bar{Y}(W_1 + W_2 + \cdots + W_n) = \bar{y}_1 W_1 + \bar{y}_2 W_2 + \cdots + \bar{y}_n W_n$$

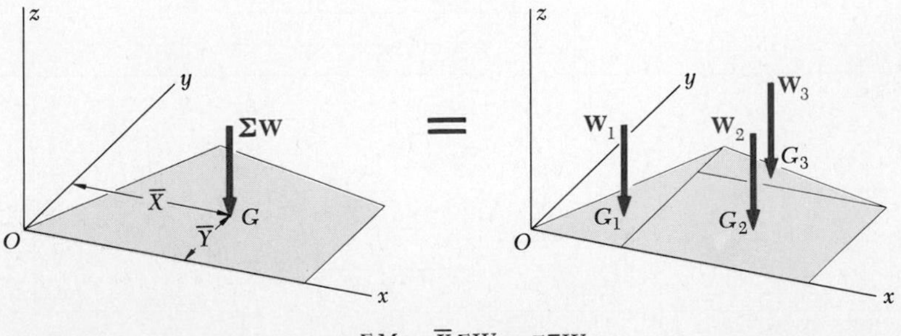

$$\Sigma M_y: \quad \bar{X}\Sigma W = \Sigma \bar{x} W$$
$$\Sigma M_x: \quad \bar{Y}\Sigma W = \Sigma \bar{y} W$$

Fig. 5.9 Center of gravity of a composite plate.

or, for short,

$$\bar{X}\Sigma W = \Sigma\bar{x}W \qquad \bar{Y}\Sigma W = \Sigma\bar{y}W \tag{5.7}$$

These equations may be solved for the coordinates $\bar{X}$ and $\bar{Y}$ of the center of gravity of the plate.

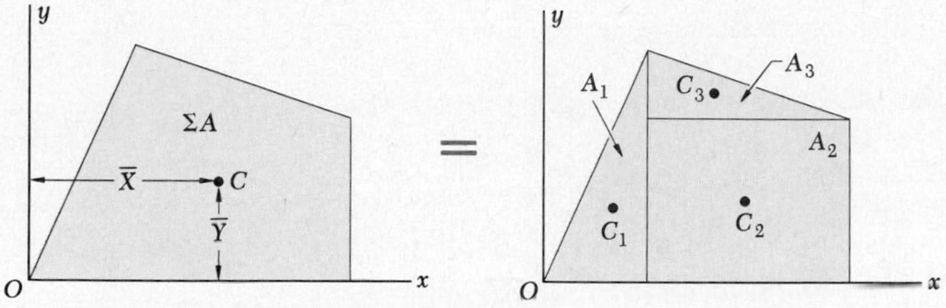

$$Q_y = \bar{X}\Sigma A = \Sigma\bar{x}A$$
$$Q_x = \bar{Y}\Sigma A = \Sigma\bar{y}A$$

Fig. 5.10 Centroid of a composite area.

If the plate is homogeneous and of uniform thickness, the center of gravity coincides with the centroid C of its area. The abscissa $\bar{X}$ of the centroid of the area may be determined by noting that the first moment Q_y of the composite area with respect to the y axis may be expressed both as the product of $\bar{X}$ and the total area and as the sum of the first moments of the elementary areas with respect to the y axis (Fig. 5.10). The ordinate $\bar{Y}$ of the centroid is found in a similar way by considering the first moment Q_x of the composite area. We have

$$Q_y = \bar{X}(A_1 + A_2 + \cdots + A_n) = \bar{x}_1A_1 + \bar{x}_2A_2 + \cdots + \bar{x}_nA_n$$
$$Q_x = \bar{Y}(A_1 + A_2 + \cdots + A_n) = \bar{y}_1A_1 + \bar{y}_2A_2 + \cdots + \bar{y}_nA_n$$

or, for short,

$$Q_y = \bar{X}\Sigma A = \Sigma\bar{x}A \qquad Q_x = \bar{Y}\Sigma A = \Sigma\bar{y}A \tag{5.8}$$

These equations yield the first moments of the composite area, or they may be solved for the coordinates $\bar{X}$ and $\bar{Y}$ of its centroid.

Care should be taken to record the moment of each area with the appropriate sign. First moments of areas, just like moments of forces, may be positive or negative. For example, an area whose centroid is located to the left of the y axis will have a negative first moment with respect to that axis. Also, the area of a hole should be recorded with a negative sign (Fig. 5.11).

Similarly, it is possible in many cases to determine the center of gravity of a composite wire or the centroid of a composite line by dividing the wire or line into simpler elements (see Sample Prob. 5.2).

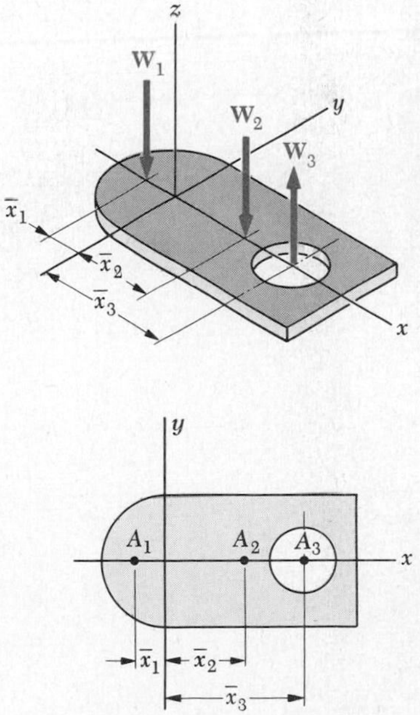

	$\bar{x}$	A	$\bar{x}A$
A_1 Semicircle	−	+	−
A_2 Full rectangle	+	+	+
A_3 Circular hole	+	−	−

Fig. 5.11

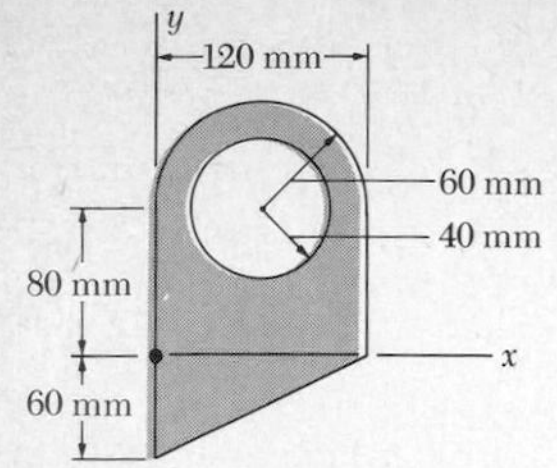

SAMPLE PROBLEM 5.1

For the plane area shown, determine (*a*) the first moments with respect to the *x* and *y* axes, (*b*) the location of the centroid.

Solution. The area is obtained by adding a rectangle, a triangle, and a semicircle and subtracting a circle. Using the coordinate axes shown, the area and the coordinates of the centroid of each of the component areas are determined and entered in the table below. The area of the circle is indicated as negative, since it is to be subtracted from the other areas. We note that the coordinate $\bar{y}$ of the centroid of the triangle is negative for the axes shown. The first moments of the component areas with respect to the coordinate axes are computed and entered in the table.

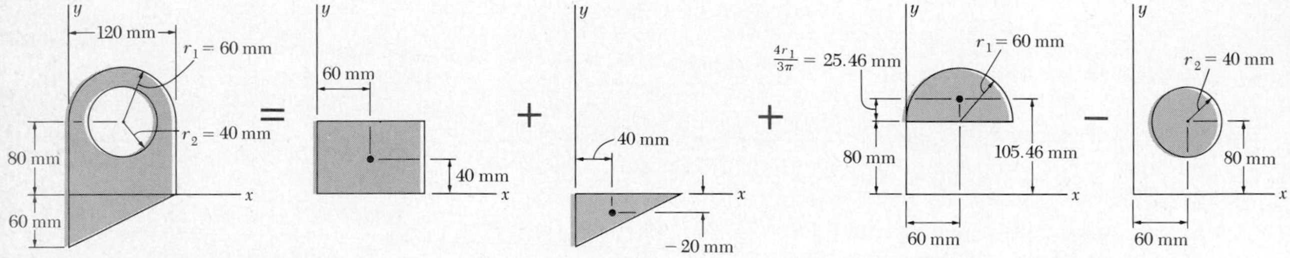

Component	A, mm^2	$\bar{x}$, mm	$\bar{y}$, mm	$\bar{x}A$, mm^3	$\bar{y}A$, mm^3
Rectangle	$(120)(80) = 9.6 \times 10^3$	60	40	$+576 \times 10^3$	$+384 \times 10^3$
Triangle	$\frac{1}{2}(120)(60) = 3.6 \times 10^3$	40	-20	$+144 \times 10^3$	-72×10^3
Semicircle	$\frac{1}{2}\pi(60)^2 = 5.655 \times 10^3$	60	105.46	$+339.3 \times 10^3$	$+596.4 \times 10^3$
Circle	$-\pi(40)^2 = -5.027 \times 10^3$	60	80	-301.6×10^3	-402.2×10^3
	$\Sigma A = 13.828 \times 10^3$			$\Sigma \bar{x}A = +757.7 \times 10^3$	$\Sigma \bar{y}A = +506.2 \times 10^3$

a. First Moments of Area. Using Eqs. (5.8), we write

$$Q_x = \Sigma \bar{y}A = 506.2 \times 10^3 \text{ mm}^3$$

$$Q_x = 506 \times 10^3 \text{ mm}^3 \quad ◄$$

$$Q_y = \Sigma \bar{x}A = 757.7 \times 10^3 \text{ mm}^3$$

$$Q_y = 758 \times 10^3 \text{ mm}^3 \quad ◄$$

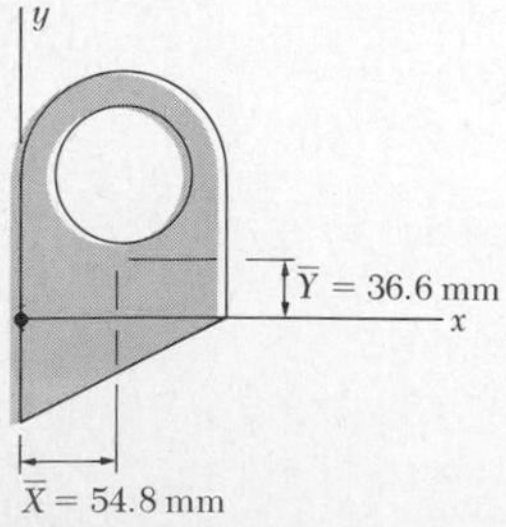

b. Location of Centroid. Substituting the values obtained from the table into the equations defining the centroid of a composite area, we obtain

$$\bar{X}\Sigma A = \Sigma \bar{x}A: \qquad \bar{X}(13.828 \times 10^3 \text{ mm}^2) = 757.7 \times 10^3 \text{ mm}^3$$

$$\bar{X} = 54.8 \text{ mm} \quad ◄$$

$$\bar{Y}\Sigma A = \Sigma \bar{y}A: \qquad \bar{Y}(13.828 \times 10^3 \text{ mm}^2) = 506.2 \times 10^3 \text{ mm}^3$$

$$\bar{Y} = 36.6 \text{ mm} \quad ◄$$

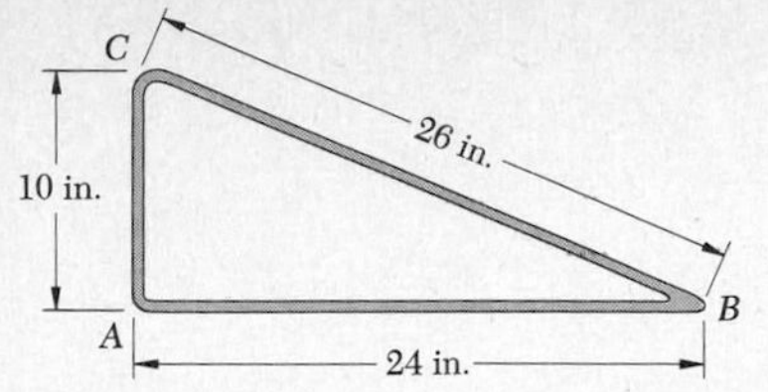

SAMPLE PROBLEM 5.2

The figure shown is made of a thin homogeneous wire. Determine its center of gravity.

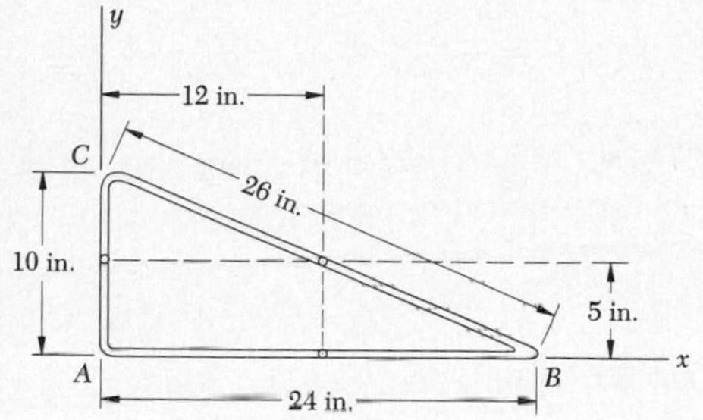

Solution. Since the figure is formed of homogeneous wire, its center of gravity may be located by determining the centroid of the corresponding line. Choosing the coordinate axes shown, with origin at A, we determine the coordinates of the centroid of each segment of line and compute its first moments with respect to the coordinate axes.

Segment	L, in.	$\bar{x}$, in.	$\bar{y}$, in.	$\bar{x}L$, in^2	$\bar{y}L$, in^2
AB	24	12	0	288	0
BC	26	12	5	312	130
CA	10	0	5	0	50
	$\Sigma L = 60$	. . .	. . .	$\Sigma \bar{x}L = 600$	$\Sigma \bar{y}L = 180$

Substituting the values obtained from the table into the equations defining the centroid of a composite line, we obtain

$\bar{X}\Sigma L = \Sigma \bar{x}L$: $\quad \bar{X}(60 \text{ in.}) = 600 \text{ in}^2 \qquad \bar{X} = 10 \text{ in.}$ ◄

$\bar{Y}\Sigma L = \Sigma \bar{y}L$: $\quad \bar{Y}(60 \text{ in.}) = 180 \text{ in}^2 \qquad \bar{Y} = 3 \text{ in.}$ ◄

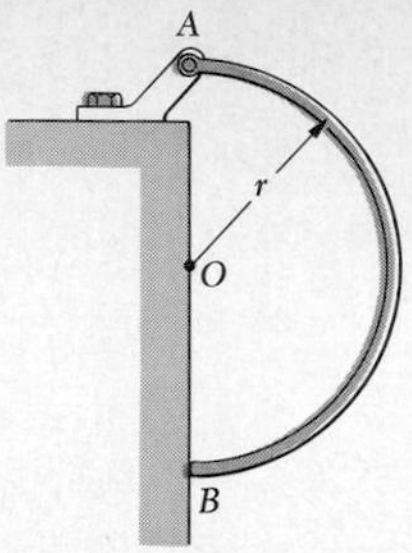

SAMPLE PROBLEM 5.3

A uniform semicircular rod of weight W and radius r is attached to a pin at A and bears against a frictionless surface at B. Determine the reactions at A and B.

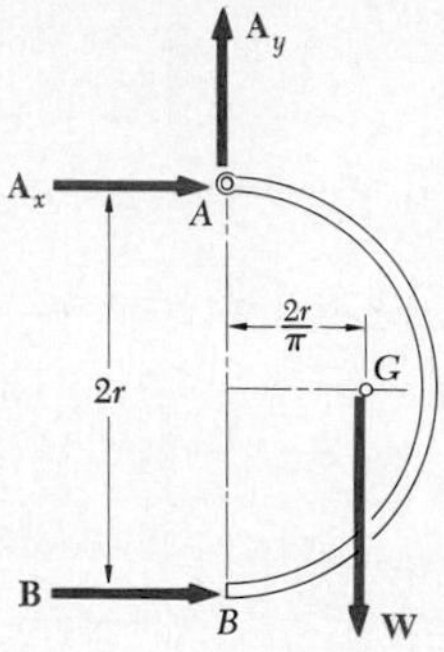

Solution. A free-body diagram of the rod is drawn. The forces acting on the rod consist of its weight **W** applied at the center of gravity G, whose position is obtained from Fig. 5.8B, of a reaction at A represented by its components $\mathbf{A}_x$ and $\mathbf{A}_y$, and of a horizontal reaction at B.

$$+\circlearrowleft \Sigma M_A = 0: \qquad B(2r) - W\left(\frac{2r}{\pi}\right) = 0$$

$$B = +\frac{W}{\pi} \qquad\qquad \mathbf{B} = \frac{W}{\pi} \rightarrow \blacktriangleleft$$

$$\xrightarrow{+} \Sigma F_x = 0: \qquad A_x + B = 0$$

$$A_x = -B = -\frac{W}{\pi} \qquad\qquad \mathbf{A}_x = \frac{W}{\pi} \leftarrow$$

$$+\uparrow \Sigma F_y = 0: \qquad A_y - W = 0 \qquad\qquad A_y = W\uparrow$$

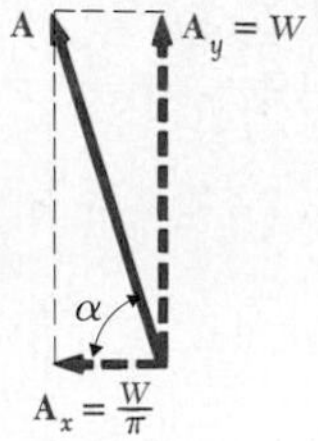

Adding the two components of the reaction at A:

$$A = \left[W^2 + \left(\frac{W}{\pi}\right)^2\right]^{1/2} \qquad\qquad A = W\left(1 + \frac{1}{\pi^2}\right)^{1/2} \blacktriangleleft$$

$$\tan\alpha = \frac{W}{W/\pi} = \pi \qquad\qquad \alpha = \tan^{-1}\pi \blacktriangleleft$$

The answers may also be expressed as follows:

$$\mathbf{A} = 1.049W \measuredangle 72.3^\circ \qquad \mathbf{B} = 0.318W\rightarrow \blacktriangleleft$$

Problems

5.1 through 5.10 Locate the centroid of the plane area shown.

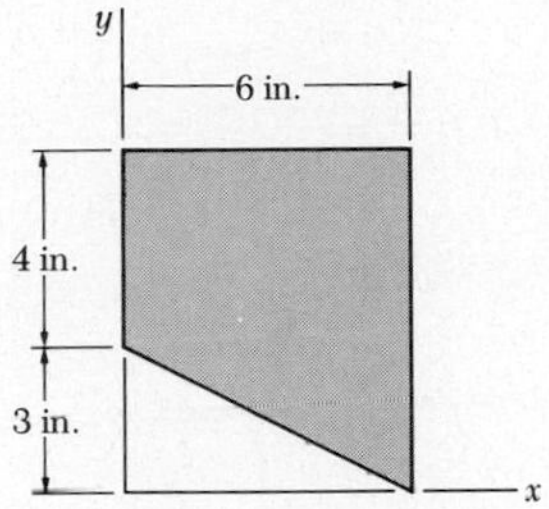

Fig. P5.1

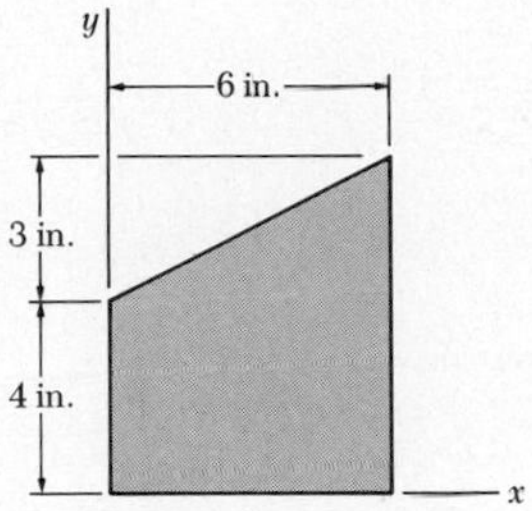

Fig. P5.2

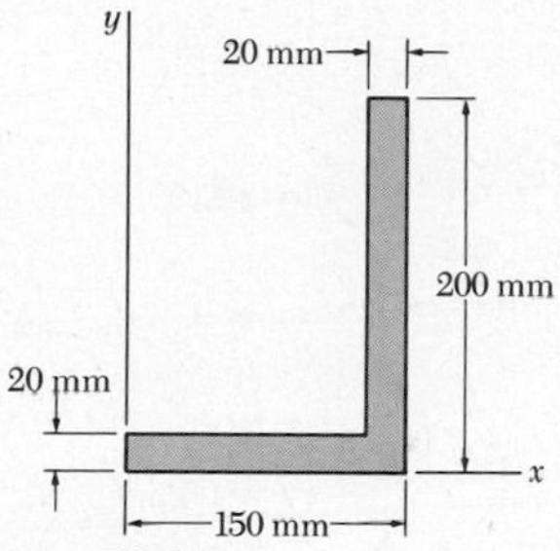

Fig. P5.3

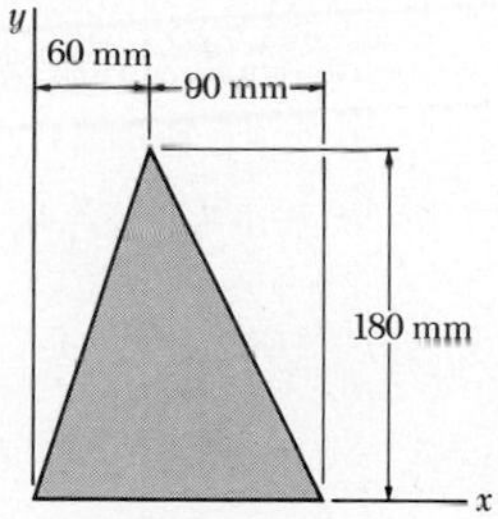

Fig. P5.4

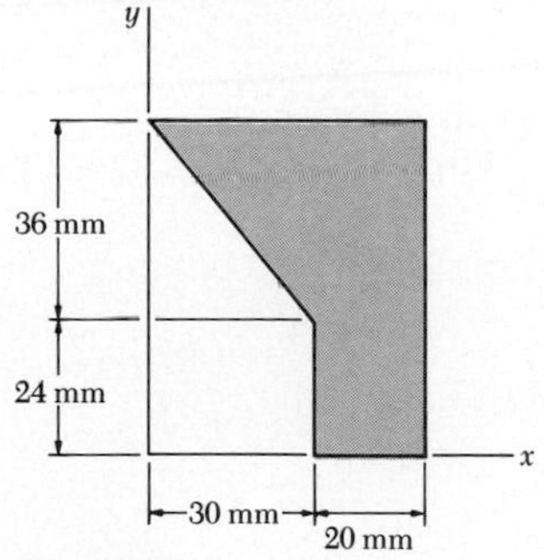

Fig. P5.5

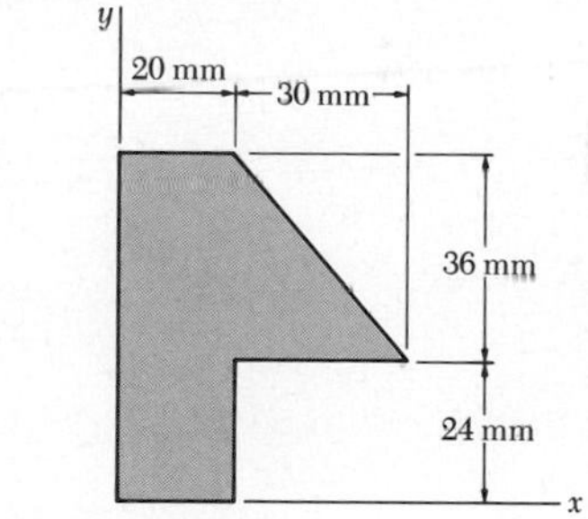

Fig. 5.6

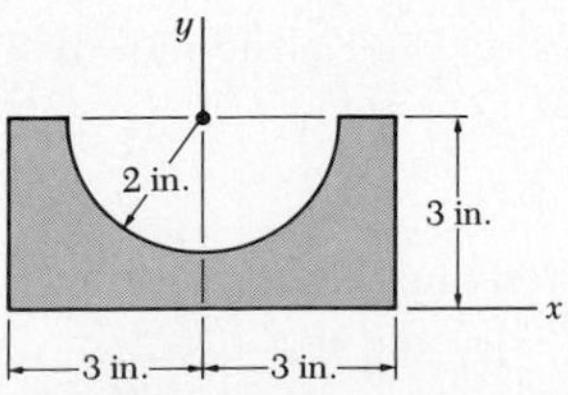

Fig. P5.7

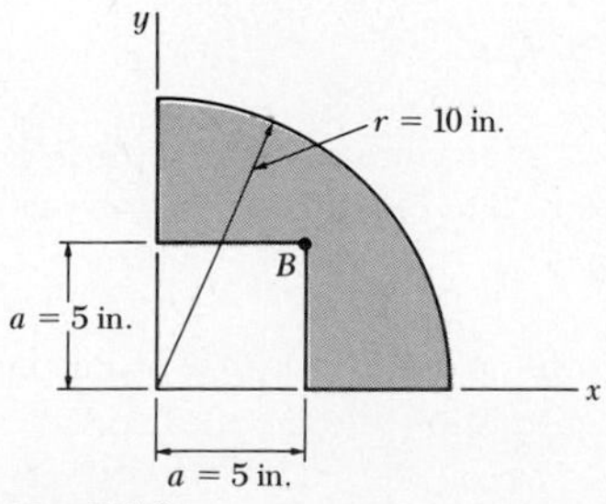

Fig. P5.8

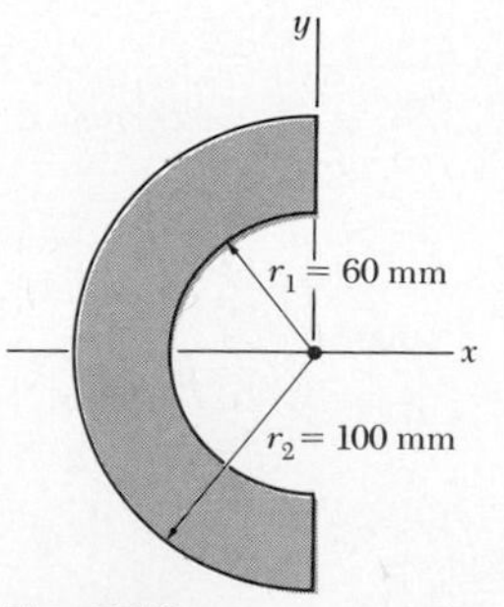

Fig. P5.9

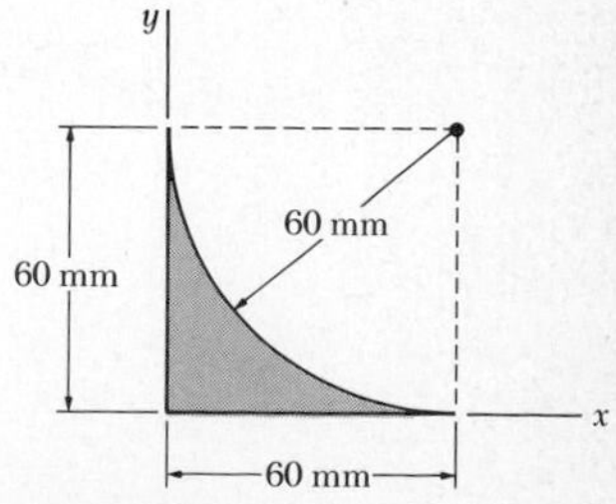

Fig. P5.10

5.11 through 5.14 Locate the centroid of the plane area shown.

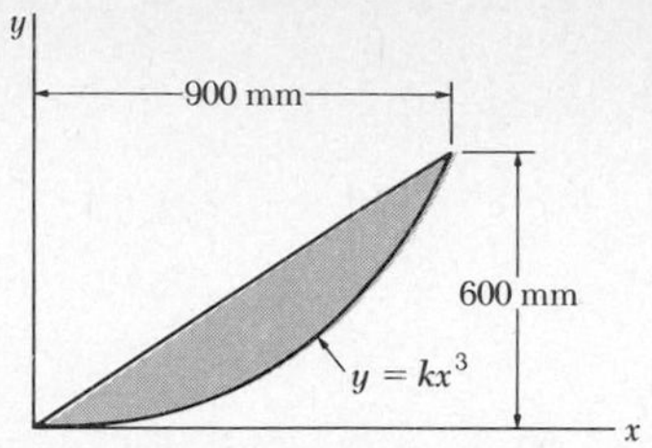

Fig. P5.11

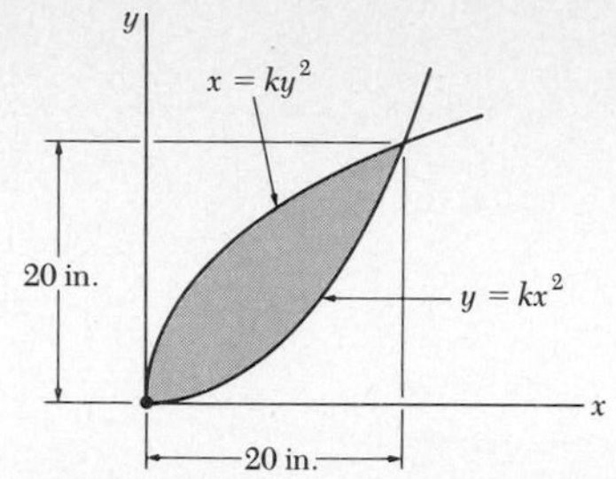

Fig. P5.12

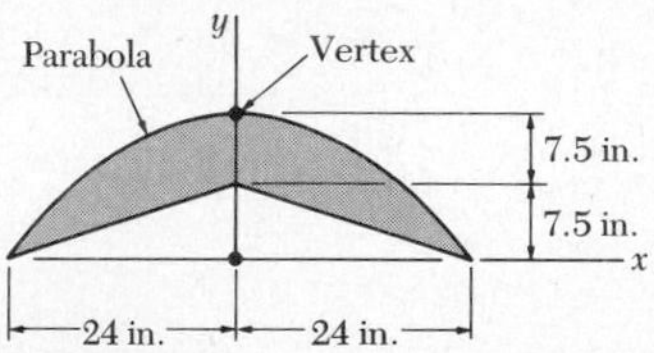

Fig. P5.13

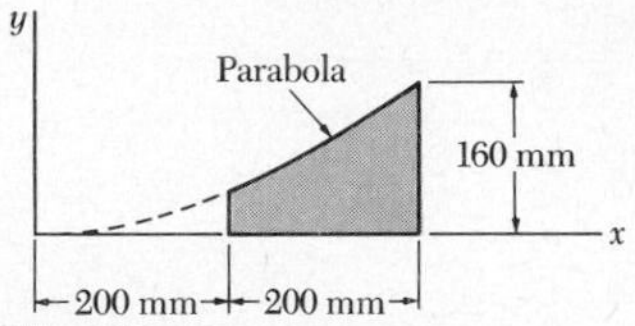

Fig. P5.14

5.15 Determine the abscissa of the centroid of the circular segment in terms of r and α.

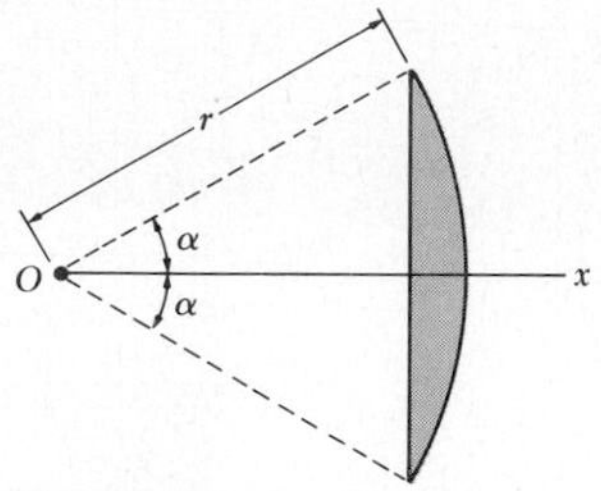

Fig. P5.15

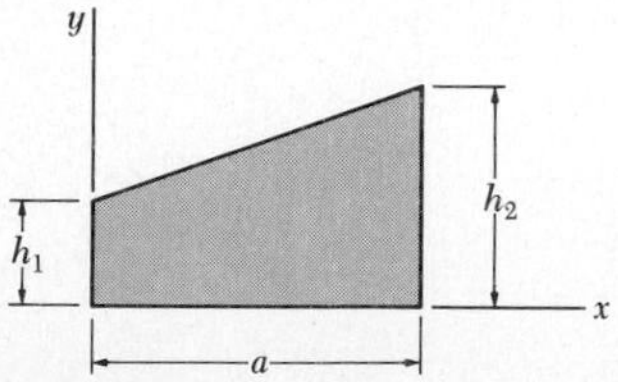

Fig. P5.16

5.16 Determine the abscissa of the centroid of the trapezoid shown in terms of h_1, h_2, and a.

5.17 Locate the centroid C in Prob. 5.9 in terms of r_1 and r_2 and show that as r_1 approaches r_2, the location of C approaches that for a semicircular arc of radius $\frac{1}{2}(r_1 + r_2)$.

5.18 For the semiannular area of Prob. 5.9, determine the ratio of r_1 to r_2 for which the centroid of the area is located at $x = -\frac{1}{2}r_2$ and $y = 0$.

5.19 For the plane area of Prob. 5.8, determine the ratio of a to r for which the centroid of the area is located at point B.

5.20 For the semiannular area of Prob. 5.9, determine the ratio of r_1 to r_2 for which the centroid of the area is located at the point of intersection of the inner circle and the x axis.

5.21 A built-up beam has been constructed by nailing together seven planks as shown. The nails are equally spaced along the beam and the beam supports a vertical load. As proved in mechanics of materials, the shearing forces exerted on the nails at A and B are proportional to the first moments with respect to the centroidal x axis of the shaded areas shown, respectively, in parts a and b of the figure. Knowing that the force exerted on the nail at A is 120 N, determine the force exerted on the nail at B.

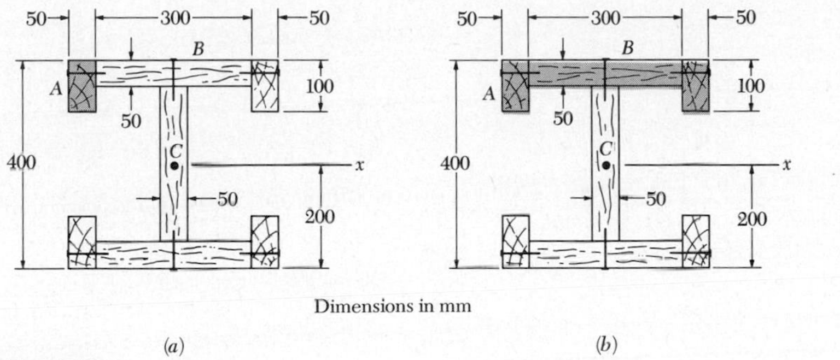

Fig. P5.21

5.22 and 5.23 The horizontal x axis is drawn through the centroid C of the area shown and divides it into two component areas A_1 and A_2. Determine the first moment of each component area with respect to the x axis and explain the result obtained.

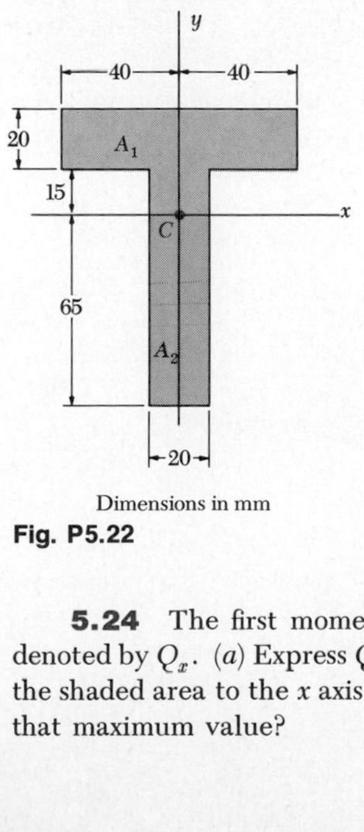

Fig. P5.22

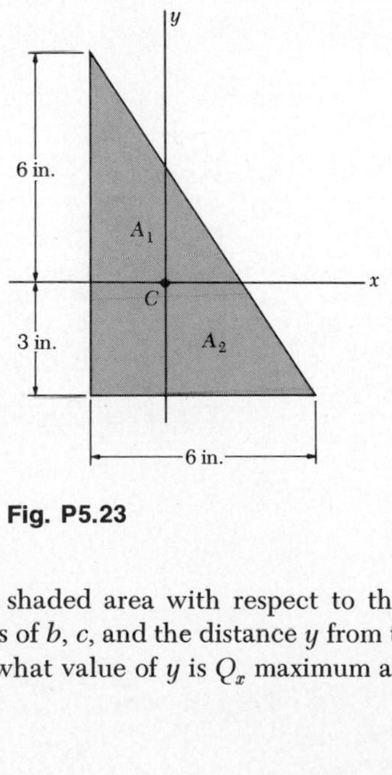

Fig. P5.23

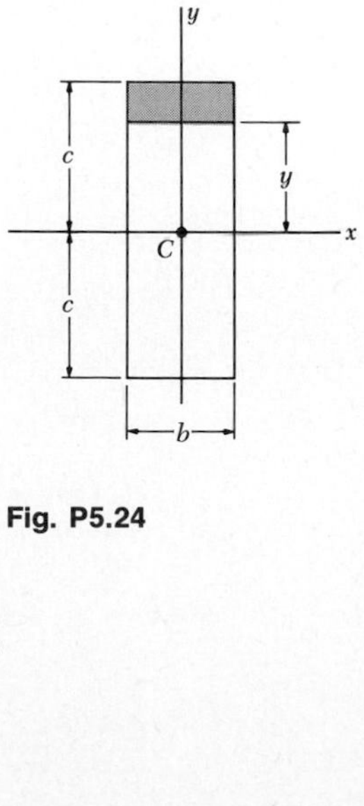

Fig. P5.24

5.24 The first moment of the shaded area with respect to the x axis is denoted by Q_x. (*a*) Express Q_x in terms of b, c, and the distance y from the base of the shaded area to the x axis. (*b*) For what value of y is Q_x maximum and what is that maximum value?

5.25 through 5.28 A thin homogeneous wire is bent to form the perimeter of the figure indicated. Locate the center of gravity of the wire figure thus formed.

5.25 Fig. P5.1
5.26 Fig. P5.2
5.27 Fig. P5.5
5.28 Fig. P5.6

5.29 A uniform circular rod of weight 20 lb and radius 18 in. is attached to a pin at B and bears against a frictionless surface at A. Determine the reactions at A and B.

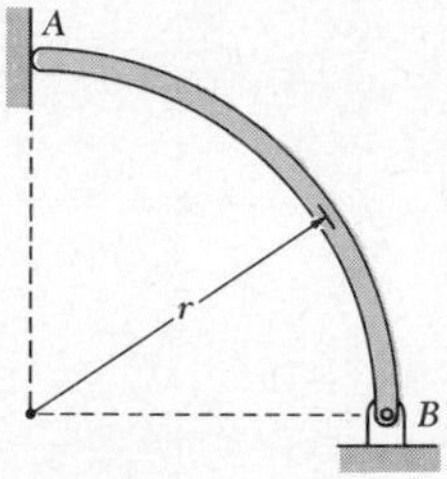

Fig. P5.29

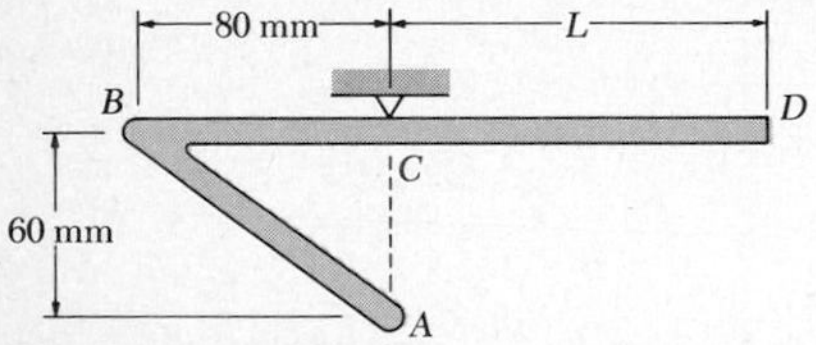

Fig. P5.30 and P5.31

5.30 The homogeneous wire $ABCD$ is bent as shown and is attached to a hinge at C. Determine the length L for which portion BCD of the wire is horizontal.

5.31 The homogeneous wire $ABCD$ is bent as shown and is attached to a hinge at C. Determine the length L for which portion AB of the wire is horizontal.

5.32 Knowing that the figure shown is formed of a thin homogeneous wire, determine the length l of portion CE of the wire for which the center of gravity of the figure is located at point C, when (a) $\theta = 30°$, (b) $\theta = 90°$.

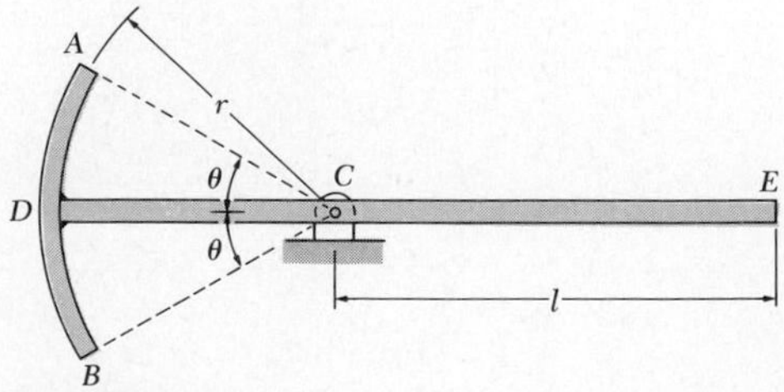

Fig. P5.32

5.33 Determine the distance h for which the centroid of the shaded area is as high above line BB' as possible when (*a*) $k = 0.25$, (*b*) $k = 0.75$.

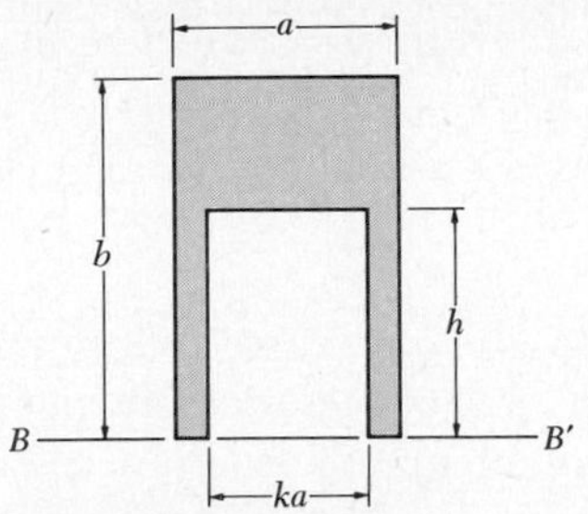

Fig. P5.33 and P5.34

5.34 Show that when the distance h is selected to maximize the distance $\bar{y}$ from line BB' to the centroid of the shaded area, we also have $\bar{y} = h$.

5.35 Determine by approximate means the x coordinate of the centroid of the area shown.

5.36 Determine by approximate means the y coordinate of the centroid of the area shown.

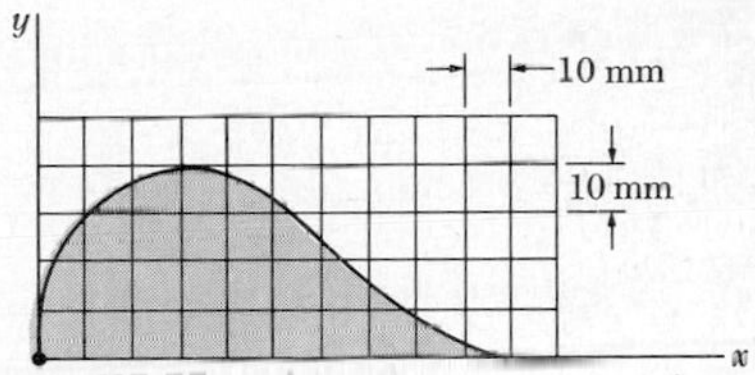

Fig. P5.35 and P5.36

5.37 Divide the parabolic spandrel shown into five vertical sections and determine by approximate means the x coordinate of its centroid; approximate the spandrel by rectangles of the form $bcc'b'$. What is the percentage error in the answer obtained? (See Fig. 5.8A for exact answer.)

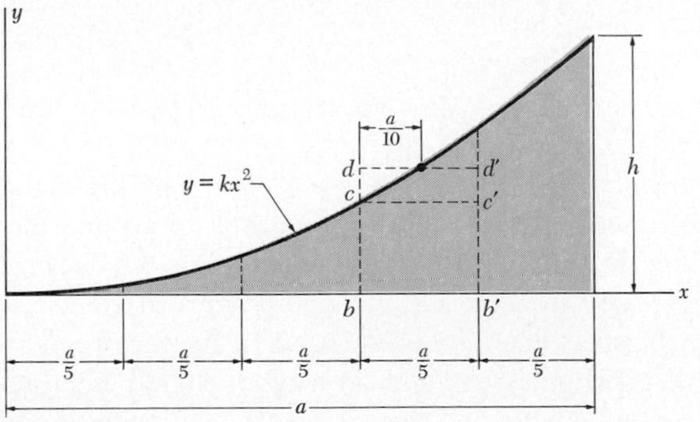

Fig. P5.37

5.38 Solve Prob. 5.37, using rectangles of the form $bdd'b'$.

5.6. Determination of Centroids by Integration. The centroid of an area bounded by analytical curves (i.e., curves defined by algebraic equations) is usually determined by computing the integrals in Eqs. (5.3) of Sec. 5.3:

$$\bar{x}A = \int x\,dA \qquad \bar{y}A = \int y\,dA \tag{5.3}$$

If the element of area dA is chosen equal to a small square of sides dx and dy, the determination of each of these integrals requires a *double integration* in x and y. A double integration is also necessary if polar coordinates are used and if dA is chosen equal to a small square of sides dr and $r\,d\theta$.

In most cases, however, it is possible to determine the coordinates of the centroid of an area by performing a single integration. This is achieved by choosing for dA a thin rectangle or strip, or a thin sector or pie-shaped element (Fig. 5.12); the centroid of the thin rectangle is located at its center, and the centroid of the thin sector at a distance $\frac{2}{3}r$ from its vertex (as for a triangle). The coordinates of the centroid of the area under consideration are then obtained by expressing that the first moment of the entire area with respect to each of the coordinate axes is equal to the sum (or integral) of the corresponding moments of the elements of area. Denoting by $\bar{x}_{el}$ and $\bar{y}_{el}$ the coordinates of the centroid of the element dA, we write

$$\begin{aligned} Q_y &= \bar{x}A = \int \bar{x}_{el}\,dA \\ Q_x &= \bar{y}A = \int \bar{y}_{el}\,dA \end{aligned} \tag{5.9}$$

If the area itself is not already known, it may also be computed from these elements.

The coordinates $\bar{x}_{el}$ and $\bar{y}_{el}$ of the centroid of the element of area should be expressed in terms of the coordinates of a point located on the curve bounding the area under consideration. Also, the element of area dA should be expressed in terms of the coordinates of the point and their differentials. This has been done in Fig. 5.12 for three common types of elements; the pie-shaped element of part c should be used when the equation of the curve bounding the area is given in polar coordinates. The appropriate expressions should be substituted in formulas (5.9), and the equation of the curve should be used to express one of the coordinates in terms of the other. The integration is thus reduced to a single integration which may be performed according to the usual rules of calculus. Once the area and the integrals in Eqs. (5.9) have been determined, these equations may be solved for the coordinates $\bar{x}$ and $\bar{y}$ of the centroid of the area.

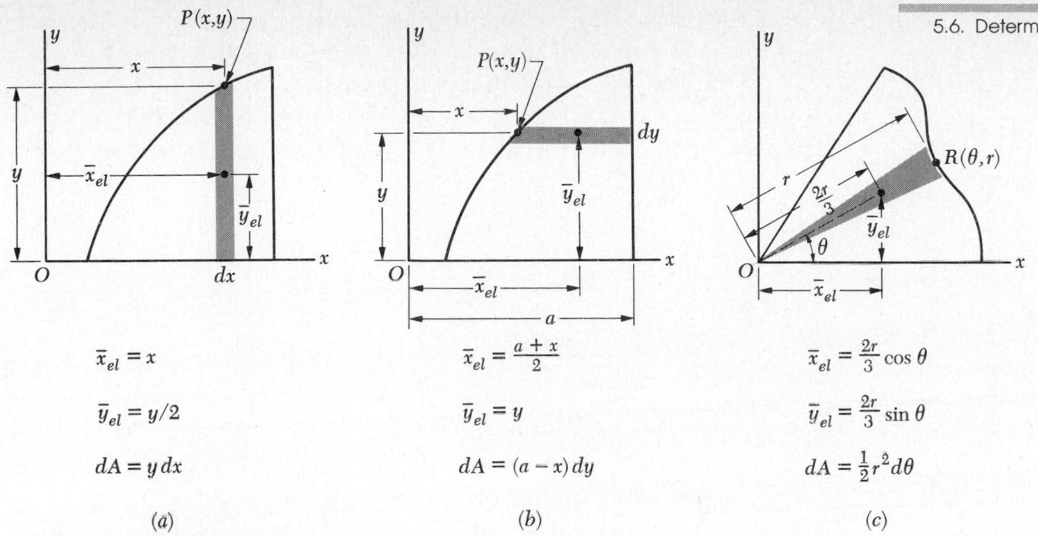

Fig. 5.12 Centroids and areas of differential elements.

The centroid of a line defined by an algebraic equation may be determined by computing the integrals in Eqs. (5.4) of Sec. 5.3:

$$\bar{x}L = \int x\,dL \qquad \bar{y}L = \int y\,dL \tag{5.4}$$

The element dL should be replaced by one of the following expressions, depending upon the type of equation used to define the line (these expressions may be derived by using the Pythagorean theorem):

$$dL = \sqrt{1 + \left(\frac{dy}{dx}\right)^2}\,dx$$

$$dL = \sqrt{1 + \left(\frac{dx}{dy}\right)^2}\,dy$$

$$dL = \sqrt{r^2 + \left(\frac{dr}{d\theta}\right)^2}\,d\theta$$

The equation of the line is then used to express one of the coordinates in terms of the other, the integration may be performed by the methods of calculus, and Eqs. (5.4) may be solved for the coordinates $\bar{x}$ and $\bar{y}$ of the centroid of the line.

5.7. Theorems of Pappus-Guldinus. These theorems, which were first formulated by the Greek geometer Pappus during the third century A.D. and later restated by the Swiss mathematician Guldinus, or Guldin (1577–1643), deal with surfaces and bodies of revolution.

A *surface of revolution* is a surface which may be generated by rotating a plane curve about a fixed axis. For example (Fig. 5.13), the surface of

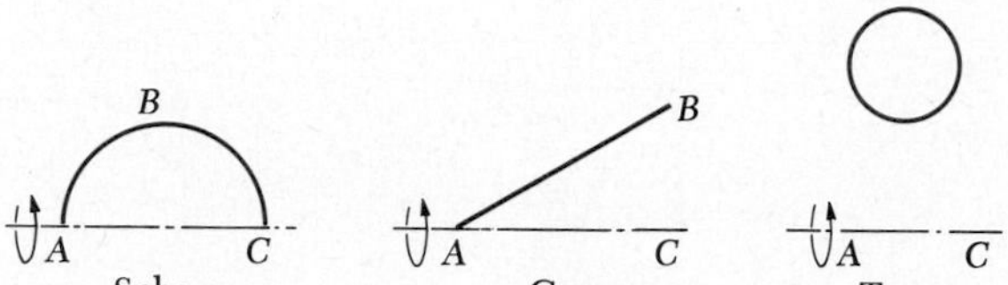

Fig. 5.13 Generating a surface of revolution.

a sphere may be obtained by rotating a semicircular arc ABC about the diameter AC; the surface of a cone by rotating a straight line AB about an axis AC; the surface of a torus or ring by rotating the circumference of a circle about a nonintersecting axis. A *body of revolution* is a body which may be generated by rotating a plane area about a fixed axis. A solid sphere may be obtained by rotating a semicircular area, a cone by rotating a triangular area, and a solid torus by rotating a full circular area (Fig. 5.14).

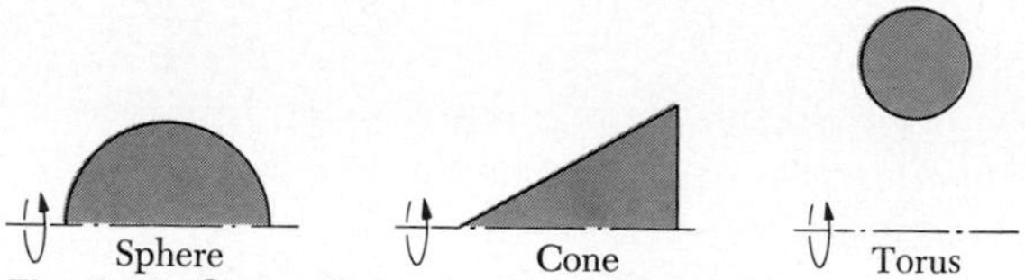

Fig. 5.14 Generating a body of revolution.

THEOREM I. *The area of a surface of revolution is equal to the length of the generating curve times the distance traveled by the centroid of the curve while the surface is being generated.*

Proof. Consider an element dL of the line L (Fig. 5.15) which is revolved about the x axis. The area dA generated by the element dL is equal to $2\pi y\, dL$. Thus, the entire area generated by L is $A = \int 2\pi y\, dL$. But we saw in Sec. 5.3 that the integral $\int y\, dL$ is equal to $\bar{y}L$. We have therefore

$$A = 2\pi \bar{y} L \tag{5.10}$$

where $2\pi\bar{y}$ is the distance traveled by the centroid of L. It should be noted

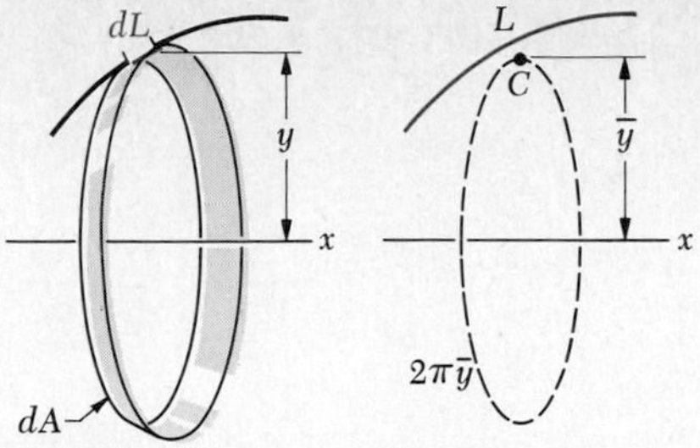

Fig. 5.15

that the generating curve should not cross the axis about which it is rotated; if it did, the two sections on either side of the axis would generate areas of opposite signs and the theorem would not apply.

THEOREM II. *The volume of a body of revolution is equal to the generating area times the distance traveled by the centroid of the area while the body is being generated.*

Proof. Consider an element dA of the area A which is revolved about the x axis (Fig. 5.16). The volume dV generated by the element dA is equal

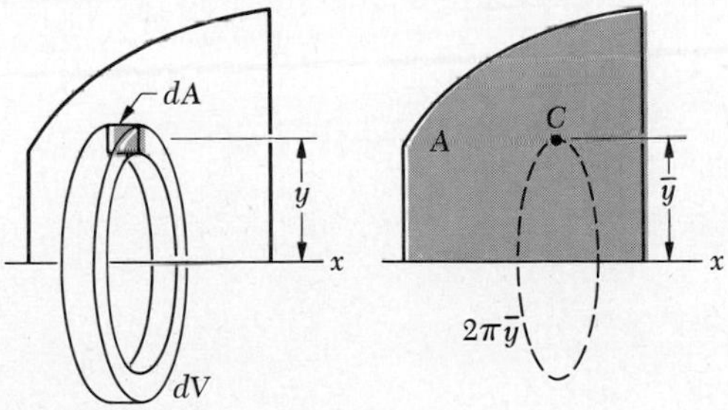

Fig. 5.16

to $2\pi y\, dA$. Thus, the entire volume generated by A is $V = \int 2\pi y\, dA$. But since the integral $\int y\, dA$ is equal to $\bar{y}A$ (Sec. 5.3), we have

$$V = 2\pi\bar{y}A \tag{5.11}$$

where $2\pi\bar{y}$ is the distance traveled by the centroid of A. Again, it should be noted that the theorem does not apply if the axis of rotation intersects the generating area.

The theorems of Pappus-Guldinus offer a simple way for computing the area of surfaces of revolution and the volume of bodies of revolution. They may also be used conversely to determine the centroid of a plane curve when the area of the surface generated by the curve is known or to determine the centroid of a plane area when the volume of the body generated by the area is known (see Sample Prob. 5.8).

SAMPLE PROBLEM 5.4

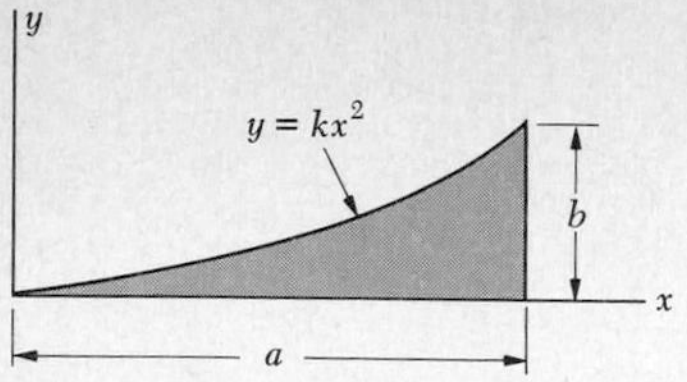

Determine by direct integration the centroid of a parabolic spandrel.

Solution. The value of k is determined by substituting $x = a$ and $y = b$ in the given equation. We have $b = ka^2$ and $k = b/a^2$. The equation of the curve is thus

$$y = \frac{b}{a^2}x^2 \qquad \text{or} \qquad x = \frac{a}{b^{1/2}}y^{1/2}$$

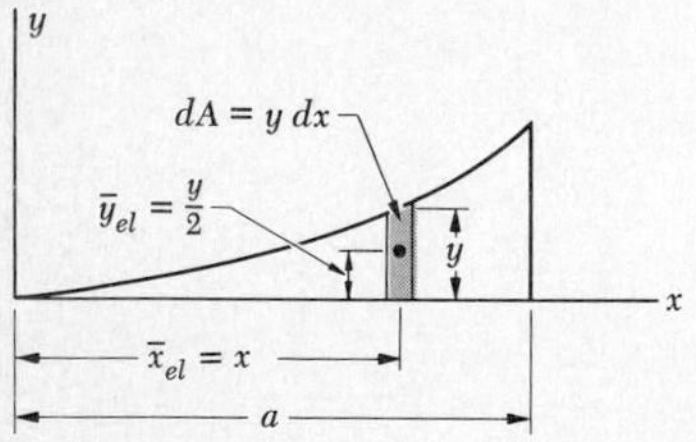

Vertical Differential Element. We choose the differential element shown and find the total area of the figure.

$$A = \int dA = \int y\,dx = \int_0^a \frac{b}{a^2}x^2\,dx = \left[\frac{b}{a^2}\frac{x^3}{3}\right]_0^a = \frac{ab}{3}$$

The first moment of the differential element with respect to the y axis is $\bar{x}_{el}\,dA$; hence, the first moment of the entire area with respect to this axis is

$$Q_y = \int \bar{x}_{el}\,dA = \int xy\,dx = \int_0^a x\left(\frac{b}{a^2}x^2\right)dx = \left[\frac{b}{a^2}\frac{x^4}{4}\right]_0^a = \frac{a^2b}{4}$$

Since $Q_y = \bar{x}A$, we have

$$\bar{x}A = \int \bar{x}_{el}\,dA \qquad \bar{x}\frac{ab}{3} = \frac{a^2b}{4} \qquad \bar{x} = \tfrac{3}{4}a \quad ◀$$

Likewise, the first moment of the differential element with respect to the x axis is $\bar{y}_{el}\,dA$, and the first moment of the entire area is

$$Q_x = \int \bar{y}_{el}\,dA = \int \frac{y}{2}y\,dx = \int_0^a \frac{1}{2}\left(\frac{b}{a^2}x^2\right)^2 dx = \left[\frac{b^2}{2a^4}\frac{x^5}{5}\right]_0^a = \frac{ab^2}{10}$$

Since $Q_x = \bar{y}A$, we have

$$\bar{y}A = \int \bar{y}_{el}\,dA \qquad \bar{y}\frac{ab}{3} = \frac{ab^2}{10} \qquad \bar{y} = \tfrac{3}{10}b \quad ◀$$

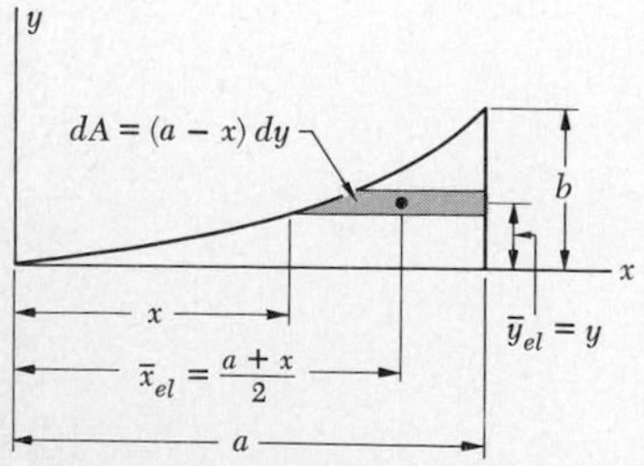

Horizontal Differential Element. The same result may be obtained by considering a horizontal element. The first moments of the area are

$$Q_y = \int \bar{x}_{el}\,dA = \int \frac{a+x}{2}(a-x)\,dy = \int_0^b \frac{a^2 - x^2}{2}\,dy$$

$$= \frac{1}{2}\int_0^b \left(a^2 - \frac{a^2}{b}y\right)dy = \frac{a^2b}{4}$$

$$Q_x = \int \bar{y}_{el}\,dA = \int y(a-x)\,dy = \int y\left(a - \frac{a}{b^{1/2}}y^{1/2}\right)dy$$

$$= \int_0^b \left(ay - \frac{a}{b^{1/2}}y^{3/2}\right)dy = \frac{ab^2}{10}$$

The expressions obtained are again substituted in the equations defining the centroid of the area to obtain $\bar{x}$ and $\bar{y}$.

SAMPLE PROBLEM 5.5

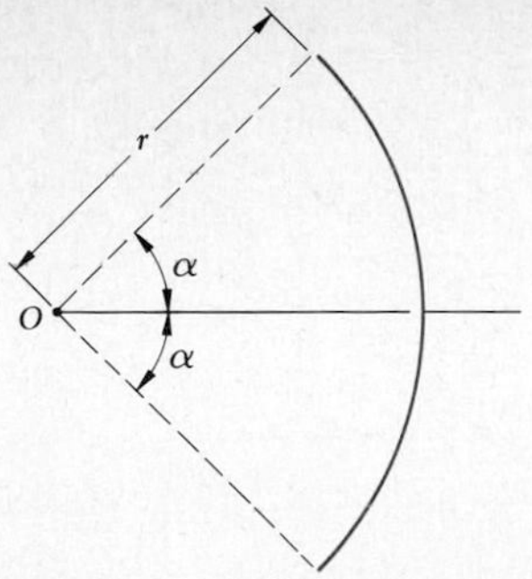

Determine the centroid of the arc of circle shown.

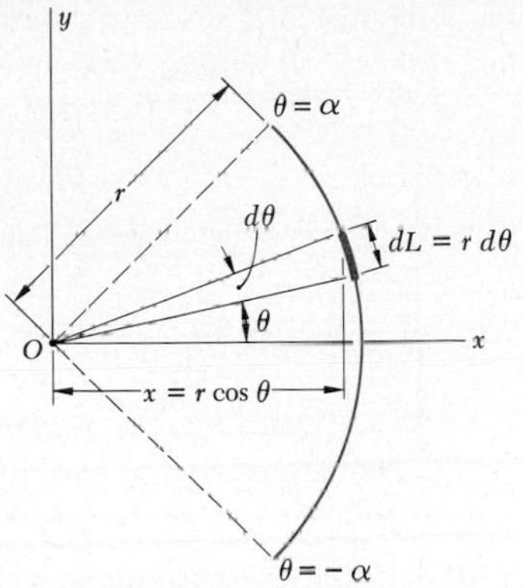

Solution. Since the arc is symmetrical with respect to the x axis, we have $\bar{y} = 0$. A differential element is chosen as shown, and the length of the arc is determined by integration.

$$L = \int dL = \int_{-\alpha}^{\alpha} r\, d\theta = r \int_{-\alpha}^{\alpha} d\theta = 2r\alpha$$

The first moment of the arc with respect to the y axis is

$$Q_y = \int x\, dL = \int_{-\alpha}^{\alpha} (r \cos\theta)(r\, d\theta) = r^2 \int_{-\alpha}^{\alpha} \cos\theta\, d\theta$$

$$= r^2[\sin\theta]_{-\alpha}^{\alpha} = 2r^2 \sin\alpha$$

Since $Q_y = \bar{x}L$, we write

$$\bar{x}(2r\alpha) = 2r^2 \sin\alpha \qquad \bar{x} = \frac{r \sin\alpha}{\alpha} \blacktriangleleft$$

SAMPLE PROBLEM 5.6

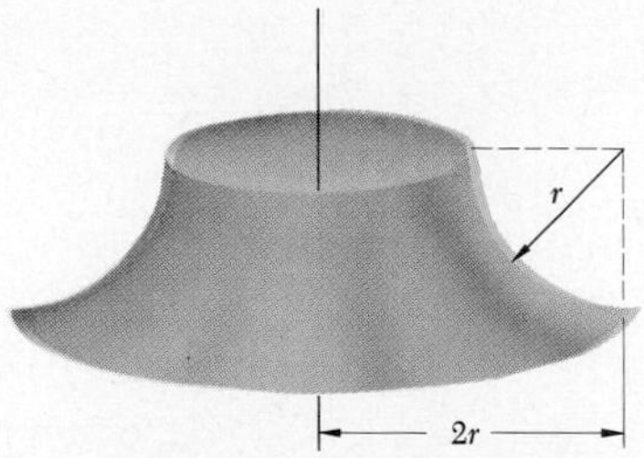

Determine the area of the surface of revolution shown, obtained by rotating a quarter-circular arc about a vertical axis.

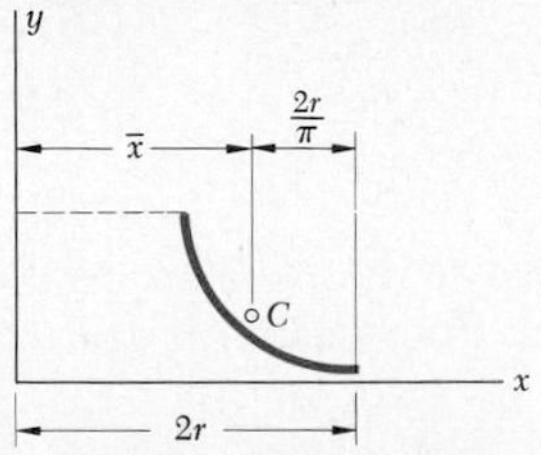

Solution. According to Theorem I of Pappus-Guldinus, the area generated is equal to the product of the length of the arc and the distance traveled by its centroid. Referring to Fig. 5.8*B*, we have

$$\bar{x} = 2r - \frac{2r}{\pi} = 2r\left(1 - \frac{1}{\pi}\right)$$

$$A = 2\pi\bar{x}L = 2\pi 2r\left(1 - \frac{1}{\pi}\right)\frac{\pi r}{2}$$

$$A = 2\pi r^2(\pi - 1) \blacktriangleleft$$

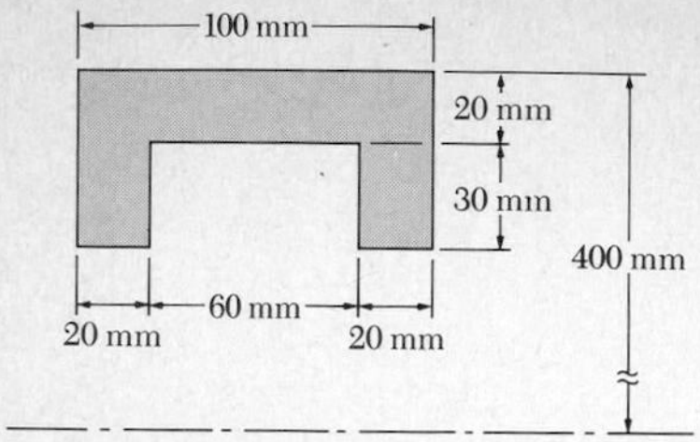

SAMPLE PROBLEM 5.7

The outside diameter of a pulley is 0.8 m, and the cross section of its rim is as shown. Determine the mass and weight of the rim, knowing that the pulley is made of steel and that the density of steel is $\rho = 7.85 \times 10^3\ \text{kg/m}^3$.

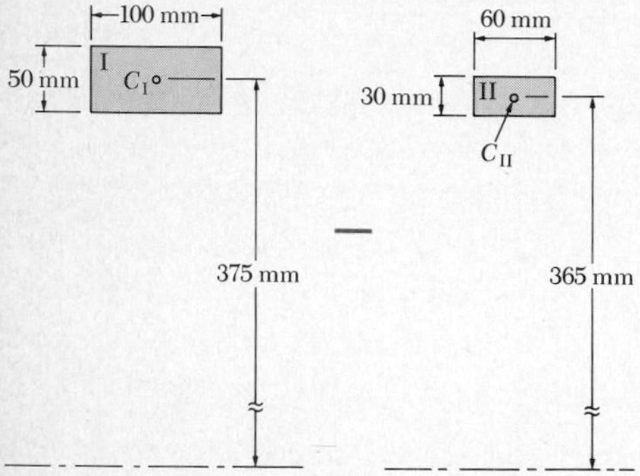

Solution. The volume of the rim may be found by applying Theorem II of Pappus-Guldinus, which states that the volume equals the product of the given cross-sectional area and of the distance traveled by its centroid in one complete revolution. However, the volume will be more easily obtained if we observe that the cross section consists of rectangle (I) with a positive area and rectangle (II) with a negative area.

	Area, mm^2	$\bar{y}$, mm	Distance Traveled by C, mm	Volume, mm^3
I	+5000	375	$2\pi(375) = 2356$	$(5000)(2356) = 11.78 \times 10^6$
II	−1800	365	$2\pi(365) = 2293$	$(-1800)(2293) = -4.13 \times 10^6$
				Volume of rim $= 7.65 \times 10^6$

Since 1 mm $= 10^{-3}$ m, we have 1 mm$^3 = (10^{-3}\ \text{m})^3 = 10^{-9}\ \text{m}^3$, and we obtain $V = 7.65 \times 10^6\ \text{mm}^3 = (7.65 \times 10^6)(10^{-9}\ \text{m}^3) = 7.65 \times 10^{-3}\ \text{m}^3$.

$$m = \rho V = (7.85 \times 10^3\ \text{kg/m}^3)(7.65 \times 10^{-3}\ \text{m}^3) \qquad m = 60.0\ \text{kg} \blacktriangleleft$$

$$W = mg = (60.0\ \text{kg})(9.81\ \text{m/s}^2) = 589\ \text{kg}\cdot\text{m/s}^2 \qquad W = 589\ \text{N} \blacktriangleleft$$

SAMPLE PROBLEM 5.8

Using the theorems of Pappus-Guldinus, determine (*a*) the centroid of a semicircular area, (*b*) the centroid of a semicircular arc. We recall that the volume of a sphere is $\frac{4}{3}\pi r^3$ and that its surface area is $4\pi r^2$.

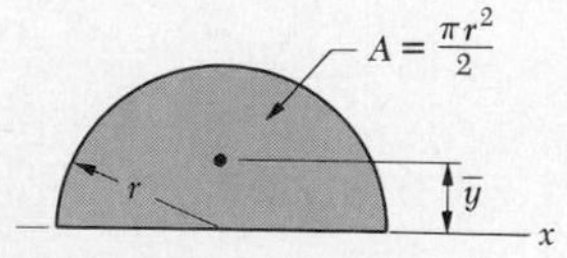

Solution. The volume of a sphere is equal to the product of the area of a semicircle and of the distance traveled by the centroid of the semicircle in one revolution about the x axis.

$$V = 2\pi\bar{y}A \qquad \tfrac{4}{3}\pi r^3 = 2\pi\bar{y}(\tfrac{1}{2}\pi r^2) \qquad \bar{y} = \frac{4r}{3\pi} \blacktriangleleft$$

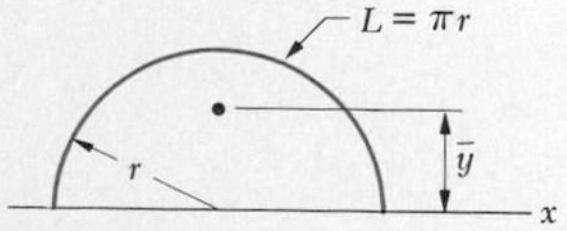

Likewise, the area of a sphere is equal to the product of the length of the generating semicircle and of the distance traveled by its centroid in one revolution.

$$A = 2\pi\bar{y}L \qquad 4\pi r^2 = 2\pi\bar{y}(\pi r) \qquad \bar{y} = \frac{2r}{\pi} \blacktriangleleft$$

Problems

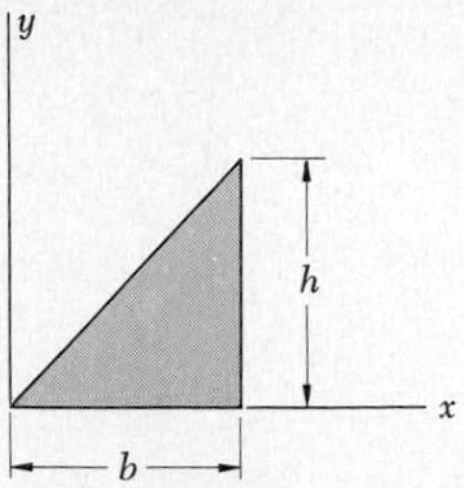

Fig. P5.39

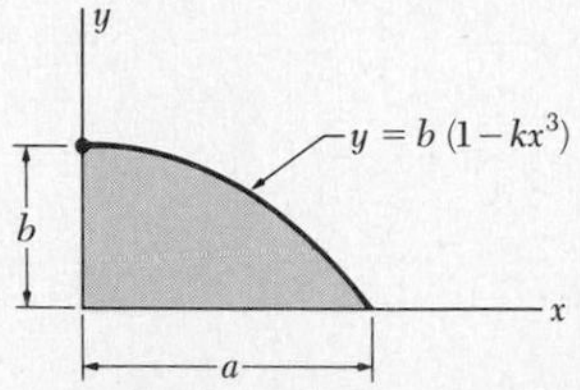

Fig. P5.40

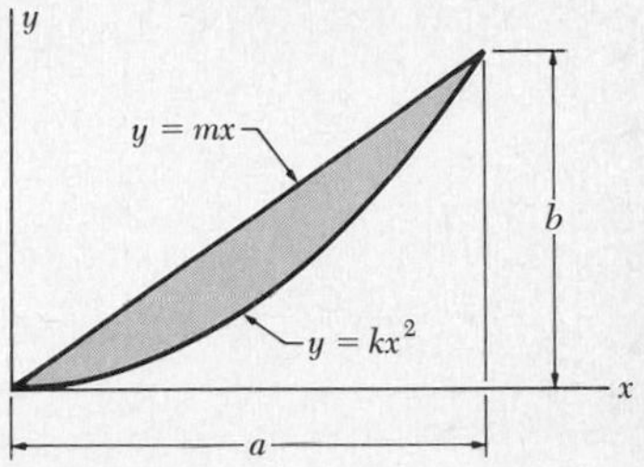

Fig. P5.41

5.39 through 5.42 Determine by direct integration the centroid of the area shown.

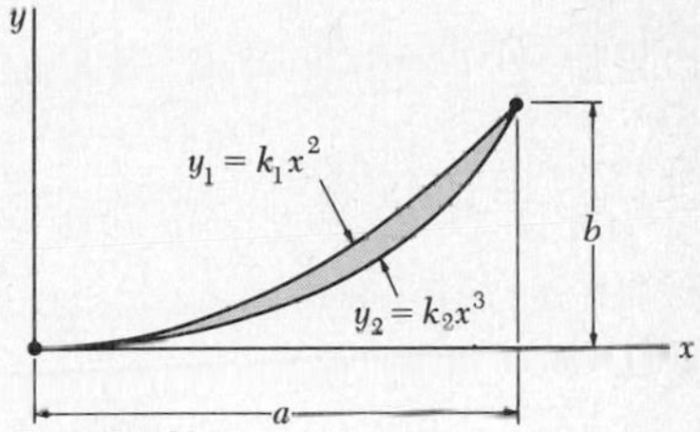

Fig. P5.42

5.43 through 5.48 Derive by direct integration the expressions for $\bar{x}$ and $\bar{y}$ given in Fig. 5.8 for:

5.43 A general spandrel ($y = kx^n$).
5.44 A quarter-elliptical area.
5.45 A semicircular area.
5.46 A semiparabolic area.
5.47 A circular sector.
5.48 A quarter-circular arc.

5.49 Determine by direct integration the x coordinate of the centroid of the area shown.

5.50 Determine by direct integration the y coordinate of the centroid of the area shown.

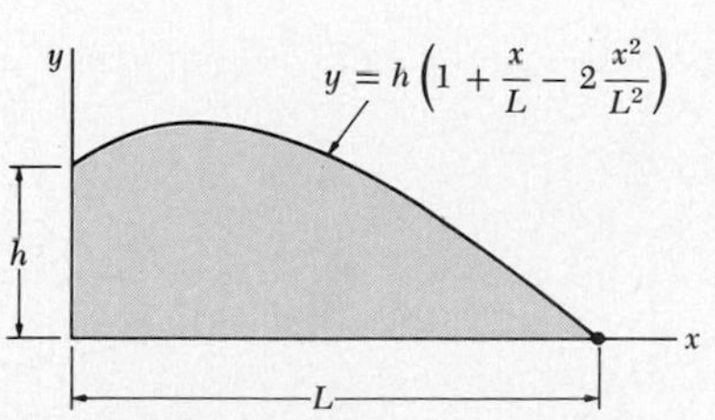

Fig. P5.49 and P5.50

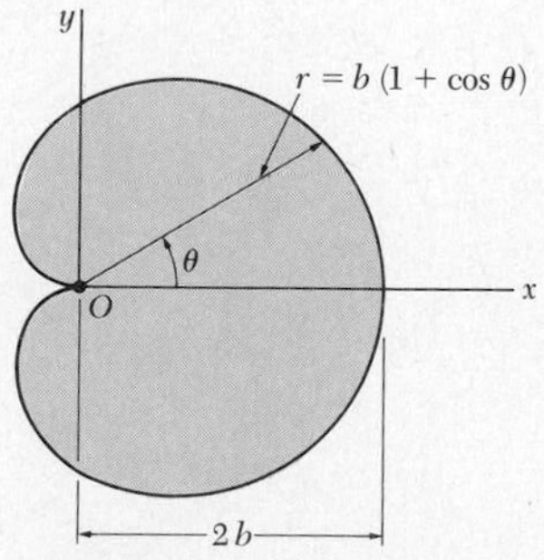

Fig. P5.51

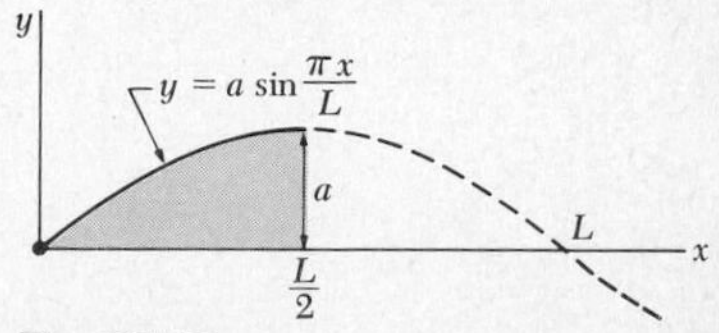

Fig. P5.52

*** 5.51 and * 5.52** Determine by direct integration the centroid of the area shown.

5.53 Determine the centroid of the area shown when $b = h$.

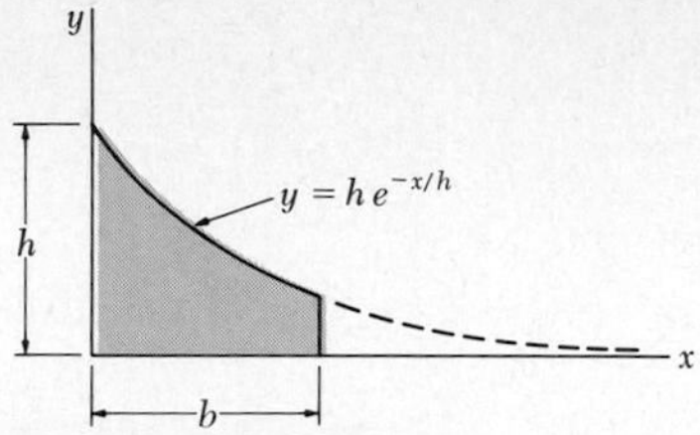

Fig. P5.53 and P5.54

5.54 Determine the centroid of the area shown when $b = \infty$.

5.55 Determine the volume of the solid obtained by rotating the area of Prob. 5.6 about (*a*) the x axis, (*b*) the y axis.

5.56 Determine the volume of the solid obtained by rotating the trapezoid of Prob. 5.2 about (*a*) the x axis, (*b*) the y axis.

5.57 Determine the volume of the solid generated by rotating the parabolic spandrel shown about (*a*) the x axis, (*b*) the axis AA'.

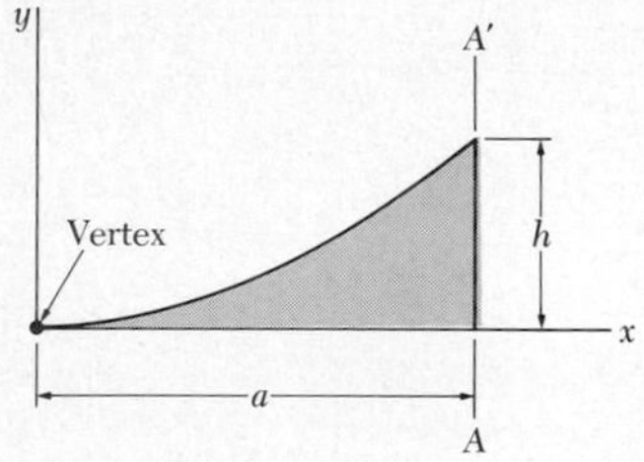

Fig. P5.57

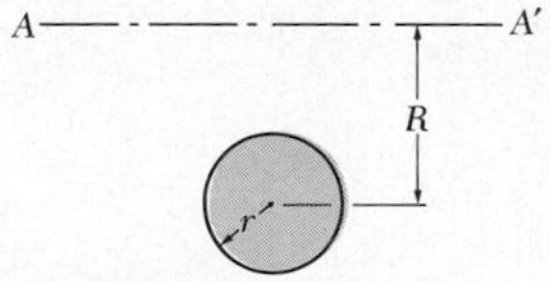

Fig. P5.58

5.58 Determine the surface area and the volume of the torus obtained by rotating the circular area shown about the axis AA'.

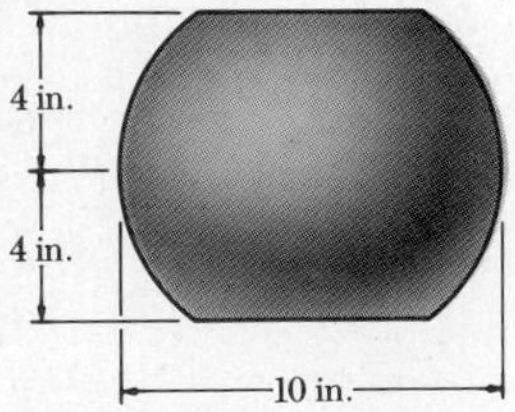

Fig. P5.59

5.59 Knowing that two equal caps have been cut from a 5-in.-diameter wooden sphere, determine the total surface area of the remaining portion.

5.60 In Prob. 5.59, determine the volume and weight of the portion remaining after the two equal caps have been removed from the 5-in.-diameter sphere. (Specific weight of wood = 40 lb/ft^3.)

5.61 A spherical pressure vessel has an inside diameter of 1.6 m. Determine (*a*) the volume of liquefied propane required to fill the vessel to a depth of 1.2 m, (*b*) the corresponding mass of the liquefied propane. (Density of liquefied propane = 580 kg/m^3.)

5.62 For the pressure vessel of Prob. 5.61, determine the area of the surface in contact with the liquefied propane.

5.63 Determine the volume and total surface area of the body shown.

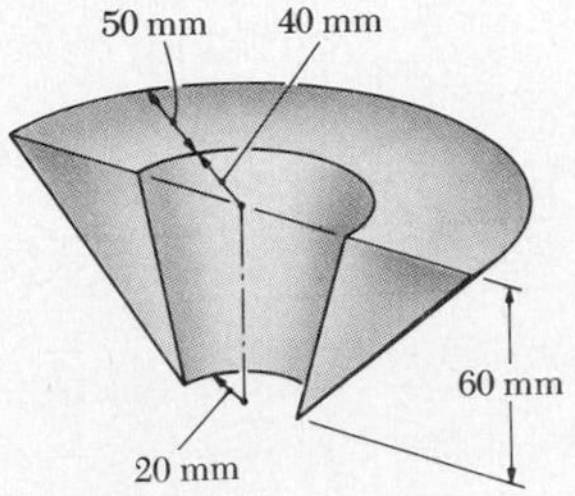

Fig. P5.63

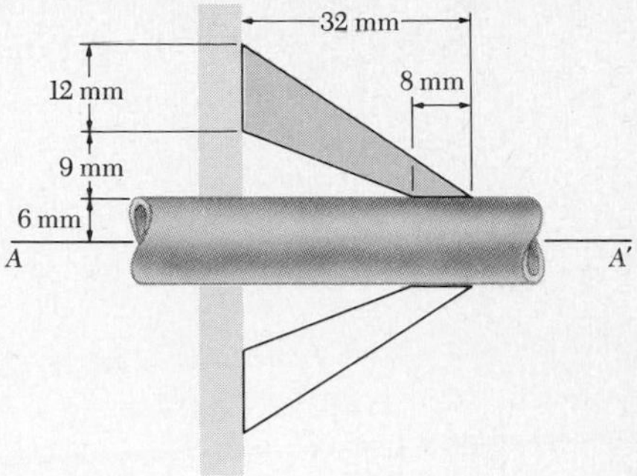

Fig. P5.64

5.64 Determine the volume and mass of the cast-iron pipe collar obtained by rotating the shaded area about the centerline of the 12-mm-diameter pipe. (Density of cast iron = 7200 kg/m^3.)

5.65 Determine the volume and weight of the solid brass knob shown. (Specific weight of brass = 0.306 lb/in^3.)

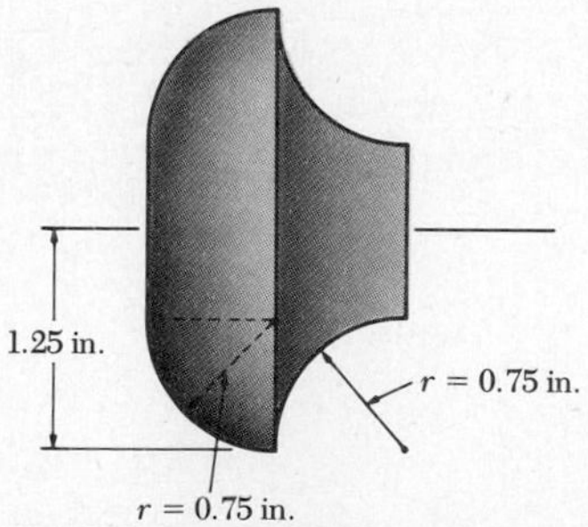

Fig. P5.65 and P5.66

5.66 Determine the total surface area of the solid brass knob shown.

5.67 The rim of a double-groove V-belt sheave of 12-in. outside diameter is made of cast iron and weighs 7.50 lb. Knowing that the cross-sectional area of the rim is 0.790 in^2, determine the distance $\overline{y}$ from the axis of rotation AA' to the centroid of the cross section of the rim. (Specific weight of cast iron = 0.264 lb/in^3.)

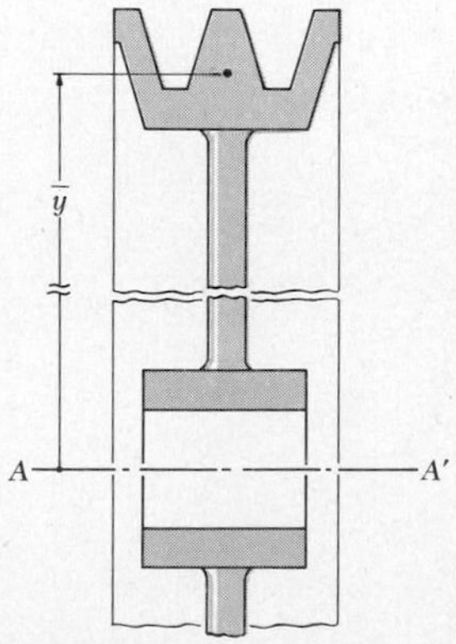

Fig. P5.67

5.68 The rotating mixing drum of a transit-mix concrete truck has the cross section shown. Determine by approximate means (*a*) the volume of the drum, (*b*) the surface area of the curved portions of the drum.

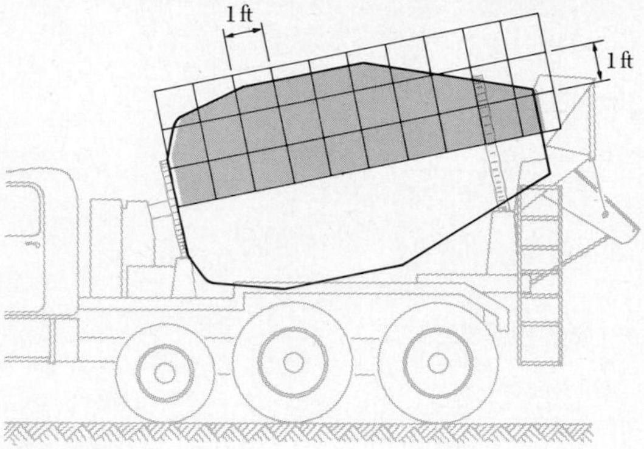

Fig. P5.68

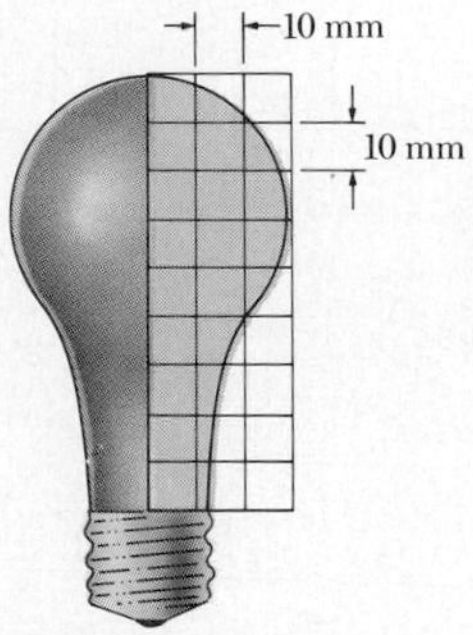

Fig. P5.69

5.69 A 60-W standard light bulb has the cross section shown. For the portion of the bulb above the base and neglecting the thickness of the glass, determine (*a*) the volume of inert gas inside the bulb, (*b*) the outside surface area of the bulb.

5.70 Determine the volume of the solid of revolution formed by revolving each of the plane areas shown about its vertical edge *AB*. Show that the volumes of the solids formed are in the ratio 6:4:3:2:1.

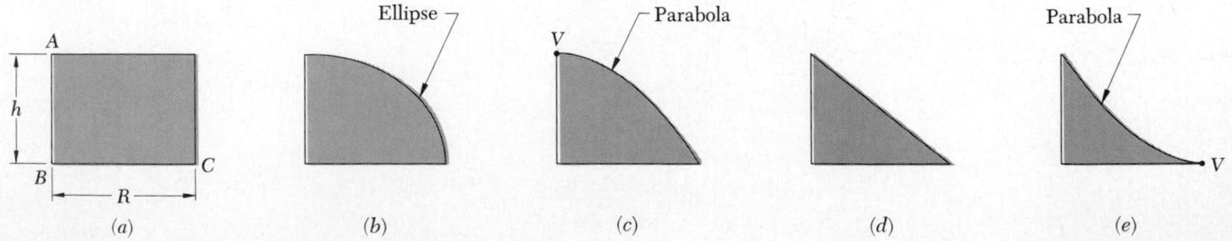

Fig. P5.70 and P5.71

5.71 Determine the volume of the solid formed by revolving each of the plane areas shown about its base *BC*. Show that the volumes of the solids formed are in the ratio 15:10:8:5:3.

***5.8. Distributed Loads on Beams.** The concept of centroid of an area may be used to solve other problems besides those dealing with the weight of flat plates. Consider, for example, a beam supporting a *distributed load;* this load may consist of the weight of materials supported directly or indirectly by the beam, or it may be caused by wind or hydrostatic pressure. The distributed load may be represented by plotting the load w supported per unit length (Fig. 5.17); this load will be expressed in N/m or in lb/ft. The magnitude of the force exerted on an element of beam of length dx is $dW = w\,dx$, and the total load supported by the beam is

$$W = \int_0^L w\,dx$$

But the product $w\,dx$ is equal in magnitude to the element of area dA shown in Fig. 5.17*a*, and W is thus equal in magnitude to the total area A under the load curve,

$$W = \int dA = A$$

We shall now determine where a *single concentrated load* **W**, of the same magnitude W as the total distributed load, should be applied on the beam if it is to produce the same reactions at the supports (Fig. 5.17*b*). This concentrated load **W**, which represents the resultant of the given distributed loading, should be equivalent to this loading as far as the free-body diagram of the entire beam is concerned. The point of application P of the equivalent concentrated load **W** will therefore be obtained by expressing that the moment of **W** about point O is equal to the sum of the moments of the elementary loads $d\mathbf{W}$ about O:

$$(OP)W = \int x\,dW$$

or, since $dW = w\,dx = dA$ and $W = A$,

$$(OP)A = \int_0^L x\,dA \tag{5.12}$$

Since the integral represents the first moment with respect to the w axis of the area under the load curve, it may be replaced by the product $\bar{x}A$. We have therefore $OP = \bar{x}$, where $\bar{x}$ is the distance from the w axis to the centroid C of the area A (this is *not* the centroid of the beam).

A distributed load on a beam may thus be replaced by a concentrated load; the magnitude of this single load is equal to the area under the load curve, and its line of action passes through the centroid of that area. It should be noted, however, that the concentrated load is equivalent to the given loading only as far as external forces are concerned. It may be used to determine reactions but should not be used to compute internal forces and deflections.

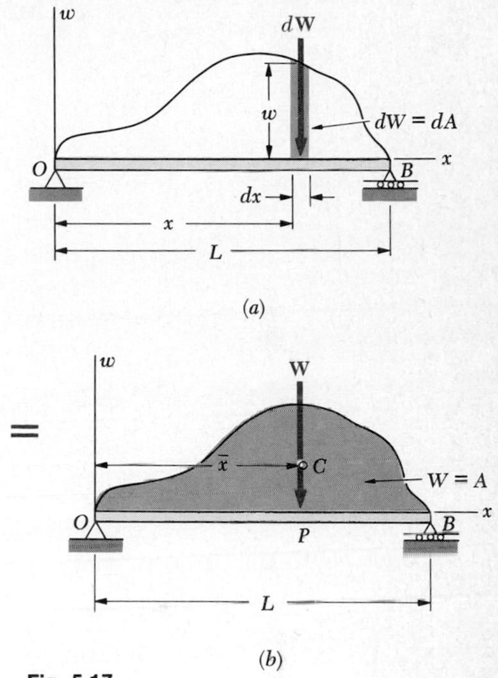

Fig. 5.17

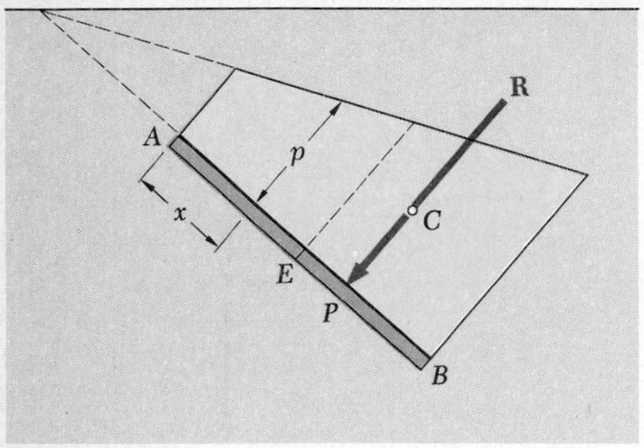

Fig. 5.18

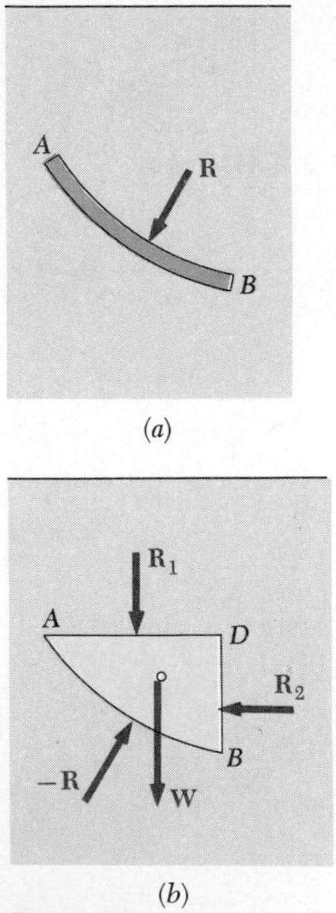

Fig. 5.19

***5.9. Forces on Submerged Surfaces.** Another example of the use of first moments and centroids of areas is obtained by considering the forces exerted on a *rectangular surface* submerged in a liquid. Consider the rectangular plate shown in Fig. 5.18; it has a length L, and its width, perpendicular to the plane of the figure, is assumed equal to unity. Since the gage pressure in a liquid is $p = \gamma h$, where γ is the specific weight of the liquid and h the vertical distance from the free surface, the pressure on the plate varies linearly with the distance x. The width of the plate being taken equal to unity, the pressure p is equal in magnitude to the load w per unit length used in Sec. 5.8.† The results obtained in that section may thus be used here, and we find that the magnitude of the resultant **R** of the forces exerted on one face of the plate is equal to the area under the pressure curve; we also find that the line of action of **R** passes through the centroid C of that area.

Noting that the area under the pressure curve is equal to $p_E L$, where p_E is the pressure at the center E of the plate and L the length (or area) of the plate, we find that the magnitude R of the resultant may be obtained by multiplying the area of the plate by the pressure at the center E of the plate. The resultant **R**, however, *should not* be applied at E; as indicated above, its line of action passes through the centroid C of the area under the pressure curve. The point of application P of the resultant **R** is known as the *center of pressure*.

We shall consider next the forces exerted by a liquid on a curved surface of constant width (Fig. 5.19*a*). Since the determination of the resultant **R** of these forces by direct integration would not be easy, we shall consider the free body obtained by detaching the volume of liquid ABD bounded by the curved surface AB and by the two plane surfaces AD and DB shown in Fig. 5.19*b*. The forces acting on the free body ABD consist of the weight **W** of the volume of liquid detached, the resultant $\mathbf{R}_1$ of the forces exerted on AD, the resultant $\mathbf{R}_2$ of the forces exerted on BD, and the resultant of the forces exerted *by the curved surface on the liquid*. This last resultant is equal and opposite to, and has the same line of action as, the resultant **R** of the forces exerted *by the liquid on the curved surface;* we shall, therefore, denote it by $-\mathbf{R}$. The forces **W**, $\mathbf{R}_1$, and $\mathbf{R}_2$ may be determined by standard methods; after their values have been found, the force $-\mathbf{R}$ will be obtained by solving the equations of equilibrium for the free body of Fig. 5.19*b*. The resultant **R** of the hydrostatic forces exerted on the curved surface will then be obtained by reversing the sense of $-\mathbf{R}$.

The methods outlined in this section may be used to determine the resultant of the hydrostatic forces exerted on the surface of dams or on rectangular gates and vanes. Resultants of forces on submerged surfaces of variable width should be determined by the methods of Chap. 9.

† We should note that, while the load w per unit length is expressed in N/m or in lb/ft, the pressure p, which represents a load per unit area, is expressed in N/m^2 or in lb/ft^2. The derived SI unit N/m^2 is called a *pascal* (Pa).

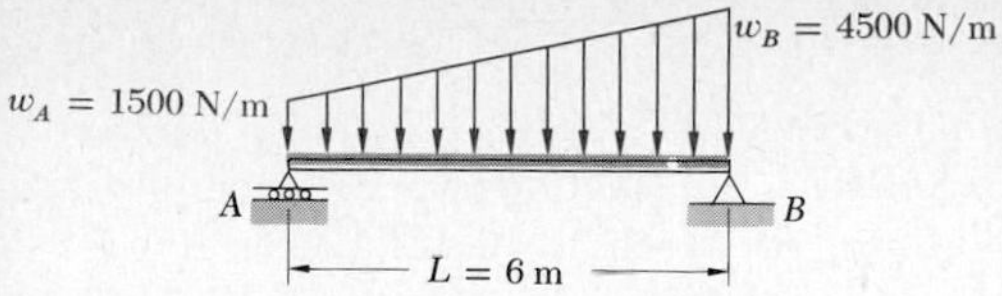

SAMPLE PROBLEM 5.9

A beam supports a distributed load as shown. (*a*) Determine the equivalent concentrated load. (*b*) Determine the reactions at the supports.

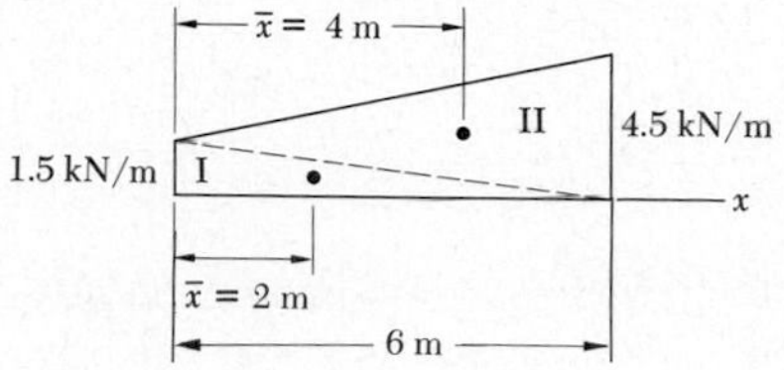

***a.* Equivalent Concentrated Load.** The magnitude of the resultant of the load is equal to the area under the load curve, and the line of action of the resultant passes through the centroid of the same area. We divide the area under the load curve into two triangles and construct the table below. To simplify the computations and tabulation, the given loads per unit length have been converted into kN/m.

Component	*A*, kN	$\bar{x}$, m	$\bar{x}A$, kN·m
Triangle I	4.5	2	9
Triangle II	13.5	4	54
	$\Sigma A = 18.0$	. . .	$\Sigma \bar{x}A = 63$

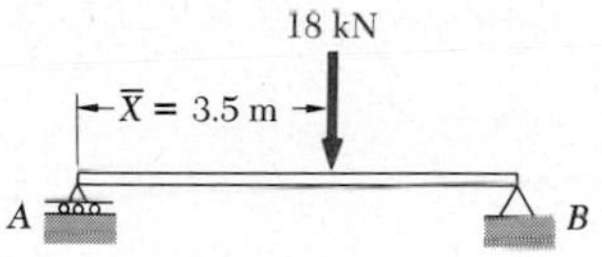

Thus, $\bar{X}\Sigma A = \Sigma \bar{x}A$: $\quad \bar{X}(18 \text{ kN}) = 63 \text{ kN}\cdot\text{m} \quad \bar{X} = 3.5 \text{ m}$

The equivalent concentrated load is

$$\mathbf{W} = 18 \text{ kN} \downarrow \blacktriangleleft$$

and its line of action is located at a distance

$$\bar{X} = 3.5 \text{ m to the right of } A \blacktriangleleft$$

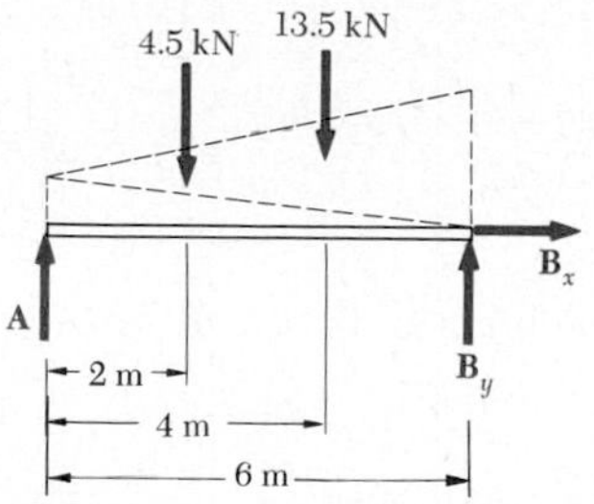

***b.* Reactions.** The reaction at *A* is vertical and is denoted by **A**; the reaction at *B* is represented by its components $\mathbf{B}_x$ and $\mathbf{B}_y$. The given load may be considered as the sum of two triangular loads as shown. The resultant of each triangular load is equal to the area of the triangle and acts at its centroid. We write the following equilibrium equations for the free body shown:

$\overset{+}{\rightarrow}\Sigma F_x = 0$: $\qquad \mathbf{B}_x = 0 \blacktriangleleft$

$+\uparrow \Sigma M_A = 0$: $\qquad -(4.5 \text{ kN})(2 \text{ m}) - (13.5 \text{ kN})(4 \text{ m}) + B_y(6 \text{ m}) = 0$

$$\mathbf{B}_y = 10.5 \text{ kN} \uparrow \blacktriangleleft$$

$+\uparrow \Sigma M_B = 0$: $\qquad +(4.5 \text{ kN})(4 \text{ m}) + (13.5 \text{ kN})(2 \text{ m}) - A(6 \text{ m}) = 0$

$$\mathbf{A} = 7.5 \text{ kN} \uparrow \blacktriangleleft$$

Alternate Solution. The given distributed load may be replaced by its resultant, which was found in part *a*. The reactions may be determined by writing the equilibrium equations $\Sigma F_x = 0$, $\Sigma M_A = 0$, and $\Sigma M_B = 0$. We again obtain

$$\mathbf{B}_x = 0 \qquad \mathbf{B}_y = 10.5 \text{ kN} \uparrow \qquad \mathbf{A} = 7.5 \text{ kN} \uparrow \blacktriangleleft$$

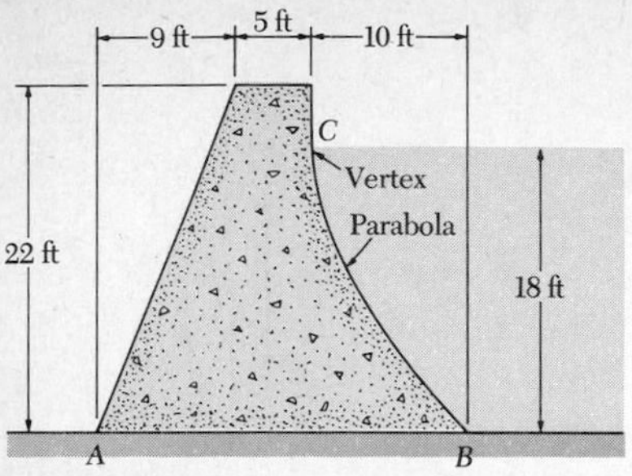

SAMPLE PROBLEM 5.10

The cross section of a concrete dam is as shown. Consider a section of the dam 1 ft thick, and determine (*a*) the resultant of the reaction forces exerted by the ground on the base of the dam *AB*, (*b*) the resultant of the pressure forces exerted by the water on the face *BC* of the dam. Specific weight of concrete = 150 lb/ft³; of water = 62.4 lb/ft³.

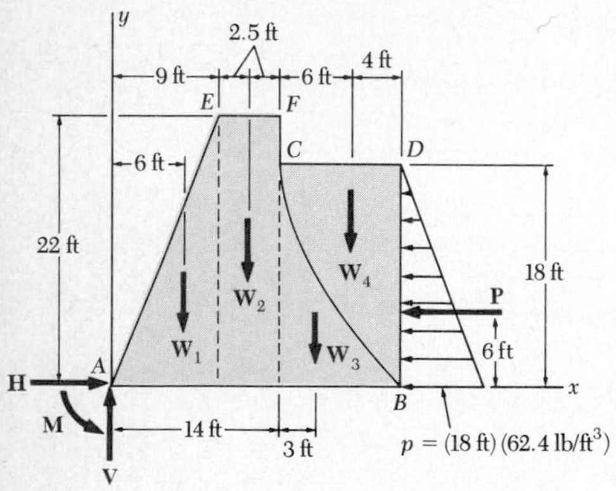

***a.* Ground Reaction.** As a free body we choose a section *AEFCDB*, 1 ft thick, of the dam and water as shown. The reaction forces exerted by the ground on the base *AB* are represented by an equivalent force-couple system at *A*. Other forces acting on the free body are the weight of the dam, represented by the weights of its components $\mathbf{W}_1$, $\mathbf{W}_2$, and $\mathbf{W}_3$, the weight of the water $\mathbf{W}_4$, and the resultant **P** of the pressure forces exerted on section *BD* by the water to the right of section *BD*. We have

$$W_1 = \tfrac{1}{2}(9 \text{ ft})(22 \text{ ft})(1 \text{ ft})(150 \text{ lb/ft}^3) = 14{,}850 \text{ lb}$$
$$W_2 = (5 \text{ ft})(22 \text{ ft})(1 \text{ ft})(150 \text{ lb/ft}^3) = 16{,}500 \text{ lb}$$
$$W_3 = \tfrac{1}{3}(10 \text{ ft})(18 \text{ ft})(1 \text{ ft})(150 \text{ lb/ft}^3) = 9000 \text{ lb}$$
$$W_4 = \tfrac{2}{3}(10 \text{ ft})(18 \text{ ft})(1 \text{ ft})(62.4 \text{ lb/ft}^3) = 7488 \text{ lb}$$
$$P = \tfrac{1}{2}(18 \text{ ft})(1 \text{ ft})(18 \text{ ft})(62.4 \text{ lb/ft}^3) = 10{,}109 \text{ lb}$$

Equilibrium Equations

$\overset{+}{\rightarrow}\Sigma F_x = 0$: $\quad H - 10{,}109 \text{ lb} = 0 \qquad \mathbf{H} = 10{,}110 \text{ lb} \rightarrow$ ◄

$+\uparrow\Sigma F_y = 0$: $\quad V - 14{,}850 \text{ lb} - 16{,}500 \text{ lb} - 9000 \text{ lb} - 7488 \text{ lb} = 0$

$\mathbf{V} = 47{,}840 \text{ lb} \uparrow$ ◄

$+\circlearrowleft \Sigma M_A = 0$: $\quad -(14{,}850 \text{ lb})(6 \text{ ft}) - (16{,}500 \text{ lb})(11.5 \text{ ft})$
$- (9000 \text{ lb})(17 \text{ ft}) - (7488 \text{ lb})(20 \text{ ft}) + (10{,}109 \text{ lb})(6 \text{ ft}) + M = 0$

$\mathbf{M} = 520{,}960 \text{ lb}\cdot\text{ft} \circlearrowleft$ ◄

We may replace the force-couple system obtained by a single force acting at a distance d to the right of A, where

$$d = \frac{520{,}960 \text{ lb}\cdot\text{ft}}{47{,}840 \text{ lb}} = 10.89 \text{ ft}$$

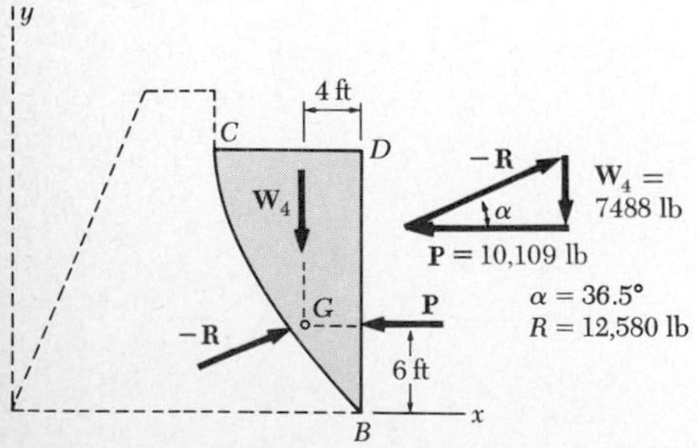

***b.* Resultant R of Water Forces.** The parabolic section of water *BCD* is chosen as a free body. The forces involved are the resultant −**R** of the forces exerted by the dam on the water, the weight $\mathbf{W}_4$, and the force **P**. Since these forces must be concurrent, −**R** passes through the point of intersection *G* of $\mathbf{W}_4$ and **P**. A force triangle is drawn from which the magnitude and direction of −**R** are determined. The resultant **R** of the forces exerted by the water on the face *BC* is equal and opposite:

$\mathbf{R} = 12{,}580 \text{ lb} \measuredangle 36.5°$ ◄

Problems

5.72 and 5.73 Determine the magnitude and location of the resultant of the distributed load shown. Also calculate the reactions at A and B.

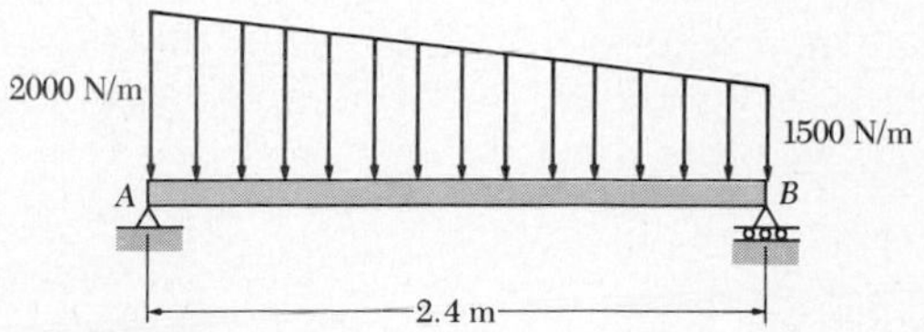

Fig. P5.72

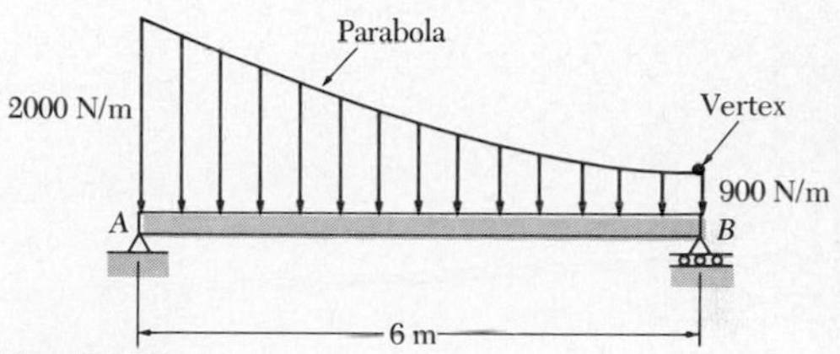

Fig. P5.73

5.74 through 5.79 Determine the reactions at the beam supports for the given loading condition.

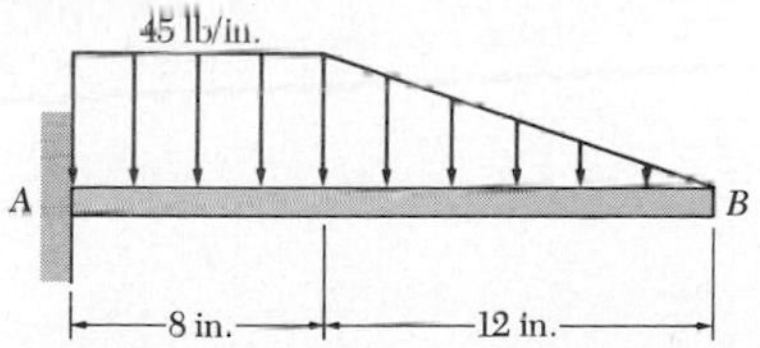

Fig. P5.74

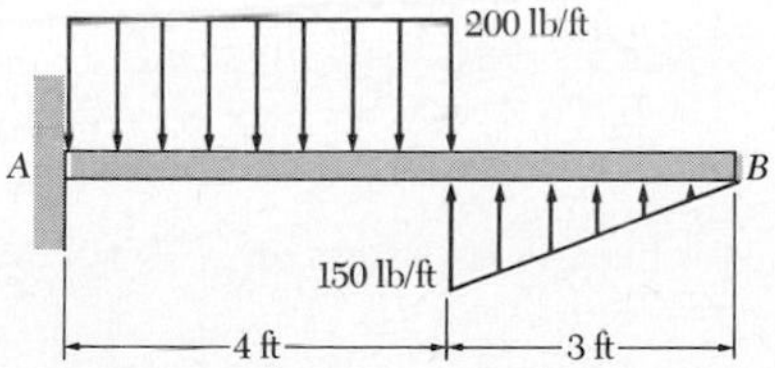

Fig. P5.75

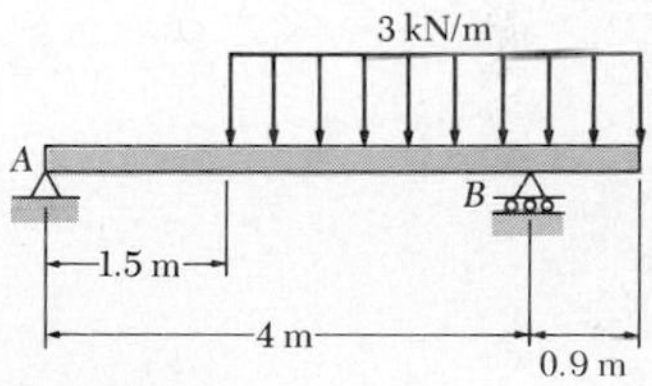

Fig. P5.76

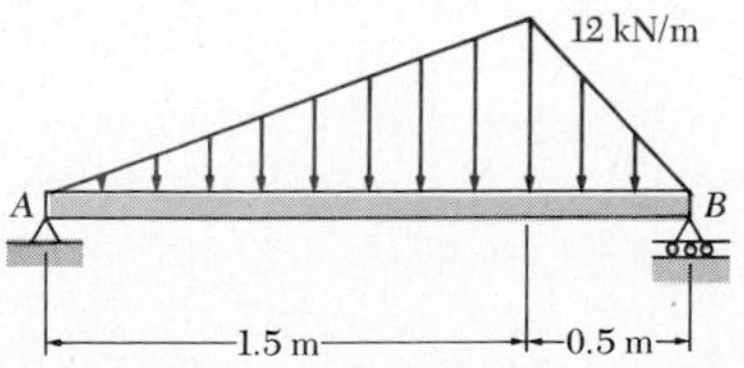

Fig. P5.77

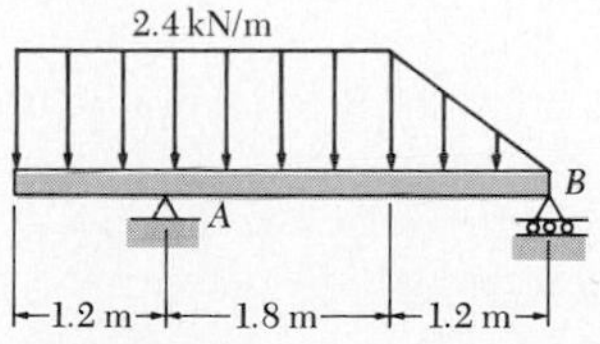

Fig. P5.78

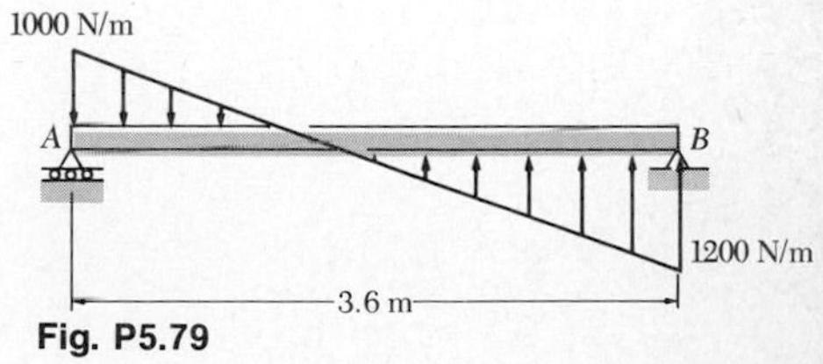

Fig. P5.79

5.80 Determine the reactions at the beam supports for the given loading condition when $w_0 = 400$ lb/ft.

5.81 Determine (*a*) the distributed load w_0 at the end A of the beam ABC for which the reaction at C is zero, (*b*) the corresponding reaction at B.

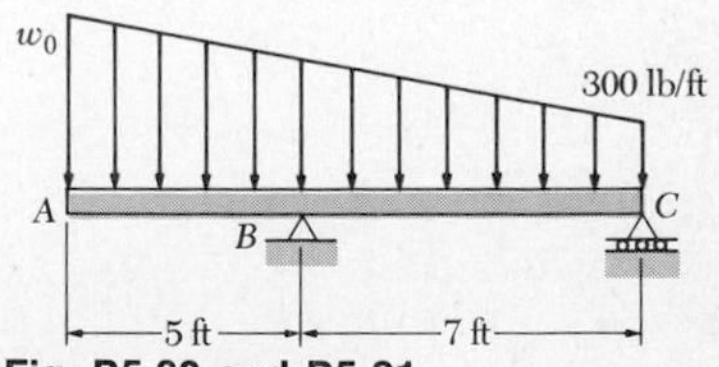

Fig. P5.80 and P5.81

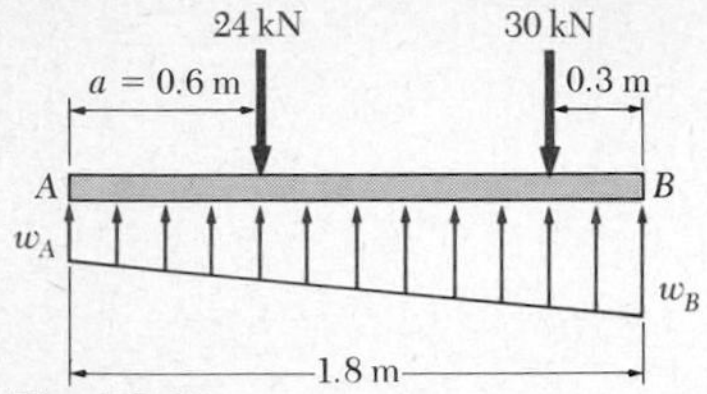

Fig. P5.82

5.82 The beam AB supports two concentrated loads and rests on soil which exerts a linearly distributed upward load as shown. Determine the values of w_A and w_B corresponding to equilibrium.

5.83 For the beam and loading of Prob. 5.82, determine (*a*) the distance a for which $w_A = 20$ kN/m, (*b*) the corresponding value of w_B.

In the following problems, use $\gamma = 62.4$ lb/ft^3 for the specific weight of fresh water and $\gamma_c = 150$ lb/ft^3 for the specific weight of concrete if U.S. customary units are used. With SI units, use $\rho = 10^3$ kg/m^3 for the density of fresh water and $\rho_c = 2.40 \times 10^3$ kg/m^3 for the density of concrete. (See footnote page 156 for the determination of the specific weight of a material from its density.)

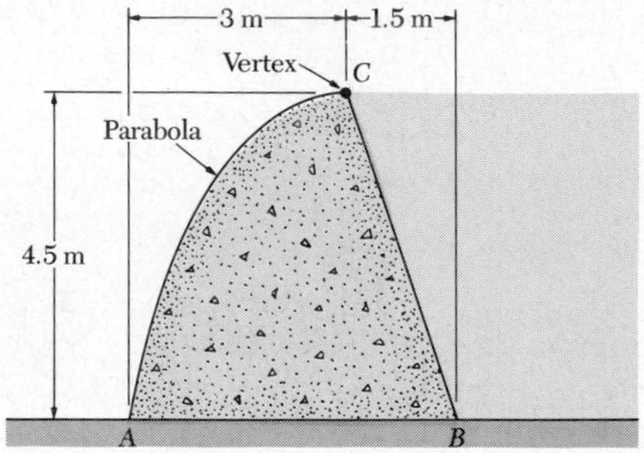

Fig. P5.84

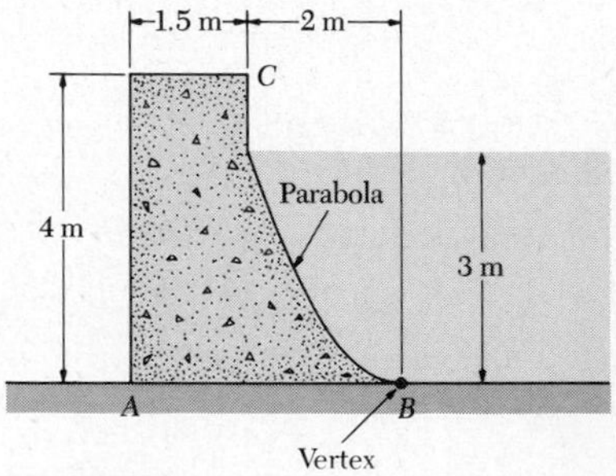

Fig. P5.85

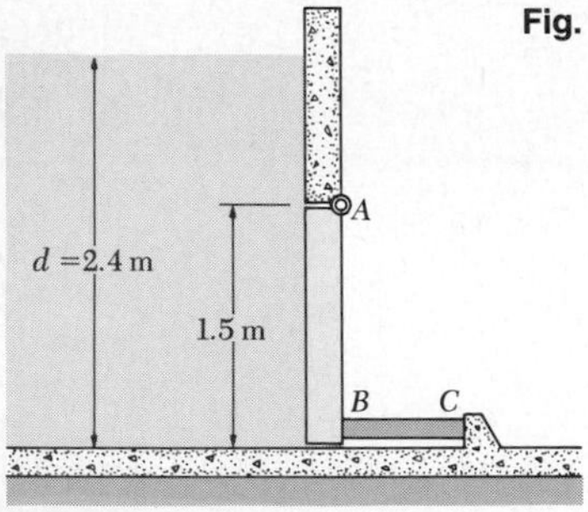

Fig. 5.86

5.84 and 5.85 The cross section of a concrete dam is as shown. Consider a section of the dam 1 ft thick, and determine (*a*) the resultant of the reaction forces exerted by the ground on the base AB of the dam, (*b*) the resultant of the pressure forces exerted by the water on the face BC of the dam.

5.86 A 1.5×1.5 m gate is located in a wall below water level. (*a*) Determine the magnitude and location of the resultant of the forces exerted by the water on the gate. (*b*) If the gate is hinged along its top edge A, determine the force exerted on the gate by the brace BC.

5.87 In Prob. 5.86, determine the depth of water d for which the force exerted on the gate by the brace BC is (*a*) 20 kN, (*b*) 9 kN.

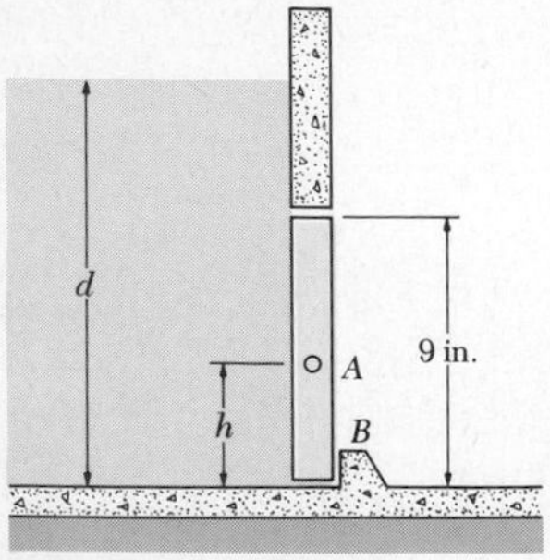

Fig. P5.88 and P5.89

5.88 An automatic valve consists of a 9×9 in. square plate which is pivoted about a horizontal axis through A located at a distance $h = 3.6$ in. above the lower edge. Determine the depth of water d for which the valve will open.

5.89 An automatic valve consists of a 9×9 in. square plate which is pivoted about a horizontal axis through A. If the valve is to open when the depth of water is $d = 18$ in., determine the distance h from the bottom of the valve to the pivot A.

5.90 The gate AB is located at the end of a 2-m-wide water channel and is supported by hinges along its top edge A. Knowing that the floor of the channel is frictionless, determine the reactions at A and B when $d = 2.4$ m.

5.91 Solve Prob. 5.90 when $d = 1.8$ m.

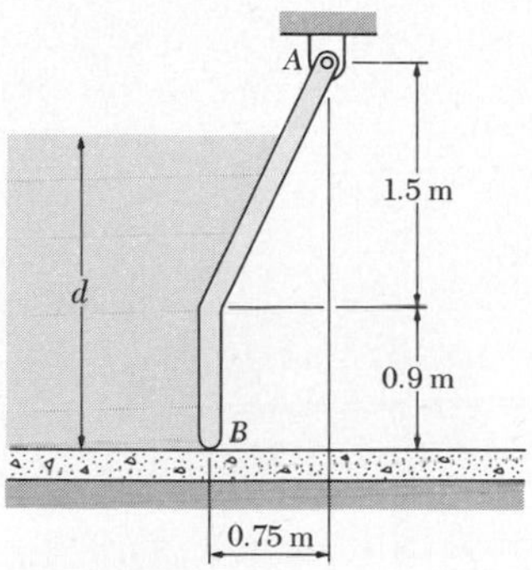

Fig. P5.90

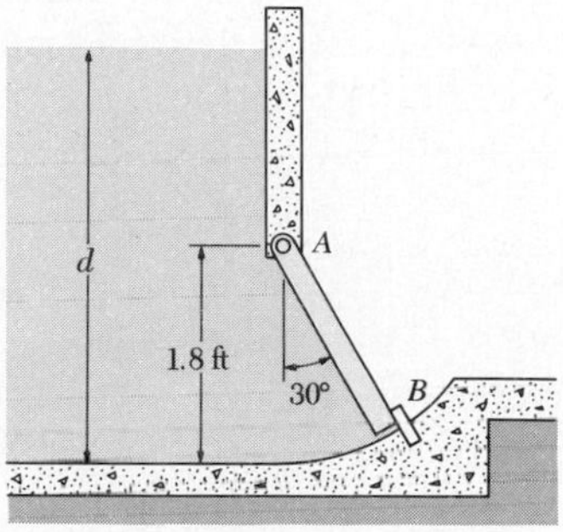

Fig. P5.92

5.92 The square gate AB is held in the position shown by hinges along its top edge A and by a shear pin at B. For a depth of water $d = 3.5$ ft, determine the force exerted on the gate by the shear pin.

5.93 Solve Prob. 5.92 when $d = 5$ ft.

5.94 A long trough is supported by a continuous hinge along its lower edge and by a series of horizontal cables attached to its upper edge. Determine the tension in each of the cables, at a time when the trough is completely full of water.

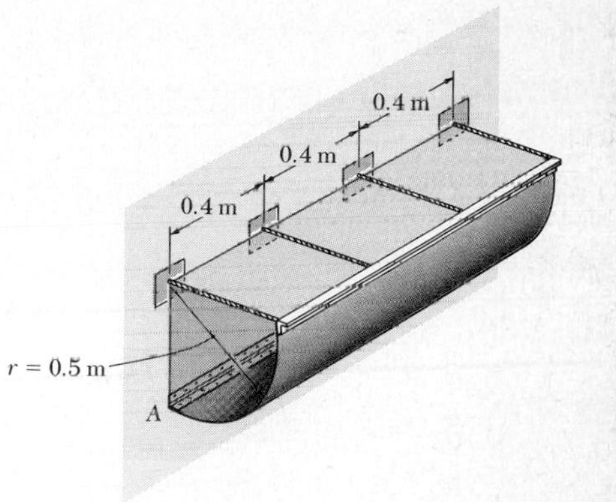

Fig. P5.94

5.95 Determine the minimum allowable value of the width a of the rectangular concrete dam if the dam is not to overturn about point A when $d = 10$ ft. Assume that a seal exists at B and that no water pressure is present under the dam.

5.96 Solve Prob. 5.95, assuming that a seal exists only at A and that the full hydrostatic pressure at B is present under the dam from A to B.

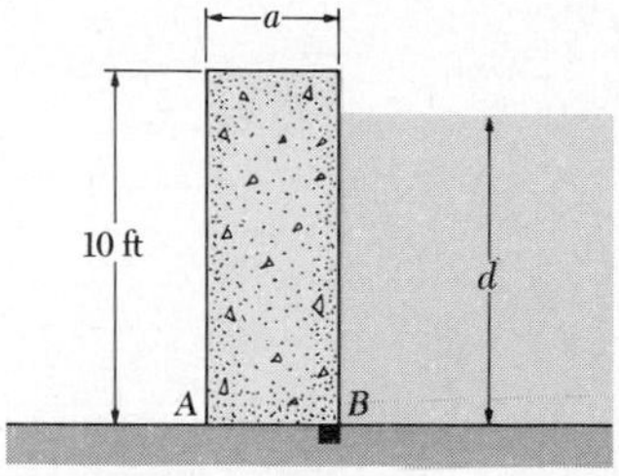

Fig. P5.95 and P5.97

5.97 Knowing that the width of the rectangular concrete dam is $a = 3$ ft, determine the maximum allowable value of the depth d of water if the dam is not to overturn about A. Assume that a seal exists at B and that no water pressure is present under the dam.

5.98 The end of a freshwater channel consists of a plate $ABCD$ which is hinged at B and is 0.4 m wide. Knowing that $a = 0.5$ m, determine the reactions at A and B.

5.99 The end of a freshwater channel consists of a plate $ABCD$ which is hinged at B and is 0.4 m wide. Determine the length a for which the reaction at A is zero.

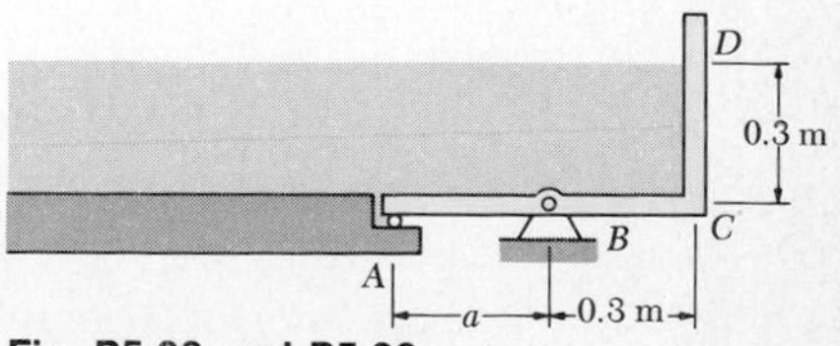

Fig. P5.98 and P5.99

VOLUMES

5.10. Center of Gravity of a Three-dimensional Body. Centroid of a Volume. The *center of gravity* G of a three-dimensional body is obtained by dividing the body into small elements and expressing that the weight $\mathbf{W}$ of the body attached at G is equivalent to the system of distributed forces $\Delta\mathbf{W}$ representing the weights of the small elements. Choosing the y axis vertical with positive sense upward (Fig. 5.20), and denoting by $\bar{\mathbf{r}}$ the position vector of G, we write that $\mathbf{W}$ is equal to the sum

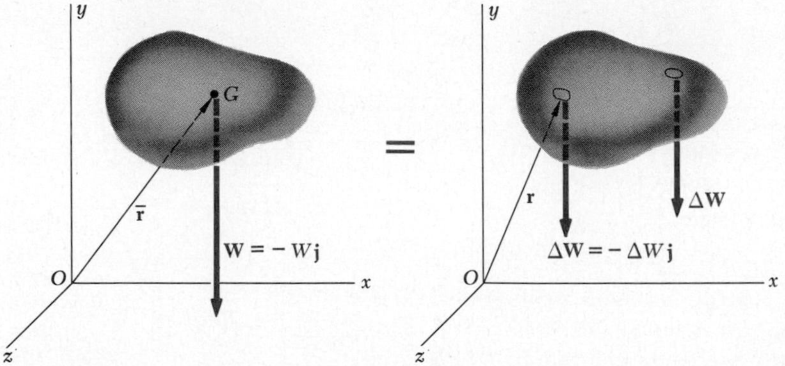

Fig. 5.20

of the elementary weights $\Delta\mathbf{W}$ and that its moment about O is equal to the sum of the moments about O of the elementary weights:

$$\Sigma\mathbf{F}: \qquad -W\mathbf{j} = \Sigma(-\Delta W\mathbf{j}) \tag{5.13}$$
$$\Delta\mathbf{M}_O: \qquad \bar{\mathbf{r}} \times (-W\mathbf{j}) = \Sigma[\mathbf{r} \times (-\Delta W\mathbf{j})]$$

Rewriting the last equation in the form

$$\bar{\mathbf{r}}W \times (-\mathbf{j}) = (\Sigma\mathbf{r}\,\Delta W) \times (-\mathbf{j}) \tag{5.14}$$

we observe that the weight $\mathbf{W}$ of the body will be equivalent to the system of the elementary weights $\Delta\mathbf{W}$ if the following conditions are satisfied:

$$W = \Sigma\,\Delta W \qquad \bar{\mathbf{r}}W = \Sigma\mathbf{r}\,\Delta W$$

Increasing the number of elements and simultaneously decreasing the size of each element, we obtain at the limit

$$W = \int dW \qquad \bar{\mathbf{r}}W = \int \mathbf{r}\,dW \tag{5.15}$$

We note that the relations obtained are independent of the orientation of the body. For example, if the body and the axes of coordinates were rotated so that the z axis pointed upward, the unit vector $-\mathbf{j}$ would be replaced by $-\mathbf{k}$ in Eqs. (5.13) and (5.14), but the relations (5.15) would remain unchanged. Resolving the vectors $\bar{\mathbf{r}}$ and $\mathbf{r}$ into rectangular components, we verify that the second of the relations (5.15) is equivalent to the three scalar equations

$$\bar{x}W = \int x\,dW \qquad \bar{y}W = \int y\,dW \qquad \bar{z}W = \int z\,dW \tag{5.16}$$

If the body is made of a homogeneous material of specific weight γ, the magnitude dW of the weight of an infinitesimal element may be expressed in terms of the volume dV of the element, and the magnitude W of the total weight in terms of the total volume V. We write

$$dW = \gamma\,dV \qquad W = \gamma V$$

Substituting for dW and W in the second of the relations (5.15), we write

$$\bar{\mathbf{r}}V = \int \mathbf{r}\,dV \tag{5.17}$$

or, in scalar form,

$$\bar{x}V = \int x\,dV \qquad \bar{y}V = \int y\,dV \qquad \bar{z}V = \int z\,dV \tag{5.18}$$

The point of coordinates $\bar{x}$, $\bar{y}$, $\bar{z}$ is also known as the *centroid C of the volume V* of the body. If the body is not homogeneous, Eqs. (5.18) cannot be used to determine the center of gravity of the body; they still define, however, the centroid of the volume.

The integral $\int x\,dV$ is known as the *first moment of the volume with respect to the yz plane*. Similarly, the integrals $\int y\,dV$ and $\int z\,dV$ define the first moments of the volume with respect to the zx plane and the xy plane, respectively. It is seen from Eqs. (5.18) that if the centroid of a volume is located in a coordinate plane, the first moment of the volume with respect to that plane is zero.

A volume is said to be symmetrical with respect to a given plane if to every point P of the volume corresponds a point P' of the same volume, such that the line PP' is perpendicular to the given plane and divided into two equal parts by that plane. The plane is said to be a *plane of symmetry* for the given volume. When a volume V possesses a plane of symmetry, the first moment of V with respect to that plane is zero and the centroid of the volume must be located in the plane of symmetry. When a volume possesses two planes of symmetry, the centroid of the volume must be located on the line of intersection of the two planes. Finally, when a volume possesses three planes of symmetry which intersect in a well-defined point (i.e., not along a common line), the point of intersection of the three planes must coincide with the centroid of the volume. This property enables us to determine immediately the centroid of the volume of spheres, ellipsoids, cubes, rectangular parallelepipeds, etc.

Centroids of unsymmetrical volumes or of volumes possessing only one or two planes of symmetry should be determined by integration (Sec. 5.12). Centroids of common shapes of volumes are shown in Fig. 5.21. It should be observed that the centroid of a volume of revolution in general *does not coincide* with the centroid of its cross section. Thus, the centroid of a hemisphere is different from that of a semicircular area, and the centroid of a cone is different from that of a triangle.

Shape		$\bar{x}$	Volume
Hemisphere	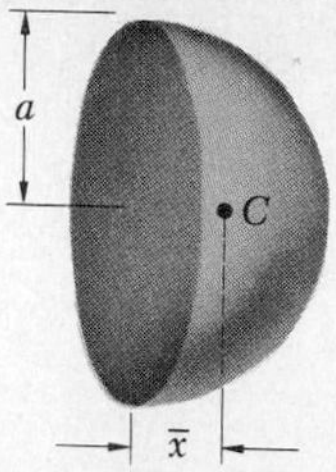	$\frac{3a}{8}$	$\frac{2}{3}\pi a^3$
Semiellipsoid of revolution	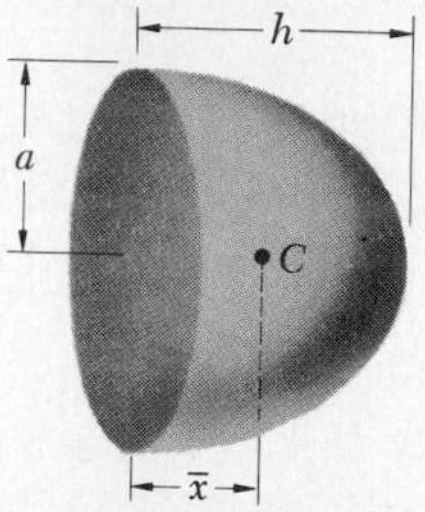	$\frac{3h}{8}$	$\frac{2}{3}\pi a^2 h$
Paraboloid of revolution	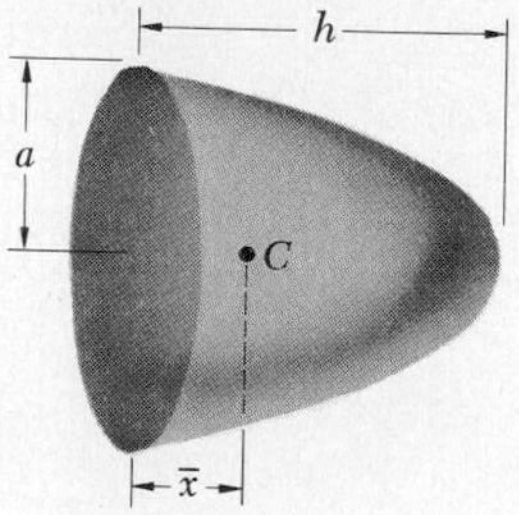	$\frac{h}{3}$	$\frac{1}{2}\pi a^2 h$
Cone	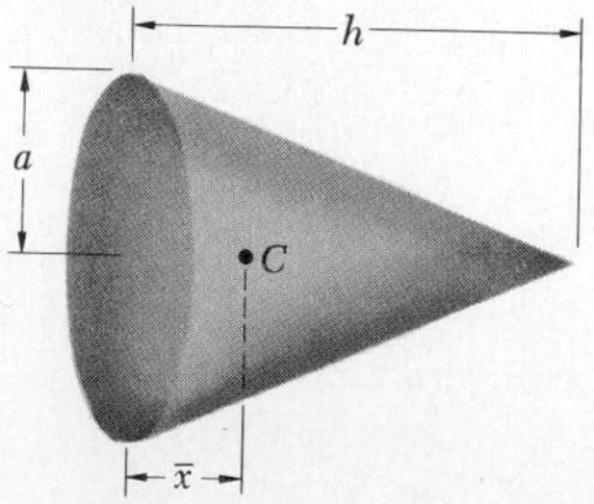	$\frac{h}{4}$	$\frac{1}{3}\pi a^2 h$
Pyramid	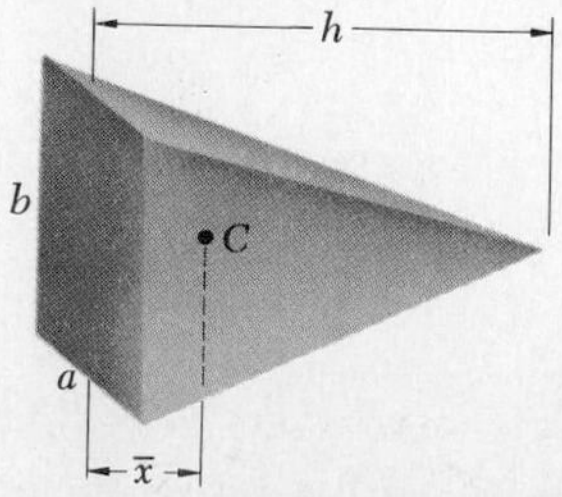	$\frac{h}{4}$	$\frac{1}{3}abh$

Fig. 5.21 Centroids of common shapes of volumes.

5.11. Composite Bodies. If a body can be divided into several of the common shapes shown in Fig. 5.21, its center of gravity G may be determined by expressing that the moment about O of its total weight is equal to the sum of the moments about O of the weights of the various component parts. Proceeding as in Sec. 5.10, we obtain the following equations defining the coordinates, $\bar{X}$, $\bar{Y}$, $\bar{Z}$ of the center of gravity G.

$$\bar{X}\Sigma W = \Sigma\bar{x}W \qquad \bar{Y}\Sigma W = \Sigma\bar{y}W \qquad \bar{Z}\Sigma W = \Sigma\bar{z}W \tag{5.19}$$

If the body is made of a homogeneous material, its center of gravity coincides with the centroid of its volume and the following equations may be used:

$$\bar{X}\Sigma V = \Sigma\bar{x}V \qquad \bar{Y}\Sigma V = \Sigma\bar{y}V \qquad \bar{Z}\Sigma V = \Sigma\bar{z}V \tag{5.20}$$

5.12. Determination of Centroids of Volumes by Integration. The centroid of a volume bounded by analytical surfaces may be determined by computing the integrals given in Sec. 5.10:

$$\bar{x}V = \int x\,dV \qquad \bar{y}V = \int y\,dV \qquad \bar{z}V = \int z\,dV \tag{5.21}$$

If the element of volume dV is chosen equal to a small cube of sides dx, dy, and dz, the determination of each of these integrals requires a *triple integration* in x, y, and z. However, it is possible to determine the coordinates of the centroid of most volumes by *double integration* if dV is chosen equal to the volume of a thin filament as shown in Fig. 5.22. The coordinates of the centroid of the volume are then obtained by writing

$$\bar{x}V = \int \bar{x}_{el}\,dV \qquad \bar{y}V = \int \bar{y}_{el}\,dV \qquad \bar{z}V = \int \bar{z}_{el}\,dV \tag{5.22}$$

and substituting for the volume dV and the coordinates $\bar{x}_{el}$, $\bar{y}_{el}$, $\bar{z}_{el}$ the expressions given in Fig. 5.22. Using the equation of the surface to express z in terms of x and y, the integration is reduced to a double integration in x and y.

If the volume under consideration possesses *two planes of symmetry*, its centroid must be located on their line of intersection. Choosing the x axis along this line, we have

$$\bar{y} = \bar{z} = 0$$

and the only coordinate to determine is $\bar{x}$. This will be done most conveniently by dividing the given volume into thin slabs parallel to the yz plane. In the particular case of a body of revolution, these slabs are circular; their volume dV is given in Fig. 5.23. Substituting for $\bar{x}_{el}$ and dV into the equation

$$\bar{x}V = \int \bar{x}_{el}\,dV \tag{5.23}$$

and expressing the radius r of the slab in terms of x, we may determine $\bar{x}$ by a single integration.

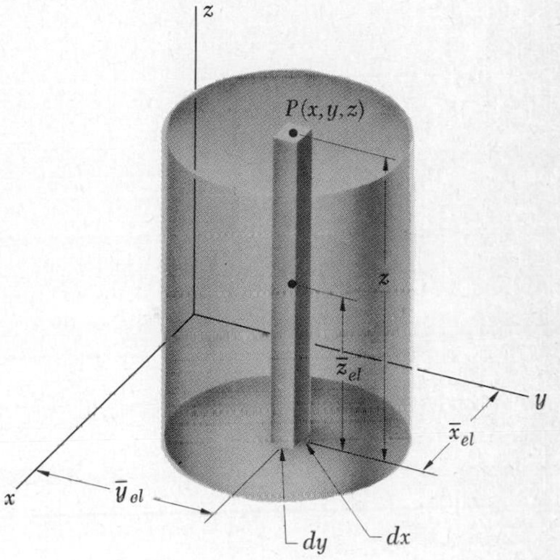

$\bar{x}_{el} = x,\ \bar{y}_{el} = y,\ \bar{z}_{el} = \frac{z}{2}$

$dV = z\,dx\,dy$

Fig. 5.22 Determination of the centroid of a volume by double integration.

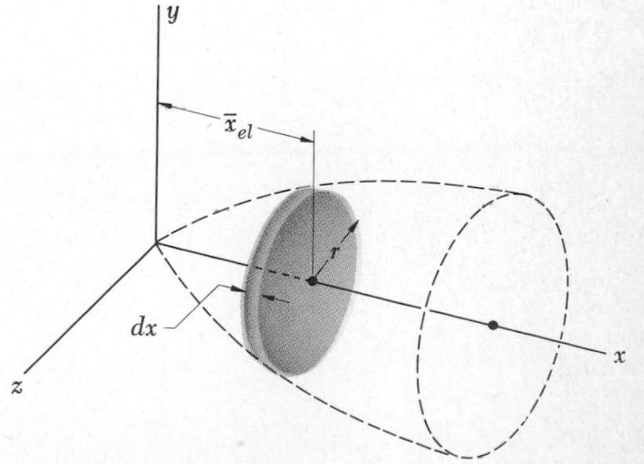

$\bar{x}_{el} = x$

$dV = \pi r^2\,dx$

Fig. 5.23 Determination of the centroid of a body of revolution.

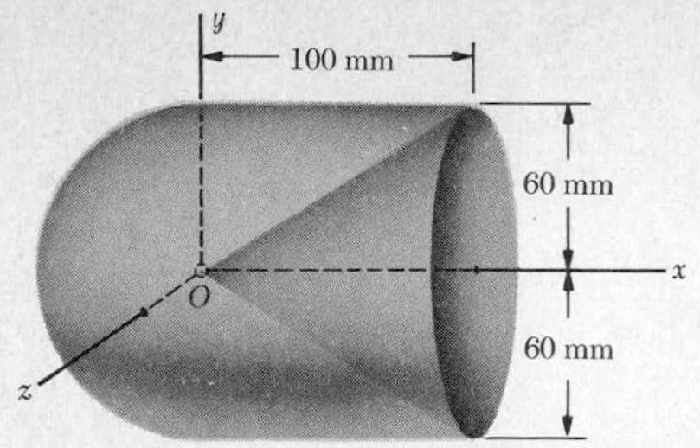

SAMPLE PROBLEM 5.11

Determine the center of gravity of the homogeneous body of revolution shown.

Solution. Because of symmetry, the center of gravity lies on the x axis. The body is seen to consist of a hemisphere, plus a cylinder, minus a cone, as shown. The volume and the abscissa of the centroid of each of these components are obtained from Fig. 5.21 and entered in the table below. The total volume of the body and the first moment of its volume with respect to the yz plane are then determined.

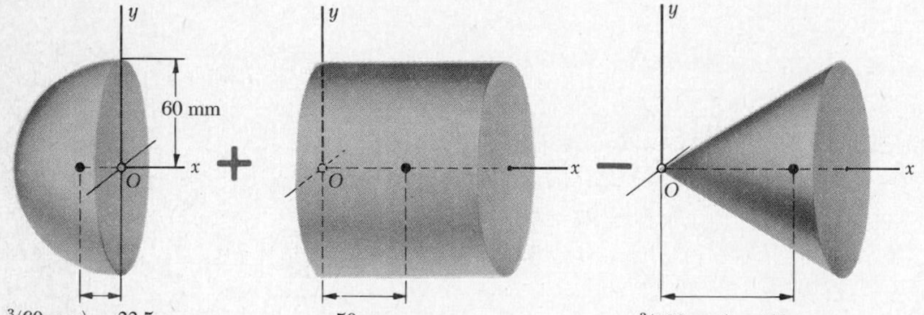

Component	Volume, mm^3	$\bar{x}$, mm	$\bar{x}V$, mm^4
Hemisphere	$\frac{1}{2}\frac{4\pi}{3}(60)^3 = 0.452 \times 10^6$	-22.5	-10.17×10^6
Cylinder	$\pi(60)^2(100) = 1.131 \times 10^6$	$+50$	$+56.55 \times 10^6$
Cone	$-\frac{\pi}{3}(60)^2(100) = -0.377 \times 10^6$	$+75$	-28.28×10^6
	$\Sigma V = 1.206 \times 10^6$	...	$\Sigma\bar{x}V = +18.10 \times 10^6$

Thus,

$\bar{X}\Sigma V = \Sigma\bar{x}V$: $\quad \bar{X}(1.206 \times 10^6 \text{ mm}^3) = 18.10 \times 10^6 \text{ mm}^4$

$\bar{X} = 15 \text{ mm}$ ◄

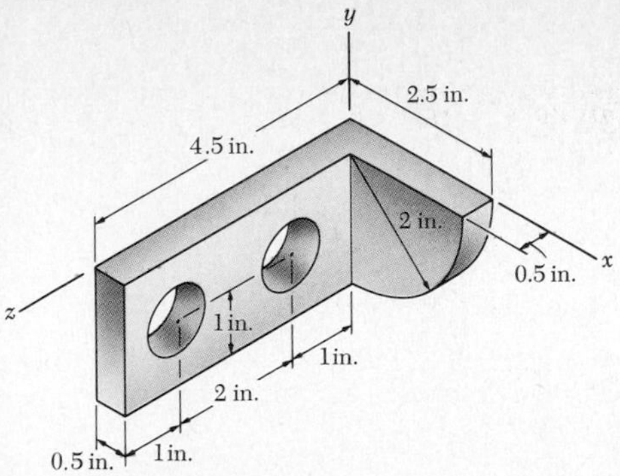

SAMPLE PROBLEM 5.12

Locate the center of gravity of the steel machine element shown. Both holes are of 1-in. diameter.

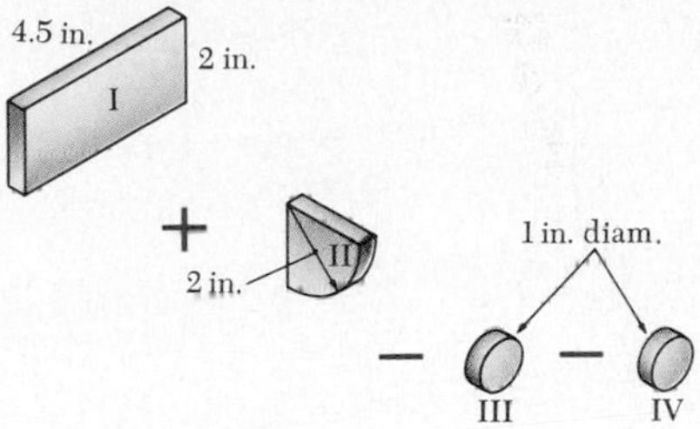

Solution. The machine element is seen to consist of a rectangular parallelepiped (I) plus a quarter cylinder (II) minus two 1-in.-diameter cylinders (III and IV). The volume and the coordinates of the centroid of each component are determined and entered in the table below. Using the data accumulated in the table, we then determine the total volume and the moments of the volume about each of the coordinate planes.

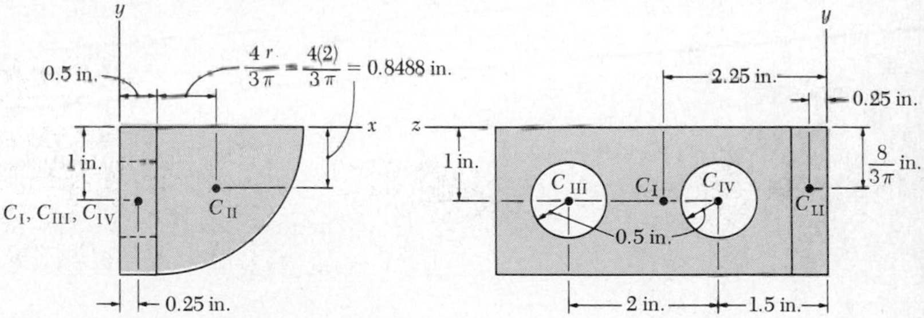

	V, in^3	$\bar{x}$, in.	$\bar{y}$, in.	$\bar{z}$, in.	$\bar{x}V$, in^4	$\bar{y}V$, in^4	$\bar{z}V$, in^4
I	$(4.5)(2)(0.5) = 4.5$	0.25	-1	2.25	1.125	-4.5	10.125
II	$\frac{1}{4}\pi(2)^2(0.5) = 1.571$	1.3488	-0.8488	0.25	2.119	-1.333	0.393
III	$-\pi(0.5)^2(0.5) = -0.3927$	0.25	-1	3.5	-0.098	0.393	-1.374
IV	$-\pi(0.5)^2(0.5) = -0.3927$	0.25	-1	1.5	-0.098	0.393	-0.589
	$\Sigma V = 5.286$				$\Sigma\bar{x}V = 3.048$	$\Sigma\bar{y}V = -5.047$	$\Sigma\bar{z}V = 8.555$

Thus,

$$\bar{X}\Sigma V = \Sigma\bar{x}V: \quad \bar{X}(5.286 \text{ in}^3) = 3.048 \text{ in}^4 \qquad \bar{X} = 0.577 \text{ in.} \blacktriangleleft$$

$$\bar{Y}\Sigma V = \Sigma\bar{y}V: \quad \bar{Y}(5.286 \text{ in}^3) = -5.047 \text{ in}^4 \qquad \bar{Y} = -0.955 \text{ in.} \blacktriangleleft$$

$$\bar{Z}\Sigma V = \Sigma\bar{z}V: \quad \bar{Z}(5.286 \text{ in}^3) = 8.555 \text{ in}^4 \qquad \bar{Z} = 1.618 \text{ in.} \blacktriangleleft$$

SAMPLE PROBLEM 5.13

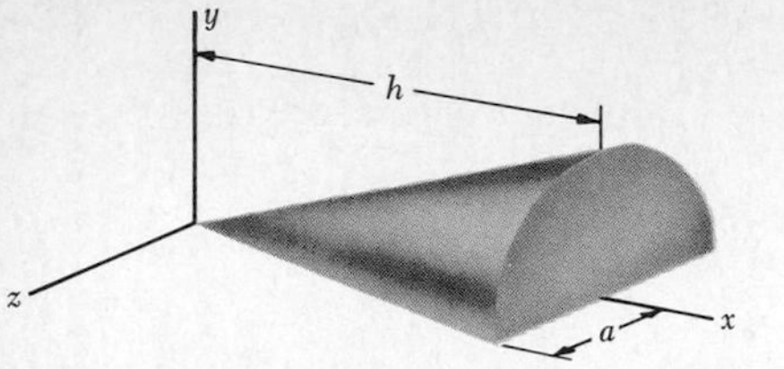

Determine the location of the centroid of the half right circular cone shown.

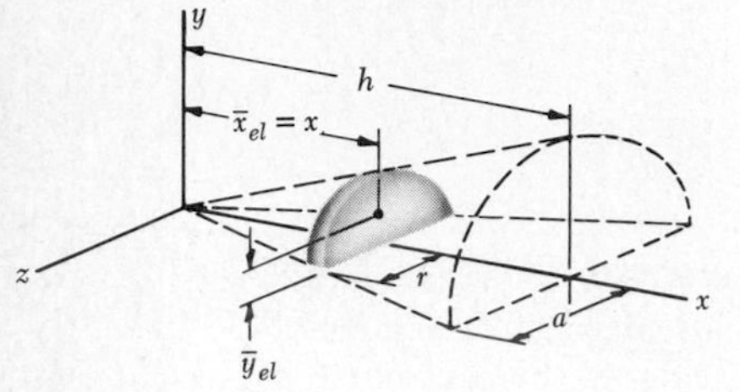

Solution. Since the xy plane is a plane of symmetry, the centroid lies in this plane and $\bar{z} = 0$. A slab of thickness dx is chosen as a differential element. The volume of this element is

$$dV = \tfrac{1}{2}\pi r^2\, dx$$

The coordinates $\bar{x}_{el}$ and $\bar{y}_{el}$ of the centroid of the element are obtained from Fig. 5.8 (semicircular area).

$$\bar{x}_{el} = x \qquad \bar{y}_{el} = \frac{4r}{3\pi}$$

We observe that r is proportional to x and write

$$\frac{r}{x} = \frac{a}{h} \qquad r = \frac{a}{h}x$$

The volume of the body is

$$V = \int dV = \int_0^h \tfrac{1}{2}\pi r^2\, dx = \int_0^h \tfrac{1}{2}\pi \left(\frac{a}{h}x\right)^2 dx = \frac{\pi a^2 h}{6}$$

The moment of the differential element with respect to the yz plane is $\bar{x}_{el}\, dV$; and the total moment of the body with respect to this plane is

$$\int \bar{x}_{el}\, dV = \int_0^h x(\tfrac{1}{2}\pi r^2)\, dx = \int_0^h x(\tfrac{1}{2}\pi)\left(\frac{a}{h}x\right)^2 dx = \frac{\pi a^2 h^2}{8}$$

Thus,

$$\bar{x}V = \int \bar{x}_{el}\, dV \qquad \bar{x}\frac{\pi a^2 h}{6} = \frac{\pi a^2 h^2}{8} \qquad \bar{x} = \tfrac{3}{4}h \blacktriangleleft$$

Likewise, the moment of the differential element with respect to the zx plane is $\bar{y}_{el}\, dV$; and the total moment is

$$\int \bar{y}_{el}\, dV = \int_0^h \frac{4r}{3\pi}(\tfrac{1}{2}\pi r^2)\, dx = \frac{2}{3}\int_0^h \left(\frac{a}{h}x\right)^3 dx = \frac{a^3 h}{6}$$

Thus,

$$\bar{y}V = \int \bar{y}_{el}\, dV \qquad \bar{y}\frac{\pi a^2 h}{6} = \frac{a^3 h}{6} \qquad \bar{y} = \frac{a}{\pi} \blacktriangleleft$$

Problems

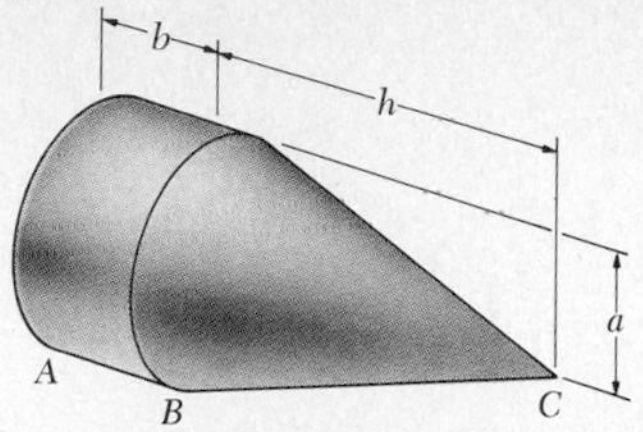

Fig. P5.100 and P5.101

5.100 Determine the location of the centroid of the composite body shown when $h = 3b$.

5.101 Determine the ratio h/b for which the centroid of the composite body is located in the plane between the cone and the cylinder.

5.102 Locate the centroid of the frustum of a right circular cone when $r_1 = 40$ mm, $r_2 = 50$ mm, and $h = 60$ mm.

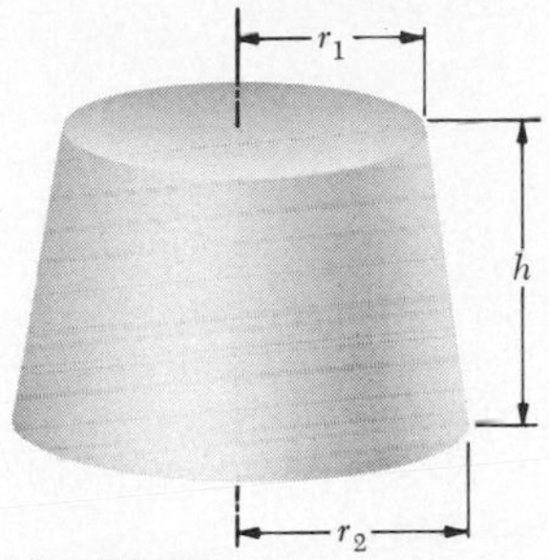

Fig. P5.102

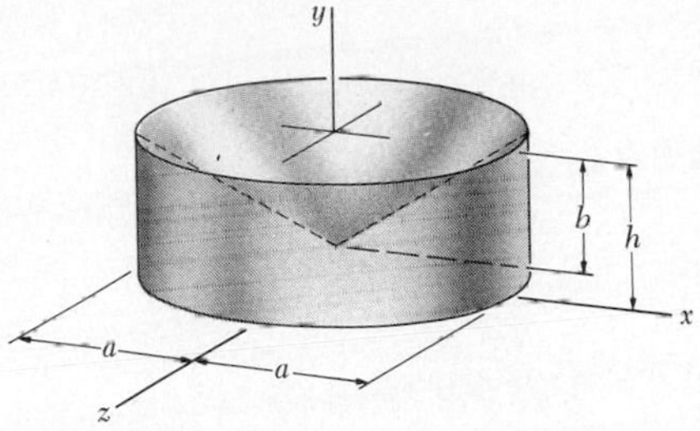

Fig. P5.103

5.103 Determine the y coordinate of the centroid of the body shown when (a) $b = \frac{1}{2}h$, (b) $b = h$.

5.104 Determine the location of the center of gravity of the parabolic reflector shown, which is formed by machining a rectangular block so that the curved surface is a paraboloid of revolution of base radius a and height h.

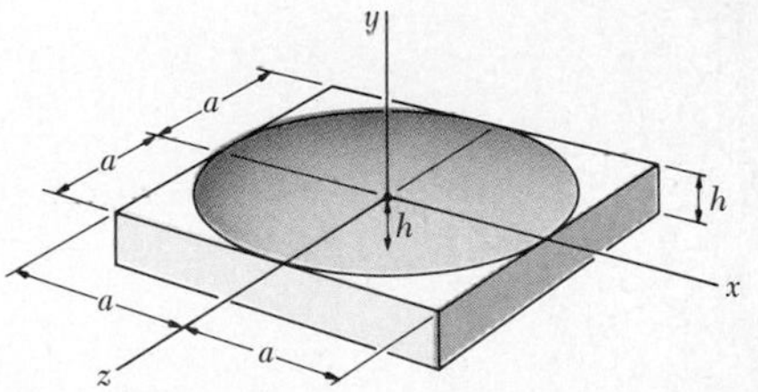

Fig. P5.104

5.105 For the machine element shown, locate the x coordinate of the center of gravity.

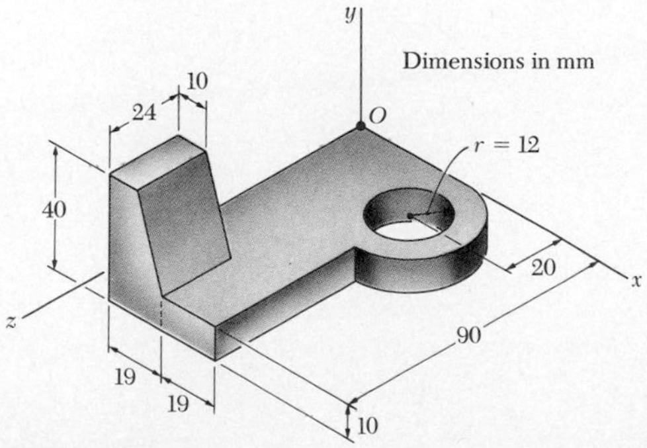

Fig. P5.105

5.106 and 5.107 For the machine element shown, locate the y coordinate of the center of gravity.

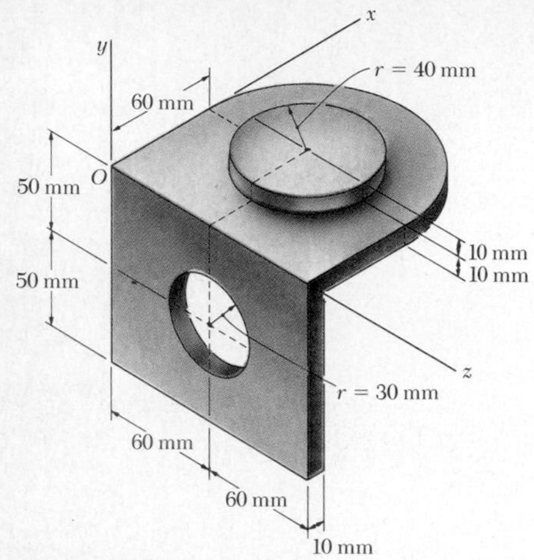

Fig. P5.106 and P5.109

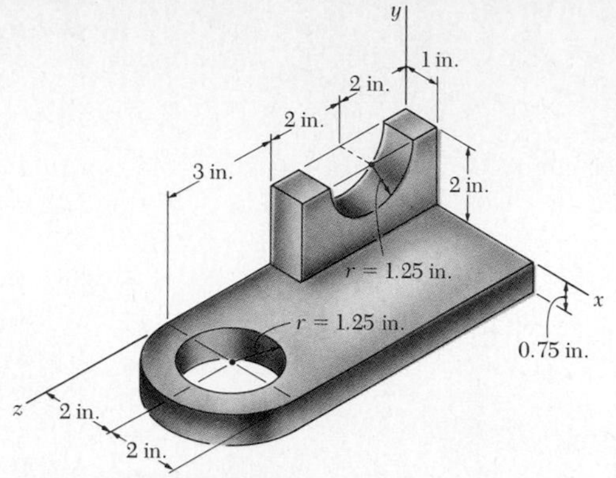

Fig. P5.107 and P5.108

5.108 For the machine element shown, locate the z coordinate of the center of gravity.

5.109 For the machine element shown, locate the x coordinate of the center of gravity.

5.110 For the machine element shown, locate the z coordinate of the center of gravity.

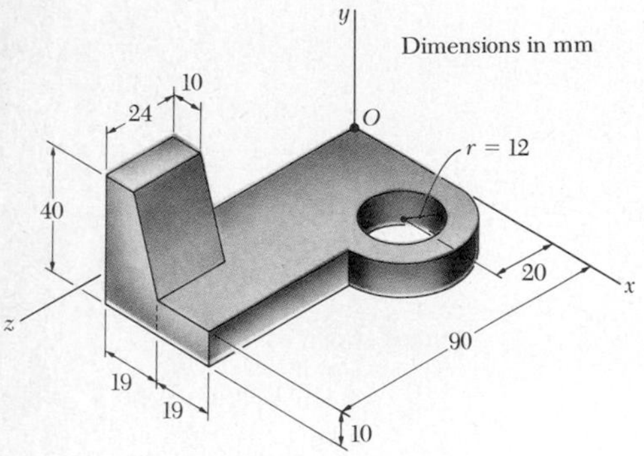

Fig. P5.110

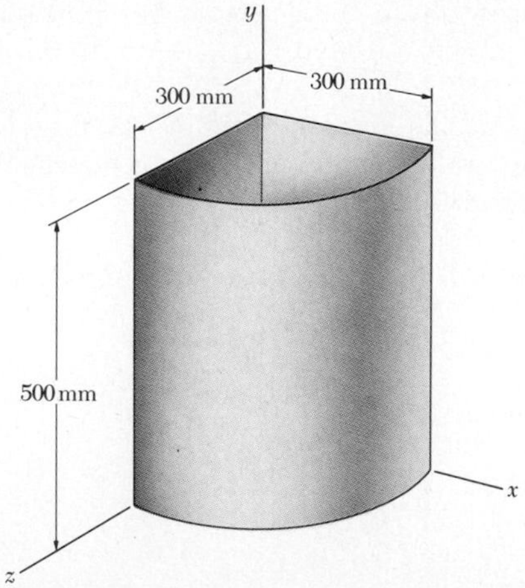

Fig. P5.111

5.111 A wastebasket, designed to fit in the corner of a room, is 500 mm high and has a base in the shape of a quarter circle of radius 300 mm. Locate the center of gravity of the wastebasket, knowing that it is made of sheet metal of uniform thickness.

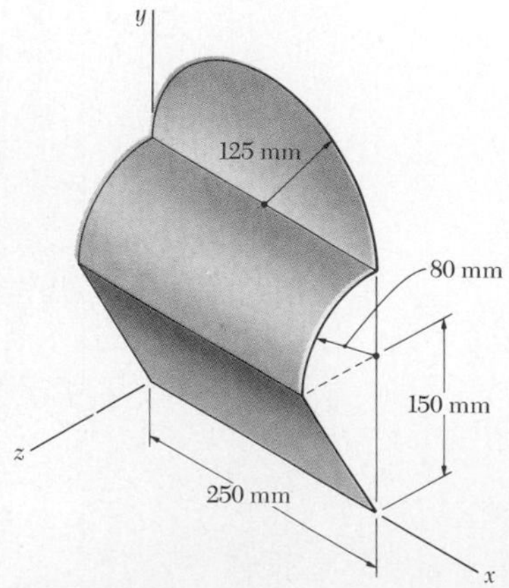

Fig. P5.112

5.112 Locate the center of gravity of the sheet-metal form shown.

5.113 Locate the center of gravity of the sheet-metal form shown.

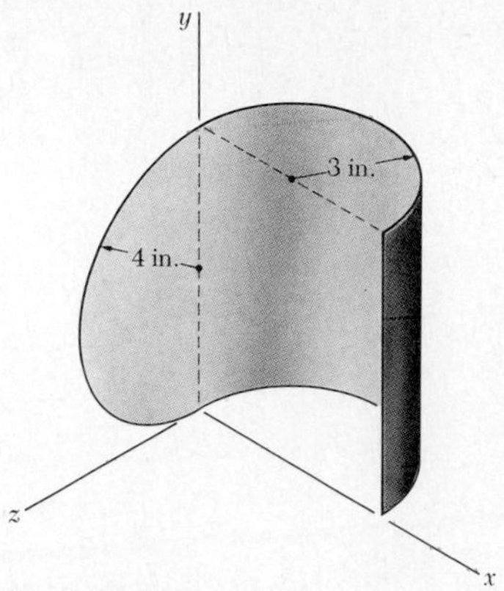

Fig. P5.113

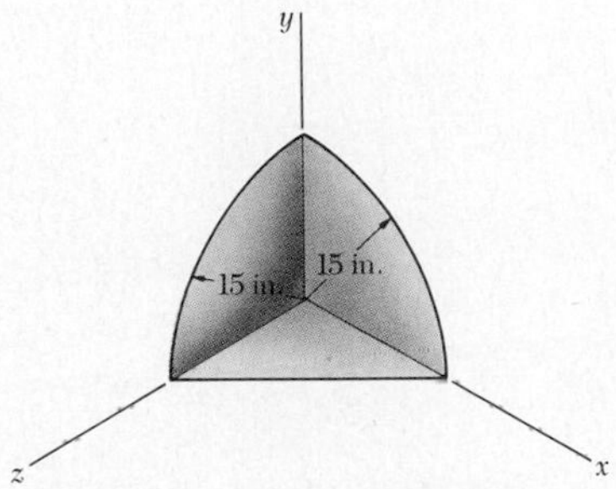

Fig. P5.114

5.114 A corner reflector for tracking by radar has two sides in the shape of a quarter circle of radius 15 in. and one side in the shape of a triangle. Locate the center of gravity of the reflector, knowing that it is made of sheet metal of uniform thickness.

5.115 and 5.116 Locate the center of gravity of the figure shown, knowing that it is made of thin brass rods of uniform diameter.

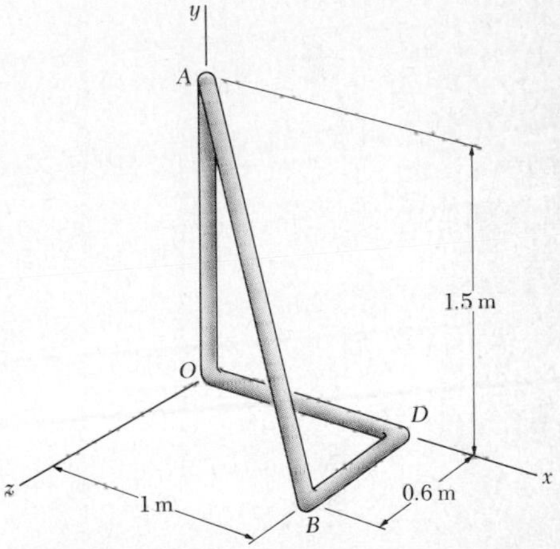

Fig. P5.115

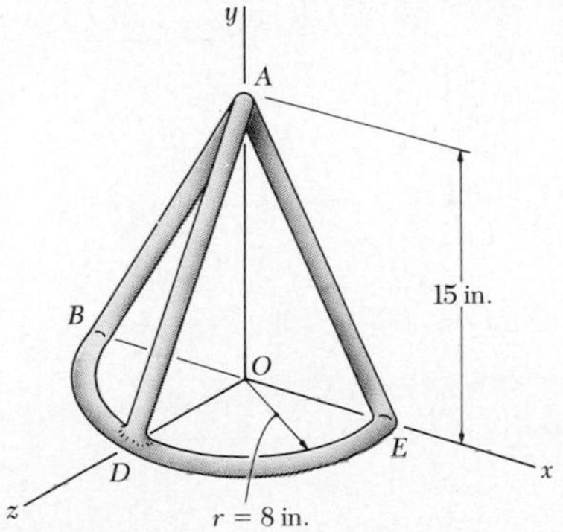

Fig. P5.116

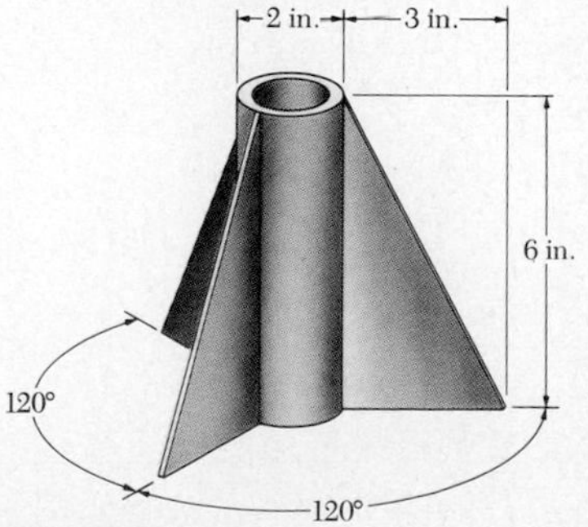

Fig. P5.117

5.117 Three brass plates are brazed to a steel pipe to form the flagpole base shown. Knowing that the pipe has a wall thickness of 0.25 in. and that each plate is 0.1875 in. thick, determine the location of the center of gravity of the base. (Specific weights: brass = $0.306\ \text{lb/in}^3$; steel = $0.284\ \text{lb/in}^3$.)

5.118 A brass collar, of length $h = 60$ mm, is mounted on an aluminum rod of length 100 mm. Locate the center of gravity of the composite body. (Densities: brass = 8470 kg/m^3; aluminum = 2800 kg/m^3.)

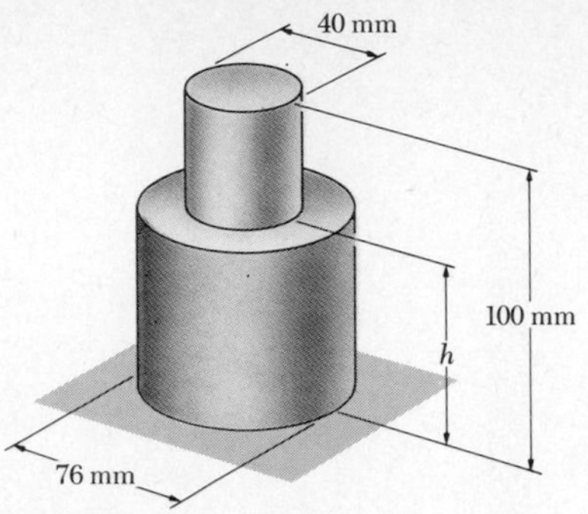

Fig. P5.118

5.119 In Prob. 5.118, determine (*a*) the length h for which the center of gravity of the composite body is as low as possible, (*b*) the corresponding location of the center of gravity.

5.120 through 5.122 Derive by direct integration the expression given for $\bar{x}$ in Fig. P5.21 for:

5.120 A hemisphere.
5.121 A semiellipsoid of revolution.
5.122 A paraboloid of revolution.

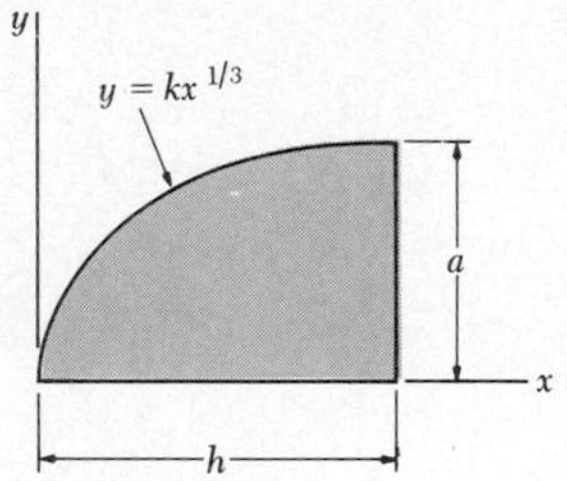

Fig. P5.124

5.123 Locate by direct integration the centroid of the frustum of the right circular cone of Prob. 5.102, expressing the result in terms of r_1, r_2, and h.

5.124 Locate the centroid of the volume obtained by rotating the shaded area about the x axis.

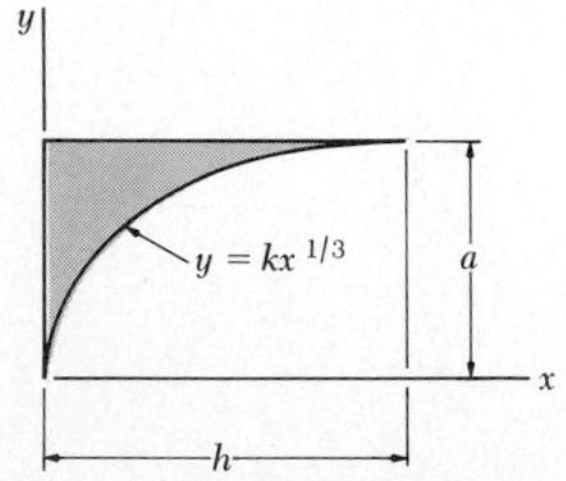

Fig. P5.125

5.125 Locate the centroid of the volume obtained by rotating the shaded area about the y axis.

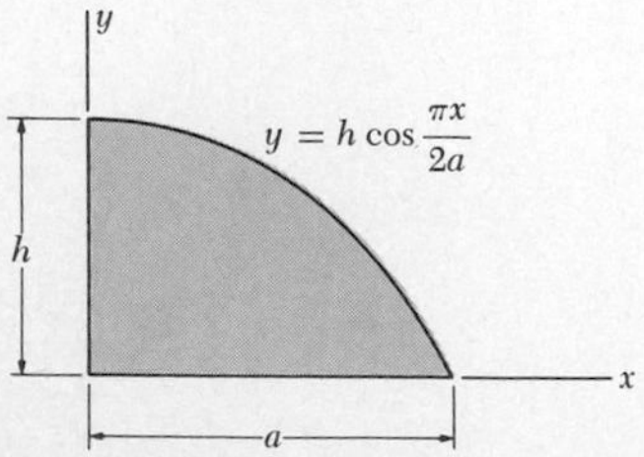

Fig. P5.126 and P5.127

* **5.126** Locate the centroid of the volume generated by revolving the portion shown of the cosine curve about the x axis.

* **5.127** Locate the centroid of the volume generated by revolving the portion shown of the cosine curve about the y axis. (*Hint.* Use as an element of volume a thin cylindrical shell of radius r and thickness dr.)

5.128 Locate the centroid of the volume of the irregular pyramid shown.

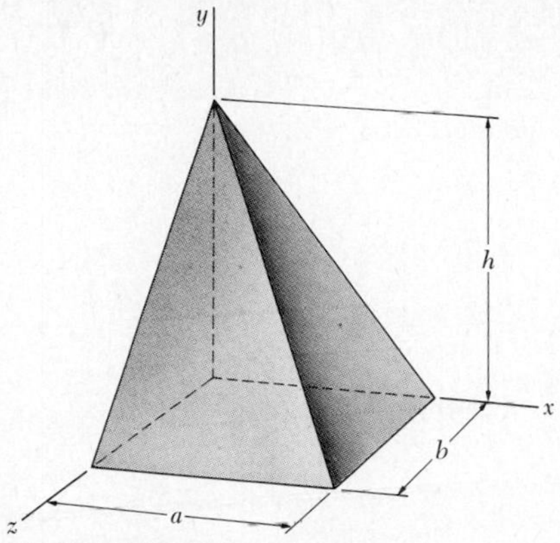

Fig. P5.128

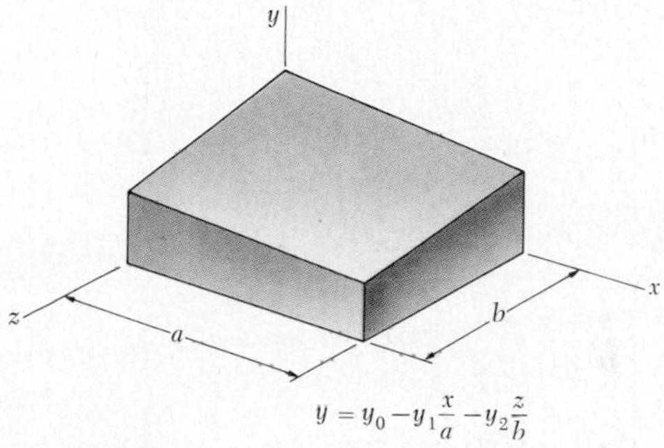

Fig. P5.129

5.129 Determine by direct integration the z coordinate of the centroid of the volume shown, which was cut from a rectangular prism by an oblique plane.

5.130 Solve Prob. 5.129, when $a = 120$ mm, $b = 90$ mm, $y_0 = 90$ mm, $y_1 = 40$ mm, and $y_2 = 20$ mm.

5.131 Locate the centroid of the section shown, which was cut from a thin circular pipe by an oblique plane.

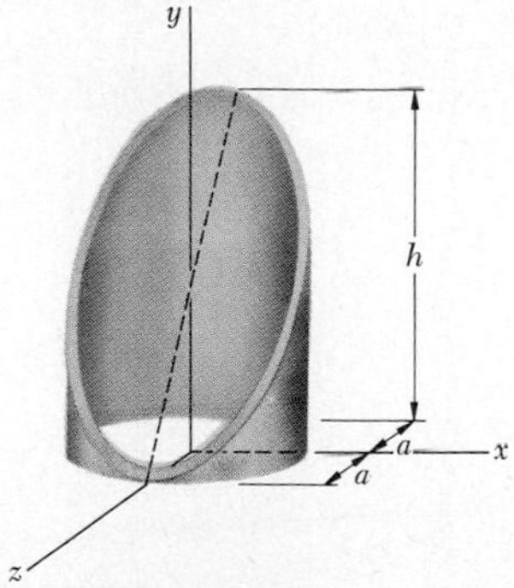

Fig. P5.131

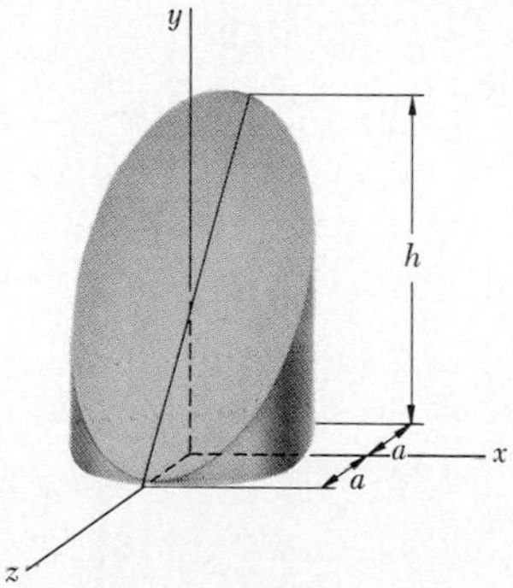

Fig. P5.132

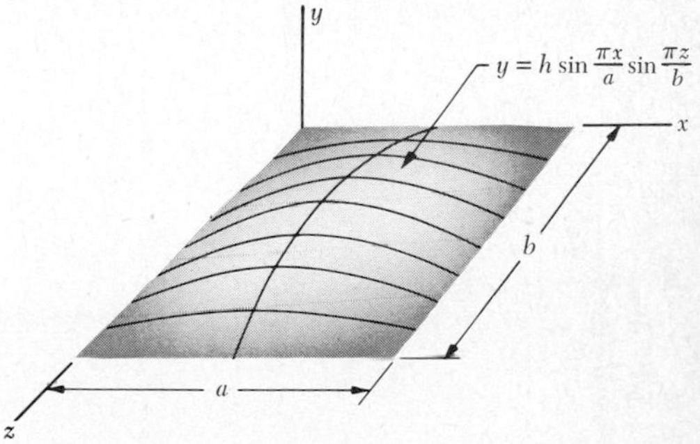

Fig. P5.133

5.132 Locate the centroid of the section shown, which was cut from a circular cylinder by an oblique plane.

5.133 Determine by direct integration the location of the centroid of the volume between the xz plane and the portion shown of the surface $y = h \sin(\pi x/a) \sin(\pi z/b)$.

Review Problems

5.134 Locate the centroid of the plane area shown.

5.135 Determine the volume and the total surface area of the solid generated by rotating the area shown about the y axis.

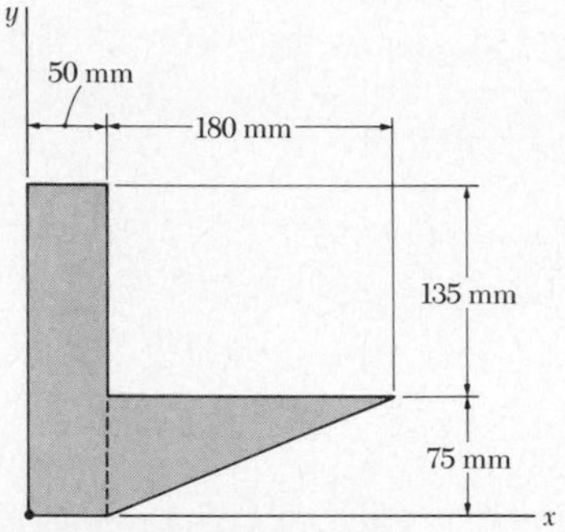

Fig. P5.134 and P5.135

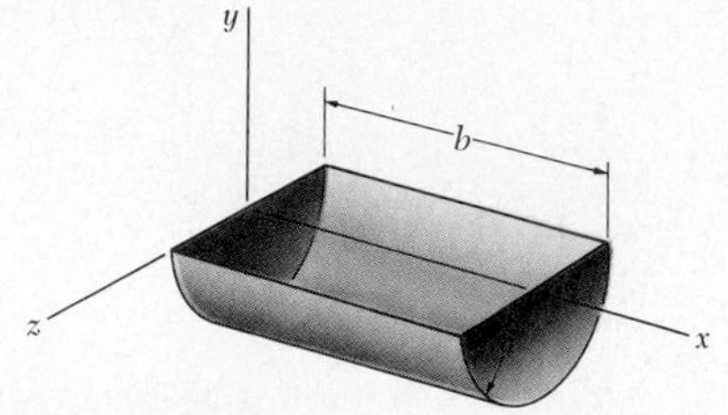

Fig. P5.136

5.136 The bin shown is made of sheet metal of uniform thickness. Determine (*a*) the center of gravity of the bin when $a = 8$ in. and $b = 24$ in., (*b*) the ratio a/b for which the y coordinate of the center of gravity is $\bar{y} = -\frac{1}{2}a$.

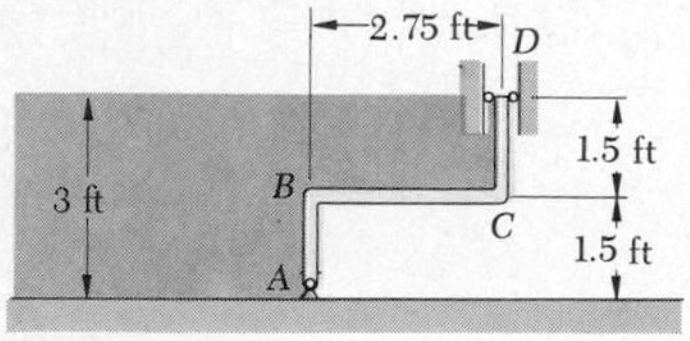

Fig. P5.137

5.137 The bent plate *ABCD* is 5 ft wide and is hinged at *A*. Determine the reactions at *A* and *D* for the water level shown.

5.138 Locate the centroid of the irregular pyramid shown. (*Hint.* See Fig. 5.21.)

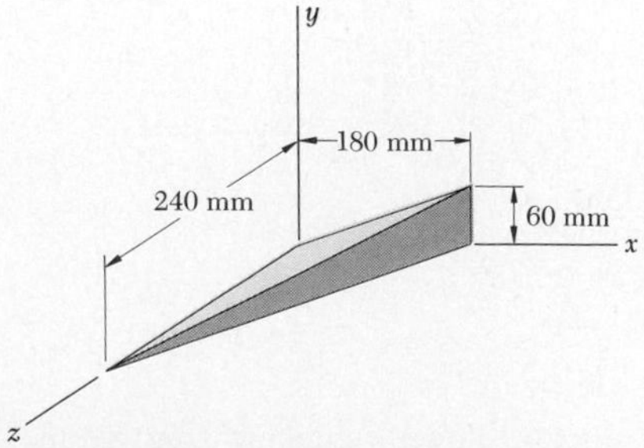

Fig. P5.138

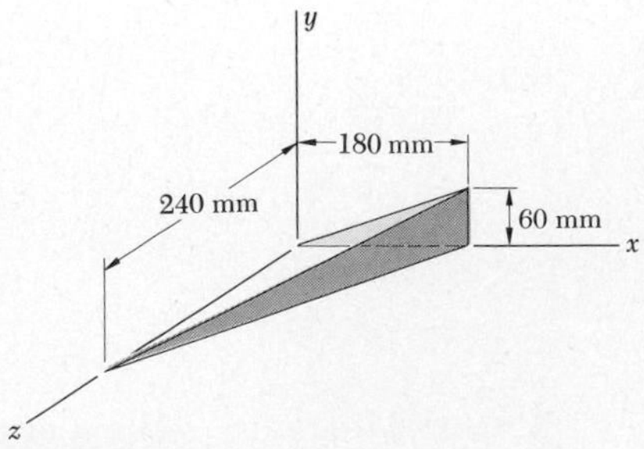

Fig. P5.139

5.139 Locate the center of gravity of the sheet-metal form shown.

5.140 The original design geometry of a plaza proposed for the corner entrance to a building is shown. Determine the change in the volume of the plaza if the dimensions used in the final design are (*a*) $a = 13$ m, $b = 22$ m, $h = 3$ m; (*b*) $a = 11$ m, $b = 22$ m, $h = 3$ m.

5.141 Solve Prob. 5.140, assuming that the final dimensions of the plaza are (*a*) $a = 9$ m, $b = 18$ m, $h = 3$ m; (*b*) $a = 9$ m, $b = 20$ m, $h = 3$ m.

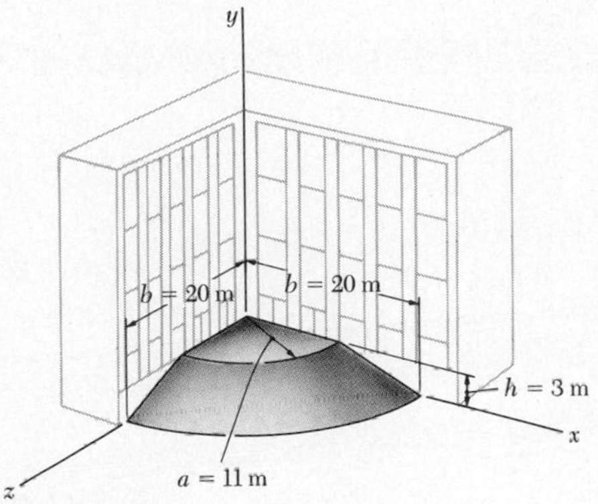

Fig. P5.140

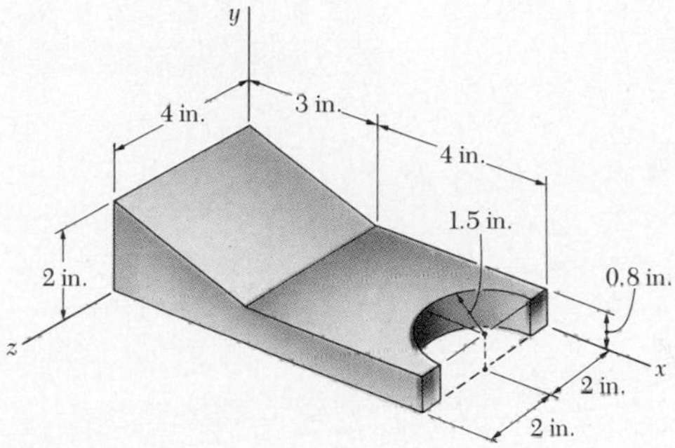

Fig. P5.142

5.142 Locate the center of gravity of the machine element shown.

5.143 Determine the reactions at the beam supports for the given loading condition.

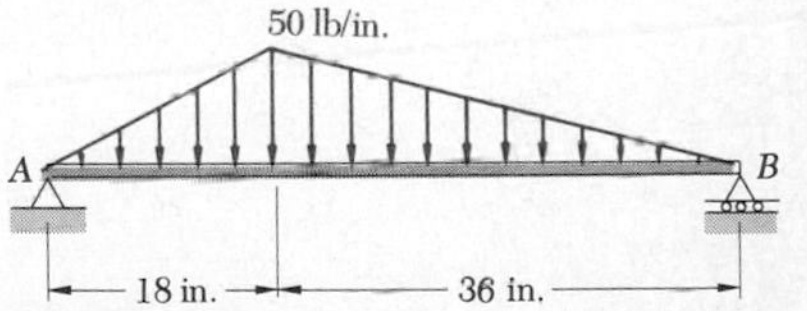

Fig. P5.143

5.144 The production-line balancing of an automobile speedometer cup is done as follows. The unbalanced cup is placed on a frictionless shaft at O and is allowed to come to rest. A hole is then punched at A; the cup rotates through an angle θ and again comes to rest. The balancing is completed by punching additional holes at B and C. Knowing that the three holes are equal in size and are at the same distance from the shaft O, determine the required angle α in terms of the angle θ.

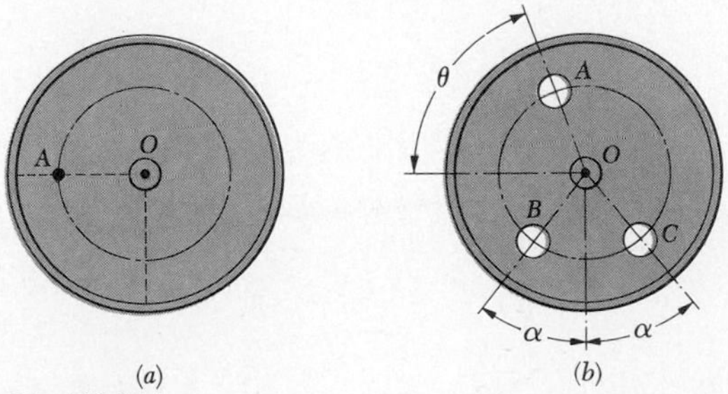

Fig. P5.144

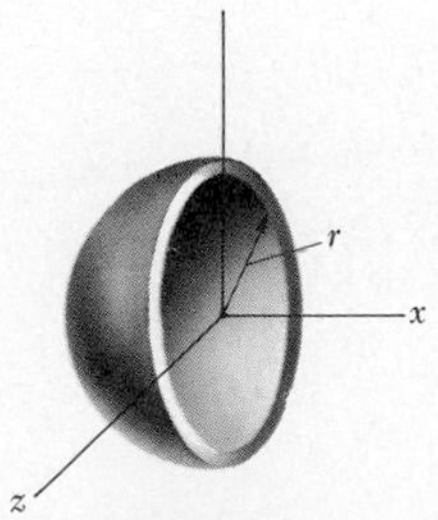

Fig. P5.145

5.145 Locate the center of gravity of a thin hemispherical shell of radius r and thickness t. (*Hint.* Consider the shell as formed by removing a hemisphere of radius r from a hemisphere of radius $r + t$; then neglect the terms containing t^2 and t^3, and keep those terms containing t.)

CHAPTER SIX

Analysis of Structures

6.1. Introduction. The problems considered in the preceding chapters concerned the equilibrium of a single rigid body, and all forces involved were external to the rigid body. We shall now consider problems dealing with the equilibrium of structures made of several connected parts. These problems call for the determination not only of the external forces acting on the structure but also of the forces which hold together the various parts of the structure. From the point of view of the structure as a whole, these forces are *internal forces*.

Consider, for example, the crane shown in Fig. 6.1*a*, which carries a load *W*. The crane consists of three beams *AD*, *CF*, and *BE* connected by frictionless pins; it is supported by a pin at *A* and by a cable *DG*. The free-body diagram of the crane has been drawn in Fig. 6.1*b*. The external

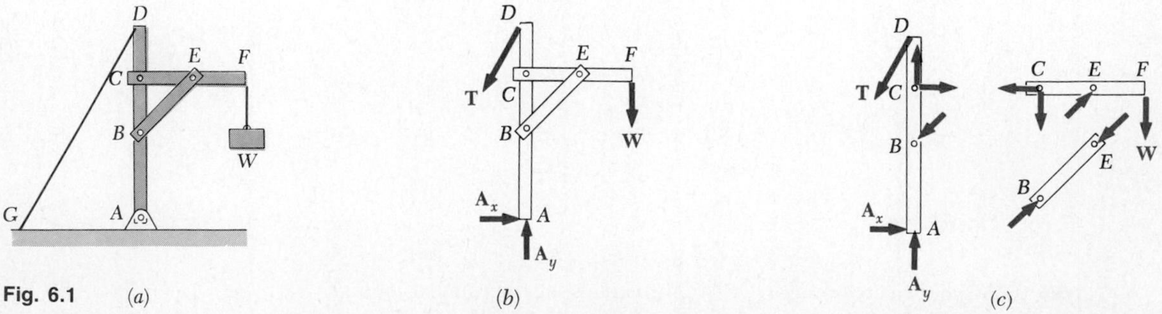

Fig. 6.1 (*a*) (*b*) (*c*)

forces are shown in the diagram and include the weight **W**, the two components $\mathbf{A}_x$ and $\mathbf{A}_y$ of the reaction at A, and the force **T** exerted by the cable at D. The internal forces holding the various parts of the crane together do not appear in the diagram. If, however, the crane is dismembered and if a free-body diagram is drawn for each of its component parts, the forces holding the three beams together must also be represented, since these forces are external forces from the point of view of each component part (Fig. 6.1*c*).

It will be noted that the force exerted at B by member BE on member AD has been represented as equal and opposite to the force exerted at the same point by member AD on member BE; similarly, the force exerted at E by BE on CF is shown equal and opposite to the force exerted by CF on BE; and the components of the force exerted at C by CF on AD are shown equal and opposite to the components of the force exerted by AD on CF. This is in conformity with Newton's third law, which states that *the forces of action and reaction between bodies in contact have the same magnitude, same line of action, and opposite sense.* As pointed out in Chap. 1, this law is one of the six fundamental principles of elementary mechanics and is based on experimental evidence. Its application is essential to the solution of problems involving connected bodies.

In this chapter, we shall consider three broad categories of engineering structures:

1. *Trusses,* which are designed to support loads and are usually stationary, fully constrained structures. Trusses consist exclusively of straight members connected at joints located at the ends of each member. Members of a truss, therefore, are *two-force members,* i.e., members acted upon by two equal and opposite forces directed along the member.

2. *Frames,* which are also designed to support loads and are also usually stationary, fully constrained structures. However, like the crane of Fig. 6.1, frames always contain at least one *multiforce member,* i.e., a member acted upon by three or more forces which, in general, are not directed along the member.

3. *Machines,* which are designed to transmit and modify forces and are structures containing moving parts. Machines, like frames, always contain at least one multiforce member.

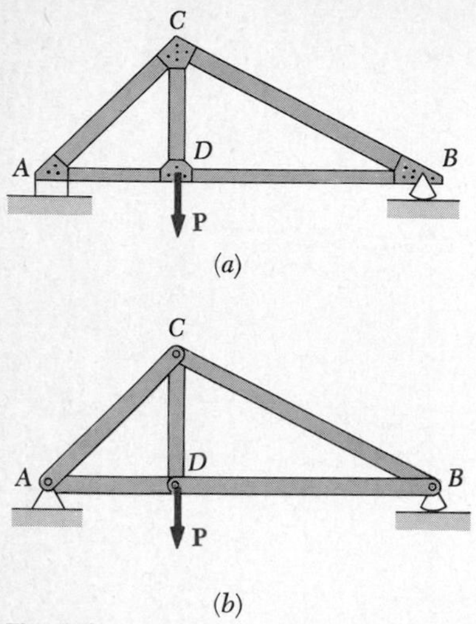

Fig. 6.2

6.2. Definition of a Truss. The truss is one of the major types of engineering structures. It provides both a practical and an economical solution to many engineering situations, especially in the design of bridges and buildings. A truss consists of straight members connected at joints; a typical truss is shown in Fig. 6.2*a*. Truss members are connected at their extremities only; thus no member is continuous through a joint. In Fig. 6.2*a*, for example, there is no member *AB*; there are instead two distinct members *AD* and *DB*. Most actual structures are made of several trusses joined together to form a space framework. Each truss is designed to carry those loads which act in its plane and thus may be treated as a two-dimensional structure.

In general, the members of a truss are slender and can support little lateral load; all loads, therefore, must be applied to the various joints, and not to the members themselves. When a concentrated load is to be applied between two joints, or when a distributed load is to be supported by the truss, as in the case of a bridge truss, a floor system must be provided which, through the use of stringers and floor beams, transmits the load to the joints (Fig. 6.3).

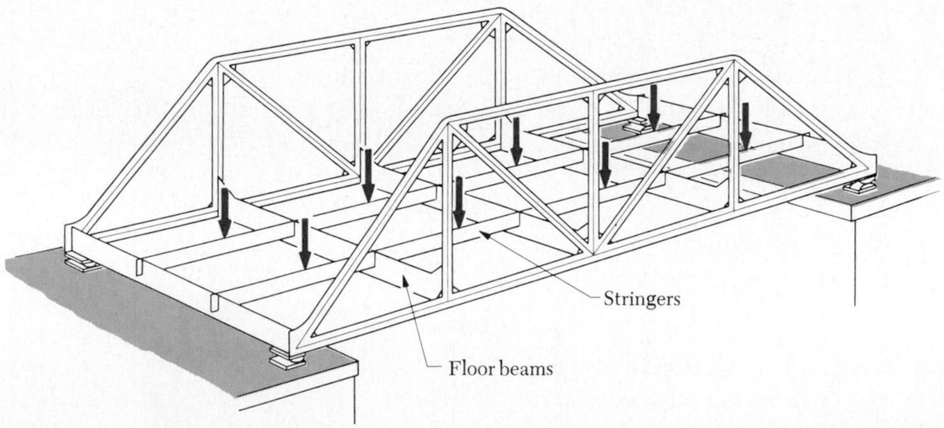

Fig. 6.3

The weights of the members of the truss are also assumed to be applied to the joints, half of the weight of each member being applied to each of the two joints the member connects. Although the members are actually joined together by means of riveted or welded connections, it is customary to assume that the members are pinned together; therefore, the forces acting at each end of a member reduce to a single force and no couple. Thus, the only forces assumed to be applied to a truss member are a single force at each end of the member. Each member may then be treated as a two-force member, and the entire truss may be considered as a group of pins and two-force members (Fig. 6.2*b*). An individual member may be acted upon as shown in either of the two sketches of Fig. 6.4. In the first sketch, the forces tend to pull the member apart, and the member is in tension, while in the second sketch, the forces tend to compress the member, and the member is in compression. Several typical trusses are shown in Fig. 6.5.

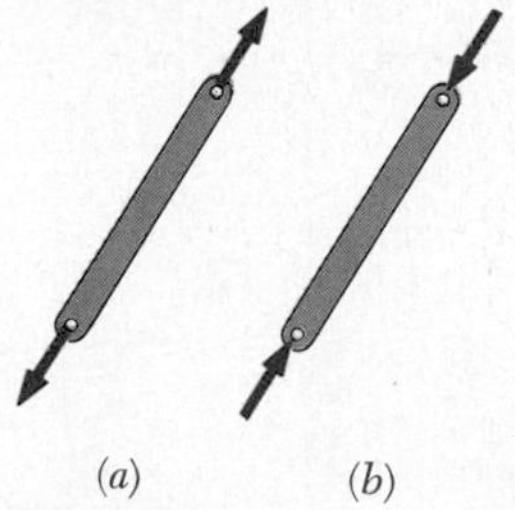

Fig. 6.4

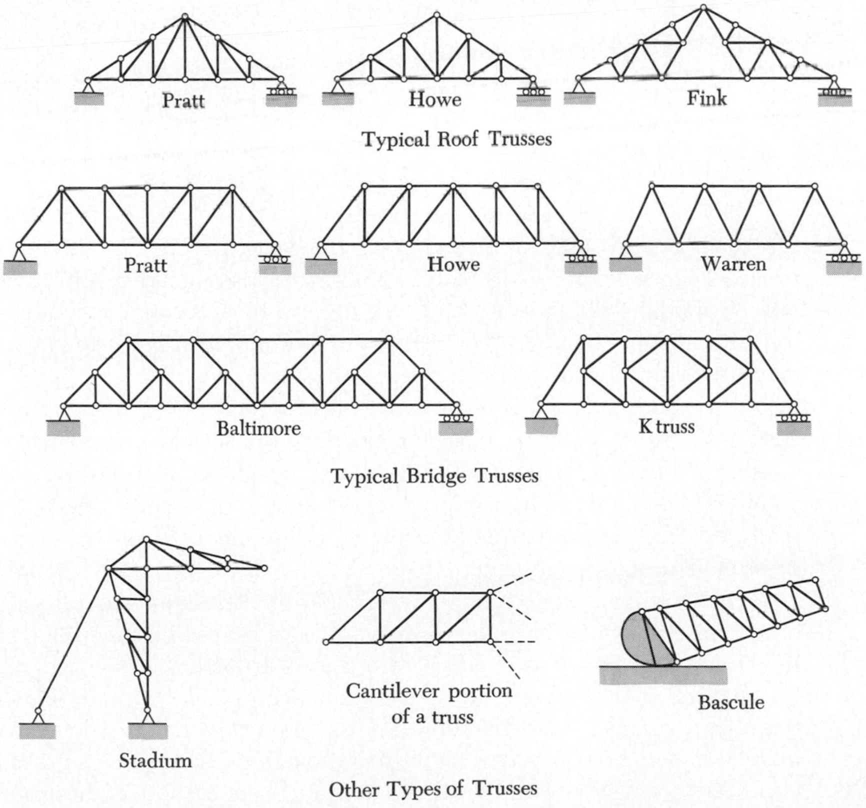

Fig. 6.5

6.3. Simple Trusses. Consider the truss of Fig. 6.6*a*, which is made of four members connected by pins at *A*, *B*, *C*, and *D*. If a load is applied at *B*, the truss will greatly deform and lose completely its original shape. On the other hand, the truss of Fig. 6.6*b*, which is made of three members connected by pins at *A*, *B*, and *C*, will deform only slightly under a load applied at *B*. The only possible deformation for this truss is one involving small changes in the length of its members. The truss of Fig. 6.6*b* is said to be a *rigid truss,* the term rigid being used here to indicate that the truss *will not collapse.*

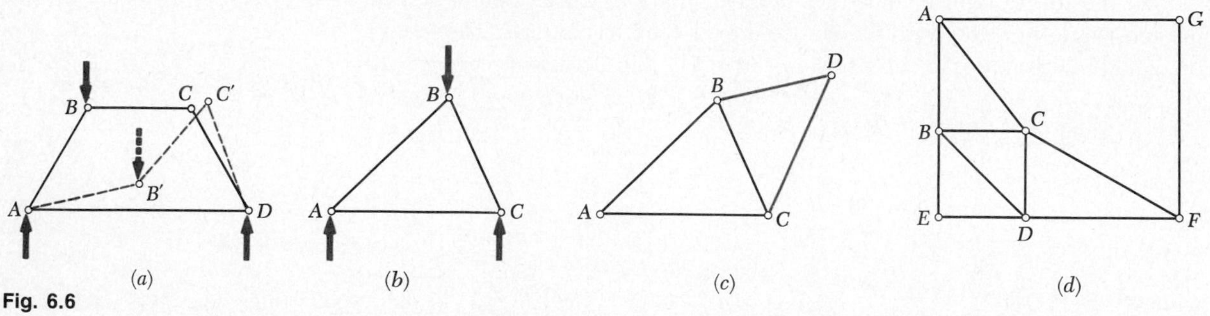

Fig. 6.6

As shown in Fig. 6.6*c*, a larger rigid truss may be obtained by adding two members *BD* and *CD* to the basic triangular truss of Fig. 6.6*b*. This procedure may be repeated as many times as desired, and the resulting truss will be rigid if each time we add two new members, we attach them to separate existing joints and connect them at a new joint.† A truss which may be constructed in this manner is called a *simple truss*.

It should be noted that a simple truss is not necessarily made only of triangles. The truss of Fig. 6.6*d*, for example, is a simple truss which was constructed from triangle *ABC* by adding successively the joints *D*, *E*, *F*, and *G*. On the other hand, rigid trusses are not always simple trusses, even when they appear to be made of triangles. The Fink and Baltimore trusses shown in Fig. 6.5, for instance, are not simple trusses, since they cannot be constructed from a single triangle in the manner described above. All the other trusses shown in Fig. 6.5 are simple trusses, as may be easily checked. (For the K truss, start with one of the central triangles.)

Returning to the basic triangular truss of Fig. 6.6*b*, we note that this truss has three members and three joints. The truss of Fig. 6.6*c* has two more members and one more joint, i.e., altogether five members and four joints. Observing that every time two new members are added, the number of joints is increased by one, we find that in a simple truss the total number of members is $m = 2n - 3$, where n is the total number of joints.

† The three joints must not be in a straight line.

6.4. Analysis of Trusses by the Method of Joints. We saw in Sec. 6.2 that a truss may be considered as a group of pins and two-force members. The truss of Fig. 6.2, whose free-body diagram is shown in Fig. 6.7*a*, may thus be dismembered, and a free-body diagram can be drawn for each pin and each member (Fig. 6.7*b*). Each member is acted upon by two forces, one at each end; these forces have the same magnitude, same line of action, and opposite sense (Sec. 4.6). Besides, Newton's third law indicates that the forces of action and reaction between a member and a pin are equal and opposite. Therefore, the forces exerted by a member on the two pins it connects must be directed along that member and be equal and opposite. The common magnitude of the forces exerted by a member on the two pins it connects is commonly referred to as the *force in the member* considered, even though this quantity is actually a scalar. Since the lines of action of all the internal forces in a truss are known, the analysis of a truss reduces to computing the forces in its various members and to determining whether each of its members is in tension or in compression.

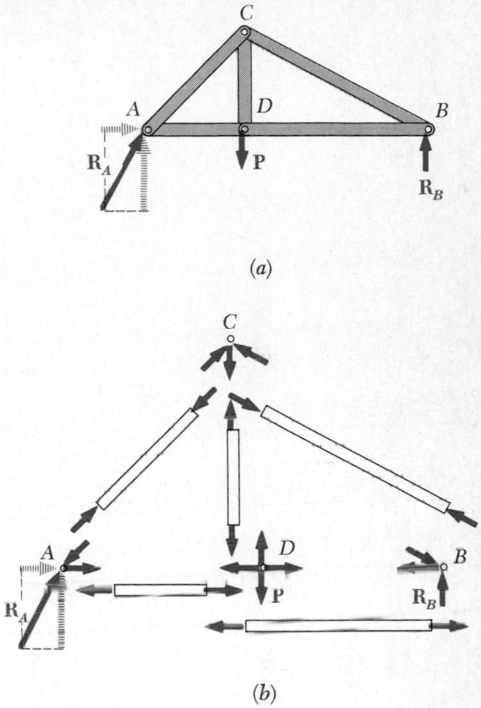

Fig. 6.7

Since the entire truss is in equilibrium, each pin must be in equilibrium. The fact that a pin is in equilibrium may be expressed by drawing its free-body diagram and writing two equilibrium equations (Sec. 2.9). If the truss contains n pins, there will be therefore $2n$ equations available, which may be solved for $2n$ unknowns. In the case of a simple truss, we have $m = 2n - 3$, that is, $2n = m + 3$, and the number of unknowns which may be determined from the free-body diagrams of the pins is thus $m + 3$. This means that the forces in all the members, as well as the two components of the reaction $\mathbf{R}_A$, and the reaction $\mathbf{R}_B$ may be found by considering the free-body diagrams of the pins.

The fact that the entire truss is a rigid body in equilibrium may be used to write three more equations involving the forces shown in the free-body diagram of Fig. 6.7*a*. Since they do not contain any new information, these equations are not independent from the equations associated with the free-body diagrams of the pins. Nevertheless, they may be used to determine immediately the components of the reactions at the supports. The arrangement of pins and members in a simple truss is such that it will then always be possible to find a joint involving only two unknown forces. These forces may be determined by the methods of Sec. 2.11 and their values transferred to the adjacent joints and treated as known quantities at these joints. This procedure may be repeated until all unknown forces have been determined.

As an example, we shall analyze the truss of Fig. 6.7 by considering successively the equilibrium of each pin, starting with a joint at which only two forces are unknown. In the truss considered, all pins are subjected to at least three unknown forces. Therefore, the reactions at the supports must first be determined by considering the entire truss as a free body and using the equations of equilibrium of a rigid body. We find in this way that $\mathbf{R}_A$ is vertical and determine the magnitudes of $\mathbf{R}_A$ and $\mathbf{R}_B$.

The number of unknown forces at joint A is thus reduced to two, and these forces may be determined by considering the equilibrium of pin A. The reaction $\mathbf{R}_A$ and the forces $\mathbf{F}_{AC}$ and $\mathbf{F}_{AD}$ exerted on pin A by members

AC and AD, respectively, must form a force triangle. First we draw $\mathbf{R}_A$ (Fig. 6.8); noting that $\mathbf{F}_{AC}$ and $\mathbf{F}_{AD}$ are directed along AC and AD, respectively, we complete the triangle and determine the magnitude and sense of $\mathbf{F}_{AC}$ and $\mathbf{F}_{AD}$. The magnitudes F_{AC} and F_{AD} represent the forces in members AC and AD. Since $\mathbf{F}_{AC}$ is directed down and to the left, that is, toward joint A, member AC pushes on pin A and is in compression. On the other hand, since $\mathbf{F}_{AD}$ is directed away from the joint, member AD pulls on pin A and is in tension.

We may now proceed to joint D, where only two forces, $\mathbf{F}_{DC}$ and $\mathbf{F}_{DB}$, are still unknown. The other forces are the load $\mathbf{P}$, which is given, and the force $\mathbf{F}_{DA}$ exerted on the pin by member AD. As indicated above, this force is equal and opposite to the force $\mathbf{F}_{AD}$ exerted by the same member on pin A. We may draw the force polygon corresponding to joint D, as shown in Fig. 6.8, and determine the forces $\mathbf{F}_{DC}$ and $\mathbf{F}_{DB}$ from that polygon. However, when more than three forces are involved, it is usually more conven-

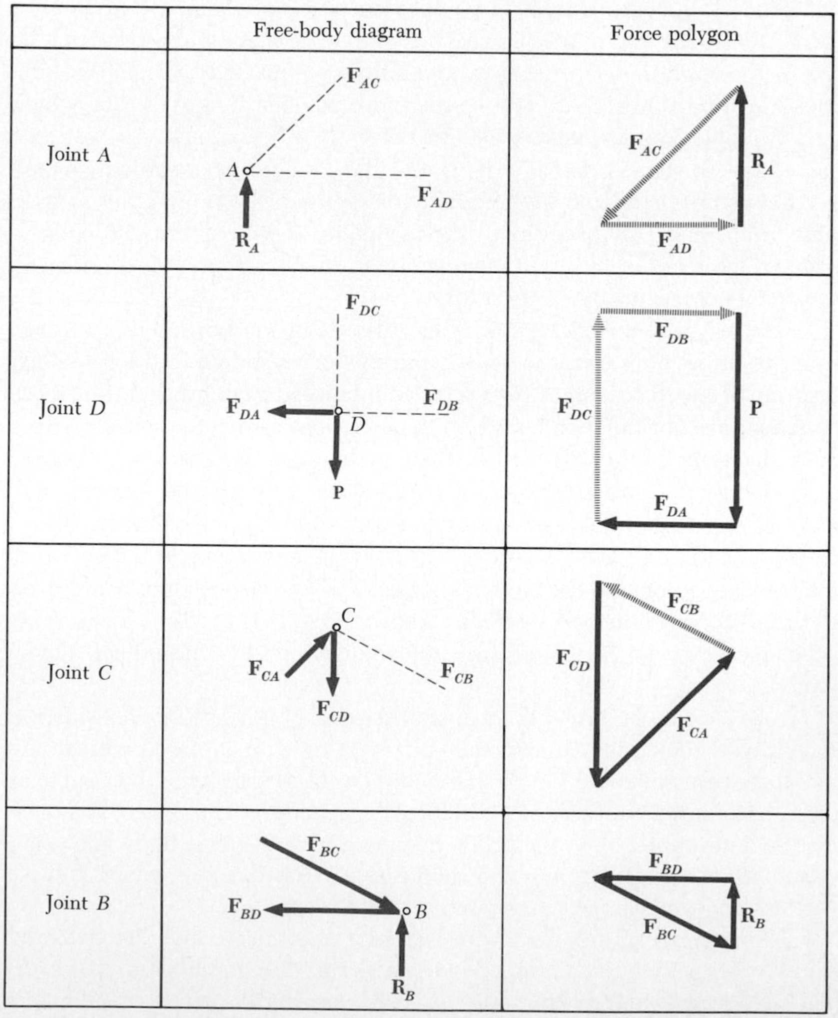

Fig. 6.8

ient to write the equations of equilibrium $\Sigma F_x = 0$, $\Sigma F_y = 0$ and solve these equations for the two unknown forces. Since both of these forces are found to be directed away from joint D, members DC and DB pull on the pin and are in tension.

Next, joint C is considered; its free-body diagram is shown in Fig. 6.8. It is noted that both $\mathbf{F}_{CD}$ and $\mathbf{F}_{CA}$ are known from the analysis of the preceding joints and that only $\mathbf{F}_{CB}$ is unknown. Since the equilibrium of each pin provides sufficient information to determine two unknowns, a check of our analysis is obtained at this joint. The force triangle is drawn, and the magnitude and sense of $\mathbf{F}_{CB}$ are determined. Since $\mathbf{F}_{CB}$ is directed toward joint C, member CB pushes on pin C and is in compression. The check is obtained by verifying that the force $\mathbf{F}_{CB}$ and member CB are parallel.

At joint B, all the forces are known. Since the corresponding pin is in equilibrium, the force triangle must close and an additional check of the analysis is obtained.

It should be noted that the force polygons shown in Fig. 6.8 are not unique. Each of them could be replaced by an alternate configuration. For example, the force triangle corresponding to joint A could be drawn as shown in Fig. 6.9. The triangle actually shown in Fig. 6.8 was obtained by drawing the three forces $\mathbf{R}_A$, $\mathbf{F}_{AC}$, and $\mathbf{F}_{AD}$ in tip-to-tail fashion in the order in which their lines of action are encountered when moving clockwise around joint A. The other force polygons in Fig. 6.8, having been drawn in the same way, can be made to fit into a single diagram, as shown in Fig. 6.10. Such a diagram, known as *Maxwell's diagram*, greatly facilitates the *graphical analysis* of truss problems.†

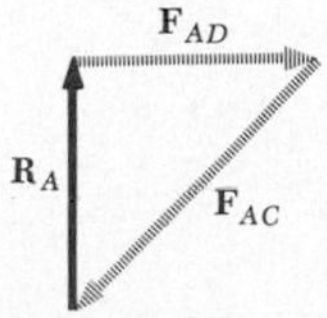

Fig. 6.9

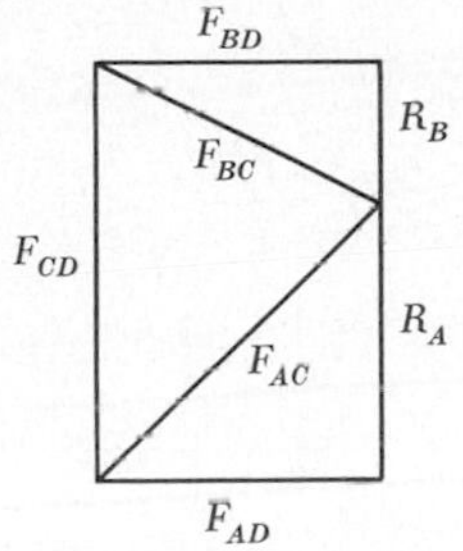

Fig. 6.10

*6.5. Joints under Special Loading Conditions.

Consider the joint shown in Fig. 6.11*a*, which connects four members lying in two intersecting straight lines. The free-body diagram of Fig. 6.11*b* shows that pin A is subjected to two pairs of directly opposite forces. The corresponding force polygon, therefore, must be a parallelogram (Fig. 6.11*c*), and *the forces in opposite members must be equal.*

† For a complete discussion of Maxwell's diagrams, see F. P. Beer and E. R. Johnston, *Mechanics for Engineers*, 3d ed., sec. 6.7, McGraw-Hill, New York, 1976.

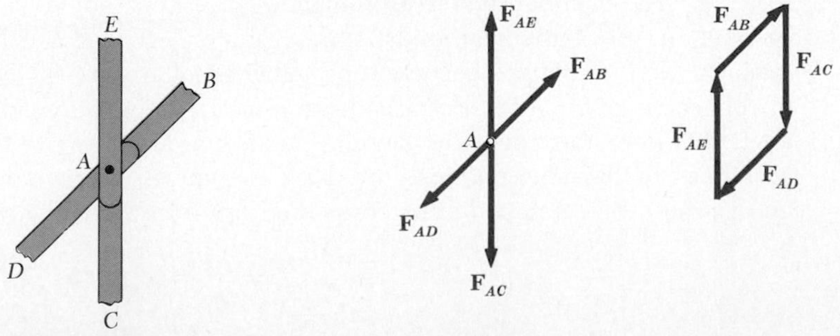

Fig. 6.11 (*a*) (*b*) (*c*)

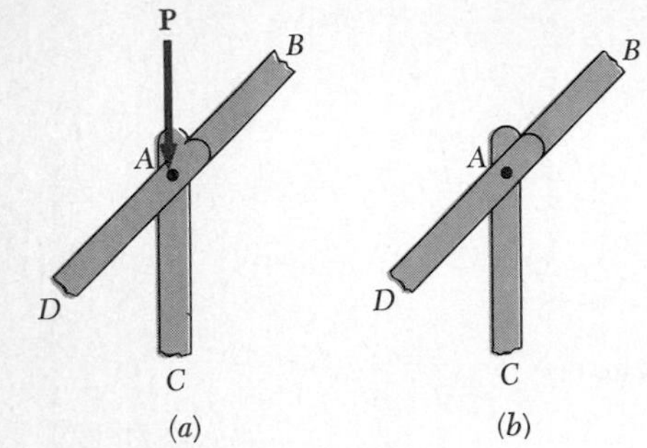

Fig. 6.12

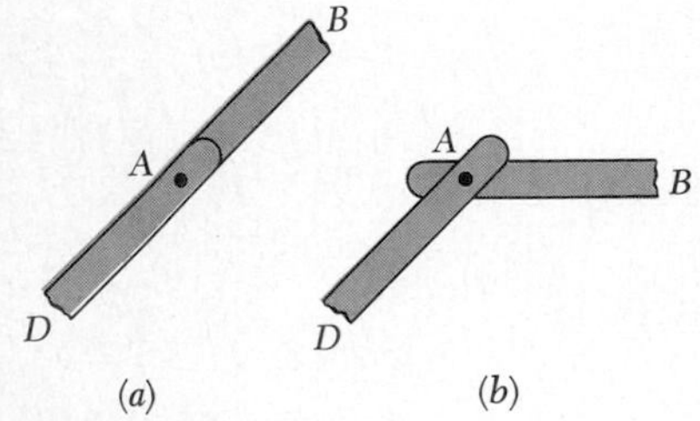

Fig. 6.13

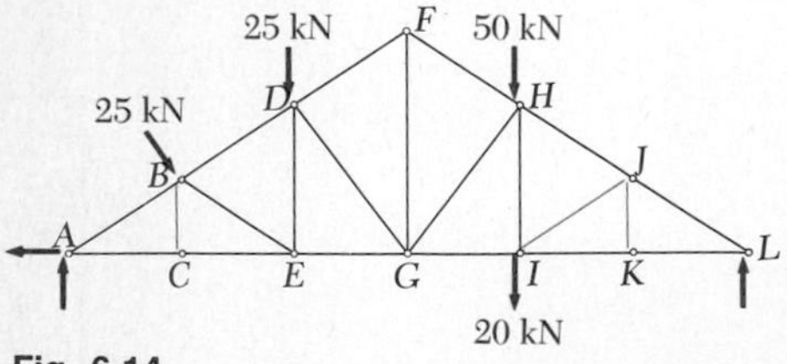

Fig. 6.14

Consider next the joint shown in Fig. 6.12*a*, which connects three members and supports a load **P**. Two of the members lie in the same line, and the load **P** acts along the third member. The free-body diagram of pin *A* and the corresponding force polygon will be as shown in Fig. 6.11*b* and *c* with $\mathbf{F}_{AE}$ replaced by the load **P**. Thus, *the forces in the two opposite members must be equal, and the force in the other member must equal P.* A particular case of special interest is shown in Fig. 6.12*b*. Since, in this case, no external load is applied to the joint, we have $P = 0$, and the force in member *AC* is zero. Member *AC* is said to be a *zero-force member.*

Consider now a joint connecting two members only. From Sec. 2.9, we know that a particle which is acted upon by two forces will be in equilibrium if the two forces have the same magnitude, same line of action, and opposite sense. In the case of the joint of Fig. 6.13*a*, which connects two members *AB* and *AD* lying in the same line, the equilibrium of pin *A* requires therefore that *the forces in the two members be equal.* In the case of the joint of Fig. 6.13*b*, the equilibrium of pin *A* is impossible unless the forces in both members are zero. Members connected as shown in Fig. 6.13*b*, therefore, must be *zero-force members*.

Spotting the joints which are under the special loading conditions listed above will expedite the analysis of a truss. Consider, for example, a Howe truss loaded as shown in Fig. 6.14. All the members represented by colored lines will be recognized as zero-force members. Joint *C* connects three members, two of which lie in the same line, and is not subjected to any external load; member *BC* is thus a zero-force member. Applying the same reasoning to joint *K*, we find that member *JK* is also a zero-force member. But joint *J* is now in the same situation as joints *C* and *K*, and member *IJ* must be a zero-force member. The examination of joints *C*, *J*, and *K* also shows that the forces in members *AC* and *CE* are equal, that the forces in members *HJ* and *JL* are equal, and that the forces in members *IK* and *KL* are equal. Furthermore, now turning our attention to joint *I*, where the 20-kN load and member *HI* are collinear, we note that the force in member *HI* is 20 kN (tension) and that the forces in members *GI* and *IK* are equal. Hence, the forces in members *GI*, *IK*, and *KL* are equal.

Students, however, should be warned against misusing the rules established in this section. For example, it would be wrong to assume that the force in member *DE* is 25 kN or that the forces in members *AB* and *BD* are equal. The conditions discussed above do not apply to joints *B* and *D*. The forces in these members and in all remaining members should be found by carrying out the analysis of joints *A*, *B*, *D*, *E*, *F*, *G*, *H*, and *L* in the usual manner. Until they have become thoroughly familiar with the conditions of application of the rules established in this section, students would be well advised to draw the free-body diagrams of all pins and to write the corresponding equilibrium equations (or draw the corresponding force polygons), whether or not the joints considered fall into the categories listed above.

A final remark concerning zero-force members: These members are not useless. While they do not carry any load under the particular loading conditions shown, the zero-force members of Fig. 6.14 will probably carry loads if the loading conditions are changed. Besides, even in the case considered, these members are needed to support the weight of the truss and to maintain the truss in the desired shape.

***6.6. Space Trusses.** When several straight members are joined together at their extremities to form a three-dimensional configuration, the structure obtained is called a *space truss*.

We recall from Sec. 6.3 that the most elementary two-dimensional rigid truss consisted of three members joined at their extremities to form the sides of a triangle; by adding two members at a time to this basic configuration, and connecting them at a new joint, it was possible to obtain a larger rigid structure which was defined as a simple truss. Similarly, the most elementary rigid space truss consists of six members joined at their extremities to form the edges of a tetrahedron $ABCD$ (Fig. 6.15*a*). By adding three members at a time to this basic configuration, such as AE, BE, and CE, attaching them at separate existing joints, and connecting them at a new joint, we can obtain a larger rigid structure which is defined as a *simple space truss* (Fig. 6.15*b*).† Observing that the basic tetrahedron has six members and four joints and that every time three members are added, the number of joints is increased by one, we conclude that in a simple space truss the total number of members is $m = 3n - 6$, where n is the total number of joints.

If a space truss is to be completely constrained and if the reactions at its supports are to be statically determinate, the supports should consist of a combination of balls, rollers, and balls and sockets which provides six unknown reactions (see Sec. 4.8). These unknown reactions may be readily determined by solving the six equations expressing that the three-dimensional truss is in equilibrium.

Although the members of a space truss are actually joined together by means of riveted or welded connections, it is assumed that each joint consists of a ball-and-socket connection. Thus, no couple will be applied to the members of the truss, and each member may be treated as a two-force member. The conditions of equilibrium for each joint will be expressed by the three equations $\Sigma F_x = 0$, $\Sigma F_y = 0$, and $\Sigma F_z = 0$. In the case of a simple space truss containing n joints, writing the conditions of equilibrium for each joint will thus yield $3n$ equations. Since $m = 3n - 6$, these equations suffice to determine all unknown forces (forces in m members and six reactions at the supports). However, to avoid solving many simultaneous equations, care should be taken to select joints in such an order that no selected joint will involve more than three unknown forces.

† The four joints must not lie in a plane.

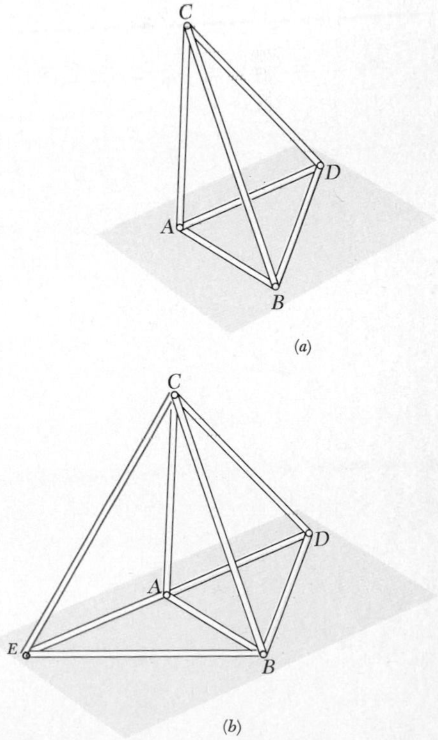

Fig. 6.15

SAMPLE PROBLEM 6.1

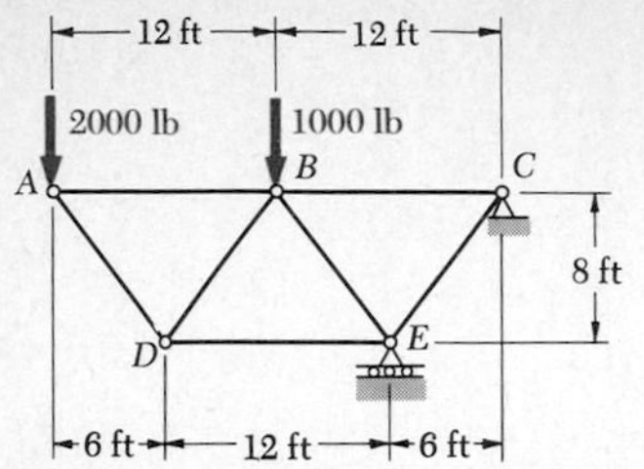

Using the method of joints, determine the force in each member of the truss shown.

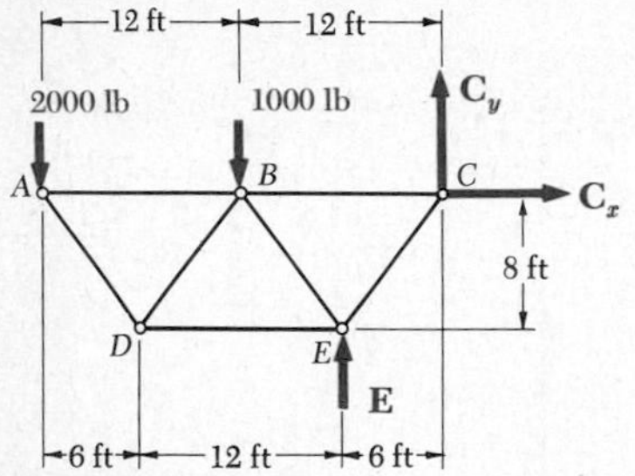

Solution. A free-body diagram of the entire truss is drawn; external forces acting on this free body consist of the applied loads and the reactions at C and E.

Equilibrium of Entire Truss.

$+\circlearrowleft \Sigma M_C = 0$: $\quad (2000 \text{ lb})(24 \text{ ft}) + (1000 \text{ lb})(12 \text{ ft}) - E(6 \text{ ft}) = 0$

$E = +10{,}000 \text{ lb}$ $\qquad\qquad \mathbf{E} = 10{,}000 \text{ lb}\uparrow$

$\xrightarrow{+} \Sigma F_x = 0$: $\qquad\qquad \mathbf{C}_x = 0$

$+\uparrow \Sigma F_y = 0$: $\quad -2000 \text{ lb} - 1000 \text{ lb} + 10{,}000 \text{ lb} + C_y = 0$

$C_y = -7000 \text{ lb}$ $\qquad\qquad \mathbf{C}_y = 7000 \text{ lb}\downarrow$

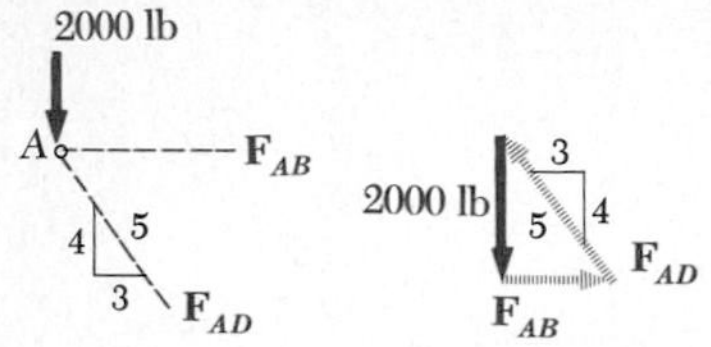

Joint A. This joint is subjected to only two unknown forces, namely, the forces exerted by members AB and AD. A force triangle is used to determine $\mathbf{F}_{AB}$ and $\mathbf{F}_{AD}$. We note that member AB pulls on the joint and thus is in tension and that member AD pushes on the joint and thus is in compression. The magnitudes of the two forces are obtained from the proportion

$$\frac{2000 \text{ lb}}{4} = \frac{F_{AB}}{3} = \frac{F_{AD}}{5}$$

$F_{AB} = 1500 \text{ lb } T$ ◄

$F_{AD} = 2500 \text{ lb } C$ ◄

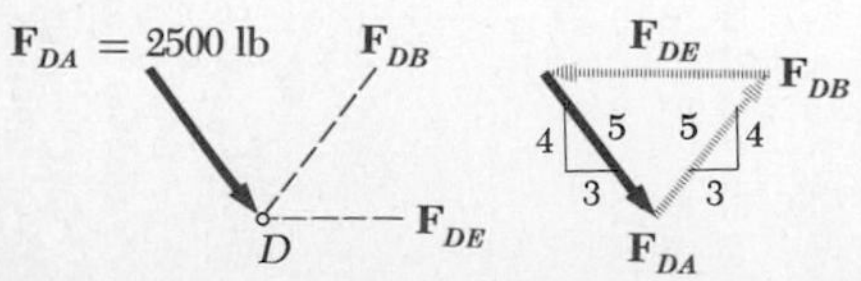

Joint D. Since the force exerted by member AD has been determined, only two unknown forces are now involved at this joint. Again, a force triangle is used to determine the unknown forces in members DB and DE.

$F_{DB} = F_{DA}$ $\qquad\qquad F_{DB} = 2500 \text{ lb } T$ ◄

$F_{DE} = 2(\frac{3}{5})F_{DA}$ $\qquad\qquad F_{DE} = 3000 \text{ lb } C$ ◄

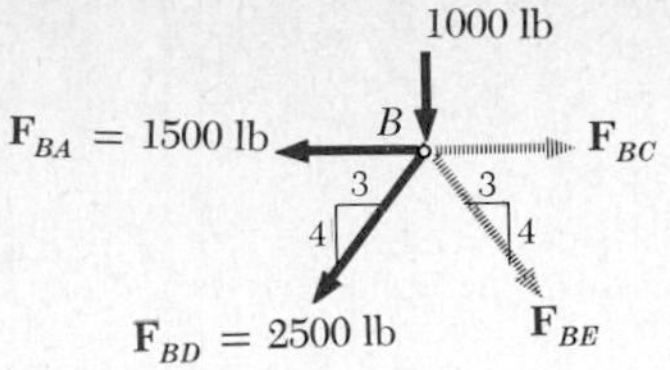

Joint B. Since more than three forces act at this joint, we determine the two unknown forces $\mathbf{F}_{BC}$ and $\mathbf{F}_{BE}$ by solving the equilibrium equations $\Sigma F_x = 0$ and $\Sigma F_y = 0$. We arbitrarily assume that both unknown forces act away from the joint, i.e., that the members are in tension. The positive value obtained for F_{BC} indicates that our assumption was correct; member BC is in tension. The negative value of F_{BE} indicates that our assumption was wrong; member BE is in compression.

$$+\uparrow\Sigma F_y = 0: \qquad -1000 - \tfrac{4}{5}(2500) - \tfrac{4}{5}F_{BE} = 0$$

$$F_{BE} = -3750 \text{ lb} \qquad F_{BE} = 3750 \text{ lb } C \blacktriangleleft$$

$$\overset{+}{\rightarrow}\Sigma F_x = 0: \qquad F_{BC} - 1500 - \tfrac{3}{5}(2500) - \tfrac{3}{5}(3750) = 0$$

$$F_{BC} = +5250 \text{ lb} \qquad F_{BC} = 5250 \text{ lb } T \blacktriangleleft$$

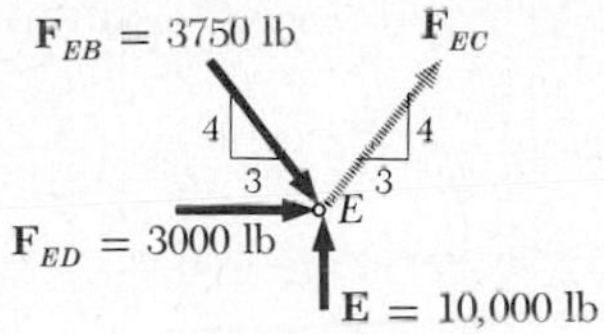

Joint E. The unknown force $\mathbf{F}_{EC}$ is assumed to act away from the joint. Summing x components, we write

$$\overset{+}{\rightarrow}\Sigma F_x = 0: \qquad \tfrac{3}{5}F_{EC} + 3000 + \tfrac{3}{5}(3750) = 0$$

$$F_{EC} = -8750 \text{ lb} \qquad F_{EC} = 8750 \text{ lb } C \blacktriangleleft$$

Summing y components, we obtain a check of our computations:

$$\begin{aligned} +\uparrow\Sigma F_y &= 10{,}000 - \tfrac{4}{5}(3750) - \tfrac{4}{5}(8750) \\ &= 10{,}000 - 3000 - 7000 = 0 \qquad \text{(checks)} \end{aligned}$$

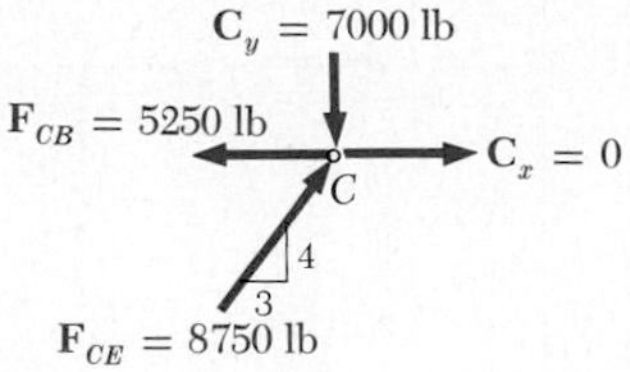

Joint C. Using the computed values of $\mathbf{F}_{CB}$ and $\mathbf{F}_{CE}$, we may determine the reactions $\mathbf{C}_x$ and $\mathbf{C}_y$ by considering the equilibrium of this joint. Since these reactions have already been determined from the equilibrium of the entire truss, we will obtain two checks of our computations. We may also merely use the computed values of all forces acting on the joint (forces in members and reactions) and check that the joint is in equilibrium:

$$\overset{+}{\rightarrow}\Sigma F_x = -5250 + \tfrac{3}{5}(8750) = -5250 + 5250 = 0 \qquad \text{(checks)}$$

$$+\uparrow\Sigma F_y = -7000 + \tfrac{4}{5}(8750) = -7000 + 7000 = 0 \qquad \text{(checks)}$$

Problems

6.1 through 6.12 Using the method of joints, determine the force in each member of the truss shown. State whether each member is in tension or compression.

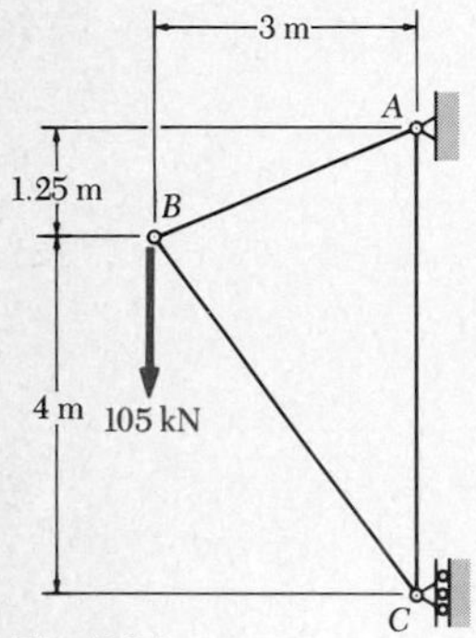

Fig. P6.1

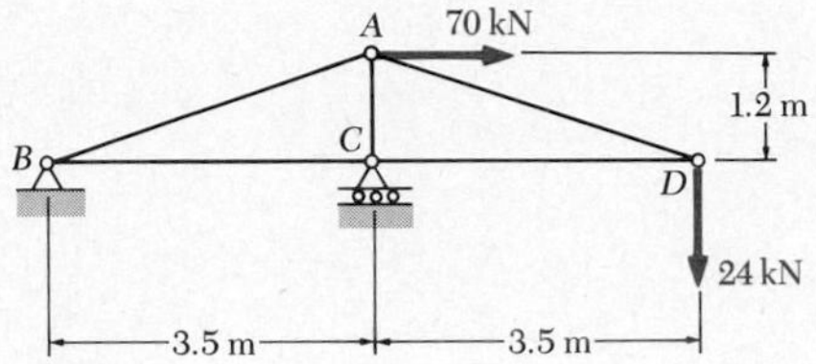

Fig. P6.2

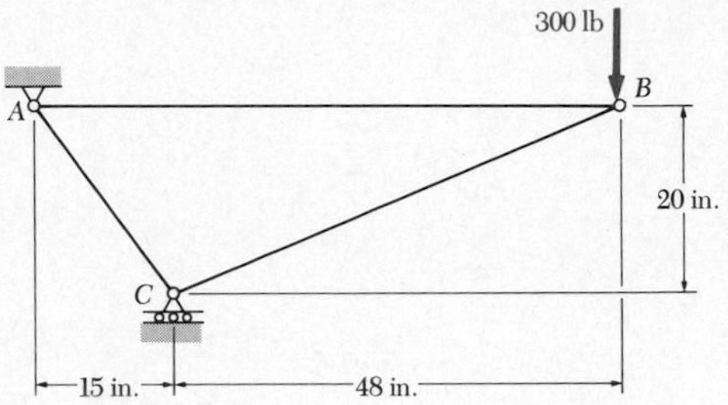

Fig. P6.3

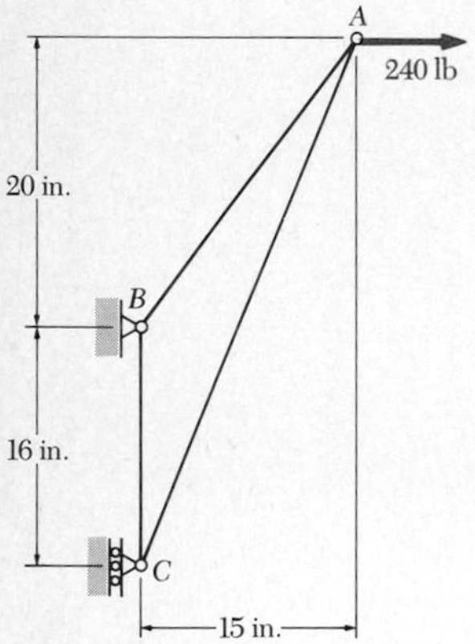

Fig. P6.4

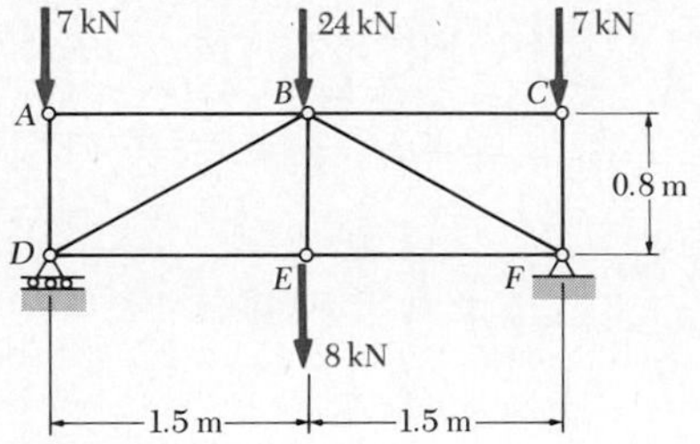

Fig. P6.5

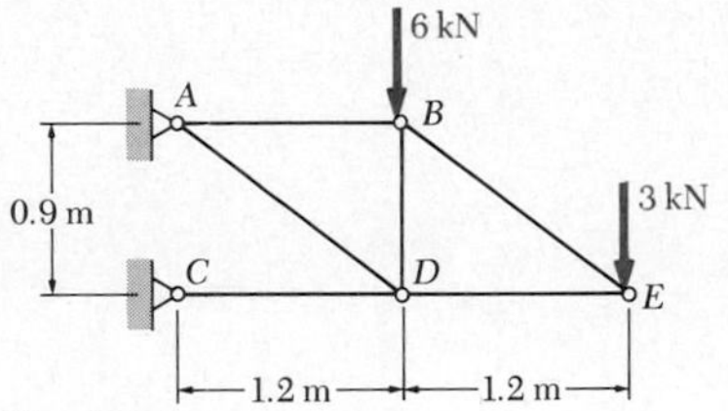

Fig. P6.6

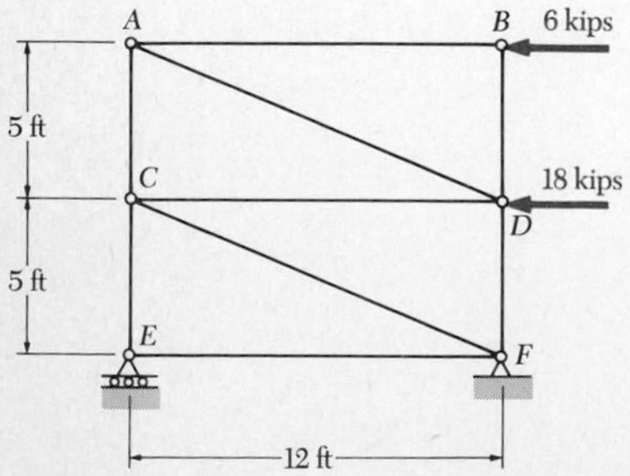

Fig. P6.7

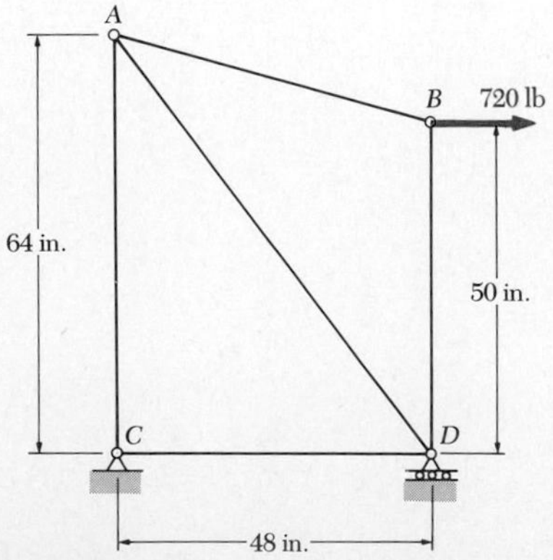

Fig. P6.8

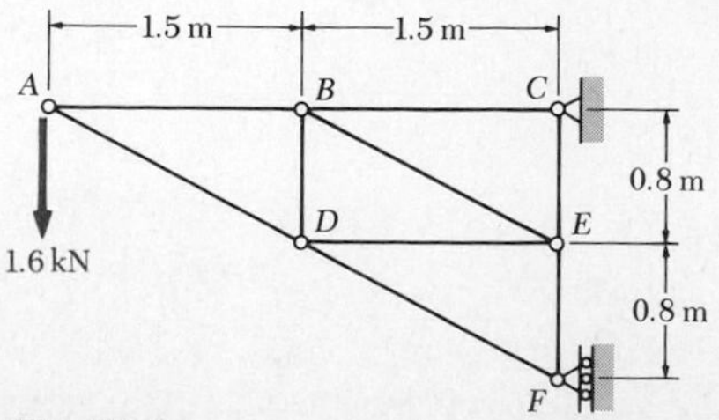

Fig. P6.9

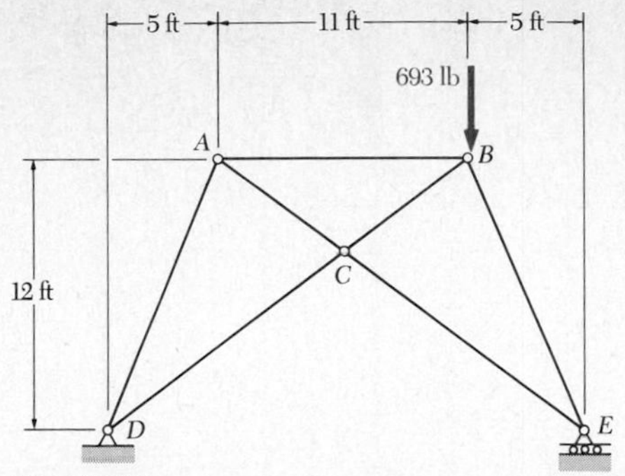

Fig. P6.10

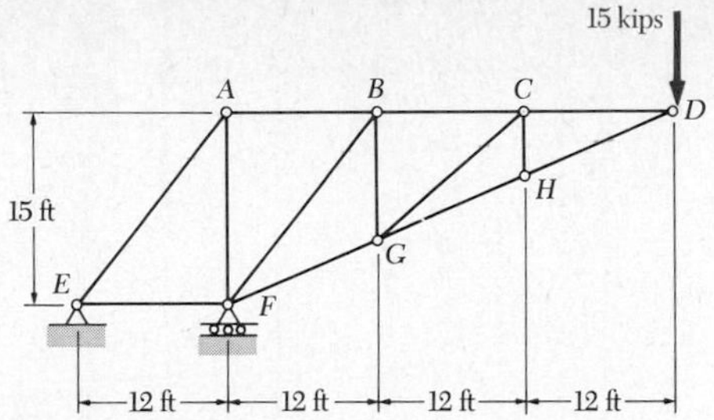

Fig. P6.11

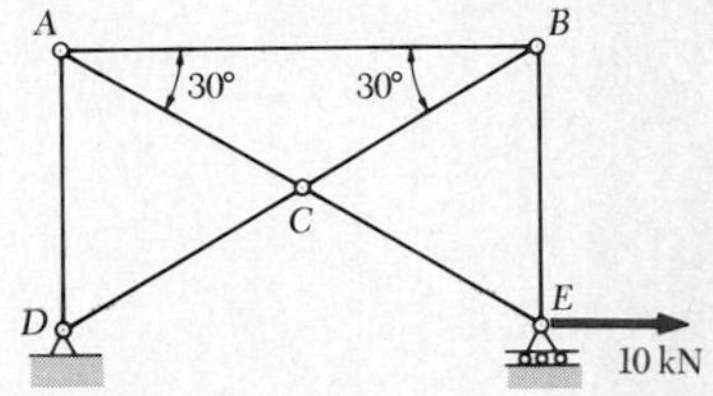

Fig. P6.12

6.13 Determine whether the trusses given in Probs. 6.10, 6.14, 6.15, and 6.16 are simple trusses.

6.14 through 6.16 Determine the zero-force members in the truss shown for the given loading.

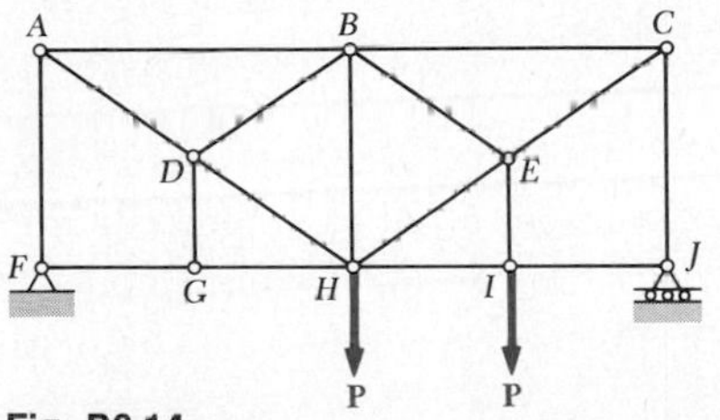

Fig. P6.14

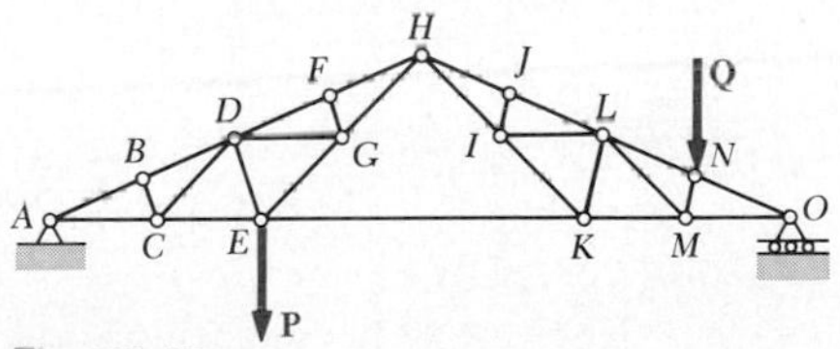

Fig. P6.15

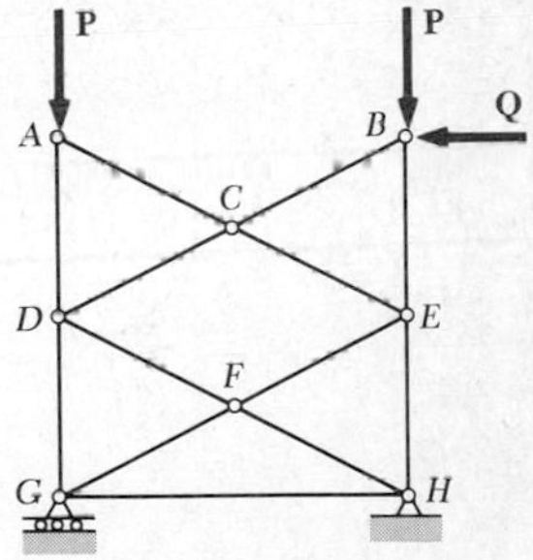

Fig. P6.16

***6.17** Six bars, each of length L, are connected to form a regular tetrahedron which rests on a frictionless horizontal surface. Determine the force in each of the six members when a vertical force **P** is applied at A.

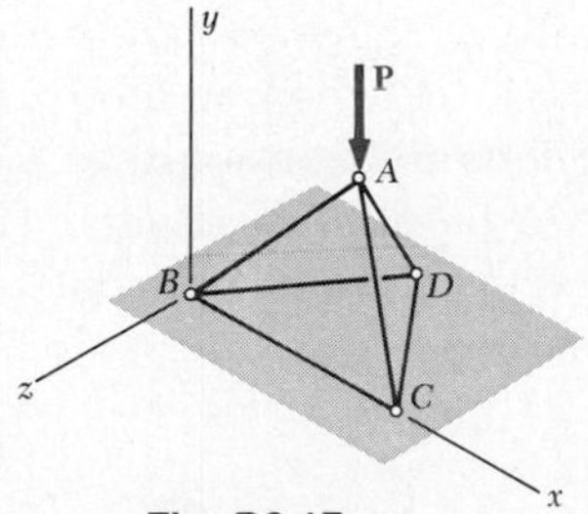

Fig. P6.17

***6.18** The three-dimensional truss is supported by the six reactions shown. If a vertical 170-lb load is applied at E, determine (a) the reactions, (b) the force in each member.

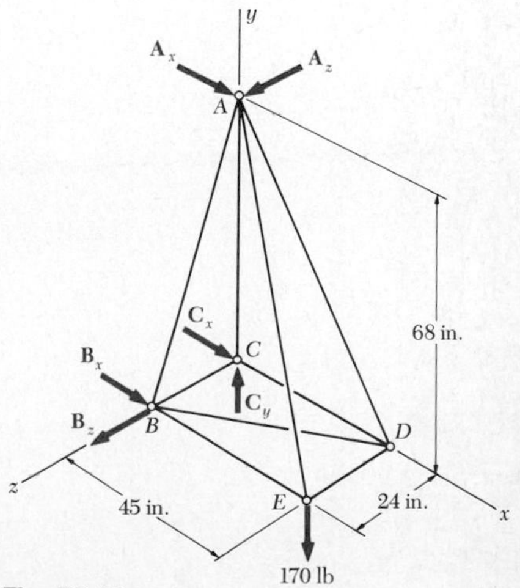

Fig. P6.18

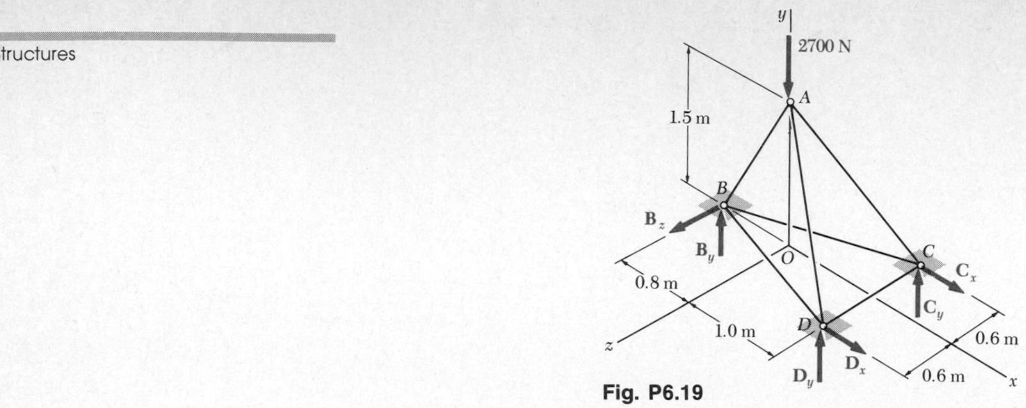

Fig. P6.19

***6.19** The three-dimensional truss is supported by the six reactions shown. If a vertical 2700-N load is applied at *A*, determine (*a*) the reactions, (*b*) the force in each member.

***6.20** Solve Prob. 6.19, assuming that the 2700-N load is replaced by a load $\mathbf{Q} = (2700 \text{ N})\mathbf{k}$ applied at *A*.

6.7. Analysis of Trusses by the Method of Sections. The method of joints is most effective when the forces in all the members of a truss are to be determined. If, however, the force in only one member or the forces in a very few members are desired, another method, the method of sections, will prove more efficient.

Assume, for example, that we want to determine the force in member *BD* of the truss shown in Fig. 6.16*a*. To do this, we must determine the force with which member *BD* acts on either joint *B* or joint *D*. If we were to use the method of joints, we would choose either joint *B* or joint *D* as a free body. However, we may also choose as a free body a larger portion of the truss, composed of several joints and members, provided that the desired force is one of the external forces acting on that portion. If, in addition, the portion of the truss is chosen so that there is a total of only three unknown forces acting upon it, the desired force may be obtained by solving the equations of equilibrium for this portion of the truss. In practice, the portion of the truss to be utilized is obtained by *passing a section* through three members of the truss, one of which is the desired member, i.e., by drawing a line which divides the truss into two completely separate parts but does not intersect more than three members. Either of the two portions of the truss obtained after the intersected members have been removed may then be used as a free body.†

† In the analysis of certain trusses, sections are passed which intersect more than three members; the forces in one, or possibly two, of the intersected members may be obtained if equilibrium equations can be found, each of which involves only one unknown (see Probs. 6.37 through 6.39).

In Fig. 6.16*a*, the section *nn* has been passed through members *BD*, *BE*, and *CE*, and the portion *ABC* of the truss is chosen as the free body (Fig. 6.16*b*). The forces acting on the free body are the loads $\mathbf{P}_1$ and $\mathbf{P}_2$ at points *A* and *B* and the three unknown forces $\mathbf{F}_{BD}$, $\mathbf{F}_{BE}$, and $\mathbf{F}_{CE}$. Since it is not known whether the members removed were in tension or compression, the three forces have been arbitrarily drawn away from the free body as if the members were in tension.

The fact that the rigid body *ABC* is in equilibrium can be expressed by writing three equations which may be solved for the three unknown forces. If only the force $\mathbf{F}_{BD}$ is desired, we need write only one equation, provided that the equation does not contain the other unknowns. Thus the equation $\Sigma M_E = 0$ yields the value of the magnitude F_{BD} of the force $\mathbf{F}_{BD}$. A positive sign in the answer will indicate that our original assumption regarding the sense of $\mathbf{F}_{BD}$ was correct and that member *BD* is in tension; a negative sign will indicate that our assumption was incorrect and that *BD* is in compression.

On the other hand, if only the force $\mathbf{F}_{CE}$ is desired, an equation which does not involve $\mathbf{F}_{BD}$ or $\mathbf{F}_{BE}$ should be written; the appropriate equation is $\Sigma M_B = 0$. Again a positive sign for the magnitude F_{CE} of the desired force indicates a correct assumption, hence tension; and a negative sign indicates an incorrect assumption, hence compression.

If only the force $\mathbf{F}_{BE}$ is desired, the appropriate equation is $\Sigma F_y = 0$. Whether the member is in tension or compression is again determined from the sign of the answer.

When the force in only one member is determined, no independent check of the computation is available. However, when all the unknown forces acting on the free body are determined, the computations can be checked by writing an additional equation. For instance, if $\mathbf{F}_{BD}$, $\mathbf{F}_{BE}$, and $\mathbf{F}_{CE}$ are determined as indicated above, the computation can be checked by verifying that $\Sigma F_x = 0$.

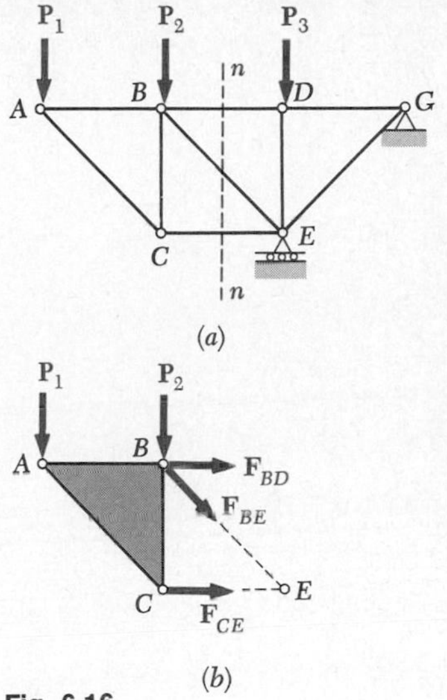

Fig. 6.16

*6.8. Trusses Made of Several Simple Trusses.

Consider two simple trusses *ABC* and *DEF*. If they are connected by three bars *BD*, *BE*, and *CE* as shown in Fig. 6.17*a*, they will form together a rigid truss *ABDF*. The trusses *ABC* and *DEF* can also be combined into a single rigid truss by joining joints *B* and *D* into a single joint *B* and by connecting joints *C* and *E* by a bar *CE* (Fig. 6.17*b*). The truss thus obtained is known as a Fink truss. It should be noted that the trusses of Fig. 6.17*a* and *b* are *not* simple trusses; they cannot be constructed from a triangular truss by adding successive pairs of members as prescribed in Sec. 6.3. They are rigid trusses, however, as we may check by comparing the systems of connections used to hold the simple trusses *ABC* and *DEF* together (three bars in Fig. 6.17*a*, one pin and one bar in Fig. 6.17*b*) with the systems of supports discussed in Secs. 4.4 and 4.5. Trusses made of several simple trusses rigidly connected are known as *compound trusses*.

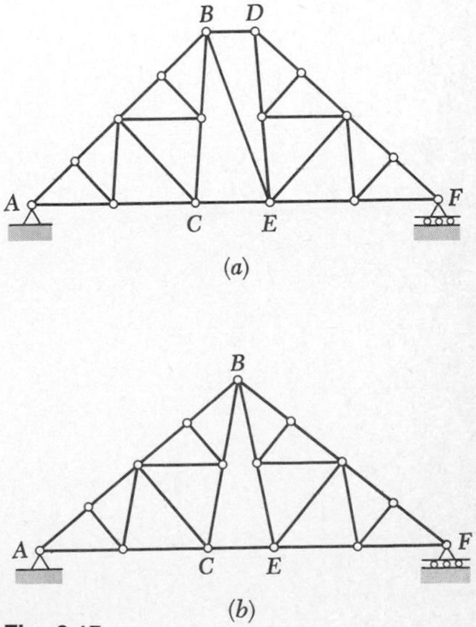

Fig. 6.17

It may be checked that in a compound truss the number of members *m* and the number of joints *n* are still related by the formula $m = 2n - 3$. If a compound truss is supported by a frictionless pin and a roller (involving three unknown reactions), the total number of unknowns is $m + 3$ and this number is therefore equal to the number $2n$ of equations obtained by expressing that the n pins are in equilibrium. Compound trusses supported by a pin and a roller, or by an equivalent system of supports, are *statically determinate, rigid,* and *completely constrained.* This means that all unknown reactions and forces in members can be determined by the methods of statics and that all equilibrium equations being satisfied, the truss will neither collapse nor move. The forces in the members, however, cannot all be determined by the method of joints, except by solving a large number of simultaneous equations. In the case of the compound truss of Fig. 6.17*a*, for example, it will be found more expeditious to pass a section through members *BD, BE,* and *CE* to determine their forces.

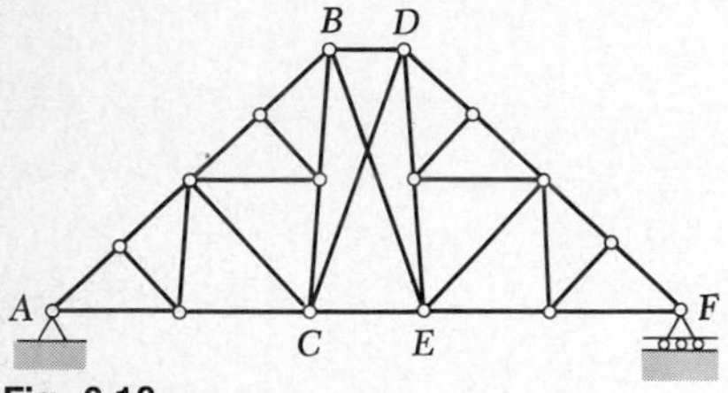

Fig. 6.18

Suppose, now, that the simple trusses *ABC* and *DEF* are connected by *four* bars *BD, BE, CD,* and *CE* (Fig. 6.18). The number of members m is now larger than $2n - 3$; the truss obtained is *overrigid,* and one of the four members *BD, BE, CD,* or *CE* is said to be *redundant.* If the truss is supported by a pin at *A* and a roller at *F,* the total number of unknowns is $m + 3$. This number is now larger than the number $2n$ of available independent equations; the truss is *statically indeterminate.*

Finally, we shall assume that the two simple trusses *ABC* and *DEF* are joined by a pin as shown in Fig. 6.19*a*. The number of members m is smaller than $2n - 3$. If the truss is supported by a pin at *A* and a roller at *F,* the total number of unknowns is $m + 3$. This number is now smaller than the number $2n$ of equilibrium equations which should be satisfied; the truss is *nonrigid* and will collapse under its own weight. However, if two pins are used to support it, the truss becomes *rigid* and will not collapse (Fig. 6.19*b*). We note that the total number of unknowns is now $m + 4$ and is equal to the number $2n$ of equations. While necessary, this condition, however, is not sufficient for the equilibrium of a structure which ceases to be rigid when detached from its supports (see Sec. 6.11).

Fig. 6.19

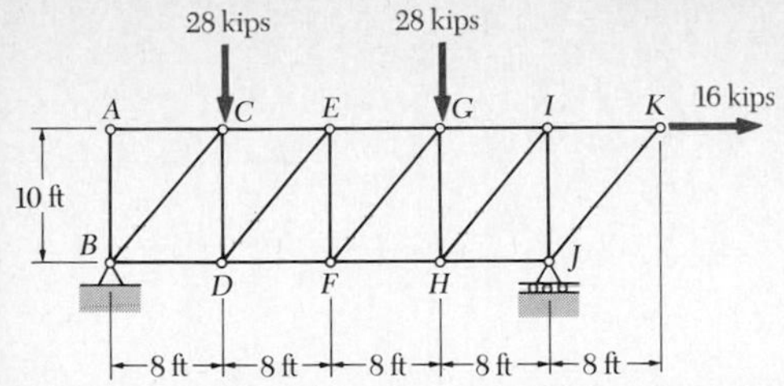

SAMPLE PROBLEM 6.2

Determine the force in members EF and GI of the truss shown.

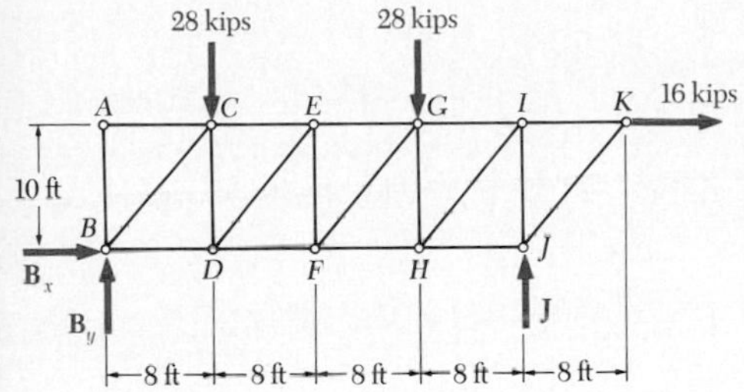

Solution. A free-body diagram of the entire truss is drawn; external forces acting on this free body consist of the applied loads and the reactions at B and J.

Equilibrium of the Entire Truss.

$+\circlearrowleft \Sigma M_B = 0$:

$$-(28 \text{ kips})(8 \text{ ft}) - (28 \text{ kips})(24 \text{ ft}) - (16 \text{ kips})(10 \text{ ft}) + J(32 \text{ ft}) = 0$$

$$J = +33 \text{ kips} \qquad \mathbf{J} = 33 \text{ kips}\uparrow$$

$\xrightarrow{+} \Sigma F_x = 0$: $\quad B_x + 16 \text{ kips} = 0$

$$B_x = -16 \text{ kips} \qquad \mathbf{B}_x = 16 \text{ kips}\leftarrow$$

$+\circlearrowleft \Sigma M_J = 0$:

$$(28 \text{ kips})(24 \text{ ft}) + (28 \text{ kips})(8 \text{ ft}) - (16 \text{ kips})(10 \text{ ft}) - B_y(32 \text{ ft}) = 0$$

$$B_y = +23 \text{ kips} \qquad \mathbf{B}_y = 23 \text{ kips}\uparrow$$

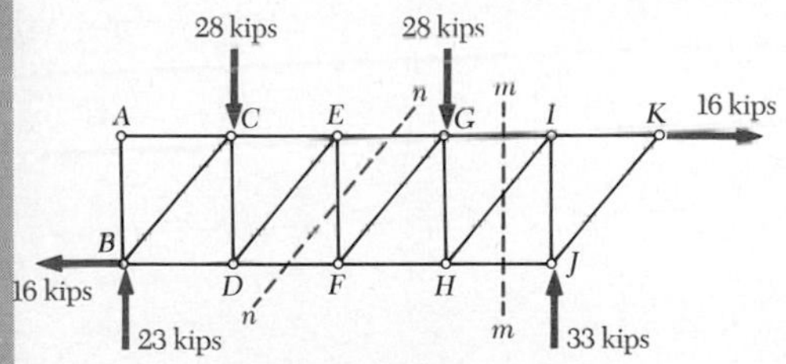

Force in Member EF. Section nn is passed through the truss so that it intersects member EF and only two additional members. After the intersected members have been removed, the left-hand portion of the truss is chosen as a free body. Three unknowns are involved; to eliminate the two horizontal forces, we write

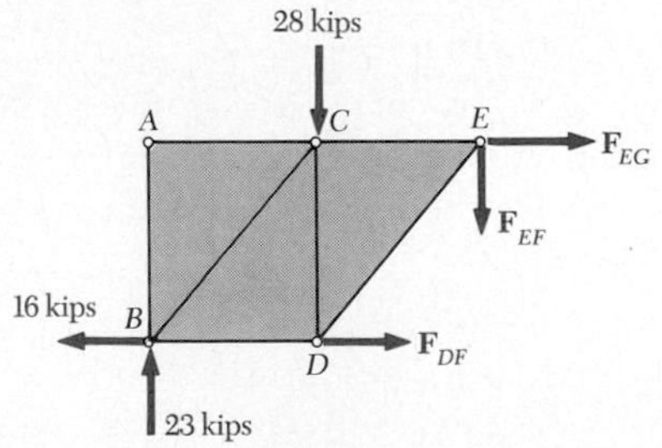

$$+\uparrow \Sigma F_y = 0: \qquad +23 \text{ kips} - 28 \text{ kips} - F_{EF} = 0$$

$$F_{EF} = -5 \text{ kips}$$

The sense of $\mathbf{F}_{EF}$ was chosen assuming member EF to be in tension; the negative sign obtained indicates that the member is in compression.

$$F_{EF} = 5 \text{ kips } C \blacktriangleleft$$

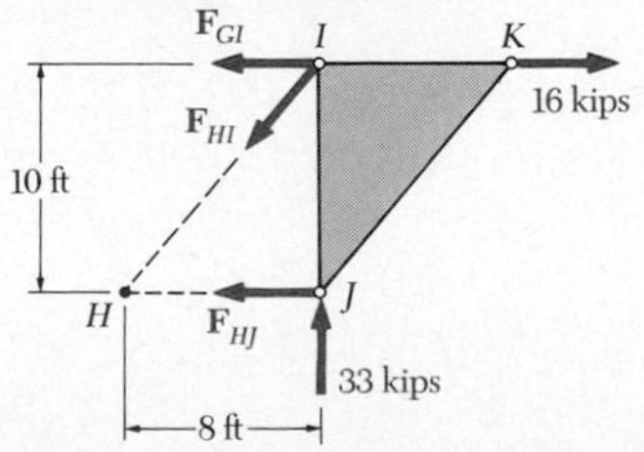

Force in Member GI. Section mm is passed through the truss so that it intersects member GI and only two additional members. After the intersected members have been removed, we choose the right-hand portion of the truss as a free body. Three unknown forces are again involved; to eliminate the two forces passing through point H, we write

$$+\circlearrowleft \Sigma M_H = 0: \qquad (33 \text{ kips})(8 \text{ ft}) - (16 \text{ kips})(10 \text{ ft}) + F_{GI}(10 \text{ ft}) = 0$$

$$F_{GI} = -10.4 \text{ kips} \qquad F_{GI} = 10.4 \text{ kips } C \blacktriangleleft$$

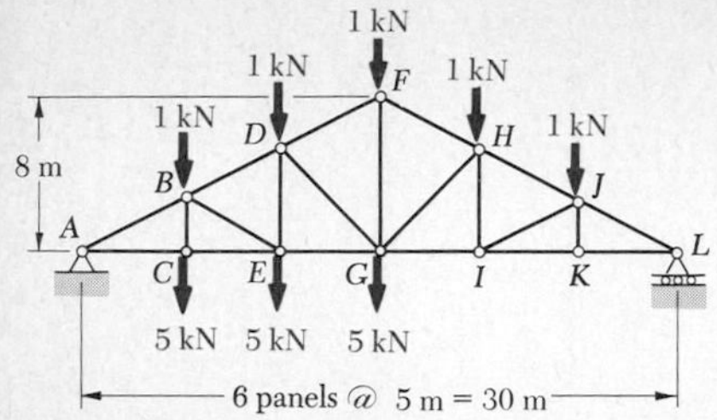

SAMPLE PROBLEM 6.3

Determine the force in members *FH*, *GH*, and *GI* of the roof truss shown.

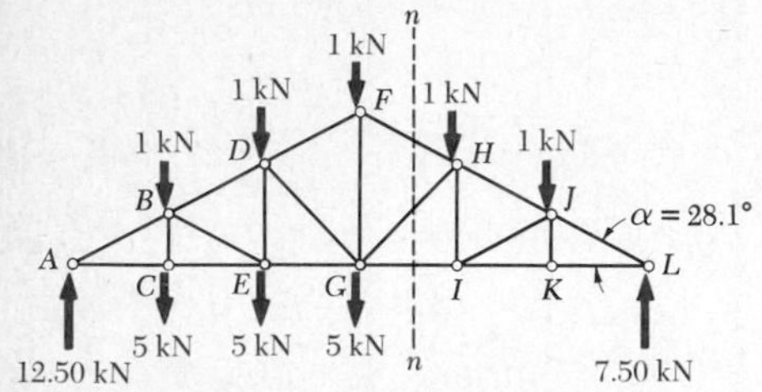

Solution. Section *nn* is passed through the truss as shown. The right-hand portion of the truss will be taken as a free body. Since the reaction at *L* acts on this free body, the value of **L** must be calculated separately, using the entire truss as a free body; the equation $\Sigma M_A = 0$ yields $\mathbf{L} = 7.50\text{ kN}\uparrow$.

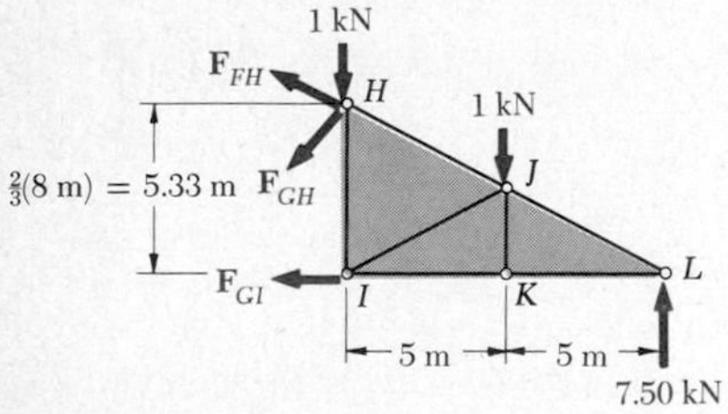

Force in Member GI. Using the portion *HLI* of the truss as a free body, the value of F_{GI} is obtained by writing

$$+\circlearrowleft \Sigma M_H = 0: \qquad (7.50\text{ kN})(10\text{ m}) - (1\text{ kN})(5\text{ m}) - F_{GI}(5.33\text{ m}) = 0$$

$$F_{GI} = +13.13\text{ kN} \qquad F_{GI} = 13.13\text{ kN } T \blacktriangleleft$$

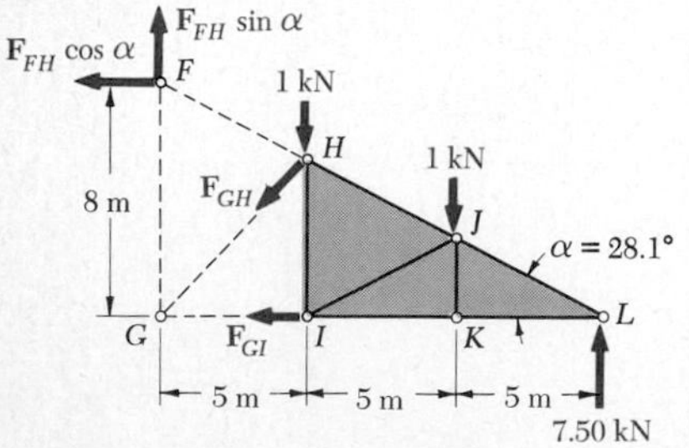

Force in Member FH. The value of F_{FH} is obtained from the equation $\Sigma M_G = 0$. We move $\mathbf{F}_{FH}$ along its line of action until it acts at point *F*, where it is resolved into its *x* and *y* components. The moment of $\mathbf{F}_{FH}$ with respect to point *G* is now equal to $(F_{FH} \cos \alpha)(8\text{ m})$.

$$+\circlearrowleft \Sigma M_G = 0:$$

$$(7.50\text{ kN})(15\text{ m}) - (1\text{ kN})(10\text{ m}) - (1\text{ kN})(5\text{ m}) + (F_{FH} \cos \alpha)(8\text{ m}) = 0$$

$$F_{FH} = -13.82\text{ kN} \qquad F_{FH} = 13.82\text{ kN } C \blacktriangleleft$$

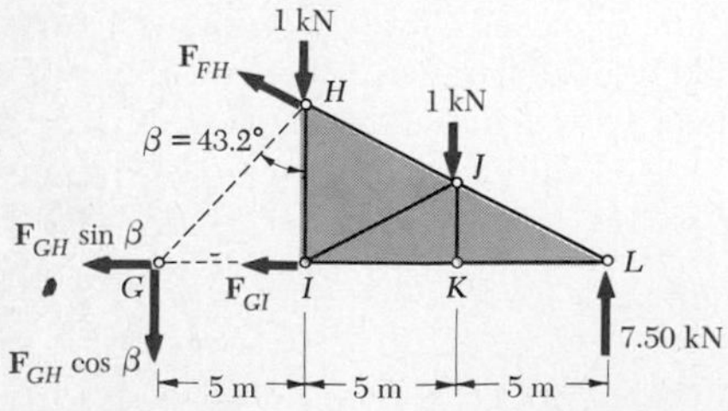

Force in Member GH. The value of F_{GH} is determined by first resolving the force $\mathbf{F}_{GH}$ into *x* and *y* components at point *G* and then solving the equation $\Sigma M_L = 0$.

$$+\circlearrowleft \Sigma M_L = 0: \qquad (1\text{ kN})(10\text{ m}) + (1\text{ kN})(5\text{ m}) + (F_{GH} \cos \beta)(15\text{ m}) = 0$$

$$F_{GH} = -1.372\text{ kN} \qquad F_{GH} = 1.372\text{ kN } C \blacktriangleleft$$

Problems

6.21 Determine the force in members *CE* and *CF* of the truss shown.

6.22 Determine the force in members *FG* and *FH* of the truss shown.

6.23 Determine the force in members *CD* and *DF* of the truss shown.

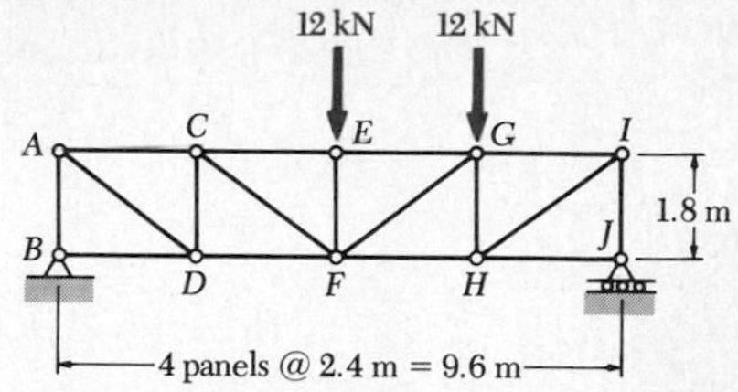

Fig. P6.21 and P6.22

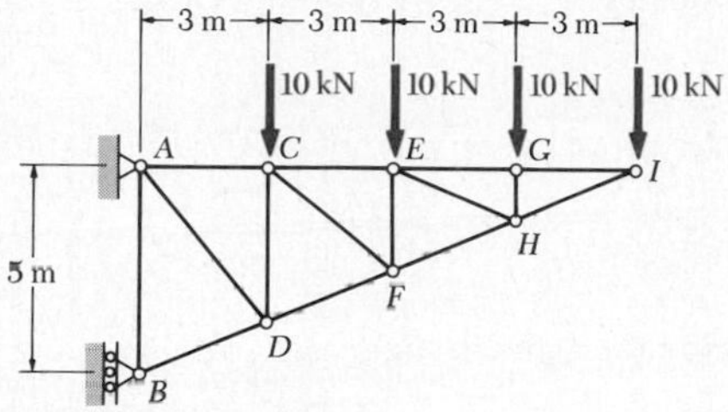

Fig. P6.23 and P6.24

6.24 Determine the force in members *CE* and *EF* of the truss shown.

6.25 Determine the force in members *DE* and *EF* of the truss shown.

6.26 Determine the force in members *FG* and *FH* of the truss shown.

6.27 Determine the force in members *CE* and *DE* of the truss shown.

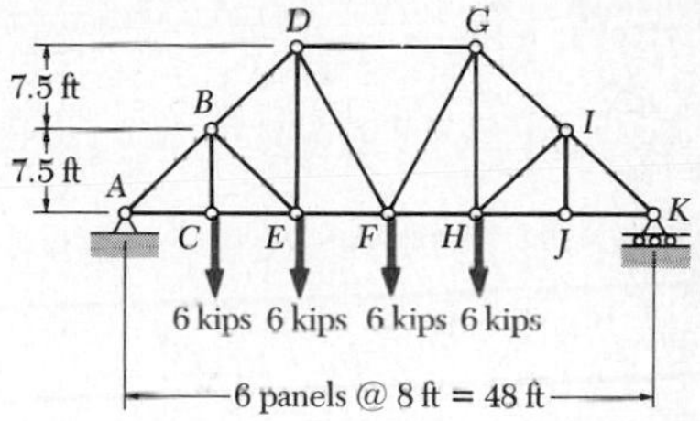

Fig. P6.25 and P6.26

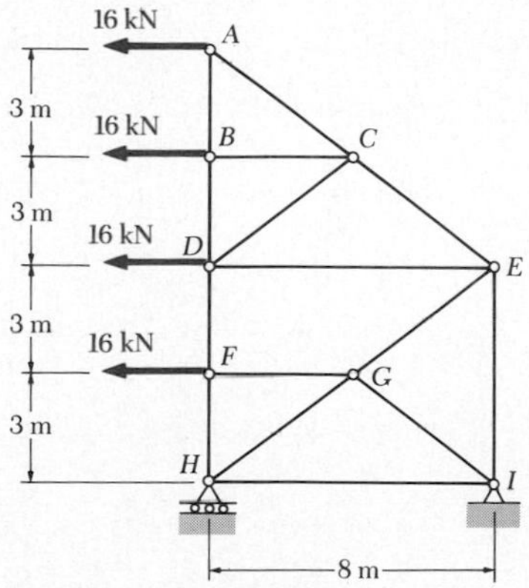

Fig. P6.27 and P6.28

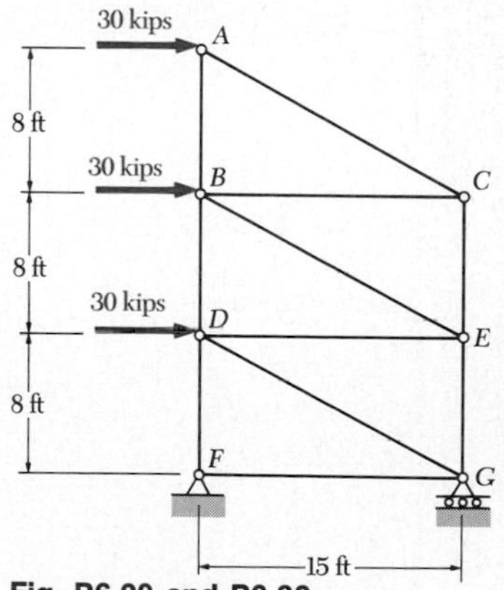

Fig. P6.29 and P6.30

6.28 Determine the force in members *EI* and *EG* of the truss shown.

6.29 Determine the force in members *BD* and *DE* of the truss shown.

6.30 Determine the force in members *DG* and *EG* of the truss shown.

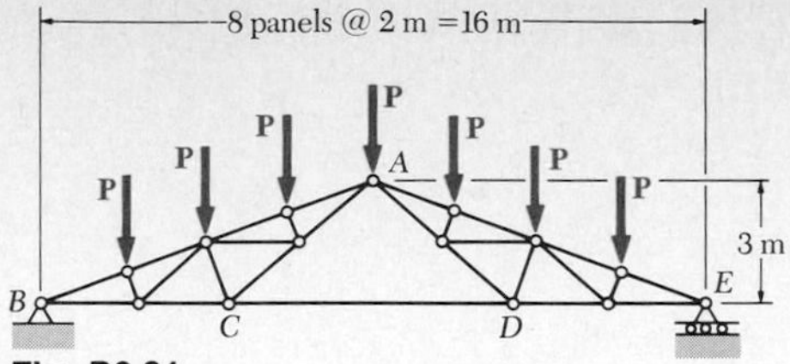

Fig. P6.31

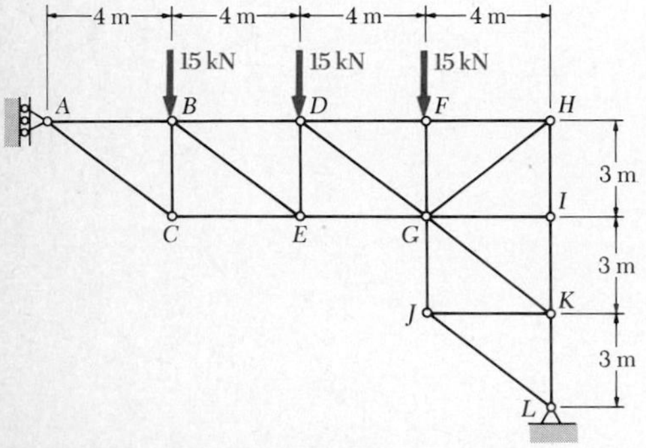

Fig. P6.32, P6.33, and P6.34

6.31 Determine the force in member *CD* of the Fink roof truss shown when $P = 15$ kN.

6.32 Determine the force in members *BD*, *BE*, and *CE* of the truss shown.

6.33 Determine the force in members *DF*, *DG*, and *EG* of the truss shown.

6.34 Determine the force in members *GJ*, *GK*, and *IK* of the truss shown.

6.35 Determine the force in members *AD*, *CD*, and *CE* of the truss shown.

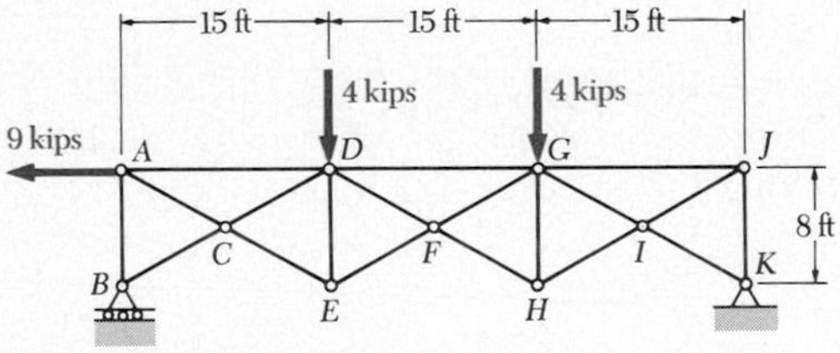

Fig. P6.35 and P6.36

6.36 Determine the force in members *DG*, *FG*, and *FH* of the truss shown.

6.37 Determine the force in members *FK* and *JO* of the truss shown. (*Hint.* Use section *a-a*.)

6.38 Determine the force in members *AC* and *BE* of the truss shown. (*Hint.* Use section *a-a*.)

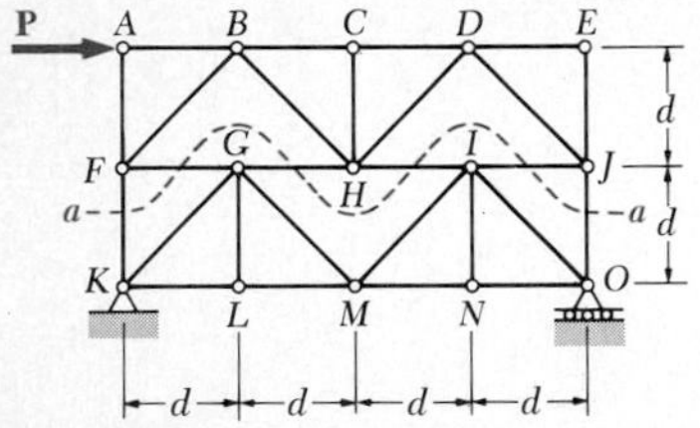

Fig. P6.37

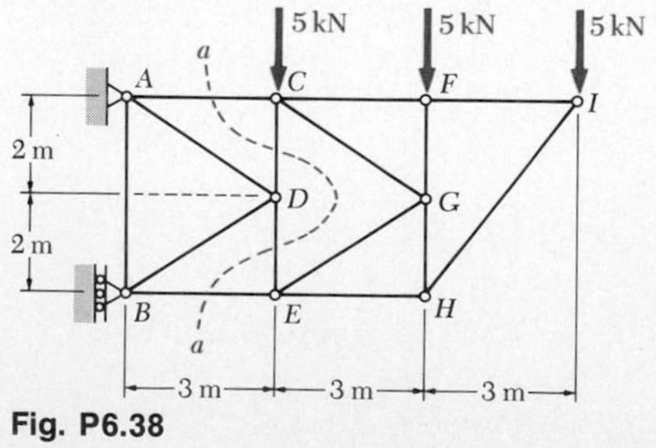

Fig. P6.38

6.39 Determine the force in member GJ of the truss shown. (*Hint.* Use section *a-a.*)

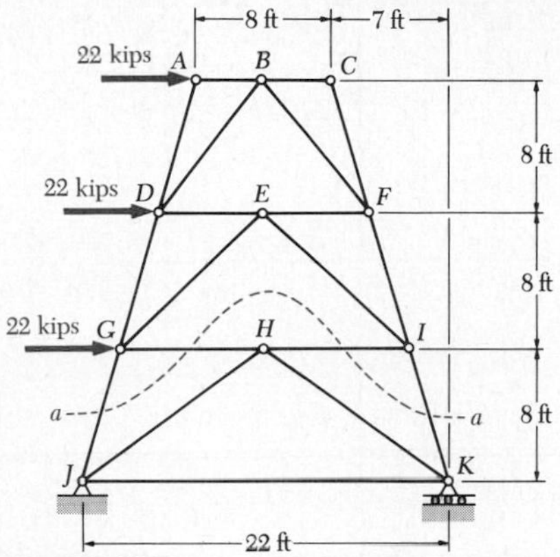

Fig. P6.39

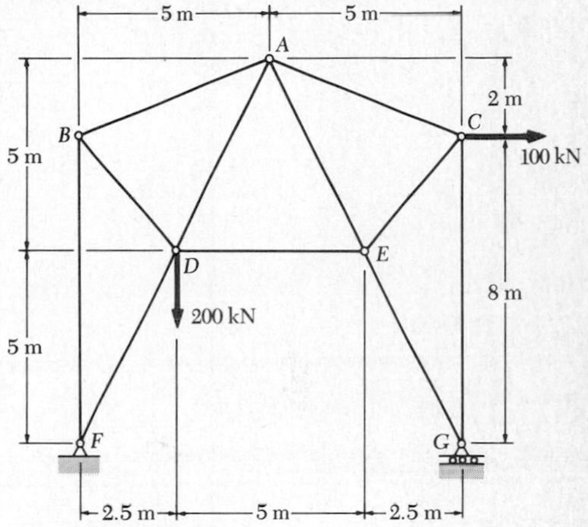

Fig. P6.40 and P6.41

6.40 Determine the force in members DE and AE of the truss shown.

6.41 Determine the force in members AB and AD of the truss shown.

6.42 The diagonal members in the center panel of the truss shown are very slender and can act only in tension; such members are known as *counters.* Determine the force in member DE and in the counters which are acting under the given loading.

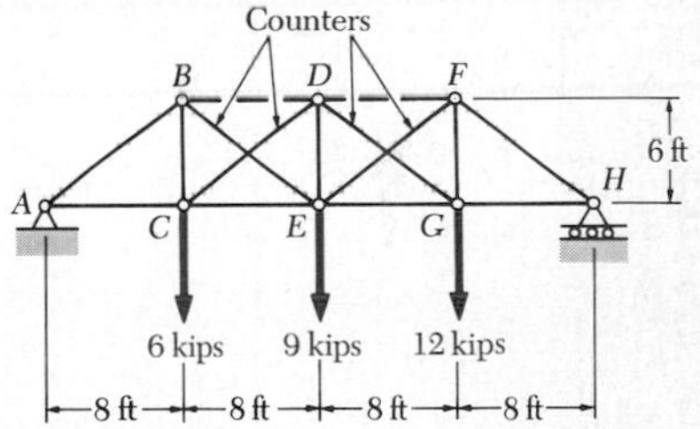

Fig. P6.42

6.43 Solve Prob. 6.42, assuming that the 6-kip load has been removed.

6.44 Solve Prob. 6.42, assuming that the 9-kip load has been removed.

6.45 and 6.46 Classify each of the given structures as completely, partially, or improperly constrained; if completely constrained, further classify as determinate or indeterminate. (All members can act both in tension and in compression.)

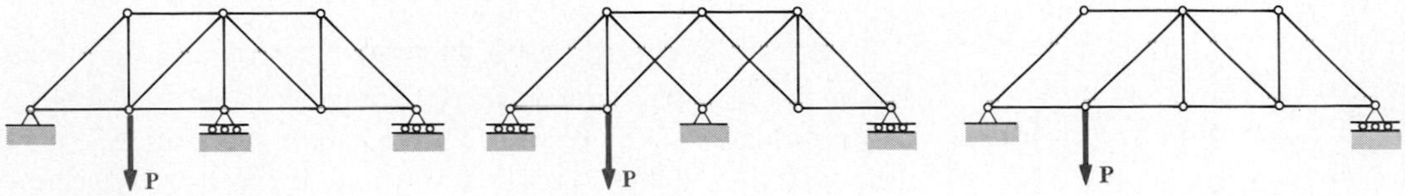

Fig. P6.45 (*a*) (*b*) (*c*)

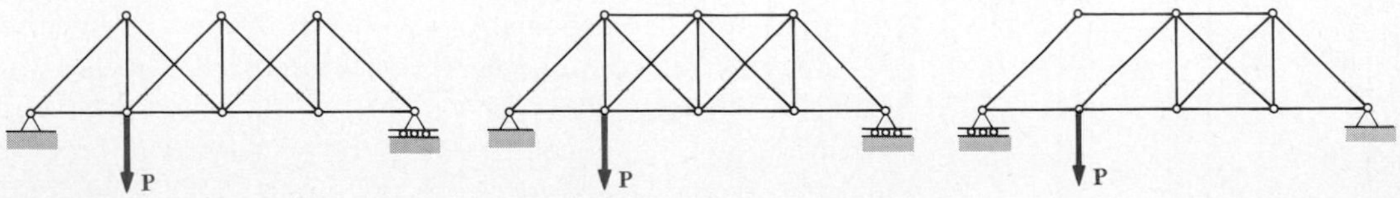

Fig. P6.46 (*a*) (*b*) (*c*)

6.47 and 6.48 Classify each of the given structures as completely, partially, or improperly constrained; if completely constrained, further classify as determinate or indeterminate. (All members can act both in tension and in compression.)

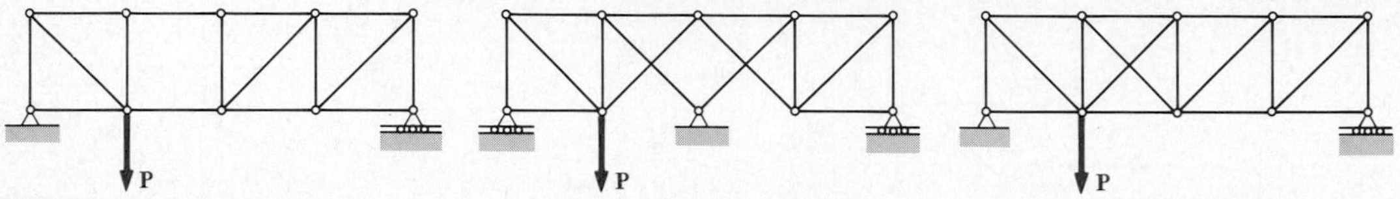

Fig. P6.47 (a) (b) (c)

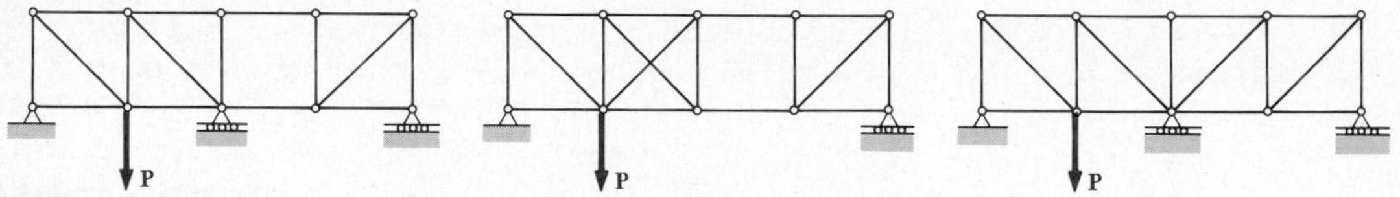

Fig. P6.48 (a) (b) (c)

FRAMES AND MACHINES

6.9. Structures Containing Multiforce Members. Under Trusses, we have considered structures consisting entirely of pins and of straight two-force members. The forces acting on the two-force members were known to be directed along the members themselves. We shall now consider structures in which at least one of the members is a *multiforce* member, i.e., a member acted upon by three or more forces. These forces will generally not be directed along the members on which they act; their direction is unknown, and they should be represented therefore by two unknown components.

Frames and machines are structures containing multiforce members. *Frames* are designed to support loads and are usually stationary, fully constrained structures. *Machines* are designed to transmit and modify forces; they may or may not be stationary and will always contain moving parts.

6.10. Analysis of a Frame. As a first example of analysis of a frame, we shall consider again the crane described in Sec. 6.1, which carries a given load W (Fig. 6.20a). The free-body diagram of the entire frame is shown in Fig. 6.20b. This diagram may be used to determine the external forces acting on the frame. Summing moments about A, we first determine the force $\mathbf{T}$ exerted by the cable; summing x and y components, we then determine the components $\mathbf{A}_x$ and $\mathbf{A}_y$ of the reaction at the pin A.

In order to determine the internal forces holding the various parts of a frame together, we must dismember the frame and draw a free-body diagram for each of its component parts (Fig. 6.20c). First, the two-force members should be considered. In this frame, member BE is the only two-force member. The forces acting at each end of this member must have the same magnitude, same line of action, and opposite sense (Sec. 4.6). They are therefore directed along BE and will be denoted, respectively, by $\mathbf{F}_{BE}$

and $-\mathbf{F}_{BE}$. Their sense will be arbitrarily assumed as shown in Fig. 6.20*c*, and the correctness of this assumption will be checked later by the sign obtained for the common magnitude F_{BE} of the two forces.

Next, we consider the multiforce members, i.e., the members which are acted upon by three or more forces. According to Newton's third law, the force exerted at *B* by member *BE* on member *AD* must be equal and opposite to the force $\mathbf{F}_{BE}$ exerted by *AD* on *BE*. Similarly, the force exerted at *E* by member *BE* on member *CF* must be equal and opposite to the force $-\mathbf{F}_{BE}$ exerted by *CF* on *BE*. The forces that the two-force member *BE* exerts on *AD* and *CF* are therefore respectively equal to $-\mathbf{F}_{BE}$ and $\mathbf{F}_{BE}$; they have the same magnitude F_{BE} and opposite sense, and should be directed as shown in Fig. 6.20*c*.

At *C* two multiforce members are connected. Since neither the direction nor the magnitude of the forces acting at *C* is known, these forces will be represented by their *x* and *y* components. The components $\mathbf{C}_x$ and $\mathbf{C}_y$ of the force acting on member *AD* will be arbitrarily directed to the right and upward. Since, according to Newton's third law, the forces exerted by member *CF* on *AD* and by member *AD* on *CF* are equal and opposite, the components of the force acting on member *CF* *must* be directed to the left and downward; they will be denoted, respectively, by $-\mathbf{C}_x$ and $-\mathbf{C}_y$. Whether the force $\mathbf{C}_x$ is actually directed to the right and the force $-\mathbf{C}_x$ is actually directed to the left will be determined later from the sign of their common magnitude C_x, a plus sign indicating that the assumption made was correct, and a minus sign that it was wrong. The free-body diagrams of the multiforce members are completed by showing the external forces acting at *A*, *D*, and *F*.†

The internal forces may now be determined by considering the free-body diagram of either of the two multiforce members. Choosing the free-body diagram of *CF*, for example, we write the equations $\Sigma M_C = 0$, $\Sigma M_E = 0$, and $\Sigma F_x = 0$, which yield the values of the magnitudes F_{BE}, C_y, and C_x, respectively. These values may be checked by verifying that member *AD* is also in equilibrium.

It should be noted that the free-body diagrams of the pins were not shown in Fig. 6.20*c*. This was because the pins were assumed to form an integral part of one of the two members they connected. This assumption can always be used to simplify the analysis of frames and machines. When a pin connects three or more members, however, or when a pin connects a support and two or more members, or when a load is applied to a pin, a clear decision must be made in choosing the member to which the pin will be assumed to belong. (If multiforce members are involved, the pin should be attached to one of these members.) The various forces exerted on the pin should then be clearly identified. This is illustrated in Sample Prob. 6.6.

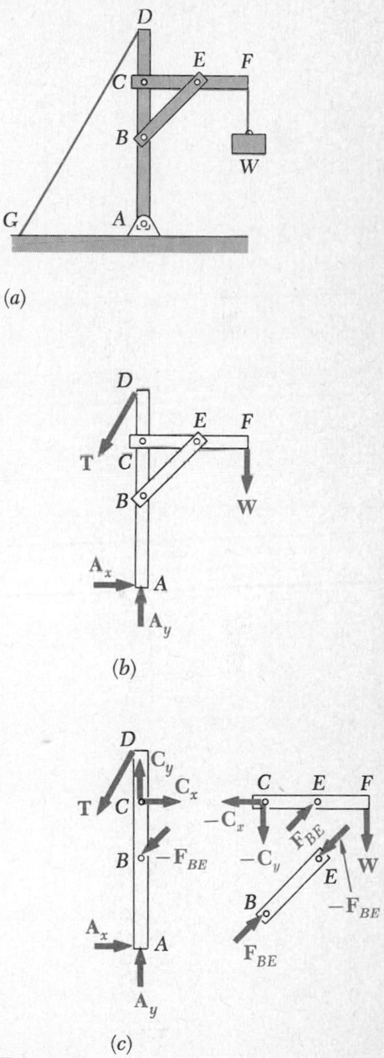

Fig. 6.20

† The use of a minus sign to distinguish the force exerted by one member on another from the equal and opposite force exerted by the second member on the first is not strictly necessary, since the two forces belong to different free-body diagrams and thus cannot easily be confused. In the Sample Problems, we shall represent by the same symbol equal and opposite forces which are applied to different free bodies. It should be noted that, under these conditions, the sign obtained for a given force component will not directly relate the sense of that component to the sense of the corresponding coordinate axis. Rather, a positive sign will indicate that *the sense assumed for that component in the free-body diagram* is correct, and a negative sign that it is wrong.

6.11. Frames Which Cease to Be Rigid When Detached from Their Supports. The crane analyzed in Sec. 6.10 was so constructed that it could keep the same shape without the help of its supports; it was therefore considered as a rigid body. Many frames, however, will collapse if detached from their supports; such frames cannot be considered as rigid bodies. Consider, for example, the frame shown in Fig. 6.21*a*, which consists of two members *AC* and *CB* carrying loads **P** and **Q** at their midpoints; the members are supported by pins at *A* and *B* and are connected by a pin at *C*. If detached from its supports, this frame will not maintain its shape; it should therefore be considered as made of *two distinct rigid parts AC* and *CB*.

The equations $\Sigma F_x = 0$, $\Sigma F_y = 0$, $\Sigma M = 0$ (about any given point) express the conditions for the *equilibrium of a rigid body* (Chap. 4); we should use them, therefore, in connection with the free-body diagrams of rigid bodies, namely, the free-body diagrams of members *AC* and *CB* (Fig. 6.21*b*). Since these members are multiforce members, and since pins are used at the supports and at the connection, the reactions at *A* and *B* and the

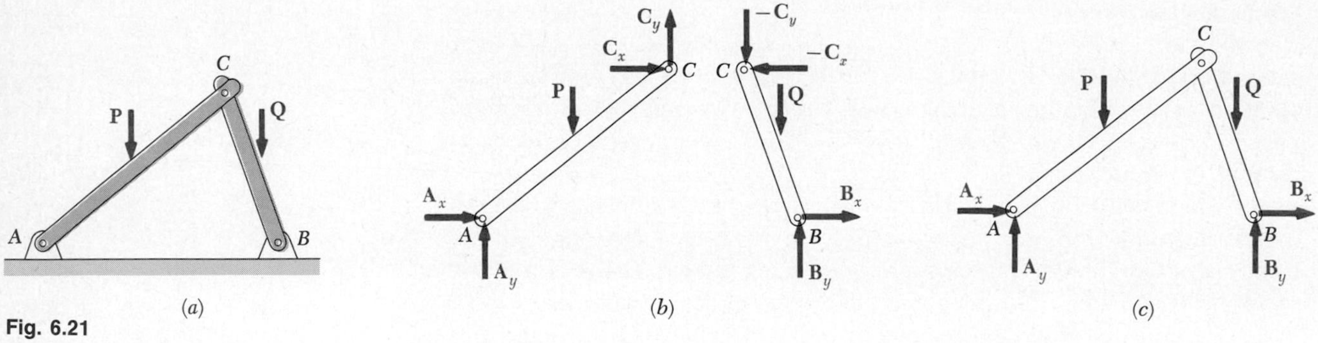

Fig. 6.21

forces at *C* will each be represented by two components. In accordance with Newton's third law, the components of the force exerted by *CB* on *AC* and the components of the force exerted by *AC* on *CB* will be represented by vectors of the same magnitude and opposite sense; thus, if the first pair of components consists of $\mathbf{C}_x$ and $\mathbf{C}_y$, the second pair will be represented by $-\mathbf{C}_x$ and $-\mathbf{C}_y$. We note that four unknown force components act on free body *AC*, while only three independent equations may be used to express that the body is in equilibrium; similarly, four unknowns, but only three equations, are associated with *CB*. However, only six different unknowns are involved in the analysis of the two members, and altogether six equations are available to express that the members are in equilibrium. Writing $\Sigma M_A = 0$ for free body *AC* and $\Sigma M_B = 0$ for *CB*, we obtain two simultaneous equations which may be solved for the common magnitude C_x of the components $\mathbf{C}_x$ and $-\mathbf{C}_x$, and for the common magnitude C_y of the components $\mathbf{C}_y$ and $-\mathbf{C}_y$. Writing, then, $\Sigma F_x = 0$ and $\Sigma F_y = 0$ for each of the two free bodies, we obtain successively the magnitudes A_x, A_y, B_x, and B_y.

We shall observe now that since the equations of equilibrium $\Sigma F_x = 0$, $\Sigma F_y = 0$, $\Sigma M = 0$ (about any given point) are satisfied by the forces acting on free body AC, and since they are also satisfied by the forces acting on free body CB, they must be satisfied when the forces acting on the two free bodies are considered simultaneously. Since the internal forces at C cancel each other, we find that the equations of equilibrium must be satisfied by the external forces shown on the free-body diagram of the frame ACB itself (Fig. 6.21*c*), although the frame is not a rigid body. These equations may be used to determine some of the components of the reactions at A and B. We shall note, however, that *the reactions cannot be completely determined from the free-body diagram of the whole frame.* It is thus necessary to dismember the frame and to consider the free-body diagrams of its component parts (Fig. 6.21*b*), even when we are interested only in finding external reactions. This may be explained by the fact that the equilibrium equations obtained for free body ACB are *necessary conditions* for the equilibrium of a nonrigid structure, *but not sufficient conditions.*

The method of solution outlined in the second paragraph of this section involved simultaneous equations. We shall now discuss a more expeditious method, which utilizes the free body ACB as well as the free bodies AC and CB. Writing $\Sigma M_A = 0$ and $\Sigma M_B = 0$ for free body ACB, we obtain B_y and A_y. Writing $\Sigma M_C = 0$, $\Sigma F_x = 0$, and $\Sigma F_y = 0$ for free body AC, we obtain successively A_x, C_x, and C_y. Finally, writing $\Sigma F_x = 0$ for ACB, we obtain B_x.

We noted above that the analysis of the frame of Fig. 6.21 involves six unknown force components and six independent equilibrium equations (the equilibrium equations for the whole frame were obtained from the original six equations and, therefore, are not independent). Moreover, we checked that all unknowns could be actually determined and that all equations could be satisfied. The frame considered is *statically determinate and rigid.*† In general, to determine whether a structure is statically determinate and rigid, we should draw a free-body diagram for each of its component parts and count the reactions and internal forces involved. We should also determine the number of independent equilibrium equations (excluding equations expressing the equilibrium of the whole structure or of groups of component parts already analyzed). If there are more unknowns than equations, the structure is *statically indeterminate.* If there are fewer unknowns than equations, the structure is *nonrigid.* If there are as many unknowns as equations, *and if all unknowns may be determined and all equations satisfied* under general loading conditions, the structure is *statically determinate and rigid;* if, however, due to an *improper arrangement* of members and supports, all unknowns cannot be determined and all equations cannot be satisfied, the structure is *statically indeterminate and nonrigid.*

† The word "rigid" is used here to indicate that the frame will maintain its shape as long as it remains attached to its supports.

SAMPLE PROBLEM 6.4

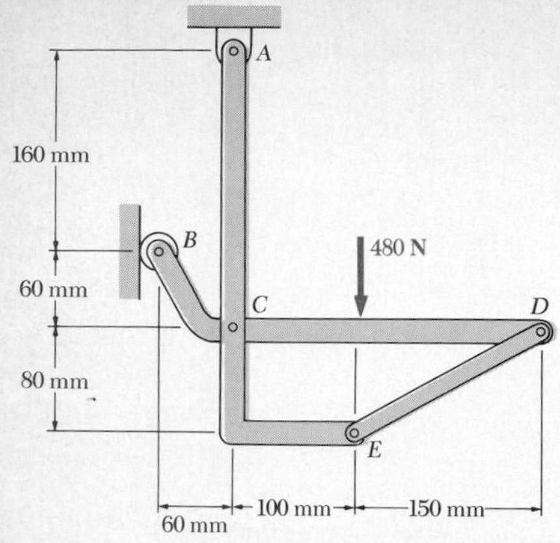

In the frame shown, members ACE and BCD are connected by a pin at C and by the link DE. For the loading shown, determine the force in link DE and the components of the force exerted at C on member BCD.

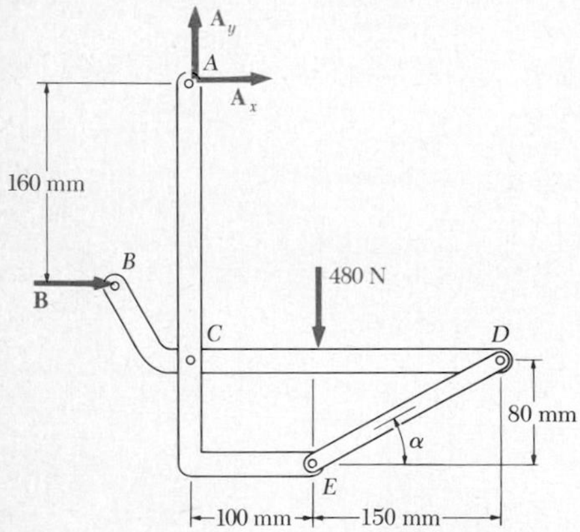

$\alpha = \tan^{-1}\dfrac{80}{150} = 28.07°$

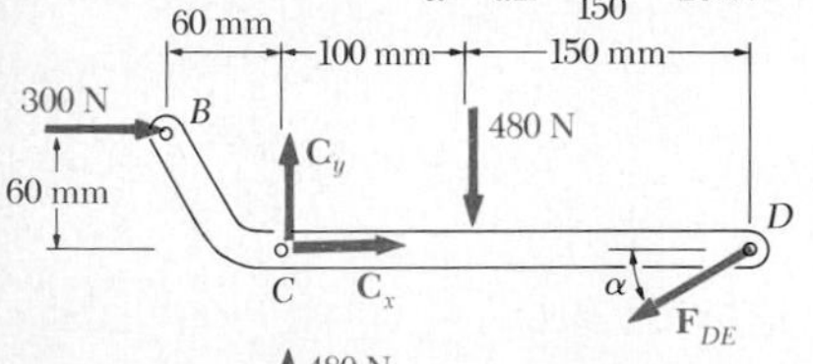

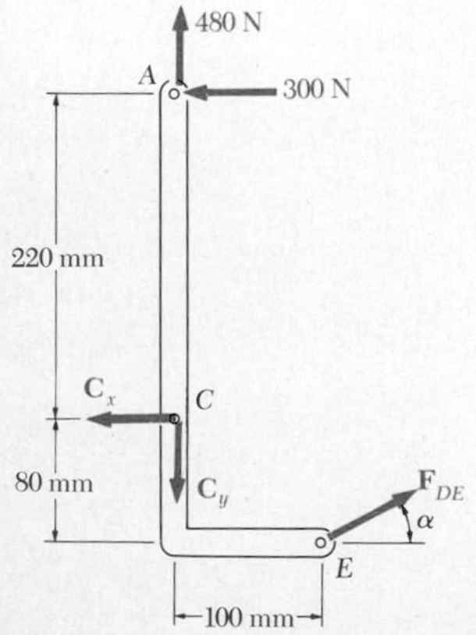

Entire Frame. Since the external reactions involve only three unknowns, we compute the reactions by considering the free-body diagram of the entire frame.

$+\uparrow\Sigma F_y = 0$: $\quad A_y - 480\text{ N} = 0 \qquad A_y = +480\text{ N} \qquad \mathbf{A}_y = 480\text{ N}\uparrow$

$+\circlearrowleft\Sigma M_A = 0$: $\quad -(480\text{ N})(100\text{ mm}) + B(160\text{ mm}) = 0$

$B = +300\text{ N} \qquad \mathbf{B} = 300\text{ N}\rightarrow$

$\overset{+}{\rightarrow}\Sigma F_x = 0$: $\quad B + A_x = 0$

$300\text{ N} + A_x = 0 \qquad A_x = -300\text{ N} \qquad \mathbf{A}_x = 300\text{ N}\leftarrow$

Members. We now dismember the frame. Since only two members are connected at C, the components of the unknown forces acting on ACE and BCD are, respectively, equal and opposite and are assumed directed as shown. We assume that link DE is in tension and exerts at D and E equal and opposite forces directed as shown.

Member *BCD*. Using the free body BCD, we write

$+\circlearrowright\Sigma M_C = 0$: $\quad (F_{DE}\sin\alpha)(250\text{ mm}) + (300\text{ N})(60\text{ mm}) + (480\text{ N})(100\text{ mm}) = 0$

$F_{DE} = -561\text{ N} \qquad F_{DE} = 561\text{ N } C$ ◄

$\overset{+}{\rightarrow}\Sigma F_x = 0$: $\quad C_x - F_{DE}\cos\alpha + 300\ N = 0$

$C_x - (-561\text{ N})\cos 28.07° + 300\text{ N} = 0 \qquad C_x = -795\text{ N}$

$+\uparrow\Sigma F_y = 0$: $\quad C_y - F_{DE}\sin\alpha - 480\text{ N} = 0$

$C_y - (-561\text{ N})\sin 28.07° - 480\text{ N} = 0 \qquad C_y = +216\text{ N}$

From the signs obtained for C_x and C_y we conclude that the force components $\mathbf{C}_x$ and $\mathbf{C}_y$ exerted on member BCD are directed respectively to the left and up. We have

$\mathbf{C}_x = 795\text{ N}\leftarrow,\ \mathbf{C}_y = 216\text{ N}\uparrow$ ◄

Member *ACE* (Check). The computations are checked by considering the free body ACE. For example,

$$+\circlearrowleft\Sigma M_A = (F_{DE}\cos\alpha)(300\text{ mm}) + (F_{DE}\sin\alpha)(100\text{ mm}) - C_x(220\text{ mm})$$
$$= (-561\cos\alpha)(300) + (-561\sin\alpha)(100) - (-795)(220) = 0$$

SAMPLE PROBLEM 6.5

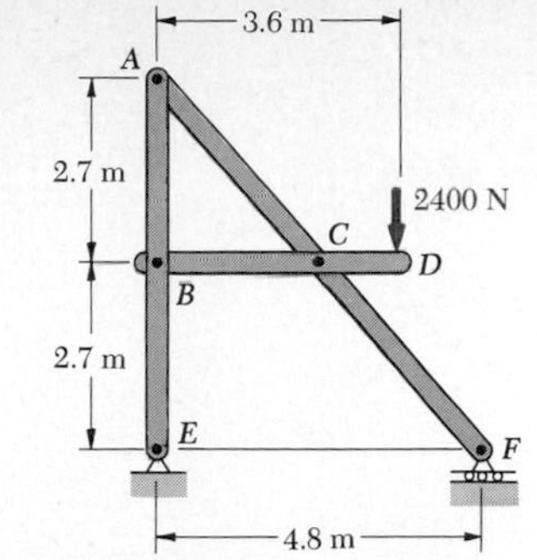

Determine the components of the forces acting on each member of the frame shown.

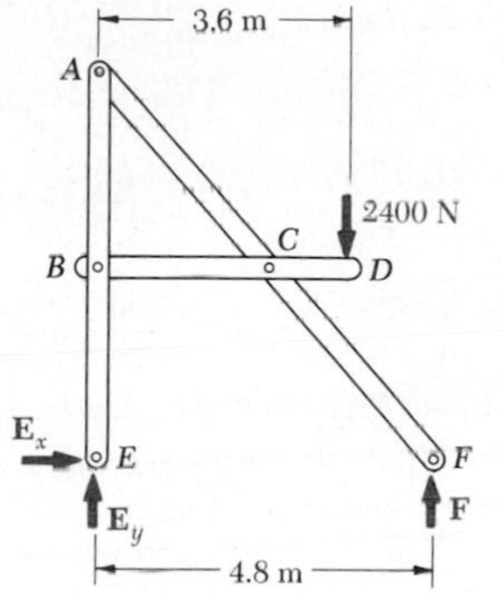

Entire Frame. Since the external reactions involve only three unknowns, we compute the reactions by considering the free-body diagram of the entire frame.

$+\circlearrowleft \Sigma M_E = 0$: $\quad -(2400\text{ N})(3.6\text{ m}) + F(4.8\text{ m}) = 0$
$F = +1800\text{ N}$ $\qquad$ $\mathbf{F} = 1800\text{ N}\uparrow$ ◀

$+\uparrow \Sigma F_y = 0$: $\quad -2400\text{ N} + 1800\text{ N} + E_y = 0$
$E_y = +600\text{ N}$ $\qquad$ $\mathbf{E}_y = 600\text{ N}\uparrow$ ◀

$\xrightarrow{+} \Sigma F_x = 0$: $\qquad$ $\mathbf{E}_x = 0$ ◀

The frame is now dismembered; since only two members are connected at each joint, equal and opposite components are shown on each member at each joint.

Member *BCD*

$+\circlearrowleft \Sigma M_B = 0$: $\quad -(2400\text{ N})(3.6\text{ m}) + C_y(2.4\text{ m}) = 0 \qquad C_y = +3600\text{ N}$ ◀
$+\circlearrowleft \Sigma M_C = 0$: $\quad -(2400\text{ N})(1.2\text{ m}) + B_y(2.4\text{ m}) = 0 \qquad B_y = +1200\text{ N}$ ◀
$\xrightarrow{+} \Sigma F_x = 0$: $\quad -B_x + C_x = 0$

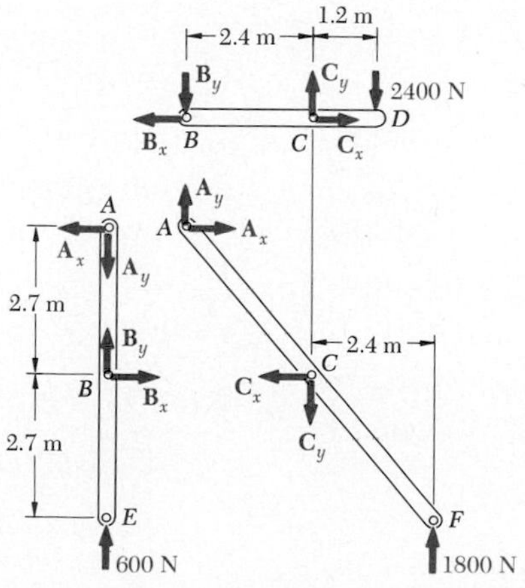

We note that neither B_x nor C_x can be obtained by considering only member *BCD*. The positive values obtained for B_y and C_y indicate that the force components $\mathbf{B}_y$ and $\mathbf{C}_y$ are directed as assumed.

Member *ABE*

$+\circlearrowleft \Sigma M_A = 0$: $\quad B_x(2.7\text{ m}) = 0 \qquad B_x = 0$ ◀
$\xrightarrow{+} \Sigma F_x = 0$: $\quad +B_x - A_x = 0 \qquad A_x = 0$ ◀
$+\uparrow \Sigma F_y = 0$: $\quad -A_y + B_y + 600\text{ N} = 0$
$-A_y + 1200\text{ N} + 600\text{ N} = 0 \qquad A_y = +1800\text{ N}$ ◀

Member *BCD*. Returning now to member *BCD*, we write

$\xrightarrow{+} \Sigma F_x = 0$: $\quad -B_x + C_x = 0 \qquad 0 + C_x = 0 \qquad C_x = 0$ ◀

Member *ACF* (Check). All unknown components have now been found; to check the results, we verify that member *ACF* is in equilibrium.

$$+\circlearrowleft \Sigma M_C = (1800\text{ N})(2.4\text{ m}) - A_y(2.4\text{ m}) - A_x(2.7\text{ m})$$
$$= (1800\text{ N})(2.4\text{ m}) - (1800\text{ N})(2.4\text{ m}) - 0 = 0 \qquad \text{(checks)}$$

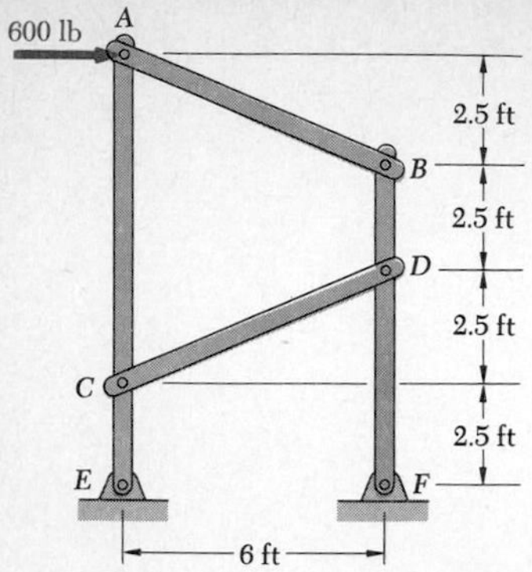

SAMPLE PROBLEM 6.6

A 600-lb horizontal force is applied to pin A of the frame shown. Determine the forces acting on the two vertical members of the frame.

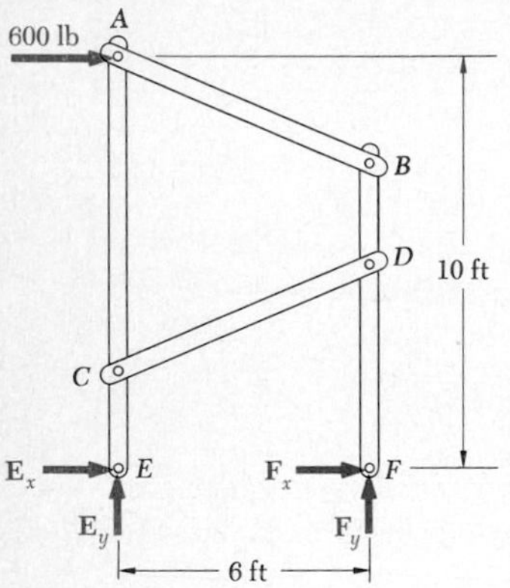

Entire Frame. The entire frame is chosen as a free body; although the reactions involve four unknowns, $\mathbf{E}_y$ and $\mathbf{F}_y$ may be determined by writing

$$+\circlearrowleft \Sigma M_E = 0: \qquad -(600 \text{ lb})(10 \text{ ft}) + F_y(6 \text{ ft}) = 0$$

$$F_y = +1000 \text{ lb} \qquad \mathbf{F}_y = 1000 \text{ lb} \uparrow \blacktriangleleft$$

$$+\uparrow \Sigma F_y = 0: \qquad E_y + F_y = 0$$

$$E_y = -1000 \text{ lb} \qquad \mathbf{E}_y = 1000 \text{ lb} \downarrow \blacktriangleleft$$

The equations of equilibrium of the entire frame are not sufficient to determine $\mathbf{E}_x$ and $\mathbf{F}_x$. The equilibrium of the various members must now be considered in order to proceed with the solution. In dismembering the frame we shall assume that pin A is attached to the multiforce member ACE and, thus, that the 600-lb force is applied to that member. We also note that AB and CD are two-force members.

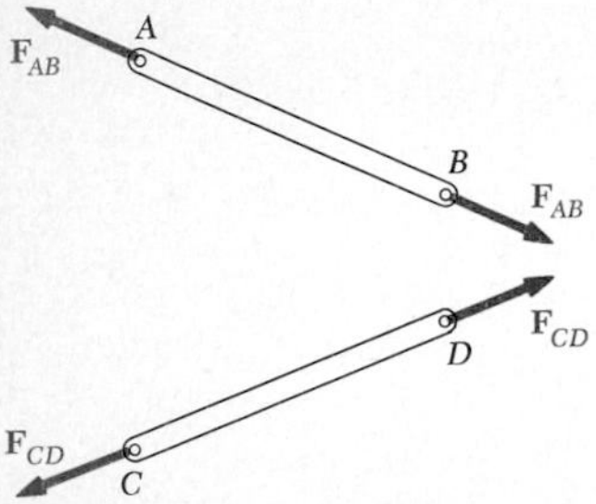

Member ACE

$$+\uparrow \Sigma F_y = 0: \qquad -\tfrac{5}{13}F_{AB} + \tfrac{5}{13}F_{CD} - 1000 \text{ lb} = 0$$

$$+\circlearrowleft \Sigma M_E = 0: \qquad -(600 \text{ lb})(10 \text{ ft}) - (\tfrac{12}{13}F_{AB})(10 \text{ ft}) - (\tfrac{12}{13}F_{CD})(2.5 \text{ ft}) = 0$$

Solving these equations simultaneously, we find

$$F_{AB} = -1040 \text{ lb} \qquad F_{CD} = +1560 \text{ lb} \blacktriangleleft$$

The signs obtained indicate that the sense assumed for F_{CD} was correct and the sense for F_{AB} incorrect. Summing now x components,

$$\overset{+}{\rightarrow} \Sigma F_x = 0: \qquad 600 \text{ lb} + \tfrac{12}{13}(-1040 \text{ lb}) + \tfrac{12}{13}(+1560 \text{ lb}) + E_x = 0$$

$$E_x = -1080 \text{ lb} \qquad \mathbf{E}_x = 1080 \text{ lb} \leftarrow \blacktriangleleft$$

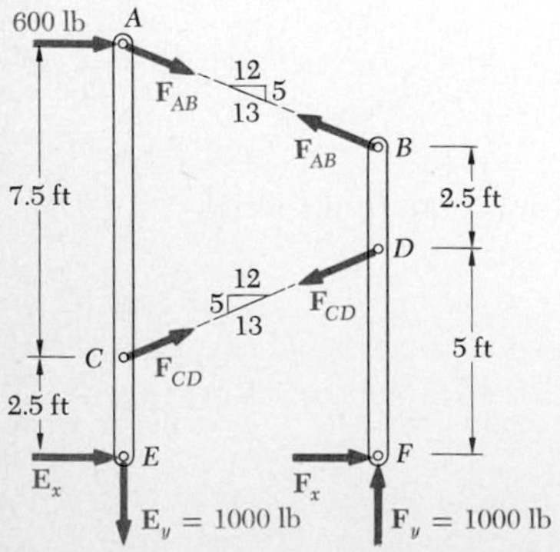

Entire Frame. Since $\mathbf{E}_x$ has been determined, we may return to the free-body diagram of the entire frame and write

$$\overset{+}{\rightarrow} \Sigma F_x = 0: \qquad 600 \text{ lb} - 1080 \text{ lb} + F_x = 0$$

$$F_x = +480 \text{ lb} \qquad \mathbf{F}_x = 480 \text{ lb} \rightarrow \blacktriangleleft$$

Member BDF (Check). We may check our computations by verifying that the equation $\Sigma M_B = 0$ is satisfied by the forces acting on member BDF.

$$\begin{aligned} +\circlearrowleft \Sigma M_B &= -(\tfrac{12}{13}F_{CD})(2.5 \text{ ft}) + (F_x)(7.5 \text{ ft}) \\ &= -\tfrac{12}{13}(1560 \text{ lb})(2.5 \text{ ft}) + (480 \text{ lb})(7.5 \text{ ft}) \\ &= -3600 \text{ lb}\cdot\text{ft} + 3600 \text{ lb}\cdot\text{ft} = 0 \qquad \text{(checks)} \end{aligned}$$

Problems

6.49 through 6.51 Determine the force in member *BD* and the components of the reaction at *C*.

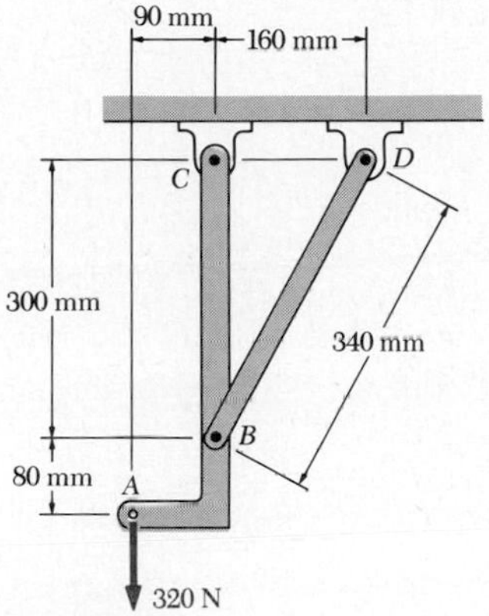

Fig. P6.49

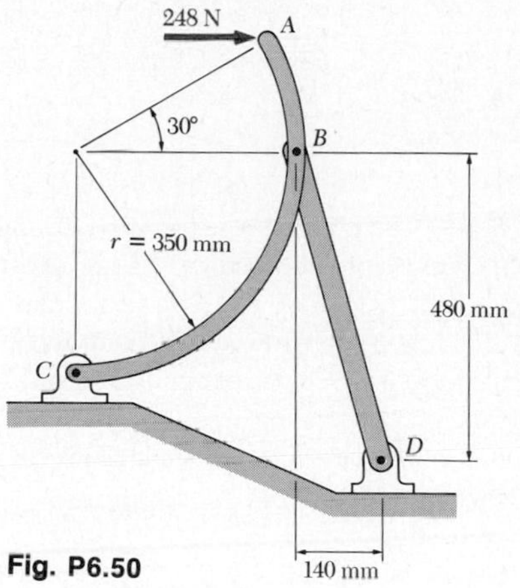

Fig. P6.50

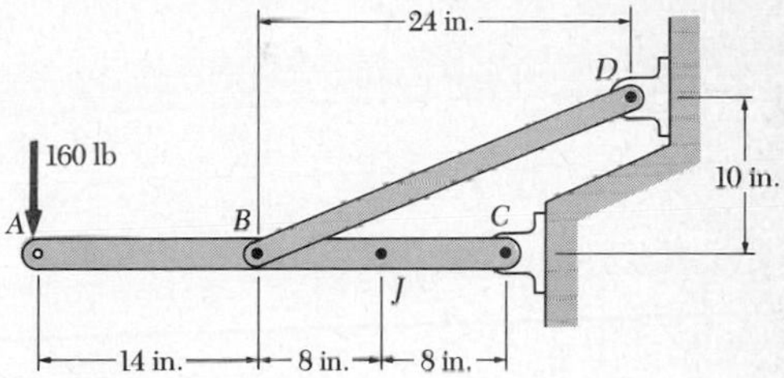

Fig. P6.51

6.52 Determine the components of all forces acting on member *ABD* of the frame shown.

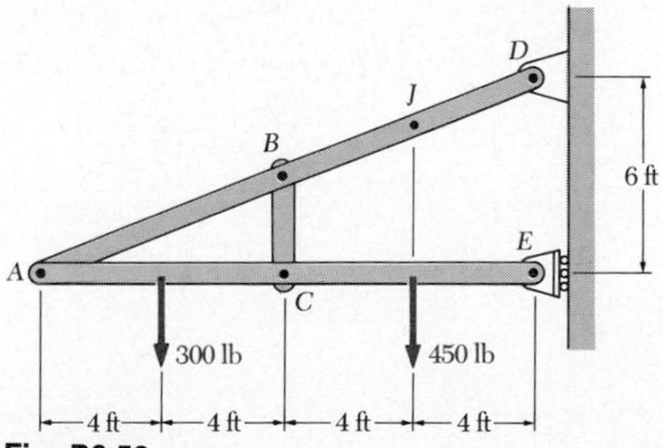

Fig. P6.52

6.53 Determine the components of all forces acting on member *ABCD* when $\theta = 0$.

6.54 Determine the components of all forces acting on member *ABCD* when $\theta = 90°$.

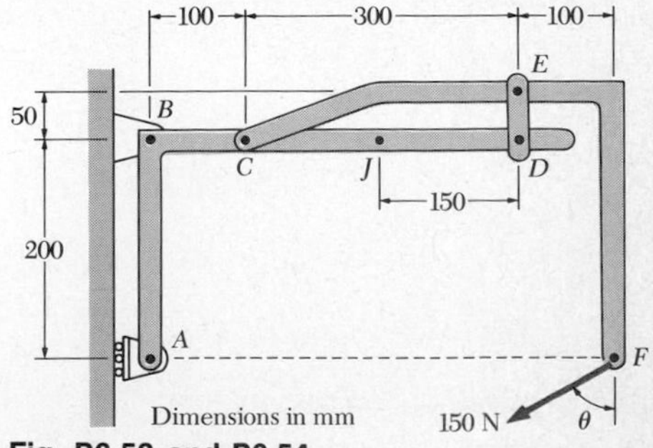

Fig. P6.53 and P6.54

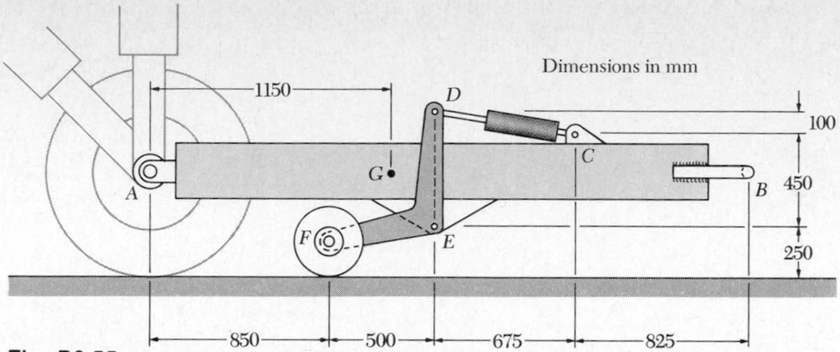

Fig. P6.55

6.55 An aircraft tow bar is positioned by means of a single hydraulic cylinder *CD* which is connected to two identical arm-and-wheel units *DEF*. The entire tow bar has a mass of 180 kg and its center of gravity is located at *G*. For the position shown, determine (*a*) the force exerted by the cylinder on bracket *C*, (*b*) the force exerted on each arm by the pin at *E*.

6.56 Solve Prob. 6.55, assuming that a 75-kg mechanic is standing on the tow bar at point *B*.

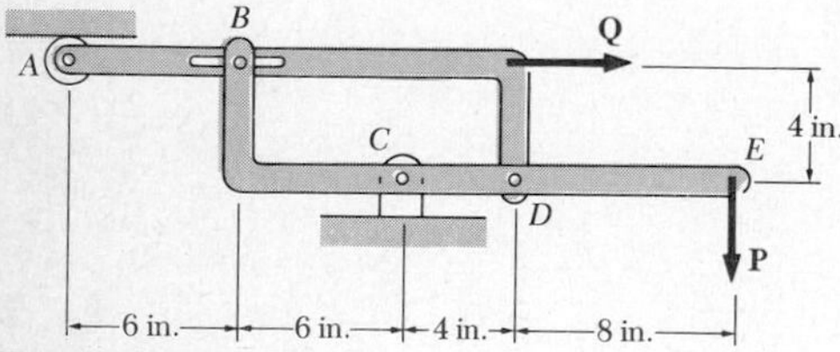

Fig. P6.57 and P6.58

6.57 Knowing that $P = 75$ lb and $Q = 45$ lb, determine the components of all forces acting on member *BCDE* of the assembly shown.

6.58 Knowing that $P = 45$ lb and $Q = 75$ lb, determine the components of all forces acting on member *BCDE* of the assembly shown.

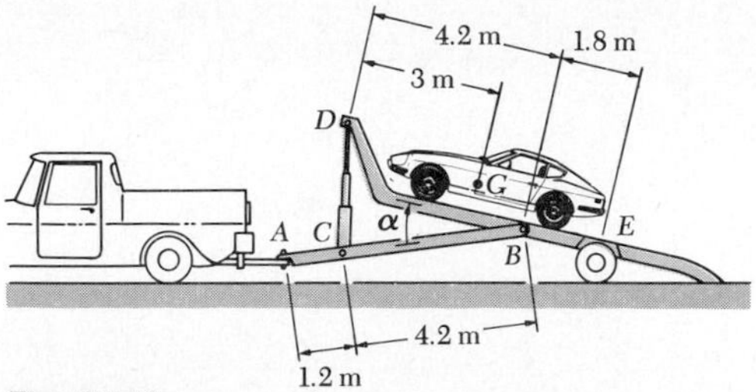

Fig. P6.59

6.59 The low-bed trailer shown is designed so that the rear end of the bed can be lowered to ground level in order to facilitate the loading of equipment or wrecked vehicles. A 1250-kg vehicle has been hauled to the position shown by a winch; the trailer is then returned to a traveling position where $\alpha = 0$ and both *AB* and *BE* are horizontal. Considering only the weight of the disabled automobile, determine the force which must be exerted by the hydraulic cylinder to maintain a position with $\alpha = 0$.

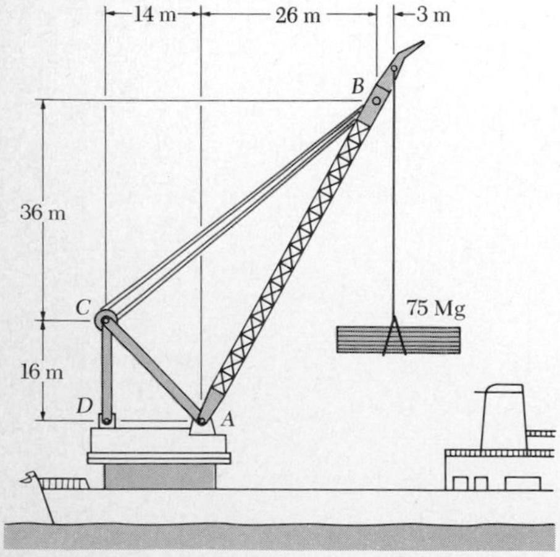

Fig. P6.60

6.60 For the marine crane shown, which is used in offshore drilling operations, determine (*a*) the force in link *CD*, (*b*) the force in the brace *AC*, (*c*) the force exerted at *A* on the boom *AB*.

6.61 Determine all the forces exerted on member AI if the frame is loaded by a clockwise couple of magnitude 1200 lb·in. applied (a) at point D, (b) at point E.

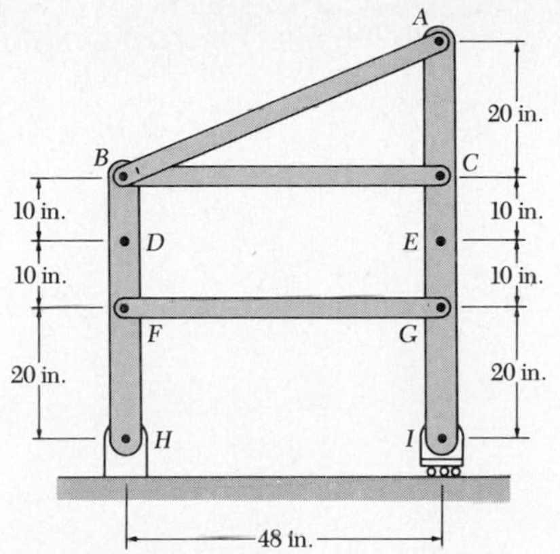

Fig. P6.61 and P6.62

6.62 Determine all the forces exerted on member AI if the frame is loaded by a 40-lb force directed horizontally to the right and applied (a) at point D, (b) at point E.

6.63 Determine the components of the reactions at A and E, (a) if the 800-N load is applied as shown, (b) if the 800-N load is moved along its line of action and is applied at point D.

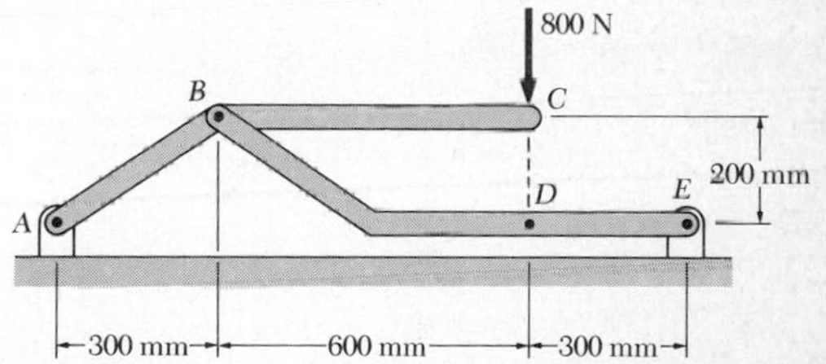

Fig. P6.63

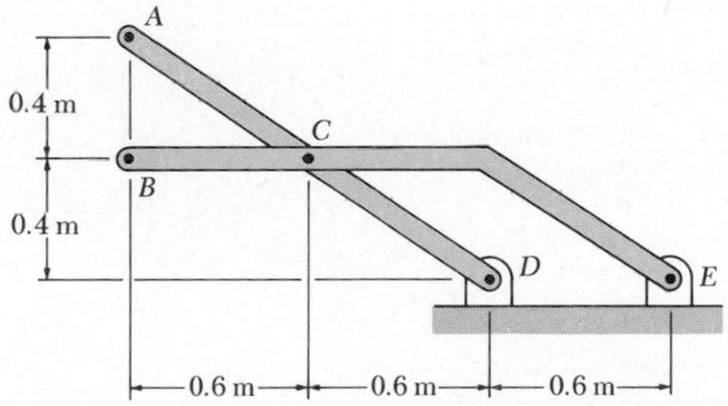

Fig. P6.64

6.64 Determine the components of the reactions at D and E if the frame is loaded by a clockwise couple of magnitude 150 N·m applied (a) at point A, (b) at point B.

6.65 A 70-kg worker stands at C on the step stool shown. Half the worker's weight is carried by the legs shown. Determine the components of the force exerted at E on leg BE, assuming that the bottom of the legs is not quite parallel to the floor and that bearing occurs only at points A and B. Neglect the weight of the stool, and assume the floor to be frictionless.

6.66 In Prob. 6.65, bearing between the legs of the step stool and the floor can occur in four ways: at A and B, A and B', A' and B, or A' and B'. Determine (a) for which combination of bearing points the force in member FG is maximum, (b) the corresponding value of the force in FG.

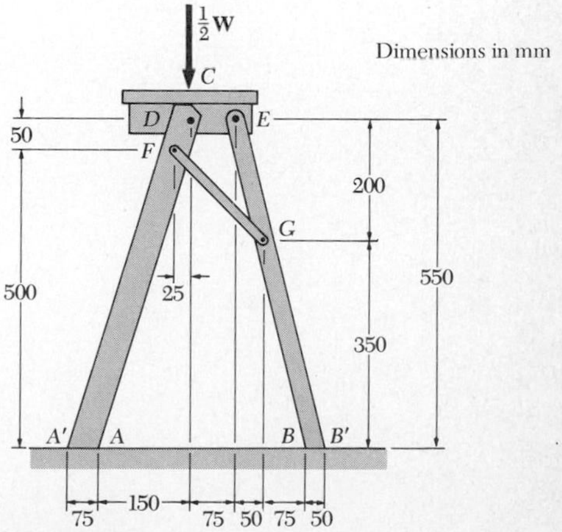

Fig. P6.65

6.67 The hydraulic cylinder *CF*, which partially controls the position of rod *DE*, has been locked in the position shown. Knowing that $\theta = 70°$, determine (*a*) the force **P** for which the tension in link *AB* is 125 lb, (*b*) the corresponding force exerted on member *BCD* at point *C*.

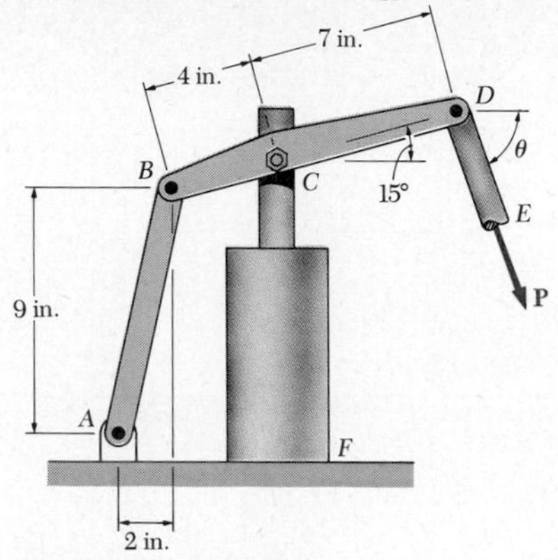

Fig. P6.67 and P6.68

6.68 The hydraulic cylinder *CF*, which partially controls the position of rod *DE*, has been locked in the position shown. Knowing that $P = 90$ lb and $\theta = 80°$, determine (*a*) the force in link *AB*, (*b*) the force exerted on member *BCD* at point *C*.

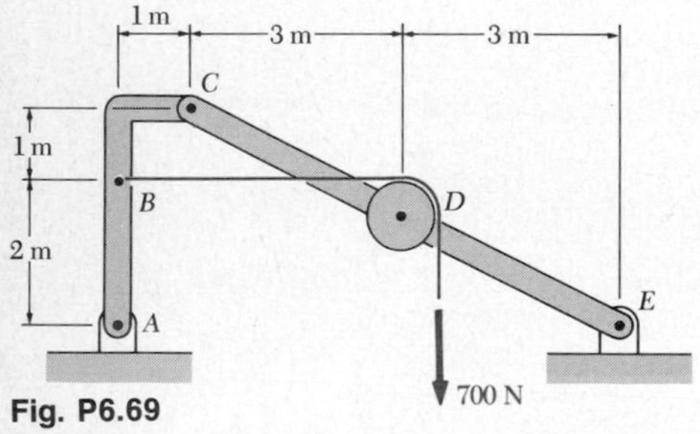

Fig. P6.69

6.69 Knowing that the pulley has a radius of 0.5 m, determine the components of the reactions at *A* and *E*.

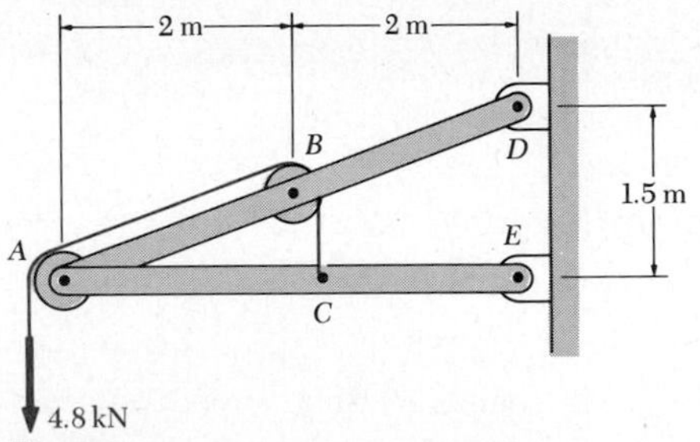

Fig. P6.70

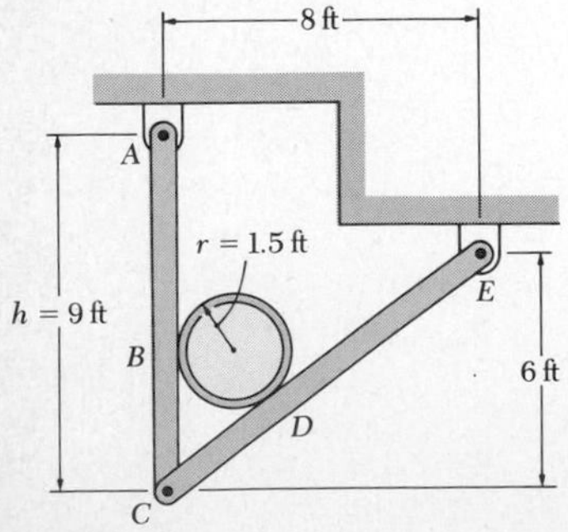

Fig. P6.71

6.70 Knowing that each pulley has a radius of 250 mm, determine the components of the reactions at *D* and *E*.

6.71 A 3-ft-diameter pipe is supported every 16 ft by a small frame; a typical frame is shown. Knowing that the combined weight of the pipe and its contents is 500 lb/ft and neglecting the effect of friction, determine the components (*a*) of the reaction at *E*, (*b*) of the force exerted at *C* on member *CDE*.

6.72 Solve Prob. 6.71, for a frame where $h = 6$ ft.

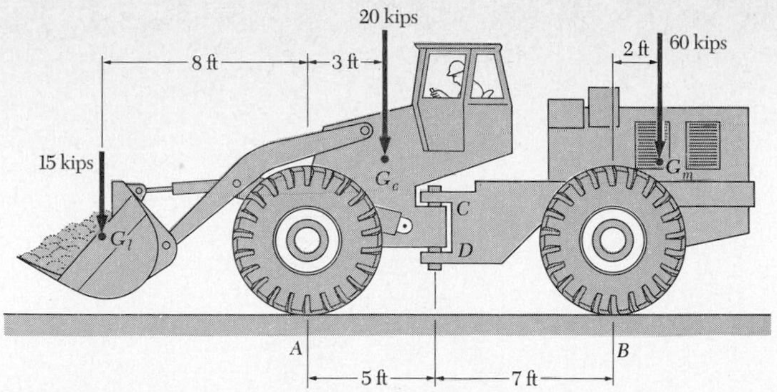

Fig. P6.73

6.73 The cab and motor units of the front-end loader shown are connected by a vertical pin located 5 ft behind the cab wheels. The distance from C to D is 30 in. The center of gravity of the 60-kip motor unit is located at G_m, while the centers of gravity of the 20-kip cab and 15-kip load are located, respectively, at G_c and G_l. Knowing that the machine is at rest with its brakes released, determine (*a*) the reactions at each of the four wheels, (*b*) the forces exerted on the motor unit at C and D.

6.74 Solve Prob. 6.73, assuming that the 15-kip load has been removed.

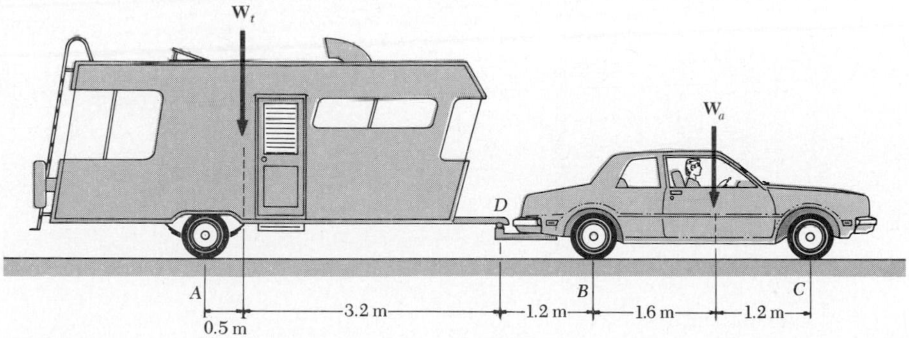

Fig. P6.75

6.75 A 1050-kg trailer is attached to a 1200-kg automobile by a ball-and-socket trailer hitch at D. Determine (*a*) the reactions at each of the six wheels when the automobile and trailer are at rest, (*b*) the additional load on each of the automobile wheels due to the trailer.

6.76 In order to obtain a better weight distribution over the four wheels of the automobile of Prob. 6.75, a compensating hitch of the type shown is used to attach the trailer to the automobile. This hitch consists of two bar springs (only one is shown in the figure) which fit into bearings inside a support rigidly attached to the automobile. The springs are also connected by chains to the trailer frame, and specially designed hooks make it possible to place both chains under tension. (*a*) Determine the tension T required in each of the two chains if the additional load due to the trailer is to be evenly distributed over the four wheels of the automobile. (*b*) What are the corresponding reactions at each of the six wheels of the trailer-automobile combination?

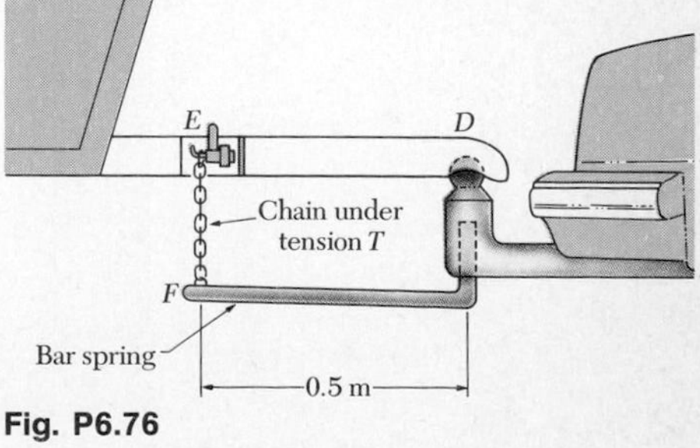

Fig. P6.76

6.77 and 6.78 For the frame and loading shown, determine the components of all forces acting on member *ABD*.

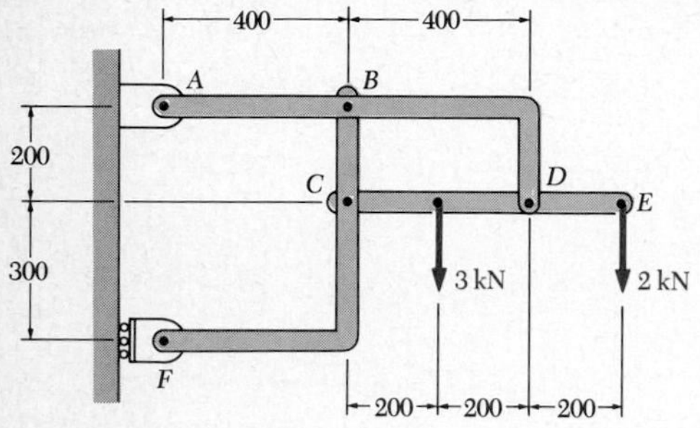

Dimensions in mm

Fig. P6.77

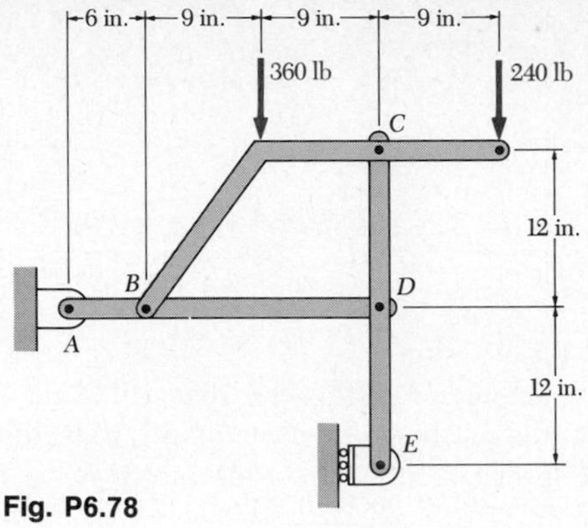

Fig. P6.78

6.79 Solve Prob. 6.78, assuming that the 360-lb load has been removed.

6.80 Solve Prob. 6.77, assuming that the 3-kN load has been removed.

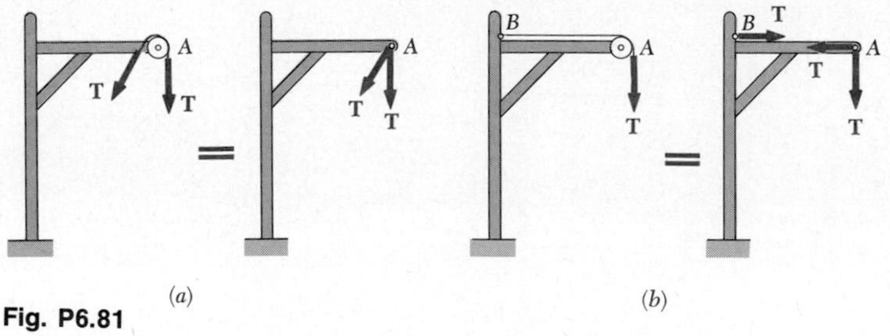

Fig. P6.81

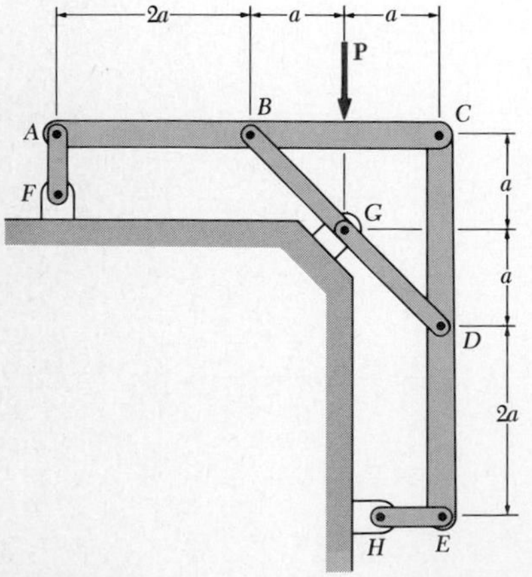

Fig. P6.82

6.81 (*a*) Show that when a frame supports a pulley at *A*, an equivalent loading of the frame and of each of its component parts may be obtained by removing the pulley and applying at *A* two forces equal and parallel to the forces of tension in the cable. (*b*) Further, show that if one end of the cable is attached to the frame at a point *B*, a force of magnitude equal to the tension should also be applied at *B*.

6.82 Members *ABC* and *CDE* are pin-connected at *C* and supported by the four links *AF*, *BG*, *GD*, and *EH*. For the loading shown, determine the force in each link.

6.83 Solve Prob. 6.82, assuming that the force **P** has been replaced by a clockwise couple of moment $\mathbf{M}_0$ applied at the same point.

6.84 through 6.86 The frame shown consists of members *AD* and *EH*, which are connected by two links. Determine the force in each link for the given loading.

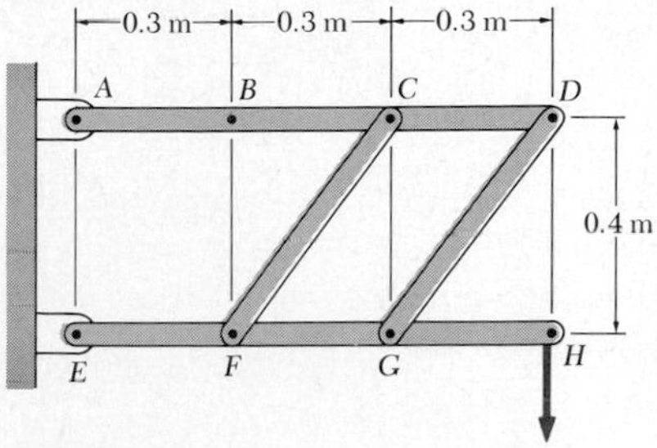

Fig. P6.84

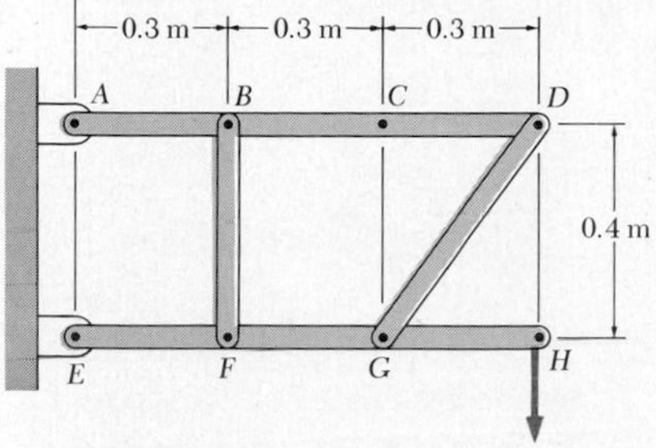

Fig. P6.85

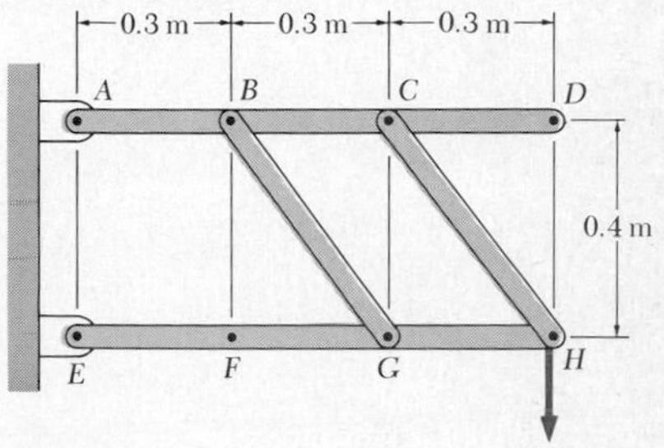

Fig. P6.86

6.87 Knowing that the surfaces at *A* and *D* are frictionless, determine the components of all forces exerted on member *ACD*.

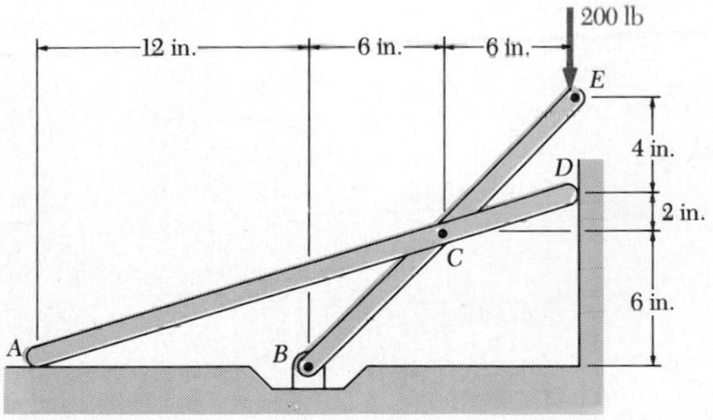

Fig. P6.87

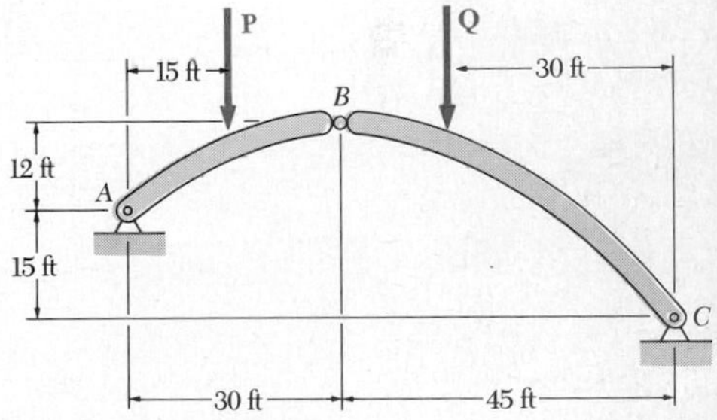

Fig. P6.89 and P6.90

6.88 Solve Prob. 6.87, assuming that the 200-lb force applied at *E* is directed horizontally to the right.

6.89 The axis of the three-hinged arch *ABC* is a parabola with vertex at *B*. Knowing that $P = 90$ kips and $Q = 60$ kips, determine (*a*) the components of the reaction at *C*, (*b*) the components of the force exerted at *B* on segment *AB*.

6.90 The axis of the three-hinged arch *ABC* is a parabola with vertex at *B*. Knowing that $P = 60$ kips and $Q = 90$ kips, determine (*a*) the components of the reaction at *C*, (*b*) the components of the force exerted at *B* on segment *AB*.

6.91 Three beams are nailed together at their midpoints to form the support system shown. Assuming that only vertical forces are exerted at the connections, determine the vertical reactions at *A*, *D*, and *F*.

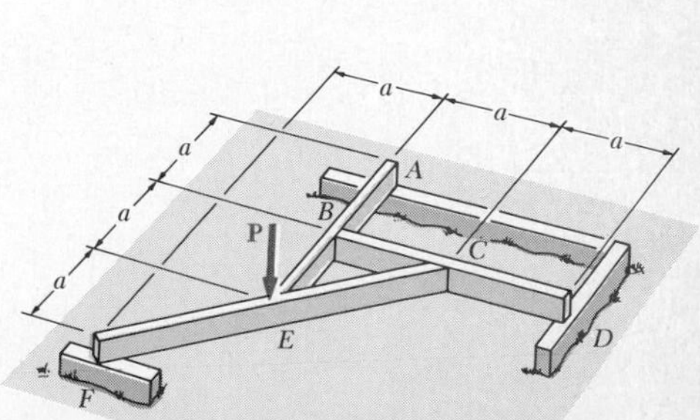

Fig. P6.91

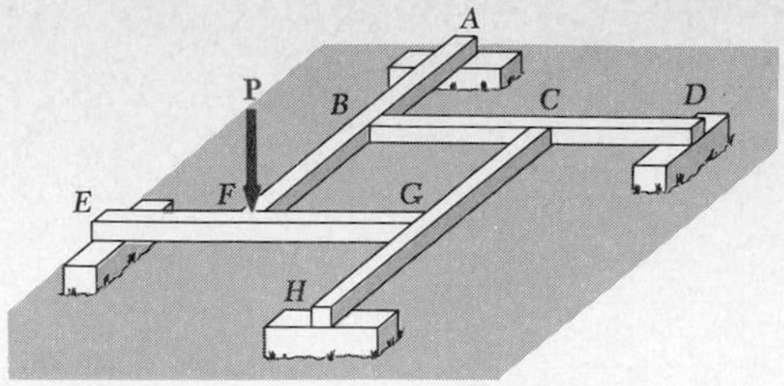

Fig. P6.92

6.92 Four beams, each of length $2a$, are nailed together at their midpoints to form the support system shown. Assuming that only vertical forces are exerted at the connections, determine the vertical reactions at A, D, E, and H.

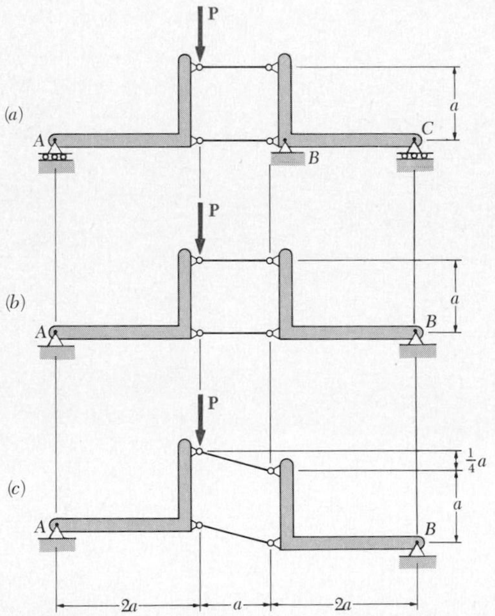
Fig. P6.93

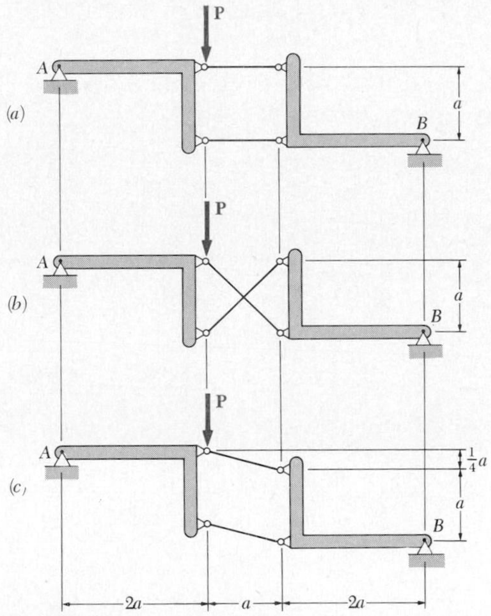
Fig. P6.94

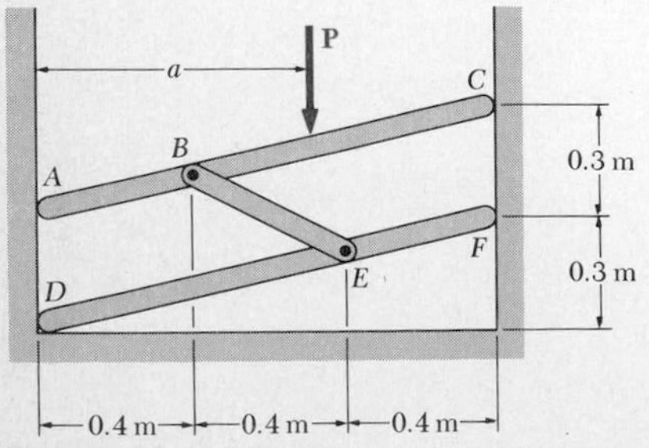

Fig. P6.95 and P6.96

6.93 and 6.94 Each of the frames shown consists of two L-shaped members connected by two rigid links. For each frame, determine the reactions at the supports and indicate whether the frame is rigid.

6.95 A vertical load **P** of magnitude 900 N is applied to member ABC. Members ABC and DEF are parallel and are placed between two frictionless walls and connected by the link BE. Knowing that $a = 0.8$ m, determine the reactions at points A, C, D, and F.

6.96 The two parallel members ABC and DEF are placed between two frictionless walls and connected by a link BE. Determine the range of values of the distance a for which the load **P** can be supported.

6.12. Machines. Machines are structures designed to transmit and modify forces. Whether they are simple tools or include complicated mechanisms, their main purpose is to transform *input forces* into *output forces*. Consider, for example, a pair of cutting pliers used to cut a wire (Fig. 6.22*a*). If we apply two equal and opposite forces **P** and −**P** on their handles, they will exert two equal and opposite forces **Q** and −**Q** on the wire (Fig. 6.22*b*).

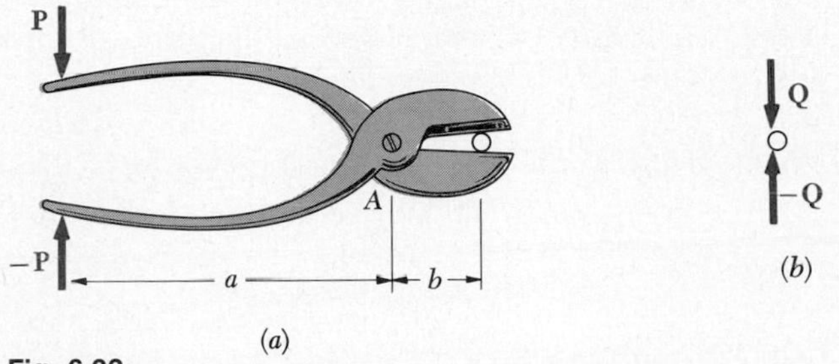

Fig. 6.22

To determine the magnitude Q of the output forces when the magnitude P of the input forces is known (or, conversely, to determine P when Q is known), we draw a free-body diagram of the pliers *alone*, showing the input forces **P** and −**P** and the *reactions* −**Q** and **Q** that the wire exerts on the pliers (Fig. 6.23). However, since a pair of pliers forms a nonrigid

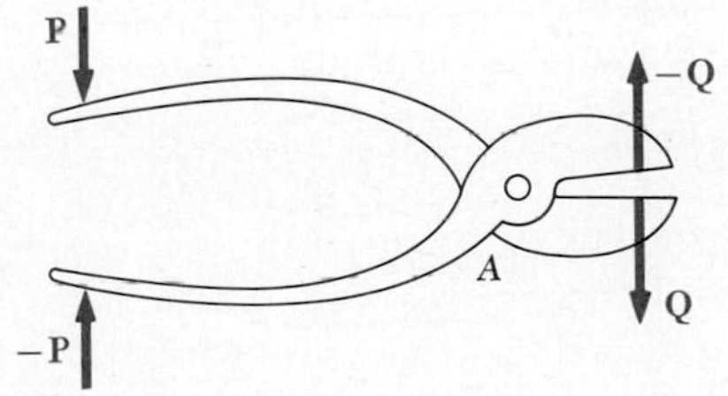

Fig. 6.23

structure, we must use one of the component parts as a free body in order to determine the unknown forces. Considering Fig. 6.24*a*, for example, and taking moments about A, we obtain the relation $Pa = Qb$, which defines the magnitude Q in terms of P or P in terms of Q. The same free-body diagram may be used to determine the components of the internal force at A; we find $A_x = 0$ and $A_y = P + Q$.

In the case of more complicated machines, it generally will be necessary to use several free-body diagrams and, possibly, to solve simultaneous equations involving various internal forces. The free bodies should be chosen to include the input forces and the reactions to the output forces, and the total number of unknown force components involved should not exceed the number of available independent equations. While it is advisable to check whether the problem is determinate before attempting to solve it, there is no point in discussing the rigidity of a machine. A machine includes moving parts and thus must be nonrigid.

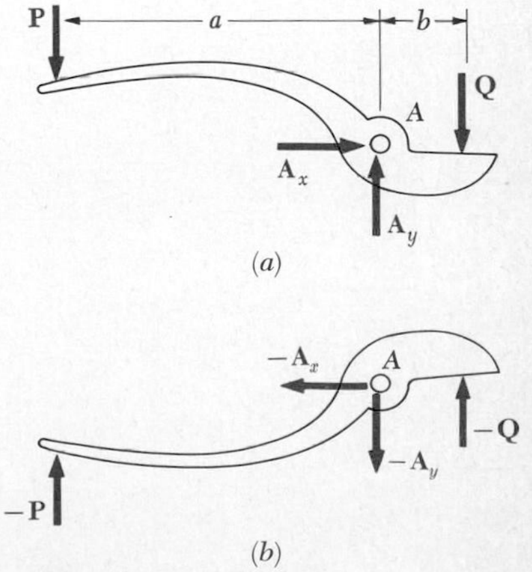

Fig. 6.24

SAMPLE PROBLEM 6.7

A hydraulic-lift table is used to raise a 1000-kg crate. It consists of a platform and two identical linkages on which hydraulic cylinders exert equal forces. (Only one linkage and one cylinder are shown.) Members *EDB* and *CG* are each of length $2a$, and member *AD* is pinned to the midpoint of *EDB*. If the crate is placed on the table, so that half of its weight is supported by the system shown, determine the force exerted by each cylinder in raising the crate for $\theta = 60°$, $a = 0.70$ m, and $L = 3.20$ m. Show that the result obtained is independent of the distance *d*.

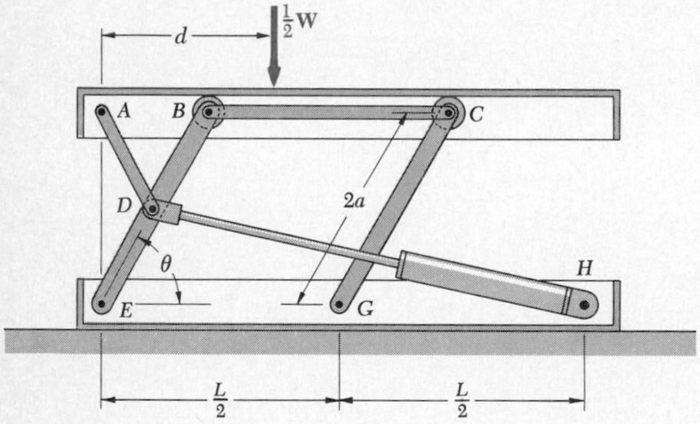

Solution. The machine considered consists of the platform and of the linkage, with an input force $\mathbf{F}_{DH}$ exerted by the cylinder and an output force equal and opposite to $\frac{1}{2}\mathbf{W}$. Since more than three unknowns would be involved, the entire mechanism is not used as a free body. The mechanism is dismembered and free-body diagrams are drawn for platform *ABC*, for roller *C*, and for member *EDB*. Since *AD*, *BC*, and *CG* are two-force members, their free-body diagrams have been omitted, and the forces they exert on the other parts of the mechanism have been drawn parallel to these members.

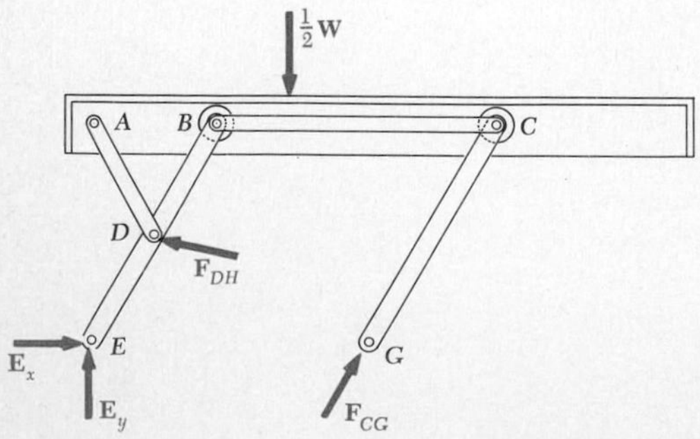

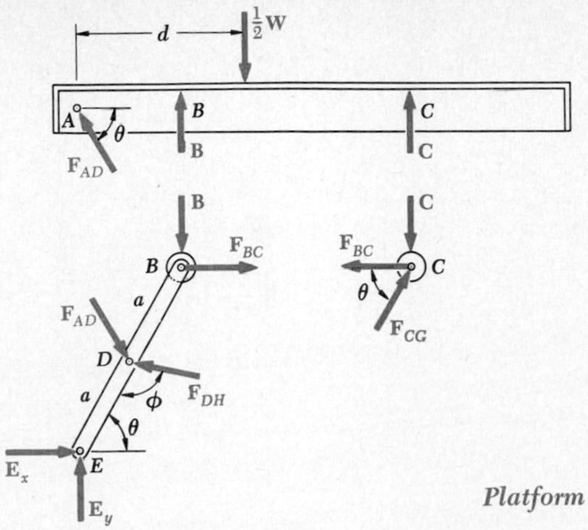

Platform ABC.

$$\overset{+}{\rightarrow}\Sigma F_x = 0: \qquad -F_{AD}\cos\theta = 0 \qquad F_{AD} = 0$$
$$+\uparrow\Sigma F_y = 0: \qquad B + C - \tfrac{1}{2}W = 0 \qquad B + C = \tfrac{1}{2}W \qquad (1)$$

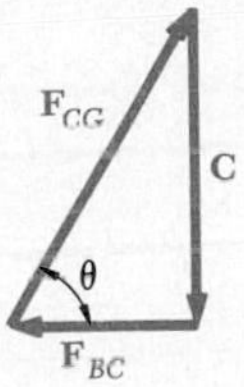

Roller C. We draw a force triangle and obtain $F_{BC} = C\cot\theta$.

Member EDB. Recalling that $F_{AD} = 0$,

$$+\circlearrowleft\Sigma M_E = 0: \qquad F_{DH}\cos(\phi - 90°)a - B(2a\cos\theta) - F_{BC}(2a\sin\theta) = 0$$
$$F_{DH}a\sin\phi - B(2a\cos\theta) - (C\cot\theta)(2a\sin\theta) = 0$$
$$F_{DH}\sin\phi - 2(B + C)\cos\theta = 0$$

Recalling Eq. (1), we have

$$F_{DH} = W\frac{\cos\theta}{\sin\phi} \qquad (2)$$

and we observe that the result obtained is independent of d. ◄

Applying first the law of sines to triangle EDH, we write

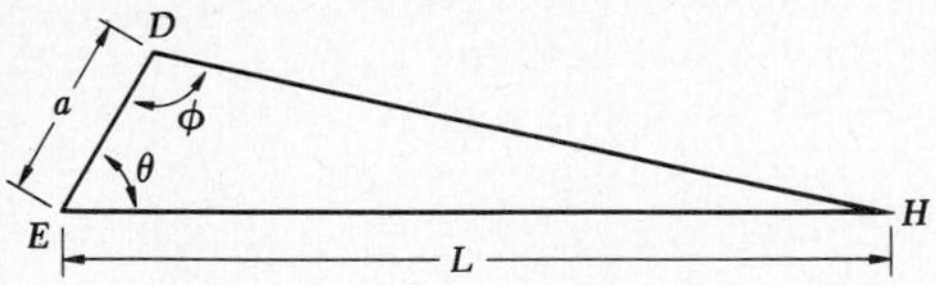

$$\frac{\sin\phi}{EH} = \frac{\sin\theta}{DH} \qquad \sin\phi = \frac{EH}{DH}\sin\theta \qquad (3)$$

Using now the law of cosines, we have

$$(DH)^2 = a^2 + L^2 - 2aL\cos\theta$$
$$= (0.70)^2 + (3.20)^2 - 2(0.70)(3.20)\cos 60°$$
$$(DH)^2 = 8.49 \qquad DH = 2.91\text{ m}$$

We also note that

$$W = mg = (1000\text{ kg})(9.81\text{ m/s}^2) = 9810\text{ N} = 9.81\text{ kN}$$

Substituting for $\sin\phi$ from (3) into (2) and using the numerical data, we write

$$F_{DH} = W\frac{DH}{EH}\cot\theta = (9.81\text{ kN})\frac{2.91\text{ m}}{3.20\text{ m}}\cot 60°$$

$$F_{DH} = 5.15\text{ kN}$$ ◄

Problems

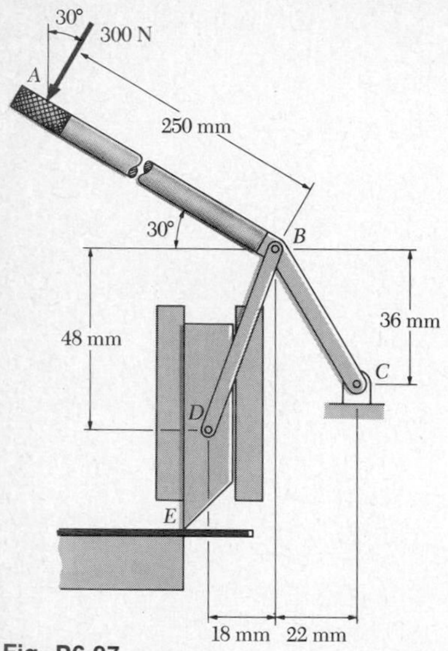

Fig. P6.97

6.97 The shear shown is used to cut and trim electronic-circuit-board laminates. For the position shown, determine (*a*) the vertical component of the force exerted on the shearing blade at *D*, (*b*) the reaction at *C*.

6.98 An 84-lb force is applied to the toggle vise at *C*. Knowing that $\theta = 90°$, determine (*a*) the vertical force exerted on the block at *D*, (*b*) the force exerted on member *ABC* at *B*.

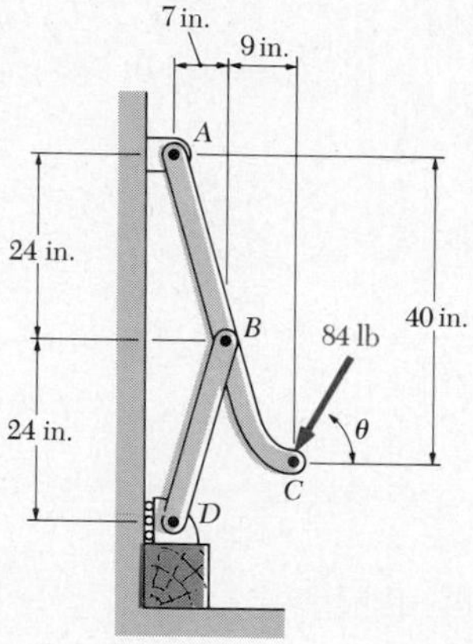

Fig. P6.98

6.99 Solve Prob. 6.98 when $\theta = 0$.

6.100 Solve Prob. 6.97, assuming that the 300-N force is applied vertically downward at *A*.

6.101 The jaw crusher shown consists of a fixed jaw and a moving jaw *AB* which is attached to a large pin at *A*. Power is delivered to shaft *C* to which a circular disk *D* is attached eccentrically. As shaft *C* rotates, the connecting rod *EF* moves toggle *GFH*. For the position shown, the magnitude of the resultant force **F** on jaw *AB* is 250 kN and the arms *GF* and *FH* of the toggle each form an angle of 6° with the horizontal. Neglecting the effect of friction, determine the force in rod *EF*. (*Note:* At the end of each crushing cycle springs, which are not shown, return the jaw and toggle to their initial positions.)

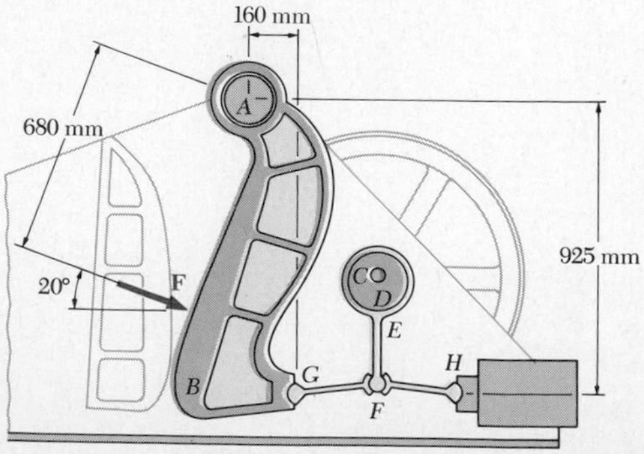

Fig. P6.101

6.102 A 9-m length of railroad track of mass 40 kg/m is lifted by the rail tongs shown. Determine the forces exerted at *D* and *F* on tong *BDF*.

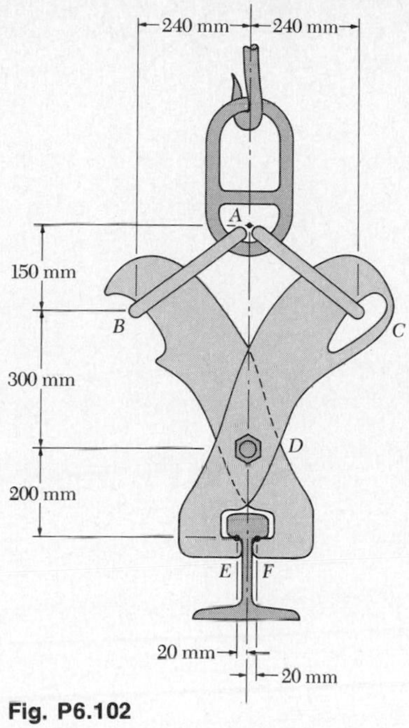

Fig. P6.102

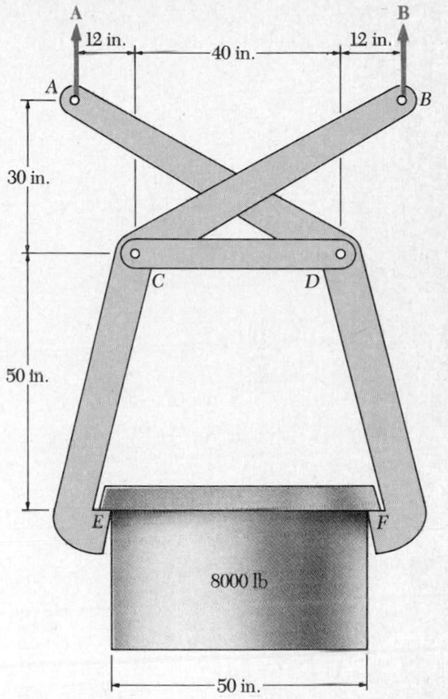

Fig. P6.103

6.103 A steel ingot weighing 8000 lb is lifted by a pair of tongs as shown. Determine the forces exerted at *C* and *E* on tong *BCE*.

6.104 If the toggle shown is added to the tongs of Prob. 6.103 and the load is lifted by applying a single force at *G*, determine the forces exerted at *C* and *E* on the tong *BCE*.

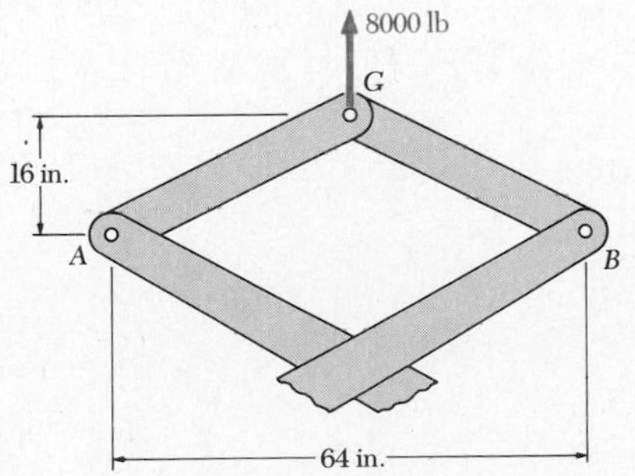

6.105 The action of the roll clamp is controlled by the two hydraulic cylinders shown. In order to hold firmly the paper roll, a vertical 1000-lb force is applied at the top of the roll by arm *CAF*. Knowing that the weight of the paper roll is 4500 lb, determine (*a*) the force exerted by each cylinder, (*b*) the force exerted at *C* on arm *BCEH*.

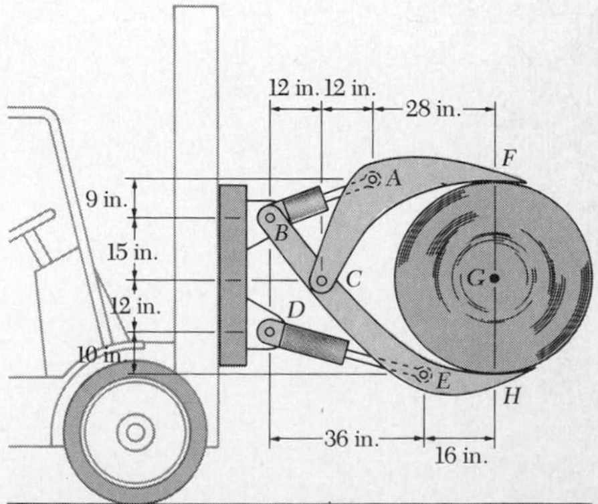

Fig. P6.105

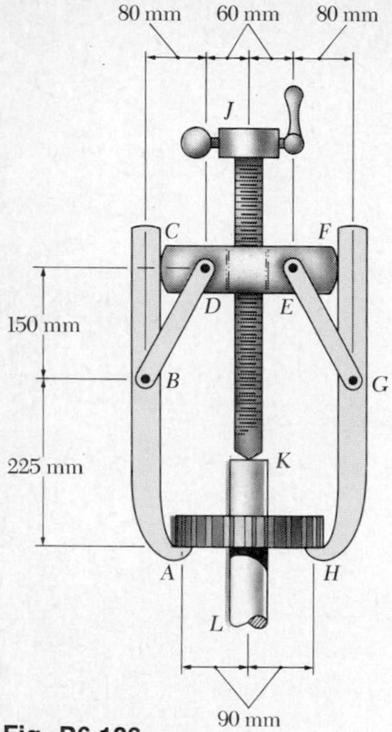

Fig. P6.106

6.106 The gear-pulling assembly shown consists of a crosshead *CF*, two grip arms *ABC* and *FGH*, two links *BD* and *EG*, and a threaded center rod *JK*. Knowing that the center rod *JK* must exert a 4500-N force on the vertical shaft *KL* in order to start the removal of the gear, determine all forces acting on grip arm *ABC*. Assume that the rounded ends of the crosshead are smooth and exert horizontal forces on the grip arms.

6.107 A force **P** of magnitude 1.8 kN is applied to the piston of the engine system shown. For each of the two positions shown, determine the couple **M** required to hold the system in equilibrium.

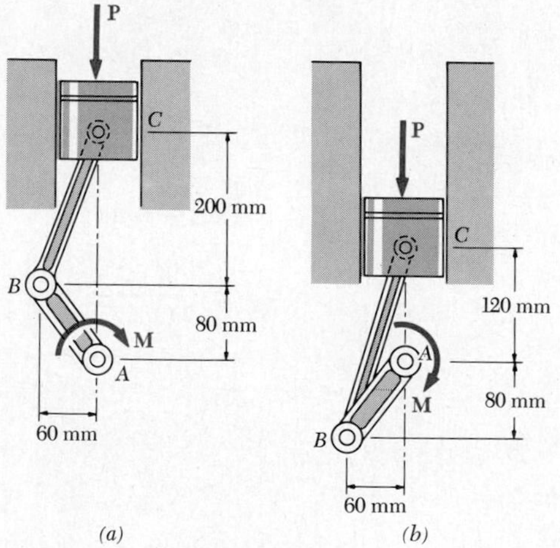

Fig. P6.107 and P6.108

6.108 A couple **M** of magnitude 180 N·m is applied to the crank of the engine system shown. For each of the two positions shown, determine the force **P** required to hold the system in equilibrium.

6.109 and 6.110 Two rods are connected by a frictionless collar *B*. Knowing that the magnitude of the couple $\mathbf{M}_A$ is 500 lb·in., determine (*a*) the couple $\mathbf{M}_C$ required for equilibrium, (*b*) the corresponding components of the reaction at *C*.

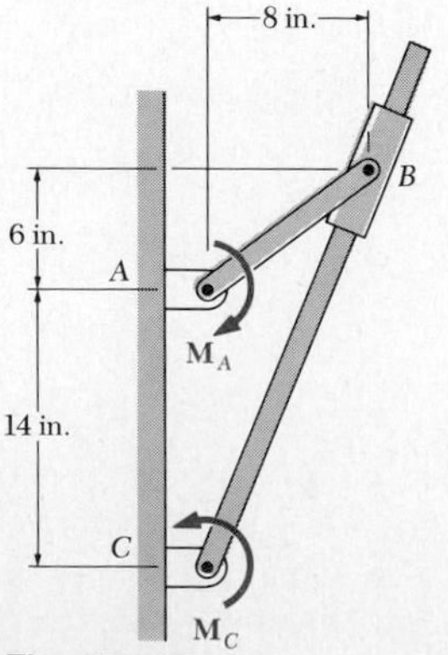

Fig. P6.109

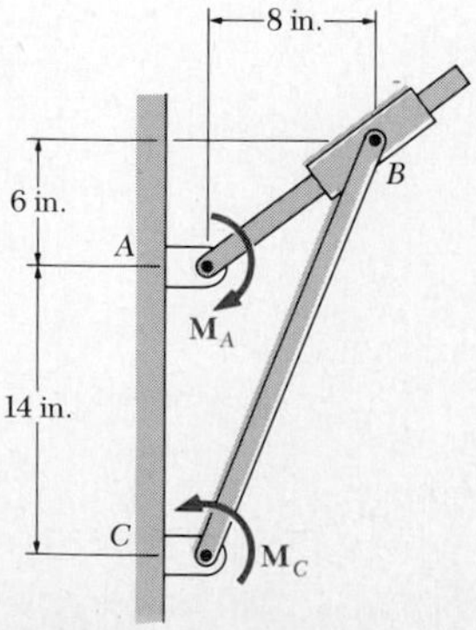

Fig. P6.110

6.111 A 48-mm-diameter pipe is gripped by the Stillson wrench shown. Portions *AB* and *DE* of the wrench are rigidly attached to each other and portion *CF* is connected by a pin at *D*. Assuming that no slipping occurs between the pipe and the wrench, determine the components of the forces exerted on the pipe at *A* and at *C*.

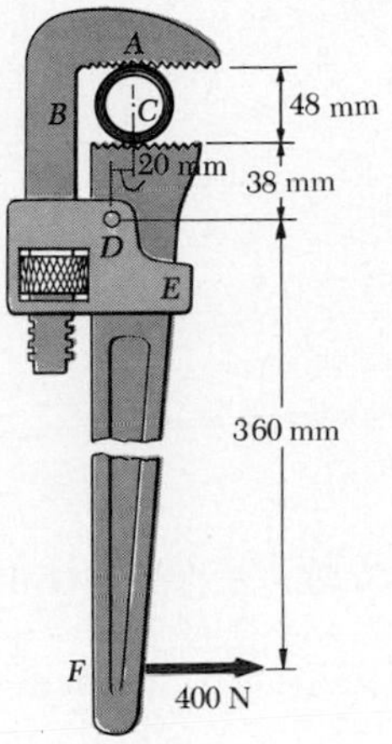

Fig. P6.111

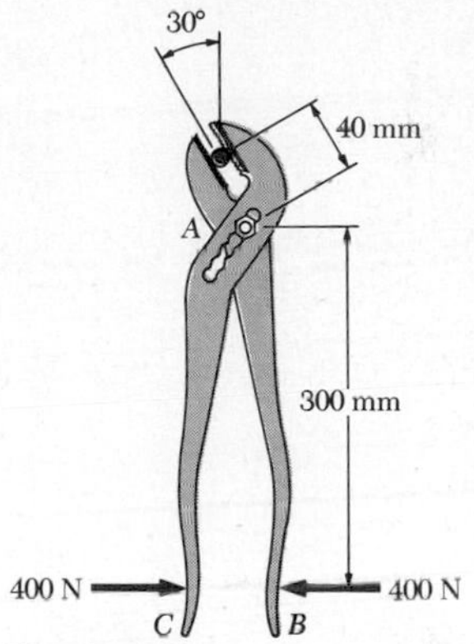

Fig. P6.112

6.112 Two 400-N forces are applied to the handles of the pliers as shown. Determine (*a*) the magnitude of the forces exerted on the rod, (*b*) the force exerted by the pin at *A* on portion *AB* of the pliers.

6.113 The tool shown is used to crimp terminals onto electric wires. Knowing that a worker will apply forces of magnitude $P = 120$ N to the handles, determine the magnitude of the crimping forces which will be exerted on the terminal.

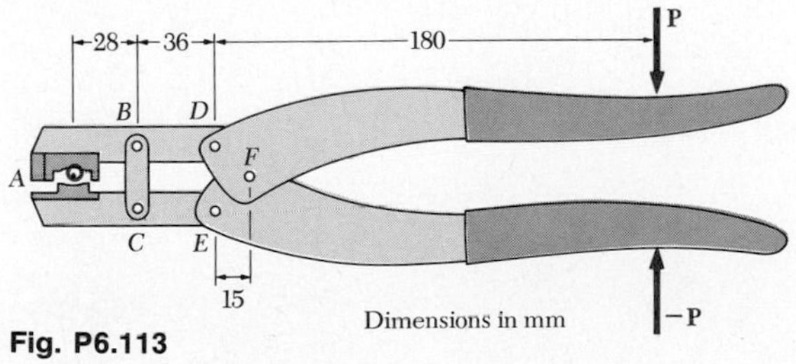

Fig. P6.113

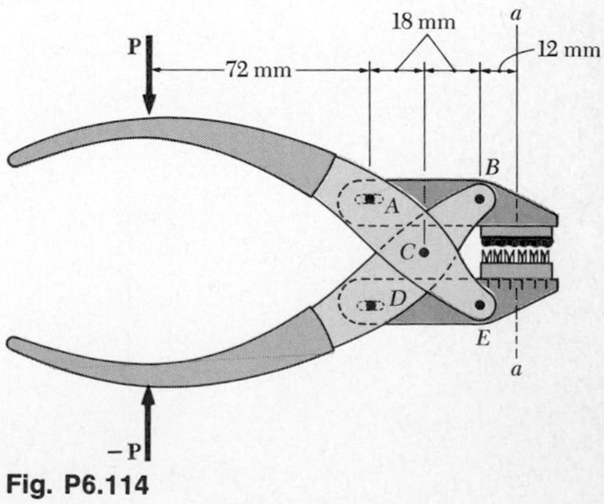

Fig. P6.114

6.114 The pliers shown are used to attach electric connectors to flat cables which carry many separate wires. The clamping jaws remain parallel as the connectors are attached. Knowing that 300-N forces directed along line *a-a* are required to complete the attachment, determine the magnitude *P* of the forces which must be applied to the handles. Assume that pins *A* and *D* slide freely in slots cut in the jaws.

6.115 A hand-operated hydraulic cylinder has been designed for use where space is severely limited. Determine the magnitude of the force exerted on the piston at D when two 80-lb forces are applied as shown.

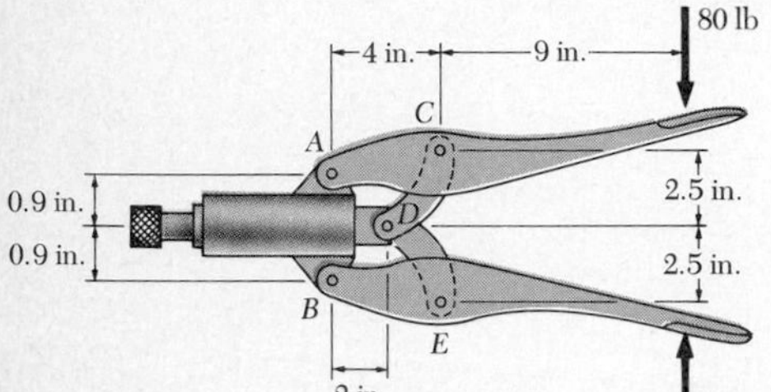

Fig. P6.115

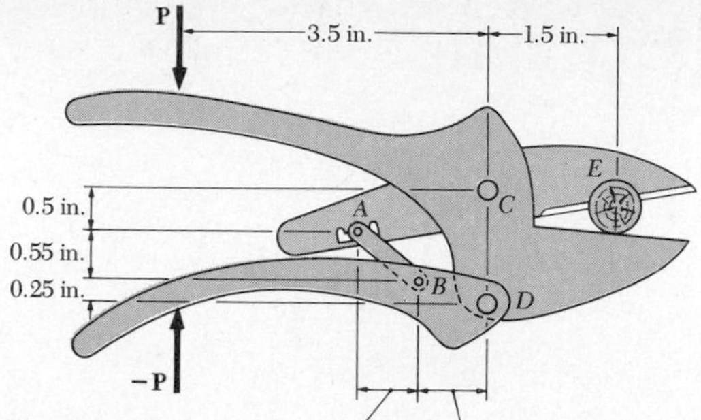

Fig. P6.116

6.116 The compound-lever pruning shears shown can be adjusted by placing pin A at various ratchet positions on blade ACE. Knowing that 300-lb vertical forces are required to complete the pruning of a twig, determine the magnitude P of the forces which must be applied to the handles when the shears are adjusted as shown.

6.117 and 6.118 Determine the force **P** which must be applied to the toggle CDE to maintain bracket ABC in the position shown.

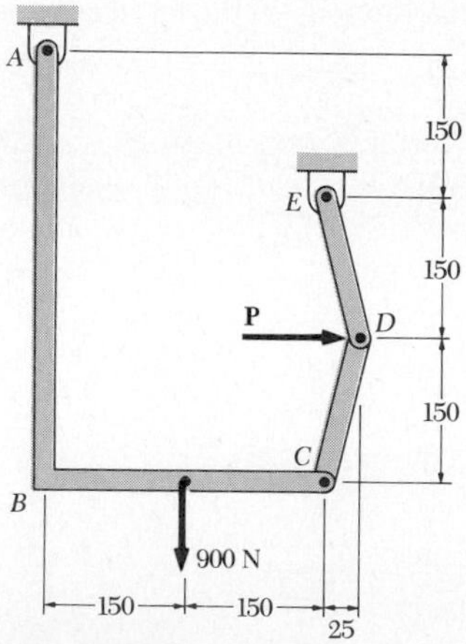

Fig. P6.117

Dimensions in mm

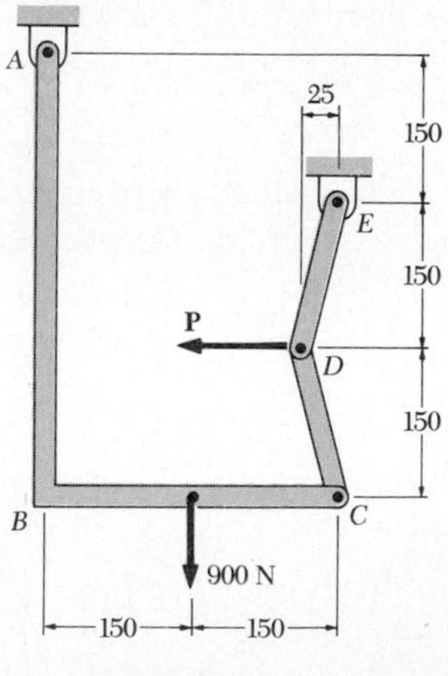

Fig. P6.118

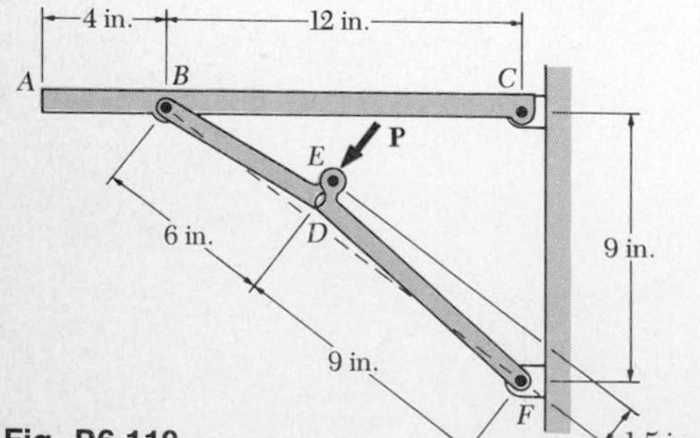

Fig. P6.119

6.119 The shelf ABC is held horizontally by a self-locking brace which consists of two parts BDE and EDF hinged at E and bearing against each other at D. Knowing that the shelf is 16 in. wide and weighs 36 lb, determine the force **P** required to release the brace. (*Hint.* To release the brace, the forces of contact at D must be zero.)

6.120 Solve Prob. 6.119, assuming that the force **P** is applied vertically downward at point E.

6.121 The telescoping arm *ABC* is used to raise a worker to the elevation of overhead electric and telephone wires. For the extension shown, the center of gravity of the 1500-lb arm is located at point *G*. The worker, the bucket, and equipment attached to the bucket together weigh 500 lb and have a combined center of gravity at *C*. For the position shown, determine the force exerted at *B* by the single hydraulic cylinder used.

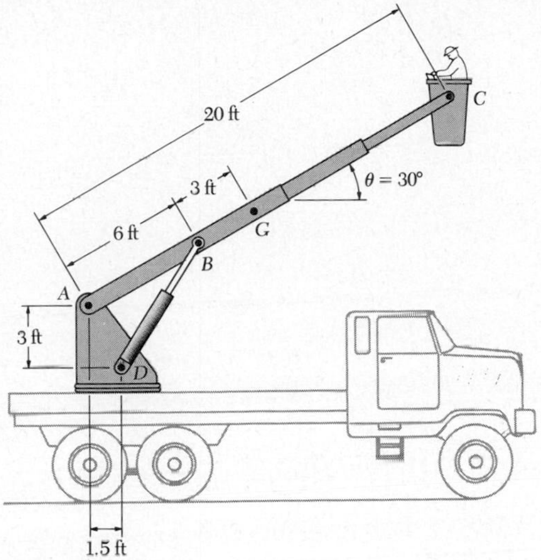

Fig. P6.121

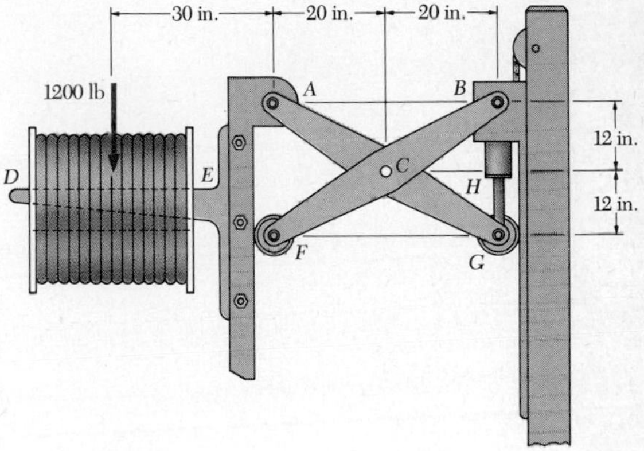

Fig. P6.122

6.122 The position of the spindle *DE* of a lift truck is partially controlled by two identical linkage-and-hydraulic-cylinder systems, only one of which is shown. A 2400-lb spool of electric cable is held by the spindle in the position shown. Knowing that the load supported by the one system shown is 1200 lb, determine (*a*) the force exerted by the hydraulic cylinder on point *G*, (*b*) the components of the force exerted on member *BCF* at point *C*.

6.123 A 400-kg concrete slab is supported by a chain and sling attached to the bucket of the front-end loader shown. The action of the bucket is controlled by two identical mechanisms, only one of which is shown. Knowing that the mechanism shown supports one-half of the 400-kg slab, determine the force exerted (*a*) by cylinder *CD*, (*b*) by cylinder *FH*.

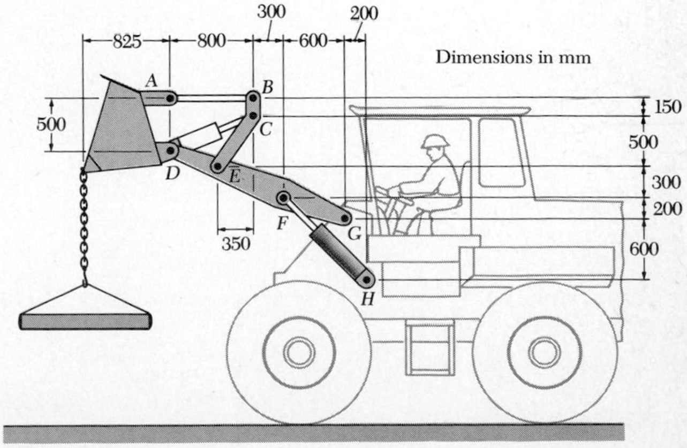

Fig. P6.123

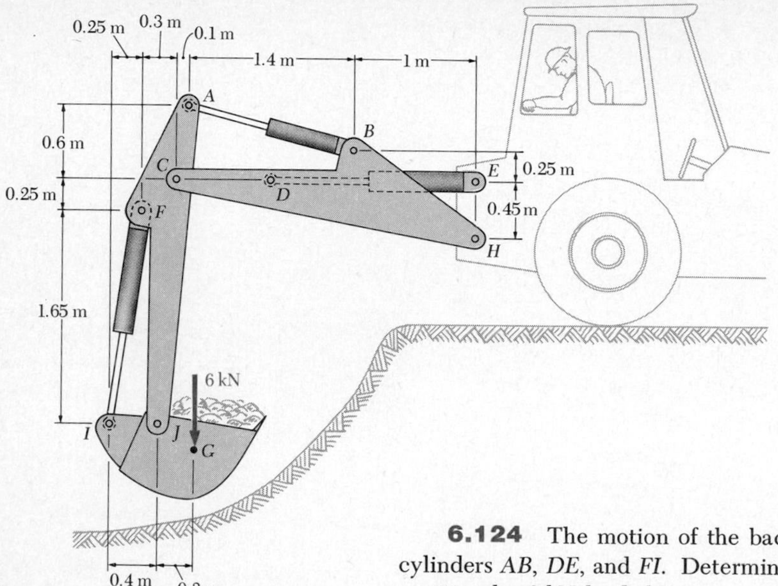

Fig. P6.124

6.124 The motion of the backhoe bucket is controlled by the hydraulic cylinders *AB*, *DE*, and *FI*. Determine the force exerted by each cylinder in supporting the 6-kN load shown.

6.125 The gears *D* and *G* are rigidly attached to shafts which are held by frictionless bearings. If $r_D = 4$ in. and $r_G = 2$ in., determine (*a*) the couple $\mathbf{M}_0$ which must be applied for equilibrium, (*b*) the reactions at *A* and *B*.

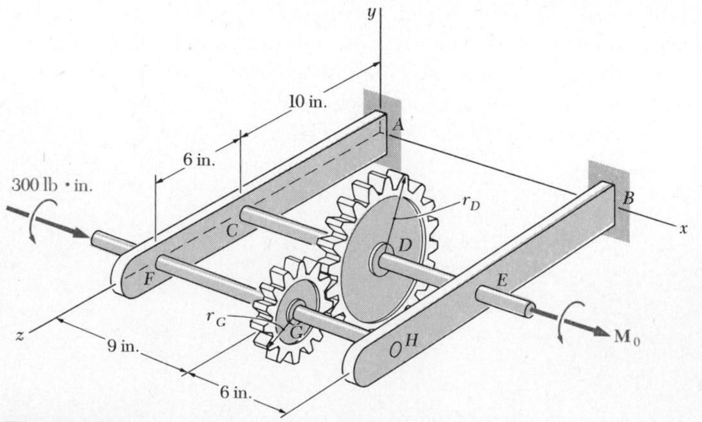

Fig. P6.125

6.126 Solve Prob. 6.125, if $r_D = 3.5$ in. and $r_G = 2.5$ in.

6.127 In the planetary-gear system shown, the radius of the central gear A is a, the radius of each planetary gear is b, and the radius of the outer gear E is $(a + 2b)$. In a particular gear system where $a = b = 40$ mm, a clockwise couple $\mathbf{M}_A$ is applied to gear A. If the system is to be in equilibrium, determine (*a*) the couple $\mathbf{M}_S$ which must be applied to the spider BCD, (*b*) the couple $\mathbf{M}_E$ which must be applied to the outer gear E.

6.128 In the planetary-gear system shown, the radius of the central gear A is a, the radius of each of the planetary gears is b, and the radius of the outer gear E is $(a + 2b)$. A clockwise couple of magnitude M_A is applied to the central gear A and a counterclockwise couple of magnitude $3.6M_A$ is applied to the spider BCD. If the system is to be in equilibrium, determine (*a*) the required ratio b/a, (*b*) the couple $\mathbf{M}_E$ which must be applied to the outer gear E.

Fig. P6.127 and P6.128

***6.129** Two shafts AC and EG, which lie in the vertical yz plane, are connected by a universal joint at D. The bearings at B and E do not exert any axial force. A couple of magnitude 25 N·m (clockwise when viewed from the positive z axis) is applied to shaft AC at A. At a time when the arm of the crosspiece attached to shaft AC is vertical, determine (*a*) the magnitude of the couple $\mathbf{M}_G$ which must be applied to shaft EG to maintain equilibrium, (*b*) the reactions at B, C, and E. (*Hint.* The sum of the couples exerted on the crosspiece must be zero.)

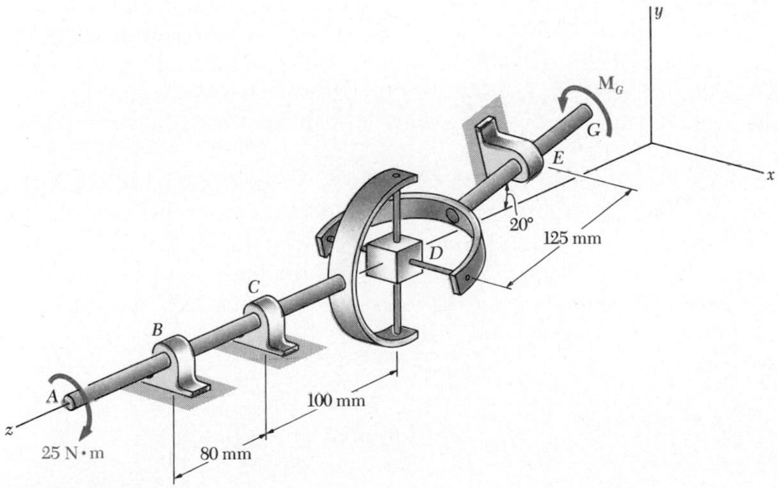

Fig. P6.129

***6.130** Solve Prob. 6.129, assuming that the arm of the crosspiece attached to shaft ABC is horizontal.

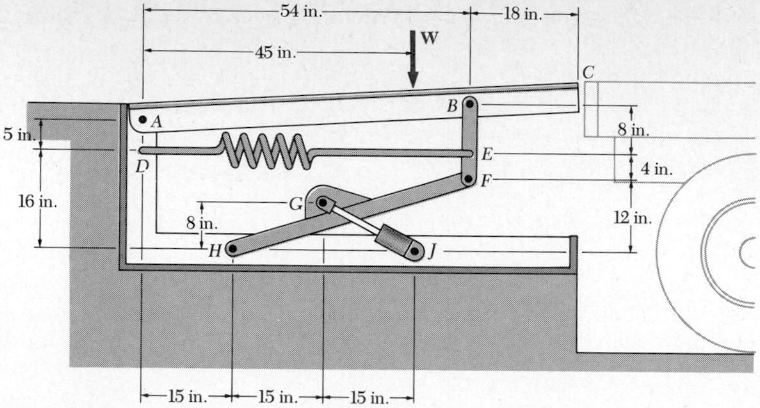

Fig. P6.131

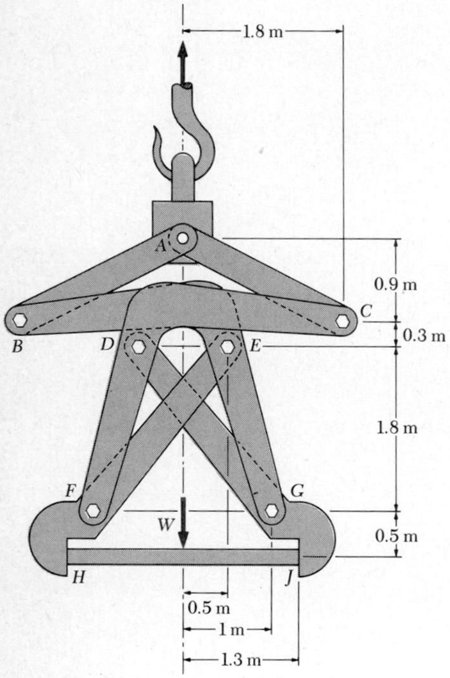

Fig. P6.133

*** 6.131** The mechanism shown is installed at some loading docks so that the dock edge C can be brought to the same elevation as the beds of various trucks. The weight of the platform and mechanism is roughly balanced by the force exerted by spring DE, while the final adjustment in elevation is controlled by the hydraulic cylinder GJ. Knowing that the weight of all moving parts is equivalent to the single force $\mathbf{W}$ of magnitude 1250 lb, determine the required tension in spring DE if the force exerted by the hydraulic cylinder is to be zero in the position shown.

*** 6.132** The weight of the platform and mechanism of Prob. 6.131 is exactly balanced by the force exerted by spring DE in the position shown and the force exerted by the hydraulic cylinder GJ is zero. A 180-lb force directed vertically downward is then applied at edge C. Determine the force which must be exerted by the hydraulic cylinder if the mechanism is to remain in the same position.

*** 6.133** The large mechanical tongs shown are used to grab and lift a thick 7500-kg steel slab HJ. Knowing that slipping does not occur between the tong grips and the slab at H and J, determine the components of all forces acting on member EFH. (*Hint.* Consider the symmetry of the tongs to establish relationships between the components of the force acting at E on EFH and the components of the force acting at D on CDF.)

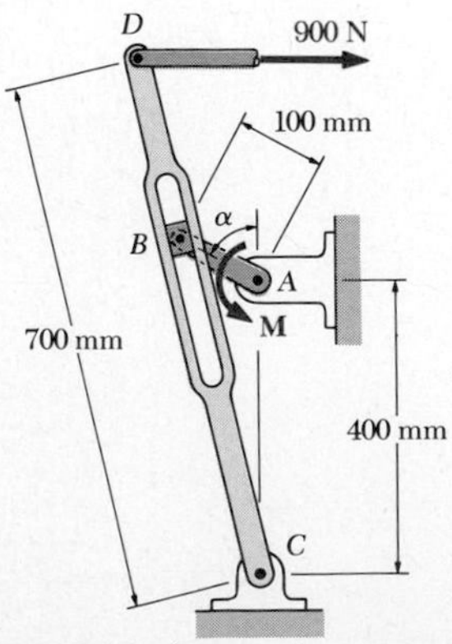

Fig. P6.134

Review Problems

6.134 The Whitworth mechanism shown is used to produce a quick-return motion of point D. The block at B is pinned to the crank AB and is free to slide in a slot cut in member CD. Determine the couple $\mathbf{M}$ which must be applied to the crank AB to hold the mechanism in equilibrium when (*a*) $\alpha = 0$, (*b*) $\alpha = 30°$.

6.135 Solve Prob. 6.134, when (*a*) $\alpha = 60°$, (*b*) $\alpha = 90°$.

6.136 Determine the reactions at the supports of the beams shown.

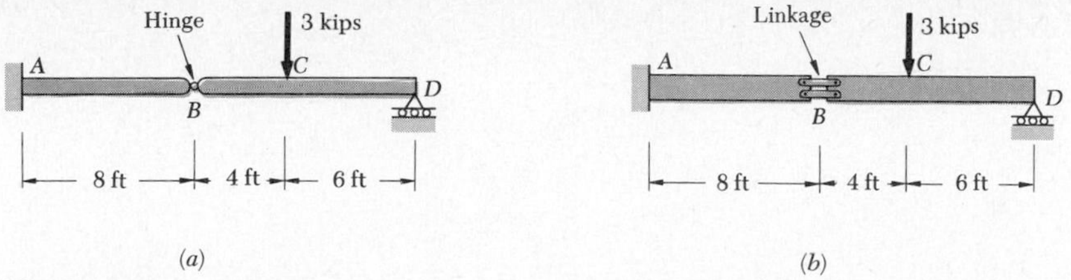

Fig. P6.136

6.137 The elevation of the platform is controlled by two identical mechanisms, only one of which is shown. A load of 1500 lb is applied to the mechanism shown. Knowing that the pin at *C* can transmit only a horizontal force, determine (*a*) the force in link *BE*, (*b*) the components of the force exerted by the hydraulic cylinder on pin *H*.

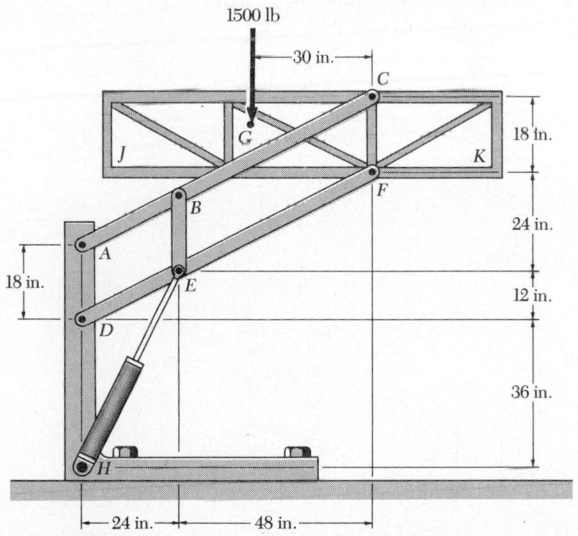

Fig. P6.137

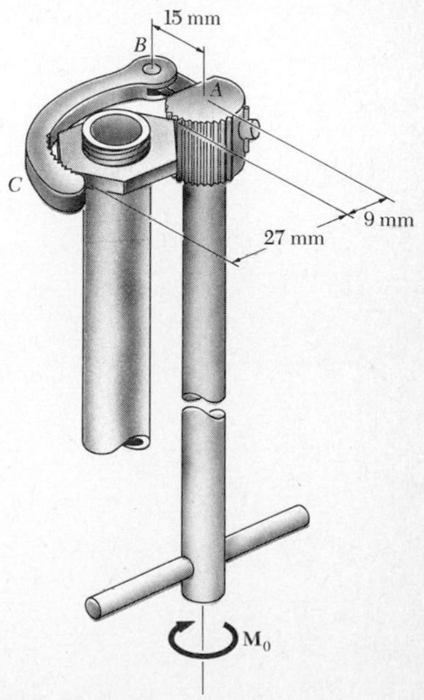

Fig. P6.138

6.138 The specialized plumbing wrench shown is used in confined areas (e.g., under a basin or sink). It consists essentially of a jaw *BC* pinned at *B* to a long rod. Knowing that the forces exerted on the nut are equivalent to a clockwise couple (when viewed from above) of magnitude 13.5 N·m, determine (*a*) the magnitude of the force exerted by pin *B* on jaw *BC*, (*b*) the couple $\mathbf{M}_0$ which is applied to the wrench.

6.139 Determine the reaction at F and the force in members AE and BD.

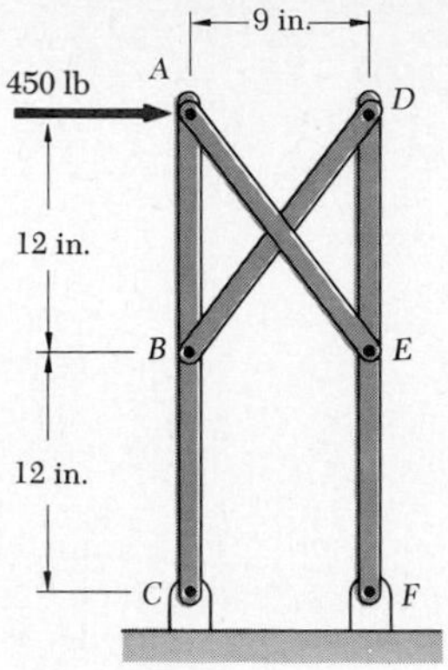

Fig. P6.139

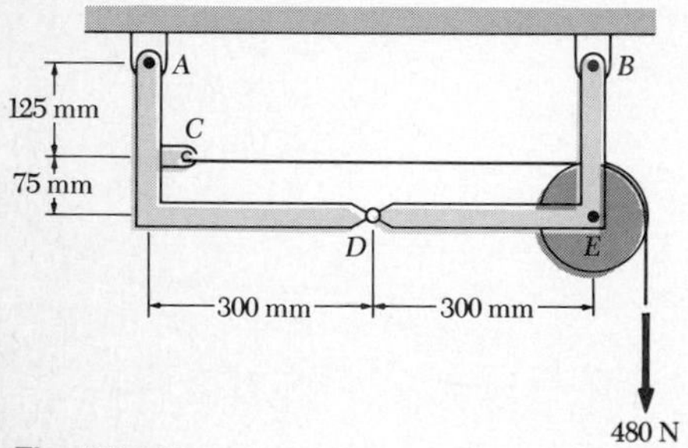

Fig. P6.140

6.140 Knowing that the pulley has a radius of 75 mm, determine the components of the reactions at A and B.

6.141 Determine the magnitude of the gripping forces produced when two 300-N forces are applied as shown.

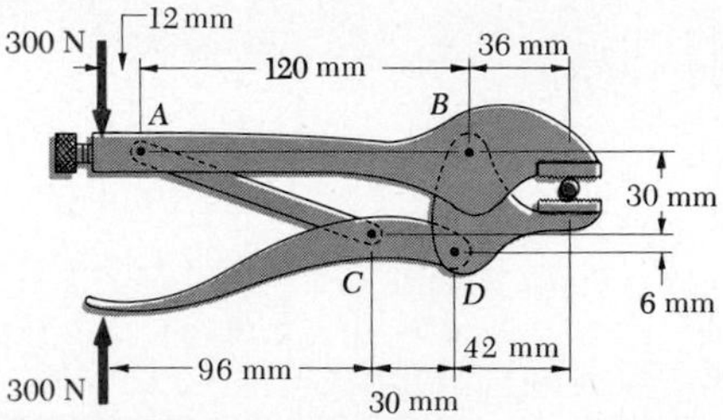

Fig. P6.141

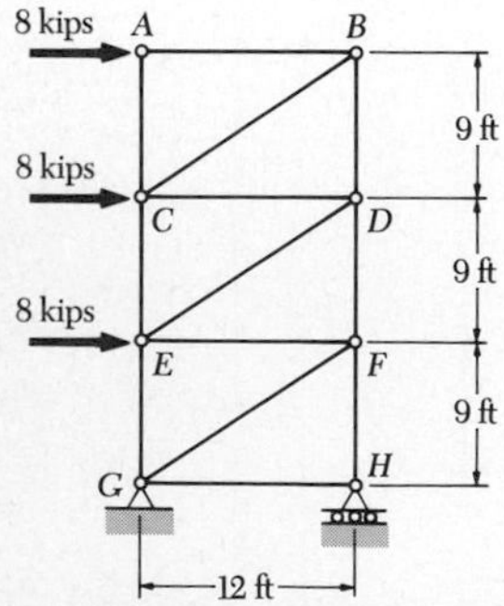

Fig. P6.142 and P6.143

6.142 Determine the force in members CE, DE, and DF of the truss shown.

6.143 Determine the force in members EG, EF, and DF of the truss shown.

6.144 The bent rod *DEF* fits into the bent pipe *ABC* as shown. Neglecting the effect of friction, determine the reactions at *A* and *F* due to the 720-N force applied at *B*.

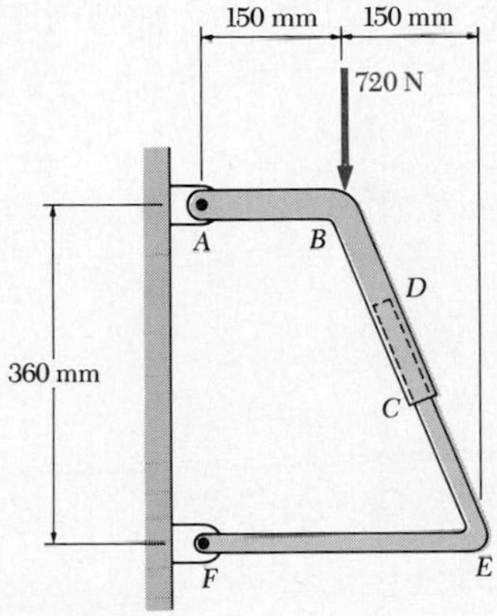

Fig. P6.144

6.145 A 300-kg block may be supported by a small frame in each of the four ways shown. The diameter of the pulley is 250 mm. For each case, determine (*a*) the force components and the couple representing the reaction at *A*, (*b*) the force exerted at *D* on the vertical member.

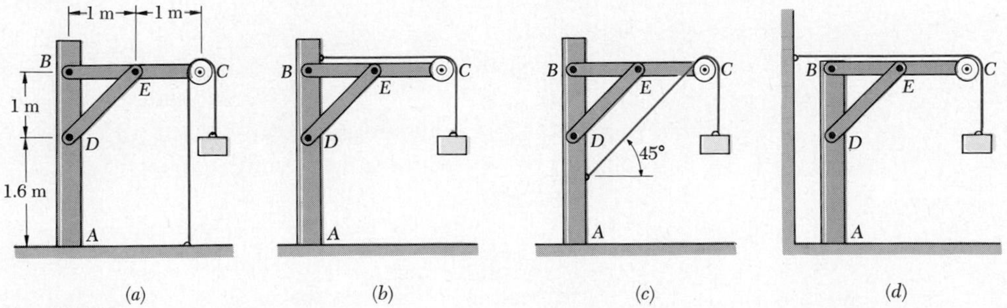

Fig. P6.145

CHAPTER SEVEN

Forces in Beams and Cables

***7.1. Introduction.** In preceding chapters, two basic problems involving structures were considered: (1) the determination of the external forces acting on a structure (Chap. 4) and (2) the determination of the forces which hold together the various members forming a structure (Chap. 6). We shall now consider the problem of determining the internal forces which hold together the various parts of a given member.

We shall first analyze the internal forces in the members of a frame, such as the crane considered earlier in Chap. 6 (Secs. 6.1 and 6.10) and note that whereas the internal forces in a straight two-force member can produce only *tension* or *compression* in that member, the internal forces in any other type of member will usually produce *shear* and *bending* as well.

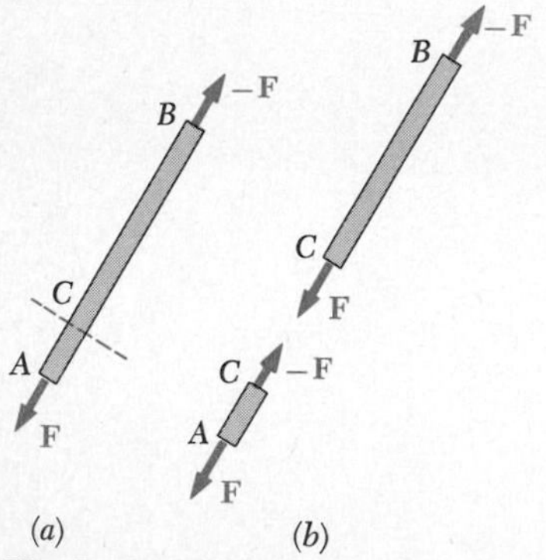

Fig. 7.1

Most of this chapter will be devoted to the analysis of the internal forces in two important types of engineering structures, namely,

1. *Beams*, which are usually long, straight prismatic members designed to support loads applied at various points along the member.
2. *Cables*, which are flexible members capable of withstanding only tension, designed to support either concentrated or distributed loads. Cables are used in many engineering applications, such as suspension bridges and transmission lines.

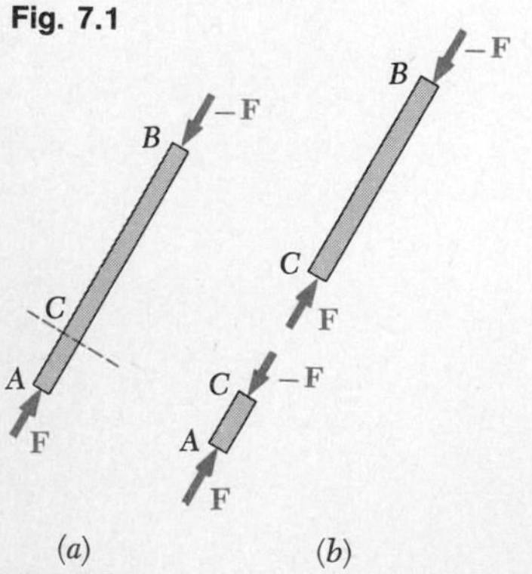

Fig. 7.2

***7.2. Internal Forces in Members.** We shall first consider a *straight two-force member AB* (Fig. 7.1*a*). From Sec. 4.6, we know that the forces **F** and −**F** acting at *A* and *B*, respectively, must be directed along *AB* in opposite sense and have the same magnitude *F*. Now, let us cut the member at *C*. To maintain the equilibrium of the free bodies *AC* and *CB* thus obtained, we must apply to *AC* a force −**F** equal and opposite to **F**, and to *CB* a force **F** equal and opposite to −**F** (Fig. 7.1*b*). These new forces are directed along *AB* in opposite sense and have the same magnitude *F*. Since the two parts *AC* and *CB* were in equilibrium before the member was cut, *internal forces* equivalent to these new forces must have

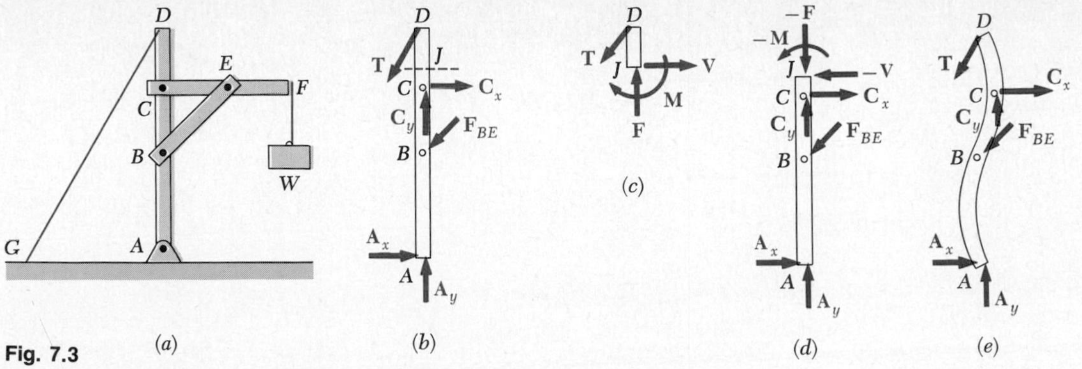

Fig. 7.3

existed in the member itself. We see that in the case of a straight two-force member, the internal forces acting on each part of the member are equivalent to an axial force. The magnitude F of this force does not depend upon the location of the section C and is referred to as the *force in member AB*. In the case considered, the member is in tension and will elongate under the action of the internal forces. In the case represented in Fig. 7.2, the member is in compression and will decrease in length under the action of the internal forces.

Next we shall consider a *multiforce member*. Take, for instance, member AD of the crane analyzed in Sec. 6.10. This crane is shown again in Fig. 7.3*a*, and the free-body diagram of member AD is drawn in Fig. 7.3*b*. We now cut member AD at J and draw a free-body diagram for each of the portions JD and AJ of the member (Fig. 7.3*c* and *d*). Considering the free body JD, we find that its equilibrium will be maintained if we apply at J a force **F** to balance the vertical component of **T**, a force **V** to balance the horizontal component of **T**, and a couple **M** to balance the moment of **T** about J. Again we conclude that internal forces must have existed at J before member AD was cut. The internal forces acting on the portion JD of member AD are equivalent to the force-couple system shown in Fig. 7.3*c*. According to Newton's third law, the internal forces acting on AJ must be equivalent to an equal and opposite force-couple system, as shown in Fig. 7.3*d*. It clearly appears that the action of the internal forces in member AD *is not limited to producing tension or compression* as in the case of straight two-force members; the internal forces *also produce shear and bending*. The force **F** is again in this case called an *axial force;* the force **V** is called a *shearing force;* and the moment **M** of the couple is known as the *bending moment at J*. We note that when determining internal forces in a member, we should clearly indicate on which portion of the member the forces are supposed to act. The deformation which will occur in member AD is sketched in Fig. 7.3*e*. The actual analysis of such a deformation is part of the study of mechanics of materials.

It should be noted that in a *two-force member which is not straight*, the internal forces are also equivalent to a force-couple system. This is shown in Fig. 7.4, where the two-force member ABC has been cut at D.

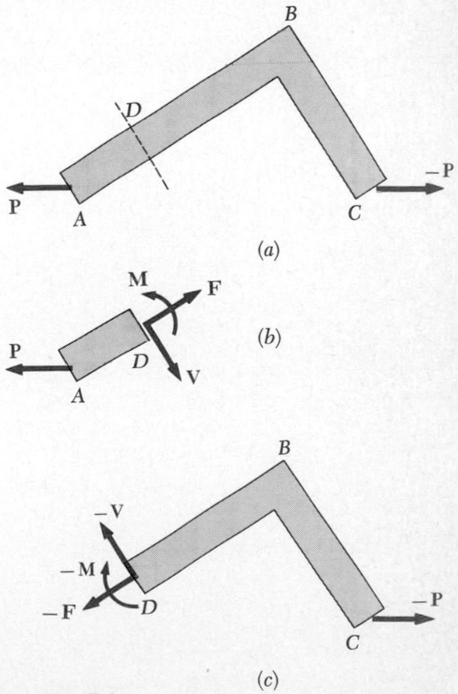

Fig. 7.4

SAMPLE PROBLEM 7.1

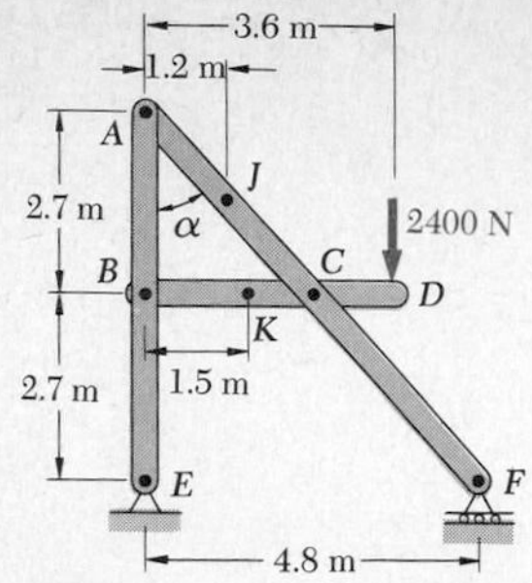

In the frame shown, determine the internal forces (*a*) in member *ACF* at point *J* and (*b*) in member *BCD* at point *K*. This frame has been previously considered in Sample Prob. 6.5.

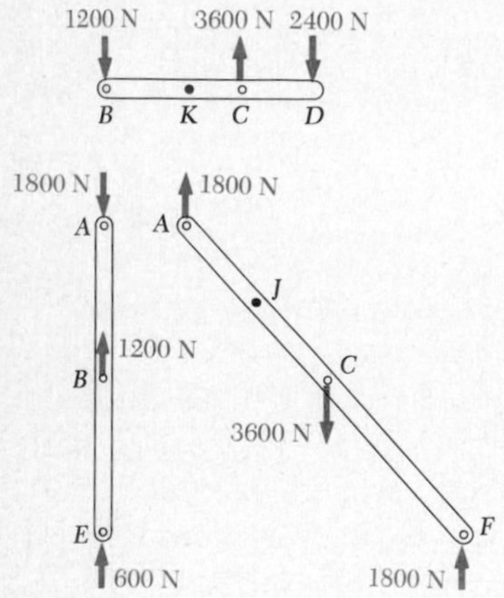

Solution. The reactions and the forces acting on each member of the frame are determined; this has been previously done in Sample Prob. 6.5, and the results are repeated here.

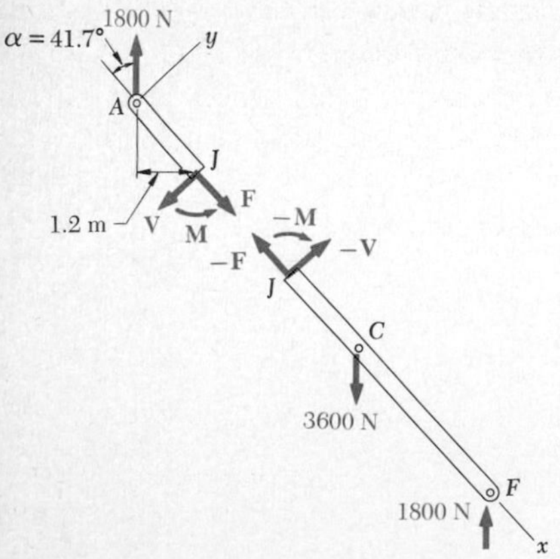

a. Internal Forces at J. Member *ACF* is cut at point *J*, and the two parts shown are obtained. The internal forces at *J* are represented by an equivalent force-couple system and may be determined by considering the equilibrium of either part. Considering the *free body AJ*, we write

$$+\circlearrowleft \Sigma M_J = 0: \quad -(1800\text{ N})(1.2\text{ m}) + M = 0$$
$$M = +2160\text{ N}\cdot\text{m} \qquad \mathbf{M} = 2160\text{ N}\cdot\text{m} \circlearrowleft \blacktriangleleft$$

$$+\searrow \Sigma F_x = 0: \quad F - (1800\text{ N}) \cos 41.7° = 0$$
$$F = +1344\text{ N} \qquad \mathbf{F} = 1344\text{ N} \searrow \blacktriangleleft$$

$$+\nearrow \Sigma F_y = 0: \quad -V + (1800\text{ N}) \sin 41.7° = 0$$
$$V = +1197\text{ N} \qquad \mathbf{V} = 1197\text{ N} \swarrow \blacktriangleleft$$

The internal forces at *J* are therefore equivalent to a couple **M**, an axial force **F**, and a shearing force **V**. The internal force-couple system acting on part *JCF* is equal and opposite.

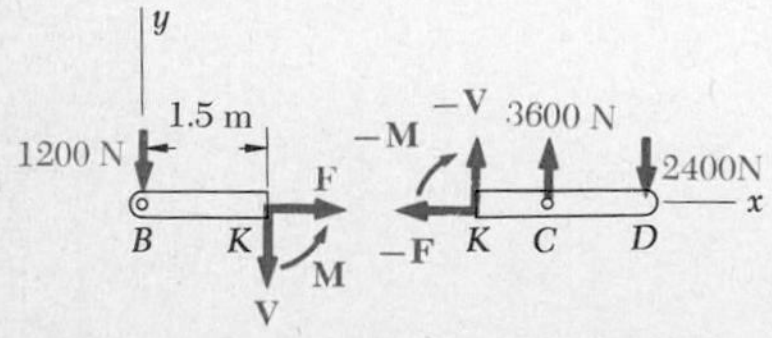

b. Internal Forces at K. We cut member *BCD* at *K* and obtain the two parts shown. Considering the *free body BK*, we write

$$+\circlearrowleft \Sigma M_K = 0: \quad (1200\text{ N})(1.5\text{ m}) + M = 0$$
$$M = -1800\text{ N}\cdot\text{m} \qquad \mathbf{M} = 1800\text{ N}\cdot\text{m} \circlearrowright \blacktriangleleft$$

$$\overset{+}{\rightarrow} \Sigma F_x = 0: \quad F = 0 \qquad \mathbf{F} = 0 \blacktriangleleft$$

$$+\uparrow \Sigma F_y = 0: \quad -1200\text{ N} - V = 0$$
$$V = -1200\text{ N} \qquad \mathbf{V} = 1200\text{ N} \uparrow \blacktriangleleft$$

Problems

7.1 through 7.4 Determine the internal forces (axial force, shearing force, and bending moment) at point J of the structure indicated:

7.1 Frame and loading of Prob. 6.51.
7.2 Frame and loading of Prob. 6.52.
7.3 Frame and loading of Prob. 6.53.
7.4 Frame and loading of Prob. 6.54.

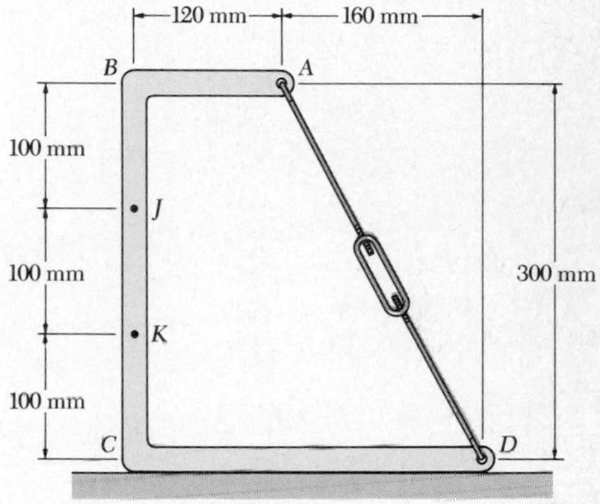

Fig. P7.5, P7.6, and P7.7

7.5 and 7.6 Knowing that the turnbuckle has been tightened until the tension in wire AD is 850 N, determine the internal forces at the point indicated:

7.5 Point J.
7.6 Point K.

7.7 It has been experimentally determined that the bending moment at point K of the frame shown is 300 N·m. Determine (*a*) the tension in wire AD, (*b*) the corresponding internal forces at point J.

7.8 The bracket ABD is supported by a pin at A and by the cable DE. Determine the internal forces just to the left of the load.

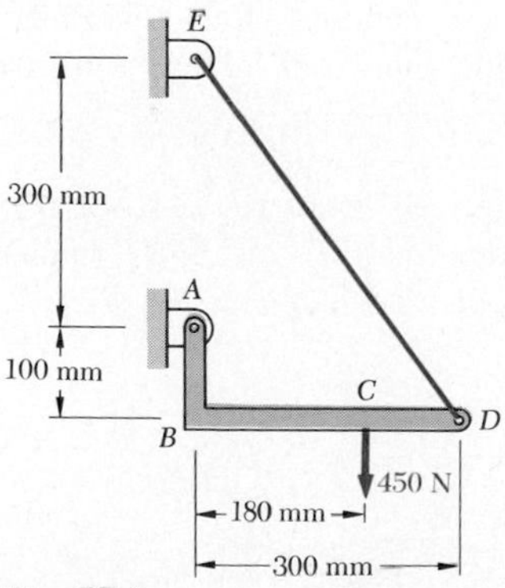

Fig. P7.8

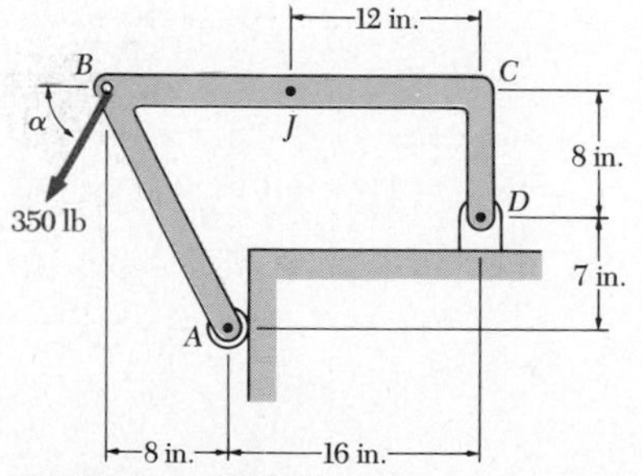

Fig. P7.9 and P7.10

7.9 Determine the internal forces at point J when $\alpha = 90°$.

7.10 Determine the internal forces at point J when $\alpha = 0$.

7.11 Knowing that the radius of each pulley is 120 mm, determine the internal forces at (*a*) point J, (*b*) point K.

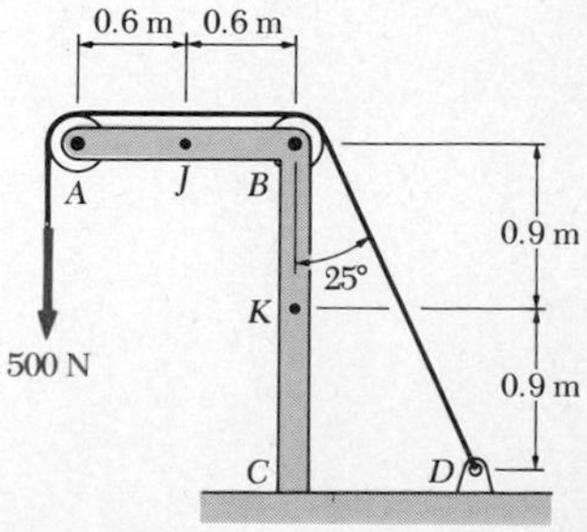

Fig. P7.11

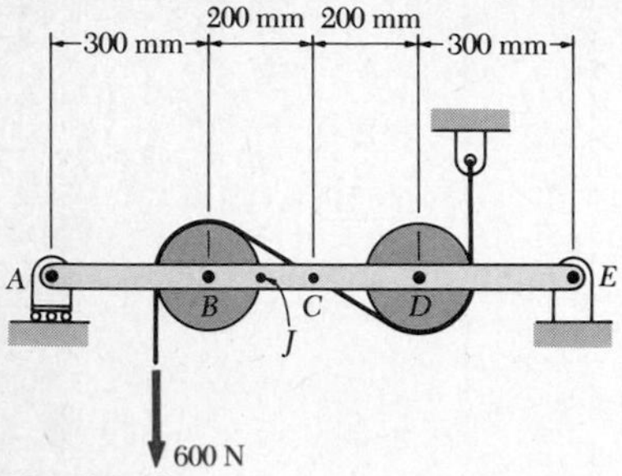

Fig. P7.12

7.12 Knowing that the radius of each pulley is 100 mm, determine the internal forces at (*a*) point *C*, (*b*) point *J*.

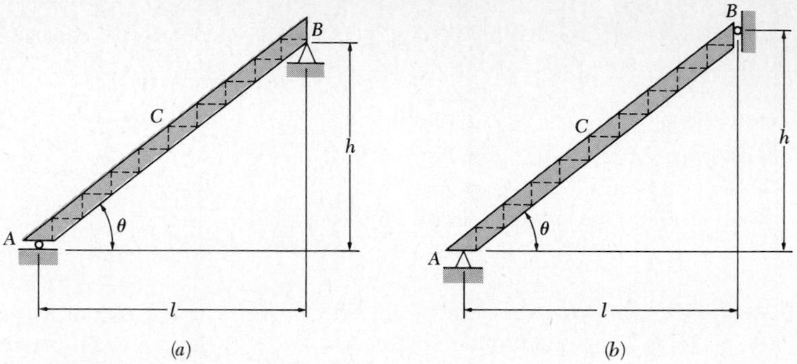

Fig. P7.13

7.13 A steel channel forms one side of a flight of stairs. If one channel weighs w lb/ft, determine the internal forces at the center of one channel due to its own weight when $w = 20$ lb/ft, $l = 12$ ft, and $h = 9$ ft. Consider each of the support conditions shown.

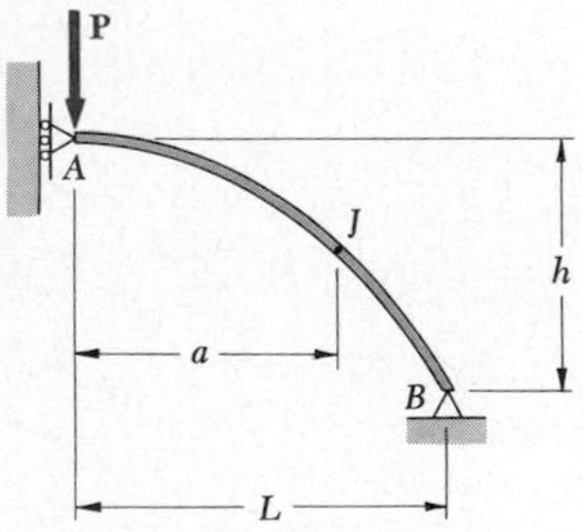

Fig. P7.14 and P7.15

7.14 The axis of the curved member *AB* is a parabola with vertex at *A*. If a vertical load **P** of magnitude 300 lb is applied at *A*, determine the internal forces at *J* when $h = 9$ in., $L = 30$ in., and $a = 20$ in.

7.15 Knowing that the axis of the curved member *AB* is a parabola with vertex at *A*, determine the magnitude and location of the maximum bending moment.

7.16 and 7.17 A half section of pipe rests on a frictionless horizontal surface as shown. If the half section of pipe has a mass of 12 kg and a diameter of 400 mm, determine the bending moment at point *J* when $\theta = 90°$.

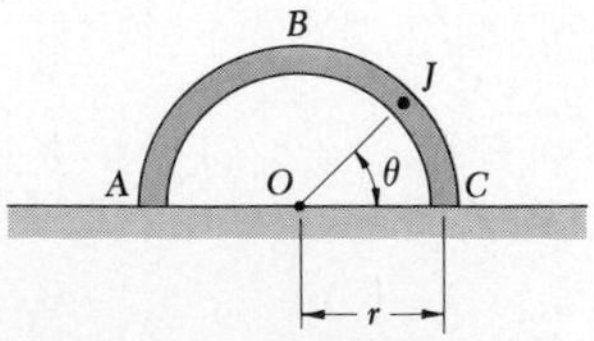

Fig. P7.16 and P7.18

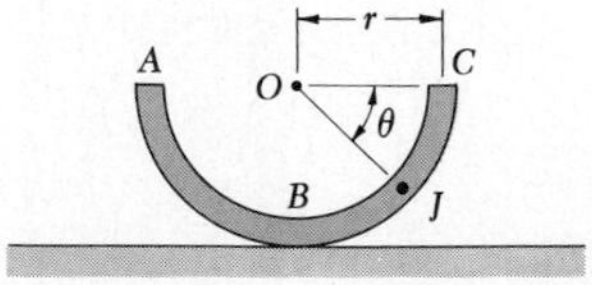

Fig. P7.17 and P7.19

***7.18 and 7.19** A half section of pipe, of weight *W* and of unit length, rests on a frictionless horizontal surface. Determine the internal force at point *J* in terms of *W*, *r*, and θ.

***7.20** In Prob. 7.19, determine the magnitude and location of the maximum internal axial force.

*** 7.3. Various Types of Loading and Support.** A structural member designed to support loads applied at various points along the member is known as a *beam*. In most cases, the loads are perpendicular to the axis of the beam and will cause only shear and bending in the beam. When the loads are not at a right angle to the beam, they will also produce axial forces in the beam.

Beams are usually long, straight prismatic bars. Designing a beam consists essentially in selecting the cross section which will provide the most effective resistance to the shear and bending produced by the applied loads. The design of the beam, therefore, includes two distinct parts. In the first part, the shearing forces and bending moments produced by the loads are determined. The second part is concerned with the selection of the cross section best suited to resist the shearing forces and bending moments determined in the first part. This portion of the chapter, Beams, deals with the first part of the problem of beam design, namely, the determination of the shearing forces and bending moments in beams subjected to various loading conditions and supported in various ways. The second part of the problem belongs to the study of mechanics of materials.

A beam may be subjected to *concentrated loads* $\mathbf{P}_1$, $\mathbf{P}_2$, . . . , expressed in newtons, pounds, or their multiples kilonewtons and kips (Fig. 7.5*a*), to a *distributed load* w, expressed in N/m, kN/m, lb/ft, or kips/ft (Fig. 7.5*b*), or to a combination of both. When the load w per unit length has a constant value over part of the beam (as between A and B in Fig. 7.5*b*), the load is said to be *uniformly distributed* over that part of the beam. The determination of the reactions at the supports may be considerably simplified if distributed loads are replaced by equivalent concentrated loads, as explained in Sec. 5.8. This substitution, however, should not be performed, or at least should be performed with care, when internal forces are being computed (see Sample Prob. 7.3).

Beams are classified according to the way in which they are supported. Several types of beams frequently used are shown in Fig. 7.6. The distance L between supports is called the *span*. It should be noted that the

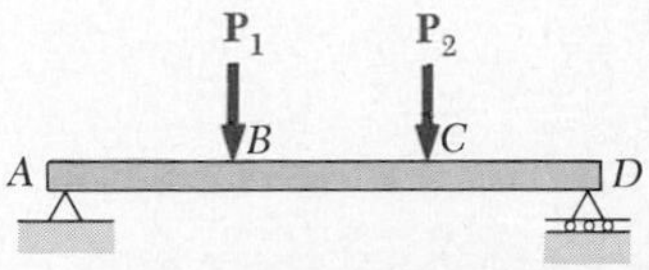

(*a*) Concentrated loads

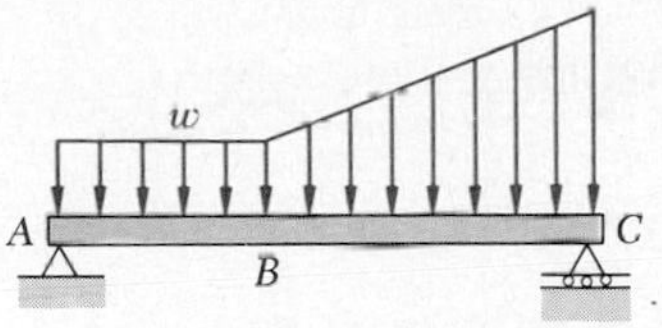

(*b*) Distributed load

Fig. 7.5 Types of loadings.

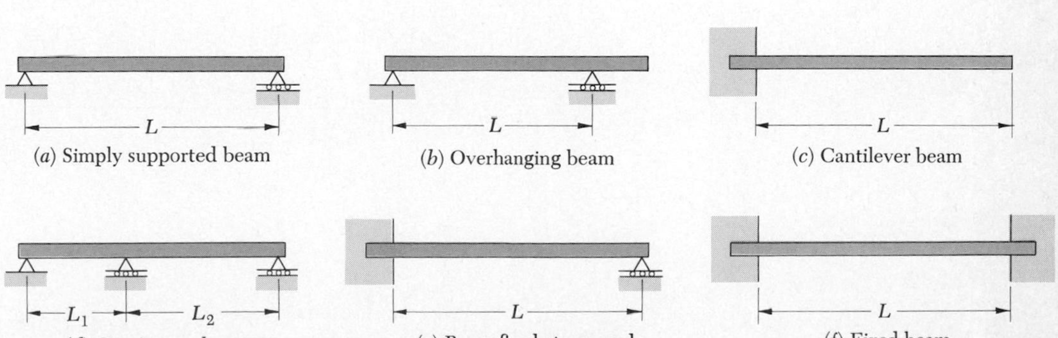

(*a*) Simply supported beam (*b*) Overhanging beam (*c*) Cantilever beam

(*d*) Continuous beam (*e*) Beam fixed at one end and simply supported at the other end (*f*) Fixed beam

Fig. 7.6 Types of beams.

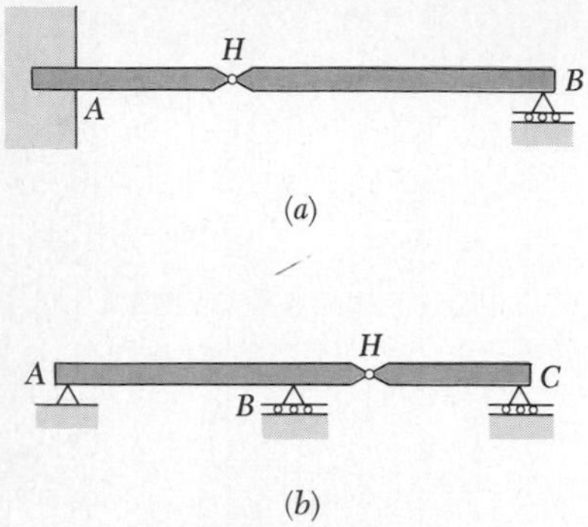

Fig. 7.7 Combined beams.

reactions will be determinate if the supports involve only three unknowns. The reactions will be statically indeterminate if more unknowns are involved; the methods of statics are not sufficient then to determine the reactions, and the properties of the beam with regard to its resistance to bending must be taken into consideration. Beams supported by two rollers are not shown here; such beams are only partially constrained and will move under certain loading conditions.

Sometimes two or more beams are connected by hinges to form a single continuous structure. Two examples of beams hinged at a point H are shown in Fig. 7.7. It will be noted that the reactions at the supports involve four unknowns and cannot be determined from the free-body diagram of the two-beam system. They can be determined, however, by considering the free-body diagram of each beam separately; six unknowns are involved (including two force components at the hinge), and six equations are available.

*** 7.4. Shear and Bending Moment in a Beam.** Consider a beam AB subjected to various concentrated and distributed loads (Fig. 7.8*a*). We propose to determine the shearing force and bending moment at any point of the beam. In the example considered here, the beam is simply supported, but the method used could be applied to any type of statically determinate beam.

First we determine the reactions at A and B by choosing the entire beam as a free body (Fig. 7.8*b*); writing $\Sigma M_A = 0$ and $\Sigma M_B = 0$, we obtain, respectively, $\mathbf{R}_B$ and $\mathbf{R}_A$.

To determine the internal forces at C, we cut the beam at C and draw the free-body diagrams of the portions AC and CB of the beam (Fig. 7.8*c*).

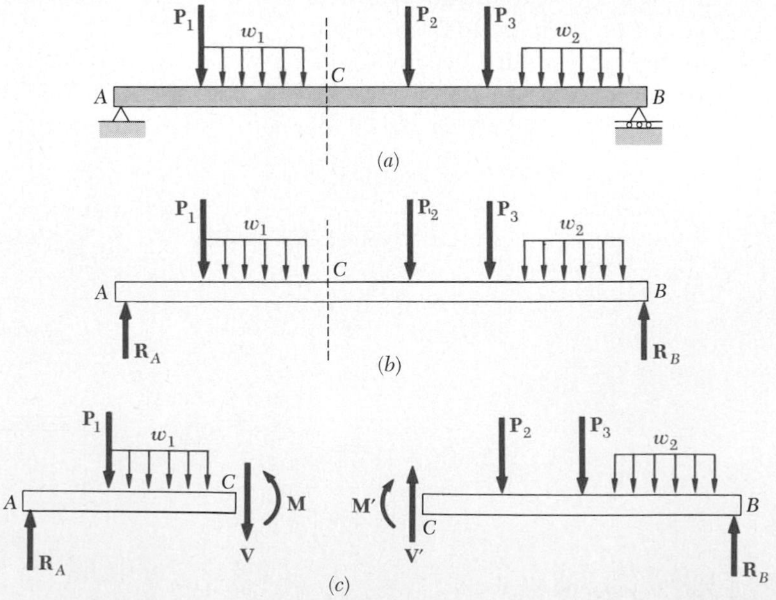

Fig. 7.8

Using the free-body diagram of AC, we may determine the shearing force $\mathbf{V}$ at C by equating to zero the sum of the vertical components of all forces acting on AC. Similarly, the bending moment $\mathbf{M}$ at C may be found by equating to zero the sum of the moments about C of all forces and couples acting on AC. We could have used just as well, however, the free-body diagram of CB† and determined the shearing force $\mathbf{V}'$ and the bending moment $\mathbf{M}'$ by equating to zero the sum of the vertical components and the sum of the moments about C of all forces and couples acting on CB. While this possible choice of alternate free bodies may facilitate the computation of the numerical values of the shearing force and bending moment, it makes it necessary to indicate on which portion of the beam the internal forces considered are acting. If the shearing force and bending moment, however, are to be computed at every point of the beam and efficiently recorded, we should not have to specify every time which portion of the beam is used as a free body. We shall adopt, therefore, the following conventions:

To determine the shearing force in a beam, *we shall always assume* that the internal forces $\mathbf{V}$ and $\mathbf{V}'$ are directed as shown in Fig. 7.8*c*. A positive value obtained for their common magnitude V will indicate that this assumption was correct and that the shearing forces are actually directed as shown. A negative value obtained for V will indicate that the assumption was wrong and that the shearing forces are directed in the opposite way. Thus, only the magnitude V, together with a plus or minus sign, needs to be recorded to define completely the shearing forces at a given point of the beam. The scalar V is commonly referred to as the *shear* at the given point of the beam.

Similarly, *we shall always assume* that the internal couples $\mathbf{M}$ and $\mathbf{M}'$ are directed as shown in Fig. 7.8*c*. A positive value obtained for their magnitude M, commonly referred to as the bending moment, will indicate that this assumption was correct, and a negative value that it was wrong. Summarizing the sign conventions we have presented, we state:

The shear V and the bending moment M at a given point of a beam are said to be positive when the internal forces and couples acting on each portion of the beam are directed as shown in Fig. 7.9a.

These conventions may be more easily remembered if we note that:

1. *The shear at C is positive when the* **external** *forces (loads and reactions) acting on the beam tend to shear off the beam at C as indicated in Fig. 7.9b.*
2. *The bending moment at C is positive when the* **external** *forces acting on the beam tend to bend the beam at C as indicated in Fig. 7.9c.*

It may also help to note that the situation described in Fig. 7.9, and corresponding to positive values of the shear and of the bending moment, is

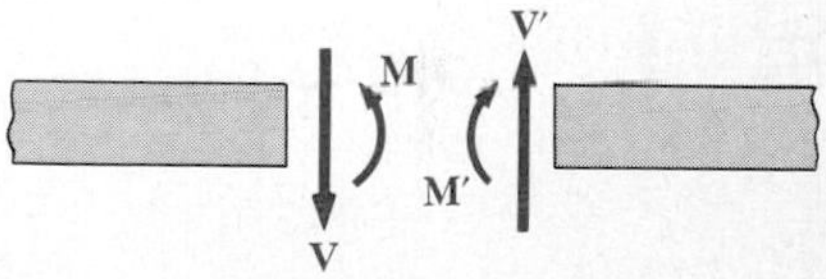

(*a*) Internal forces at section (positive shear and positive bending moment)

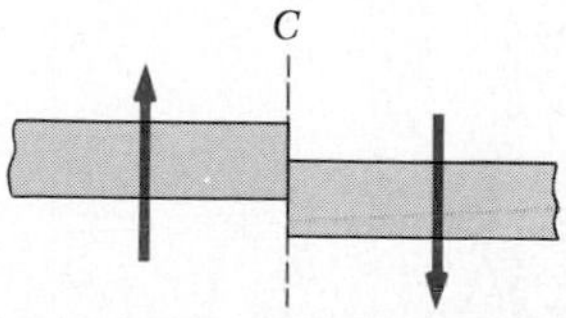

(*b*) Effect of external forces (positive shear)

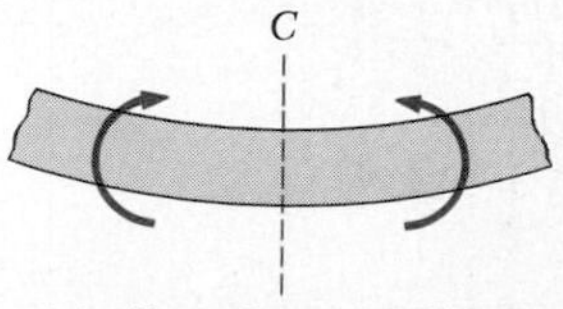

(*c*) Effect of external forces (positive bending moment)

Fig. 7.9

†The force and couple representing the internal forces acting on CB will now be denoted by $\mathbf{V}'$ and $\mathbf{M}'$, rather than by $-\mathbf{V}$ and $-\mathbf{M}$ as done earlier, in order to avoid confusion when applying the sign convention which we are about to introduce.

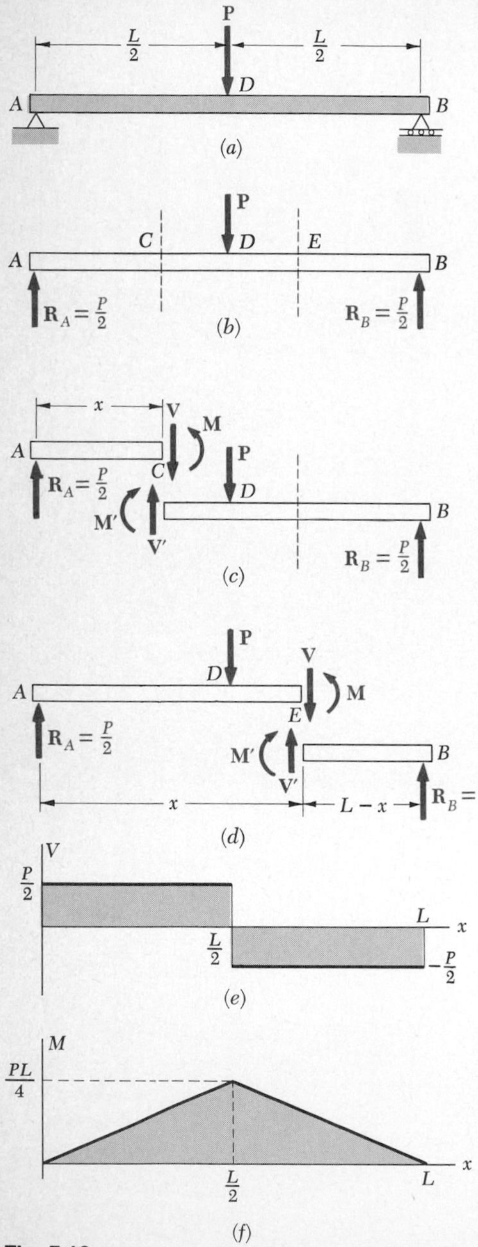

Fig. 7.10

precisely the situation which occurs in the left half of a simply supported beam carrying a single concentrated load at its midpoint. This particular example is fully discussed in the following section.

***7.5. Shear and Bending-Moment Diagrams.** Now that shear and bending moment have been clearly defined in sense as well as in magnitude, we may easily record their values at any point of a beam by plotting these values against the distance x measured from one end of the beam. The graphs obtained in this way are called, respectively, the *shear diagram* and the *bending-moment diagram*. As an example, consider a simply supported beam AB of span L subjected to a single concentrated load $\mathbf{P}$ applied at its midpoint D (Fig. 7.10*a*). We first determine the reactions at the supports from the free-body diagram of the entire beam (Fig. 7.10*b*); we find that the magnitude of each reaction is equal to $P/2$.

Next we cut the beam at a point C between A and D and draw the free-body diagrams of AC and CB (Fig. 7.10*c*). *Assuming that shear and bending moment are positive*, we direct the internal forces $\mathbf{V}$ and $\mathbf{V}'$ and the internal couples $\mathbf{M}$ and $\mathbf{M}'$ as indicated in Fig. 7.9*a*. Considering the free body AC and writing that the sum of the vertical components and the sum of the moments about C of the forces acting on the free body are zero, we find $V = +P/2$ and $M = +Px/2$. Both shear and bending moment are therefore positive; this may be checked by observing that the reaction at A tends to shear off and to bend the beam at C as indicated in Fig. 7.9*b* and *c*. We may plot V and M between A and D (Fig. 7.10*e* and *f*); the shear has a constant value $V = P/2$, while the bending moment increases linearly from $M = 0$ at $x = 0$ to $M = PL/4$ at $x = L/2$.

Cutting, now, the beam at a point E between D and B and considering the free body EB (Fig. 7.10*d*), we write that the sum of the vertical components and the sum of the moments about E of the forces acting on the free body are zero. We obtain $V = -P/2$ and $M = P(L - x)/2$. The shear is therefore negative and the bending moment positive; this may be checked by observing that the reaction at B bends the beam at E as indicated in Fig. 7.9*c* but tends to shear it off in a manner opposite to that shown in Fig. 7.9*b*. We can complete, now, the shear and bending-moment diagrams of Fig. 7.10*e* and *f*; the shear has a constant value $V = -P/2$ between D and B, while the bending moment decreases linearly from $M = PL/4$ at $x = L/2$ to $M = 0$ at $x = L$.

We shall note that when a beam is subjected to concentrated loads only, the shear is of constant value between loads and the bending moment varies linearly between loads. On the other hand, when a beam is subjected to distributed loads, the shear and bending moment vary quite differently (see Sample Prob. 7.3).

SAMPLE PROBLEM 7.2

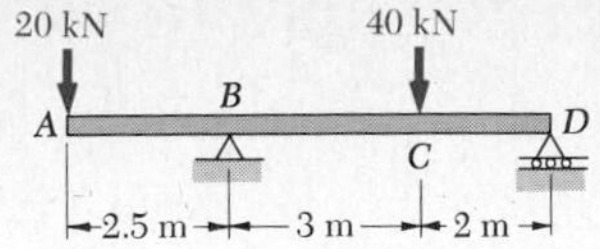

Draw the shear and bending-moment diagram for the beam and loading shown.

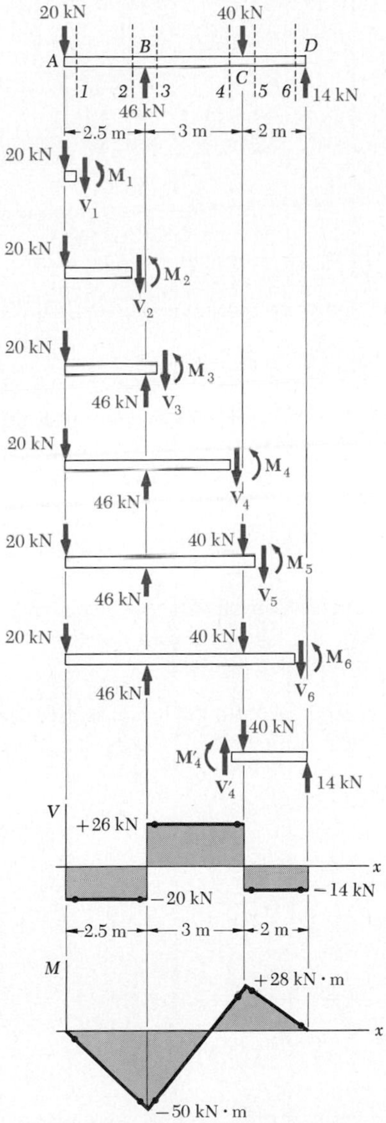

Solution. The reactions are determined by considering the entire beam as a free body; they are

$$\mathbf{R}_B = 46\ \text{kN}\uparrow \qquad \mathbf{R}_D = 14\ \text{kN}\uparrow$$

We first determine the internal forces just to the right of the 20-kN load at A. Considering the stub of beam to the left of section *1* as a free body and assuming V and M to be positive (according to the standard convention), we write

$$+\uparrow\Sigma F_y = 0: \qquad -20\ \text{kN} - V_1 = 0 \qquad V_1 = -20\ \text{kN}$$

$$+\circlearrowleft\ \Sigma M_1 = 0: \qquad (20\ \text{kN})(0\ \text{m}) + M_1 = 0 \qquad M_1 = 0$$

We next consider as a free body the portion of beam to the left of section *2* and write

$$+\uparrow\Sigma F_y = 0: \qquad -20\ \text{kN} - V_2 = 0 \qquad V_2 = -20\ \text{kN}$$

$$+\circlearrowleft\ \Sigma M_2 = 0: \qquad (20\ \text{kN})(2.5\ \text{m}) + M_2 = 0 \qquad M_2 = -50\ \text{kN}\cdot\text{m}$$

The shear and bending moment at sections *3*, *4*, *5*, and *6* are determined in a similar way from the free-body diagrams shown. We obtain

$$V_3 = +26\ \text{kN} \qquad M_3 = -50\ \text{kN}\cdot\text{m}$$

$$V_4 = +26\ \text{kN} \qquad M_4 = +28\ \text{kN}\cdot\text{m}$$

$$V_5 = -14\ \text{kN} \qquad M_5 = +28\ \text{kN}\cdot\text{m}$$

$$V_6 = -14\ \text{kN} \qquad M_6 = 0$$

For several of the latter sections, the results may be more easily obtained by considering as a free body the portion of the beam to the right of the section. For example, considering the portion of the beam to the right of section *4*, we write

$$+\uparrow\Sigma F_y = 0: \qquad V_4 - 40\ \text{kN} + 14\ \text{kN} = 0 \qquad V_4 = +26\ \text{kN}$$

$$+\circlearrowleft\ \Sigma M_4 = 0: \qquad -M_4 + (14\ \text{kN})(2\ \text{m}) = 0 \qquad M_4 = +28\ \text{kN}\cdot\text{m}$$

We may now plot the six points shown on the shear and bending-moment diagrams. As indicated in Sec. 7.5, the shear is of constant value between concentrated loads, and the bending moment varies linearly; we obtain therefore the shear and bending-moment diagrams shown.

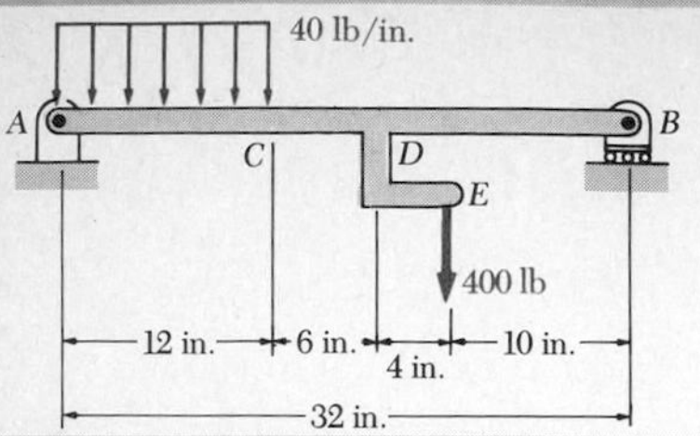

SAMPLE PROBLEM 7.3

Draw the shear and bending-moment diagrams for the beam AB. The distributed load of 40 lb/in. extends over 12 in. of the beam, from A to C, and the 400-lb load is applied at E.

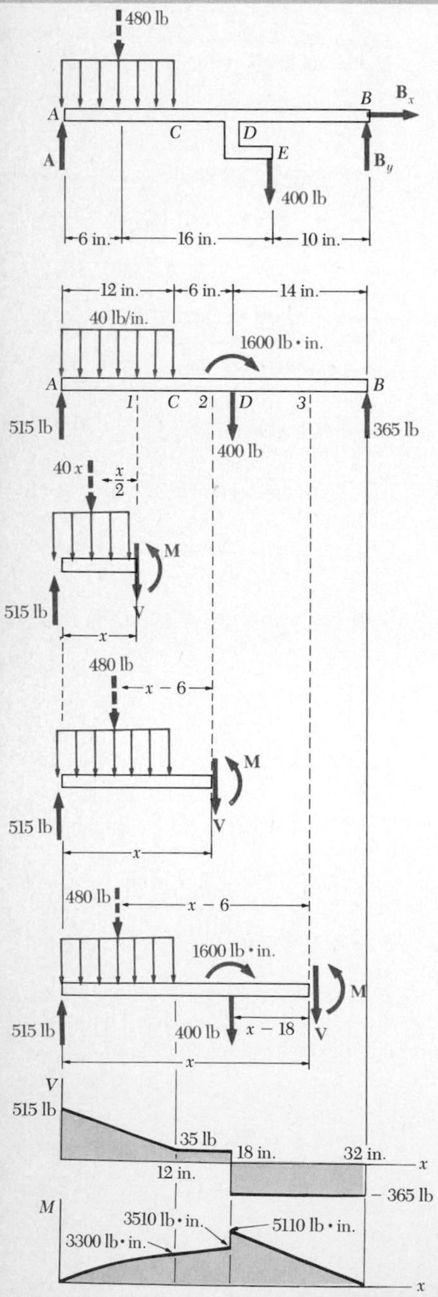

Solution. The reactions are determined by considering the entire beam as a free body.

$+\circlearrowleft \Sigma M_A = 0$: $\quad B_y(32\text{ in.}) - (480\text{ lb})(6\text{ in.}) - (400\text{ lb})(22\text{ in.}) = 0$

$B_y = +365\text{ lb} \qquad \mathbf{B}_y = 365\text{ lb}\uparrow$

$+\circlearrowleft \Sigma M_B = 0$: $\quad (480\text{ lb})(26\text{ in.}) + (400\text{ lb})(10\text{ in.}) - A(32\text{ in.}) = 0$

$A = +515\text{ lb} \qquad \mathbf{A} = 515\text{ lb}\uparrow$

$\xrightarrow{+} \Sigma F_x = 0$: $\quad B_x = 0 \qquad \mathbf{B}_x = 0$

The 400-lb load is now replaced by an equivalent force-couple system acting on the beam at point D.

From A to C. We determine the internal forces at a distance x from point A by considering the portion of beam to the left of section *1*. That part of the distributed load acting on the free body is replaced by its resultant, and we write

$+\uparrow \Sigma F_y = 0$: $\quad 515 - 40x - V = 0 \qquad V = 515 - 40x$

$+\circlearrowleft \Sigma M_1 = 0$: $\quad -515x + 40x(\tfrac{1}{2}x) + M = 0 \qquad M = 515x - 20x^2$

Since the free-body diagram shown may be used for all values of x smaller than 12 in., the expressions obtained for V and M are valid in the region $0 < x < 12$ in.

From C to D. Considering the portion of beam to the left of section *2* and again replacing the distributed load by its resultant, we obtain

$+\uparrow \Sigma F_y = 0$: $\quad 515 - 480 - V = 0 \qquad V = 35\text{ lb}$

$+\circlearrowleft \Sigma M_2 = 0$: $\quad -515x + 480(x - 6) + M = 0 \qquad M = (2880 + 35x)\text{ lb}\cdot\text{in.}$

These expressions are valid in the region 12 in. $< x <$ 18 in.

From D to B. Using the portion of beam to the left of section *3*, we obtain for the region 18 in. $< x <$ 32 in.

$+\uparrow \Sigma F_y = 0$: $\quad 515 - 480 - 400 - V = 0 \qquad V = -365\text{ lb}$

$+\circlearrowleft \Sigma M_3 = 0$: $\quad -515x + 480(x - 6) - 1600 + 400(x - 18) + M = 0$

$M = (11{,}680 - 365x)\text{ lb}\cdot\text{in.}$

The shear and bending-moment diagrams for the entire beam may now be plotted. We note that the couple of moment 1600 lb·in. applied at point D introduces a discontinuity into the bending-moment diagram.

Problems

7.21 through 7.26 Draw the shear and bending-moment diagrams for the beam and loading shown.

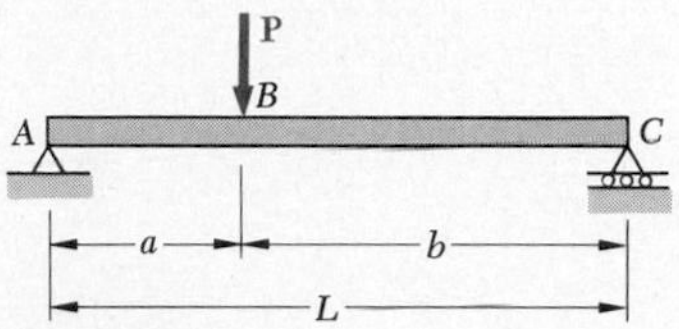

Fig. P7.21

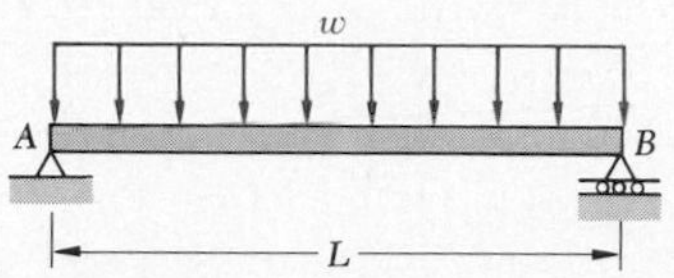

Fig. P7.22

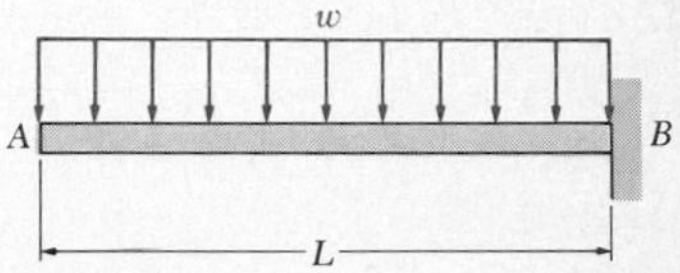

Fig. P7.23

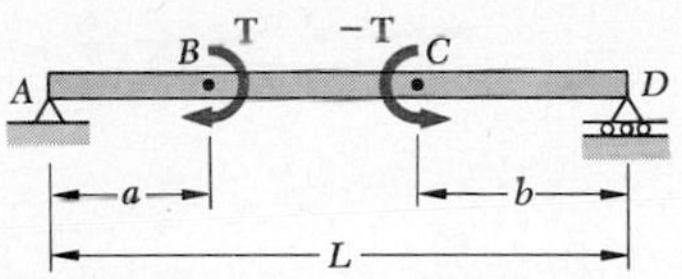

Fig. P7.24

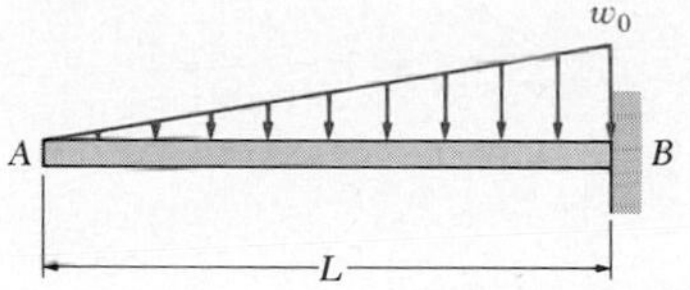

Fig. P7.25

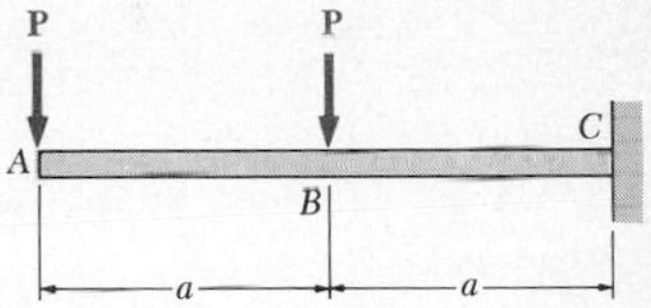

Fig. P7.26

7.27 and 7.28 Draw the shear and bending-moment diagrams for the beam *AB*.

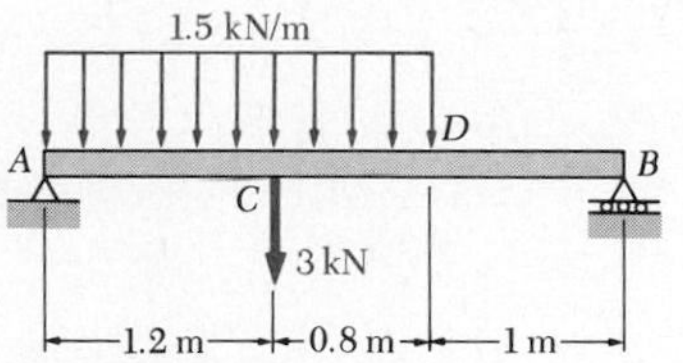

Fig. P7.27

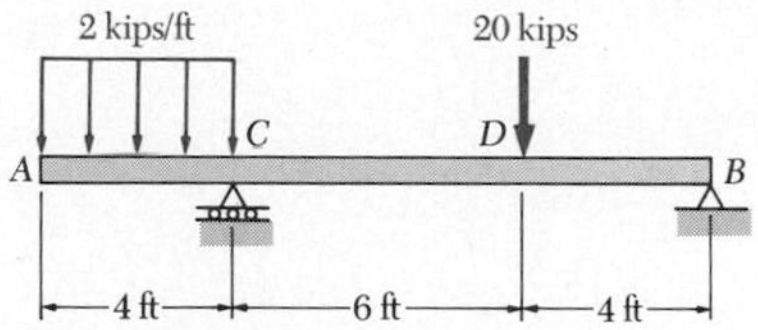

Fig. P7.28

7.29 Draw the shear and bending-moment diagrams for the beam *AB* if the magnitude of the force **P** is 75 lb.

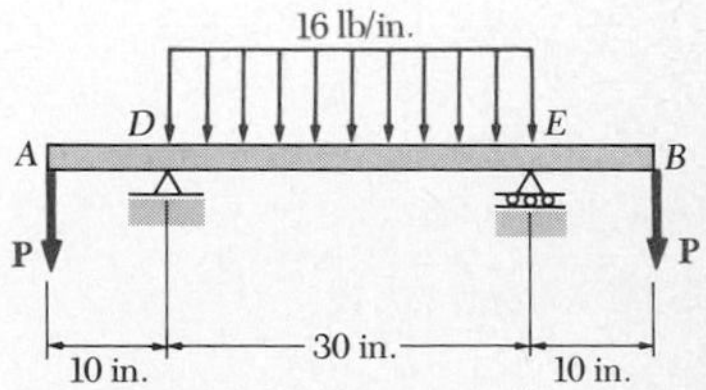

Fig. P7.29 and P7.31

7.30 Draw the shear and bending-moment diagrams for the beam *AB* if $a = 0.3$ m.

***7.31** Determine the magnitude of the force **P** for which the maximum absolute value of the bending moment in the beam is as small as possible. (*Hint.* Draw the bending-moment diagram and then equate the absolute values of the largest positive and negative bending moments obtained.)

***7.32** Determine the distance *a* for which the maximum absolute value of the bending moment in the beam is as small as possible. (See hint of Prob. 7.31.)

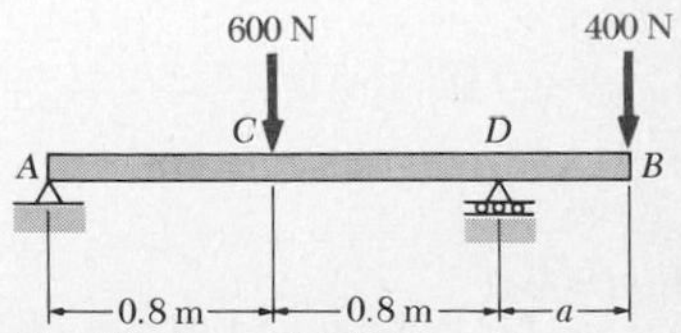

Fig. P7.30 and P7.32

7.33 through 7.35 Assuming the upward reaction of the ground to be uniformly distributed, draw the shear and bending-moment diagrams for the beam AB.

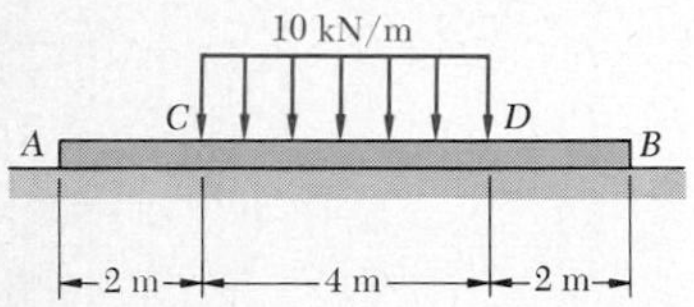

Fig. P7.33

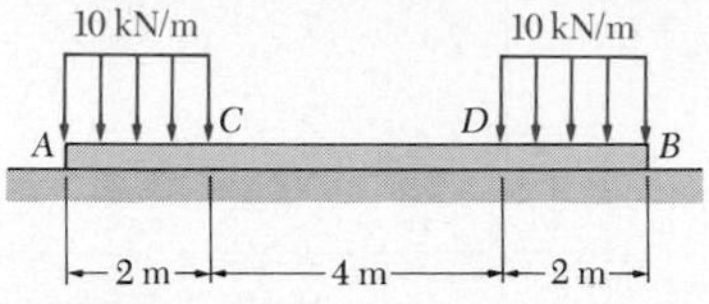

Fig. P7.34

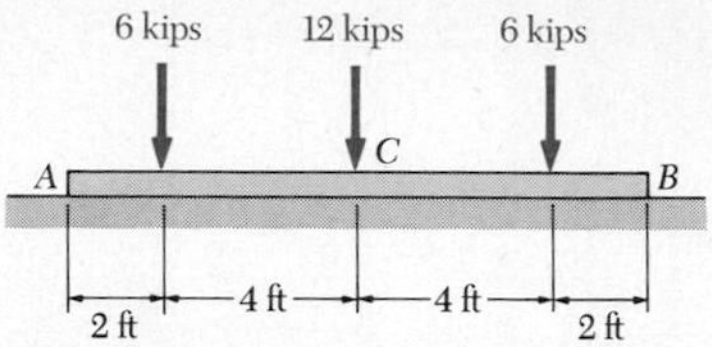

Fig. P7.35

7.36 Solve Prob. 7.35, assuming that the 12-kip load has been removed.

7.37 through 7.39 Draw the shear and bending-moment diagrams for the beam AB.

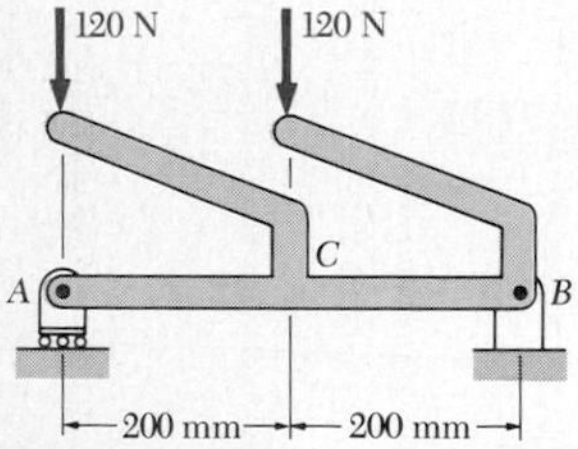

Fig. P7.37

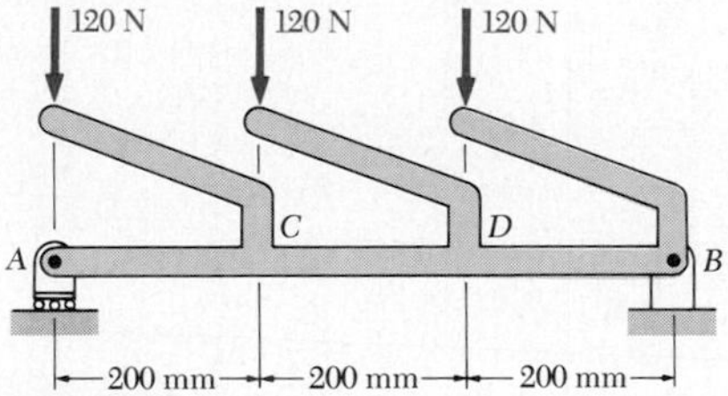

Fig. P7.38

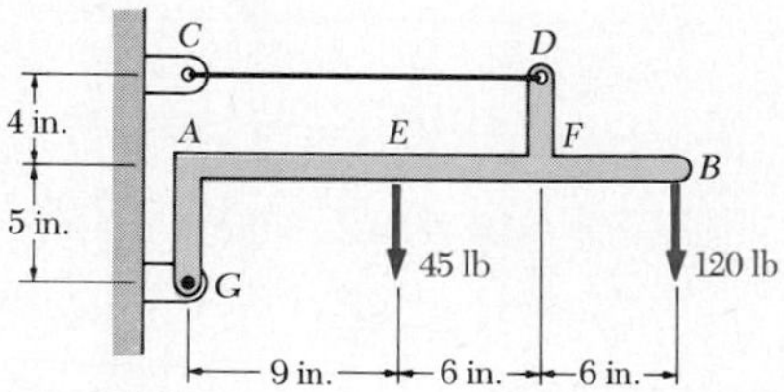

Fig. P7.39

7.40 Solve Prob. 7.39, assuming that the 45-lb load has been removed.

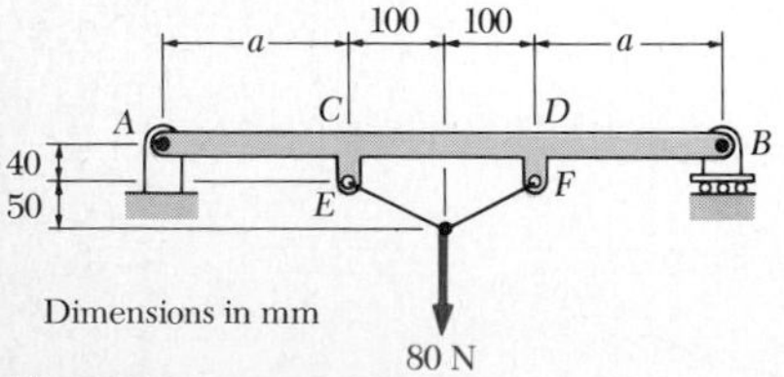

Fig. P7.41 and P7.42

7.41 For the beam and loading shown, (a) determine the dimension a for which the bending moment is zero from C to D, (b) draw the corresponding shear and bending-moment diagrams.

7.42 Draw the shear and bending-moment diagrams for the beam AB if $a = 200$ mm.

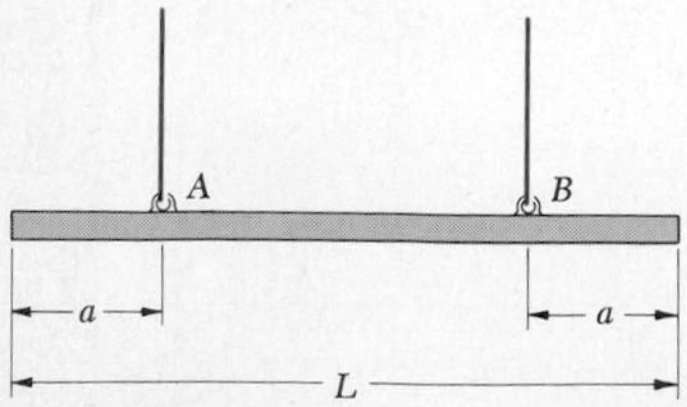

Fig. P7.43

***7.43** A uniform beam is to be picked up by crane cables attached at A and B. Determine the distance a from the ends of the beam to the points where the cables should be attached if the maximum absolute value of the bending moment in the beam is to be as small as possible. (*Hint.* Draw the bending-moment diagram in terms of a, L, and the weight w per unit length, and then equate the absolute values of the largest positive and negative bending moments obtained.)

*** 7.44** In order to reduce the bending moment in the cantilever beam AB, a cable and counterweight are permanently attached at end B. Determine the magnitude of the counterweight for which the maximum absolute value of the bending moment in the beam is as small as possible. (*a*) Consider only the case when the force **P** is actually applied at C. (*b*) Consider the more general case when the force **P** may either be applied at C or removed.

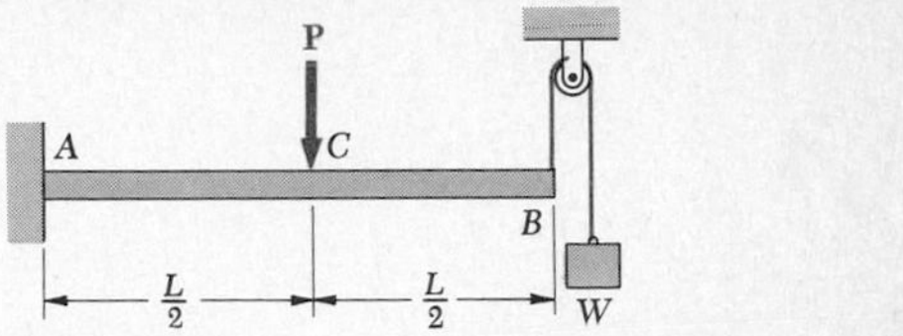

Fig. P7.44

*** 7.6. Relations among Load, Shear, and Bending Moment.** When a beam carries more than two or three concentrated loads, or when it carries distributed loads, the method outlined in Sec. 7.5 for plotting shear and bending moment may prove quite cumbersome. The construction of the shear diagram and, especially, of the bending-moment diagram will be greatly facilitated if certain relations existing among load, shear, and bending moment are taken into consideration.

Let us consider a simply supported beam AB carrying a distributed load w per unit length (Fig. 7.11*a*), and let C and C' be two points of the beam at a distance Δx from each other. The shear and bending moment at C will be denoted by V and M, respectively, and will be assumed positive; the shear and bending moment at C' will be denoted by $V + \Delta V$ and $M + \Delta M$.

We shall now detach the portion of beam CC' and draw its free-body diagram (Fig. 7.11*b*). The forces exerted on the free body include a load of magnitude $w\,\Delta x$ and internal forces and couples at C and C'. Since shear and bending moment have been assumed positive, the forces and couples will be directed as shown in the figure.

Relations between Load and Shear. We write that the sum of the vertical components of the forces acting on the free body CC' is zero:

$$V - (V + \Delta V) - w\,\Delta x = 0$$
$$\Delta V = -w\,\Delta x$$

Dividing both members of the equation by Δx and then letting Δx approach zero, we obtain

$$\frac{dV}{dx} = -w \qquad (7.1)$$

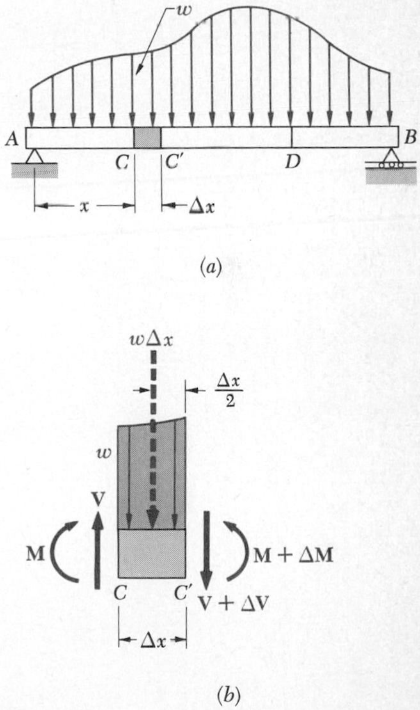

Fig. 7.11

Formula (7.1) indicates that for a beam loaded as shown in Fig. 7.11*a*, the slope dV/dx of the shear curve is negative; the numerical value of the slope at any point is equal to the load per unit length at that point.

Integrating (7.1) between points C and D, we obtain

$$V_D - V_C = -\int_{x_C}^{x_D} w\,dx \qquad (7.2)$$

$$V_D - V_C = -(\text{area under load curve between } C \text{ and } D) \qquad (7.2')$$

Note that this result could also have been obtained by considering the equilibrium of the portion of beam CD, since the area under the load curve represents the total load applied between C and D.

It should be observed that formula (7.1) *is not valid* at a point where a concentrated load is applied; the shear curve is discontinuous at such a point, as seen in Sec. 7.5. Similarly, formulas (7.2) and (7.2′) cease to be valid when concentrated loads are applied between C and D, since they do not take into account the sudden change in shear caused by a concentrated load. Formulas (7.2) and (7.2′), therefore, should be applied only between successive concentrated loads.

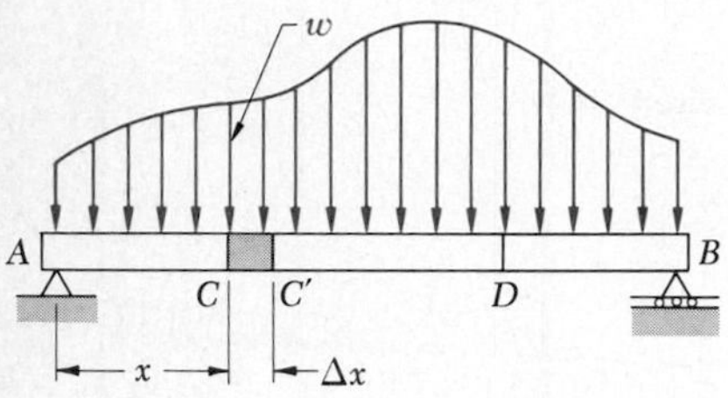

(a)

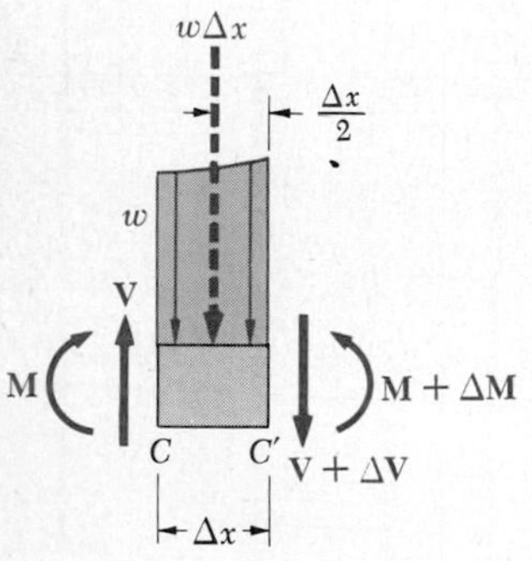

(b)

Fig. 7.11 (*repeated*)

Relations between Shear and Bending Moment. Returning to the free-body diagram of Fig. 7.11*b*, and writing now that the sum of the moments about C' is zero, we obtain

$$(M + \Delta M) - M - V\,\Delta x + w\,\Delta x \frac{\Delta x}{2} = 0$$

$$\Delta M = V\,\Delta x - \tfrac{1}{2}w(\Delta x)^2$$

Dividing both members of the equation by Δx and then letting Δx approach zero, we obtain

$$\frac{dM}{dx} = V \tag{7.3}$$

Formula (7.3) indicates that the slope dM/dx of the bending-moment curve is equal to the value of the shear. This is true at any point where the shear has a well-defined value, i.e., at any point where no concentrated load is applied. Formula (7.3) also shows that the shear is zero at points where the bending moment is maximum. This property facilitates the determination of the points where the beam is likely to fail under bending.

Integating (7.3) between points C and D, we obtain

$$M_D - M_C = \int_{x_C}^{x_D} V\,dx \tag{7.4}$$

$$M_D - M_C = \text{area under shear curve between } C \text{ and } D \tag{7.4′}$$

Note that the area under the shear curve should be considered positive where the shear is positive and negative where the shear is negative. Formulas (7.4) and (7.4′) are valid even when concentrated loads are applied between C and D, as long as the shear curve has been correctly drawn. The formulas cease to be valid, however, if a *couple* is applied at a point between C and D, since they do not take into account the sudden change in bending moment caused by a couple (see Sample Prob. 7.7).

Example. Let us consider a simply supported beam AB of span L carrying a uniformly distributed load w (Fig. 7.12a). From the free-body diagram of the entire beam we determine the magnitude of the reactions at the supports: $R_A = R_B = wL/2$ (Fig. 7.12b). Next, we draw the shear diagram. Close to the end A of the beam, the shear is equal to R_A, that is, to $wL/2$, as we may check by considering as a free body a very small portion of the beam. Using formula (7.2), we may then determine the shear V at any distance x from A; we write

$$V - V_A = -\int_0^x w\,dx = -wx$$

$$V = V_A - wx = \frac{wL}{2} - wx = w\left(\frac{L}{2} - x\right)$$

The shear curve is thus an oblique straight line which crosses the x axis at $x = L/2$ (Fig. 7.12c). Considering, now, the bending moment, we first observe that $M_A = 0$. The value M of the bending moment at any distance x from A may then be obtained from formula (7.4); we have

$$M - M_A = \int_0^x V\,dx$$

$$M = \int_0^x w\left(\frac{L}{2} - x\right)dx = \frac{w}{2}(Lx - x^2)$$

The bending-moment curve is a parabola. The maximum value of the bending moment occurs when $x = L/2$, since V (and thus dM/dx) is zero for that value of x. Substituting $x = L/2$ in the last equation, we obtain $M_{max} = wL^2/8$.

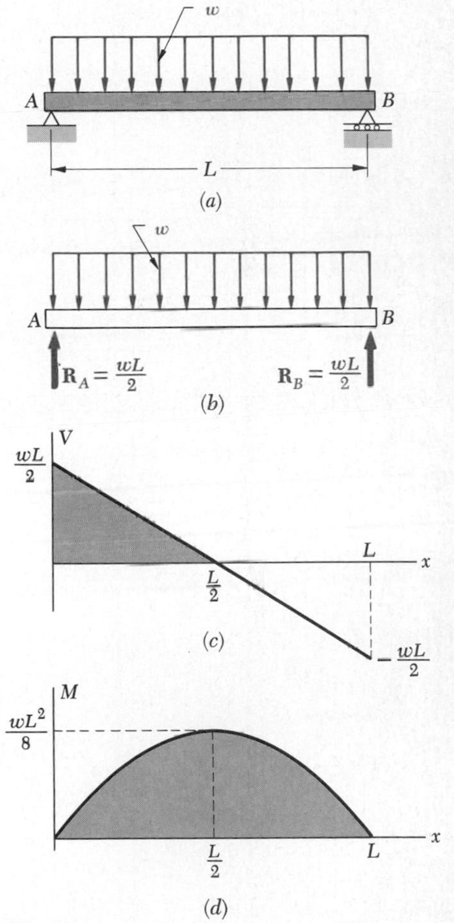

Fig. 7.12

In most engineering applications, the value of the bending moment needs to be known only at a few specific points. Once the shear diagram has been drawn, and after M has been determined at one of the ends of the beam, the value of the bending moment may then be obtained at any given point by computing the area under the shear curve and using formula (7.4′). For instance, since $M_A = 0$ for the beam of Fig. 7.12, the maximum value of the bending moment for that beam may be obtained simply by measuring the area of the shaded triangle in the shear diagram:

$$M_{max} = \frac{1}{2}\frac{L}{2}\frac{wL}{2} = \frac{wL^2}{8}$$

We note that in this example, the load curve is a horizontal straight line, the shear curve an oblique straight line, and the bending-moment curve a parabola. If the load curve had been an oblique straight line (first degree), the shear curve would have been a parabola (second degree) and the bending-moment curve a cubic (third degree). The shear and bending-moment curves will always be, respectively, one and two degrees higher than the load curve. With this in mind, we should be able to sketch the shear and bending-moment diagrams without actually determining the functions $V(x)$ and $M(x)$, once a few values of the shear and bending moment have been computed. The sketches obtained will be more accurate if we make use of the fact that at any point where the curves are continuous, the slope of the shear curve is equal to $-w$ and the slope of the bending-moment curve is equal to V.

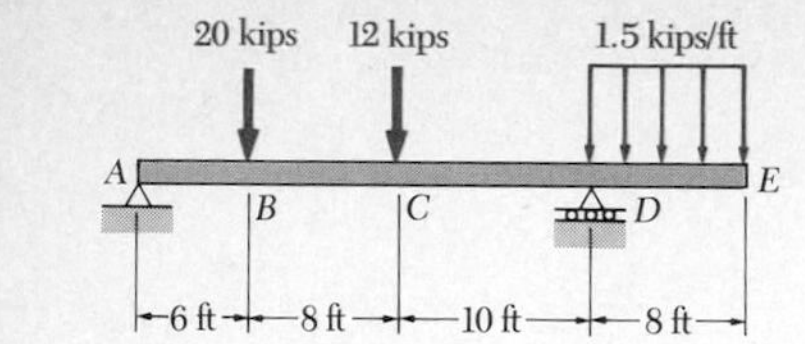

SAMPLE PROBLEM 7.4

Draw the shear and bending-moment diagrams for the beam and loading shown.

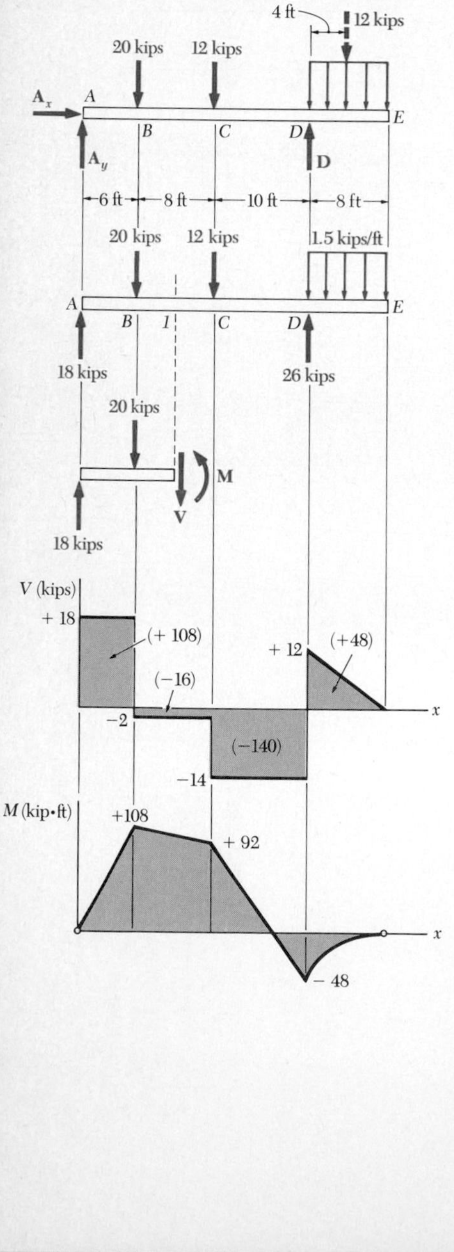

Solution. Considering the entire beam as a free body, we determine the reactions:

$+\Sigma \circlearrowleft M_A = 0$:

$$D(24\text{ ft}) - (20\text{ kips})(6\text{ ft}) - (12\text{ kips})(14\text{ ft}) - (12\text{ kips})(28\text{ ft}) = 0$$

$$D = +26\text{ kips} \qquad \mathbf{D} = 26\text{ kips}\uparrow$$

$+\uparrow\Sigma F_y = 0$: $\quad A_y - 20\text{ kips} - 12\text{ kips} + 26\text{ kips} - 12\text{ kips} = 0$

$$A_y = +18\text{ kips} \qquad \mathbf{A}_y = 18\text{ kips}\uparrow$$

$\overset{+}{\rightarrow}\Sigma F_x = 0$: $\quad A_x = 0 \qquad \mathbf{A}_x = 0$

We also note that at both A and E the bending moment is zero; thus two points (indicated by dots) are obtained on the bending-moment diagram.

Shear Diagram. Since $dV/dx = -w$, we find that between concentrated loads and reactions the slope of the shear diagram is zero (i.e., the shear is constant). The shear at any point is determined by dividing the beam into two parts and considering either part as a free body. For example, using the portion of beam to the left of section *1*, we obtain the shear between B and C:

$+\uparrow\Sigma F_y = 0$: $\quad +18\text{ kips} - 20\text{ kips} - V = 0 \qquad V = -2\text{ kips}$

We also find that the shear is $+12$ kips just to the right of D and zero at end E. Since the slope $dV/dx = -w$ is constant between D and E, the shear diagram between these two points is a straight line.

Bending-Moment Diagram. We recall that the area under the shear curve between two points is equal to the change in bending moment between the same two points. For convenience, the area of each portion of the shear diagram is computed and is indicated on the diagram. Since the bending moment M_A at the left end is known to be zero, we write

$$\begin{aligned} M_B - M_A &= +108 & M_B &= +108\text{ kip}\cdot\text{ft} \\ M_C - M_B &= -\ 16 & M_C &= +\ 92\text{ kip}\cdot\text{ft} \\ M_D - M_C &= -140 & M_D &= -\ 48\text{ kip}\cdot\text{ft} \\ M_E - M_D &= +\ 48 & M_E &= 0 \end{aligned}$$

Since M_E is known to be zero, a check of the computations is obtained.

Between the concentrated loads and reactions the shear is constant, thus the slope dM/dx is constant and the bending-moment diagram is drawn by connecting the known points with straight lines. Between D and E where the shear diagram is an oblique straight line, the bending-moment diagram is a parabola.

From the V and M diagrams we note that $V_{max} = 18$ kips and $M_{max} = 108$ kip·ft.

SAMPLE PROBLEM 7.5

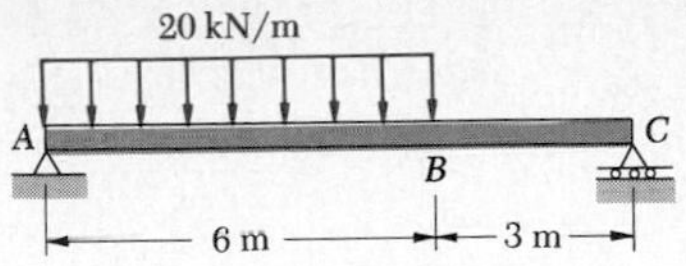

Draw the shear and bending-moment diagrams for the beam and loading shown.

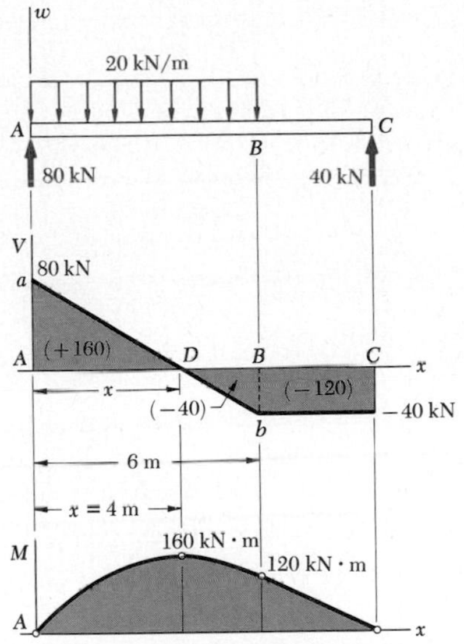

Solution. Considering the entire beam as a free body, we obtain the reactions

$$\mathbf{R}_A = 80\text{ kN}\uparrow \qquad \mathbf{R}_C = 40\text{ kN}\uparrow$$

Shear Diagram. The shear just to the right of A is $V_A = +80$ kN. Since the change in shear between two points is equal to *minus* the area under the load curve between the same two points, we obtain V_B by writing

$$V_B - V_A = -(20\text{ kN/m})(6\text{ m}) = -120\text{ kN}$$
$$V_B = -120 + V_A = -120 + 80 = -40\text{ kN}$$

The slope $dV/dx = -w$ being constant between A and B, the shear diagram between these two points is represented by a straight line. Between B and C, the area under the load curve is zero; therefore,

$$V_C - V_B = 0 \qquad V_C = V_B = -40\text{ kN}$$

and the shear is constant between B and C.

Bending-Moment Diagram. We note that the bending moment at each end of the beam is zero. In order to determine the maximum bending moment, we locate the section D of the beam where $V = 0$. Considering the portion of the shear diagram between A and B, we note that the triangles DAa and DBb are similar; thus,

$$\frac{x}{80\text{ kN}} = \frac{6 - x}{40\text{ kN}} \qquad x = 4\text{ m}$$

The maximum bending moment occurs at point D, where we have $dM/dx = V = 0$. The areas of the various portions of the shear diagram are computed and are given (in parentheses) on the diagram. Since the area of the shear diagram between two points is equal to the change in bending moment between the same two points, we write

$$M_D - M_A = +160\text{ kN}\cdot\text{m} \qquad M_D = +160\text{ kN}\cdot\text{m}$$
$$M_B - M_D = -\ 40\text{ kN}\cdot\text{m} \qquad M_B = +120\text{ kN}\cdot\text{m}$$
$$M_C - M_B = -120\text{ kN}\cdot\text{m} \qquad M_C = 0$$

The bending-moment diagram consists of an arc of parabola followed by a segment of straight line; the slope of the parabola at A is equal to the value of V at that point.

SAMPLE PROBLEM 7.6

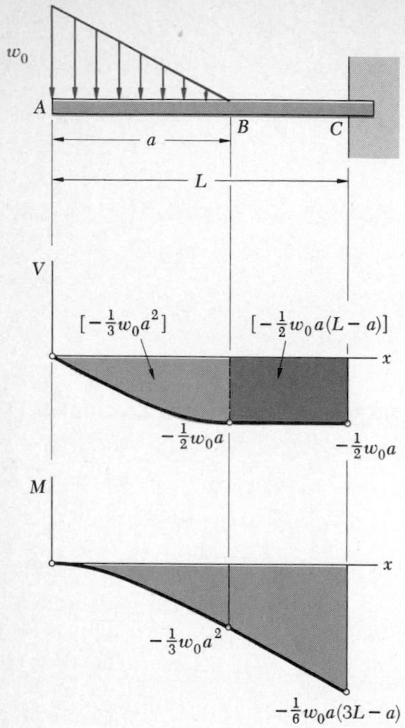

Sketch the shear and bending-moment diagrams for the cantilever beam shown.

Solution. ***Shear Diagram.*** At the free end of the beam, we find $V_A = 0$. Between A and B, the area under the load curve is $\frac{1}{2}w_0a$; we find V_B by writing

$$V_B - V_A = -\tfrac{1}{2}w_0a \qquad V_B = -\tfrac{1}{2}w_0a$$

Between B and C, the beam is not loaded; thus $V_C = V_B$. At A, we have $w = w_0$ and, according to Eq. (7.1), the slope of the shear curve is $dV/dx = -w_0$, while at B the slope is $dV/dx = 0$. Between A and B, the loading decreases linearly, and the shear diagram is parabolic. Between B and C, $w = 0$, and the shear diagram is a horizontal line.

Bending-Moment Diagram. The bending moment M_A at the free end of the beam is zero. We compute the area under the shear curve and write

$$M_B - M_A = -\tfrac{1}{3}w_0a^2 \qquad M_B = -\tfrac{1}{3}w_0a^2$$
$$M_C - M_B = -\tfrac{1}{2}w_0a(L-a)$$
$$M_C = -\tfrac{1}{6}w_0a(3L-a)$$

The sketch of the bending-moment diagram is completed by recalling that $dM/dx = V$. We find that between A and B the diagram is represented by a cubic curve with zero slope at A, and between B and C by a straight line.

SAMPLE PROBLEM 7.7

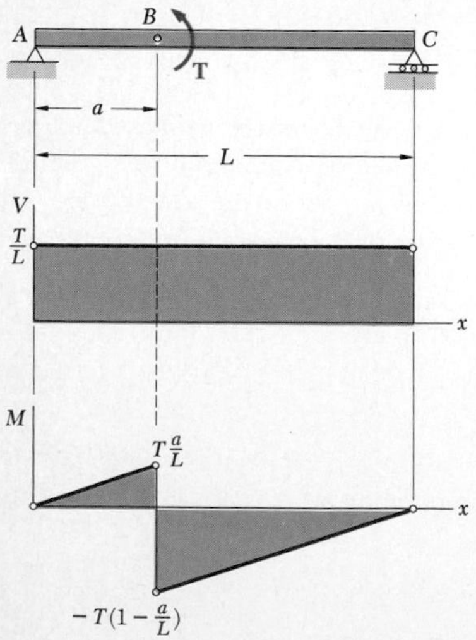

The simple beam AC is loaded by a couple of magnitude T applied at point B. Draw the shear and bending-moment diagrams of the beam.

Solution. The entire beam is taken as a free body, and we obtain

$$\mathbf{R}_A = \frac{T}{L}\uparrow \qquad \mathbf{R}_C = \frac{T}{L}\downarrow$$

The shear at any section is constant and equal to T/L. Since a couple is applied at B, the bending-moment diagram is discontinuous at B; the bending moment decreases suddenly by an amount equal to T.

Problems

7.45 Using the methods of Sec. 7.6, solve Prob. 7.21.
7.46 Using the methods of Sec. 7.6, solve Prob. 7.22.
7.47 Using the methods of Sec. 7.6, solve Prob. 7.25.
7.48 Using the methods of Sec. 7.6, solve Prob. 7.26.
7.49 Using the methods of Sec. 7.6, solve Prob. 7.27.
7.50 Using the methods of Sec. 7.6, solve Prob. 7.28.
7.51 Using the methods of Sec. 7.6, solve Prob. 7.29.
7.52 Using the methods of Sec. 7.6, solve Prob. 7.30.
7.53 Using the methods of Sec. 7.6, solve Prob. 7.33.
7.54 Using the methods of Sec. 7.6, solve Prob. 7.34.
7.55 Using the methods of Sec. 7.6, solve Prob. 7.35.
7.56 Using the methods of Sec. 7.6, solve Prob. 7.38.

7.57 through 7.60 Draw the shear and bending-moment diagrams for the beam and loading shown.

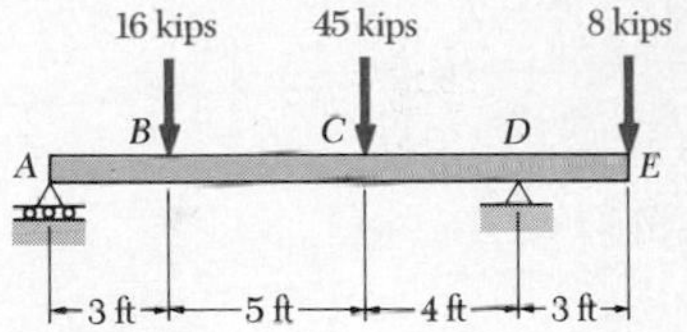

Fig. P7.57

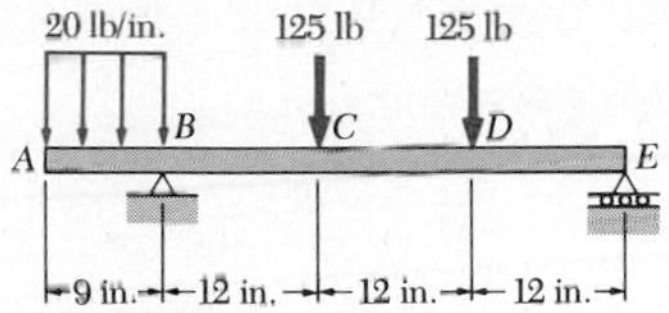

Fig. P7.58

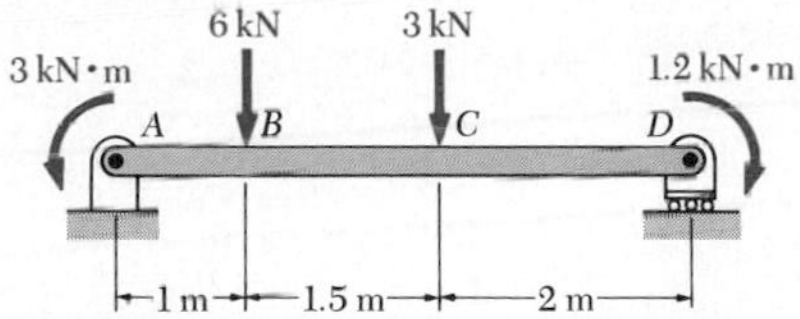

Fig. P7.59

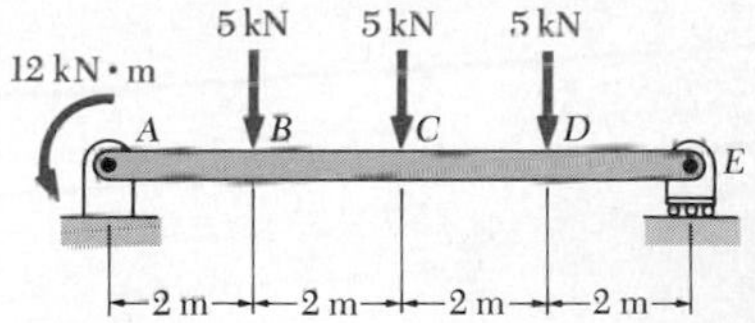

Fig. P7.60

7.61 through 7.66 Draw the shear and bending-moment diagrams for the beam and loading shown, and determine the location and magnitude of the maximum bending moment.

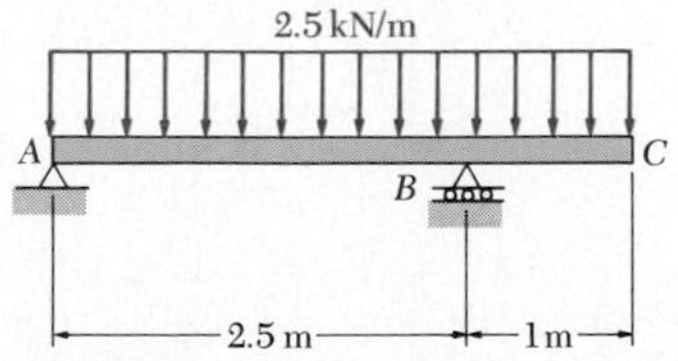

Fig. P7.61

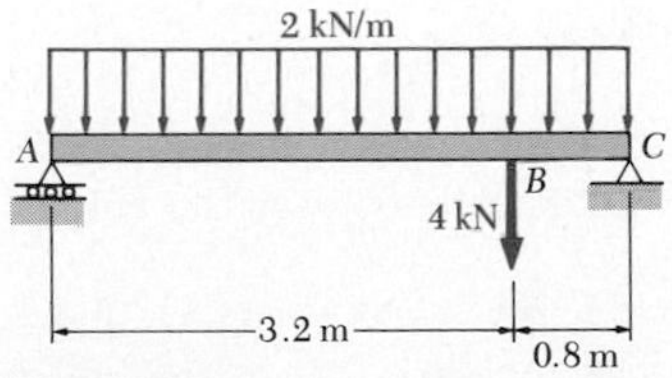

Fig. P7.62

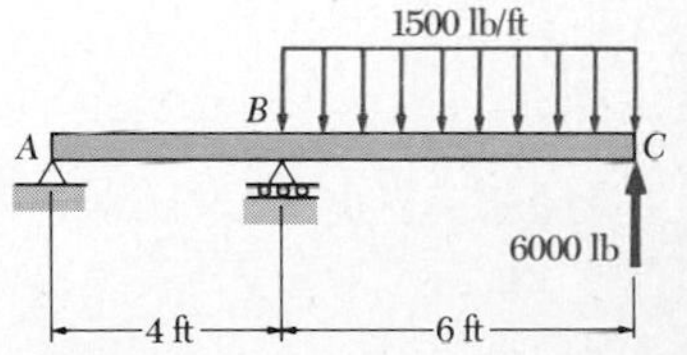

Fig. P7.63

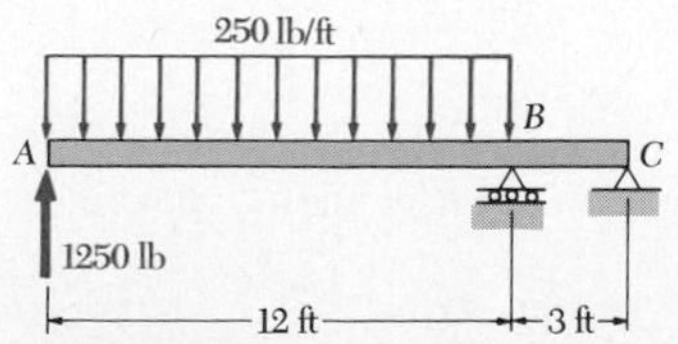

Fig. P7.64

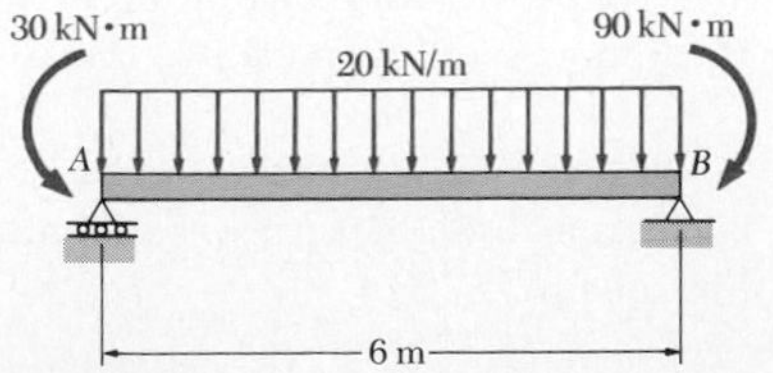

Fig. P7.65

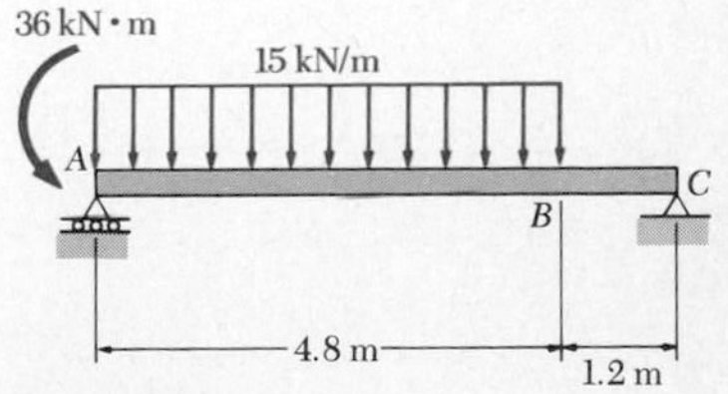

Fig. 7.66

7.67 Determine the equations of the shear and bending-moment curves for the beam and loading of Prob. 7.25. (Place the origin at point A.)

7.68 through 7.70 Determine the equations of the shear and bending-moment curves for the given beam and loading. Also determine the magnitude and location of the maximum bending moment in the beam.

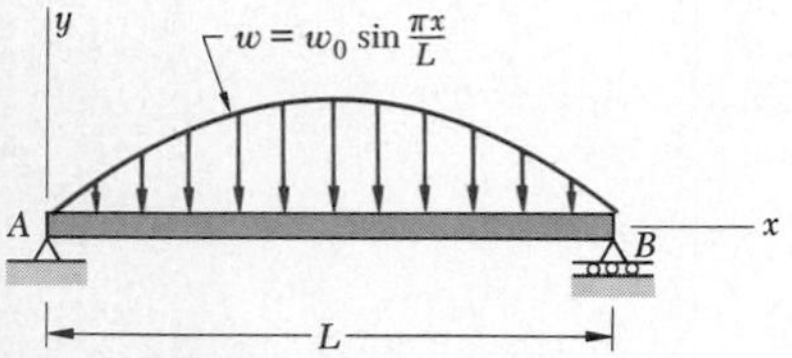

Fig. P7.68

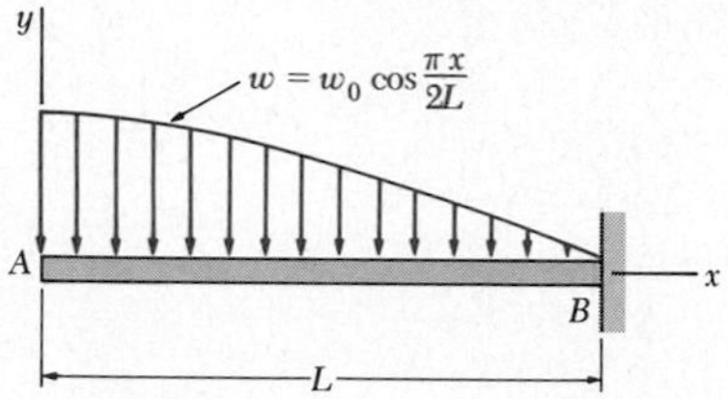

Fig. P7.69

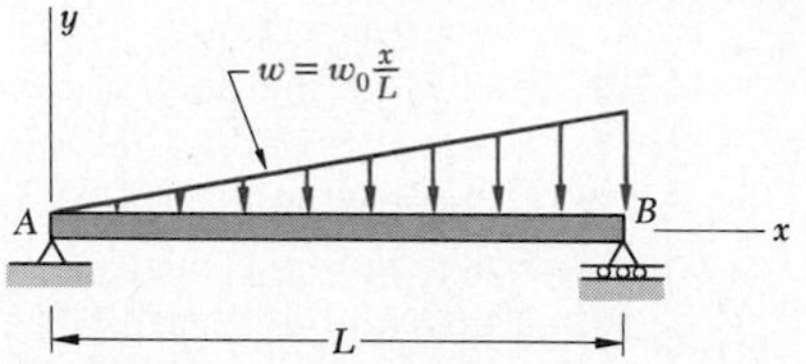

Fig. P7.70

***7.71** The beam AB is acted upon by the uniformly distributed load of 1.5 kN/m and by two forces **P** and **Q**. It has been experimentally determined that the bending moment is +360 N·m at point D and +960 N·m at point E. Determine **P** and **Q**, and draw the shear and bending-moment diagrams for the beam.

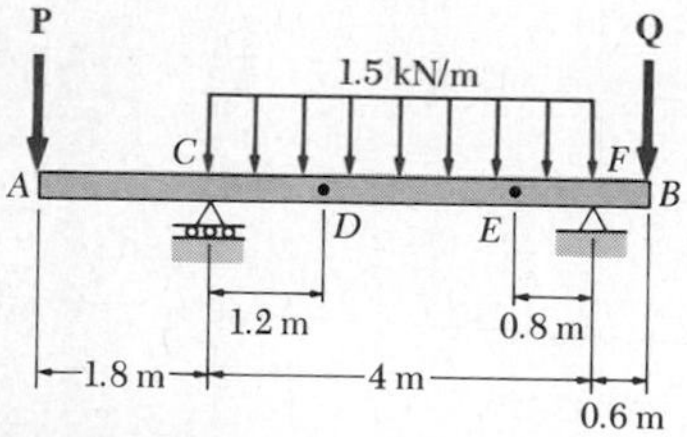

Fig. P7.71

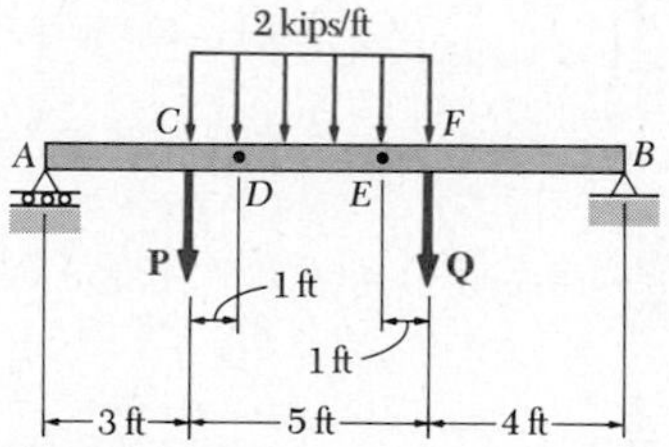

Fig. P7.72

***7.72** The beam AB is acted upon by the uniformly distributed load of 2 kips/ft and by two forces **P** and **Q**. It has been experimentally determined that the bending moment is +172 kip·ft at point D and +235 kip·ft at point E. Determine **P** and **Q**, and draw the shear and bending-moment diagrams for the beam.

***7.73** Solve Prob. 7.72, assuming that the uniformly distributed load of 2 kips/ft extends over the entire beam AB.

***7.74** Solve Prob. 7.71, assuming that the uniformly distributed load of 1.5 kN/m extends over the entire beam AB.

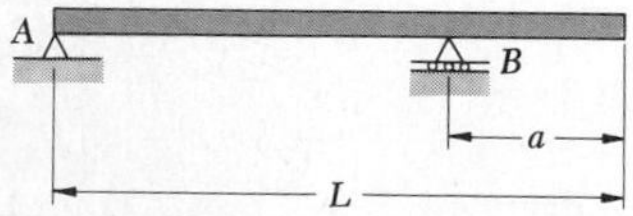

Fig. P7.75

***7.75** A uniform beam of weight w per unit length is supported as shown. Determine the distance a if the maximum absolute value of the bending moment is to be as small as possible. (See hint of Prob. 7.43.)

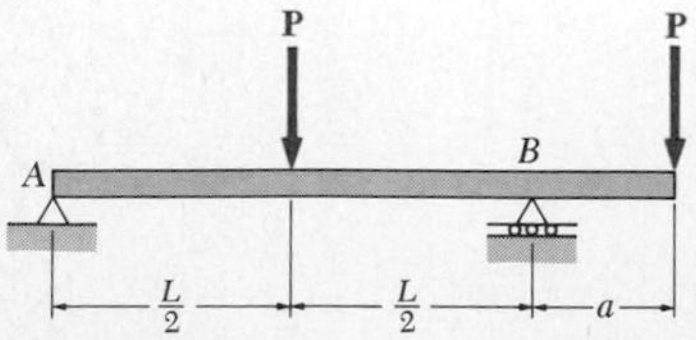

Fig. P7.76

***7.76** For the beam and loading shown, determine the distance a if the maximum absolute value of the bending moment is to be as small as possible. (See hint of Prob. 7.31.)

*** 7.7. Cables with Concentrated Loads.** Cables are used in many engineering applications, such as suspension bridges, transmission lines, aerial tramways, guy wires for high towers, etc. Cables may be divided into two categories, according to their loading: (1) cables supporting concentrated loads, (2) cables supporting distributed loads. In this section, we shall examine cables of the first category.

Consider a cable attached to two fixed points A and B and supporting n given vertical concentrated loads $\mathbf{P}_1, \mathbf{P}_2, \ldots, \mathbf{P}_n$ (Fig. 7.13*a*). We assume that the cable is *flexible*, i.e., that its resistance to bending is small and may be neglected. We further assume that the *weight of the cable is negligible* compared with the loads supported by the cable. Any portion of cable between successive loads may therefore be considered as a two-force member, and the internal forces at any point in the cable reduce to a *force of tension directed along the cable*.

We assume that each of the loads lies in a given vertical line, i.e., that the horizontal distance from support A to each of the loads is known; we also assume that the horizontal and vertical distances between the supports are known. We propose to determine the shape of the cable, i.e., the vertical distance from support A to each of the points $C_1, C_2, \ldots, C_n$, and also the tension T in each portion of the cable.

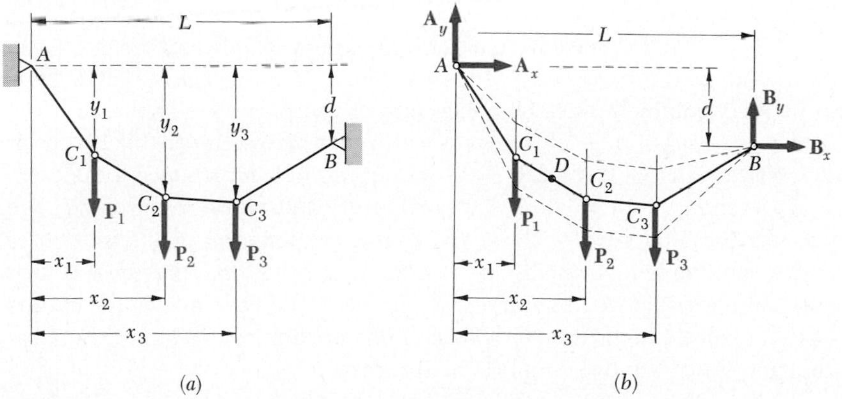

Fig. 7.13

We first draw the free-body diagram of the entire cable (Fig. 7.13*b*). Since the slope of the portions of cable attached at A and B is not known, the reactions at A and B must be represented by two components each. Thus, four unknowns are involved, and the three equations of equilibrium are not sufficient to determine the reactions at A and B.† We must therefore obtain an additional equation by considering the equilibrium of a por-

† Clearly, the cable is not a rigid body; the equilibrium equations represent, therefore, *necessary but not sufficient conditions* (see Sec. 6.11).

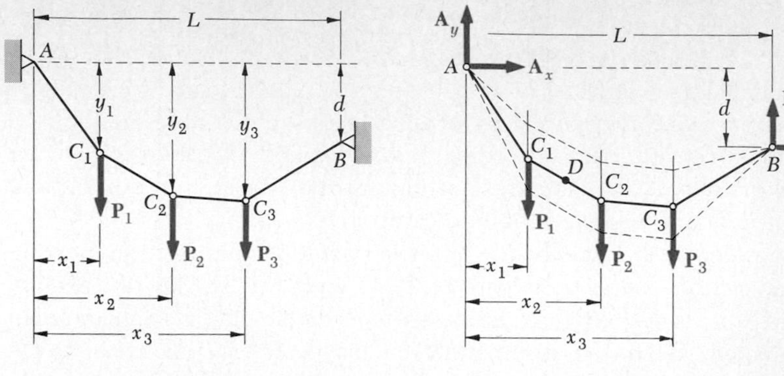

(a) (b)

Fig. 7.13 (*repeated*)

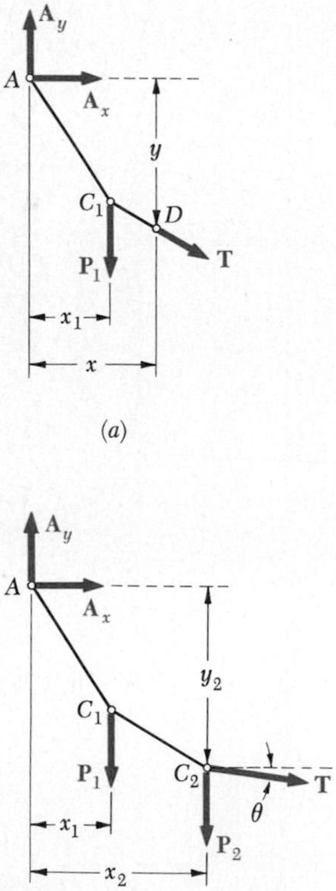

(b)

Fig. 7.14

tion of the cable. This is possible if we know the coordinates x and y of a point D of the cable. Drawing the free-body diagram of the portion of cable AD (Fig. 7.14*a*) and writing $\Sigma M_D = 0$, we obtain an additional relation between the scalar components A_x and A_y and may determine the reactions at A and B. The problem would remain indeterminate, however, if we did not know the coordinates of D, unless some other relation between A_x and A_y (or between B_x and B_y) were given. The cable might hang in any of various possible ways, as indicated by the dashed lines in Fig. 7.13*b*.

Once A_x and A_y have been determined, the vertical distance from A to any point of the cable may be easily found. Considering point C_2, for example, we draw the free-body diagram of the portion of cable AC_2 (Fig. 7.14*b*). Writing $\Sigma M_{C_2} = 0$, we obtain an equation which may be solved for y_2. Writing $\Sigma F_x = 0$ and $\Sigma F_y = 0$, we obtain the components of the force **T** representing the tension in the portion of cable to the right of C_2. We observe that $T \cos \theta = -A_x$; *the horizontal component of the tension force is the same at any point of the cable.* It follows that the tension T is maximum when $\cos \theta$ is minimum, i.e., in the portion of cable which has the largest angle of inclination θ. Clearly, this portion of cable must be adjacent to one of the two supports of the cable.

***7.8. Cables with Distributed Loads.** Consider a cable attached to two fixed points A and B and carrying a *distributed load* (Fig. 7.15*a*). We saw in the preceding section that for a cable supporting concentrated loads, the internal force at any point is a force of tension directed along the cable. In the case of a cable carrying a distributed load, the cable hangs in the shape of a curve, and the internal force at a point D is a force of tension **T** *directed along the tangent to the curve.* Given a certain distributed load, we propose in this section to determine the tension at any point of the cable. We shall also see in the following sections how the shape of the cable may be determined for two particular types of distributed loads.

Considering the most general case of distributed load, we draw the free-body diagram of the portion of cable extending from the lowest point

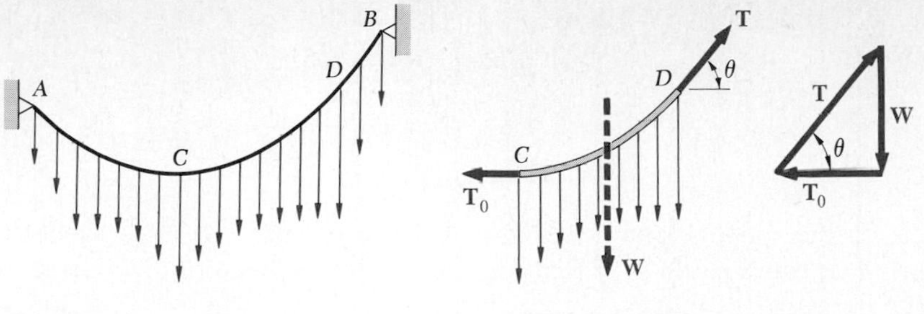

Fig. 7.15 (a) (b) (c)

C to a given point D of the cable (Fig. 7.15b). The forces acting on the free body are the tension force $\mathbf{T}_0$ at C, which is horizontal, the tension force **T** at D, directed along the tangent to the cable at D, and the resultant **W** of the distributed load supported by the portion of cable CD. Drawing the corresponding force triangle (Fig. 7.15c), we obtain the following relations:

$$T\cos\theta = T_0 \qquad T\sin\theta = W \tag{7.5}$$

$$T = \sqrt{T_0^2 + W^2} \qquad \tan\theta = \frac{W}{T_0} \tag{7.6}$$

From the relations (7.5), it appears that the horizontal component of the tension force **T** is the same at any point and that the vertical component of **T** is equal to the magnitude W of the load measured from the lowest point. Relations (7.6) show that the tension T is minimum at the lowest point and maximum at one of the two points of support.

*** 7.9. Parabolic Cable.** Let us assume, now, that the cable AB carries a load *uniformly distributed along the horizontal* (Fig. 7.16a). Cables of suspension bridges may be assumed loaded in this way, since the weight of the cables is small compared with the weight of the roadway. We denote by w the load per unit length (*measured horizontally*) and express it in N/m or in lb/ft. Choosing coordinate axes with origin at the lowest point C of the cable, we find that the magnitude W of the total load carried by the portion of cable extending from C to the point D of coordinates x and y is $W = wx$. The relations (7.6) defining the magnitude and direction of the tension force at D become

$$T = \sqrt{T_0^2 + w^2x^2} \qquad \tan\theta = \frac{wx}{T_0} \tag{7.7}$$

Moreover, the distance from D to the line of action of the resultant **W** is equal to half the horizontal distance from C to D (Fig. 7.16b). Summing moments about D, we write

$$+\circlearrowleft \Sigma M_D = 0: \qquad wx\frac{x}{2} - T_0 y = 0$$

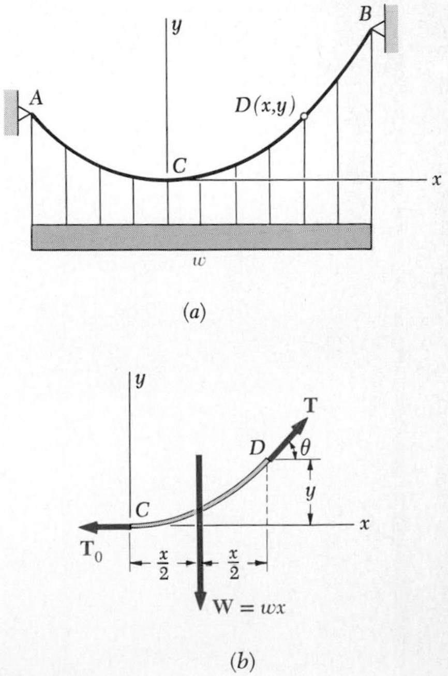

Fig. 7.16

and, solving for y,

$$y = \frac{wx^2}{2T_0} \tag{7.8}$$

This is the equation of a *parabola* with a vertical axis and its vertex at the origin of coordinates. The curve formed by cables loaded uniformly along the horizontal is thus a parabola.†

When the supports A and B of the cable have the same elevation, the distance L between the supports is called the *span* of the cable and the vertical distance h from the supports to the lowest point is called the *sag* of the cable (Fig. 7.17*a*). If the span and sag of a cable are known, and if the load w per unit horizontal length is given, the minimum tension T_0 may be found by substituting $x = L/2$ and $y = h$ in formula (7.8). Formulas (7.7) and (7.8) will then define the tension at any point and the shape of the cable.

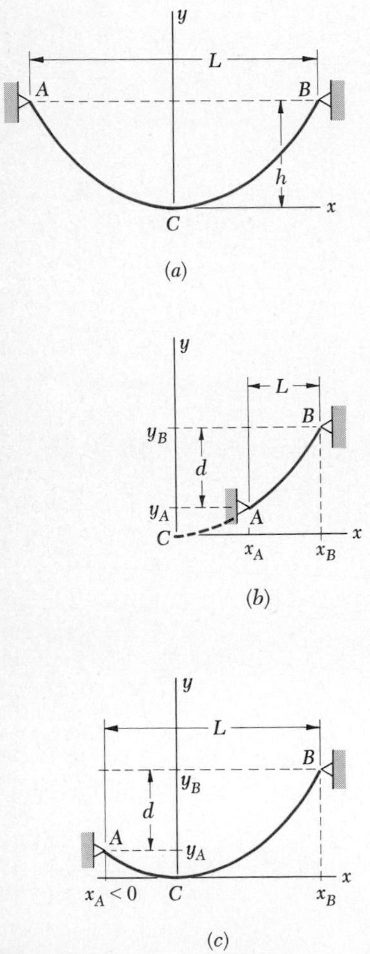

Fig. 7.17

When the supports have different elevations, the position of the lowest point of the cable is not known and the coordinates x_A, y_A and x_B, y_B of the supports must be determined. To this effect, we express that the coordinates of A and B satisfy Eq. (7.8) and that $x_B - x_A = L$, $y_B - y_A = d$, where L and d denote, respectively, the horizontal and vertical distances between the two supports (Fig. 7.17*b* and *c*).

The length of the cable from its lowest point C to its support B may be obtained from the formula

$$s_B = \int_0^{x_B} \sqrt{1 + \left(\frac{dy}{dx}\right)^2}\, dx \tag{7.9}$$

Differentiating (7.8), we obtain the derivative $dy/dx = wx/T_0$; substituting into (7.9) and using the binomial theorem to expand the radical in an infinite series, we have

$$s_B = \int_0^{x_B} \sqrt{1 + \frac{w^2x^2}{T_0^2}}\, dx = \int_0^{x_B} \left(1 + \frac{w^2x^2}{2T_0^2} - \frac{w^4x^4}{8T_0^4} + \cdots\right) dx$$

$$s_B = x_B\left(1 + \frac{w^2x_B^2}{6T_0^2} - \frac{w^4x_B^4}{40T_0^4} + \cdots\right)$$

and, since $wx_B^2/2T_0 = y_B$,

$$s_B = x_B\left[1 + \frac{2}{3}\left(\frac{y_B}{x_B}\right)^2 - \frac{2}{5}\left(\frac{y_B}{x_B}\right)^4 + \cdots\right] \tag{7.10}$$

The series converges for values of the ratio y_B/x_B less than 0.5; in most cases, this ratio is much smaller, and only the first two terms of the series need be computed.

† Cables hanging under their own weight are not loaded uniformly along the horizontal, and they do not form a parabola. The error introduced by assuming a parabolic shape for cables hanging under their own weight, however, is small when the cable is sufficiently taut. A complete discussion of cables hanging under their own weight is given in the next section.

SAMPLE PROBLEM 7.8

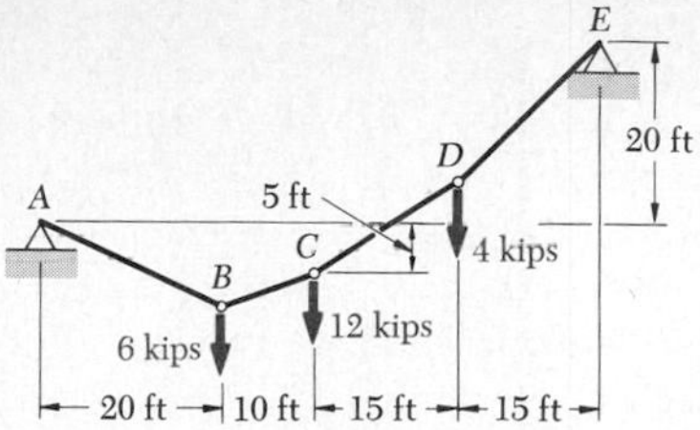

The cable AE supports three vertical loads from the points indicated. If point C is 5 ft below the left support, determine (*a*) the elevations of points B and D, (*b*) the maximum slope and the maximum tension in the cable.

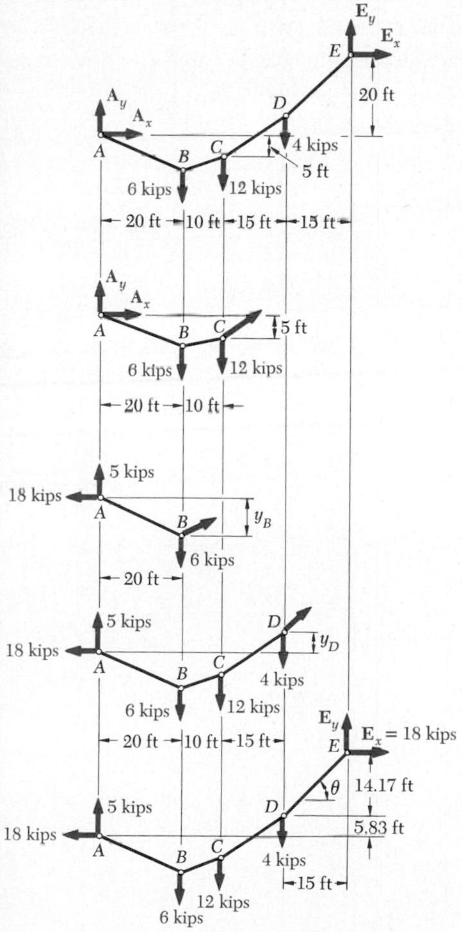

Solution. The reaction components $\mathbf{A}_x$ and $\mathbf{A}_y$ are determined as follows:

Free Body: Entire Cable

$+\circlearrowleft \Sigma M_E = 0:$

$$A_x(20 \text{ ft}) - A_y(60 \text{ ft}) + (6 \text{ kips})(40 \text{ ft}) + (12 \text{ kips})(30 \text{ ft}) + (4 \text{ kips})(15 \text{ ft}) = 0$$

$$20A_x - 60A_y + 660 = 0$$

Free Body: ABC

$$+\circlearrowleft \Sigma M_C = 0: \quad -A_x(5 \text{ ft}) - A_y(30 \text{ ft}) + (6 \text{ kips})(10 \text{ ft}) = 0$$

$$-5A_x - 30A_y + 60 = 0$$

Solving the two equations simultaneously, we obtain

$$A_x = -18 \text{ kips} \qquad \mathbf{A}_x = 18 \text{ kips} \leftarrow$$

$$A_y = +5 \text{ kips} \qquad \mathbf{A}_y = 5 \text{ kips} \uparrow$$

a. Elevation of Point B. Considering the portion of cable AB *as a free body,* we write

$$+\circlearrowleft \Sigma M_B = 0: \quad (18 \text{ kips})y_B - (5 \text{ kips})(20 \text{ ft}) = 0$$

$$y_B = 5.56 \text{ ft below } A \quad ◀$$

Elevation of Point D. Using the portion of cable $ABCD$ *as a free body,* we write

$+\circlearrowleft \Sigma M_D = 0:$

$$-(18 \text{ kips})y_D - (5 \text{ kips})(45 \text{ ft}) + (6 \text{ kips})(25 \text{ ft}) + (12 \text{ kips})(15 \text{ ft}) = 0$$

$$y_D = 5.83 \text{ ft above } A \quad ◀$$

b. Maximum Slope and Maximum Tension. We observe that the maximum slope occurs in portion DE. Since the horizontal component of the tension is constant and equal to 18 kips, we write

$$\tan\theta = \frac{14.17}{15 \text{ ft}} \qquad \theta = 43.4° \quad ◀$$

$$T_{max} = \frac{18 \text{ kips}}{\cos\theta} \qquad T_{max} = 24.8 \text{ kips} \quad ◀$$

SAMPLE PROBLEM 7.9

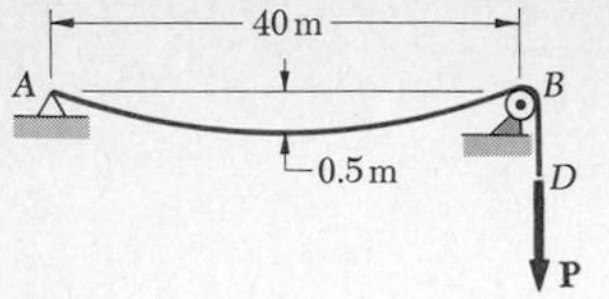

A light cable is attached to a support at A, passes over a small pulley at B, and supports a load $\mathbf{P}$. Knowing that the sag of the cable is 0.5 m and that the mass per unit length of the cable is 0.75 kg/m, determine (*a*) the magnitude of the load $\mathbf{P}$, (*b*) the slope of the cable at B, and (*c*) the total length of the cable from A to B. Since the ratio of the sag to the span is small, assume the cable to be parabolic. Also, neglect the weight of the portion of cable from B to D.

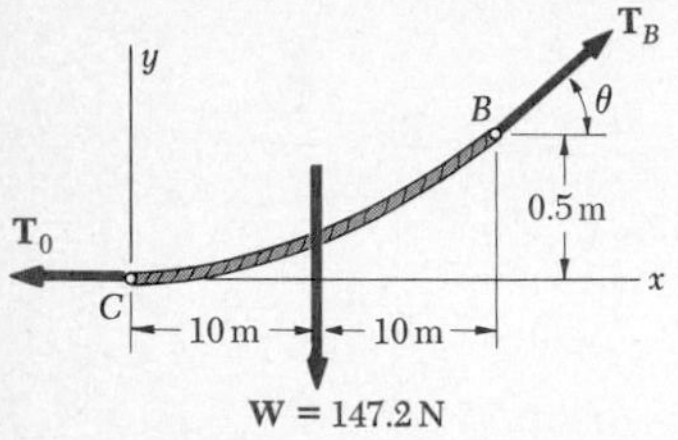

***a.* Load P.** We denote by C the lowest point of the cable and draw the free-body diagram of the portion CB of cable. Assuming the load to be uniformly distributed along the horizontal, we write

$$w = (0.75 \text{ kg/m})(9.81 \text{ m/s}^2) = 7.36 \text{ N/m}$$

The total load for the portion CB of the cable is

$$W = wx_B = (7.36 \text{ N/m})(20 \text{ m}) = 147.2 \text{ N}$$

and is applied halfway between C and B. Summing moments about B, we write

$$+\circlearrowleft \Sigma M_B = 0: \qquad (147.2 \text{ N})(10 \text{ m}) - T_0(0.5 \text{ m}) = 0 \qquad T_0 = 2944 \text{ N}$$

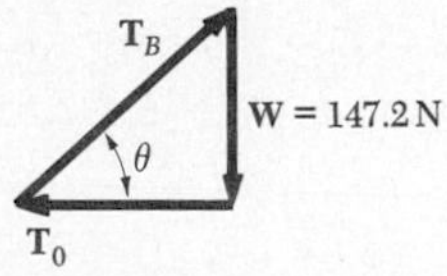

From the force triangle we obtain

$$T_B = \sqrt{T_0^2 + W^2}$$
$$= \sqrt{(2944 \text{ N})^2 + (147.2 \text{ N})^2} = 2948 \text{ N}$$

Since the tension on each side of the pulley is the same, we find

$$P = T_B = 2948 \text{ N} \quad ◀$$

***b.* Slope of Cable at *B*.** We also obtain from the force triangle

$$\tan\theta = \frac{W}{T_0} = \frac{147.2 \text{ N}}{2944 \text{ N}} = 0.05$$

$$\theta = 2.9° \quad ◀$$

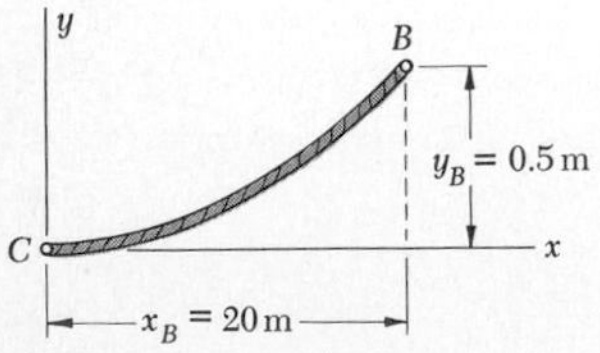

***c.* Length of Cable.** Applying Eq. (7.10) between C and B, we write

$$s_B = x_B\left[1 + \frac{2}{3}\left(\frac{y_B}{x_B}\right)^2 + \cdots\right]$$
$$= (20 \text{ m})\left[1 + \frac{2}{3}\left(\frac{0.5 \text{ m}}{20 \text{ m}}\right)^2 + \cdots\right] = 20.00833 \text{ m}$$

The total length of the cable between A and B is twice this value,

$$\text{Length} = 2s_B = 40.0167 \text{ m} \quad ◀$$

Problems

7.77 Three loads are suspended as shown from the cable. Knowing that $h_C = 3$ m, determine (*a*) the components of the reaction at *E*, (*b*) the maximum value of the tension in the cable.

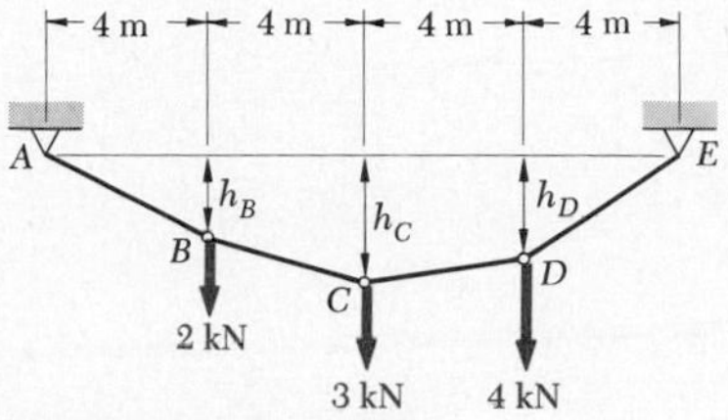

Fig. P7.77 and P7.78

7.78 Determine the sag at point *C* if the maximum tension in the cable is 13 kN.

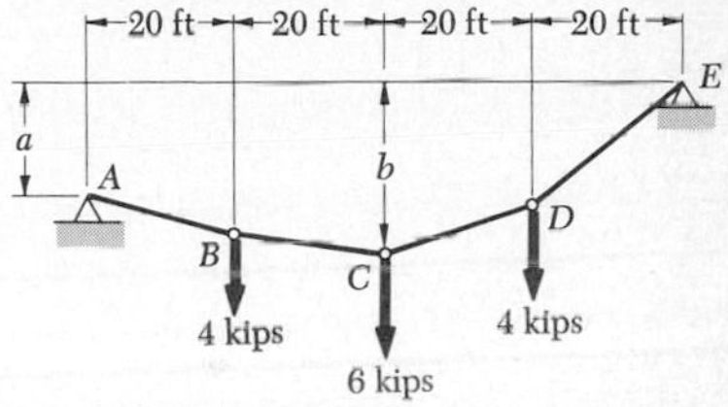

Fig. P7.79 and P7.80

7.79 If $a = 16$ ft and $b = 18$ ft, determine the components of the reaction at *E* for the cable and loading shown.

7.80 If $a = b = 16$ ft, determine (*a*) the components of the reaction at *E*, (*b*) the maximum tension in the cable.

7.81 If $h_C = 7$ m, determine (*a*) the components of the reaction at *E*, (*b*) the maximum tension in the cable.

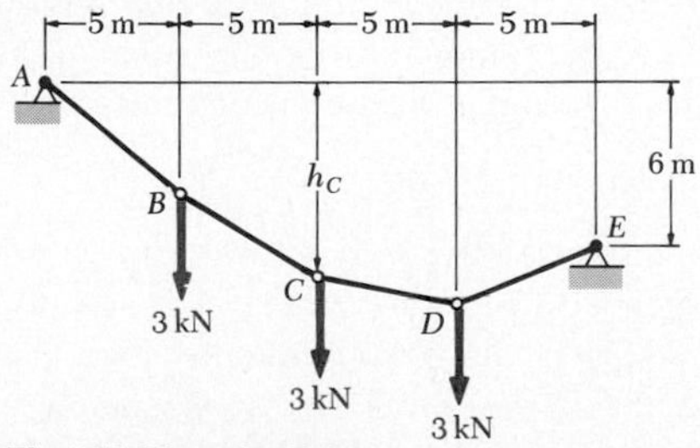

Fig. P7.81 and P7.82

7.82 Determine (*a*) the distance h_C for which the portion *CD* of the cable is horizontal, (*b*) the corresponding components of the reaction at *E*.

7.83 Cable *ABC* supports two loads as shown. Knowing that the distance *b* is to be 7 m, determine (*a*) the required magnitude of the force **P**, (*b*) the corresponding distance *a*.

7.84 Cable *ABC* supports two loads as shown. Determine the distances *a* and *b* when a horizontal force **P** of magnitude 1000 N is applied at *C*.

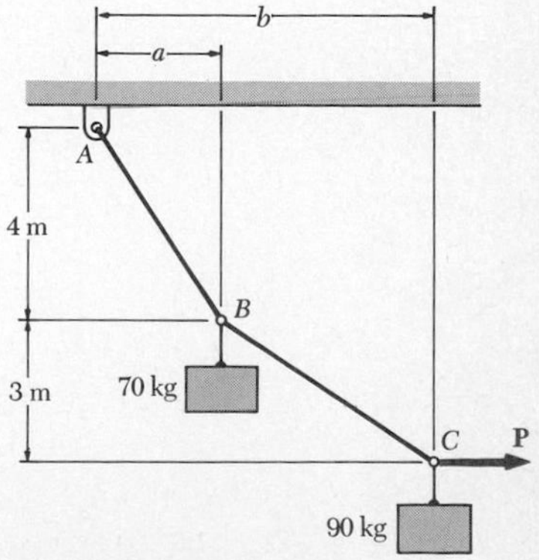

Fig. P7.83 and P7.84

7.85 A 400-lb load is attached to a small pulley which may roll on the cable *ACB*. The pulley and load are held in the position shown by a second cable *DE* which is parallel to the portion *CB* of the main cable. Determine (*a*) the reactions at *A* and *B*, (*b*) the tension in cable *ACB*, (*c*) the tension in cable *DE*. Neglect the radius of the pulleys and the weight of the cables.

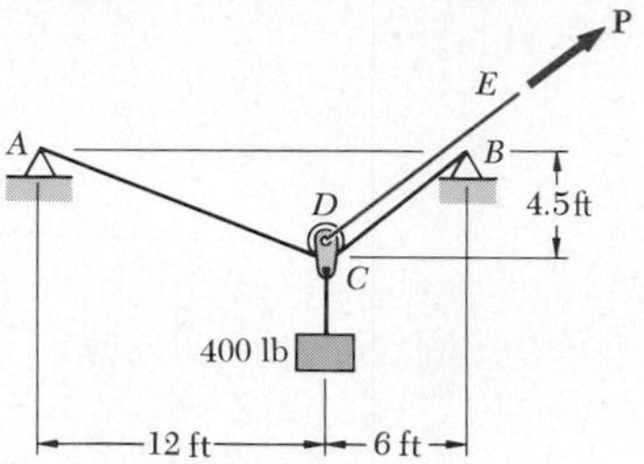

Fig. P7.85

7.86 Knowing that the 440-lb force applied at *C* and the block attached at *B* maintain the cable *ABCD* in the position shown, determine (*a*) the reaction at *D*, (*b*) the required weight *W* of the block, (*c*) the tension in each portion of the cable.

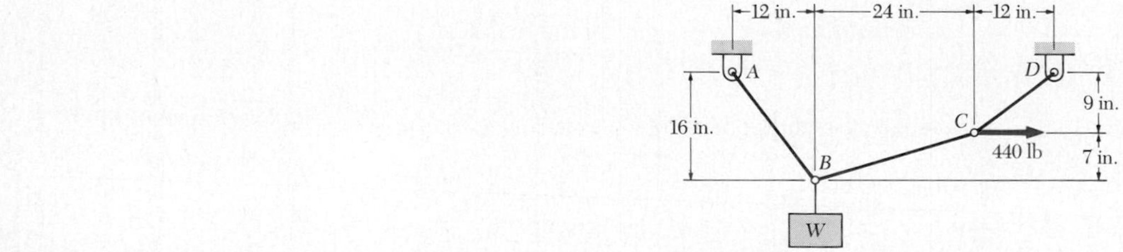

Fig. P7.86

7.87 An electric wire having a mass per unit length of 0.4 kg/m is strung between two insulators at the same elevation and 40 m apart. Knowing that the sag of the wire is 1.2 m, determine (*a*) the maximum tension in the wire, (*b*) the length of the wire.

7.88 The center span of the Verrazano-Narrows Bridge consists of two uniform roadways suspended from four cables. The uniform load supported by each cable is $w = 10.8$ kips/ft along the horizontal. Knowing that the span L is 4260 ft and the sag h is 390 ft, determine (*a*) the maximum tension in each cable. (*b*) the length of each cable.

7.89 The design of the Verrazano-Narrows Bridge included the effect of extreme temperature changes which cause the sag of the center span to vary from $h_w = 386$ ft in winter to $h_s = 394$ ft in summer. Knowing that the span is $L =$ 4260 ft, determine the change in length of the cables due to extreme temperature variation.

7.90 A 61-m length of wire having a mass per unit length of 1.8 kg/m is used to span a horizontal distance of 60 m. Determine (*a*) the approximate sag of the wire, (*b*) the maximum tension in the wire. [*Hint.* Use only the first two terms of Eq. (7.10).]

7.91 The total mass of cable ACB is 10 kg. Assuming that the mass of the cable is distributed uniformly along the horizontal, determine (*a*) the sag h, (*b*) the slope of the cable at A.

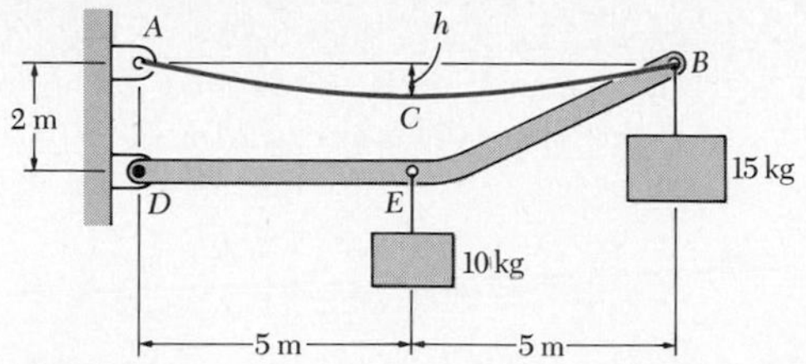

Fig. P7.91

7.92 Solve Prob. 7.91, assuming that the 10-kg load attached at E is removed.

7.93 Cable AB supports a load distributed uniformly along the horizontal as shown. The lowest point of the cable is located at a distance a = 12 ft below the support A. Determine the maximum and minimum values of the tension in the cable.

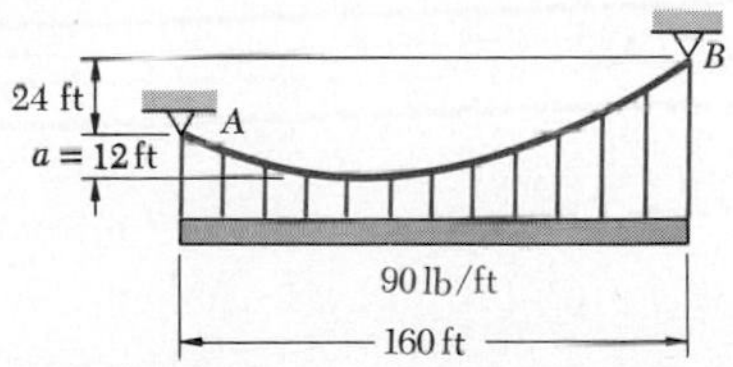

Fig. P7.93

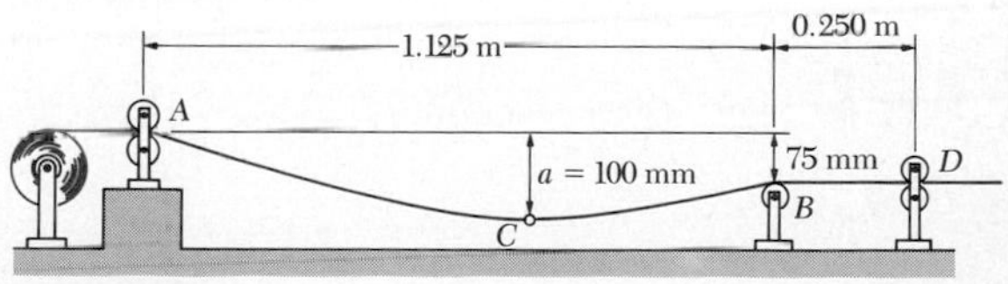

Fig. P7.94

7.94 Before being fed into a printing press located to the right of D, a continuous sheet of paper having a mass per unit length of 300 g/m passes over rollers at A and B. Assuming that the curve formed by the sheet is parabolic, determine (*a*) the location of the lowest point C, (*b*) the maximum tension in the sheet.

7.95 Solve Prob. 7.94, assuming that a = 90 mm.

7.96 Solve Prob. 7.93, assuming that a = 9 ft.

***7.97** A cable AB of span L and a simple beam $A'B'$ of the same span are subjected to identical vertical loadings as shown. Show that the magnitude of the bending moment at a point C' in the beam is equal to the product of $T_0 h$, where T_0 is the magnitude of the horizontal component of the tension force in the cable and h is the vertical distance between point C of the cable and the chord joining the points of support A and B.

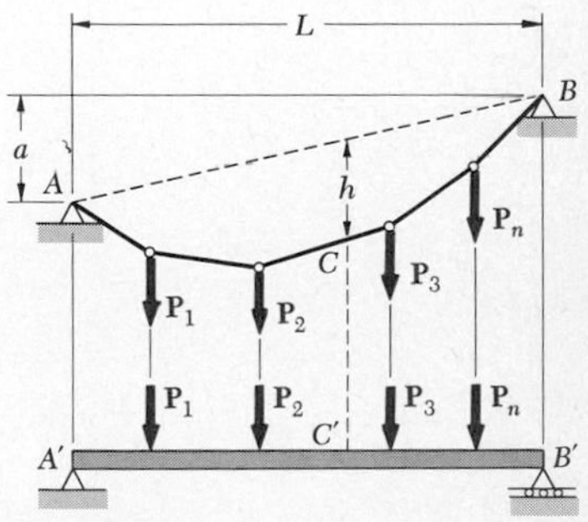

Fig. P7.97

***7.98** Show that the curve assumed by a cable which carries a distributed load $w(x)$ is defined by the differential equation $d^2y/dx^2 = w(x)/T_0$, where T_0 is the tension at the lowest point.

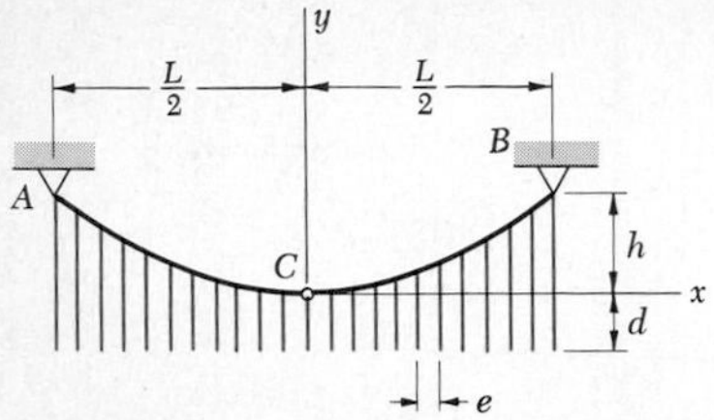

Fig. P7.100

*** 7.99** Using the property indicated in Prob. 7.98, determine the curve assumed by a cable of span L and sag h carrying a distributed load per unit length $w = w_0 \cos(\pi x/L)$, where x is measured from mid-span. Also determine the maximum and minimum values of the tension.

*** 7.100** A large number of ropes are tied to a light wire which is suspended from two points A and B at the same level. Show that, if the lower ends of the ropes have been cut so that they lie in the same horizontal line and if the ropes are kept uniformly spaced, the curve assumed by the wire ACB is

$$y + d = d \cosh \left(\frac{w_r}{T_0 e}\right)^{1/2} x$$

where w_r is the weight per unit length of the ropes, d is the length of the shortest rope, and e is the horizontal distance between adjacent ropes. (*Hint.* Use the property indicated in Prob. 7.98.)

* 7.10. Catenary.

We shall consider now a cable AB carrying a load *uniformly distributed along the cable itself* (Fig. 7.18*a*). Cables hanging under their own weight are loaded in this way. We denote by w the load per unit length (*measured along the cable*) and express it in N/m or in lb/ft. The magnitude W of the total load carried by a portion of cable of length s extending from the lowest point C to a point D is $W = ws$. Substituting this value for W in formula (7.6), we obtain the tension at D:

$$T = \sqrt{T_0^2 + w^2 s^2}$$

In order to simplify the subsequent computations, we shall introduce the constant $c = T_0/w$. We thus write

$$T_0 = wc \qquad W = ws \qquad T = w\sqrt{c^2 + s^2} \tag{7.11}$$

The free-body diagram of the portion of cable CD is shown in Fig. 7.18*b*. This diagram, however, cannot be used to obtain directly the equation of the curve assumed by the cable, since we do not know the horizontal distance from D to the line of action of the resultant $\mathbf{W}$ of the load. To obtain this equation, we shall write first that the horizontal projection of a

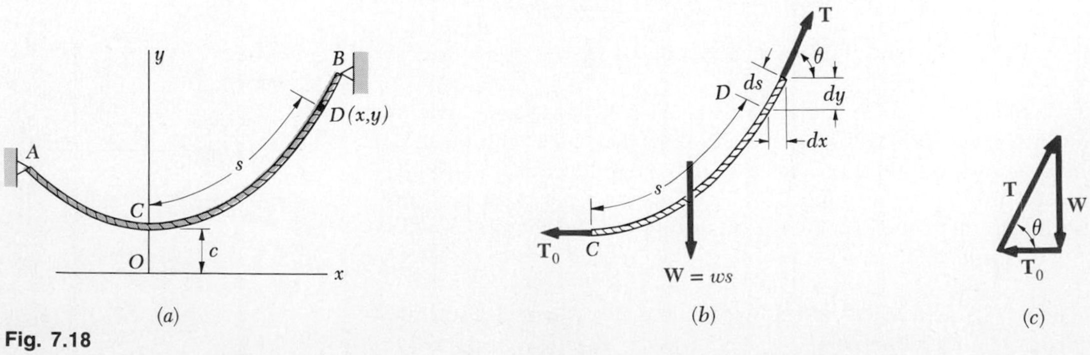

Fig. 7.18

small element of cable of length ds is $dx = ds \cos \theta$. Observing from Fig. 7.18c that $\cos \theta = T_0/T$ and using (7.11), we write

$$dx = ds \cos \theta = \frac{T_0}{T} ds = \frac{wc\, ds}{w\sqrt{c^2 + s^2}} = \frac{ds}{\sqrt{1 + s^2/c^2}}$$

Selecting the origin O of the coordinates at a distance c directly below C (Fig. 7.18a) and integrating from $C(0, c)$ to $D(x, y)$, we obtain†

$$x = \int_0^s \frac{ds}{\sqrt{1 + s^2/c^2}} = c\left[\sinh^{-1}\frac{s}{c}\right]_0^s = c \sinh^{-1}\frac{s}{c}$$

This equation, which relates the length s of the portion of cable CD and the horizontal distance x, may be written in the form

$$s = c \sinh \frac{x}{c} \tag{7.15}$$

The relation between the coordinates x and y may now be obtained by writing $dy = dx \tan \theta$. Observing from Fig. 7.18c that $\tan \theta = W/T_0$ and using (7.11) and (7.15), we write

$$dy = dx \tan \theta = \frac{W}{T_0} dx = \frac{s}{c} dx = \sinh \frac{x}{c} dx$$

Integrating from $C(0, c)$ to $D(x, y)$ and using (7.12) and (7.13), we obtain

$$y - c = \int_0^x \sinh \frac{x}{c} dx = c\left[\cosh \frac{x}{c}\right]_0^x = c\left(\cosh \frac{x}{c} - 1\right)$$

$$y - c = c \cosh \frac{x}{c} - c$$

†This integral may be found in all standard integral tables. The function

$$z = \sinh^{-1} u$$

(read "arc hyperbolic sine u") is the *inverse* of the function $u = \sinh z$ (read "hyperbolic sine z"). This function and the function $v = \cosh z$ (read "hyperbolic cosine z") are defined as follows:

$$u = \sinh z = \tfrac{1}{2}(e^z - e^{-z}) \qquad v = \cosh z = \tfrac{1}{2}(e^z + e^{-z})$$

Numerical values of the functions $\sinh z$ and $\cosh z$ are found in *tables of hyperbolic functions*. They may also be computed on most calculators either directly or from the above definitions. The student is referred to any calculus text for a complete description of the properties of these functions. In this section, we shall make use only of the following properties, which may be easily derived from the above definitions:

$$\frac{d \sinh z}{dz} = \cosh z \qquad \frac{d \cosh z}{dz} = \sinh z \tag{7.12}$$

$$\sinh 0 = 0 \qquad \cosh 0 = 1 \tag{7.13}$$

$$\cosh^2 z - \sinh^2 z = 1 \tag{7.14}$$

which reduces to

$$y = c \cosh \frac{x}{c} \tag{7.16}$$

This is the equation of a *catenary* with vertical axis. The ordinate c of the lowest point C is called the *parameter* of the catenary. Squaring both sides of Eqs. (7.15) and (7.16), subtracting, and taking (7.14) into account, we obtain the following relation between y and s:

$$y^2 - s^2 = c^2 \tag{7.17}$$

Solving (7.17) for s^2 and carrying into the last of the relations (7.11), we write these relations as follows:

$$T_0 = wc \qquad W = ws \qquad T = wy \tag{7.18}$$

The last relation indicates that the tension at any point P of the cable is proportional to the vertical distance from D to the horizontal line representing the x axis.

When the supports A and B of the cable have the same elevation, the distance L between the supports is called the *span* of the cable and the vertical distance h from the supports to the lowest point C is called the *sag* of the cable. These definitions are the same that were given in the case of parabolic cables, but it should be noted that because of our choice of coordinate axes, the sag h is now

$$h = y_A - c \tag{7.19}$$

It should also be observed that certain catenary problems involve transcendental equations which must be solved by successive approximations (see Sample Prob. 7.10). When the cable is fairly taut, however, the load may be assumed uniformly distributed *along the horizontal* and the catenary may be replaced by a parabola. The solution of the problem is thus greatly simplified, while the error introduced is small.

When the supports A and B have different elevations, the position of the lowest point of the cable is not known. The problem may be solved then in a manner similar to that indicated for parabolic cables, by expressing that the cable must pass through the supports and that $x_B - x_A = L$, $y_B - y_A = d$, where L and d denote, respectively, the horizontal and vertical distances between the two supports.

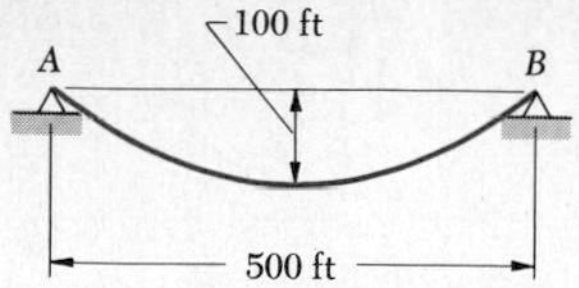

SAMPLE PROBLEM 7.10

A uniform cable weighing 3 lb/ft is suspended between two points A and B as shown. Determine (a) the maximum and minimum values of the tension in the cable, (b) the length of the cable.

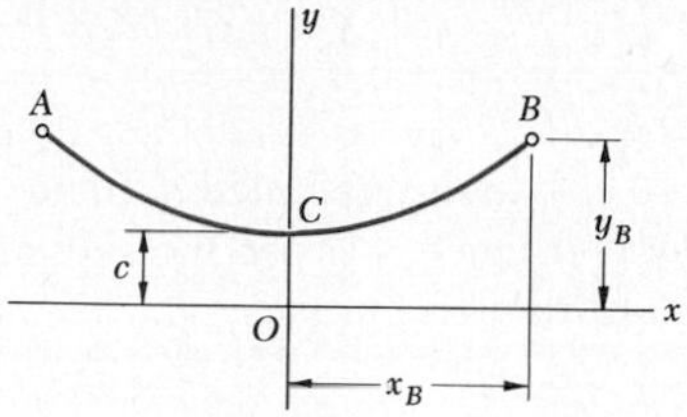

Solution. ***Equation of Cable.*** The origin of coordinates is placed at a distance c below the lowest point of the cable. The equation of the cable is given by Eq. (7.16),

$$y = c \cosh \frac{x}{c}$$

The coordinates of point B are

$$x_B = 250 \text{ ft} \qquad y_B = 100 + c$$

Substituting these coordinates into the equation of the cable, we obtain

$$100 + c = c \cosh \frac{250}{c}$$

$$\frac{100}{c} + 1 = \cosh \frac{250}{c}$$

The value of c is determined by assuming successive trial values, as shown in the following table:

c	$\frac{250}{c}$	$\frac{100}{c}$	$\frac{100}{c} + 1$	$\cosh \frac{250}{c}$
300	0.833	0.333	1.333	1.367
350	0.714	0.286	1.286	1.266
330	0.758	0.303	1.303	1.301
328	0.762	0.305	1.305	1.305

Taking $c = 328$, we have

$$y_B = 100 + c = 428 \text{ ft}$$

a. Maximum and Minimum Values of the Tension. Using Eqs. (7.18), we obtain

$$T_{min} = T_0 = wc = (3 \text{ lb/ft})(328 \text{ ft}) \qquad T_{min} = 984 \text{ lb} \blacktriangleleft$$
$$T_{max} = T_B = wy_B = (3 \text{ lb/ft})(428 \text{ ft}) \qquad T_{max} = 1284 \text{ lb} \blacktriangleleft$$

b. Length of Cable. One-half the length of the cable is found by solving Eq. (7.17):

$$y_B^2 - s_{CB}^2 = c^2 \qquad s_{CB}^2 = y_B^2 - c^2 = (428)^2 - (328)^2 \qquad s_{CB} = 275 \text{ ft}$$

The total length of the cable is therefore

$$s_{AB} = 2s_{CB} = 2(275 \text{ ft}) \qquad s_{AB} = 550 \text{ ft} \blacktriangleleft$$

Problems

7.101 A 30-m rope is strung between the roofs of two buildings, each 9 m high. The maximum tension is found to be 250 N and the lowest point of the cable is observed to be 3 m above the ground. Determine (*a*) the horizontal distance between the buildings, (*b*) the total mass of the rope.

7.102 A 50-m steel surveying tape has a mass of 1.6 kg. If the tape is stretched between two points at the same elevation and pulled until the tension at each end is 60 N, determine the horizontal distance between the ends of the tape. Neglect the elongation of the tape due to the tension.

7.103 An aerial tramway cable of length 600 ft and weighing 3.0 lb/ft is suspended between two points at the same elevation. Knowing that the sag is 150 ft, find (*a*) the horizontal distance between the supports, (*b*) the maximum tension in the cable.

7.104 A 20-ft chain of weight 30 lb is suspended between two points at the same elevation. Knowing that the sag is observed to be 8 ft, determine (*a*) the distance between the supports, (*b*) the maximum tension in the chain.

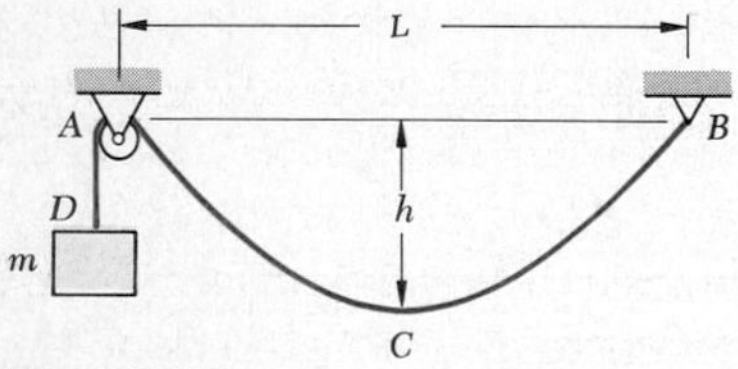

Fig. P7.105

7.105 A counterweight D of mass 50 kg is attached to a cable which passes over a small pulley at A and is attached to a support at B. Knowing that $L = 12$ m and $h = 4$ m, determine (*a*) the length of the cable from A to B, (*b*) the mass per unit length of the cable. Neglect the mass of the cable from A to D.

7.106 Solve Prob. 7.105, assuming that the mass of the counterweight D is 40 kg.

7.107 A wire of length 60 ft and total weight 20 lb is suspended between two points at the same elevation and 30 ft apart. Determine the sag and the maximum tension.

7.108 A 240-ft wire is suspended between two points at the same elevation and 160 ft apart. Knowing that the maximum tension is 50 lb, determine the sag and the total weight of the wire.

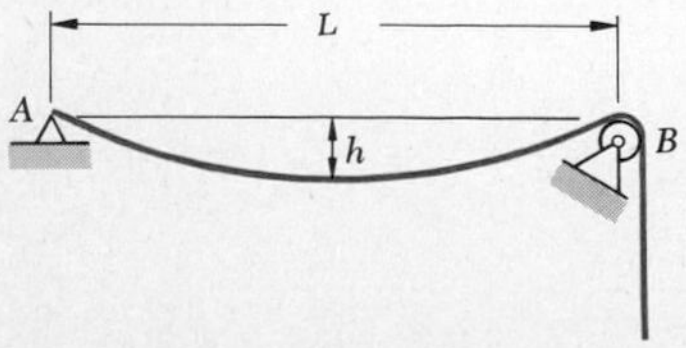

Fig. P7.109 and P7.110

7.109 A uniform cord 750 mm long passes over a frictionless pulley at B and is attached to a rigid support at A. Knowing that $L = 250$ mm, determine the smaller of the two values of h for which the cord is in equilibrium.

7.110 A uniform cord 50 in. long passes over a frictionless pulley at B and is attached to a rigid support at A. Knowing that $L = 20$ in., determine the smaller of the two values of h for which the cord is in equilibrium.

7.111 To the left of point B the long cable $ABDE$ rests on the rough horizontal surface shown. Knowing that the cable weighs 1.5 lb/ft, determine the force $\mathbf{F}$ required when $a = 6$ ft.

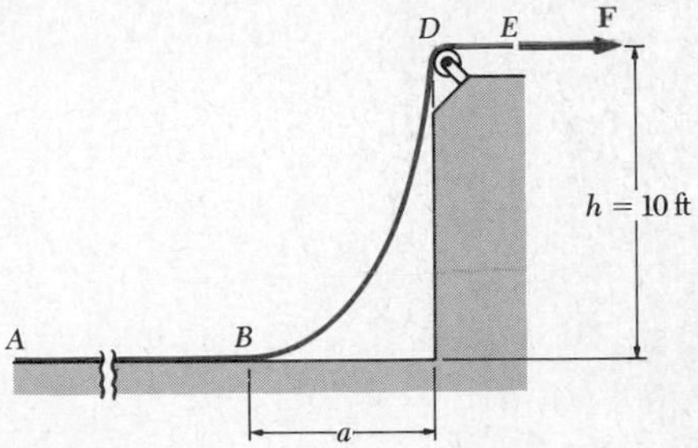

Fig. P7.111

7.112 Solve Prob. 7.111 when $a = 10$ ft.

7.113 Denoting by θ the angle formed by a uniform cable and the horizontal, show that at any point $y = c/\cos\theta$.

***7.114** (*a*) Determine the maximum allowable horizontal span for a uniform cable of weight w per unit length if the tension in the cable is not to exceed the value T_m. (*b*) Using the result of part *a*, find the maximum span of a drawn-steel wire for which $w = 0.25$ lb/ft and $T_m = 8000$ lb.

***7.115** A cable has a mass per unit length of 4 kg/m and is supported as shown. Knowing that the span L is 6 m, determine the *two* values of the sag h for which the maximum tension is 500 N.

***7.116** Determine the sag-to-span ratio for which the maximum tension in the cable is equal to the total weight of the entire cable AB.

***7.117** A cable, of weight w per unit length, is suspended between two points at the same elevation and a distance L apart. Determine the sag-to-span ratio for which the maximum tension is as small as possible. What are the corresponding values of θ_B and T_m?

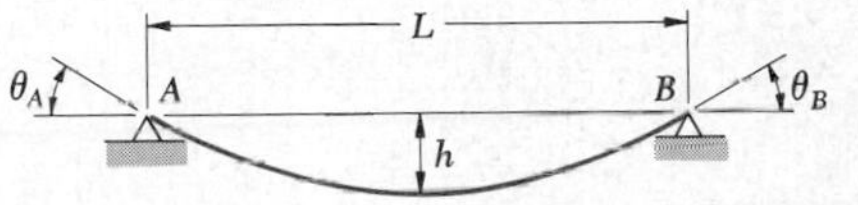

Fig. P7.115, P7.116, and P7.117

Review Problems

7.118 A 14-mm-diameter wire rope has a mass per unit length of 0.7 kg/m and is suspended from two supports at the same elevation and 120 m apart. If the sag is 30 m, determine (*a*) the total length of the cable, (*b*) the maximum tension.

7.119 Two cables of the same gage are attached to a transmission tower at B. Since the tower is slender, the horizontal component of the resultant of the forces exerted by the cables at B is to be zero. Assuming the cables to be parabolic, determine the required sag h of cable AB.

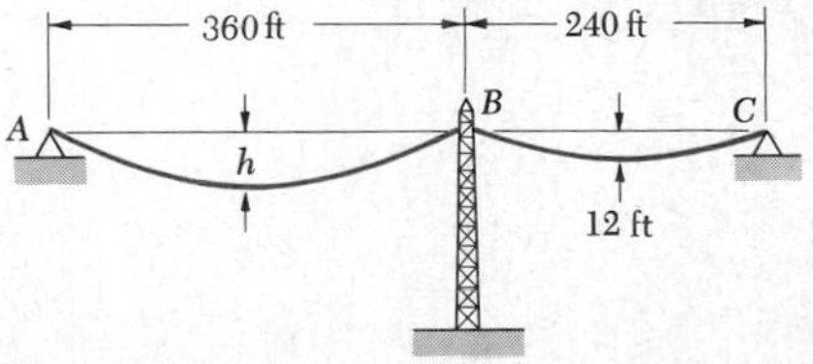

Fig. P7.119

7.120 and 7.121 Draw the shear and bending-moment diagrams for the beam and loading shown.

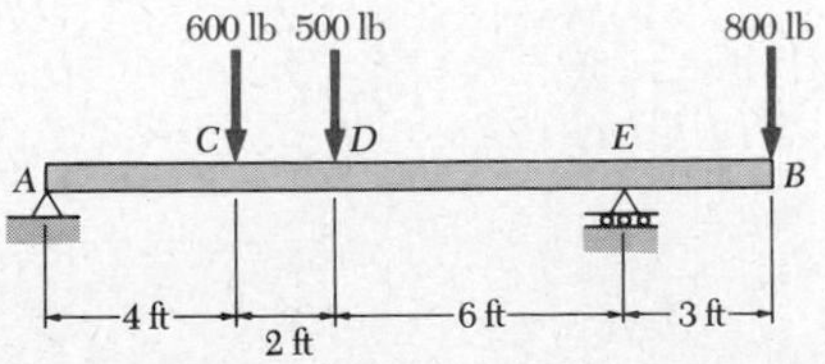

Fig. P7.120

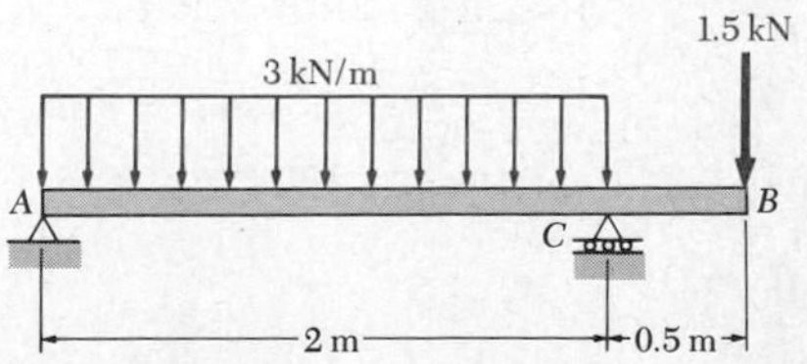

Fig. P7.121

7.122 A 5-kN load is applied at point F of plank EFG, which is suspended as shown from cable $ABCD$. If $h_B = 1.5$ m, determine (*a*) the distance a, (*b*) the corresponding maximum tension in cable $ABCD$.

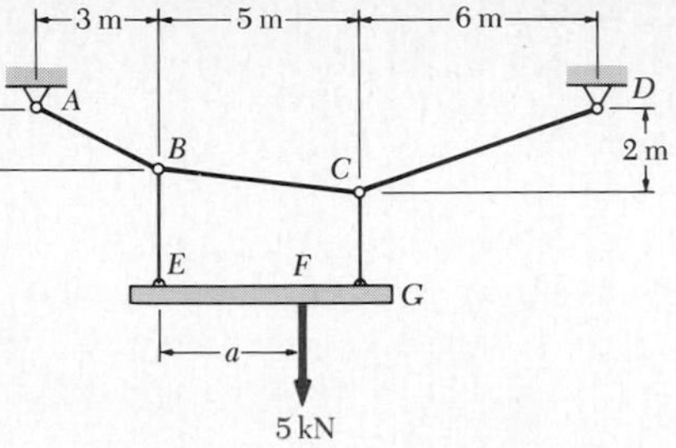

Fig. P7.122 and P7.123

7.123 A 5-kN load is applied at point F of plank EFG, which is suspended as shown from cable $ABCD$. If $a = 2.5$ m, determine (*a*) the sag h_B, (*b*) the corresponding maximum tension in cable $ABCD$.

7.124 For each of the arrangements shown, determine the internal forces at point F of beam AB.

7.125 For arrangements a and b, draw the shear and bending-moment diagrams for beam AB.

7.126 For arrangements c and d, draw the shear and bending-moment diagrams for beam AB.

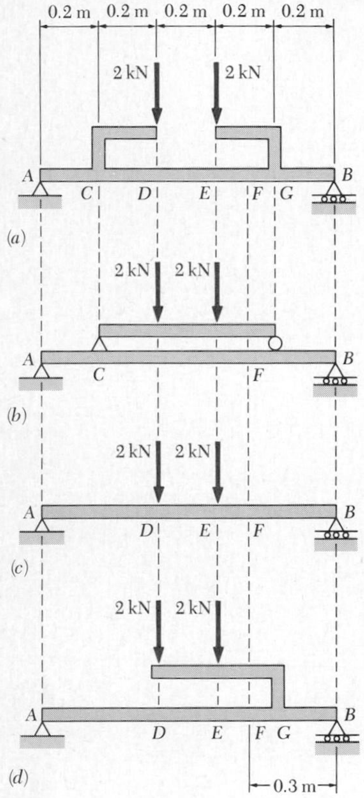

Fig. P7.124, P7.125, and P7.126

7.127 A cable of length $L + \Delta$ is suspended between two points which are at the same elevation and a distance L apart. (*a*) Assuming that Δ is small compared with L and that the cable is parabolic, determine the approximate sag in terms of L and Δ. (*b*) If $L = 30$ m and $\Delta = 0.8$ m and the mass of the wire is 0.35 kg/m, use the result of part a to determine approximate values for the sag and the maximum tension in the wire. [*Hint.* Use only the first two terms of Eq. (7.10).]

7.128 A uniform rod $ABCDE$ has been bent and is held by a wire sling as shown. Knowing that the rod weighs 3.6 lb, draw the shear and bending-moment diagrams for portion BCD of the rod.

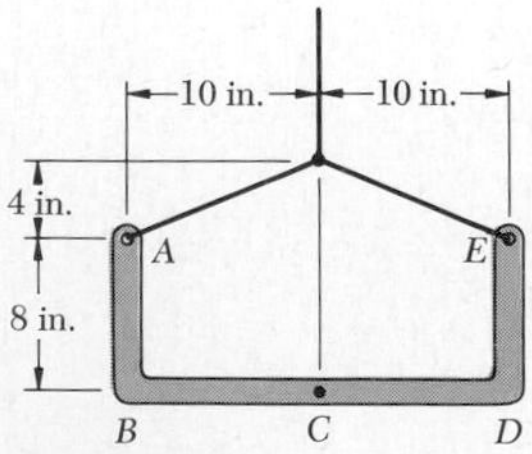

Fig. P7.128

7.129 The rod AB is attached to a hinge at A and is free at B. Determine (*a*) the ratio of w_2 to w_1 for which the rod is in equilibrium, (*b*) the equations of the shear and bending-moment curves for the rod, (*c*) the magnitude and location of the maximum bending moment.

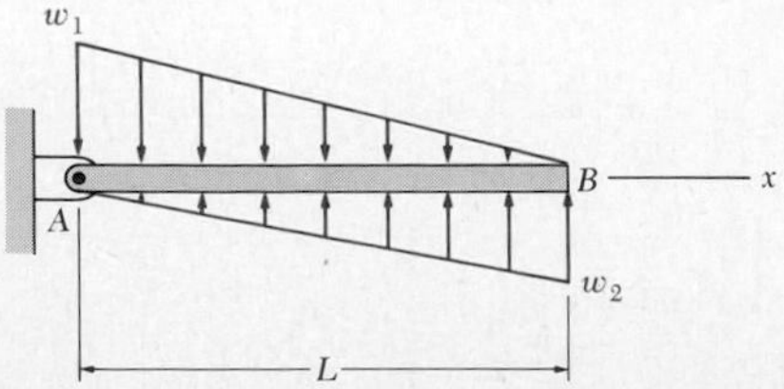

Fig. P7.129

CHAPTER EIGHT
Friction

8.1. Introduction. In the preceding chapters, it was assumed that surfaces in contact were either *frictionless* or *rough*. If they were frictionless, the force each surface exerted on the other was normal to the surfaces and the two surfaces could move freely with respect to each other. If they were rough, it was assumed that tangential forces could develop to prevent the motion of one surface with respect to the other.

This view was a simplified one. Actually, no perfectly frictionless surface exists. When two surfaces are in contact, tangential forces, called *friction forces*, will always develop if one attempts to move one surface with respect to the other. On the other hand, these friction forces are limited in magnitude and will not prevent motion if sufficiently large forces are applied. The distinction between frictionless and rough surfaces is thus a matter of degree. This will be seen more clearly in the present chapter, which is devoted to the study of friction and of its applications to common engineering situations.

There are two types of friction: *dry friction*, sometimes called *Coulomb friction*, and *fluid friction*. Fluid friction develops between layers of fluid moving at different velocities. Fluid friction is of great importance in problems involving the flow of fluids through pipes and orifices or dealing with bodies immersed in moving fluids. It is also basic in the analysis of the motion of *lubricated mechanisms*. Such problems are considered in texts on fluid mechanics. We shall limit our present study to dry friction, i.e., to problems involving rigid bodies which are in contact along *nonlubricated* surfaces.

In the first part of this chapter, we shall analyze the equilibrium of various rigid bodies and structures, assuming dry friction at the surfaces of contact. Later we shall consider a number of specific engineering applications where dry friction plays an important role. These are wedges, square-threaded screws, journal bearings, thrust bearings, rolling resistance, and belt friction.

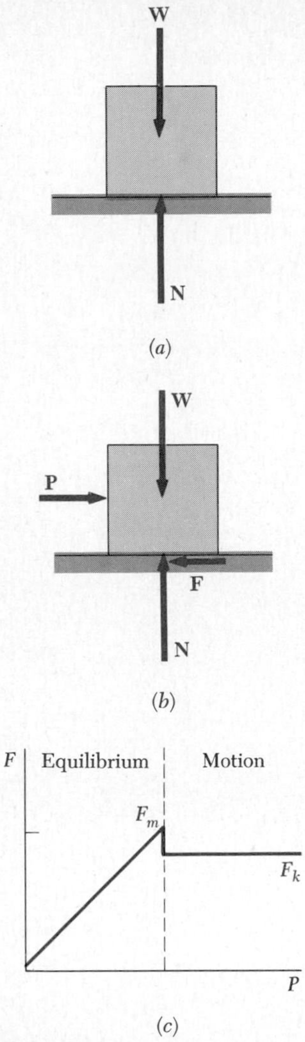

Fig. 8.1

8.2. The Laws of Dry Friction. Coefficients of Friction. The laws of dry friction are best understood by the following experiment. A block of weight **W** is placed on a horizontal plane surface (Fig. 8.1*a*). The forces acting on the block are its weight **W** and the reaction of the surface. Since the weight has no horizontal component, the reaction of the surface also has no horizontal component; the reaction is therefore *normal* to the surface and is represented by **N** in Fig. 8.1*a*. Suppose, now, that a horizontal force **P** is applied to the block (Fig. 8.1*b*). If **P** is small, the block will not move; some other horizontal force must therefore exist, which balances **P**. This other force is the *static-friction force* **F**, which is actually the resultant of a great number of forces acting over the entire surface of contact between the block and the plane. The nature of these forces is not known exactly, but it is generally assumed that these forces are due to the irregularities of the surfaces in contact and also, to a certain extent, to molecular attraction.

If the force **P** is increased, the friction force **F** also increases, continuing to oppose **P**, until its magnitude reaches a certain *maximum value* F_m (Fig. 8.1*c*). If **P** is further increased, the friction force cannot balance it any more and the block starts sliding. As soon as the block has been set in motion, the magnitude of **F** drops from F_m to a lower value F_k. This is because there is less interpenetration between the irregularities of the surfaces in contact when these surfaces move with respect to each other. From then on, the block keeps sliding with increasing velocity while the friction force, denoted by $\mathbf{F}_k$ and called the *kinetic-friction force*, remains approximately constant.

Experimental evidence shows that the maximum value F_m of the static-friction force is proportional to the normal component N of the reaction of the surface. We have

$$F_m = \mu_s N \tag{8.1}$$

where μ_s is a constant called the *coefficient of static friction*. Similarly, the magnitude F_k of the kinetic-friction force may be put in the form

$$F_k = \mu_k N \tag{8.2}$$

where μ_k is a constant called the *coefficient of kinetic friction*. The coefficients of friction μ_s and μ_k do not depend upon the area of the surfaces in contact. Both coefficients, however, depend strongly on the *nature* of the surfaces in contact. Since they also depend upon the exact condition of the surfaces, their value is seldom known with an accuracy greater than 5 percent. Approximate values of coefficients of static friction are given in Table 8.1 for various dry surfaces. The corresponding values of the coefficient of kinetic friction would be about 25 percent smaller. Since coeffi-

Table 8.1 Approximate Values of Coefficient of Static Friction for Dry Surfaces

Metal on metal	0.15–0.60
Metal on wood	0.20–0.60
Metal on stone	0.30–0.70
Metal on leather	0.30–0.60
Wood on wood	0.25–0.50
Wood on leather	0.25–0.50
Stone on stone	0.40–0.70
Earth on earth	0.20–1.00
Rubber on concrete	0.60–0.90

cients of friction are dimensionless quantities, the values given in Table 8.1 may be used with both SI units and U.S. customary units.

From the description given above, it appears that four different situations may occur when a rigid body is in contact with a horizontal surface:

1. The forces applied to the body do no tend to move it along the surface of contact; there is no friction force (Fig. 8.2*a*).

2. The applied forces tend to move the body along the surface of contact but are not large enough to set it in motion. The friction force **F** which has developed may be found by solving the equations of equilibrium for the body. Since there is no evidence that the maximum value of the static-friction force has been reached, the equation $F_m = \mu_s N$ *cannot be used* to determine the friction force (Fig. 8.2*b*).

3. The applied forces are such that the body is just about to slide. We say that *motion is impending*. The friction force **F** has reached its maximum value F_m and, together with the normal force **N**, balances the applied forces. Both the equations of equilibrium and the equation $F_m = \mu_s N$ *may be used*. We also note that the friction force has a sense opposite to the sense of impending motion (Fig. 8.2*c*).

4. The body is sliding under the action of the applied forces, and the equations of equilibrium do not apply any more. However, **F** is now equal to the kinetic-friction force $\mathbf{F}_k$ and the equation $F_k = \mu_k N$ may be used. The sense of $\mathbf{F}_k$ is opposite to the sense of motion (Fig. 8.2*d*).

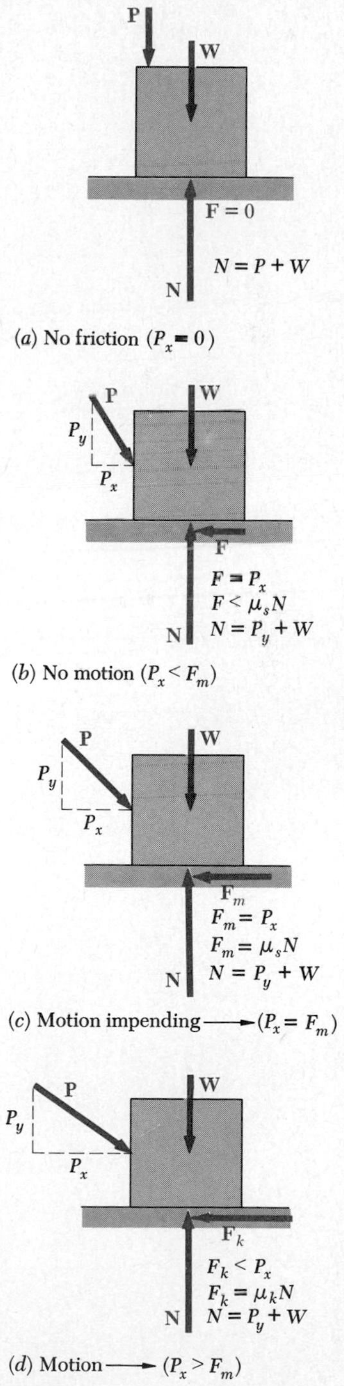

Fig. 8.2

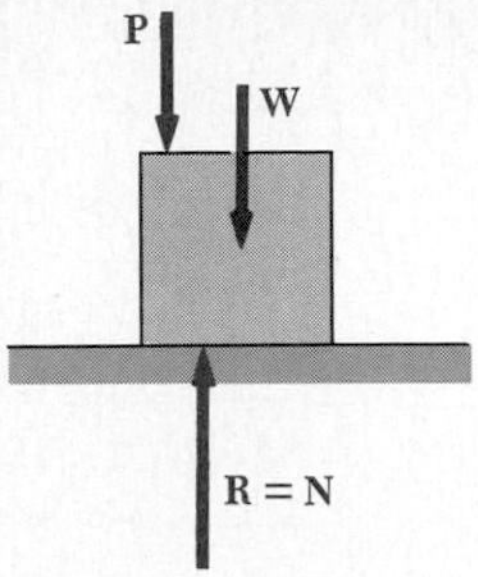

(*a*) No friction

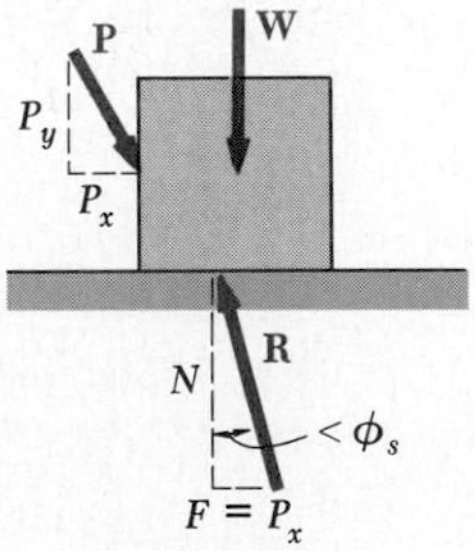

(*b*) No motion

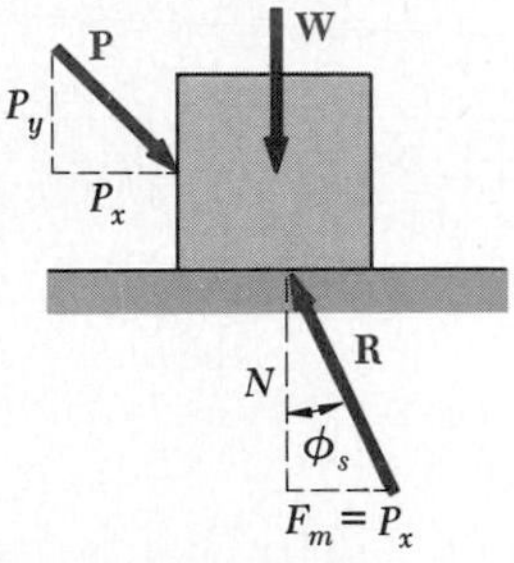

(*c*) Motion impending ⟶

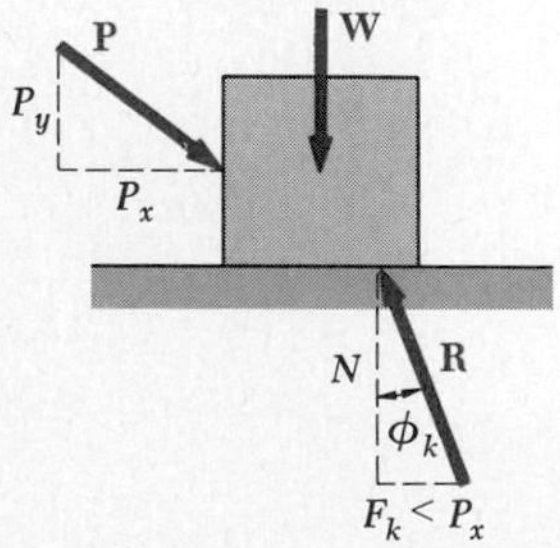

(*d*) Motion ⟶

Fig. 8.3

8.3. Angles of Friction. It is sometimes found convenient to replace the normal force **N** and the friction force **F** by their resultant **R**. Let us consider again a block of weight **W** resting on a horizontal plane surface. If no horizontal force is applied to the block, the resultant **R** reduces to the normal force **N** (Fig. 8.3*a*). However, if the applied force **P** has a horizontal component $\mathbf{P}_x$ which tends to move the block, the force **R** will have a horizontal component **F** and, thus, will form a certain angle with the vertical (Fig. 8.3*b*). If $\mathbf{P}_x$ is increased until motion becomes impending, the angle between **R** and the vertical grows and reaches a maximum value (Fig. 8.3*c*). This value is called the *angle of static friction* and is denoted by ϕ_s. From the geometry of Fig. 8.3*c*, we note that

$$\tan \phi_s = \frac{F_m}{N} = \frac{\mu_s N}{N}$$

$$\tan \phi_s = \mu_s \tag{8.3}$$

If motion actually takes place, the magnitude of the friction force drops to F_k; similarly, the angle between **R** and **N** drops to a lower value ϕ_k, called the *angle of kinetic friction* (Fig. 8.3*d*). From the geometry of Fig. 8.3*d*, we write

$$\tan \phi_k = \frac{F_k}{N} = \frac{\mu_k N}{N}$$

$$\tan \phi_k = \mu_k \tag{8.4}$$

Another example will show how the angle of friction may be used to advantage in the analysis of certain types of problems. Consider a block resting on a board which may be given any desired inclination; the block is subjected to no other force than its weight **W** and the reaction **R** of the board. If the board is horizontal, the force **R** exerted by the board on the block is perpendicular to the board and balances the weight **W** (Fig. 8.4*a*). If the board is given a small angle of inclination θ, the force **R** will deviate from the perpendicular to the board by the angle θ and will keep balancing **W** (Fig. 8.4*b*); it will then have a normal component **N** of magnitude $N = W \cos \theta$ and a tangential component **F** of magnitude $F = W \sin \theta$.

If we keep increasing the angle of inclination, motion will soon become impending. At that time, the angle between **R** and the normal will have reached its maximum value ϕ_s (Fig. 8.4*c*). The value of the angle of inclination corresponding to impending motion is called the *angle of repose*. Clearly, the angle of repose is equal to the angle of static friction ϕ_s. If the angle of inclination θ is further increased, motion starts and the angle between **R** and the normal drops to the lower value ϕ_k (Fig. 8.4*d*). The reaction **R** is not vertical any more, and the forces acting on the block are unbalanced.

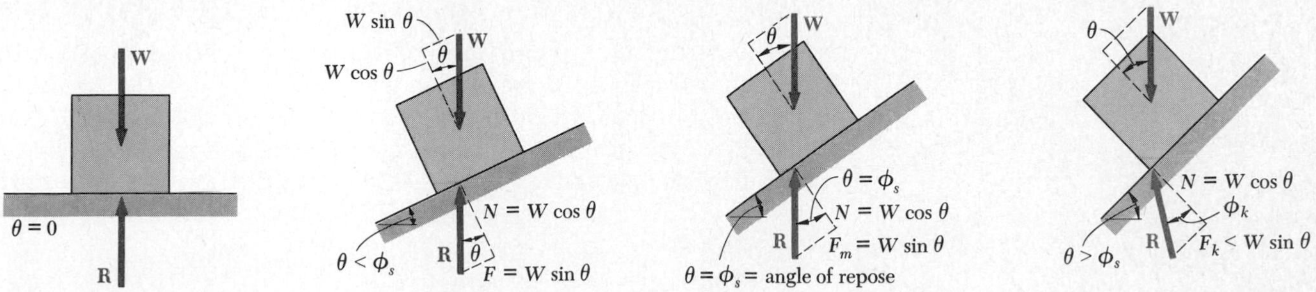

(*a*) No friction (*b*) No motion (*c*) Motion impending (*d*) Motion

Fig. 8.4

8.4. Problems Involving Dry Friction. Problems involving dry friction are found in many engineering applications. Some deal with simple situations such as the block sliding on a plane described in the preceding sections. Others involve more complicated situations as in Sample Prob. 8.3; many deal with the stability of rigid bodies in accelerated motion and will be studied in dynamics. Also, a number of common machines and mechanisms may be analyzed by applying the laws of dry friction. These include wedges, screws, journal and thrust bearings, and belt transmissions. They will be studied in the following sections.

The *methods* which should be used to solve problems involving dry friction are the same that were used in the preceding chapters. If a problem involves only a motion of translation, with no possible rotation, the body under consideration may usually be treated as a particle, and the methods of Chap. 2 may be used. If the problem involves a possible rotation, the body must be considered as a rigid body and the methods of Chap. 4 should be used. If the structure considered is made of several parts, the principle of action and reaction must be used as was done in Chap. 6.

If the body considered is acted upon by more than three forces (including the reactions at the surfaces of contact), the reaction at each surface will be represented by its components **N** and **F** and the problem will be solved from the equations of equilibrium. If only three forces act on the body under consideration, it may be found more convenient to represent each reaction by the single force **R** and to solve the problem by drawing a force triangle.

Most problems involving friction fall into one of the following *three groups:* In the *first group* of problems, all applied forces are given, and the coefficients of friction are known; we are to determine whether the body considered will remain at rest or slide. The friction force **F** *required to maintain equilibrium* is unknown (its magnitude is *not* equal to $\mu_s N$) and should be determined, together with the normal force **N**, by drawing a free-body diagram and *solving the equations of equilibrium* (Fig. 8.5*a*).

The value found for the magnitude F of the friction force is then compared with the maximum value $F_m = \mu_s N$. If F is smaller than or equal to F_m, the body remains at rest. If the value found for F is larger than F_m, equilibrium cannot be maintained and motion takes place; the actual magnitude of the friction force is then $F_k = \mu_k N$.

In problems of the *second group*, all applied forces are given, and the motion is known to be impending; we are to determine the value of the coefficient of static friction. Here again, we determine the friction force and the normal force by drawing a free-body diagram and solving the equations of equilibrium (Fig. 8.5*b*). Since we know that the value found for F is the maximum value F_m, the coefficient of friction may be found by writing and solving the equation $F_m = \mu_s N$.

In problems of the *third group*, the coefficient of static friction is given, and it is known that the motion is impending in a given direction; we are to determine the magnitude or the direction of one of the applied forces. The friction force should be shown in the free-body diagram with a *sense opposite to that of the impending motion* and with a magnitude $F_m = \mu_s N$ (Fig. 8.5*c*). The equations of equilibrium may then be written, and the desired force may be determined.

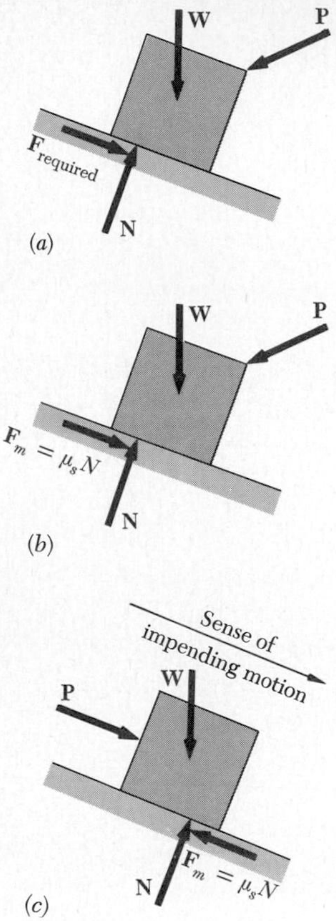

Fig. 8.5

As noted above, it may be more convenient when only three forces are involved to represent the reaction of the surface by a single force **R** and to solve the problem by drawing a force triangle. Such a solution is used in Sample Prob. 8.2.

When two bodies A and B are in contact (Fig. 8.6*a*), the forces of friction exerted, respectively, by A on B and by B on A are equal and opposite (Newton's third law). It is important in drawing the free-body diagram of one of the bodies, to include the appropriate friction force with its correct sense. The following rule should then be observed: *The sense of the friction force acting on A is opposite to that of the motion* (*or impending motion*) *of A as observed from B* (Fig. 8.6*b*).† The sense of the friction force acting on B is determined in a similar way (Fig. 8.6*c*). Note that the motion of A as observed from B is a *relative motion*. Body A may be fixed; yet it will have a relative motion with respect to B if B itself moves. Also, A may actually move down yet be observed from B to move up if B moves down faster than A.

†It is therefore *the same as that of the motion of B as observed from A*.

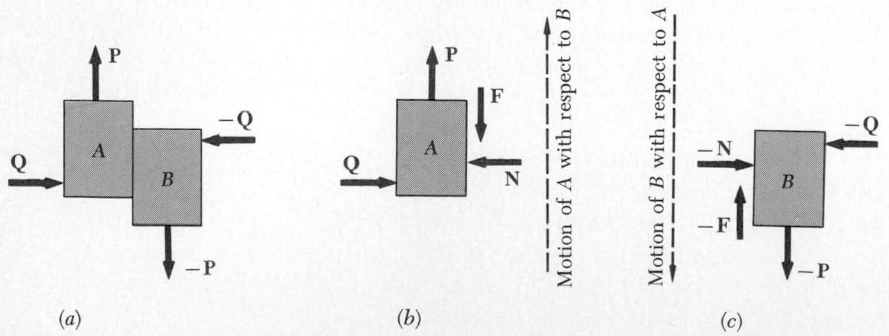

Fig. 8.6

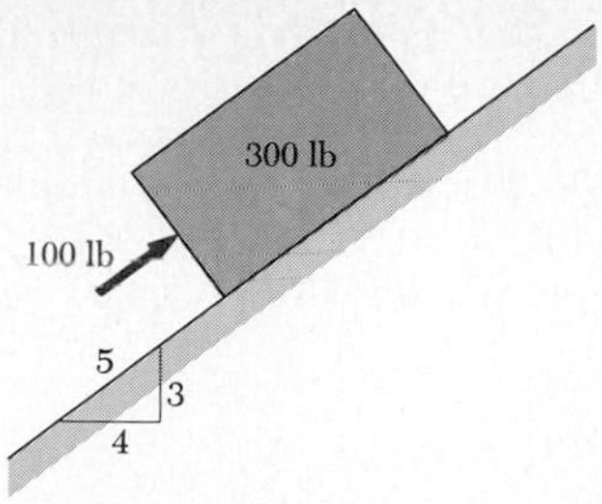

SAMPLE PROBLEM 8.1

A 100-lb force acts as shown on a 300-lb block placed on an inclined plane. The coefficients of friction between the block and the plane are $\mu_s = 0.25$ and $\mu_k = 0.20$. Determine whether the block is in equilibrium, and find the value of the friction force.

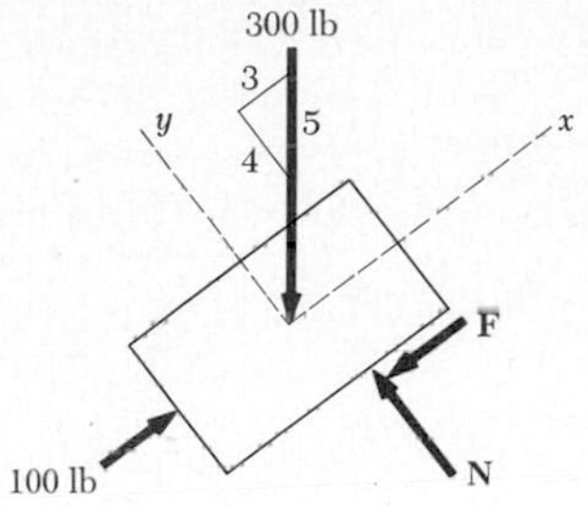

Force Required for Equilibrium. We first determine the value of the friction force *required to maintain equilibrium*. Assuming that **F** is directed down and to the left, we draw the free-body diagram of the block and write

$$+\nearrow \Sigma F_x = 0: \quad 100 \text{ lb} - \tfrac{3}{5}(300 \text{ lb}) - F = 0$$
$$F = -80 \text{ lb} \qquad \mathbf{F} = 80 \text{ lb} \nearrow$$

$$+\nwarrow \Sigma F_y = 0: \quad N - \tfrac{4}{5}(300 \text{ lb}) = 0$$
$$N = +240 \text{ lb} \qquad \mathbf{N} = 240 \text{ lb} \nwarrow$$

The force **F** required to maintain equilibrium is an 80-lb force directed up and to the right; the tendency of the block is thus to move down the plane.

Maximum Friction Force. The magnitude of the maximum friction force which may be developed is

$$F_m = \mu_s N \qquad F_m = 0.25(240 \text{ lb}) = 60 \text{ lb}$$

Since the value of the force required to maintain equilibrium (80 lb) is larger than the maximum value which may be obtained (60 lb), equilibrium will not be maintained and *the block will slide down the plane.*

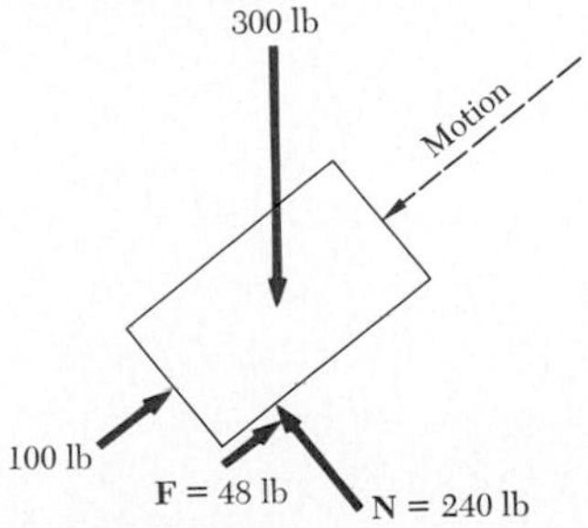

Actual Value of Friction Force. The magnitude of the actual friction force is obtained as follows:

$$F_{\text{actual}} = F_k = \mu_k N$$
$$= 0.20(240 \text{ lb}) = 48 \text{ lb}$$

The sense of this force is opposite to the sense of motion; the force is thus directed up and to the right:

$$\mathbf{F}_{\text{actual}} = 48 \text{ lb} \nearrow \quad ◄$$

It should be noted that the forces acting on the block are not balanced; the resultant is

$$\tfrac{3}{5}(300 \text{ lb}) - 100 \text{ lb} - 48 \text{ lb} = 32 \text{ lb} \swarrow$$

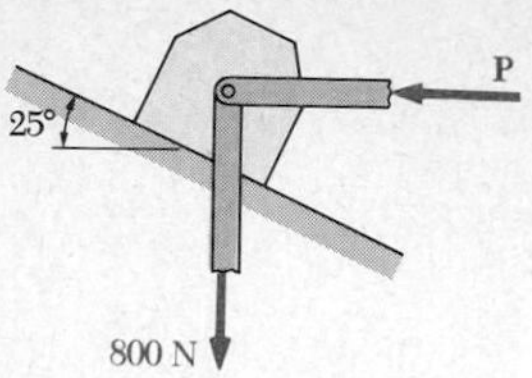

SAMPLE PROBLEM 8.2

A support block is acted upon by two forces as shown. Knowing that the coefficients of friction between the block and the incline are $\mu_s = 0.35$ and $\mu_k = 0.25$, determine the force **P** required (*a*) to start the block moving up the incline, (*b*) to keep it moving up, (*c*) to prevent it from sliding down.

Solution. For each part of the problem we draw a free-body diagram of the block and a force triangle including the 800-N vertical force, the horizontal force **P**, and the force **R** exerted on the block by the incline. The direction of **R** must be determined in each separate case. We note that since **P** is perpendicular to the 800-N force, the force triangle is a right triangle, which may easily be solved for **P**. In most other problems, however, the force triangle will be an oblique triangle and should be solved by applying the law of sines.

a. Force P *to Start Block Moving Up*

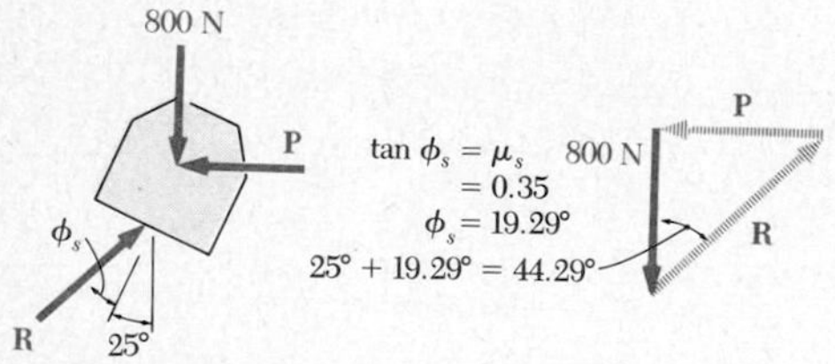

$$P = (800\text{ N}) \tan 44.29°$$

P = 780 N ← ◀

b. Force P *to Keep Block Moving Up*

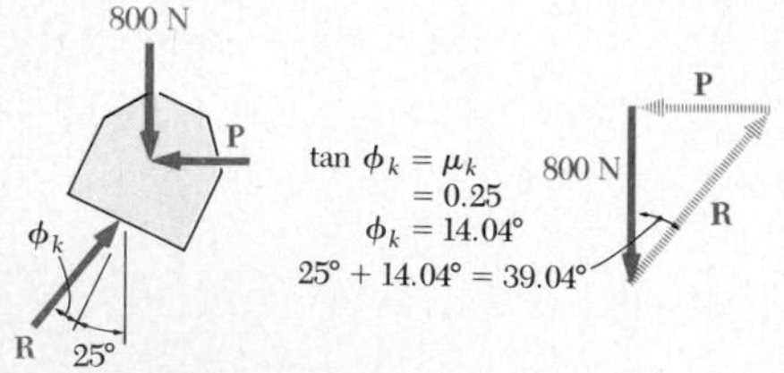

$$P = (800\text{ N}) \tan 39.04°$$

P = 649 N ← ◀

c. Force P *to Prevent Block from Sliding Down*

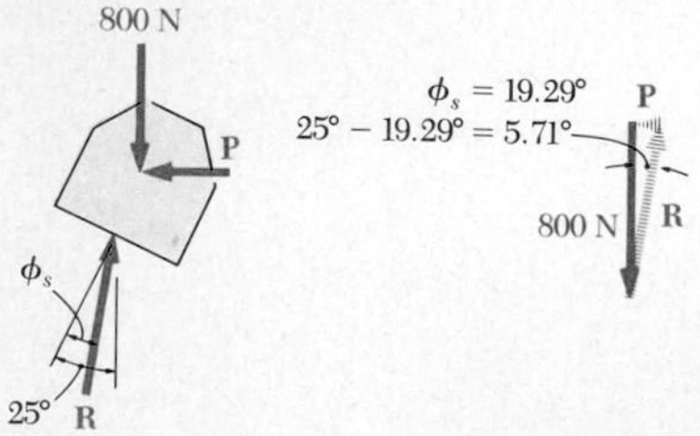

$$P = (800\text{ N}) \tan 5.71°$$

P = 80.0 N ← ◀

SAMPLE PROBLEM 8.3

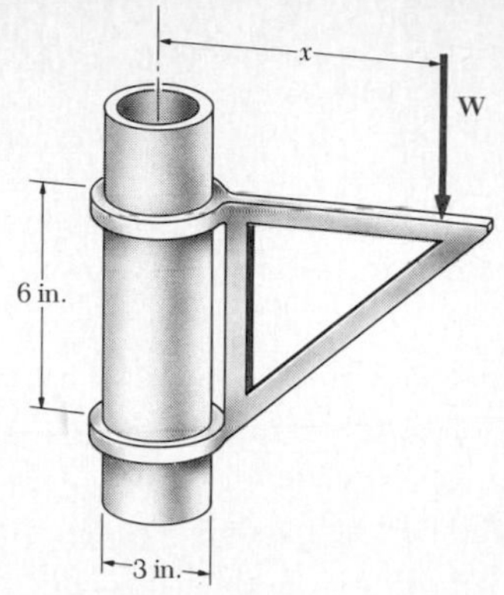

The movable bracket shown may be placed at any height on the 3-in.-diameter pipe. If the coefficient of static friction between the pipe and bracket is 0.25, determine the minimum distance x at which the load **W** can be supported. Neglect the weight of the bracket.

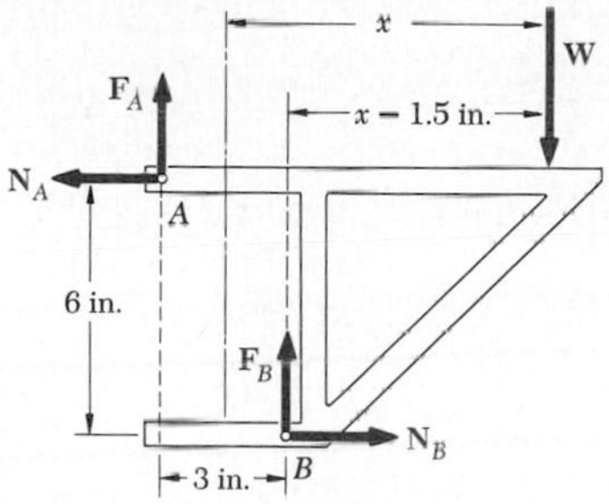

Solution. We draw the free-body diagram of the bracket. When **W** is placed at the minimum distance x from the axis of the pipe, the bracket is just about to slip, and the forces of friction at A and B have reached their maximum values:

$$F_A = \mu_s N_A = 0.25\,N_A$$
$$F_B = \mu_s N_B = 0.25\,N_B$$

Equilibrium Equations

$\overset{+}{\rightarrow}\Sigma F_x = 0$: $\quad N_B - N_A = 0$

$$N_B = N_A$$

$+\uparrow\Sigma F_y = 0$: $\quad F_A + F_B - W = 0$

$$0.25N_A + 0.25N_B = W$$

And, since N_B has been found equal to N_A,

$$0.50N_A = W$$
$$N_A = 2W$$

$+\circlearrowleft\Sigma M_B = 0$: $\quad N_A(6\text{ in.}) - F_A(3\text{ in.}) - W(x - 1.5\text{ in.}) = 0$

$$6N_A - 3(0.25N_A) - Wx + 1.5W = 0$$
$$6(2W) - 0.75(2W) - Wx + 1.5W = 0$$

Dividing through by W and solving for x,

$$x = 12\text{ in.} \quad ◀$$

Problems

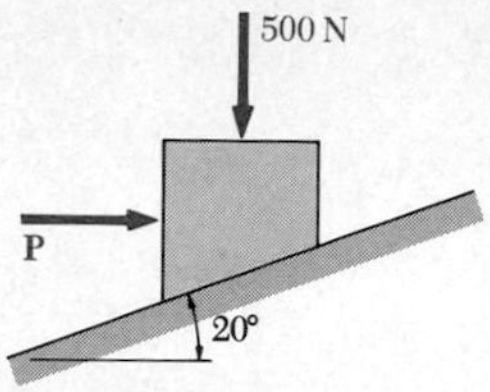

Fig. P8.1

8.1 The coefficients of friction between the block and the incline are $\mu_s = 0.30$ and $\mu_k = 0.25$. Determine whether the block is in equilibrium and find the magnitude and direction of the friction force when $P = 150$ N.

8.2 Solve Prob. 8.1, assuming that $P = 400$ N.

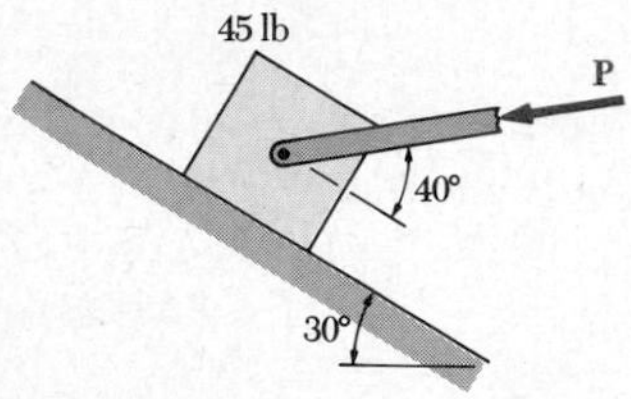

Fig. P8.3

8.3 The coefficients of friction between the 45-lb block and the incline are $\mu_s = 0.40$ and $\mu_k = 0.30$. Determine whether the block is in equilibrium and find the magnitude and direction of the friction force when $P = 100$ lb.

8.4 Solve Prob. 8.3, assuming that $P = 60$ lb.

8.5 For the block of Prob. 8.3, find the smallest value of P required (*a*) to start the block up the incline, (*b*) to keep it moving up, (*c*) to prevent it from moving down.

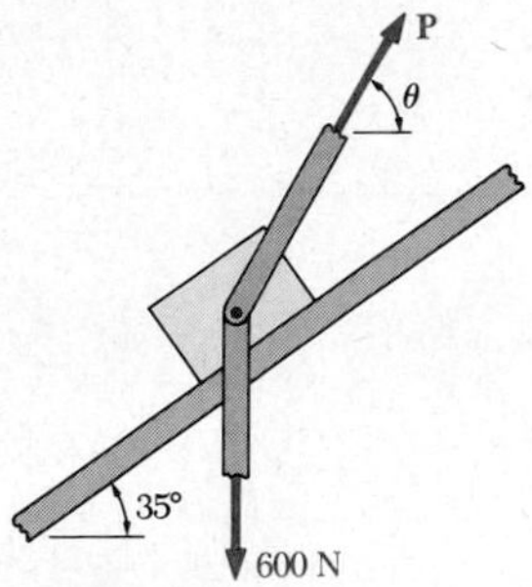

Fig. P8.6 and P8.7

8.6 The coefficients of friction between the block and the rail are $\mu_s = 0.25$ and $\mu_k = 0.20$. Knowing that $\theta = 60°$, determine the smallest value of P required (*a*) to start the block up the rail, (*b*) to keep it moving up, (*c*) to prevent it from moving down.

8.7 The coefficients of friction between the block and the rail are $\mu_s = 0.25$ and $\mu_k = 0.20$. Find the magnitude and direction of the smallest force **P** required (*a*) to start the block up the rail, (*b*) to keep the block from moving down.

8.8 A block of weight $W = 60$ lb rests on a rough plane as shown. Knowing that $\alpha = 20°$ and $\mu_s = 0.30$, determine the magnitude and direction of the smallest force **P** required (*a*) to start the block up the plane, (*b*) to prevent the block from moving down.

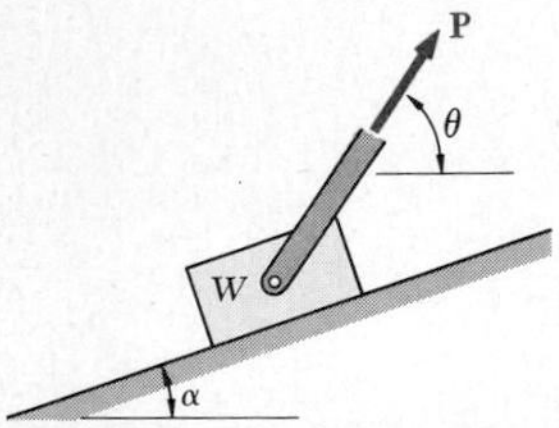

Fig. P8.8

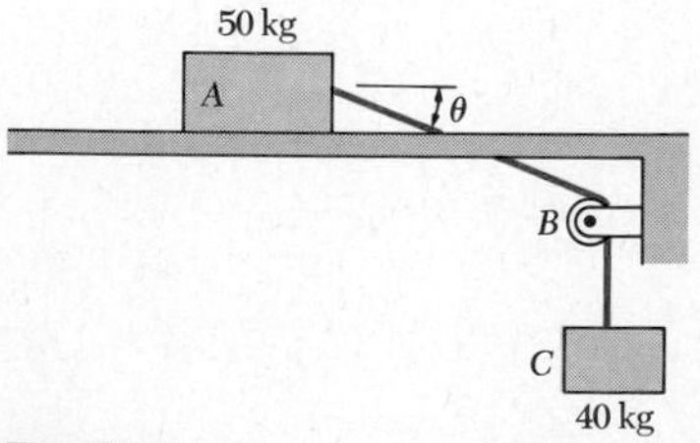

Fig. P8.9

8.9 Two blocks are connected by a cable as shown. Knowing that the coefficient of static friction between block A and the horizontal surface is 0.50, determine the value of θ for which motion is impending.

8.10 The 40-kg block A hangs from a cable as shown. Pulley C is connected by a short link to block E, which rests on a horizontal rail. Knowing that the coefficient of static friction between block E and the rail is 0.30, and neglecting the weight of block E and the friction in the pulleys, determine the maximum allowable value of θ if the system is to remain in equilibrium.

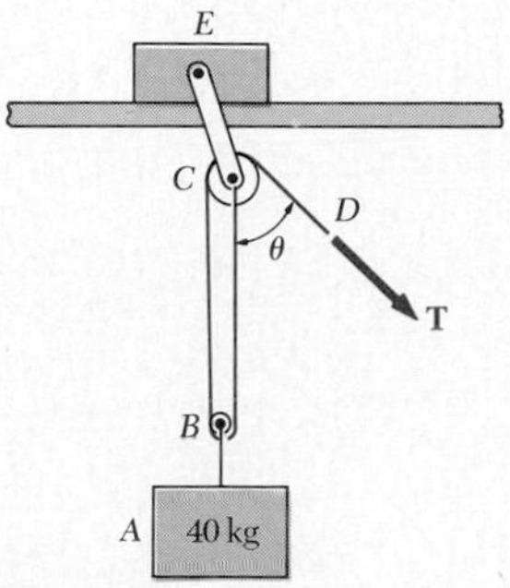

Fig. P8.10

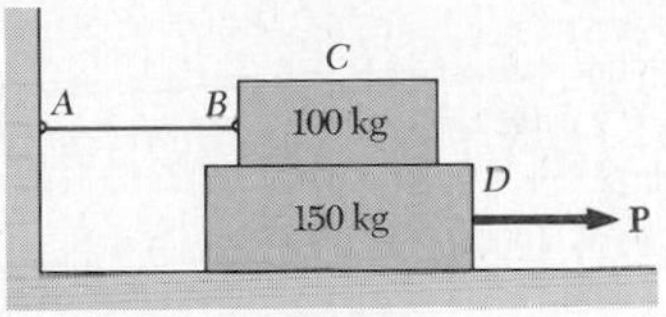

Fig. P8.11 and P8.12

8.11 The coefficients of friction are $\mu_s = 0.25$ and $\mu_k = 0.20$ between all surfaces of contact. Determine the smallest force **P** required to start block D moving if (a) block C is restrained by cable AB as shown, (b) cable AB is removed.

8.12 The coefficients of friction are $\mu_s = 0.25$ and $\mu_k = 0.20$ between all surfaces of contact. Knowing that $P = 800$ N, determine whether block D moves and find the resultant of the friction forces exerted on block D if (a) block C is restrained as shown, (b) cable AB is removed.

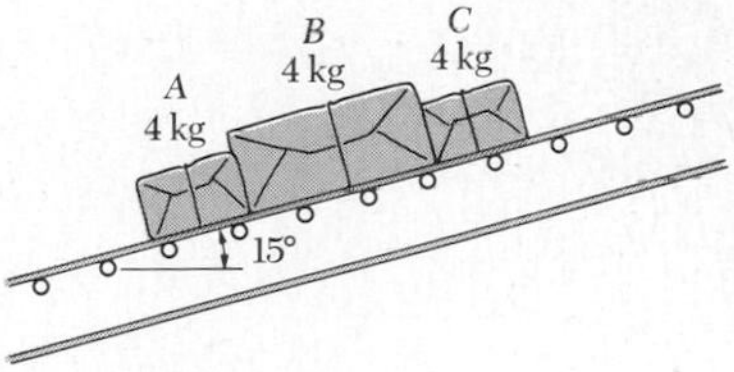

Fig. P8.13

8.13 Three 4-kg packages A, B, and C are placed on a conveyor belt which is at rest. Between the belt and both packages A and C the coefficients of friction are $\mu_s = 0.30$ and $\mu_k = 0.20$; between package B and the belt the coefficients are $\mu_s = 0.10$ and $\mu_k = 0.08$. The packages are placed on the belt so that they are in contact with each other and at rest. Determine which, if any, of the packages will move and the friction force acting on each package.

8.14 Solve Prob. 8.13, assuming that package B is placed to the right of both packages A and C.

8.15 Block A weighs 50 lb and block B weighs 25 lb. Knowing that the coefficient of static friction is 0.15 between all surfaces of contact, determine the value of θ for which motion is impending.

8.16 Solve Prob. 8.15, assuming that the coefficient of static friction is 0.15 between the two blocks, and zero between block B and the incline.

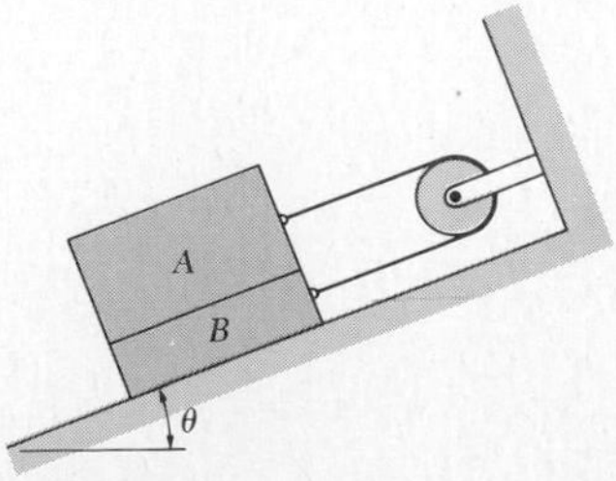

Fig. P8.15

8.17 For the system of Prob. 8.15, and assuming that the coefficients of friction are $\mu_s = 0.15$ and $\mu_k = 0.12$ between the two blocks, and zero between block B and the incline, determine the magnitude of the friction force between the two blocks when (a) $\theta = 25°$, (b) $\theta = 35°$.

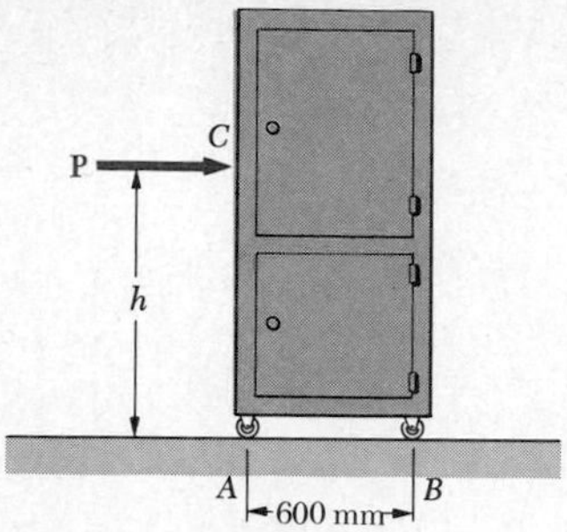

Fig. P8.18 and P8.19

8.18 A 60-kg cabinet is mounted on casters which can be locked to prevent their rotation. The coefficient of friction is 0.30. If $h = 800$ mm, determine the magnitude of the force **P** required to move the cabinet to the right (*a*) if all casters are locked, (*b*) if the casters at *B* are locked and the casters at *A* are free to rotate, (*c*) if the casters at *A* are locked and the casters at *B* are free to rotate.

8.19 A 60-kg cabinet is mounted on casters which can be locked to prevent their rotation. The coefficient of friction between the floor and each caster is 0.30. Assuming that the casters at both *A* and *B* are locked, determine (*a*) the force **P** required to move the cabinet to the right, (*b*) the largest allowable value of *h* if the cabinet is not to tip over.

8.20 A half section of pipe, weighing 200 lb, is to be moved to the right along the floor without tipping. Knowing that the coefficient of friction between the pipe and the floor is 0.40, determine (*a*) the largest allowable value of α, (*b*) the corresponding tension *T*.

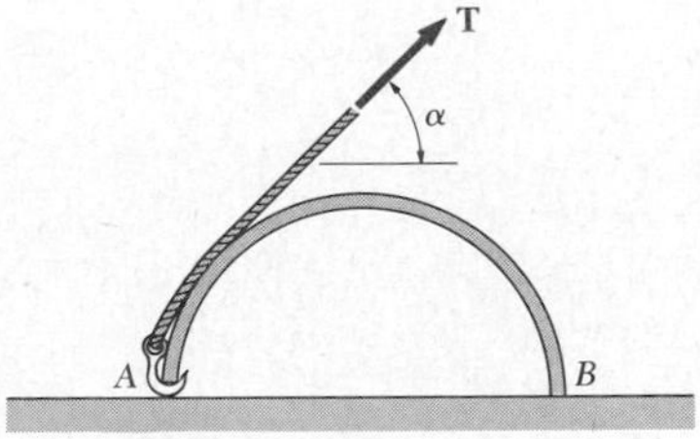

Fig. P8.20 and P8.21

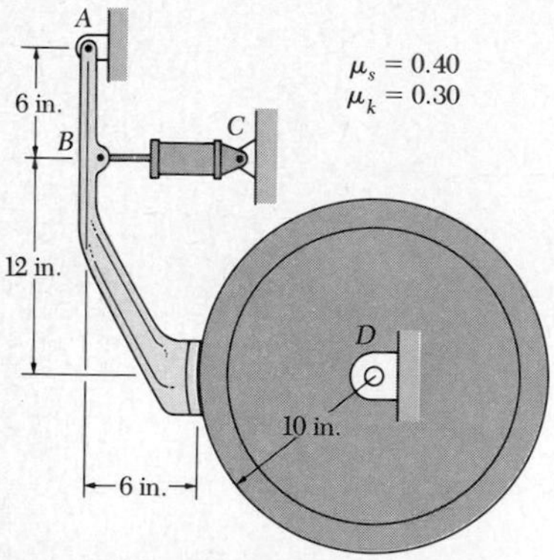

Fig. P8.22 and P8.23

8.21 A half section of pipe, weighing 200 lb, is pulled by a cable as shown. The coefficient of friction between the pipe and the floor is 0.40. If $\alpha = 30°$, determine (*a*) the tension *T* required to move the pipe, (*b*) whether the pipe will slide or tip.

8.22 A couple of magnitude 60 lb·ft is applied to the drum. Determine the smallest force which must be exerted by the hydraulic cylinder if the drum is not to rotate, when the applied couple is directed (*a*) clockwise, (*b*) counterclockwise.

8.23 The hydraulic cylinder exerts on point *B* a force of 500 lb directed to the right. Determine the moment of the friction force about the axle of the drum when the drum is rotating (*a*) clockwise, (*b*) counterclockwise.

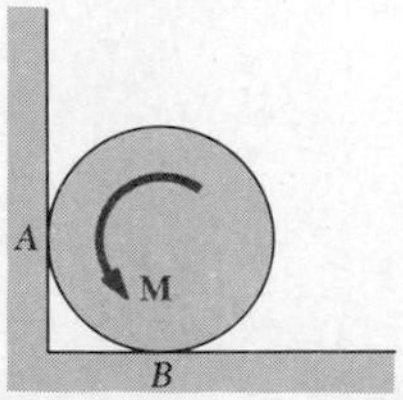

Fig. P8.24 and P8.25

8.24 Find the magnitude of the largest couple **M** which may be applied to the cylinder if it is not to spin. The cylinder has a weight *W* and a radius *r*, and the coefficient of static friction μ_s is the same at *A* and *B*.

8.25 The cylinder has a weight *W* and a radius *r*. Express in terms of *W* and *r* the magnitude of the largest couple **M** which may be applied to the cylinder if it is not to spin, assuming the coefficient of static friction to be (*a*) zero at *A* and 0.40 at *B*, (*b*) 0.30 at *A* and 0.40 at *B*.

*** 8.26** A cord is attached to and partially wound around a cylinder of weight W and radius r which rests on an incline as shown. Knowing that the coefficient of static friction between the cylinder and the incline is 0.30, find (*a*) the smallest allowable value of θ if the cylinder is to remain in equilibrium, (*b*) the corresponding value of the tension in the cord.

*** 8.27** A cord is attached to and partially wound around a cylinder of weight W and radius r which rests on an incline as shown. Knowing that $\theta = 40°$, find (*a*) the tension in the cord, (*b*) the smallest value of the coefficient of static friction between the cylinder and the incline for which equilibrium is maintained.

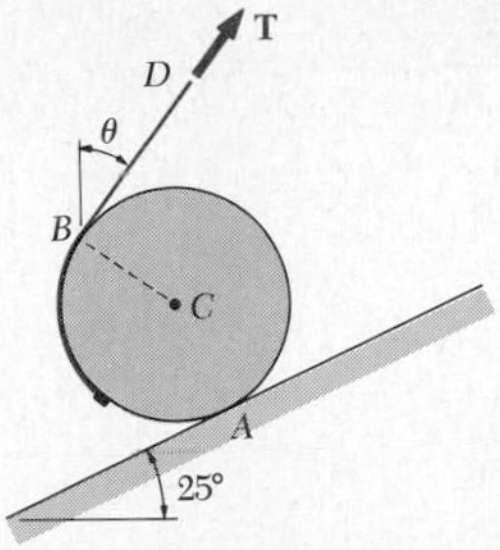

Fig. P8.26 and P8.27

8.28 A car is stopped with its front wheels resting against a curb when its driver starts the engine and tries to drive over the curb. Knowing that the radius of the wheels is 280 mm, that the coefficient of static friction between the tires and the pavement is 0.90, and that 60 percent of the weight of the car is distributed over its front wheels and 40 percent over its rear wheels, determine the largest curb height h that the car can negotiate, assuming (*a*) front-wheel drive, (*b*) rear-wheel drive.

Fig. P8.28

8.29 Solve Prob. 8.28, assuming that the weight of the car is equally distributed over its front and rear wheels.

8.30 The end A of a slender, uniform rod of length L and weight W bears on a horizontal surface, while its end B is supported by a cord BC. Knowing that the coefficients of friction are $\mu_s = 0.35$ and $\mu_k = 0.25$, determine (*a*) the maximum value of θ for which equilibrium is maintained, (*b*) the corresponding value of the tension in the cord.

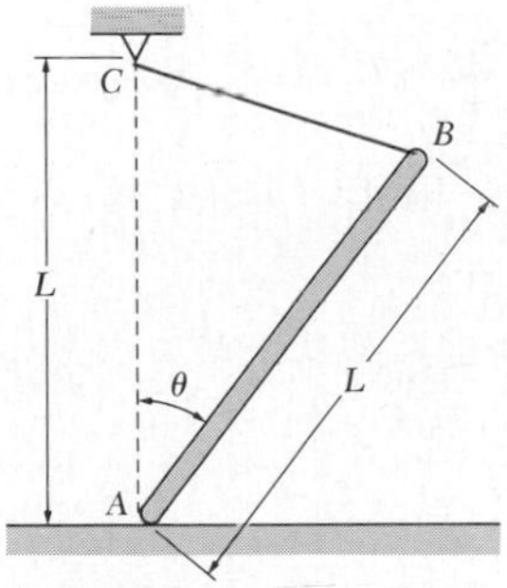

Fig. P8.30

8.31 Determine whether the rod of Prob. 8.30 is in equilibrium when $\theta = 30°$ and find the magnitude and direction of the friction force exerted on the rod.

8.32 A slender rod of length L is lodged between peg C and the vertical wall and supports a load **P** at end A. Knowing that $\theta = 30°$ and that the coefficient of static friction is 0.20 at both B and C, find the range of values of the ratio L/a for which equilibrium is maintained.

8.33 Solve Prob. 8.32, assuming that the coefficient of static friction is 0.20 at B and zero at C.

8.34 Solve Prob. 8.32, assuming that the coefficient of static friction is 0.20 at C and zero at B.

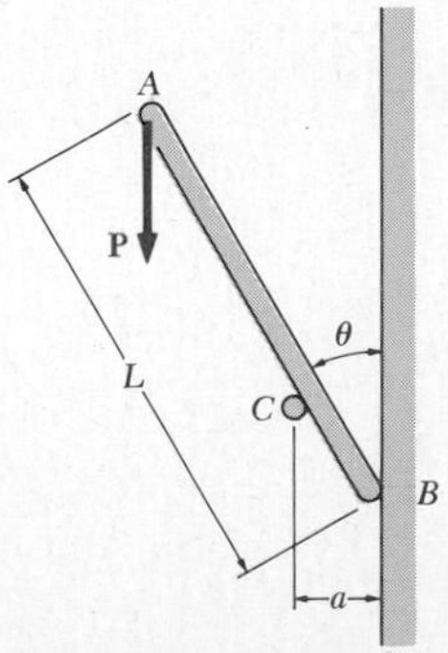

Fig. P8.32 and P8.35

8.35 A slender rod of length L is lodged between peg C and the vertical wall and supports a load **P** at end A. Knowing that $L = 10a$ and $\theta = 30°$ and that the coefficients of friction are $\mu_s = 0.20$ and $\mu_k = 0.15$ at C, and zero at B, determine whether the rod is in equilibrium.

8.36 Solve Prob. 8.35, assuming that $L = 5a$ and $\theta = 30°$ and that the coefficients of friction are $\mu_s = 0.20$ and $\mu_k = 0.15$ at B, and zero at C.

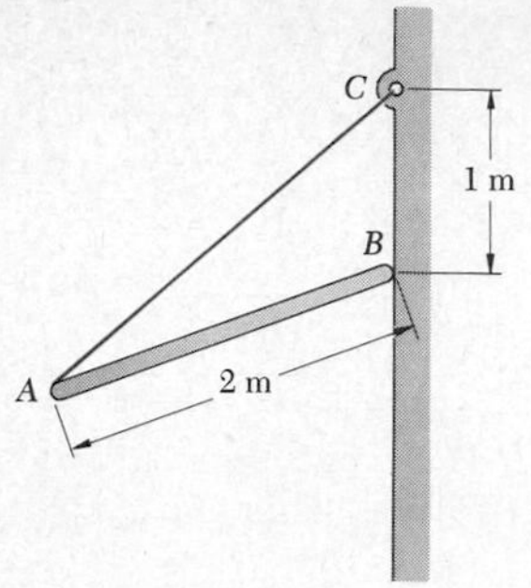

Fig. P8.37

8.37 A 10-kg rod AB, of length 2 m and of uniform cross section, is held in equilibrium as shown, with one end against a vertical wall and the other attached to a cord. Knowing that the coefficients of friction between the rod and the wall are $\mu_s = 0.35$ and $\mu_k = 0.25$, determine the range of values of the length L of the cord for which equilibrium is maintained.

8.38 Assuming that the length of cord AC in Prob. 8.37 is $L = 2.50$ m, determine whether the rod is in equilibrium and find the magnitude and direction of the friction force exerted on the rod.

8.39 Solve Prob. 8.37, assuming that the coefficients of friction between the rod and the wall are $\mu_s = 0.60$ and $\mu_k = 0.45$.

8.40 The shear shown is used to cut and trim electronic-circuit-board laminates. Knowing that the coefficient of kinetic friction between the blade and the vertical guide is 0.25, determine the force exerted by the edge E of the blade on the laminate.

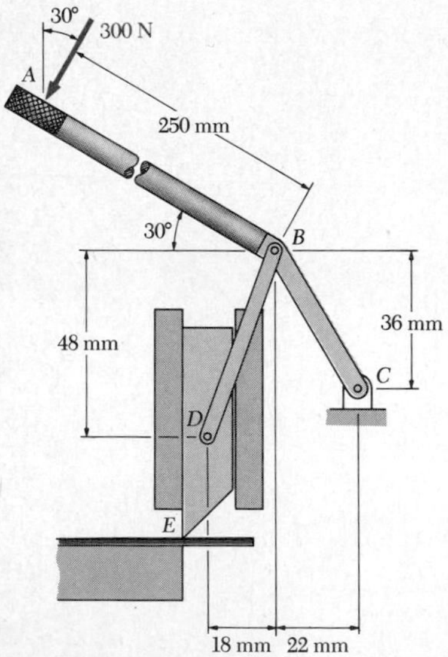

Fig. P8.40

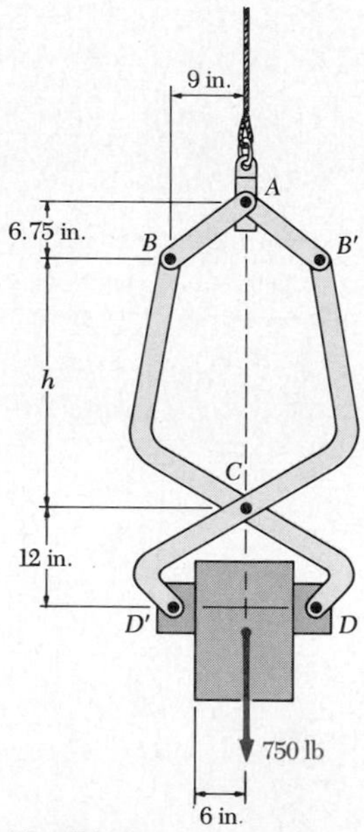

Fig. P8.41

8.41 The friction tongs shown are used to lift a 750-lb casting. Knowing that $h = 30$ in., determine the smallest allowable value of the coefficient of static friction between the casting and blocks D and D'.

8.42 For the friction tongs of Prob. 8.41, determine the smallest allowable value of h if the coefficient of static friction between the casting and blocks D and D' is 0.30.

8.43 Two large cylinders, each of radius $r = 600$ mm, rotate in opposite directions and form the main elements of a crusher for stone aggregate. The distance d is set equal to the maximum desired size of the crushed aggregate. If $d = 25$ mm and $\mu_s = 0.30$, determine the size s of the largest stones which will be pulled through the crusher by friction alone.

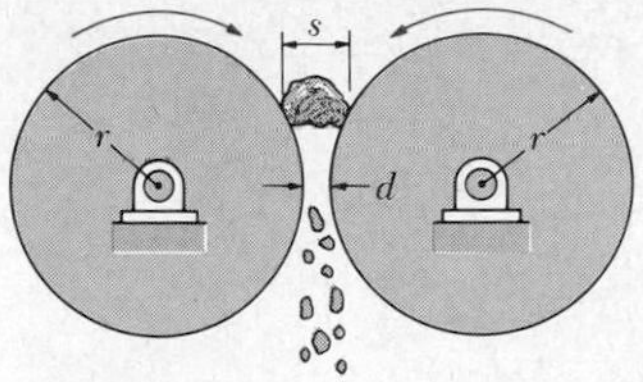

Fig. P8.43

8.44 A safety device used by workers climbing ladders fixed to high structures consists of a rail attached to the ladder and a sleeve which may slide on the flange of the rail. A chain connects the worker's belt to the end of an eccentric cam which may rotate about an axle attached to the sleeve at C. Determine the smallest allowable common value of the coefficient of static friction between the flange of the rail, the pins at A and B, and the eccentric cam if the sleeve is not to slide down when the chain is pulled vertically downward.

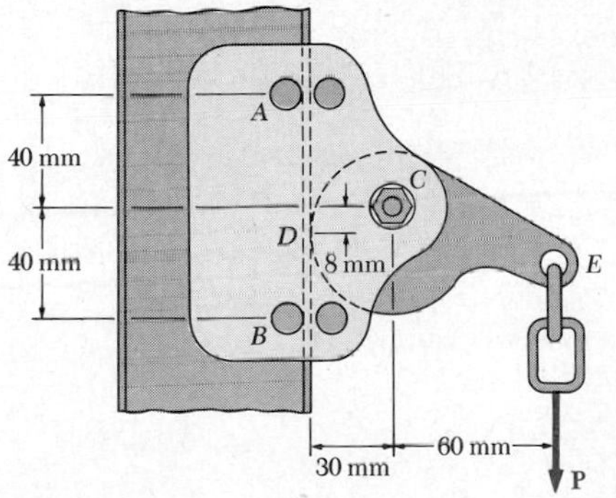

Fig. P8.44

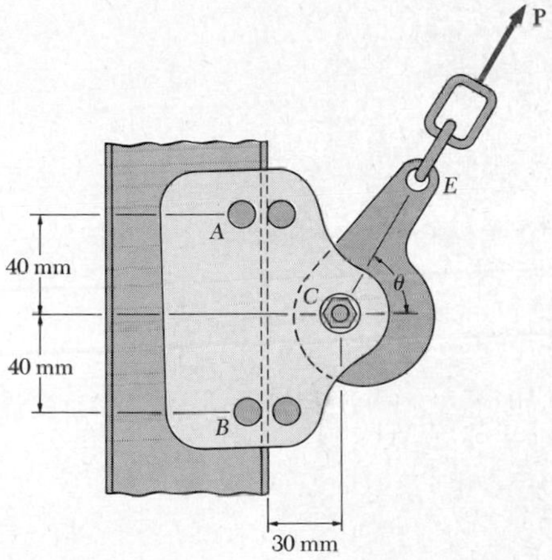

Fig. P8.45

8.45 To be of practical use, the safety sleeve described in the preceding problem must be free to slide along the rail when pulled upward. Determine the largest allowable value of the coefficient of static friction between the flange of the rail and the pins at A and B if the sleeve is to be free to slide when pulled as shown in the figure, assuming (*a*) $\theta = 70°$, (*b*) $\theta = 60°$, (*c*) $\theta = 50°$.

8.46 The uniform rod AB lies in a vertical plane with its ends resting against the surfaces AC and BC. Knowing that the coefficient of static friction is 0.30 between the rod and each of the surfaces, determine the range of values of θ corresponding to equilibrium when (*a*) $\alpha = 45°$, (*b*) $\alpha = 30°$.

8.47 The uniform rod AB lies in a vertical plane with its ends resting against the surfaces AC and BC. Determine the range of values of θ corresponding to equilibrium in terms of α and the angle of static friction ϕ_s between the rod and the two surfaces. Consider the case when (*a*) $\alpha > 2\phi_s$, (*b*) $\alpha \leq 2\phi_s$.

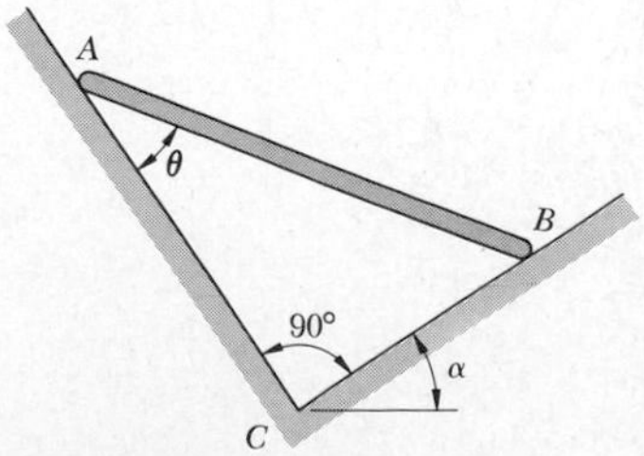

Fig. P8.46 and P8.47

8.48 The slender rod AB of length $l = 300$ mm is attached to a collar at A and rests on a wheel located at a vertical distance $a = 50$ mm from the horizontal rod on which the collar slides. Neglecting the friction at C and knowing that the coefficient of static friction between the collar and the horizontal rod is 0.25, determine the range of values of Q for which equilibrium is maintained when $P = 60$ N and $\theta = 20°$.

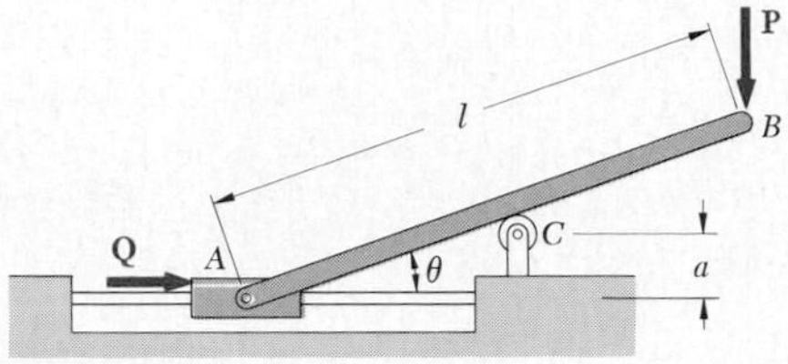

Fig. P8.48

8.49 Solve Prob. 8.48, assuming that the wheel at C is frozen in its bearings and that the coefficient of static friction between the wheel and rod AB is also 0.25.

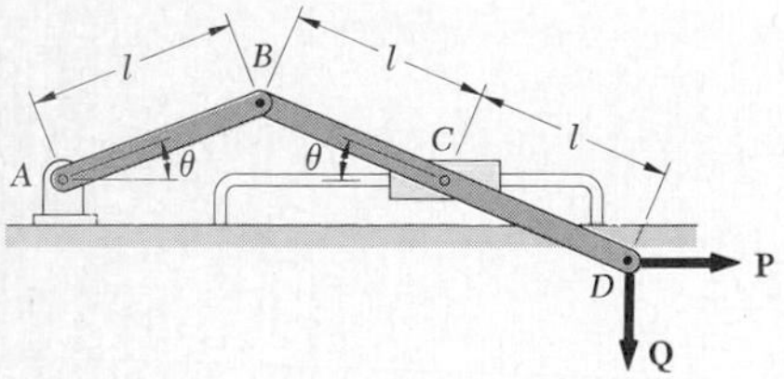

Fig. P8.50

8.50 The mechanism shown is acted upon by the force **P** of magnitude $P = 20$ lb. Knowing that $\theta = 25°$ and that the coefficient of static friction between collar C and the horizontal rod is 0.30, determine the range of values of Q for which equilibrium is maintained.

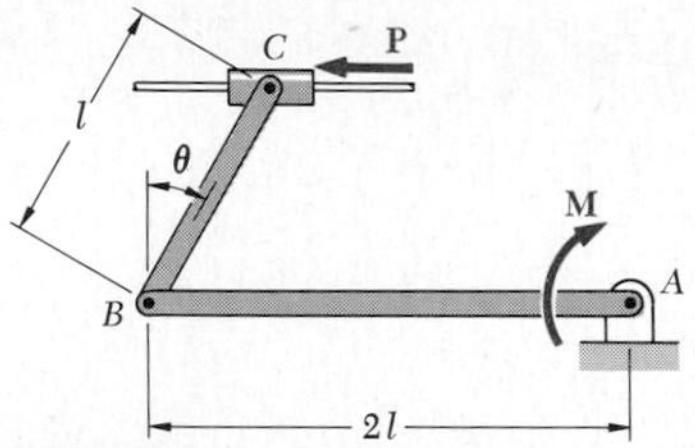

Fig. P8.51

8.51 Denoting by μ_s the coefficient of static friction between the collar C and the horizontal rod, determine the largest magnitude of the couple **M** for which equilibrium is maintained. Explain what happens if $\mu_s \geq \tan\theta$.

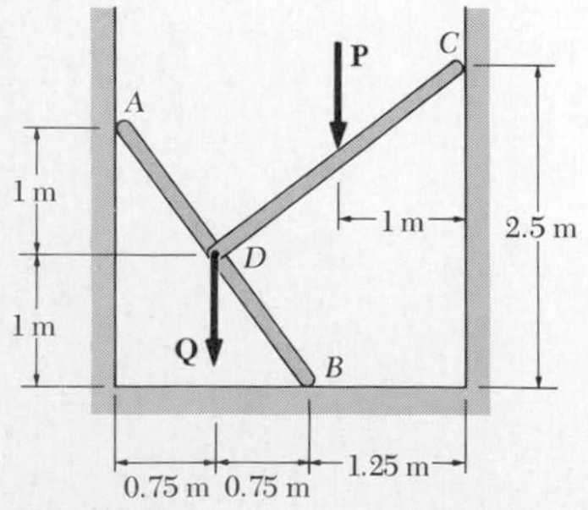

Fig. P8.52

***8.52** Two 2.5-m beams are pin-connected at D and support two loads **P** and **Q** as shown. Knowing that the coefficient of static friction is zero at A and 0.25 at B and C, determine the smallest value of P for which equilibrium is maintained when $Q = 550$ N. (*Hint.* Note that C moves up when B moves to the right.)

***8.53** Solve Prob. 8.52, assuming that the coefficient of static friction is 0.25 at all surfaces of contact.

***8.54** Two slender rods of negligible weight are pin-connected at A and attached to the 20-lb block B and the 60-lb block C as shown. The coefficient of static friction is 0.60 between all surfaces of contact. Determine the range of values of P for which equilibrium is maintained.

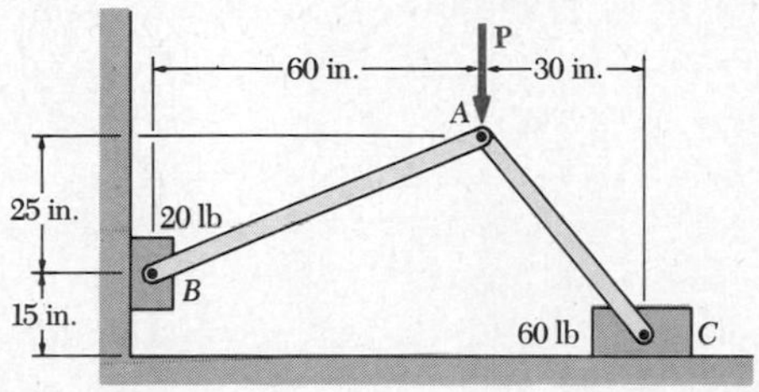

Fig. P8.54 and P8.55

***8.55** Two slender rods of negligible weight are pin-connected at A and attached to the 20-lb block B and the 60-lb block C as shown. A vertical load of magnitude $P = 300$ lb is applied at A. Determine (*a*) the minimum value of the coefficient of static friction (at both B and C) for which equilibrium is maintained, (*b*) whether sliding is impending at B or at C.

***8.56** Solve Prob. 8.55, assuming that $P = 200$ lb.

***8.57** A 3-m plank lies on two rollers A and B, located 1.5 m apart on a 10° incline. Roller A rotates slowly clockwise at a constant speed; roller B is idle. Initially, the plank is placed on the rollers with its center of gravity at A ($x = 0$) and its end C at B. Knowing that the coefficients of friction between the board and roller A are $\mu_s = 0.40$ and $\mu_k = 0.30$, (*a*) determine the value reached by x when the plank starts slipping on roller A, (*b*) describe what happens next.

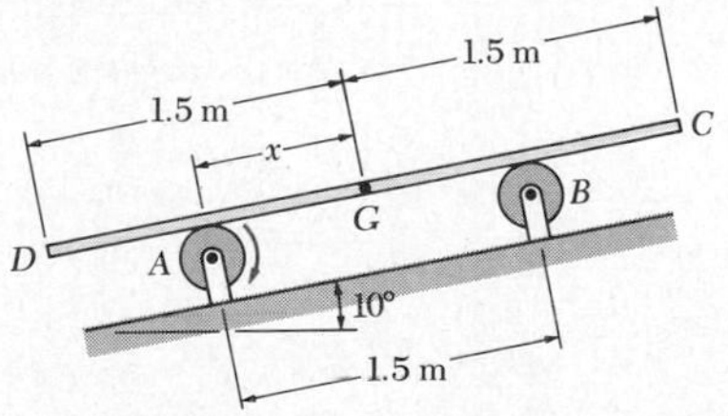

Fig. P8.57

***8.58** Solve Prob. 8.57, assuming that roller B is frozen in its bearings and that the coefficients of friction are the same at B as at A.

***8.59** The mathematical model shown, consisting of blocks connected by springs, has been developed for the analysis of a certain structure. The weight of each block is $W = 5$ N, the constant of each spring is $k = 20$ N/m, and the coefficient of friction between the base and each block is 0.40. Knowing that the initial tension in each spring is zero, construct a graph showing the magnitude of the force **P** versus the position of block A as P increases from zero to 5 N and then decreases to zero.

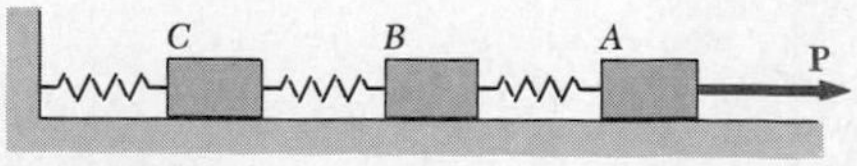

Fig. P8.59

8.5. Wedges.

Wedges are simple machines used to raise large stone blocks and other heavy loads. These loads may be raised by applying to the wedge a force usually considerably smaller than the weight of the load. Besides, because of the friction existing between the surfaces in contact, a wedge, if properly shaped, will remain in place after being forced under the load. Wedges may thus be used advantageously to make small adjustments in the position of heavy pieces of machinery.

Consider the block A shown in Fig. 8.7*a*. This block rests against a vertical wall B and is to be raised slightly by forcing a wedge C between block A and a second wedge D. We want to find the minimum value of the force **P** which must be applied to the wedge C to move the block. We shall assume that the weight **W** of the block is known. It may have been given in pounds or determined in newtons from the mass of the block expressed in kilograms.

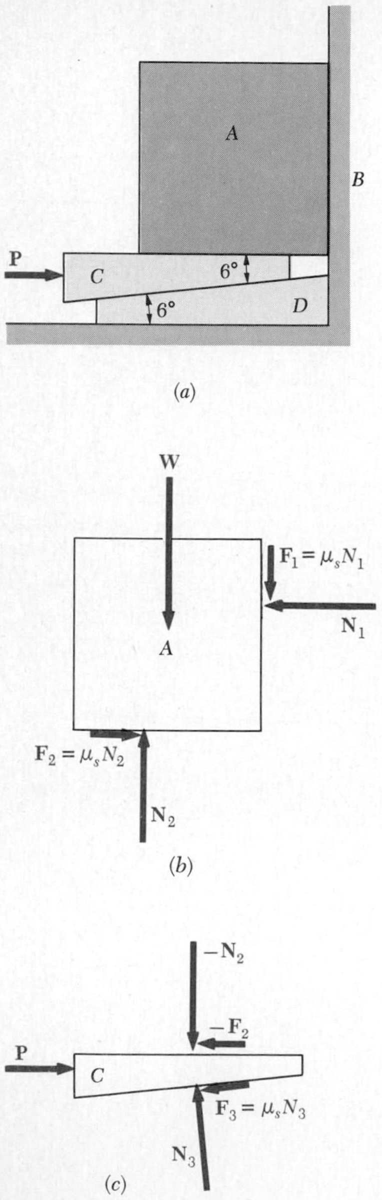

Fig. 8.7

The free-body diagrams of block A and of wedge C have been drawn in Fig. 8.7*b* and *c*. The forces acting on the block include its weight and the normal and friction forces at the surfaces of contact with wall B and wedge C. The magnitudes of the friction forces $\mathbf{F}_1$ and $\mathbf{F}_2$ are equal, respectively, to $\mu_s N_1$ and $\mu_s N_2$ since the motion of the block must be started. It is important to show the friction forces with their correct sense. Since the block will move upward, the force $\mathbf{F}_1$ exerted by the wall on the block must be directed downward. On the other hand, since the wedge C moves to the right, the relative motion of A with respect to C is to the left and the force $\mathbf{F}_2$ exerted by C on A must be directed to the right.

Considering now the free body C in Fig. 8.7*c*, we note that the forces acting on C include the applied force **P** and the normal and friction forces at the surfaces of contact with A and D. The weight of the wedge is small compared with the other forces involved and may be neglected. The forces acting on C are equal and opposite to the forces $\mathbf{N}_2$ and $\mathbf{F}_2$ acting on A and are denoted, respectively, by $-\mathbf{N}_2$ and $-\mathbf{F}_2$; the friction force $-\mathbf{F}_2$ must therefore be directed to the left. We check that the force $\mathbf{F}_3$ is also directed to the left.

The total number of unknowns involved in the two free-body diagrams may be reduced to four if the friction forces are expressed in terms of the normal forces. Expressing that block A and wedge C are in equilibrium will provide four equations which may be solved to obtain the magnitude of **P**. It should be noted that in the example considered here, it will be more convenient to replace each pair of normal and friction forces by their resultant. Each free body is then subjected to only three forces, and the problem may be solved by drawing the corresponding force triangles (see Sample Prob. 8.4).

8.6. Square-threaded Screws.

Square-threaded screws are frequently used in jacks, presses, and other mechanisms. Their analysis is similar to that of a block sliding along an inclined plane.

Consider the jack shown in Fig. 8.8. The screw carries a load **W** and is supported by the base of the jack. Contact between screw and base takes place along a portion of their threads. By applying a force **P** on the handle, the screw may be made to turn and to raise the load **W**.

The thread of the base has been unwrapped and shown as a straight line in Fig. 8.9*a*. The correct slope was obtained by plotting horizontally the product $2\pi r$, where r is the mean radius of the thread, and vertically the *lead* L of the screw, i.e., the distance through which the screw advances in one turn. The angle θ this line forms with the horizontal is the *lead angle*. Since the force of friction between two surfaces in contact does not depend upon the area of contact, the two threads may be assumed to be in contact over a much smaller area than they actually are and the screw may be represented by the block shown in Fig. 8.9*a*. It should be noted, however, that in this analysis of the jack, the friction between cap and screw is neglected.

The free-body diagram of the block should include the load **W**, the reaction **R** of the base thread, and a horizontal force **Q** having the same effect as the force **P** exerted on the handle. The force **Q** should have the same moment as **P** about the axis of the screw and its magnitude should thus be $Q = Pa/r$. The force **Q**, and thus the force **P** required to raise the load **W**, may be obtained from the free-body diagram shown in Fig. 8.9*a*. The friction angle is taken equal to ϕ_s since the load will presumably be raised through a succession of short strokes. In mechanisms providing for the continuous rotation of a screw, it may be desirable to distinguish between the force required to start motion (using ϕ_s) and that required to maintain motion (using ϕ_k).

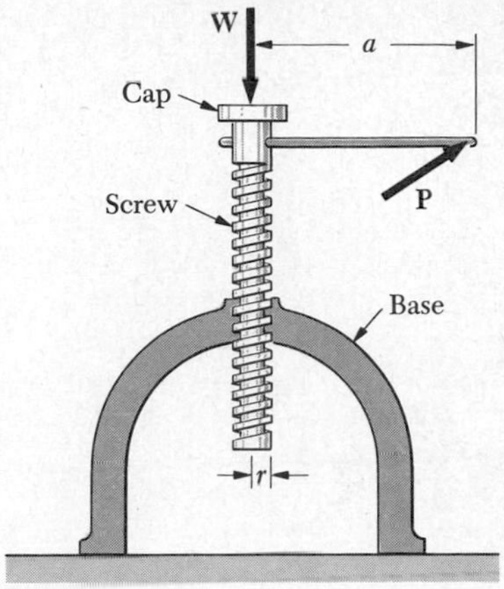

Fig. 8.8

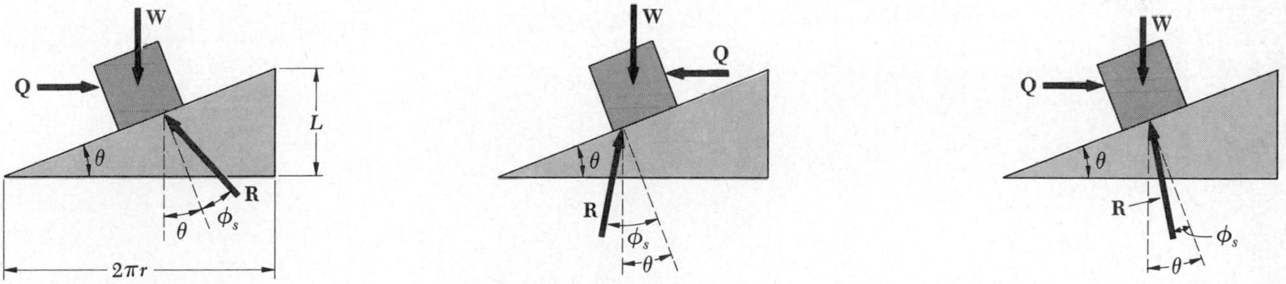

(*a*) Impending motion upward (*b*) Impending motion downward with $\phi_s > \theta$ (*c*) Impending motion downward with $\phi_s < \theta$

Fig. 8.9 Block-and-incline analysis of a screw.

If the friction angle ϕ_s is larger than the lead angle θ, the screw is said to be *self-locking;* it will remain in place under the load. To lower the load, we must then apply the force shown in Fig. 8.9*b*. If ϕ_s is smaller than θ, the screw will unwind under the load; it is then necessary to apply the force shown in Fig. 8.9*c* to maintain equilibrium.

The lead of a screw should not be confused with its *pitch*. The lead was defined as the distance through which the screw advances in one turn; the pitch is the distance measured between two consecutive threads. While lead and pitch are equal in the case of *single-threaded* screws, they are different in the case of *multiple-threaded* screws, i.e., screws having several independent threads. It is easily verified that for double-threaded screws, the lead is twice as large as the pitch; for triple-threaded screws, it is three times as large as the pitch; etc.

SAMPLE PROBLEM 8.4

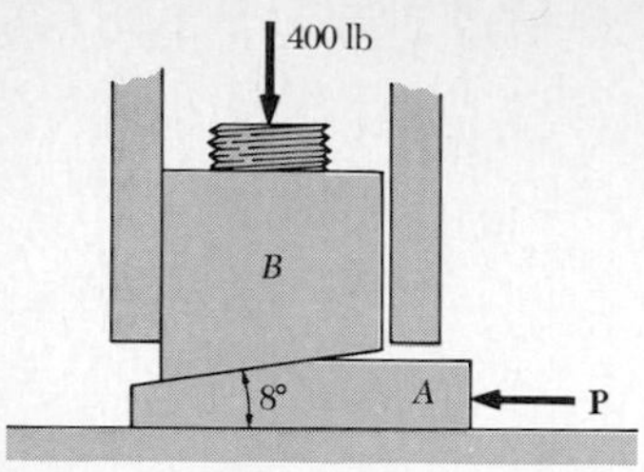

The position of the machine block B is adjusted by moving the wedge A. Knowing that the coefficient of static friction is 0.35 between all surfaces of contact, determine the force **P** required (*a*) to raise block B, (*b*) to lower block B.

Solution. For each part, the free-body diagrams of block B and wedge A are drawn, together with the corresponding force triangles, and the law of sines is used to find the desired forces. We note that since $\mu_s = 0.35$, the angle of friction is

$$\phi_s = \tan^{-1} 0.35 = 19.3°$$

a. Force P to Raise Block

Free Body: Block B

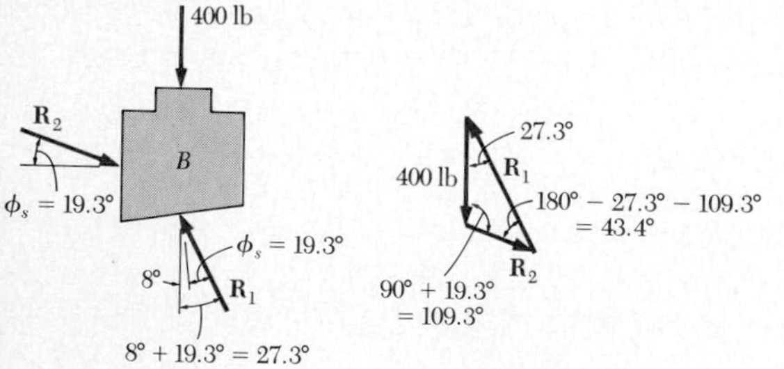

$$\frac{R_1}{\sin 109.3°} = \frac{400 \text{ lb}}{\sin 43.4°}$$

$$R_1 = 549 \text{ lb}$$

Free Body: Wedge A

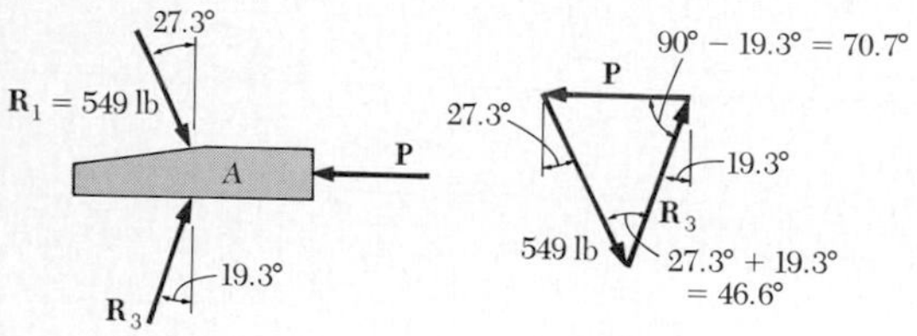

$$\frac{P}{\sin 46.6°} = \frac{549 \text{ lb}}{\sin 70.7°}$$

$$P = 423 \text{ lb} \qquad \mathbf{P} = 423 \text{ lb} \leftarrow$$

b. Force P to Lower Block

Free Body: Block B

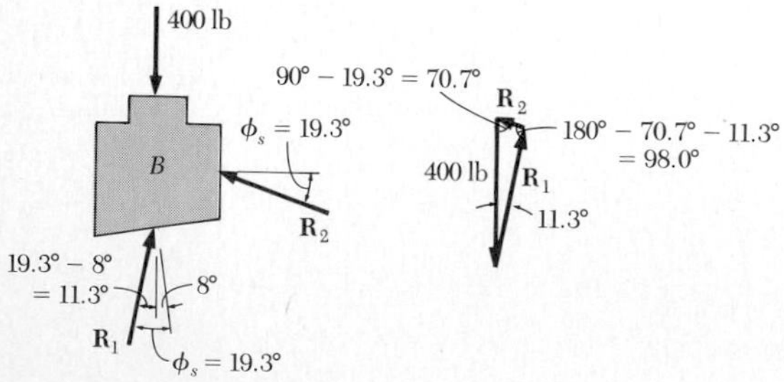

$$\frac{R_1}{\sin 70.7°} = \frac{400 \text{ lb}}{\sin 98.0°}$$

$$R_1 = 381 \text{ lb}$$

Free Body: Wedge A

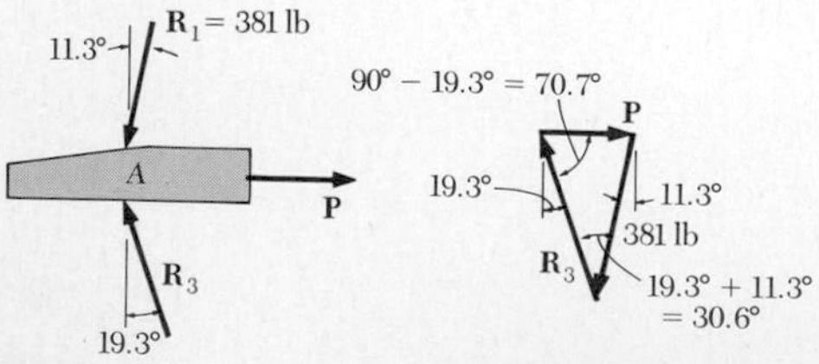

$$\frac{P}{\sin 30.6°} = \frac{381 \text{ lb}}{\sin 70.7°}$$

$$P = 206 \text{ lb} \qquad \mathbf{P} = 206 \text{ lb} \rightarrow$$

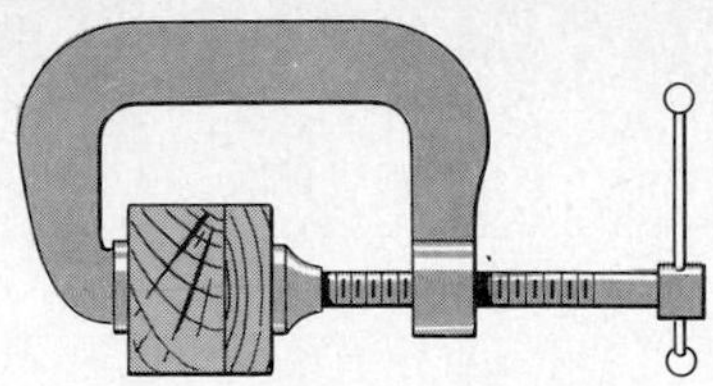

SAMPLE PROBLEM 8.5

A clamp is used to hold two pieces of wood together as shown. The clamp has a double square thread of mean diameter equal to 10 mm and with a pitch of 2 mm. The coefficient of friction between threads is $\mu_s = 0.30$. If a maximum torque of 40 N·m is applied in tightening the clamp, determine (*a*) the force exerted on the pieces of wood, (*b*) the torque required to loosen the clamp.

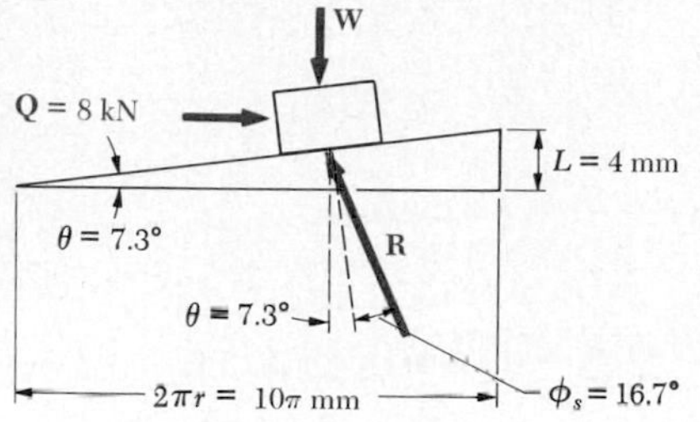

a. **Force Exerted by Clamp.** The mean radius of the screw is $r = 5$ mm. Since the screw is double-threaded, the lead L is equal to twice the pitch: $L = 2(2 \text{ mm}) = 4$ mm. The lead angle θ and the friction angle ϕ_s are obtained by writing

$$\tan\theta = \frac{L}{2\pi r} = \frac{4 \text{ mm}}{10\pi \text{ mm}} = 0.1273 \qquad \theta = 7.3°$$

$$\tan\phi_s = \mu_s = 0.30 \qquad \phi_s = 16.7°$$

The force **Q** which should be applied to the block representing the screw is obtained by expressing that its moment Qr about the axis of the screw is equal to the applied torque.

$$Q(5 \text{ mm}) = 40 \text{ N·m}$$

$$Q = \frac{40 \text{ N·m}}{5 \text{ mm}} = \frac{40 \text{ N·m}}{5 \times 10^{-3} \text{ m}} = 8000 \text{ N} = 8 \text{ kN}$$

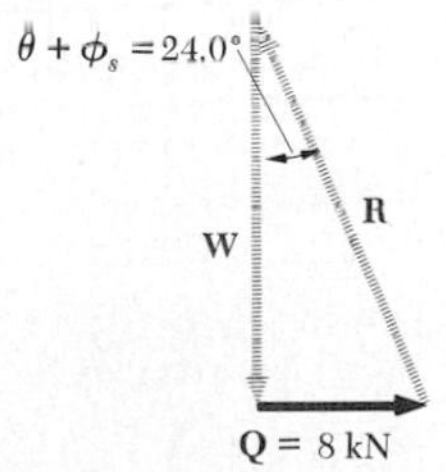

The free-body diagram and the corresponding force triangle may now be drawn for the block; the magnitude of the force **W** exerted on the pieces of wood is obtained by solving the triangle.

$$W = \frac{Q}{\tan(\theta + \phi_s)} = \frac{8 \text{ kN}}{\tan 24.0°}$$

$$W = 17.97 \text{ kN} \quad ◄$$

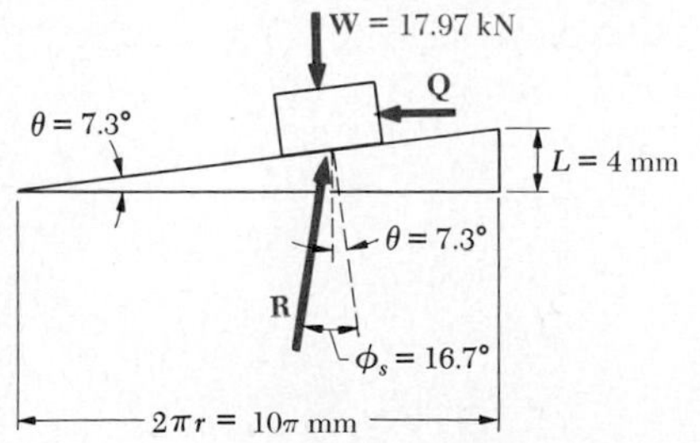

b. **Torque Required to Loosen Clamp.** The force **Q** required to loosen the clamp and the corresponding torque are obtained from the free-body diagram and force triangle shown.

$$Q = W\tan(\phi_s - \theta) = (17.97 \text{ kN}) \tan 9.4°$$
$$= 2.975 \text{ kN}$$

$$\text{Torque} = Qr = (2.975 \text{ kN})(5 \text{ mm})$$
$$= (2.975 \times 10^3 \text{ N})(5 \times 10^{-3} \text{ m}) = 14.87 \text{ N·m}$$

$$\text{Torque} = 14.87 \text{ N·m} \quad ◄$$

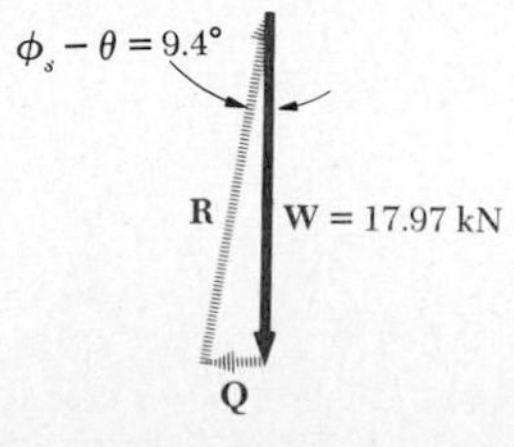

Problems

8.60 The machine part *ABC* is supported by a frictionless hinge at *B* and a 10° wedge at *C*. Knowing that the coefficient of static friction is 0.20 at both surfaces of the wedge, determine (*a*) the force **P** required to move the wedge to the left, (*b*) the components of the corresponding reaction at *B*.

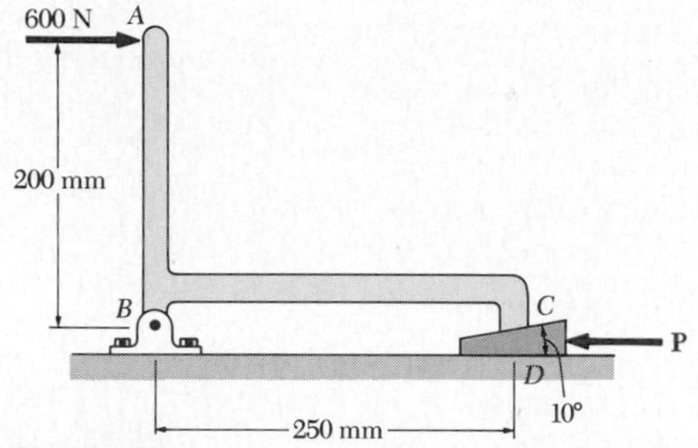

Fig. P8.60

8.61 Solve Prob. 8.60, assuming that the wedge is to be moved to the right.

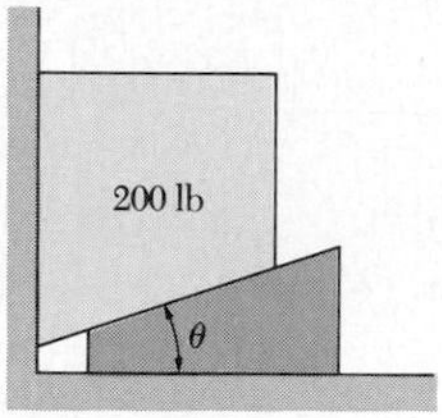

Fig. P8.62

8.62 A 200-lb block rests as shown on a wedge of negligible weight. Knowing that the coefficient of static friction is 0.30 at all surfaces of contact, determine the angle θ for which sliding is impending and compute the corresponding value of the normal force exerted on the block by the vertical wall.

8.63 Solve Prob. 8.62, assuming (*a*) that the coefficient of static friction is zero between the block and the vertical wall and 0.30 at the other two surfaces of contact, (*b*) that the coefficient of static friction is zero between the block and the wedge and 0.30 at the other two surfaces of contact.

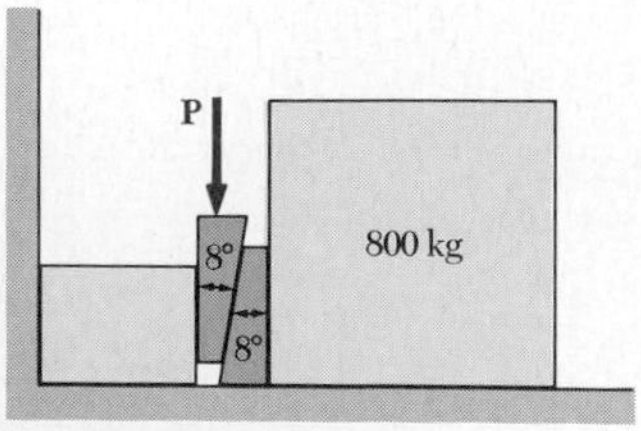

Fig. P8.64

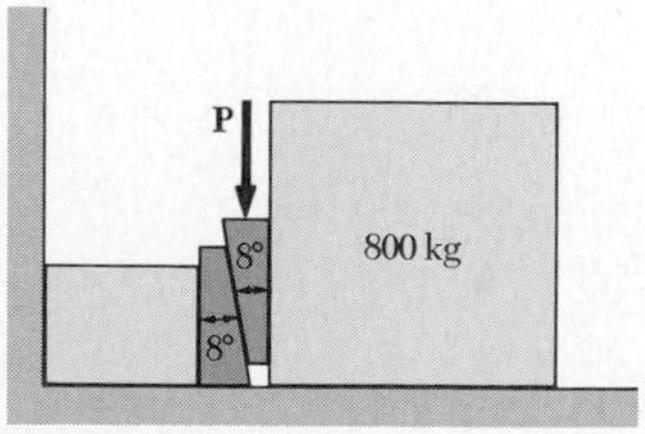

Fig. P8.65

8.64 and 8.65 Two 8° wedges of negligible weight are used to move and position the 800-kg block. Knowing that the coefficient of static friction is 0.25 at all surfaces of contact, determine the smallest force **P** which should be applied as shown to one of the wedges.

8.66 The elevation of the end of the steel floor beam is adjusted by means of the steel wedges *E* and *F*. The steel base plate *CD* has been welded to the lower flange of the beam; the end reaction of the beam is known to be 15,000 lb. The coefficient of static friction is 0.30 between two steel surfaces and 0.60 between steel and concrete. If horizontal motion of the base plate is prevented by the force **Q**, determine (*a*) the force **P** required to raise the beam, (*b*) the corresponding force **Q**.

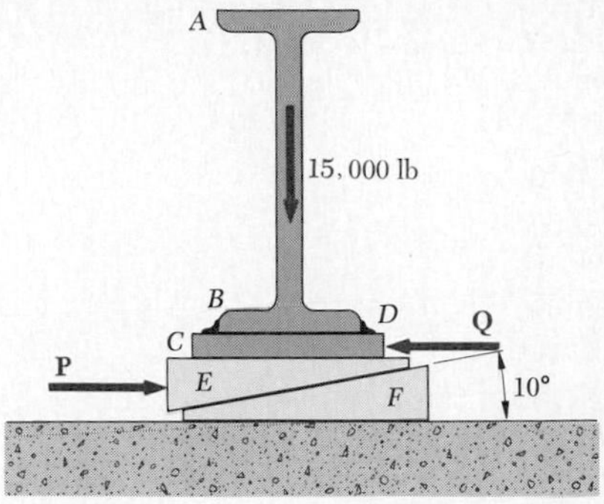

Fig. P8.66

8.67 Solve Prob. 8.66, assuming that the end of the beam is to be lowered.

8.68 A wedge *A* of negligible weight is to be driven between two 40-kg plates *B* and *C*. The coefficient of static friction between all surfaces of contact is 0.35. Determine the magnitude of the force **P** required to start moving the wedge (*a*) if the plates are equally free to move, (*b*) if plate *C* is securely bolted to the surface.

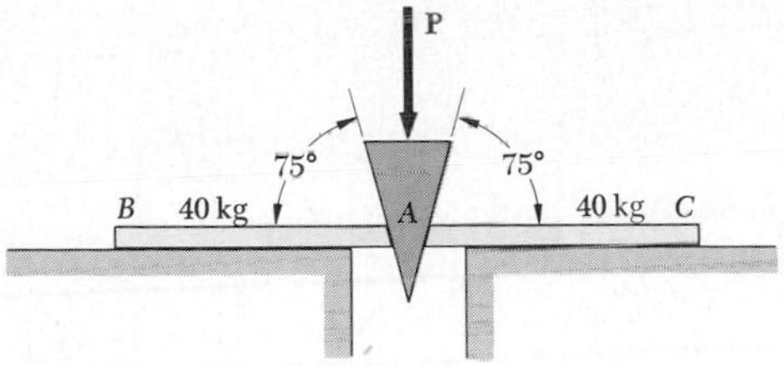

Fig. P8.68

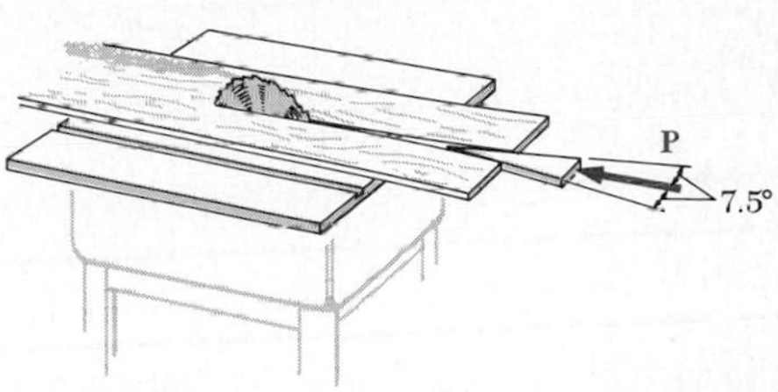

Fig. P8.69

8.69 A 15° wedge is forced into a saw cut to prevent binding of the circular saw. The coefficient of static friction between the wedge and the wood is 0.25. Knowing that a horizontal force **P** of magnitude 140 N was required to insert the wedge, determine the magnitude of the forces exerted on the board by the wedge after it has been inserted.

8.70 The spring of the door latch has a constant of 300 N/m and in the position shown exerts a 2.4-N force on the bolt. The coefficient of static friction between the bolt and the guide plate is 0.40; all other surfaces are well lubricated and may be assumed frictionless. Determine the magnitude of the force **P** required to start closing the door.

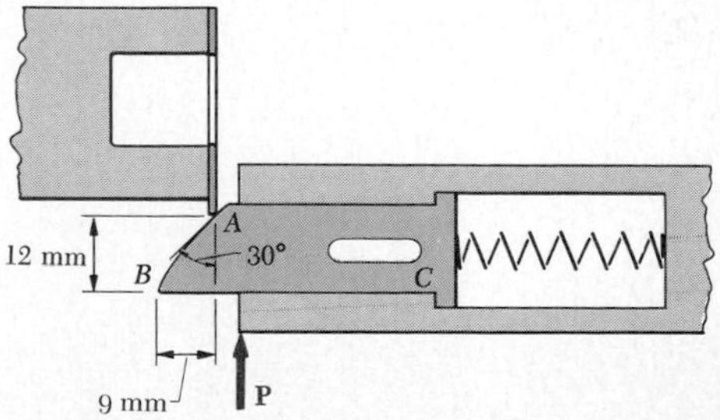

Fig. P8.70

8.71 In Prob. 8.70, determine the angle which the face of the bolt should form with the line *BC* if the force **P** required to close the door is to be the same for both the position shown and the position when *B* is almost at the guide plate.

8.72 A 5° wedge is to be forced under a 1400-lb machine base at *A*. Knowing that the coefficient of static friction is 0.20 at all surfaces, (*a*) determine the force **P** required to move the wedge, (*b*) indicate whether the machine will move.

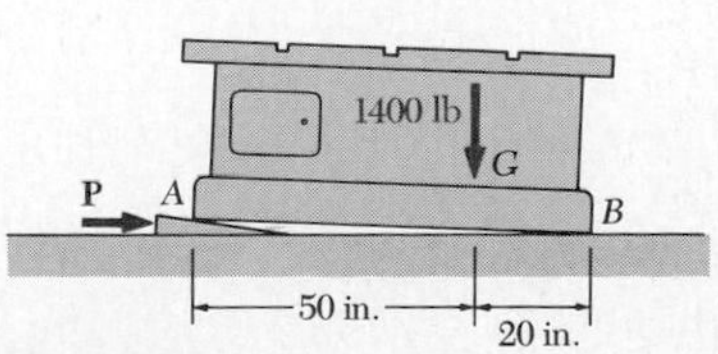

Fig. P8.72

8.73 Solve Prob. 8.72, assuming that the wedge is to be forced under the machine base at *B* instead of *A*.

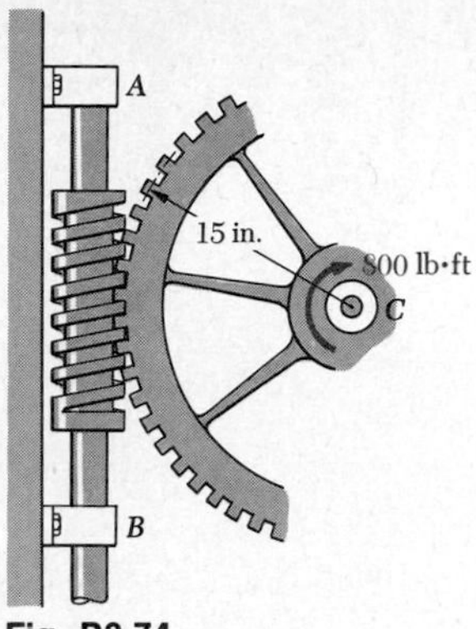

Fig. P8.74

8.74 The square-threaded worm gear shown has a mean radius of 2 in. and a pitch of $\frac{1}{2}$ in. The large gear is subjected to a constant clockwise torque of 800 lb·ft. Knowing that the coefficient of static friction between gear teeth is 0.10, determine the torque which must be applied to shaft *AB* in order to rotate the large gear counterclockwise. Neglect friction in the bearings at *A*, *B*, and *C*.

8.75 In Prob. 8.74, determine the torque which must be applied to shaft *AB* in order to rotate the large gear clockwise.

8.76 The main features of a screw-luffing crane are shown. Distances *AD* and *CD* are each 9 m. The position of the 6-Mg boom *CDE* is controlled by the screw *ABC*, which is double-threaded at each end (left-handed thread at *A*, right-handed thread at *C*). Each thread has a pitch of 16 mm and a mean diameter of 200 mm. If the coefficient of static friction is 0.08, determine the magnitude of the couple which must be applied to the screw (*a*) to raise the boom, (*b*) to lower the boom.

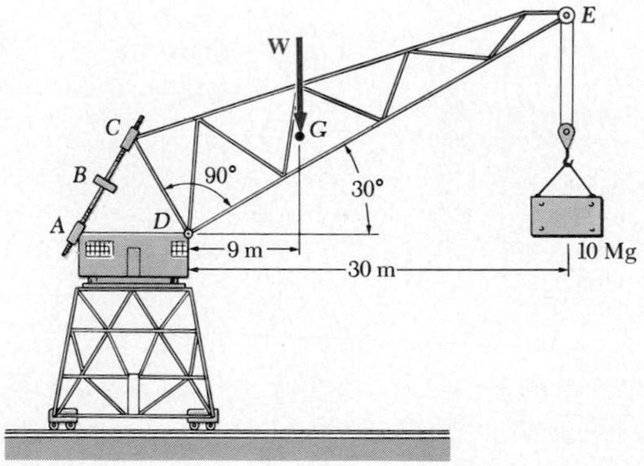

Fig. P8.76

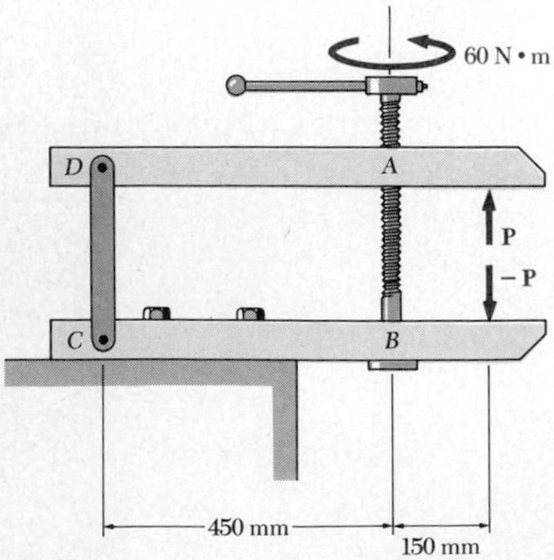

Fig. P8.78

8.77 Derive the following formulas relating the load **W** and the force **P** exerted on the handle of the jack discussed in Sec. 8.6: (*a*) $P = (Wr/a)\tan(\theta + \phi_s)$, to raise the load; (*b*) $P = (Wr/a)\tan(\phi_s - \theta)$, to lower the load if the screw is self-locking; (*c*) $P = (Wr/a)\tan(\theta - \phi_s)$, to hold the load if the screw is not self-locking.

8.78 In the vise shown, the screw is single-threaded in the upper member; it passes through the lower member and is held by a frictionless washer. The pitch of the screw is 3 mm, its mean radius is 12 mm, and the coefficient of static friction is 0.15. Determine the magnitude *P* of the forces exerted by the jaws when a 60-N·m torque is applied to the screw.

8.79 Solve Prob. 8.78, assuming that the screw is single-threaded at both *A* and *B* (right-handed thread at *A* and left-handed thread at *B*).

8.80 The vise shown consists of two members connected by two double-threaded screws of mean radius 0.20 in. and pitch 0.06 in. The lower member is threaded at A and B ($\mu_s = 0.35$), but the upper member is not threaded. It is desired to apply two equal and opposite forces of 100 lb on the blocks held between the jaws. (*a*) What screw should be adjusted first? (*b*) What is the maximum torque applied in tightening the second screw?

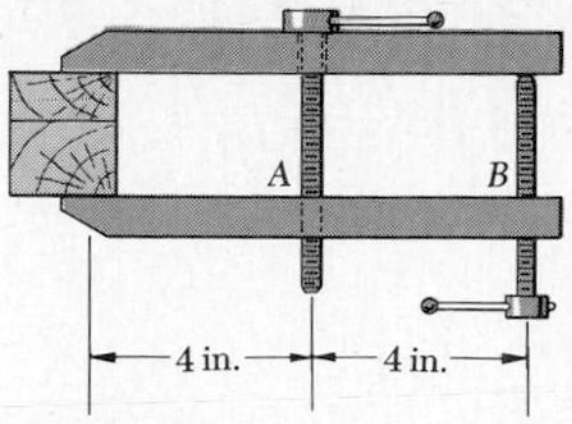

Fig. P8.80

8.81 Solve part *b* of Prob. 8.80, assuming that the wrong screw is adjusted first.

8.82 In the machinist's vise shown, the movable jaw D is rigidly attached to the tongue AB which fits loosely into the fixed body of the vise. The screw is single-threaded into the fixed base and has a mean diameter of 15 mm and a pitch of 6 mm. The coefficient of static friction is 0.25 between the threads and also between the tongue and the body. Neglecting bearing friction between the screw and the movable head, determine the torque which must be applied to the handle in order to produce a clamping force of 4 kN.

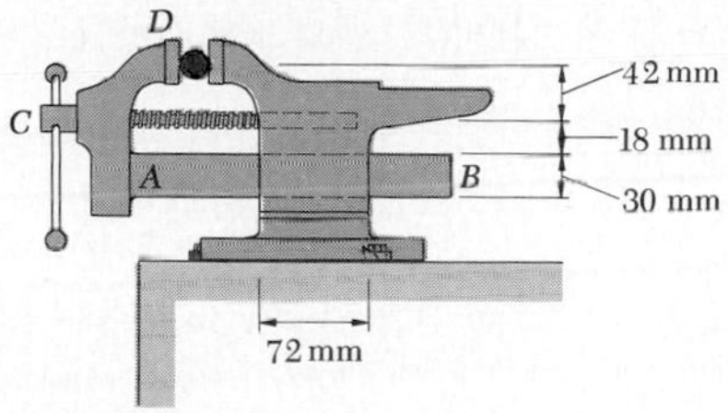

Fig. P8.82

8.83 In Prob. 8.82, a clamping force of 4 kN was obtained by tightening the vise. Determine the torque which must be applied to the screw to loosen the vise.

*** 8.7. Journal Bearings. Axle Friction.** Journal bearings are used to provide lateral support to rotating shafts and axles. Thrust bearings, which will be studied in the next section, are used to provide axial support to shafts and axles. If the journal bearing is fully lubricated, the frictional resistance depends upon the speed of rotation, the clearance between axle and bearing, and the viscosity of the lubricant. As indicated in Sec. 8.1, such problems are studied in fluid mechanics. The methods of this chapter, however, may be applied to the study of axle friction when the bearing is not lubricated or only partially lubricated. We may then assume that the axle and the bearing are in direct contact along a single straight line.

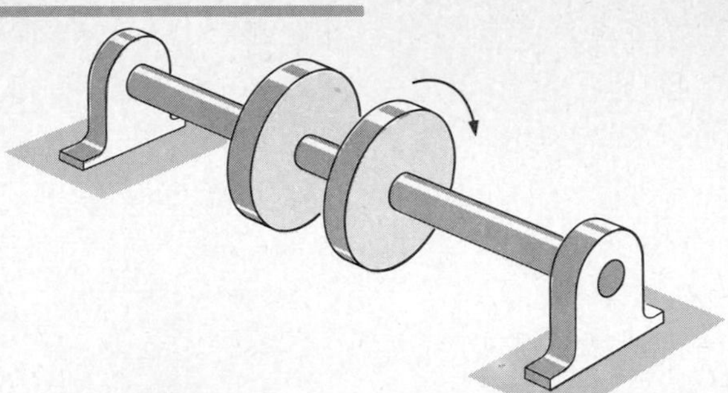

(a)

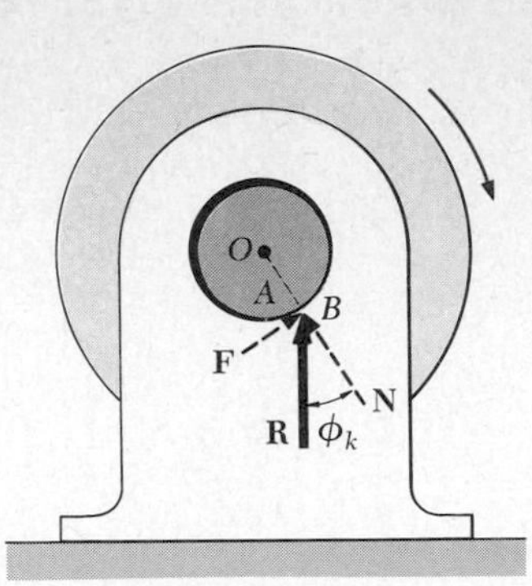

(b)

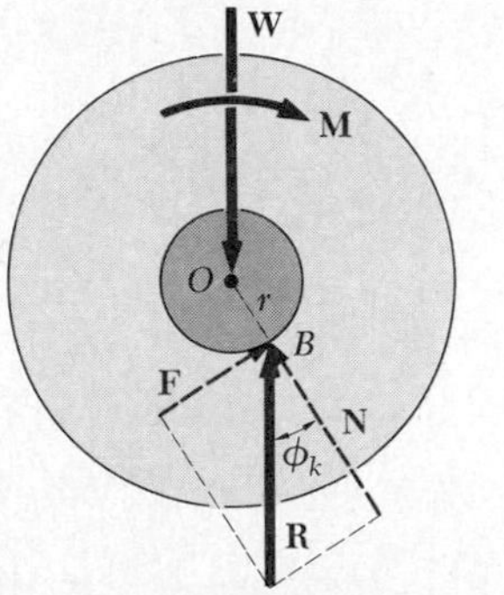

(c)

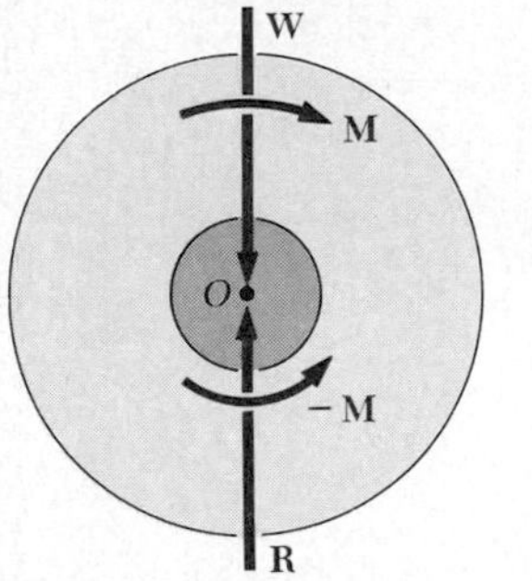

(d)

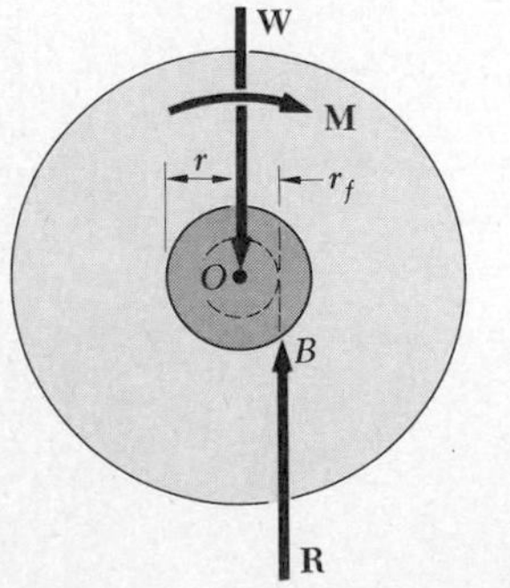

(e)

Fig. 8.10

Consider two wheels, each of weight **W**, rigidly mounted on an axle supported symmetrically by two journal bearings (Fig. 8.10*a*). If the wheels rotate, we find that to keep them rotating at constant speed, it is necessary to apply to each of them a couple **M**. A free-body diagram has been drawn in Fig. 8.10*c*, which represents one of the wheels and the corresponding half axle in projection on a plane perpendicular to the axle. The forces acting on the free body include the weight **W** of the wheel, the couple **M** required to maintain its motion, and a force **R** representing the reaction of the bearing. This force is vertical, equal, and opposite to **W** but does not pass through the center O of the axle; **R** is located to the right of O at a distance such that its moment about O balances the moment **M** of the couple. Contact between the axle and bearing, therefore, does not take place at the lowest point A when the axle rotates. It takes place at point B (Fig. 8.10*b*) or, rather, along a straight line intersecting the plane of the figure at B. Physically, this is explained by the fact that when the wheels are set in motion, the axle "climbs" in the bearings until slippage occurs. After sliding back slightly, the axle settles more or less in the position shown. This position is such that the angle between the reaction **R** and the normal to the surface of the bearing is equal to the angle of kinetic friction ϕ_k. The distance from O to the line of action of **R** is thus $r \sin \phi_k$, where r is the radius of the axle. Writing that $\Sigma M_O = 0$ for the forces acting on the free body considered, we obtain the magnitude of the couple **M** required to overcome the frictional resistance of one of the bearings:

$$M = Rr \sin \phi_k \tag{8.5}$$

Observing that, for small values of the angle of friction, $\sin \phi_k$ may be replaced by $\tan \phi_k$, that is, by μ_k, we write the approximate formula

$$M \approx Rr\mu_k \tag{8.6}$$

In the solution of certain problems, it may be more convenient to let the line of action of **R** pass through O, as it does when the axle does not rotate. A couple $-$**M** of the same magnitude as the couple **M** but of opposite sense must then be added to the reaction **R** (Fig. 8.10*d*). This couple represents the frictional resistance of the bearing.

In case a graphical solution is preferred, the line of action of **R** may be readily drawn (Fig. 8.10*e*) if we note that it must be tangent to a circle centered at O and of radius

$$r_f = r \sin \phi_k \approx r\mu_k \tag{8.7}$$

This circle is called the *circle of friction* of the axle and bearing and is independent of the loading conditions of the axle.

*8.8. Thrust Bearings. Disk Friction.

Thrust bearings are used to provide axial support to rotating shafts and axles. They are of two types: (1) *end bearings* and (2) *collar bearings* (Fig. 8.11). In the case of collar bearings, friction forces develop between the two ring-shaped areas which are in contact. In the case of end bearings, friction takes place over full circular areas, or over ring-shaped areas when the end of the shaft is hollow. Friction between circular areas, called *disk friction*, also occurs in other mechanisms, such as *disk clutches*.

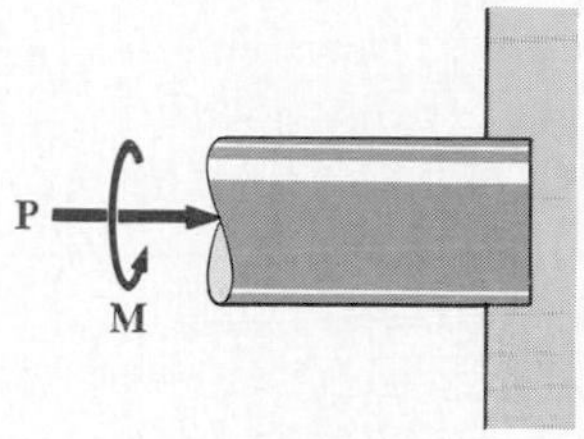

(*a*) End bearing

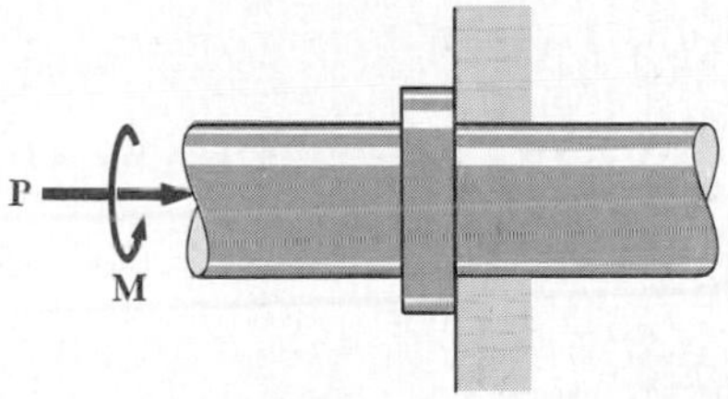

(*b*) Collar bearing

Fig. 8.11 Thrust bearings.

To obtain a formula which is valid in the most general case of disk friction, we shall consider a rotating hollow shaft. A couple **M** keeps the shaft rotating at constant speed while a force **P** maintains it in contact with a fixed bearing (Fig. 8.12). Contact between the shaft and the bearing takes place over a ring-shaped area of inner radius R_1 and outer radius R_2. Assuming that the pressure between the two surfaces in contact is uniform, we find that the magnitude of the normal force $\Delta\mathbf{N}$ exerted on an element of area ΔA is $\Delta N = P\,\Delta A/A$, where $A = \pi(R_2^2 - R_1^2)$, and that the magnitude of the friction force $\Delta\mathbf{F}$ acting on ΔA is $\Delta F = \mu_k\,\Delta N$. Denoting by r the distance from the axis of the shaft to the element of area ΔA, we express as follows the moment ΔM of $\Delta\mathbf{F}$ about the axis of the shaft:

$$\Delta M = r\,\Delta F = \frac{r\mu_k P\,\Delta A}{\pi(R_2^2 - R_1^2)}$$

The equilibrium of the shaft requires that the moment **M** of the couple applied to the shaft be equal in magnitude to the sum of the moments of the friction forces $\Delta\mathbf{F}$. Replacing ΔA by the infinitesimal element $dA = r\,d\theta\,dr$ used with polar coordinates, and integrating over the area of contact, we thus obtain the following expression for the magnitude of the couple **M**

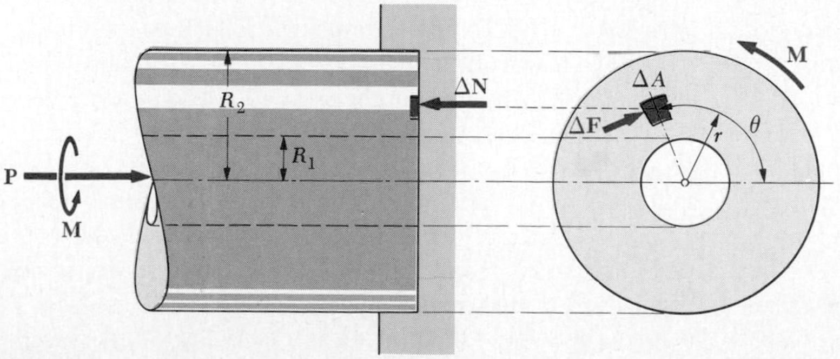

Fig. 8.12

required to overcome the frictional resistance of the bearing:

$$M = \frac{\mu_k P}{\pi(R_2^2 - R_1^2)} \int_0^{2\pi} \int_{R_1}^{R_2} r^2\, dr\, d\theta$$

$$= \frac{\mu_k P}{\pi(R_2^2 - R_1^2)} \int_0^{2\pi} \tfrac{1}{3}(R_2^3 - R_1^3)\, d\theta$$

$$M = \tfrac{2}{3}\mu_k P \frac{R_2^3 - R_1^3}{R_2^2 - R_1^2} \tag{8.8}$$

When contact takes place over a full circle of radius R, formula (8.8) reduces to

$$M = \tfrac{2}{3}\mu_k PR \tag{8.9}$$

The value of M is then the same as would be obtained if contact between shaft and bearing took place at a single point located at a distance $2R/3$ from the axis of the shaft.

The largest torque which may be transmitted by a disk clutch without causing slippage is given by a formula similar to (8.9), where μ_k has been replaced by the coefficient of static friction μ_s.

***8.9. Wheel Friction. Rolling Resistance.** The wheel is one of the most important inventions of our civilization. Its use makes it possible to move heavy loads with relatively little effort. Because the point of the wheel in contact with the ground at any given instant has no relative motion with respect to the ground, the wheel eliminates the large friction forces which would arise if the load were in direct contact with the ground. In practice, however, the wheel is not perfect, and some resistance to its motion exists. This resistance has two distinct causes. It is due (1) to a combined effect of axle friction and friction at the rim and (2) to the fact that the wheel and the ground deform, with the result that contact between wheel and ground takes place, not at a single point, but over a certain area.

To understand better the first cause of resistance to the motion of a wheel, we shall consider a railroad car supported by eight wheels mounted on axles and bearings. The car is assumed to be moving to the right at constant speed along a straight horizontal track. The free-body diagram of one of the wheels is shown in Fig. 8.13*a*. The forces acting on the free body include the load **W** supported by the wheel and the normal reaction **N** of the track. Since **W** is drawn through the center O of the axle, the frictional resistance of the bearing should be represented by a counterclockwise couple **M** (see Sec. 8.7). To keep the free body in equilibrium, we must add two equal and opposite forces **P** and **F**, forming a clockwise couple of moment $-$**M**. The force **F** is the friction force exerted by the track on the wheel, and **P** represents the force which should be applied to the wheel to keep it rolling at constant speed. Note that the forces **P** and **F** would not exist if there were no friction between wheel and track. The couple **M** representing the axle friction would then be zero; the wheel would slide on the track without turning in its bearing.

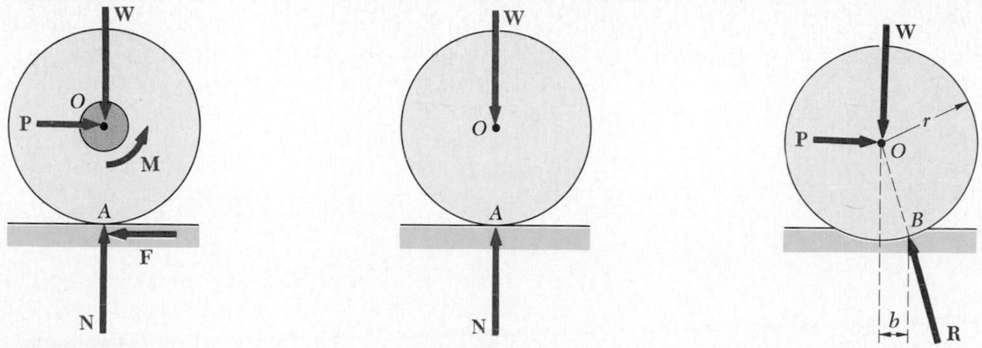

(*a*) Effect of axle friction (*b*) Free wheel (*c*) Rolling resistance
Fig. 8.13

The couple **M** and the forces **P** and **F** also reduce to zero when there is no axle friction. For example, a wheel which is not held in bearings and rolls freely and at constant speed on horizontal ground (Fig. 8.13*b*) will be subjected to only two forces: its own weight **W** and the normal reaction **N** of the ground. No friction force will act on the wheel, regardless of the value of the coefficient of friction between wheel and ground. A wheel rolling freely on horizontal ground should thus keep rolling indefinitely.

Experience, however, indicates that the wheel will slow down and eventually come to rest. This is due to the second type of resistance mentioned at the beginning of this section, known as the *rolling resistance*. Under the load **W**, both the wheel and the ground deform slightly, causing the contact between wheel and ground to take place over a certain area. Experimental evidence shows that the resultant of the forces exerted by the ground on the wheel over this area is a force **R** applied at a point B, which is not located directly under the center O of the wheel, but slightly in front of it (Fig. 8.13*c*). To balance the moment of **W** about B and to keep the wheel rolling at constant speed, it is necessary to apply a horizontal force **P** at the center of the wheel. Writing $\Sigma M_B = 0$, we obtain

$$Pr = Wb \tag{8.10}$$

where r = radius of wheel
b = horizontal distance between O and B

The distance b is commonly called the *coefficient of rolling resistance*. It should be noted that b is not a dimensionless coefficient since it represents a length; b is usually expressed in inches or in millimeters. The value of b depends upon several parameters in a manner which has not yet been clearly established. Values of the coefficient of rolling resistance vary from about 0.01 in. or 0.25 mm for a steel wheel on a steel rail to 5.0 in. or 125 mm for the same wheel on soft ground.

SAMPLE PROBLEM 8.6

A pulley of diameter 4 in. can rotate about a fixed shaft of diameter 2 in. The coefficients of static and kinetic friction between the pulley and shaft are both assumed equal to 0.20. Determine (*a*) the smallest vertical force **P** required to raise a 500-lb load, (*b*) the smallest vertical force **P** required to hold the load, (*c*) the smallest horizontal force **P** required to raise the same load.

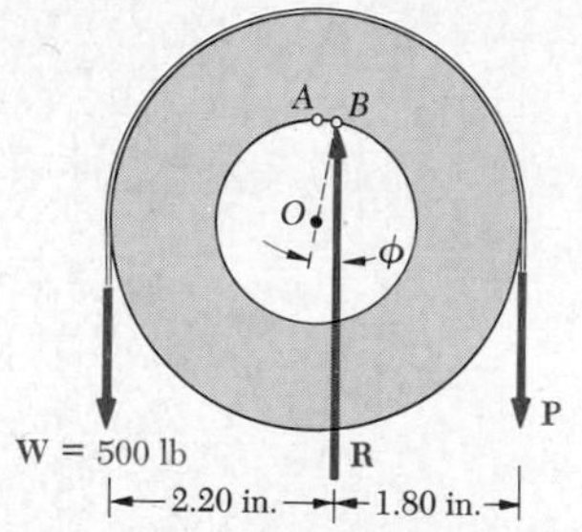

***a.* Vertical Force P Required to Raise the Load.** When the forces in both parts of the rope are equal, contact between the pulley and shaft takes place at *A*. When **P** is increased, the pulley rolls around the shaft slightly and contact takes place at *B*. The free-body diagram of the pulley when motion is impending is drawn. The perpendicular distance from the center *O* of the pulley to the line of action of **R** is

$$r_f = r \sin\phi \approx r\mu \qquad r_f \approx (1\text{ in.})0.20 = 0.20\text{ in.}$$

Summing moments about *B*, we write

$$+\circlearrowleft \Sigma M_B = 0: \qquad (2.20\text{ in.})(500\text{ lb}) - (1.80\text{ in.})P = 0$$
$$P = 611\text{ lb} \qquad\qquad \mathbf{P} = 611\text{ lb}\downarrow \blacktriangleleft$$

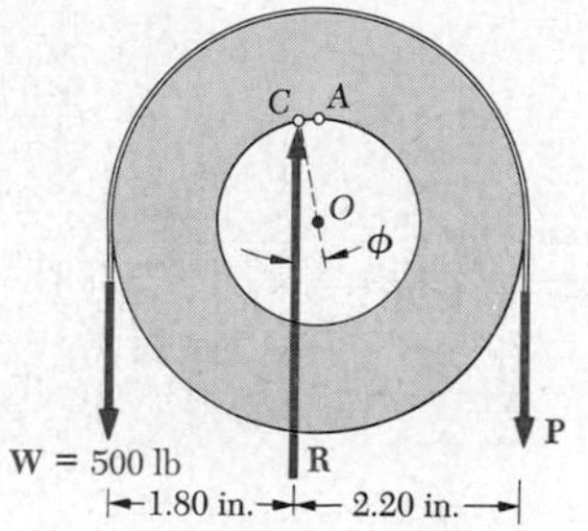

***b.* Vertical Force P to Hold the Load.** As the force **P** is decreased, the pulley rolls around the shaft and contact takes place at *C*. Considering the pulley as a free body and summing moments about *C*, we write

$$+\circlearrowleft \Sigma M_C = 0: \qquad (1.80\text{ in.})(500\text{ lb}) - (2.20\text{ in.})P = 0$$
$$P = 409\text{ lb} \qquad\qquad \mathbf{P} = 409\text{ lb}\downarrow \blacktriangleleft$$

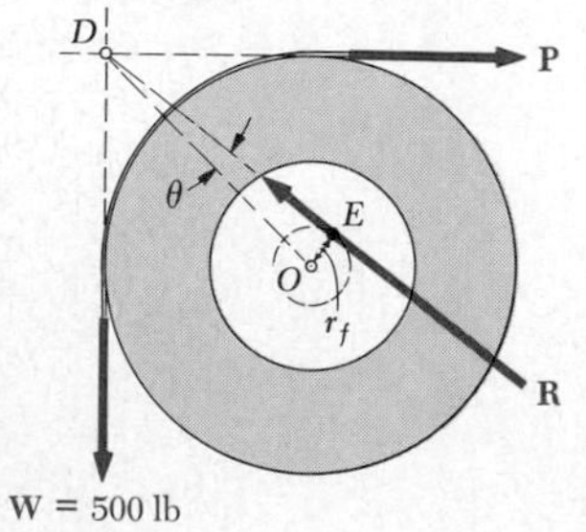

***c.* Horizontal Force P to Raise the Load.** Since the three forces **W**, **P**, and **R** are not parallel, they must be concurrent. The direction of **R** is thus determined from the fact that its line of action must pass through the point of intersection *D* of **W** and **P**, and must be tangent to the circle of friction. Recalling that the radius of the circle of friction is $r_f = 0.20$ in., we write

$$\sin\theta = \frac{OE}{OD} = \frac{0.20\text{ in.}}{(2\text{ in.})\sqrt{2}} = 0.0707 \qquad \theta = 4.1°$$

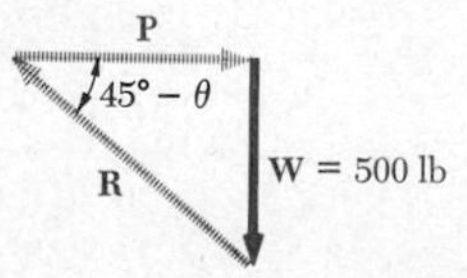

From the force triangle, we obtain

$$P = W \cot(45° - \theta) = (500\text{ lb}) \cot 40.9°$$
$$= 577\text{ lb} \qquad\qquad \mathbf{P} = 577\text{ lb}\rightarrow \blacktriangleleft$$

Problems

8.84 A hot-metal ladle and its contents have a mass of 60 Mg. Knowing that the coefficient of static friction between the hooks and the pinion is 0.30, determine the tension in cable AB required to start tipping the ladle.

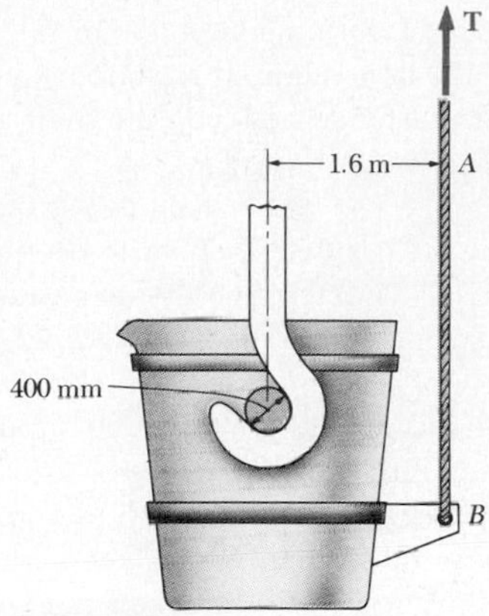

Fig. P8.84

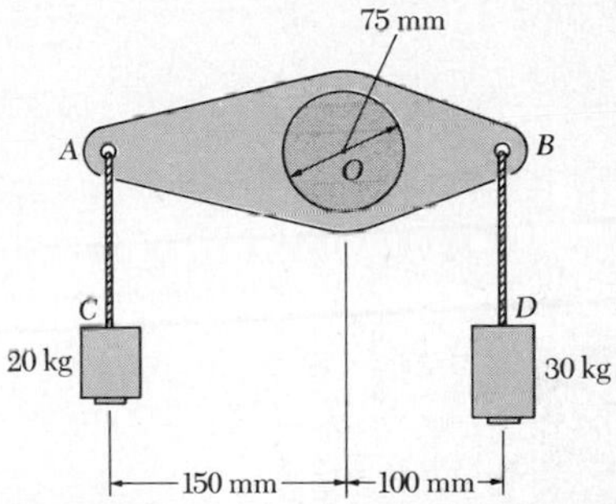

Fig. P8.85

8.85 A lever of negligible weight is loosely fitted onto a 75-mm-diameter fixed shaft. It is observed that the lever will just start rotating if a 2-kg mass is added at C. Determine the coefficient of static friction between the shaft and the lever.

8.86 A windlass of diameter 6 in. is used to raise or lower a 100-lb load. It is supported by two poorly lubricated bearings ($\mu_s = 0.40$) of diameter 2 in. For each of the two positions shown, determine the magnitude of the force **P** required to start raising the load.

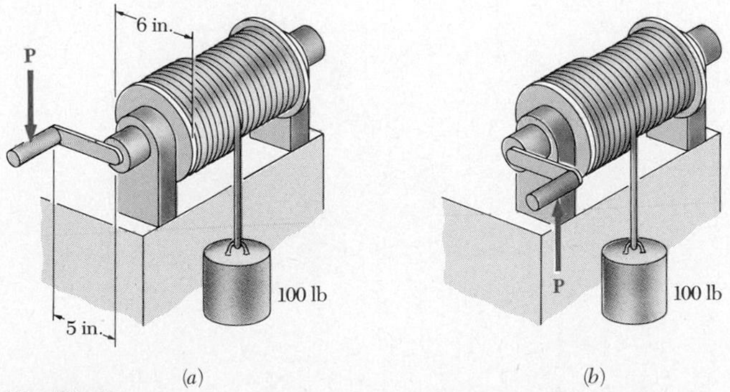

Fig. P8.86

8.87 For the windlass of Prob. 8.86, and for each of the two positions shown, determine the magnitude of the smallest force **P** which will hold the load in place.

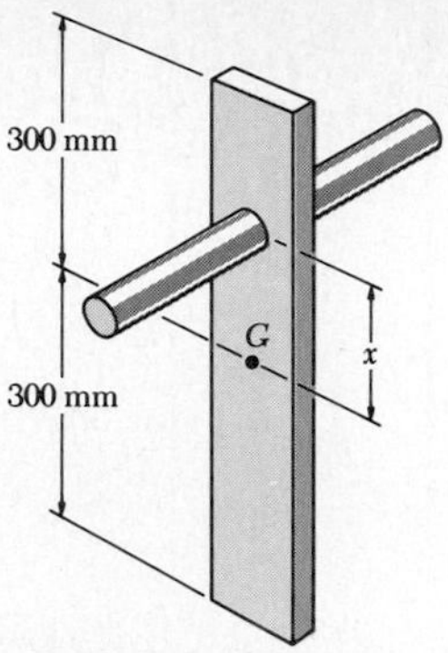

Fig. P8.88

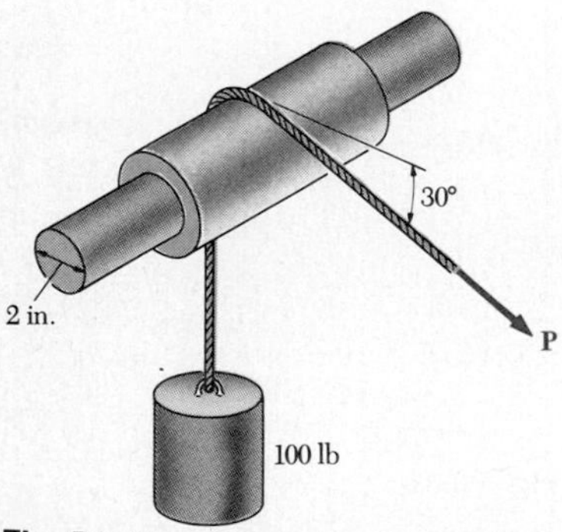

Fig. P8.90

8.88 A uniform bar of length 600 mm is supported by a horizontal shaft of diameter 50 mm. The shaft fits loosely in a circular hole in the bar and the coefficient of static friction is 0.40. If $x = 150$ mm and the shaft is slowly rotated, determine the angle at which the bar will slip.

8.89 For the bar of Prob. 8.88, determine the maximum value of x for which the bar will not slip when the shaft is slowly rotated through a full turn.

8.90 A bushing of 3-in. outside diameter fits loosely on a horizontal 2-in.-diameter shaft. If a force **P** forming an angle of 30° with the horizontal and having a magnitude $P = 150$ lb is required to start raising the 100-lb load, determine the coefficient of static friction between the shaft and the bushing. Assume that the rope does not slip on the bushing.

8.91 Knowing that the coefficient of static friction between the shaft and the bushing of Prob. 8.90 is 0.30, determine the magnitude of the smallest force **P** required (*a*) to start raising the 100-lb load, (*b*) to hold the 100-lb load in place.

8.92 A loaded railroad car has a mass of 30 Mg and is supported by eight wheels of 800-mm diameter with 125-mm-diameter axles. Knowing that the coefficients of friction are $\mu_s = 0.020$ and $\mu_k = 0.015$, determine the horizontal force required (*a*) to get the car in motion, (*b*) to keep it moving with a constant velocity. Neglect the rolling resistance between the wheels and the track.

8.93 A scooter is to be designed to roll down a 2 percent slope at a constant speed. Assuming that the coefficient of kinetic friction between the 25-mm-diameter axles and the bearings is 0.10, determine the required diameter of the wheels. Neglect the rolling resistance between the wheels and the ground.

8.94 A couple of magnitude $M = 100$ N·m is required to start the vertical shaft rotating. Determine the coefficient of static friction.

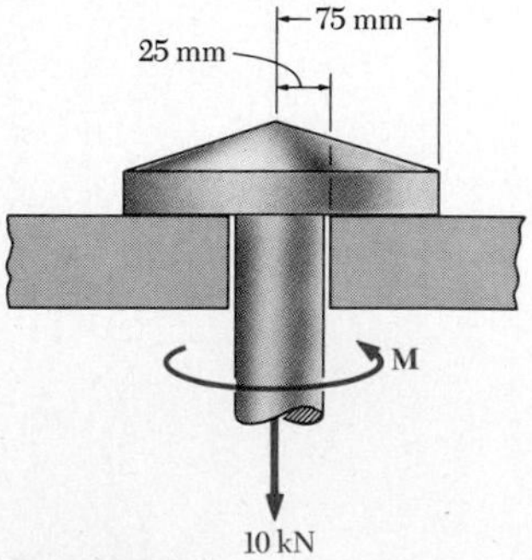

Fig. P8.94

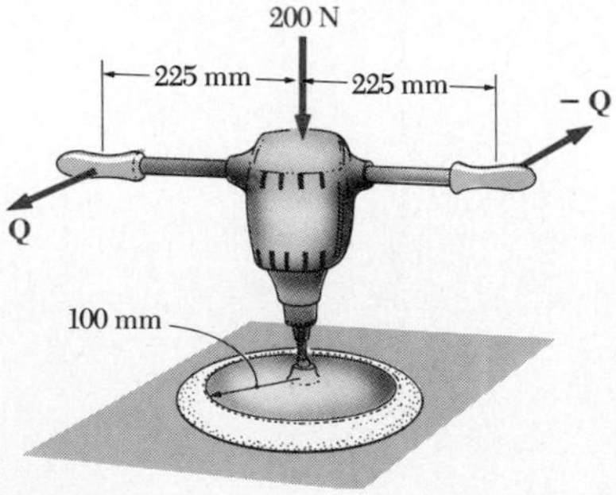

Fig. P8.95

8.95 A vertical force of 200 N is applied to a 6-kg electric polisher as it is operated on a horizontal surface. Knowing that the coefficient of kinetic friction is 0.25, determine the magnitude Q of the forces required to prevent rotation of the handles. Assume the force between the disk and the surface to be uniformly distributed.

8.96 Four springs, each of constant 30 lb/in., are used to press plate AB against the head on the vertical rod. In the position shown each spring has been compressed 2 in. The plate AB may move vertically, but it is constrained against rotation by bolts which pass through holes cut in the plate. Knowing that $\mathbf{P} = 0$ and that the coefficient of friction is 0.30, determine the magnitude of the couple $\mathbf{M}$ required to rotate the rod.

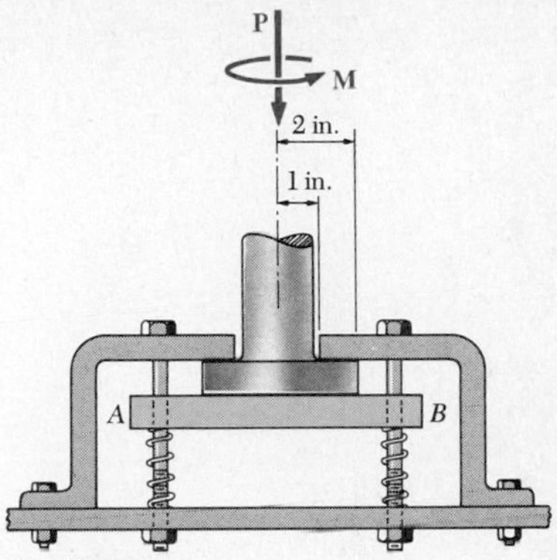

Fig. P8.96

8.97 In Prob. 8.96, determine the required magnitude of the force $\mathbf{P}$ if the rod is to rotate when the magnitude of the couple $\mathbf{M}$ is 180 lb·in. For what magnitude of the force $\mathbf{P}$ is the rod most easily rotated?

* **8.98** As the surfaces of shaft and bearing wear out, the frictional resistance of a thrust bearing decreases. It is generally assumed that the wear is directly proportional to the distance traveled by any given point of the shaft, and thus to the distance r from the point to the axis of the shaft. Assuming, then, that the normal force per unit area is inversely proportional to r, show that the magnitude M of the couple required to overcome the frictional resistance of a worn-out end bearing (with contact over the full circular area) is equal to 75 percent of the value given by formula (8.9) for a new bearing.

* **8.99** Assuming that bearings wear out as indicated in Prob. 8.98, show that the magnitude M of the couple required to overcome the frictional resistance of a worn-out collar bearing is

$$M = \tfrac{1}{2}\mu_k P(R_1 + R_2)$$

where P = magnitude of the total axial force
R_1, R_2 = inner and outer radii of collar

* **8.100** Assuming that the pressure between the surfaces of contact is uniform, show that the magnitude M of the couple required to overcome frictional resistance for the conical pivot shown is

$$M = \frac{2}{3}\frac{\mu_k P}{\sin\theta}\frac{R_2^3 - R_1^3}{R_2^2 - R_1^2}$$

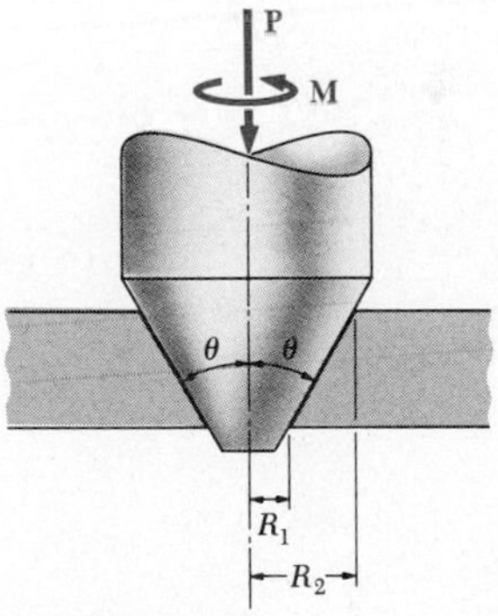

Fig. P8.100

8.101 Solve Prob. 8.95, assuming that the normal force per unit area between the disk and the surface varies linearly from a maximum at the center to zero at the circumference of the disk.

8.102 A circular disk of diameter 6 in. rolls at a constant velocity down a 2 percent incline. Determine the coefficient of rolling resistance.

8.103 Determine the horizontal force required to move a 2500-lb automobile along a horizontal road at a constant velocity. Neglect all forms of friction except rolling resistance, and assume the coefficient of rolling resistance to be 0.05 in. The diameter of each tire is 23 in.

8.104 Solve Prob. 8.92, including the effect of a coefficient of rolling resistance of 0.5 mm.

8.105 Solve Prob. 8.93, including the effect of a coefficient of rolling resistance of 1.75 mm.

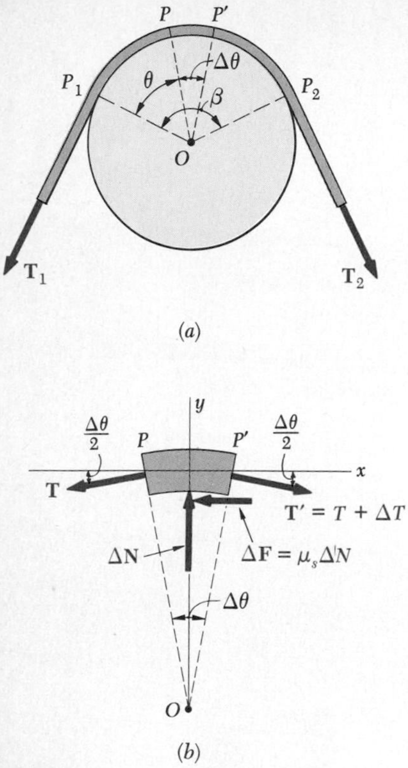

Fig. 8.14

8.10. Belt Friction. Consider a flat belt passing over a fixed cylindrical drum (Fig. 8.14*a*). We propose to determine the relation existing between the values T_1 and T_2 of the tension in the two parts of the belt when the belt is just about to slide toward the right.

Let us detach from the belt a small element PP' subtending an angle $\Delta\theta$. Denoting by T the tension at P and by $T + \Delta T$ the tension at P', we draw the free-body diagram of the element of the belt (Fig. 8.14*b*). Besides the two forces of tension, the forces acting on the free body are the normal component $\Delta\mathbf{N}$ of the reaction of the drum and the friction force $\Delta\mathbf{F}$. Since motion is assumed to be impending, we have $\Delta F = \mu_s\,\Delta N$. It should be noted that if $\Delta\theta$ is made to approach zero, the magnitudes ΔN, ΔF, and the *difference* ΔT between the tension at P and the tension at P' will also approach zero; the value T of the tension at P, however, will remain unchanged. This observation helps in understanding our choice of notations.

Choosing the coordinate axes shown in Fig. 8.14*b*, we write the equations of equilibrium for the element PP':

$$\Sigma F_x = 0: \qquad (T + \Delta T)\cos\frac{\Delta\theta}{2} - T\cos\frac{\Delta\theta}{2} - \mu_s\,\Delta N = 0 \tag{8.11}$$

$$\Sigma F_y = 0: \qquad \Delta N - (T + \Delta T)\sin\frac{\Delta\theta}{2} - T\sin\frac{\Delta\theta}{2} = 0 \tag{8.12}$$

Solving Eq. (8.12) for ΔN and substituting into (8.11), we obtain after reductions

$$\Delta T\cos\frac{\Delta\theta}{2} - \mu_s(2T + \Delta T)\sin\frac{\Delta\theta}{2} = 0$$

We shall now divide both terms by $\Delta\theta$; as far as the first term is concerned, this will be simply done by dividing ΔT by $\Delta\theta$. The division of the second term is carried out by dividing the terms in the parentheses by 2 and the sine by $\Delta\theta/2$. We write

$$\frac{\Delta T}{\Delta\theta}\cos\frac{\Delta\theta}{2} - \mu_s\left(T + \frac{\Delta T}{2}\right)\frac{\sin(\Delta\theta/2)}{\Delta\theta/2} = 0$$

If we now let $\Delta\theta$ approach 0, the cosine approaches 1 and $\Delta T/2$ approaches zero as noted above. On the other hand, the quotient of $\sin(\Delta\theta/2)$ over $\Delta\theta/2$ approaches 1, according to a lemma derived in all calculus textbooks. Since the limit of $\Delta T/\Delta\theta$ is by definition equal to the derivative $dT/d\theta$, we write

$$\frac{dT}{d\theta} - \mu_s T = 0 \qquad \frac{dT}{T} = \mu_s\,d\theta$$

We shall now integrate both members of the last equation obtained from P_1 to P_2 (Fig. 8.14*a*). At P_1, we have $\theta = 0$ and $T = T_1$; at P_2, we have $\theta = \beta$ and $T = T_2$. Integrating between these limits, we write

$$\int_{T_1}^{T_2}\frac{dT}{T} = \int_0^{\beta}\mu_s\,d\theta$$

$$\ln T_2 - \ln T_1 = \mu_s\beta$$

or, noting that the left-hand member is equal to the natural logarithm of the quotient of T_2 and T_1,

$$\ln \frac{T_2}{T_1} = \mu_s \beta \tag{8.13}$$

This relation may also be written in the form

$$\frac{T_2}{T_1} = e^{\mu_s \beta} \tag{8.14}$$

The formulas we have derived apply equally well to problems involving flat belts passing over fixed cylindrical drums and to problems involving ropes wrapped around a post or capstan. They may also be used to solve problems involving band brakes. In such problems, it is the drum which is about to rotate, while the band remains fixed. The formulas may also be applied to problems involving belt drives. In these problems, both the pulley and the belt rotate; our concern is then to find whether the belt will slip, i.e., whether it will move *with respect* to the pulley.

Formulas (8.13) and (8.14) should be used only if the belt, rope, or brake is *about to slip*. Formula (8.14) will be used if T_1 or T_2 is desired; formula (8.13) will be preferred if either μ_s or the angle of contact β is desired. We should note that T_2 is always larger than T_1; T_2 therefore represents the tension in that part of the belt or rope which *pulls*, while T_1 is the tension in the part which *resists*. We should also observe that the angle of contact β must be expressed in *radians*. The angle β may be larger than 2π; for example, if a rope is wrapped n times around a post, β is equal to $2\pi n$.

If the belt, rope, or brake is actually slipping, formulas similar to (8.13) and (8.14), but involving the coefficient of kinetic friction μ_k, should be used. If the belt, rope, or brake does not slip and is not about to slip, none of these formulas may be used.

The belts used in belt drives are often V-shaped. Such a belt, called a *V belt*, is shown in Fig. 8.15*a*. It is seen that contact between belt and pulley takes place along the sides of the groove. The relation existing between the values T_1 and T_2 of the tension in the two parts of the belt when the belt is just about to slip may again be obtained by drawing the free-body diagram of an element of belt (Fig. 8.15*b* and *c*). Equations similar to (8.11) and (8.12) are derived, but the magnitude of the total friction force acting on the element is now $2\,\Delta F$, and the sum of the y components of the normal forces is $2\,\Delta N \sin(\alpha/2)$. Proceeding as above, we obtain

$$\frac{T_2}{T_1} = e^{\mu_s \beta / \sin(\alpha/2)} \tag{8.15}$$

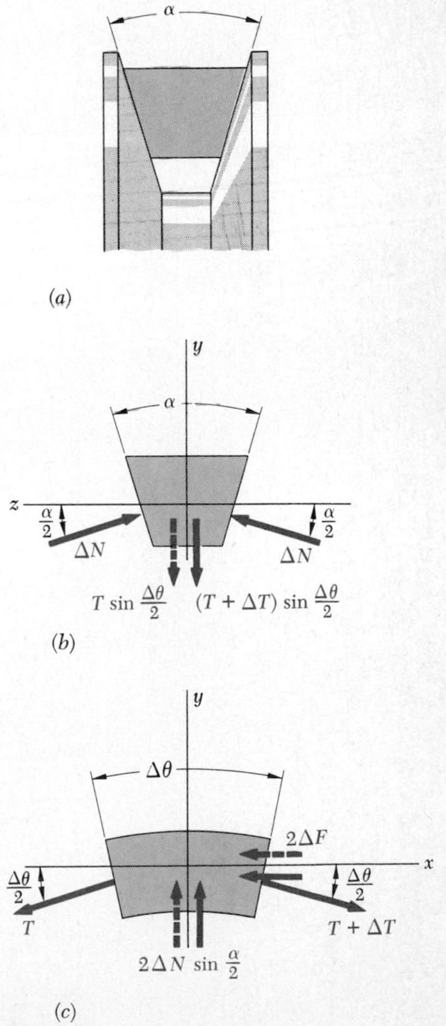

Fig. 8.15

SAMPLE PROBLEM 8.7

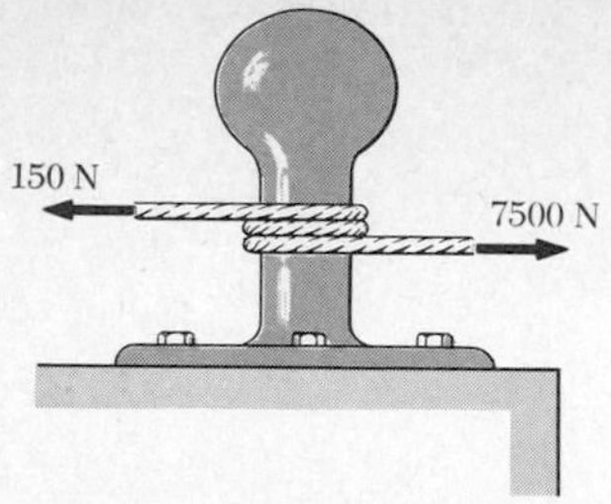

A hawser thrown from a ship to a pier is wrapped two full turns around a capstan. The tension in the hawser is 7500 N; by exerting a force of 150 N on its free end, a dock worker can just keep the hawser from slipping. (*a*) Determine the coefficient of friction between the hawser and the capstan. (*b*) Determine the tension in the hawser that could be resisted by the 150-N force if the hawser were wrapped three full turns around the capstan.

***a.* Coefficient of Friction.** Since slipping of the hawser is impending, we use Eq. (8.13):

$$\ln \frac{T_2}{T_1} = \mu_s \beta$$

Since the hawser is wrapped two full turns around the capstan, we have

$$\beta = 2(2\pi \text{ rad}) = 12.6 \text{ rad}$$
$$T_1 = 150 \text{ N} \qquad T_2 = 7500 \text{ N}$$

Therefore,

$$\mu_s \beta = \ln \frac{T_2}{T_1}$$

$$\mu_s(12.6 \text{ rad}) = \ln \frac{7500 \text{ N}}{150 \text{ N}} = \ln 50 = 3.91$$

$$\mu_s = 0.31 \blacktriangleleft$$

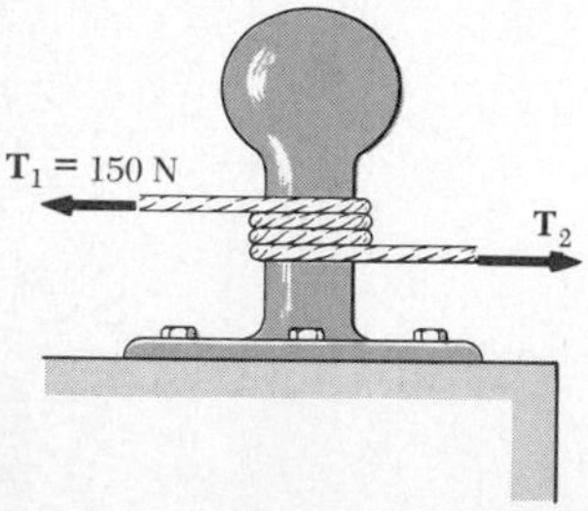

***b.* Hawser Wrapped Three Turns around Capstan.** Using the value of μ_s obtained in part *a*, we have now

$$\beta = 3(2\pi \text{ rad}) = 18.9 \text{ rad}$$
$$T_1 = 150 \text{ N} \qquad \mu_s = 0.31$$

Substituting these values into Eq. (8.14), we obtain

$$\frac{T_2}{T_1} = e^{\mu_s \beta}$$

$$\frac{T_2}{150 \text{ N}} = e^{(0.31)(18.9)} = e^{5.86} = 350$$

$$T_2 = 52\,500 \text{ N} \qquad T_2 = 52.5 \text{ kN} \blacktriangleleft$$

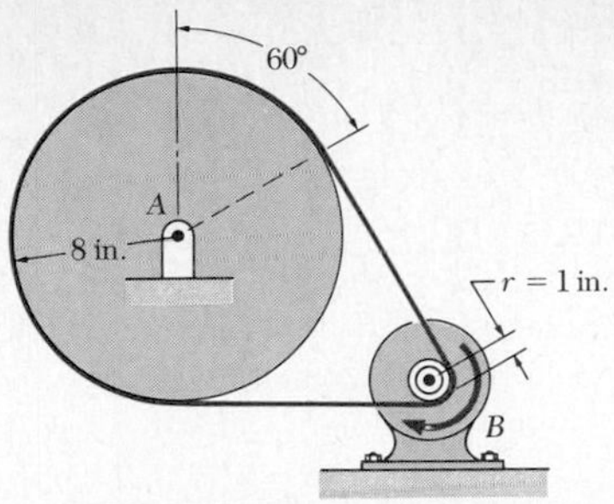

SAMPLE PROBLEM 8.8

A flat belt connects pulley A, which drives a machine tool, to pulley B, which is attached to the shaft of an electric motor. The coefficients of friction are $\mu_s = 0.25$ and $\mu_k = 0.20$ between both pulleys and the belt. Knowing that the maximum allowable tension in the belt is 600 lb, determine the largest torque which can be exerted by the belt on pulley A.

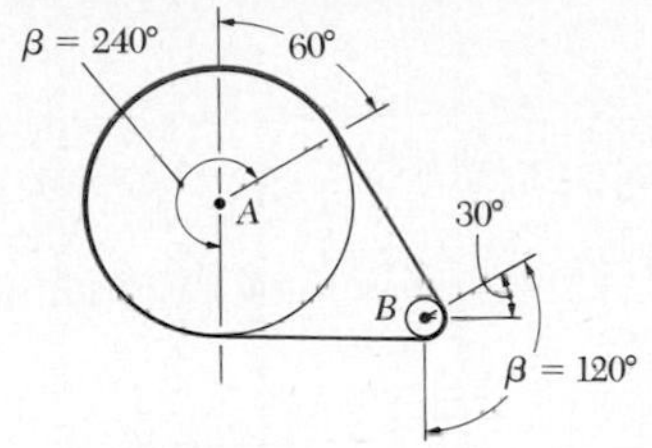

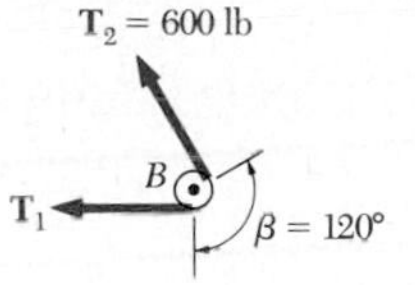

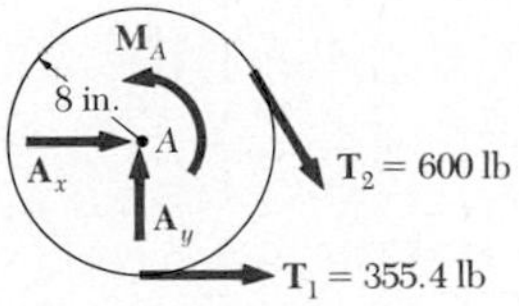

Solution. Since the resistance to slippage depends upon the angle of contact β between pulley and belt, as well as upon the coefficient of static friction μ_s, and since μ_s is the same for both pulleys, slippage will occur first on pulley B, for which β is smaller.

Pulley B. Using Eq. (8.14) with $T_2 = 600$ lb, $\mu_s = 0.25$, and $\beta = 120° = 2\pi/3$ rad, we write

$$\frac{T_2}{T_1} = e^{\mu_s \beta} \qquad \frac{600 \text{ lb}}{T_1} = e^{0.25(2\pi/3)} = 1.688$$

$$T_1 = \frac{600 \text{ lb}}{1.688} = 355.4 \text{ lb}$$

Pulley A. We draw the free-body diagram of pulley A. The couple $\mathbf{M}_A$ is applied to the pulley by the machine tool to which it is attached and is equal and opposite to the torque exerted by the belt. We write

$$+\circlearrowleft \Sigma M_A = 0: \qquad M_A - (600 \text{ lb})(8 \text{ in.}) + (355.4 \text{ lb})(8 \text{ in.}) = 0$$

$$M_A = 1957 \text{ lb}\cdot\text{in.} \qquad M_A = 163.1 \text{ lb}\cdot\text{ft} \blacktriangleleft$$

Note. We may check that the belt does not slip on pulley A by computing the value of μ_s required to prevent slipping at A and verifying that it is smaller than the actual value of μ_s. From Eq. (8.13) we have

$$\mu_s \beta = \ln \frac{T_2}{T_1} = \ln \frac{600 \text{ lb}}{355.4 \text{ lb}} = 0.524$$

and, since $\beta = 240° = 4\pi/3$ rad,

$$\frac{4\pi}{3}\mu_s = 0.524 \qquad \mu_s = 0.125 < 0.25$$

Problems

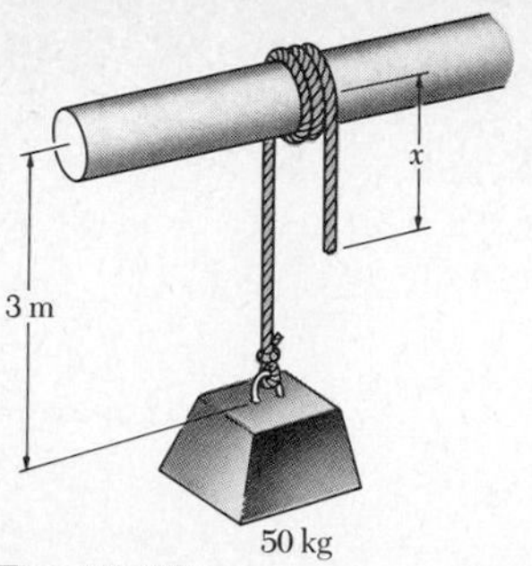

Fig. P8.107

8.106 A hawser is wrapped two full turns around a capstan head. By exerting a force of 800 N on the free end of the hawser, a seaman can resist a force of 60 kN on the other end of the hawser. Determine (*a*) the coefficient of static friction, (*b*) the number of times the hawser should be wrapped around the capstan if a 200-kN force is to be resisted by the same 800-N force.

8.107 A rope having a mass per unit length of 0.8 kg/m is wound $2\frac{1}{2}$ times around a horizontal rod. What length x of rope should be left hanging if a 50-kg load is to be supported? The coefficient of static friction between the rope and the rod is 0.25.

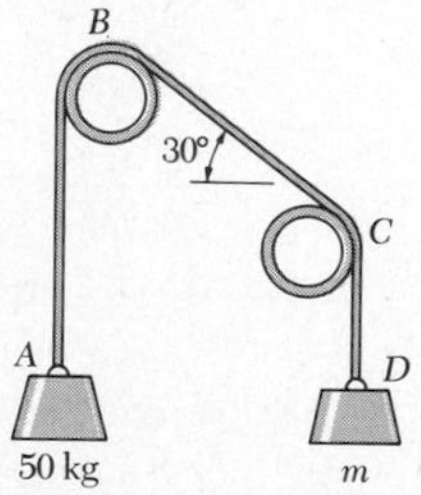

Fig. P8.110 and P8.111

8.108 Assume that the bushing of Prob. 8.90 has become frozen to the shaft and cannot rotate. Determine the coefficient of static friction between the bushing and the rope if a force **P** of magnitude 150 lb is required to start raising the 100-lb load.

8.109 Assume that the bushing of Prob. 8.90 has become frozen to the shaft and cannot rotate. If the coefficient of static friction between the bushing and the rope is 0.30, determine the magnitude of the smallest force **P** required (*a*) to start raising the 100-lb load, (*b*) to hold the 100-lb load in place.

8.110 A rope *ABCD* is looped over two pipes as shown. Knowing that the coefficient of static friction is 0.25, determine (*a*) the smallest value of the mass m for which equilibrium is possible, (*b*) the corresponding tension in portion *BC* of the rope.

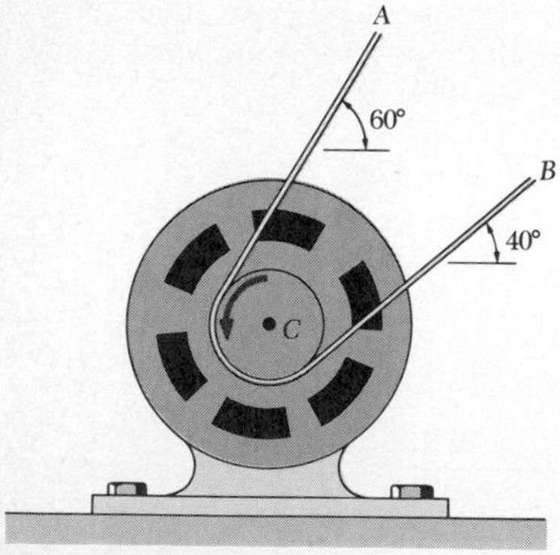

Fig. P8.112

8.111 A rope *ABCD* is looped over two pipes as shown. Knowing that the coefficient of static friction is 0.25, determine (*a*) the largest value of the mass m for which equilibrium is possible, (*b*) the corresponding tension in portion *BC* of the rope.

8.112 A flat belt is used to transmit the 25-lb·ft torque developed by an electric motor. The drum in contact with the belt has a diameter of 6 in., and the coefficient of static friction between the belt and the drum is 0.30. Determine the minimum allowable value of the tension in each part of the belt if the belt is not to slip.

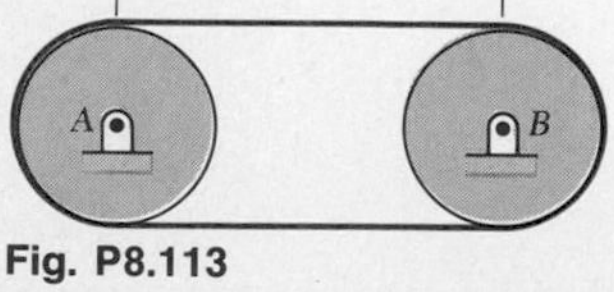

Fig. P8.113

8.113 A flat belt is used to transmit a torque from pulley *A* to pulley *B*. The radius of each pulley is 50 mm and the coefficient of static friction is 0.30. Determine the largest torque which can be transmitted if the allowable belt tension is 3 kN.

8.114 Solve Prob. 8.113, assuming that the belt is looped around the pulleys in a figure 8.

8.115 A flat belt passes over two idler pulleys and under a rotating drum of diameter 8 in. The axle of the drum is free to move vertically in a slot, and a spring keeps the drum in contact with the belt. What is the minimum force which should be exerted by the spring if slippage is not to occur when a 30-lb·ft torque is applied to the drum? The coefficient of static friction between belt and drum is 0.25.

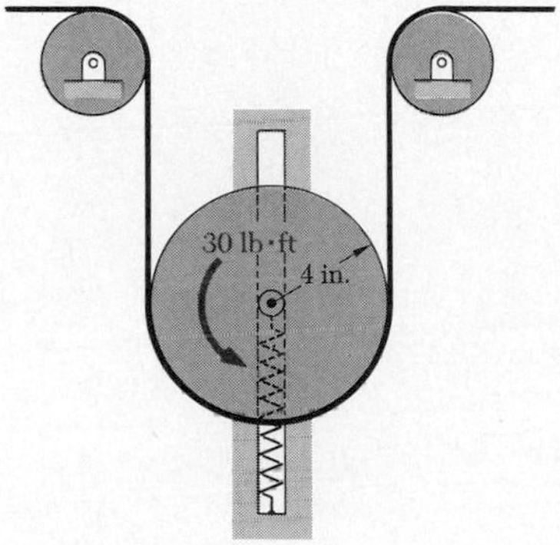

Fig. P8.115

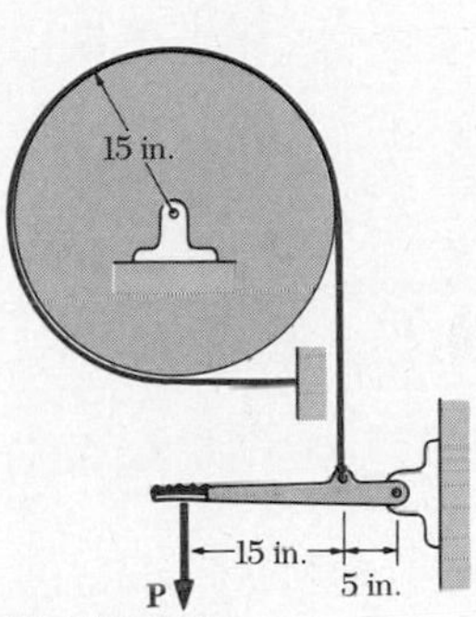

Fig. P8.116

8.116 A band brake is used to control the speed of a flywheel as shown. The coefficients of friction are $\mu_s = 0.30$ and $\mu_k = 0.25$. What torque should be applied to the flywheel to keep it rotating clockwise at a constant speed when $P = 10$ lb?

8.117 Solve Prob. 8.116, assuming that the flywheel rotates counterclockwise.

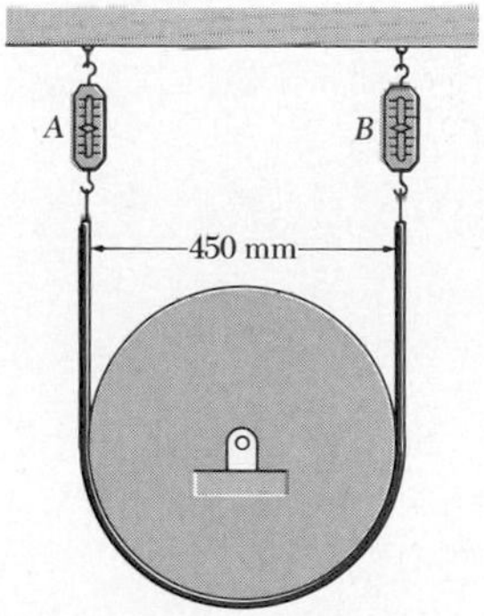

Fig. P8.118

8.118 The setup shown is used to measure the power output of a small turbine. The coefficient of kinetic friction is 0.15. When the flywheel is at rest, the reading of each spring scale is 75 N. What will be the reading of each scale when the flywheel is rotating clockwise? Assume that the belt is of constant length.

8.119 A brake drum of radius $r = 125$ mm is rotating counterclockwise when a force **P** of magnitude 50 N is applied at A. Knowing that the coefficient of kinetic friction is 0.40, determine the moment about O of the friction forces applied to the drum when $a = 200$ mm and $b = 250$ mm.

8.120 Knowing that $r = 125$ mm and $a = 200$ mm, determine the maximum value of the coefficient of kinetic friction for which the brake is not self-locking when the drum rotates counterclockwise.

8.121 Knowing that the coefficient of kinetic friction is 0.40, determine the minimum value of the ratio a/r for which the brake is not self-locking. Assume that $a > r$ and that the drum revolves counterclockwise.

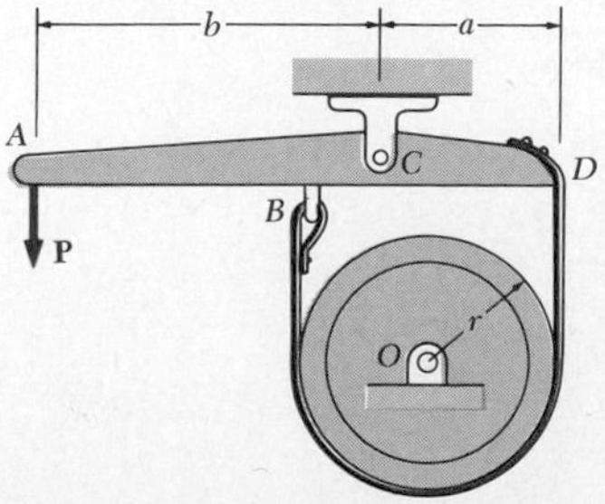

Fig. P8.119, P8.120, and P8.121

8.122 A cable is placed around three pipes, each of 4-in. outside diameter, located in the same horizontal plane. Two of the pipes are fixed and do not rotate; the third pipe is rotated slowly. Knowing that the coefficients of friction are $\mu_s = 0.30$ and $\mu_k = 0.25$ for each pipe, determine the largest weight W which can be raised (*a*) if only pipe A is rotated, (*b*) if only pipe B is rotated, (*c*) if only pipe C is rotated.

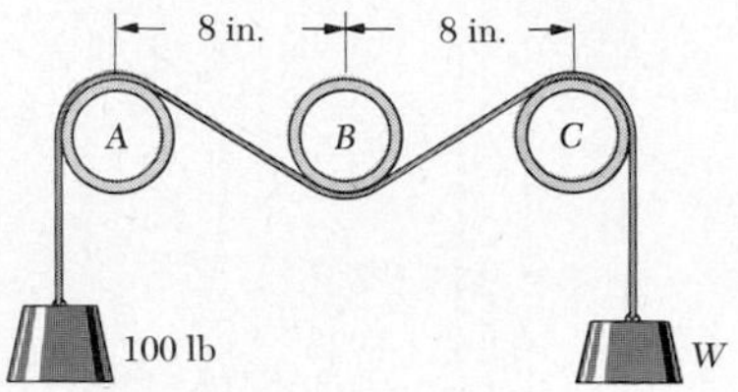

Fig. P8.122

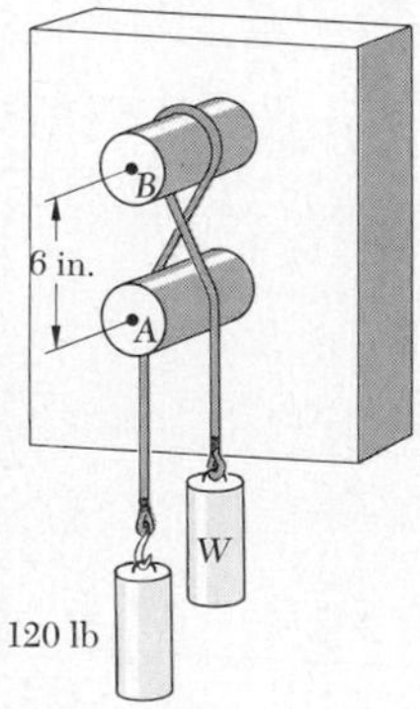

Fig. P8.123

8.123 A cord is placed over two cylinders, each of 3-in. diameter. Knowing that the coefficients of friction are $\mu_s = 0.25$ and $\mu_k = 0.20$, determine the largest weight W which can be raised when cylinder B is rotated slowly and cylinder A is kept fixed.

8.124 The strap wrench shown is used to grip the pipe firmly and at the same time not mar the external surface of the pipe. Knowing that $a = 250$ mm and $r = 50$ mm, determine the minimum coefficient of static friction for which the wrench will be self-locking.

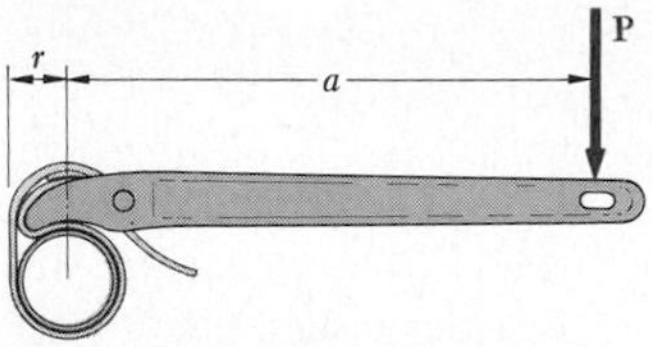

Fig. P8.124 and P8.125

8.125 Denoting by μ_s the coefficient of static friction between the strap and the pipe, determine the minimum value of the ratio r/a for which the strap wrench will be self-locking.

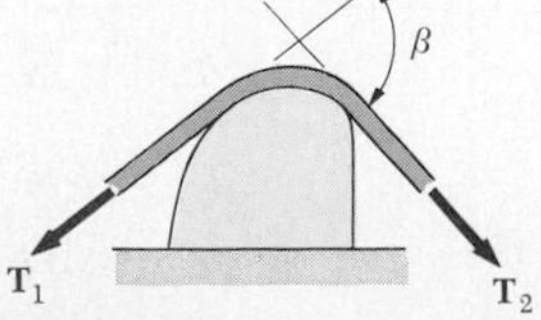

Fig. P8.126

8.126 Prove that Eqs. (8.13) and (8.14) are valid for any shape of surface provided that the coefficient of friction is the same at all points of contact.

8.127 Complete the derivation of Eq. (8.15), which relates the tension in both parts of a V belt.

8.128 Solve Prob. 8.112, assuming that the flat belt and pulley are replaced by a V belt and V pulley with $\alpha = 28°$. (The angle α is as shown in Fig. 8.15*a*.)

Review Problems

8.129 The axle of the pulley is frozen and cannot rotate with respect to the block. Knowing that the coefficient of static friction between cable *ABCD* and the pulley is 0.30, determine (*a*) the maximum allowable value of θ if the system is to remain in equilibrium, (*b*) the corresponding reactions at *A* and *D*. (Assume that the straight portions of the cable meet at point *E*.)

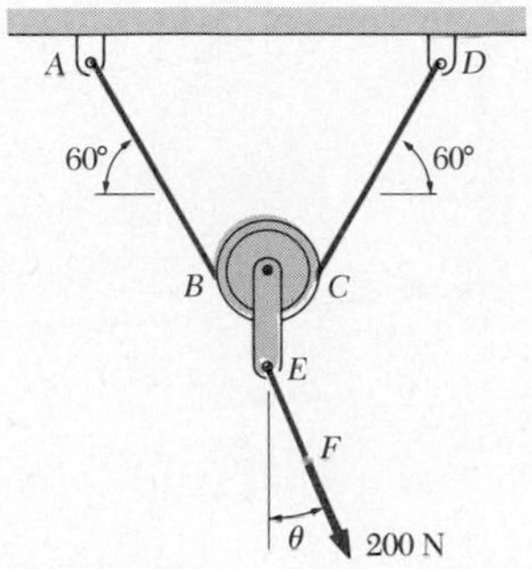

Fig. P8.129

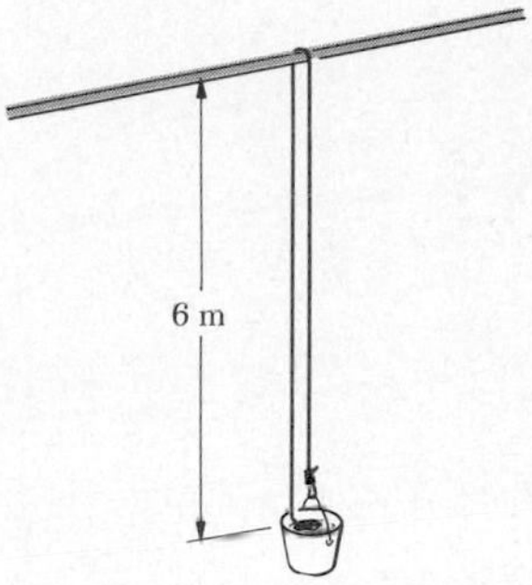

Fig. P8.130

8.130 A 15-m rope passes over a small horizontal shaft; one end of the rope is attached to a 2.4-kg bucket and the excess rope is coiled inside the bucket. The coefficient of static friction between the rope and the shaft is 0.30, and the rope has a mass per unit length of 800 g/m. If the shaft is slowly rotated, determine how far the bucket (*a*) can be raised before the rope slips on the shaft, (*b*) can be lowered before slipping occurs.

8.131 The 50-lb plate *ABCD* is attached to collars at *A* and *D* which may slide on the vertical rod. Knowing that the coefficient of static friction is 0.40 between both collars and the rod, determine whether the plate is in equilibrium in the position shown when the magnitude of the vertical force applied at *E* is (*a*) $P = 0$, (*b*) $P = 20$ lb.

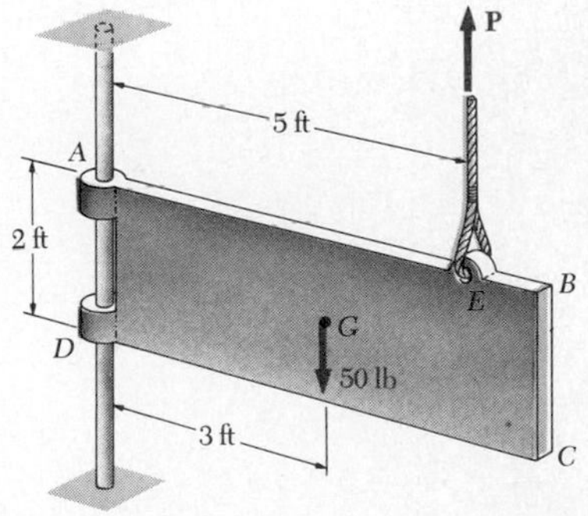

Fig. P8.131

8.132 In Prob. 8.131, determine the range of values of the magnitude *P* of the vertical force applied at *E* for which the plate will move downward.

8.133 Knowing that the coefficient of friction between the joists and the plank *AB* is 0.30, determine the smallest distance *a* for which the plank will slip at *C*.

8.134 A 1.2-m plank of mass 3 kg rests on two joists. Knowing that the coefficient of static friction between the plank and the joists is 0.30, determine the magnitude of the horizontal force required to move the plank when (*a*) $a = 750$ mm, (*b*) $a = 900$ mm.

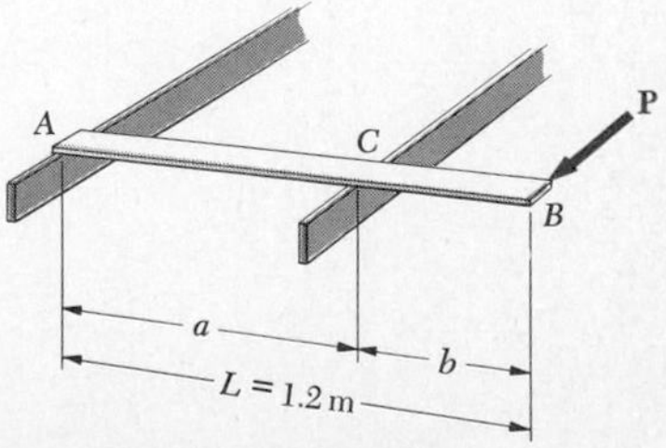

Fig. P8.133 and P8.134

8.135 A 12° wedge is used to split a log. The coefficient of friction between the wedge and the wood is 0.35. Knowing that a force **P** of magnitude 2.5 kN was required to insert the wedge, determine the magnitude of the forces exerted on the log by the wedge after it has been inserted.

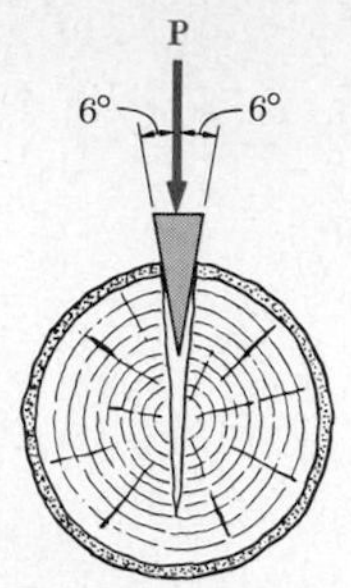

Fig. P8.135

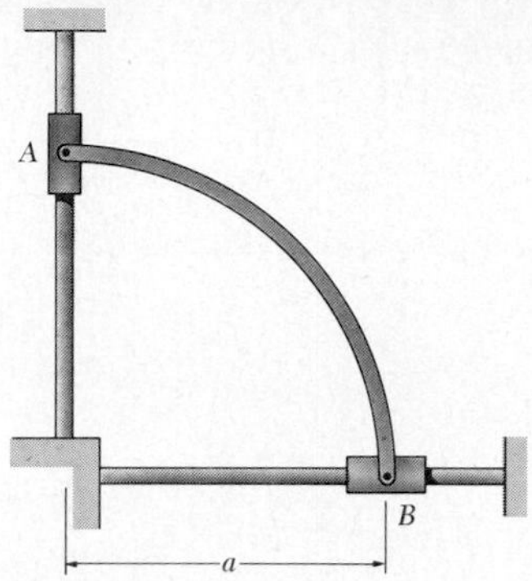

Fig. P8.136

8.136 The uniform rod AB of radius a and weight W is attached to collars at A and B which may slide on the rods shown. Denoting by μ_s the coefficient of static friction between each collar and the rod upon which it may slide, determine the smallest value of μ_s for which the rod will remain in equilibrium in the position shown.

8.137 Two blocks A and B are connected by a cable as shown. Knowing that the coefficient of static friction at all surfaces of contact is 0.30 and neglecting the friction of the pulleys, determine the magnitude P of the smallest force required to move the blocks.

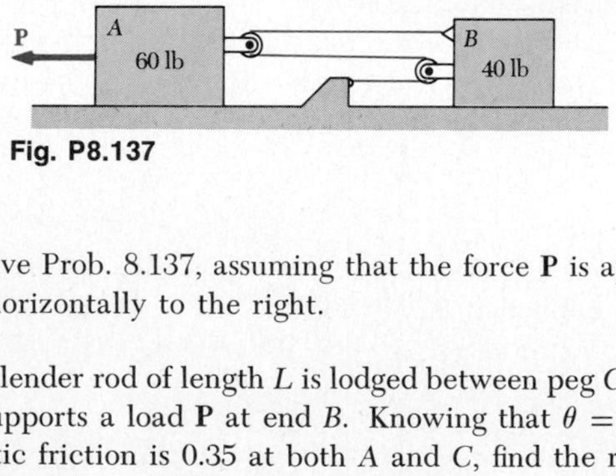

Fig. P8.137

8.138 Solve Prob. 8.137, assuming that the force **P** is applied to block B and is directed horizontally to the right.

8.139 A slender rod of length L is lodged between peg C and the horizontal ceiling and supports a load **P** at end B. Knowing that $\theta = 30°$ and that the coefficient of static friction is 0.35 at both A and C, find the range of values of the ratio L/a for which equilibrium is maintained.

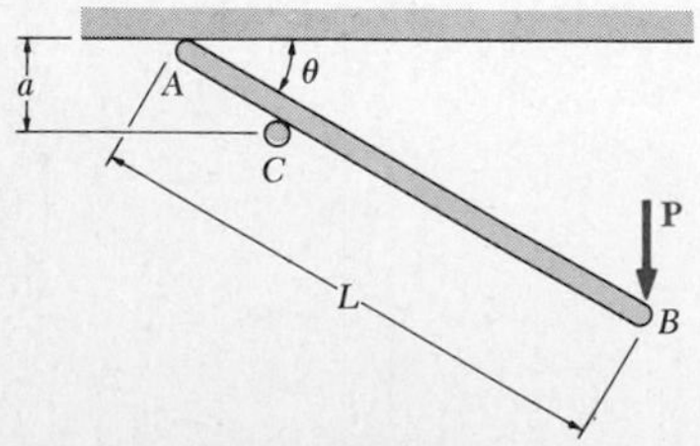

Fig. P8.139 and P8.140

8.140 A slender rod of length L is lodged between peg C and the horizontal ceiling and supports a load **P** at end B. Knowing that $L = 8a$, $\theta = 15°$, and that the coefficients of friction are $\mu_s = 0.60$ and $\mu_k = 0.50$ at A, and zero at C, determine whether the rod is in equilibrium and find the value of the friction force at A.

CHAPTER NINE

Distributed Forces: Moments of Inertia

9.1. Introduction. In Chap. 5, we analyzed various systems of forces distributed over an area or volume. The three main types of forces considered were (1) weights of homogeneous plates of uniform thickness (Secs. 5.3 through 5.6), (2) distributed loads on beams and hydrostatic forces (Secs. 5.8 and 5.9), and (3) weights of homogeneous three-dimensional bodies (Secs. 5.10 and 5.11). In the case of homogeneous plates, the magnitude ΔW of the weight of an element of plate was proportional to the area ΔA of the element. In the case of distributed loads on beams, the magnitude ΔW of each elementary weight was represented by an element of area $\Delta A = \Delta W$ under the load curve; in the case of hydrostatic forces on submerged rectangular surfaces, a similar procedure was followed. In the case of homogeneous three-dimensional bodies, the magnitude ΔW of the weight of an element of the body was proportional to the volume ΔV of the element. Thus, in all cases considered in Chap. 5, the distributed forces were proportional to the elementary areas or volumes associated with them. The resultant of these forces, therefore, could be obtained by summing the corresponding areas or volumes and the moment of the resultant about any given axis could be determined by computing the first moments of the areas or volumes about that axis.

In the first part of this chapter, we shall consider distributed forces $\Delta \mathbf{F}$ whose magnitudes depend not only upon the element of area ΔA on which they act but also upon the distance from ΔA to some given axis. More precisely, the magnitude of the force per unit area $\Delta F/\Delta A$ will vary linearly with the distance to the axis. As we shall see in the next section, forces of this type are found in the study of the bending of beams and in problems involving submerged nonrectangular surfaces. Assuming that the elementary forces involved are distributed over an area A and vary linearly with the distance y to the x axis, we shall find that while the magnitude of their resultant $\mathbf{R}$ depends upon the first moment $Q_x = \int y\, dA$ of the area A, the location of the point where $\mathbf{R}$ should be applied depends upon the *second moment*, or *moment of inertia*, $I_x = \int y^2\,dA$ of the same area with respect to the x axis. We shall learn to compute the moments of inertia of various areas with respect to given x and y axes. We shall also introduce the *polar moment of inertia* $J_O = \int r^2\,dA$ of an area, where r is the distance from the element of area dA to a point O. To facilitate our computations, we shall establish a relation between the moment of inertia I_x of an area A with respect to a given x axis and its moment of inertia $I_{x'}$ with respect to a parallel centroidal x' axis (parallel-axis theorem). We shall also study the transformation of the moments of inertia of a given area under a rotation of axes (Secs. 9.9 and 9.10).

In the second part of the chapter, we shall determine the moments of inertia of various *masses* with respect to a given axis. As we shall see in Sec. 9.11, the moment of inertia of a given mass about an axis AA' is defined as the integral $I = \int r^2\,dm$, where r is the distance from the axis AA' to the element of mass dm. Moments of inertia of masses are encountered in dynamics in problems involving the rotation of a rigid body about an axis. To facilitate the computation of mass moments of inertia, we shall introduce the parallel-axis theorem (Sec. 9.12) and analyze the transformation of moments of inertia of masses under a rotation of axes (Secs. 9.16 and 9.17).

MOMENTS OF INERTIA OF AREAS

9.2. Second Moment, or Moment of Inertia, of an Area.

In the first part of this chapter, we shall consider distributed forces $\Delta\mathbf{F}$ of magnitude ΔF proportional to the element of area ΔA on which they act and varying linearly with the distance from ΔA to some given axis.

Consider, for example, a beam of uniform cross section, subjected to two equal and opposite couples applied at each end of the beam. Such a beam is said to be in *pure bending*, and it is shown in mechanics of materials that the internal forces in any section of the beam are distributed forces whose magnitudes $\Delta F = ky\,\Delta A$ vary linearly with the distance y from an axis passing through the centroid of the section. This axis, represented by the x axis in Fig. 9.1, is known as the *neutral axis* of the section. The forces on one side of the neutral axis are forces of compression and on the other side, forces of tension, while on the neutral axis itself the forces are zero.

The magnitude of the resultant $\mathbf{R}$ of the elementary forces $\Delta\mathbf{F}$ over the entire section is

$$R = \int ky\,dA = k\int y\,dA$$

The last integral obtained is recognized as the *first moment* Q_x of the section about the x axis; it is equal to $\bar{y}A$ and to zero, since the centroid of the section is located on the x axis. The system of the forces $\Delta\mathbf{F}$ thus reduces to a couple. The magnitude M of this couple (bending moment) must be equal to the sum of the moments $\Delta M_x = y\,\Delta F = ky^2\,\Delta A$ of the elementary forces. Integrating over the entire section, we obtain

$$M = \int ky^2\,dA = k\int y^2\,dA$$

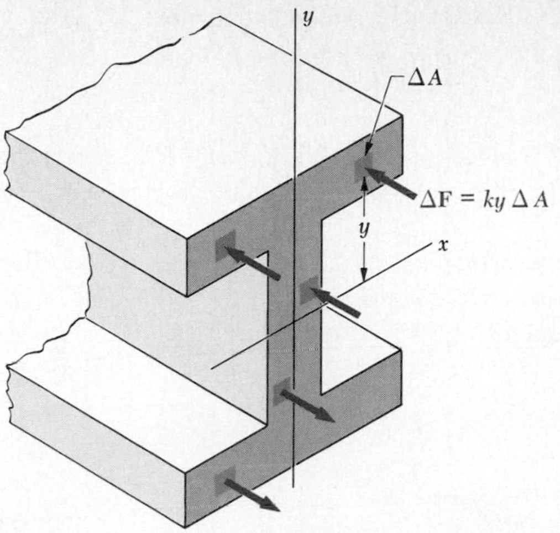

Fig. 9.1

The last integral is known as the *second moment*, or *moment of inertia*,† of the beam section with respect to the x axis and is denoted by I_x. It is obtained by multiplying each element of area dA by the *square of its distance* from the x axis and integrating over the beam section. Since each product $y^2\,dA$ is positive, whether y is itself positive or negative (or zero if y is zero), the integral I_x will always be different from zero and positive.

Another example of second moment, or moment of inertia, of an area is provided by the following problem of hydrostatics: A vertical circular gate used to close the outlet of a large reservoir is submerged under water as shown in Fig. 9.2. What is the resultant of the forces exerted by the water on the gate, and what is the moment of the resultant about the line of intersection of the plane of the gate with the water surface (x axis)?

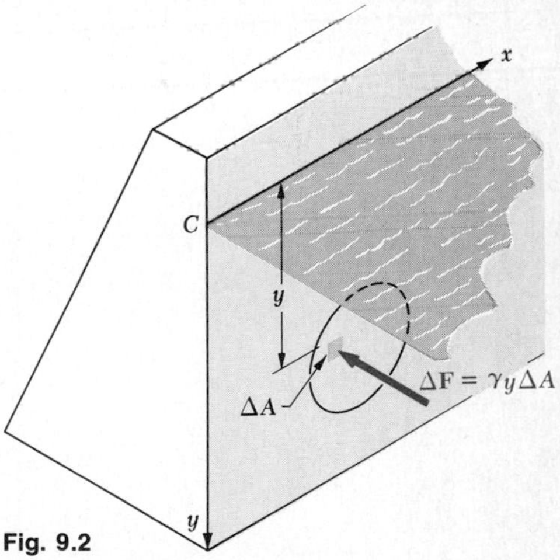

Fig. 9.2

If the gate were rectangular, the resultant of the forces of pressure could be determined from the pressure curve, as was done in Sec. 5.9. Since the gate is circular, however, a more general method must be used. Denoting by y the depth of an element of area ΔA and by γ the specific weight of water, the pressure at the element is $p = \gamma y$, and the magnitude of the elementary force exerted on ΔA is $\Delta F = p\,\Delta A = \gamma y\,\Delta A$. The magnitude of the resultant of the elementary forces is thus

$$R = \int \gamma y\,dA = \gamma \int y\,dA$$

and may be obtained by computing the first moment $Q_x = \int y\,dA$ of the area of the gate with respect to the x axis. The moment M_x of the resultant must be equal to the sum of the moments $\Delta M_x = y\,\Delta F = \gamma y^2\,\Delta A$ of the

† The term second moment is more proper than the term moment of inertia, since, logically, the latter should be used only to denote integrals of mass (see Sec. 9.11). In common engineering practice, however, moment of inertia is used in connection with areas as well as masses.

elementary forces. Integrating over the area of the gate, we have

$$M_x = \int \gamma y^2\, dA = \gamma \int y^2\, dA$$

Here again, the integral obtained represents the second moment, or moment of inertia, I_x of the area with respect to the x axis.

9.3. Determination of the Moment of Inertia of an Area by Integration. We have defined in the preceding section the second moment, or moment of inertia, of an area A with respect to the x axis.

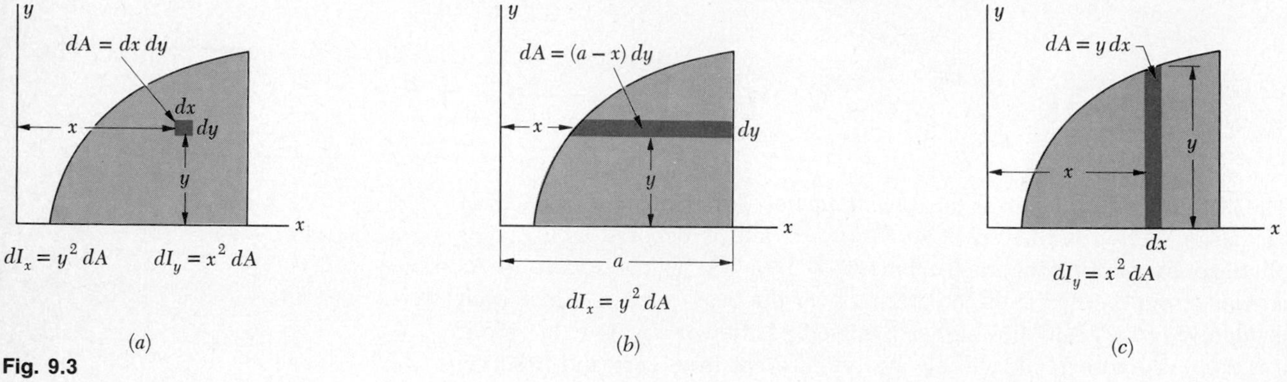

Fig. 9.3

Defining in a similar way the moment of inertia I_y of the area A with respect to the y axis, we write (Fig. 9.3a)

$$I_x = \int y^2\, dA \qquad I_y = \int x^2\, dA \tag{9.1}$$

These integrals, known as the *rectangular moments of inertia* of the area A, may be more easily computed if we choose for dA a thin strip parallel to one of the axes of coordinates. To compute I_x, the strip is chosen parallel to the x axis, so that all the points forming the strip are at the same distance y from the x axis (Fig. 9.3b); the moment of inertia dI_x of the strip is then obtained by multiplying the area dA of the strip by y^2. To compute I_y, the strip is chosen parallel to the y axis so that all the points forming the strip are at the same distance x from the y axis (Fig. 9.3c); the moment of inertia dI_y of the strip is $x^2\, dA$.

Moment of Inertia of a Rectangular Area. As an example, we shall determine the moment of inertia of a rectangle with respect to its base (Fig. 9.4). Dividing the rectangle into strips parallel to the x axis, we obtain

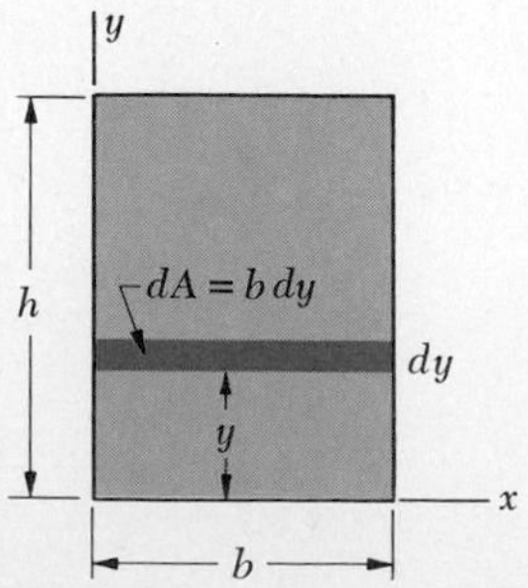

Fig. 9.4

$$dA = b\, dy \qquad dI_x = y^2 b\, dy \qquad I_x = \int_0^h by^2\, dy = \tfrac{1}{3}bh^3 \tag{9.2}$$

Computing I_x and I_y from the Same Elementary Strips. The formula just derived may be used to determine the moment of inertia dI_x with respect to the x axis of a rectangular strip parallel to the y axis such as the one shown in Fig. 9.3*c*. Making $b = dx$ and $h = y$ in formula (9.2), we write

$$dI_x = \tfrac{1}{3}y^3\,dx$$

On the other hand, we have

$$dI_y = x^2\,dA = x^2 y\,dx$$

The same element may thus be used to compute the moments of inertia I_x and I_y of a given area (Fig. 9.5).

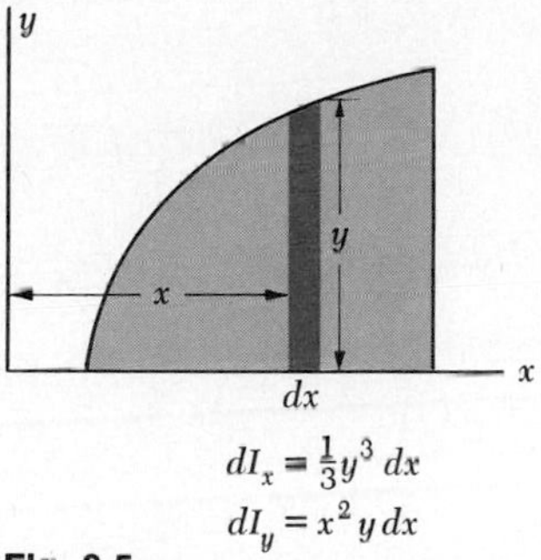

Fig. 9.5

9.4. Polar Moment of Inertia. An integral of great importance in problems concerning the torsion of cylindrical shafts and in problems dealing with the rotation of slabs is

$$J_O = \int r^2\,dA \tag{9.3}$$

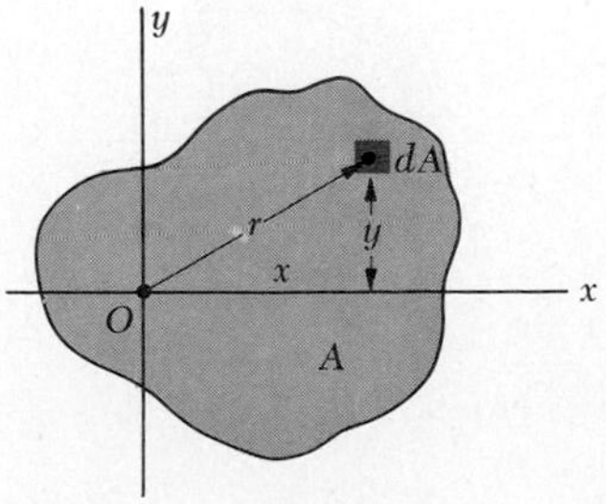

Fig. 9.6

where r is the distance from the element of area dA to the pole O (Fig. 9.6). This integral is the *polar moment of inertia* of the area A with respect to O.

The polar moment of inertia of a given area may be computed from the rectangular moments of inertia I_x and I_y of the area if these integrals are already known. Indeed, noting that $r^2 = x^2 + y^2$, we write

$$J_O = \int r^2\,dA = \int (x^2 + y^2)\,dA = \int y^2\,dA + \int x^2\,dA$$

that is,

$$J_O = I_x + I_y \tag{9.4}$$

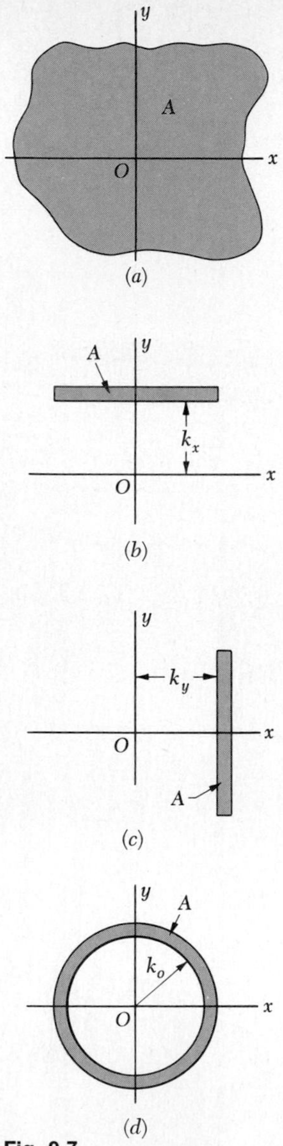

Fig. 9.7

9.5. Radius of Gyration of an Area. Consider an area A which has a moment of inertia I_x with respect to the x axis (Fig. 9.7*a*). Let us imagine that we concentrate this area into a thin strip parallel to the x axis (Fig. 9.7*b*). If the area A, thus concentrated, is to have the same moment of inertia with respect to the x axis, the strip should be placed at a distance k_x from the x axis, defined by the relation

$$I_x = k_x^2 A$$

Solving for k_x, we write

$$k_x = \sqrt{\frac{I_x}{A}} \tag{9.5}$$

The distance k_x is referred to as the *radius of gyration* of the area with respect to the x axis. We may define in a similar way the radii of gyration k_y and k_O (Fig. 9.7*c* and *d*); we write

$$I_y = k_y^2 A \qquad k_y = \sqrt{\frac{I_y}{A}} \tag{9.6}$$

$$J_O = k_O^2 A \qquad k_O = \sqrt{\frac{J_O}{A}} \tag{9.7}$$

Substituting for J_O, I_x, and I_y in terms of the radii of gyration in the relation (9.4), we observe that

$$k_O^2 = k_x^2 + k_y^2 \tag{9.8}$$

Example. As an example, let us compute the radius of gyration k_x of the rectangle shown in Fig. 9.8. Using formulas (9.5) and (9.2), we write

$$k_x^2 = \frac{I_x}{A} = \frac{\frac{1}{3}bh^3}{bh} = \frac{h^2}{3} \qquad k_x = \frac{h}{\sqrt{3}}$$

The radius of gyration k_x of the rectangle is shown in Fig. 9.8. It should not be confused with the ordinate $\bar{y} = h/2$ of the centroid of the area. While k_x depends upon the *second moment*, or moment of inertia, of the area, the ordinate $\bar{y}$ is related to the *first moment* of the area.

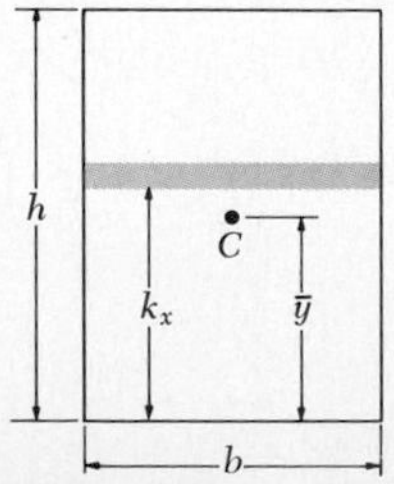

Fig. 9.8

SAMPLE PROBLEM 9.1

Determine the moment of inertia of a triangle with respect to its base.

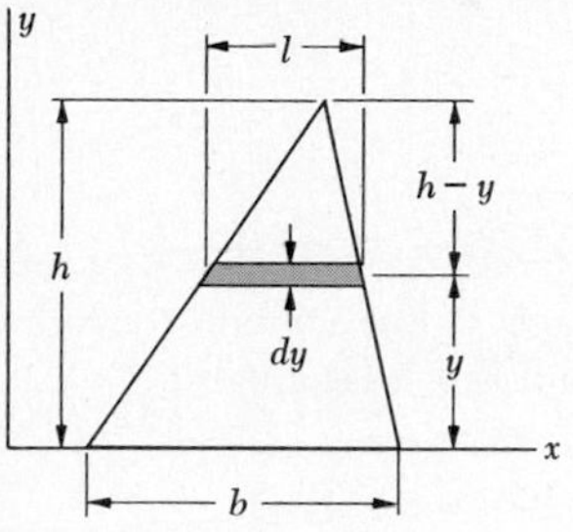

Solution. A triangle of base b and height h is drawn; the x axis is chosen to coincide with the base. A differential strip parallel to the x axis is chosen. Since all portions of the strip are at the same distance from the x axis, we write

$$dI_x = y^2\,dA \qquad dA = l\,dy$$

From the similar triangles, we have

$$\frac{l}{b} = \frac{h-y}{h} \qquad l = b\frac{h-y}{h} \qquad dA = b\frac{h-y}{h}\,dy$$

Integrating dI_x from $y = 0$ to $y = h$, we obtain

$$I_x = \int y^2\,dA = \int_0^h y^2 b\frac{h-y}{h}\,dy = \frac{b}{h}\int_0^h (hy^2 - y^3)\,dy$$

$$= \frac{b}{h}\left[h\frac{y^3}{3} - \frac{y^4}{4}\right]_0^h \qquad I_x = \frac{bh^3}{12} \blacktriangleleft$$

SAMPLE PROBLEM 9.2

(*a*) Determine the centroidal polar moment of inertia of a circular area by direct integration. (*b*) Using the result of part *a*, determine the moment of inertia of a circular area with respect to a diameter.

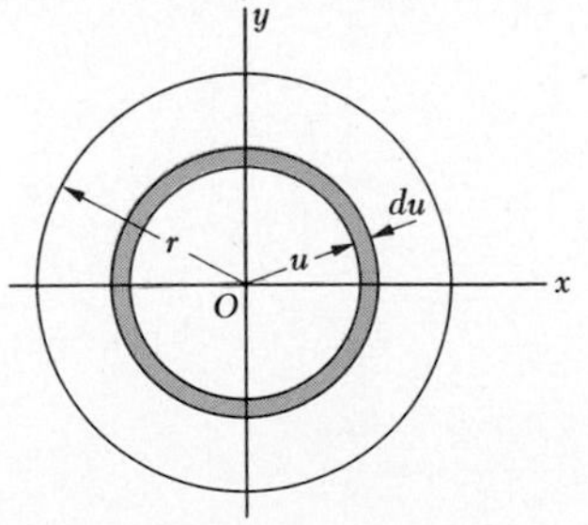

***a.* Polar Moment of Inertia.** An annular differential element of area is chosen. Since all portions of this differential area are at the same distance from the origin, we write

$$dJ_O = u^2\,dA \qquad dA = 2\pi u\,du$$

$$J_O = \int dJ_O = \int_0^r u^2(2\pi u\,du) = 2\pi\int_0^r u^3\,du$$

$$J_O = \frac{\pi}{2}r^4 \blacktriangleleft$$

***b.* Moment of Inertia with Respect to a Diameter.** Because of the symmetry of the circular area we have $I_x = I_y$. We then write

$$J_O = I_x + I_y = 2I_x \qquad \frac{\pi}{2}r^4 = 2I_x \qquad I_{\text{diameter}} = I_x = \frac{\pi}{4}r^4 \blacktriangleleft$$

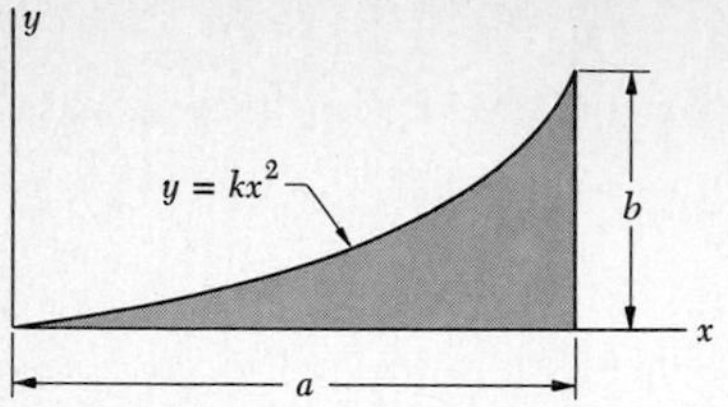

SAMPLE PROBLEM 9.3

(a) Determine the moment of inertia of the shaded area shown with respect to each of the coordinate axes. This area has also been considered in Sample Prob. 5.4. (b) Using the results of part a, determine the radius of gyration of the shaded area with respect to each of the coordinate axes.

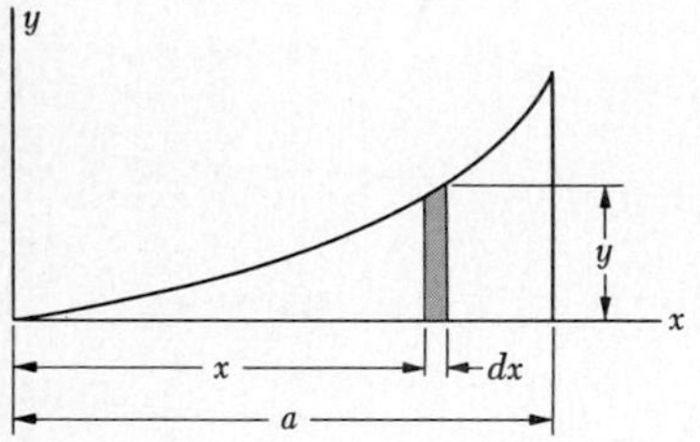

Solution. Referring to Sample Prob. 5.4, we obtain the following expressions for the equation of the curve and the total area:

$$y = \frac{b}{a^2}x^2 \qquad A = \tfrac{1}{3}ab$$

Moment of Inertia I_x. A vertical differential element of area is chosen. Since all portions of this element are *not* at the same distance from the x axis, we must treat the element as a thin rectangle. The moment of inertia of the element with respect to the x axis is

$$dI_x = \tfrac{1}{3}y^3\,dx = \frac{1}{3}\left(\frac{b}{a^2}x^2\right)^3 dx = \frac{1}{3}\frac{b^3}{a^6}x^6\,dx$$

$$I_x = \int dI_x = \int_0^a \frac{1}{3}\frac{b^3}{a^6}x^6\,dx = \left[\frac{1}{3}\frac{b^3}{a^6}\frac{x^7}{7}\right]_0^a$$

$$I_x = \frac{ab^3}{21} \quad ◄$$

Moment of Inertia I_y. The same vertical differential element of area is used. Since all portions of the element are at the same distance from the y axis, we write

$$dI_y = x^2\,dA = x^2(y\,dx) = x^2\left(\frac{b}{a^2}x^2\right)dx = \frac{b}{a^2}x^4\,dx$$

$$I_y = \int dI_y = \int_0^a \frac{b}{a^2}x^4\,dx = \left[\frac{b}{a^2}\frac{x^5}{5}\right]_0^a$$

$$I_y = \frac{a^3b}{5} \quad ◄$$

Radii of Gyration k_x and k_y

$$k_x^2 = \frac{I_x}{A} = \frac{ab^3/21}{ab/3} = \frac{b^2}{7} \qquad k_x = \sqrt{\tfrac{1}{7}}\,b \quad ◄$$

$$k_y^2 = \frac{I_y}{A} = \frac{a^3b/5}{ab/3} = \tfrac{3}{5}a^2 \qquad k_y = \sqrt{\tfrac{3}{5}}\,a \quad ◄$$

Problems

9.1 through 9.4 Determine by direct integration the moment of inertia of the shaded area with respect to the y axis.

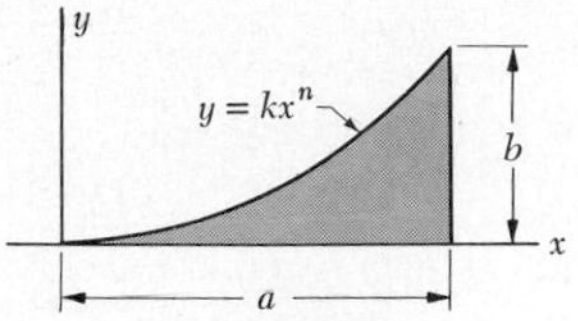

Fig. P9.1 and P9.5

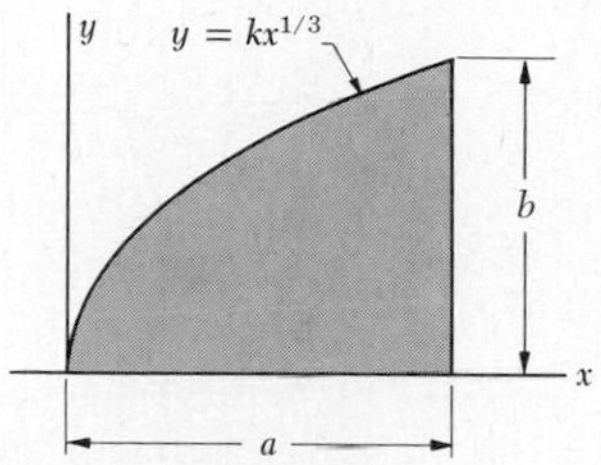

Fig. P9.2 and P9.6

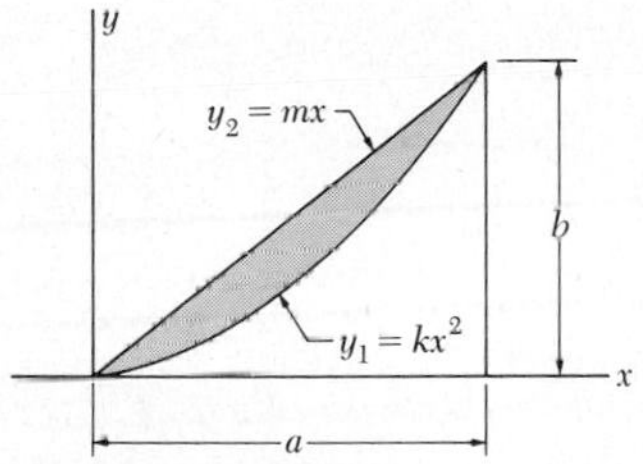

Fig. P9.3 and P9.7

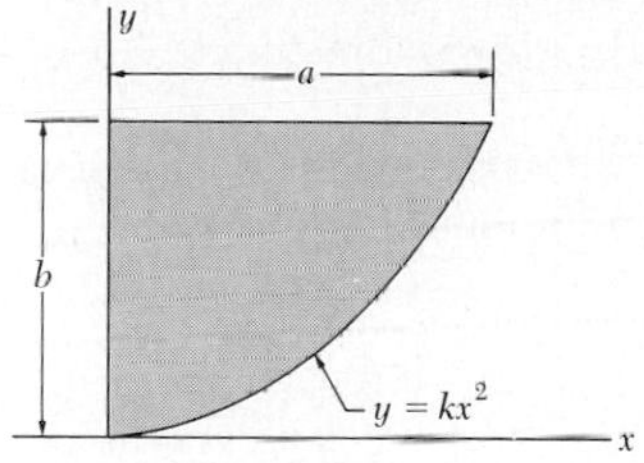

Fig. P9.4 and P9.8

9.5 through 9.8 Determine by direct integration the moment of inertia of the shaded area with respect to the x axis.

9.9 and 9.10 Determine the moment of inertia and radius of gyration of the shaded area shown with respect to the x axis.

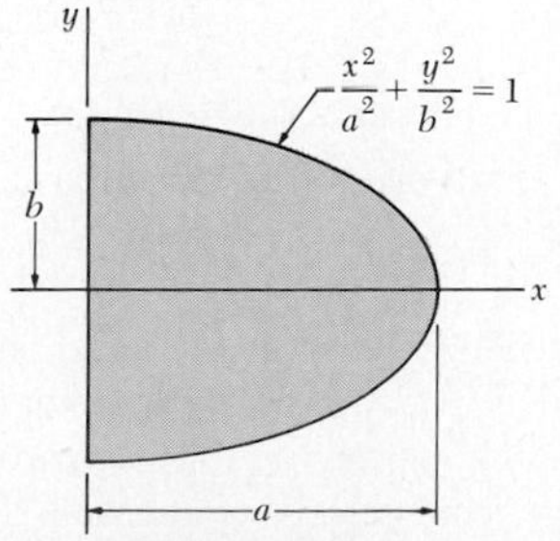

Fig. P9.9 and P9.11

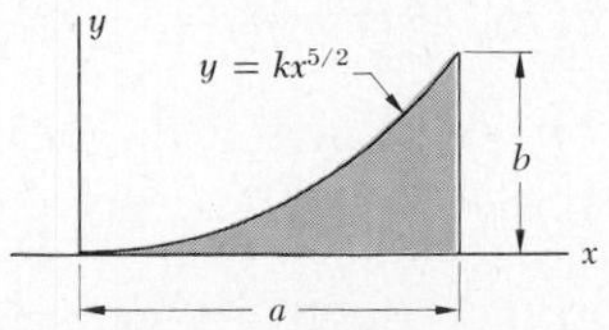

Fig. P9.10 and P9.12

9.11 and 9.12 Determine the moment of inertia and radius of gyration of the shaded area shown with respect to the y axis.

9.13 Determine the moment of inertia and radius of gyration of the shaded area shown with respect to the x axis.

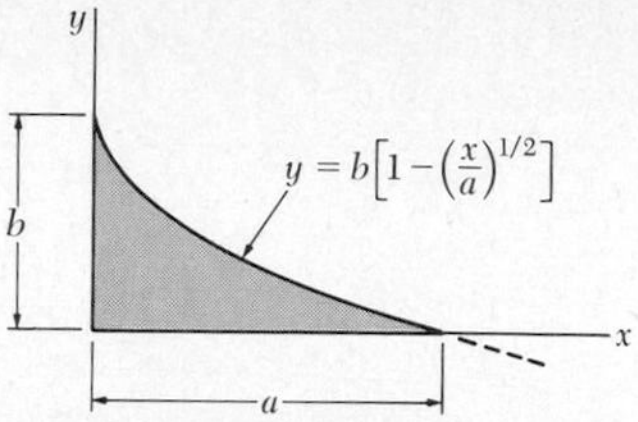

Fig. P9.13 and P9.14

9.14 Determine the moment of inertia and radius of gyration of the shaded area shown with respect to the y axis.

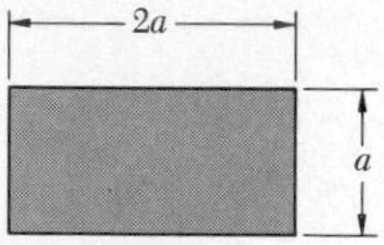

Fig. P9.15

9.15 Determine the polar moment of inertia and the polar radius of gyration of the rectangle shown with respect to the midpoint of one of its (a) longer sides, (b) shorter sides.

9.16 Determine the polar moment of inertia and the polar radius of gyration of the trapezoid shown with respect to point P_1.

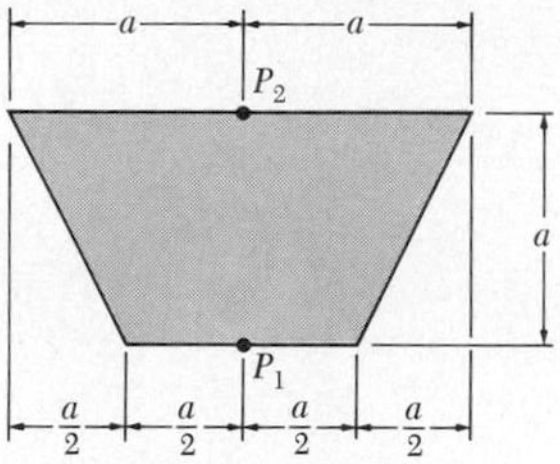

Fig. P9.16 and P9.17

9.17 Determine the polar moment of inertia and the polar radius of gyration of the trapezoid shown with respect to point P_2.

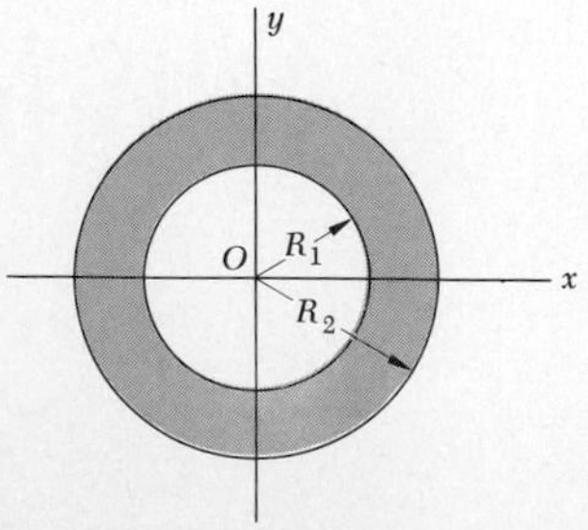

Fig. P9.18 and P9.19

9.18 (a) Determine by direct integration the polar moment of inertia of the annular area shown. (b) Using the result of part a, determine the moment of inertia of the given area with respect to the x axis.

9.19 (a) Show that the polar radius of gyration k_O of the annular area shown is approximately equal to the mean radius $R_m = (R_1 + R_2)/2$ for small values of the thickness $t = R_2 - R_1$. (b) Determine the percentage error introduced by using R_m in place of k_O for values of t/R_m respectively equal to 1, $\frac{1}{2}$, and $\frac{1}{10}$.

***9.20** Prove that the centroidal polar moment of inertia of a given area A cannot be smaller than $A^2/2\pi$. (*Hint.* Compare the moment of inertia of the given area with the moment of inertia of a circle of the same area and same centroid.)

9.6. Parallel-Axis Theorem. Consider the moment of inertia I of an area A with respect to an axis AA' (Fig. 9.9). Denoting by y the distance from an element of area dA to AA', we write

$$I = \int y^2\, dA$$

Let us now draw an axis BB' parallel to AA' through the centroid C of the area; this axis is called a *centroidal axis*. Denoting by y' the distance from

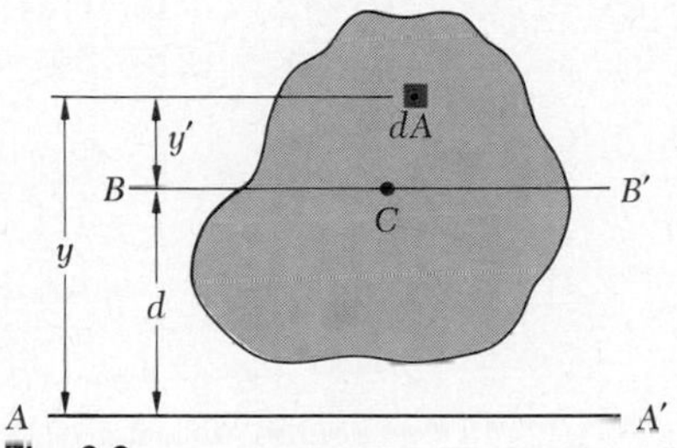

Fig. 9.9

the element dA to BB', we write $y = y' + d$, where d is the distance between the axes AA' and BB'. Substituting for y in the integral representing I, we write

$$I = \int y^2\, dA = \int (y' + d)^2\, dA$$
$$= \int y'^2\, dA + 2d \int y'\, dA + d^2 \int dA$$

The first integral represents the moment of inertia $\bar{I}$ of the area with respect to the centroidal axis BB'. The second integral represents the first moment of the area with respect to BB'; since the centroid C of the area is located on that axis, the second integral must be zero. Finally, we observe that the last integral is equal to the total area A. We write therefore

$$I = \bar{I} + Ad^2 \qquad (9.9)$$

This formula expresses that the moment of inertia I of an area with respect to any given axis AA' is equal to the moment of inertia $\bar{I}$ of the area with respect to a centroidal axis BB' parallel to AA' *plus* the product Ad^2 of the area A and of the square of the distance d between the two axes. This theorem is known as the *parallel-axis theorem*. Substituting k^2A for I and $\bar{k}^2A$ for $\bar{I}$, the theorem may also be expressed in the following way:

$$k^2 = \bar{k}^2 + d^2 \qquad (9.10)$$

A similar theorem may be used to relate the polar moment of inertia J_O of an area about a point O and the polar moment of inertia $\bar{J}_C$ of the same area about its centroid C. Denoting by d the distance between O and C, we write

$$J_O = \bar{J}_C + Ad^2 \qquad \text{or} \qquad k_O^2 = \bar{k}_C^2 + d^2 \qquad (9.11)$$

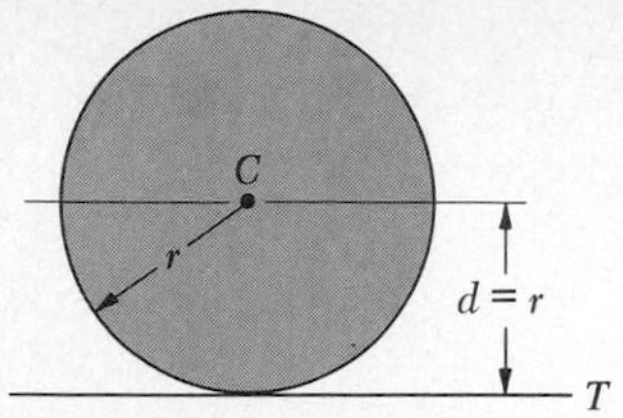

Fig. 9.10

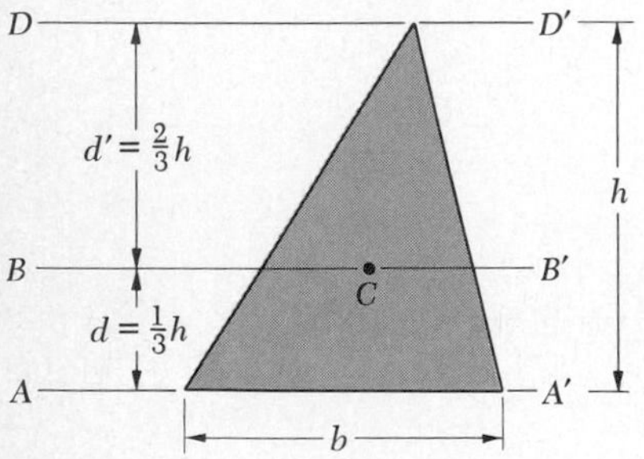

Fig. 9.11

Example 1. As an application of the parallel-axis theorem, we shall determine the moment of inertia I_T of a circular area with respect to a line tangent to the circle (Fig. 9.10). We found in Sample Prob. 9.2 that the moment of inertia of a circular area about a centroidal axis is $\bar{I} = \frac{1}{4}\pi r^4$. We may write, therefore,

$$I_T = \bar{I} + Ad^2 = \tfrac{1}{4}\pi r^4 + \pi r^2 r^2 = \tfrac{5}{4}\pi r^4$$

Example 2. The parallel-axis theorem may also be used to determine the centroidal moment of inertia of an area when the moment of inertia of this area with respect to some parallel axis is known. Consider, for instance, a triangular area (Fig. 9.11). We found in Sample Prob. 9.1 that the moment of inertia of a triangle with respect to its base AA' is equal to $\frac{1}{12}bh^3$. Using the parallel-axis theorem, we write

$$I_{AA'} = \bar{I}_{BB'} + Ad^2$$
$$\bar{I}_{BB'} = I_{AA'} - Ad^2 = \tfrac{1}{12}bh^3 - \tfrac{1}{2}bh(\tfrac{1}{3}h)^2 = \tfrac{1}{36}bh^3$$

It should be observed that the product Ad^2 was *subtracted* from the given moment of inertia in order to obtain the centroidal moment of inertia of the triangle. While this product is *added* in transferring *from* a centroidal axis to a parallel axis, it should be *subtracted* in transferring *to* a centroidal axis. In other words, the moment of inertia of an area is always smaller with respect to a centroidal axis than with respect to any other parallel axis.

Returning to Fig. 9.11, we observe that the moment of inertia of the triangle with respect to a line DD' drawn through a vertex may be obtained by writing

$$I_{DD'} = \bar{I}_{BB'} + Ad'^2 = \tfrac{1}{36}bh^3 + \tfrac{1}{2}bh(\tfrac{2}{3}h)^2 = \tfrac{1}{4}bh^3$$

Note that $I_{DD'}$ *could not* have been obtained directly from $I_{AA'}$. The parallel-axis theorem can be applied only if one of the two parallel axes passes through the centroid of the area.

9.7. Moments of Inertia of Composite Areas. Consider a composite area A made of several component areas A_1, A_2, etc. Since the integral representing the moment of inertia of A may be subdivided into integrals computed over A_1, A_2, etc., the moment of inertia of A with respect to a given axis will be obtained by adding the moments of inertia of the areas A_1, A_2, etc., with respect to the same axis. The moment of inertia of an area made of several of the common shapes shown in Fig. 9.12 may thus be obtained from the formulas given in that figure. Before adding the moments of inertia of the component areas, however, the parallel-axis theorem should be used to transfer each moment of inertia to the desired axis. This is shown in Sample Probs. 9.4 and 9.5.

The properties of the cross sections of various structural shapes are given in Fig. 9.13. As noted in Sec. 9.2, the moment of inertia of a beam section about its neutral axis is closely related to the value of the internal forces. The determination of moments of inertia is thus a prerequisite to the analysis and design of structural members.

It should be noted that the radius of gyration of a composite area is *not* equal to the sum of the radii of gyration of the component areas. In order to determine the radius of gyration of a composite area, it is necessary first to compute the moment of inertia of the area.

Rectangle	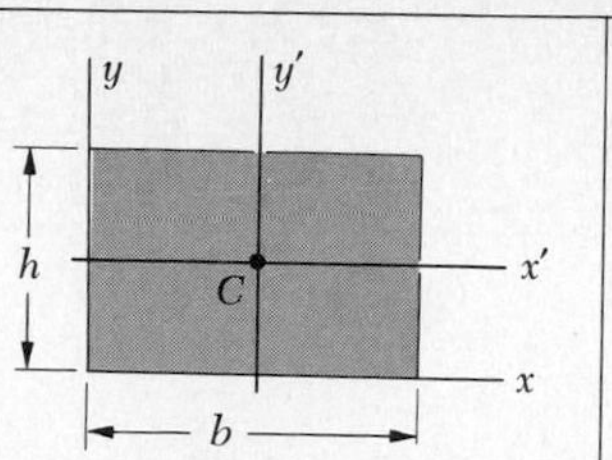	$\bar{I}_{x'} = \frac{1}{12}bh^3$ $\bar{I}_{y'} = \frac{1}{12}b^3h$ $I_x = \frac{1}{3}bh^3$ $I_y = \frac{1}{3}b^3h$ $J_C = \frac{1}{12}bh(b^2 + h^2)$
Triangle	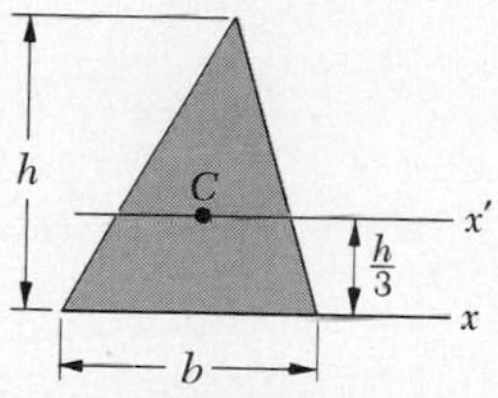	$\bar{I}_{x'} = \frac{1}{36}bh^3$ $I_x = \frac{1}{12}bh^3$
Circle	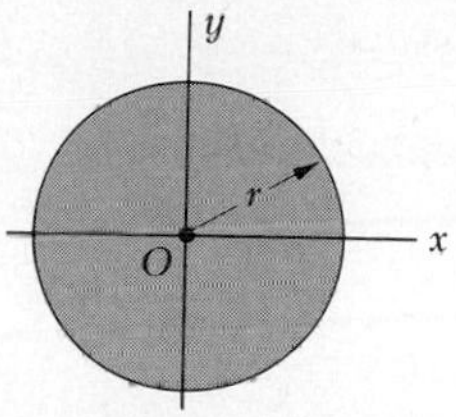	$\bar{I}_x = \bar{I}_y = \frac{1}{4}\pi r^4$ $J_O = \frac{1}{2}\pi r^4$
Semicircle	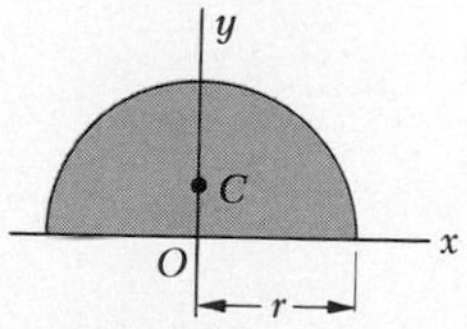	$I_x = I_y = \frac{1}{8}\pi r^4$ $J_O = \frac{1}{4}\pi r^4$
Quarter circle	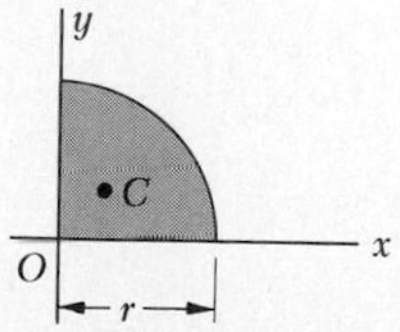	$I_x = I_y = \frac{1}{16}\pi r^4$ $J_O = \frac{1}{8}\pi r^4$
Ellipse	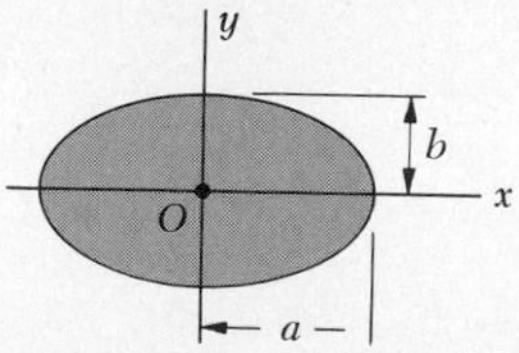	$\bar{I}_x = \frac{1}{4}\pi ab^3$ $\bar{I}_y = \frac{1}{4}\pi a^3b$ $J_O = \frac{1}{4}\pi ab(a^2 + b^2)$

Fig. 9.12 Moments of inertia of common geometric shapes.

Fig. 9.13A Properties of Rolled-Steel Shapes (U.S. Customary Units).*

	Designation	Area in^2	Depth in.	Width in.	Axis X-X			Axis Y-Y		
					$\bar{I}_x$, in^4	$\bar{k}_x$, in.	$\bar{y}$, in.	$\bar{I}_y$, in^4	$\bar{k}_y$, in.	$\bar{x}$, in.
W Shapes (Wide-Flange Shapes) 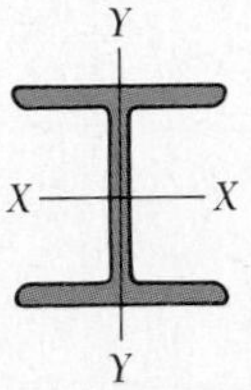	W18 × 76†	22.3	18.21	11.035	1330	7.73		152	2.61	
	W16 × 57	16.8	16.43	7.120	758	6.72		43.1	1.60	
	W14 × 38	11.2	14.10	6.770	385	5.88		26.7	1.55	
	W8 × 31	9.13	8.00	7.995	110	3.47		37.1	2.02	
S Shapes (American Standard Shapes) 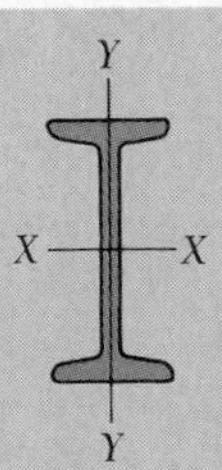	S18 × 55.7†	16.1	18.00	6.001	804	7.07		20.8	1.14	
	S12 × 31.8	9.35	12.00	5.000	218	4.83		9.36	1.00	
	S10 × 25.4	7.46	10.00	4.661	124	4.07		6.79	0.954	
	S6 × 12.5	3.67	6.00	3.332	22.1	2.45		1.82	0.705	
C Shapes (American Standard Channels) 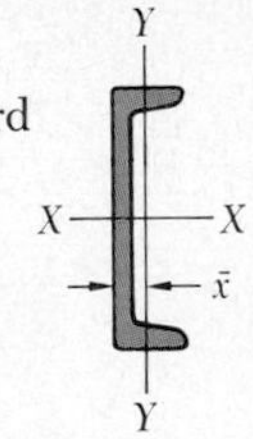	C12 × 20.7†	6.09	12.00	2.942	129	4.61		3.88	0.799	0.698
	C10 × 15.3	4.49	10.00	2.600	67.4	3.87		2.28	0.713	0.634
	C8 × 11.5	3.38	8.00	2.260	32.6	3.11		1.32	0.625	0.571
	C6 × 8.2	2.40	6.00	1.920	13.1	2.34		0.692	0.537	0.512
Angles 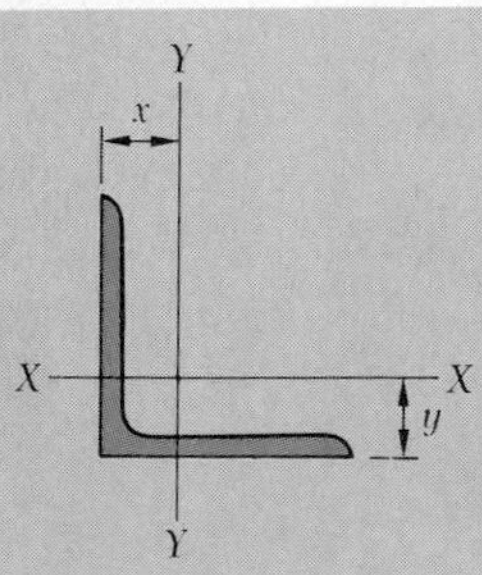	L6 × 6 × 1‡	11.00			35.5	1.80	1.86	35.5	1.80	1.86
	L4 × 4 × $\frac{1}{2}$	3.75			5.56	1.22	1.18	5.56	1.22	1.18
	L3 × 3 × $\frac{1}{4}$	1.44			1.24	0.930	0.842	1.24	0.930	0.842
	L6 × 4 × $\frac{1}{2}$	4.75			17.4	1.91	1.99	6.27	1.15	0.987
	L5 × 3 × $\frac{1}{2}$	3.75			9.45	1.59	1.75	2.58	0.829	0.750
	L3 × 2 × $\frac{1}{4}$	1.19			1.09	0.957	0.993	0.392	0.574	0.493

*Courtesy of the American Institute of Steel Construction, Chicago, Illinois

†Nominal depth in inches and weight in pounds per foot.

‡Depth, width, and thickness in inches.

Fig. 9.13B Properties of Rolled-Steel Shapes (SI Units).

					Axis X-X			Axis Y-Y		
	Designation	Area mm^2	Depth mm	Width mm	$\bar{I}_x$ 10^6 mm^4	$\bar{k}_x$ mm	$\bar{y}$ mm	$\bar{I}_y$ 10^6 mm^4	$\bar{k}_y$ mm	$\bar{x}$ mm
W Shapes (Wide-Flange Shapes) 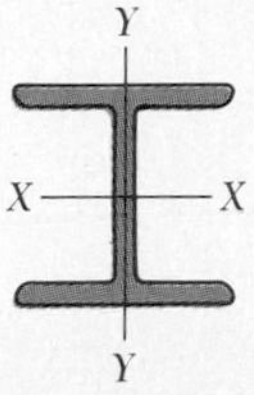	W460 × 113†	14400	463	280	554	196.3		63.3	66.3	
	W410 × 85	10800	417	181	316	170.7		17.94	40.6	
	W360 × 57	7230	358	172	160.2	149.4		11.11	39.4	
	W200 × 46.1	5890	203	203	45.8	88.1		15.44	51.3	
S Shapes (American Standard Shapes) 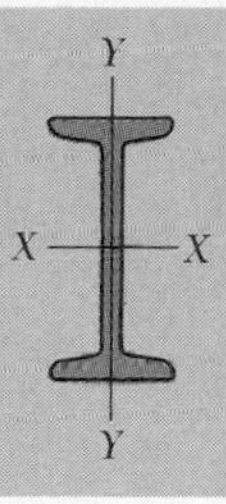	S460 × 81.4†	10390	457	152	335	179.6		8.66	29.0	
	W310 × 47.3	6032	305	127	90.7	122.7		3.90	25.4	
	S250 × 37.8	4806	254	118	51.6	103.4		2.83	24.2	
	S150 × 18.6	2362	152	84	9.2	62.2		0.758	17.91	
C Shapes (American Standard Channels) 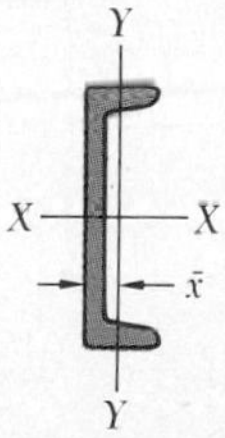	C310 × 30.8†	3929	305	74	53.7	117.1		1.615	20.29	17.73
	C250 × 22.8	2897	254	65	28.1	98.3		0.949	18.11	16.10
	C200 × 17.1	2181	203	57	13.57	79.0		0.549	15.88	14.50
	C150 × 12.2	1548	152	48	5.45	59.4		0.288	13.64	13.00
Angles 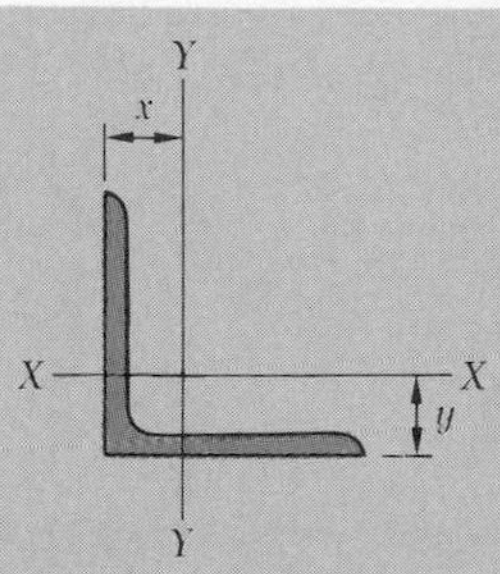	L152 × 152 × 25.4‡	7100			14.78	45.6	47.2	14.78	45.6	47.2
	L102 × 102 × 12.7	2420			2.31	30.9	30.0	2.31	30.9	30.0
	L76 × 76 × 6.4	929			0.516	23.6	21.4	0.516	23.6	21.4
	L152 × 102 × 12.7	3060			7.24	48.6	50.5	2.61	29.2	25.1
	L127 × 76 × 12.7	2420			3.93	40.3	44.5	1.074	21.1	19.05
	L76 × 51 × 6.4	768			0.454	24.3	25.2	0.163	14.58	12.52

†Nominal depth in millimeters and mass in kilograms per meter.

‡Depth, width, and thickness in millimeters.

SAMPLE PROBLEM 9.4

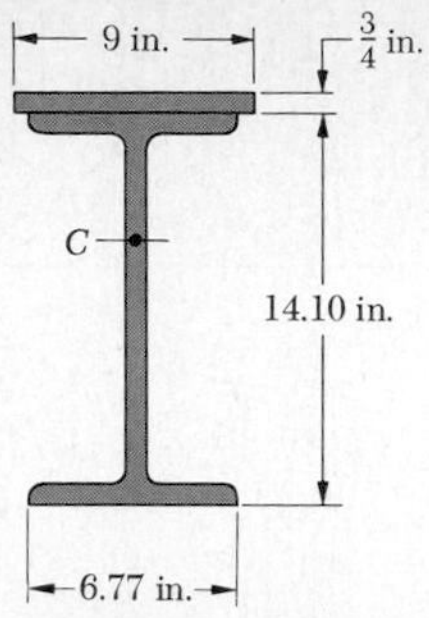

The strength of a W14 × 38 rolled-steel beam is increased by attaching a 9 × $\frac{3}{4}$ in. plate to its upper flange as shown. Determine the moment of inertia and the radius of gyration of the composite section with respect to an axis through its centroid C and parallel to the plate.

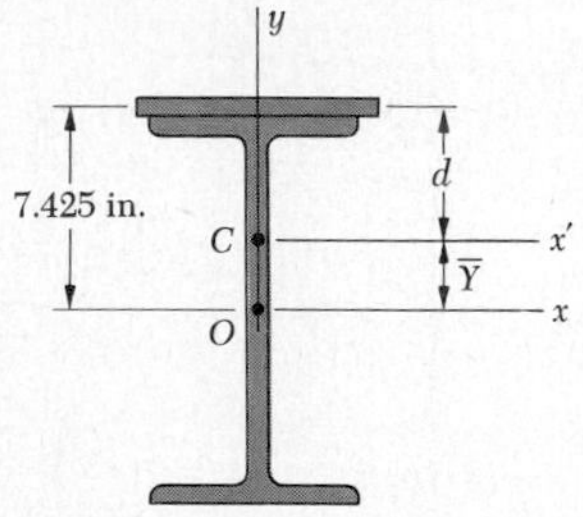

Solution. The origin of coordinates O is placed at the centroid of the wide-flange shape, and the distance $\overline{Y}$ to the centroid of the composite section is computed by the methods of Chap. 5. The area of the wide-flange shape is found by referring to Fig. 9.13A. The area and y coordinate of the centroid of the plate are

$$A = (9 \text{ in.})(0.75 \text{ in.}) = 6.75 \text{ in}^2$$
$$\overline{y} = \tfrac{1}{2}(14.10 \text{ in.}) + \tfrac{1}{2}(0.75 \text{ in.}) = 7.425 \text{ in.}$$

Section	Area, in²	$\overline{y}$, in.	$\overline{y}A$, in³
Plate	6.75	7.425	50.12
Wide-flange shape	11.20	0	0
	17.95	. . .	50.12

$$\overline{Y}\Sigma A = \Sigma\overline{y}A \qquad \overline{Y}(17.95) = 50.12 \qquad \overline{Y} = 2.792 \text{ in.}$$

Moment of Inertia. The parallel-axis theorem is used to determine the moments of inertia of the wide-flange shape and of the plate with respect to the x' axis. This axis is a centroidal axis for the composite section, but *not* for either of the elements considered separately. The value of $\overline{I}_x$ for the wide-flange shape is obtained from Fig. 9.13A.

For the wide-flange shape,

$$I_{x'} = \overline{I}_x + A\overline{Y}^2 = 385 + (11.20)(2.792)^2 = 472.3 \text{ in}^4$$

For the plate,

$$I_{x'} = \overline{I}_x + Ad^2 = (\tfrac{1}{12})(9)(\tfrac{3}{4})^3 + (6.75)(7.425 - 2.792)^2 = 145.2 \text{ in}^4$$

For the composite area,

$$I_{x'} = 472.3 + 145.2 = 617.5 \text{ in}^4$$

$$I_{x'} = 618 \text{ in}^4 \blacktriangleleft$$

Radius of Gyration

$$k_{x'}^2 = \frac{I_{x'}}{A} = \frac{617.5 \text{ in}^4}{17.95 \text{ in}^2}$$

$$k_{x'} = 5.87 \text{ in.} \blacktriangleleft$$

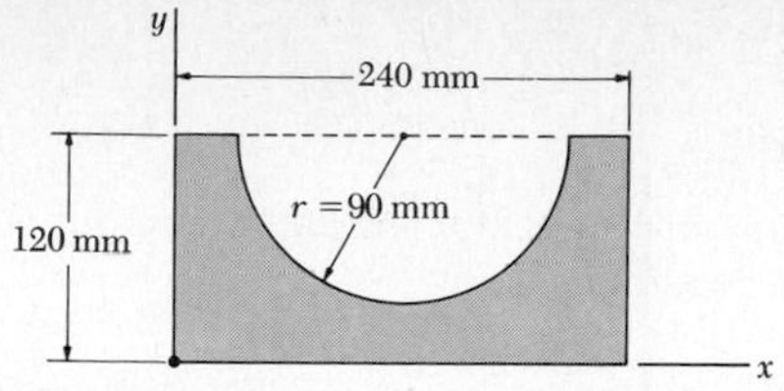

SAMPLE PROBLEM 9.5

Determine the moment of inertia of the shaded area with respect to the x axis.

Solution. The given area may be obtained by subtracting a half circle from a rectangle. The moments of inertia of the rectangle and of the half circle will be computed separately.

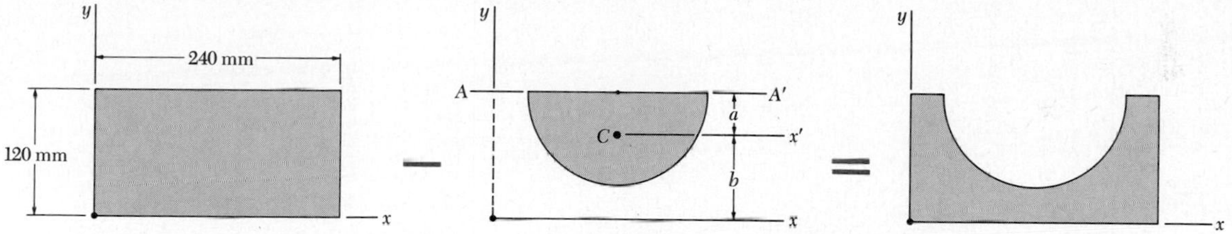

I_x for Rectangle. Referring to Fig. 9.12, we obtain

$$I_x = \tfrac{1}{3}bh^3 = \tfrac{1}{3}(240 \text{ mm})(120 \text{ mm})^3 = 138.2 \times 10^6 \text{ mm}^4$$

I_x for the Half Circle. Referring to Fig. 5.8, we locate the centroid C of the half circle with respect to diameter AA'.

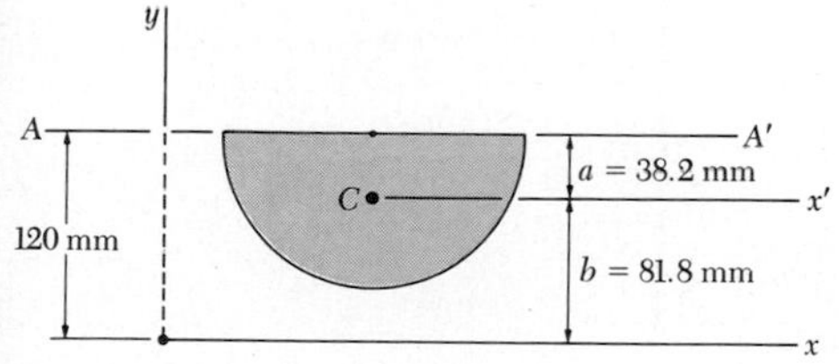

$$a = \frac{4r}{3\pi} = \frac{(4)(90 \text{ mm})}{3\pi} = 38.2 \text{ mm}$$

The distance b from the centroid C to the x axis is

$$b = 120 \text{ mm} - a = 120 \text{ mm} - 38.2 \text{ mm} = 81.8 \text{ mm}$$

Referring now to Fig. 9.12, we compute the moment of inertia of the half circle with respect to diameter AA'; we also compute the area of the half circle.

$$I_{AA'} = \tfrac{1}{8}\pi r^4 = \tfrac{1}{8}\pi(90 \text{ mm})^4 = 25.76 \times 10^6 \text{ mm}^4$$
$$A = \tfrac{1}{2}\pi r^2 = \tfrac{1}{2}\pi(90 \text{ mm})^2 = 12.72 \times 10^3 \text{ mm}^2$$

Using the parallel-axis theorem, we obtain the value of $\bar{I}_{x'}$:

$$I_{AA'} = \bar{I}_{x'} + Aa^2$$
$$25.76 \times 10^6 \text{ mm}^4 = \bar{I}_{x'} + (12.72 \times 10^3 \text{ mm}^2)(38.2 \text{ mm})^2$$
$$\bar{I}_{x'} = 7.20 \times 10^6 \text{ mm}^4$$

Again using the parallel-axis theorem, we obtain the value of I_x:

$$I_x = \bar{I}_{x'} + Ab^2 = 7.20 \times 10^6 \text{ mm}^4 + (12.72 \times 10^3 \text{ mm}^2)(81.8 \text{ mm})^2$$
$$= 92.3 \times 10^6 \text{ mm}^4$$

I_x for Given Area. Subtracting the moment of inertia of the half circle from that of the rectangle, we obtain

$$I_x = 138.2 \times 10^6 \text{ mm}^4 - 92.3 \times 10^6 \text{ mm}^4$$
$$I_x = 45.9 \times 10^6 \text{ mm}^4 \blacktriangleleft$$

Problems

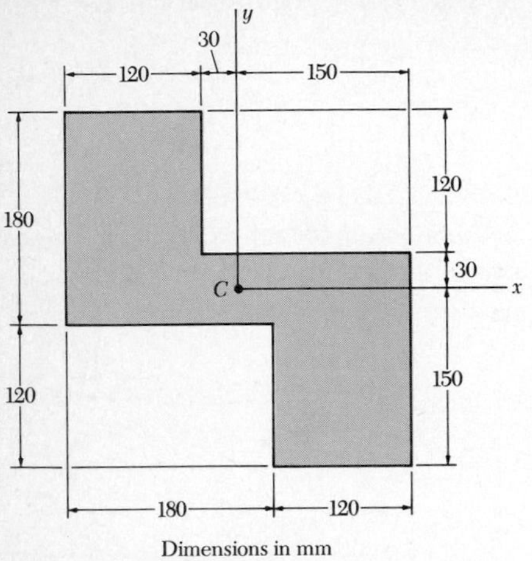

Fig. P9.21 and P9.22

9.21 and 9.23 Determine the moment of inertia and the radius of gyration of the shaded area with respect to the x axis.

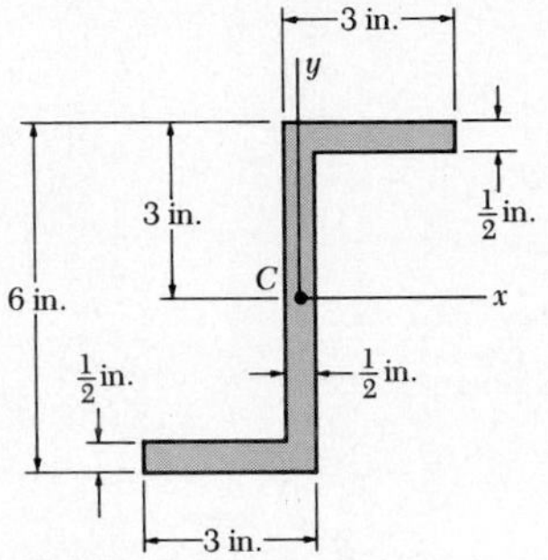

Fig. P9.23 and P9.24

9.22 and 9.24 Determine the moment of inertia and the radius of gyration of the shaded area with respect to the y axis.

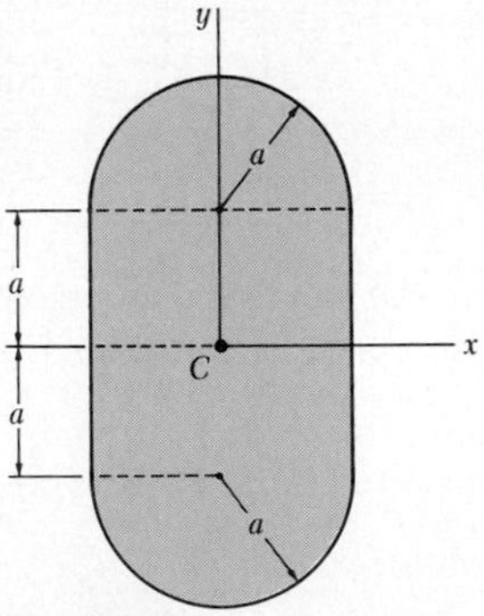

Fig. P9.25

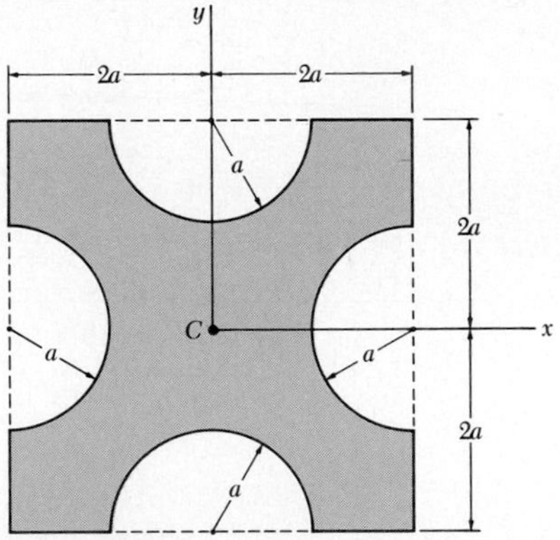

Fig. P9.26

9.25 and 9.26 Determine the moments of inertia of the shaded area shown with respect to the x and y axes when $a = 20$ mm.

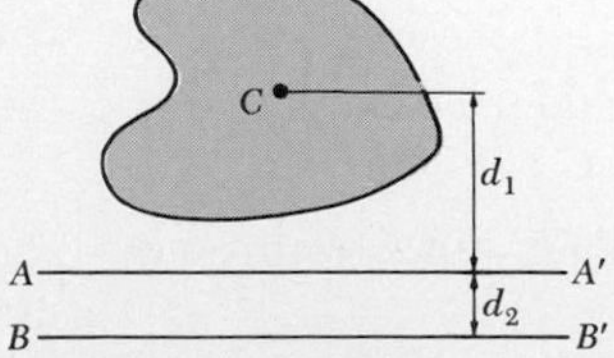

Fig. P9.27 and P9.28

9.27 Determine the shaded area and its moment of inertia with respect to a centroidal axis parallel to AA', knowing that its moments of inertia with respect to AA' and BB' are respectively 4.1×10^6 mm^4 and 6.9×10^6 mm^4, and that $d_1 = 30$ mm and $d_2 = 10$ mm.

9.28 Knowing that the shaded area is equal to 7500 mm^2 and that its moment of inertia with respect to AA' is 31×10^6 mm^4, determine its moment of inertia with respect to BB', for $d_1 = 60$ mm and $d_2 = 15$ mm.

9.29 The shaded area shown is equal to 20 in^2, and for $d_1 = 6$ in. the polar moment of inertia of the area with respect to point B is $J_B = 800$ in^4. Knowing that $\bar{I}_x = 4\bar{I}_y$, determine (*a*) $\bar{I}_x$, (*b*) J_A for $d_2 = 5$ in.

9.30 For the shaded area shown, determine the polar moment of inertia with respect to point D, knowing that the polar moments of inertia with respect to points A and B are, respectively, $J_A = 4000$ in^4 and $J_B = 6240$ in^4, and that $d_1 = 8$ in. and $d_2 = 6$ in.

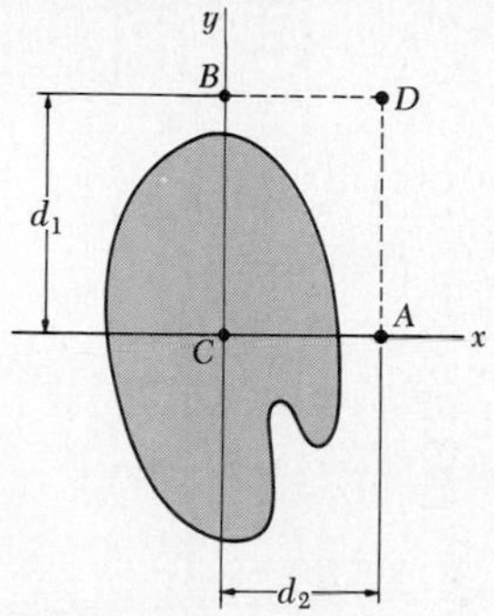

Fig. P9.29 and P9.30

9.31 and 9.32 Determine the moments of inertia $\bar{I}_x$ and $\bar{I}_y$ of the area shown with respect to centroidal axes respectively parallel and perpendicular to the side AB.

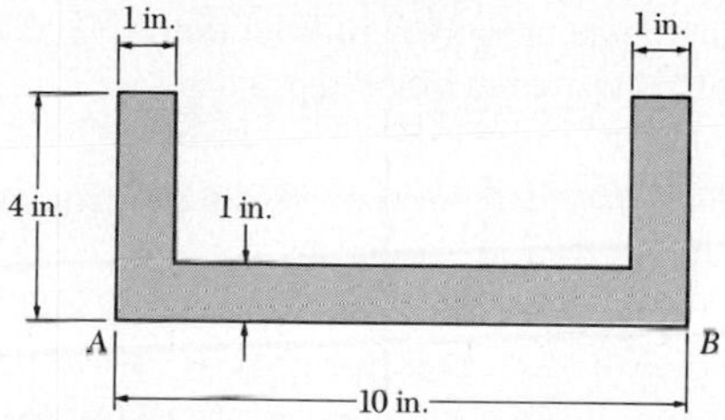

Fig. P9.31

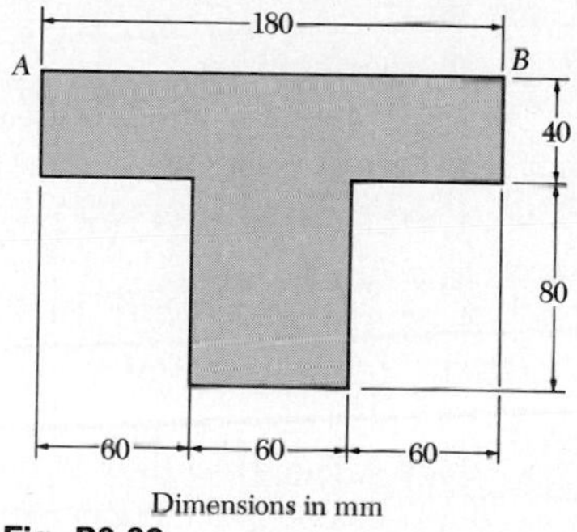

Fig. P9.32

9.33 and 9.34 Determine the polar moment of inertia of the area shown with respect to (*a*) point O, (*b*) the centroid of the area.

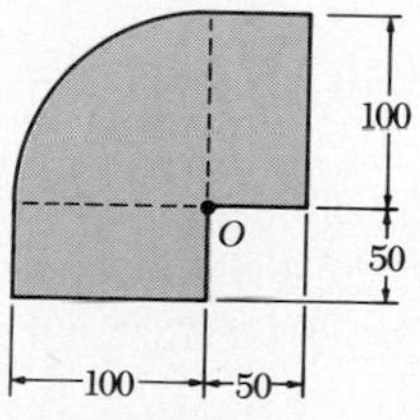

Fig. P9.33

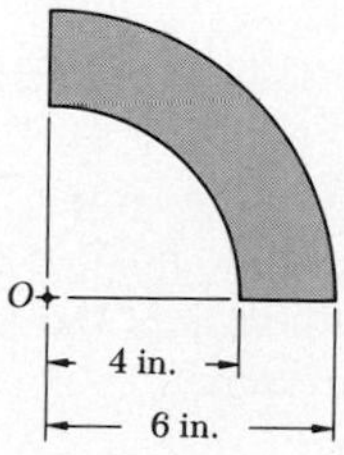

Fig. P9.34

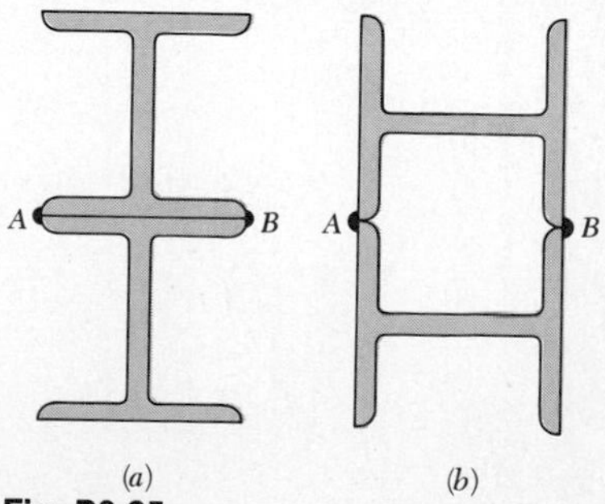

Fig. P9.35

9.35 Two W8 × 31 rolled sections may be welded at A and B in either of the two ways shown. For each arrangement, determine the moment of inertia of the section with respect to the horizontal centroidal axis.

9.36 Two $6 \times 4 \times \frac{1}{2}$ in. angles are welded together to form the section shown. Determine the moments of inertia and the radii of gyration of the section with respect to the centroidal axes shown.

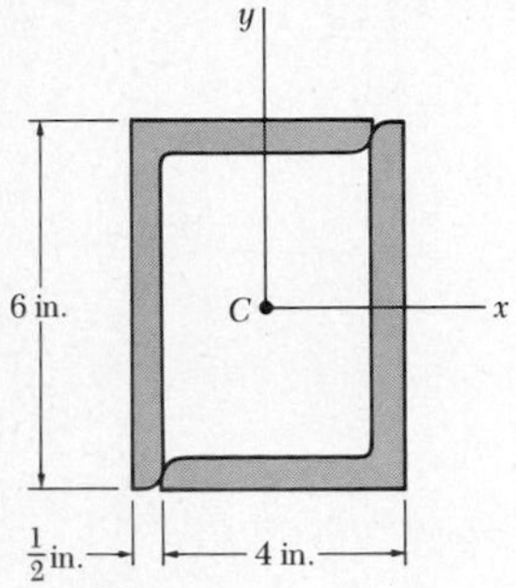

Fig. P9.36

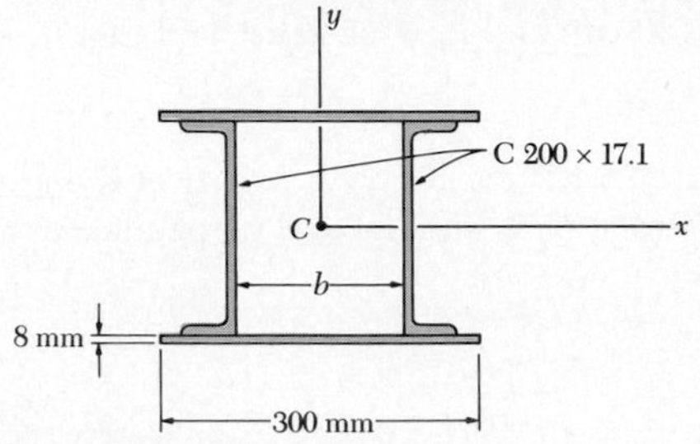

Fig. P9.37

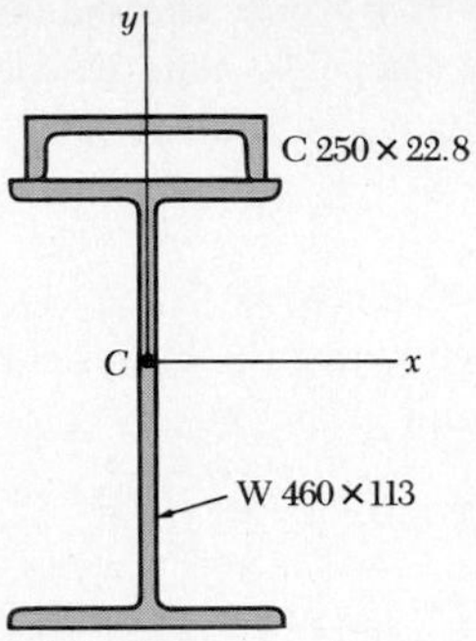

Fig. P9.39

9.37 Two channels and two plates are used to form the column section shown. For $b = 160$ mm, determine the moments of inertia and the radii of gyration of the combined section with respect to the centroidal axes.

9.38 In Prob. 9.37, determine the distance b for which the centroidal moments of inertia $\bar{I}_x$ and $\bar{I}_y$ of the column section are equal.

9.39 The strength of the wide-flange rolled section shown is increased by welding a channel to its upper flange. Determine the moments of inertia of the combined section with respect to its centroidal x and y axes.

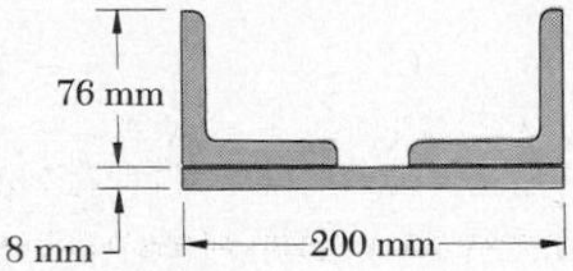

Fig. P9.40

9.40 Two $76 \times 76 \times 6.4$ mm angles are welded to an 8×200 mm steel plate as shown. Determine the moments of inertia of the combined section with respect to centroidal axes respectively parallel and perpendicular to the plate.

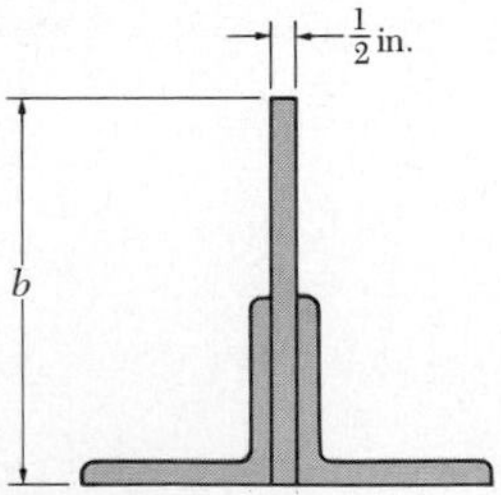

Fig. P9.41

9.41 Two $4 \times 4 \times \frac{1}{2}$ in. angles are welded to a steel plate as shown. For $b = 8$ in., determine the moments of inertia of the combined section with respect to centroidal axes respectively parallel and perpendicular to the plate.

9.42 Solve Prob. 9.41, assuming that $b = 10$ in.

9.43 through 9.45 The panel shown forms the end of a trough which is filled with water to the line AA'. Referring to Sec. 9.2, determine the depth of the point of application of the resultant of the hydrostatic forces acting on the panel (center of pressure).

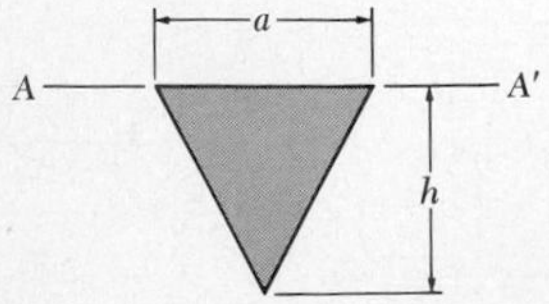

Fig. P9.43

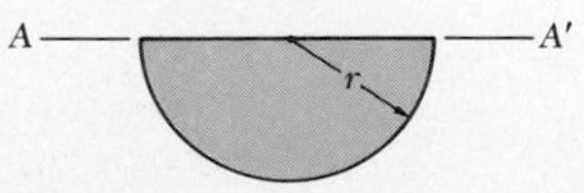

Fig. P9.44

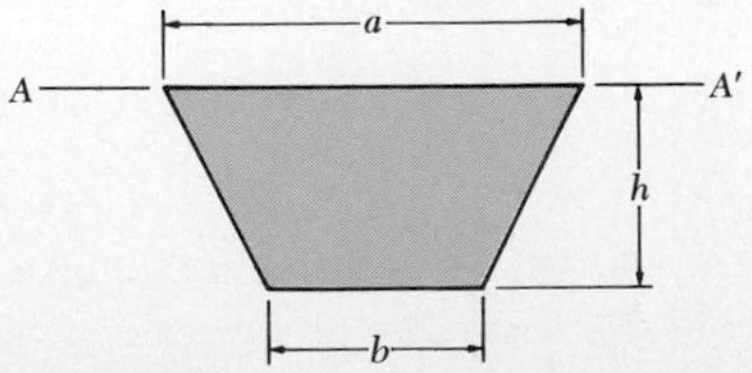

Fig. P9.45

9.46 The center of a vertical circular gate, 1.6 m in diameter, is located 2.4 m below the water surface. The gate is held by three bolts equally spaced as shown. Determine the force on each bolt.

9.47 Solve Prob. 9.46, assuming that the center of the gate is located 2 m below the water surface.

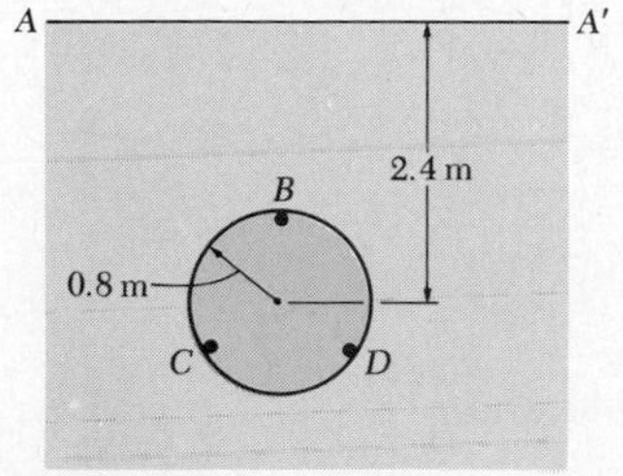

Fig. P9.46

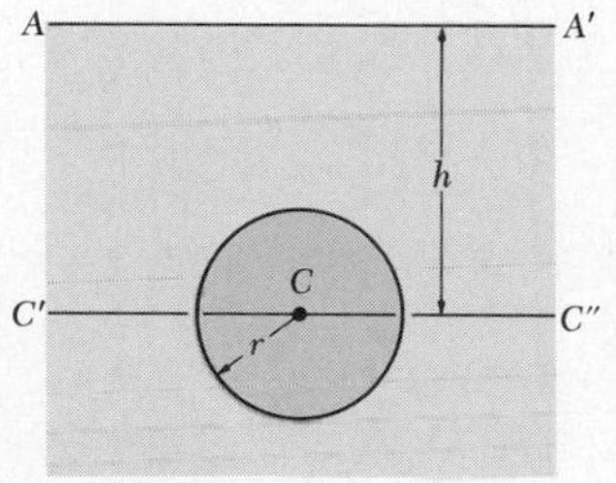

Fig. P9.48

9.48 A vertical circular gate of radius r is hinged about its diameter $C'C''$. Denoting the specific weight of water by γ, determine (*a*) the reaction at each of the two hinges, (*b*) the moment of the couple required to keep the gate closed.

***9.49** Determine the x coordinate of the centroid of the volume shown; this volume was obtained by intersecting a circular cylinder by an oblique plane. (*Hint.* The height of the volume is proportional to the x coordinate; consider an analogy between this height and the water pressure on a submerged surface.)

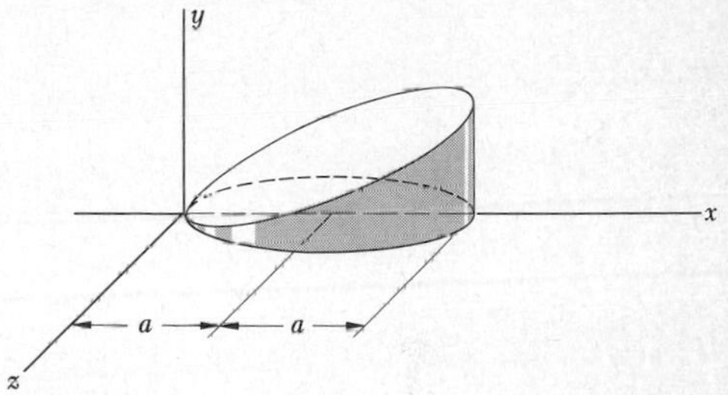

Fig. P9.49

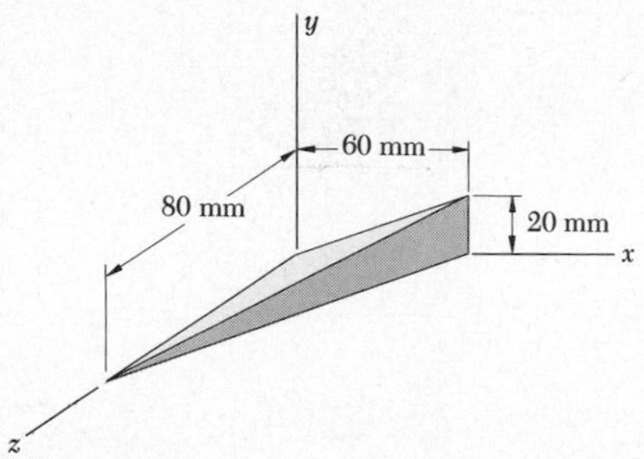

Fig. P9.50

***9.50** Determine the x coordinate of the centroid of the volume shown. (See hint of Prob. 9.49.)

***9.51** Show that the system of hydrostatic forces acting on a submerged plane area A may be reduced to a force **P** at the centroid C of the area and two couples. The force **P** is perpendicular to the area and of magnitude $P = \gamma A\bar{y} \sin\theta$, where γ is the specific weight of the liquid, and the couples are represented by vectors directed as shown and of magnitude $M_{x'} = \gamma \bar{I}_{x'} \sin\theta$ and $M_{y'} = \gamma \bar{I}_{x'y'} \sin\theta$, where $\bar{I}_{x'y'} = \int x'y'\, dA$ (see Sec. 9.8). Note that the couples are independent of the depth at which the area is submerged.

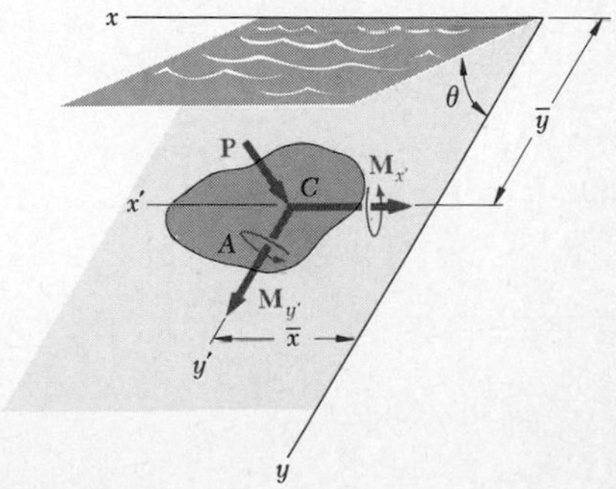

Fig. P9.51

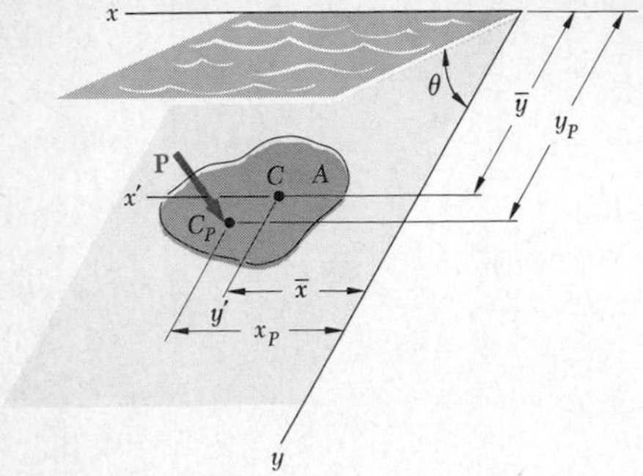

Fig. P9.52

***9.52** Show that the resultant of the hydrostatic forces acting on a submerged plane area A is a force **P** perpendicular to the area and of magnitude $P = \gamma A\bar{y} \sin\theta = \bar{p}A$, where γ is the specific weight of the liquid and $\bar{p}$ the pressure at the centroid C of the area. Show that **P** is applied at a point C_P, called the center of pressure, of coordinates $x_P = I_{xy}/A\bar{y}$ and $y_P = I_x/A\bar{y}$, where $I_{xy} = \int xy\,dA$ (see Sec. 9.8). Show also that the difference of ordinates $y_P - \bar{y}$ is equal to $k_{x'}^2/\bar{y}$ and thus depends upon the depth at which the area is submerged.

*9.8. Product of Inertia.

The integral

$$I_{xy} = \int xy\,dA \tag{9.12}$$

obtained by multiplying each element dA of an area A by its coordinates x and y and integrating over the area (Fig. 9.14), is known as the *product of inertia* of the area A with respect to the x and y axes. Unlike the moments of inertia I_x and I_y, the product of inertia I_{xy} may be either positive or negative.

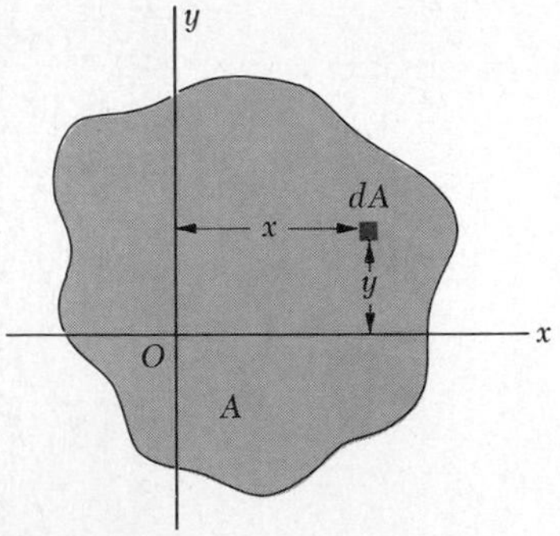

Fig. 9.14

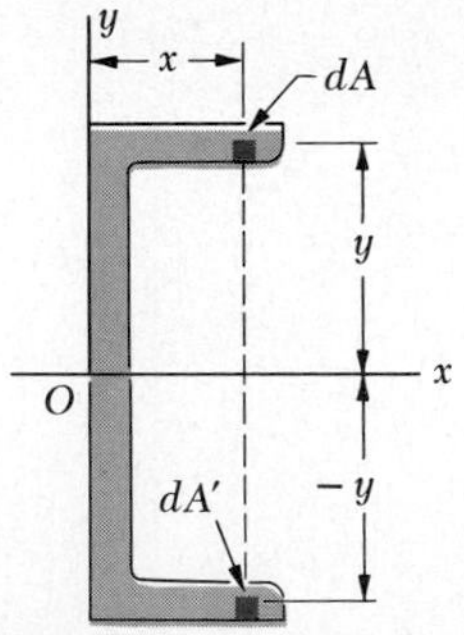

Fig. 9.15

When one or both of the x and y axes are axes of symmetry for the area A, the product of inertia I_{xy} is zero. Consider, for example, the channel section shown in Fig. 9.15. Since this section is symmetrical with respect to the x axis, we can associate to each element dA of coordinates x and y an element dA' of coordinates x and $-y$. Clearly, the contributions of any pair of elements chosen in this way cancel out, and the integral (9.12) reduces to zero.

A parallel-axis theorem similar to the one established in Sec. 9.6 for moments of inertia may be derived for products of inertia. Consider an area A and a system of rectangular coordinates x and y (Fig. 9.16). Through the centroid C of the area, of coordinates $\bar{x}$ and $\bar{y}$, we draw two *centroidal axes* x' and y' parallel, respectively, to the x and y axes. Denoting by x and y the coordinates of an element of area dA with respect to the original axes, and by x' and y' the coordinates of the same element with respect to the centroidal axes, we write $x = x' + \bar{x}$ and $y = y' + \bar{y}$. Substituting into (9.12), we obtain the following expression for the product of inertia I_{xy}:

$$\begin{aligned} I_{xy} &= \int xy\, dA = \int (x' + \bar{x})(y' + \bar{y})\, dA \\ &= \int x'y'\, dA + \bar{y}\int x'\, dA + \bar{x}\int y'\, dA + \bar{x}\bar{y}\int dA \end{aligned}$$

The first integral represents the product of inertia $\bar{I}_{x'y'}$ of the area A with respect to the centroidal axes x' and y'. The next two integrals represent first moments of the area with respect to the centroidal axes; they reduce to zero, since the centroid C is located on these axes. Finally, we observe that the last integral is equal to the total area A. We write therefore

$$I_{xy} = \bar{I}_{x'y'} + \bar{x}\bar{y}A \tag{9.13}$$

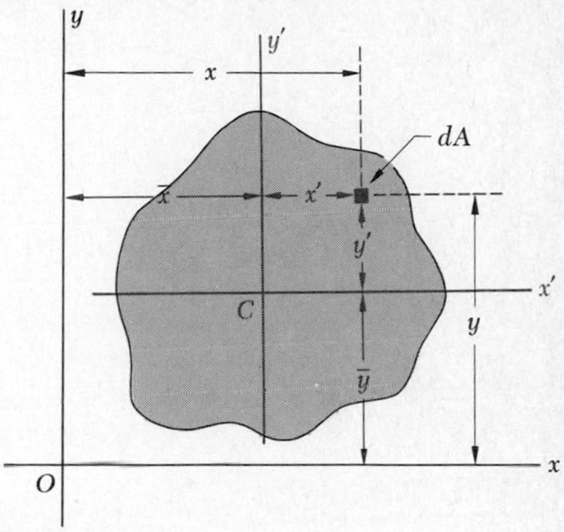

Fig. 9.16

* 9.9. Principal Axes and Principal Moments of Inertia.

Consider the area A and the coordinate axes x and y (Fig. 9.17). We assume that the moments and product of inertia

$$I_x = \int y^2\, dA \qquad I_y = \int x^2\, dA \qquad I_{xy} = \int xy\, dA \tag{9.14}$$

of the area A are known, and we propose to determine the moments and product of inertia $I_{x'}$, $I_{y'}$, and $I_{x'y'}$ of A with respect to new axes x' and y' obtained by rotating the original axes about the origin through an angle θ.

We first note the following relations between the coordinates x', y' and x, y of an element of area dA:

$$x' = x\cos\theta + y\sin\theta \qquad y' = y\cos\theta - x\sin\theta$$

Substituting for y' in the expression for $I_{x'}$, we write

$$\begin{aligned} I_{x'} &= \int (y')^2\, dA = \int (y\cos\theta - x\sin\theta)^2\, dA \\ &= \cos^2\theta \int y^2\, dA - 2\sin\theta\cos\theta \int xy\, dA + \sin^2\theta \int x^2\, dA \end{aligned}$$

Taking the relations (9.14) into account, we write

$$I_{x'} = I_x\cos^2\theta - 2I_{xy}\sin\theta\cos\theta + I_y\sin^2\theta \tag{9.15}$$

Similarly, we obtain for $I_{y'}$ and $I_{x'y'}$ the expressions

$$I_{y'} = I_x\sin^2\theta + 2I_{xy}\sin\theta\cos\theta + I_y\cos^2\theta \tag{9.16}$$
$$I_{x'y'} = (I_x - I_y)\sin\theta\cos\theta + I_{xy}(\cos^2\theta - \sin^2\theta) \tag{9.17}$$

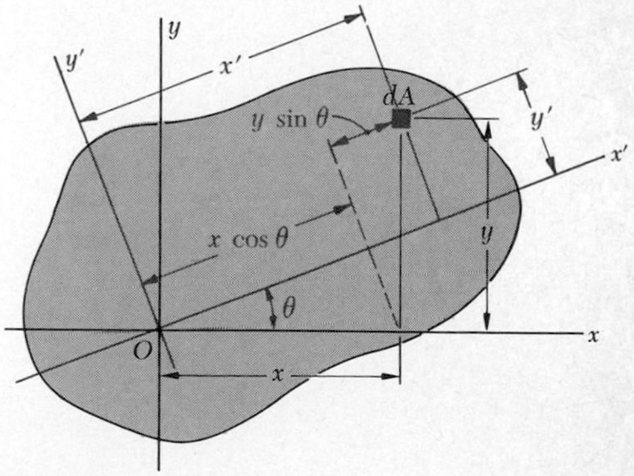

Fig. 9.17

Recalling the trigonometric relations

$$\sin 2\theta = 2 \sin \theta \cos \theta \qquad \cos 2\theta = \cos^2 \theta - \sin^2 \theta$$

and

$$\cos^2 \theta = \frac{1 + \cos 2\theta}{2} \qquad \sin^2 \theta = \frac{1 - \cos 2\theta}{2}$$

we may write (9.15), (9.16), and (9.17) as follows:

$$I_{x'} = \frac{I_x + I_y}{2} + \frac{I_x - I_y}{2} \cos 2\theta - I_{xy} \sin 2\theta \tag{9.18}$$

$$I_{y'} = \frac{I_x + I_y}{2} - \frac{I_x - I_y}{2} \cos 2\theta + I_{xy} \sin 2\theta \tag{9.19}$$

$$I_{x'y'} = \frac{I_x - I_y}{2} \sin 2\theta + I_{xy} \cos 2\theta \tag{9.20}$$

We observe, by adding (9.18) and (9.19) member by member, that

$$I_{x'} + I_{y'} = I_x + I_y \tag{9.21}$$

This result could have been anticipated, since both members of (9.21) are equal to the polar moment of inertia J_O.

Equations (9.18) and (9.20) are the parametric equations of a circle. This means that if we choose a set of rectangular axes and plot a point M of abscissa $I_{x'}$ and ordinate $I_{x'y'}$ for any given value of the parameter θ, all the points thus obtained will lie on a circle. To establish this property we shall eliminate θ from Eqs. (9.18) and (9.20); this is done by transposing $(I_x + I_y)/2$ in Eq. (9.18), squaring both members of Eqs. (9.18) and (9.20), and adding. We write

$$\left(I_{x'} - \frac{I_x + I_y}{2}\right)^2 + I_{x'y'}^2 = \left(\frac{I_x - I_y}{2}\right)^2 + I_{xy}^2 \tag{9.22}$$

Setting

$$I_{\text{ave}} = \frac{I_x + I_y}{2} \qquad \text{and} \qquad R = \sqrt{\left(\frac{I_x - I_y}{2}\right)^2 + I_{xy}^2} \tag{9.23}$$

we write the identity (9.22) in the form

$$(I_{x'} - I_{\text{ave}})^2 + I_{x'y'}^2 = R^2 \tag{9.24}$$

which is the equation of a circle of radius R centered at the point C of abscissa I_{ave} and ordinate 0 (Fig. 9.18*a*). It may be observed that Eqs. (9.19) and (9.20) are the parametric equations of the same circle. Furthermore, due to the symmetry of the circle about the horizontal axis, the same result would have been obtained if instead of plotting M, we had plotted a point N of abscissa $I_{y'}$ and ordinate $-I_{x'y'}$ (Fig. 9.18*b*). This property will be used in Sec. 9.10.

The two points A and B where the circle obtained intersects the axis of abscissas are of special interest: Point A corresponds to the maximum value of the moment of inertia $I_{x'}$, while point B corresponds to its minimum

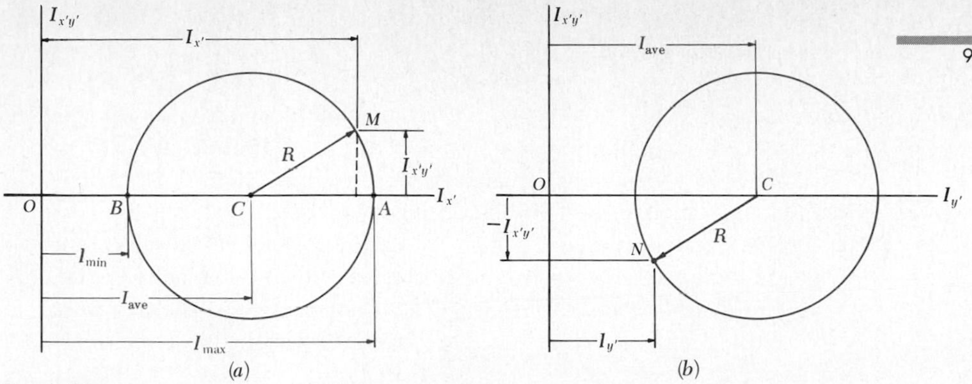

Fig. 9.18

value. Besides, both points correspond to a zero value of the product of inertia $I_{x'y'}$. Thus, the values θ_m of the parameter θ which correspond to the points A and B may be obtained by setting $I_{x'y'} = 0$ in Eq. (9.20). We obtain†

$$\tan 2\theta_m = -\frac{2I_{xy}}{I_x - I_y} \tag{9.25}$$

This equation defines two values $2\theta_m$ which are 180° apart and thus two values θ_m which are 90° apart. One of them corresponds to point A in Fig. 9.18*a* and to an axis through O in Fig. 9.17 with respect to which the moment of inertia of the given area is maximum; the other value corresponds to point B and an axis through O with respect to which the moment of inertia of the area is minimum. The two axes thus defined, which are perpendicular to each other, are called the *principal axes of the area about O*, and the corresponding values I_{max} and I_{min} of the moment of inertia are called the *principal moments of inertia of the area about O*. Since the two values θ_m defined by Eq. (9.25) were obtained by setting $I_{x'y'} = 0$ in Eq. (9.20), it is clear that the product of inertia of the given area with respect to its principal axes is zero.

We observe from Fig. 9.18*a* that

$$I_{max} = I_{ave} + R \qquad \text{and} \qquad I_{min} = I_{ave} - R \tag{9.26}$$

Substituting for I_{ave} and R from formulas (9.23), we write

$$I_{max,min} = \frac{I_x + I_y}{2} \pm \sqrt{\left(\frac{I_x - I_y}{2}\right)^2 + I_{xy}^2} \tag{9.27}$$

Unless it is possible to tell by inspection which of the two principal axes corresponds to I_{max} and which corresponds to I_{min}, it is necessary to substitute one of the values of θ_m into Eq. (9.18) in order to determine which of

†This relation may also be obtained by differentiating $I_{x'}$ in Eq. (9.18) and setting $dI_{x'}/d\theta = 0$.

the two corresponds to the maximum value of the moment of inertia of the area about O.

Referring to Sec. 9.8, we note that if an area possesses an axis of symmetry through a point O, this axis must be a principal axis of the area about O. On the other hand, a principal axis does not need to be an axis of symmetry; whether or not an area possesses properties of symmetry, it will have two principal axes of inertia about any point O.

The properties established here hold for any point O located inside or outside the given area. If the point O is chosen to coincide with the centroid of the area, any axis through O is a centroidal axis; the two principal axes of the area about its centroid are referred to as the *principal centroidal axes of the area.*

***9.10. Mohr's Circle for Moments and Products of Inertia.** The circle used in the preceding section to illustrate the relations existing between the moments and products of inertia of a given area with respect to axes passing through a fixed point O was first introduced by the German engineer Otto Mohr (1835–1918) and is known as *Mohr's circle*. We shall see that if the moments and product of inertia of an area A are known with respect to two rectangular x and y axes through a point O, Mohr's circle may be used to determine graphically (*a*) the principal axes and principal moments of inertia of the area about O, or (*b*) the moments and product of inertia of the area with respect to any other pair of rectangular axes x' and y' through O.

Consider a given area A and two rectangular coordinate axes x and y (Fig. 9.19*a*). We shall assume that the moments of inertia I_x and I_y and the product of inertia I_{xy} are known, and we shall represent them on a diagram by plotting a point X of coordinates I_x and I_{xy} and a point Y of coordinates I_y and $-I_{xy}$ (Fig. 9.19*b*). If I_{xy} is positive, as assumed in Fig. 9.19*a*, point X is located above the horizontal axis and point Y below, as shown in Fig. 9.19*b*. If I_{xy} is negative, X is located below the horizontal axis and Y above. Joining X and Y by a straight line, we define the point C of intersection of line XY with the horizontal axis and draw the circle of center C and diameter XY. Noting that the abscissa of C and the radius of the circle are respectively equal to the quantities I_{ave} and R defined by the formulas (9.23), we conclude that the circle obtained is Mohr's circle for the given area about point O. Thus the abscissas of the points A and B where the circle intersects the horizontal axes represent respectively the principal moments of inertia $I_{\max}$ and $I_{\min}$ of the area.

We also note that, since $\tan(XCA) = 2I_{xy}/(I_x - I_y)$, the angle XCA is equal in magnitude to one of the angles $2\theta_m$ which satisfy Eq. (9.25); thus the angle θ_m which defines in Fig. 9.19*a* the principal axis Oa corresponding to point A in Fig. 9.19*b* may be obtained by dividing in half the angle XCA measured on Mohr's circle. We further observe that if $I_x > I_y$ and $I_{xy} > 0$, as in the case considered here, the rotation which brings CX into CA is clockwise. But, in that case, the angle θ_m obtained from Eq. (9.25) and defining the principal axis Oa in Fig. 9.19*a* is negative; thus the rota-

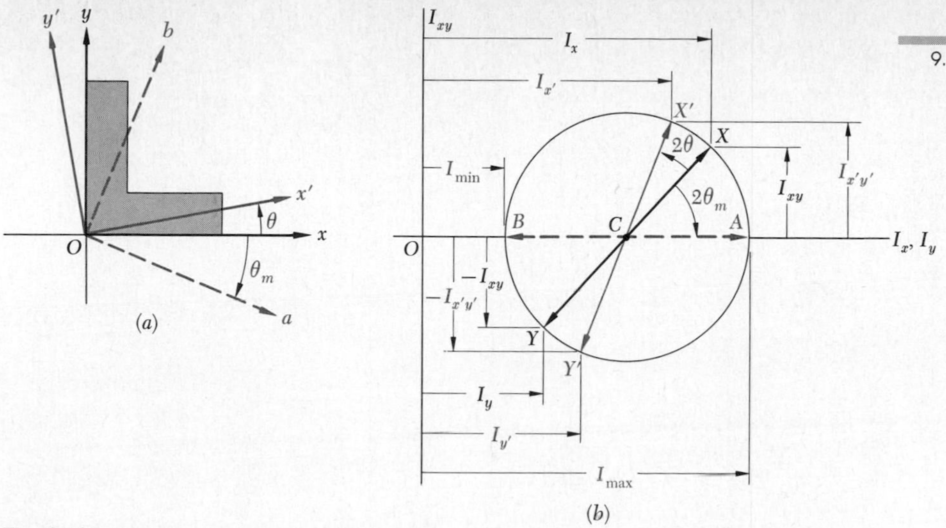

Fig. 9.19

tion bringing Ox into Oa is also clockwise. We conclude that the senses of rotation in both parts of Fig. 9.19 are the same; if a clockwise rotation through $2\theta_m$ is required to bring CX into CA on Mohr's circle, a clockwise rotation through θ_m will bring Ox into the corresponding principal axis Oa in Fig. 9.19*a*.

Since Mohr's circle is uniquely defined, the same circle may be obtained by considering the moments and product of inertia of the area A with respect to the rectangular axes x' and y' (Fig. 9.19*a*). The point X' of coordinates $I_{x'}$ and $I_{x'y'}$ and the point Y' of coordinates $I_{y'}$ and $-I_{x'y'}$ are therefore located on Mohr's circle, and the angle $X'CA$ in Fig. 9.19*b* must be equal to twice the angle $x'Oa$ in Fig. 9.19*a*. Since, as noted above, the angle XCA is twice the angle xOa, it follows that the angle XCX' in Fig. 9.19*b* is twice the angle xOx' in Fig. 9.19*a*. The diameter $X'Y'$ defining the moments and product of inertia $I_{x'}$, $I_{y'}$, and $I_{x'y'}$ of the given area with respect to rectangular axes x' and y' forming an angle θ with the x and y axes may be obtained by rotating through an angle 2θ the diameter XY corresponding to the moments and product of inertia I_x, I_y, and I_{xy}. We note that the rotation which brings the diameter XY into the diameter $X'Y'$ in Fig. 9.19*b* has the same sense as the rotation which brings the xy axes into the $x'y'$ axes in Fig. 9.19*a*.

It should be noted that the use of Mohr's circle is not limited to graphical solutions, i.e., to solutions based on the careful drawing and measuring of the various parameters involved. By merely sketching Mohr's circle and using trigonometry, one may easily derive the various relations required for a numerical solution of a given problem. Actual computations may then be carried out on a calculator (see Sample Prob. 9.8).

SAMPLE PROBLEM 9.6

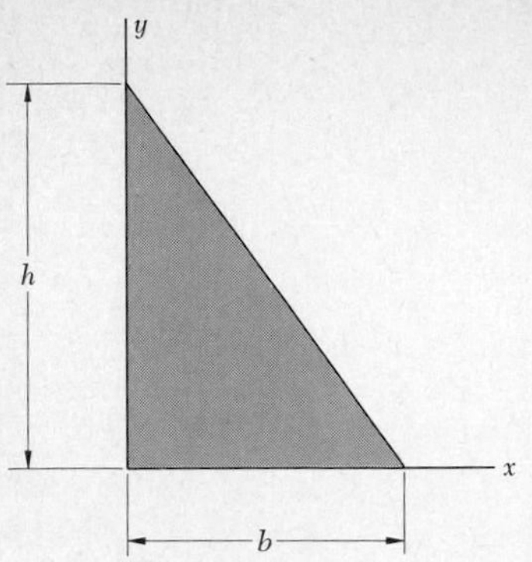

Determine the product of inertia of the right triangle shown (*a*) with respect to the x and y axes and (*b*) with respect to centroidal axes parallel to the x and y axes.

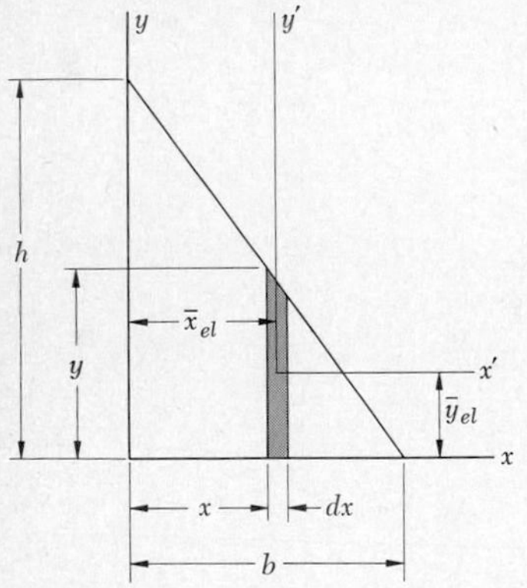

***a*. Product of Inertia I_{xy}.** A vertical rectangular strip is chosen as the differential element of area. Using the parallel-axis theorem, we write

$$dI_{xy} = dI_{x'y'} + \bar{x}_{el}\bar{y}_{el}\,dA$$

Since the element is symmetrical with respect to the x' and y' axes, we note that $dI_{x'y'} = 0$. From the geometry of the triangle, we obtain

$$y = h\left(1 - \frac{x}{b}\right) \qquad dA = y\,dx = h\left(1 - \frac{x}{b}\right)dx$$

$$\bar{x}_{el} = x \qquad \bar{y}_{el} = \tfrac{1}{2}y = \tfrac{1}{2}h\left(1 - \frac{x}{b}\right)$$

Integrating dI_{xy} from $x = 0$ to $x = b$, we obtain

$$I_{xy} = \int dI_{xy} = \int \bar{x}_{el}\bar{y}_{el}\,dA = \int_0^b x(\tfrac{1}{2})h^2\left(1 - \frac{x}{b}\right)^2 dx$$

$$= h^2 \int_0^b \left(\frac{x}{2} - \frac{x^2}{b} + \frac{x^3}{2b^2}\right)dx = h^2\left[\frac{x^2}{4} - \frac{x^3}{3b} + \frac{x^4}{8b^2}\right]_0^b$$

$$I_{xy} = \tfrac{1}{24}b^2h^2 \quad ◄$$

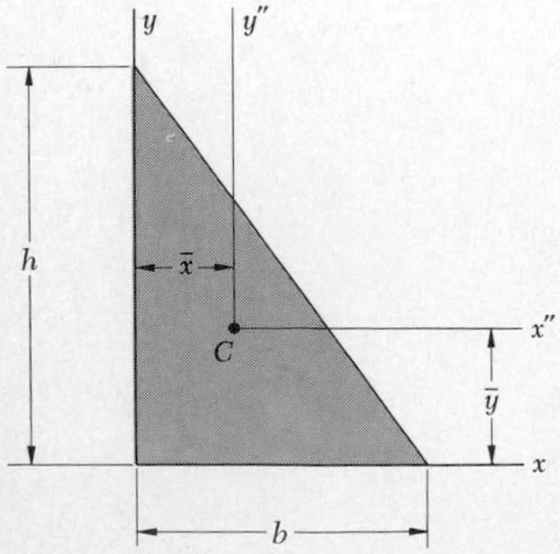

***b*. Product of Inertia $\bar{I}_{x''y''}$.** The coordinates of the centroid of the triangle are

$$\bar{x} = \tfrac{1}{3}b \qquad \bar{y} = \tfrac{1}{3}h$$

Using the expression for I_{xy} obtained in part *a*, we apply the parallel-axis theorem and write

$$I_{xy} = \bar{I}_{x''y''} + \bar{x}\bar{y}A$$

$$\tfrac{1}{24}b^2h^2 = \bar{I}_{x''y''} + (\tfrac{1}{3}b)(\tfrac{1}{3}h)(\tfrac{1}{2}bh)$$

$$\bar{I}_{x''y''} = \tfrac{1}{24}b^2h^2 - \tfrac{1}{18}b^2h^2$$

$$\bar{I}_{x''y''} = -\tfrac{1}{72}b^2h^2 \quad ◄$$

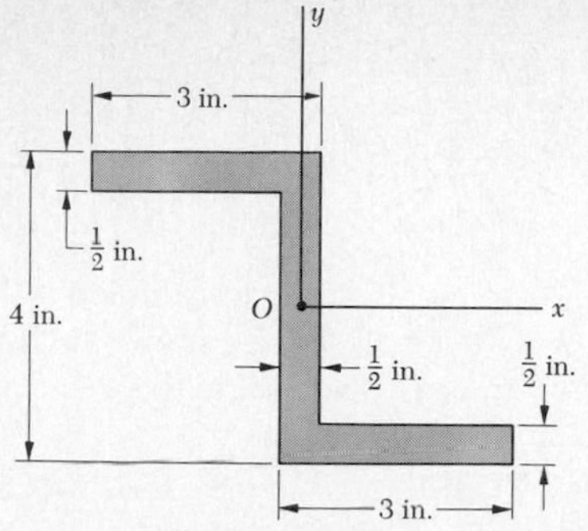

SAMPLE PROBLEM 9.7

For the section shown, the moments of inertia with respect to the x and y axes have been computed and are known to be

$$I_x = 10.38 \text{ in}^4 \qquad I_y = 6.97 \text{ in}^4$$

Determine (*a*) the principal axes of the section about O, (*b*) the values of the principal moments of inertia of the section about O.

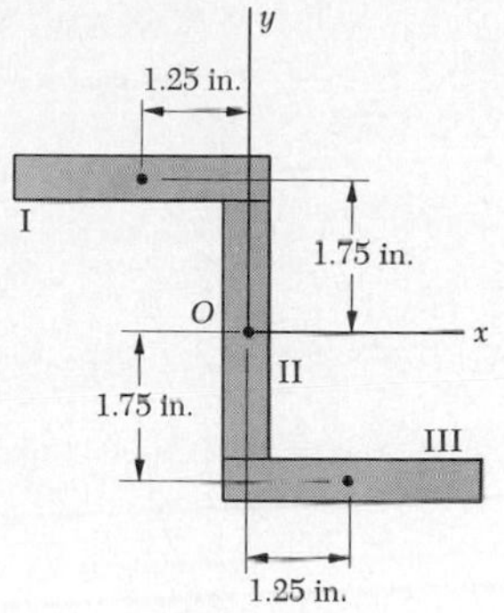

Solution. We first compute the product of inertia with respect to the x and y axes. The area is divided into three rectangles as shown. We note that the product of inertia $\bar{I}_{x'y'}$ with respect to centroidal axes parallel to the x and y axes is zero for each rectangle. Using the parallel-axis theorem $I_{xy} = \bar{I}_{x'y'} + \bar{x}\bar{y}A$, we thus find that for each rectangle, I_{xy} reduces to $\bar{x}\bar{y}A$.

Rectangle	Area, in^2	$\bar{x}$, in.	$\bar{y}$, in.	$\bar{x}\bar{y}A$, in^4
I	1.5	−1.25	+1.75	−3.28
II	1.5	0	0	0
III	1.5	+1.25	−1.75	−3.28
				−6.56

$$I_{xy} = \Sigma\bar{x}\bar{y}A = -6.56 \text{ in}^4$$

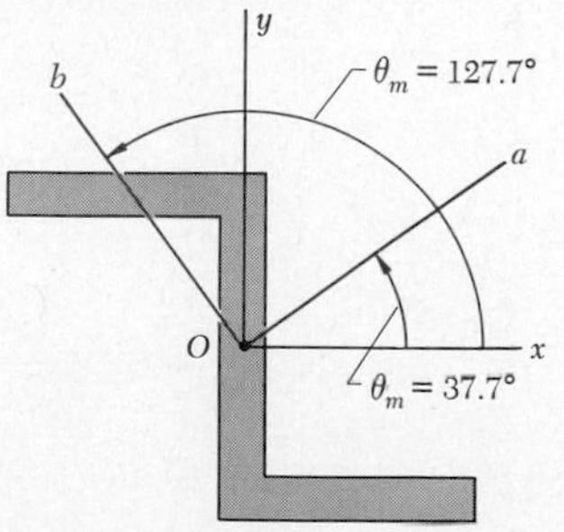

a. Principal Axes. Since the magnitudes of I_x, I_y, and I_{xy} are known, Eq. (9.25) is used to determine the values of θ_m:

$$\tan 2\theta_m = -\frac{2I_{xy}}{I_x - I_y} = -\frac{2(-6.56)}{10.38 - 6.97} = +3.85$$

$$2\theta_m = 75.4° \text{ and } 255.4°$$

$$\theta_m = 37.7° \qquad \text{and} \qquad \theta_m = 127.7° \quad ◄$$

b. Principal Moments of Inertia. Using Eq. (9.27), we write

$$I_{\max,\min} = \frac{I_x + I_y}{2} \pm \sqrt{\left(\frac{I_x - I_y}{2}\right)^2 + I_{xy}^2}$$

$$= \frac{10.38 + 6.97}{2} \pm \sqrt{\left(\frac{10.38 - 6.97}{2}\right)^2 + (-6.56)^2}$$

$$I_{\max} = 15.45 \text{ in}^4 \qquad I_{\min} = 1.897 \text{ in}^4 \quad ◄$$

Noting that the area of the section is farther away from the a axis than from the b axis, we conclude that $I_a = I_{\max} = 15.45$ in^4 and $I_b = I_{\min} = 1.897$ in^4. This conclusion may be verified by substituting $\theta = 37.7°$ into Eqs. (9.18) and (9.19).

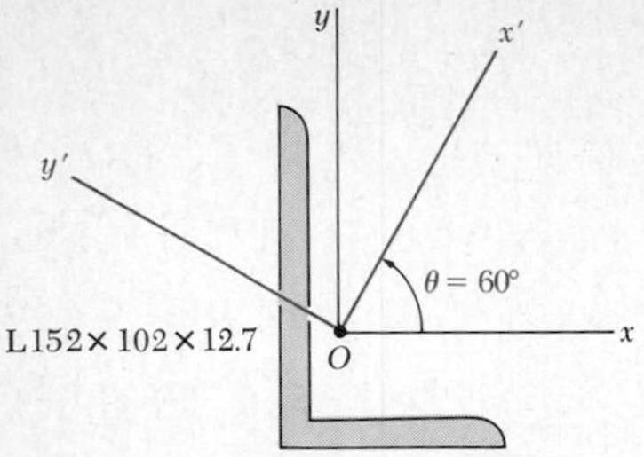

SAMPLE PROBLEM 9.8

For the section shown, the moments and product of inertia with respect to the x and y axes have been computed and are known to be

$$I_x = 7.24 \times 10^6 \text{ mm}^4 \qquad I_y = 2.61 \times 10^6 \text{ mm}^4 \qquad I_{xy} = -2.54 \times 10^6 \text{mm}^4$$

Using Mohr's circle, determine (*a*) the principal axes of the section about O, (*b*) the values of the principal moments of inertia of the section about O, (*c*) the moments and product of inertia of the section with respect to the x' and y' axes forming an angle of 60° with the x and y axes.

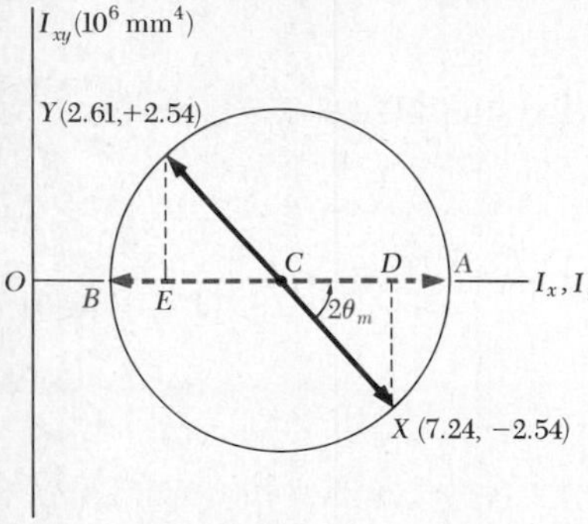

Solution. We first plot point X of coordinates $I_x = 7.24$, $I_{xy} = -2.54$, and point Y of coordinates $I_y = 2.61$, $-I_{xy} = +2.54$. Joining X and Y by a straight line, we define the center C of Mohr's circle. The abscissa of C, which represents I_{ave}, and the radius R of the circle may be measured directly or calculated as follows:

$$I_{ave} = OC = \tfrac{1}{2}(I_x + I_y) = \tfrac{1}{2}(7.24 \times 10^6 + 2.61 \times 10^6) = 4.925 \times 10^6 \text{ mm}^4$$
$$CD = \tfrac{1}{2}(I_x - I_y) = \tfrac{1}{2}(7.24 \times 10^6 - 2.61 \times 10^6) = 2.315 \times 10^6 \text{ mm}^4$$
$$R = \sqrt{(CD)^2 + (DX)^2} = \sqrt{(2.315 \times 10^6)^2 + (2.54 \times 10^6)^2} = 3.437 \times 10^6 \text{ mm}^4$$

a. Principal Axes. The principal axes of the section correspond to points A and B on Mohr's circle and the angle through which we should rotate CX to bring it into CA defines $2\theta_m$. We have

$$\tan 2\theta_m = \frac{DX}{CD} = \frac{2.54}{2.315} = 1.097 \qquad 2\theta_m = 47.6°\circlearrowleft \qquad \theta_m = 23.8°\circlearrowleft \quad ◄$$

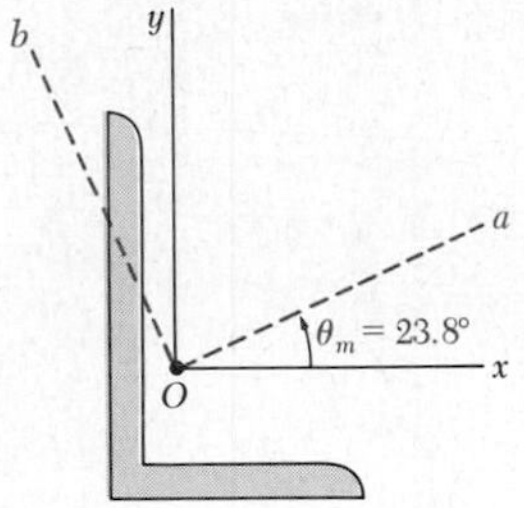

Thus the principal axis Oa corresponding to the maximum value of the moment of inertia is obtained by rotating the x axis through 23.8° counterclockwise; the principal axis corresponding to the minimum value of the moment of inertia may be obtained by rotating the y axis through the same angle.

b. Principal Moments of Inertia. The principal moments of inertia are represented by the abscissas of A and B. We have

$$I_{max} = OA = OC + CA = I_{ave} + R = (4.925 + 3.437)10^6 \text{ mm}^4$$
$$I_{max} = 8.36 \times 10^6 \text{ mm}^4 \quad ◄$$
$$I_{min} = OB = OC - BC = I_{ave} - R = (4.925 - 3.437)10^6 \text{ mm}^4$$
$$I_{min} = 1.49 \times 10^6 \text{ mm}^4 \quad ◄$$

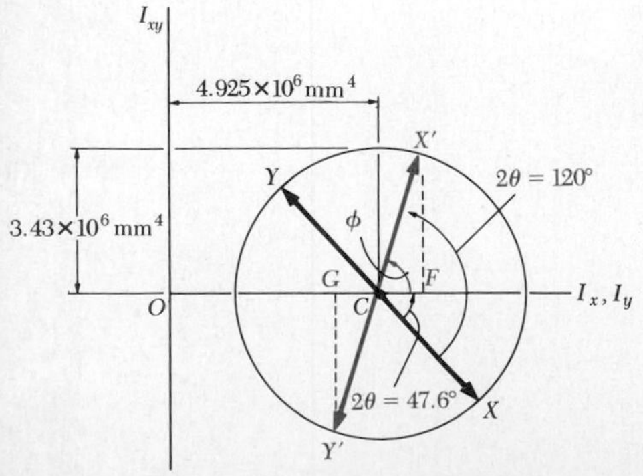

c. Moments and Product of Inertia with Respect to x'y' Axes. The points X' and Y' on Mohr's circle which correspond to the x' and y' axes are obtained by rotating CX and CY through an angle $2\theta = 2(60°) = 120°$ counterclockwise. The coordinates of X' and Y' yield the desired moments and product of inertia. Noting that the angle that CX' forms with the horizontal axis is $\phi = 120° - 47.6° = 72.4°$, we write

$$I_{x'} = OF = OC + CF = 4.925 \times 10^6 \text{ mm}^4 + (3.437 \times 10^6 \text{ mm}^4) \cos 72.4°$$
$$I_{x'} = 5.96 \times 10^6 \text{ mm}^4 \quad ◄$$
$$I_{y'} = OG = OC - GC = 4.925 \times 10^6 \text{ mm}^4 - (3.437 \times 10^6 \text{ mm}^4) \cos 72.4°$$
$$I_{y'} = 3.89 \times 10^6 \text{ mm}^4 \quad ◄$$
$$I_{x'y'} = FX' = (3.437 \times 10^6 \text{ mm}^4) \sin 72.4°$$
$$I_{x'y'} = 3.28 \times 10^6 \text{ mm}^4 \quad ◄$$

Problems

9.53 through 9.56 Determine by direct integration the product of inertia of the given area with respect to the x and y axes.

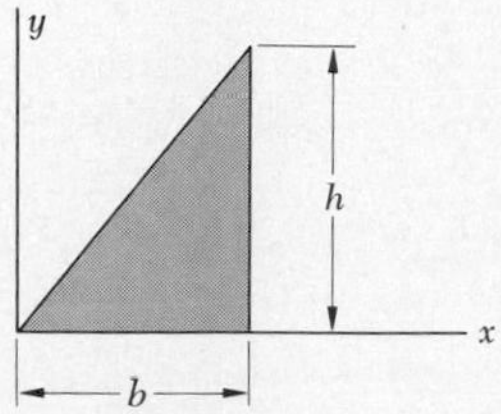

Fig. P9.53

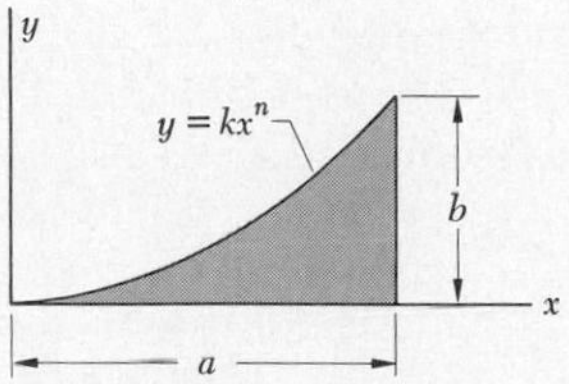

Fig. P9.54

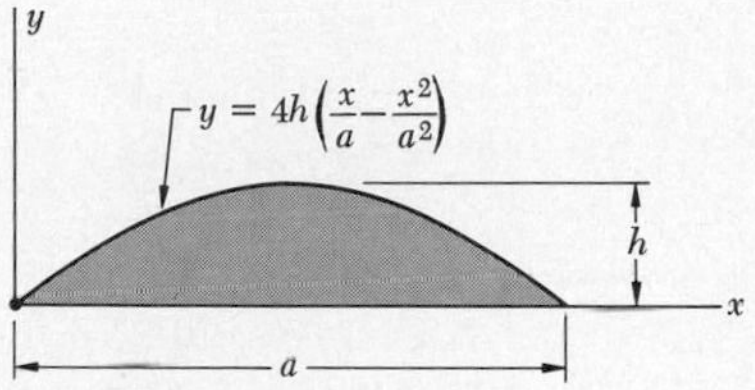

Fig. P9.55

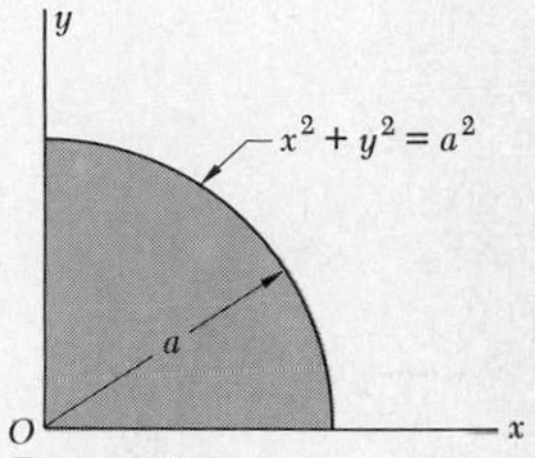

Fig. P9.56

9.57 through 9.62 Using the parallel-axis theorem, determine the product of inertia of the area shown with respect to the centroidal x and y axes.

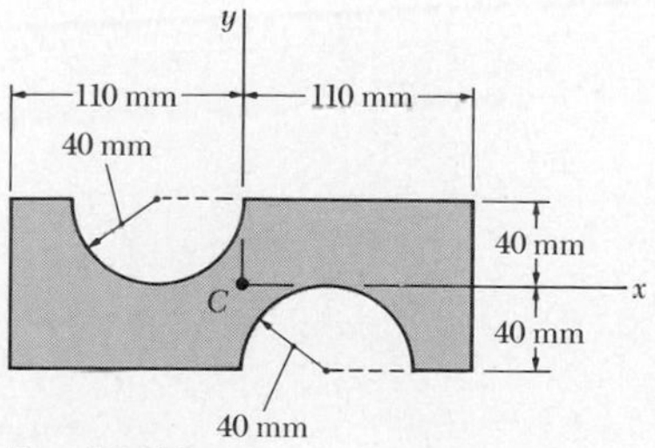

Fig. P9.57

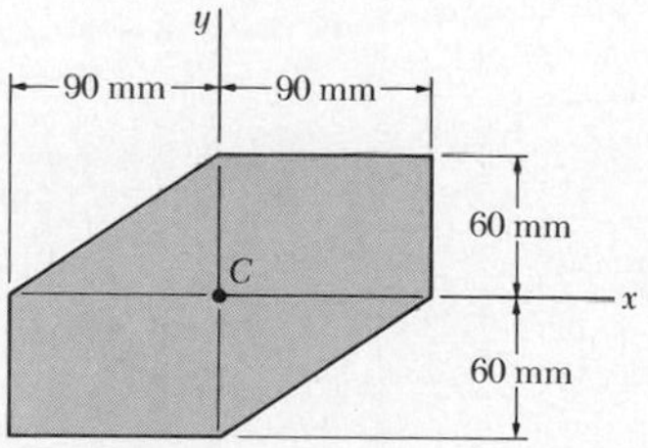

Fig. P9.58

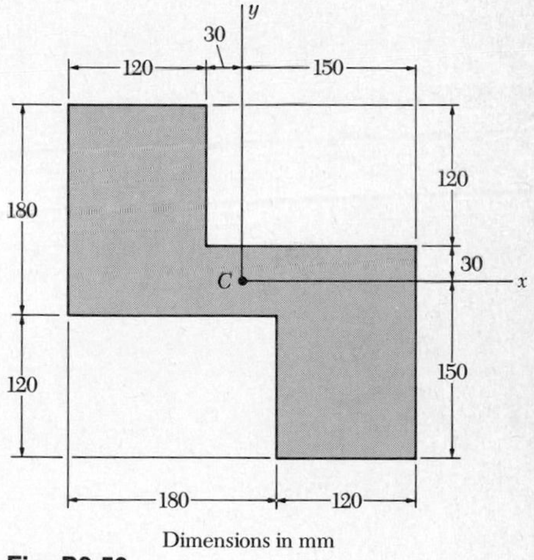

Fig. P9.59

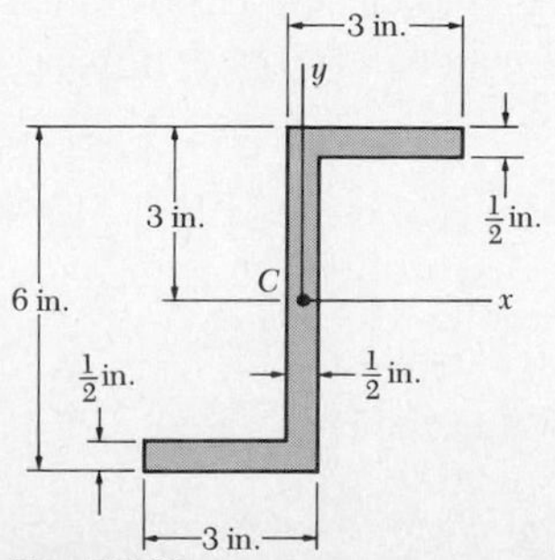

Fig. P9.60

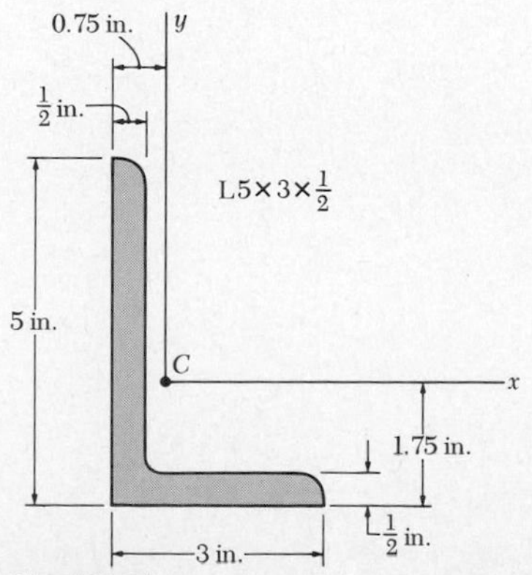

Fig. P9.61

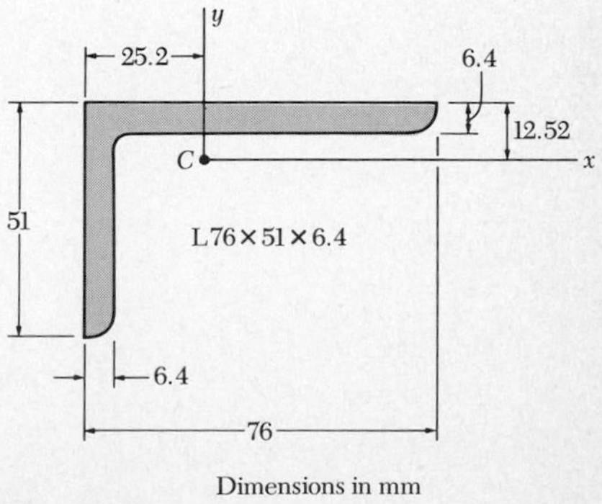

Fig. P9.62

9.63 Determine the moments of inertia and the product of inertia of the quarter circle of Prob. 9.56 with respect to new axes obtained by rotating the x and y axes about O (a) through 30° counterclockwise, (b) through 45° counterclockwise.

9.64 Determine the moments of inertia and the product of inertia of the area of Prob. 9.60 with respect to new centroidal axes obtained by rotating the x and y axes through 30° counterclockwise.

9.65 Determine the moments of inertia and the product of inertia of the L $5 \times 3 \times \frac{1}{2}$ in. angle cross section of Prob. 9.61 with respect to new centroidal axes obtained by rotating the x and y axes through 30° clockwise. (The moments of inertia $\bar{I}_x$ and $\bar{I}_y$ of the section are given in Fig. 9.13*A*.)

9.66 Determine the moments of inertia and the product of inertia of the L $76 \times 51 \times 6.4$ mm angle cross section of Prob. 9.62 with respect to new centroidal axes obtained by rotating the x and y axes through 30° clockwise. (The moments of inertia $\bar{I}_x$ and $\bar{I}_y$ of the section are given in Fig. 9.13*B*.)

9.67 Determine the orientation of the principal axes through the centroid and the corresponding values of the moment of inertia for the area of Prob. 9.57. (*Given:* $\bar{I}_x = 6.16 \times 10^6 \text{ mm}^4$ and $\bar{I}_y = 60.9 \times 10^6 \text{ mm}^4$.)

9.68 through 9.70 For the area indicated, determine the orientation of the principal axes through the centroid and the corresponding values of the moment of inertia.

9.68 Area of Prob. 9.58.
9.69 Area of Prob. 9.59.
9.70 Area of Prob. 9.60.

9.71 and 9.72 For the angle cross section indicated, determine the orientation of the principal axes through the centroid and the corresponding values of the moment of inertia. (The moments of inertia $\bar{I}_x$ and $\bar{I}_y$ of the cross sections are given in Fig. 9.13.)

9.71 The L $5 \times 3 \times \frac{1}{2}$ in. angle of Prob. 9.61.
9.72 The L $76 \times 51 \times 6.4$ mm angle of Prob. 9.62.

9.73 Using Mohr's circle, determine the moments of inertia and the product of inertia of the quarter circle of Prob. 9.56 with respect to new axes obtained by rotating the x and y axes about O (a) through 30° counterclockwise, (b) through 45° counterclockwise.

9.74 Using Mohr's circle, determine the moments of inertia and the product of inertia of the area of Prob. 9.60 with respect to new centroidal axes obtained by rotating the x and y axes through 30° counterclockwise.

9.75 Using Mohr's circle, determine the moments of inertia and the product of inertia of the L $5 \times 3 \times \frac{1}{2}$ in. angle cross section of Prob. 9.61 with respect to new centroidal axes obtained by rotating the x and y axes through 30° clockwise. (The moments of inertia $\bar{I}_x$ and $\bar{I}_y$ of the section are given in Fig. 9.13*A*.)

9.76 Using Mohr's circle, determine the moment of inertia and the product of inertia of the L 76 × 51 × 6.4 mm angle cross section of Prob. 9.62 with respect to new centroidal axes obtained by rotating the x and y axes through 30° clockwise. (The moments of inertia $\bar{I}_x$ and $\bar{I}_y$ of the section are given in Fig. 9.13*B*.)

9.77 Using Mohr's circle, determine the orientation of the principal axes through the centroid and the corresponding values of the moment of inertia for the area of Prob. 9.57. (*Given:* $\bar{I}_x = 6.16 \times 10^6\ \text{mm}^4$ and $\bar{I}_y = 60.9 \times 10^6\ \text{mm}^4$.)

9.78 through 9.80 Using Mohr's circle, determine for the area indicated the orientation of the principal axes through the centroid and the corresponding values of the moment of inertia.

9.78 Area of Prob. 9.58.
9.79 Area of Prob. 9.59.
9.80 Area of Prob. 9.60.

9.81 and 9.82 Using Mohr's circle, determine for the angle cross section indicated the orientation of the principal axes through the centroid and the corresponding values of the moment of inertia. (The moments of inertia $\bar{I}_x$ and $\bar{I}_y$ of the cross sections are given in Fig. 9.13.)

9.81 The L 5 × 3 × $\frac{1}{2}$ in. angle of Prob. 9.61.
9.82 The L 76 × 51 × 6.4 mm angle of Prob. 9.62.

9.83 Using Mohr's circle show that for any regular polygon (such as a pentagon), (*a*) the moment of inertia with respect to every axis through the centroid is the same, (*b*) the product of inertia with respect to any pair of rectangular axes through the centroid is zero.

9.84 The moments and product of inertia of an L 127 × 76 × 12.7 mm angle cross section with respect to two rectangular axes x and y through C are, respectively, $\bar{I}_x = 3.93 \times 10^6\ \text{mm}^4$, $\bar{I}_y = 1.074 \times 10^6\ \text{mm}^4$, and $\bar{I}_{xy} < 0$, with the minimum value of the moment of inertia of the area with respect to any axis through C being $\bar{I}_{min} = 0.655 \times 10^6\ \text{mm}^4$. Using Mohr's circle, determine (*a*) the product of inertia $\bar{I}_{xy}$ of the area, (*b*) the orientation of the principal axes, (*c*) the value of $\bar{I}_{max}$.

*** 9.85** Prove that the expression $I_{x'}I_{y'} - I_{x'y'}^2$, where $I_{x'}$, $I_{y'}$, and $I_{x'y'}$ represent, respectively, the moments and product of inertia of a given area with respect to two rectangular axes x' and y' through a given point O, is independent of the orientation of the x' and y' axes. Considering the particular case when the x' and y' axes correspond to the maximum value of $I_{x'y'}$, show that the given expression represents the square of the tangent drawn from the origin of the coordinates to Mohr's circle.

*** 9.86** Using the invariance property established in the preceding problem, express the product of inertia I_{xy} of an area A with respect to two rectangular axes through O in terms of the moments of inertia I_x and I_y of A and of the principal moments of inertia I_{min} and I_{max} of A about O. Apply the formula obtained to calculate the product of inertia $\bar{I}_{xy}$ of the L 6 × 4 × $\frac{1}{2}$ in. angle cross section shown in Fig. 9.13*A*, knowing that its minimum moment of inertia is 3.60 in^4.

MOMENTS OF INERTIA OF MASSES

9.11. Moment of Inertia of a Mass. Consider a small mass Δm mounted on a rod of negligible mass which may rotate freely about an axis AA' (Fig. 9.20*a*). If a couple is applied to the system, the rod and mass, assumed initially at rest, will start rotating about AA'. The details of this motion will be studied later in dynamics. At present, we wish only to indicate that the time required for the system to reach a given speed of rotation is proportional to the mass Δm and to the square of the distance r. The product $r^2 \Delta m$ provides, therefore, a measure of the *inertia* of the system, i.e., of the resistance the system offers when we try to set it in motion. For this reason, the product $r^2 \Delta m$ is called the *moment of inertia* of the mass Δm with respect to the axis AA'.

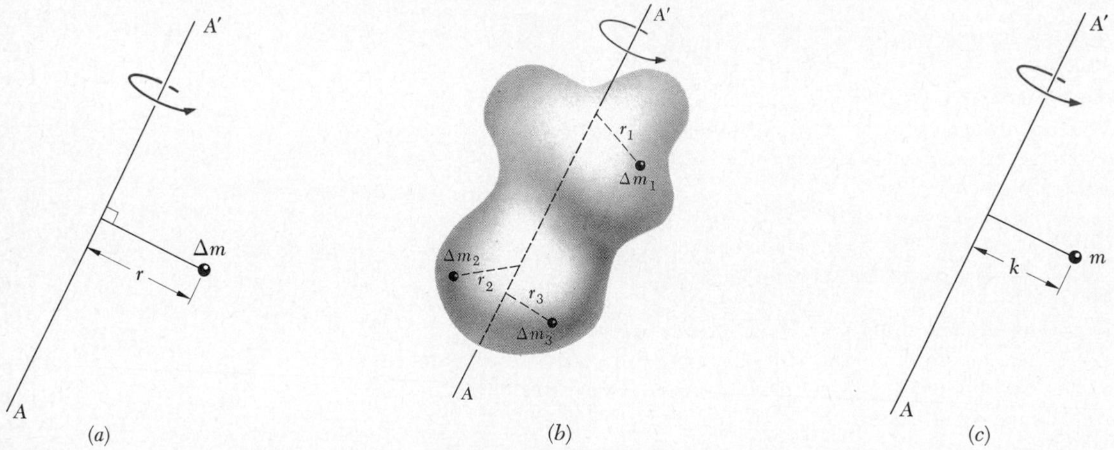

Fig. 9.20

Consider now a body of mass m which is to be rotated about an axis AA' (Fig. 9.20*b*). Dividing the body into elements of mass Δm_1, Δm_2, etc., we find that the resistance offered by the body is measured by the sum $r_1^2 \Delta m_1 + r_2^2 \Delta m_2 + \cdots$. This sum defines, therefore, the moment of inertia of the body with respect to the axis AA'. Increasing the number of elements, we find that the moment of inertia is equal, at the limit, to the integral

$$I = \int r^2 \, dm \tag{9.28}$$

The *radius of gyration* k of the body with respect to the axis AA' is defined by the relation

$$I = k^2 m \qquad \text{or} \qquad k = \sqrt{\frac{I}{m}} \tag{9.29}$$

The radius of gyration k represents, therefore, the distance at which the entire mass of the body should be concentrated if its moment of inertia with respect to AA' is to remain unchanged (Fig. 9.20*c*). Whether it is kept in its original shape (Fig. 9.20*b*) or whether it is concentrated as shown in Fig. 9.20*c*, the mass m will react in the same way to a rotation, or *gyration*, about AA'.

If SI units are used, the radius of gyration k is expressed in meters and the mass m in kilograms. The moment of inertia of a mass, therefore, will be expressed in $kg \cdot m^2$. If U.S. customary units are used, the radius of gyration is expressed in feet and the mass in slugs, i.e., in $lb \cdot s^2/ft$. The moment of inertia of a mass, then, will be expressed in $lb \cdot ft \cdot s^2$.†

The moment of inertia of a body with respect to a coordinate axis may easily be expressed in terms of the coordinates x, y, z of the element of mass dm (Fig. 9.21). Noting, for example, that the square of the distance r from

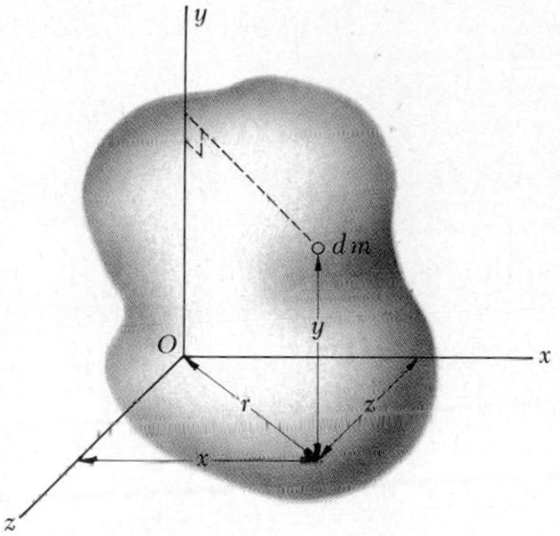

Fig. 9.21

the element dm to the y axis is $z^2 + x^2$, we express the moment of inertia of the body with respect to the y axis as

$$I_y = \int r^2\, dm = \int (z^2 + x^2)\, dm$$

Similar expressions may be obtained for the moments of inertia with respect to the x and z axes. We write

$$\begin{aligned} I_x &= \int (y^2 + z^2)\, dm \\ I_y &= \int (z^2 + x^2)\, dm \\ I_z &= \int (x^2 + y^2)\, dm \end{aligned} \qquad (9.30)$$

†It should be kept in mind when converting the moment of inertia of a mass from U.S. customary units to SI units that the base unit pound used in the derived unit $lb \cdot ft \cdot s^2$ is a unit of force (*not* of mass) and should therefore be converted into newtons. We have

$$1\ lb \cdot ft \cdot s^2 = (4.45\ N)(0.3048\ m)(1\ s)^2 = 1.356\ N \cdot m \cdot s^2$$

or, since $N = kg \cdot m/s^2$,

$$1\ lb \cdot ft \cdot s^2 = 1.356\ kg \cdot m^2$$

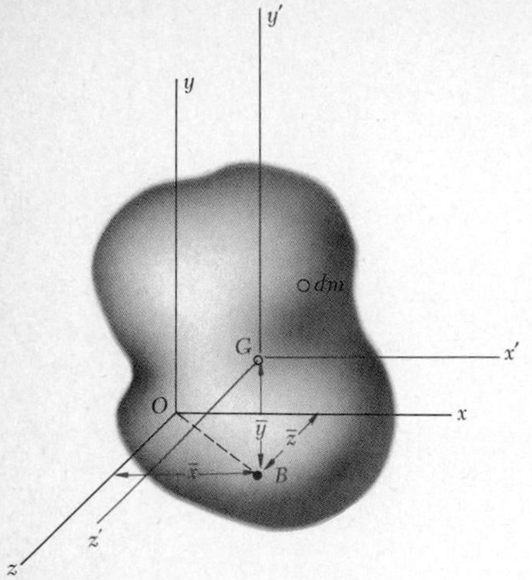

Fig. 9.22

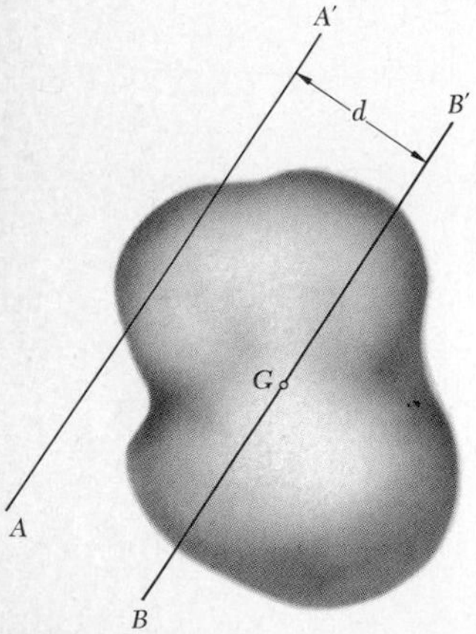

Fig. 9.23

9.12. Parallel-Axis Theorem. Consider a body of mass m. Let $Oxyz$ be a system of rectangular coordinates with origin at an arbitrary point O, and $Gx'y'z'$ a system of parallel *centroidal axes*, i.e., a system with origin at the center of gravity G of the body† and with axes x', y', z', respectively, parallel to x, y, z (Fig. 9.22). Denoting by $\bar{x}$, $\bar{y}$, $\bar{z}$ the coordinates of G with respect to $Oxyz$, we write the following relations between the coordinates x, y, z of the element dm with respect to $Oxyz$ and its coordinates x', y', z' with respect to the centroidal axes $Gx'y'z'$:

$$x = x' + \bar{x} \qquad y = y' + \bar{y} \qquad z = z' + \bar{z} \tag{9.31}$$

Referring to Eqs. (9.30), we may express the moment of inertia of the body with respect to the x axis as follows:

$$\begin{aligned} I_x &= \int (y^2 + z^2)\, dm = \int [(y' + \bar{y})^2 + (z' + \bar{z})^2]\, dm \\ &= \int (y'^2 + z'^2)\, dm + 2\bar{y}\int y'\, dm + 2\bar{z}\int z'\, dm + (\bar{y}^2 + \bar{z}^2)\int dm \end{aligned}$$

The first integral in the expression obtained represents the moment of inertia $\bar{I}_{x'}$ of the body with respect to the centroidal axis x'; the second and third integrals represent the first moment of the body with respect to the $z'x'$ and $x'y'$ planes, respectively, and, since both planes contain G, the two integrals are zero; the last integral is equal to the total mass m of the body. We write, therefore,

$$I_x = \bar{I}_{x'} + m(\bar{y}^2 + \bar{z}^2) \tag{9.32}$$

and, similarly,

$$I_y = \bar{I}_{y'} + m(\bar{z}^2 + \bar{x}^2) \qquad I_z = \bar{I}_{z'} + m(\bar{x}^2 + \bar{y}^2) \tag{9.32$'$}$$

We easily verify from Fig. 9.22 that the sum $\bar{z}^2 + \bar{x}^2$ represents the square of the distance OB between the y and y' axis. Similarly, $\bar{y}^2 + \bar{z}^2$ and $\bar{x}^2 + \bar{y}^2$ represent the squares of the distances between the x and x' axes, and the z and z' axes, respectively. Denoting by d the distance between an arbitrary axis AA' and a parallel centroidal axis BB' (Fig. 9.23), we may, therefore, write the following general relation between the moment of inertia I of the body with respect to AA' and its moment of inertia $\bar{I}$ with respect to BB':

$$I = \bar{I} + md^2 \tag{9.33}$$

Expressing the moments of inertia in terms of the corresponding radii of gyration, we may also write

$$k^2 = \bar{k}^2 + d^2 \tag{9.34}$$

where k and $\bar{k}$ represent the radii of gyration about AA' and BB', respectively.

† Note that the term centroidal is used to define an axis passing through the center of gravity G of the body, whether or not G coincides with the centroid of the volume of the body.

9.13. Moments of Inertia of Thin Plates. Consider a thin plate of uniform thickness t, made of a homogeneous material of density ρ (density = mass per unit volume). The mass moment of inertia of the plate with respect to an axis AA' *contained in the plane* of the plate (Fig. 9.24*a*) is

$$I_{AA',\text{mass}} = \int r^2\, dm$$

Since $dm = \rho t\, dA$, we write

$$I_{AA',\text{mass}} = \rho t \int r^2\, dA$$

But r represents the distance of the element of area dA to the axis AA'; the

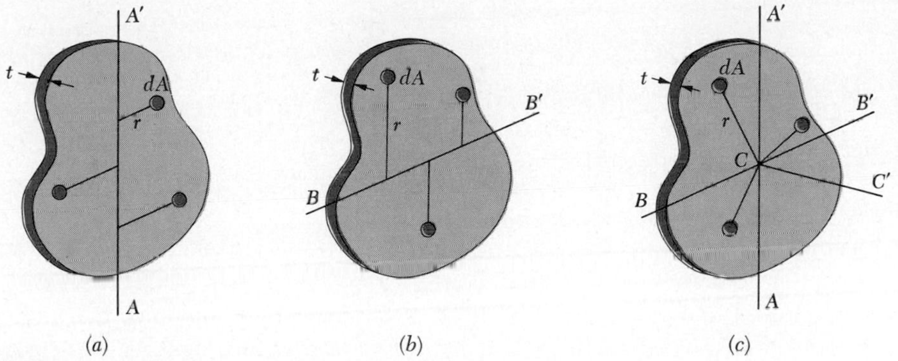

Fig. 9.24

integral is therefore equal to the moment of inertia of the area of the plate with respect to AA'. We have

$$I_{AA',\text{mass}} = \rho t I_{AA',\text{area}} \tag{9.35}$$

Similarly, we have with respect to an axis BB' perpendicular to AA' (Fig. 9.24*b*)

$$I_{BB',\text{mass}} = \rho t I_{BB',\text{area}} \tag{9.36}$$

Considering now the axis CC' *perpendicular* to the plate through the point of intersection C of AA' and BB' (Fig. 9.24*c*), we write

$$I_{CC',\text{mass}} = \rho t J_{C,\text{area}} \tag{9.37}$$

where J_C is the *polar* moment of inertia of the area of the plate with respect to point C.

Recalling the relation $J_C = I_{AA'} + I_{BB'}$ existing between polar and rectangular moments of inertia of an area, we write the following relation between the mass moments of inertia of a thin plate:

$$I_{CC'} = I_{AA'} + I_{BB'} \tag{9.38}$$

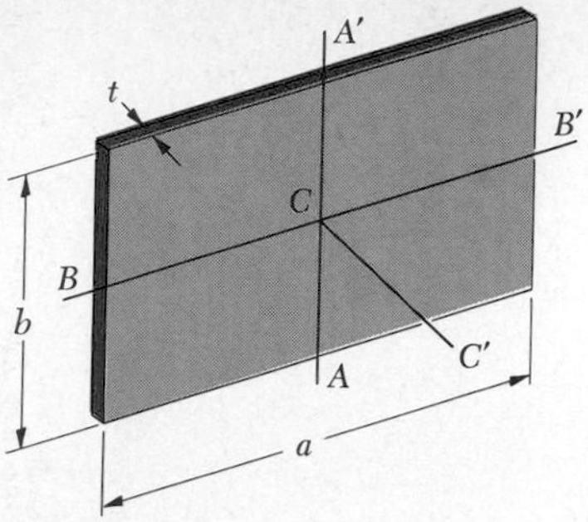

Fig. 9.25

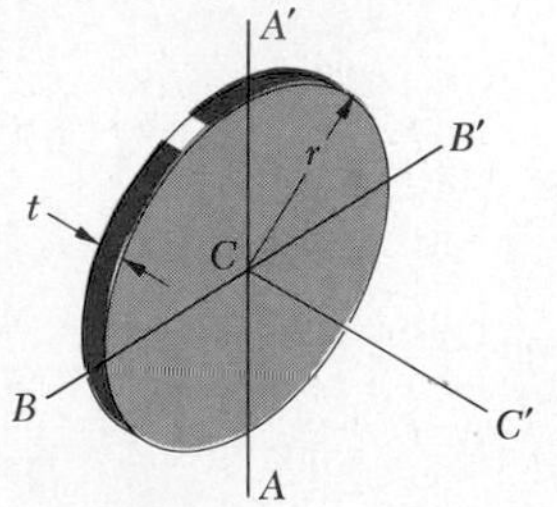

Fig. 9.26

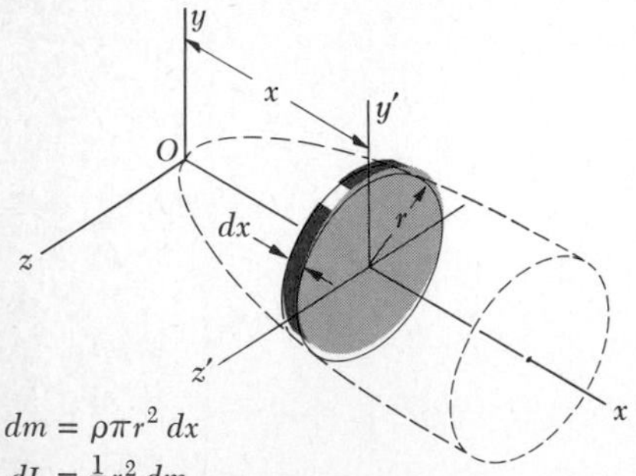

$dm = \rho\pi r^2\, dx$

$dI_x = \frac{1}{2}r^2\, dm$

$dI_y = dI_{y'} + x^2\, dm = (\frac{1}{4}r^2 + x^2)\, dm$

$dI_z = dI_{z'} + x^2\, dm = (\frac{1}{4}r^2 + x^2)\, dm$

Fig. 9.27 Determination of the moment of inertia of a body of revolution.

Rectangular Plate. In the case of a reactangular plate of sides a and b (Fig. 9.25), we obtain the following mass moments of inertia with respect to axes through the center of gravity of the plate:

$$I_{AA',\text{mass}} = \rho t I_{AA',\text{area}} = \rho t(\tfrac{1}{12}a^3 b)$$
$$I_{BB',\text{mass}} = \rho t I_{BB',\text{area}} = \rho t(\tfrac{1}{12}ab^3)$$

Observing that the product ρabt is equal to the mass m of the plate, we write the mass moments of inertia of a thin rectangular plate as follows:

$$I_{AA'} = \tfrac{1}{12}ma^2 \qquad I_{BB'} = \tfrac{1}{12}mb^2 \tag{9.39}$$
$$I_{CC'} = I_{AA'} + I_{BB'} = \tfrac{1}{12}m(a^2 + b^2) \tag{9.40}$$

Circular Plate. In the case of a circular plate, or disk, of radius r (Fig. 9.26), we write

$$I_{AA',\text{mass}} = \rho t I_{AA',\text{area}} = \rho t(\tfrac{1}{4}\pi r^4)$$

Observing that the product $\rho\pi r^2 t$ is equal to the mass m of the plate and that $I_{AA'} = I_{BB'}$, we write the mass moments of inertia of a circular plate as follows:

$$I_{AA'} = I_{BB'} = \tfrac{1}{4}mr^2 \tag{9.41}$$
$$I_{CC'} = I_{AA'} + I_{BB'} = \tfrac{1}{2}mr^2 \tag{9.42}$$

9.14. Determination of the Moment of Inertia of a Three-dimensional Body by Integration. The moment of inertia of a three-dimensional body is obtained by computing the integral $I = \int r^2\, dm$. If the body is made of a homogeneous material of density ρ, we have $dm = \rho\, dV$ and write $I = \rho\int r^2\, dV$. This integral depends only upon the shape of the body. Thus, in order to compute the moment of inertia of a three-dimensional body, it will generally be necessary to perform a triple, or at least a double, integration.

However, if the body possesses two planes of symmetry, it is usually possible to determine the body's moment of inertia through a single integration by choosing as an element of mass dm the mass of a thin slab perpendicular to the planes of symmetry. In the case of bodies of revolution, for example, the element of mass should be a thin disk (Fig. 9.27). Using formula (9.42), the moment of inertia of the disk with respect to the axis of revolution may be readily expressed as indicated in Fig. 9.27. Its moment of inertia with respect to each of the other two axes of coordinates will be obtained by using formula (9.41) and the parallel-axis theorem. Integration of the expressions obtained will yield the desired moment of inertia of the body of revolution.

9.15. Moments of Inertia of Composite Bodies. The moments of inertia of a few common shapes are shown in Fig. 9.28. The moment of inertia with respect to a given axis of a body made of several of these simple shapes may be obtained by computing the moments of inertia of its component parts about the desired axis and adding them together. We should note, as we have already noted in the case of areas, that the radius of gyration of a composite body *cannot* be obtained by adding the radii of gyration of its component parts.

Slender rod	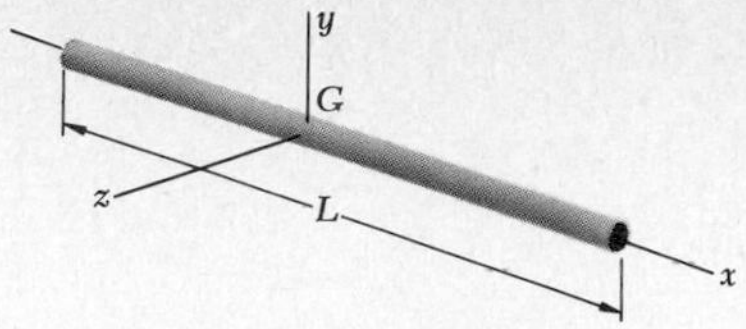	$I_y = I_z = \frac{1}{12}mL^2$
Thin rectangular plate	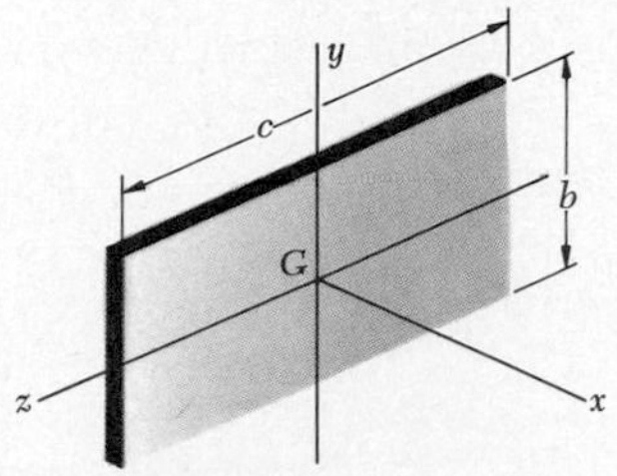	$I_x = \frac{1}{12}m(b^2 + c^2)$ $I_y = \frac{1}{12}mc^2$ $I_z = \frac{1}{12}mb^2$
Rectangular prism	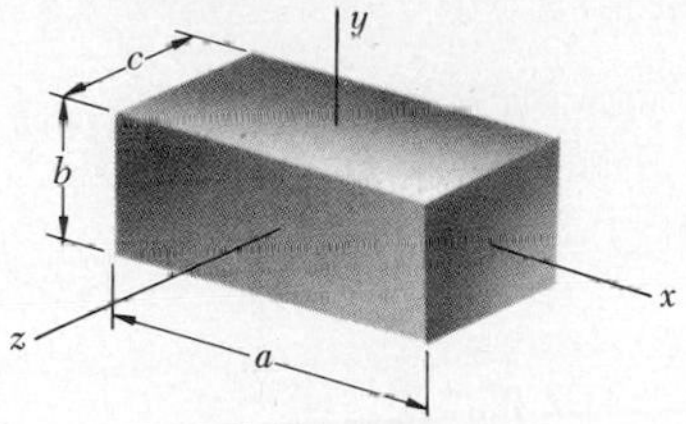	$I_x = \frac{1}{12}m(b^2 + c^2)$ $I_y = \frac{1}{12}m(c^2 + a^2)$ $I_z = \frac{1}{12}m(a^2 + b^2)$
Thin disk	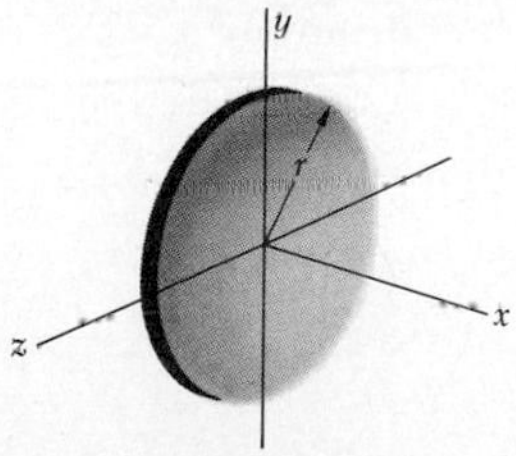	$I_x = \frac{1}{2}mr^2$ $I_y = I_z = \frac{1}{4}mr^2$
Circular cylinder	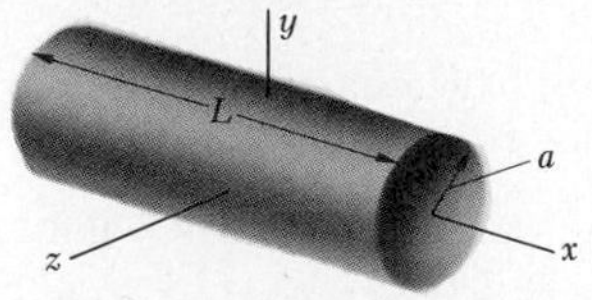	$I_x = \frac{1}{2}ma^2$ $I_y = I_z = \frac{1}{12}m(3a^2 + L^2)$
Circular cone	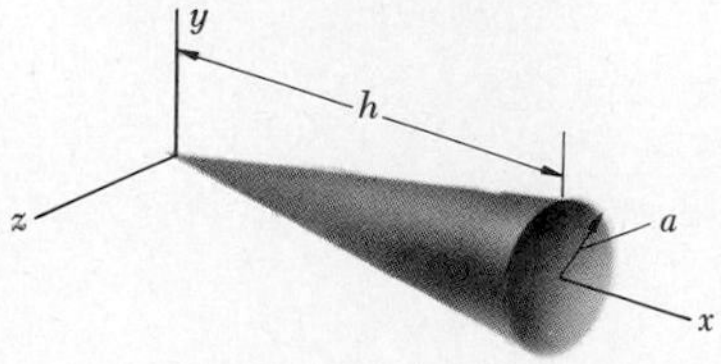	$I_x = \frac{3}{10}ma^2$ $I_y = I_z = \frac{3}{5}m(\frac{1}{4}a^2 + h^2)$
Sphere	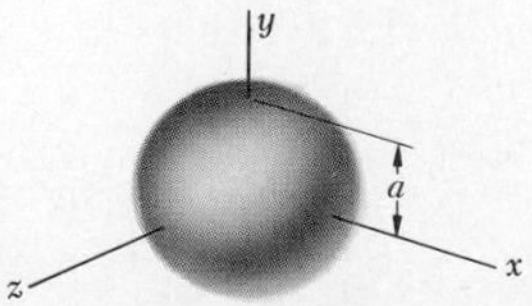	$I_x = I_y = I_z = \frac{2}{5}ma^2$

Fig. 9.28 Mass moments of inertia of common geometric shapes.

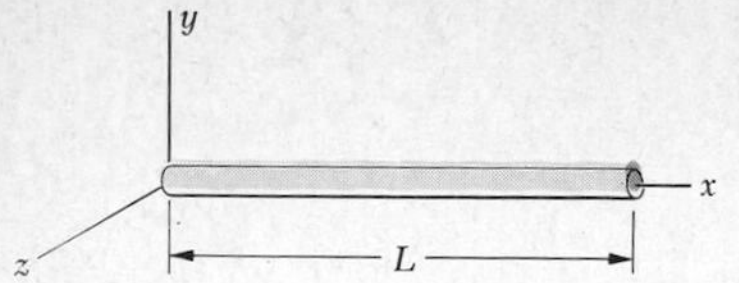

SAMPLE PROBLEM 9.9

Determine the mass moment of inertia of a slender rod of length L and mass m with respect to an axis perpendicular to the rod and passing through one end of the rod.

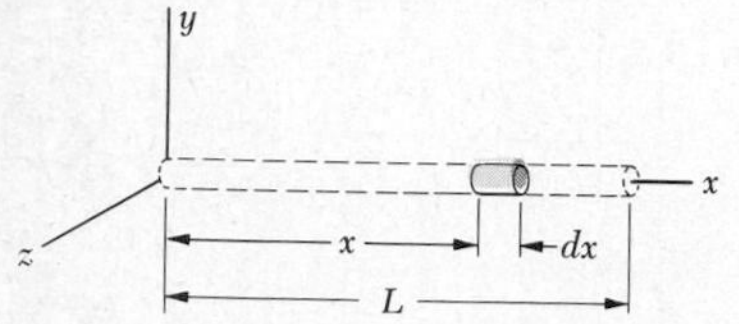

Solution. Choosing the differential element of mass shown, we write

$$dm = \frac{m}{L}\,dx$$

$$I_y = \int x^2\,dm = \int_0^L x^2 \frac{m}{L}\,dx = \left[\frac{m}{L}\frac{x^3}{3}\right]_0^L \qquad I_y = \tfrac{1}{3}mL^2 \blacktriangleleft$$

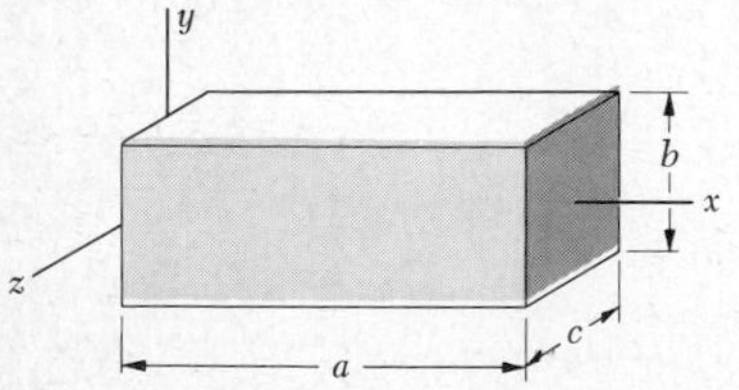

SAMPLE PROBLEM 9.10

Determine the mass moment of inertia of the homogeneous rectangular prism shown with respect to the z axis.

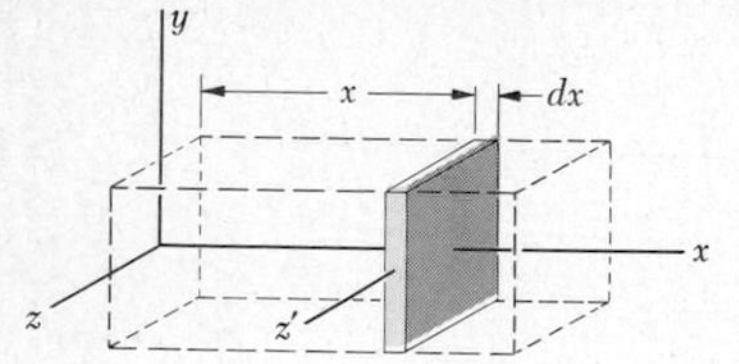

Solution. We choose as a differential element of mass the thin slab shown for which

$$dm = \rho bc\,dx$$

Referring to Sec. 9.13, we find that the moment of inertia of the element with respect to the z' axis is

$$dI_{z'} = \tfrac{1}{12}b^2\,dm$$

Applying the parallel-axis theorem, we obtain the mass moment of inertia of the slab with respect to the z axis.

$$dI_z = dI_{z'} + x^2\,dm = \tfrac{1}{12}b^2\,dm + x^2\,dm = (\tfrac{1}{12}b^2 + x^2)\,\rho bc\,dx$$

Integrating from $x = 0$ to $x = a$, we obtain

$$I_z = \int dI_z = \int_0^a (\tfrac{1}{12}b^2 + x^2)\,\rho bc\,dx = \rho abc\,(\tfrac{1}{12}b^2 + \tfrac{1}{3}a^2)$$

Since the total mass of the prism is $m = \rho abc$, we may write

$$I_z = m(\tfrac{1}{12}b^2 + \tfrac{1}{3}a^2) \qquad I_z = \tfrac{1}{12}m(4a^2 + b^2) \blacktriangleleft$$

We note that if the prism is slender, b is small compared to a and the expression for I_z reduces to $\frac{1}{3}ma^2$, which is the result obtained in Sample Prob. 9.9 when $L = a$.

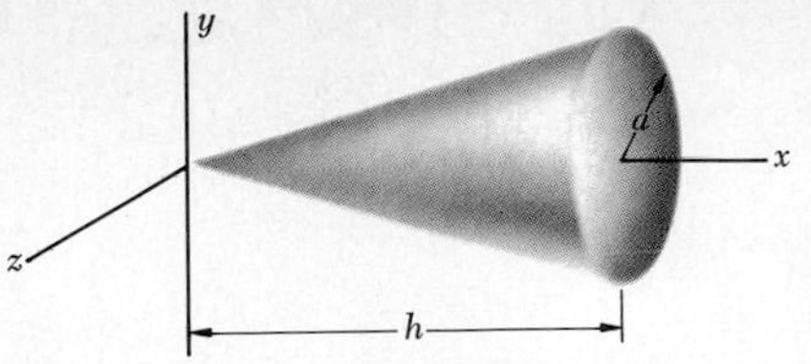

SAMPLE PROBLEM 9.11

Determine the mass moment of inertia of a right circular cone with respect to (*a*) its longitudinal axis, (*b*) an axis through the apex of the cone and perpendicular to its longitudinal axis, (*c*) an axis through the centroid of the cone and perpendicular to its longitudinal axis.

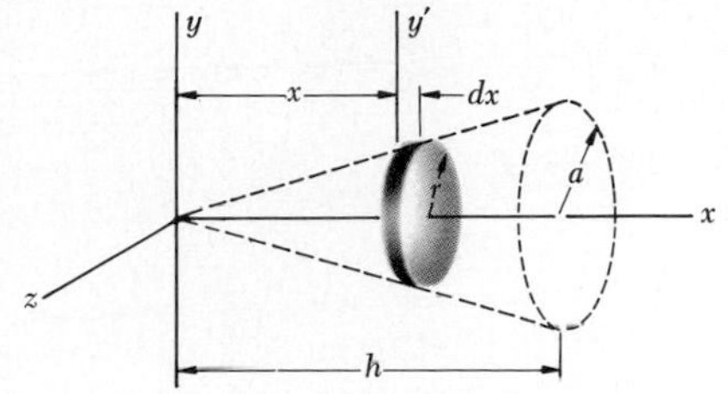

Solution. We choose the differential element of mass shown.

$$r = a\frac{x}{h} \qquad dm = \rho\pi r^2\,dx = \rho\pi\frac{a^2}{h^2}x^2\,dx$$

a. Moment of Inertia I_x. Using the expression derived in Sec. 9.13 for a thin disk, we compute the mass moment of inertia of the differential element with respect to the x axis.

$$dI_x = \tfrac{1}{2}r^2\,dm = \tfrac{1}{2}\left(a\frac{x}{h}\right)^2\left(\rho\pi\frac{a^2}{h^2}x^2\,dx\right) = \tfrac{1}{2}\rho\pi\frac{a^4}{h^4}x^4\,dx$$

Integrating from $x = 0$ to $x = h$, we obtain

$$I_x = \int dI_x = \int_0^h \tfrac{1}{2}\rho\pi\frac{a^4}{h^4}x^4\,dx = \tfrac{1}{2}\rho\pi\frac{a^4}{h^4}\frac{h^5}{5} = \tfrac{1}{10}\rho\pi a^4 h$$

Since the total mass of the cone is $m = \tfrac{1}{3}\rho\pi a^2 h$, we may write

$$I_x = \tfrac{1}{10}\rho\pi a^4 h = \tfrac{3}{10}a^2(\tfrac{1}{3}\rho\pi a^2 h) = \tfrac{3}{10}ma^2 \qquad I_x = \tfrac{3}{10}ma^2 \quad ◄$$

b. Moment of Inertia I_y. The same differential element will be used. Applying the parallel-axis theorem and using the expression derived in Sec. 9.13 for a thin disk, we write

$$dI_y = dI_{y'} + x^2\,dm = \tfrac{1}{4}r^2\,dm + x^2\,dm = (\tfrac{1}{4}r^2 + x^2)\,dm$$

Substituting the expressions for r and dm, we obtain

$$dI_y = \left(\frac{1}{4}\frac{a^2}{h^2}x^2 + x^2\right)\left(\rho\pi\frac{a^2}{h^2}x^2\,dx\right) = \rho\pi\frac{a^2}{h^2}\left(\frac{a^2}{4h^2} + 1\right)x^4\,dx$$

$$I_y = \int dI_y = \int_0^h \rho\pi\frac{a^2}{h^2}\left(\frac{a^2}{4h^2} + 1\right)x^4\,dx = \rho\pi\frac{a^2}{h^2}\left(\frac{a^2}{4h^2} + 1\right)\frac{h^5}{5}$$

Introducing the total mass of the cone m, we rewrite I_y as follows:

$$I_y = \tfrac{3}{5}(\tfrac{1}{4}a^2 + h^2)\tfrac{1}{3}\rho\pi a^2 h \qquad I_y = \tfrac{3}{5}m(\tfrac{1}{4}a^2 + h^2) \quad ◄$$

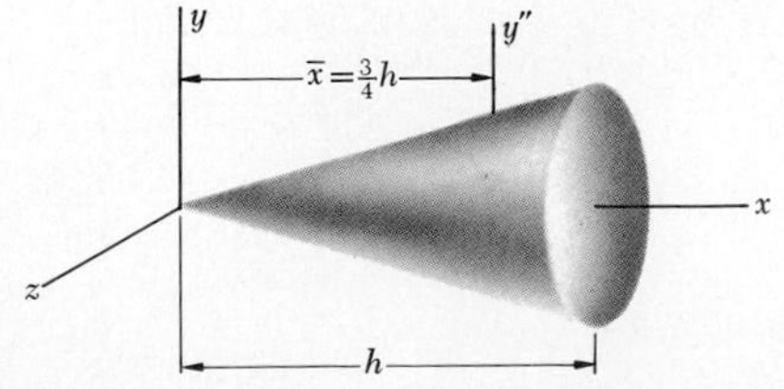

c. Moment of Inertia $\bar{I}_{y''}$. We apply the parallel-axis theorem and write

$$I_y = \bar{I}_{y''} + m\bar{x}^2$$

Solving for $\bar{I}_{y''}$ and recalling that $\bar{x} = \tfrac{3}{4}h$, we have

$$\bar{I}_{y''} = I_y - m\bar{x}^2 = \tfrac{3}{5}m(\tfrac{1}{4}a^2 + h^2) - m(\tfrac{3}{4}h)^2$$

$$\bar{I}_{y''} = \tfrac{3}{20}m(a^2 + \tfrac{1}{4}h^2) \quad ◄$$

SAMPLE PROBLEM 9.12

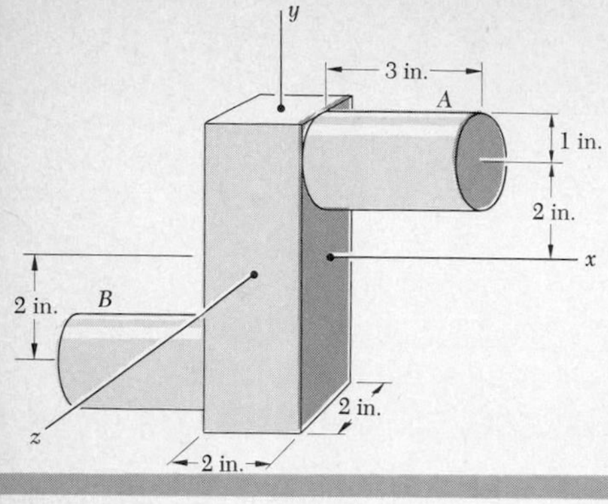

A steel forging consists of a rectangular prism $6 \times 2 \times 2$ in. and of two cylinders of diameter 2 in. and length 3 in., as shown. Determine the mass moments of inertia with respect to the coordinate axes. (Specific weight of steel = 490 lb/ft³.)

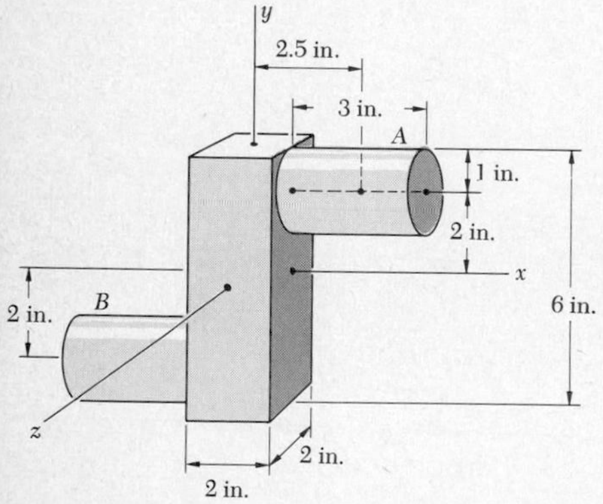

Computation of Masses
Prism

$$V = (2\text{ in.})(2\text{ in.})(6\text{ in.}) = 24\text{ in}^3$$

$$W = \frac{(24\text{ in}^3)(490\text{ lb/ft}^3)}{1728\text{ in}^3/\text{ft}^3} = 6.81\text{ lb}$$

$$m = \frac{6.81\text{ lb}}{32.2\text{ ft/s}^2} = 0.211\text{ lb}\cdot\text{s}^2/\text{ft}$$

Each Cylinder

$$V = \pi(1\text{ in.})^2(3\text{ in.}) = 9.42\text{ in}^3$$

$$W = \frac{(9.42\text{ in}^3)(490\text{ lb/ft}^3)}{1728\text{ in}^3/\text{ft}^3} = 2.67\text{ lb}$$

$$m = \frac{2.67\text{ lb}}{32.2\text{ ft/s}^2} = 0.0829\text{ lb}\cdot\text{s}^2/\text{ft}$$

Mass Moments of Inertia. The mass moments of inertia of each component are computed from Fig. 9.28, using the parallel-axis theorem when necessary. Note that all lengths should be expressed in feet.

Prism

$$I_x = I_z = \tfrac{1}{12}(0.211\text{ lb}\cdot\text{s}^2/\text{ft})[(\tfrac{6}{12}\text{ ft})^2 + (\tfrac{2}{12}\text{ ft})^2] = 4.88 \times 10^{-3}\text{ lb}\cdot\text{ft}\cdot\text{s}^2$$
$$I_y = \tfrac{1}{12}(0.211\text{ lb}\cdot\text{s}^2/\text{ft})[(\tfrac{2}{12}\text{ ft})^2 + (\tfrac{2}{12}\text{ ft})^2] = 0.977 \times 10^{-3}\text{ lb}\cdot\text{ft}\cdot\text{s}^2$$

Each Cylinder

$$I_x = \tfrac{1}{2}ma^2 + m\bar{y}^2 = \tfrac{1}{2}(0.0829\text{ lb}\cdot\text{s}^2/\text{ft})(\tfrac{1}{12}\text{ ft})^2 + (0.0829\text{ lb}\cdot\text{s}^2/\text{ft})(\tfrac{2}{12}\text{ ft})^2 = 2.59 \times 10^{-3}\text{ lb}\cdot\text{ft}\cdot\text{s}^2$$
$$I_y = \tfrac{1}{12}m(3a^2 + L^2) + m\bar{x}^2 = \tfrac{1}{12}(0.0829\text{ lb}\cdot\text{s}^2/\text{ft})[3(\tfrac{1}{12}\text{ ft})^2 + (\tfrac{3}{12}\text{ ft})^2] + (0.0829\text{ lb}\cdot\text{s}^2/\text{ft})(\tfrac{2.5}{12}\text{ ft})^2 = 4.17 \times 10^{-3}\text{ lb}\cdot\text{ft}\cdot\text{s}^2$$
$$I_z = \tfrac{1}{12}m(3a^2 + L^2) + m(\bar{x}^2 + \bar{y}^2) = \tfrac{1}{12}(0.0829)[3(\tfrac{1}{12})^2 + (\tfrac{3}{12})^2] + (0.0829)[(\tfrac{2.5}{12})^2 + (\tfrac{2}{12})^2] = 6.48 \times 10^{-3}\text{ lb}\cdot\text{ft}\cdot\text{s}^2$$

Entire Body. Adding the values obtained,

$$I_x = 4.88 \times 10^{-3} + 2(2.59 \times 10^{-3}) \qquad I_x = 10.06 \times 10^{-3}\text{ lb}\cdot\text{ft}\cdot\text{s}^2 \blacktriangleleft$$
$$I_y = 0.977 \times 10^{-3} + 2(4.17 \times 10^{-3}) \qquad I_y = 9.32 \times 10^{-3}\text{ lb}\cdot\text{ft}\cdot\text{s}^2 \blacktriangleleft$$
$$I_z = 4.88 \times 10^{-3} + 2(6.48 \times 10^{-3}) \qquad I_z = 17.84 \times 10^{-3}\text{ lb}\cdot\text{ft}\cdot\text{s}^2 \blacktriangleleft$$

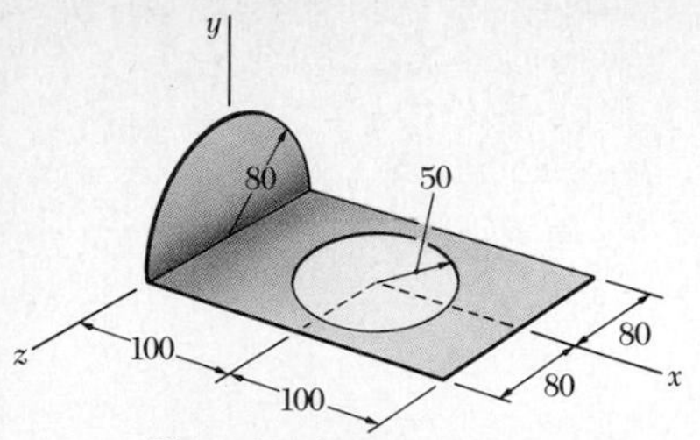

Dimensions in mm

SAMPLE PROBLEM 9.13

A thin steel plate 4 mm thick is cut and bent to form the machine part shown. Knowing that the density of steel is 7850 kg/m^3, determine the mass moment of inertia of the machine part with respect to the coordinate axes.

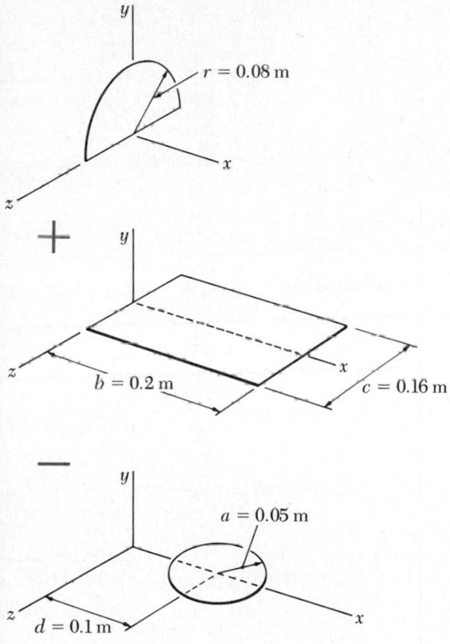

Solution. We observe that the machine part consists of a semicircular plate, plus a rectangular plate, minus a circular plate.

Computation of Masses. ***Semicircular Plate***

$$V_1 = \tfrac{1}{2}\pi r^2 t = \tfrac{1}{2}\pi(0.08\text{ m})^2(0.004\text{ m}) = 40.21 \times 10^{-6}\text{ m}^3$$
$$m_1 = \rho V_1 = (7.85 \times 10^3\text{ kg/m}^3)(40.21 \times 10^{-6}\text{ m}^3) = 0.3156\text{ kg}$$

Rectangular Plate

$$V_2 = (0.200\text{ m})(0.160\text{ m})(0.004\text{ m}) = 128 \times 10^{-6}\text{ m}^3$$
$$m_2 = \rho V_2 = (7.85 \times 10^3\text{ kg/m}^3)(128 \times 10^{-6}\text{ m}^3) = 1.005\text{ kg}$$

Circular Plate

$$V_3 = \pi a^2 t = \pi(0.050\text{ m})^2(0.004\text{ m}) = 31.42 \times 10^{-6}\text{ m}^3$$
$$m_3 = \rho V_3 = (7.85 \times 10^3\text{ kg/m}^3)(31.42 \times 10^{-6}\text{ m}^3) = 0.2466\text{ kg}$$

Mass Moments of Inertia. Using the method presented in Sec. 9.13, we compute the mass moment of inertia of each component.

Semicircular Plate. From Fig. 9.28, we observe that for a circular plate of mass m and radius r

$$I_x = \tfrac{1}{2}mr^2 \qquad I_y = I_z = \tfrac{1}{4}mr^2$$

Because of symmetry, we note that for a semicircular plate

$$I_x = \tfrac{1}{2}(\tfrac{1}{2}mr^2) \qquad I_y = I_z = \tfrac{1}{2}(\tfrac{1}{4}mr^2)$$

Since the mass of the semicircular plate is $m_1 = \tfrac{1}{2}m$, we have

$$I_x = \tfrac{1}{2}m_1r^2 = \tfrac{1}{2}(0.3156\text{ kg})(0.08\text{ m})^2 = 1.010 \times 10^{-3}\text{ kg}\cdot\text{m}^2$$
$$I_y = I_z = \tfrac{1}{4}(\tfrac{1}{2}mr^2) = \tfrac{1}{4}m_1r^2 = \tfrac{1}{4}(0.3156\text{ kg})(0.08\text{ m})^2 = 0.505 \times 10^{-3}\text{ kg}\cdot\text{m}^2$$

Rectangular Plate

$$I_x = \tfrac{1}{12}m_2c^2 = \tfrac{1}{12}(1.005\text{ kg})(0.16\text{ m})^2 = 2.144 \times 10^{-3}\text{ kg}\cdot\text{m}^2$$
$$I_z = \tfrac{1}{3}m_2b^2 = \tfrac{1}{3}(1.005\text{ kg})(0.2\text{ m})^2 = 13.400 \times 10^{-3}\text{ kg}\cdot\text{m}^2$$
$$I_y = I_x + I_z = (2.144 + 13.400)(10^{-3}) = 15.544 \times 10^{-3}\text{ kg}\cdot\text{m}^2$$

Circular Plate

$$I_x = \tfrac{1}{4}m_3a^2 = \tfrac{1}{4}(0.2466\text{ kg})(0.05\text{ m})^2 = 0.154 \times 10^{-3}\text{ kg}\cdot\text{m}^2$$
$$I_y = \tfrac{1}{2}m_3a^2 + m_3d^2$$
$$= \tfrac{1}{2}(0.2466\text{ kg})(0.05\text{ m})^2 + (0.2466\text{ kg})(0.1\text{ m})^2 = 2.774 \times 10^{-3}\text{ kg}\cdot\text{m}^2$$
$$I_z = \tfrac{1}{4}m_3a^2 + m_3d^2 = \tfrac{1}{4}(0.2466\text{ kg})(0.05\text{ m})^2 + (0.2466\text{ kg})(0.1\text{ m})^2$$
$$= 2.620 \times 10^{-3}\text{ kg}\cdot\text{m}^2$$

Entire Machine Part

$$I_x = (1.010 + 2.144 - 0.154)(10^{-3})\text{ kg}\cdot\text{m}^2 \qquad I_x = 3.00 \times 10^{-3}\text{ kg}\cdot\text{m}^2 \blacktriangleleft$$
$$I_y = (0.505 + 15.544 - 2.774)(10^{-3})\text{ kg}\cdot\text{m}^2 \qquad I_y = 13.28 \times 10^{-3}\text{ kg}\cdot\text{m}^2 \blacktriangleleft$$
$$I_z = (0.505 + 13.400 - 2.620)(10^{-3})\text{ kg}\cdot\text{m}^2 \qquad I_z = 11.29 \times 10^{-3}\text{ kg}\cdot\text{m}^2 \blacktriangleleft$$

Problems

9.87 A thin semicircular plate has a radius a and a mass m. Determine the mass moment of inertia of the plate with respect to (*a*) the centroidal axis BB', (*b*) the centroidal axis CC'.

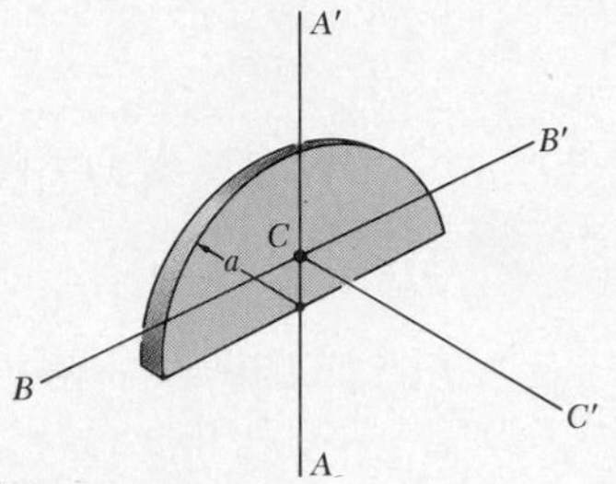

Fig. P9.87

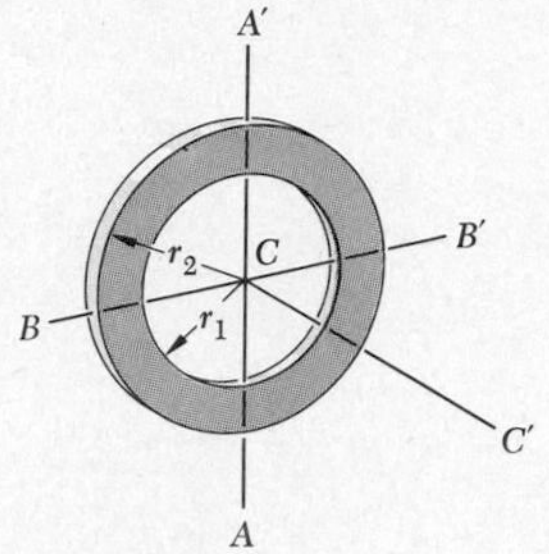

Fig. P9.88

9.88 Determine the mass moment of inertia of a ring of mass m, cut from a thin uniform plate, with respect to (*a*) the diameter AA' of the ring, (*b*) the axis CC' perpendicular to the plane of the ring.

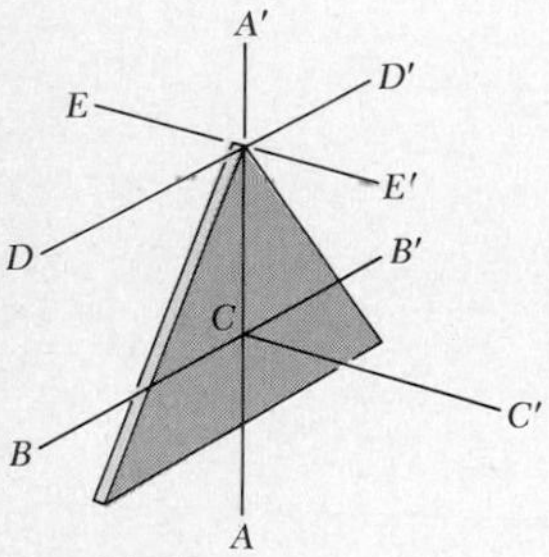

Fig. P9.89

9.89 A thin plate of mass m is cut in the shape of an equilateral triangle of side a. Determine the mass moment of inertia of the plate with respect to (*a*) the centroidal axes AA' and BB' in the plane of the plate, (*b*) the centroidal axis CC' perpendicular to the plate.

9.90 Determine the mass moments of inertia of the plate of Prob. 9.89 with respect to the axes DD' and EE' parallel to the centroidal axes BB' and CC', respectively.

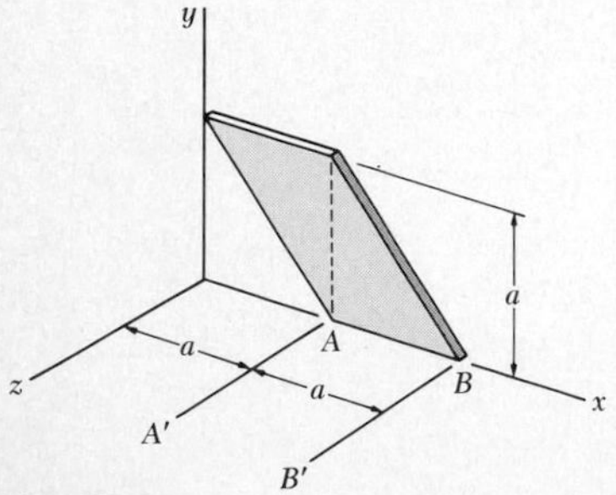

Fig. P9.91

9.91 A thin plate of mass m is cut in the shape of a parallelogram as shown. Determine the mass moment of inertia of the plate with respect to (*a*) the x axis, (*b*) the axis BB' perpendicular to the plate.

9.92 Determine the mass moment of inertia of the plate of Prob. 9.91 with respect to (*a*) the y axis, (*b*) the axis AA' perpendicular to the plate.

9.93 The area shown is revolved about the x axis to form a homogeneous solid of revolution of mass m. Express the mass moment of inertia of the solid with respect to the x axis in terms of m, a, and n. The expression obtained may be used to verify (*a*) the value given in Fig. 9.28 for a cone (with $n = 1$), (*b*) the answer to Prob. 9.95 (with $n = \frac{1}{2}$).

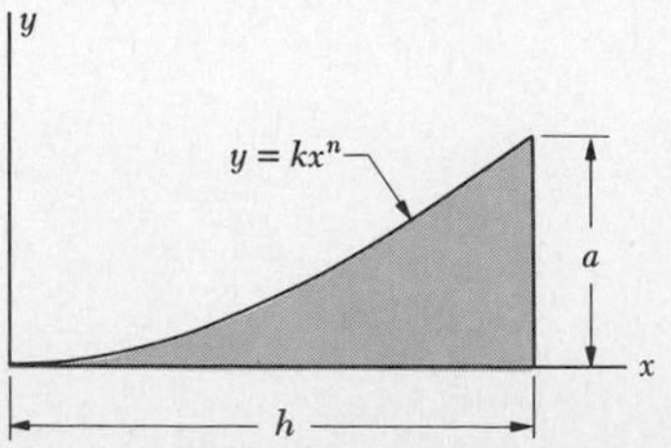

Fig. P9.93

9.94 Determine by direct integration the mass moment of inertia with respect to the y axis of the right circular cylinder shown, assuming a uniform density and a mass m.

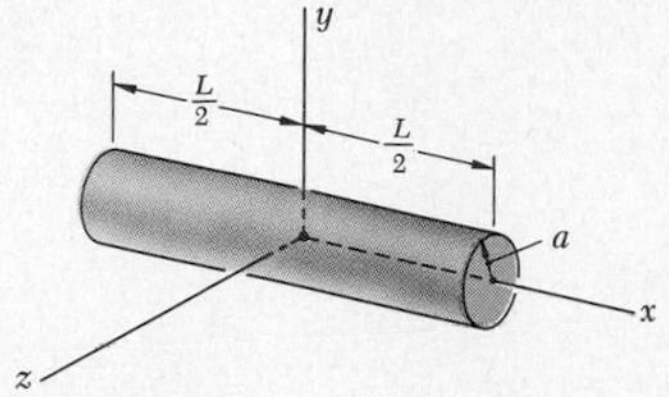

Fig. P9.94

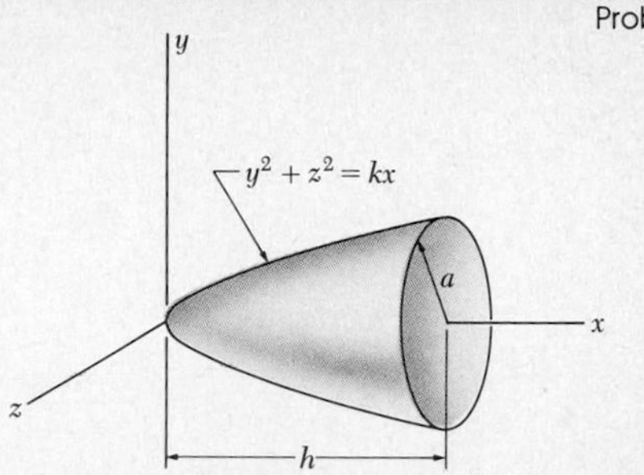

Fig. P9.95 and P9.96

9.95 Determine by direct integration the mass moment of inertia and the radius of gyration with respect to the x axis of the paraboloid shown, assuming a uniform density and a mass m.

9.96 Determine by direct integration the mass moment of inertia and the radius of gyration with respect to the y axis of the paraboloid shown, assuming a uniform density and a mass m.

9.97 Determine by direct integration the mass moment of inertia with respect to the x axis of the pyramid shown, assuming a uniform density and a mass m.

9.98 Determine by direct integration the mass moment of inertia with respect to the y axis of the pyramid shown, assuming a uniform density and a mass m.

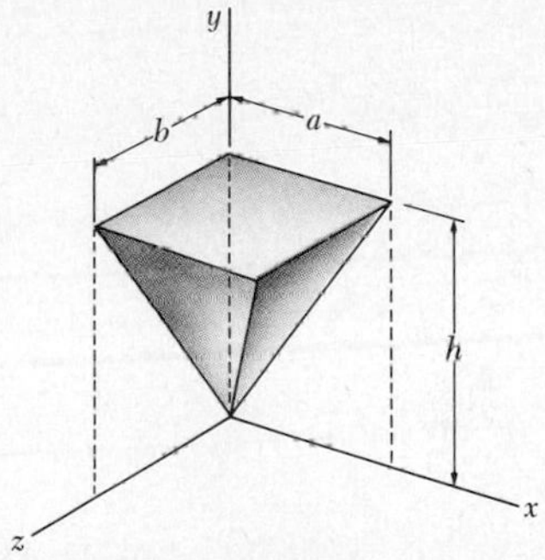

Fig. P9.97 and P9.98

9.99 A thin steel wire is bent into the shape of a circular arc as shown. Denoting by m' the mass per unit length of the wire, determine the mass moments of inertia of the wire with respect to the coordinate axes.

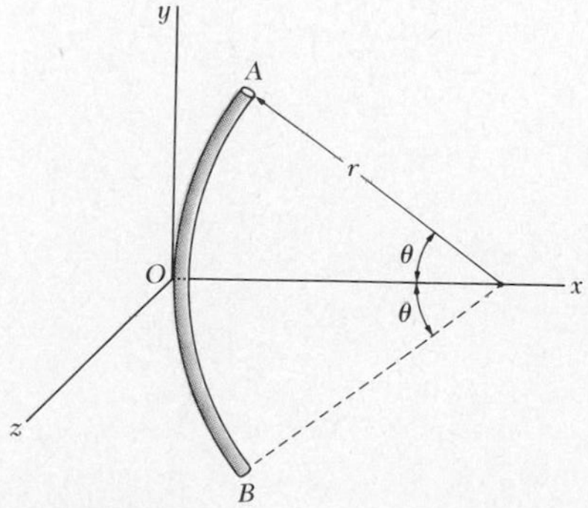

Fig. P9.99

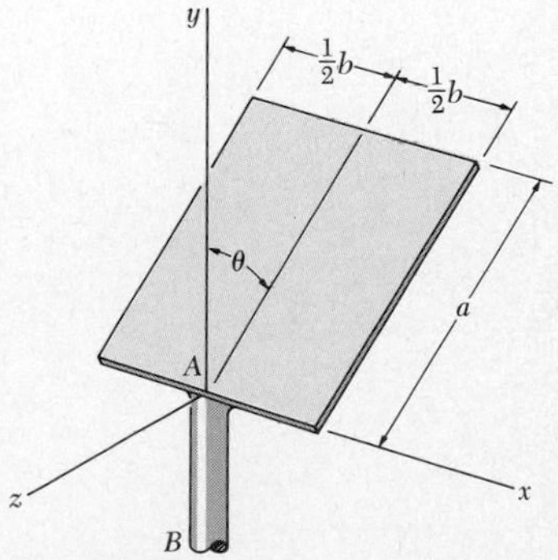

Fig. P9.100

9.100 A thin rectangular plate of mass m is welded to a vertical shaft AB with which it forms an angle θ. Determine by direct integration the mass moment of inertia with respect to (*a*) the y axis, (*b*) the z axis.

9.101 The cross section of a small flywheel is shown. The rim and hub are connected by eight spokes (two of which are shown in the cross section). Each spoke has a cross-sectional area of 160 mm^2. Determine the mass moment of inertia and radius of gyration of the flywheel with respect to the axis of rotation. (Density of steel = 7850 kg/m^3.)

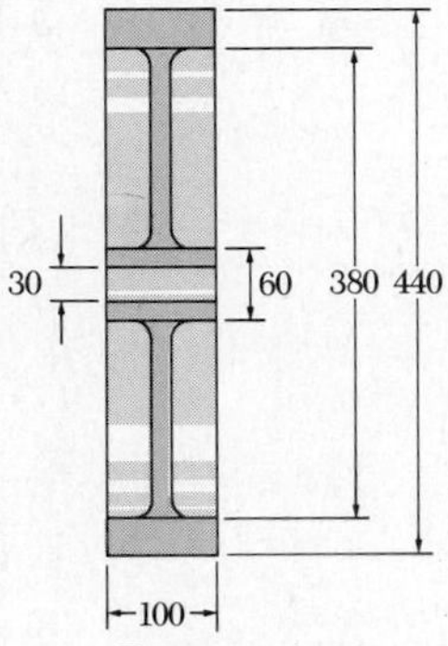

Fig. P9.101

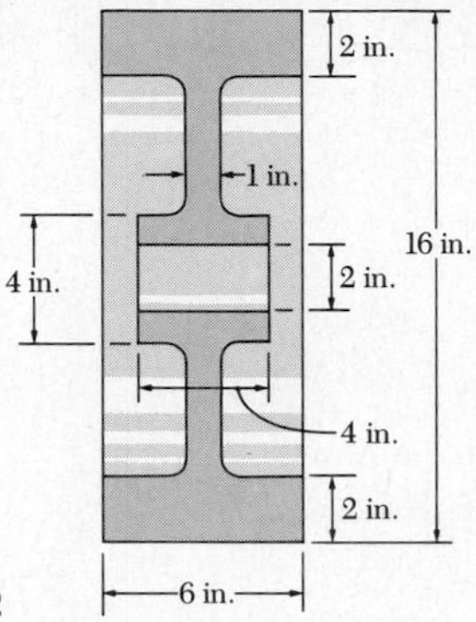

Fig. P9.102

9.102 Determine the mass moment of inertia and the radius of gyration of the steel flywheel shown with respect to the axis of rotation. The web of the flywheel consists of a solid plate 1 in. thick. (Specific weight of steel = 490 lb/ft^3.)

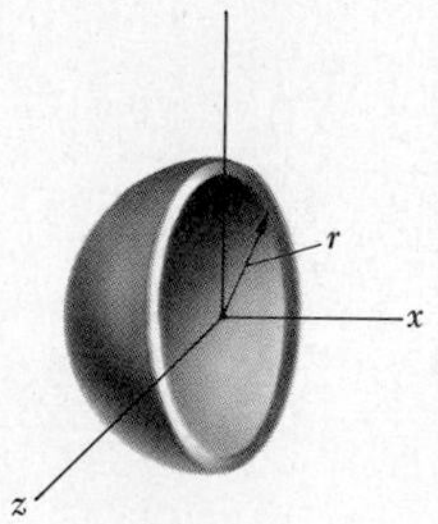

Fig. P9.103

9.103 Knowing that the thin hemispherical shell shown is of mass m and thickness t, determine the mass moment of inertia and the radius of gyration of the shell with respect to the x axis. (*Hint.* Consider the shell as formed by removing a hemisphere of radius r from a hemisphere of radius $r + t$; then neglect the terms containing t^2 and t^3 and keep those terms containing t.)

9.104 The machine part shown is formed by machining a conical surface into a circular cylinder. For $b = \frac{1}{2}h$, determine the mass moment of inertia and the radius of gyration of the machine part with respect to the y axis.

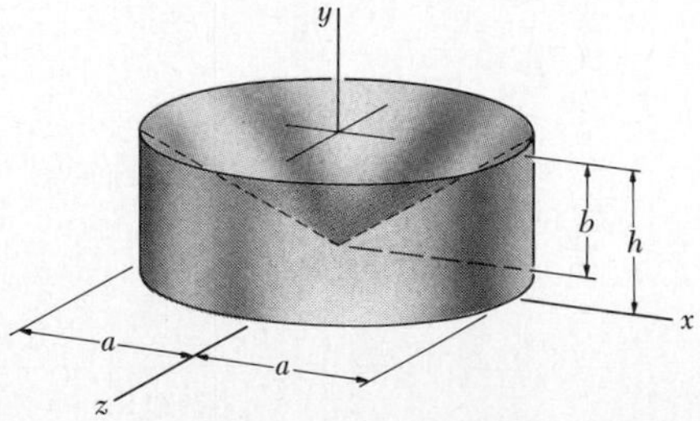

Fig. P9.104

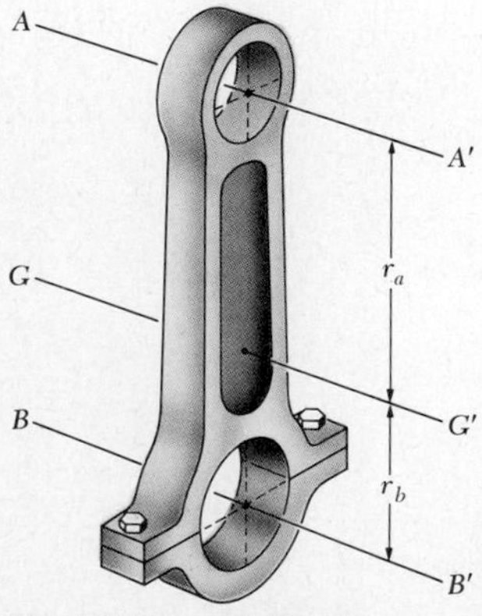

Fig. P9.105 and P9.106

9.105 For the 1.8-kg connecting rod shown, it has been experimentally determined that the mass moments of inertia of the rod with respect to the centerline axes of the bearings AA' and BB' are, respectively, $I_{AA'} = 65\ \text{g}\cdot\text{m}^2$ and $I_{BB'} = 33\ \text{g}\cdot\text{m}^2$. Knowing that $r_a + r_b = 275$ mm, determine (*a*) the location of the centroidal axis GG', (*b*) the radius of gyration with respect to axis GG'.

9.106 Knowing that for the 3.1-kg connecting rod shown the mass moment of inertia with respect to axis AA' is 160 g·m^2, determine the mass moment of inertia with respect to axis BB', for $r_a = 204$ mm and $r_b = 126$ mm.

9.107 A circular hole of radius r is to be drilled through the center of a rectangular steel plate to form the machine component shown. Denoting the density of the steel by ρ, determine (*a*) the mass moment of inertia of the component with respect to the axis BB', (*b*) the value of r for which, given a and h, $I_{BB'}$ is maximum, (*c*) the corresponding value of $I_{BB'}$ and of the radius of gyration $k_{BB'}$.

9.108 Determine the mass moment of inertia and the radius of gyration of the steel machine component shown with respect to the axis BB' when $a = 8$ in., $h = 4$ in., and $r = 3$ in. (Specific weight of steel $= 490$ lb/ft^3.)

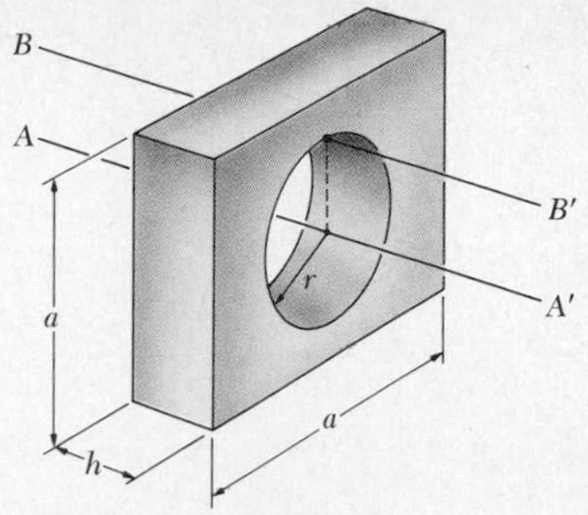

Fig. P9.107 and P9.108

9.109 and 9.110 A section of sheet steel 2 mm thick is cut and bent into the machine component shown. Knowing that the density of steel is 7850 kg/m^3, determine the mass moment of inertia of the component with respect to (*a*) the x axis, (*b*) the y axis, (*c*) the z axis.

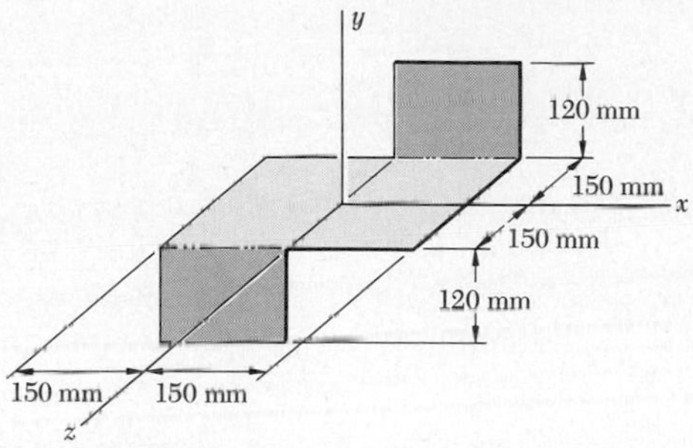

Fig. P9.109

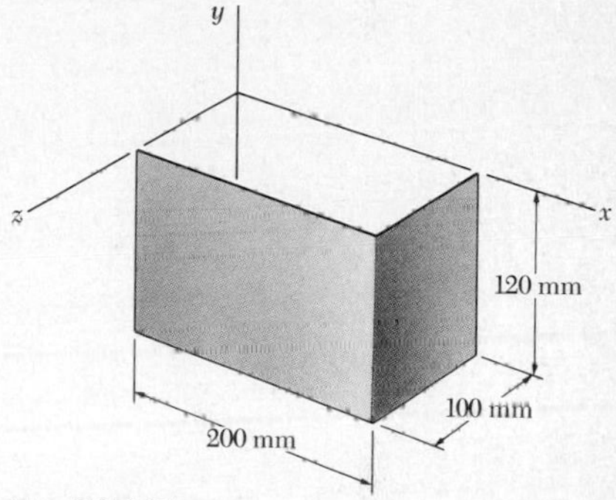

Fig. P9.110

9.111 A corner reflector for tracking by radar has two sides in the shape of a quarter circle of radius 15 in. and one side in the shape of a triangle. Each part of the reflector is formed from aluminum plate of uniform 0.05-in. thickness. Knowing that the specific weight of the aluminum used is 170 lb/ft^3, determine the mass moment of inertia of the reflector with respect to each of the coordinate axes.

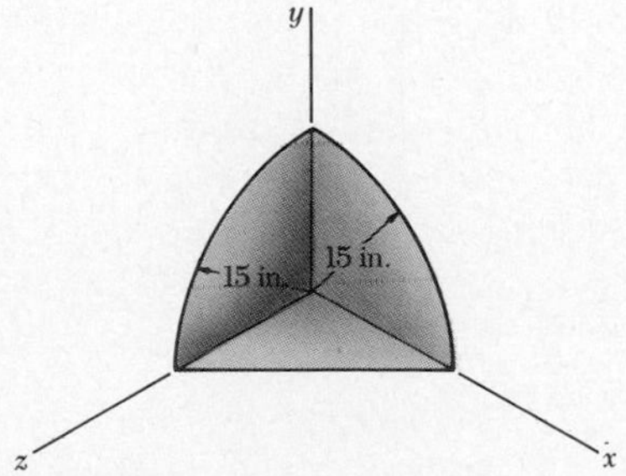

Fig. P9.111

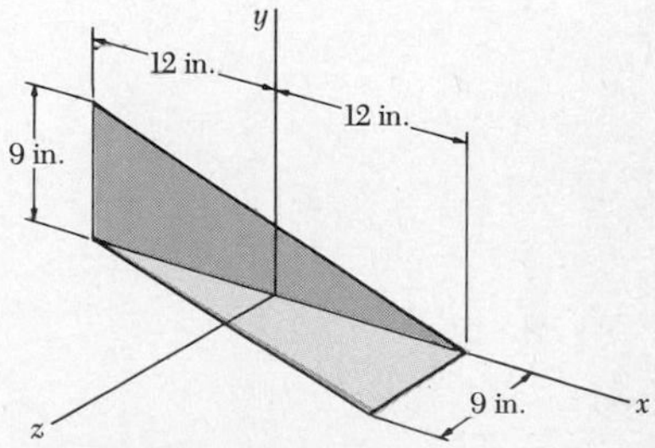

Fig. P9.112

9.112 A section of sheet steel 0.03 in. thick is cut and bent into the sheet-metal machine component shown. Knowing that the specific weight of steel is 490 lb/ft^3, determine the mass moment of inertia of the component with respect to each of the coordinate axes.

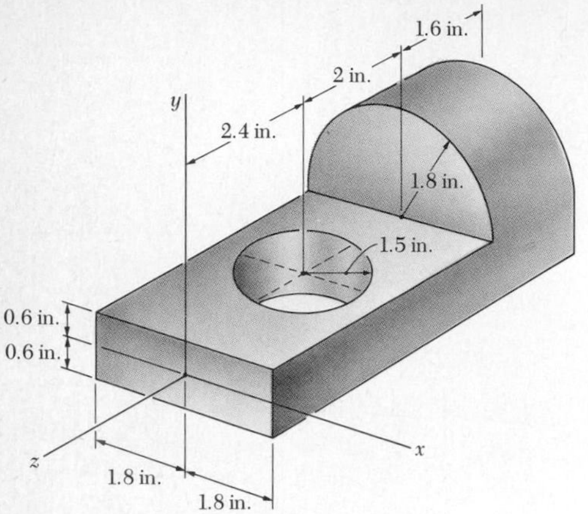

Fig. P9.114 and P9.115

9.113 and 9.114 Determine the mass moment of inertia and the radius of gyration of the steel machine element shown with respect to the x axis. (Specific weight of steel = 490 lb/ft^3; density of steel = 7850 kg/m^3.)

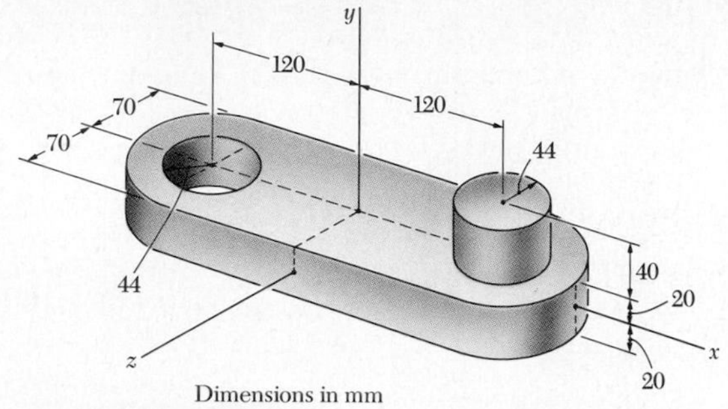

Fig. P9.113 and P9.116

9.115 and 9.116 Determine the mass moment of inertia and the radius of gyration of the steel machine element shown with respect to the y axis. (Specific weight of steel = 490 lb/ft^3; density of steel = 7850 kg/m^3.)

9.117 A homogenous wire with a mass per unit length of 1.5 kg/m is used to form the figure shown. Determine the mass moment of inertia of the wire figure with respect to (a) the x axis, (b) the y axis.

9.118 In Prob. 9.117, determine the mass moment of inertia of the wire figure with respect to the z axis.

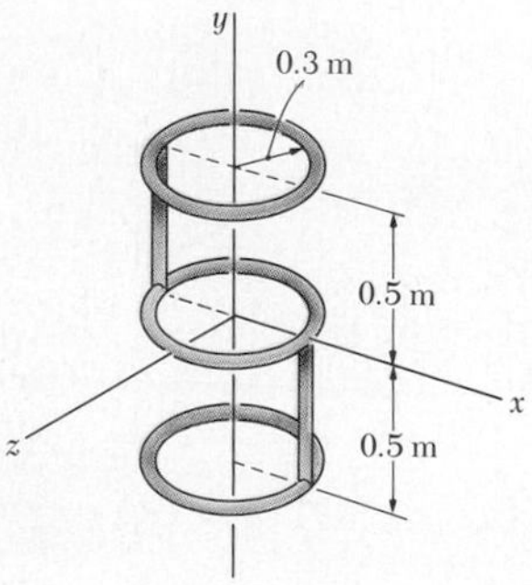

Fig. P9.117

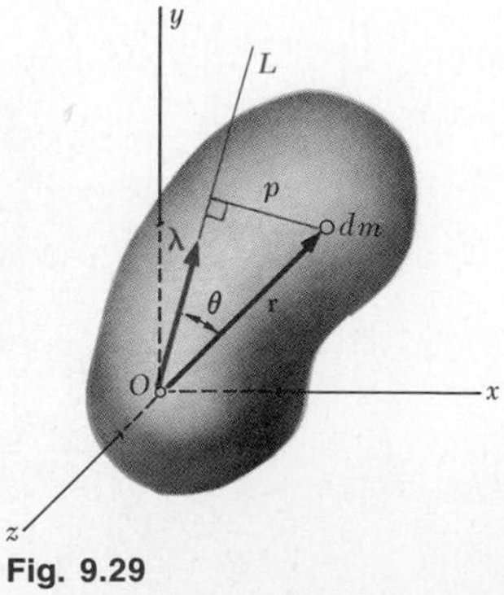

Fig. 9.29

***9.16. Moment of Inertia of a Body with Respect to an Arbitrary Axis through *O*. Mass Products of Inertia.** We shall see in this section how the moment of inertia of a body may be determined with respect to an arbitrary axis OL through the origin (Fig. 9.29) if we have computed beforehand its moments of inertia with respect to the three coordinate axes, as well as certain other quantities to be defined below.

The moment of inertia of the body with respect to OL is represented by the integral $I_{OL} = \int p^2\,dm$, where p denotes the perpendicular distance from the element of mass dm to the axis OL. But, denoting by $\boldsymbol{\lambda}$ the unit vector along OL and by $\mathbf{r}$ the position vector of the element dm, we observe that the perpendicular distance p is equal to the magnitude $r \sin\theta$ of the vector product $\boldsymbol{\lambda} \times \mathbf{r}$. We therefore write

$$I_{OL} = \int p^2\,dm = \int |\boldsymbol{\lambda} \times \mathbf{r}|^2\,dm \tag{9.43}$$

Expressing the square in terms of the rectangular components of the vector product, we have

$$I_{OL} = \int [(\lambda_x y - \lambda_y x)^2 + (\lambda_y z - \lambda_z y)^2 + (\lambda_z x - \lambda_x z)^2]\,dm$$

where the components $\lambda_x, \lambda_y, \lambda_z$ of the unit vector $\boldsymbol{\lambda}$ represent the direction cosines of the axis OL, and the components x, y, z of $\mathbf{r}$ represent the coordinates of the element of mass dm. Expanding the squares in the expression obtained and rearranging the terms, we write

$$I_{OL} = \lambda_x^2 \int (y^2 + z^2)\,dm + \lambda_y^2 \int (z^2 + x^2)\,dm + \lambda_z^2 \int (x^2 + y^2)\,dm$$
$$-2\lambda_x\lambda_y \int xy\,dm - 2\lambda_y\lambda_z \int yz\,dm - 2\lambda_z\lambda_x \int zx\,dm \qquad (9.44)$$

Referring to Eqs. (9.30), we note that the first three integrals in (9.44) represent, respectively, the moments of inertia I_x, I_y, and I_z of the body with respect to the coordinate axes. The last three integrals in (9.44), which involve products of coordinates, are called the *products of inertia* of the body with respect to the x and y axes, the y and z axes, and the z and x axes, respectively. We write

$$I_{xy} = \int xy\,dm \qquad I_{yz} = \int yz\,dm \qquad I_{zx} = \int zx\,dm \qquad (9.45)$$

Substituting for the various integrals from (9.30) and (9.45) into (9.44), we have

$$I_{OL} = I_x\lambda_x^2 + I_y\lambda_y^2 + I_z\lambda_z^2 - 2I_{xy}\lambda_x\lambda_y - 2I_{yz}\lambda_y\lambda_z - 2I_{zx}\lambda_z\lambda_x \qquad (9.46)$$

We note that the definition of the products of inertia of a mass given in Eqs. (9.45) is an extension of the definition of the product of inertia of an area (Sec. 9.8). Mass products of inertia reduce to zero under the same conditions of symmetry as products of inertia of areas, and the parallel-axis theorem for mass products of inertia is expressed by relations similar to the formula derived for the product of inertia of an area. Substituting for x, y, z from Eqs. (9.31) into Eqs. (9.45), we verify that

$$\begin{aligned} I_{xy} &= \bar{I}_{x'y'} + m\bar{x}\bar{y} \\ I_{yz} &= \bar{I}_{y'z'} + m\bar{y}\bar{z} \\ I_{zx} &= \bar{I}_{z'x'} + m\bar{z}\bar{x} \end{aligned} \qquad (9.47)$$

where $\bar{x}$, $\bar{y}$, $\bar{z}$ are the coordinates of the center of gravity G of the body, and $\bar{I}_{x'y'}$, $\bar{I}_{y'z'}$, $\bar{I}_{z'x'}$ denote the products of inertia with respect to the centroidal axes x', y', z' (Fig. 9.22).

*9.17. Ellipsoid of Inertia. Principal Axes of Inertia.

Let us assume that the moment of inertia of the body considered in the preceding section has been determined with respect to a large number of axes OL through the fixed point O, and that a point Q has been plotted on each axis OL at a distance $OQ = 1/\sqrt{I_{OL}}$ from O. The locus of the points Q thus obtained forms a surface (Fig. 9.30). The equation of that surface may be obtained by substituting $1/(OQ)^2$ for I_{OL} in (9.46) and multiplying both sides of the equation by $(OQ)^2$. Observing that

$$(OQ)\lambda_x = x \qquad (OQ)\lambda_y = y \qquad (OQ)\lambda_z = z$$

where x, y, z denote the rectangular coordinates of a point Q of the surface, we write

$$I_x x^2 + I_y y^2 + I_z z^2 - 2I_{xy}xy - 2I_{yz}yz - 2I_{zx}zx = 1 \qquad (9.48)$$

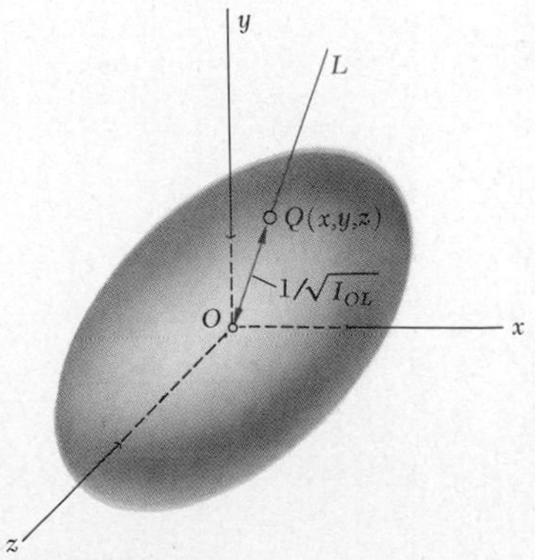

Fig. 9.30

The equation obtained is that of a *quadric*. Since the moment of inertia I_{OL} is different from zero for every axis OL, no point Q may be at an infinite distance from O. Thus, the quadric obtained is an *ellipsoid*. This ellipsoid, which defines the moment of inertia of the body with respect to any axis through O, is known as the *ellipsoid of inertia* of the body at O.

We observe that if the axes in Fig. 9.30 are rotated, the coefficients of the equation defining the ellipsoid change, since these are equal to the moments and products of inertia of the body with respect to the rotated coordinate axes. However, the *ellipsoid itself remains unaffected*, since its shape depends only upon the distribution of mass in the body considered. Suppose now that we choose as coordinate axes the principal axes x', y', z' of the ellipsoid of inertia (Fig. 9.31). The equation of the ellipsoid with respect to these coordinate axes will be of the form

$$I_{x'}x'^2 + I_{y'}y'^2 + I_{z'}z'^2 = 1 \tag{9.49}$$

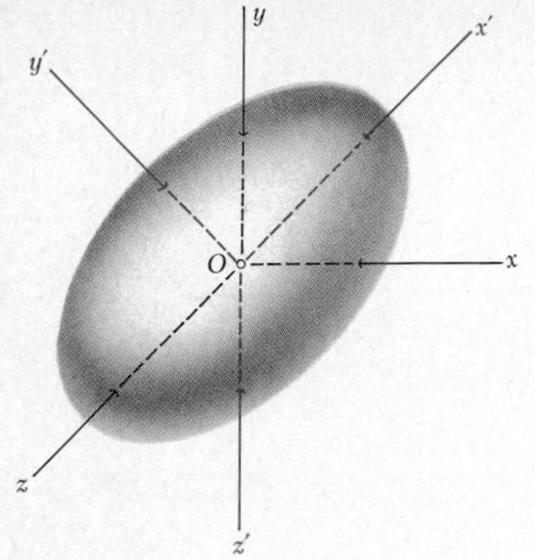

Fig. 9.31

which does not contain any product of coordinates. Thus, the products of inertia of the body with respect to the x', y', z' axes are zero. The x', y', z' axes are known as the *principal axes of inertia* of the body at O, and the coefficients $I_{x'}$, $I_{y'}$, $I_{z'}$ as the *principal moments of inertia* of the body at O. Note that, given a body of arbitrary shape and a point O, it is always possible to find axes which are the principal axes of inertia of the body at O, that is, axes with respect to which the products of inertia of the body are zero. Indeed, no matter how odd or irregular the shape of the body may be, the moments of inertia of the body with respect to axes through O will define an ellipsoid, and this ellipsoid will have principal axes which, by definition, are the principal axes of the body at O.

If the principal axes of inertia x', y', z' are used as coordinate axes, the expression obtained in Eq. (9.46) for the moment of inertia of a body with respect to an arbitrary axis through O reduces to

$$I_{OL} = I_{x'}\lambda_{x'}^2 + I_{y'}\lambda_{y'}^2 + I_{z'}\lambda_{z'}^2 \tag{9.50}$$

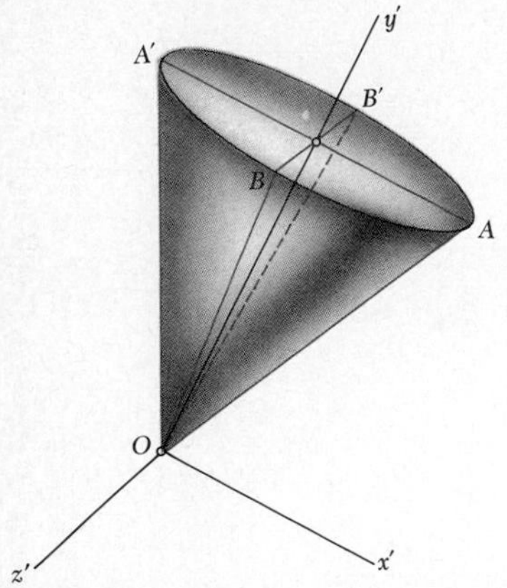

Fig. 9.32

While the determination of the principal axes of inertia of a body of arbitrary shape is somewhat involved and would require solving a cubic equation, there are many cases when these axes may be spotted immediately. Consider, for instance, the homogeneous cone of elliptical base shown in Fig. 9.32; this cone possesses two mutually perpendicular planes of symmetry OAA' and OBB'. We check from the definition (9.45) that if the $x'y'$ and $y'z'$ planes are chosen to coincide with the two planes of symmetry, all the products of inertia are zero. The x', y', and z' axes thus selected are therefore the principal axes of inertia of the cone at O. In the case of the homogeneous regular tetrahedron $OABC$ shown in Fig. 9.33, the line joining the corner O to the center D of the opposite face is a principal axis of inertia at O and any line through O perpendicular to OD is also a principal axis of inertia at O. This property may be recognized if we observe that a rotation through 120° about OD leaves the shape and the mass distribution of the tetrahedron unchanged. It follows that the ellipsoid of inertia at O also remains unchanged under this rotation. The ellipsoid, therefore, is of revolution about OD, and the line OD, as well as any perpendicular line through O, must be a principal axis of the ellipsoid.

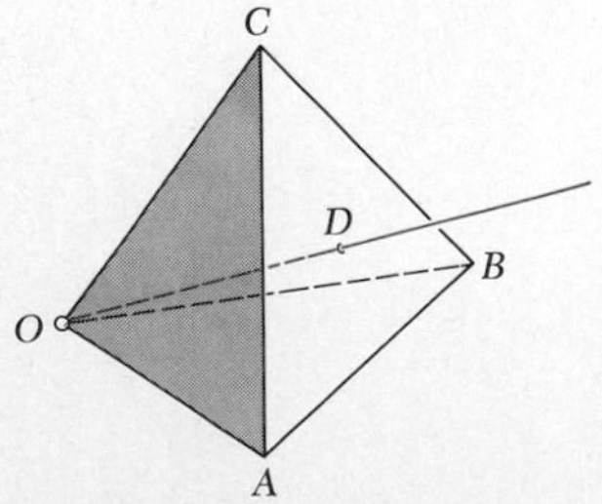

Fig. 9.33

SAMPLE PROBLEM 9.14

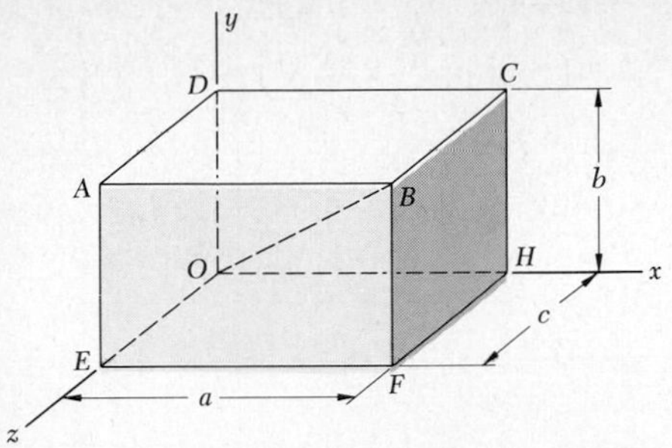

Consider a rectangular prism of mass m and sides a, b, c. Determine (a) the mass moments and products of inertia of the prism with respect to the coordinate axes shown, (b) its moment of inertia with respect to the diagonal OB.

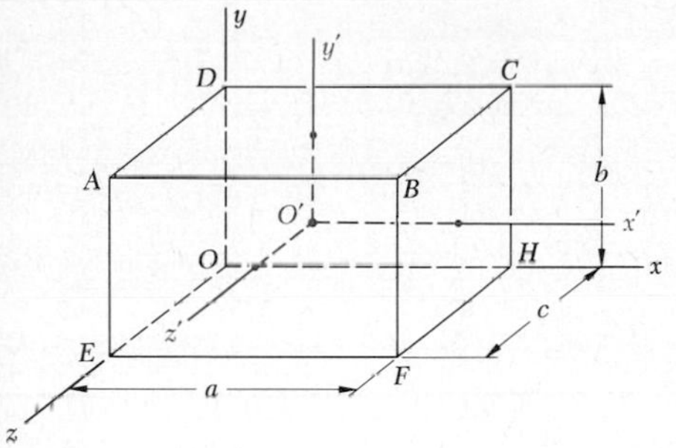

a. Moments and Products of Inertia with Respect to the Coordinate Axes. ***Moments of Inertia.*** Introducing the centroidal axes x', y', z', with respect to which the moments of inertia are given in Fig. 9.28, we apply the parallel-axis theorem:

$$I_x = \bar{I}_{x'} + m(\bar{y}^2 + \bar{z}^2) = \tfrac{1}{12}m(b^2 + c^2) + m(\tfrac{1}{4}b^2 + \tfrac{1}{4}c^2)$$

$$I_x = \tfrac{1}{3}m(b^2 + c^2) \blacktriangleleft$$

Similarly, $$I_y = \tfrac{1}{3}m(c^2 + a^2) \qquad I_z = \tfrac{1}{3}m(a^2 + b^2) \blacktriangleleft$$

Products of Inertia. Because of symmetry, the products of inertia with respect to the centroidal axes x', y', z' are zero and these axes are principal axes of inertia. Using the parallel-axis theorem, we have

$$I_{xy} = \bar{I}_{x'y'} + m\bar{x}\bar{y} = 0 + m(\tfrac{1}{2}a)(\tfrac{1}{2}b) \qquad I_{xy} = \tfrac{1}{4}mab \blacktriangleleft$$

Similarly, $$I_{yz} = \tfrac{1}{4}mbc \qquad I_{zx} = \tfrac{1}{4}mca \blacktriangleleft$$

b. Moment of Inertia with Respect to OB. We recall Eq. (9.46):

$$I_{OB} = I_x\lambda_x^2 + I_y\lambda_y^2 + I_z\lambda_z^2 - 2I_{xy}\lambda_x\lambda_y - 2I_{yz}\lambda_y\lambda_z - 2I_{zx}\lambda_z\lambda_x$$

where the direction cosines of OB are

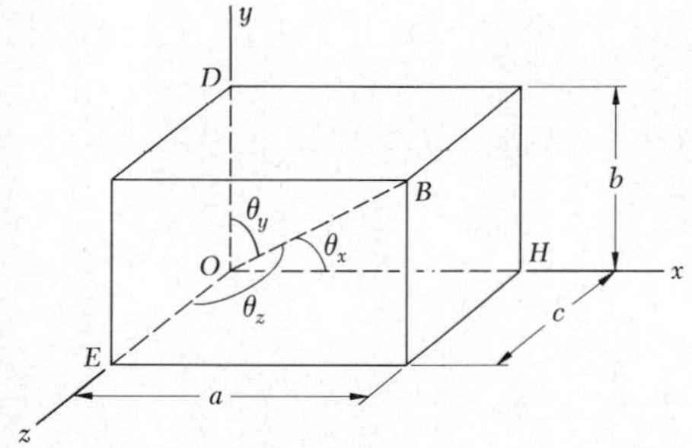

$$\lambda_x = \cos\theta_x = \frac{OH}{OB} = \frac{a}{(a^2 + b^2 + c^2)^{1/2}}$$

$$\lambda_y = \frac{b}{(a^2 + b^2 + c^2)^{1/2}} \qquad \lambda_z = \frac{c}{(a^2 + b^2 + c^2)^{1/2}}$$

Substituting the values obtained for the moments and products of inertia and for the direction cosines:

$$I_{OB} = \frac{1}{a^2 + b^2 + c^2}[\tfrac{1}{3}m(b^2 + c^2)a^2 + \tfrac{1}{3}m(c^2 + a^2)b^2 + \tfrac{1}{3}m(a^2 + b^2)c^2$$
$$- \tfrac{1}{2}ma^2b^2 - \tfrac{1}{2}mb^2c^2 - \tfrac{1}{2}mc^2a^2]$$

$$I_{OB} = \frac{m}{6}\,\frac{a^2b^2 + b^2c^2 + c^2a^2}{a^2 + b^2 + c^2} \blacktriangleleft$$

Alternate Solution. The moment of inertia I_{OB} may be obtained directly from the principal moments of inertia $\bar{I}_{x'}$, $\bar{I}_{y'}$, $\bar{I}_{z'}$, since the line OB passes through the centroid O'. Since the x', y', z' axes are principal axes of inertia, we use Eq. (9.50) and write

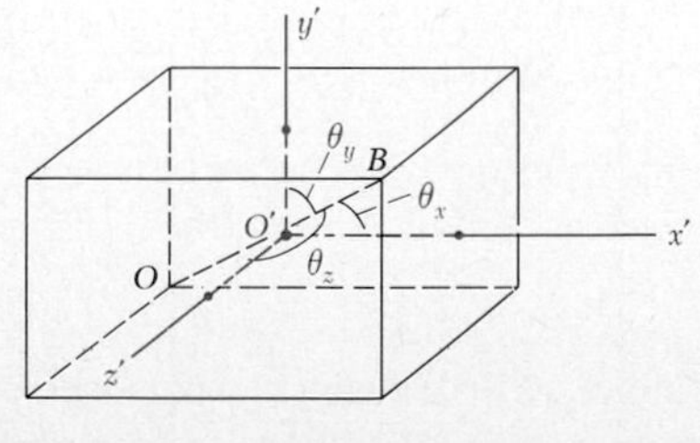

$$I_{OB} = \bar{I}_{x'}\lambda_x^2 + \bar{I}_{y'}\lambda_y^2 + \bar{I}_{z'}\lambda_z^2$$

$$= \frac{1}{a^2 + b^2 + c^2}\left[\frac{m}{12}(b^2 + c^2)a^2 + \frac{m}{12}(c^2 + a^2)b^2 + \frac{m}{12}(a^2 + b^2)c^2\right]$$

$$I_{OB} = \frac{m}{6}\,\frac{a^2b^2 + b^2c^2 + c^2a^2}{a^2 + b^2 + c^2} \blacktriangleleft$$

Problems

9.119 and 9.120 Determine the mass products of inertia I_{xy}, I_{yz}, and I_{zx} of the steel machine element shown. (Specific weight of steel $=$ 490 lb/ft^3; density of steel $=$ 7850 kg/m^3.)

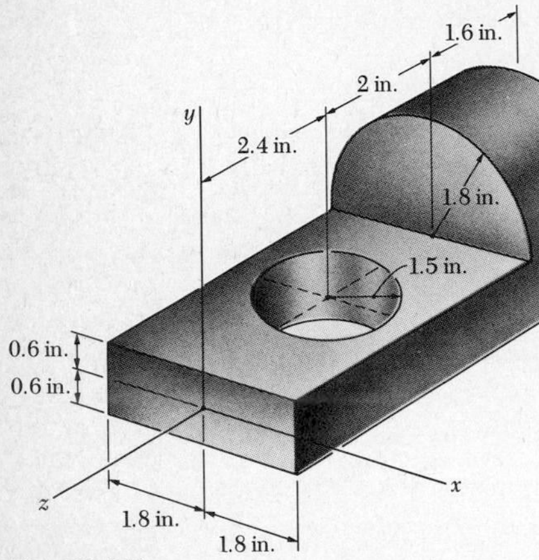

Fig. P9.119

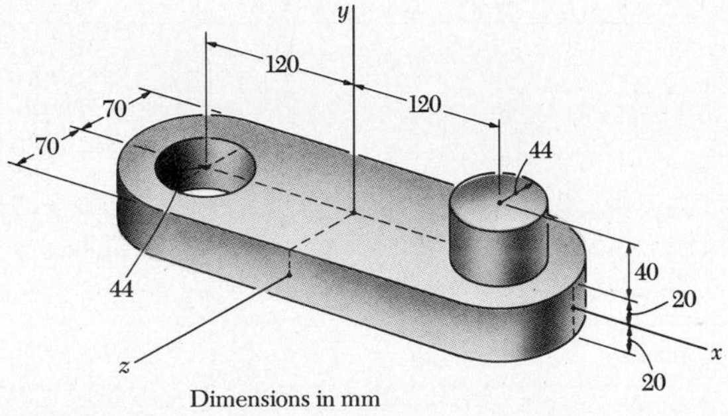

Fig. P9.120

9.121 A section of sheet steel, 2 mm thick, is cut and bent into the machine component shown. Knowing that the density of steel is 7850 kg/m^3, determine the mass products of inertia I_{xy}, I_{yz}, and I_{zx} of the component.

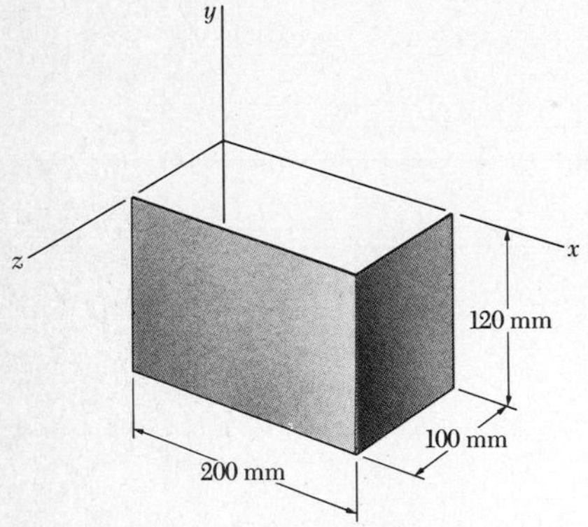

Fig. P9.121

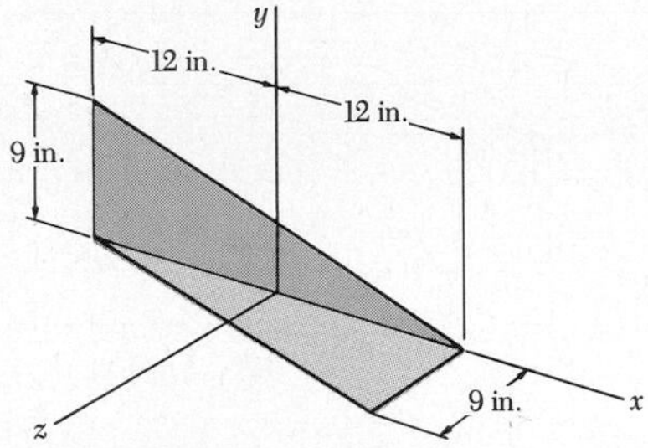

Fig. P9.122

9.122 A section of sheet steel, 0.03 in. thick, is cut and bent into the sheet-metal machine component shown. Knowing that the specific weight of steel is 490 lb/ft^3, determine the mass products of inertia I_{xy}, I_{yz}, and I_{zx} of the component. (*Hint.* See results of Sample Prob. 9.6.)

9.123 For the corner reflector of Prob. 9.111, determine the mass products of inertia I_{xy}, I_{yz}, and I_{zx}. (*Hint.* See results of Sample Prob. 9.6 and Prob. 9.56.)

9.124 A homogeneous wire with a mass per unit length of 1.5 kg/m is used to form the figure shown. Determine the mass products of inertia I_{xy}, I_{yz}, and I_{zx} of the wire figure.

9.125 Complete the derivation of Eqs. (9.47), which express the parallel-axis theorem for mass products of inertia.

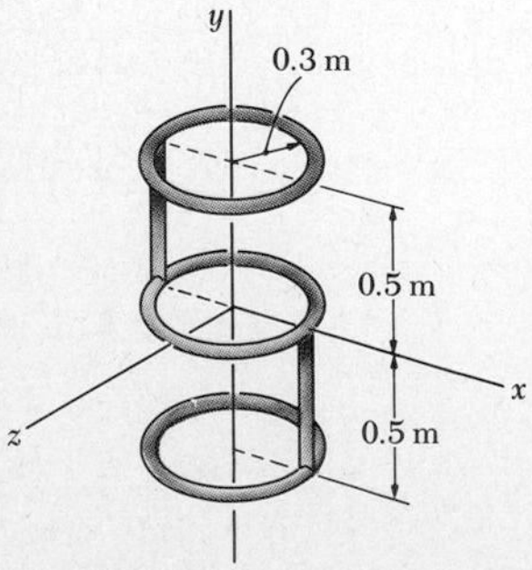

Fig. P9.124

9.126 For the homogeneous tetrahedron of mass m which is shown, (*a*) determine by direct integration the mass product of inertia I_{zx}, (*b*) deduce I_{yz} and I_{xy} from the result obtained in part *a*.

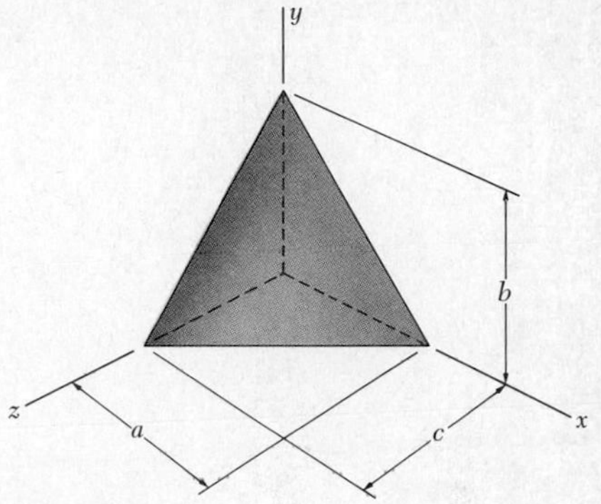

Fig. P9.126

9.127 Determine the mass moment of inertia of the right circular cone of Sample Prob. 9.11 with respect to a generator of the cone.

9.128 Determine the mass moment of inertia of the rectangular prism of Sample Prob. 9.14 with respect to the diagonal OF of its base.

9.129 Determine the mass moment of inertia of the sheet-metal machine component of Probs. 9.112 and 9.122 with respect to the axis through the origin which forms equal angles with the x, y, and z axes.

9.130 Determine the mass moment of inertia of the machine part of Sample Prob. 9.13 with respect to an axis through the origin characterized by the unit vector $\boldsymbol{\lambda} = \frac{2}{3}\mathbf{i} + \frac{1}{3}\mathbf{j} + \frac{2}{3}\mathbf{k}$.

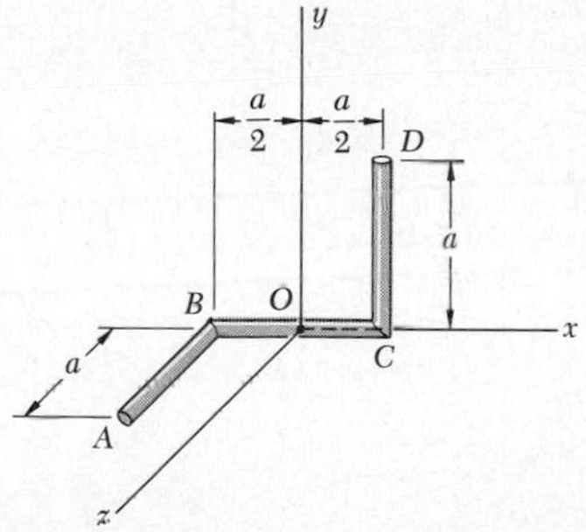

Fig. P9.131

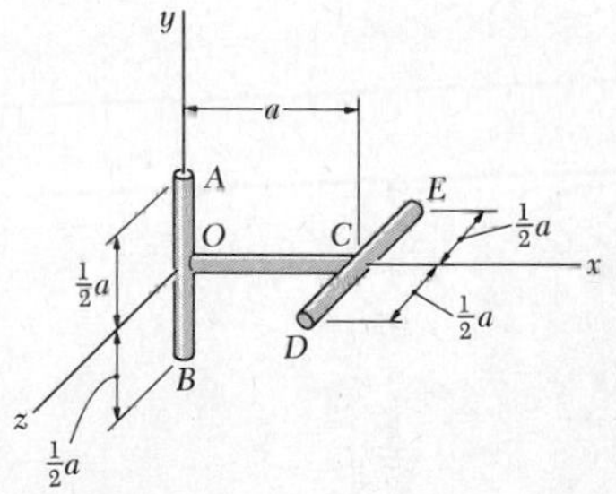

Fig. P9.132

9.131 and 9.132 Three uniform rods, each of mass m, are welded together as shown. Determine (*a*) the mass moments of inertia and the mass products of inertia with respect to the coordinate axes, (*b*) the mass moment of inertia with respect to a line joining the origin O and point D.

9.133 The thin bent plate shown is of uniform density and mass m. Determine its mass moment of inertia with respect to a line joining points B and C.

9.134 The thin bent plate shown is of uniform density and mass m. Determine its mass moment of inertia with respect to a line joining the origin O and point A.

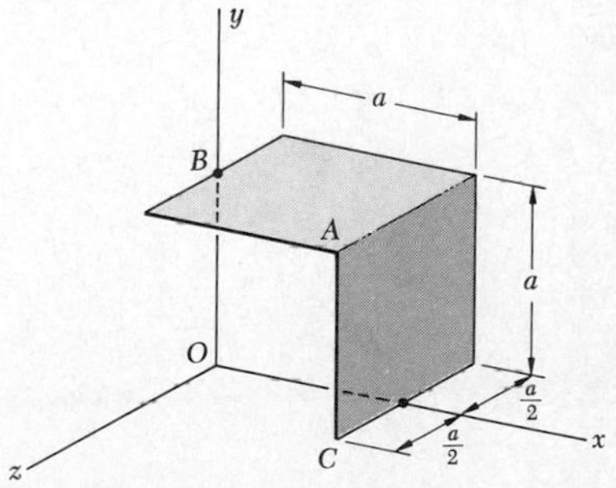

Fig. P9.133 and P9.134

9.135 Determine the value of the ratio a/h for which the ellipsoid of inertia of the right circular cone of Sample Prob. 9.11 is a sphere when computed (*a*) at the apex of the cone, (*b*) at the centroid of the cone.

9.136 Consider a homogeneous circular cylinder of radius a and length L. Determine the value of the ratio a/L for which the ellipsoid of inertia of the cylinder is a sphere when computed (*a*) at the centroid of the cylinder, (*b*) at the center of one of its bases.

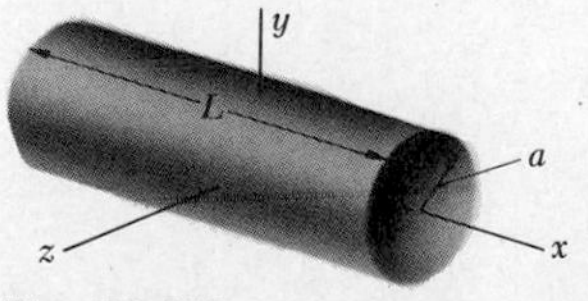

Fig. P9.136

9.137 Given an arbitrary solid and three rectangular axes x, y, and z, prove that the mass moment of inertia of the solid with respect to any one of the three axes cannot be larger than the sum of the moments of inertia of the solid with respect to the other two axes; i.e., prove that the inequality $I_x \leq I_y + I_z$ is satisfied, as well as two similar inequalities. Further prove that, if the solid is homogeneous and of revolution, and if x is the axis of revolution and y a transverse axis, then $I_y \geq \frac{1}{2}I_x$.

9.138 Consider a cube of mass m and side a. (a) Show that the ellipsoid of inertia at the center of the cube is a sphere, and use this property to determine the mass moment of inertia of the cube with respect to one of its diagonals. (b) Show that the ellipsoid of inertia at one of the corners of the cube is an ellipsoid of revolution, and determine the principal moments of inertia of the cube at that point.

9.139 Given a homogeneous solid of mass m and of arbitrary shape, and three rectangular axes x, y, and z of origin O, prove that the sum $I_x + I_y + I_z$ of the mass moments of inertia of the solid cannot be smaller than the similar sum computed for a sphere of the same mass and same material centered at O. Further prove, using the result of Prob. 9.137, that if the solid is of revolution and if x is the axis of revolution, then its moment of inertia I_y about a transverse axis y must satisfy the inequality

$$I_y \geq \tfrac{3}{10} ma^2$$

where a is the radius of the sphere of the same mass and same material.

Review Problems

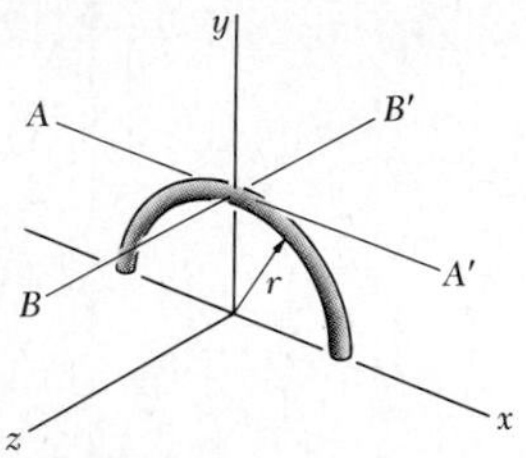

Fig. P9.140

9.140 A thin wire has been bent to form a half ring of radius r. Denoting the mass of the wire by m, determine the mass moment of inertia of the wire with respect to (a) the axis AA', (b) the axis BB'.

9.141 Determine the moment of inertia of the shaded area with respect to (a) the x axis, (b) the y axis.

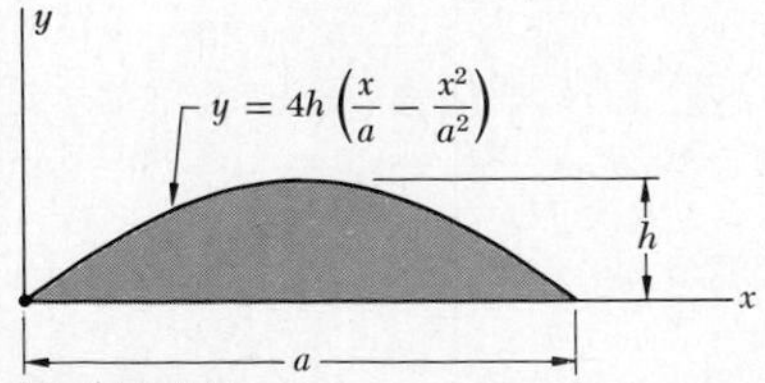

Fig. P9.141

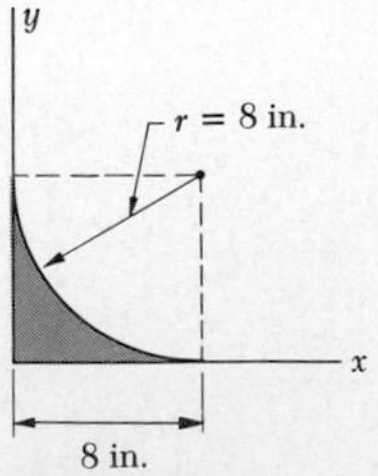

Fig. P9.142 and P9.143

9.142 Determine the moment of inertia of the shaded area with respect to the y axis.

9.143 Determine the product of inertia of the shaded area with respect to the x and y axes. (*Hint.* Use the answer to Prob. 9.56.)

9.144 Determine the centroidal moments of inertia $\bar{I}_x$ and $\bar{I}_y$ and the centroidal radii of gyration $\bar{k}_x$ and $\bar{k}_y$ for the shaded area shown.

9.145 Determine the orientation of the principal axes through the centroid C and the corresponding values of the moment of inertia for the shaded area shown.

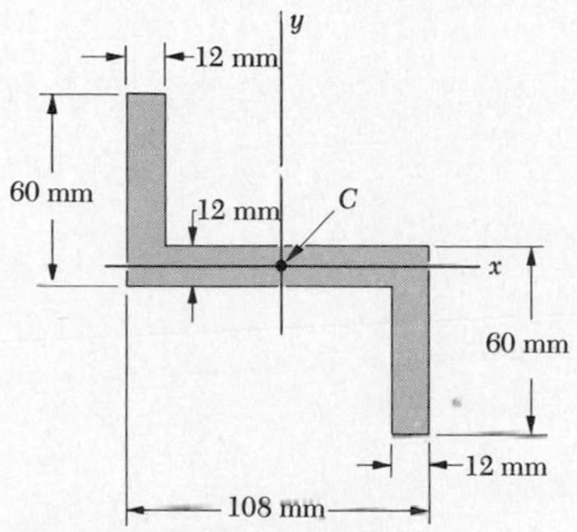

Fig. P9.144 and P9.145

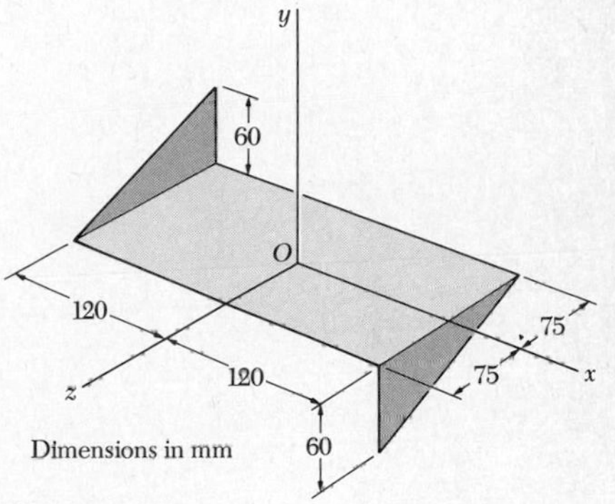

Fig. P9.146

9.146 A section of sheet steel, 2 mm thick, is cut and bent into the machine component shown. Knowing that the density of steel is 7850 kg/m^3, determine the mass moment of inertia and the radius of gyration of the component with respect to the x axis.

9.147 For the machine component of Prob. 9.146, determine the mass moment of inertia and the radius of gyration with respect to the y axis.

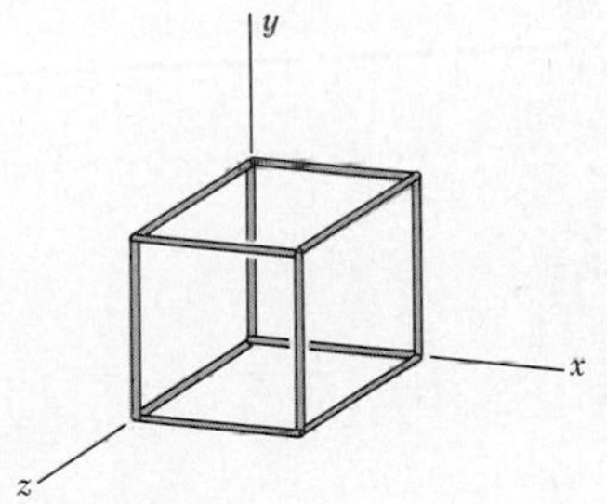

Fig. P9.148

9.148 Twelve uniform slender rods, each of length l, are welded together to form the cubical figure shown. Denoting by m the total mass of the twelve rods, determine the mass moment of inertia and the radius of gyration of the figure with respect to the x axis.

9.149 Determine the polar moment of inertia and the polar radius of gyration of an equilateral triangle of side a with respect to one of its vertices.

9.150 A thin plate of mass m is cut in the shape of a regular hexagon of side a. Determine the mass moment of inertia and the radius of gyration of the plate with respect to the axis CC' perpendicular to the plate.

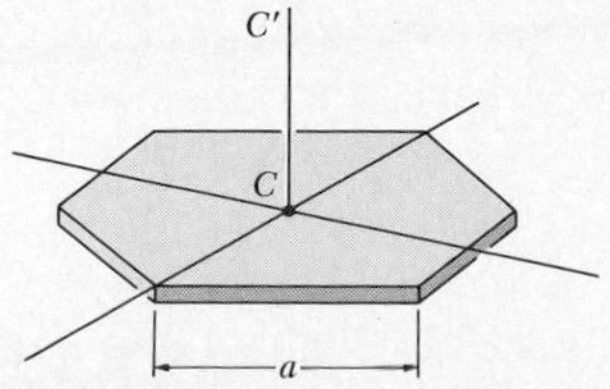

Fig. P9.150

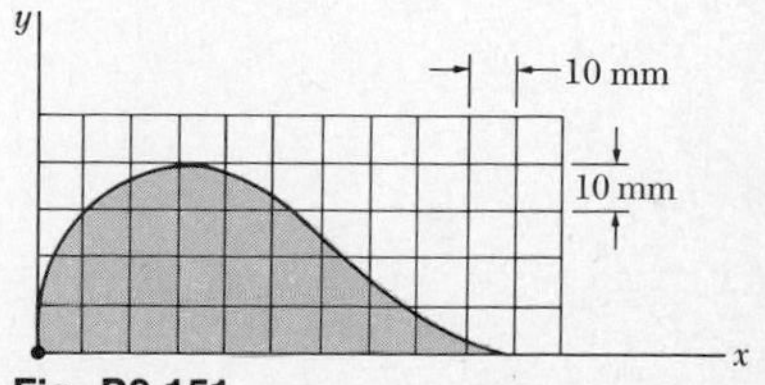

Fig. P9.151

9.151 Determine by approximate means the moment of inertia of the shaded area with respect to (*a*) the x axis, (*b*) the y axis.

CHAPTER TEN
Method of Virtual Work

***10.1. Introduction.** In the preceding chapters, problems involving the equilibrium of rigid bodies were solved by expressing that the external forces acting on the bodies were balanced. The equations of equilibrium $\Sigma F_x = 0$, $\Sigma F_y = 0$, $\Sigma M_A = 0$ were written and solved for the desired unknowns. We shall now consider a different method, which will prove more effective for solving certain types of equilibrium problems. This method is based on the *principle of virtual work* and was first formally used by the Swiss mathematician Jean Bernoulli in the eighteenth century.

As we shall see in Sec. 10.3, the principle of virtual work states that if a particle or rigid body, or, more generally, if a system of connected rigid bodies is in equilibrium under various external forces, and if the system is given an arbitrary displacement from that position of equilibrium, the total work done by the external forces during the displacement is zero. This principle may be applied particularly effectively to the solution of problems involving the equilibrium of machines or mechanisms consisting of several connected members.

In the second part of the chapter, the method of virtual work will be applied in an alternate form based on the concept of *potential energy*. As we shall see in Sec. 10.8, if a particle, rigid body, or system of rigid bodies is in equilibrium, then the derivative of its potential energy with respect to a variable defining its position must be zero.

In this chapter, we shall also learn to evaluate the mechanical efficiency of a machine (Sec. 10.5), and to determine whether a given position of equilibrium is stable, unstable, or neutral (Sec. 10.9).

***10.2. Work of a Force.** We shall first define the terms *displacement* and *work* as they are used in mechanics. Consider a particle which moves from a point A to a neighboring point A' (Fig. 10.1). If $\mathbf{r}$ denotes the position vector corresponding to point A, the small vector joining A and A' may be denoted by the differential $d\mathbf{r}$; the vector $d\mathbf{r}$ is called the *displacement* of the particle. Now let us assume that a force $\mathbf{F}$ is acting on the particle. The *work of the force* $\mathbf{F}$ *corresponding to the displacement* $d\mathbf{r}$ is defined as the quantity

$$dU = \mathbf{F} \cdot d\mathbf{r} \tag{10.1}$$

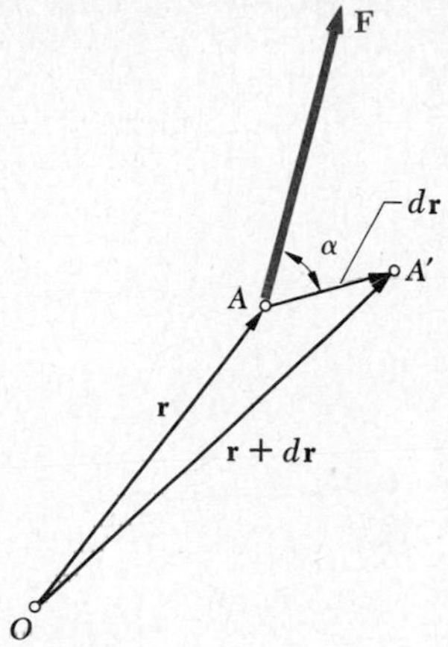

Fig. 10.1

obtained by forming the scalar product of the force $\mathbf{F}$ and of the displacement $d\mathbf{r}$. Denoting respectively by F and ds the magnitudes of the force and of the displacement, and by α the angle formed by $\mathbf{F}$ and $d\mathbf{r}$, and recalling the definition of the scalar product of two vectors (Sec. 3.9), we write

$$dU = F\,ds \cos \alpha \tag{10.1'}$$

Being a *scalar quantity*, work has a magnitude and a sign, but no direction. We also note that work should be expressed in units obtained by multiplying units of length by units of force. Thus, if U.S. customary units are used, work should be expressed in ft·lb or in·lb. If SI units are used, work should be expressed in N·m. The unit of work N·m is called a *joule* (J).†

It follows from (10.1′) that the work dU is positive if the angle α is acute and negative if α is obtuse. Three particular cases are of special interest. If the force $\mathbf{F}$ has the same direction as $d\mathbf{r}$, the work dU reduces to $F\,ds$. If $\mathbf{F}$ has a direction opposite to that of $d\mathbf{r}$, the work is $dU = -F\,ds$. Finally, if $\mathbf{F}$ is perpendicular to $d\mathbf{r}$, the work dU is zero.

The work dU of a force $\mathbf{F}$ during a displacement $d\mathbf{r}$ may also be considered as the product of F and of the component $ds \cos \alpha$ of the displacement $d\mathbf{r}$ along $\mathbf{F}$ (Fig. 10.2*a*). This view is particularly useful in the computation of the work done by the weight $\mathbf{W}$ of a body (Fig. 10.2*b*). The work of $\mathbf{W}$ is equal to the product of W and of the vertical displacement dy of the center of gravity G of the body. If the displacement is downward, the work is positive; if it is upward, the work is negative.

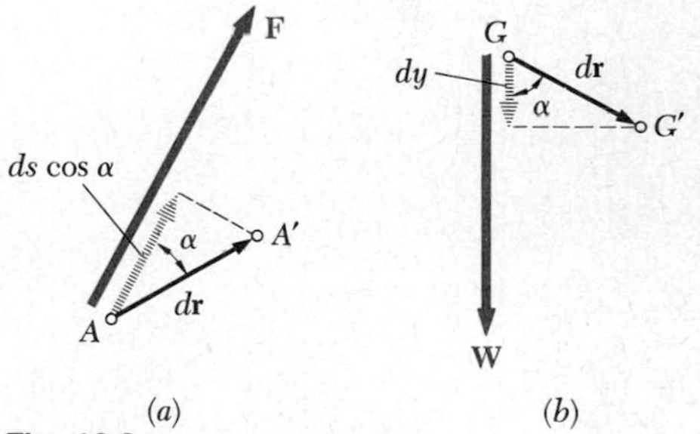

Fig. 10.2

A number of forces frequently encountered in statics *do no work*. They are forces applied to fixed points ($ds = 0$) or acting in a direction perpendicular to the displacement ($\cos \alpha = 0$). Among the forces which do no work are the following: the reaction at a frictionless pin when the body

† The joule is the SI unit of *energy*, whether in mechanical form (work, potential energy, kinetic energy) or in chemical, electrical, or thermal form. We should note that even though N·m = J, the moment of a force must be expressed in N·m, and not in joules, since the moment of a force is not a form of energy.

supported rotates about the pin, the reaction at a frictionless surface when the body in contact moves along the surface, the reaction at a roller moving along its track, the weight of a body when its center of gravity moves horizontally, the friction force acting on a wheel rolling without slipping (since at any instant the point of contact does not move). Examples of forces which *do work* are the weight of a body (except in the case considered above), the friction force acting on a body sliding on a rough surface, and most forces applied on a moving body.

In certain cases, the sum of the work done by several forces is zero. Consider, for example, two rigid bodies AC and BC connected at C by a *frictionless pin* (Fig. 10.3*a*). Among the forces acting on AC is the force **F**

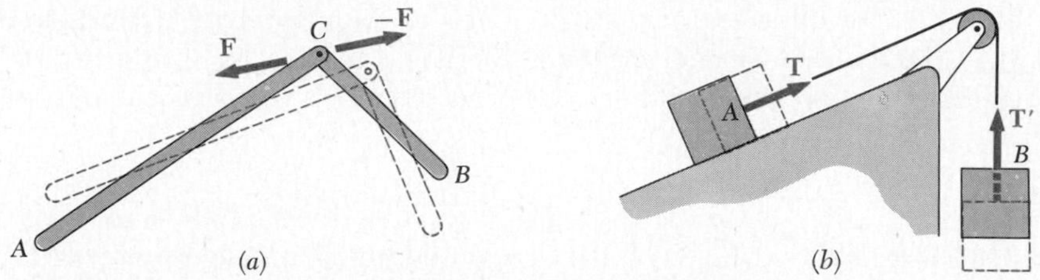

Fig. 10.3

exerted at C by BC. In general, the work of this force will not be zero, but it will be equal in magnitude and opposite in sign to the work of the force −**F** exerted by AC on BC, since these forces are equal and opposite and are applied to the same particle. Thus, when the total work done by all the forces acting on AB and BC is considered, the work of the two internal forces at C cancels out. A similar result is obtained if we consider a system consisting of two blocks connected by an *inextensible cord AB* (Fig. 10.3*b*). The work of the tension force **T** at A is equal in magnitude to the work of the tension force **T**′ at B, since these forces have the same magnitude and the points A and B move through the same distance; but in one case the work is positive, and in the other it is negative. Thus, the work of the internal forces again cancels out.

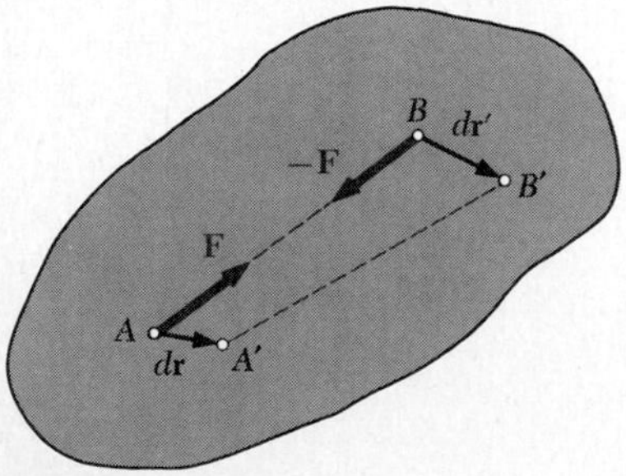

Fig. 10.4

It may be shown that the total work of the internal forces holding together the particles of a rigid body is zero. Consider two particles A and B of a rigid body and the two equal and opposite forces **F** and −**F** they exert on each other (Fig. 10.4). While, in general, small displacements $d\mathbf{r}$ and $d\mathbf{r}'$ of the two particles are different, the components of these displacements along AB must be equal; otherwise, the particles would not remain at the same distance from each other, and the body would not be rigid. Therefore, the work of **F** is equal in magnitude and opposite in sign to the work of −**F**, and their sum is zero.

In computing the work of the external forces acting on a rigid body, it is often convenient to determine the work of a couple without considering separately the work of each of the two forces forming the couple. Consider the two forces **F** and −**F** forming a couple of moment **M** and acting on a rigid body (Fig. 10.5). Any small displacement of the rigid body bringing A

and B, respectively, into A' and B'' may be divided into two parts, one in which points A and B undergo equal displacements $d\mathbf{r}_1$, the other in which A' remains fixed while B' moves into B'' through a displacement $d\mathbf{r}_2$ of magnitude $ds_2 = r\, d\theta$. In the first part of the motion, the work of **F** is equal in magnitude and opposite in sign to the work of $-\mathbf{F}$, and their sum is zero. In the second part of the motion, only force **F** works, and its work is $dU = F\, ds_2 = Fr\, d\theta$. But the product Fr is equal to the magnitude M of the moment of the couple. Thus, the work of a couple of moment **M** acting on a rigid body is

$$dU = M\, d\theta \qquad (10.2)$$

where $d\theta$ is the small angle expressed in radians through which the body rotates. We again note that work should be expressed in units obtained by multiplying units of force by units of length.

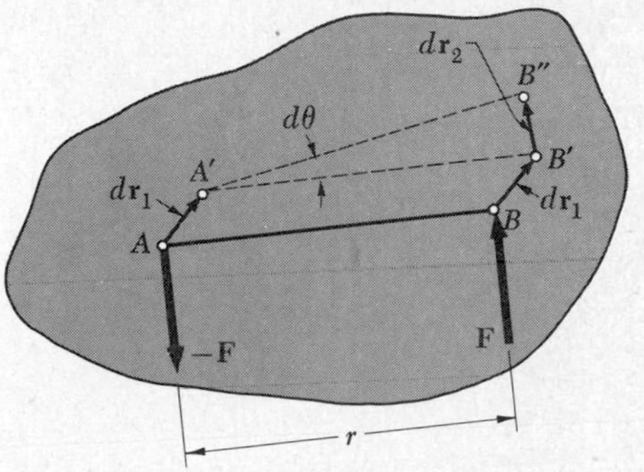

Fig. 10.5

***10.3. Principle of Virtual Work.** Consider a particle acted upon by several forces $\mathbf{F}_1, \mathbf{F}_2, \ldots, \mathbf{F}_n$ (Fig. 10.6). We shall assume that the particle undergoes a small displacement from A to A'. This displacement is possible, but it will not necessarily take place. The forces may be balanced and the particle at rest, or the particle may move under the action of the given forces in a direction different from that of AA'. The displacement considered is therefore an imaginary displacement; it is called a *virtual displacement* and is denoted by $\delta\mathbf{r}$. The symbol $\delta\mathbf{r}$ represents a differential of the first order; it is used to distinguish the virtual displacement from the displacement $d\mathbf{r}$ which would take place under actual motion. As we shall see, virtual displacements may be used to determine whether the conditions of equilibrium of a particle are satisfied.

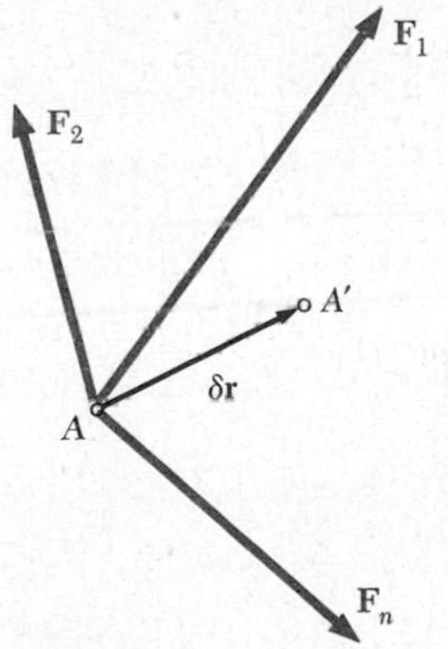

Fig. 10.6

The work of each of the forces $\mathbf{F}_1, \mathbf{F}_2, \ldots, \mathbf{F}_n$ during the virtual displacement $\delta\mathbf{r}$ is called *virtual work*. The virtual work of all the forces acting on the particle of Fig. 10.6 is

$$\begin{aligned}\delta U &= \mathbf{F}_1 \cdot \delta\mathbf{r} + \mathbf{F}_2 \cdot \delta\mathbf{r} + \cdots + \mathbf{F}_n \cdot \delta\mathbf{r} \\ &= (\mathbf{F}_1 + \mathbf{F}_2 + \cdots + \mathbf{F}_n) \cdot \delta\mathbf{r}\end{aligned}$$

or

$$\delta U = \mathbf{R} \cdot \delta\mathbf{r} \qquad (10.3)$$

where **R** is the resultant of the given forces. Thus, the total virtual work of the forces $\mathbf{F}_1, \mathbf{F}_2, \ldots, \mathbf{F}_n$ is equal to the virtual work of their resultant **R**.

The principle of virtual work for a particle states that, *if a particle is in equilibrium, the total virtual work of the forces acting on the particle is zero for any virtual displacement of the particle.* This condition is necessary: if the particle is in equilibrium, the resultant **R** of the forces is zero, and it follows from (10.3) that the total virtual work δU is zero. The condition is also sufficient: if the total virtual work δU is zero for any virtual displacement, the scalar product $\mathbf{R} \cdot \delta\mathbf{r}$ is zero for any $\delta\mathbf{r}$, and the resultant **R** must be zero.

In the case of a rigid body, the principle of virtual work states that *if a rigid body is in equilibrium, the total virtual work of the external forces acting on the rigid body is zero for any virtual displacement of the body.* The condition is necessary: if the body is in equilibrium, all the particles forming the body are in equilibrium and the total virtual work of the forces acting on all the particles must be zero; but we have seen in the preceding section that the total work of the internal forces is zero; the total work of the external forces must therefore also be zero. The condition may also be proved to be sufficient.

The principle of virtual work may be extended to the case of a *system of connected rigid bodies.* If the system remains connected during the virtual displacement, *only the work of the forces external to the system need be considered,* since the total work of the internal forces at the various connections is zero.

***10.4. Applications of the Principle of Virtual Work.** The principle of virtual work is particularly effective when applied to the solution of problems involving machines or mechanisms consisting of several connected rigid bodies. Consider, for instance, the toggle vise ACB of Fig. 10.7*a*, used to compress a wooden block. We wish to determine the

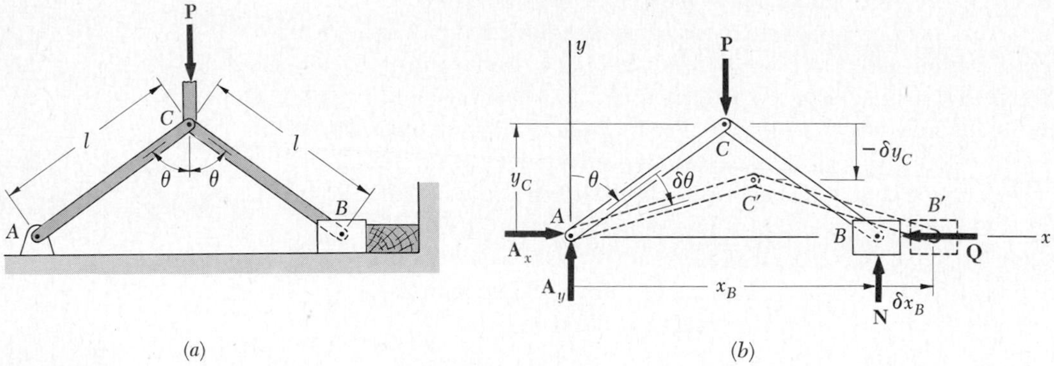

Fig. 10.7

force exerted by the vise on the block when a given force **P** is applied at C, assuming that there is no friction. Denoting by **Q** the reaction of the block on the vise, we draw the free-body diagram of the vise and consider the virtual displacement obtained by giving to the angle θ a positive increment $\delta\theta$ (Fig. 10.7*b*). Choosing a system of coordinate axes with origin at A, we note that x_B increases while y_C decreases. This is indicated in the figure by means of the positive increment δx_B and the negative increment $-\delta y_C$. The reactions $\mathbf{A}_x$, $\mathbf{A}_y$, and **N** will do no work during the virtual displacement considered, and we need only compute the work of **P** and **Q**. Since **Q** and δx_B have opposite senses, the virtual work of **Q** is $\delta U_Q = -Q\,\delta x_B$. Since **P** and the increment shown $(-\delta y_C)$ have the same sense, the virtual work of **P** is $\delta U_P = +P(-\delta y_C) = -P\,\delta y_C$. The minus signs obtained could have been predicted by simply noting that the forces **Q** and **P** are directed opposite to the positive x and y axes, respectively. Expressing the coordinates x_B and y_C in terms of the angle θ and differentiating, we obtain

$$\begin{aligned} x_B &= 2l \sin\theta & y_C &= l\cos\theta \\ \delta x_B &= 2l\cos\theta\,\delta\theta & \delta y_C &= -l\sin\theta\,\delta\theta \end{aligned} \tag{10.4}$$

The total virtual work of the forces **Q** and **P** is thus

$$\begin{aligned} \delta U &= \delta U_Q + \delta U_P = -Q\,\delta x_B - P\,\delta y_C \\ &= -2Ql\cos\theta\,\delta\theta + Pl\sin\theta\,\delta\theta \end{aligned}$$

Making $\delta U = 0$, we obtain

$$2Ql\cos\theta\,\delta\theta = Pl\sin\theta\,\delta\theta \tag{10.5}$$

$$Q = \tfrac{1}{2}P\tan\theta \tag{10.6}$$

The superiority of the method of virtual work over the conventional equilibrium equations in the problem considered here is clear: by using the method of virtual work, we were able to eliminate all unknown reactions, while the equation $\Sigma M_A = 0$ would have eliminated only two of the unknown reactions. We may take advantage of this characteristic of the method of virtual work to solve many problems involving machines and mechanisms. *If the virtual displacement considered is consistent with the constraints imposed by the supports and connections, all reactions and internal forces are eliminated and only the work of the loads, applied forces, and friction forces need be considered.*

We shall observe that the method of virtual work may also be used to solve problems involving completely constrained structures, although the virtual displacements considered will never actually take place. Consider, for example, the frame ACB shown in Fig. 10.8*a*. If point A is kept fixed, while B is given a horizontal virtual displacement (Fig. 10.8*b*), we need consider only the work of **P** and $\mathbf{B}_x$. We may thus determine the reaction

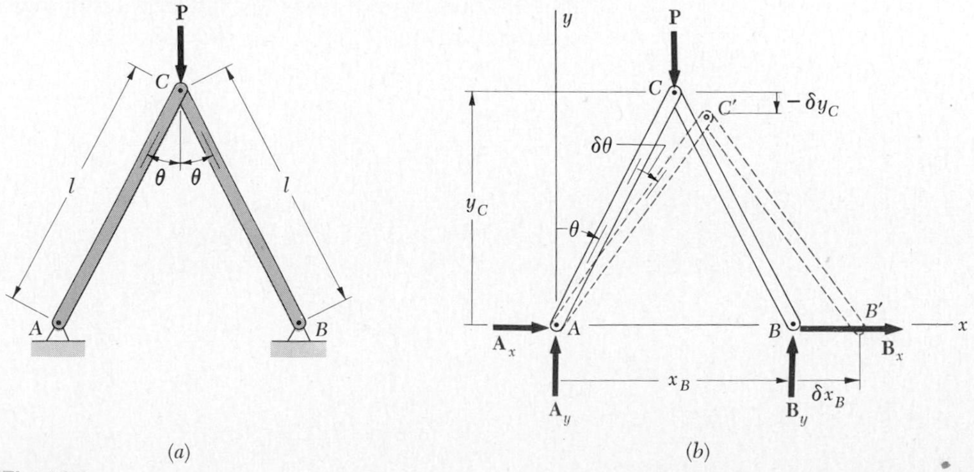

Fig. 10.8

component $\mathbf{B}_x$ in the same way as the force $\mathbf{Q}$ of the preceding example (Fig. 10.7*b*); we have

$$B_x = -\tfrac{1}{2}P \tan\theta$$

Keeping B fixed and giving to A a horizontal virtual displacement, we may similarly determine the reaction component $\mathbf{A}_x$. The components $\mathbf{A}_y$ and $\mathbf{B}_y$ may be determined by rotating the frame ACB as a rigid body about B and A, respectively.

The method of virtual work may also be used to determine the configuration of a system in equilibrium under given forces. For example, the value of the angle θ for which the linkage of Fig. 10.7 is in equilibrium under two given forces $\mathbf{P}$ and $\mathbf{Q}$ may be obtained by solving Eq. (10.6) for $\tan\theta$.

It should be noted, however, that the attractiveness of the method of virtual work depends to a large extent upon the existence of simple geometric relations between the various virtual displacements involved in the solution of a given problem. When no such simple relations exist, it is usually advisable to revert to the conventional method of Chap. 6.

* 10.5. Real Machines. Mechanical Efficiency.

In analyzing the toggle vise in the preceding section, we assumed that no friction forces were involved. Thus, the virtual work consisted only of the work of the applied force $\mathbf{P}$ and of the reaction $\mathbf{Q}$. But the work of the reaction $\mathbf{Q}$ is equal in magnitude and opposite in sign to the work of the force exerted by the vise on the block. Equation (10.5), therefore, expresses that the *output work* $2Ql \cos\theta\, \delta\theta$ is equal to the *input work* $Pl \sin\theta\, \delta\theta$. A machine in which input and output work are equal is said to be an "ideal" machine. In a "real" machine, friction forces will always do some work, and the output work will be smaller than the input work.

Consider, for example, the toggle vise of Fig. 10.7*a*, and assume now that a friction force $\mathbf{F}$ develops between the sliding block B and the horizontal plane (Fig. 10.9). Using the conventional methods of statics and

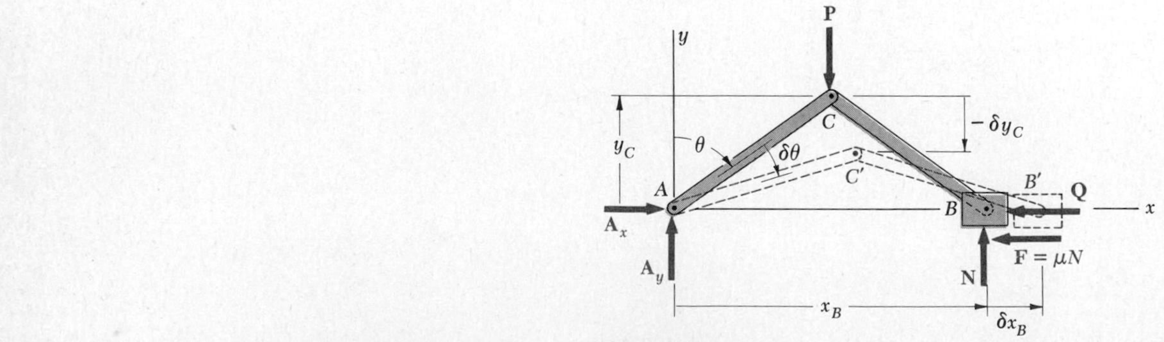

Fig. 10.9

summing moments about A, we find $N = P/2$. Denoting by μ the coefficient of friction between block B and the horizontal plane, we have $F = \mu N = \mu P/2$. Recalling formulas (10.4), we find that the total virtual work of the forces **Q**, **P**, and **F** during the virtual displacement shown in Fig. 10.9 is

$$\begin{aligned}\delta U &= -Q\,\delta x_B - P\,\delta y_C - F\,\delta x_B \\ &= -2Ql\cos\theta\,\delta\theta + Pl\sin\theta\,\delta\theta - \mu Pl\cos\theta\,\delta\theta\end{aligned}$$

Making $\delta U = 0$, we obtain

$$2Ql\cos\theta\,\delta\theta = Pl\sin\theta\,\delta\theta - \mu Pl\cos\theta\,\delta\theta \qquad (10.7)$$

which expresses that the output work is equal to the input work minus the work of the friction force. Solving for Q, we have

$$Q = \tfrac{1}{2}P(\tan\theta - \mu) \qquad (10.8)$$

We note that $Q = 0$ when $\tan\theta = \mu$, that is, when θ is equal to the angle of friction ϕ, and that $Q < 0$ when $\theta < \phi$. The toggle vise may thus be used only for values θ larger than the angle of friction.

The *mechanical efficiency* of a machine is defined as the ratio

$$\eta = \frac{\text{output work}}{\text{input work}} \qquad (10.9)$$

Clearly, the mechanical efficiency of an ideal machine is $\eta = 1$, since input and output work are then equal, while the mechanical efficiency of a real machine will always be less than 1.

In the case of the toggle vise we have just analyzed, we write

$$\eta = \frac{\text{output work}}{\text{input work}} = \frac{2Ql\cos\theta\,\delta\theta}{Pl\sin\theta\,\delta\theta}$$

Substituting from (10.8) for Q, we obtain

$$\eta = \frac{P(\tan\theta - \mu)l\cos\theta\,\delta\theta}{Pl\sin\theta\,\delta\theta} = 1 - \mu\cot\theta \qquad (10.10)$$

We check that in the absence of friction forces, we would have $\mu = 0$ and $\eta = 1$. In the general case, when μ is different from zero, the efficiency η becomes zero for $\mu\cot\theta = 1$, that is, for $\tan\theta = \mu$, or $\theta = \tan^{-1}\mu = \phi$. We check again that the toggle vise may be used only for values of θ larger than the angle of friction ϕ.

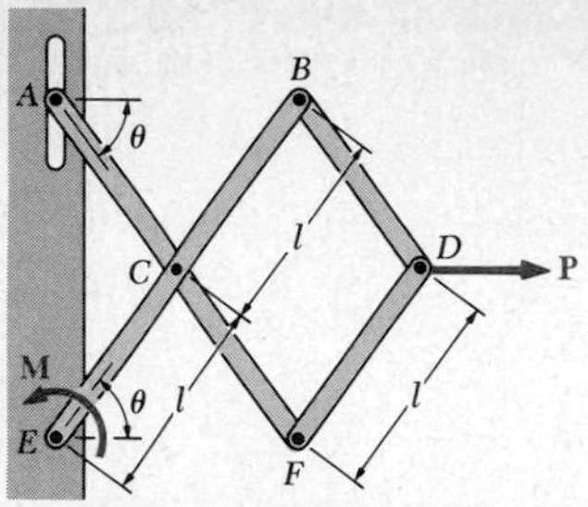

SAMPLE PROBLEM 10.1

Using the method of virtual work, determine the magnitude of the couple **M** required to maintain the equilibrium of the mechanism shown.

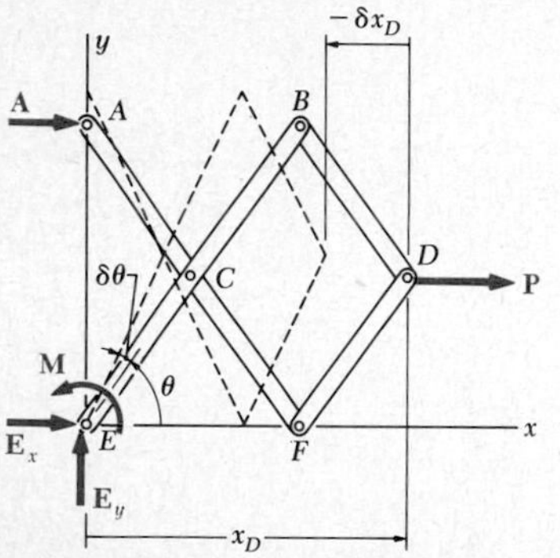

Solution. Choosing a coordinate system with origin at E, we write

$$x_D = 3l \cos \theta \qquad \delta x_D = -3l \sin \theta \, \delta\theta$$

Principle of Virtual Work. Since the reactions **A**, $\mathbf{E}_x$, and $\mathbf{E}_y$ will do no work during the virtual displacement, the total virtual work done by **M** and **P** must be zero. Noting that **P** acts in the positive x direction and **M** acts in the positive θ direction, we write

$$\delta U = 0: \qquad +M\,\delta\theta + P\,\delta x_D = 0$$
$$+M\,\delta\theta + P(-3l \sin\theta\,\delta\theta) = 0$$

$$M = 3Pl \sin \theta \quad ◄$$

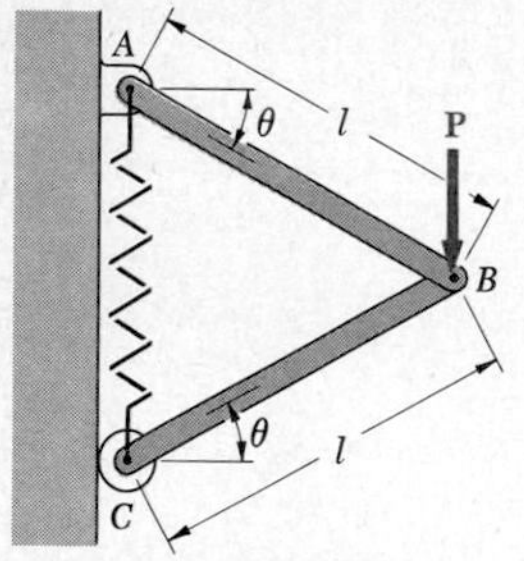

SAMPLE PROBLEM 10.2

Determine the expressions for θ and for the tension in the spring which correspond to the equilibrium position of the mechanism. The unstretched length of the spring is h, and the constant of the spring is k. Neglect the weight of the mechanism.

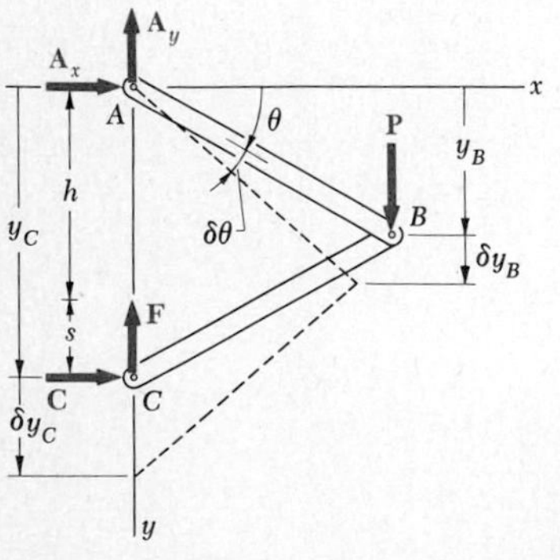

Solution. With the coordinate system shown

$$y_B = l \sin\theta \qquad y_C = 2l \sin\theta$$
$$\delta y_B = l \cos\theta\,\delta\theta \qquad \delta y_C = 2l \cos\theta\,\delta\theta$$

The elongation of the spring is $\quad s = y_C - h = 2l \sin\theta - h$

The magnitude of the force exerted at C by the spring is

$$F = ks = k(2l \sin\theta - h) \tag{1}$$

Principle of Virtual Work. Since the reactions $\mathbf{A}_x$, $\mathbf{A}_y$, and **C** do no work, the total virtual work done by **P** and **F** must be zero.

$$\delta U = 0: \qquad P\,\delta y_B - F\,\delta y_C = 0$$
$$P(l \cos\theta\,\delta\theta) - k(2l \sin\theta - h)(2l \cos\theta\,\delta\theta) = 0$$

$$\sin\theta = \frac{P + 2kh}{4kl} \quad ◄$$

Substituting this expression into (1), we obtain $\quad F = \frac{1}{2}P \quad ◄$

SAMPLE PROBLEM 10.3

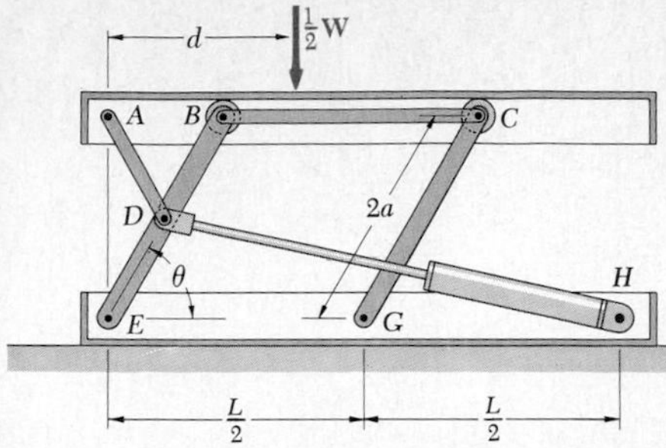

A hydraulic-lift table is used to raise a 1000-kg crate. It consists of a platform and of two identical linkages on which hydraulic cylinders exert equal forces. (Only one linkage and one cylinder are shown.) Members *EDB* and *CG* are each of length $2a$, and member *AD* is pinned to the midpoint of *EDB*. If the crate is placed on the table, so that half of its weight is supported by the system shown, determine the force exerted by each cylinder in raising the crate for $\theta = 60°$, $a = 0.70$ m, and $L = 3.20$ m. This mechanism has been previously considered in Sample Prob. 6.7.

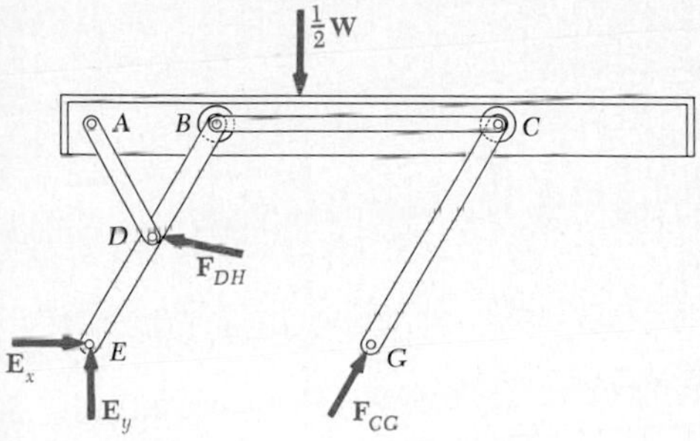

Solution. The machine considered consists of the platform and of the linkage, with an input force $\mathbf{F}_{DH}$ exerted by the cylinder and an output force equal and opposite to $\frac{1}{2}\mathbf{W}$.

Principle of Virtual Work. We first observe that the reactions at *E* and *G* do no work. Denoting by y the elevation of the platform above the base, and by s the length *DH* of the cylinder-and-piston assembly, we write

$$\delta U = 0: \qquad -\tfrac{1}{2}W\,\delta y + F_{DH}\,\delta s = 0 \qquad (1)$$

The vertical displacement δy of the platform is expressed in terms of the angular displacement $\delta\theta$ of *EDB* as follows:

$$y = (EB)\sin\theta = 2a\sin\theta$$
$$\delta y = 2a\cos\theta\,\delta\theta$$

To express δs similarly in terms of $\delta\theta$, we first note that by the law of cosines,

$$s^2 = a^2 + L^2 - 2aL\cos\theta$$

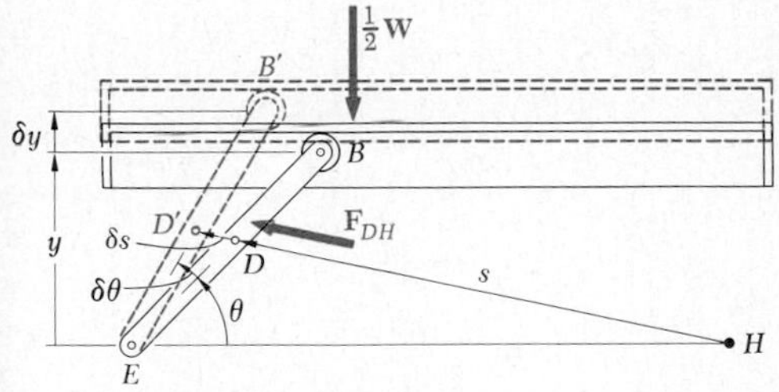

Differentiating,

$$2s\,\delta s = -2aL(-\sin\theta)\,\delta\theta$$

$$\delta s = \frac{aL\sin\theta}{s}\,\delta\theta$$

Substituting for δy and δs into (1), we write

$$(-\tfrac{1}{2}W)2a\cos\theta\,\delta\theta + F_{DH}\frac{aL\sin\theta}{s}\,\delta\theta = 0$$

$$F_{DH} = W\frac{s}{L}\cot\theta$$

With the given numerical data, we have

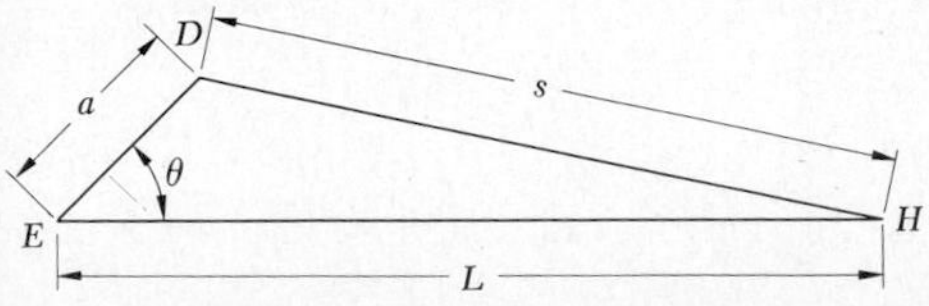

$$W = mg = (1000 \text{ kg})(9.81 \text{ m/s}^2) = 9810 \text{ N} = 9.81 \text{ kN}$$
$$s^2 = a^2 + L^2 - 2aL\cos\theta$$
$$= (0.70)^2 + (3.20)^2 - 2(0.70)(3.20)\cos 60° = 8.49$$
$$s = 2.91 \text{ m}$$

$$F_{DH} = W\frac{s}{L}\cot\theta = (9.81 \text{ kN})\frac{2.91 \text{ m}}{3.20 \text{ m}}\cot 60°$$

$$F_{DH} = 5.15 \text{ kN} \blacktriangleleft$$

Problems

10.1 Determine the mass m which balances the 5-kg block.

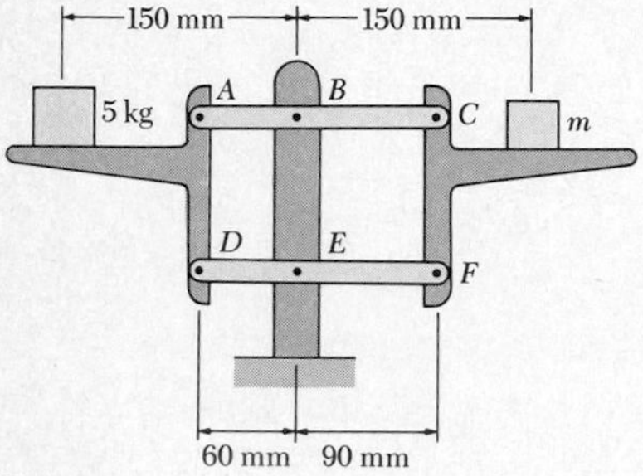

Fig. P10.1

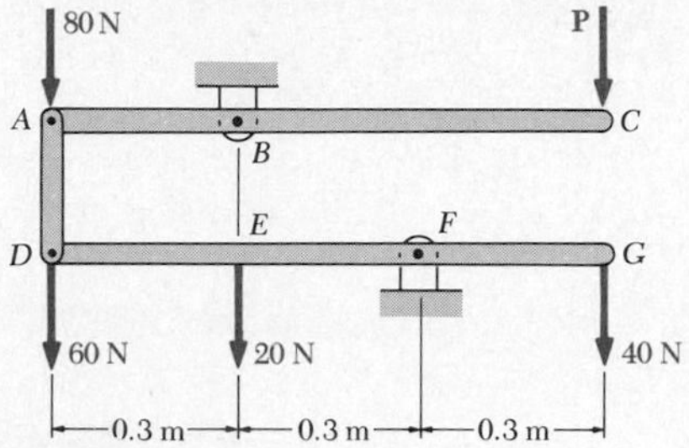

Fig. P10.2

10.2 Determine the magnitude of the force **P** required to maintain the equilibrium of the linkage shown.

10.3 Determine the horizontal force **P** which must be applied at A to maintain the equilibrium of the linkage.

10.4 Determine the couple **M** which must be applied to member ABC to maintain the equilibrium of the linkage.

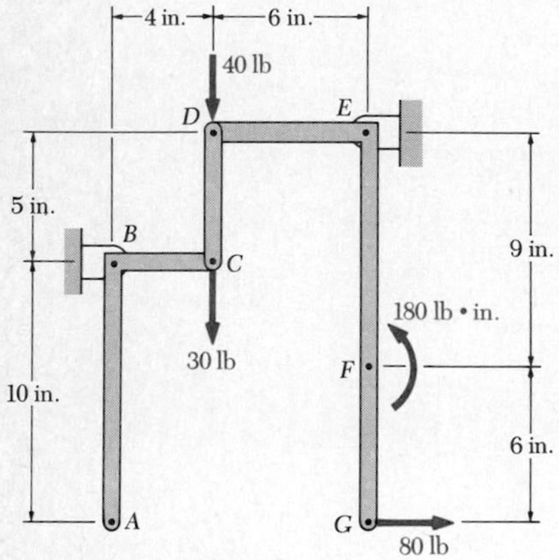

Fig. P10.3 and P10.4

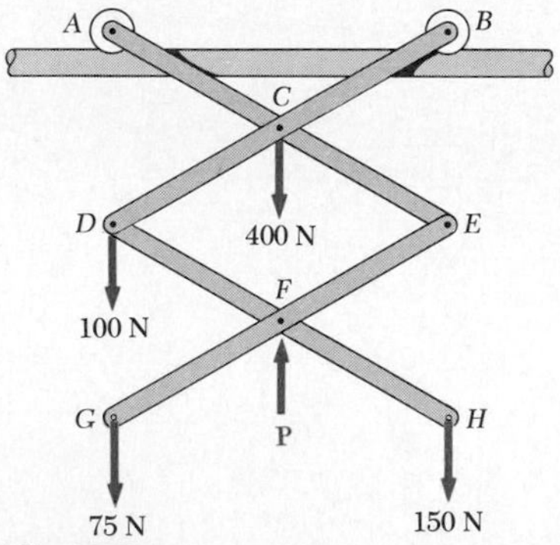

Fig. P10.5

10.5 Determine the force **P** required to maintain the equilibrium of the linkage shown. All members are of the same length and the wheels at A and B roll freely on the horizontal rod.

10.6 Solve Prob. 10.5, assuming that the vertical force **P** is applied at point E.

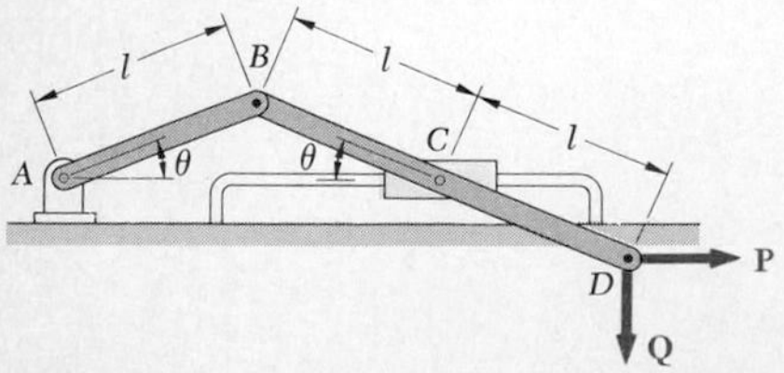

Fig. P10.7

10.7 The mechanism shown is acted upon by the force **P**; derive an expression for the magnitude of the force **Q** required to maintain equilibrium.

10.8 The mechanism shown is acted upon by the force **P**; derive an expression for the magnitude of the force **Q** required to maintain equilibrium.

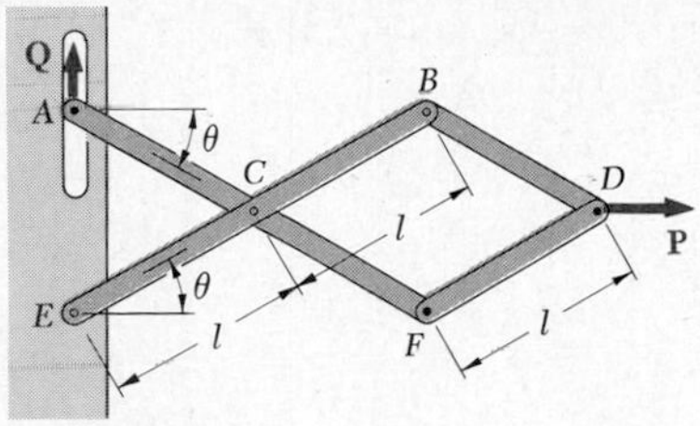

Fig. P10.8

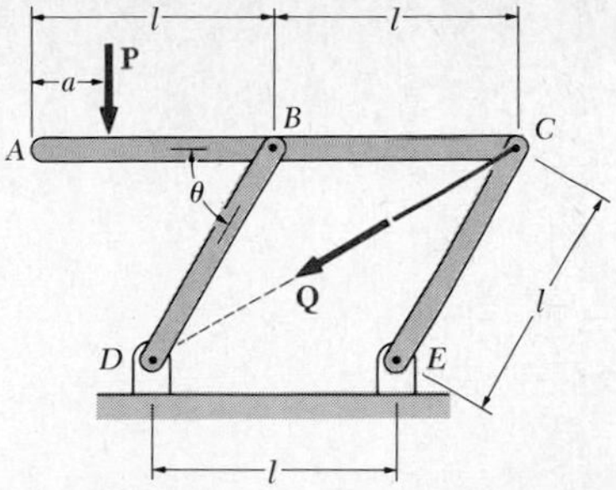

Fig. P10.9

10.9 Knowing that the line of action of the force **Q** passes through point *D*, derive an expression for the magnitude of **Q** required to maintain equilibrium.

10.10 Rod *AD* is acted upon by a vertical force **P** at end *A* and by two equal and opposite horizontal forces of magnitude *Q* at points *B* and *C*. Derive an expression for the magnitude *Q* of the horizontal forces required for equilibrium.

10.11 The slender rod *AB* is attached to a collar at *A* and rests on a wheel at *C*. Neglecting the effect of friction, derive an expression for the magnitude of the force **Q** required to maintain equilibrium.

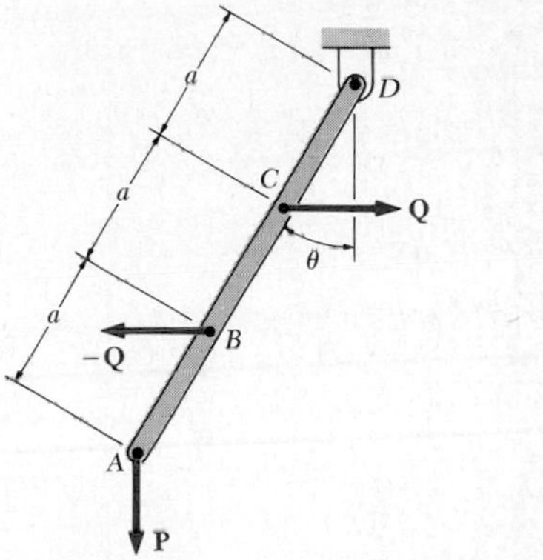

Fig. P10.10

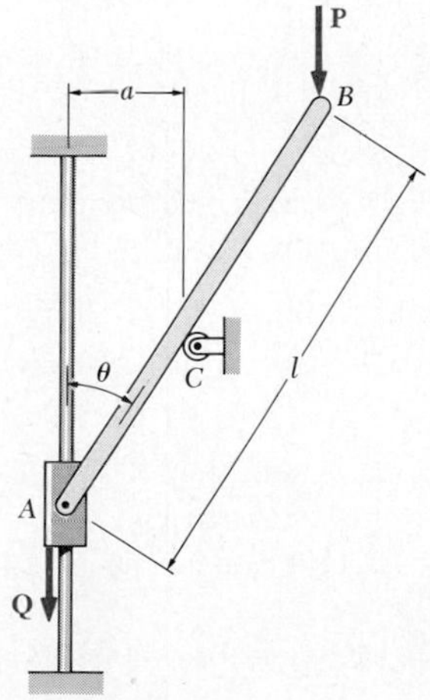

Fig. P10.11

10.12 Solve Prob. 10.11, assuming that the force **P** applied at *B* is horizontal and directed to the right.

10.13 Each of the uniform rods *AB* and *BC* is of weight *W*. Derive an expression for the magnitude of the couple **M** required to maintain equilibrium.

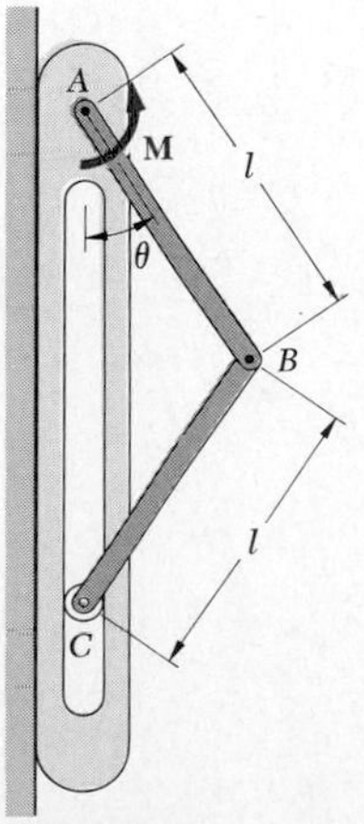

Fig. P10.13

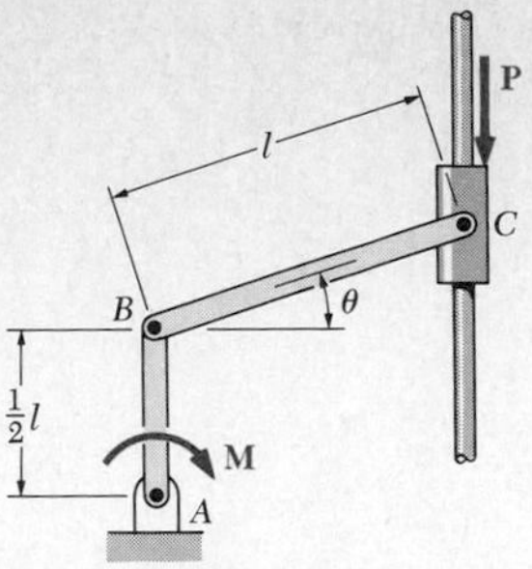

Fig. P10.14

10.14 Derive an expression for the magnitude of the couple **M** required to maintain the equilibrium of the linkage in the position shown.

10.15 and 10.16 Knowing that rod AB is of length $2l$, derive an expression for the magnitude of the couple **M** required to maintain equilibrium.

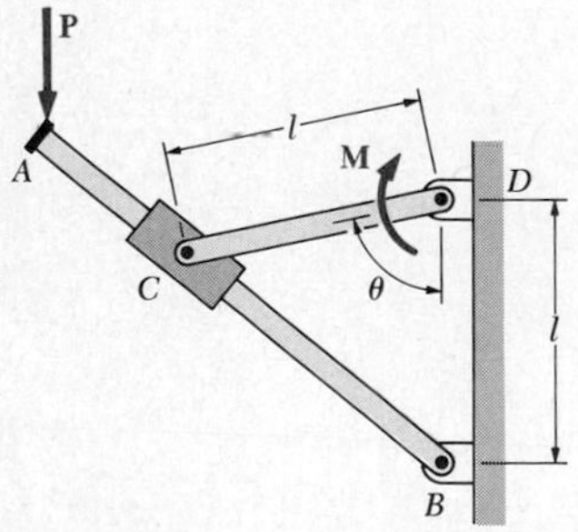

Fig. P10.15

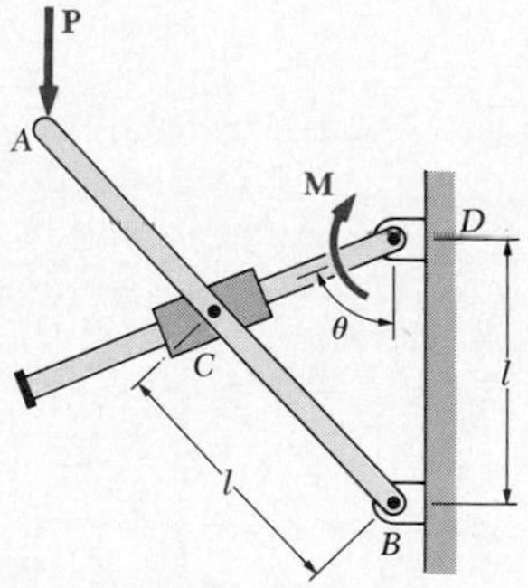

Fig. P10.16

10.17 Solve Prob. 10.15, assuming that the force **P** applied at A is horizontal and directed to the left.

10.18 Solve Prob. 10.16, assuming that the force **P** applied at A is horizontal and directed to the left.

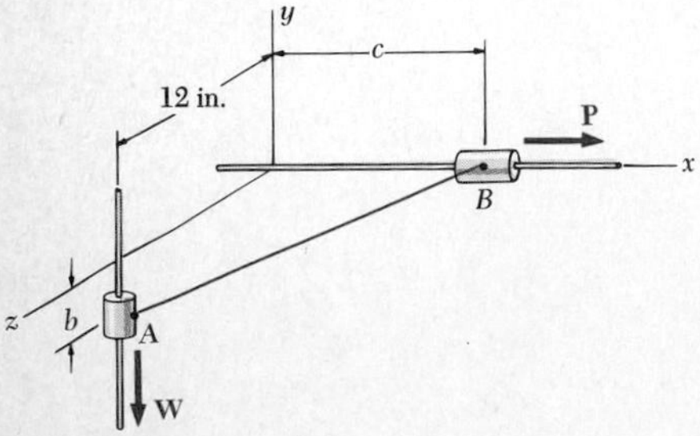

Fig. P10.19

10.19 Collar A weighs 12.6 lb and may slide freely on the frictionless vertical rod. Knowing that the length of the wire AB is 22 in., determine the magnitude of **P** required for equilibrium when (*a*) $c = 4$ in., (*b*) $c = 12$ in.

10.20 Solve Prob. 10.19 when (*a*) $c = 14$ in., (*b*) $c = 18$ in.

10.21 In Prob. 10.9, determine the magnitude of the force **Q** required for equilibrium when $l = 500$ mm, $\theta = 70°$, and $P = 350$ N.

10.22 In Prob. 10.14, determine the magnitude of the couple **M** required for equilibrium when $l = 500$ mm, $\theta = 35°$ and, $P = 500$ N.

10.23 Determine the value of θ corresponding to the equilibrium position of the mechanism of Prob. 10.7 when $P = 45$ lb and $Q = 60$ lb.

10.24 Determine the value of θ corresponding to the equilibrium position of the mechanism of Prob. 10.8 when $P = 50$ lb and $Q = 160$ lb.

10.25 Determine the value of θ corresponding to the equilibrium position of the mechanism of Prob. 10.11 when $P = 200$ N, $Q = 400$ N, $l = 600$ mm, and $a = 100$ mm.

10.26 Determine the value of θ corresponding to the equilibrium position of the mechanism of Prob. 10.10 when $P = 500$ N, $Q = 400$ N, and $a = 100$ mm.

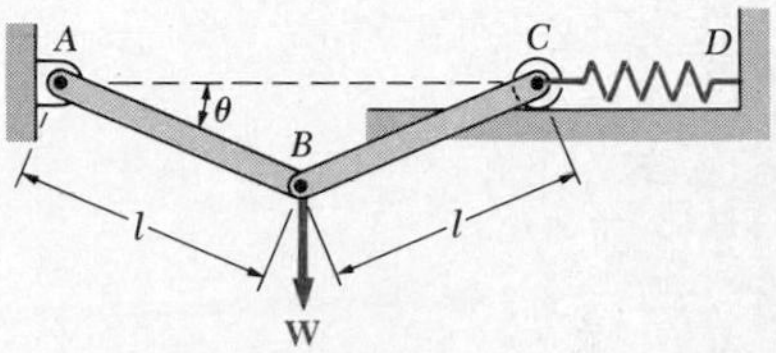

Fig. P10.27 and P10.28

10.27 A vertical load **W** is applied to the linkage at B. The constant of the spring is k, and the spring is unstretched when AB and BC are horizontal. Neglecting the weight of the linkage, derive an equation in θ, W, l, and k, which must be satisfied when the linkage is in equilibrium.

10.28 A load **W** of magnitude 900 N is applied to the linkage at B. Neglecting the weight of the linkage and knowing that $l = 225$ mm, determine the value of θ corresponding to equilibrium. The constant of the spring is $k = 2$ kN/m, and the spring is unstretched when AB and BC are horizontal. (*Hint.* Solve the equation obtained for θ by trial.)

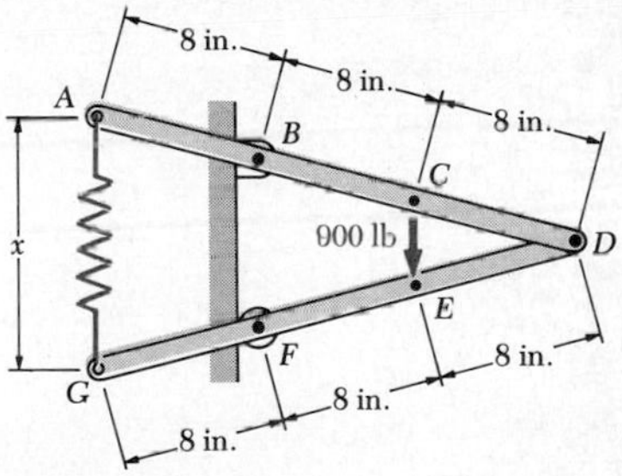

Fig. P10.29

10.29 Two bars AD and DG are connected by a pin at D and by a spring AG. Knowing that the spring is 12 in. long when unstretched and that the constant of the spring is 125 lb/in., determine the value of x corresponding to equilibrium when a 900-lb load is applied at E as shown.

10.30 Solve Prob. 10.29, assuming that the 900-lb vertical force is applied at C instead of E.

10.31 A load **W** of magnitude 80 lb is applied to the mechanism at C. Neglecting the weight of the mechanism, determine the value of θ corresponding to equilibrium. The constant of the spring is $k = 20$ lb/in., and the spring is unstretched when $\theta = 0$.

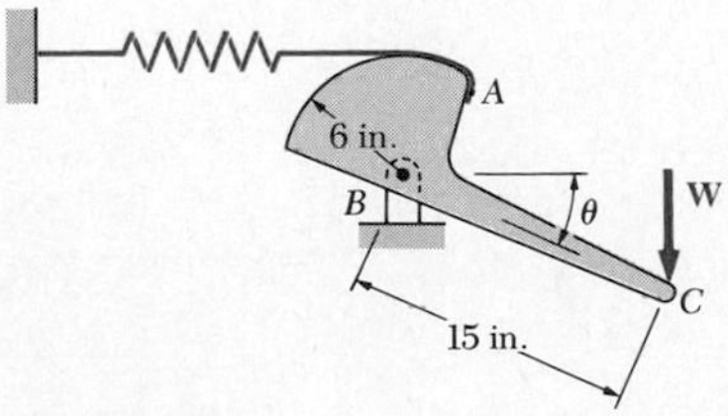

Fig. P10.31

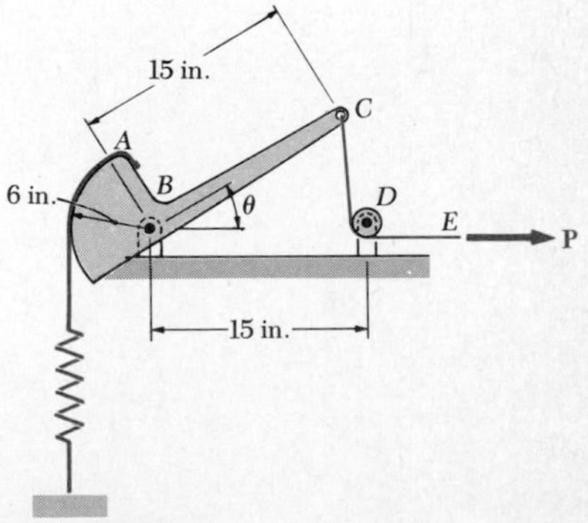

Fig. P10.32

10.32 A force **P** of magnitude 60 lb is applied to end E of cable CDE, which passes under pulley D and is attached to the mechanism at C. Neglecting the weight of the mechanism and the radius of the pulley, determine the value of θ corresponding to equilibrium. The constant of the spring is $k = 20$ lb/in., and the spring is unstretched when $\theta = 90°$.

10.33 The lever AB is attached to the horizontal shaft BC which passes through a bearing and is welded to a fixed support at C. The torsional spring constant of the shaft BC is K; that is, a couple of magnitude K is required to rotate end B through one radian. Knowing that the shaft is untwisted when AB is horizontal, determine the value of θ corresponding to the position of equilibrium, if $P = 120$ N, $l = 300$ mm, and $K = 16$ N·m/rad.

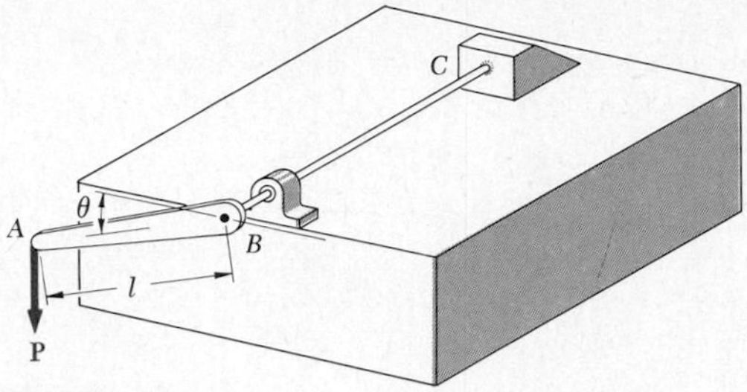

Fig. P10.33

10.34 Solve Prob. 10.33, if $P = 400$ N, $l = 300$ mm, and $K = 16$ N·m/rad. Obtain an answer in each of the following quadrants: $0 < \theta < 90°$, $270° < \theta < 360°$, and $360° < \theta < 450°$. (It is assumed that K remains constant for the angles of twist considered.)

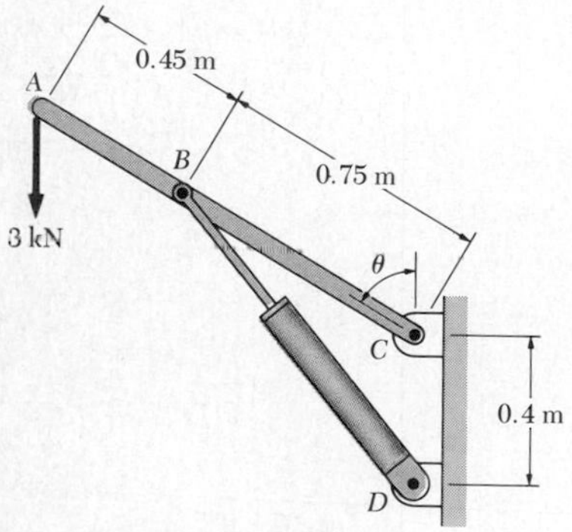

Fig. P10.35 and P10.36

10.35 The position of boom ABC is controlled by the hydraulic cylinder BD. For the loading shown, determine the force exerted by the hydraulic cylinder on pin B when $\theta = 60°$.

10.36 The position of boom ABC is controlled by the hydraulic cylinder BD. For the loading shown, (*a*) express the force exerted by the hydraulic cylinder on pin B as a function of the length BD, (*b*) determine the largest possible value of the angle θ if the maximum force that the cylinder can exert on pin B is 12.5 kN.

10.37 The position of member ABC is controlled by the hydraulic cylinder CD. For the loading shown, determine the force exerted by the hydraulic cylinder on pin C when $\theta = 65°$.

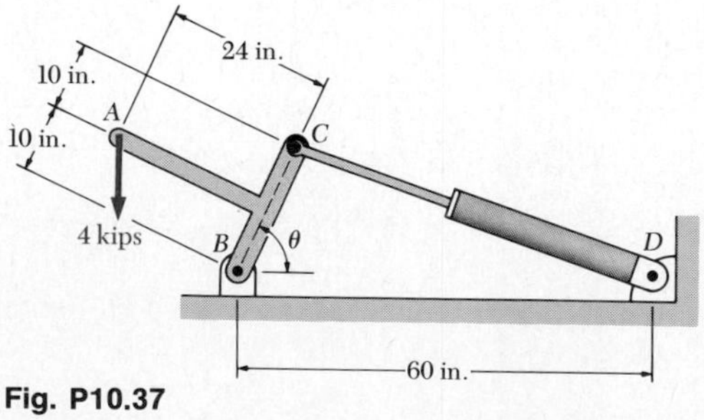

Fig. P10.37

10.38 Solve Prob. 10.37 when $\theta = 115°$.

10.39 A block of weight W is pulled up a plane forming an angle α with the horizontal by a force **P** directed along the plane. If μ is the coefficient of friction between the block and the plane, derive an expression for the mechanical efficiency of the system. Show that the mechanical efficiency cannot exceed $\frac{1}{2}$ if the block is to remain in place when the force **P** is removed.

10.40 Derive an expression for the mechanical efficiency of the jack discussed in Sec. 8.6. Show that if the jack is to be self-locking, the mechanical efficiency cannot exceed $\frac{1}{2}$.

10.41 Denoting by μ the coefficient of friction between collar C and the vertical rod, derive an expression for the magnitude of the largest couple **M** for which equilibrium is maintained in the position shown. Explain what happens if $\mu \geq \tan\theta$.

10.42 Knowing that the coefficient of friction between collar C and the vertical rod is 0.30, determine the magnitude of the largest and smallest couple **M** for which equilibrium is maintained in the position shown, when $\theta = 30°$, $l = 250$ mm, and $P = 400$ N.

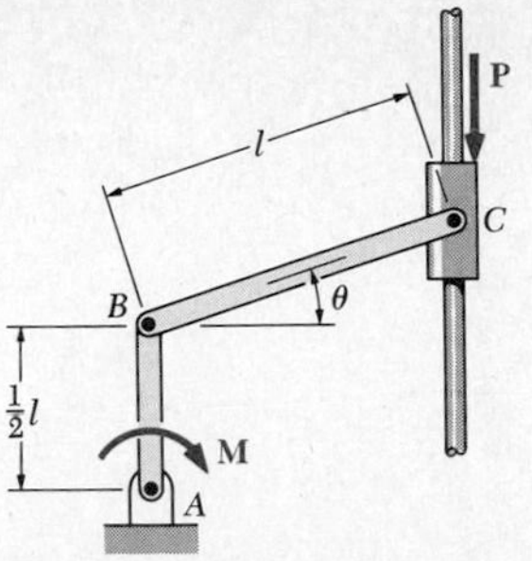

Fig. P10.41 and P10.42

10.43 Using the method of virtual work, determine the reaction at E.

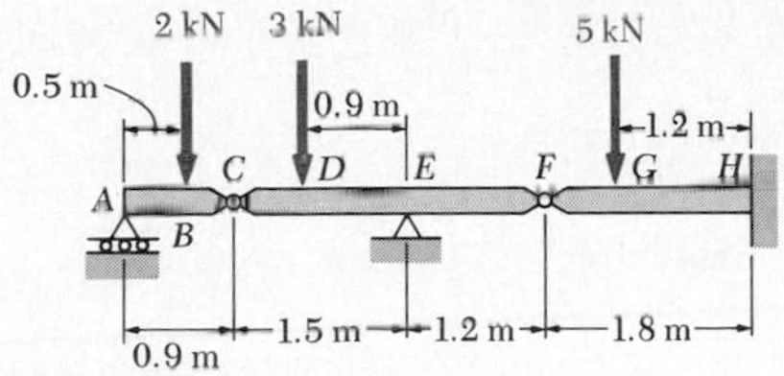

Fig. P10.43 and P10.44

10.44 Using the method of virtual work, determine separately the force and couple representing the reaction at H.

10.45 In Prob. 10.5 the force **P** is removed and the linkage is maintained in equilibrium by a cord which is attached to pins E and H. Determine the tension in the cord.

10.46 A slender rod AB, of weight W, is attached to blocks A and B which move freely in the guides shown. The blocks are connected by an elastic cord which passes over a pulley at C. (*a*) Express the tension in the cord in terms of W and θ. (*b*) Determine the value of θ for which the tension in the cord is equal to $2W$.

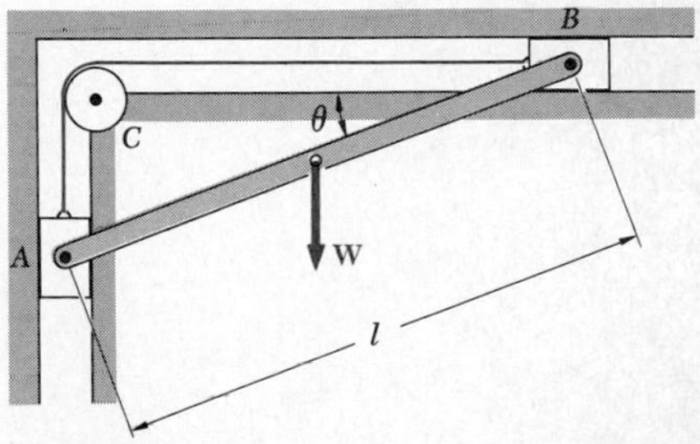

Fig. P10.46

10.47 Determine the vertical movement of joint C if member FG is lengthened by 1.5 in. (*Hint.* Apply a vertical load at joint C, and using the methods of Chap. 6, compute the force exerted by member FG on joints F and G. Then apply the method of virtual work for a virtual displacement making member FG longer. This method should be used only for small changes in the length of members.)

10.48 Determine the horizontal movement of joint C if member FG is lengthened by 1.5 in. (Follow hint of Prob. 10.47, applying a *horizontal* force at joint C.)

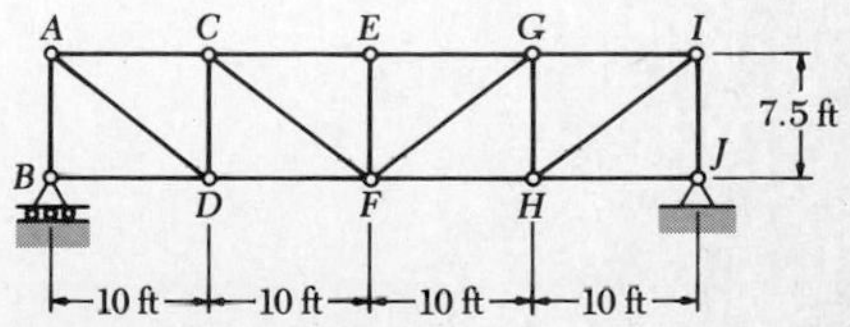

Fig. P10.47 and P10.48

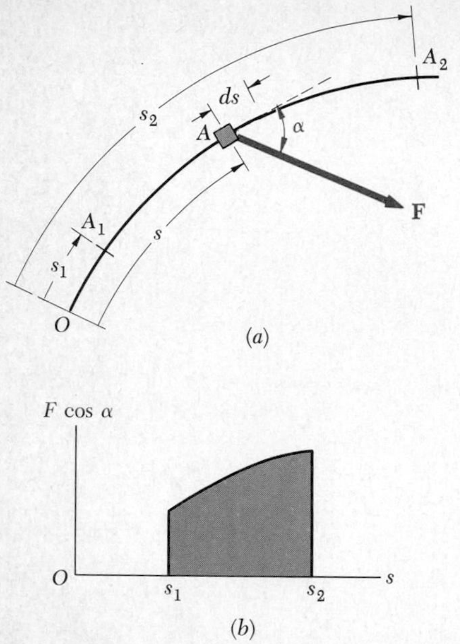

Fig. 10.10

* 10.6. Work of a Force during a Finite Displacement.

Consider a force **F** acting on a particle. The work of **F** corresponding to an infinitesimal displacement $d\mathbf{r}$ of the particle was defined in Sec. 10.2 as

$$dU = \mathbf{F} \cdot d\mathbf{r} \tag{10.1}$$

The work of **F** corresponding to a finite displacement of the particle from A_1 to A_2 (Fig. 10.10*a*) is denoted by $U_{1\to2}$ and is obtained by integrating (10.1) along the curve described by the particle:

$$U_{1\to2} = \int_{A_1}^{A_2} \mathbf{F} \cdot d\mathbf{r} \tag{10.11}$$

Using the alternate expression

$$dU = F\,ds \cos\alpha \tag{10.1'}$$

given in Sec. 10.2 for the elementary work dU, we may also express the work $U_{1\to2}$ as

$$U_{1\to2} = \int_{s_1}^{s_2} (F\cos\alpha)\,ds \tag{10.11'}$$

where the variable of integration s measures the distance along the path traveled by the particle. The work $U_{1\to2}$ is represented by the area under the curve obtained by plotting $F \cos\alpha$ against s (Fig. 10.10*b*). In the case of a force **F** of constant magnitude acting in the direction of motion, formula (10.11′) yields $U_{1\to2} = F(s_2 - s_1)$.

Recalling from Sec. 10.2 that the work of a couple of moment **M** during an infinitesimal rotation $d\theta$ of a rigid body is

$$dU = M\,d\theta \tag{10.2}$$

we express as follows the work of the couple during a finite rotation of the body:

$$U_{1\to2} = \int_{\theta_1}^{\theta_2} M\,d\theta \tag{10.12}$$

In the case of a constant couple, formula (10.12) yields

$$U_{1\to2} = M(\theta_2 - \theta_1)$$

Work of a Weight. It was stated in Sec. 10.2 that the work of the weight **W** of a body during an infinitesimal displacement of the body is equal to the product of W and of the vertical displacement of the center of gravity of the body. With the y axis pointing upward, the work of **W** during a finite displacement of the body (Fig. 10.11) is obtained by writing

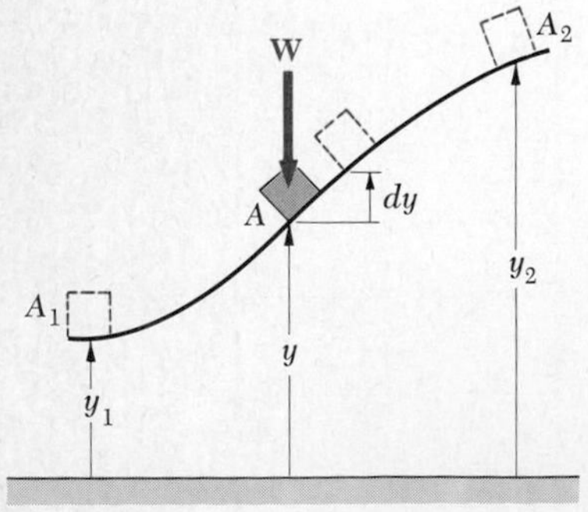

Fig. 10.11

$$dU = -W\,dy$$

Integrating from A_1 to A_2, we have

$$U_{1\to2} = -\int_{y_1}^{y_2} W\,dy = Wy_1 - Wy_2 \tag{10.13}$$

or

$$U_{1\to2} = -W(y_2 - y_1) = -W\,\Delta y \tag{10.13'}$$

where Δy is the vertical displacement from A_1 to A_2. The work of the weight **W** is thus equal to *the product of W and of the vertical displacement of the center of gravity of the body.* The work is *positive* when $\Delta y < 0$, that is, *when the body moves down.*

Work of the Force Exerted by a Spring. Consider a body A attached to a fixed point B by a spring; it is assumed that the spring is undeformed when the body is at A_0 (Fig. 10.12*a*). Experimental evidence shows that the magnitude of the force **F** exerted by the spring on a body A is proportional to the deflection x of the spring measured from the position A_0. We have

$$F = kx \tag{10.14}$$

where k is the *spring constant,* expressed in N/m if SI units are used, and in lb/ft or lb/in. if U.S. customary units are used. The work of the force **F** exerted by the spring during a finite displacement of the body from $A_1(x = x_1)$ to $A_2(x = x_2)$ is obtained by writing

$$dU = -F\,dx = -kx\,dx$$

$$U_{1\to2} = -\int_{x_1}^{x_2} kx\,dx = \tfrac{1}{2}kx_1^2 - \tfrac{1}{2}kx_2^2 \tag{10.15}$$

Care should be taken to express k and x in consistent units. For example, if U.S. customary units are used, k should be expressed in lb/ft and x in feet, or k in lb/in. and x in inches; in the first case, the work is obtained in ft·lb; in the second case, in in·lb. We note that the work of the force **F** exerted by the spring on the body is *positive* when $x_2 < x_1$, that is, *when the spring is returning to its undeformed position.*

Since Eq. (10.14) is the equation of a straight line of slope k passing through the origin, the work $U_{1\to2}$ of **F** during the displacement from A_1 to A_2 may be obtained by evaluating the area of the trapezoid shown in Fig. 10.12*b*. This is done by computing the values F_1 and F_2 and multiplying the base Δx of the trapezoid by its mean height $\frac{1}{2}(F_1 + F_2)$. Since the work of the force **F** exerted by the spring is positive for a negative value of Δx, we write

$$U_{1\to2} = -\tfrac{1}{2}(F_1 + F_2)\,\Delta x \tag{10.16}$$

Formula (10.16) is usually more convenient to use than (10.15) and affords fewer chances of confusing the units involved.

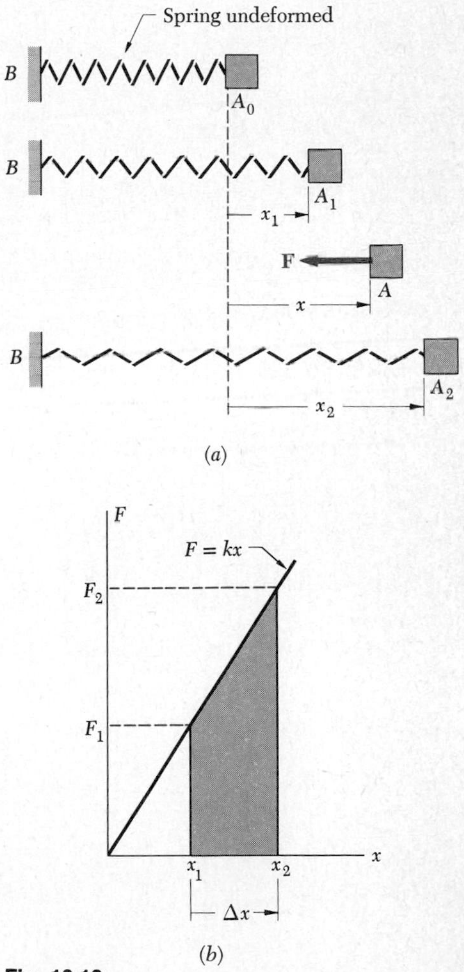

Fig. 10.12

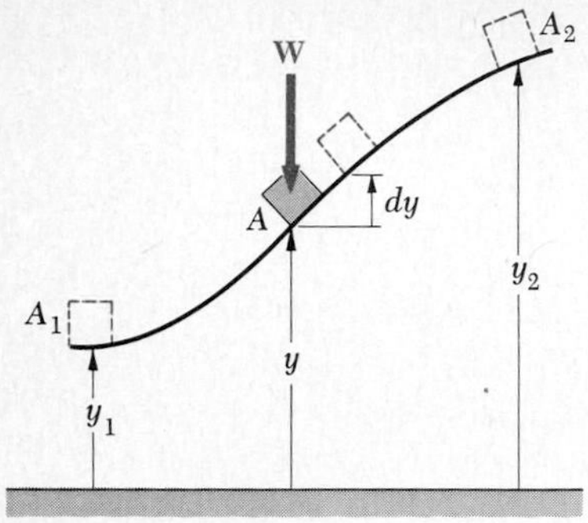

Fig. 10.11 ***(repeated)***

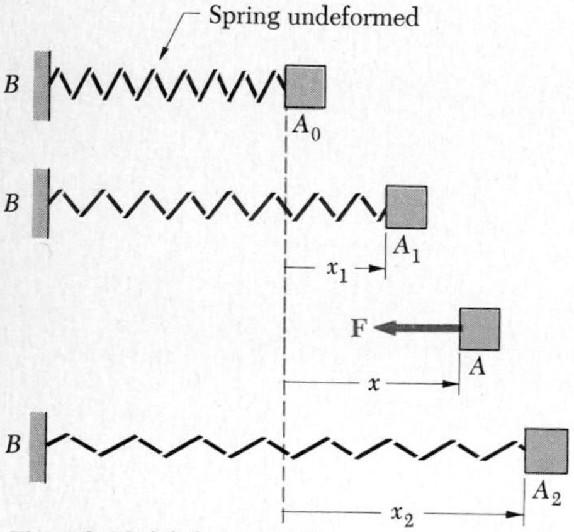

Fig. 10.12 (*a*) ***(repeated)***

***10.7. Potential Energy.** Considering again the body of Fig. 10.11, we note from (10.13) that the work of the weight **W** during a finite displacement is obtained by subtracting the value of the function Wy corresponding to the second position of the body from its value corresponding to the first position. The work of **W** is thus independent of the actual path followed; it depends only upon the initial and final values of the function Wy. This function is called the *potential energy* of the body with respect to the *force of gravity* **W** and is denoted by V_g. We write

$$U_{1\rightarrow 2} = (V_g)_1 - (V_g)_2 \qquad \text{with } V_g = Wy \tag{10.17}$$

We note that if $(V_g)_2 > (V_g)_1$, that is, *if the potential energy increases* during the displacement (as in the case considered here), *the work* $U_{1\rightarrow 2}$ *is negative.* If, on the other hand, the work of **W** is positive, the potential energy decreases. Therefore, the potential energy V_g of the body provides a measure of *the work which may be done* by its weight **W**. Since only the *change* in potential energy, and not the actual value of V_g, is involved in formula (10.17), an arbitrary constant may be added to the expression obtained for V_g. In other words, the level from which the elevation y is measured may be chosen arbitrarily. Note that potential energy is expressed in the same units as work, i.e., in joules (J) if SI units are used,† and in ft·lb or in·lb if U.S. customary units are used.

Considering now the body of Fig. 10.12*a*, we note from formula (10.15) that the work of the elastic force **F** is obtained by subtracting the value of the function $\frac{1}{2}kx^2$ corresponding to the second position of the body from its value corresponding to the first position. This function is denoted by V_e and is called the *potential energy* of the body with respect to the *elastic force* **F**. We write

$$U_{1\rightarrow 2} = (V_e)_1 - (V_e)_2 \qquad \text{with } V_e = \tfrac{1}{2}kx^2 \tag{10.18}$$

and observe that during the displacement considered, the work of the force **F** exerted by the spring on the body is negative and the potential energy V_e increases. We should note that the expression obtained for V_e is valid only if the deflection of the spring is measured from its undeformed position.

The concept of potential energy may be used when forces other than gravity forces and elastic forces are involved. It remains valid as long as the elementary work dU of the force considered is an *exact differential.* It is then possible to find a function V, called potential energy, such that

$$dU = -dV \tag{10.19}$$

Integrating (10.19) over a finite displacement, we obtain the general formula

$$U_{1\rightarrow 2} = V_1 - V_2 \tag{10.20}$$

†See footnote, page 389.

which expresses that *the work of the force is independent of the path followed and is equal to minus the change in potential energy.* A force which satisfies Eq. (10.20) is said to be a *conservative force.*†

*** 10.8. Potential Energy and Equilibrium.** The application of the principle of virtual work is considerably simplified when the potential energy of a system is known. In the case of a virtual displacement, formula (10.19) becomes $\delta U = -\delta V$. Besides, if the position of the system is defined by a single independent variable θ, we may write $\delta V = (dV/d\theta)\,\delta\theta$. Since $\delta\theta$ must be different from zero, the condition $\delta U = 0$ for the equilibrium of the system becomes

$$\frac{dV}{d\theta} = 0 \tag{10.21}$$

In terms of potential energy, the principle of virtual work states therefore that *if a system is in equilibrium, the derivative of its total potential energy is zero.* If the position of the system depends upon several independent variables (the system is then said to possess *several degrees of freedom*), the partial derivatives of V with respect to each of the independent variables must be zero.

Consider, for example, a structure made of two members AC and CB and carrying a load W at C. The structure is supported by a pin at A and a roller at B, and a spring BD connects B to a fixed point D (Fig. 10.13*a*). The constant of the spring is k, and it is assumed that the natural length of the spring is equal to AD and thus that the spring is undeformed when B coincides with A. Neglecting the friction forces and the weight of the members, we find that the only forces which work during a displacement of the structure are the weight $\mathbf{W}$ and the force $\mathbf{F}$ exerted by the spring at point B (Fig. 10.13*b*). The total potential energy of the system will thus be obtained by adding the potential energy V_g corresponding to the gravity force $\mathbf{W}$ and the potential energy V_e corresponding to the elastic force $\mathbf{F}$.

Choosing a coordinate system with origin at A and noting that the deflection of the spring, measured from its undeformed position, is $AB = x_B$, we write

$$V_e = \tfrac{1}{2}kx_B^2 \qquad V_g = Wy_C$$

Expressing the coordinates x_B and y_C in terms of the angle θ, we have

$$\begin{aligned} x_B &= 2l\sin\theta & y_C &= l\cos\theta \\ V_e &= \tfrac{1}{2}k(2l\sin\theta)^2 & V_g &= W(l\cos\theta) \\ V = V_e + V_g &= 2kl^2\sin^2\theta + Wl\cos\theta \end{aligned} \tag{10.22}$$

The positions of equilibrium of the system are obtained by equating to zero the derivative of the potential energy V. We write

$$\frac{dV}{d\theta} = 4kl^2\sin\theta\cos\theta - Wl\sin\theta = 0$$

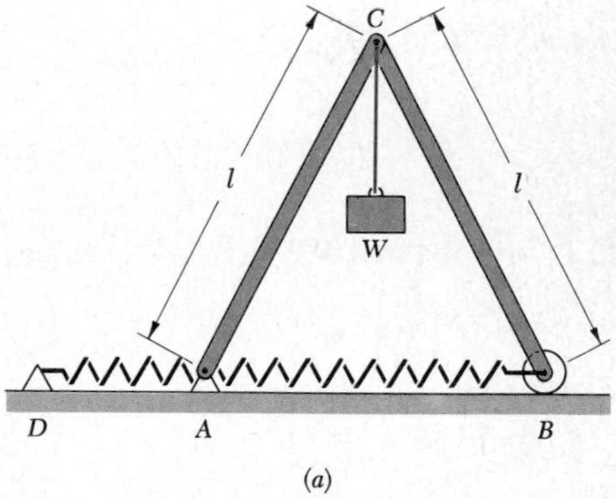

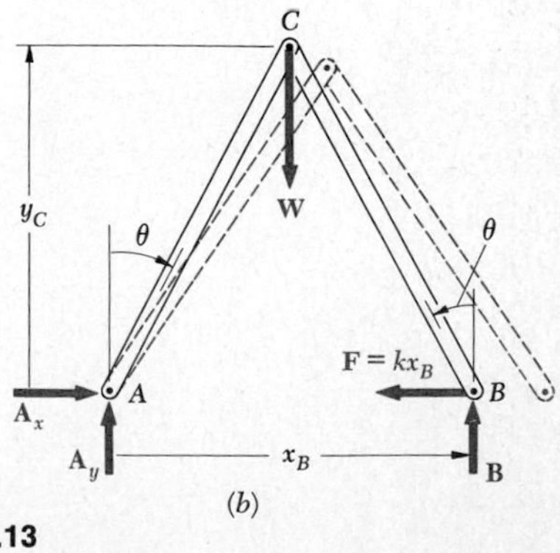

Fig. 10.13

†A detailed discussion of conservative forces is given in Sec. 13.7 of *Dynamics*.

or, factoring $l \sin\theta$,

$$\frac{dV}{d\theta} = l \sin\theta(4kl \cos\theta - W) = 0$$

There are therefore two positions of equilibrium, corresponding respectively to the values $\theta = 0$ and $\theta = \cos^{-1}(W/4kl)$.†

*** 10.9. Stability of Equilibrium.** Consider the three uniform rods of length $2a$ and weight **W** shown in Fig. 10.14. While each rod is in equilibrium, there is an important difference between the three cases considered. Suppose that each rod is slightly disturbed from its position of equilibrium and then released: rod a will move back toward its original position, rod b will keep moving away from its original position, and rod c will remain in its new position. In case a, the equilibrium of the rod is said to be *stable;* in case b, the equilibrium is said to be *unstable;* and, in case c, to be *neutral.*

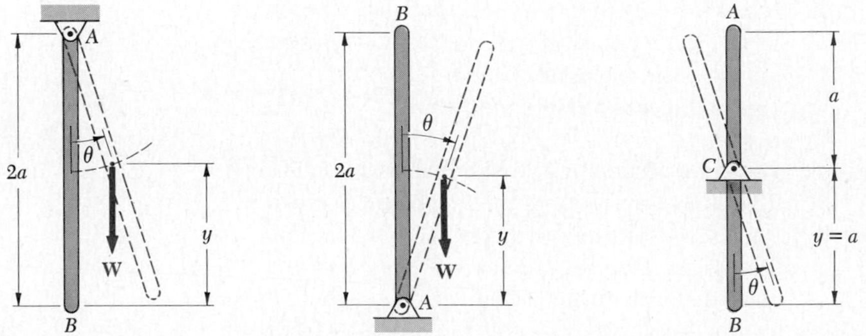

Fig. 10.14 (a) Stable equilibrium (b) Unstable equilibrium (c) Neutral equilibrium

Recalling from Sec. 10.7 that the potential energy V_g with respect to gravity is equal to Wy, where y is the elevation of the point of application of **W** measured from an arbitrary level, we observe that the potential energy of rod a is minimum in the position of equilibrium considered, that the potential energy of rod b is maximum, and that the potential energy of rod c is constant. Equilibrium is thus *stable, unstable,* or *neutral* according to whether the potential energy is *minimum, maximum,* or *constant* (Fig. 10.15).

That the result obtained is quite general may be seen as follows: We first observe that a force always tends to do positive work and thus to decrease the potential energy of the system on which it is applied. Therefore, when a system is disturbed from its position of equilibrium, the forces acting on the system will tend to bring it back to its original position if V is minimum (Fig. 10.15a) and to move it farther away if V is maximum (Fig. 10.15b). If V is constant (Fig. 10.15c), the forces will not tend to move the system either way.

† The second position does not exist if $W > 4kl$ (see Prob. 10.81 for a further discussion of the equilibrium of this system).

Recalling from calculus that a function is minimum or maximum according to whether its second derivative is positive or negative, we may summarize as follows the conditions for the equilibrium of a system with one degree of freedom (i.e., a system the position of which is defined by a single independent variable θ):

$$\frac{dV}{d\theta} = 0 \qquad \frac{d^2V}{d\theta^2} > 0\text{: stable equilibrium}$$
$$\frac{dV}{d\theta} = 0 \qquad \frac{d^2V}{d\theta^2} < 0\text{: unstable equilibrium} \tag{10.23}$$

If both the first and the second derivatives of V are zero, it is necessary to examine derivatives of a higher order to determine whether the equilibrium is stable, unstable, or neutral. The equilibrium will be neutral if all

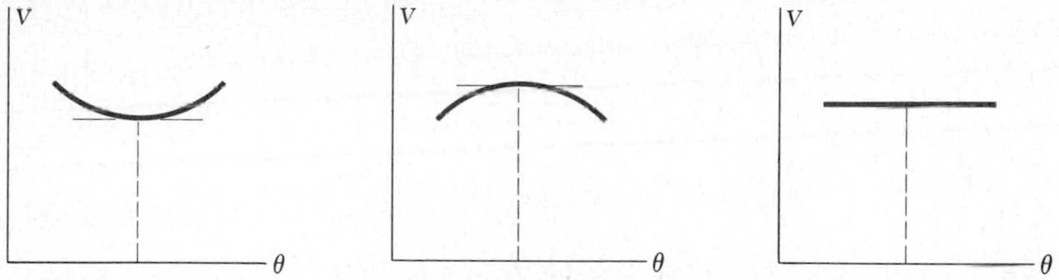

Fig. 10.15 (*a*) Stable equilibrium (*b*) Unstable equilibrium (*c*) Neutral equilibrium

derivatives are zero, since the potential energy V is then a constant. The equilibrium will be stable if the first derivative found to be different from zero is of even order, and if that derivative is positive. In all other cases the equilibrium will be unstable.

If the system considered possesses *several degrees of freedom,* the potential energy V depends upon several variables and it is thus necessary to apply the theory of functions of several variables to determine whether V is minimum. It may be verified that a system with two degrees of freedom will be stable, and the corresponding potential energy $V(\theta_1, \theta_2)$ will be minimum, if the following relations are satisfied simultaneously:

$$\frac{\partial V}{\partial \theta_1} = \frac{\partial V}{\partial \theta_2} = 0$$
$$\left(\frac{\partial^2 V}{\partial \theta_1\, \partial \theta_2}\right)^2 - \frac{\partial^2 V}{\partial \theta_1^2}\frac{\partial^2 V}{\partial \theta_2^2} < 0 \tag{10.24}$$
$$\frac{\partial^2 V}{\partial \theta_1^2} > 0 \quad \text{or} \quad \frac{\partial^2 V}{\partial \theta_2^2} > 0$$

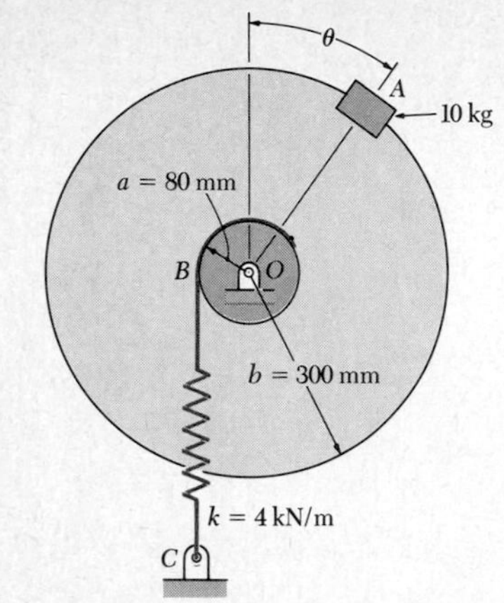

SAMPLE PROBLEM 10.4

A 10-kg block is attached to the rim of a 300-mm-radius disk as shown. Knowing that spring BC is unstretched when $\theta = 0$, determine the position or positions of equilibrium, and state in each case whether the equilibrium is stable, unstable, or neutral.

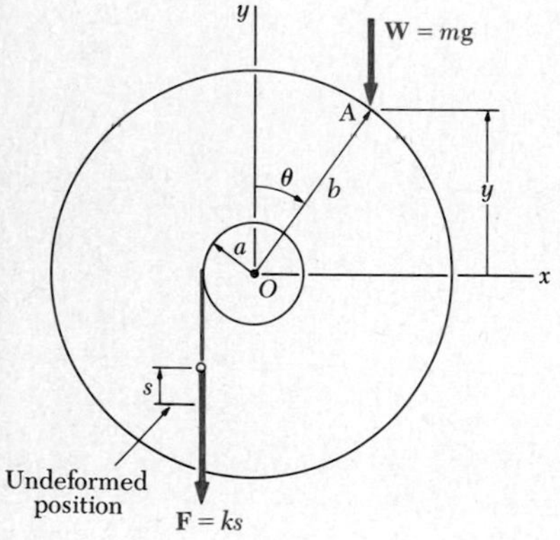

Potential Energy. Denoting by s the deflection of the spring from its undeformed position and placing the origin of coordinates at O, we obtain

$$V_e = \tfrac{1}{2}ks^2 \qquad V_g = Wy = mgy$$

Measuring θ in radians, we have

$$s = a\theta \qquad y = b\cos\theta$$

Substituting for s and y in the expressions for V_e and V_g, we write

$$V_e = \tfrac{1}{2}ka^2\theta^2 \qquad V_g = mgb\cos\theta$$
$$V = V_e + V_g = \tfrac{1}{2}ka^2\theta^2 + mgb\cos\theta$$

Positions of Equilibrium. Setting $dV/d\theta = 0$, we write

$$\frac{dV}{d\theta} = ka^2\theta - mgb\sin\theta = 0$$

$$\sin\theta = \frac{ka^2}{mgb}\theta$$

Substituting $a = 0.08$ m, $b = 0.3$ m, $k = 4$ kN/m, and $m = 10$ kg, we obtain

$$\sin\theta = \frac{(4\text{ kN/m})(0.08\text{ m})^2}{(10\text{ kg})(9.81\text{ m/s}^2)(0.3\text{ m})}\theta$$
$$\sin\theta = 0.8699\,\theta$$

where θ is expressed in radians. Solving by trial for θ, we find

$$\theta = 0 \quad \text{and} \quad \theta = 0.902\text{ rad}$$
$$\theta = 0 \quad \text{and} \quad \theta = 51.7° \blacktriangleleft$$

Stability of Equilibrium. The second derivative of the potential energy V with respect to θ is

$$\frac{d^2V}{d\theta^2} = ka^2 - mgb\cos\theta$$
$$= (4\text{ kN/m})(0.08\text{ m})^2 - (10\text{ kg})(9.81\text{ m/s}^2)(0.3\text{ m})\cos\theta$$
$$= 25.6 - 29.43\cos\theta$$

For $\theta = 0$: $\quad \dfrac{d^2V}{d\theta^2} = 25.6 - 29.43\cos 0° = -3.83 < 0$

The equilibrium is unstable for $\theta = 0°$ ◀

For $\theta = 51.7°$: $\quad \dfrac{d^2V}{d\theta^2} = 25.6 - 29.43\cos 51.7° = +7.36 > 0$

The equilibrium is stable for $\theta = 51.7°$ ◀

Problems

10.49 Using the method of Sec. 10.8, solve Prob. 10.27.

10.50 Using the method of Sec. 10.8, solve Prob. 10.28.

10.51 Using the method of Sec. 10.8, solve Prob. 10.29.

10.52 Using the method of Sec. 10.8, solve Prob. 10.30.

10.53 Show that the equilibrium is neutral in Prob. 10.5.

10.54 Show that the equilibrium is neutral in Prob. 10.2.

10.55 Two uniform rods, each of mass m, are attached to gears of equal radii as shown. Determine the positions of equilibrium of the system and state in each case whether the equilibrium is stable, unstable, or neutral.

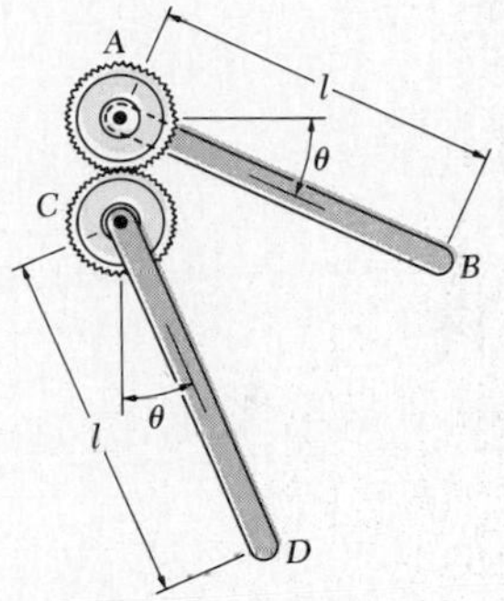

Fig. P10.55 and P10.56

10.56 Two uniform rods, AB and CD, are attached to gears of equal radii as shown. Knowing that $m_{AB} = 400$ g and $m_{CD} = 600$ g, determine the positions of equilibrium of the system and state in each case whether the equilibrium is stable, unstable, or neutral.

10.57 Two uniform rods, AB and CD, are attached to gears as shown. Knowing that rod AB weighs 2 lb and that rod CD weighs 3 lb, determine the positions of equilibrium of the system and state in each case whether the equilibrium is stable, unstable, or neutral.

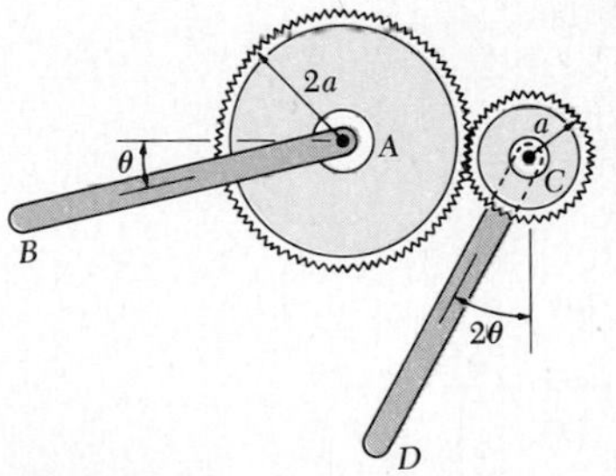

Fig. P10.57

10.58 Solve Prob. 10.57, assuming that rods AB and CD each weighs 3 lb.

10.59 Using the method of Sec. 10.8, solve Prob. 10.33. Determine whether the equilibrium is stable, unstable, or neutral. (*Hint.* The potential energy corresponding to the couple exerted by a torsion spring is $\frac{1}{2}K\theta^2$, where K is the torsional spring constant and θ is the angle of twist.)

10.60 In Prob. 10.34, determine whether each of the positions of equilibrium is stable, unstable, or neutral. (See hint of Prob. 10.59.)

10.61 A load **W** of magnitude 120 lb is applied to the mechanism at C. Neglecting the weight of the mechanism, determine the value of θ corresponding to equilibrium and check that the equilibrium is stable. The constant of the spring is $k = 20$ lb/in., and the spring is unstretched when $\theta = 0$.

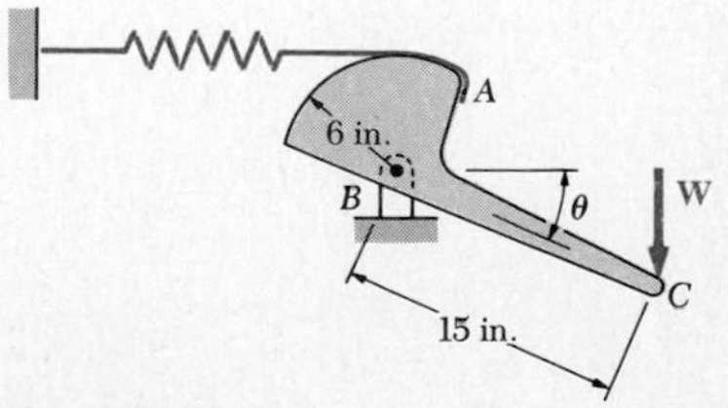

Fig. P10.61

10.62 Solve Prob. 10.61, assuming that the spring is unstretched when $\theta = 30°$.

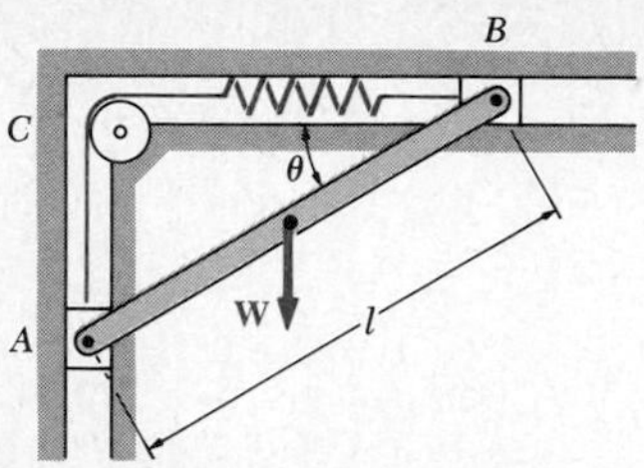

Fig. P10.63

10.63 A slender rod AB, of weight W, is attached to blocks A and B which may move freely in the guides shown. The constant of the spring is k and the spring is unstretched when AB is horizontal. Neglecting the weight of the blocks, derive an equation in θ, W, l, and k which must be satisfied when the rod is in equilibrium.

10.64 In Prob. 10.63, determine three values of θ corresponding to equilibrium when $W = 30$ N, $k = 200$ N/m, and $l = 0.5$ m. State in each case whether the equilibrium is stable, unstable, or neutral.

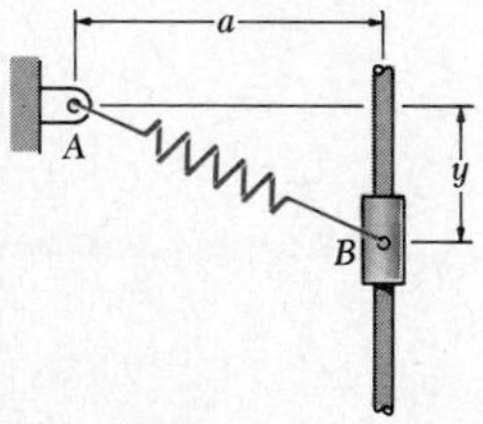

Fig. P10.65

10.65 A collar B, of weight W, may move freely along the vertical rod shown. The constant of the spring is k and the spring is unstretched when $y = 0$. (*a*) Derive an equation in y, W, a, and k which must be satisfied when the collar is in equilibrium. (*b*) Determine the value of y corresponding to equilibrium when $W = 60$ N, $a = 400$ mm, and $k = 1$ kN/m, and check that the equilibrium is stable.

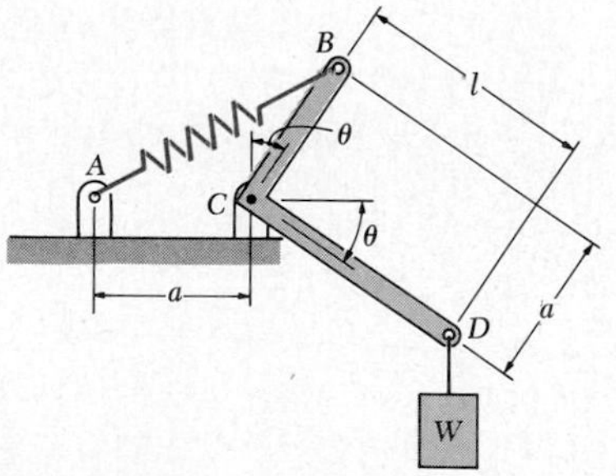

Fig. P10.66

10.66 The constant of spring AB is k and the spring is unstretched when $\theta = 0$. (*a*) Neglecting the weight of the rigid arm BCD, derive an equation in θ, k, a, l, and W which must be satisfied when the arm is in equilibrium. (*b*) Determine three values of θ corresponding to equilibrium when $k = 60$ lb/in., $a = 10$ in., $l = 15$ in., and $W = 80$ lb. State in each case whether the equilibrium is stable, unstable, or neutral.

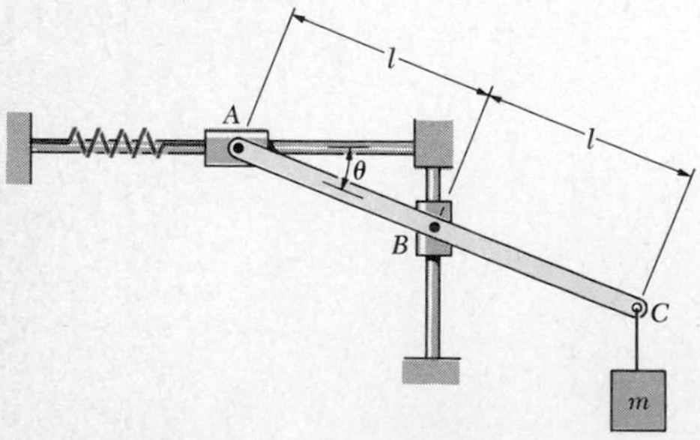

Fig. P10.67

10.67 Bar ABC is attached to collars A and B which may move freely on the rods shown. The constant of the spring is k and the spring is unstretched when $\theta = 0$. (*a*) Neglecting the weight of bar ABC, derive an equation in θ, m, k, and l which must be satisfied when bar ABC is in equilibrium. (*b*) Determine the value of θ corresponding to equilibrium when $m = 5$ kg, $k = 1.2$ kN/m, and $l = 250$ mm, and check that the equilibrium is stable.

10.68 Solve Prob. 10.67, assuming that the spring is unstretched when $\theta = 30°$.

10.69 A vertical load **P** is applied at end B of rod BC. The constant of the spring is k and the spring is unstretched when $\theta = 60°$. (*a*) Neglecting the weight of the rod, express the angle θ corresponding to equilibrium in terms of P, k, and l. (*b*) Determine the values of θ corresponding to equilibrium when $P = 40$ lb, $k = 20$ lb/in., and $l = 8$ in. State in each case whether the equilibrium is stable, unstable, or neutral.

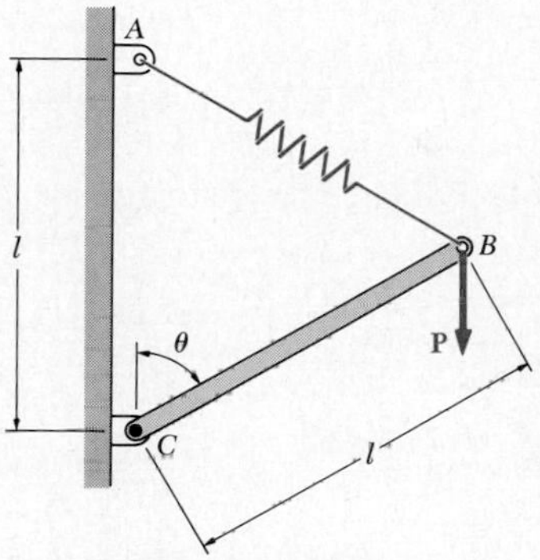

Fig. P10.69 and P10.70

10.70 A vertical load **P** of magnitude 100 lb is applied at end B of rod BC. The constant of the spring is $k = 20$ lb/in. and the spring is unstretched when $\theta = 60°$. Neglecting the weight of the rod and knowing that $l = 8$ in., determine the value of θ corresponding to equilibrium. State whether the equilibrium is stable, unstable, or neutral.

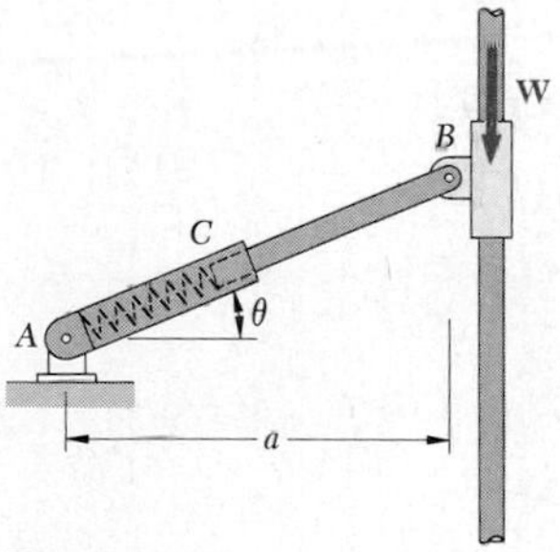

Fig. P10.71

10.71 The internal spring AC is of constant k and is undeformed when $\theta = 45°$. Derive an equation defining the values of θ corresponding to equilibrium positions.

10.72 In Prob. 10.71, determine the positive values of θ corresponding to equilibrium positions when $W = 100$ N, $k = 3$ kN/m, and $a = 500$ mm. State in each case whether the equilibrium is stable, unstable, or neutral.

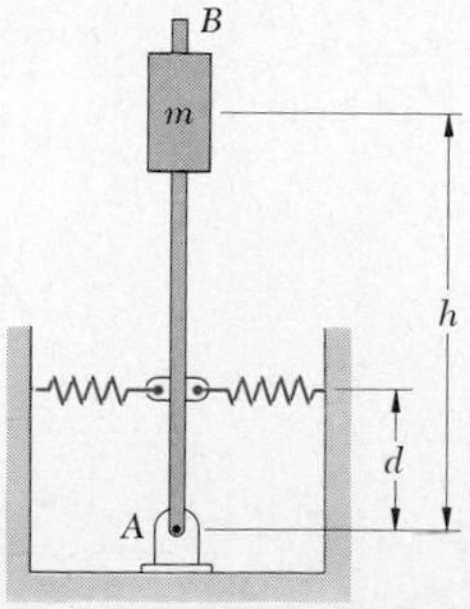

Fig. P10.73 and P10.74

10.73 The rod AB is attached to a hinge at A and to two springs, each of constant k. If $h = 750$ mm, $d = 400$ mm, and $m = 80$ kg, determine the range of values of k for which the equilibrium of rod AB is stable in the position shown. Each spring can act in either tension or compression.

10.74 If $m = 75$ kg, $h = 650$ mm, and the constant of each spring is $k = 2.5$ kN/m, determine the range of values of the distance d for which the equilibrium of rod AB is stable in the position shown. Each spring can act in either tension or compression.

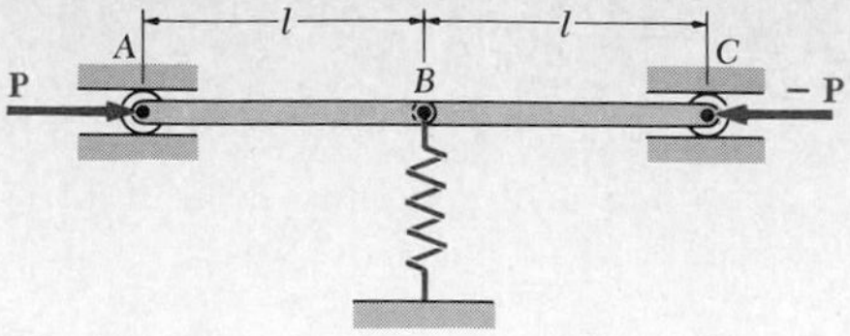

Fig. P10.75

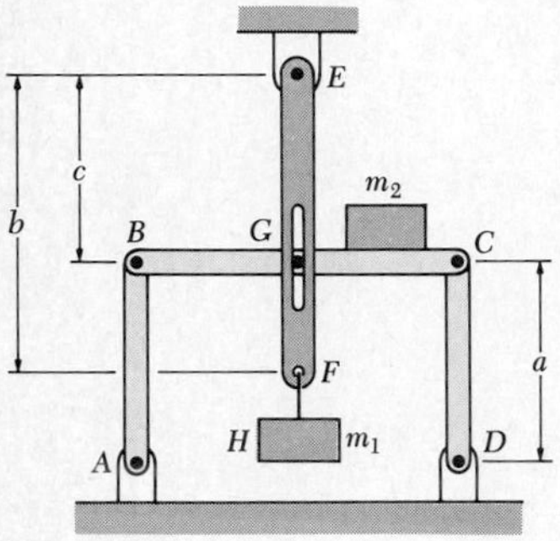

Fig. P10.76

10.75 Two bars AB and BC of negligible weight are attached to a single spring of constant k which is undeformed when the bars are horizontal. Determine the range of values of the magnitude P of the two equal and opposite forces **P** and $-$**P** for which the equilibrium of the system is stable in the position shown.

10.76 Bars BC and EF are connected at G by a pin which is attached to BC and may slide freely in a slot cut in EF. Determine the smallest mass m_1 for which the equilibrium of the mechanism is stable in the position shown.

*** 10.77 and 10.78** The bars shown, each of length l and of negligible weight, are attached to springs each of constant k. The springs are undeformed and the system is in equilibrium when $\theta_1 = \theta_2 = 0$. Determine the range of values of P for which the equilibrium position is stable.

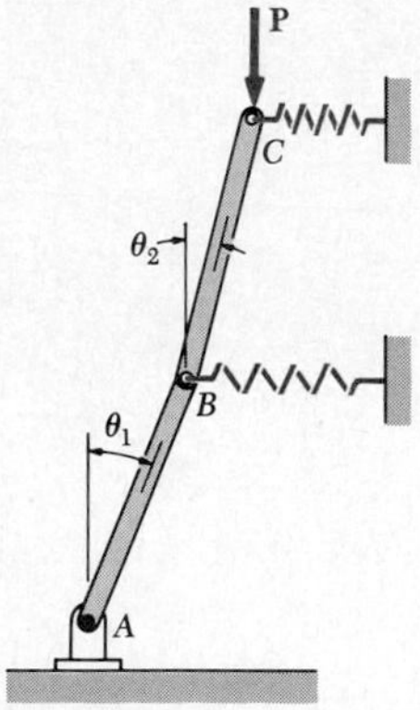

Fig. P10.77

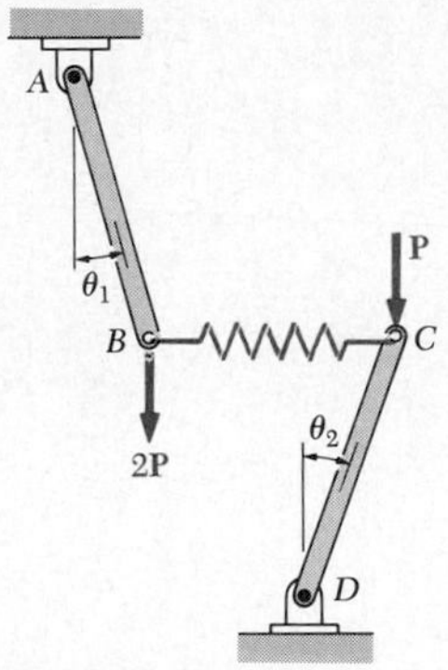

Fig. P10.78

*** 10.79** Solve Prob. 10.78, assuming that the vertical force applied at B is increased to 5**P**.

*** 10.80** Solve Prob. 10.77, knowing that $l = 25$ in. and $k = 45$ lb/in.

10.81 In Sec. 10.8, two positions of equilibrium were obtained for the system shown in Fig. 10.13, namely, $\theta = 0$ and $\theta = \cos^{-1}(W/4kl)$. Show that (*a*) if $W < 4kl$, the equilibrium is stable in the first position ($\theta = 0$) and unstable in the second, (*b*) if $W = 4kl$, the two positions coincide and the equilibrium is unstable, (*c*) if $W > 4kl$, the equilibrium is unstable in the first position ($\theta = 0$) and the second position does not exist. (*Note.* It is assumed that the system must deform as shown and that the system cannot rotate as a single rigid body about A when A and B coincide.)

Review Problems

10.82 The pliers shown are used to attach electric connectors to flat cables which carry many separate wires. Knowing that 300-N forces directed along line *a-a* are required to complete the attachment, determine the magnitude *P* of the forces which must be applied to the handles.

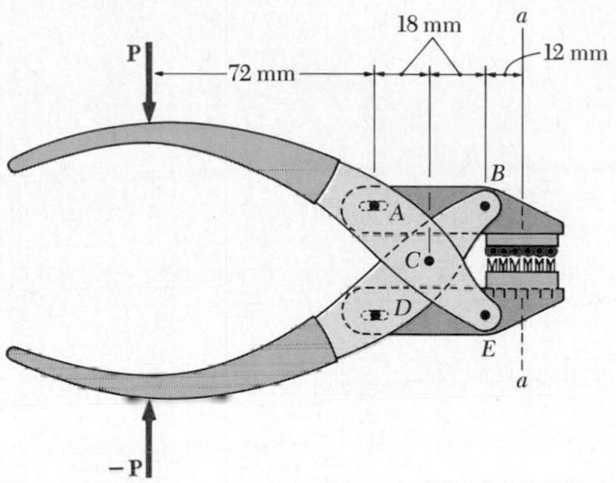

Fig. P10.82

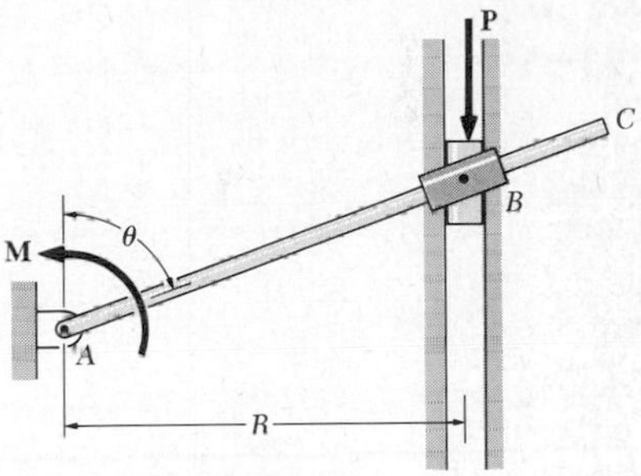

Fig. P10.83 and P10.84

10.83 Collar *B* may slide along rod *AC* and is attached by a pin to a block which may slide in the vertical slot. Derive an expression for the magnitude of the couple **M** required to maintain equilibrium.

10.84 Determine the value of θ corresponding to the equilibrium position of rod *AC* when $R = 10$ in., $P = 50$ lb, and $M = 2000$ lb·in.

10.85 A spring *AB* of constant *k* is attached to two identical gears as shown. A uniform bar *CD* of mass *m* is supported by cords wrapped around drums of radius *b* which are attached to the gears. If the spring is undeformed when $\theta = 0$, obtain an equation defining the angle θ corresponding to the equilibrium position.

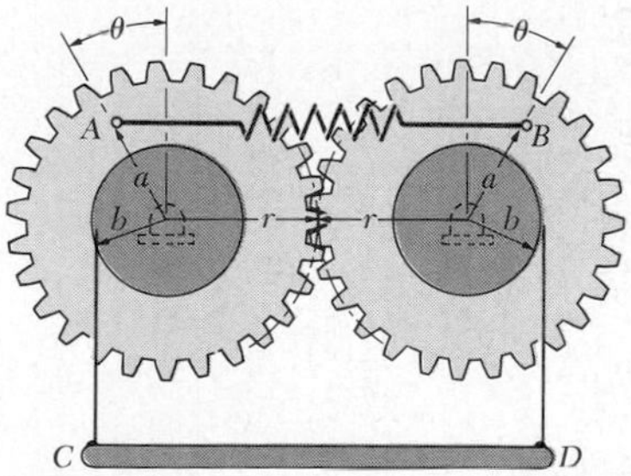

Fig. P10.85

10.86 For the mechanism of Prob. 10.85 the following numerical values are given: $k = 6$ kN/m, $a = 80$ mm, $b = 60$ mm, $r = 120$ mm, and $m = 50$ kg. Determine the values of θ corresponding to equilibrium positions and state in each case whether the equilibrium is stable, unstable, or neutral.

10.87 Collar *A* may slide freely on the balance beam *DCE* which is supported by a frictionless pin at *C*. The position of collar *A* on the beam is controlled by a cord which is attached to a fixed cylindrical support of radius *a*. Knowing that $l_A = l$ when beam *DCE* is horizontal, derive an equation in θ, m_A, m_B, *l*, and *a* which must be satisfied when the balance beam is in equilibrium.

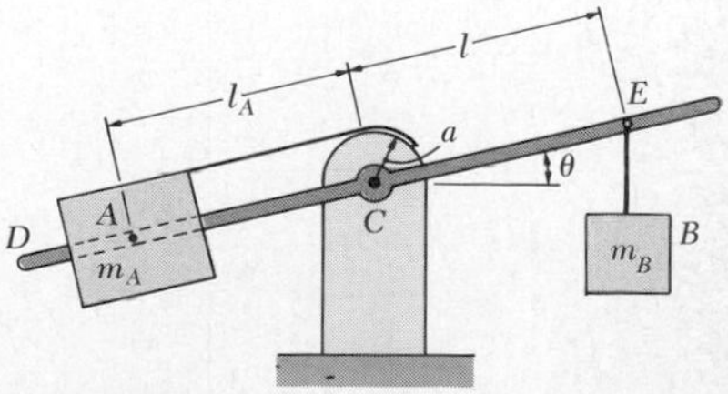

Fig. P10.87

10.88 Determine the value of θ corresponding to the equilibrium position of the balance beam of Prob. 10.87, when $m_A = 500$ g, $m_B = 450$ g, $l = 200$ mm, and $a = 40$ mm.

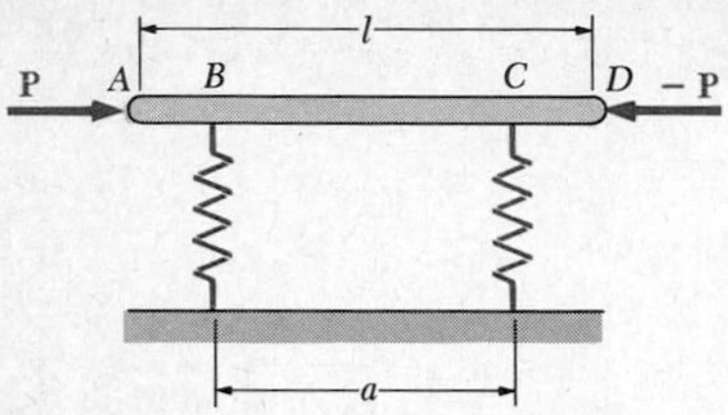

Fig. P10.89

10.89 The horizontal bar AD is attached to two springs of constant k and is in equilibrium in the position shown. Determine the range of values of the magnitude P of the two equal and opposite *horizontal* forces **P** and **−P** for which the equilibrium position is stable (*a*) if $AB = CD$, (*b*) if $AB = 2(CD)$.

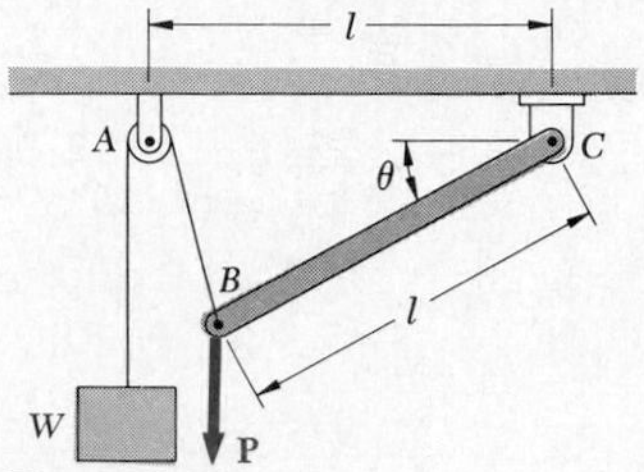

Fig. P10.90

10.90 (*a*) Derive an equation defining the angle θ corresponding to the equilibrium position. (*b*) Determine the angle θ corresponding to equilibrium if $P = 2W$.

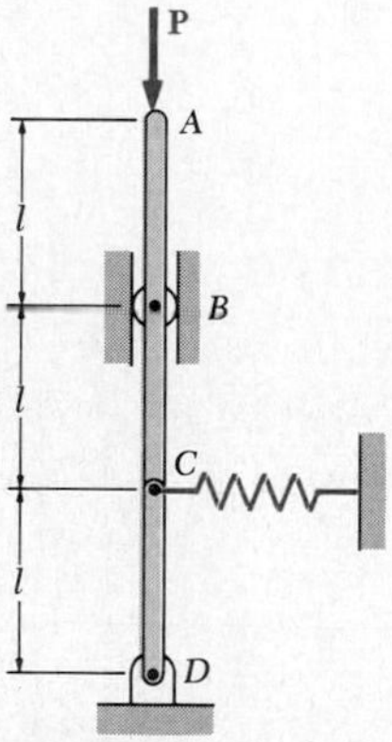

Fig. P10.91

10.91 Two bars ABC and CD are attached to a single spring of constant k which is unstretched when the bars are vertical. Determine the range of values of P for which the equilibrium of the system is stable in the position shown.

10.92 Blocks A and B, of mass m_A and m_B, respectively, are connected by a cable AB which passes over a pulley at C. Neglecting the effect of friction, derive an equation in θ, β, b, m_A, and m_B which must be satisfied when the system is in equilibrium.

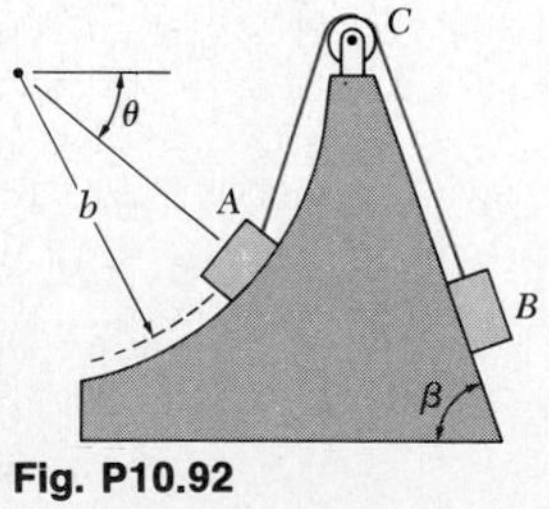

Fig. P10.92

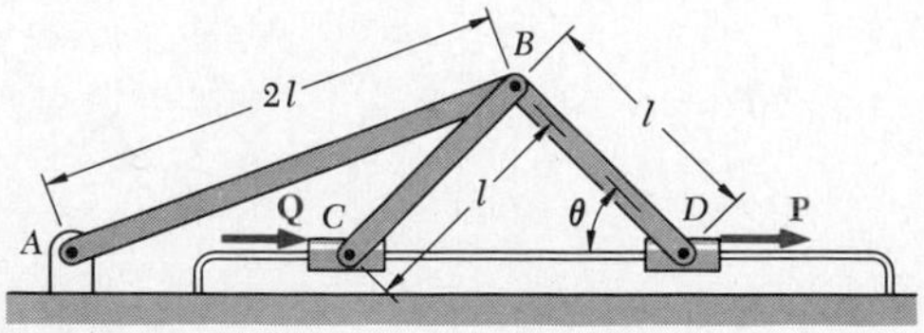

Fig. P10.93

10.93 The mechanism shown is acted upon by the force **P**; derive an expression for the magnitude of the force **Q** required to maintain equilibrium.

CHAPTER ELEVEN

Kinematics of Particles

11.1. Introduction to Dynamics. Chapters 1 to 10 were devoted to *statics*, i.e., to the analysis of bodies at rest. We shall now begin the study of *dynamics*, which is the part of mechanics that deals with the analysis of bodies in motion.

While the study of statics goes back to the time of the Greek philosophers, the first significant contribution to dynamics was made by Galileo (1564–1642). Galileo's experiments on uniformly accelerated bodies led Newton (1642–1727) to formulate his fundamental laws of motion.

Dynamics is divided into two parts: (1) *Kinematics*, which is the study of the geometry of motion; kinematics is used to relate displacement, velocity, acceleration, and time, without reference to the cause of the motion. (2) *Kinetics*, which is the study of the relation existing between the forces acting on a body, the mass of the body, and the motion of the body; kinetics is used to predict the motion caused by given forces or to determine the forces required to produce a given motion.

Chapters 11 to 14 are devoted to the *dynamics of particles*, and Chap. 11 more particularly to the *kinematics of particles*. The use of the word particles does not imply that we shall restrict our study to that of small corpuscles; it rather indicates that in these first chapters we shall study the motion of bodies—possibly as large as cars, rockets, or airplanes—without regard to their size. By saying that the bodies are analyzed as particles, we mean that only their motion as an entire unit will be considered; any rotation about their own mass center will be neglected. There are cases, however, when such a rotation is not negligible; the bodies may not then be considered as particles. The analysis of such motions will be carried out in later chapters dealing with the *dynamics of rigid bodies*.

In the first part of Chap. 11, we shall analyze the rectilinear motion of a particle; that is, we shall determine at every instant the position, velocity, and acceleration of a particle as it moves along a straight line. After first studying the motion of a particle by general methods of analysis, we shall consider two important particular cases, namely, the uniform motion and the uniformly accelerated motion of a particle (Secs. 11.4 and 11.5). We shall then consider in Sec. 11.6 the simultaneous motion of several particles and introduce the concept of the relative motion of one particle with respect to another. The first part of this chapter concludes with a study of graphical methods of analysis and their application to the solution of various problems involving the rectilinear motion of particles (Secs. 11.7 and 11.8).

In the second part of this chapter, we shall analyze the motion of a particle as it moves along a curved path. Since the position, velocity, and acceleration of a particle will be defined as vector quantities, the concept of the derivative of a vector function will be introduced in Sec. 11.10 and added to our mathematical tools. We shall then consider applications where the motion of a particle is defined by the rectangular components of its velocity and acceleration; at this point, the motion of a projectile will be analyzed (Sec. 11.11). In Sec. 11.12, we shall consider the motion of a particle relative to a reference frame in translation. Finally, we shall analyze the curvilinear motion of a particle in terms of components other than rectangular. In Sec. 11.13, we shall introduce the tangential and normal components of the velocity and acceleration of a particle; and in Sec. 11.14, the radial and transverse components of its velocity and acceleration.

RECTILINEAR MOTION OF PARTICLES

11.2. Position, Velocity, and Acceleration. A particle moving along a straight line is said to be in *rectilinear motion.* At any given instant t, the particle will occupy a certain position on the straight line. To define the position P of the particle, we choose a fixed origin O on the straight line and a positive direction along the line. We measure the distance x from O to P and record it with a plus or minus sign, according to whether P is reached from O by moving along the line in the positive or the negative direction. The distance x, with the appropriate sign, completely defines the position of the particle; it is called the *position coordinate* of the particle considered. For example, the position coordinate corresponding to P in Fig. 11.1a is $x = +5$ m, while the coordinate corresponding to P' in Fig. 11.1b is $x' = -2$ m.

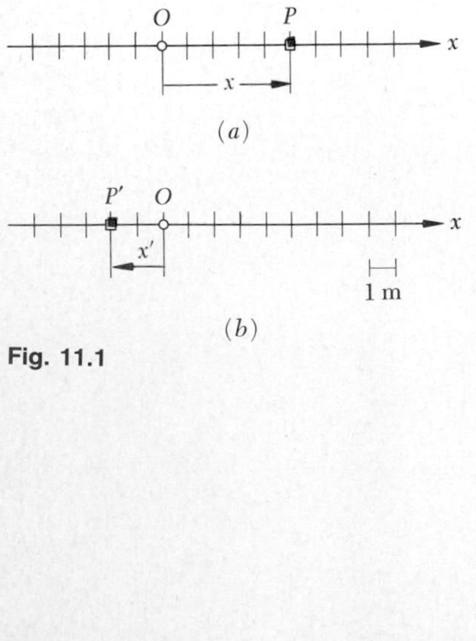

Fig. 11.1

When the position coordinate x of a particle is known for every value of time t, we say that the motion of the particle is known. The "timetable" of the motion may be given in the form of an equation in x and t, such as $x = 6t^2 - t^3$, or in the form of a graph of x versus t as shown in Fig. 11.6. The units most generally used to measure the position coordinate x are the meter (m) in the SI system of units,† and the foot (ft) in the U.S. customary

†Cf. Sec. 1.3.

system of units. Time t will generally be measured in seconds (s).

Consider the position P occupied by the particle at time t and the corresponding coordinate x (Fig. 11.2). Consider also the position P' occupied by the particle at a later time $t + \Delta t$; the position coordinate of P' may be obtained by adding to the coordinate x of P the small displacement Δx, which will be positive or negative according to whether P' is to the right or to the left of P. The *average velocity* of the particle over the time interval Δt is defined as the quotient of the displacement Δx and the time interval Δt:

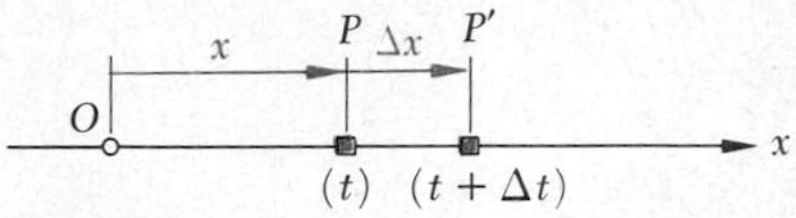

Fig. 11.2

$$\text{Average velocity} = \frac{\Delta x}{\Delta t}$$

If SI units are used, Δx is expressed in meters and Δt in seconds; the average velocity will thus be expressed in meters per second (m/s). If U.S. customary units are used, Δx is expressed in feet and Δt in seconds; the average velocity will then be expressed in feet per second (ft/s).

The *instantaneous velocity* v of the particle at the instant t is obtained from the average velocity by choosing shorter and shorter time intervals Δt and displacements Δx:

$$\text{Instantaneous velocity} = v = \lim_{\Delta t \to 0} \frac{\Delta x}{\Delta t}$$

The instantaneous velocity will also be expressed in m/s or ft/s. Observing that the limit of the quotient is equal, by definition, to the derivative of x with respect to t, we write

$$v = \frac{dx}{dt} \tag{11.1}$$

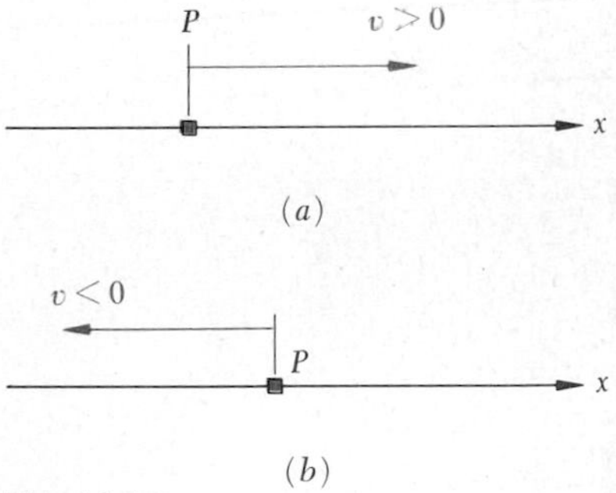

Fig. 11.3

The velocity v is represented by an algebraic number which may be positive or negative.† A positive value of v indicates that x increases, i.e., that the particle moves in the positive direction (Fig. 11.3a); a negative value of v indicates that x decreases, i.e., that the particle moves in the negative direction (Fig. 11.3b). The magnitude of v is known as the *speed* of the particle.

Consider the velocity v of the particle at time t and also its velocity $v + \Delta v$ at a later time $t + \Delta t$ (Fig. 11.4). The *average acceleration* of the particle over the time interval Δt is defined as the quotient of Δv and Δt:

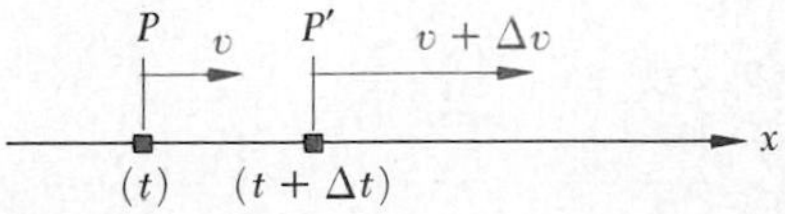

Fig. 11.4

$$\text{Average acceleration} = \frac{\Delta v}{\Delta t}$$

† As we shall see in Sec. 11.9, the velocity is actually a vector quantity. However, since we are considering here the rectilinear motion of a particle, where the velocity of the particle has a known and fixed direction, we need only specify the sense and magnitude of the velocity; this may be conveniently done by using a scalar quantity with a plus or minus sign. The same remark will apply to the acceleration of a particle in rectilinear motion.

If SI units are used, Δv is expressed in m/s and Δt in seconds; the average acceleration will thus be expressed in m/s^2. If U.S. customary units are used, Δv is expressed in ft/s and Δt in seconds; the average acceleration will then be expressed in ft/s^2.

The *instantaneous acceleration* a of the particle at the instant t is obtained from the average acceleration by choosing smaller and smaller values for Δt and Δv:

$$\text{Instantaneous acceleration} = a = \lim_{\Delta t \to 0} \frac{\Delta v}{\Delta t}$$

The instantaneous acceleration will also be expressed in m/s^2 or ft/s^2. The limit of the quotient is by definition the derivative of v with respect to t and measures the rate of change of the velocity. We write

$$a = \frac{dv}{dt} \tag{11.2}$$

or, substituting for v from (11.1),

$$a = \frac{d^2x}{dt^2} \tag{11.3}$$

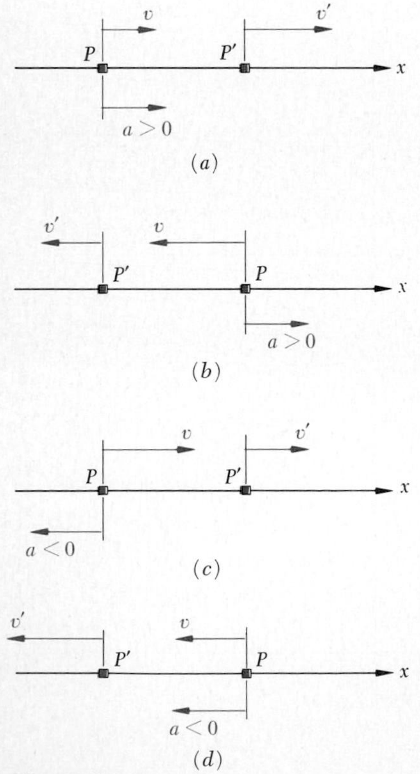

Fig. 11.5

The acceleration a is represented by an algebraic number which may be positive or negative.† A positive value of a indicates that the velocity (i.e., the algebraic number v) increases. This may mean that the particle is moving faster in the positive direction (Fig. 11.5*a*) or that it is moving more slowly in the negative direction (Fig. 11.5*b*); in both cases, Δv is positive. A negative value of a indicates that the velocity decreases; either the particle is moving more slowly in the positive direction (Fig. 11.5*c*) or it is moving faster in the negative direction (Fig. 11.5*d*).

The term *deceleration* is sometimes used to refer to a when the speed of the particle (i.e., the magnitude of v) decreases; the particle is then moving more slowly. For example, the particle of Fig. 11.5 is decelerated in parts b and c, while it is truly accelerated (i.e., moves faster) in parts a and d.

Another expression may be obtained for the acceleration by eliminating the differential dt in Eqs. (11.1) and (11.2). Solving (11.1) for dt, we obtain $dt = dx/v$; carrying into (11.2), we write

$$a = v\frac{dv}{dx} \tag{11.4}$$

†See footnote, page 419.

Example. Consider a particle moving in a straight line, and assume that its position is defined by the equation

$$x = 6t^2 - t^3$$

where t is expressed in seconds and x in meters. The velocity v at any time t is obtained by differentiating x with respect to t:

$$v = \frac{dx}{dt} = 12t - 3t^2$$

The acceleration a is obtained by differentiating again with respect to t:

$$a = \frac{dv}{dt} = 12 - 6t$$

The position coordinate, the velocity, and the acceleration have been plotted against t in Fig. 11.6. The curves obtained are known as *motion curves*. It should be kept in mind, however, that the particle does not move along any of these curves; the particle moves in a straight line. Since the derivative of a function measures the slope of the corresponding curve, the slope of the x–t curve at any given time is equal to the value of v at that time and the slope of the v–t curve is equal to the value of a. Since $a = 0$ at $t = 2$ s, the slope of the v–t curve must be zero at $t = 2$ s; the velocity reaches a maximum at this instant. Also, since $v = 0$ at $t = 0$ and at $t = 4$ s, the tangent to the x–t curve must be horizontal for both these values of t.

A study of the three motion curves of Fig. 11.6 shows that the motion of the particle from $t = 0$ to $t = \infty$ may be divided into four phases:

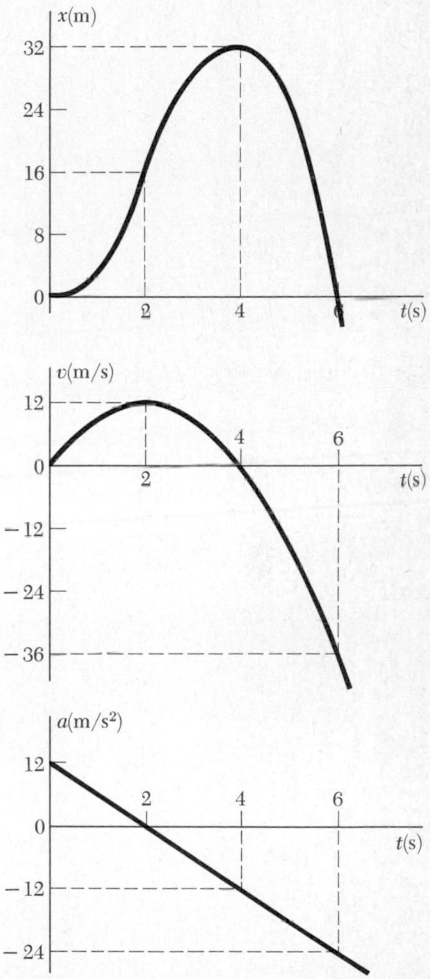

Fig. 11.6

1. The particle starts from the origin, $x = 0$, with no velocity but with a positive acceleration. Under this acceleration, the particle gains a positive velocity and moves in the positive direction. From $t = 0$ to $t = 2$ s, x, v, and a are all positive.

2. At $t = 2$ s, the acceleration is zero; the velocity has reached its maximum value. From $t = 2$ s to $t = 4$ s, v is positive, but a is negative; the particle still moves in the positive direction but more and more slowly; the particle is decelerated.

3. At $t = 4$ s, the velocity is zero; the position coordinate x has reached its maximum value. From then on, both v and a are negative; the particle is accelerated and moves in the negative direction with increasing speed.

4. At $t = 6$ s, the particle passes through the origin; its coordinate x is then zero, while the total distance traveled since the beginning of the motion is 64 m. For values of t larger than 6 s, x, v, and a will all be negative. The particle keeps moving in the negative direction, away from O, faster and faster.

11.3. Determination of the Motion of a Particle. We saw in the preceding section that the motion of a particle is said to be known if the position of the particle is known for every value of the time t. In practice, however, a motion is seldom defined by a relation between x and t. More often, the conditions of the motion will be specified by the type of acceleration that the particle possesses. For example, a freely falling body will have a constant acceleration, directed downward and equal to $9.81\ \mathrm{m/s^2}$, or $32.2\ \mathrm{ft/s^2}$; a mass attached to a spring which has been stretched will have an acceleration proportional to the instantaneous elongation of the spring measured from the equilibrium position; etc. In general, the acceleration of the particle may be expressed as a function of one or more of the variables x, v, and t. In order to determine the position coordinate x in terms of t, it will thus be necessary to perform two successive integrations.

We shall consider three common classes of motion:

1. $a = f(t)$. *The Acceleration Is a Given Function of* t. Solving (11.2) for dv and substituting $f(t)$ for a, we write

$$dv = a\,dt$$
$$dv = f(t)\,dt$$

Integrating both members, we obtain the equation

$$\int dv = \int f(t)\,dt$$

which defines v in terms of t. It should be noted, however, that an arbitrary constant will be introduced as a result of the integration. This is due to the fact that there are many motions which correspond to the given acceleration $a = f(t)$. In order to uniquely define the motion of the particle, it is necessary to specify the *initial conditions* of the motion, i.e., the value v_0 of the velocity and the value x_0 of the position coordinate at $t = 0$. Replacing the indefinite integrals by *definite integrals* with lower limits corresponding to the initial conditions $t = 0$ and $v = v_0$ and upper limits corresponding to $t = t$ and $v = v$, we write

$$\int_{v_0}^{v} dv = \int_0^t f(t)\,dt$$

$$v - v_0 = \int_0^t f(t)\,dt$$

which yields v in terms of t.

We shall now solve (11.1) for dx:

$$dx = v\,dt$$

and substitute for v the expression just obtained. Both members are then integrated, the left-hand member with respect to x from $x = x_0$ to $x = x$, and the right-hand member with respect to t from $t = 0$ to $t = t$. The position coordinate x is thus obtained in terms of t; the motion is completely determined.

Two important particular cases will be studied in greater detail in Secs. 11.4 and 11.5: the case when $a = 0$, corresponding to a *uniform motion,* and the case when $a =$ constant, corresponding to a *uniformly accelerated motion.*

2. $a = f(x)$. *The Acceleration Is a Given Function of* x. Rearranging Eq. (11.4) and substituting $f(x)$ for a, we write

$$v\,dv = a\,dx$$
$$v\,dv = f(x)\,dx$$

Since each member contains only one variable, we may integrate the equation. Denoting again by v_0 and x_0, respectively, the initial values of the velocity and of the position coordinate, we obtain

$$\int_{v_0}^{v} v\,dv = \int_{x_0}^{x} f(x)\,dx$$

$$\tfrac{1}{2}v^2 - \tfrac{1}{2}v_0^2 = \int_{x_0}^{x} f(x)\,dx$$

which yields v in terms of x. We now solve (11.1) for dt:

$$dt = \frac{dx}{v}$$

and substitute for v the expression just obtained. Both members may be integrated, and the desired relation between x and t is obtained.

3. $a = f(v)$. *The Acceleration Is a Given Function of* v. We may then substitute $f(v)$ for a either in (11.2) or in (11.4) to obtain either of the following relations:

$$f(v) = \frac{dv}{dt} \qquad\qquad f(v) = v\frac{dv}{dx}$$

$$dt = \frac{dv}{f(v)} \qquad\qquad dx = \frac{v\,dv}{f(v)}$$

Integration of the first equation will yield a relation between v and t; integration of the second equation will yield a relation between v and x. Either of these relations may be used in conjunction with Eq. (11.1) to obtain the relation between x and t which characterizes the motion of the particle.

SAMPLE PROBLEM 11.1

The position of a particle which moves along a straight line is defined by the relation $x = t^3 - 6t^2 - 15t + 40$, where x is expressed in feet and t in seconds. Determine (*a*) the time at which the velocity will be zero, (*b*) the position and distance traveled by the particle at that time, (*c*) the acceleration of the particle at that time, (*d*) the distance traveled by the particle from $t = 4$ s to $t = 6$ s.

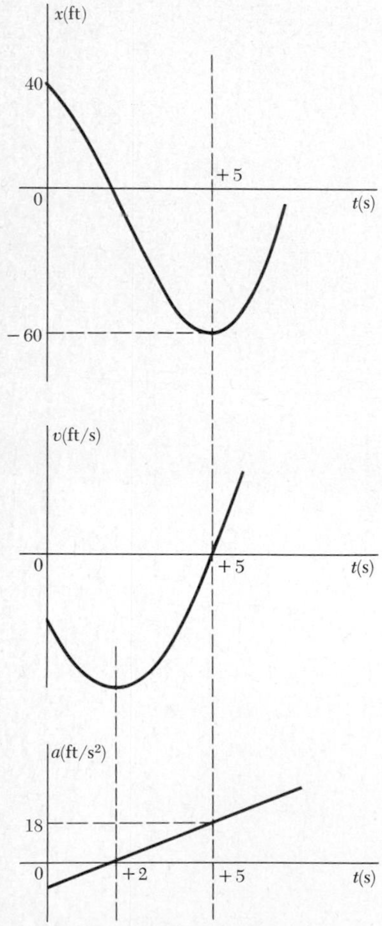

Solution. The equations of motion are

$$x = t^3 - 6t^2 - 15t + 40 \qquad (1)$$

$$v = \frac{dx}{dt} = 3t^2 - 12t - 15 \qquad (2)$$

$$a = \frac{dv}{dt} = 6t - 12 \qquad (3)$$

a. Time at Which $v = 0$. We make $v = 0$ in (2):

$$3t^2 - 12t - 15 = 0 \qquad t = -1\text{ s} \quad \text{and} \quad t = +5\text{ s} \quad ◄$$

Only the root $t = +5$ s corresponds to a time after the motion has begun: for $t < 5$ s, $v < 0$, the particle moves in the negative direction; for $t > 5$ s, $v > 0$, the particle moves in the positive direction.

b. Position and Distance Traveled When $v = 0$. Carrying $t = +5$ s into (1), we have

$$x_5 = (5)^3 - 6(5)^2 - 15(5) + 40 \qquad x_5 = -60\text{ ft} \quad ◄$$

The initial position at $t = 0$ was $x_0 = +40$ ft. Since $v \neq 0$ during the interval $t = 0$ to $t = 5$ s, we have

$$\text{Distance traveled} = x_5 - x_0 = -60\text{ ft} - 40\text{ ft} = -100\text{ ft}$$

$$\text{Distance traveled} = 100\text{ ft in the negative direction} \quad ◄$$

c. Acceleration When $v = 0$. We carry $t = +5$ s into (3):

$$a_5 = 6(5) - 12 \qquad a_5 = +18\text{ ft/s}^2 \quad ◄$$

d. Distance Traveled from $t = 4$ s to $t = 6$ s. Since the particle moves in the negative direction from $t = 4$ s to $t = 5$ s and in the positive direction from $t = 5$ s to $t = 6$ s, we shall compute separately the distance traveled during each of these time intervals.

From $t = 4$ s to $t = 5$ s: $\quad x_5 = -60$ ft

$$x_4 = (4)^3 - 6(4)^2 - 15(4) + 40 = -52\text{ ft}$$

$$\text{Distance traveled} = x_5 - x_4 = -60\text{ ft} - (-52\text{ ft}) = -8\text{ ft}$$
$$= 8\text{ ft in the negative direction}$$

From $t = 5$ s to $t = 6$ s: $\quad x_5 = -60$ ft

$$x_6 = (6)^3 - 6(6)^2 - 15(6) + 40 = -50\text{ ft}$$

$$\text{Distance traveled} = x_6 - x_5 = -50\text{ ft} - (-60\text{ ft}) = +10\text{ ft}$$
$$= 10\text{ ft in the positive direction}$$

Total distance traveled from $t = 4$ s to $t = 6$ s is

$$8\text{ ft} + 10\text{ ft} = 18\text{ ft} \quad ◄$$

SAMPLE PROBLEM 11.2

A ball is tossed with a velocity of 10 m/s directed vertically upward from a window located 20 m above the ground. Knowing that the acceleration of the ball is constant and equal to 9.81 m/s^2 downward, determine (*a*) the velocity v and elevation y of the ball above the ground at any time t, (*b*) the highest elevation reached by the ball and the corresponding value of t, (*c*) the time when the ball will hit the ground and the corresponding velocity. Draw the v–t and y–t curves.

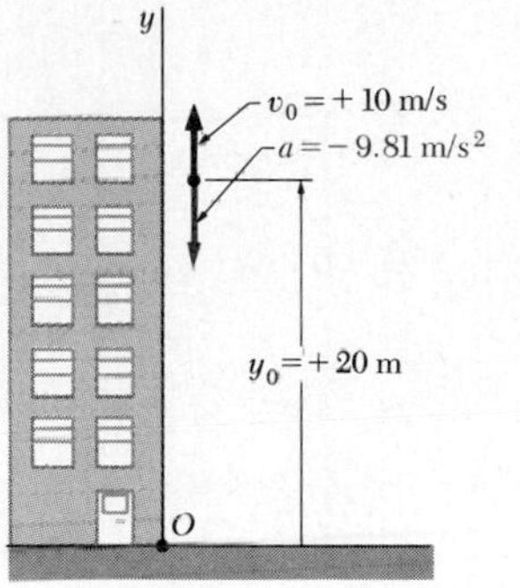

a. Velocity and Elevation. The y axis measuring the position coordinate (or elevation) is chosen with its origin O on the ground and its positive sense upward. The value of the acceleration and the initial values of v and y are as indicated. Substituting for a in $a = dv/dt$ and noting that at $t = 0$, $v_0 = +10$ m/s, we have

$$\frac{dv}{dt} = a = -9.81 \text{ m/s}^2$$

$$\int_{v_0=10}^{v} dv = -\int_0^t 9.81\, dt$$

$$[v]_{10}^{v} = -[9.81t]_0^t$$

$$v - 10 = -9.81t$$

$$v = 10 - 9.81t \quad (1) \quad ◄$$

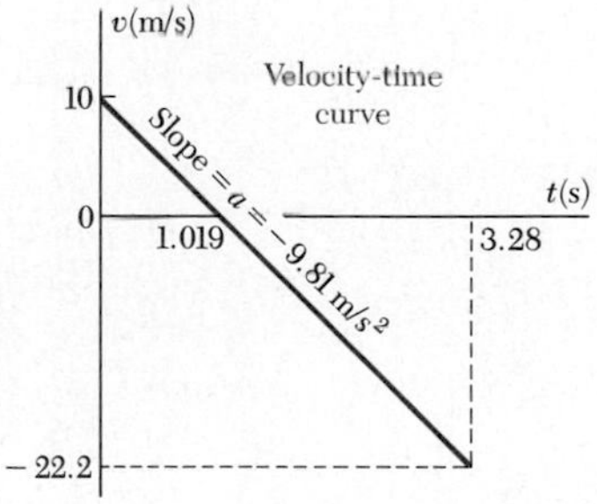

Substituting for v in $v = dy/dt$ and noting that at $t = 0$, $y_0 = 20$ m, we have

$$\frac{dy}{dt} = v = 10 - 9.81t$$

$$\int_{y_0=20}^{y} dy = \int_0^t (10 - 9.81t)\, dt$$

$$[y]_{20}^{y} = [10t - 4.905t^2]_0^t$$

$$y - 20 = 10t - 4.905t^2$$

$$y = 20 + 10t - 4.905t^2 \quad (2) \quad ◄$$

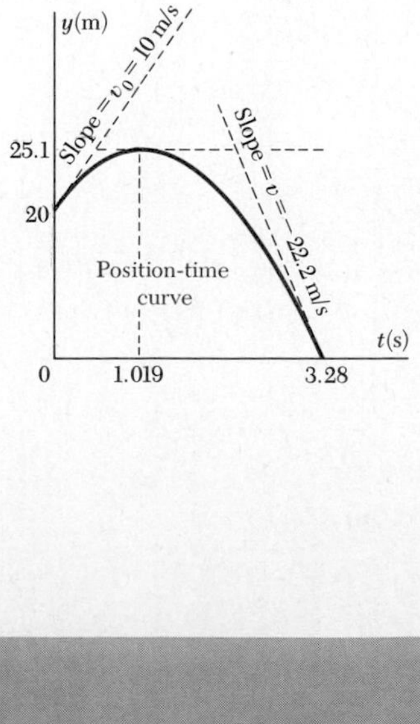

b. Highest Elevation. When the ball reaches its highest elevation, we have $v = 0$. Substituting into (1), we obtain

$$10 - 9.81t = 0 \qquad t = 1.019 \text{ s} \quad ◄$$

Carrying $t = 1.019$ s into (2), we have

$$y = 20 + 10(1.019) - 4.905(1.019)^2 \qquad y = 25.1 \text{ m} \quad ◄$$

c. Ball Hits the Ground. When the ball hits the ground, we have $y = 0$. Substituting into (2), we obtain

$$20 + 10t - 4.905t^2 = 0 \qquad t = -1.243 \text{ s} \quad \text{and} \quad t = +3.28 \text{ s} \quad ◄$$

Only the root $t = +3.28$ s corresponds to a time after the motion has begun. Carrying this value of t into (1), we have

$$v = 10 - 9.81(3.28) = -22.2 \text{ m/s} \qquad v = 22.2 \text{ m/s} \downarrow \quad ◄$$

SAMPLE PROBLEM 11.3

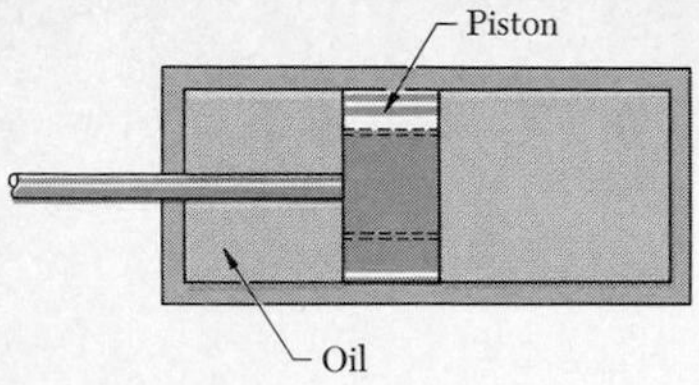

The brake mechanism used to reduce recoil in certain types of guns consists essentially of a piston which is attached to the barrel and may move in a fixed cylinder filled with oil. As the barrel recoils with an initial velocity v_0, the piston moves and oil is forced through orifices in the piston, causing the piston and the barrel to decelerate at a rate proportional to their velocity, that is, $a = -kv$. Express (*a*) v in terms of t, (*b*) x in terms of t, (*c*) v in terms of x. Draw the corresponding motion curves.

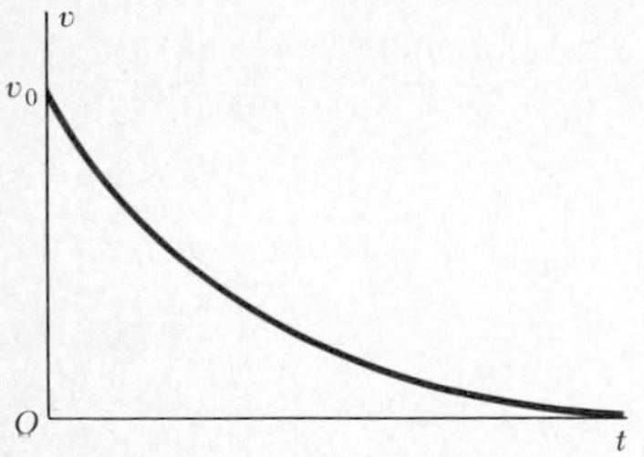

a. v in Terms of t. Substituting $-kv$ for a in the fundamental formula defining acceleration, $a = dv/dt$, we write

$$-kv = \frac{dv}{dt} \qquad \frac{dv}{v} = -k\,dt \qquad \int_{v_0}^{v} \frac{dv}{v} = -k\int_0^t dt$$

$$\ln\frac{v}{v_0} = -kt \qquad\qquad v = v_0 e^{-kt} \blacktriangleleft$$

b. x in Terms of t. Substituting the expression just obtained for v into $v = dx/dt$, we write

$$v_0 e^{-kt} = \frac{dx}{dt}$$

$$\int_0^x dx = v_0\int_0^t e^{-kt}\,dt$$

$$x = -\frac{v_0}{k}[e^{-kt}]_0^t = -\frac{v_0}{k}(e^{-kt} - 1)$$

$$x = \frac{v_0}{k}(1 - e^{-kt}) \blacktriangleleft$$

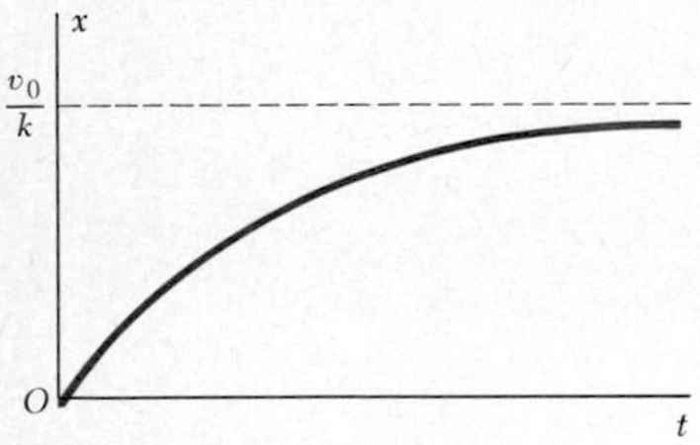

c. v in Terms of x. Substituting $-kv$ for a in $a = v\,dv/dx$, we write

$$-kv = v\frac{dv}{dx}$$

$$dv = -k\,dx$$

$$\int_{v_0}^{v} dv = -k\int_0^x dx$$

$$v - v_0 = -kx \qquad\qquad v = v_0 - kx \blacktriangleleft$$

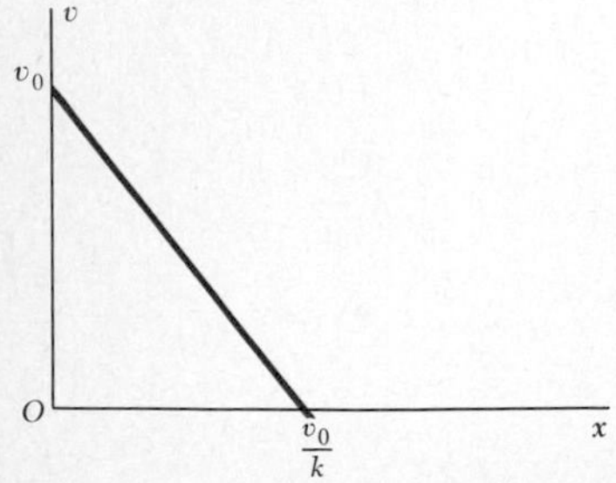

Check. Part *c* could have been solved by eliminating t from the answers obtained for parts *a* and *b*. This alternate method may be used as a check. From part *a* we obtain $e^{-kt} = v/v_0$; substituting in the answer of part *b*, we obtain

$$x = \frac{v_0}{k}(1 - e^{-kt}) = \frac{v_0}{k}\left(1 - \frac{v}{v_0}\right) \qquad v = v_0 - kx \qquad \text{(checks)}$$

Problems

11.1 The motion of a particle is defined by the relation $x = 2t^3 - 15t^2 + 36t - 10$, where x is expressed in meters and t in seconds. Determine the position, velocity, and acceleration when $t = 4$ s.

11.2 The motion of a particle is defined by the relation $x = t^4 - 3t^3 + t$, where x is expressed in inches and t in seconds. Determine the position, velocity, and acceleration when $t = 3$ s.

11.3 The motion of a particle is defined by the relation $x = t^3 - 3t^2 + 6$, where x is expressed in feet and t in seconds. Determine the time, position, and acceleration when $v = 0$.

11.4 The motion of a particle is defined by the relation $x = t^3 - 9t^2 + 15t + 18$, where x is expressed in meters and t in seconds. Determine the time, position, and acceleration when $v = 0$.

11.5 The motion of a particle is defined by the relation $x = t^3 - 12t^2 + 45t - 20$, where x is expressed in meters and t in seconds. Determine (*a*) when the velocity is zero, (*b*) the position, acceleration, and total distance traveled when $t = 5$ s.

11.6 The motion of a particle is defined by the relation $x = 2t^3 - 15t^2 + 24t + 4$, where x is expressed in meters and t in seconds. Determine (*a*) when the velocity is zero, (*b*) the position and the total distance traveled when the acceleration is zero.

11.7 The acceleration of a particle is defined by the relation $a = -5 \text{ ft/s}^2$. If $v = +30$ ft/s and $x = 0$ when $t = 0$, determine the velocity, position, and total distance traveled when $t = 8$ s.

11.8 The acceleration of a particle is directly proportional to the time t. At $t = 0$, the velocity of the particle is $v = -9$ m/s. Knowing that $v = 0$ and $x = 12$ m when $t = 3$ s, write the equations of motion of the particle.

11.9 The acceleration of a particle is defined by the relation $a = 32 - 6t^2$. The particle starts at $t = 0$ with $v = 0$ and $x = 50$ m. Determine (*a*) the time when the velocity is again zero, (*b*) the position and velocity when $t = 6$ s, (*c*) the total distance traveled by the particle from $t = 0$ to $t = 6$ s.

11.10 The acceleration of a particle is defined by the relation $a = kt^2$. (*a*) Knowing that $v = -24$ ft/s when $t = 0$ and that $v = +40$ ft/s when $t = 4$ s, determine the constant k. (*b*) Write the equations of motion, knowing also that $x = 6$ ft when $t = 2$ s.

11.11 The acceleration of a particle is defined by the relation $a = -kx^{-2}$. The particle starts with no initial velocity at $x = 900$ mm, and it is observed that its velocity is 10 m/s when $x = 300$ mm. Determine (*a*) the value of k, (*b*) the velocity of the particle when $x = 500$ mm.

11.12 The acceleration of a particle is defined by the relation $a = -k/x$. It has been experimentally determined that $v = 4$ m/s when $x = 250$ mm and that $v = 3$ m/s when $x = 500$ mm. Determine (*a*) the velocity of the particle when $x = 750$ mm, (*b*) the position of the particle at which its velocity is zero.

11.13 The acceleration of an oscillating particle is defined by the relation $a = -kx$. Find the value of k such that $v = 24$ in./s when $x = 0$ and $x = 6$ in. when $v = 0$.

11.14 The acceleration of a particle is defined by the relation $a = 90 - 6x^2$, where a is expressed in in./s^2 and x in inches. The particle starts with no initial velocity at the position $x = 0$. Determine (*a*) the velocity when $x = 5$ in., (*b*) the position where the velocity is again zero, (*c*) the position where the velocity is maximum.

11.15 The acceleration of a particle is defined by the relation $a = -16x(1 + kx^2)$, where a is expressed in m/s^2 and x in meters. Knowing that $v = 20$ m/s when $x = 0$, determine the velocity when $x = 4$ m, for (*a*) $k = 0$, (*b*) $k = 0.002$ m/s^2, (*c*) $k = -0.002$ m/s^2.

11.16 The acceleration of a particle is defined by the relation $a = -60x^{-1.5}$, where a is expressed in m/s^2 and x in meters. Knowing that the particle starts with no initial velocity at $x = 4$ m, determine the velocity of the particle when (*a*) $x = 2$ m, (*b*) $x = 1$ m, (*c*) $x = 100$ mm.

11.17 The acceleration of a particle is defined by the relation $a = -3v$, where a is expressed in in./s^2 and v in in./s. Knowing that at $t = 0$ the velocity is 60 in./s, determine (*a*) the distance the particle will travel before coming to rest, (*b*) the time required for the particle to come to rest, (*c*) the time required for the velocity of the particle to be reduced to 1 percent of its initial value.

11.18 The acceleration of a particle is defined by the relation $a = -kv^2$, where a is expressed in ft/s^2 and v in ft/s. The particle starts at $x = 0$ with a velocity of 20 ft/s and when $x = 100$ ft the velocity is found to be 15 ft/s. Determine the distance the particle will travel (*a*) before its velocity drops to 10 ft/s, (*b*) before it comes to rest.

11.19 Solve Prob. 11.18, assuming that the acceleration of the particle is defined by the relation $a = -kv^3$.

11.20 The magnitude in m/s^2 of the deceleration due to air resistance of the nose cone of a small experimental rocket is known to be $0.0005v^2$, where v is expressed in m/s. If the nose cone is projected vertically from the ground with an initial velocity of 100 m/s, determine the maximum height that it will reach. [*Hint.* The total acceleration is $-(g + 0.0005v^2)$, where $g = 9.81$ m/s^2.]

11.21 Determine the velocity of the nose cone of Prob. 11.20 when it returns to the ground. (*Hint.* The total acceleration is $g - 0.0005v^2$, where $g = 9.81$ m/s^2.)

11.22 The acceleration of a particle is defined by the relation $a = -kv^{1.5}$. The particle starts at $t = 0$ and $x = 0$ with an initial velocity v_0. (*a*) Show that the velocity and position coordinate at any time t are related by the equation $x/t = \sqrt{v_0 v}$. (*b*) Knowing that for $v_0 = 36$ m/s the particle comes to rest after traveling 3 m, determine the velocity of the particle and the time when $x = 2$ m.

11.23 The velocity of a particle is defined by the relation $v = 40 - 0.2x$, where v is expressed in m/s and x in meters. Knowing that $x = 0$ at $t = 0$, determine (*a*) the distance traveled before the particle comes to rest, (*b*) the acceleration at $t = 0$, (*c*) the time when $x = 50$ m.

11.24 A projectile enters a resisting medium at $x = 0$ with an initial velocity $v_0 = 900$ ft/s and travels 4 in. before coming to rest. Assuming that the velocity of the projectile was defined by the relation $v = v_0 - kx$, where v is expressed in ft/s and x in feet, determine (*a*) the initial acceleration of the projectile, (*b*) the time required for the projectile to penetrate 3.9 in. into the resisting medium.

11.25 The acceleration due to gravity at an altitude y above the surface of the earth may be expressed as

$$a = \frac{-9.81}{\left(1 + \dfrac{y}{6.37 \times 10^6}\right)^2}$$

where a is measured in m/s^2 and y in meters. Using this expression, compute the height reached by a bullet fired vertically upward from the surface of the earth with the following initial velocities: (*a*) 300 m/s, (*b*) 3000 m/s, (*c*) 11.18 km/s.

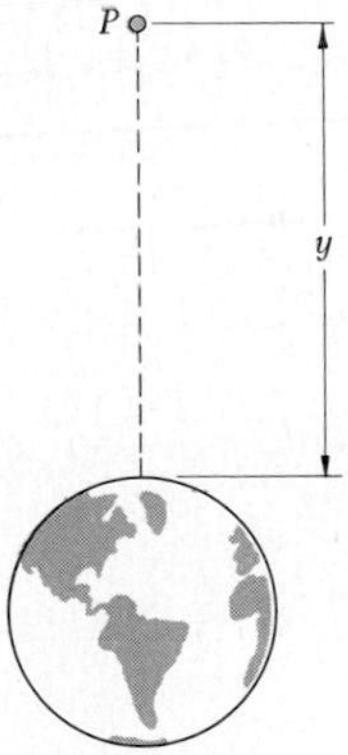

Fig. P11.25

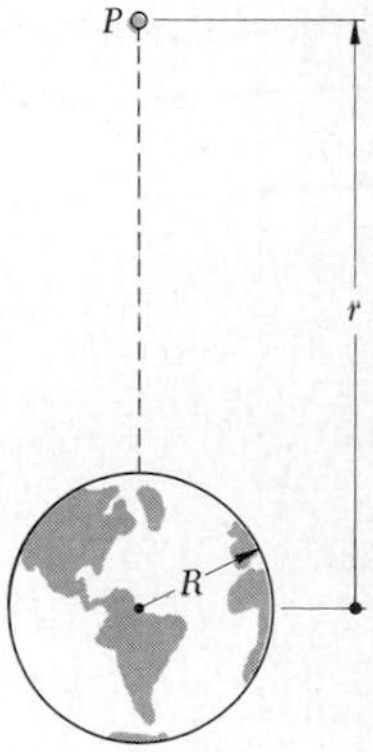

Fig. P11.26

11.26 The acceleration due to gravity of a particle falling toward the earth is $a = -gR^2/r^2$, where r is the distance from the *center* of the earth to the particle, R is the radius of the earth, and g is the acceleration due to gravity at the surface of the earth. Derive an expression for the *escape velocity*, i.e., for the minimum velocity with which a particle should be projected vertically upward from the surface of the earth if it is not to return to the earth. (*Hint.* $v = 0$ for $r = \infty$.)

11.27 The acceleration of a particle is $a = k \sin(\pi t/T)$. Knowing that both the velocity and the position coordinate of the particle are zero when $t = 0$, determine (*a*) the equations of motion, (*b*) the maximum velocity, (*c*) the position at $t = 2T$, (*d*) the average velocity during the interval $t = 0$ to $t = 2T$.

11.28 The position of an oscillating particle is defined by the relation $x = A \sin(pt + \phi)$. Denoting the velocity and position coordinate when $t = 0$ by v_0 and x_0, respectively, show (*a*) that $\tan\phi = x_0 p/v_0$, (*b*) that the maximum value of the position coordinate is

$$A = \sqrt{x_0^2 + \left(\frac{v_0}{p}\right)^2}$$

11.4. Uniform Rectilinear Motion. Uniform rectilinear motion is a type of straight-line motion which is frequently encountered in practical applications. In this motion, the acceleration a of the particle is zero for every value of t. The velocity v is therefore constant, and Eq. (11.1) becomes

$$\frac{dx}{dt} = v = \text{constant}$$

The position coordinate x is obtained by integrating this equation. Denoting by x_0 the initial value of x, we write

$$\int_{x_0}^{x} dx = v \int_0^t dt$$

$$x - x_0 = vt$$

$$x = x_0 + vt \qquad (11.5)$$

This equation may be used *only if the velocity of the particle is known to be constant.*

11.5 Uniformly Accelerated Rectilinear Motion. Uniformly accelerated rectilinear motion is another common type of motion. In this motion, the acceleration a of the particle is constant, and Eq. (11.2) becomes

$$\frac{dv}{dt} = a = \text{constant}$$

The velocity v of the particle is obtained by integrating this equation:

$$\int_{v_0}^{v} dv = a \int_0^t dt$$

$$v - v_0 = at$$

$$v = v_0 + at \qquad (11.6)$$

where v_0 is the initial velocity. Substituting for v into (11.1), we write

$$\frac{dx}{dt} = v_0 + at$$

Denoting by x_0 the initial value of x and integrating, we have

$$\int_{x_0}^{x} dx = \int_0^t (v_0 + at)\, dt$$

$$x - x_0 = v_0 t + \tfrac{1}{2}at^2$$

$$x = x_0 + v_0 t + \tfrac{1}{2}at^2 \qquad (11.7)$$

We may also use Eq. (11.4) and write

$$v\frac{dv}{dx} = a = \text{constant}$$

$$v\, dv = a\, dx$$

Integrating both sides, we obtain

$$\int_{v_0}^{v} v\,dv = a\int_{x_0}^{x} dx$$

$$\tfrac{1}{2}(v^2 - v_0^2) = a(x - x_0)$$

$$v^2 = v_0^2 + 2a(x - x_0) \tag{11.8}$$

The three equations we have derived provide useful relations among position coordinate, velocity, and time in the case of a uniformly accelerated motion, as soon as appropriate values have been substituted for a, v_0, and x_0. The origin O of the x axis should first be defined and a positive direction chosen along the axis; this direction will be used to determine the signs of a, v_0, and x_0. Equation (11.6) relates v and t and should be used when the value of v corresponding to a given value of t is desired, or inversely. Equation (11.7) relates x and t; Eq. (11.8) relates v and x. An important application of uniformly accelerated motion is the motion of a *freely falling body.* The acceleration of a freely falling body (usually denoted by g) is equal to 9.81 m/s², or 32.2 ft/s².

It is important to keep in mind that the three equations above may be used *only when the acceleration of the particle is known to be constant.* If the acceleration of the particle is variable, its motion should be determined from the fundamental equations (11.1) to (11.4), according to the methods outlined in Sec. 11.3.

11.6. Motion of Several Particles. When several particles move independently along the same line, independent equations of motion may be written for each particle. Whenever possible, time should be recorded from the same initial instant for all particles, and displacements should be measured from the same origin and in the same direction. In other words, a single clock and a single measuring tape should be used.

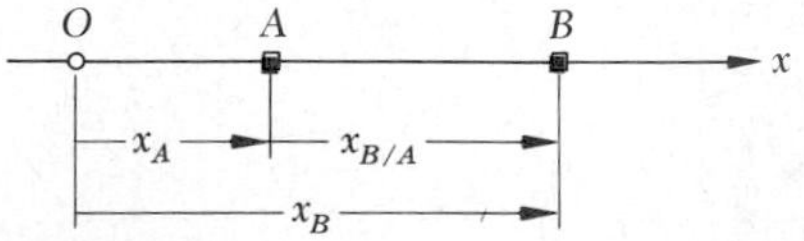

Fig. 11.7

Relative Motion of Two Particles. Consider two particles A and B moving along the same straight line (Fig. 11.7). If the position coordinates x_A and x_B are measured from the same origin, the difference $x_B - x_A$ defines the *relative position coordinate of B with respect to* A and is denoted by $x_{B/A}$. We write

$$x_{B/A} = x_B - x_A \qquad \text{or} \qquad x_B = x_A + x_{B/A} \tag{11.9}$$

A positive sign for $x_{B/A}$ means that B is to the right of A; a negative sign means that B is to the left of A, regardless of the position of A and B with respect to the origin.

The rate of change of $x_{B/A}$ is known as the *relative velocity of B with respect to* A and is denoted by $v_{B/A}$. Differentiating (11.9), we write

$$v_{B/A} = v_B - v_A \qquad \text{or} \qquad v_B = v_A + v_{B/A} \tag{11.10}$$

A positive sign for $v_{B/A}$ means that B is *observed from* A to move in the positive direction; a negative sign, that it is observed to move in the negative direction.

The rate of change of $v_{B/A}$ is known as the *relative acceleration of* B *with respect to* A and is denoted by $a_{B/A}$. Differentiating (11.10), we obtain†

$$a_{B/A} = a_B - a_A \quad \text{or} \quad a_B = a_A + a_{B/A} \tag{11.11}$$

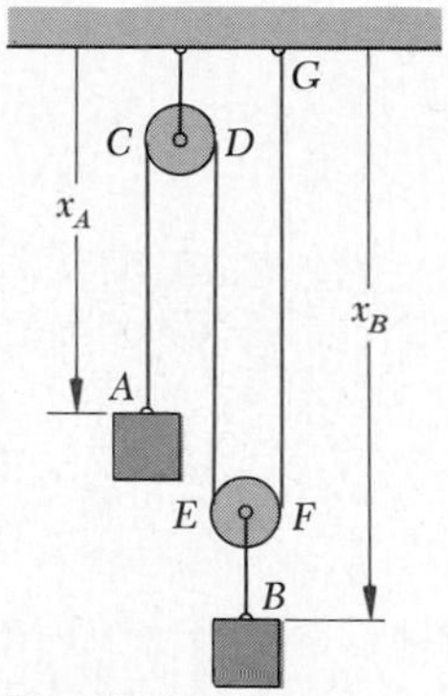

Fig. 11.8

Dependent Motions. Sometimes, the position of a particle will depend upon the position of another or of several other particles. The motions are then said to be dependent. For example, the position of block B in Fig. 11.8 depends upon the position of block A. Since the rope $ACDEFG$ is of constant length, and since the lengths of the portions of rope CD and EF wrapped around the pulleys remain constant, it follows that the sum of the lengths of the segments AC, DE, and FG is constant. Observing that the length of the segment AC differs from x_A only by a constant, and that, similarly, the lengths of the segments DE and FG differ from x_B only by a constant, we write

$$x_A + 2x_B = \text{constant}$$

Since only one of the two coordinates x_A and x_B may be chosen arbitrarily, we say that the system shown in Fig. 11.8 has *one degree of freedom*. From the relation between the position coordinates x_A and x_B, it follows that if x_A is given an increment Δx_A, that is, if block A is lowered by an amount Δx_A, the coordinate x_B will receive an increment $\Delta x_B = -\frac{1}{2}\Delta x_A$; that is, block B will rise by half the same amount; this may easily be checked directly from Fig. 11.8.

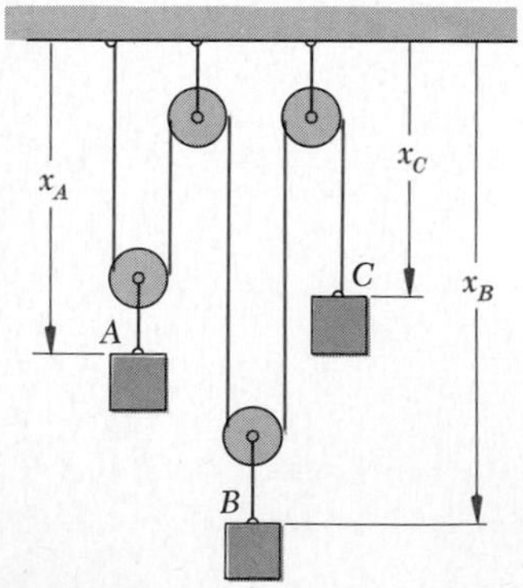

Fig. 11.9

In the case of the three blocks of Fig. 11.9, we may again observe that the length of the rope which passes over the pulleys is constant, and thus that the following relation must be satisfied by the position coordinates of the three blocks:

$$2x_A + 2x_B + x_C = \text{constant}$$

Since two of the coordinates may be chosen arbitrarily, we say that the system shown in Fig. 11.9 has *two degrees of freedom.*

When the relation existing between the position coordinates of several particles is *linear*, a similar relation holds between the velocities and between the accelerations of the particles. In the case of the blocks of Fig. 11.9, for instance, we differentiate twice the equation obtained and write

$$2\frac{dx_A}{dt} + 2\frac{dx_B}{dt} + \frac{dx_C}{dt} = 0 \quad \text{or} \quad 2v_A + 2v_B + v_C = 0$$

$$2\frac{dv_A}{dt} + 2\frac{dv_B}{dt} + \frac{dv_C}{dt} = 0 \quad \text{or} \quad 2a_A + 2a_B + a_C = 0$$

†Note that the product of the subscripts A and B/A used in the right-hand member of Eqs. (11.9), (11.10), and (11.11) is equal to the subscript B used in their left-hand member.

SAMPLE PROBLEM 11.4

A ball is thrown vertically upward from the 12-m level in an elevator shaft, with an initial velocity of 18 m/s. At the same instant an open-platform elevator passes the 5-m level, moving upward with a constant velocity of 2 m/s. Determine (*a*) when and where the ball will hit the elevator, (*b*) the relative velocity of the ball with respect to the elevator when the ball hits the elevator.

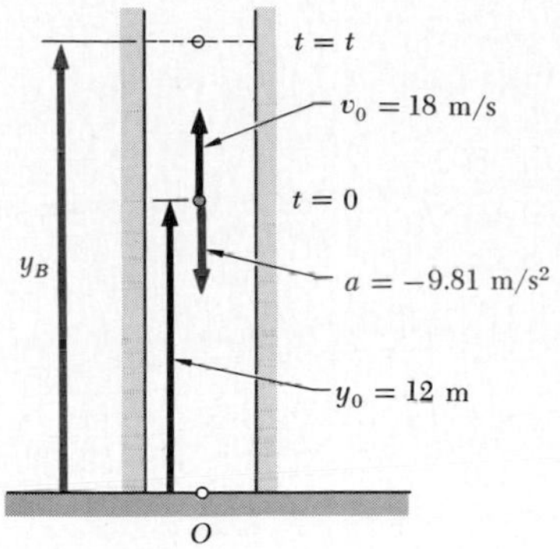

Motion of Ball. Since the ball has a constant acceleration, its motion is *uniformly accelerated.* Placing the origin O of the y axis at ground level and choosing its positive direction upward, we find that the initial position is $y_0 = +12$ m, the initial velocity is $v_0 = +18$ m/s, and the acceleration is $a = -9.81$ m/s^2. Substituting these values in the equations for uniformly accelerated motion, we write

$$v_B = v_0 + at \qquad v_B = 18 - 9.81t \tag{1}$$

$$y_B = y_0 + v_0t + \tfrac{1}{2}at^2 \qquad y_B = 12 + 18t - 4.905t^2 \tag{2}$$

Motion of Elevator. Since the elevator has a constant velocity, its motion is *uniform.* Again placing the origin O at the ground level and choosing the positive direction upward, we note that $y_0 = +5$ m and write

$$v_E = +2 \text{ m/s} \tag{3}$$

$$y_E = y_0 + v_Et \qquad y_E = 5 + 2t \tag{4}$$

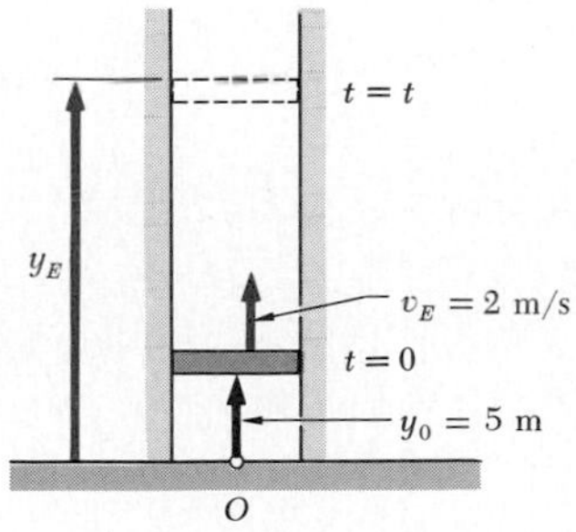

Ball Hits Elevator. We first note that the same time t and the same origin O were used in writing the equations of motion of both the ball and the elevator. We see from the figure that when the ball hits the elevator,

$$y_E = y_B \tag{5}$$

Substituting for y_E and y_B from (2) and (4) into (5), we have

$$5 + 2t = 12 + 18t - 4.905t^2$$

$$t = -0.39 \text{ s} \quad \text{and} \quad t = 3.65 \text{ s} \blacktriangleleft$$

Only the root $t = 3.65$ s corresponds to a time after the motion has begun. Substituting this value into (4), we have

$$y_E = 5 + 2(3.65) = 12.30 \text{ m}$$

Elevation from ground = 12.30 m ◄

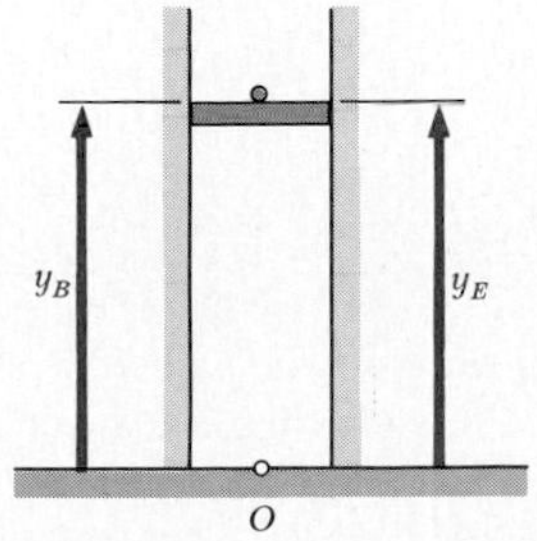

The relative velocity of the ball with respect to the elevator is

$$v_{B/E} = v_B - v_E = (18 - 9.81t) - 2 = 16 - 9.81t$$

When the ball hits the elevator at time $t = 3.65$ s, we have

$$v_{B/E} = 16 - 9.81(3.65) \qquad v_{B/E} = -19.81 \text{ m/s} \blacktriangleleft$$

The negative sign means that the ball is observed from the elevator to be moving in the negative sense (downward).

SAMPLE PROBLEM 11.5

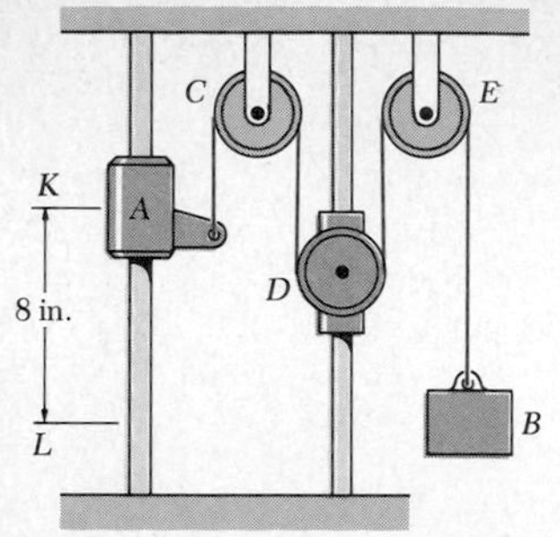

Collar A and block B are connected by a cable passing over three pulleys C, D, and E as shown. Pulleys C and E are fixed, while D is attached to a collar which is pulled downward with a constant velocity of 3 in./s. At $t = 0$, collar A starts moving downward from position K with a constant acceleration and no initial velocity. Knowing that the velocity of collar A is 12 in./s as it passes through point L, determine the change in elevation, the velocity, and the acceleration of block B when collar A passes through L.

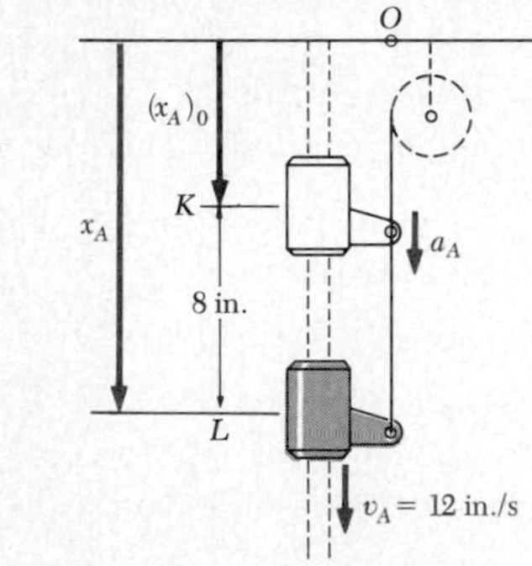

Motion of Collar A. We place the origin O at the upper horizontal surface and choose the positive direction downward. We observe that when $t = 0$, collar A is at position K and $(v_A)_0 = 0$. Since $v_A = 12$ in./s and $x_A - (x_A)_0 = 8$ in. when the collar passes through L, we write

$$v_A^2 = (v_A)_0^2 + 2a_A[x_A - (x_A)_0] \qquad (12)^2 = 0 + 2a_A(8)$$

$$a_A = 9 \text{ in./s}$$

The time at which collar A reaches point L is obtained by writing

$$v_A = (v_A)_0 + a_A t \qquad 12 = 0 + 9t \qquad t = 1.333 \text{ s}$$

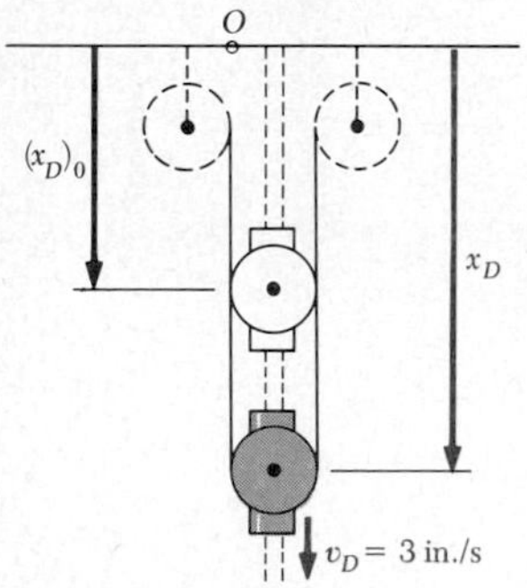

Motion of Pulley D. Recalling that the positive direction is downward, we write

$$a_D = 0 \qquad v_D = 3 \text{ in./s} \qquad x_D = (x_D)_0 + v_D t = (x_D)_0 + 3t$$

When collar A reaches L, at $t = 1.333$ s, we have

$$x_D = (x_D)_0 + 3(1.333) = (x_D)_0 + 4$$

Thus,

$$x_D - (x_D)_0 = 4 \text{ in.}$$

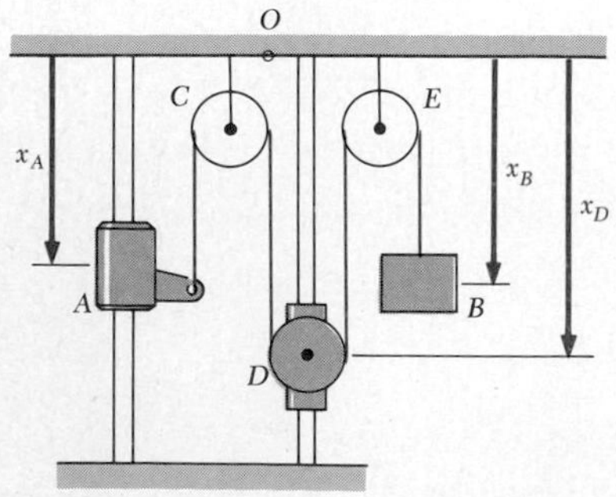

Motion of Block B. We note that the total length of cable $ACDEB$ differs from the quantity $(x_A + 2x_D + x_B)$ only by a constant. Since the cable length is constant during the motion, this quantity must also remain constant. Thus considering the times $t = 0$ and $t = 1.333$ s, we write

$$x_A + 2x_D + x_B = (x_A)_0 + 2(x_D)_0 + (x_B)_0 \qquad (1)$$

$$[x_A - (x_A)_0] + 2[x_D - (x_D)_0] + [x_B - (x_B)_0] = 0 \qquad (2)$$

But we know that $x_A - (x_A)_0 = 8$ in. and $x_D - (x_D)_0 = 4$ in.; substituting these values in (2), we find

$$8 + 2(4) + [x_B - (x_B)_0] = 0 \qquad x_B - (x_B)_0 = -16 \text{ in.}$$

Thus: Change in elevation of $B = 16$ in. ↑ ◄

Differentiating (1) twice, we obtain equations relating the velocities and the accelerations of A, B, and D. Substituting for the velocities and accelerations of A and D at $t = 1.333$ s, we have

$v_A + 2v_D + v_B = 0$: $\quad 12 + 2(3) + v_B = 0$

$$v_B = -18 \text{ in./s} \qquad v_B = 18 \text{ in./s} \uparrow \ ◄$$

$a_A + 2a_D + a_B = 0$: $\quad 9 + 2(0) + a_B = 0$

$$a_B = -9 \text{ in./s}^2 \qquad a_B = 9 \text{ in./s}^2 \uparrow \ ◄$$

Problems

11.29 A ball is thrown vertically upward from a point on a tower located 25 m above the ground. Knowing that the ball strikes the ground 3 s after release, determine the speed with which the ball (*a*) was thrown upward, (*b*) strikes the ground.

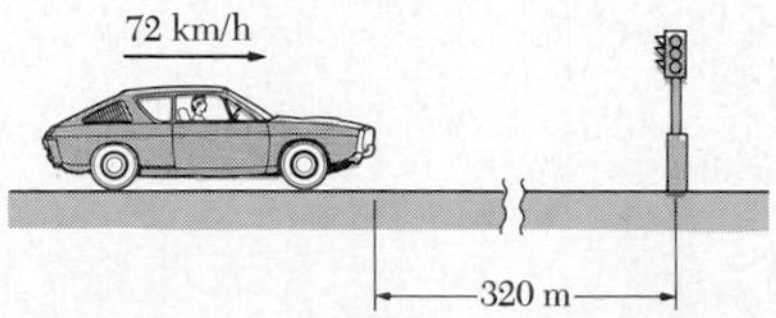

Fig. P11.30

11.30 A motorist is traveling at 72 km/h when she observes that a traffic light 320 m ahead of her turns red. The traffic light is timed to stay red for 22 s. If the motorist wishes to pass the light without stopping just as it turns green again, determine (*a*) the required uniform deceleration of the car, (*b*) the speed of the car as it passes the light.

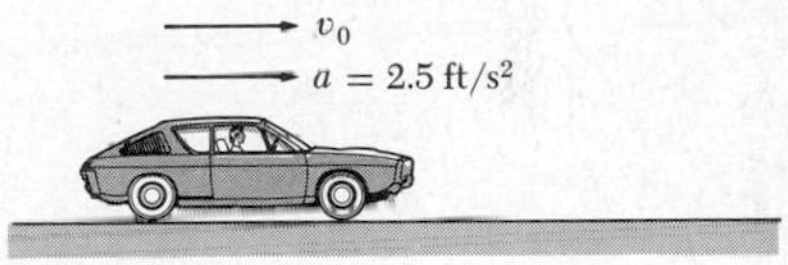

Fig. P11.31

11.31 An automobile travels 800 ft in 20 s while being accelerated at a constant rate of 2.5 ft/s^2. Determine (*a*) its initial velocity, (*b*) its final velocity, (*c*) the distance traveled during the first 10 s.

11.32 A stone is released from an elevator moving up at a speed of 12 ft/s and reaches the bottom of the shaft in 2.5 s. (*a*) How high was the elevator when the stone was released? (*b*) With what speed does the stone strike the bottom of the shaft?

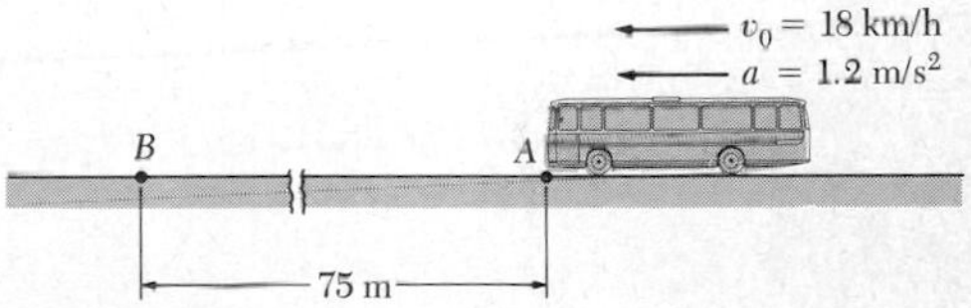

Fig. P11.33

11.33 A bus is accelerated at the rate of 1.2 m/s^2 as it travels from *A* to *B*. Knowing that the speed of the bus was $v_0 = 18$ km/h as it passed *A*, determine (*a*) the time required for the bus to reach *B*, (*b*) the corresponding speed as it passes *B*.

11.34 Solve Prob. 11.33, assuming that the velocity of the bus as it passed *A* was 36 km/h.

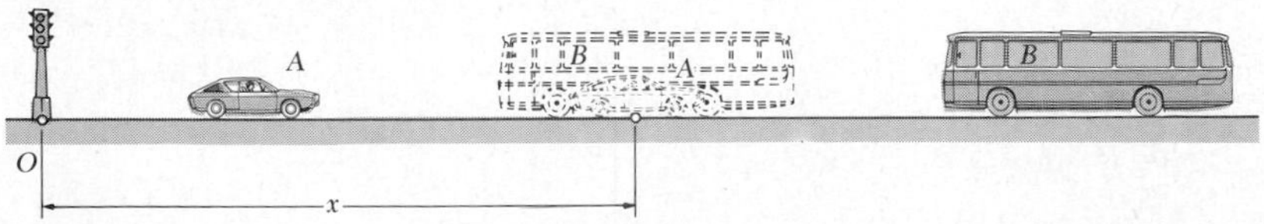

Fig. P11.35

11.35 Automobile *A* starts from *O* and accelerates at the constant rate of 0.8 m/s^2. A short time later it is passed by bus *B* which is traveling in the opposite direction at a constant speed of 5 m/s. Knowing that bus *B* passes point *O* 22 s after automobile *A* started from there, determine when and where the vehicles passed each other.

11.36 An open-platform elevator is moving down a mine shaft at a constant velocity v_e when the elevator platform hits and dislodges a stone. (*a*) Assuming that the stone starts falling with no initial velocity, show that the stone will hit the platform with a relative velocity of magnitude v_e. (*b*) If $v_e = 7.5$ m/s, determine when and where the stone will hit the elevator platform.

11.37 A freight elevator moving upward with a constant velocity of 6 ft/s passes a passenger elevator which is stopped. Four seconds later the passenger elevator starts upward with a constant acceleration of 2.4 ft/s². Determine (*a*) when and where the elevators will be at the same height, (*b*) the speed of the passenger elevator at that time.

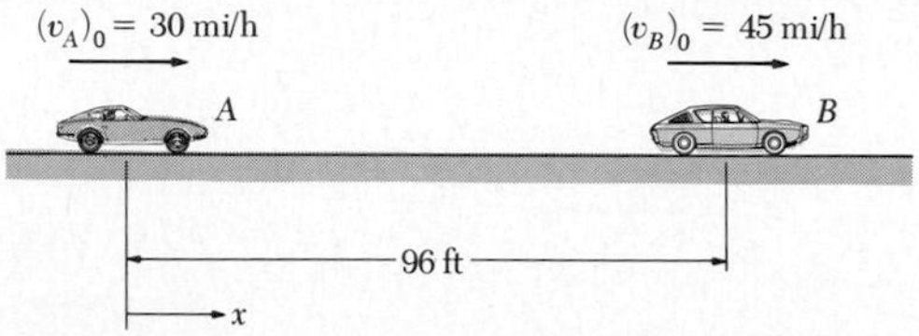

Fig. P11.38

11.38 Automobiles A and B are traveling in adjacent highway lanes and at $t = 0$ have the positions and speeds shown. Knowing that automobile A has a constant acceleration of 2 ft/s² and that B has a constant deceleration of 1.5 ft/s², determine (*a*) when and where A will overtake B, (*b*) the speed of each automobile at that time.

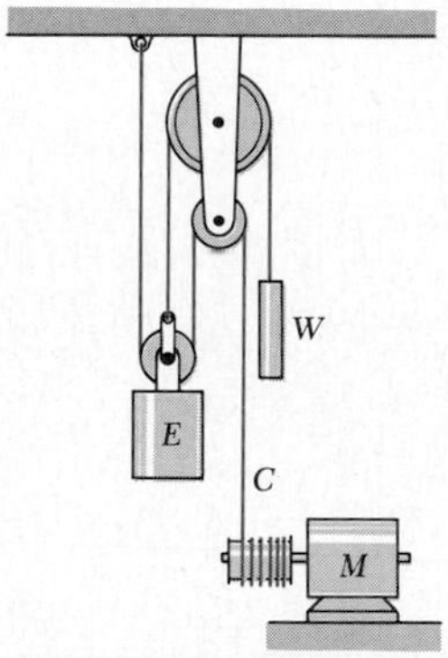

Fig. P11.39 and P11.40

11.39 The elevator shown in the figure moves upward at the constant velocity of 4 m/s. Determine (*a*) the velocity of the cable C, (*b*) the velocity of the counterweight W, (*c*) the relative velocity of the cable C with respect to the elevator, (*d*) the relative velocity of the counterweight W with respect to the elevator.

11.40 The elevator shown starts from rest and moves upward with a constant acceleration. If the counterweight W moves through 10 m in 4 s, determine (*a*) the acceleration of the elevator, (*b*) the acceleration of cable C, (*c*) the velocity of the elevator after 6 s.

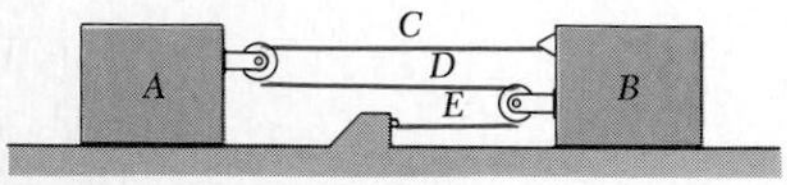

Fig. P11.41 and P11.42

11.41 The slider block B moves to the right with a constant velocity of 20 in./s. Determine (*a*) the velocity of block A, (*b*) the velocity of portion D of the cable, (*c*) the relative velocity of A with respect to B, (*d*) the relative velocity of portion C of the cable with respect to portion D.

11.42 The slider block A starts from rest and moves to the left with a constant acceleration. Knowing that the velocity of block B is 12 in./s after moving 24 in., determine (*a*) the accelerations of A and B, (*b*) the velocity and position of A after 5 s.

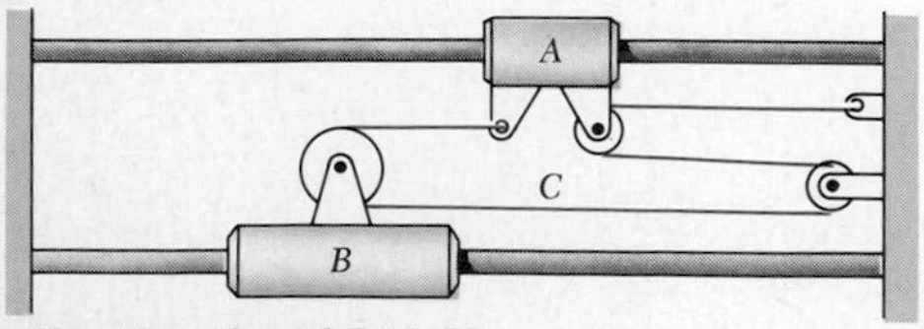

Fig. P11.43 and P11.44

11.43 Collar A starts from rest and moves to the left with a constant acceleration. Knowing that after 4 s the relative velocity of collar B with respect to collar A is 300 mm/s, determine (*a*) the accelerations of A and B, (*b*) the position and velocity of B after 5 s.

11.44 In the position shown collar B moves to the left with a velocity of 150 mm/s. Determine (*a*) the velocity of collar A, (*b*) the velocity of portion C of the cable, (*c*) the relative velocity of portion C of the cable with respect to collar B.

11.45 Under normal operating conditions, tape is transferred between the reels shown at a speed of 600 mm/s. At $t = 0$, portion A of the tape is moving to the right at a speed of 480 mm/s and has a constant acceleration. Knowing that portion B of the tape has a constant speed of 600 mm/s and that the speed of portion A reaches 600 mm/s at $t = 5$ s, determine (a) the acceleration and velocity of the compensator C at $t = 3$ s, (b) the distance through which C will have moved at $t = 5$ s.

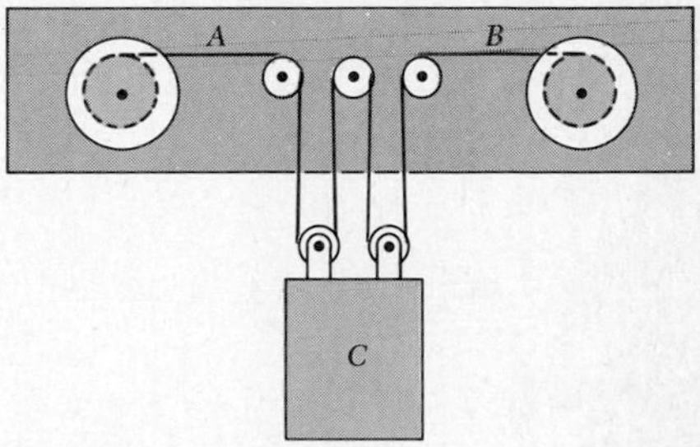

Fig. P11.45 and P11.46

11.46 Portions A and B of the tape shown start from rest at $t = 0$ and are each accelerated uniformly until a speed of 600 mm/s is reached; each portion of tape then moves at a constant speed of 600 mm/s. Knowing that portions A and B reached the speed of 600 mm/s in, respectively, 15 s and 12 s, determine (a) the acceleration and velocity of the compensator C at $t = 8$ s, (b) the distance through which C will have moved when both portions of tape reach their final speed.

11.47 Collar A starts from rest at $t = 0$ and moves upward with a constant acceleration of 2.5 in./s^2. Knowing that collar B moves downward with a constant velocity of 15 in./s, determine (a) the time at which the velocity of block C is zero, (b) the corresponding position of block C.

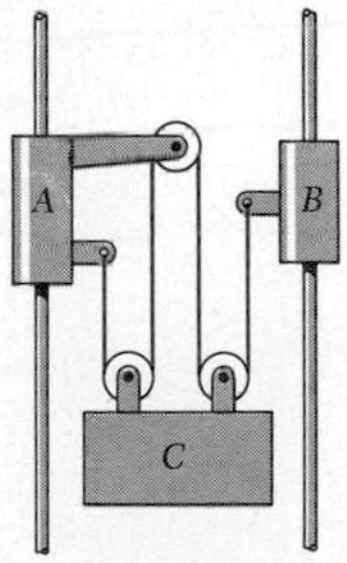

Fig. P11.47 and P11.48

11.48 Collars A and B start from rest and move with the following accelerations: $a_A = 3t$ in./s^2 upward and $a_B = 9$ in./s^2 downward. Determine (a) the time at which the velocity of block C is again zero, (b) the distance through which block C will have moved at that time.

11.49 (a) Choosing the positive sense downward for each block, express the velocity of A in terms of the velocities of B and C. (b) Knowing that both blocks A and C start from rest and move downward with the respective accelerations $a_A = 50$ mm/s^2 and $a_C = 110$ mm/s^2, determine the velocity and position of B after 3 s.

11.50 The three blocks shown move with constant velocities. Find the velocity of each block, knowing that the relative velocity of A with respect to C is 300 mm/s upward and that the relative velocity of B with respect to A is 200 mm/s downward.

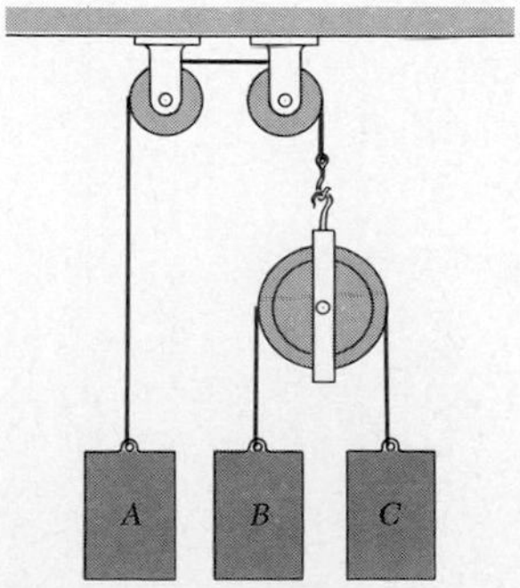

Fig. P11.49 and P11.50

***11.7. Graphical Solution of Rectilinear-Motion Problems.** It was observed in Sec. 11.2 that the fundamental formulas

$$v = \frac{dx}{dt} \quad \text{and} \quad a = \frac{dv}{dt}$$

have a geometrical significance. The first formula expresses that the velocity at any instant is equal to the slope of the x–t curve at the same instant (Fig. 11.10). The second formula expresses that the acceleration is equal to

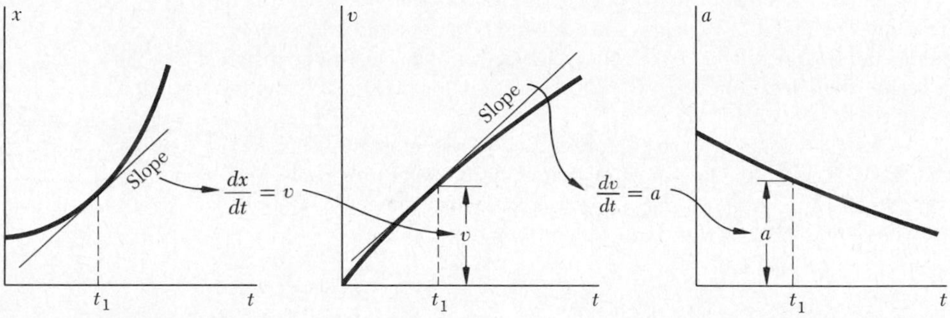

Fig. 11.10

the slope of the v–t curve. These two properties may be used to derive graphically the v–t and a–t curves of a motion when the x–t curve is known.

Integrating the two fundamental formulas from a time t_1 to a time t_2, we write

$$x_2 - x_1 = \int_{t_1}^{t_2} v\,dt \quad \text{and} \quad v_2 - v_1 = \int_{t_1}^{t_2} a\,dt \tag{11.12}$$

The first formula expresses that the area measured under the v–t curve from t_1 to t_2 is equal to the change in x during that time interval (Fig. 11.11). The second formula expresses similarly that the area measured under the a–t curve from t_1 to t_2 is equal to the change in v during that time interval. These two properties may be used to determine graphically the x–t curve of a motion when its v–t curve or its a–t curve is known (see Sample Prob. 11.6).

Graphical solutions are particularly useful when the motion considered is defined from experimental data and when x, v, and a are not analytical functions of t. They may also be used to advantage when the motion consists of distinct parts and when its analysis requires writing a different equation for each of its parts. When using a graphical solution, however, one should be careful to note (1) that the area under the v–t curve measures the *change in x*, not x itself, and, similarly, that the area under the a–t curve measures the change in v; (2) that while an area above the t axis corresponds to an increase in x or v, an area located below the t axis measures a decrease in x or v.

It will be useful to remember in drawing motion curves that if the velocity is constant, it will be represented by a horizontal straight line; the position coordinate x will then be a linear function of t and will be repre-

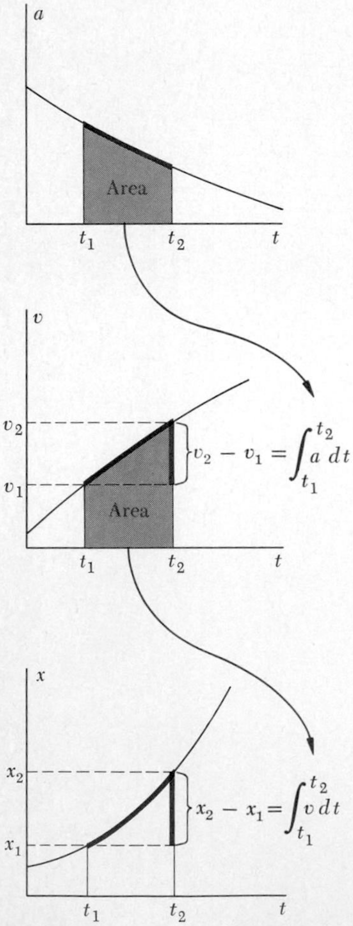

Fig. 11.11

sented by an oblique straight line. If the acceleration is constant and different from zero, it will be represented by a horizontal straight line; v will then be a linear function of t, represented by an oblique straight line; and x will be expressed as a second-degree polynomial in t, represented by a parabola. If the acceleration is a linear function of t, the velocity and the position coordinate will be equal, respectively, to second-degree and third-degree polynomials; a will then be represented by an oblique straight line, v by a parabola, and x by a cubic. In general, if the acceleration is a polynomial of degree n in t, the velocity will be a polynomial of degree $n + 1$ and the position coordinate a polynomial of degree $n + 2$; these polynomials are represented by motion curves of a corresponding degree.

***11.8. Other Graphical Methods.** An alternative graphical solution may be used to determine directly from the a–t curve the position of a particle at a given instant. Denoting, respectively, by x_0 and v_0 the values of x and v at $t = 0$ and by x_1 and v_1 their values at $t = t_1$, and observing that the area under the v–t curve may be divided into a rectangle of area $v_0 t_1$ and horizontal differential elements of area $(t_1 - t)dv$ (Fig. 11.12*a*), we write

$$x_1 - x_0 = \text{area under } v\text{–}t \text{ curve} = v_0 t_1 + \int_{v_0}^{v_1} (t_1 - t)\, dv$$

Substituting $dv = a\, dt$ in the integral, we obtain

$$x_1 - x_0 = v_0 t_1 + \int_0^{t_1} (t_1 - t) a\, dt$$

Referring to Fig. 11.12*b*, we note that the integral represents the first moment of the area under the a–t curve with respect to the line $t = t_1$ bounding the area on the right. This method of solution is known, therefore, as the *moment-area method.* If the abscissa $\bar{t}$ of the centroid C of the area is known, the position coordinate x_1 may be obtained by writing

$$x_1 = x_0 + v_0 t_1 + (\text{area under } a\text{–}t \text{ curve})(t_1 - \bar{t}) \tag{11.13}$$

If the area under the a–t curve is a composite area, the last term in (11.13) may be obtained by multiplying each component area by the distance from its centroid to the line $t = t_1$. Areas above the t axis should be considered as positive and areas below the t axis, as negative.

Another type of motion curve, the v–x curve, is sometimes used. If such a curve has been plotted (Fig. 11.13), the acceleration a may be obtained at any time by drawing the normal to the curve and *measuring the subnormal BC.* Indeed, observing that the angle between AC and AB is equal to the angle θ between the horizontal and the tangent at A (the slope of which is $\tan\theta = dv/dx$), we write

$$BC = AB \tan\theta = v\frac{dv}{dx}$$

and thus, recalling formula (11.4),

$$BC = a$$

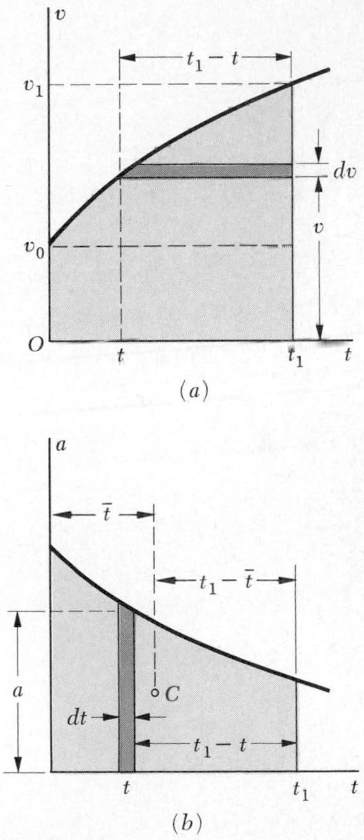

Fig. 11.12

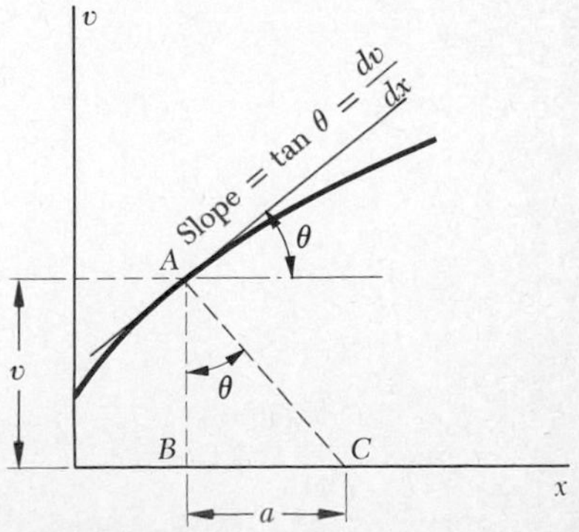

Fig. 11.13

SAMPLE PROBLEM 11.6

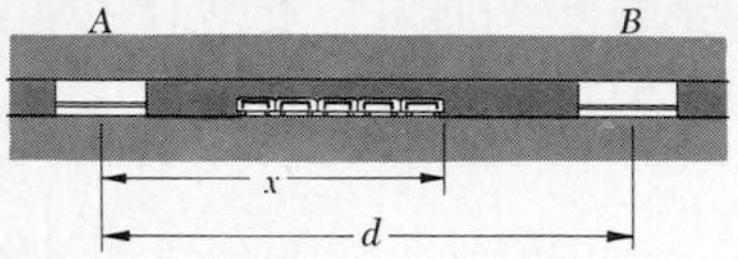

A subway train leaves station A; it gains speed at the rate of 4 ft/s^2 for 6 s and then at the rate of 6 ft/s^2 until it has reached the speed of 48 ft/s. The train maintains the same speed until it approaches station B; brakes are then applied, giving the train a constant deceleration and bringing it to a stop in 6 s. The total running time from A to B is 40 s. Draw the a–t, v–t, and x–t curves, and determine the distance between stations A and B.

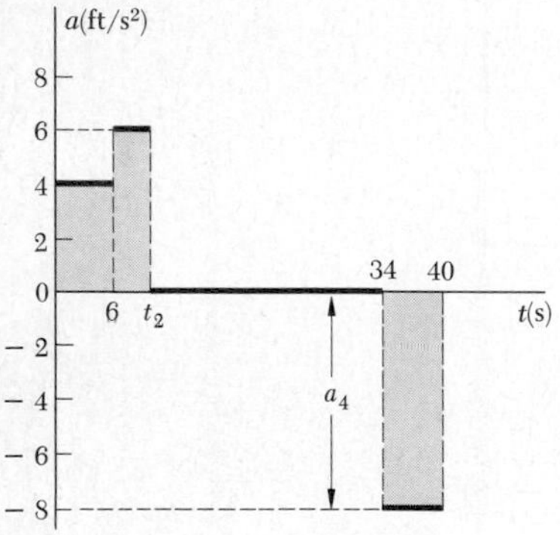

Acceleration-Time Curve. Since the acceleration is either constant or zero, the a–t curve is made of horizontal straight-line segments. The values of t_2 and a_4 are determined as follows:

$0 < t < 6$: Change in v = area under a–t curve

$$v_6 - 0 = (6 \text{ s})(4 \text{ ft/s}^2) = 24 \text{ ft/s}$$

$6 < t < t_2$: Since the velocity increases from 24 to 48 ft/s,

Change in v = area under a–t curve

$$48 - 24 = (t_2 - 6)(6 \text{ ft/s}^2) \qquad t_2 = 10 \text{ s}$$

$t_2 < t < 34$: Since the velocity is constant, the acceleration is zero.

$34 < t < 40$: Change in v = area under a–t curve

$$0 - 48 = (6 \text{ s})a_4 \qquad a_4 = -8 \text{ ft/s}^2$$

The acceleration being negative, the corresponding area is below the t axis; this area represents a decrease in velocity.

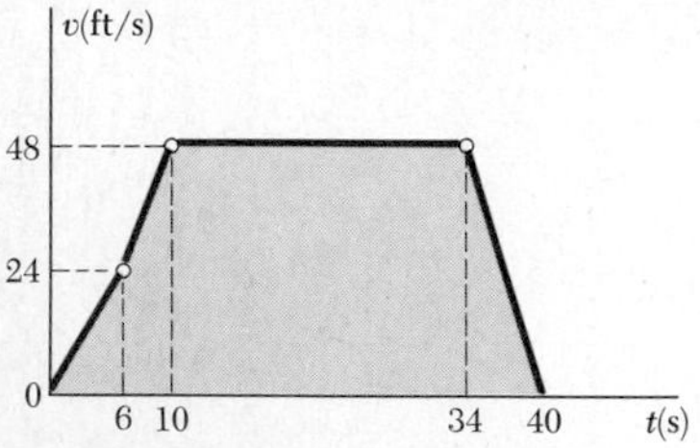

Velocity-Time Curve. Since the acceleration is either constant or zero, the v–t curve is made of segments of straight line connecting the points determined above.

Change in x = area under v–t curve

$0 < t < 6$: $\quad x_6 - 0 = \frac{1}{2}(6)(24) = 72 \text{ ft}$

$6 < t < 10$: $\quad x_{10} - x_6 = \frac{1}{2}(4)(24 + 48) = 144 \text{ ft}$

$10 < t < 34$: $\quad x_{34} - x_{10} = (24)(48) = 1152 \text{ ft}$

$34 < t < 40$: $\quad x_{40} - x_{34} = \frac{1}{2}(6)(48) = 144 \text{ ft}$

Adding the changes in x, we obtain the distance from A to B:

$$d = x_{40} - 0 = 1512 \text{ ft}$$

$$d = 1512 \text{ ft} \blacktriangleleft$$

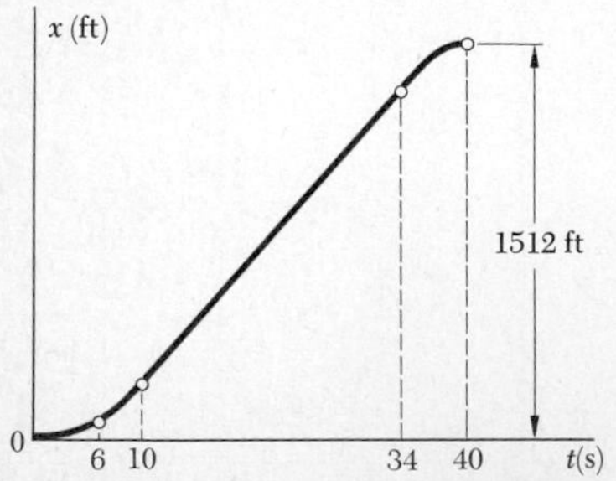

Position-Time Curve. The points determined above should be joined by three arcs of parabola and one segment of straight line. The construction of the x–t curve will be performed more easily and more accurately if we keep in mind that for any value of t the slope of the tangent to the x–t curve is equal to the value of v at that instant.

Problems

11.51 A particle moves in a straight line with the acceleration shown in the figure. Knowing that it starts from the origin with $v_0 = -14$ ft/s, plot the v–t and x–t curves for $0 < t < 15$ s and determine (*a*) the maximum value of the velocity of the particle, (*b*) the maximum value of its position coordinate.

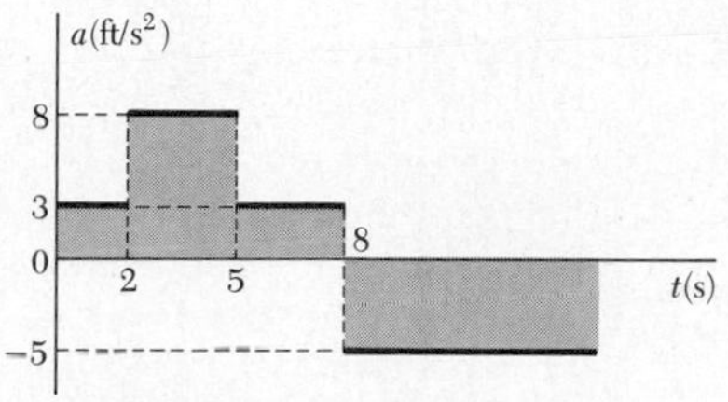

Fig. P11.51

11.52 For the particle and motion of Prob. 11.51, plot the v–t and x–t curves for $0 < t < 15$ s and determine the velocity of the particle, its position, and the total distance traveled after 10 s.

11.53 A particle moves in a straight line with the velocity shown in the figure. Knowing that $x = -8$ m at $t = 0$, draw the a–t and x–t curves for $0 < t < 20$ s and determine (*a*) the maximum value of the position coordinate of the particle, (*b*) the values of t for which the particle is at a distance of 18 m from the origin.

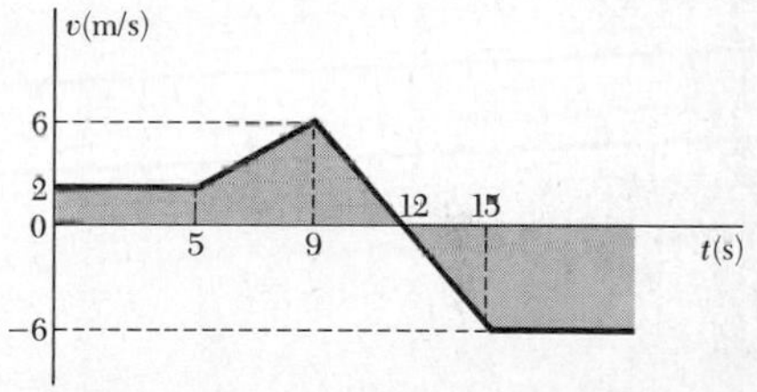

Fig. P11.53

11.54 For the particle and motion of Prob. 11.53, plot the a–t and x–t curves for $0 < t < 20$ s and determine (*a*) the total distance traveled by the particle during the period $t = 0$ to $t = 15$ s, (*b*) the two values of t for which the particle passes through the origin.

11.55 A bus starts from rest at point A and accelerates at the rate of 0.75 m/s^2 until it reaches a speed of 9 m/s. It then proceeds at 9 m/s until the brakes are applied; it comes to rest at point B, 27 m beyond the point where the brakes were applied. Assuming uniform deceleration and knowing that the distance between A and B is 180 m, determine the time required for the bus to travel from A to B.

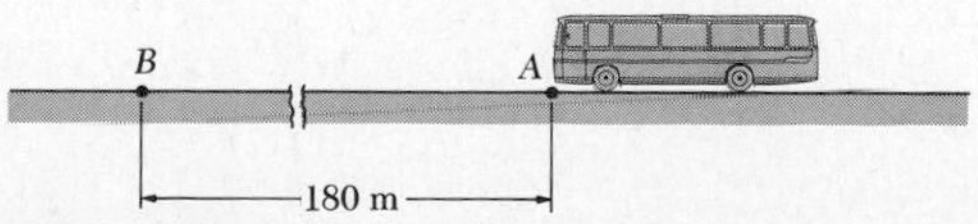

Fig. P11.55

11.56 A series of city traffic signals is timed so that an automobile traveling at a constant speed of 54 km/h will reach each signal just as it turns green. A motorist misses a signal and is stopped at signal A. Knowing that the next signal B is 300 m ahead and that the maximum acceleration of the automobile is 1.8 m/s^2, determine what the motorist should do to keep the maximum speed as small as possible, yet reach signal B just as it turns green. What is the maximum speed reached?

11.57 Firing a howitzer causes the barrel to recoil 42 in. before a braking mechanism brings it to rest. From a high-speed photographic record, it is found that the maximum value of the recoil velocity is 250 in./s and that this is reached 0.02 s after firing. Assuming that the recoil period consists of two phases during which the acceleration has, respectively, a constant positive value a_1 and a constant negative value a_2, determine (*a*) the values of a_1 and a_2, (*b*) the position of the barrel 0.02 s after firing, (*c*) the time at which the velocity of the barrel is zero.

11.58 During a finishing operation the bed of an industrial planer moves alternately 36 in. to the right and 36 in. to the left. The velocity of the bed is limited to a maximum value of 6 in./s to the right and 9 in./s to the left; the acceleration is successively equal to 3 in./s^2 to the right, zero, 3 in./s^2 to the left, zero, etc. Determine the time required for the bed to complete a full cycle, and draw the v–t and x–t curves.

11.59 A motorist is traveling at 72 km/h when she observes that a traffic signal 320 m ahead of her turns red. She knows that the signal is timed to stay red for 22 s. What should she do to pass the signal at 72 km/h just as it turns green again? Draw the v–t curve, selecting the solution which calls for the smallest possible deceleration and acceleration, and determine (*a*) the deceleration and acceleration in m/s^2, (*b*) the minimum speed reached in km/h.

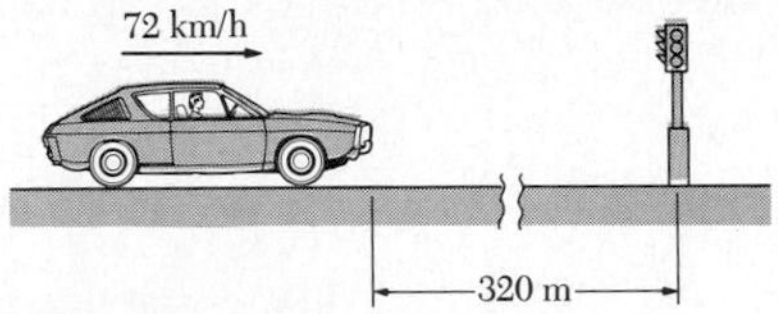

Fig. P11.59

11.60 Solve Prob. 11.59, knowing that the acceleration of the automobile cannot exceed 0.75 m/s^2.

11.61 An automobile at rest is passed by a truck traveling at a constant speed of 54 km/h. The automobile starts and accelerates for 10 s at a constant rate until it reaches a speed of 81 km/h. If the automobile then maintains a constant speed of 81 km/h, determine when and where it will overtake the truck, assuming that the automobile starts (*a*) just as the truck passes it, (*b*) 2 s after the truck has passed it.

11.62 A motorcycle and an automobile are both traveling at the constant speed of 40 mi/h; the motorcycle is 50 ft behind the automobile. The motorcyclist wants to pass the automobile, i.e., he wishes to place his motorcycle at *B*, 50 ft in front of the automobile, and then resume the speed of 40 mi/h. The maximum acceleration of the motorcycle is 6 ft/s^2 and the maximum deceleration obtained by applying the brakes is 18 ft/s^2. What is the shortest time in which the motorcyclist can complete the passing operation if he does not at any time exceed a speed of 55 mi/h? Draw the v–t curve.

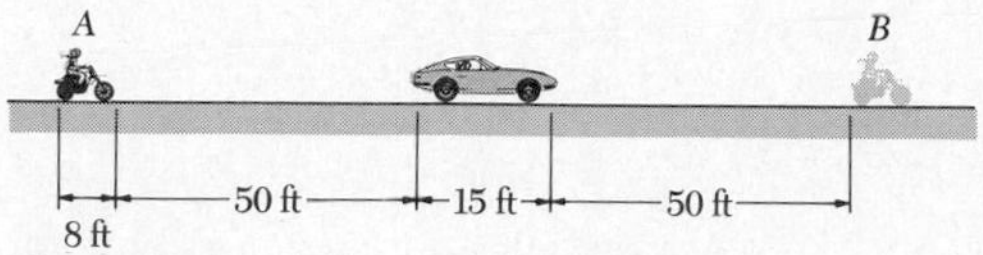

Fig. P11.62

11.63 Solve Prob. 11.62, assuming that the motorcyclist does not pay any attention to the speed limit while passing and concentrates on reaching position *B* and resuming a speed of 40 mi/h in the shortest possible time. What is the maximum speed reached? Draw the v–t curve.

11.64 A car and a truck are both traveling at a constant speed v_0; the car is 35 ft behind the truck. The truck driver suddenly applies his brakes, causing the truck to decelerate at the constant rate of 10 ft/s^2. Two seconds later the driver of the car applies her brakes and barely manages to avoid a rear-end collision. Determine the constant rate at which the car decelerated if (*a*) v_0 = 60 mi/h, (*b*) v_0 = 30 mi/h.

11.65 Car *A* is traveling at the constant speed v_A. It approaches car *B*, which is traveling in the same direction at the constant speed of 63 km/h. The driver of car *B* notices car *A* when it is still 50 m behind him and then accelerates at the constant rate of 0.8 m/s^2 to avoid being passed or struck by car *A*. Knowing that the closest that *A* comes to *B* is 10 m, determine the speed v_A of car *A*.

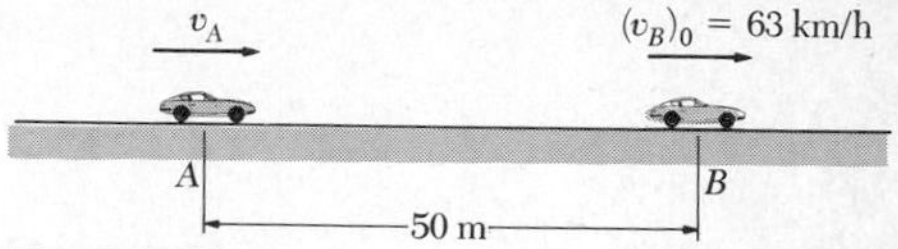

Fig. P11.65

11.66 A fighter plane flying horizontally in a straight line at 900 km/h is overtaking a bomber flying in the same straight line at 720 km/h. The pilot of the fighter plane fires an air-to-air missile at the bomber when his plane is 1150 m behind the bomber. The missile accelerates at a constant rate of 400 m/s^2 for 1 s and then travels at a constant speed. (*a*) How many seconds after firing will the missile reach the bomber? (*b*) If both planes continue at constant speeds, what will be the distance between the planes when the missile strikes the bomber?

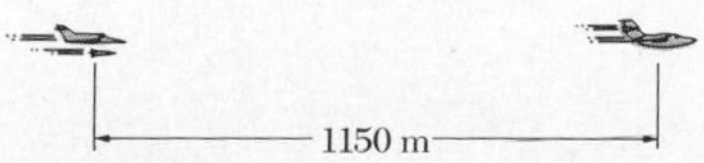

Fig. P11.66

11.67 The acceleration record shown was obtained for a small airplane traveling along a straight course. Knowing that $x = 0$ and $v = 50$ m/s when $t = 0$, determine (*a*) the velocity and position of the plane at $t = 20$ s, (*b*) its average velocity during the interval $6 \text{ s} < t < 14$ s.

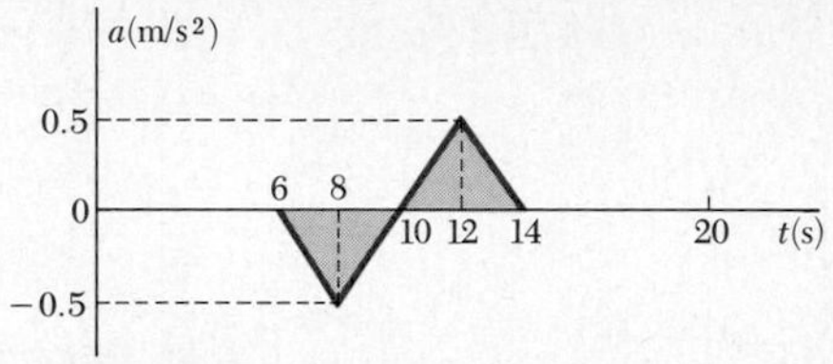

Fig. P11.67

11.68 A train starts at a station and accelerates uniformly at a rate of 1 ft/s^2 until it reaches a speed of 20 ft/s; it then proceeds at the constant speed of 20 ft/s. Determine the time and the distance traveled if its *average* velocity is (*a*) 12 ft/s, (*b*) 18 ft/s.

11.69 The rate of change of acceleration is known as the *jerk;* large or abrupt rates of change of acceleration cause discomfort to elevator passengers. If the jerk, or rate of change of the acceleration, of an elevator is limited to ± 1.5 ft/s^2 per second, determine (*a*) the shortest time required for an elevator, starting from rest, to rise 24 ft and stop, (*b*) the corresponding average velocity of the elevator.

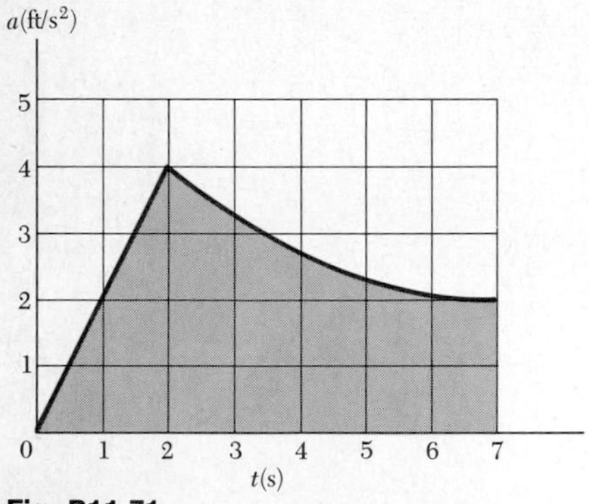

Fig. P11.71

11.70 In order to maintain passenger comfort, the acceleration of an elevator is limited to ± 1.2 m/s^2 and the jerk, or rate of change of acceleration, is limited to ± 0.4 m/s^2 per second. If the elevator starts from rest, determine (*a*) the shortest time required for it to attain a constant velocity of 6 m/s, (*b*) the distance traveled in that time, (*c*) the corresponding average velocity of the elevator.

11.71 The acceleration record shown was obtained for an automobile traveling on a straight highway. Knowing that the initial velocity of the automobile was 15 mi/h, determine the velocity and distance traveled when (*a*) $t = 3$ s, (*b*) $t = 6$ s.

11.72 A test sled is brought to rest in 5 s after the completion of an experiment. An accelerometer attached to the sled provides the acceleration record shown. Determine by approximate means (*a*) the initial velocity of the sled, (*b*) the distance that the sled travels during the interval $0 < t < 5$ s.

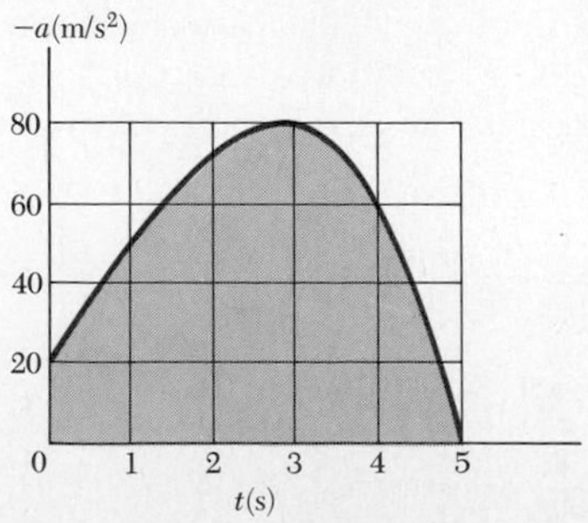

Fig. P11.72

11.73 The maximum possible deceleration of a passenger train under emergency conditions was determined experimentally; the results are shown (solid curve) in the figure. If the brakes are applied when the train is traveling at 108 km/h, determine by approximate means (*a*) the time required for the train to come to rest, (*b*) the distance traveled in that time.

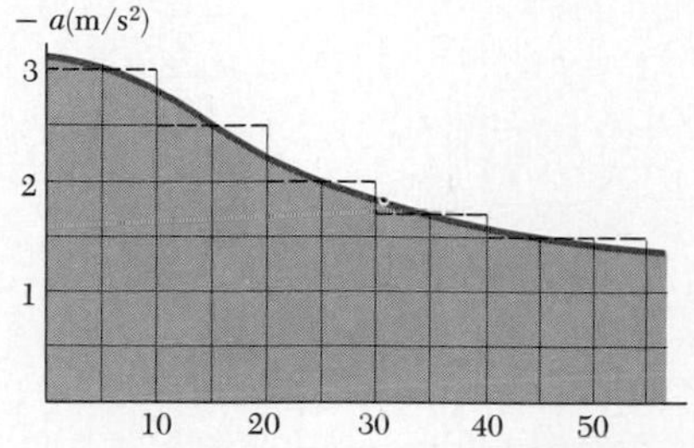

Fig. P11.73

11.74 The v–x curve shown was obtained experimentally during the motion of the bed of an industrial planer. Determine by approximate means the acceleration (*a*) when $x = 3$ in., (*b*) when $v = 4$ in./s.

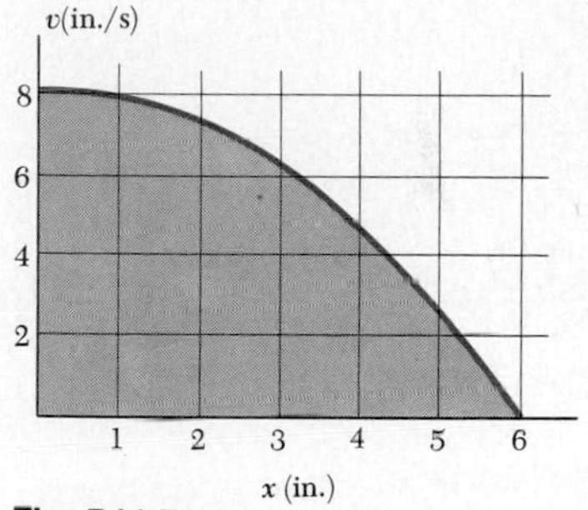

Fig. P11.74

11.75 Using the method of Sec. 11.8, derive the formula $x = x_0 + v_0t + \frac{1}{2}at^2$ for the position coordinate of a particle in uniformly accelerated rectilinear motion.

11.76 The acceleration of an object subjected to the pressure wave of a large explosion is defined approximately by the curve shown. The object is initially at rest and is again at rest at time t_1. Using the method of Sec. 11.8, determine (*a*) the time t_1, (*b*) the distance through which the object is moved by the pressure wave.

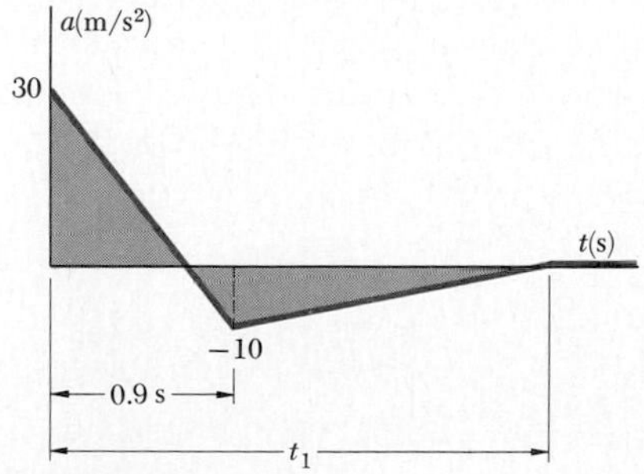

Fig. P11.76

11.77 For the particle of Prob. 11.53, draw the a–t curve and, using the method of Sec. 11.8, determine (*a*) the position of the particle when $t = 10$ s, (*b*) the maximum value of its position coordinate.

11.78 Using the method of Sec. 11.8, determine the position of the particle of Prob. 11.51 when $t = 8$ s.

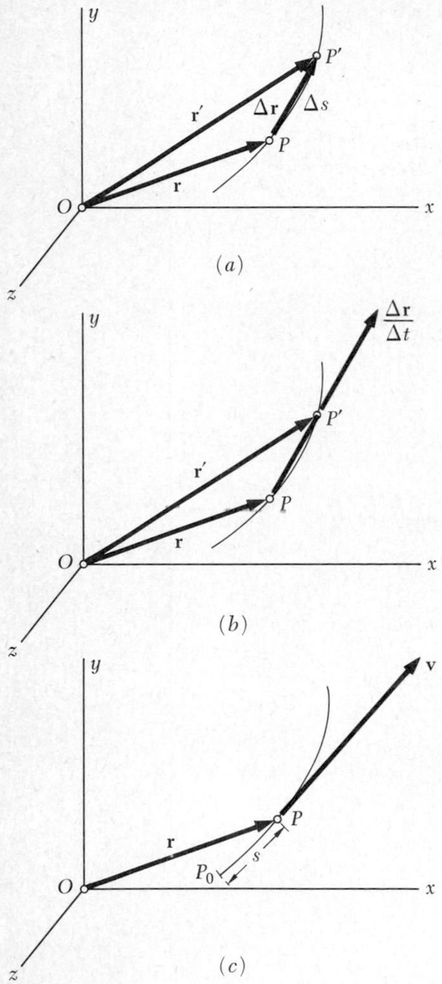

Fig. 11.14

11.9. Position Vector, Velocity, and Acceleration. When a particle moves along a curve other than a straight line, we say that the particle is in *curvilinear motion*. To define the position P occupied by the particle at a given time t, we select a fixed reference system, such as the x, y, z axes shown in Fig. 11.14a, and draw the vector $\mathbf{r}$ joining the origin O and point P. Since the vector $\mathbf{r}$ is characterized by its magnitude r and its direction with respect to the reference axes, it completely defines the position of the particle with respect to those axes; the vector $\mathbf{r}$ is referred to as the *position vector* of the particle at time t.

Consider now the vector $\mathbf{r}'$ defining the position P' occupied by the same particle at a later time $t + \Delta t$. The vector $\Delta\mathbf{r}$ joining P and P' represents the change in the position vector during the time interval Δt since, as we may easily check from Fig. 11.14a, the vector $\mathbf{r}'$ is obtained by adding the vectors $\mathbf{r}$ and $\Delta\mathbf{r}$ according to the triangle rule. We note that $\Delta\mathbf{r}$ represents a change in *direction* as well as a change in *magnitude* of the position vector $\mathbf{r}$. The *average velocity* of the particle over the time interval Δt is defined as the quotient of $\Delta\mathbf{r}$ and Δt. Since $\Delta\mathbf{r}$ is a vector and Δt is a scalar, the quotient $\Delta\mathbf{r}/\Delta t$ is a vector attached at P, of the same direction as $\Delta\mathbf{r}$, and of magnitude equal to the magnitude of $\Delta\mathbf{r}$ divided by Δt (Fig. 11.14b).

The *instantaneous velocity* of the particle at time t is obtained by choosing shorter and shorter time intervals Δt and, correspondingly, shorter and shorter vector increments $\Delta\mathbf{r}$. The instantaneous velocity is thus represented by the vector

$$\mathbf{v} = \lim_{\Delta t \to 0} \frac{\Delta\mathbf{r}}{\Delta t} \tag{11.14}$$

As Δt and $\Delta\mathbf{r}$ become shorter, the points P and P' get closer; the vector $\mathbf{v}$ obtained at the limit must therefore be tangent to the path of the particle (Fig. 11.14c).

Since the position vector $\mathbf{r}$ depends upon the time t, we may refer to it as a *vector function* of the scalar variable t and denote it by $\mathbf{r}(t)$. Extending the concept of derivative of a scalar function introduced in elementary calculus, we shall refer to the limit of the quotient $\Delta\mathbf{r}/\Delta t$ as the *derivative* of the vector function $\mathbf{r}(t)$. We write

$$\mathbf{v} = \frac{d\mathbf{r}}{dt} \tag{11.15}$$

The magnitude v of the vector $\mathbf{v}$ is called the *speed* of the particle. It may be obtained by substituting for the vector $\Delta\mathbf{r}$ in formula (11.14) its magnitude represented by the straight-line segment PP'. But the length of the segment PP' approaches the length Δs of the arc PP' as Δt decreases (Fig. 11.14a), and we may write

$$v = \lim_{\Delta t \to 0} \frac{PP'}{\Delta t} = \lim_{\Delta t \to 0} \frac{\Delta s}{\Delta t} \qquad v = \frac{ds}{dt} \tag{11.16}$$

The speed v may thus be obtained by differentiating with respect to t the length s of the arc described by the particle.

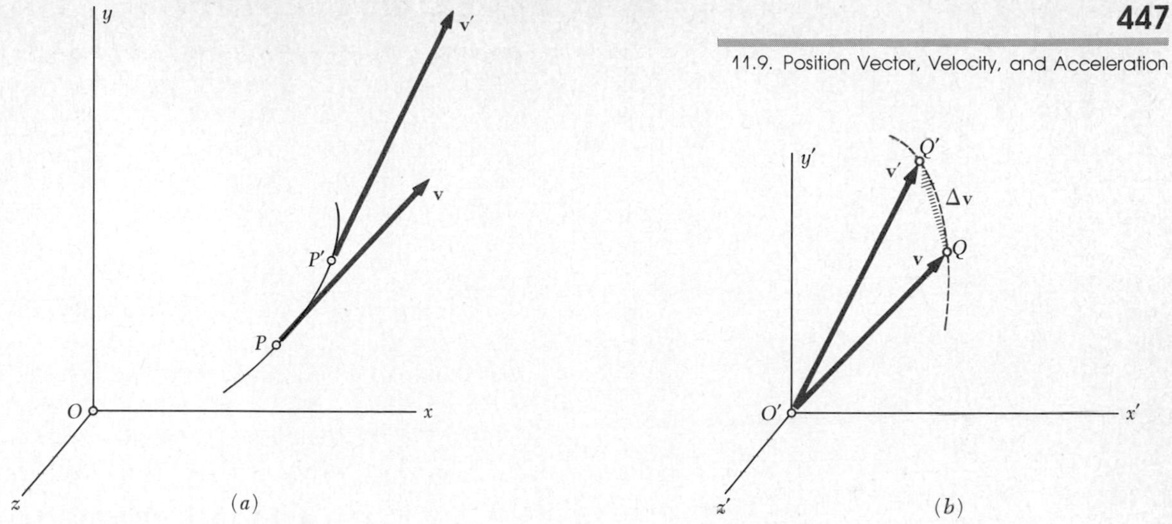

Consider the velocity **v** of the particle at time t and also its velocity **v**′ at a later time $t + \Delta t$ (Fig. 11.15*a*). Let us draw both vectors **v** and **v**′ from the same origin O' (Fig. 11.15*b*). The vector $\Delta\mathbf{v}$ joining Q and Q' represents the change in the velocity of the particle during the time interval Δt, since the vector **v**′ may be obtained by adding the vectors **v** and $\Delta\mathbf{v}$. We should note that $\Delta\mathbf{v}$ represents a change in the *direction* of the velocity as well as a change in *speed*. The *average acceleration* of the particle over the time interval Δt is defined as the quotient of $\Delta\mathbf{v}$ and Δt. Since $\Delta\mathbf{v}$ is a vector and Δt a scalar, the quotient $\Delta\mathbf{v}/\Delta t$ is a vector of the same direction as $\Delta\mathbf{v}$.

The *instantaneous acceleration* of the particle at time t is obtained by choosing smaller and smaller values for Δt and $\Delta\mathbf{v}$. The instantaneous acceleration is thus represented by the vector

$$\mathbf{a} = \lim_{\Delta t \to 0} \frac{\Delta\mathbf{v}}{\Delta t} \tag{11.17}$$

Noting that the velocity **v** is a vector function $\mathbf{v}(t)$ of the time t, we may refer to the limit of the quotient $\Delta\mathbf{v}/\Delta t$ as the derivative of **v** with respect to t. We write

$$\mathbf{a} = \frac{d\mathbf{v}}{dt} \tag{11.18}$$

We observe that the acceleration **a** is tangent to the curve described by the tip Q of the vector **v** when the latter is drawn from a fixed origin O' (Fig. 11.15*c*) and that, in general, the acceleration is *not* tangent to the path of the particle (Fig. 11.15*d*). The curve described by the tip of **v** and shown in Fig. 11.15*c* is called the *hodograph* of the motion.

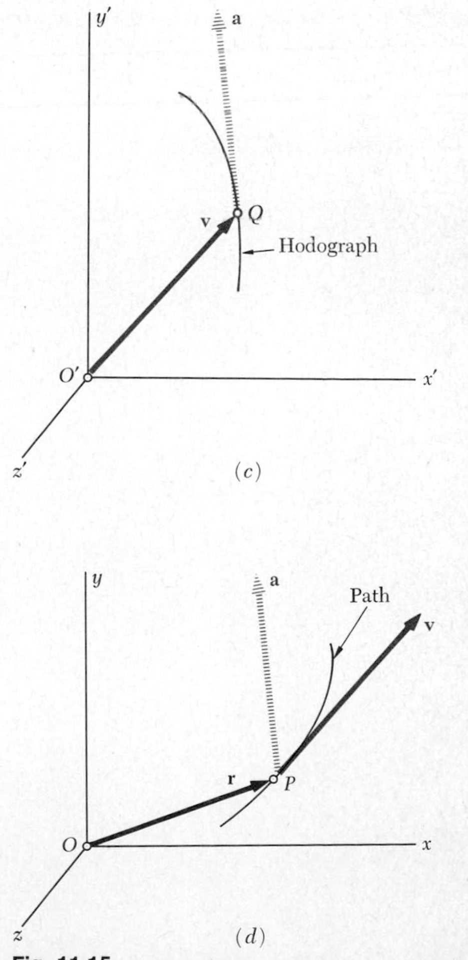

Fig. 11.15

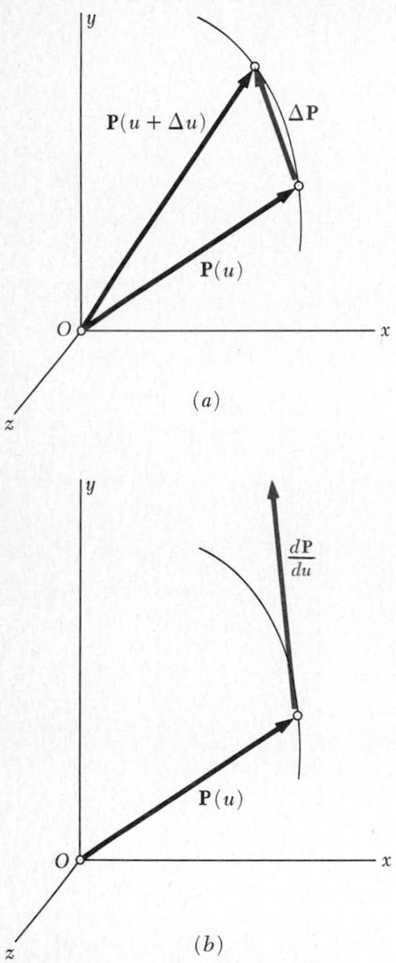

Fig. 11.16

11.10. Derivatives of Vector Functions. We saw in the preceding section that the velocity **v** of a particle in curvilinear motion may be represented by the derivative of the vector function $\mathbf{r}(t)$ characterizing the position of the particle. Similarly, the acceleration **a** of the particle may be represented by the derivative of the vector function $\mathbf{v}(t)$. In this section, we shall give a formal definition of the derivative of a vector function and establish a few rules governing the differentiation of sums and products of vector functions.

Let $\mathbf{P}(u)$ be a vector function of the scalar variable u. By that we mean that the scalar u completely defines the magnitude and direction of the vector **P**. If the vector **P** is drawn from a fixed origin O and the scalar u is allowed to vary, the tip of **P** will describe a given curve in space. Consider the vectors **P** corresponding, respectively, to the values u and $u + \Delta u$ of the scalar variable (Fig. 11.16*a*). Let $\Delta\mathbf{P}$ be the vector joining the tips of the two given vectors; we write

$$\Delta\mathbf{P} = \mathbf{P}(u + \Delta u) - \mathbf{P}(u)$$

Dividing through by Δu and letting Δu approach zero, *we define the derivative of the vector function* $\mathbf{P}(u)$:

$$\frac{d\mathbf{P}}{du} = \lim_{\Delta u \to 0} \frac{\Delta\mathbf{P}}{\Delta u} = \lim_{\Delta u \to 0} \frac{\mathbf{P}(u + \Delta u) - \mathbf{P}(u)}{\Delta u} \tag{11.19}$$

As Δu approaches zero, the line of action of $\Delta\mathbf{P}$ becomes tangent to the curve of Fig. 11.16*a*. Thus, the derivative $d\mathbf{P}/du$ of the vector function $\mathbf{P}(u)$ *is tangent to the curve described by the tip of* $\mathbf{P}(u)$ (Fig. 11.16*b*).

We shall now show that the standard rules for the differentiation of the sums and products of scalar functions may be extended to vector functions. Consider first the *sum of two vector functions* $\mathbf{P}(u)$ and $\mathbf{Q}(u)$ of the same scalar variable u. According to the definition given in (11.19), the derivative of the vector $\mathbf{P} + \mathbf{Q}$ is

$$\frac{d(\mathbf{P} + \mathbf{Q})}{du} = \lim_{\Delta u \to 0} \frac{\Delta(\mathbf{P} + \mathbf{Q})}{\Delta u} = \lim_{\Delta u \to 0} \left(\frac{\Delta\mathbf{P}}{\Delta u} + \frac{\Delta\mathbf{Q}}{\Delta u}\right)$$

or since the limit of a sum is equal to the sum of the limits of its terms,

$$\frac{d(\mathbf{P} + \mathbf{Q})}{du} = \lim_{\Delta u \to 0} \frac{\Delta\mathbf{P}}{\Delta u} + \lim_{\Delta u \to 0} \frac{\Delta\mathbf{Q}}{\Delta u}$$

$$\frac{d(\mathbf{P} + \mathbf{Q})}{du} = \frac{d\mathbf{P}}{du} + \frac{d\mathbf{Q}}{du} \tag{11.20}$$

Next we shall consider the *product of a scalar function* $f(u)$ *and of a vector function* $\mathbf{P}(u)$ of the same scalar variable u. The derivative of the vector $f\mathbf{P}$ is

$$\frac{d(f\mathbf{P})}{du} = \lim_{\Delta u \to 0} \frac{(f + \Delta f)(\mathbf{P} + \Delta\mathbf{P}) - f\mathbf{P}}{\Delta u} = \lim_{\Delta u \to 0} \left(\frac{\Delta f}{\Delta u}\mathbf{P} + f\frac{\Delta\mathbf{P}}{\Delta u}\right)$$

or recalling the properties of the limits of sums and products,

$$\frac{d(f\mathbf{P})}{du} = \frac{df}{du}\mathbf{P} + f\frac{d\mathbf{P}}{du} \tag{11.21}$$

The derivatives of the *scalar product* and of the *vector product* of two vector functions $\mathbf{P}(u)$ and $\mathbf{Q}(u)$ may be obtained in a similar way. We have

$$\frac{d(\mathbf{P}\cdot\mathbf{Q})}{du} = \frac{d\mathbf{P}}{du}\cdot\mathbf{Q} + \mathbf{P}\cdot\frac{d\mathbf{Q}}{du} \tag{11.22}$$

$$\frac{d(\mathbf{P}\times\mathbf{Q})}{du} = \frac{d\mathbf{P}}{du}\times\mathbf{Q} + \mathbf{P}\times\frac{d\mathbf{Q}}{du} \tag{11.23}$$†

We shall use the properties established above to determine the *rectangular components of the derivative of a vector function* $\mathbf{P}(u)$. Resolving $\mathbf{P}$ into components along fixed rectangular axes x, y, z, we write

$$\mathbf{P} = P_x\mathbf{i} + P_y\mathbf{j} + P_z\mathbf{k} \tag{11.24}$$

where P_x, P_y, P_z are the rectangular scalar components of the vector $\mathbf{P}$, and $\mathbf{i}$, $\mathbf{j}$, $\mathbf{k}$ the unit vectors corresponding, respectively, to the x, y, and z axes (Sec. 2.12). By (11.20), the derivative of $\mathbf{P}$ is equal to the sum of the derivatives of the terms in the right-hand member. Since each of these terms is the product of a scalar and a vector function, we should use (11.21). But the unit vectors $\mathbf{i}$, $\mathbf{j}$, $\mathbf{k}$ have a constant magnitude (equal to 1) and fixed directions. Their derivatives are therefore zero, and we write

$$\frac{d\mathbf{P}}{du} = \frac{dP_x}{du}\mathbf{i} + \frac{dP_y}{du}\mathbf{j} + \frac{dP_z}{du}\mathbf{k} \tag{11.25}$$

Noting that the coefficients of the unit vectors are, by definition, the scalar components of the vector $d\mathbf{P}/du$, we conclude that *the rectangular scalar components of the derivative* $d\mathbf{P}/du$ *of the vector function* $\mathbf{P}(u)$ are obtained by differentiating the corresponding scalar components of $\mathbf{P}$.

Rate of Change of a Vector. When the vector $\mathbf{P}$ is a function of the time t, its derivative $d\mathbf{P}/dt$ represents the *rate of change* of $\mathbf{P}$ with respect to the frame $Oxyz$. Resolving $\mathbf{P}$ into rectangular components, we have, by (11.25),

$$\frac{d\mathbf{P}}{dt} = \frac{dP_x}{dt}\mathbf{i} + \frac{dP_y}{dt}\mathbf{j} + \frac{dP_z}{dt}\mathbf{k}$$

or, using dots to indicate differentiation with respect to t,

$$\dot{\mathbf{P}} = \dot{P}_x\mathbf{i} + \dot{P}_y\mathbf{j} + \dot{P}_z\mathbf{k} \tag{11.25'}$$

† Since the vector product is not commutative (Sec. 3.4), the order of the factors in (11.23) must be maintained.

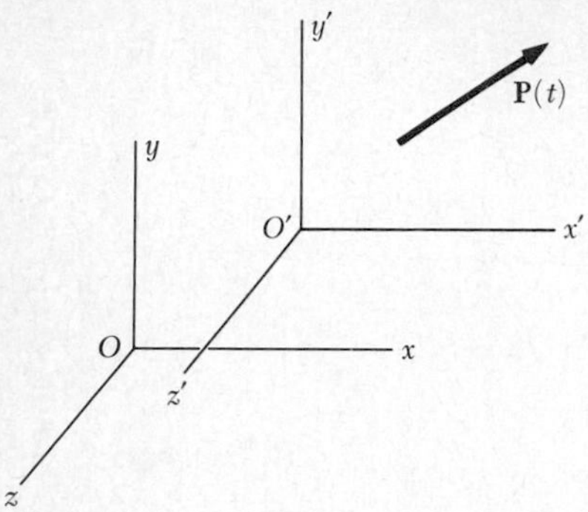

Fig. 11.17

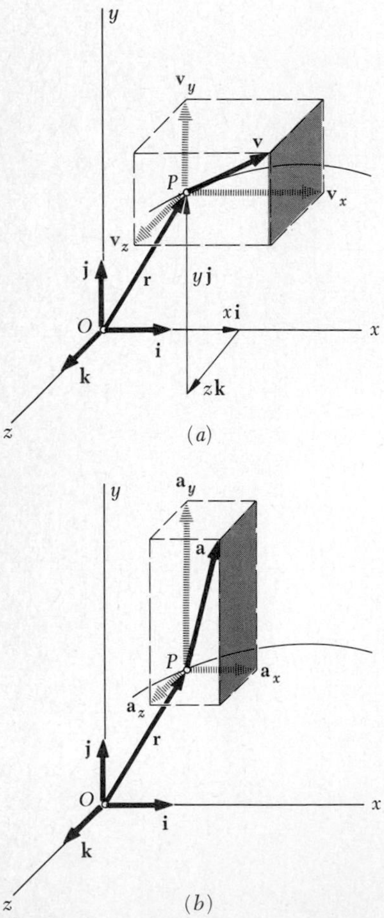

Fig. 11.18

As we shall see in Sec. 15.10, the rate of change of a vector, as observed from a *moving frame of reference,* is, in general, different from its rate of change as observed from a fixed frame of reference. However, if the moving frame $O'x'y'z'$ is in *translation,* i.e., if its axes remain parallel to the corresponding axes of the fixed frame $Oxyz$ (Fig. 11.17), the same unit vectors, **i**, **j**, **k** are used in both frames, and the vector **P** has, at any given instant, the same components P_x, P_y, P_z in both frames. It follows from (11.25′) that the rate of change $\dot{\mathbf{P}}$ is the same with respect to the frames $Oxyz$ and $O'x'y'z'$. We state, therefore: *The rate of change of a vector is the same with respect to a fixed frame and with respect to a frame in translation.* This property will greatly simplify our work, since we shall deal mainly with frames in translation.

11.11. Rectangular Components of Velocity and Acceleration. When the position of a particle P is defined at any instant by its rectangular coordinates x, y, and z, it is convenient to resolve the velocity **v** and the acceleration **a** of the particle into rectangular components (Fig. 11.18).

Resolving the position vector **r** of the particle into rectangular components, we write

$$\mathbf{r} = x\mathbf{i} + y\mathbf{j} + z\mathbf{k} \tag{11.26}$$

where the coordinates x, y, z are functions of t. Differentiating twice, we obtain

$$\mathbf{v} = \frac{d\mathbf{r}}{dt} = \dot{x}\mathbf{i} + \dot{y}\mathbf{j} + \dot{z}\mathbf{k} \tag{11.27}$$

$$\mathbf{a} = \frac{d\mathbf{v}}{dt} = \ddot{x}\mathbf{i} + \ddot{y}\mathbf{j} + \ddot{z}\mathbf{k} \tag{11.28}$$

where $\dot{x}$, $\dot{y}$, $\dot{z}$ and $\ddot{x}$, $\ddot{y}$, $\ddot{z}$ represent, respectively, the first and second derivatives of x, y, and z with respect to t. It follows from (11.27) and (11.28) that the scalar components of the velocity and acceleration are

$$v_x = \dot{x} \qquad v_y = \dot{y} \qquad v_z = \dot{z} \tag{11.29}$$

$$a_x = \ddot{x} \qquad a_y = \ddot{y} \qquad a_z = \ddot{z} \tag{11.30}$$

A positive value for v_x indicates that the vector component $\mathbf{v}_x$ is directed to the right, a negative value that it is directed to the left; the sense of each of the other vector components may be determined in a similar way from the sign of the corresponding scalar component. If desired, the magnitudes and directions of the velocity and acceleration may be obtained from their scalar components by the methods of Secs. 2.7 and 2.12.

The use of rectangular components to describe the position, the velocity, and the acceleration of a particle is particularly effective when the component a_x of the acceleration depends only upon t, x, and/or v_x, and when, similarly, a_y depends only upon t, y, and/or v_y, and a_z upon t, z, and/or v_z. Equations (11.30) may then be integrated independently, and so may Eqs. (11.29). In other words, the motion of the particle in the x direction, its motion in the y direction, and its motion in the z direction may be considered separately.

In the case of the *motion of a projectile*, for example, it may be shown (see Sec. 12.5) that the components of the acceleration are

$$a_x = \ddot{x} = 0 \qquad a_y = \ddot{y} = -g \qquad a_z = \ddot{z} = 0$$

if the resistance of the air is neglected. Denoting by x_0, y_0, z_0 the coordinates of the gun, and by $(v_x)_0$, $(v_y)_0$, $(v_z)_0$ the components of the initial velocity $\mathbf{v}_0$ of the projectile, we integrate twice in t and obtain

$$\begin{aligned} v_x &= \dot{x} = (v_x)_0 & v_y &= \dot{y} = (v_y)_0 - gt & v_z &= \dot{z} = (v_z)_0 \\ x &= x_0 + (v_x)_0 t & y &= y_0 + (v_y)_0 t - \tfrac{1}{2}gt^2 & z &= z_0 + (v_z)_0 t \end{aligned}$$

If the projectile is fired in the xy plane from the origin O, we have $x_0 = y_0 = z_0 = 0$ and $(v_z)_0 = 0$, and the equations of motion reduce to

$$\begin{aligned} v_x &= (v_x)_0 & v_y &= (v_y)_0 - gt & v_z &= 0 \\ x &= (v_x)_0 t & y &= (v_y)_0 t - \tfrac{1}{2}gt^2 & z &= 0 \end{aligned}$$

These equations show that the projectile remains in the xy plane and that its motion in the horizontal direction is uniform, while its motion in the vertical direction is uniformly accelerated. The motion of a projectile may thus be replaced by two independent rectilinear motions, which are easily visualized if we assume that the projectile is fired vertically with an initial velocity $(\mathbf{v}_y)_0$ from a platform moving with a constant horizontal velocity $(\mathbf{v}_x)_0$ (Fig. 11.19). The coordinate x of the projectile is equal at any instant to the distance traveled by the platform, while its coordinate y may be computed as if the projectile were moving along a vertical line.

It may be observed that the equations defining the coordinates x and y of a projectile at any instant are the parametric equations of a parabola. Thus, the trajectory of a projectile is *parabolic*. This result, however, ceases to be valid when the resistance of the air or the variation with altitude of the acceleration of gravity is taken into account.

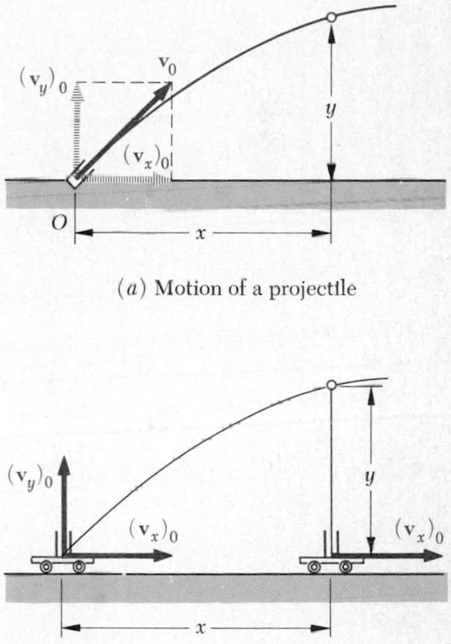

(*a*) Motion of a projectile

(*b*) Equivalent rectilinear motions

Fig. 11.19

11.12. Motion Relative to a Frame in Translation.

In the preceding section, a single frame of reference was used to describe the motion of a particle. In most cases this frame was attached to the earth and was considered as fixed. We shall now analyze situations in which it is convenient to use simultaneously several frames of reference. If one of the frames is attached to the earth, we shall call it a *fixed frame of reference*, and the other frames will be referred to as *moving frames of reference*. It should be understood, however, that the selection of a fixed frame of reference is purely arbitrary. Any frame may be designated as "fixed"; all other frames not rigidly attached to this frame will then be described as "moving."

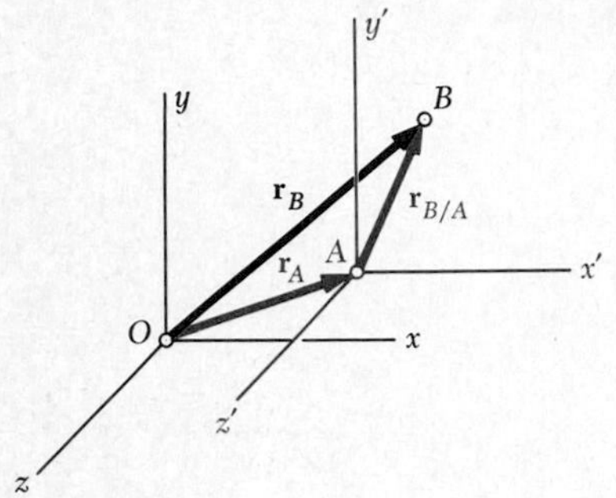

Fig. 11.20

Consider two particles A and B moving in space (Fig. 11.20); the vectors $\mathbf{r}_A$ and $\mathbf{r}_B$ define their positions at any given instant with respect to the fixed frame of reference $Oxyz$. Consider now a system of axes x', y', z' centered at A and parallel to the x, y, z axes. While the origin of these axes moves, their orientation remains the same; the frame of reference $Ax'y'z'$ is in *translation* with respect to $Oxyz$. The vector $\mathbf{r}_{B/A}$ joining A and B defines *the position of B relative to the moving frame* $Ax'y'z'$ (or, for short, *the position of B relative to A*).

We note from Fig. 11.20 that the position vector $\mathbf{r}_B$ of particle B is the sum of the position vector $\mathbf{r}_A$ of particle A and of the position vector $\mathbf{r}_{B/A}$ of B relative to A; we write

$$\mathbf{r}_B = \mathbf{r}_A + \mathbf{r}_{B/A} \tag{11.31}$$

Differentiating (11.31) with respect to t within the fixed frame of reference, and using dots to indicate time derivatives, we have

$$\dot{\mathbf{r}}_B = \dot{\mathbf{r}}_A + \dot{\mathbf{r}}_{B/A} \tag{11.32}$$

The derivatives $\dot{\mathbf{r}}_A$ and $\dot{\mathbf{r}}_B$ represent, respectively, the velocities $\mathbf{v}_A$ and $\mathbf{v}_B$ of the particles A and B. The derivative $\dot{\mathbf{r}}_{B/A}$ represents the rate of change of $\mathbf{r}_{B/A}$ with respect to the frame $Ax'y'z'$, as well as with respect to the fixed frame, since $Ax'y'z'$ is in translation (Sec. 11.10). This derivative therefore defines *the velocity* $\mathbf{v}_{B/A}$ *of B relative to the frame* $Ax'y'z'$ (or, for short, *the velocity* $\mathbf{v}_{B/A}$ *of B relative to A*). We write

$$\mathbf{v}_B = \mathbf{v}_A + \mathbf{v}_{B/A} \tag{11.33}$$

Differentiating Eq. (11.33) with respect to t, and using the derivative $\dot{\mathbf{v}}_{B/A}$ to define *the acceleration* $\mathbf{a}_{B/A}$ *of B relative to the frame* $Ax'y'z'$ (or, for short, *the acceleration* $\mathbf{a}_{B/A}$ *of B relative to A*), we write

$$\mathbf{a}_B = \mathbf{a}_A + \mathbf{a}_{B/A} \tag{11.34}$$

The motion of B with respect to the fixed frame $Oxyz$ is referred to as the *absolute motion of B*. The equations derived in this section show that *the absolute motion of B may be obtained by combining the motion of A and the relative motion of B with respect to the moving frame attached to A*. Equation (11.33), for example, expresses that the absolute velocity $\mathbf{v}_B$ of particle B may be obtained by adding vectorially the velocity of A and the velocity of B relative to the frame $Ax'y'z'$. Equation (11.34) expresses a similar property in terms of the accelerations.† We should keep in mind, however, that *the frame* $Ax'y'z'$ *is in translation;* that is, while it moves with A, it maintains the same orientation. As we shall see later (Sec. 15.14), different relations must be used in the case of a rotating frame of reference.

†Note that the product of the subscripts A and B/A used in the right-hand member of Eqs. (11.31) through (11.34) is equal to the subscript B used in their left-hand member.

SAMPLE PROBLEM 11.7

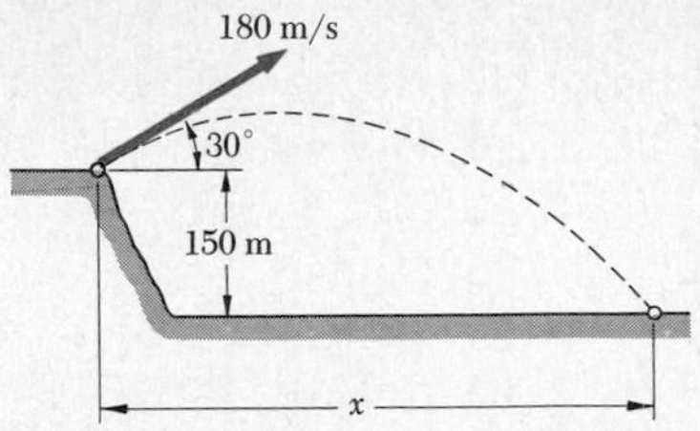

A projectile is fired from the edge of a 150-m cliff with an initial velocity of 180 m/s, at an angle of 30° with the horizontal. Neglecting air resistance, find (*a*) the horizontal distance from the gun to the point where the projectile strikes the ground, (*b*) the greatest elevation above the ground reached by the projectile.

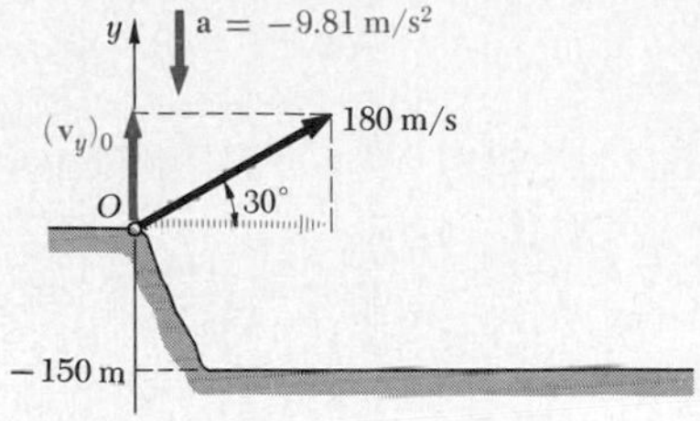

Solution. We shall consider separately the vertical and the horizontal motion.

Vertical Motion. Uniformly accelerated motion. Choosing the positive sense of the y axis upward and placing the origin O at the gun, we have

$$(v_y)_0 = (180\text{ m/s}) \sin 30° = +90\text{ m/s}$$
$$a = -9.81\text{ m/s}^2$$

Substituting into the equations of uniformly accelerated motion, we have

$$v_y = (v_y)_0 + at \qquad v_y = 90 - 9.81t \qquad (1)$$
$$y = (v_y)_0 t + \tfrac{1}{2}at^2 \qquad y = 90t - 4.90t^2 \qquad (2)$$
$$v_y^2 = (v_y)_0^2 + 2ay \qquad v_y^2 = 8100 - 19.62y \qquad (3)$$

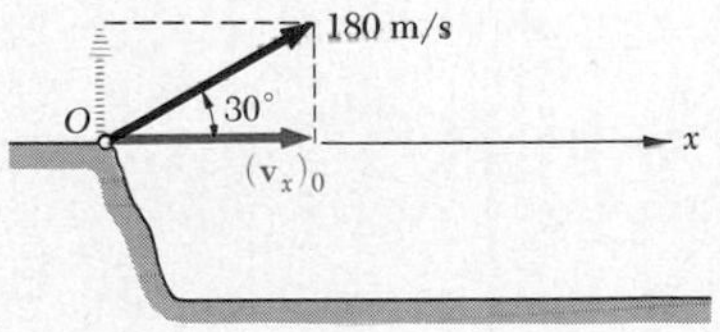

Horizontal Motion. Uniform motion. Choosing the positive sense of the x axis to the right, we have

$$(v_x)_0 = (180\text{ m/s}) \cos 30° = +155.9\text{ m/s}$$

Substituting into the equation of uniform motion, we obtain

$$x = (v_x)_0 t \qquad x = 155.9t \qquad (4)$$

a. **Horizontal Distance.** When the projectile strikes the ground, we have

$$y = -150\text{ m}$$

Carrying this value into Eq. (2) for the vertical motion, we write

$$-150 = 90t - 4.90t^2 \qquad t^2 - 18.37t - 30.6 = 0 \qquad t = 19.91\text{ s}$$

Carrying $t = 19.91$ s into Eq. (4) for the horizontal motion, we obtain

$$x = 155.9(19.91) \qquad x = 3100\text{ m} \blacktriangleleft$$

b. **Greatest Elevation.** When the projectile reaches its greatest elevation, we have $v_y = 0$; carrying this value into Eq. (3) for the vertical motion, we write

$$0 = 8100 - 19.62y \qquad y = 413\text{ m}$$
$$\text{Greatest elevation above ground} = 150\text{ m} + 413\text{ m}$$
$$= 563\text{ m} \blacktriangleleft$$

SAMPLE PROBLEM 11.8

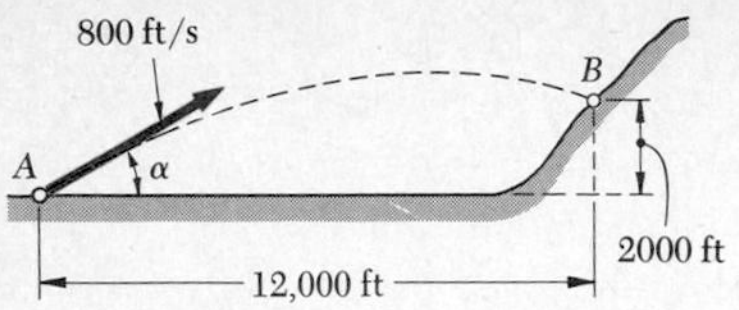

A projectile is fired with an initial velocity of 800 ft/s at a target B located 2000 ft above the gun A and at a horizontal distance of 12,000 ft. Neglecting air resistance, determine the value of the firing angle α.

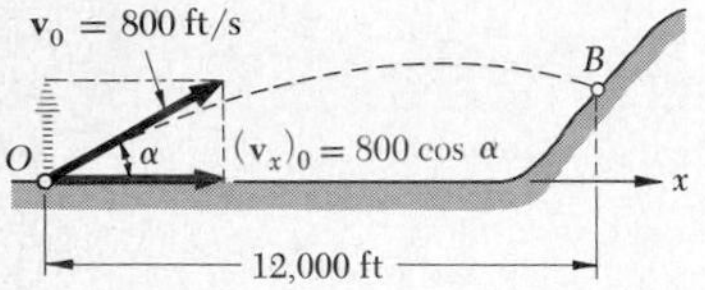

Solution. We shall consider separately the horizontal and the vertical motion.

Horizontal Motion. Placing the origin of coordinates at the gun, we have

$$(v_x)_0 = 800 \cos \alpha$$

Substituting into the equation of uniform horizontal motion, we obtain

$$x = (v_x)_0 t \qquad x = (800 \cos \alpha)t$$

The time required for the projectile to move through a horizontal distance of 12,000 ft is obtained by making x equal to 12,000 ft.

$$12{,}000 = (800 \cos \alpha)t$$

$$t = \frac{12{,}000}{800 \cos \alpha} = \frac{15}{\cos \alpha}$$

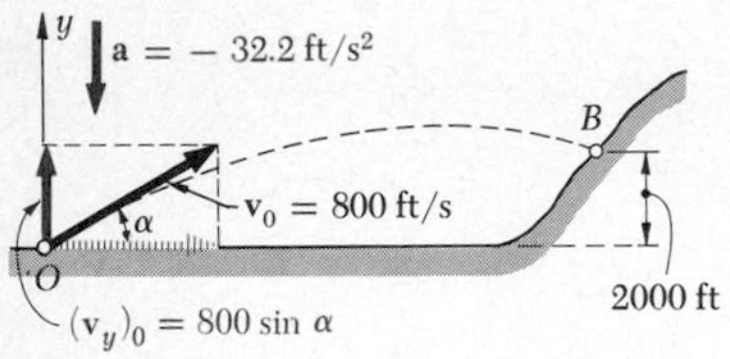

Vertical Motion

$$(v_y)_0 = 800 \sin \alpha \qquad a = -32.2 \text{ ft/s}^2$$

Substituting into the equation of uniformly accelerated vertical motion, we obtain

$$y = (v_y)_0 t + \tfrac{1}{2}at^2 \qquad y = (800 \sin \alpha)t - 16.1t^2$$

Projectile Hits Target. When $x = 12{,}000$ ft, we must have $y = 2000$ ft. Substituting for y and making t equal to the value found above, we write

$$2000 = 800 \sin \alpha \frac{15}{\cos \alpha} - 16.1\left(\frac{15}{\cos \alpha}\right)^2$$

Since $1/\cos^2 \alpha = \sec^2 \alpha = 1 + \tan^2 \alpha$, we have

$$2000 = 800(15) \tan \alpha - 16.1(15^2)(1 + \tan^2 \alpha)$$
$$3622 \tan^2 \alpha - 12{,}000 \tan \alpha + 5622 = 0$$

Solving this quadratic equation for $\tan \alpha$, we have

$$\tan \alpha = 0.565 \qquad \text{and} \qquad \tan \alpha = 2.75$$

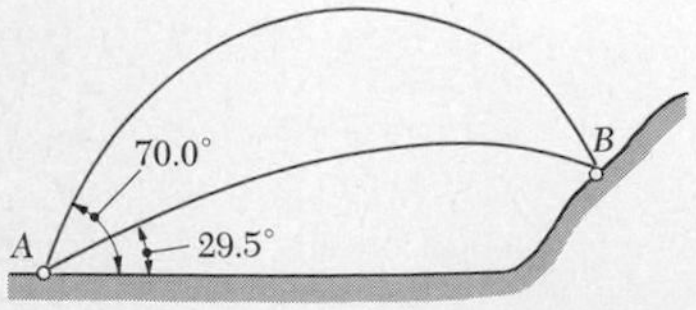

$$\alpha = 29.5° \qquad \text{and} \qquad \alpha = 70.0° \blacktriangleleft$$

The target will be hit if either of these two firing angles is used (see figure).

SAMPLE PROBLEM 11.9

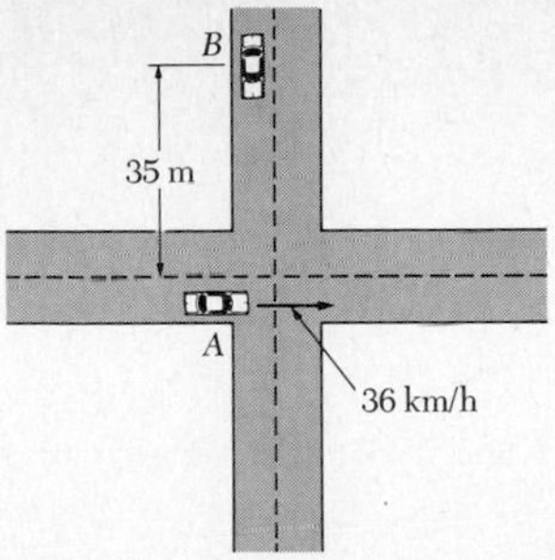

Automobile A is traveling east at the constant speed of 36 km/h. As automobile A crosses the intersection shown, automobile B starts from rest 35 m north of the intersection and moves south with a constant acceleration of 1.2 m/s². Determine the position, velocity, and acceleration of B relative to A five seconds after A crosses the intersection.

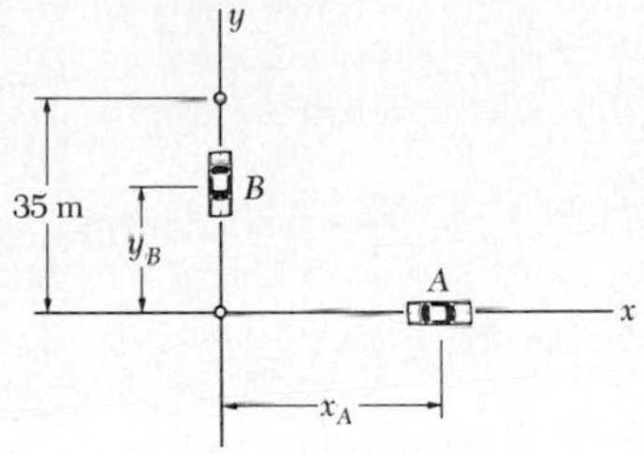

Solution. We choose x and y axes with origin at the intersection of the two streets and with positive senses directed respectively east and north.

Motion of Automobile A. First the speed is expressed in m/s:

$$v_A = \left(36\frac{\text{km}}{\text{h}}\right)\left(\frac{1000\text{ m}}{1\text{ km}}\right)\left(\frac{1\text{ h}}{3600\text{ s}}\right) = 10\text{ m/s}$$

Noting that the motion of A is uniform, we write, for any time t,

$$a_A = 0$$
$$v_A = +10\text{ m/s}$$
$$x_A = (x_A)_0 + v_A t = 0 + 10t$$

For $t = 5$ s, we have

$$a_A = 0 \qquad \mathbf{a}_A = 0$$
$$v_A = +10\text{ m/s} \qquad \mathbf{v}_A = 10\text{ m/s} \rightarrow$$
$$x_A = +(10\text{ m/s})(5\text{ s}) = +50\text{ m} \qquad \mathbf{r}_A = 50\text{ m} \rightarrow$$

Motion of Automobile B. We note that the motion of B is uniformly accelerated, and write

$$a_B = -1.2\text{ m/s}^2$$
$$v_B = (v_B)_0 + at = 0 - 1.2t$$
$$y_B = (y_B)_0 + (v_B)_0 t + \tfrac{1}{2}a_B t^2 = 35 + 0 - \tfrac{1}{2}(1.2)t^2$$

For $t = 5$ s, we have

$$a_B = -1.2\text{ m/s}^2 \qquad \mathbf{a}_B = 1.2\text{ m/s}^2 \downarrow$$
$$v_B = -(1.2\text{ m/s}^2)(5\text{ s}) = -6\text{ m/s} \qquad \mathbf{v}_B = 6\text{ m/s} \downarrow$$
$$y_B = 35 - \tfrac{1}{2}(1.2\text{ m/s}^2)(5\text{ s})^2 = +20\text{ m} \qquad \mathbf{r}_B = 20\text{ m} \uparrow$$

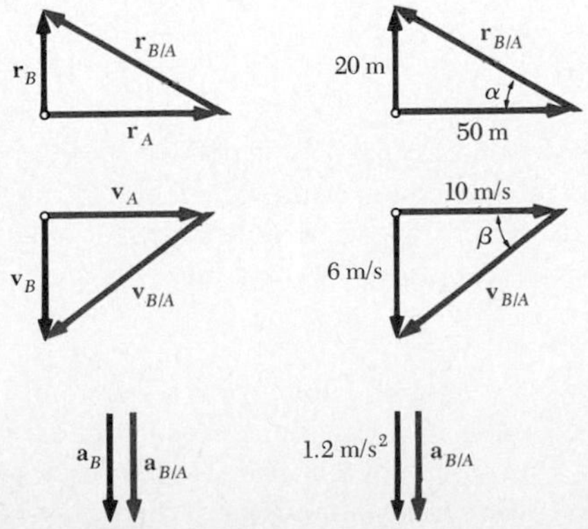

Motion of B Relative to A. We draw the triangle corresponding to the vector equation $\mathbf{r}_B = \mathbf{r}_A + \mathbf{r}_{B/A}$ and obtain the magnitude and direction of the position vector of B relative to A.

$$r_{B/A} = 53.9\text{ m} \qquad \alpha = 21.8^\circ \qquad \mathbf{r}_{B/A} = 53.9\text{ m} \measuredangle 21.8^\circ \quad ◀$$

Proceeding in a similar fashion, we find the velocity and acceleration of B relative to A.

$$\mathbf{v}_B = \mathbf{v}_A + \mathbf{v}_{B/A}$$
$$v_{B/A} = 11.66\text{ m/s} \qquad \beta = 31.0^\circ \qquad \mathbf{v}_{B/A} = 11.66\text{ m/s} \measuredangle 31.0^\circ \quad ◀$$
$$\mathbf{a}_B = \mathbf{a}_A + \mathbf{a}_{B/A} \qquad \mathbf{a}_{B/A} = 1.2\text{ m/s}^2 \downarrow \quad ◀$$

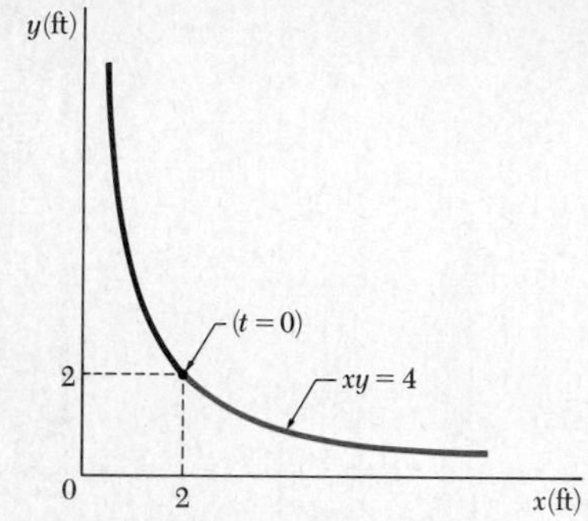

Fig. P11.80

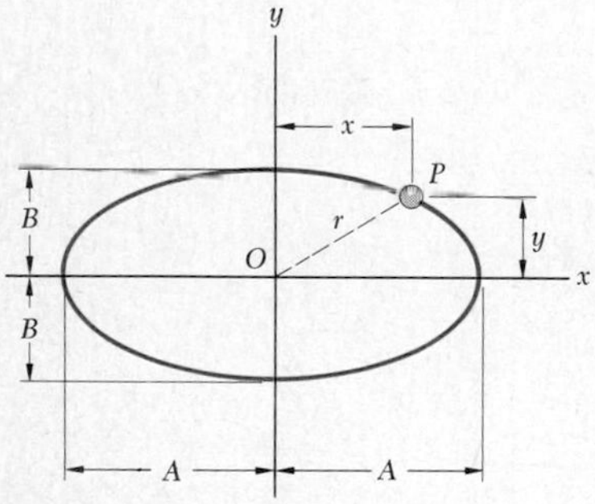

Fig. P11.83

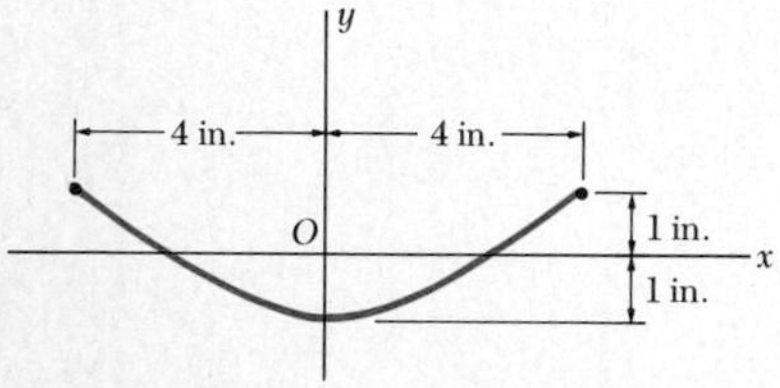

Fig. P11.84

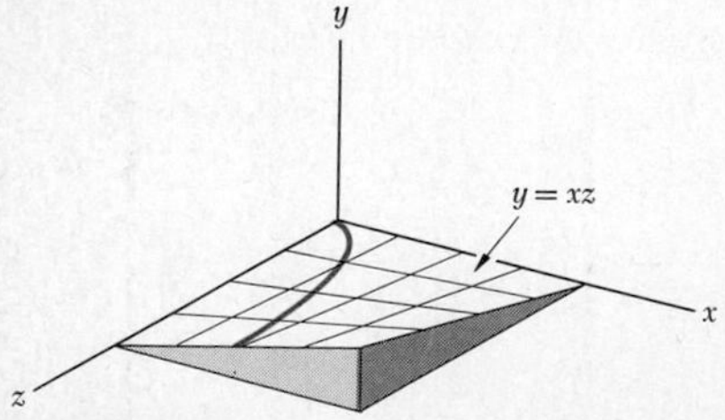

Fig. P11.86

Problems

Note. **Neglect air resistance in problems concerning projectiles.**

11.79 The motion of a particle is defined by the equations $x = 1.5t^2 - 6t$ and $y = 6t^2 - 2t^3$, where x and y are expressed in inches and t in seconds. Determine the velocity and acceleration when (*a*) $t = 1$ s, (*b*) $t = 2$ s, (*c*) $t = 3$ s.

11.80 The motion of a particle is defined by the equations $x = 2(t + 1)^2$ and $y = 2(t + 1)^{-2}$, where x and y are expressed in feet and t in seconds. Show that the path of the particle is part of the rectangular hyperbola shown and determine the velocity and acceleration when (*a*) $t = 0$, (*b*) $t = \frac{1}{2}$ s.

11.81 The motion of a particle is defined by the equations $x = 2t^2 - 4t$ and $y = 2(t - 1)^2 - 4(t - 1)$, where x and y are expressed in meters and t in seconds. Determine (*a*) the magnitude of the smallest velocity reached by the particle, (*b*) the corresponding time, position, and direction of the velocity.

11.82 The motion of a particle is defined by the equations $x = 120 \sin \frac{1}{2}\pi t$ and $y = 40t^2$, where x and y are expressed in millimeters and t in seconds. Determine the velocity and acceleration of the particle when (*a*) $t = 1$ s, (*b*) $t = 2$ s.

11.83 A particle moves in an elliptic path defined by the position vector $\mathbf{r} = (A \cos pt)\mathbf{i} + (B \sin pt)\mathbf{j}$. Show that the acceleration (*a*) is directed toward the origin, (*b*) is proportional to the distance from the origin to the particle.

11.84 The motion of a vibrating particle is defined by the position vector $\mathbf{r} = (4 \sin \pi t)\mathbf{i} - (\cos 2\pi t)\mathbf{j}$, where r is expressed in inches and t in seconds. (*a*) Determine the velocity and acceleration when $t = 1$ s. (*b*) Show that the path of the particle is parabolic.

11.85 The three-dimensional motion of a particle is defined by the position vector $\mathbf{r} = (R \sin pt)\mathbf{i} + ct\mathbf{j} + (R \cos pt)\mathbf{k}$. Determine the magnitudes of the velocity and acceleration of the particle. (The space curve described by the particle is a helix.)

11.86 The three-dimensional motion of a particle is defined by the position vector $\mathbf{r} = At\mathbf{i} + ABt^3\mathbf{j} + Bt^2\mathbf{k}$, where r is expressed in meters and t in seconds. Show that the space curve described by the particle lies on the hyperbolic paraboloid $y = xz$. For $A = B = 1$, determine the magnitudes of the velocity and acceleration when (*a*) $t = 0$, (*b*) $t = 2$ s.

11.87 A man standing at the 18-m level of a tower throws a stone in a horizontal direction. Knowing that the stone hits the ground 25 m from the bottom of the tower, determine (*a*) the initial velocity of the stone, (*b*) the distance at which a stone would hit the ground if it were thrown horizontally with the same velocity from the 22-m level of the tower.

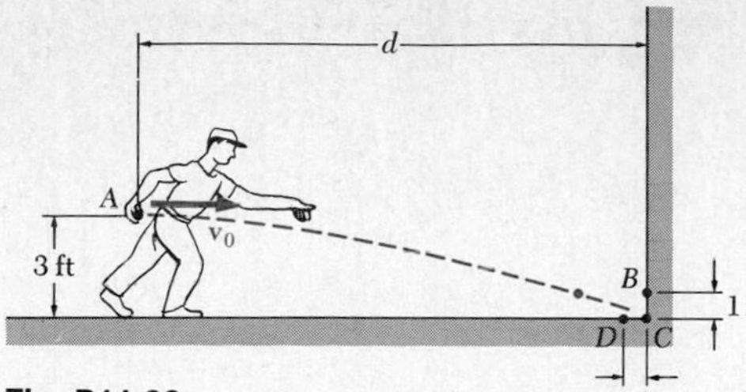

Fig. P11.88

11.88 A handball player throws a ball from A with a horizontal velocity $\mathbf{v}_0$. Knowing that $d = 20$ ft, determine (*a*) the value of v_0 for which the ball will strike the corner C, (*b*) the range of values of v_0 for which the ball will strike the corner region BCD.

11.89 Solve Prob. 11.88, assuming that the distance from point A to the wall is $d = 30$ ft.

11.90 Sand is discharged at A from a horizontal conveyor belt with an initial velocity $\mathbf{v}_0$. Determine the range of values of v_0 for which the sand will enter the vertical chute shown.

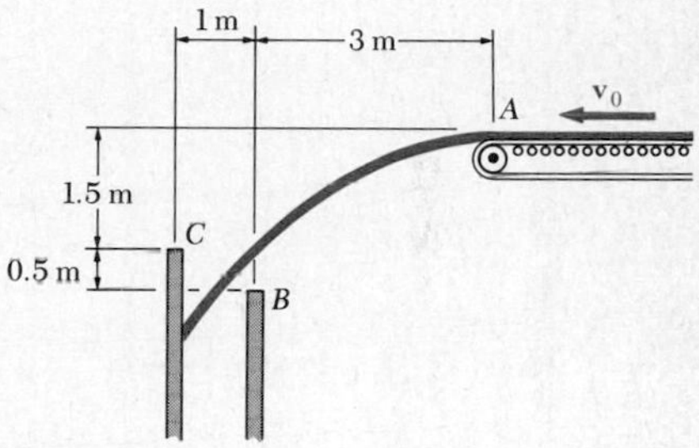

Fig. P11.90

11.91 A projectile is fired with an initial velocity of 1000 ft/s at a target balloon located 1000 ft above the ground. Knowing that the projectile hits the balloon, determine (*a*) the horizontal distance d, (*b*) the time of flight of the projectile.

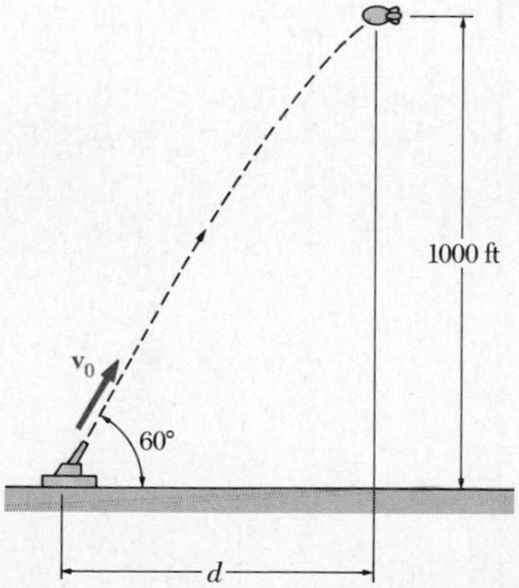

Fig. P11.91

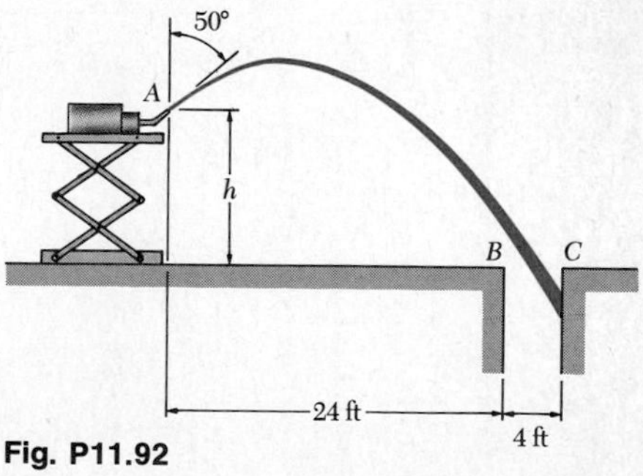

Fig. P11.92

11.92 A pump is located near the edge of the horizontal platform shown. The nozzle at A discharges water with an initial velocity of 25 ft/s at an angle of 50° with the vertical. Determine the range of values of the height h for which the water enters the opening BC.

11.93 An oscillating water sprinkler is operated at point A on an incline which forms an angle α with the horizontal. The sprinkler discharges water with an initial velocity $\mathbf{v}_0$ at an angle ϕ with the vertical which varies from $-\phi_0$ to $+\phi_0$. Knowing that $v_0 = 10$ m/s, $\phi_0 = 40°$, and $\alpha = 10°$, determine the horizontal distance between the sprinkler and points B and C which define the watered area.

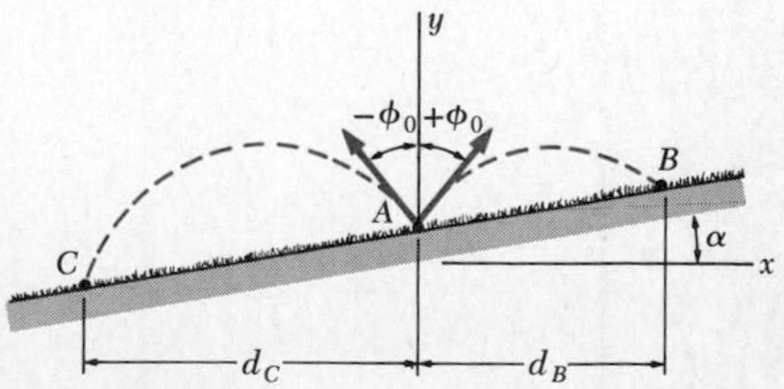

Fig. P11.93

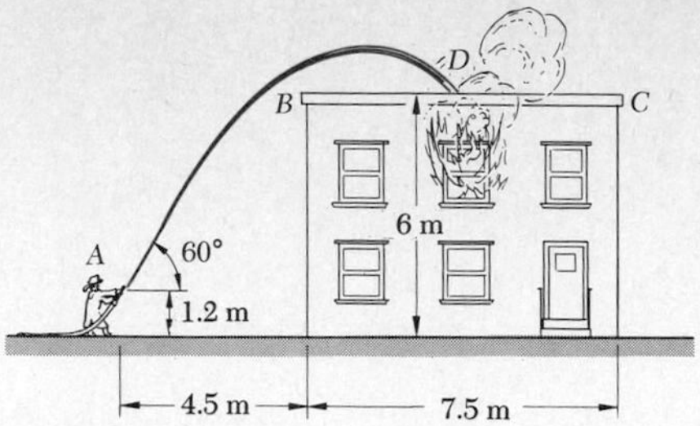

Fig. P11.94

11.94 A nozzle at A discharges water with an initial velocity of 12 m/s at an angle of 60° with the horizontal. Determine where the stream of water strikes the roof. Check that the stream will clear the edge of the roof.

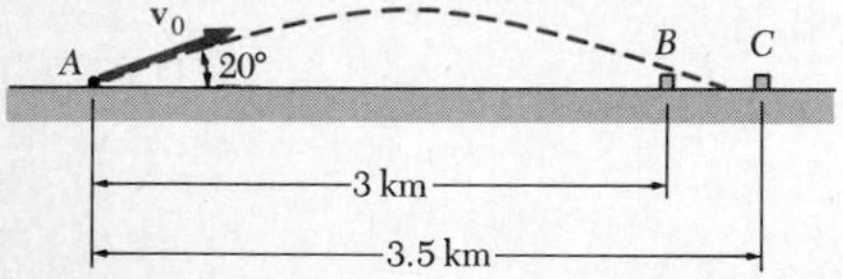

Fig. P11.96

11.95 In Prob. 11.94, determine the range of values of the initial velocity for which the water will fall on the roof.

11.96 A projectile is fired with an initial velocity $\mathbf{v}_0$ at an angle of 20° with the horizontal. Determine the required value of v_0 if the projectile is to hit (*a*) point B, (*b*) point C.

11.97 Sand is discharged at A from a conveyor belt and falls onto the top of a stockpile at B. Knowing that the conveyor belt forms an angle $\alpha = 20°$ with the horizontal, determine the speed v_0 of the belt.

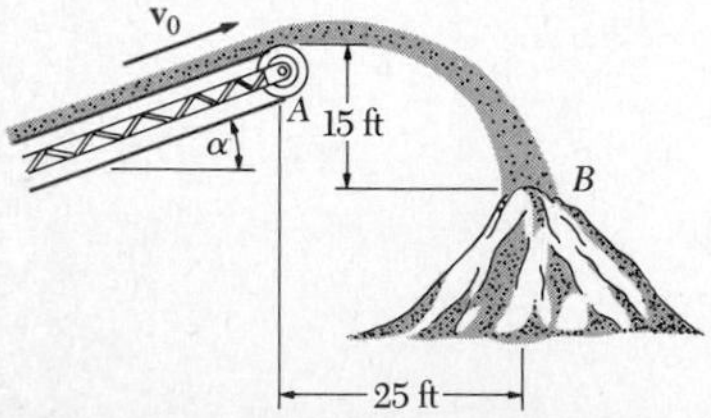

Fig. P11.97, P11.99, and P11.100

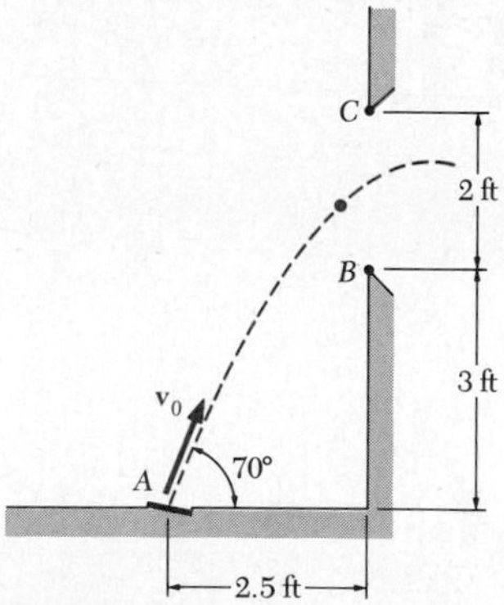

Fig. P11.98

11.98 A ball is dropped onto a pad at A and rebounds with a velocity $\mathbf{v}_0$ at an angle of 70° with the horizontal. Determine the range of values of v_0 for which the ball will enter the opening BC.

11.99 Knowing that the conveyor belt moves at the constant speed $v_0 = 24$ ft/s, determine the angle α for which the sand is deposited on the stockpile at B.

***11.100** Knowing that the conveyor belt moves at the constant speed v_0, determine (*a*) the smallest value of v_0 for which sand can be deposited on the stockpile at B, (*b*) the corresponding value of α.

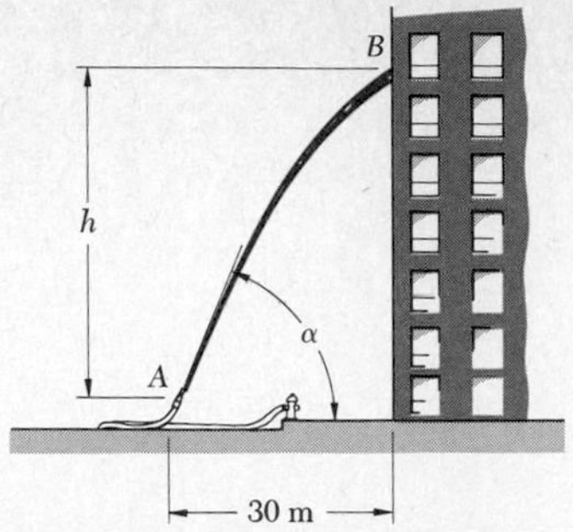

Fig. P11.101 and P11.102

*11.101 A fire nozzle discharges water with an initial velocity $\mathbf{v}_0$ of 24 m/s. Knowing that the nozzle is located 30 m from a building, determine (*a*) the maximum height h that can be reached by the water, (*b*) the corresponding angle α.

11.102 A fire nozzle located at A discharges water with an initial velocity $\mathbf{v}_0$ of 24 m/s. Knowing that the stream of water strikes the building at a height $h = 18$ m above the ground, determine the angle α.

11.103 A player throws a ball with an initial velocity $\mathbf{v}_0$ of 15 m/s from a point A located 1.5 m above the floor. Knowing that $h = 3$ m, determine the angle α for which the ball will strike the wall at point B.

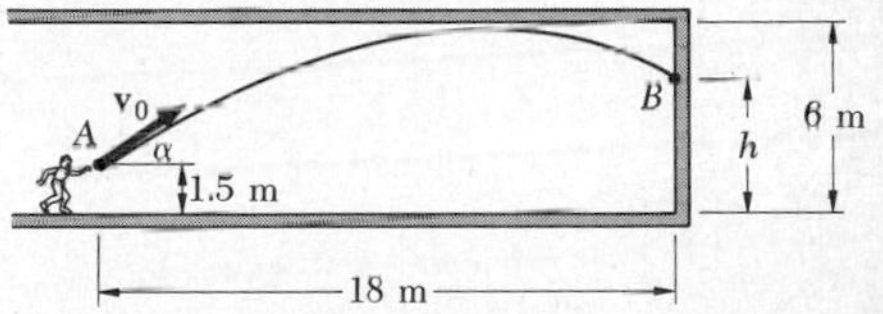

Fig. P11.103 and P11.104

11.104 A player throws a ball with an initial velocity $\mathbf{v}_0$ of 15 m/s from a point A located 1.5 m above the floor. Knowing that the ceiling of the gymnasium is 6 m high, determine the highest point B at which the ball can strike the wall 18 m away.

11.105 The velocities of boats A and C are as shown and the relative velocity of boat B with respect to A is $\mathbf{v}_{B/A} = 4$ m/s ∠ 50°. Determine (*a*) $v_{A/C}$, (*b*) $v_{C/B}$, (*c*) the change in position of B with respect to C during a 10-s interval. Also show that for any motion, $\mathbf{v}_{B/A} + \mathbf{v}_{C/B} + \mathbf{v}_{A/C} = 0$.

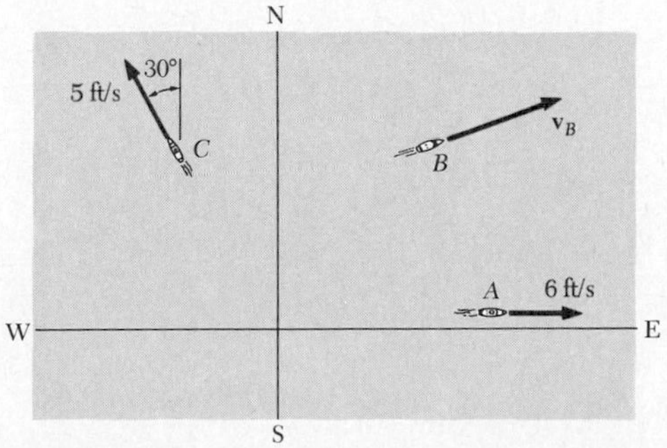

Fig. P11.105

11.106 Instruments in an airplane indicate that with respect to the air, the plane is moving north at a speed of 500 km/h. At the same time ground-based radar indicates that the plane is moving at a speed of 530 km/h in a direction 5° east of north. Determine the magnitude and direction of the velocity of the air.

11.107 Two airplanes A and B are flying at the same altitude; plane A is flying due east at a constant speed of 900 km/h, while plane B is flying southwest at a constant speed of 600 km/h. Determine the change in position of plane B relative to plane A, which takes place during a 2-min interval.

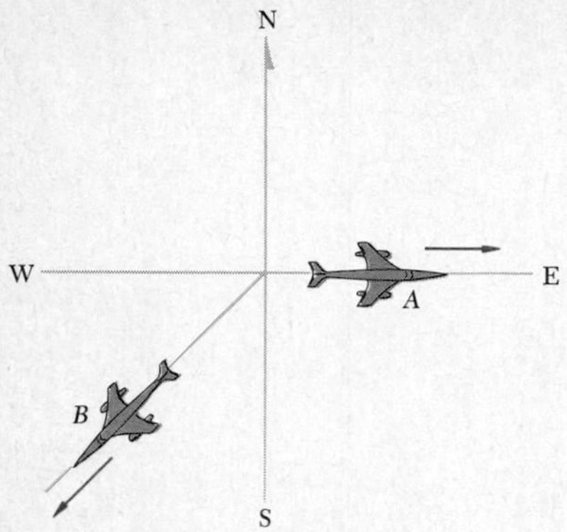

Fig. P11.107

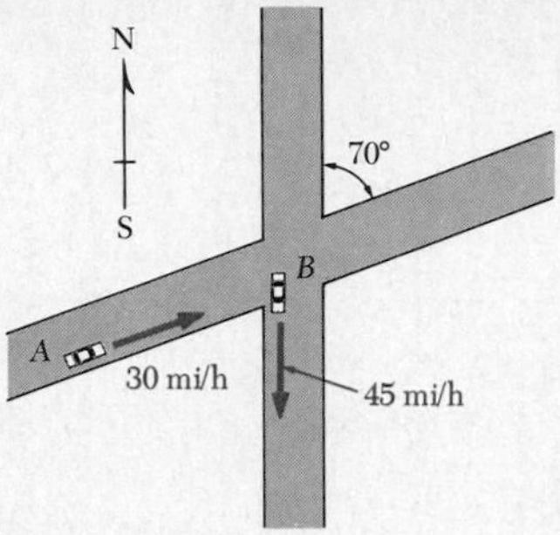

Fig. P11.108

11.108 Four seconds after automobile B passes through the intersection shown, automobile A passes through the same intersection. Knowing that the speed of each automobile is constant, determine (a) the relative velocity of B with respect to A, (b) the change in position of B with respect to A during a 3-s interval, (c) the distance between the two automobiles 2 s after A has passed through the intersection.

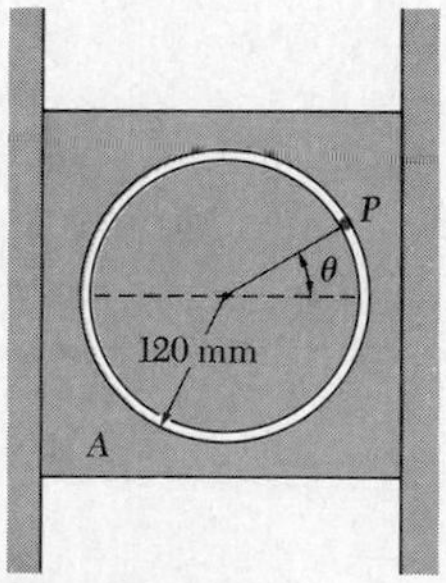

Fig. P11.109

11.109 Pin P moves with a constant speed of 90 mm/s in a counterclockwise sense along a circular slot which has been milled in the slider block A shown. Knowing that the block moves downward at a constant speed of 60 mm/s, determine the velocity of pin P when (a) $\theta = 30°$, (b) 120°.

11.110 At $t = 0$, wedge A starts moving to the right with a constant acceleration of 100 mm/s^2 and block B starts moving along the wedge toward the left with a constant acceleration of 150 mm/s^2 relative to the wedge. Determine (a) the acceleration of block B, (b) the velocity of block B when $t = 4$ s.

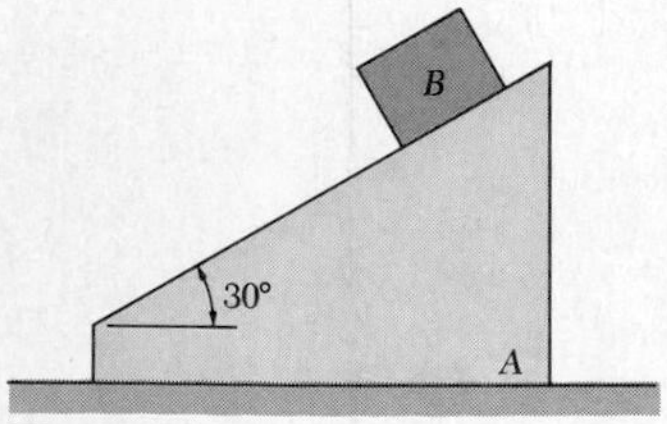

Fig. P11.110

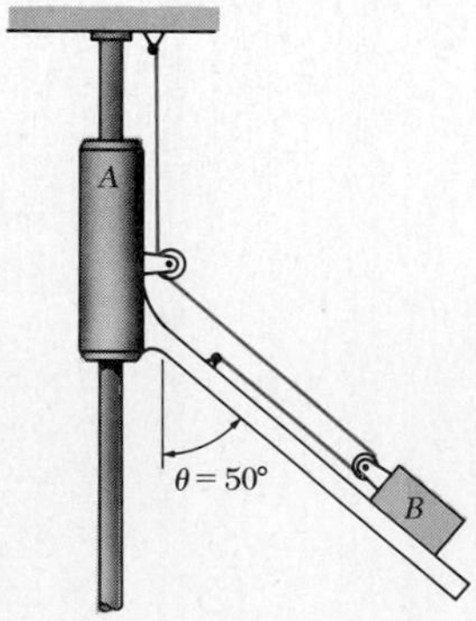

Fig. P11.111

11.111 Knowing that at the instant shown assembly A has a velocity of 16 in./s and an acceleration of 24 in./s^2 both directed downward, determine (a) the velocity of block B, (b) the acceleration of block B.

11.112 Solve Prob. 11.111, assuming that $\theta = 30°$.

11.113 Water is discharged at A with an initial velocity of 12 m/s and strikes a series of vanes at B. Knowing that the vanes move downward with a constant speed of 1.5 m/s, determine the velocity and acceleration of the water relative to the vane at B.

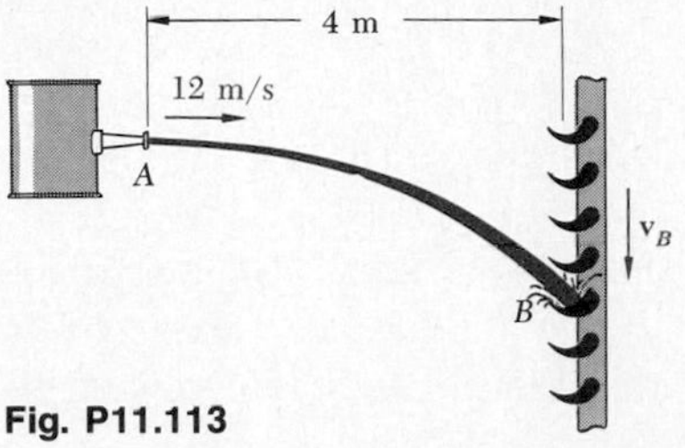

Fig. P11.113

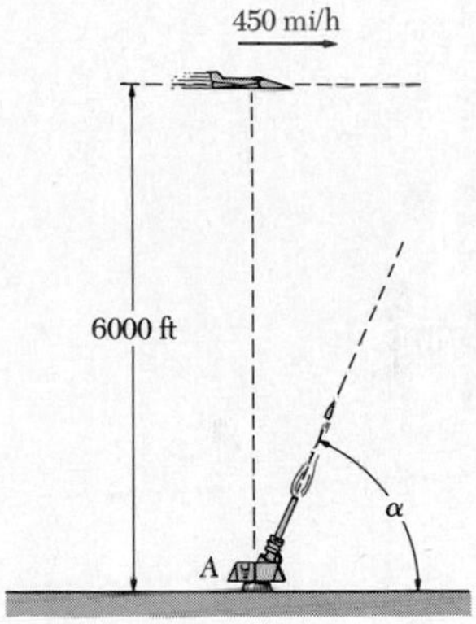

Fig. P11.114

11.114 An antiaircraft gun fires a shell as a plane passes directly over the position of the gun, at an altitude of 6000 ft. The muzzle velocity of the shell is 1500 ft/s. Knowing that the plane is flying horizontally at 450 mi/h, determine (*a*) the required firing angle if the shell is to hit the plane, (*b*) the velocity and acceleration of the shell relative to the plane at the time of impact.

11.115 As observed from a ship moving due east at 8 km/h, the wind appears to blow from the south. After the ship has changed course, and as it is moving due north at 8 km/h, the wind appears to blow from the southwest. Assuming that the wind velocity is constant during the period of observation, determine the magnitude and direction of the true wind velocity.

11.116 A small motorboat maintains a constant speed of 2 m/s relative to the water as it is maneuvering in a tidal current. When the boat is directed due east, it is observed from shore to move due south and when it is directed toward the northeast it is observed to move due west. Determine the speed and direction of the current.

11.117 During a rainstorm the paths of the raindrops appear to form an angle of 30° with the vertical and to be directed to the left when observed from a side window of a train moving at a speed of 10 mi/h. A short time later, after the speed of the train has increased to 15 mi/h, the angle between the vertical and the paths of the drops appears to be 45°. If the train were stopped, at what angle and with what velocity would the drops be observed to fall?

11.118 As the speed of the train of Prob. 11.117 increases, the angle between the vertical and the paths of the drops becomes equal to 60°. Determine the speed of the train at that time.

11.13. Tangential and Normal Components. We saw in Sec. 11.9 that the velocity of a particle is a vector tangent to the path of the particle but that, in general, the acceleration is not tangent to the path. It is sometimes convenient to resolve the acceleration into components directed, respectively, along the tangent and the normal to the path of the particle.

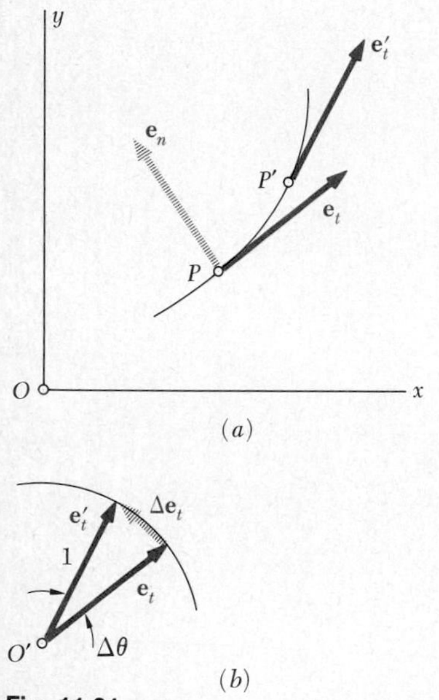

Fig. 11.21

Plane Motion of a Particle. We shall first consider a particle which moves along a curve contained in the plane of the figure. Let P be the position of the particle at a given instant. We attach at P a unit vector $\mathbf{e}_t$ tangent to the path of the particle and pointing toward the direction of motion (Fig. 11.21*a*). Let $\mathbf{e}_t'$ be the unit vector corresponding to the position P' of the particle at a later instant. Drawing both vectors from the same origin O', we define the vector $\Delta\mathbf{e}_t = \mathbf{e}_t' - \mathbf{e}_t$ (Fig. 11.21*b*). Since $\mathbf{e}_t$ and $\mathbf{e}_t'$ are of unit length, their tips lie on a circle of radius 1. Denoting by $\Delta\theta$ the angle formed by $\mathbf{e}_t$ and $\mathbf{e}_t'$, we find that the magnitude of $\Delta\mathbf{e}_t$ is $2 \sin (\Delta\theta/2)$. Considering now the vector $\Delta\mathbf{e}_t/\Delta\theta$, we note that as $\Delta\theta$ approaches zero, this vector becomes tangent to the unit circle of Fig. 11.21*b*, i.e., perpendicular to $\mathbf{e}_t$, and that its magnitude approaches

$$\lim_{\Delta\theta\to 0} \frac{2 \sin (\Delta\theta/2)}{\Delta\theta} = \lim_{\Delta\theta\to 0} \frac{\sin (\Delta\theta/2)}{\Delta\theta/2} = 1$$

Thus, the vector obtained at the limit is a unit vector along the normal to the path of the particle, in the direction toward which $\mathbf{e}_t$ turns. Denoting this vector by $\mathbf{e}_n$, we write

$$\mathbf{e}_n = \lim_{\Delta\theta\to 0} \frac{\Delta\mathbf{e}_t}{\Delta\theta}$$

$$\mathbf{e}_n = \frac{d\mathbf{e}_t}{d\theta} \tag{11.35}$$

Since the velocity $\mathbf{v}$ of the particle is tangent to the path, we may express it as the product of the scalar v and the unit vector $\mathbf{e}_t$. We have

$$\mathbf{v} = v\mathbf{e}_t \tag{11.36}$$

To obtain the acceleration of the particle, we shall differentiate (11.36) with respect to t. Applying the rule for the differentiation of the product of a scalar and a vector function (Sec. 11.10), we write

$$\mathbf{a} = \frac{d\mathbf{v}}{dt} = \frac{dv}{dt}\mathbf{e}_t + v\frac{d\mathbf{e}_t}{dt} \tag{11.37}$$

But

$$\frac{d\mathbf{e}_t}{dt} = \frac{d\mathbf{e}_t}{d\theta}\frac{d\theta}{ds}\frac{ds}{dt}$$

Recalling from (11.16) that $ds/dt = v$, from (11.35) that $d\mathbf{e}_t/d\theta = \mathbf{e}_n$, and from elementary calculus that $d\theta/ds$ is equal to $1/\rho$, where ρ is the radius of curvature of the path at P (Fig. 11.22), we have

$$\frac{d\mathbf{e}_t}{dt} = \frac{v}{\rho}\mathbf{e}_n \qquad (11.38)$$

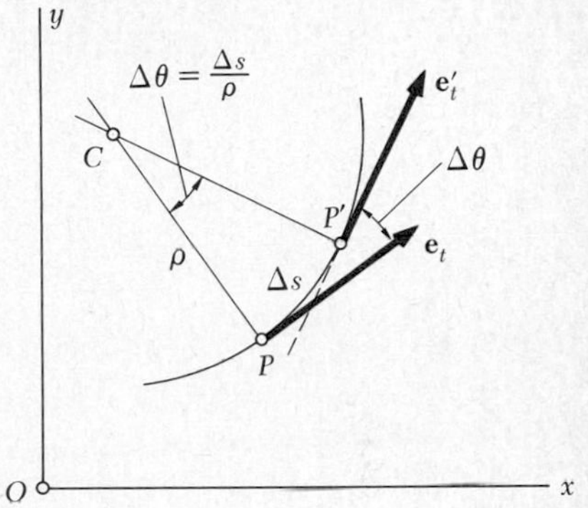

Fig. 11.22

Substituting into (11.37), we obtain

$$\mathbf{a} = \frac{dv}{dt}\mathbf{e}_t + \frac{v^2}{\rho}\mathbf{e}_n \qquad (11.39)$$

Thus, the scalar components of the acceleration are

$$a_t = \frac{dv}{dt} \qquad a_n = \frac{v^2}{\rho} \qquad (11.40)$$

The relations obtained express that the *tangential component* of the acceleration is equal to the *rate of change of the speed of the particle,* while the *normal component* is equal to the *square of the speed divided by the radius of curvature of the path at P.* Depending upon whether the speed of the particle increases or decreases, a_t is positive or negative, and the vector component $\mathbf{a}_t$ points in the direction of motion or against the direction of motion. The vector component $\mathbf{a}_n$, on the other hand, *is always directed toward the center of curvature C of the path* (Fig. 11.23).

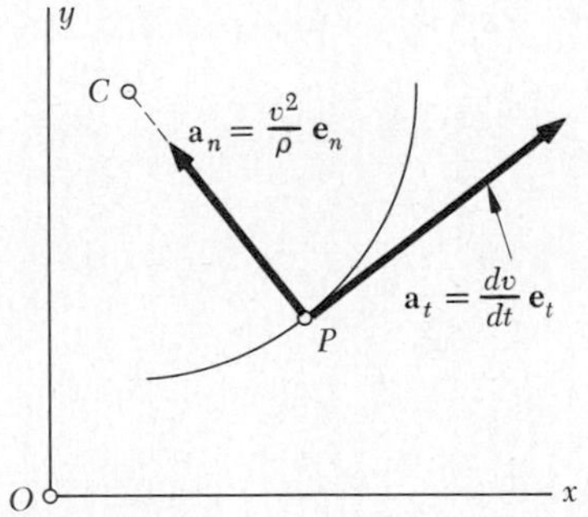

Fig. 11.23

It appears from the above that the tangential component of the acceleration reflects a change in the speed of the particle, while its normal component reflects a change in the direction of motion of the particle. The acceleration of a particle will be zero only if both its components are zero. Thus, the acceleration of a particle moving with constant speed along a curve will not be zero, unless the particle happens to pass through a point of inflection of the curve (where the radius of curvature is infinite) or unless the curve is a straight line.

The fact that the normal component of the acceleration depends upon the radius of curvature of the path followed by the particle is taken into account in the design of structures or mechanisms as widely different as airplane wings, railroad tracks, and cams. In order to avoid sudden changes in the acceleration of the air particles flowing past a wing, wing profiles are designed without any sudden change in curvature. Similar care is taken in designing railroad curves, to avoid sudden changes in the acceleration of the cars (which would be hard on the equipment and unpleasant for the passengers). A straight section of track, for instance, is never directly followed by a circular section. Special transition sections are used, to help pass smoothly from the infinite radius of curvature of the straight section to the finite radius of the circular track. Likewise, in the design of high-speed cams, abrupt changes in acceleration are avoided by using transition curves which produce a continuous change in acceleration.

Motion of a Particle in Space. The relations (11.39) and (11.40) still hold in the case of a particle moving along a space curve. However, since there is an infinite number of straight lines which are perpendicular to the tangent at a given point P of a space curve, it is then necessary to define more precisely the direction of the unit vector $\mathbf{e}_n$.

Let us consider again the unit vectors $\mathbf{e}_t$ and $\mathbf{e}'_t$ tangent to the path of the particle at two neighboring points P and P' (Fig. 11.24*a*) and the vector $\Delta\mathbf{e}_t$ representing the difference between $\mathbf{e}_t$ and $\mathbf{e}'_t$ (Fig. 11.24*b*). Let us now

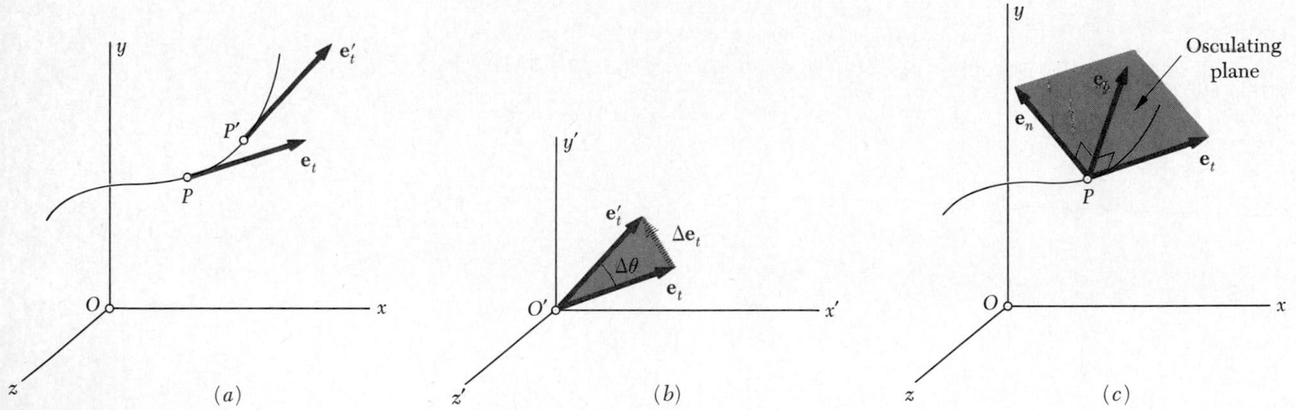

Fig. 11.24

imagine a plane through P (Fig. 11.24*a*) parallel to the plane defined by the vectors $\mathbf{e}_t$, $\mathbf{e}'_t$, and $\Delta\mathbf{e}_t$ (Fig. 11.24*b*). This plane contains the tangent to the curve at P and is parallel to the tangent at P'. If we let P' approach P, we obtain at the limit the plane which fits the curve most closely in the neighborhood of P. This plane is called the *osculating plane* at P.† It follows from this definition that the osculating plane contains the unit vector $\mathbf{e}_n$, since this vector represents the limit of the vector $\Delta\mathbf{e}_t/\Delta\theta$. The normal defined by $\mathbf{e}_n$ is thus contained in the osculating plane; it is called the *principal normal* at P. The unit vector $\mathbf{e}_b = \mathbf{e}_t \times \mathbf{e}_n$ which completes the right-handed triad $\mathbf{e}_t$, $\mathbf{e}_n$, $\mathbf{e}_b$ (Fig. 11.24*c*) defines the *binormal* at P. The binormal is thus perpendicular to the osculating plane. We conclude that as stated in (11.39), the acceleration of the particle at P may be resolved into two components, one along the tangent, the other along the principal normal at P. The acceleration has no component along the binormal.

†From the Latin *osculari*, to embrace.

11.14. Radial and Transverse Components. In certain problems of plane motion, the position of the particle P is defined by its polar coordinates r and θ (Fig. 11.25*a*). It is then convenient to resolve the velocity and acceleration of the particle into components parallel and perpendicular, respectively, to the line OP. These components are called *radial* and *transverse components*.

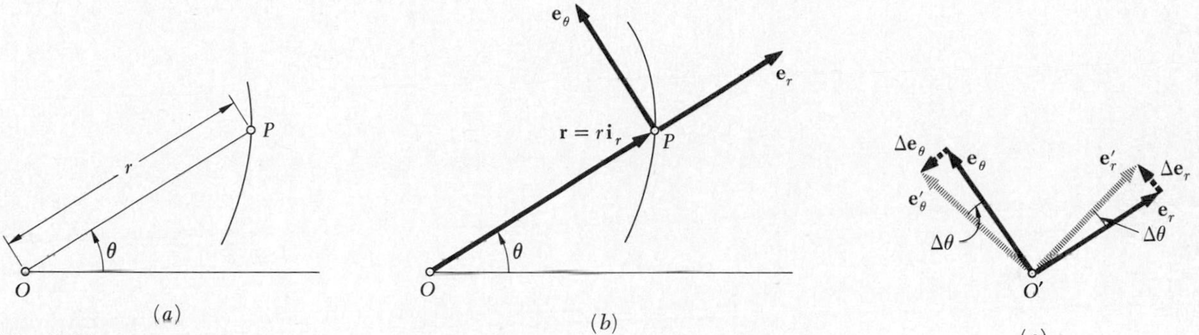

Fig. 11.25

We attach at P two unit vectors, $\mathbf{e}_r$ and $\mathbf{e}_\theta$ (Fig. 11.25*b*). The vector $\mathbf{e}_r$ is directed along OP and the vector $\mathbf{e}_\theta$ is obtained by rotating $\mathbf{e}_r$ through 90° counterclockwise. The unit vector $\mathbf{e}_r$ defines the *radial* direction, i.e., the direction in which P would move if r were increased and θ kept constant; the unit vector $\mathbf{e}_\theta$ defines the *transverse* direction, i.e., the direction in which P would move if θ were increased and r kept constant. A derivation similar to the one we used in Sec. 11.13 to determine the derivative of the unit vector $\mathbf{e}_t$ leads to the relations

$$\frac{d\mathbf{e}_r}{d\theta} = \mathbf{e}_\theta \qquad \frac{d\mathbf{e}_\theta}{d\theta} = -\mathbf{e}_r \tag{11.41}$$

where $-\mathbf{e}_r$ denotes a unit vector of sense opposite to that of $\mathbf{e}_r$ (Fig. 11.25*c*). Using the chain rule of differentiation, we express the time derivatives of the unit vectors $\mathbf{e}_r$ and $\mathbf{e}_\theta$ as follows:

$$\frac{d\mathbf{e}_r}{dt} = \frac{d\mathbf{e}_r}{d\theta}\frac{d\theta}{dt} = \mathbf{e}_\theta\frac{d\theta}{dt} \qquad \frac{d\mathbf{e}_\theta}{dt} = \frac{d\mathbf{e}_\theta}{d\theta}\frac{d\theta}{dt} = -\mathbf{e}_r\frac{d\theta}{dt}$$

or, using dots to indicate differentiation with respect to t,

$$\dot{\mathbf{e}}_r = \dot{\theta}\mathbf{e}_\theta \qquad \dot{\mathbf{e}}_\theta = -\dot{\theta}\mathbf{e}_r \tag{11.42}$$

To obtain the velocity $\mathbf{v}$ of the particle P, we express the position vector $\mathbf{r}$ of P as the product of the scalar r and the unit vector $\mathbf{e}_r$ and differentiate with respect to t:

$$\mathbf{v} = \frac{d}{dt}(r\mathbf{e}_r) = \dot{r}\mathbf{e}_r + r\dot{\mathbf{e}}_r$$

or, recalling the first of the relations (11.42),

$$\mathbf{v} = \dot{r}\mathbf{e}_r + r\dot{\theta}\mathbf{e}_\theta \tag{11.43}$$

Differentiating again with respect to t to obtain the acceleration, we write

$$\mathbf{a} = \frac{d\mathbf{v}}{dt} = \ddot{r}\mathbf{e}_r + \dot{r}\dot{\mathbf{e}}_r + \dot{r}\dot{\theta}\mathbf{e}_\theta + r\ddot{\theta}\mathbf{e}_\theta + r\dot{\theta}\dot{\mathbf{e}}_\theta$$

or, substituting for $\dot{\mathbf{e}}_r$ and $\dot{\mathbf{e}}_\theta$ from (11.42) and factoring $\mathbf{e}_r$ and $\mathbf{e}_\theta$,

$$\mathbf{a} = (\ddot{r} - r\dot{\theta}^2)\mathbf{e}_r + (r\ddot{\theta} + 2\dot{r}\dot{\theta})\mathbf{e}_\theta \tag{11.44}$$

The scalar components of the velocity and acceleration in the radial and transverse directions are therefore

$$v_r = \dot{r} \qquad v_\theta = r\dot{\theta} \tag{11.45}$$

$$a_r = \ddot{r} - r\dot{\theta}^2 \qquad a_\theta = r\ddot{\theta} + 2\dot{r}\dot{\theta} \tag{11.46}$$

It is important to note that a_r is *not* equal to the time derivative of v_r, and that a_θ is *not* equal to the time derivative of v_θ.

In the case of a particle moving along a circle of center O, we have $r =$ constant, $\dot{r} = \ddot{r} = 0$, and the formulas (11.43) and (11.44) reduce, respectively, to

$$\mathbf{v} = r\dot{\theta}\mathbf{e}_\theta \qquad \mathbf{a} = -r\dot{\theta}^2\mathbf{e}_r + r\ddot{\theta}\mathbf{e}_\theta \tag{11.47}$$

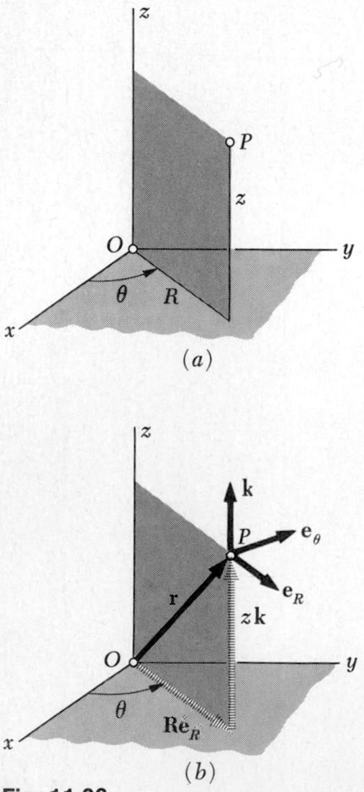

Fig. 11.26

Extension to the Motion of a Particle in Space: Cylindrical Coordinates. The position of a particle P in space is sometimes defined by its cylindrical coordinates R, θ, and z (Fig. 11.26*a*). It is then convenient to use the unit vectors $\mathbf{e}_R$, $\mathbf{e}_\theta$, and $\mathbf{k}$ shown in Fig. 11.26*b*. Resolving the position vector $\mathbf{r}$ of the particle P into components along the unit vectors, we write

$$\mathbf{r} = R\mathbf{e}_R + z\mathbf{k} \tag{11.48}$$

Observing that $\mathbf{e}_R$ and $\mathbf{e}_\theta$ define, respectively, the radial and transverse direction in the horizontal xy plane, and that the vector $\mathbf{k}$, which defines the *axial* direction, is constant in direction as well as in magnitude, we easily verify that

$$\mathbf{v} = \frac{d\mathbf{r}}{dt} = \dot{R}\mathbf{e}_R + R\dot{\theta}\mathbf{e}_\theta + \dot{z}\mathbf{k} \tag{11.49}$$

$$\mathbf{a} = \frac{d\mathbf{v}}{dt} = (\ddot{R} - R\dot{\theta}^2)\mathbf{e}_R + (R\ddot{\theta} + 2\dot{R}\dot{\theta})\mathbf{e}_\theta + \ddot{z}\mathbf{k} \tag{11.50}$$

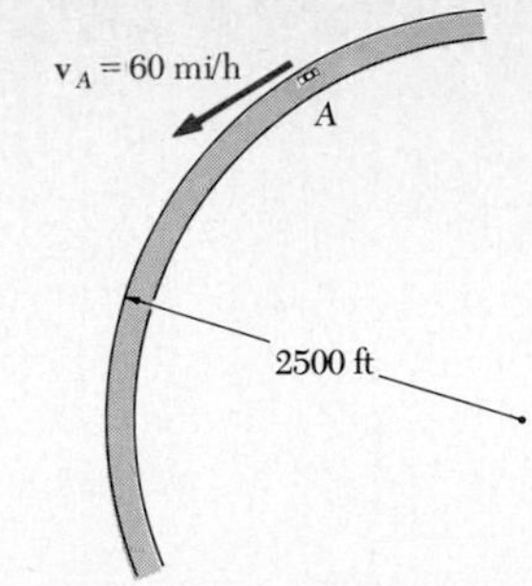

SAMPLE PROBLEM 11.10

A motorist is traveling on a curved section of highway of radius 2500 ft at the speed of 60 mi/h. The motorist suddenly applies the brakes, causing the automobile to slow down at a constant rate. Knowing that after 8 s the speed has been reduced to 45 mi/h, determine the acceleration of the automobile immediately after the brakes have been applied.

Tangential Component of Acceleration. First the speeds are expressed in ft/s.

$$60 \text{ mi/h} = \left(60 \frac{\text{mi}}{\text{h}}\right)\left(\frac{5280 \text{ ft}}{1 \text{ mi}}\right)\left(\frac{1 \text{ h}}{3600 \text{ s}}\right) = 88 \text{ ft/s}$$

$$45 \text{ mi/h} = 66 \text{ ft/s}$$

Since the automobile slows down at a constant rate, we have

$$a_t = \text{average } a_t = \frac{\Delta v}{\Delta t} = \frac{66 \text{ ft/s} - 88 \text{ ft/s}}{8 \text{ s}} = -2.75 \text{ ft/s}^2$$

Normal Component of Acceleration. Immediately after the brakes have been applied, the speed is still 88 ft/s, and we have

$$a_n = \frac{v^2}{\rho} = \frac{(88 \text{ ft/s})^2}{2500 \text{ ft}} = 3.10 \text{ ft/s}^2$$

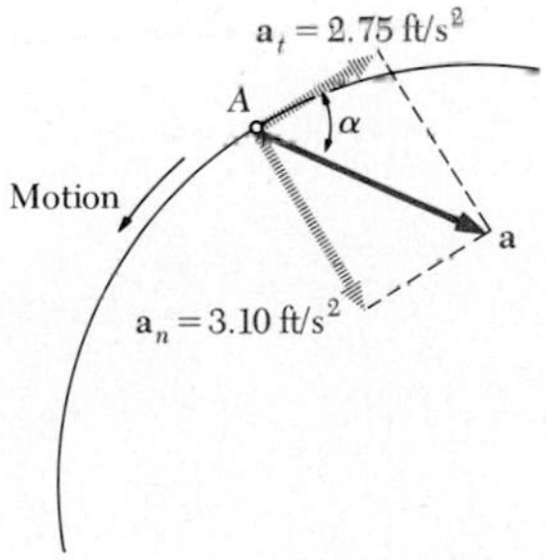

Magnitude and Direction of Acceleration. The magnitude and direction of the resultant **a** of the components $\mathbf{a}_n$ and $\mathbf{a}_t$ are

$$\tan \alpha = \frac{a_n}{a_t} = \frac{3.10 \text{ ft/s}^2}{2.75 \text{ ft/s}^2} \qquad \alpha = 48.4° \blacktriangleleft$$

$$a = \frac{a_n}{\sin \alpha} = \frac{3.10 \text{ ft/s}^2}{\sin 48.4°} \qquad a = 4.14 \text{ ft/s}^2 \blacktriangleleft$$

SAMPLE PROBLEM 11.11

Determine the minimum radius of curvature of the trajectory described by the projectile considered in Sample Prob. 11.7.

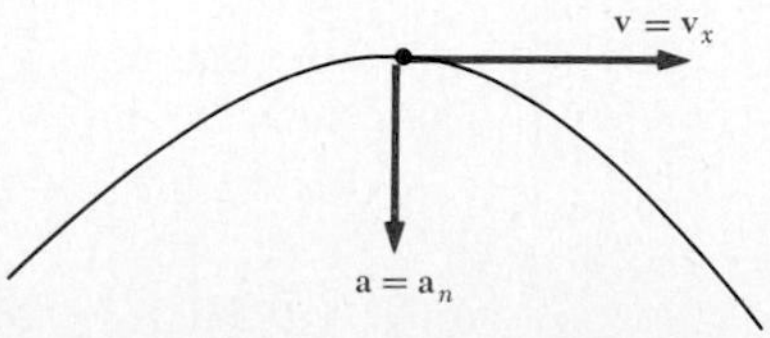

Solution. Since $a_n = v^2/\rho$, we have $\rho = v^2/a_n$. The radius will be small when v is small or when a_n is large. The speed v is minimum at the top of the trajectory since $v_y = 0$ at that point; a_n is maximum at that same point, since the direction of the vertical coincides with the direction of the normal. Therefore, the minimum radius of curvature occurs at the top of the trajectory. At this point, we have

$$v = v_x = 155.9 \text{ m/s} \qquad a_n = a = 9.81 \text{ m/s}^2$$

$$\rho = \frac{v^2}{a_n} = \frac{(155.9 \text{ m/s})^2}{9.81 \text{ m/s}^2} \qquad \rho = 2480 \text{ m} \blacktriangleleft$$

SAMPLE PROBLEM 11.12

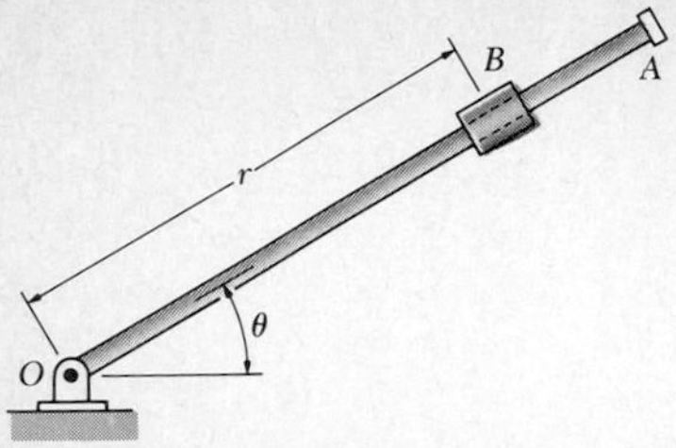

The rotation of the 0.9-m arm OA about O is defined by the relation $\theta = 0.15t^2$, where θ is expressed in radians and t in seconds. Collar B slides along the arm in such a way that its distance from O is $r = 0.9 - 0.12t^2$, where r is expressed in meters and t in seconds. After the arm OA has rotated through 30°, determine (*a*) the total velocity of the collar, (*b*) the total acceleration of the collar, (*b*) the relative acceleration of the collar with respect to the arm.

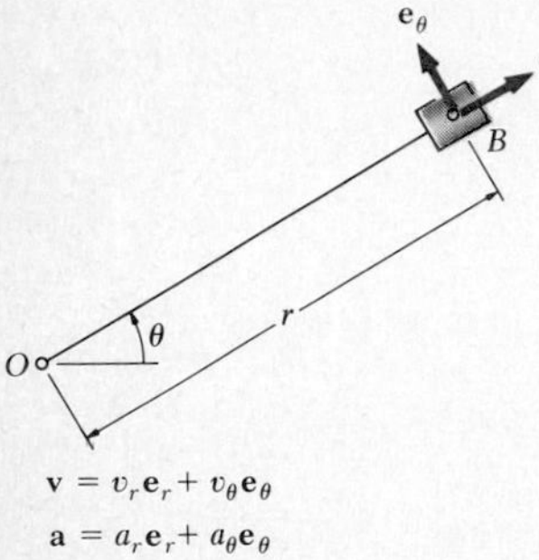

Solution. We first find the time t at which $\theta = 30°$. Substituting $\theta = 30° = 0.524$ rad into the expression for θ, we obtain

$$\theta = 0.15t^2 \qquad 0.524 = 0.15t^2 \qquad t = 1.869 \text{ s}$$

Equations of Motion. Substituting $t = 1.869$ s in the expressions for r, θ, and their first and second derivatives, we have

$$r = 0.9 - 0.12t^2 = 0.481 \text{ m} \qquad \theta = 0.15t^2 = 0.524 \text{ rad}$$
$$\dot{r} = -0.24t = -0.449 \text{ m/s} \qquad \dot{\theta} = 0.30t = 0.561 \text{ rad/s}$$
$$\ddot{r} = -0.24 = -0.240 \text{ m/s}^2 \qquad \ddot{\theta} = 0.30 = 0.300 \text{ rad/s}^2$$

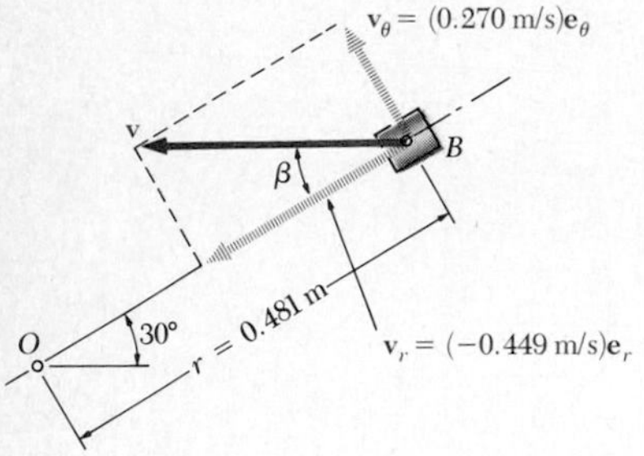

a. Velocity of B. Using Eqs. (11.45), we obtain the values of v_r and v_θ when $t = 1.869$ s.

$$v_r = \dot{r} = -0.449 \text{ m/s}$$
$$v_\theta = r\dot{\theta} = 0.481(0.561) = 0.270 \text{ m/s}$$

Solving the right triangle shown, we obtain the magnitude and direction of the velocity,

$$v = 0.524 \text{ m/s} \qquad \beta = 31.0° \quad ◀$$

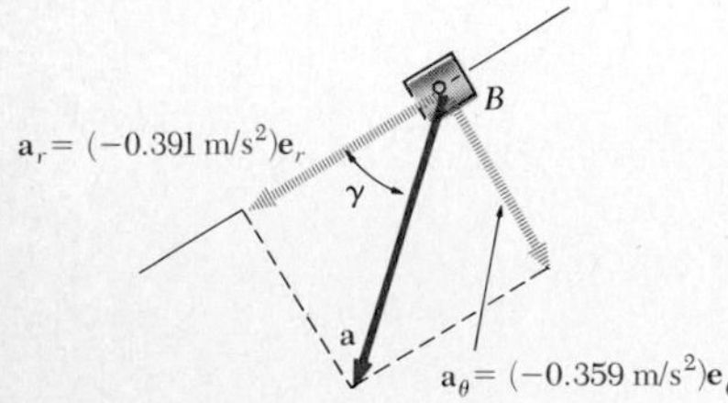

b. Acceleration of B. Using Eqs. (11.46), we obtain

$$a_r = \ddot{r} - r\dot{\theta}^2$$
$$= -0.240 - 0.481(0.561)^2 = -0.391 \text{ m/s}^2$$
$$a_\theta = r\ddot{\theta} + 2\dot{r}\dot{\theta}$$
$$= 0.481(0.300) + 2(-0.449)(0.561) = -0.359 \text{ m/s}^2$$
$$a = 0.531 \text{ m/s}^2 \qquad \gamma = 42.6° \quad ◀$$

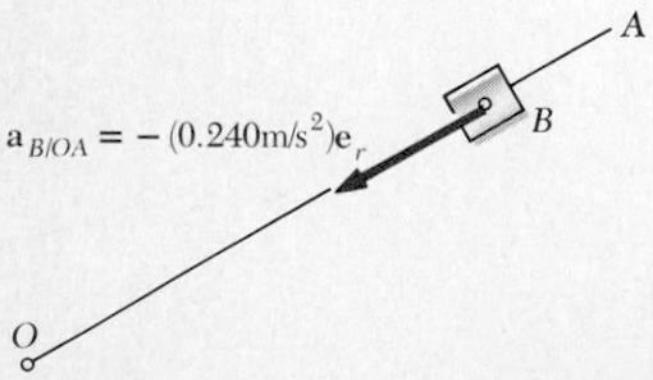

c. Acceleration of B with Respect to Arm OA. We note that the motion of the collar with respect to the arm is rectilinear and defined by the coordinate r. We write

$$a_{B/OA} = \ddot{r} = -0.240 \text{ m/s}^2$$
$$a_{B/OA} = 0.240 \text{ m/s}^2 \text{ toward } O. \quad ◀$$

Problems

11.119 What is the smallest radius which should be used for a highway curve if the normal component of the acceleration of a car traveling at 60 mi/h is not to exceed 2.5 ft/s²?

11.120 A motorist drives along the circular exit ramp of a turnpike at the constant speed v_0. Knowing that the odometer indicates a distance of 0.6 km between point A where the automobile is going due south and B where it is going due north, determine the speed v_0 for which the normal component of the acceleration is $0.08g$.

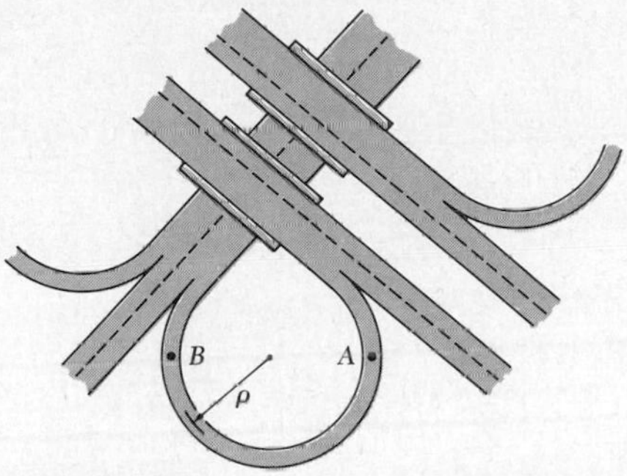

Fig. P11.120

11.121 During the test of an airplane it is desired that the normal component of the acceleration be equal to $6g$. Determine the radius of the horizontal circle along which the pilot should fly if the speed of the airplane is 1800 km/h.

11.122 A computer tape moves over two drums at a constant speed v_0. Knowing that the normal component of the acceleration of the portion of tape in contact with drum B is 400 ft/s², determine (*a*) the speed v_0, (*b*) the normal component of the acceleration of the portion of tape in contact with drum A.

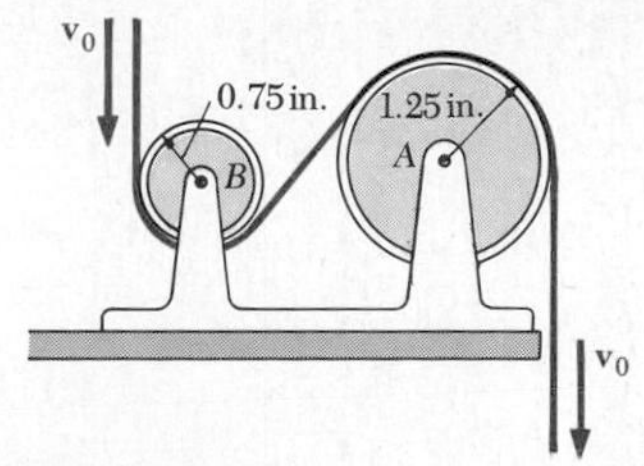

Fig. P11.122

11.123 A motorist is traveling on a curved portion of highway of radius 400 m at a speed of 90 km/h. The brakes are suddenly applied, causing the speed to decrease at a constant rate of 1.2 m/s². Determine the magnitude of the total acceleration of the automobile (*a*) immediately after the brakes have been applied, (*b*) 5 s later.

11.124 A bus starts from rest on a curve of 250-m radius and accelerates at the constant rate $a_t = 0.6$ m/s². Determine the distance and time that the bus will travel before the magnitude of its total acceleration is 0.75 m/s².

11.125 A motorist decreases the speed of an automobile at a constant rate from 45 to 30 mi/h over a distance of 750 ft along a curve of 1500-ft radius. Determine the magnitude of the total acceleration of the automobile after it has traveled 500 ft along the curve.

11.126 As a train enters a curve of radius 4000 ft at a speed of 60 mi/h, the brakes are applied sufficiently to cause the magnitude of the *total* acceleration of the train to be 2.5 ft/s^2. After 10 s the brakes are more fully applied so that the magnitude of the total acceleration is again 2.5 ft/s^2. If this second brake setting is then maintained, what is the total time required to bring the train to a stop?

11.127 Automobile A is traveling along a straight highway, while B is moving along a circular exit ramp of 80-m radius. The speed of A is being increased at the rate of 2 m/s^2 and the speed of B is being decreased at the rate of 1.2 m/s^2. For the position shown, determine (a) the velocity of A relative to B, (b) the acceleration of A relative to B.

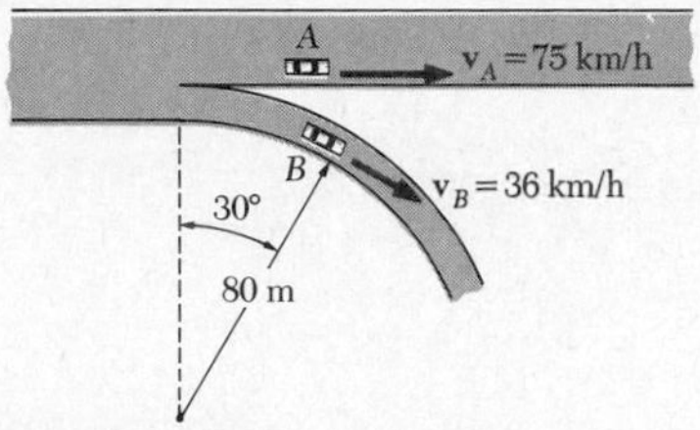

Fig. P11.127

11.128 Solve Prob. 11.127, assuming that the velocity of B is 18 km/h and is being decreased at the rate of 1.2 m/s^2.

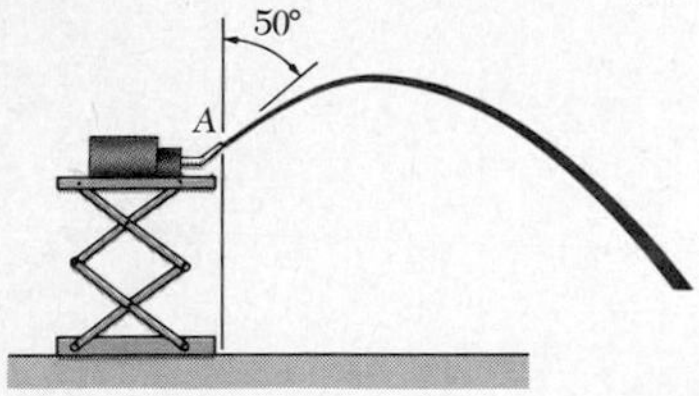

Fig. P11.129

11.129 A nozzle discharges a stream of water in the direction shown with an initial velocity of 25 ft/s. Determine the radius of curvature of the stream (a) as it leaves the nozzle, (b) at the maximum height of the stream.

11.130 For each of the two firing angles obtained in Sample Prob. 11.8, determine the radius of curvature of the trajectory described by the projectile as it leaves the gun.

11.131 Determine the radius of curvature of the trajectory described by the projectile of Sample Prob. 11.7 (a) as the projectile leaves the gun, (b) at the maximum elevation of the projectile.

11.132 From measurements of a photograph it has been found that as the stream of water shown left the nozzle at A, it had a radius of curvature of 25 m. Determine (a) the initial velocity $\mathbf{v}_A$ of the stream, (b) the radius of curvature of the stream at its maximum height.

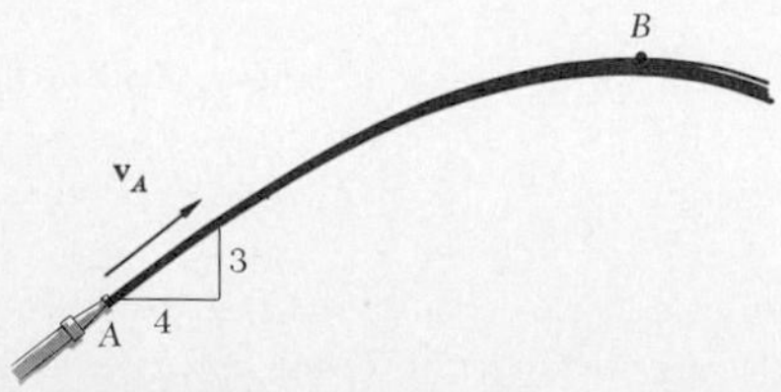

Fig. P11.132

***11.133** Determine the radius of curvature of the helix of Prob. 11.85.

***11.134** Determine the radius of curvature of the path described by the particle of Prob. 11.86 when (*a*) $t = 0$, (*b*) $t = 2$ s.

11.135 A satellite will travel indefinitely in a circular orbit around the earth if the normal component of its acceleration is equal to $g(R/r)^2$, where $g = 32.2 \text{ ft/s}^2$, R = radius of the earth = 3960 mi, and r = distance from the center of the earth to the satellite. Determine the height above the surface of the earth at which a satellite will travel indefinitely around the earth at a speed of 16,500 mi/h.

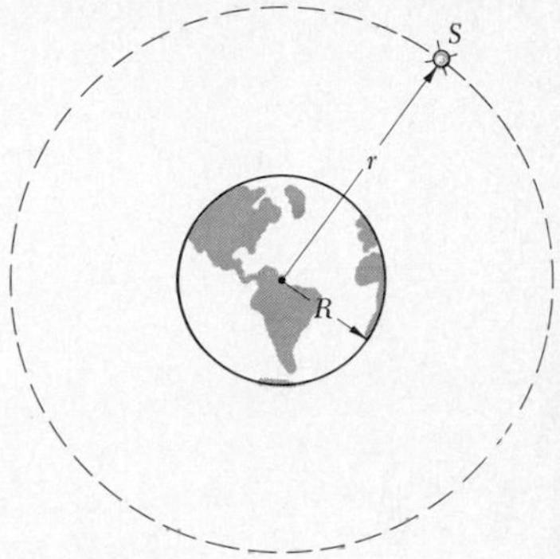

Fig. P11.135

11.136 Determine the speed of a space shuttle traveling in a circular orbit 140 mi above the surface of the earth. (See information given in Prob. 11.135.)

11.137 Show that the speed of an earth satellite traveling in a circular orbit is inversely proportional to the square root of the radius of its orbit. Knowing that the radius of the earth is 6370 km and that $g = 9.81 \text{ m/s}^2$, determine the minimum time required for a satellite to circle the earth. (See information given in Prob. 11.135.)

11.138 Assuming that the orbit of the moon is a circle of radius 384×10^3 km, determine the speed of the moon relative to the earth. (See information given in Probs. 11.135 and 11.137.)

11.139 The rotation of rod OA about O is defined by the relation $\theta = t^3 - 4t$, where θ is expressed in radians and t in seconds. Collar B slides along the rod in such a way that its distance from O is $r = 25t^3 - 50t^2$, where r is expressed in millimeters and t in seconds. When $t = 1$ s, determine (*a*) the velocity of the collar, (*b*) the total acceleration of the collar, (*c*) the acceleration of the collar relative to the rod.

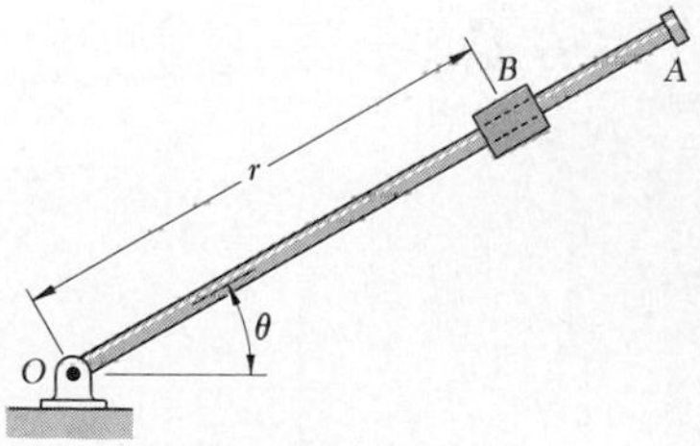

Fig. P11.139 and P11.140

11.140 The rotation of rod OA about O is defined by the relation $\theta = \frac{1}{2}\pi(4t - 3t^2)$, where θ is expressed in radians and t in seconds. Collar B slides along the rod in such a way that its distance from O is $r = 1.25t^2 - 0.9t^3$, where r is expressed in meters and t in seconds. When $t = 1$ s, determine (*a*) the velocity of the collar, (*b*) the total acceleration of the collar, (*c*) the acceleration of the collar relative to the rod.

11.141 The two-dimensional motion of a particle is defined by the relations $r = 2b \cos \omega t$ and $\theta = \omega t$, where b and ω are constants. Determine (*a*) the velocity and acceleration of the particle at any instant, (*b*) the radius of curvature of its path. What conclusion can you draw regarding the path of the particle?

11.142 The path of a particle P is an Archimedean spiral. The motion of the particle is defined by the relations $r = 10t$ and $\theta = 2\pi t$, where r is expressed in inches, t in seconds, and θ in radians. Determine the velocity and acceleration of the particle when (*a*) $t = 0$, (*b*) $t = 0.25$ s.

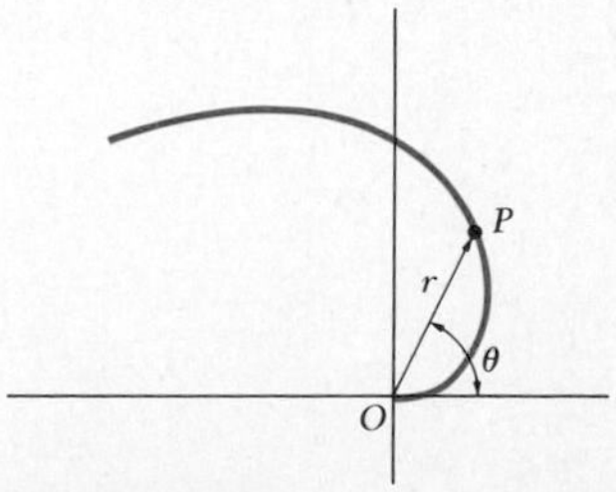

Fig. P11.142

* **11.143** The motion of particle P, along the elliptic path shown, is defined by the relations $r = 35/(1 - 0.75 \cos \pi t)$ and $\theta = \pi t$, where r is expressed in millimeters, t in seconds, and θ in radians. Determine the velocity and acceleration of the particle when (*a*) $t = 0$, (*b*) $t = 0.5$ s.

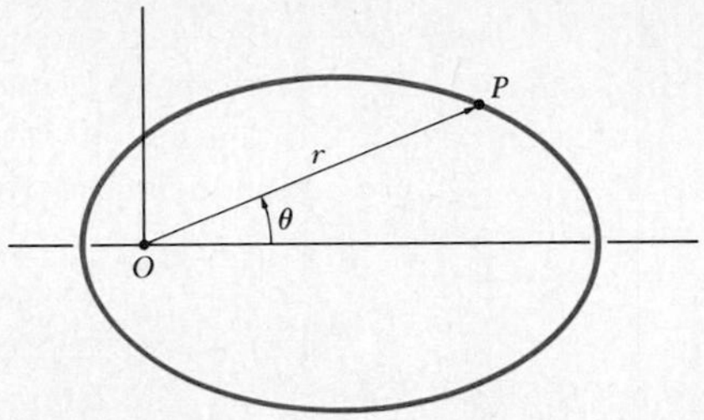

Fig. P11.143

* **11.144** Solve Prob. 11.143, when (*a*) $t = 1$ s, (*b*) $t = 1.5$ s.

11.145 A rocket is fired vertically from a launching pad at B. Its flight is tracked by radar from point A. Determine the velocity of the rocket in terms of b, θ, and $\dot{\theta}$.

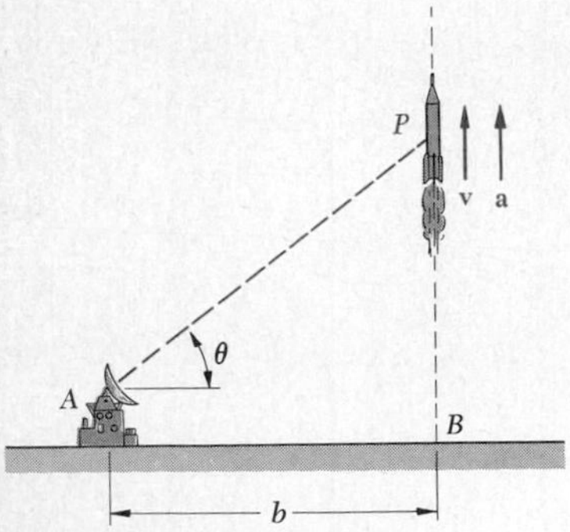

Fig. P11.145 and P11.149

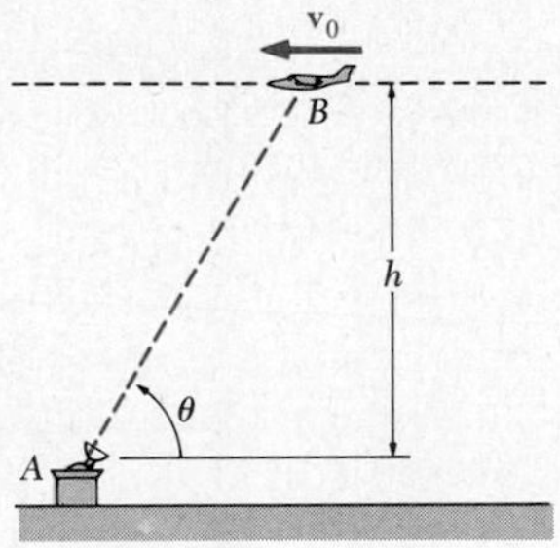

Fig. P11.146

11.146 The flight path of airplane B is a horizontal straight line and passes directly over a radar tracking station at A. Knowing that the airplane moves to the left with the constant velocity $\mathbf{v}_0$, determine $d\theta/dt$ in terms of v_0, h, and θ.

11.147 Determine the acceleration of the rocket of Prob. 11.145 in terms of b, θ, $\dot{\theta}$, and $\ddot{\theta}$.

11.148 In Prob. 11.146, determine $d^2\theta/dt^2$ in terms of v_0, h, and θ.

11.149 A test rocket is fired vertically from a launching pad at B. When the rocket is at P the angle of elevation is $\theta = 42.0°$, and 0.5 s later it is $\theta = 43.2°$. Knowing that $b = 3$ km, determine approximately the speed of the rocket during the 0.5-s interval.

11.150 An airplane passes over a radar tracking station at A and continues to fly due east. When the airplane is at P, the distance and angle of elevation of the plane are, respectively, $r = 11{,}200$ ft and $\theta = 26.5°$. Two seconds later the radar station sights the plane at $r = 12{,}300$ ft and $\theta = 23.3°$. Determine approximately the speed and the angle of dive α of the plane during the 2-s interval.

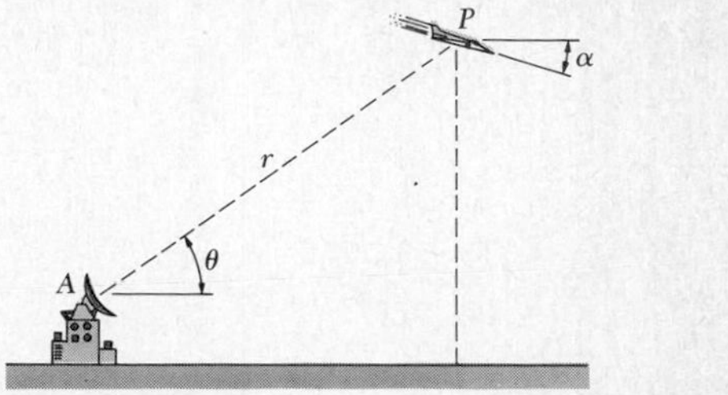

Fig. P11.150

11.151 and 11.152 A particle moves along the spiral shown; determine the magnitude of the velocity of the particle in terms of b, θ, and $\dot{\theta}$.

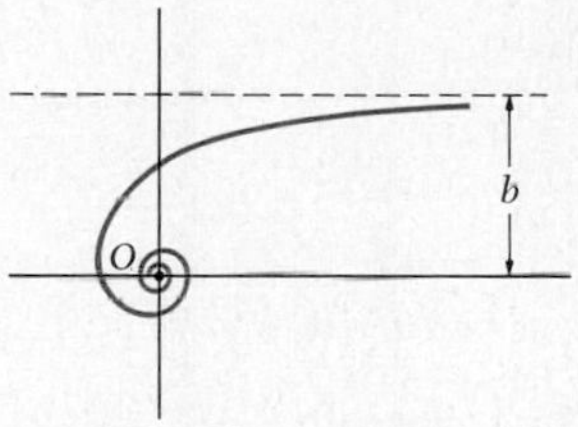

Hyperbolic spiral $r\theta = b$
Fig. P11.151 and P11.153

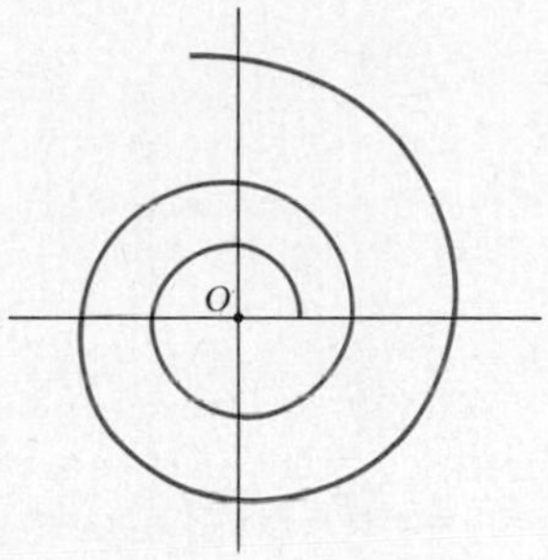

Logarithmic spiral $r = e^{b\theta}$
Fig. P11.152 and P11.154

11.153 and 11.154 A particle moves along the spiral shown. Knowing that $\dot{\theta}$ is constant and denoting this constant by ω, determine the magnitude of the acceleration of the particle in terms of b, θ, and ω.

11.155 As rod OA rotates, pin P moves along the parabola BCD. Knowing that the equation of the parabola is $r = 2b/(1 + \cos\theta)$ and that $\theta = kt$, determine the velocity and acceleration of P when (*a*) $\theta = 0$, (*b*) $\theta = 90°$.

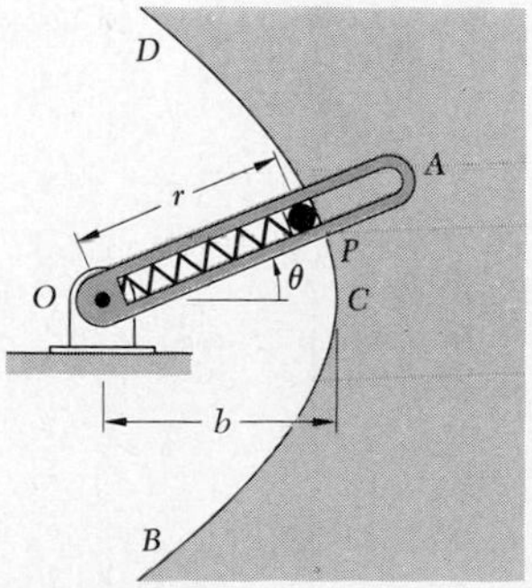

Fig. P11.155

11.156 The pin at B is free to slide along the circular slot DE and along the rotating rod OC. Assuming that the rod OC rotates at a constant rate $\dot{\theta}$, (*a*) show that the acceleration of pin B is of constant magnitude, (*b*) determine the direction of the acceleration of pin B.

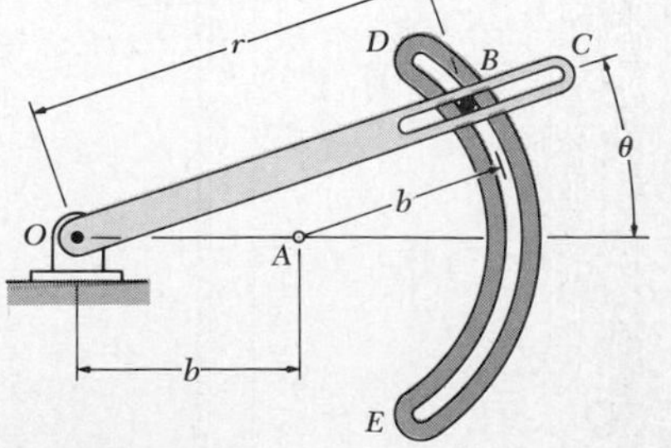

Fig. P11.156

11.157 The motion of a particle on the surface of a right circular cylinder is defined by the relations $R = A$, $\theta = 2\pi t$, and $z = B \sin 2\pi nt$, where A and B are constants and n is an integer. Determine the magnitudes of the velocity and acceleration of the particle at any time t.

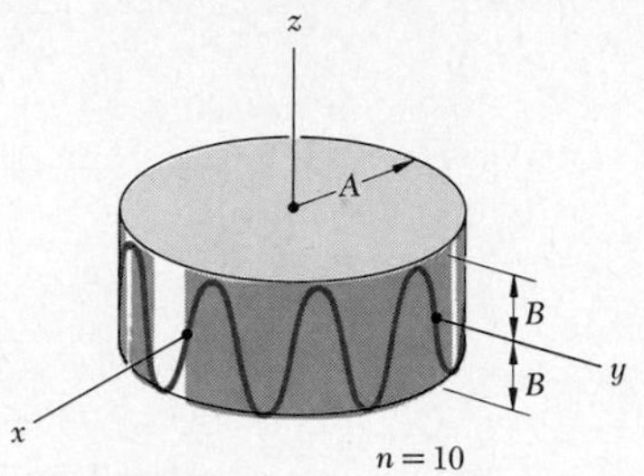

Fig. P11.157

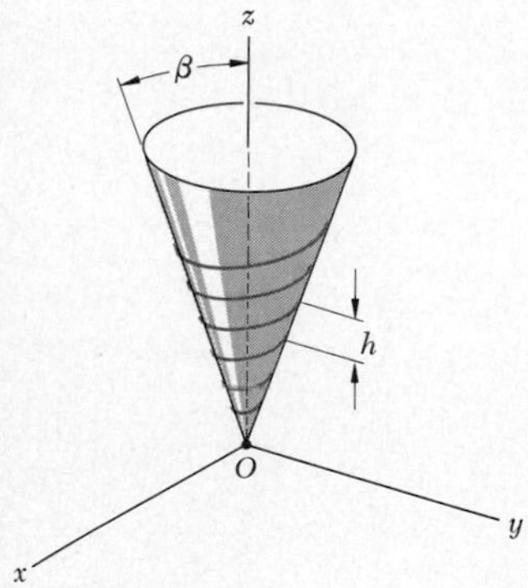

Fig. P11.159

11.158 For the case when $n = 1$ in Prob. 11.157, (*a*) show that the path of the particle is contained in a plane, (*b*) determine the maximum and minimum radii of curvature of the path.

11.159 The motion of a particle on the surface of a right circular cone is defined by the relations $R = ht \tan \beta$, $\theta = 2\pi t$, and $z = ht$, where β is the apex angle of the cone and h is the distance the particle rises in one passage around the cone. Determine the magnitudes of the velocity and acceleration at any time t.

11.160 The three-dimensional motion of a particle is defined by the relations $R = A(1 - e^{-t})$, $\theta = 2\pi t$, and $z = B(1 - e^{-t})$. Determine the magnitudes of the velocity and acceleration when (*a*) $t = 0$, (*b*) $t = \infty$.

11.161 The three-dimensional motion of a particle is defined by the relations $R = A$, $\theta = 2\pi t$, and $z = A \sin^2 2\pi t$. Determine the magnitudes of the velocity and acceleration at any time t.

11.162 The three-dimensional motion of a particle is defined by the relations $R = 2k \cos t$, $\theta = t$, and $z = pt$. Determine (*a*) the path of the particle, (*b*) the magnitudes of the velocity and acceleration at any time t, (*c*) the radius of curvature of the path at any time t.

***11.163** For the helix of Prob. 11.85, determine the angle that the osculating plane forms with the y axis.

***11.164** Determine the direction of the binormal of the path described by the particle of Prob. 11.86 when (*a*) $t = 0$, (*b*) $t = 2$ s.

Review Problems

11.165 The speed of a racing car is increased at a constant rate from 72 km/h to 108 km/h over a distance of 120 m along a curve of 200-m radius. Determine the magnitude of the total acceleration of the car after it has traveled 80 m along the curve.

11.166 The acceleration of a collar moving in a straight line is defined by the relation $a = 50 \sin \frac{1}{2}\pi t$, where a is expressed in mm/s² and t in seconds. Knowing that $x = 0$ and $v = 0$ when $t = 0$, determine (*a*) the maximum velocity of the collar, (*b*) its position at $t = 4$ s, (*c*) its average velocity over the interval $0 < t < 4$ s. Sketch the a–t, v–t, and x–t curves for the motion.

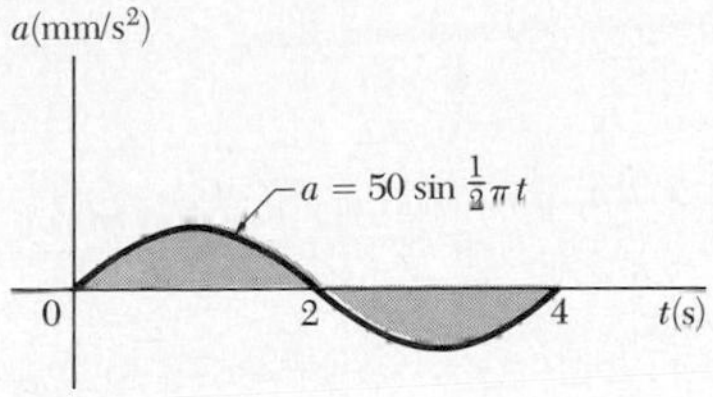

Fig. P11.166

11.167 The acceleration of a particle moving along the x axis is directed toward the origin O and is inversely proportional to the cube of the distance of the particle from O. The particle starts from rest at $x = 3$ ft, and it is observed that its velocity is 5 ft/s when $x = 2$ ft. Determine the velocity of the particle when (*a*) $x = 1$ ft, (*b*) $x = 0.5$ ft.

11.168 A nozzle discharges a stream of water with an initial velocity $\mathbf{v}_0$ of 75 ft/s into the end of a horizontal pipe of inside diameter $d = 6$ ft. Determine the largest distance x that the stream can reach.

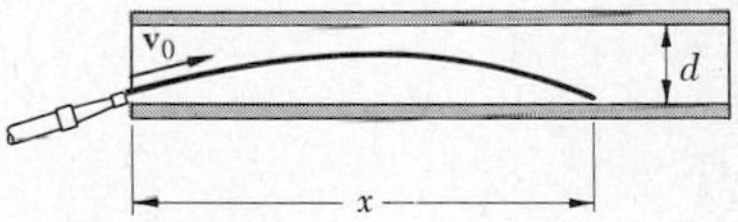

Fig. P11.168

11.169 A ball is projected from point A with a velocity $\mathbf{v}_0$ which is perpendicular to the incline shown. Knowing that the ball strikes the incline at B, determine the range R in terms of v_0 and β.

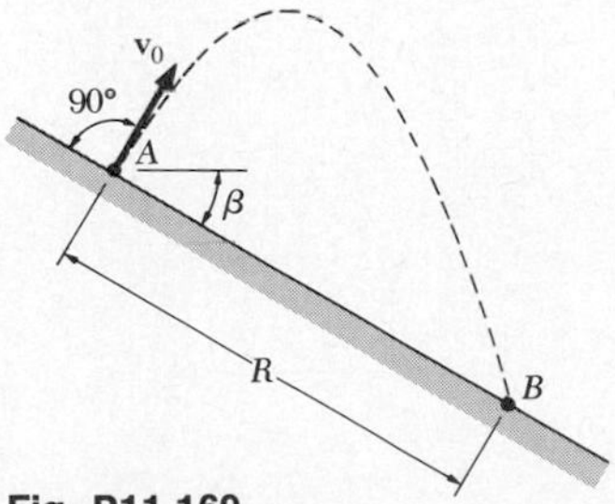

Fig. P11.169

11.170 In Prob. 11.169, determine the range R when $v_0 = 10$ m/s and $\beta = 30°$.

11.171 Airplane A is flying due east at 700 km/h, while airplane B is flying at 500 km/h at the same altitude and in a direction to the west of south. Knowing that the speed of B with respect to A is 1125 km/h, determine the direction of the flight path of B.

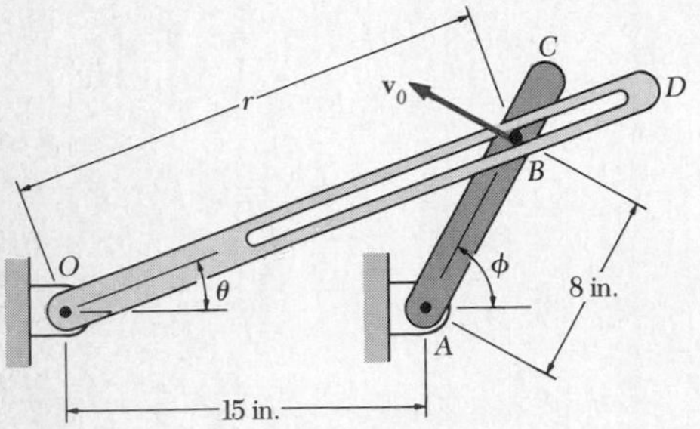

Fig. P11.173

11.172 A stone is dropped from ground level into a mine shaft as an ore bucket passes a point 120 m below ground level. Determine where and when the stone will strike the bucket if the speed of the bucket is 8 m/s (*a*) upward, (*b*) downward.

11.173 Pin B is attached to the rotating arm AC and moves at a constant speed $v_0 = 115$ in./s. Knowing that pin B slides freely in a slot cut in arm OD, determine the rate $\dot{\theta}$ at which arm OD rotates and the radial component v_r of the velocity of pin B (*a*) when $\phi = 0$, (*b*) when $\phi = 90°$.

11.174 A police officer on a motorcycle is escorting a motorcade which is traveling at 30 mi/h. The police officer suddenly decides to take a new position in the motorcade, 200 ft ahead. Assuming that he accelerates and decelerates at the rate of 11 ft/s^2 and that he does not exceed at any time a speed of 45 mi/h draw the a–t and v–t curves for his motion and determine (*a*) the shortest time in which he can occupy his new position in the motorcade, (*b*) the distance he will travel in that time.

11.175 In the position shown, collar A moves downward with a velocity of 80 mm/s and the relative velocity of block C with respect to collar B is 40 mm/s downward. Determine (*a*) the velocity of collar B, (*b*) the velocity of block C.

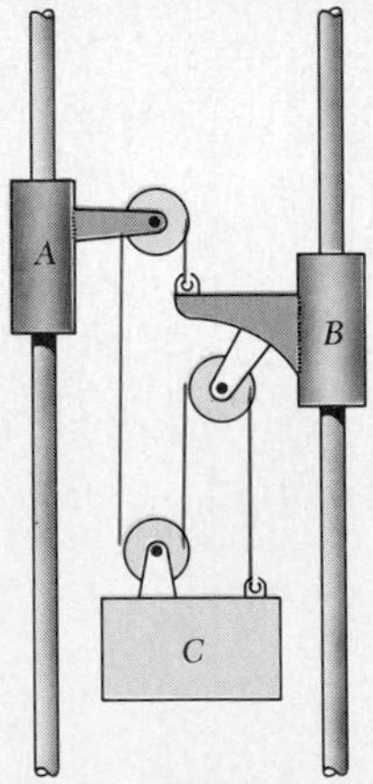

Fig. P11.175 and P11.176

11.176 Collars A and B start from rest and move with the following upward accelerations: $a_A = 160$ mm/s^2 and $a_B = 100$ mm/s^2. Determine the relative velocity of block C with respect to collar A after 4 s.

CHAPTER TWELVE

Kinetics of Particles: Newton's Second Law

12.1. Introduction. Newton's first and third laws of motion were used extensively in statics to study bodies at rest and the forces acting upon them. These two laws are also used in dynamics; in fact, they are sufficient for the study of the motion of bodies which have no acceleration. However, when bodies are accelerated, i.e., when the magnitude or the direction of their velocity changes, it is necessary to use Newton's second law of motion to relate the motion of the body with the forces acting on it.

In this chapter we shall discuss Newton's second law and apply it to the analysis of the motion of particles. As we state in Sec. 12.2, if the resultant of the forces acting on a particle is not zero, the particle will have an acceleration proportional to the magnitude of the resultant and in the direction of this resultant force. Moreover, the ratio of the magnitudes of the resultant force and of the acceleration may be used to define the *mass* of the particle.

In Sec. 12.3, we shall define the *linear momentum* of a particle as the product $\mathbf{L} = m\mathbf{v}$ of the mass m and velocity $\mathbf{v}$ of the particle and show that Newton's second law may be expressed in an alternative form relating the rate of change of the linear momentum with the resultant of the forces acting on that particle.

Section 12.4 stresses the need for consistent units in the solution of dynamics problems and provides a review of the two systems used in this text, namely, the International System of Units (SI units) and the U.S. customary units.

In Secs. 12.5 and 12.6 and in the Sample Problems which follow, Newton's second law is applied to the solution of engineering problems, using either rectangular components or tangential and normal components of the forces and accelerations involved. We recall that an actual body—possibly as large as a car, a rocket, or an airplane—may be considered as a particle for the purpose of analyzing its motion, as long as the effect of a rotation of the body about its mass center may be ignored.

The second part of the chapter is devoted to the solution of problems in terms of radial and transverse components, with particular emphasis on the motion of a particle under a central force. In Sec. 12.7, we shall define the *angular momentum* $\mathbf{H}_O$ of a particle about a point O as the moment about O of the linear momentum of the particle: $\mathbf{H}_O = \mathbf{r} \times m\mathbf{v}$. It then follows from Newton's second law that the rate of change of the angular momentum $\mathbf{H}_O$ of a particle is equal to the sum of the moments about O of the forces acting on that particle.

Section 12.9 deals with the motion of a particle under a *central force*, i.e., under a force directed toward or away from a fixed point O. Since such a force has zero moment about O, it follows that the angular momentum of the particle about O is conserved. This property greatly simplifies the analysis of the motion of a particle under a central force and will be applied to the solution of problems involving the orbital motion of bodies under gravitational attraction (Sec. 12.10).

Sections 12.11 through 12.13 are optional. They present a more extensive discussion of orbital motion and contain a number of problems related to space mechanics.

12.2. Newton's Second Law of Motion. Newton's second law may be stated as follows:

If the resultant force acting on a particle is not zero, the particle will have an acceleration proportional to the magnitude of the resultant and in the direction of this resultant force.

Newton's second law of motion may best be understood if we imagine the following experiment: A particle is subjected to a force $\mathbf{F}_1$ of constant direction and constant magnitude F_1. Under the action of that force, the particle will be observed to move in a straight line and *in the direction of the force* (Fig. 12.1*a*). By determining the position of the particle at various instants, we find that its acceleration has a constant magnitude a_1. If the experiment is repeated with forces $\mathbf{F}_2$, $\mathbf{F}_3$, etc., of different magnitude or direction (Fig. 12.1*b* and *c*), we find each time that the particle moves in the direction of the force acting on it and that the magnitudes a_1, a_2, a_3, etc., of the accelerations are proportional to the magnitudes F_1, F_2, F_3, etc., of the corresponding forces,

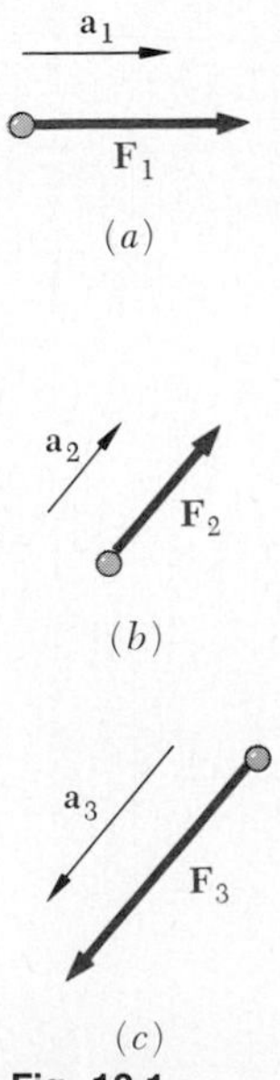

Fig. 12.1

$$\frac{F_1}{a_1} = \frac{F_2}{a_2} = \frac{F_3}{a_3} = \cdots = \text{constant}$$

The constant value obtained for the ratio of the magnitudes of the forces and accelerations is a characteristic of the particle under consideration. It is called the *mass* of the particle and is denoted by m. When a particle of mass m is acted upon by a force $\mathbf{F}$, the force $\mathbf{F}$ and the acceleration $\mathbf{a}$ of the particle must therefore satisfy the relation

$$\mathbf{F} = m\mathbf{a} \tag{12.1}$$

This relation provides a complete formulation of Newton's second law; it expresses not only that the magnitudes of **F** and **a** are proportional but also (since m is a positive scalar) that the vectors **F** and **a** have the same direction (Fig. 12.2). We should note that Eq. (12.1) still holds when **F** is not constant but varies with t in magnitude or direction. The magnitudes of **F** and **a** remain proportional, and the two vectors have the same direction at any given instant. However, they will not, in general, be tangent to the path of the particle.

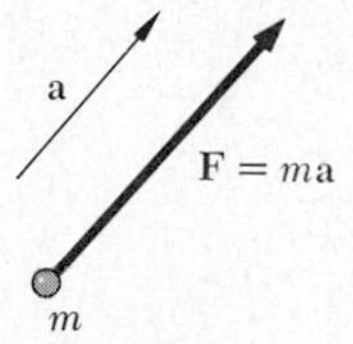

Fig. 12.2

When a particle is subjected simultaneously to several forces, Eq. (12.1) should be replaced by

$$\Sigma \mathbf{F} = m\mathbf{a} \tag{12.2}$$

where $\Sigma\mathbf{F}$ represents the sum, or resultant, of all the forces acting on the particle.

It should be noted that the system of axes with respect to which the acceleration **a** is determined is not arbitrary. These axes must have a constant orientation with respect to the stars, and their origin must either be attached to the sun† or move with a constant velocity with respect to the sun. Such a system of axes is called a *newtonian frame of reference.*‡ A system of axes attached to the earth does *not* constitute a newtonian frame of reference, since the earth rotates with respect to the stars and is accelerated with respect to the sun. However, in most engineering applications, the acceleration **a** may be determined with respect to axes attached to the earth and Eqs. (12.1) and (12.2) used without any appreciable error. On the other hand, these equations do not hold if **a** represents a relative acceleration measured with respect to moving axes, such as axes attached to an accelerated car or to a rotating piece of machinery.

We may observe that if the resultant $\Sigma\mathbf{F}$ of the forces acting on the particle is zero, it follows from Eq. (12.2) that the acceleration **a** of the particle is also zero. If the particle is initially at rest ($\mathbf{v}_0 = 0$) with respect to the newtonian frame of reference used, it will thus remain at rest ($\mathbf{v} = 0$). If originally moving with a velocity $\mathbf{v}_0$, the particle will maintain a constant velocity $\mathbf{v} = \mathbf{v}_0$; that is, it will move with the constant speed v_0 in a straight line. This, we recall, is the statement of Newton's first law (Sec. 2.10). Thus, Newton's first law is a particular case of Newton's second law and may be omitted from the fundamental principles of mechanics.

†More accurately, to the mass center of the solar system.

‡Since the stars are not actually fixed, a more rigorous definition of a newtonian frame of reference (also called *inertial system*) is *one with respect to which Eq.* (12.2) *holds.*

12.3. Linear Momentum of a Particle. Rate of Change of Linear Momentum. Replacing the acceleration $\mathbf{a}$ by the derivative $d\mathbf{v}/dt$ in Eq. (12.2), we write

$$\Sigma \mathbf{F} = m\frac{d\mathbf{v}}{dt}$$

or, since the mass m of the particle is constant,

$$\Sigma \mathbf{F} = \frac{d}{dt}(m\mathbf{v}) \qquad (12.3)$$

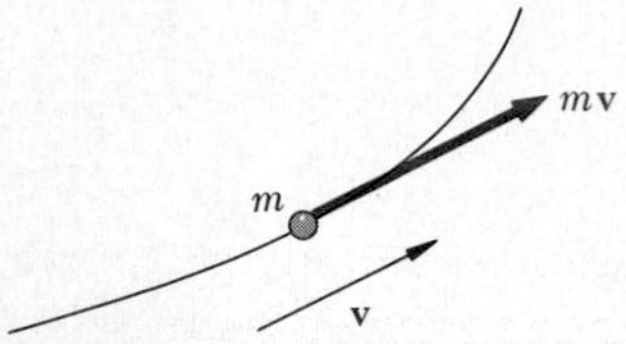

Fig. 12.3

The vector $m\mathbf{v}$ is called the *linear momentum*, or simply the *momentum*, of the particle. It has the same direction as the velocity of the particle and its magnitude is equal to the product of the mass m and the speed v of the particle (Fig. 12.3). Equation (12.3) expresses that *the resultant of the forces acting on the particle is equal to the rate of change of the linear momentum of the particle.* It is in this form that the second law of motion was originally stated by Newton. Denoting by $\mathbf{L}$ the linear momentum of the particle,

$$\mathbf{L} = m\mathbf{v} \qquad (12.4)$$

and by $\dot{\mathbf{L}}$ its derivative with respect to t, we may write Eq. (12.3) in the alternative form

$$\Sigma \mathbf{F} = \dot{\mathbf{L}} \qquad (12.5)$$

It should be noted that the mass m of the particle is assumed to be constant in Eqs. (12.3) to (12.5). Equation (12.3) or (12.5) should therefore not be used to solve problems involving the motion of bodies, such as rockets, which gain or lose mass. Problems of that type will be considered in Sec. 14.12.†

It follows from Eq. (12.3) that the rate of change of the linear momentum $m\mathbf{v}$ is zero when $\Sigma \mathbf{F} = 0$. Thus, *if the resultant force acting on a particle is zero, the linear momentum of the particle remains constant, both in magnitude and direction.* This is the principle of *conservation of linear momentum* for a particle, which we may recognize as just an alternative statement of Newton's first law (Sec. 2.10).

12.4. Systems of Units. In using the fundamental equation $\mathbf{F} = m\mathbf{a}$, the units of force, mass, length, and time cannot be chosen arbitrarily. If they are, the magnitude of the force $\mathbf{F}$ required to give an acceleration $\mathbf{a}$ to the mass m will *not* be numerically equal to the product ma; it will be only proportional to this product. Thus, we may choose three of the four units arbitrarily but must choose the fourth unit so that the equation $\mathbf{F} = m\mathbf{a}$ is satisfied. The units are then said to form a system of consistent kinetic units.

†On the other hand, Eqs. (12.3) and (12.5) do hold in *relativistic mechanics*, where the mass m of the particle is assumed to vary with the speed of the particle.

Two systems of consistent kinetic units are currently used by American engineers, the International System of Units (SI units†), and the U.S. customary units. Since both systems have been discussed in detail in Sec. 1.3, we shall describe them only briefly in this section.

International System of Units (SI Units). In this system, the base units are the units of length, mass, and time, and are called, respectively, the *meter* (m), the *kilogram* (kg), and the *second* (s). All three are arbitrarily defined (Sec. 1.3). The unit of force is a derived unit. It is called the *newton* (N) and is defined as the force which gives an acceleration of 1 m/s² to a mass of 1 kg (Fig. 12.4) From Eq. (12.1) we write

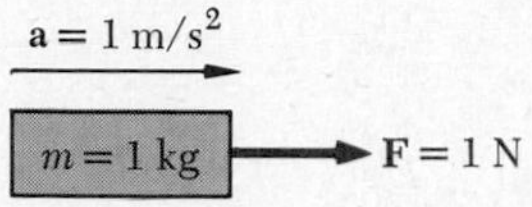

Fig. 12.4

$$1 \text{ N} = (1 \text{ kg})(1 \text{ m/s}^2) = 1 \text{ kg}\cdot\text{m/s}^2$$

The SI units are said to form an *absolute* system of units. This means that the three base units chosen are independent of the location where measurements are made. The meter, the kilogram, and the second may be used anywhere on the earth; they may even be used on another planet. They will always have the same significance.

The *weight* **W** of a body, or *force of gravity* exerted on that body, should, like any other force, be expressed in newtons. Since a body subjected to its own weight acquires an acceleration equal to the acceleration of gravity g, it follows from Newton's second law that the magnitude W of the weight of a body of mass m is

$$W = mg \tag{12.6}$$

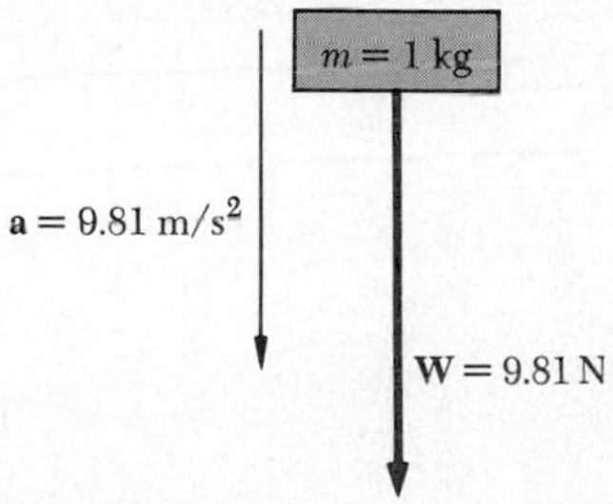

Fig. 12.5

Recalling that $g = 9.81 \text{ m/s}^2$, we find that the weight of a body of mass 1 kg (Fig. 12.5) is

$$W = (1 \text{ kg})(9.81 \text{ m/s}^2) = 9.81 \text{ N}$$

Multiples and submultiples of the units of length, mass, and force are frequently used in engineering practice. They are, respectively, the *kilometer* (km) and the *millimeter* (mm); the *megagram*‡ (Mg) and the *gram* (g); and the *kilonewton* (kN). By definition,

$$1 \text{ km} = 1000 \text{ m} \qquad 1 \text{ mm} = 0.001 \text{ m}$$
$$1 \text{ Mg} = 1000 \text{ kg} \qquad 1 \text{ g} = 0.001 \text{ kg}$$
$$1 \text{ kN} = 1000 \text{ N}$$

The conversion of these units to meters, kilograms, and newtons, respectively, can be effected by simply moving the decimal point three places to the right or to the left.

Units other than the units of mass, length, and time may all be expressed in terms of these three base units. For example, the unit of linear momentum may be obtained by recalling the definition of linear momentum and writing

$$mv = (\text{kg})(\text{m/s}) = \text{kg}\cdot\text{m/s}$$

† SI stands for *Système International d'Unités* (French).

‡ Also known as a *metric ton*.

U.S. Customary Units. Most practicing American engineers still commonly use a system in which the base units are the units of length, force, and time. These units are, respectively, the *foot* (ft), the *pound* (lb), and the *second* (s). The second is the same as the corresponding SI unit. The foot is defined as 0.3048 m. The pound is defined as the *weight* of a platinum standard, called the *standard pound* and kept at the National Bureau of Standards in Washington, the mass of which is 0.453 592 43 kg. Since the weight of a body depends upon the gravitational attraction of the earth, which varies with location, it is specified that the standard pound should be placed at sea level and at the latitude of 45° to properly define a force of 1 lb. Clearly the U.S. customary units do not form an absolute system of units. Because of their dependence upon the gravitational attraction of the earth, they are said to form a *gravitational* system of units.

While the standard pound also serves as the unit of mass in commercial transactions in the United States, it cannot be so used in engineering computations since such a unit would not be consistent with the base units defined in the preceding paragraph. Indeed, when acted upon by a force of 1 lb, that is, when subjected to its own weight, the standard pound receives the acceleration of gravity, $g = 32.2\ \text{ft/s}^2$ (Fig. 12.6), not the unit acceleration required by Eq. (12.1). The unit of mass consistent with the foot, the pound, and the second is the mass which receives an acceleration of $1\ \text{ft/s}^2$ when a force of 1 lb is applied to it (Fig. 12.7). This unit, sometimes called a *slug*, can be derived from the equation $F = ma$ after substituting 1 lb and $1\ \text{ft/s}^2$ for F and a, respectively. We write

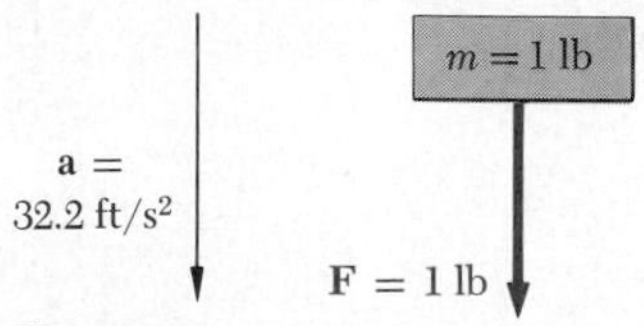

Fig. 12.6

$$F = ma \qquad 1\ \text{lb} = (1\ \text{slug})(1\ \text{ft/s}^2)$$

and obtain

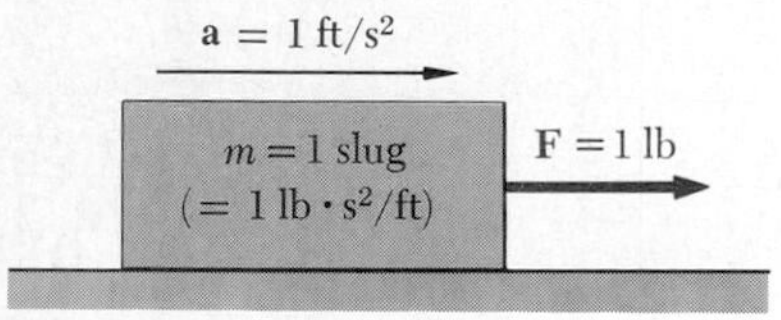

Fig. 12.7

$$1\ \text{slug} = \frac{1\ \text{lb}}{1\ \text{ft/s}^2} = 1\ \text{lb}\cdot\text{s}^2/\text{ft}$$

Comparing Figs. 12.6 and 12.7, we conclude that the slug is a mass 32.2 times larger than the mass of the standard pound.

The fact that bodies are characterized in the U.S. customary system of units by their weight in pounds rather than by their mass in slugs was a convenience in the study of statics, where we were dealing constantly with weights and other forces and only seldom with masses. However, in the study of kinetics, where forces, masses, and accelerations are involved, we repeatedly shall have to express in slugs the mass m of a body, the weight W of which has been given in pounds. Recalling Eq. (12.6), we shall write

$$m = \frac{W}{g} \tag{12.7}$$

where g is the acceleration of gravity ($g = 32.2\ \text{ft/s}^2$).

Units other than the units of force, length, and time may all be expressed in terms of these three base units. For example, the unit of linear momentum may be obtained by recalling the definition of linear momentum and writing

$$mv = (\text{lb}\cdot\text{s}^2/\text{ft})(\text{ft/s}) = \text{lb}\cdot\text{s}$$

The conversion from U.S. customary units to SI units, and vice versa, has been discussed in Sec. 1.4. We shall recall the conversion factors obtained, respectively, for the units of length, force, and mass:

Length:	1 ft = 0.3048 m
Force:	1 lb = 4.448 N
Mass:	1 slug = 1 lb·s²/ft = 14.59 kg

Although it cannot be used as a consistent unit of mass, we also recall that the mass of the standard pound is, by definition,

$$1 \text{ pound-mass} = 0.4536 \text{ kg}$$

This constant may be used to determine the *mass* in SI units (kilograms) of a body which has been characterized by its *weight* in U.S. customary units (pounds).

12.5. Equations of Motion. Consider a particle of mass m acted upon by several forces. We recall from Sec. 12.2 that Newton's second law may be expressed by writing the equation

$$\Sigma \mathbf{F} = m\mathbf{a} \tag{12.2}$$

which relates the forces acting on the particle and the vector $m\mathbf{a}$ (Fig. 12.8). In order to solve problems involving the motion of a particle, however, it will be found more convenient to replace Eq. (12.2) by equivalent equations involving scalar quantities.

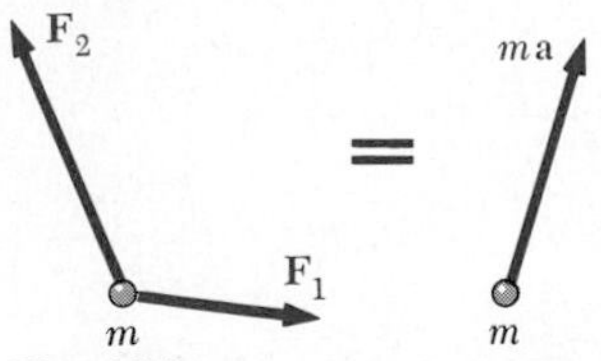

Fig. 12.8

Rectangular Components. Resolving each force $\mathbf{F}$ and the acceleration $\mathbf{a}$ into rectangular components, we write

$$\Sigma(F_x\mathbf{i} + F_y\mathbf{j} + F_z\mathbf{k}) = m(a_x\mathbf{i} + a_y\mathbf{j} + a_z\mathbf{k})$$

from which it follows that

$$\Sigma F_x = ma_x \qquad \Sigma F_y = ma_y \qquad \Sigma F_z = ma_z \tag{12.8}$$

Recalling from Sec. 11.11 that the components of the acceleration are equal to the second derivatives of the coordinates of the particle, we have

$$\Sigma F_x = m\ddot{x} \qquad \Sigma F_y = m\ddot{y} \qquad \Sigma F_z = m\ddot{z} \tag{12.8'}$$

Consider, as an example, the motion of a projectile. If the resistance of the air is neglected, the only force acting on the projectile after it has been fired is its weight $\mathbf{W} = -W\mathbf{j}$. The equations defining the motion of the projectile are therefore

$$m\ddot{x} = 0 \qquad m\ddot{y} = -W \qquad m\ddot{z} = 0$$

and the components of the acceleration of the projectile are

$$\ddot{x} = 0 \qquad \ddot{y} = -\frac{W}{m} = -g \qquad \ddot{z} = 0$$

where g is 9.81 m/s^2, or 32.2 ft/s^2. The equations obtained may be integrated independently, as was shown in Sec. 11.11, to obtain the velocity and displacement of the projectile at any instant.

When a problem involves two or more bodies, equations of motion should be written for each of the bodies (see Sample Probs. 12.3 and 12.4). We also recall from Sec. 12.2 that all accelerations should be measured with respect to a newtonian frame of reference. In most engineering applications accelerations may be determined with respect to axes attached to the earth, but relative accelerations measured with respect to moving axes, such as axes attached to an accelerated body, cannot be substituted for **a** in the equations of motion.

Tangential and Normal Components. Resolving the forces and the acceleration of the particle into components along the tangent to the path (in the direction of motion) and the normal (toward the inside of the path) (Fig. 12.9), and substituting into Eq. (12.2), we obtain the two scalar equations

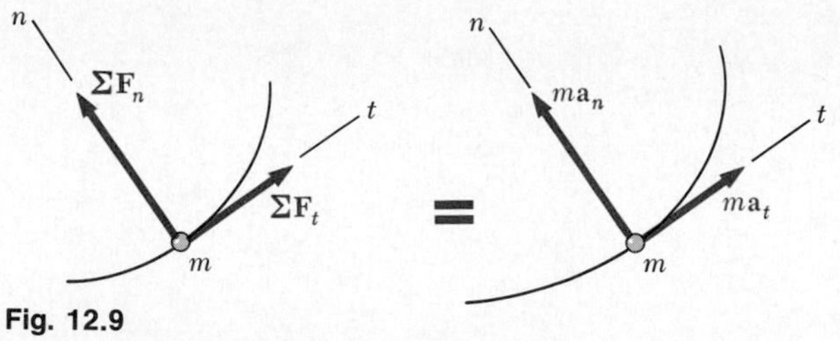

Fig. 12.9

$$\Sigma F_t = ma_t \qquad \Sigma F_n = ma_n \tag{12.9}$$

Substituting for a_t and a_n from Eqs. (11.40), we have

$$\Sigma F_t = m\frac{dv}{dt} \qquad \Sigma F_n = m\frac{v^2}{\rho} \tag{12.9'}$$

The equations obtained may be solved for two unknowns.

12.6. Dynamic Equilibrium. Returning to Eq. (12.2) and transposing the right-hand member, we write Newton's second law in the alternative form

$$\Sigma \mathbf{F} - m\mathbf{a} = 0 \tag{12.10}$$

which expresses that if we add the vector $-m\mathbf{a}$ to the forces acting on the particle, *we obtain a system of vectors equivalent to zero* (Fig. 12.10). The vector $-m\mathbf{a}$, of magnitude ma and of *direction opposite* to that of the acceleration, is called an *inertia vector.* The particle may thus be considered to be in equilibrium under the given forces and the inertia vector. The particle is said to be in *dynamic equilibrium,* and the problem under consideration may be solved by the methods developed earlier in statics.

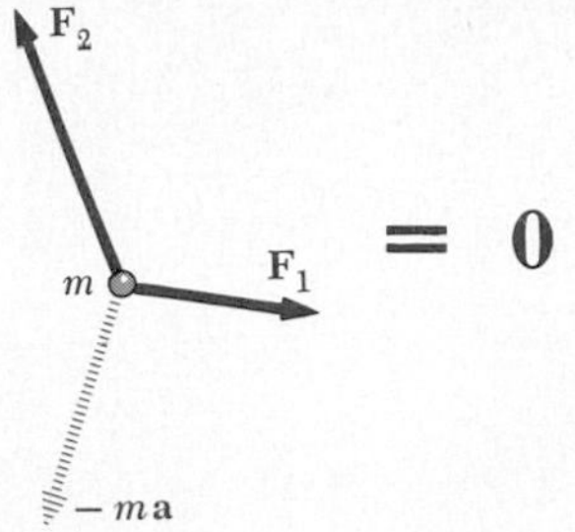

Fig. 12.10

In the case of coplanar forces, we may draw in tip-to-tail fashion all the vectors shown in Fig. 12.10, *including the inertia vector,* to form a closed-vector polygon. Or we may write that the sums of the components of all the vectors in Fig. 12.10, including again the inertia vector, are zero. Using rectangular components, we therefore write

$$\Sigma F_x = 0 \qquad \Sigma F_y = 0 \qquad \textit{including inertia vector} \tag{12.11}$$

When tangential and normal components are used, it is more convenient to represent the inertia vector by its two components $-m\mathbf{a}_t$ and $-m\mathbf{a}_n$ in the sketch itself (Fig. 12.11). The tangential component of the inertia vector provides a measure of the resistance the particle offers to a change in speed, while its normal component (also called *centrifugal force*) represents the tendency of the particle to leave its curved path. We should note that either of these two components may be zero under special conditions: (1) if the particle starts from rest, its initial velocity is zero and the normal component of the inertia vector is zero at $t = 0$; (2) if the particle moves at constant speed along its path, the tangential component of the inertia vector is zero and only its normal component needs to be considered.

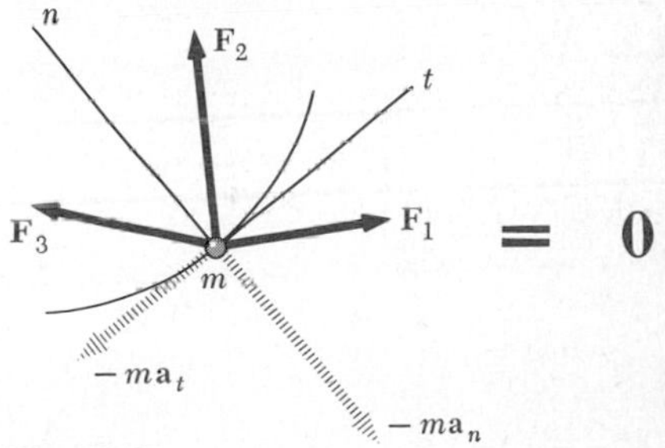

Fig. 12.11

Because they measure the resistance that particles offer when we try to set them in motion or when we try to change the conditions of their motion, inertia vectors are often called *inertia forces.* The inertia forces, however, are not forces like the forces found in statics, which are either contact forces or gravitational forces (weights). Many people, therefore, object to the use of the word "force" when referring to the vector $-m\mathbf{a}$ or even avoid altogether the concept of dynamic equilibrium. Others point out that inertia forces and actual forces, such as gravitational forces, affect our senses in the same way and cannot be distinguished by physical measurements. A man riding in an elevator which is accelerated upward will have the feeling that his weight has suddenly increased; and no measurement made within the elevator could establish whether the elevator is truly accelerated or whether the force of attraction exerted by the earth has suddenly increased.

Sample problems have been solved in this text by the direct application of Newton's second law, as illustrated in Figs. 12.8 and 12.9, rather than by the method of dynamic equilibrium.

SAMPLE PROBLEM 12.1

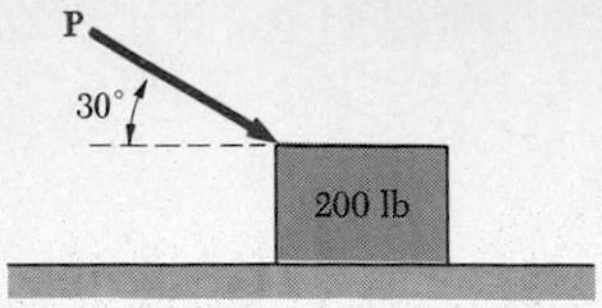

A 200-lb block rests on a horizontal plane. Find the magnitude of the force **P** required to give the block an acceleration of 10 ft/s² to the right. The coefficient of kinetic friction between the block and the plane is $\mu_k = 0.25$.

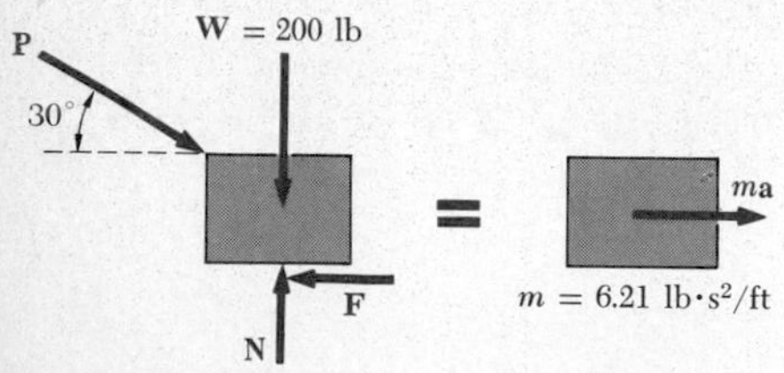

Solution. The mass of the block is

$$m = \frac{W}{g} = \frac{200 \text{ lb}}{32.2 \text{ ft/s}^2} = 6.21 \text{ lb} \cdot \text{s}^2/\text{ft}$$

We note that $F = \mu_k N = 0.25N$ and that $a = 10 \text{ ft/s}^2$. Expressing that the forces acting on the block are equivalent to the vector $m\mathbf{a}$, we write

$$\xrightarrow{+} \Sigma F_x = ma: \quad P \cos 30° - 0.25N = (6.21 \text{ lb} \cdot \text{s}^2/\text{ft})(10 \text{ ft/s}^2)$$

$$P \cos 30° - 0.25N = 62.1 \text{ lb} \quad (1)$$

$$+\uparrow \Sigma F_y = 0: \quad N - P \sin 30° - 200 \text{ lb} = 0 \quad (2)$$

Solving (2) for N and carrying the result into (1), we obtain

$$N = P \sin 30° + 200 \text{ lb}$$

$$P \cos 30° - 0.25(P \sin 30° + 200 \text{ lb}) = 62.1 \text{ lb} \qquad P = 151 \text{ lb} \blacktriangleleft$$

SAMPLE PROBLEM 12.2

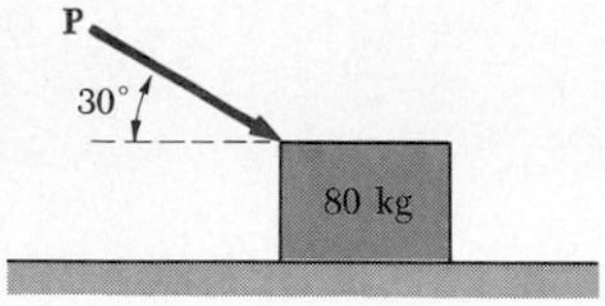

An 80-kg block rests on a horizontal plane. Find the magnitude of the force **P** required to give the block an acceleration of 2.5 m/s² to the right. The coefficient of kinetic friction between the block and the plane is $\mu_k = 0.25$.

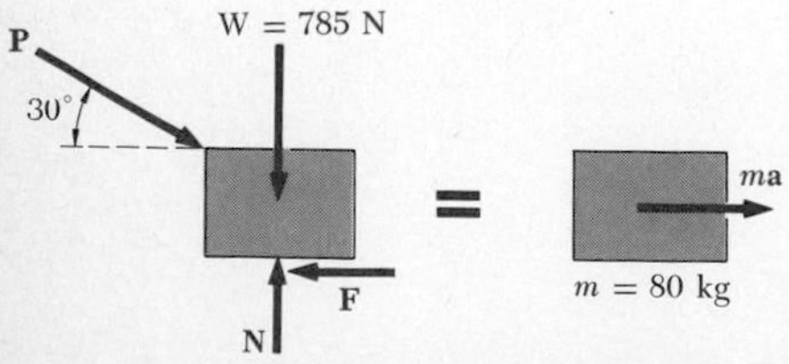

Solution. The weight of the block is

$$W = mg = (80 \text{ kg})(9.81 \text{ m/s}^2) = 785 \text{ N}$$

We note that $F = \mu_k N = 0.25N$ and that $a = 2.5 \text{ m/s}^2$. Expressing that the forces acting on the block are equivalent to the vector $m\mathbf{a}$, we write

$$\xrightarrow{+} \Sigma F_x = ma: \quad P \cos 30° - 0.25N = (80 \text{ kg})(2.5 \text{ m/s}^2)$$

$$P \cos 30° - 0.25N = 200 \text{ N} \quad (1)$$

$$+\uparrow \Sigma F_y = 0: \quad N - P \sin 30° - 785 \text{ N} = 0 \quad (2)$$

Solving (2) for N and carrying the result into (1), we obtain

$$N = P \sin 30° + 785 \text{ N}$$

$$P \cos 30° - 0.25(P \sin 30° + 785 \text{ N}) = 200 \text{ N} \qquad P = 535 \text{ N} \blacktriangleleft$$

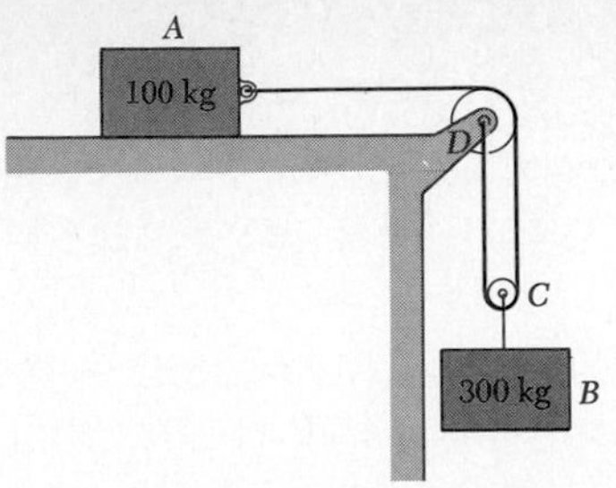

SAMPLE PROBLEM 12.3

The two blocks shown start from rest. The horizontal plane and the pulley are frictionless, and the pulley is assumed to be of negligible mass. Determine the acceleration of each block and the tension in each cord.

Kinematics. We note that if block A moves through x_A to the right, block B moves down through

$$x_B = \tfrac{1}{2}x_A$$

Differentiating twice with respect to t, we have

$$a_B = \tfrac{1}{2}a_A \qquad (1)$$

Kinetics. We shall apply Newton's second law successively to block A, block B, and pulley C.

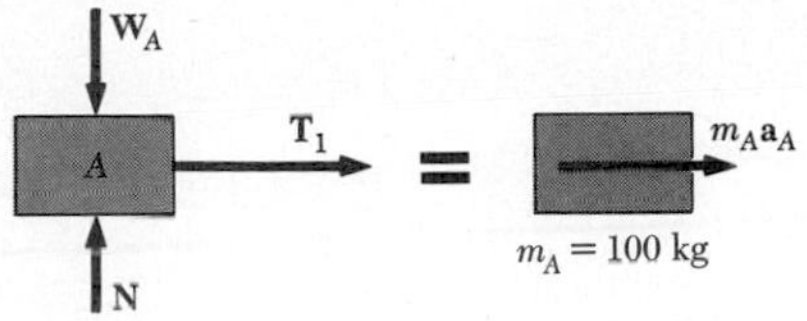

Block A. Denoting by T_1 the tension in cord ACD, we write

$$\xrightarrow{+} \Sigma F_x = m_A a_A: \qquad T_1 = 100a_A \qquad (2)$$

Block B. Observing that the weight of block B is

$$W_B = m_B g = (300\text{ kg})(9.81\text{ m/s}^2) = 2940\text{ N}$$

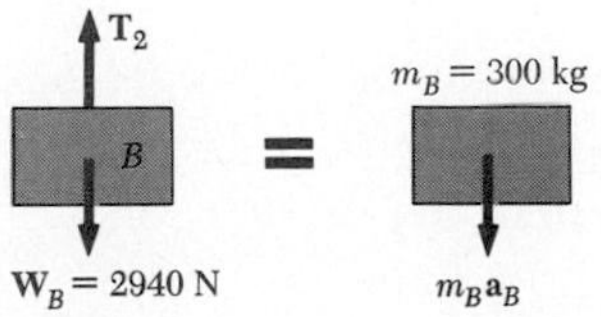

and denoting by T_2 the tension in cord BC, we write

$$+\downarrow \Sigma F_y = m_B a_B: \qquad 2940 - T_2 = 300a_B$$

or, substituting for a_B from (1),

$$2940 - T_2 = 300(\tfrac{1}{2}a_A)$$
$$T_2 = 2940 - 150a_A \qquad (3)$$

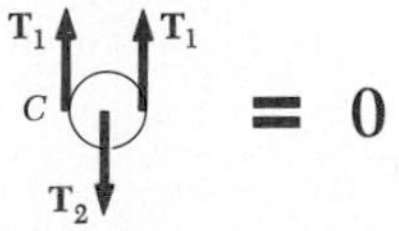

Pulley C. Since m_C is assumed to be zero, we have

$$+\downarrow \Sigma F_y = m_C a_C = 0: \qquad T_2 - 2T_1 = 0 \qquad (4)$$

Substituting for T_1 and T_2 from (2) and (3), respectively, into (4) we write

$$2940 - 150a_A - 2(100a_A) = 0$$
$$2940 - 350a_A = 0 \qquad a_A = 8.40\text{ m/s}^2 \blacktriangleleft$$

Substituting the value obtained for a_A into (1) and (2), we have

$$a_B = \tfrac{1}{2}a_A = \tfrac{1}{2}(8.40\text{ m/s}^2) \qquad a_B = 4.20\text{ m/s}^2 \blacktriangleleft$$
$$T_1 = 100a_A = (100\text{ kg})(8.40\text{ m/s}^2) \qquad T_1 = 840\text{ N} \blacktriangleleft$$

Recalling (4), we write

$$T_2 = 2T_1 \qquad T_2 = 2(840\text{ N}) \qquad T_2 = 1680\text{ N} \blacktriangleleft$$

We note that the value obtained for T_2 is *not* equal to the weight of block B.

SAMPLE PROBLEM 12.4

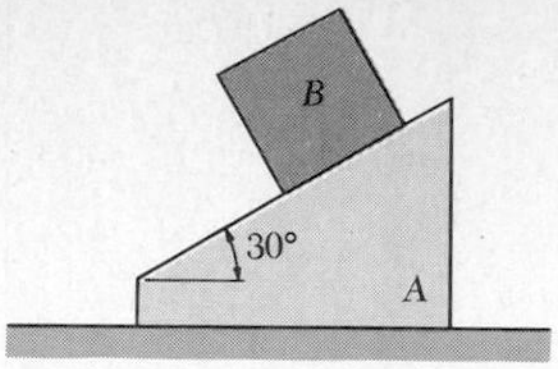

The 12-lb block B starts from rest and slides on the 30-lb wedge A, which is supported by a horizontal surface. Neglecting friction, determine (*a*) the acceleration of the wedge, (*b*) the acceleration of the block relative to the wedge.

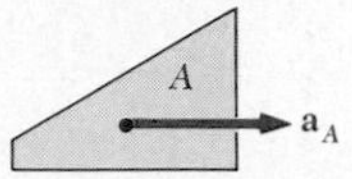

Kinematics. We shall first examine the accelerations of the wedge and of the block.

Wedge A. Since the wedge is constrained to move on the horizontal surface, its acceleration $\mathbf{a}_A$ is horizontal. We shall assume that it is directed to the right.

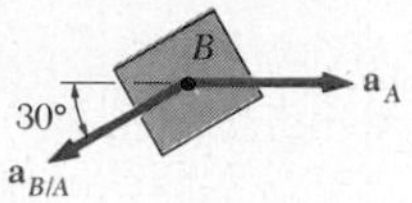

Block B. The acceleration $\mathbf{a}_B$ of block B may be expressed as the sum of the acceleration of A and of the acceleration of B relative to A. We have

$$\mathbf{a}_B = \mathbf{a}_A + \mathbf{a}_{B/A}$$

where $\mathbf{a}_{B/A}$ is directed along the sloping surface of the wedge.

Kinetics. We shall draw the free-body diagrams of the wedge and of the block and apply Newton's second law.

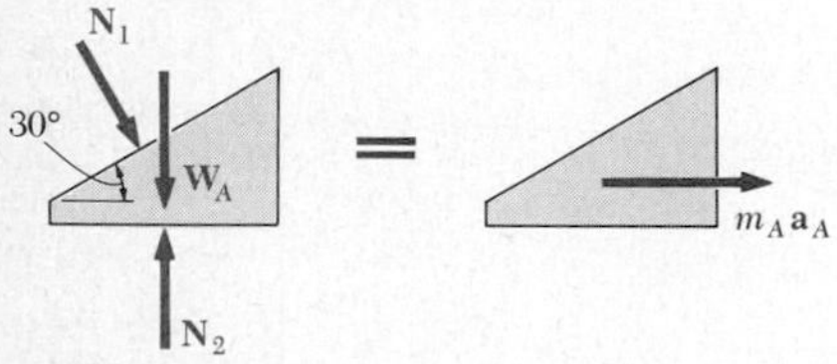

Wedge A. We denote respectively by $\mathbf{N}_1$ and $\mathbf{N}_2$ the forces exerted by the block and the horizontal surface on wedge A.

$$\overset{+}{\rightarrow} \Sigma F_x = m_A a_A: \qquad N_1 \sin 30° = m_A a_A$$

$$0.5N_1 = (W_A/g)a_A \tag{1}$$

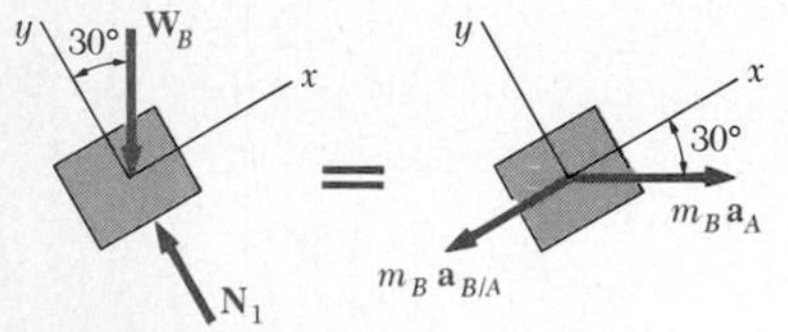

Block B. Using the coordinate axes shown and resolving $\mathbf{a}_B$ into its components $\mathbf{a}_A$ and $\mathbf{a}_{B/A}$, we write

$$+\nearrow \Sigma F_x = m_B a_x: \qquad -W_B \sin 30° = m_B a_A \cos 30° - m_B a_{B/A}$$

$$-W_B \sin 30° = (W_B/g)(a_A \cos 30° - a_{B/A})$$

$$a_{B/A} = a_A \cos 30° + g \sin 30° \tag{2}$$

$$+\nwarrow \Sigma F_y = m_B a_y: \qquad N_1 - W_B \cos 30° = -m_B a_A \sin 30°$$

$$N_1 - W_B \cos 30° = -(W_B/g)a_A \sin 30° \tag{3}$$

a. Acceleration of Wedge A. Substituting for N_1 from Eq. (1) into Eq. (3), we have

$$2(W_A/g)a_A - W_B \cos 30° = -(W_B/g)a_A \sin 30°$$

Solving for a_A and substituting the numerical data, we write

$$a_A = \frac{W_B \cos 30°}{2W_A + W_B \sin 30°}g = \frac{(12 \text{ lb}) \cos 30°}{2(30 \text{ lb}) + (12 \text{ lb}) \sin 30°}(32.2 \text{ ft/s}^2)$$

$$a_A = +5.07 \text{ ft/s}^2 \qquad \mathbf{a}_A = 5.07 \text{ ft/s}^2 \rightarrow \quad ◄$$

b. Acceleration of Block *B* relative to A. Substituting the value obtained for a_A into Eq. (2), we have

$$a_{B/A} = (5.07 \text{ ft/s}^2) \cos 30° + (32.2 \text{ ft/s}^2) \sin 30°$$

$$a_{B/A} = +20.5 \text{ ft/s}^2 \qquad \mathbf{a}_{B/A} = 20.5 \text{ ft/s}^2 \swarrow 30° \quad ◄$$

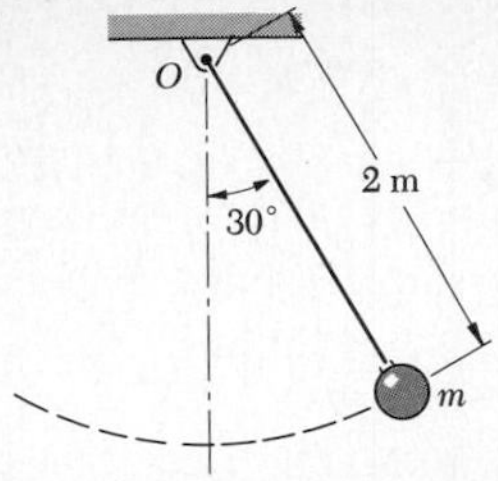

SAMPLE PROBLEM 12.5

The bob of a 2-m pendulum describes an arc of circle in a vertical plane. If the tension in the cord is 2.5 times the weight of the bob for the position shown, find the velocity and acceleration of the bob in that position.

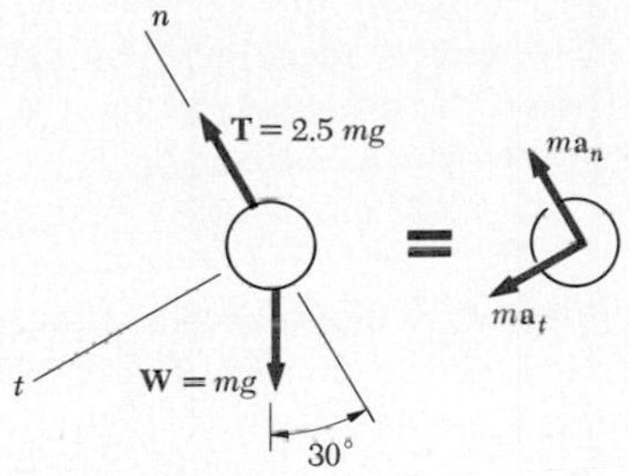

Solution. The weight of the bob is $W = mg$; the tension in the cord is thus $2.5\,mg$. Recalling that $\mathbf{a}_n$ is directed toward O and assuming $\mathbf{a}_t$ as shown, we apply Newton's second law and obtain

$+\swarrow \Sigma F_t = ma_t$: $\quad mg \sin 30° = ma_t$

$$a_t = g \sin 30° = +4.90 \text{ m/s}^2 \qquad \mathbf{a}_t = 4.90 \text{ m/s}^2 \swarrow \blacktriangleleft$$

$+\nwarrow \Sigma F_n = ma_n$: $\quad 2.5\,mg - mg \cos 30° = ma_n$

$$a_n = 1.634\,g = +16.03 \text{ m/s}^2 \qquad \mathbf{a}_n = 16.03 \text{ m/s}^2 \nwarrow \blacktriangleleft$$

Since $a_n = v^2/\rho$, we have $v^2 = \rho a_n = (2 \text{ m})(16.03 \text{ m/s}^2)$

$$v = \pm 5.66 \text{ m/s} \qquad \mathbf{v} = 5.66 \text{ m/s} \nearrow \text{(up or down)} \blacktriangleleft$$

SAMPLE PROBLEM 12.6

Determine the rated speed of a highway curve of radius $\rho = 400$ ft banked through an angle $\theta = 18°$. The rated speed of a banked curved road is the speed at which a car should travel if no lateral friction force is to be exerted on its wheels.

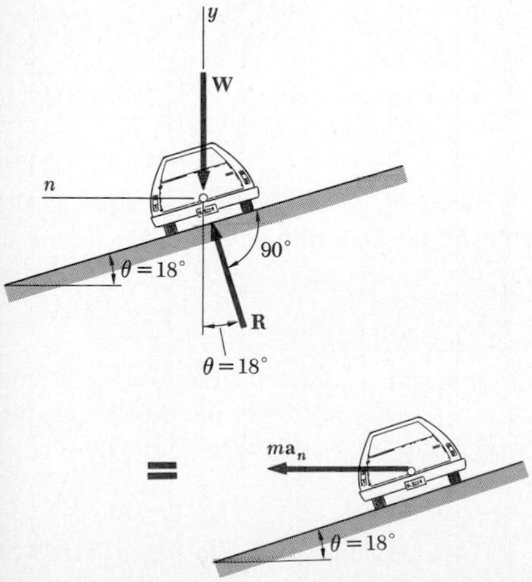

Solution. The car travels in a *horizontal* circular path of radius ρ. The normal component $\mathbf{a}_n$ of the acceleration is directed toward the center of the path; its magnitude is $a_n = v^2/\rho$, where v is the speed of the car in ft/s. The mass m of the car is W/g, where W is the weight of the car. Since no lateral friction force is to be exerted on the car, the reaction $\mathbf{R}$ of the road is shown perpendicular to the roadway. Applying Newton's second law, we write

$$+\uparrow \Sigma F_y = 0: \qquad R \cos\theta - W = 0 \qquad R = \frac{W}{\cos\theta} \tag{1}$$

$$\overset{+}{\leftarrow} \Sigma F_n = ma_n: \qquad R \sin\theta = \frac{W}{g} a_n \tag{2}$$

Substituting for R from (1) into (2), and recalling that $a_n = v^2/\rho$,

$$\frac{W}{\cos\theta} \sin\theta = \frac{W}{g}\frac{v^2}{\rho} \qquad v^2 = g\rho \tan\theta$$

Substituting the given data, $\rho = 400$ ft and $\theta = 18°$, into this equation, we obtain

$$v^2 = (32.2 \text{ ft/s}^2)(400 \text{ ft}) \tan 18°$$
$$v = 64.7 \text{ ft/s} \qquad v = 44.1 \text{ mi/h} \blacktriangleleft$$

Problems

12.1 The acceleration due to gravity on the moon is 1.62 m/s^2. Determine (*a*) the weight in newtons, (*b*) the mass in kilograms, on the moon, of a gold bar, the mass of which has been officially designated as 2 kg.

12.2 The value of g at any latitude ϕ may be obtained from the formula

$$g = 32.09(1 + 0.0053 \sin^2 \phi) \text{ ft/s}^2$$

which takes into account the effect of the rotation of the earth, as well as the fact that the earth is not truly spherical. Determine to four significant figures (*a*) the weight in pounds, (*b*) the mass in pounds, (*c*) the mass in lb·s^2/ft, at the latitudes of 0°, 45°, and 90°, of a silver bar, the mass of which has been officially designated as 5 lb.

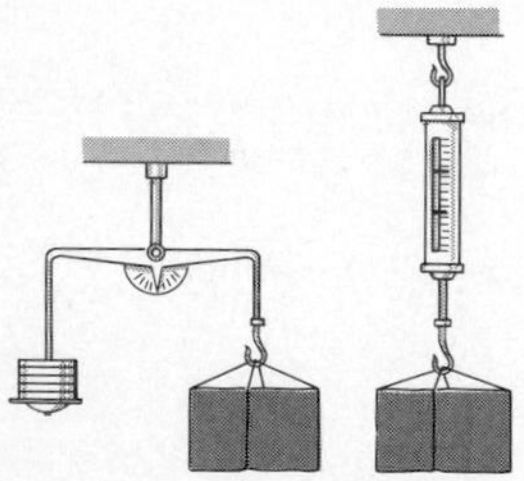

Fig. P12.3

12.3 Two boxes are weighed on the scales shown: scale a is a lever scale; scale b is a spring scale. The scales are attached to the roof of an elevator. When the elevator is at rest, each scale indicates a mass of 12 kg. If the spring scale indicates a mass of 10 kg, determine (*a*) the acceleration of the elevator, (*b*) the mass indicated by the lever scale.

12.4 A 400-kg satellite has been placed in a circular orbit 1500 km above the surface of the earth. The acceleration of gravity at this elevation is 6.43 m/s^2. Determine the linear momentum of the satellite, knowing that its orbital speed is 25.6×10^3 km/h.

12.5 Determine the maximum theoretical speed that may be achieved over a distance of 50 m by a car starting from rest, knowing that the coefficient of static friction is 0.80 between the tires and the pavement. Assume four-wheel drive.

12.6 Solve Prob. 12.5, knowing that 60 percent of the weight of the car is distributed over its front wheels and 40 percent over its rear wheels, and assuming (*a*) front-wheel drive, (*b*) rear-wheel drive.

12.7 A motorist traveling at a speed of 90 km/h suddenly applies the brakes and comes to a stop after skidding 50 m. Determine (*a*) the time required for the car to stop, (*b*) the coefficient of friction between the tires and the pavement.

Fig. P12.8

12.8 A car has been traveling up a long 2 percent grade at a constant speed of 55 mi/h. If the driver does not change the setting of the throttle or shift gears as the car reaches the top of the hill, what will be the acceleration of the car as it starts moving down the 3 percent grade?

12.9 A 3000-lb automobile is driven down a 5° incline at a speed of 50 mi/h when the brakes are applied, causing a total braking force of 1200 lb to be applied to the automobile. Determine the distance traveled by the automobile before it comes to a stop.

12.10 The two blocks shown are originally at rest. Neglecting the masses of the pulleys and the effect of friction in the pulleys and between block A and the incline, determine (*a*) the acceleration of each block, (*b*) the tension in the cable.

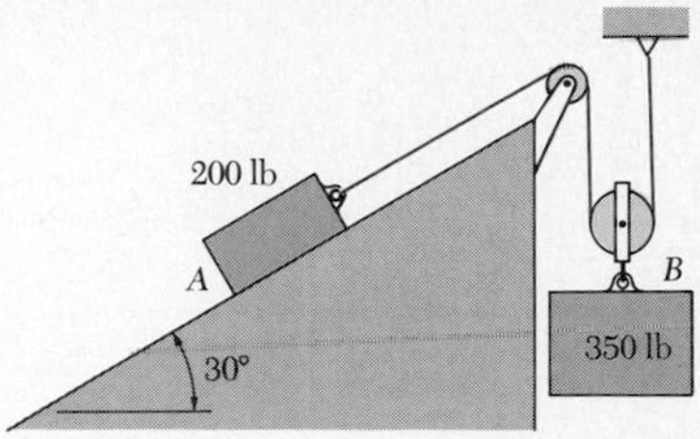

Fig. P12.10

12.11 Solve Prob. 12.10, assuming that the coefficients of friction between block A and the incline are $\mu_s = 0.25$ and $\mu_k = 0.20$.

12.12 A 1050-kg trailer is hitched to a 1200-kg car. The car and trailer are traveling at 90 km/h when the driver applies the brakes on both the car and the trailer. Knowing that the braking forces exerted on the car and the trailer are 4500 N and 3600 N, respectively, determine (*a*) the deceleration of the car and trailer, (*b*) the horizontal component of the force exerted by the trailer hitch on the car.

Fig. P12.12

12.13 Solve Prob. 12.12, assuming that the trailer brakes are inoperative.

12.14 Block A has a mass of 30 kg and block B a mass of 15 kg. The coefficients of friction between all surfaces of contact are $\mu_s = 0.15$ and $\mu_k = 0.10$. Knowing that $\theta = 30°$ and that the magnitude of the force **P** applied to block A is 250 N, determine (*a*) the acceleration of block A, (*b*) the tension in the cord.

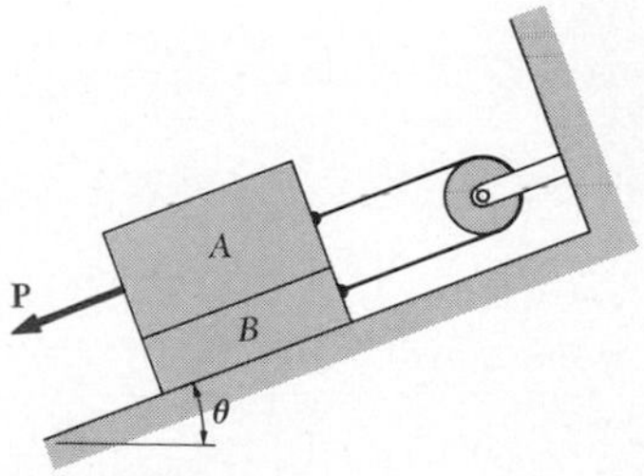

Fig. P12.14

12.15 Solve Prob. 12.14, assuming that the 250-N force **P** is applied to block B instead of block A.

12.16 Knowing that the system shown starts from rest, find the velocity at $t = 1.2$ s of (*a*) collar A, (*b*) collar B. Neglect the masses of the pulleys and the effect of friction.

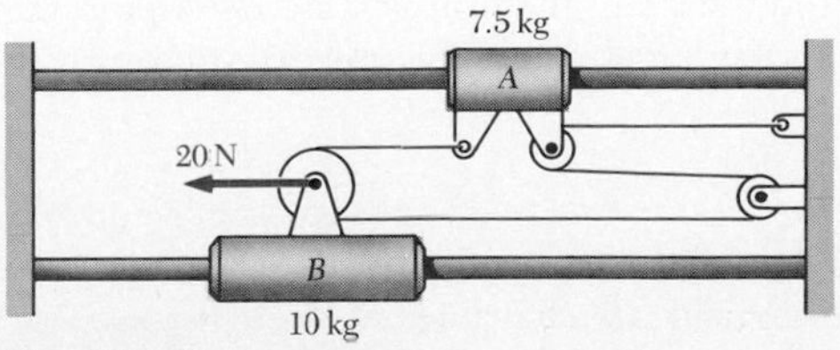

Fig. P12.16

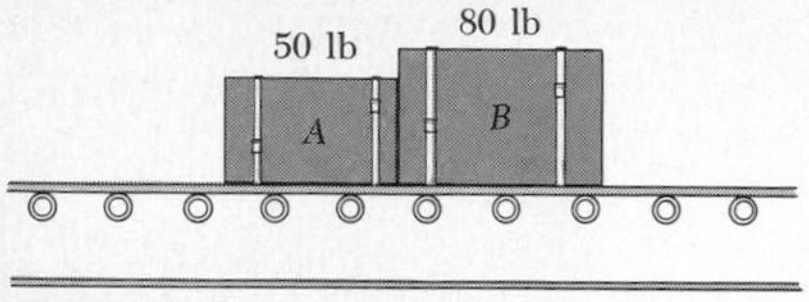

Fig. P12.17

12.17 Two packages are placed on a conveyor belt which is at rest. The coefficient of kinetic friction is 0.20 between the belt and package A, and 0.10 between the belt and package B. If the belt is suddenly started to the right and slipping occurs between the belt and the packages, determine (a) the acceleration of the packages, (b) the force exerted by package A on package B.

12.18 Each of the systems shown is initially at rest. Assuming the pulleys to be weightless and neglecting axle friction, determine for each system (a) the acceleration of block A, (b) the velocity of block A after 2 s, (c) the velocity of block A after it has moved through 8 ft.

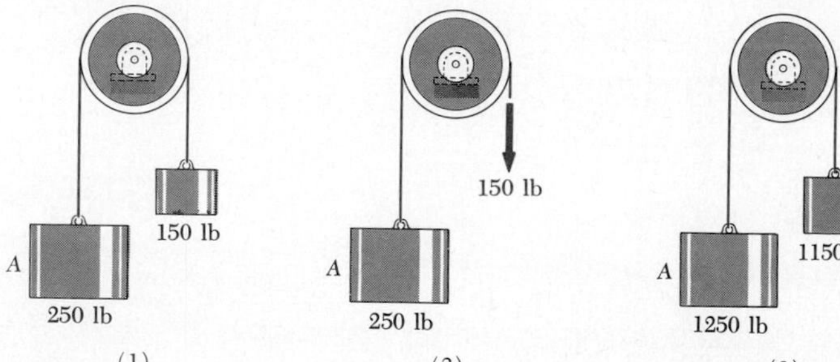

Fig. P12.18

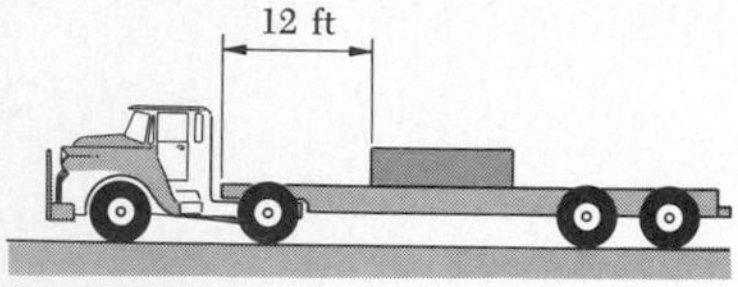

Fig. P12.19

12.19 The coefficients of friction between the load and the flat-bed trailer shown are $\mu_s = 0.50$ and $\mu_k = 0.40$. Knowing that the speed of the rig is 45 mi/h, determine the shortest distance in which the rig can be brought to a stop if the load is not to shift.

12.20 The tractor-trailer of Prob. 12.19 is traveling at 60 mi/h when the driver makes an emergency stop, causing the rig to skid to rest in 5 s. Determine (a) whether the load will shift, (b) if the load does shift, the relative velocity with which it will reach the forward edge of the trailer.

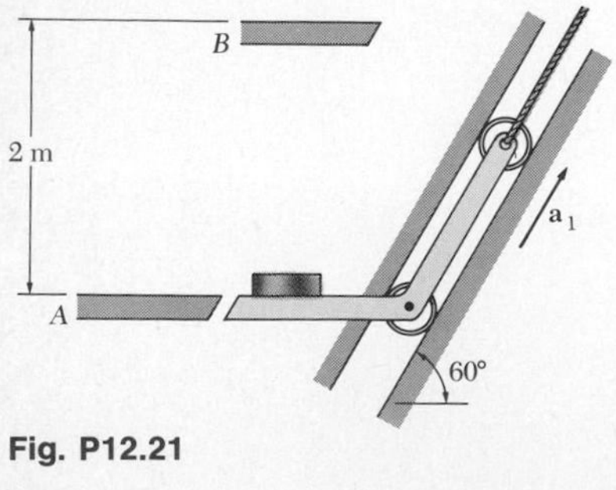

Fig. P12.21

12.21 In a manufacturing process, disks are moved from level A to level B by the lifting arm shown. The arm starts from level A with no initial velocity, moves first with a constant acceleration $\mathbf{a}_1$ as shown, then with a constant deceleration $\mathbf{a}_2$, and comes to a stop at level B. Knowing that the coefficient of static friction between the disks and the arm is 0.25, determine the largest allowable acceleration $\mathbf{a}_1$ and the largest allowable deceleration $\mathbf{a}_2$ if the disks are not to slide.

12.22 If the disks of Prob. 12.21 are to be moved from level A to level B in the shortest possible time without sliding, determine (a) the maximum velocity reached by the lifting arm, (b) the time required to move each disk.

12.23 A chain of mass m' per unit length rests over a small pulley of negligible mass. If the chain is released in the position shown ($h > \frac{1}{2}l$), determine (*a*) the velocity v_1 of the chain as the end of the chain leaves the pulley, (*b*) the maximum possible value of v_1 as h approaches $\frac{1}{2}l$.

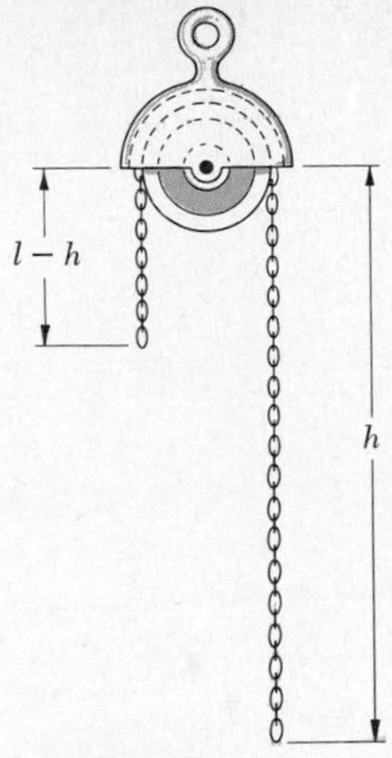

Fig. P12.23

12.24 An airplane has a mass of 30 Mg and its engines develop a total thrust of 50 kN during take-off. If the drag **D** exerted on the plane has a magnitude $D = 2.50\,v^2$, where v is expressed in meters per second and D in newtons, and if the plane becomes airborne at a speed of 270 km/h, determine the length of runway required for the plane to take off.

12.25 For the airplane of Prob. 12.24, determine the time required to take off.

12.26 A constant force **P** is applied to a piston and rod of total mass m in order to make them move in a cylinder filled with oil. As the piston moves, the oil is forced through orifices in the piston and exerts on the piston an additional force of magnitude kv, proportional to the speed v of the piston and in a direction opposite to its motion. Express the distance x traveled by the piston as a function of the time t, assuming that the piston starts from rest at $t = 0$ and $x = 0$.

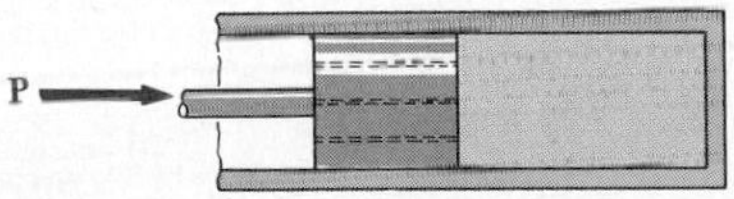

Fig. P12.26

12.27 Block A weighs 20 lb, and blocks B and C weigh 10 lb each. Knowing that $P = 2.5$ lb and that the blocks are initially at rest, determine at $t = 3$ s the velocity (*a*) of B relative to A, (*b*) of C relative to A. Neglect the weights of the pulleys and the effect of friction.

12.28 Block A weighs 20 lb, and blocks B and C weigh 10 lb each. Knowing that the blocks are initially at rest and that B moves through 8 ft in 2 s, determine (*a*) the magnitude of the force **P**. (*b*) the tension in the cord AD. Neglect the weights of the pulleys and the effect of friction.

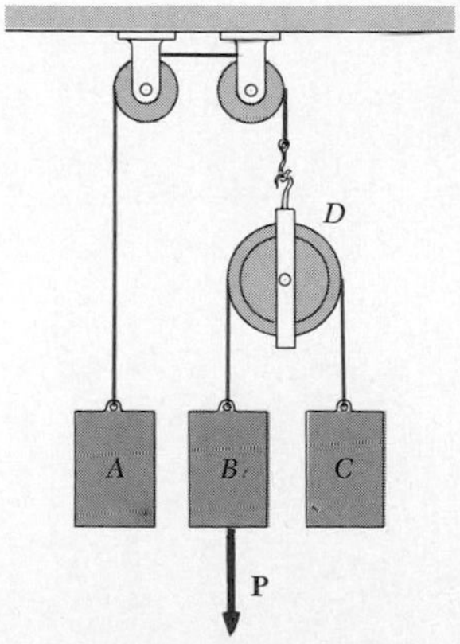

Fig. P12.27 and P12.28

12.29 An 8-kg block B rests as shown on a 12-kg bracket A. The coefficients of friction are $\mu_s = 0.40$ and $\mu_k = 0.30$ between block B and bracket A, and there is no friction in the pulley or between the bracket and the horizontal surface. If $P = 30$ N, determine the velocity of B relative to A after B has moved 400 mm with respect to A.

12.30 For the system of Prob. 12.29, determine (*a*) the maximum force **P** which may be exerted on the cord if block B is not to slide on bracket A, (*b*) the corresponding acceleration of the bracket.

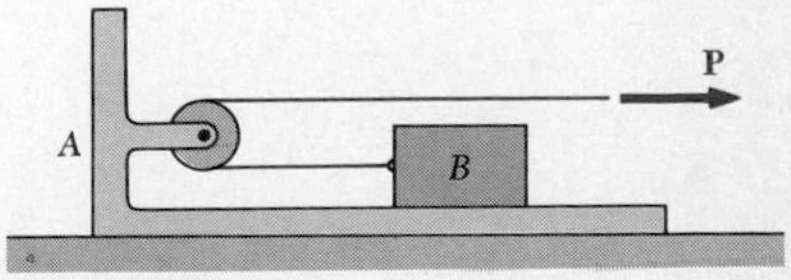

Fig. P12.29

12.31 A 12-lb block B rests as shown on the upper surface of a 30-lb wedge A. Neglecting friction, determine (*a*) the acceleration of A, (*b*) the acceleration of B relative to A, immediately after the system is released from rest.

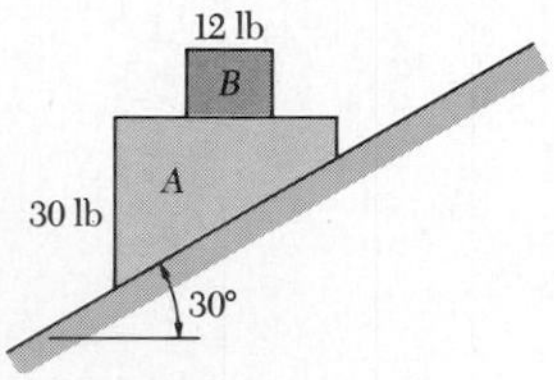

Fig. P12.31

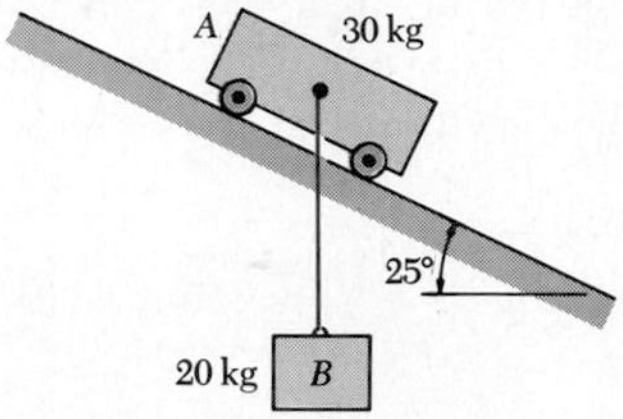

Fig. P12.32

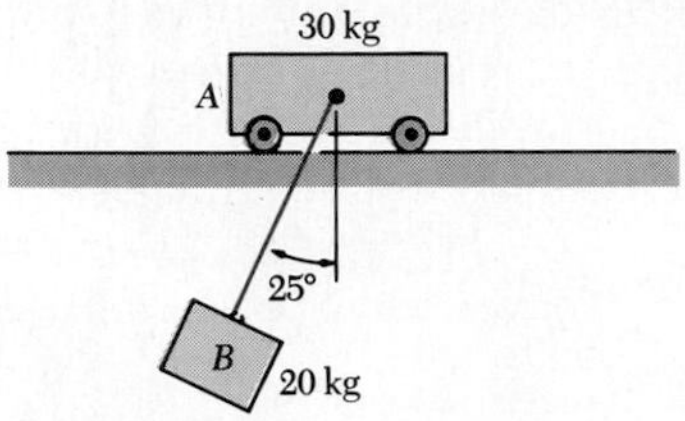

Fig. P12.33

12.32 and 12.33 A 20-kg block B is suspended from a 2-m cord attached to a 30-kg cart A. Neglecting friction, determine (*a*) the acceleration of the cart, (*b*) the tension in the cord, immediately after the system is released from rest in the position shown.

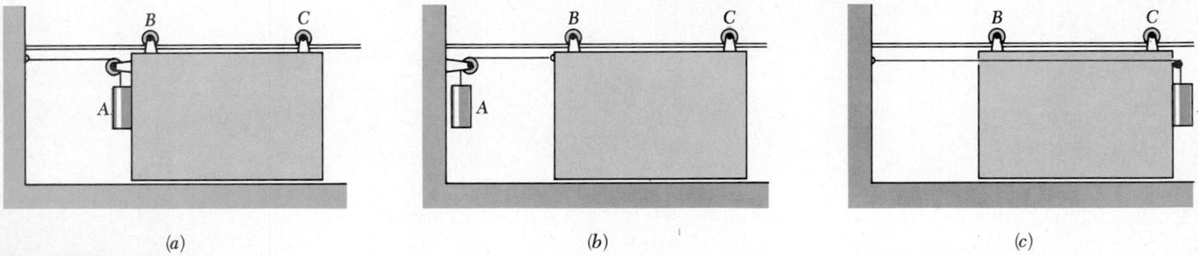

Fig. P12.34

12.34 A 40-lb sliding panel is supported by rollers at B and C. A 25-lb counterweight A is attached to a cable as shown and, in cases *a* and *c*, is initially in contact with a vertical edge of the panel. Neglecting friction, determine in each case shown the acceleration of the panel and the tension in the cord immediately after the system is released from rest.

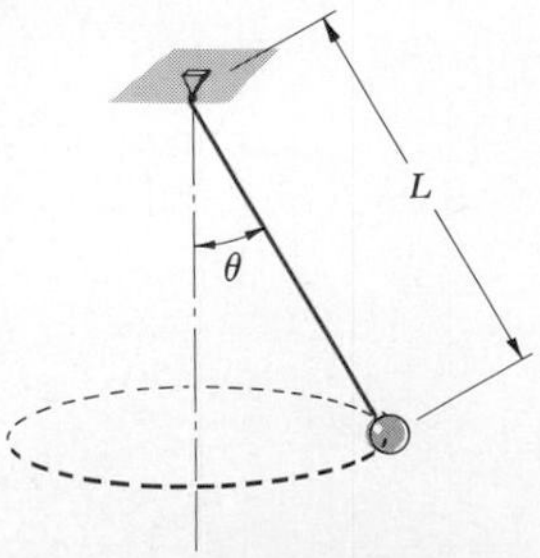

Fig. P12.35 and P12.36

12.35 A 5-lb ball revolves in a horizontal circle as shown. Knowing that $L = 4$ ft and that the maximum allowable tension in the cord is 12 lb, determine (*a*) the maximum allowable speed, (*b*) the corresponding value of the angle θ.

12.36 A 3-kg ball revolves in a horizontal circle as shown at a constant speed of 1.2 m/s. Knowing that $L = 800$ mm, determine (*a*) the angle θ that the cord forms with the vertical, (*b*) the tension in the cord.

12.37 and 12.38 A single wire ACB passes through a ring at C attached to a sphere which revolves at the constant speed v in the horizontal circle shown. Knowing that the tension is the same in both portions of the wire, determine the speed v.

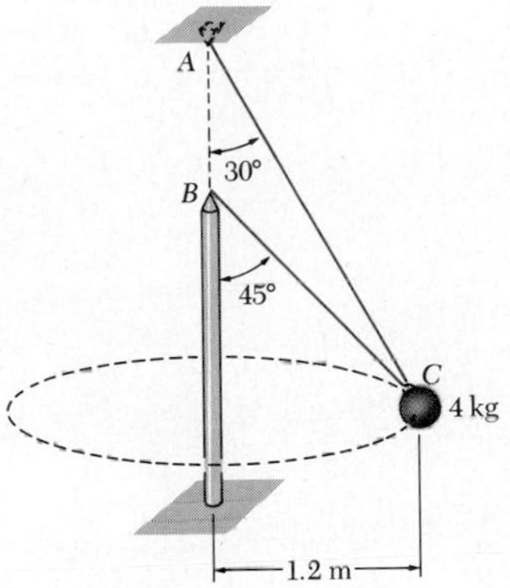

Fig. P12.37, P12.39, P12.40 and P12.42

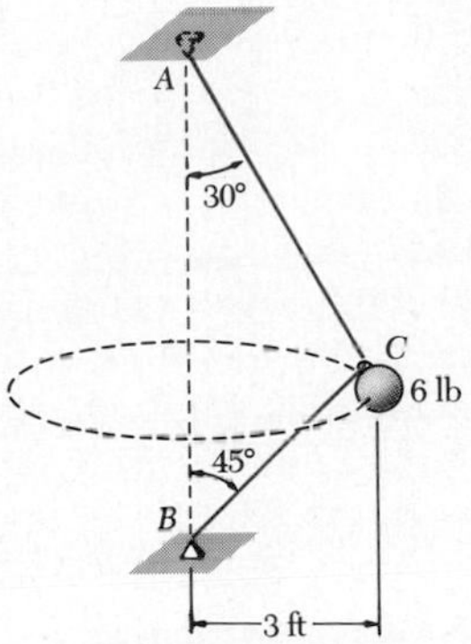

Fig. P12.38 and P12.41

12.39 Two wires AC and BC are tied at C to a sphere which revolves at the constant speed v in the horizontal circle shown. Determine the range of values of v for which both wires remain taut.

12.40 Two wires AC and BC are tied at C to a sphere which revolves at the constant speed v in the horizontal circle shown. Determine the range of the allowable values of v if the tension in either of the wires is not to exceed 35 N.

*** 12.41** Two wires AC and BC are tied at C to a sphere which revolves at the constant speed v in the horizontal circle shown. Determine the range of the allowable values of v if both wires are to remain taut and if the tension in either of the wires is not to exceed 12 lb.

*** 12.42** Two wires AC and BC are tied at C to a sphere which revolves at the constant speed v in the horizontal circle shown. Determine the range of the allowable values of v if both wires are to remain taut and if the tension in either of the wires is not to exceed 50 N.

12.43 A small sphere of weight W is held as shown by two wires AB and CD. If wire AB is cut, determine the tension in the other wire (a) before AB is cut, (b) immediately after AB has been cut.

12.44 Solve Prob. 12.43, assuming that wire CD is cut instead of wire AB.

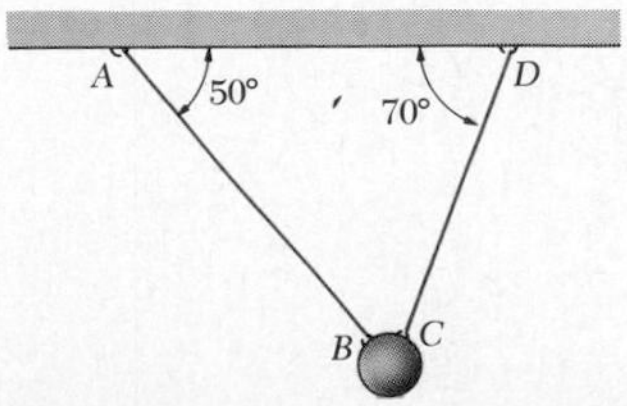

Fig. P12.43

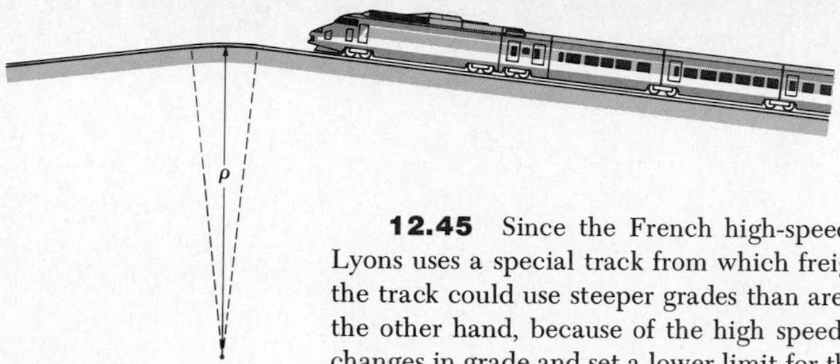

Fig. P12.45

12.45 Since the French high-speed (TGV) train running from Paris to Lyons uses a special track from which freight trains are barred, the designers of the track could use steeper grades than are usually permissible for railroads. On the other hand, because of the high speeds involved, they had to avoid sudden changes in grade and set a lower limit for the radius of curvature ρ of the vertical profile of the track. (*a*) Knowing that the train's maximum speed in test runs was 382 km/h, determine the smallest allowable value of ρ if the train is not to leave the track at that speed. (*b*) Assuming the value of ρ found in part *a*, determine the force exerted by his seat on an 80-kg passenger at the top and at the bottom of a hill when the train is traveling at 270 km/h.

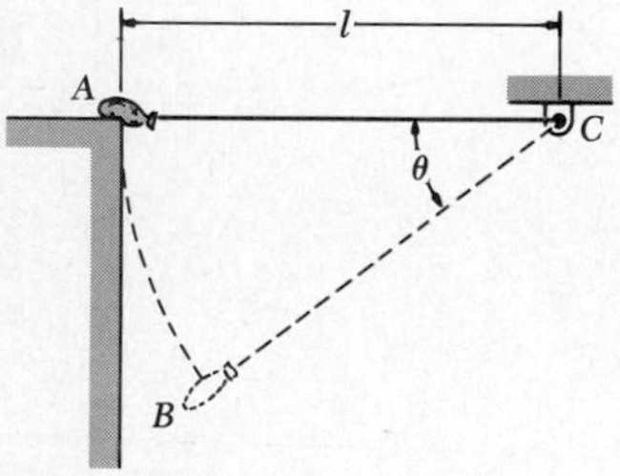

Fig. P12.46

12.46 A bag is gently pushed off the top of a wall at A and swings in a vertical plane at the end of a rope of length l. (*a*) For any position B of the bag, determine the tangential component a_t of its acceleration and obtain its velocity v by integration. (*b*) Determine the value of θ for which the rope will break, knowing that it can withstand a maximum tension equal to twice the weight of the bag.

12.47 The roller-coaster track shown is contained in a vertical plane. The portion of track between A and B is straight and horizontal, while the portions to the left of A and to the right of B have radii of curvature as indicated. A car is traveling at a speed of 45 mi/h when the brakes are suddenly applied, causing the wheels of the car to slide on the track ($\mu_k = 0.25$). Determine the initial deceleration of the car if the brakes are applied as the car (*a*) has almost reached A, (*b*) is traveling between A and B, (*c*) has just passed B.

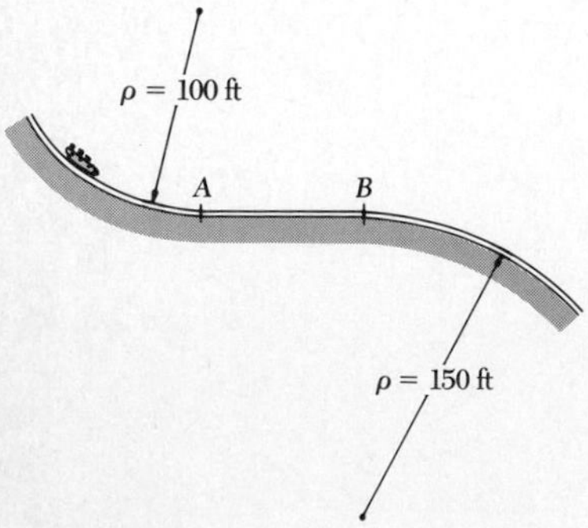

Fig. P12.47

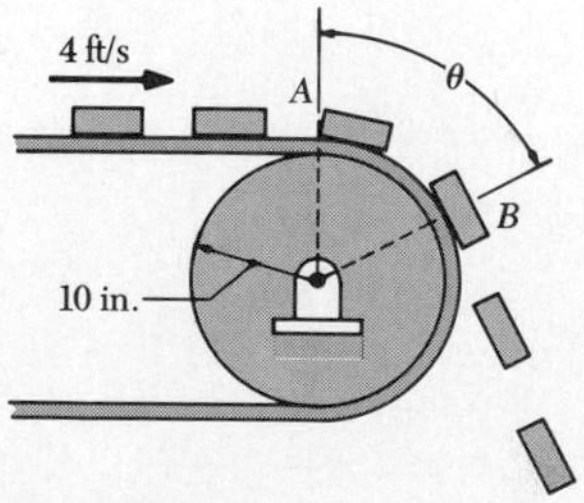

Fig. P12.48

12.48 A series of small packages, each weighing $\frac{3}{4}$ lb, are discharged from a conveyor belt as shown. Knowing that the coefficient of static friction between each package and the conveyor belt is 0.40, determine (*a*) the force exerted by the belt on a package just after it has passed point A, (*b*) the angle θ defining the point B where the packages first *slip* relative to the belt.

12.49 A 200-g block B fits inside a small cavity cut in arm OA, which rotates in the vertical plane at a constant rate such that $v = 2$ m/s. Knowing that the spring exerts on block B a force of magnitude $P = 0.8$ N and neglecting the effect of friction, determine the range of values of θ for which block B is in contact with the face of the cavity closest to the axis of rotation O.

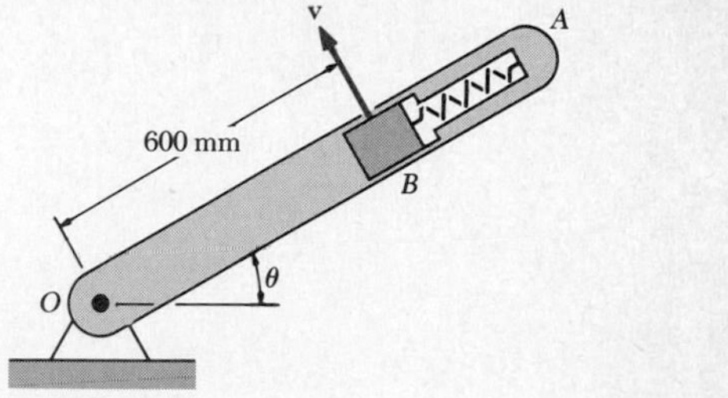

Fig. P12.49

12.50 Solve Prob. 12.49, assuming that the spring exerts on block B a force of magnitude $P = 3$ N.

12.51 Express the minimum and maximum safe speeds with respect to skidding of a car traveling on a banked road, in terms of the radius r of the curve, the banking angle θ, and the friction angle ϕ_s between the tires and the pavement.

12.52 A stunt driver proposes to drive a small automobile at the speed of 40 mi/h on the vertical wall of a circular pit of radius 50 ft. Knowing that the mass center of the automobile and driver is 2 ft from the wall, determine the minimum required value of the coefficient of static friction between the tires and the wall.

12.53 For the highway curve of Sample Prob. 12.6, determine the maximum safe speed, assuming that the coefficient of static friction between the tires and the pavement is 0.60.

12.54 A curve in a speed track has a radius of 150 m and a rated speed of 135 km/h. (See Sample Prob. 12.6 for the definition of rated speed.) Knowing that a racing car starts skidding on the curve when traveling at the speed of 300 km/h, determine (*a*) the banking angle θ, (*b*) the coefficient of static friction between the tires and the track under the prevailing conditions, (*c*) the minimum speed at which the same car could negotiate that curve.

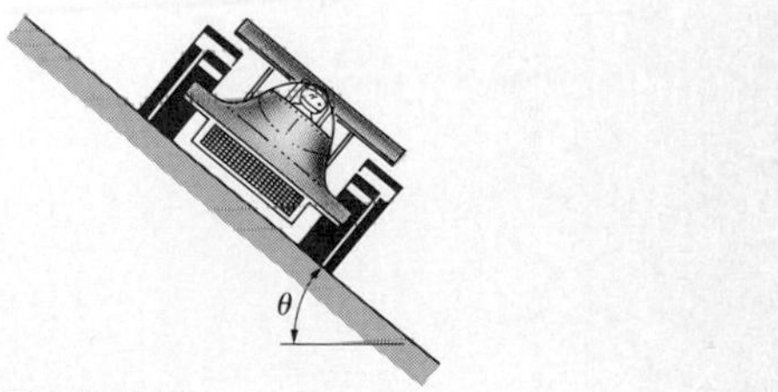

Fig. P12.54

12.55 A small 200-g collar C may slide on a semicircular rod which is made to rotate about the vertical AB at the constant rate of 6 rad/s. Determine the minimum required value of the coefficient of static friction between the collar and the rod if the collar is not to slide when (*a*) $\theta = 90°$, (*b*) $\theta = 75°$, (*c*) $\theta = 45°$. Indicate in each case the direction of the impending motion.

12.56 For the collar and rod of Prob. 12.55, and assuming that the coefficients of friction are $\mu_s = 0.25$ and $\mu_k = 0.20$, determine the magnitude and direction of the friction force exerted on the collar immediately after it has been released on the rotating rod in the position corresponding to (*a*) $\theta = 70°$, (*b*) $\theta = 30°$. Indicate also in each case whether the collar will slide on the rod.

***12.57** For the collar and rod of Prob. 12.55, determine (*a*) the three values of θ for which the collar will not slide on the rod, assuming no friction between the collar and the rod, (*b*) the range of values of θ for which the collar will not slide on the rod, assuming a coefficient of static friction of 0.25. (*Hint.* In part *b* solve by trial the equation obtained for θ.)

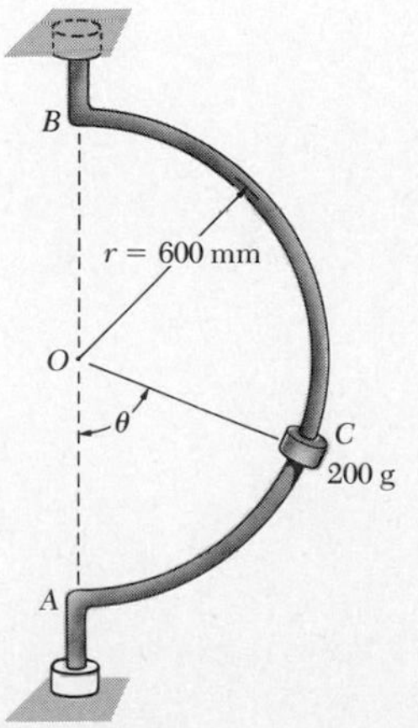

Fig. P12.55

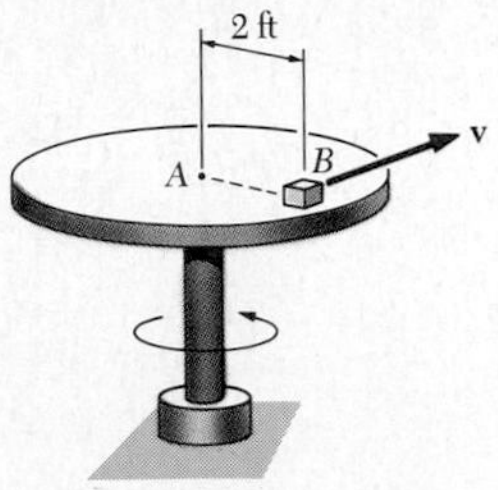

Fig. P12.58 and P12.59

12.58 A small block B is supported by a turntable which starting from rest is rotated in such a way that the block undergoes a constant tangential acceleration $a_t = 6\ \text{ft/s}^2$. Knowing that the coefficient of static friction between the block and the turntable is 0.60, determine (*a*) how long it will take for the block to start slipping on the turntable, (*b*) the speed v of the block at that instant.

12.59 A small block B is supported by a turntable which starting from rest is rotated in such a way that the block undergoes a constant tangential acceleration. Knowing that the coefficient of static friction between the block and the turntable is 0.50, determine the smallest interval of time in which the block can reach the speed $v = 5\ \text{ft/s}$ without slipping.

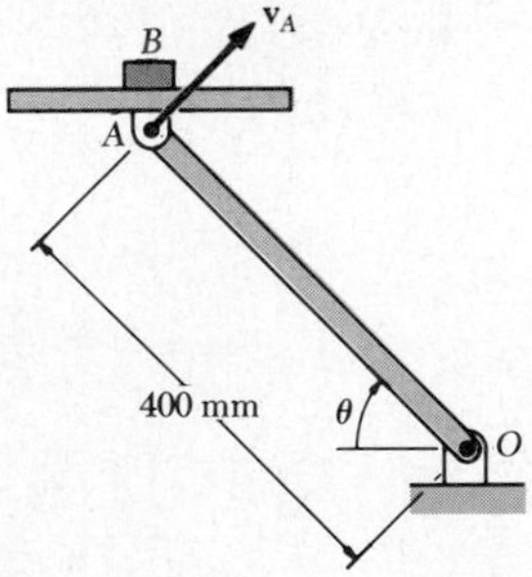

Fig. P12.60

12.60 A small block B is supported by a platform connected at A to rod OA. Point A describes a circle in a vertical plane at the constant speed v_A, while the platform is constrained to remain horizontal throughout its motion by the use of a special linkage (not shown in the figure). The coefficients of friction between the block and the platform are $\mu_s = 0.40$ and $\mu_k = 0.30$. Determine (*a*) the maximum allowable speed v_A if the block is not to slide on the platform, (*b*) the values of θ for which sliding is impending.

***12.61** For the system of Prob. 12.60, it is observed that block B slides intermittently on the platform when point A moves in a vertical circle at the constant speed $v_A = 1.22\ \text{m/s}$. Determine the values of θ for which the block starts sliding.

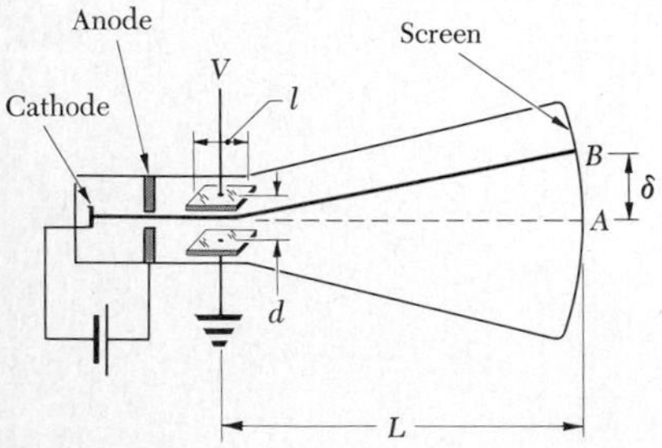

Fig. P12.62

12.62 In the cathode-ray tube shown, electrons emitted by the cathode and attracted by the anode pass through a small hole in the anode and keep traveling in a straight line with a speed v_0 until they strike the screen at A. However, if a difference of potential V is established between the two parallel plates, each electron will be subjected to a force **F** perpendicular to the plates while it travels between the plates and will strike the screen at point B at a distance δ from A. The magnitude of the force **F** is $F = eV/d$, where $-e$ is the charge of the electron and d is the distance between the plates. Derive an expression for the deflection δ in terms of V, v_0, the charge $-e$ of the electron, its mass m, and the dimensions d, l, and L.

12.63 In Prob. 12.62, determine the smallest allowable value of the ratio d/l in terms of e, m, v_0, and V if the electrons are not to strike the positive plate.

12.64 A manufacturer wishes to design a new cathode-ray tube which will be only half as long as his current model. If the size of the screen is to remain the same, how should the length l of the plates be modified if all the other characteristics of the circuit are to remain unchanged? (See Prob. 12.62 for description of cathode-ray tube.)

12.7. Angular Momentum of a Particle. Rate of Change of Angular Momentum. Consider a particle P of mass m moving with respect to a newtonian frame of reference $Oxyz$. As we saw in Sec. 12.3, the linear momentum of the particle at a given instant is defined as the vector $m\mathbf{v}$ obtained by multiplying the velocity $\mathbf{v}$ of the particle by its mass m. The moment about O of the vector $m\mathbf{v}$ is called the *moment of momentum*, or the *angular momentum*, of the particle about O at that instant and is denoted by $\mathbf{H}_O$. Recalling the definition of the moment of a vector (Sec. 3.6), and denoting by $\mathbf{r}$ the position vector of P, we write

$$\mathbf{H}_O = \mathbf{r} \times m\mathbf{v} \tag{12.12}$$

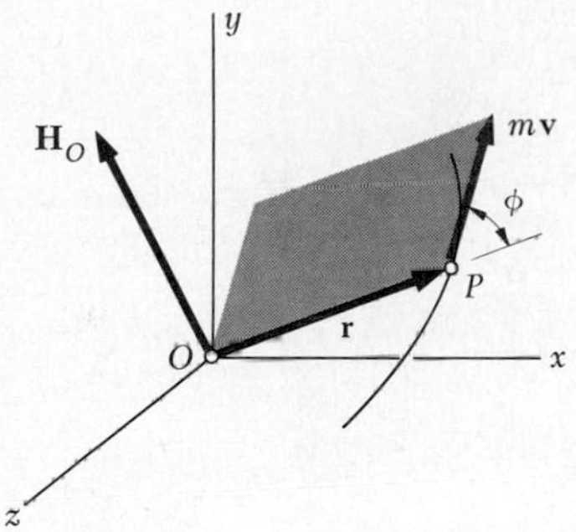

Fig. 12.12

and note that $\mathbf{H}_O$ is a vector perpendicular to the plane containing $\mathbf{r}$ and $m\mathbf{v}$, and of magnitude

$$H_O = rmv \sin \phi \tag{12.13}$$

where ϕ is the angle between $\mathbf{r}$ and $m\mathbf{v}$ (Fig. 12.12). The sense of $\mathbf{H}_O$ may be determined from the sense of $m\mathbf{v}$ by applying the right-hand rule. The unit of angular momentum is obtained by multiplying the units of length and of linear momentum (Sec. 12.4). With SI units we have

$$(\text{m})(\text{kg}\cdot\text{m/s}) = \text{kg}\cdot\text{m}^2/\text{s}$$

while, with U.S. customary units, we write

$$(\text{ft})(\text{lb}\cdot\text{s}) = \text{ft}\cdot\text{lb}\cdot\text{s}$$

Resolving the vectors $\mathbf{r}$ and $m\mathbf{v}$ into components, and applying formula (3.10), we write

$$\mathbf{H}_O = \begin{vmatrix} \mathbf{i} & \mathbf{j} & \mathbf{k} \\ x & y & z \\ mv_x & mv_y & mv_z \end{vmatrix} \tag{12.14}$$

The components of $\mathbf{H}_O$, which also represent the moments of the linear momentum $m\mathbf{v}$ about the coordinate axes, may be obtained by expanding the determinant in (12.14). We have

$$\begin{aligned} H_x &= m(yv_z - zv_y) \\ H_y &= m(zv_x - xv_z) \\ H_z &= m(xv_y - yv_x) \end{aligned} \tag{12.15}$$

In the case of a particle moving in the xy plane, we have $z = v_z = 0$ and the components H_x and H_y reduce to zero. The angular momentum is thus perpendicular to the xy plane; it is then completely defined by the scalar

$$H_O = H_z = m(xv_y - yv_x) \tag{12.16}$$

which will be positive or negative, according to the sense in which the particle is observed to move from O. If polar coordinates are used, we resolve the linear momentum of the particle into radial and transverse components (Fig. 12.13) and write

$$H_O = rmv \sin \phi = rmv_\theta \tag{12.17}$$

or, recalling from (11.45) that $v_\theta = r\dot{\theta}$,

$$H_O = mr^2\dot{\theta} \tag{12.18}$$

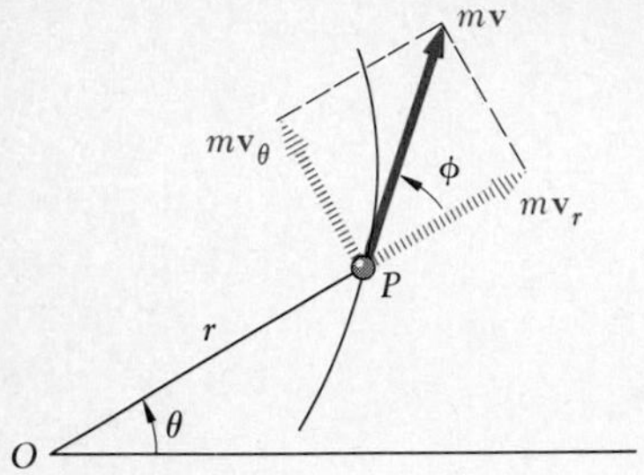

Fig. 12.13

We shall now compute the derivative with respect to t of the angular momentum $\mathbf{H}_O$ of a particle P moving in space. Differentiating both members of Eq. (12.12), and recalling the rule for the differentiation of a vector product (Sec. 11.10), we write

$$\dot{\mathbf{H}}_O = \dot{\mathbf{r}} \times m\mathbf{v} + \mathbf{r} \times m\dot{\mathbf{v}} = \mathbf{v} \times m\mathbf{v} + \mathbf{r} \times m\mathbf{a}$$

Since the vectors $\mathbf{v}$ and $m\mathbf{v}$ are collinear, the first term of the expression obtained is zero; and, by Newton's second law, $m\mathbf{a}$ is equal to the sum $\Sigma\mathbf{F}$ of the forces acting on P. Noting that $\mathbf{r} \times \Sigma\mathbf{F}$ represents the sum $\Sigma\mathbf{M}_O$ of the moments about O of these forces, we write

$$\Sigma\mathbf{M}_O = \dot{\mathbf{H}}_O \tag{12.19}$$

Equation (12.19), which results directly from Newton's second law, states that *the sum of the moments about O of the forces acting on the particle is equal to the rate of change of the moment of momentum, or angular momentum, of the particle about O.*

12.8. Equations of Motion in Terms of Radial and Transverse Components. Consider a particle P, of polar coordinates r and θ, which moves in a plane under the action of several forces. Resolving the forces and the acceleration of the particle into radial and transverse components (Fig. 12.14), and substituting into Eq. (12.2), we obtain the two scalar equations

$$\Sigma F_r = ma_r \qquad \Sigma F_\theta = ma_\theta \tag{12.20}$$

Substituting for a_r and a_θ from Eqs. (11.46), we have

$$\Sigma F_r = m(\ddot{r} - r\dot{\theta}^2) \tag{12.21}$$

$$\Sigma F_\theta = m(r\ddot{\theta} + 2\dot{r}\dot{\theta}) \tag{12.22}$$

The equations obtained may be solved for two unknowns.

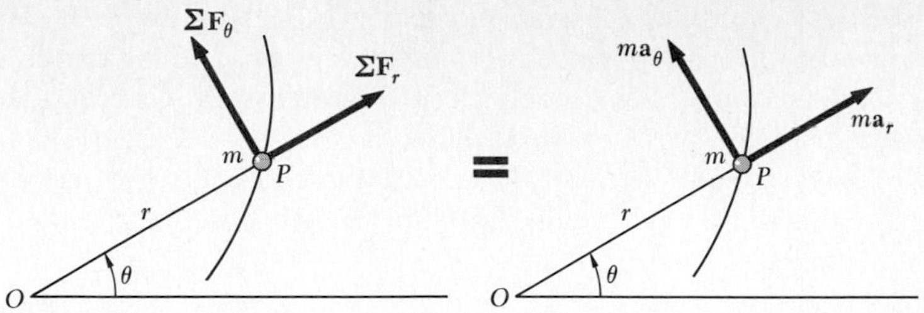

Fig. 12.14

Equation (12.22) could have been derived from Eq. (12.19). Recalling (12.18) and noting that $\Sigma M_O = r\Sigma F_\theta$, Eq. (12.19) yields

$$r\Sigma F_\theta = \frac{d}{dt}(mr^2\dot\theta)$$
$$= m(r^2\ddot\theta + 2r\dot r\dot\theta)$$

and, after dividing both members by r,

$$\Sigma F_\theta = m(r\ddot\theta + 2\dot r\dot\theta) \qquad (12.22)$$

12.9. Motion under a Central Force. Conservation of Angular Momentum. When the only force acting on a particle P is a force $\mathbf{F}$ directed toward or away from a fixed point O, the particle is said to be moving *under a central force*, and the point O is referred to as the *center of force* (Fig. 12.15). Since the line of action of $\mathbf{F}$ passes through O, we must have $\Sigma \mathbf{M}_O = 0$ at any given instant. Substituting into Eq. (12.19), we therefore obtain

$$\dot{\mathbf{H}}_O = 0$$

for all values of t or, integrating in t,

$$\mathbf{H}_O = \text{constant} \qquad (12.23)$$

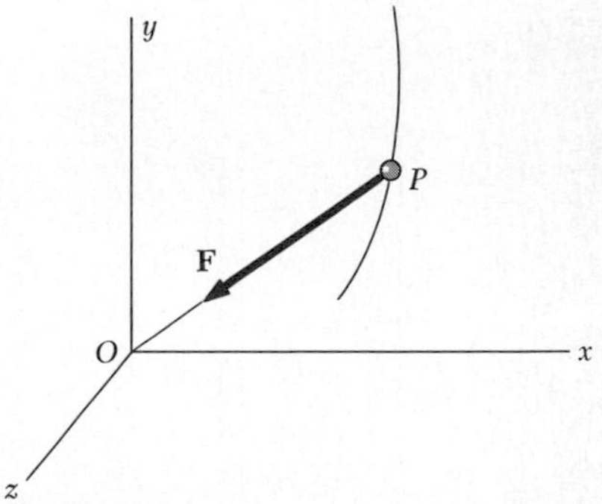

Fig. 12.15

We thus conclude that *the angular momentum of a particle moving under a central force is constant, both in magnitude and direction.*

Recalling the definition of the angular momentum of a particle (Sec. 12.7). we write

$$\mathbf{r} \times m\mathbf{v} = \mathbf{H}_O = \text{constant} \qquad (12.24)$$

from which it follows that the position vector $\mathbf{r}$ of the particle P must be perpendicular to the constant vector $\mathbf{H}_O$. Thus, a particle under a central

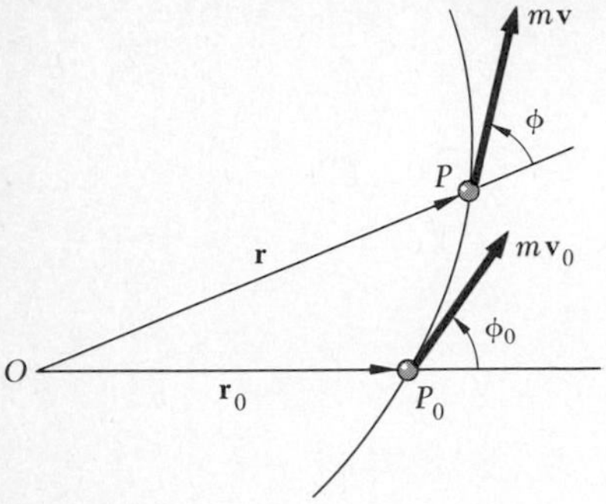

Fig. 12.16

force moves in a fixed plane perpendicular to $\mathbf{H}_O$. The vector $\mathbf{H}_O$ and the fixed plane are defined by the initial position vector $\mathbf{r}_0$ and the initial velocity $\mathbf{v}_0$ of the particle. For convenience, we shall assume that the plane of the figure coincides with the fixed plane of motion (Fig. 12.16).

Since the magnitude H_O of the angular momentum of the particle P is constant, the right-hand member in Eq. (12.13) must be constant. We therefore write

$$rmv \sin \phi = r_0 m v_0 \sin \phi_0 \tag{12.25}$$

This relation applies to the motion of any particle under a central force. Since the gravitational force exerted by the sun on a planet is a central force directed toward the center of the sun, Eq. (12.25) is fundamental to the study of planetary motion. For a similar reason, it is also fundamental to the study of the motion of space vehicles in orbit about the earth.

Recalling Eq. (12.18), we may alternatively express the fact that the magnitude H_O of the angular momentum of the particle P is constant by writing

$$mr^2\dot{\theta} = H_O = \text{constant} \tag{12.26}$$

or, dividing by m and denoting by h the angular momentum per unit mass H_O/m,

$$r^2\dot{\theta} = h \tag{12.27}$$

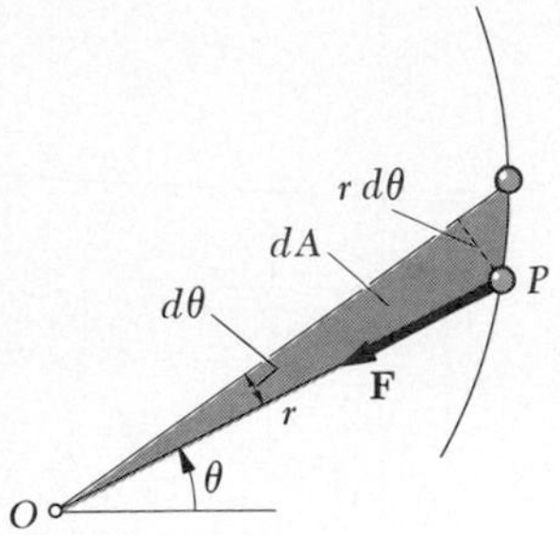

Fig. 12.17

Equation (12.27) may be given an interesting geometric interpretation. Observing from Fig. 12.17 that the radius vector OP sweeps an infinitesimal area $dA = \frac{1}{2}r^2\, d\theta$ as it rotates through an angle $d\theta$, and defining the *areal velocity* of the particle as the quotient dA/dt, we note that the left-hand member of Eq. (12.27) represents twice the areal velocity of the particle. We thus conclude that *when a particle moves under a central force, its areal velocity is constant.*

12.10. Newton's Law of Gravitation.

As we saw in the preceding section, the gravitational force exerted by the sun on a planet, or by the earth on an orbiting satellite, is an important example of a central force. In this section we shall learn how to determine the magnitude of a gravitational force.

In his *law of universal gravitation,* Newton states that two particles at a distance r from each other and, respectively, of mass M and m attract each other with equal and opposite forces $\mathbf{F}$ and $-\mathbf{F}$ directed along the line joining the particles (Fig. 12.18). The common magnitude F of the two forces is

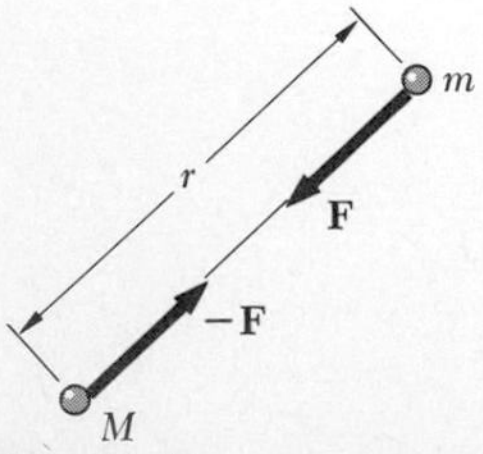

Fig. 12.18

$$F = G\frac{Mm}{r^2} \tag{12.28}$$

where G is a universal constant, called the *constant of gravitation*. Experiments show that the value of G is $(66.73 \pm 0.03) \times 10^{-12}\ m^3/kg \cdot s^2$ in SI units, or approximately $34.4 \times 10^{-9}\ ft^4/lb \cdot s^4$ in U.S. customary units. While gravitational forces exist between any pair of bodies, their effect is appreciable only when one of the bodies has a very large mass. The effect of gravitational forces is apparent in the case of the motion of a planet about the sun, of satellites orbiting about the earth, or of bodies falling on the surface of the earth.

Since the force exerted by the earth on a body of mass m located on or near its surface is defined as the weight $\mathbf{W}$ of the body, we may substitute the magnitude $W = mg$ of the weight for F, and the radius R of the earth for r, in Eq. (12.28). We obtain

$$W = mg = \frac{GM}{R^2}m \qquad \text{or} \qquad g = \frac{GM}{R^2} \tag{12.29}$$

where M is the mass of the earth. Since the earth is not truly spherical, the distance R from the center of the earth depends upon the point selected on its surface, and the values of W and g will thus vary with the altitude and latitude of the point considered. Another reason for the variation of W and g with the latitude is that a system of axes attached to the earth does not constitute a newtonian frame of reference (see Sec. 12.2). A more accurate definition of the weight of a body should therefore include a component representing the centrifugal force due to the rotation of the earth. Values of g at sea level vary from $9.781\ m/s^2$, or $32.09\ ft/s^2$, at the equator to $9.833\ m/s^2$, or $32.26\ ft/s^2$, at the poles.†

The force exerted by the earth on a body of mass m located in space at a distance r from its center may be found from Eq. (12.28). The computations will be somewhat simplified if we note that according to Eq. (12.29), the product of the constant of gravitation G and of the mass M of the earth may be expressed as

$$GM = gR^2 \tag{12.30}$$

where g and the radius R of the earth will be given their average values $g = 9.81\ m/s^2$ and $R = 6.37 \times 10^6$ m in SI units,‡ or $g = 32.2\ ft/s^2$ and $R = (3960\ mi)(5280\ ft/mi)$ in U.S. customary units.

The discovery of the law of universal gravitation has often been attributed to the fact that, after observing an apple falling from a tree, Newton had reflected that the earth must attract an apple and the moon in much the same way. While it is doubtful that this incident actually took place, it may be said that Newton would not have formulated his law if he had not first perceived that the acceleration of a falling body must have the same cause as the acceleration which keeps the moon in its orbit. This basic concept of continuity of the gravitational attraction is more easily understood now, when the gap between the apple and the moon is being filled with long-range ballistic missiles and artificial earth satellites.

†A formula expressing g in terms of the latitude ϕ was given in Prob. 12.2.

‡The value of R is easily found if one recalls that the circumference of the earth is $2\pi R = 40 \times 10^6$ m.

SAMPLE PROBLEM 12.7

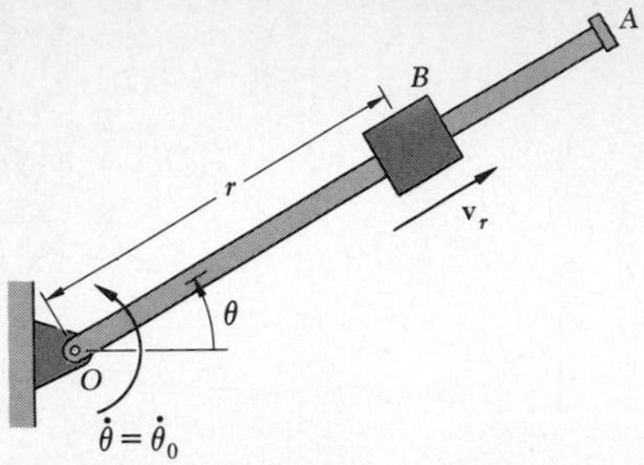

A block B of mass m may slide freely on a frictionless arm OA which rotates in a horizontal plane at a constant rate $\dot\theta_0$. Knowing that B is released at a distance r_0 from O, express as a function of r, (*a*) the component v_r of the velocity of B along OA, (*b*) the magnitude of the horizontal force **F** exerted on B by the arm OA.

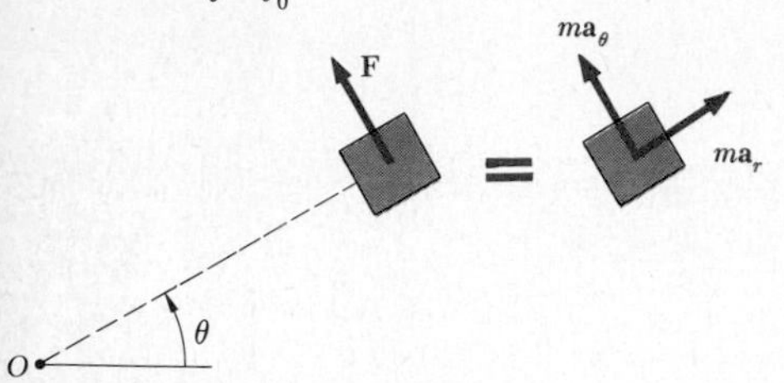

Solution. Since all other forces are perpendicular to the plane of the figure, the only force shown acting on B is the force **F** perpendicular to OA.

Equations of Motion. Using radial and transverse components,

$$+\nearrow \Sigma F_r = ma_r: \qquad 0 = m(\ddot r - r\dot\theta^2) \qquad (1)$$

$$+\nwarrow \Sigma F_\theta = ma_\theta: \qquad F = m(r\ddot\theta + 2\dot r\dot\theta) \qquad (2)$$

***a.* Component $\mathbf{v}_r$ of Velocity.** Since $v_r = \dot r$, we have

$$\ddot r = \dot v_r = \frac{dv_r}{dt} = \frac{dv_r}{dr}\frac{dr}{dt} = v_r\frac{dv_r}{dr}$$

Substituting for $\ddot r$ into (1), recalling that $\dot\theta = \dot\theta_0$, and separating the variables,

$$v_r\,dv_r = \dot\theta_0^2 r\,dr$$

Multiplying by 2, and integrating form 0 to v_r and from r_0 to r,

$$v_r^2 = \dot\theta_0^2(r^2 - r_0^2) \qquad v_r = \dot\theta_0(r^2 - r_0^2)^{1/2} \blacktriangleleft$$

***b.* Horizontal Force F.** Making $\dot\theta = \dot\theta_0$, $\ddot\theta = 0$, $\dot r = v_r$ in Eq. (2), and substituting for v_r the expression obtained in part *a*,

$$F = 2m\dot\theta_0(r^2 - r_0^2)^{1/2}\dot\theta_0 \qquad F = 2m\dot\theta_0^2(r^2 - r_0^2)^{1/2} \blacktriangleleft$$

SAMPLE PROBLEM 12.8

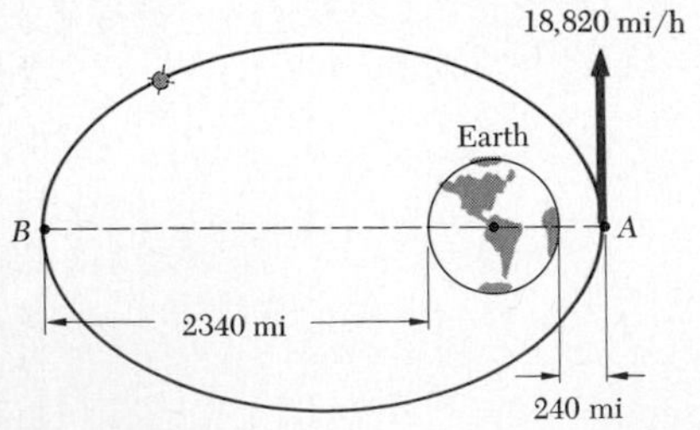

A satellite is launched in a direction parallel to the surface of the earth with a velocity of 18,820 mi/h from an altitude of 240 mi. Determine the velocity of the satellite as it reaches its maximum altitude of 2340 mi. It is recalled that the radius of the earth is 3960 mi.

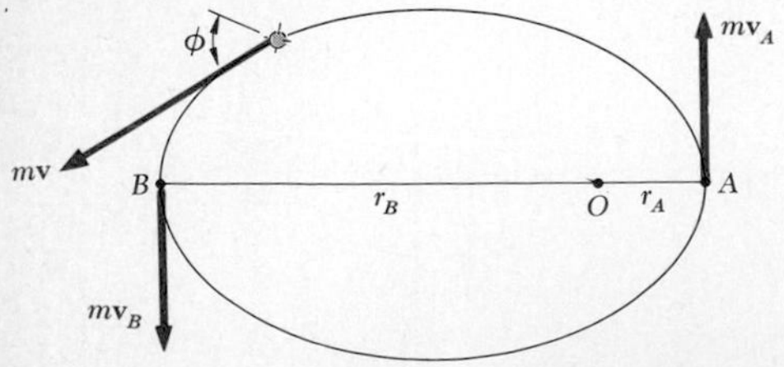

Solution. Since the satellite is moving under a central force directed toward the center O of the earth, its angular momentum $\mathbf{H}_O$ is constant. From Eq. (12.13) we have

$$rmv \sin\phi = H_O = \text{constant}$$

which shows that v is minimum at B, where both r and $\sin\phi$ are maximum. Expressing conservation of angular momentum between A and B,

$$r_A m v_A = r_B m v_B$$

$$v_B = v_A\frac{r_A}{r_B} = (18{,}820 \text{ mi/h})\frac{3960 \text{ mi} + 240 \text{ mi}}{3960 \text{ mi} + 2340 \text{ mi}}$$

$$v_B = 12{,}550 \text{ mi/h} \blacktriangleleft$$

Problems

12.65 The two-dimensional motion of particle B is defined by the relations $r = 25\,t^3 - 50\,t^2$ and $\theta = t^3 - 4\,t$, where r is expressed in millimeters, t in seconds, and θ in radians. If the particle has a mass of 2 kg and moves in a horizontal plane, determine the radial and transverse components of the force acting on the particle when (*a*) $t = 0$, (*b*) $t = 1$ s.

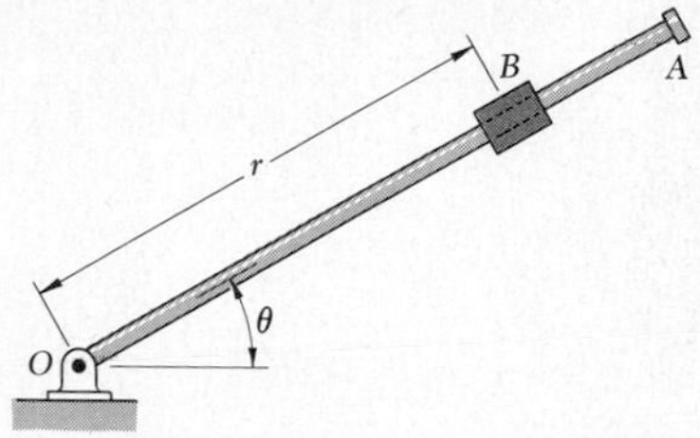

Fig. P12.65

12.66 For the motion defined in Prob. 12.65, determine the radial and transverse components of the force acting on the 2-kg particle as it passes again through the origin at $t = 2$ s.

12.67 For the motion defined in Sample Prob. 11.12, determine the force exerted by the arm OA on collar B and the radial force $\mathbf{Q}$ which must be applied to the collar (*a*) when $\theta = 0°$, (*b*) when $\theta = 30°$. Assume that collar B has a mass of 2 kg, and that the arm rotates in a vertical plane.

12.68 Pin B weighs 4 oz and is free to slide along the rotating arm OC and along the circular slot DE of radius $b = 20$ in. Neglecting friction and assuming that rod OC is made to rotate at the constant rate $\dot{\theta}_0 = 15$ rad/s in a horizontal plane, determine for any given value of θ (*a*) the radial and transverse components of the resultant force exerted on pin B, (*b*) the forces $\mathbf{P}$ and $\mathbf{Q}$ exerted on pin B, respectively, by rod OC and the wall of slot DE.

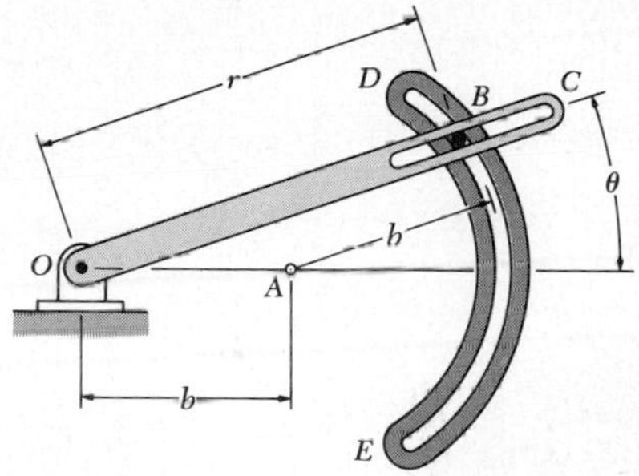

Fig. P12.68

12.69 Solve Prob. 12.68 for the position $\theta = 20°$, assuming that at that instant $\dot{\theta} = 15$ rad/s and $\ddot{\theta} = 250$ rad/s^2.

12.70 Slider C has a mass of 200 g and may move in a slot cut in arm AB, which rotates at the constant rate $\dot{\theta}_0 = 12$ rad/s in a horizontal plane. The slider is attached to a spring of constant $k = 36$ N/m, which is unstretched when $r = 0$. Knowing that the slider passes through the position $r = 400$ mm with a radial velocity $v_r = +1.8$ m/s, determine at that instant (*a*) the radial and transverse components of its acceleration, (*b*) its acceleration relative to arm AB, (*c*) the horizontal force exerted on the slider by arm AB.

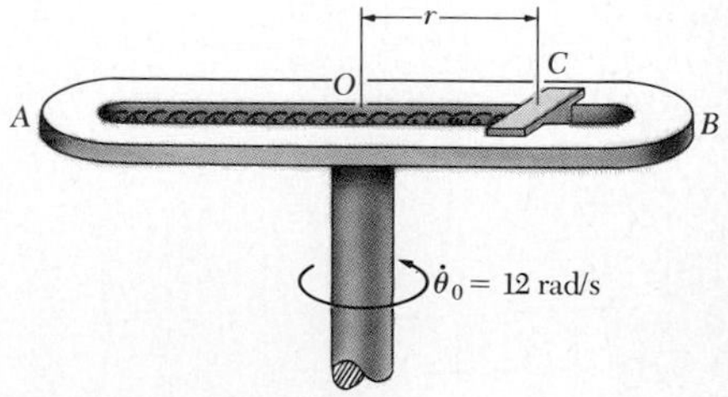

Fig. P12.70 and P12.71

***12.71** Slider C has a mass of 200 g and may move in a slot cut in arm AB, which rotates at the constant rate $\dot{\theta}_0 = 12$ rad/s in a horizontal plane. The slider is attached to a spring of constant $k = 36$ N/m, which is unstretched when $r = 0$. Knowing that the slider is released with no radial velocity in the position $r = 500$ mm and neglecting friction, determine for the position $r = 300$ mm (*a*) the radial and transverse components of the velocity of the slider, (*b*) the radial and transverse components of its acceleration, (*c*) the horizontal force exerted on the slider by arm AB.

***12.72** Solve Prob. 12.71, assuming that the spring is unstretched when slider C is located 45 mm to the left of the midpoint O of arm AB ($r = -45$ mm).

12.73 A particle of mass m is projected from point A with an initial velocity $\mathbf{v}_0$ perpendicular to the line OA and moves under a central force $\mathbf{F}$ along a semicircular path of diameter OA. Observing that $r = r_0 \cos\theta$ and using Eq. (12.27), show that the speed v of the particle is inversely proportional to the square of the distance r from the particle to the center of force O.

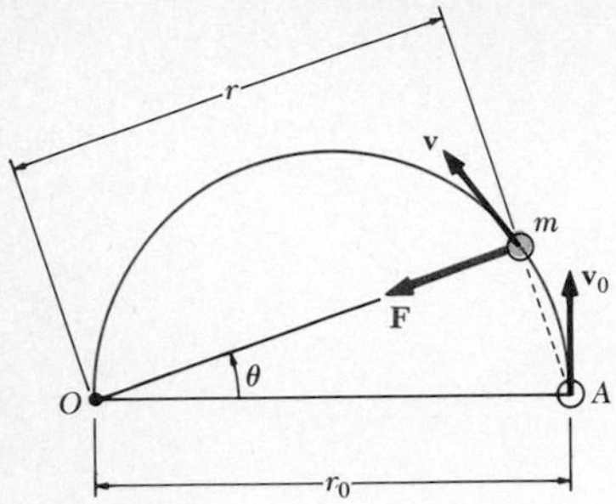

Fig. P12.73

12.74 A particle of mass m is projected from point A with an initial velocity $\mathbf{v}_0$ perpendicular to OA and moves under a central force $\mathbf{F}$ directed away from the center of force O. Knowing that the particle follows a path defined by the equation $r = r_0/\cos 2\theta$, and using Eq. (12.27), express the radial and transverse components of the velocity $\mathbf{v}$ of the particle as functions of θ.

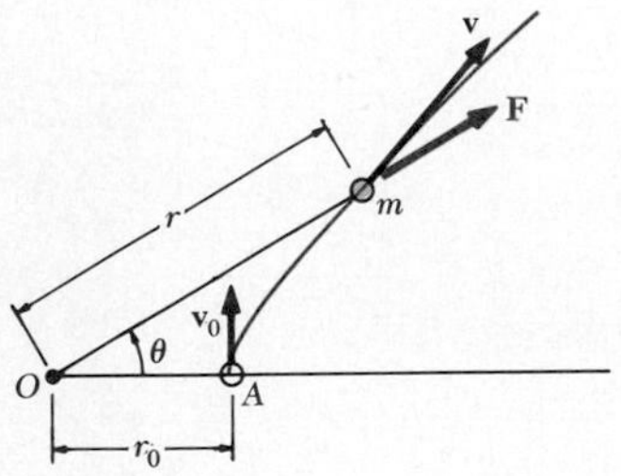

Fig. P12.74

12.75 For the particle of Prob. 12.73, (a) show that the central force $\mathbf{F}$ is inversely proportional to the fifth power of the distance r from the particle to the center of force O, (b) determine the magnitude of $\mathbf{F}$ for $\theta = 0$.

12.76 For the particle of Prob. 12.74, (a) show that the central force $\mathbf{F}$ is inversely proportional to the cube of the distance r from the particle to the center of force O, (b) determine the magnitude of $\mathbf{F}$ for $\theta = 0$.

12.77 Denoting by ρ the mean density of a planet, show that the minimum time required by a satellite to complete one full revolution about the planet is $(3\pi/G\rho)^{1/2}$, where G is the constant of gravitation.

12.78 Show that the radius r of the moon's orbit may be determined from the radius R of the earth, the acceleration of gravity g at the surface of the earth, and the time τ required by the moon to complete one full revolution about the earth. Compute r knowing that $\tau = 27.3$ days, giving the answer in both SI and U.S. customary units.

12.79 Communication satellites have been placed in a geosynchronous orbit, i.e., in a circular orbit such that they complete one full revolution about the earth in one sidereal day (23 h 56 min), and thus appear stationary with respect to the ground. Determine (a) the altitude of the satellites above the surface of the earth, (b) the velocity with which they describe their orbit. Give the answers in both SI and U.S. customary units.

12.80 The periodic times of two of the planet Jupiter's satellites, Io and Callisto, have been observed to be, respectively, 1 day 18 h 28 min and 16 days 16 h 32 min. Knowing that the radius of Callisto's orbit is 1.884×10^6 km, determine (a) the mass of the planet Jupiter, (b) the radius of Io's orbit. (The periodic time of a satellite is the time it requires to complete one full revolution about the planet.)

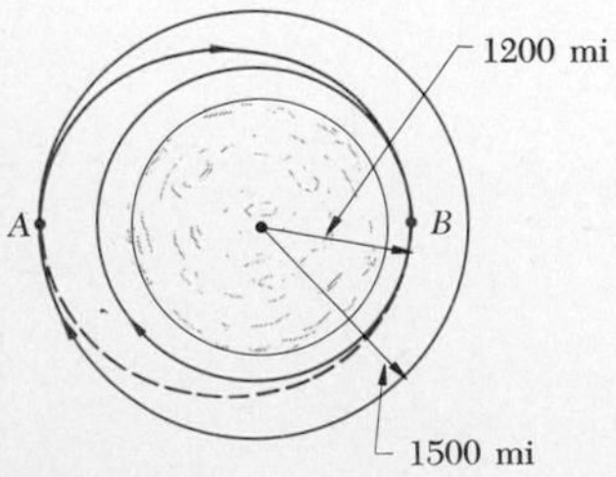

Fig. P12.81

12.81 An Apollo spacecraft describes a circular orbit with a 1500-mi radius around the moon with a velocity of 3190 mi/h. In order to transfer it to a smaller circular orbit with a 1200-mi radius, the spacecraft is first placed on an elliptic path AB by reducing its velocity to 3000 mi/h as it passes through A. Determine (a) the velocity of the spacecraft as it approaches B on the elliptic path, (b) the value to which its velocity must be reduced at B to insert it into the smaller circular orbit.

12.82 A space probe is to be placed in a circular orbit of 8000-km radius about the planet Venus in a specified plane. As the probe reaches A, the point of its original trajectory closest to Venus, it is inserted in a first elliptic transfer orbit by reducing its speed by Δv_A. This orbit brings it to point B with a much reduced velocity. There the probe is inserted in a second transfer orbit located in the specified plane by changing the direction of its velocity and further reducing its speed by Δv_B. Finally, as the probe reaches point C, it is inserted in the desired circular orbit by reducing its speed by Δv_C. Knowing that the mass of Venus is 0.82 times the mass of the earth, that $r_A = 12 \times 10^3$ km and $r_B = 96 \times 10^3$ km, that the probe reaches A with a velocity of 7400 m/s and B with a velocity of 869 m/s, and that its speed is further reduced by 146 m/s at B, determine by how much the velocity of the probe should be reduced (*a*) at A, (*b*) at C.

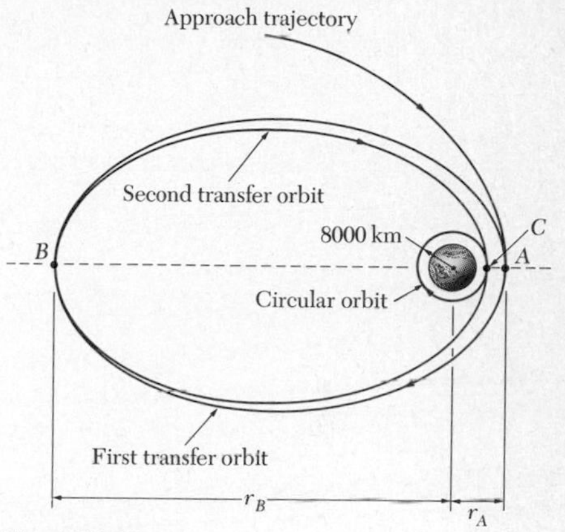

Fig. P12.82

12.83 For the space probe of Prob. 12.82, it is known that $r_A = 10 \times 10^3$ km and $r_B = 100 \times 10^3$ km, that the probe reaches A with a velocity of 8100 m/s, and that its speed is reduced by 400 m/s at A and by 2300 m/s at C. Determine (*a*) the velocity of the probe as it reaches B, (*b*) by how much its velocity should be reduced at B.

12.84 A space tug is used to place communication satellites into a geosynchronous orbit (see Prob. 12.79) at an altitude of 22,230 mi above the surface of the earth. The tug initially describes a circular orbit at an altitude of 220 mi and is inserted in an elliptic transfer orbit by firing its engine as it passes through A, thus increasing its velocity by 7910 ft/s. By how much should its velocity be increased as it reaches B to insert it in the geosynchronous orbit?

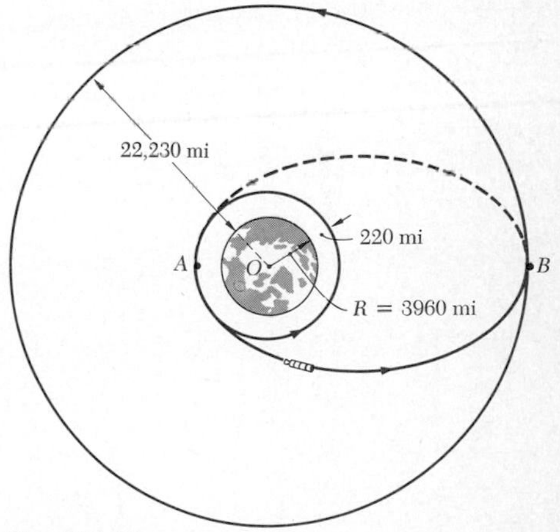

Fig. P12.84

12.85 A 5-lb ball is mounted on a horizontal rod which is free to rotate about a vertical shaft. In the position shown, the speed of the ball is $v_1 = 24$ in./s and the ball is held by a cord attached to the shaft. The cord is suddenly cut and the ball moves to position B as the rod rotates. Neglecting the mass of the rod, determine (*a*) the radial and transverse components of the acceleration of the ball immediately after the cord has been cut, (*b*) the acceleration of the ball relative to the rod at the same instant, (*c*) the speed of the ball after it has reached the stop B.

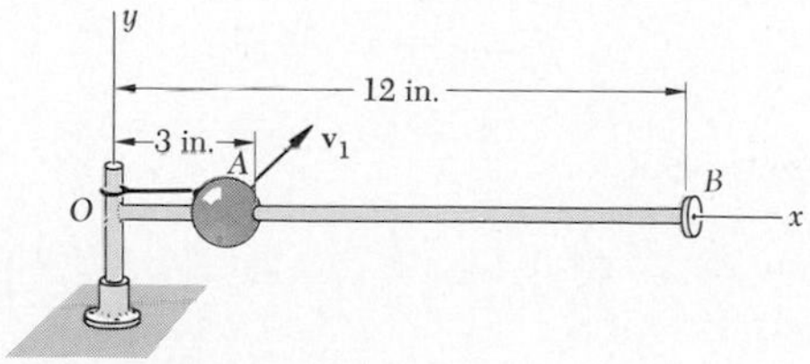

Fig. P12.85

12.86 A 4-oz ball slides on a smooth horizontal table at the end of a string which passes through a small hole in the table at O. When the length of string above the table is $r_1 = 18$ in., the speed of the ball is $v_1 = 5$ ft/s. Knowing that the breaking strength of the string is 4.00 lb, determine (*a*) the smallest distance r_2 which can be achieved by slowly drawing the string through the hole, (*b*) the corresponding speed v_2.

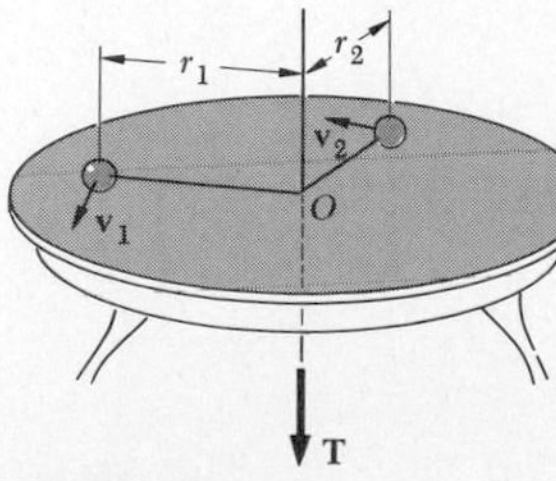

Fig. P12.86

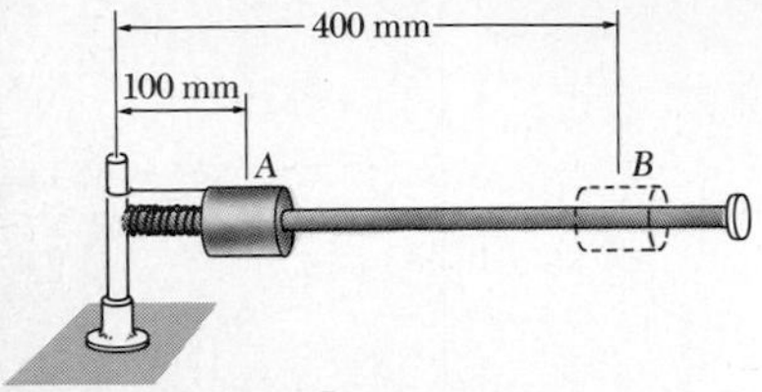

Fig. P12.87

12.87 A 250-g collar may slide on a horizontal rod which is free to rotate about a vertical shaft. The collar is initially held at A by a cord attached to the shaft and compresses a spring of constant 6 N/m, which is undeformed when the collar is located 500 mm from the shaft. As the rod rotates at the rate $\dot{\theta}_0 =$ 16 rad/s, the cord is cut and the collar moves out along the rod. Neglecting friction and the mass of the rod, determine for the position B of the collar (a) the transverse component of the velocity of the collar. (b) the radial and transverse components of its acceleration, (c) the acceleration of the collar relative to the rod.

***12.88** In Prob. 12.87, determine for the position B of the collar, (a) the radial component of the velocity of the collar, (b) the value of $\ddot{\theta}$.

*12.11. Trajectory of a Particle under a Central Force.

Consider a particle P moving under a central force **F**. We propose to obtain the differential equation which defines its trajectory.

Assuming that the force **F** is directed toward the center of force O, we note that ΣF_r and ΣF_θ reduce, respectively, to $-F$ and zero in Eqs. (12.21) and (12.22). We therefore write

$$m(\ddot{r} - r\dot{\theta}^2) = -F \tag{12.31}$$

$$m(r\ddot{\theta} + 2\dot{r}\dot{\theta}) = 0 \tag{12.32}$$

These equations define the motion of P. We shall, however, replace Eq. (12.32) by Eq. (12.27), which is more convenient to use and which is equivalent to Eq. (12.32), as we may easily check by differentiating it with respect to t. We write

$$r^2\dot{\theta} = h \qquad \text{or} \qquad r^2\frac{d\theta}{dt} = h \tag{12.33}$$

Equation (12.33) may be used to eliminate the independent variable t from Eq. (12.31). Solving Eq. (12.33) for $\dot{\theta}$ or $d\theta/dt$, we have

$$\dot{\theta} = \frac{d\theta}{dt} = \frac{h}{r^2} \tag{12.34}$$

from which it follows that

$$\dot{r} = \frac{dr}{dt} = \frac{dr}{d\theta}\frac{d\theta}{dt} = \frac{h}{r^2}\frac{dr}{d\theta} = -h\frac{d}{d\theta}\left(\frac{1}{r}\right) \tag{12.35}$$

$$\ddot{r} = \frac{d\dot{r}}{dt} = \frac{d\dot{r}}{d\theta}\frac{d\theta}{dt} = \frac{h}{r^2}\frac{d\dot{r}}{d\theta}$$

or, substituting for $\dot{r}$ from (12.35),

$$\ddot{r} = \frac{h}{r^2}\frac{d}{d\theta}\left[-h\frac{d}{d\theta}\left(\frac{1}{r}\right)\right]$$

$$\ddot{r} = -\frac{h^2}{r^2}\frac{d^2}{d\theta^2}\left(\frac{1}{r}\right) \tag{12.36}$$

Substituting for $\dot{\theta}$ and $\ddot{r}$ from (12.34) and (12.36), respectively, into Eq. (12.31), and introducing the function $u = 1/r$, we obtain after reductions

$$\frac{d^2u}{d\theta^2} + u = \frac{F}{mh^2u^2} \tag{12.37}$$

In deriving Eq. (12.37), the force **F** was assumed directed toward O. The magnitude F should therefore be positive if **F** is actually directed toward O (attractive force) and negative if **F** is directed away from O (repulsive force). If F is a known function of r and thus of u, Eq. (12.37) is a differential equation in u and θ. This differential equation defines the trajectory followed by the particle under the central force **F**. The equation of the trajectory will be obtained by solving the differential equation (12.37) for u as a function of θ and determining the constants of integration from the initial conditions.

***12.12. Application to Space Mechanics.** After the last stage of their launching rockets has burned out, earth satellites and other space vehicles are subjected to only the gravitational pull of the earth. Their motion may therefore be determined from Eqs. (12.33) and (12.37), which govern the motion of a particle under a central force, after F has been replaced by the expression obtained for the force of gravitational attraction.† Setting in Eq. (12.37)

$$F = \frac{GMm}{r^2} = GMmu^2$$

where M = mass of earth
m = mass of space vehicle
r = distance from center of earth to vehicle
$u = 1/r$

we obtain the differential equation

$$\frac{d^2u}{d\theta^2} + u = \frac{GM}{h^2} \tag{12.38}$$

where the right-hand member is observed to be a constant.

The solution of the differential equation (12.38) is obtained by adding the particular solution $u = GM/h^2$ to the general solution $u = C\cos(\theta - \theta_0)$ of the corresponding homogeneous equation (i.e., the equation obtained by setting the right-hand member equal to zero). Choosing

† It is assumed that the space vehicles considered here are attracted by only the earth and that their mass is negligible compared with the mass of the earth. If a vehicle moves very far from the earth, its path may be affected by the attraction of the sun, the moon, or another planet.

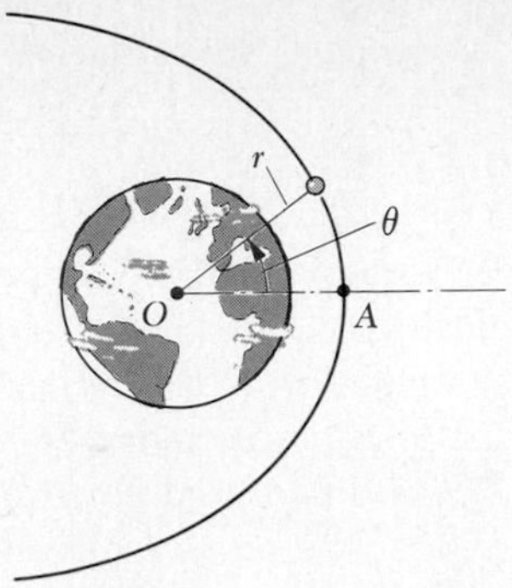

Fig. 12.19

the polar axis so that $\theta_0 = 0$, we write

$$\frac{1}{r} = u = \frac{GM}{h^2} + C\cos\theta \tag{12.39}$$

Equation (12.39) is the equation of a *conic section* (ellipse, parabola, or hyperbola) in the polar coordinates r and θ. The origin O of the coordinates, which is located at the center of the earth, is a *focus* of this conic section, and the polar axis is one of its axes of symmetry (Fig. 12.19).

The ratio of the constants C and GM/h^2 defines the *eccentricity* ε of the conic section; setting

$$\varepsilon = \frac{C}{GM/h^2} = \frac{Ch^2}{GM} \tag{12.40}$$

we may write Eq. (12.39) in the form

$$\frac{1}{r} = \frac{GM}{h^2}(1 + \varepsilon\cos\theta) \tag{12.39'}$$

Three cases may be distinguished:

1. $\varepsilon > 1$, or $C > GM/h^2$: There are two values θ_1 and $-\theta_1$ of the polar angle, defined by $\cos\theta_1 = -GM/Ch^2$, for which the right-hand member of Eq. (12.39) becomes zero. For both these values, the radius vector r becomes infinite; the conic section is a *hyperbola* (Fig. 12.20).
2. $\varepsilon = 1$, or $C = GM/h^2$: The radius vector becomes infinite for $\theta = 180°$; the conic section is a *parabola*.
3. $\varepsilon < 1$, or $C < GM/h^2$: The radius vector remains finite for every value of θ; the conic section is an *ellipse*. In the particular case when $\varepsilon = C = 0$, the length of the radius vector is constant; the conic section is a circle.

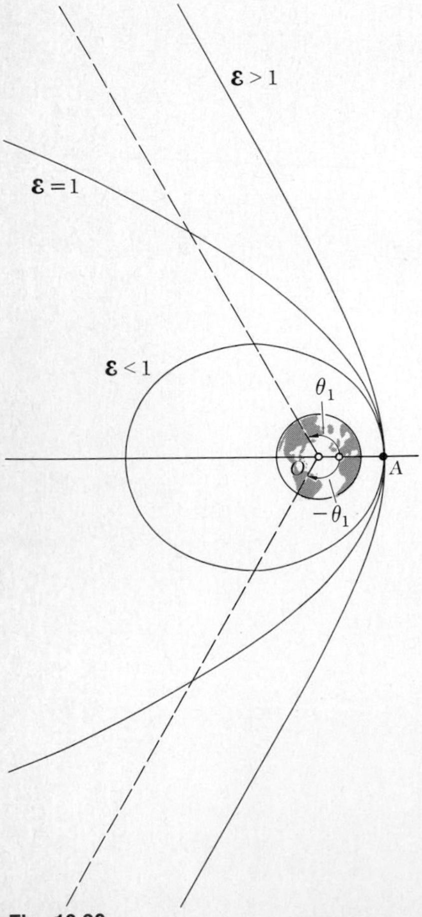

Fig. 12.20

We shall see now how the constants C and GM/h^2 which characterize the trajectory of a space vehicle may be determined from the position and the velocity of the space vehicle at the beginning of its free flight. We shall assume, as is generally the case, that the powered phase of its flight has been programmed in such a way that as the last stage of the launching rocket burns out, the vehicle has a velocity parallel to the surface of the earth (Fig. 12.21). In other words, we shall assume that the space vehicle begins its free flight at the vertex A of its trajectory.†

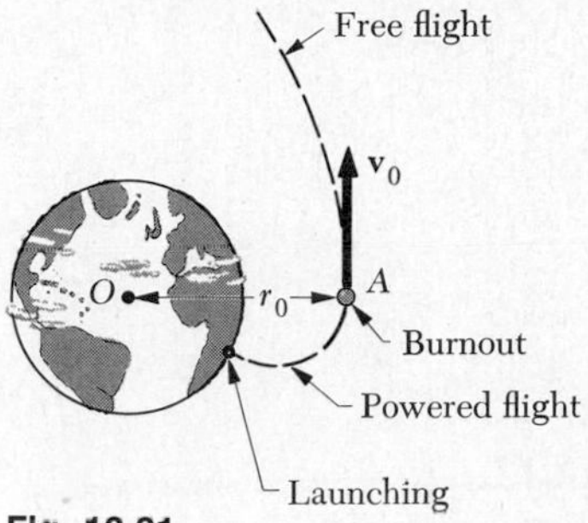

Fig. 12.21

Denoting, respectively, by r_0 and v_0 the radius vector and speed of the vehicle at the beginning of its free flight, we observe, since the velocity reduces to its transverse component, that $v_0 = r_0\dot{\theta}_0$. Recalling Eq. (12.27), we express the angular momentum per unit mass h as

$$h = r_0^2\dot{\theta}_0 = r_0 v_0 \tag{12.41}$$

The value obtained for h may be used to determine the constant GM/h^2. We also note that the computation of this constant will be simplified if we use the relation indicated in Sec. 12.10:

$$GM = gR^2 \tag{12.30}$$

where R is the radius of the earth ($R = 6.37 \times 10^6$ m or 3960 mi) and g the acceleration of gravity at the surface of the earth.

The constant C will be determined by setting $\theta = 0$, $r = r_0$ in Eq. (12.39); we obtain

$$C = \frac{1}{r_0} - \frac{GM}{h^2} \tag{12.42}$$

Substituting for h from (12.41), we may then easily express C in terms of r_0 and v_0.

Let us now determine the initial conditions corresponding to each of the three fundamental trajectories indicated above. Considering first the parabolic trajectory, we set C equal to GM/h^2 in Eq. (12.42) and eliminate h between Eqs. (12.41) and (12.42). Solving for v_0, we obtain

$$v_0 = \sqrt{\frac{2GM}{r_0}}$$

We may easily check that a larger value of the initial velocity corresponds to a hyperbolic trajectory and a smaller value, to an elliptic orbit. Since the value of v_0 obtained for the parabolic trajectory is the smallest value for which the space vehicle does not return to its starting point, it is called the *escape velocity*. We write therefore

$$v_{esc} = \sqrt{\frac{2GM}{r_0}} \quad \text{or} \quad v_{esc} = \sqrt{\frac{2gR^2}{r_0}} \tag{12.43}$$

if we make use of Eq. (12.30). We note that the trajectory will be (1) hyperbolic if $v_0 > v_{esc}$; (2) parabolic if $v_0 = v_{esc}$; (3) elliptic if $v_0 < v_{esc}$.

† Problems involving oblique launchings will be considered in Sec. 13.9.

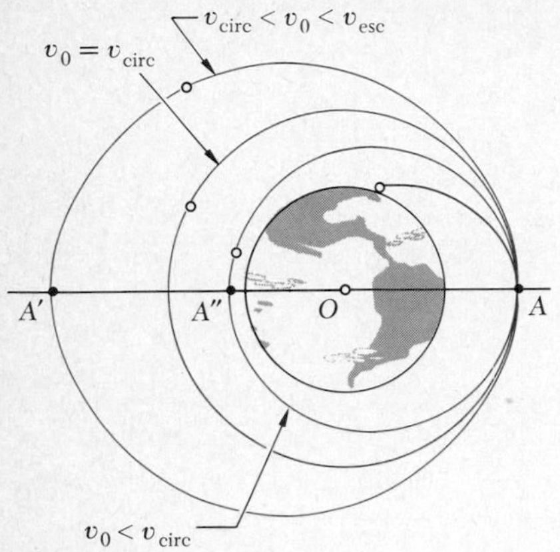

Fig. 12.22

Among the various possible elliptic orbits, one is of special interest, the *circular orbit,* which is obtained when $C = 0$. The value of the initial velocity corresponding to a circular orbit is easily found to be

$$v_{circ} = \sqrt{\frac{GM}{r_0}} \qquad \text{or} \qquad v_{circ} = \sqrt{\frac{gR^2}{r_0}} \tag{12.44}$$

if Eq. (12.30) is taken into account. We may note from Fig. 12.22 that for values of v_0 comprised between v_{circ} and v_{esc}, point A where free flight begins is the point of the orbit closest to the earth; this point is called the *perigee,* while point A', which is farthest away from the earth, is known as the *apogee.* For values of v_0 smaller than v_{circ}, point A becomes the apogee, while point A'', on the other side of the orbit, becomes the perigee. For values of v_0 much smaller than v_{circ}, the trajectory of the space vehicle intersects the surface of the earth; in such a case, the vehicle does not go into orbit.

Ballistic missiles, which are designed to hit the surface of the earth, also travel along elliptic trajectories. In fact, we should now realize that any object projected in vacuum with an initial velocity v_0 smaller than v_{esc} will move along an elliptic path. It is only when the distances involved are small that the gravitational field of the earth may be assumed uniform, and that the elliptic path may be approximated by a parabolic path, as was done earlier (Sec. 11.11) in the case of conventional projectiles.

Periodic Time. An important characteristic of the motion of an earth satellite is the time required by the satellite to describe its orbit. This time is known as the *periodic time* of the satellite and is denoted by τ. We first observe, in view of the definition of the areal velocity (Sec. 12.9), that τ may be obtained by dividing the area inside the orbit by the areal velocity. Since the area of an ellipse is equal to πab, where a and b denote, respectively, the semimajor and semiminor axes, and since the areal velocity is equal to $h/2$, we write

$$\tau = \frac{2\pi ab}{h} \tag{12.45}$$

While h may be readily determined from r_0 and v_0 in the case of a satellite launched in a direction parallel to the surface of the earth, the semiaxes a and b are not directly related to the initial conditions. Since, on the other hand, the values r_0 and r_1 of r corresponding to the perigee and apogee of the orbit may easily be determined from Eq. (12.39), we shall express the semiaxes a and b in terms of r_0 and r_1.

Consider the elliptic orbit shown in Fig. 12.23. The earth's center is located at O and coincides with one of the two foci of the ellipse, while the points A and A' represent, respectively, the perigee and apogee of the orbit. We easily check that

$$r_0 + r_1 = 2a$$

and thus

$$a = \tfrac{1}{2}(r_0 + r_1) \tag{12.46}$$

Recalling that the sum of the distances from each of the foci to any point of the ellipse is constant, we write

$$O'B + BO = O'A + OA = 2a \qquad \text{or} \qquad BO = a$$

On the other hand, we have $CO = a - r_0$. We may therefore write

$$b^2 = (BC)^2 = (BO)^2 - (CO)^2 = a^2 - (a - r_0)^2$$
$$b^2 = r_0(2a - r_0) = r_0 r_1$$

and thus

$$b = \sqrt{r_0 r_1} \tag{12.47}$$

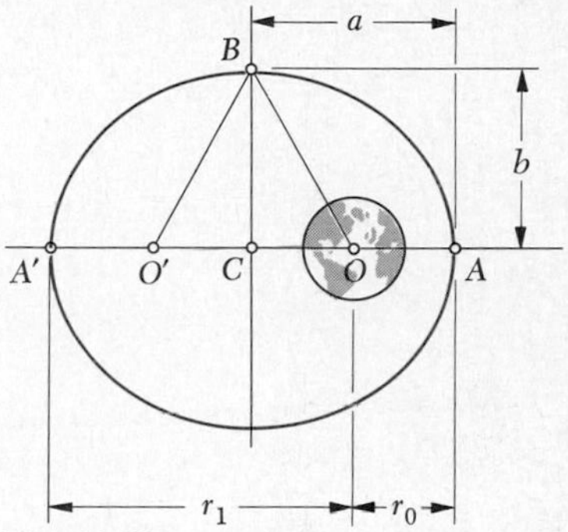

Fig. 12.23

Formulas (12.46) and (12.47) indicate that the semimajor and semiminor axes of the orbit are equal, respectively, to the arithmetic and geometric means of the maximum and minimum values of the radius vector. Once r_0 and r_1 have been determined, the lengths of the semiaxes may thus be easily computed and substituted for a and b in formula (12.45).

*** 12.13. Kepler's Laws of Planetary Motion.** The equations governing the motion of an earth satellite may be used to describe the motion of the moon around the earth. In that case, however, the mass of the moon is not negligible compared with the mass of the earth, and the results obtained are not entirely accurate.

The theory developed in the preceding sections may also be applied to the study of the motion of the planets around the sun. While another error is introduced by neglecting the forces exerted by the planets on one another, the approximation obtained is excellent. Indeed, the properties expressed by Eq. (12.39), where M now represents the mass of the sun, and by Eq. (12.33) had been discovered by the German astronomer Johann Kepler (1571–1630) from astronomical observations of the motion of the planets, even before Newton had formulated his fundamental theory.

Kepler's three *laws of planetary motion* may be stated as follows:

1. Each planet describes an ellipse, with the sun located at one of its foci.
2. The radius vector drawn from the sun to a planet sweeps equal areas in equal times.
3. The squares of the periodic times of the planets are proportional to the cubes of the semimajor axes of their orbits.

The first law states a particular case of the result established in Sec. 12.12, while the second law expresses that the areal velocity of each planet is constant (see Sec. 12.9). Kepler's third law may also be derived from the results obtained in Sec. 12.12.†

†See Prob. 12.115.

SAMPLE PROBLEM 12.9

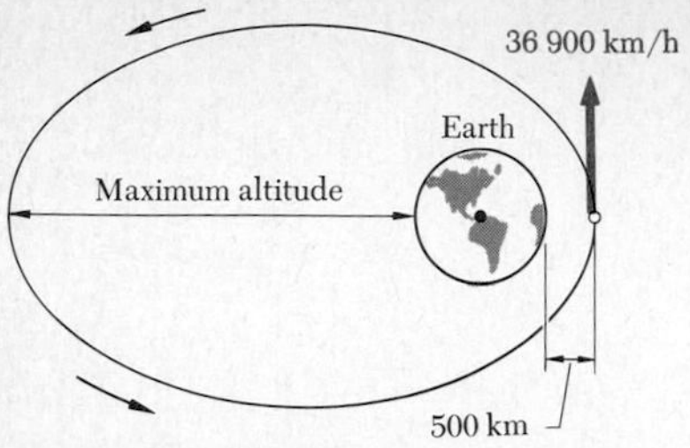

A satellite is launched in a direction parallel to the surface of the earth with a velocity of 36 900 km/h from an altitude of 500 km. Determine (*a*) the maximum altitude reached by the satellite, (*b*) the periodic time of the satellite.

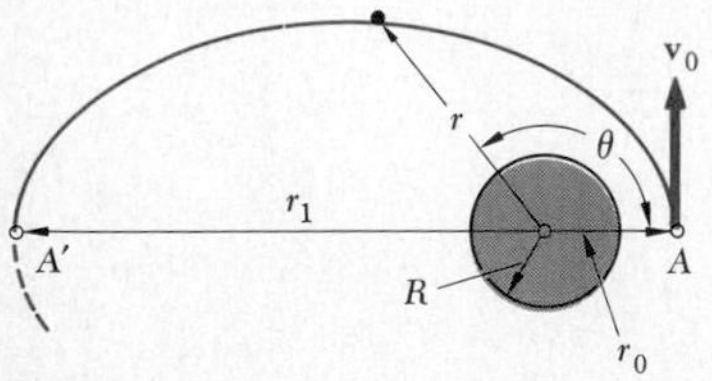

***a.* Maximum Altitude.** After launching, the satellite is subjected only to the gravitational attraction of the earth; its motion is thus governed by Eq. (12.39),

$$\frac{1}{r} = \frac{GM}{h^2} + C \cos \theta \qquad (1)$$

Since the radial component of the velocity is zero at the point of launching A, we have $h = r_0 v_0$. Recalling that the radius of the earth is $R = 6370$ km, we compute

$$r_0 = 6370 \text{ km} + 500 \text{ km} = 6870 \text{ km} = 6.87 \times 10^6 \text{ m}$$

$$v_0 = 36\,900 \text{ km/h} = \frac{36.9 \times 10^6 \text{ m}}{3.6 \times 10^3 \text{ s}} = 10.25 \times 10^3 \text{ m/s}$$

$$h = r_0 v_0 = (6.87 \times 10^6 \text{ m})(10.25 \times 10^3 \text{ m/s}) = 70.4 \times 10^9 \text{ m}^2/\text{s}$$

$$h^2 = 4.96 \times 10^{21} \text{ m}^4/\text{s}^2$$

Since $GM = gR^2$, where R is the radius of the earth, we have

$$GM = gR^2 = (9.81 \text{ m/s}^2)(6.37 \times 10^6 \text{ m})^2 = 398 \times 10^{12} \text{ m}^3/\text{s}^2$$

$$\frac{GM}{h^2} = \frac{398 \times 10^{12} \text{ m}^3/\text{s}^2}{4.96 \times 10^{21} \text{ m}^4/\text{s}^2} = 80.3 \times 10^{-9} \text{ m}^{-1}$$

Substituting this value into (1), we obtain

$$\frac{1}{r} = 80.3 \times 10^{-9} \text{ m}^{-1} + C \cos \theta \qquad (2)$$

Noting that at point A we have $\theta = 0$ and $r = r_0 = 6.87 \times 10^6$ m, we compute the constant C:

$$\frac{1}{6.87 \times 10^6 \text{ m}} = 80.3 \times 10^{-9} \text{ m}^{-1} + C \cos 0° \qquad C = 65.3 \times 10^{-9} \text{ m}^{-1}$$

At A', the point on the orbit farthest from the earth, we have $\theta = 180°$. Using (2), we compute the corresponding distance r_1:

$$\frac{1}{r_1} = 80.3 \times 10^{-9} \text{ m}^{-1} + (65.3 \times 10^{-9} \text{ m}^{-1}) \cos 180°$$

$$r_1 = 66.7 \times 10^6 \text{ m} = 66\,700 \text{ km}$$

Maximum altitude = 66 700 km − 6370 km = 60 300 km ◄

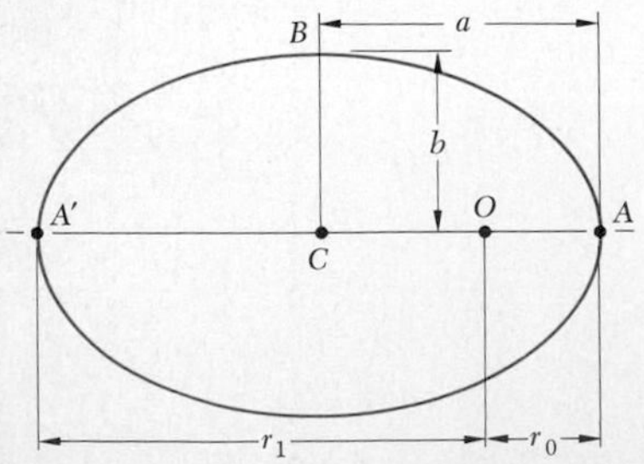

***b.* Periodic Time.** Since A and A' are the perigee and apogee, respectively, of the elliptic orbit, we use Eqs. (12.46) and (12.47) and compute the semimajor and semiminor axes of the orbit:

$$a = \tfrac{1}{2}(r_0 + r_1) = \tfrac{1}{2}(6.87 + 66.7)(10^6) \text{ m} = 36.8 \times 10^6 \text{ m}$$

$$b = \sqrt{r_0 r_1} = \sqrt{(6.87)(66.7)} \times 10^6 \text{ m} = 21.4 \times 10^6 \text{ m}$$

$$\tau = \frac{2\pi ab}{h} = \frac{2\pi(36.8 \times 10^6 \text{ m})(21.4 \times 10^6 \text{ m})}{70.4 \times 10^9 \text{ m}^2/\text{s}}$$

$$\tau = 70.3 \times 10^3 \text{ s} = 1171 \text{ min} = 19 \text{ h } 31 \text{ min}$$ ◄

Problems

12.89 For the particle of Prob. 12.73, and using Eq. (12.37), show that the central force **F** is inversely proportional to the fifth power of the distance r from the particle to the center of force O.

12.90 For the particle of Prob. 12.74, and using Eq. (12.37), show that the central force **F** is inversely proportional to the cube of the distance r from the particle to the center of force O.

12.91 A particle of mass m describes the hyperbolic spiral $r = b/\theta$ under a central force **F** directed toward the center of force O. Using Eq. (12.37), show that **F** is inversely proportional to the cube of the distance r from the particle to O.

12.92 A particle of mass m describes the logarithmic spiral $r = r_0 e^{b\theta}$ under a central force **F** directed toward the center of force O. Using Eq. (12.37), show that **F** is inversely proportional to the cube of the distance r from the particle to O.

12.93 It was observed that as the spacecraft Voyager 1 reached the point on its trajectory closest to the planet Jupiter, it was at a distance of 350×10^3 km from the center of the planet and had a velocity of 26.9 km/s. Determine the mass of Jupiter, assuming that the trajectory of the spacecraft was parabolic.

12.94 It was observed that as the spacecraft Voyager 2 reached the point on its trajectory closest to the planet Jupiter, it was at a distance of 715×10^3 km from the center of the planet. Assuming the trajectory of the spacecraft to be parabolic and using the data given in Prob. 12.93 for Voyager 1, determine the maximum velocity of Voyager 2 on its approach to Jupiter.

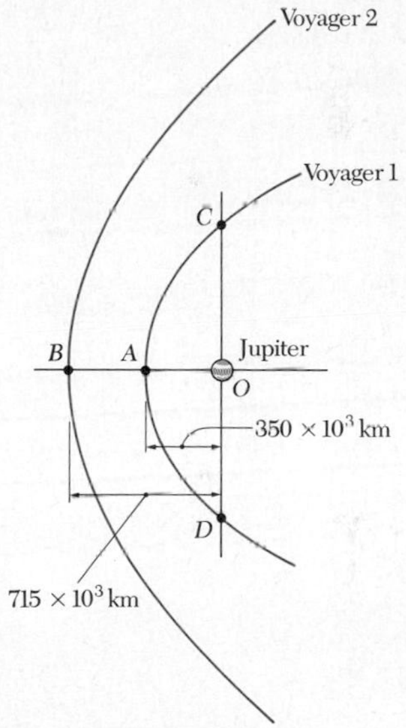

Fig. P12.93 and P12.94

12.95 Solve Prob. 12.94, assuming that the eccentricity of the trajectory of the spacecraft Voyager 2 was $\varepsilon = 1.20$.

12.96 It was observed that as the spacecraft Voyager 1 reached the point of its trajectory closest to the planet Saturn, it was at a distance of 115×10^3 mi from the center of the planet and had a velocity of 68.8×10^3 ft/s. Knowing that Tethys, one of Saturn's satellites, describes a circular orbit of radius 183×10^3 mi at a speed of 37.2×10^3 ft/s, determine the eccentricity of the trajectory of Voyager 1 on its approach to Saturn.

12.97 A satellite describes an elliptic orbit about a planet of mass M. Denoting by r_0 and r_1, respectively, the minimum and maximum values of the distance r from the satellite to the center of the planet, derive the relation

$$\frac{1}{r_0} + \frac{1}{r_1} = \frac{2GM}{h^2}$$

where h is the angular momentum per unit mass of the satellite.

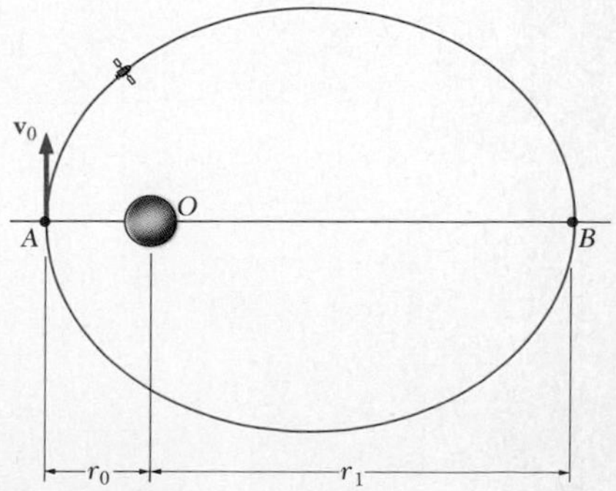

Fig. P12.97

12.98 At engine burnout on its second mission, the space shuttle Columbia had reached point A at an altitude of 40 mi and had a horizontal velocity $\mathbf{v}_0$. Knowing that its first orbit was elliptic and that the shuttle was transferred to a circular orbit as it passed through point B at an altitude of 150 mi above the surface of the earth, determine (*a*) the speed v_0 of the shuttle at engine burnout, (*b*) the increase in speed required at B to insert the shuttle on the circular orbit.

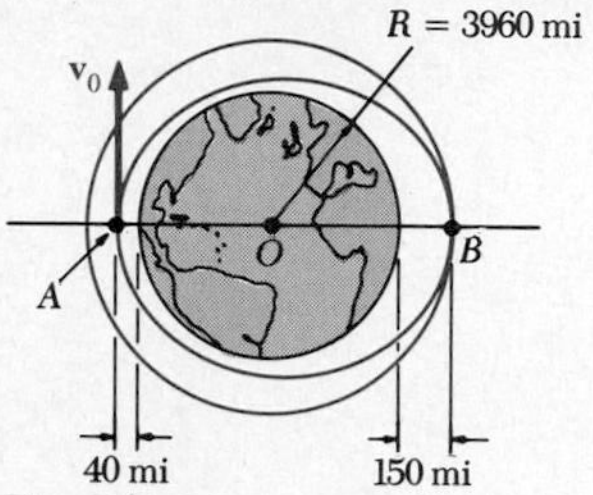

Fig. P12.98

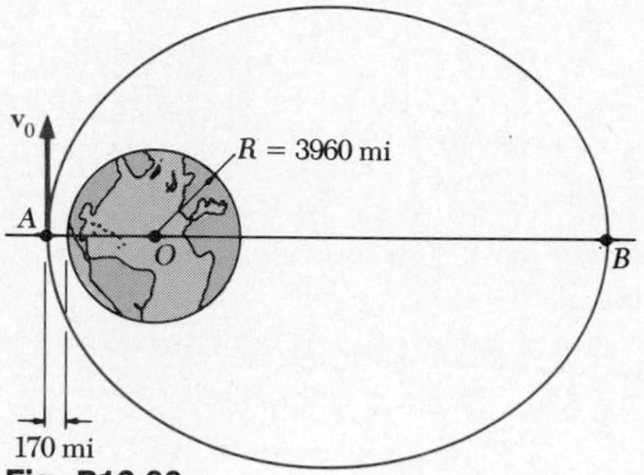

Fig. P12.99

12.99 At engine burnout an Explorer satellite was 170 mi above the surface of the earth and had a horizontal velocity $\mathbf{v}_0$ of magnitude 32.6×10^3 ft/s. Determine (*a*) the highest altitude reached by the satellite, (*b*) its velocity at its apogee B.

12.100 A space probe is describing a circular orbit about a planet of radius R. The altitude of the probe above the surface of the planet is αR and its velocity is v_0. In order to place the probe on an elliptic orbit which will bring it closer to the planet, its velocity is reduced from v_0 to βv_0, where $\beta < 1$, by firing its engine for a short interval of time. Determine the smallest permissible value of β if the probe is not to crash on the surface of the planet.

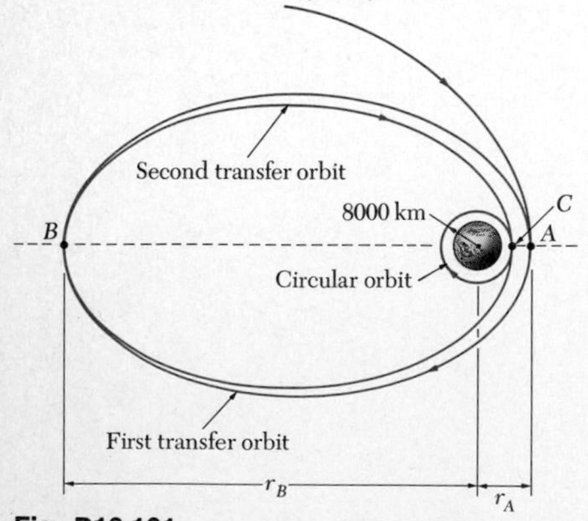

Fig. P12.101

12.101 A space probe is to be placed in a circular orbit of 8000-km radius about the planet Venus in a specified plane. As the probe reaches A, the point of its original trajectory closest to Venus, it is inserted in a first elliptic transfer orbit by reducing its speed by Δv_A. This orbit brings it to point B with a much reduced velocity. There the probe is inserted in a second transfer orbit located in the specified plane by changing the direction of its velocity and further reducing its speed by Δv_B. Finally, as the probe reaches point C, it is inserted in the desired circular orbit by reducing its speed by Δv_C. Knowing that the mass of Venus is 0.82 times the mass of the earth, that $r_A = 12.5 \times 10^3$ km and $r_B = 100 \times 10^3$ km, and that the probe approaches A on a parabolic trajectory, determine by how much the velocity of the probe should be reduced (*a*) at A, (*b*) at B, (*c*) at C.

12.102 For the space probe of Prob. 12.101, it is known that $r_A = 12.5 \times 10^3$ km and that the velocity of the probe is reduced to 6890 m/s as it passes through A. Determine (*a*) the distance from the center of Venus to point B, (*b*) the amounts by which the velocity of the probe should be reduced at B and C, respectively.

12.103 Determine the time needed for the space probe of Prob. 12.101 to travel from A to B on its first transfer orbit.

12.104 Determine the time needed for the space probe of Prob. 12.101 to travel from B to C on its second transfer orbit.

12.105 Determine the periodic time of the Explorer satellite of Prob. 12.99.

12.106 For the shuttle Columbia of Prob. 12.98, determine (*a*) the periodic time of the shuttle on its final circular orbit, (*b*) the time needed for the shuttle to travel from A to B on its original elliptic orbit.

12.107 Determine the time needed for the spacecraft Voyager 1 of Prob. 12.93 to travel from C to D on its parabolic trajectory.

12.108 Halley's comet travels in an elongated elliptic orbit for which the minimum distance from the sun is approximately $\frac{1}{2}r_E$, where $r_E = 92.9 \times 10^6$ mi is the mean distance from the sun to the earth. Knowing that the periodic time of Halley's comet is about 76 years, determine the maximum distance from the sun reached by the comet.

12.109 A space probe is describing a circular orbit of radius nR with a velocity v_0 about a planet of radius R and center O. As the probe passes through point A, its velocity is reduced from v_0 to βv_0, where $\beta < 1$, to place the probe on a crash trajectory. Show that the angle AOB, where B denotes the point of impact of the probe on the planet, depends only upon the values of n and β.

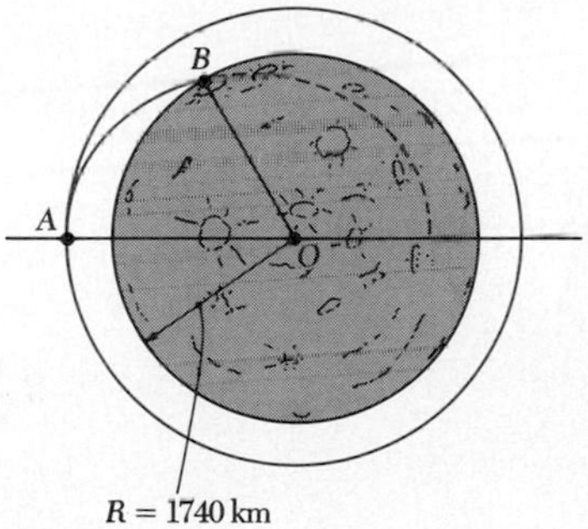

Fig. P12.110

12.110 Upon completion of their moon-exploration mission, the two astronauts forming the crew of an Apollo lunar excursion module (LEM) would rejoin the command module which had remained in a circular orbit around the moon. Before their return to earth, the astronauts would position their craft so that the LEM faced to the rear. As the command module passed through A, the LEM would be cast adrift and crash on the moon's surface at point B. Knowing that the command module was orbiting the moon at an altitude of 150 km and that the angle AOB was 60°, determine the velocity of the LEM relative to the command module as it was cast adrift. (*Hint.* Point A is the apogee of the elliptic crash trajectory. It is also recalled that the mass of the moon is 0.01230 times the mass of the earth.)

12.111 In Prob. 12.110, determine the angle AOB defining the point of impact of the LEM on the surface of the moon, assuming that the command module was orbiting the moon at an altitude of 120 km and that the LEM was cast adrift with a velocity of 40 m/s relative to the command module.

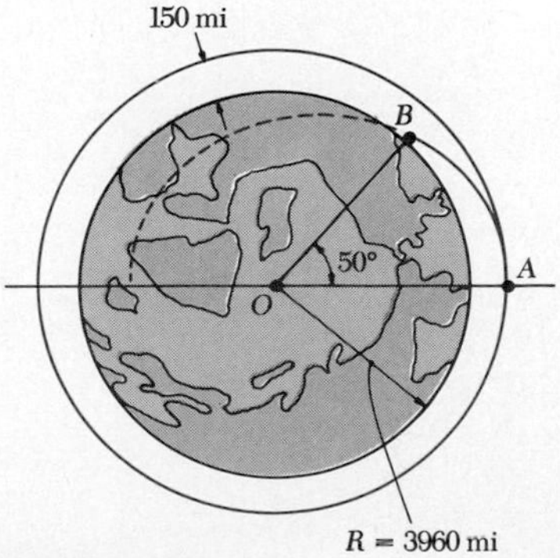

Fig. P12.112

12.112 A space shuttle is describing a circular orbit at an altitude of 150 mi above the surface of the earth. As it passes through A it fires its engine for a short interval of time to reduce its speed by 4 percent and begin its descent toward the earth. Determine the altitude of the shuttle at point B, knowing that the angle AOB is equal to 50°. (*Hint.* Point A is the apogee of the elliptic descent trajectory.)

12.113 A satellite describes an elliptic orbit about a planet. Denoting by r_0 and r_1 the distances corresponding, respectively, to the perigee and apogee of the orbit, show that the curvature of the orbit at each of these two points may be expressed as

$$\frac{1}{\rho} = \frac{1}{2}\left(\frac{1}{r_0} + \frac{1}{r_1}\right)$$

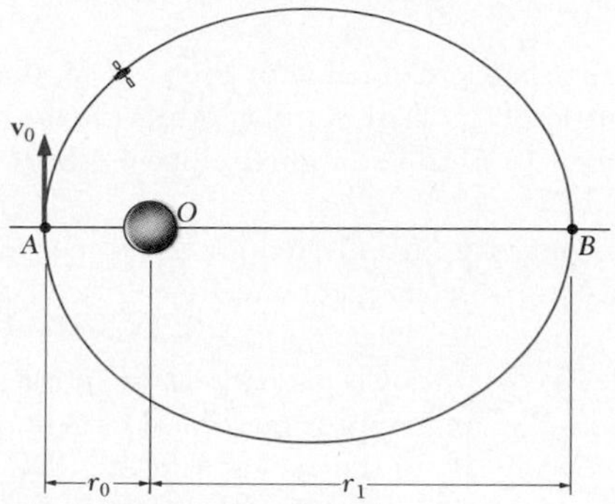

Fig. P12.113 and P12.114

12.114 (*a*) Express the eccentricity ε of the elliptic orbit described by a satellite about a planet in terms of the distances r_0 and r_1 corresponding, respectively, to the perigee and apogee of the orbit. (*b*) Use the result obtained in part *a* to determine the eccentricities of the two transfer orbits described in Prob. 12.101.

12.115 Derive Kepler's third law of planetary motion from Eqs. (12.39) and (12.45).

12.116 Show that the angular momentum per unit mass h of a satellite describing an elliptic orbit of semimajor axis a and eccentricity ε about a planet of mass M may be expressed as

$$h = \sqrt{GMa(1 - \varepsilon^2)}$$

Review Problems

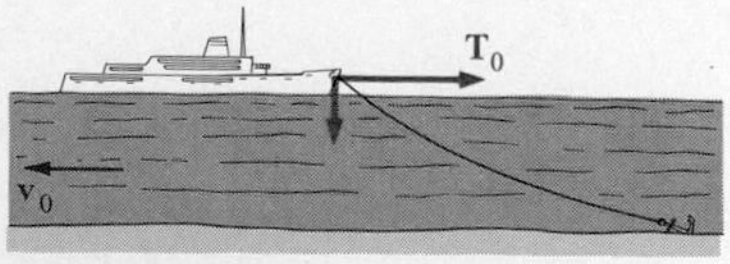

Fig. P12.117

12.117 A ship of total mass m is anchored in the middle of a river which is flowing with a constant velocity $\mathbf{v}_0$. The horizontal component of the force exerted on the ship by the anchor chain is $\mathbf{T}_0$. If the anchor chain suddenly breaks, determine the time required for the ship to attain a velocity equal to $\frac{1}{2}\mathbf{v}_0$. Assume that the frictional resistance of the water is proportional to the velocity of the ship relative to the water.

12.118 A ball is attached to a cord and swung through a full circle in a vertical plane. Knowing that the radius of the circle is 1.25 m, determine the smallest velocity that the ball should have at the top of the circle if the cord is to remain taut.

12.119 The coefficients of friction between the 45-lb crate and the 30-lb cart are $\mu_s = 0.25$ and $\mu_k = 0.20$. If a force **P** of magnitude 20 lb is applied to the cart, determine the acceleration (*a*) of the cart, (*b*) of the crate, (*c*) of the crate with respect to the cart.

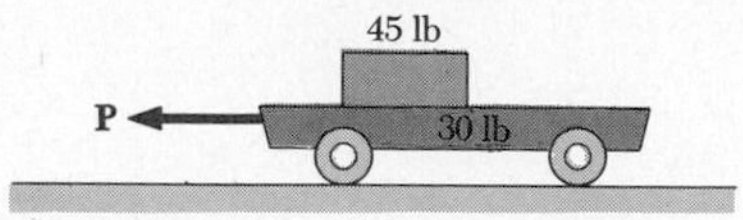

Fig. P12.119

12.120 For the cart and crate of Prob. 12.119, determine the acceleration of the cart as a function of the magnitude of the force **P**.

12.121 The assembly shown rotates about a vertical axis at a constant rate. Knowing that the coefficient of static friction between the small block *A* and the cylindrical wall is 0.30, determine the lowest speed *v* for which the block will remain in contact with the wall.

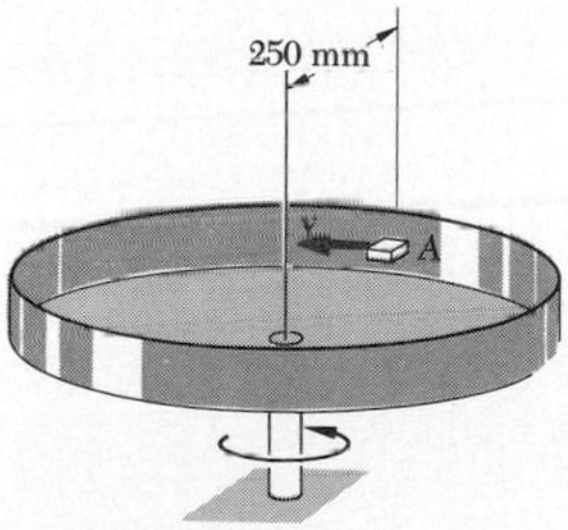

Fig. P12.121

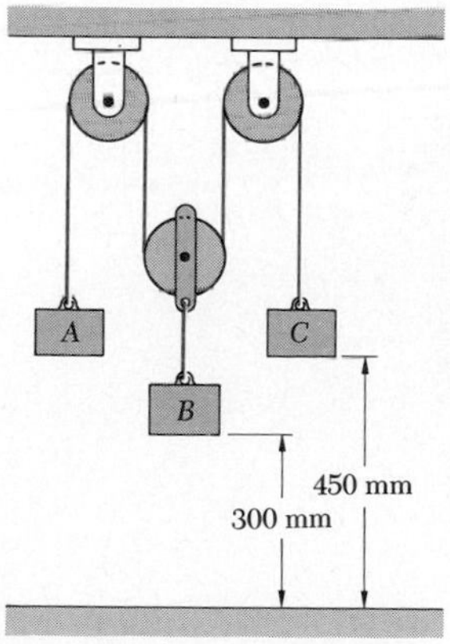

Fig. P12.122 and P12.123

12.122 Knowing that blocks *B* and *C* strike the ground simultaneously and exactly 1 s after the system is released from rest, determine m_B and m_C in terms of m_A.

12.123 Determine the acceleration of each block when $m_A = 5$ kg, $m_B = 15$ kg, and $m_C = 10$ kg. Which block strikes the ground first?

12.124 Two solid steel spheres, each of radius 100 mm, are placed so that their surfaces are in contact. (*a*) Determine the force of gravitational attraction between the spheres, knowing that the density of steel is 7850 kg/m³. (*b*) If the spheres are moved 2 mm apart and released with zero velocity, determine the approximate time required for their gravitational attraction to bring them back into contact. (*Hint.* Assume that the gravitational forces remain constant.)

12.125 A small ball swings in a horizontal circle at the end of a cord of length l_1, which forms an angle θ_1 with the vertical. The cord is then slowly drawn through the support at *O* until the length of the free end is l_2. (*a*) Derive a relation between l_1, l_2, θ_1, and θ_2. (*b*) If the ball is set in motion so that initially $l_1 = 40$ in. and $\theta_1 = 30°$, determine the angle θ_2 when $l_2 = 30$ in.

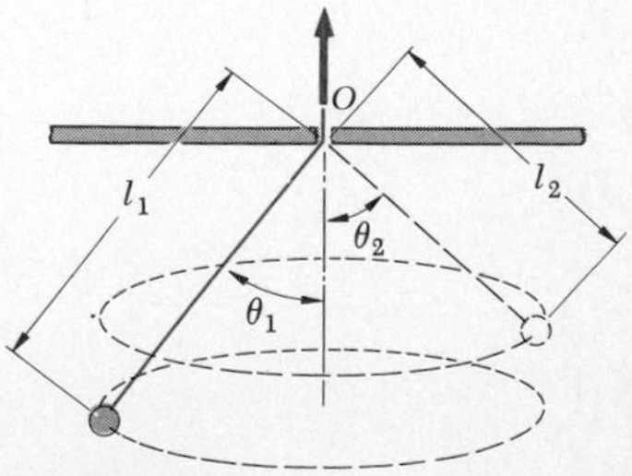

Fig. P12.125

12.126 A spacecraft is describing an elliptic orbit of minimum altitude $h_A = 1040$ mi and maximum altitude $h_B = 6040$ mi above the surface of the earth. (*a*) Determine the speed of the spacecraft at *A*. (*b*) If its engine is fired as the spacecraft passes through point *A* and its speed is increased by 10 percent, determine the maximum altitude reached by the spacecraft on its new orbit.

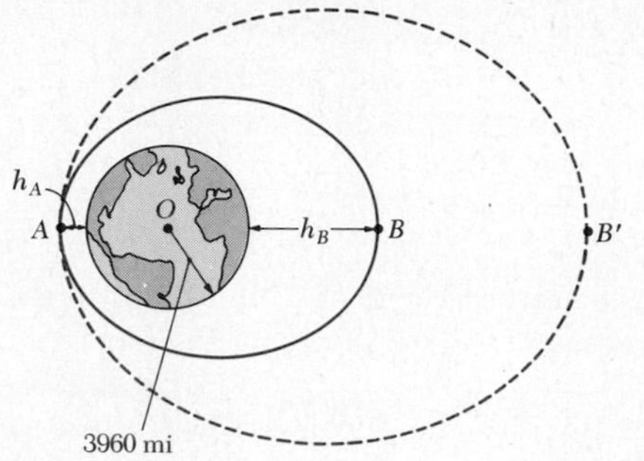

Fig. P12.126

12.127 Determine the approximate time required for an object to fall to the surface of the earth after being released with no velocity from a distance equal to the radius of the orbit of the moon, namely, 384×10^3 km. (*Hint.* Assume that the object is given a very small initial velocity in a transverse direction, say, $v_\theta = 1$ m/s, and determine the periodic time τ of the object on the resulting orbit. An examination of the orbit will show that the time of fall must be approximately equal to $\frac{1}{2}\tau$.)

12.128 Neglecting the effect of friction, determine (*a*) the acceleration of each block, (*b*) the tension in the cable.

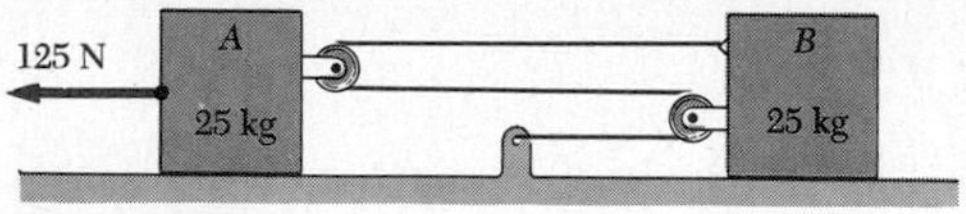

Fig. P12.128

CHAPTER THIRTEEN

Kinetics of Particles: Energy and Momentum Methods

13.1. Introduction. In the preceding chapter, most problems dealing with the motion of particles were solved through the use of the fundamental equation of motion $\mathbf{F} = m\mathbf{a}$. Given a particle acted upon by a force $\mathbf{F}$, we could solve this equation for the acceleration $\mathbf{a}$; then, by applying the principles of kinematics, we could determine from $\mathbf{a}$ the velocity and position of the particle at any time.

If the equation $\mathbf{F} = m\mathbf{a}$ and the principles of kinematics are combined, two additional methods of analysis may be obtained, the *method of work and energy* and the *method of impulse and momentum.* The advantage of these methods lies in the fact that they make the determination of the acceleration unnecessary. Indeed, the method of work and energy relates directly force, mass, velocity, and displacement, while the method of impulse and momentum relates force, mass, velocity, and time.

The method of work and energy will be considered first. In Secs. 13.2 through 13.4, we shall discuss the *work of a force* and the *kinetic energy of a particle* and apply the principle of work and energy to the solution of engineering problems. The concepts of *power* and *efficiency* of a machine will be introduced in Sec. 13.5.

Sections 13.6 through 13.8 are devoted to the concept of *potential energy* of a conservative force and to the application of the principle of conservation of energy to various problems of practical interest. In Sec. 13.9, we shall see how the principles of conservation of energy and of conservation of angular momentum may be used jointly to solve problems of space mechanics.

The second part of the chapter is devoted to the *principle of impulse and momentum* and to its application to the study of the motion of a particle. As we shall see in Sec. 13.11, this principle is particularly effective in the study of the *impulsive motion* of a particle, where very large forces are applied for a very short time interval.

In Secs. 13.12 through 13.14, we shall consider the *central impact* of two bodies. It will be shown that a certain relation exists between the relative velocities of the two colliding bodies before and after impact. This relation may be used together with the fact that the total momentum of the two bodies is conserved to solve a number of problems of practical interest.

Finally, in Sec. 13.15, we shall learn to select from the three fundamental methods presented in Chaps. 12 and 13 the method best suited for the solution of a given problem. We shall also see how the principle of conservation of energy and the method of impulse and momentum may be combined to solve problems involving only conservative forces, except for a short impact phase during which impulsive forces must also be taken into consideration.

13.2. Work of a Force. We shall first define the terms *displacement* and *work* as they are used in mechanics.† Consider a particle which moves from a point A to a neighboring point A' (Fig. 13.1). If $\mathbf{r}$ denotes the position vector corresponding to point A, the small vector joining A and A' may be denoted by the differential $d\mathbf{r}$; the vector $d\mathbf{r}$ is called the *displacement* of the particle. Now, let us assume that a force $\mathbf{F}$ is acting on the particle. The *work of the force* $\mathbf{F}$ *corresponding to the displacement* $d\mathbf{r}$ is defined as the quantity

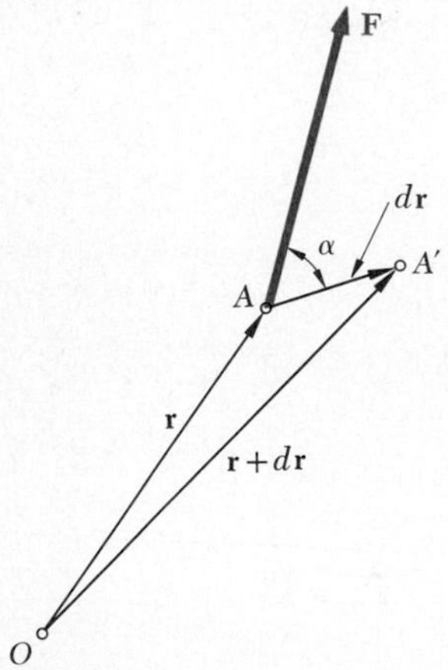

Fig. 13.1

$$dU = \mathbf{F} \cdot d\mathbf{r} \tag{13.1}$$

obtained by forming the scalar product of the force $\mathbf{F}$ and of the displacement $d\mathbf{r}$. Denoting, respectively, by F and ds the magnitudes of the force and of the displacement, and by α the angle formed by $\mathbf{F}$ and $d\mathbf{r}$, and recalling the definition of the scalar product of two vectors (Sec. 3.9), we write

$$dU = F\, ds \cos \alpha \tag{13.1$'$}$$

Using formula (3.30), we may also express the work dU in terms of the rectangular components of the force and of the displacement:

$$dU = F_x\, dx + F_y\, dy + F_z\, dz \tag{13.1$''$}$$

Being a *scalar quantity*, work has a magnitude and a sign but no direction. We also note that work should be expressed in units obtained by multiplying units of length by units of force. Thus, if U.S. customary units are used, work should be expressed in ft · lb or in · lb. If SI units are used, work should

† The definition of work was given in Sec. 10.2, and the basic properties of the work of a force were outlined in Secs. 10.2 and 10.6. For convenience, we repeat here the portions of this material which relate to the kinetics of particles.

be expressed in N·m. The unit of work N·m is called a *joule* (J).† Recalling the conversion factors indicated in Sec. 12.4, we write

$$1 \text{ ft}\cdot\text{lb} = (1 \text{ ft})(1 \text{ lb}) = (0.3048 \text{ m})(4.448 \text{ N}) = 1.356 \text{ J}$$

It follows from (13.1′) that the work dU is positive if the angle α is acute and negative if α is obtuse. Three particular cases are of special interest. If the force **F** has the same direction as $d\mathbf{r}$, the work dU reduces to $F\,ds$. If **F** has a direction opposite to that of $d\mathbf{r}$, the work is $dU = -F\,ds$. Finally, if **F** is perpendicular to $d\mathbf{r}$, the work dU is zero.

The work of **F** during a *finite* displacement of the particle from A_1 to A_2 (Fig. 13.2*a*) is obtained by integrating Eq. (13.1) along the path described by the particle. This work, denoted by $U_{1\to2}$, is

$$U_{1\to2} = \int_{A_1}^{A_2} \mathbf{F}\cdot d\mathbf{r} \tag{13.2}$$

Using the alternate expression (13.1′) for the elementary work dU, and observing that $F\cos\alpha$ represents the tangential component F_t of the force, we may also express the work $U_{1\to2}$ as

$$U_{1\to2} = \int_{s_1}^{s_2} (F\cos\alpha)\,ds = \int_{s_1}^{s_2} F_t\,ds \tag{13.2′}$$

where the variable of integration s measures the distance traveled by the particle along the path. The work $U_{1\to2}$ is represented by the area under the curve obtained by plotting $F_t = F\cos\alpha$ against s (Fig. 13.2*b*).

When the force **F** is defined by its rectangular components, the expression (13.1″) may be used for the elementary work. We write then

$$U_{1\to2} = \int_{A_1}^{A_2} (F_x\,dx + F_y\,dy + F_z\,dz) \tag{13.2″}$$

where the integration is to be performed along the path described by the particle.

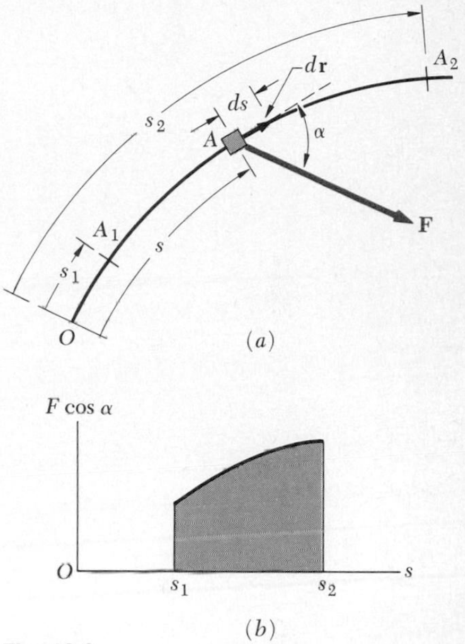

Fig. 13.2

Work of a Constant Force in Rectilinear Motion. When a particle moving in a straight line is acted upon by a force **F** of constant magnitude and of constant direction (Fig. 13.3), formula (13.2′) yields

$$U_{1\to2} = (F\cos\alpha)\,\Delta x \tag{13.3}$$

where α = angle the force forms with direction of motion
Δx = displacement from A_1 to A_2

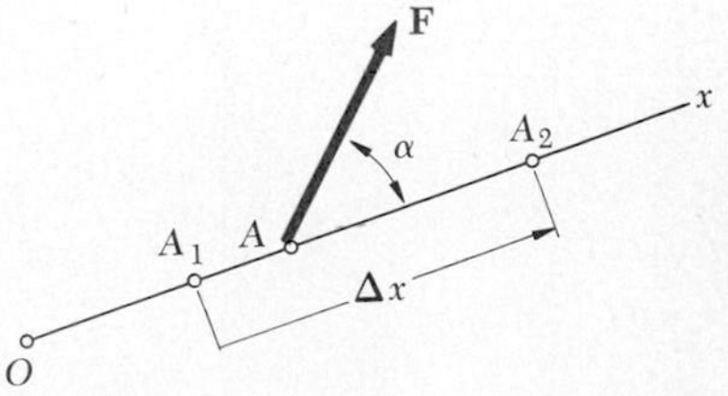

Fig. 13.3

† The joule (J) is the SI unit of *energy,* whether in mechanical form (work, potential energy, kinetic energy) or in chemical, electrical, or thermal form. We should note that even though N·m = J, the moment of a force must be expressed in N·m and not in joules, since the moment of a force is not a form of energy.

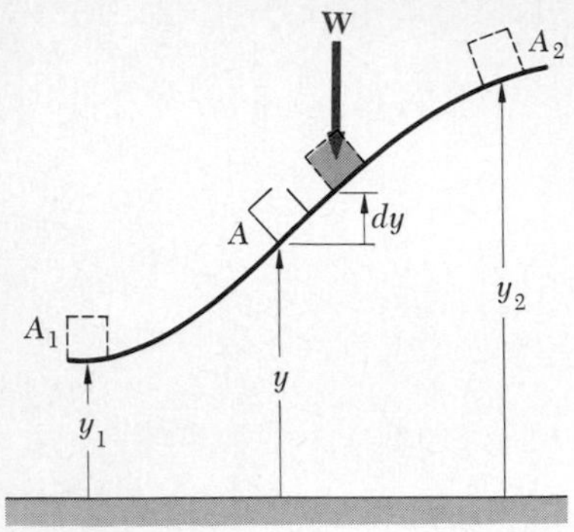

Fig. 13.4

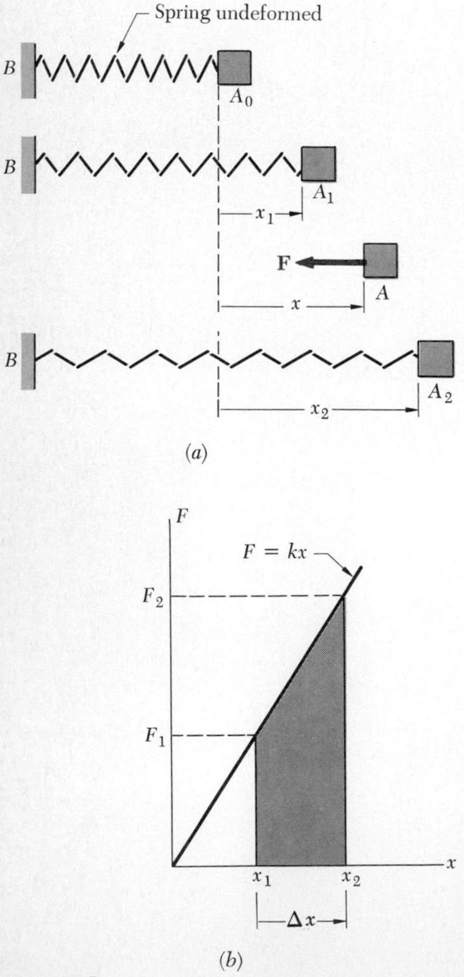

Fig. 13.5

Work of the Force of Gravity. The work of the weight **W** of a body, i.e., of the force of gravity exerted on that body, is obtained by substituting the components of **W** into (13.1″) and (13.2″). With the y axis chosen upward (Fig. 13.4), we have $F_x = 0$, $F_y = -W$, $F_z = 0$, and we write

$$dU = -W\,dy$$

$$U_{1\to 2} = -\int_{y_1}^{y_2} W\,dy = Wy_1 - Wy_2 \tag{13.4}$$

or

$$U_{1\to 2} = -W(y_2 - y_1) = -W\,\Delta y \tag{13.4'}$$

where Δy is the vertical displacement from A_1 to A_2. The work of the weight **W** is thus equal to *the product of W and of the vertical displacement of the center of gravity of the body*. The work is *positive* when $\Delta y < 0$, that is, *when the body moves down.*

Work of the Force Exerted by a Spring. Consider a body A attached to a fixed point B by a spring; it is assumed that the spring is undeformed when the body is at A_0 (Fig. 13.5*a*). Experimental evidence shows that the magnitude of the force **F** exerted by the spring on body A is proportional to the deflection x of the spring measured from the position A_0. We have

$$F = kx \tag{13.5}$$

where k is the *spring constant,* expressed in N/m or kN/m if SI units are used and in lb/ft or lb/in. if U.S. customary units are used.†

The work of the force **F** exerted by the spring during a finite displacement of the body from $A_1(x = x_1)$ to $A_2(x = x_2)$ is obtained by writing

$$dU = -F\,dx = -kx\,dx$$

$$U_{1\to 2} = -\int_{x_1}^{x_2} kx\,dx = \tfrac{1}{2}kx_1^2 - \tfrac{1}{2}kx_2^2 \tag{13.6}$$

Care should be taken to express k and x in consistent units. For example, if U.S. customary units are used, k should be expressed in lb/ft and x in feet, or k in lb/in. and x in inches; in the first case, the work is obtained in ft·lb, in the second case, in in·lb. We note that the work of the force **F** exerted by the spring on the body is *positive* when $x_2 < x_1$, that is, *when the spring is returning to its undeformed position.*

Since Eq.(13.5) is the equation of a straight line of slope k passing through the origin, the work $U_{1\to 2}$ of **F** during the displacement from A_1 to

†The relation $F = kx$ is correct under static conditions only. Under dynamic conditions, formula (13.5) should be modified to take the inertia of the spring into account. However, the error introduced by using the relation $F = kx$ in the solution of kinetics problems is small if the mass of the spring is small compared with the other masses in motion.

A_2 may be obtained by evaluating the area of the trapezoid shown in Fig. 13.5*b*. This is done by computing F_1 and F_2 and multiplying the base Δx of the trapezoid by its mean height $\frac{1}{2}(F_1 + F_2)$. Since the work of the force **F** exerted by the spring is positive for a negative value of Δx, we write

$$U_{1\to 2} = -\tfrac{1}{2}(F_1 + F_2)\,\Delta x \qquad (13.6')$$

Formula (13.6′) is usually more convenient to use than (13.6) and affords fewer chances of confusing the units involved.

Work of a Gravitational Force. We saw in Sec. 12.10 that two particles at distance r from each other and, respectively, of mass M and m, attract each other with equal and opposite forces **F** and −**F** directed along the line joining the particles, and of magnitude

$$F = G\frac{Mm}{r^2}$$

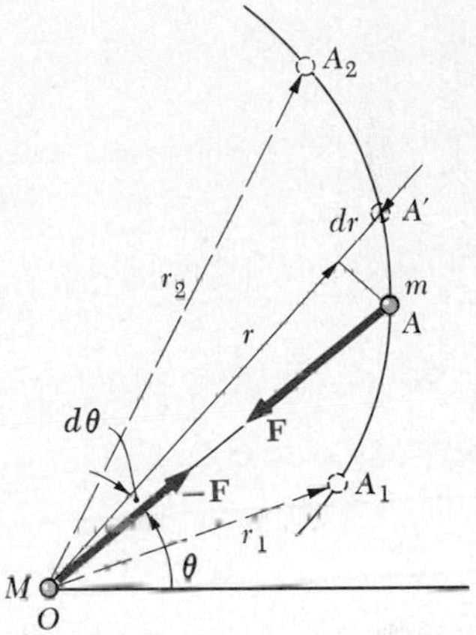

Fig. 13.6

Let us assume that the particle M occupies a fixed position O while the particle m moves along the path shown in Fig. 13.6. The work of the force **F** exerted on the particle m during an infinitesimal displacement of the particle from A to A' may be obtained by multiplying the magnitude F of the force by the radial component dr of the displacement. Since **F** is directed toward O, the work is negative and we write

$$dU = -F\,dr = -G\frac{Mm}{r^2}\,dr$$

The work of the gravitational force **F** during a finite displacement from $A_1(r = r_1)$ to $A_2(r = r_2)$ is therefore

$$U_{1\to 2} = -\int_{r_1}^{r_2} \frac{GMm}{r^2}\,dr = \frac{GMm}{r_2} - \frac{GMm}{r_1} \qquad (13.7)$$

The formula obtained may be used to determine the work of the force exerted by the earth on a body of mass m at a distance r from the center of the earth, when r is larger than the radius R of the earth. The letter M represents then the mass of the earth; recalling the first of the relations (12.29), we may thus replace the product GMm in Eq. (13.7) by WR^2, where R is the radius of the earth ($R = 6.37 \times 10^6$ m or 3960 mi) and W the value of the weight of the body at the surface of the earth.

A number of forces frequently encountered in problems of kinetics *do no work*. They are forces applied to fixed points ($ds = 0$) or acting in a direction perpendicular to the displacement ($\cos\alpha = 0$). Among the forces which do no work are the following: the reaction at a frictionless pin when the body supported rotates about the pin, the reaction at a frictionless surface when the body in contact moves along the surface, the reaction at a roller moving along its track, and the weight of a body when its center of gravity moves horizontally.

13.3. Kinetic Energy of a Particle. Principle of Work and Energy. Consider a particle of mass m acted upon by a force $\mathbf{F}$ and moving along a path which is either rectilinear or curved (Fig. 13.7). Expressing Newton's second law in terms of the tangential components of the force and of the acceleration (see Sec. 12.5), we write

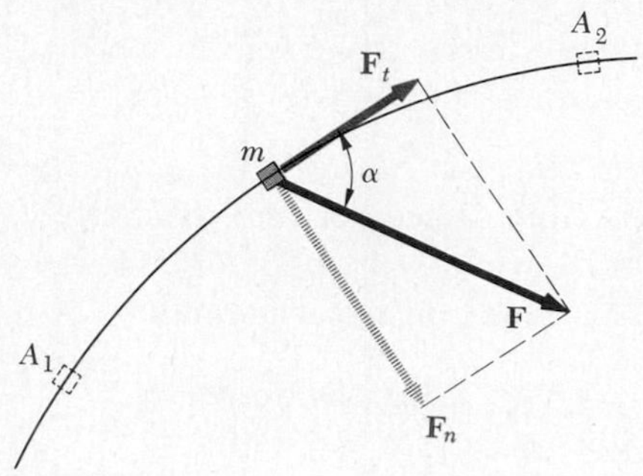

Fig. 13.7

$$F_t = ma_t \qquad \text{or} \qquad F_t = m\frac{dv}{dt}$$

where v is the speed of the particle. Recalling from Sec. 11.9 that $v = ds/dt$, we obtain

$$F_t = m\frac{dv}{ds}\frac{ds}{dt} = mv\frac{dv}{ds}$$

$$F_t\,ds = mv\,dv$$

Integrating from A_1, where $s = s_1$ and $v = v_1$, to A_2, where $s = s_2$ and $v = v_2$, we write

$$\int_{s_1}^{s_2} F_t\,ds = m\int_{v_1}^{v_2} v\,dv = \tfrac{1}{2}mv_2^2 - \tfrac{1}{2}mv_1^2 \qquad (13.8)$$

The left-hand member of Eq. (13.8) represents the work $U_{1\to2}$ of the force $\mathbf{F}$ exerted on the particle during the displacement from A_1 to A_2; as indicated in Sec. 13.2, the work $U_{1\to2}$ is a scalar quantity. The expression $\frac{1}{2}mv^2$ is also a scalar quantity; it is defined as the kinetic energy of the particle and is denoted by T. We write

$$T = \tfrac{1}{2}mv^2 \qquad (13.9)$$

Substituting into (13.8), we have

$$U_{1\to2} = T_2 - T_1 \tag{13.10}$$

which expresses that, when a particle moves from A_1 to A_2 under the action of a force **F**, *the work of the force* **F** *is equal to the change in kinetic energy of the particle.* This is known as the *principle of work and energy.* Rearranging the terms in (13.10), we write

$$T_1 + U_{1\to2} = T_2 \tag{13.11}$$

Thus, *the kinetic energy of the particle at* A_2 *may be obtained by adding to its kinetic energy at* A_1 *the work done during the displacement from* A_1 *to* A_2 *by the force* **F** *exerted on the particle.* As Newton's second law from which it is derived, the principle of work and energy applies only with respect to a newtonian frame of reference (Sec. 12.2). The speed v used to determine the kinetic energy T should therefore be measured with respect to a newtonian frame of reference.

Since both work and kinetic energy are scalar quantities, their sum may be computed as an ordinary algebraic sum, the work $U_{1\to2}$ being considered as positive or negative according to the direction of **F**. When several forces act on the particle, the expression $U_{1\to2}$ represents the total work of the forces acting on the particle; it is obtained by adding algebraically the work of the various forces.

As noted above, the kinetic energy of a particle is a scalar quantity. It further appears from the definition $T = \frac{1}{2}mv^2$ that the kinetic energy is always positive, regardless of the direction of motion of the particle. Considering the particular case when $v_1 = 0$, $v_2 = v$, and substituting $T_1 = 0$, $T_2 = T$ into (13.10), we observe that the work done by the forces acting on the particle is equal to T. Thus, the kinetic energy of a particle moving with a speed v represents the work which must be done to bring the particle from rest to the speed v. Substituting $T_1 = T$ and $T_2 = 0$ into (13.10), we also note that when a particle moving with a speed v is brought to rest, the work done by the forces acting on the particle is $-T$. Assuming that no energy is dissipated into heat, we conclude that the work done by the forces exerted *by the particle* on the bodies which cause it to come to rest is equal to T. Thus, the kinetic energy of a particle also represents *the capacity to do work associated with the speed of the particle.*

The kinetic energy is measured in the same units as work, i.e., in joules if SI units are used, and in ft·lb if U.S. customary units are used. We check that, in SI units,

$$T = \tfrac{1}{2}mv^2 = \text{kg}(\text{m/s})^2 = (\text{kg}\cdot\text{m/s}^2)\text{m} = \text{N}\cdot\text{m} = \text{J}$$

while, in customary units,

$$T = \tfrac{1}{2}mv^2 = (\text{lb}\cdot\text{s}^2/\text{ft})(\text{ft/s})^2 = \text{lb}\cdot\text{ft}$$

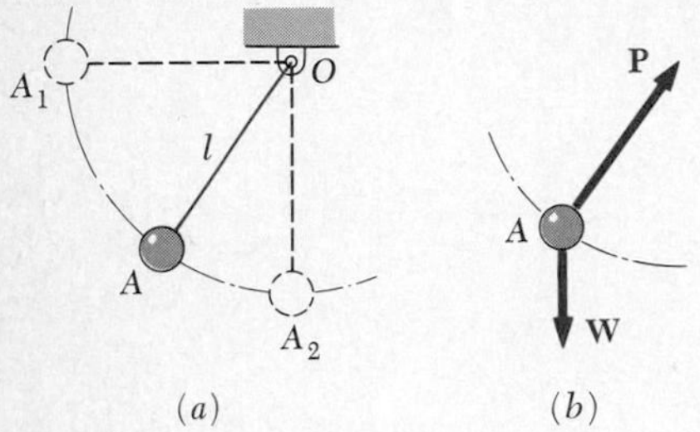

Fig. 13.8

13.4. Applications of the Principle of Work and Energy. The application of the principle of work and energy greatly simplifies the solution of many problems involving forces, displacements, and velocities. Consider, for example, the pendulum OA consisting of a bob A of weight W attached to a cord of length l (Fig. 13.8a). The pendulum is released with no initial velocity from a horizontal position OA_1 and allowed to swing in a vertical plane. We wish to determine the speed of the bob as it passes through A_2, directly under O.

We first determine the work done during the displacement from A_1 to A_2 by the forces acting on the bob. We draw a free-body diagram of the bob, showing all the *actual* forces acting on it, i.e., the weight **W** and the force **P** exerted by the cord (Fig. 13.8b). (An inertia vector is not an actual force and *should not* be included in the free-body diagram.) We note that the force **P** does no work, since it is normal to the path; the only force which does work is thus the weight **W**. The work of **W** is obtained by multiplying its magnitude W by the vertical displacement l (Sec. 13.2); since the displacement is downward, the work is positive. We therefore write $U_{1\rightarrow 2} = Wl$.

Considering, now, the kinetic energy of the bob, we find $T_1 = 0$ at A_1 and $T_2 = \frac{1}{2}(W/g)v_2^2$ at A_2. We may now apply the principle of work and energy; recalling formula (13.11), we write

$$T_1 + U_{1\rightarrow 2} = T_2 \qquad 0 + Wl = \frac{1}{2}\frac{W}{g}v_2^2$$

Solving for v_2, we find $v_2 = \sqrt{2gl}$. We note that the speed obtained is that of a body falling freely from a height l.

The example we have considered illustrates the following advantages of the method of work and energy:

1. In order to find the speed at A_2, there is no need to determine the acceleration in an intermediate position A and to integrate the expression obtained from A_1 to A_2.
2. All quantities involved are scalars and may be added directly, without using x and y components.
3. Forces which do no work are eliminated from the solution of the problem.

What is an advantage in one problem, however, may become a disadvantage in another. It is evident, for instance, that the method of work and energy cannot be used to directly determine an acceleration. We also note that it should be supplemented by the direct application of Newton's second law in order to determine a force which is normal to the path of the particle, since such a force does no work. Suppose, for example, that we wish to determine the tension in the cord of the pendulum of Fig. 13.8a as the bob passes through A_2. We draw a free-body diagram of the bob in that position (Fig. 13.9) and express Newton's second law in terms of tangential and normal components. The equations $\Sigma F_t = ma_t$ and $\Sigma F_n = ma_n$ yield, respectively, $a_t = 0$ and

$$P - W = ma_n = \frac{W}{g}\frac{v_2^2}{l}$$

But the speed at A_2 was determined earlier by the method of work and energy. Substituting $v_2^2 = 2gl$ and solving for P, we write

$$P = W + \frac{W}{g}\frac{2gl}{l} = 3W$$

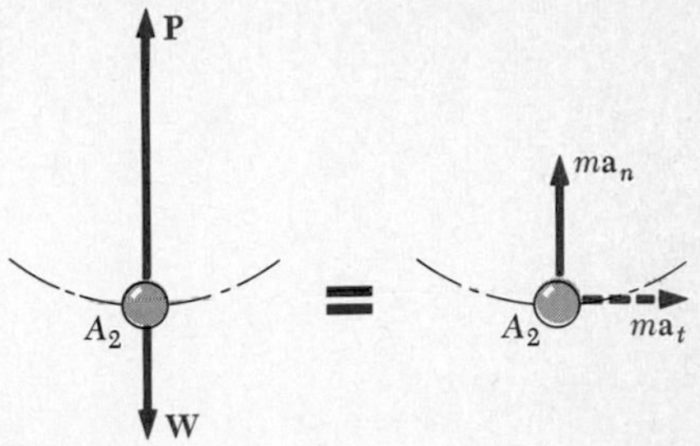

Fig. 13.9

When a problem involves two particles or more, each particle may be considered separately and the principle of work and energy may be applied to each particle. Adding the kinetic energies of the various particles, and considering the work of all the forces acting on them, we may also write a single equation of work and energy for all the particles involved. We have

$$T_1 + U_{1\to2} = T_2 \tag{13.11}$$

where T represents the arithmetic sum of the kinetic energies of the particles involved (all terms are positive) and $U_{1\to2}$ the work of all the forces acting on the particles, *including the forces of action and reaction exerted by the particles on each other.* In problems involving bodies connected by *inextensible cords or links,* however, the work of the forces exerted by a given cord or link on the two bodies it connects cancels out, since the points of application of these forces move through equal distances (see Sample Prob. 13.2).†

Since friction forces have a direction opposite to that of the displacement of the body on which they act, *the work of friction forces is always negative.* This work represents energy dissipated into heat and always results in a decrease in the kinetic energy of the body involved (see Sample Prob. 13.3).

13.5. Power and Efficiency. *Power* is defined as the time rate at which work is done. In the selection of a motor or engine, power is a much more important criterion than is the actual amount of work to be performed. A small motor or a large power plant may both be used to do a given amount of work; but the small motor may require a month to do the work done by the power plant in a matter of minutes. If ΔU is the work done during the time interval Δt, then the average power during this time interval is

$$\text{Average power} = \frac{\Delta U}{\Delta t}$$

Letting Δt approach zero, we obtain at the limit

$$\text{Power} = \frac{dU}{dt} \tag{13.12}$$

† The application of the method of work and energy to a system of particles is discussed in detail in Chap. 14.

Substituting the scalar product $\mathbf{F} \cdot d\mathbf{r}$ for dU, we may also write

$$\text{Power} = \frac{dU}{dt} = \frac{\mathbf{F} \cdot d\mathbf{r}}{dt}$$

and, recalling that $d\mathbf{r}/dt$ represents the velocity $\mathbf{v}$ of the point of application of $\mathbf{F}$,

$$\text{Power} = \mathbf{F} \cdot \mathbf{v} \tag{13.13}$$

Since power was defined as the time rate at which work is done, it should be expressed in units obtained by dividing units of work by the unit of time. Thus, if SI units are used, power should be expressed in J/s; this unit is called a *watt* (W). We have

$$1\text{ W} = 1\text{ J/s} = 1\text{ N}\cdot\text{m/s}$$

If U.S. customary units are used, power should be expressed in ft·lb/s or in *horsepower* (hp), with the latter defined as

$$1\text{ hp} = 550\text{ ft}\cdot\text{lb/s}$$

Recalling from Sec. 13.2 that 1 ft·lb = 1.356 J, we verify that

$$1\text{ ft}\cdot\text{lb/s} = 1.356\text{ J/s} = 1.356\text{ W}$$
$$1\text{ hp} = 550(1.356\text{ W}) = 746\text{ W} = 0.746\text{ kW}$$

The *mechanical efficiency* of a machine was defined in Sec. 10.5 as the ratio of the output work to the input work:

$$\eta = \frac{\text{output work}}{\text{input work}} \tag{13.14}$$

This definition is based on the assumption that work is done at a constant rate. The ratio of the output to the input work is therefore equal to the ratio of the rates at which output and input work are done, and we have

$$\eta = \frac{\text{power output}}{\text{power input}} \tag{13.15}$$

Because of energy losses due to friction, the output work is always smaller than the input work, and, consequently, the power output is always smaller than the power input. The mechanical efficiency of a machine is therefore always less than 1.

When a machine is used to transform mechanical energy into electric energy, or thermal energy into mechanical energy, its *overall efficiency* may be obtained from formula (13.15). The overall efficiency of a machine is always less than 1; it provides a measure of all the various energy losses involved (losses of electric or thermal energy as well as frictional losses). We should note that it is necessary, before using formula (13.15), to express the power output and the power input in the same units.

SAMPLE PROBLEM 13.1

An automobile weighing 4000 lb is driven down a 5° incline at a speed of 60 mi/h when the brakes are applied, causing a constant total braking force (applied by the road on the tires) of 1500 lb. Determine the distance traveled by the automobile as it comes to a stop.

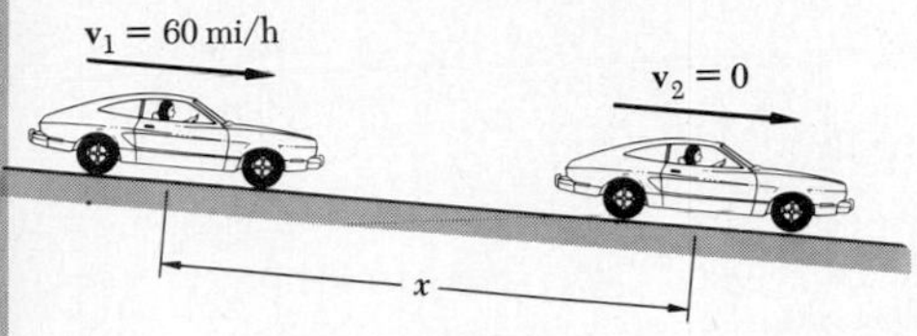

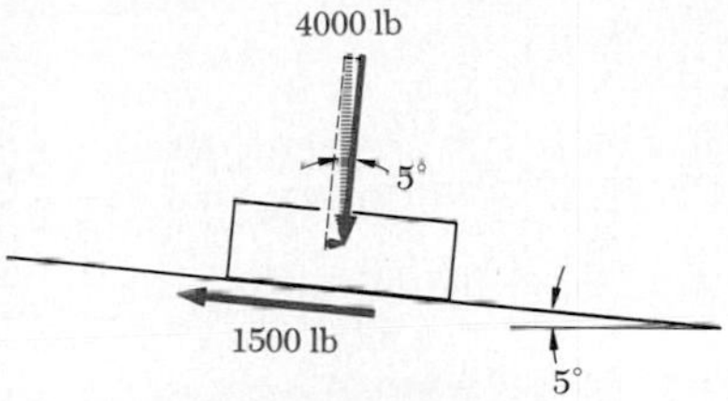

Solution. ***Kinetic Energy***

Position 1:
$$v_1 = \left(60\frac{\text{mi}}{\text{h}}\right)\left(\frac{5280\text{ ft}}{1\text{ mi}}\right)\left(\frac{1\text{ h}}{3600\text{ s}}\right) = 88\text{ ft/s}$$
$$T_1 = \tfrac{1}{2}mv_1^2 = \tfrac{1}{2}(4000/32.2)(88)^2 = 481{,}000\text{ ft}\cdot\text{lb}$$

Position 2:
$$v_2 = 0 \qquad T_2 = 0$$

Work
$$U_{1\to2} = -1500x + (4000\sin 5°)x = -1151x$$

Principle of Work and Energy
$$T_1 + U_{1\to2} = T_2$$
$$481{,}000 - 1151x = 0 \qquad x = 418\text{ ft} \blacktriangleleft$$

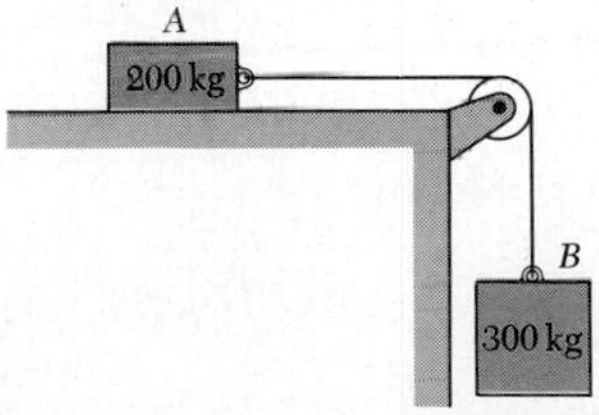

SAMPLE PROBLEM 13.2

Two blocks are joined by an inextensible cable as shown. If the system is released from rest, determine the velocity of block A after it has moved 2 m. Assume that the coefficient of kinetic friction between block A and the plane is $\mu_k = 0.25$ and that the pulley is weightless and frictionless.

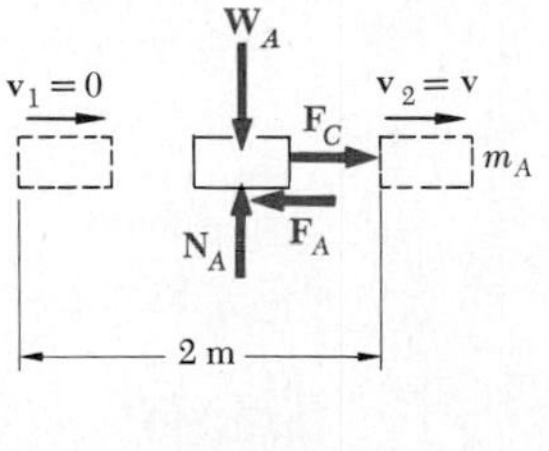

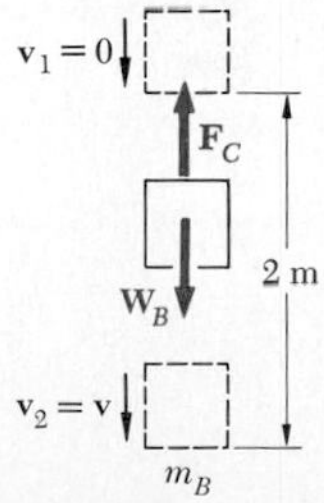

Solution. ***Work and Energy for Block A.*** We denote by $\mathbf{F}_A$ the friction force, by $\mathbf{F}_C$ the force exerted by the cable, and write

$$m_A = 200\text{ kg} \qquad W_A = (200\text{ kg})(9.81\text{ m/s}^2) = 1962\text{ N}$$
$$F_A = \mu_k N_A = \mu_k W_A = 0.25(1962\text{ N}) = 490\text{ N}$$
$$T_1 + U_{1\to2} = T_2: \quad 0 + F_C(2\text{ m}) - F_A(2\text{ m}) = \tfrac{1}{2}m_A v^2$$
$$F_C(2\text{ m}) - (490\text{ N})(2\text{ m}) = \tfrac{1}{2}(200\text{ kg})v^2 \qquad (1)$$

Work and Energy for Block B. We write

$$m_B = 300\text{ kg} \qquad W_B = (300\text{ kg})(9.81\text{ m/s}^2) = 2940\text{ N}$$
$$T_1 + U_{1\to2} = T_2: \quad 0 + W_B(2\text{ m}) - F_C(2\text{ m}) = \tfrac{1}{2}m_B v^2$$
$$(2940\text{ N})(2\text{ m}) - F_C(2\text{ m}) = \tfrac{1}{2}(300\text{ kg})v^2 \qquad (2)$$

Adding the left-hand and right-hand members of (1) and (2), we observe that the work of the forces exerted by the cable on A and B cancels out:

$$(2940\text{ N})(2\text{ m}) - (490\text{ N})(2\text{ m}) = \tfrac{1}{2}(200\text{ kg} + 300\text{ kg})v^2$$
$$4900\text{ J} = \tfrac{1}{2}(500\text{ kg})v^2 \qquad v = 4.43\text{ m/s} \blacktriangleleft$$

SAMPLE PROBLEM 13.3

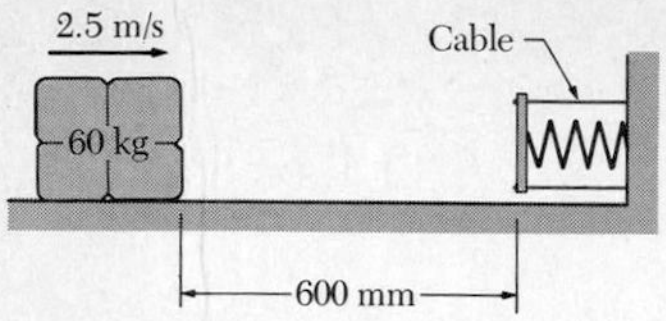

A spring is used to stop a 60-kg package which is sliding on a horizontal surface. The spring has a constant $k = 20$ kN/m and is held by cables so that it is initially compressed 120 mm. Knowing that the package has a velocity of 2.5 m/s in the position shown and that the maximum additional deflection of the spring is 40 mm, determine (*a*) the coefficient of kinetic friction between the package and the surface, (*b*) the velocity of the package as it passes again through the position shown.

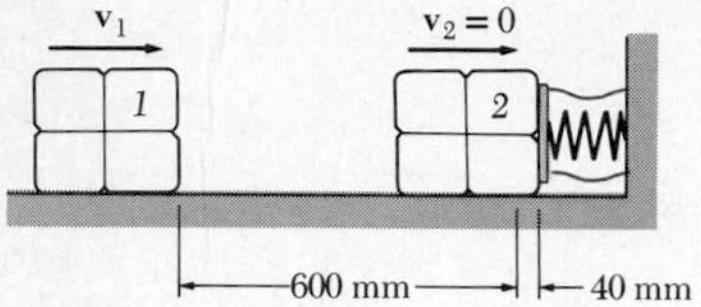

a.* Motion from Position *1* to Position *2

Kinetic Energy. *Position 1:* $\quad v_1 = 2.5$ m/s

$$T_1 = \tfrac{1}{2}mv_1^2 = \tfrac{1}{2}(60 \text{ kg})(2.5 \text{ m/s})^2 = 187.5 \text{ N}\cdot\text{m} = 187.5 \text{ J}$$

Position 2 (maximum spring deflection): $\quad v_2 = 0 \qquad T_2 = 0$

Work

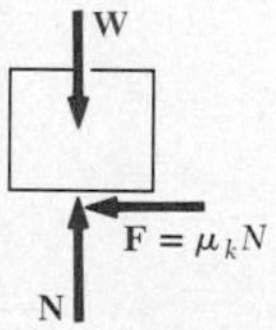

Friction Force **F**. We have

$$F = \mu_k N = \mu_k W = \mu_k mg = \mu_k(60 \text{ kg})(9.81 \text{ m/s}^2) = 588.6\,\mu_k$$

The work of **F** is negative and equal to

$$(U_{1\to 2})_f = -Fx = -(588.6\,\mu_k)(0.600 \text{ m} + 0.040 \text{ m}) = -377\,\mu_k$$

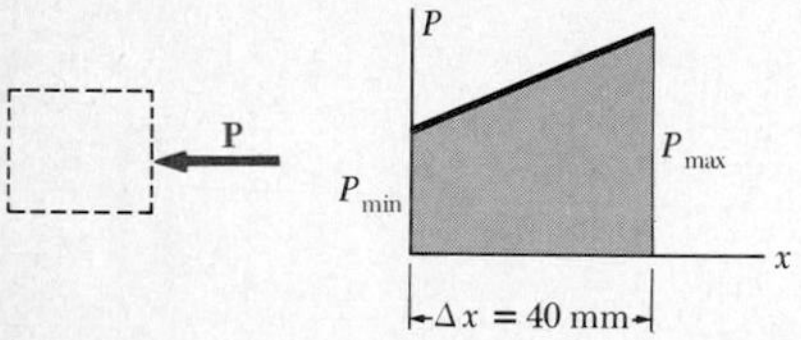

Spring Force **P**. The variable force **P** exerted by the spring does an amount of negative work equal to the area under the force-deflection curve of the spring force. We have

$$P_{\text{min}} = kx_0 = (20 \text{ kN/m})(120 \text{ m}) = (20\,000 \text{ N/m})(0.120 \text{ m}) = 2400 \text{ N}$$
$$P_{\text{max}} = P_{\text{min}} + k\,\Delta x = 2400 \text{ N} + (20 \text{ kN/m})(40 \text{ mm}) = 3200 \text{ N}$$
$$(U_{1\to 2})_e = -\tfrac{1}{2}(P_{\text{min}} + P_{\text{max}})\,\Delta x = -\tfrac{1}{2}(2400 \text{ N} + 3200 \text{ N})(0.040 \text{ m}) = -112.0 \text{ J}$$

The total work is thus

$$U_{1\to 2} = (U_{1\to 2})_f + (U_{1\to 2})_e = -377\,\mu_k - 112.0 \text{ J}$$

Principle of Work and Energy

$$T_1 + U_{1\to 2} = T_2: \qquad 187.5 \text{ J} - 377\,\mu_k - 112.0 \text{ J} = 0 \qquad\qquad \mu_k = 0.20 \blacktriangleleft$$

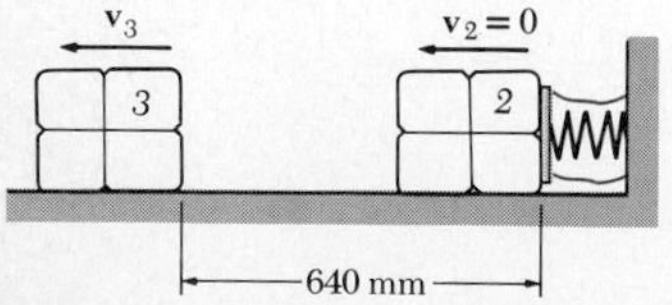

b.* Motion from Position *2* to Position *3

Kinetic Energy. *Position 2:* $\quad v_2 = 0 \qquad T_2 = 0$

Position 3: $\qquad T_3 = \tfrac{1}{2}mv_3^2 = \tfrac{1}{2}(60 \text{ kg})v_3^2$

Work. Since the distances involved are the same, the numerical values of the work of the friction force **F** and of the spring force **P** are the same as above. However, while the work of **F** is still negative, the work of **P** is now positive.

$$U_{2\to 3} = -377\mu_k + 112.0 \text{ J} = -75.5 \text{ J} + 112.0 \text{ J} = +36.5 \text{ J}$$

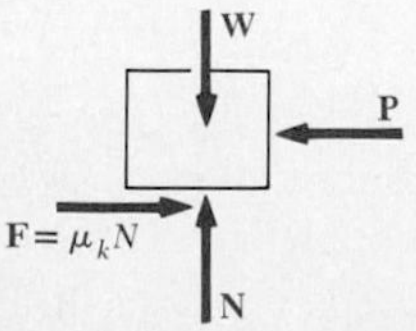

Principle of Work and Energy

$$T_2 + U_{2\to 3} = T_3: \qquad 0 + 36.5 \text{ J} = \tfrac{1}{2}(60 \text{ kg})v_3^2$$
$$v_3 = 1.103 \text{ m/s} \qquad\qquad \mathbf{v}_3 = 1.103 \text{ m/s} \leftarrow \blacktriangleleft$$

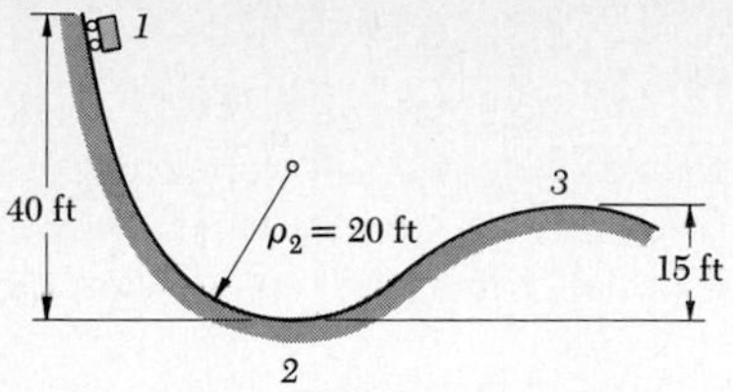

SAMPLE PROBLEM 13.4

A 2000-lb car starts from rest at point *1* and moves without friction down the track shown. (*a*) Determine the force exerted by the track on the car at point 2, where the radius of curvature of the track is 20 ft. (*b*) Determine the minimum safe value of the radius of curvature at point *3*.

***a.* Force Exerted by the Track at Point 2.** The principle of work and energy is used to determine the velocity of the car as it passes through point *2*.

Kinetic Energy: $T_1 = 0 \qquad T_2 = \frac{1}{2}mv_2^2 = \frac{1}{2}\frac{W}{g}v_2^2$

Work. The only force which does work is the weight **W**. Since the vertical displacement from point *1* to point *2* is 40 ft downward, the work of the weight is

$$U_{1\to2} = +W(40 \text{ ft})$$

Principle of Work and Energy

$$T_1 + U_{1\to2} = T_2 \qquad 0 + W(40 \text{ ft}) = \frac{1}{2}\frac{W}{g}v_2^2$$

$$v_2^2 = 80g = 80(32.2) \qquad v_2 = 50.8 \text{ ft/s}$$

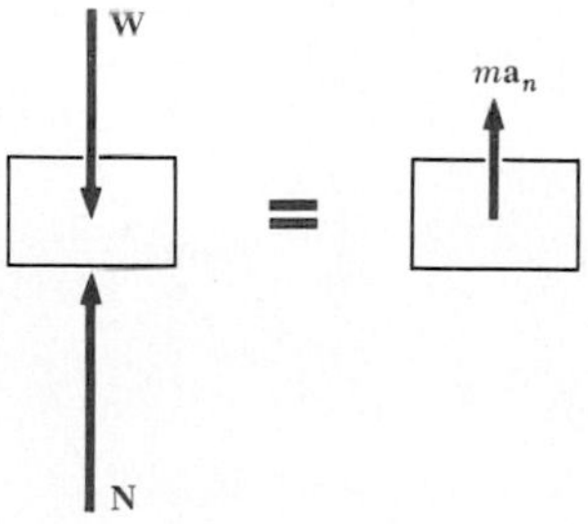

Newton's Second Law at Point 2. The acceleration $\mathbf{a}_n$ of the car at point *2* has a magnitude $a_n = v_2^2/\rho$ and is directed upward. Since the external forces acting on the car are **W** and **N**, we write

$$+\uparrow\Sigma F_n = ma_n: \qquad -W + N = ma_n$$

$$= \frac{W}{g}\frac{v_2^2}{\rho}$$

$$= \frac{W}{g}\frac{80g}{20}$$

$$N = 5W \qquad \mathbf{N} = 10{,}000 \text{ lb} \uparrow \blacktriangleleft$$

b.* Minimum Value of ρ at Point *3*.** ***Principle of Work and Energy. Applying the principle of work and energy between point *1* and point *3*, we obtain

$$T_1 + U_{1\to3} = T_3 \qquad 0 + W(25 \text{ ft}) = \frac{1}{2}\frac{W}{g}v_3^2$$

$$v_3^2 = 50g = 50(32.2) \qquad v_3 = 40.1 \text{ ft/s}$$

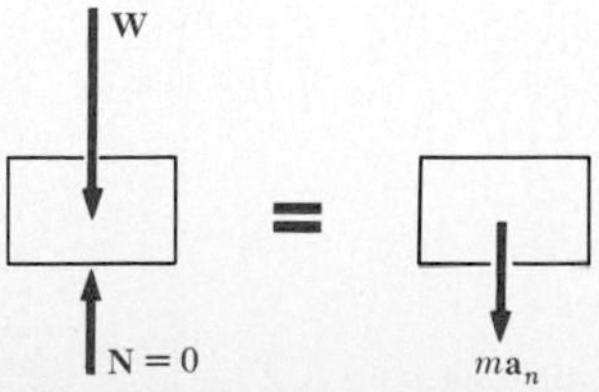

Newton's Second Law at Point 3. The minimum safe value of ρ occurs when $\mathbf{N} = 0$. In this case, the acceleration $\mathbf{a}_n$, of magnitude $a_n = v_3^2/\rho$, is directed downward, and we write

$$+\downarrow\Sigma F_n = ma_n: \qquad W = \frac{W}{g}\frac{v_3^2}{\rho}$$

$$= \frac{W}{g}\frac{50g}{\rho} \qquad \rho = 50 \text{ ft} \blacktriangleleft$$

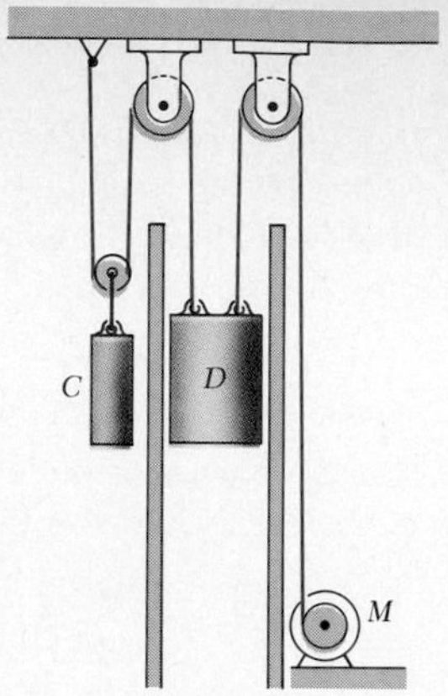

SAMPLE PROBLEM 13.5

The dumb-waiter D and its load have a combined weight of 600 lb, while the counterweight C weighs 800 lb. Determine the power delivered by the electric motor M when the dumb-waiter (*a*) is moving up at a constant speed of 8 ft/s, (*b*) has an instantaneous velocity of 8 ft/s and an acceleration of 2.5 ft/s², both directed upward.

Solution. Since the force **F** exerted by the motor cable has the same direction as the velocity $\mathbf{v}_D$ of the dumb-waiter, the power is equal to Fv_D, where $v_D = 8$ ft/s. To obtain the power, we must first determine **F** in each of the two given situations.

a. Uniform Motion. We have $\mathbf{a}_C = \mathbf{a}_D = 0$; both bodies are in equilibrium.

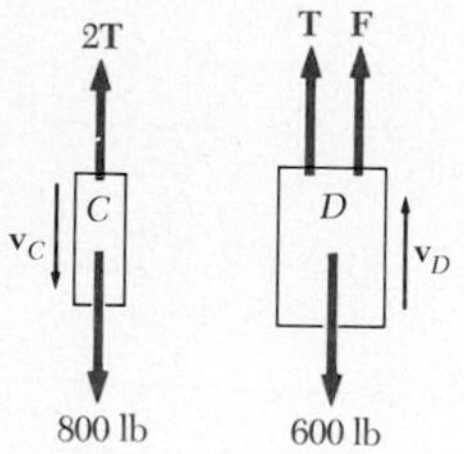

Free Body C: $+\uparrow\Sigma F_y = 0$: $\quad 2T - 800 \text{ lb} = 0 \qquad T = 400 \text{ lb}$

Free Body D: $+\uparrow\Sigma F_y = 0$: $\quad F + T - 600 \text{ lb} = 0$

$$F = 600 \text{ lb} - T = 600 \text{ lb} - 400 \text{ lb} = 200 \text{ lb}$$

$$Fv_D = (200 \text{ lb})(8 \text{ ft/s}) = 1600 \text{ ft}\cdot\text{lb/s}$$

$$\text{Power} = (1600 \text{ ft}\cdot\text{lb/s})\frac{1 \text{ hp}}{550 \text{ ft}\cdot\text{lb/s}} = 2.91 \text{ hp} \quad ◄$$

b. Accelerated Motion. We have

$$\mathbf{a}_D = 2.5 \text{ ft/s}^2\uparrow \qquad \mathbf{a}_C = -\tfrac{1}{2}\mathbf{a}_D = 1.25 \text{ ft/s}^2\downarrow$$

The equations of motion are

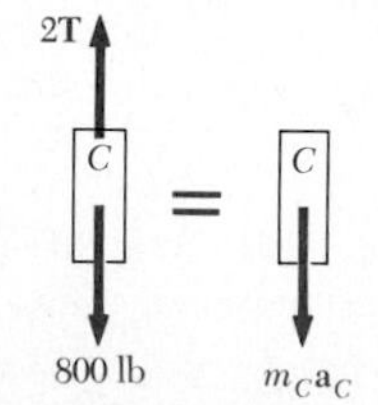

Free Body C: $+\downarrow\Sigma F_y = m_C a_C$: $\quad 800 - 2T = \dfrac{800}{32.2}(1.25) \qquad T = 384.5 \text{ lb}$

Free Body D: $+\uparrow\Sigma F_y = m_D a_D$: $\quad F + T - 600 = \dfrac{600}{32.2}(2.5)$

$$F + 384.5 - 600 = 46.6 \qquad F = 262.1 \text{ lb}$$

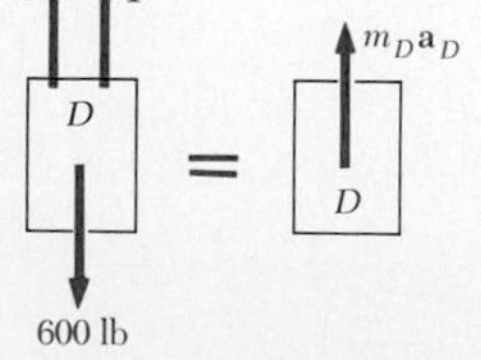

$$Fv_D = (262.1 \text{ lb})(8 \text{ ft/s}) = 2097 \text{ ft}\cdot\text{lb/s}$$

$$\text{Power} = (2097 \text{ ft}\cdot\text{lb/s})\frac{1 \text{ hp}}{550 \text{ ft}\cdot\text{lb/s}} = 3.81 \text{ hp} \quad ◄$$

Problems

13.1 A 400-kg satellite was placed in a circular orbit 1500 km above the surface of the earth. At this elevation the acceleration of gravity is 6.43 m/s^2. Determine the kinetic energy of the satellite, knowing that its orbital speed is 25.6×10^3 km/h.

13.2 A stone which weighs 4 lb is dropped from a height h and strikes the ground with a velocity of 60 ft/s. (*a*) Find the kinetic energy of the stone as it strikes the ground and the height h from which it was dropped. (*b*) Solve part *a*, assuming that the same stone is dropped on the moon. (Acceleration of gravity on the moon = 5.31 ft/s^2.)

13.3 Determine the maximum theoretical speed that may be achieved over a distance of 50 m by a car starting from rest, knowing that the coefficient of static friction is 0.80 between the tires and the pavement and that 60 percent of the weight of the car is distributed over its front wheels and 40 percent over its rear wheels. Assume (*a*) front-wheel drive, (*b*) rear-wheel drive.

13.4 Solve Prob. 13.3, assuming four-wheel drive.

13.5 In an iron-ore mixing operation, a bucket full of ore is suspended from a traveling crane which is moving with a speed v = 3.75 m/s along a stationary bridge. If the crane suddenly stops, determine the additional horizontal distance through which the bucket will move.

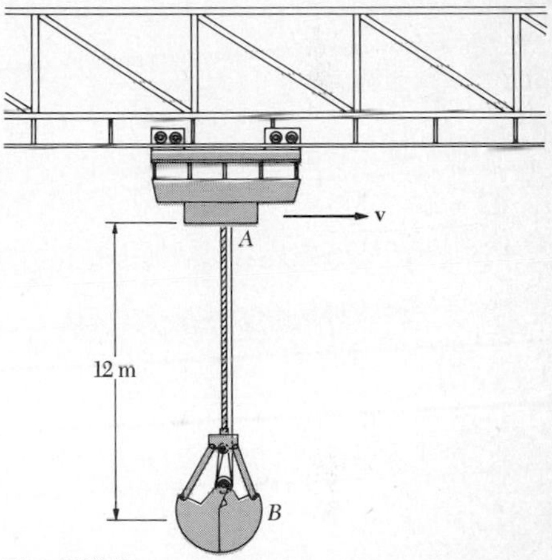

Fig. P13.5

13.6 Determine the speed v with which the crane of Prob. 13.5 was moving if the cables supporting the bucket B swing through an angle of 18° after the crane is brought to a sudden stop.

13.7 A 24-lb package is placed with no initial velocity at the top of a chute. Knowing that the coefficient of kinetic friction between the package and the chute is 0.25, determine (*a*) how far the package will slide on the horizontal portion of the chute, (*b*) the maximum velocity reached by the package, (*c*) the amount of energy dissipated due to friction between A and B.

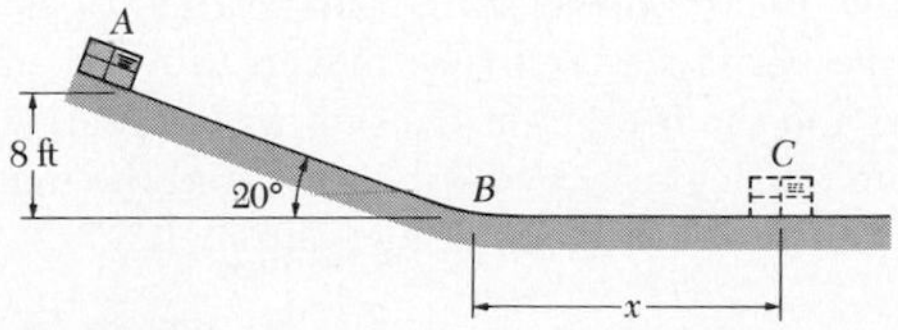

Fig. P13.7

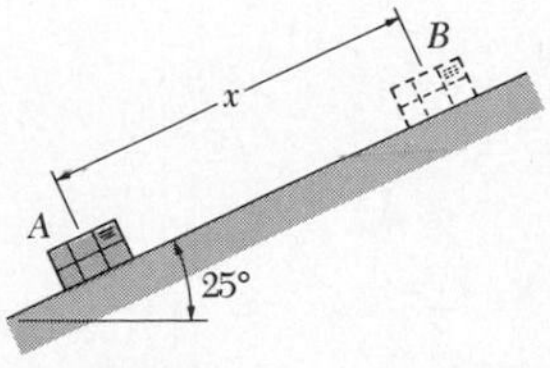

Fig. P13.8

13.8 A 20-lb package is projected up a 25° incline with an initial velocity of 24 ft/s. Knowing that the coefficient of kinetic friction between the package and the incline is 0.20, determine (*a*) the maximum distance x that the package will move up the incline, (*b*) the velocity of the package as it returns to its original position, (*c*) the total amount of energy dissipated due to friction.

13.9 A 1050-kg trailer is hitched to a 1200-kg car. The car and trailer are traveling at 90 km/h when the driver applies the brakes on both the car and the trailer. Knowing that the braking forces exerted on the car and the trailer are 4500 N and 3600 N, respectively, determine (*a*) the distance traveled by the car and trailer before they come to a stop, (*b*) the horizontal component of the force exerted by the trailer hitch on the car.

Fig. P13.9

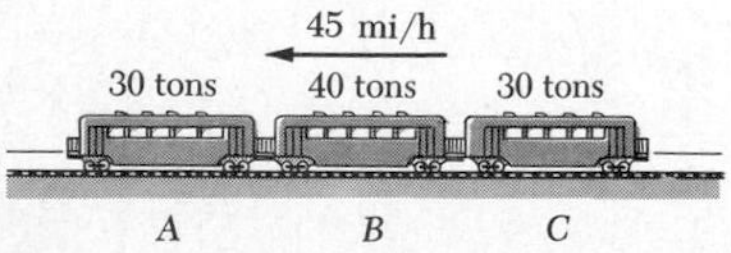

Fig. P13.10

13.10 The subway train shown is traveling at a speed of 45 mi/h when the brakes are fully applied on the wheels of cars *A* and *B*, causing them to slide on the track, but are not applied on the wheels of car *C*. Knowing that the coefficient of kinetic friction is 0.30 between the wheels and the track, determine (*a*) the distance required to bring the train to a stop, (*b*) the force in each coupling.

13.11 Solve Prob. 13.10, assuming that the brakes are applied only on the wheels of car *A*.

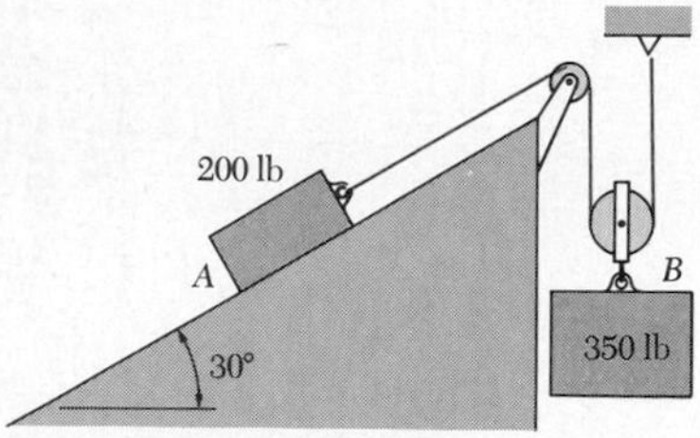

Fig. P13.13 and P13.15

13.12 Solve Prob. 13.9, assuming that the trailer brakes are inoperative.

13.13 The two blocks shown are originally at rest. Neglecting the masses of the pulleys and the effect of friction in the pulleys and between block *A* and the incline, determine (*a*) the velocity of block *A* after it has moved through 5 ft, (*b*) the tension in the cable.

13.14 Solve Prob. 12.18*c*, using the method of work and energy.

13.15 The two blocks shown start from rest and it is observed that the velocity of block *A* is 6.75 ft/s after it has moved through 5 ft. Neglecting the masses of the pulleys and the friction in the pulleys, determine (*a*) the amount of energy dissipated due to friction between block *A* and the incline, (*b*) the coefficient of kinetic friction between block *A* and the incline.

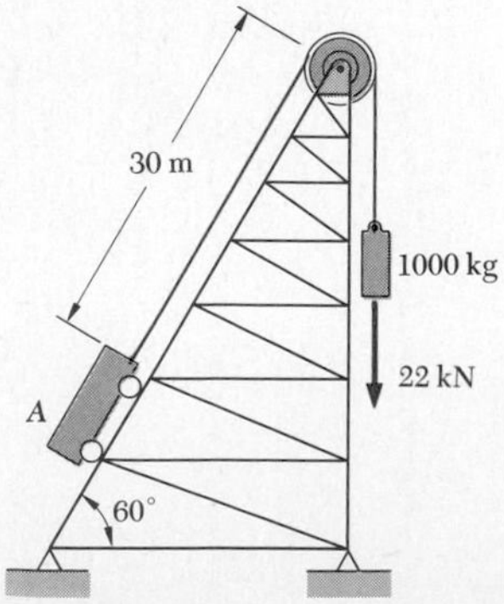

Fig. P13.16

13.16 The total mass of loading car *A* and its load is 3500 kg. The car is connected to a 1000-kg counterweight and is at rest when a constant 22-kN force is applied as shown. (*a*) If the force acts through the entire motion, what is the speed of the car after it has traveled 30 m? (*b*) If after the car has moved through a distance *x* the 22-kN force is removed, the car will coast to rest. After what distance *x* should the force be removed if the car is to come to rest after a total movement of 30 m?

13.17 Knowing that the system shown starts from rest, determine (*a*) the velocity of collar *A* after it has moved through 320 mm, (*b*) the corresponding velocity of collar *B*, (*c*) the tension in the cable. Neglect the masses of the pulleys and the effect of friction.

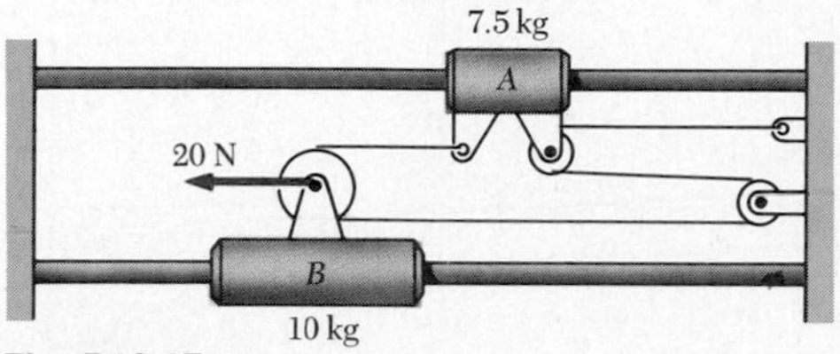

Fig. P13.17

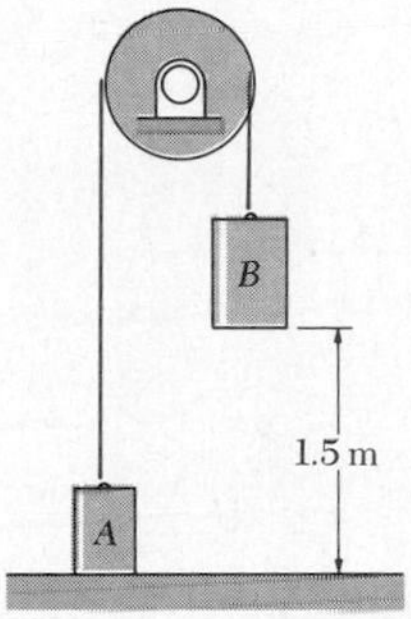

Fig. P13.18 and P13.19

13.18 Two blocks *A* and *B*, of mass 8 kg and 12 kg, respectively, hang from a cable which passes over a pulley of negligible mass. Knowing that the blocks are released from rest and that the energy dissipated by axle friction in the pulley is 10 J, determine (*a*) the velocity of block *B* as it strikes the ground, (*b*) the force exerted by the cable on each of the two blocks during the motion.

13.19 Two blocks *A* and *B*, of mass 12 kg and 15 kg, respectively, hang from a cable which passes over a pulley of negligible mass. The blocks are released from rest in the positions shown and block *B* is observed to strike the ground with a velocity of 1.6 m/s. Determine (*a*) the energy dissipated due to axle friction in the pulley, (*b*) the force exerted by the cable on each of the two blocks during the motion.

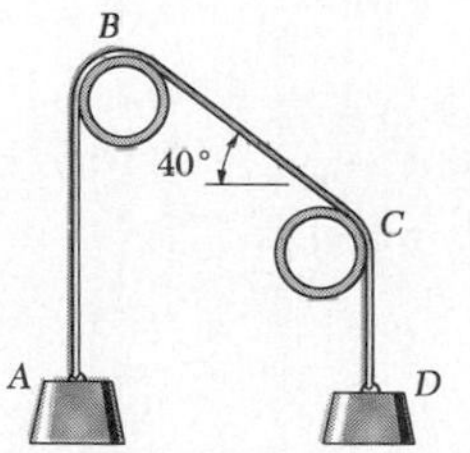

Fig. P13.20 and P13.21

13.20 Two blocks *A* and *D*, weighing, respectively, 125 lb and 300 lb, are attached to a rope which passes over two fixed pipes *B* and *C* as shown. It is observed that when the system is released from rest, block *A* acquires a velocity of 8 ft/s after moving 5 ft up. Determine (*a*) the force exerted by the rope on each of the two blocks during the motion, (*b*) the coefficient of kinetic friction between the rope and the pipes, (*c*) the energy dissipated due to friction.

13.21 Two blocks *A* and *D* are attached to a rope which passes over two fixed pipes *B* and *C* as shown. The coefficients of friction between the rope and the pipes are $\mu_s = 0.25$ and $\mu_k = 0.20$. Knowing that the masses of blocks *A* and *B* are, respectively, 50 kg and 125 kg and that the system is released from rest, determine (*a*) the velocity of *A* after it has moved 1.2 m up, (*b*) the force exerted by the rope on each of the two blocks during the motion, (*c*) the energy dissipated due to friction.

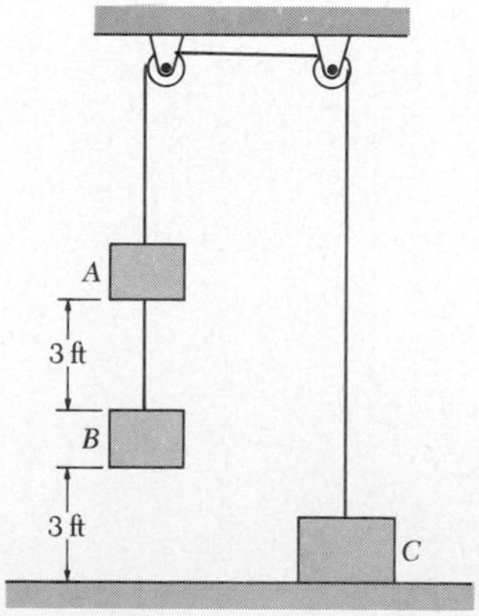

Fig. P13.22 and P13.23

13.22 Blocks *A* and *B* weigh 10 lb each, and block *C* weighs 12 lb. Knowing that the blocks are released from rest in the positions shown and neglecting friction, determine (*a*) the velocity of block *B* just before it strikes the ground, (*b*) the velocity of block *A* just before it strikes block *B*.

13.23 Blocks *A* and *B* weigh 10 lb each. Neglecting friction, determine the weight of block *C* so that when released from rest in the position shown, the system will come to rest again with block *A* just touching block *B*.

13.24 Four packages, each having a mass of 20 kg, are placed as shown on a conveyor belt which is disengaged from its drive motor. Package *1* is just to the right of the horizontal portion of the belt. If the system is released from rest, determine the velocities of packages *1* and *2* as they fall off the belt at point *A*. Assume that the mass of the belt and rollers is small compared with the mass of the packages.

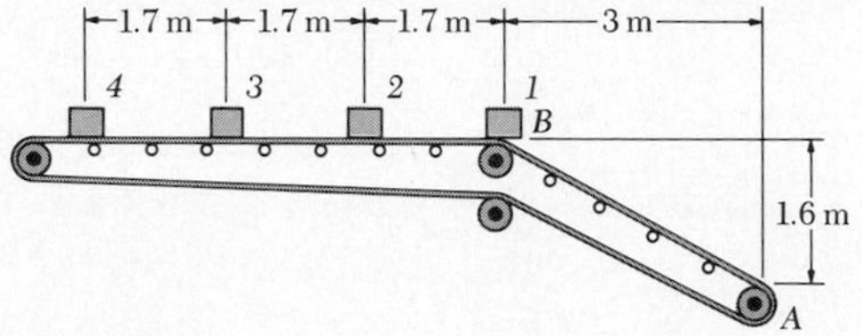

Fig. P13.24

13.25 Knowing that package *2* in Prob. 13.24 has a velocity of 4.72 m/s as it falls off the belt, determine the velocities of packages *3* and *4* as they fall off the belt at point *A*.

13.26 Two blocks *A* and *B*, of mass 5 kg and 6 kg, respectively, are connected by a cord which passes over pulleys as shown. A collar *C*, of mass 4 kg, is placed on block *A* and the system is released from rest. After the blocks have moved through 0.9 m, collar *C* is removed and the blocks continue to move. Determine the velocity of block *A* just before it strikes the ground.

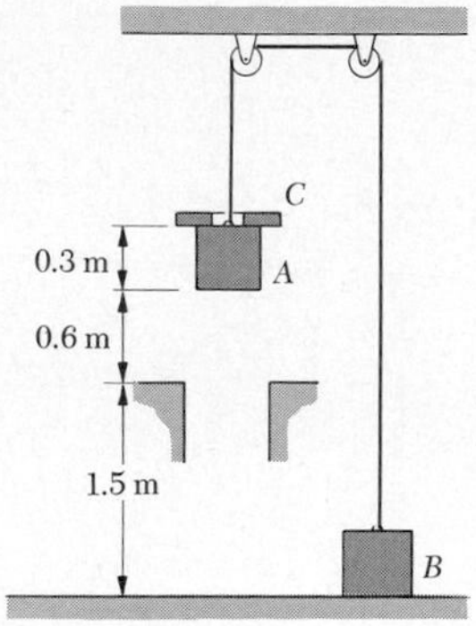

Fig. P13.26

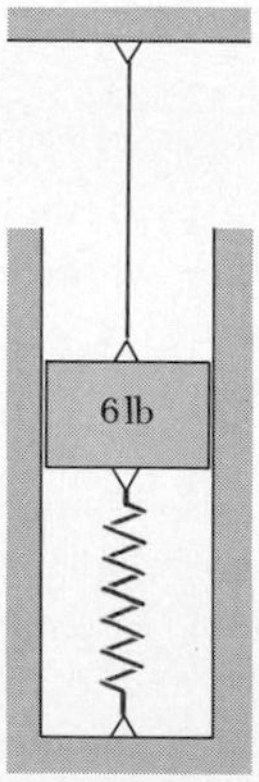

Fig. P13.27

13.27 A 6-lb block is attached to a cable and to a spring as shown. The constant of the spring is $k = 5$ lb/in. and the tension in the cable is 4 lb. If the cable is cut, determine (*a*) the maximum displacement of the block, (*b*) the maximum velocity of the block.

13.28 Solve Prob. 13.27, assuming that the initial tension in the cable is 8 lb.

13.29 An 8-kg plunger is released from rest in the position shown and is stopped by two nested springs; the constant of the outer spring is $k_1 = 4$ kN/m and the constant of the inner spring is $k_2 = 12$ kN/m. If the maximum deflection of the outer spring is observed to be 125 mm, determine the height h from which the plunger was released.

13.30 An 8-kg plunger is released from rest in the position shown and is stopped by two nested springs; the constant of the outer spring is $k_1 = 4$ kN/m and the constant of the inner spring is $k_2 = 12$ kN/m. If the plunger is released from the height $h = 600$ mm, determine the maximum deflection of the outer spring.

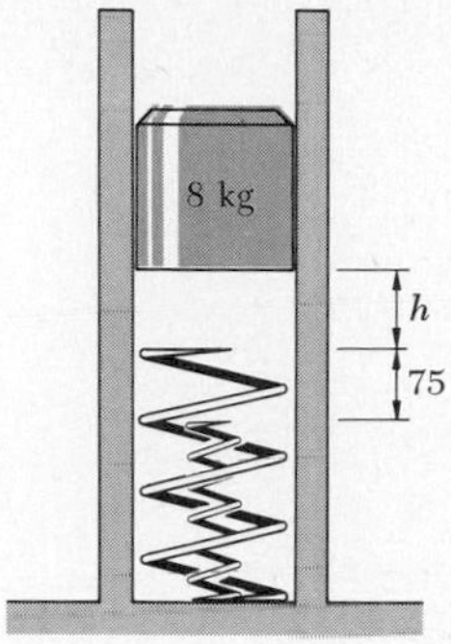

Fig. P13.29 and P13.30

13.31 It is shown in mechanics of materials that when an elastic beam AB supports a block of weight W at a given point D, the deflection y_{st} of point D (called the static deflection) is proportional to W. Show that if the same block is dropped from a height h onto the beam and hits it at D, the maximum deflection y_m of point D in the ensuing motion may be expressed as

$$y_m = y_{st}\left(1 + \sqrt{1 + \frac{2h}{y_{st}}}\right)$$

Note that this formula is approximate, since it is based on the assumption that the block does not bounce off the beam and that no energy is dissipated in the impact.

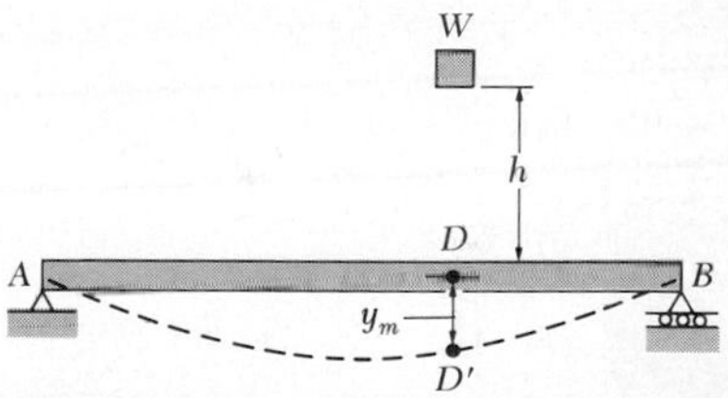

Fig. P13.31

13.32 Two types of energy-absorbing fenders designed to be used on a pier are statically loaded. The force-deflection curve for each type of fender is given in the graph. Determine the maximum deflection of each fender when a 100-ton ship moving at 1 mi/h strikes the fender and is brought to rest.

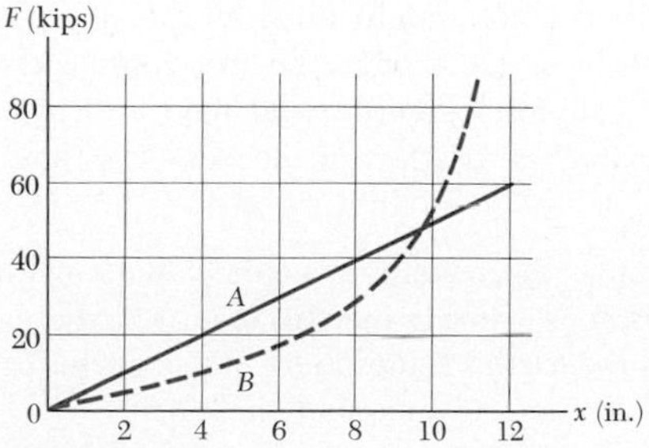

Fig. P13.32

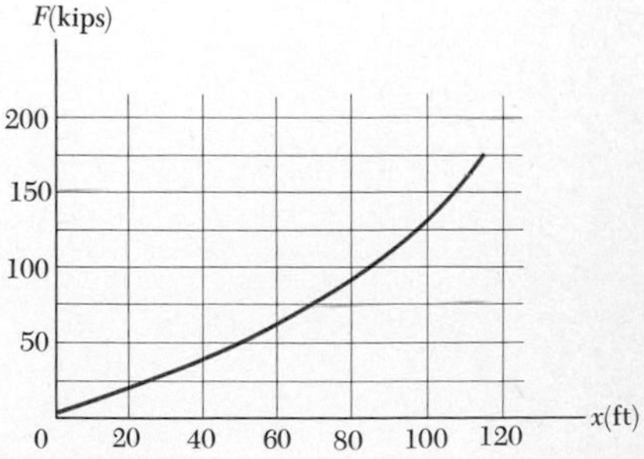

Fig. P13.33

13.33 A 12,000-lb airplane lands on an aircraft carrier and is caught by an arresting cable which is characterized by the force-deflection diagram shown. Knowing that the landing speed of the plane is 100 mi/h, determine (*a*) the distance required for the plane to come to rest, (*b*) the maximum rate of deceleration of the plane.

13.34 Nonlinear springs are classified as hard or soft, depending upon the curvature of their force-deflection curves (see figure). If a delicate instrument having a mass of 5 kg is placed on a spring of length l so that its base is just touching the undeformed spring and then inadvertently released from that position, determine the maximum deflection x_m of the spring and the maximum force F_m exerted by the spring, assuming (*a*) a linear spring of constant $k = 3$ kN/m, (*b*) a hard, nonlinear spring, for which $F = (3 \text{ kN/m})x(1 + 160x^2)$.

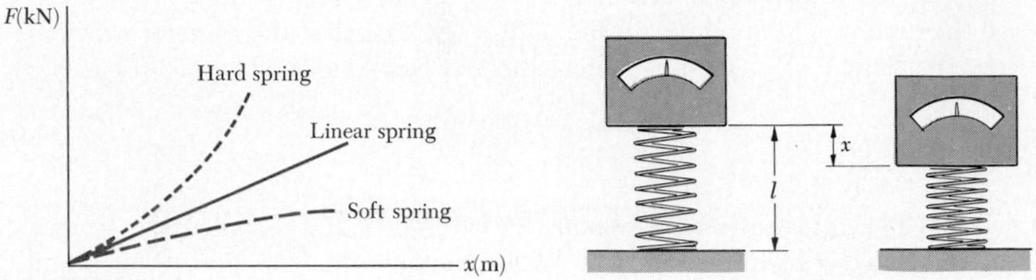

Fig. P13.34

13.35 Solve Prob. 13.34, assuming that in part *b*, the hard spring has been replaced by a soft, nonlinear spring, for which $F = (3 \text{ kN/m})x(1 - 160x^2)$.

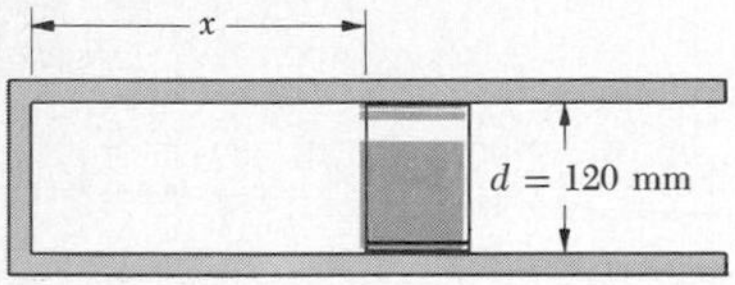

Fig. P13.36

13.36 A 120-mm-diameter piston having a mass of 5 kg slides without friction in a cylinder. The pressure p within the cylinder varies inversely as the volume of the cylinder and is equal to the atmospheric pressure $p_a = 101.3 \text{ kN/m}^2$ when $x = 200$ mm. If the piston is moved to the left and released with no velocity when $x = 80$ mm, determine the maximum velocity reached by the piston in the ensuing motion.

13.37 In Prob. 13.36, determine the maximum value of the coordinate x after the piston has been released with no velocity in the position $x = 80$ mm.

13.38 A bullet is fired straight up from the surface of the moon with an initial velocity of 500 m/s. Determine the maximum elevation reached by the bullet, (*a*) assuming a uniform gravitational field with $g = 1.62 \text{ m/s}^2$, (*b*) using Newton's law of gravitation. (Radius of moon = 1740 km.)

13.39 Determine the maximum velocity with which an object dropped from a very large distance will hit the surface (*a*) of the earth, (*b*) of the moon. Neglect the effect of the earth's atmosphere. (The radius of the moon is 1740 km and its mass is 0.01230 times the mass of the earth.)

13.40 A rocket is fired vertically from the ground. Knowing that at burnout the rocket is 50 mi above the ground and has a velocity of 15,000 ft/s, determine the highest altitude it will reach.

13.41 A rocket is fired vertically from the ground. What should be its velocity $\mathbf{v}_B$ at burnout, 50 mi above the ground, if it is to reach an altitude of 800 mi?

13.42 Sphere C and block A are both moving to the left with a velocity $\mathbf{v}_0$ when the block is suddenly stopped by the wall. Determine the smallest velocity $\mathbf{v}_0$ for which the sphere C will swing in a full circle about the pivot B (a) if BC is a slender rod of negligible mass, (b) if BC is a cord.

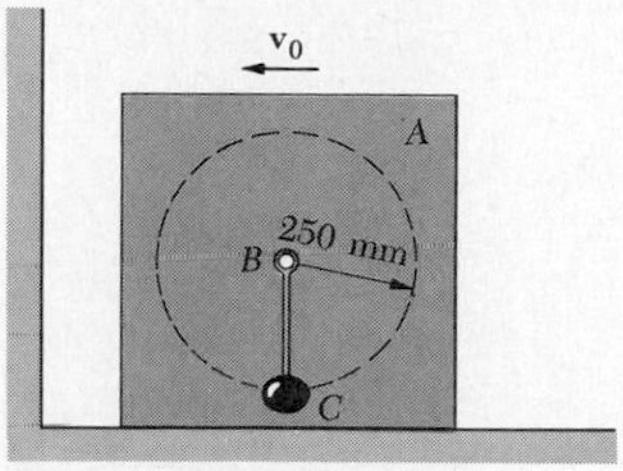

Fig. P13.42

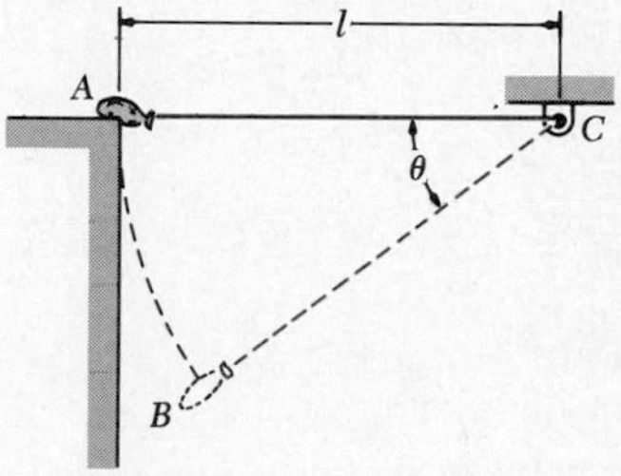

Fig. P13.43

13.43 A bag is gently pushed off the top of a wall at A and swings in a vertical plane at the end of a rope of length l. Determine the angle θ for which the rope will break, knowing that it can withstand a maximum tension 50 percent larger than the weight of the bag.

13.44 A roller coaster starts from rest at A, rolls down the track to B, describes a circular loop of 40-ft diameter, and moves up and down past point E. Knowing that $h = 60$ ft and assuming no energy loss due to friction, determine (a) the force exerted by his seat on a 160-lb rider at B and D, (b) the minimum value of the radius of curvature at E if the roller coaster is not to leave the track at that point.

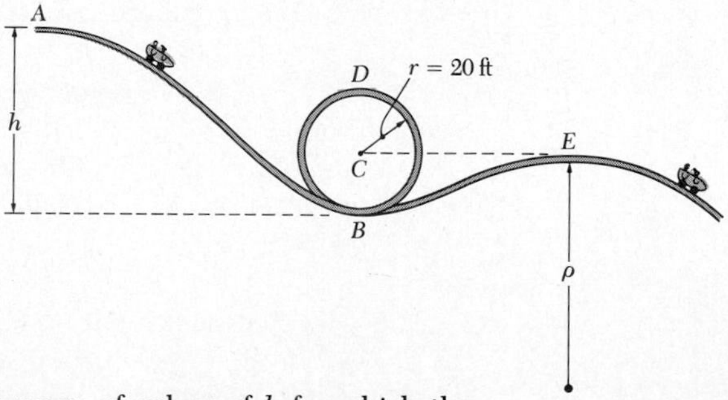

Fig. P13.44

13.45 In Prob. 13.44, determine the range of values of h for which the roller coaster will not leave the track at D or E, knowing that the radius of curvature at E is $\rho = 75$ ft. Assume no energy loss due to friction.

13.46 A small block slides with velocity $\mathbf{v}_0$ on the horizontal surface AB. Neglecting friction and knowing that $v_0 = 0.5\sqrt{gr}$, express in terms of r (a) the elevation h of point C where the block will leave the cylindrical surface BD, (b) the distance d from D to point E where it will hit the ground.

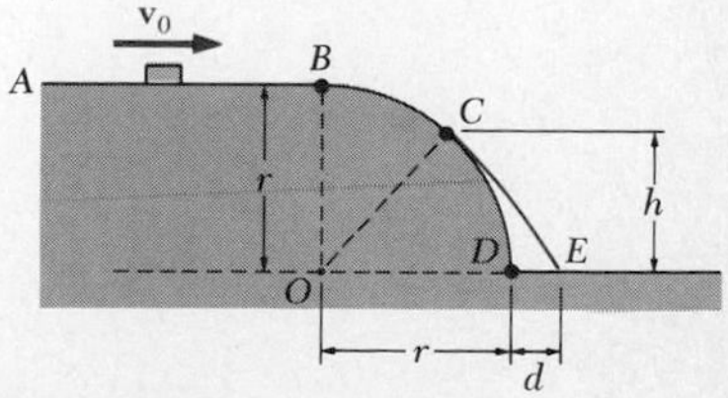

Fig. P13.46

13.47 For the block of Prob. 13.46, and knowing that $r = 600$ mm, determine (a) the smallest value of v_0 for which the block leaves the surface ABD at point B and the corresponding value of d, (b) the smallest possible values of h and d (when v_0 approaches zero).

13.48 A 15-g bullet leaves a fixed rifle barrel 2 ms after being fired. Knowing that the muzzle velocity is 800 m/s and neglecting friction, determine the average power developed by the rifle.

13.49 A 75-kg man runs up a 6-m-high flight of stairs in 8 s. (*a*) What is the average power developed by the man? (*b*) If a 60-kg woman can develop 80 percent as much power, how long will it take her to run up a 4-m-high flight of stairs?

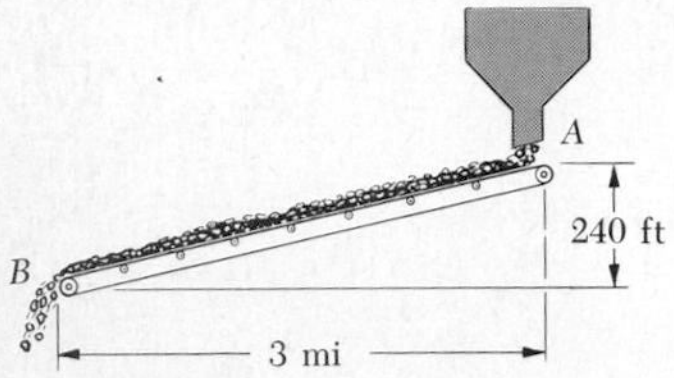

Fig. P13.50

13.50 Crushed stone is being moved from a quarry at A to a loading dock at B at the rate of 500 tons/h. An electric generator is attached to the system in order to maintain a constant belt speed. Knowing that the efficiency of the belt-generator system is 0.70, determine the average power in kilowatts developed by the generator if the belt speed is (*a*) 6 ft/s, (*b*) 10 ft/s.

13.51 The escalator shown is designed to transport 6000 persons per hour at a constant speed of 1.5 ft/s. Assuming an average weight of 150 lb per person, determine (*a*) the average power required, (*b*) the required capacity of the motor if the mechanical efficiency is 85 percent and if a 300 percent overload is to be allowed.

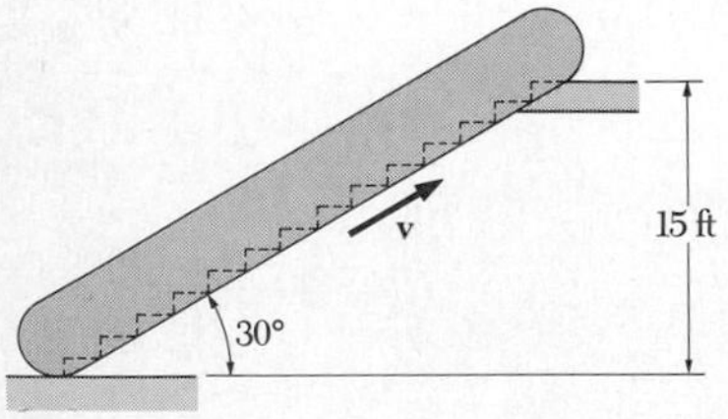

Fig. P13.51

13.52 A train of total mass equal to 500 Mg starts from rest and accelerates uniformly to a speed of 90 km/h in 50 s. After reaching this speed, the train travels with a constant velocity. The track is horizontal and axle friction and rolling resistance result in a total force of 15 kN in a direction opposite to the direction of motion. Determine the power required as a function of time.

13.53 Solve Prob. 13.52, assuming that during the entire motion, the train is traveling up a 2.5 percent grade.

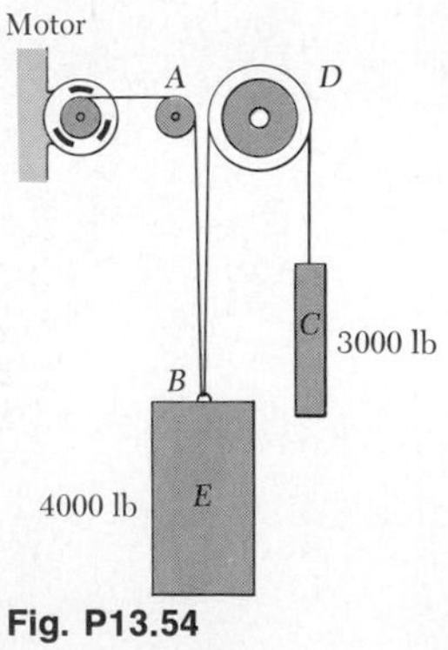

Fig. P13.54

13.54 Elevator E weighs 4000 lb when fully loaded and is connected to a 3000-lb counterweight C. Determine the power delivered by the electric motor when the elevator (*a*) is moving up at a constant speed of 20 ft/s, (*b*) has an instantaneous velocity of 20 ft/s and an acceleration of 3 ft/s^2, both directed upward.

13.55 For the dumb-waiter system of Sample Prob. 13.5, it is known that the power delivered by the motor is 3 hp at the instant the speed of the dumb-waiter is 6 ft/s upward. Determine the acceleration of the dumb-waiter at that instant.

13.56 The fluid transmission of a 12-Mg truck permits the engine to deliver an essentially constant power of 40 kW to the driving wheels. Determine the time required and the distance traveled as the speed of the truck is increased (*a*) from 24 km/h to 48 km/h, (*b*) from 48 km/h to 72 km/h.

13.57 The frictional resistance of a ship is known to vary directly as the 1.75 power of the speed v of the ship. A single tugboat at full power can tow the ship at a constant speed of 5.4 km/h by exerting a constant force of 250 kN. Determine (*a*) the power developed by the tugboat, (*b*) the maximum speed at which two tugboats, capable of delivering the same power, can tow the ship.

13.6. Potential Energy.† Let us consider again a body of weight **W** which moves along a curved path from a point A_1 of elevation y_1 to a point A_2 of elevation y_2 (Fig. 13.4). We recall from Sec. 13.2 that the work of the force of gravity **W** during this displacement is

$$U_{1\to2} = Wy_1 - Wy_2 \tag{13.4}$$

The work of **W** may thus be obtained by subtracting the value of the function Wy corresponding to the second position of the body from its value corresponding to the first position. The work of **W** is independent of the actual path followed; it depends only upon the initial and final values of the function Wy. This function is called the *potential energy* of the body with respect to the *force of gravity* **W** and is denoted by V_g. We write

$$U_{1\to2} = (V_g)_1 - (V_g)_2 \qquad \text{with } V_g = Wy \tag{13.16}$$

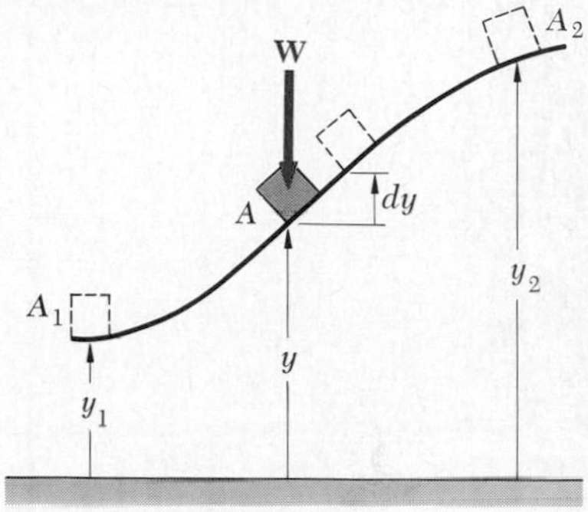

Fig. 13.4 (*repeated*)

We note that if $(V_g)_2 > (V_g)_1$, that is, *if the potential energy increases* during the displacement (as in the case considered here), *the work $U_{1\to2}$ is negative.* If, on the other hand, the work of **W** is positive, the potential energy decreases. Therefore, the potential energy V_g of the body provides a measure of the work which may be done by its weight **W**. Since only the *change* in potential energy, and not the actual value of V_g, is involved in formula (13.16), an arbitrary constant may be added to the expression obtained for V_g. In other words, the level, or datum, from which the elevation y is measured may be chosen arbitrarily. Note that potential energy is expressed in the same units as work, i.e., in joules if SI units are used, and in ft·lb or in·lb if U.S. customary units are used.

It should be noted that the expression just obtained for the potential energy of a body with respect to gravity is valid only as long as the weight **W** of the body may be assumed to remain constant, i.e., as long as the displacements of the body are small compared with the radius of the earth. In the case of a space vehicle, however, we should take into consideration the variation of the force of gravity with the distance r from the center of the earth. Using the expression obtained in Sec. 13.2 for the work of a gravitational force, we write (Fig. 13.6)

$$U_{1\to2} = \frac{GMm}{r_2} - \frac{GMm}{r_1} \tag{13.7}$$

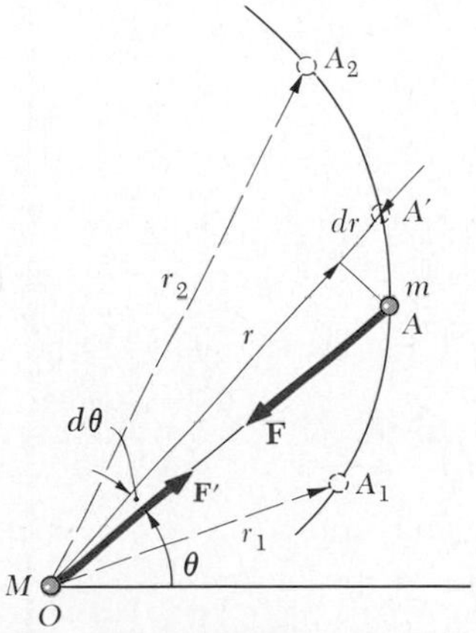

Fig. 13.6 (*repeated*)

The work of the force of gravity may therefore be obtained by subtracting the value of the function $-GMm/r$ corresponding to the second position of the body from its value corresponding to the first position. Thus, the expression which should be used for the potential energy V_g when the variation in the force of gravity cannot be neglected is

$$V_g = -\frac{GMm}{r} \tag{13.17}$$

†Some of the material in this section has already been considered in Sec. 10.7.

Taking the first of the relations (12.29) into account, we write V_g in the alternate form

$$V_g = -\frac{WR^2}{r} \tag{13.17'}$$

where R is the radius of the earth and W the value of the weight of the body at the surface of the earth. When either of the relations (13.17) or (13.17′) is used to express V_g, the distance r should, of course, be measured from the center of the earth.† Note that V_g is always negative and that it approaches zero for very large values of r.

Consider, now, a body attached to a spring and moving from a position A_1, corresponding to a deflection x_1 of the spring, to a position A_2, corresponding to a deflection x_2 (Fig. 13.5). We recall from Sec. 13.2 that the work of the force **F** exerted by the spring on the body is

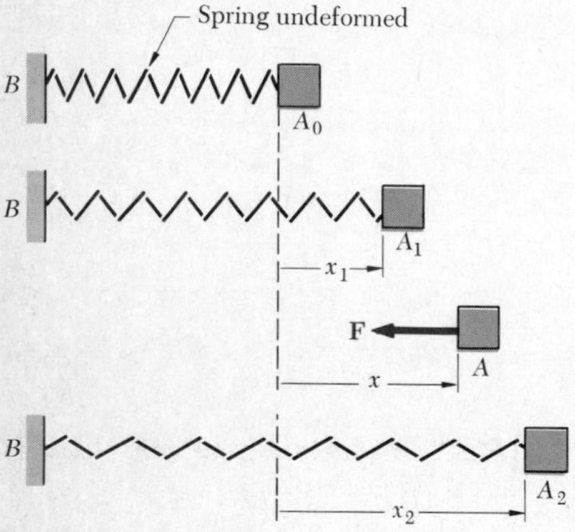

Fig. 13.5 (*repeated*)

$$U_{1\to2} = \tfrac{1}{2}kx_1^2 - \tfrac{1}{2}kx_2^2 \tag{13.6}$$

The work of the elastic force is thus obtained by subtracting the value of the function $\frac{1}{2}kx^2$ corresponding to the second position of the body from its value corresponding to the first position. This function is denoted by V_e and is called the *potential energy* of the body with respect to the *elastic force* **F**. We write

$$U_{1\to2} = (V_e)_1 - (V_e)_2 \qquad \text{with } V_e = \tfrac{1}{2}kx^2 \tag{13.18}$$

and observe that during the displacement considered, the work of the force **F** exerted by the spring on the body is negative and the potential energy V_e increases. We should note that the expression obtained for V_e is valid only if the deflection of the spring is measured from its undeformed position. On the other hand, formula (13.18) may be used even when the spring is rotated about its fixed end (Fig. 13.10*a*). The work of the elastic force depends only upon the initial and final deflections of the spring (Fig. 13.10*b*).

†The expressions given for V_g in (13.17) and (13.17′) are valid only when $r \geqq R$, that is, when the body considered is above the surface of the earth.

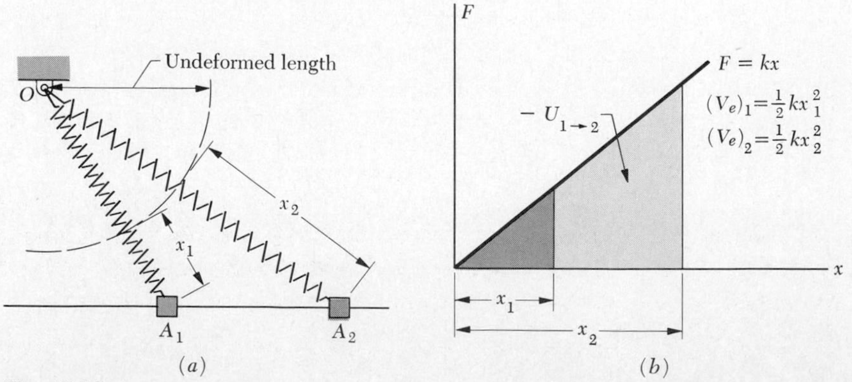

Fig. 13.10

The concept of potential energy may be used when forces other than gravity forces and elastic forces are involved. Indeed, it remains valid as long as the work of the force considered is independent of the path followed by its point of application as this point moves from a given position A_1 to a given position A_2. Such forces are said to be *conservative forces;* the general properties of conservative forces are studied in the following section.

*** 13.7. Conservative Forces.** As indicated in the preceding section, a force **F** acting on a particle A is said to be conservative *if its work* $U_{1\to2}$ *is independent of the path followed by the particle A as it moves from* A_1 *to* A_2 (Fig. 13.11*a*). We may then write

$$U_{1\to2} = V(x_1, y_1, z_1) - V(x_2, y_2, z_2) \qquad (13.19)$$

or, for short,

$$U_{1\to2} = V_1 - V_2 \qquad (13.19')$$

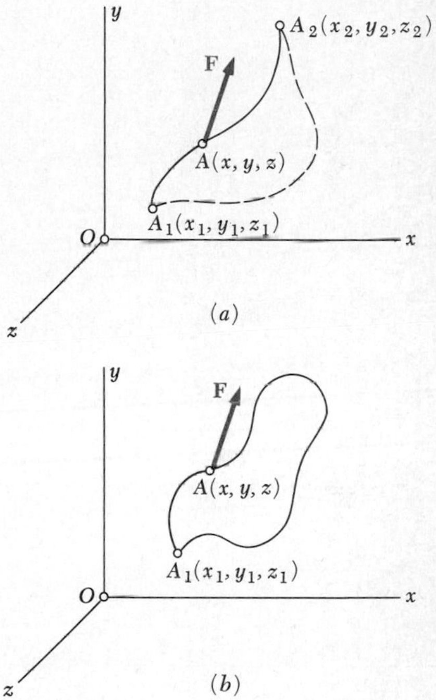

Fig. 13.11

The function $V(x, y, z)$ is called the potential energy, or *potential function,* of **F**.

We note that if A_2 is chosen to coincide with A_1, that is, if the particle describes a closed path (Fig. 13.11*b*), we have $V_1 = V_2$ and the work is zero. We may thus write for any conservative force **F**

$$\oint \mathbf{F} \cdot d\mathbf{r} = 0 \qquad (13.20)$$

where the circle on the integral sign indicates that the path is closed.

Let us now apply (13.19) between two neighboring points $A(x, y, z)$ and $A'(x + dx, y + dy, z + dz)$. The elementary work dU corresponding to the displacement $d\mathbf{r}$ from A to A' is

$$dU = V(x, y, z) - V(x + dx, y + dy, z + dz)$$

or

$$dU = -dV(x, y, z) \qquad (13.21)$$

Thus, the elementary work of a conservative force is an *exact differential.*

Substituting for dU in (13.21) the expression obtained in (13.1″), and recalling the definition of the differential of a function of several variables, we write

$$F_x\,dx + F_y\,dy + F_z\,dz = -\left(\frac{\partial V}{\partial x}dx + \frac{\partial V}{\partial y}dy + \frac{\partial V}{\partial z}dz\right)$$

from which it follows that

$$F_x = -\frac{\partial V}{\partial x} \qquad F_y = -\frac{\partial V}{\partial y} \qquad F_z = -\frac{\partial V}{\partial z} \tag{13.22}$$

It is clear that the components of **F** must be functions of the coordinates x, y, z. Thus, a *necessary* condition for a conservative force is that it depend only upon the position of its point of application. The relations (13.22) may be expressed more concisely if we write

$$\mathbf{F} = F_x\mathbf{i} + F_y\mathbf{j} + F_z\mathbf{k} = -\left(\frac{\partial V}{\partial x}\mathbf{i} + \frac{\partial V}{\partial y}\mathbf{j} + \frac{\partial V}{\partial z}\mathbf{k}\right)$$

The vector in parentheses is known as the *gradient of the scalar function* V and is denoted by **grad** V. We thus write for any conservative force

$$\mathbf{F} = -\mathbf{grad}\ V \tag{13.23}$$

The relations (13.19) to (13.23) were shown to be satisfied by any conservative force. It may also be shown that if a force **F** satisfies one of these relations, **F** must be a conservative force.

13.8. Conservation of Energy. We saw in the preceding two sections that the work of a conservative force, such as the weight of a particle or the force exerted by a spring, may be expressed as a change in potential energy. When a particle moves under the action of conservative forces, the principle of work and energy stated in Sec. 13.3 may be expressed in a modified form. Substituting for $U_{1\rightarrow 2}$ from (13.19′) into (13.10), we write

$$V_1 - V_2 = T_2 - T_1$$

$$T_1 + V_1 = T_2 + V_2 \tag{13.24}$$

Formula (13.24) indicates that when a particle moves under the action of conservative forces, *the sum of the kinetic energy and of the potential energy of the particle remains constant*. The sum $T + V$ is called the *total mechanical energy* of the particle and is denoted by E.

Consider, for example, the pendulum analyzed in Sec. 13.4, which is released with no velocity from A_1 and allowed to swing in a vertical plane (Fig. 13.12). Measuring the potential energy from the level of A_2, we have, at A_1,

$$T_1 = 0 \qquad V_1 = Wl \qquad T_1 + V_1 = Wl$$

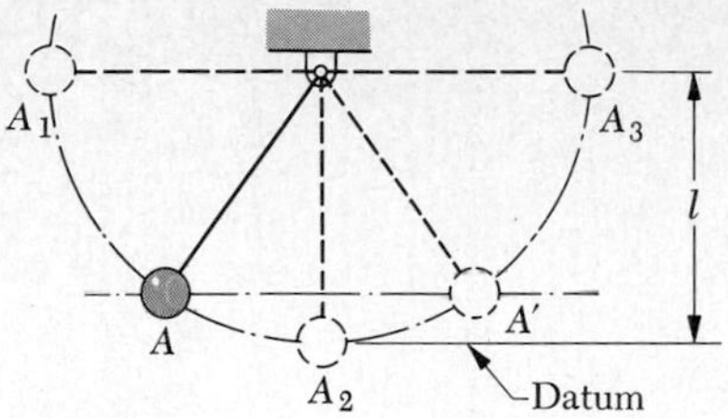

Fig. 13.12

Recalling that at A_2, the speed of the pendulum is $v_2 = \sqrt{2gl}$, we have

$$T_2 = \tfrac{1}{2}mv_2^2 = \frac{1}{2}\frac{W}{g}(2gl) = Wl \qquad V_2 = 0$$

$$T_2 + V_2 = Wl$$

We thus check that the total mechanical energy $E = T + V$ of the pendulum is the same at A_1 and A_2. While the energy is entirely potential at A_1, it becomes entirely kinetic at A_2 and, as the pendulum keeps swinging to the right, the kinetic energy is transformed back into potential energy. At A_3, we shall have $T_3 = 0$ and $V_3 = Wl$.

Since the total mechanical energy of the pendulum remains constant and since its potential energy depends only upon its elevation, the kinetic energy of the pendulum will have the same value at any two points located on the same level. Thus, the speed of the pendulum is the same at A and at A' (Fig. 13.12). This result may be extended to the case of a particle moving along any given path, regardless of the shape of the path, as long as the only forces acting on the particle are its weight and the normal reaction of the path. The particle of Fig. 13.13, for example, which slides in a vertical plane along a frictionless track, will have the same speed at A, A', and A''.

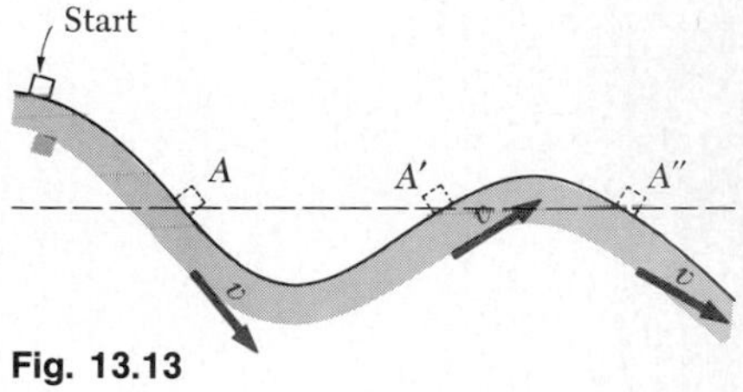

Fig. 13.13

While the weight of a particle and the force exerted by a spring are conservative forces, *friction forces are nonconservative forces.* In other words, *the work of a friction force cannot be expressed as a change in potential energy.* The work of a friction force depends upon the path followed by its point of application; and while the work $U_{1\to2}$ defined by (13.19) is positive or negative according to the sense of motion, *the work of a friction force,* as we noted in Sec. 13.4, *is always negative.* It follows that when a mechanical system involves friction, its total mechanical energy does not remain constant but decreases. The energy of the system, however, is not lost; it is transformed into heat, and the sum of the *mechanical energy* and of the *thermal energy* of the system remains constant.

Other forms of energy may also be involved in a system. For instance, a generator converts mechanical energy into *electric energy;* a gasoline engine converts *chemical energy* into mechanical energy; a nuclear reactor converts *mass* into thermal energy. If all forms of energy are considered, the energy of any system may be considered as constant and the principle of conservation of energy remains valid under all conditions.

13.9. Motion under a Conservative Central Force. Application to Space Mechanics. We saw in Sec. 12.9 that when a particle P moves under a central force $\mathbf{F}$, the angular momentum $\mathbf{H}_O$ of the particle about the center of force O is constant. If the force $\mathbf{F}$ is also conservative, there exists a potential energy V associated with $\mathbf{F}$, and the total energy $E = T + V$ of the particle is constant (Sec. 13.8). Thus, when a particle moves under a conservative central force, both the principle of conservation of angular momentum and the principle of conservation of energy may be used to study its motion.

Consider, for example, a space vehicle moving under the earth's gravitational force. We shall assume that it begins its free flight at point P_0 at a distance r_0 from the center of the earth, with a velocity $\mathbf{v}_0$ forming an angle ϕ_0 with the radius vector OP_0 (Fig. 13.14). Let P be a point of the trajectory described by the vehicle; we denote by r the distance from O to P, by $\mathbf{v}$ the velocity of the vehicle at P, and by ϕ the angle formed by $\mathbf{v}$ and the radius vector OP. Applying the principle of conservation of angular momentum about O between P_0 and P (Sec. 12.9), we write

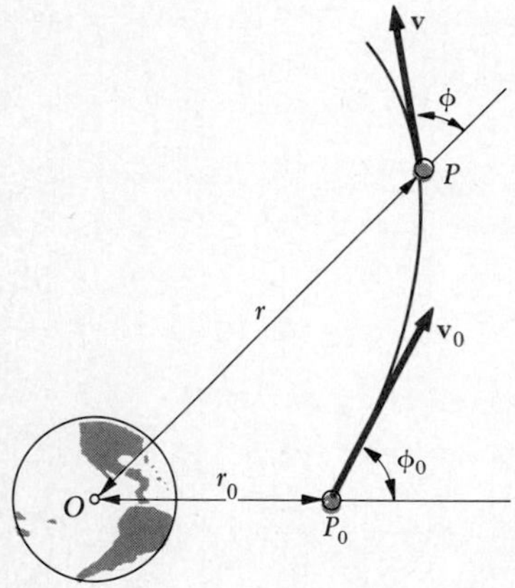

Fig. 13.14

$$r_0 m v_0 \sin \phi_0 = r m v \sin \phi \qquad (13.25)$$

Recalling the expression (13.17) obtained for the potential energy due to a gravitational force, we apply the principle of conservation of energy between P_0 and P and write

$$T_0 + V_0 = T + V$$

$$\tfrac{1}{2} m v_0^2 - \frac{GMm}{r_0} = \tfrac{1}{2} m v^2 - \frac{GMm}{r} \qquad (13.26)$$

where M is the mass of the earth.

Equation (13.26) may be solved for the magnitude v of the velocity of the vehicle at P when the distance r from O to P is known; Eq. (13.25) may then be used to determine the angle ϕ that the velocity forms with the radius vector OP.

Equations (13.25) and (13.26) may also be used to determine the maximum and minimum values of r in the case of a satellite launched from P_0 in a direction forming an angle ϕ_0 with the vertical OP_0 (Fig. 13.15). The desired values of r are obtained by making $\phi = 90°$ in (13.25) and eliminating v between Eqs. (13.25) and (13.26).

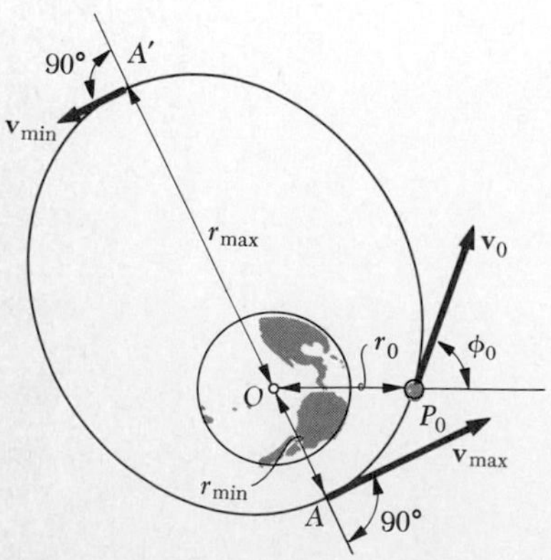

Fig. 13.15

It should be noted that the application of the principles of conservation of energy and of conservation of angular momentum leads to a more fundamental formulation of the problems of space mechanics than does the method indicated in Sec. 12.12. In all cases involving oblique launchings, it will also result in much simpler computations. And while the method of Sec. 12.12 must be used when the actual trajectory or the periodic time of a space vehicle is to be determined, the calculations will be simplified if the conservation principles are first used to compute the maximum and minimum values of the radius vector r.

SAMPLE PROBLEM 13.6

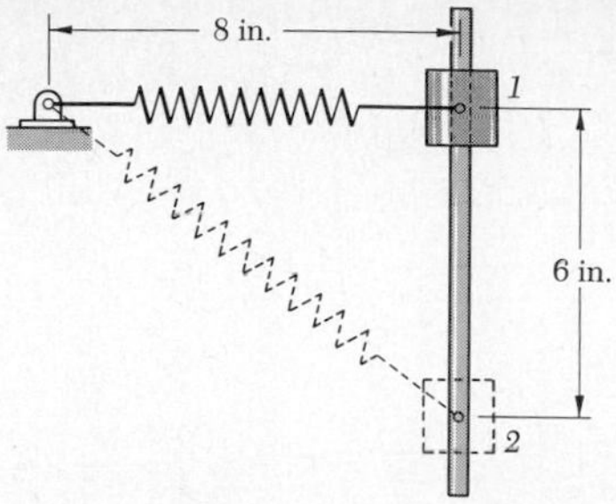

A 20-lb collar slides without friction along a vertical rod as shown. The spring attached to the collar has an undeformed length of 4 in. and a constant of 3 lb/in. If the collar is released from rest in position *1*, determine its velocity after it has moved 6 in. to position 2.

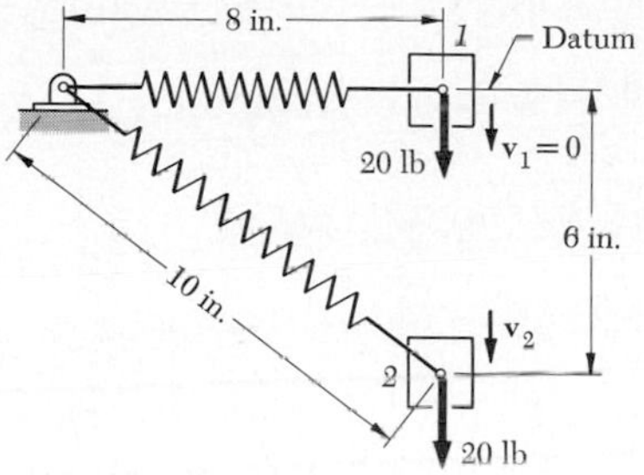

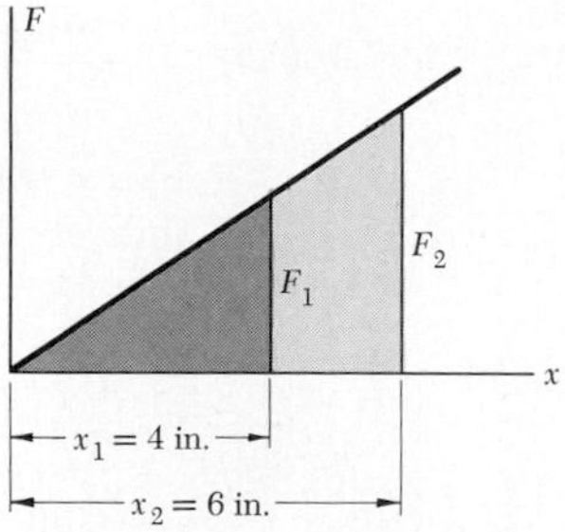

Position ***1.*** ***Potential Energy.*** The elongation of the spring is

$$x_1 = 8 \text{ in.} - 4 \text{ in.} = 4 \text{ in.}$$

and we have

$$V_e = \tfrac{1}{2}kx_1^2 = \tfrac{1}{2}(3 \text{ lb/in.})(4 \text{ in.})^2 = 24 \text{ in}\cdot\text{lb}$$

Choosing the datum as shown, we have $V_g = 0$. Therefore,

$$V_1 = V_e + V_g = 24 \text{ in}\cdot\text{lb} = 2 \text{ ft}\cdot\text{lb}$$

Kinetic Energy. Since the velocity in position *1* is zero, $T_1 = 0$.

Position 2. ***Potential Energy.*** The elongation of the spring is

$$x_2 = 10 \text{ in.} - 4 \text{ in.} = 6 \text{ in.}$$

and we have

$$V_e = \tfrac{1}{2}kx_2^2 = \tfrac{1}{2}(3 \text{ lb/in.})(6 \text{ in.})^2 = 54 \text{ in}\cdot\text{lb}$$
$$V_g = Wy = (20 \text{ lb})(-6 \text{ in.}) = -120 \text{ in}\cdot\text{lb}$$

Therefore,

$$V_2 = V_e + V_g = 54 - 120 = -66 \text{ in}\cdot\text{lb} = -5.5 \text{ ft}\cdot\text{lb}$$

Kinetic Energy

$$T_2 = \tfrac{1}{2}mv_2^2 = \frac{1}{2}\frac{20}{32.2}v_2^2 = 0.311v_2^2$$

Conservation of Energy. Applying the principle of conservation of energy between positions *1* and 2, we write

$$T_1 + V_1 = T_2 + V_2$$
$$0 + 2 \text{ ft}\cdot\text{lb} = 0.311v_2^2 - 5.5 \text{ ft}\cdot\text{lb}$$
$$v_2 = \pm 4.91 \text{ ft/s}$$

$$\mathbf{v}_2 = 4.91 \text{ ft/s}\downarrow$$ ◄

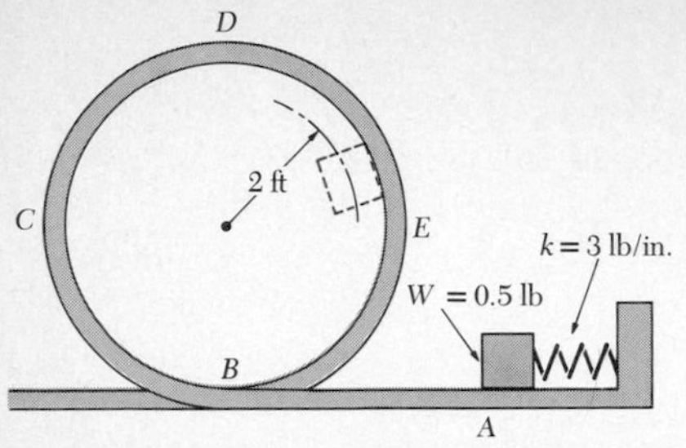

SAMPLE PROBLEM 13.7

The 0.5-lb pellet is pushed against the spring at A and released from rest. Neglecting friction, determine the smallest deflection of the spring for which the pellet will travel around the loop $ABCDE$ and remain at all times in contact with the loop.

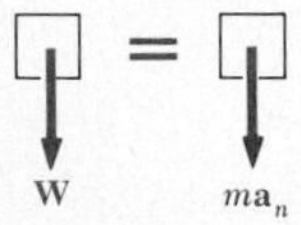

Required Speed at Point *D*. As the pellet passes through the highest point D, its potential energy with respect to gravity is maximum; thus, at the same point its kinetic energy and its speed are minimum. Since the pellet must remain in contact with the loop, the force **N** exerted on the pellet by the loop must be equal to, or greater than, zero. Setting $\mathbf{N} = 0$, we compute the smallest possible speed v_D.

$$+\downarrow \Sigma F_n = ma_n: \qquad W = ma_n \qquad mg = ma_n \qquad a_n = g$$

$$a_n = \frac{v_D^2}{r}: \qquad v_D^2 = ra_n = rg = (2\text{ ft})(32.2\text{ ft/s}^2) = 64.4\text{ ft}^2/\text{s}^2$$

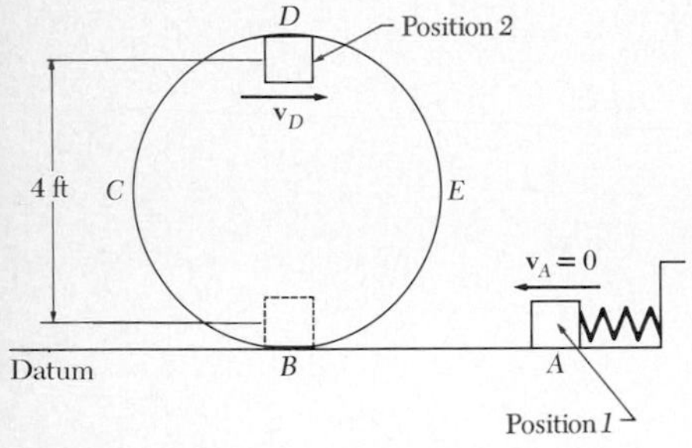

Position *1*. ***Potential Energy.*** Denoting by x the deflection of the spring and noting that $k = 3\text{ lb/in.} = 36\text{ lb/ft}$, we write

$$V_e = \tfrac{1}{2}kx^2 = \tfrac{1}{2}(36\text{ lb/ft})x^2 = 18x^2$$

Choosing the datum at A, we have $V_g = 0$; therefore

$$V_1 = V_e + V_g = 18x^2$$

Kinetic Energy. Since the pellet is released from rest, $v_A = 0$ and we have $T_1 = 0$.

Position 2. ***Potential Energy.*** The spring is now undeformed; thus $V_e = 0$. Since the pellet is 4 ft above the datum, we have

$$V_g = Wy = (0.5\text{ lb})(4\text{ ft}) = 2\text{ ft}\cdot\text{lb}$$
$$V_2 = V_e + V_g = 2\text{ ft}\cdot\text{lb}$$

Kinetic Energy. Using the value of v_D^2 obtained above, we write

$$T_2 = \tfrac{1}{2}mv_D^2 = \frac{1}{2}\frac{0.5\text{ lb}}{32.2\text{ ft/s}^2}(64.4\text{ ft}^2/\text{s}^2) = 0.5\text{ ft}\cdot\text{lb}$$

Conservation of Energy. Applying the principle of conservation of energy between positions *1* and *2*, we write

$$T_1 + V_1 = T_2 + V_2$$
$$0 + 18x^2 = 0.5\text{ ft}\cdot\text{lb} + 2\text{ ft}\cdot\text{lb}$$
$$x = 0.3727\text{ ft} \qquad\qquad x = 4.47\text{ in.} \blacktriangleleft$$

SAMPLE PROBLEM 13.8

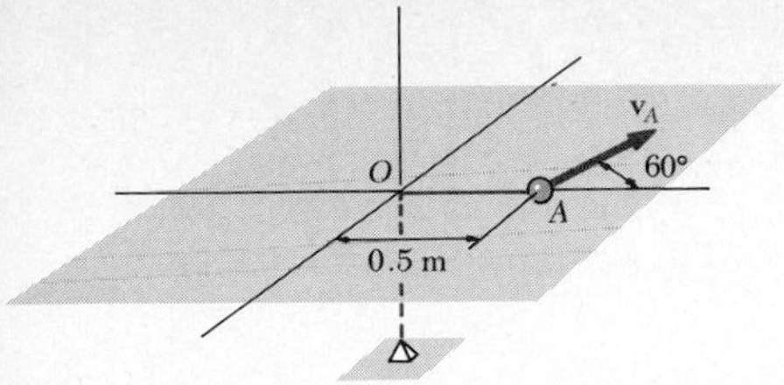

A sphere of mass $m = 0.6$ kg is attached to an elastic cord of constant $k = 100$ N/m, which is undeformed when the sphere is located at the origin O. Knowing that the sphere may slide without friction on the horizontal surface and that in the position shown its velocity $\mathbf{v}_A$ has a magnitude of 20 m/s, determine (*a*) the maximum and minimum distances from the sphere to the origin O, (*b*) the corresponding values of its speed.

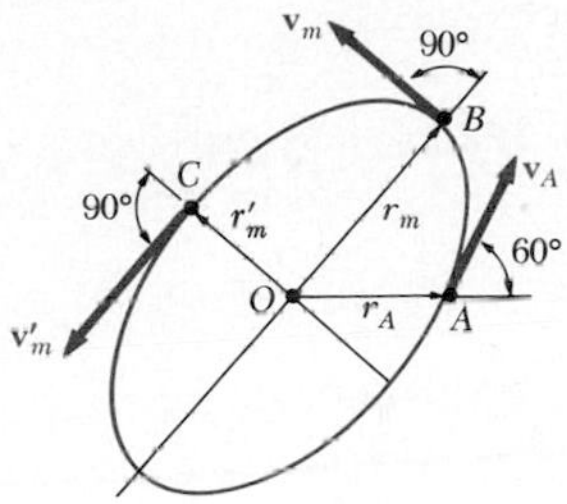

Solution. The force exerted by the cord on the sphere passes through the fixed point O, and its work may be expressed as a change in potential energy. It is therefore a conservative central force, and both the total energy of the sphere and its angular momentum about O are conserved.

Conservation of Angular Momentum About O. At point B, where the distance from O is maximum, the velocity of the sphere is perpendicular to OB and the angular momentum is $r_m v_m$. A similar property holds at point C, where the distance from O is minimum. Expressing conservation of angular momentum between A and B, we write

$$r_A m v_A \sin 60° = r_m m v_m$$
$$(0.5\text{ m})(0.6\text{ kg})(20\text{ m/s})\sin 60° = r_m(0.6\text{ kg})v_m$$
$$v_m = \frac{8.66}{r_m} \qquad (1)$$

Conservation of Energy

At point A:
$$T_A = \tfrac{1}{2}mv_A^2 = \tfrac{1}{2}(0.6\text{ kg})(20\text{ m/s})^2 = 120\text{ J}$$
$$V_A = \tfrac{1}{2}kr_A^2 = \tfrac{1}{2}(100\text{ N/m})(0.5\text{ m})^2 = 12.5\text{ J}$$

At point B:
$$T_B = \tfrac{1}{2}mv_m^2 = \tfrac{1}{2}(0.6\text{ kg})v_m^2 = 0.3v_m^2$$
$$V_B = \tfrac{1}{2}kr_m^2 = \tfrac{1}{2}(100\text{ N/m})r_m^2 = 50r_m^2$$

Applying the principle of conservation of energy between points A and B, we write

$$T_A + V_A = T_B + V_B$$
$$120 + 12.5 = 0.3v_m^2 + 50r_m^2 \qquad (2)$$

***a.* Maximum and Minimum Values of Distance.** Substituting for v_m from Eq. (1) into Eq. (2) and solving for r_m^2, we obtain

$$r_m^2 = 2.468, \text{ or } 0.1824 \qquad r_m = 1.571\text{ m},\ r'_m = 0.427\text{ m} \blacktriangleleft$$

***b.* Corresponding Values of Speed.** Substituting the values obtained for r_m and r'_m into Eq. (1), we have

$$v_m = \frac{8.66}{1.571} \qquad v_m = 5.51\text{ m/s} \blacktriangleleft$$

$$v'_m = \frac{8.66}{0.427} \qquad v'_m = 20.3\text{ m/s} \blacktriangleleft$$

Note. It may be shown that the path of the sphere is an ellipse of center O.

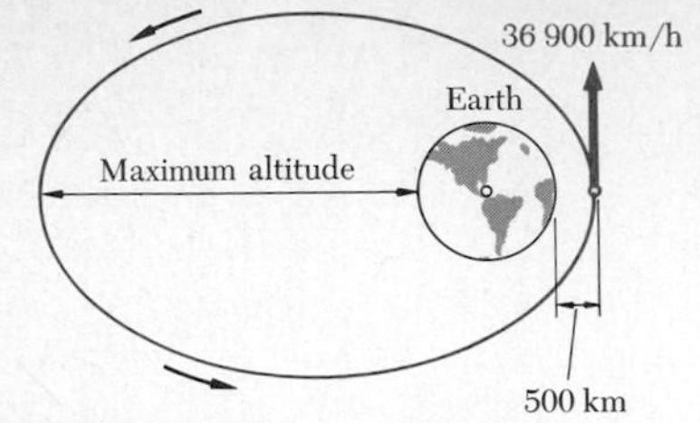

SAMPLE PROBLEM 13.9

A satellite is launched in a direction parallel to the surface of the earth with a velocity of 36 900 km/h from an altitude of 500 km. Determine (*a*) the maximum altitude reached by the satellite, (*b*) the maximum allowable error in the direction of launching if the satellite is to go into orbit and come no closer than 200 km to the surface of the earth.

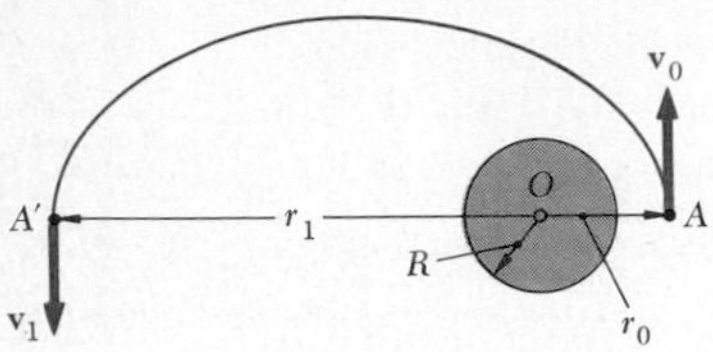

***a.* Maximum Altitude.** We denote by A' the point of the orbit farthest from the earth and by r_1 the corresponding distance from the center of the earth. Since the satellite is in free flight between A and A', we apply the principle of conservation of energy:

$$T_A + V_A = T_{A'} + V_{A'}$$

$$\tfrac{1}{2}mv_0^2 - \frac{GMm}{r_0} = \tfrac{1}{2}mv_1^2 - \frac{GMm}{r_1} \qquad (1)$$

Since the only force acting on the satellite is the force of gravity, which is a central force, the angular momentum of the satellite about O is conserved. Considering points A and A', we write

$$r_0 m v_0 = r_1 m v_1 \qquad v_1 = v_0 \frac{r_0}{r_1} \qquad (2)$$

Substituting this expression for v_1 into Eq. (1) and dividing each term by the mass m, we obtain after rearranging the terms,

$$\tfrac{1}{2}v_0^2\left(1 - \frac{r_0^2}{r_1^2}\right) = \frac{GM}{r_0}\left(1 - \frac{r_0}{r_1}\right) \qquad 1 + \frac{r_0}{r_1} = \frac{2GM}{r_0 v_0^2} \qquad (3)$$

Recalling that the radius of the earth is $R = 6370$ km, we compute

$$r_0 = 6370 \text{ km} + 500 \text{ km} = 6870 \text{ km} = 6.87 \times 10^6 \text{ m}$$
$$v_0 = 36\,900 \text{ km/h} = (36.9 \times 10^6 \text{ m})/(3.6 \times 10^3 \text{ s}) = 10.25 \times 10^3 \text{ m/s}$$
$$GM = gR^2 = (9.81 \text{ m/s}^2)(6.37 \times 10^6 \text{ m})^2 = 398 \times 10^{12} \text{ m}^3/\text{s}^2$$

Substituting these values into (3), we obtain $r_1 = 66.8 \times 10^6$ m.

$$\textit{Maximum altitude} = 66.8 \times 10^6 \text{ m} - 6.37 \times 10^6 \text{ m} = 60.4 \times 10^6 \text{ m}$$
$$= 60\,400 \text{ km} \quad ◄$$

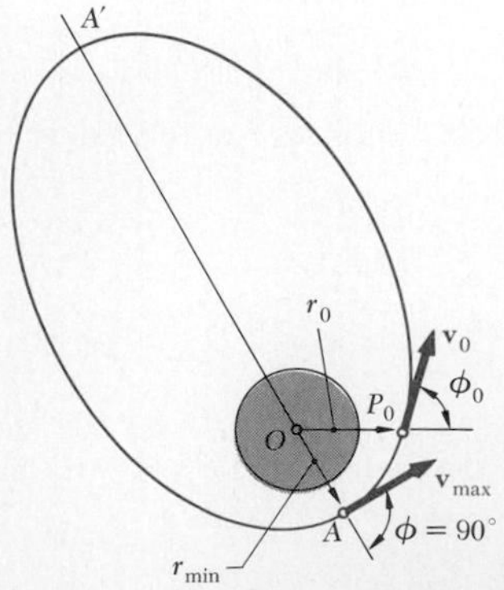

***b.* Allowable Error in Direction of Launching.** The satellite is launched from P_0 in a direction forming an angle ϕ_0 with the vertical OP_0. The value of ϕ_0 corresponding to $r_{min} = 6370$ km + 200 km = 6570 km is obtained by applying the principles of conservation of energy and of conservation of angular momentum between P_0 and A:

$$\tfrac{1}{2}mv_0^2 - \frac{GMm}{r_0} = \tfrac{1}{2}mv_{max}^2 - \frac{GMm}{r_{min}} \qquad (4)$$

$$r_0 m v_0 \sin\phi_0 = r_{min} m v_{max} \qquad (5)$$

Solving (5) for v_{max} and then substituting for v_{max} into (4), we may solve (4) for $\sin\phi_0$. Using the values of v_0 and GM computed in part *a* and noting that $r_0/r_{min} = 6870/6570 = 1.0457$, we find

$$\sin\phi_0 = 0.9801 \qquad \phi_0 = 90° \pm 11.5° \qquad \text{Allowable error} = \pm 11.5° \quad ◄$$

Problems

13.58 A collar C of mass m slides without friction on a horizontal rod between springs A and B. If the collar is pushed to the left until spring A is compressed 60 mm and released, determine the distance through which the collar will travel and the maximum velocity it will reach (*a*) if $m = 0.5$ kg, (*b*) if $m = 2$ kg.

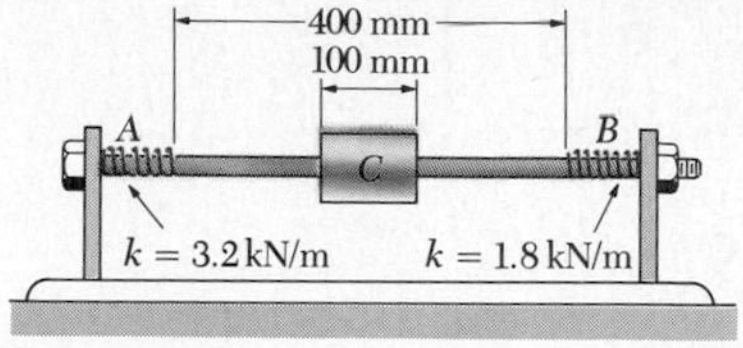

Fig. P13.58

13.59 A 0.8-kg collar C may slide without friction along a horizontal rod. It is attached to a spring of constant $k = 300$ N/m and 500-mm undeformed length. Knowing that the collar is released from rest in the position shown, determine the maximum velocity it will reach in the ensuing motion.

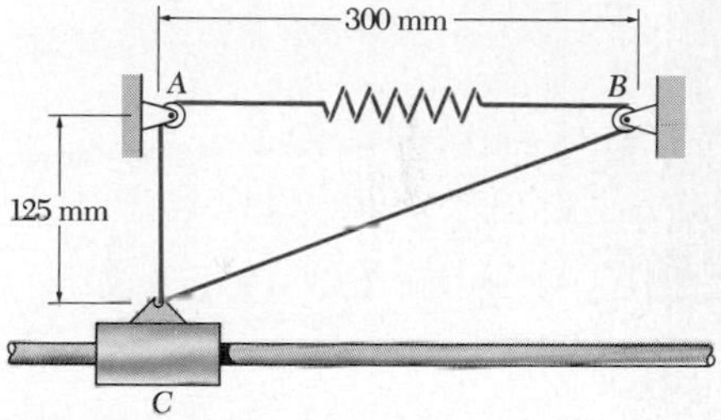

Fig. P13.59

13.60 A 2-lb collar is attached to a spring and slides without friction along a circular rod which lies in a *horizontal* plane. The spring has a constant $k = 3$ lb/in. and is undeformed when the collar is at B. If the collar is released from rest at C, determine the speed of the collar as it passes through point B.

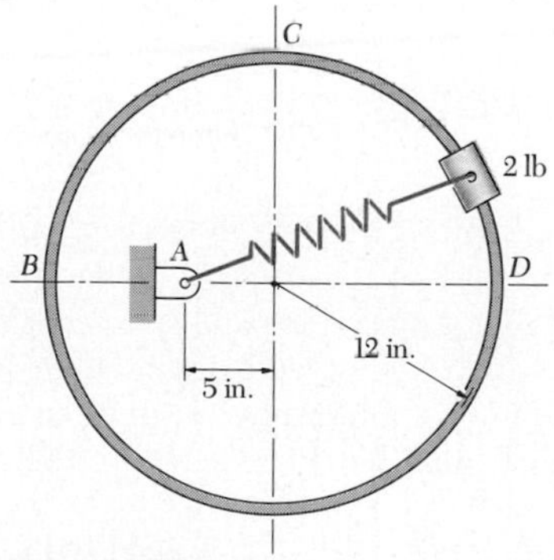

Fig. P13.60

13.61 It is possible for the collar of Prob. 13.60 to have a continuous, although nonuniform, motion along the rod. If the speed of the collar at B is to be twice the speed of the collar at D, determine (*a*) the required speed at D, (*b*) the corresponding speed at C.

13.62 An elastic cord is stretched between two points A and B, located 12 in. apart in the same horizontal plane. When stretched directly between A and B, the tension in the cord is 5 lb. The cord is then stretched as shown until its midpoint C has moved through 4.5 in. to C'; a force of 30 lb is required to hold the cord at C'. A 2-oz pellet is placed at C', and the cord is released. Determine the speed of the pellet as it passes through C.

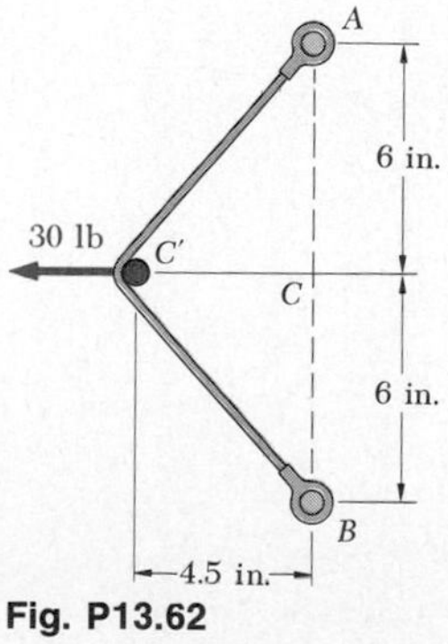

Fig. P13.62

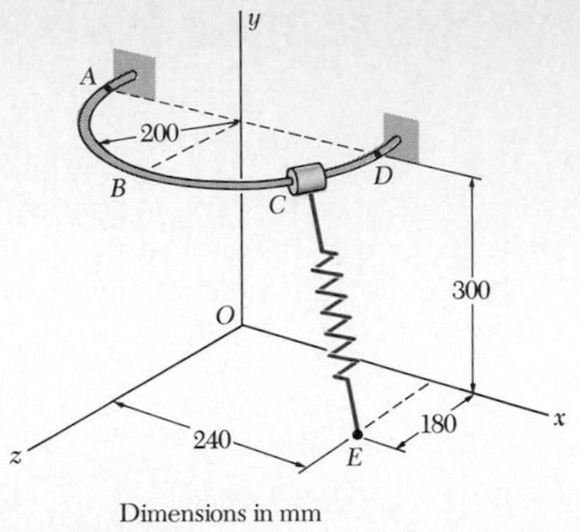

Fig. P13.63

13.63 A 600-g collar C may slide along a horizontal, semicircular rod ABD. The spring CE has an undeformed length of 250 mm and a spring constant of 135 N/m. Knowing that the collar is released from rest at A and neglecting friction, determine the speed of the collar (a) at B, (b) at D.

13.64 In Prob. 13.63, determine (a) the maximum speed of the collar, (b) the coordinates of the point where the collar reaches its maximum speed.

13.65 A 30-Mg railroad car starts from rest and coasts down a 1 percent incline for a distance of 20 m. It is stopped by a bumper having a spring constant of 2400 kN/m. (a) What is the speed of the car at the bottom of the incline? (b) How many millimeters will the spring be compressed?

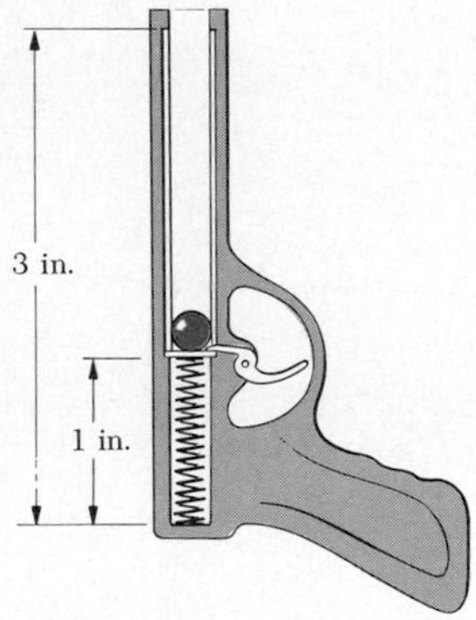

Fig. P13.66

13.66 A toy spring gun is used to shoot 1-oz bullets vertically upward. The undeformed length of the spring is 6 in.; it is compressed to a length of 1 in. when the gun is ready to be shot and expands to a length of 3 in. as the bullet leaves the gun. A force of 10 lb is required to maintain the spring in firing position when the length of the spring is 1 in. Determine (a) the velocity of the bullet as it leaves the gun, (b) the maximum height reached by the bullet.

13.67 A spring is used to stop a 150-lb package which is moving down a 20° incline. The spring has a constant $k = 150$ lb/in. and is held by cables so that it is initially compressed 4 in. Knowing that the velocity of the package is 10 ft/s when it is 30 ft from the spring and neglecting friction, determine the maximum additional deformation of the spring in bringing the package to rest.

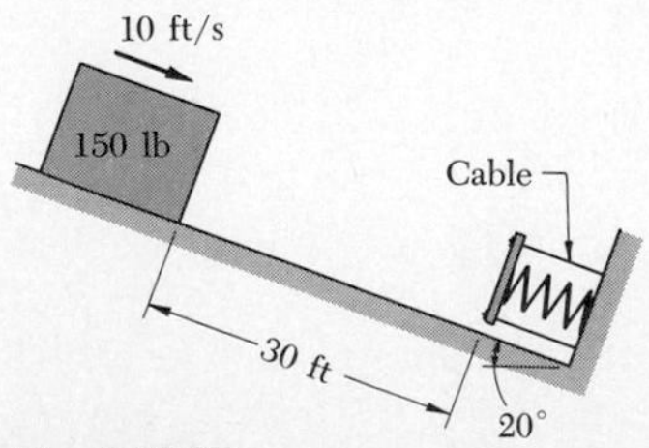

Fig. P13.67

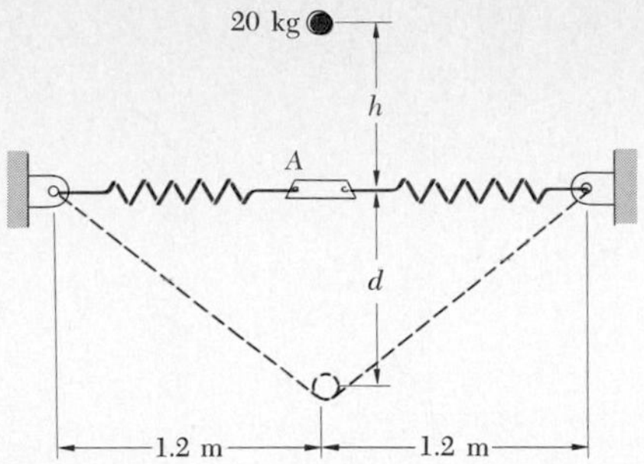

Fig. P13.68

13.68 Two springs are attached to a piece of cloth A of negligible mass, as shown. The initial tension in each spring is 500 N, and the spring constant of each spring is $k = 2$ kN/m. A 20-kg ball is released from a height h above A; the ball hits the cloth, causing it to move through a maximum distance $d = 0.9$ m. Determine the height h.

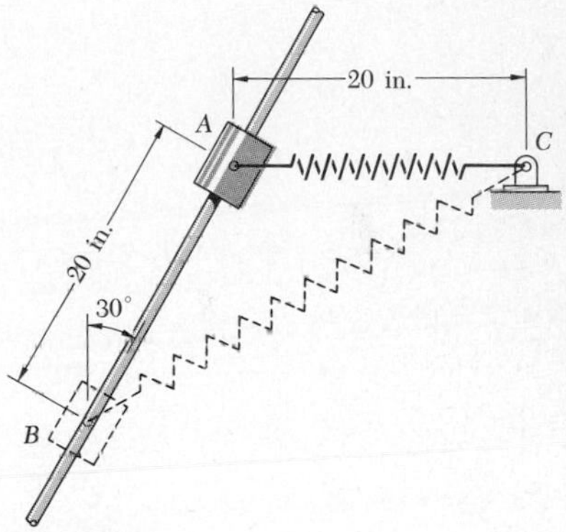

Fig. P13.69

13.69 A 12-lb collar slides without friction along a rod which forms an angle of 30° with the vertical. The spring is unstretched when the collar is at A. If the collar is released from rest at A, determine the value of the spring constant k for which the collar has zero velocity at B.

13.70 A 2-kg collar is attached to a spring and slides without friction in a vertical plane along the curved rod ABC. The spring is undeformed when the collar is at C and its constant is 600 N/m. If the collar is released at A with no initial velocity, determine its velocity (a) as it passes through B, (b) as it reaches C.

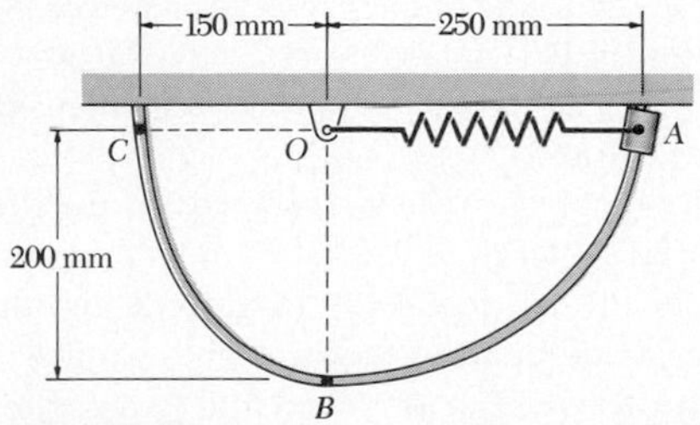

Fig. P13.70

13.71 Solve Prob. 13.70, assuming that the undeformed length of the spring is 120 mm and that its constant is 800 N/m.

13.72 Knowing that the rod of Prob. 13.70 has a radius of curvature of 200 mm at point B, determine the force exerted by the rod on the collar as it passes through point B.

13.73 A 200-g pellet is released from rest at A and slides without friction along the surface shown. Determine the force exerted by the surface on the pellet as it passes through (a) point B, (b) point C.

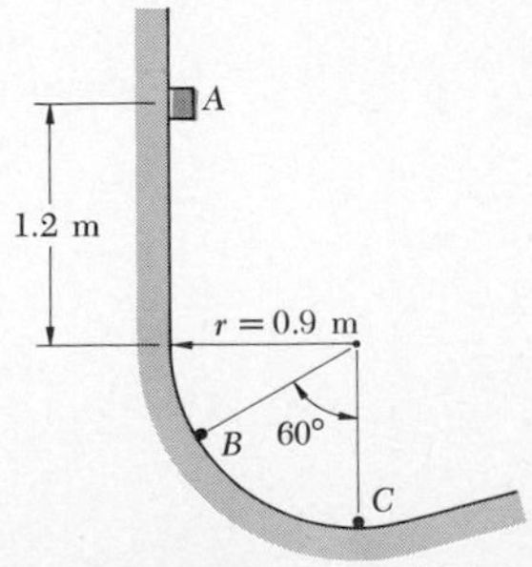

Fig. P13.73

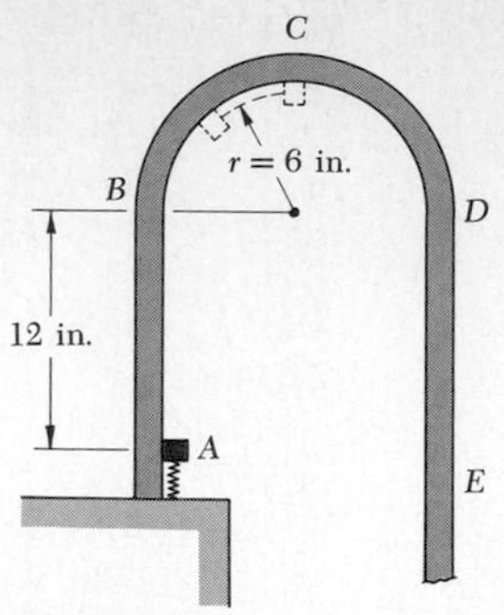

Fig. P13.74

13.74 A $\frac{1}{2}$-lb pellet is released from rest at A when the spring is compressed 3 in. and travels around the loop $ABCDE$. Determine the smallest value of the spring constant for which the pellet will travel around the loop and will at all times remain in contact with the loop.

13.75 In Prob. 13.63, determine the force exerted by the rod on collar C as the collar passes through point B.

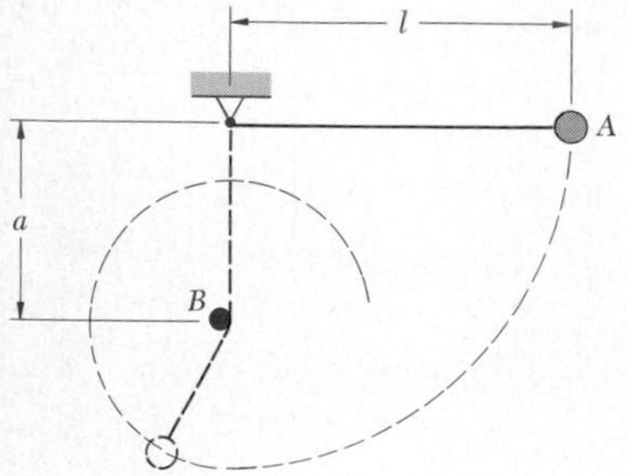

Fig. P13.76

13.76 The pendulum shown is released from rest at A and swings through 90° before the cord touches the fixed peg B. Determine the smallest value of a for which the pendulum bob will describe a circle about the peg.

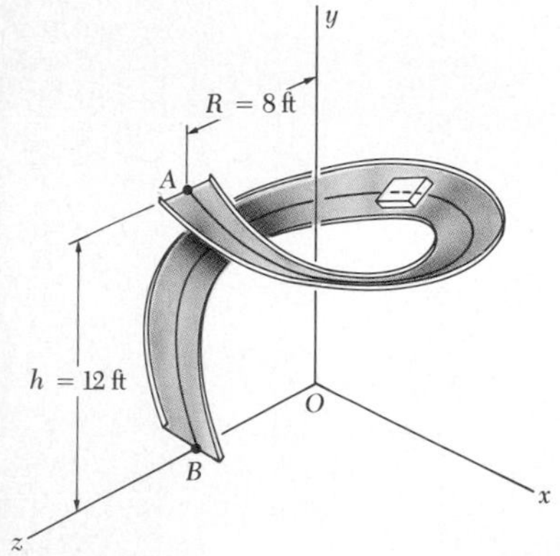

Fig. P13.77

***13.77** Packages are moved from point A on the upper floor of a warehouse to point B on the lower floor, 12 ft directly below A, by means of a chute, the centerline of which is in the shape of a helix of vertical axis y and radius $R = 8$ ft. The cross section of the chute is to be banked in such a way that each package, after being released at A with no velocity, will slide along the centerline of the chute, without ever touching its edges. Neglecting friction, (*a*) express as a function of the elevation y of a given point P of the centerline the angle ϕ formed by the normal to the surface of the chute at P and the principal normal of the centerline at that point, (*b*) determine the magnitude and direction of the force exerted by the chute on a 20-lb package as it reaches point B. *Hint.* The principal normal to the helix at any point P is horizontal and directed toward the y axis, and the radius of curvature of the helix is $\rho = R[1 + (h/2\pi R)^2]$.

***13.78** Prove that a force $\mathbf{F}(x, y, z)$ is conservative if, and only if, the following relations are satisfied:

$$\frac{\partial F_x}{\partial y} = \frac{\partial F_y}{\partial x} \qquad \frac{\partial F_y}{\partial z} = \frac{\partial F_z}{\partial y} \qquad \frac{\partial F_z}{\partial x} = \frac{\partial F_x}{\partial z}$$

***13.79** A force $\mathbf{F} = (x\mathbf{i} + y\mathbf{j} + z\mathbf{k})(x^2 + y^2 + z^2)$ acts on a particle $P(x, y, z)$ which moves in space. (*a*) Using the relations derived in Prob. 13.78, prove that $\mathbf{F}$ is a conservative force. (*b*) Determine the potential function $V(x, y, z)$ associated with $\mathbf{F}$.

***13.80** A force **F** acts on a particle $P(x, y)$ which moves in the xy plane. Determine whether **F** is a conservative force and compute the work of **F** when P describes in counterclockwise sense the square of vertices A, B, C, and D, (a) when $\mathbf{F} = ky\mathbf{i}$, ($b$) when $\mathbf{F} = k(y\mathbf{i} + x\mathbf{j})$.

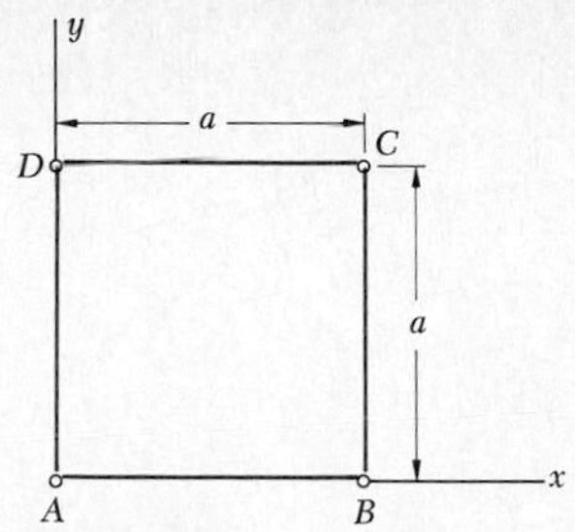

Fig. P13.80

***13.81** A force $\mathbf{F} = (x\mathbf{i} + y\mathbf{j})/(x^2 + y^2)^2$ acts on a particle $P(x, y)$ which moves in the xy plane. (a) Using the first of the relations derived in Prob. 13.78, prove that **F** is a conservative force. (b) Determine the potential function $V(x, y)$ associated with **F**.

13.82 (a) Determine the kinetic energy per unit mass which a missile must have after being fired from the surface of the earth if it is to reach an infinite distance from the earth. (b) What is the initial velocity of the missile (called *escape velocity*)? Give your answers in SI units and show that the answer to part b is independent of the firing angle.

13.83 A satellite of mass m describes a circular orbit of radius r around the earth. Express (a) its potential energy, (b) its kinetic energy, (c) its total energy, as a function of r. Denote by R the radius of the earth, by g the acceleration of gravity at the surface of the earth, and assume that the potential energy of the satellite is zero on its launching pad.

13.84 (a) Show by setting $r = R + y$ in formula (13.17′) and expanding in a power series in y/R that the expression obtained in (13.16) for the potential energy V_g due to gravity is a first-order approximation for the expression given in (13.17′). (b) Using the same expansion, derive a second-order approximation for V_g.

13.85 A lunar excursion module (LEM) was used in the Apollo moon-landing missions to save fuel by making it unnecessary to launch the entire Apollo spacecraft from the moon's surface on its return trip to the earth. Check the effectiveness of this approach by computing the energy per kilogram required for a spacecraft to escape the gravitational field of the moon if the spacecraft starts (a) from the moon's surface, (b) from a circular orbit 100 km above the moon's surface. Neglect the effect of the earth's gravitational field. (The radius of the moon is 1740 km and its mass is 0.01230 times the mass of the earth.)

13.86 During its fifth mission, the space shuttle Columbia ejected two communication satellites while describing a circular orbit, 185 mi above the surface of the earth. Knowing that one of these satellites weighed 8000 lb, determine (a) the additional energy required to place the satellite in a geosynchronous orbit (see Prob. 12.79) at an altitude of 22,230 mi above the surface of the earth, (b) the energy required to place it in the same orbit by launching it from the surface of the earth.

13.87 Show that the ratio of the potential and kinetic energies of an electron as it enters the plates of the cathode-ray tube of Prob. 12.62 is equal to $d\delta/lL$. (Place the datum at the surface of the positive plate.)

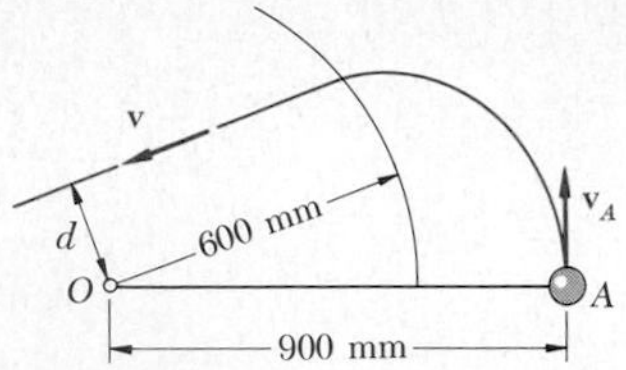

Fig. P13.88

13.88 A 200-g ball may slide on a horizontal frictionless surface and is attached to a fixed point O by means of an elastic cord of constant $k = 150$ N/m and undeformed length equal to 600 mm. The ball is placed at point A, 900 mm from O, and is given an initial velocity $\mathbf{v}_A$ in a direction perpendicular to OA. Knowing that the ball passes at a distance $d = 100$ mm from point O, determine (*a*) the initial speed v_A of the ball, (*b*) its speed v after the cord has become slack.

13.89 Knowing that the initial speed of the ball of Prob. 13.88 is $v_A = 3$ m/s, determine (*a*) the speed v of the ball after the cord has become slack, (*b*) the closest distance d that the ball will come to O.

13.90 For the ball of Prob. 13.88, determine the smallest magnitude of the initial velocity $\mathbf{v}_A$ for which the elastic cord remains taut at all times.

13.91 In Sample Prob. 13.8, determine the required magnitude of the velocity $\mathbf{v}_A$ if the maximum distance from the origin O reached by the sphere is to be 1.6 m.

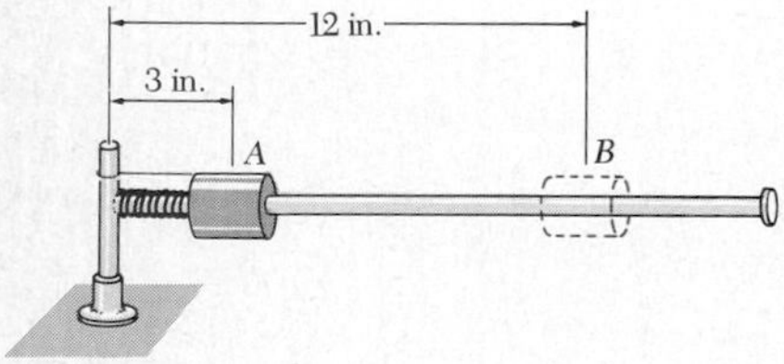

Fig. P13.92

13.92 A $\frac{1}{2}$-lb collar may slide on a horizontal rod which is free to rotate about a vertical shaft. The collar is initially held at A by a cord attached to the shaft and compresses a spring of constant 2.5 lb/ft, which is undeformed when the collar is located 9 in. from the shaft. As the rod rotates at the rate $\dot{\theta}_0 = 12$ rad/s, the cord is cut and the collar moves along the rod. Neglecting friction and the mass of the rod, and assuming that the spring is attached to the collar and to the shaft, determine the radial and transverse components of the velocity of the collar as it passes through B.

13.93 In Prob. 13.92, determine the maximum distance r from the vertical shaft that the collar will reach. (*Hint.* Solve by trial the equation obtained for r.)

13.94 through 13.96 Using the principles of conservation of energy and conservation of angular momentum, solve the following problems:

13.94 Prob. 12.98.

13.95 Prob. 12.99.

13.96 Prob. 12.102*a*.

13.97 Using the principles of conservation of energy and conservation of angular momentum, determine the velocity of the probe of Prob. 12.101 (*a*) just after it has been placed on its second transfer orbit at B, (*b*) as it reaches point C.

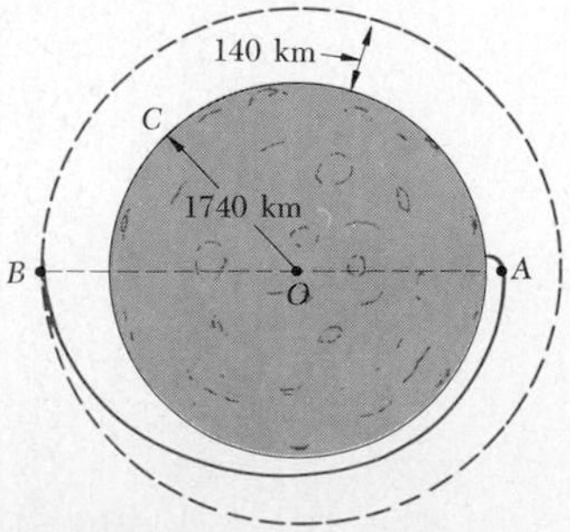

Fig. P13.98

13.98 After completing their moon-exploration mission, the two astronauts forming the crew of an Apollo lunar excursion module (LEM) prepared to rejoin the command module which was orbiting the moon at an altitude of 140 km. They fired the LEM's engine, brought it along a curved path to a point A, 8 km above the moon's surface, and shut off the engine. Knowing that the LEM was moving at that time in a direction parallel to the moon's surface and that it then coasted along an elliptic path to a rendezvous at B with the command module, determine (*a*) the speed of the LEM at engine shutoff, (*b*) the relative velocity with which the command module approached the LEM at B. (The radius of the moon is 1740 km and its mass is 0.01230 times the mass of the earth.)

13.99 While describing a circular orbit, 185 mi above the surface of the earth, a space shuttle ejects at point A an inertial upper stage (IUS) carrying a communication satellite to be placed in a geosynchronous orbit (see Prob. 12.79) at an altitude of 22,230 mi above the surface of the earth. Determine (*a*) the velocity of the IUS relative to the shuttle after its engine has been fired at A, (*b*) the increase in velocity required at B to place the satellite in its final orbit.

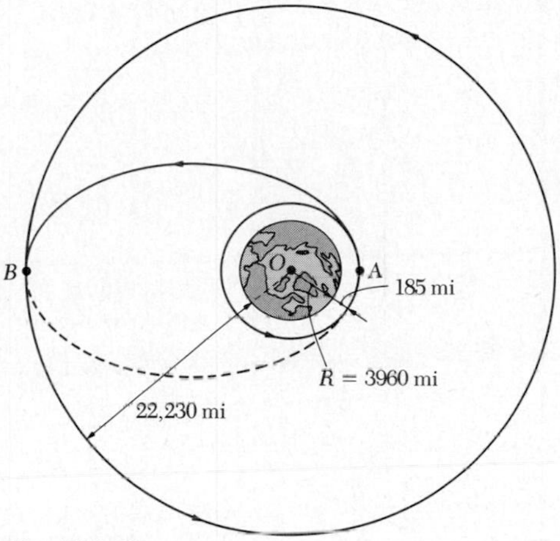

Fig. P13.99

13.100 Upon the LEM's return to the command module, the Apollo spacecraft of Prob. 13.98 was turned around so that the LEM faced to the rear. The LEM was then cast adrift with a velocity of 200 m/s relative to the command module. Determine the magnitude and direction (angle ϕ formed with the vertical OC) of the velocity $\mathbf{v}_C$ of the LEM just before it crashed at C on the moon's surface.

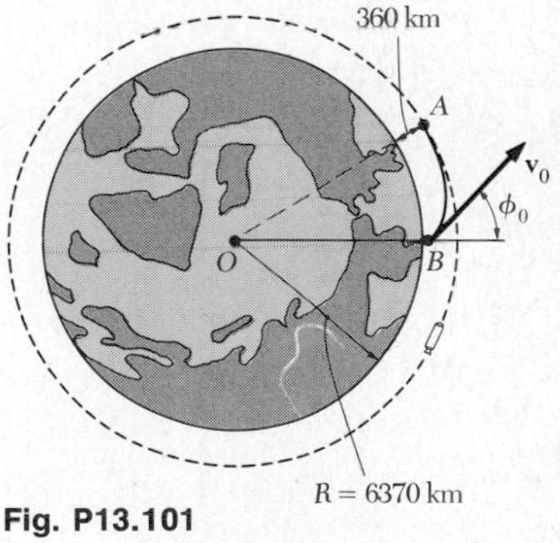

Fig. P13.101

13.101 A space shuttle is to rendezvous with an orbiting laboratory which circles the earth at the constant altitude of 360 km. The shuttle has reached an altitude of 60 km when its engine is shut off, and its velocity $\mathbf{v}_0$ forms an angle $\phi_0 = 45°$ with the vertical OB at that time. What magnitude should $\mathbf{v}_0$ have if the shuttle's trajectory is to be tangent at A to the orbit of the laboratory?

13.102 A space shuttle is describing a circular orbit at an altitude of 150 mi above the surface of the earth. As it passes through A, it fires its engine for a short interval of time to reduce its speed by 4 percent and begin its descent toward the earth. Determine the magnitude and direction (angle ϕ formed with the vertical OB) of the velocity $\mathbf{v}_B$ of the shuttle as it reaches point B at an altitude of 30 mi.

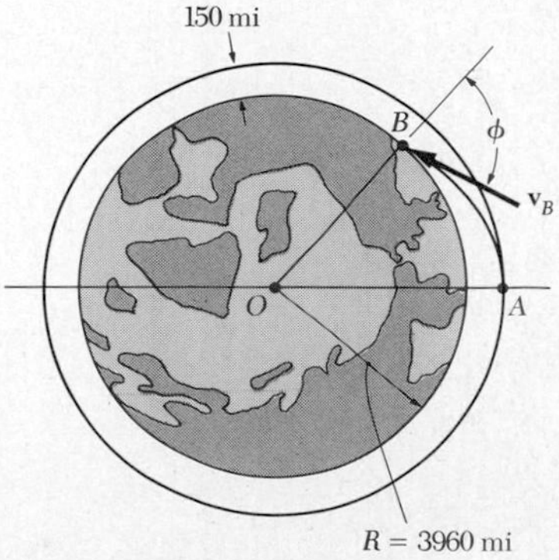

Fig. P13.102 and P13.103

13.103 A space shuttle is describing a circular orbit at an altitude of 150 mi above the surface of the earth. As it passes through A, it fires its engine to reduce its speed and begin its descent toward the earth. As it reaches point B at an altitude of 30 mi, its velocity $\mathbf{v}_B$ is observed to form an angle $\phi = 75°$ with the vertical. What is the magnitude of its velocity at that instant?

13.104 At engine burnout, an experimental rocket has reached an altitude of 400 km and has a velocity $\mathbf{v}_0$ of magnitude 7500 m/s forming an angle of 35° with the vertical. Determine the maximum altitude reached by the rocket.

13.105 At engine burnout, an experimental rocket has reached an altitude of 400 km and has a velocity $\mathbf{v}_0$ of magnitude 8000 m/s. What angle should $\mathbf{v}_0$ form with the vertical if the rocket is to reach a maximum altitude of 4000 km?

13.106 Show that the values v_1 and v_2 of the speed of an earth satellite at the perigee A and the apogee A' of an elliptic orbit are defined by the relations

$$v_1^2 = \frac{2GM}{r_1 + r_2}\frac{r_2}{r_1} \qquad v_2^2 = \frac{2GM}{r_1 + r_2}\frac{r_1}{r_2}$$

where M is the mass of the earth, and r_1 and r_2 represent, respectively, the minimum and maximum distance of the orbit to the center of the earth.

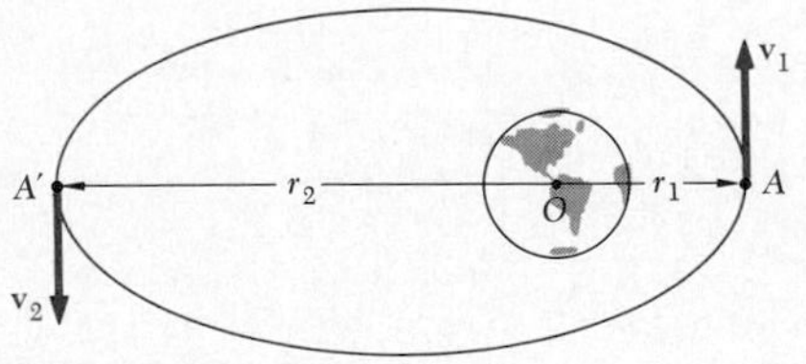

Fig. P13.106 and P13.107

13.107 Show that the total energy E of an earth satellite of mass m describing an elliptic orbit is $E = -GMm/(r_1 + r_2)$, where M is the mass of the earth, and r_1 and r_2 represent, respectively, the minimum and maximum distance of the orbit to the center of the earth. (It is recalled that the gravitational potential energy of a satellite was defined as being zero at an infinite distance from the earth.)

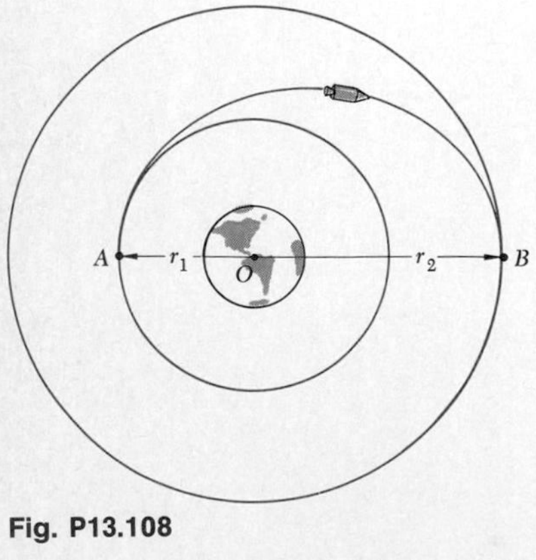

Fig. P13.108

13.108 A spacecraft of mass m describes a circular orbit of radius r_1 around the earth. (*a*) Show that the additional energy ΔE which must be imparted to the spacecraft to transfer it to a circular orbit of larger radius r_2 is

$$\Delta E = \frac{GMm(r_2 - r_1)}{2r_1 r_2}$$

where M is the mass of the earth. (*b*) Further show that if the transfer from one circular orbit to the other is executed by placing the spacecraft on a transitional semielliptic path AB, the amounts of energy ΔE_A and ΔE_B which must be imparted at A and B are, respectively, proportional to r_2 and r_1:

$$\Delta E_A = \frac{r_2}{r_1 + r_2}\Delta E \qquad \Delta E_B = \frac{r_1}{r_1 + r_2}\Delta E$$

13.109 A missile is fired from the ground with an initial velocity $\mathbf{v}_0$ forming an angle ϕ_0 with the vertical. If the missile is to reach a maximum altitude equal to αR, where R is the radius of the earth, (*a*) show that the required angle ϕ_0 is defined by the relation

$$\sin \phi_0 = (1 + \alpha)\sqrt{1 - \frac{\alpha}{1 + \alpha}\left(\frac{v_{esc}}{v_0}\right)^2}$$

where v_{esc} is the escape velocity, (*b*) determine the range of allowable values of v_0.

13.110 A satellite is projected into space with a velocity $\mathbf{v}_0$ at a distance r_0 from the center of the earth by the last stage of its launching rocket. The velocity $\mathbf{v}_0$ was designed to send the satellite into a circular orbit of radius r_0. However, owing to a malfunction of control, the satellite is not projected horizontally but at an angle α with the horizontal and, as a result, is propelled into an elliptic orbit. Determine the maximum and minimum values of the distance from the center of the earth to the satellite.

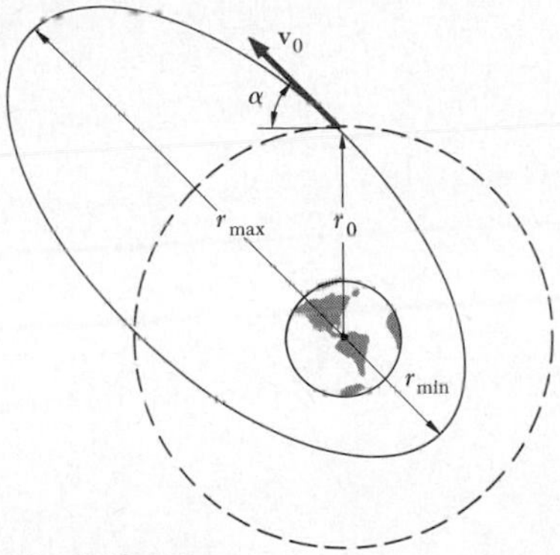

Fig. P13.110

***13.111** Using the answers obtained in Prob. 13.110, show that the intended circular orbit and the resulting elliptic orbit intersect at the ends of the minor axis of the elliptic orbit.

***13.112** (*a*) Express in terms of r_{min} and v_{max} the angular momentum per unit mass, h, and the total energy per unit mass, E/m, of a space vehicle moving under the gravitational attraction of a planet of mass M (Fig. 13.15). (*b*) Eliminating v_{max} between the equations obtained, derive the formula

$$\frac{1}{r_{min}} = \frac{GM}{h^2}\left[1 + \sqrt{1 + \frac{2E}{m}\left(\frac{h}{GM}\right)^2}\right]$$

(*c*) Show that the eccentricity ε of the trajectory of the vehicle may be expressed as

$$\varepsilon = \sqrt{1 + \frac{2E}{m}\left(\frac{h}{GM}\right)^2}$$

(*d*) Further show that the trajectory of the vehicle is a hyperbola, an ellipse, or a parabola, depending on whether E is positive, negative, or zero.

13.10. Principle of Impulse and Momentum.

A third basic method for the solution of problems dealing with the motion of particles will be considered now. This method is based on the principle of impulse and momentum and may be used to solve problems involving force, mass, velocity, and time. It is of particular interest in the solution of problems involving impulsive motion or impact (Secs. 13.11 and 13.12).

Consider a particle of mass m acted upon by a force $\mathbf{F}$. As we saw in Sec. 12.3, Newton's second law may be expressed in the form

$$\mathbf{F} = \frac{d}{dt}(m\mathbf{v}) \tag{13.27}$$

where $m\mathbf{v}$ is the linear momentum of the particle. Multiplying both sides of Eq. (13.27) by dt and integrating from a time t_1 to a time t_2, we write

$$\mathbf{F}\,dt = d(m\mathbf{v})$$

$$\int_{t_1}^{t_2} \mathbf{F}\,dt = m\mathbf{v}_2 - m\mathbf{v}_1$$

or, transposing the last term,

$$m\mathbf{v}_1 + \int_{t_1}^{t_2} \mathbf{F}\,dt = m\mathbf{v}_2 \tag{13.28}$$

The integral in Eq. (13.28) is a vector known as the *linear impulse,* or simply the *impulse,* of the force $\mathbf{F}$ during the interval of time considered. Resolving $\mathbf{F}$ into rectangular components, we write

$$\mathbf{Imp}_{1\to 2} = \int_{t_1}^{t_2} \mathbf{F}\,dt$$

$$= \mathbf{i}\int_{t_1}^{t_2} F_x\,dt + \mathbf{j}\int_{t_1}^{t_2} F_y\,dt + \mathbf{k}\int_{t_1}^{t_2} F_z\,dt \tag{13.29}$$

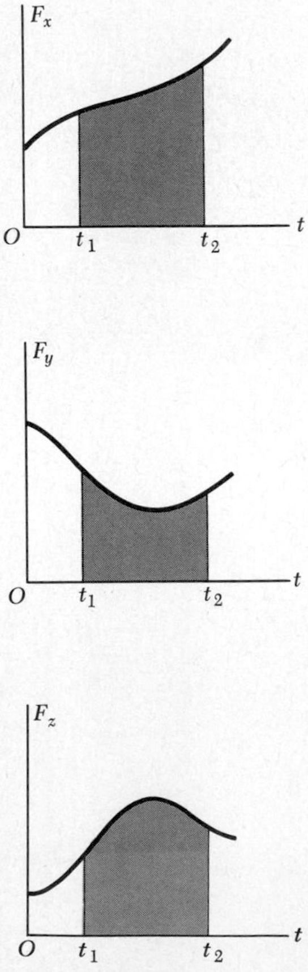

Fig. 13.16

and note that the components of the impulse of the force $\mathbf{F}$ are, respectively, equal to the areas under the curves obtained by plotting the components F_x, F_y, and F_z against t (Fig. 13.16). In the case of a force $\mathbf{F}$ of constant magnitude and direction, the impulse is represented by the vector $\mathbf{F}(t_2 - t_1)$, which has the same direction as $\mathbf{F}$.

If SI units are used, the magnitude of the impulse of a force is expressed in N·s. But, recalling the definition of the newton, we have

$$\text{N}\cdot\text{s} = (\text{kg}\cdot\text{m/s}^2)\cdot\text{s} = \text{kg}\cdot\text{m/s}$$

which is the unit obtained in Sec. 12.4 for the linear momentum of a particle. We thus check that Eq. (13.28) is dimensionally correct. If U.S. customary units are used, the impulse of a force is expressed in lb·s, which is also the unit obtained in Sec. 12.4 for the linear momentum of a particle.

Equation (13.28) expresses that when a particle is acted upon by a force **F** during a given time interval, *the final momentum* $m\mathbf{v}_2$ *of the particle may be obtained by adding vectorially its initial momentum* $m\mathbf{v}_1$ *and the impulse of the force* **F** *during the time interval considered* (Fig. 13.17).

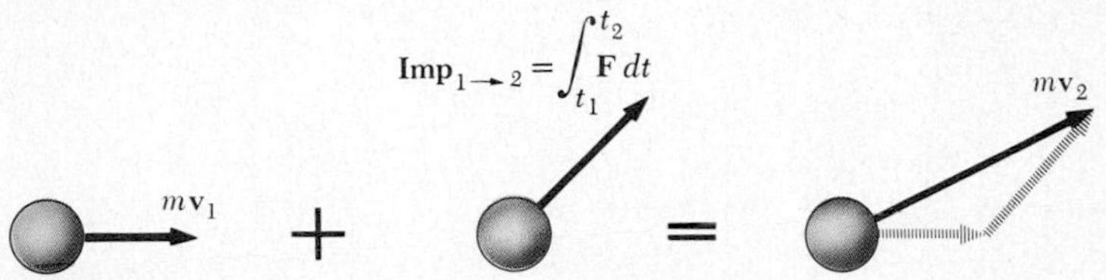

Fig. 13.17

We write

$$m\mathbf{v}_1 + \mathbf{Imp}_{1\to 2} = m\mathbf{v}_2 \tag{13.30}$$

We note that while kinetic energy and work are scalar quantities, momentum and impulse are vector quantities. To obtain an analytic solution, it is thus necessary to replace Eq. (13.30) by the equivalent component equations

$$\begin{aligned}(mv_x)_1 + \int_{t_1}^{t_2} F_x\,dt &= (mv_x)_2 \\ (mv_y)_1 + \int_{t_1}^{t_2} F_y\,dt &= (mv_y)_2 \\ (mv_z)_1 + \int_{t_1}^{t_2} F_z\,dt &= (mv_z)_2\end{aligned} \tag{13.31}$$

When several forces act on a particle, the impulse of each of the forces must be considered. We have

$$m\mathbf{v}_1 + \Sigma\,\mathbf{Imp}_{1\to 2} = m\mathbf{v}_2 \tag{13.32}$$

Again, the equation obtained represents a relation between vector quantities; in the actual solution of a problem, it should be replaced by the corresponding component equations.

When a problem involves two particles or more, each particle may be considered separately and Eq. (13.32) may be written for each particle. We may also add vectorially the momenta of all the particles and the impulses of all the forces involved. We write then

$$\Sigma m\mathbf{v}_1 + \Sigma\,\mathbf{Imp}_{1\to 2} = \Sigma m\mathbf{v}_2 \tag{13.33}$$

Since the forces of action and reaction exerted by the particles on each other form pairs of equal and opposite forces, and since the time interval from t_1 to t_2 is common to all the forces involved, the impulses of the forces of action and reaction cancel out, and only the impulses of the external forces need be considered.†

If no external force is exerted on the particles or, more generally, if the sum of the external forces is zero, the second term in Eq. (13.33) vanishes, and Eq. (13.33) reduces to

$$\Sigma m\mathbf{v}_1 = \Sigma m\mathbf{v}_2 \tag{13.34}$$

which expresses that *the total momentum of the particles is conserved.* Consider, for example, two boats, of mass m_A and m_B, initially at rest, which are being pulled together (Fig. 13.18). If the resistance of the water

Fig. 13.18

is neglected, the only external forces acting on the boats are their weights and the buoyant forces exerted on them. Since these forces are balanced, we write

$$\Sigma m\mathbf{v}_1 = \Sigma m\mathbf{v}_2$$
$$0 = m_A\mathbf{v}'_A + m_B\mathbf{v}'_B$$

where $\mathbf{v}'_A$ and $\mathbf{v}'_B$ represent the velocities of the boats after a finite interval of time. The equation obtained indicates that the boats move in opposite directions (toward each other) with velocities inversely proportional to their masses.‡

13.11. Impulsive Motion. In some problems, a very large force may act during a very short time interval on a particle and produce a definite change in momentum. Such a force is called an *impulsive force* and the resulting motion an *impulsive motion.* For example, when a baseball is struck, the contact between bat and ball takes place during a very short time interval Δt. But the average value of the force $\mathbf{F}$ exerted by the bat on

† We should note the difference between this statement and the corresponding statement made in Sec. 13.4 regarding the work of the forces of action and reaction between several particles. While the sum of the impulses of these forces is always zero, the sum of their work is zero only under special circumstances, e.g., when the various bodies involved are connected by inextensible cords or links and are thus constrained to move through equal distances.

‡ The application of the method of impulse and momentum to a system of particles and the concept of conservation of momentum for a system of particles are discussed in detail in Chap. 14.

the ball is very large, and the resulting impulse $\mathbf{F}\,\Delta t$ is large enough to change the sense of motion of the ball (Fig. 13.19).

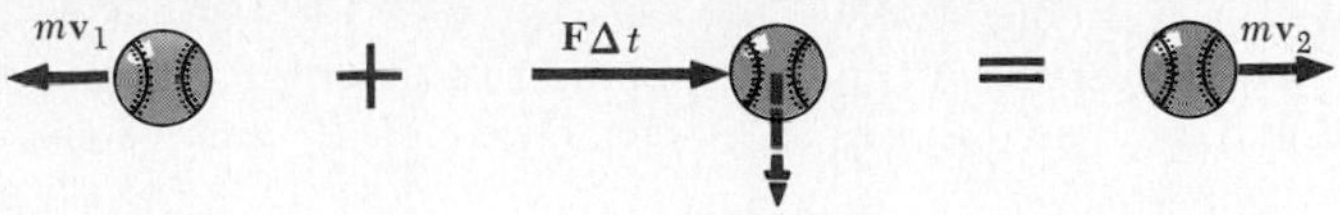

Fig. 13.19

When impulsive forces act on a particle, Eq. (13.32) becomes

$$m\mathbf{v}_1 + \Sigma\mathbf{F}\,\Delta t = m\mathbf{v}_2 \tag{13.35}$$

Any force which is not an impulsive force may be neglected, since the corresponding impulse $\mathbf{F}\,\Delta t$ is very small. *Nonimpulsive forces* include the weight of the body, the force exerted by a spring, or any other force which is *known* to be small compared with an impulsive force. Unknown reactions may or may not be impulsive; their impulse should therefore be included in Eq. (13.35) as long as it has not been proved negligible. The impulse of the weight of the baseball considered above, for example, may be neglected. If the motion of the bat is analyzed, the impulse of the weight of the bat may also be neglected. The impulses of the reactions of the player's hands on the bat, however, should be included; these impulses will not be negligible if the ball is incorrectly hit.

We note that the method of impulse and momentum is particularly effective in the analysis of the impulsive motion of a particle, since it involves only the initial and final velocities of the particle and the impulses of the forces exerted on the particle. The direct application of Newton's second law, on the other hand, would require the determination of the forces as functions of the time and the integration of the equations of motion over the time interval Δt.

In the case of the impulsive motion of several particles, Eq. (13.33) may be used. It reduces to

$$\Sigma m\mathbf{v}_1 + \Sigma\mathbf{F}\,\Delta t = \Sigma m\mathbf{v}_2 \tag{13.36}$$

where the second term involves only impulsive, external forces. If all the external forces acting on the various particles are nonimpulsive, the second term in Eq. (13.36) vanishes and this equation reduces to Eq. (13.34). We write

$$\Sigma m\mathbf{v}_1 = \Sigma m\mathbf{v}_2 \tag{13.34}$$

which expresses that the total momentum of the particles is conserved. This situation occurs, for example, when two particles which are moving freely collide with one another. We should note, however, that while the total momentum of the particles is conserved, their total energy is generally *not* conserved. Problems involving the collision or *impact* of two particles will be discussed in detail in Secs. 13.12 through 13.14.

SAMPLE PROBLEM 13.10

An automobile weighing 4000 lb is driven down a 5° incline at a speed of 60 mi/h when the brakes are applied, causing a constant total braking force (applied by the road on the tires) of 1500 lb. Determine the time required for the automobile to come to a stop.

Solution. We apply the principle of impulse and momentum. Since each force is constant in magnitude and direction, each corresponding impulse is equal to the product of the force and of the time interval t.

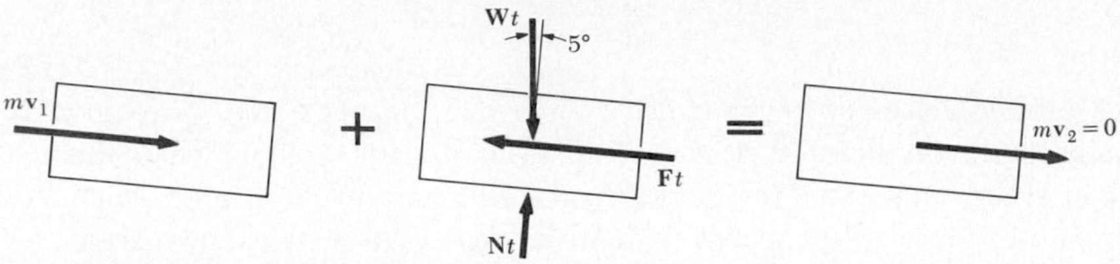

$$m\mathbf{v}_1 + \Sigma\, \mathbf{Imp}_{1\to 2} = m\mathbf{v}_2$$

$+\searrow x$ components: $mv_1 + (W \sin 5°)t - Ft = 0$

$(4000/32.2)(88 \text{ ft/s}) + (4000 \sin 5°)t - 1500t = 0$ $\qquad t = 9.49 \text{ s}$ ◄

SAMPLE PROBLEM 13.11

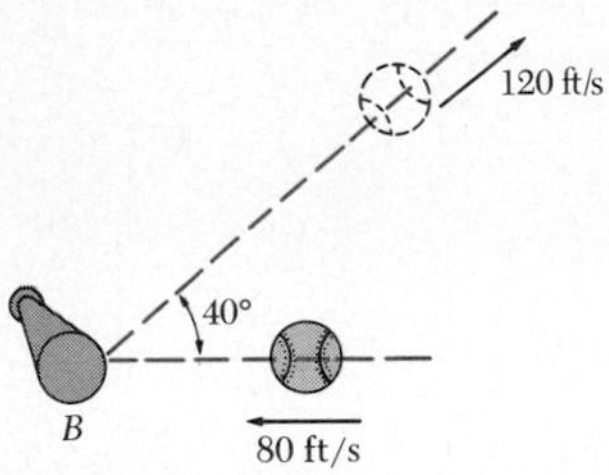

A 4-oz baseball is pitched with a velocity of 80 ft/s toward a batter. After the ball is hit by the bat B, it has a velocity of 120 ft/s in the direction shown. If the bat and ball are in contact 0.015 s, determine the average impulsive force exerted on the ball during the impact.

Solution. We apply the principle of impulse and momentum to the ball. Since the weight of the ball is a nonimpulsive force, we shall neglect it.

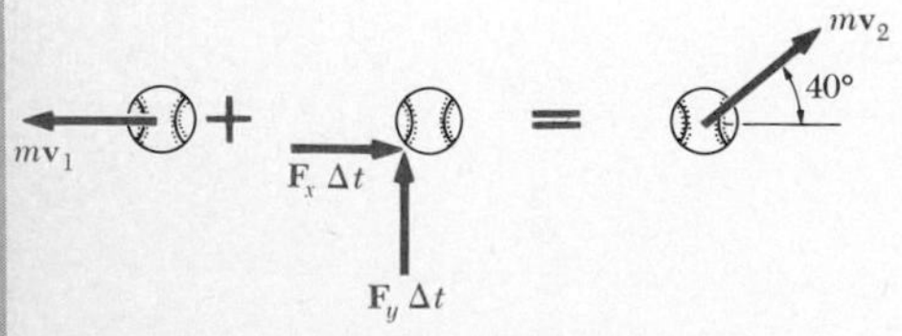

$$m\mathbf{v}_1 + \Sigma\, \mathbf{Imp}_{1\to 2} = m\mathbf{v}_2$$

$\xrightarrow{+}$ x components: $\qquad -mv_1 + F_x\,\Delta t = mv_2 \cos 40°$

$$-\frac{\frac{4}{16}}{32.2}(80 \text{ ft/s}) + F_x(0.015 \text{ s}) = \frac{\frac{4}{16}}{32.2}(120 \text{ ft/s}) \cos 40°$$

$$F_x = +89.0 \text{ lb}$$

$+\uparrow y$ components: $\qquad 0 + F_y\,\Delta t = mv_2 \sin 40°$

$$F_y(0.015 \text{ s}) = \frac{\frac{4}{16}}{32.2}(120 \text{ ft/s}) \sin 40°$$

$$F_y = +39.9 \text{ lb}$$

From its components F_x and F_y we determine the magnitude and direction of the force $\mathbf{F}$:

$\mathbf{F} = 97.5 \text{ lb} \measuredangle 24.2°$ ◄

SAMPLE PROBLEM 13.12

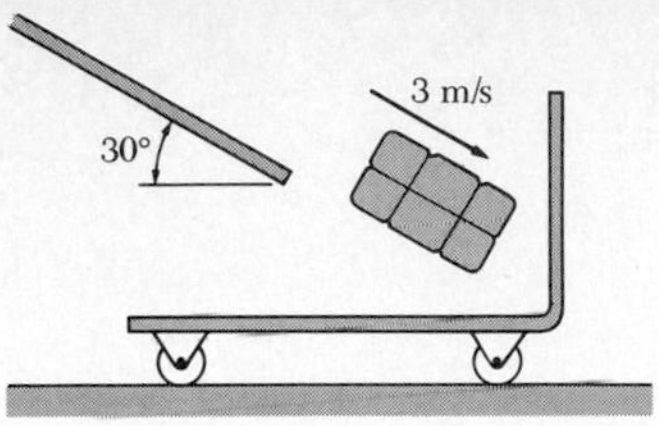

A 10-kg package drops from a chute into a 25-kg cart with a velocity of 3 m/s. Knowing that the cart is initially at rest and may roll freely, determine (*a*) the final velocity of the cart, (*b*) the impulse exerted by the cart on the package, (*c*) the fraction of the initial energy lost in the impact.

Solution. We first apply the principle of impulse and momentum to the package-cart system to determine the velocity $\mathbf{v}_2$ of the cart and package. We then apply the same principle to the package alone to determine the impulse $\mathbf{F}\,\Delta t$ exerted on it.

a. Impulse-Momentum Principle: Package and Cart

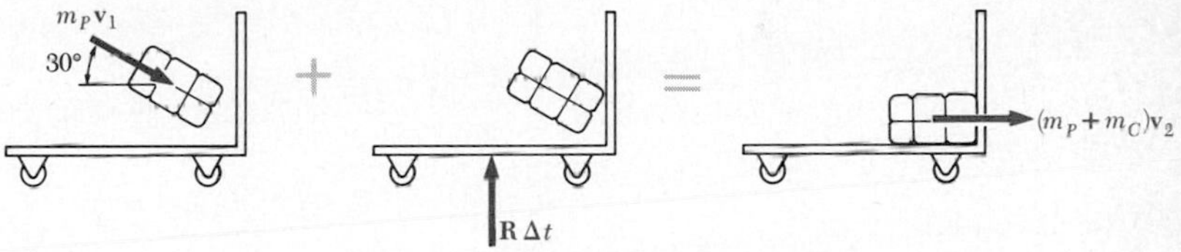

$$m_P\mathbf{v}_1 + \Sigma\,\mathbf{Imp}_{1\to 2} = (m_P + m_C)\mathbf{v}_2$$

$\xrightarrow{+}$ x components: $\quad m_P v_1 \cos 30° + 0 = (m_P + m_C)v_2$

$$(10\text{ kg})(3\text{ m/s})\cos 30° = (10\text{ kg} + 25\text{ kg})v_2$$

$$\mathbf{v}_2 = 0.742\text{ m/s} \rightarrow \blacktriangleleft$$

We note that the equation used expresses conservation of momentum in the x direction.

b. Impulse-Momentum Principle: Package

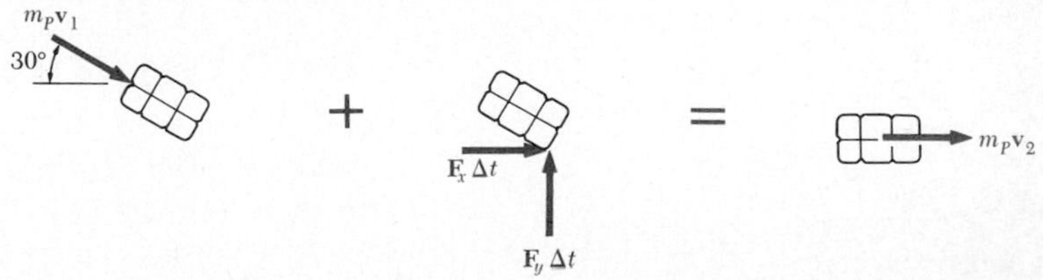

$$m_P\mathbf{v}_1 + \Sigma\,\mathbf{Imp}_{1\to 2} = m_P\mathbf{v}_2$$

$\xrightarrow{+}$ x components: $\quad m_P v_1 \cos 30° + F_x\,\Delta t = (10\text{ kg})(0.742\text{ m/s})$

$$F_x\,\Delta t = -18.56\text{ N}\cdot\text{s}$$

$+\uparrow y$ components: $\quad -m_P v_1 \sin 30° + F_y\,\Delta t = 0$

$$-(10\text{ kg})(3\text{ m/s})\sin 30° + F_y\,\Delta t = 0$$

$$F_y\,\Delta t = +15\text{ N}\cdot\text{s}$$

The impulse exerted on the package is $\quad \mathbf{F}\,\Delta t = 23.9\text{ N}\cdot\text{s} \measuredangle 38.9° \blacktriangleleft$

c. Fraction of Energy Lost. The initial and final energies are

$$T_1 = \tfrac{1}{2}m_P v_1^2 = \tfrac{1}{2}(10\text{ kg})(3\text{ m/s})^2 = 45\text{ J}$$

$$T_2 = \tfrac{1}{2}(m_P + m_C)v_2^2 = \tfrac{1}{2}(10\text{ kg} + 25\text{ kg})(0.742\text{ m/s})^2 = 9.63\text{ J}$$

The fraction of energy lost is $\quad \dfrac{T_1 - T_2}{T_1} = \dfrac{45\text{ J} - 9.63\text{ J}}{45\text{ J}} = 0.786 \blacktriangleleft$

Problems

13.113 A 50,000-ton ocean liner has an initial velocity of 3 mi/h. Neglecting the frictional resistance of the water, determine the time required to bring the liner to rest by using a single tugboat which exerts a constant force of 45 kips.

13.114 A 1200-kg automobile is moving at a speed of 90 km/h when the brakes are fully applied, causing all four wheels to skid. Determine the time required to stop the automobile (*a*) on dry pavement ($\mu_k = 0.75$), (*b*) on an icy road ($\mu_k = 0.10$).

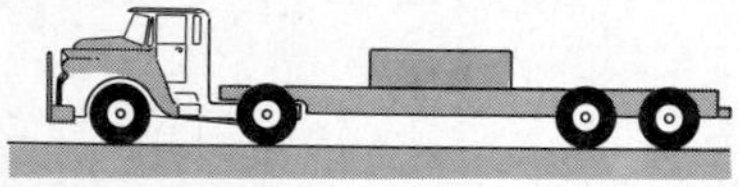

Fig. P13.115

13.115 The coefficients of friction between the load and the flatbed trailer shown are $\mu_s = 0.50$ and $\mu_k = 0.40$. Knowing that the speed of the rig is 72 km/h, determine the shortest time in which the rig can be brought to a stop if the load is not to shift.

13.116 A 10-lb particle is acted upon by the force, expressed in pounds, $\mathbf{F} = (4 - 3t)\mathbf{i} + (2 - t^2)\mathbf{j} + (2 + t)\mathbf{k}$. Knowing that at $t = 0$ the velocity of the particle is $\mathbf{v} = (75\text{ ft/s})\mathbf{i} + (50\text{ ft/s})\mathbf{j} - (125\text{ ft/s})\mathbf{k}$, determine (*a*) the time at which the velocity of the particle is parallel to the *yz* plane, (*b*) the corresponding velocity of the particle.

13.117 A 5-kg particle is acted upon by the force, expressed in newtons, $\mathbf{F} = 10t\,\mathbf{i} + (15 - 6t)\mathbf{j} + 4t^3\mathbf{k}$. Knowing that the velocity of the particle at $t = 0$ is $\mathbf{v} = -(8\text{ m/s})\mathbf{i} + (5\text{ m/s})\mathbf{j} - (20\text{ m/s})\mathbf{k}$, determine the velocity of the particle at $t = 5$ s.

13.118 Using the principle of impulse and momentum, solve Prob. 12.16.

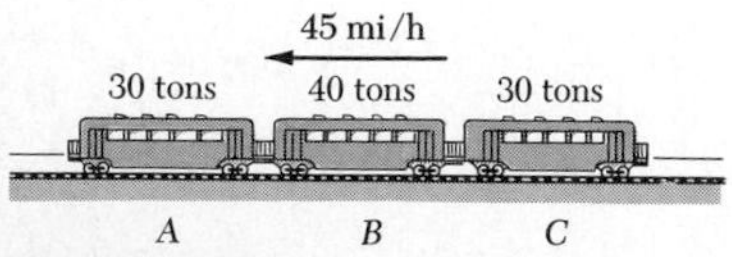

Fig. P13.119

13.119 The subway train shown is traveling at a speed of 45 mi/h when the brakes are fully applied on the wheels of cars *A* and *B*, causing them to slide on the track, but are not applied on the wheels of car *C*. Knowing that the coefficient of kinetic friction is 0.30 between the wheels and the track, determine (*a*) the time required to bring the train to a stop, (*b*) the force in each coupling.

13.120 Solve Prob. 13.119, assuming that the brakes are applied only on the wheels of car *A*.

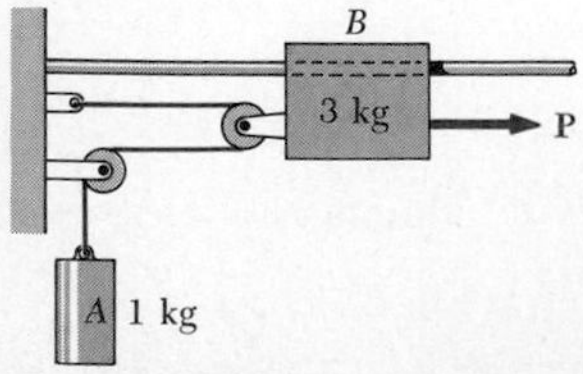

Fig. P13.121

13.121 The system shown is initially at rest. Neglecting friction, determine (*a*) the force **P** required if the velocity of collar *B* is to be 5 m/s after 2 s, (*b*) the corresponding tension in the cable.

13.122 Using the principle of impulse and momentum, solve Prob. 12.18*b*.

13.123 Two packages are placed on an incline as shown. The coefficients of friction are $\mu_s = 0.25$ and $\mu_k = 0.20$ between the incline and package A, and $\mu_s = 0.15$ and $\mu_k = 0.12$ between the incline and package B. Knowing that the packages are in contact when released, determine (*a*) the velocity of each package after 3 s, (*b*) the force exerted by package A on package B.

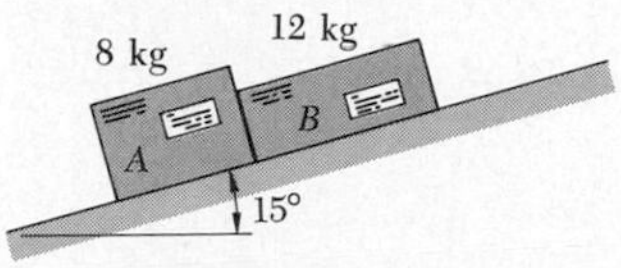

Fig. P13.123

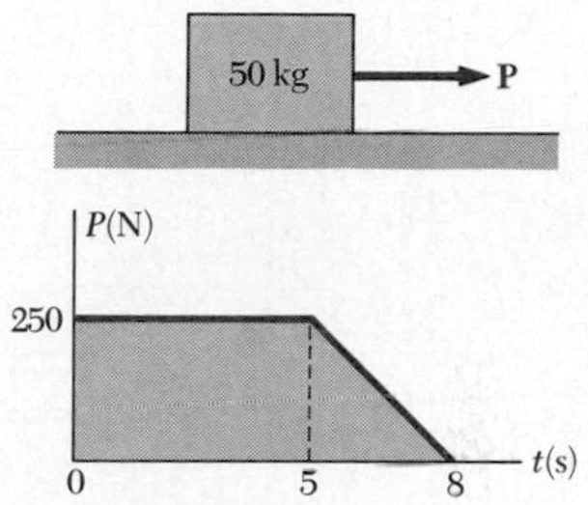

Fig. P13.124

13.124 A 50-kg block initially at rest is acted upon by a force **P** which varies as shown. Knowing that the coefficient of kinetic friction between the block and the horizontal surface is 0.20, determine the velocity of the block (*a*) at $t = 5$ s, (*b*) at $t = 8$ s.

13.125 In Prob. 13.124, determine (*a*) the maximum velocity reached by the block and the corresponding time, (*b*) the time at which the block comes to rest.

13.126 The pressure wave produced by an explosion exerts on a 50-lb block a force **P** whose variation with time may be approximated as shown. Knowing that the block was initially at rest and neglecting the effect of friction, determine (*a*) the maximum velocity reached by the block, (*b*) the velocity of the block at $t = 0.75$ s.

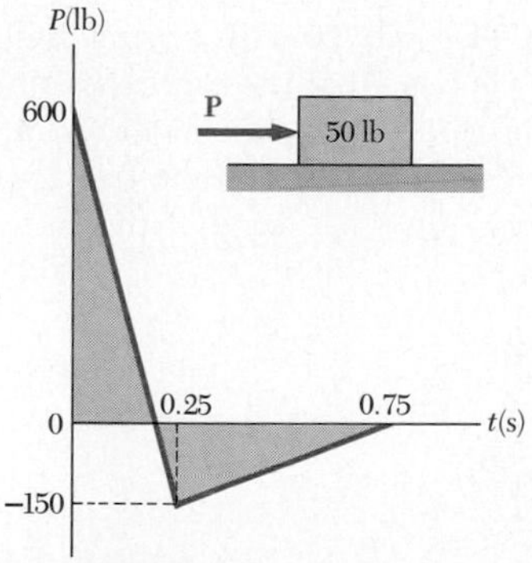

Fig. P13.126

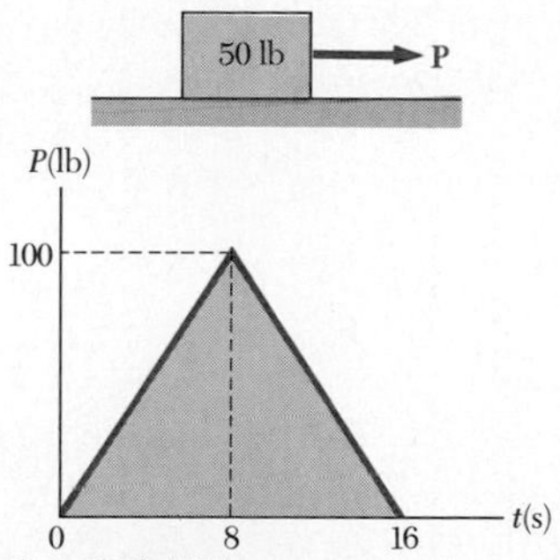

Fig. P13.127

13.127 A 50-block initially at rest is acted upon by a force **P** which varies as shown. Knowing that the coefficients of friction between the block and the horizontal surface are $\mu_s = 0.50$ and $\mu_k = 0.40$, determine the time at which the block will (*a*) start moving, (*b*) stop moving.

13.128 A 30-g bullet is fired with a velocity of 640 m/s into a wooden block which rests against a solid vertical wall. Knowing that the bullet is brought to rest in 0.8 ms, determine the average impulsive force exerted by the bullet on the block.

13.129 A 2500-lb car moving with a velocity of 3 mi/h hits a garage wall and is brought to rest in 0.05 s. Determine the average impulsive force exerted by the wall on the car bumper.

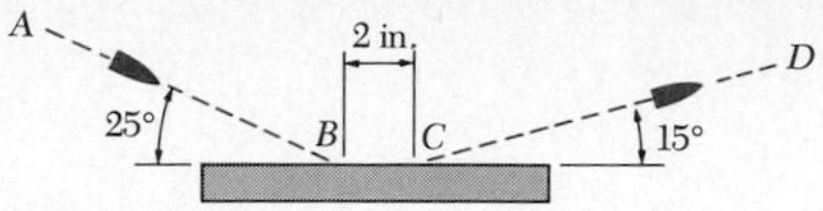

Fig. P13.130

13.130 A 1-oz steel-jacketed bullet is fired with a velocity of 2200 ft/s toward a steel plate and ricochets along the path CD with a velocity of 1800 ft/s. Knowing that the bullet leaves a 2-in. scratch on the surface of the plate and assuming that it has an average speed of 2000 ft/s while in contact with the plate, determine the magnitude and direction of the impulsive force exerted by the plate on the bullet.

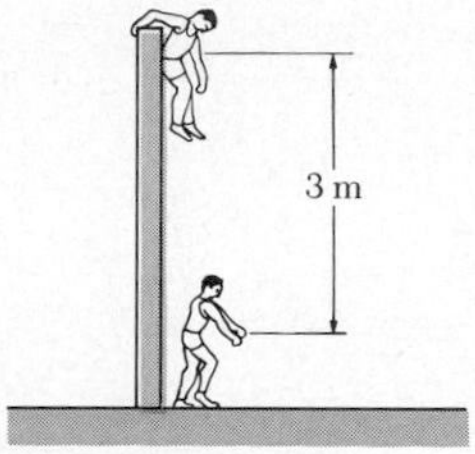

Fig. P13.131

13.131 After scaling a wall, a man lets himself drop 3 m to the ground. If his body comes to a complete stop 0.100 s after his feet first touch the ground, determine the vertical component of the average impulsive force exerted by the ground on his feet.

13.132 A 50-g bullet is fired with a horizontal velocity of 500 m/s into a 4-kg wooden block which is at rest on a frictionless, horizontal surface. Determine (*a*) the final velocity of the block, (*b*) the ratio of the final kinetic energy of the block and bullet to the initial kinetic energy of the bullet.

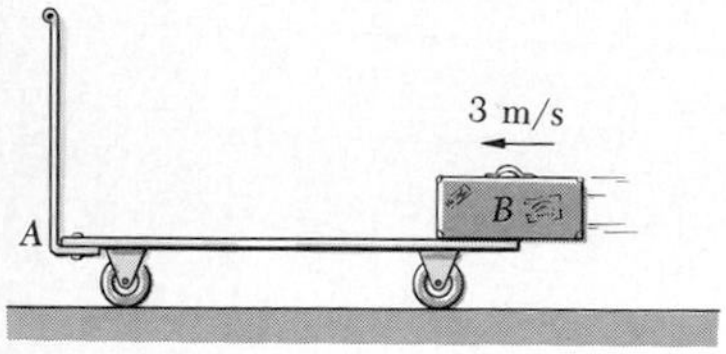

Fig. P13.133

13.133 An airline employee tosses a 15-kg suitcase with a horizontal velocity of 3 m/s onto a 35-kg baggage carrier. Knowing that the carrier is initially at rest and can roll freely, determine (*a*) the velocity of the carrier after the suitcase has slid to a relative stop on the carrier, (*b*) the ratio of the final kinetic energy of the carrier and suitcase to the initial kinetic energy of the suitcase.

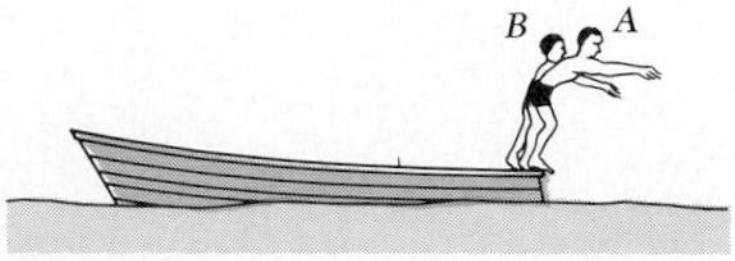

Fig. P13.134

13.134 Two swimmers A and B, of mass 75 kg and 50 kg, respectively, dive off the end of a 250-kg boat. Each swimmer dives so that his relative horizontal velocity with respect to the boat is 4 m/s. If the boat is initially at rest, determine its final velocity, assuming that (*a*) the two swimmers dive simultaneously, (*b*) swimmer A dives first, (*c*) swimmer B dives first.

Fig. P13.135

13.135 A 50-Mg railroad car moving with a velocity of 3 km/h is to be coupled to a 30-Mg car which is at rest. Determine (*a*) the final velocity of the coupled cars, (*b*) the average impulsive force acting on each car if the coupling is completed in 0.4 s.

13.136 Car A was traveling due north through an intersection when it was hit broadside by car B which was traveling due east. While both drivers admitted having ignored the four-way stop signs at the intersection, each claimed that he was traveling at the 35-mi/h speed limit and that the other was traveling much faster. Knowing that car A weighs 2000 lb, car B 3600 lb, and that inspection of the scene of the accident showed that as a result of the impact, the two cars got stuck together and skidded in a direction 40° north of east, determine (*a*) which of the two cars was actually traveling at 35 mi/h, (*b*) how fast the other car was moving.

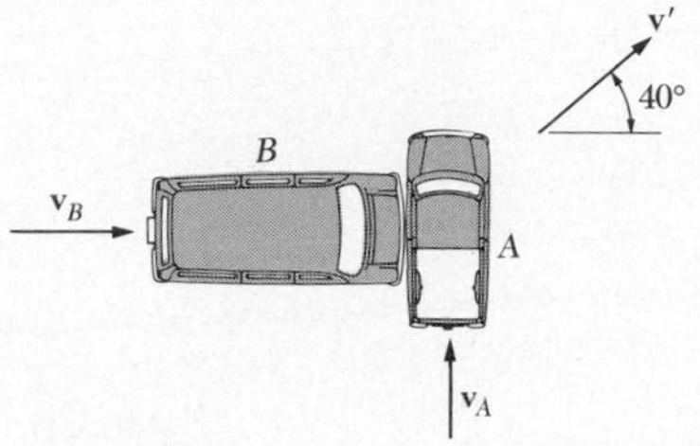

Fig. P13.136

13.137 An old 4000-lb gun fires a 20-lb shell with an initial velocity of 2000 ft/s at an angle of 30°. The gun rests on a horizontal surface and is free to move horizontally. Assuming that the barrel of the gun is rigidly attached to the frame (no recoil mechanism) and that the shell leaves the barrel 6 ms after firing, determine (*a*) the recoil velocity of the gun, (*b*) the resultant **R** of the vertical impulsive forces exerted by the ground on the gun.

Fig. P13.137

13.138 A small rivet connecting two pieces of sheet metal is being clinched by hammering. Determine the impulse exerted on the rivet and the energy absorbed by the rivet under each blow, knowing that the head of the hammer has a mass of 800 g and that it strikes the rivet with a velocity of 6 m/s. Assume that the hammer does not rebound and that the anvil is supported by springs and (*a*) has an infinite mass (rigid support), (*b*) has a mass of 4 kg.

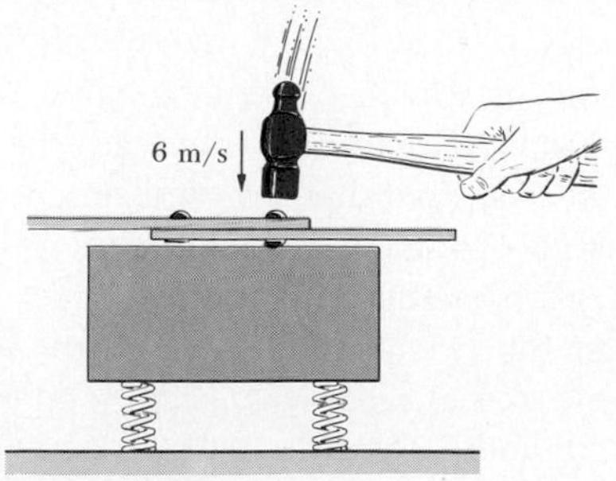

Fig. P13.138

13.139 In order to test the resistance of a chain to impact, the chain is suspended from an 80-kg block supported by two columns. A rod attached to the last link of the chain is then hit by a 20-kg cylinder dropped from a 1.2-m height. Determine the initial impulse exerted on the chain and the energy absorbed by the chain, assuming that the cylinder does not rebound and that the columns supporting the 80-kg block (*a*) are perfectly rigid, (*b*) are equivalent to two perfectly elastic springs.

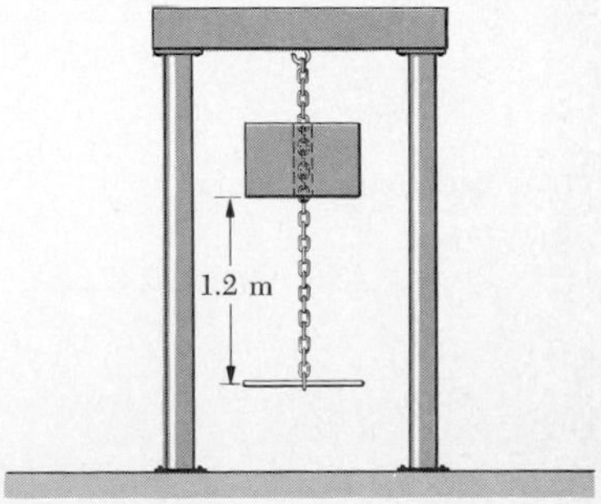

Fig. P13.139

13.12. Impact. A collision between two bodies which occurs in a very small interval of time, and during which the two bodies exert on each other relatively large forces, is called an *impact*. The common normal to the surfaces in contact during the impact is called the *line of impact*. If the mass centers of the two colliding bodies are located on this line, the impact is a *central impact*. Otherwise, the impact is said to be *eccentric*. We shall limit our present study to that of the central impact of two particles and postpone until later the analysis of the eccentric impact of two rigid bodies (Sec. 17.12).

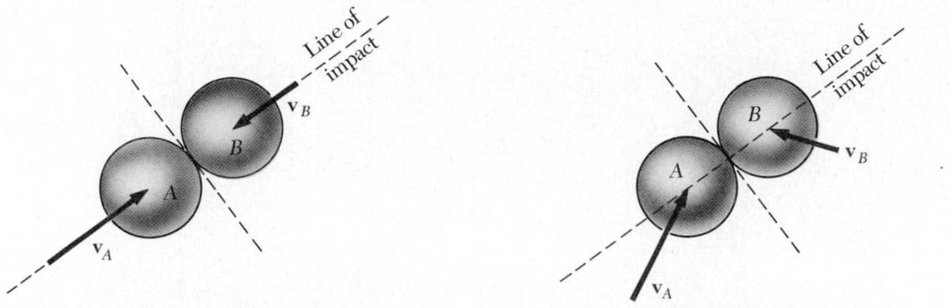

Fig. 13.20 (*a*) Direct central impact (*b*) Oblique central impact

If the velocities of the two particles are directed along the line of impact, the impact is said to be a *direct impact* (Fig. 13.20*a*). If, on the other hand, either or both particles move along a line other than the line of impact, the impact is said to be an *oblique impact* (Fig. 13.20*b*).

13.13. Direct Central Impact. Consider two particles A and B, of mass m_A and m_B, which are moving in the same straight line and to the right with known velocities $\mathbf{v}_A$ and $\mathbf{v}_B$ (Fig. 13.21*a*). If $\mathbf{v}_A$ is larger than $\mathbf{v}_B$, particle A will eventually strike particle B. Under the impact, the two particles, will *deform* and, at the end of the period of deformation, they will have the same velocity $\mathbf{u}$ (Fig. 13.21*b*). A period of *restitution* will then take place, at the end of which, depending upon the magnitude of the impact forces and upon the materials involved, the two particles either will have regained their original shape or will stay permanently deformed. Our purpose here is to determine the velocities $\mathbf{v}'_A$ and $\mathbf{v}'_B$ of the particles at the end of the period of restitution (Fig. 13.21*c*).

Considering first the two particles as a single system, we note that there is no impulsive, external force. Thus, the total momentum of the two particles is conserved, and we write

$$m_A\mathbf{v}_A + m_B\mathbf{v}_B = m_A\mathbf{v}'_A + m_B\mathbf{v}'_B$$

Since all the velocities considered are directed along the same axis, we may replace the equation obtained by the following relation involving only scalar components:

$$m_A v_A + m_B v_B = m_A v'_A + m_B v'_B \tag{13.37}$$

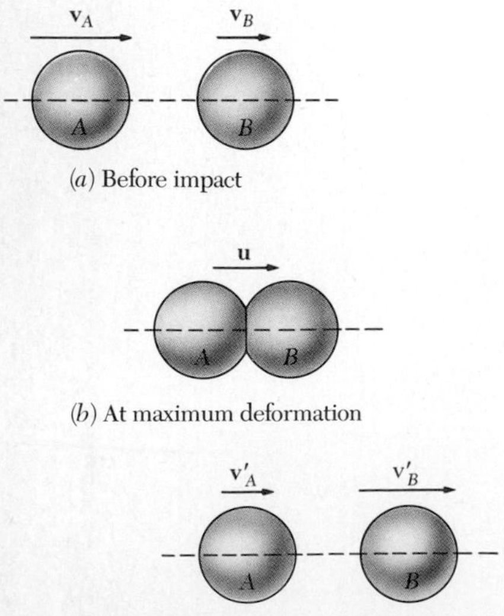

(*a*) Before impact

(*b*) At maximum deformation

(*c*) After impact

Fig. 13.21

A positive value for any of the scalar quantities v_A, v_B, v'_A, or v'_B means that the corresponding vector is directed to the right; a negative value indicates that the corresponding vector is directed to the left.

To obtain the velocities $\mathbf{v}'_A$ and $\mathbf{v}'_B$, it is necessary to establish a second relation between the scalars v'_A and v'_B. For this purpose, we shall consider now the motion of particle A during the period of deformation and apply the principle of impulse and momentum. Since the only impulsive force acting on A during this period is the force $\mathbf{P}$ exerted by B (Fig. 13.22a), we write, using again scalar components,

$$m_A v_A - \int P\,dt = m_A u \tag{13.38}$$

where the integral extends over the period of deformation. Considering now the motion of A during the period of restitution, and denoting by $\mathbf{R}$ the force exerted by B on A during this period (Fig. 13.22b), we write

$$m_A u - \int R\,dt = m_A v'_A \tag{13.39}$$

where the integral extends over the period of restitution.

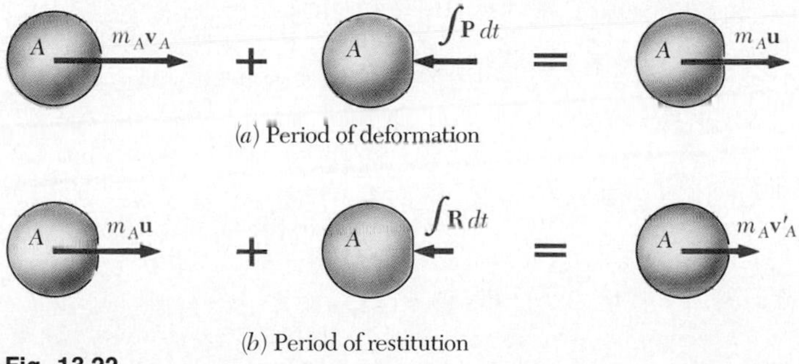

(a) Period of deformation

(b) Period of restitution

Fig. 13.22

In general, the force $\mathbf{R}$ exerted on A during the period of restitution differs from the force $\mathbf{P}$ exerted during the period of deformation, and the magnitude $\int R\,dt$ of its impulse is smaller than the magnitude $\int P\,dt$ of the impulse of $\mathbf{P}$. The ratio of the magnitudes of the impulses corresponding, respectively, to the period of restitution and to the period of deformation is called the *coefficient of restitution* and is denoted by e. We write

$$e = \frac{\int R\,dt}{\int P\,dt} \tag{13.40}$$

The value of the coefficient e is always between 0 and 1 and depends to a large extent on the two materials involved. However, it also varies considerably with the impact velocity and the shape and size of the two colliding bodies.

Solving Eqs. (13.38) and (13.39) for the two impulses and substituting into (13.40), we write

$$e = \frac{u - v'_A}{v_A - u} \tag{13.41}$$

A similar analysis of particle B leads to the relation

$$e = \frac{v'_B - u}{u - v_B} \tag{13.42}$$

Since the quotients in (13.41) and (13.42) are equal, they are also equal to the quotient obtained by adding, respectively, their numerators and their denominators. We have, therefore,

$$e = \frac{(u - v'_A) + (v'_B - u)}{(v_A - u) + (u - v_B)} = \frac{v'_B - v'_A}{v_A - v_B}$$

and

$$v'_B - v'_A = e(v_A - v_B) \tag{13.43}$$

Since $v'_B - v'_A$ represents the relative velocity of the two particles after impact and $v_A - v_B$ their relative velocity before impact, formula (13.43) expresses that *the relative velocity of the two particles after impact may be obtained by multiplying their relative velocity before impact by the coefficient of restitution.* This property is used to determine experimentally the value of the coefficient of restitution of two given materials.

The velocities of the two particles after impact may now be obtained by solving Eqs. (13.37) and (13.43) simultaneously for v'_A and v'_B. It is recalled that the derivation of Eqs. (13.37) and (13.43) was based on the assumption that particle B is located to the right of A, and that both particles are initially moving to the right. If particle B is initially moving to the left, the scalar v_B should be considered negative. The same sign convention holds for the velocities after impact: a positive sign for v'_A will indicate that particle A moves to the right after impact and a negative sign, that it moves to the left.

Two particular cases of impact are of special interest:

1. $e = 0$, *Perfectly Plastic Impact.* When $e = 0$, Eq. (13.43) yields $v'_B = v'_A$. There is no period of restitution, and both particles stay together after impact. Substituting $v'_B = v'_A = v'$ into Eq. (13.37), which expresses that the total momentum of the particles is conserved, we write

$$m_A v_A + m_B v_B = (m_A + m_B)v' \tag{13.44}$$

This equation may be solved for the common velocity v' of the two particles after impact.

2. $e = 1$, *Perfectly Elastic Impact.* When $e = 1$, Eq. (13.43) reduces to

$$v'_B - v'_A = v_A - v_B \tag{13.45}$$

which expresses that the relative velocities before and after impact are equal. The impulses received by each particle during the period of deformation and during the period of restitution are equal.

The particles move away from each other after impact with the same velocity with which they approached each other before impact. The velocities v'_A and v'_B may be obtained by solving Eqs. (13.37) and (13.45) simultaneously.

It is worth noting that *in the case of a perfectly elastic impact, the total energy of the two particles,* as well as their total momentum, *is conserved.* Equations (13.37) and (13.45) may be written as follows:

$$m_A(v_A - v'_A) = m_B(v'_B - v_B) \qquad (13.37')$$

$$v_A + v'_A = v_B + v'_B \qquad (13.45')$$

Multiplying (13.37′) and (13.45′) member by member, we have

$$m_A(v_A - v'_A)(v_A + v'_A) = m_B(v'_B - v_B)(v'_B + v_B)$$

$$m_A v_A^2 - m_A(v'_A)^2 = m_B(v'_B)^2 - m_B v_B^2$$

Rearranging the terms in the equation obtained, and multiplying by $\frac{1}{2}$, we write

$$\tfrac{1}{2}m_A v_A^2 + \tfrac{1}{2}m_B v_B^2 = \tfrac{1}{2}m_A(v'_A)^2 + \tfrac{1}{2}m_B(v'_B)^2 \qquad (13.46)$$

which expresses that the kinetic energy of the particles is conserved. It should be noted, however, that *in the general case of impact,* i.e., when e is not equal to 1, *the total energy of the particles is not conserved.* This may be shown in any given case by comparing the kinetic energies before and after impact. The lost kinetic energy is in part transformed into heat and in part spent in generating elastic waves within the two colliding bodies.

13.14. Oblique Central Impact. Let us now consider the case when the velocities of the two colliding particles are *not* directed along the line of impact (Fig. 13.23). As indicated in Sec. 13.12, the impact is said to

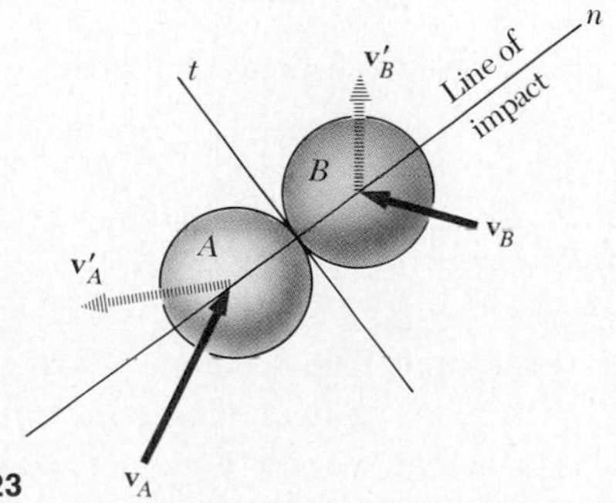

Fig. 13.23

be *oblique.* Since the velocities $\mathbf{v}'_A$ and $\mathbf{v}'_B$ of the particles after impact are unknown in direction as well as in magnitude, their determination will require the use of four independent equations.

We choose as coordinate axes the n axis along the line of impact, i.e., along the common normal to the surfaces in contact, and the t axis along their common tangent. Assuming that the particles are perfectly *smooth and frictionless,* we observe that the only impulses exerted on the particles

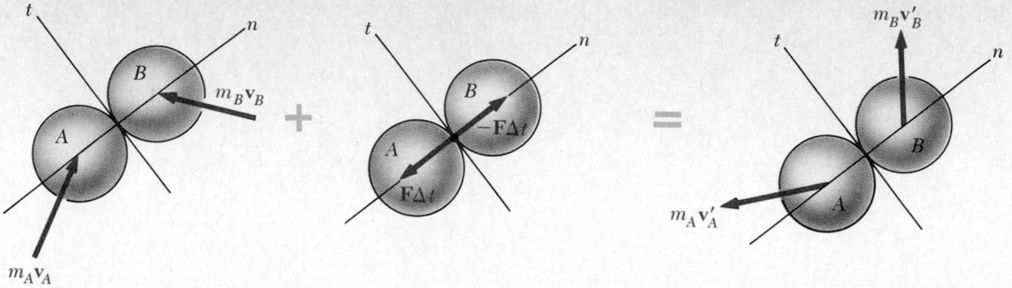

Fig. 13.24

during the impact are due to internal forces directed along the line of impact, i.e., along the n axis (Fig. 13.24). It follows that

1. The component along the t axis of the momentum of each particle, considered separately, is conserved; hence the t component of the velocity of each particle remains unchanged. We write

$$(v_A)_t = (v'_A)_t \qquad (v_B)_t = (v'_B)_t \tag{13.37}$$

2. The component along the n axis of the total momentum of the two particles is conserved. We write

$$m_A(v_A)_n + m_B(v_B)_n = m_A(v'_A)_n + m_B(v'_B)_n \tag{13.48}$$

3. The component along the n axis of the relative velocity of the two particles after impact is obtained by multiplying the n component of their relative velocity before impact by the coefficient of restitution. Indeed, a derivation similar to that given in Sec. 13.13 for direct central impact yields

$$(v'_B)_n - (v'_A)_n = e[(v_A)_n - (v_B)_n] \tag{13.49}$$

We have thus obtained four independent equations which may be solved for the components of the velocities of A and B after impact. This method of solution is illustrated in Sample Prob. 13.15.

Our analysis of the oblique central impact of two particles has been based so far on the assumption that both particles moved freely before and after the impact. We shall now examine the case when one or both of the colliding particles is constrained in its motion. Consider, for instance, the collision between block A, which is constrained to move on a horizontal surface, and ball B, which is free to move in the plane of the figure (Fig. 13.25). Assuming no friction between the block and the ball, or between the block and the horizontal surface, we note that the impulses exerted on the system consist of the impulses of the internal forces $\mathbf{F}$ and $-\mathbf{F}$ directed along the line of impact, i.e., along the n axis, and of the impulse of the external force $\mathbf{F}_{\text{ext}}$ exerted by the horizontal surface on block A and directed along the vertical (Fig. 13.26).

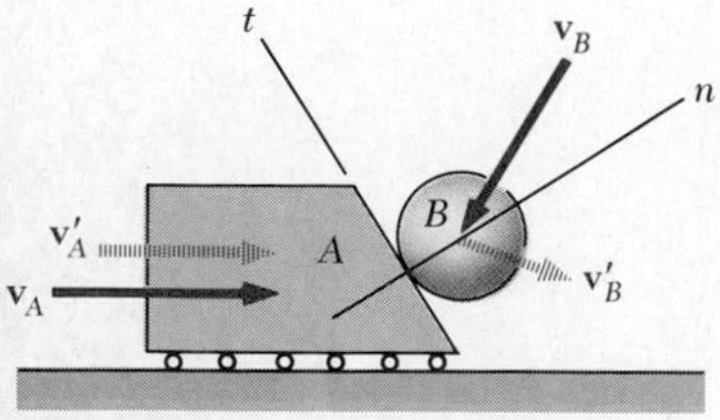

Fig. 13.25

The velocities of block A and ball B immediately after the impact are represented by three unknowns, namely, the magnitude of the velocity $\mathbf{v}'_A$ of block A, which is known to be horizontal, and the magnitude and direc-

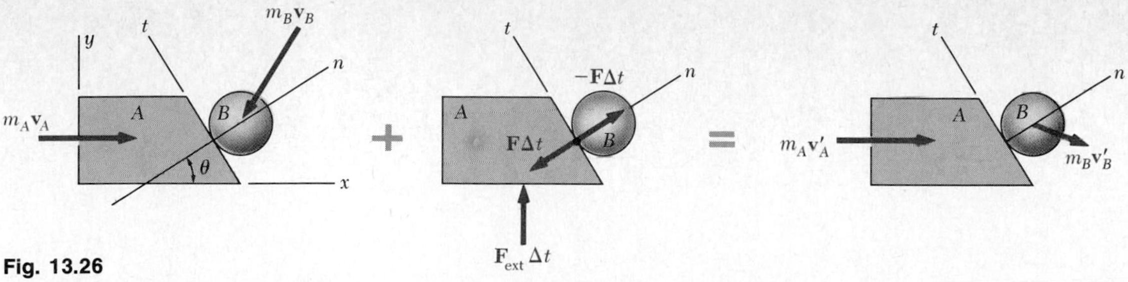

Fig. 13.26

tion of the velocity $\mathbf{v}'_B$ of ball B. We shall therefore write three equations by expressing that

1. The component along the t axis of the momentum of ball B is conserved; hence the t component of the velocity of ball B remains unchanged. We write

$$(v_B)_t = (v'_B)_t \tag{13.50}$$

2. The component along the horizontal x axis of the total momentum of block A and ball B is conserved. We write

$$m_A v_A + m_B(v_B)_x = m_A v'_A + m_B(v'_B)_x \tag{13.51}$$

3. The component along the n axis of the relative velocity of block A and ball B after impact is obtained by multiplying the n component of their relative velocity before impact by the coefficient of restitution. We write again

$$(v'_B)_n - (v'_A)_n = e[(v_A)_n - (v_B)_n] \tag{13.49}$$

We should note, however, that in the case considered here, the validity of Eq. (13.49) cannot be established through a mere extension of the derivation given in Sec. 13.13 for the direct central impact of two particles moving in a straight line. Indeed, these particles were not subjected to any external impulse, while block A in the present analysis is subjected to the impulse exerted by the horizontal surface. To prove that Eq. (13.49) is still valid, we shall first apply the principle of impulse and momentum to block A over the period of deformation (Fig. 13.27). Considering only the horizontal components, we write

$$m_A v_A - (\textstyle\int P\,dt)\cos\theta = m_A u \tag{13.52}$$

where the integral extends over the period of deformation and where $\mathbf{u}$ represents the velocity of block A at the end of that period. Considering

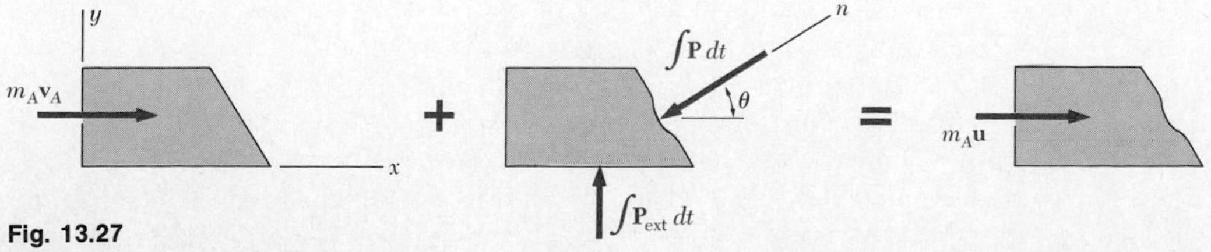

Fig. 13.27

now the period of restitution, we write in a similar way

$$m_A u - (\int R\, dt) \cos\theta = m_A v'_A \tag{13.53}$$

where the integral extends over the period of restitution.

Recalling from Sec. 13.13 the definition of the coefficient of restitution, we write

$$e = \frac{\int R\, dt}{\int P\, dt} \tag{13.40}$$

Solving Eqs. (13.52) and (13.53) for the integrals $\int P\, dt$ and $\int R\, dt$, and substituting into Eq. (13.40), we have, after reductions,

$$e = \frac{u - v'_A}{v_A - u}$$

or, multiplying all velocities by $\cos\theta$ to obtain their projections on the line of impact,

$$e = \frac{u_n - (v'_A)_n}{(v_A)_n - u_n} \tag{13.54}$$

We note that Eq. (13.54) is identical to Eq. (13.41) of Sec. 13.13, except for the subscripts n which are used here to indicate that we are considering velocity components along the line of impact. Since the motion of ball B is unconstrained, the proof of Eq. (13.49) may be completed in the same manner as the derivation of Eq. (13.43) of Sec. 13.13. Thus, we conclude that the relation (13.49) between the components along the line of impact of the relative velocities of two colliding particles remains valid when one of the particles is constrained in its motion. The validity of this relation may easily be extended to the case when both particles are constrained in their motion.

13.15. Problems Involving Energy and Momentum. We have now at our disposal three different methods for the solution of kinetics problems: the direct application of Newton's second law, $\Sigma\mathbf{F} = m\mathbf{a}$, the method of work and energy, and the method of impulse and momentum. To derive maximum benefit from these three methods, we should be able to choose the method best suited for the solution of a given problem. We should also be prepared to use different methods for solving the various parts of a problem when such a procedure seems advisable.

We have already seen that the method of work and energy is in many cases more expeditious than the direct application of Newton's second law. As indicated in Sec. 13.4, however, the method of work and energy has limitations, and it must sometimes be supplemented by the use of $\Sigma\mathbf{F} = m\mathbf{a}$. This is the case, for example, when we wish to determine an acceleration or a normal force.

There is generally no great advantage in using the method of impulse and momentum for the solution of problems involving no impulsive forces. It will usually be found that the equation $\Sigma\mathbf{F} = m\mathbf{a}$ yields a solution just as fast and that the method of work and energy, if it applies, is more rapid and more convenient. However, the method of impulse and momentum is the only practicable method in problems of impact. A solution based on the direct application of $\Sigma\mathbf{F} = m\mathbf{a}$ would be unwieldy, and the method of work and energy cannot be used since impact (unless perfectly elastic) involves a loss of mechanical energy.

Many problems involve only conservative forces, except for a short impact phase during which impulsive forces act. The solution of such problems may be divided into several parts. While the part corresponding to the impact phase calls for the use of the method of impulse and momentum and of the relation between relative velocities, the other parts may usually be solved by the method of work and energy. The use of the equation $\Sigma\mathbf{F} = m\mathbf{a}$ will be necessary, however, if the problem involves the determination of a normal force.

Consider, for example, a pendulum A, of mass m_A and length l, which is released with no velocity from a position A_1 (Fig. 13.28*a*). The pendulum swings freely in a vertical plane and hits a second pendulum B, of mass m_B and same length l, which is initially at rest. After the impact (with coefficient of restitution e), pendulum B swings through an angle θ that we wish to determine.

The solution of the problem may be divided into three parts:

1. *Pendulum A Swings from A_1 to A_2.* The principle of conservation of energy may be used to determine the velocity $(\mathbf{v}_A)_2$ of the pendulum at A_2 (Fig. 13.28*b*).
2. *Pendulum A Hits Pendulum B.* Using the fact that the total momentum of the two pendulums is conserved and the relation between their relative velocities, we determine the velocities $(\mathbf{v}_A)_3$ and $(\mathbf{v}_B)_3$ of the two pendulums after impact (Fig. 13.28*c*).
3. *Pendulum B Swings from B_3 to B_4.* Applying the principle of conservation of energy to pendulum B, we determine the maximum elevation y_4 reached by that pendulum (Fig. 13.28*d*). The angle θ may then be determined by trigonometry.

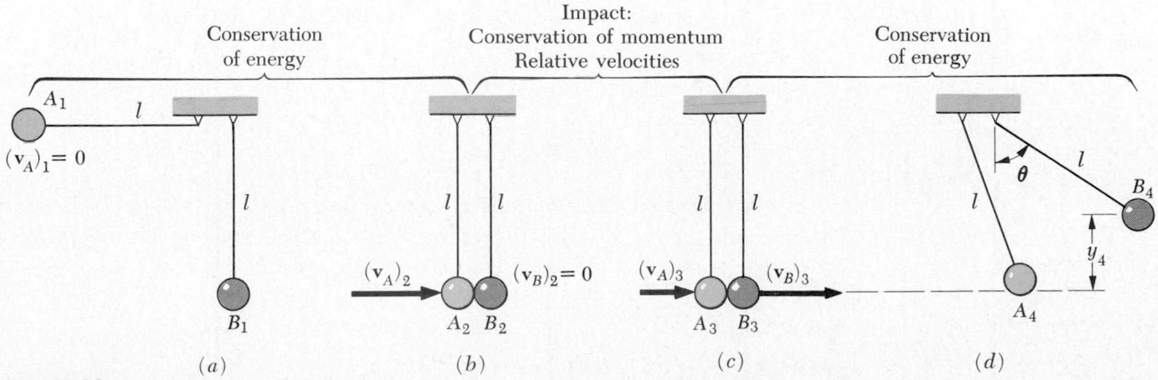

Fig. 13.28

We note that the method of solution just described should be supplemented by the use of $\Sigma\mathbf{F} = m\mathbf{a}$ if the tensions in the cords holding the pendulums are to be determined.

SAMPLE PROBLEM 13.13

A 20-Mg railroad car moving at a speed of 0.5 m/s to the right collides with a 35-Mg car which is at rest. If after the collision the 35-Mg car is observed to move to the right at a speed of 0.3 m/s, determine the coefficient of restitution between the two cars.

Solution. We express that the total momentum of the two cars is conserved.

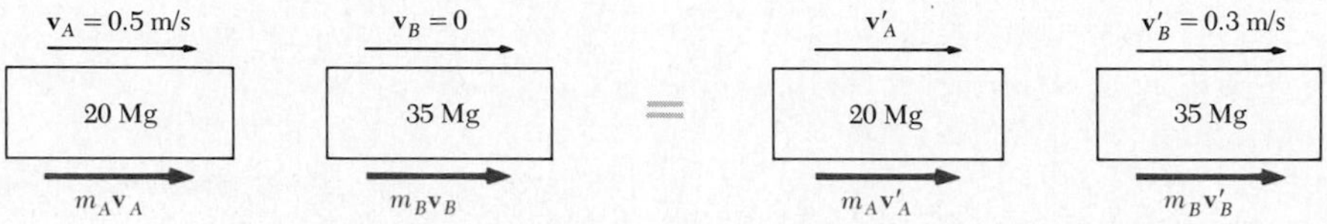

$$m_A \mathbf{v}_A + m_B \mathbf{v}_B = m_A \mathbf{v}'_A + m_B \mathbf{v}'_B$$
$$(20 \text{ Mg})(+0.5 \text{ m/s}) + (35 \text{ Mg})(0) = (20 \text{ Mg})v'_A + (35 \text{ Mg})(+0.3 \text{ m/s})$$
$$v'_A = -0.025 \text{ m/s} \qquad \mathbf{v}'_A = 0.025 \text{ m/s} \leftarrow$$

The coefficient of restitution is obtained by writing

$$e = \frac{v'_B - v'_A}{v_A - v_B} = \frac{+0.3 - (-0.025)}{+0.5 - 0} = \frac{0.325}{0.5} \qquad e = 0.65 \blacktriangleleft$$

SAMPLE PROBLEM 13.14

A ball is thrown against a frictionless, vertical wall. Immediately before the ball strikes the wall, its velocity has a magnitude v and forms an angle of 30° with the horizontal. Knowing that $e = 0.90$, determine the magnitude and direction of the velocity of the ball as it rebounds from the wall.

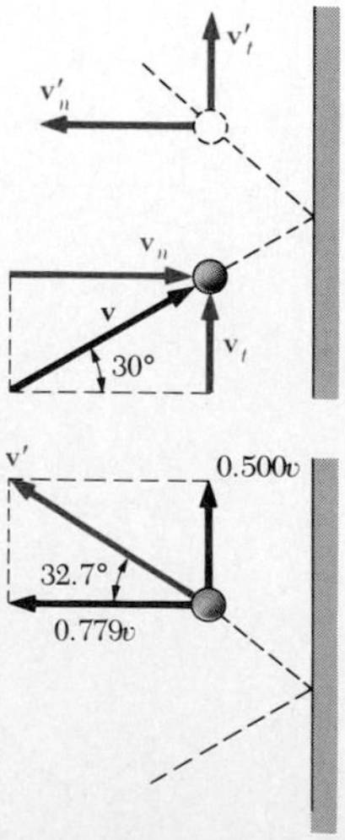

Solution. We resolve the initial velocity of the ball into components, respectively, perpendicular and parallel to the wall:

$$v_n = v \cos 30° = 0.866v \qquad v_t = v \sin 30° = 0.500v$$

Motion Parallel to the Wall. Since the wall is frictionless, the impulse it exerts on the ball is perpendicular to the wall. Thus, the component parallel to the wall of the momentum of the ball is conserved and we have

$$\mathbf{v}'_t = \mathbf{v}_t = 0.500v \uparrow$$

Motion Perpendicular to the Wall. Since the mass of the wall (and earth) is essentially infinite, expressing that the total momentum of the ball and wall is conserved would yield no useful information. Using the relation (13.49) between relative velocities, we write

$$0 - v'_n = e(v_n - 0)$$
$$v'_n = -0.90(0.866v) = -0.779v \qquad \mathbf{v}'_n = 0.779v \leftarrow$$

Resultant Motion. Adding vectorially the components $\mathbf{v}'_n$ and $\mathbf{v}'_t$,

$$\mathbf{v}' = 0.926v \measuredangle 32.7° \blacktriangleleft$$

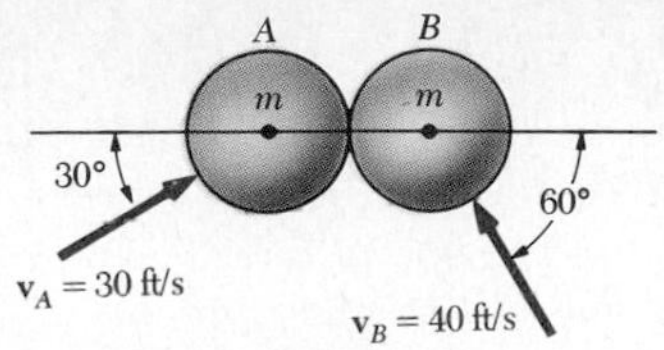

SAMPLE PROBLEM 13.15

The magnitude and direction of the velocities of two identical frictionless balls before they strike each other are as shown. Assuming $e = 0.90$, determine the magnitude and direction of the velocity of each ball after the impact.

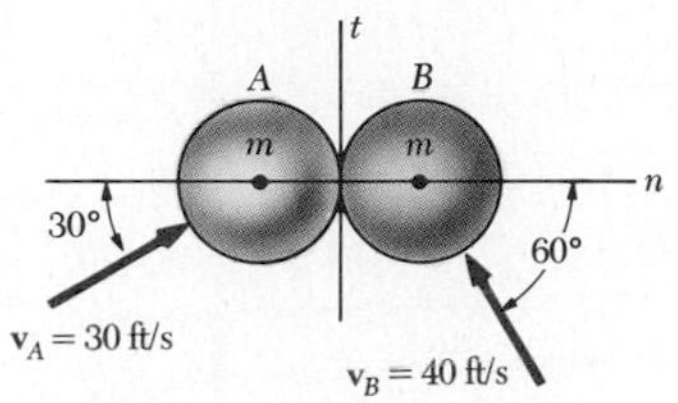

Solution. The impulsive forces acting between the balls during the impact are directed along a line joining the centers of the balls called the *line of impact*. Resolving the velocities into components directed, respectively, along the line of impact and along the common tangent to the surfaces in contact, we write

$$(v_A)_n = v_A \cos 30° = +26.0 \text{ ft/s}$$
$$(v_A)_t = v_A \sin 30° = +15.0 \text{ ft/s}$$
$$(v_B)_n = -v_B \cos 60° = -20.0 \text{ ft/s}$$
$$(v_B)_t = v_B \sin 60° = +34.6 \text{ ft/s}$$

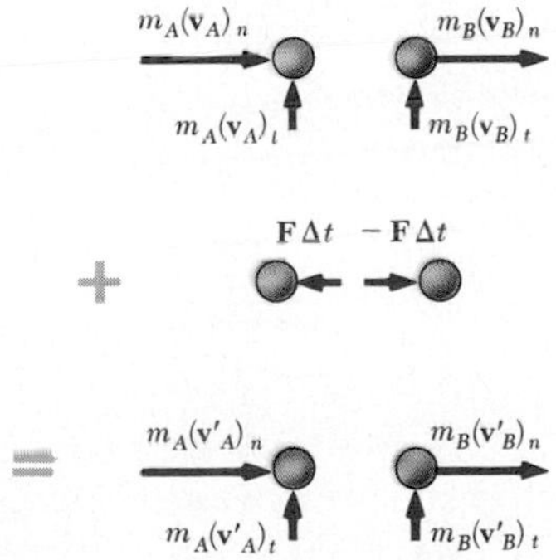

Principle of Impulse and Momentum. In the adjoining sketches we show in turn the initial momenta, the impulses, and the final momenta.

Motion along the Common Tangent. Considering only the t components, we apply the principle of impulse and momentum to each ball *separately*. Since the impulsive forces are directed along the line of impact, the t component of the momentum, and hence the t component of the velocity of each ball, is unchanged. We have

$$(\mathbf{v}'_A)_t = 15.0 \text{ ft/s} \uparrow \qquad (\mathbf{v}'_B)_t = 34.6 \text{ ft/s} \uparrow$$

Motion along the Line of Impact. In the n direction, we consider the two balls as a single system and note that by Newton's third law, the internal impulses are, respectively, $\mathbf{F}\,\Delta t$ and $-\mathbf{F}\,\Delta t$ and cancel. We thus write that the total momentum of the balls is conserved:

$$m_A(v_A)_n + m_B(v_B)_n = m_A(v'_A)_n + m_B(v'_B)_n$$
$$m(26.0) + m(-20.0) = m(v'_A)_n + m(v'_B)_n$$
$$(v'_A)_n + (v'_B)_n = 6.0 \quad (1)$$

Using the relation (13.49) between relative velocities, we write

$$(v'_B)_n - (v'_A)_n = e[(v_A)_n - (v_B)_n]$$
$$(v'_B)_n - (v'_A)_n = (0.90)[26.0 - (-20.0)]$$
$$(v'_B)_n - (v'_A)_n = 41.4 \quad (2)$$

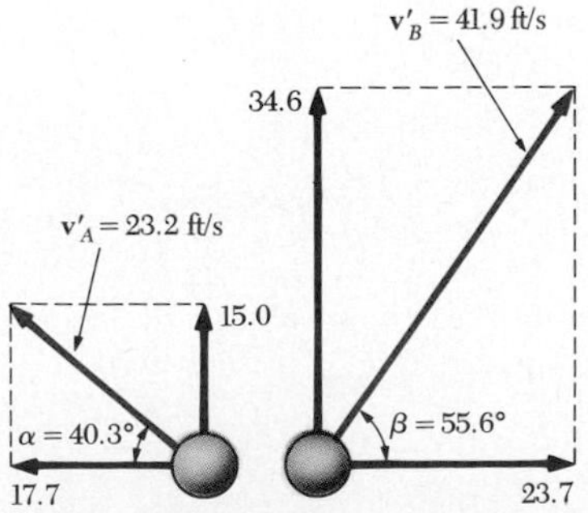

Solving Eqs. (1) and (2) simultaneously, we obtain

$$(v'_A)_n = -17.7 \qquad (v'_B)_n = +23.7$$
$$(\mathbf{v}'_A)_n = 17.7 \text{ ft/s} \leftarrow \qquad (\mathbf{v}'_B)_n = 23.7 \text{ ft/s} \rightarrow$$

Resultant Motion. Adding vectorially the velocity components of each ball, we obtain

$$\mathbf{v}'_A = 23.2 \text{ ft/s} \nwarrow 40.3° \qquad \mathbf{v}'_B = 41.9 \text{ ft/s} \nearrow 55.6° \blacktriangleleft$$

SAMPLE PROBLEM 13.16

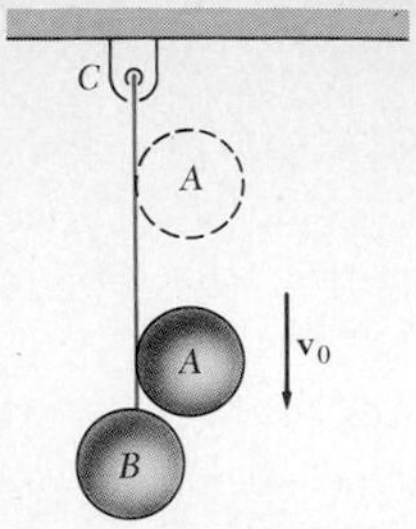

Ball B is hanging from an inextensible cord BC. An identical ball A is released from rest when it is just touching the cord and acquires a velocity $\mathbf{v}_0$ before striking ball B. Assuming perfectly elastic impact ($e = 1$) and no friction, determine the velocity of each ball immediately after impact.

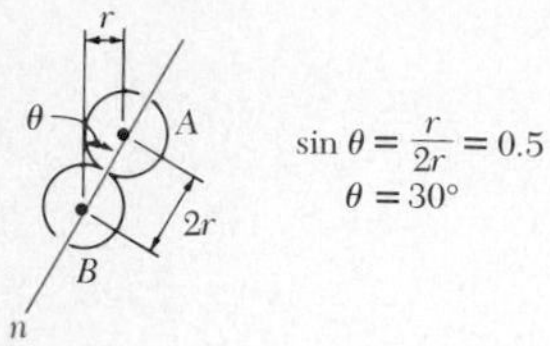

Solution. Since ball B is constrained to move in a circle of center C, its velocity $\mathbf{v}'_B$ after impact must be horizontal. Thus the problem involves three unknowns: the magnitude v'_B of the velocity of B, and the magnitude and direction of the velocity $\mathbf{v}'_A$ of A after impact.

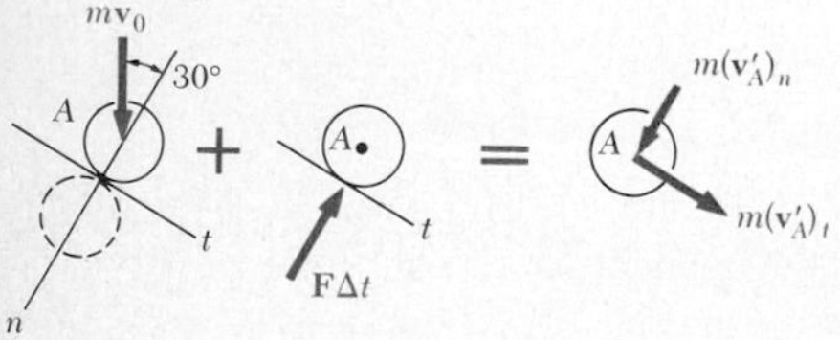

Impulse-Momentum Principle: Ball A

$$m\mathbf{v}_A + \mathbf{F}\,\Delta t = m\mathbf{v}'_A$$

$+\searrow t$ components: $\quad mv_0 \sin 30° + 0 = m(v'_A)_t$

$$(v'_A)_t = 0.5v_0 \tag{1}$$

We note that the equation used expresses conservation of the momentum of ball A along the common tangent to balls A and B.

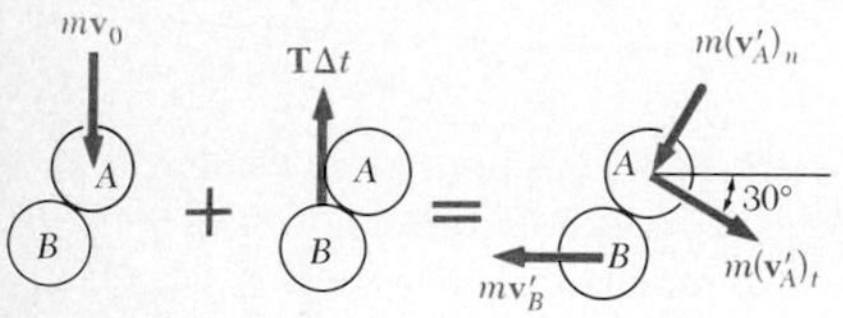

Impulse-Momentum Principle: Balls A and B

$$m\mathbf{v}_A + \mathbf{T}\,\Delta t = m\mathbf{v}'_A + m\mathbf{v}'_B$$

$\overset{+}{\rightarrow} x$ components: $\quad 0 = m(v'_A)_t \cos 30° - m(v'_A)_n \sin 30° - mv'_B$

We note that the equation obtained expresses conservation of the total momentum in the x direction. Substituting for $(v'_A)_t$ from Eq. (1) and rearranging terms, we write

$$0.5(v'_A)_n + v'_B = 0.433v_0 \tag{2}$$

Relative Velocities along the Line of Impact. Since $e = 1$, Eq. (13.49) yields

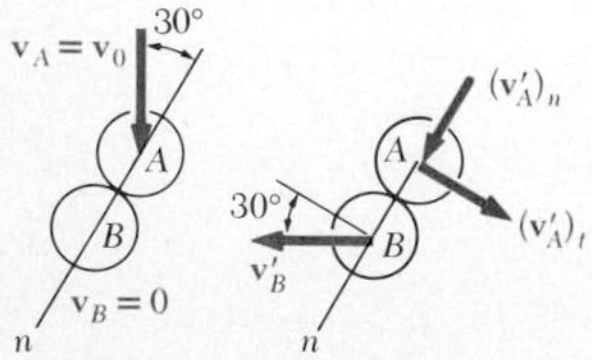

$$(v'_B)_n - (v'_A)_n = (v_A)_n - (v_B)_n$$
$$v'_B \sin 30° - (v'_A)_n = v_0 \cos 30° - 0$$
$$0.5v'_B - (v'_A)_n = 0.866v_0 \tag{3}$$

Solving Eqs. (2) and (3) simultaneously, we obtain

$$(v'_A)_n = -0.520v_0 \qquad v'_B = 0.693v_0$$

$$\mathbf{v}'_B = 0.693v_0 \leftarrow \blacktriangleleft$$

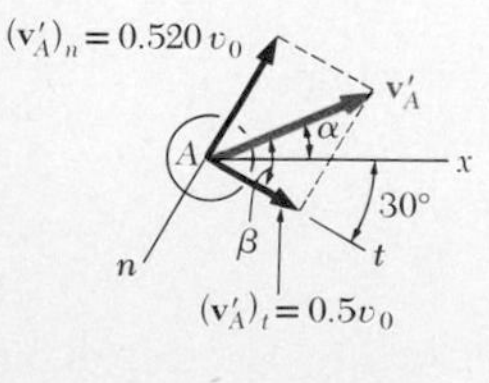

Recalling Eq. (1) we draw the adjoining sketch and obtain by trigonometry

$$v'_A = 0.721v_0 \qquad \beta = 46.1° \qquad \alpha = 46.1° - 30° = 16.1°$$

$$\mathbf{v}'_A = 0.721v_0 \measuredangle 16.1° \blacktriangleleft$$

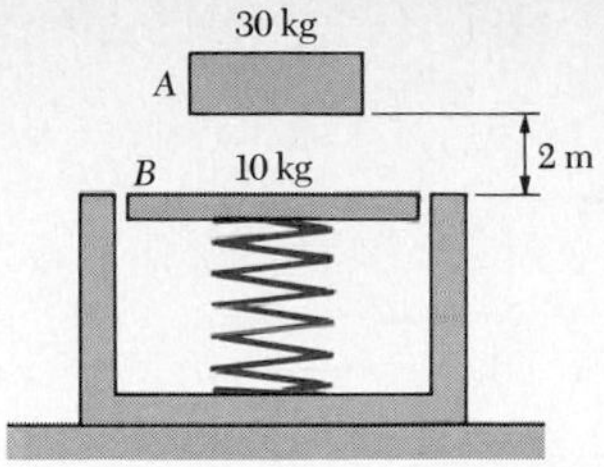

SAMPLE PROBLEM 13.17

A 30-kg block is dropped from a height of 2 m onto the 10-kg pan of a spring scale. Assuming the impact to be perfectly plastic, determine the maximum deflection of the pan. The constant of the spring is $k = 20$ kN/m.

Solution. The impact between the block and the pan *must* be treated separately; therefore we divide the solution into three parts.

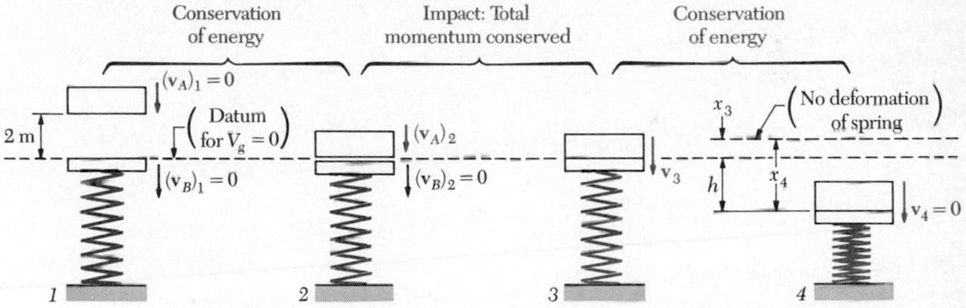

Conservation of Energy. Block: $W_A = (30 \text{ kg})(9.81 \text{ m/s}^2) = 294$ N

$$T_1 = \tfrac{1}{2}m_A(v_A)_1^2 = 0 \qquad V_1 = W_A y = (294 \text{ N})(2 \text{ m}) = 588 \text{ J}$$

$$T_2 = \tfrac{1}{2}m_A(v_A)_2^2 = \tfrac{1}{2}(30 \text{ kg})(v_A)_2^2 \qquad V_2 = 0$$

$T_1 + V_1 = T_2 + V_2$: $\qquad 0 + 588 \text{ J} = \tfrac{1}{2}(30 \text{ kg})(v_A)_2^2 + 0$

$$(v_A)_2 = +6.26 \text{ m/s} \qquad (\mathbf{v}_A)_2 = 6.26 \text{ m/s} \downarrow$$

Impact: Conservation of Momentum. Since the impact is perfectly plastic, $e = 0$; the block and pan move together after the impact.

$$m_A(v_A)_2 + m_B(v_B)_2 = (m_A + m_B)v_3$$

$$(30 \text{ kg})(6.26 \text{ m/s}) + 0 = (30 \text{ kg} + 10 \text{ kg})v_3$$

$$v_3 = +4.70 \text{ m/s} \qquad \mathbf{v}_3 = 4.70 \text{ m/s} \downarrow$$

Conservation of Energy. Initially the spring supports the weight W_B of the pan; thus the initial deflection of the spring is

$$x_3 = \frac{W_B}{k} = \frac{(10 \text{ kg})(9.81 \text{ m/s}^2)}{20 \times 10^3 \text{ N/m}} = \frac{98.1 \text{ N}}{20 \times 10^3 \text{ N/m}} = 4.91 \times 10^{-3} \text{ m}$$

Denoting by x_4 the total maximum deflection of the spring, we write

$$T_3 = \tfrac{1}{2}(m_A + m_B)v_3^2 = \tfrac{1}{2}(30 \text{ kg} + 10 \text{ kg})(4.70 \text{ m/s})^2 = 442 \text{ J}$$

$$V_3 = V_g + V_e = 0 + \tfrac{1}{2}kx_3^2 = \tfrac{1}{2}(20 \times 10^3)(4.91 \times 10^{-3})^2 = 0.241 \text{ J}$$

$$T_4 = 0$$

$$V_4 = V_g + V_e = (W_A + W_B)(-h) + \tfrac{1}{2}kx_4^2 = -(392)h + \tfrac{1}{2}(20 \times 10^3)x_4^2$$

Noting that the displacement of the pan is $h = x_4 - x_3$, we write

$T_3 + V_3 = T_4 + V_4$:

$$442 + 0.241 = 0 - 392(x_4 - 4.91 \times 10^{-3}) + \tfrac{1}{2}(20 \times 10^3)x_4^2$$

$$x_4 = 0.230 \text{ m} \qquad h = x_4 - x_3 = 0.230 \text{ m} - 4.91 \times 10^{-3} \text{ m}$$

$$h = 0.225 \text{ m} \qquad h = 225 \text{ mm} \blacktriangleleft$$

Problems

13.140 The velocities of two collars before impact are as shown. If after impact the velocity of collar A is observed to be 5.4 m/s to the left, determine (*a*) the velocity of collar B after impact, (*b*) the coefficient of restitution between the two collars.

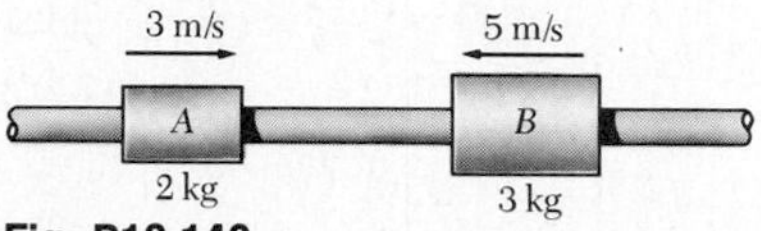

Fig. P13.140

13.141 Solve Prob. 13.140, assuming that the velocity of collar A after impact is observed to be 3 m/s to the left.

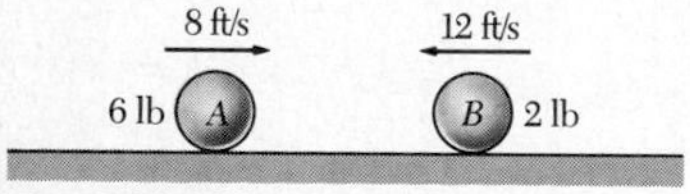

Fig. P13.142

13.142 Two small spheres A and B are made of different materials and have the masses indicated. They are moving on a frictionless, horizontal surface with the velocities shown when they hit each other. Knowing that the coefficient of restitution between the spheres is $e = 0.80$, determine (*a*) the velocity of each sphere after impact, (*b*) the energy loss due to the impact.

13.143 Solve Prob. 13.142, assuming that sphere A is moving to the left at the speed of 8 ft/s.

13.144 Two identical cars B and C are at rest on a loading dock with their brakes released. Car A of the same model, which has been pushed by dock workers, hits car B with a velocity of 1.6 m/s, causing a series of collisions among the three cars. Assuming a coefficient of restitution $e = 0.70$ between the bumpers, determine the velocity of each car after *all* collisions have taken place.

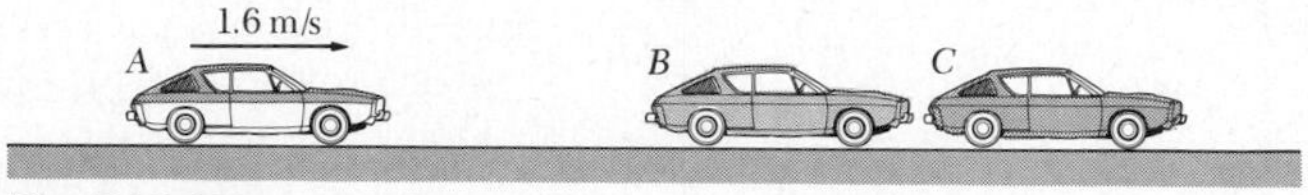

Fig. P13.144

13.145 In Prob. 13.144, it is observed that the final velocity of car C is 0.9 m/s to the right. Determine (*a*) the coefficient of restitution e between the bumpers, (*b*) the velocities of cars A and B after *all* collisions have taken place.

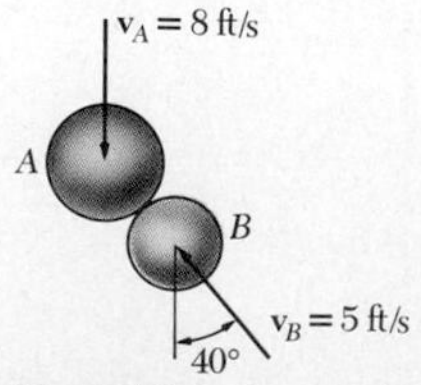

Fig. P13.146

13.146 A 2.5-lb ball A is falling vertically with a velocity of magnitude $v_A = 8$ ft/s when it is hit as shown by a 1.5-lb ball B which has a velocity of magnitude $v_B = 5$ ft/s. Knowing that the coefficient of restitution between the two balls is $e = 0.75$ and assuming no friction, determine the velocity of each ball immediately after impact.

13.147 Two identical billiard balls may move freely on a horizontal table. Ball A has a velocity $\mathbf{v}_0$ as shown and hits ball B, which is at rest, at a point C defined by $\theta = 30°$. Knowing that the coefficient of restitution between the two balls is $e = 0.90$ and assuming no friction, determine the velocity of each ball after impact.

13.148 Two identical billiard balls may move freely on a horizontal table. Ball A has a velocity $\mathbf{v}_0$ as shown and hits ball B which is at rest. Denoting by e the coefficient of restitution between the two balls and assuming no friction, determine (*a*) the value of θ for which the component of the velocity of B in a direction perpendicular to the incoming path of A will be maximum after impact, (*b*) the magnitude of that component.

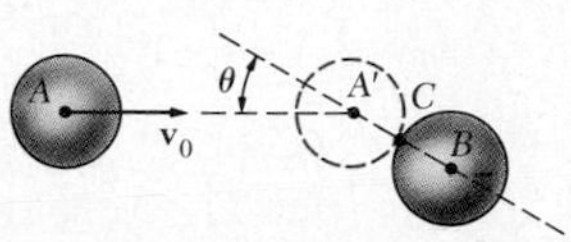

Fig. P13.147 and P13.148

13.149 A billiard player wishes to have ball A hit ball B obliquely and then ball C squarely. Assuming perfectly elastic impact ($e = 1$) and denoting by r the radius of the balls and by d the distance between the centers of B and C, determine (*a*) the angle θ defining point D where ball B should be hit, (*b*) the ranges of values of angles ABC and ACB for which this play is possible.

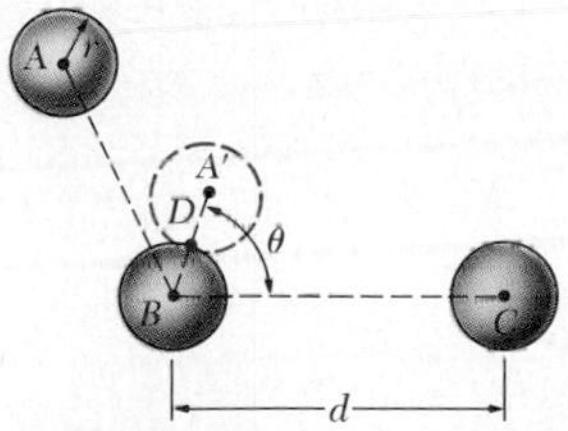

Fig. P13.149

13.150 A ball falling vertically hits a 90° corner at B with a velocity $\mathbf{v}_0$. Show that after hitting the corner again at C, it will rebound vertically with a velocity $-e\mathbf{v}_0$, where e is the coefficient of restitution between the ball and the corner. Neglect friction and assume that the short intermediate path BC is a straight line.

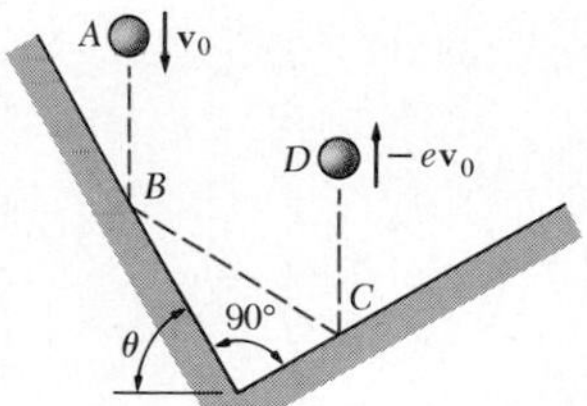

Fig. P13.150

13.151 In Prob. 13.150, determine the required value of θ if the path BC is to be horizontal.

13.152 A small ball A is dropped from a height h onto a rigid, frictionless plate at B and bounces to point C at the same elevation as B. Knowing that $\theta = 20°$ and that the coefficient of restitution between the ball and the plate is $e = 0.40$, determine the distance d.

13.153 A small ball A is dropped from a height h onto a rigid, frictionless plate at B and bounces to point C at the same elevation as B. Determine the value of θ for which the distance d is maximum and the corresponding value of d, assuming that the coefficient of restitution between the ball and the plate is (*a*) $e = 1$, (*b*) $e = 0.30$.

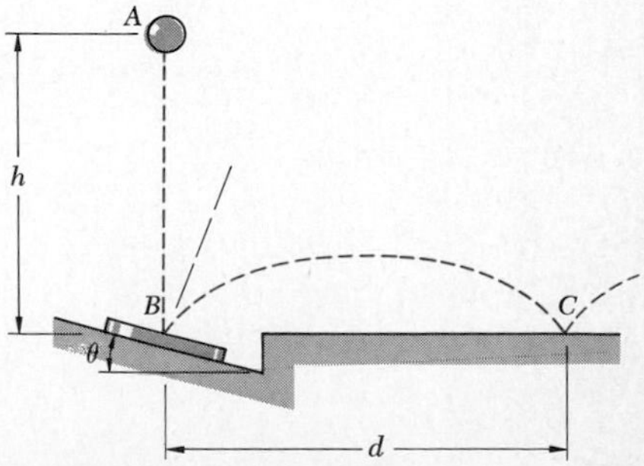

Fig. P13.152 and P13.153

13.154 A ball moving with the horizontal velocity $\mathbf{v}_0$ drops from A through the vertical distance $h_0 = 10$ in. to a frictionless floor. Knowing that the ball hits the floor at a distance $d_0 = 4$ in. from B and that the coefficient of restitution between the ball and the floor is $e = 0.80$, determine (*a*) the height h_1 and length d_1 of the first bounce, (*b*) the height h_2 and length d_2 of the second bounce.

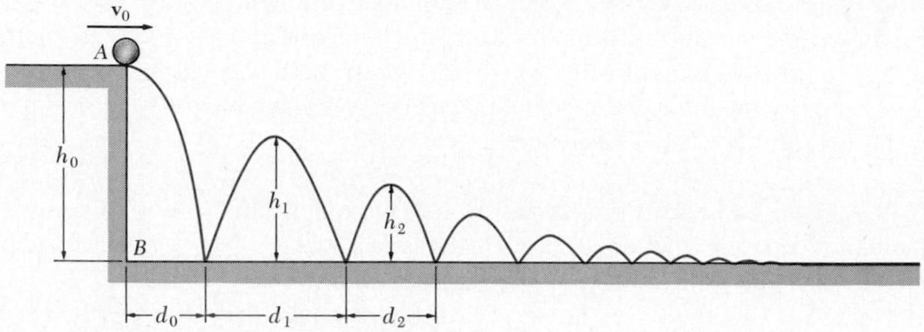

Fig. P13.154, P13.155, and P13.156

13.155 A ball moving with the horizontal velocity $\mathbf{v}_0$ drops from A through the vertical distance $h_0 = 16$ in. to a frictionless floor. Knowing that the ball hits the floor at a distance $d_0 = 4$ in. from B and that the height of its first bounce is $h_1 = 9$ in., determine (*a*) the coefficient of restitution between the ball and the floor, (*b*) the length d_1 of the first bounce.

13.156 A ball moving with a horizontal velocity of magnitude $v_0 = 0.2$ m/s drops from A to a frictionless floor. Knowing that the ball hits the floor at a distance $d_0 = 60$ mm from B and that the length of its first bounce is $d_1 = 96$ mm, determine (*a*) the coefficient of restitution between the ball and the floor, (*b*) the height h_1 of the first bounce.

13.157 In Prob. 13.156, determine how long the ball will keep bouncing after first hitting the floor.

13.158 A sphere A of mass m hits squarely with the velocity $\mathbf{v}_0$ a sphere B of the same mass m which is hanging from an inextensible wire BC. Knowing that $\theta = 30°$ and that the coefficient of restitution between the two spheres is $e = 0.75$, determine the velocity of each sphere immediately after impact.

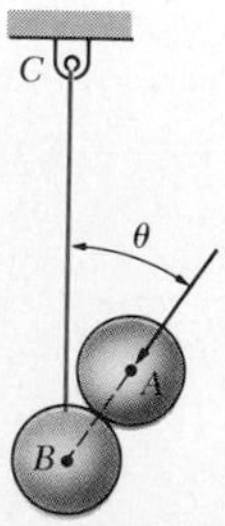

Fig. P13.158 and P13.159

13.159 A sphere A of mass m hits squarely with the velocity $\mathbf{v}_0$ a sphere B of the same mass m which is hanging from an inextensible wire BC. Denoting by e the coefficient of restitution between the two spheres, determine (*a*) the required value of the angle θ if the velocity of sphere A is to be zero immediately after impact, (*b*) the corresponding velocity of sphere B.

13.160 A sphere A of mass $m_A = 2$ kg is released from rest in the position shown and strikes the frictionless, inclined surface of a wedge B of mass $m_B = 6$ kg with a velocity of magnitude $v_0 = 3$ m/s. The wedge, which is supported by rollers and may move freely in the horizontal direction, is initially at rest. Knowing that $\theta = 30°$ and that the coefficient of restitution between the sphere and the wedge is $e = 0.80$, determine the velocities of the sphere and of the wedge immediately after impact.

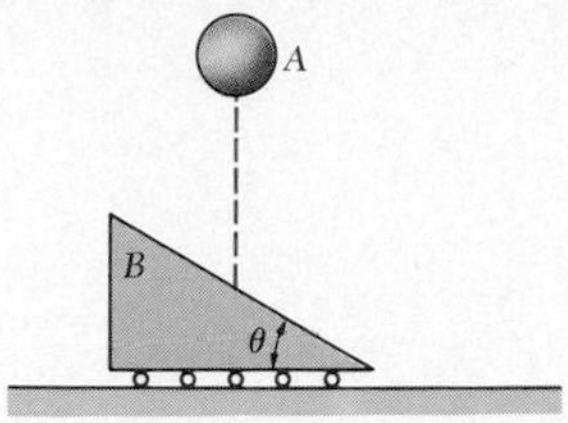

Fig. P13.160 and P13.161

13.161 A sphere A of mass m_A is released from rest in the position shown and strikes the frictionless, inclined surface of a wedge B of mass m_B. The wedge, which is supported by rollers and may move freely in the horizontal direction, is initially at rest. Knowing that the sphere is observed to bounce off horizontally to the right, express the coefficient of restitution e between the sphere and the surface of the incline in terms of m_A, m_B, and the angle θ of the wedge.

13.162 In Prob. 13.161, knowing that $m_A = 1.5$ kg, $m_B = 6$ kg, and $e = 0.75$, determine the angle θ of the wedge for which the sphere bounces off horizontally to the right, assuming that the wedge (*a*) is free to move in the horizontal direction, (*b*) is rigidly attached to the ground.

13.163 A 2-lb sphere A is released from rest when $\theta_A = 50°$ and strikes a 4-lb sphere B which is at rest. Knowing that the coefficient of restitution is $e = 0.80$, determine the values of θ_A and θ_B corresponding to the highest positions to which the spheres will rise after impact.

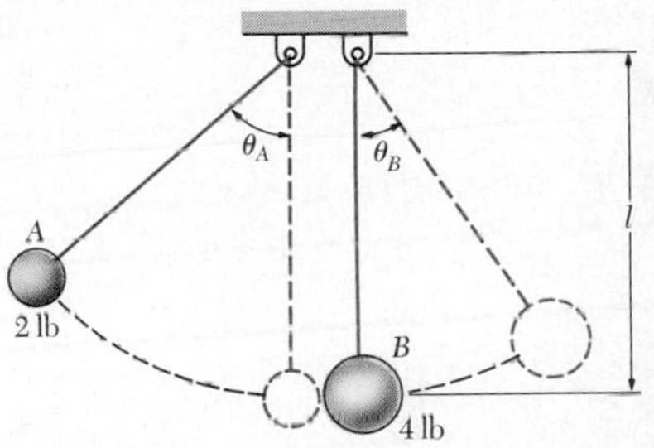

Fig. P13.163 and P13.164

13.164 A 2-lb sphere A is released from rest when $\theta_A = 60°$ and strikes a 4-lb sphere B which is at rest. Knowing that the velocity of sphere A is zero after impact, determine (*a*) the coefficient of restitution e, (*b*) the value of θ_B corresponding to the highest position to which sphere B will rise.

13.165 A 2-kg sphere A is released from rest when $\theta_A = 60°$ and strikes a 2.5-kg block B which is at rest. It is observed that the velocity of the sphere is zero after impact and that the block moves through 1.5 m before coming to rest. Determine (*a*) the coefficient of restitution between the sphere and the block, (*b*) the coefficient of kinetic friction between the block and the floor.

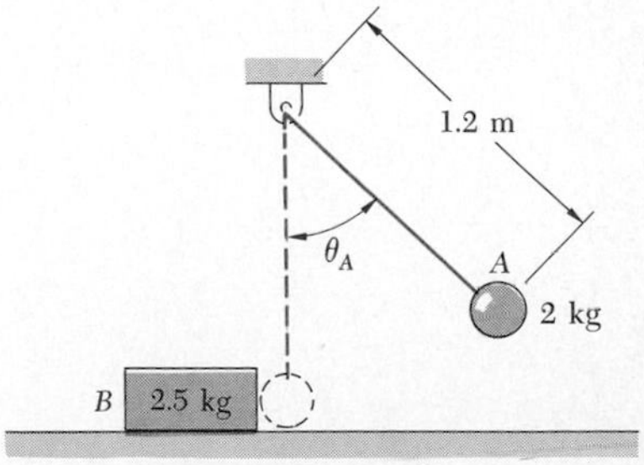

Fig. P13.165 and P13.166

13.166 A 2-kg sphere A is released from rest when $\theta_A = 90°$ and strikes a 2.5-kg block B which is at rest. Knowing that the coefficient of restitution between the sphere and the block is 0.75 and that the coefficient of kinetic friction between the block and the floor is 0.25, determine (*a*) how far block B will move, (*b*) the percentage of the initial energy lost in friction between the block and the floor.

13.167 A 25-g bullet is fired with a velocity of magnitude $v_0 = 550$ m/s into a 5-kg block of wood. Knowing that the coefficient of kinetic friction between the block and the floor is 0.30, determine (*a*) how far the block will move, (*b*) the percentage of the initial energy lost in friction between the block and the floor.

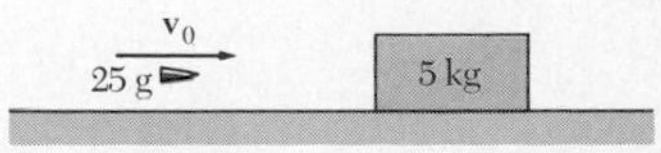

Fig. P13.167

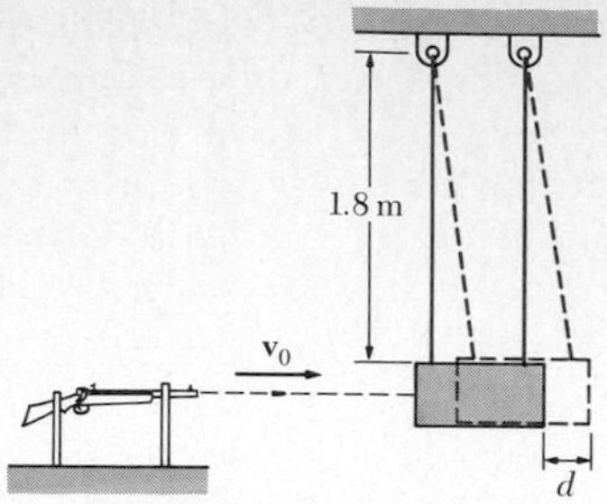

Fig. P13.168

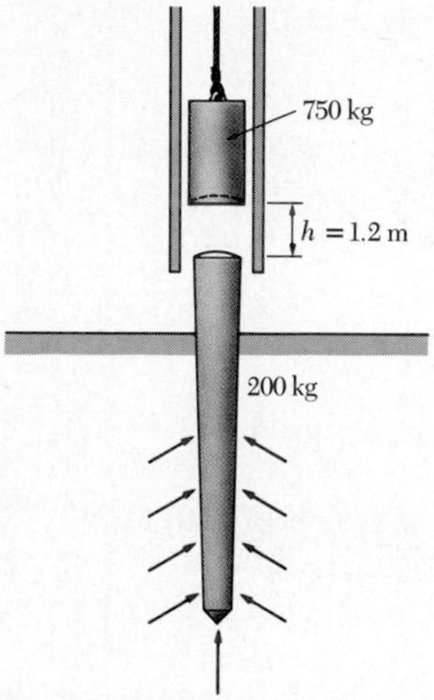

Fig. P13.169 and P13.170

13.168 A ballistic pendulum consisting of a 30-kg block suspended as shown from two wires 1.8-m long is used to measure the muzzle velocity of a rifle. If the pendulum swings through a horizontal distance $d = 250$ mm when a 40-g bullet is fired into it, determine the muzzle velocity v_0 of the rifle.

13.169 It is desired to drive a 200-kg pile into the ground until the resistance to its penetration is 100 kN. Each blow of the 750-kg hammer is the result of a 1.2-m free fall onto the top of the pile. Determine how far the pile will be driven into the ground by a single blow when the 100-kN resistance is achieved. Assume that the impact is perfectly plastic.

13.170 The 750-kg hammer of a drop-hammer pile driver falls from a height of 1.2 m onto the top of a 200-kg pile. The pile is driven 100 mm into the ground. Assuming perfectly plastic impact, determine the average resistance of the ground to penetration.

13.171 A 2-oz ball is dropped from a height h_0 onto an 8-oz plate. The ball is observed to rebound to a height $h_1 = 22.5$ in. when the plate rests directly on hard ground and to a height $h_2 = 6.40$ in. when a foam-rubber mat is placed between the plate and the ground. Determine (*a*) the coefficient of restitution between the ball and the plate, (*b*) the height h_0 from which the ball was dropped.

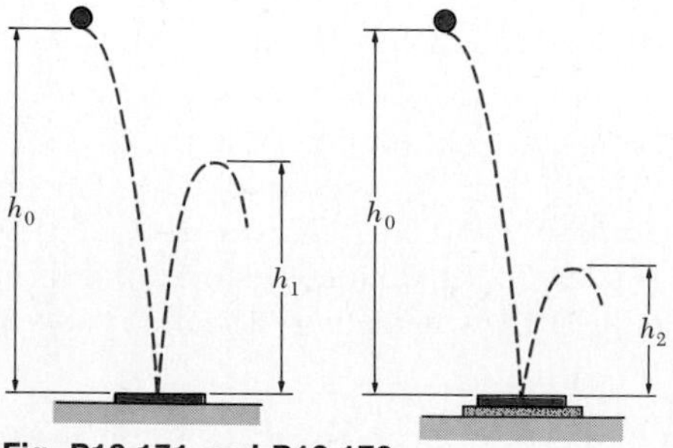

Fig. P13.171 and P13.172

13.172 A 1.5-oz ball is dropped from a height $h_0 = 25$ in. onto a small plate. The ball is observed to rebound to a height $h_1 = 16$ in. when the plate rests directly on hard ground and to a height $h_2 = 9$ in. when a foam-rubber mat is placed between the plate and the ground. Determine (*a*) the coefficient of restitution between the ball and the plate, (*b*) the weight of the plate.

13.173 Ball B is hanging from an inextensible cord. An identical ball A is released from rest when it is just touching the cord and drops through the vertical distance $h_A = 150$ mm before striking ball B. Assuming perfectly elastic impact ($e = 1$) and no friction, determine the resulting maximum vertical displacement h_B of ball B.

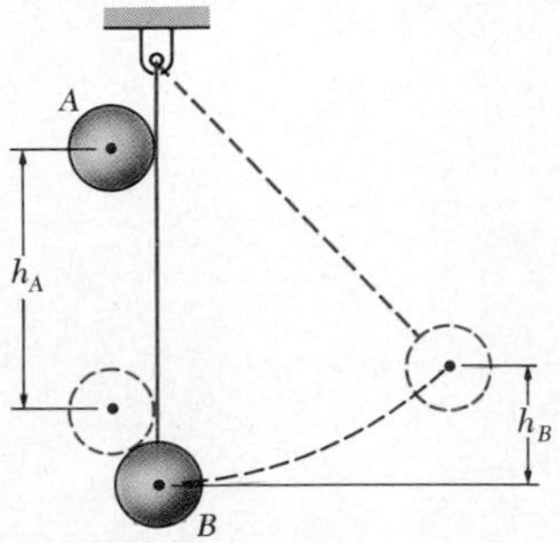

Fig. P13.173

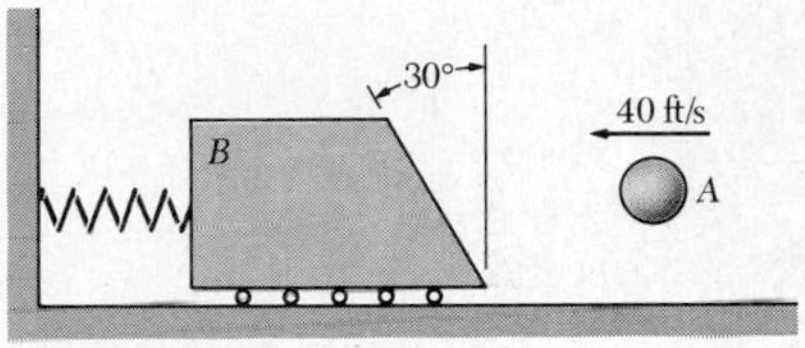

Fig. P13.174

13.174 A 2-lb sphere A is moving to the left with a velocity of 40 ft/s when it strikes the inclined surface of a 5-lb block B which is at rest. The block is supported by rollers and is attached to a spring of constant $k = 12$ lb/in. Knowing that the coefficient of restitution between the sphere and the block is $e = 0.75$ and neglecting friction, determine the maximum deflection of the spring.

13.175 A bumper is designed to protect a 2500-lb automobile from damage when it hits a rigid wall at speeds up to 9 mi/h. Assuming perfectly plastic impact, determine (a) the energy absorbed by the bumper during the impact, (b) the speed at which the automobile can hit another 2500-lb automobile without incurring any damage if the other automobile is similarly protected and is at rest with its brakes released.

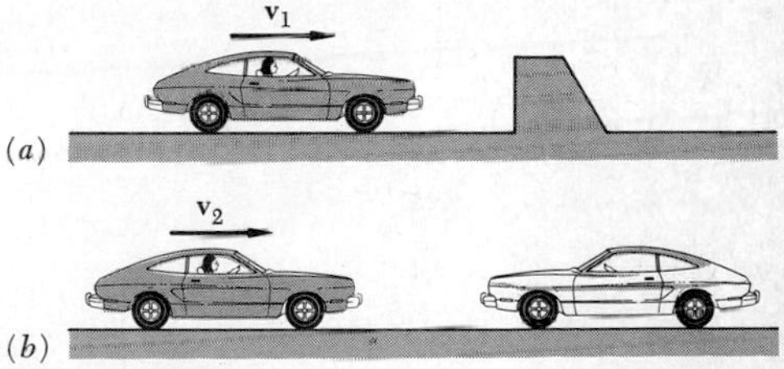

Fig. P13.175

***13.176** A ball of mass m_A moving with a velocity $\mathbf{v}_A$ strikes squarely a second ball of mass m_B which is at rest. Denoting by e the coefficient of restitution between the two balls, show that the percentage of energy lost during the impact is

$$\frac{100(1 - e^2)}{1 + (m_A/m_B)}$$

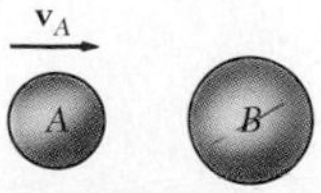

Fig. P13.176

Review Problems

13.177 Three 15-kg packages rest on a belt which passes over a pulley and is attached to a 40-kg block. Knowing that the coefficients of friction between the belt and the horizontal surface and also between the belt and the packages are $\mu_s = 0.50$ and $\mu_k = 0.40$, determine the speed of package B as it falls off the belt at E.

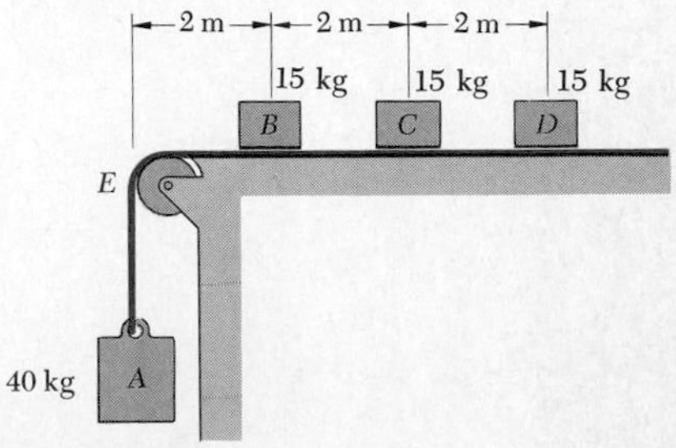

Fig. P13.177

13.178 In Prob. 13.177, determine the speed of package C as it falls off the belt at E.

13.179 A 500-g block P rests on a frictionless horizontal table at a distance of 400 mm from a fixed pin O. The block is attached to pin O by an elastic cord of constant $k = 100$ N/m and of undeformed length 900 mm. If the block is set in motion to the right as shown, determine (a) the speed v_1 for which the distance from O to the block P will reach a maximum value of 1.2 m, (b) the speed v_2 when $OP = 1.2$ m, (c) the radius of curvature of the path of the block when $OP = 1.2$ m.

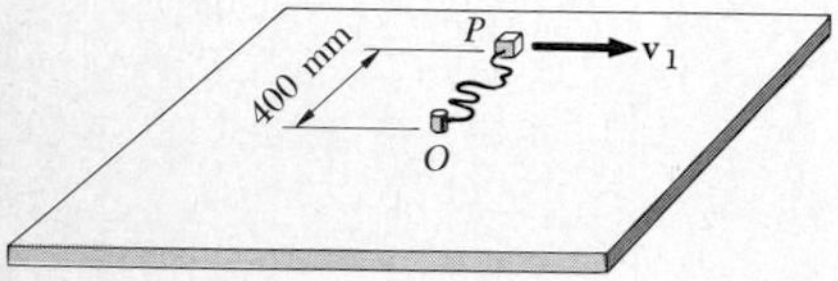

Fig. P13.179

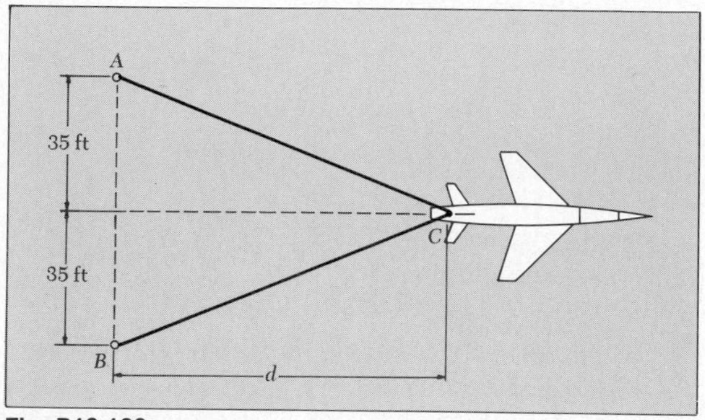

Fig. P13.180

13.180 A 36,000-lb airplane lands on an aircraft carrier and is caught by an arresting cable. The cable is inextensible and is paid out at A and B from mechanisms located below deck and consisting of pistons moving in long oil-filled cylinders. Knowing that the piston-cylinder system maintains a constant tension of 90 kips in the cable during the entire landing, determine the landing speed of the airplane if it travels a distance $d = 100$ ft after being caught by the cable.

13.181 For the airplane and aircraft carrier of Prob. 13.180, determine the value of the tension in the arresting cable, knowing that the airplane lands with a speed of 120 mi/h and travels a distance $d = 120$ ft after being caught by the cable.

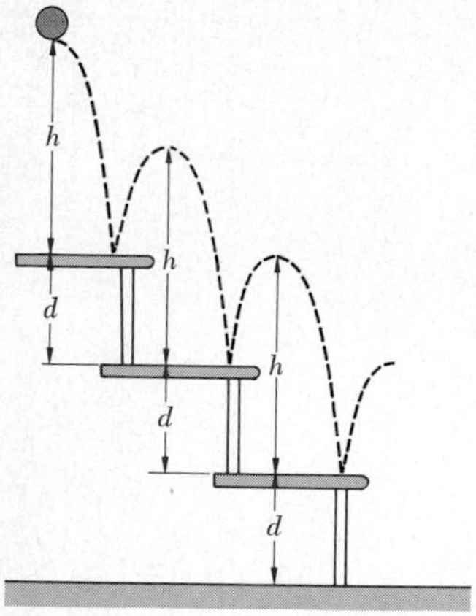

Fig. P13.182

13.182 A ball is dropped from a height h above the landing and bounces down a flight of stairs. Denoting by e the coefficient of restitution, determine the value of h for which the ball will bounce the same height above each step.

13.183 A 15-Mg truck and a 45-Mg railroad flatcar are both at rest with their brakes released. An engine bumps the flatcar and causes the flatcar alone to start moving with a velocity of 0.8 m/s to the right. Assuming $e = 1$ between the truck and the ends of the flatcar and neglecting friction, determine the velocities of the truck and of the flatcar after (a) end A strikes the truck, (b) the truck strikes end B.

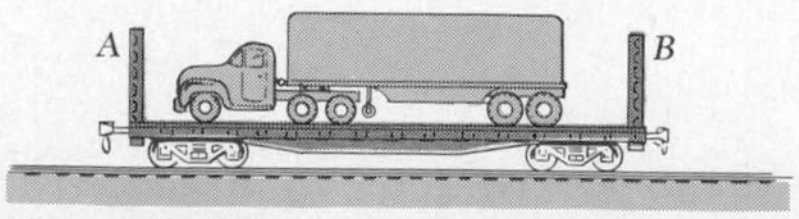

Fig. P13.183

13.184 A machine part is forged in a small drop forge. The hammer has a mass of 200 kg and is dropped from a height of 1.4 m. Determine the initial impulse exerted on the machine part and the energy absorbed by it, assuming that $e = 0$ and that the 500-kg anvil (*a*) is resting directly on hard ground, (*b*) is supported by springs.

13.185 An elevator travels upward at a constant speed of 2 m/s. A child riding the elevator throws a 0.6-kg stone upward with a speed of 5 m/s *relative* to the elevator. Determine (*a*) the work done by the child in throwing the stone, (*b*) the difference in the values of the kinetic energy of the stone before and after it was thrown. (*c*) Why are the values obtained in parts *a* and *b* not the same?

13.186 A dime which is at rest on a *rough* surface is struck squarely by a half dollar moving to the right. After the impact, each coin slides and comes to rest; the dime slides 480 mm to the right, and the half dollar slides 95 mm to the right. Assuming that the coefficient of friction is the same for each coin, determine the value of the coefficient of restitution between the coins. (*Masses:* half-dollar, 12.50 g; quarter dollar, 6.25 g; dime, 2.50 g.)

13.187 A light train made of two cars travels at 45 mi/h. Car *A* weighs 20 tons and car *B* weighs 25 tons. When the brakes are applied, a constant braking force of 4000 lb is applied to each car. Determine (*a*) the time required for the train to stop after the brakes are applied, (*b*) the force in the coupling between the cars while the train is slowing down.

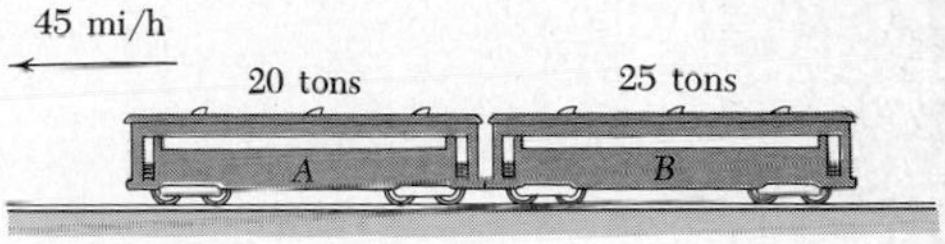

Fig. P13.187

13.188 Collar *A* is dropped 30 in. onto collar *B*, which is resting on a spring of constant $k = 4$ lb/in. Assuming perfectly plastic impact, determine (*a*) the maximum deflection of collar *B*, (*b*) the energy lost during the impact.

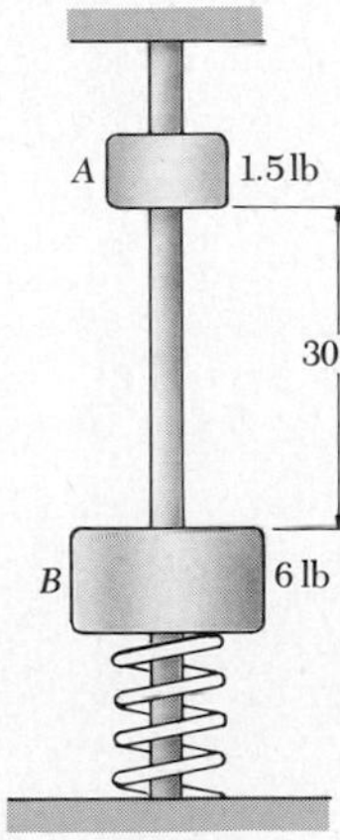

Fig. P13.188

CHAPTER FOURTEEN
Systems of Particles

14.1. Introduction. We shall be concerned in this chapter with the motion of *systems of particles*, i.e., with the motion of a large number of particles considered together. In the first part of the chapter, we shall discuss systems consisting of well-defined particles, while in the second part we shall analyze the motion of variable systems, i.e., systems which are continually gaining or losing particles, or doing both at the same time.

In Sec. 14.2, we shall first apply Newton's second law to each particle of the system. Defining the *effective force* of a particle as the product $m_i\mathbf{a}_i$ of its mass m_i and its acceleration $\mathbf{a}_i$, we shall show that the *external forces* acting on the various particles form a system equipollent to the system of the effective forces, i.e., both systems have the same resultant and same moment resultant about any given point. In Sec. 14.3, we shall further show that the resultant and moment resultant of the external forces are equal, respectively, to the rate of change of the total linear momentum and of the total angular momentum of the particles of the system.

In Sec. 14.4, we shall define the *mass center* of a system of particles and describe the motion of that point, while in Sec. 14.5 we shall discuss the motion of the particles about their mass center. We shall be concerned in Sec. 14.6 with the conditions under which the linear momentum and the angular momentum of a system of particles are conserved, and we shall apply the results obtained to the solution of various problems.

Sections 14.7 and 14.8 deal with the application of the work-energy principle to a system of particles, and Sec. 14.9 with the application of the impulse-momentum principle. These sections also contain a number of problems of practical interest.

It should be noted that while the derivations given in the first part of this chapter are carried out for a system of independent particles, they remain valid when the particles of the system are rigidly connected, i.e., when they form a rigid body. In fact, the results obtained here will form the foundation of our discussion of the kinetics of rigid bodies in Chaps. 16 through 18.

The second part of this chapter is devoted to the study of variable systems of particles. In Sec. 14.11 we shall consider steady streams of particles, such as a stream of water diverted by a fixed vane, or the flow of air through a jet engine, and we shall learn to determine the force exerted by the stream on the vane and the thrust developed by the engine. Finally, in Sec. 14.12, we shall analyze systems which gain mass by continually absorbing particles or lose mass by continually expelling particles. Among the various practical applications of this analysis will be the determination of the thrust developed by a rocket engine.

14.2. Application of Newton's Laws to the Motion of a System of Particles. Effective Forces. In order to derive the equations of motion for a system of n particles, we shall begin by writing Newton's second law for each individual particle of the system. Consider the particle P_i, where $1 \leq i \leq n$. Let m_i be the mass of P_i and $\mathbf{a}_i$ its acceleration with respect to the newtonian frame of reference $Oxyz$. We shall denote by $\mathbf{f}_{ij}$ the force exerted on P_i by another particle P_j of the system (Fig. 14.1); this force is called an *internal force*. The resultant of the

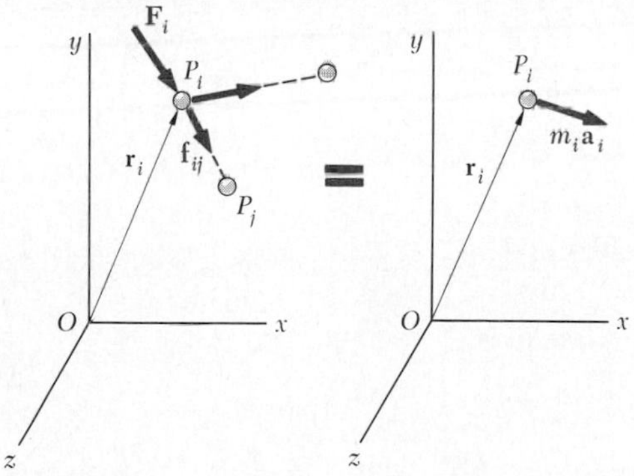

Fig. 14.1

internal forces exerted on P_i by all the other particles of the system is thus $\sum_{j=1}^{n} \mathbf{f}_{ij}$ (where $\mathbf{f}_{ii}$ has no meaning and is assumed to be equal to zero). Denoting, on the other hand, by $\mathbf{F}_i$ the resultant of all the *external forces* acting on P_i, we write Newton's second law for the particle P_i as follows:

$$\mathbf{F}_i + \sum_{j=1}^{n} \mathbf{f}_{ij} = m_i \mathbf{a}_i \tag{14.1}$$

Denoting by $\mathbf{r}_i$ the position vector of P_i and taking the moments about O of the various terms in Eq. (14.1), we also write

$$\mathbf{r}_i \times \mathbf{F}_i + \sum_{j=1}^{n} (\mathbf{r}_i \times \mathbf{f}_{ij}) = \mathbf{r}_i \times m_i \mathbf{a}_i \tag{14.2}$$

Repeating this procedure for each particle P_i of the system, we obtain n equations of the type (14.1) and n equations of the type (14.2), where i takes successively the values 1, 2, . . . , n. The vectors $m_i\mathbf{a}_i$ are referred to as the *effective forces* of the particles.† Thus the equations obtained express the fact that the external forces $\mathbf{F}_i$ and the internal forces $\mathbf{f}_{ij}$ acting on the various particles form a system equivalent to the system of the effective forces $m_i\mathbf{a}_i$ (i.e., one system may be replaced by the other) (Fig. 14.2).

Before proceeding further with our derivation, let us examine the internal forces $\mathbf{f}_{ij}$. We note that these forces occur in pairs $\mathbf{f}_{ij}$, $\mathbf{f}_{ji}$, where $\mathbf{f}_{ij}$ represents the force exerted by the particle P_j on the particle P_i and $\mathbf{f}_{ji}$ the force exerted by P_i on P_j (Fig. 14.2). Now, according to Newton's third law

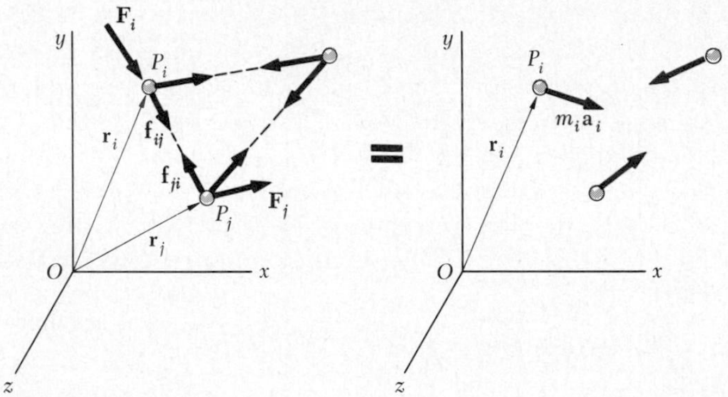

Fig. 14.2

(Sec. 6.1), as extended by Newton's law of gravitation to particles acting at a distance (Sec. 12.10), the forces $\mathbf{f}_{ij}$ and $\mathbf{f}_{ji}$ are equal and opposite and have the same line of action. Their sum is therefore $\mathbf{f}_{ij} + \mathbf{f}_{ji} = 0$ and the sum of their moments about O is

$$\mathbf{r}_i \times \mathbf{f}_{ij} + \mathbf{r}_j \times \mathbf{f}_{ji} = \mathbf{r}_i \times (\mathbf{f}_{ij} + \mathbf{f}_{ji}) + (\mathbf{r}_j - \mathbf{r}_i) \times \mathbf{f}_{ji} = 0$$

since the vectors $\mathbf{r}_j - \mathbf{r}_i$ and $\mathbf{f}_{ji}$ in the last term are collinear. Adding all the internal forces of the system, and summing their moments about O, we obtain the equations

$$\sum_{i=1}^{n}\sum_{j=1}^{n} \mathbf{f}_{ij} = 0 \qquad \sum_{i=1}^{n}\sum_{j=1}^{n} (\mathbf{r}_i \times \mathbf{f}_{ij}) = 0 \tag{14.3}$$

which express the fact that the resultant and the moment resultant of the internal forces of the system are zero.

Returning now to the n equations (14.1), where $i = 1, 2, \ldots, n$, we add them member by member. Taking into account the first of Eqs. (14.3), we obtain

$$\sum_{i=1}^{n} \mathbf{F}_i = \sum_{i=1}^{n} m_i\mathbf{a}_i \tag{14.4}$$

† Since these vectors represent the resultants of the forces acting on the various particles of the system, they may truly be considered as forces.

Proceeding similarly with Eqs. (14.2) and taking into account the second of Eqs. (14.3), we have

$$\sum_{i=1}^{n} (\mathbf{r}_i \times \mathbf{F}_i) = \sum_{i=1}^{n} (\mathbf{r}_i \times m_i \mathbf{a}_i) \tag{14.5}$$

Equations (14.4) and (14.5) express the fact that the system of the external forces $\mathbf{F}_i$ and the system of the effective forces $m_i\mathbf{a}_i$ have the same resultant and the same moment resultant. Referring to the definition given in Sec. 3.19 for two equipollent systems of vectors, we may therefore state that *the system of the external forces acting on the particles and the system of the effective forces of the particles are equipollent*† (Fig. 14.3).

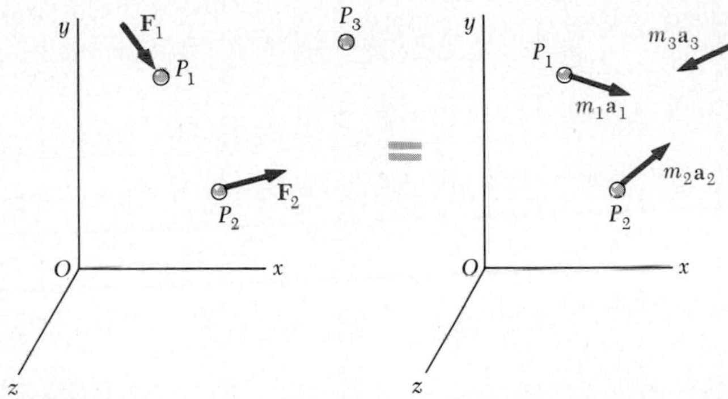

Fig. 14.3

We may note that Eqs. (14.3) express the fact that the system of the internal forces $\mathbf{f}_{ij}$ is equipollent to zero. It does *not* follow, however, that the internal forces have no effect on the particles under consideration. Indeed, the gravitational forces that the sun and the planets exert on one another are internal to the solar system and equipollent to zero. Yet these forces are alone responsible for the motion of the planets about the sun.

Similarly, it does not follow from Eqs. (14.4) and (14.5) that two systems of external forces which have the same resultant and the same moment resultant will have the same effect on a given system of particles. Clearly, the systems shown in Figs. 14.4*a* and 14.4*b* have the same resultant and the same moment resultant; yet the first system accelerates particle A and leaves particle B unaffected, while the second accelerates B and does not affect A. It is important to recall that when we stated in Sec. 3.19 that

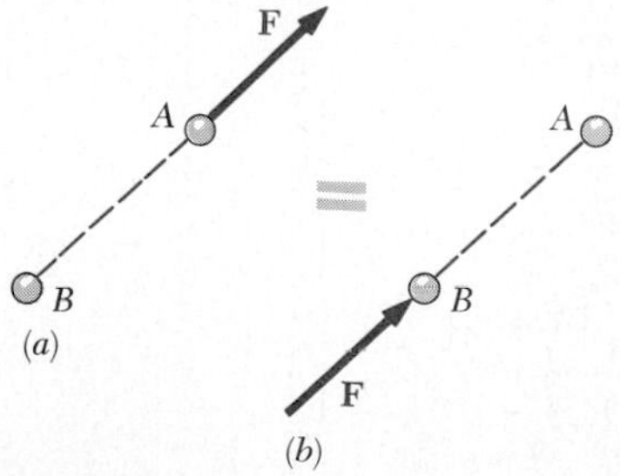

Fig. 14.4

†The result just obtained is often referred to as *d'Alembert's principle*, after the French mathematician Jean le Rond d'Alembert (1717–1783). However, d'Alembert's original statement refers to the motion of a system of connected bodies, with $\mathbf{f}_{ij}$ representing constraint forces which if applied by themselves, will not cause the system to move. Since, as it will now be shown, this is in general not the case for the internal forces acting on a system of free particles, we shall postpone the consideration of d'Alembert's principle until the study of the motion of rigid bodies (Chap. 16).

two equipollent systems of forces acting on a rigid body are also equivalent, we specifically noted that this property could *not* be extended to a system of forces acting on a set of independent particles such as those considered in this chapter.

In order to avoid any confusion, we shall use gray equals signs to connect equipollent systems of vectors, such as those shown in Figs. 14.3 and 14.4. These signs will indicate that the two systems of vectors have the same resultant and the same moment resultant. Colored equals signs will continue to be used to indicate that two systems of vectors are equivalent, i.e., that one system may actually be replaced by the other (Fig. 14.2).

14.3. Linear and Angular Momentum of a System of Particles. Equations (14.4) and (14.5), obtained in the preceding section for the motion of a system of particles, may be expressed in a more condensed form if we introduce the linear and the angular momentum of the system of particles. Defining the linear momentum **L** of the system of particles as the sum of the linear momenta of the various particles of the system (Sec. 12.3), we write

$$\mathbf{L} = \sum_{i=1}^{n} m_i \mathbf{v}_i \tag{14.6}$$

Defining the angular momentum $\mathbf{H}_O$ about O of the system of particles in a similar way (Sec. 12.7), we have

$$\mathbf{H}_O = \sum_{i=1}^{n} (\mathbf{r}_i \times m_i \mathbf{v}_i) \tag{14.7}$$

Differentiating both members of Eqs. (14.6) and (14.7) with respect to t, we write

$$\dot{\mathbf{L}} = \sum_{i=1}^{n} m_i \dot{\mathbf{v}}_i = \sum_{i=1}^{n} m_i \mathbf{a}_i \tag{14.8}$$

and

$$\begin{aligned} \dot{\mathbf{H}}_O &= \sum_{i=1}^{n} (\dot{\mathbf{r}}_i \times m_i \mathbf{v}_i) + \sum_{i=1}^{n} (\mathbf{r}_i \times m_i \dot{\mathbf{v}}_i) \\ &= \sum_{i=1}^{n} (\mathbf{v}_i \times m_i \mathbf{v}_i) + \sum_{i=1}^{n} (\mathbf{r}_i \times m_i \mathbf{a}_i) \end{aligned}$$

which reduces to

$$\dot{\mathbf{H}}_O = \sum_{i=1}^{n} (\mathbf{r}_i \times m_i \mathbf{a}_i) \tag{14.9}$$

since the vectors $\mathbf{v}_i$ and $m_i \mathbf{v}_i$ are collinear.

We observe that the right-hand members of Eqs. (14.8) and (14.9) are, respectively, identical with the right-hand members of Eqs. (14.4) and (14.5). It follows that the left-hand members of these equations are respectively equal. Recalling that the left-hand member of Eq. (14.5) represents the sum of the moments $\mathbf{M}_O$ about O of the external forces acting on the particles of the system, and omitting the subscript i from the sums, we write

$$\Sigma \mathbf{F} = \dot{\mathbf{L}} \tag{14.10}$$

$$\Sigma \mathbf{M}_O = \dot{\mathbf{H}}_O \tag{14.11}$$

These equations express that *the resultant and the moment resultant about the fixed point O of the external forces are respectively equal to the rates of change of the linear momentum and of the angular momentum about O of the system of particles.*

14.4. Motion of the Mass Center of a System of Particles. Equation (14.10) may be written in an alternate form if the *mass center* of the system of particles is considered. The mass center of the system is the point G defined by the position vector $\bar{\mathbf{r}}$, which satisfies the relation

$$m\bar{\mathbf{r}} = \sum_{i=1}^{n} m_i \mathbf{r}_i \tag{14.12}$$

where m represents the total mass $\sum_{i=1}^{n} m_i$ of the particles. Resolving the position vectors $\bar{\mathbf{r}}$ and $\mathbf{r}_i$ into rectangular components, we obtain the following three scalar equations, which may be used to determine the coordinates $\bar{x}$, $\bar{y}$, $\bar{z}$ of the mass center:

$$m\bar{x} = \sum_{i=1}^{n} m_i x_i \qquad m\bar{y} = \sum_{i=1}^{n} m_i y_i \qquad m\bar{z} = \sum_{i=1}^{n} m_i z_i \tag{14.12$'$}$$

Since $m_i g$ represents the weight of the particle P_i, and mg the total weight of the particles, we note that G is also the center of gravity of the system of particles. However, in order to avoid any confusion, we shall call G the *mass center* of the system of particles when discussing properties of the system associated with the *mass* of the particles, and we shall refer to it as the *center of gravity* of the system when considering properties associated with the *weight* of the particles. Particles located outside the gravitational field of the earth, for example, have a mass but no weight. We may then properly refer to their mass center, but obviously not to their center of gravity.†

†It may also be pointed out that the mass center and the center of gravity of a system of particles do not exactly coincide, since the weights of the particles are directed toward the center of the earth and thus do not truly form a system of parallel forces.

Differentiating both members of Eq. (14.12) with respect to t, we write

$$m\dot{\bar{\mathbf{r}}} = \sum_{i=1}^{n} m_i \dot{\mathbf{r}}_i$$

or

$$m\bar{\mathbf{v}} = \sum_{i=1}^{n} m_i \mathbf{v}_i \tag{14.13}$$

where $\bar{\mathbf{v}}$ represents the velocity of the mass center G of the system of particles. But the right-hand member of Eq. (14.13) is, by definition, the linear momentum $\mathbf{L}$ of the system (Sec. 14.3). We therefore have

$$\mathbf{L} = m\bar{\mathbf{v}} \tag{14.14}$$

and, differentiating both members with respect to t,

$$\dot{\mathbf{L}} = m\bar{\mathbf{a}} \tag{14.15}$$

where $\bar{\mathbf{a}}$ represents the acceleration of the mass center G. Substituting for $\dot{\mathbf{L}}$ from (14.15) into (14.10), we write the equation

$$\Sigma\mathbf{F} = m\bar{\mathbf{a}} \tag{14.16}$$

which defines the motion of the mass center G of the system of particles.

We note that Eq. (14.16) is identical with the equation we would obtain for a particle of mass m equal to the total mass of the particles of the system, acted upon by all the external forces. We state therefore: *The mass center of a system of particles moves as if the entire mass of the system and all the external forces were concentrated at that point.*

This principle is best illustrated by the motion of an exploding shell. We know that if the resistance of the air is neglected, a shell may be assumed to travel along a parabolic path. After the shell has exploded, the mass center G of the fragments of shell will continue to travel along the same path. Indeed, point G must move as if the mass and the weight of all fragments were concentrated at G; it must move, therefore, as if the shell had not exploded.

It should be noted that the preceding derivation does not involve the moments of the external forces. Therefore, *it would be wrong to assume* that the external forces are equipollent to a vector $m\bar{\mathbf{a}}$ attached at the mass center G. This is, in general, not the case, since, as we shall see in the next section, the sum of the moments about G of the external forces is, in general, not equal to zero.

14.5. Angular Momentum of a System of Particles about Its Mass Center. In some applications (for example, in the analysis of the motion of a rigid body) it is convenient to consider the motion of the particles of the system with respect to a centroidal frame of reference $Gx'y'z'$ which translates with respect to the newtonian frame of reference $Oxyz$ (Fig. 14.5). While such a frame is not, in general, a newtonian frame of reference, we shall see that the fundamental relation (14.11) still holds when the frame $Oxyz$ is replaced by $Gx'y'z'$.

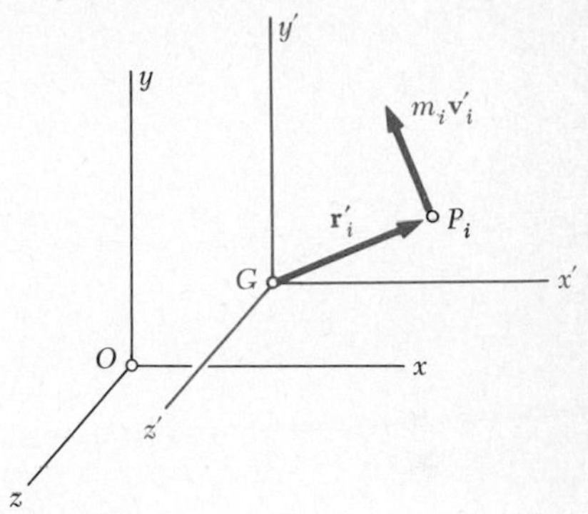

Fig. 14.5

Denoting, respectively, by $\mathbf{r}'_i$ and $\mathbf{v}'_i$ the position vector and the velocity of the particle P_i relative to the moving frame of reference $Gx'y'z'$, we define the *angular momentum* $\mathbf{H}'_G$ of the system of particles *about the mass center* G as follows:

$$\mathbf{H}'_G = \sum_{i=1}^{n} (\mathbf{r}'_i \times m_i\mathbf{v}'_i) \tag{14.17}$$

We now differentiate both members of Eq. (14.17) with respect to t. This operation being similar to that performed in Sec. 14.3 on Eq. (14.7), we write immediately

$$\dot{\mathbf{H}}'_G = \sum_{i=1}^{n} (\mathbf{r}'_i \times m_i\mathbf{a}'_i) \tag{14.18}$$

where $\mathbf{a}'_i$ denotes the acceleration of P_i relative to the moving frame of reference. Referring to Sec. 11.12, we write

$$\mathbf{a}_i = \overline{\mathbf{a}} + \mathbf{a}'_i$$

where $\mathbf{a}_i$ and $\overline{\mathbf{a}}$ denote, respectively, the accelerations of P_i and G relative to the frame $Oxyz$. Solving for $\mathbf{a}'_i$ and substituting into (14.18), we have

$$\dot{\mathbf{H}}'_G = \sum_{i=1}^{n} (\mathbf{r}'_i \times m_i\mathbf{a}_i) - \left(\sum_{i=1}^{n} m_i\mathbf{r}'_i\right) \times \overline{\mathbf{a}} \tag{14.19}$$

But, by (14.12), the second sum in Eq. (14.19) is equal to $m\overline{\mathbf{r}}'$ and, thus, to zero, since the position vector $\overline{\mathbf{r}}'$ of G relative to the frame $Gx'y'z'$ is clearly zero. On the other hand, since $\mathbf{a}_i$ represents the acceleration of P_i relative to a newtonian frame, we may use Eq. (14.1) and replace $m_i\mathbf{a}_i$ by the sum of the internal forces $\mathbf{f}_{ij}$ and of the resultant $\mathbf{F}_i$ of the external forces acting on P_i. But a reasoning similar to that used in Sec. 14.2 shows that the moment resultant about G of the internal forces $\mathbf{f}_{ij}$ of the entire system is zero. The first sum in Eq. (14.19) reduces therefore to the moment resultant about G of the external forces acting on the particles of the system, and we write

$$\Sigma\mathbf{M}_G = \dot{\mathbf{H}}'_G \tag{14.20}$$

which expresses that *the moment resultant about G of the external forces is equal to the rate of change of the angular momentum about G of the system of particles.*

It should be noted that in Eq. (14.17), we defined the angular momentum $\mathbf{H}'_G$ as the sum of the moments about G of the momenta of the particles $m_i\mathbf{v}'_i$ *in their motion relative to the centroidal frame of reference* $Gx'y'z'$. We may sometimes want to compute the sum $\mathbf{H}_G$ of the moments about G of the momenta of the particles $m_i\mathbf{v}_i$ *in their absolute motion,* i.e., in their motion as observed from the newtonian frame of reference $Oxyz$ (Fig. 14.6):

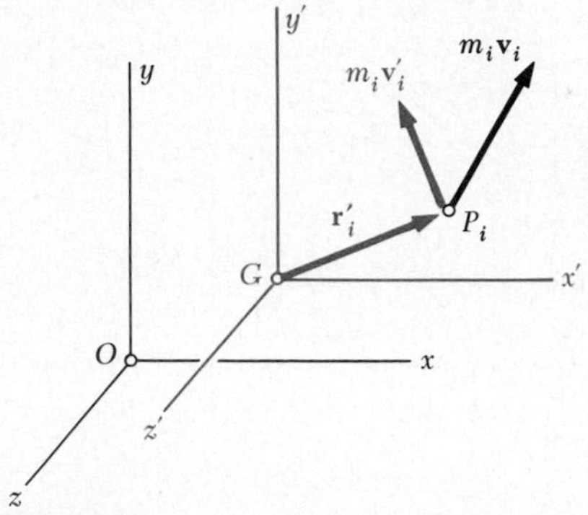

Fig. 14.6

$$\mathbf{H}_G = \sum_{i=1}^{n} (\mathbf{r}'_i \times m_i\mathbf{v}_i) \tag{14.21}$$

Remarkably, the angular momenta $\mathbf{H}'_G$ and $\mathbf{H}_G$ are identically equal. This may be verified by referring to Sec. 11.12 and writing

$$\mathbf{v}_i = \overline{\mathbf{v}} + \mathbf{v}'_i \tag{14.22}$$

Substituting for $\mathbf{v}_i$ from (14.22) into Eq. (14.21), we have

$$\mathbf{H}_G = \left(\sum_{i=1}^{n} m_i\mathbf{r}'_i\right) \times \overline{\mathbf{v}} + \sum_{i=1}^{n} (\mathbf{r}'_i \times m_i\mathbf{v}'_i)$$

But, as observed earlier, the first sum is equal to zero. Thus $\mathbf{H}_G$ reduces to the second sum which, by definition, is equal to $\mathbf{H}'_G$.†

†Note that this property is peculiar to the centroidal frame $Gx'y'z'$ and does not hold, in general, for other frames of reference (see Prob. 14.23).

Taking advantage of the property we have just established, we shall simplify our notation by dropping the prime (′) from Eq. (14.20). We therefore write

$$\Sigma \mathbf{M}_G = \dot{\mathbf{H}}_G \tag{14.23}$$

where it is understood that the angular momentum $\mathbf{H}_G$ may be computed by forming the moments about G of the momenta of the particles in their motion with respect to either the newtonian frame $Oxyz$ or the centroidal frame $Gx'y'z'$:

$$\mathbf{H}_G = \sum_{i=1}^{n} (\mathbf{r}'_i \times m_i\mathbf{v}_i) = \sum_{i=1}^{n} (\mathbf{r}'_i \times m_i\mathbf{v}'_i) \tag{14.24}$$

14.6. Conservation of Momentum for a System of Particles. If no external force acts on the particles of a system, the left-hand members of Eqs. (14.10) and (14.11) are equal to zero and these equations reduce to $\dot{\mathbf{L}} = 0$ and $\dot{\mathbf{H}}_O = 0$. We conclude that

$$\mathbf{L} = \text{constant} \qquad \mathbf{H}_O = \text{constant} \tag{14.25}$$

The equations obtained express that the linear momentum of the system of particles and its angular momentum about the fixed point O are conserved.

In some applications, such as problems involving central forces, the moment about a fixed point O of each of the external forces may be zero, without any of the forces being zero. In such cases, the second of Eqs. (14.25) still holds; the angular momentum of the system of particles about O is conserved.

The concept of conservation of momentum may also be applied to the analysis of the motion of the mass center G of a system of particles and to the analysis of the motion of the system about G. For example, if the sum of the external forces is zero, the first of Eqs. (14.25) applies. Recalling Eq. (14.14), we write

$$\overline{\mathbf{v}} = \text{constant} \tag{14.26}$$

which expresses that the mass center G of the system moves in a straight line and at a constant speed. On the other hand, if the sum of the moments about G of the external forces is zero, it follows from Eq. (14.23) that the angular momentum of the system about its mass center is conserved:

$$\mathbf{H}_G = \text{constant} \tag{14.27}$$

SAMPLE PROBLEM 14.1

A 200-kg space vehicle is observed at $t = 0$ to pass through the origin of a newtonian reference frame $Oxyz$ with the velocity $\mathbf{v}_0 = (150 \text{ m/s})\mathbf{i}$ relative to the frame. Following the detonation of explosive charges, the vehicle separates into three parts, A, B, and C, of mass 100 kg, 60 kg, and 40 kg, respectively. Knowing that, at $t = 2.5$ s, the positions of parts A and B are observed to be $A(555, -180, 240)$ and $B(255, 0, -120)$, where the coordinates are expressed in meters, determine the position of part C at that time.

Solution. Since there is no external force, the mass center G of the system moves with the constant velocity $\mathbf{v}_0 = (150 \text{ m/s})\mathbf{i}$. At $t = 2.5$ s, its position is

$$\bar{\mathbf{r}} = \mathbf{v}_0 t = (150 \text{ m/s})\mathbf{i}(2.5 \text{ s}) = (375 \text{ m})\mathbf{i}$$

Recalling Eq. (14.12), we write

$$m\bar{\mathbf{r}} = m_A\mathbf{r}_A + m_B\mathbf{r}_B + m_C\mathbf{r}_C$$

$$(200 \text{ kg})(375 \text{ m})\mathbf{i} = (100 \text{ kg})[(555 \text{ m})\mathbf{i} - (180 \text{ m})\mathbf{j} + (240 \text{ m})\mathbf{k}] + (60 \text{ kg})[(255 \text{ m})\mathbf{i} - (120 \text{ m})\mathbf{k}] + (40 \text{ kg})\mathbf{r}_C$$

$$\mathbf{r}_C = (105 \text{ m})\mathbf{i} + (450 \text{ m})\mathbf{j} - (420 \text{ m})\mathbf{k} \quad ◄$$

SAMPLE PROBLEM 14.2

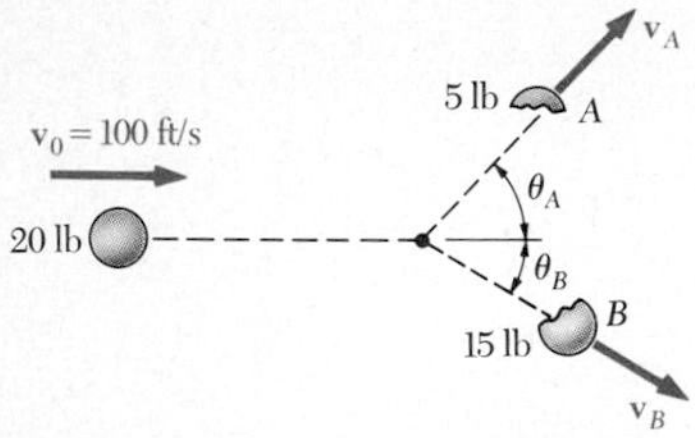

A 20-lb projectile is moving with a velocity of 100 ft/s when it explodes into two fragments A and B, weighing 5 lb and 15 lb, respectively. Knowing that immediately after the explosion fragments A and B travel in directions defined, respectively, by $\theta_A = 45°$ and $\theta_B = 30°$, determine the velocity of each fragment.

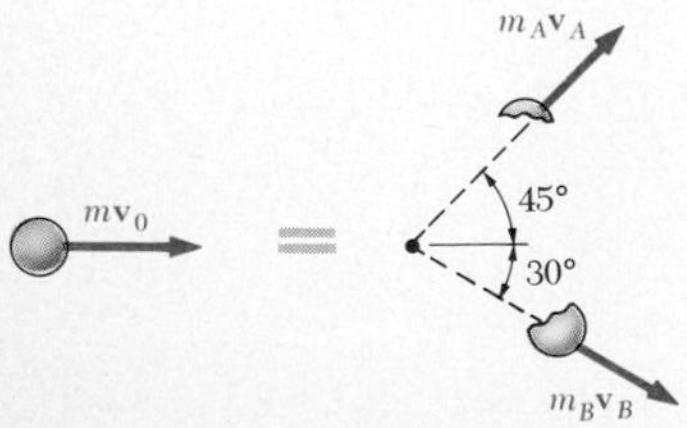

Solution. Since there is no external force, the linear momentum of the system is conserved, and we write

$$m_A\mathbf{v}_A + m_B\mathbf{v}_B = m\mathbf{v}_0$$

$$(5/g)\mathbf{v}_A + (15/g)\mathbf{v}_B = (20/g)\mathbf{v}_0$$

$\xrightarrow{+}$ x components: $\quad 5v_A \cos 45° + 15v_B \cos 30° = 20(100)$

$+\uparrow y$ components: $\quad 5v_A \sin 45° - 15v_B \sin 30° = 0$

Solving simultaneously the two equations for v_A and v_B, we have

$$v_A = 207 \text{ ft/s} \qquad v_B = 97.6 \text{ ft/s}$$

$$\mathbf{v}_A = 207 \text{ ft/s} \measuredangle 45° \qquad \mathbf{v}_B = 97.6 \text{ ft/s} \searrow 30° \quad ◄$$

Problems

14.1 A 25-g bullet is fired in a horizontal direction through block A and becomes embedded in block B. The bullet causes A and B to start moving with velocities of 2.4 and 1.8 m/s, respectively. Determine (*a*) the initial velocity $\mathbf{v}_0$ of the bullet, (*b*) the velocity of the bullet as it travels from block A to block B.

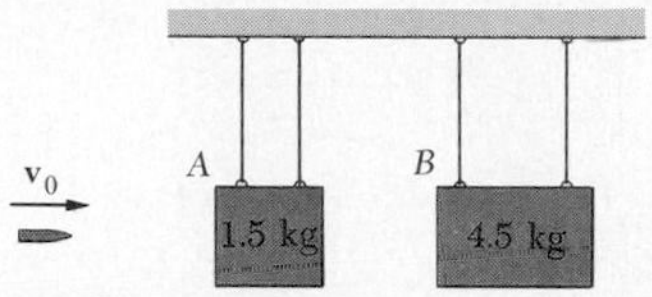

Fig. P14.1 and P14.2

14.2 A 25-g bullet is fired in a horizontal direction through block A and becomes embedded in block B. Knowing that $v_0 = 600$ m/s and that block B starts moving with a velocity of 2 m/s, determine (*a*) the velocity with which block A starts moving, (*b*) the velocity of the bullet as it travels from block A to block B.

14.3 A 40-ton boxcar A is moving with a velocity of 4.5 mi/h in a railroad switchyard when it strikes, and is automatically coupled with, a 20-ton flatcar B which carries a 30-ton trailer truck C. Both the flatcar and the truck are at rest with their brakes released. Determine the velocity of the cars and of the truck (*a*) immediately after the coupling of the two cars, (*b*) immediately after the end of the flatcar hits the truck. Neglect friction and assume the impact between the flatcar and the truck to be perfectly plastic ($e = 0$).

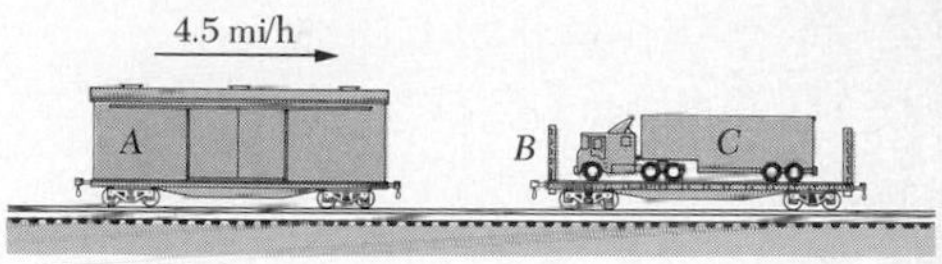

Fig. P14.3

14.4 Three identical freight cars have the velocities indicated. Assuming that car B is first hit by car A, determine the velocity of each car after all collisions have taken place, if (*a*) all three cars get automatically coupled, (*b*) cars A and B get automatically and tightly coupled, while cars B and C bounce off each other with a coefficient of restitution $e = 1$.

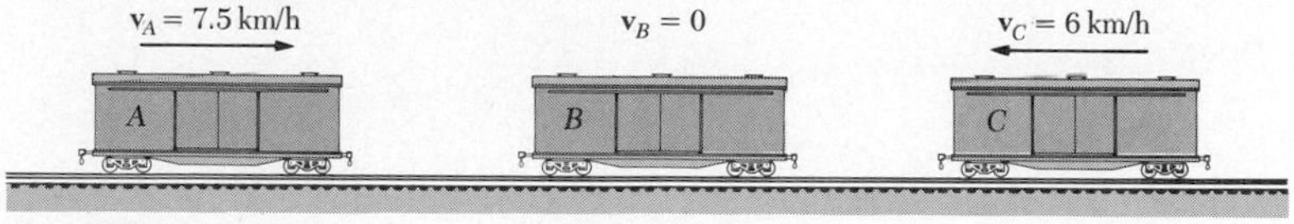

Fig. P14.4

14.5 Solve Prob. 14.4, assuming that car B is first hit by car C.

14.6 Solve Prob. 14.3, assuming the impact between the flatcar and the truck to be perfectly elastic ($e = 1$).

14.7 A system consists of three particles A, B, and C. We know that $m_A = 3$ kg, $m_B = 4$ kg, and $m_C = 5$ kg and that the velocities of the particles expressed in m/s are, respectively, $\mathbf{v}_A = 2\mathbf{i} + 3\mathbf{j} - 2\mathbf{k}$, $\mathbf{v}_B = v_x\mathbf{i} + 2\mathbf{j} + v_z\mathbf{k}$, and $\mathbf{v}_C = -3\mathbf{i} - 2\mathbf{j} + \mathbf{k}$. Determine (*a*) the components v_x and v_z of the velocity of particle B for which the angular momentum $\mathbf{H}_O$ of the system about O is parallel to the x axis, (*b*) the corresponding value of $\mathbf{H}_O$.

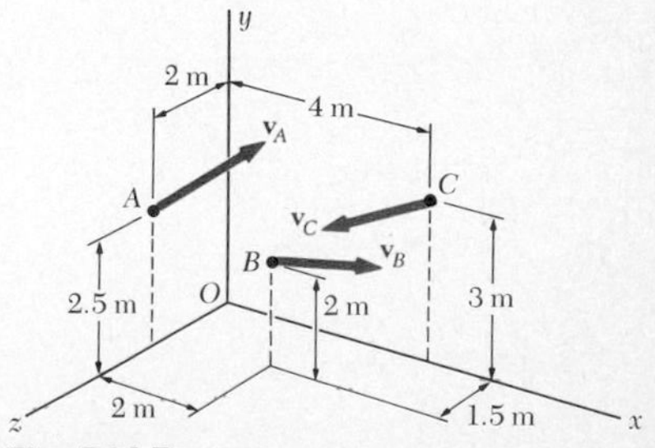

Fig. P14.7

14.8 For the system of particles of Prob. 14.7, determine (*a*) the components v_x and v_z of the velocity of particle B for which the angular momentum $\mathbf{H}_O$ of the system about O is parallel to the z axis, (*b*) the corresponding value of $\mathbf{H}_O$.

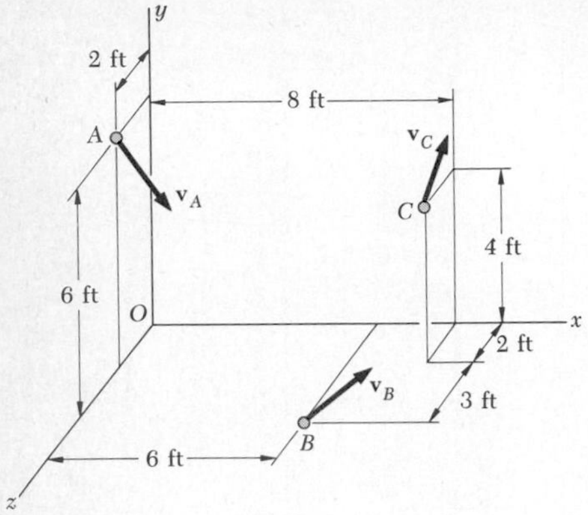

Fig. P14.9

14.9 A system consists of three particles A, B, and C. We know that $W_A = 2$ lb, $W_B = 4$ lb, and $W_C = 6$ lb and that the velocities of the particles expressed in ft/s are, respectively, $\mathbf{v}_A = 6\mathbf{i} - 4\mathbf{j} + 8\mathbf{k}$, $\mathbf{v}_B = 8\mathbf{i} + 6\mathbf{j}$, and $\mathbf{v}_C = 4\mathbf{i} + 10\mathbf{j} - 6\mathbf{k}$. (*a*) Determine the angular momentum $\mathbf{H}_O$ of the system about O. (*b*) Using the result of part *a* and the answers to Prob. 14.10, check that the relation given in Prob. 14.21 is satisfied.

14.10 For the system of particles of Prob. 14.9, determine (*a*) the position vector $\bar{\mathbf{r}}$ of the mass center G of the system, (*b*) the linear momentum $m\bar{\mathbf{v}}$ of the system, (*c*) the angular momentum $\mathbf{H}_G$ of the system about G.

14.11 A 600-lb space vehicle traveling with a velocity $\mathbf{v}_0 = (1200 \text{ ft/s})\mathbf{k}$ passes through the origin O at $t = 0$. Explosive charges then separate the vehicle into three parts A, B, and C, weighing, respectively, 100 lb, 200 lb, and 300 lb. Knowing that at $t = 2$ s, the positions of parts B and C are observed to be B (600, 1320, 3240) and C (−480, −960, 1920), where the coordinates are expressed in feet, determine the corresponding position of part A. Neglect the effect of gravity.

14.12 An airliner weighing 50 tons and flying due east at an altitude of 28,000 ft and a speed of 540 mi/h suddenly explodes, breaking into three fragments A, B, and C. Fragment A, weighing 25 tons, and fragment B, weighing 15 tons, are found in a wooded area at the points shown in the figure. Knowing that the explosion occurred when the airliner was directly above point O and assuming that all three fragments hit the ground at the same time, determine the location of fragment C. Neglect the resistance of the air.

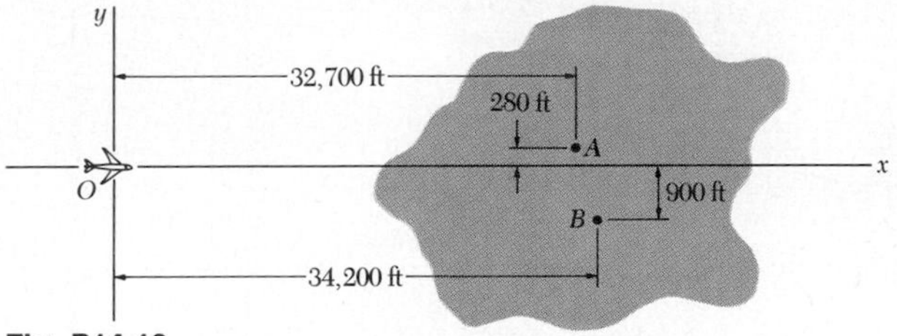

Fig. P14.12

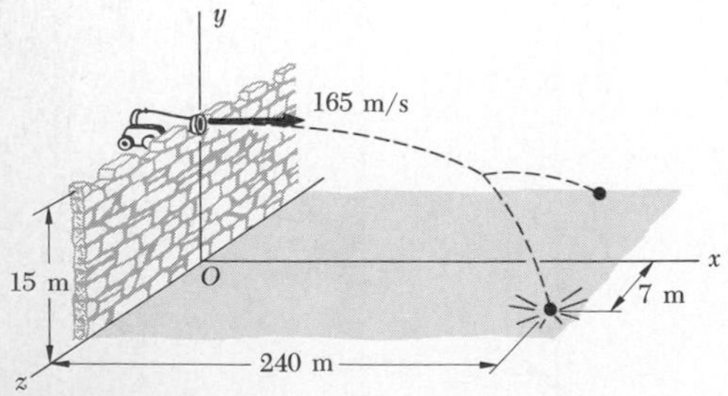

Fig. P14.13

14.13 Two 15-kg cannon balls are chained together and fired horizontally with a velocity of 165 m/s from the top of a 15-m wall. The chain breaks during the flight of the cannon balls and one of them strikes the ground at $t = 1.5$ s, at a distance of 240 m from the foot of the wall, and 7 m to the right of the line of fire. Determine the position of the other cannon ball at that instant. Neglect the resistance of the air.

14.14 Solve Prob. 14.13, if the cannon ball which first strikes the ground has a mass of 12 kg, and the other a mass of 18 kg. Assume that the time of flight and the point of impact of the first cannon ball remain the same.

14.15 Knowing that fragment A of Sample Prob. 14.2 has a velocity of magnitude $v_A = 150$ ft/s and travels in the direction $\theta_A = 55°$, determine the magnitude and direction of the velocity of fragment B immediately after the explosion.

14.16 A 500-lb space vehicle is traveling with a velocity $\mathbf{v}_0 = (1500 \text{ ft/s})\mathbf{i}$ when explosive charges separate it into three parts A, B, and C, weighing, respectively, 120 lb, 180 lb, and 200 lb. Knowing that immediately after the explosion, the velocity of part A is $\mathbf{v}_A = (2000 \text{ ft/s})\mathbf{i} - (200 \text{ ft/s})\mathbf{j} + (300 \text{ ft/s})\mathbf{k}$ and that the velocity of part B is $\mathbf{v}_B = (1000 \text{ ft/s})\mathbf{i} + (300 \text{ ft/s})\mathbf{j} - (500 \text{ ft/s})\mathbf{k}$, determine the corresponding velocity of part C.

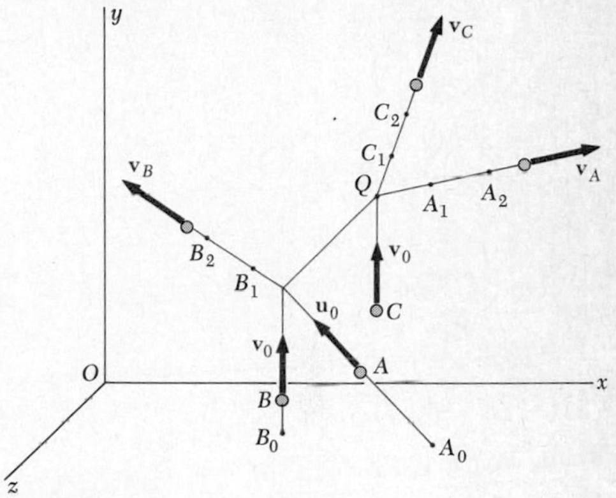

Fig. P14.17

14.17 In a scattering experiment, an alpha particle A is projected with the velocity $\mathbf{u}_0 = -(480 \text{ m/s})\mathbf{i} + (600 \text{ m/s})\mathbf{j} - (640 \text{ m/s})\mathbf{k}$ into a stream of oxygen nuclei moving with the common velocity $\mathbf{v}_0 = (480 \text{ m/s})\mathbf{j}$. After colliding successively with the nuclei B and C, particle A is observed to move along the path defined by the points A_1 (240, 220, 160), A_2 (320, 300, 200), while nuclei B and C are observed to move along paths defined, respectively, by B_1 (107, 200, 170), B_2 (74, 270, 160) and by C_1 (200, 212, 130), C_2 (200, 260, 115). All paths are along straight lines and all coordinates are expressed in millimeters. Knowing that the mass of an oxygen nucleus is four times that of an alpha particle, determine the speed of each of the three particles after the collisions.

14.18 A 1-kg shell is moving in a direction perpendicular to a wall when at point D, it explodes into three fragments A, B, and C, respectively, of mass 300 g, 300 g, and 400 g. Knowing that the fragments hit the wall at the points indicated and that $v_A = 490$ m/s, determine the velocity $\mathbf{v}_0$ of the shell before it exploded.

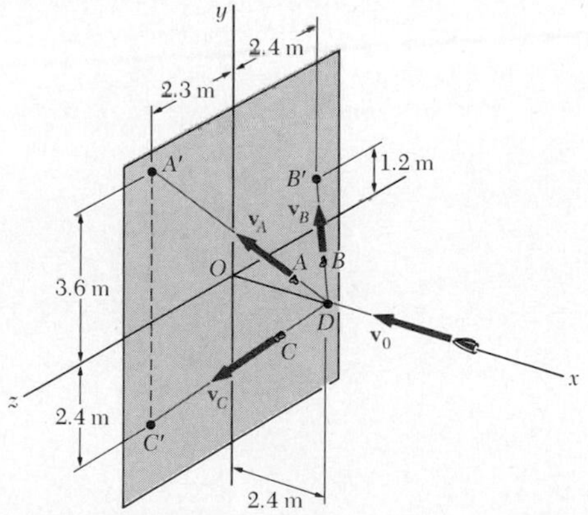

Fig. P14.18

14.19 A 3000-kg helicopter A was traveling due east at a speed of 120 km/h and an altitude of 800 m when it was hit by a 5000-kg helicopter B. As a result of the collision, both helicopters lost their lift, and their entangled wreckage fell to the ground in 12 s at a point located 525 m east and 135 m south of the point of impact. Neglecting air resistance, determine the velocity components of helicopter B just before the collision.

14.20 An archer hits a game bird flying in a horizontal straight line 9 m above the ground with a 32-g wooden arrow. Knowing that the arrow strikes the bird from behind with a velocity of 110 m/s at an angle of 30° with the vertical, and that the bird falls to the ground in 1.5 s and 15 m beyond the point where it was hit, determine (a) the mass of the bird, (b) the speed at which it was flying when it was hit.

14.21 Derive the relation

$$\mathbf{H}_O = \bar{\mathbf{r}} \times m\bar{\mathbf{v}} + \mathbf{H}_G$$

between the angular momenta $\mathbf{H}_O$ and $\mathbf{H}_G$ defined in Eqs. (14.7) and (14.24), respectively. The vectors $\bar{\mathbf{r}}$ and $\bar{\mathbf{v}}$ define, respectively, the position and velocity of the mass center G of the system of particles relative to the newtonian frame of reference $Oxyz$, and m represents the total mass of the system.

14.22 Show that Eq. (14.23) may be derived directly from Eq. (14.11) by substituting for $\mathbf{H}_O$ the expression given in Prob. 14.21.

14.23 Consider the frame of reference $Ax'y'z'$ in translation with respect to the newtonian frame of reference $Oxyz$. We define the angular momentum $\mathbf{H}'_A$ of a system of n particles about A as the sum

$$\mathbf{H}'_A = \sum_{i=1}^{n} \mathbf{r}'_i \times m_i \mathbf{v}'_i \tag{1}$$

of the moments about A of the momenta $m_i\mathbf{v}'_i$ of the particles in their motion relative to the frame $Ax'y'z'$. Denoting by $\mathbf{H}_A$ the sum

$$\mathbf{H}_A = \sum^{n} \mathbf{r}'_i \times m_i \mathbf{v}_i \tag{2}$$

of the moments about A of the momenta $m_i\mathbf{v}_i$ of the particles in their motion relative to the newtonian frame $Oxyz$, show that $\mathbf{H}_A = \mathbf{H}'_A$ at a given instant if, and only if, one of the following conditions is satisfied at that instant: (*a*) A has zero velocity with respect to the frame $Oxyz$, (*b*) A coincides with the mass center G of the system, (*c*) the velocity $\mathbf{v}_A$ relative to $Oxyz$ is directed along the line AG.

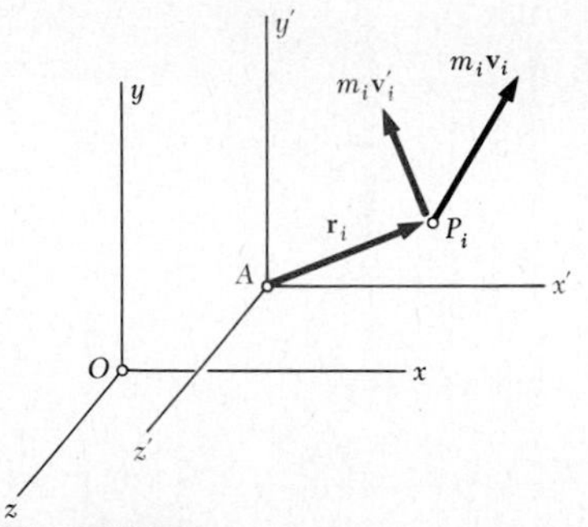

Fig. P14.23

14.24 Show that the relation

$$\Sigma\mathbf{M}_A = \dot{\mathbf{H}}'_A$$

where $\mathbf{H}'_A$ is defined by Eq. (1) of Prob. 14.23 and where $\Sigma\mathbf{M}_A$ represents the sum of the moments about A of the external forces acting on the system of particles, is valid if, and only if, one of the following conditions is satisfied: (*a*) the frame $Ax'y'z'$ is itself a newtonian frame of reference, (*b*) A coincides with the mass center G, (*c*) the acceleration $\mathbf{a}_A$ of A relative to $Oxyz$ is directed along the line AG.

14.7. Kinetic Energy of a System of Particles. The kinetic energy T of a system of particles is defined as the sum of the kinetic energies of the various particles of the system. Referring to Sec. 13.3, we therefore write

$$T = \frac{1}{2}\sum_{i=1}^{n} m_i v_i^2 \tag{14.28}$$

Using a Centroidal Frame of Reference. It is often convenient when computing the kinetic energy of a system comprising a large number of particles (as in the case of a rigid body) to consider separately the motion of the mass center G of the system and the motion of the system relative to a moving frame of reference attached to G.

Let P_i be a particle of the system, $\mathbf{v}_i$ its velocity relative to the newtonian frame of reference $Oxyz$, and $\mathbf{v}'_i$ its velocity relative to the moving frame $Gx'y'z'$ which is in translation with respect to $Oxyz$ (Fig. 14.7). We recall from the preceding section that

$$\mathbf{v}_i = \overline{\mathbf{v}} + \mathbf{v}'_i \tag{14.22}$$

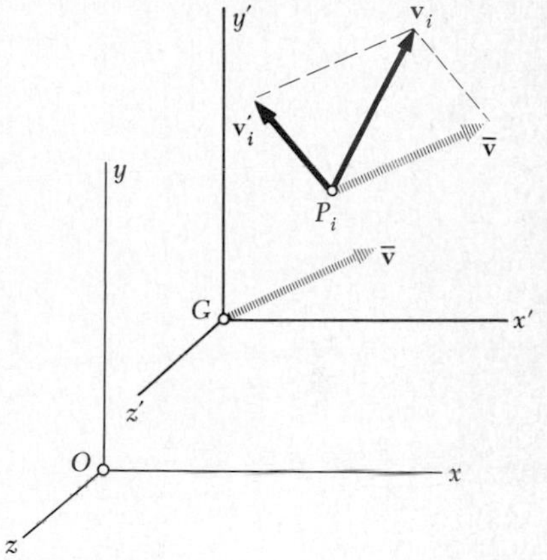

Fig. 14.7

where $\overline{\mathbf{v}}$ denotes the velocity of the mass center G relative to the newtonian frame $Oxyz$. Observing that v_i^2 is equal to the scalar product $\mathbf{v}_i \cdot \mathbf{v}_i$, we express as follows the kinetic energy T of the system relative to the newtonian frame $Oxyz$:

$$T = \frac{1}{2}\sum_{i=1}^{n} m_i v_i^2 = \frac{1}{2}\sum_{i=1}^{n} (m_i \mathbf{v}_i \cdot \mathbf{v}_i)$$

or, substituting for $\mathbf{v}_i$ from (14.22),

$$T = \frac{1}{2}\sum_{i=1}^{n} [m_i(\overline{\mathbf{v}} + \mathbf{v}'_i)\cdot(\overline{\mathbf{v}} + \mathbf{v}'_i)]$$

$$= \frac{1}{2}\left(\sum_{i=1}^{n} m_i\right)\bar{v}^2 + \overline{\mathbf{v}}\cdot\sum_{i=1}^{n} m_i \mathbf{v}'_i + \frac{1}{2}\sum_{i=1}^{n} m_i v_i'^2$$

The first sum represents the total mass m of the system. Recalling Eq. (14.13), we note that the second sum is equal to $m\overline{\mathbf{v}}'$ and thus to zero, since $\overline{\mathbf{v}}'$, which represents the velocity of G relative to the frame $Gx'y'z'$, is clearly zero. We therefore write

$$T = \tfrac{1}{2}m\bar{v}^2 + \frac{1}{2}\sum_{i=1}^{n} m_i v_i'^2 \tag{14.29}$$

This equation shows that the kinetic energy T of a system of particles may be obtained *by adding the kinetic energy of the mass center G* (assuming the entire mass concentrated at G) *and the kinetic energy of the system in its motion relative to the frame Gx′y′z′.*

14.8. Work-Energy Principle. Conservation of Energy for a System of Particles. The principle of work and energy may be applied to each particle P_i of a system of particles. We write

$$T_1 + U_{1\to2} = T_2 \tag{14.30}$$

for each particle P_i, where $U_{1\to2}$ represents the work done by the internal forces $\mathbf{f}_{ij}$ and the resultant external force $\mathbf{F}_i$ acting on P_i. Adding the kinetic energies of the various particles of the system, and considering the work of all the forces involved, we may apply Eq. (14.30) to the entire system. The quantities T_1 and T_2 now represent the kinetic energy of the entire system and may be computed from either Eq. (14.28) or Eq. (14.29). The quantity $U_{1\to2}$ represents the work of all the forces acting on the particles of the system. We should note that while the internal forces $\mathbf{f}_{ij}$ and $\mathbf{f}_{ji}$ are equal and opposite, the work of these forces, in general, will not cancel out, since the particles P_i and P_j on which they act will, in general, undergo different displacements. Therefore, in computing $U_{1\to2}$, *we should consider the work of the internal forces* $\mathbf{f}_{ij}$ *as well as the work of the external forces* $\mathbf{F}_i$.

If all the forces acting on the particles of the system are conservative, Eq. (14.30) may be replaced by

$$T_1 + V_1 = T_2 + V_2 \tag{14.31}$$

where V represents the potential energy associated with the internal and external forces acting on the particles of the system. Equation (14.31) expresses the principle of *conservation of energy* for the system of particles.

14.9. Principle of Impulse and Momentum for a System of Particles. Integrating Eqs. (14.10) and (14.11) in t from a time t_1 to a time t_2, we write

$$\sum \int_{t_1}^{t_2} \mathbf{F}\,dt = \mathbf{L}_2 - \mathbf{L}_1 \tag{14.32}$$

$$\sum \int_{t_1}^{t_2} \mathbf{M}_O\,dt = (\mathbf{H}_O)_2 - (\mathbf{H}_O)_1 \tag{14.33}$$

Recalling the definition of the linear impulse of a force given in Sec. 13.10, we observe that the integrals in Eq. (14.32) represent the linear impulses of the external forces acting on the particles of the system. We shall refer in a similar way to the integrals in Eq. (14.33) as the *angular impulses* about O of the external forces. Thus, Eq. (14.32) expresses that the sum of the linear impulses of the external forces acting on the system is equal to the change in linear momentum of the system. Similarly, Eq. (14.33) expresses that the sum of the angular impulses about O of the external forces is equal to the change in angular momentum about O of the system.

In order to understand the physical significance of Eqs. (14.32) and (14.33), we shall rearrange the terms in these equations and write

$$\mathbf{L}_1 + \sum \int_{t_1}^{t_2} \mathbf{F}\, dt = \mathbf{L}_2 \tag{14.34}$$

$$(\mathbf{H}_O)_1 + \sum \int_{t_1}^{t_2} \mathbf{M}_O\, dt = (\mathbf{H}_O)_2 \tag{14.35}$$

We have sketched in parts *a* and *c* of Fig. 14.8 the momenta of the particles of the system at times t_1 and t_2, respectively, and we have shown in part *b* of the same figure a vector equal to the sum of the linear impulses of the external forces and a couple of moment equal to the sum of the angular impulses about *O* of the external forces. For simplicity, the particles have

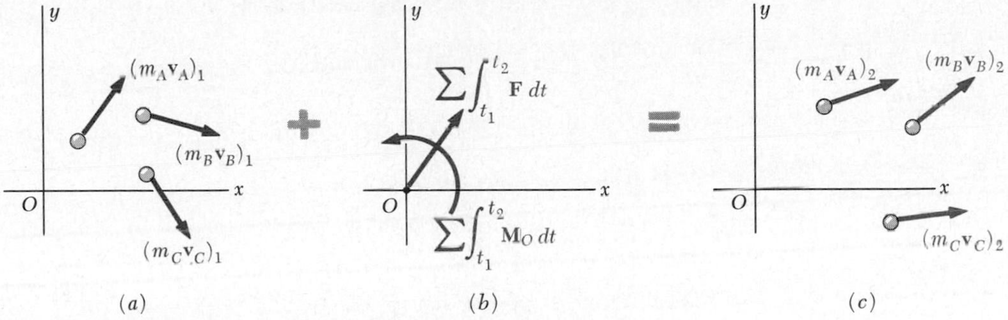

Fig. 14.8

been assumed to move in the plane of the figure, but the present discussion remains valid in the case of particles moving in space. Recalling from Eq. (14.6) that **L**, by definition, is the resultant of the momenta $m_i\mathbf{v}_i$, we note that Eq. (14.34) expresses that the resultant of the vectors shown in parts *a* and *b* of Fig. 14.8 is equal to the resultant of the vectors shown in part *c* of the same figure. Recalling from Eq. (14.7) that $\mathbf{H}_O$ is the moment resultant of the momenta $m_i\mathbf{v}_i$, we note that Eq. (14.35) similarly expresses that the moment resultant of the vectors in parts *a* and *b* of Fig. 14.8 is equal to the moment resultant of the vectors in part *c*. Together, Eqs. (14.34) and (14.35) thus express that *the momenta of the particles at time t_1 and the impulses of the external forces from t_1 to t_2 form a system of vectors equipollent to the system of the momenta of the particles at time t_2*. This has been indicated in Fig. 14.8 by the use of gray plus and equals signs.

If no external force acts on the particles of the system, the integrals in Eqs. (14.34) and (14.35) are zero, and these equations reduce to

$$\mathbf{L}_1 = \mathbf{L}_2 \tag{14.36}$$

$$(\mathbf{H}_O)_1 = (\mathbf{H}_O)_2 \tag{14.37}$$

We thus check the result obtained in Sec. 14.6: If no external force acts on the particles of a system, the linear momentum and the angular momentum about *O* of the system of particles are conserved. The system of the initial momenta is equipollent to the system of the final momenta, and it follows that the angular momentum of the system of particles about *any* fixed point is conserved.

SAMPLE PROBLEM 14.3

For the 200-kg space vehicle of Sample Prob. 14.1, it is known that at $t = 2.5$ s, the velocity of part A is $\mathbf{v}_A = (270 \text{ m/s})\mathbf{i} - (120 \text{ m/s})\mathbf{j} + (160 \text{ m/s})\mathbf{k}$ and the velocity of part B is parallel to the xz plane. Determine the velocity of part C.

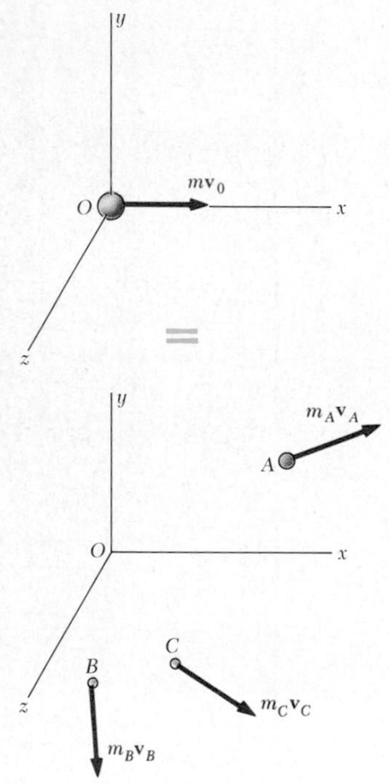

Solution. Since there is no external force, the initial momentum $m\mathbf{v}_0$ is equipollent to the system of the final momenta. Equating first the sums of the vectors in both parts of the adjoining sketch, and then the sums of their moments about O, we write

$$\mathbf{L}_1 = \mathbf{L}_2: \qquad m\mathbf{v}_0 = m_A\mathbf{v}_A + m_B\mathbf{v}_B + m_C\mathbf{v}_C \tag{1}$$

$$(\mathbf{H}_O)_1 = (\mathbf{H}_O)_2: \quad 0 = \mathbf{r}_A \times m_A\mathbf{v}_A + \mathbf{r}_B \times m_B\mathbf{v}_B + \mathbf{r}_C \times m_C\mathbf{v}_C \tag{2}$$

Recalling from Sample Prob. 14.1 that $\mathbf{v}_0 = (150 \text{ m/s})\mathbf{i}$,

$$m_A = 100 \text{ kg} \qquad m_B = 60 \text{ kg} \qquad m_C = 40 \text{ kg}$$
$$\mathbf{r}_A = (555 \text{ m})\mathbf{i} - (180 \text{ m})\mathbf{j} + (240 \text{ m})\mathbf{k}$$
$$\mathbf{r}_B = (255 \text{ m})\mathbf{i} - (120 \text{ m})\mathbf{k}$$
$$\mathbf{r}_C = (105 \text{ m})\mathbf{i} + (450 \text{ m})\mathbf{j} - (420 \text{ m})\mathbf{k}$$

and using the information given in the statement of this problem, we rewrite Eqs. (1) and (2) as follows:

$$200(150\mathbf{i}) = 100(270\mathbf{i} - 120\mathbf{j} + 160\mathbf{k}) + 60[(v_B)_x\mathbf{i} + (v_B)_z\mathbf{k}] + 40[(v_C)_x\mathbf{i} + (v_C)_y\mathbf{j} + (v_C)_z\mathbf{k}] \tag{1'}$$

$$0 = 100\begin{vmatrix} \mathbf{i} & \mathbf{j} & \mathbf{k} \\ 555 & -180 & 240 \\ 270 & -120 & 160 \end{vmatrix} + 60\begin{vmatrix} \mathbf{i} & \mathbf{j} & \mathbf{k} \\ 255 & 0 & -120 \\ (v_B)_x & 0 & (v_B)_z \end{vmatrix} + 40\begin{vmatrix} \mathbf{i} & \mathbf{j} & \mathbf{k} \\ 105 & 450 & -420 \\ (v_C)_x & (v_C)_y & (v_C)_z \end{vmatrix} \tag{2'}$$

Equating to zero the coefficient of $\mathbf{j}$ in (1′) and the coefficients of $\mathbf{i}$ and $\mathbf{k}$ in (2′), we write, after reductions, the three scalar equations

$$(v_C)_y - 300 = 0$$
$$450(v_C)_z + 420(v_C)_y = 0$$
$$105(v_C)_y - 450(v_C)_x - 45\,000 = 0$$

which yield, respectively,

$$(v_C)_y = 300 \qquad (v_C)_z = -280 \qquad (v_C)_x = -30$$

The velocity of part C is thus

$$\mathbf{v}_C = -(30 \text{ m/s})\mathbf{i} + (300 \text{ m/s})\mathbf{j} - (280 \text{ m/s})\mathbf{k} \quad ◄$$

SAMPLE PROBLEM 14.4

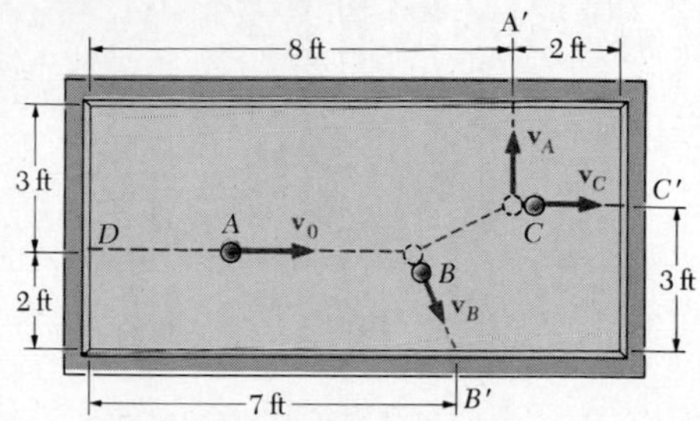

In a game of billiards, ball A is given an initial velocity $\mathbf{v}_0$ of magnitude $v_0 = 10$ ft/s along line DA parallel to the axis of the table. It hits ball B and then ball C, which are both at rest. Knowing that A and C hit the sides of the table squarely at points A' and C', respectively, that B hits the side obliquely at B', and assuming frictionless surfaces and perfectly elastic impacts, determine the velocities $\mathbf{v}_A$, $\mathbf{v}_B$, and $\mathbf{v}_C$ with which the balls hit the sides of the table. (*Remark.* In this sample problem and in several of the problems which follow, the billiard balls are assumed to be particles moving freely in a horizontal plane, rather than the rolling and sliding spheres they actually are.)

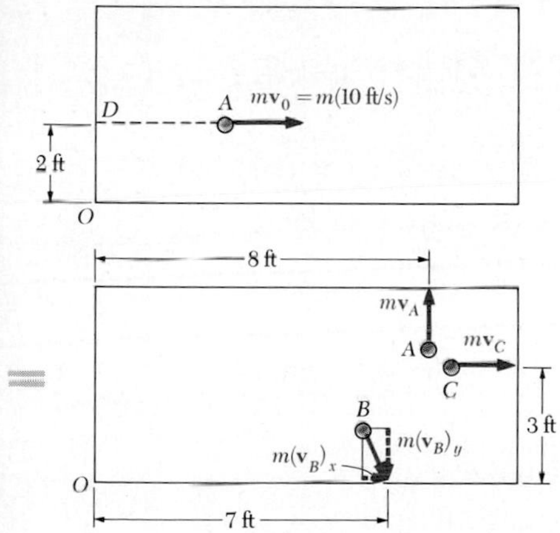

Solution. ***Conservation of Momentum.*** Since there is no external force, the initial momentum $m\mathbf{v}_0$ is equipollent to the system of momenta after the two collisions (and before any of the balls hits the side of the table). Referring to the adjoining sketch, we write

$\rightarrow x$ components: $$m(10 \text{ ft/s}) = m(v_B)_x + mv_C \quad (1)$$

$+\uparrow y$ components: $$0 = mv_A - m(v_B)_y \quad (2)$$

$+\circlearrowleft$ moments about O: $$-(2 \text{ ft})m(10 \text{ ft/s}) = (8 \text{ ft})mv_A - (7 \text{ ft})m(v_B)_y - (3 \text{ ft})mv_C \quad (3)$$

Solving the three equations for v_A, $(v_B)_x$, and $(v_B)_y$ in terms of v_C,

$$v_A = (v_B)_y = 3v_C - 20 \qquad (v_B)_x = 10 - v_C \quad (4)$$

Conservation of Energy. Since the surfaces are frictionless and the impacts are perfectly elastic, the initial kinetic energy $\frac{1}{2}mv_0^2$ is equal to the final kinetic energy of the system:

$$\tfrac{1}{2}mv_0^2 = \tfrac{1}{2}m_A v_A^2 + \tfrac{1}{2}m_B v_B^2 + \tfrac{1}{2}m_C v_C^2$$
$$v_A^2 + (v_B)_x^2 + (v_B)_y^2 + v_C^2 = (10 \text{ ft/s})^2 \quad (5)$$

Substituting for v_A, $(v_B)_x$, and $(v_B)_y$ from (4) into (5), we have

$$2(3v_C - 20)^2 + (10 - v_C)^2 + v_C^2 = 100$$
$$20v_C^2 - 260v_C + 800 = 0$$

Solving for v_C, we find $v_C = 5$ ft/s and $v_C = 8$ ft/s. Since only the second root yields a positive value for v_A after substitution into Eqs. (4), we conclude that $v_C = 8$ ft/s and

$$v_A = (v_B)_y = 3(8) - 20 = 4 \text{ ft/s} \qquad (v_B)_x = 10 - 8 = 2 \text{ ft/s}$$

$$\mathbf{v}_A = 4 \text{ ft/s} \uparrow \qquad \mathbf{v}_B = 4.47 \text{ ft/s} ⦪ 63.4° \qquad \mathbf{v}_C = 8 \text{ ft/s} \rightarrow \blacktriangleleft$$

Problems

14.25 In Prob. 14.1, determine the energy lost due to friction as the bullet (*a*) moves through block *A*, (*b*) penetrates block *B*.

14.26 In Prob. 14.3, determine the energy lost (*a*) in the coupling of the two cars, (*b*) as the end of the flatcar hits the truck.

14.27 Each of the three freight cars of Prob. 14.4 has a mass of 40 Mg. Assuming that car *B* is first hit by car *A* and gets automatically coupled with it and then is hit by car *C* and gets automatically coupled with it, determine the energy lost (*a*) in the first coupling, (*b*) in the second coupling.

14.28 In Prob. 14.18, determine the work done by the internal forces during the explosion of the shell.

Fig. P14.29

14.29 Two automobiles *A* and *B*, of mass m_A and m_B, respectively, are traveling in opposite directions when they collide head on. The impact is assumed perfectly plastic, and it is further assumed that the energy absorbed by each automobile is equal to its loss of kinetic energy with respect to a moving frame of reference attached to the mass center of the two-vehicle system. Denoting by E_A and E_B, respectively, the energy absorbed by automobile *A* and by automobile *B*, (*a*) show that $E_A/E_B = m_B/m_A$, that is, the amount of energy absorbed by each vehicle is inversely proportional to its mass, (*b*) compute E_A and E_B, knowing that automobile *A* weighs 3600 lb and had a speed of 45 mi/h before the crash, while automobile *B* weighs 2000 lb and had a speed of 30 mi/h.

14.30 It is assumed that each of the two automobiles involved in the collision described in Prob. 14.29 had been designed to safely withstand a test in which it crashed into a solid, immovable wall at the speed v_0. The severity of the collision of Prob. 14.29 may then be measured for each vehicle by the ratio of the energy it absorbed in the collision to the energy it absorbed in the test. On that basis, show that the collision described in Prob. 14.29 is $(m_A/m_B)^2$ times more severe for automobile *B* than for automobile *A*.

14.31 A space vehicle which is drifting in space consists of two parts *A* and *B* connected by explosive bolts. Parts *A* and *B* compress four springs, each of which has a potential energy of 120 ft·lb. When the bolts are exploded the springs expand, causing parts *A* and *B* to move away from each other. Determine the relative velocity of *B* with respect to *A*, knowing that (*a*) each part weighs 2000 lb, (*b*) part *A* weighs 3000 lb and part *B* 1000 lb.

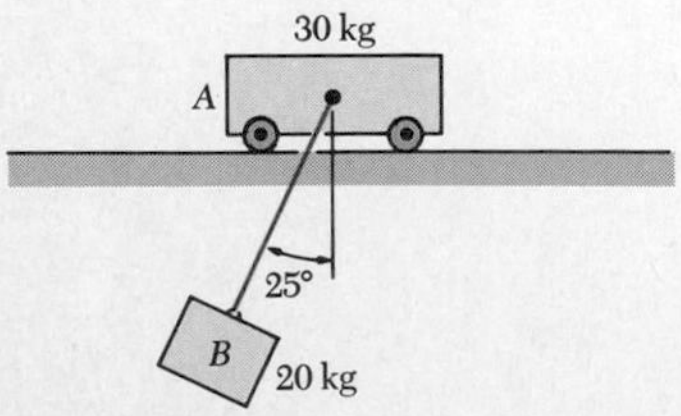

Fig. P14.32

14.32 A 20-kg block *B* is suspended from a 2-m cord attached to a 30-kg cart *A*, which may roll freely on a frictionless, horizontal track. If the system is released from rest in the position shown, determine the velocities of *A* and *B* as *B* passes directly under *A*.

14.33 Ball B, of mass m_B, is suspended from a cord of length l attached to a cart A, of mass m_A, which may roll freely on a frictionless, horizontal track. If the ball is given an initial horizontal velocity $\mathbf{v}_0$ while the cart is at rest, determine (*a*) the maximum vertical distance h through which B will rise, (*b*) the velocity of B as it reaches its maximum elevation. (*Hint.* At that instant, the velocity of B relative to A is zero.) Discuss the particular case when $m_A \gg m_B$ and when $m_A \ll m_B$.

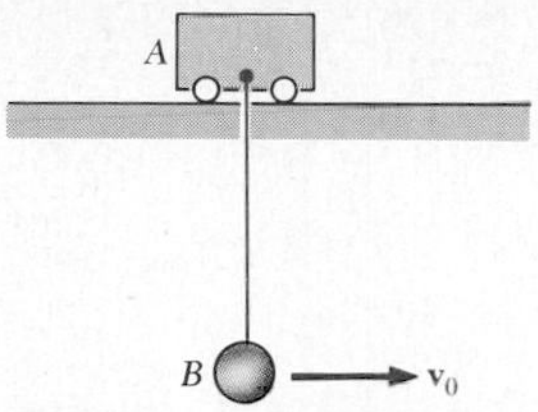

Fig. P14.33 and P14.34

14.34 Ball B is suspended from a cord of length l attached to a cart A, which may roll freely on a frictionless, horizontal track. The ball and the cart have the same mass m. If the ball is given an initial horizontal velocity $\mathbf{v}_0$ while the cart is at rest, describe the subsequent motion of the system, specifying the velocities of A and B for the following successive values of the angle θ (assumed positive counterclockwise) that the cord will form with the vertical: (*a*) $\theta = \theta_{max}$, (*b*) $\theta = 0$, (*c*) $\theta = \theta_{min}$.

14.35 In a game of billiards, ball A is moving with the velocity $\mathbf{v}_0 = (4 \text{ m/s})\mathbf{i}$ when it strikes balls B and C which are at rest side by side. After the collision, the three balls are observed to move in the directions shown, with $\theta = 25°$. Assuming frictionless surfaces and perfectly elastic impacts (i.e., conservation of energy), determine the magnitudes of the velocities v_A, v_B, and v_C.

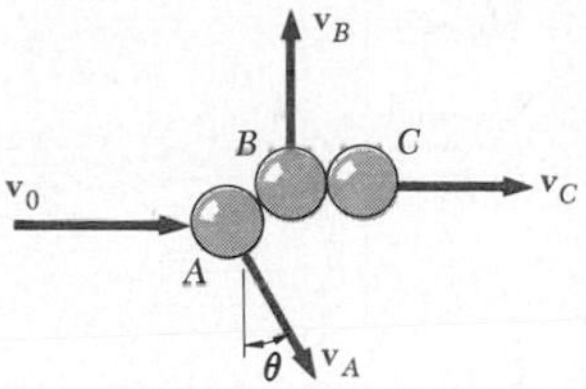

Fig. P14.35

14.36 A 12-lb block B starts from rest and slides on the 30-lb wedge A, which is supported by a horizontal surface. Neglecting friction, determine (*a*) the velocity of B relative to A after it has slid 3 ft down the inclined surface of the wedge, (*b*) the corresponding velocity of A.

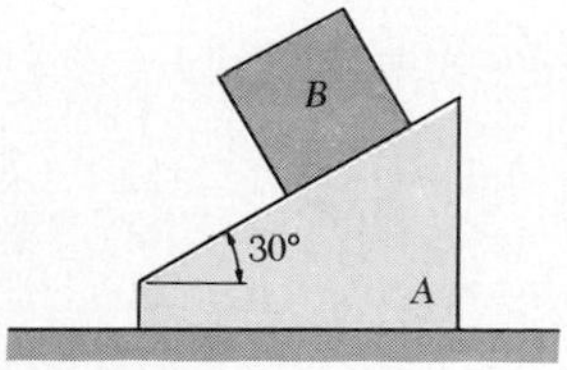

Fig. P14.36

14.37 Three spheres, each of mass m, may slide freely on a frictionless, horizontal surface. Spheres A and B are attached to an inextensible, inelastic cord of length l and are at rest in the position shown when sphere B is struck squarely by sphere C which is moving to the right with a velocity $\mathbf{v}_0$. Knowing that the cord is taut when sphere B is struck by sphere C and assuming perfectly elastic impact between B and C, and thus conservation of energy for the entire system, determine the velocity of each sphere immediately after impact.

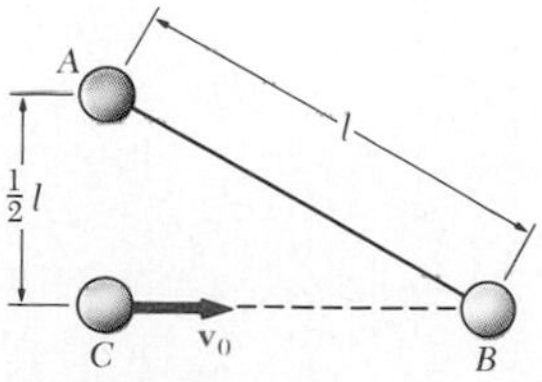

Fig. P14.37

14.38 Three spheres, each of mass m, may slide freely on a frictionless, horizontal surface. Spheres A and B are attached to an inextensible, inelastic cord of length l and are at rest in the position shown when sphere B is struck squarely by sphere C which is moving to the right with a velocity $\mathbf{v}_0$. Knowing that the cord is slack when sphere B is struck by sphere C and assuming perfectly elastic impact between B and C, determine (*a*) the velocity of each sphere immediately after the cord becomes taut, (*b*) the fraction of the initial kinetic energy of the system which is dissipated when the cord becomes taut.

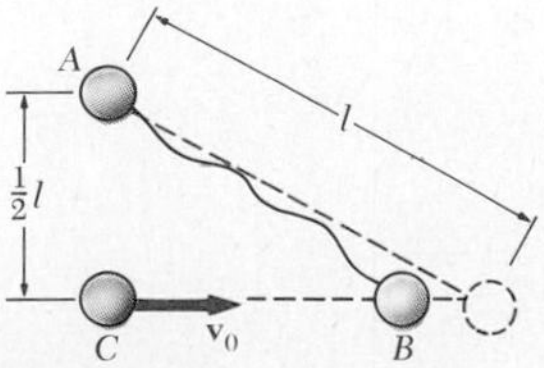

Fig. P14.38

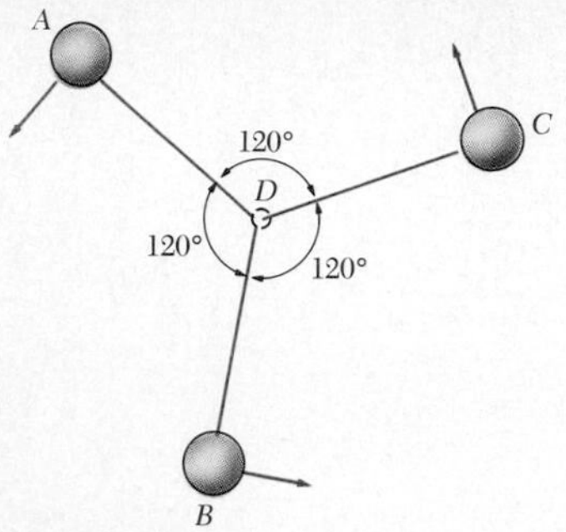

Fig. P14.39

14.39 Three small spheres A, B, and C, each of mass m, are connected to a small ring D by means of three inextensible, inelastic cords of length l which are equally spaced. The spheres may slide freely on a frictionless horizontal surface and are rotating initially with a speed v_0 about ring D which is at rest. Suddenly cord CD breaks. After the other two cords have again become taut, determine (*a*) the speed of ring D, (*b*) the relative speed with which spheres A and B rotate about D, (*c*) the fraction of the energy of the original system which is dissipated when cords AD and BD again become taut.

14.40 A 600-lb space vehicle traveling with a velocity $\mathbf{v}_0 = (1200 \text{ ft/s})\mathbf{k}$ passes through the origin O. Explosive charges then separate the vehicle into three parts A, B, and C, weighing, respectively, 100 lb, 200 lb, and 300 lb. Knowing that shortly thereafter the positions of the three parts are, respectively, $A(240, 240, 2160)$, $B(600, 1320, 3240)$, and $C(-480, -960, 1920)$, where the coordinates are expressed in feet, that the velocity of B is $\mathbf{v}_B = (400 \text{ ft/s})\mathbf{i} + (880 \text{ ft/s})\mathbf{j} + (1760 \text{ ft/s})\mathbf{k}$, and that the x component of the velocity of C is -320 ft/s, determine the velocity of part A.

14.41 In the scattering experiment of Prob. 14.17, it is known that the alpha particle is projected from A_0 (260, -20, 340) and that it collides with the oxygen nucleus C at $Q(200, 180, 140)$, where all coordinates are expressed in millimeters. Determine the coordinates of point B_0 where the original path of nucleus B intersects the xz plane. (*Hint.* Express that the angular momentum of the three particles about Q is conserved.)

14.42 Three small identical spheres A, B, and C are attached to three strings, 150 mm long, which are tied to a ring G. The ring has a velocity $\mathbf{v}_0 = (0.3 \text{ m/s})\mathbf{i}$ and moves along the x axis. The three spheres are rotating counterclockwise about G with a relative velocity of 0.6 m/s. Suddenly ring G breaks and the three spheres move freely in the horizontal xy plane, with A and B following paths parallel to the y axis and C a path parallel to the x axis. Determine (*a*) the velocity of each sphere, (*b*) the distance d.

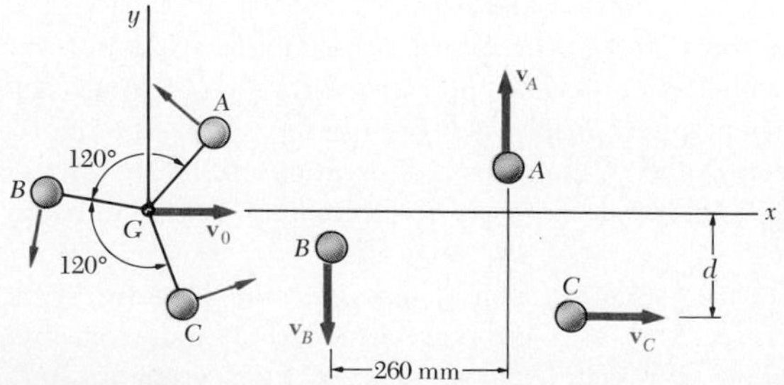

Fig. P14.42

14.43 Two small disks A and B, of mass 3 kg and 1.5 kg, respectively, may slide on a horizontal, frictionless surface. They are connected by a cord, 600 mm long, and spin counterclockwise about their mass center G at the rate of 10 rad/s. At $t = 0$, the coordinates of G are $\bar{x}_0 = 0$, $\bar{y}_0 = 2$ m, and its velocity is $\bar{\mathbf{v}}_0 = (1.2 \text{ m/s})\mathbf{i} + (0.96 \text{ m/s})\mathbf{j}$. Shortly thereafter the cord breaks; disk A is then observed to move along a path parallel to the y axis and disk B along a path which intersects the x axis at a distance $b = 7.5$ m from O. Determine (*a*) the velocities of A and B after the cord breaks, (*b*) the distance a from the y axis to the path of A.

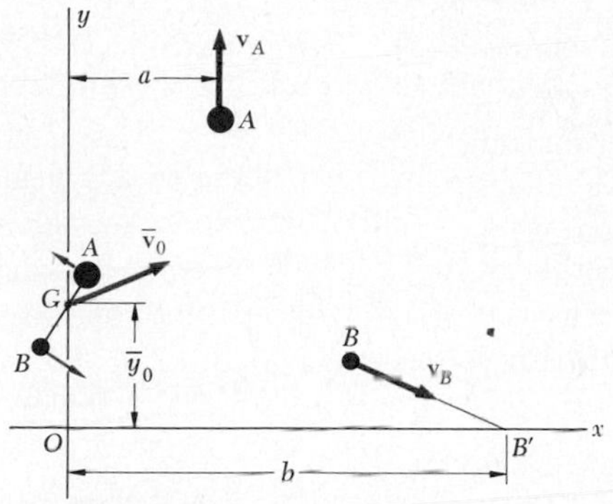

Fig. P14.43

14.44 In a game of billiards, ball A is given an initial velocity $\mathbf{v}_0$ along line DA parallel to the axis of the table. It hits ball B and then ball C, which are both at rest. Balls A and C are observed to hit the sides of the table squarely at points A' and C', respectively, and ball B to hit the side obliquely at B'. Knowing that $v_0 = 12$ ft/s, $v_C = 7$ ft/s, and $c = 3$ ft., determine (*a*) the velocities $\mathbf{v}_A$ and $\mathbf{v}_B$ of balls A and B, (*b*) the point A' where ball A hits the side of the table. Assume frictionless surfaces and perfectly elastic impacts (i.e., conservation of energy).

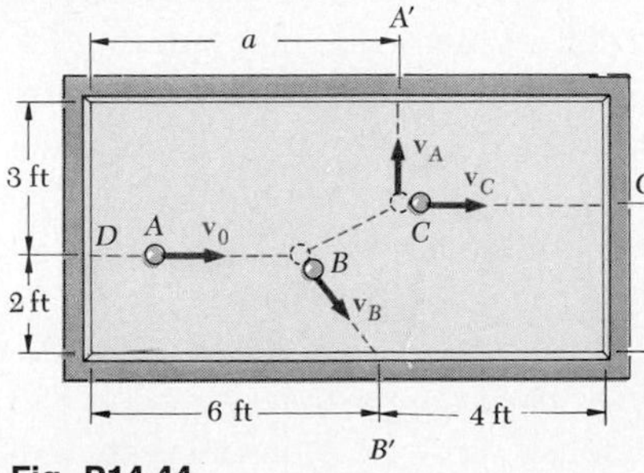

Fig. P14.44

14.45 For the game of billiards of Prob. 14.44, it is now assumed that $v_0 = 10$ ft/s, $v_A = 4$ ft/s, and $a = 7$ ft. Determine (*a*) the velocities $\mathbf{v}_B$ and $\mathbf{v}_C$ of balls B and C, (*b*) the point C' where ball C hits the side of the table.

***14.10. Variable Systems of Particles.** All the systems of particles considered so far consisted of well-defined particles. These systems did not gain or lose any particles during their motion. In a large number of engineering applications, however, it is necessary to consider *variable systems of particles,* i.e., systems which are continually gaining or losing particles, or doing both at the same time. Consider, for example, a hydraulic turbine. Its analysis involves the determination of the forces exerted by a stream of water on rotating blades, and we note that the particles of water in contact with the blades form an everchanging system which continually acquires and loses particles. Rockets furnish another example of variable systems, since their propulsion depends upon the continual ejection of fuel particles.

We recall that all the kinetics principles established so far were derived for constant systems of particles, which neither gain nor lose particles. We must therefore find a way to reduce the analysis of a variable system of particles to that of an auxiliary constant system. The procedure to follow is indicated in Secs. 14.11 and 14.12 for two broad categories of applications.

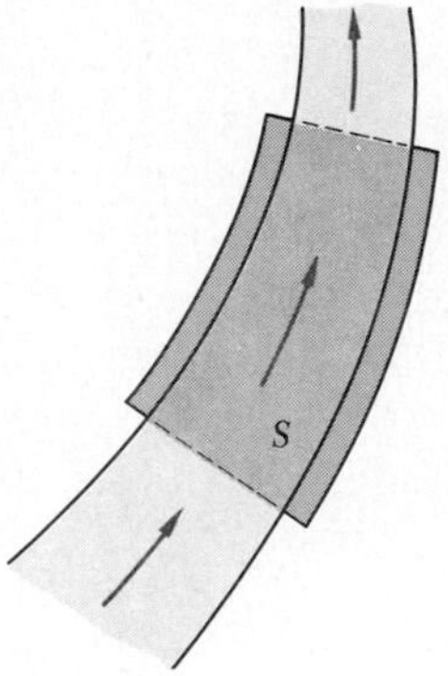

Fig. 14.9

***14.11. Steady Stream of Particles.** Consider a steady stream of particles, such as a stream of water diverted by a fixed vane or a flow of air through a duct or through a blower. In order to determine the resultant of the forces exerted on the particles in contact with the vane, duct, or blower, we isolate these particles and denote by S the system thus defined (Fig. 14.9). We observe that S is a variable system of particles, since it continually gains particles flowing in and loses an equal number of particles flowing out. Therefore, the kinetics principles that have been established so far cannot be directiy applied to S.

However, we may easily define an auxiliary system of particles which does remain constant for a short interval of time Δt. Consider at time t the system S *plus* the particles which will enter S during the interval of time Δt (Fig. 14.10*a*). Next, consider at time $t + \Delta t$ the system S *plus* the particles which have left S during the interval Δt (Fig. 14.10*c*). Clearly, *the same particles are involved in both cases,* and we may apply to these particles the principle of impulse and momentum. Since the total mass m of the system S remains constant, the particles entering the system and those leaving the system in the time Δt must have the same mass Δm. Denoting by $\mathbf{v}_A$ and $\mathbf{v}_B$, respectively, the velocities of the particles entering S at A and leaving S at B, we represent the momentum of the particles entering S by $(\Delta m)\mathbf{v}_A$ (Fig. 14.10*a*) and the momentum of the particles leaving S by $(\Delta m)\mathbf{v}_B$ (Fig. 14.10*c*). We also represent the momenta $m_i\mathbf{v}_i$ of the particles forming S and the impulses of the forces exerted on S by the appropriate vectors, and indicate by gray plus and equals signs that the system of the momenta and impulses in parts a and b of Fig. 14.10 is equipollent to the system of the momenta in part c of the same figure.

Since the resultant $\Sigma m_i\mathbf{v}_i$ of the momenta of the particles of S is found on both sides of the equals sign, it may be omitted. We conclude that *the system formed by the momentum $(\Delta m)\mathbf{v}_A$ of the particles entering S in the*

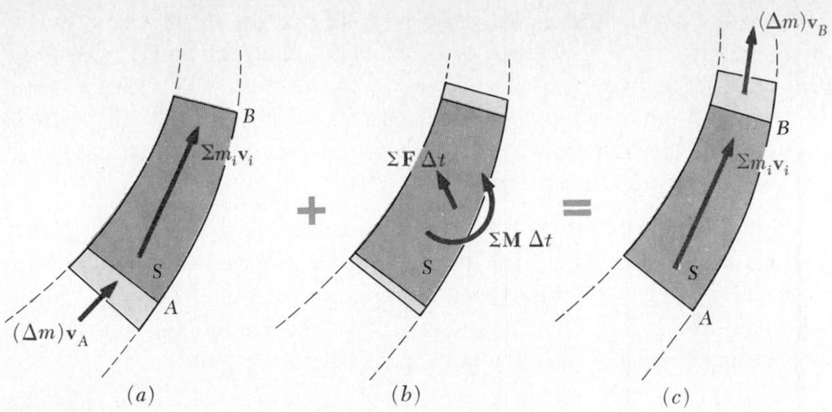

Fig. 14.10

time Δt and the impulses of the forces exerted on S during that time is equipollent to the momentum $(\Delta m)\mathbf{v}_B$ of the particles leaving S in the same time Δt. We may therefore write

$$(\Delta m)\mathbf{v}_A + \Sigma\mathbf{F}\,\Delta t = (\Delta m)\mathbf{v}_B \tag{14.38}$$

A similar equation may be obtained by taking the moments of the vectors involved (see Sample Prob. 14.5). Dividing all terms of Eq. (14.38) by Δt and letting Δt approach zero, we obtain at the limit

$$\Sigma\mathbf{F} = \frac{dm}{dt}(\mathbf{v}_B - \mathbf{v}_A) \tag{14.39}$$

where $\mathbf{v}_B - \mathbf{v}_A$ represents the difference between the *vectors* $\mathbf{v}_B$ and $\mathbf{v}_A$.

If SI units are used, dm/dt is expressed in kg/s and the velocities in m/s; we check that both members of Eq. (14.39) are expressed in the same units (newtons). If U.S. customary units are used, dm/dt must be expressed in slugs/s and the velocities in ft/s; we check again that both members of the equation are expressed in the same units (pounds).†

The principle we have established may be used to analyze a large number of engineering applications. Some of the most common are indicated below.

Fluid Stream Diverted by a Vane. If the vane is fixed, the method of analysis given above may be applied directly to find the force **F** exerted by the vane on the stream. We note that **F** is the only force which needs to be considered since the pressure in the stream is constant (atmospheric pressure). The force exerted by the stream on the vane will be equal and

†It is often convenient to express the mass rate of flow dm/dt as the product ρQ, where ρ is the density of the stream (mass per unit volume) and Q its volume rate of flow (volume per unit time). If SI units are used, ρ is expressed in kg/m^3 (for instance, $\rho = 1000$ kg/m^3 for water) and Q in m^3/s. However, if U.S. customary units are used, ρ will generally have to be computed from the corresponding specific weight γ (weight per unit volume), $\rho = \gamma/g$. Since γ is expressed in lb/ft^3 (for instance, $\gamma = 62.4$ lb/ft^3 for water), ρ is obtained in slugs/ft^3. The volume rate of flow Q is expressed in ft^3/s.

opposite to **F**. If the vane moves with a constant velocity, the stream is not steady. However, it will appear steady to an observer moving with the vane. We should therefore choose a system of axes moving with the vane. Since this system of axes is not accelerated, Eq. (14.38) may still be used, but $\mathbf{v}_A$ and $\mathbf{v}_B$ must be replaced by the *relative velocities* of the stream with respect to the vane (see Sample Prob. 14.6).

Fluid Flowing through a Pipe. The force exerted by the fluid on a pipe transition such as a bend or a contraction may be determined by considering the system of particles S in contact with the transition. Since, in general, the pressure in the flow will vary, we should also consider the forces exerted on S by the adjoining portions of the fluid.

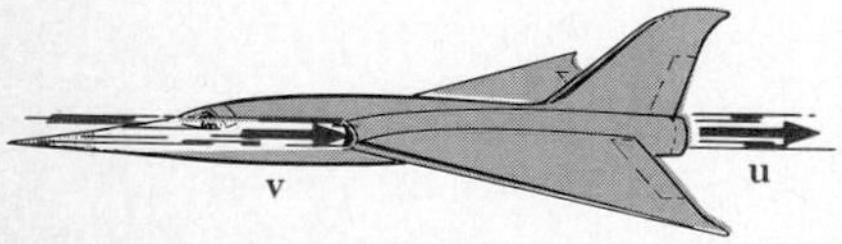

Fig. 14.11

Jet Engine. In a jet engine, air enters with no velocity through the front of the engine and leaves through the rear with a high velocity. The energy required to accelerate the air particles is obtained by burning fuel. While the exhaust gases contain burned fuel, the mass of the fuel is small compared with the mass of the air flowing through the engine and usually may be neglected. Thus, the analysis of a jet engine reduces to that of an air stream. This stream may be considered as a steady stream if all velocities are measured with respect to the airplane. The air stream shall be assumed, therefore, to enter the engine with a velocity **v** of magnitude equal to the speed of the airplane and to leave with a velocity **u** equal to the relative velocity of the exhaust gases (Fig. 14.11). Since the intake and exhaust pressures are nearly atmospheric, the only external force which needs to be considered is the force exerted by the engine on the air stream. This force is equal and opposite to the thrust.†

Fan. We consider the system of particles S shown in Fig. 14.12. The velocity $\mathbf{v}_A$ of the particles entering the system is assumed equal to zero,

† Note that if the airplane is accelerated, it cannot be used as a newtonian frame of reference. The same result will be obtained for the thrust, however, by using a reference frame at rest with respect to the atmosphere, since the air particles will then be observed to enter the engine with no velocity and to leave it with a velocity of magnitude $u - v$.

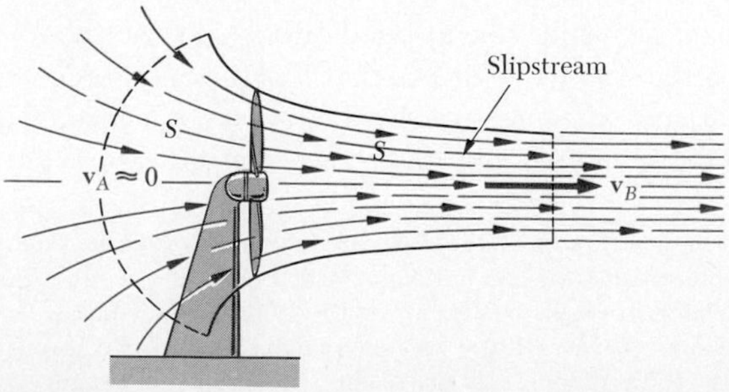

Fig. 14.12

and the velocity $\mathbf{v}_B$ of the particles leaving the system is the velocity of the *slipstream*. The rate of flow may be obtained by multiplying v_B by the cross-sectional area of the slipstream. Since the pressure all around S is atmospheric, the only external force acting on S is the thrust of the fan.

Airplane Propeller. In order to obtain a steady stream of air, velocities should be measured with respect to the airplane. Thus, the air particles will be assumed to enter the system with a velocity **v** of magnitude equal to the speed of the airplane and to leave with a velocity **u** equal to the relative velocity of the slipstream.

*** 14.12. Systems Gaining or Losing Mass.** We shall now analyze a different type of variable system of particles, namely, a system which gains mass by continually absorbing particles or loses mass by continually expelling particles. Consider the system S shown in Fig. 14.13. Its mass, equal to m at the instant t, increases by Δm in the interval of time Δt.

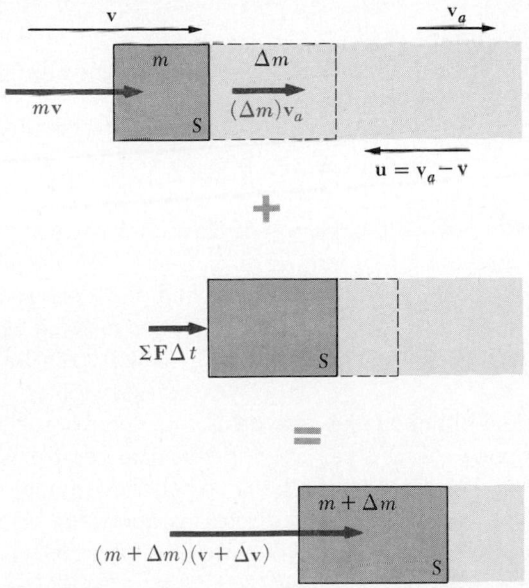

Fig. 14.13

In order to apply the principle of impulse and momentum to the analysis of this system, we must consider at time t the system S *plus* the particles of mass Δm which S absorbs during the time interval Δt. The velocity of S at time t is denoted by **v**, and its velocity at time $t + \Delta t$ is denoted by $\mathbf{v} + \Delta\mathbf{v}$, while the absolute velocity of the particles which are absorbed is denoted by $\mathbf{v}_a$. Applying the principle of impulse and momentum, we write

$$m\mathbf{v} + (\Delta m)\mathbf{v}_a + \Sigma\mathbf{F}\,\Delta t = (m + \Delta m)(\mathbf{v} + \Delta\mathbf{v})$$

Solving for the sum $\Sigma\mathbf{F}\,\Delta t$ of the impulses of the external forces acting on S (excluding the forces exerted by the particles being absorbed), we have

$$\Sigma\mathbf{F}\,\Delta t = m\Delta\mathbf{v} + \Delta m(\mathbf{v} - \mathbf{v}_a) + (\Delta m)(\Delta\mathbf{v}) \tag{14.40}$$

Introducing the *relative velocity* $\mathbf{u}$ with respect to S of the particles which are absorbed, we write $\mathbf{u} = \mathbf{v}_a - \mathbf{v}$ and note, since $v_a < v$, that the relative velocity $\mathbf{u}$ is directed to the left, as shown in Fig. 14.13. Neglecting the last term in Eq. (14.40), which is of the second order, we write

$$\Sigma\mathbf{F}\,\Delta t = m\,\Delta\mathbf{v} - (\Delta m)\mathbf{u}$$

Dividing through by Δt and letting Δt approach zero, we have at the limit†

$$\Sigma\mathbf{F} = m\frac{d\mathbf{v}}{dt} - \frac{dm}{dt}\mathbf{u} \tag{14.41}$$

Rearranging the terms and recalling that $d\mathbf{v}/dt = \mathbf{a}$, where $\mathbf{a}$ is the acceleration of the system S, we write

$$\Sigma\mathbf{F} + \frac{dm}{dt}\mathbf{u} = m\mathbf{a} \tag{14.42}$$

which shows that the action on S of the particles being absorbed is equivalent to a thrust

$$\mathbf{P} = \frac{dm}{dt}\mathbf{u} \tag{14.43}$$

which tends to slow down the motion of S, since the relative velocity $\mathbf{u}$ of the particles is directed to the left. If SI units are used, dm/dt is expressed in kg/s, the relative velocity u in m/s, and the corresponding thrust in newtons. If U.S. customary units are used, dm/dt must be expressed in slugs/s and u in ft/s; the corresponding thrust will then be expressed in pounds.‡

The equations obtained may also be used to determine the motion of a system S losing mass. In this case, the rate of change of mass is negative, and the action on S of the particles being expelled is equivalent to a thrust in the direction of $-\mathbf{u}$, that is, in the direction opposite to that in which the particles are being expelled. A *rocket* represents a typical case of a system continually losing mass (see Sample Prob. 14.7).

†When the absolute velocity $\mathbf{v}_a$ of the particles absorbed is zero, we have $\mathbf{u} = -\mathbf{v}$, and formula (14.41) becomes

$$\Sigma\mathbf{F} = \frac{d}{dt}(m\mathbf{v})$$

Comparing the formula obtained to Eq. (12.3) of Sec. 12.3, we observe that Newton's second law may be applied to a system gaining mass. *provided that the particles absorbed are initially at rest.* It may also be applied to a system losing mass, *provided that the velocity of the particles expelled is zero* with respect to the frame of reference selected.

‡See footnote on page 617.

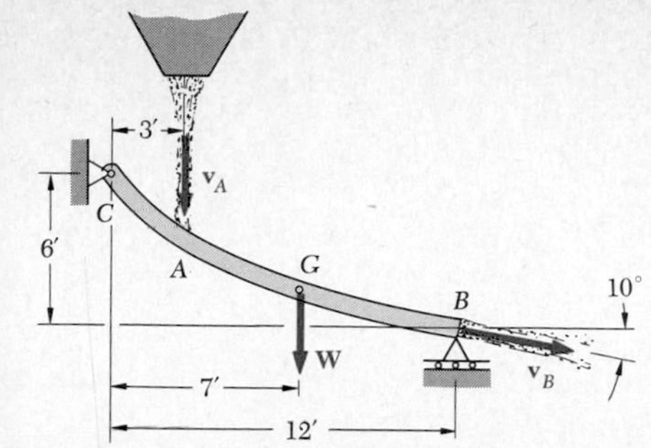

SAMPLE PROBLEM 14.5

Grain falls from a hopper onto a chute CB at the rate of 240 lb/s. It hits the chute at A with a velocity of 20 ft/s and leaves at B with a velocity of 15 ft/s, forming an angle of 10° with the horizontal. Knowing that the combined weight of the chute and of the grain it supports is a force $\mathbf{W}$ of magnitude 600 lb applied at G, determine the reaction at the roller-support B and the components of the reaction at the hinge C.

Solution. We apply the principle of impulse and momentum for the time interval Δt to the system consisting of the chute, the grain it supports, and the amount of grain which hits the chute in the interval Δt. Since the chute does not move, it has no momentum. We also note that the sum $\Sigma m_i \mathbf{v}_i$ of the momenta of the particles supported by the chute is the same at t and $t + \Delta t$ and thus may be omitted.

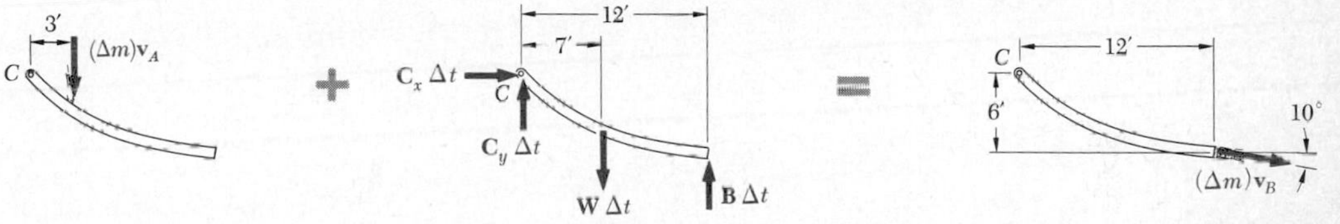

Since the system formed by the momentum $(\Delta m)\mathbf{v}_A$ and the impulses is equipollent to the momentum $(\Delta m)\mathbf{v}_B$, we write

$\xrightarrow{+}$ x components: $\quad C_x\,\Delta t = (\Delta m)v_B \cos 10° \qquad (1)$

$+\uparrow y$ components: $\quad -(\Delta m)v_A + C_y\,\Delta t - W\,\Delta t + B\,\Delta t$
$$= -(\Delta m)v_B \sin 10° \qquad (2)$$

$+\circlearrowleft$ moments about C: $\quad -3(\Delta m)v_A - 7(W\,\Delta t) + 12(B\,\Delta t)$
$$= 6(\Delta m)v_B \cos 10° - 12(\Delta m)v_B \sin 10° \qquad (3)$$

Using the given data, $W = 600$ lb, $v_A = 20$ ft/s, $v_B = 15$ ft/s, $\Delta m/\Delta t = 240/32.2 = 7.45$ slugs/s, and solving Eq. (3) for B and Eq. (1) for C_x,

$$12B = 7(600) + 3(7.45)(20) + 6(7.45)(15)(\cos 10° - 2 \sin 10°)$$
$$12B = 5075 \qquad B = 423 \text{ lb} \qquad \mathbf{B} = 423 \text{ lb} \uparrow \blacktriangleleft$$
$$C_x = (7.45)(15) \cos 10° = 110.1 \text{ lb} \qquad \mathbf{C}_x = 110.1 \text{ lb} \rightarrow \blacktriangleleft$$

Substituting for B and solving Eq. (2) for C_y,

$$C_y = 600 - 423 + (7.45)(20 - 15 \sin 10°) = 307 \text{ lb}$$
$$\mathbf{C}_y = 307 \text{ lb} \uparrow \blacktriangleleft$$

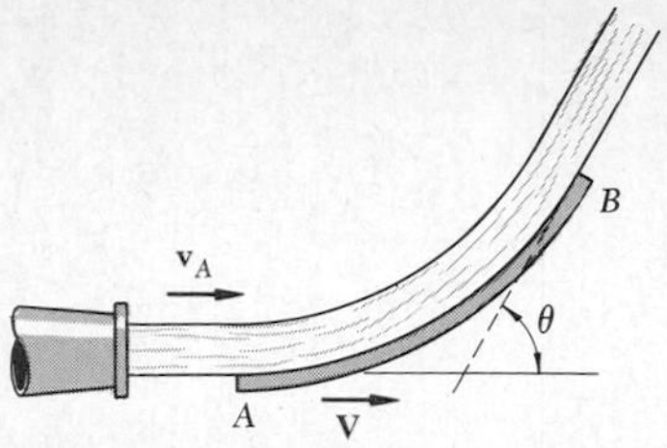

SAMPLE PROBLEM 14.6

A nozzle discharges a stream of water of cross-sectional area A with a velocity $\mathbf{v}_A$. The stream is deflected by a *single* blade which moves to the right with a constant vleocity $\mathbf{V}$. Assuming that the water moves along the blade at constant speed, determine (*a*) the components of the force $\mathbf{F}$ exerted by the blade on the stream, (*b*) the velocity $\mathbf{V}$ for which maximum power is developed.

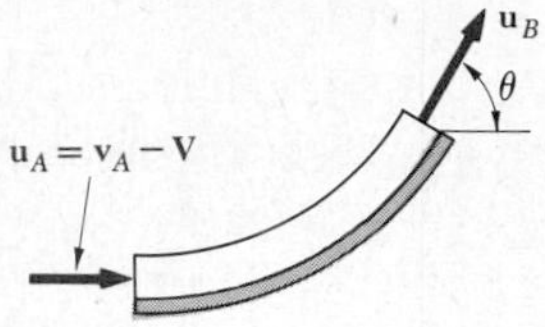

***a.* Components of Force Exerted on Stream.** We choose a coordinate system which moves with the blade at a constant velocity $\mathbf{V}$. The particles of water strike the blade with a relative velocity $\mathbf{u}_A = \mathbf{v}_A - \mathbf{V}$ and leave the blade with a relative velocity $\mathbf{u}_B$. Since the particles move along the blade at a constant speed, the relative velocities $\mathbf{u}_A$ and $\mathbf{u}_B$ have the same magnitude u. Denoting the density of water by ρ, the mass of the particles striking the blade during the time interval Δt is $\Delta m = A\rho(v_A - V)\,\Delta t$; an equal mass of particles leaves the blade during Δt. We apply the principle of impulse and momentum to the system formed by the particles in contact with the blade and by those striking the blade in the time Δt.

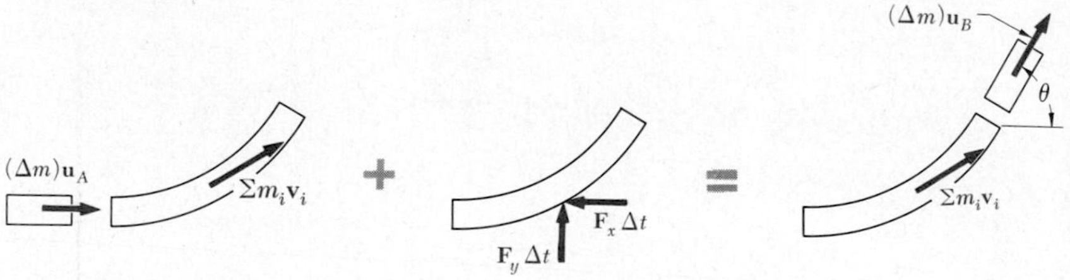

Recalling that $\mathbf{u}_A$ and $\mathbf{u}_B$ have the same magnitude u, and omitting the momentum $\Sigma m_i \mathbf{v}_i$ which appears on both sides, we write

$\xrightarrow{+}$ x components: $\quad (\Delta m)u - F_x\,\Delta t = (\Delta m)u\cos\theta$

$+\uparrow y$ components: $\quad +F_y\,\Delta t = (\Delta m)u\sin\theta$

Substituting $\Delta m = A\rho(v_A - V)\,\Delta t$ and $u = v_A - V$, we obtain

$$\mathbf{F}_x = A\rho(v_A - V)^2(1 - \cos\theta)\leftarrow \qquad \mathbf{F}_y = A\rho(v_A - V)^2\sin\theta\uparrow \quad ◀$$

***b.* Velocity of Blade for Maximum Power.** The power is obtained by multiplying the velocity V of the blade by the component F_x of the force exerted by the stream on the blade.

$$\text{Power} = F_x V = A\rho(v_A - V)^2(1 - \cos\theta)V$$

Differentiating the power with respect to V and setting the derivative equal to zero, we obtain

$$\frac{d(\text{power})}{dV} = A\rho(v_A^2 - 4v_A V + 3V^2)(1 - \cos\theta) = 0$$

$$V = v_A \qquad V = \tfrac{1}{3}v_A \qquad \text{For maximum power } \mathbf{V} = \tfrac{1}{3}v_A\rightarrow \quad ◀$$

Note. These results are valid only when a *single* blade deflects the stream. Different results are obtained when a series of blades deflects the stream, as in a Pelton-wheel turbine. (See Prob. 14.70.)

SAMPLE PROBLEM 14.7

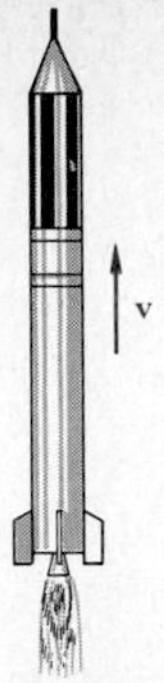

A rocket of initial mass m_0 (including shell and fuel) is fired vertically at time $t = 0$. The fuel is consumed at a constant rate $q = dm/dt$ and is expelled at a constant speed u relative to the rocket. Derive an expression for the velocity of the rocket at time t, neglecting the resistance of the air.

Solution. At time t, the mass of the rocket shell and remaining fuel is $m = m_0 - qt$, and the velocity is $\mathbf{v}$. During the time interval Δt, a mass of fuel $\Delta m = q\,\Delta t$ is expelled with a speed u relative to the rocket. Denoting by $\mathbf{v}_e$ the absolute velocity of the expelled fuel, we apply the principle of impulse and momentum between time t and time $t + \Delta t$.

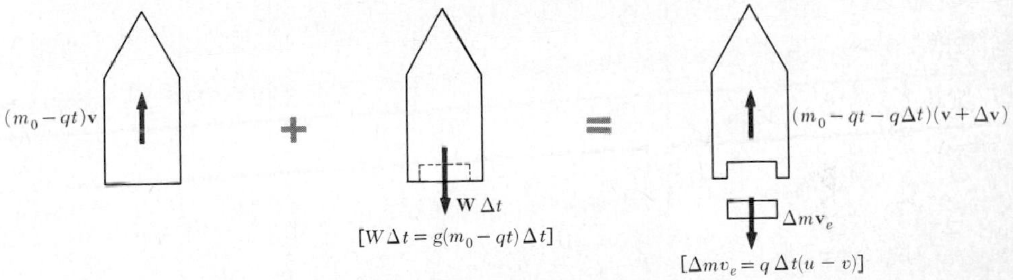

We write

$$(m_0 - qt)v - g(m_0 - qt)\,\Delta t = (m_0 - qt - q\,\Delta t)(v + \Delta v) - q\,\Delta t(u - v)$$

Dividing through by Δt, and letting Δt approach zero, we obtain

$$-g(m_0 - qt) = (m_0 - qt)\frac{dv}{dt} - qu$$

Separating variables and integrating from $t = 0$, $v = 0$ to $t = t$, $v = v$,

$$dv = \left(\frac{qu}{m_0 - qt} - g\right)dt \qquad \int_0^v dv = \int_0^t \left(\frac{qu}{m_0 - qt} - g\right)dt$$

$$v = [-u \ln (m_0 - qt) - gt]_0^t \qquad v = u \ln \frac{m_0}{m_0 - qt} - gt \quad ◄$$

Remark. The mass remaining at time t_f, after all the fuel has been expended, is equal to the mass of the rocket shell $m_s = m_0 - qt_f$, and the maximum velocity attained by the rocket is $v_m = u \ln (m_0/m_s) - gt_f$. Assuming that the fuel is expelled in a relatively short period of time, the term gt_f is small and we have $v_m \approx u \ln (m_0/m_s)$. In order to escape the gravitational field of the earth, a rocket must reach a velocity of 11.18 km/s. Assuming $u = 2200$ m/s and $v_m = 11.18$ km/s, we obtain $m_0/m_s = 161$. Thus, to project each kilogram of the rocket shell into space, it is necessary to consume more than 161 kg of fuel if a propellant yielding $u = 2200$ m/s is used.

Problems

Note. In the following problems use $\rho = 1000$ kg/m^3 for the density of water in SI units, and $\gamma = 62.4$ lb/ft^3 for its specific weight in U.S. customary units. (See footnote on page 617.)

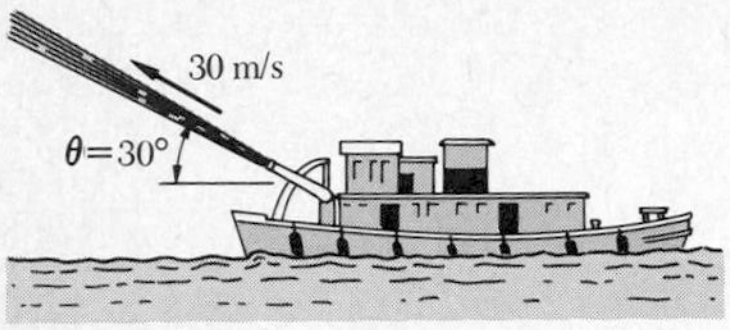

Fig. P14.46

14.46 A hose discharges water at the rate of 9 m^3/min from the stern of a 20-Mg fireboat. If the velocity of the water stream is 30 m/s, determine the reaction on the boat.

14.47 Tree limbs and branches are being fed at D at the rate of 5 kg/s into a shredder which spews the resulting wood chips at C with a velocity of 25 m/s. Determine the horizontal component of the force exerted by the shredder on the truck hitch at A.

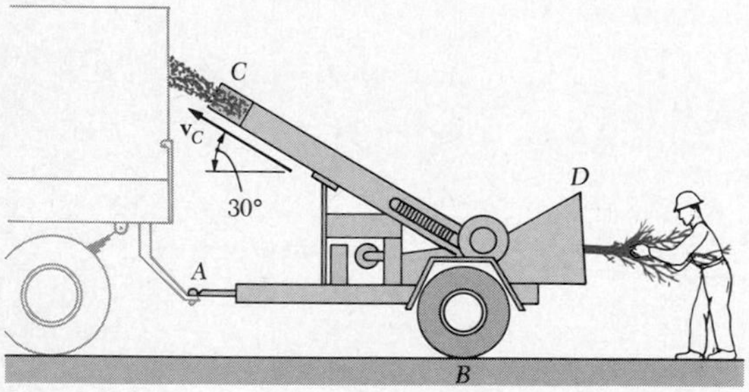

Fig. P14.47

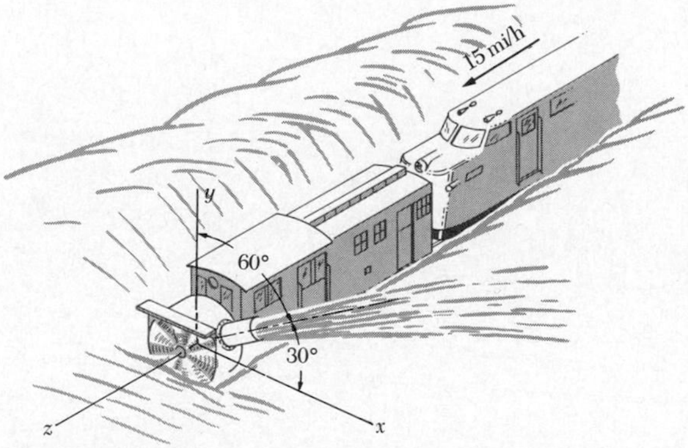

Fig. P14.48

14.48 A rotary power plow is used to remove snow from a level section of railroad track. The plow car is placed ahead of an engine which propels it at a constant speed of 15 mi/h. The plow clears 180 tons of snow per minute, projecting it in the direction shown with a velocity of 40 ft/s relative to the plow car. Neglecting rolling friction, determine (*a*) the magnitude of the force **P** exerted by the engine on the plow car, (*b*) the lateral force exerted on the plow car by the track.

14.49 For the rotary power plow of Prob. 14.48, determine the angle that the discharge duct should form with the z axis if at the speed considered, the force **P** exerted by the engine on the plow car is to be zero.

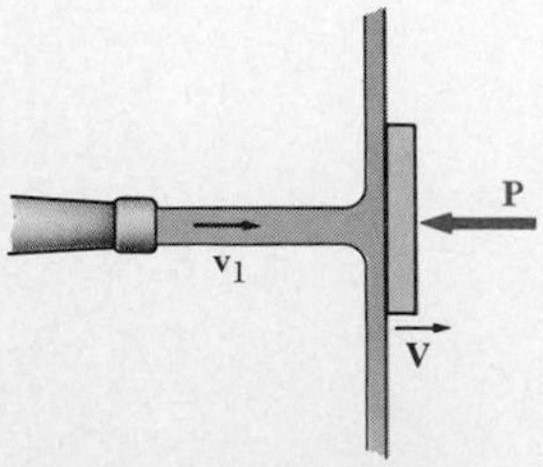

Fig. P14.50 and P14.51

14.50 A stream of water of cross-sectional area A and velocity $\mathbf{v}_1$ strikes a plate which is held motionless by a force **P**. Determine the magnitude of **P**, knowing that $A = 600$ mm^2, $v_1 = 30$ m/s, and $V = 0$.

14.51 A stream of water of cross-sectional area A and velocity $\mathbf{v}_1$ strikes a plate which moves to the right with a velocity **V**. Determine the magnitude of **V**, knowing that $A = 400$ mm^2, $v_1 = 35$ m/s, and $P = 360$ N.

14.52 Water flows in a continuous sheet from between two plates A and B with a velocity **v** of magnitude 120 ft/s. The stream is split into two parts by a smooth horizontal plate C. Knowing that the rates of flow in each of the two resulting streams are, respectively, $Q_1 = 40$ gal/min and $Q_2 = 160$ gal/min, determine (*a*) the angle θ, (*b*) the total force exerted by the streams on the plate ($1 \text{ ft}^3 = 7.48$ gal).

14.53 Water flows in a continuous sheet from between two plates A and B with a velocity **v** of magnitude 80 ft/s. The stream is split into two parts by a smooth horizontal plate C. Determine the rates of flow Q_1 and Q_2 in each of the two resulting streams, knowing that $\theta = 30°$ and that the total force exerted by the streams on the plate is a 75-lb vertical force ($1 \text{ ft}^3 = 7.48$ gal).

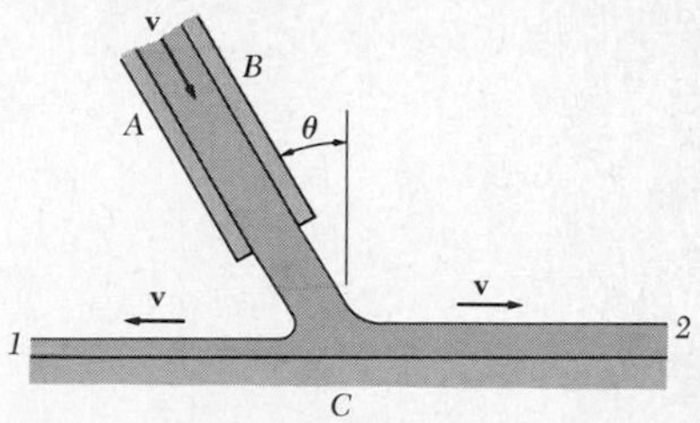

Fig. P14.52 and P14.53

14.54 The nozzle shown discharges water at the rate of 300 gal/min. Knowing that at both A and B the stream of water moves with a velocity of magnitude 80 ft/s and neglecting the weight of the vane, determine the components of the reactions at C and D ($1 \text{ ft}^3 = 7.48$ gal).

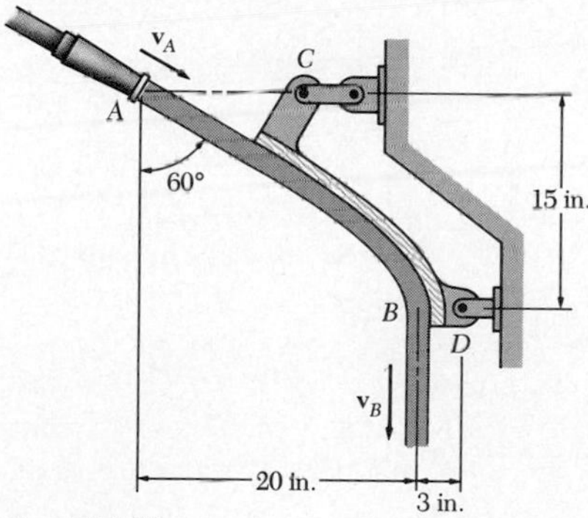

Fig. P14.54

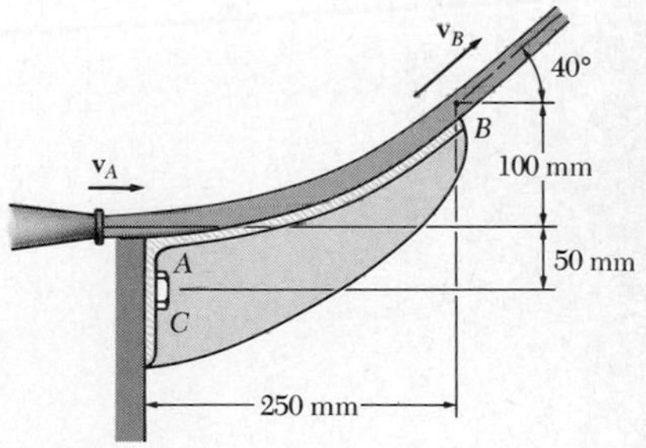

Fig. P14.55

14.55 The nozzle shown discharges water at the rate of $0.95 \text{ m}^3/\text{min}$. Knowing that at both A and B the water moves with a velocity of magnitude 40 m/s and neglecting the weight of the vane, determine the force-couple system which must be applied at C to hold the vane in place.

14.56 Knowing that the blade AB of Sample Prob. 14.6 is in the shape of an arc of circle, show that the resultant force **F** exerted by the blade on the stream is applied at the midpoint C of the arc AB. (*Hint.* First show that the line of action of **F** must pass through the center O of the circle.)

14.57 A jet of water having a cross-sectional area $A = 1.2 \text{ in}^2$ and moving with a velocity of magnitude $v_A = v_B = 60$ ft/s is deflected by the two vanes shown, which are welded to a vertical plate. Knowing that the combined weight of the plate and vanes is 10 lb, determine the reactions at C and D.

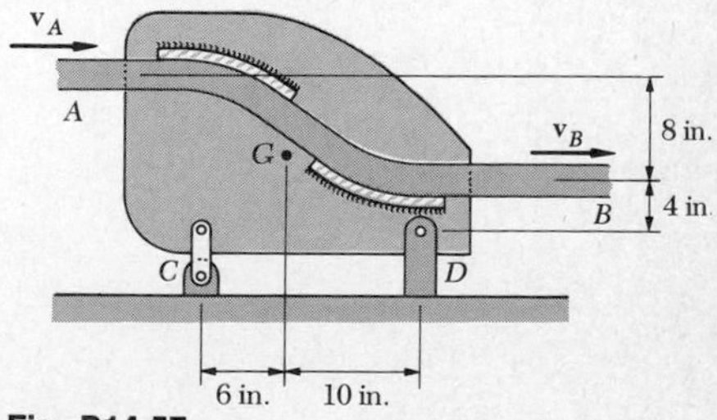

Fig. P14.57

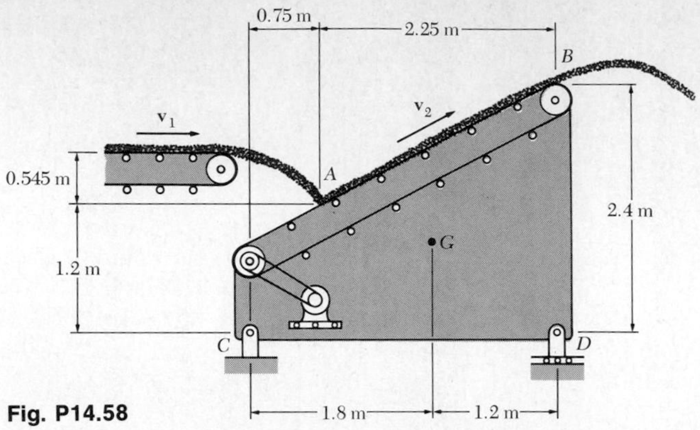

Fig. P14.58

14.58 Coal is being discharged from a first conveyor belt at the rate of 120 kg/s. It is received at A by a second belt which discharges it again at B. Knowing that $v_1 = 3$ m/s and $v_2 = 4.25$ m/s and that the second conveyor belt assembly and the coal it supports have a total mass of 472 kg, determine the reactions at C and D.

14.59 Solve Prob. 14.58, assuming that the velocity of the second conveyor belt is decreased to $v_2 = 2.55$ m/s and that the mass of coal it supports is increased by 48 kg.

14.60 The total drag due to air friction on a jet airplane traveling at 900 km/h is 30 kN. Knowing that the exhaust velocity is 650 m/s relative to the airplane, determine the mass of air which must pass through the engine per second to maintain the speed of 900 km/h in level flight.

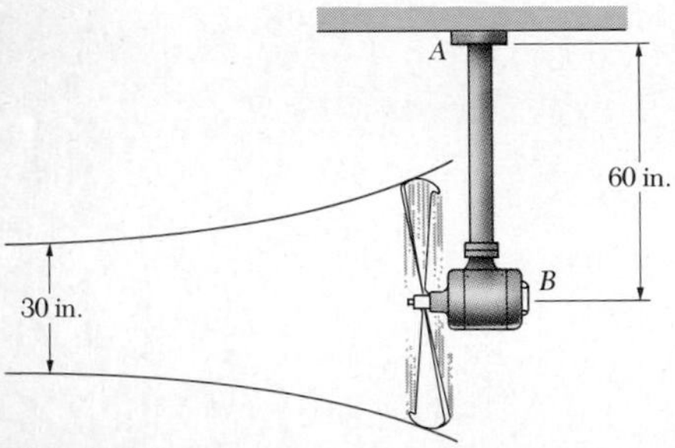

Fig. P14.62

14.61 While cruising in level flight at a speed of 600 mi/h, a jet airplane scoops in air at the rate of 170 lb/s and discharges it with a velocity of 2000 ft/s relative to the airplane. Determine the total drag due to air friction on the airplane.

14.62 For the ceiling-mounted fan shown, determine the maximum allowable air velocity in the slipstream if the bending moment in the supporting rod AB is not to exceed 80 lb·ft. Assume $\gamma = 0.076$ lb/ft^3 for air and neglect the approach velocity of the air.

14.63 The jet engine shown scoops in air at A at the rate of 75 kg/s and discharges it at B with a velocity of 800 m/s relative to the airplane. Determine the magnitude and line of action of the propulsive thrust developed by the engine when the speed of the airplane is (*a*) 500 km/h, (*b*) 1000 km/h.

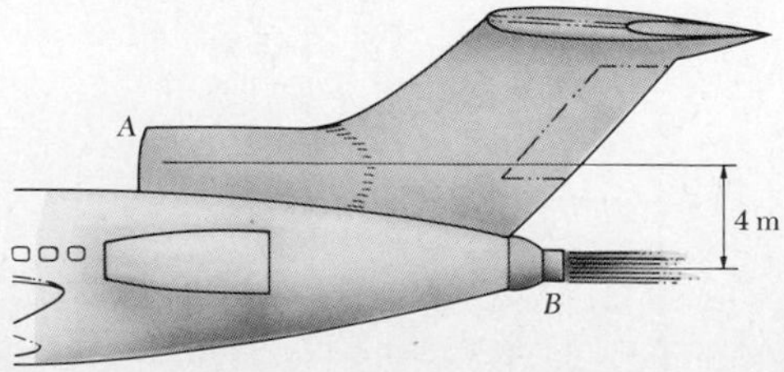

Fig. P14.63

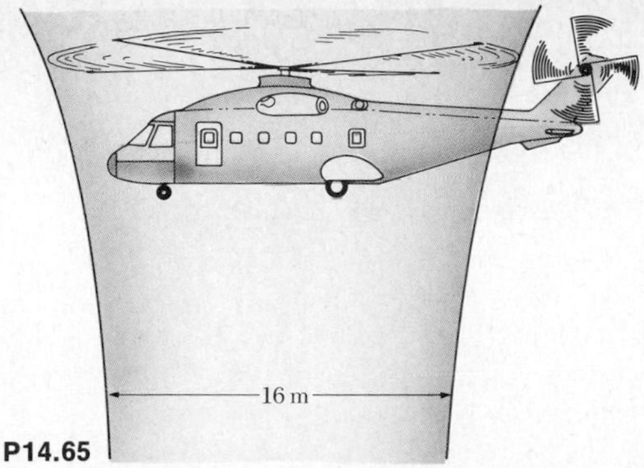

Fig. P14.64 and P14.65

14.64 The helicopter shown has a mass of 10 Mg when empty and can produce a maximum downward air speed of 25 m/s in its 16-m-diameter slipstream. Assuming $\rho = 1.21 \text{ kg/m}^3$ for air, determine the maximum combined payload and fuel load the helicopter can carry while hovering in midair.

14.65 The helicopter shown has a mass of 10 Mg. Assuming $\rho = 1.21 \text{ kg/m}^3$ for air, determine the downward air speed the helicopter produces in its 16-m-diameter slipstream while hovering in midair with a combined payload and fuel load of 4000 kg.

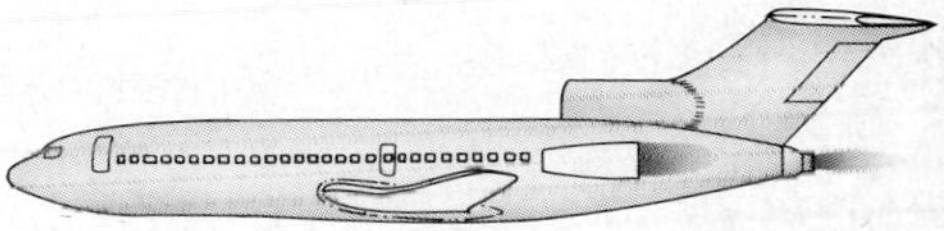

Fig. P14.66

14.66 A jet airliner is cruising at the speed of 990 km/h with each of its three engines discharging air with a velocity of 675 m/s relative to the plane. Determine the speed of the airliner after it has lost the use of (*a*) one of its engines, (*b*) two of its engines. Assume that the drag due to air friction is proportional to the square of the speed and that the remaining engines keep operating at the same rate.

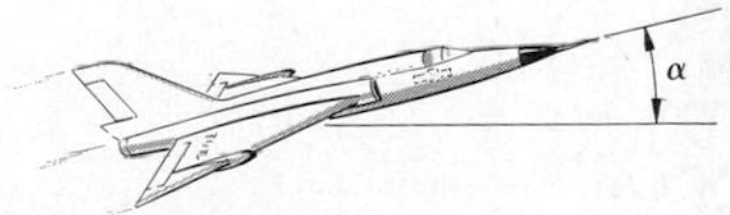

Fig. P14.67

14.67 A 30,000-lb jet airplane maintains a constant speed of 540 mi/h while climbing at an angle $\alpha = 10°$. The airplane scoops in air at the rate of 500 lb/s and discharges it with a velocity of 2250 ft/s relative to the airplane. If the pilot changes to a horizontal flight and the same engine conditions are maintained, determine (*a*) the initial acceleration of the plane, (*b*) the maximum horizontal speed attained. Assume that the drag due to air friction is proportional to the square of the speed.

14.68 The wind turbine-generator shown has an output-power rating of 15 kW for a wind speed of 43.2 km/h. For the given wind speed, determine (*a*) the kinetic energy of the air particles entering the 7.60-m-diameter circle per second, (*b*) the efficiency of this energy-conversion system. Assume $\rho = 1.21 \text{ kg/m}^3$ for air.

14.69 For a given wind speed, the wind turbine-generator shown produces 20 kW of electric power and, as an energy-conversion system, has an efficiency of 0.30. Determine (*a*) the kinetic energy of the air particles entering the 7.60-m-diameter circle per second, (*b*) the wind speed.

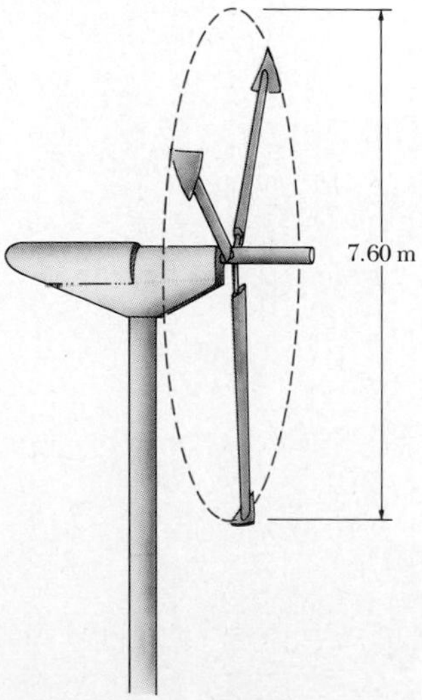

Fig. P14.68 and P14.69

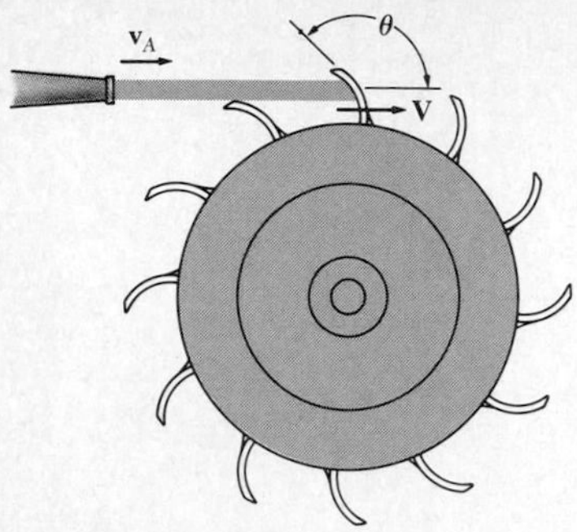

Fig. P14.70

14.70 In a Pelton-wheel turbine, a stream of water is deflected by a series of blades so that the rate at which water is deflected by the blades is equal to the rate at which water issues from the nozzle ($\Delta m/\Delta t = A\rho v_A$). Using the same notation as in Sample Prob. 14.6, (*a*) determine the velocity **V** of the blades for which maximum power is developed, (*b*) derive an expression for the maximum power, (*c*) derive an expression for the mechanical efficiency.

14.71 While cruising in level flight at a speed of 600 mi/h, a jet airplane scoops in air at the rate of 200 lb/s and discharges it with a velocity of 2200 ft/s relative to the airplane. Determine (*a*) the power actually used to propel the airplane, (*b*) the total power developed by the engine, (*c*) the mechanical efficiency of the airplane.

14.72 A circular reentrant orifice (also called Borda's mouthpiece) of diameter D is placed at a depth h below the surface of a tank. Knowing that the speed of the issuing stream is $v = \sqrt{2gh}$ and assuming that the speed of approach v_1 is zero, show that the diameter of the stream is $d = D/\sqrt{2}$. (*Hint.* Consider the section of water indicated, and note that P is equal to the pressure at a depth h multiplied by the area of the orifice.)

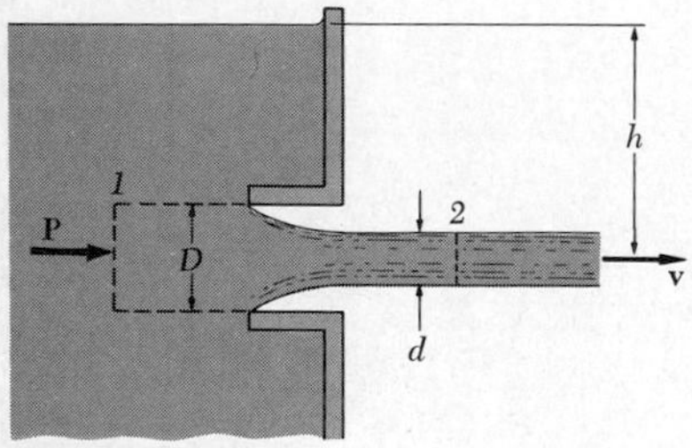

Fig. P14.72

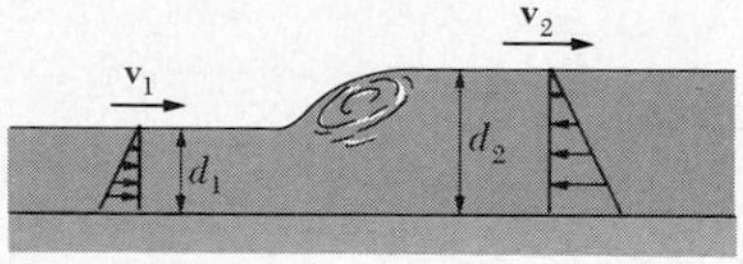

Fig. P14.73

***14.73** The depth of water flowing in a rectangular channel of width b at a speed v_1 and depth d_1 increases to a depth d_2 at a *hydraulic jump*. Express the rate of flow Q in terms of b, d_1, and d_2.

***14.74** Determine the rate of flow in the channel of Prob. 14.73, knowing that $d_1 = 0.8$ m, $d_2 = 1.2$ m, and the rectangular channel is 3 m wide.

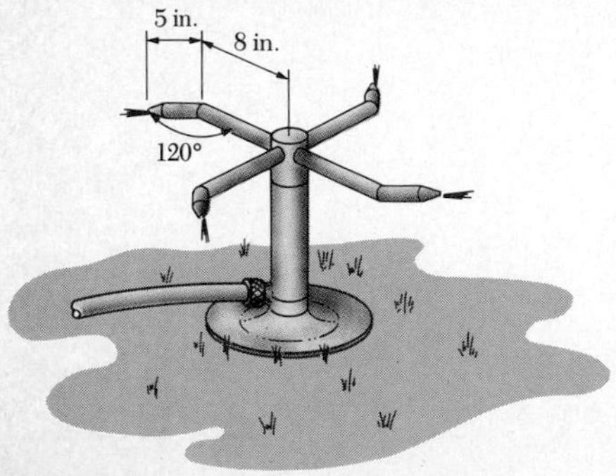

Fig. P14.75

14.75 A garden sprinkler has four rotating arms, each of which consists of two horizontal straight sections of pipe forming an angle of 120°. Each arm discharges water at the rate of 4 gal/min with a velocity of 50 ft/s relative to the arm. Knowing that the friction between the moving and stationary parts of the sprinkler is equivalent to a couple of magnitude $M = 0.250$ lb·ft, determine the constant rate at which the sprinkler rotates (1 ft^3 = 7.48 gal).

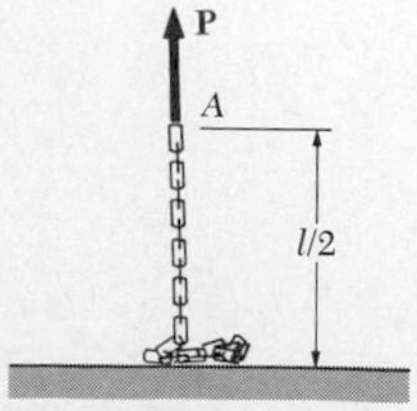

Fig. P14.76

14.76 A chain of length l and total mass m lies in a pile on the floor. If its end A is raised vertically at a constant speed v, determine (*a*) the force **P** applied to A at the time when half the chain is off the floor, (*b*) the reaction exerted by the floor at that time.

14.77 Solve Prob. 14.76, assuming that the end A of the chain is being *lowered* to the floor at a constant speed v.

14.78 The ends of a chain lie in piles at A and C. When given an initial speed v, the chain keeps moving freely at that speed over the pulley at B. Neglecting friction, determine the required value of h.

14.79 The ends of a chain lie in piles at A and C. When released from rest at time $t = 0$, the chain moves over the pulley at B, which has a negligible mass. Denoting by L the length of chain connecting the two piles and neglecting friction, determine the speed v of the chain at time t.

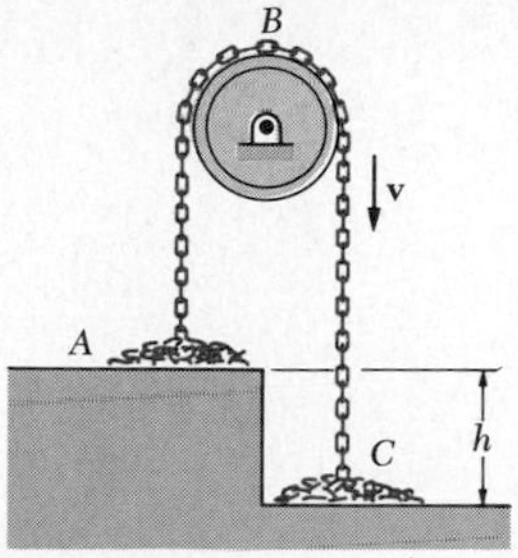

Fig. P14.78 and P14.79

14.80 For the chain of Prob. 14.79, it is known that $h = 300$ mm, and that the length of chain connecting the two piles is $L = 4$ m. Neglecting friction, determine at $t = 2$ s (*a*) the speed v of the chain, (*b*) the length of chain which has been transferred from pile A to pile C.

14.81 Determine the speed v of the chain of Prob. 14.79, after a length x of chain has been transferred from pile A to pile C.

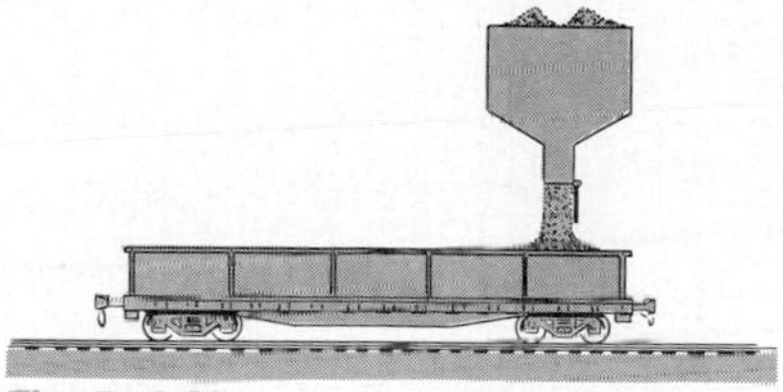
Fig. P14.82

14.82 A railroad car of length L and mass m_0 when empty is moving freely on a horizontal track while being loaded with sand from a stationary chute at the rate $q = dm/dt$. Knowing that the car was approaching the chute with a speed v_0, determine after the car has cleared the chute (*a*) the mass of the car and its load, (*b*) the speed of the car.

14.83 A possible method for reducing the speed of a training plane as it lands on an aircraft carrier consists in having the tail of the plane hook into the end of a heavy chain of length l which lies in a pile below deck. Denoting by m the mass of the plane and by v_0 its speed at touchdown, and assuming no other retarding force, determine (*a*) the required mass of the chain if the speed of the plane is to be reduced to βv_0, where $\beta < 1$, (*b*) the maximum value of the force exerted by the chain on the plane.

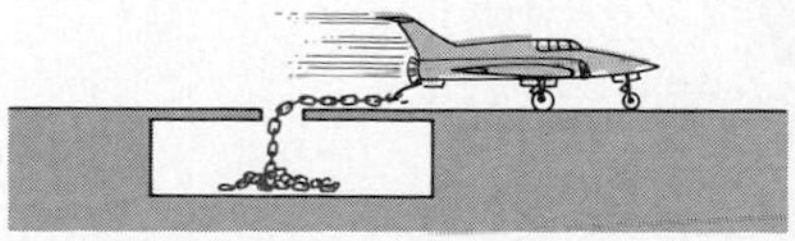
Fig. P14.83 and P14.84

14.84 As a 15,000-lb training plane lands on an aircraft carrier at a speed of 120 mi/h, its tail hooks into the end of a long chain which lies in a pile below deck. Knowing that the chain weighs 25 lb/ft and assuming no other retarding force, determine the maximum deceleration of the plane.

14.85 For the railroad car of Prob. 14.82, express in terms of t the magnitude of the horizontal force **P** which should be applied to the car to keep it moving, (*a*) at the constant speed v_0, (*b*) with a constant acceleration a, while being loaded. Assume that $t = 0$ when the loading operation begins.

14.86 The main propulsion system of a space shuttle consists of three identical rocket engines, each of which burns the hydrogen-oxygen propellant at the rate of 800 lb/s and ejects it with a relative velocity of 12,000 ft/s. Determine the total thrust provided by the three engines.

Fig. P14.86 and P14.87

14.87 The main propulsion system of a space shuttle consists of three identical rocket engines which provide a total thrust of 1250 kips. Determine the rate at which the hydrogen-oxygen propellant is burned by each of the three engines, knowing that it is ejected with a relative velocity of 12,000 ft/s.

14.88 A rocket has a mass of 1200 kg, including 1000 kg of fuel, which is consumed at the rate of 12.5 kg/s and ejected with a relative velocity of 4000 m/s. Knowing that the rocket is fired vertically from the ground, determine its acceleration (*a*) as it is fired, (*b*) as the last particle of fuel is being consumed.

14.89 The acceleration of a rocket is observed to be 30 m/s^2 at $t = 0$, as it is fired vertically from the ground, and 350 m/s^2 at $t = 80$ s. Knowing that the fuel is consumed at the rate of 10 kg/s, determine (*a*) the initial mass of the rocket, (*b*) the relative velocity with which the fuel is ejected.

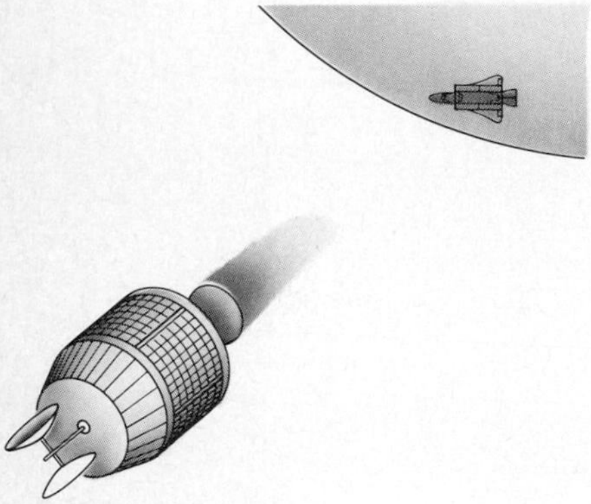

Fig. P14.90 and P14.91

14.90 A weather satellite of mass 4000 kg, including fuel, has been ejected from a space shuttle describing a low circular orbit around the earth. After the satellite has slowly drifted to a safe distance from the shuttle, its engine is fired to increase its velocity by 2430 m/s as a first step to its transfer to a geosynchronous orbit. Knowing that the fuel is ejected with a relative velocity of 3800 m/s, determine the mass of fuel consumed in this maneuver.

14.91 Determine the increase in velocity of the weather satellite of Prob. 14.90 after 1200 kg of fuel has been consumed.

14.92 An 800-lb spacecraft is mounted on top of a rocket weighing 38,400 lb, including 36,000 lb of fuel. Knowing that the fuel is consumed at the rate of 400 lb/s and ejected with a relative velocity of 12,000 ft/s, determine the maximum speed imparted to the spacecraft when the rocket is fired vertically from the ground.

14.93 The rocket used to launch the 800-lb spacecraft of Prob. 14.92 is redesigned to include two stages A and B, each weighing 19,200 lb, including 18,000 lb of fuel. The fuel is again consumed at the rate of 400 lb/s and ejected with a relative velocity of 12,000 ft/s. Knowing that when stage A expels its last particle of fuel, its casing is released and jettisoned, determine (*a*) the speed of the rocket at that instant, (*b*) the maximum speed imparted to the spacecraft.

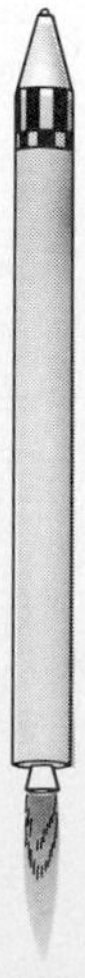

Fig. P14.92

Fig. P14.93

14.94 Determine the altitude reached by the spacecraft of Prob. 14.92 when all the fuel of its launching rocket has been consumed.

14.95 For the spacecraft and the two-stage launching rocket of Prob. 14.93, determine the altitude at which (*a*) stage *A* of the rocket is released, (*b*) the fuel of both stages has been consumed.

14.96 Knowing that the engine of the weather satellite of Prob. 14.90 was fired for 126 s, determine the distance separating the weather satellite from the space shuttle as the engine was turned off.

14.97 For the rocket of Prob. 14.88, determine (*a*) the altitude at which all the fuel has been consumed, (*b*) the velocity of the rocket at that time.

14.98 In a rocket, the kinetic energy imparted to the consumed and ejected fuel is wasted as far as propelling the rocket is concerned. The useful power is equal to the product of the force available to propel the rocket and the speed of the rocket. If v is the speed of the rocket and u is the relative speed of the expelled fuel, show that the mechanical efficiency of the rocket is $\eta = 2uv/(u^2 + v^2)$. Explain why $\eta = 1$ when $u = v$.

14.99 In a jet airplane, the kinetic energy imparted to the exhaust gases is wasted as far as propelling the airplane is concerned. The useful power is equal to the product of the force available to propel the airplane and the speed of the airplane. If v is the speed of the airplane and u is the relative speed of the expelled gases, show that the mechanical efficiency of the airplane is $\eta = 2v/(u + v)$. Explain why $\eta = 1$ when $u = v$.

Review Problems

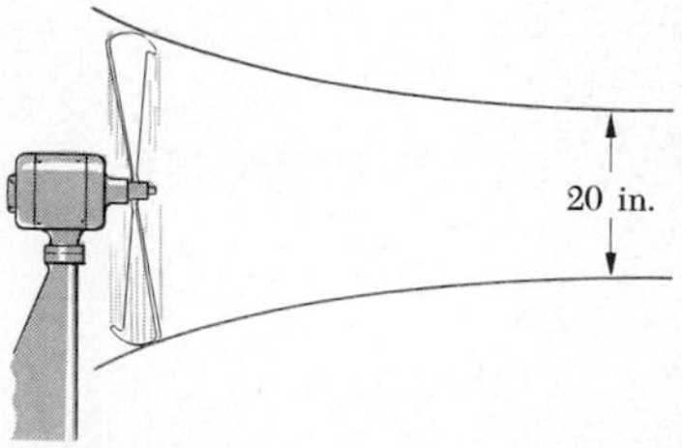

Fig. P14.100

14.100 The slipstream of a fan has a diameter of 20 in. and a velocity of 35 ft/s relative to the fan. Assuming $\gamma = 0.076$ lb/ft^3 for air and neglecting the velocity of approach of the air, determine the force required to hold the fan motionless.

14.101 A chain of length l and mass m falls through a small hole in a plate. Initially, when y is very small, the chain is at rest. In each case shown, determine (*a*) the acceleration of the first link *A* as a function of y, (*b*) the velocity of the chain as the last link passes through the hole. In case *1* assume that the individual links are at rest until they fall through the hole; in case 2 assume that at any instant all links have the same speed. Ignore the effect of friction.

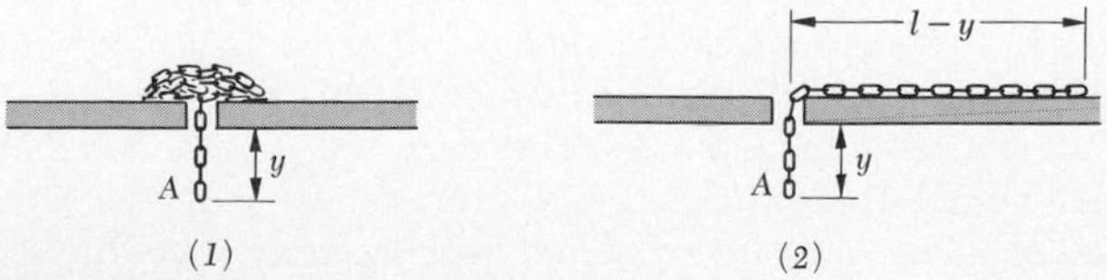

Fig. P14.101

14.102 In a game of billiards, ball A is moving with the velocity $\mathbf{v}_0 = (3\text{ m/s})\mathbf{i}$ when it strikes balls B and C which are at rest side by side. After the collision, A is observed to move with the velocity $\mathbf{v}_A = (1.176\text{ m/s})\mathbf{i} - (1.368\text{ m/s})\mathbf{j}$, while B and C move in the directions shown. Determine (*a*) the magnitudes of the velocities $\mathbf{v}_B$ and $\mathbf{v}_C$, (*b*) the percentage of the initial energy lost in the collision.

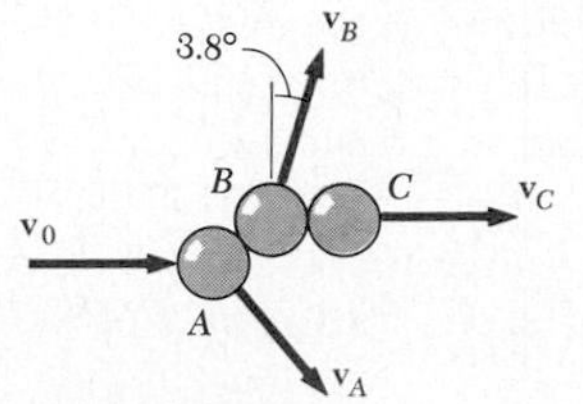

Fig. P14.102

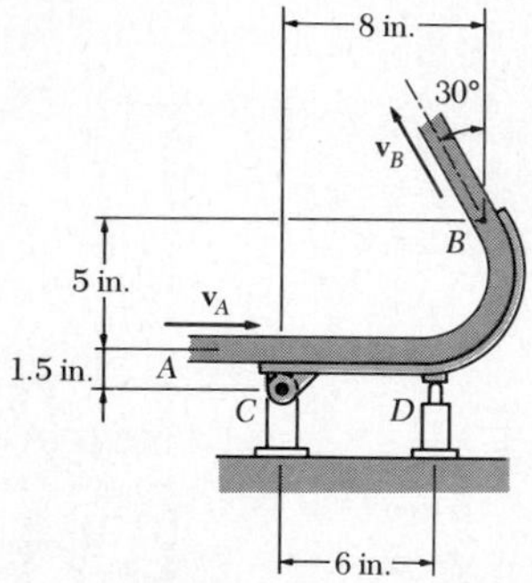

Fig. P14.103

14.103 The stream of water shown flows at the rate of 250 gal/min and moves with a velocity of magnitude 120 ft/s at both A and B. The vane is supported by a pin connection at C and by a load cell at D which can exert only a vertical force. Neglecting the weight of the vane, determine the reactions at C and D ($1\text{ ft}^3 = 7.48$ gal).

14.104 Three identical freight cars are located on the same track. Car A is moving to the right with a velocity $\mathbf{v}_0$, while cars B and C are at rest. Determine the velocity of each car after all collisions have taken place, assuming that cars A and B bounce off each other with a coefficient of restitution $e = 1$ and that cars B and C (*a*) also bounce off each other with $e = 1$, (*b*) get automatically coupled as they hit each other, (*c*) were already tightly coupled when A hit B.

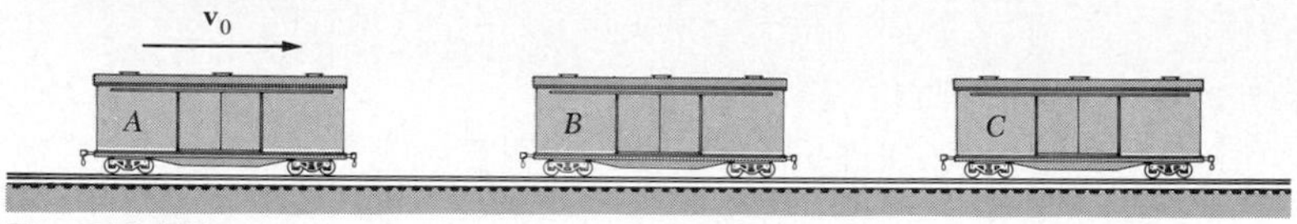

Fig. P14.104

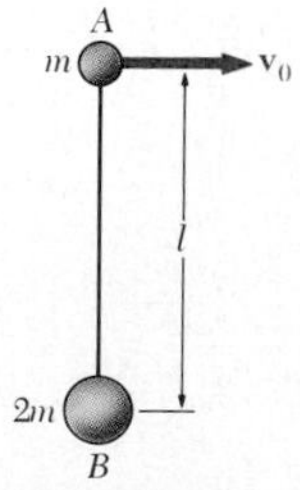

Fig. P14.105

14.105 Two small spheres A and B, respectively of mass m and $2m$, are connected by a rigid rod of length l and negligible mass. The two spheres are resting on a horizontal, frictionless surface when A is suddenly given the velocity $\mathbf{v}_0 = v_0\mathbf{i}$. Determine (*a*) the linear momentum of the system and its angular momentum about its mass center G, (*b*) the velocities of A and B after the rod AB has rotated through 90°, (*c*) the velocities of A and B after the rod AB has rotated through 180°.

14.106 A spacecraft is launched vertically by a two-stage rocket. When the speed is 16×10^3 km/h, the first-stage casing is released and the second-stage rocket is fired. Fuel is consumed at the rate of 100 kg/s and ejected with a relative velocity of 2500 m/s. Knowing that the combined mass of the second-stage rocket and spacecraft is 10 Mg, including 8.5 Mg of fuel, determine (*a*) the maximum speed attained by the spacecraft, (*b*) the distance between the spacecraft and the first-stage casing as the last particle of fuel is being expelled by the second-stage rocket.

14.107 When the cord connecting particles A and B is severed, the compressed spring causes the particles to fly apart (the spring is not connected to the particles). The potential energy of the compressed spring is known to be 60 J and the assembly has an initial velocity $\mathbf{v}_0$ as shown. If the cord is severed when $\theta = 25°$, determine the resulting velocity of each particle.

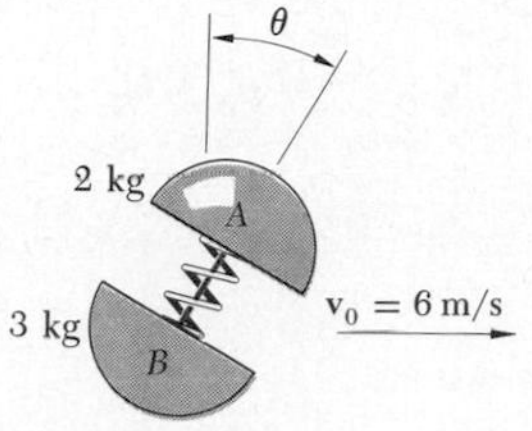

Fig. P14.107

14.108 Gravel falls with practically zero velocity onto a conveyor belt at the constant rate $q = dm/dt$. (*a*) Determine the magnitude of the force **P** required to maintain a constant belt speed v. (*b*) Show that the kinetic energy acquired by the gravel in a given time interval is equal to half the work done in that interval by the force **P**. Explain what happens to the other half of the work done by **P**.

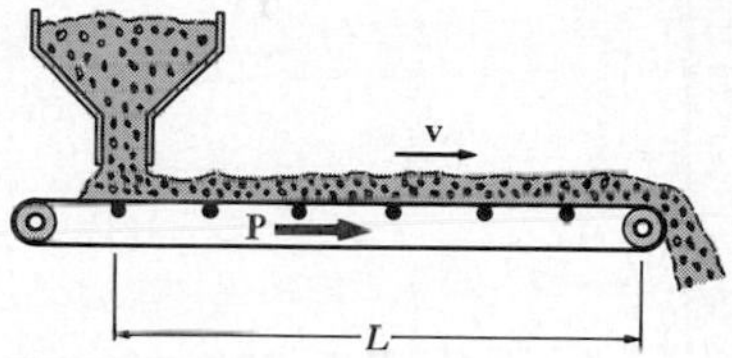

Fig. P14.108

14.109 A space vehicle describing a circular orbit at a speed of 24×10^3 km/h releases at its front end a capsule which has a gross mass of 500 kg, including 375 kg of fuel. If the fuel is consumed at the constant rate of 15 kg/s and ejected with a relative velocity of 2500 m/s, determine (*a*) the tangential acceleration of the capsule as the engine is fired, (*b*) the maximum speed attained by the capsule.

Fig. P14.109

14.110 In a game of billiards, ball A is given an initial velocity $\mathbf{v}_0$ along line DA parallel to the axis of the table. It hits ball B and then ball C, which are at rest. Knowing that A and C hit the sides of the table squarely at points A' and C', respectively, with velocities of magnitude $v_A = 4$ ft/s and $v_C = 6$ ft/s, and assuming frictionless surfaces and perfectly elastic impacts (i.e., conservation of energy), determine (*a*) the initial velocity $\mathbf{v}_0$ of ball A, (*b*) the velocity $\mathbf{v}_B$ of ball B, (*c*) the point B' where B hits the side of the table.

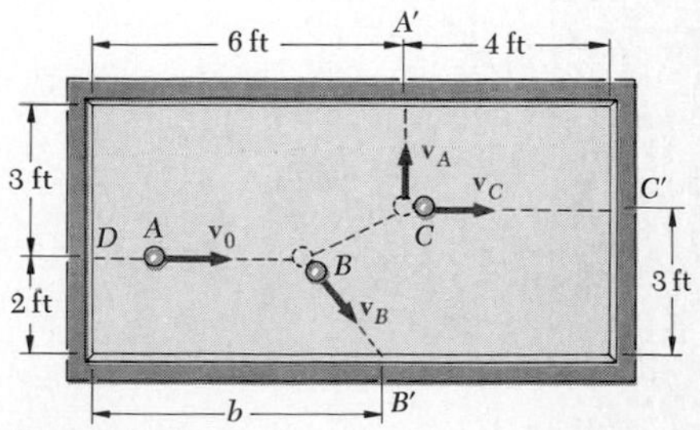

Fig. P14.110

14.111 A 15-kg projectile is passing through the origin O with a velocity $\mathbf{v}_0 = (40 \text{ m/s})\mathbf{i}$ when it explodes into two fragments A and B, of mass 6 kg and 9 kg, respectively. Knowing that 3 s later the position of the first fragment is $A(150$ m, 12 m, -24 m$)$, determine the position of fragment B at the same instant. Assume $a_y = -g = -9.81$ m/s^2 and neglect the resistance of the air.

CHAPTER FIFTEEN

Kinematics of Rigid Bodies

15.1. Introduction. In this chapter, we shall study the kinematics of *rigid bodies*. We shall investigate the relations existing between the time, the positions, the velocities, and the accelerations of the various particles forming a rigid body. As we shall see, the various types of rigid-body motion may be conveniently grouped as follows:

1. *Translation.* A motion is said to be a translation if any straight line inside the body keeps the same direction during the motion. It may also be observed that in a translation all the particles forming the body move along parallel paths. If these paths are straight lines, the motion is said to be a *rectilinear translation* (Fig. 15.1); if the paths are curved lines, the motion is a *curvilinear translation* (Fig. 15.2).
2. *Rotation about a Fixed Axis.* In this motion, the particles forming the rigid body move in parallel planes along circles centered on the same fixed axis (Fig. 15.3). If this axis, called the *axis of rotation*, intersects the rigid body, the particles located on the axis have zero velocity and zero acceleration.

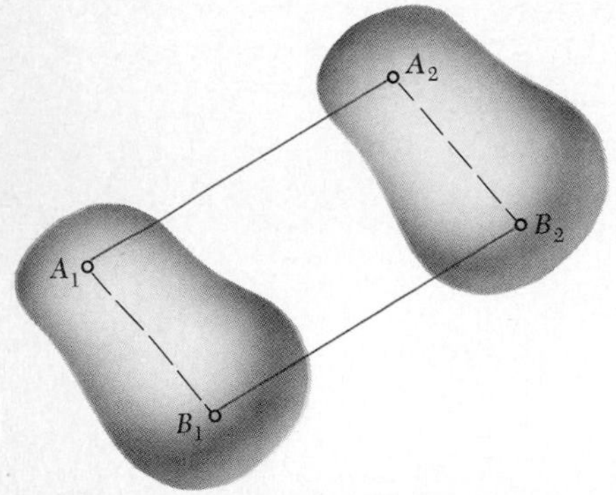

Fig. 15.1

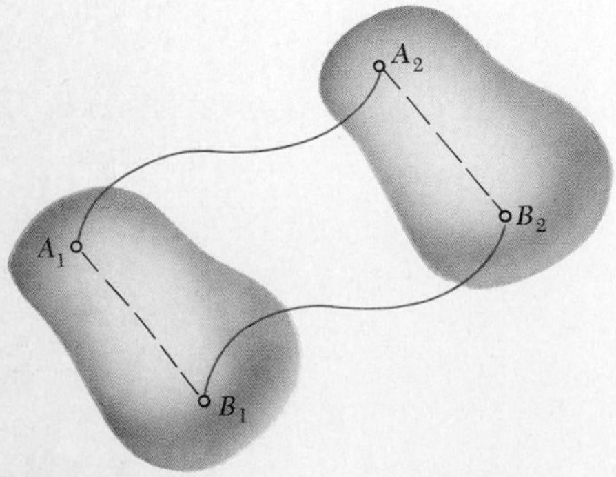

Fig. 15.2

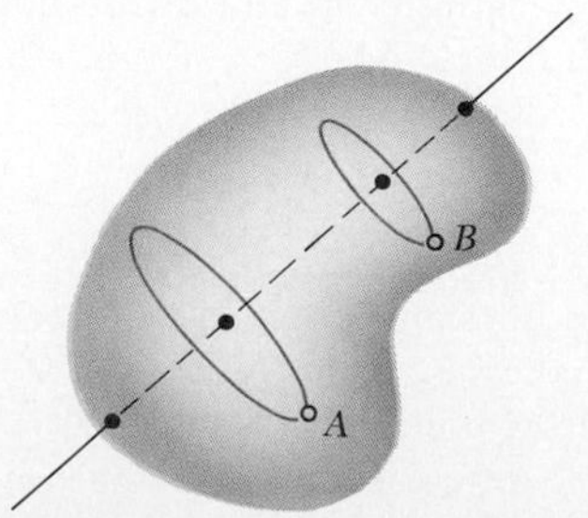

Fig. 15.3

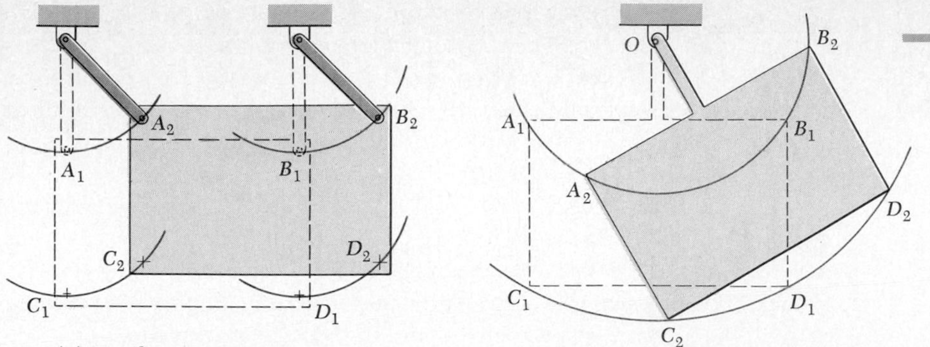

(a) Curvilinear translation

Fig. 15.4

(b) Rotation

Rotation should not be confused with certain types of curvilinear translation. For example, the plate shown in Fig. 15.4*a* is in curvilinear translation, with all its particles moving along *parallel* circles, while the plate shown in Fig. 15.4*b* is in rotation, with all its particles moving along *concentric* circles. In the first case, any given straight line drawn on the plate will maintain the same direction, whereas in the second case, point O remains fixed.

Because each particle moves in a given plane, the rotation of a body about a fixed axis is said to be a *plane motion*.

3. *General Plane Motion.* There are many other types of plane motion, i.e., motions in which all the particles of the body move in parallel planes. Any plane motion which is neither a rotation nor a translation is referred to as a general plane motion. Two examples of general plane motion are given in Fig. 15.5.

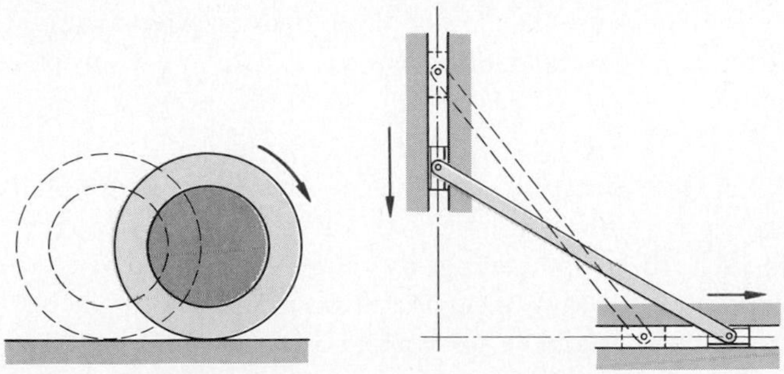

(a) Rolling wheel

Fig. 15.5

(b) Sliding rod

4. *Motion about a Fixed Point.* This is the three-dimensional motion of a rigid body attached at a fixed point O. An example of motion about a fixed point is provided by the motion of a top on a rough floor (Fig. 15.6).
5. *General Motion.* Any motion of a rigid body which does not fall in any of the categories above is referred to as a general motion.

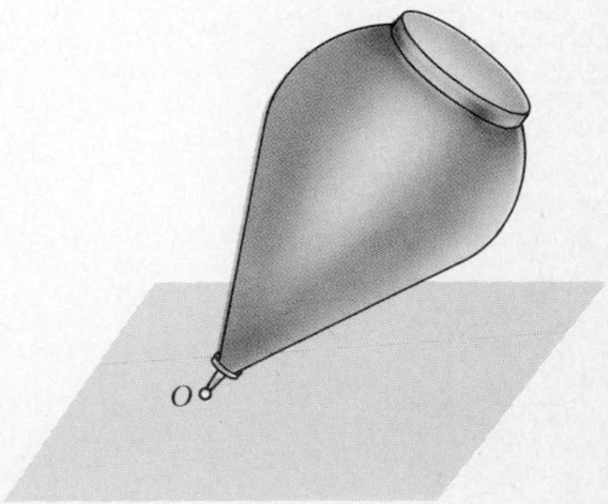

Fig. 15.6

After a brief discussion in Sec. 15.2 of the motion of translation, we shall consider in Sec. 15.3 the rotation of a rigid body about a fixed axis. We shall define the *angular velocity* and the *angular acceleration* of the body and learn to express the velocity and the acceleration of a given point of the body in terms of its position vector and the angular velocity and angular acceleration of the body.

The following sections are devoted to the study of the general plane motion of a rigid body and to its application to the analysis of mechanisms such as gears, connecting rods, and pin-connected linkages. Resolving the plane motion of a slab into a translation and a rotation (Secs. 15.5 and 15.6), we shall then express the velocity of a point B of the slab as the sum of the velocity of a reference point A and of the velocity of B relative to a frame of reference translating with A (i.e., moving with A but not rotating). The same approach is used later in Sec. 15.8 to express the acceleration of B in terms of the acceleration of A and of the acceleration of B relative to a frame translating with A.

An alternative method for the analysis of velocities in plane motion, based on the concept of *instantaneous center of rotation*, is given in Sec. 15.7; and still another method of analysis, based on the use of parametric expressions for the coordinates of a given point, is presented in Sec. 15.9.

The motion of a particle relative to a rotating frame of reference and the concept of *Coriolis acceleration* are discussed in Secs. 15.10 and 15.11, and the results obtained are applied to the analysis of the plane motion of mechanisms containing parts which slide on each other.

The remaining part of the chapter is devoted to the analysis of the three-dimensional motion of a rigid body, namely, the motion of a rigid body with a fixed point and the general motion of a rigid body. In Secs. 15.12 and 15.13, a fixed frame of reference or a frame of reference in translation will be used to carry out this analysis; whereas in Secs. 15.14 and 15.15, we shall consider the motion of the body relative to a rotating frame or to a frame in general motion and use again the concept of Coriolis acceleration.

15.2. Translation. Consider a rigid body in translation (either rectilinear or curvilinear translation), and let A and B be any two of its particles (Fig. 15.7*a*). Denoting, respectively, by $\mathbf{r}_A$ and $\mathbf{r}_B$ the position vectors of A and B with respect to a fixed frame of reference and by $\mathbf{r}_{B/A}$ the vector joining A and B, we write

$$\mathbf{r}_B = \mathbf{r}_A + \mathbf{r}_{B/A} \tag{15.1}$$

Let us differentiate this relation with respect to t. We note that from the very definition of a translation, the vector $\mathbf{r}_{B/A}$ must maintain a constant direction; its magnitude must also be constant, since A and B belong to the same rigid body. Thus, the derivative of $\mathbf{r}_{B/A}$ is zero and we have

$$\mathbf{v}_B = \mathbf{v}_A \tag{15.2}$$

Differentiating once more, we write

$$\mathbf{a}_B = \mathbf{a}_A \qquad (15.3)$$

Thus, *when a rigid body is in translation, all the points of the body have the same velocity and the same acceleration at any given instant* (Fig. 15.7*b* and *c*). In the case of curvilinear translation, the velocity and acceleration change in direction as well as in magnitude at every instant. In the case of rectilinear translation, all particles of the body move along parallel straight lines, and their velocity and acceleration keep the same direction during the entire motion.

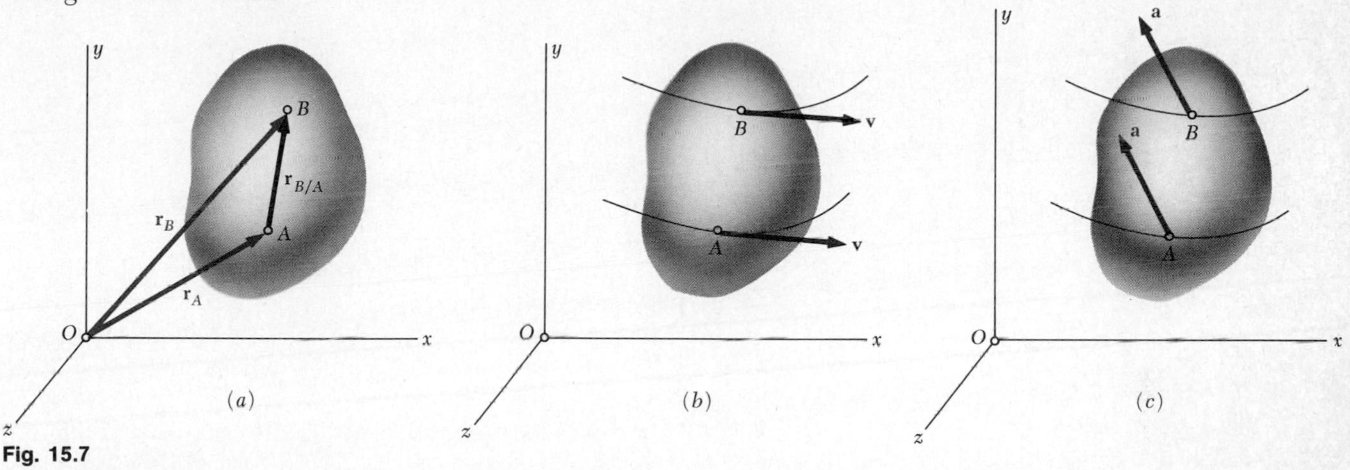

Fig. 15.7

15.3. Rotation about a Fixed Axis. Consider a rigid body which rotates about a fixed axis AA'. Let P be a point of the body and $\mathbf{r}$ its position vector with respect to a fixed frame of reference. For convenience, we shall assume that the frame is centered at point O on AA' and that the z axis coincides with AA' (Fig. 15.8). Let B be the projection of P on AA'; since P must remain at a constant distance from B, it will describe a circle of center B and of radius $r \sin \phi$, where ϕ denotes the angle formed by $\mathbf{r}$ and AA'.

The position of P and of the entire body is completely defined by the angle θ the line BP forms with the zx plane. The angle θ is known as the *angular coordinate* of the body. The angular coordinate is defined as positive when counterclockwise as viewed from A' and will be expressed in radians (rad) or, occasionally, in degrees (°) or revolutions (rev). We recall that

$$1 \text{ rev} = 2\pi \text{ rad} = 360°$$

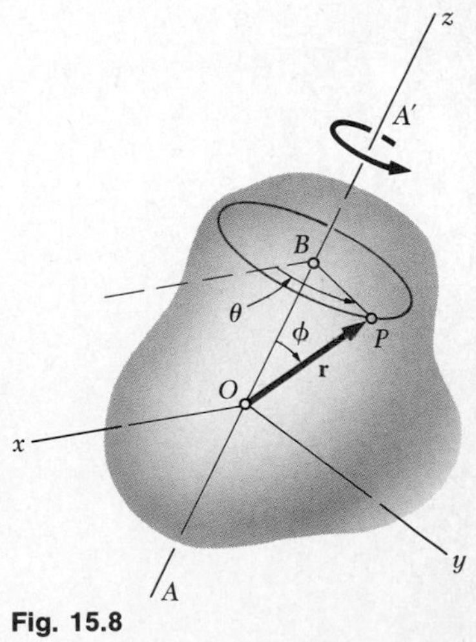

Fig. 15.8

We recall from Sec. 11.9 that the velocity $\mathbf{v} = d\mathbf{r}/dt$ of a particle P is a vector tangent to the path of P and of magnitude $v = ds/dt$. Observing that the length Δs of the arc described by P when the body rotates through $\Delta\theta$ is

$$\Delta s = (BP)\,\Delta\theta = (r \sin \phi)\,\Delta\theta$$

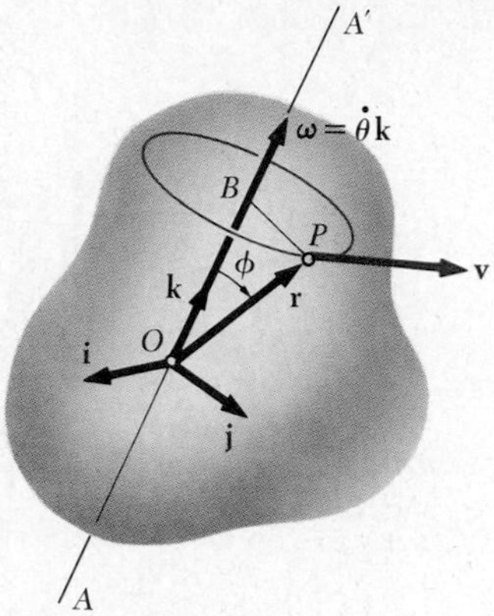

Fig. 15.9

and dividing both members by Δt, we obtain at the limit, as Δt approaches zero,

$$v = \frac{ds}{dt} = r\dot{\theta} \sin \phi \tag{15.4}$$

where $\dot{\theta}$ denotes the time derivative of θ. (Note that while the angle θ depends on the position of P within the body, the rate of change $\dot{\theta}$ is itself independent of P.) We conclude that the velocity $\mathbf{v}$ of P is a vector perpendicular to the plane containing AA' and $\mathbf{r}$, and of magnitude v defined by (15.4). But this is precisely the result we would obtain if we drew along AA' a vector $\boldsymbol{\omega} = \dot{\theta}\mathbf{k}$ and formed the vector product $\boldsymbol{\omega} \times \mathbf{r}$ (Fig. 15.9). We thus write

$$\mathbf{v} = \frac{d\mathbf{r}}{dt} = \boldsymbol{\omega} \times \mathbf{r} \tag{15.5}$$

The vector

$$\boldsymbol{\omega} = \omega\mathbf{k} = \dot{\theta}\mathbf{k} \tag{15.6}$$

is called the *angular velocity* of the body. It is directed along the axis of rotation, it is equal in magnitude to the rate of change $\dot{\theta}$ of the angular coordinate, and its sense may be obtained by the right-hand rule (Sec. 3.6) from the sense of rotation of the body.†

We shall now determine the acceleration $\mathbf{a}$ of the particle P. Differentiating (15.5) and recalling the rule for the differentiation of a vector product (Sec. 11.10), we write

$$\begin{aligned} \mathbf{a} &= \frac{d\mathbf{v}}{dt} = \frac{d}{dt}(\boldsymbol{\omega} \times \mathbf{r}) \\ &= \frac{d\boldsymbol{\omega}}{dt} \times \mathbf{r} + \boldsymbol{\omega} \times \frac{d\mathbf{r}}{dt} \\ &= \frac{d\boldsymbol{\omega}}{dt} \times \mathbf{r} + \boldsymbol{\omega} \times \mathbf{v} \end{aligned} \tag{15.7}$$

The vector $d\boldsymbol{\omega}/dt$ is denoted by $\boldsymbol{\alpha}$ and called the *angular acceleration* of the body. Substituting also for $\mathbf{v}$ from (15.5), we have

$$\mathbf{a} = \boldsymbol{\alpha} \times \mathbf{r} + \boldsymbol{\omega} \times (\boldsymbol{\omega} \times \mathbf{r}) \tag{15.8}$$

Differentiating (15.6), and recalling that $\mathbf{k}$ is constant in magnitude and direction, we have

$$\boldsymbol{\alpha} = \alpha\mathbf{k} = \dot{\omega}\mathbf{k} = \ddot{\theta}\mathbf{k} \tag{15.9}$$

† It will be shown in Sec. 15.12 in the more general case of a rigid body rotating simultaneously about axes having different directions that angular velocities obey the parallelogram law of addition and, thus, are actually vector quantities.

Thus, the angular acceleration of a body rotating about a fixed axis is a vector directed along the axis of rotation, and equal in magnitude to the rate of change $\dot{\omega}$ of the angular velocity. Returning to (15.8), we note that the acceleration of P is the sum of two vectors. The first vector is equal to the vector product $\boldsymbol{\alpha} \times \mathbf{r}$; it is tangent to the circle described by P and represents, therefore, the tangential component of the acceleration. The second vector is equal to the *vector triple product* $\boldsymbol{\omega} \times (\boldsymbol{\omega} \times \mathbf{r})$ obtained by forming the vector product of $\boldsymbol{\omega}$ and $\boldsymbol{\omega} \times \mathbf{r}$; since $\boldsymbol{\omega} \times \mathbf{r}$ is tangent to the circle described by P, the vector triple product is directed toward the center B of the circle and represents, therefore, the normal component of the acceleration.

Rotation of a Representative Slab. The rotation of a rigid body about a fixed axis may be defined by the motion of a representative slab in a reference plane perpendicular to the axis of rotation. Let us choose the xy plane as the reference plane and assume that it coincides with the plane of the figure, with the z axis pointing out of the paper (Fig. 15.10). Recalling from (15.6) that $\boldsymbol{\omega} = \omega\mathbf{k}$, we note that a positive value of the scalar ω corresponds to a counterclockwise rotation of the representative slab, and a negative value to a clockwise rotation. Substituting $\omega\mathbf{k}$ for $\boldsymbol{\omega}$ into Eq. (15.5), we express the velocity of any given point P of the slab as

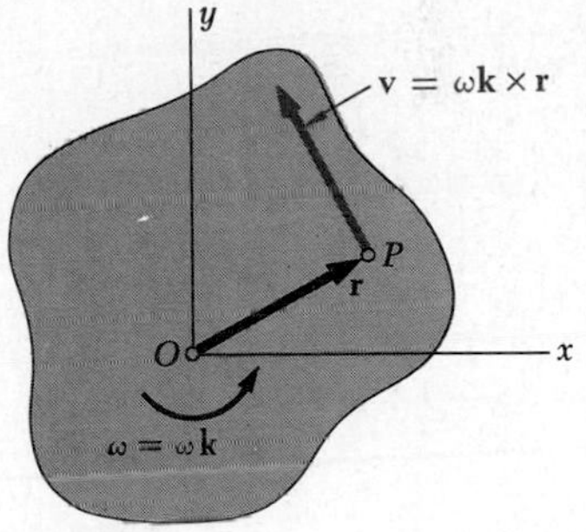

Fig. 15.10

$$\mathbf{v} = \omega\mathbf{k} \times \mathbf{r} \tag{15.10}$$

Since the vectors $\mathbf{k}$ and $\mathbf{r}$ are mutually perpendicular, the magnitude of the velocity $\mathbf{v}$ is

$$v = r\omega \tag{15.10$'$}$$

and its direction may be obtained by rotating $\mathbf{r}$ through 90° in the sense of rotation of the slab.

Substituting $\boldsymbol{\omega} = \omega\mathbf{k}$ and $\boldsymbol{\alpha} = \alpha\mathbf{k}$ into Eq. (15.8), and observing that cross-multiplying $\mathbf{r}$ twice by $\mathbf{k}$ results in a 180° rotation of the vector $\mathbf{r}$, we express the acceleration of point P as

$$\mathbf{a} = \alpha\mathbf{k} \times \mathbf{r} - \omega^2\mathbf{r} \tag{15.11}$$

Resolving $\mathbf{a}$ into tangential and normal components (Fig. 15.11), we write

$$\begin{aligned} \mathbf{a}_t &= \alpha\mathbf{k} \times \mathbf{r} \qquad & a_t &= r\alpha \\ \mathbf{a}_n &= -\omega^2\mathbf{r} \qquad & a_n &= r\omega^2 \end{aligned} \tag{15.11$'$}$$

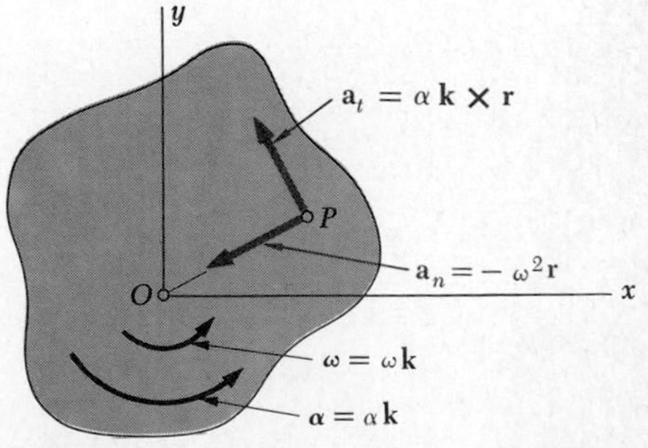

Fig. 15.11

The tangential component $\mathbf{a}_t$ points in the counterclockwise direction if the scalar α is positive, and in the clockwise direction if α is negative. The normal component $\mathbf{a}_n$ always points in the direction opposite to that of $\mathbf{r}$, that is, toward O.

15.4. Equations Defining the Rotation of a Rigid Body about a Fixed Axis. The motion of a rigid body rotating about a fixed axis AA' is said to be *known* when its angular coordinate θ may be expressed as a known function of t. In practice, however, the rotation of a rigid body is seldom defined by a relation between θ and t. More often, the conditions of motion will be specified by the type of angular acceleration that the body possesses. For example, α may be given as a function of t, or as a function of θ, or as a function of ω. Recalling the relations (15.6) and (15.9), we write

$$\omega = \frac{d\theta}{dt} \tag{15.12}$$

$$\alpha = \frac{d\omega}{dt} = \frac{d^2\theta}{dt^2} \tag{15.13}$$

or, solving (15.12) for dt and substituting into (15.13),

$$\alpha = \omega\frac{d\omega}{d\theta} \tag{15.14}$$

Since these equations are similar to those obtained in Chap. 11 for the rectilinear motion of a particle, their integration may be performed by following the procedure outlined in Sec. 11.3.

Two particular cases of rotation are frequently encountered:

1. *Uniform Rotation.* This case is characterized by the fact that the angular acceleration is zero. The angular velocity is thus constant, and the angular coordinate is given by the formula

$$\theta = \theta_0 + \omega t \tag{15.15}$$

2. *Uniformly Accelerated Rotation.* In this case, the angular acceleration is constant. The following formulas relating angular velocity, angular coordinate, and time may then be derived in a manner similar to that described in Sec. 11.5. The similitude between the formulas derived here and those obtained for the rectilinear uniformly accelerated motion of a particle is easily noted.

$$\begin{aligned} \omega &= \omega_0 + \alpha t \\ \theta &= \theta_0 + \omega_0 t + \tfrac{1}{2}\alpha t^2 \\ \omega^2 &= \omega_0^2 + 2\alpha(\theta - \theta_0) \end{aligned} \tag{15.16}$$

It should be emphasized that formula (15.15) may be used only when $\alpha = 0$, and formulas (15.16) only when $\alpha =$ constant. In any other case, the general formulas (15.12) to (15.14) should be used.

SAMPLE PROBLEM 15.1

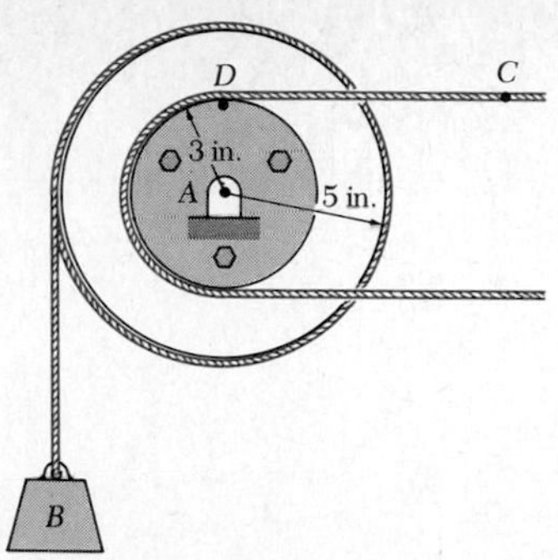

Load B is connected to a double pulley by one of the two inextensible cables shown. The motion of the pulley is controlled by cable C, which has a constant acceleration of 9 in./s^2 and an initial velocity of 12 in./s, both directed to the right. Determine (a) the number of revolutions executed by the pulley in 2 s, (b) the velocity and change in position of the load B after 2 s, and (c) the acceleration of point D on the rim of the inner pulley at $t = 0$.

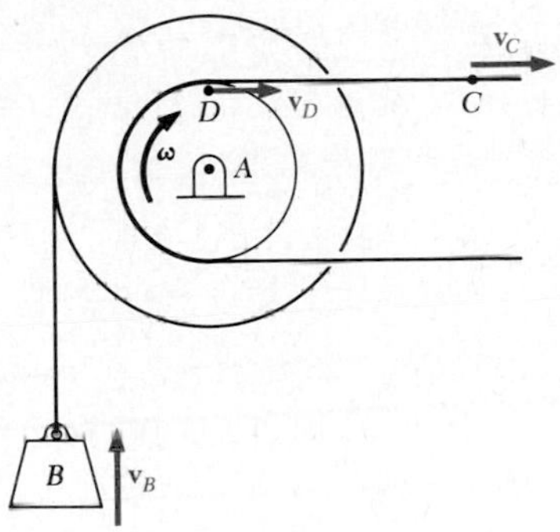

***a.* Motion of Pulley.** Since the cable is inextensible, the velocity of point D is equal to the velocity of point C and the tangential component of the acceleration of D is equal to the acceleration of C.

$$(\mathbf{v}_D)_0 = (\mathbf{v}_C)_0 = 12 \text{ in./s} \rightarrow \qquad (\mathbf{a}_D)_t = \mathbf{a}_C = 9 \text{ in./s}^2 \rightarrow$$

Noting that the distance from D to the center of the pulley is 3 in., we write

$$(v_D)_0 = r\omega_0 \qquad 12 \text{ in./s} = (3 \text{ in.})\omega_0 \qquad \boldsymbol{\omega}_0 = 4 \text{ rad/s} \curvearrowright$$

$$(a_D)_t = r\alpha \qquad 9 \text{ in./s}^2 = (3 \text{ in.})\alpha \qquad \boldsymbol{\alpha} = 3 \text{ rad/s}^2 \curvearrowright$$

Using the equations of uniformly accelerated motion, we obtain, for $t = 2$ s,

$$\omega = \omega_0 + \alpha t = 4 \text{ rad/s} + (3 \text{ rad/s}^2)(2 \text{ s}) = 10 \text{ rad/s}$$
$$\boldsymbol{\omega} = 10 \text{ rad/s} \curvearrowright$$
$$\theta = \omega_0 t + \tfrac{1}{2}\alpha t^2 = (4 \text{ rad/s})(2 \text{ s}) + \tfrac{1}{2}(3 \text{ rad/s}^2)(2 \text{ s})^2 = 14 \text{ rad}$$
$$\theta = 14 \text{ rad} \curvearrowright$$

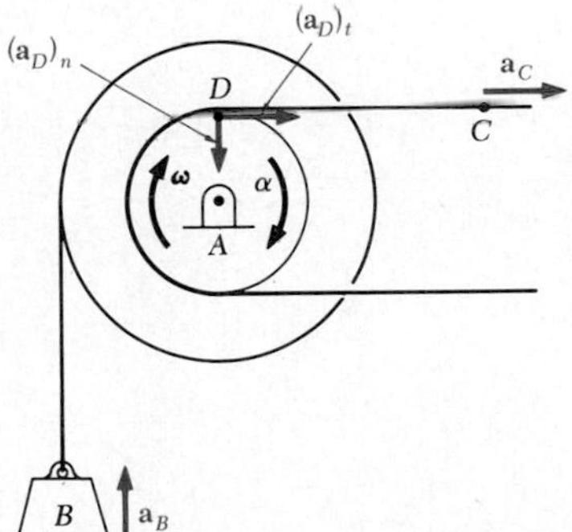

$$\text{Number of revolutions} = (14 \text{ rad})\left(\frac{1 \text{ rev}}{2\pi \text{ rad}}\right) = 2.23 \text{ rev} \quad ◀$$

***b.* Motion of Load *B*.** Using the following relations between linear and angular motion, with $r = 5$ in., we write

$$v_B = r\omega = (5 \text{ in.})(10 \text{ rad/s}) = 50 \text{ in./s} \qquad \mathbf{v}_B = 50 \text{ in./s} \uparrow \quad ◀$$
$$\Delta y_B = r\theta = (5 \text{ in.})(14 \text{ rad}) = 70 \text{ in.} \qquad \Delta y_B = 70 \text{ in. upward} \quad ◀$$

***c.* Acceleration of Point *D* at $t = 0$.** The tangential component of the acceleration is

$$(\mathbf{a}_D)_t = \mathbf{a}_C = 9 \text{ in./s}^2 \rightarrow$$

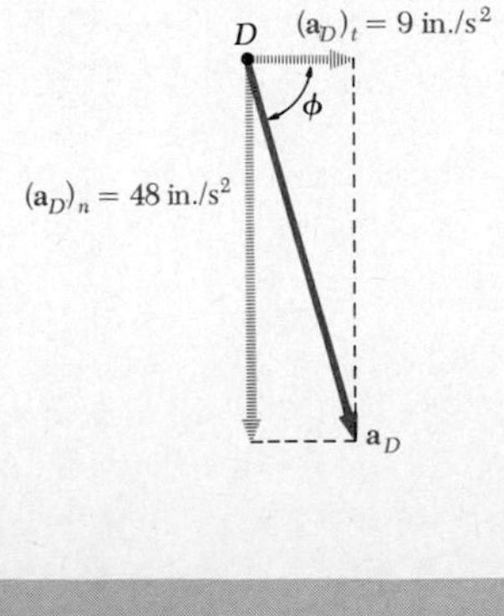

Since, at $t = 0$, $\omega_0 = 4$ rad/s, the normal component of the acceleration is

$$(a_D)_n = r_D\omega_0^2 = (3 \text{ in.})(4 \text{ rad/s})^2 = 48 \text{ in./s}^2 \qquad (\mathbf{a}_D)_n = 48 \text{ in./s}^2 \downarrow$$

The magnitude and direction of the total acceleration may be obtained by writing

$$\tan\phi = (48 \text{ in./s}^2)/(9 \text{ in./s}^2) \qquad \phi = 79.4^\circ$$
$$a_D \sin 79.4^\circ = 48 \text{ in./s}^2 \qquad a_D = 48.8 \text{ in./s}^2$$
$$\mathbf{a}_D = 48.8 \text{ in./s}^2 \measuredangle 79.4^\circ \quad ◀$$

Problems

15.1 The motion of an oscillating crank is defined by the relation $\theta = \theta_0 \sin(2\pi t/T)$, where θ is expressed in radians and t in seconds. Knowing that $\theta_0 = 1.2$ rad and $T = 0.5$ s, determine the maximum angular velocity and the maximum angular acceleration of the crank.

15.2 The motion of a disk rotating in an oil bath is defined by the relation $\theta = \theta_0(1 - e^{-t/4})$, where θ is expressed in radians and t in seconds. Knowing that $\theta_0 = 0.80$ rad, determine the angular coordinate, velocity, and acceleration of the disk when (*a*) $t = 0$, (*b*) $t = 4$ s, (*c*) $t = \infty$.

15.3 The rotor of an electric motor has a speed of 1800 rpm when the power is cut off. The rotor is then observed to come to rest after executing 625 revolutions. Assuming uniformly accelerated motion, determine (*a*) the angular acceleration, (*b*) the time required for the rotor to come to rest.

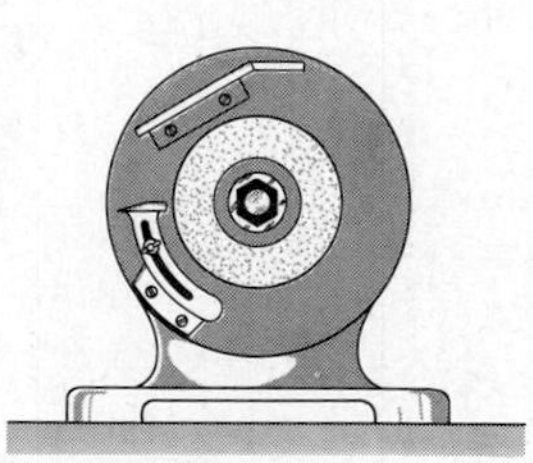

Fig. P15.4

15.4 A small grinding wheel is attached to the shaft of an electric motor which has a rated speed of 1800 rpm. When the power is turned on, the unit reaches its rated speed in 5 s, and when the power is turned off, the unit coasts to rest in 90 s. Assuming uniformly accelerated motion, determine the number of revolutions that the motor executes (*a*) in reaching its rated speed, (*b*) in coasting to rest.

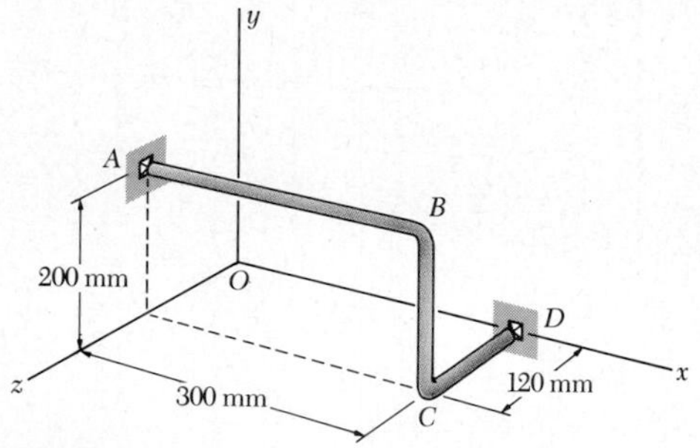

Fig. P15.5

15.5 The bent rod *ABCD* rotates about a line joining points *A* and *D* with a constant angular velocity of 95 rad/s. Knowing that at the instant considered the velocity of corner *C* is downward, determine the velocity and acceleration of corner *B*.

15.6 In Prob. 15.5, determine the velocity and acceleration of corner *B*, assuming that the angular velocity is 95 rad/s and decreases at the rate of 380 rad/s^2.

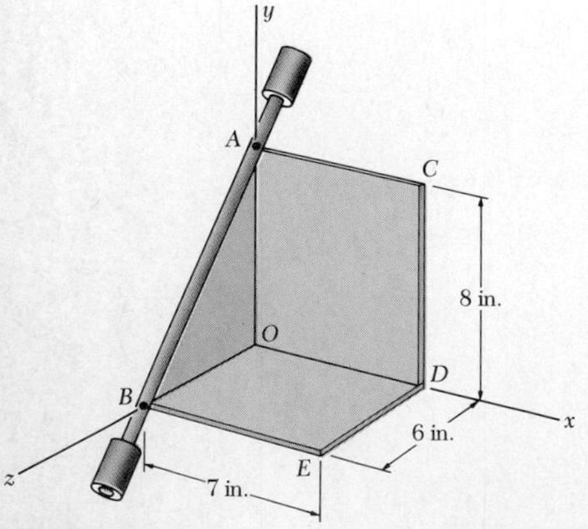

Fig. P15.7

15.7 A triangular plate and two rectangular plates are welded to each other and to the straight rod *AB*. The entire welded unit rotates about the axis *AB* with a constant angular velocity of 5 rad/s. Knowing that at the instant considered the velocity of corner *E* is downward, determine the velocity and acceleration of corner *D*.

15.8 The assembly shown consists of the straight rod ABC which passes through and is welded to the rectangular plate $DEFH$. The assembly rotates about the axis AC with a constant angular velocity of 9 rad/s. Knowing that the motion when viewed from C is counterclockwise, determine the velocity and acceleration of corner F.

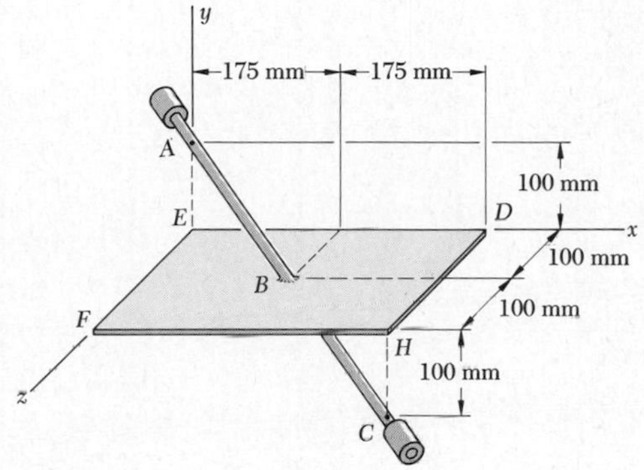

Fig. P15.8

15.9 In Prob. 15.8, assuming that the angular velocity is 9 rad/s and decreases at the rate of 13.5 rad/s^2, determine the velocity and acceleration of corner H.

15.10 In Prob. 15.7, determine the velocity and acceleration of corner D, assuming that the angular velocity is 5 rad/s and increases at the rate of 20 rad/s^2.

15.11 The earth makes one complete revolution about the sun in 365.24 days. Assuming that the orbit of the earth is circular and has a radius of 93,000,000 mi, determine the velocity and acceleration of the earth.

15.12 The earth makes one complete revolution on its axis in 23.93 h. Knowing that the mean radius of the earth is 3960 mi, determine the linear velocity and acceleration of a point on the surface of the earth (*a*) at the equator, (*b*) at Philadelphia, latitude 40° north, (*c*) at the North Pole.

15.13 A series of small machine components being moved by a conveyor belt pass over the 180-mm-radius idler pulley shown. At the instant shown, the velocity of point A is 450 mm/s to the left and its acceleration is 315 mm/s^2 to the right. Determine (*a*) the angular velocity and angular acceleration of the idler pulley, (*b*) the total acceleration of the machine component at B.

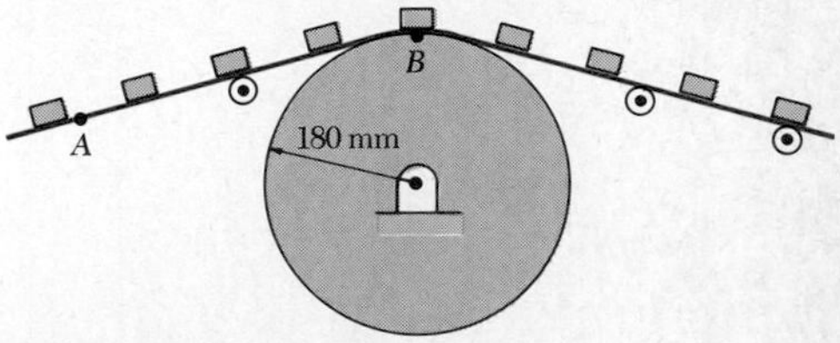

Fig. P15.13 and P15.14

15.14 A series of small machine components being moved by a conveyor belt pass over the 180-mm-radius idler pulley shown. At the instant shown, the angular velocity of the idler pulley is 3 rad/s clockwise. Determine the angular acceleration of the pulley for which the total acceleration of the machine component at B is 2 m/s^2.

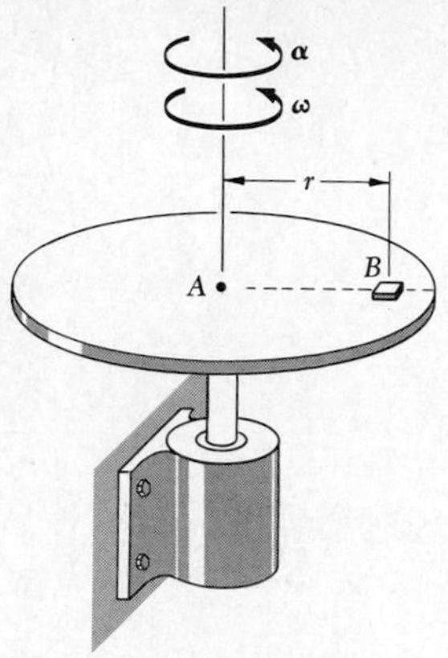

Fig. P15.15 and P15.16

15.15 It is known that the static-friction force between the small block B and the plate will be exceeded and that the block will start sliding on the plate when the total acceleration of the block reaches 3 m/s^2. If the plate starts from rest at $t = 0$ and is accelerated at the constant rate of 4 rad/s^2, determine the time t and the angular velocity of the plate when the block starts sliding, assuming $r = 200$ mm.

15.16 A small block B rests on a horizontal plate which rotates about a fixed axis. The plate starts from rest at $t = 0$ and is accelerated at the constant rate of 0.5 rad/s^2. Knowing that $r = 200$ mm, determine the magnitude of the total acceleration of the block when (a) $t = 0$, (b) $t = 1$ s, (c) $t = 2$ s.

15.17 The belt sander shown is initially at rest. If the driving drum A has a constant angular acceleration of 120 rad/s^2 counterclockwise, determine the acceleration of the belt at point C and at the work station D, two seconds after the sander has been turned on.

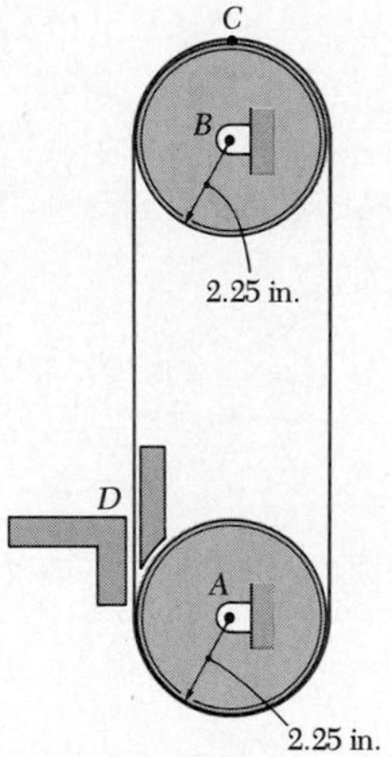

Fig. P15.17 and P15.18

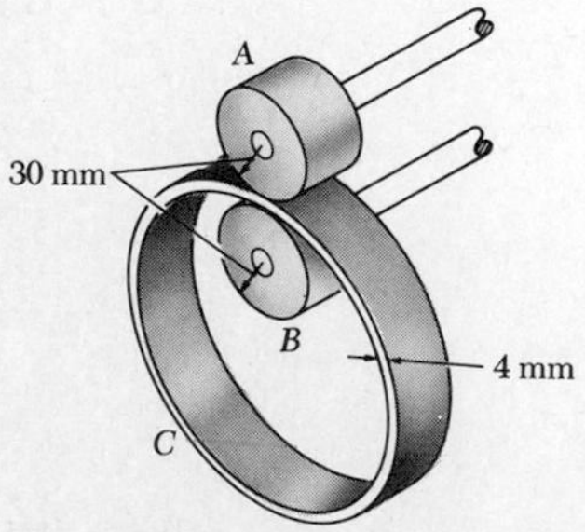

Fig. P15.19

15.18 The rated speed of drum A of the belt sander shown is 3450 rpm. When the power is turned off it is observed that the sander coasts from its rated speed to rest in 5 s. Assuming uniformly decelerated motion, determine the velocity and acceleration of point C of the belt (a) immediately before the power is turned off, (b) 4.5 s later.

15.19 Ring C has an inside radius of 72 mm and an outside radius of 76 mm and is positioned between two wheels A and B, each of 30-mm outside radius. Knowing that wheel A rotates with a constant angular velocity of 400 rpm and that no slipping occurs, determine (a) the angular velocity of ring C and of wheel B, (b) the acceleration of the points of A and B which are in contact with C.

15.20 A computer tape moves over two drums. During a 3-s interval the speed of the tape is increased uniformly from $v_0 = 2$ ft/s to $v_1 = 5$ ft/s. Knowing that the tape does not slip on the drums, determine (*a*) the angular acceleration of drum A, (*b*) the number of revolutions executed by drum A during the 3-s interval.

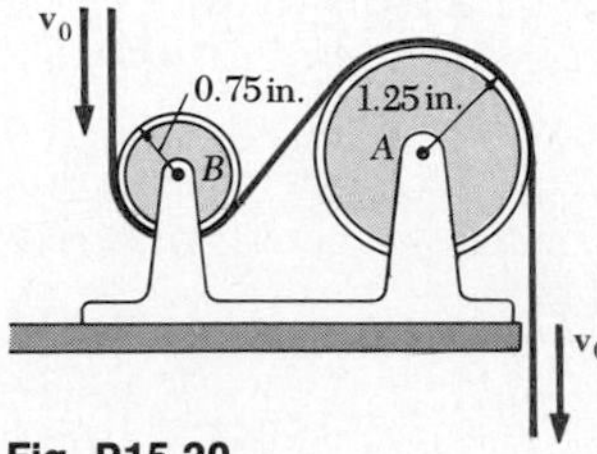

Fig. P15.20

15.21 Solve Prob. 15.20, considering drum *B* instead of drum *A*.

15.22 A mixing drum of 125-mm outside radius rests on two casters, each of 25-mm radius. The drum executes 12 rev during the time interval *t*, while its angular velocity is being increased uniformly from 25 to 45 rpm. Knowing that no slipping occurs between the drum and the casters, determine (*a*) the angular acceleration of the casters, (*b*) the time interval *t*.

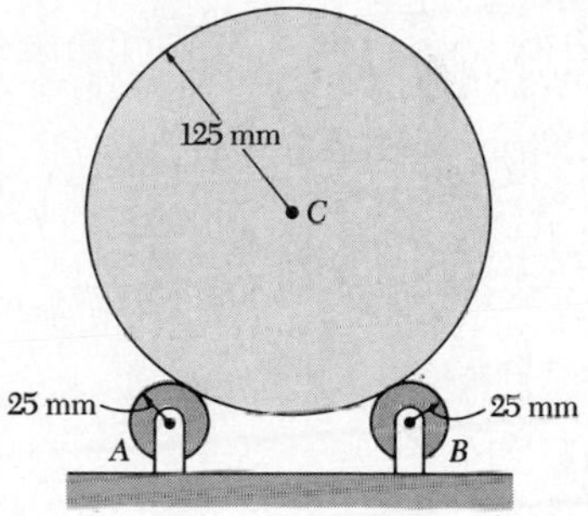

Fig. P15.22

15.23 A pulley and two loads are connected by inextensible cords as shown. Load *A* has a constant acceleration of 2.5 m/s^2 and an initial velocity of 3.75 m/s, both directed upward. Determine (*a*) the number of revolutions executed by the pulley in 3 s, (*b*) the velocity and position of load *B* after 3 s, (*c*) the acceleration of point *C* on the rim of the pulley at $t = 0$.

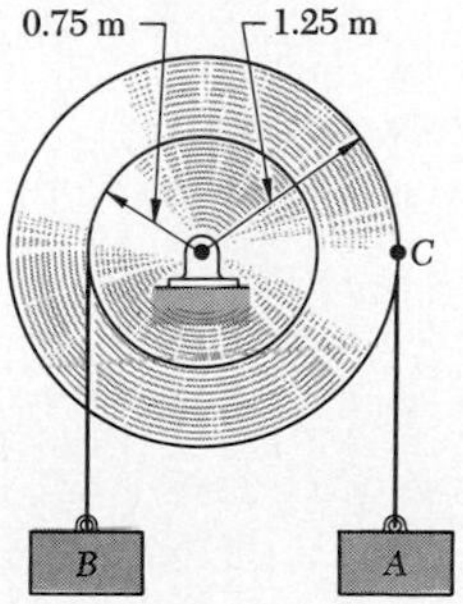

Fig. P15.23

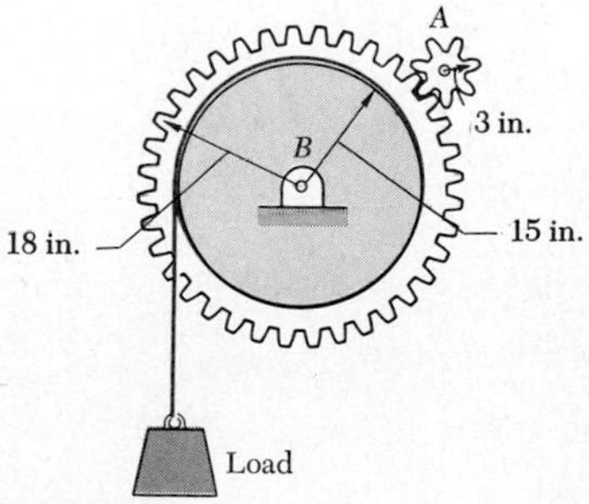

Fig. P15.24

15.24 A load is to be raised 20 ft by the hoisting system shown. Assuming gear *A* is initially at rest, accelerates uniformly to a speed of 120 rpm in 5 s, and then maintains a constant speed of 120 rpm, determine (*a*) the number of revolutions executed by gear *A* in raising the load, (*b*) the time required to raise the load.

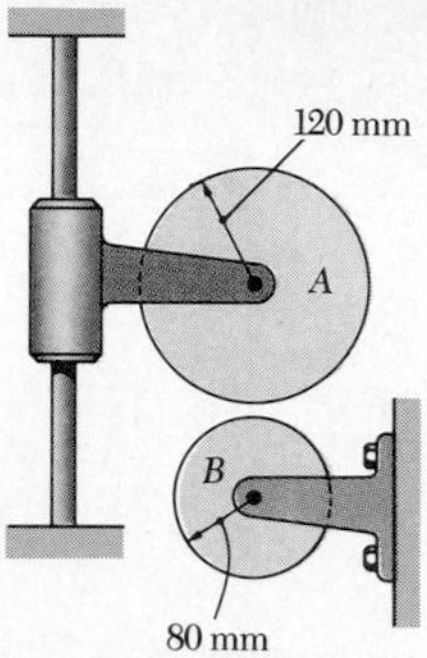

Fig. P15.25 and P15.26

15.25 Disk A is at rest when it is brought into contact with disk B, which is rotating freely at 450 rpm clockwise. After 6 s of slippage, during which each disk has a constant angular acceleration, disk B reaches a final angular velocity of 140 rpm clockwise. Determine the angular acceleration of each disk during the period of slippage.

15.26 and 15.27 A simple friction drive consists of two disks A and B. Initially, disk B has a clockwise angular velocity of 500 rpm, and disk A is at rest. It is known that disk B will coast to rest in 60 s. However, rather than waiting until both disks are at rest to bring them together, disk A is given a constant angular acceleration of 3 rad/s^2 counterclockwise. Determine (*a*) at what time the disks may be brought together if they are not to slip, (*b*) the angular velocity of each disk as contact is made.

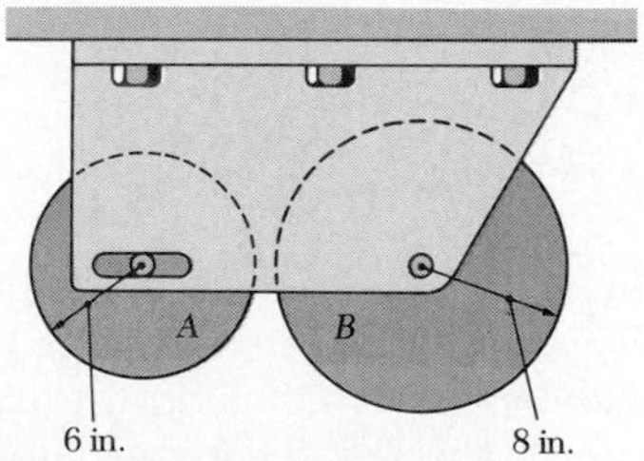

Fig. P15.27 and P15.28

15.28 Two friction wheels A and B are both rotating freely at 240 rpm counterclockwise when they are brought into contact. After 8 s of slippage, during which each wheel has a constant angular acceleration, wheel B reaches a final angular velocity of 60 rpm counterclockwise. Determine (*a*) the angular acceleration of each wheel during the period of slippage, (*b*) the time at which the angular velocity of wheel A is equal to zero.

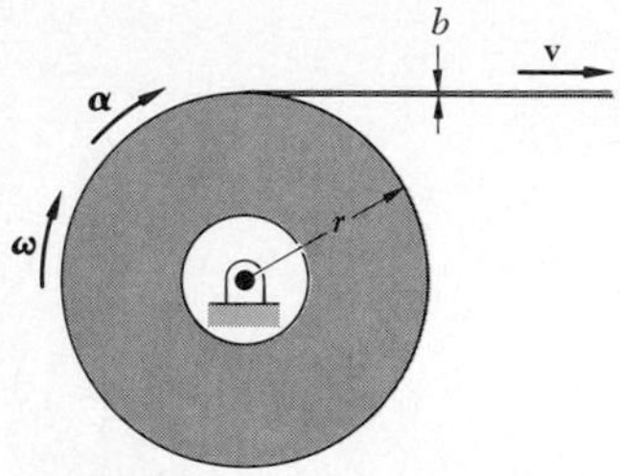

Fig. P15.29

* **15.29** In a continuous printing process, paper is drawn into the presses at a constant speed v. Denoting by r the radius of paper on the roll at any given time and by b the thickness of the paper, derive an expression for the angular acceleration of the paper roll.

15.5. General Plane Motion. As indicated in Sec. 15.1, we understand by general plane motion a plane motion which is neither a translation nor a rotation. As we shall presently see, however, *a general plane motion may always be considered as the sum of a translation and a rotation.*

Consider, for example, a wheel rolling on a straight track (Fig. 15.12). Over a certain interval of time, two given points A and B will have moved, respectively, from A_1 to A_2 and from B_1 to B_2. The same result could be obtained through a translation which would bring A and B into A_2 and B_1' (the line AB remaining vertical), followed by a rotation about A bringing B into B_2. Although the original rolling motion differs from the combination of translation and rotation when these motions are taken in succession, the original motion may be completely duplicated by a combination of simultaneous translation and rotation.

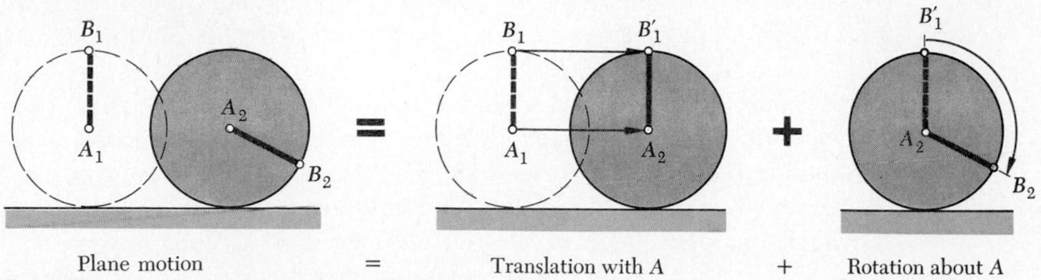

Fig. 15.12

Another example of plane motion is given in Fig. 15.13, which represents a rod whose extremities slide, respectively, along a horizontal and a vertical track. This motion may be replaced by a translation in a horizontal direction and a rotation about A (Fig. 15.13*a*) or by a translation in a vertical direction and a rotation about B (Fig. 15.13*b*).

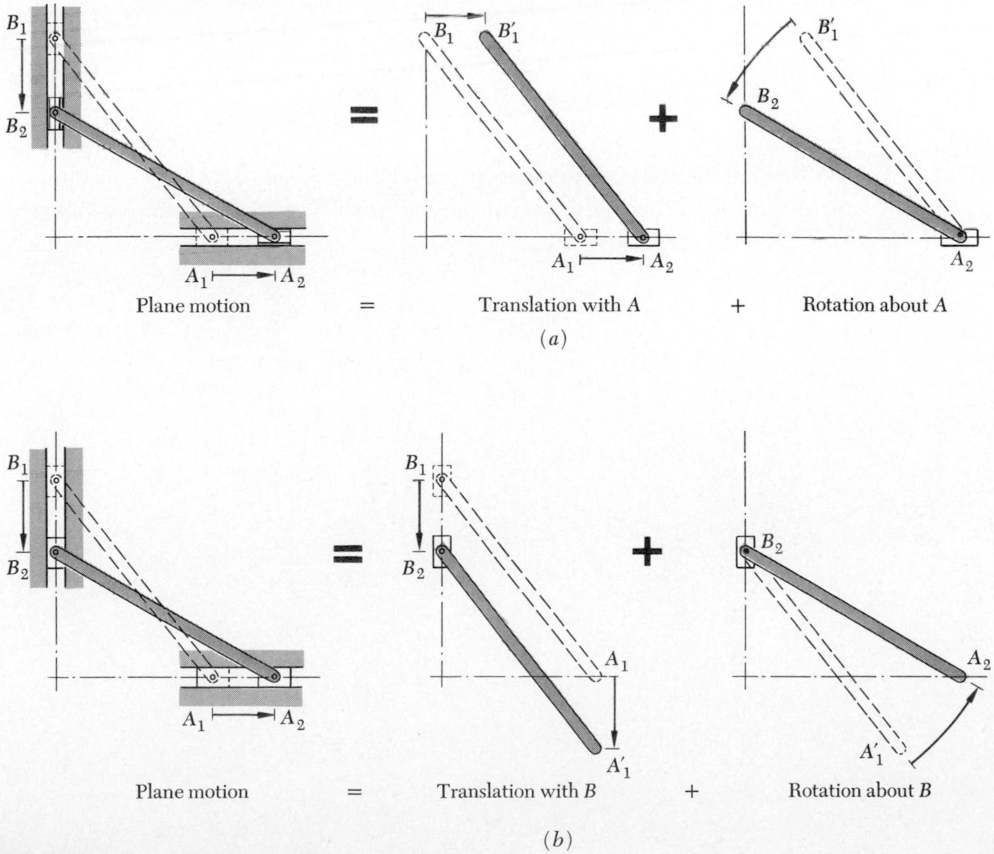

Fig. 15.13

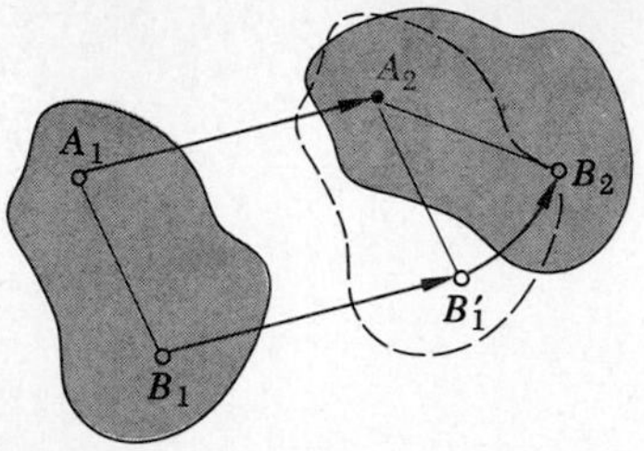

Fig. 15.14

In general, we shall consider a small displacement which brings two particles A and B of a representative slab, respectively, from A_1 and B_1 into A_2 and B_2 (Fig. 15.14). This displacement may be divided into two parts, one in which the particles move into A_2 and B_1' while the line AB maintains the same direction, the other in which B moves into B_2 while A remains fixed. The first part of the motion is clearly a translation and the second part a rotation about A.

Recalling from Sec. 11.12 the definition of the "relative motion" of a particle with respect to a moving frame of reference—as opposed to its "absolute motion" with respect to a fixed frame of reference—we may restate as follows the result obtained above: Given two particles A and B of a rigid slab in plane motion, the relative motion of B with respect to a frame attached to A and of fixed orientation is a rotation. To an observer moving with A but not rotating, particle B will appear to describe an arc of circle centered at A.

15.6. Absolute and Relative Velocity in Plane Motion. We saw in the preceding section that any plane motion of a slab may be replaced by a translation defined by the motion of an arbitrary reference point A, and a simultaneous rotation about A. The absolute velocity $\mathbf{v}_B$ of a particle B of the slab is obtained from the relative-velocity formula derived in Sec. 11.12,

$$\mathbf{v}_B = \mathbf{v}_A + \mathbf{v}_{B/A} \tag{15.17}$$

where the right-hand member represents a vector sum. The velocity $\mathbf{v}_A$ corresponds to the translation of the slab with A, while the relative velocity

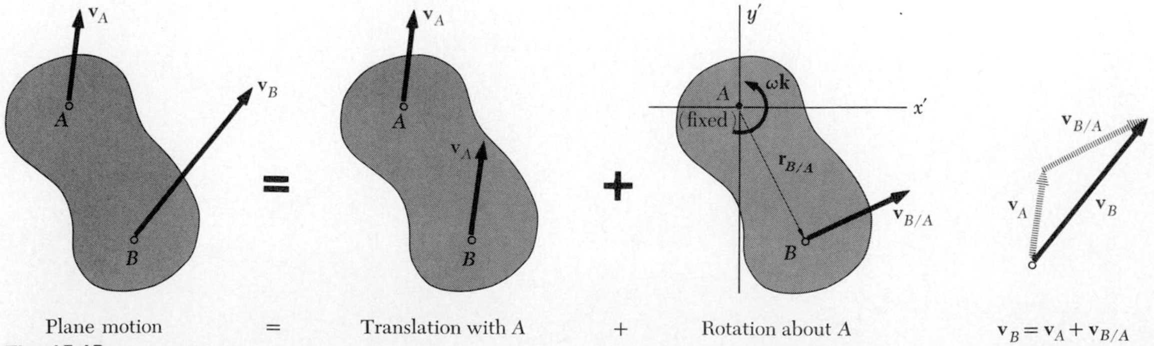

Fig. 15.15

$\mathbf{v}_{B/A}$ is associated with the rotation of the slab about A and is measured with respect to axes centered at A and of fixed orientation (Fig. 15.15). Denoting by $\mathbf{r}_{B/A}$ the position vector of B relative to A, and by $\omega\mathbf{k}$ the angular velocity of the slab with respect to axes of fixed orientation, we have from (15.10) and (15.10′)

$$\mathbf{v}_{B/A} = \omega\mathbf{k} \times \mathbf{r}_{B/A} \qquad v_{B/A} = r\omega \tag{15.18}$$

where r is the distance from A to B. Substituting for $\mathbf{v}_{B/A}$ from (15.18) into (15.17), we may also write

$$\mathbf{v}_B = \mathbf{v}_A + \omega\mathbf{k} \times \mathbf{r}_{B/A} \tag{15.17′}$$

As an example, we shall consider again the rod AB of Fig. 15.13. Assuming that the velocity $\mathbf{v}_A$ of end A is known, we propose to find the velocity $\mathbf{v}_B$ of end B and the angular velocity $\boldsymbol{\omega}$ of the rod, in terms of the velocity $\mathbf{v}_A$, the length l, and the angle θ. Choosing A as a reference point, we express that the given motion is equivalent to a translation with A and a simultaneous rotation about A (Fig. 15.16). The absolute velocity of B must therefore be equal to the vector sum

$$\mathbf{v}_B = \mathbf{v}_A + \mathbf{v}_{B/A} \tag{15.17}$$

We note that while the direction of $\mathbf{v}_{B/A}$ is known, its magnitude $l\omega$ is unknown. However, this is compensated for by the fact that the direction of $\mathbf{v}_B$ is known. We may therefore complete the diagram of Fig. 15.16. Solving for the magnitudes v_B and ω, we write

$$v_B = v_A \tan\theta \qquad \omega = \frac{v_{B/A}}{l} = \frac{v_A}{l\cos\theta} \tag{15.19}$$

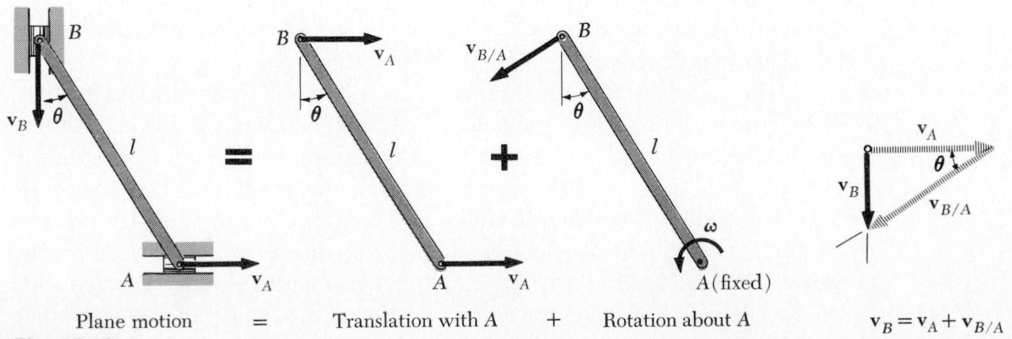

Fig. 15.16

The same result may be obtained by using B as a point of reference. Resolving the given motion into a translation with B and a simultaneous rotation about B (Fig. 15.17), we write the equation

$$\mathbf{v}_A = \mathbf{v}_B + \mathbf{v}_{A/B} \tag{15.20}$$

which is represented graphically in Fig. 15.17. We note that $\mathbf{v}_{A/B}$ and $\mathbf{v}_{B/A}$ have the same magnitude $l\omega$ but opposite sense. The sense of the relative velocity depends, therefore, upon the point of reference which has been selected and should be carefully ascertained from the appropriate diagram (Fig. 15.16 or 15.17).

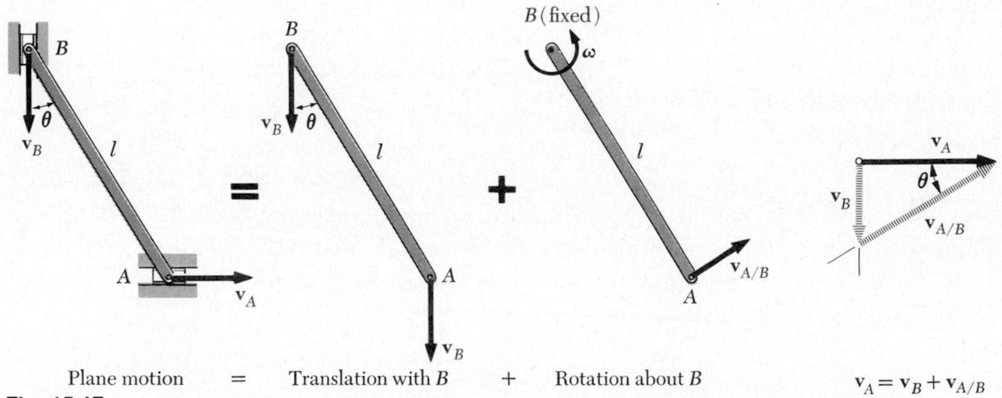

Fig. 15.17

Finally, we observe that the angular velocity ω of the rod in its rotation about B is the same as in its rotation about A. It is measured in both cases by the rate of change of the angle θ. This result is quite general; we should therefore bear in mind that *the angular velocity* ω *of a rigid body in plane motion is independent of the reference point.*

Most mechanisms consist not of one but of *several* moving parts. When the various parts of a mechanism are pin-connected, its analysis may be carried out by considering each part as a rigid body, while keeping in mind that the points where two parts are connected must have the same absolute velocity (see Sample Prob. 15.3). A similar analysis may be used when gears are involved, since the teeth in contact must also have the same absolute velocity. However, when a mechanism contains parts which slide on each other, the relative velocity of the part in contact must be taken into account (see Secs. 15.10 and 15.11).

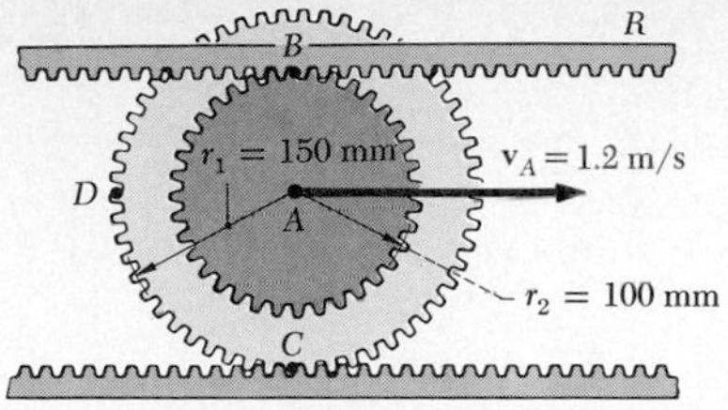

SAMPLE PROBLEM 15.2

The double gear shown rolls on the stationary lower rack; the velocity of its center A is 1.2 m/s directed to the right. Determine (a) the angular velocity of the gear, (b) the velocities of the upper rack R and of point D of the gear.

***a.* Angular Velocity of the Gear.** Since the gear rolls on the lower rack, its center A moves through a distance equal to the outer circumference $2\pi r_1$ for each full revolution of the gear. Noting that 1 rev $= 2\pi$ rad, and that when A moves to the right ($x_A > 0$) the gear rotates clockwise ($\theta < 0$), we write

$$\frac{x_A}{2\pi r_1} = -\frac{\theta}{2\pi} \qquad x_A = -r_1\theta$$

Differentiating with respect to the time t and substituting the known values $v_A = 1.2$ m/s and $r_1 = 150$ mm $= 0.150$ m, we obtain

$$v_A = -r_1\omega \qquad 1.2 \text{ m/s} = -(0.150 \text{ m})\omega \qquad \omega = -8 \text{ rad/s}$$

$$\boldsymbol{\omega} = \omega\mathbf{k} = -(8 \text{ rad/s})\mathbf{k} \quad ◀$$

where $\mathbf{k}$ is a unit vector pointing out of the paper.

***b.* Velocities.** The rolling motion is resolved into two component motions: a translation with the center A and a rotation about the center A. In the translation, all points of the gear move with the same velocity $\mathbf{v}_A$. In the rotation, each point P of the gear moves about A with a relative velocity $\mathbf{v}_{P/A} = \omega\mathbf{k} \times \mathbf{r}_{P/A}$, where $\mathbf{r}_{P/A}$ is the position vector of P relative to A.

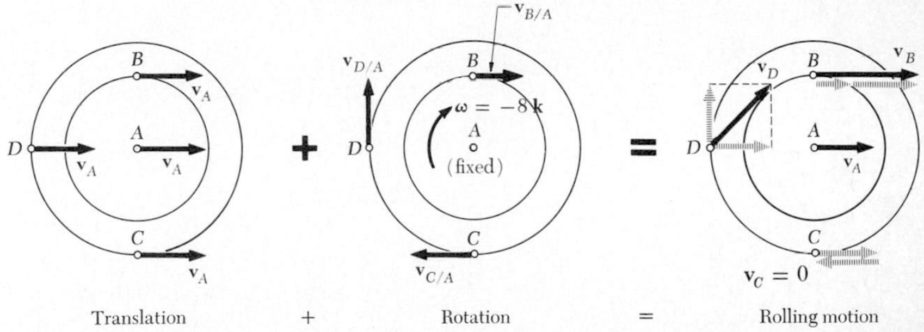

Velocity of Upper Rack. The velocity of the upper rack is equal to the velocity of point B; we write

$$\begin{aligned}\mathbf{v}_R = \mathbf{v}_B = \mathbf{v}_A + \mathbf{v}_{B/A} &= \mathbf{v}_A + \omega\mathbf{k} \times \mathbf{r}_{B/A} \\ &= (1.2 \text{ m/s})\mathbf{i} - (8 \text{ rad/s})\mathbf{k} \times (0.100 \text{ m})\mathbf{j} \\ &= (1.2 \text{ m/s})\mathbf{i} + (0.8 \text{ m/s})\mathbf{i} = (2 \text{ m/s})\mathbf{i}\end{aligned}$$

$$\mathbf{v}_R = 2 \text{ m/s} \rightarrow \quad ◀$$

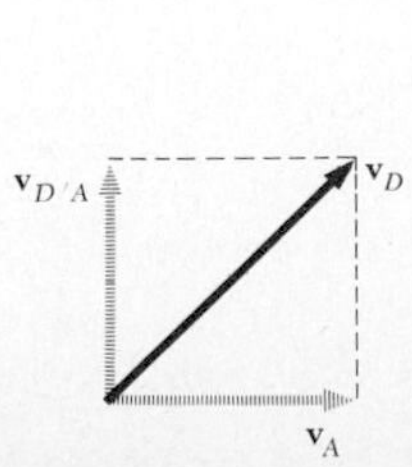

Velocity of Point D:

$$\begin{aligned}\mathbf{v}_D = \mathbf{v}_A + \mathbf{v}_{D/A} &= \mathbf{v}_A + \omega\mathbf{k} \times \mathbf{r}_{D/A} \\ &= (1.2 \text{ m/s})\mathbf{i} - (8 \text{ rad/s})\mathbf{k} \times (-0.150 \text{ m})\mathbf{i} \\ &= (1.2 \text{ m/s})\mathbf{i} + (1.2 \text{ m/s})\mathbf{j}\end{aligned}$$

$$\mathbf{v}_D = 1.697 \text{ m/s} \measuredangle 45° \quad ◀$$

SAMPLE PROBLEM 15.3

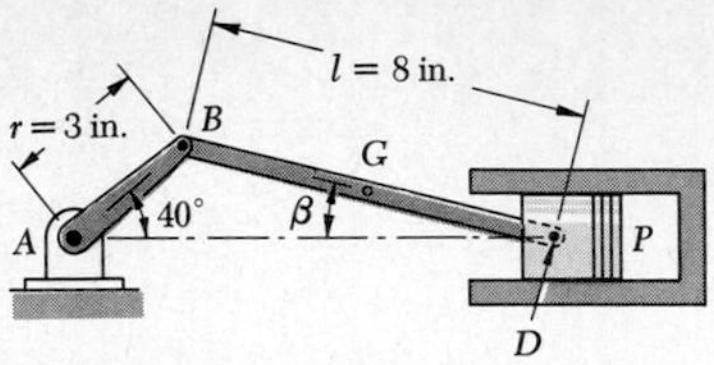

In the engine system shown, the crank AB has a constant clockwise angular velocity of 2000 rpm. For the crank position indicated, determine (*a*) the angular velocity of the connecting rod BD, (*b*) the velocity of the piston P.

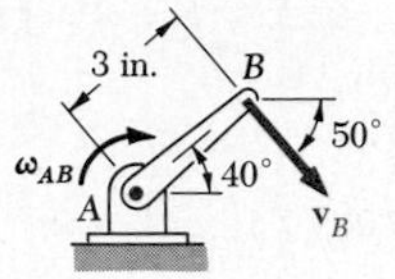

Motion of Crank *AB*. The crank AB rotates about point A. Expressing ω_{AB} in rad/s and writing $v_B = r\omega_{AB}$, we obtain

$$\omega_{AB} = \left(2000\,\frac{\text{rev}}{\text{min}}\right)\left(\frac{1\text{ min}}{60\text{ s}}\right)\left(\frac{2\pi\text{ rad}}{1\text{ rev}}\right) = 209\text{ rad/s}$$

$$v_B = (AB)\omega_{AB} = (3\text{ in.})(209\text{ rad/s}) = 627\text{ in./s}$$

$$\mathbf{v}_B = 627\text{ in./s} \searrow 50°$$

Motion of Connecting Rod *BD*. We consider this motion as a general plane motion. Using the law of sines, we compute the angle β between the connecting rod and the horizontal:

$$\frac{\sin 40°}{8\text{ in.}} = \frac{\sin \beta}{3\text{ in.}} \qquad \beta = 13.9°$$

The velocity $\mathbf{v}_D$ of the point D where the rod is attached to the piston must be horizontal, while the velocity of point B is equal to the velocity $\mathbf{v}_B$ obtained above. Resolving the motion of BD into a translation with B and a rotation about B, we obtain

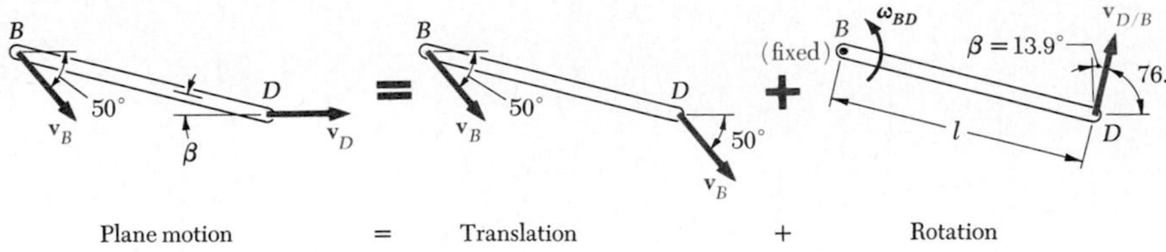

Expressing the relation between the velocities $\mathbf{v}_D$, $\mathbf{v}_B$, and $\mathbf{v}_{D/B}$, we write

$$\mathbf{v}_D = \mathbf{v}_B + \mathbf{v}_{D/B}$$

We draw the vector diagram corresponding to this equation. Recalling that $\beta = 13.9°$, we determine the angles of the triangle and write

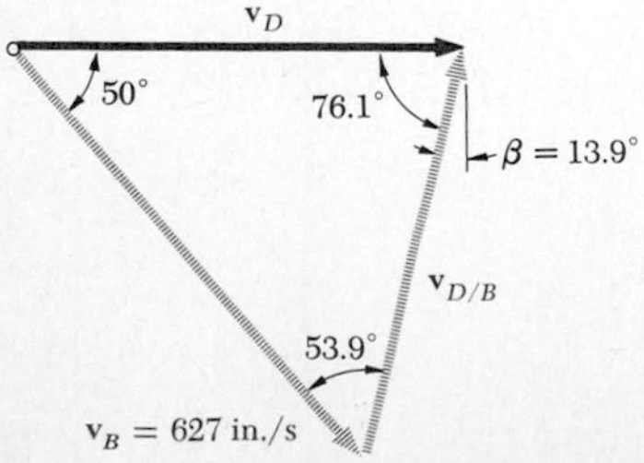

$$\frac{v_D}{\sin 53.9°} = \frac{v_{D/B}}{\sin 50°} = \frac{627\text{ in./s}}{\sin 76.1°}$$

$$v_{D/B} = 495\text{ in./s} \qquad \mathbf{v}_{D/B} = 495\text{ in./s} \measuredangle 76.1°$$

$$v_D = 522\text{ in./s} = 43.5\text{ ft/s} \qquad \mathbf{v}_D = 43.5\text{ ft/s} \rightarrow$$

$$\mathbf{v}_P = \mathbf{v}_D = 43.5\text{ ft/s} \rightarrow \quad ◄$$

Since $v_{D/B} = l\omega_{BD}$, we have

$$495\text{ in./s} = (8\text{ in.})\omega_{BD} \qquad \omega_{BD} = 61.9\text{ rad/s} ↺ \quad ◄$$

Problems

15.30 An automobile travels to the right at a constant speed of 45 mi/h. If the diameter of a wheel is 22 in., determine the velocities of points B, C, D, and E on the rim of the wheel.

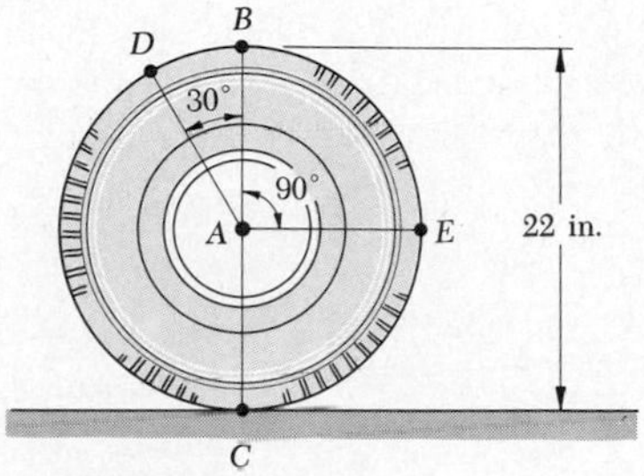

Fig. P15.30

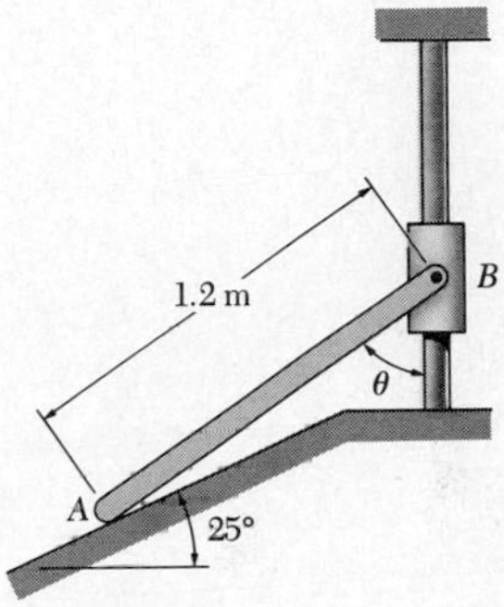

Fig. P15.31

15.31 Collar B moves upward with a constant velocity of 1.8 m/s. At the instant when $\theta = 50°$, determine (*a*) the angular velocity of rod AB, (*b*) the velocity of end A of the rod.

15.32 Solve Prob. 15.31, assuming that $\theta = 35°$.

15.33 Rod AB is 30 in. long and slides with its ends in contact with the floor and the inclined plane. End A moves with a constant velocity of 25 in./s to the right. At the instant when $\theta = 25°$, determine (*a*) the angular velocity of the rod, (*b*) the velocity of end B.

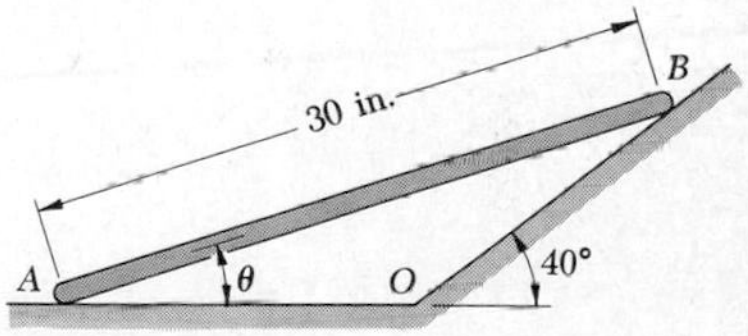

Fig. P15.33

15.34 The plate shown moves in the xy plane. Knowing that $(v_A)_x = 300$ mm/s, $(v_B)_y = -180$ mm/s, and $(v_C)_x = -150$ mm/s, determine (*a*) the angular velocity of the plate, (*b*) the velocity of point A.

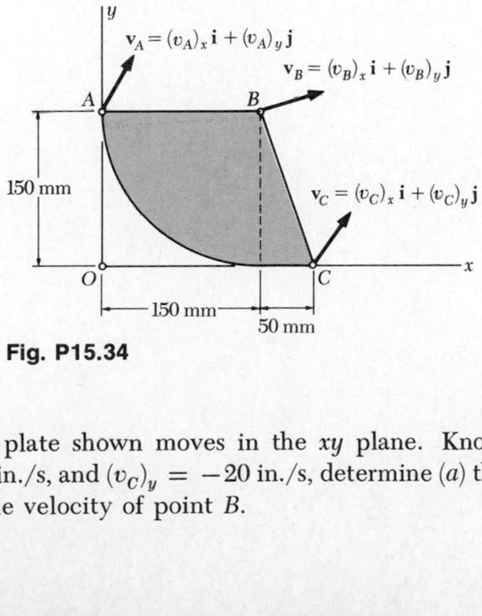

Fig. P15.34

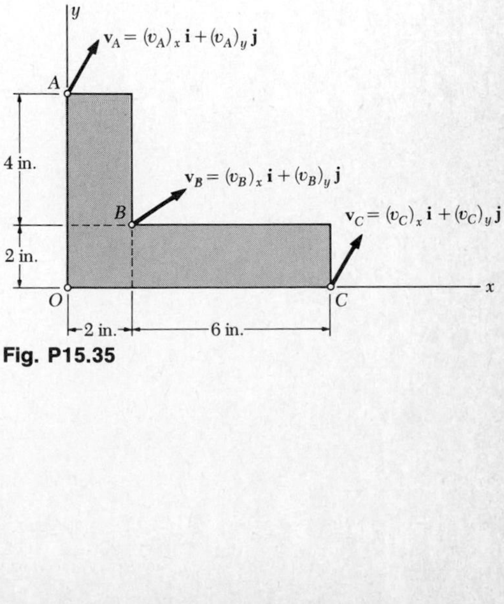

Fig. P15.35

15.35 The plate shown moves in the xy plane. Knowing that $(v_A)_x = 24$ in./s, $(v_B)_x = 8$ in./s, and $(v_C)_y = -20$ in./s, determine (*a*) the angular velocity of the plate, (*b*) the velocity of point B.

Fig. P15.38 and P15.39

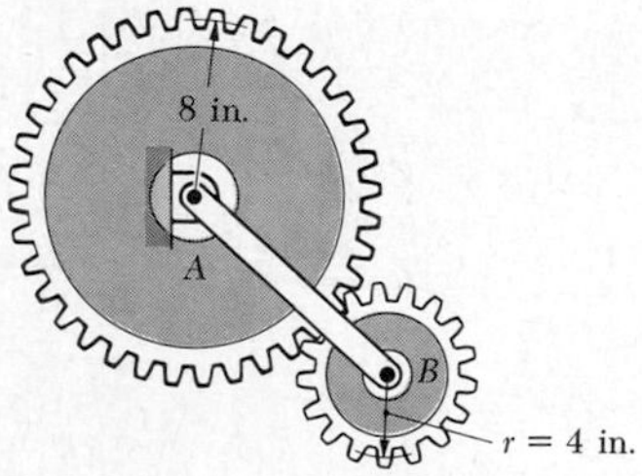

Fig. P15.40 and P15.41

15.36 In Prob. 15.35, determine (*a*) the velocity of point *A*, (*b*) the point of the plate with zero velocity.

15.37 In Prob. 15.34, determine (*a*) the velocity of point *B*, (*b*) the point of the plate with zero velocity.

15.38 In the planetary gear system shown, the radius of gears *A*, *B*, *C*, and *D* is *a* and the radius of the outer gear *E* is 3*a*. Knowing that the angular velocity of gear *A* is ω_A clockwise and that the outer gear *E* is stationary, determine (*a*) the angular velocity of each planetary gear, (*b*) the angular velocity of the spider connecting the planetary gears.

15.39 In the planetary gear system shown, the radius of gears *A*, *B*, *C*, and *D* is 45 mm and the radius of the outer gear *E* is 135 mm. Knowing that gear *E* has an angular velocity of 120 rpm clockwise and that the central gear *A* has an angular velocity of 150 rpm clockwise, determine (*a*) the angular velocity of each planetary gear, (*b*) the angular velocity of the spider connecting the planetary gears.

15.40 Gear *A* rotates clockwise with a constant angular velocity of 60 rpm. Determine the angular velocity of gear *B* if the angular velocity of arm *AB* is (*a*) 40 rpm counterclockwise, (*b*) 40 rpm clockwise.

15.41 Gear *A* rotates clockwise with a constant angular velocity of 150 rpm. Knowing that at the same time arm *AB* rotates clockwise with a constant angular velocity of 125 rpm, determine the angular velocity of gear *B*.

15.42 Three gears *A*, *B*, and *C* are pinned at their centers to rod *ABC*. Knowing that gear *A* does not rotate, determine the angular velocity of gears *B* and *C* when the rod *ABC* rotates clockwise with a constant angular velocity of 48 rpm.

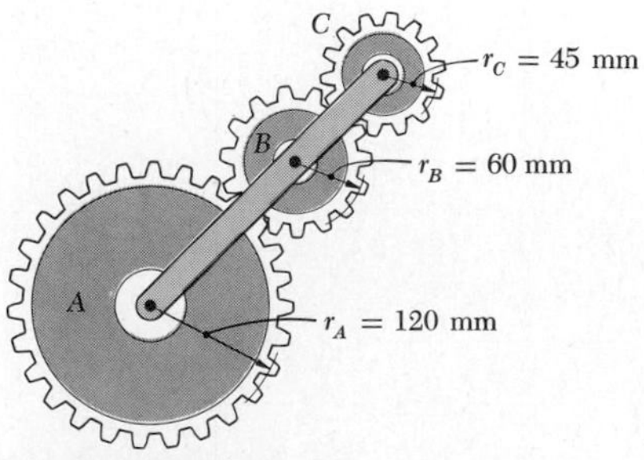

Fig. P15.42 and P15.43

15.43 Three gears *A*, *B*, and *C* are pinned at their centers to rod *ABC*. Knowing that gear *C* does not rotate, determine the angular velocity of gears *A* and *B* when the rod *ABC* rotates clockwise with an angular velocity of 48 rpm.

15.44 Crank AB has a constant angular velocity of 200 rpm counterclockwise. Determine the angular velocity of rod BD and the velocity of collar D when (a) $\theta = 0$, (b) $\theta = 90°$, (c) $\theta = 180°$.

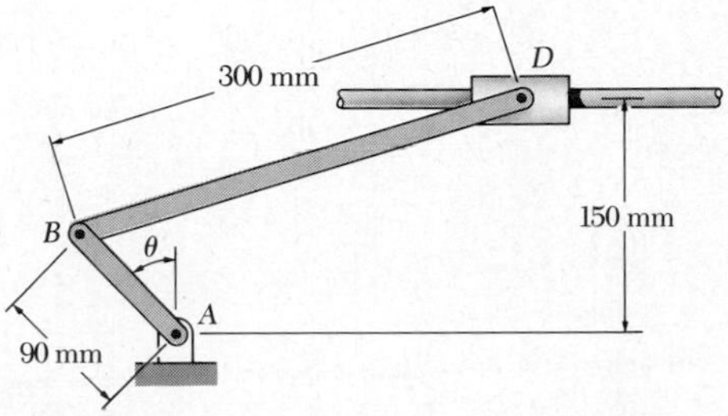

Fig. P15.44

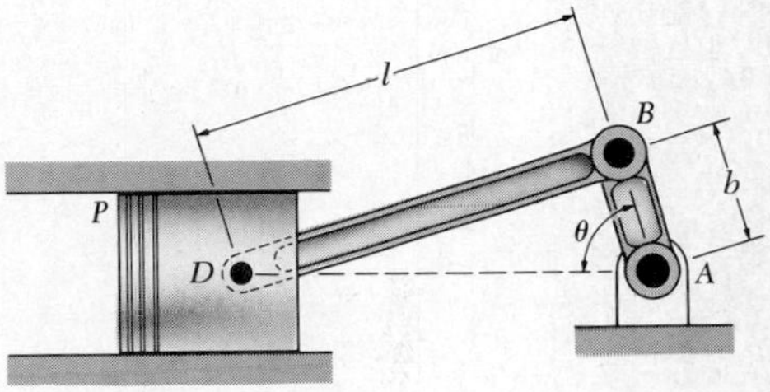

Fig. P15.45

15.45 In the engine system shown, $l = 10$ in. and $b = 3$ in.; the crank AB rotates with a constant angular velocity of 750 rpm clockwise. Determine the velocity of the piston P and the angular velocity of the connecting rod for the position corresponding to (a) $\theta = 0$, (b) $\theta = 90°$, (c) $\theta = 180°$.

15.46 Solve Prob. 15.45 for the position corresponding to $\theta = 30°$.

15.47 Solve Prob. 15.44 for the position corresponding to $\theta = 60°$.

15.48 through 15.51 In the position shown, bar AB has a constant angular velocity of 3 rad/s counterclockwise. Determine the angular velocity of bars BD and DE.

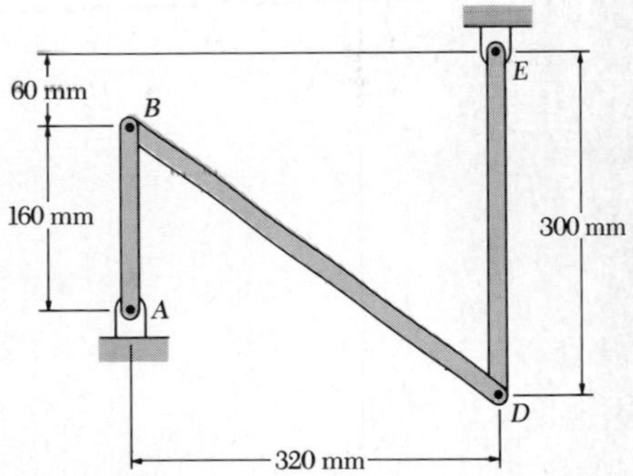

Fig. P15.48

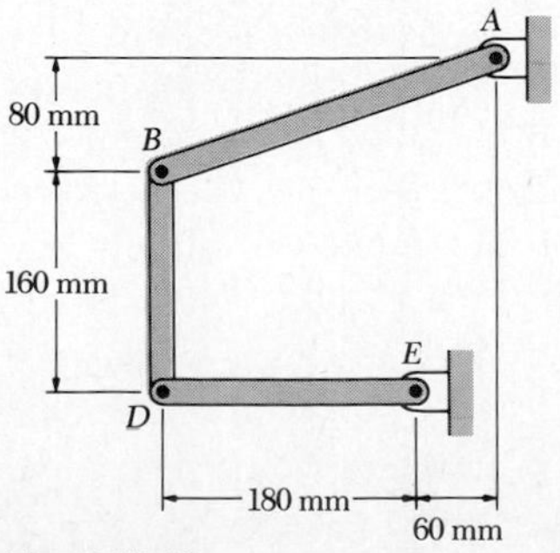

Fig. P15.49

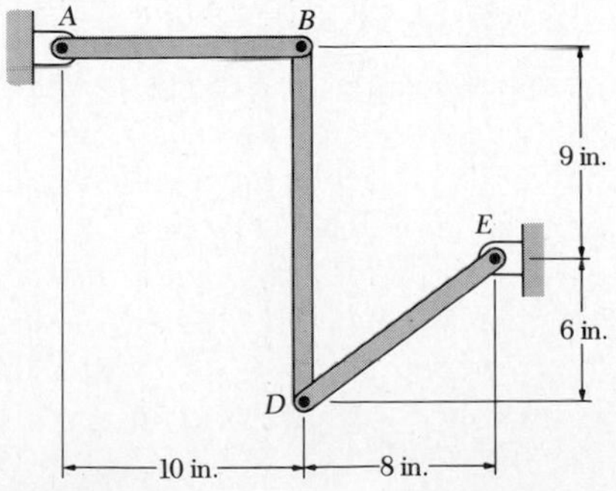

Fig. P15.50

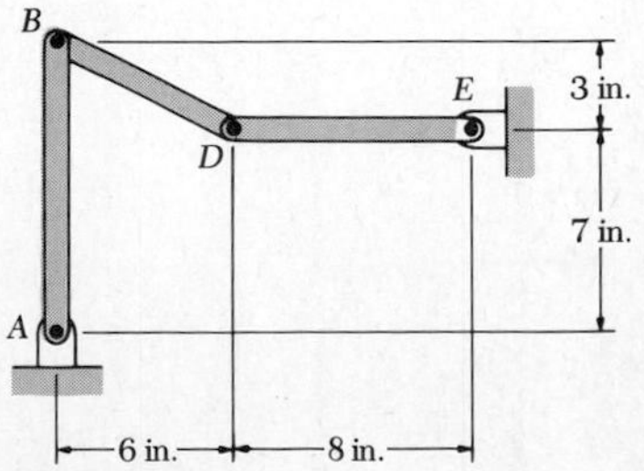

Fig. P15.51

15.52 The flanged wheel shown rolls to the right with a constant velocity of 1.5 m/s. Knowing that rod AB is 1.2 m long, determine the velocity of A and the angular velocity of the rod when (*a*) $\beta = 0$, (*b*) $\beta = 90°$.

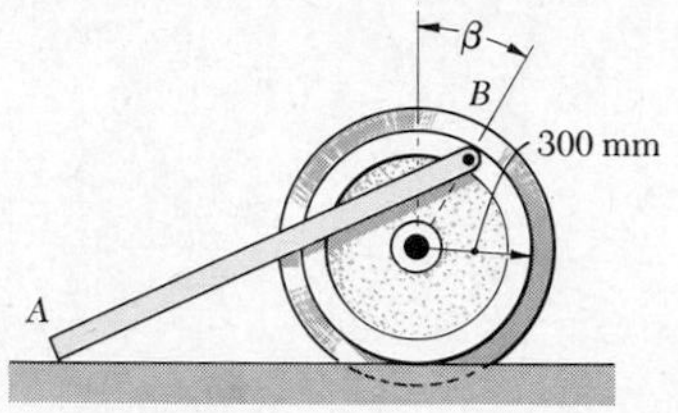

Fig. P15.52

15.53 Solve Prob. 15.52, assuming that $\beta = 30°$.

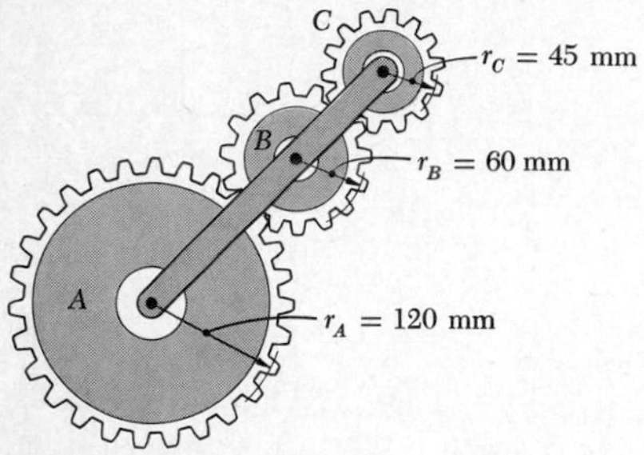

Fig. P15.54

***15.54** For the gearing system shown, derive an expression for the angular velocity ω_C of gear C and show that ω_C is independent of the radius of gear B. Assume that point A is fixed and denote the angular velocities of rod ABC and gear A by ω_{ABC} and ω_A, respectively.

15.7. Instantaneous Center of Rotation in Plane Motion. Consider the general plane motion of a slab. We shall show that at any given instant the velocities of the various particles of the slab are the same as if the slab were rotating about a certain axis perpendicular to the plane of the slab, called the *instantaneous axis of rotation.* This axis intersects the plane of the slab at a point C, called the *instantaneous center of rotation* of the slab.

To prove our statement, we first recall that the plane motion of a slab may always be replaced by a translation defined by the motion of an arbitrary reference point A, and by a rotation about A. As far as the velocities are concerned, the translation is characterized by the velocity $\mathbf{v}_A$ of the reference point A and the rotation is characterized by the angular velocity $\boldsymbol{\omega}$ of the slab (which is independent of the choice of A). Thus, the velocity $\mathbf{v}_A$ of point A and the angular velocity $\boldsymbol{\omega}$ of the slab define completely the velocities of all the other particles of the slab (Fig. 15.18*a*). Now let us assume that $\mathbf{v}_A$ and $\boldsymbol{\omega}$ are known and that they are both different from zero. (If $\mathbf{v}_A = 0$, point A is itself the instantaneous center of rotation, and if $\boldsymbol{\omega} = 0$, all the particles have the same velocity $\mathbf{v}_A$.) These velocities could be obtained by letting the slab rotate with the angular velocity $\boldsymbol{\omega}$ about a point C located on the perpendicular to $\mathbf{v}_A$ at a distance $r = v_A/\omega$ from A

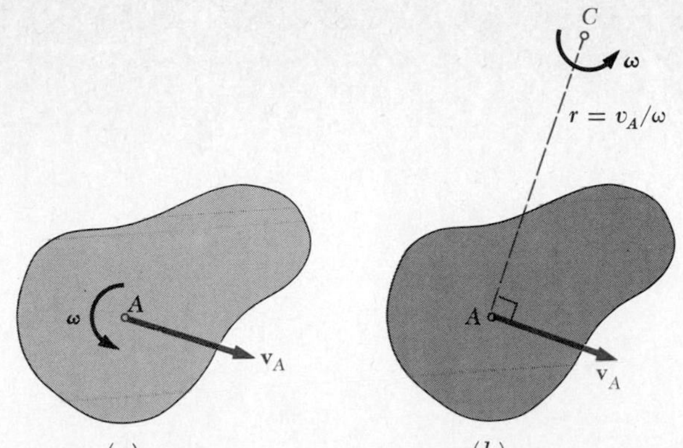

(a) (b)

Fig. 15.18

as shown in Fig. 15.18*b*. We check that the velocity of *A* would be perpendicular to *AC* and that its magnitude would be $r\omega = (v_A/\omega)\omega = v_A$. Thus the velocities of all the other particles of the slab would be the same as originally defined. Therefore, *as far as the velocities are concerned, the slab seems to rotate about the instantaneous center C* at the instant considered.

The position of the instantaneous center may be defined in two other ways. If the directions of the velocities of two particles *A* and *B* of the slab are known and if they are different, the instantaneous center *C* is obtained by drawing the perpendicular to $\mathbf{v}_A$ through *A* and the perpendicular to $\mathbf{v}_B$ through *B* and determining the point in which these two lines intersect (Fig. 15.19*a*). If the velocities $\mathbf{v}_A$ and $\mathbf{v}_B$ of two particles *A* and *B* are perpendicular to the line *AB* and if their magnitudes are known, the instantaneous center may be found by intersecting the line *AB* with the line joining the extremities of the vectors $\mathbf{v}_A$ and $\mathbf{v}_B$ (Fig. 15.19*b*). Note that if $\mathbf{v}_A$ and $\mathbf{v}_B$ were parallel in Fig. 15.19*a* or if $\mathbf{v}_A$ and $\mathbf{v}_B$ had the same magnitude in Fig. 15.19*b*, the instantaneous center *C* would be at an infinite distance and $\boldsymbol{\omega}$ would be zero; all points of the slab would have the same velocity.

(a) (b)

Fig. 15.19

To see how the concept of instantaneous center of rotation may be put to use, let us consider again the rod of Sec. 15.6. Drawing the perpendicular to $\mathbf{v}_A$ through A and the perpendicular to $\mathbf{v}_B$ through B (Fig. 15.20), we obtain the instantaneous center C. At the instant considered, the velocities

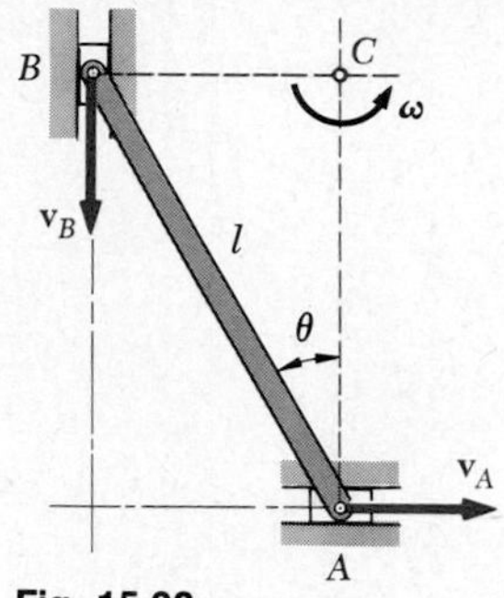

Fig. 15.20

of all the particles of the rod are thus the same as if the rod rotated about C. Now, if the magnitude v_A of the velocity of A is known, the magnitude ω of the angular velocity of the rod may be obtained by writing

$$\omega = \frac{v_A}{AC} = \frac{v_A}{l \cos \theta}$$

The magnitude of the velocity of B may then be obtained by writing

$$v_B = (BC)\omega = l \sin \theta \frac{v_A}{l \cos \theta} = v_A \tan \theta$$

Note that only *absolute* velocities are involved in the computation.

The instantaneous center of a slab in plane motion may be located either on the slab or outside the slab. If it is located on the slab, the particle C coinciding with the instantaneous center at a given instant t must have zero velocity at that instant. However, it should be noted that the instantaneous center of rotation is valid only at a given instant. Thus, the particle C of the slab which coincides with the instantaneous center at time t will generally not coincide with the instantaneous center at time $t + \Delta t$; while its velocity is zero at time t, it will probably be different from zero at time $t + \Delta t$. This means that, in general, the particle C *does not have zero acceleration* and, therefore, that the *accelerations* of the various particles of the slab *cannot* be determined as if the slab were rotating about C.

As the motion of the slab proceeds, the instantaneous center moves in space. But it was just pointed out that the position of the instantaneous center on the slab keeps changing. Thus, the instantaneous center describes one curve in space, called the *space centrode*, and another curve on the slab, called the *body centrode* (Fig. 15.21). It may be shown that at any instant, these two curves are tangent at C and that as the slab moves, the body centrode appears to *roll* on the space centrode.

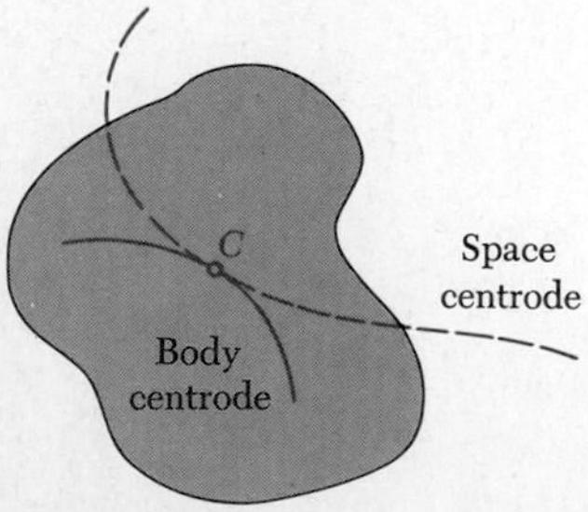

Fig. 15.21

SAMPLE PROBLEM 15.4

Solve Sample Prob. 15.2, using the method of the instantaneous center of rotation.

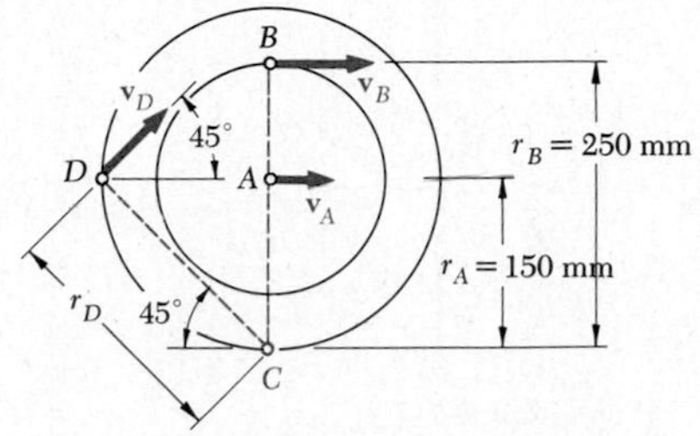

a. **Angular Velocity of the Gear.** Since the gear rolls on the stationary lower rack, the point of contact C of the gear with the rack has no velocity; point C is therefore the instantaneous center of rotation. We write

$$v_A = r_A\omega \qquad 1.2\text{ m/s} = (0.150\text{ m})\omega \qquad \omega = 8\text{ rad/s} \downarrow \quad ◄$$

b. **Velocities.** All points of the gear seem to rotate about the instantaneous center as far as velocities are concerned.

Velocity of Upper Rack. Recalling that $v_R = v_B$, we write

$$v_R = v_B = r_B\omega \qquad v_R = (0.250\text{ m})(8\text{ rad/s}) = 2\text{ m/s}$$

$$\mathbf{v}_R = 2\text{ m/s} \rightarrow \quad ◄$$

Velocity of Point D. Since $r_D = (0.150\text{ m})\sqrt{2} = 0.2121$ m, we write

$$v_D = r_D\omega \qquad v_D = (0.2121\text{ m})(8\text{ rad/s}) = 1.697\text{ m/s}$$

$$\mathbf{v}_D = 1.697\text{ m/s} \measuredangle 45° \quad ◄$$

SAMPLE PROBLEM 15.5

Solve Sample Prob. 15.3, using the method of the instantaneous center of rotation.

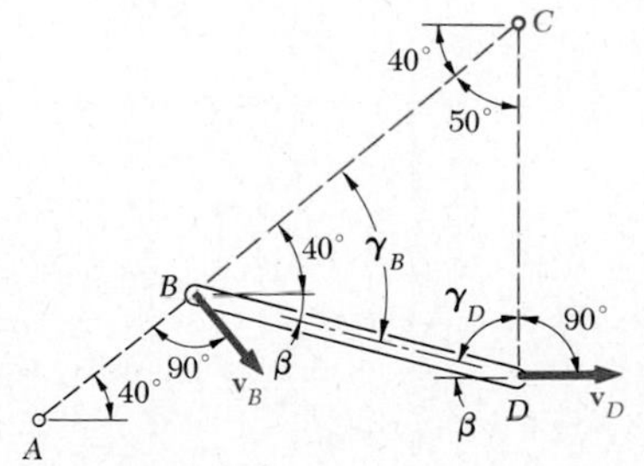

Motion of Crank *AB*. Referring to Sample Prob. 15.3, we obtain the velocity of point B; $\mathbf{v}_B = 627$ in./s $\searrow$ 50°.

Motion of the Connecting Rod *BD*. We first locate the instantaneous center C by drawing lines perpendicular to the absolute velocities $\mathbf{v}_B$ and $\mathbf{v}_D$. Recalling from Sample Prob. 15.3 that $\beta = 13.9°$ and that $BD = 8$ in., we solve the triangle BCD.

$$\gamma_B = 40° + \beta = 53.9° \qquad \gamma_D = 90° - \beta = 76.1°$$

$$\frac{BC}{\sin 76.1°} = \frac{CD}{\sin 53.9°} = \frac{8\text{ in.}}{\sin 50°}$$

$$BC = 10.14\text{ in.} \qquad CD = 8.44\text{ in.}$$

Since the connecting rod BD seems to rotate about point C, we write

$$v_B = (BC)\omega_{BD}$$

$$627\text{ in./s} = (10.14\text{ in.})\omega_{BD}$$

$$\omega_{BD} = 61.9\text{ rad/s} \circlearrowleft \quad ◄$$

$$v_D = (CD)\omega_{BD} = (8.44\text{ in.})(61.9\text{ rad/s})$$

$$= 522\text{ in./s} = 43.5\text{ ft/s}$$

$$\mathbf{v}_P = \mathbf{v}_D = 43.5\text{ ft/s} \rightarrow \quad ◄$$

Problems

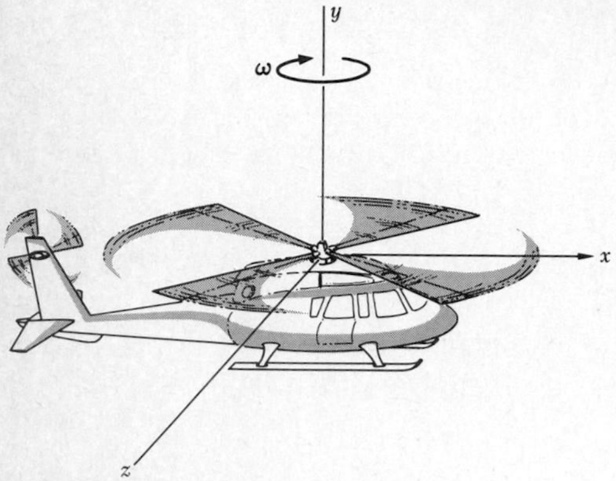

Fig. P15.55

15.55 A helicopter moves horizontally in the x direction at a speed of 105 mi/h. Knowing that the main blades rotate clockwise at an angular velocity of 160 rpm, determine the instantaneous axis of rotation of the main blades.

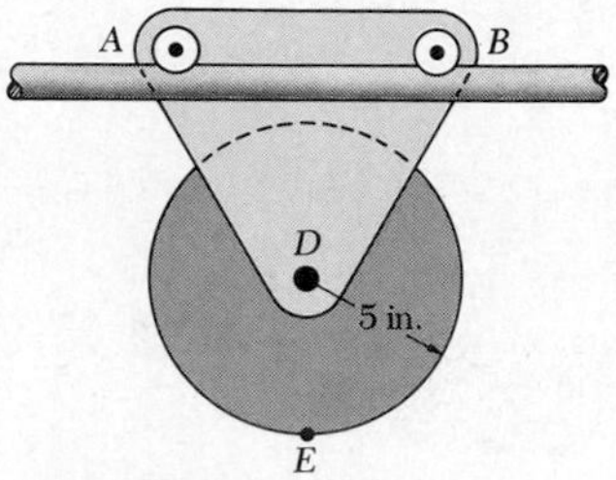

Fig. P15.56

15.56 The trolley shown moves to the left along a horizontal pipe support at a speed of 24 in./s. Knowing that the 5-in.-radius disk has an angular velocity of 8 rad/s counterclockwise, determine (*a*) the instantaneous center of rotation of the disk, (*b*) the velocity of point E.

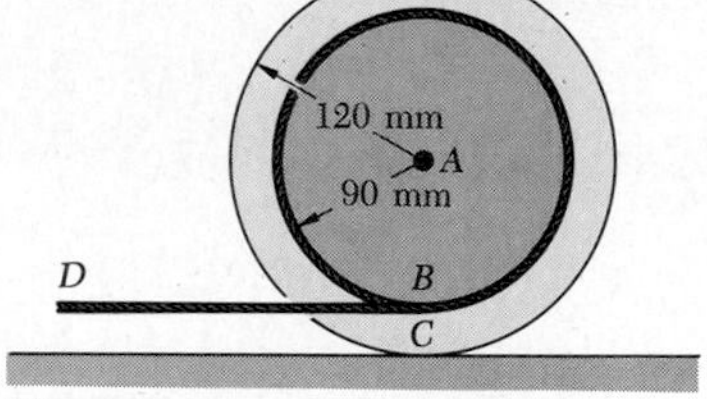

Fig. P15.57

15.57 A drum of radius 90 mm is mounted on a cylinder of radius 120 mm. A cord is wound around the drum, and its extremity D is pulled to the left at a constant velocity of 150 mm/s, causing the cylinder to roll without sliding. Determine (*a*) the angular velocity of the cylinder, (*b*) the velocity of the center of the cylinder, (*c*) the length of cord which is wound or unwound per second.

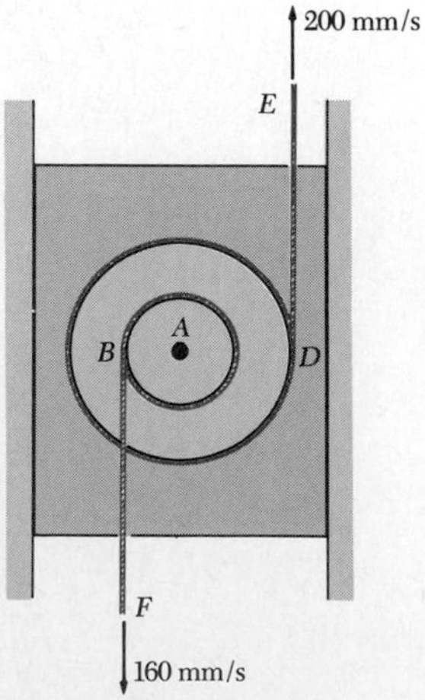

Fig. P15.58

15.58 A double pulley is attached to a slider block by a pin at A. The 30-mm-radius inner pulley is rigidly attached to the 60-mm-radius outer pulley. Knowing that each of the two cords is pulled at a constant speed as shown, determine (*a*) the instantaneous center of rotation of the double pulley, (*b*) the velocity of the slider block, (*c*) the number of millimeters of cord wrapped or unwrapped on each pulley per second.

15.59 Knowing that at the instant shown the angular velocity of rod *BE* is 3 rad/s counterclockwise, determine (*a*) the angular velocity of rod *AD*, (*b*) the velocity of collar *D*, (*c*) the velocity of point *A*.

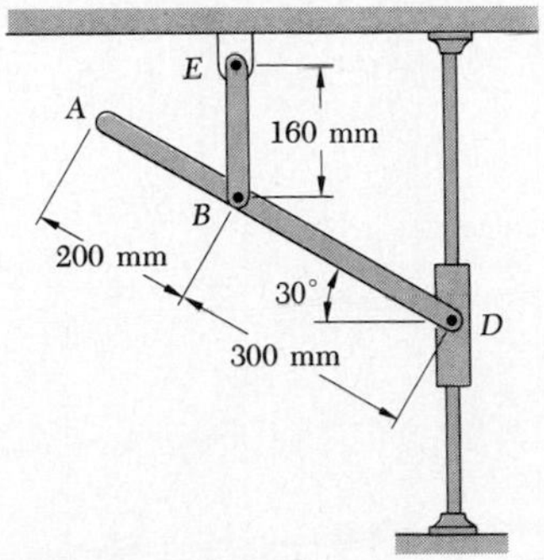

Fig. P15.59 and P15.60

15.60 Knowing that at the instant shown the velocity of collar *D* is 1.8 m/s upward, determine (*a*) the angular velocity of rod *AD*, (*b*) the velocity of point *B*, (*c*) the velocity of point *A*.

15.61 Knowing that at the instant shown the velocity of collar *D* is 48 in./s upward, determine (*a*) the instantaneous center of rotation of link *BD*, (*b*) the angular velocities of crank *AB* and link *BD*, (*c*) the velocity of the midpoint of link *BD*.

15.62 Knowing that at the instant shown the angular velocity of crank *AB* is 2.7 rad/s clockwise, determine (*a*) the angular velocity of link *BD*, (*b*) the velocity of collar *D*, (*c*) the velocity of the midpoint of link *BD*.

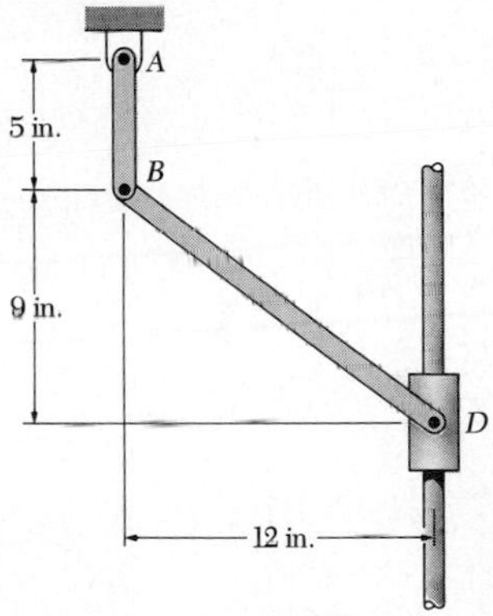

Fig. P15.61 and P15.62

15.63 The rod *BDE* is partially guided by a wheel at *D* which rolls in a vertical track. Knowing that at the instant shown the angular velocity of crank *AB* is 5 rad/s clockwise and that $\beta = 30°$, determine (*a*) the angular velocity of the rod, (*b*) the velocity of point *E*.

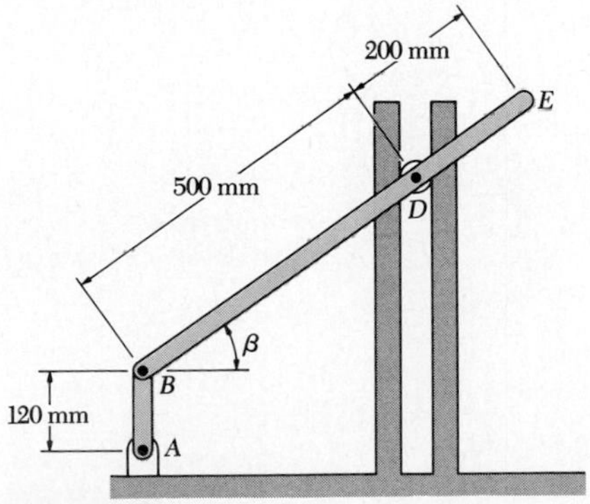

Fig. P15.63

15.64 Solve Prob. 15.63, assuming that $\beta = 40°$.

15.65 Two links AB and BD, each 500 mm long, are connected at B and guided by hydraulic cylinders attached at A and D. Knowing that D is stationary and that the velocity of A is 1.5 m/s to the right, determine at the instant shown (*a*) the angular velocity of each link, (*b*) the velocity of B.

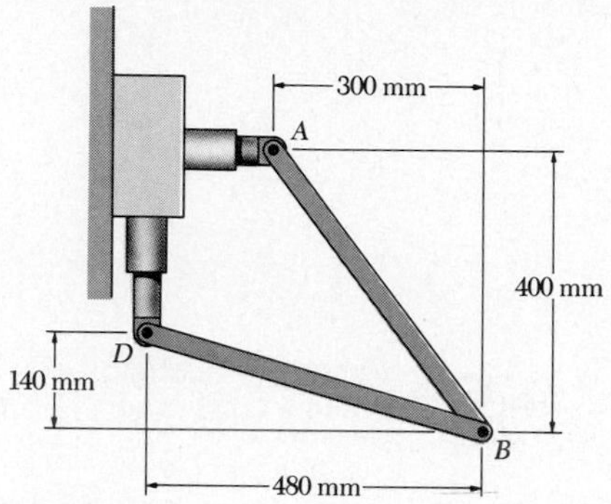

Fig. P15.65

15.66 Solve Prob. 15.65, assuming that A is stationary and that the velocity of D is 1.5 m/s downward.

15.67 Two slots have been cut in the plate FG and the plate has been placed so that the slots fit two fixed pins A and B. Knowing that at the instant shown the angular velocity of crank DE is 8 rad/s clockwise, determine (*a*) the velocity of point F, (*b*) the velocity of point G.

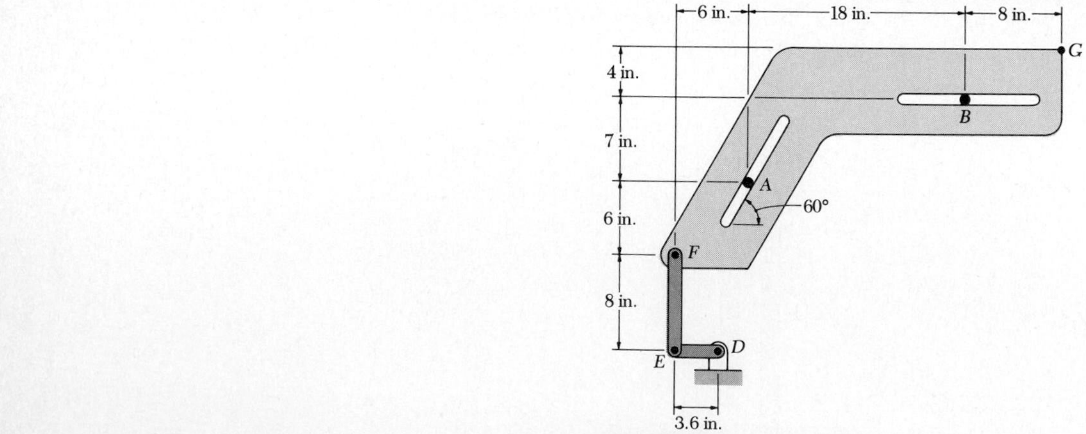

Fig. P15.67

15.68 In Prob. 15.67, determine the velocity of the point of the plate which is in contact with (*a*) pin A, (*b*) pin B.

15.69 Two links AB and BD, each of length 30 in., are connected to three collars. At the instant shown $\beta = 70°$ and the velocity of collar A is 90 in./s to the right. Determine (*a*) the corresponding value of γ, (*b*) the velocity of collar D.

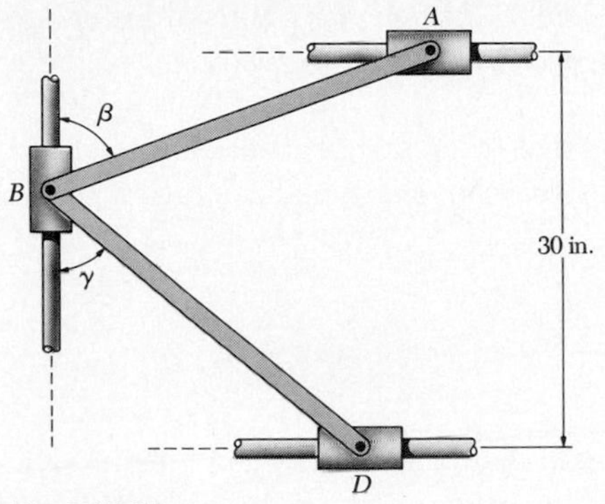

Fig. P15.69

15.70 Solve Prob. 15.69, assuming that $\beta = 55°$.

15.71 and 15.72 Two 500-mm rods are pin-connected at D as shown. Knowing that B moves to the right with a constant velocity of 480 mm/s, determine at the instant shown (*a*) the angular velocity of each rod, (*b*) the velocity of E.

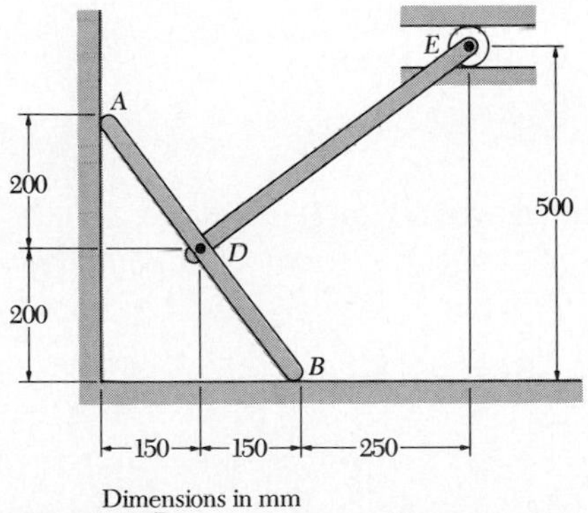

Fig. P15.71

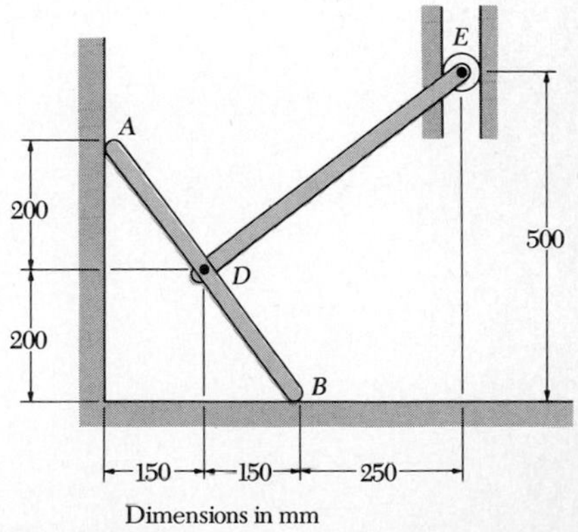

Fig. P15.72

15.73 Describe the space centrode and the body centrode of link AB of Prob. 15.69 as collar B moves upward. (*Note.* The body centrode need not lie on a physical portion of the link.)

15.74 Describe the space centrode and the body centrode of the drum of Prob. 15.57 as the drum rolls on the horizontal surface.

15.75 Using the method of Sec. 15.7, solve Prob. 15.49.
15.76 Using the method of Sec. 15.7, solve Prob. 15.50.
15.77 Using the method of Sec. 15.7, solve Prob. 15.51.
15.78 Using the method of Sec. 15.7, solve Prob. 15.52.

15.8. Absolute and Relative Acceleration in Plane Motion. We saw in Sec. 15.5 that any plane motion may be replaced by a translation defined by the motion of an arbitrary reference point A, and a simultaneous rotation about A. This property was used in Sec. 15.6 to determine the velocity of the various points of a moving slab. We shall now use the same property to determine the acceleration of the points of the slab.

We first recall that the absolute acceleration $\mathbf{a}_B$ of a particle of the slab may be obtained from the relative-acceleration formula derived in Sec. 11.12,

$$\mathbf{a}_B = \mathbf{a}_A + \mathbf{a}_{B/A} \tag{15.21}$$

where the right-hand member represents a vector sum. The acceleration $\mathbf{a}_A$ corresponds to the translation of the slab with A, while the relative acceleration $\mathbf{a}_{B/A}$ is associated with the rotation of the slab about A and is measured with respect to axes centered at A and of fixed orientation. We recall from Sec. 15.3 that the relative acceleration $\mathbf{a}_{B/A}$ may be resolved into two components, a *tangential component* $(\mathbf{a}_{B/A})_t$ perpendicular to the line AB, and a *normal component* $(\mathbf{a}_{B/A})_n$ directed toward A (Fig. 15.22). Denoting by $\mathbf{r}_{B/A}$ the position vector of B relative to A and, respectively, by $\omega\mathbf{k}$ and $\alpha\mathbf{k}$ the angular velocity and angular acceleration of the slab with respect to axes of fixed orientation, we have

$$\begin{aligned} (\mathbf{a}_{B/A})_t &= \alpha\mathbf{k} \times \mathbf{r}_{B/A} & \qquad (a_{B/A})_t &= r\alpha \\ (\mathbf{a}_{B/A})_n &= -\omega^2\mathbf{r}_{B/A} & \qquad (a_{B/A})_n &= r\omega^2 \end{aligned} \tag{15.22}$$

where r is the distance from A to B. Substituting into (15.21) the expressions obtained for the tangential and normal components of $\mathbf{a}_{B/A}$, we may also write

$$\mathbf{a}_B = \mathbf{a}_A + \alpha\mathbf{k} \times \mathbf{r}_{B/A} - \omega^2\mathbf{r}_{B/A} \tag{15.21'}$$

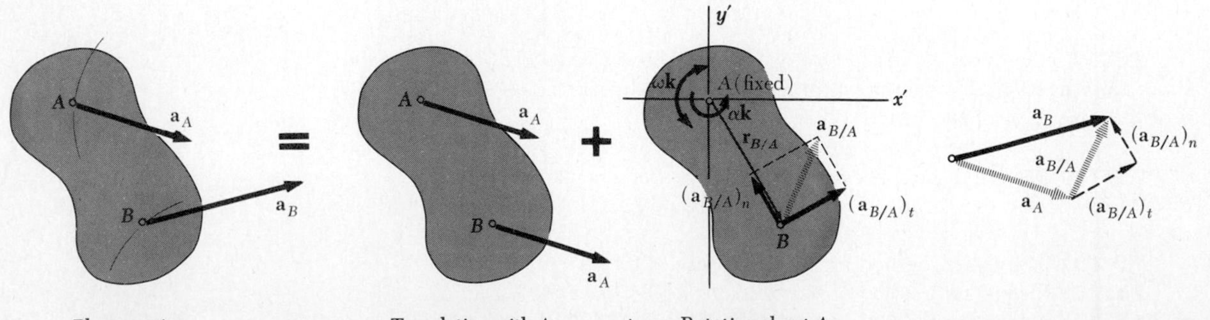

Fig. 15.22

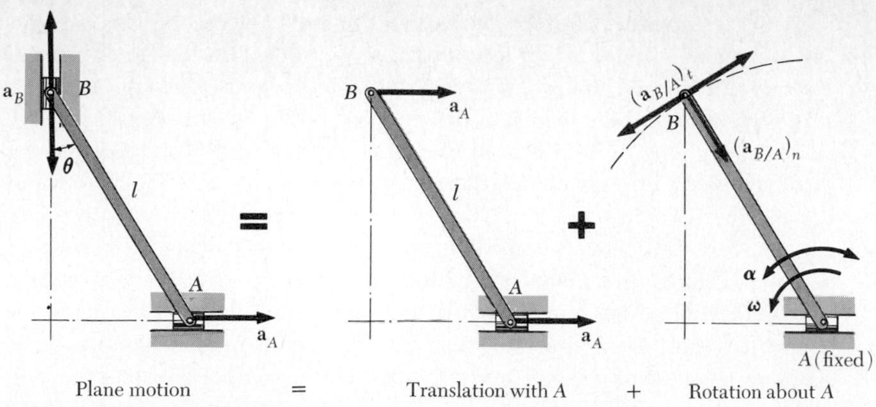

Fig. 15.23

As an example, we shall consider again the rod AB whose extremities slide, respectively, along a horizontal and a vertical track (Fig. 15.23). Assuming that the velocity $\mathbf{v}_A$ and the acceleration $\mathbf{a}_A$ of A are known, we propose to determine the acceleration $\mathbf{a}_B$ of B and the angular acceleration $\boldsymbol{\alpha}$ of the rod. Choosing A as a reference point, we express that the given motion is equivalent to a translation with A and a rotation about A. The absolute acceleration of B must be equal to the sum

$$\begin{aligned}\mathbf{a}_B &= \mathbf{a}_A + \mathbf{a}_{B/A} \\ &= \mathbf{a}_A + (\mathbf{a}_{B/A})_n + (\mathbf{a}_{B/A})_t \qquad (15.23)\end{aligned}$$

where $(\mathbf{a}_{B/A})_n$ has the magnitude $l\omega^2$ and is *directed toward A*, while $(\mathbf{a}_{B/A})_t$ has the magnitude $l\alpha$ and is perpendicular to AB. There is no way of telling at the present time whether the tangential component $(\mathbf{a}_{B/A})_t$ is directed to the left or to the right, and students should not rely on their "intuition" in this matter. We shall therefore indicate both possible directions for this component in Fig. 15.23. Similarly, we indicate both possible senses for $\mathbf{a}_B$, since we do not know whether point B is accelerated upward or downward.

Equation (15.23) has been expressed geometrically in Fig. 15.24. Four different vector polygons may be obtained, depending upon the sense of $\mathbf{a}_A$ and the relative magnitude of a_A and $(a_{B/A})_n$. If we are to determine a_B and α from one of these diagrams, we must know not only a_A and θ but also ω. The angular velocity of the rod should therefore be separately determined by one of the methods indicated in Secs. 15.6 and 15.7. The values of a_B and α may then be obtained by considering successively the x and y components of the vectors shown in Fig. 15.24. In the case of polygon a, for example, we write

$\overset{+}{\rightarrow}$ x components: $\quad 0 = a_A + l\omega^2 \sin\theta - l\alpha\cos\theta$

$+\uparrow y$ components: $\quad -a_B = -l\omega^2\cos\theta - l\alpha\sin\theta$

and solve for a_B and α. The two unknowns may also be obtained by direct measurement on the vector polygon. In that case, care should be taken to draw first the known vectors $\mathbf{a}_A$ and $(\mathbf{a}_{B/A})_n$.

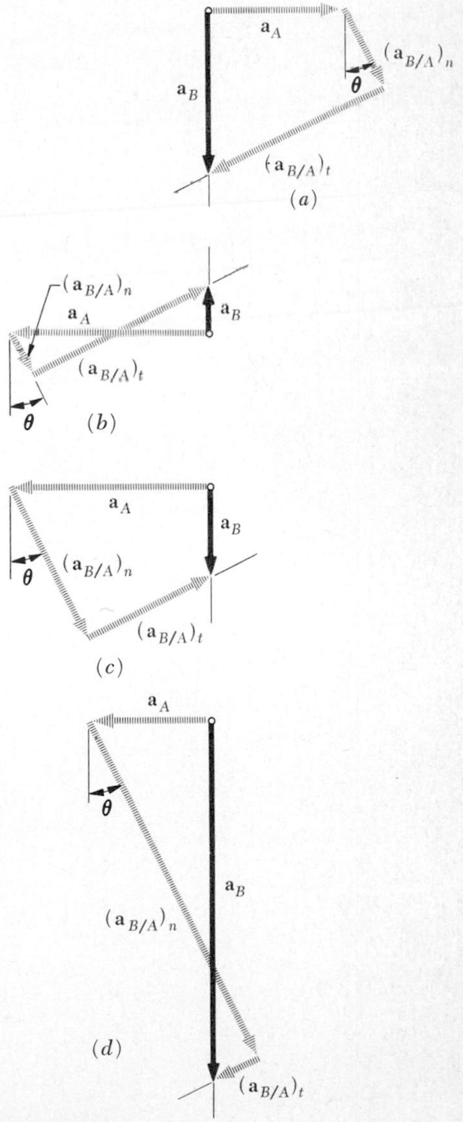

Fig. 15.24

It is quite evident that the determination of accelerations is considerably more involved than the determination of velocities. Yet in the example considered here, the extremities A and B of the rod were moving along straight tracks, and the diagrams drawn were relatively simple. If A and B had moved along curved tracks, the accelerations $\mathbf{a}_A$ and $\mathbf{a}_B$ should have been resolved into normal and tangential components and the solution of the problem would have involved six different vectors.

When a mechanism consists of several moving parts which are pin-connected, its analysis may be carried out by considering each part as a rigid body, while keeping in mind that the points where two parts are connected must have the same absolute acceleration (see Sample Prob. 15.7). In the case of meshed gears, the tangential components of the accelerations of the teeth in contact are equal, but their normal components are different.

***15.9. Analysis of Plane Motion in Terms of a Parameter.** In the case of certain mechanisms, it is possible to express the coordinates x and y of all the significant points of the mechanism by means of simple analytic expressions containing a single parameter. It may be advantageous in such a case to determine directly the absolute velocity and the absolute acceleration of the various points of the mechanism, since the components of the velocity and of the acceleration of a given point may be obtained by differentiating the coordinates x and y of that point.

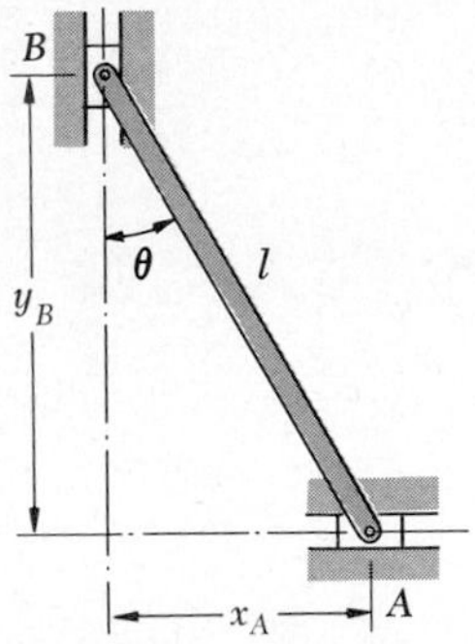

Fig. 15.25

Let us consider again the rod AB whose extremities slide, respectively, in a horizontal and a vertical track (Fig. 15.25). The coordinates x_A and y_B of the extremities of the rod may be expressed in terms of the angle θ the rod forms with the vertical:

$$x_A = l \sin\theta \qquad y_B = l\cos\theta \tag{15.24}$$

Differentiating Eqs. (15.24) twice with respect to t, we write

$$\begin{aligned} v_A &= \dot{x}_A = l\dot{\theta}\cos\theta \\ a_A &= \ddot{x}_A = -l\dot{\theta}^2\sin\theta + l\ddot{\theta}\cos\theta \end{aligned}$$

$$\begin{aligned} v_B &= \dot{y}_B = -l\dot{\theta}\sin\theta \\ a_B &= \ddot{y}_B = -l\dot{\theta}^2\cos\theta - l\ddot{\theta}\sin\theta \end{aligned}$$

Recalling that $\dot{\theta} = \omega$ and $\ddot{\theta} = \alpha$, we obtain

$$v_A = l\omega\cos\theta \qquad v_B = -l\omega\sin\theta \tag{15.25}$$

$$a_A = -l\omega^2\sin\theta + l\alpha\cos\theta \qquad a_B = -l\omega^2\cos\theta - l\alpha\sin\theta \tag{15.26}$$

We note that a positive sign for v_A or a_A indicates that the velocity $\mathbf{v}_A$ or the acceleration $\mathbf{a}_A$ is directed to the right; a positive sign for v_B or a_B indicates that $\mathbf{v}_B$ or $\mathbf{a}_B$ is directed upward. Equations (15.25) may be used, for example, to determine v_B and ω when v_A and θ are known. Substituting for ω in (15.26), we may then determine a_B and α if a_A is known.

SAMPLE PROBLEM 15.6

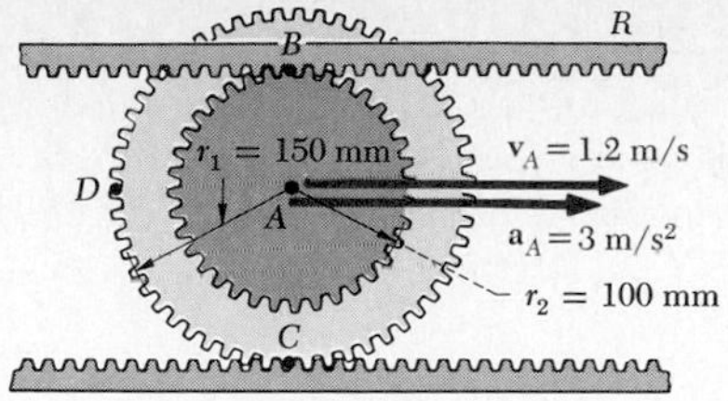

The center of the double gear of Sample Prob. 15.2 has a velocity of 1.2 m/s to the right and an acceleration of 3 m/s² to the right. Recalling that the lower rack is stationary, determine (*a*) the angular acceleration of the gear, (*b*) the acceleration of points *B*, *C*, and *D* of the gear.

***a.* Angular Acceleration of the Gear.** In Sample Prob. 15.2, we found that $x_A = -r_1\theta$ and $v_A = -r_1\omega$. Differentiating the latter with respect to time, we obtain $a_A = -r_1\alpha$.

$$v_A = -r_1\omega \qquad 1.2 \text{ m/s} = -(0.150 \text{ m})\omega \qquad \omega = -8 \text{ rad/s}$$
$$a_A = -r_1\alpha \qquad 3 \text{ m/s}^2 = -(0.150 \text{ m})\alpha \qquad \alpha = -20 \text{ rad/s}^2$$
$$\boldsymbol{\alpha} = \alpha\mathbf{k} = -(20 \text{ rad/s}^2)\mathbf{k} \quad ◄$$

***b.* Accelerations.** The rolling motion of the gear is resolved into a translation with *A* and a rotation about *A*.

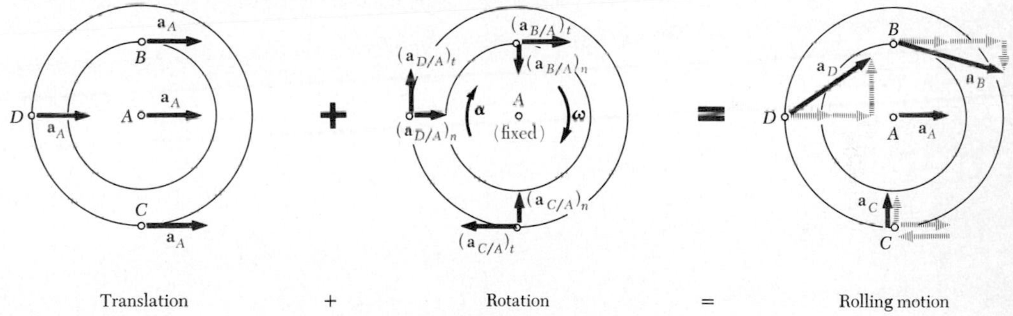

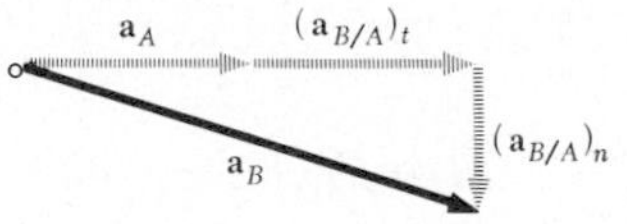

Acceleration of Point B. Adding vectorially the accelerations corresponding to the translation and to the rotation, we obtain

$$\begin{aligned}\mathbf{a}_B &= \mathbf{a}_A + \mathbf{a}_{B/A} = \mathbf{a}_A + (\mathbf{a}_{B/A})_t + (\mathbf{a}_{B/A})_n \\ &= \mathbf{a}_A + \alpha\mathbf{k} \times \mathbf{r}_{B/A} - \omega^2\mathbf{r}_{B/A} \\ &= (3 \text{ m/s}^2)\mathbf{i} - (20 \text{ rad/s}^2)\mathbf{k} \times (0.100 \text{ m})\mathbf{j} - (8 \text{ rad/s})^2(0.100 \text{ m})\mathbf{j} \\ &= (3 \text{ m/s}^2)\mathbf{i} + (2 \text{ m/s}^2)\mathbf{i} - (6.40 \text{ m/s}^2)\mathbf{j}\end{aligned}$$
$$\mathbf{a}_B = 8.12 \text{ m/s}^2 \searrow 52.0° \quad ◄$$

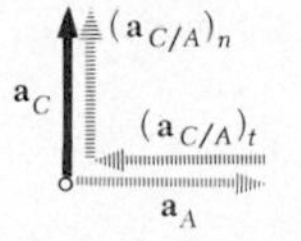

Acceleration of Point C

$$\begin{aligned}\mathbf{a}_C &= \mathbf{a}_A + \mathbf{a}_{C/A} = \mathbf{a}_A + \alpha\mathbf{k} \times \mathbf{r}_{C/A} - \omega^2\mathbf{r}_{C/A} \\ &= (3 \text{ m/s}^2)\mathbf{i} - (20 \text{ rad/s}^2)\mathbf{k} \times (-0.150 \text{ m})\mathbf{j} - (8 \text{ rad/s})^2(-0.150 \text{ m})\mathbf{j} \\ &= (3 \text{ m/s}^2)\mathbf{i} - (3 \text{ m/s}^2)\mathbf{i} + (9.60 \text{ m/s}^2)\mathbf{j}\end{aligned}$$
$$\mathbf{a}_C = 9.60 \text{ m/s}^2 \uparrow \quad ◄$$

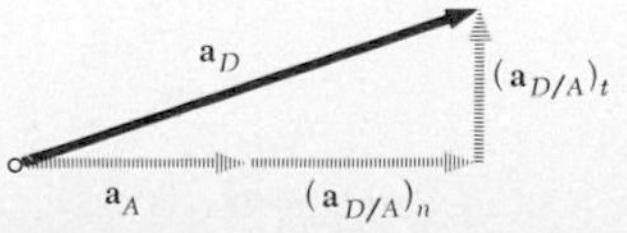

Acceleration of Point D

$$\begin{aligned}\mathbf{a}_D &= \mathbf{a}_A + \mathbf{a}_{D/A} = \mathbf{a}_A + \alpha\mathbf{k} \times \mathbf{r}_{D/A} - \omega^2\mathbf{r}_{D/A} \\ &= (3 \text{ m/s}^2)\mathbf{i} - (20 \text{ rad/s}^2)\mathbf{k} \times (-0.150 \text{ m})\mathbf{i} - (8 \text{ rad/s})^2(-0.150 \text{ m})\mathbf{i} \\ &= (3 \text{ m/s}^2)\mathbf{i} + (3 \text{ m/s}^2)\mathbf{j} + (9.60 \text{ m/s}^2)\mathbf{i}\end{aligned}$$
$$\mathbf{a}_D = 12.95 \text{ m/s}^2 \angle 13.4° \quad ◄$$

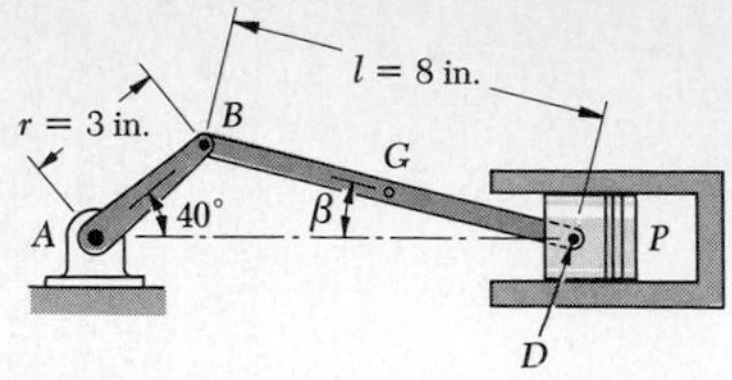

SAMPLE PROBLEM 15.7

Crank AB of the engine system of Sample Prob. 15.3 has a constant clockwise angular velocity of 2000 rpm. For the crank position shown, determine the angular acceleration of the connecting rod BD and the acceleration of point D.

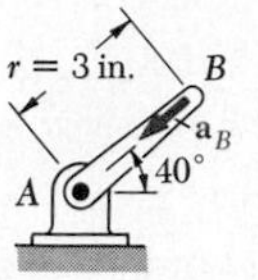

Motion of Crank *AB*. Since the crank rotates about A with constant $\omega_{AB} = 2000 \text{ rpm} = 209 \text{ rad/s}$, we have $\alpha_{AB} = 0$. The acceleration of B is therefore directed toward A and has a magnitude

$$a_B = r\omega_{AB}^2 = (\tfrac{3}{12}\text{ ft})(209 \text{ rad/s})^2 = 10{,}920 \text{ ft/s}^2$$
$$\mathbf{a}_B = 10{,}920 \text{ ft/s}^2 \measuredangle 40°$$

Motion of the Connecting Rod *BD*. The angular velocity $\boldsymbol{\omega}_{BD}$ and the value of β were obtained in Sample Prob. 15.3.

$$\boldsymbol{\omega}_{BD} = 61.9 \text{ rad/s} \circlearrowleft \qquad \beta = 13.9°$$

The motion of BD is resolved into a translation with B and a rotation about B. The relative acceleration $\mathbf{a}_{D/B}$ is resolved into normal and tangential components:

$$(a_{D/B})_n = (BD)\omega_{BD}^2 = (\tfrac{8}{12}\text{ ft})(61.9 \text{ rad/s})^2 = 2550 \text{ ft/s}^2$$
$$(\mathbf{a}_{D/B})_n = 2550 \text{ ft/s}^2 \measuredangle 13.9°$$
$$(a_{D/B})_t = (BD)\alpha_{BD} = (\tfrac{8}{12})\alpha_{BD} = 0.667\alpha_{BD}$$
$$(\mathbf{a}_{D/B})_t = 0.667\alpha_{BD} \measuredangle 76.1°$$

While $(\mathbf{a}_{D/B})_t$ must be perpendicular to BD, its sense is not known.

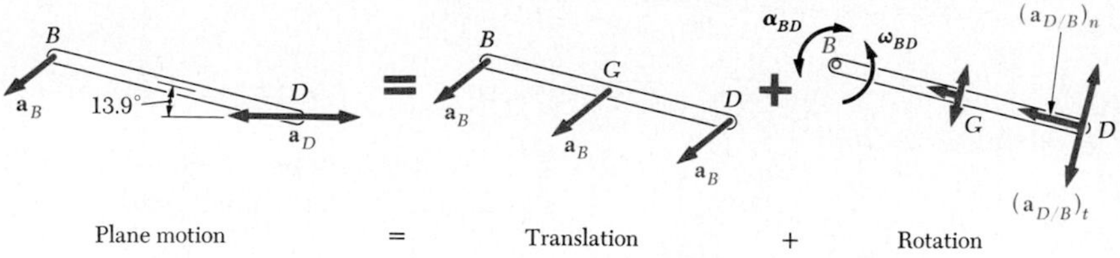

Noting that the acceleration $\mathbf{a}_D$ must be horizontal, we write

$$\mathbf{a}_D = \mathbf{a}_B + \mathbf{a}_{D/B} = \mathbf{a}_B + (\mathbf{a}_{D/B})_n + (\mathbf{a}_{D/B})_t$$
$$[a_D \leftrightarrow] = [10{,}920 \measuredangle 40°] + [2560 \measuredangle 13.9°] + [0.667\alpha_{BD} \measuredangle 76.1°]$$

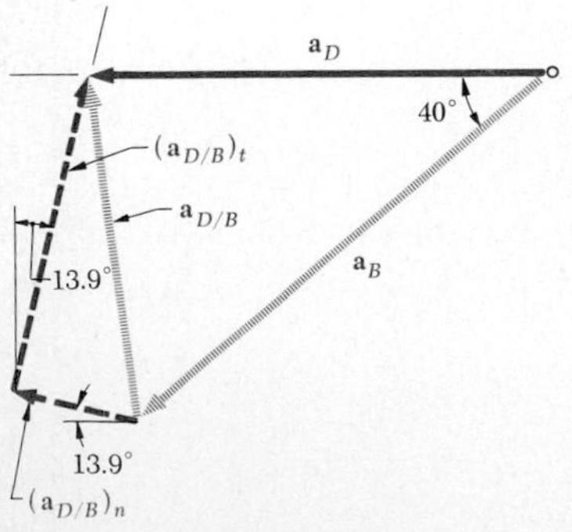

Equating x and y components, we obtain the following scalar equations:

$\xrightarrow{+}$ x components:

$$-a_D = -10{,}920 \cos 40° - 2550 \cos 13.9° + 0.667\alpha_{BD} \sin 13.9°$$

$+\uparrow y$ components:

$$0 = -10{,}920 \sin 40° + 2550 \sin 13.9° + 0.667\alpha_{BD} \cos 13.9°$$

Solving the equations simultaneously, we obtain $\alpha_{BD} = +9890 \text{ rad/s}^2$ and $a_D = +9260 \text{ ft/s}^2$. The positive signs indicate that the senses shown on the vector polygon are correct; we write

$$\boldsymbol{\alpha}_{BD} = 9890 \text{ rad/s}^2 \circlearrowleft \quad ◄$$
$$\mathbf{a}_D = 9260 \text{ ft/s}^2 \leftarrow \quad ◄$$

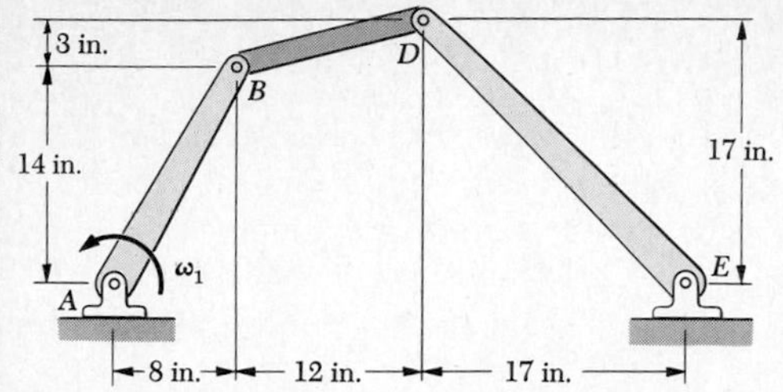

SAMPLE PROBLEM 15.8

The linkage *ABDE* moves in the vertical plane. Knowing that in the position shown crank *AB* has a constant angular velocity $\boldsymbol{\omega}_1$ of 20 rad/s counterclockwise, determine the angular velocities and angular accelerations of the connecting rod *BD* and of the crank *DE*.

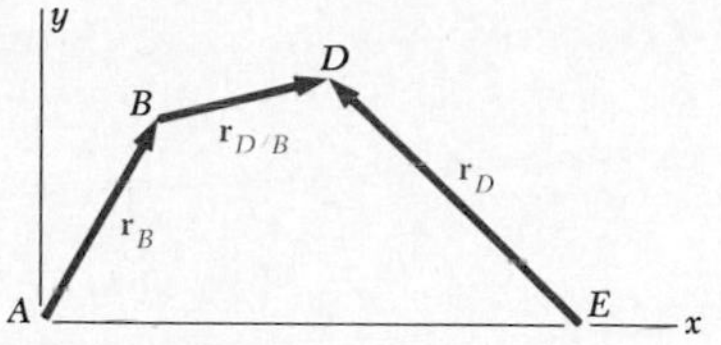

$$\mathbf{r}_B = 8\mathbf{i} + 14\mathbf{j}$$
$$\mathbf{r}_D = -17\mathbf{i} + 17\mathbf{j}$$
$$\mathbf{r}_{D/B} = 12\mathbf{i} + 3\mathbf{j}$$

Solution. While this problem could be solved by the method used in Sample Prob. 15.7, we shall make full use of the vector approach in the present case. The position vectors $\mathbf{r}_B$, $\mathbf{r}_D$, and $\mathbf{r}_{D/B}$ are chosen as shown in the sketch.

Velocities. Since the motion of each element of the linkage is contained in the plane of the figure, we have

$$\boldsymbol{\omega}_{AB} = \omega_{AB}\mathbf{k} = (20 \text{ rad/s})\mathbf{k} \qquad \boldsymbol{\omega}_{BD} = \omega_{BD}\mathbf{k} \qquad \boldsymbol{\omega}_{DE} = \omega_{DE}\mathbf{k}$$

where $\mathbf{k}$ is a unit vector pointing out of the paper. We now write

$$\mathbf{v}_D = \mathbf{v}_B + \mathbf{v}_{D/B}$$
$$\omega_{DE}\mathbf{k} \times \mathbf{r}_D = \omega_{AB}\mathbf{k} \times \mathbf{r}_B + \omega_{BD}\mathbf{k} \times \mathbf{r}_{D/B}$$
$$\omega_{DE}\mathbf{k} \times (-17\mathbf{i} + 17\mathbf{j}) = 20\mathbf{k} \times (8\mathbf{i} + 14\mathbf{j}) + \omega_{BD}\mathbf{k} \times (12\mathbf{i} + 3\mathbf{j})$$
$$-17\omega_{DE}\mathbf{j} - 17\omega_{DE}\mathbf{i} = 160\mathbf{j} - 280\mathbf{i} + 12\omega_{BD}\mathbf{j} - 3\omega_{BD}\mathbf{i}$$

Equating the coefficients of the unit vectors $\mathbf{i}$ and $\mathbf{j}$, we obtain the following two scalar equations:

$$-17\omega_{DE} = -280 - 3\omega_{BD}$$
$$-17\omega_{DE} = +160 + 12\omega_{BD}$$
$$\boldsymbol{\omega}_{BD} = -(29.33 \text{ rad/s})\mathbf{k} \qquad \boldsymbol{\omega}_{DE} = (11.29 \text{ rad/s})\mathbf{k} \quad ◄$$

Accelerations. Noting that at the instant considered crank *AB* has a constant angular velocity, we write

$$\boldsymbol{\alpha}_{AB} = 0 \qquad \boldsymbol{\alpha}_{BD} = \alpha_{BD}\mathbf{k} \qquad \boldsymbol{\alpha}_{DE} = \alpha_{DE}\mathbf{k}$$
$$\mathbf{a}_D = \mathbf{a}_B + \mathbf{a}_{D/B} \qquad (1)$$

Each term of Eq. (1) is evaluated separately:

$$\mathbf{a}_D = \alpha_{DE}\mathbf{k} \times \mathbf{r}_D - \omega_{DE}^2\mathbf{r}_D$$
$$= \alpha_{DE}\mathbf{k} \times (-17\mathbf{i} + 17\mathbf{j}) - (11.29)^2(-17\mathbf{i} + 17\mathbf{j})$$
$$= -17\alpha_{DE}\mathbf{j} - 17\alpha_{DE}\mathbf{i} + 2170\mathbf{i} - 2170\mathbf{j}$$
$$\mathbf{a}_B = \alpha_{AB}\mathbf{k} \times \mathbf{r}_B - \omega_{AB}^2\mathbf{r}_B = 0 - (20)^2(8\mathbf{i} + 14\mathbf{j})$$
$$= -3200\mathbf{i} - 5600\mathbf{j}$$
$$\mathbf{a}_{D/B} = \alpha_{BD}\mathbf{k} \times \mathbf{r}_{D/B} - \omega_{BD}^2\mathbf{r}_{D/B}$$
$$= \alpha_{BD}\mathbf{k} \times (12\mathbf{i} + 3\mathbf{j}) - (29.33)^2(12\mathbf{i} + 3\mathbf{j})$$
$$= 12\alpha_{BD}\mathbf{j} - 3\alpha_{BD}\mathbf{i} - 10{,}320\mathbf{i} - 2580\mathbf{j}$$

Substituting into Eq. (1) and equating the coefficients of $\mathbf{i}$ and $\mathbf{j}$, we obtain

$$-17\alpha_{DE} + 3\alpha_{BD} = -15{,}690$$
$$-17\alpha_{DE} - 12\alpha_{BD} = -6010$$
$$\boldsymbol{\alpha}_{BD} = -(645 \text{ rad/s}^2)\mathbf{k} \qquad \boldsymbol{\alpha}_{DE} = (809 \text{ rad/s}^2)\mathbf{k} \quad ◄$$

Problems

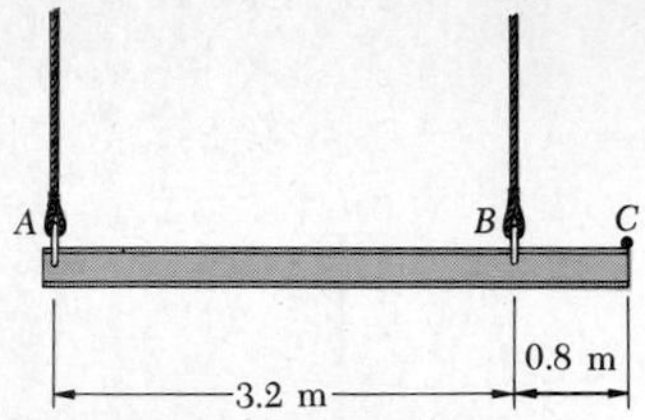

Fig. P15.79 and P15.80

15.79 A 4-m steel beam is lowered by means of two cables unwinding at the same speed from overhead cranes. As the beam approaches the ground, the crane operators apply brakes to slow down the unwinding motion. At the instant considered the deceleration of the cable attached at A is 5 m/s^2, while that of the cable attached at B is 2 m/s^2. Determine (a) the angular acceleration of the beam, (b) the acceleration of point C.

15.80 The acceleration of point C is 0.4 m/s^2 downward and the angular acceleration of the beam is 0.75 rad/s^2 clockwise. Knowing that the angular velocity of the beam is zero at the instant considered, determine the acceleration of each cable.

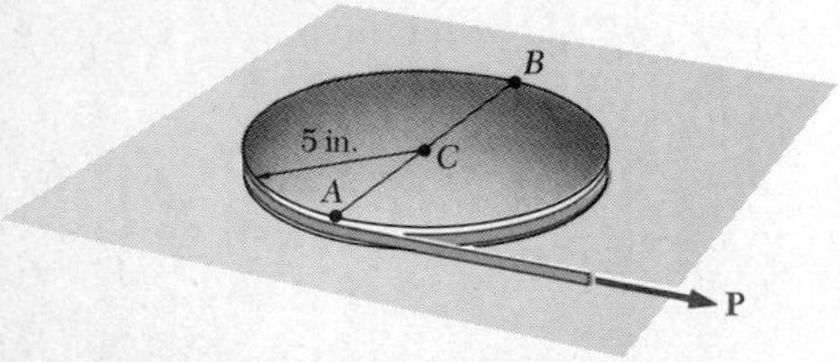

Fig. P15.81

15.81 A tape is wrapped around a 10-in.-diameter disk which is at rest on a horizontal table. A force **P** applied as shown produces the following accelerations: $\mathbf{a}_A = 30$ in./s^2 to the right, $\boldsymbol{\alpha} = 4$ rad/s^2 counterclockwise as viewed from above. Determine the acceleration (a) of point B, (b) of point C.

15.82 In Prob. 15.81, determine the point of the plate which (a) has no acceleration, (b) has an acceleration of 22 in./s^2 to the right.

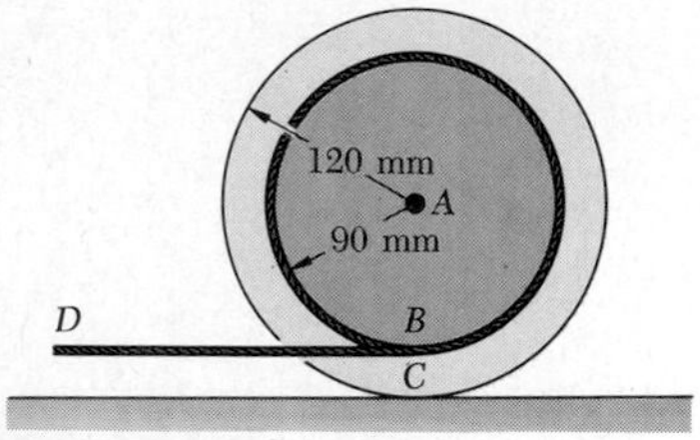

Fig. P15.83

15.83 A drum of radius 90 mm is mounted on a cylinder of radius 120 mm. A cord is wound around the drum and is pulled in such a way that point D has a velocity of 90 mm/s and an acceleration of 450 mm/s^2, both directed to the left. Assuming that the cylinder rolls without slipping, determine the acceleration (a) of point A, (b) of point B, (c) of point C.

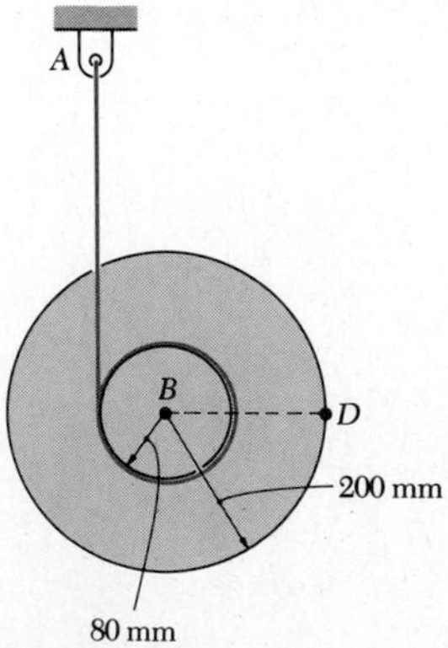

Fig. P15.84

15.84 At the instant shown the center B of the double pulley has a velocity of 0.6 m/s and an acceleration of 2.4 m/s^2, both directed downward. Knowing that the cord wrapped around the inner pulley is attached to a fixed support at A, determine the acceleration of point D.

15.85 An automobile travels to the right at a constant speed of 45 mi/h. Knowing that the diameter of a wheel is 22 in., determine the acceleration (*a*) of point *C*, (*b*) of point *D*.

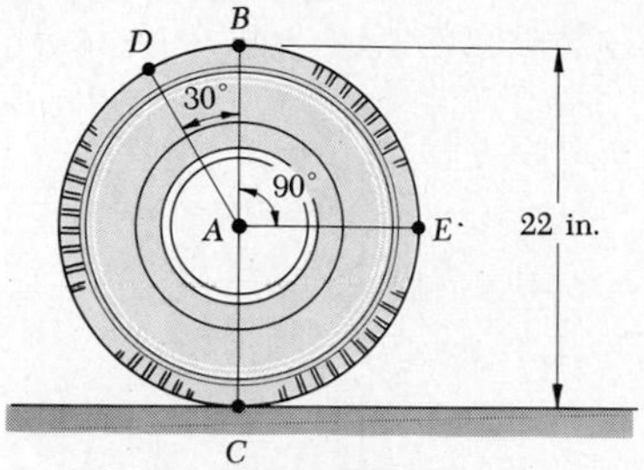

Fig. P15.85 and P15.86

15.86 An automobile is started from rest and has a constant acceleration of 7 ft/s² to the right. Knowing that the wheel shown rolls without sliding, determine the speed of the automobile at which the magnitude of the acceleration of point *B* of the wheel is 50 ft/s².

15.87 The 125-mm-radius drum rolls without slipping on a portion of a belt which moves downward to the left with a constant velocity of 150 mm/s. Knowing that at a given instant the velocity and acceleration of the center *A* of the drum are as shown, determine the acceleration of point *D*.

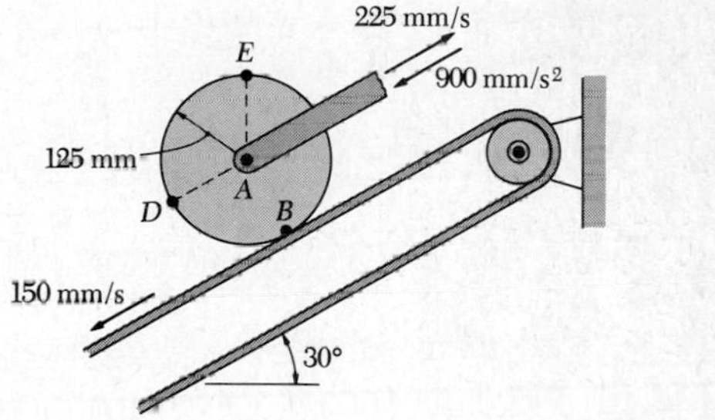

Fig. P15.87

15.88 In Prob. 15.87, determine the acceleration of point *E*.

15.89 Crank *AB* has a constant angular velocity of 200 rpm counterclockwise. Determine the acceleration of collar *D* when (*a*) $\theta = 0$, (*b*) $\theta = 90°$.

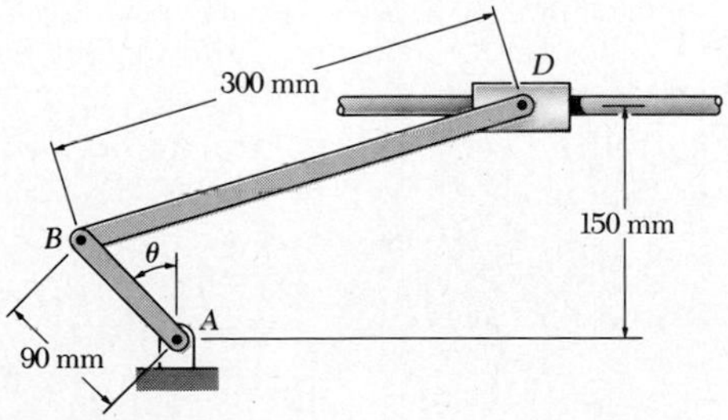

Fig. P15.89

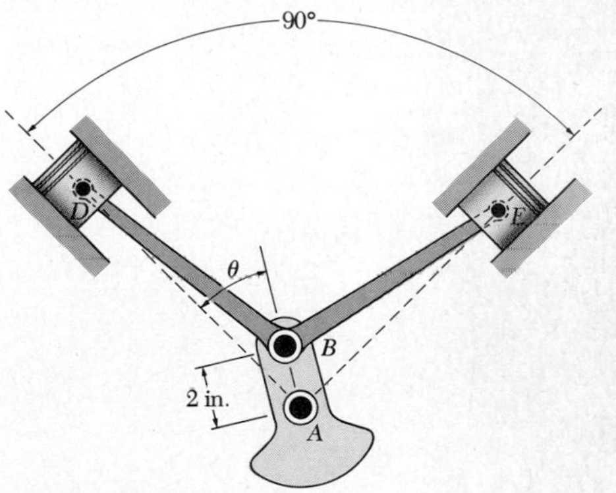

Fig. P15.90

15.90 In the two-cylinder air compressor shown the connecting rods *BD* and *BE* are each 8 in. long and the crank *AB* rotates about the fixed point *A* with a constant angular velocity of 1800 rpm clockwise. Determine the acceleration of each piston when $\theta = 0°$.

15.91 In Prob. 15.90, determine the acceleration of piston *E* when $\theta = 45°$.

15.92 Arm AB rotates with a constant angular velocity of 90 rpm clockwise. Knowing that gear A does not rotate, determine the acceleration of the tooth of gear B which is in contact with gear A.

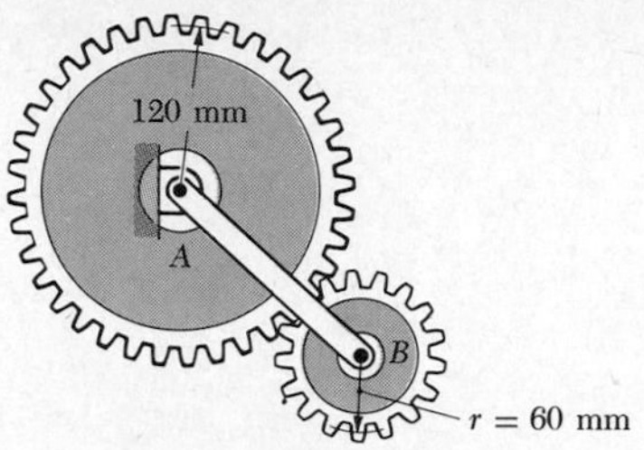

Fig. P15.92

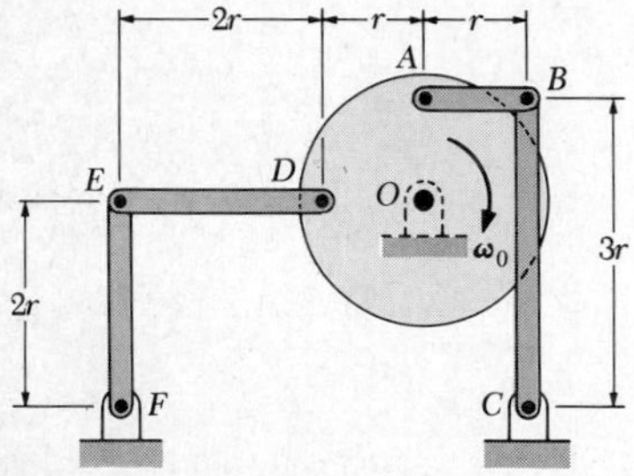

Fig. P15.93 and P15.94

15.93 At the instant shown, the disk rotates with a constant angular velocity ω_0 clockwise. Determine the angular velocities and the angular accelerations of the rods AB and BC.

15.94 At the instant shown, the disk rotates with a constant angular velocity ω_0 clockwise. Determine the angular velocities and the angular accelerations of the rods DE and EF.

15.95 and 15.96 For the linkage indicated, determine the angular acceleration (*a*) of bar BD, (*b*) of bar DE.

15.95 Linkage of Prob. 15.51.
15.96 Linkage of Prob. 15.48.

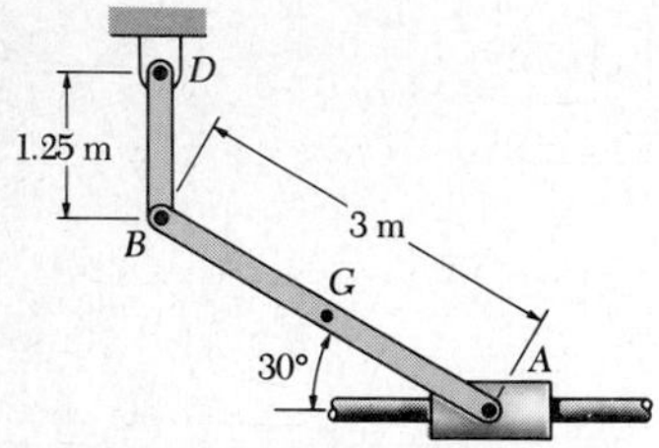

Fig. P15.97 and P15.98

15.97 End A of rod AB moves to the right with a constant velocity of 2.5 m/s. For the position shown, determine (*a*) the angular acceleration of rod AB, (*b*) the acceleration of the midpoint G of rod AB.

15.98 In the position shown, end A of rod AB has a velocity of 2.5 m/s and an acceleration of 1.5 m/s^2, both directed to the right. Determine (*a*) the angular acceleration of rod AB, (*b*) the acceleration of the midpoint G of rod AB.

15.99 In the position shown, end A of rod AB has a velocity of 3 ft/s and an acceleration of 2.5 ft/s^2, both directed to the left. Determine (*a*) the angular acceleration of rod AB, (*b*) the acceleration of the midpoint G of rod AB.

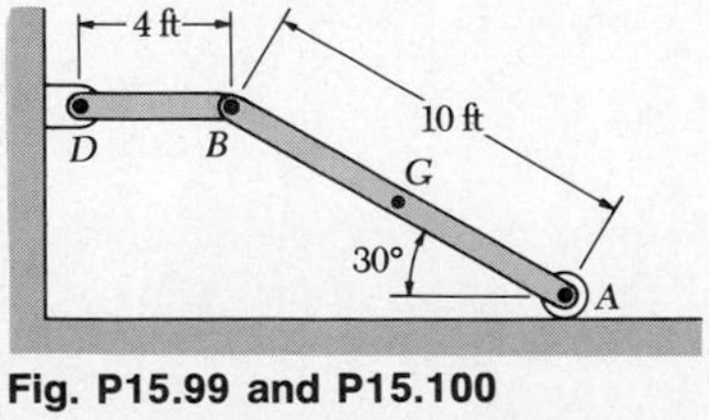

Fig. P15.99 and P15.100

15.100 End A of rod AB moves to the left with a constant velocity of 3 ft/s. For the position shown, determine (*a*) the angular acceleration of rod AB, (*b*) the acceleration of the midpoint G of rod AB.

15.101 Denoting by $\mathbf{r}_A$ the position vector of a point A of a rigid slab which moves in plane motion, show that (*a*) the position vector $\mathbf{r}_C$ of the instantaneous center of rotation is

$$\mathbf{r}_C = \mathbf{r}_A + \frac{\boldsymbol{\omega} \times \mathbf{v}_A}{\omega^2}$$

where $\boldsymbol{\omega}$ is the angular velocity of the slab and $\mathbf{v}_A$ the velocity of point A, (*b*) the acceleration of the instantaneous center of rotation is zero if, and only if,

$$\mathbf{a}_A = \frac{\alpha}{\omega}\mathbf{v}_A + \boldsymbol{\omega} \times \mathbf{v}_A$$

where $\boldsymbol{\alpha} = \alpha\mathbf{k}$ is the angular acceleration of the slab.

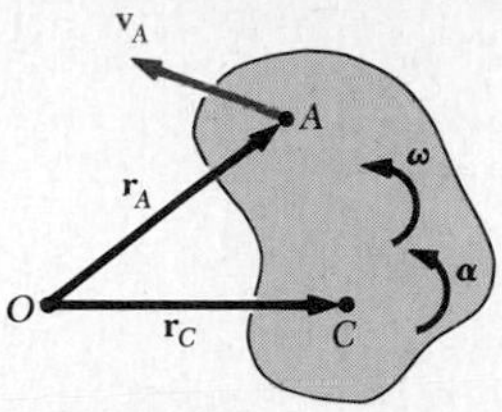

Fig. P15.101

***15.102** Rod AB slides with its ends in contact with the floor and the inclined plane. Using the method of Sec. 15.9, derive an expression for the angular velocity of the rod in terms of v_B, θ, l, and β.

***15.103** Derive an expression for the angular acceleration of the rod AB in terms of v_B, θ, l, and β, knowing that the acceleration of point B is zero.

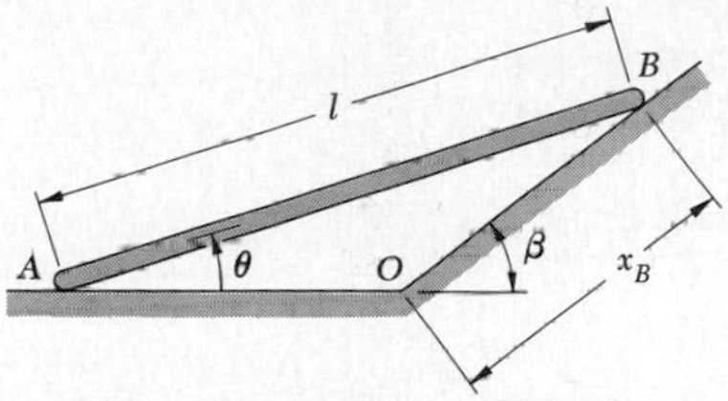

Fig. P15.102 and P15.103

***15.104** Blocks A and B are connected by a cable AB which passes over a pulley at C. Knowing that block A moves downward along the curved surface with a constant speed v_0, determine (*a*) the velocity of block B, (*b*) the acceleration of block B.

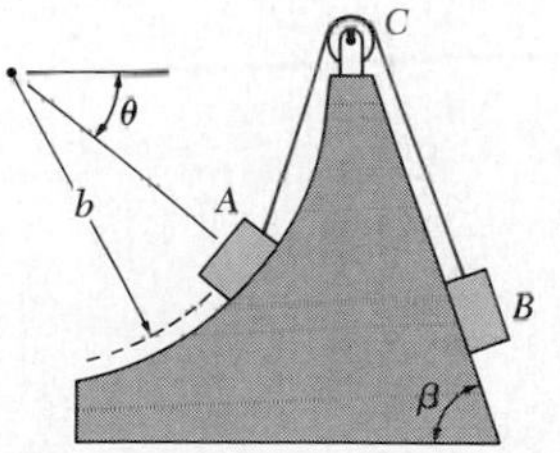

Fig. P15.104

***15.105** The drive disk of the Scotch crosshead mechanism shown has an angular velocity $\boldsymbol{\omega}$ and an angular acceleration $\boldsymbol{\alpha}$, both directed clockwise. Using the method of Sec. 15.9, derive an expression for (*a*) the velocity of point A, (*b*) the acceleration of point A.

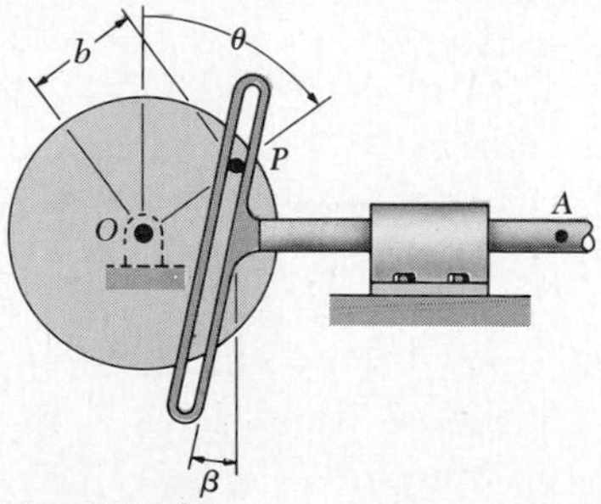

Fig. P15.105

***15.106** Solve Prob. 15.105, assuming that $\beta = 0$.

***15.107** Knowing that rod AB rotates with an angular velocity $\boldsymbol{\omega}$ and with an angular acceleration $\boldsymbol{\alpha}$, both counterclockwise, derive expressions for the velocity and acceleration of collar C.

***15.108** Knowing that collar C moves upward with a constant velocity $\mathbf{v}_0$, derive expressions for the angular velocity and angular acceleration of rod AB in the position shown.

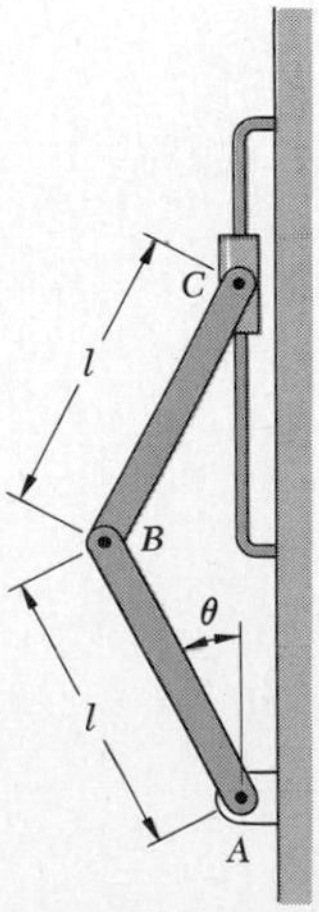

Fig. P15.107 and P15.108

*** 15.109** The crank AB rotates with a constant clockwise angular velocity ω, and $\theta = 0$ at $t = 0$. Using the method of Sec. 15.9, derive an expression for the velocity of the piston P in terms of the time t.

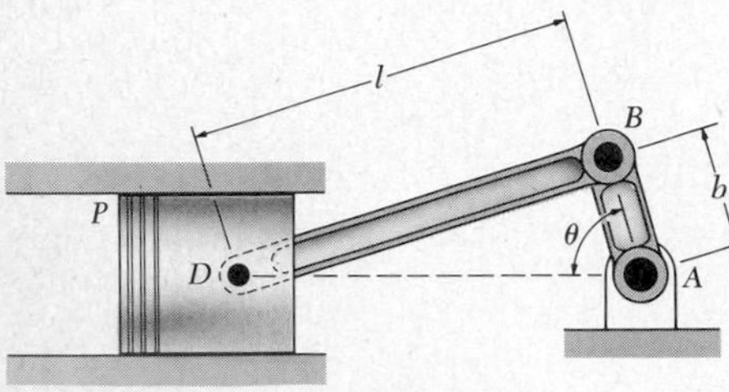

Fig. P15.109

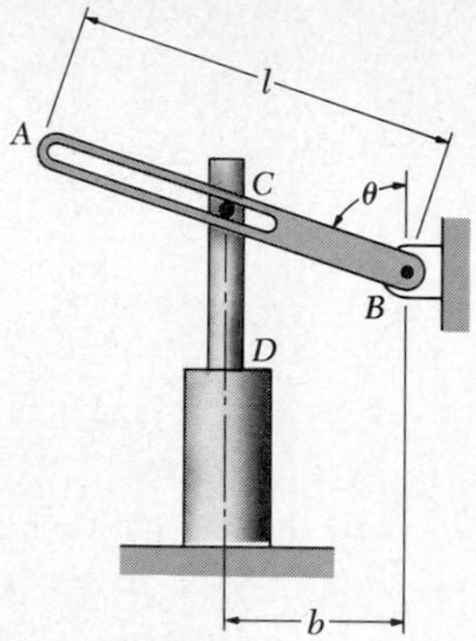

Fig. P15.110

*** 15.110** Pin C is attached to rod CD and slides in a slot cut in arm AB. Knowing that rod CD moves vertically upward with a constant velocity $\mathbf{v}_0$, derive an expression for (*a*) the angular velocity of arm AB, (*b*) the components of the velocity of point A.

*** 15.111** In Prob. 15.110, derive an expression for the angular acceleration of arm AB.

*** 15.112** A disk of radius r rolls to the right with a constant velocity $\mathbf{v}$. Denoting by P the point of the rim in contact with the ground at $t = 0$, derive expressions for the horizontal and vertical components of the velocity of P at any time t. (The curve described by point P is called a *cycloid*.)

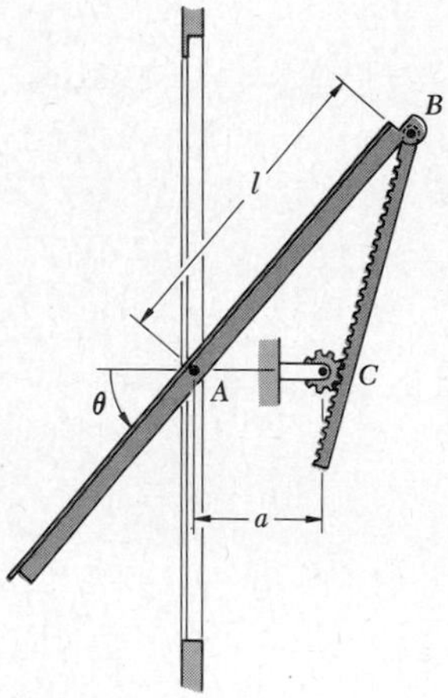

Fig. P15.113

*** 15.113** The position of a factory window is controlled by the rack and pinion shown. Knowing that the pinion C has a radius r and rotates counterclockwise at a constant rate ω, derive an expression for the angular velocity of the window.

15.10. Rate of Change of a Vector with Respect to a Rotating Frame. We saw in Sec. 11.10 that the rate of change of a vector is the same with respect to a fixed frame and with respect to a frame in translation. In this section, we shall compare the rates of change of a vector $\mathbf{Q}$ with respect to a fixed frame and with respect to a rotating frame of reference.† We shall also learn to determine the rate of change of $\mathbf{Q}$ with respect to one frame of reference when $\mathbf{Q}$ is defined by its components in another frame.

Consider two frames of reference centered at O, a fixed frame $OXYZ$ and a frame $Oxyz$ which rotates about the fixed axis OA; let $\mathbf{\Omega}$ denote the angular velocity of the frame $Oxyz$ at a given instant (Fig. 15.26). Consider now a vector function $\mathbf{Q}(t)$ represented by the vector $\mathbf{Q}$ attached at O; as the time t varies, both the direction and the magnitude of $\mathbf{Q}$ change. Since

† It is recalled that the selection of a fixed frame of reference is arbitrary. Any frame may be designated as "fixed"; all others will then be considered as moving.

the variation of $\mathbf{Q}$ is viewed differently by an observer using $OXYZ$ as a frame of reference and by an observer using $Oxyz$, we should expect the rate of change of $\mathbf{Q}$ to depend upon the frame of reference which has been selected. Therefore, we shall denote by $(\dot{\mathbf{Q}})_{OXYZ}$ the rate of change of $\mathbf{Q}$ with respect to the fixed frame $OXYZ$, and by $(\dot{\mathbf{Q}})_{Oxyz}$ its rate of change with respect to the rotating frame of $Oxyz$. We propose to determine the relation existing between these two rates of change.

Let us first resolve the vector $\mathbf{Q}$ into components along the x, y, and z axes of the rotating frame. Denoting by $\mathbf{i}$, $\mathbf{j}$, and $\mathbf{k}$ the corresponding unit vectors, we write

$$\mathbf{Q} = Q_x\mathbf{i} + Q_y\mathbf{j} + Q_z\mathbf{k} \tag{15.27}$$

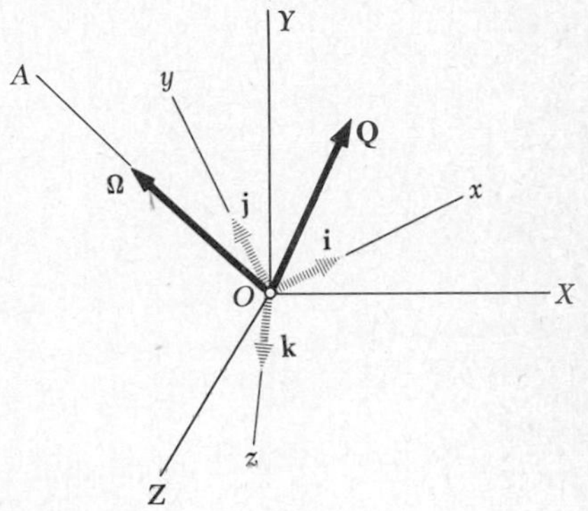

Fig. 15.26

Differentiating (15.27) with respect to t and considering the unit vectors $\mathbf{i}$, $\mathbf{j}$, $\mathbf{k}$ as fixed, we obtain the rate of change of $\mathbf{Q}$ *with respect to the rotating frame Oxyz:*

$$(\dot{\mathbf{Q}})_{Oxyz} = \dot{Q}_x\mathbf{i} + \dot{Q}_y\mathbf{j} + \dot{Q}_z\mathbf{k} \tag{15.28}$$

To obtain the rate of change of $\mathbf{Q}$ *with respect to the fixed frame* $OXYZ$, we must consider the unit vectors $\mathbf{i}$, $\mathbf{j}$, $\mathbf{k}$ as variable when differentiating (15.27). We therefore write

$$(\dot{\mathbf{Q}})_{OXYZ} = \dot{Q}_x\mathbf{i} + \dot{Q}_y\mathbf{j} + \dot{Q}_z\mathbf{k} + Q_x\frac{d\mathbf{i}}{dt} + Q_y\frac{d\mathbf{j}}{dt} + Q_z\frac{d\mathbf{k}}{dt} \tag{15.29}$$

Recalling (15.28), we observe that the sum of the first three terms in the right-hand member of (15.29) represents the rate of change $(\dot{\mathbf{Q}})_{Oxyz}$. We note, on the other hand, that the rate of change $(\dot{\mathbf{Q}})_{OXYZ}$ would reduce to the last three terms in (15.29) if the vector $\mathbf{Q}$ were fixed within the frame $Oxyz$, since $(\dot{\mathbf{Q}})_{Oxyz}$ would then be zero. But in that case, $(\dot{\mathbf{Q}})_{OXYZ}$ would represent the velocity of a particle located at the tip of $\mathbf{Q}$ and belonging to a body rigidly attached to the frame $Oxyz$. Thus, the last three terms in (15.29) represent the velocity of that particle; since the frame $Oxyz$ has an angular velocity $\boldsymbol{\Omega}$ with respect to $OXYZ$ at the instant considered, we write, by (15.5),

$$Q_x\frac{d\mathbf{i}}{dt} + Q_y\frac{d\mathbf{j}}{dt} + Q_z\frac{d\mathbf{k}}{dt} = \boldsymbol{\Omega} \times \mathbf{Q} \tag{15.30}$$

Substituting from (15.28) and (15.30) into (15.29), we obtain the fundamental relation

$$(\dot{\mathbf{Q}})_{OXYZ} = (\dot{\mathbf{Q}})_{Oxyz} + \boldsymbol{\Omega} \times \mathbf{Q} \tag{15.31}$$

We conclude that the rate of change of the vector $\mathbf{Q}$ with respect to the fixed frame $OXYZ$ is made of two parts: The first part represents the rate of change of $\mathbf{Q}$ with respect to the rotating frame $Oxyz$; the second part, $\boldsymbol{\Omega} \times \mathbf{Q}$, is induced by the rotation of the frame $Oxyz$.

The use of relation (15.31) simplifies the determination of the rate of change of a vector $\mathbf{Q}$ with respect to a fixed frame of reference $OXYZ$ when the vector **Q** is defined by its components along the axes of a rotating frame *Oxyz*, since this relation does not require the separate computation of the derivatives of the unit vectors defining the orientation of the rotating frame.

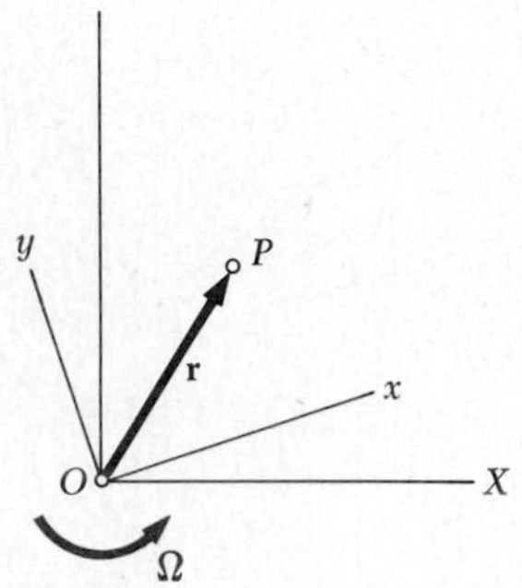

Fig. 15.27

15.11. Plane Motion of a Particle Relative to a Rotating Frame. Coriolis Acceleration. Consider two frames of reference, both centered at O and both in the plane of the figure, a fixed frame OXY, and a rotating frame Oxy (Fig. 15.27). Let P be a particle moving in the plane of the figure. While the position vector **r** of P is the same in both frames, its rate of change depends upon the frame of reference which has been selected.

The absolute velocity $\mathbf{v}_P$ of the particle is defined as the velocity observed from the fixed frame OXY and is equal to the rate of change $(\dot{\mathbf{r}})_{OXY}$ of **r** with respect to that frame. We may, however, express $\mathbf{v}_P$ in terms of the rate of change $(\dot{\mathbf{r}})_{Oxy}$ observed from the rotating frame if we make use of Eq. (15.31). Denoting by $\boldsymbol{\Omega}$ the angular velocity of the frame Oxy with respect to OXY at the instant considered, we write

$$\mathbf{v}_P = (\dot{\mathbf{r}})_{OXY} = \boldsymbol{\Omega} \times \mathbf{r} + (\dot{\mathbf{r}})_{Oxy} \tag{15.32}$$

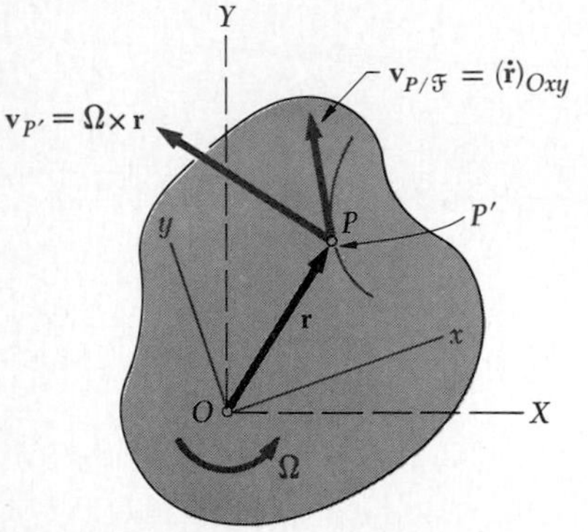

Fig. 15.28

But $(\dot{\mathbf{r}})_{Oxy}$ defines the velocity of the particle P relative to the rotating frame Oxy. Denoting the rotating frame by $\mathcal{F}$ for short, we shall represent the velocity $(\dot{\mathbf{r}})_{Oxy}$ of P relative to the rotating frame by the symbol $\mathbf{v}_{P/\mathcal{F}}$. If we imagine that a rigid slab has been attached to the rotating frame, $v_{P/\mathcal{F}}$ will represent the velocity of P along the path that it describes on that slab (Fig. 15.28). On the other hand, the term $\boldsymbol{\Omega} \times \mathbf{r}$ in (15.32) will represent the velocity $\mathbf{v}_{P'}$ of the point P' of the slab—or rotating frame—which coincides with P at the instant considered. Thus, we have

$$\mathbf{v}_P = \mathbf{v}_{P'} + \mathbf{v}_{P/\mathcal{F}} \tag{15.33}$$

where $\mathbf{v}_P$ = absolute velocity of particle P
$\mathbf{v}_{P'}$ = velocity of point P' of moving frame $\mathcal{F}$ coinciding with P
$\mathbf{v}_{P/\mathcal{F}}$ = velocity of P relative to moving frame $\mathcal{F}$

The absolute acceleration $\mathbf{a}_P$ of the particle is defined as the rate of change of $\mathbf{v}_P$ with respect to the fixed frame OXY. Computing the rates of change with respect to OXY of the terms in (15.32), we write

$$\mathbf{a}_P = \dot{\mathbf{v}}_P = \dot{\boldsymbol{\Omega}} \times \mathbf{r} + \boldsymbol{\Omega} \times \dot{\mathbf{r}} + \frac{d}{dt}[(\dot{\mathbf{r}})_{Oxy}] \tag{15.34}$$

where all derivatives are defined with respect to OXY, except where indicated otherwise. Referring to Eq. (15.31), we note that the last term in (15.34) may be expressed as

$$\frac{d}{dt}[(\dot{\mathbf{r}})_{Oxy}] = (\ddot{\mathbf{r}})_{Oxy} + \boldsymbol{\Omega} \times (\dot{\mathbf{r}})_{Oxy}$$

On the other hand, $\dot{\mathbf{r}}$ represents the velocity $\mathbf{v}_P$ and may be replaced by the right-hand member of Eq. (15.32). After completing these two substitutions into (15.34), we write

$$\mathbf{a}_P = \dot{\mathbf{\Omega}} \times \mathbf{r} + \mathbf{\Omega} \times (\mathbf{\Omega} \times \mathbf{r}) + 2\mathbf{\Omega} \times (\dot{\mathbf{r}})_{Oxy} + (\ddot{\mathbf{r}})_{Oxy} \tag{15.35}$$

Referring to the expression (15.8) obtained in Sec. 15.3 for the acceleration of a particle in a rigid body rotating about a fixed axis, we note that the sum of the first two terms represents the acceleration $\mathbf{a}_{P'}$ of the point P' of the rotating frame which coincides with P at the instant considered. On the other hand, the last term defines the acceleration $\mathbf{a}_{P/\mathcal{F}}$ of P relative to the rotating frame. If it were not for the third term, which has not been accounted for, a relation similar to (15.33) could be written for the accelerations, and $\mathbf{a}_P$ could be expressed as the sum of $\mathbf{a}_{P'}$ and $\mathbf{a}_{P/\mathcal{F}}$. However, it is clear that *such a relation would be incorrect* and that we must include the additional term. This term, which we shall denote by $\mathbf{a}_c$, is called the *complementary acceleration*, or *Coriolis acceleration*, after the French mathematician De Coriolis (1792–1843). We write

$$\mathbf{a}_P = \mathbf{a}_{P'} + \mathbf{a}_{P/\mathcal{F}} + \mathbf{a}_c \tag{15.36}$$

where $\mathbf{a}_P$ = absolute acceleration of particle P
$\mathbf{a}_{P'}$ = acceleration of point P' of moving frame $\mathcal{F}$ coinciding with P
$\mathbf{a}_{P/\mathcal{F}}$ = acceleration of P relative to moving frame $\mathcal{F}$
$\mathbf{a}_c = 2\mathbf{\Omega} \times (\dot{\mathbf{r}})_{Oxy} = 2\mathbf{\Omega} \times \mathbf{v}_{P/\mathcal{F}}$
= complementary, or Coriolis, acceleration†

We note that since point P' moves in a circle about the origin O, its acceleration $\mathbf{a}_{P'}$ has, in general, two components: a component $(\mathbf{a}_{P'})_t$ tangent to the circle, and a component $(\mathbf{a}_{P'})_n$ directed toward O. Similarly, the acceleration $\mathbf{a}_{P/\mathcal{F}}$ generally has two components: a component $(\mathbf{a}_{P/\mathcal{F}})_t$ tangent to the path that P describes on the rotating slab, and a component $(\mathbf{a}_{P/\mathcal{F}})_n$ directed toward the center of curvature of that path. We further note that since the vector $\mathbf{\Omega}$ is perpendicular to the plane of motion, and thus to $\mathbf{v}_{P/\mathcal{F}}$, the magnitude of the Coriolis acceleration $\mathbf{a}_c = 2\mathbf{\Omega} \times \mathbf{v}_{P/\mathcal{F}}$ is equal to $2\Omega v_{P/\mathcal{F}}$, and its direction may be obtained by rotating the vector $\mathbf{v}_{P/\mathcal{F}}$ through 90° in the sense of rotation of the moving frame (Fig. 15.29). The Coriolis acceleration reduces to zero when either $\mathbf{\Omega}$ or $\mathbf{v}_{P/\mathcal{F}}$ is zero.

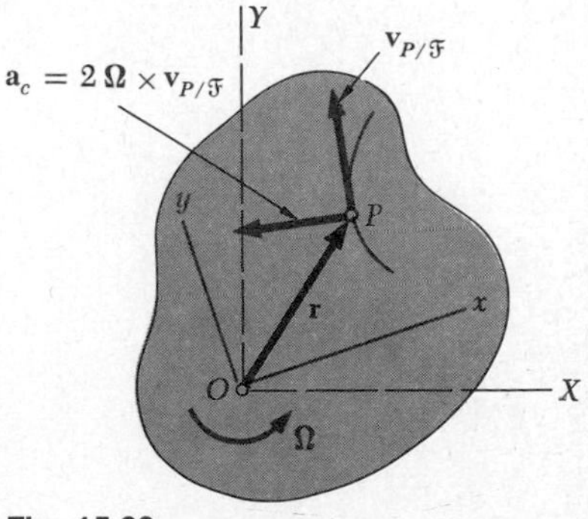

Fig. 15.29

†It is important to note the difference between Eq. (15.36) and Eq. (15.21) of Sec. 15.8. When we wrote

$$\mathbf{a}_B = \mathbf{a}_A + \mathbf{a}_{B/A} \tag{15.21}$$

in Sec. 15.8, we were expressing the absolute acceleration of point B as the sum of its acceleration $\mathbf{a}_{B/A}$ relative to a *frame in translation* and of the acceleration $\mathbf{a}_A$ of a point of that frame. We are now trying to relate the absolute acceleration of point P to its acceleration $\mathbf{a}_{P/\mathcal{F}}$ relative to a *rotating frame* $\mathcal{F}$ and to the acceleration $\mathbf{a}_{P'}$ of the point P' of that frame which coincides with P; Eq. (15.36) shows that because the frame is rotating, it is necessary to include an additional term representing the Coriolis acceleration $\mathbf{a}_c$.

The following example will help in understanding the physical meaning of the Coriolis acceleration. Consider a collar P which is made to slide at a constant relative speed u along a rod OB rotating at a constant angular velocity $\boldsymbol{\omega}$ about O (Fig. 15.30a). According to formula (15.36), the absolute acceleration of P may be obtained by adding vectorially the acceleration $\mathbf{a}_A$ of the point A of the rod coinciding with P, the relative acceleration $\mathbf{a}_{P/OB}$ of P with respect to the rod, and the Coriolis acceleration $\mathbf{a}_c$. Since the angular velocity $\boldsymbol{\omega}$ of the rod is constant, $\mathbf{a}_A$ reduces to its normal component $(\mathbf{a}_A)_n$ of magnitude $r\omega^2$; and since u is constant, the relative acceleration $\mathbf{a}_{P/OB}$ is zero. According to the definition given above, the Coriolis acceleration is a vector perpendicular to OB, of magnitude $2\omega u$, and directed as shown in the figure. The acceleration of the collar P consists, therefore, of the two vectors shown in Fig. 15.30a. Note that the result obtained may be checked by applying the relation (11.44).

To understand better the significance of the Coriolis acceleration, we shall consider the absolute velocity of P at time t and at time $t + \Delta t$ (Fig. 15.30b). At time t, the velocity may be resolved into its components $\mathbf{u}$ and $\mathbf{v}_A$, and at time $t + \Delta t$ into its components $\mathbf{u}'$ and $\mathbf{v}_{A'}$. Drawing these components from the same origin (Fig. 15.30c), we note that the change in velocity during the time Δt may be represented by the sum of three vectors, $\overrightarrow{RR'}$, $\overrightarrow{TT''}$, and $\overrightarrow{T''T'}$. The vector $\overrightarrow{TT''}$ measures the change in direction of the velocity $\mathbf{v}_A$, and the quotient $\overrightarrow{TT''}/\Delta t$ represents the acceleration $\mathbf{a}_A$ when Δt approaches zero. We check that the direction of $\overrightarrow{TT''}$ is that of $\mathbf{a}_A$ when Δt approaches zero and that

$$\lim_{\Delta t \to 0} \frac{TT''}{\Delta t} = \lim_{\Delta t \to 0} v_A \frac{\Delta\theta}{\Delta t} = r\omega\omega = r\omega^2 = a_A$$

The vector $\overrightarrow{RR'}$ measures the change in direction of $\mathbf{u}$ due to the rotation of the rod; the vector $\overrightarrow{T''T'}$ measures the change in magnitude of $\mathbf{v}_A$ due to the motion of P on the rod. The vectors $\overrightarrow{RR'}$ and $\overrightarrow{T''T'}$ result from the *combined effect* of the relative motion of P and of the rotation of the rod; they would vanish if *either* of these two motions stopped. We may easily verify that the sum of these two vectors defines the Coriolis acceleration. Their direction is that of $\mathbf{a}_c$ when Δt approaches zero, and since $RR' = u\,\Delta\theta$ and $T''T' = v_{A'} - v_A = (r + \Delta r)\omega - r\omega = \omega\,\Delta r$, we check that

$$\lim_{\Delta t \to 0} \left(\frac{RR'}{\Delta t} + \frac{T''T'}{\Delta t} \right) = \lim_{\Delta t \to 0} \left(u\frac{\Delta\theta}{\Delta t} + \omega\frac{\Delta r}{\Delta t} \right) = u\omega + \omega u = 2\omega u = a_c$$

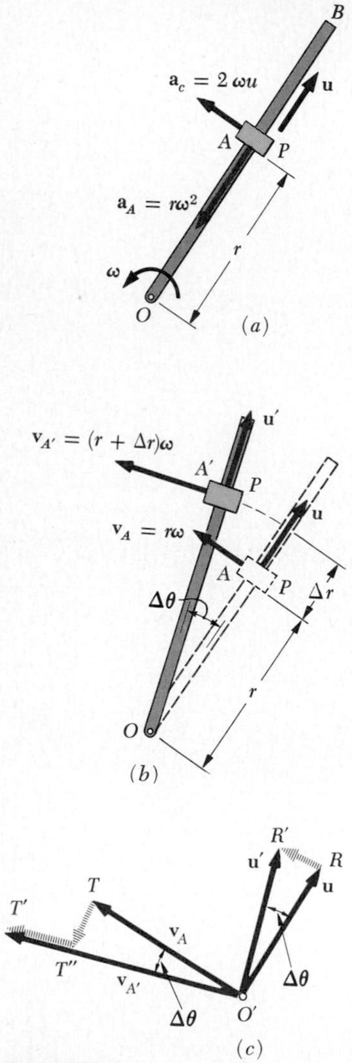

Fig. 15.30

Formulas (15.33) and (15.36) may be used to analyze the motion of mechanisms which contain parts sliding on each other. They make it possible, for example, to relate the absolute and relative motions of sliding pins and collars (see Sample Probs. 15.9 and 15.10). The concept of Coriolis acceleration is also very useful in the study of long-range projectiles and of other bodies whose motions are appreciably affected by the rotation of the earth. As was pointed out in Sec. 12.2, a system of axes attached to the earth does not truly constitute a newtonian frame of reference; such a system of axes should actually be considered as rotating. The formulas derived in this section will therefore facilitate the study of the motion of bodies with respect to axes attached to the earth.

SAMPLE PROBLEM 15.9

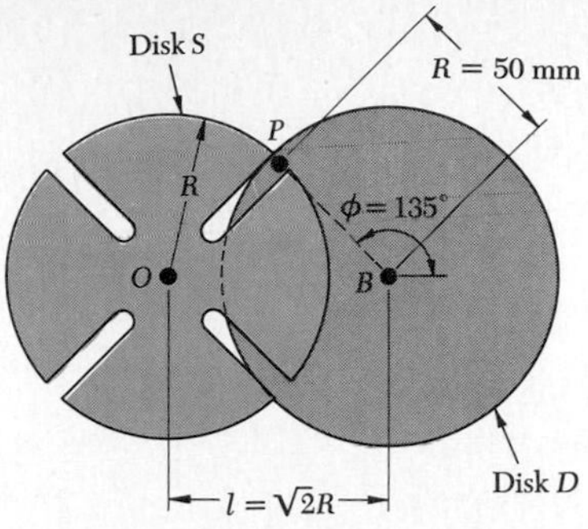

The Geneva mechanism shown is used in many counting instruments and in other applications where an intermittent rotary motion is required. Disk D rotates with a constant counterclockwise angular velocity ω_D of 10 rad/s. A pin P is attached to disk D and slides along one of several slots cut in disk S. It is desirable that the angular velocity of disk S be zero as the pin enters and leaves each slot; in the case of four slots, this will occur if the distance between the centers of the disks is $l = \sqrt{2}\,R$.

At the instant when $\phi = 150°$, determine (*a*) the angular velocity of disk S, (*b*) the velocity of pin P relative to disk S.

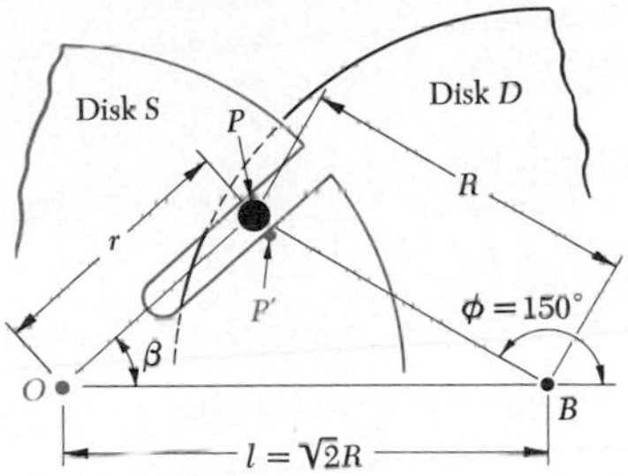

Solution. We solve triangle OPB, which corresponds to the position $\phi = 150°$. Using the law of cosines, we write

$$r^2 = R^2 + l^2 - 2Rl\cos 30° = 0.551R^2 \qquad r = 0.742R = 37.1 \text{ mm}$$

From the law of sines,

$$\frac{\sin\beta}{R} = \frac{\sin 30°}{r} \qquad \sin\beta = \frac{\sin 30°}{0.742} \qquad \beta = 42.4°$$

Since pin P is attached to disk D, and since disk D rotates about point B, the magnitude of the absolute velocity of P is

$$v_P = R\omega_D = (50 \text{ mm})(10 \text{ rad/s}) = 500 \text{ mm/s}$$
$$\mathbf{v}_P = 500 \text{ mm/s} \measuredangle 60°$$

We consider now the motion of pin P along the slot in disk S. Denoting by P' the point of disk S which coincides with P at the instant considered and selecting a rotating frame $\mathcal{S}$ attached to disk S, we write

$$\mathbf{v}_P = \mathbf{v}_{P'} + \mathbf{v}_{P/\mathcal{S}}$$

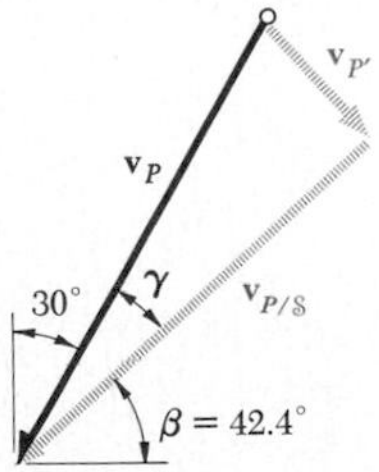

Noting that $\mathbf{v}_{P'}$ is perpendicular to the radius OP and that $\mathbf{v}_{P/\mathcal{S}}$ is directed along the slot, we draw the velocity triangle corresponding to the equation above. From the triangle, we compute

$$\gamma = 90° - 42.4° - 30° = 17.6°$$
$$v_{P'} = v_P \sin\gamma = (500 \text{ mm/s}) \sin 17.6°$$
$$\mathbf{v}_{P'} = 151.2 \text{ mm/s} \searrow 42.4°$$
$$v_{P/\mathcal{S}} = v_P \cos\gamma = (500 \text{ mm/s}) \cos 17.6°$$
$$\mathbf{v}_{P/S} = \mathbf{v}_{P/\mathcal{S}} = 477 \text{ mm/s} \measuredangle 42.4° \quad ◄$$

Since $\mathbf{v}_{P'}$ is perpendicular to the radius OP, we write

$$v_{P'} = r\omega_{\mathcal{S}} \qquad 151.2 \text{ mm/s} = (37.1 \text{ mm})\omega_{\mathcal{S}}$$
$$\omega_S = \omega_{\mathcal{S}} = 4.08 \text{ rad/s} \curvearrowright \quad ◄$$

SAMPLE PROBLEM 15.10

In the Geneva mechanism of Sample Prob. 15.9, disk D rotates with a constant counterclockwise angular velocity $\boldsymbol{\omega}_D$ of 10 rad/s. At the instant when $\phi = 150°$, determine the angular acceleration of disk S.

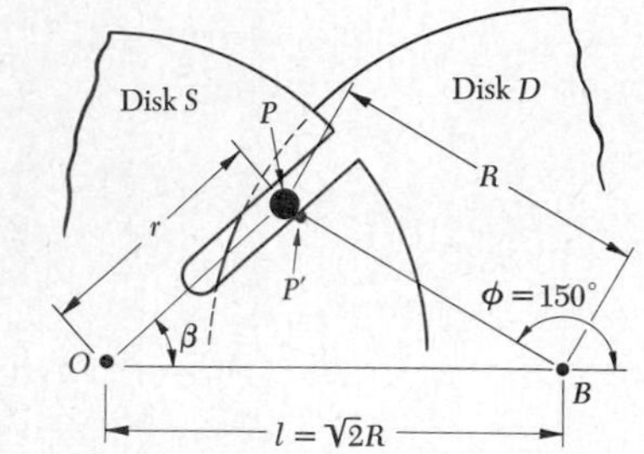

Solution. Referring to Sample Prob. 15.9, we obtain the angular velocity of the frame $\mathcal{S}$ attached to disk S and the velocity of the pin relative to $\mathcal{S}$:

$$\omega_{\mathcal{S}} = 4.08 \text{ rad/s} \circlearrowright$$

$$\beta = 42.4° \qquad \mathbf{v}_{P/\mathcal{S}} = 477 \text{ mm/s} \nearrow 42.4°$$

Since pin P moves with respect to the rotating frame $\mathcal{S}$, we write

$$\mathbf{a}_P = \mathbf{a}_{P'} + \mathbf{a}_{P/\mathcal{S}} + \mathbf{a}_c \tag{1}$$

Each term of this vector equation is investigated separately.

Absolute Acceleration $\mathbf{a}_P$**.** Since disk D rotates with a constant angular velocity, the absolute acceleration $\mathbf{a}_P$ is directed toward B. We have

$$a_P = R\omega_D^2 = (50 \text{ mm})(10 \text{ rad/s})^2 = 5000 \text{ mm/s}^2$$

$$\mathbf{a}_P = 5000 \text{ mm/s}^2 \searrow 30°$$

Acceleration $\mathbf{a}_{P'}$ ***of the Coinciding Point P′.*** The acceleration $\mathbf{a}_{P'}$ of the point P' of the frame $\mathcal{S}$ which coincides with P at the instant considered is resolved into normal and tangential components. (We recall from Sample Prob. 15.9 that $r = 37.1$ mm.)

$$(a_{P'})_n = r\omega_{\mathcal{S}}^2 = (37.1 \text{ mm})(4.08 \text{ rad/s})^2 = 618 \text{ mm/s}^2$$

$$(\mathbf{a}_{P'})_n = 618 \text{ mm/s}^2 \nearrow 42.4°$$

$$(a_{P'})_t = r\alpha_{\mathcal{S}} = 37.1\alpha_{\mathcal{S}} \qquad (\mathbf{a}_{P'})_t = 37.1\alpha_{\mathcal{S}} \nwarrow 42.4°$$

Relative Acceleration $\mathbf{a}_{P/\mathcal{S}}$**.** Since the pin P moves in a straight slot cut in disk S, the relative acceleration $\mathbf{a}_{P/\mathcal{S}}$ must be parallel to the slot; i.e., its direction must be $\swarrow$ 42.4°.

Coriolis Acceleration $\mathbf{a}_c$**.** Rotating the relative velocity $\mathbf{v}_{P/\mathcal{S}}$ through 90° in the sense of $\boldsymbol{\omega}_{\mathcal{S}}$, we obtain the direction of the Coriolis component of the acceleration: $\nwarrow$ 42.4°. We write

$$a_c = 2\omega_{\mathcal{S}}v_{P/\mathcal{S}} = 2(4.08 \text{ rad/s})(477 \text{ mm/s}) = 3890 \text{ mm/s}^2$$

$$\mathbf{a}_c = 3890 \text{ mm/s}^2 \nwarrow 42.4°$$

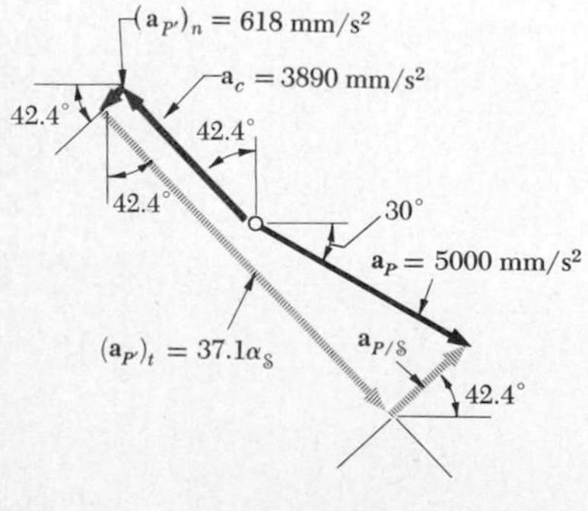

We rewrite Eq. (1) and substitute the accelerations found above:

$$\mathbf{a}_P = (\mathbf{a}_{P'})_n + (\mathbf{a}_{P'})_t + \mathbf{a}_{P/\mathcal{S}} + \mathbf{a}_c$$

$$[5000 \searrow 30°] = [618 \nearrow 42.4°] + [37.1\alpha_{\mathcal{S}} \nwarrow 42.4°]$$

$$+ [a_{P/\mathcal{S}} \swarrow 42.4°] + [3890 \nwarrow 42.4°]$$

Equating components in a direction perpendicular to the slot,

$$5000 \cos 17.6° = 37.1\alpha_{\mathcal{S}} - 3890$$

$$\boldsymbol{\alpha}_S = \boldsymbol{\alpha}_{\mathcal{S}} = 233 \text{ rad/s}^2 \circlearrowright \blacktriangleleft$$

Problems

15.114 and 15.115 Two rotating rods are connected by a slider block *P*. The rod attached at *A* rotates with a constant clockwise angular velocity ω_A. For the given data, determine for the position shown (*a*) the angular velocity of the rod attached at *B*, (*b*) the relative velocity of the slider block *P* with respect to the rod on which it slides.

15.114 $b = 8$ in., $\omega_A = 6$ rad/s.
15.115 $b = 250$ mm, $\omega_A = 8$ rad/s.

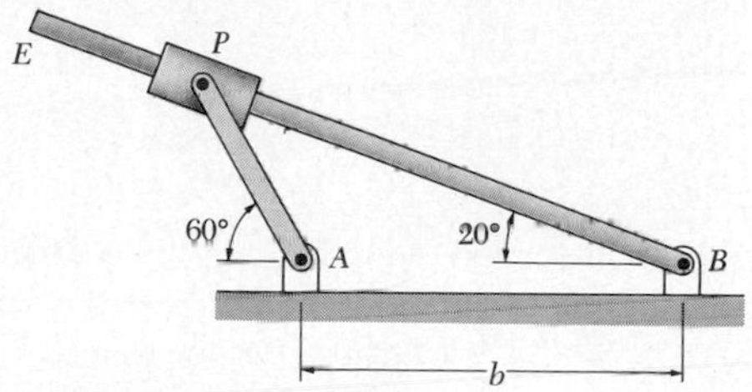

Fig. P15.114 and P15.116

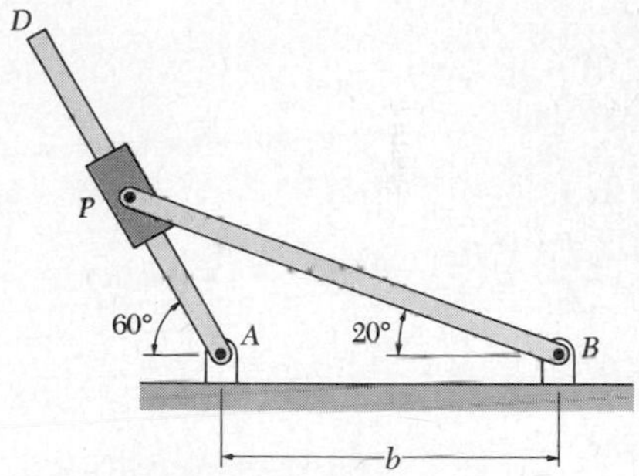

Fig. P15.115 and P15.117

15.116 and 15.117 Two rotating rods are connected by a slider block *P*. The velocity $\mathbf{v}_0$ of the slider block relative to the rod on which it slides is constant and is directed outward. For the given data, determine the angular velocity of each rod for the position shown.

15.116 $b = 250$ mm, $v_0 = 360$ mm/s.
15.117 $b = 8$ in., $v_0 = 12$ in./s.

15.118 The motion of pin *P* is guided by slots cut in rods *AE* and *BD*. Knowing that the rods rotate with the constant angular velocities $\boldsymbol{\omega}_A = 4$ rad/s ↻ and $\boldsymbol{\omega}_B = 3$ rad/s ↻, determine the velocity of pin *P* for the position shown.

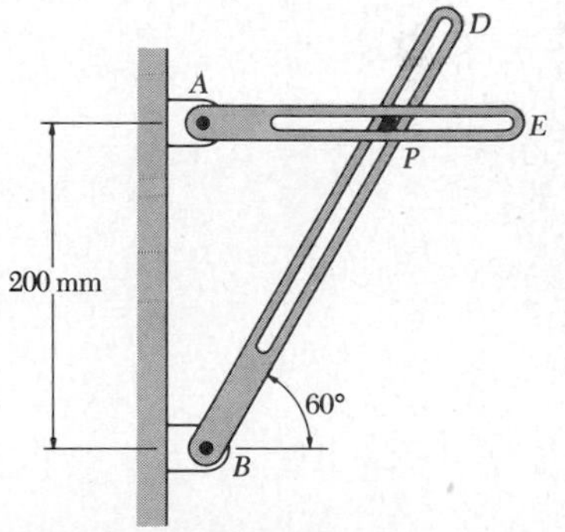

Fig. P15.118 and P15.119

15.119 The motion of pin *P* is guided by slots cut in rods *AE* and *BD*. Knowing that the rods rotate with the constant angular velocities $\boldsymbol{\omega}_A = 4$ rad/s ↺ and $\boldsymbol{\omega}_B = 3$ rad/s ↻, determine the velocity of pin *P* for the position shown.

15.120 Four pins slide in four separate slots cut in a circular plate as shown. When the plate is at rest, each pin has a velocity directed as shown and of the same constant magnitude *u*. If each pin maintains the same velocity in relation to the plate when the plate rotates about *O* with a constant *counterclockwise* angular velocity $\boldsymbol{\omega}$, determine the acceleration of each pin.

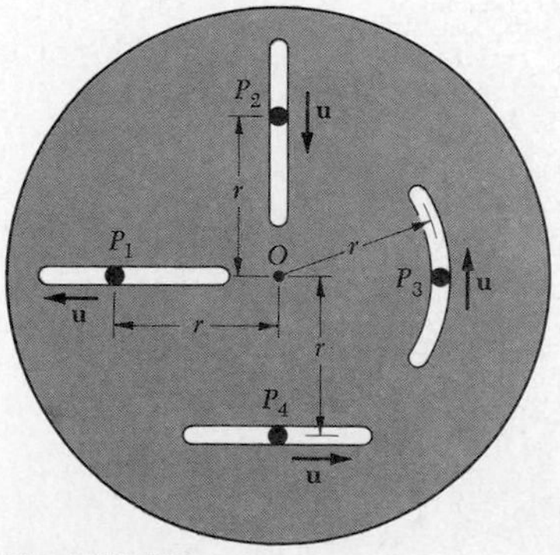

Fig. P15.120

15.121 Solve Prob. 15.120, assuming that the plate rotates about *O* with a constant *clockwise* angular velocity $\boldsymbol{\omega}$.

15.122 At the instant shown the length of the boom is being decreased at the constant rate of 8 in./s and the boom is being lowered at the constant rate of 0.08 rad/s. Knowing that $\theta = 30°$, determine (*a*) the velocity of point *B*, (*b*) the acceleration of point *B*.

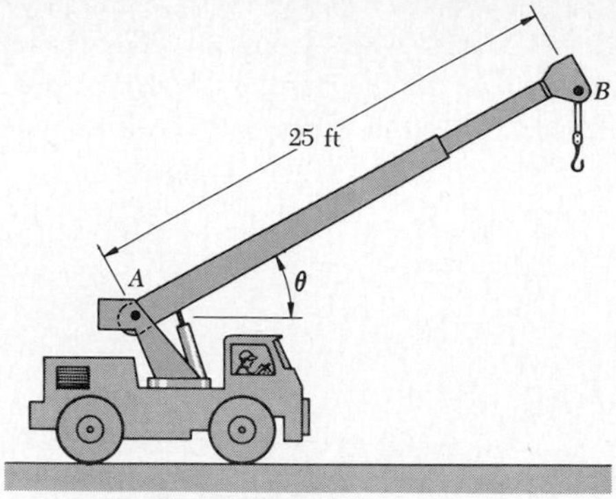

Fig. P15.122 and P15.123

15.123 At the instant shown the length of the boom is being increased at the constant rate of 8 in./s and the boom is being lowered at the constant rate of 0.08 rad/s. Knowing that $\theta = 30°$, determine (*a*) the velocity of point *B*, (*b*) the acceleration of point *B*.

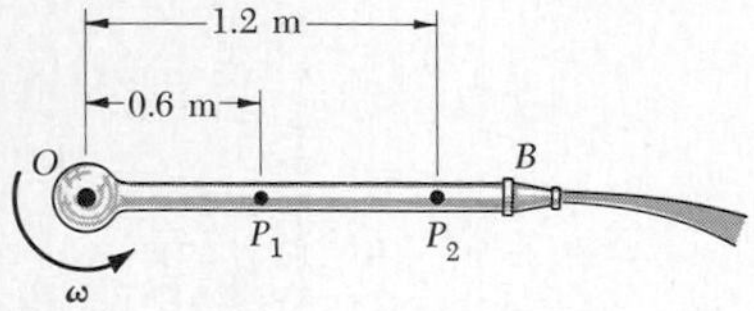

Fig. P15.124

15.124 Water flows at a constant rate through a straight pipe *OB* which rotates counterclockwise with a constant angular velocity of 150 rpm. If the velocity of the water relative to the pipe is 6 m/s, determine the total acceleration (*a*) of the particle of water P_1, (*b*) of the particle of water P_2.

15.125 In the automated welding setup shown, the position of the two welding tips *G* and *H* is controlled by the hydraulic cylinder *D* and the rod *BC*. The cylinder is bolted to the vertical plate which at the instant shown rotates counterclockwise about *A* with a constant angular velocity of 1.2 rad/s. Knowing that at the same instant the length *EF* of the welding asssembly is increasing at the constant rate of 300 mm/s, determine (*a*) the velocity of tip *H*, (*b*) the acceleration of tip *H*.

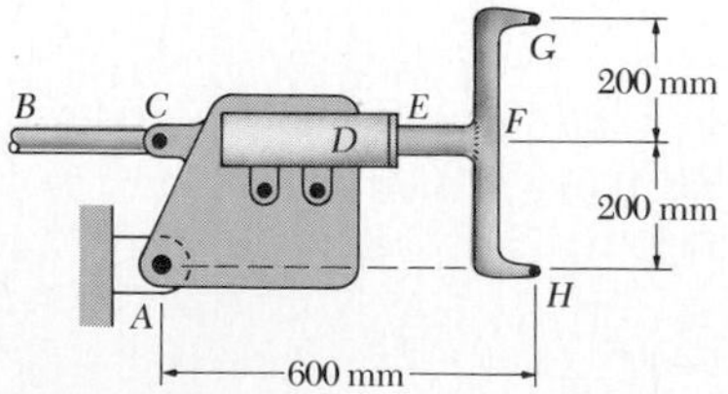

Fig. P15.125

15.126 In Prob. 15.125, determine (*a*) the velocity of tip *G*, (*b*) the acceleration of tip *G*.

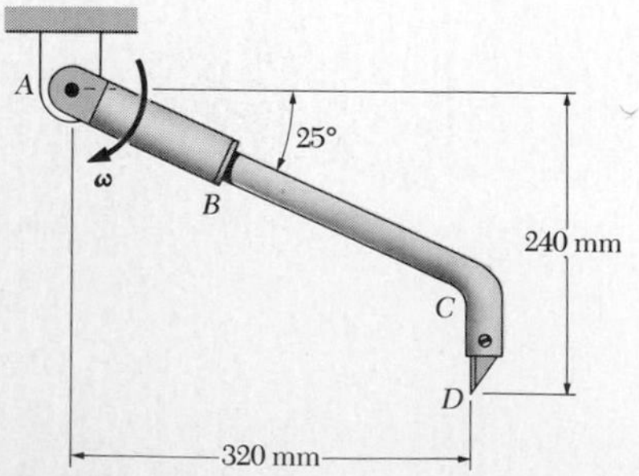

Fig. P15.127

15.127 The motion of blade *D* is controlled by the robot arm *ABC*. At the instant shown the arm is rotating clockwise at the constant rate $\omega = 1.5$ rad/s and the length of portion *BC* of the arm is being decreased at the constant rate of 180 mm/s. Determine (*a*) the velocity of *D*, (*b*) the acceleration of *D*.

15.128 The hydraulic cylinder CD is welded to an arm which rotates clockwise about A at the constant rate $\omega = 3$ rad/s. Knowing that in the position shown the rod BE is being moved to the right at the constant rate of 12 in./s with respect to the cylinder, determine (*a*) the velocity of point B, (*b*) the acceleration of point B.

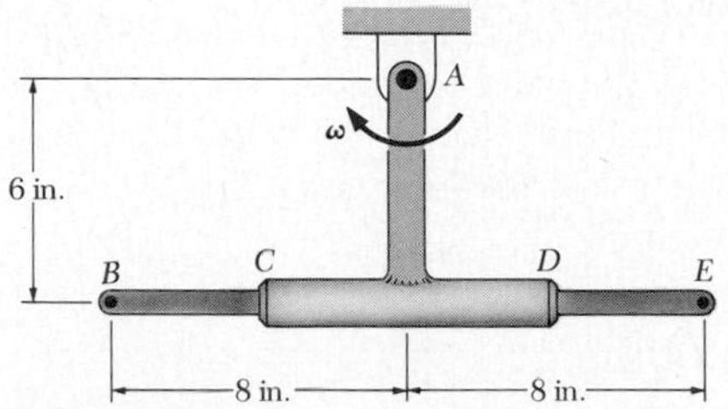

Fig. P15.128

15.129 In Prob. 15.128, determine (*a*) the velocity of point E, (*b*) the acceleration of point E.

15.130 The Geneva mechanism shown is used to provide an intermittent rotary motion of disk S. Disk D rotates with a constant counterclockwise angular velocity ω_D of 10 rad/s. A pin P is attached to disk D and may slide in one of the six equally spaced slots cut in disk S. It is desirable that the angular velocity of disk S be zero as the pin enters and leaves each of the six slots; this will occur if the distance between the centers of the disks and the radii of the disks are related as shown. Determine the angular velocity and angular acceleration of disk S at the instant when $\phi = 150°$.

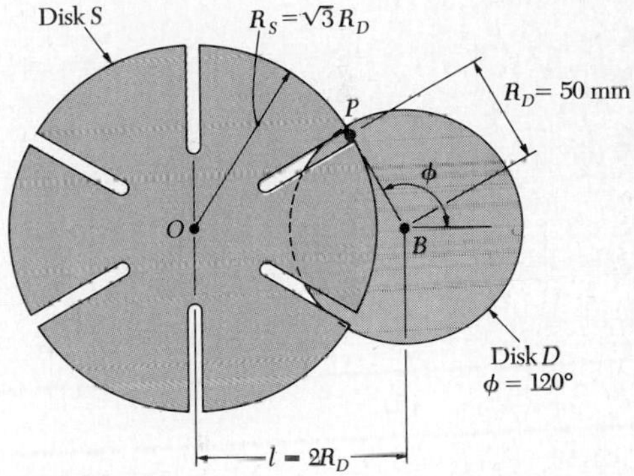

Fig. P15.130

15.131 In Prob. 15.130, determine the angular velocity and angular acceleration of disk S at the instant when $\phi = 135°$.

15.132 In Prob. 15.114, determine the angular acceleration of the rod attached at B.

15.133 In Prob. 15.117, determine the angular acceleration of the rod attached at B.

15.134 Water flows through the curved pipe OB, which has a uniform radius of 15 in. and which rotates with a constant counterclockwise angular velocity of 120 rpm. If the velocity of the water relative to the pipe is 40 ft/s, determine the total acceleration of the particle of water P.

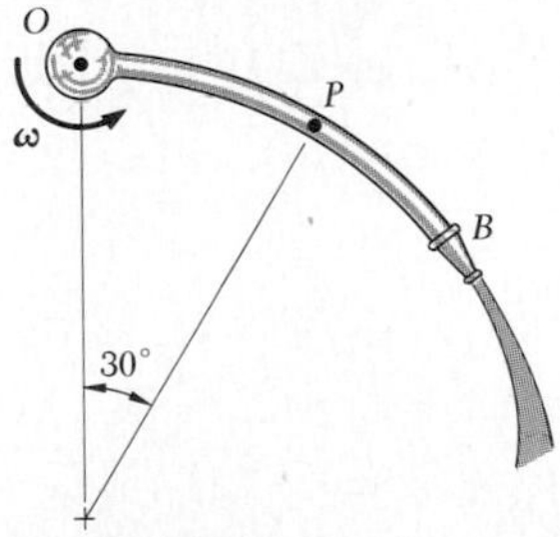

Fig. P15.134

15.135 Solve Prob. 15.134, assuming that the curved pipe rotates with a constant clockwise angular velocity of 120 rpm.

15.136 Rod AD is bent in the shape of an arc of circle of radius $b = 120$ mm. The position of the rod is controlled by pin B which slides in a horizontal slot and also slides along the rod. Knowing that at the instant shown pin B moves to the right with a constant speed of 40 mm/s, determine (*a*) the angular velocity of the rod, (*b*) the angular acceleration of the rod.

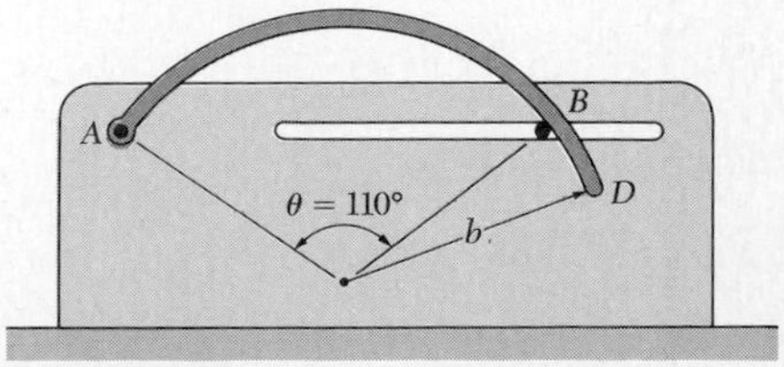

Fig. P15.136

15.137 Solve Prob. 15.136 when $\theta = 90°$.

*** 15.138** In Prob. 15.118, determine the acceleration of pin P.

*** 15.139** In Prob. 15.119, determine the acceleration of pin P.

*** 15.140** Rod AB passes through a collar which is welded to link DE. Knowing that at the instant shown block A moves to the right with a constant speed of 75 in./s, determine (*a*) the angular velocity of rod AB, (*b*) the velocity relative to the collar of the point of the rod in contact with the collar, (*c*) the acceleration of the point of the rod in contact with the collar. (*Hint.* Rod AB and link DE have the same $\boldsymbol{\omega}$ and the same $\boldsymbol{\alpha}$.)

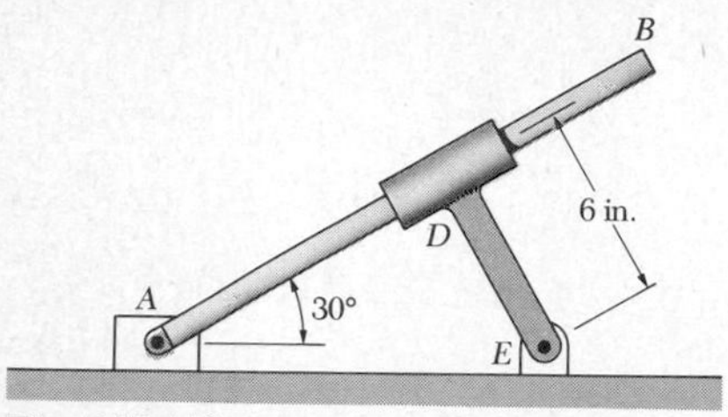

Fig. P15.140

*** 15.141** Solve Prob. 15.140, assuming that block A moves to the left with a constant speed of 75 in./s.

* 15.12. Motion about a Fixed Point.

We have studied in Sec. 15.3 the motion of a rigid body constrained to rotate about a fixed axis. We shall now consider the more general case of the motion of a rigid body which has a fixed point O.

First, we shall prove that *the most general displacement of a rigid body with a fixed point O is equivalent to a rotation of the body about an axis through O.*† Instead of considering the rigid body itself, we may detach a sphere of center O from the body and analyze the motion of that sphere. Clearly, the motion of the sphere completely characterizes the motion of the given body. Since three points define the position of a solid in space, the center O and two points A and B on the surface of the sphere will define the position of the sphere and thus the position of the body. Let A_1 and B_1 characterize the position of the sphere at one instant, and A_2 and B_2 its position at a later instant (Fig. 15.31*a*). Since the sphere is rigid, the lengths of the arcs of great circle A_1B_1 and A_2B_2 must be equal, but except for this requirement, the positions of A_1, A_2, B_1, and B_2 are arbitrary. We propose to prove that the points A and B may be brought, respectively, from A_1 and B_1 into A_2 and B_2 by a single rotation of the sphere about an axis.

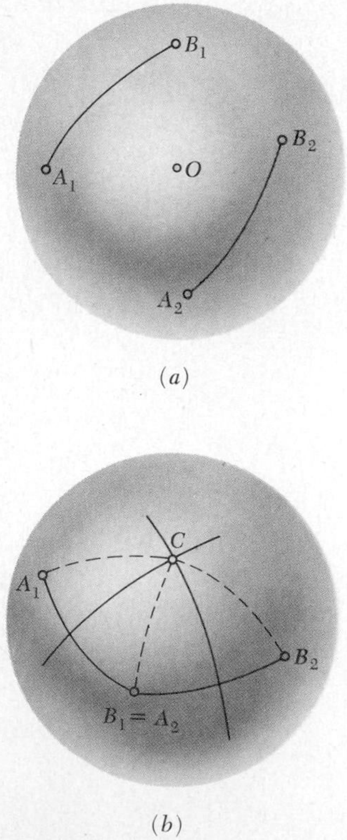

Fig. 15.31

For convenience, and without loss of generality, we may select point B so that its initial position coincides with the final position of A; thus, $B_1 = A_2$ (Fig. 15.31*b*). We draw the arcs of great circle A_1A_2, A_2B_2 and the arcs bisecting, respectively, A_1A_2 and A_2B_2. Let C be the point of intersection of these last two arcs; we complete the construction by drawing A_1C, A_2C, and B_2C. As pointed out above, $A_1B_1 = A_2B_2$ on account of the rigidity of the sphere; on the other hand, since C is by construction equidistant from A_1, A_2, and B_2, we have $A_1C = A_2C = B_2C$. As a result, the spherical triangles A_1CA_2 and B_1CB_2 are congruent and the angles A_1CA_2 and B_1CB_2 are equal. Denoting by θ the common value of these angles, we conclude that the sphere may be brought from its initial position into its final position by a single rotation through θ about the axis OC.

† This is known as Euler's theorem.

It follows that the motion during a time interval Δt of a rigid body with a fixed point O may be considered as a rotation through $\Delta\theta$ about a certain axis. Drawing along that axis a vector of magnitude $\Delta\theta/\Delta t$ and letting Δt approach zero, we obtain at the limit the *instantaneous axis of rotation* and the angular velocity $\boldsymbol{\omega}$ of the body at the instant considered (Fig. 15.32). The velocity of a particle P of the body may then be obtained, as in Sec. 15.3, by forming the vector product of $\boldsymbol{\omega}$ and of the position vector $\mathbf{r}$ of the particle:

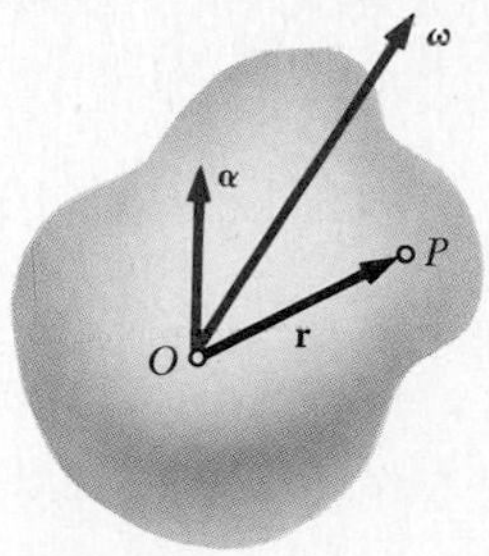

Fig. 15.32

$$\mathbf{v} = \frac{d\mathbf{r}}{dt} = \boldsymbol{\omega} \times \mathbf{r} \tag{15.37}$$

The acceleration of the particle is obtained by differentiating (15.37) with respect to t. As in Sec. 15.3 we have

$$\mathbf{a} = \boldsymbol{\alpha} \times \mathbf{r} + \boldsymbol{\omega} \times (\boldsymbol{\omega} \times \mathbf{r}) \tag{15.38}$$

where the angular acceleration $\boldsymbol{\alpha}$ is defined as the derivative

$$\boldsymbol{\alpha} = \frac{d\boldsymbol{\omega}}{dt} \tag{15.39}$$

of the angular velocity $\boldsymbol{\omega}$.

In the case of the motion of a rigid body with a fixed point, the direction of $\boldsymbol{\omega}$ and of the instantaneous axis of rotation changes from one instant to the next. The angular acceleration $\boldsymbol{\alpha}$ therefore reflects the change in direction of $\boldsymbol{\omega}$ as well as its change in magnitude and, in general, *is not directed along the instantaneous axis of rotation.* While the particles of the body located on the instantaneous axis of rotation have zero velocity at the instant considered, they do not have zero acceleration. Also, the accelerations of the various particles of the body *cannot* be determined as if the body were rotating permanently about the instantaneous axis.

Recalling the definition of the velocity of a particle with position vector $\mathbf{r}$, we note that the angular acceleration $\boldsymbol{\alpha}$, as expressed in (15.39), represents the velocity of the tip of the vector $\boldsymbol{\omega}$. This property may be useful in the determination of the angular acceleration of a rigid body. For example, it follows that the vector $\boldsymbol{\alpha}$ is tangent to the curve described in space by the tip of the vector $\boldsymbol{\omega}$.

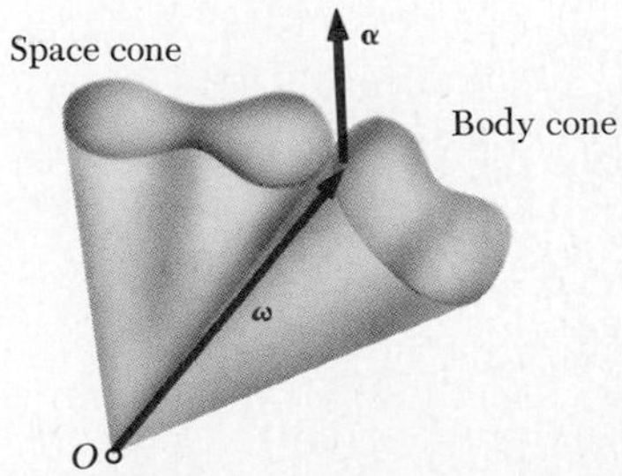

Fig. 15.33

We should note that the vector $\boldsymbol{\omega}$ moves within the body, as well as in space. It thus generates two cones called, respectively, the *body cone* and the *space cone* (Fig. 15.33).† It may be shown that at any given instant, the two cones are tangent along the instantaneous axis of rotation and that as the body moves, the body cone appears to *roll* on the space cone.

†It is recalled that a *cone* is, by definition, a surface generated by a straight line passing through a fixed point. In general, the cones considered here *will not be circular cones.*

Before concluding our analysis of the motion of a rigid body with a fixed point, we should prove that angular velocities are actually vectors. As it was indicated in Sec. 2.3, some quantities, such as the *finite rotations* of a rigid body, have magnitude and direction but do not obey the parallelogram law of addition; these quantities cannot be considered as vectors. We shall see presently that angular velocities (and also *infinitesimal rotations*) *do obey* the parallelogram law and thus are truly vector quantities.

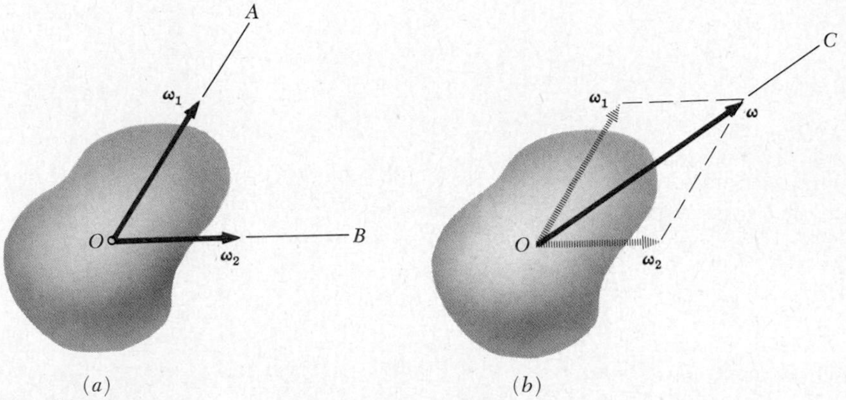

Fig. 15.34

Consider a rigid body with a fixed point O which at a given instant rotates simultaneously about the axes OA and OB with angular velocities $\boldsymbol{\omega}_1$ and $\boldsymbol{\omega}_2$ (Fig. 15.34*a*). We know that this motion must be equivalent at the instant considered to a single rotation of angular velocity $\boldsymbol{\omega}$. We propose to show that

$$\boldsymbol{\omega} = \boldsymbol{\omega}_1 + \boldsymbol{\omega}_2 \tag{15.40}$$

i.e., that the resulting angular velocity may be obtained by adding $\boldsymbol{\omega}_1$ and $\boldsymbol{\omega}_2$ by the parallelogram law (Fig. 15.34*b*).

Consider a particle P of the body, defined by the position vector $\mathbf{r}$. Denoting, respectively, by $\mathbf{v}_1$, $\mathbf{v}_2$, and $\mathbf{v}$ the velocity of P when the body rotates about OA only, about OB only, and about both axes simultaneously, we write

$$\mathbf{v} = \boldsymbol{\omega} \times \mathbf{r} \qquad \mathbf{v}_1 = \boldsymbol{\omega}_1 \times \mathbf{r} \qquad \mathbf{v}_2 = \boldsymbol{\omega}_2 \times \mathbf{r} \tag{15.41}$$

But the vectorial characters of *linear* velocities is well established (since they represent the derivatives of position vectors). We have therefore

$$\mathbf{v} = \mathbf{v}_1 + \mathbf{v}_2$$

where the plus sign indicates vector addition. Substituting from (15.41), we write

$$\boldsymbol{\omega} \times \mathbf{r} = \boldsymbol{\omega}_1 \times \mathbf{r} + \boldsymbol{\omega}_2 \times \mathbf{r}$$
$$\boldsymbol{\omega} \times \mathbf{r} = (\boldsymbol{\omega}_1 + \boldsymbol{\omega}_2) \times \mathbf{r}$$

where the plus sign still indicates vector addition. Since the relation obtained holds for an arbitrary $\mathbf{r}$, we conclude that (15.40) must be true.

*** 15.13. General Motion.** We shall now consider the most general motion of a rigid body in space. Let A and B be two particles of the body. We recall from Sec. 11.12 that the velocity of B with respect to the fixed frame of reference $OXYZ$ may be expressed as

$$\mathbf{v}_B = \mathbf{v}_A + \mathbf{v}_{B/A} \tag{15.42}$$

where $\mathbf{v}_{B/A}$ is the velocity of B relative to a frame $AX'Y'Z'$ attached to A and of fixed orientation (Fig. 15.35). Since A is fixed in this frame, the motion of the body relative to $AX'Y'Z'$ is the motion of a body with a fixed point. The relative velocity $\mathbf{v}_{B/A}$ may, therefore, be obtained from (15.37), after $\mathbf{r}$ has been replaced by the position vector $\mathbf{r}_{B/A}$ of B relative to A. Substituting for $\mathbf{v}_{B/A}$ into (15.42), we write

$$\mathbf{v}_B = \mathbf{v}_A + \boldsymbol{\omega} \times \mathbf{r}_{B/A} \tag{15.43}$$

where $\boldsymbol{\omega}$ is the angular velocity of the body at the instant considered.

The acceleration of B is obtained by a similar reasoning. We first write

$$\mathbf{a}_B = \mathbf{a}_A + \mathbf{a}_{B/A}$$

and, recalling Eq. (15.38),

$$\mathbf{a}_B = \mathbf{a}_A + \boldsymbol{\alpha} \times \mathbf{r}_{B/A} + \boldsymbol{\omega} \times (\boldsymbol{\omega} \times \mathbf{r}_{B/A}) \tag{15.44}$$

where $\boldsymbol{\alpha}$ is the angular acceleration of the body at the instant considered.

Equations (15.43) and (15.44) show that *the most general motion of a rigid body is equivalent, at any given instant, to the sum of a translation,* in which all the particles of the body have the same velocity and acceleration as a reference particle A, *and of a motion in which particle A is assumed to be fixed.*†

It may easily be shown by solving (15.43) and (15.44) for $\mathbf{v}_A$ and $\mathbf{a}_A$ that the motion of the body with respect to a frame attached to B would be characterized by the same vectors $\boldsymbol{\omega}$ and $\boldsymbol{\alpha}$ as its motion relative to $AX'Y'Z'$. The angular velocity and angular acceleration of a rigid body at a given instant are thus independent of the choice of the reference point. On the other hand, one should keep in mind that whether it is attached to A or to B, the moving frame should maintain a fixed orientation; that is, it should remain parallel to the fixed reference frame $OXYZ$ throughout the motion of the rigid body. In many problems it is found more convenient to use a moving frame which is allowed to rotate as well as to translate. The use of such moving frames will be discussed in Secs. 15.14 and 15.15.

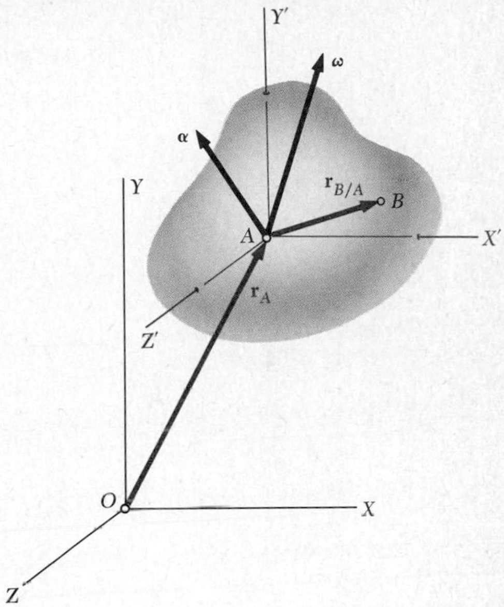

Fig. 15.35

† It is recalled from Sec. 15.12 that, in general, the vectors $\boldsymbol{\omega}$ and $\boldsymbol{\alpha}$ are not collinear, and that the accelerations of the particles of the body in their motion relative to the frame $AX'Y'Z'$ cannot be determined as if the body were rotating permanently about the instantaneous axis through A.

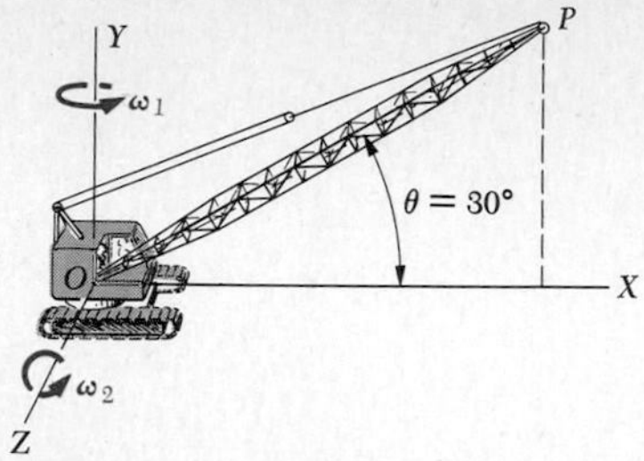

SAMPLE PROBLEM 15.11

The crane shown rotates with a constant angular velocity $\boldsymbol{\omega}_1$ of 0.30 rad/s. Simultaneously, the boom is being raised with a constant angular velocity $\boldsymbol{\omega}_2$ of 0.50 rad/s relative to the cab. Knowing that the length of the boom OP is $l = 12$ m, determine (*a*) the angular velocity $\boldsymbol{\omega}$ of the boom, (*b*) the angular acceleration $\boldsymbol{\alpha}$ of the boom, (*c*) the velocity $\mathbf{v}$ of the tip of the boom, (*d*) the acceleration $\mathbf{a}$ of the tip of the boom.

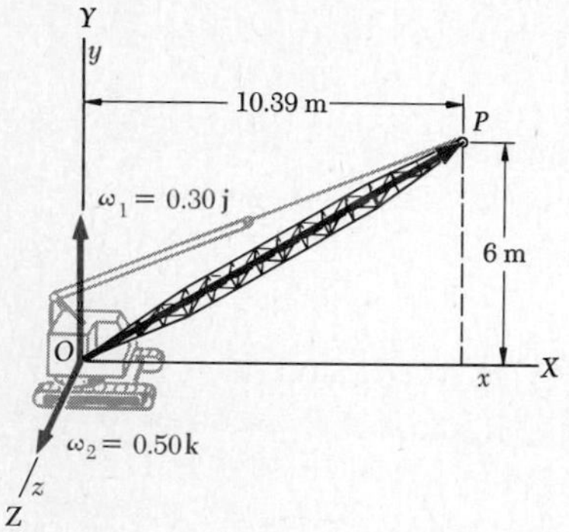

***a.* Angular Velocity of Boom.** Adding the angular velocity $\boldsymbol{\omega}_1$ of the cab and the angular velocity $\boldsymbol{\omega}_2$ of the boom relative to the cab, we obtain the angular velocity $\boldsymbol{\omega}$ of the boom at the instant considered:

$$\boldsymbol{\omega} = \boldsymbol{\omega}_1 + \boldsymbol{\omega}_2 \qquad \boldsymbol{\omega} = (0.30 \text{ rad/s})\mathbf{j} + (0.50 \text{ rad/s})\mathbf{k} \quad ◄$$

***b.* Angular Acceleration of Boom.** The angular acceleration $\boldsymbol{\alpha}$ of the boom is obtained by differentiating $\boldsymbol{\omega}$. Since the vector $\boldsymbol{\omega}_1$ is constant in magnitude and direction, we have

$$\boldsymbol{\alpha} = \dot{\boldsymbol{\omega}} = \dot{\boldsymbol{\omega}}_1 + \dot{\boldsymbol{\omega}}_2 = 0 + \dot{\boldsymbol{\omega}}_2$$

where the rate of change $\dot{\boldsymbol{\omega}}_2$ is to be computed with respect to the fixed frame $OXYZ$. However, it is more convenient to use a frame $Oxyz$ attached to the cab and rotating with it, since the vector $\boldsymbol{\omega}_2$ also rotates with the cab and therefore has zero rate of change with respect to that frame. Using Eq. (15.31) with $\mathbf{Q} = \boldsymbol{\omega}_2$ and $\boldsymbol{\Omega} = \boldsymbol{\omega}_1$, we write

$$(\dot{\mathbf{Q}})_{OXYZ} = (\dot{\mathbf{Q}})_{Oxyz} + \boldsymbol{\Omega} \times \mathbf{Q}$$
$$(\dot{\boldsymbol{\omega}}_2)_{OXYZ} = (\dot{\boldsymbol{\omega}}_2)_{Oxyz} + \boldsymbol{\omega}_1 \times \boldsymbol{\omega}_2$$
$$\boldsymbol{\alpha} = (\dot{\boldsymbol{\omega}}_2)_{OXYZ} = 0 + (0.30 \text{ rad/s})\mathbf{j} \times (0.50 \text{ rad/s})\mathbf{k}$$
$$\boldsymbol{\alpha} = (0.15 \text{ rad/s}^2)\mathbf{i} \quad ◄$$

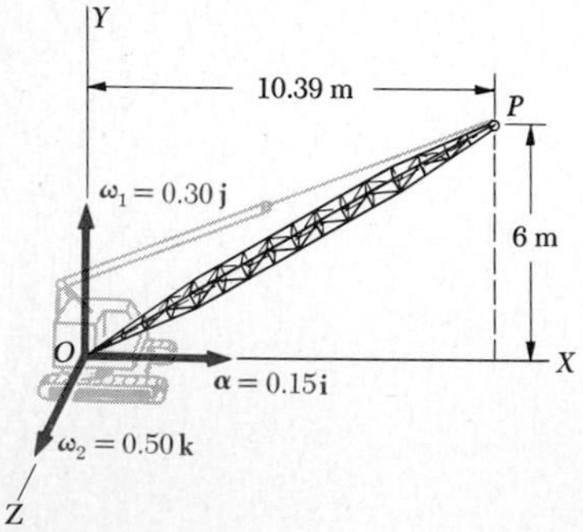

***c.* Velocity of Tip of Boom.** Noting that the position vector of point P is $\mathbf{r} = (10.39 \text{ m})\mathbf{i} + (6 \text{ m})\mathbf{j}$ and using the expression found for $\boldsymbol{\omega}$ in part *a*, we write

$$\mathbf{v} = \boldsymbol{\omega} \times \mathbf{r} = \begin{vmatrix} \mathbf{i} & \mathbf{j} & \mathbf{k} \\ 0 & 0.30 \text{ rad/s} & 0.50 \text{ rad/s} \\ 10.39 \text{ m} & 6 \text{ m} & 0 \end{vmatrix}$$

$$\mathbf{v} = -(3 \text{ m/s})\mathbf{i} + (5.20 \text{ m/s})\mathbf{j} - (3.12 \text{ m/s})\mathbf{k} \quad ◄$$

***d.* Acceleration of Tip of Boom.** Recalling that $\mathbf{v} = \boldsymbol{\omega} \times \mathbf{r}$, we write

$$\mathbf{a} = \boldsymbol{\alpha} \times \mathbf{r} + \boldsymbol{\omega} \times (\boldsymbol{\omega} \times \mathbf{r}) = \boldsymbol{\alpha} \times \mathbf{r} + \boldsymbol{\omega} \times \mathbf{v}$$

$$\mathbf{a} = \begin{vmatrix} \mathbf{i} & \mathbf{j} & \mathbf{k} \\ 0.15 & 0 & 0 \\ 10.39 & 6 & 0 \end{vmatrix} + \begin{vmatrix} \mathbf{i} & \mathbf{j} & \mathbf{k} \\ 0 & 0.30 & 0.50 \\ -3 & 5.20 & -3.12 \end{vmatrix}$$

$$= 0.90\mathbf{k} - 0.94\mathbf{i} - 2.60\mathbf{i} - 1.50\mathbf{j} + 0.90\mathbf{k}$$

$$\mathbf{a} = -(3.54 \text{ m/s}^2)\mathbf{i} - (1.50 \text{ m/s}^2)\mathbf{j} + (1.80 \text{ m/s}^2)\mathbf{k} \quad ◄$$

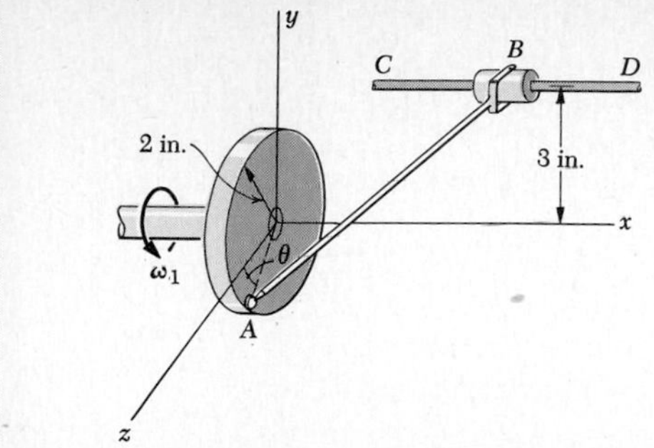

SAMPLE PROBLEM 15.12

The rod AB, of length 7 in., is attached to the disk by a ball-and-socket connection and to the collar B by a clevis. The disk rotates in the yz plane at a constant rate $\omega_1 = 12$ rad/s, while the collar is free to slide along the horizontal rod CD. For the position $\theta = 0$, determine (a) the velocity of the collar, (b) the angular velocity of the rod.

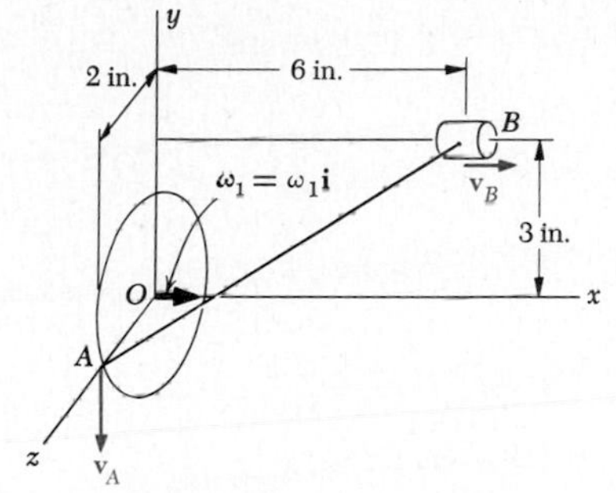

$\boldsymbol{\omega}_1 = 12\mathbf{i}$
$\mathbf{r}_A = 2\mathbf{k}$
$\mathbf{r}_B = 6\mathbf{i} + 3\mathbf{j}$
$\mathbf{r}_{B/A} = 6\mathbf{i} + 3\mathbf{j} - 2\mathbf{k}$

***a.* Velocity of Collar.** Since point A is attached to the disk and since collar B moves parallel to the x axis, we have

$$\mathbf{v}_A = \boldsymbol{\omega}_1 \times \mathbf{r}_A = 12\mathbf{i} \times 2\mathbf{k} = -24\mathbf{j} \qquad \mathbf{v}_B = v_B\mathbf{i}$$

Denoting by $\boldsymbol{\omega}$ the angular velocity of the rod, we write

$$\mathbf{v}_B = \mathbf{v}_A + \mathbf{v}_{B/A} = \mathbf{v}_A + \boldsymbol{\omega} \times \mathbf{r}_{B/A}$$

$$v_B\mathbf{i} = -24\mathbf{j} + \begin{vmatrix} \mathbf{i} & \mathbf{j} & \mathbf{k} \\ \omega_x & \omega_y & \omega_z \\ 6 & 3 & -2 \end{vmatrix}$$

$$v_B\mathbf{i} = -24\mathbf{j} + (-2\omega_y - 3\omega_z)\mathbf{i} + (6\omega_z + 2\omega_x)\mathbf{j} + (3\omega_x - 6\omega_y)\mathbf{k}$$

Equating the coefficients of the unit vectors, we obtain

$$v_B = \quad -2\omega_y \quad -3\omega_z \qquad (1)$$
$$24 = 2\omega_x \quad +6\omega_z \qquad (2)$$
$$0 = 3\omega_x \quad -6\omega_y \qquad (3)$$

Multiplying Eqs. (1), (2), (3), respectively, by 6, 3, -2 and adding, we write

$$6v_B + 72 = 0 \qquad v_B = -12 \qquad \mathbf{v}_B = -(12 \text{ in./s})\mathbf{i} \quad ◄$$

***b.* Angular Velocity of Rod *AB*.** We note that the angular velocity cannot be determined from Eqs. (1), (2), and (3), since the determinant formed by the coefficients of ω_x, ω_y, and ω_z is zero. We must therefore obtain an additional equation by considering the constraint imposed by the clevis at B.

The collar-clevis connection at B permits rotation of AB about the rod CD and also about an axis perpendicular to the plane containing AB and CD. It prevents rotation of AB about the axis EB, which is perpendicular to CD and lies in the plane containing AB and CD. Thus the projection of $\boldsymbol{\omega}$ on $\mathbf{r}_{E/B}$ must be zero and we write†

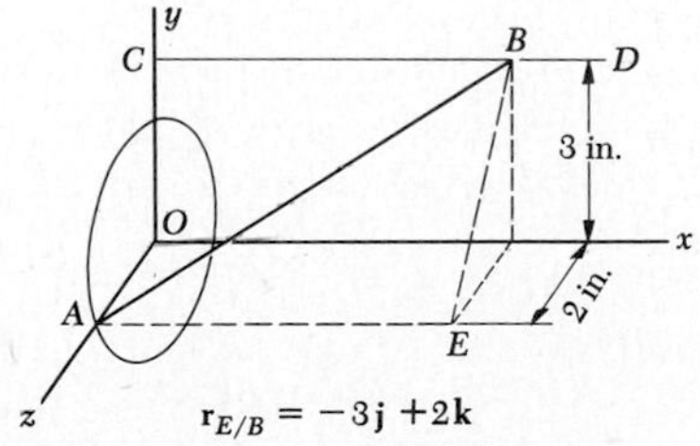

$\mathbf{r}_{E/B} = -3\mathbf{j} + 2\mathbf{k}$

$$\boldsymbol{\omega} \cdot \mathbf{r}_{E/B} = 0 \qquad (\omega_x\mathbf{i} + \omega_y\mathbf{j} + \omega_z\mathbf{k}) \cdot (-3\mathbf{j} + 2\mathbf{k}) = 0$$
$$-3\omega_y + 2\omega_z = 0 \qquad (4)$$

Solving Eqs. (1) through (4) simultaneously, we obtain

$$v_B = -12 \qquad \omega_x = 3.69 \qquad \omega_y = 1.846 \qquad \omega_z = 2.77$$

$$\boldsymbol{\omega} = (3.69 \text{ rad/s})\mathbf{i} + (1.846 \text{ rad/s})\mathbf{j} + (2.77 \text{ rad/s})\mathbf{k} \quad ◄$$

† We could also note that the direction of EB is that of the vector triple product $\mathbf{r}_{B/C} \times (\mathbf{r}_{B/C} \times \mathbf{r}_{B/A})$ and write $\boldsymbol{\omega} \cdot [\mathbf{r}_{B/C} \times (\mathbf{r}_{B/C} \times \mathbf{r}_{B/A})] = 0$. This formulation would be particularly useful if the rod CD were skew.

Problems

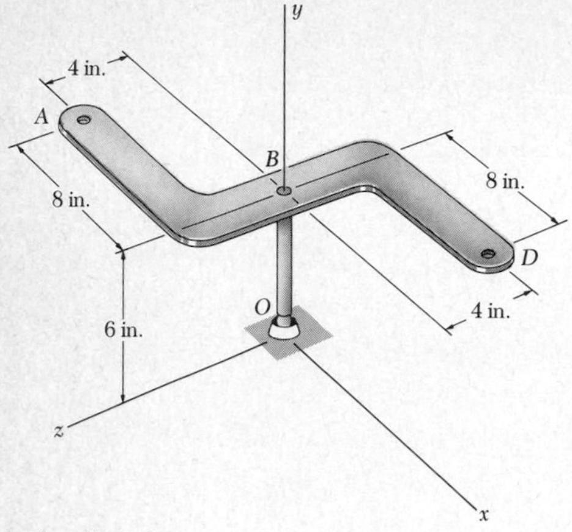

Fig. 15.142

15.142 Plate ABD and rod OB are rigidly connected and rotate about the ball-and-socket joint O with an angular velocity $\boldsymbol{\omega} = \omega_x\mathbf{i} + \omega_y\mathbf{j} + \omega_z\mathbf{k}$. Knowing that $\mathbf{v}_A = (14 \text{ in./s})\mathbf{i} + (16 \text{ in./s})\mathbf{j} + (v_A)_z\mathbf{k}$ and $\omega_x = 2$ rad/s, determine (*a*) the angular velocity of the assembly, (*b*) the velocity of point D.

15.143 Solve Prob. 15.142, assuming that $\omega_x = -2$ rad/s.

15.144 The rigid body shown rotates about the origin of coordinates with an angular velocity $\boldsymbol{\omega}$. Denoting the velocity of point A by $\mathbf{v}_A = (v_A)_x\mathbf{i} + (v_A)_y\mathbf{j} + (v_A)_z\mathbf{k}$, and knowing that $(v_A)_x = 150$ mm/s and $(v_A)_y = -75$ mm/s, determine the velocity component $(v_A)_z$.

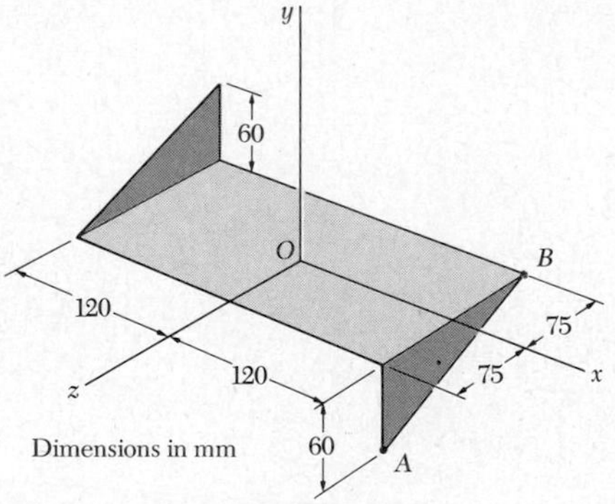

Fig. P15.144 and P15.145

15.145 The rigid body shown rotates about the origin of coordinates with an angular velocity $\boldsymbol{\omega} = \omega_x\mathbf{i} + \omega_y\mathbf{j} + \omega_z\mathbf{k}$. Knowing that $(v_A)_y = 165$ mm/s, $(v_B)_y = 315$ mm/s, and $\omega_y = -3$ rad/s, determine (*a*) the angular velocity of the body, (*b*) the velocities of points A and B.

15.146 The disk of a portable sander rotates at the constant rate $\omega_1 =$ 4400 rpm as shown. Determine the angular acceleration of the disk as a worker rotates the sander about the z axis with an angular velocity of 0.5 rad/s and an angular acceleration of 2.5 rad/s^2, both clockwise when viewed from the positive z axis.

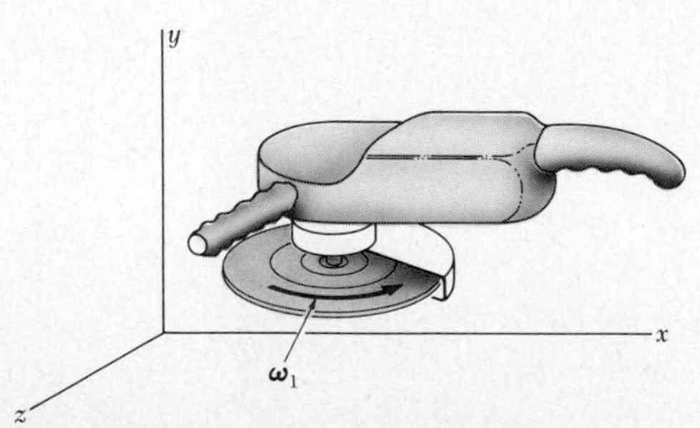

Fig. P15.146

15.147 The rotor of an electric motor rotates at the constant rate $\omega_1 = 3600$ rpm. Determine the angular acceleration of the rotor as the motor is rotated about the y axis with a constant angular velocity of 5 rpm clockwise when viewed from the positive y axis.

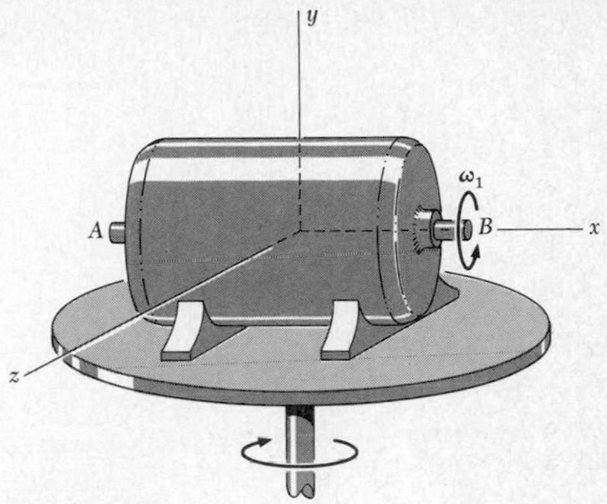

Fig. P15.147

15.148 The fan of an automobile engine rotates about a horizontal axis at the rate of 2500 rpm in a clockwise sense when viewed from the rear of the engine. Knowing that the automobile is turning right along a path of radius 40 ft at a constant speed of 9 mi/h, determine the angular acceleration of the fan at the instant the automobile is moving due north.

15.149 Two disks A and B are mounted on an axle of length $2R$ and roll without sliding on a horizontal floor. Knowing that the axle rotates with a constant angular velocity $\boldsymbol{\omega}_1$, determine (*a*) the angular velocity of disk A, (*b*) the angular acceleration of disk A.

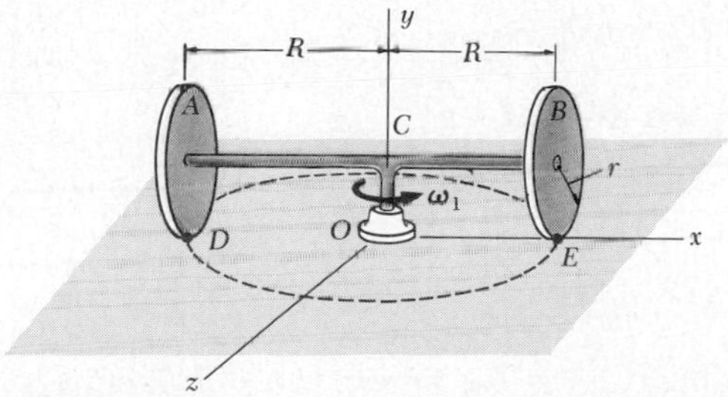

Fig. P15.149

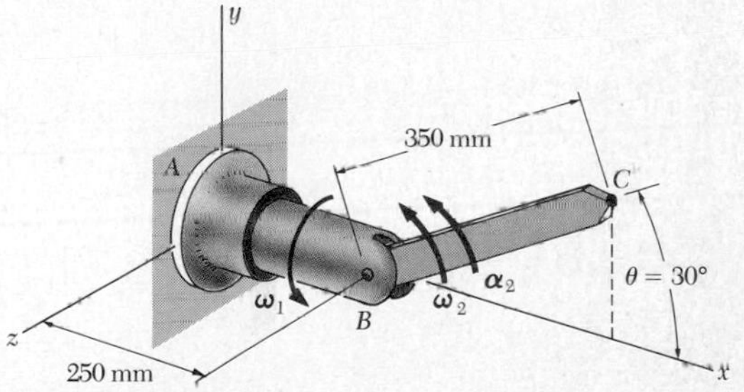

Fig. P15.151

15.150 In Prob. 15.149, determine (*a*) the angular velocity of disk B, (*b*) the angular acceleration of disk B.

15.151 The robot component shown rotates with a constant angular velocity $\boldsymbol{\omega}_1$ of 2.5 rad/s about the x axis. At the instant shown, arm BC is rotating about the z axis with an angular velocity $\boldsymbol{\omega}_2$ of 3 rad/s and an angular acceleration α_2 of 4 rad/s^2. Determine the angular acceleration of arm BC.

15.152 In Prob. 15.151, determine (*a*) the velocity of point C, (*b*) the acceleration of point C.

15.153 A disk of radius r spins at the constant rate ω_2 about a horizontal axle held by a fork-ended vertical rod which rotates at the constant rate ω_1. For the position shown, determine the acceleration of point P for an arbitrary value of the angle θ.

15.154 A disk of radius r spins at the constant rate ω_2 about a horizontal axle held by a fork-ended vertical rod which rotates at the constant rate ω_1. For the position shown, determine (*a*) the angular acceleration of the disk, (*b*) the acceleration of point P on the rim of the disk if $\theta = 0$, (*c*) the acceleration of P if $\theta = 90°$.

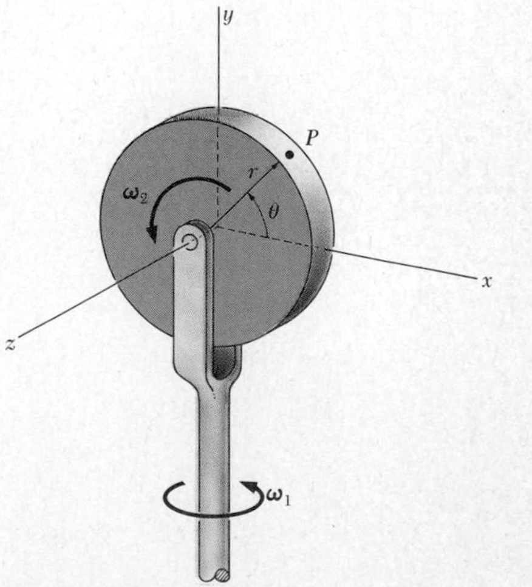

Fig. P15.153 and P15.154

15.155 At the instant shown the telescope CD is being rotated simultaneously at the constant rate $\omega_1 = 0.15$ rad/s about the y axis, and at the constant rate $\omega_2 = 0.25$ rad/s about the z axis. Determine the velocity and acceleration of the center D of the objective lens.

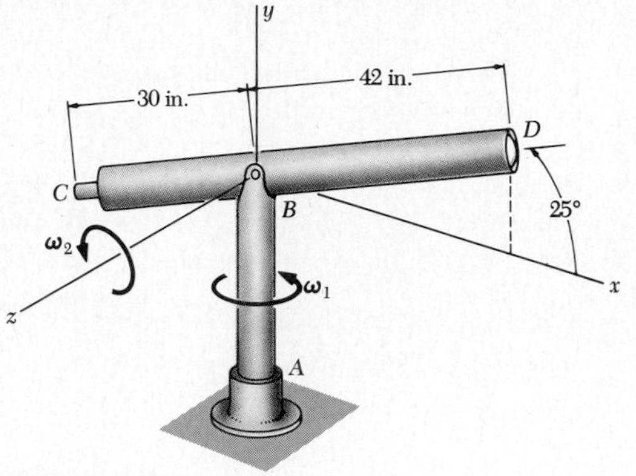

Fig. P15.155

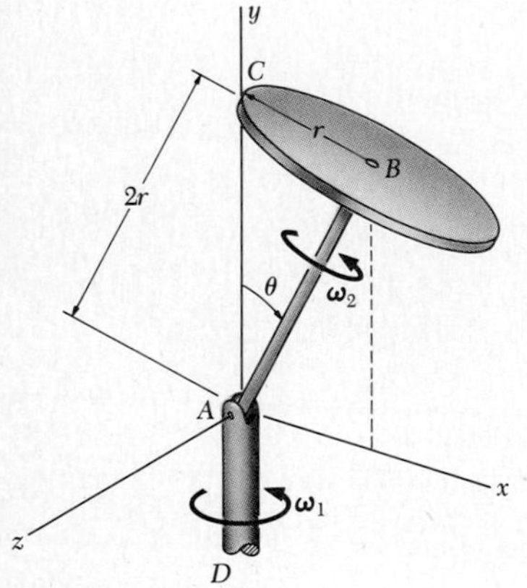

Fig. P15.157

15.156 In Prob. 15.155, determine the velocity and acceleration of the eyepiece C.

15.157 A disk of radius r is mounted on an axle of length $2r$. The axle is attached to a vertical shaft AD which rotates at the constant rate ω_1 and the disk rotates about the axle AB at the constant rate ω_2. Knowing that the angle θ remains constant and that the rim of the disk touches the y axis, determine (a) the angular acceleration of the disk, (b) the velocity of point C of the disk, (c) the acceleration of point C of the disk.

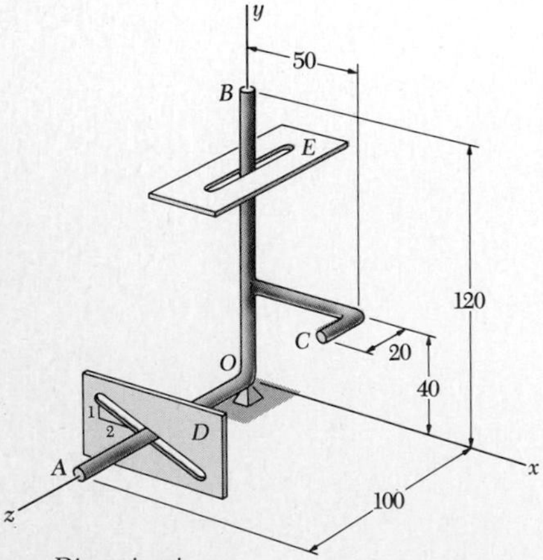

Dimensions in mm

Fig. P15.158

15.158 Several rods are brazed together to form the robotic guide arm shown which is attached to a ball-and-socket joint at O. Rod OA slides in a straight inclined slot while rod OB slides in a slot parallel to the z axis. Knowing that at the instant shown $\mathbf{v}_B = (180 \text{ mm/s})\mathbf{k}$, determine ($a$) the angular velocity of the guide arm, (b) the velocity of point A, (c) the velocity of point C.

15.159 In Prob. 15.158 the speed of point B is known to be constant. For the position shown, determine (a) the angular acceleration of the guide arm, (b) the acceleration of point C.

15.160 In Prob. 15.158 the speed of point B is being decreased at the rate of 90 mm/s². For the position shown, determine (a) the angular acceleration of the assembly, (b) the acceleration of point C.

15.161 In the planetary gear system shown gears A and B are rigidly connected to each other and rotate as a unit about the inclined shaft. Gears C and D rotate with constant angular velocities of 20 rad/s and 30 rad/s, respectively (both counterclockwise when viewed from the right). Choosing the x axis to the right, the y axis upward, and the z axis pointing out of the plane of the figure, determine (a) the common angular velocity of gears A and B, (b) the angular velocity of shaft FH, which is rigidly attached to the inclined shaft.

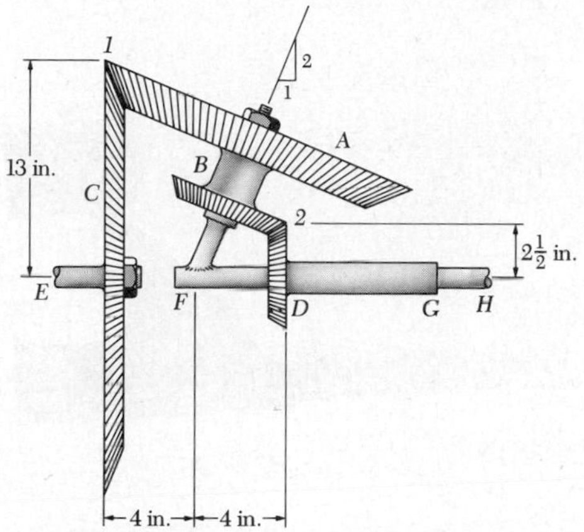

Fig. P15.161

15.162 In Prob. 15.161, determine (a) the common angular acceleration of gears A and B, (b) the acceleration of the tooth of gear A which is in contact with gear C at point 1.

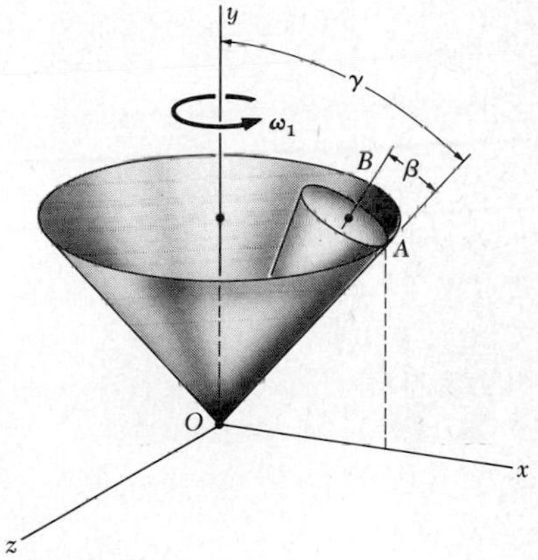

Fig. P15.163

15.163 The small cone shown rolls without slipping on the inside surface of the large fixed cone. Denoting by $\boldsymbol{\omega}_1$ the constant angular velocity of the axis OB about the y axis, determine in terms of ω_1, β, and γ, (a) the rate of spin of the cone about the axis OB, (b) the total angular velocity of the cone, (c) the angular acceleration of the cone.

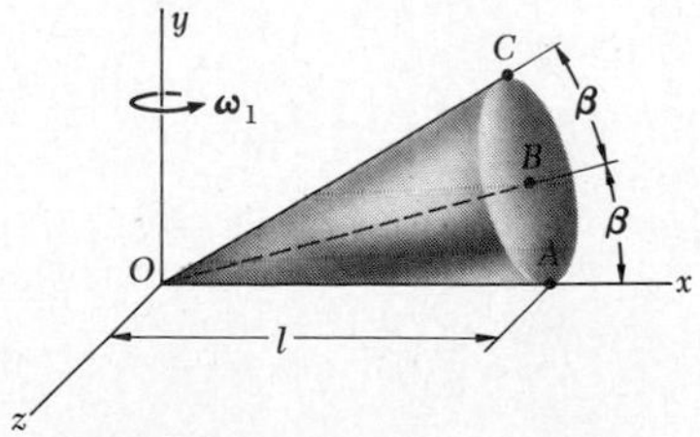

Fig. P15.164

15.164 The cone shown rolls on the zx plane with its apex at the origin of coordinates. Denoting by $\boldsymbol{\omega}_1$ the constant angular velocity of the axis OB of the cone about the y axis, determine (a) the rate of spin of the cone about the axis OB, (b) the total angular velocity of the cone, (c) the angular acceleration of the cone.

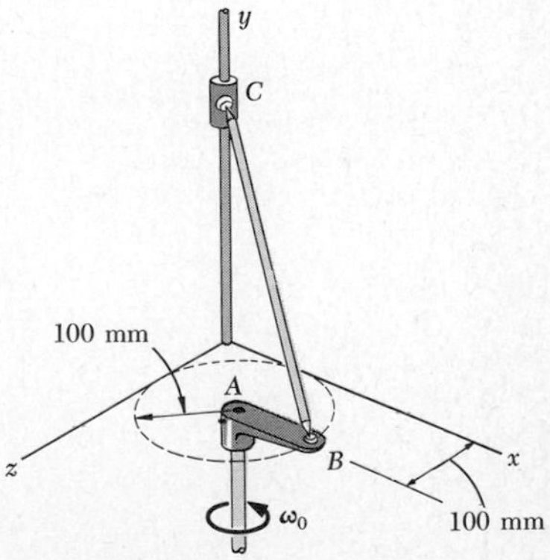

Fig. P15.165

15.165 Rod BC, of length 525 mm, is connected by ball-and-socket joints to the collar C and to the rotating arm AB. Knowing that arm AB rotates in the zx plane at the constant rate $\omega_0 = 19$ rad/s, determine the velocity of collar C.

15.166 Rod AB, of length 11 in., is connected by ball-and-socket joints to collars A and B, which slide along the two rods shown. Knowing that collar B moves downward with a constant speed of 54 in./s, determine the velocity of collar A when $c = 2$ in.

15.167 Solve Prob. 15.166 when $c = 7$ in.

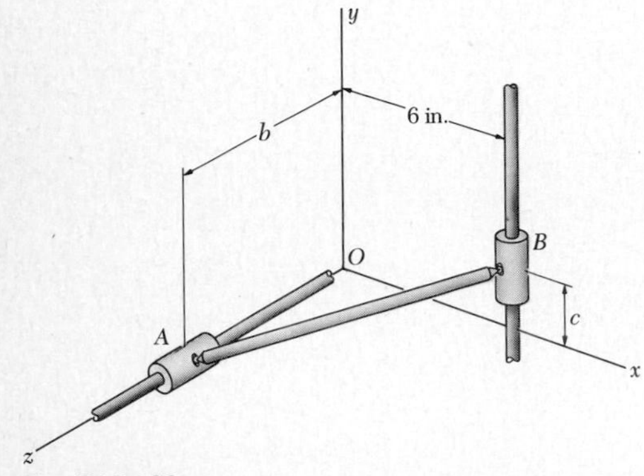

Fig. P15.166

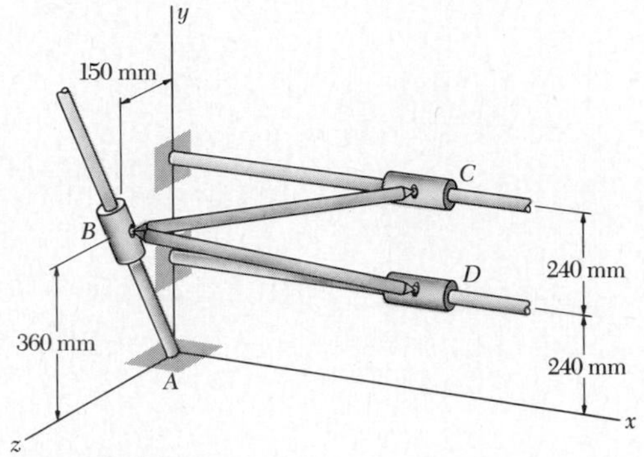

Fig. P15.168

15.168 Rods BC and BD are each 630 mm long and are connected by ball-and-socket joints to collars which may slide on the fixed rods shown. Knowing that at the instant shown, collar B moves toward A with a constant speed of 260 mm/s, determine the velocity of collar C.

15.169 In Prob. 15.168, determine the velocity of collar D.

15.170 Rod BC, of length 420 mm, is connected by ball-and-socket joints to the rotating arm AB and to the collar C. Arm AB is of length 60 mm and rotates in a horizontal plane at the constant rate $\omega_0 = 20$ rad/s. At the instant shown, when arm AB is parallel to the x axis, determine the velocity of collar C.

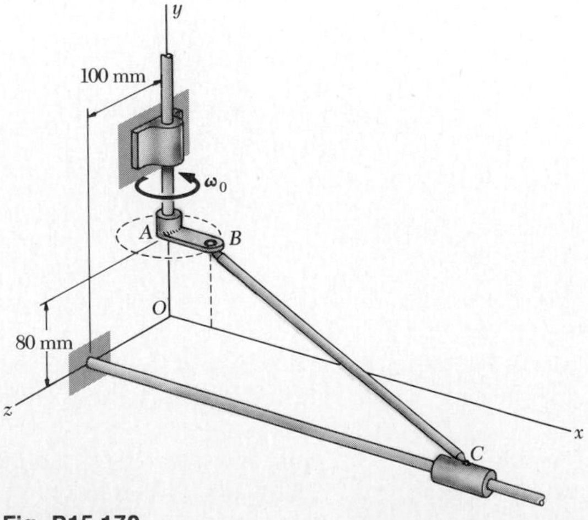

Fig. P15.170

15.171 Two shafts AC and EG, which lie in the vertical yz plane, are connected by a universal joint at D. Shaft AC rotates with a constant angular velocity ω_1 as shown. At a time when the arm of the crosspiece attached to shaft AC is vertical, determine the angular velocity of shaft EG.

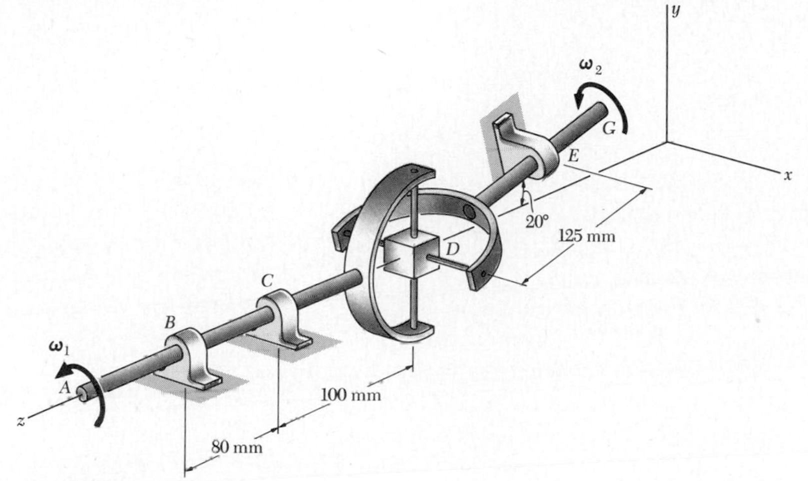

Fig. P15.171

15.172 Solve Prob. 15.171, assuming that the arm of the crosspiece attached to shaft AC is horizontal.

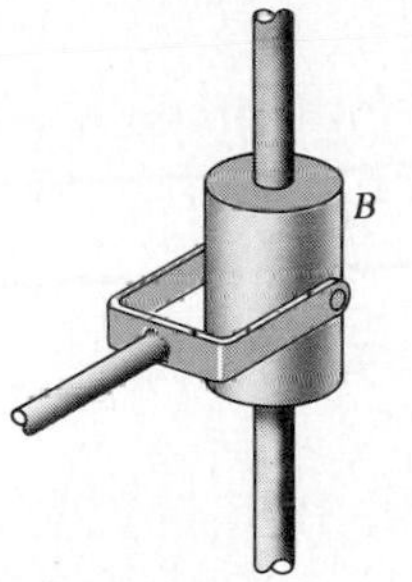

Fig. P15.173

15.173 In Prob. 15.166, the ball-and-socket joint between the rod and collar B is replaced by the clevis connection shown. Determine (*a*) the angular velocity of the rod, (*b*) the velocity of collar A.

15.174 In Prob. 15.170, the ball-and-socket joint between the rod and collar C is replaced by the clevis connection shown. Determine (*a*) the angular velocity of the rod, (*b*) the velocity of collar C.

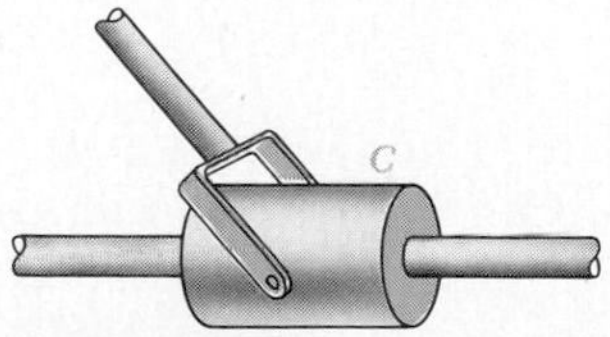

Fig. P15.174

***15.175** In Prob. 15.166, determine the acceleration of collar A when $c = 7$ in.

***15.176** In Prob. 15.166, determine the acceleration of collar A when $c = 2$ in.

***15.177** In Prob. 15.168, determine the acceleration of collar C.

***15.178** In Prob. 15.170, determine the acceleration of collar C.

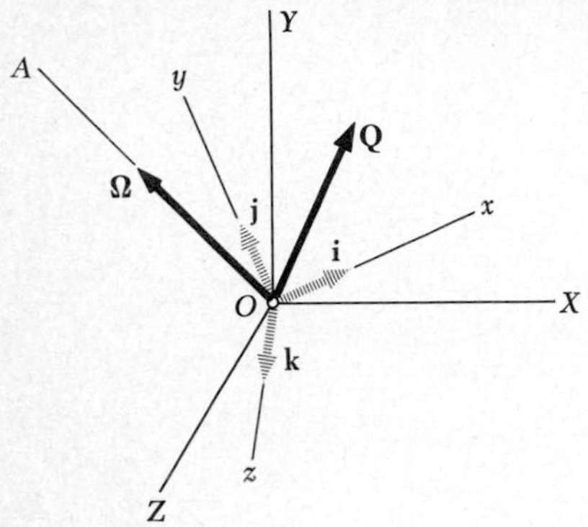

Fig. 15.36

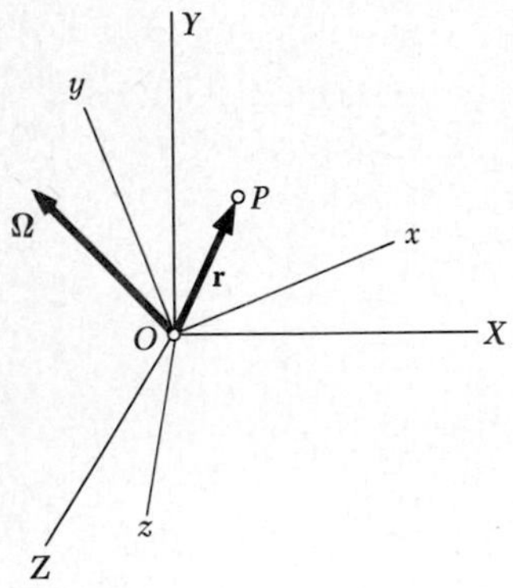

Fig. 15.37

***15.14. Three-dimensional Motion of a Particle Relative to a Rotating Frame. Coriolis Acceleration.** We saw in Sec. 15.10 that given a vector function $\mathbf{Q}(t)$ and two frames of reference centered at O—a fixed frame $OXYZ$ and a rotating frame $Oxyz$—the rates of change of $\mathbf{Q}$ with respect to the two frames satisfy the relation

$$(\dot{\mathbf{Q}})_{OXYZ} = (\dot{\mathbf{Q}})_{Oxyz} + \mathbf{\Omega} \times \mathbf{Q} \tag{15.31}$$

We had assumed at the time that the frame $Oxyz$ was constrained to rotate about a fixed axis OA. However, the derivation given in Sec. 15.10 remains valid when the frame $Oxyz$ is constrained only to have a fixed point O. Under this more general assumption, the axis OA represents the *instantaneous* axis of rotation of the frame $Oxyz$ (Sec. 15.12) and the vector $\mathbf{\Omega}$, its angular velocity at the instant considered (Fig. 15.36).

We shall now consider the three-dimensional motion of a particle P relative to a rotating frame $Oxyz$ constrained to have a fixed origin O. Let $\mathbf{r}$ be the position vector of P at a given instant and $\mathbf{\Omega}$ be the angular velocity of the frame $Oxyz$ with respect to the fixed frame $OXYZ$ at the same instant (Fig. 15.37). The derivations given in Sec. 15.11 for the two-dimensional motion of a particle may readily be extended to the three-dimensional case, and we may express the absolute velocity $\mathbf{v}_P$ of P (i.e., its velocity with respect to the fixed frame $OXYZ$) as

$$\mathbf{v}_P = \mathbf{\Omega} \times \mathbf{r} + (\dot{\mathbf{r}})_{Oxyz} \tag{15.45}$$

Denoting by $\mathfrak{F}$ the rotating frame $Oxyz$, we write this relation in the alternative form

$$\mathbf{v}_P = \mathbf{v}_{P'} + \mathbf{v}_{P/\mathfrak{F}} \tag{15.46}$$

where $\mathbf{v}_P$ = absolute velocity of particle P
$\mathbf{v}_{P'}$ = velocity of point P' of moving frame $\mathfrak{F}$ coinciding with P
$\mathbf{v}_{P/\mathfrak{F}}$ = velocity of P relative to moving frame $\mathfrak{F}$

The absolute acceleration $\mathbf{a}_P$ of P may be expressed as

$$\mathbf{a}_P = \dot{\mathbf{\Omega}} \times \mathbf{r} + \mathbf{\Omega} \times (\mathbf{\Omega} \times \mathbf{r}) + 2\mathbf{\Omega} \times (\dot{\mathbf{r}})_{Oxyz} + (\ddot{\mathbf{r}})_{Oxyz} \tag{15.47}$$

We may also use the alternative form

$$\mathbf{a}_P = \mathbf{a}_{P'} + \mathbf{a}_{P/\mathfrak{F}} + \mathbf{a}_c \tag{15.48}$$

where $\mathbf{a}_P$ = absolute acceleration of particle P
$\mathbf{a}_{P'}$ = acceleration of point P' of moving frame $\mathcal{F}$ coinciding with P
$\mathbf{a}_{P/\mathcal{F}}$ = acceleration of P relative to moving frame $\mathcal{F}$
$\mathbf{a}_c = 2\boldsymbol{\Omega} \times (\dot{\mathbf{r}})_{Oxyz} = 2\boldsymbol{\Omega} \times \mathbf{v}_{P/\mathcal{F}}$
= complementary, or Coriolis, acceleration†

We note that the Coriolis acceleration is perpendicular to the vectors $\boldsymbol{\Omega}$ and $\mathbf{v}_{P/\mathcal{F}}$. However, since these vectors are usually not perpendicular to each other, the magnitude of $\mathbf{a}_c$ is in general *not* equal to $2\Omega v_{P/\mathcal{F}}$, as was the case for the plane motion of a particle. We further note that the Coriolis acceleration reduces to zero when the vectors $\boldsymbol{\Omega}$ and $\mathbf{v}_{P/\mathcal{F}}$ are parallel, or when either of them is zero.

Rotating frames of reference are particularly useful in the study of the three-dimensional motion of rigid bodies. If a rigid body has a fixed point O, as was the case for the crane of Sample Prob. 15.11, we may use a frame $Oxyz$ which is neither fixed nor rigidly attached to the rigid body. Denoting by $\boldsymbol{\Omega}$ the angular velocity of the frame $Oxyz$, we then resolve the angular velocity $\boldsymbol{\omega}$ of the body into the components $\boldsymbol{\Omega}$ and $\boldsymbol{\omega}_{B/\mathcal{F}}$, where the second component represents the angular velocity of the body relative to the frame $Oxyz$ (see Sample Prob. 15.14). An appropriate choice of the rotating frame will often lead to a simpler analysis of the motion of the rigid body than would be possible with axes of fixed orientation. This is especially true in the case of the general three-dimensional motion of a rigid body, i.e., when the rigid body under consideration has no fixed point (see Sample Prob. 15.15).

*** 15.15. Frame of Reference in General Motion.** Consider a fixed frame of reference $OXYZ$ and a frame $Axyz$ which moves in a known, but arbitrary, fashion with respect to $OXYZ$ (Fig. 15.38). Let P be a particle moving in space. The position of P is defined at any instant by the vector $\mathbf{r}_P$ in the fixed frame, and by the vector $\mathbf{r}_{P/A}$ in the moving frame. Denoting by $\mathbf{r}_A$ the position vector of A in the fixed frame, we have

$$\mathbf{r}_P = \mathbf{r}_A + \mathbf{r}_{P/A} \tag{15.49}$$

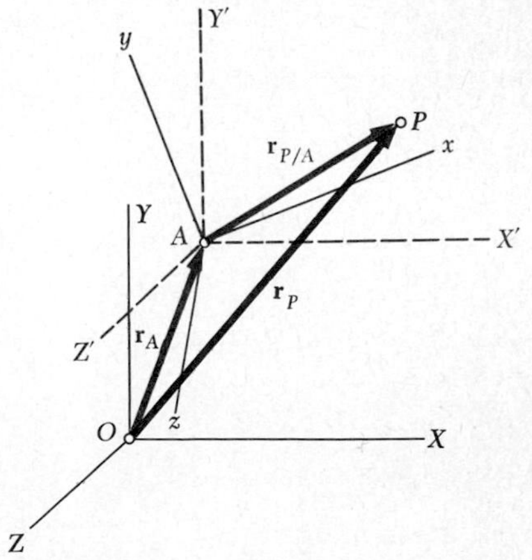

Fig. 15.38

The absolute velocity $\mathbf{v}_P$ of the particle is obtained by writing

$$\mathbf{v}_P = \dot{\mathbf{r}}_P = \dot{\mathbf{r}}_A + \dot{\mathbf{r}}_{P/A} \tag{15.50}$$

where the derivatives are defined with respect to the fixed frame $OXYZ$. The first term in the right-hand member of (15.50) thus represents the velocity $\mathbf{v}_A$ of the origin A of the moving axes. On the other hand, since the rate of change of a vector is the same with respect to a fixed frame and with respect to a frame in translation (Sec. 11.10), the second term may be regarded as the velocity $\mathbf{v}_{P/A}$ of P relative to the frame $AX'Y'Z'$ of the same orientation as $OXYZ$ and the same origin as $Axyz$. We therefore have

$$\mathbf{v}_P = \mathbf{v}_A + \mathbf{v}_{P/A} \tag{15.51}$$

† It is important to note the difference between Eq. (15.48) and Eq. (15.21) of Sec. 15.8. See the footnote on page 677.

But the velocity $\mathbf{v}_{P/A}$ of P relative to $AX'Y'Z'$ may be obtained from (15.45) by substituting $\mathbf{r}_{P/A}$ for $\mathbf{r}$ in that equation. We write

$$\mathbf{v}_P = \mathbf{v}_A + \mathbf{\Omega} \times \mathbf{r}_{P/A} + (\dot{\mathbf{r}}_{P/A})_{Axyz} \qquad (15.52)$$

where $\mathbf{\Omega}$ is the angular velocity of the frame $Axyz$ at the instant considered.

The absolute acceleration $\mathbf{a}_P$ of the particle is obtained by differentiating (15.51) and writing

$$\mathbf{a}_P = \dot{\mathbf{v}}_P = \dot{\mathbf{v}}_A + \dot{\mathbf{v}}_{P/A} \qquad (15.53)$$

where the derivatives are defined with respect to either of the frames $OXYZ$ or $AX'Y'Z'$. Thus, the first term in the right-hand member of (15.53) represents the acceleration $\mathbf{a}_A$ of the origin A of the moving axes and the second term, the acceleration $\mathbf{a}_{P/A}$ of P relative to the frame $AX'Y'Z'$. This acceleration may be obtained from (15.47) by substituting $\mathbf{r}_{P/A}$ for $\mathbf{r}$. We therefore write

$$\begin{aligned}\mathbf{a}_P = \mathbf{a}_A + \dot{\mathbf{\Omega}} \times \mathbf{r}_{P/A} + \mathbf{\Omega} \times (\mathbf{\Omega} \times \mathbf{r}_{P/A})\\ + 2\mathbf{\Omega} \times (\dot{\mathbf{r}}_{P/A})_{Axyz} + (\ddot{\mathbf{r}}_{P/A})_{Axyz}\end{aligned} \qquad (15.54)$$

Formulas (15.52) and (15.54) make it possible to determine the velocity and acceleration of a given particle with respect to a fixed frame of reference, when the motion of the particle is known with respect to a moving frame. These formulas become more significant, and considerably easier to remember, if we note that the sum of the first two terms in (15.52) represents the velocity of the point P' of the moving frame which coincides with P at the instant considered, and that the sum of the first three terms in (15.54) represents the acceleration of the same point. Thus, the relations (15.46) and (15.48) of the preceding section are still valid in the case of a reference frame in general motion, and we write

$$\mathbf{v}_P = \mathbf{v}_{P'} + \mathbf{v}_{P/\mathcal{F}} \qquad (15.46)$$
$$\mathbf{a}_P = \mathbf{a}_{P'} + \mathbf{a}_{P/\mathcal{F}} + \mathbf{a}_c \qquad (15.48)$$

where the various vectors involved have been defined in Sec. 15.14.

We may note that if the moving reference frame $\mathcal{F}$ (or $Axyz$) is in translation, the velocity and acceleration of the point P' of the frame which coincides with P become, respectively, equal to the velocity and acceleration of the origin A of the frame. On the other hand, since the frame maintains a fixed orientation, $\mathbf{a}_c$ is zero, and the relations (15.46) and (15.48) reduce, respectively, to the relations (11.33) and (11.34) derived in Sec. 11.12.

SAMPLE PROBLEM 15.13

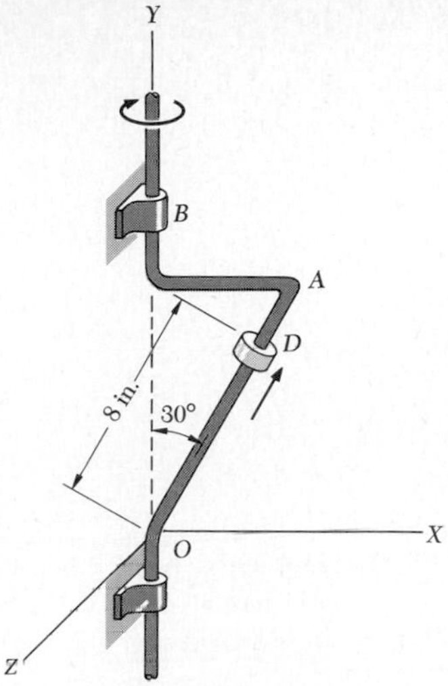

The bent rod OAB rotates about the vertical OB. At the instant considered, its angular velocity and angular acceleration are, respectively, 20 rad/s and 200 rad/s^2, both clockwise when viewed from the positive Y axis. The collar D moves along the rod and at the instant considered, $OD = 8$ in., and the velocity and acceleration of the collar relative to the rod are, respectively, 50 in./s and 600 in./s^2, both upward. Determine (*a*) the velocity of the collar, (*b*) the acceleration of the collar.

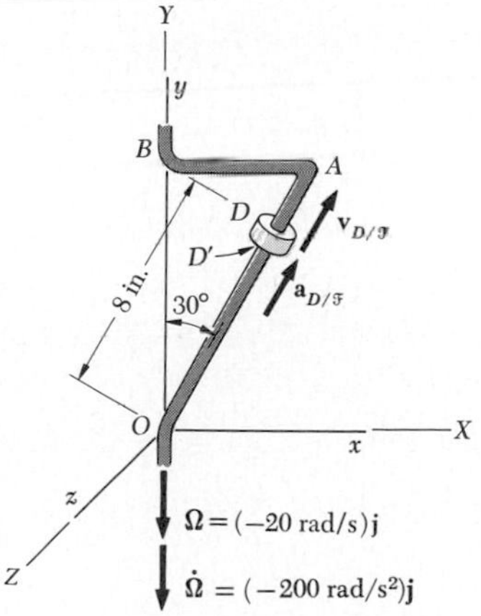

Frames of Reference. The frame $OXYZ$ is fixed. We attach the rotating frame $Oxyz$ to the bent rod. Its angular velocity and angular acceleration relative to $OXYZ$ are therefore $\mathbf{\Omega} = (-20 \text{ rad/s})\mathbf{j}$ and $\dot{\mathbf{\Omega}} = (-200 \text{ rad/s}^2)\mathbf{j}$, respectively. The position vector of D is

$$\mathbf{r} = (8 \text{ in.})(\sin 30°\mathbf{i} + \cos 30°\mathbf{j}) = (4 \text{ in.})\mathbf{i} + (6.93 \text{ in.})\mathbf{j}$$

a. Velocity $\mathbf{v}_D$. Denoting by D' the point of the rod which coincides with D and by $\mathcal{F}$ the rotating frame $Oxyz$, we write from Eq. (15.46)

$$\mathbf{v}_D = \mathbf{v}_{D'} + \mathbf{v}_{D/\mathcal{F}} \tag{1}$$

where

$$\mathbf{v}_{D'} = \mathbf{\Omega} \times \mathbf{r} = (-20 \text{ rad/s})\mathbf{j} \times [(4 \text{ in.})\mathbf{i} + (6.93 \text{ in.})\mathbf{j}] = (80 \text{ in./s})\mathbf{k}$$
$$\mathbf{v}_{D/\mathcal{F}} = (50 \text{ in./s})(\sin 30°\mathbf{i} + \cos 30°\mathbf{j}) = (25 \text{ in./s})\mathbf{i} + (43.3 \text{ in./s})\mathbf{j}$$

Substituting the values obtained for $\mathbf{v}_{D'}$ and $\mathbf{v}_{D/\mathcal{F}}$ into (1), we find

$$\mathbf{v}_D = (25 \text{ in./s})\mathbf{i} + (43.3 \text{ in./s})\mathbf{j} + (80 \text{ in./s})\mathbf{k} \quad ◄$$

b. Acceleration $\mathbf{a}_D$. From Eq. (15.48) we write

$$\mathbf{a}_D = \mathbf{a}_{D'} + \mathbf{a}_{D/\mathcal{F}} + \mathbf{a}_c \tag{2}$$

where

$$\begin{aligned}
\mathbf{a}_{D'} &= \dot{\mathbf{\Omega}} \times \mathbf{r} + \mathbf{\Omega} \times (\mathbf{\Omega} \times \mathbf{r}) \\
&= (-200 \text{ rad/s}^2)\mathbf{j} \times [(4 \text{ in.})\mathbf{i} + (6.93 \text{ in.})\mathbf{j}] - (20 \text{ rad/s})\mathbf{j} \times (80 \text{ in./s})\mathbf{k} \\
&= +(800 \text{ in./s}^2)\mathbf{k} - (1600 \text{ in./s}^2)\mathbf{i} \\
\mathbf{a}_{D/\mathcal{F}} &= (600 \text{ in./s}^2)(\sin 30°\mathbf{i} + \cos 30°\mathbf{j}) = (300 \text{ in./s}^2)\mathbf{i} + (520 \text{ in./s}^2)\mathbf{j} \\
\mathbf{a}_c &= 2\mathbf{\Omega} \times \mathbf{v}_{D/\mathcal{F}} \\
&= 2(-20 \text{ rad/s})\mathbf{j} \times [(25 \text{ in./s})\mathbf{i} + (43.3 \text{ in./s})\mathbf{j}] = (1000 \text{ in./s}^2)\mathbf{k}
\end{aligned}$$

Substituting the values obtained for $\mathbf{a}_{D'}$, $\mathbf{a}_{D/\mathcal{F}}$, and $\mathbf{a}_c$ into (2),

$$\mathbf{a}_D = -(1300 \text{ in./s}^2)\mathbf{i} + (520 \text{ in./s}^2)\mathbf{j} + (1800 \text{ in./s}^2)\mathbf{k} \quad ◄$$

SAMPLE PROBLEM 15.14

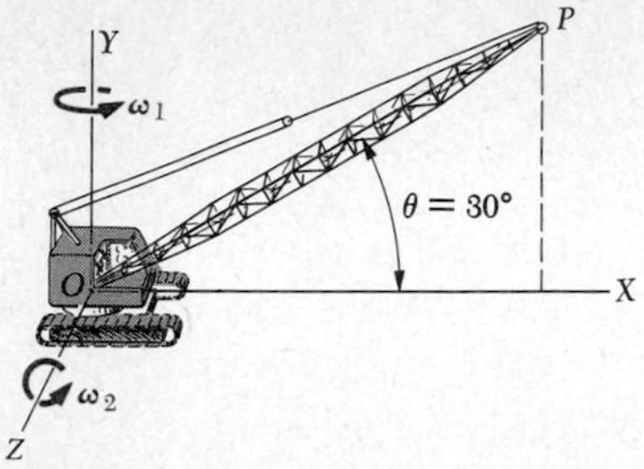

The crane shown rotates with a constant angular velocity $\boldsymbol{\omega}_1$ of 0.30 rad/s. Simultaneously, the boom is being raised with a constant angular velocity $\boldsymbol{\omega}_2$ of 0.50 rad/s relative to the cab. Knowing that the length of the boom OP is $l = 12$ m, determine (*a*) the velocity of the tip of the boom, (*b*) the acceleration of the tip of the boom.

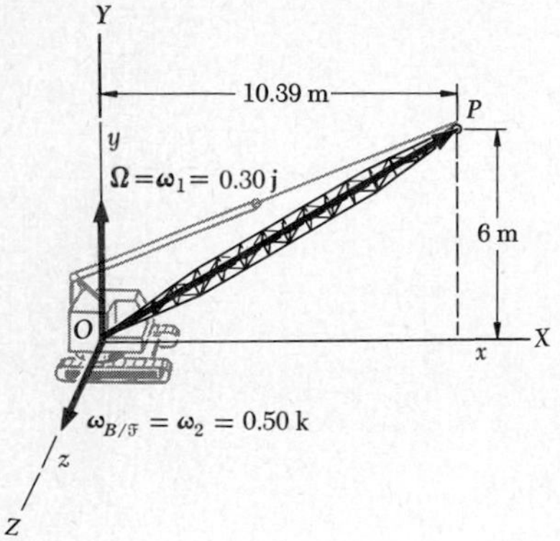

Frames of Reference. The frame $OXYZ$ is fixed. We attach the rotating frame $Oxyz$ to the cab. Its angular velocity with respect to the frame $OXYZ$ is therefore $\boldsymbol{\Omega} = \boldsymbol{\omega}_1 = (0.30\text{ rad/s})\mathbf{j}$. The angular velocity of the boom relative to the cab and the rotating frame $Oxyz$ (or $\mathcal{F}$, for short) is $\boldsymbol{\omega}_{B/\mathcal{F}} = \boldsymbol{\omega}_2 = (0.50\text{ rad/s})\mathbf{k}$.

***a.* Velocity $\mathbf{v}_P$.** From Eq. (15.46) we write

$$\mathbf{v}_P = \mathbf{v}_{P'} + \mathbf{v}_{P/\mathcal{F}} \qquad (1)$$

where $\mathbf{v}_{P'}$ is the velocity of the point P' of the rotating frame which coincides with P:

$$\mathbf{v}_{P'} = \boldsymbol{\Omega} \times \mathbf{r} = (0.30\text{ rad/s})\mathbf{j} \times [(10.39\text{ m})\mathbf{i} + (6\text{ m})\mathbf{j}] = -(3.12\text{ m/s})\mathbf{k}$$

and where $\mathbf{v}_{P/\mathcal{F}}$ is the velocity of P relative to the rotating frame $Oxyz$. But the angular velocity of the boom relative to $Oxyz$ was found to be $\boldsymbol{\omega}_{B/\mathcal{F}} = (0.50\text{ rad/s})\mathbf{k}$. The velocity of its tip P relative to $Oxyz$ is therefore

$$\begin{aligned}\mathbf{v}_{P/\mathcal{F}} &= \boldsymbol{\omega}_{B/\mathcal{F}} \times \mathbf{r} = (0.50\text{ rad/s})\mathbf{k} \times [(10.39\text{ m})\mathbf{i} + (6\text{ m})\mathbf{j}] \\ &= -(3\text{ m/s})\mathbf{i} + (5.20\text{ m/s})\mathbf{j}\end{aligned}$$

Substituting the values obtained for $\mathbf{v}_{P'}$ and $\mathbf{v}_{P/\mathcal{F}}$ into (1), we find

$$\mathbf{v}_P = -(3\text{ m/s})\mathbf{i} + (5.20\text{ m/s})\mathbf{j} - (3.12\text{ m/s})\mathbf{k} \quad ◀$$

***b.* Acceleration $\mathbf{a}_P$.** From Eq. (15.48) we write

$$\mathbf{a}_P = \mathbf{a}_{P'} + \mathbf{a}_{P/\mathcal{F}} + \mathbf{a}_c \qquad (2)$$

Since $\boldsymbol{\Omega}$ and $\boldsymbol{\omega}_{B/\mathcal{F}}$ are both constant, we have

$$\begin{aligned}\mathbf{a}_{P'} &= \boldsymbol{\Omega} \times (\boldsymbol{\Omega} \times \mathbf{r}) = (0.30\text{ rad/s})\mathbf{j} \times (-3.12\text{ m/s})\mathbf{k} = -(0.94\text{ m/s}^2)\mathbf{i} \\ \mathbf{a}_{P/\mathcal{F}} &= \boldsymbol{\omega}_{B/\mathcal{F}} \times (\boldsymbol{\omega}_{B/\mathcal{F}} \times \mathbf{r}) \\ &= (0.50\text{ rad/s})\mathbf{k} \times [-(3\text{ m/s})\mathbf{i} + (5.20\text{ m/s})\mathbf{j}] \\ &= -(1.50\text{ m/s}^2)\mathbf{j} - (2.60\text{ m/s}^2)\mathbf{i} \\ \mathbf{a}_c &= 2\boldsymbol{\Omega} \times \mathbf{v}_{P/\mathcal{F}} \\ &= 2(0.30\text{ rad/s})\mathbf{j} \times [-(3\text{ m/s})\mathbf{i} + (5.20\text{ m/s})\mathbf{j}] = (1.80\text{ m/s}^2)\mathbf{k}\end{aligned}$$

Substituting for $\mathbf{a}_{P'}$, $\mathbf{a}_{P/\mathcal{F}}$, and $\mathbf{a}_c$ into (2), we find

$$\mathbf{a}_P = -(3.54\text{ m/s}^2)\mathbf{i} - (1.50\text{ m/s}^2)\mathbf{j} + (1.80\text{ m/s}^2)\mathbf{k} \quad ◀$$

SAMPLE PROBLEM 15.15

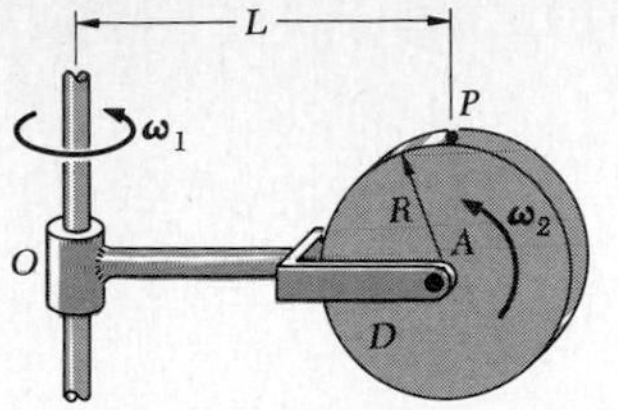

Disk D, of radius R, is pinned to end A of the arm OA of length L located in the plane of the disk. The arm rotates about a vertical axis through O at the constant rate ω_1, and the disk rotates about A at the constant rate ω_2. Determine (*a*) the velocity of point P located directly above A, (*b*) the acceleration of P, (*c*) the angular velocity and angular acceleration of the disk.

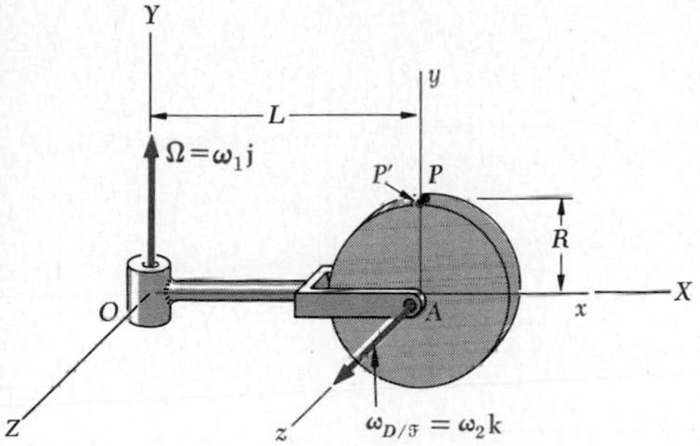

Frames of Reference. The frame $OXYZ$ is fixed. We attach the moving frame $Axyz$ to the arm OA. Its angular velocity with respect to the frame $OXYZ$ is therefore $\boldsymbol{\Omega} = \omega_1 \mathbf{j}$. The angular velocity of disk D relative to the moving frame $Axyz$ (or $\mathfrak{F}$, for short) is $\boldsymbol{\omega}_{D/\mathfrak{F}} = \omega_2 \mathbf{k}$. The position vector of P relative to O is $\mathbf{r} = L\mathbf{i} + R\mathbf{j}$ and its position vector relative to A is $\mathbf{r}_{P/A} = R\mathbf{j}$.

a. **Velocity $\mathbf{v}_P$.** Denoting by P' the point of the moving frame which coincides with P, we write from Eq. (15.46)

$$\mathbf{v}_P = \mathbf{v}_{P'} + \mathbf{v}_{P/\mathfrak{F}} \tag{1}$$

where $\mathbf{v}_{P'} = \boldsymbol{\Omega} \times \mathbf{r} = \omega_1 \mathbf{j} \times (L\mathbf{i} + R\mathbf{j}) = -\omega_1 L\mathbf{k}$

$$\mathbf{v}_{P/\mathfrak{F}} = \boldsymbol{\omega}_{D/\mathfrak{F}} \times \mathbf{r}_{P/A} = \omega_2 \mathbf{k} \times R\mathbf{j} = -\omega_2 R\mathbf{i}$$

Substituting the values obtained for $\mathbf{v}_{P'}$ and $\mathbf{v}_{P/\mathfrak{F}}$ into (1), we find

$$\mathbf{v}_P = -\omega_2 R\mathbf{i} - \omega_1 L\mathbf{k} \quad ◀$$

b. **Acceleration $\mathbf{a}_P$.** From Eq. (15.48) we write

$$\mathbf{a}_P = \mathbf{a}_{P'} + \mathbf{a}_{P/\mathfrak{F}} + \mathbf{a}_c \tag{2}$$

Since $\boldsymbol{\Omega}$ and $\boldsymbol{\omega}_{D/\mathfrak{F}}$ are both constant, we have

$$\mathbf{a}_{P'} = \boldsymbol{\Omega} \times (\boldsymbol{\Omega} \times \mathbf{r}) = \omega_1 \mathbf{j} \times (-\omega_1 L\mathbf{k}) = -\omega_1^2 L\mathbf{i}$$
$$\mathbf{a}_{P/\mathfrak{F}} = \boldsymbol{\omega}_{D/\mathfrak{F}} \times (\boldsymbol{\omega}_{D/\mathfrak{F}} \times \mathbf{r}_{P/A}) = \omega_2 \mathbf{k} \times (-\omega_2 R\mathbf{i}) = -\omega_2^2 R\mathbf{j}$$
$$\mathbf{a}_c = 2\boldsymbol{\Omega} \times \mathbf{v}_{P/\mathfrak{F}} = 2\omega_1 \mathbf{j} \times (-\omega_2 R\mathbf{i}) = 2\omega_1\omega_2 R\mathbf{k}$$

Substituting the values obtained into (2), we find

$$\mathbf{a}_P = -\omega_1^2 L\mathbf{i} - \omega_2^2 R\mathbf{j} + 2\omega_1\omega_2 R\mathbf{k} \quad ◀$$

c. **Angular Velocity and Angular Acceleration of Disk.**

$$\boldsymbol{\omega} = \boldsymbol{\Omega} + \boldsymbol{\omega}_{D/\mathfrak{F}} \qquad \boldsymbol{\omega} = \omega_1 \mathbf{j} + \omega_2 \mathbf{k} \quad ◀$$

Using Eq. (15.31) with $\mathbf{Q} = \boldsymbol{\omega}$, we write

$$\boldsymbol{\alpha} = (\dot{\boldsymbol{\omega}})_{OXYZ} = (\dot{\boldsymbol{\omega}})_{Axyz} + \boldsymbol{\Omega} \times \boldsymbol{\omega}$$
$$= 0 + \omega_1 \mathbf{j} \times (\omega_1 \mathbf{j} + \omega_2 \mathbf{k})$$

$$\boldsymbol{\alpha} = \omega_1 \omega_2 \mathbf{i} \quad ◀$$

Problems

15.179 The bent rod ABC rotates at the constant rate $\omega_1 = 6$ rad/s. Knowing that the collar D moves toward end C at a constant relative speed $u = 100$ in./s, determine for the position shown (*a*) the velocity of D, (*b*) the acceleration of D.

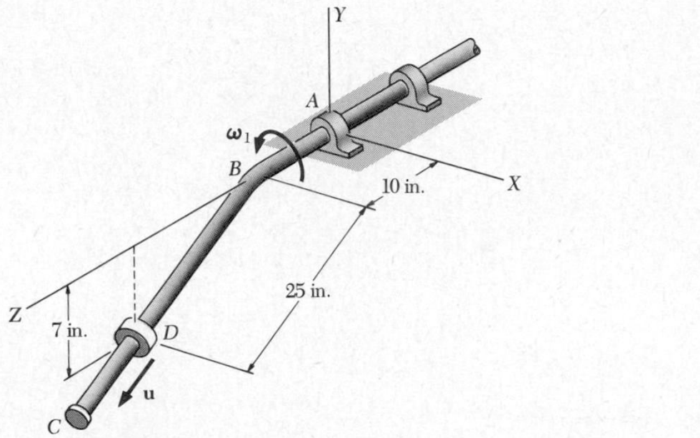

Fig. P15.179

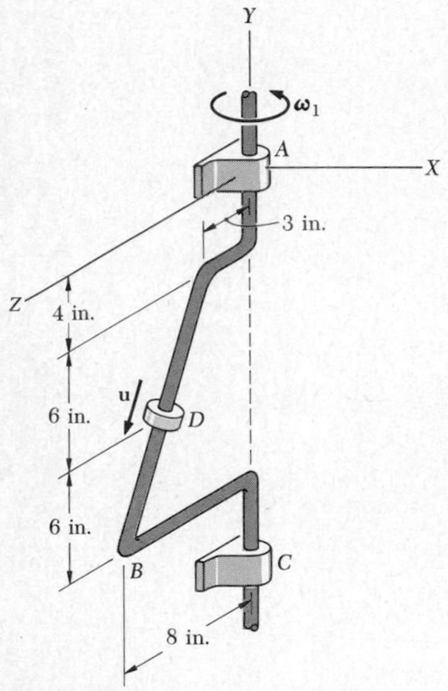

Fig. P15.180

15.180 The bent rod ABC rotates at the constant rate $\omega_1 = 4$ rad/s. Knowing that the collar D moves downward along the rod at a constant relative speed $u = 65$ in./s, determine for the position shown (*a*) the velocity of D, (*b*) the acceleration of D.

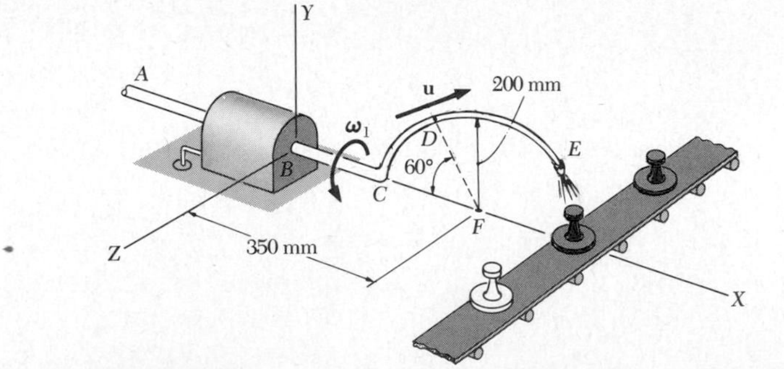

Fig. P15.181

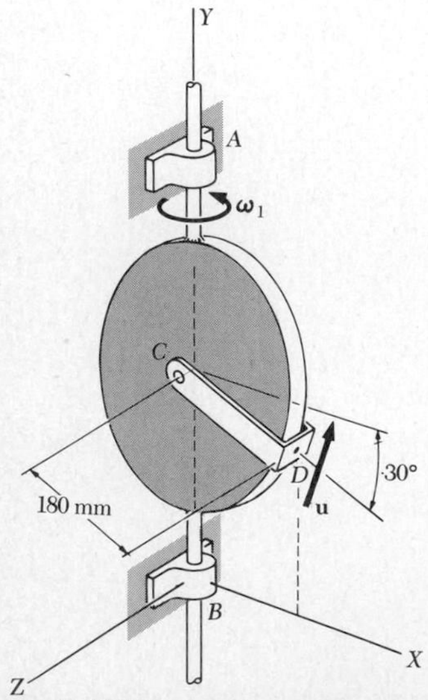

Fig. P15.182

15.181 Manufactured items are spray-painted as they pass through the automated work station shown. Knowing that the bent pipe ACE rotates at the constant rate $\omega_1 = 0.6$ rad/s and that at point D the paint moves through the pipe at a constant relative speed $u = 150$ mm/s, determine for the position shown (*a*) the velocity of the paint at D, (*b*) the acceleration of the paint at D.

15.182 The circular plate shown rotates about its vertical diameter at the constant rate $\omega_1 = 8$ rad/s. Knowing that in the position shown the disk lies in the XY plane and point D of strap CD moves upward at a constant relative speed $u = 1.2$ m/s, determine (*a*) the velocity of D, (*b*) the acceleration of D.

15.183 Solve Prob. 15.180, assuming that at the instant shown the angular velocity $\boldsymbol{\omega}_1$ of the rod is 4 rad/s and is decreasing at the rate of 10 rad/s^2, while the relative speed u of the collar is 65 in./s and is decreasing at the rate of 208 in./s^2.

15.184 Solve Prob. 15.182, assuming that at the instant shown the angular velocity $\boldsymbol{\omega}_1$ of the plate is 8 rad/s and is decreasing at the rate of 25 rad/s^2 while the relative speed u of point D of strap CD is 1.2 m/s and is decreasing at the rate of 3 m/s^2.

15.185 Using the method of Sec. 15.14, solve Prob. 15.157.
15.186 Using the method of Sec. 15.14, solve Prob. 15.152.
15.187 Using the method of Sec. 15.14, solve Prob. 15.155.
15.188 Using the method of Sec. 15.14, solve Prob. 15.156.

15.189 The body AB and rod BC of the robot component shown rotate at the constant rate $\omega_1 = 0.60$ rad/s about the Y axis. Simultaneously a wire-and-pulley control causes arm CD to rotate about C at the constant rate $\omega_2 = d\beta/dt = 0.45$ rad/s. Knowing that $\beta = 120°$, determine (*a*) the angular acceleration of arm CD, (*b*) the velocity of D, (*c*) the acceleration of D.

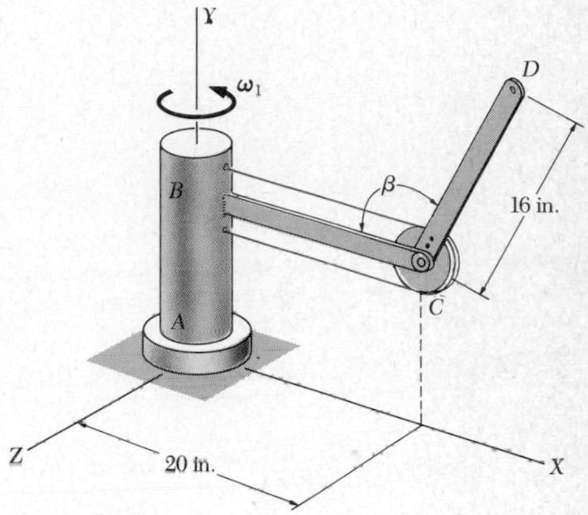

Fig. P15.189

15.190 Solve Prob. 15.189, assuming that $\beta = 60°$.

15.191 A disk of radius r rotates at the constant rate ω_2 with respect to the arm ABC, which itself rotates at the constant rate ω_1 about the X axis. Determine (*a*) the angular velocity and angular acceleration of the disk, (*b*) the velocity and acceleration of point D on the rim of the disk.

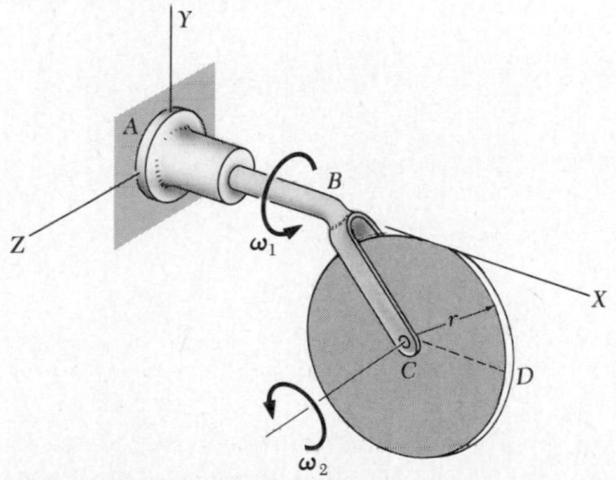

Fig. P15.191 and P15.192

15.192 A disk of radius $r = 120$ mm rotates at the constant rate $\omega_2 =$ 5 rad/s with respect to the arm ABC, which itself rotates at the constant rate $\omega_1 = 4$ rad/s about the X axis. Determine the velocity and acceleration of point D on the rim of the disk.

15.193 The remote manipulator system (RMS) shown is used to deploy payloads from the cargo bay of space shuttles. At the instant shown, the whole RMS is rotating at the constant rate $\omega_1 = 0.02$ rad/s about the axis AB. At the same time, portion BCD rotates as a rigid body at the constant rate $\omega_2 = d\beta/dt = 0.03$ rad/s about an axis through B parallel to the X axis. Knowing that $\beta = 30°$, determine (*a*) the angular acceleration of BCD, (*b*) the velocity of D, (*c*) the acceleration of D.

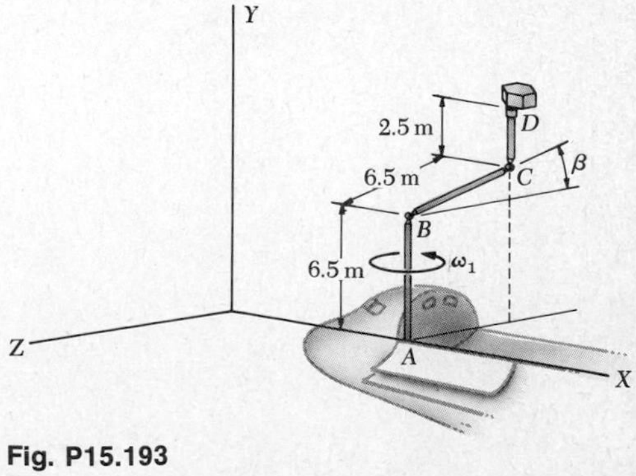

Fig. P15.193

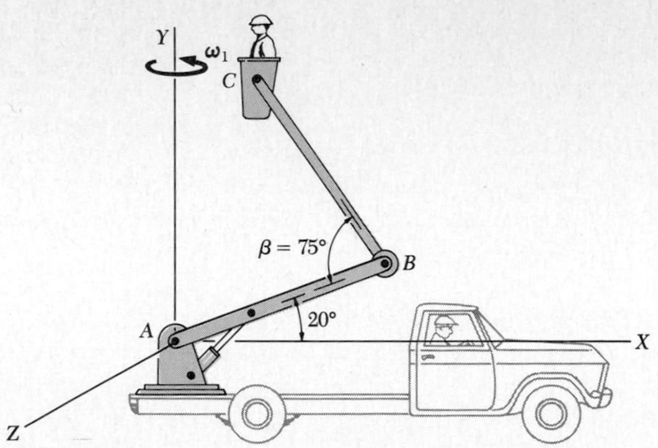

Fig. P15.194

15.194 The mechanism shown is used to raise a worker to the elevation of overhead electric and telephone wires. The entire mechanism rotates at the constant rate $\omega_1 = 0.15$ rad/s about the Y axis. The angle between arm AB and the horizontal is constant, while arm BC is being lowered at the constant rate $\omega_2 = d\beta/dt = -0.25$ rad/s. Knowing that AB and BC are each 4 m long, determine the acceleration of C at the instant shown.

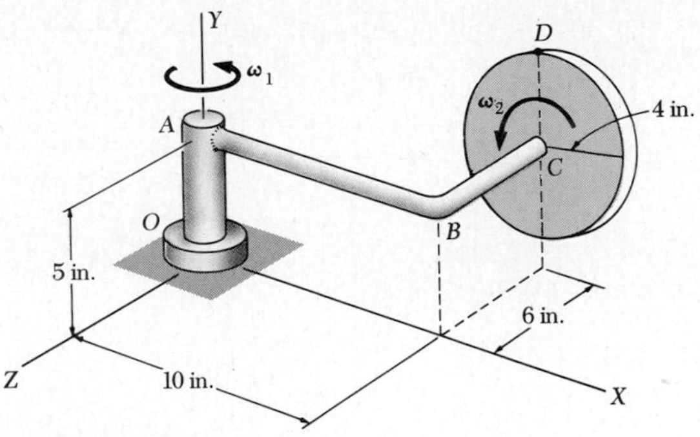

Fig. P15.195

15.195 A disk of 4-in. radius rotates at the constant rate $\omega_2 = 5$ rad/s with respect to the arm ABC, which itself rotates at the constant rate $\omega_1 = 3$ rad/s about the Y axis. Determine (*a*) the angular acceleration of the disk, (*b*) the acceleration of point D on the rim of the disk.

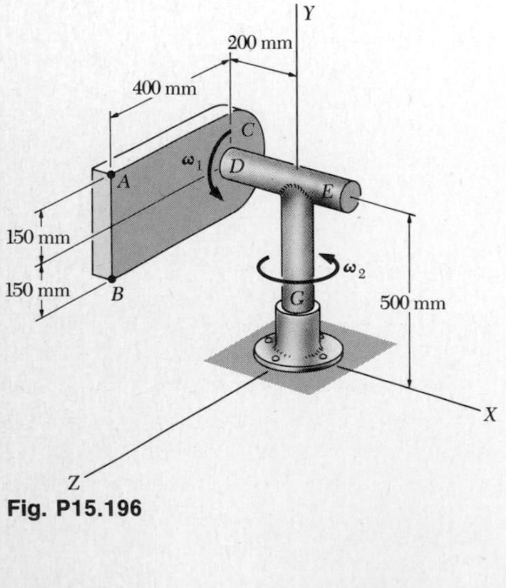

Fig. P15.196

15.196 In the portion of an industrial robot shown plate ABC rotates at the constant rate $\omega_1 = 1.8$ rad/s with respect to component DEG. At the same time, the entire unit rotates about the Y axis at the constant rate $\omega_2 = 1.5$ rad/s. Knowing that at the instant shown DE is parallel to the X axis and plate ABC is parallel to the YZ plane, determine (*a*) the velocity of point A, (*b*) the acceleration of point A.

15.197 In Prob. 15.196, determine (*a*) the velocity of point B, (*b*) the acceleration of point B.

15.198 The 120-ft blades of a wind-turbine generator rotate at the constant rate $\omega_1 = 25$ rpm. Knowing that at the instant shown the entire unit is being rotated about the Y axis at the constant rate $\omega_2 = 0.1$ rad/s, determine (*a*) the angular acceleration of the blades, (*b*) the velocity and acceleration of blade tip B.

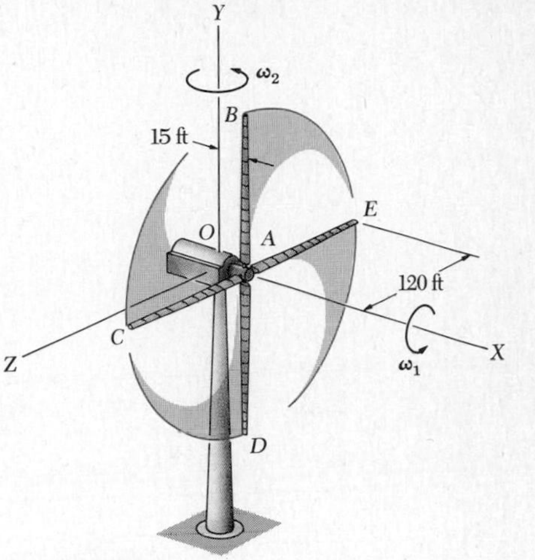

Fig. P15.198

15.199 In Prob. 15.198, determine the velocity and acceleration of (*a*) blade tip C, (*b*) blade tip D.

15.200 The position of the stylus tip A is controlled by the robot shown. In the position shown the stylus moves at a constant speed $u = 150$ mm/s relative to the solenoid BC. At the same time, arm CD rotates at the constant rate $\omega_2 = 1.6$ rad/s with respect to component DEG. Knowing that the entire robot rotates about the X axis at the constant rate $\omega_1 = 1.2$ rad/s, determine (*a*) the velocity of A, (*b*) the acceleration of A.

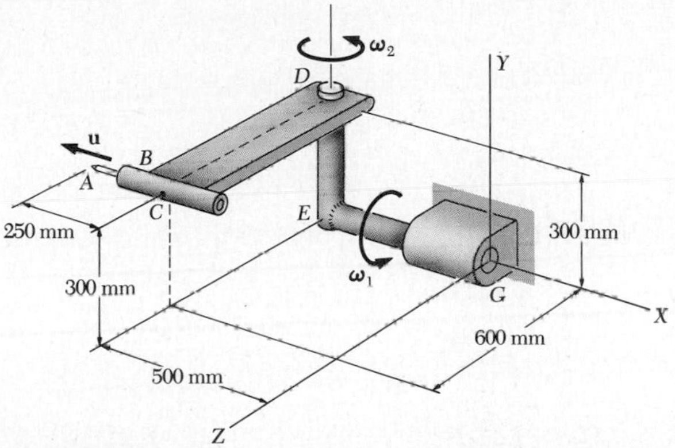

Fig. P15.200

15.201 A square plate of side $2r$ is welded to a vertical shaft which rotates with a constant angular velocity $\boldsymbol{\omega}_1$. At the same time, rod AB of length r rotates about the center of the plate with a constant angular velocity $\boldsymbol{\omega}_2$ with respect to the plate. For the position of the plate shown, determine the acceleration of end B of the rod if (*a*) $\theta = 0$, (*b*) $\theta = 90°$, (*c*) $\theta = 180°$.

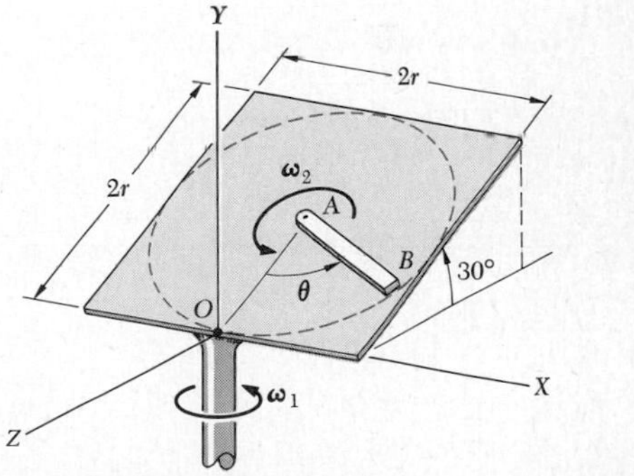

Fig. P15.201

15.202 Solve Prob. 15.201, assuming $\omega_1 = 3$ rad/s, $\omega_2 = 2$ rad/s, and $r = 120$ mm.

Review Problems

15.203 The flanged wheel rolls without slipping on the horizontal rail. If at a given instant the velocity and acceleration of the center of the wheel are as shown, determine the acceleration (*a*) of point B, (*b*) of point C, (*c*) of point D.

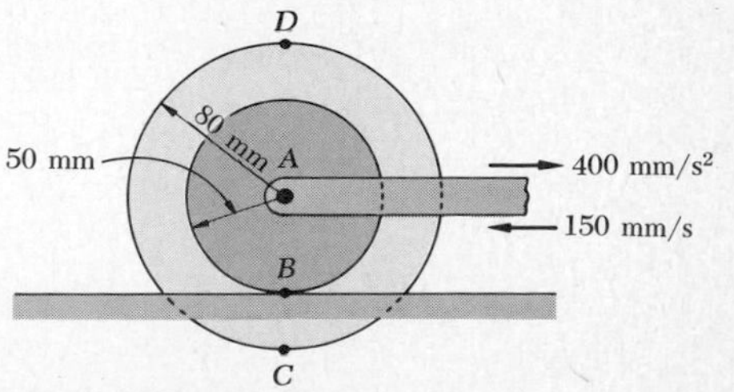

Fig. P15.203

15.204 The propeller of a small airplane rotates at a constant rate of 1800 rpm in a clockwise sense when viewed by the pilot. Knowing that the airplane is turning left along a horizontal circular path of radius 1200 m, and that the speed of the airplane is 270 km/h, determine the angular acceleration of the propeller at the instant the airplane is moving due west.

15.205 Rod AC of length $2b$ is attached to a collar at A and passes through a pivoted collar at B. Knowing that the collar A moves upward with a constant velocity $\mathbf{v}_A$, derive an expression for (*a*) the angular velocity of rod AC, (*b*) the velocity of the point of the rod in contact with the pivoted collar B, (*c*) the angular acceleration of rod AC.

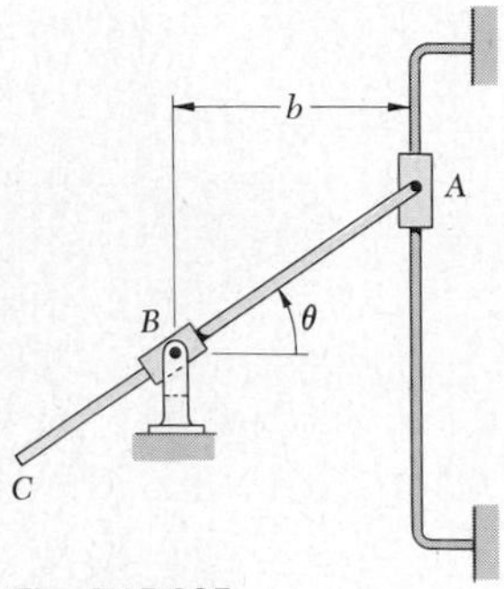

Fig. P15.205

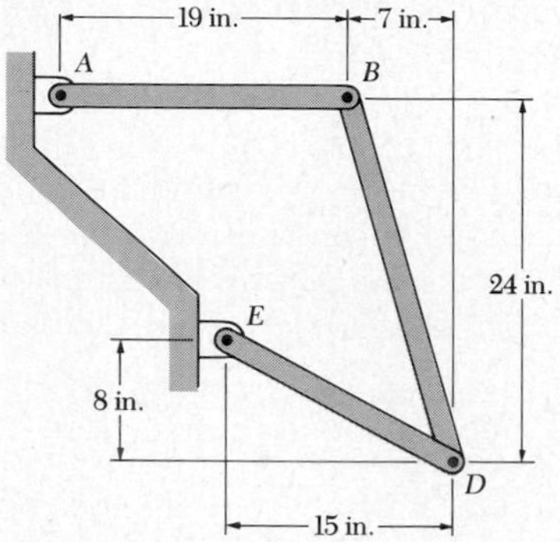

Fig. P15.206

15.206 In the position shown, bar AB has a constant angular velocity of 4 rad/s counterclockwise. Determine the angular velocity (*a*) of bar BD, (*b*) of bar DE.

15.207 For the linkage of Prob. 15.206, determine the angular acceleration (*a*) of bar BD, (*b*) of bar DE.

15.208 The telescoping arm AB is used to raise a worker to the elevation of electric and telephone wires. Knowing that the length AB increases at the constant rate $dl/dt = 0.25$ m/s and that the arm rotates at the constant rate $\omega_1 = 0.30$ rad/s about a vertical axis, while the angle θ it forms with the horizontal maintains a constant value, determine the acceleration of point B when $l = 6$ m.

15.209 Solve Prob. 15.208, assuming that the angle θ increases at the constant rate $d\theta/dt = 0.40$ rad/s.

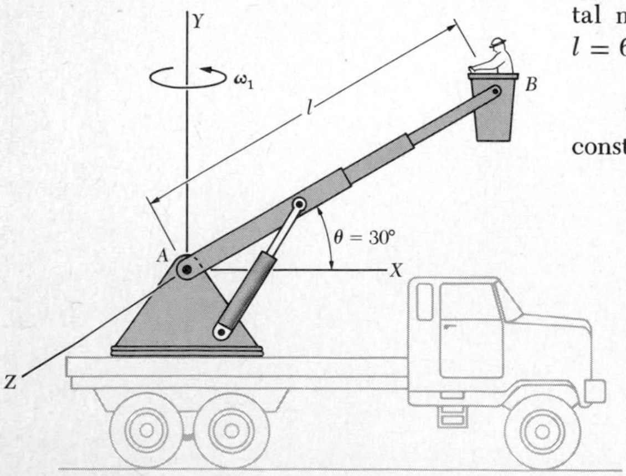

Fig. P15.208

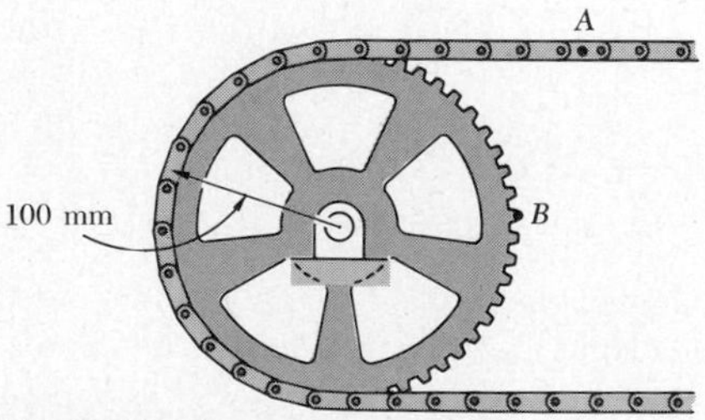

Fig. P15.210

15.210 The sprocket wheel and chain are initially at rest. If the acceleration of point A of the chain has a constant magnitude of 120 mm/s^2 and is directed to the left, determine the acceleration of sprocket B as the wheel is completing a full revolution.

15.211 Gear A rolls on the fixed gear B and rotates about the axle AD which is rigidly attached at D to the vertical shaft DE. Knowing that shaft DE rotates with a constant angular velocity $\boldsymbol{\omega}_1$, determine (*a*) the rate of spin of gear A about the axle AD, (*b*) the angular acceleration of gear A, (*c*) the acceleration of tooth C of gear A.

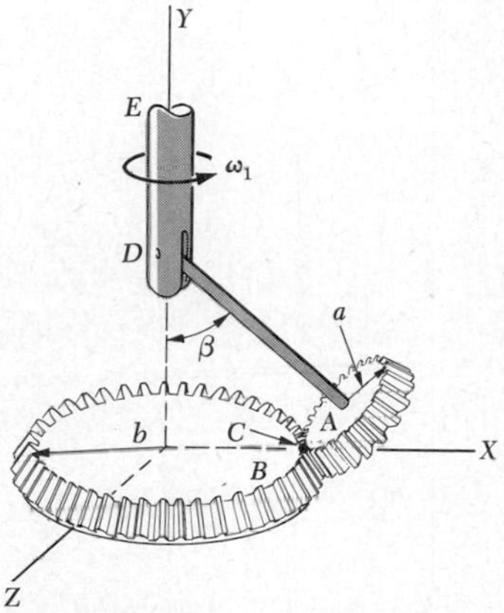

Fig. P15.211

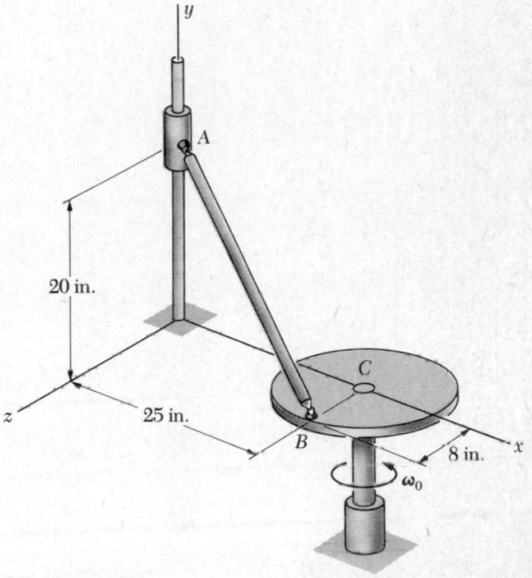

Fig. P15.212

15.212 Rod AB is connected by ball-and-socket joints to the collar A and to the 16-in.-diameter rotating disk C. Knowing that the disk rotates counterclockwise in the zx plane at the constant rate $\omega_0 = 3$ rad/s, determine the velocity of collar A.

15.213 Two collars C and D move along the vertical rod shown. Knowing that the velocity of collar C is 660 mm/s downward, determine (*a*) the velocity of collar D, (*b*) the angular velocity of member AB.

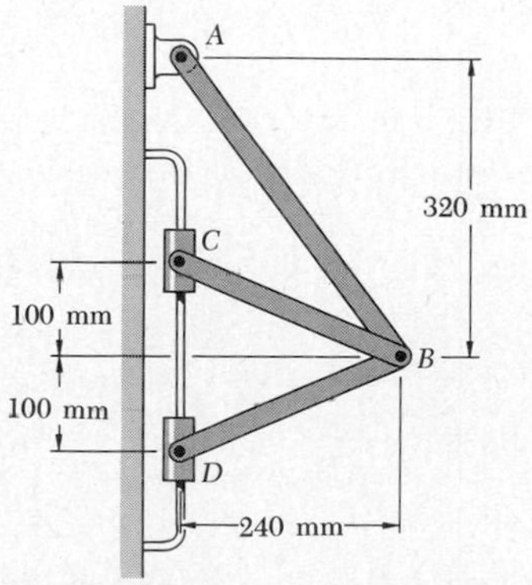

Fig. P15.213

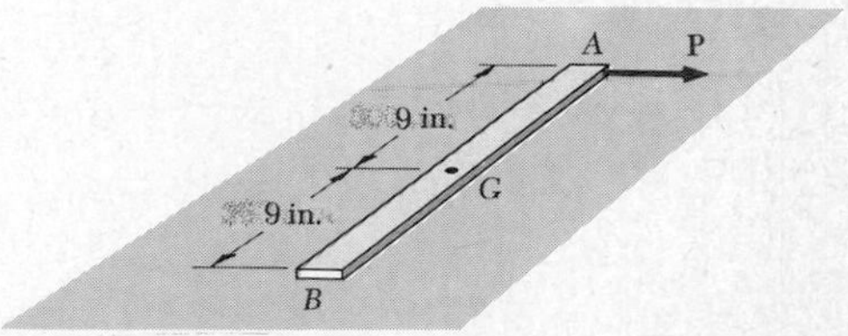

Fig. P15.214

15.214 An 18-in. rod rests on a frictionless horizontal table. A force **P** applied as shown produces the following accelerations: $\mathbf{a}_A = 30$ in./s² to the right, $\boldsymbol{\alpha} = 2.5$ rad/s² clockwise as viewed from above. Determine (*a*) the acceleration of point B, (*b*) the point of the rod which has no acceleration.

CHAPTER SIXTEEN

Plane Motion of Rigid Bodies: Forces and Accelerations

16.1. Introduction. In this chapter and in Chaps. 17 and 18, we shall study the *kinetics of rigid bodies,* i.e., the relations existing between the forces acting on a rigid body, the shape and mass of the body, and the motion produced. In Chaps. 12 and 13, we studied similar relations, assuming then that the body could be considered as a particle, i.e., that its mass could be concentrated in one point and that all forces acted at that point. We shall now take the shape of the body into account, as well as the exact location of the points of application of the forces. We shall also be concerned not only with the motion of the body as a whole but also with the motion of the body about its mass center.

Our approach will be to consider rigid bodies as made of large numbers of particles and to use the results obtained in Chap. 14 for the motion of systems of particles. In this chapter, we shall use specifically Eq. (14.16), $\Sigma \mathbf{F} = m\bar{\mathbf{a}}$, which relates the resultant of the external forces and the acceleration of the mass center G of the system of particles, and Eq. (14.23), $\Sigma \mathbf{M}_G = \dot{\mathbf{H}}_G$, which relates the moment resultant of the external forces and the angular momentum of the system of particles about G.

Except for Sec. 16.2, which applies to the most general case of the motion of a rigid body, the results derived in this chapter will be limited in two ways: (1) They will be restricted to the *plane motion* of rigid bodies, i.e., to a motion in which each particle of the body remains at a constant distance from a fixed reference plane. (2) The rigid bodies considered will consist only of plane slabs and of bodies which are symmetrical with respect to the reference plane.† The study of the plane motion of non-

† Or, more generally, bodies which have a principal centroidal axis of inertia perpendicular to the reference plane.

symmetrical three-dimensional bodies and, more generally, the motion of rigid bodies in three-dimensional space will be postponed until Chap. 18.

In Sec. 16.3, we define the angular momentum of a rigid body in plane motion and show that the rate of change of the angular momentum $\dot{\mathbf{H}}_G$ about the mass center is equal to the product $\bar{I}\boldsymbol{\alpha}$ of the centroidal mass moment of inertia $\bar{I}$ and the angular acceleration $\boldsymbol{\alpha}$ of the body. D'Alembert's principle, introduced in Sec. 16.4, is used to prove that the external forces acting on a rigid body are equivalent to a vector $m\bar{\mathbf{a}}$ attached at the mass center and a couple of moment $\bar{I}\boldsymbol{\alpha}$.

In Sec. 16.5, we derive the principle of transmissibility using only the parallelogram law and Newton's laws of motion. The principle of transmissibility may therefore be removed from the list of axioms (Sec. 1.2) required for the study of the statics and dynamics of rigid bodies.

The concept of free-body-diagram equations is introduced in Sec. 16.6. Emphasis will be placed on the use of free-body-diagram equations in the solution of all problems involving the plane motion of rigid bodies.

After considering the plane motion of connected rigid bodies in Sec. 16.7, we shall be prepared to solve a variety of problems involving the translation, centroidal rotation, or unconstrained motion of rigid bodies. In Sec. 16.8 and in the remaining part of the chapter, we shall consider the solution of problems involving the noncentroidal rotation, rolling motion, or other partially constrained plane motions of rigid bodies.

16.2. Equations of Motion for a Rigid Body. Consider a rigid body acted upon by several external forces $\mathbf{F}_1$, $\mathbf{F}_2$, $\mathbf{F}_3$, etc. (Fig. 16.1). We may assume that the body is made of a large number n of particles of mass Δm_i ($i = 1, 2, \ldots, n$) and apply the results obtained in Chap. 14 for a system of particles (Fig. 16.2). Considering first the motion of the mass center G of the body with respect to the newtonian frame of reference $Oxyz$, we recall Eq. (14.16) and write

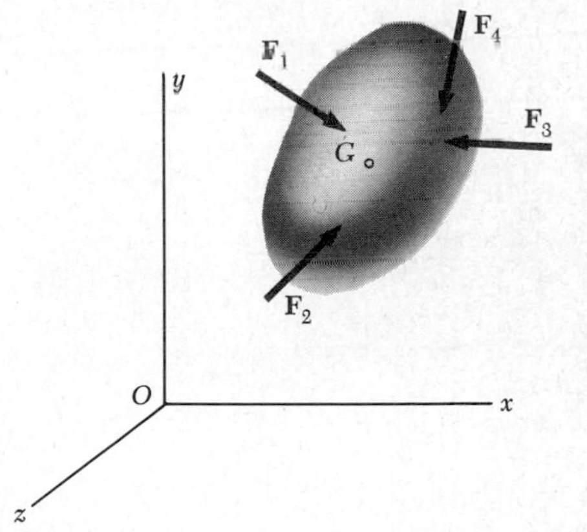

Fig. 16.1

$$\Sigma\mathbf{F} = m\bar{\mathbf{a}} \tag{16.1}$$

where m is the mass of the body and $\bar{\mathbf{a}}$ the acceleration of the mass center G. Turning now to the motion of the body relative to the centroidal frame of reference $Gx'y'z'$, we recall Eq. (14.23) and write

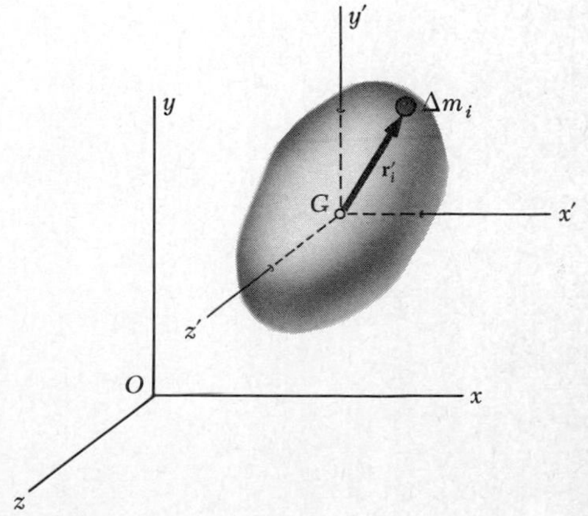

Fig. 16.2

$$\Sigma\mathbf{M}_G = \dot{\mathbf{H}}_G \tag{16.2}$$

where $\dot{\mathbf{H}}_G$ represents the rate of change of $\mathbf{H}_G$, the angular momentum about G of the system of particles forming the rigid body. In the following we shall simply refer to $\mathbf{H}_G$ as the *angular momentum of the rigid body about its mass center G.* Together Eqs. (16.1) and (16.2) express that *the*

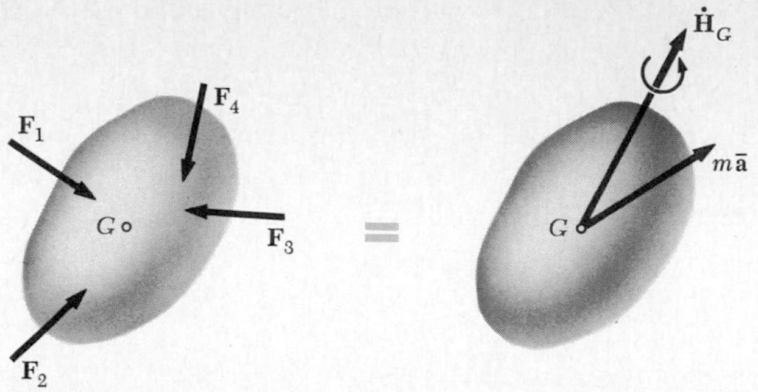

Fig. 16.3

system of the external forces is equipollent to the system consisting of the vector $m\bar{\mathbf{a}}$ attached at G and the couple of moment $\dot{\mathbf{H}}_G$ (Fig. 16.3).†

Equations (16.1) and (16.2) apply in the most general case of the motion of a rigid body. In the rest of this chapter, however, we shall limit our analysis to the *plane motion* of rigid bodies, i.e., to a motion in which each particle remains at a constant distance from a fixed reference plane, and we shall assume that the rigid bodies considered consist only of plane slabs and of bodies which are symmetrical with respect to the reference plane. Further study of the plane motion of nonsymmetrical three-dimensional bodies and of the motion of rigid bodies in three-dimensional space will be postponed until Chap. 18.

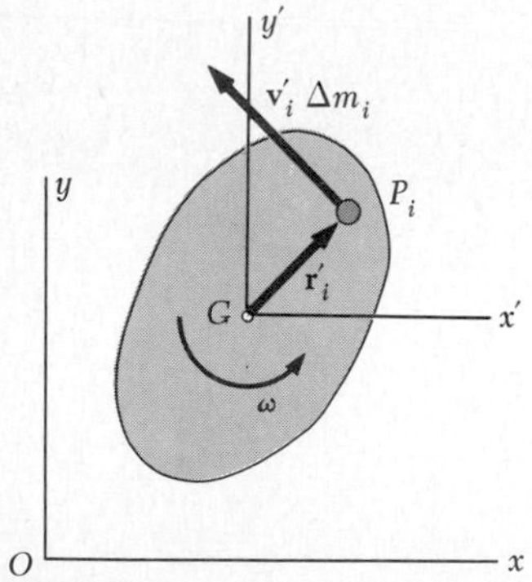

Fig. 16.4

16.3. Angular Momentum of a Rigid Body in Plane Motion. Consider a rigid slab in plane motion. Assuming that the slab is made of a large number n of particles P_i of mass Δm_i and recalling Eq. (14.24) of Sec. 14.5, we note that the angular momentum $\mathbf{H}_G$ of the slab about its mass center G may be computed by taking the moments about G of the momenta of the particles of the slab in their motion with respect to either of the frames Oxy or $Gx'y'$ (Fig. 16.4). Choosing the latter course, we write

$$\mathbf{H}_G = \sum_{i=1}^{n} (\mathbf{r}'_i \times \mathbf{v}'_i\,\Delta m_i) \tag{16.3}$$

where $\mathbf{r}'_i$ and $\mathbf{v}'_i\,\Delta m_i$ denote, respectively, the position vector and the linear momentum of the particle P_i relative to the centroidal frame of reference $Gx'y'$. But since the particle belongs to the slab, we have $\mathbf{v}'_i = \boldsymbol{\omega} \times \mathbf{r}'_i$, where $\boldsymbol{\omega}$ is the angular velocity of the slab at the instant considered. We write

$$\mathbf{H}_G = \sum_{i=1}^{n} [\mathbf{r}'_i \times (\boldsymbol{\omega} \times \mathbf{r}'_i)\Delta m_i]$$

† Since the systems involved act on a rigid body, we could conclude at this point, by referring to Sec. 3.19, that the two systems are *equivalent* as well as equipollent and use colored rather than gray equals signs in Fig. 16.3. However, by postponing this conclusion, we shall be able to arrive at it independently (Secs. 16.4 and 18.5), thereby eliminating the necessity of including the principle of transmissibility among the axioms of mechanics (Sec. 16.5).

Referring to Fig. 16.4, we easily verify that the expression obtained represents a vector of the same direction as $\boldsymbol{\omega}$ (that is, perpendicular to the slab) and of magnitude equal to $\omega \Sigma r_i'^2 \, \Delta m_i$. Recalling that the sum $\Sigma r_i'^2 \, \Delta m_i$ represents the moment of inertia $\bar{I}$ of the slab about a centroidal axis perpendicular to the slab, we conclude that the angular momentum $\mathbf{H}_G$ of the slab about its mass center is

$$\mathbf{H}_G = \bar{I}\boldsymbol{\omega} \tag{16.4}$$

Differentiating both members of Eq. (16.4) we obtain

$$\dot{\mathbf{H}}_G = \bar{I}\dot{\boldsymbol{\omega}} = \bar{I}\boldsymbol{\alpha} \tag{16.5}$$

Thus the rate of change of the angular momentum of the slab is represented by a vector of the same direction as $\boldsymbol{\alpha}$ (that is, perpendicular to the slab) and of magnitude $\bar{I}\alpha$.

It should be kept in mind that the results obtained in this section have been derived for a rigid slab in plane motion. As we shall see in Chap. 18, they remain valid in the case of the plane motion of rigid bodies which are symmetrical with respect to the reference plane.† However, they do not apply in the case of nonsymmetrical bodies or in the case of three-dimensional motion.

16.4. Plane Motion of a Rigid Body. D'Alembert's Principle. Consider a rigid slab of mass m moving under the action of several external forces $\mathbf{F}_1$, $\mathbf{F}_2$, $\mathbf{F}_3$, etc., contained in the plane of the slab (Fig. 16.5). Substituting for $\dot{\mathbf{H}}_G$ from Eq. (16.5) into Eq. (16.2) and writing the fundamental equations of motion (16.1) and (16.2) in scalar form, we have

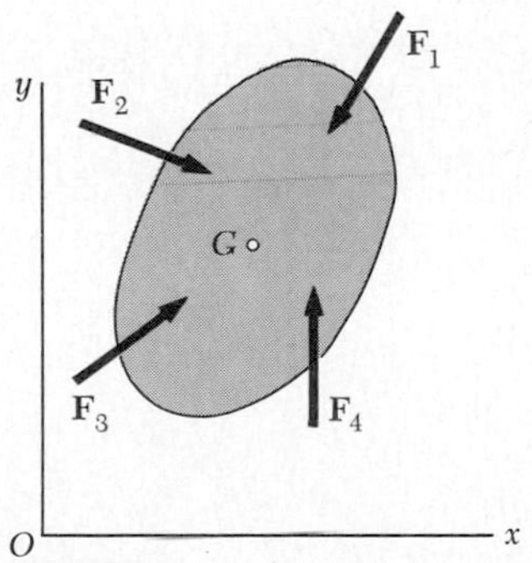

Fig. 16.5

$$\Sigma F_x = m\bar{a}_x \qquad \Sigma F_y = m\bar{a}_y \qquad \Sigma M_G = \bar{I}\alpha \tag{16.6}$$

Equations (16.6) show that the acceleration of the mass center G of the slab and its angular acceleration $\boldsymbol{\alpha}$ may easily be obtained, once the resultant of the external forces acting on the slab and their moment resultant about G have been determined. Given appropriate initial conditions, the coordinates $\bar{x}$ and $\bar{y}$ of the mass center and the angular coordinate θ of the slab may then be obtained at any instant t by integration. Thus *the motion of the slab is completely defined by the resultant and moment resultant about G of the external forces acting on it.*

This property, which will be extended in Chap. 18 to the case of the three-dimensional motion of a rigid body, is characteristic of the motion of a rigid body. Indeed, as we saw in Chap. 14, the motion of a system of particles which are not rigidly connected will in general depend upon the

† Or, more generally, bodies which have a principal centroidal axis of inertia perpendicular to the reference plane.

specific external forces, as well as upon the internal forces, acting on the various particles.

Since the motion of a rigid body depends only upon the resultant and moment resultant of the external forces acting on it, it follows that *two systems of forces which are equipollent,* i.e., which have the same resultant and the same moment resultant, *are also equivalent;* that is, they have exactly the same effect on a given rigid body.†

Consider in particular the system of the external forces acting on a rigid body (Fig. 16.6*a*) and the system of the effective forces associated with the particles forming the rigid body (Fig. 16.6*b*). It was shown in Sec. 14.2 that the two systems thus defined are equipollent. But since the particles considered now form a rigid body, it follows from the discussion above that the two systems are also equivalent. We may thus state that *the external forces acting on a rigid body are equivalent to the effective forces of the various particles forming the body.* This statement is referred to as *d'Alembert's principle* after the French mathematician Jean le Rond d'Alembert (1717–1783), even though d'Alembert's original statement was written in a somewhat different form.

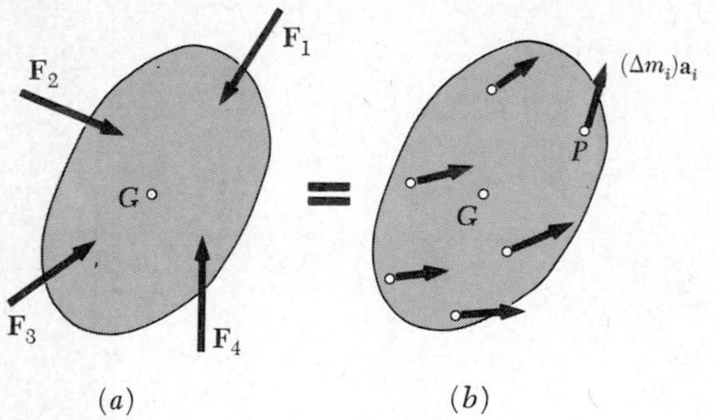

Fig. 16.6

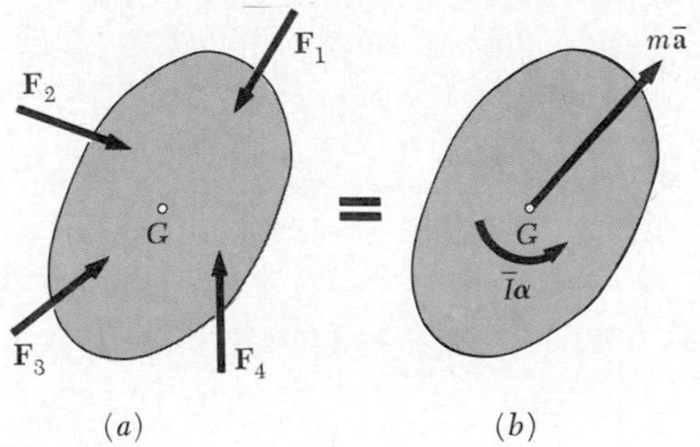

Fig. 16.7

The significance of d'Alembert's principle has been emphasized by the use of a colored equals sign in Fig. 16.6 and also in Fig. 16.7, where using results obtained earlier in this section, we have replaced the effective forces by a vector $m\bar{\mathbf{a}}$ attached at the mass center G of the slab and a couple of moment $\bar{I}\boldsymbol{\alpha}$.

† This result has already been derived in Sec. 3.19 from the principle of transmissibility (Sec. 3.3). The present derivation is independent of that principle, however, and will make possible its elimination from the axioms of mechanics (Sec. 16.5).

Translation. In the particular case of a body in translation, the angular acceleration of the body is identically equal to zero and its effective forces reduce to the vector $m\bar{\mathbf{a}}$ attached at G (Fig. 16.8). Thus, the resultant of the external forces acting on a rigid body in translation passes through the mass center of the body and is equal to $m\bar{\mathbf{a}}$.

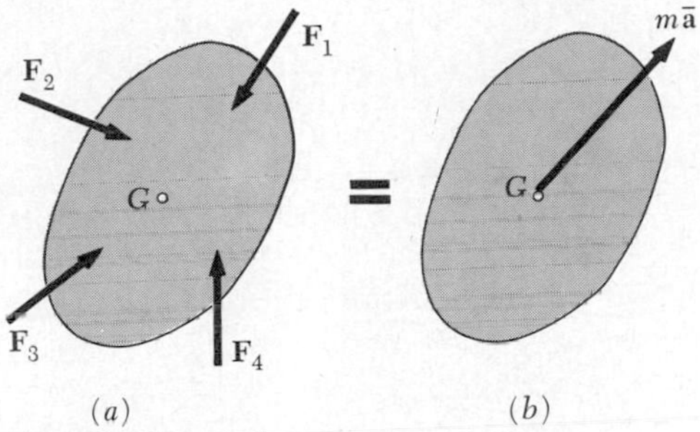

Fig. 16.8 Translation.

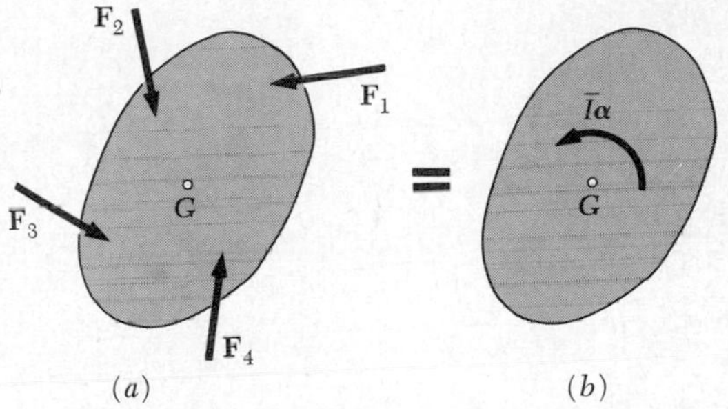

Fig. 16.9 Centroidal rotation.

Centroidal Rotation. When a slab, or, more generally, a body symmetrical with respect to the reference plane, rotates about a fixed axis perpendicular to the reference plane and passing through its mass center G, we say that the body is in *centroidal rotation*. Since the acceleration $\bar{\mathbf{a}}$ is identically equal to zero, the effective forces of the body reduce to the couple $\bar{I}\boldsymbol{\alpha}$ (Fig. 16.9). Thus, the external forces acting on a body in centroidal rotation are equivalent to a couple of moment $\bar{I}\boldsymbol{\alpha}$.

General Plane Motion. Comparing Fig. 16.7 with Figs. 16.8 and 16.9, we observe that from the point of view of *kinetics*, the most general plane motion of a rigid body symmetrical with respect to the reference plane may be replaced by the sum of a translation and a centroidal rotation. We should note that this statement is more restrictive than the similar statement made earlier from the point of view of *kinematics* (Sec. 15.5), since we now require that the mass center of the body be selected as the reference point.

Referring to Eqs. (16.6), we observe that the first two equations are identical with the equations of motion of a particle of mass m acted upon by the given forces $\mathbf{F}_1$, $\mathbf{F}_2$, $\mathbf{F}_3$, etc. We thus check that *the mass center G of a rigid body in plane motion moves as if the entire mass of the body were concentrated at that point, and as if all the external forces acted on it.* We recall that this result has already been obtained in Sec. 14.4 in the general case of a system of particles, the particles being not necessarily rigidly connected. We also note, as we did in Sec. 14.4, that the system of the external forces does not, in general, reduce to a single vector $m\bar{\mathbf{a}}$ attached at G. Therefore, in the general case of the plane motion of a rigid body, *the resultant of the external forces acting on the body does not pass through the mass center of the body.*

Finally, we may observe that the last of Eqs. (16.6) would still be valid if the rigid body while subjected to the same applied forces were constrained to rotate about a fixed axis through G. Thus, *a rigid body in plane motion rotates about its mass center as if this point were fixed.*

***16.5. A Remark on the Axioms of the Mechanics of Rigid Bodies.** The fact that two equipollent systems of external forces acting on a rigid body are also equivalent, i.e., have the same effect on that rigid body, has already been established in Sec. 3.19. But there it was derived from the *principle of transmissibility,* one of the axioms used in our study of the statics of rigid bodies. It should be noted that this axiom has not been used in the present chapter, because Newton's second and third laws of motion make its use unnecessary in the study of the dynamics of rigid bodies.

In fact, the principle of transmissibility may now be *derived* from the other axioms used in the study of mechanics. This principle stated, without proof (Sec. 3.3), that the conditions of equilibrium or motion of a rigid body remain unchanged if a force $\mathbf{F}$ acting at a given point of the rigid body is replaced by a force $\mathbf{F}'$ of the same magnitude and same direction, but acting at a different point, provided that the two forces have the same line of action. But since $\mathbf{F}$ and $\mathbf{F}'$ have the same moment about any given point, it is clear that they form two equipollent systems of external forces. Thus, we may now *prove,* as a result of what we established in the preceding section, that $\mathbf{F}$ and $\mathbf{F}'$ have the same effect on the rigid body (Fig. 3.3).

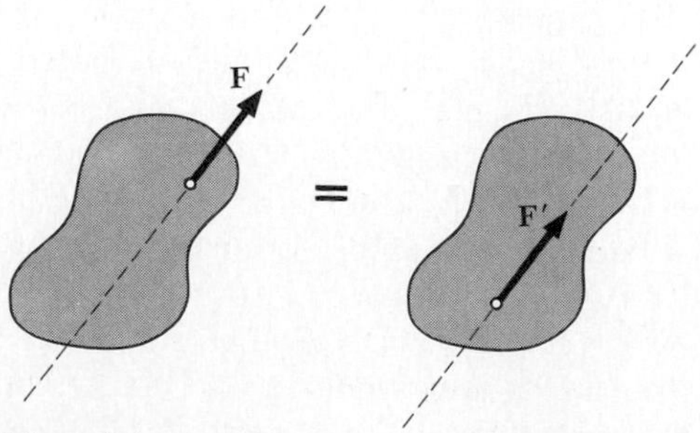

Fig. 3.3 ***(repeated)***

The principle of transmissibility may therefore be removed from the list of axioms required for the study of the mechanics of rigid bodies. These axioms are reduced to the parallelogram law of addition of vectors and to Newton's laws of motion.

16.6. Solution of Problems Involving the Motion of a Rigid Body. We saw in Sec. 16.4 that when a rigid body is in plane motion, there exists a fundamental relation between the forces $\mathbf{F}_1$, $\mathbf{F}_2$, $\mathbf{F}_3$, etc., acting on the body, the acceleration $\bar{\mathbf{a}}$ of its mass center, and the angular acceleration $\boldsymbol{\alpha}$ of the body. This relation, which is represented in Fig. 16.7 in the form of a *free-body-diagram equation,* may be used to determine the acceleration $\bar{\mathbf{a}}$ and the angular acceleration $\boldsymbol{\alpha}$ produced by a given system of forces acting on a rigid body or, conversely, to determine the forces which produce a given motion of the rigid body.

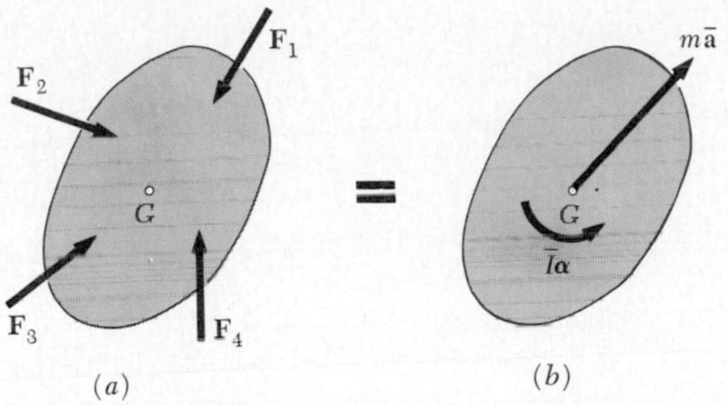

Fig. 16.7 *(repeated)*

While the three algebraic equations (16.6) may be used to solve problems of plane motion,† our experience in statics suggests that the solution of many problems involving rigid bodies could be simplified by an appropriate choice of the point about which the moments of the forces are computed. It is therefore preferable to remember the relation existing between the forces and the accelerations in the pictorial form shown in Fig. 16.7, and to derive from this fundamental relation the component or moment equations which fit best the solution of the problem under consideration.

The fundamental relation shown in Fig. 16.7 may be presented in an alternative form if we add to the external forces an inertia vector $-m\bar{\mathbf{a}}$ of sense opposite to that of $\bar{\mathbf{a}}$, attached at G, and an inertia couple $-\bar{I}\boldsymbol{\alpha}$ of moment equal in magnitude to $\bar{I}\alpha$ and of sense opposite to that of $\boldsymbol{\alpha}$ (Fig. 16.10). The system obtained is equivalent to zero, and the rigid body is said to be in dynamic equilibrium.

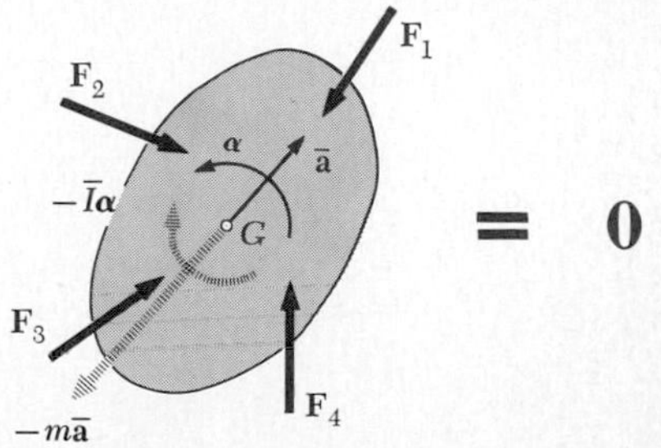

Fig. 16.10

Whether the principle of equivalence of external and effective forces is directly applied, as in Fig. 16.7, or whether the concept of dynamic equilibrium is introduced, as in Fig. 16.10, the use of free-body-diagram equations showing vectorially the relationship existing between the forces applied on the rigid body and the resulting linear and angular accelerations presents considerable advantages over the blind application of formulas (16.6). These advantages may be summarized as follows:

† We recall that the last of Eqs. (16.6) is valid only in the case of the plane motion of a rigid body symmetrical with respect to the reference plane. In all other cases, the methods of Chap. 18 should be used.

1. A much clearer understanding of the effect of the forces on the motion of the body will result from the use of a pictorial representation.
2. This approach makes it possible to divide the solution of a dynamics problem into two parts: In the first part, the analysis of the kinematic and kinetic characteristics of the problem leads to the free-body diagrams of Fig. 16.7 or 16.10; in the second part, the diagram obtained is used to analyze by the methods of Chap. 3 the various forces and vectors involved.
3. A unified approach is provided for the analysis of the plane motion of a rigid body, regardless of the particular type of motion involved. While the kinematics of the various motions considered may vary from one case to the other, the approach to the kinetics of the motion is consistently the same. In every case we shall draw a diagram showing the external forces, the vector $m\bar{\mathbf{a}}$ associated with the motion of G, and the couple $\bar{I}\alpha$ associated with the rotation of the body about G.
4. The resolution of the plane motion of a rigid body into a translation and a centroidal rotation, which is used here, is a basic concept which may be applied effectively throughout the study of mechanics. We shall use it again in Chap. 17 with the method of work and energy and the method of impulse and momentum.
5. As we shall see in Chap. 18, this approach may be extended to the study of the general three-dimensional motion of a rigid body. The motion of the body will again be resolved into a translation and a rotation about the mass center, and free-body-diagram equations will be used to indicate the relationship existing between the external forces and the rates of change of the linear and angular momentum of the body.

16.7. Systems of Rigid Bodies. The method described in the preceding section may also be used in problems involving the plane motion of several connected rigid bodies. A diagram similar to Fig. 16.7 or Fig. 16.10 may be drawn for each part of the system. The equations of motion obtained from these diagrams are solved simultaneously.

In some cases, as in Sample Prob. 16.3, a single diagram may be drawn for the entire system. This diagram should include all the external forces, as well as the vectors $m\bar{\mathbf{a}}$ and the couples $\bar{I}\alpha$ associated with the various parts of the system. However, internal forces, such as the forces exerted by connecting cables, may be omitted since they occur in pairs of equal and opposite forces and are thus equipollent to zero. The equations obtained by expressing that the system of the external forces is equipollent to the system of the effective forces may be solved for the remaining unknowns.†

This second approach may not be used in problems involving more than three unknowns, since only three equations of motion are available when a single diagram is used. We shall not elaborate upon this point, since the discussion involved would be completely similar to that given in Sec. 6.11 in the case of the equilibrium of a system of rigid bodies.

†Note that we cannot speak of *equivalent* systems since we are not dealing with a single rigid body.

SAMPLE PROBLEM 16.1

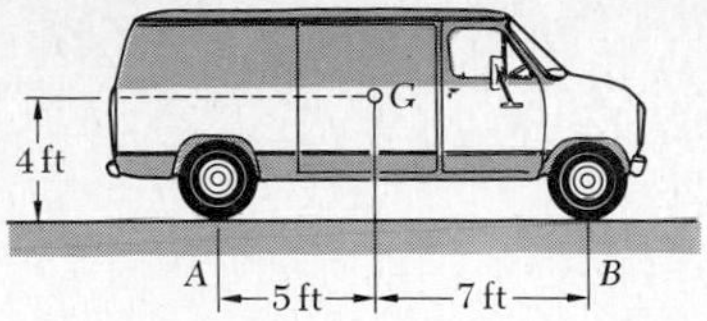

When the forward speed of the truck shown was 30 ft/s, the brakes were suddenly applied, causing all four wheels to stop rotating. It was observed that the truck skidded to rest in 20 ft. Determine the magnitude of the normal reaction and of the friction force at each wheel as the truck skidded to rest.

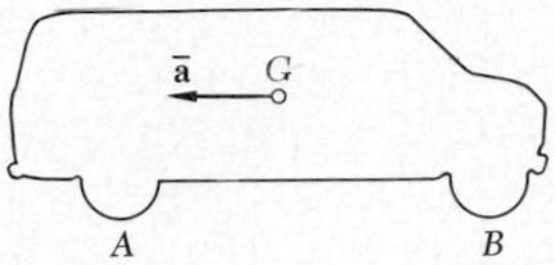

Kinematics of Motion. Choosing the positive sense to the right and using the equations of uniformly accelerated motion, we write

$$\bar{v}_0 = +30 \text{ ft/s} \qquad \bar{v}^2 = \bar{v}_0^2 + 2\bar{a}\bar{x} \qquad 0 = (30)^2 + 2\bar{a}(20)$$

$$\bar{a} = -22.5 \text{ ft/s}^2 \qquad \bar{\mathbf{a}} = 22.5 \text{ ft/s}^2 \leftarrow$$

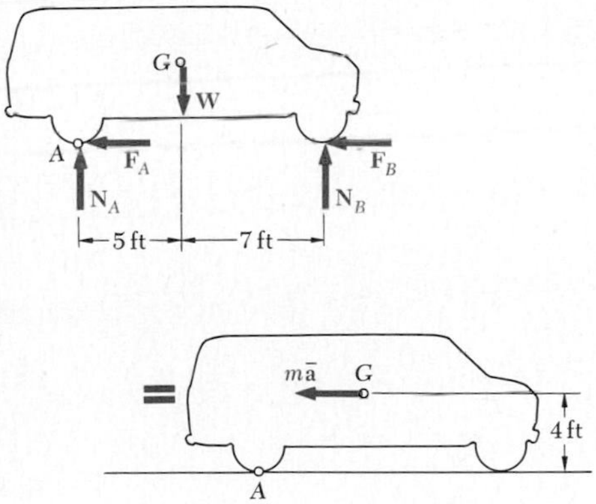

Equations of Motion. The external forces consist of the weight $\mathbf{W}$ of the truck and of the normal reactions and friction forces at the wheels. (The vectors $\mathbf{N}_A$ and $\mathbf{F}_A$ represent the sum of the reactions at the rear wheels, while $\mathbf{N}_B$ and $\mathbf{F}_B$ represent the sum of the reactions at the front wheels.) Since the truck is in translation, the effective forces reduce to the vector $m\bar{\mathbf{a}}$ attached at G. Three equations of motion are obtained by expressing that the system of the external forces is equivalent to the system of the effective forces.

$$+\uparrow \Sigma F_y = \Sigma (F_y)_{\text{eff}}: \qquad N_A + N_B - W = 0$$

Since $F_A = \mu_k N_A$ and $F_B = \mu_k N_B$ where μ_k is the coefficient of kinetic friction, we find that

$$F_A + F_B = \mu_k (N_A + N_B) = \mu_k W$$

$$\overset{+}{\rightarrow} \Sigma F_x = \Sigma (F_x)_{\text{eff}}: \qquad -(F_A + F_B) = -m\bar{a}$$

$$-\mu_k W = -\frac{W}{32.2 \text{ ft/s}^2}(22.5 \text{ ft/s}^2)$$

$$\mu_k = 0.699$$

$$+\circlearrowleft \Sigma M_A = \Sigma (M_A)_{\text{eff}}: \; -W(5 \text{ ft}) + N_B(12 \text{ ft}) = m\bar{a}(4 \text{ ft})$$

$$-W(5 \text{ ft}) + N_B(12 \text{ ft}) = \frac{W}{32.2 \text{ ft/s}^2}(22.5 \text{ ft/s}^2)(4 \text{ ft})$$

$$N_B = 0.650W$$

$$F_B = \mu_k N_B = (0.699)(0.650W) \qquad F_B = 0.454W$$

$$+\uparrow \Sigma F_y = \Sigma (F_y)_{\text{eff}}: \qquad N_A + N_B - W = 0$$

$$N_A + 0.650W - W = 0$$

$$N_A = 0.350W$$

$$F_A = \mu_k N_A = (0.699)(0.350W) \qquad F_A = 0.245W$$

Reactions at Each Wheel. Recalling that the values computed above represent the sum of the reactions at the two front wheels or the two rear wheels, we obtain the magnitude of the reactions at each wheel by writing

$$N_{\text{front}} = \tfrac{1}{2}N_B = 0.325W \qquad N_{\text{rear}} = \tfrac{1}{2}N_A = 0.175W \quad \blacktriangleleft$$

$$F_{\text{front}} = \tfrac{1}{2}F_B = 0.227W \qquad F_{\text{rear}} = \tfrac{1}{2}F_A = 0.122W \quad \blacktriangleleft$$

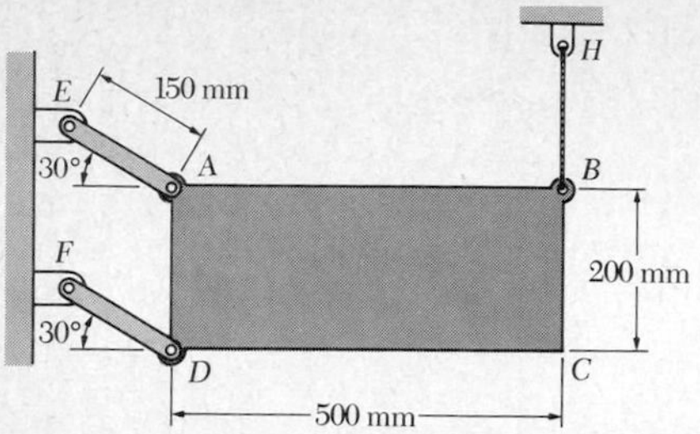

SAMPLE PROBLEM 16.2

The thin plate $ABCD$ has a mass of 8 kg and is held in the position shown by the wire BH and two links AE and DF. Neglecting the mass of the links, determine immediately after wire BH has been cut (*a*) the acceleration of the plate, (*b*) the force in each link.

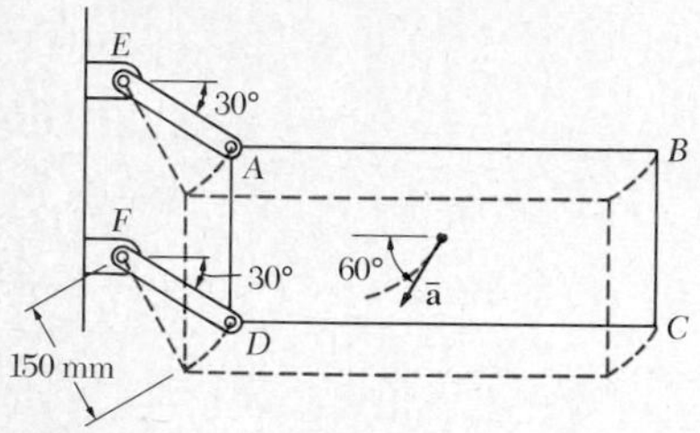

Kinematics of Motion. After wire BH has been cut, we observe that corners A and D move along parallel circles of radius 150 mm centered, respectively, at E and F. The motion of the plate is thus a curvilinear translation; the particles forming the plate move along parallel circles of radius 150 mm.

At the instant wire BH is cut, the velocity of the plate is zero. Thus the acceleration $\bar{\mathbf{a}}$ of the mass center G of the plate is tangent to the circular path which will be described by G.

Equations of Motion. The external forces consist of the weight $\mathbf{W}$ and the forces $\mathbf{F}_{AE}$ and $\mathbf{F}_{DF}$ exerted by the links. Since the plate is in translation, the effective forces reduce to the vector $m\bar{\mathbf{a}}$ attached at G and directed along the t axis. We shall now express that the system of the external forces is equivalent to the system of the effective forces.

a. Acceleration of the Plate.

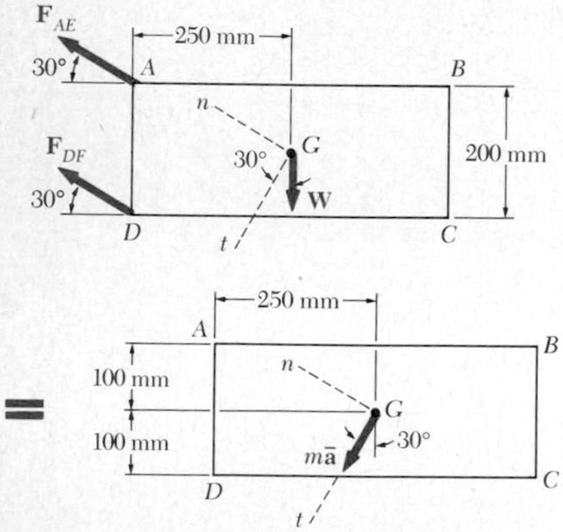

$+\swarrow \Sigma F_t = \Sigma (F_t)_{\text{eff}}$:

$$W \cos 30° = m\bar{a}$$

$$mg \cos 30° = m\bar{a}$$

$$\bar{a} = g \cos 30° = (9.81 \text{ m/s}^2) \cos 30° \qquad (1)$$

$$\bar{\mathbf{a}} = 8.50 \text{ m/s}^2 \measuredangle 60° \quad ◀$$

b. Forces in Links AE and DF.

$+\nwarrow \Sigma F_n = \Sigma (F_n)_{\text{eff}}$: $\quad F_{AE} + F_{DF} - W \sin 30° = 0 \qquad (2)$

$+\circlearrowright \Sigma M_G = \Sigma (M_G)_{\text{eff}}$:

$$(F_{AE} \sin 30°)(250 \text{ mm}) - (F_{AE} \cos 30°)(100 \text{ mm})$$

$$+(F_{DF} \sin 30°)(250 \text{ mm}) + (F_{DF} \cos 30°)(100 \text{ mm}) = 0$$

$$38.4F_{AE} + 211.6F_{DF} = 0$$

$$F_{DF} = -0.1815F_{AE} \qquad (3)$$

Substituting for F_{DF} from (3) into (2), we write

$$F_{AE} - 0.1815F_{AE} - W \sin 30° = 0$$

$$F_{AE} = 0.6109W$$

$$F_{DF} = -0.1815(0.6109W) = -0.1109W$$

Noting that $W = mg = (8 \text{ kg})(9.81 \text{ m/s}^2) = 78.48$ N, we have

$$F_{AE} = 0.6109(78.48 \text{ N}) \qquad F_{AE} = 47.9 \text{ N } T \quad ◀$$

$$F_{DF} = -0.1109(78.48 \text{ N}) \qquad F_{DF} = 8.70 \text{ N } C \quad ◀$$

SAMPLE PROBLEM 16.3

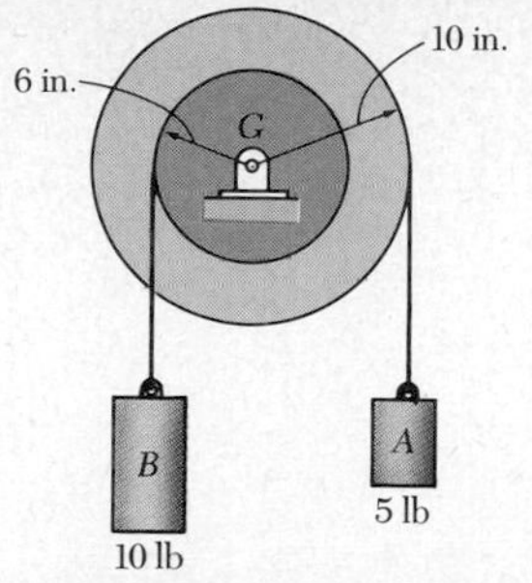

A pulley weighing 12 lb and having a radius of gyration of 8 in. is connected to two cylinders as shown. Assuming no axle friction, determine the angular acceleration of the pulley and the acceleration of each cylinder.

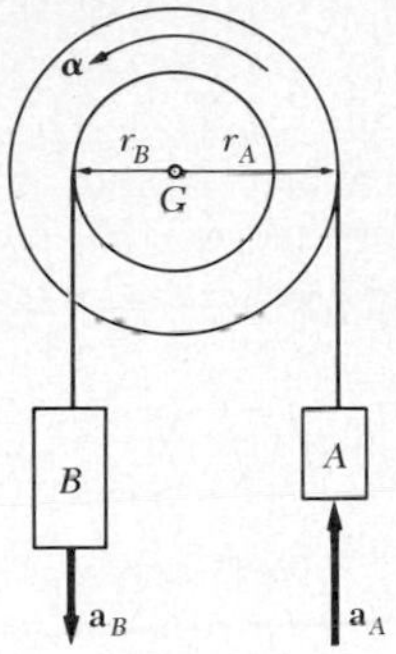

Sense of Motion. Although an arbitrary sense of motion may be assumed (since no friction forces are involved) and later checked by the sign of the answer, we may prefer first to determine the actual sense of rotation of the pulley. We first find the weight of cylinder B required to maintain the equilibrium of the pulley when it is acted upon by the 5-lb cylinder A. We write

$$+\circlearrowleft \Sigma M_G = 0: \qquad W_B(6 \text{ in.}) - (5 \text{ lb})(10 \text{ in.}) = 0 \qquad W_B = 8.33 \text{ lb}$$

Since cylinder B actually weighs 10 lb, the pulley will rotate counterclockwise.

Kinematics of Motion. Assuming $\boldsymbol{\alpha}$ counterclockwise and noting that $a_A = r_A\alpha$ and $a_B = r_B\alpha$, we obtain

$$\mathbf{a}_A = (\tfrac{10}{12} \text{ ft})\alpha \uparrow \qquad \mathbf{a}_B = (\tfrac{6}{12} \text{ ft})\alpha \downarrow$$

Equations of Motion. A single system consisting of the pulley and the two cylinders is considered. Forces external to this system consist of the weights of the pulley and the two cylinders and of the reaction at G. (The forces exerted by the cables on the pulley and on the cylinders are internal to the system considered and cancel out.) Since the motion of the pulley is a centroidal rotation and the motion of each cylinder is a translation, the effective forces reduce to the couple $\bar{I}\boldsymbol{\alpha}$ and the two vectors $m\mathbf{a}_A$ and $m\mathbf{a}_B$. The centroidal moment of inertia of the pulley is

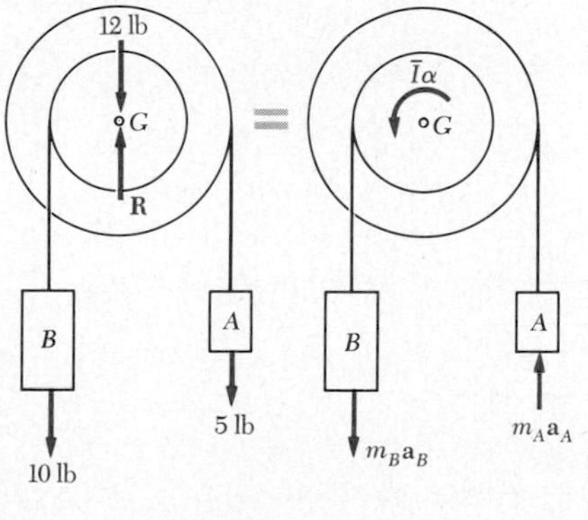

$$\bar{I} = m\bar{k}^2 = \frac{W}{g}\bar{k}^2 = \frac{12 \text{ lb}}{32.2 \text{ ft/s}^2}(\tfrac{8}{12} \text{ ft})^2 = 0.1656 \text{ lb}\cdot\text{ft}\cdot\text{s}^2$$

Since the system of the external forces is equipollent to the system of the effective forces, we write

$$+\circlearrowleft \Sigma M_G = \Sigma(M_G)_{\text{eff}}:$$

$$(10 \text{ lb})(\tfrac{6}{12} \text{ ft}) - (5 \text{ lb})(\tfrac{10}{12} \text{ ft}) = +\bar{I}\alpha + m_B a_B(\tfrac{6}{12} \text{ ft}) + m_A a_A(\tfrac{10}{12} \text{ ft})$$

$$(10)(\tfrac{6}{12}) - (5)(\tfrac{10}{12}) = 0.1656\alpha + \tfrac{10}{32.2}(\tfrac{6}{12}\alpha)(\tfrac{6}{12}) + \tfrac{5}{32.2}(\tfrac{10}{12}\alpha)(\tfrac{10}{12})$$

$$\alpha = +2.374 \text{ rad/s}^2 \qquad \boldsymbol{\alpha} = 2.37 \text{ rad/s}^2 \circlearrowleft \blacktriangleleft$$

$$a_A = r_A\alpha = (\tfrac{10}{12} \text{ ft})(2.374 \text{ rad/s}^2) \qquad \mathbf{a}_A = 1.978 \text{ ft/s}^2 \uparrow \blacktriangleleft$$

$$a_B = r_B\alpha = (\tfrac{6}{12} \text{ ft})(2.374 \text{ rad/s}^2) \qquad \mathbf{a}_B = 1.187 \text{ ft/s}^2 \downarrow \blacktriangleleft$$

SAMPLE PROBLEM 16.4

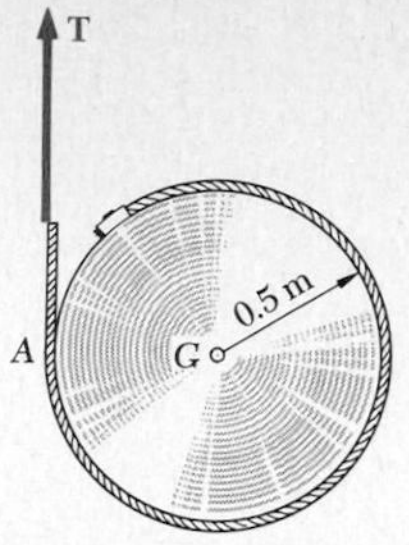

A cord is wrapped around a homogeneous disk of radius $r = 0.5$ m and mass $m = 15$ kg. If the cord is pulled upward with a force **T** of magnitude 180 N, determine (*a*) the acceleration of the center of the disk, (*b*) the angular acceleration of the disk, (*c*) the acceleration of the cord.

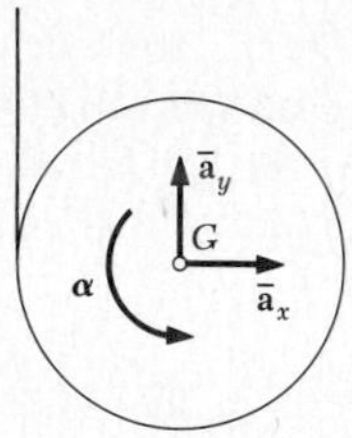

Equations of Motion. We assume that the components $\bar{\mathbf{a}}_x$ and $\bar{\mathbf{a}}_y$ of the acceleration of the center are directed, respectively, to the right and upward and that the angular acceleration of the disk is counterclockwise. The external forces acting on the disk consist of the weight **W** and the force **T** exerted by the cord. This system is equivalent to the system of the effective forces, which consists of a vector of components $m\bar{\mathbf{a}}_x$ and $m\bar{\mathbf{a}}_y$ attached at G and a couple $\bar{I}\boldsymbol{\alpha}$. We write

$$\overset{+}{\rightarrow} \Sigma F_x = \Sigma(F_x)_{\text{eff}}: \qquad 0 = m\bar{a}_x \qquad \bar{\mathbf{a}}_x = 0 \blacktriangleleft$$

$$+\uparrow \Sigma F_y = \Sigma(F_y)_{\text{eff}}: \qquad T - W = m\bar{a}_y$$

$$\bar{a}_y = \frac{T - W}{m}$$

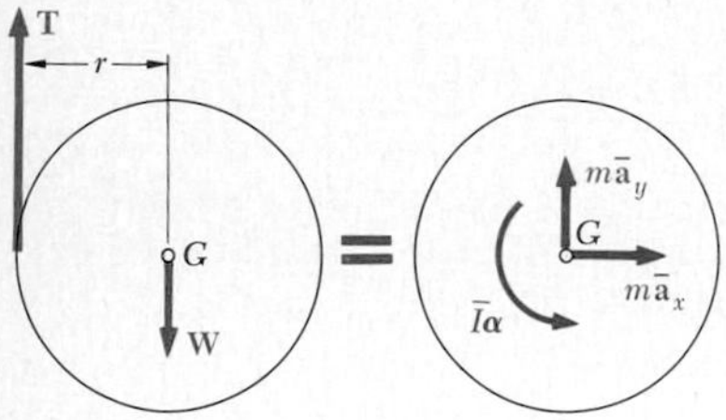

Since $T = 180$ N, $m = 15$ kg, and $W = (15 \text{ kg})(9.81 \text{ m/s}^2) = 147.1$ N, we have

$$\bar{a}_y = \frac{180 \text{ N} - 147.1 \text{ N}}{15 \text{ kg}} = +2.19 \text{ m/s}^2 \qquad \bar{\mathbf{a}}_y = 2.19 \text{ m/s}^2 \uparrow \blacktriangleleft$$

$$+\circlearrowleft \Sigma M_G = \Sigma(M_G)_{\text{eff}}: \qquad -Tr = \bar{I}\alpha$$

$$-Tr = (\tfrac{1}{2}mr^2)\alpha$$

$$\alpha = -\frac{2T}{mr} = -\frac{2(180 \text{ N})}{(15 \text{ kg})(0.5 \text{ m})} = -48.0 \text{ rad/s}^2$$

$$\boldsymbol{\alpha} = 48.0 \text{ rad/s}^2 \circlearrowright \blacktriangleleft$$

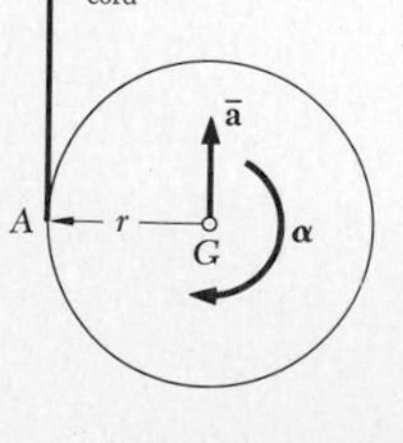

Acceleration of Cord. Since the acceleration of the cord is equal to the tangential component of the acceleration of point A on the disk, we write

$$\mathbf{a}_{\text{cord}} = (\mathbf{a}_A)_t = \bar{\mathbf{a}} + (\mathbf{a}_{A/G})_t$$
$$= [2.19 \text{ m/s}^2 \uparrow] + [(0.5 \text{ m})(48 \text{ rad/s}^2) \uparrow]$$
$$\mathbf{a}_{\text{cord}} = 26.2 \text{ m/s}^2 \uparrow \blacktriangleleft$$

SAMPLE PROBLEM 16.5

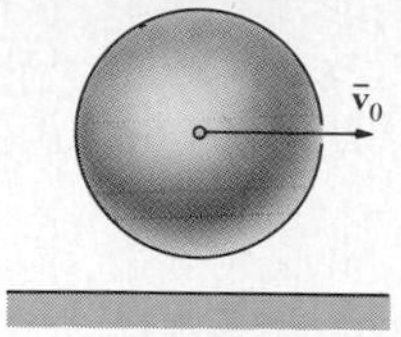

A uniform sphere of mass m and radius r is projected along a rough horizontal surface with a linear velocity $\bar{\mathbf{v}}_0$ and no angular velocity. Denoting by μ_k the coefficient of kinetic friction between the sphere and the floor, determine (*a*) the time t_1 at which the sphere will start rolling without sliding, (*b*) the linear velocity and angular velocity of the sphere at time t_1.

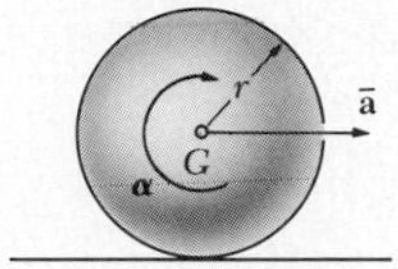

Equations of Motion. The positive sense is chosen to the right for $\bar{\mathbf{a}}$ and clockwise for $\boldsymbol{\alpha}$. The external forces acting on the sphere consist of the weight **W**, the normal reaction **N**, and the friction force **F**. Since the point of the sphere in contact with the surface is sliding to the right, the friction force **F** is directed to the left. While the sphere is sliding, the magnitude of the friction force is $F = \mu_k N$. The effective forces consist of the vector $m\bar{\mathbf{a}}$ attached at G and the couple $\bar{I}\boldsymbol{\alpha}$. Expressing that the system of the external forces is equivalent to the system of the effective forces, we write

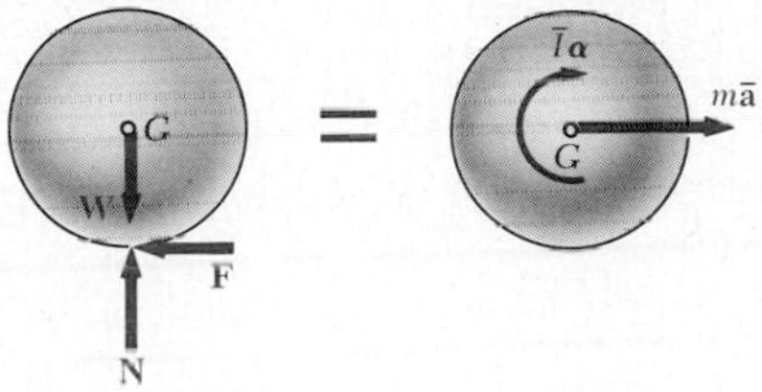

$$+\uparrow \Sigma F_y = \Sigma (F_y)_{\text{eff}}: \qquad N - W = 0$$

$$N = W = mg \qquad F = \mu_k N = \mu_k mg$$

$$\xrightarrow{+} \Sigma F_x = \Sigma (F_x)_{\text{eff}}: \qquad -F = m\bar{a} \qquad -\mu_k mg = m\bar{a} \qquad \bar{a} = -\mu_k g$$

$$+\circlearrowright \Sigma M_G = \Sigma (M_G)_{\text{eff}}: \qquad Fr = \bar{I}\alpha$$

Noting that $\bar{I} = \frac{2}{5}mr^2$ and substituting the value obtained for F, we write

$$(\mu_k mg)r = \tfrac{2}{5}mr^2\alpha \qquad \alpha = \frac{5}{2}\frac{\mu_k g}{r}$$

Kinematics of Motion. As long as the sphere both rotates and slides, its linear and angular motions are uniformly accelerated.

$$t = 0,\ \bar{v} = \bar{v}_0 \qquad \bar{v} = \bar{v}_0 + \bar{a}t = \bar{v}_0 - \mu_k g t \tag{1}$$

$$t = 0,\ \omega_0 = 0 \qquad \omega = \omega_0 + \alpha t = 0 + \left(\frac{5}{2}\frac{\mu_k g}{r}\right)t \tag{2}$$

The sphere will start rolling without sliding when the velocity $\mathbf{v}_C$ of the point of contact C is zero. At that time, $t = t_1$, point C becomes the instantaneous center of rotation, and we have

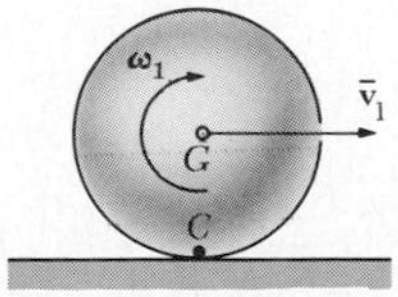

$$\bar{v}_1 = r\omega_1 \tag{3}$$

Substituting in (3) the values obtained for $\bar{v}_1$ and ω_1 by making $t = t_1$ in (1) and (2), respectively, we write

$$\bar{v}_0 - \mu_k g t_1 = r\left(\frac{5}{2}\frac{\mu_k g}{r}t_1\right) \qquad t_1 = \frac{2}{7}\frac{\bar{v}_0}{\mu_k g} \blacktriangleleft$$

Substituting for t_1 into (2), we have

$$\omega_1 = \frac{5}{2}\frac{\mu_k g}{r}t_1 = \frac{5}{2}\frac{\mu_k g}{r}\left(\frac{2}{7}\frac{\bar{v}_0}{\mu_k g}\right) \qquad \omega_1 = \frac{5}{7}\frac{\bar{v}_0}{r} \qquad \omega_1 = \frac{5}{7}\frac{\bar{v}_0}{r} \circlearrowright \blacktriangleleft$$

$$\bar{v}_1 = r\omega_1 = r\left(\frac{5}{7}\frac{\bar{v}_0}{r}\right) \qquad \bar{v}_1 = \tfrac{5}{7}\bar{v}_0 \qquad \mathbf{v}_1 = \tfrac{5}{7}\bar{v}_0 \rightarrow \blacktriangleleft$$

Problems

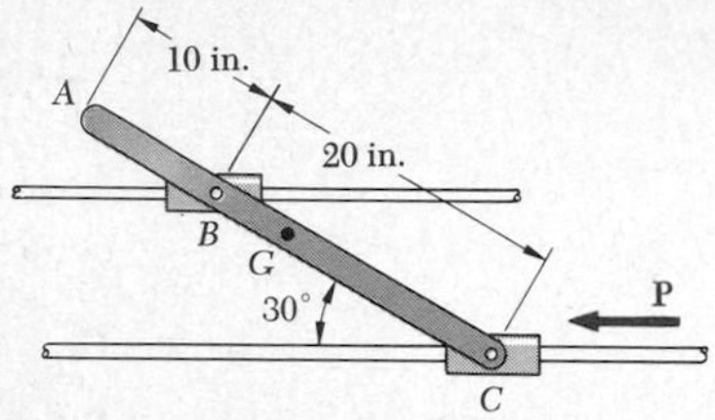

Fig. P16.1

16.1 A uniform rod ABC weighs 16 lb and is connected to two collars of negligible weight which slide on horizontal, frictionless rods located in the same vertical plane. If a force **P** of magnitude 4 lb is applied at C, determine (*a*) the acceleration of the rod, (*b*) the reactions at B and C.

16.2 In Prob. 16.1, determine (*a*) the required magnitude of **P** if the reaction at B is to be 8 lb upward, (*b*) the corresponding acceleration of the rod.

16.3 and 16.4 The motion of the 1.5-kg rod AB is guided by two small wheels which roll freely in horizontal slots. If a force **P** of magnitude 5 N is applied at B, determine (*a*) the acceleration of the rod, (*b*) the reactions at A and B.

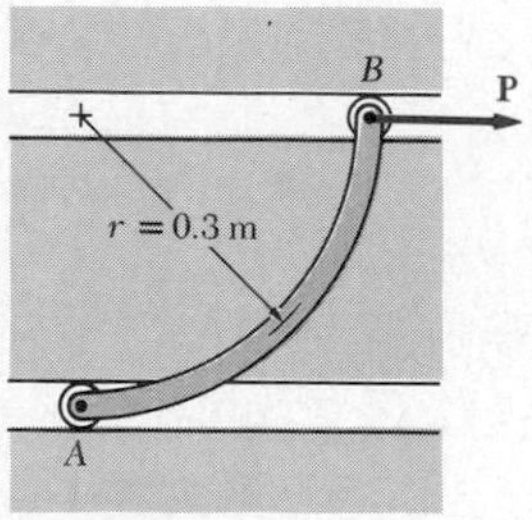

Fig. P16.3

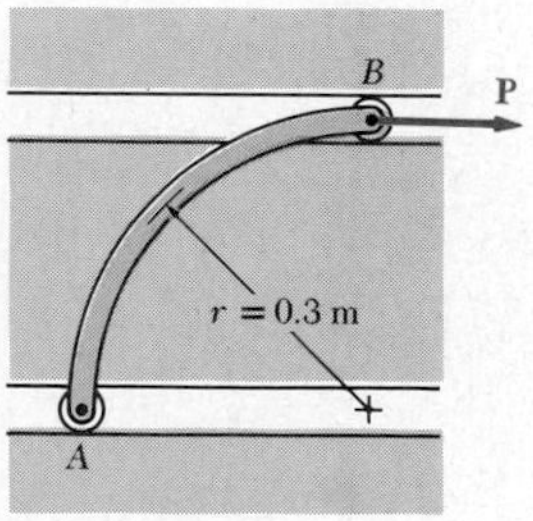

Fig. P16.4

16.5 A uniform rod BC of mass 3 kg is connected to a collar A by a 0.2-m cord AB. Neglecting the mass of the collar and cord, determine (*a*) the smallest constant acceleration $\mathbf{a}_A$ for which the cord and the rod lie in a straight line, (*b*) the corresponding tension in the cord.

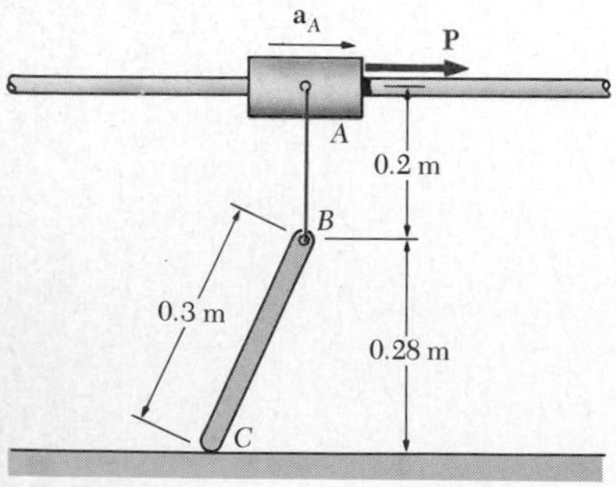

Fig. P16.5

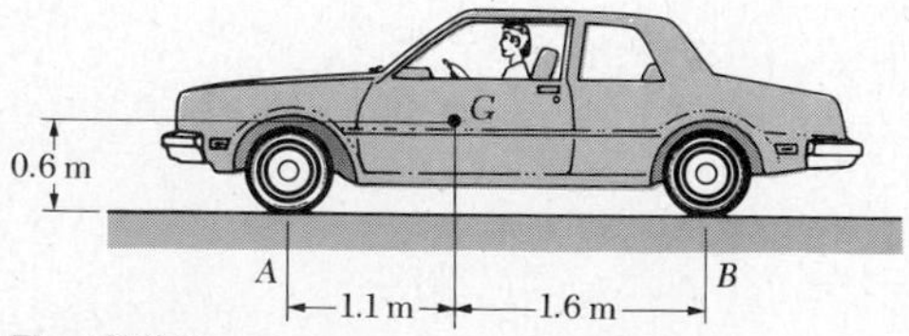

Fig. P16.6

16.6 Knowing that the coefficient of static friction between the tires and road is 0.80 for the car shown, determine the maximum possible acceleration on a level road, assuming (*a*) four-wheel drive, (*b*) conventional rear-wheel drive, (*c*) front-wheel drive.

16.7 Determine the distance through which the truck of Sample Prob. 16.1 will skid if (*a*) the rear-wheel brakes fail to operate, (*b*) the front-wheel brakes fail to operate.

16.8 A 20-kg rectangular panel is suspended from two skids *A* and *B* and is maintained in the position shown by a wire *CD*. Knowing that the coefficient of kinetic friction between each skid and the inclined track is $\mu_k = 0.15$, determine the reaction at *B* immediately after wire *CD* has been cut.

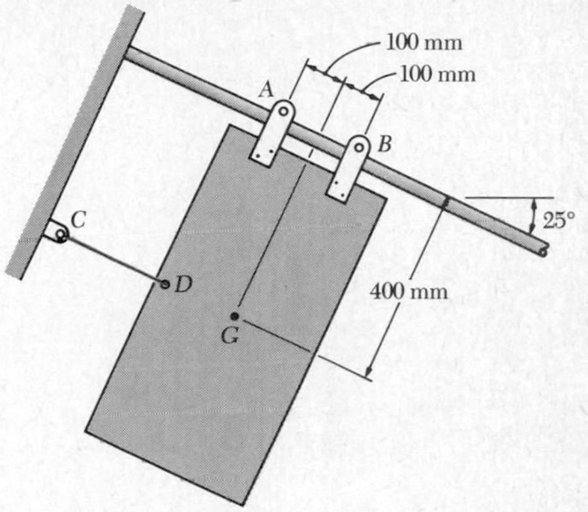

Fig. P16.8

16.9 In Prob. 16.8, determine (*a*) the maximum value of μ_k for which the skid at *B* remains in contact with the track after wire *CD* has been cut, (*b*) the corresponding acceleration of the panel.

16.10 Cylindrical cans are transported from one elevation to another by the moving horizontal arms shown. Assuming that $\mu_s = 0.25$ between the cans and the arms, determine (*a*) the magnitude of the upward acceleration **a** for which the cans slide on the horizontal arms, (*b*) the smallest ratio h/d for which the cans tip before they slide.

16.11 Solve Prob. 16.10, assuming that the acceleration **a** of the horizontal arms is directed downward.

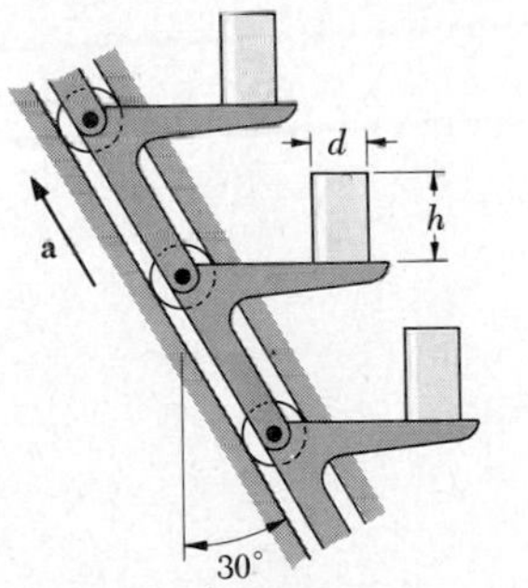

Fig. P16.10

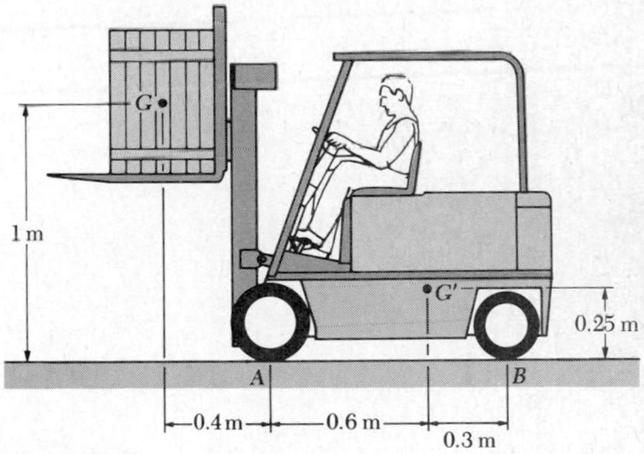

Fig. P16.12

16.12 A 2500-kg fork-lift truck carries the 1200-kg crate at the height shown. The truck is moving to the left when the brakes are applied, causing a deceleration of 3 m/s^2. Knowing that the coefficient of static friction between the crate and the fork lift is 0.60, determine the vertical component of the reaction at each wheel.

16.13 In Prob. 16.12, determine the maximum allowable deceleration of the truck if the crate is not to slide forward and if the truck is not to tip forward.

16.14 A 70-lb cabinet is mounted on casters which allow it to move freely ($\mu = 0$) on the floor. If a 35-lb force is applied as shown, determine (*a*) the acceleration of the cabinet, (*b*) the range of values of *h* for which the cabinet will not tip.

16.15 Solve Prob. 16.14, assuming that the casters are locked and slide along the rough floor ($\mu_k = 0.20$).

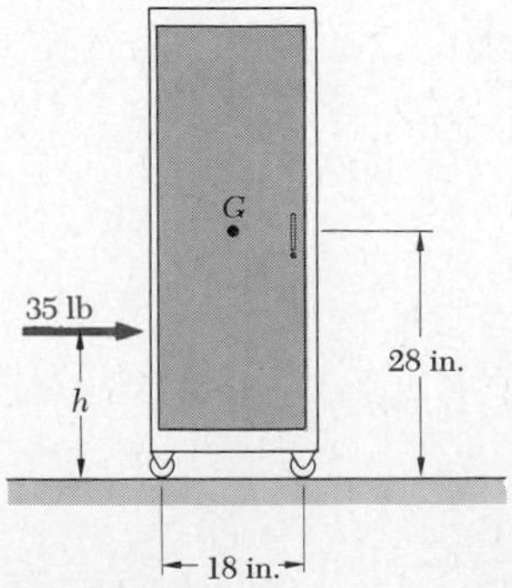

Fig. P16.14

16.16 A completely filled barrel and its contents have a combined mass of 125 kg. A 60-kg cylinder is connected to the barrel as shown. Knowing that the coefficients of friction between the barrel and the floor are $\mu_s = 0.35$ and $\mu_k = 0.30$, determine (*a*) the acceleration of the barrel, (*b*) the range of values of h for which the barrel will not tip.

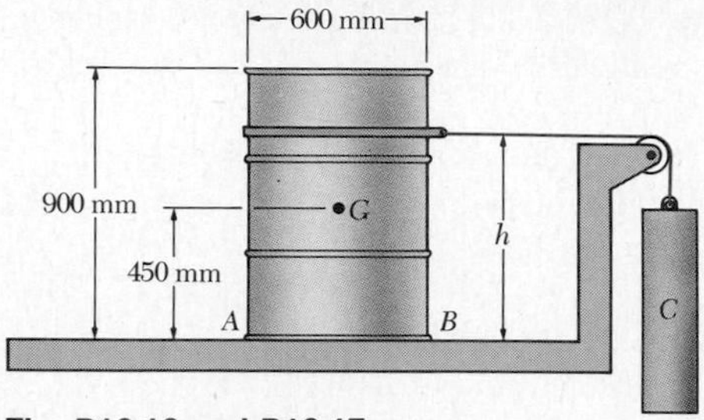

Fig. P16.16 and P16.17

16.17 A completely filled barrel and its contents have a combined mass of 160 kg. A cylinder C of mass m_C is connected to the barrel at a height $h = 700$ mm. Knowing that the coefficients of friction between the barrel and the floor are $\mu_s = 0.40$ and $\mu_k = 0.35$, determine the range of values of the mass of cylinder C for which the barrel will not tip.

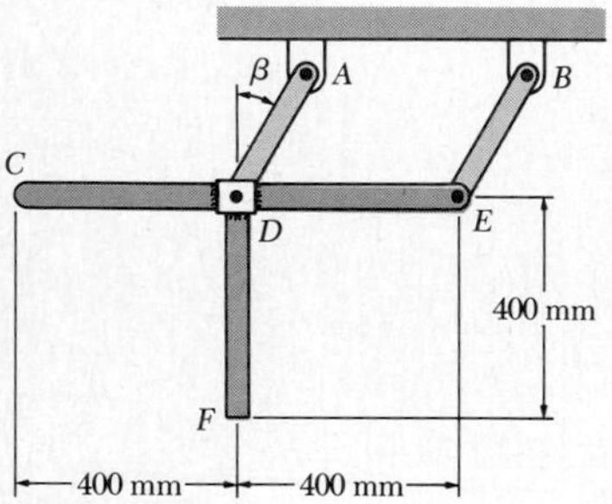

Fig. P16.18

16.18 Three uniform rods CD, DE, and DF, each of mass 1.8 kg, are welded together and are pin-connected to two links AD and BE. Neglecting the mass of the links, determine the force in each link immediately after the system has been released from rest with $\beta = 30°$.

16.19 Solve Prob. 16.18, knowing that the system has been released from rest with $\beta = 60°$.

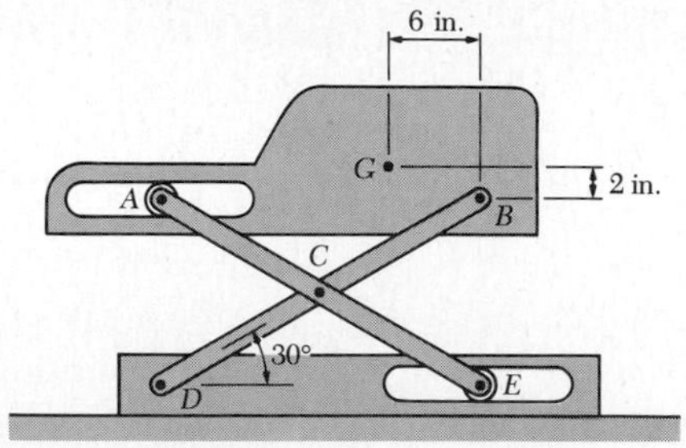

Fig. P16.20

16.20 Members ACE and DCB are each 24 in. long and are connected by a pin at C. The mass center of the 15-lb member AB is located at G. Determine (*a*) the acceleration of AB immediately after the system has been released from rest in the position shown, (*b*) the corresponding force exerted by roller A on member AB. Neglect the weight of members ACE and DCB.

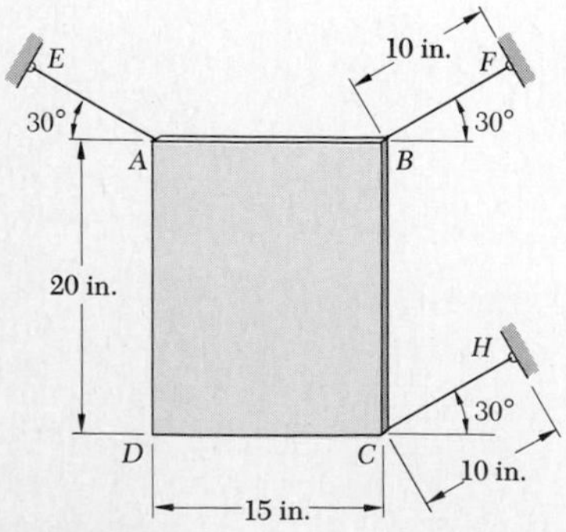

Fig. P16.21

16.21 The thin plate $ABCD$ weighs 18 lb and is held in position by the three inextensible wires AE, BF, and CH. Wire AE is then cut. Determine (*a*) the acceleration of the plate, (*b*) the tension in wires BF and CH immediately after wire AE has been cut.

16.22 A uniform semicircular plate of mass 5 kg is attached to two links *AB* and *DE*, each 250 mm long, and moves under its own weight. Neglecting the mass of the links and knowing that in the position shown the velocity of the plate is 1.8 m/s, determine the force in each link.

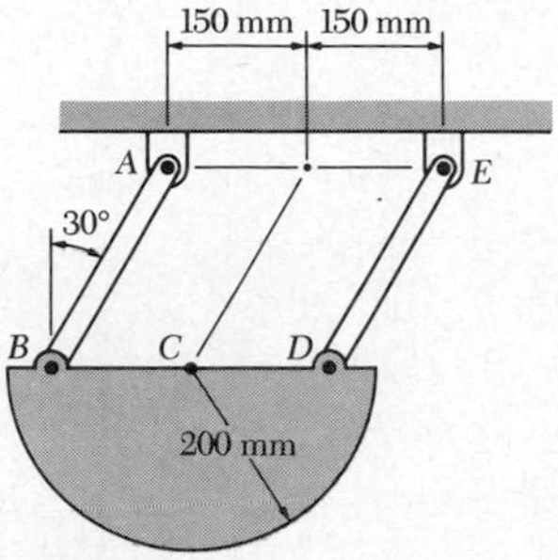

Fig. P16.22

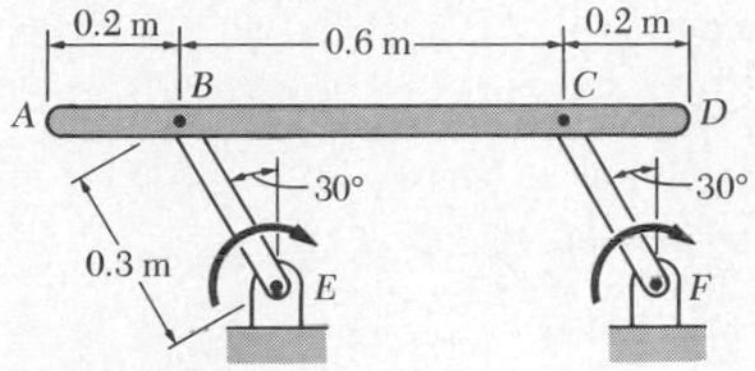

Fig. P16.23

16.23 The cranks *BE* and *CF* are made to rotate at a constant speed of 90 rpm. For the position shown, determine the vertical components of the forces exerted on the 6-kg uniform rod *ABCD* by the pins *B* and *C*.

16.24 The T-shaped rod *ABCD* is guided by two pins which slide freely in parallel curved slots of radius 5 in. The rod weighs 4 lb, and its mass center is located at point *G*. Knowing that for the position shown the *horizontal* component of the velocity of *D* is 3 ft/s to the right and the *horizontal* component of the acceleration of *D* is 20 ft/s² to the right, determine the magnitude of the force **P**.

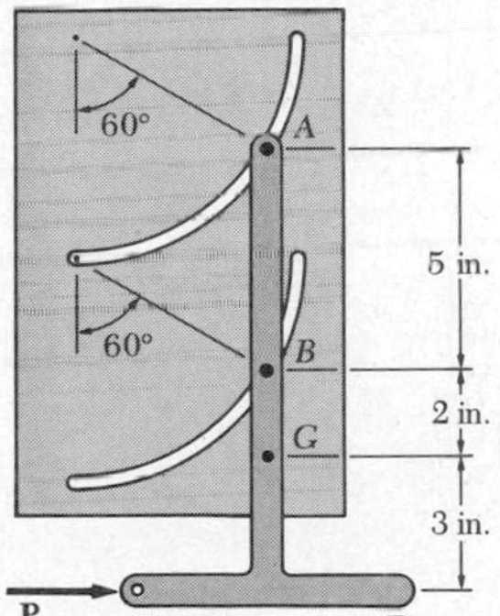

Fig. P16.24

16.25 Solve Prob. 16.24, knowing that for the position shown the *horizontal* component of the velocity of *D* is 3 ft/s to the right and the *horizontal* component of the acceleration of *D* is zero.

***16.26** A 16-lb block is placed on a 4-lb platform *BD* which is held in the position shown by three wires. Determine the accelerations of the block and of the platform immediately after wire *AB* has been cut. Assume that the block (*a*) is rigidly attached to the platform, (*b*) can slide without friction on the platform.

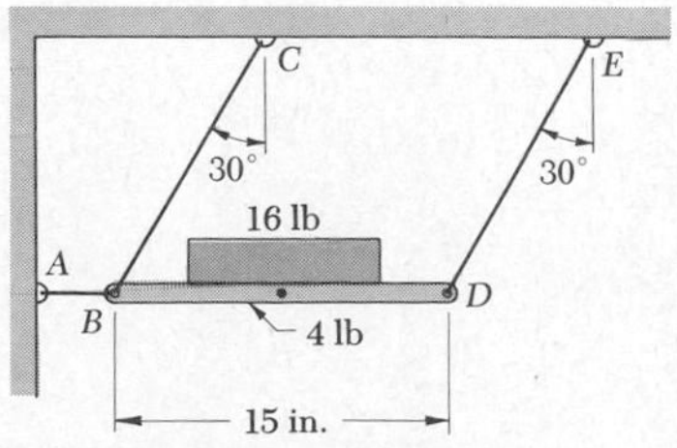

Fig. P16.26 and P16.27

***16.27** The coefficients of friction between the 16-lb block and the 4-lb platform *BD* are $\mu_s = 0.50$ and $\mu_k = 0.40$. Determine the accelerations of the block and of the platform immediately after wire *AB* has been cut.

***16.28** Draw the shear and bending-moment diagrams for the horizontal rod *CD* of Prob. 16.18.

***16.29** Draw the shear and bending-moment diagrams for the horizontal rod *ABCD* of Prob. 16.23.

16.30 For a rigid slab in translation, show that the system of the effective forces consists of vectors $(\Delta m_i)\bar{\mathbf{a}}$ attached to the various particles of the slab, where $\bar{\mathbf{a}}$ is the acceleration of the mass center G of the slab. Further show, by computing their sum and the sum of their moments about G, that the effective forces reduce to a single vector $m\bar{\mathbf{a}}$ attached at G.

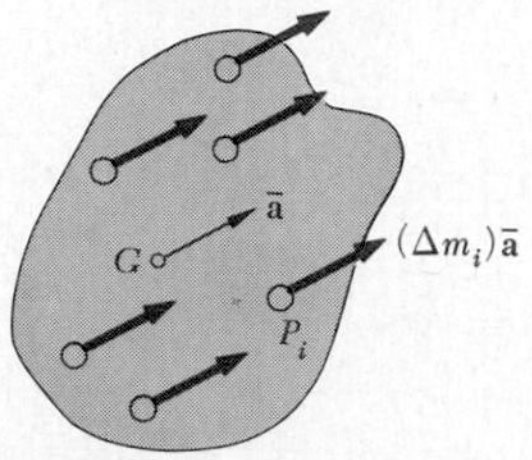

Fig. P16.30

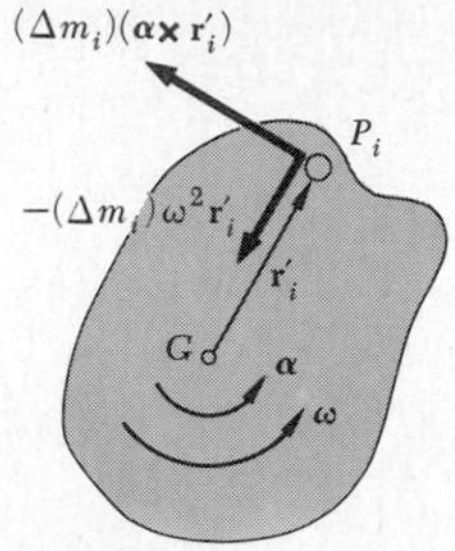

Fig. P16.31

16.31 For a rigid slab in centroidal rotation, show that the system of the effective forces consists of vectors $-(\Delta m_i)\omega^2\mathbf{r}'_i$ and $(\Delta m_i)(\boldsymbol{\alpha} \times \mathbf{r}'_i)$ attached to the various particles P_i of the slab, where $\boldsymbol{\omega}$ and $\boldsymbol{\alpha}$ are the angular velocity and angular acceleration of the slab, and where $\mathbf{r}'_i$ denotes the position vector of the particle P_i relative to the mass center G of the slab. Further show, by computing their sum and the sum of their moments about G, that the effective forces reduce to a couple $\bar{I}\boldsymbol{\alpha}$.

16.32 An electric motor is rotating at 1800 rpm when the load and power are cut off. The rotor weighs 125 lb and has a radius of gyration of 6 in. If the kinetic friction of the rotor produces a couple of moment 9 lb·in., how many revolutions will the rotor execute before stopping?

16.33 A turbine-generator unit is shut off when its rotor is rotating at 3600 rpm; it is observed that the rotor coasts to rest in 8.4 min. Knowing that the 1600-kg rotor has a radius of gyration of 220 mm, determine the average magnitude of the couple due to bearing friction.

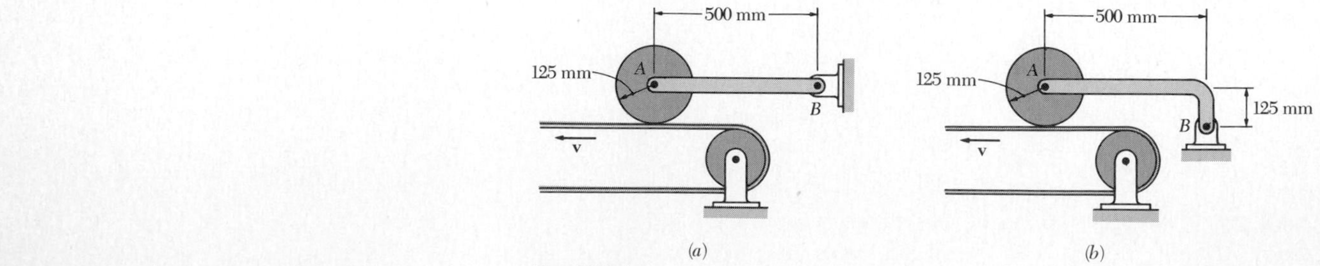

Fig. P16.34

16.34 The 4-kg disk is at rest when it is placed in contact with a conveyor belt moving at a constant speed. The link AB connecting the center of the disk to the support at B is of negligible weight. Knowing that the coefficient of kinetic friction between the disk and the belt is 0.40, determine for each of the arrangements shown the angular acceleration of the disk while slipping occurs.

16.35 Solve Prob. 16.34, assuming that the direction of motion of the conveyor belt is reversed.

16.36 Three disks of the same thickness and same material are attached to a shaft as shown. Disks A and B are each of mass 3 kg and radius $r = 250$ mm. A couple $\mathbf{M}$ of magnitude 35 N·m is applied to disk A when the system is at rest. Determine the radius nr of disk C if the angular acceleration of the system is to be 50 rad/s^2.

16.37 Three disks of the same thickness and same material are attached to a shaft as shown. Disks A and B each have a radius r; disk C has a radius nr. A couple $\mathbf{M}$ of constant magnitude is applied when the system is at rest. Determine the radius of disk C which results in the largest tangential acceleration of a point on the rim of disk C.

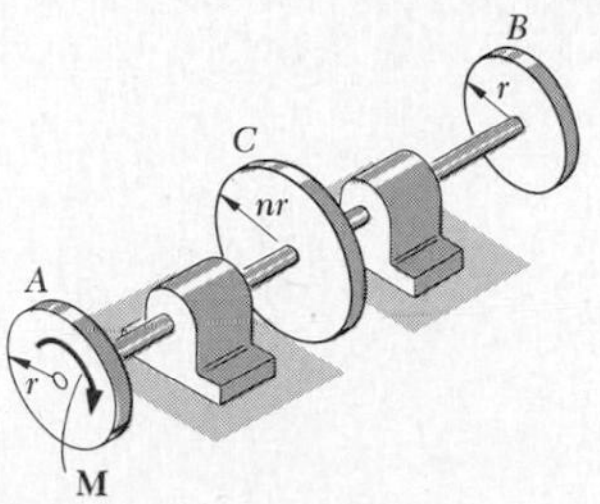

Fig. P16.36 and P16.37

16.38 The 10-in.-radius brake drum is attached to a larger flywheel which is not shown. The total mass moment of inertia of the flywheel and drum is 13.5 lb·ft·s^2. Knowing that the initial angular velocity is 180 rpm clockwise, determine the force which must be exerted by the hydraulic cylinder if the system is to stop in 50 revolutions.

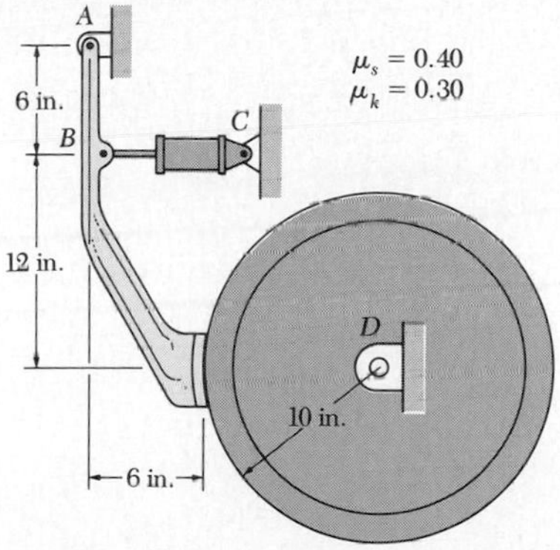

Fig. P16.38

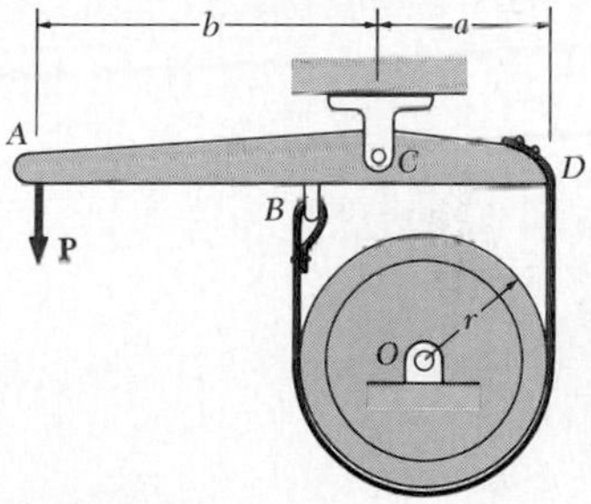

Fig. P16.40

16.39 Solve Prob. 16.38, assuming that the initial angular velocity of the flywheel is 180 rpm counterclockwise.

16.40 The 125-mm-radius brake drum is attached to a flywheel which is not shown. The drum and flywheel together have a mass of 325 kg and a radius of gyration of 725 mm. The coefficient of kinetic friction between the brake band and the drum is 0.40. Knowing that a force $\mathbf{P}$ of magnitude 50 N is applied at A when the angular velocity is 240 rpm counterclockwise, determine the time required to stop the flywheel when $a = 200$ mm and $b = 250$ mm.

16.41 Solve Prob. 16.40, assuming that the coefficient of kinetic friction is 0.30 between the brake band and the drum.

16.42 The double pulley shown has a total mass of 6 kg and a centroidal radius of gyration of 135 mm. Five collars, each of mass 1.2 kg, are attached to cords A and B as shown. When the system is at rest and in equilibrium, one collar is removed from cord A. Neglecting friction, determine (a) the angular acceleration of the pulley, (b) the velocity of cord A at $t = 2.5$ s.

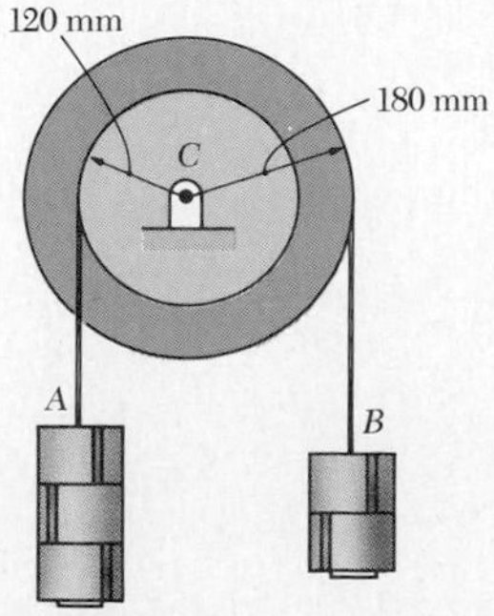

Fig. P16.42

16.43 Solve Prob. 16.42, assuming that one collar is removed from cord B.

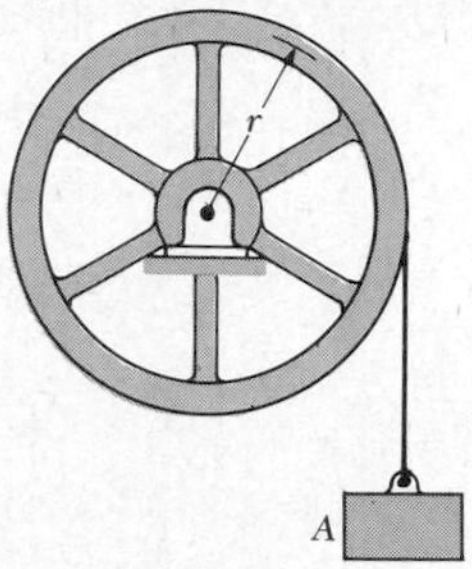

Fig. P16.44, P16.45, and P16.46

16.44 In order to determine the mass moment of inertia of a flywheel of radius $r = 2$ ft, a block A of weight 20 lb is attached to a cord wrapped around the rim of the flywheel. The block is released from rest and is observed to fall 10 ft in 4.60 s. To eliminate bearing friction from the computation, a second block of weight 40 lb is used and is observed to fall 10 ft in 3.10 s. Assuming that the moment of the couple due to bearing friction is constant, determine the mass moment of inertia of the flywheel.

16.45 The flywheel shown has a mass m_F and a radius of gyration $\bar{k}$. A block A of mass m is attached to a wire wrapped around the rim of radius r. Neglecting the effect of friction, derive a formula for the acceleration of the block in terms of m_F, m, r, and $\bar{k}$.

16.46 The flywheel shown has a mass of 100 kg and a radius of gyration of 300 mm. A block A of mass 15 kg is attached to a wire wrapped around the rim of radius $r = 400$ mm. The system is released from rest. Neglecting the effect of friction, determine (a) the acceleration of block A, (b) the speed of block A after it has moved 3 m.

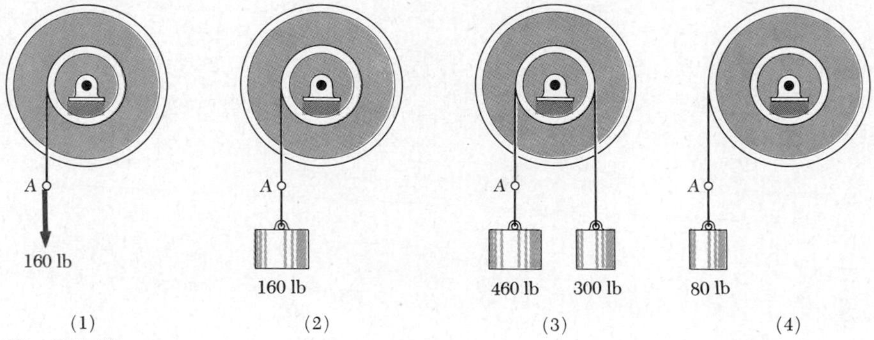

Fig. P16.47

16.47 Each of the double pulleys shown has a mass moment of inertia of 15 lb·ft·s^2 and is initially at rest. The outside radius is 18 in., and the inner radius is 9 in. Determine (a) the angular acceleration of each pulley, (b) the angular velocity of each pulley at $t = 3$ s, (c) the angular velocity of each pulley after point A on the cord has moved 10 ft.

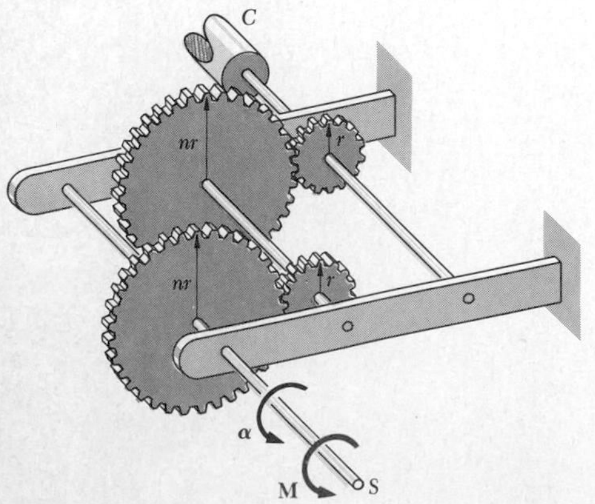

Fig. P16.48

16.48 A coder C, used to record in digital form the rotation of a shaft S, is connected to the shaft by means of the gear train shown, which consists of four gears of the same thickness and of the same material. Two of the gears have a radius r and the other two a radius nr. Let I_R denote the ratio M/α of the moment M of the couple applied to the shaft S and of the resulting angular acceleration α of S. (I_R is sometimes called the "reflected moment of inertia" of the coder and gear train.) Determine I_R in terms of the gear ratio n, the moment of inertia I_0 of the first gear, and the moment of inertia I_C of the coder. Neglect the moments of inertia of the shafts.

16.49 Each of the gears A and B has a mass of 2 kg and a radius of gyration of 75 mm; gear C has a mass of 10 kg and a radius of gyration of 225 mm. If a couple $\mathbf{M}$ of constant magnitude 6 N·m is applied to gear C, determine (a) the angular acceleration of gear A, (b) the tangential force which gear C exerts on gear A.

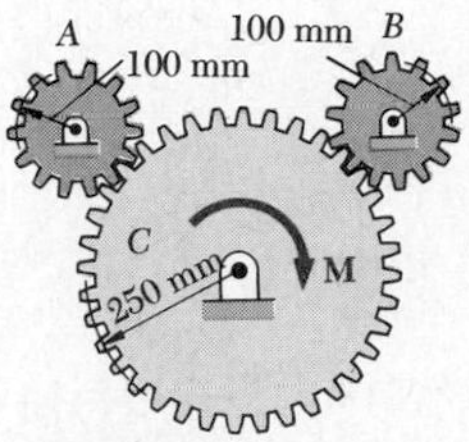

Fig. P16.49

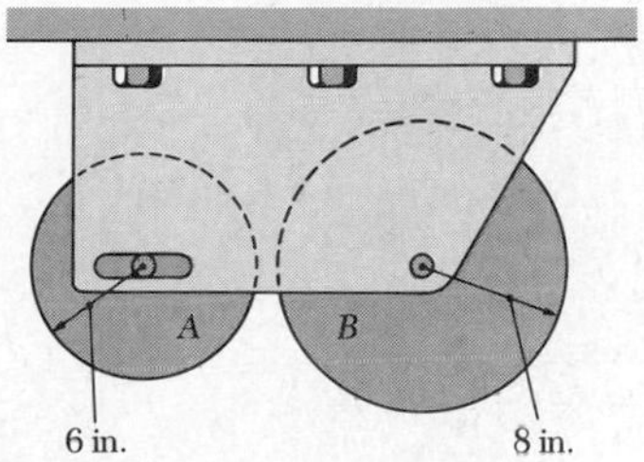

Fig. P16.50

16.50 Disk A weighs 5 lb and has an initial angular velocity of 900 rpm clockwise; disk B weighs 10 lb and is initially at rest. The disks are brought together by applying, as shown, a horizontal force of magnitude $P = 3$ lb to the axle of disk A. Knowing that $\mu_k = 0.25$ and neglecting bearing friction, determine (a) the angular acceleration of each disk, (b) the final angular velocity of each disk.

16.51 Solve Prob. 16.50, assuming that disk A was initially at rest and disk B had an angular velocity of 900 rpm clockwise.

16.52 Disk A has a mass $m_A = 3$ kg, a radius $r_A = 200$ mm, and an initial angular velocity $\omega_0 = 240$ rpm clockwise. Disk B has a mass $m_B = 1.2$ kg, a radius $r_B = 120$ mm, and is at rest when it is brought into contact with disk A. Knowing that $\mu_k = 0.30$ between the disks and neglecting bearing friction, determine (a) the angular acceleration of each disk, (b) the reaction at the support C.

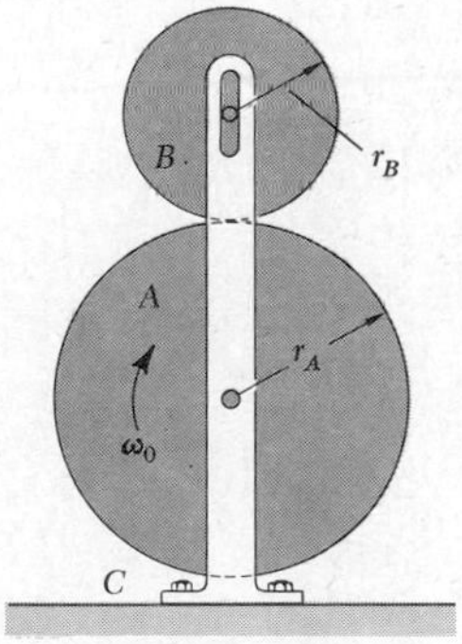

Fig. P16.52 and P16.53

16.53 Disk B is at rest when it is brought into contact with disk A, which has an initial angular velocity ω_0. Show that (a) the final angular velocities of the disks are independent of the coefficient of friction μ_k between the disks as long as $\mu_k \neq 0$, (b) the final angular velocity of disk A depends only upon ω_0 and the ratio of the masses m_A and m_B of the two disks.

16.54 A cylinder of radius r and mass m rests on two small casters A and B as shown. Initially, the cylinder is at rest and is set in motion by rotating caster B clockwise at high speed so that slipping occurs between the cylinder and caster B. Denoting by μ_k the coefficient of kinetic friction and neglecting the moment of inertia of the free caster A, derive an expression for the angular acceleration of the cylinder.

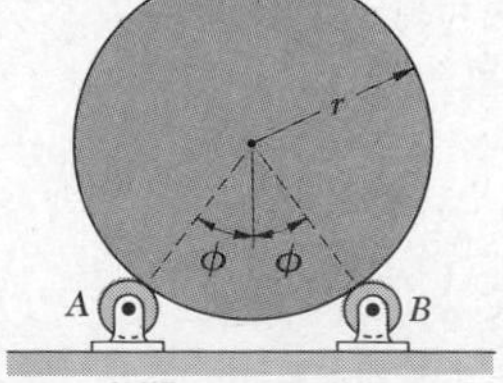

Fig. P16.54

16.55 In Prob. 16.54, assume that no slipping can occur between caster B and the cylinder (such a case would exist if the cylinder and caster had gear teeth along their rims). Derive an expression for the maximum allowable counterclockwise acceleration $\boldsymbol{\alpha}$ of the cylinder if it is not to lose contact with the caster at A.

16.56 Show that the system of the effective forces for a rigid slab in plane motion reduces to a single vector, and express the distance from the mass center G of the slab to the line of action of this vector in terms of the centroidal radius of gyration $\bar{k}$ of the slab, the magnitude $\bar{a}$ of the acceleration of G, and the angular acceleration α.

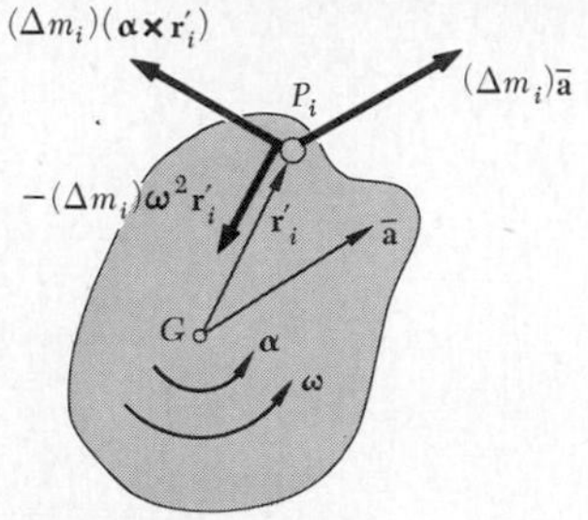

Fig. P16.57

16.57 For a rigid slab in plane motion, show that the system of the effective forces consists of vectors $(\Delta m_i)\bar{\mathbf{a}}$, $-(\Delta m_i)\omega^2\mathbf{r}'_i$, and $(\Delta m_i)(\boldsymbol{\alpha} \times \mathbf{r}'_i)$ attached to the various particles P_i of the slab, where $\bar{\mathbf{a}}$ is the acceleration of the mass center G of the slab, $\boldsymbol{\omega}$ is the angular velocity of the slab, $\boldsymbol{\alpha}$ is its angular acceleration, and $\mathbf{r}'_i$ denotes the position vector of the particle P_i relative to G. Further show, by computing their sum and the sum of their moments about G, that the effective forces reduce to a vector $m\bar{\mathbf{a}}$ attached at G and a couple $\bar{I}\boldsymbol{\alpha}$.

16.58 A 10-in.-diameter disk weighing 15 lb rests on a frictionless horizontal surface. A force $\mathbf{P}$ of magnitude 3 lb is applied to a tape wrapped around the disk. Determine the acceleration of (*a*) point A, (*b*) point B.

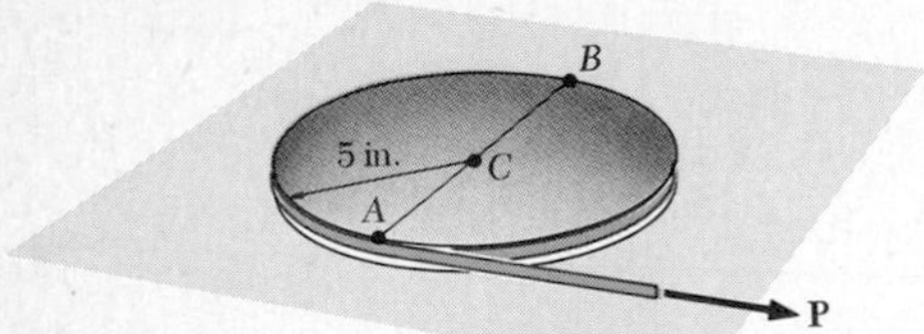

Fig. P16.58

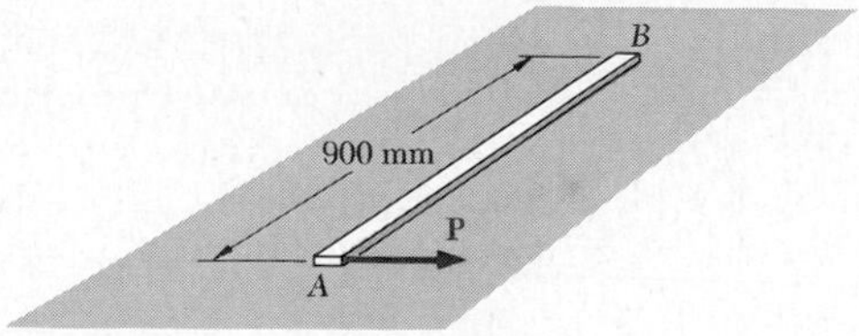

Fig. P16.59

16.59 The uniform slender rod AB of mass 1.25 kg is at rest on a frictionless horizontal surface. A force $\mathbf{P}$ of magnitude 3 N is applied at A in a horizontal direction perpendicular to the rod. Determine (*a*) the angular acceleration of the rod, (*b*) the acceleration of the center of the rod, (*c*) the point of the rod which has no acceleration.

16.60 (*a*) In Prob. 16.59, determine the point of the rod AB at which the force $\mathbf{P}$ should be applied if the acceleration of point B is to be zero. (*b*) Knowing that the magnitude of $\mathbf{P}$ is 3 N, determine the corresponding angular acceleration of the rod and the acceleration of the center of the rod.

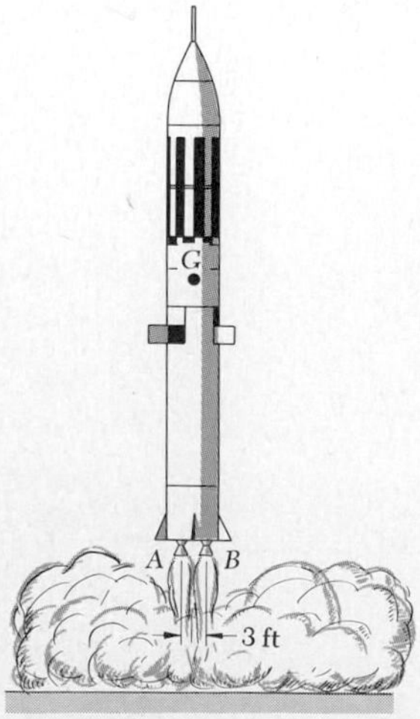

Fig. P16.61

16.61 Shortly after being fired, the experimental rocket shown weighs 25,000 lb and is moving upward with an acceleration of 45 ft/s². If at that instant the rocket engine A fails, while rocket engine B continues to operate, determine (*a*) the acceleration of the mass center of the rocket, (*b*) the angular acceleration of the rocket. Assume that the rocket is a uniform slender rod 48 ft long.

16.62 A 4-m beam of mass 250 kg is lowered from a considerable height by means of two cables unwinding from overhead cranes. As the beam approaches the ground, the crane operators apply brakes to slow the unwinding motion. Determine the acceleration of each cable at that instant, knowing that $T_A = 1000$ N and $T_B = 1800$ N.

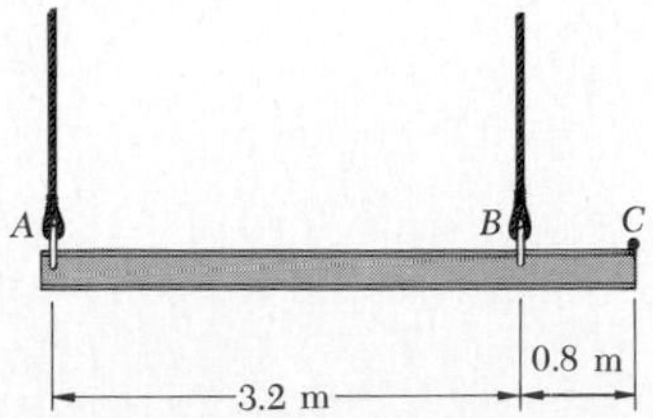

Fig. P16.62 and P16.63

16.63 A 4-m beam of mass 250 kg is lowered from a considerable height by means of two cables unwinding from overhead cranes. As the beam approaches the ground, the crane operators apply brakes to slow the unwinding motion. The deceleration of cable A is 2 m/s^2 and that of cable B is 0.5 m/s^2. Determine the tension in each cable.

16.64 The 800-lb crate shown is being lowered by means of two overhead cranes. Knowing that at the instant shown the deceleration of cable A is 21 ft/s^2 and that of cable B is 3 ft/s^2, determine the tension in each cable.

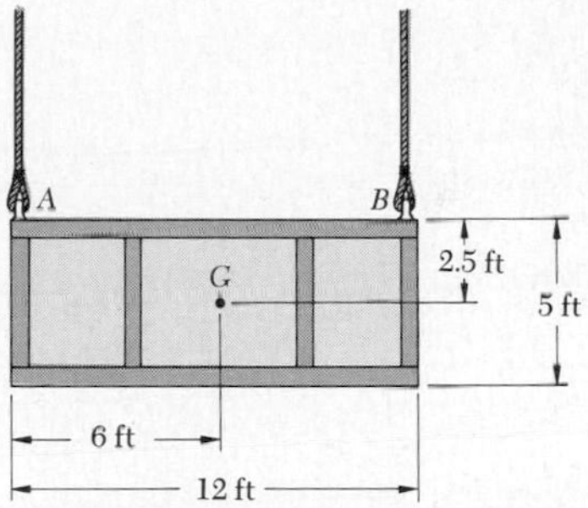

Fig. P16.64 and P16.65

16.65 The 800-lb crate shown is being lowered by means of two overhead cranes. As the crate approaches the ground, the crane operators apply brakes to slow the motion. Determine the acceleration of each cable at that instant, knowing that $T_A = 650$ lb and $T_B = 550$ lb.

16.66 By pulling on the cord of a yo-yo just fast enough, a person manages to make the yo-yo spin counterclockwise, while remaining at a constant height above the floor. Denoting the weight of the yo-yo by W, the radius of the inner drum on which the cord is wound by r, and the radius of gyration of the yo-yo by $\bar{k}$, determine (*a*) the tension in the cord, (*b*) the angular acceleration of the yo-yo.

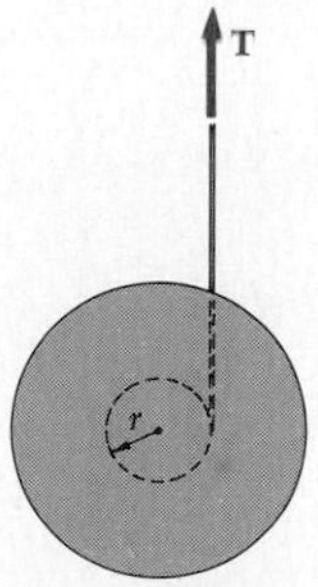

Fig. P16.66

16.67 For the disk and cord of Sample Prob. 16.4, determine the magnitude of the force **T** and the corresponding angular acceleration of the disk, knowing that the acceleration of the center of the disk is 0.8 m/s^2 (*a*) downward, (*b*) upward.

16.68 Structural steel beams are formed by passing them through successive pairs of rolls. The roll shown weighs 2800 lb, has a centroidal radius of gyration of 6 in., and is lifted by two cables looped around its shaft. Knowing that for each cable $T_A = 700$ lb and $T_B = 725$ lb, determine (*a*) the angular acceleration of the roll, (*b*) the acceleration of its mass center.

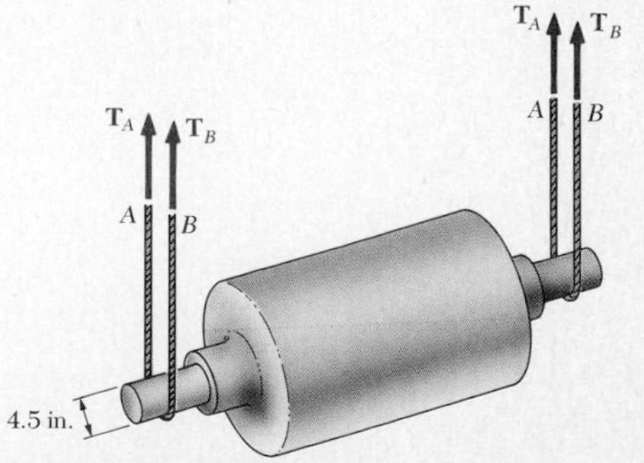

Fig. P16.68 and P16.69

16.69 The steel roll shown weighs 2800 lb, has a centroidal radius of gyration of 6 in., and is being lowered by two cables looped around its shaft. Knowing that at the instant shown the acceleration of the roll is 5 in./s^2 downward and that for each cable $T_A = 680$ lb, determine (*a*) the corresponding value of the tension T_B, (*b*) the angular acceleration of the roll.

16.70 through 16.73 A uniform slender bar AB of mass m is suspended from two springs as shown. If spring 2 breaks, determine at that instant (*a*) the angular acceleration of the bar, (*b*) the acceleration of point A, (*c*) the acceleration of point B.

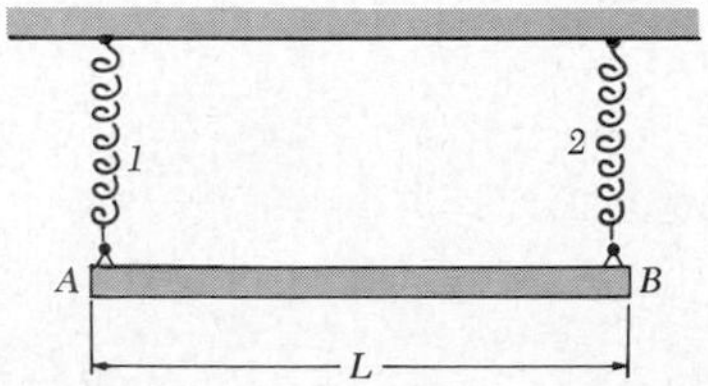

Fig. P16.70

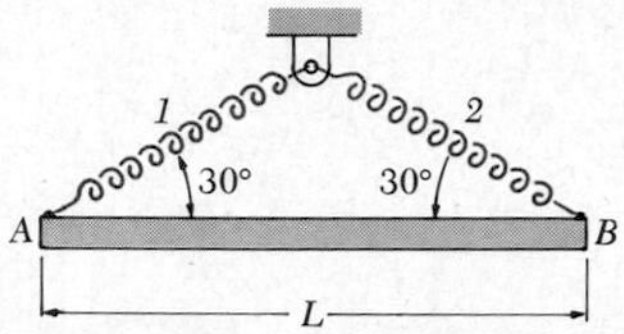

Fig. P16.71

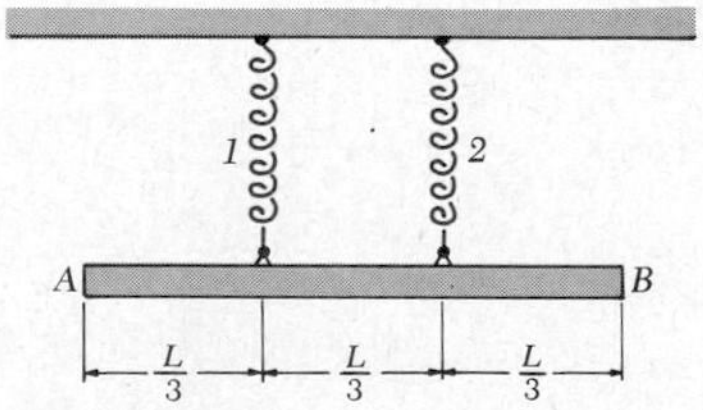

Fig. P16.72

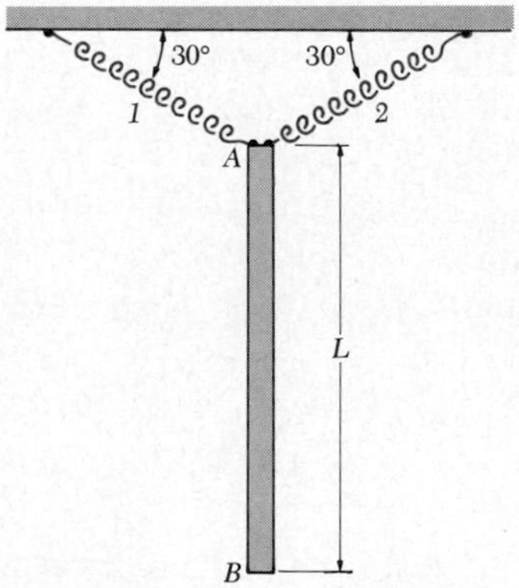

Fig. P16.73

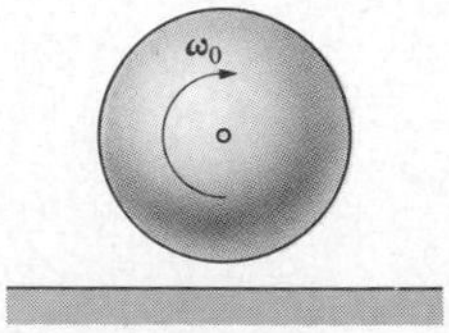

Fig. P16.74

16.74 A sphere of radius r and mass m is placed on a horizontal surface with no linear velocity but with a clockwise angular velocity ω_0. Denoting by μ_k the coefficient of kinetic friction between the sphere and the floor, determine (*a*) the time t_1 at which the sphere will start rolling without sliding, (*b*) the linear and angular velocities of the sphere at time t_1.

16.75 Solve Prob. 16.74, assuming that the sphere is replaced by a uniform disk of radius r and mass m.

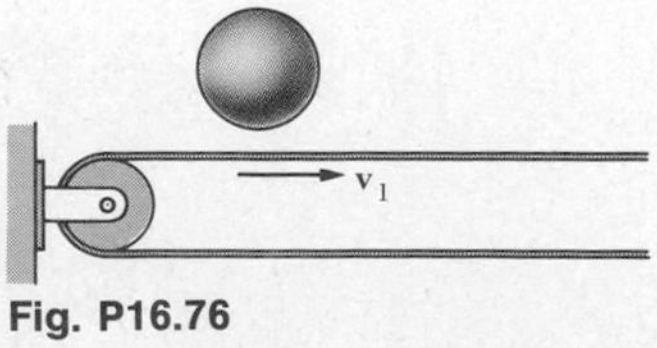

Fig. P16.76

16.76 A uniform sphere of radius r and mass m is placed with no initial velocity on a belt which moves to the right with a constant velocity $\mathbf{v}_1$. Denoting by μ_k the coefficient of kinetic friction between the sphere and the belt, determine (*a*) the time t_1 at which the sphere will start rolling without sliding, (*b*) the linear velocity and the angular velocity of the sphere at time t_1.

16.77 Solve Prob. 16.76, assuming that the sphere is replaced by a wheel of radius r, mass m, and centroidal radius of gyration $\bar{k}$.

16.8. Constrained Plane Motion. Most engineering applications deal with rigid bodies which are moving under given constraints. Cranks, for example, must rotate about a fixed axis, wheels must roll without sliding, connecting rods must describe certain prescribed motions. In all such cases, definite relations exist between the components of the acceleration $\bar{\mathbf{a}}$ of the mass center G of the body considered and its angular acceleration $\boldsymbol{\alpha}$; the corresponding motion is said to be a *constrained motion.*

The solution of a problem involving a constrained plane motion calls first for a *kinematic analysis* of the problem. Consider, for example, a slender rod AB of length l and mass m whose extremities are connected to blocks of negligible mass which slide along horizontal and vertical frictionless tracks. The rod is pulled by a force **P** applied at A (Fig. 16.11). We know from Sec. 15.8 that the acceleration $\bar{\mathbf{a}}$ of the mass center G of the rod may be determined at any given instant from the position of the rod, its angular velocity, and its angular acceleration at that instant. Suppose, for instance, that the values of θ, ω, and α are known at a given instant and that we wish to determine the corresponding value of the force **P**, as well as the reactions at A and B. We should first *determine the components* $\bar{a}_x$ *and* $\bar{a}_y$ *of the acceleration of the mass center* G by the method of Sec. 15.8. We next apply d'Alembert's principle (Fig. 16.12), using the expressions obtained for $\bar{a}_x$ and $\bar{a}_y$. The unknown forces **P**, $\mathbf{N}_A$, and $\mathbf{N}_B$ may then be determined by writing and solving the appropriate equations.

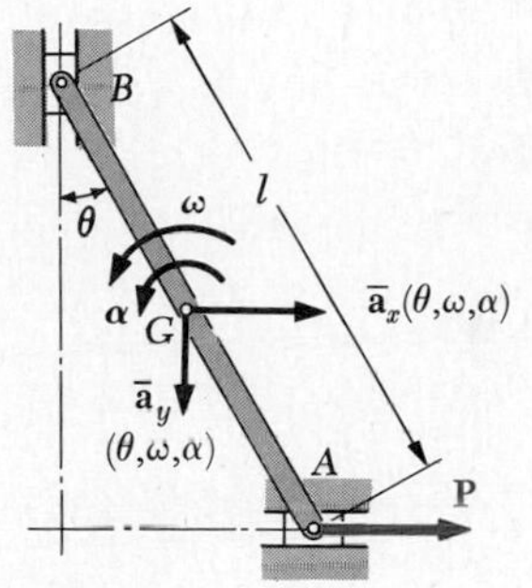

Fig. 16.11

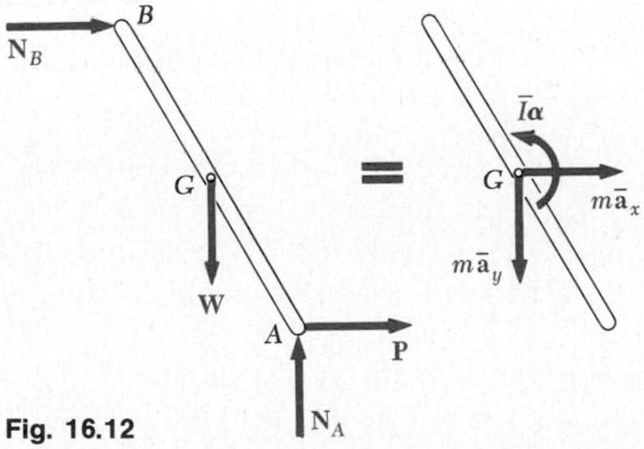

Fig. 16.12

Suppose now that the applied force **P**, the angle θ, and the angular velocity ω of the rod are known at a given instant and that we wish to find the angular acceleration α of the rod and the components $\bar{a}_x$ and $\bar{a}_y$ of the acceleration of its mass center at that instant, as well as the reactions at A and B. The preliminary kinematic study of the problem will have for its object *to express the components* $\bar{a}_x$ *and* $\bar{a}_y$ *of the acceleration of* G *in terms of the angular acceleration* α *of the rod.* This will be done by first expressing the acceleration of a suitable reference point such as A in terms of the angular acceleration α. The components $\bar{a}_x$ and $\bar{a}_y$ of the acceleration of G may then be determined in terms of α, and the expressions obtained carried into Fig. 16.12. Three equations may then be derived in terms of α, N_A, and N_B, and solved for the three unknowns (see Sample Prob. 16.10). Note that the method of dynamic equilibrium may also be used to carry out the solution of the two types of problems we have considered (Fig. 16.13).

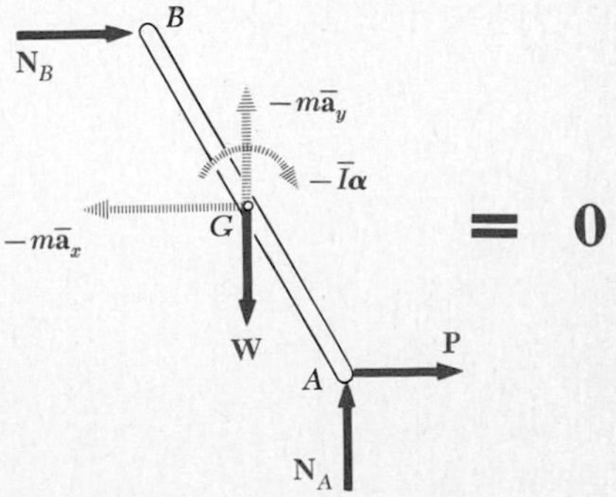

Fig. 16.13

When a mechanism consists of *several moving parts,* the approach just described may be used with each part of the mechanism. The procedure required to determine the various unknowns is then similar to the procedure followed in the case of the equilibrium of a system of connected rigid bodies (Sec. 6.11).

We have analyzed earlier two particular cases of constrained plane motion, the translation of a rigid body, in which the angular acceleration of the body is constrained to be zero, and the centroidal rotation, in which the acceleration $\bar{\mathbf{a}}$ of the mass center of the body is constrained to be zero. Two other particular cases of constrained plane motion are of special interest, the *noncentroidal rotation* of a rigid body and the *rolling motion* of a disk or wheel. These two cases should be analyzed by one of the general methods described above. However, in view of the range of their applications, they deserve a few special comments.

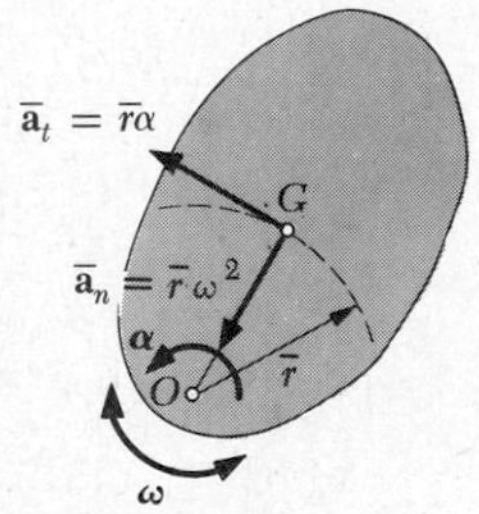

Fig. 16.14

Noncentroidal Rotation. This is the motion of a rigid body constrained to rotate about a fixed axis which does not pass through its mass center. Such a motion is called a *noncentroidal rotation*. The mass center G of the body moves along a circle of radius $\bar{r}$ centered at the point O, where the axis of rotation intersects the plane of reference (Fig. 16.14). Denoting, respectively, by $\boldsymbol{\omega}$ and $\boldsymbol{\alpha}$ the angular velocity and the angular acceleration of the line OG, we obtain the following expressions for the tangential and normal components of the acceleration of G:

$$\bar{a}_t = \bar{r}\alpha \qquad \bar{a}_n = \bar{r}\omega^2 \tag{16.7}$$

Since line OG belongs to the body, its angular velocity $\boldsymbol{\omega}$ and its angular acceleration $\boldsymbol{\alpha}$ also represent the angular velocity and the angular acceleration of the body in its motion relative to G. Equations (16.7) define, therefore, the kinematic relation existing between the motion of the mass center G and the motion of the body about G. They should be used to eliminate $\bar{a}_t$ and $\bar{a}_n$ from the equations obtained by applying d'Alembert's principle (Fig. 16.15) or the method of dynamic equilibrium (Fig. 16.16).

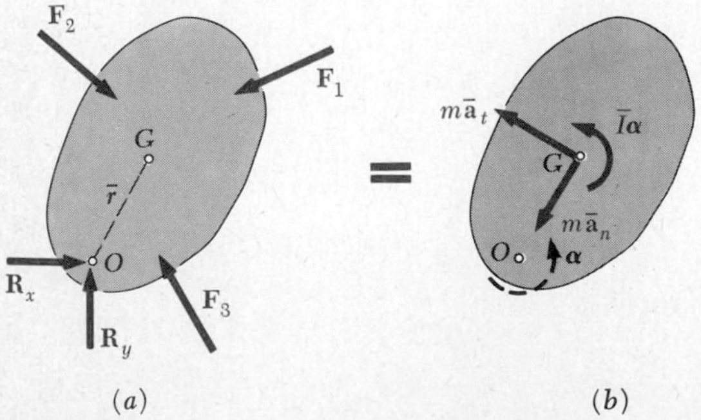

Fig. 16.15

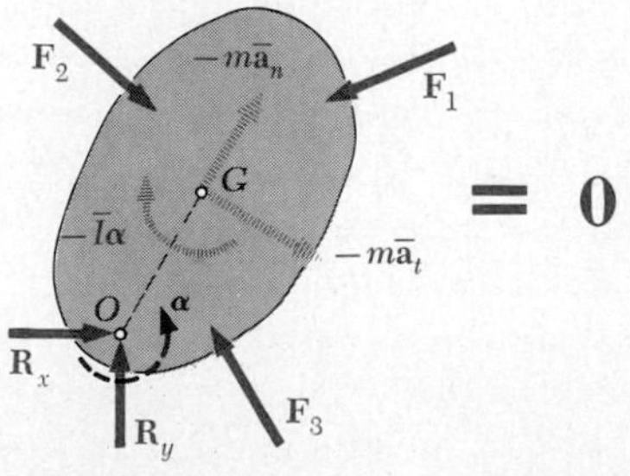

Fig. 16.16

An interesting relation may be obtained by equating the moments about the fixed point O of the forces and vectors shown, respectively, in parts a and b of Fig. 16.15. We write

$$+\circlearrowleft \Sigma M_O = \bar{I}\alpha + (m\bar{r}\alpha)\bar{r} = (\bar{I} + m\bar{r}^2)\alpha$$

But according to the parallel-axis theorem, we have $\bar{I} + m\bar{r}^2 = I_O$, where I_O denotes the moment of inertia of the rigid body about the fixed axis. We therefore write

$$\Sigma M_O = I_O\alpha \tag{16.8}$$

Although formula (16.8) expresses an important relation between the sum of the moments of the external forces about the fixed point O and the product $I_O\alpha$, it should be clearly understood that this formula *does not mean* that the system of the external forces is equivalent to a couple of moment $I_O\alpha$. The system of the effective forces, and thus the system of the external forces, reduces to a couple only when O coincides with G—that is, *only when the rotation is centroidal* (Sec. 16.4). In the more general case of noncentroidal rotation, the system of the external forces does not reduce to a couple.

A particular case of noncentroidal rotation is of special interest—the case of *uniform rotation*, in which the angular velocity $\boldsymbol{\omega}$ is constant. Since $\boldsymbol{\alpha}$ is zero, the inertia couple in Fig. 16.16 vanishes and the inertia vector reduces to its normal component. This component (also called *centrifugal force*) represents the tendency of the rigid body to break away from the axis of rotation.

Rolling Motion. Another important case of plane motion is the motion of a disk or wheel rolling on a plane surface. If the disk is constrained to roll without sliding, the acceleration $\bar{\mathbf{a}}$ of its mass center G and its angular acceleration $\boldsymbol{\alpha}$ are not independent. Assuming that the disk is balanced, so that its mass center and its geometric center coincide, we first write that the distance $\bar{x}$ traveled by G during a rotation θ of the disk is $\bar{x} = r\theta$, where r is the radius of the disk. Differentiating this relation twice, we write

$$\bar{a} = r\alpha \tag{16.9}$$

Recalling that the system of the effective forces in plane motion reduces to a vector $m\bar{\mathbf{a}}$ and a couple $\bar{I}\boldsymbol{\alpha}$, we find that in the particular case of the rolling motion of a balanced disk, the effective forces reduce to a vector of magnitude $mr\alpha$ attached at G and to a couple of magnitude $\bar{I}\alpha$. We may thus express that the external forces are equivalent to the vector and couple shown in Fig. 16.17.

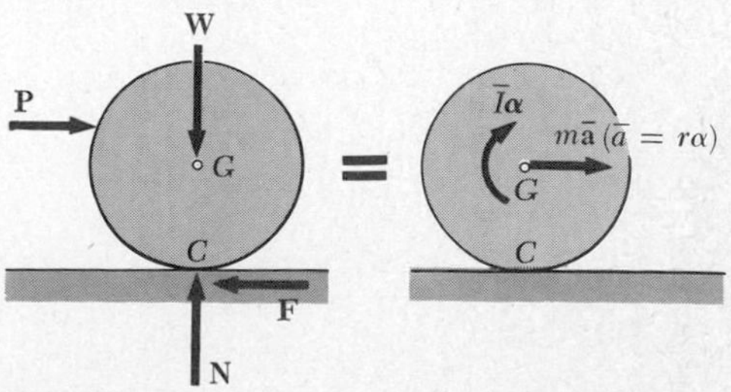

Fig. 16.17

When a disk *rolls without sliding*, there is no relative motion between the point of the disk in contact with the ground and the ground itself. As far as the computation of the friction force **F** is concerned, a rolling disk may thus be compared with a block at rest on a surface. The magnitude F of the friction force may have any value, as long as this value does not exceed the maximum value $F_m = \mu_s N$, where μ_s is the coefficient of static friction and N the magnitude of the normal force. In the case of a rolling disk, the magnitude F of the friction force should therefore be determined independently of N by solving the equation obtained from Fig. 16.17.

When *sliding is impending*, the friction force reaches its maximum value $F_m = \mu_s N$ and may be obtained from N.

When the disk *rotates and slides* at the same time, a relative motion exists between the point of the disk which is in contact with the ground and the ground itself, and the force of friction has the magnitude $F_k = \mu_k N$, where μ_k is the coefficient of kinetic friction. In this case, however, the motion of the mass center G of the disk and the rotation of the disk about G are independent, and $\bar{a}$ is not equal to $r\alpha$.

These three different cases may be summarized as follows:

Rolling, no sliding:	$F \leq \mu_s N$	$\bar{a} = r\alpha$
Rolling, sliding impending:	$F = \mu_s N$	$\bar{a} = r\alpha$
Rotating and sliding:	$F = \mu_s N$	$\bar{a}$ *and* α *independent*

When it is not known whether or not a disk slides, it should first be assumed that the disk rolls without sliding. If F is found smaller than, or equal to, $\mu_s N$, the assumption is proved correct. If F is found larger than $\mu_s N$, the assumption is incorrect and the problem should be started again, assuming rotating and sliding.

When a disk is *unbalanced*, i.e., when its mass center G does not coincide with its geometric center O, the relation (16.9) does not hold between $\bar{a}$ and α. However, a similar relation holds between the magnitude a_O of the acceleration of the geometric center and the angular acceleration α of an unbalanced disk which rolls without sliding. We have

$$a_O = r\alpha \tag{16.10}$$

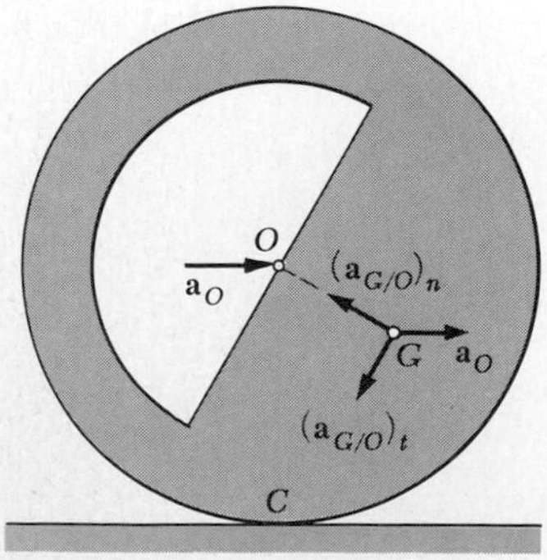

Fig. 16.18

To determine $\bar{a}$ in terms of the angular acceleration α and the angular velocity ω of the disk, we may use the relative-acceleration formula

$$\begin{aligned} \bar{\mathbf{a}} = \mathbf{a}_G &= \mathbf{a}_O + \mathbf{a}_{G/O} \\ &= \mathbf{a}_O + (\mathbf{a}_{G/O})_t + (\mathbf{a}_{G/O})_n \end{aligned} \tag{16.11}$$

where the three component accelerations obtained have the directions indicated in Fig. 16.18 and the magnitudes $a_O = r\alpha$, $(a_{G/O})_t = (OG)\alpha$, and $(a_{G/O})_n = (OG)\omega^2$.

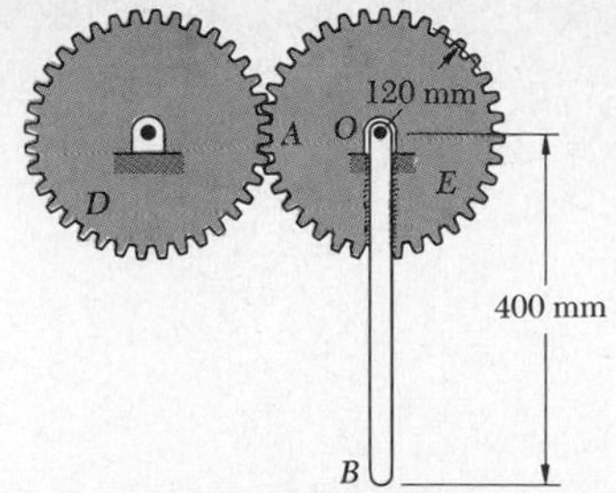

SAMPLE PROBLEM 16.6

The portion AOB of a mechanism consists of a 400-mm steel rod OB welded to a gear E of radius 120 mm which may rotate about a horizontal shaft O. It is actuated by a gear D and, at the instant shown, has a clockwise angular velocity of 8 rad/s and a counterclockwise angular acceleration of 40 rad/s^2. Knowing that rod OB has a mass of 3 kg and gear E a mass of 4 kg and a radius of gyration of 85 mm, determine (*a*) the tangential force exerted by gear D on gear E, (*b*) the components of the reaction at shaft O.

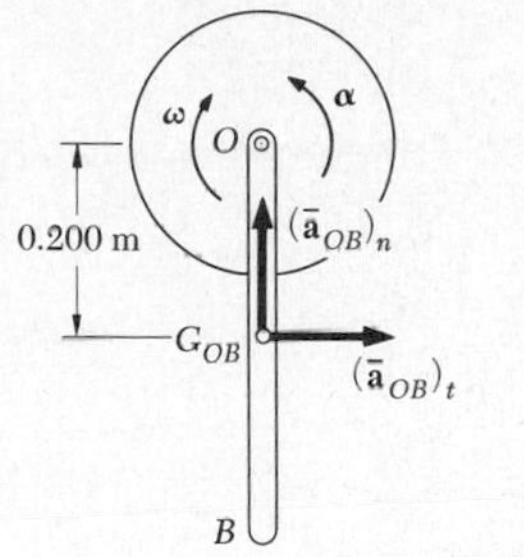

Solution. In determining the effective forces of the rigid body AOB we shall consider separately gear E and rod OB. Therefore, we shall first determine the components of the acceleration of the mass center G_{OB} of the rod:

$$(\bar{a}_{OB})_t = \bar{r}\alpha = (0.200 \text{ m})(40 \text{ rad/s}^2) = 8 \text{ m/s}^2$$
$$(\bar{a}_{OB})_n = \bar{r}\omega^2 = (0.200 \text{ m})(8 \text{ rad/s})^2 = 12.8 \text{ m/s}^2$$

Equations of Motion. Two sketches of the rigid body AOB have been drawn. The first shows the external forces consisting of the weight $\mathbf{W}_E$ of gear E, the weight $\mathbf{W}_{OB}$ of the rod OB, the force $\mathbf{F}$ exerted by gear D, and the components $\mathbf{R}_x$ and $\mathbf{R}_y$ of the reaction at O. The magnitudes of the weights are, respectively,

$$W_E = m_E g = (4 \text{ kg})(9.81 \text{ m/s}^2) = 39.2 \text{ N}$$
$$W_{OB} = m_{OB} g = (3 \text{ kg})(9.81 \text{ m/s}^2) = 29.4 \text{ N}$$

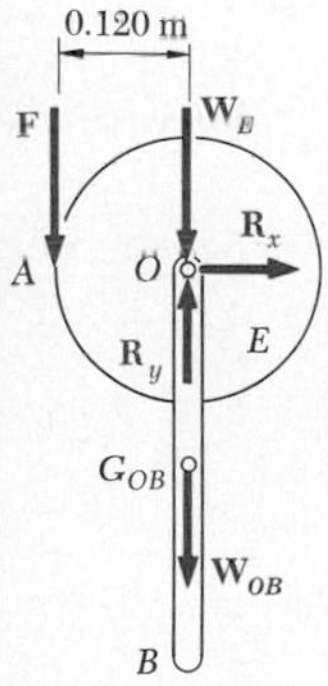

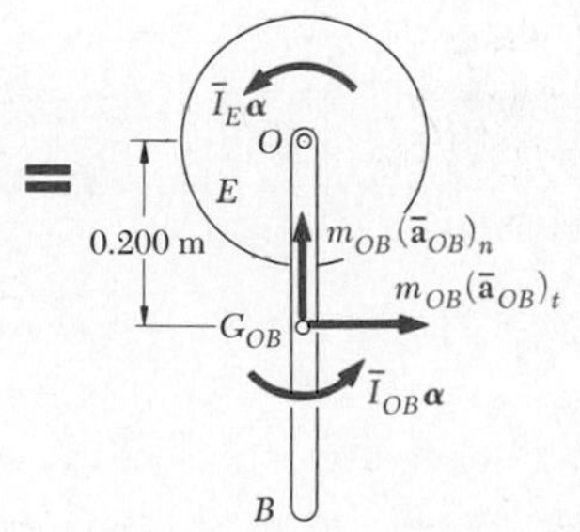

The second sketch shows the effective forces, which consist of a couple $\bar{I}_E\boldsymbol{\alpha}$ (since gear E is in centroidal rotation) and of a couple and two vector components at the mass center of OB. Since the accelerations are known, we compute the magnitudes of these components and couples:

$$\bar{I}_E\alpha = m_E\bar{k}_E^2\alpha = (4 \text{ kg})(0.085 \text{ m})^2(40 \text{ rad/s}^2) = 1.156 \text{ N}\cdot\text{m}$$
$$m_{OB}(\bar{a}_{OB})_t = (3 \text{ kg})(8 \text{ m/s}^2) = 24.0 \text{ N}$$
$$m_{OB}(\bar{a}_{OB})_n = (3 \text{ kg})(12.8 \text{ m/s}^2) = 38.4 \text{ N}$$
$$\bar{I}_{OB}\alpha = (\tfrac{1}{12}m_{OB}L^2)\alpha = \tfrac{1}{12}(3 \text{ kg})(0.400 \text{ m})^2(40 \text{ rad/s}^2) = 1.600 \text{ N}\cdot\text{m}$$

Expressing that the system of the external forces is equivalent to the system of the effective forces, we write the following equations:

$+\circlearrowleft \Sigma M_O = \Sigma(M_O)_{\text{eff}}$:

$$F(0.120 \text{ m}) = \bar{I}_E\alpha + m_{OB}(\bar{a}_{OB})_t(0.200 \text{ m}) + \bar{I}_{OB}\alpha$$
$$F(0.120 \text{ m}) = 1.156 \text{ N}\cdot\text{m} + (24.0 \text{ N})(0.200 \text{ m}) + 1.600 \text{ N}\cdot\text{m}$$

$$F = 63.0 \text{ N} \qquad \mathbf{F} = 63.0 \text{ N} \downarrow \blacktriangleleft$$

$\xrightarrow{+} \Sigma F_x = \Sigma(F_x)_{\text{eff}}$: $\quad R_x = m_{OB}(\bar{a}_{OB})_t$

$$R_x = 24.0 \text{ N} \qquad \mathbf{R}_x = 24.0 \text{ N} \rightarrow \blacktriangleleft$$

$+\uparrow \Sigma F_y = \Sigma(F_y)_{\text{eff}}$: $\quad R_y - F - W_E - W_{OB} = m_{OB}(\bar{a}_{OB})_n$

$$R_y - 63.0 \text{ N} - 39.2 \text{ N} - 29.4 \text{ N} = 38.4 \text{ N}$$

$$R_y = 170.0 \text{ N} \qquad \mathbf{R}_y = 170.0 \text{ N} \uparrow \blacktriangleleft$$

SAMPLE PROBLEM 16.7

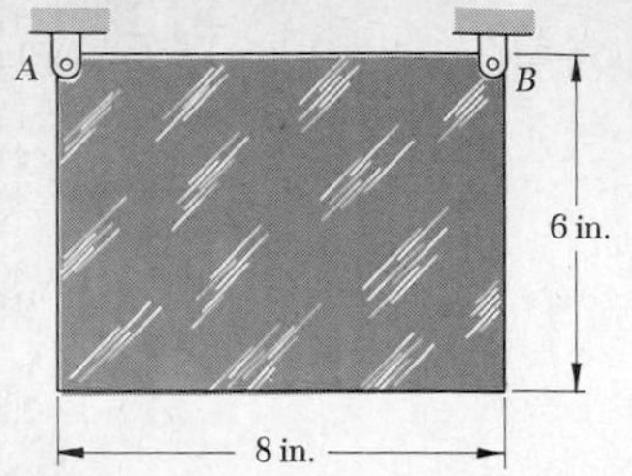

A 6×8 in. rectangular plate weighing 60 lb is suspended from two pins A and B. If pin B is suddenly removed, determine (a) the angular acceleration of the plate, (b) the components of the reaction at pin A, immediately after pin B has been removed.

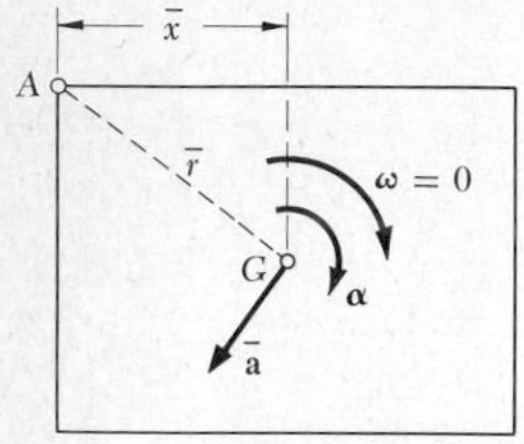

***a.* Angular Acceleration.** We observe that as the plate rotates about point A, its mass center G describes a circle of radius $\bar{r}$ with center at A.

Since the plate is released from rest ($\omega = 0$), the normal component of the acceleration of G is zero. The magnitude of the acceleration $\bar{\mathbf{a}}$ of the mass center G is thus $\bar{a} = \bar{r}\alpha$. We draw the diagram shown to express that the external forces are equivalent to the effective forces:

$$+\circlearrowright \Sigma M_A = \Sigma(M_A)_{\text{eff}}: \qquad W\bar{x} = (m\bar{a})\bar{r} + \bar{I}\alpha$$

Since $\bar{a} = \bar{r}\alpha$, we have

$$W\bar{x} = m(\bar{r}\alpha)\bar{r} + \bar{I}\alpha \qquad \alpha = \frac{W\bar{x}}{\dfrac{W}{g}\bar{r}^2 + \bar{I}} \tag{1}$$

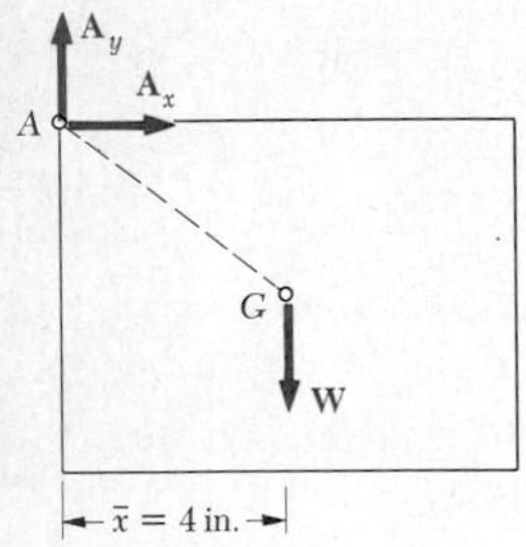

The centroidal moment of inertia of the plate is

$$\bar{I} = \frac{m}{12}(a^2 + b^2) = \frac{60 \text{ lb}}{12(32.2 \text{ ft/s}^2)}[(\tfrac{8}{12} \text{ ft})^2 + (\tfrac{6}{12} \text{ ft})^2]$$
$$= 0.1078 \text{ lb}\cdot\text{ft}\cdot\text{s}^2$$

Substituting this value of $\bar{I}$ together with $W = 60$ lb, $\bar{r} = \frac{5}{12}$ ft, and $\bar{x} = \frac{4}{12}$ ft into Eq. (1), we obtain

$$\alpha = +46.4 \text{ rad/s}^2 \qquad \boldsymbol{\alpha} = 46.4 \text{ rad/s}^2 \circlearrowright \blacktriangleleft$$

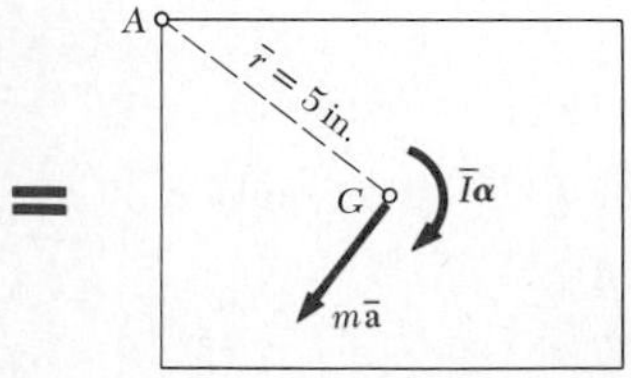

***b.* Reaction at *A*.** Using the computed value of α, we determine the magnitude of the vector $m\bar{\mathbf{a}}$ attached at G,

$$m\bar{a} = m\bar{r}\alpha = \frac{60 \text{ lb}}{32.2 \text{ ft/s}^2}(\tfrac{5}{12} \text{ ft})(46.4 \text{ rad/s}^2) = 36.0 \text{ lb}$$

Showing this result on the diagram, we write the equations of motion

$$\overset{+}{\rightarrow} \Sigma F_x = \Sigma(F_x)_{\text{eff}}: \qquad A_x = -\tfrac{3}{5}(36 \text{ lb})$$
$$= -21.6 \text{ lb} \qquad \mathbf{A}_x = 21.6 \text{ lb} \leftarrow \blacktriangleleft$$

$$+\uparrow \Sigma F_y = \Sigma(F_y)_{\text{eff}}: \qquad A_y - 60 \text{ lb} = -\tfrac{4}{5}(36 \text{ lb})$$
$$A_y = +31.2 \text{ lb} \qquad \mathbf{A}_y = 31.2 \text{ lb} \uparrow \blacktriangleleft$$

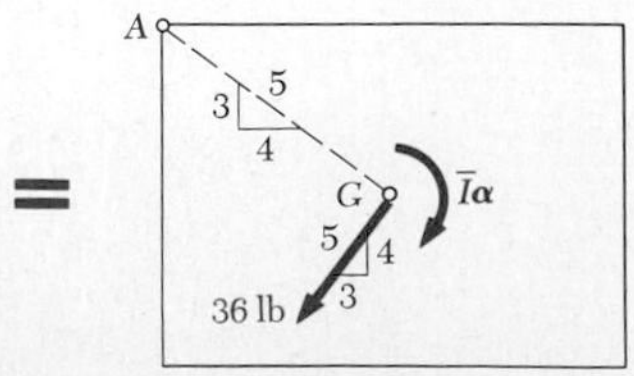

The couple $\bar{I}\boldsymbol{\alpha}$ is not involved in the last two equations; nevertheless, it should be indicated on the diagram.

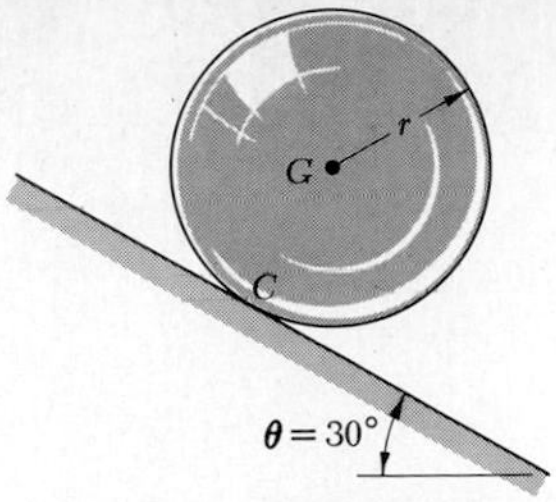

SAMPLE PROBLEM 16.8

A sphere of radius r and weight W is released with no initial velocity on the incline and rolls without slipping. Determine (*a*) the minimum value of the coefficient of static friction compatible with the rolling motion, (*b*) the velocity of the center G of the sphere after the sphere has rolled 10 ft, (*c*) the velocity of G if the sphere were to move 10 ft down a frictionless 30° incline.

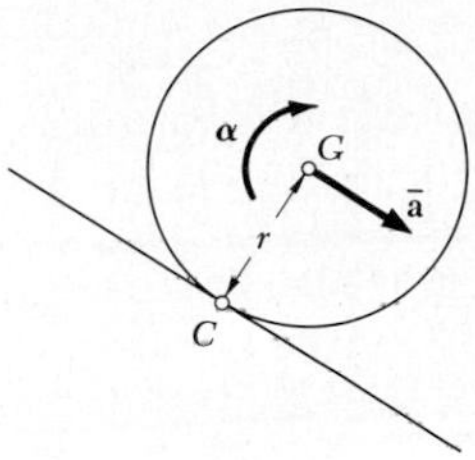

***a.* Minimum μ_s for Rolling Motion.** The external forces **W**, **N**, and **F** form a system equivalent to the system of effective forces represented by the vector $m\bar{\mathbf{a}}$ and the couple $\bar{I}\boldsymbol{\alpha}$. Since the sphere rolls without sliding, we have $\bar{a} = r\alpha$.

$$+\circlearrowright \Sigma M_C = \Sigma (M_C)_{\text{eff}}: \qquad (W \sin\theta)r = (m\bar{a})r + \bar{I}\alpha$$
$$(W \sin\theta)r = (mr\alpha)r + \bar{I}\alpha$$

Noting that $m = W/g$ and $\bar{I} = \frac{2}{5}mr^2$, we write

$$(W\sin\theta)r = \left(\frac{W}{g}r\alpha\right)r + \frac{2}{5}\frac{W}{g}r^2\alpha \qquad \alpha = +\frac{5g\sin\theta}{7r}$$

$$\bar{a} = r\alpha = \frac{5g\sin\theta}{7} = \frac{5(32.2 \text{ ft/s}^2)\sin 30°}{7} = 11.50 \text{ ft/s}^2$$

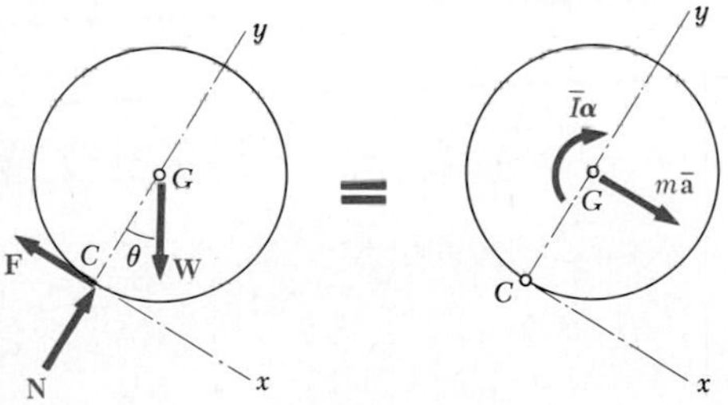

$$+\searrow \Sigma F_x = \Sigma (F_x)_{\text{eff}}: \qquad W\sin\theta - F = m\bar{a}$$
$$W\sin\theta - F = \frac{W}{g}\frac{5g\sin\theta}{7}$$
$$F = +\tfrac{2}{7}W\sin\theta = \tfrac{2}{7}W\sin 30° \qquad \mathbf{F} = 0.143W \searrow 30°$$

$$+\nearrow \Sigma F_y = \Sigma (F_y)_{\text{eff}}: \qquad N - W\cos\theta = 0$$
$$N = W\cos\theta = 0.866W \qquad \mathbf{N} = 0.866W \angle 60°$$

$$\mu_s = \frac{F}{N} = \frac{0.143W}{0.866W} \qquad \mu_s = 0.165 \blacktriangleleft$$

***b.* Velocity of Rolling Sphere.** We have uniformly accelerated motion:

$$\bar{v}_0 = 0 \qquad \bar{a} = 11.50 \text{ ft/s}^2 \qquad \bar{x} = 10 \text{ ft} \qquad \bar{x}_0 = 0$$
$$\bar{v}^2 = \bar{v}_0^2 + 2\bar{a}(\bar{x} - \bar{x}_0) \qquad \bar{v}^2 = 0 + 2(11.50 \text{ ft/s}^2)(10 \text{ ft})$$
$$\bar{v} = 15.17 \text{ ft/s} \qquad \bar{\mathbf{v}} = 15.17 \text{ ft/s} \searrow 30° \blacktriangleleft$$

***c.* Velocity of Sliding Sphere.** Assuming now no friction, we have $F = 0$ and obtain

$$+\circlearrowright \Sigma M_G = \Sigma (M_G)_{\text{eff}}: \qquad 0 = \bar{I}\alpha \qquad \alpha = 0$$

$$+\searrow \Sigma F_x = \Sigma (F_x)_{\text{eff}}: \qquad W\sin 30° = m\bar{a} \qquad 0.50W = \frac{W}{g}\bar{a}$$

$$\bar{a} = +16.1 \text{ ft/s}^2 \qquad \bar{\mathbf{a}} = 16.1 \text{ ft/s}^2 \searrow 30°$$

Substituting $\bar{a} = 16.1 \text{ ft/s}^2$ into the equations for uniformly accelerated motion, we obtain

$$\bar{v}^2 = \bar{v}_0^2 + 2\bar{a}(\bar{x} - \bar{x}_0) \qquad \bar{v}^2 = 0 + 2(16.1 \text{ ft/s}^2)(10 \text{ ft})$$
$$\bar{v} = 17.94 \text{ ft/s} \qquad \bar{\mathbf{v}} = 17.94 \text{ ft/s} \searrow 30° \blacktriangleleft$$

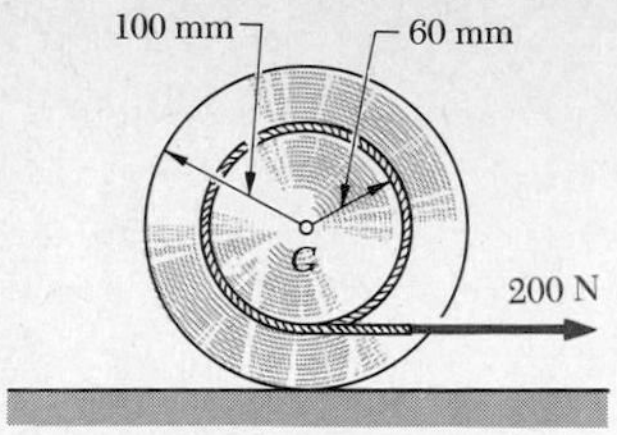

SAMPLE PROBLEM 16.9

A cord is wrapped around the inner drum of a wheel and pulled horizontally with a force of 200 N. The wheel has a mass of 50 kg and a radius of gyration of 70 mm. Knowing that $\mu_s = 0.20$ and $\mu_k = 0.15$, determine the acceleration of G and the angular acceleration of the wheel.

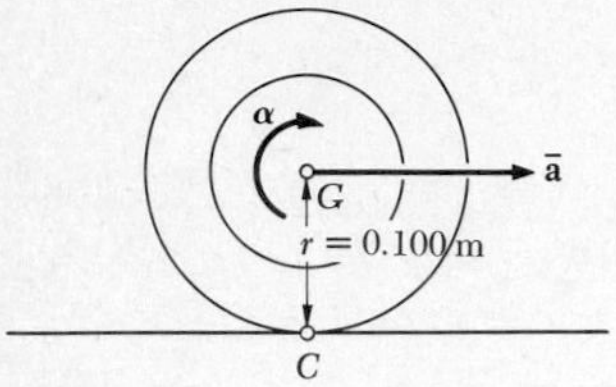

***a.* Assume Rolling without Sliding.** In this case, we have

$$\bar{a} = r\alpha = (0.100 \text{ m})\alpha$$

By comparing the friction force obtained with the maximum available friction force, we shall determine whether this assumption is justified. The moment of inertia of the wheel is

$$\bar{I} = m\bar{k}^2 = (50 \text{ kg})(0.070 \text{ m})^2 = 0.245 \text{ kg}\cdot\text{m}^2$$

Equations of Motion

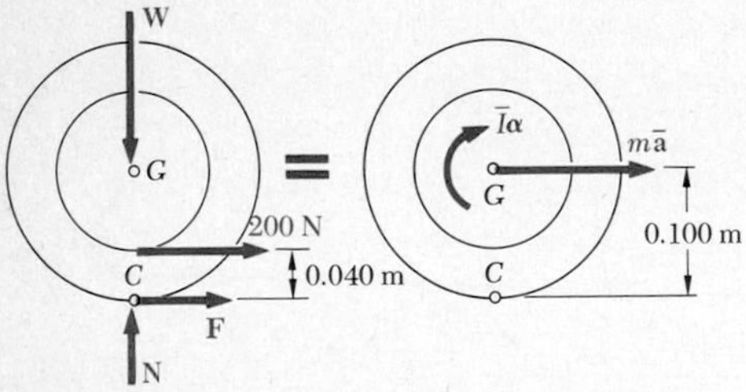

$+\circlearrowright\Sigma M_C = \Sigma(M_C)_{\text{eff}}$: $\quad (200 \text{ N})(0.040 \text{ m}) = m\bar{a}(0.100 \text{ m}) + \bar{I}\alpha$

$$8.00 \text{ N}\cdot\text{m} = (50 \text{ kg})(0.100 \text{ m})\alpha(0.100 \text{ m}) + (0.245 \text{ kg}\cdot\text{m}^2)\alpha$$

$$\alpha = +10.74 \text{ rad/s}^2$$

$$\bar{a} = r\alpha = (0.100 \text{ m})(10.74 \text{ rad/s}^2) = 1.074 \text{ m/s}^2$$

$\overset{+}{\rightarrow}\Sigma F_x = \Sigma(F_x)_{\text{eff}}$: $\quad F + 200 \text{ N} = m\bar{a}$

$$F + 200 \text{ N} = (50 \text{ kg})(1.074 \text{ m/s}^2)$$

$$F = -146.3 \text{ N} \qquad \mathbf{F} = 146.3 \text{ N} \leftarrow$$

$+\uparrow\Sigma F_y = \Sigma(F_y)_{\text{eff}}$:

$N - W = 0 \qquad N - W = mg = (50 \text{ kg})(9.81 \text{ m/s}^2) = 490.5 \text{ N}$

$$\mathbf{N} = 490.5 \text{ N} \uparrow$$

Maximum Available Friction Force

$$F_{\text{max}} = \mu_s N = 0.20(490.5 \text{ N}) = 98.1 \text{ N}$$

Since $F > F_{\text{max}}$, the assumed motion is impossible.

***b.* Rotating and Sliding.** Since the wheel must rotate and slide at the same time, we draw a new diagram, where $\bar{\mathbf{a}}$ and $\boldsymbol{\alpha}$ are independent and where

$$F = F_k = \mu_k N = 0.15(490.5 \text{ N}) = 73.6 \text{ N}$$

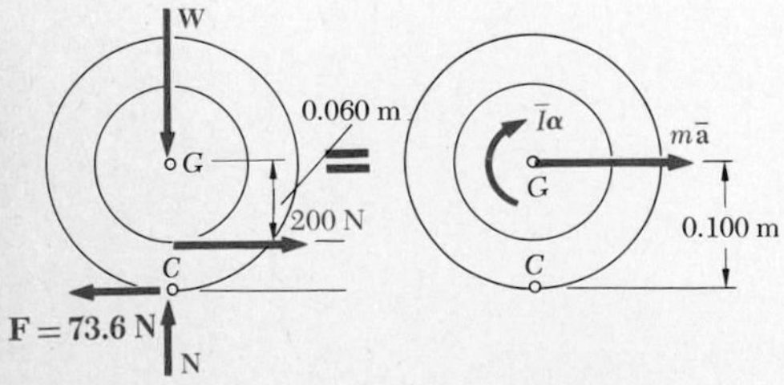

From the computation of part *a*, it appears that **F** should be directed to the left. We write the following equations of motion:

$\overset{+}{\rightarrow}\Sigma F_x = \Sigma(F_x)_{\text{eff}}$: $\quad 200 \text{ N} - 73.6 \text{ N} = (50 \text{ kg})\bar{a}$

$$\bar{a} = +2.53 \text{ m/s}^2 \qquad \bar{\mathbf{a}} = 2.53 \text{ m/s}^2 \rightarrow \blacktriangleleft$$

$+\circlearrowright\Sigma M_G = \Sigma(M_G)_{\text{eff}}$:

$$(73.6 \text{ N})(0.100 \text{ m}) - (200 \text{ N})(0.060 \text{ m}) = (0.245 \text{ kg}\cdot\text{m}^2)\alpha$$

$$\alpha = -18.94 \text{ rad/s}^2 \qquad \boldsymbol{\alpha} = 18.94 \text{ rad/s}^2 \circlearrowleft \blacktriangleleft$$

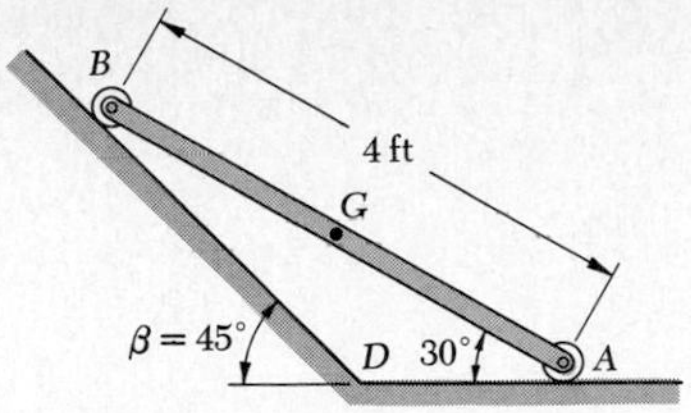

SAMPLE PROBLEM 16.10

The extremities of a 4-ft rod, weighing 50 lb, may move freely and with no friction along two straight tracks as shown. If the rod is released with no velocity from the position shown, determine (*a*) the angular acceleration of the rod, (*b*) the reactions at A and B.

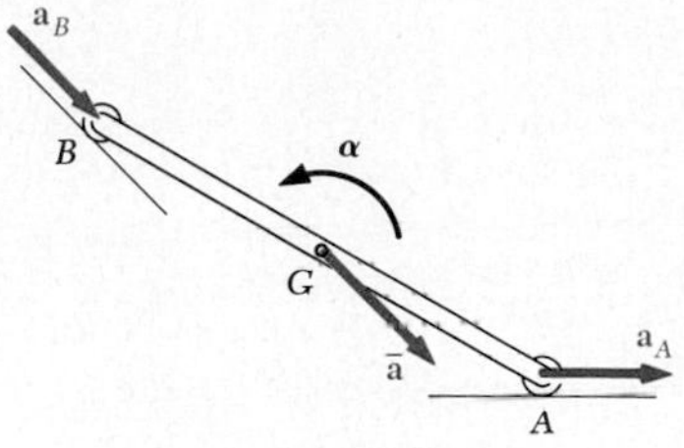

Kinematics of Motion. Since the motion is constrained, the acceleration of G must be related to the angular acceleration $\boldsymbol{\alpha}$. To obtain this relation, we shall first determine the magnitude of the acceleration $\mathbf{a}_A$ of point A in terms of α; assuming $\boldsymbol{\alpha}$ directed counterclockwise and noting that $a_{B/A} = 4\alpha$, we write

$$\mathbf{a}_B = \mathbf{a}_A + \mathbf{a}_{B/A}$$

$$[a_B \searrow 45°] = [a_A \rightarrow] + [4\alpha \swarrow 60°]$$

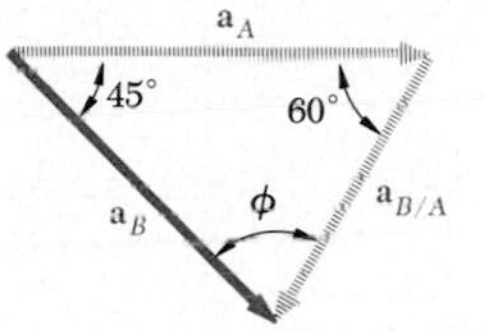

Noting that $\phi = 75°$ and using the law of sines, we obtain

$$a_A = 5.46\alpha \qquad a_B = 4.90\alpha$$

The acceleration of G is now obtained by writing

$$\bar{\mathbf{a}} = \mathbf{a}_G = \mathbf{a}_A + \mathbf{a}_{G/A}$$

$$\bar{\mathbf{a}} = [5.46\alpha \rightarrow] + [2\alpha \swarrow 60°]$$

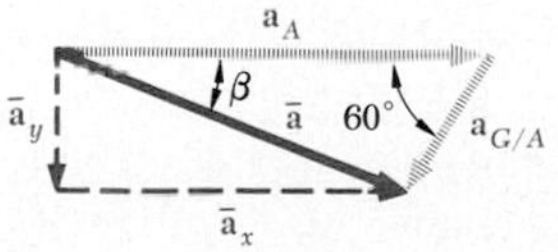

Resolving $\bar{\mathbf{a}}$ into x and y components, we obtain

$$\bar{a}_x = 5.46\alpha - 2\alpha \cos 60° = 4.46\alpha \qquad \bar{\mathbf{a}}_x = 4.46\alpha \rightarrow$$

$$\bar{a}_y = -2\alpha \sin 60° = -1.732\alpha \qquad \bar{\mathbf{a}}_y = 1.732\alpha \downarrow$$

Kinetics of Motion. We draw the two sketches shown to express that the system of external forces is equivalent to the system of effective forces represented by the vector of components $m\bar{\mathbf{a}}_x$ and $m\bar{\mathbf{a}}_y$ attached at G and the couple $\bar{I}\boldsymbol{\alpha}$. We compute the following magnitudes:

$$\bar{I} = \tfrac{1}{12}ml^2 = \frac{1}{12}\frac{50\text{ lb}}{32.2\text{ ft/s}^2}(4\text{ ft})^2 = 2.07\text{ lb}\cdot\text{ft}\cdot\text{s}^2 \qquad \bar{I}\alpha = 2.07\alpha$$

$$m\bar{a}_x = \frac{50}{32.2}(4.46\alpha) = 6.93\alpha \qquad m\bar{a}_y = -\frac{50}{32.2}(1.732\alpha) = -2.69\alpha$$

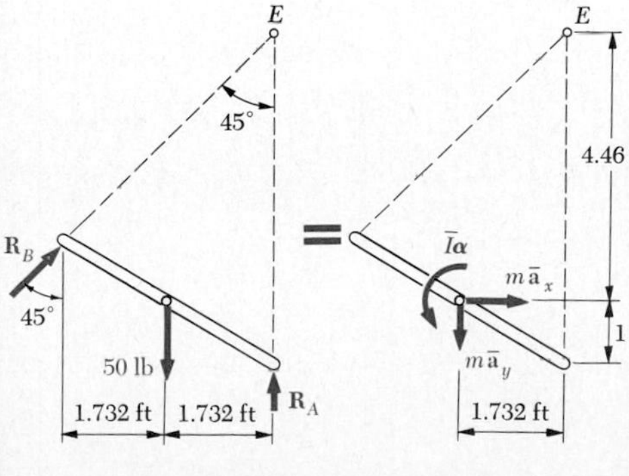

Equations of Motion

$+\circlearrowleft \Sigma M_E = \Sigma(M_E)_{\text{eff}}$:

$$(50)(1.732) = (6.93\alpha)(4.46) + (2.69\alpha)(1.732) + 2.07\alpha$$

$$\alpha = +2.30\text{ rad/s}^2 \qquad \boldsymbol{\alpha} = 2.30\text{ rad/s}^2 \circlearrowleft \blacktriangleleft$$

$\xrightarrow{+} \Sigma F_x = \Sigma(F_x)_{\text{eff}}$:

$$R_B \sin 45° = (6.93)(2.30) = 15.94$$

$$R_B = 22.5\text{ lb} \qquad \mathbf{R}_B = 22.5\text{ lb} \measuredangle 45° \blacktriangleleft$$

$+\uparrow \Sigma F_y = \Sigma(F_y)_{\text{eff}}$:

$$R_A + R_B \cos 45° - 50 = -(2.69)(2.30)$$

$$R_A = -6.19 - 15.94 + 50 = 27.9\text{ lb} \qquad \mathbf{R}_A = 27.9\text{ lb} \uparrow \blacktriangleleft$$

Problems

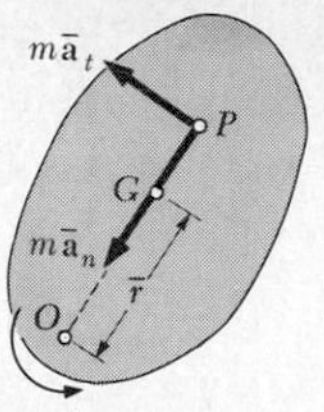

Fig. P16.78

16.78 Show that the couple $\bar{I}\boldsymbol{\alpha}$ of Fig. 16.15 may be eliminated by attaching the vectors $m\bar{\mathbf{a}}_t$ and $m\bar{\mathbf{a}}_n$ at a point P called the *center of percussion*, located on line OG at a distance $GP = \bar{k}^2/\bar{r}$ from the mass center of the body.

16.79 A uniform slender rod, of length $L = 900$ mm and mass $m = 2.5$ kg, hangs freely from a hinge at A. If a force $\mathbf{P}$ of magnitude 15 N is applied at B horizontally to the left ($h = L$), determine (a) the angular acceleration of the rod, (b) the components of the reaction at A.

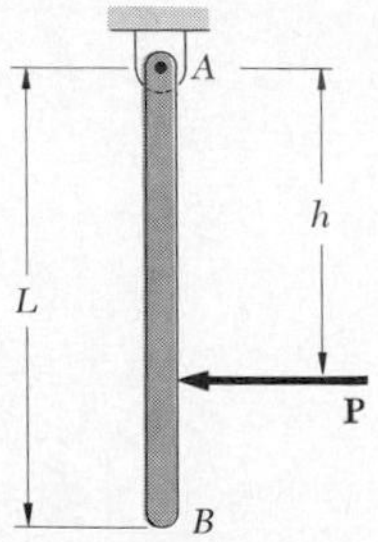

Fig. P16.79

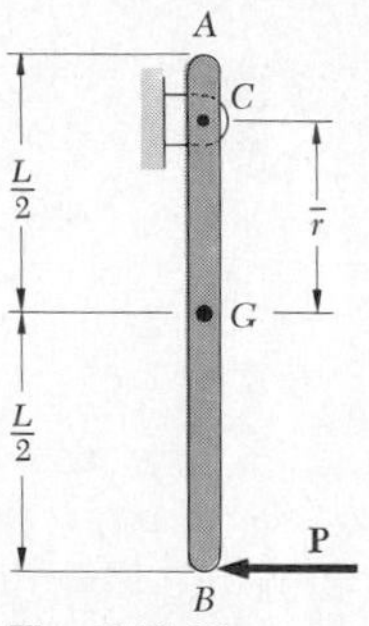

Fig. P16.80

16.80 A uniform slender rod, of length $L = 36$ in. and weight $W = 7.5$ lb, is supported as shown. A horizontal force $\mathbf{P}$ of magnitude 2.5 lb is applied at end B. For $\bar{r} = \frac{1}{4}L = 9$ in., determine (a) the angular acceleration of the rod, (b) the components of the reaction at C.

16.81 In Prob. 16.80, determine (a) the distance $\bar{r}$ for which the horizontal component of the reaction at C is zero, (b) the corresponding angular acceleration of the rod.

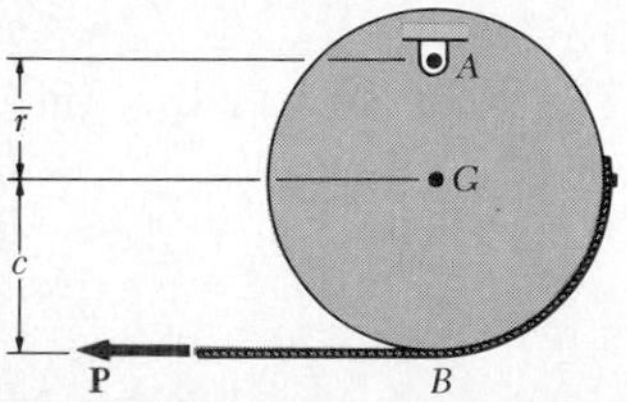

Fig. P16.82

16.82 A uniform disk, of radius $c = 160$ mm and mass $m = 6$ kg, hangs freely from a pin support at A. A force $\mathbf{P}$ of magnitude 20 N is applied as shown to a cord wrapped around the disk. For $\bar{r} = \frac{3}{4}c = 120$ mm, determine (a) the angular acceleration of the disk, (b) the components of the reaction at A.

16.83 In Prob. 16.82, determine (a) the distance $\bar{r}$ for which the horizontal component of the reaction at A is zero, (b) the corresponding angular acceleration of the disk.

16.84 In Prob. 16.79, determine (a) the distance h for which the horizontal component of the reaction at A is zero, (b) the corresponding angular acceleration of the rod.

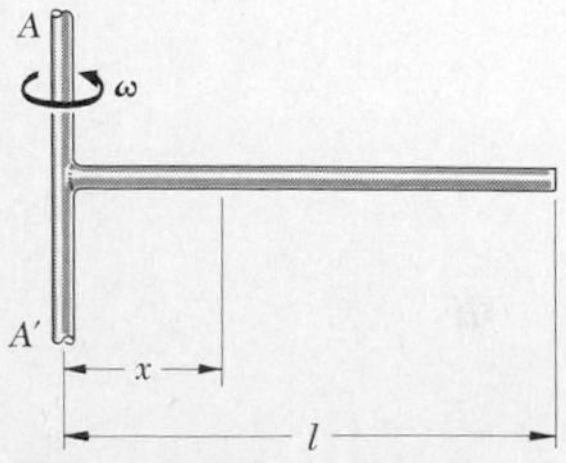

Fig. P16.85

16.85 A uniform slender rod of length l and mass m rotates about a vertical axis AA' at a constant angular velocity $\boldsymbol{\omega}$. Determine the tension in the rod at a distance x from the axis of rotation.

16.86 The rim of a flywheel has a mass of 1800 kg and a mean radius of 600 mm. As the flywheel rotates at a constant angular velocity of 360 rpm, radial forces are exerted on the rim by the spokes and internal forces are developed within the rim. Neglecting the weight of the spokes, determine (*a*) the internal forces in the rim, assuming the radial forces exerted by the spokes to be zero, (*b*) the radial force exerted by each spoke, assuming the tangential forces in the rim to be zero.

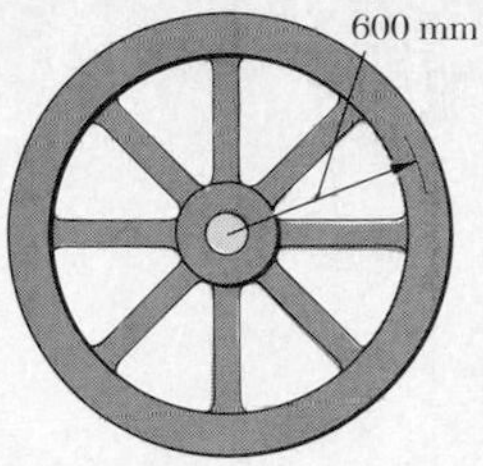

Fig. P16.86

16.87 A large flywheel is mounted on a horizontal shaft and rotates at a constant rate of 900 rpm. Experimental data show that the total force exerted by the flywheel on the shaft varies from 6000 lb upward to 9000 lb downward. Determine (*a*) the weight of the flywheel, (*b*) the distance from the center of the shaft to the mass center of the flywheel.

16.88 A portion of a circular cylindrical shell forms a small vane which is welded to the vertical shaft AB. The vane and shaft rotate about the y axis with a constant angular velocity of 180 rpm counterclockwise. Knowing that the vane weighs 4 lb, determine the horizontal components of the reaction at A.

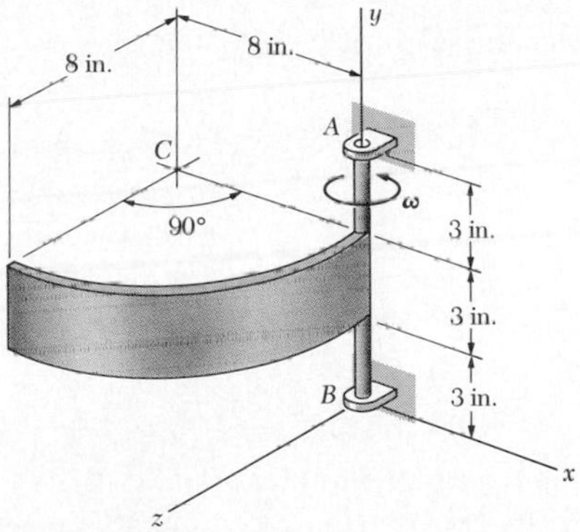

Fig. P16.88

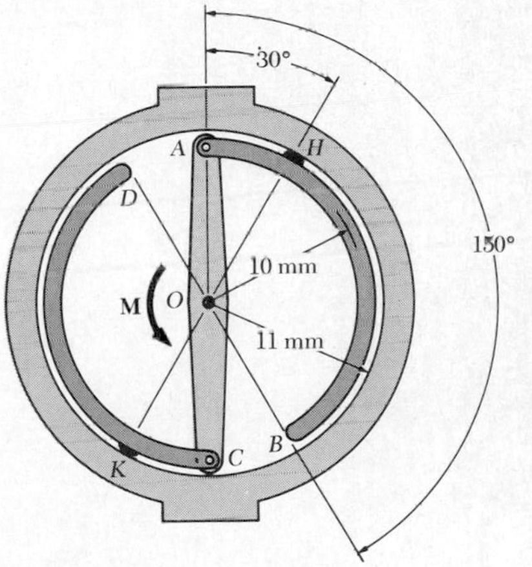

Fig. P16.89

16.89 Centrifugal clutches of the type shown are used to control the operating speed of equipment such as movie cameras and dial telephones. Thin curved members AB and CD are connected by pins at A and C to the arm AC which may rotate about a fixed point O. Each of the members AB and CD has a mass of 4.5 g and a radius of 10 mm. As the clutch rotates counterclockwise, knobs H and K slide on the inside of a fixed cylindrical surface of radius 11 mm. Knowing that the coefficient of kinetic friction at H and K is 0.35 and that the clutch is to have a constant angular velocity of 3000 rpm, determine the couple **M** which must be applied to arm AC.

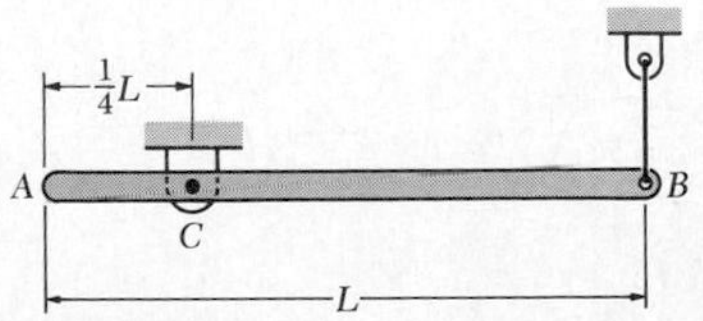

Fig. P16.91

16.90 For the centrifugal clutch of Prob. 16.89, determine the constant angular velocity which will result from the application of a couple **M** of constant magnitude 50 N·mm.

16.91 and 16.92 A uniform beam of length L and weight W is supported as shown. If the cable suddenly breaks, determine (*a*) the acceleration of end B, (*b*) the reaction at the pin support.

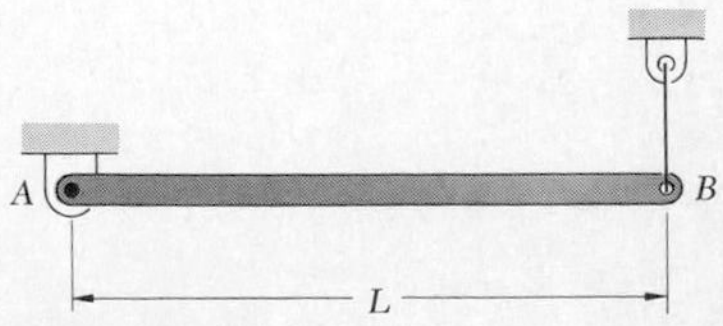

Fig. P16.92

16.93 A uniform square plate of weight W is supported as shown. If the cable suddenly breaks, determine (*a*) the angular acceleration of the plate, (*b*) the acceleration of corner C, (*c*) the reaction at A.

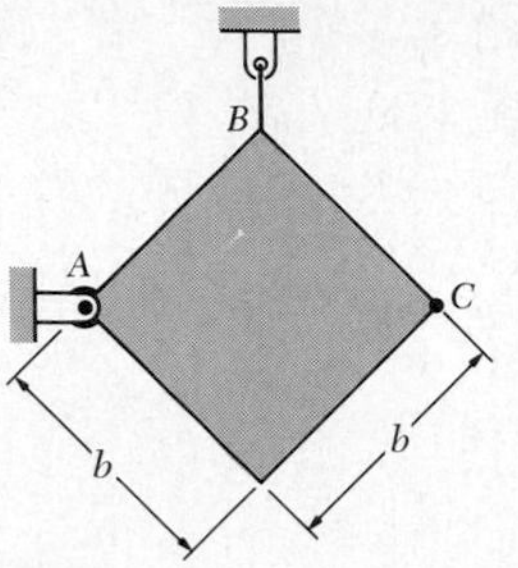

Fig. P16.93

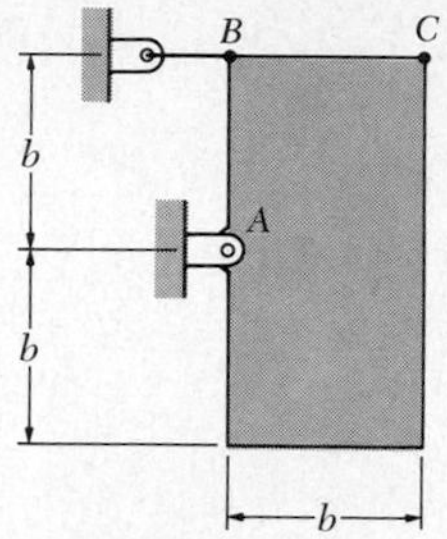

Fig. P16.94

16.94 A uniform rectangular plate of weight W is supported as shown. If the cable suddenly breaks, determine (*a*) the angular acceleration of the plate, (*b*) the acceleration of corner C, (*c*) the reaction at A.

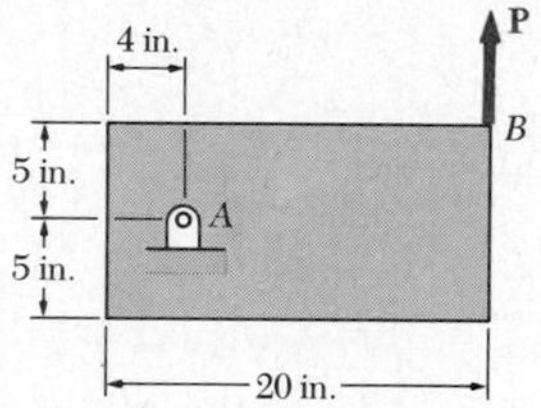

Fig. P16.95 and P16.96

16.95 An 8-lb uniform plate swings freely about A in a vertical plane. Knowing that $\mathbf{P} = 0$ and that in the position shown the plate has an angular velocity of 15 rad/s counterclockwise, determine (*a*) the angular acceleration of the plate, (*b*) the components of the reaction at A.

16.96 An 8-lb uniform plate rotates about A in a vertical plane under the combined effect of gravity and of the vertical force $\mathbf{P}$. Knowing that at the instant shown the plate has an angular velocity of 25 rad/s and an angular acceleration of 20 rad/s^2 both counterclockwise, determine (*a*) the force $\mathbf{P}$, (*b*) the components of the reaction at A.

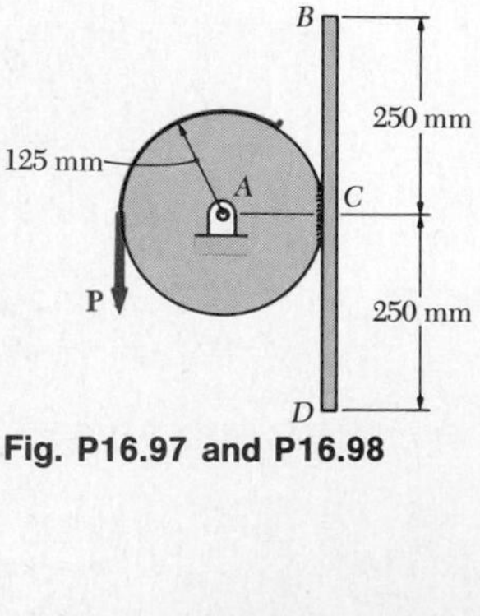

Fig. P16.97 and P16.98

16.97 A 3-kg slender rod is welded to the edge of a 2-kg uniform disk as shown. The assembly rotates about A in a vertical plane under the combined effect of gravity and of the vertical force $\mathbf{P}$. Knowing that at the instant shown the assembly has an angular velocity of 12 rad/s and an angular acceleration of 24 rad/s^2 both counterclockwise, determine (*a*) the force $\mathbf{P}$, (*b*) the components of the reaction at A.

16.98 A 3-kg slender rod is welded to the edge of a 2-kg uniform disk as shown. The assembly swings freely about A in a vertical plane. Knowing that $\mathbf{P} = 0$ and that in the position shown the assembly has an angular velocity of 16 rad/s counterclockwise, determine (*a*) the angular acceleration of the assembly, (*b*) the components of the reaction at A.

16.99 Two uniform rods, ABC of mass 3 kg and DCE of mass 4 kg, are connected by a pin at C and by two cords BD and BE. The T-shaped assembly rotates in a vertical plane under the combined effect of gravity and of a couple **M** which is applied to rod ABC. Knowing that at the instant shown the tension is 8 N in cord BD and zero in cord BE, determine (*a*) the angular acceleration of the assembly, (*b*) the couple **M**.

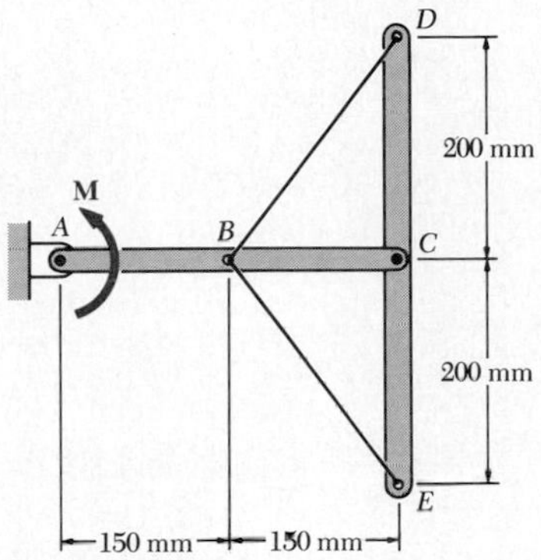

Fig. P16.99

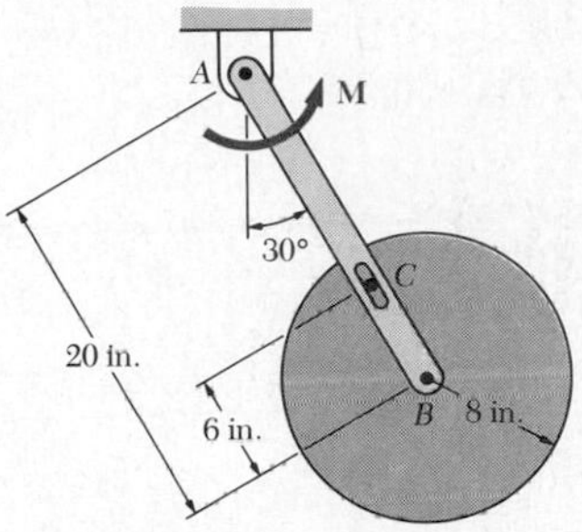

Fig. P16.100

16.100 A 16-lb uniform disk is attached to the 9-lb slender rod AB by means of frictionless pins at B and C. The assembly rotates in a vertical plane under the combined effect of gravity and of a couple **M** which is applied to rod AB. Knowing that at the instant shown the assembly has an angular velocity of 6 rad/s and an angular acceleration of 15 rad/s^2, both counterclockwise, determine (*a*) the couple **M**, (*b*) the force exerted by pin C on member AB.

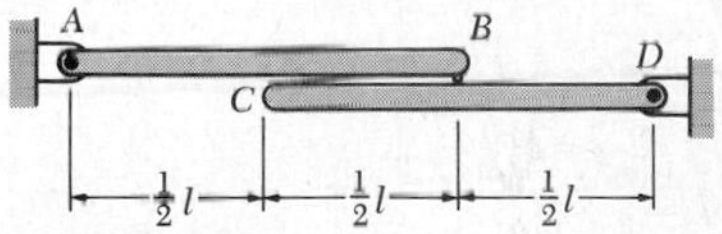

Fig. P16.101

16.101 Two slender rods, each of length l and mass m, are released from rest in the position shown. Knowing that a small frictionless knob at end B of rod AB bears on rod CD, determine immediately after release, (*a*) the acceleration of end C of rod CD, (*b*) the force exerted on knob B.

16.102 A 3.5-m plank of mass 30 kg rests on two horizontal pipes AB and CD of a scaffolding. The pipes are 2.5 m apart, and the plank overhangs 0.5 m at each end. A 70-kg worker is standing on the plank when pipe CD suddenly breaks. Determine the initial acceleration of the worker, knowing that he was standing (*a*) in the middle of the plank, (*b*) just above pipe CD.

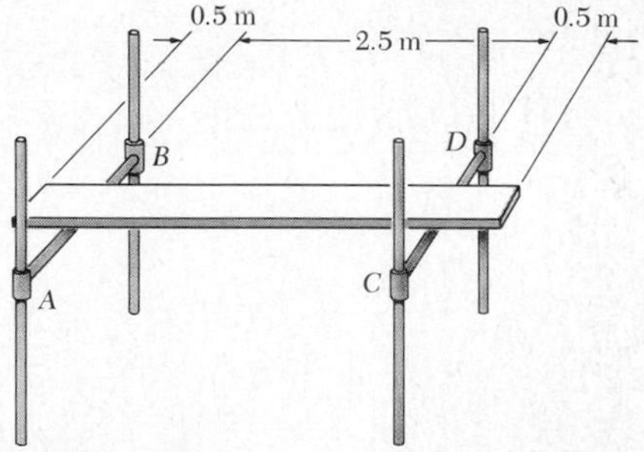

Fig. P16.102

16.103 The uniform rod AB of mass m is released from rest when $\beta = 60°$. Assuming that the friction force between end A and the surface is large enough to prevent sliding, determine (*a*) the angular acceleration of the rod just after release, (*b*) the normal reaction and the friction force at A, (*c*) the minimum value of μ_s compatible with the described motion.

***16.104** Knowing that the coefficient of static friction between the rod and the floor is 0.30, determine the range of values of β for which the rod *will* slip immediately after being released from rest.

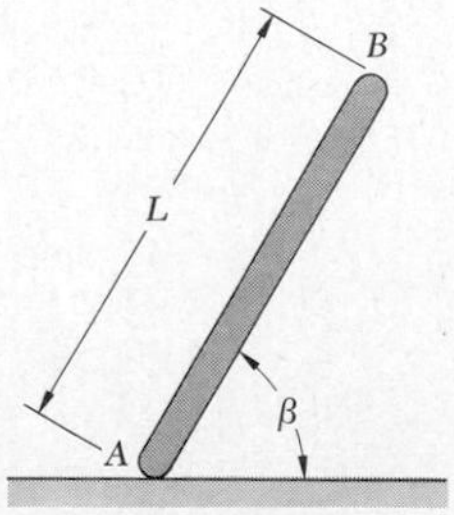

Fig. P16.103 and P16.104

16.105 Derive the equation $\Sigma M_C = I_C\alpha$ for the rolling disk of Fig. 16.17, where ΣM_C represents the sum of the moments of the external forces about the instantaneous center C and I_C the moment of inertia of the disk about C.

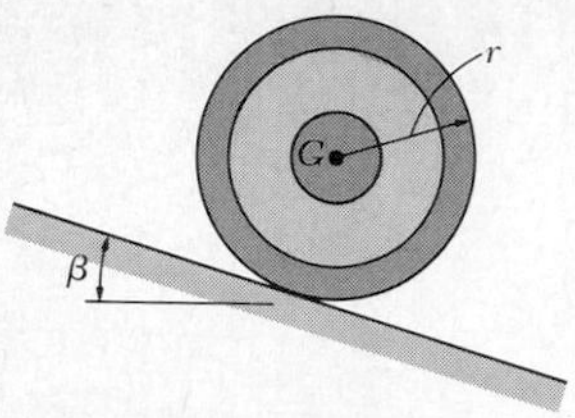

Fig. P16.107

16.106 Show that in the case of an unbalanced disk, the equation derived in Prob. 16.105 is valid only when the mass center G, the geometric center O, and the instantaneous center C happen to lie in a straight line.

16.107 A wheel of radius r and centroidal radius of gyration $\bar{k}$ is released from rest on the incline and rolls without slipping. Derive an expression for the acceleration of the center of the wheel in terms of r, $\bar{k}$, β, and g.

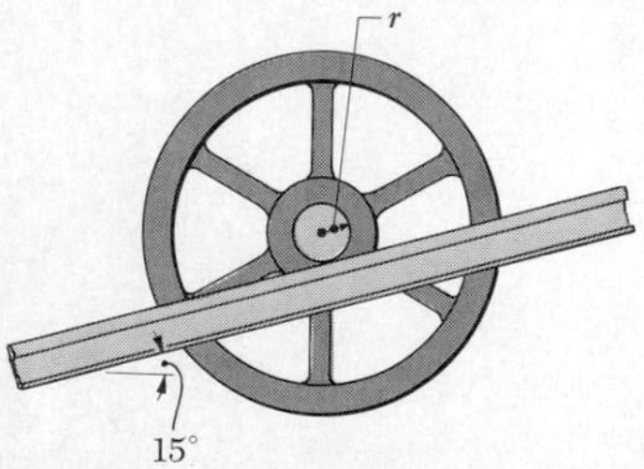

Fig. P16.108 and P16.109

16.108 A flywheel is rigidly attached to a shaft of 50-mm radius which may roll along parallel rails as shown. When released from rest, the system rolls through a distance of 3.5 m in 20 s. Determine the centroidal radius of gyration of the system.

16.109 A flywheel of centroidal radius of gyration $\bar{k} = 400$ mm is rigidly attached to a shaft of radius $r = 25$ mm which may roll along parallel rails. Knowing that the system is released from rest, determine the distance through which it will roll in 15 s.

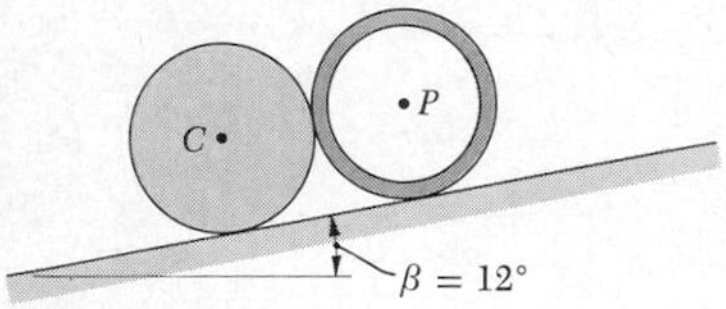

Fig. P16.110

16.110 A homogeneous cylinder C and a section of pipe P are in contact when they are released from rest. Knowing that both the cylinder and the pipe roll without slipping, determine the clear distance between them after 3 s.

16.111 through 16.114 A drum of 4-in. radius is attached to a disk of 8-in. radius. The disk and drum have a total weight of 10 lb and a radius of gyration of 6 in. A cord is attached as shown and pulled with a force **P** of magnitude 5 lb. Knowing that the disk rolls without sliding, determine (*a*) the angular acceleration of the disk and the acceleration of G, (*b*) the minimum value of the coefficient of static friction compatible with this motion.

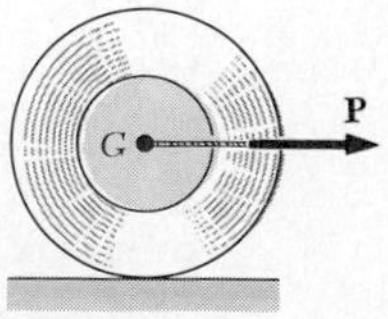

Fig. P16.111 and P16.115

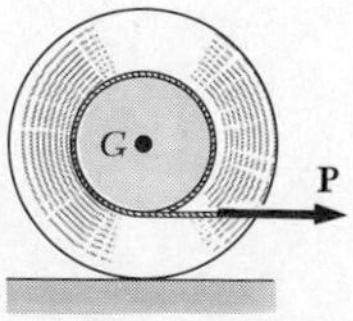

Fig. P16.112 and P16.116

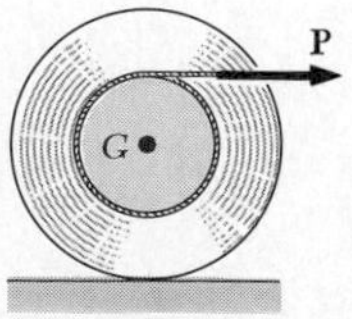

Fig. P16.113 and P16.117

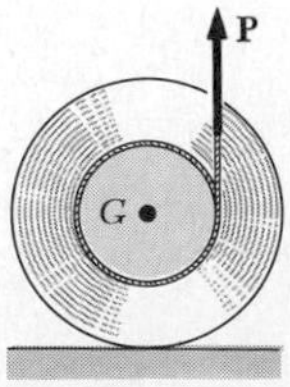

Fig. P16.114 and P16.118

16.115 through 16.118 A drum of 80-mm radius is attached to a disk of 160-mm radius. The disk and drum have a total mass of 5 kg and a radius of gyration of 120 mm. A cord is attached as shown and pulled with a force **P** of magnitude 18 N. Knowing that the coefficients of static and kinetic friction are $\mu_s = 0.20$ and $\mu_k = 0.15$, determine (*a*) whether or not the disk slides, (*b*) the angular acceleration of the disk and the acceleration of G.

16.119 and 16.120 The 9-kg carriage is supported as shown by two uniform disks each of mass 6 kg and radius 150 mm. Knowing that the disks roll without sliding, determine the acceleration of the carriage when a force of 30 N is applied to it.

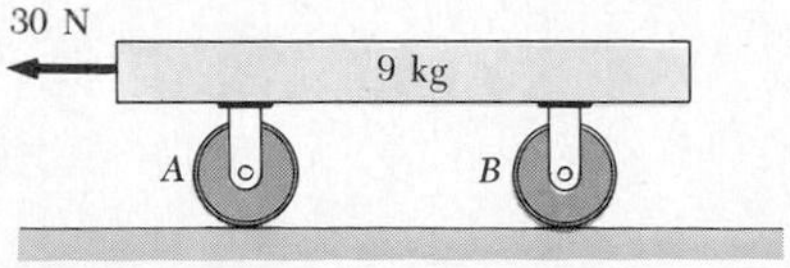

Fig. P16.119

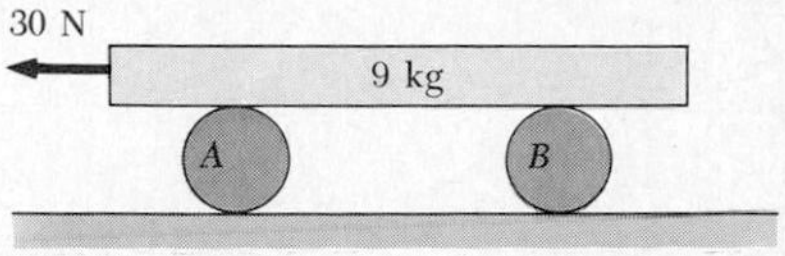

Fig. P16.120

16.121 and 16.122 Gear C has a mass of 3 kg and a centroidal radius of gyration of 75 mm. The uniform bar AB has a mass of 2.5 kg and gear D is stationary. If the system is released from rest in the position shown, determine (a) the angular acceleration of gear C, (b) the acceleration of point B.

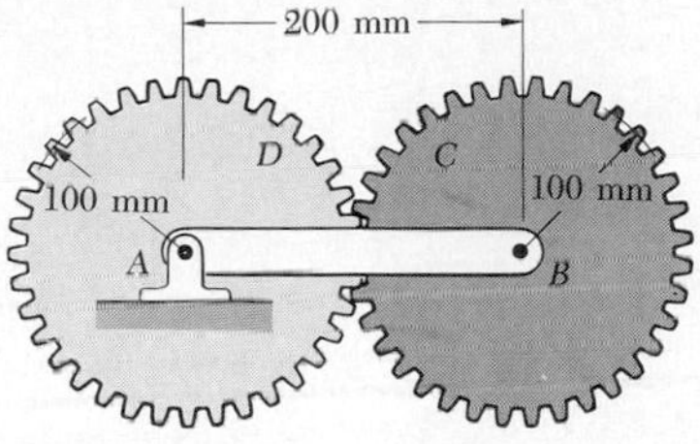

Fig. P16.121

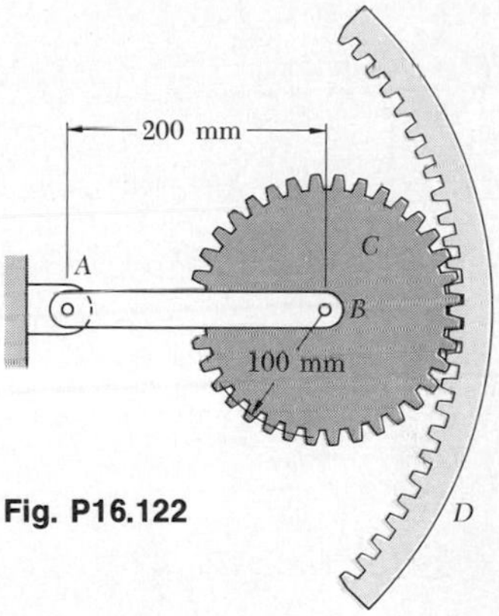

Fig. P16.122

16.123 A block B of mass m is attached to a cord wrapped around a cylinder of the same mass m and of radius r. The cylinder rolls without sliding on a horizontal surface. Determine the components of the accelerations of the center A of the cylinder and of the block B immediately after the system has been released from rest if (a) the block hangs freely, (b) the motion of the block is guided by a rigid member DAE, frictionless and of negligible mass, which is hinged to the cylinder at A.

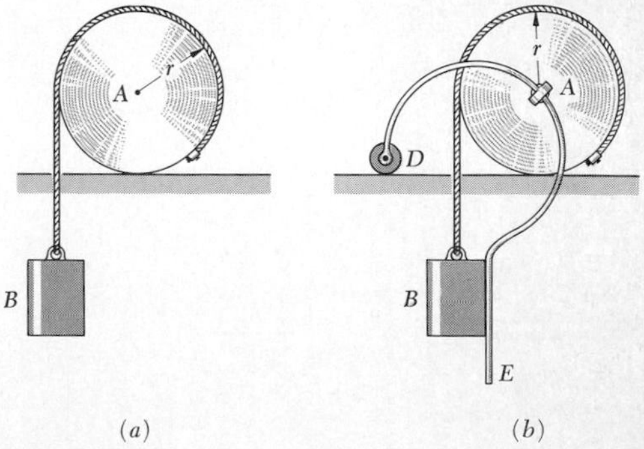

Fig. P16.123

16.124 The disk-and-drum assembly A has a total weight of 15 lb and a centroidal radius of gyration of 6 in. A cord is attached as shown to the assembly A and to the 10-lb uniform disk B. Knowing that the disks roll without sliding, determine the acceleration of the center of each disk for $P = 4$ lb and $Q = 0$.

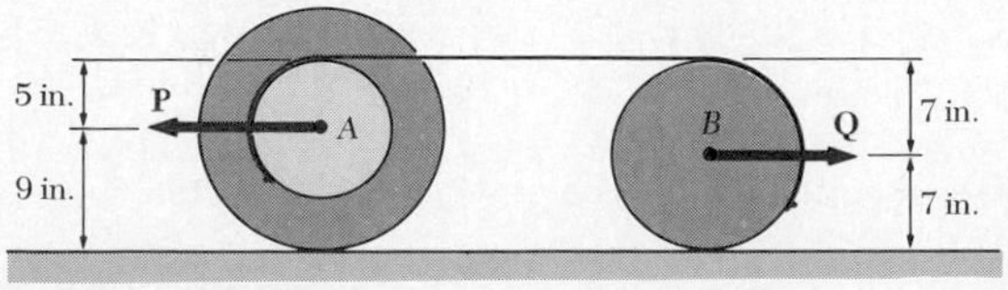

Fig. P16.124

16.125 Solve Prob. 16.124 for $P = 0$ and $Q = 4$ lb.

16.126 Two uniform disks A and B, each of weight 4 lb, are connected by a 3-lb rod CD as shown. A counterclockwise couple **M** of magnitude 1.5 lb·ft is applied to disk A. Knowing that the disks roll without sliding, determine (*a*) the acceleration of the center of each disk, (*b*) the horizontal component of the force exerted on disk B by pin D.

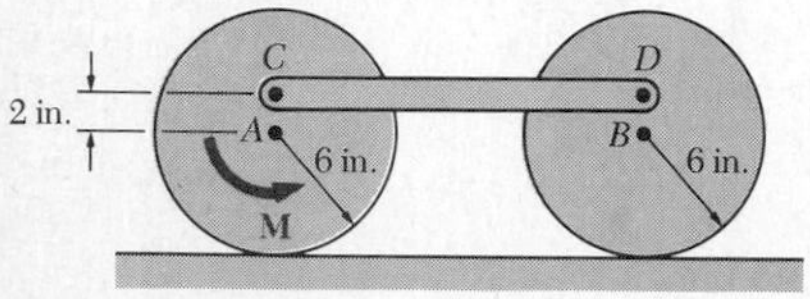

Fig. P16.126

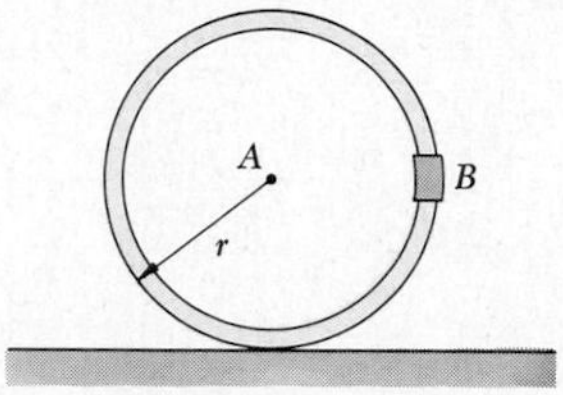

Fig. P16.127

16.127 A small block of mass m is attached at B to a hoop of mass m and radius r. The system is released from rest when B is directly above A and rolls without sliding. Knowing that at the instant shown the system has a clockwise angular velocity of magnitude $\omega = \sqrt{g/2r}$, determine (*a*) the angular acceleration of the hoop, (*b*) the acceleration of B.

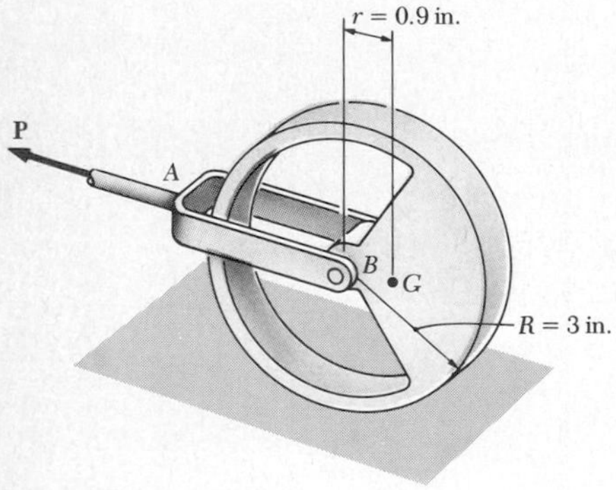

Fig. P16.128

16.128 The center of gravity G of a 2.5-lb unbalanced tracking wheel is located at a distance $r = 0.9$ in. from its geometric center B. The radius of the wheel is $R = 3$ in. and its centroidal radius of gyration is $\bar{k} = 2.2$ in. At the instant shown the center B of the wheel has a velocity of 1.5 ft/s and an acceleration of 5 ft/s^2, both directed to the left. Knowing that the wheel rolls without sliding and neglecting the weight of the driving yoke AB, determine the horizontal force **P** applied to the yoke.

16.129 For the wheel of Prob. 16.128, determine the horizontal force **P** applied to the yoke, knowing that at the instant shown the center B has a velocity of 1.5 ft/s and an acceleration of 5 ft/s^2, both directed to the right.

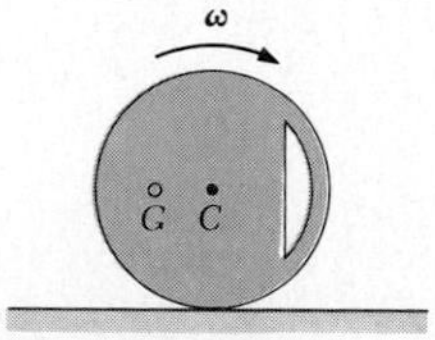

Fig. P16.130

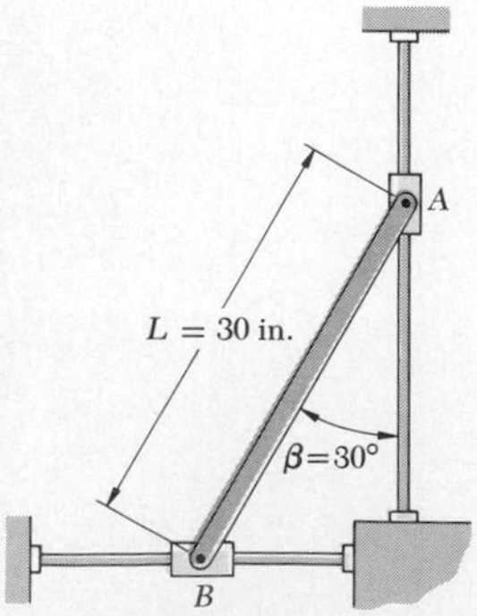

Fig. P16.131

16.130 The mass center G of a 5-kg wheel of radius $R = 300$ mm is located at a distance $r = 100$ mm from its geometric center C. The centroidal radius of gyration is $\bar{k} = 150$ mm. As the wheel rolls without sliding, its angular velocity varies and it is observed that $\omega = 8$ rad/s in the position shown. Determine the corresponding angular acceleration of the wheel.

16.131 The ends of the 6-lb slender rod AB are attached to collars of negligible weight which slide without friction along fixed rods. If rod AB is released from rest in the position shown, determine immediately after release (*a*) the angular acceleration of the rod, (*b*) the reaction at A, (*c*) the reaction at B.

16.132 The motion of the 3-kg uniform rod ACB is guided by two blocks of negligible mass which slide without friction in the slots shown. If the rod is released from rest in the position shown, determine immediately after release (*a*) the angular acceleration of the rod, (*b*) the reaction at A.

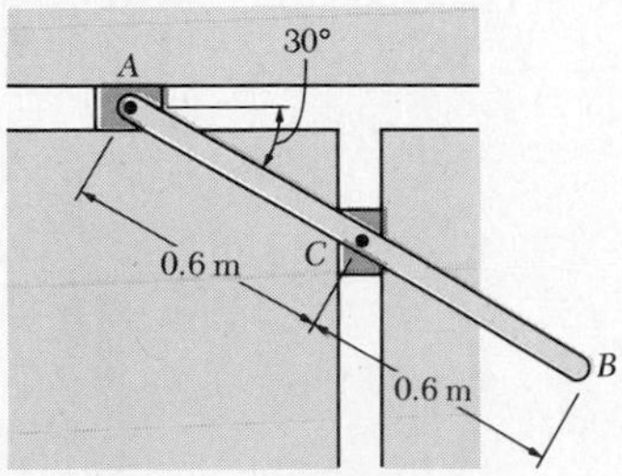

Fig. P16.132 and P16.133

16.133 The motion of the 3-kg uniform rod ACB is guided by two blocks of negligible mass which slide without friction in the slots shown. A horizontal force **P** is applied to block A, causing the rod to start from rest with a counterclockwise angular acceleration of 12 rad/s^2. Determine (*a*) the required force **P**, (*b*) the corresponding reaction at A.

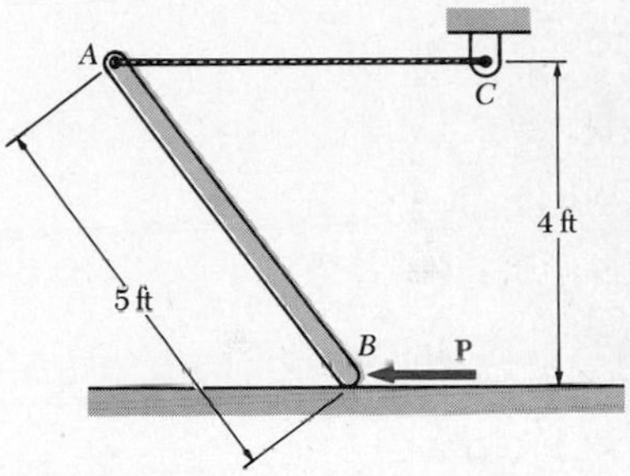

Fig. P16.134

16.134 End B of the 15-lb uniform rod AB rests on a frictionless floor, while end A is attached to a horizontal cable AC. Knowing that at the instant shown the force **P** causes end B of the rod to start from rest with an acceleration of 9 ft/s^2 to the left, determine (*a*) the force **P**, (*b*) the corresponding tension in cable AC.

16.135 End A of the 5-kg uniform rod AB rests on the inclined surface, while end B is attached to a collar of negligible mass which may slide along the vertical rod shown. Knowing that the rod is released from rest when $\theta = 35°$ and neglecting the effect of friction, determine immediately after release (*a*) the angular acceleration of the rod, (*b*) the reaction at B.

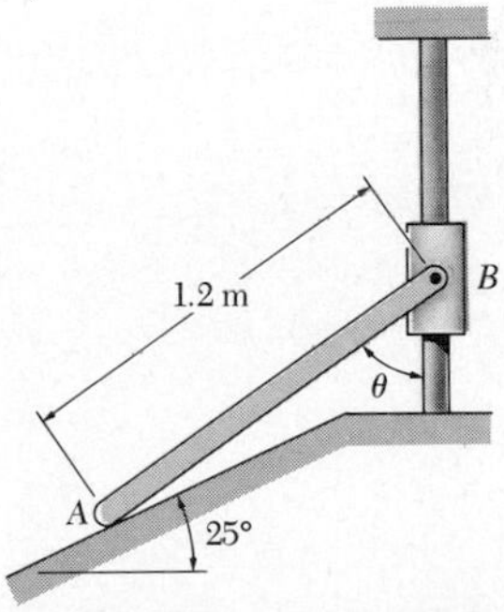

Fig. P16.135 and P16.138

16.136 Rod AB weighs 3 lb and is released from rest in the position shown. Assuming that the ends of the rod slide without friction, determine immediately after release (*a*) the angular acceleration of the rod, (*b*) the reaction at B.

16.137 Ends A and B of the 3-lb uniform rod slide without friction along the surfaces shown. When the rod is at rest a horizontal force **P** is applied at A, causing end A of the rod to start moving to the right with an acceleration of 8 ft/s^2. Determine (*a*) the force **P**, (*b*) the corresponding reaction at B.

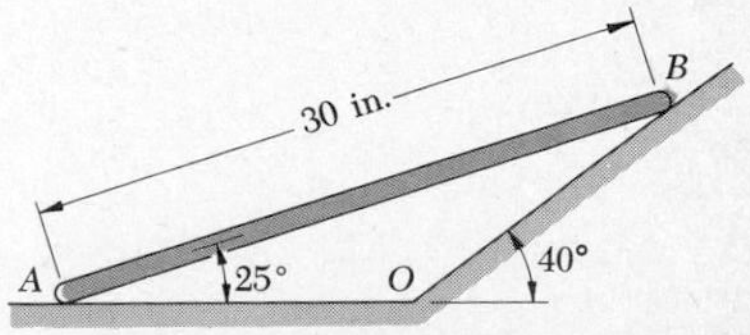

Fig. P16.136 and P16.137

16.138 End A of the 5-kg uniform rod AB rests on the inclined surface, while end B is attached to a collar of negligible mass which may slide along the vertical rod shown. When the rod is at rest a vertical force **P** is applied at B, causing end B of the rod to start moving upward with an acceleration of 4 m/s^2. Knowing that $\theta = 35°$, determine the force **P**.

16.139 The 5-kg uniform rod BD is connected as shown to crank AB and to a collar of negligible mass. Crank AB has a constant angular velocity of 20 rad/s counterclockwise. Neglecting the effect of friction, determine the reaction at D when $\theta = 0$.

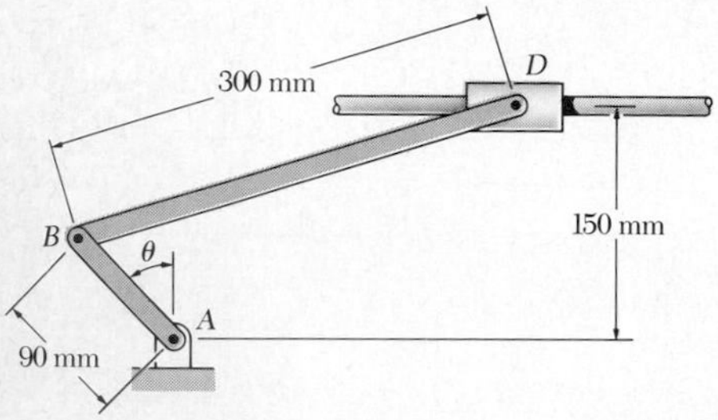

Fig. P16.139

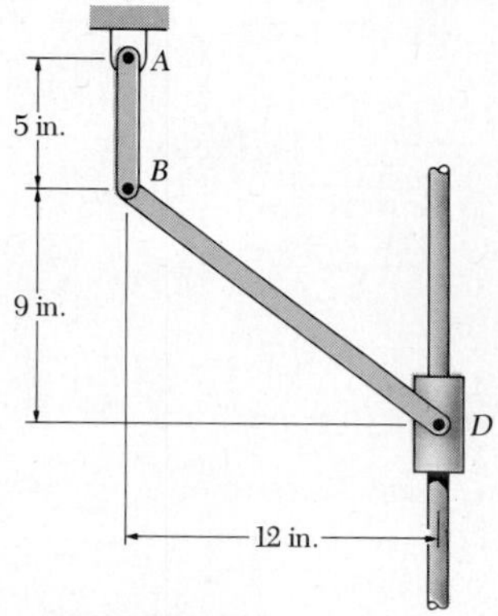

Fig. P16.140

16.140 The 1.5-lb uniform rod BD is connected to crank AB and to a collar of negligible weight. In the position shown crank AB has an angular velocity of 9 rad/s counterclockwise and an angular acceleration of 45 rad/s^2 clockwise. Neglecting the effect of friction, determine the reaction at D.

16.141 In Prob. 16.140, determine the reaction at D, knowing that in the position shown crank AB has an angular velocity of 9 rad/s and an angular acceleration of 45 rad/s^2, both counterclockwise.

16.142 Solve Prob. 16.139 when $\theta = 180°$.

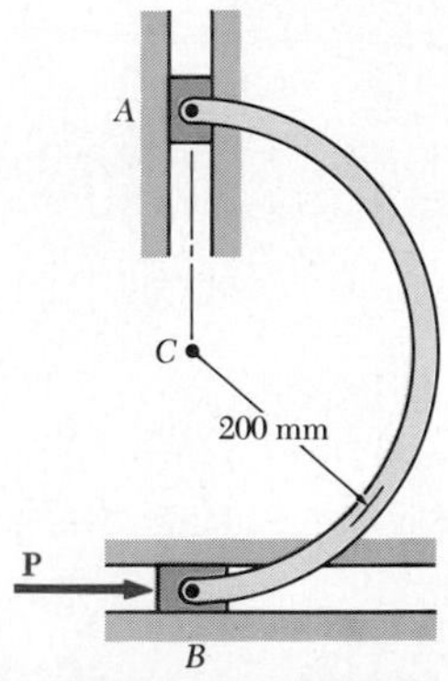

Fig. P16.143 and P16.144

16.143 The motion of a 1.5-kg semicircular rod is guided by two blocks of negligible mass which slide without friction in the slots shown. A horizontal and variable force **P** is applied at B, causing B to move to the right with a constant speed of 5 m/s. For the position shown, determine (*a*) the force **P**, (*b*) the reaction at B.

16.144 The motion of a 1.5-kg semicircular rod is guided by two blocks of negligible mass which slide without friction in the slots shown. Knowing that for the position shown $\mathbf{P} = 0$ and the speed of B is 5 m/s to the right, determine (*a*) the acceleration of B, (*b*) the reaction at B.

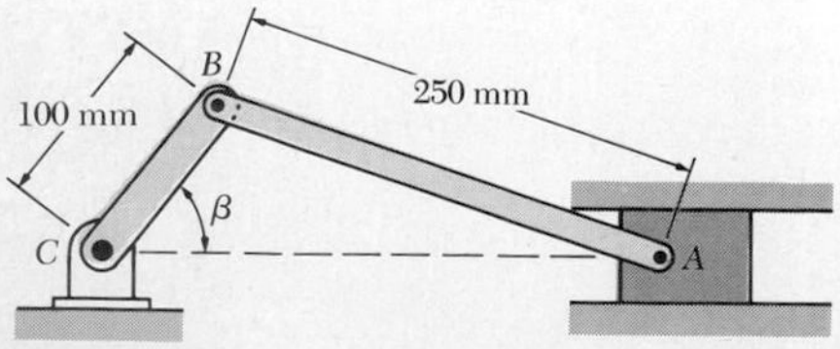

Fig. P16.145

16.145 The 1.8-kg sliding block is connected to crank BC by the uniform rod AB of mass 1.2 kg. Knowing that the crank has a constant angular velocity of 600 rpm, determine the forces exerted on the rod at A and B when $\beta = 0$. Assume that the motion takes place in a horizontal plane.

16.146 Solve Prob. 16.145 when $\beta = 180°$.

16.147 Two rods AB and BC, of mass m' per unit length, are connected as shown to a disk which is made to rotate in a vertical plane at a constant angular velocity ω_0. For the position shown, determine the components of the forces exerted at A and B on rod AB.

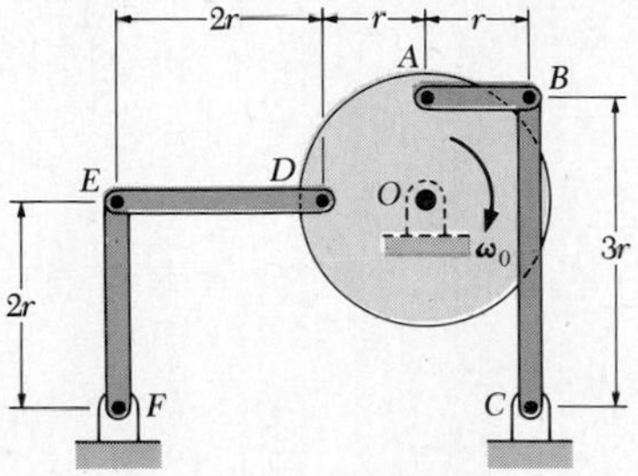

Fig. P16.147 and P16.148

16.148 Two rods DE and EF, of mass m' per unit length, are connected as shown to a disk which is made to rotate in a vertical plane at a constant angular velocity ω_0. For the position shown, determine the components of the forces exerted at D and E on rod DE.

***16.149 and *16.150** The 3-kg cylinder B and the 2-kg wedge A are held at rest in the position shown by the cord C. Assuming that the cylinder rolls without sliding on the wedge and neglecting friction between the wedge and the ground, determine (a) the acceleration of the wedge, (b) the angular acceleration of the cylinder, immediately after the cord C has been cut.

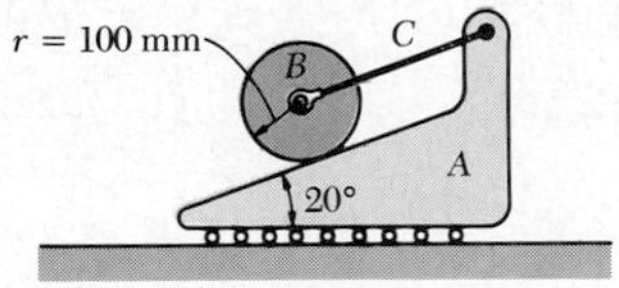

Fig. P16.149

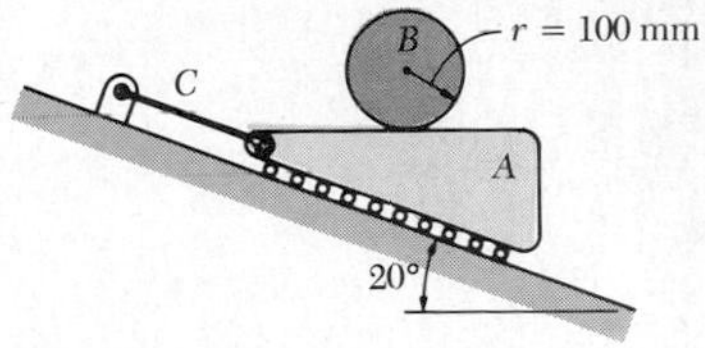

Fig. P16.150

***16.151 and *16.152** The uniform rod AB, of weight 20 lb and length 3 ft, is attached to the 30-lb cart C. Neglecting friction, determine immediately after the system has been released from rest, (a) the acceleration of the cart, (b) the angular acceleration of the rod.

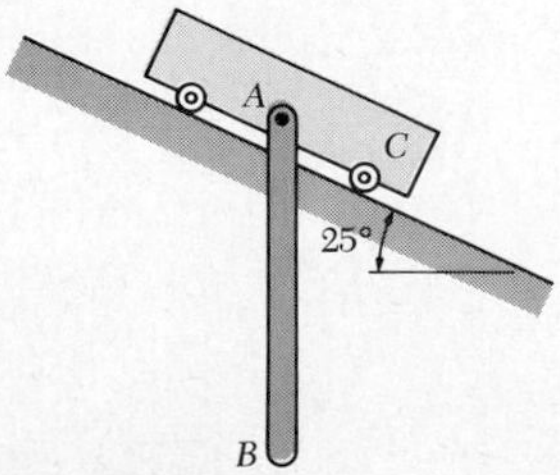

Fig. P16.151

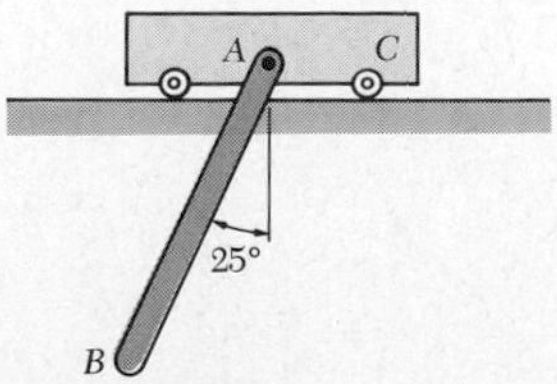

Fig. P16.152

***16.153** A uniform slender rod of length L and mass m is released from rest in the position shown. Derive an expression for (*a*) the angular acceleration of the rod, (*b*) the acceleration of end A, (*c*) the reaction at A, immediately after release. Neglect the mass and friction of the roller at A.

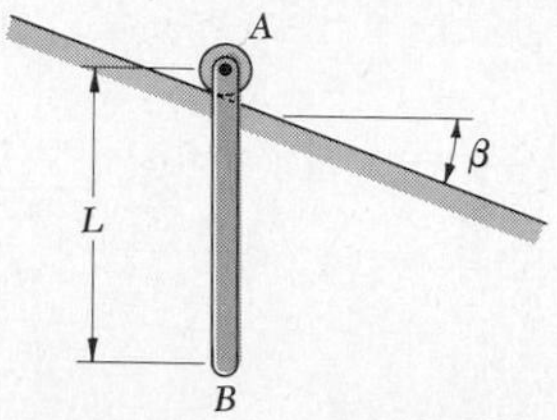

Fig. P16.153 and P16.154

***16.154** A uniform rod AB, of mass 3 kg and length $L = 1.2$ m, is released from rest in the position shown. Knowing that $\beta = 30°$, determine the values immediately after release of (*a*) the angular acceleration of the rod, (*b*) the acceleration of end A, (*c*) the reaction at A. Neglect the mass and friction of the roller at A.

***16.155** Each of the bars AB and BC is of length $L = 18$ in. and weight 3 lb. A horizontal force **P** of magnitude 4 lb is applied at C. Determine the angular acceleration of each bar.

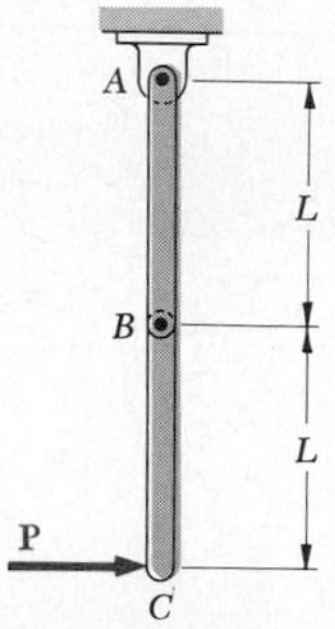

Fig. P16.155

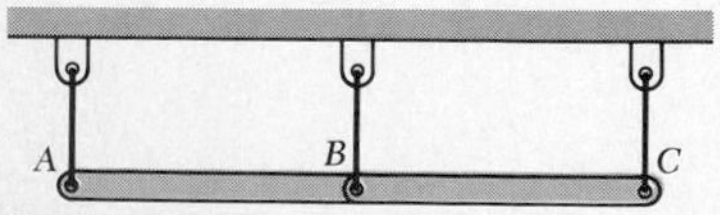

Fig. P16.156

***16.156** Two identical uniform rods are connected by a pin at B and are held in a horizontal position by three wires as shown. If the wires attached at A and B are cut simultaneously, determine at that instant the acceleration of (*a*) point A, (*b*) point B.

***16.157** (*a*) Determine the magnitude and the location of the maximum bending moment in the rod of Prob. 16.79. (*b*) Show that the answer to part *a* is independent of the mass of the rod.

***16.158** Draw the shear and bending-moment diagrams for the beam of Prob. 16.92 immediately after the cable at B breaks.

Review Problems

16.159 A uniform disk of mass $m = 3$ kg and radius $r = 125$ mm is supported by the belt *ABCD* which is bolted to the disk at *B* and *C*. If the belt suddenly breaks at a point between *A* and *B*, determine at that instant (*a*) the acceleration of the center of the disk, (*b*) the tension in portion *CD* of the belt.

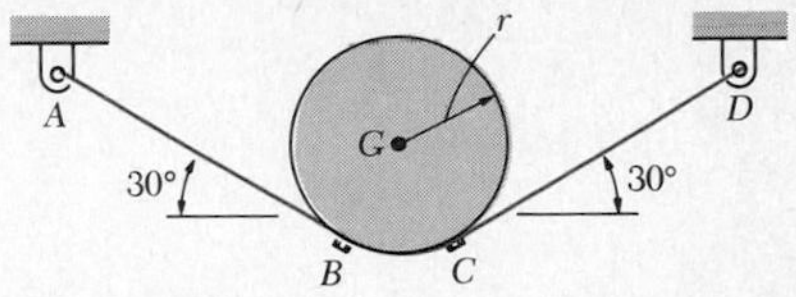

Fig. P16.159

16.160 and 16.161 A uniform plate of mass *m* is suspended in each of the ways shown. For each case determine the acceleration of the center of the plate immediately after the connection at *B* has been released.

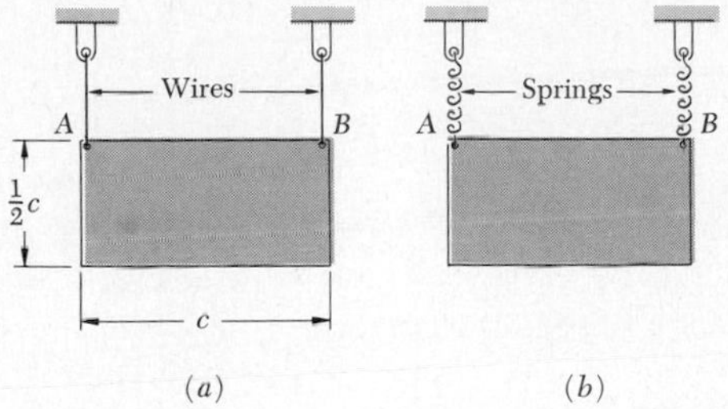

Fig. P16.160

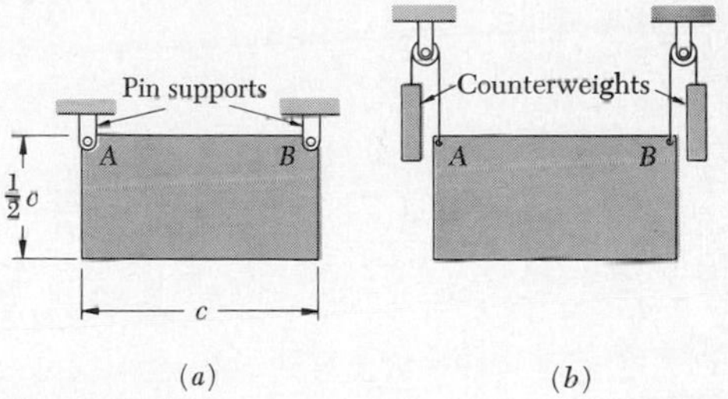

Fig. P16.161

16.162 A cyclist rides a bicycle at a speed of 20 mi/h. The distance between axles is 42 in., and the mass center of the cyclist and bicycle is located 26 in. behind the front axle and 40 in. above the ground. If the cyclist applies the brakes on the front wheel only, determine the shortest distance in which she can stop without being thrown over the front wheel.

16.163 Two uniform rods, each of mass *m*, are attached as shown to small gears of negligible mass. If the rods are released from rest in the position shown, determine the angular acceleration of rod *AB* immediately after release, assuming (*a*) $\theta = 0$, (*b*) $\theta = 30°$.

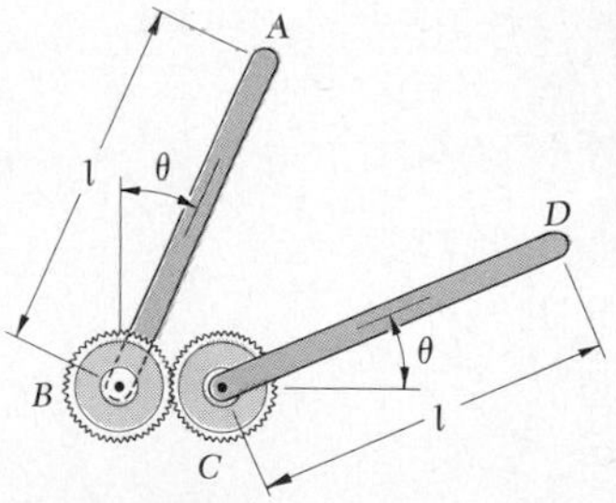

Fig. P16.163

16.164 A section of pipe rests on a plate. The plate is then given a constant acceleration **a** directed to the right. Assuming that the pipe rolls on the plate, determine (*a*) the acceleration of the pipe, (*b*) the distance through which the plate will move before the pipe reaches end *A*.

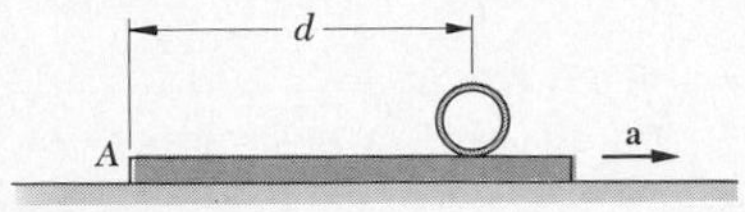

Fig. P16.164

16.165 Solve Prob. 16.164, assuming that the pipe is replaced (1) by a solid cylinder, (2) by a sphere.

16.166 A 3-kg roll of paper of radius $r = 150$ mm is attached by a frictionless pin C to a 5-kg trolley which may roll freely along a horizontal pipe. Knowing that a horizontal force of magnitude $P = 12$ N is applied to the paper at D, determine (*a*) the acceleration of the trolley, (*b*) the acceleration of point D.

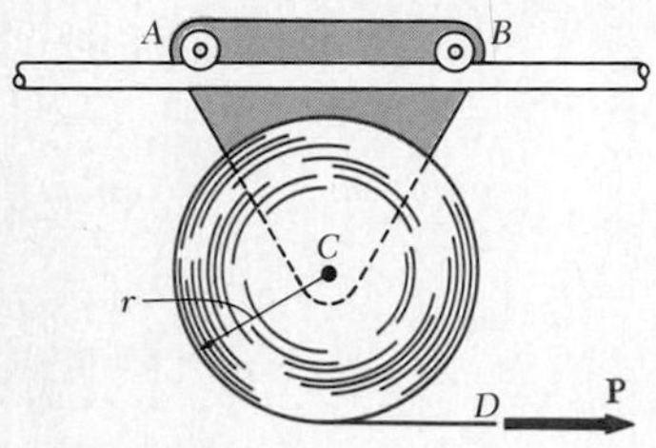

Fig. P16.166

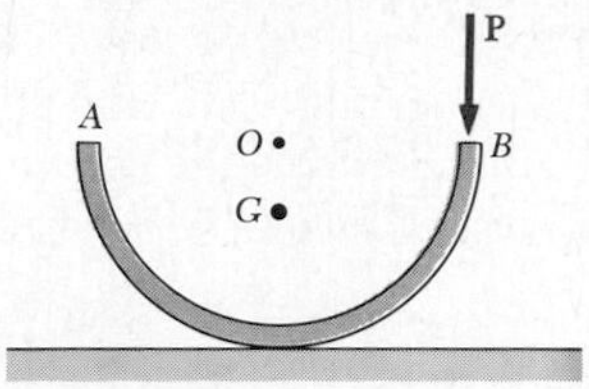

Fig. P16.167

16.167 A half section of pipe of mass m and radius r rests on a rough horizontal surface. A vertical force **P** is applied as shown. Assuming that the section rolls without sliding, derive an expression (*a*) for its angular acceleration, (*b*) for the minimum value of μ_s compatible with this motion. [*Hint.* Note that $OG = 2r/\pi$ and that, by the parallel-axis theorem, $\bar{I} = mr^2 - m(OG)^2$.]

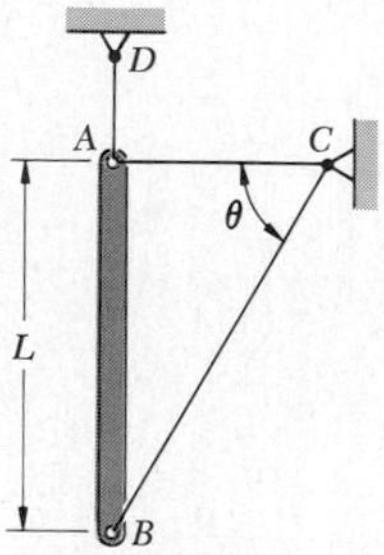

Fig. P16.168

16.168 A uniform slender rod AB, of length $L = 1.2$ m and mass 3 kg, is held in the position shown by three wires. If $\theta = 60°$, determine the tension in wires AC and BC immediately after wire AD has been cut.

16.169 Identical cylinders of mass m and radius r are pushed by a series of moving arms. Knowing that the coefficients of friction between all surfaces are $\mu_s = 0.40$ and $\mu_k = 0.30$, and denoting by a the magnitude of the acceleration of the arms, determine (*a*) the maximum allowable value of a if each cylinder is to roll without sliding, (*b*) the minimum allowable value of a if each cylinder is to move to the right without rotating.

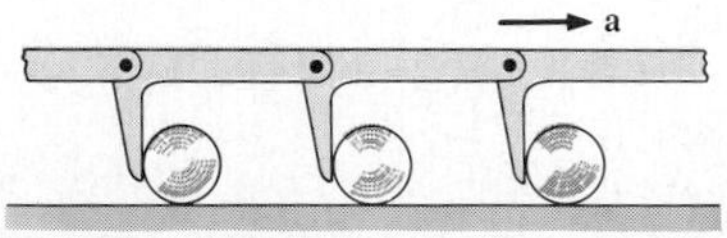

Fig. P16.169

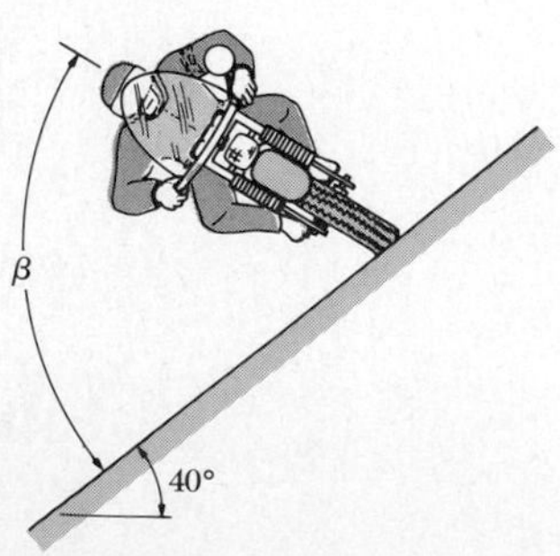

Fig. P16.170

16.170 A motorcyclist drives around a banked racing track at a constant speed v. The coefficient of static friction between the motorcycle tires and the track is $\mu_s = 0.50$. Assuming that the combined mass center of the motorcycle and driver travels in a horizontal circle of radius $r = 60$ ft, determine (*a*) the maximum and minimum safe values of the speed v with respect to skidding, (*b*) the corresponding values of the angle β.

CHAPTER SEVENTEEN

Plane Motion of Rigid Bodies: Energy and Momentum Methods

17.1. Introduction. In this chapter the method of work and energy and the method of impulse and momentum will be used to analyze the plane motion of rigid bodies and of systems of rigid bodies.

The method of work and energy will be considered first. In Secs. 17.2 through 17.5, we shall define the work of a force and of a couple and then obtain an expression for the kinetic energy of a rigid body in plane motion. The principle of work and energy will then be used to solve problems involving displacements and velocities. In Sec. 17.6, the principle of conservation of energy will be applied to the solution of a variety of engineering problems.

In the second part of the chapter, we shall discuss in Secs. 17.8 and 17.9 the principle of impulse and momentum and apply it to the solution of problems involving velocities and time. In Sec. 17.10, the concept of conservation of angular momentum will be introduced and discussed.

In the last part of the chapter (Secs. 17.11 and 17.12), we shall consider problems involving the eccentric impact of rigid bodies. As was done in Chap. 13, where we considered the impact of particles, the coefficient of restitution between the colliding bodies will be used together with the principle of impulse and momentum in the solution of impact problems. It will also be shown that the method used is applicable not only when the colliding bodies move freely after the impact but also when the bodies are partially constrained in their motion.

17.2. Principle of Work and Energy for a Rigid Body. The principle of work and energy will now be used to analyze the plane motion of rigid bodies. As was pointed out in Chap. 13, the method of work and energy is particularly well adapted to the solution of problems involving velocities and displacements. Its main advantage resides in the fact that the work of forces and the kinetic energy of particles are scalar quantities.

In order to apply the principle of work and energy to the analysis of the motion of a rigid body, we shall again assume that the rigid body is made of a large number n of particles of mass Δm_i. Recalling Eq. (14.30) of Sec. 14.8, we write

$$T_1 + U_{1\to 2} = T_2 \tag{17.1}$$

where T_1, T_2 = initial and final values of total kinetic energy of the particles forming the rigid body
$U_{1\to 2}$ = work of all forces acting on the various particles of the body

The total kinetic energy

$$T = \frac{1}{2}\sum_{i=1}^{n} \Delta m_i\, v_i^2 \tag{17.2}$$

is obtained by adding positive scalar quantities and is itself a positive scalar quantity. We shall see later how T may be determined for various types of motion of a rigid body.

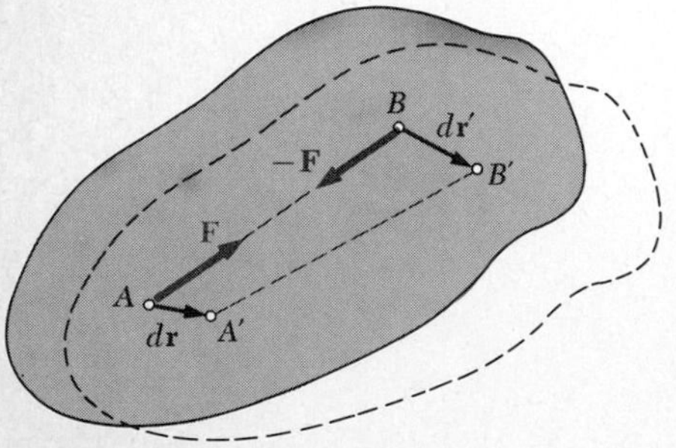

Fig. 17.1

The expression $U_{1\to 2}$ in (17.1) represents the work of all the forces acting on the various particles of the body, whether these forces are internal or external. However, as we shall see presently, the total work of the internal forces holding together the particles of a rigid body is zero. Consider two particles A and B of a rigid body and the two equal and opposite forces $\mathbf{F}$ and $-\mathbf{F}$ they exert on each other (Fig. 17.1). While, in general, small displacements $d\mathbf{r}$ and $d\mathbf{r}'$ of the two particles are different, the components of these displacements along AB must be equal; otherwise, the particles would not remain at the same distance from each other, and the body would not be rigid. Therefore, the work of $\mathbf{F}$ is equal in magnitude and opposite in sign to the work of $-\mathbf{F}$, and their sum is zero. Thus, the total work of the internal forces acting on the particles of a rigid body is zero, and *the expression* $U_{1\to 2}$ *in Eq.* (17.1) *reduces to the work of the external forces* acting on the body during the displacement considered.

17.3. Work of Forces Acting on a Rigid Body. We saw in Sec. 13.2 that the work of a force **F** during a displacement of its point of application from A_1 to A_2 is

$$U_{1\to 2} = \int_{A_1}^{A_2} \mathbf{F} \cdot d\mathbf{r} \tag{17.3}$$

or

$$U_{1\to 2} = \int_{s_1}^{s_2} (F \cos \alpha)\, ds \tag{17.3'}$$

where F is the magnitude of the force, α the angle it forms with the direction of motion of its point of application A, and s the variable of integration which measures the distance traveled by A along its path.

In computing the work of the external forces acting on a rigid body, it is often convenient to determine the work of a couple without considering separately the work of each of the two forces forming the couple. Consider the two forces **F** and $-$**F** forming a couple of moment **M** and acting on a rigid body (Fig. 17.2). Any small displacement of the rigid body bringing A and B, respectively, into A' and B'' may be divided into two parts, one in which points A and B undergo equal displacements $d\mathbf{r}_1$, the other in which A' remains fixed while B' moves into B'' through a displacement $d\mathbf{r}_2$ of magnitude $ds_2 = r\, d\theta$. In the first part of the motion, the work of **F** is equal in magnitude and opposite in sign to the work of $-$**F** and their sum is zero. In the second part of the motion, only force **F** works, and its work is $dU = F\, ds_2 = Fr\, d\theta$. But the product Fr is equal to the magnitude M of the moment of the couple. Thus, the work of a couple of moment **M** acting on a rigid body is

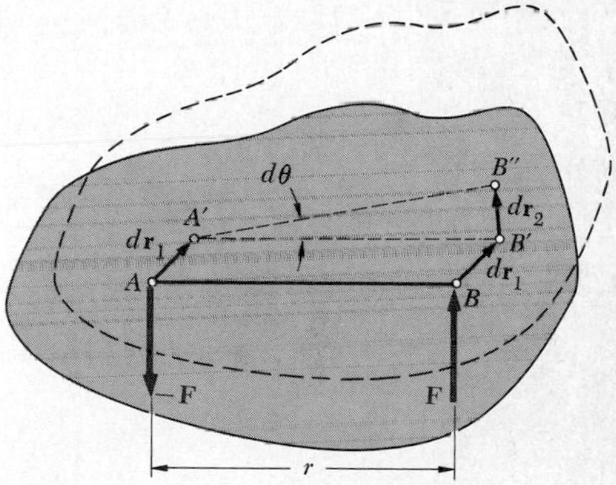

Fig. 17.2

$$dU = M\, d\theta \tag{17.4}$$

where $d\theta$ is the small angle expressed in radians through which the body rotates. We again note that work should be expressed in units obtained by multiplying units of force by units of length. The work of the couple during a finite rotation of the rigid body is obtained by integrating both members of (17.4) from the initial value θ_1 of the angle θ to its final value θ_2. We write

$$U_{1\to 2} = \int_{\theta_1}^{\theta_2} M\, d\theta \tag{17.5}$$

When the moment **M** *of the couple is constant,* formula (17.5) reduces to

$$U_{1\to 2} = M(\theta_2 - \theta_1) \tag{17.6}$$

It was pointed out in Sec. 13.2 that a number of forces encountered in problems of kinetics *do no work.* They are forces applied to fixed points or acting in a direction perpendicular to the displacement of their point of application. Among the forces which do no work the following have been listed: the reaction at a frictionless pin when the body supported rotates about the pin, the reaction at a frictionless surface when the body in contact moves along the surface, the weight of a body when its center of gravity moves horizontally. We should also indicate now that *when a rigid body rolls without sliding on a fixed surface, the friction force* **F** *at the point of contact C does no work.* The velocity $\mathbf{v}_C$ of the point of contact C is zero, and the work of the friction force **F** during a small displacement of the rigid body is

$$dU = F\,ds_C = F(v_C\,dt) = 0$$

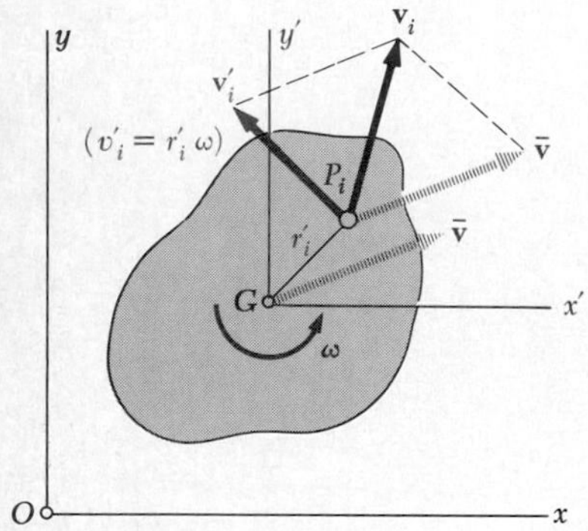

Fig. 17.3

17.4. Kinetic Energy of a Rigid Body in Plane Motion. Consider a rigid body of mass m in plane motion. We recall from Sec. 14.7 that, if the absolute velocity $\mathbf{v}_i$ of each particle P_i of the body is expressed as the sum of the velocity $\bar{\mathbf{v}}$ of the mass center G of the body and of the velocity $\mathbf{v}'_i$ of the particle relative to a frame $Gx'y'$ attached to G and of fixed orientation (Fig. 17.3), the kinetic energy of the system of particles forming the rigid body may be written in the form

$$T = \tfrac{1}{2}m\bar{v}^2 + \frac{1}{2}\sum_{i=1}^{n} \Delta m_i\, v_i'^2 \tag{17.7}$$

But the magnitude v'_i of the relative velocity of P_i is equal to the product $r'_i\omega$ of the distance r'_i of P_i from the axis through G perpendicular to the plane of motion and of the magnitude ω of the angular velocity of the body at the instant considered. Substituting into (17.7), we have

$$T = \tfrac{1}{2}m\bar{v}^2 + \frac{1}{2}\left(\sum_{i=1}^{n} r_i'^2\,\Delta m_i\right)\omega^2 \tag{17.8}$$

or, since the sum represents the moment of inertia $\bar{I}$ of the body about the axis through G,

$$T = \tfrac{1}{2}m\bar{v}^2 + \tfrac{1}{2}\bar{I}\omega^2 \tag{17.9}$$

We note that in the particular case of a body in translation ($\omega = 0$), the expression obtained reduces to $\frac{1}{2}m\bar{v}^2$, while in the case of a centroidal rotation ($\bar{v} = 0$), it reduces to $\frac{1}{2}\bar{I}\omega^2$. We conclude that the kinetic energy of a rigid body in plane motion may be separated into two parts: (1) the kinetic energy $\frac{1}{2}m\bar{v}^2$ associated with the motion of the mass center G of the body, and (2) the kinetic energy $\frac{1}{2}\bar{I}\omega^2$ associated with the rotation of the body about G.

Noncentroidal Rotation. The relation (17.9) is valid for any type of plane motion and may, therefore, be used to express the kinetic energy of a rigid body rotating with an angular velocity $\boldsymbol{\omega}$ about a fixed axis through O (Fig. 17.4). In that case, however, the kinetic energy of the body may be expressed more directly by noting that the speed v_i of the particle P_i is equal to the product $r_i\omega$ of the distance r_i of P_i from the fixed axis and of the magnitude ω of the angular velocity of the body at the instant considered. Substituting into (17.2), we write

$$T = \frac{1}{2}\sum_{i=1}^{n} \Delta m_i\,(r_i\omega)^2 = \frac{1}{2}\left(\sum_{i=1}^{n} r_i^2\,\Delta m_i\right)\omega^2$$

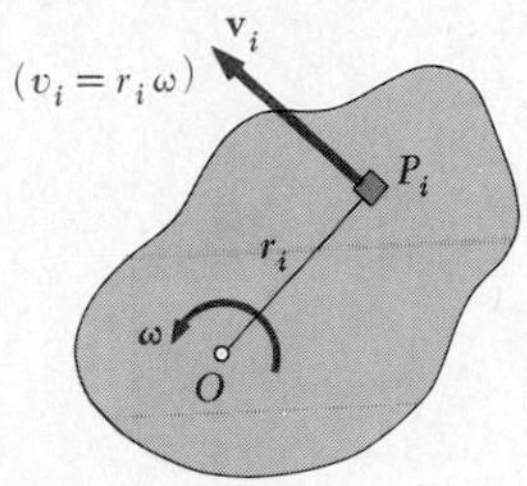

Fig. 17.4

or, since the last sum represents the moment of inertia I_O of the body about the fixed axis through O,

$$T = \tfrac{1}{2}I_O\omega^2 \tag{17.10}$$

We note that the results obtained are not limited to the motion of plane slabs or to the motion of bodies which are symmetrical with respect to the reference plane. They may be applied to the study of the plane motion of any rigid body, regardless of its shape. However, since Eq. (17.9) is applicable to any plane motion, we shall use it in the solution of sample problems, rather than Eq. (17.10) which may be used only for a noncentroidal rotation.

17.5. Systems of Rigid Bodies. When a problem involves several rigid bodies, each rigid body may be considered separately, and the principle of work and energy may be applied to each body. Adding the kinetic energies of all the particles and considering the work of all the forces involved, we may also write the equation of work and energy for the entire system. We have

$$T_1 + U_{1\to 2} = T_2 \tag{17.11}$$

where T represents the arithmetic sum of the kinetic energies of the rigid bodies forming the system (all terms are positive) and $U_{1\to 2}$ the work of all the forces acting on the various bodies, whether these forces are *internal* or *external* from the point of view of the system as a whole.

The method of work and energy is particularly useful in solving problems involving pin-connected members, or blocks and pulleys connected by inextensible cords, or meshed gears. In all these cases, the internal forces occur by pairs of equal and opposite forces, and the points of application of the forces in each pair *move through equal distances* during a small displacement of the system. As a result, the work of the internal forces is zero, and $U_{1\to 2}$ reduces to the work of the *forces external to the system.*

17.6. Conservation of Energy. We saw in Sec. 13.6 that the work of conservative forces, such as the weight of a body or the force exerted by a spring, may be expressed as a change in potential energy. When a rigid body, or a system of rigid bodies, moves under the action of conservative forces, the principle of work and energy stated in Sec. 17.2 may be expressed in a modified form. Substituting for $U_{1\rightarrow 2}$ from (13.19′) into (17.1), we write

$$T_1 + V_1 = T_2 + V_2 \tag{17.12}$$

Formula (17.12) indicates that when a rigid body, or a system of rigid bodies, moves under the action of conservative forces, *the sum of the kinetic energy and of the potential energy of the system remains constant.* It should be noted that in the case of the plane motion of a rigid body, the kinetic energy of the body should include both the *translational* term $\frac{1}{2}m\bar{v}^2$ and the *rotational* term $\frac{1}{2}\bar{I}\omega^2$.

As an example of application of the principle of conservation of energy, we shall consider a slender rod AB, of length l and mass m, whose extremities are connected to blocks of negligible mass sliding along horizontal and vertical tracks. We assume that the rod is released with no initial velocity from a horizontal position (Fig. 17.5*a*), and we wish to determine its angular velocity after it has rotated through an angle θ (Fig. 17.5*b*).

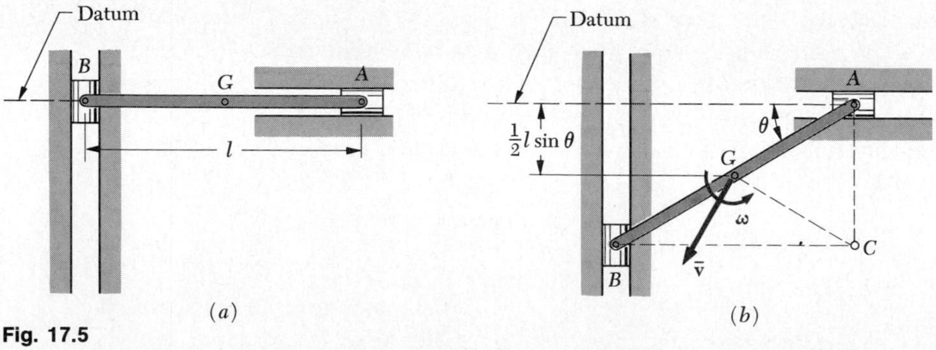

Fig. 17.5

Since the initial velocity is zero, we have $T_1 = 0$. Measuring the potential energy from the level of the horizontal track, we write $V_1 = 0$. After the rod has rotated through θ, the center of gravity G of the rod is at a distance $\frac{1}{2}l \sin\theta$ below the reference level and we have

$$V_2 = -\tfrac{1}{2}Wl\sin\theta = -\tfrac{1}{2}mgl\sin\theta$$

Observing that in this position, the instantaneous center of the rod is located at C, and that $CG = \frac{1}{2}l$, we write $\bar{v}_2 = \frac{1}{2}l\omega$ and obtain

$$T_2 = \tfrac{1}{2}m\bar{v}_2^2 + \tfrac{1}{2}\bar{I}\omega_2^2 = \tfrac{1}{2}m(\tfrac{1}{2}l\omega)^2 + \tfrac{1}{2}(\tfrac{1}{12}ml^2)\omega^2$$
$$= \frac{1}{2}\frac{ml^2}{3}\omega^2$$

Applying the principle of conservation of energy, we write

$$T_1 + V_1 = T_2 + V_2$$
$$0 = \frac{1}{2}\frac{ml^2}{3}\omega^2 - \tfrac{1}{2}mgl\sin\theta$$
$$\omega = \left(\frac{3g}{l}\sin\theta\right)^{1/2}$$

We recall that the advantages of the method of work and energy, as well as its shortcomings, were indicated in Sec. 13.4. In this connection, we wish to mention that the method of work and energy must be supplemented by the application of d'Alembert's principle when reactions at fixed axles, at rollers, or at sliding blocks are to be determined. For example, in order to compute the reactions at the extremities A and B of the rod of Fig. 17.5*b*, a diagram should be drawn to express that the system of the external forces applied to the rod is equivalent to the vector $m\bar{\mathbf{a}}$ and the couple $\bar{I}\boldsymbol{\alpha}$. The angular velocity $\boldsymbol{\omega}$ of the rod, however, is determined by the method of work and energy before the equations of motion are solved for the reactions. The complete analysis of the motion of the rod and of the forces exerted on the rod requires, therefore, the combined use of the method of work and energy and of the principle of equivalence of the external and effective forces.

17.7. Power. *Power* was defined in Sec. 13.5 as the time rate at which work is done. In the case of a body acted upon by a force **F**, and moving with a velocity **v**, the power was expressed as follows:

$$\text{Power} = \frac{dU}{dt} = \mathbf{F}\cdot\mathbf{v} \tag{13.13}$$

In the case of a rigid body rotating with an angular velocity $\boldsymbol{\omega}$ and acted upon by a couple of moment **M** parallel to the axis of rotation, we have, by (17.4),

$$\text{Power} = \frac{dU}{dt} = \frac{M\,d\theta}{dt} = M\omega \tag{17.13}$$

The various units used to measure power, such as the watt and the horsepower, were defined in Sec. 13.5.

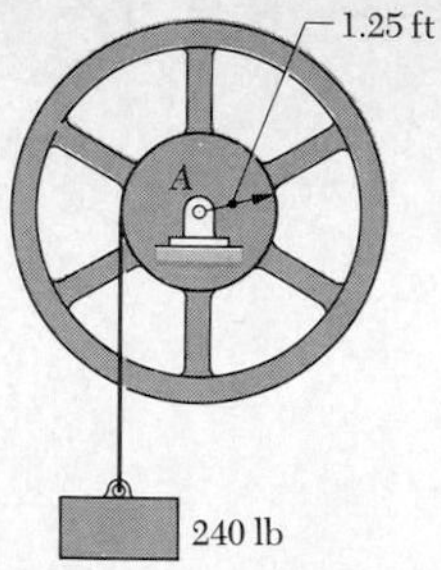

SAMPLE PROBLEM 17.1

A 240-lb block is suspended from an inextensible cable which is wrapped around a drum of 1.25-ft radius rigidly attached to a flywheel. The drum and flywheel have a combined centroidal moment of inertia $\bar{I} = 10.5\ \text{lb}\cdot\text{ft}\cdot\text{s}^2$. At the instant shown, the velocity of the block is 6 ft/s directed downward. Knowing that the bearing at A is poorly lubricated and that the bearing friction is equivalent to a couple **M** of magnitude 60 lb·ft, determine the velocity of the block after it has moved 4 ft downward.

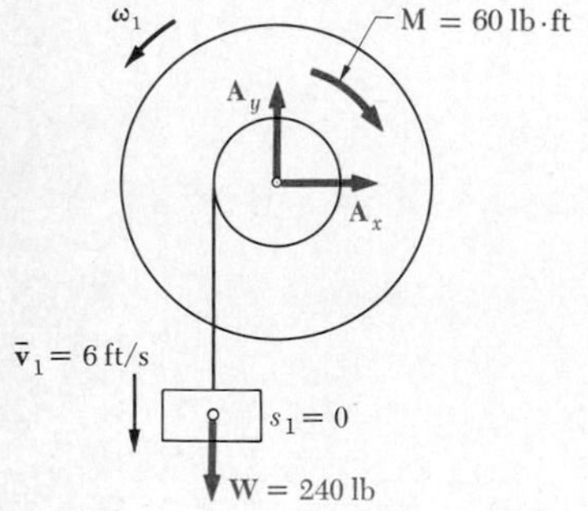

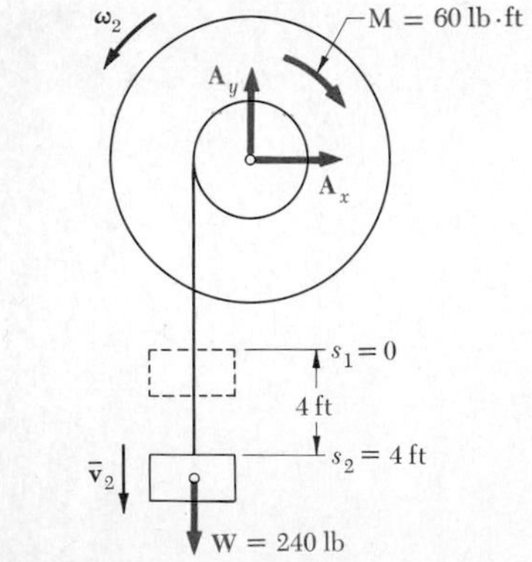

Solution. We consider the system formed by the flywheel and the block. Since the cable is inextensible, the work done by the internal forces exerted by the cable cancels. The initial and final positions of the system and the external forces acting on the system are as shown.

Kinetic Energy. *Position 1.*

Block: $\bar{v}_1 = 6\ \text{ft/s}$

Flywheel: $\omega_1 = \dfrac{\bar{v}_1}{r} = \dfrac{6\ \text{ft/s}}{1.25\ \text{ft}} = 4.80\ \text{rad/s}$

$$T_1 = \tfrac{1}{2}m\bar{v}_1^2 + \tfrac{1}{2}\bar{I}\omega_1^2$$
$$= \frac{1}{2}\frac{240\ \text{lb}}{32.2\ \text{ft/s}^2}(6\ \text{ft/s})^2 + \tfrac{1}{2}(10.5\ \text{lb}\cdot\text{ft}\cdot\text{s}^2)(4.80\ \text{rad/s})^2$$
$$= 255\ \text{ft}\cdot\text{lb}$$

Position 2. Noting that $\omega_2 = \bar{v}_2/1.25$, we write

$$T_2 = \tfrac{1}{2}m\bar{v}_2^2 + \tfrac{1}{2}\bar{I}\omega_2^2$$
$$= \frac{1}{2}\frac{240}{32.2}(\bar{v}_2)^2 + (\tfrac{1}{2})(10.5)\left(\frac{\bar{v}_2}{1.25}\right)^2 = 7.09\bar{v}_2^2$$

Work. During the motion, only the weight **W** of the block and the friction couple **M** do work. Noting that **W** does positive work and that the friction couple **M** does negative work, we write

$$s_1 = 0 \qquad s_2 = 4\ \text{ft}$$
$$\theta_1 = 0 \qquad \theta_2 = \frac{s_2}{r} = \frac{4\ \text{ft}}{1.25\ \text{ft}} = 3.20\ \text{rad}$$
$$U_{1\to2} = W(s_2 - s_1) - M(\theta_2 - \theta_1)$$
$$= (240\ \text{lb})(4\ \text{ft}) - (60\ \text{lb}\cdot\text{ft})(3.20\ \text{rad})$$
$$= 768\ \text{ft}\cdot\text{lb}$$

Principle of Work and Energy

$$T_1 + U_{1\to2} = T_2$$
$$255\ \text{ft}\cdot\text{lb} + 768\ \text{ft}\cdot\text{lb} = 7.09\bar{v}_2^2$$
$$\bar{v}_2 = 12.01\ \text{ft/s} \qquad \bar{\mathbf{v}}_2 = 12.01\ \text{ft/s}\downarrow \quad ◄$$

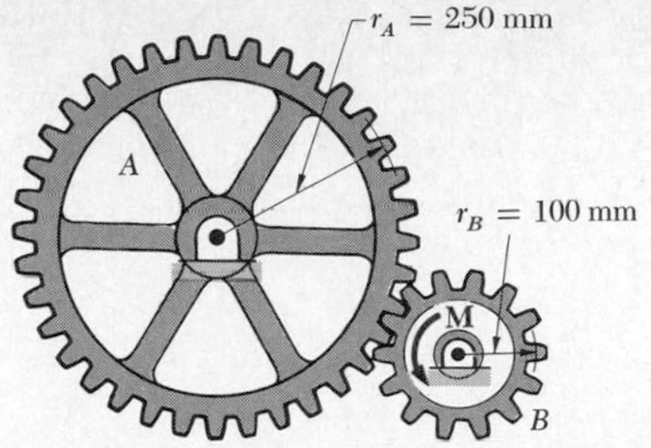

SAMPLE PROBLEM 17.2

Gear A has a mass of 10 kg and a radius of gyration of 200 mm, while gear B has a mass of 3 kg and a radius of gyration of 80 mm. The system is at rest when a couple **M** of magnitude 6 N·m is applied to gear B. Neglecting friction, determine (*a*) the number of revolutions executed by gear B before its angular velocity reaches 600 rpm, (*b*) the tangential force which gear B exerts on gear A.

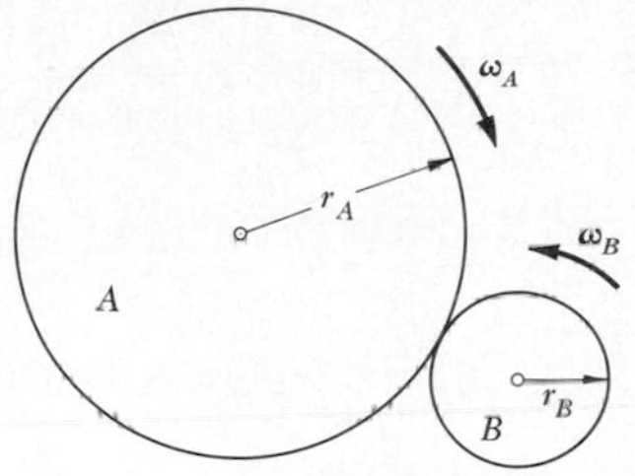

Motion of Entire System. Noting that the peripheral speeds of the gears are equal, we write

$$r_A\omega_A = r_B\omega_B \qquad \omega_A = \omega_B\frac{r_B}{r_A} = \omega_B\frac{100\text{ mm}}{250\text{ mm}} = 0.40\omega_B$$

For $\omega_B = 600$ rpm, we have

$$\omega_B = 62.8\text{ rad/s} \qquad \omega_A = 0.40\omega_B = 25.1\text{ rad/s}$$
$$\bar{I}_A = m_A\bar{k}_A^2 = (10\text{ kg})(0.200\text{ m})^2 = 0.400\text{ kg}\cdot\text{m}^2$$
$$\bar{I}_B = m_B\bar{k}_B^2 = (3\text{ kg})(0.080\text{ m})^2 = 0.0192\text{ kg}\cdot\text{m}^2$$

Kinetic Energy. Since the system is initially at rest, $T_1 = 0$. Adding the kinetic energies of the two gears when $\omega_B = 600$ rpm, we obtain

$$\begin{aligned}T_2 &= \tfrac{1}{2}\bar{I}_A\omega_A^2 + \tfrac{1}{2}\bar{I}_B\omega_B^2\\ &= \tfrac{1}{2}(0.400\text{ kg}\cdot\text{m}^2)(25.1\text{ rad/s})^2 + \tfrac{1}{2}(0.0192\text{ kg}\cdot\text{m}^2)(62.8\text{ rad/s})^2\\ &= 163.9\text{ J}\end{aligned}$$

Work. Denoting by θ_B the angular displacement of gear B, we have

$$U_{1\to2} = M\theta_B = (6\text{ N}\cdot\text{m})(\theta_B\text{ rad}) = (6\,\theta_B)\text{ J}$$

Principle of Work and Energy

$$T_1 + U_{1\to2} = T_2$$
$$0 + (6\,\theta_B)\text{ J} = 163.9\text{ J}$$
$$\theta_B = 27.32\text{ rad} \qquad \theta_B = 4.35\text{ rev} \blacktriangleleft$$

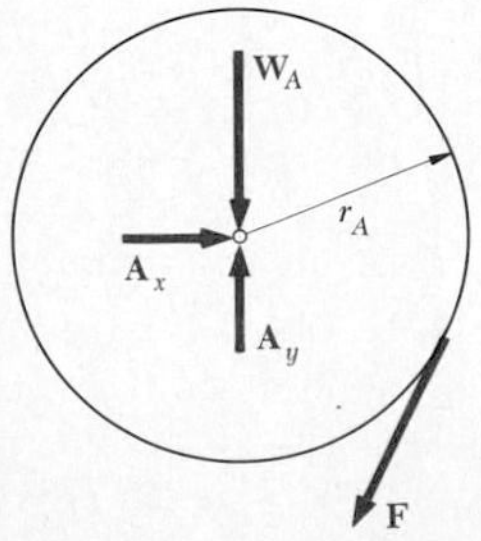

Motion of Gear A. ***Kinetic Energy.*** Initially, gear A is at rest, $T_1 = 0$. When $\omega_B = 600$ rpm, the kinetic energy of gear A is

$$T_2 = \tfrac{1}{2}\bar{I}_A\omega_A^2 = \tfrac{1}{2}(0.400\text{ kg}\cdot\text{m}^2)(25.1\text{ rad/s})^2 = 126.0\text{ J}$$

Work. The forces acting on gear A are as shown. The tangential force **F** does work equal to the product of its magnitude and of the length $\theta_A r_A$ of the arc described by the point of contact. Since $\theta_A r_A = \theta_B r_B$, we have

$$U_{1\to2} = F(\theta_B r_B) = F(27.3\text{ rad})(0.100\text{ m}) = F(2.73\text{ m})$$

Principle of Work and Energy

$$T_1 + U_{1\to2} = T_2$$
$$0 + F(2.73\text{ m}) = 126.0\text{ J}$$
$$F = +46.2\text{ N} \qquad \mathbf{F} = 46.2\text{ N} \swarrow \blacktriangleleft$$

SAMPLE PROBLEM 17.3

A sphere, a cylinder, and a hoop, each having the same mass and the same radius, are released from rest on an incline. Determine the velocity of each body after it has rolled through a distance corresponding to a change in elevation h.

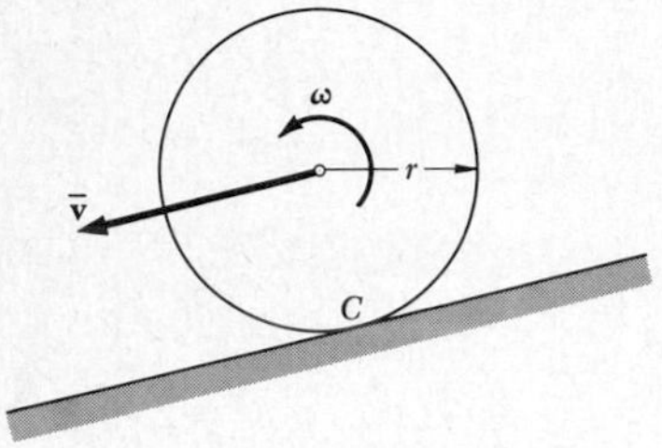

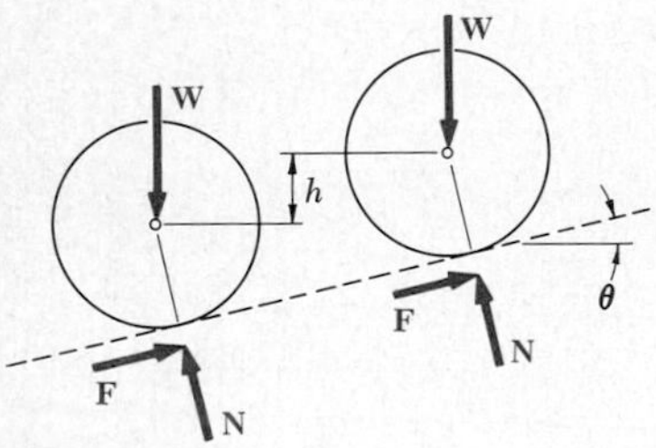

Solution. We shall first solve the problem in general terms and then find particular results for each body. We denote the mass by m, the centroidal moment of inertia by $\bar{I}$, the weight by W, and the radius by r.

Since each body rolls, the instantaneous center of rotation is located at C and we write

$$\omega = \frac{\bar{v}}{r}$$

Kinetic Energy

$$T_1 = 0$$
$$T_2 = \tfrac{1}{2}m\bar{v}^2 + \tfrac{1}{2}\bar{I}\omega^2$$
$$= \tfrac{1}{2}m\bar{v}^2 + \tfrac{1}{2}\bar{I}\left(\frac{\bar{v}}{r}\right)^2 = \tfrac{1}{2}\left(m + \frac{\bar{I}}{r^2}\right)\bar{v}^2$$

Work. Since the friction force **F** in rolling motion does no work,

$$U_{1\to2} = Wh$$

Principle of Work and Energy

$$T_1 + U_{1\to2} = T_2$$
$$0 + Wh = \tfrac{1}{2}\left(m + \frac{\bar{I}}{r^2}\right)\bar{v}^2 \qquad \bar{v}^2 = \frac{2Wh}{m + \bar{I}/r^2}$$

Noting that $W = mg$, we rearrange the result and obtain

$$\bar{v}^2 = \frac{2gh}{1 + \bar{I}/mr^2}$$

Velocities of Sphere, Cylinder, and Hoop. Introducing successively the particular expression for $\bar{I}$, we obtain

Sphere:	$\bar{I} = \frac{2}{5}mr^2$	$\bar{v} = 0.845\sqrt{2gh}$ ◄
Cylinder:	$\bar{I} = \frac{1}{2}mr^2$	$\bar{v} = 0.816\sqrt{2gh}$ ◄
Hoop:	$\bar{I} = mr^2$	$\bar{v} = 0.707\sqrt{2gh}$ ◄

Remark. We may compare the results with the velocity attained by a frictionless block sliding through the same distance. The solution is identical to the above solution except that $\omega = 0$; we find $\bar{v} = \sqrt{2gh}$.

Comparing the results, we note that the velocity of the body is independent of both its mass and radius. However, the velocity does depend upon the quotient $\bar{I}/mr^2 = \bar{k}^2/r^2$, which measures the ratio of the rotational kinetic energy to the translational kinetic energy. Thus the hoop, which has the largest $\bar{k}$ for a given radius r, attains the smallest velocity, while the sliding block, which does not rotate, attains the largest velocity.

SAMPLE PROBLEMS 17.4

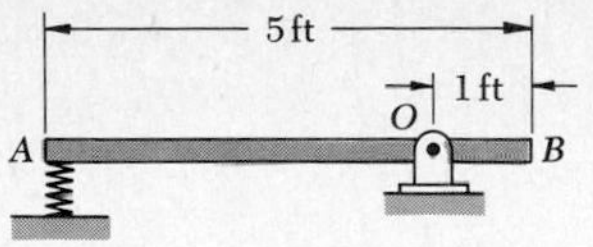

A 30-lb slender rod AB is 5 ft long and is pivoted about a point O which is 1 ft from end B. The other end is pressed against a spring of constant $k = 1800$ lb/in. until the spring is compressed 1 in. The rod is then in a horizontal position. If the rod is released from this position, determine its angular velocity and the reaction at the pivot O as the rod passes through a vertical position.

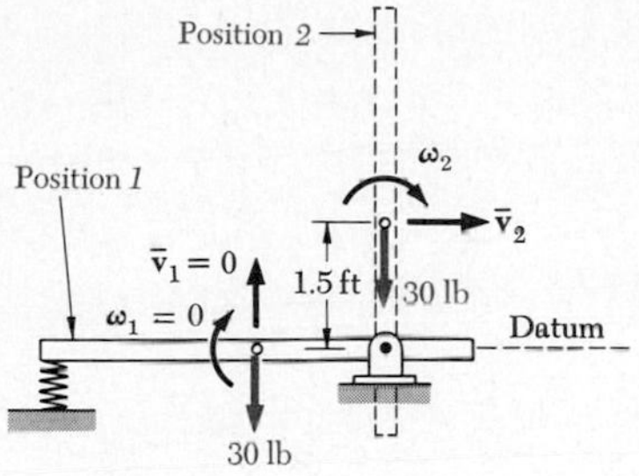

Position 1. ***Potential Energy.*** Since the spring is compressed 1 in., we have $x_1 = 1$ in.

$$V_e = \tfrac{1}{2}kx_1^2 = \tfrac{1}{2}(1800 \text{ lb/in.})(1 \text{ in.})^2 = 900 \text{ in}\cdot\text{lb}$$

Choosing the datum as shown, we have $V_g = 0$; therefore,

$$V_1 = V_e + V_g = 900 \text{ in}\cdot\text{lb} = 75 \text{ ft}\cdot\text{lb}$$

Kinetic Energy. Since the velocity in position *1* is zero, we have $T_1 = 0$.

Position 2. ***Potential Energy.*** The elongation of the spring is zero, and we have $V_e = 0$. Since the center of gravity of the rod is now 1.5 ft above the datum,

$$V_g = (30 \text{ lb})(+1.5 \text{ ft}) = 45 \text{ ft}\cdot\text{lb}$$
$$V_2 = V_e + V_g = 45 \text{ ft}\cdot\text{lb}$$

Kinetic Energy. Denoting by $\boldsymbol{\omega}_2$ the angular velocity of the rod in position 2, we note that the rod rotates about O and write $\bar{v}_2 = \bar{r}\omega_2 = 1.5\omega_2$.

$$\bar{I} = \tfrac{1}{12}ml^2 = \frac{1}{12}\frac{30 \text{ lb}}{32.2 \text{ ft/s}^2}(5 \text{ ft})^2 = 1.941 \text{ lb}\cdot\text{ft}\cdot\text{s}^2$$

$$T_2 = \tfrac{1}{2}m\bar{v}_2^2 + \tfrac{1}{2}\bar{I}\omega_2^2 = \frac{1}{2}\frac{30}{32.2}(1.5\omega_2)^2 + \tfrac{1}{2}(1.941)\omega_2^2 = 2.019\omega_2^2$$

Conservation of Energy

$$T_1 + V_1 = T_2 + V_2$$
$$0 + 75 \text{ ft}\cdot\text{lb} = 2.019\omega_2^2 + 45 \text{ ft}\cdot\text{lb}$$

$$\omega_2 = 3.86 \text{ rad/s} \circlearrowright \blacktriangleleft$$

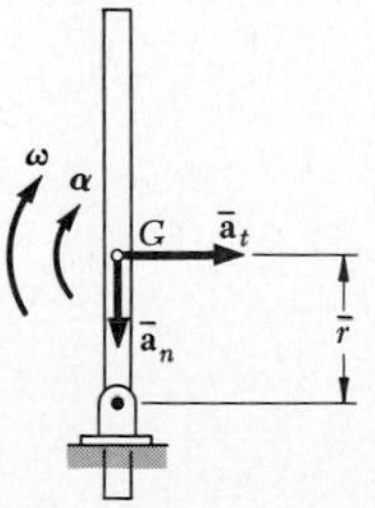

Reaction in Position 2. Since $\omega_2 = 3.86$ rad/s, the components of the acceleration of G as the rod passes through position *2* are

$$\bar{a}_n = \bar{r}\omega_2^2 = (1.5 \text{ ft})(3.86 \text{ rad/s})^2 = 22.3 \text{ ft/s}^2 \qquad \bar{\mathbf{a}}_n = 22.3 \text{ ft/s}^2 \downarrow$$
$$\bar{a}_t = \bar{r}\alpha \qquad \bar{\mathbf{a}}_t = \bar{r}\alpha \rightarrow$$

We express that the system of external forces is equivalent to the system of effective forces represented by the vector of components $m\bar{\mathbf{a}}_t$ and $m\bar{\mathbf{a}}_n$ attached at G and the couple $\bar{I}\boldsymbol{\alpha}$.

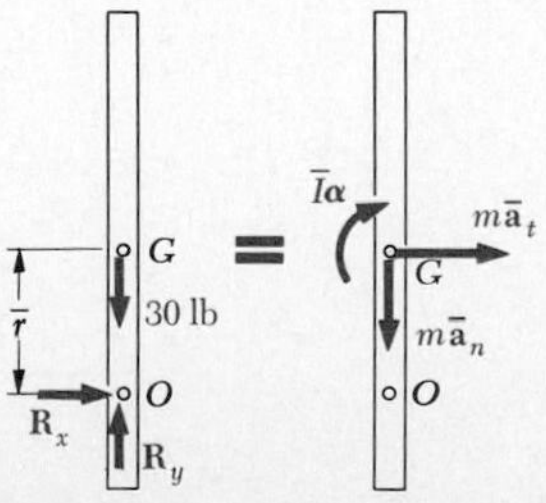

$+\circlearrowright\Sigma M_O = \Sigma(M_O)_{\text{eff}}$: $\quad 0 = \bar{I}\alpha + m(\bar{r}\alpha)\bar{r} \qquad \alpha = 0$

$\xrightarrow{+} \Sigma F_x = \Sigma(F_x)_{\text{eff}}$: $\quad R_x = m(\bar{r}\alpha) \qquad R_x = 0$

$+\uparrow\Sigma F_y = \Sigma(F_y)_{\text{eff}}$: $\quad R_y - 30 \text{ lb} = -m\bar{a}_n$

$$R_y - 30 \text{ lb} = -\frac{30 \text{ lb}}{32.2 \text{ ft/s}^2}(22.3 \text{ ft/s}^2)$$

$$R_y = +9.22 \text{ lb} \qquad \mathbf{R} = 9.22 \text{ lb} \uparrow \blacktriangleleft$$

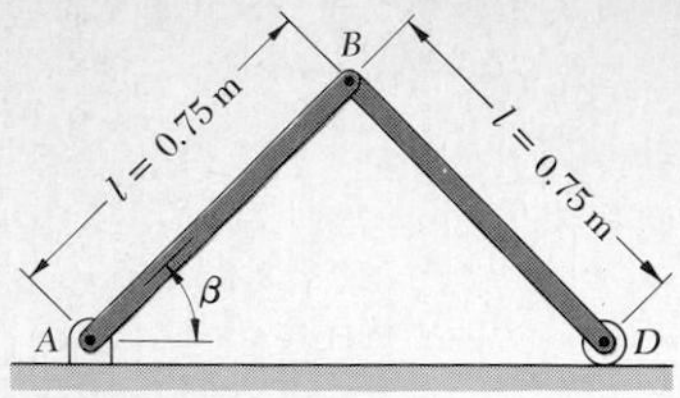

SAMPLE PROBLEM 17.5

Each of the two slender rods shown is 0.75 m long and has a mass of 6 kg. If the system is released from rest when $\beta = 60°$, determine (*a*) the angular velocity of rod *AB* when $\beta = 20°$, (*b*) the velocity of point *D* at the same instant.

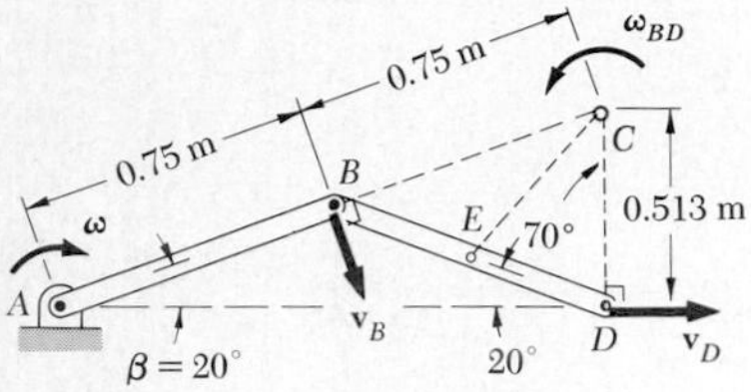

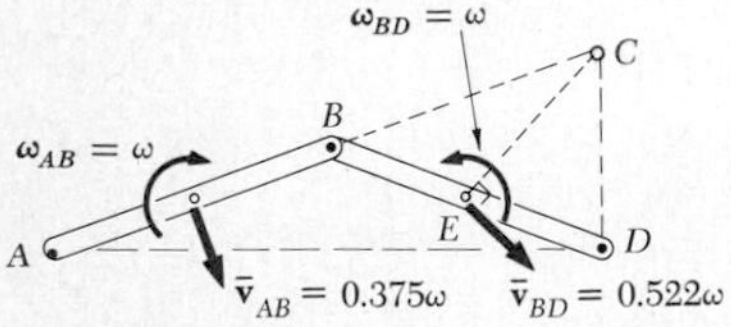

Kinematics of Motion When $\beta = 20°$. Since $\mathbf{v}_B$ is perpendicular to the rod *AB* and $\mathbf{v}_D$ is horizontal, the instantaneous center of rotation of rod *BD* is located at *C*. Considering the geometry of the figure, we obtain

$$BC = 0.75 \text{ m} \qquad CD = 2(0.75 \text{ m}) \sin 20° = 0.513 \text{ m}$$

Applying the law of cosines to triangle *CDE*, where *E* is located at the mass center of rod *BD*, we find $EC = 0.522$ m. Denoting by ω the angular velocity of rod *AB*, we have

$$\bar{v}_{AB} = (0.375 \text{ m})\omega \qquad \bar{\mathbf{v}}_{AB} = 0.375\omega \searrow$$
$$v_B = (0.75 \text{ m})\omega \qquad \mathbf{v}_B = 0.75\omega \searrow$$

Since rod *BD* seems to rotate about point *C*, we may write

$$v_B = (BC)\omega_{BD} \qquad (0.75 \text{ m})\omega = (0.75 \text{ m})\omega_{BD} \qquad \omega_{BD} = \omega \circlearrowleft$$
$$\bar{v}_{BD} = (EC)\omega_{BD} = (0.522 \text{ m})\omega \qquad \bar{\mathbf{v}}_{BD} = 0.522\omega \searrow$$

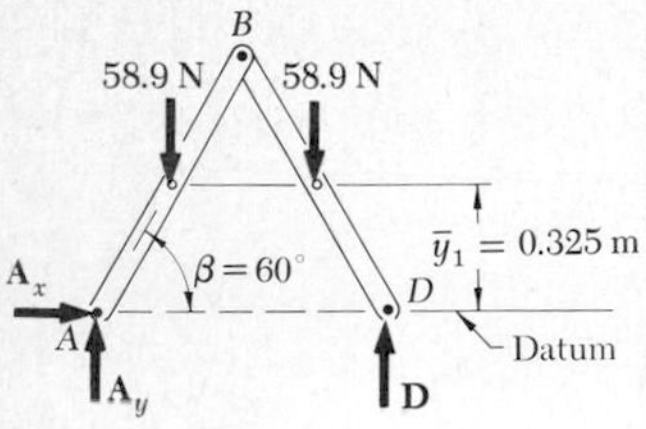

Position *1*

Position 1. ***Potential Energy.*** Choosing the datum as shown, and observing that $W = (6 \text{ kg})(9.81 \text{ m/s}^2) = 58.9$ N, we have

$$V_1 = 2W\bar{y}_1 = 2(58.9 \text{ N})(0.325 \text{ m}) = 38.3 \text{ J}$$

Kinetic Energy. Since the system is at rest, $T_1 = 0$.

Position 2. ***Potential Energy***

$$V_2 = 2W\bar{y}_2 = 2(58.9 \text{ N})(0.128 \text{ m}) = 15.1 \text{ J}$$

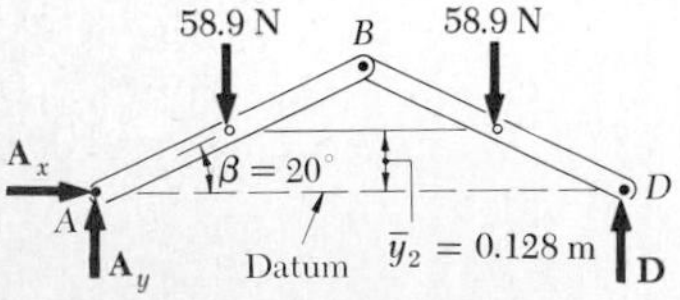

Position 2

Kinetic Energy

$$\bar{I}_{AB} = \bar{I}_{BD} = \tfrac{1}{12}ml^2 = \tfrac{1}{12}(6 \text{ kg})(0.75 \text{ m})^2 = 0.281 \text{ kg}\cdot\text{m}^2$$
$$T_2 = \tfrac{1}{2}m\bar{v}_{AB}^2 + \tfrac{1}{2}\bar{I}_{AB}\omega_{AB}^2 + \tfrac{1}{2}m\bar{v}_{BD}^2 + \tfrac{1}{2}\bar{I}_{BD}\omega_{BD}^2$$
$$= \tfrac{1}{2}(6)(0.375\omega)^2 + \tfrac{1}{2}(0.281)\omega^2 + \tfrac{1}{2}(6)(0.522\omega)^2 + \tfrac{1}{2}(0.281)\omega^2$$
$$= 1.520\omega^2$$

Conservation of Energy

$$T_1 + V_1 = T_2 + V_2$$
$$0 + 38.3 \text{ J} = 1.520\omega^2 + 15.1 \text{ J}$$
$$\omega = 3.91 \text{ rad/s} \qquad \omega_{AB} = 3.91 \text{ rad/s} \circlearrowright \quad ◀$$

Velocity of Point *D*

$$v_D = (CD)\omega = (0.513 \text{ m})(3.91 \text{ rad/s}) = 2.01 \text{ m/s}$$
$$\mathbf{v}_D = 2.01 \text{ m/s} \rightarrow \quad ◀$$

Problems

17.1 A large flywheel weighs 6000 lb and has a radius of gyration of 36 in. It is observed that 1500 revolutions are required for the flywheel to coast from an angular velocity of 300 rpm to rest. Determine the average magnitude of the couple due to kinetic friction in the bearings.

17.2 The rotor of an electric motor has an angular velocity of 3600 rpm when the load and power are cut off. The 50-kg rotor, which has a centroidal radius of gyration of 180 mm, then coasts to rest. Knowing that the kinetic friction of the rotor produces a couple of magnitude 3.5 N·m, determine the number of revolutions that the rotor executes before coming to rest.

17.3 Three disks of the same thickness and same material are attached to a shaft as shown. Disks A and B each have a radius r; disk C has a radius nr. A couple **M** of constant magnitude is applied when the system is at rest and is removed after the system has executed one revolution. Determine the radius of disk C which results in the largest final speed of a point on the rim of disk C.

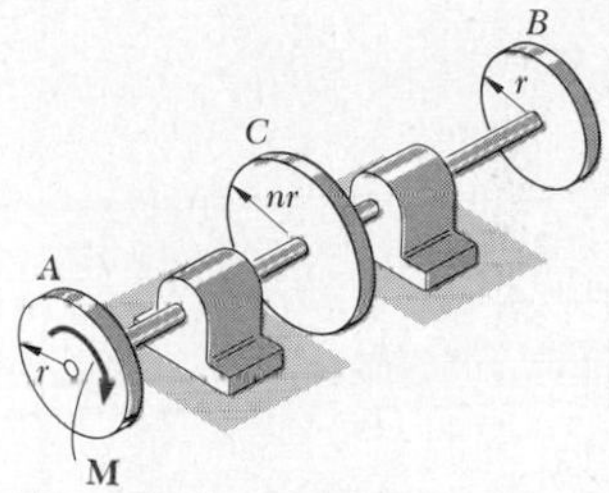

Fig. P17.3 and P17.4

17.4 Three disks of the same thickness and same material are attached to a shaft as shown. Disks A and B each have a mass of 8 kg and a radius $r = 240$ mm. A couple **M** of magnitude 40 N·m is applied to disk A when the system is at rest. Determine the radius nr of disk C if the angular velocity of the system is to be 900 rpm after 25 revolutions.

17.5 The flywheel of a punching machine has a mass of 300 kg and a radius of gyration of 600 mm. Each punching operation requires 2500 J of work. (*a*) Knowing that the speed of the flywheel is 300 rpm just before a punching, determine the speed immediately after the punching. (*b*) If a constant 25-N·m couple is applied to the shaft of the flywheel, determine the number of revolutions executed before the speed is again 300 rpm.

17.6 The flywheel of a small punch rotates at 300 rpm. It is known that 1800 ft·lb of work must be done each time a hole is punched. It is desired that the speed of the flywheel after one punching be not less than 90 percent of the original speed of 300 rpm. (*a*) Determine the required moment of inertia of the flywheel. (*b*) If a constant 25-lb·ft couple is applied to the shaft of the flywheel, determine the number of revolutions which must occur between each punching, knowing that the initial velocity is to be 300 rpm at the start of each punching.

17.7 A disk of uniform thickness and initially at rest is placed in contact with the belt, which moves with a constant velocity **v**. Denoting by μ_k the coefficient of kinetic friction between the disk and the belt, derive an expression for the number of revolutions executed by the disk before it reaches a constant angular velocity.

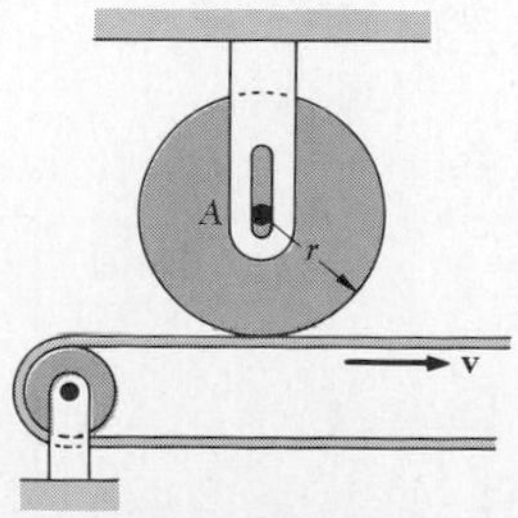

Fig. P17.7 and P17.8

17.8 Disk A has a mass of 4 kg and a radius $r = 90$ mm; it is at rest when it is placed in contact with the belt, which moves with a constant speed $v = 15$ m/s. Knowing that $\mu_k = 0.25$ between the disk and the belt, determine the number of revolutions executed by the disk before it reaches a constant angular velocity.

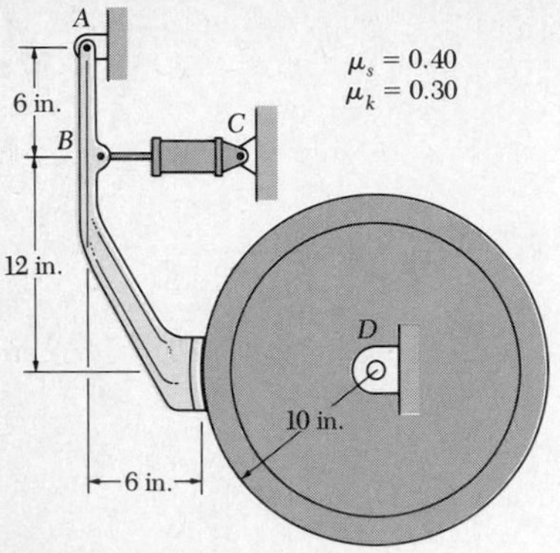

Fig. P17.9 and P17.10

17.9 The 10-in.-radius brake drum is attached to a larger flywheel which is not shown. The total mass moment of inertia of the flywheel and drum is 13.5 lb·ft·s^2. Knowing that the initial angular velocity is 180 rpm clockwise, determine the force which must be exerted by the hydraulic cylinder if the system is to stop in 50 revolutions.

17.10 Solve Prob. 17.9, assuming that the initial angular velocity of the flywheel is 180 rpm counterclockwise.

17.11 Using the principle of work and energy, solve Prob. 16.46*b*.

17.12 Gear *G* weighs 8 lb and has a radius of gyration of 2.4 in., while gear *D* weighs 30 lb and has a radius of gyration of 4.8 in. A couple $\mathbf{M} = (20 \text{ lb}\cdot\text{in.})\mathbf{i}$ is applied to shaft *CE*. Neglecting the mass of shafts *CE* and *FH*, determine (*a*) the number of revolutions of gear *G* required for its angular velocity to increase from 240 to 720 rpm, (*b*) the corresponding tangential force acting on gear *G*.

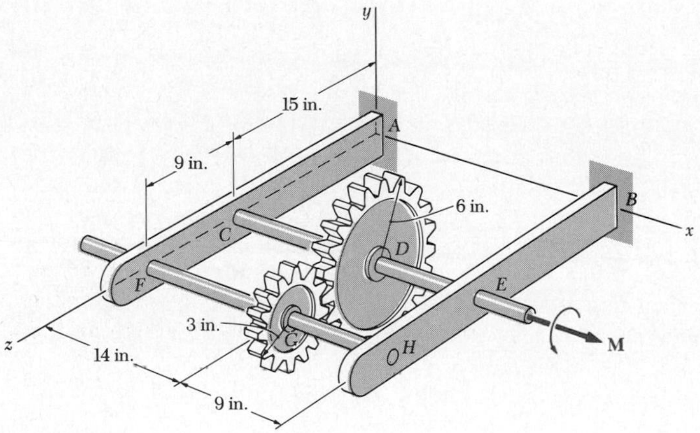

Fig. P17.12

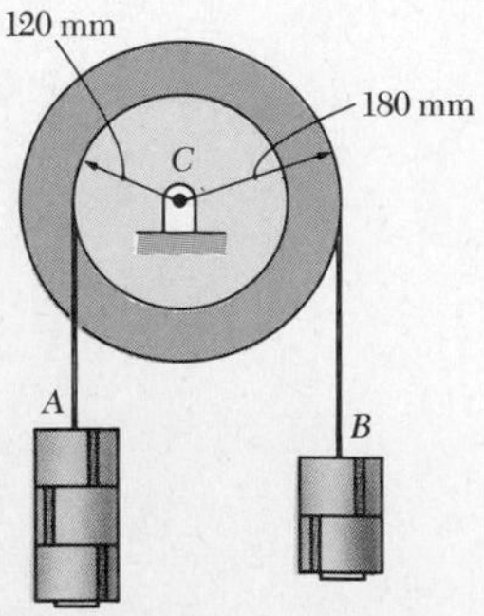

Fig. P17.14

17.13 Solve Prob. 17.12, assuming that the couple $\mathbf{M} = (20 \text{ lb}\cdot\text{in.})\mathbf{i}$ is applied to shaft *FH*.

17.14 The double pulley shown has a total mass of 6 kg and a centroidal radius of gyration of 135 mm. Five collars, each of mass 1.2 kg, are attached to cords *A* and *B* as shown. When the system is at rest and in equilibrium, one collar is removed from cord *A*. Knowing that the bearing friction is equivalent to a couple **M** of magnitude 0.5 N·m, determine the velocity of cord *A* after it has moved 600 mm.

17.15 Solve Prob. 17.14, assuming that one collar is removed from cord *B*.

17.16 The pulley shown has a mass of 5 kg and a radius of gyration of 150 mm. The 3-kg cylinder C is attached to a cord which is wrapped around the pulley as shown. A 1.8-kg collar B is then placed on the cylinder and the system is released from rest. After the cylinder has moved 300 mm, the collar is removed and the cylinder continues to move downward into a pit. Determine the velocity of cylinder C just before it strikes the bottom D of the pit.

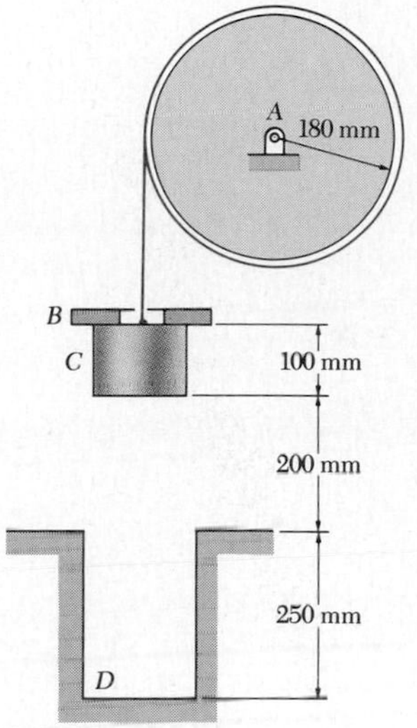

Fig. P17.16

17.17 and 17.18 A slender rod of length L and weight W is supported as shown. After the cable is cut the rod swings freely. (*a*) Determine the angular velocity of the rod as it first passes through a vertical position and the corresponding reaction at the pin support. (*b*) Solve part *a* for $W = 5$ lb and $L = 3$ ft.

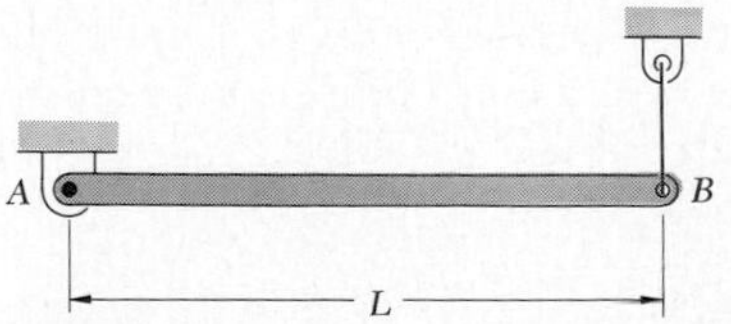

Fig. P17.17

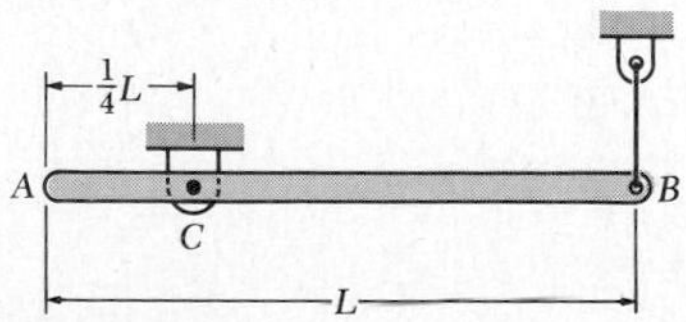

Fig. P17.18

17.19 A 4-kg slender rod rotates in a *vertical* plane about a pivot at B. A spring of constant $k = 200$ N/m and of unstretched length 150 mm is attached to the rod as shown. Knowing that in the position shown the rod has an angular velocity of 3 rad/s counterclockwise, determine the angular velocity of the rod after it has rotated through (*a*) 90°, (*b*) 180°.

17.20 Solve Prob. 17.19, assuming that the rod rotates in a *horizontal* plane.

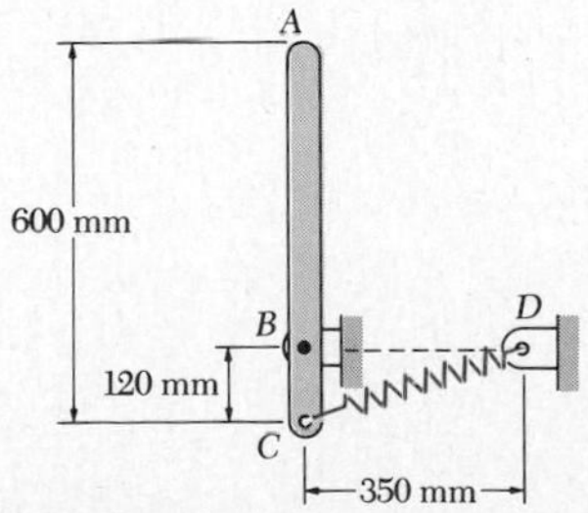

Fig. P17.19

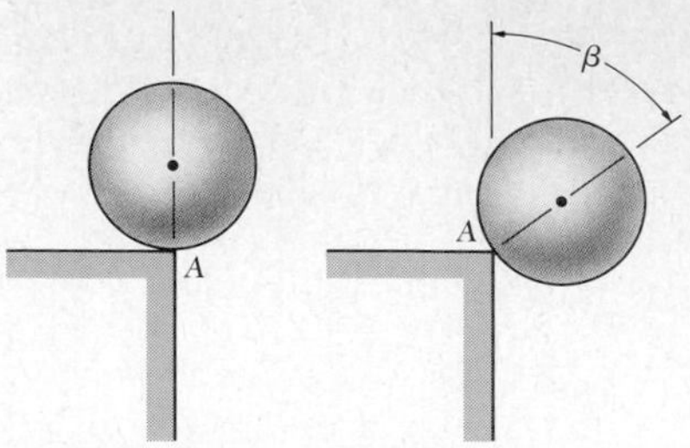

Fig. P17.21

17.21 A uniform sphere of radius r is placed at corner A and is given a slight clockwise motion. Assuming that the corner is sharp and becomes slightly embedded in the sphere, so that the coefficient of static friction at A is very large, determine (*a*) the angle β through which the sphere will have rotated when it loses contact with the corner, (*b*) the corresponding velocity of the center of the sphere.

17.22 Solve Prob. 17.21, assuming that the sphere is replaced by a uniform cylinder of radius r.

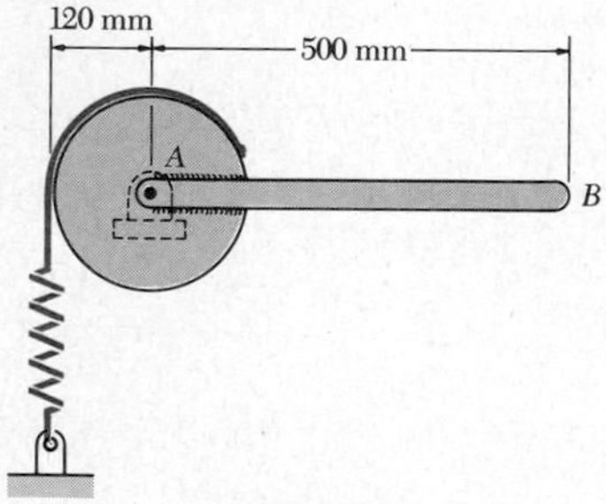

Fig. P17.23 and P17.24

17.23 The 5-kg slender rod AB is welded to the 3-kg uniform disk which rotates about a pivot at A. A spring of constant 75 N/m is attached to the disk and is unstretched when rod AB is horizontal. Knowing that the assembly is released from rest in the position shown, determine its angular velocity after it has rotated through 90°.

17.24 The 5-kg slender rod AB is welded to the 3-kg uniform disk which rotates about a pivot at A. A spring of constant 75 N/m is attached to the disk and is unstretched when rod AB is horizontal. Determine the required angular velocity of the assembly when it is in the position shown, if its angular velocity is to be 8 rad/s after it has rotated through 90° clockwise.

17.25 Two 6-lb slender rods are welded to the edge of an 8-lb uniform disk as shown. Knowing that in the position shown the assembly has an angular velocity of 9 rad/s clockwise, determine the largest and the smallest angular velocity of the assembly as it rotates about the pivot C.

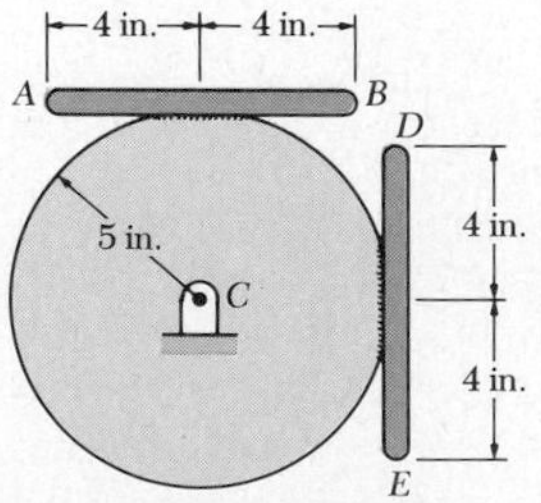

Fig. 17.25 and P17.26

17.26 Two 6-lb slender rods are welded to the edge of an 8-lb uniform disk as shown. The assembly is released from rest in the position shown and swings freely about the pivot C. Determine (*a*) the angular velocity of the assembly after it has rotated through 90°, (*b*) the maximum velocity attained by the assembly as it swings freely.

17.27 A flywheel of centroidal radius of gyration $\bar{k} = 25$ in. is rigidly attached to a shaft of radius $r = 1.25$ in. which may roll along parallel rails. Knowing that the system is released from rest, determine the velocity of the center of the shaft after it has moved 10 ft.

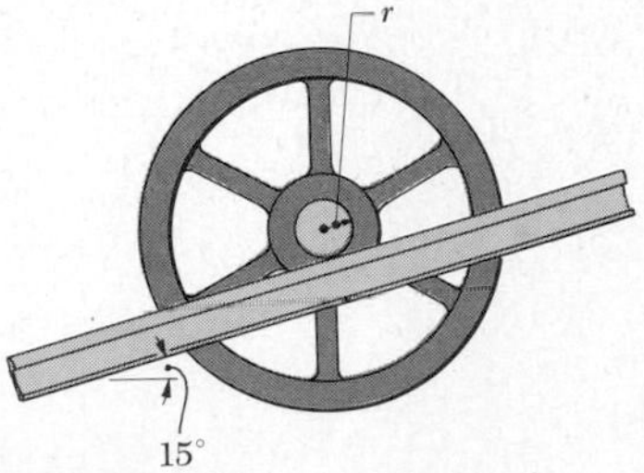

Fig. P17.27 and P17.28

17.28 A flywheel is rigidly attached to a 40-mm-radius shaft which rolls without sliding along parallel rails. The system is released from rest and attains a speed of 160 mm/s after moving 1.5 m along the rails. Determine the centroidal radius of gyration of the system.

17.29 The mass center G of a 5-kg wheel of radius $R = 300$ mm is located at a distance $r = 100$ mm from its geometric center C. The centroidal radius of gyration of the wheel is $\bar{k} = 150$ mm. As the wheel rolls without sliding, its angular velocity is observed to vary. Knowing that $\omega = 6$ rad/s in the position shown, determine (a) the angular velocity of the wheel when the mass center G is directly above the geometric center C, (b) the reaction at the horizontal surface at the same instant.

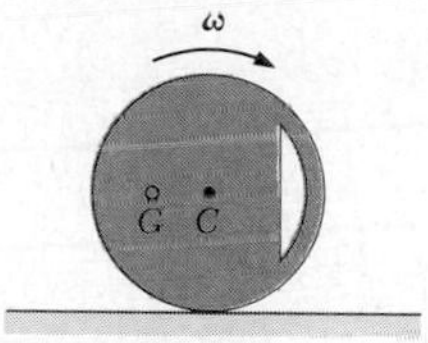

Fig. P17.29

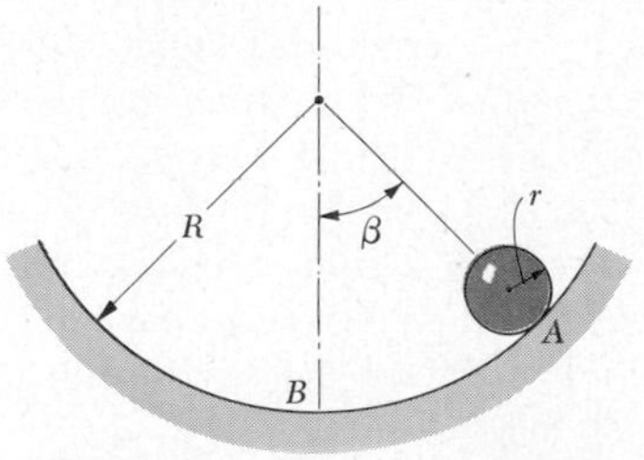

Fig. P17.30

17.30 A sphere of weight W and radius r rolls without slipping inside a curved surface of radius R. Knowing that the sphere is released from rest in the position shown, derive an expression (a) for the linear velocity of the sphere as it passes through B, (b) for the magnitude of the vertical reaction at that instant.

17.31 Solve Prob. 17.30, assuming that the sphere is replaced by a uniform cylinder of weight W and radius r.

17.32 A half section of pipe of mass m and radius r is released from rest in the position shown. Assuming that the pipe rolls without sliding, determine (a) its angular velocity after it has rolled through 90°, (b) the reaction at the horizontal surface at the same instant. [*Hint.* Note that $GO = 2r/\pi$ and that by the parallel-axis theorem, $\bar{I} = mr^2 - m(GO)^2$.]

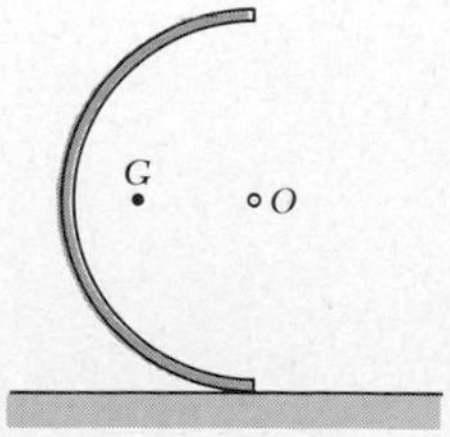

Fig. P17.32

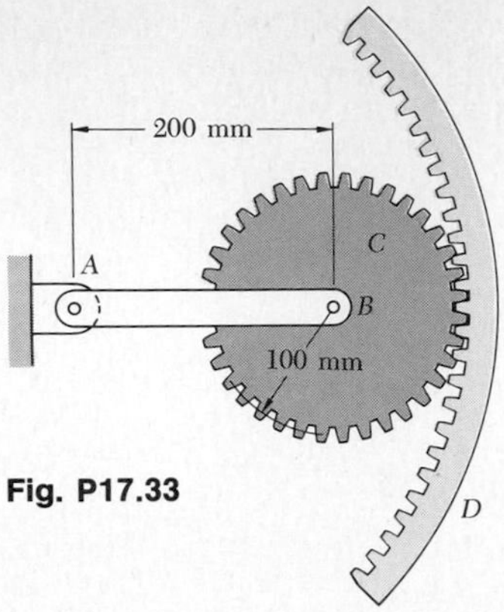

Fig. P17.33

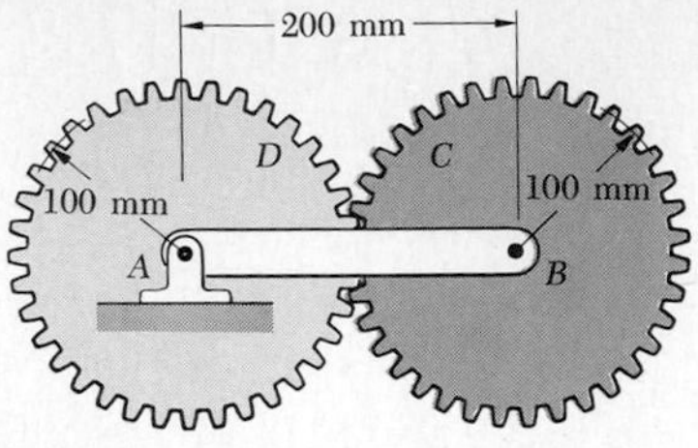

Fig. P17.34

17.33 and 17.34 Gear C has a mass of 3 kg and a centroidal radius of gyration of 75 mm. The uniform bar AB has a mass of 2.5 kg, and gear D is stationary. If the system is released from rest in the position shown, determine the velocity of point B after bar AB has rotated through 90°.

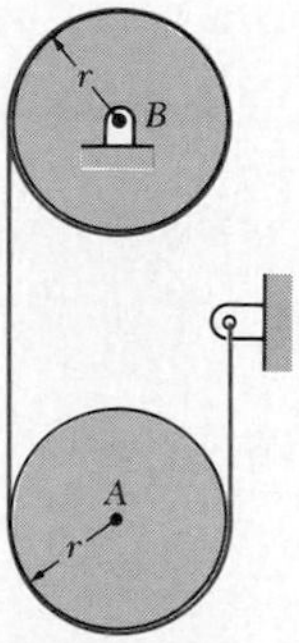

Fig. P17.35 and P17.36

17.35 Two uniform cylinders, each of weight $W = 14$ lb and radius $r = 5$ in., are connected by a belt as shown. Knowing that at the instant shown the angular velocity of cylinder B is 30 rad/s clockwise, determine (*a*) the distance through which cylinder A will rise before the angular velocity of cylinder B is reduced to 5 rad/s, (*b*) the tension in the portion of belt connecting the two cylinders.

17.36 Two uniform cylinders, each of weight $W = 14$ lb and radius $r = 5$ in., are connected by a belt as shown. If the system is released from rest, determine (*a*) the velocity of the center of cylinder A after it has moved through 3 ft, (*b*) the tension in the portion of belt connecting the two cylinders.

17.37 The 8-kg rod AB is attached by pins to two 5-kg uniform disks as shown. The assembly rolls without sliding on a horizontal surface. If the assembly is released from rest when $\theta = 60°$, determine (*a*) the angular velocity of the disks when $\theta = 180°$, (*b*) the force exerted by the surface on each disk at that instant.

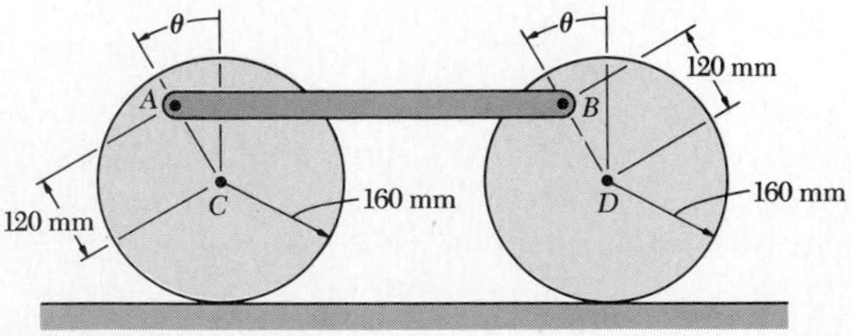

Fig. P17.37 and P17.38

17.38 The 8-kg rod AB is attached by pins to two 5-kg uniform disks as shown. The assembly rolls without sliding on a horizontal surface. Knowing that the velocity of rod AB is 700 mm/s to the left when $\theta = 0$, determine (*a*) the velocity of rod AB when $\theta = 180°$, (*b*) the force exerted by the surface on each disk at that instant.

17.39 The motion of the 5-lb slender bar AB is guided by collars of negligible weight which slide freely on the vertical and horizontal rods shown. Knowing that the bar is released from rest when $\theta = 30°$, determine the velocity of collars A and B when $\theta = 60°$.

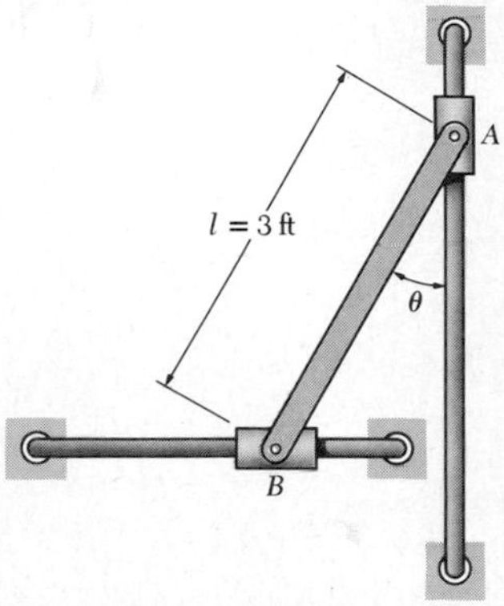

Fig. P17.39 and P17.40

17.40 The motion of the 5-lb slender bar AB is guided by collars of negligible weight which slide freely on the vertical and horizontal rods shown. Knowing that the bar is released from rest when $\theta = 45°$, determine the velocity of collar A when $\theta = 90°$.

17.41 The ends of a 4-kg rod AB are constrained to move along slots cut in a vertical plate as shown. A spring of constant 500 N/m is attached to end A in such a way that its tension is zero when $\theta = 0$. If the rod is released from rest when $\theta = 0$, determine the angular velocity of the rod and the velocity of end B when $\theta = 30°$.

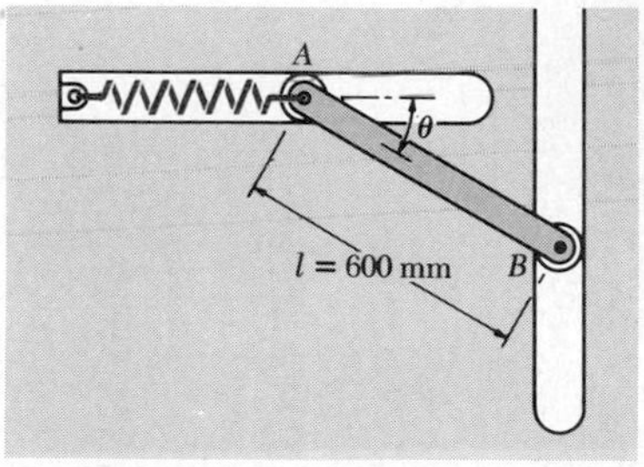

Fig. P17.41 and P17.42

17.42 The ends of a 4-kg rod AB are constrained to move along slots cut in a vertical plate as shown. A spring of constant 500 N/m is attached to end A in such a way that its tension is zero when $\theta = 0$. If the rod is released from rest when $\theta = 50°$, determine the angular velocity of the rod and the velocity of end B when $\theta = 0$.

17.43 The motion of a slender rod of length R is guided by pins at A and B which slide freely in slots cut in a vertical plate as shown. If end B is moved slightly to the right and then released, determine the angular velocity of the rod and the velocity of its mass center (*a*) as end B passes through point D, (*b*) as end A passes through point D.

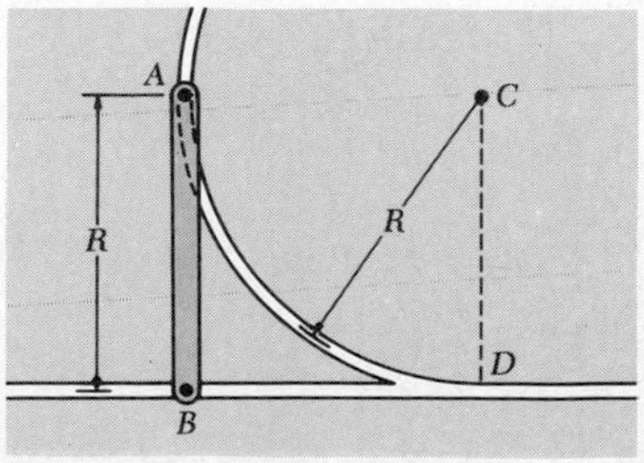

Fig. P17.43 and P17.44

17.44 The motion of a slender rod of length R is guided by pins at A and B which slide freely in slots cut in a vertical plate as shown. If end B is moved slightly to the left and then released, determine the angular velocity of the rod and the velocity of its mass center (*a*) at the instant when the velocity of end B is zero, (*b*) as end B passes through point D.

17.45 The uniform rods AB and BC are of mass 2 kg and 5 kg, respectively, and collar C has a mass of 3 kg. If the system is released from rest in the position shown, determine the velocity of point B after rod AB has rotated through 90°.

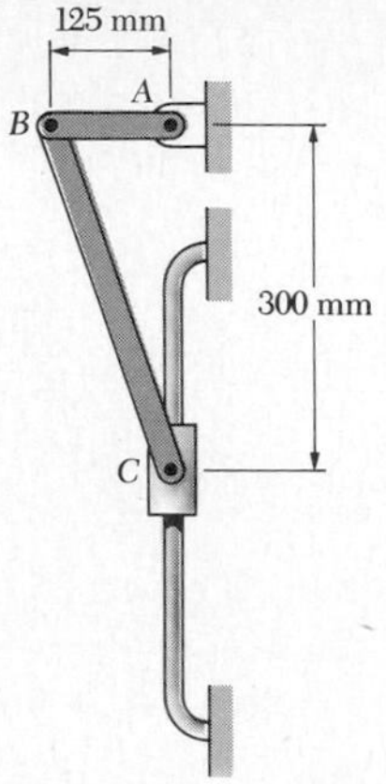

Fig. P17.45 and P17.46

17.46 The uniform rods AB and BC are of mass 2 kg and 5 kg, respectively, and collar C has a mass of 3 kg. Knowing that at the instant shown the velocity of collar C is 0.8 m/s downward, determine the velocity of point B after rod AB has rotated through 90°.

17.47 Two uniform rods, each of mass m and length l, are connected to form the linkage shown. End D of rod BD may slide freely in the vertical slot, while end A of rod AB is attached to a fixed pin support. If the system is released from rest in the position shown, determine the velocity of end D at the instant when (a) ends A and D are at the same elevation, (b) rod AB is horizontal.

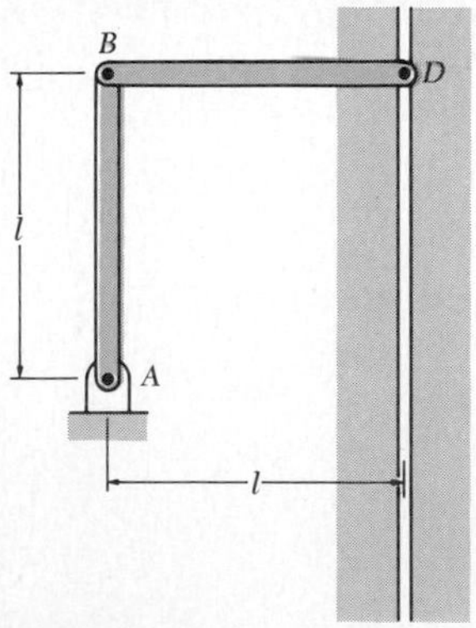

Fig. P17.47

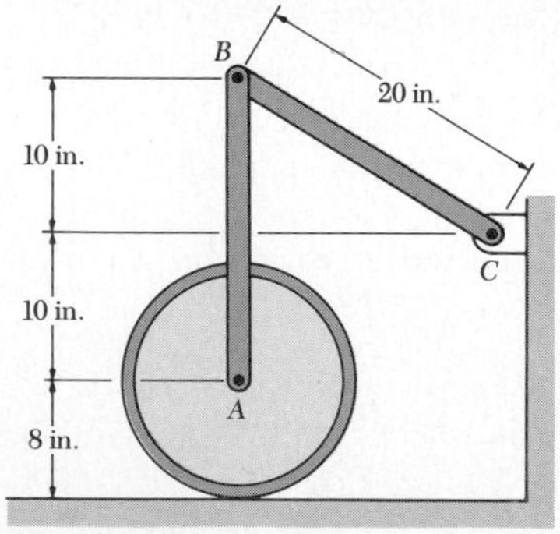

Fig. P17.48

17.48 Wheel A weighs 9 lb, has a centroidal radius of gyration of 6 in., and rolls without sliding on the horizontal surface. Each of the uniform rods AB and BC is 20 in. long and weighs 5 lb. If point A is moved slightly to the left and released, determine the velocity of point A as rod BC passes through a horizontal position.

17.49 The motor shown runs a machine attached to the shaft at A. The motor develops 5 hp and runs at a constant speed of 360 rpm. Determine the magnitude of the couple exerted (a) by the shaft on pulley A, (b) by the motor on pulley B.

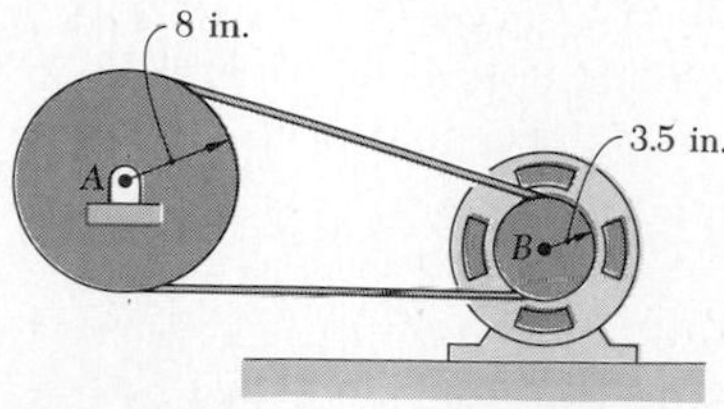

Fig. P17.49

17.50 Three shafts and four gears are used to form a gear train which will transmit 7.5 kW from the motor at A to a machine tool at F. (Bearings for the shafts are omitted in the sketch.) Knowing that the frequency of the motor is 30 Hz, determine the magnitude of the couple which is applied to shaft (a) AB, (b) CD, (c) EF.

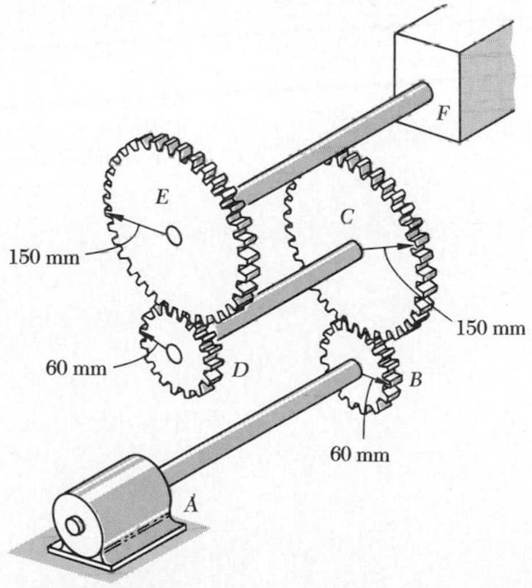

Fig. P17.50

17.51 Determine the moment of the couple which must be exerted by a motor to develop 250 W at a speed of (a) 3450 rpm, (b) 1150 rpm.

17.52 Knowing that the maximum allowable couple which can be applied to a shaft is 8 kip·in., determine the maximum horsepower which can be transmitted by the shaft (a) at 60 rpm, (b) at 300 rpm.

17.8. Principle of Impulse and Momentum for the Plane Motion of a Rigid Body. We shall now apply the principle of impulse and momentum to the analysis of the plane motion of rigid bodies and of systems of rigid bodies. As was pointed out in Chap. 13, the method of impulse and momentum is particularly well adapted to the solution of problems involving time and velocities. Moreover, the principle of impulse and momentum provides the only practicable method for the solution of problems involving impulsive motion or impact (Secs. 17.11 and 17.12).

Considering again a rigid body as made of a large number of particles P_i, we recall from Sec. 14.9 that the system formed by the momenta of the particles at time t_1 and the system of the impulses of the external forces applied from t_1 to t_2 are together equipollent to the system formed by the momenta of the particles at time t_2. Since the vectors associated with a rigid body may be considered as sliding vectors, it follows (Sec. 3.19) that the systems of vectors shown in Fig. 17.6 are not only equipollent but truly *equivalent* in the sense that the vectors on the left-hand side of the equals sign may be transformed into the vectors on the right-hand side through the use of the fundamental operations listed in Sec. 3.13. We therefore write

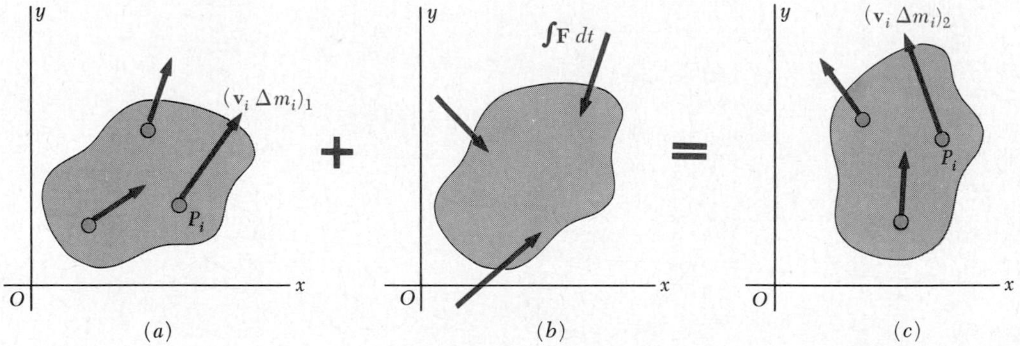

Fig. 17.6

$$\text{Syst Momenta}_1 + \text{Syst Ext Imp}_{1\to 2} = \text{Syst Momenta}_2 \tag{17.14}$$

But the momenta $\mathbf{v}_i\,\Delta m_i$ of the particles may be reduced to a vector attached at G, equal to their sum

$$\mathbf{L} = \sum_{i=1}^{n} \mathbf{v}_i\,\Delta m_i$$

and to a couple of moment equal to the sum of their moments about G

$$\mathbf{H}_G = \sum_{i=1}^{n} \mathbf{r}'_i \times \mathbf{v}_i\,\Delta m_i$$

We recall from Sec. 14.3 that $\mathbf{L}$ and $\mathbf{H}_G$ define, respectively, the linear momentum and the angular momentum about G of the system of particles forming the rigid body. We also note from Eq. (14.14) that $\mathbf{L} = m\overline{\mathbf{v}}$. On the other hand, restricting the present analysis to the plane motion of a rigid slab or of a rigid body symmetrical with respect to the reference plane, we recall from Eq. (16.4) that $\mathbf{H}_G = \bar{I}\boldsymbol{\omega}$. We thus conclude that the system of the momenta $\mathbf{v}_i\,\Delta m_i$ is equivalent to the *linear momentum vector* $m\overline{\mathbf{v}}$ attached at G and to the *angular momentum couple* $\bar{I}\boldsymbol{\omega}$ (Fig. 17.7).

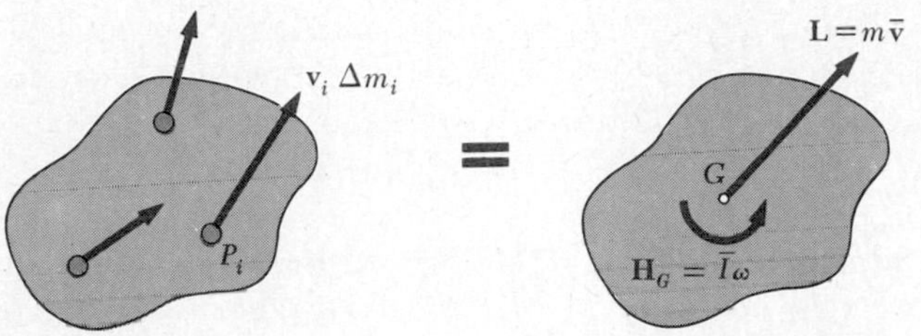

Fig. 17.7

Observing that the system of momenta reduces to the vector $m\overline{\mathbf{v}}$ in the particular case of a translation ($\boldsymbol{\omega} = 0$) and to the couple $\bar{I}\boldsymbol{\omega}$ in the particular case of a centroidal rotation ($\overline{\mathbf{v}} = 0$), we verify once more that the plane motion of a rigid body symmetrical with respect to the reference plane may be resolved into a translation with the mass center G and a rotation about G.

Replacing the system of momenta in parts *a* and *c* of Fig. 17.6 by the equivalent linear momentum vector and angular momentum couple, we obtain the three diagrams shown in Fig. 17.8. This figure expresses as a

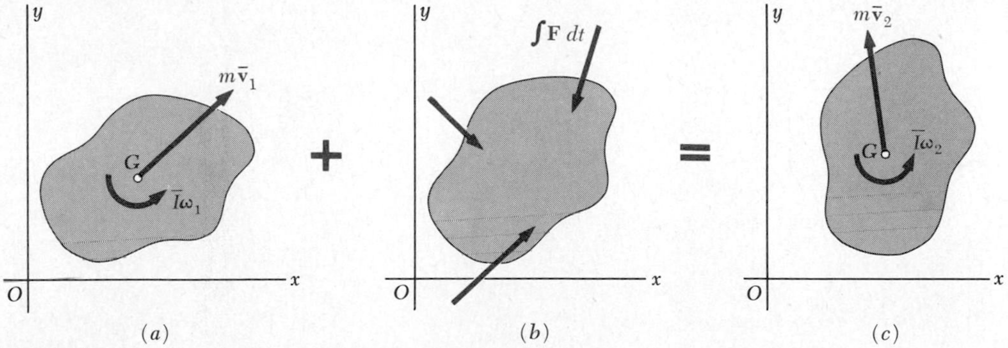

Fig. 17.8

free-body-diagram equation the fundamental relation (17.14) in the case of the plane motion of a rigid slab or of a rigid body symmetrical with respect to the reference plane.

Three equations of motion may be derived from Fig. 17.8. Two equations are obtained by summing and equating the x and y *components* of the momenta and impulses, and the third by summing and equating the *moments* of these vectors *about any given point*. The coordinate axes may be chosen fixed in space, or they may be allowed to move with the mass center of the body while maintaining a fixed direction. In either case, the point about which moments are taken should keep the same position relative to the coordinate axes during the interval of time considered.

In deriving the three equations of motion for a rigid body, care should be taken not to add indiscriminately linear and angular momenta. Confusion will be avoided if it is kept in mind that $m\bar{v}_x$ and $m\bar{v}_y$ represent the *components of a vector,* namely, the linear momentum vector $m\bar{\mathbf{v}}$, while $\bar{I}\omega$ represents the *magnitude of a couple,* namely, the angular momentum couple $\bar{I}\boldsymbol{\omega}$. Thus the quantity $\bar{I}\omega$ should be added only to the *moment* of the linear momentum $m\bar{\mathbf{v}}$, never to this vector itself nor to its components. All quantities involved will then be expressed in the same units, namely N·m·s or lb·ft·s.

Noncentroidal Rotation. In this particular case of plane motion, the magnitude of the velocity of the mass center of the body is $\bar{v} = \bar{r}\omega$, where $\bar{r}$ represents the distance from the mass center to the fixed axis of rotation and $\boldsymbol{\omega}$ the angular velocity of the body at the instant considered; the magnitude of the momentum vector attached at G is thus $m\bar{v} = m\bar{r}\omega$. Summing the moments about O of the momentum vector and momentum couple (Fig. 17.9) and using the parallel-axis theorem for moments of inertia, we find that the angular momentum $\mathbf{H}_O$ of the body about O has the magnitude†

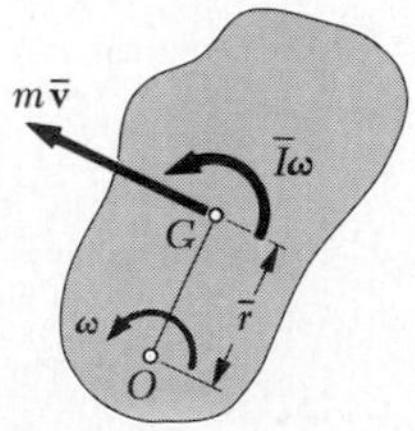

Fig. 17.9

$$\bar{I}\omega + (m\bar{r}\omega)\bar{r} = (\bar{I} + m\bar{r}^2)\omega = I_O\omega \tag{17.15}$$

Equating the moments about O of the momenta and impulses in (17.14), we write

$$I_O\omega_1 + \sum\int_{t_1}^{t_2} M_O\,dt = I_O\omega_2 \tag{17.16}$$

In the general case of plane motion of a rigid body symmetrical with respect to the reference plane, Eq. (17.16) may be used with respect to the instantaneous axis of rotation under certain conditions. It is recommended, however, that all problems of plane motion be solved by the general method described earlier in this section.

17.9. Systems of Rigid Bodies. The motion of several rigid bodies may be analyzed by applying the principle of impulse and momentum to each body separately (Sample Prob. 17.6).

However, in solving problems involving no more than three unknowns (including the impulses of unknown reactions), it is often found convenient

†Note that the sum $\mathbf{H}_A$ of the moments about an arbitrary point A of the momenta of the particles of a rigid slab is, in general, *not* equal to $I_A\omega$. (See Prob. 17.67.)

to apply the principle of impulse and momentum to the system as a whole. The momentum and impulse diagrams are drawn for the entire system of bodies. The diagrams of momenta should include a momentum vector, a momentum couple, or both, for each moving part of the system. Impulses of forces internal to the system may be omitted from the impulse diagram since they occur in pairs of equal and opposite vectors. Summing and equating successively the x components, y components, and moments of all vectors involved, one obtains three relations which express that the momenta at time t_1 and the impulses of the external forces form a system equipollent to the system of the momenta at time t_2.† Again, care should be taken not to add indiscriminately linear and angular momenta; each equation should be checked to make sure that consistent units have been used. This approach has been used in Sample Prob. 17.8 and, further on, in Sample Probs. 17.9 and 17.10.

17.10. Conservation of Angular Momentum. When no external force acts on a rigid body or a system of rigid bodies, the impulses of the external forces are zero and the system of the momenta at time t_1 is equipollent to the system of the momenta at time t_2. Summing and equating successively the x components, y components, and moments of the momenta at times t_1 and t_2, we conclude that the total linear momentum of the system is conserved in any direction and that its total angular momentum is conserved about any point.

There are many engineering applications, however, in which *the linear momentum is not conserved* yet *the angular momentum* $\mathbf{H}_O$ of the system about a given point O is *conserved:*

$$(\mathbf{H}_O)_1 = (\mathbf{H}_O)_2 \tag{17.17}$$

Such cases occur when the lines of action of all external forces pass through O or, more generally, when the sum of the angular impulses of the external forces about O is zero.

Problems involving *conservation of angular momentum* about a point O may be solved by the general method of impulse and momentum, i.e., by drawing momentum and impulse diagrams as described in Secs. 17.8 and 17.9. Equation (17.17) is then obtained by summing and equating moments about O (Sample Prob. 17.8). As we shall see later in Sample Prob. 17.9, two additional equations may be written by summing and equating x and y components; these equations may be used to determine two unknown linear impulses, such as the impulses of the reaction components at a fixed point.

†Note that as in Sec. 16.7, we cannot speak of *equivalent* systems since we are not dealing with a single rigid body.

SAMPLE PROBLEM 17.6

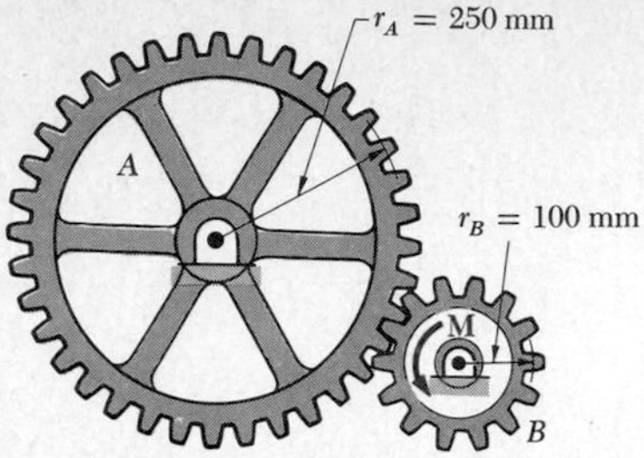

Gear A has a mass of 10 kg and a radius of gyration of 200 mm, while gear B has a mass of 3 kg and a radius of gyration of 80 mm. The system is at rest when a couple **M** of magnitude 6 N·m is applied to gear B. Neglecting friction, determine (*a*) the time required for the angular velocity of gear B to reach 600 rpm, (*b*) the tangential force which gear B exerts on gear A. These gears have been previously considered in Sample Prob. 17.2.

Solution. We apply the principle of impulse and momentum to each gear separately. Since all forces and the couple are constant, their impulses are obtained by multiplying them by the unknown time t. We recall from Sample Prob. 17.2 that the centroidal moments of inertia and the final angular velocities are

$$\bar{I}_A = 0.400\ \text{kg}\cdot\text{m}^2 \qquad \bar{I}_B = 0.0192\ \text{kg}\cdot\text{m}^2$$
$$(\omega_A)_2 = 25.1\ \text{rad/s} \qquad (\omega_B)_2 = 62.8\ \text{rad/s}$$

Principle of Impulse and Momentum for Gear A. The systems of initial momenta, impulses, and final momenta are shown in three separate sketches.

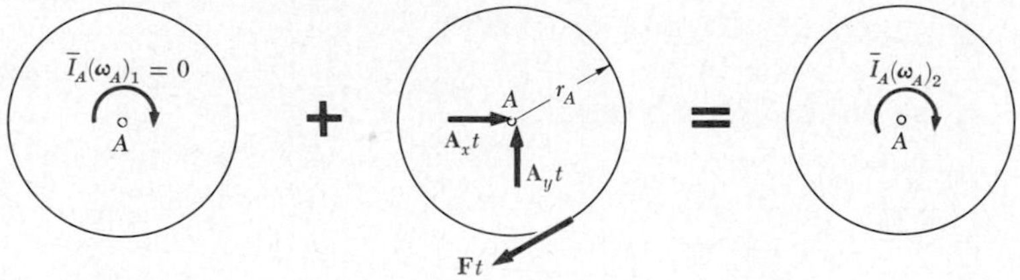

$$\textbf{Syst Momenta}_1 + \textbf{Syst Ext Imp}_{1\to 2} = \textbf{Syst Momenta}_2$$

$+\circlearrowleft$ moments about A: $\quad 0 - Ftr_A = -\bar{I}_A(\omega_A)_2$

$$Ft(0.250\ \text{m}) = (0.400\ \text{kg}\cdot\text{m}^2)(25.1\ \text{rad/s})$$
$$Ft = 40.2\ \text{N}\cdot\text{s}$$

Principle of Impulse and Momentum for Gear B.

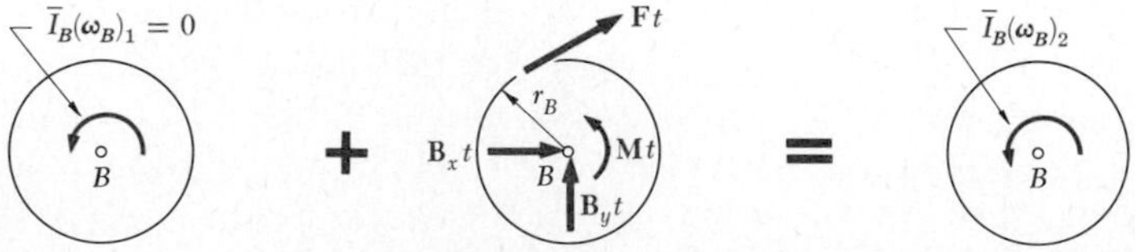

$$\textbf{Syst Momenta}_1 + \textbf{Syst Ext Imp}_{1\to 2} = \textbf{Syst Momenta}_2$$

$+\circlearrowleft$ moments about B: $\quad 0 + Mt - Ftr_B = \bar{I}_B(\omega_B)_2$

$$+(6\ \text{N}\cdot\text{m})t - (40.2\ \text{N}\cdot\text{s})(0.100\ \text{m}) = (0.0192\ \text{kg}\cdot\text{m}^2)(62.8\ \text{rad/s})$$

$$t = 0.871\ \text{s} \quad ◄$$

Recalling that $Ft = 40.2\ \text{N}\cdot\text{s}$, we write

$$F(0.871\ \text{s}) = 40.2\ \text{N}\cdot\text{s} \qquad F = +46.2\ \text{N}$$

Thus, the force exerted by gear B on gear A is $\quad \mathbf{F} = 46.2\ \text{N} \swarrow \quad ◄$

SAMPLE PROBLEM 17.7

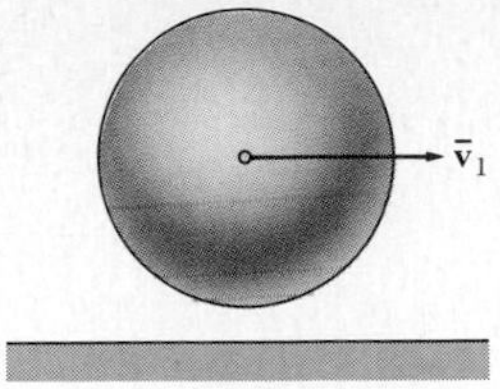

A uniform sphere of mass m and radius r is projected along a rough horizontal surface with a linear velocity $\bar{\mathbf{v}}_1$ and no angular velocity. Denoting by μ_k the coefficient of kinetic friction between the sphere and the surface, determine (*a*) the time t_2 at which the sphere will start rolling without sliding, (*b*) the linear and angular velocities of the sphere at time t_2.

Solution. While the sphere is sliding relative to the surface, it is acted upon by the normal force **N**, the friction force **F**, and its weight **W** of magnitude $W = mg$.

Principle of Impulse and Momentum. We apply the principle of impulse and momentum to the sphere from the time $t_1 = 0$ when it is placed on the surface until the time $t_2 = t$ when it starts rolling without sliding.

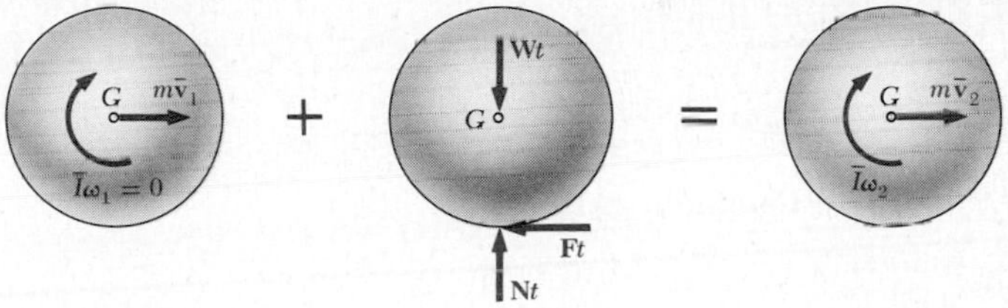

Syst Momenta$_1$ + Syst Ext Imp$_{1\to 2}$ = Syst Momenta$_2$

$+\uparrow y$ components: $\quad Nt - Wt = 0 \qquad (1)$

$\xrightarrow{+}$ x components: $\quad m\bar{v}_1 - Ft = m\bar{v}_2 \qquad (2)$

$+\downarrow$moments about G: $\quad Ftr = \bar{I}\omega_2 \qquad (3)$

From (1) we obtain $N = W = mg$. During the entire time interval considered, sliding occurs at point C and we have $F = \mu_k N = \mu_k mg$. Substituting for F into (2), we write

$$m\bar{v}_1 - \mu_k mgt = m\bar{v}_2 \qquad \bar{v}_2 = \bar{v}_1 - \mu_k gt \qquad (4)$$

Substituting $F = \mu_k mg$ and $\bar{I} = \frac{2}{5}mr^2$ into (3),

$$\mu_k mgtr = \tfrac{2}{5}mr^2\omega_2 \qquad \omega_2 = \frac{5}{2}\frac{\mu_k g}{r}t \qquad (5)$$

The sphere will start rolling without sliding when the velocity $\mathbf{v}_C$ of the point of contact is zero. At that time, point C becomes the instantaneous center of rotation, and we have $\bar{v}_2 = r\omega_2$. Substituting from (4) and (5), we write

$$\bar{v}_2 = r\omega_2 \qquad \bar{v}_1 - \mu_k gt = r\left(\frac{5}{2}\frac{\mu_k g}{r}t\right) \qquad t = \frac{2}{7}\frac{\bar{v}_1}{\mu_k g} \quad ◄$$

Substituting this expression for t into (5),

$$\omega_2 = \frac{5}{2}\frac{\mu_k g}{r}\left(\frac{2}{7}\frac{\bar{v}_1}{\mu_k g}\right) \qquad \omega_2 = \frac{5}{7}\frac{\bar{v}_1}{r} \qquad \omega_2 = \frac{5}{7}\frac{\bar{v}_1}{r} \downarrow \quad ◄$$

$$\bar{v}_2 = r\omega_2 \qquad \bar{v}_2 = r\left(\frac{5}{7}\frac{v_1}{r}\right) \qquad \bar{\mathbf{v}}_2 = \tfrac{5}{7}\bar{v}_1 \rightarrow \quad ◄$$

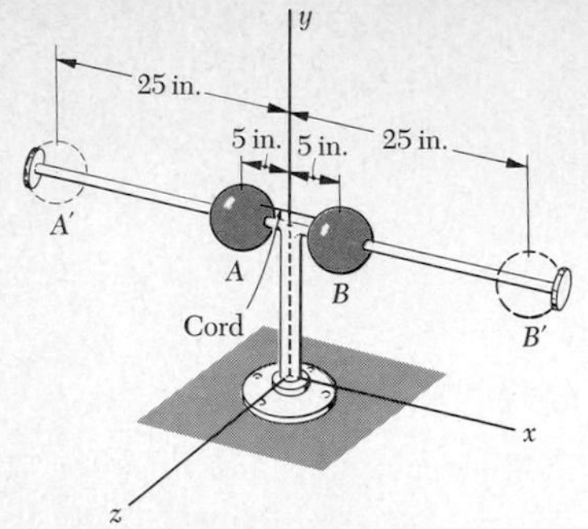

SAMPLE PROBLEM 17.8

Two solid spheres of radius 3 in., weighing 2 lb each, are mounted at A and B on the horizontal rod $A'B'$, which rotates freely about the vertical with a counterclockwise angular velocity of 6 rad/s. The spheres are held in position by a cord which is suddenly cut. Knowing that the centroidal moment of inertia of the rod and pivot is $\bar{I}_R = 0.25 \text{ lb}\cdot\text{ft}\cdot\text{s}^2$, determine (*a*) the angular velocity of the rod after the spheres have moved to positions A' and B', (*b*) the energy lost due to the plastic impact of the spheres and the stops at A' and B'.

***a.* Principle of Impulse and Momentum.** In order to determine the final angular velocity of the rod, we shall express that the initial momenta of the various parts of the system and the impulses of the external forces are together equipollent to the final momenta of the system.

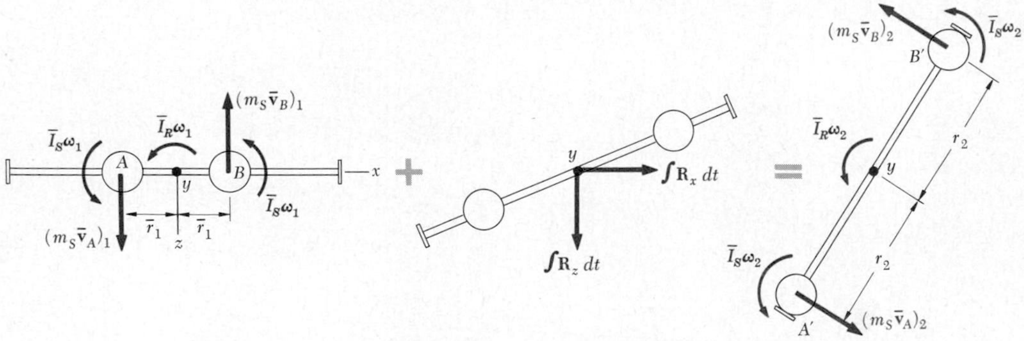

Syst Momenta$_1$ + Syst Ext Imp$_{1\to2}$ = Syst Momenta$_2$

Observing that the external forces consist of the weights and the reaction at the pivot, which have no moment about the y axis, and noting that $\bar{v}_A = \bar{v}_B = \bar{r}\omega$, we write

$+\circlearrowleft$moments about y axis:

$$2(m_S\bar{r}_1\omega_1)\bar{r}_1 + 2\bar{I}_S\omega_1 + \bar{I}_R\omega_1 = 2(m_S\bar{r}_2\omega_2)\bar{r}_2 + 2\bar{I}_S\omega_2 + \bar{I}_R\omega_2$$

$$(2m_S\bar{r}_1^2 + 2\bar{I}_S + \bar{I}_R)\omega_1 = (2m_S\bar{r}_2^2 + 2\bar{I}_S + \bar{I}_R)\omega_2 \qquad (1)$$

which expresses that *the angular momentum of the system about the y axis is conserved.* We now compute

$$\bar{I}_S = \tfrac{2}{5}m_Sa^2 = \frac{2}{5}\left(\frac{2 \text{ lb}}{32.2 \text{ ft/s}^2}\right)(\tfrac{3}{12}\text{ ft})^2 = 0.00155 \text{ lb}\cdot\text{ft}\cdot\text{s}^2$$

$$m_S\bar{r}_1^2 = \frac{2}{32.2}\left(\frac{5}{12}\right)^2 = 0.0108 \qquad m_S\bar{r}_2^2 = \frac{2}{32.2}\left(\frac{25}{12}\right)^2 = 0.2696$$

Substituting these values and $\bar{I}_R = 0.25$, $\omega_1 = 6$ rad/s into (1):

$$0.275(6 \text{ rad/s}) = 0.792\omega_2 \qquad \omega_2 = 2.08 \text{ rad/s} \circlearrowleft \blacktriangleleft$$

***b.* Energy Lost.** The kinetic energy of the system at any instant is

$$T = 2(\tfrac{1}{2}m_S\bar{v}^2 + \tfrac{1}{2}\bar{I}_S\omega^2) + \tfrac{1}{2}\bar{I}_R\omega^2 = \tfrac{1}{2}(2m_S\bar{r}^2 + 2\bar{I}_S + \bar{I}_R)\omega^2$$

Recalling the numerical values found above, we have

$$T_1 = \tfrac{1}{2}(0.275)(6)^2 = 4.95 \text{ ft}\cdot\text{lb} \qquad T_2 = \tfrac{1}{2}(0.792)(2.08)^2 = 1.713 \text{ ft}\cdot\text{lb}$$

$$\Delta T = T_2 - T_1 = 1.71 - 4.95 \qquad \Delta T = -3.24 \text{ ft}\cdot\text{lb} \blacktriangleleft$$

Problems

17.53 The rotor of a steam turbine has an angular velocity of 5400 rpm when the steam supply is suddenly cut off. The rotor weighs 200 lb and has a radius of gyration of 6 in. If kinetic friction produces a couple of magnitude 24 lb·in., determine the time required for the rotor to coast to rest.

17.54 A large flywheel has a mass of 2500 kg and a radius of gyration of 1.06 m. It is observed that 8.7 min is required for the flywheel to coast to rest from an angular velocity of 300 rpm. Determine the average magnitude of the couple due to kinetic friction in the bearings of the flywheel.

17.55 Three disks of the same thickness and same material are attached to a shaft as shown. Disks A and B each have a mass of 8 kg and a radius $r =$ 240 mm. A couple **M** of magnitude 40 N·m is applied to disk A when the system is at rest. Determine the radius nr of disk C if the angular velocity of the system is to be 900 rpm after 3 s.

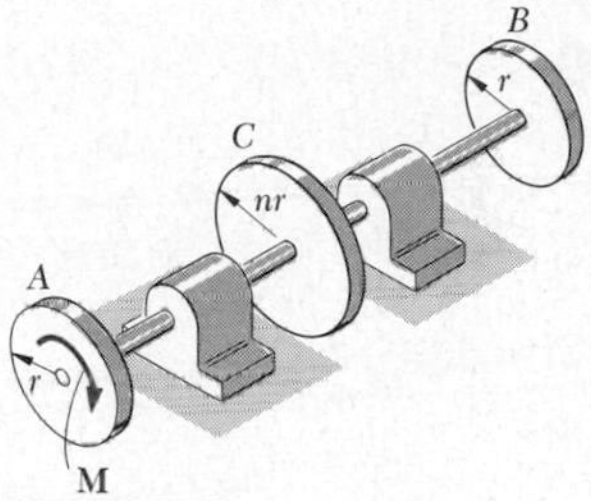

Fig. P17.55

17.56 A bolt located 2 in. from the center of an automobile wheel is tightened by applying the couple shown for 0.10 s. Assuming that the wheel is free to rotate and is initially at rest, determine the resulting angular velocity of the wheel. The wheel weighs 42 lb and has a radius of gyration of 11 in.

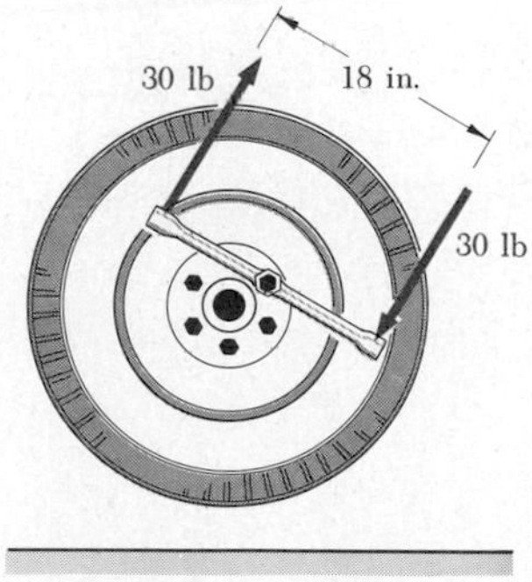

Fig. P17.56

17.57 A disk of constant thickness, initially at rest, is placed in contact with a belt which moves with a constant velocity **v**. The link AB connecting the center of the disk to the support at B has a negligible weight. Denoting by μ_k the coefficient of kinetic friction between the disk and the belt, derive an expression for the time required for the disk to reach a constant angular velocity.

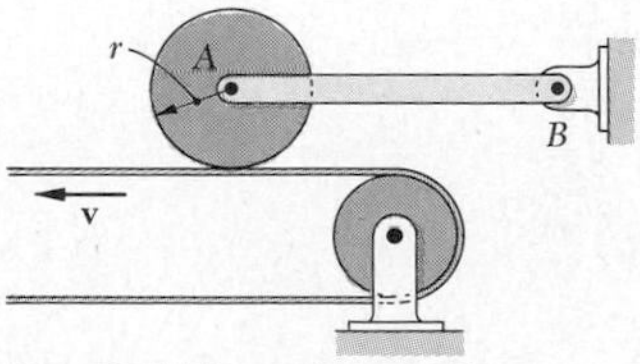

Fig. P17.57 and P17.58

17.58 Disk A has a mass of 3 kg and a radius $r = 90$ mm; it is placed in contact with the belt, which moves with a constant speed $v = 18$ m/s. The link AB connecting the center of the disk to the support at B has a negligible weight. Knowing that $\mu_k = 0.20$ between the disk and the belt, determine the time required for the disk to reach a constant angular velocity.

17.59 Solve Prob. 17.58, assuming that the link AB has a mass of 750 g.

17.60 Each of the double pulleys shown has a centroidal mass moment of inertia of $0.10\ \mathrm{kg \cdot m^2}$, an inner radius of 80 mm, and an outer radius of 120 mm. Neglecting bearing friction, determine (*a*) the velocity of the cylinder 3 s after the system is released from rest, (*b*) the tension in the cord connecting the pulleys.

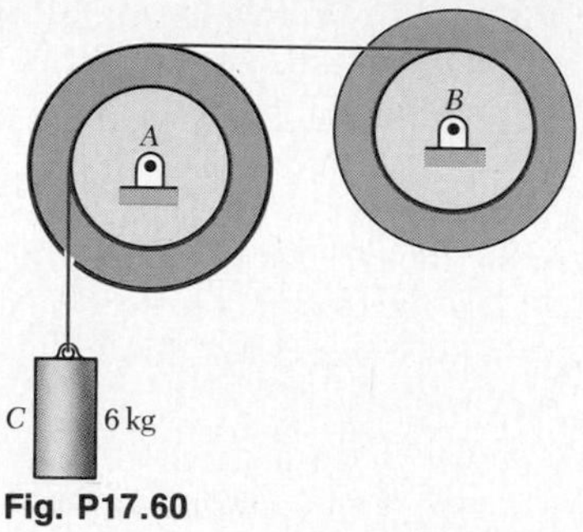

Fig. P17.60

17.61 Solve Prob. 17.60, assuming that the bearing friction at *A* and at *B* is equivalent to a couple of magnitude $0.25\ \mathrm{N \cdot m}$.

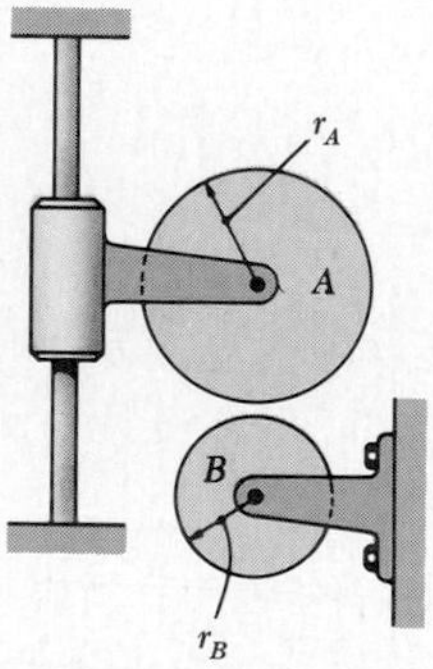

Fig. P17.62 and P17.63

17.62 Disk *B* has a mass $m_B = 3$ kg, a radius $r_B = 80$ mm, and an initial angular velocity $\omega_0 = 600$ rpm clockwise. Disk *A* has a mass $m_A = 5$ kg, a radius $r_A = 120$ mm, and is at rest when it is brought into contact with disk *B*. Neglecting friction in the bearings, determine (*a*) the final angular velocity of each disk, (*b*) the total impulse of the friction force which acted on disk *A*.

17.63 Disk *A* is at rest when it is brought into contact with disk *B*, which has an initial angular velocity $\boldsymbol{\omega}_0$. Show that the final angular velocity of disk *B* depends only on ω_0 and the ratio of the masses m_A and m_B of the two disks.

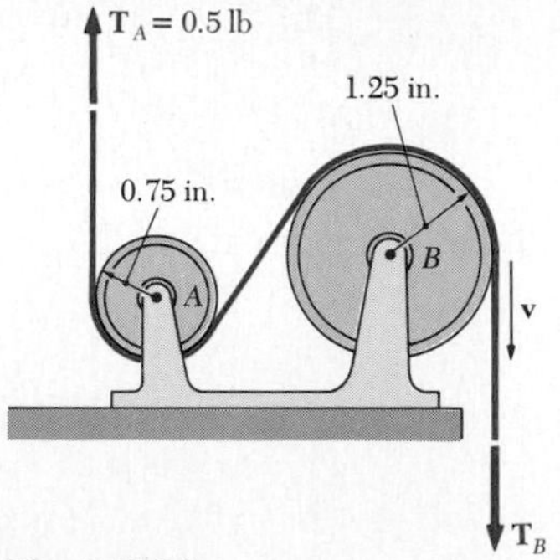

Fig. P17.64

17.64 A computer tape moves over the two drums shown. Drum *A* weighs 0.90 lb and has a radius of gyration of 0.70 in., while drum *B* weighs 1.8 lb and has a radius of gyration of 1.10 in. In the portion of tape above drum *A* the tension is constant and equal to $T_A = 0.5$ lb. Knowing that the tape is initially at rest, determine (*a*) the required constant tension T_B if the velocity of the tape is to be $v = 8$ ft/s after 0.2 s, (*b*) the corresponding tension in the portion of tape between the drums.

17.65 Show that the system of momenta for a rigid slab in plane motion reduces to a single vector, and express the distance from the mass center G to the line of action of this vector in terms of the centroidal radius of gyration $\bar{k}$ of the slab, the magnitude $\bar{v}$ of the velocity of G, and the angular velocity ω.

17.66 Show that, when a rigid slab rotates about a fixed axis through O perpendicular to the slab, the system of momenta of its particles is equivalent to a single vector of magnitude $m\bar{r}\omega$, perpendicular to the line OG, and applied to a point P on this line, called the *center of percussion*, at a distance $GP = \bar{k}^2/\bar{r}$ from the mass center of the slab.

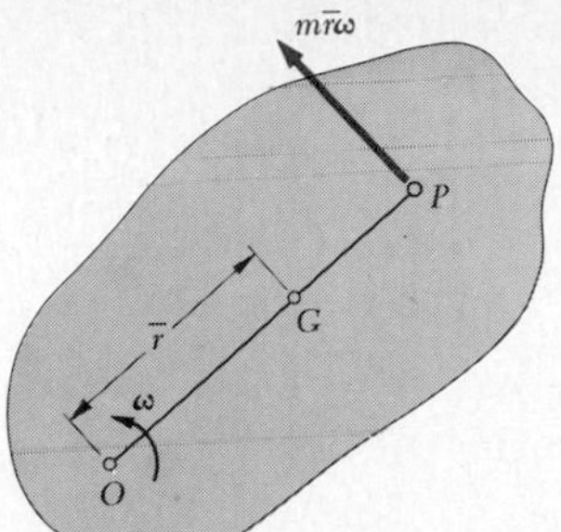

Fig. P17.66

17.67 Show that the sum $\mathbf{H}_A$ of the moments about a point A of the momenta of the particles of a rigid slab in plane motion is equal to $I_A\omega$, where ω is the angular velocity of the slab at the instant considered and I_A the moment of inertia of the slab about A, if and only if one of the following conditions is satisfied: (*a*) A is the mass center of the slab, (*b*) A is the instantaneous center of rotation, (*c*) the velocity of A is directed along a line joining point A and the mass center G.

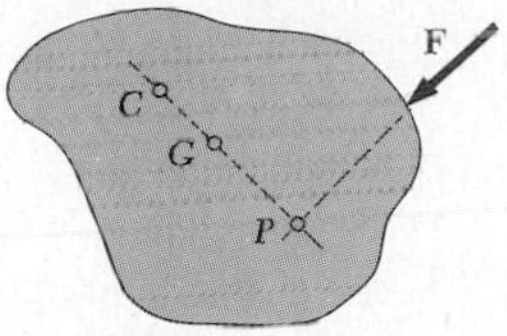

Fig. P17.68

17.68 Consider a rigid slab initially at rest and subjected to an impulsive force $\mathbf{F}$ contained in the plane of the slab. We define the *center of percussion* P as the point of intersection of the line of action of $\mathbf{F}$ with the perpendicular drawn from G. (*a*) Show that the instantaneous center of rotation C of the slab is located on line GP at a distance $GC = \bar{k}^2/GP$ on the opposite side of G. (*b*) Show that if the center of percussion were located at C, the instantaneous center of rotation would be located at P.

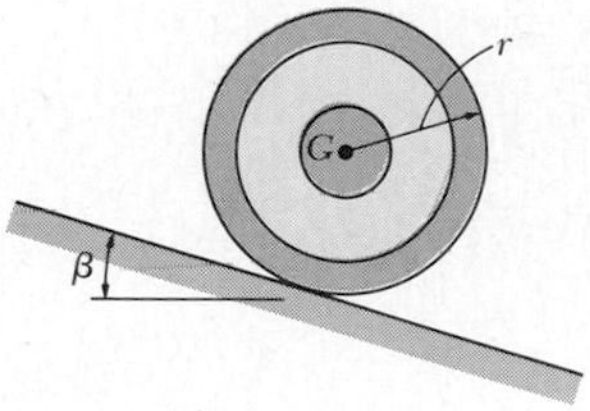

Fig. P17.69

17.69 A wheel of radius r and centroidal radius of gyration $\bar{k}$ is placed on an incline and released from rest at time $t = 0$. Assuming that the wheel rolls without slipping, determine (*a*) the velocity of the center at time t, (*b*) the coefficient of static friction required to prevent slipping.

17.70 A drum of 60-mm radius is attached to a disk of 120-mm radius. The disk and the drum have a total mass of 4 kg and a radius of gyration of 90 mm. A cord is wrapped around the drum and pulled with a force $\mathbf{P}$ of magnitude 15 N. Knowing that the disk is initially at rest, determine (*a*) the velocity of the center G after 2.5 s, (*b*) the friction force required to prevent slipping.

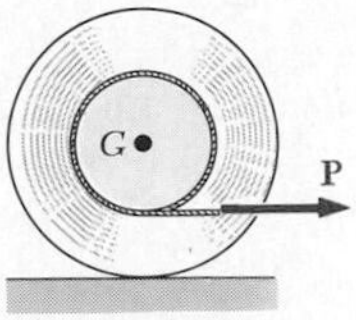

Fig. P17.70

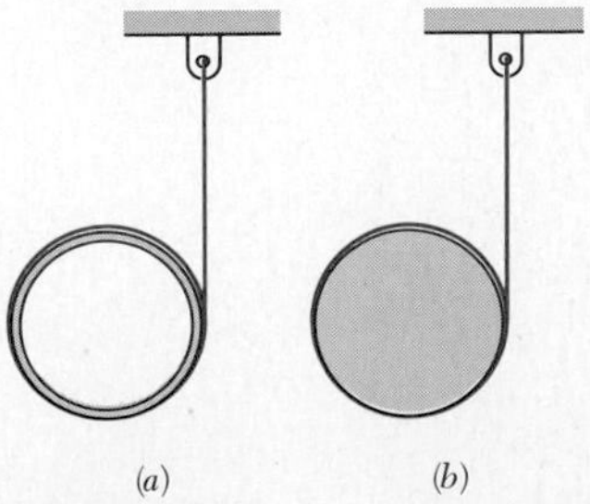

Fig. P17.71

17.71 Cords are wrapped around a thin-walled pipe and a solid cylinder as shown. Knowing that the pipe and the cylinder are each released from rest at time $t = 0$, determine at time t the velocity of the center of (*a*) the pipe, (*b*) the cylinder.

17.72 The double pulley shown weighs 6 lb and has a centroidal radius of gyration of 3 in. The 2-in.-radius inner pulley is rigidly attached to the 4-in.-radius outer pulley, and the motion of the center A is guided by a smooth pin at A which slides in a vertical slot. When the pulley is at rest, a force $\mathbf{P}$ of magnitude 8 lb is applied to cord C. Determine the velocity of the center of the pulley after 5 s.

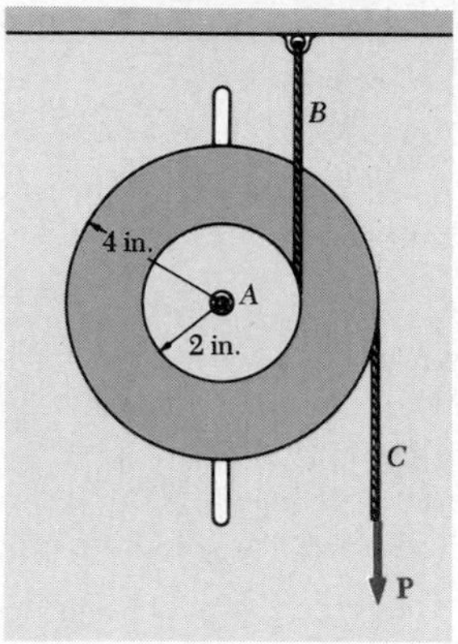

Fig. P17.72

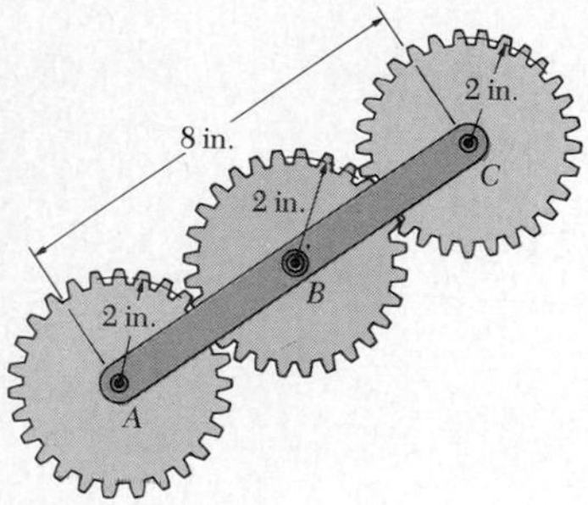

Fig. P17.74

17.73 Solve Prob. 17.72, assuming that the force $\mathbf{P}$ is replaced by an 8-lb weight attached to cord C.

17.74 In the gear arrangement shown, gears A and C are attached to the rod ABC, which is free to rotate about B, while the inner gear B is fixed. Knowing that the system is at rest, determine the moment of the couple $\mathbf{M}$ which must be applied to rod ABC, if 3 s later the angular velocity of the rod is to be 300 rpm clockwise. Gears A and C weigh 3 lb each and may be considered as disks of radius 2 in.; rod ABC weighs 5 lb.

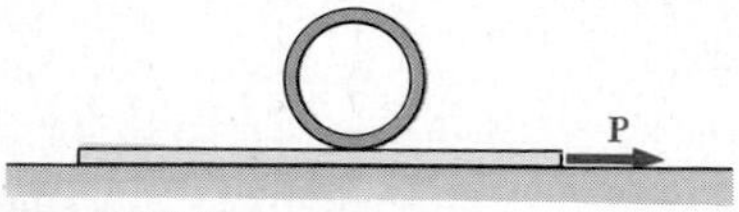

Fig. P17.75

17.75 A 240-mm-diameter pipe of mass 16 kg rests on a 4-kg plate. The pipe and plate are initially at rest when a force $\mathbf{P}$ of magnitude 80 N is applied for 0.50 s. Knowing that $\mu_s = 0.25$ and $\mu_k = 0.20$ between the plate and both the pipe and the floor, determine (*a*) whether the pipe slides with respect to the plate, (*b*) the resulting velocities of the pipe and of the plate.

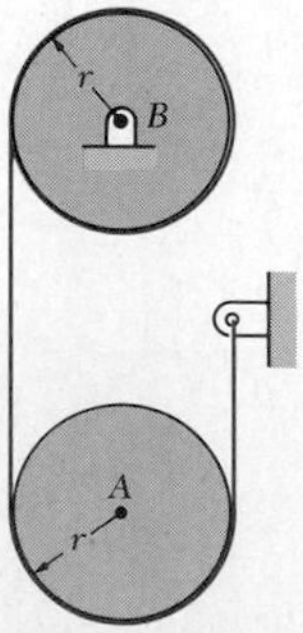

Fig. P17.76

17.76 Two uniform cylinders, each of weight $W = 14$ lb and radius $r = 5$ in., are connected by a belt as shown. If the system is released from rest, determine (*a*) the velocity of the center of cylinder A after 3 s, (*b*) the tension in the portion of belt connecting the two cylinders.

17.77 and 17.78 The 6-kg carriage is supported as shown by two uniform disks, each having a mass of 4 kg and a radius of 75 mm. Knowing that the carriage is initially at rest, determine the velocity of the carriage 2.5 s after the 10-N force has been applied. Assume that the disks roll without sliding.

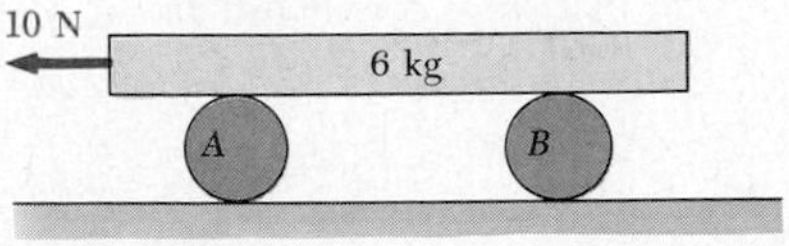

Fig. P17.77

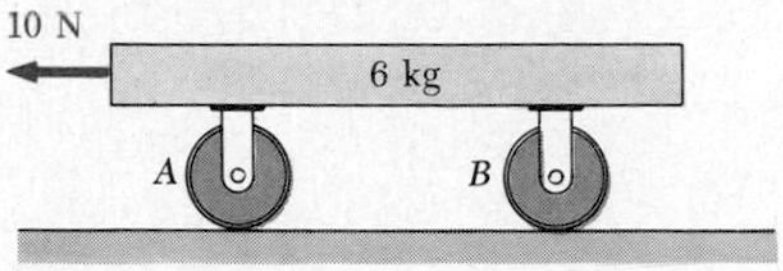

Fig. P17.78

17.79 Solve Sample Prob. 17.7, assuming that the sphere is replaced by a uniform cylinder of mass m and radius r.

17.80 A 4-kg bar AB is attached by a pin at D to a 5-kg circular plate which may rotate freely about a vertical axis. Knowing that when the bar is vertical the angular velocity of the plate is 120 rpm, determine (a) the angular velocity of the plate after the bar has swung into a horizontal position and has come to rest against pin C, (b) the energy lost due to the plastic impact at C.

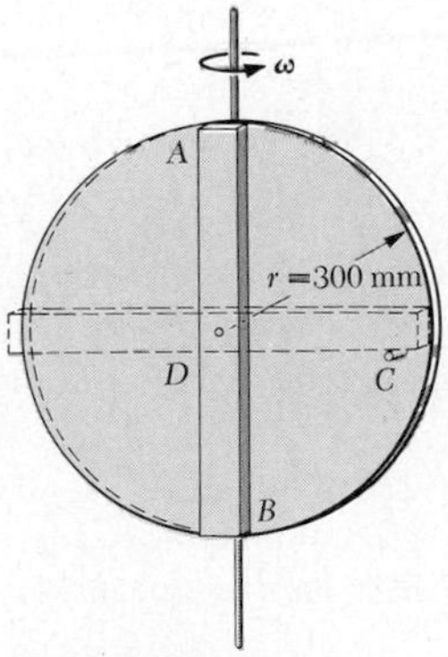

Fig. P17.80

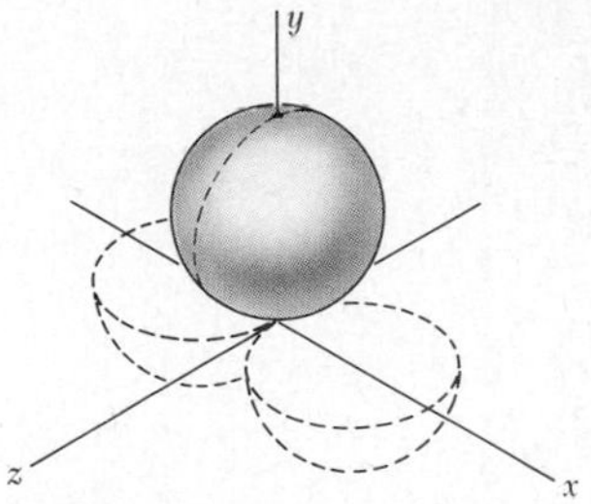

Fig. P17.81

17.81 A thin-walled spherical shell of mass m and radius r is used as a protective cover for a space shuttle experiment. In order to remove the cover, it is separated into two hemispherical shells which move to the final position shown. Knowing that during a test of the cover the initial angular velocity of the shell is $\boldsymbol{\omega} = \omega_0 \mathbf{j}$, determine the angular velocity of the cover after it has reached its final position. (*Hint.* For a thin-walled shell, $I_{\text{diameter}} = \frac{2}{3}mr^2$.)

17.82 The rod AB is of mass m and slides freely inside the tube CD which is also of mass m. The angular velocity of the assembly was ω_1 when the rod was entirely inside the tube ($x = 0$). Neglecting the effect of friction, determine the angular velocity of the assembly when $x = \frac{2}{3}L$.

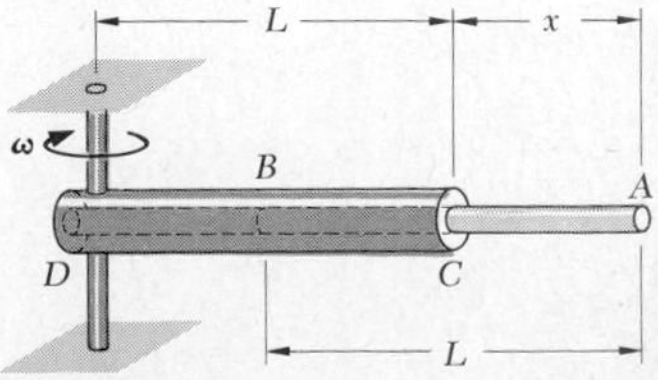

Fig. P17.82

17.83 A 1.5-kg tube AB may slide freely on rod DE which in turn may rotate freely in a horizontal plane. Initially the assembly is rotating with an angular velocity $\omega = 6$ rad/s and the tube is held in position by a cord. The moment of inertia of the rod and bracket about the vertical axis of rotation is $0.20 \text{ kg} \cdot \text{m}^2$ and the centroidal moment of inertia of the tube about a vertical axis is $0.0015 \text{ kg} \cdot \text{m}^2$. If the cord suddenly breaks, determine (*a*) the angular velocity of the assembly after the tube has moved to end E, (*b*) the energy lost during the plastic impact at E.

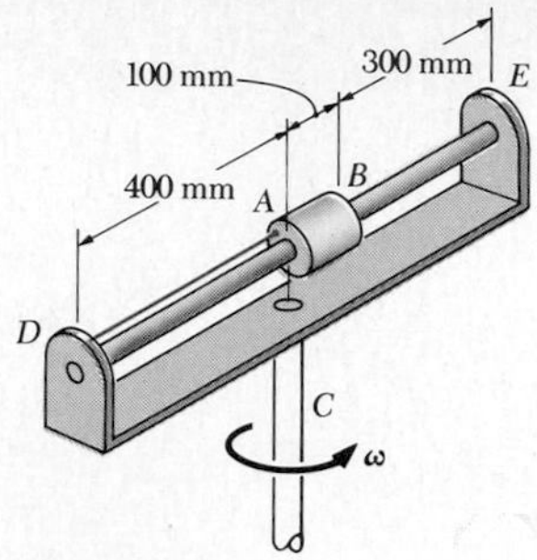

Fig. P17.83

17.84 Two square panels, each of side b, are attached with hinges to a rectangular plate and held by a wire in the position shown. The plate and the panels are made of the same material and have the same thickness. The entire assembly is rotating with an angular velocity ω_0 when the wire breaks. Determine the angular velocity of the assembly after the panels have come to rest in a horizontal position.

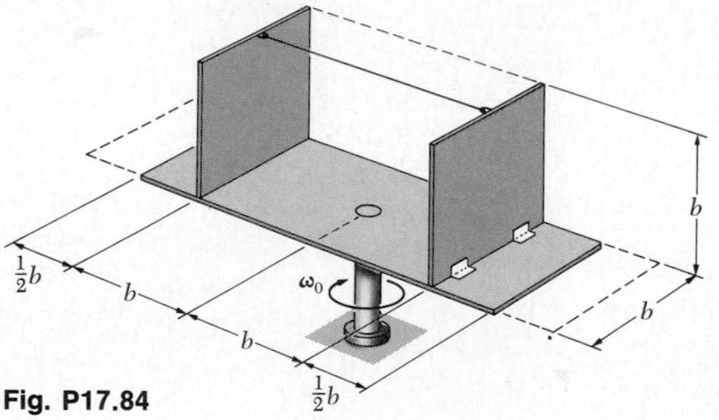

Fig. P17.84

17.85 Two 4-kg disks and a small motor are mounted on a 6-kg rectangular platform which is free to rotate about a central vertical spindle. The normal operating speed of the motor is 180 rpm. If the motor is started when the system is at rest, determine the angular velocity of all elements of the system after the motor has attained its normal operating speed. Neglect the mass of the motor and of the belt.

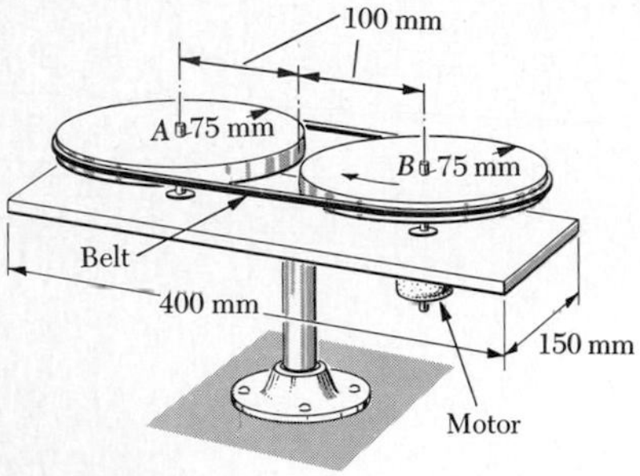

Fig. P17.85

17.86 Solve Prob. 17.85, assuming that the belt is removed and that the bearing between disk A and the plate is (*a*) perfectly lubricated, (*b*) frozen.

17.87 In the helicopter shown, a vertical tail propeller is used to prevent rotation of the cab as the speed of the main blades is changed. Assuming that the tail propeller is not operating, determine the final angular velocity of the cab after the speed of the main blades has been changed from 200 to 300 rpm. The speed of the main blades is measured relative to the cab, which has a centroidal moment of inertia of $800 \text{ lb} \cdot \text{ft} \cdot \text{s}^2$. Each of the four main blades is assumed to be a 15-ft slender rod weighing 60 lb.

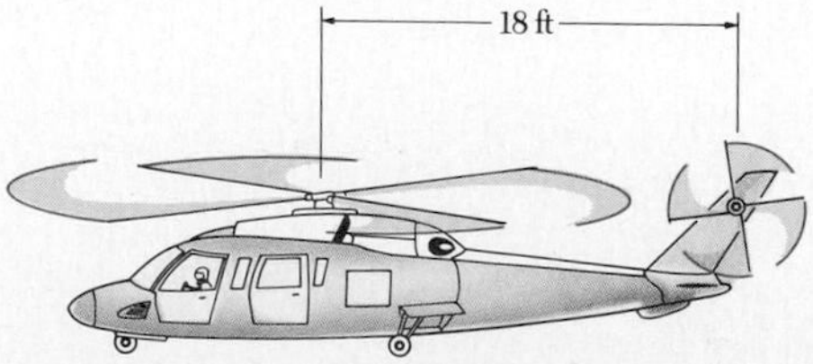

Fig. P17.87

17.88 Assuming that the tail propeller in Prob. 17.87 is operating and that the angular velocity of the cab remains zero, determine the final horizontal velocity of the cab when the speed of the main blades is changed from 200 to 300 rpm. The cab weighs 1500 lb and is initially at rest. Also determine the force exerted by the tail propeller if this change in speed takes place uniformly in 12 s.

17.89 Collar B has a mass of 2 kg and may slide freely on rod OA, which in turn may rotate freely in the horizontal plane. The assembly is rotating with an angular velocity $\omega = 1.5$ rad/s when a spring located between A and B is released, projecting the collar along the rod with an initial relative speed $v_r = 1.5$ m/s. Knowing that the moment of inertia about O of the rod and spring is $0.3\ \text{kg}\cdot\text{m}^2$, determine (*a*) the minimum distance between the collar and point O in the ensuing motion, (*b*) the corresponding angular velocity of the assembly.

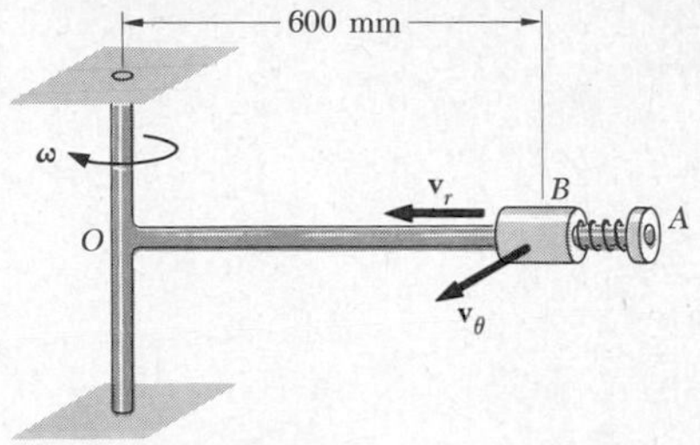

Fig. P17.89

17.90 In Prob. 17.89, determine the required magnitude of the initial relative velocity $\mathbf{v}_r$ if during the ensuing motion the minimum distance between collar B and point O is to be 300 mm.

17.91 In Prob. 17.83, determine the velocity of the tube relative to the rod as the tube strikes end E of the assembly.

17.92 In Prob. 17.82, determine the velocity of the rod relative to the tube when $x = \frac{2}{3}L$.

17.93 In Prob. 17.82, determine the velocity of the rod relative to the tube when $x = \frac{1}{2}L$.

***17.94** The uniform rod AB, of weight 20 lb and length 3 ft, is attached to the 30-lb cart C. Knowing that the system is released from rest in the position shown and neglecting friction, determine (*a*) the velocity of point B as the rod AB passes through a vertical position, (*b*) the corresponding velocity of the cart C.

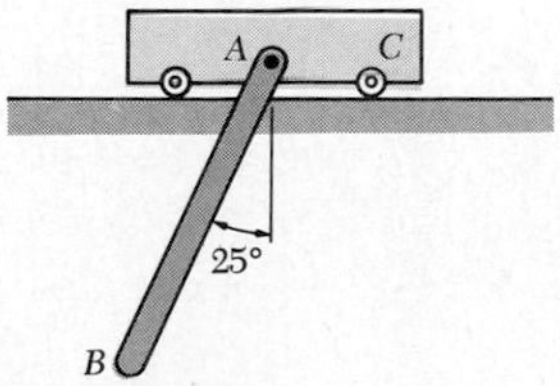

Fig. P17.94

***17.95** The 3-kg cylinder B and the 2-kg wedge A are at rest in the position shown. The cord C connecting the cylinder and the wedge is then cut and the cylinder rolls without sliding on the wedge. Neglecting friction between the wedge and the ground, determine (*a*) the angular velocity of the cylinder after it has rolled 150 mm down the wedge, (*b*) the corresponding velocity of the wedge.

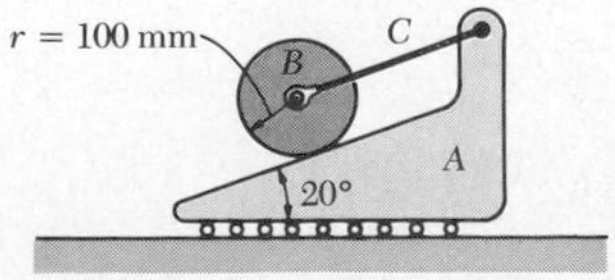

Fig. P17.95

17.11. Impulsive Motion. We saw in Chap. 13 that the method of impulse and momentum is the only practicable method for the solution of problems involving impulsive motion. Now we shall also find that compared with the various problems considered in the preceding sections, problems involving impulsive motion are particularly well adapted to a solution by the method of impulse and momentum. The computation of linear impulses and angular impulses is quite simple, since the time interval considered being very short, the bodies involved may be assumed to occupy the same position during that time interval.

17.12. Eccentric Impact. In Secs. 13.13 and 13.14, we learned to solve problems of *central impact*, i.e., problems in which the mass centers of the two colliding bodies are located on the line of impact. We shall now analyze the *eccentric impact* of two rigid bodies. Consider two bodies which collide, and denote by $\mathbf{v}_A$ and $\mathbf{v}_B$ the velocities before impact *of the two points of contact* A and B (Fig. 17.10*a*). Under the impact, the two

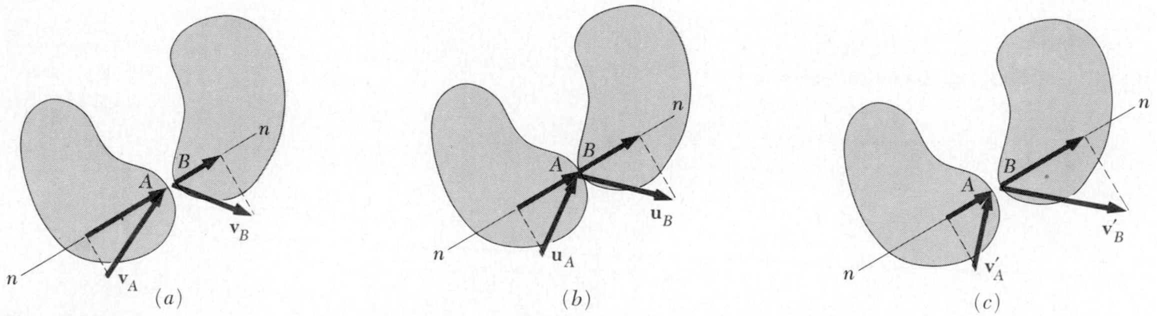

Fig. 17.10

bodies will *deform*, and at the end of the period of deformation, the velocities $\mathbf{u}_A$ and $\mathbf{u}_B$ of A and B will have equal components along the line of impact nn (Fig. 17.10*b*). A period of *restitution* will then take place, at the end of which A and B will have velocities $\mathbf{v}'_A$ and $\mathbf{v}'_B$ (Fig. 17.10*c*). Assuming that the bodies are frictionless, we find that the forces they exert on each other are directed along the line of impact. Denoting, respectively, by $\int P\,dt$ and $\int R\,dt$ the magnitude of the impulse of one of these forces during the period of deformation and during the period of restitution, we recall that the coefficient of restitution e is defined as the ratio

$$e = \frac{\int R\,dt}{\int P\,dt} \tag{17.18}$$

We propose to show that the relation established in Sec. 13.13 between the relative velocities of two particles before and after impact also holds between the components along the line of impact of the relative velocities of the two points of contact A and B. We propose to show, therefore, that

$$(v'_B)_n - (v'_A)_n = e[(v_A)_n - (v_B)_n] \qquad (17.19)$$

We shall first assume that the motion of each of the two colliding bodies of Fig. 17.10 is unconstrained. Thus the only impulsive forces exerted on the bodies during the impact are applied at A and B, respectively. Consider the body to which point A belongs and draw the three momentum and impulse diagrams corresponding to the period of deformation (Fig. 17.11). We denote by $\bar{\mathbf{v}}$ and $\bar{\mathbf{u}}$, respectively, the velocity of the mass center

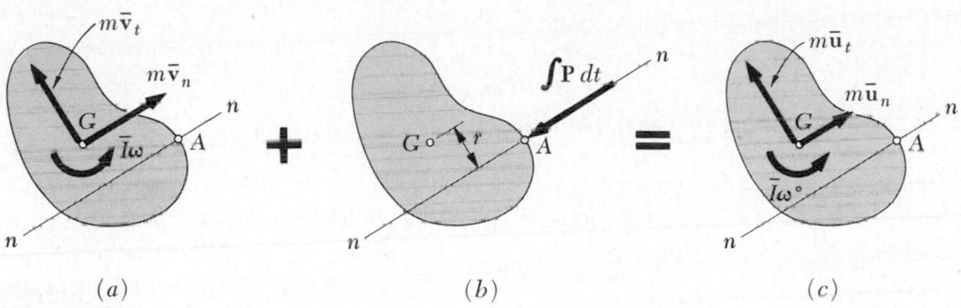

Fig. 17.11

at the beginning and at the end of the period of deformation, and by $\boldsymbol{\omega}$ and $\boldsymbol{\omega}^*$ the angular velocity of the body at the same instants. Summing and equating the components of the momenta and impulses along the line of impact nn, we write

$$m\bar{v}_n - \int P\,dt = m\bar{u}_n \qquad (17.20)$$

Summing and equating the moments about G of the momenta and impulses, we also write

$$\bar{I}\omega - r\int P\,dt = \bar{I}\omega^* \qquad (17.21)$$

where r represents the perpendicular distance from G to the line of impact. Considering now the period of restitution, we obtain in a similar way

$$m\bar{u}_n - \int R\,dt = m\bar{v}'_n \qquad (17.22)$$

$$\bar{I}\omega^* - r\int R\,dt = \bar{I}\omega' \qquad (17.23)$$

where $\bar{\mathbf{v}}'$ and $\boldsymbol{\omega}'$ represent, respectively, the velocity of the mass center and the angular velocity of the body after impact. Solving (17.20) and (17.22) for the two impulses and substituting into (17.18), and then solving (17.21) and (17.23) for the same two impulses and substituting again into (17.18), we obtain the following two alternative expressions for the coefficient of restitution:

$$e = \frac{\bar{u}_n - \bar{v}'_n}{\bar{v}_n - \bar{u}_n} \qquad e = \frac{\omega^* - \omega'}{\omega - \omega^*} \tag{17.24}$$

Multiplying by r the numerator and denominator of the second expression obtained for e, and adding respectively to the numerator and denominator of the first expression, we have

$$e = \frac{\bar{u}_n + r\omega^* - (\bar{v}'_n + r\omega')}{\bar{v}_n + r\omega - (\bar{u}_n + r\omega^*)} \tag{17.25}$$

Observing that $\bar{v}_n + r\omega$ represents the component $(v_A)_n$ along nn of the velocity of the point of contact A and that, similarly, $\bar{u}_n + r\omega^*$ and $\bar{v}'_n + r\omega'$ represent, respectively, the components $(u_A)_n$ and $(v'_A)_n$, we write

$$e = \frac{(u_A)_n - (v'_A)_n}{(v_A)_n - (u_A)_n} \tag{17.26}$$

The analysis of the motion of the second body leads to a similar expression for e in terms of the components along nn of the successive velocities of point B. Recalling that $(u_A)_n = (u_B)_n$, and eliminating these two velocity components by a manipulation similar to the one used in Sec. 13.13, we obtain relation (17.19).

If one or both of the colliding bodies is constrained to rotate about a fixed point O, as in the case of a compound pendulum (Fig. 17.12a), an impulsive reaction will be exerted at O (Fig. 17.12b). We shall verify that while their derivation must be modified, Eqs. (17.26) and (17.19) remain valid. Applying formula (17.16) to the period of deformation and to the period of restitution, we write

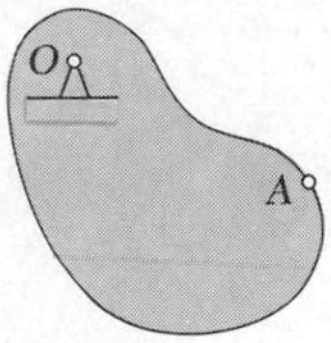

(a)

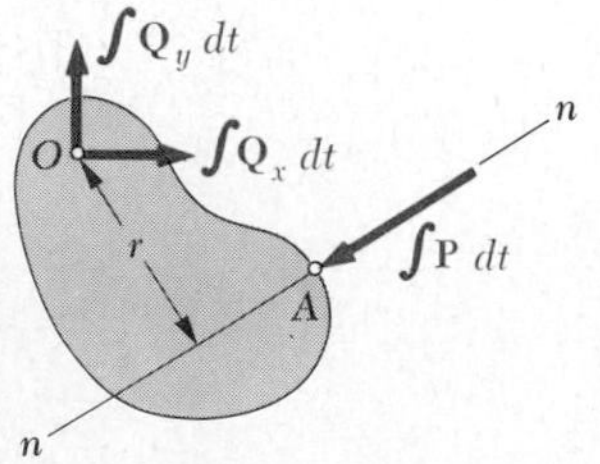

(b)

Fig. 17.12

$$I_O\omega - r\int P\,dt = I_O\omega^* \tag{17.27}$$
$$I_O\omega^* - r\int R\,dt = I_O\omega' \tag{17.28}$$

where r represents the perpendicular distance from the fixed point O to the line of impact. Solving (17.27) and (17.28) for the two impulses and substituting into (17.18), and then observing that $r\omega$, $r\omega^*$, and $r\omega'$ represent the components along nn of the successive velocities of point A, we write

$$e = \frac{\omega^* - \omega'}{\omega - \omega^*} = \frac{r\omega^* - r\omega'}{r\omega - r\omega^*} = \frac{(u_A)_n - (v'_A)_n}{(v_A)_n - (u_A)_n}$$

and check that Eq. (17.26) still holds. Thus Eq. (17.19) remains valid when one or both of the colliding bodies is constrained to rotate about a fixed point O.

In order to determine the velocities of the two colliding bodies after impact, relation (17.19) should be used in conjunction with one or several other equations obtained by applying the principle of impulse and momentum (Sample Prob. 17.10).

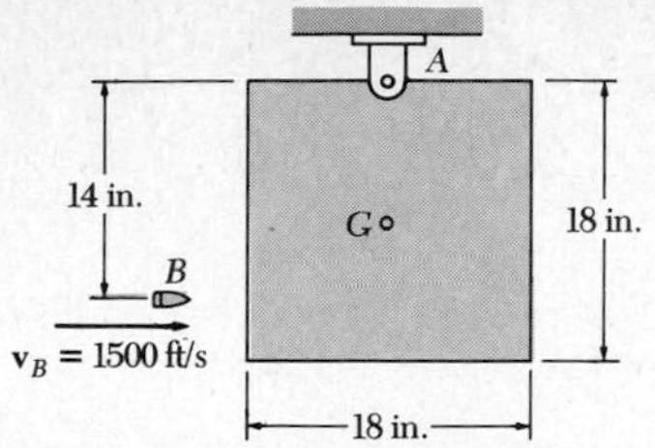

SAMPLE PROBLEM 17.9

A 0.05-lb bullet B is fired with a horizontal velocity of 1500 ft/s into the side of a 20-lb square panel suspended from a hinge at A. Knowing that the panel is initially at rest, determine (*a*) the angular velocity of the panel immediately after the bullet becomes embedded, (*b*) the impulsive reaction at A, assuming that the bullet becomes embedded in 0.0006 s.

Solution. ***Principle of Impulse and Momentum.*** We consider the bullet and the panel as a single system and express that the initial momenta of the bullet and panel and the impulses of the external forces are together equipollent to the final momenta of the system. Since the time interval $\Delta t = 0.0006$ s is very short, we neglect all nonimpulsive forces and consider only the external impulses $\mathbf{A}_x \Delta t$ and $\mathbf{A}_y \Delta t$.

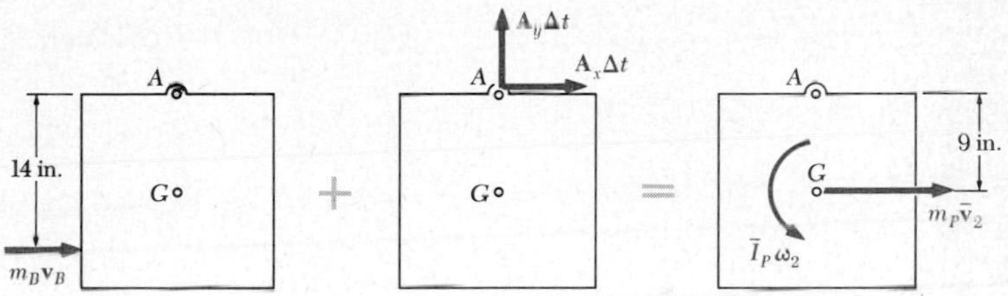

$$\textbf{Syst Momenta}_1 + \textbf{Syst Ext Imp}_{1\to 2} = \textbf{Syst Momenta}_2$$

+↻ moments about A: $\quad m_Bv_B(\frac{14}{12}\text{ ft}) + 0 = m_P\bar{v}_2(\frac{9}{12}\text{ ft}) + \bar{I}_P\omega_2 \qquad (1)$

$\xrightarrow{+}$ x components: $\quad m_Bv_B + A_x\,\Delta t = m_P\bar{v}_2 \qquad (2)$

$+\uparrow y$ components: $\quad 0 + A_y\,\Delta t = 0 \qquad (3)$

The centroidal mass moment of inertia of the square panel is

$$\bar{I}_P = \tfrac{1}{6}m_Pb^2 = \frac{1}{6}\left(\frac{20\text{ lb}}{32.2}\right)(\tfrac{18}{12}\text{ ft})^2 = 0.2329\text{ lb}\cdot\text{ft}\cdot\text{s}^2$$

Substituting this value as well as the given data into (1) and noting that

$$\bar{v}_2 = (\tfrac{9}{12}\text{ ft})\omega_2$$

we write

$$\left(\frac{0.05}{32.2}\right)(1500)(\tfrac{14}{12}) = 0.2329\omega_2 + \left(\frac{20}{32.2}\right)(\tfrac{9}{12}\omega_2)(\tfrac{9}{12})$$

$$\omega_2 = 4.67\text{ rad/s} \qquad \omega_2 = 4.67\text{ rad/s}\circlearrowleft \blacktriangleleft$$

$$\bar{v}_2 = (\tfrac{9}{12}\text{ ft})\omega_2 = (\tfrac{9}{12}\text{ ft})(4.67\text{ rad/s}) = 3.50\text{ ft/s}$$

Substituting $\bar{v}_2 = 3.50$ ft/s and $\Delta t = 0.0006$ s into Eq. (2), we have

$$\left(\frac{0.05}{32.2}\right)(1500) + A_x(0.0006) = \left(\frac{20}{32.2}\right)(3.50)$$

$$A_x = -259\text{ lb} \qquad \mathbf{A}_x = 259\text{ lb} \leftarrow \blacktriangleleft$$

From Eq. (3), we find $\quad A_y = 0 \qquad \mathbf{A}_y = 0 \blacktriangleleft$

SAMPLE PROBLEM 17.10

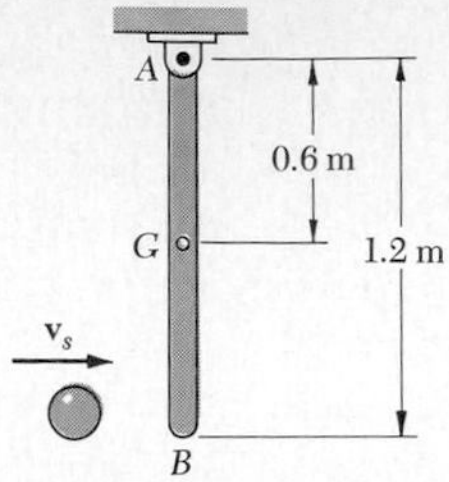

A 2-kg sphere moving horizontally to the right with an initial velocity of 5 m/s strikes the lower end of an 8-kg rigid rod AB. The rod is suspended from a hinge at A and is initially at rest. Knowing that the coefficient of restitution between the rod and sphere is 0.80, determine the angular velocity of the rod and the velocity of the sphere immediately after the impact.

Principle of Impulse and Momentum. We consider the rod and sphere as a single system and express that the initial momenta of the rod and sphere and the impulses of the external forces are together equipollent to the final momenta of the system. We note that the only impulsive force external to the system is the impulsive reaction at A.

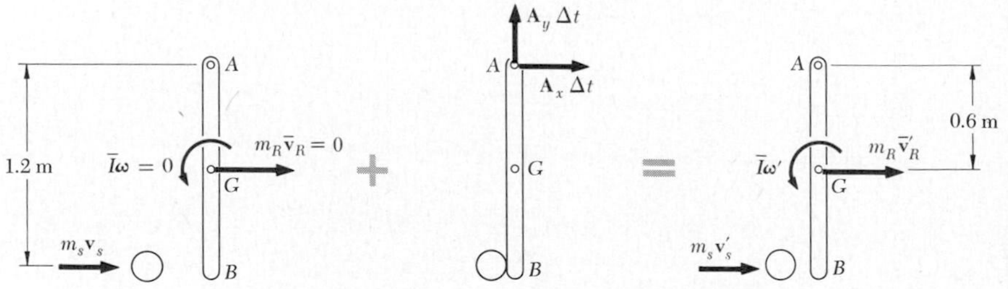

$$\textbf{Syst Momenta}_1 + \textbf{Syst Ext Imp}_{1\to 2} = \textbf{Syst Momenta}_2$$

$+\circlearrowleft$moments about A:

$$m_s v_s(1.2\text{ m}) = m_s v'_s(1.2\text{ m}) + m_R \bar{v}'_R(0.6\text{ m}) + \bar{I}\omega' \qquad (1)$$

Since the rod rotates about A, we have $\bar{v}'_R = \bar{r}\omega' = (0.6\text{ m})\omega'$. Also,

$$\bar{I} = \tfrac{1}{12}mL^2 = \tfrac{1}{12}(8\text{ kg})(1.2\text{ m})^2 = 0.96\text{ kg}\cdot\text{m}^2$$

Substituting these values and the given data into Eq. (1), we have

$$(2\text{ kg})(5\text{ m/s})(1.2\text{ m}) = (2\text{ kg})v'_s(1.2\text{ m}) + (8\text{ kg})(0.6\text{ m})\omega'(0.6\text{ m}) + (0.96\text{ kg}\cdot\text{m}^2)\omega'$$
$$12 = 2.4v'_s + 3.84\omega' \qquad (2)$$

Relative Velocities. Choosing positive to the right, we write

$$v'_B - v'_s = e(v_s - v_B)$$

Substituting $v_s = 5$ m/s, $v_B = 0$, and $e = 0.80$, we obtain

$$v'_B - v'_s = 0.80(5\text{ m/s}) \qquad (3)$$

Again noting that the rod rotates about A, we write

$$v'_B = (1.2\text{ m})\omega' \qquad (4)$$

Solving Eqs. (2) to (4) simultaneously, we obtain

$$\omega' = 3.21\text{ rad/s} \qquad \omega' = 3.21\text{ rad/s}\circlearrowleft \blacktriangleleft$$
$$v'_s = -0.143\text{ m/s} \qquad \mathbf{v}'_s = 0.143\text{ m/s} \leftarrow \blacktriangleleft$$

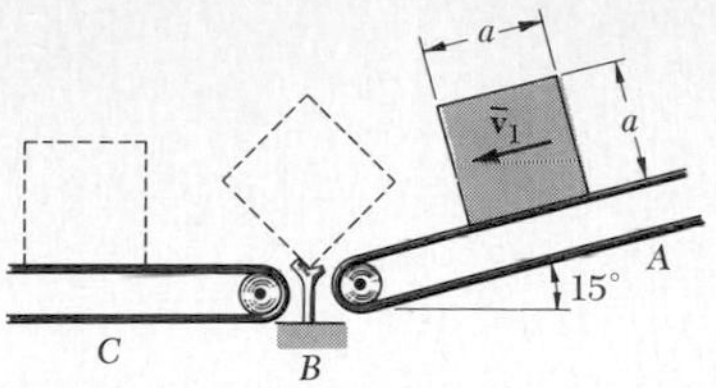

SAMPLE PROBLEM 17.11

A square package of side a and mass m moves down a conveyor belt A with a constant velocity $\overline{\mathbf{v}}_1$. At the end of the conveyor belt, the corner of the package strikes a rigid support at B. Assuming that the impact at B is perfectly plastic, derive an expression for the smallest magnitude of the velocity $\overline{\mathbf{v}}_1$ for which the package will rotate about B and reach conveyor belt C.

Principle of Impulse and Momentum. Since the impact between the package and the support is perfectly plastic, the package rotates about B during the impact. We apply the principle of impulse and momentum to the package and note that the only impulsive force external to the package is the impulsive reaction at B.

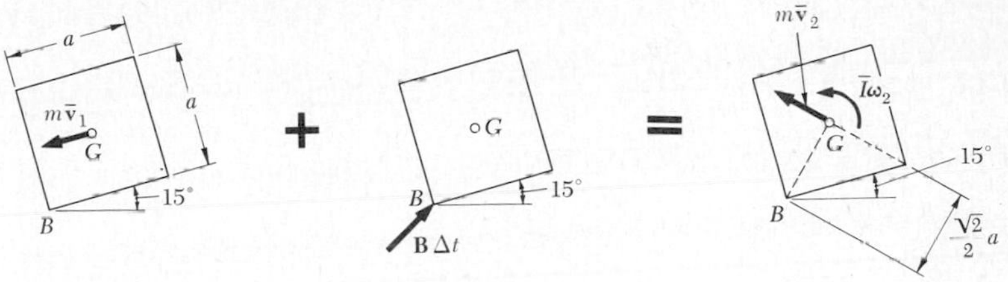

Syst Momenta$_1$ + Syst Ext Imp$_{1\to 2}$ = Syst Momenta$_2$

$+\circlearrowleft$moments about B: $\quad (m\overline{v}_1)(\frac{1}{2}a) + 0 = (m\overline{v}_2)(\frac{1}{2}\sqrt{2}a) + \overline{I}\omega_2 \qquad (1)$

Since the package rotates about B, we have $\overline{v}_2 = (GB)\omega_2 = \frac{1}{2}\sqrt{2}a\omega_2$. We substitute this expression, together with $\overline{I} = \frac{1}{6}ma^2$, into Eq. (1):

$$(m\overline{v}_1)(\tfrac{1}{2}a) = m(\tfrac{1}{2}\sqrt{2}a\omega_2)(\tfrac{1}{2}\sqrt{2}a) + \tfrac{1}{6}ma^2\omega_2 \qquad \overline{v}_1 = \tfrac{4}{3}a\omega_2 \qquad (2)$$

Position 2

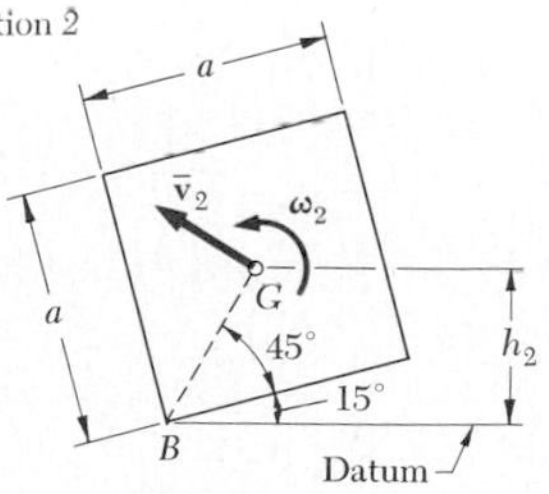

$GB = \frac{1}{2}\sqrt{2}a = 0.707a$
$h_2 = GB\sin(45° + 15°)$
$= 0.612a$

Principle of Conservation of Energy. We apply the principle of conservation of energy between position 2 and position 3.

Position 2. $V_2 = Wh_2$. Recalling that $\overline{v}_2 = \frac{1}{2}\sqrt{2}a\omega_2$, we write

$$T_2 = \tfrac{1}{2}m\overline{v}_2^2 + \tfrac{1}{2}\overline{I}\omega_2^2 = \tfrac{1}{2}m(\tfrac{1}{2}\sqrt{2}a\omega_2)^2 + \tfrac{1}{2}(\tfrac{1}{6}ma^2)\omega_2^2 = \tfrac{1}{3}ma^2\omega_2^2$$

Position 3. Since the package must reach conveyor belt C, it must pass through position 3 where G is directly above B. Also, since we wish to determine the smallest velocity for which the package will reach this position, we choose $\overline{v}_3 = \omega_3 = 0$. Therefore $T_3 = 0$ and $V_3 = Wh_3$.

Position 3

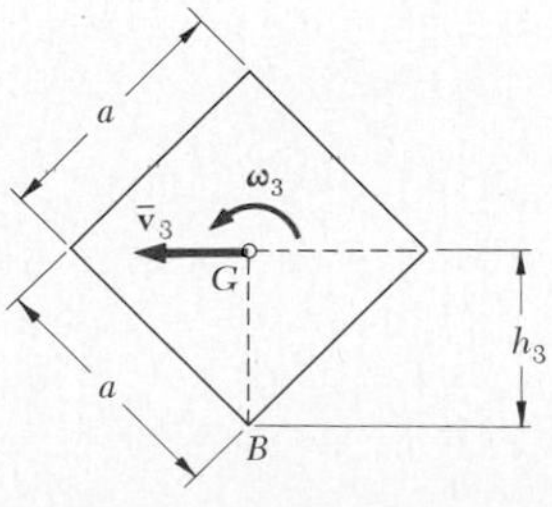

$h_3 = GB = 0.707a$

Conservation of Energy

$$T_2 + V_2 = T_3 + V_3$$
$$\tfrac{1}{3}ma^2\omega_2^2 + Wh_2 = 0 + Wh_3$$
$$\omega_2^2 = \frac{3W}{ma^2}(h_3 - h_2) = \frac{3g}{a^2}(h_3 - h_2) \qquad (3)$$

Substituting the computed values of h_2 and h_3 into Eq. (3), we obtain

$$\omega_2^2 = \frac{3g}{a^2}(0.707a - 0.612a) = \frac{3g}{a^2}(0.095a) \qquad \omega_2 = \sqrt{0.285g/a}$$

$$\overline{v}_1 = \tfrac{4}{3}a\omega_2 = \tfrac{4}{3}a\sqrt{0.285g/a} \qquad \overline{v}_1 = 0.712\sqrt{ga} \blacktriangleleft$$

Problems

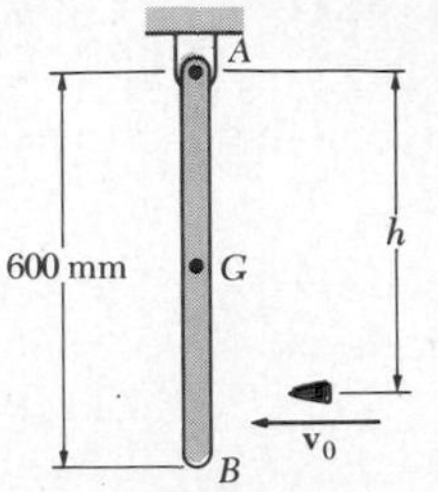

Fig. P17.96

17.96 A 32-g bullet is fired with a horizontal velocity of 450 m/s into a 5-kg wooden beam *AB* suspended from a pin support at *A*. Knowing that $h = 500$ mm and that the beam is initially at rest, determine (*a*) the velocity of the mass center *G* of the beam immediately after the bullet becomes embedded, (*b*) the impulsive reaction at *A*, assuming that the bullet becomes embedded in 1 ms.

17.97 In Prob. 17.96, determine (*a*) the required distance *h* if the impulsive reaction at *A* is to be zero, (*b*) the corresponding velocity of the mass center *G* of the beam after the bullet becomes embedded.

17.98 A 0.10-lb bullet is fired with a horizontal velocity of 1200 ft/s into a 20-lb wooden disk suspended from a pin support at *A*. Knowing that the disk is initially at rest, determine (*a*) the required distance *h* if the impulsive reaction at *A* is to be zero, (*b*) the corresponding velocity of the center *G* of the disk immediately after the bullet becomes embedded.

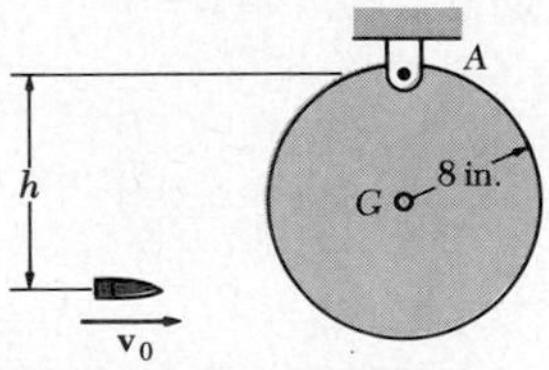

Fig. P17.98 and P17.99

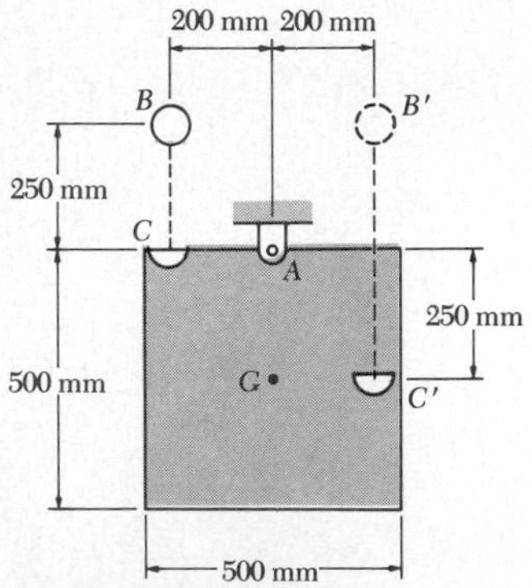

Fig. P17.100 and P17.101

17.99 A 0.10-lb bullet is fired with a horizontal velocity of 1200 ft/s into a 20-lb wooden disk suspended from a pin support at *A*. Knowing that $h = 10$ in. and that the disk is initially at rest, determine (*a*) the velocity of the center *G* of the disk immediately after the bullet becomes embedded, (*b*) the impulsive reaction at *A*, assuming that the bullet becomes embedded in 0.001 s.

17.100 An 8-kg wooden panel is suspended from a pin support at *A* and is initially at rest. A 2-kg metal sphere is released from rest at *B* and falls into a hemispherical cup *C* attached to the panel at a point located on its top edge. Assuming that the impact is perfectly plastic, determine the velocity of the mass center *G* of the panel immediately after the impact.

17.101 An 8-kg wooden panel is suspended from a pin support at *A* and is initially at rest. A 2-kg metal sphere is released from rest at *B′* and falls into a hemispherical cup *C′* attached to the panel at the same level as the mass center *G*. Assuming that the impact is perfectly plastic, determine the velocity of the mass center *G* of the panel immediately after the impact.

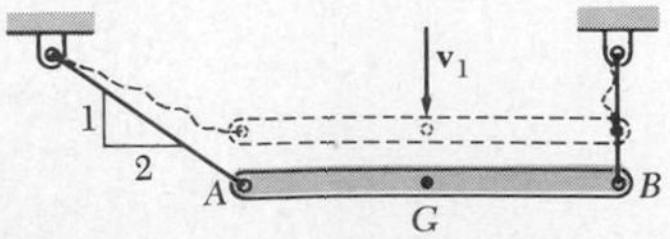

Fig. P17.102

17.102 A slender rod of length *L* is falling with a velocity v_1 at the instant when the cords simultaneously become taut. Assuming that the impacts are perfectly plastic, determine the angular velocity of the rod and the velocity of its mass center immediately after the cords become taut.

17.103 A uniform square panel of side L and mass m is supported by a frictionless horizontal table. The panel is moving to the right with a velocity $\mathbf{v}_1$ when a hook located at the midpoint E of side AB engages the fixed pin F. Assuming that the impact is perfectly plastic, determine the angular velocity of the panel in its subsequent rotation about F.

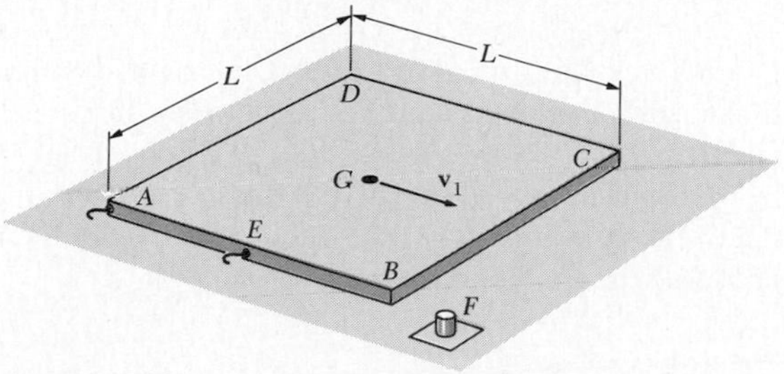

Fig. P17.103

17.104 Solve Prob. 17.103, assuming that the hook at E is removed and that a hook located at corner A engages the fixed pin F.

17.105 A uniform disk of radius r and mass m is supported by a frictionless horizontal table. Initially the disk is spinning freely about its mass center G with a constant angular velocity ω_1. Suddenly a latch B is moved to the right and is struck by a small stop A welded to the edge of the disk. Assuming that the impact of A and B is perfectly plastic, determine the angular velocity of the disk and the velocity of its mass center immediately after impact.

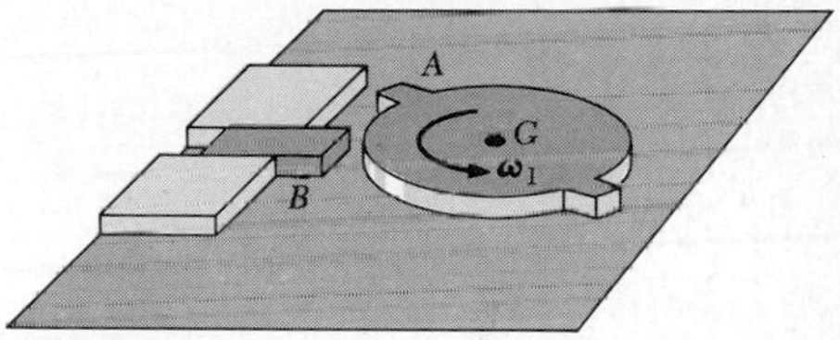

Fig. P17.105

17.106 A uniform slender rod AB is equipped at both ends with the hooks shown and is supported by a frictionless horizontal table. Initially the rod is hooked at A to a fixed pin C about which it rotates with the constant angular velocity ω_1. Suddenly end B of the rod hits and gets hooked to the pin D, causing end A to be released. Determine the magnitude of the angular velocity ω_2 of the rod in its subsequent rotation about D.

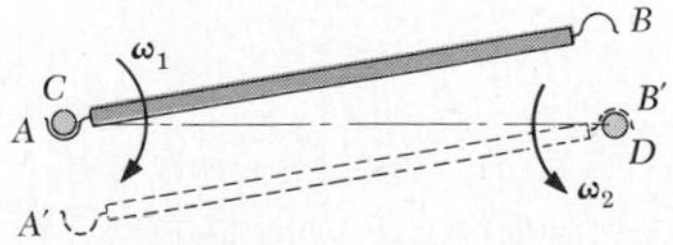

Fig. P17.106

17.107 A slender rod of length L and mass m is released from rest in the position shown and hits the edge D. Assuming perfectly plastic impact at D, determine for $b = 0.6L$, (a) the angular velocity of the rod immediately after the impact, (b) the maximum angle through which the rod will rotate after the impact.

17.108 A slender rod of mass m and length L is released from rest in the position shown and hits the edge D. Assuming perfectly elastic impact ($e = 1$) at D, determine the distance b for which the rod will rebound with no angular velocity.

17.109 Solve Prob. 17.108, assuming $e = 0.50$.

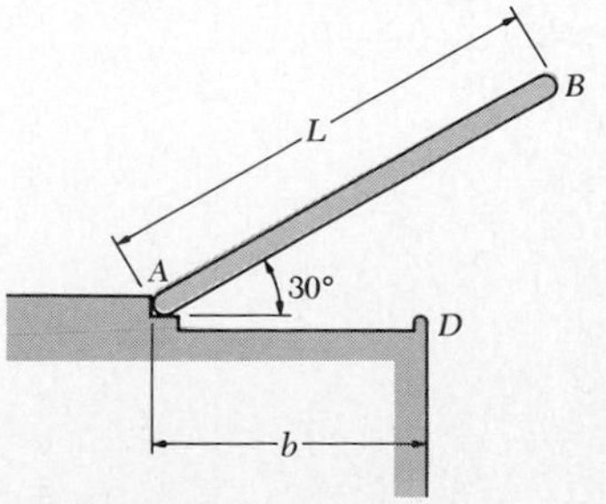

Fig. P17.107 and P17.108

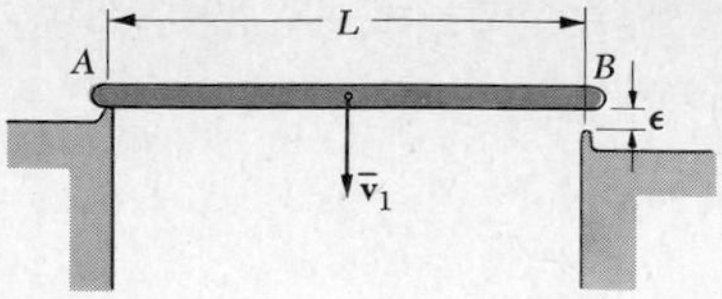

Fig. P17.110

17.110 A uniform slender rod of length L is dropped onto rigid supports at A and B. Immediately before striking A the velocity of the rod is $\bar{\mathbf{v}}_1$. Since support B is slightly lower than support A, the rod strikes A before it strikes B. Assuming perfectly elastic impact at both A and B, determine the angular velocity of the rod and the velocity of its mass center immediately after the rod (*a*) strikes support A, (*b*) strikes support B, (*c*) again strikes support A.

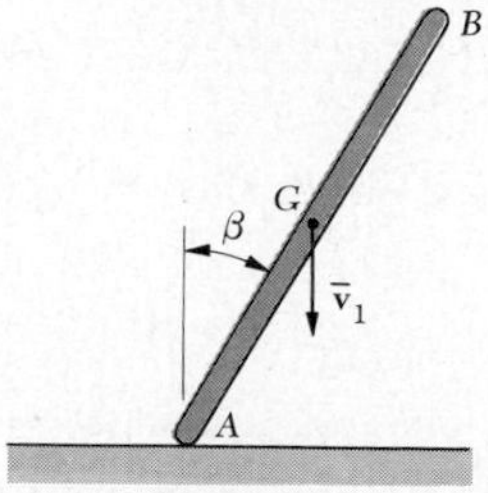

Fig. P17.111

17.111 A slender rod of length L forming an angle β with the vertical strikes a frictionless floor at A with a vertical velocity $\bar{\mathbf{v}}_1$ and no angular velocity. Assuming that the impact at A is perfectly elastic, derive an expression for the angular velocity of the rod immediately after impact.

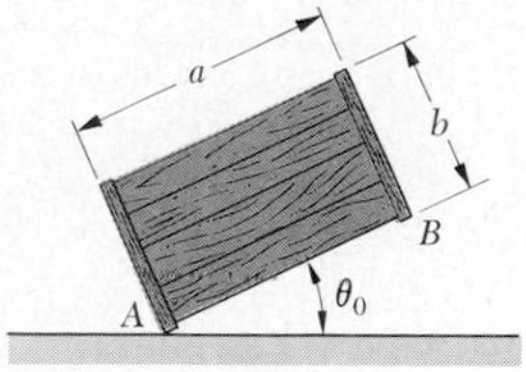

Fig. P17.112

17.112 A uniformly loaded rectangular crate is released from rest in the position shown. Assuming that the floor is sufficiently rough to prevent slipping and that the impact at B is perfectly plastic, determine the smallest value of the ratio a/b for which corner A will remain in contact with the floor.

17.113 A uniformly loaded square crate is released from rest with its corner D directly above A; it rotates about A until its corner B strikes the floor, and then rotates about B. The floor is sufficiently rough to prevent slipping and the impact at B is perfectly plastic. Denoting by ω_0 the angular velocity of the crate immediately before B strikes the floor, determine (*a*) the angular velocity of the crate immediately after B strikes the floor, (*b*) the fraction of the kinetic energy of the crate lost during the impact, (*c*) the angle θ through which the crate will rotate after B strikes the floor.

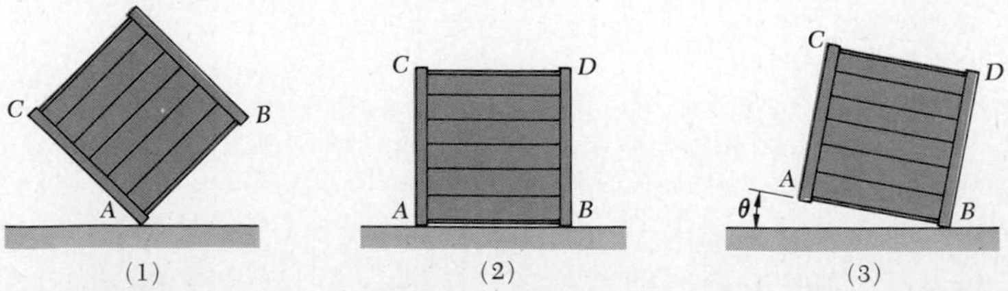

Fig. P17.113

17.114 Rod AB, of mass 1.5 kg and length $L = 0.5$ m, is suspended from a pin A which may move freely along a horizontal guide. If an impulse $\mathbf{Q}\,\Delta t = (1.8\ \text{N}\cdot\text{s})\mathbf{i}$ is applied at B, determine the maximum angle θ_m through which the rod will rotate during its subsequent motion.

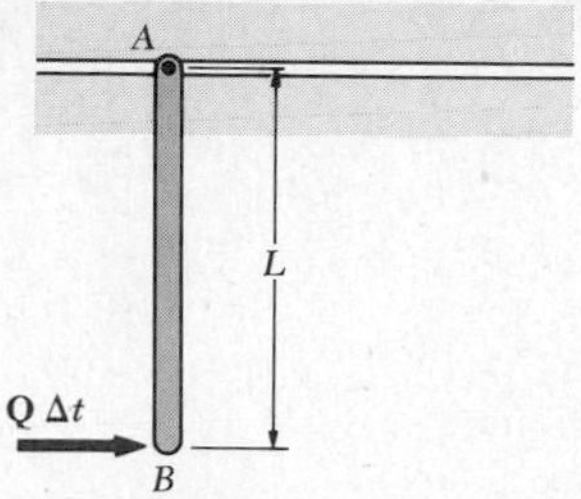

Fig. P17.114

17.115 For the rod of Prob. 17.114, determine the magnitude $Q\,\Delta t$ of the impulse for which the maximum angle through which the rod will rotate is (*a*) 90°, (*b*) 120°.

17.116 The plank CDE of mass m_P rests on top of a small pivot at D. A gymnast A of mass m stands on the plank at end C; a second gymnast B of the same mass m jumps from a height h and strikes the plank at E. Assuming perfectly plastic impact, determine the height to which gymnast A will rise. (Assume that gymnast A stands completely rigid.)

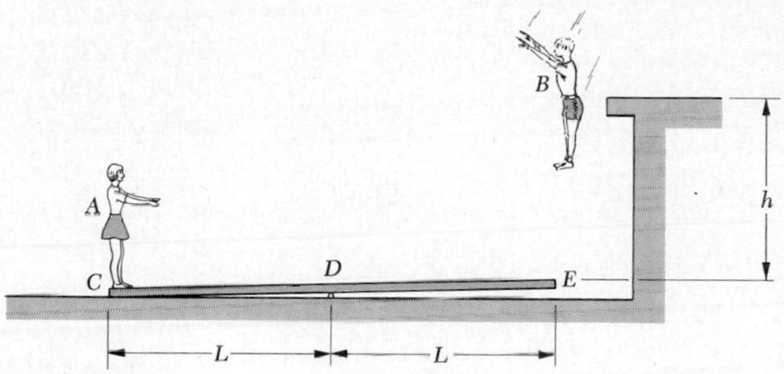

Fig. P17.116

17.117 A slender rod CD, of length L and mass m, is placed upright as shown on a frictionless horizontal surface. A second and identical rod AB is moving along the surface with a velocity $\bar{\mathbf{v}}_1$ when its end B strikes squarely end C of rod CD. Assuming that the impact is perfectly elastic, determine immediately after the impact, (*a*) the velocity of rod AB, (*b*) the angular velocity of rod CD, (*c*) the velocity of the mass center of rod CD.

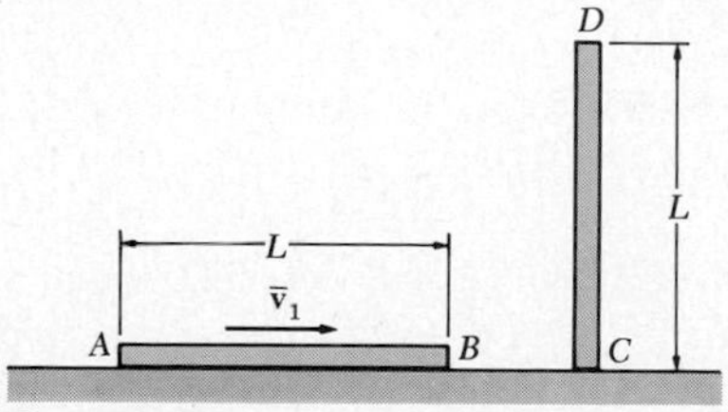

Fig. P17.117

17.118 Solve Prob. 17.117, assuming $e = 0.50$.

17.119 A sphere of mass m is dropped onto the end of a slender rod BD of length L and mass $2m$. The rod is attached to a pin support at C which is located at a distance $c = \frac{1}{4}L$ from end B. Denoting by $\mathbf{v}_1$ the velocity of the sphere just before it strikes the rod and assuming perfectly elastic impact, determine immediately after the impact (*a*) the velocity of the sphere, (*b*) the angular velocity of the rod.

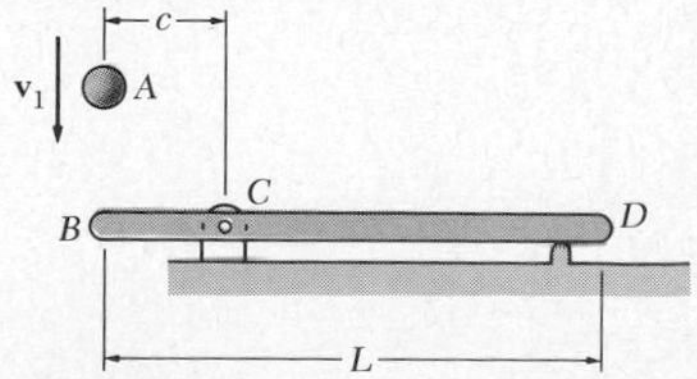

Fig. P17.119

17.120 Solve Prob. 17.119, assuming that $c = \frac{1}{2}L$.

17.121 A sphere of weight W_S is dropped from a height h and strikes at A the uniform slender plank AB of weight W which is held by two inextensible cords. Knowing that the impact is perfectly plastic and that the sphere remains attached to the plank, determine the velocity of the sphere immediately after impact.

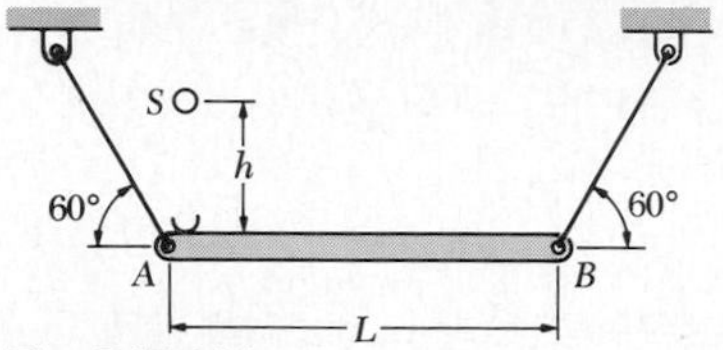

Fig. P17.121

17.122 Solve Prob. 17.121 when $W = 6$ lb, $W_S = 3$ lb, $L = 3$ ft, and $h = 4$ ft.

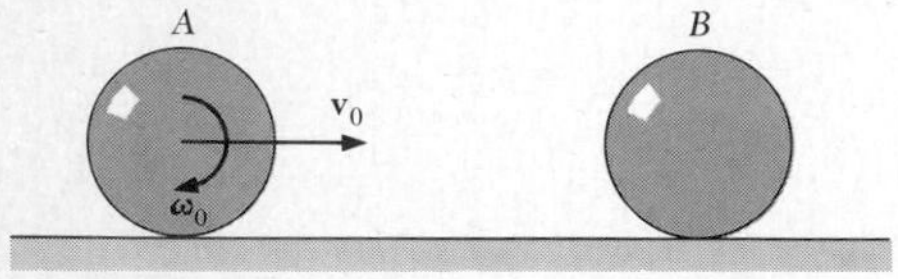

Fig. P17.123

17.123 A sphere A of mass m and radius r rolls without slipping with a velocity $\mathbf{v}_0$ on a horizontal plane. It hits squarely an identical sphere B which is at rest. Denoting by μ_k the coefficient of kinetic friction between the spheres and the plane, neglecting friction between the spheres, and assuming perfectly elastic impact ($e = 1$), determine (*a*) the linear and angular velocity of each sphere immediately after impact, (*b*) the velocity of each sphere after it has started rolling uniformly. (*c*) Discuss the special case when $\mu_k = 0$.

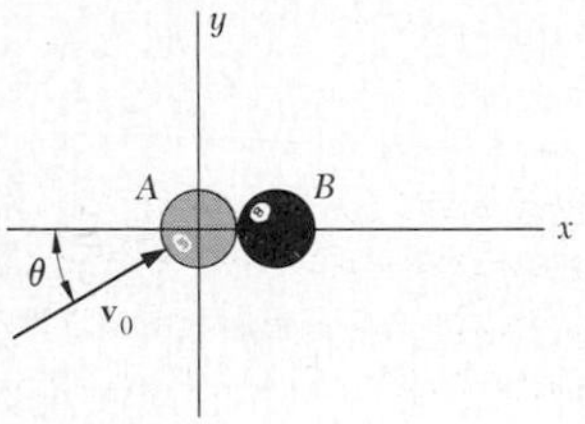

Fig. P17.124

17.124 In a game of billiards, ball A is rolling without slipping with a velocity $\mathbf{v}_0$ as it hits obliquely ball B which is at rest. Denoting by r the radius of each ball, by μ_k the coefficient of kinetic friction between the balls and the table, neglecting friction between the balls, and assuming perfectly elastic impact ($e = 1$), determine (*a*) the linear and angular velocity of each ball immediately after impact, (*b*) the velocity of B after it has started rolling uniformly.

17.125 In Prob. 17.124, determine the equation of the path described by the center of ball A while the ball is slipping.

17.126 For the billiard balls of Prob. 17.124, determine (*a*) the velocity of ball A after it has started rolling again without slipping, (*b*) the angle ϕ formed by the velocities of balls A and B after they have finished slipping.

17.127 The uniform plate $ABCD$ is falling with a velocity $\mathbf{v}_1$ when wire BE becomes taut. Assuming that the impact is perfectly plastic, determine the angular velocity of the plate and the velocity of its mass center immediately after the impact.

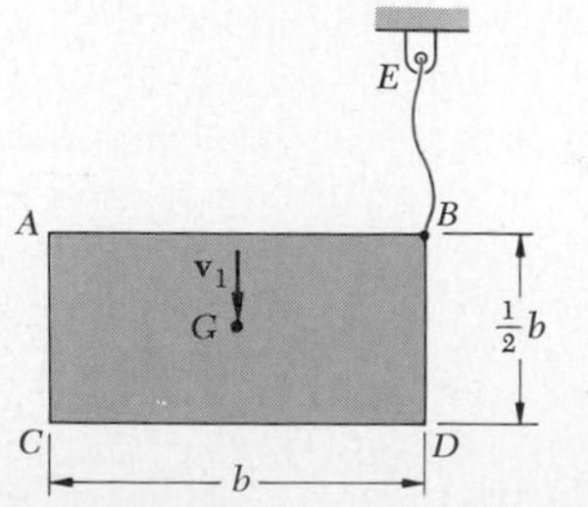

Fig. P17.127

17.128 A small rubber ball of radius r is thrown against a rough floor with a velocity $\mathbf{v}_A$ of magnitude v_0 and a "backspin" $\boldsymbol{\omega}_A$ of magnitude ω_0. It is observed that the ball bounces from A to B, then from B to A, then from A to B, etc. Assuming perfectly elastic impact, determine the required magnitude ω_0 of the backspin in terms of v_0 and r.

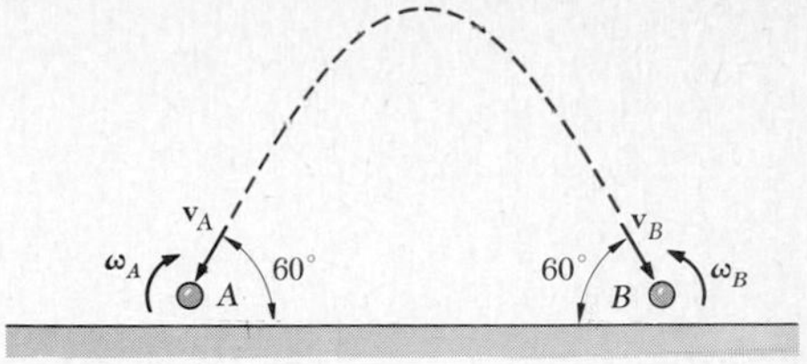

Fig. P17.128

17.129 A rectangular slab of mass m_S moves across a series of rollers, each of which is equivalent to a uniform disk of mass m_R and is initially at rest. Since the length of the slab is slightly less than three times the distance b between two adjacent rollers, the slab leaves a roller just before it reaches another one. Each time a new roller enters into contact with the slab, slipping occurs between the roller and the slab for a short period of time (less than the time needed for the slab to move through the distance b). Denoting by $\mathbf{v}_0$ the velocity of the slab in the position shown, determine the velocity of the slab after it has moved (a) a distance b, (b) a distance nb.

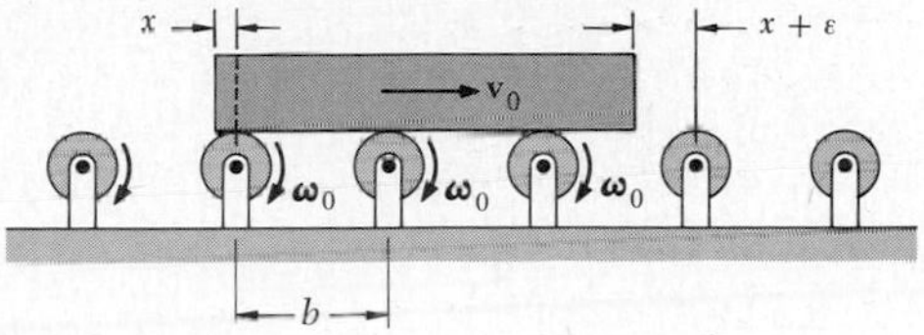

Fig. P17.129

17.130 Solve Prob. 17.129, assuming $v_0 = 6$ m/s, $m_S = 9$ kg, $m_R = 2$ kg, and $n = 8$.

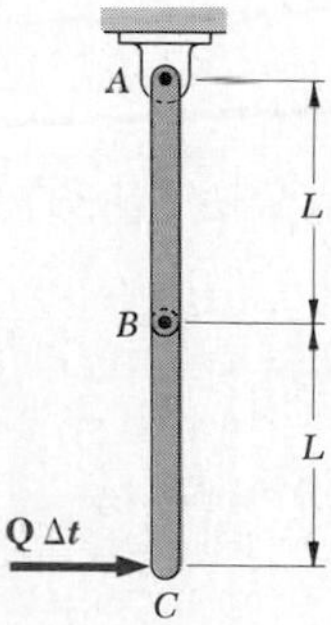

Fig. P17.131

***17.131** Each of the bars AB and BC is of length $L = 15$ in. and weight $W = 2.5$ lb. Determine the angular velocity of each bar immediately after the impulse $\mathbf{Q}\,\Delta t = (0.30\text{ lb}\cdot\text{s})\mathbf{i}$ is applied at C.

Review Problems

17.132 A uniform sphere of radius R rolls to the right without slipping on a horizontal surface and strikes a step of height $\frac{1}{2}R$. Assuming that the impact is perfectly plastic and that no slipping occurs between the sphere and the corner of the step, determine the angular velocity of the sphere and the velocity of its center immediately after the impact.

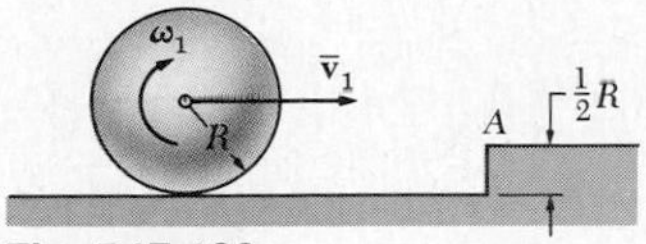

Fig. P17.132

17.133 A 50-lb uniform cylindrical roller, initially at rest, is acted upon by a 15-lb force as shown. Assuming that the body rolls without slipping, determine (a) the velocity of the center G after it has moved 4 ft, (b) the friction force required to prevent slipping.

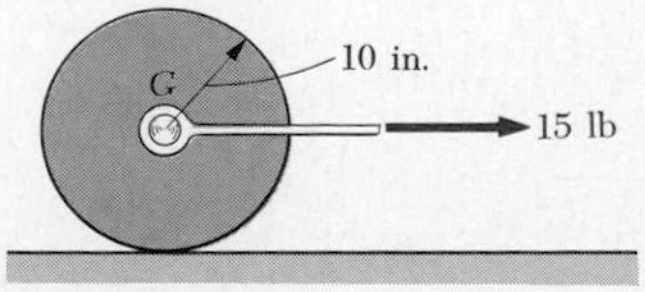

Fig. P17.133

17.134 A uniform rod of length L and weight W is attached to two wires, each of length b. The rod is released from rest when $\theta = 0$ and swings to the position $\theta = 90°$, at which time wire BD suddenly breaks. Determine the tension in wire AC (a) immediately before wire BD breaks, (b) immediately after wire BD breaks.

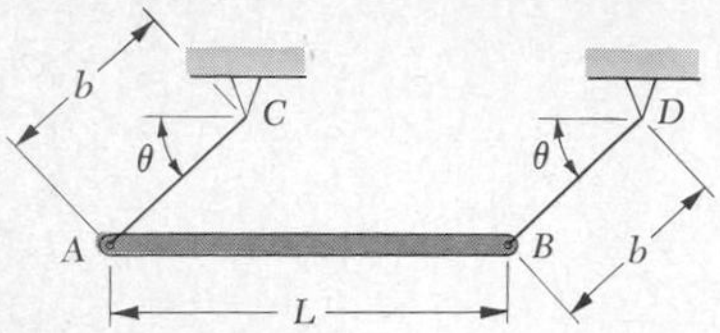

Fig. P17.134

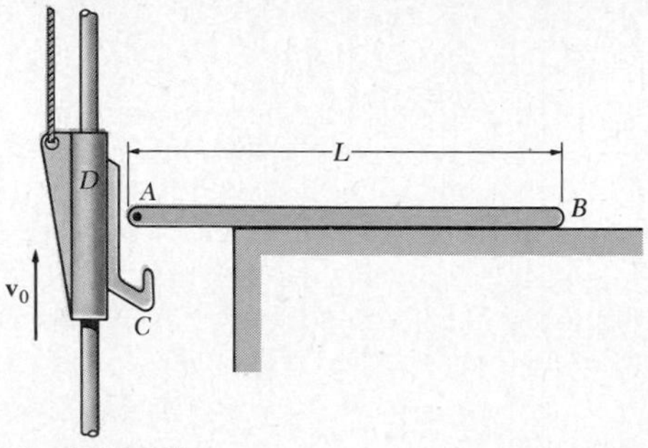

Fig. P17.135

17.135 The uniform rod AB of mass m is at rest on a frictionless horizontal surface when hook C of slider D engages a small pin at A. Knowing that the slider is pulled upward with a constant velocity $\mathbf{v}_0$, determine the impulse exerted on the rod (a) at A, (b) at B. Assume that the velocity of the slider is unchanged and that the impact is perfectly plastic.

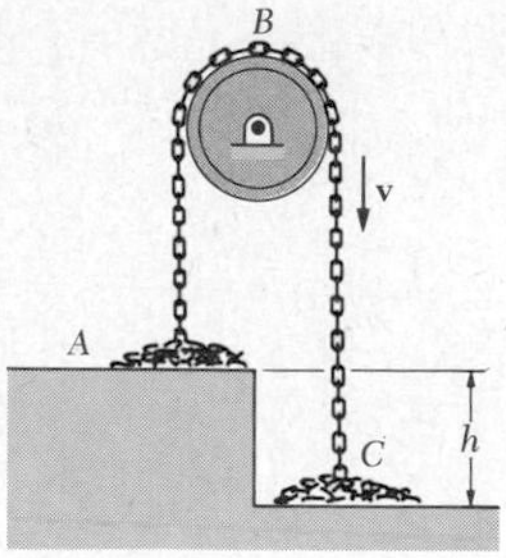

Fig. P17.136

17.136 The ends of a chain having a mass per unit length of 1.6 kg/m lie in piles at A and C. When released from rest at time $t = 0$, the chain moves over the pulley at B which has a mass of 2.5 kg, a radius of gyration of 160 mm, and an outside radius of 200 mm. It is known that $h = 300$ mm, and that the length of the chain connecting the two piles is $L = 4$ m. Neglecting friction, determine at $t = 2$ s (a) the speed v of the chain, (b) the length of chain which has been transferred from pile A to pile C. (Compare with Prob. 14.80 where the mass of the pulley was neglected.)

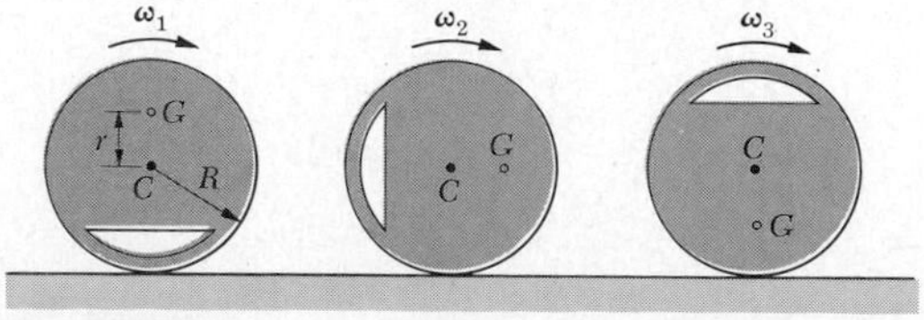

Fig. P17.137

17.137 The mass center G of a wheel of radius R is located at a distance r from its geometric center C. The centroidal radius of gyration of the wheel is denoted by $\bar{k}$. As the wheel rolls freely and without sliding on a horizontal plane, its angular velocity is observed to vary. Denoting by ω_1, ω_2, and ω_3, respectively, the angular velocity of the wheel when G is directly above C, level with C, and directly below C, show that ω_1, ω_2, and ω_3 satisfy the relation

$$\frac{\omega_2^2 - \omega_1^2}{\omega_3^2 - \omega_2^2} = \frac{g/R + \omega_1^2}{g/R + \omega_3^2}$$

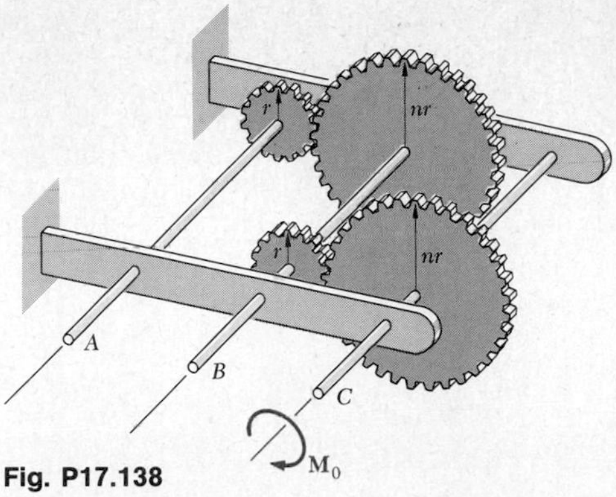

Fig. P17.138

17.138 The gear train shown consists of four gears of the same thickness and of the same material; two gears are of radius r, and the other two are of radius nr. The system is at rest when the couple $\mathbf{M}_0$ is applied to shaft C. Denoting by I_0 the moment of inertia of a gear of radius r, determine the angular velocity of shaft A if the couple $\mathbf{M}_0$ is applied for one revolution of shaft C.

17.139 For the gear train of Prob. 17.138, determine the angular velocity of shaft A if the couple $\mathbf{M}_0$ is applied for t seconds to shaft C.

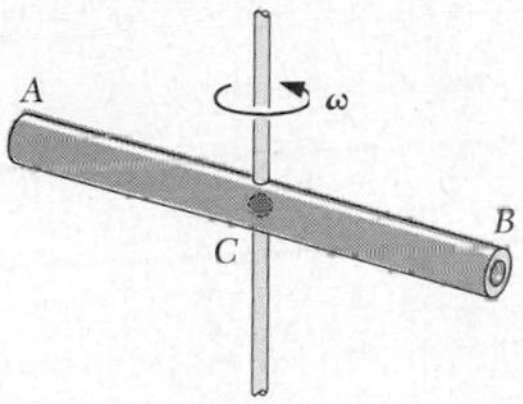

Fig. P17.140

17.140 A small 400-g ball may slide in a slender tube of length 1 m and of mass 1.2 kg which rotates freely about a vertical axis passing through its center C. If the angular velocity of the tube is 8 rad/s as the ball passes through C, determine the angular velocity of the tube (a) just before the ball leaves the tube, (b) just after the ball has left the tube.

17.141 Knowing that in Prob. 17.140 the speed of the ball is 1.5 m/s as it passes through C, determine the radial and transverse components of the velocity of the ball as it leaves the tube at B.

17.142 A small matchbox is placed on top of the rod AB. End B of the rod is given a slight horizontal push, causing it to slide on the horizontal floor. Assuming no friction and neglecting the weight of the matchbox, determine the angle θ through which the rod will have rotated when the matchbox loses contact with the rod.

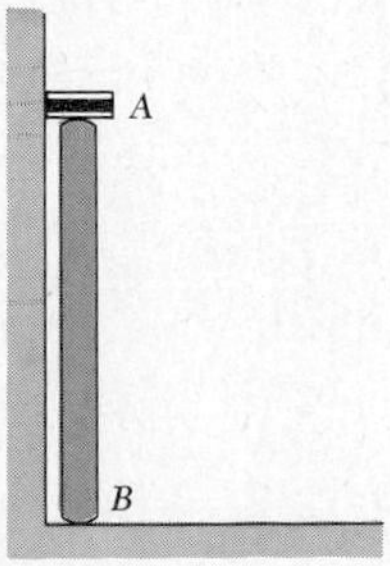

Fig. P17.142

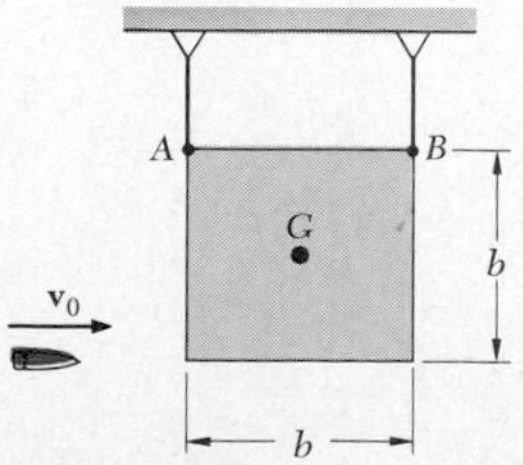

Fig. P17.143

17.143 A bullet of mass m is fired with a horizontal velocity $\mathbf{v}_0$ into the lower corner of a square panel of much larger mass M. The panel is held by two vertical wires as shown. Determine the velocity of the center G of the panel immediately after the bullet becomes embedded.

CHAPTER EIGHTEEN

Kinetics of Rigid Bodies in Three Dimensions

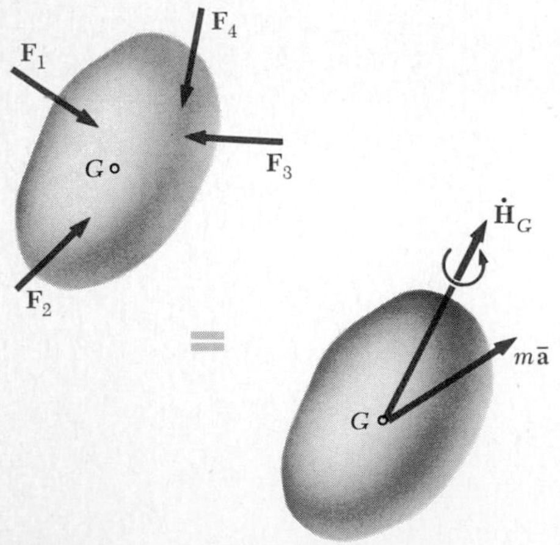

Fig. 18.1

***18.1. Introduction.** In Chaps. 16 and 17 we were concerned with the plane motion of rigid bodies and of systems of rigid bodies. In Chap. 16 and in the second half of Chap. 17 (momentum method), our study was further restricted to that of plane slabs and of bodies symmetrical with respect to the reference plane. However, many of the fundamental results obtained in these two chapters remain valid in the case of the motion of a rigid body in three dimensions.

For example, the two fundamental equations

$$\Sigma \mathbf{F} = m\bar{\mathbf{a}} \tag{18.1}$$

$$\Sigma \mathbf{M}_G = \dot{\mathbf{H}}_G \tag{18.2}$$

on which the analysis of the plane motion of a rigid body was based, remain valid in the most general case of motion of a rigid body. As it was indicated in Sec. 16.2, these equations express that the system of the external forces is equipollent to the system consisting of the vector $m\bar{\mathbf{a}}$ attached at G and the couple of moment $\dot{\mathbf{H}}_G$ (Fig. 18.1). However, the relation $\mathbf{H}_G = \bar{I}\boldsymbol{\omega}$, which enabled us to determine the angular momentum of a rigid slab and which played an important part in the solution of problems involving the plane motion of slabs and bodies symmetrical with respect to the reference plane, ceases to be valid in the case of nonsymmetrical bodies or three-dimensional motion. It will thus be necessary for us to develop in Sec. 18.2 a more general method for computing the angular momentum $\mathbf{H}_G$ of a rigid body in three dimensions.

Similarly, the main feature of the impulse-momentum method discussed in Sec. 17.7, namely, the reduction of the momenta of the particles of a rigid body to a linear momentum vector $m\overline{\mathbf{v}}$ attached at the mass center G of the body and an angular momentum couple $\mathbf{H}_G$, remains valid. Here again, however, the relation $\mathbf{H}_G = \overline{I}\boldsymbol{\omega}$ will have to be discarded and replaced by the more general relation to be developed in Sec. 18.2.

We also note that the work-energy principle (Sec. 17.2) and the principle of conservation of energy (Sec. 17.6) still apply in the case of the motion of a rigid body in three dimensions. However, the expression obtained in Sec. 17.4 for the kinetic energy of a rigid body in plane motion will be replaced by a new expression to be developed in Sec. 18.4 for a rigid body in three-dimensional motion.

In the second part of the chapter, we shall first learn to determine the rate of change $\dot{\mathbf{H}}_G$ of the angular momentum $\mathbf{H}_G$ of a three-dimensional rigid body, using a rotating frame of reference with respect to which the moments and products of inertia of the body remain constant (Sec. 18.5). We shall then express Eqs. (18.1) and (18.2) in the form of free-body-diagram equations, which may be used to solve various problems involving the three-dimensional motion of rigid bodies (Secs. 18.6 through 18.8).

The last part of the chapter (Secs. 18.9 through 18.11) is devoted to the study of the motion of the gyroscope or, more generally, of an axisymmetrical body with a fixed point located on its axis of symmetry. In Sec. 18.10, we shall consider the particular case of the steady precession of a gyroscope and, in Sec. 18.11, the motion of an axisymmetrical body subjected to no force, except its own weight.

*** 18.2. Angular Momentum of a Rigid Body in Three Dimensions.** We shall see in this section how the angular momentum $\mathbf{H}_G$ of the body about its mass center G may be determined from the angular velocity $\boldsymbol{\omega}$ of the body in the case of three-dimensional motion.

According to Eq. (14.24), the angular momentum of the body about G may be expressed as

$$\mathbf{H}_G = \sum_{i=1}^{n} (\mathbf{r}'_i \times \mathbf{v}'_i \,\Delta m_i) \tag{18.3}$$

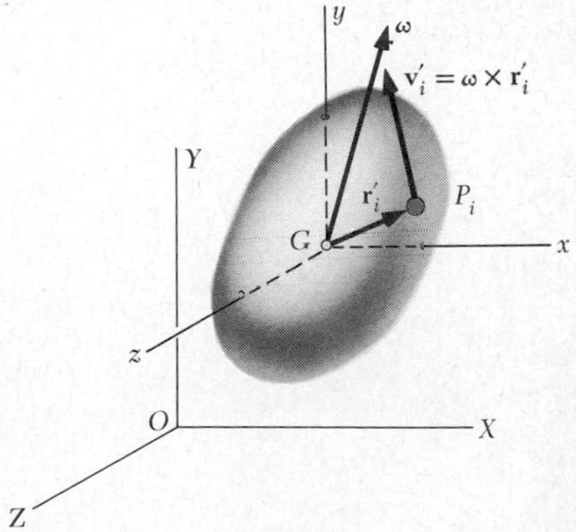

Fig. 18.2

where $\mathbf{r}'_i$ and $\mathbf{v}'_i$ denote, respectively, the position vector and the velocity of the particle P_i, of mass Δm_i, relative to the centroidal frame $Gxyz$ (Fig. 18.2). But $\mathbf{v}'_i = \boldsymbol{\omega} \times \mathbf{r}'_i$, where $\boldsymbol{\omega}$ is the angular velocity of the body at the instant considered. Substituting into (18.3), we have

$$\mathbf{H}_G = \sum_{i=1}^{n} [\mathbf{r}'_i \times (\boldsymbol{\omega} \times \mathbf{r}'_i) \,\Delta m_i]$$

Recalling the rule for determining the rectangular components of a vector product (Sec. 3.5), we obtain the following expression for the x component of the angular momentum:

$$H_x = \sum_{i=1}^{n} [y_i(\boldsymbol{\omega} \times \mathbf{r}'_i)_z - z_i(\boldsymbol{\omega} \times \mathbf{r}'_i)_y]\,\Delta m_i$$

$$= \sum_{i=1}^{n} [y_i(\omega_x y_i - \omega_y x_i) - z_i(\omega_z x_i - \omega_x z_i)]\,\Delta m_i$$

$$= \omega_x \sum_i (y_i^2 + z_i^2)\,\Delta m_i - \omega_y \sum_i x_i y_i\,\Delta m_i - \omega_z \sum_i z_i x_i\,\Delta m_i$$

Replacing the sums by integrals in this expression and in the two similar expressions which are obtained for H_y and H_z, we have

$$\begin{aligned} H_x &= \omega_x \int (y^2 + z^2)\,dm - \omega_y \int xy\,dm - \omega_z \int zx\,dm \\ H_y &= -\omega_x \int xy\,dm + \omega_y \int (z^2 + x^2)\,dm - \omega_z \int yz\,dm \\ H_z &= -\omega_x \int zx\,dm - \omega_y \int yz\,dm + \omega_z \int (x^2 + y^2)\,dm \end{aligned} \tag{18.4}$$

We note that the integrals containing squares represent the *centroidal mass moments of inertia* of the body about the x, y, and z axes, respectively (Sec. 9.11); we have

$$\bar{I}_x = \int (y^2 + z^2)\,dm \qquad \bar{I}_y = \int (z^2 + x^2)\,dm \qquad \bar{I}_z = \int (x^2 + y^2)\,dm \tag{18.5}$$

Similarly, the integrals containing products of coordinates represent the *centroidal mass products of inertia* of the body (Sec. 9.16); we have

$$\bar{I}_{xy} = \int xy\,dm \qquad \bar{I}_{yz} = \int yz\,dm \qquad \bar{I}_{zx} = \int zx\,dm \tag{18.6}$$

Substituting from (18.5) and (18.6) into (18.4), we obtain the components of the angular momentum $\mathbf{H}_G$ of the body about its mass center G:

$$\begin{aligned} H_x &= +\bar{I}_x\,\omega_x - \bar{I}_{xy}\omega_y - \bar{I}_{xz}\omega_z \\ H_y &= -\bar{I}_{yx}\omega_x + \bar{I}_y\,\omega_y - \bar{I}_{yz}\omega_z \\ H_z &= -\bar{I}_{zx}\omega_x - \bar{I}_{zy}\omega_y + \bar{I}_z\,\omega_z \end{aligned} \tag{18.7}$$

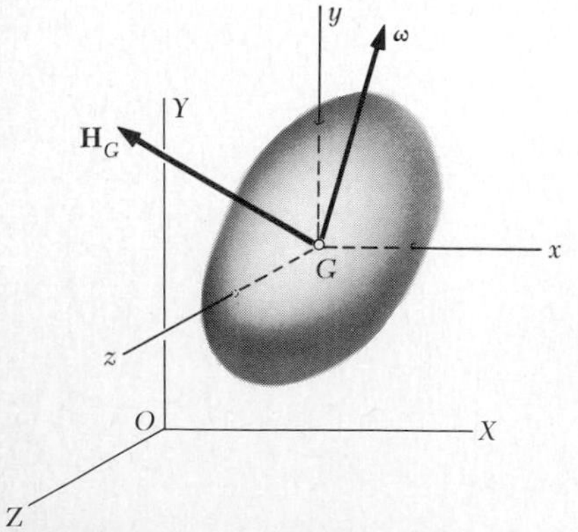

Fig. 18.3

The relations (18.7) show that the operation which transforms the vector $\boldsymbol{\omega}$ into the vector $\mathbf{H}_G$ (Fig. 18.3) is characterized by the array of moments and products of inertia

$$\begin{pmatrix} \bar{I}_x & -\bar{I}_{xy} & -\bar{I}_{xz} \\ -\bar{I}_{yx} & \bar{I}_y & -\bar{I}_{yz} \\ -\bar{I}_{zx} & -\bar{I}_{zy} & \bar{I}_z \end{pmatrix} \tag{18.8}$$

The array (18.8) defines the *inertia tensor* of the body at its mass center G.† A new array of moments and products of inertia would be obtained if a different system of axes were used. The transformation characterized by this new array, however, would still be the same. Clearly, the angular momentum $\mathbf{H}_G$ corresponding to a given angular velocity $\boldsymbol{\omega}$ is independent of the choice of the coordinate axes. As was shown in Sec. 9.17, it is always possible to select a system of axes $Gx'y'z'$, called *principal axes of inertia,* with respect to which all the products of inertia of a given body are zero. The array (18.8) takes then the diagonalized form

$$\begin{pmatrix} \bar{I}_{x'} & 0 & 0 \\ 0 & \bar{I}_{y'} & 0 \\ 0 & 0 & \bar{I}_{z'} \end{pmatrix} \qquad (18.9)$$

where $\bar{I}_{x'}$, $\bar{I}_{y'}$, $\bar{I}_{z'}$ represent the *principal centroidal moments of inertia* of the body, and the relations (18.7) reduce to

$$H_{x'} = \bar{I}_{x'}\omega_{x'} \qquad H_{y'} = \bar{I}_{y'}\omega_{y'} \qquad H_{z'} = \bar{I}_{z'}\omega_{z'} \qquad (18.10)$$

We note that if the three principal centroidal moments of inertia $\bar{I}_{x'}$, $\bar{I}_{y'}$, $\bar{I}_{z'}$ are equal, the components $H_{x'}$, $H_{y'}$, $H_{z'}$ of the angular momentum about G are proportional to the components $\omega_{x'}$, $\omega_{y'}$, $\omega_{z'}$ of the angular velocity, and the vectors $\mathbf{H}_G$ and $\boldsymbol{\omega}$ are collinear. In general, however, the principal moments of inertia will be different, and the vectors $\mathbf{H}_G$ and $\boldsymbol{\omega}$ *will have different directions,* except when two of the three components of $\boldsymbol{\omega}$ happen to be zero, i.e., when $\boldsymbol{\omega}$ is directed along one of the coordinate axes. Thus, *the angular momentum $\mathbf{H}_G$ of a rigid body and its angular velocity $\boldsymbol{\omega}$ have the same direction if, and only if, $\boldsymbol{\omega}$ is directed along a principal axis of inertia.*‡ Since this condition is satisfied in the case of the plane motion of a rigid body symmetrical with respect to the reference plane, we were able in Secs. 16.3 and 17.8 to represent the angular momentum $\mathbf{H}_G$ of such a body by the vector $\bar{I}\boldsymbol{\omega}$. We must realize, however, that this result cannot be extended to the case of the plane motion of a nonsymmetrical body, or to the case of the three-dimensional motion of a

†Setting $\bar{I}_x = I_{11}$, $\bar{I}_y = I_{22}$, $\bar{I}_z = I_{33}$, and $-\bar{I}_{xy} = I_{12}$, $-\bar{I}_{xz} = I_{13}$, etc., we may write the inertia tensor (18.8) in the standard form

$$\begin{pmatrix} I_{11} & I_{12} & I_{13} \\ I_{21} & I_{22} & I_{23} \\ I_{31} & I_{32} & I_{33} \end{pmatrix}$$

Denoting by H_1, H_2, H_3 the components of the angular momentum $\mathbf{H}_G$ and by ω_1, ω_2, ω_3 the components of the angular velocity $\boldsymbol{\omega}$, we may write the relations (18.7) in the form

$$H_i = \sum_j I_{ij}\omega_j$$

where i and j take the values 1, 2, 3. The quantities I_{ij} are said to be the *components* of the inertia tensor. Since $I_{ij} = I_{ji}$, the inertia tensor is a *symmetric tensor of the second order.*

‡In the particular case when $\bar{I}_{x'} = \bar{I}_{y'} = \bar{I}_{z'}$, any line through G may be considered as a principal axis of inertia, and the vectors $\mathbf{H}_G$ and $\boldsymbol{\omega}$ are always collinear.

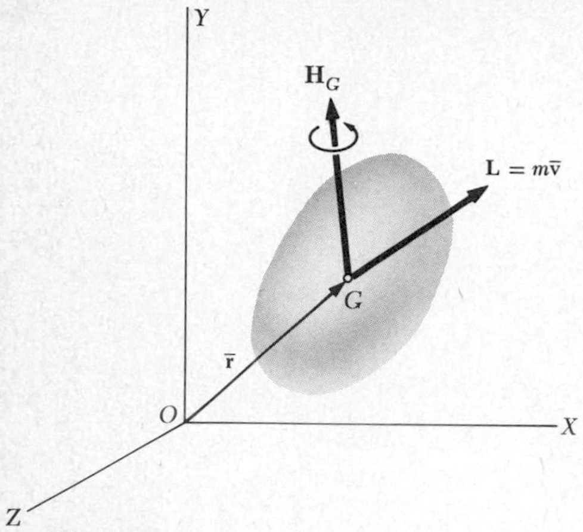

Fig. 18.4

rigid body. Except when $\boldsymbol{\omega}$ happens to be directed along a principal axis of inertia, the angular momentum and angular velocity of a rigid body have different directions, and the relation (18.7) or (18.10) must be used to determine $\mathbf{H}_G$ from $\boldsymbol{\omega}$.

Reduction of the Momenta of the Particles of a Rigid Body to a Momentum Vector and a Couple at G. We saw in Sec. 17.8 that the system formed by the momenta of the various particles of a rigid body may be reduced to a vector $\mathbf{L}$ attached at the mass center G of the body, representing the linear momentum of the body, and to a couple $\mathbf{H}_G$, representing the angular momentum of the body about G (Fig. 18.4). We are now in a position to determine the vector $\mathbf{L}$ and the couple $\mathbf{H}_G$ in the most general case of three-dimensional motion of a rigid body. As in the case of the two-dimensional motion considered in Sec. 17.8, the linear momentum $\mathbf{L}$ of the body is equal to the product $m\overline{\mathbf{v}}$ of its mass m and of the velocity $\overline{\mathbf{v}}$ of its mass center G. The angular momentum $\mathbf{H}_G$, however, cannot be obtained any more by simply multiplying the angular velocity $\boldsymbol{\omega}$ of the body by the scalar $\overline{I}$; it must now be obtained from the components of $\boldsymbol{\omega}$ and from the centroidal moments and products of inertia of the body through the use of Eq. (18.7) or (18.10).

We should also note that once the linear momentum $m\overline{\mathbf{v}}$ and the angular momentum $\mathbf{H}_G$ of a rigid body have been determined, its angular momentum $\mathbf{H}_O$ about any given point O may be obtained by adding the moments about O of the vector $m\overline{\mathbf{v}}$ and of the couple $\mathbf{H}_G$. We write

$$\mathbf{H}_O = \overline{\mathbf{r}} \times m\overline{\mathbf{v}} + \mathbf{H}_G \tag{18.11}$$

Angular Momentum of a Rigid Body Constrained to Rotate about a Fixed Point. In the particular case of a rigid body constrained to rotate in three-dimensional space about a fixed point O (Fig. 18.5*a*), it is sometimes convenient to determine the angular momentum $\mathbf{H}_O$ of the body about the fixed point O. While $\mathbf{H}_O$ could be obtained by first computing $\mathbf{H}_G$ as indicated above and then using Eq. (18.11), it is often advantageous to determine $\mathbf{H}_O$ directly from the angular velocity $\boldsymbol{\omega}$ of the body and its moments and products of inertia with respect to a frame $Oxyz$ centered at the fixed point O. Recalling Eq. (14.7), we write

$$\mathbf{H}_O = \sum_{i=1}^{n} (\mathbf{r}_i \times \mathbf{v}_i \, \Delta m_i) \tag{18.12}$$

where $\mathbf{r}_i$ and $\mathbf{v}_i$ denote, respectively, the position vector and the velocity of the particle P_i with respect to the fixed frame $Oxyz$. Substituting $\mathbf{v}_i = \boldsymbol{\omega} \times \mathbf{r}_i$, and after manipulations similar to those used in the earlier part of this section, we find that the components of the angular momentum $\mathbf{H}_O$ (Fig. 18.5*b*) are given by the relations

$$\begin{aligned} H_x &= +I_x\,\omega_x - I_{xy}\omega_y - I_{xz}\omega_z \\ H_y &= -I_{yx}\omega_x + I_y\,\omega_y - I_{yz}\omega_z \\ H_z &= -I_{zx}\omega_x - I_{zy}\omega_y + I_z\,\omega_z \end{aligned} \tag{18.13}$$

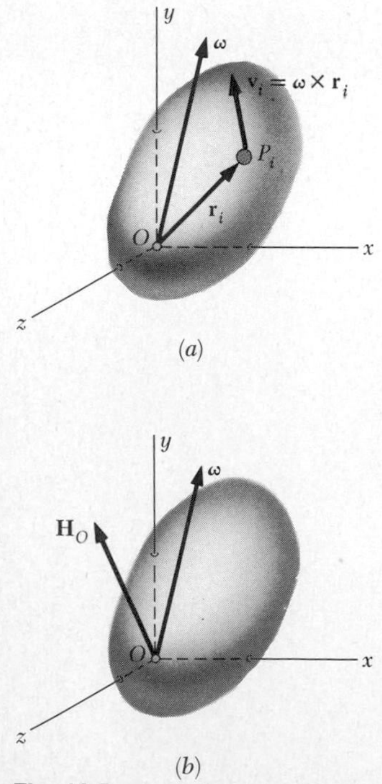

Fig. 18.5

where the moments of inertia I_x, I_y, I_z and the products of inertia I_{xy}, I_{yz}, I_{zx} are computed with respect to the frame $Oxyz$ centered at the fixed point O.

*** 18.3. Application of the Principle of Impulse and Momentum to the Three-dimensional Motion of a Rigid Body.** Before we can apply the fundamental equation (18.2) to the solution of problems involving the three-dimensional motion of a rigid body, we shall have to learn to compute the derivative of the vector $\mathbf{H}_G$. This will be done in Sec. 18.5. We may, however, immediately use the results obtained in the preceding section to solve problems by the impulse-momentum method.

Recalling that the system formed by the momenta of the particles of a rigid body reduces to a linear momentum vector $m\overline{\mathbf{v}}$ attached at the mass center G of the body and an angular momentum couple $\mathbf{H}_G$, we represent graphically the fundamental relation

$$\textbf{Syst Momenta}_1 + \textbf{Syst Ext Imp}_{1\to 2} = \textbf{Syst Momenta}_2 \qquad (17.4)$$

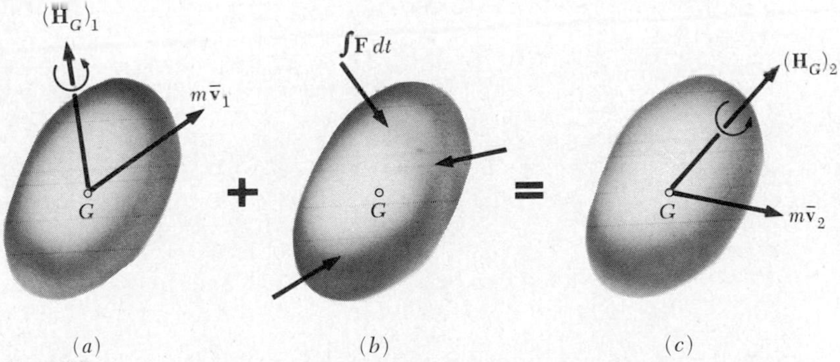

Fig. 18.6

by means of the three sketches shown in Fig. 18.6. To solve a given problem, we may use these sketches to write appropriate component and moment equations, keeping in mind that the components of the angular momentum $\mathbf{H}_G$ are related to the components of the angular velocity $\boldsymbol{\omega}$ by Eqs. (18.7) of the preceding section.

In solving problems dealing with the motion of a body rotating about a fixed point O, it will be convenient to eliminate the impulse of the reaction at O by writing an equation involving the moments of the momenta and impulses about O. We recall that the angular momentum $\mathbf{H}_O$ of the body about the fixed point O may be obtained either directly from Eqs. (18.13), or by first computing its linear momentum $m\overline{\mathbf{v}}$ and its angular momentum $\mathbf{H}_G$ and then using Eq. (18.11).

***18.4. Kinetic Energy of a Rigid Body in Three Dimensions.** Consider a rigid body of mass m in three-dimensional motion. We recall from Sec. 14.6 that if the absolute velocity $\mathbf{v}_i$ of each particle P_i of the body is expressed as the sum of the velocity $\overline{\mathbf{v}}$ of the mass center G of the body and of the velocity $\mathbf{v}'_i$ of the particle relative to a frame $Gxyz$ attached to G and of fixed orientation (Fig. 18.7), the kinetic energy of the

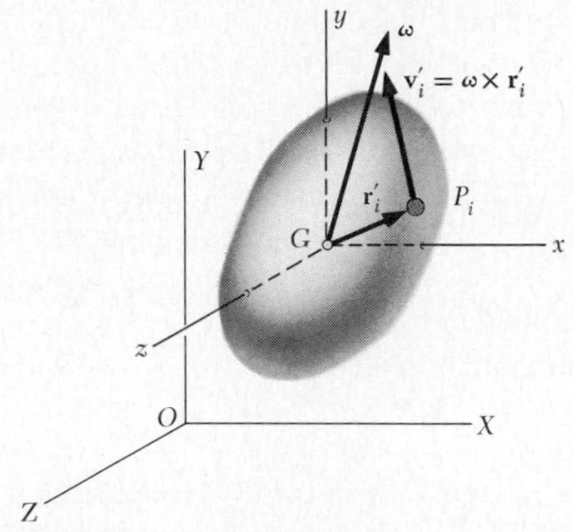

Fig. 18.7

system of particles forming the rigid body may be written in the form

$$T = \tfrac{1}{2}m\overline{v}^2 + \frac{1}{2}\sum_{i=1}^{n}\Delta m_i v_i'^2 \tag{18.14}$$

where the last term represents the kinetic energy T' of the body relative to the centroidal frame $Gxyz$. Since $v'_i = |\mathbf{v}'_i| = |\boldsymbol{\omega} \times \mathbf{r}'_i|$, we write

$$T' = \frac{1}{2}\sum_{i=1}^{n}\Delta m_i v_i'^2 = \frac{1}{2}\sum_{i=1}^{n}|\boldsymbol{\omega} \times \mathbf{r}'_i|^2\,\Delta m_i$$

Expressing the square in terms of the rectangular components of the vector product, and replacing the sums by integrals, we have

$$\begin{aligned} T' &= \int[(\omega_x y - \omega_y x)^2 + (\omega_y z - \omega_z y)^2 + (\omega_z x - \omega_x z)^2]\,dm \\ &= \omega_x^2\int(y^2 + z^2)\,dm + \omega_y^2\int(z^2 + x^2)\,dm + \omega_z^2\int(x^2 + y^2)\,dm \\ &\qquad - 2\omega_x\omega_y\int xy\,dm - 2\omega_y\omega_z\int yz\,dm - 2\omega_z\omega_x\int zx\,dm \end{aligned}$$

or, recalling the relations (18.5) and (18.6),

$$T' = \tfrac{1}{2}(\overline{I}_x\omega_x^2 + \overline{I}_y\omega_y^2 + \overline{I}_z\omega_z^2 - 2\overline{I}_{xy}\omega_x\omega_y - 2\overline{I}_{yz}\omega_y\omega_z - 2\overline{I}_{zx}\omega_z\omega_x) \tag{18.15}$$

Substituting into (18.14) the expression (18.15) we have just obtained for the kinetic energy of the body relative to centroidal axes, we write

$$\begin{aligned} T = \tfrac{1}{2}m\overline{v}^2 + \tfrac{1}{2}(\overline{I}_x\omega_x^2 + \overline{I}_y\omega_y^2 + \overline{I}_z\omega_z^2 &- 2\overline{I}_{xy}\omega_x\omega_y \\ &- 2\overline{I}_{yz}\omega_y\omega_z - 2\overline{I}_{zx}\omega_z\omega_x) \end{aligned} \tag{18.16}$$

If the axes of coordinates are chosen so that they coincide at the instant considered with the principal axes x', y', z' of the body, the relation obtained reduces to

$$T = \tfrac{1}{2}m\bar{v}^2 + \tfrac{1}{2}(\bar{I}_{x'}\omega_{x'}^2 + \bar{I}_{y'}\omega_{y'}^2 + \bar{I}_{z'}\omega_{z'}^2) \tag{18.17}$$

where $\bar{\mathbf{v}}$ = velocity of mass center
$\boldsymbol{\omega}$ = angular velocity
m = mass of rigid body
$\bar{I}_{x'}, \bar{I}_{y'}, \bar{I}_{z'}$ = principal centroidal moments of inertia

The results we have obtained enable us to extend to the three-dimensional motion of a rigid body the application of the principle of work and energy (Sec. 17.2) and of the principle of conservation of energy (Sec. 17.6).

Kinetic Energy of a Rigid Body with a Fixed Point. In the particular case of a rigid body rotating in three-dimensional space about a fixed point O, the kinetic energy of the body may be expressed in terms of its moments and products of inertia with respect to axes attached at O (Fig. 18.8). Recalling the definition of the kinetic energy of a system of particles,

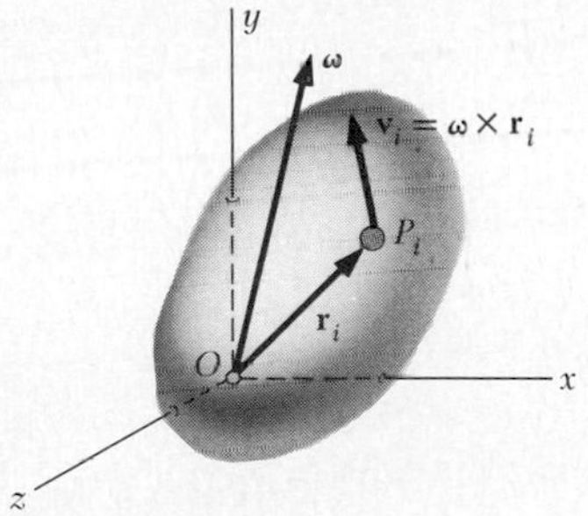

Fig. 18.8

and substituting $v_i = |\mathbf{v}_i| = |\boldsymbol{\omega} \times \mathbf{r}_i|$, we write

$$T = \frac{1}{2}\sum_{i=1}^{n} \Delta m_i v_i^2 = \frac{1}{2}\sum_{i=1}^{n} |\boldsymbol{\omega} \times \mathbf{r}_i|^2 \, \Delta m_i \tag{18.18}$$

Manipulations similar to those used to derive Eq. (18.15) yield

$$T = \tfrac{1}{2}(I_x\omega_x^2 + I_y\omega_y^2 + I_z\omega_z^2 - 2I_{xy}\omega_x\omega_y - 2I_{yz}\omega_y\omega_z - 2I_{zx}\omega_z\omega_x) \tag{18.19}$$

or, if the principal axes x', y', z' of the body at the origin O are chosen as coordinate axes,

$$T = \tfrac{1}{2}(I_{x'}\omega_{x'}^2 + I_{y'}\omega_{y'}^2 + I_{z'}\omega_{z'}^2) \tag{18.20}$$

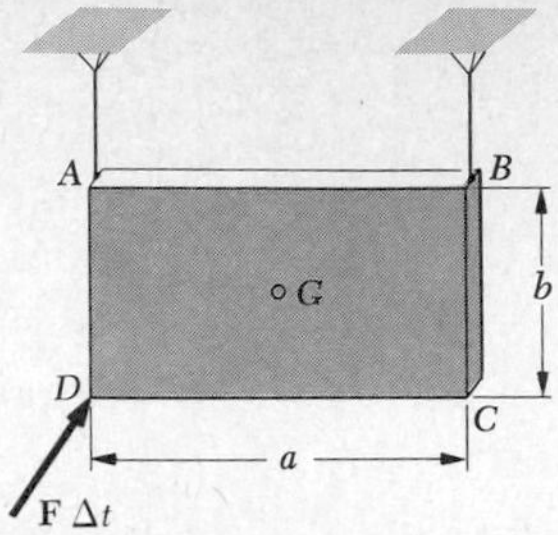

SAMPLE PROBLEM 18.1

A rectangular plate of mass m suspended from two wires at A and B is hit at D in a direction perpendicular to the plate. Denoting by $\mathbf{F}\,\Delta t$ the impulse applied at D, determine immediately after impact (*a*) the velocity of the mass center G, (*b*) the angular velocity of the plate.

Solution. We shall assume that the wires remain taut and thus that the components $\bar{v}_y$ of $\bar{\mathbf{v}}$ and ω_z of $\boldsymbol{\omega}$ are zero after impact. We therefore have

$$\bar{\mathbf{v}} = \bar{v}_x\mathbf{i} + \bar{v}_z\mathbf{k} \qquad \boldsymbol{\omega} = \omega_x\mathbf{i} + \omega_y\mathbf{j}$$

and since the x, y, z axes are principal axes of inertia,

$$\mathbf{H}_G = \bar{I}_x\omega_x\mathbf{i} + \bar{I}_y\omega_y\mathbf{j} \qquad \mathbf{H}_G = \tfrac{1}{12}mb^2\omega_x\mathbf{i} + \tfrac{1}{12}ma^2\omega_y\mathbf{j} \qquad (1)$$

Principle of Impulse and Momentum. Since the initial momenta are zero, the system of the impulses must be equivalent to the system of the final momenta:

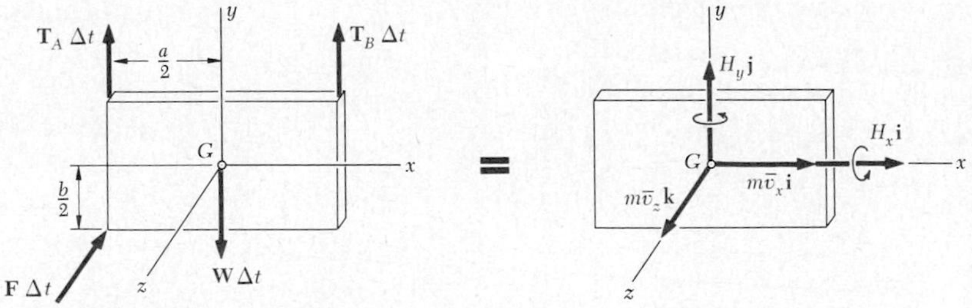

a. Velocity of Mass Center. Equating the components of the impulses and momenta in the x and z directions:

x components: $\quad 0 = m\bar{v}_x \qquad \bar{v}_x = 0$

z components: $\quad -F\,\Delta t = m\bar{v}_z \qquad \bar{v}_z = -F\,\Delta t/m$

$$\bar{\mathbf{v}} = \bar{v}_x\mathbf{i} + \bar{v}_z\mathbf{k} \qquad \bar{\mathbf{v}} = -(F\,\Delta t/m)\mathbf{k} \quad ◀$$

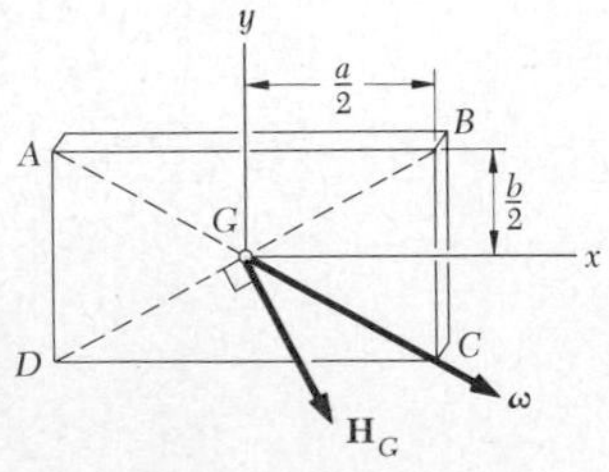

b. Angular Velocity. Equating the moments of the impulses and momenta about the x and y axes:

About x axis: $\quad \tfrac{1}{2}bF\,\Delta t = H_x$

About y axis: $\quad -\tfrac{1}{2}aF\,\Delta t = H_y$

$$\mathbf{H}_G = H_x\mathbf{i} + H_y\mathbf{j} \qquad \mathbf{H}_G = \tfrac{1}{2}bF\,\Delta t\mathbf{i} - \tfrac{1}{2}aF\,\Delta t\mathbf{j} \qquad (2)$$

Comparing Eqs. (1) and (2), we conclude that

$$\omega_x = 6F\,\Delta t/mb \qquad \omega_y = -6F\,\Delta t/ma$$

$$\boldsymbol{\omega} = \omega_x\mathbf{i} + \omega_y\mathbf{j} \qquad \boldsymbol{\omega} = (6F\,\Delta t/mab)(a\mathbf{i} - b\mathbf{j}) \quad ◀$$

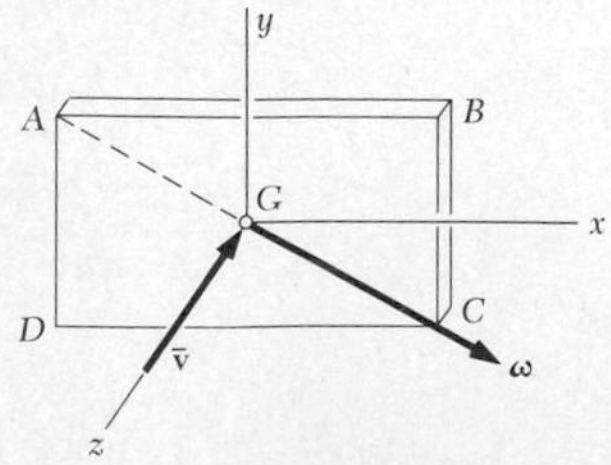

We note that $\boldsymbol{\omega}$ is directed along the diagonal AC.

Remark: Equating the y components of the impulses and momenta, and their moments about the z axis, we obtain two additional equations which yield $T_A = T_B = \tfrac{1}{2}W$. We thus verify that the wires remain taut and that our assumption was correct.

SAMPLE PROBLEM 18.2

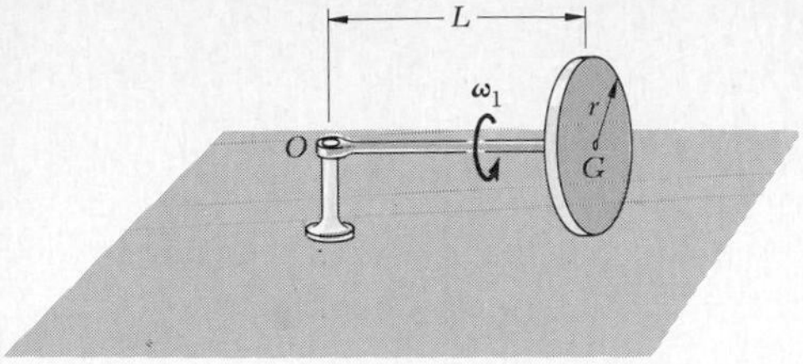

A homogeneous disk of radius r and mass m is mounted on an axle OG of length L and negligible mass. The axle is pivoted at the fixed point O, and the disk is constrained to roll on a horizontal floor. Knowing that the disk rotates counterclockwise at the rate ω_1 about the axle OG, determine (a) the angular velocity of the disk, (b) its angular momentum about O, (c) its kinetic energy, (d) the vector and couple at G equivalent to the momenta of the particles of the disk.

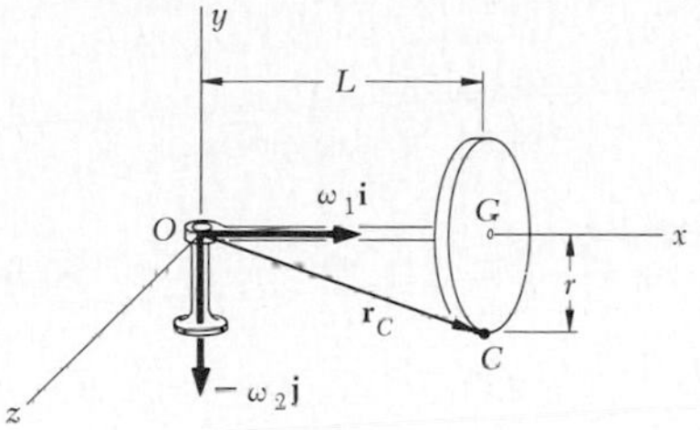

***a.* Angular Velocity.** As the disk rotates about the axle OG it also rotates with the axle about the y axis at a rate ω_2 clockwise. The total angular velocity of the disk is therefore

$$\boldsymbol{\omega} = \omega_1\mathbf{i} - \omega_2\mathbf{j} \qquad (1)$$

To determine ω_2 we write that the velocity of C is zero:

$$\mathbf{v}_C = \boldsymbol{\omega} \times \mathbf{r}_C = 0$$
$$(\omega_1\mathbf{i} - \omega_2\mathbf{j}) \times (L\mathbf{i} - r\mathbf{j}) = 0$$
$$(L\omega_2 - r\omega_1)\mathbf{k} = 0 \qquad \omega_2 = r\omega_1/L$$

Substituting into (1) for ω_2: $\boldsymbol{\omega} = \omega_1\mathbf{i} - (r\omega_1/L)\mathbf{j}$ ◄

***b.* Angular Momentum About *O*.** Assuming the axle to be part of the disk, we may consider the disk to have a fixed point at O. Since the x, y, and z axes are principal axes of inertia for the disk,

$$H_x = I_x\omega_x = (\tfrac{1}{2}mr^2)\omega_1$$
$$H_y = I_y\omega_y = (mL^2 + \tfrac{1}{4}mr^2)(-r\omega_1/L)$$
$$H_z = I_z\omega_z = (mL^2 + \tfrac{1}{4}mr^2)\,0 = 0$$
$$\mathbf{H}_O = \tfrac{1}{2}mr^2\omega_1\mathbf{i} - m(L^2 + \tfrac{1}{4}r^2)(r\omega_1/L)\mathbf{j} \quad ◄$$

***c.* Kinetic Energy.** Using the values obtained for the moments of inertia and the components of $\boldsymbol{\omega}$, we have

$$T = \tfrac{1}{2}(I_x\omega_x^2 + I_y\omega_y^2 + I_z\omega_z^2) = \tfrac{1}{2}[\tfrac{1}{2}mr^2\omega_1^2 + m(L^2 + \tfrac{1}{4}r^2)(-r\omega_1/L)^2]$$
$$T = \tfrac{1}{8}mr^2\left(6 + \frac{r^2}{L^2}\right)\omega_1^2 \quad ◄$$

***d.* Momentum Vector and Couple at *G*.** The linear momentum vector $m\bar{\mathbf{v}}$ and the angular momentum couple $\mathbf{H}_G$ are

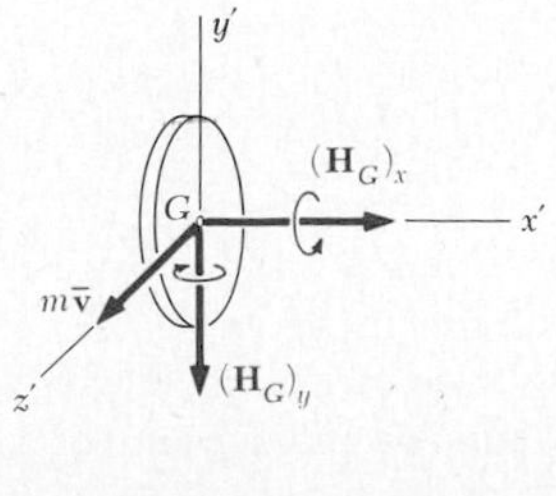

$$m\bar{\mathbf{v}} = mr\omega_1\mathbf{k} \quad ◄$$

and

$$\mathbf{H}_G = \bar{I}_{x'}\omega_x\mathbf{i} + \bar{I}_{y'}\omega_y\mathbf{j} + \bar{I}_{z'}\omega_z\mathbf{k} = \tfrac{1}{2}mr^2\omega_1\mathbf{i} + \tfrac{1}{4}mr^2(-r\omega_1/L)\mathbf{j}$$
$$\mathbf{H}_G = \tfrac{1}{2}mr^2\omega_1\left(\mathbf{i} - \frac{r}{2L}\mathbf{j}\right) \quad ◄$$

Problems

18.1 Two uniform rods CD and DE, each weighing 2.5 lb, are welded to the shaft AB. Knowing that the angular velocity of the shaft is constant and equal to 12 rad/s, determine the angle that the angular momentum of the assembly about D forms with the shaft AB.

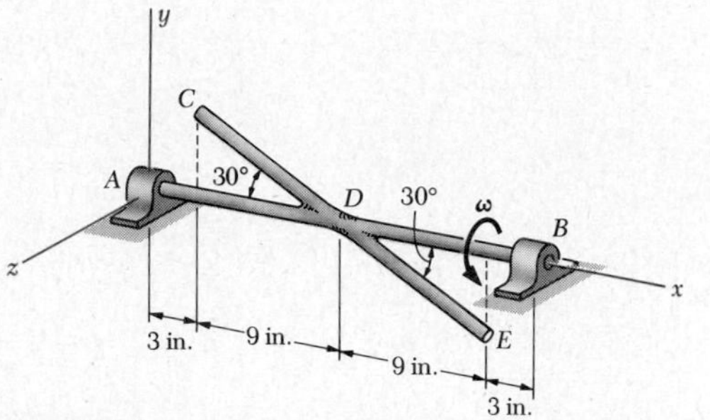

Fig. P18.1

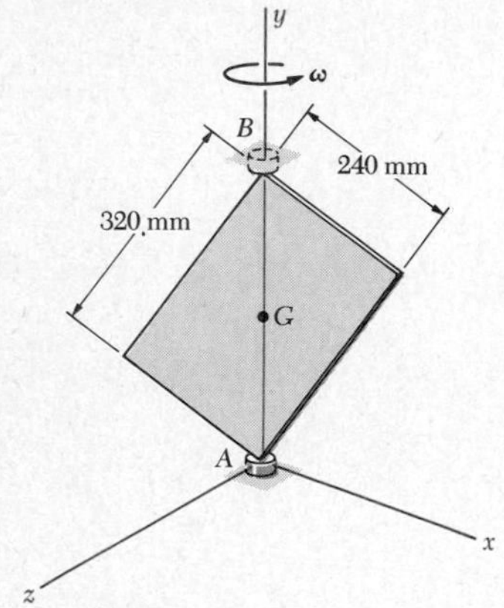

Fig. P18.2

18.2 A thin rectangular plate of mass 6 kg rotates about its vertical diagonal AB with an angular velocity ω. Knowing that the z axis is perpendicular to the plate and that ω is constant and equal to 5 rad/s, determine the angular momentum of the plate about its mass center G.

18.3 A thin homogeneous square plate of mass m and side a is welded to a vertical shaft AB with which it forms an angle of 45°. Knowing that the shaft rotates with an angular velocity ω, determine the angular momentum of the plate about A.

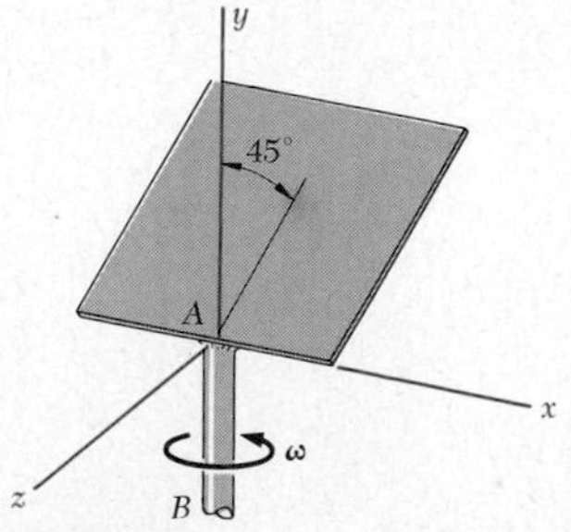

Fig. P18.3

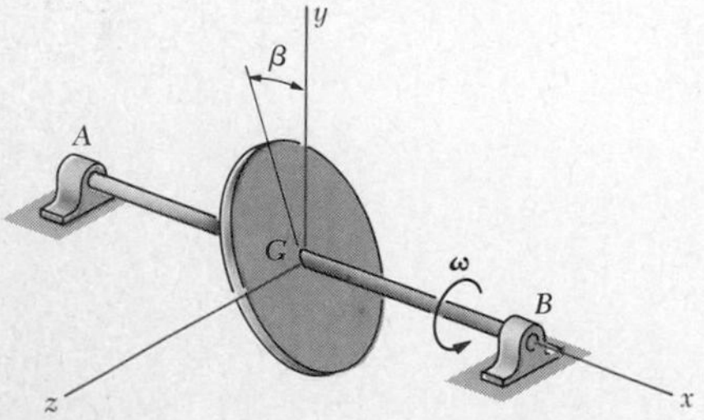

Fig. P18.4

18.4 A thin homogeneous disk of mass m and radius r is mounted on the horizontal axle AB. The plane of the disk forms an angle $\beta = 25°$ with the vertical. Knowing that the axle rotates with an angular velocity ω, determine the angle θ formed by the axle and the angular momentum of the disk about G.

18.5 In Prob. 18.4, determine (*a*) the value of β for which the angle θ formed by the axle and the angular momentum $\mathbf{H}_G$ is maximum, (*b*) the corresponding maximum value of θ.

18.6 A thin disk of weight $W = 8$ lb rotates at the constant rate $\omega_2 = 12$ rad/s with respect to the arm OA, which itself rotates at the constant rate $\omega_1 = 4$ rad/s about the y axis. Determine the angular momentum of the disk about its center A.

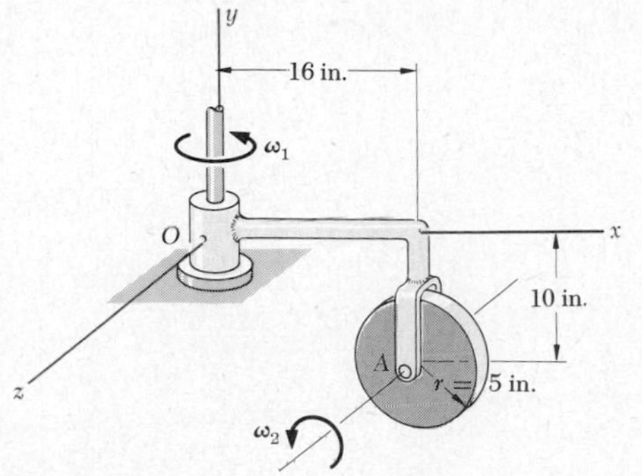

Fig. P18.6

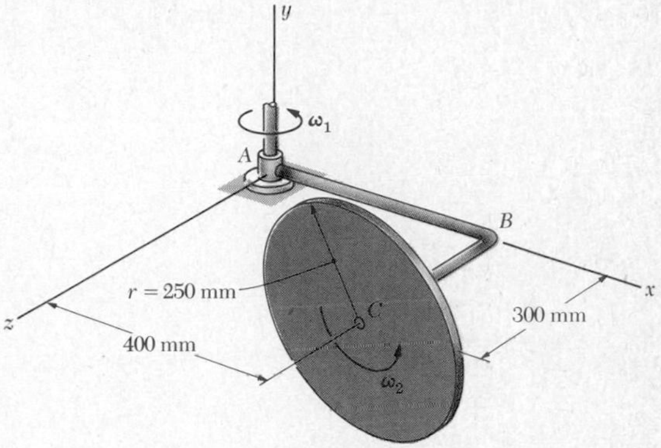

Fig. P18.7

18.7 A thin disk of mass $m = 5$ kg rotates at the constant rate $\omega_2 = 8$ rad/s with respect to the bent axle ABC, which itself rotates at the constant rate $\omega_1 = 3$ rad/s about the y axis. Determine the angular momentum of the disk about its center C.

18.8 The 25-kg projectile shown has a radius of gyration of 60 mm about its axis of symmetry Gx and of 250 mm about its transverse axis Gy. Its angular velocity $\boldsymbol{\omega}$ may be resolved into two components: one component, directed along the axis of symmetry Gx, measures the *rate of spin* of the projectile, while the other, directed along GD, measures its *rate of precession*. Knowing that $\theta = 5°$ and that the angular momentum of the projectile about its mass center G is $\mathbf{H}_G = (320\ \text{g}\cdot\text{m}^2/\text{s})\mathbf{i} + (9\ \text{g}\cdot\text{m}^2/\text{s})\mathbf{j}$, determine (*a*) the rate of spin, (*b*) the rate of precession.

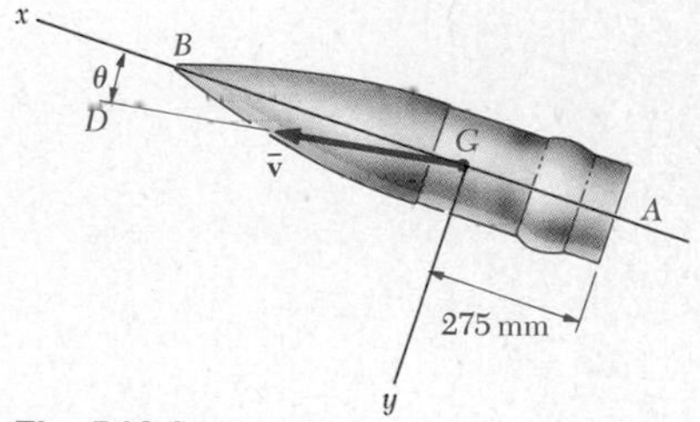

Fig. P18.8

18.9 Determine the angular momentum $\mathbf{H}_A$ of the projectile of Prob. 18.8 about the center A of its base, knowing that its mass center G has a velocity $\overline{\mathbf{v}}$ of 650 m/s. Give your answer in terms of components, respectively, parallel to the x and y axes shown and to a third axis z pointing toward you.

18.10 Determine the angular momentum of the disk of Prob. 18.7 about point A.

18.11 Determine the angular momentum of the disk of Prob. 18.6 about point O.

18.12 Determine the angular momentum $\mathbf{H}_O$ of the disk of Sample Prob. 18.2 from the expressions obtained for its linear momentum $m\overline{\mathbf{v}}$ and its angular momentum $\mathbf{H}_G$, using Eq. (18.11). Verify that the result obtained is the same that was obtained by direct computation.

18.13 Show that when a rigid body rotates about a fixed axis, its angular momentum is the same about any two points A and B on the fixed axis ($\mathbf{H}_A = \mathbf{H}_B$) if, and only if, the mass center G of the body is located on the fixed axis.

18.14 A homogeneous rod of mass m' per unit length is used to form the shaft shown. Knowing that the shaft rotates with a constant angular velocity ω, determine (*a*) the angular momentum of the shaft about G, (*b*) the angle formed by the angular momentum and the axis AB.

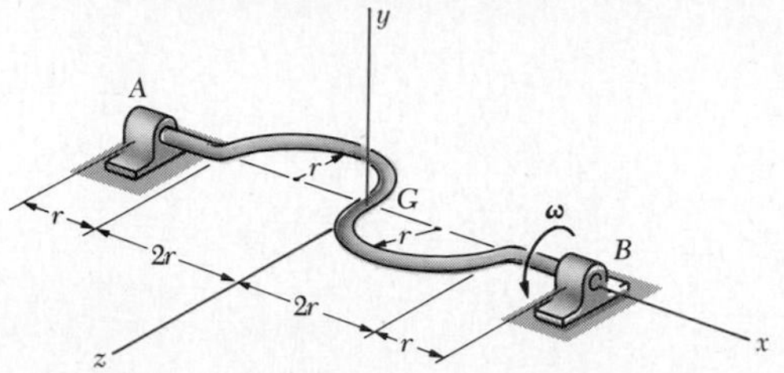

Fig. P18.14

18.15 Determine the angular momentum of the shaft of Prob. 18.14 about (*a*) point A, (*b*) point B.

18.16 Two triangular plates, each of mass 5 kg, are welded to a vertical shaft AB. Knowing that the system rotates at the constant rate $\omega = 8$ rad/s, determine its angular momentum about G.

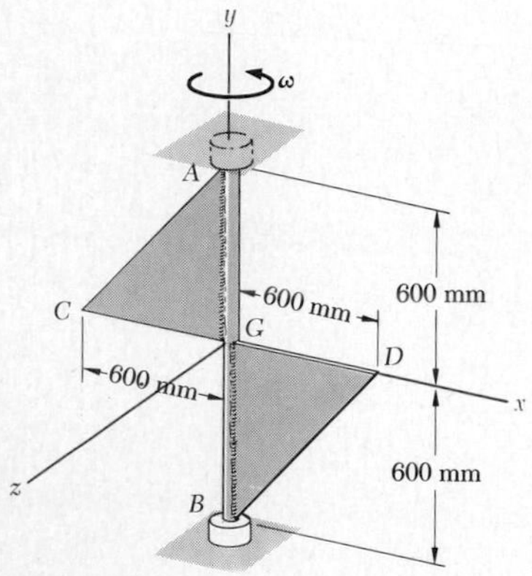

Fig. P18.16

18.17 Two L-shaped arms, each weighing 4 lb, are welded at the third points of the 2-ft shaft AB. Knowing that shaft AB rotates at the constant rate $\omega = 240$ rpm, determine (*a*) the angular momentum of the body about A, (*b*) the angle formed by the angular momentum and the shaft AB.

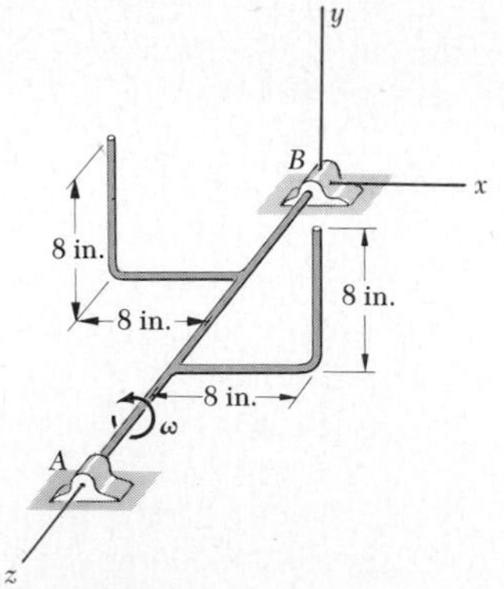

Fig. P18.17

18.18 For the body of Prob. 18.17, determine (*a*) the angular momentum about B, (*b*) the angle formed by the angular momentum and the shaft BA.

18.19 Two circular plates, each of mass 5 kg, are rigidly connected by a rod AB of negligible mass and suspended from point A as shown. If an impulse $\mathbf{F}\,\Delta t = -(3\ \mathrm{N\cdot s})\mathbf{k}$ is applied at point D, determine (*a*) the velocity of the mass center G of the assembly, (*b*) the angular velocity of the assembly.

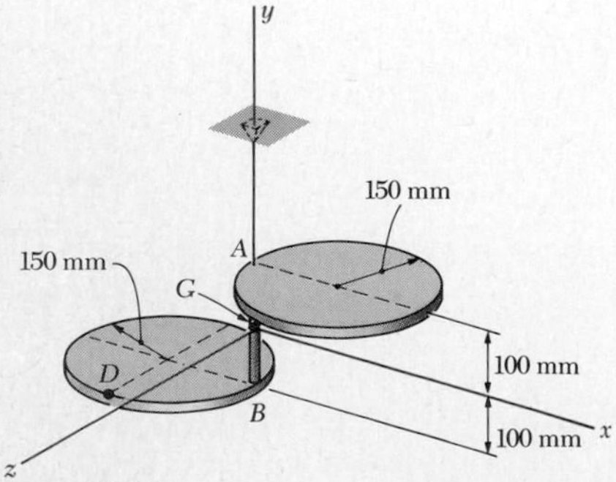

Fig. P18.19

18.20 A uniform rod of total mass m is bent into the shape shown and is suspended by a wire attached at its mass center G. The bent rod is hit at C in a direction perpendicular to the plane containing the rod (in the negative z direction). Denoting the corresponding impulse by $\mathbf{F}\,\Delta t$, determine immediately after the impact (*a*) the velocity of the mass center G, (*b*) the angular velocity of the rod.

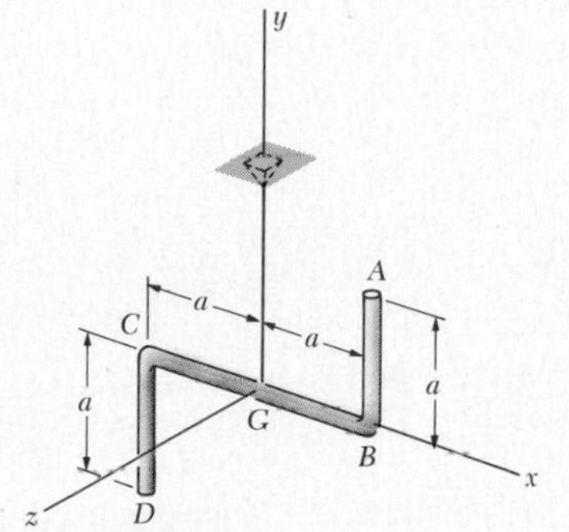

Fig. P18.20

18.21 Solve Prob. 18.20, assuming that the bent rod is hit at D.

18.22 A uniform rod of total mass m is bent into the shape shown and is suspended by a wire attached at B. The bent rod is hit at C in the negative z direction. Denoting the corresponding impulse by $\mathbf{F}\,\Delta t$, determine immediately after the impact (*a*) the velocity of the mass center G, (*b*) the angular velocity of the rod.

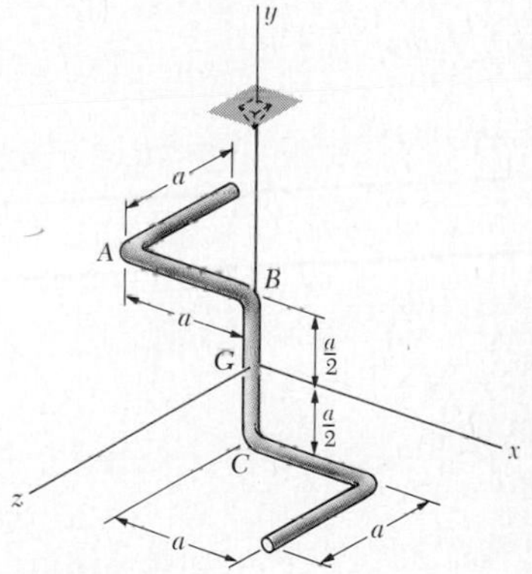

Fig. P18.22

18.23 Solve Prob. 18.22, assuming that the bent rod is hit at A in a vertical downward direction.

18.24 and 18.25 Two uniform rods AB and CD are welded together at B to form a T-shaped assembly of total mass m. The assembly is suspended from a ball-and-socket joint at A and is hit at C in a direction perpendicular to its plane (in the negative z direction). Denoting the corresponding impulse by $\mathbf{F}\,\Delta t$, determine immediately after the impact (*a*) the angular velocity of the assembly, (*b*) its instantaneous axis of rotation.

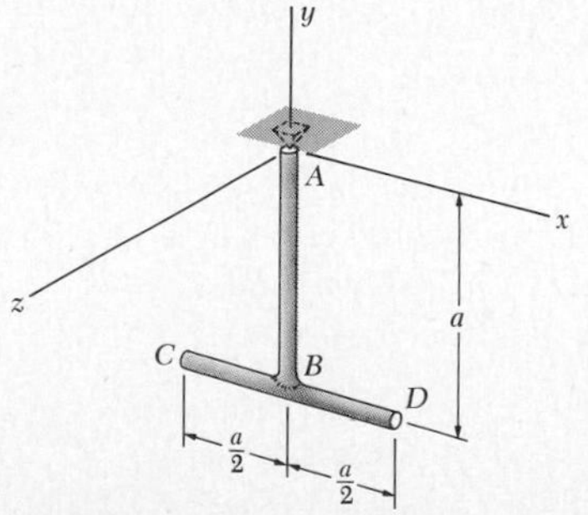

Fig. P18.24

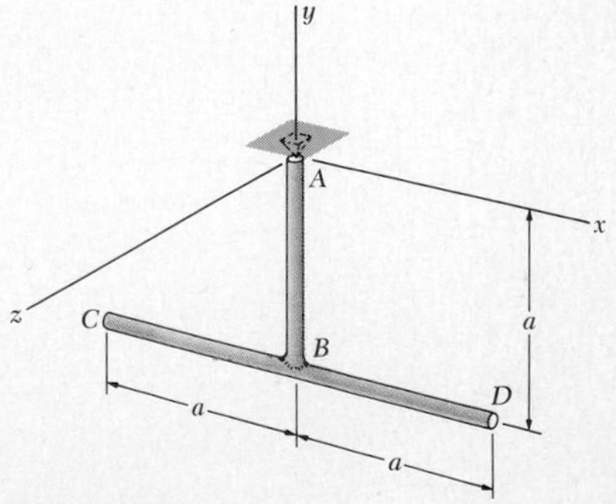

Fig. P18.25

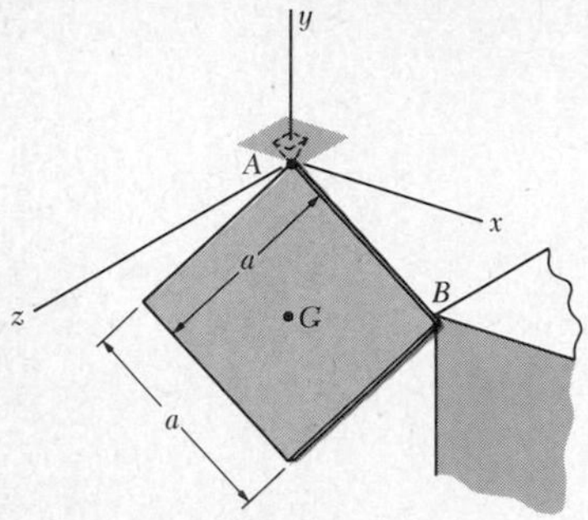

Fig. P18.26

18.26 A square plate of side a and mass m supported by a ball-and-socket joint at A is rotating about the y axis with a constant angular velocity $\boldsymbol{\omega} = \omega_0 \mathbf{j}$ when an obstruction is suddenly introduced at B. Assuming the impact at B to be perfectly plastic, determine immediately after the impact (*a*) the angular velocity of the plate (*b*) the velocity of its mass center G.

18.27 Determine the impulse exerted on the plate of Prob. 18.26 during the impact by (*a*) the obstruction at B, (*b*) the support at A.

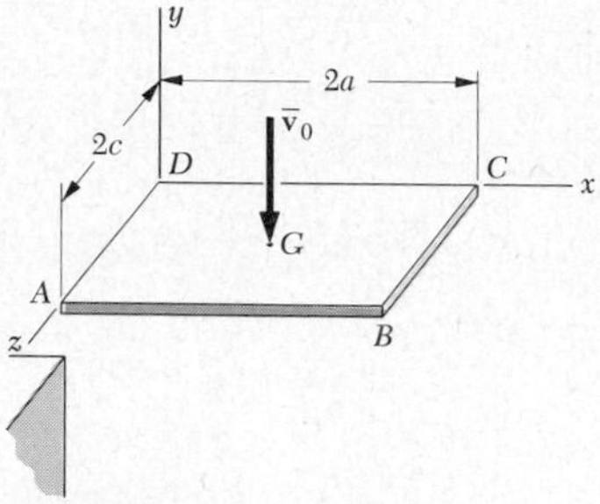

Fig. P18.28

18.28 A rectangular plate of mass m is falling with a velocity $\bar{\mathbf{v}}_0$ and no angular velocity when its corner A strikes an obstruction. Assuming the impact at A to be perfectly plastic, determine immediately after the impact (*a*) the angular velocity of the plate, (*b*) the velocity of its mass center G.

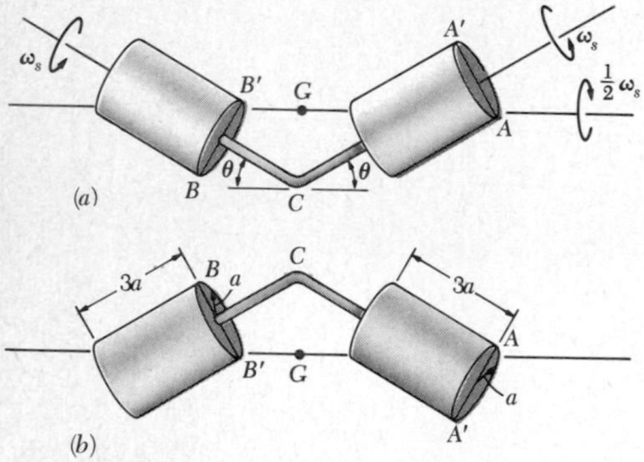

Fig. P18.29

18.29 In order to analyze the twist performed by cats to land on their paws after being dropped upside down, the model shown may be used. It consists of two homogeneous cylinders, each of mass m, radius a, and length $3a$, mounted on a bent shaft. Small motors located inside the cylinders make it possible to spin each cylinder about the shaft at the rate ω_s after the assembly has been released from rest in position *a*. Observing that the entire assembly must rotate at the rate $\frac{1}{2}\omega_s$ about a horizontal axis through its mass center G if it is to complete a half turn about that axis when each cylinder has spun through 360° with respect to the shaft (position *b*), determine the required angle θ that each portion of shaft should form with the horizontal. Neglect the mass of the shaft and of the motors. (*Hint.* Since no external force except its weight acts on the assembly, the angular momentum of the assembly about G must remain equal to zero).

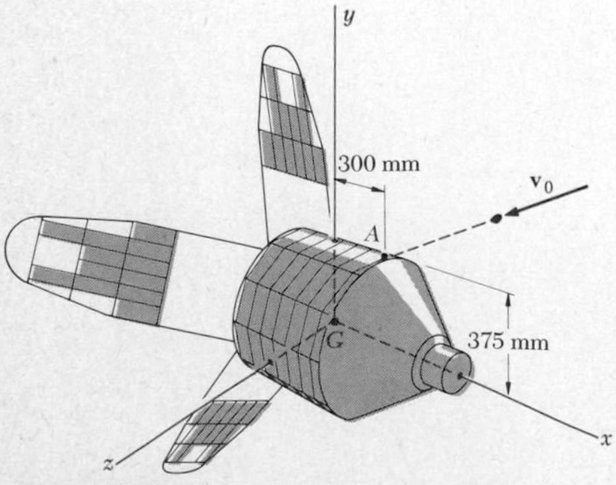

Fig. P18.30

18.30 A satellite of mass 160 kg has no angular velocity when it is struck at A by a 20-g meteorite traveling with a velocity $\mathbf{v}_0 = -(720 \text{ m/s})\mathbf{i} - (540 \text{ m/s})\mathbf{j} + (1200 \text{ m/s})\mathbf{k}$ relative to the satellite. Knowing that the radii of gyration of the satellite are $\bar{k}_x = 300$ mm and $\bar{k}_y = \bar{k}_z = 400$ mm, determine the angular velocity of the satellite immediately after the meteorite has become imbedded.

18.31 Solve Prob. 18.30, assuming that the satellite was initially spinning about its axis of symmetry with an angular velocity of 1.25 rad/s clockwise as viewed from the positive x axis.

18.32 An 800-lb geostationary satellite is spinning with the angular velocity $\boldsymbol{\omega}_0 = (1.5 \text{ rad/s})\mathbf{j}$ when it is hit at A by a 6-oz meteorite traveling at a speed $v_0 = 4500$ ft/s relative to the satellite. Knowing that the radii of gyration of the satellite are $\bar{k}_x = \bar{k}_z = 28.8$ in. and $\bar{k}_y = 32.4$ in., and that immediately after the meteorite has become imbedded the angular velocity of the satellite is observed to be $\boldsymbol{\omega} = (0.6 \text{ rad/s})\mathbf{j} + (0.37 \text{ rad/s})\mathbf{k}$, determine the components of the velocity $\mathbf{v}_0$ of the meteorite relative to the satellite.

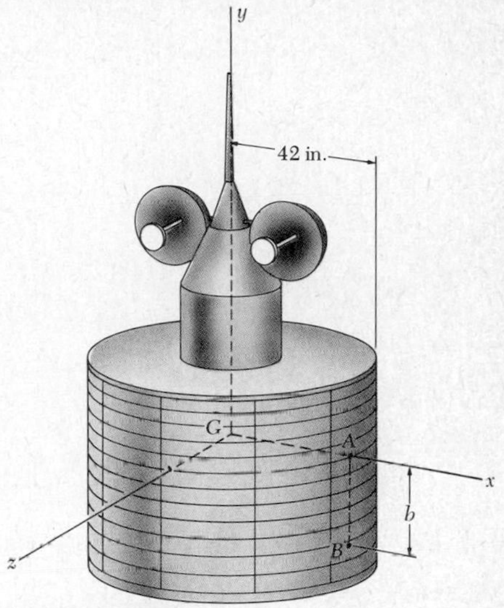

Fig. P18.32 and P18.33

18.33 An 800-lb geostationary satellite is spinning with the angular velocity $\boldsymbol{\omega}_0 = (1.5 \text{ rad/s})\mathbf{j}$ when it is hit at B by a 6-oz meteorite traveling with the velocity $\mathbf{v}_0 = -(2800 \text{ ft/s})\mathbf{i} + (2900 \text{ ft/s})\mathbf{j} - (2000 \text{ ft/s})\mathbf{k}$ relative to the satellite. Knowing that the radii of gyration of the satellite are $\bar{k}_x = \bar{k}_z = 28.8$ in. and $\bar{k}_y = 32.4$ in., and that immediately after the meteorite has become imbedded the x component of the angular velocity $\boldsymbol{\omega}$ of the satellite is observed to be $\omega_x = 0.3$ rad/s, determine (*a*) the distance b, (*b*) the y and z components of $\boldsymbol{\omega}$.

18.34 Denoting, respectively, by $\boldsymbol{\omega}$, $\mathbf{H}_O$, and T the angular velocity, the angular momentum, and the kinetic energy of a rigid body with a fixed point O, (*a*) prove that

$$\mathbf{H}_O \cdot \boldsymbol{\omega} = 2T$$

(*b*) show that the angle θ between $\boldsymbol{\omega}$ and $\mathbf{H}_O$ will always be acute.

18.35 Show that the kinetic energy of a rigid body with a fixed point O may be expressed as

$$T = \tfrac{1}{2} I_{OL} \omega^2$$

where $\boldsymbol{\omega}$ is the instantaneous angular velocity of the body and I_{OL} its moment of inertia about the line of action OL of $\boldsymbol{\omega}$. Derive this expression (*a*) from Eqs. (9.46) and (18.19), (*b*) by considering T as the sum of the kinetic energies of particles P_i describing circles of radius ρ_i about line OL.

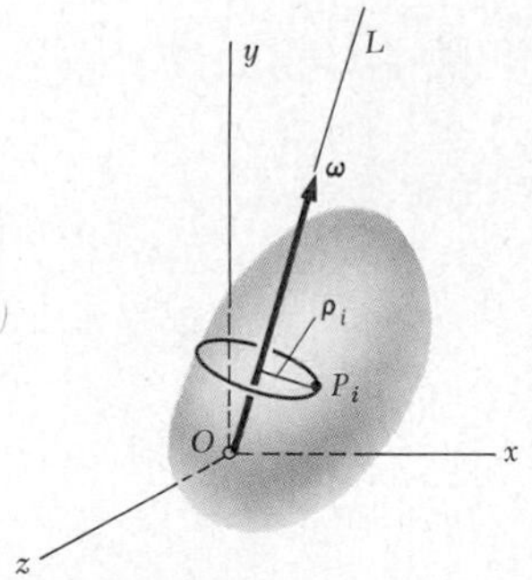

Fig. P18.35

18.36 Determine the kinetic energy of the plate of Prob. 18.2.

18.37 Determine the kinetic energy of the plate of Prob. 18.3.

18.38 Determine the kinetic energy of the disk of Prob. 18.4.

18.39 Determine the kinetic energy of the disk of Prob. 18.7.

18.40 Determine the kinetic energy of the disk of Prob. 18.6.

18.41 Determine the kinetic energy of the body of Prob. 18.17.

18.42 Determine the kinetic energy lost when the plate of Prob. 18.26 hits the obstruction at B.

18.43 Determine the kinetic energy lost when the plate of Prob. 18.28 hits the obstruction at A.

18.44 Determine the kinetic energy of the satellite of Prob. 18.30 in its motion about its mass center after the collision with the meteorite.

18.45 Determine the change in the kinetic energy of the satellite of Prob. 18.32 in its motion about its mass center due to the collision with the meteorite.

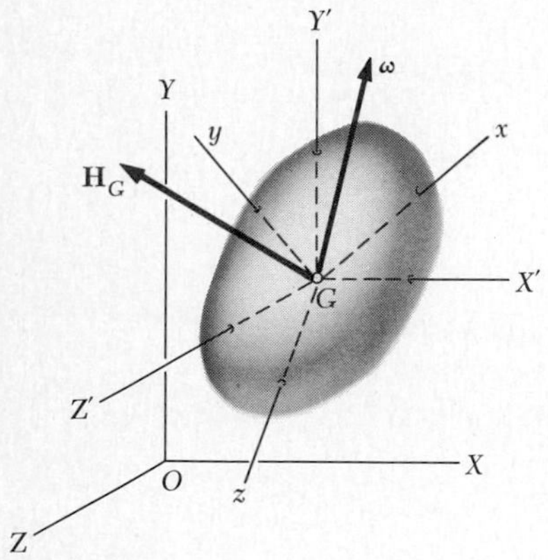

Fig. P18.9

* 18.5. Motion of a Rigid Body in Three Dimensions.

As was indicated in Sec. 18.2, the fundamental equations

$$\Sigma \mathbf{F} = m\bar{\mathbf{a}} \tag{18.1}$$
$$\Sigma \mathbf{M}_G = \dot{\mathbf{H}}_G \tag{18.2}$$

remain valid in the most general case of the motion of a rigid body. Before Eq. (18.2) could be applied to the three-dimensional motion of a rigid body, however, it was necessary to derive Eqs. (18.7), which relate the components of the angular momentum $\mathbf{H}_G$ and those of the angular velocity $\boldsymbol{\omega}$. It still remains for us to find an effective and convenient way for computing the components of the derivative $\dot{\mathbf{H}}_G$ of the angular momentum.

Since $\mathbf{H}_G$ represents the angular momentum of the body in its motion relative to centroidal axes $GX'Y'Z'$ of fixed orientation (Fig. 18.9), and since $\dot{\mathbf{H}}_G$ represents the rate of change of $\mathbf{H}_G$ with respect to the same axes, it would seem natural to use components of $\boldsymbol{\omega}$ and $\mathbf{H}_G$ along the axes X', Y', Z' in writing the relations (18.7). But since the body rotates, its moments and products of inertia would change continually, and it would be necessary to determine their values as functions of the time. It is therefore more convenient to use axes x, y, z attached to the body, thus making sure that its moments and products of inertia will maintain the same values during the motion. This is permissible since, as indicated earlier, the transformation of $\boldsymbol{\omega}$ into $\mathbf{H}_G$ is independent of the system of coordinate axes which has been selected. The angular velocity $\boldsymbol{\omega}$, however, should still be *defined* with respect to the frame $GX'Y'Z'$ of fixed orientation. The vector $\boldsymbol{\omega}$ may then be *resolved* into components along the rotating x, y, and z axes. Applying the relations (18.7), we obtain the *components* of the vector $\mathbf{H}_G$ along the rotating axes. The vector $\mathbf{H}_G$, however, represents the angular momentum about G of the body *in its motion relative to the frame $GX'Y'Z'$*.

Differentiating with respect to t the components of the angular momentum in (18.7), we define the rate of change of the vector $\mathbf{H}_G$ with respect to the rotating frame $Gxyz$:

$$(\dot{\mathbf{H}}_G)_{Gxyz} = \dot{H}_x\mathbf{i} + \dot{H}_y\mathbf{j} + \dot{H}_z\mathbf{k} \tag{18.21}$$

where $\mathbf{i}$, $\mathbf{j}$, $\mathbf{k}$ are the unit vectors along the rotating axes. Recalling from Sec. 15.10 that the rate of change $\dot{\mathbf{H}}_G$ of the vector $\mathbf{H}_G$ with respect to the frame $GX'Y'Z'$ may be obtained by adding to $(\dot{\mathbf{H}}_G)_{Gxyz}$ the vector product $\boldsymbol{\Omega} \times \mathbf{H}_G$, where $\boldsymbol{\Omega}$ denotes the angular velocity of the rotating frame, we write

$$\dot{\mathbf{H}}_G = (\dot{\mathbf{H}}_G)_{Gxyz} + \boldsymbol{\Omega} \times \mathbf{H}_G \tag{18.22}$$

where $\mathbf{H}_G$ = angular momentum of the body with respect to the frame $GX'Y'Z'$ of fixed orientation

$(\dot{\mathbf{H}}_G)_{Gxyz}$ = rate of change of $\mathbf{H}_G$ with respect to the rotating frame $Gxyz$, to be computed from the relations (18.7) and (18.21)

$\boldsymbol{\Omega}$ = angular velocity of the rotating frame $Gxyz$

Substituting for $\dot{\mathbf{H}}_G$ from (18.22) into (18.2), we have

$$\Sigma \mathbf{M}_G = (\dot{\mathbf{H}}_G)_{Gxyz} + \boldsymbol{\Omega} \times \mathbf{H}_G \tag{18.23}$$

If the rotating frame is attached to the body, as has been assumed in this discussion, its angular velocity $\boldsymbol{\Omega}$ is identically equal to the angular velocity $\boldsymbol{\omega}$ of the body. There are many applications, however, where it is advantageous to use a frame of reference which is not actually attached to the body but rotates in an independent manner. For example, if the body considered is axisymmetrical, as in Sample Prob. 18.5 or Sec. 18.9, it is possible to select a frame of reference with respect to which the moments and products of inertia of the body remain constant, but which rotates less than the body itself.† As a result, simpler expressions may be obtained for the angular velocity $\boldsymbol{\omega}$ and the angular momentum $\mathbf{H}_G$ of the body than would have been possible if the frame of reference had actually been attached to the body. It is clear that in such cases the angular velocity $\boldsymbol{\Omega}$ of the rotating frame and the angular velocity $\boldsymbol{\omega}$ of the body are different.

*** 18.6. Euler's Equations of Motion. Extension of d'Alembert's Principle to the Motion of a Rigid Body in Three Dimensions.** If the x, y, and z axes are chosen to coincide with the principal axes of inertia of the body, the simplified relations (18.10) may be used to determine the components of the angular momentum $\mathbf{H}_G$. Omitting the primes from the subscripts, we write

$$\mathbf{H}_G = \bar{I}_x \omega_x \mathbf{i} + \bar{I}_y \omega_y \mathbf{j} + \bar{I}_z \omega_z \mathbf{k} \tag{18.24}$$

where $\bar{I}_x$, $\bar{I}_y$, and $\bar{I}_z$ denote the principal centroidal moments of inertia of the body. Substituting for $\mathbf{H}_G$ from (18.24) into (18.23) and setting $\boldsymbol{\Omega} = \boldsymbol{\omega}$, we obtain the three scalar equations

$$\begin{aligned} \Sigma M_x &= \bar{I}_x \dot{\omega}_x - (\bar{I}_y - \bar{I}_z)\omega_y \omega_z \\ \Sigma M_y &= \bar{I}_y \dot{\omega}_y - (\bar{I}_z - \bar{I}_x)\omega_z \omega_x \\ \Sigma M_z &= \bar{I}_z \dot{\omega}_z - (\bar{I}_x - \bar{I}_y)\omega_x \omega_y \end{aligned} \tag{18.25}$$

These equations, called *Euler's equations of motion* after the Swiss mathematician Leonhard Euler (1707–1783), may be used to analyze the motion of a rigid body about its mass center. In the following sections, however, we shall use Eq. (18.23) in preference to Eqs. (18.25), since the former is more general, and the compact vectorial form in which it is expressed is easier to remember.

Writing Eq. (18.1) in scalar form, we obtain the three additional equations

$$\Sigma F_x = m\bar{a}_x \qquad \Sigma F_y = m\bar{a}_y \qquad \Sigma F_z = m\bar{a}_z \tag{18.26}$$

which together with Euler's equations form a system of six differential equations. Given appropriate initial conditions, these differential equa-

† More specifically, the frame of reference will have no spin (see Sec. 18.9).

tions have a unique solution. Thus, the motion of a rigid body in three dimensions is completely defined by the resultant and the moment resultant of the external forces acting on it. This result will be recognized as a generalization of a similar result obtained in Sec. 16.4 in the case of the plane motion of a rigid slab. It follows that in three as well as two dimensions, two systems of forces which are equipollent are also equivalent; that is, they have the same effect on a given rigid body.

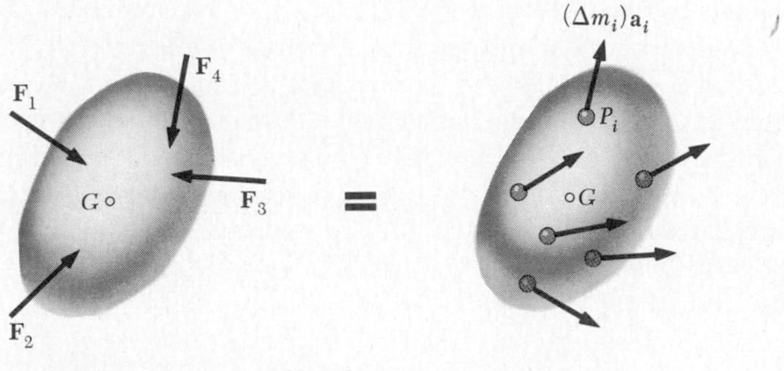

Fig. 18.10

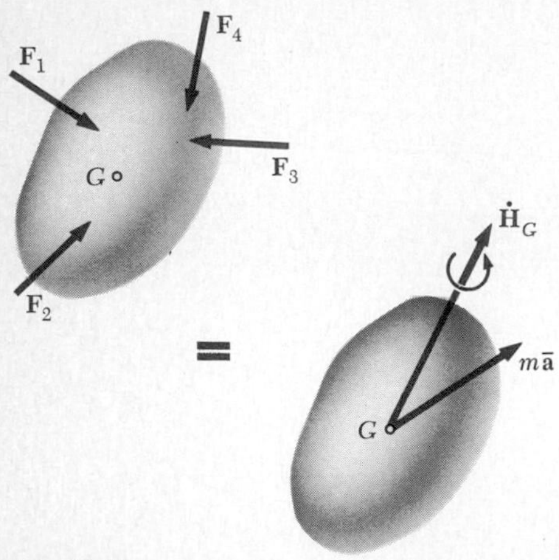

Fig. 18.11

Considering in particular the system of the external forces acting on a rigid body (Fig. 18.10*a*) and the system of the effective forces associated with the particles forming the rigid body (Fig. 18.10*b*), we may state that the two systems—which were shown in Sec. 14.2 to be equipollent—are also equivalent. This is the extension of d'Alembert's principle to the three-dimensional motion of a rigid body. Replacing the effective forces in Fig. 18.10*b* by an equivalent force-couple system, we verify that the system of the external forces acting on a rigid body in three-dimensional motion is equivalent to the system consisting of the vector $m\bar{\mathbf{a}}$ attached at the mass center G of the body and the couple of moment $\dot{\mathbf{H}}_G$ (Fig. 18.11), where $\dot{\mathbf{H}}_G$ is obtained from the relations (18.7) and (18.22). Problems involving the three-dimensional motion of a rigid body may be solved by considering the free-body-diagram equation represented in Fig. 18.11 and writing appropriate scalar equations relating the components or moments of the external and effective forces (see Sample Prob. 18.3).

*** 18.7. Motion of a Rigid Body about a Fixed Point.** When a rigid body is constrained to rotate about a fixed point O, it is desirable to write an equation involving the moments about O of the external and effective forces, since this equation will not contain the unknown reaction at O. While such an equation may be obtained from Fig. 18.11, it may be more convenient to write it by considering the rate of change of the angular momentum $\mathbf{H}_O$ of the body about the fixed point O (Fig. 18.12). Recalling Eq. (14.11), we write

$$\Sigma \mathbf{M}_O = \dot{\mathbf{H}}_O \tag{18.27}$$

where $\dot{\mathbf{H}}_O$ denotes the rate of change of the vector $\mathbf{H}_O$ with respect to the fixed frame $OXYZ$. A derivation similar to that used in Sec. 18.5 enables us to relate $\dot{\mathbf{H}}_O$ to the rate of change $(\dot{\mathbf{H}}_O)_{Oxyz}$ of $\mathbf{H}_O$ with respect to the rotating frame $Oxyz$. Substitution into (18.27) leads to the equation

$$\Sigma \mathbf{M}_O = (\dot{\mathbf{H}}_O)_{Oxyz} + \boldsymbol{\Omega} \times \mathbf{H}_O \tag{18.28}$$

where $\Sigma \mathbf{M}_O$ = sum of the moments about O of the forces applied to the rigid body
$\mathbf{H}_O$ = angular momentum of the body with respect to the fixed frame $OXYZ$
$(\dot{\mathbf{H}}_O)_{Oxyz}$ = rate of change of $\mathbf{H}_O$ with respect to the rotating frame $Oxyz$, to be computed from the relations (18.13)
$\boldsymbol{\Omega}$ = angular velocity of the rotating frame $Oxyz$

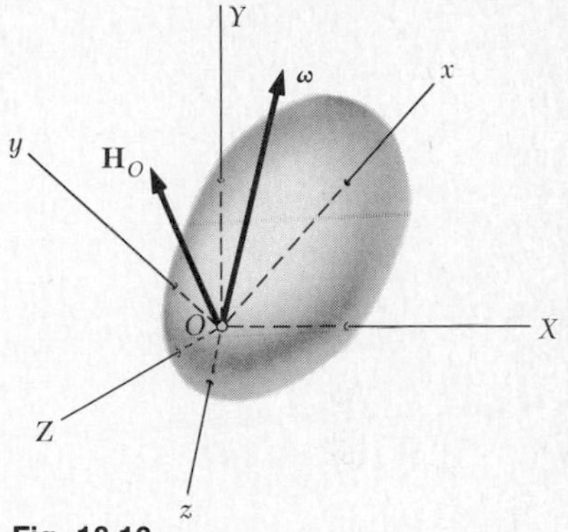

Fig. 18.12

If the rotating frame is attached to the body, its angular velocity $\boldsymbol{\Omega}$ is identically equal to the angular velocity $\boldsymbol{\omega}$ of the body. However, as indicated in the last paragraph of Sec. 18.5, there are many applications where it is advantageous to use a frame of reference which is not actually attached to the body but rotates in an independent manner.

*** 18.8. Rotation of a Rigid Body about a Fixed Axis.** We shall use Eq. (18.28), which was derived in the preceding section, to analyze the motion of a rigid body constrained to rotate about a fixed axis AB (Fig. 18.13). First, we note that the angular velocity of the body with respect to the fixed frame $OXYZ$ is represented by the vector $\boldsymbol{\omega}$ directed along the axis of rotation. Attaching the moving frame of reference $Oxyz$ to the body, with the z axis along AB, we have $\boldsymbol{\omega} = \omega\mathbf{k}$. Substituting $\omega_x = 0$, $\omega_y = 0$, $\omega_z = \omega$ into the relations (18.13), we obtain the components along the rotating axes of the angular momentum $\mathbf{H}_O$ of the body about O:

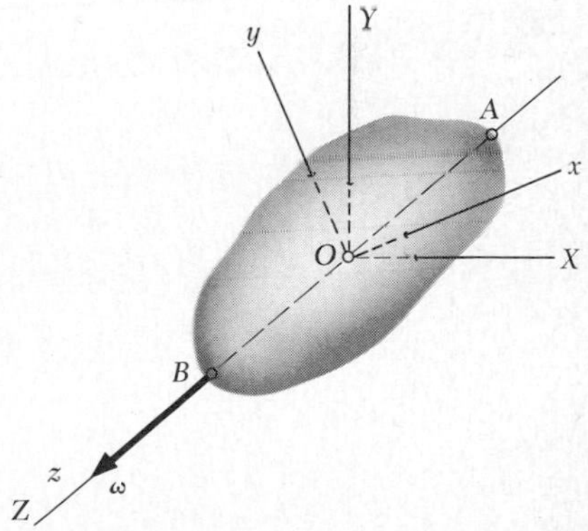

Fig. 18.13

$$H_x = -I_{xz}\omega \qquad H_y = -I_{yz}\omega \qquad H_z = I_z\omega$$

Since the frame $Oxyz$ is attached to the body, we have $\boldsymbol{\Omega} = \boldsymbol{\omega}$ and Eq. (18.28) yields

$$\begin{aligned}
\Sigma \mathbf{M}_O &= (\dot{\mathbf{H}}_O)_{Oxyz} + \boldsymbol{\omega} \times \mathbf{H}_O \\
&= (-I_{xz}\mathbf{i} - I_{yz}\mathbf{j} + I_z\mathbf{k})\dot{\omega} + \omega\mathbf{k} \times (-I_{xz}\mathbf{i} - I_{yz}\mathbf{j} + I_z\mathbf{k})\omega \\
&= (-I_{xz}\mathbf{i} - I_{yz}\mathbf{j} + I_z\mathbf{k})\alpha + (-I_{xz}\mathbf{j} + I_{yz}\mathbf{i})\omega^2
\end{aligned}$$

The result obtained may be expressed by the three scalar equations

$$\begin{aligned}
\Sigma M_x &= -I_{xz}\alpha + I_{yz}\omega^2 \\
\Sigma M_y &= -I_{yz}\alpha - I_{xz}\omega^2 \\
\Sigma M_z &= I_z\alpha
\end{aligned} \tag{18.29}$$

When the forces applied to the body are known, the angular acceleration α may be obtained from the last of Eqs. (18.29). The angular velocity ω is then determined by integration and the values obtained for α and ω may be substituted into the first two equations (18.29). These equations plus the three equations (18.26) which define the motion of the mass center of the body may then be used to determine the reactions at the bearings A and B.

It should be noted that axes other than the ones shown in Fig. 18.13 may be selected to analyze the rotation of a rigid body about a fixed axis. In many cases, the principal axes of inertia of the body will be found more advantageous. It is therefore wise to revert to Eq. (18.28) and to select the system of axes which best fits the problem under consideration.

If the rotating body is symmetrical with respect to the xy plane, the products of inertia I_{xz} and I_{yz} are equal to zero and Eqs. (18.29) reduce to

$$\Sigma M_x = 0 \qquad \Sigma M_y = 0 \qquad \Sigma M_z = I_z \alpha \tag{18.30}$$

which is in accord with the results obtained in Chap. 16. If, on the other hand, the products of inertia I_{xz} and I_{yz} are different from zero, the sum of the moments of the external forces about the x and y axes will also be different from zero, even when the body rotates at a constant rate ω. Indeed, in the latter case, Eqs. (18.29) yield

$$\Sigma M_x = I_{yz}\omega^2 \qquad \Sigma M_y = -I_{xz}\omega^2 \qquad \Sigma M_z = 0 \tag{18.31}$$

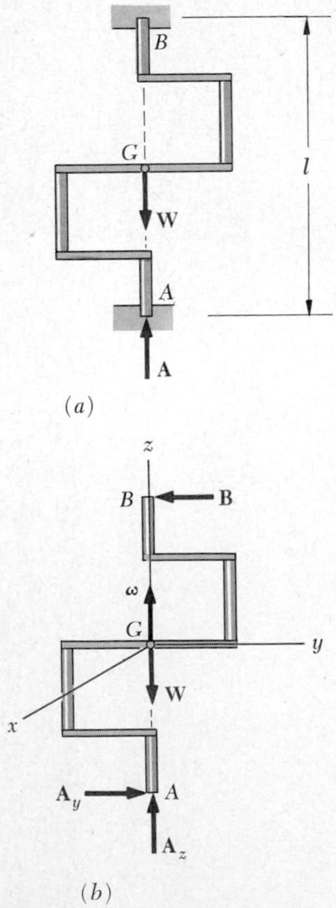

(a)
(b)
Fig. 18.14

This last observation leads us to discuss the *balancing of rotating shafts*. Consider, for instance, the crankshaft shown in Fig. 18.14*a*, which is symmetrical about its mass center G. We first observe that when the crankshaft is at rest, it exerts no lateral thrust on its supports, since its center of gravity G is located directly above A. The shaft is said to be *statically balanced*. The reaction at A, often referred to as a *static reaction*, is vertical and its magnitude is equal to the weight W of the shaft. Let us now assume that the shaft rotates with a constant angular velocity $\boldsymbol{\omega}$. Attaching our frame of reference to the shaft, with its origin at G, the z axis along AB, and the y axis in the plane of symmetry of the shaft (Fig. 18.14*b*), we note that I_{xz} is zero and that I_{yz} is positive. According to Eqs. (18.31), the external forces must include a couple of moment $I_{yz}\omega^2\mathbf{i}$. Since this couple is formed by the reaction at B and the horizontal component of the reaction at A, we have

$$\mathbf{A}_y = \frac{I_{yz}\omega^2}{l}\mathbf{j} \qquad \mathbf{B} = -\frac{I_{yz}\omega^2}{l}\mathbf{j} \tag{18.32}$$

Since the bearing reactions are proportional to ω^2, the shaft will have a tendency to tear away from its bearings when rotating at high speeds. Moreover, since the bearing reactions $\mathbf{A}_y$ and $\mathbf{B}$, called *dynamic reactions*, are contained in the yz plane, they rotate with the shaft and cause the structure supporting it to vibrate. These undesirable effects will be avoided if by rearranging the distribution of mass around the shaft or by adding corrective masses, we let I_{yz} become equal to zero. The dynamic reactions $\mathbf{A}_y$ and $\mathbf{B}$ will vanish and the reactions at the bearings will reduce to the static reaction $\mathbf{A}_z$, the direction of which is fixed. The shaft will then be *dynamically as well as statically balanced.*

SAMPLE PROBLEM 18.3

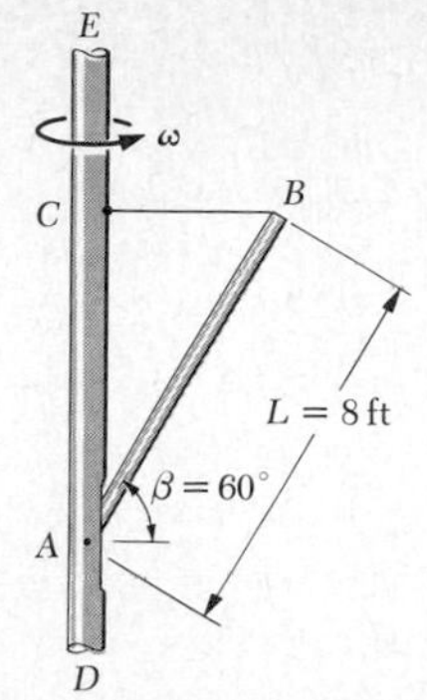

A slender rod AB of length $L = 8$ ft and weight $W = 40$ lb is pinned at A to a vertical axle DE which rotates with a constant angular velocity $\boldsymbol{\omega}$ of 15 rad/s. The rod is maintained in position by means of a horizontal wire BC attached to the axle and to the end B of the rod. Determine the tension in the wire and the reaction at A.

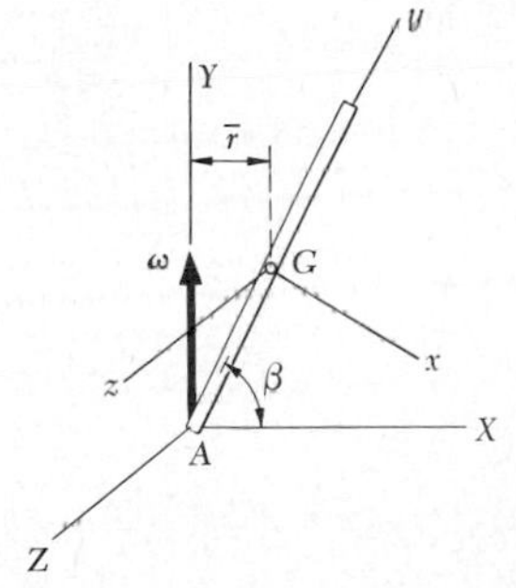

Solution. The effective forces reduce to the vector $m\bar{\mathbf{a}}$ attached at G and the couple $\dot{\mathbf{H}}_G$. Since G describes a horizontal circle of radius $\bar{r} = \frac{1}{2}L \cos \beta$ at the constant rate ω, we have

$$\bar{\mathbf{a}} = \mathbf{a}_n = -\bar{r}\omega^2\mathbf{I} = -(\tfrac{1}{2}L \cos \beta)\omega^2\mathbf{I} = -(450 \text{ ft/s}^2)\mathbf{I}$$

$$m\bar{\mathbf{a}} = \frac{40}{g}(-450\mathbf{I}) = -(559 \text{ lb})\mathbf{I}$$

Determination of $\dot{\mathbf{H}}_G$. We first compute the angular momentum $\mathbf{H}_G$. Using the principal centroidal axes of inertia x, y, z, we write

$$\bar{I}_x = \tfrac{1}{12}mL^2 \qquad \bar{I}_y = 0 \qquad \bar{I}_z = \tfrac{1}{12}mL^2$$

$$\omega_x = -\omega \cos \beta \qquad \omega_y = \omega \sin \beta \qquad \omega_z = 0$$

$$\mathbf{H}_G = \bar{I}_x\omega_x\mathbf{i} + \bar{I}_y\omega_y\mathbf{j} + \bar{I}_z\omega_z\mathbf{k}$$

$$\mathbf{H}_G = -\tfrac{1}{12}mL^2\omega \cos \beta\, \mathbf{i}$$

The rate of change $\dot{\mathbf{H}}_G$ of $\mathbf{H}_G$ with respect to axes of fixed orientation is obtained from Eq. (18.22). Observing that the rate of change $(\dot{\mathbf{H}}_G)_{Gxyz}$ of $\mathbf{H}_G$ with respect to the rotating frame $Gxyz$ is zero, and that the angular velocity $\boldsymbol{\Omega}$ of that frame is equal to the angular velocity $\boldsymbol{\omega}$ of the rod, we have

$$\dot{\mathbf{H}}_G = (\dot{\mathbf{H}}_G)_{Gxyz} + \boldsymbol{\omega} \times \mathbf{H}_G$$

$$\dot{\mathbf{H}}_G = 0 + (-\omega \cos \beta\, \mathbf{i} + \omega \sin \beta\, \mathbf{j}) \times (-\tfrac{1}{12}mL^2\omega \cos \beta\, \mathbf{i})$$

$$\dot{\mathbf{H}}_G = \tfrac{1}{12}mL^2\omega^2 \sin \beta \cos \beta\, \mathbf{k} = (645 \text{ lb}\cdot\text{ft})\mathbf{k}$$

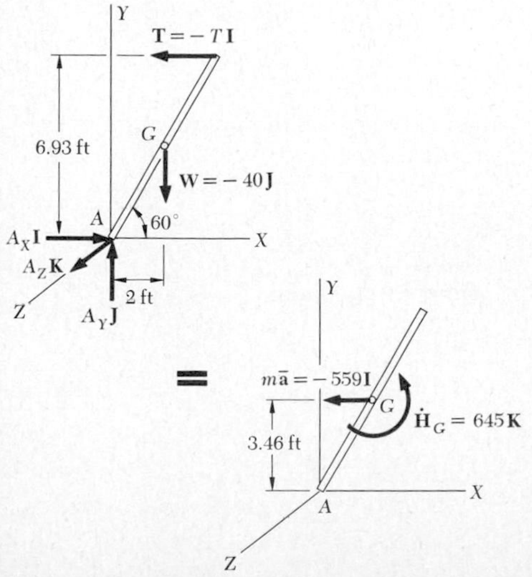

Equations of Motion. Expressing that the system of the external forces is equivalent to the system of the effective forces, we write

$\Sigma\mathbf{M}_A = \Sigma(\mathbf{M}_A)_{\text{eff}}$:

$$6.93\mathbf{J} \times (-T\mathbf{I}) + 2\mathbf{I} \times (-40\mathbf{J}) = 3.46\mathbf{J} \times (-559\mathbf{I}) + 645\,\mathbf{K}$$

$$(6.93T - 80)\mathbf{K} = (1934 + 645)\mathbf{K} \qquad T = 384 \text{ lb} \blacktriangleleft$$

$\Sigma\mathbf{F} = \Sigma\mathbf{F}_{\text{eff}}$: $\quad A_X\mathbf{I} + A_Y\mathbf{J} + A_Z\mathbf{K} - 384\mathbf{I} - 40\mathbf{J} = -559\mathbf{I}$

$$\mathbf{A} = -(175 \text{ lb})\mathbf{I} + (40 \text{ lb})\mathbf{J} \blacktriangleleft$$

Remark. The value of T could have been obtained from $\mathbf{H}_A$ and Eq. (18.28). However, the method used here also yields the reaction at A. Moreover, it draws attention to the effect of the asymmetry of the rod on the solution of the problem by clearly showing that both the vector $m\bar{\mathbf{a}}$ and the couple $\dot{\mathbf{H}}_G$ must be used to represent the effective forces.

SAMPLE PROBLEM 18.4

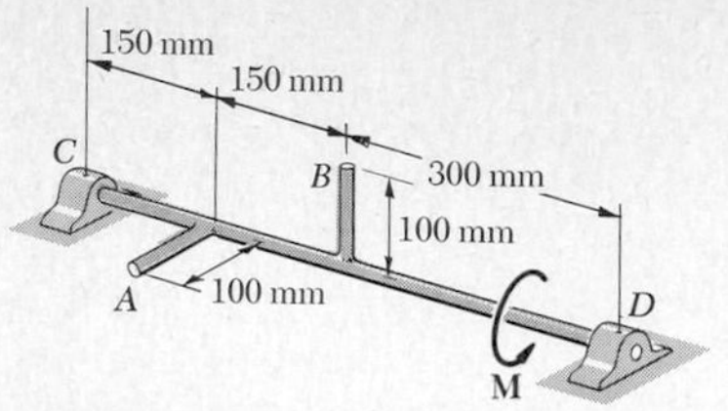

Two 100-mm rods A and B, each of mass 300 g, are welded to the shaft CD which is supported by bearings at C and D. If a couple $\mathbf{M}$ of magnitude equal to 6 N·m is applied to the shaft, determine the components of the dynamic reactions at C and D at the instant when the shaft has reached an angular velocity of 1200 rpm. Neglect the moment of inertia of the shaft itself.

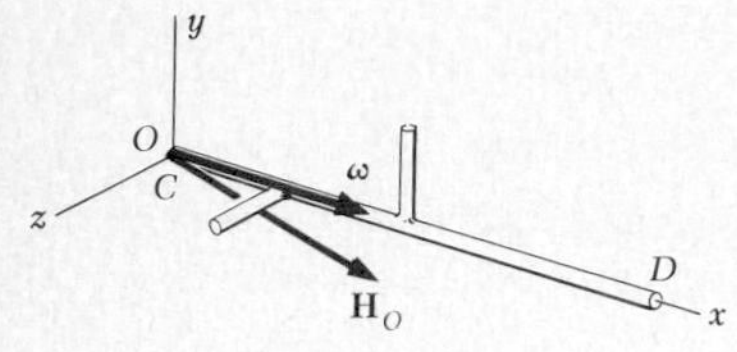

Angular Momentum about O. We attach to the body the frame of reference $Oxyz$ and note that the axes chosen are not principal axes of inertia for the body. Since the body rotates about the x axis, we have $\omega_x = \omega$ and $\omega_y = \omega_z = 0$. Substituting into Eqs. (18.13),

$$H_x = I_x\omega \qquad H_y = -I_{xy}\omega \qquad H_z = -I_{xz}\omega$$
$$\mathbf{H}_O = (I_x\mathbf{i} - I_{xy}\mathbf{j} - I_{xz}\mathbf{k})\omega$$

Moments of the External Forces about O. Since the frame of reference rotates with the angular velocity $\boldsymbol{\omega}$, Eq. (18.28) yields

$$\begin{aligned}\Sigma\mathbf{M}_O &= (\dot{\mathbf{H}}_O)_{Oxyz} + \boldsymbol{\omega} \times \mathbf{H}_O \\ &= (I_x\mathbf{i} - I_{xy}\mathbf{j} - I_{xz}\mathbf{k})\alpha + \omega\mathbf{i} \times (I_x\mathbf{i} - I_{xy}\mathbf{j} - I_{xz}\mathbf{k})\omega \\ &= I_x\alpha\mathbf{i} - (I_{xy}\alpha - I_{xz}\omega^2)\mathbf{j} - (I_{xz}\alpha + I_{xy}\omega^2)\mathbf{k} \qquad (1)\end{aligned}$$

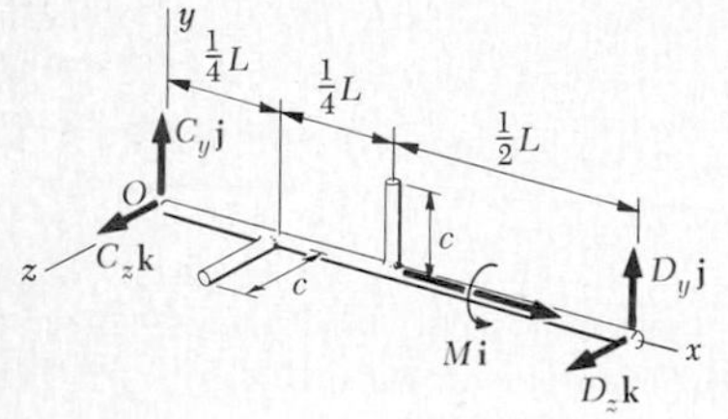

Dynamic Reaction at D. The external forces consist of the weights of the shaft and rods, the couple $\mathbf{M}$, the static reactions at C and D, and the dynamic reactions at C and D. Since the weights and static reactions are balanced, the external forces reduce to the couple $\mathbf{M}$ and the dynamic reactions $\mathbf{C}$ and $\mathbf{D}$ as shown in the figure. Taking moments about O, we have

$$\Sigma\mathbf{M}_O = L\mathbf{i} \times (D_y\mathbf{j} + D_z\mathbf{k}) + M\mathbf{i} = M\mathbf{i} - D_zL\mathbf{j} + D_yL\mathbf{k} \qquad (2)$$

Equating the coefficients of the unit vector $\mathbf{i}$ in (1) and (2),

$$M = I_x\alpha \qquad M = 2(\tfrac{1}{3}mc^2)\alpha \qquad \alpha = 3M/2mc^2$$

Equating the coefficients of $\mathbf{k}$ and $\mathbf{j}$ in (1) and (2):

$$D_y = -(I_{xz}\alpha + I_{xy}\omega^2)/L \qquad D_z = (I_{xy}\alpha - I_{xz}\omega^2)/L \qquad (3)$$

Using the parallel-axis theorem, and noting that the product of inertia of each rod is zero with respect to centroidal axes, we have

$$I_{xy} = \Sigma m\bar{x}\bar{y} = m(\tfrac{1}{2}L)(\tfrac{1}{2}c) = \tfrac{1}{4}mLc$$
$$I_{xz} = \Sigma m\bar{x}\bar{z} = m(\tfrac{1}{4}L)(\tfrac{1}{2}c) = \tfrac{1}{8}mLc$$

Substituting into (3) the values found for I_{xy}, I_{xz}, and α:

$$D_y = -\tfrac{3}{16}(M/c) - \tfrac{1}{4}mc\omega^2 \qquad D_z = \tfrac{3}{8}(M/c) - \tfrac{1}{8}mc\omega^2$$

Substituting $\omega = 1200$ rpm $= 125.7$ rad/s, $c = 0.100$ m, $M = 6$ N·m, and $m = 0.300$ kg, we have

$$D_y = -129.8 \text{ N} \qquad D_z = -36.8 \text{ N} \blacktriangleleft$$

Dynamic Reaction at C. Using a frame of reference attached at D, we obtain equations similar to Eqs. (3), which yield

$$C_y = -152.2 \text{ N} \qquad C_z = -155.2 \text{ N} \blacktriangleleft$$

SAMPLE PROBLEM 18.5

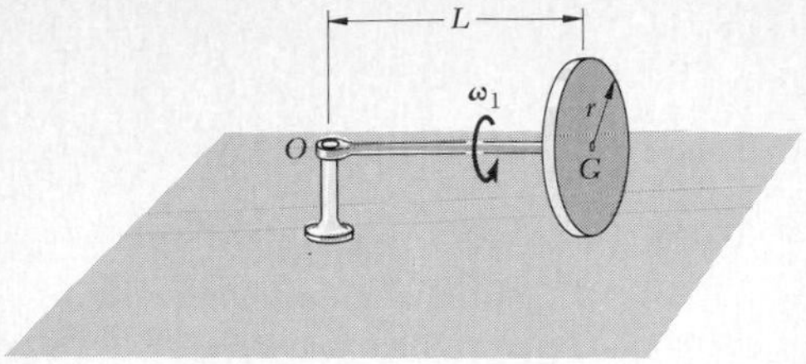

A homogeneous disk of radius r and mass m is mounted on an axle OG of length L and negligible mass. The axle is pivoted at the fixed point O and the disk is constrained to roll on a horizontal floor. Knowing that the disk rotates counterclockwise at the constant rate ω_1 about the axle, determine (*a*) the force (assumed vertical) exerted by the floor on the disk, (*b*) the reaction at the pivot O.

Solution. The effective forces reduce to the vector $m\bar{\mathbf{a}}$ attached at G and the couple $\dot{\mathbf{H}}_G$. Recalling from Sample Prob. 18.2 that the axle rotates about the y axis at the rate $\omega_2 = r\omega_1/L$, we write

$$m\bar{\mathbf{a}} = -mL\omega_2^2\mathbf{i} = -mL(r\omega_1/L)^2\mathbf{i} = -(mr^2\omega_1^2/L)\mathbf{i} \tag{1}$$

Determination of $\dot{\mathbf{H}}_G$. We recall from Sample Prob. 18.2 that the angular momentum of the disk about G is

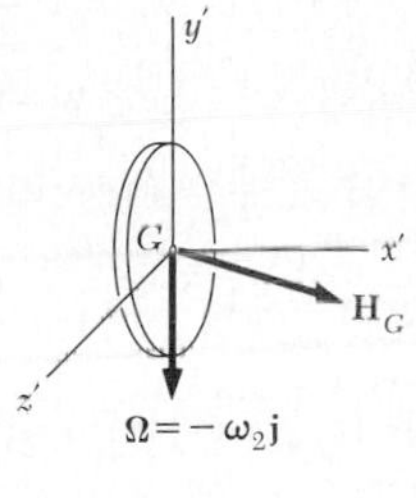

$$\mathbf{H}_G = \tfrac{1}{2}mr^2\omega_1\left(\mathbf{i} - \frac{r}{2L}\mathbf{j}\right)$$

where $\mathbf{H}_G$ is resolved into components along the rotating axes x', y', z', with x' along OG and y' vertical. The rate of change $\dot{\mathbf{H}}_G$ of $\mathbf{H}_G$ with respect to axes of fixed orientation is obtained from Eq. (18.22). Noting that the rate of change $(\dot{\mathbf{H}}_G)_{Gx'y'z'}$ of $\mathbf{H}_G$ with respect to the rotating frame is zero, and that the angular velocity $\boldsymbol{\Omega}$ of that frame is

$$\boldsymbol{\Omega} = -\omega_2\mathbf{j} = -(r\omega_1/L)\mathbf{j}$$

we have

$$\begin{aligned}\dot{\mathbf{H}}_G &= (\dot{\mathbf{H}}_G)_{Gx'y'z'} + \boldsymbol{\Omega} \times \mathbf{H}_G \\ &= 0 - \frac{r\omega_1}{L}\mathbf{j} \times \tfrac{1}{2}mr^2\omega_1\left(\mathbf{i} - \frac{r}{2L}\mathbf{j}\right) \\ &= \tfrac{1}{2}mr^2(r/L)\omega_1^2\mathbf{k} \end{aligned} \tag{2}$$

Equations of Motion. Expressing that the system of the external forces is equivalent to the system of the effective forces, we write

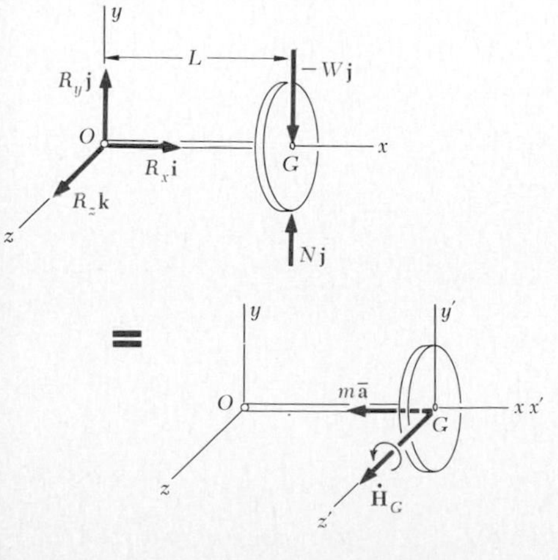

$\Sigma\mathbf{M}_O = \Sigma(\mathbf{M}_O)_{\text{eff}}$: $\quad L\mathbf{i} \times (N\mathbf{j} - W\mathbf{j}) = \dot{\mathbf{H}}_G$

$$(N - W)L\mathbf{k} = \tfrac{1}{2}mr^2(r/L)\omega_1^2\mathbf{k}$$

$$N = W + \tfrac{1}{2}mr(r/L)^2\omega_1^2 \qquad \mathbf{N} = [W + \tfrac{1}{2}mr(r/L)^2\omega_1^2]\mathbf{j} \tag{3}$$ ◄

$\Sigma\mathbf{F} = \Sigma\mathbf{F}_{\text{eff}}$: $\quad \mathbf{R} + N\mathbf{j} - W\mathbf{j} = m\bar{\mathbf{a}}$

Substituting for N from (3), for $m\bar{\mathbf{a}}$ from (1), and solving for $\mathbf{R}$, we have

$$\mathbf{R} = -(mr^2\omega_1^2/L)\mathbf{i} - \tfrac{1}{2}mr(r/L)^2\omega_1^2\mathbf{j}$$

$$\mathbf{R} = -\frac{mr^2\omega_1^2}{L}\left(\mathbf{i} + \frac{r}{2L}\mathbf{j}\right)$$ ◄

Problems

18.46 Determine the rate of change $\dot{\mathbf{H}}_G$ of the angular momentum $\mathbf{H}_G$ of the plate of Prob. 18.2.

18.47 Determine the rate of change $\dot{\mathbf{H}}_A$ of the angular momentum $\mathbf{H}_A$ of the plate of Prob. 18.3, knowing that its angular velocity $\boldsymbol{\omega}$ remains constant.

18.48 Determine the rate of change $\dot{\mathbf{H}}_G$ of the angular momentum $\mathbf{H}_G$ of the disk of Prob. 18.4, knowing that its angular velocity $\boldsymbol{\omega}$ remains constant.

18.49 Determine the rate of change $\dot{\mathbf{H}}_D$ of the angular momentum $\mathbf{H}_D$ of the assembly of Prob. 18.1.

18.50 Determine the rate of change $\dot{\mathbf{H}}_A$ of the angular momentum $\mathbf{H}_A$ of the disk of Prob. 18.6.

18.51 Determine the rate of change $\dot{\mathbf{H}}_C$ of the angular momentum $\mathbf{H}_C$ of the disk of Prob. 18.7.

18.52 Determine the rate of change $\dot{\mathbf{H}}_G$ of the angular momentum $\mathbf{H}_G$ of the disk of Prob. 18.4, knowing that it has an angular velocity $\boldsymbol{\omega} = \omega\mathbf{i}$ and an angular acceleration $\boldsymbol{\alpha} = \alpha\mathbf{i}$.

18.53 Determine the rate of change $\dot{\mathbf{H}}_A$ of the angular momentum $\mathbf{H}_A$ of the plate of Prob. 18.3, knowing that it has an angular velocity $\boldsymbol{\omega} = \omega\mathbf{j}$ and an angular acceleration $\boldsymbol{\alpha} = \alpha\mathbf{j}$.

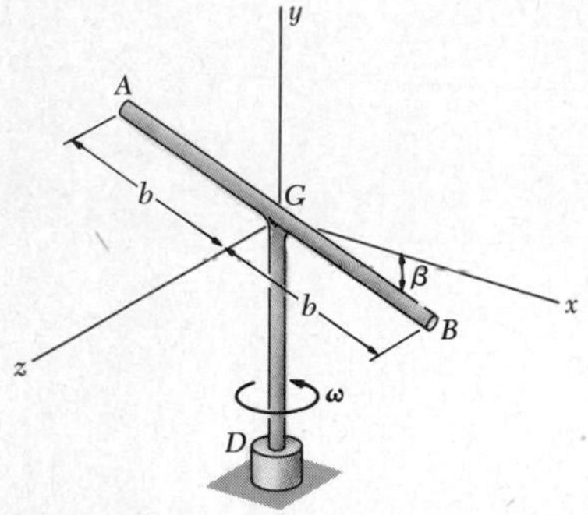

Fig. P18.54

18.54 A thin homogeneous rod AB of mass m and length $2b$ is welded at its midpoint G to a vertical shaft GD. Knowing that the shaft rotates with a constant angular velocity $\boldsymbol{\omega}$, determine the couple exerted by the shaft on rod AB.

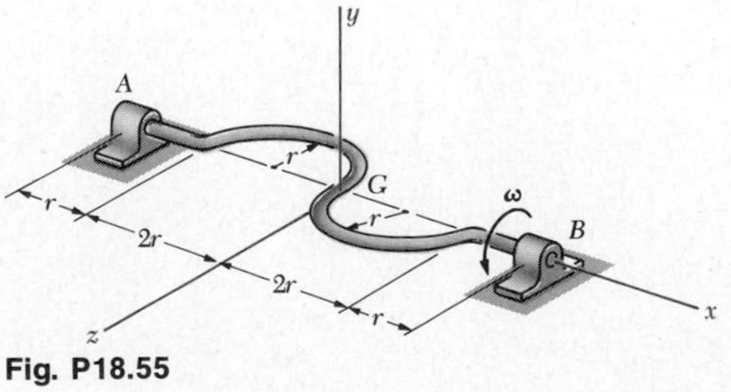

Fig. P18.55

18.55 A thin homogeneous rod of weight w per unit length is used to form the shaft shown. Knowing that the shaft rotates with a constant angular velocity $\boldsymbol{\omega}$, determine the dynamic reactions at A and B.

18.56 A thin homogeneous triangular plate of mass 2.5 kg is welded to a light vertical axle supported by bearings at A and B. Knowing that the plate rotates at the constant rate $\omega = 8$ rad/s, determine the dynamic reactions at A and B.

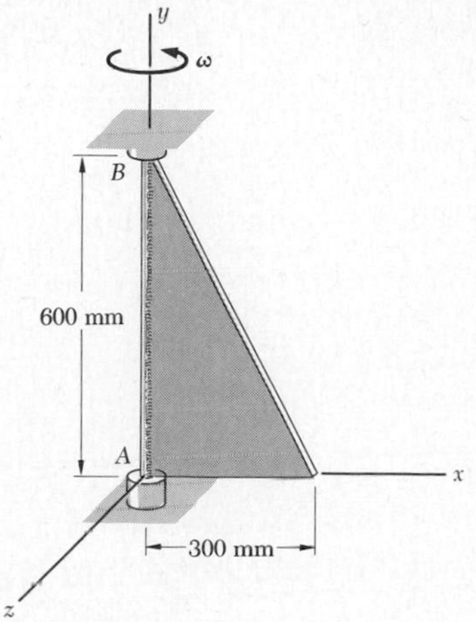

Fig. P18.56

18.57 Two L-shaped arms, each weighing 4 lb, are welded at the third points of a 2-ft horizontal shaft supported by bearings at A and B. Knowing that the shaft rotates at the constant rate $\omega = 240$ rpm, determine the dynamic reactions at A and B.

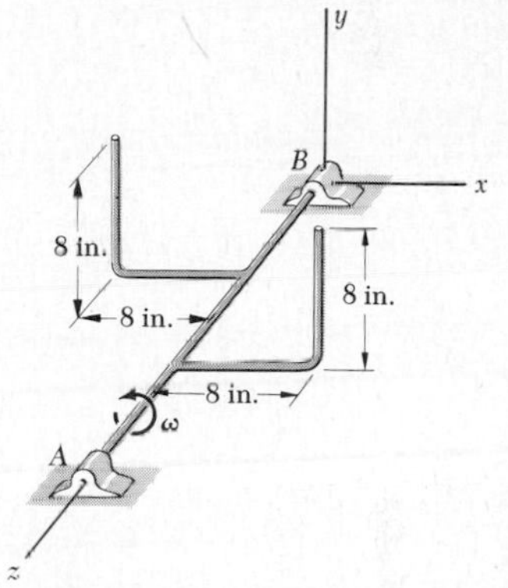

Fig. P18.57 and P18.58

18.58 Two L-shaped arms, each weighing 4 lb, are welded at the third points of a 2-ft horizontal shaft supported by bearings at A and B. The assembly is a rest ($\omega = 0$) when a couple of moment $\mathbf{M}_0 = (20 \text{ lb}\cdot\text{in.})\mathbf{k}$ is applied to the shaft. Determine (*a*) the resulting angular acceleration of the shaft, (*b*) the dynamic reactions at A and B immediately after the couple has been applied.

18.59 Knowing that the plate of Prob. 18.56 is initially at rest ($\omega = 0$) when a couple of moment $\mathbf{M}_0 = (0.45 \text{ N}\cdot\text{m})\mathbf{j}$ is applied to it, determine (*a*) the resulting angular acceleration of the plate, (*b*) the dynamic reactions at A and B immediately after the couple has been applied.

18.60 The sheet-metal component shown is of uniform thickness and has a mass of 400 g. It is attached to a light axle supported by bearings at A and B located 120 mm apart. The component is at rest when it is subjected to a couple $\mathbf{M}_0$ as shown. If the resulting angular acceleration is $\boldsymbol{\alpha} = (12 \text{ rad/s}^2)\mathbf{k}$, determine (*a*) the couple $\mathbf{M}_0$, (*b*) the dynamic reactions at A and B immediately after the couple has been applied.

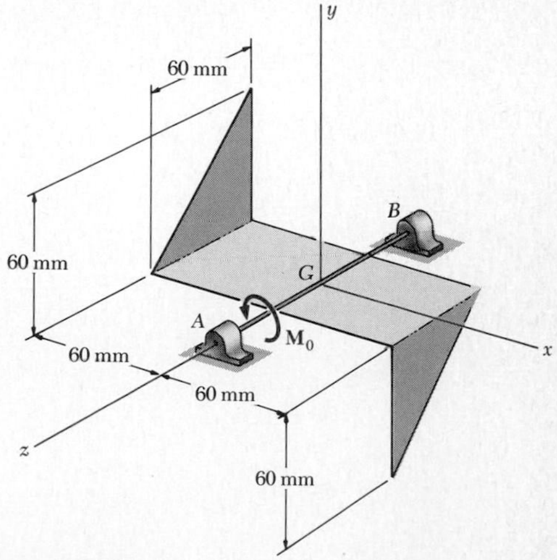

Fig. P18.60

18.61 Knowing that the shaft of Prob. 18.55 is originally at rest ($\omega = 0$), determine (*a*) the couple $\mathbf{M}_0$ required to impart to the shaft an acceleration $\boldsymbol{\alpha} = \alpha\mathbf{i}$, (*b*) the dynamic reactions at A and B immediately after the couple $\mathbf{M}_0$ has been applied.

18.62 For the sheet-metal component of Prob. 18.60, determine (*a*) the angular velocity of the component 0.5 s after the couple $\mathbf{M}_0$ has been applied to it, (*b*) the dynamic reactions at A and B at that time.

18.63 Knowing that the shaft GD in Prob. 18.54 has an angular velocity $\boldsymbol{\omega} = \omega\mathbf{j}$ and an angular acceleration $\boldsymbol{\alpha} = \alpha\mathbf{j}$ at the instant shown, determine the components of the couple exerted by the shaft on rod AB.

18.64 The blade of an oscillating fan and the rotor of its motor have a total mass of 200 g and a combined radius of gyration of 75 mm. They are supported by bearings located at A and B, 120 mm apart, and have an angular velocity $\boldsymbol{\omega}_1$ of 2400 rpm at the high-speed setting. Determine the dynamic reactions at A and B when the motor casing has an angular velocity $\boldsymbol{\omega}_2$ of 0.5 rad/s as shown.

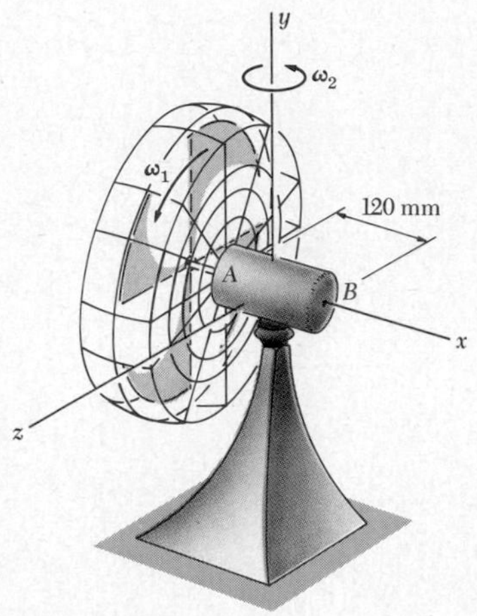

Fig. P18.64

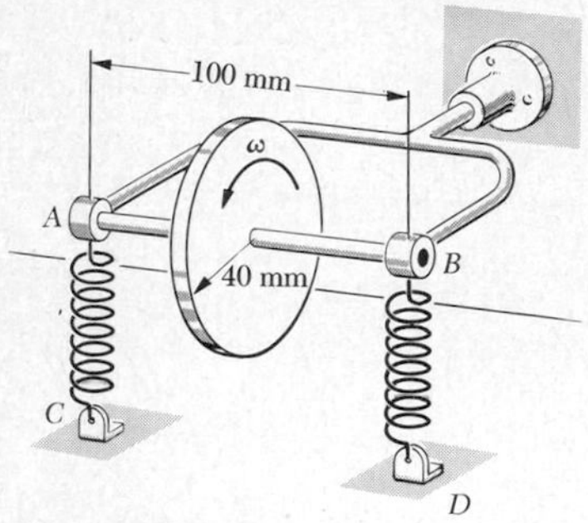

Fig. P18.65

18.65 The essential structure of a certain type of aircraft turn indicator is shown. Springs AC and BD are initially stretched and exert equal vertical forces at A and B when the airplane is traveling in a straight path. Each spring has a constant of 500 N/m and the uniform disk has a mass of 200 g and spins at the rate of 10 000 rpm. Determine the angle through which the yoke will rotate when the pilot executes a horizontal turn of 750-m radius to the right at a speed of 800 km/h. Indicate whether point A will move up or down.

18.66 Each wheel of an automobile weighs 45 lb and has a radius of gyration of 10 in. The automobile travels around an unbanked curve of radius 600 ft at a speed of 60 mi/h. Knowing that each wheel is of 24-in. diameter and that the transverse distance between the wheels is 58 in., determine the additional normal reaction at each outside wheel due to the rotation of the car.

18.67 An airplane has a single four-bladed propeller which weighs 250 lb and has a radius of gyration of 32 in. Knowing that the propeller rotates at 1800 rpm clockwise as seen from the front, determine the magnitude of the couple exerted by the propeller on its shaft and the resulting effect on the airplane when the pilot executes a horizontal turn of 1500-ft radius to the left at a speed of 375 mi/h.

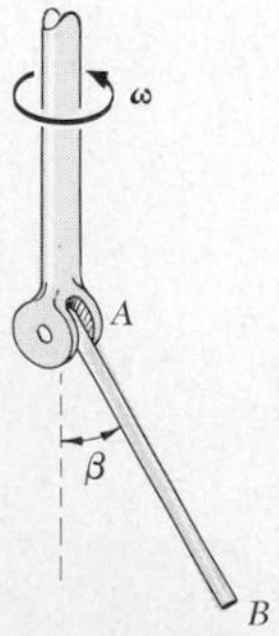

Fig. P18.68

18.68 A uniform rod AB of length l and mass m is attached to the pin of a clevis which rotates with a constant angular velocity $\boldsymbol{\omega}$. Determine (*a*) the constant angle β that the rod forms with the vertical, (*b*) the range of values of ω for which the rod remains vertical ($\beta = 0$).

18.69 A uniform square plate of side a and mass m is hinged at A and B to a clevis which rotates with a constant angular velocity $\boldsymbol{\omega}$. Determine (a) the value of ω for which the angle β maintains the constant value $\beta = 30°$, (b) the range of values of ω for which the plate remains vertical ($\beta = 90°$).

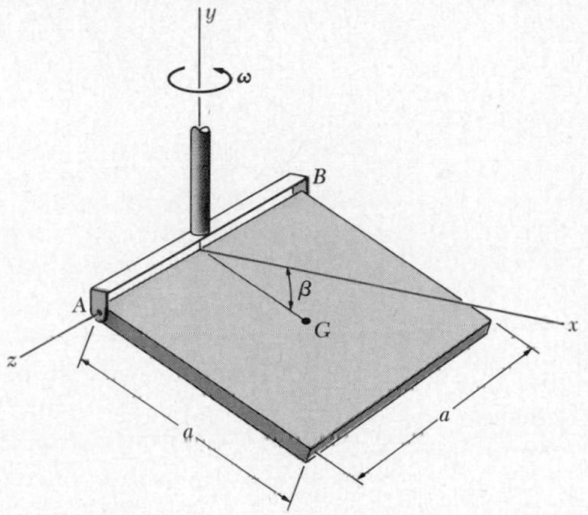

Fig. P18.69

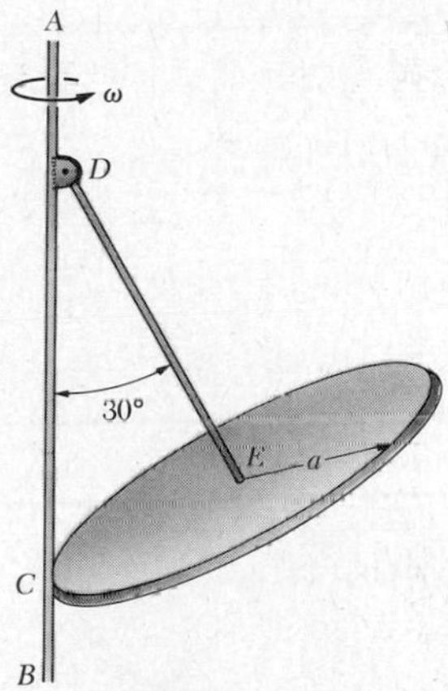

Fig. P18.70

18.70 A disk of mass m and radius a is rigidly attached to a rod DE of negligible mass. The rod DE is attached to a vertical shaft AB by a clevis at D and the disk leans against the shaft at C. Noting that when the shaft AB is made to rotate, the same point of the disk will remain in contact with the shaft at C, determine the magnitude of the angular velocity $\boldsymbol{\omega}$ for which the reaction at C will be zero.

18.71 A slender rod AB of length L is attached to the arm BC by a clevis at B and is made to rotate about the vertical with an angular velocity $\boldsymbol{\omega}$. Knowing that the rod forms an angle $\theta = 25°$ with the horizontal, derive an expression for the corresponding magnitude of the angular velocity $\boldsymbol{\omega}$.

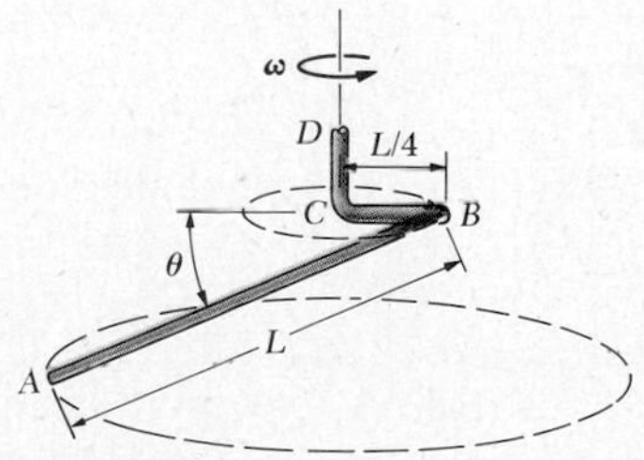

Fig. P18.71

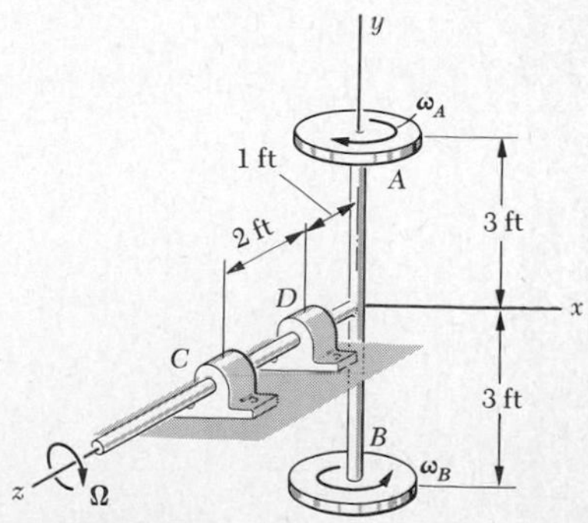

Fig. P18.72

18.72 Two disks, each of weight 10 lb, and radius 12 in., spin as shown at 1200 rpm about the rod AB, which is attached to shaft CD. The entire system is made to rotate about the z axis with an angular velocity $\boldsymbol{\Omega}$ of 60 rpm. (a) Determine the dynamic reactions at C and D due to gyroscopic action as the system passes through the position shown. (b) Solve part a assuming that the direction of spin of disk B is reversed.

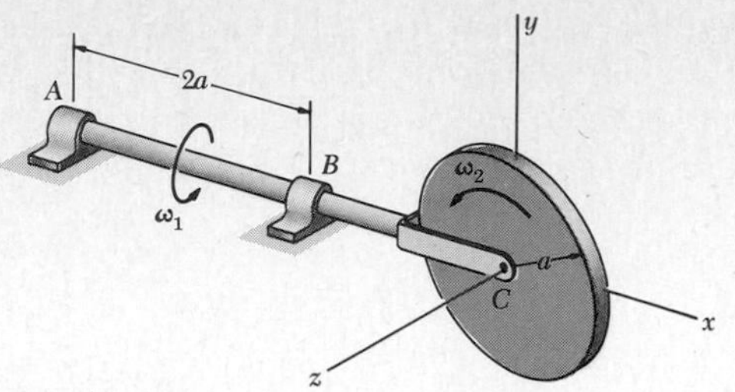

Fig. P18.73

18.73 A thin homogeneous disk of mass m and radius a is held by a fork-ended horizontal rod ABC. The disk and the rod rotate with the angular velocities shown. Assuming that both ω_1 and ω_2 are constant, determine the dynamic reactions at A and B.

18.74 A thin disk of mass $m = 5$ kg rotates with an angular velocity $\boldsymbol{\omega}_2$ with respect to the bent axle ABC, which itself rotates with an angular velocity $\boldsymbol{\omega}_1$ about the y axis. Knowing that $\omega_1 = 3$ rad/s and $\omega_2 = 8$ rad/s and that both are constant, determine the force-couple system representing the dynamic reaction at the support at A.

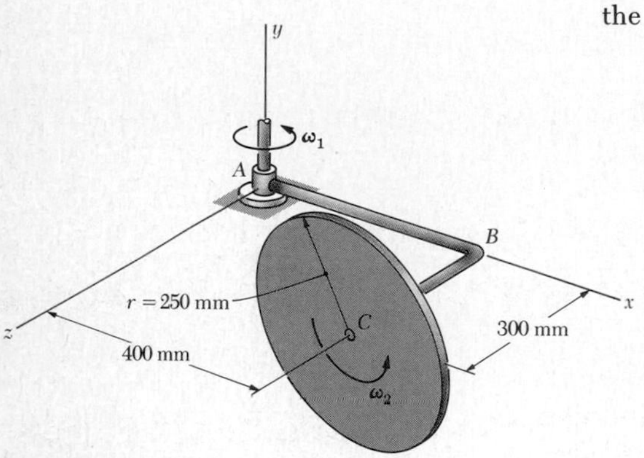

Fig. P18.74

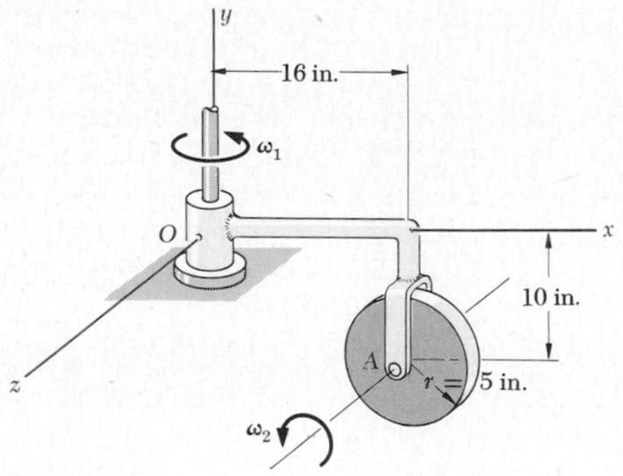

Fig. P18.75

18.75 A thin disk of weight $W = 8$ lb rotates with an angular velocity $\boldsymbol{\omega}_2$ with respect to the arm OA, which itself rotates with an angular velocity $\boldsymbol{\omega}_1$ about the y axis. Knowing that $\omega_1 = 4$ rad/s and $\omega_2 = 12$ rad/s and that both are constant, determine the force-couple system representing the dynamic reaction at the support at O.

18.76 For the disk of Prob. 18.74, determine (a) the couple $M_1\mathbf{j}$ which should be applied to the bent axle ABC to give it an angular acceleration $\boldsymbol{\alpha}_1 = (6 \text{ rad/s}^2)\mathbf{j}$ when $\omega_1 = 3$ rad/s, knowing that the disk rotates at the constant rate $\omega_2 = 8$ rad/s, (b) the force-couple system representing the dynamic reaction at A at that instant. Assume that the bent axle ABC has a negligible mass.

18.77 For the disk of Prob. 18.74, determine (a) the couple $M_1\mathbf{j}$ which should be applied to the bent axle ABC to keep it rotating at the constant rate $\omega_1 = 3$ rad/s at an instant when the angular velocity $\boldsymbol{\omega}_2$ of the disk has a magnitude of 8 rad/s and decreases at the rate of 2 rad/s^2 due to the axle friction at C, (b) the force-couple representing the dynamic reaction at A at that instant.

18.78 For the disk of Prob. 18.73, determine (*a*) the couple $M_1\mathbf{i}$ which should be applied to rod *AB* to give it an angular acceleration $\alpha_1\mathbf{i}$ when the rod has an angular velocity $\omega_1\mathbf{i}$, knowing that the disk rotates at the constant rate ω_2, (*b*) the dynamic reactions at *A* and *B* at that instant.

18.79 For the disk of Prob. 18.75, determine (*a*) the couple $M_1\mathbf{j}$ which should be applied to arm *OA* to give it an angular acceleration $\boldsymbol{\alpha}_1 = (6\ \text{rad/s}^2)\mathbf{j}$ when $\omega_1 = 4$ rad/s, knowing that the disk rotates at the constant rate $\omega_2 =$ 12 rad/s, (*b*) the force-couple system representing the dynamic reaction at *O* at that instant. Assume that arm *OA* has a negligible mass.

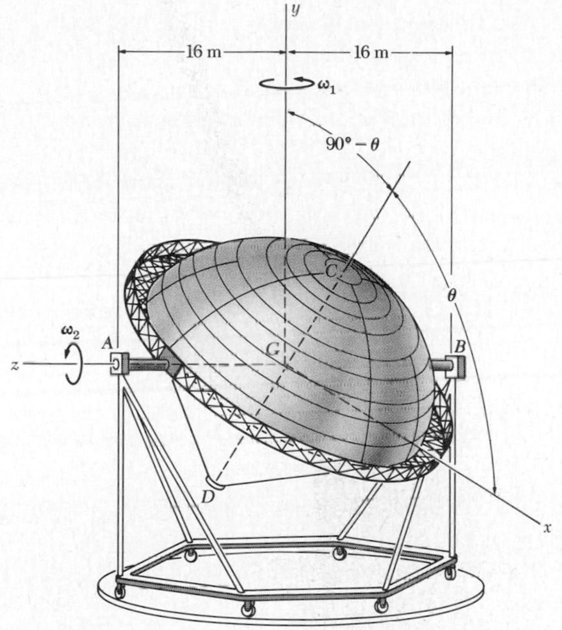

Fig. P18.80

***18.80** An experimental Fresnel-lens solar-energy concentrator may rotate about the horizontal axis *AB* which passes through its mass center *G*. It is supported at *A* and *B* by a steel framework which may rotate about the vertical *y* axis. The concentrator has a mass of 30 Mg, a radius of gyration of 12 m about its axis of symmetry *CD*, and a radius of gyration of 10 m about any transverse axis through *G*. Knowing that the angular velocities $\boldsymbol{\omega}_1$ and $\boldsymbol{\omega}_2$ have constant magnitudes equal to 0.20 rad/s and 0.15 rad/s, respectively, determine for the position $\theta = 30°$ (*a*) the forces exerted on the concentrator at *A* and *B*, (*b*) the couple $M_2\mathbf{k}$ applied to the concentrator at that instant.

***18.81** The slender homogeneous rod *AB* of mass *m* and length *L* is made to rotate at the constant rate ω_2 about the horizontal *z* axis, while the frame *CD* is made to rotate at the constant rate ω_1 about the vertical *y* axis. Express as a function of the angle θ (*a*) the couple $\mathbf{M}_1$ required to maintain the rotation of the frame, (*b*) the couple $\mathbf{M}_2$ required to maintain the rotation of the rod, (*c*) the dynamic reactions at the supports *C* and *D*.

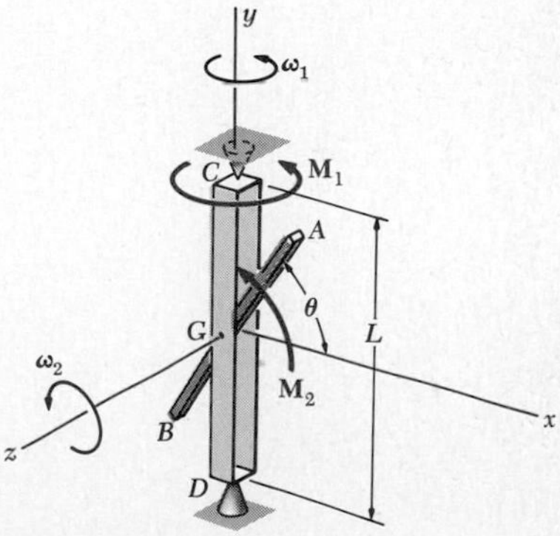

Fig. P18.81

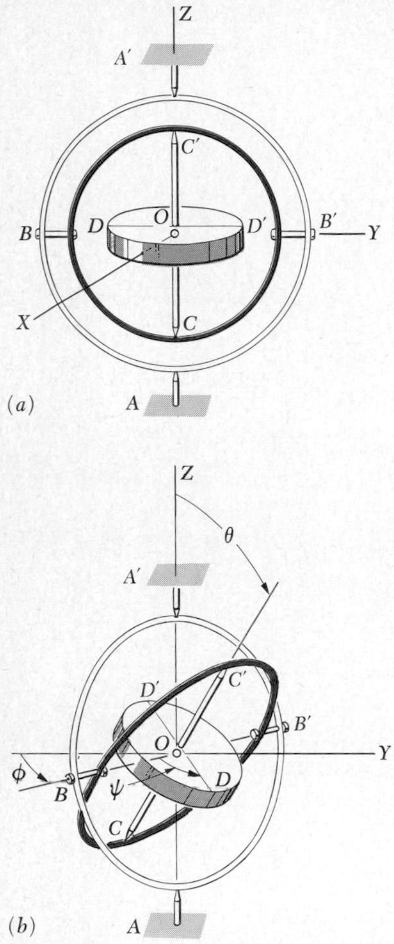

Fig. 18.15

***18.9. Motion of a Gyroscope. Eulerian Angles.** A *gyroscope* consists essentially of a rotor which may spin freely about its geometric axis. When mounted in a Cardan's suspension (Fig. 18.15), a gyroscope may assume any orientation, but its mass center must remain fixed in space. In order to define the position of a gyroscope at a given instant, we shall select a fixed frame of reference $OXYZ$, with the origin O located at the mass center of the gyroscope and the Z axis directed along the line defined by the bearings A and A' of the outer gimbal, and we shall consider a reference position of the gyroscope in which the two gimbals and a given diameter DD' of the rotor are located in the fixed YZ plane (Fig. 18.15*a*). The gyroscope may be brought from this reference position into any arbitrary position (Fig. 18.15*b*) by means of the following steps: (1) a rotation of the outer gimbal through an angle ϕ about the axis AA', (2) a rotation of the inner gimbal through θ about BB', (3) a rotation of the rotor through ψ about CC'. The angles ϕ, θ, and ψ are called the *Eulerian angles;* they completely characterize the position of the gyroscope at any given instant. Their derivatives $\dot{\phi}$, $\dot{\theta}$, and $\dot{\psi}$ define, respectively, the rate of *precession,* the rate of *nutation,* and the rate of *spin* of the gyroscope at the instant considered.

In order to compute the components of the angular velocity and of the angular momentum of the gyroscope, we shall use a rotating system of axes *Oxyz attached to the inner gimbal,* with the y axis along BB' and the z axis along CC' (Fig. 18.16). These axes are principal axes of inertia for the gyroscope but while they follow it in its precession and nutation, they do not spin. For that reason, they are more convenient to use than axes actually attached to the gyroscope. We shall now express the angular velocity $\boldsymbol{\omega}$ of the gyroscope with respect to the fixed frame of reference $OXYZ$ as the sum of three partial angular velocities corresponding, respectively, to the precession, the nutation, and the spin of the gyroscope. Denoting by $\mathbf{i}$, $\mathbf{j}$, $\mathbf{k}$ the unit vectors along the rotating axes, and by $\mathbf{K}$ the unit vector along the fixed Z axis, we have

$$\boldsymbol{\omega} = \dot{\phi}\mathbf{K} + \dot{\theta}\mathbf{j} + \dot{\psi}\mathbf{k} \tag{18.33}$$

Since the vector components obtained for $\boldsymbol{\omega}$ in (18.33) are not orthogonal (Fig. 18.16), we shall resolve the unit vector $\mathbf{K}$ into components along the x and z axes; we write

$$\mathbf{K} = -\sin\theta\,\mathbf{i} + \cos\theta\,\mathbf{k} \tag{18.34}$$

and, substituting for $\mathbf{K}$ into (18.33),

$$\boldsymbol{\omega} = -\dot{\phi}\sin\theta\,\mathbf{i} + \dot{\theta}\mathbf{j} + (\dot{\psi} + \dot{\phi}\cos\theta)\mathbf{k} \tag{18.35}$$

Since the coordinate axes are principal axes of inertia, the components of the angular momentum $\mathbf{H}_O$ may be obtained by multiplying the components of $\boldsymbol{\omega}$ by the moments of inertia of the rotor about the x, y, and z axes, respectively. Denoting by I the moment of inertia of the rotor about its spin axis, by I' its moment of inertia about a transverse axis through O, and neglecting the mass of the gimbals, we write

$$\mathbf{H}_O = -I'\dot{\phi}\sin\theta\,\mathbf{i} + I'\dot{\theta}\mathbf{j} + I(\dot{\psi} + \dot{\phi}\cos\theta)\mathbf{k} \qquad (18.36)$$

Recalling that the rotating axes are attached to the inner gimbal, and thus do not spin, we express their angular velocity as the sum

$$\boldsymbol{\Omega} = \dot{\phi}\mathbf{K} + \dot{\theta}\mathbf{j} \qquad (18.37)$$

or, substituting for $\mathbf{K}$ from (18.34),

$$\boldsymbol{\Omega} = -\dot{\phi}\sin\theta\,\mathbf{i} + \dot{\theta}\mathbf{j} + \dot{\phi}\cos\theta\,\mathbf{k} \qquad (18.38)$$

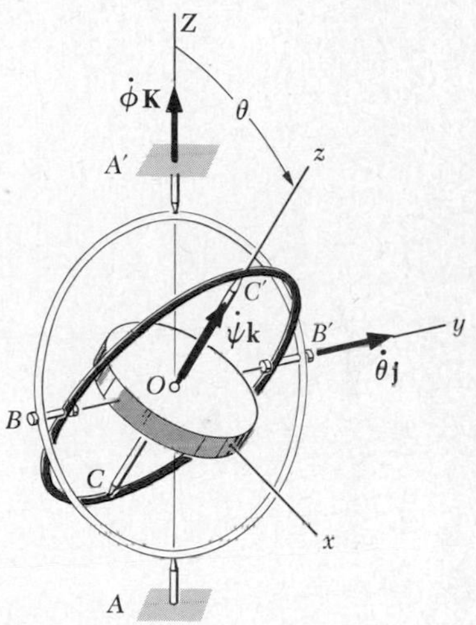

Fig. 18.16

Substituting for $\mathbf{H}_O$ and $\boldsymbol{\Omega}$ from (18.36) and (18.38) into the equation

$$\Sigma\mathbf{M}_O = (\dot{\mathbf{H}}_O)_{Oxyz} + \boldsymbol{\Omega} \times \mathbf{H}_O \qquad (18.28)$$

we obtain the three differential equations

$$\begin{aligned}
\Sigma M_x &= -I'(\ddot{\phi}\sin\theta + 2\dot{\theta}\dot{\phi}\cos\theta) + I\dot{\theta}(\dot{\psi} + \dot{\phi}\cos\theta)\\
\Sigma M_y &= I'(\ddot{\theta} - \dot{\phi}^2\sin\theta\cos\theta) + I\dot{\phi}\sin\theta(\dot{\psi} + \dot{\phi}\cos\theta) \qquad (18.39)\\
\Sigma M_z &= I\frac{d}{dt}(\dot{\psi} + \dot{\phi}\cos\theta)
\end{aligned}$$

The equations (18.39) define the motion of a gyroscope subjected to a given system of forces when the mass of its gimbals is neglected. They may also be used to define the motion of an *axisymmetrical body* (or body of revolution) attached at a point on its axis of symmetry, or the motion of an axisymmetrical body about its mass center. While the gimbals of the gyroscope helped us visualize the Eulerian angles, it is clear that these angles may be used to define the position of any rigid body with respect to axes centered at a point of the body, regardless of the way in which the body is actually supported.

Since the equations (18.39) are nonlinear, it will not be possible, in general, to express the Eulerian angles ϕ, θ, and ψ as analytical functions of the time t, and numerical methods of solution may have to be used. However, as we shall see in the following sections, there are several particular cases of interest which may be analyzed easily.

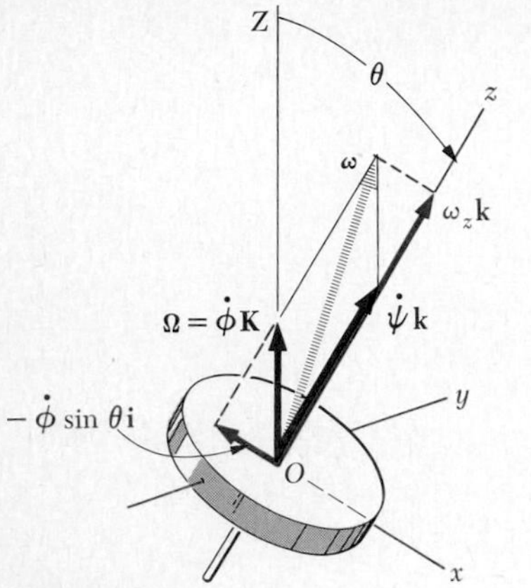

Fig. 18.17

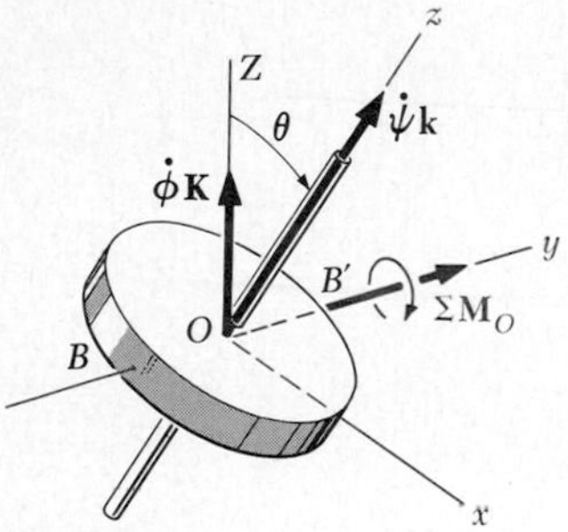

Fig. 18.18

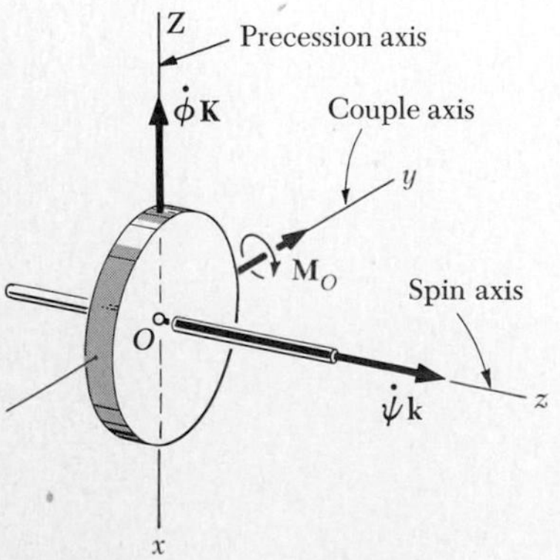

Fig. 18.19

* 18.10. Steady Precession of a Gyroscope.

We shall consider in this section the particular case of gyroscopic motion in which the angle θ, the rate of precession $\dot\phi$, and the rate of spin $\dot\psi$ remain constant. We propose to determine the forces which must be applied to the gyroscope to maintain this motion, known as the *steady precession* of a gyroscope.

Instead of applying the general equations (18.39), we shall determine the sum of the moments of the required forces by computing the rate of change of the angular momentum of the gyroscope in the particular case considered. We first note that the angular velocity $\boldsymbol{\omega}$ of the gyroscope, its angular momentum $\mathbf{H}_O$, and the angular velocity $\boldsymbol{\Omega}$ of the rotating frame of reference (Fig. 18.17) reduce, respectively, to

$$\boldsymbol{\omega} = -\dot\phi \sin\theta\, \mathbf{i} + \omega_z \mathbf{k} \tag{18.40}$$

$$\mathbf{H}_O = -I'\dot\phi \sin\theta\, \mathbf{i} + I\omega_z \mathbf{k} \tag{18.41}$$

$$\boldsymbol{\Omega} = -\dot\phi \sin\theta\, \mathbf{i} + \dot\phi \cos\theta\, \mathbf{k} \tag{18.42}$$

where $\omega_z = \dot\psi + \dot\phi\cos\theta$ = component along the spin axis of the total angular velocity of the gyroscope

Since θ, $\dot\phi$, and $\dot\psi$ are constant, the vector $\mathbf{H}_O$ is constant in magnitude and direction with respect to the rotating frame of reference, and its rate of change $(\dot{\mathbf{H}}_O)_{Oxyz}$ with respect to that frame is zero. Thus Eq. (18.28) reduces to

$$\Sigma\mathbf{M}_O = \boldsymbol{\Omega} \times \mathbf{H}_O \tag{18.43}$$

which yields, after substitutions from (18.41) and (18.42),

$$\Sigma\mathbf{M}_O = (I\omega_z - I'\dot\phi\cos\theta)\dot\phi\sin\theta\,\mathbf{j} \tag{18.44}$$

Since the mass center of the gyroscope is fixed in space, we have, by (18.1), $\Sigma\mathbf{F} = 0$; thus, the forces which must be applied to the gyroscope to maintain its steady precession reduce to a couple of moment equal to the right-hand member of Eq. (18.44). We note that *this couple should be applied about an axis perpendicular to the precession axis and to the spin axis of the gyroscope* (Fig. 18.18).

In the particular case when the precession axis and the spin axis are at a right angle to each other, we have $\theta = 90°$ and Eq. (18.44) reduces to

$$\Sigma\mathbf{M}_O = I\dot\psi\dot\phi\mathbf{j} \tag{18.45}$$

Thus, if we apply to the gyroscope a couple $\mathbf{M}_O$ about an axis perpendicular to its axis of spin, the gyroscope will precess about an axis perpendicular to both the spin axis and the couple axis, in a sense such that the vectors representing, respectively, the spin, the couple, and the precession form a right-handed triad (Fig. 18.19).

Because of the relatively large couples required to change the orientation of their axles, gyroscopes are used as stabilizers in torpedoes and ships.

Spinning bullets and shells remain tangent to their trajectory because of gyroscopic action. And a bicycle is easier to keep balanced at high speeds because of the stabilizing effect of its spinning wheels. However, gyroscopic action is not always welcome and must be taken into account in the design of bearings supporting rotating shafts subjected to forced precession. The reactions exerted by its propellers on an airplane which changes its direction of flight must also be taken into consideration and compensated for whenever possible.

*** 18.11. Motion of an Axisymmetrical Body under No Force.** We shall consider in this section the motion about its mass center of an axisymmetrical body under no force, except its own weight. Examples of such a motion are furnished by projectiles, if air resistance is neglected, and by artificial satellites and space vehicles after burnout of their launching rockets.

Since the sum of the moments of the external forces about the mass center G of the body is zero, Eq. (18.2) yields $\dot{\mathbf{H}}_G = 0$. It follows that the angular momentum $\mathbf{H}_G$ of the body about G is constant. Thus, the direction of $\mathbf{H}_G$ is fixed in space and may be used to define the Z axis, or axis of precession (Fig. 18.20). Selecting a rotating system of axes $Gxyz$ with the z axis along the axis of symmetry of the body and the x axis in the plane defined by the Z and z axes, we have

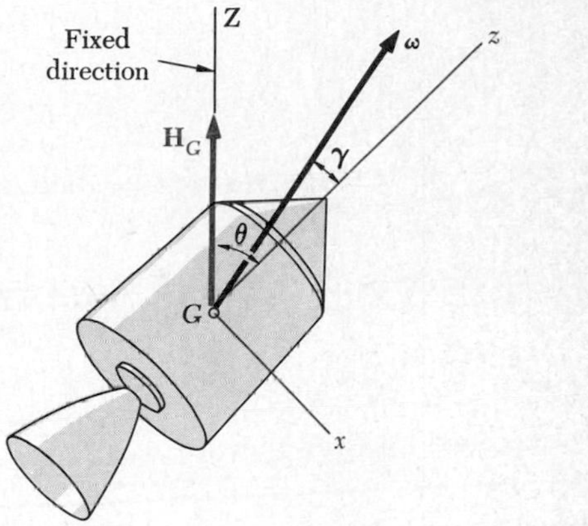

Fig. 18.20

$$H_x = -H_G \sin\theta \qquad H_y = 0 \qquad H_z = H_G \cos\theta \tag{18.46}$$

where θ represents the angle formed by the Z and z axes, and H_G denotes the constant magnitude of the angular momentum of the body about G. Since the x, y, and z axes are principal axes of inertia for the body considered, we may write

$$H_x = I'\omega_x \qquad H_y = I'\omega_y \qquad H_z = I\omega_z \tag{18.47}$$

where I denotes the moment of inertia of the body about its axis of symmetry, and I' its moment of inertia about a transverse axis through G. It follows from Eqs. (18.46) and (18.47) that

$$\omega_x = -\frac{H_G \sin\theta}{I'} \qquad \omega_y = 0 \qquad \omega_z = \frac{H_G \cos\theta}{I} \tag{18.48}$$

The second of the relations obtained shows that the angular velocity $\boldsymbol{\omega}$ has no component along the y axis, i.e., along an axis perpendicular to the Zz plane. Thus, the angle θ formed by the Z and z axes remains constant and *the body is in steady precession about the Z axis.*

Dividing the first and third of the relations (18.48) member by member, and observing from Fig. 18.21 that $-\omega_x/\omega_z = \tan\gamma$, we obtain the following relation between the angles γ and θ that the vectors $\boldsymbol{\omega}$ and $\mathbf{H}_G$, respectively, form with the axis of symmetry of the body:

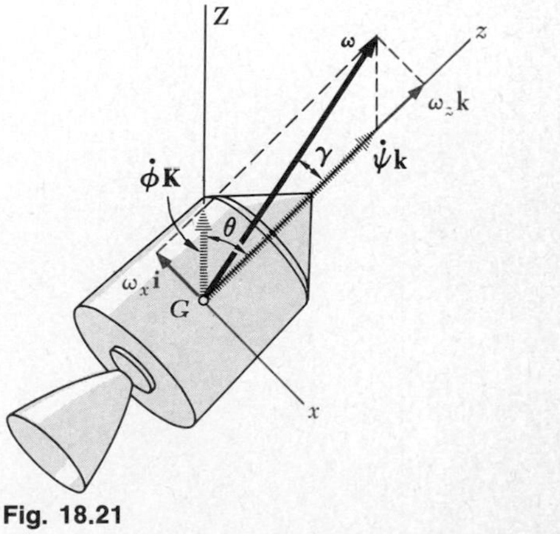

Fig. 18.21

$$\tan\gamma = \frac{I}{I'}\tan\theta \tag{18.49}$$

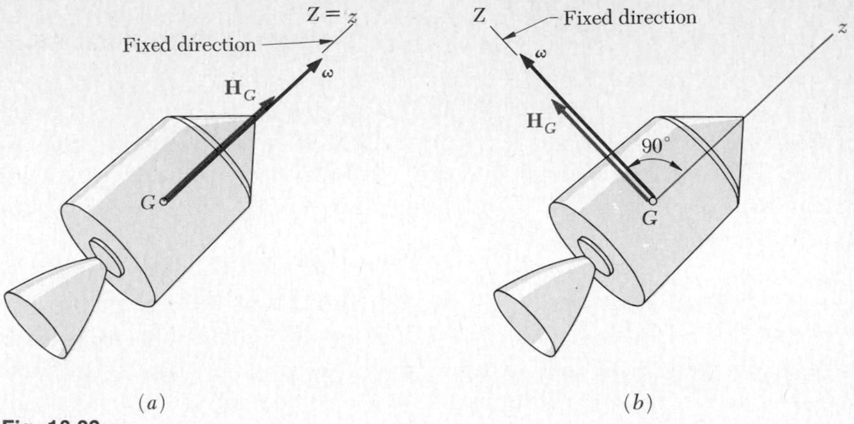

Fig. 18.22

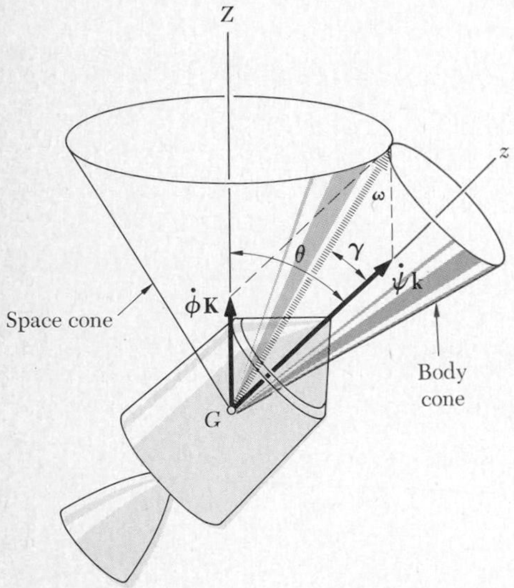

Fig. 18.23

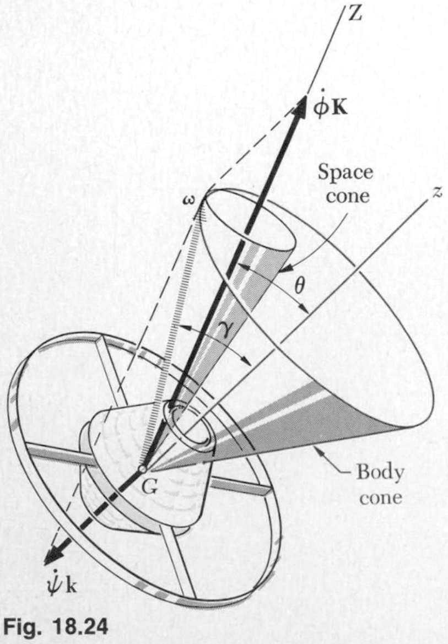

Fig. 18.24

There are two particular cases of motion of an axisymmetrical body under no force which involve no precession: (1) If the body is set to spin about its axis of symmetry, we have $\omega_x = 0$ and, by (18.47), $H_x = 0$; the vectors $\boldsymbol{\omega}$ and $\mathbf{H}_G$ have the same orientation and the body keeps spinning about its axis of symmetry (Fig. 18.22*a*). (2) If the body is set to spin about a transverse axis, we have $\omega_z = 0$ and, by (18.47), $H_z = 0$; again $\boldsymbol{\omega}$ and $\mathbf{H}_G$ have the same orientation and the body keeps spinning about the given transverse axis (Fig. 18.22*b*).

Considering now the general case represented in Fig. 18.21, we recall from Sec. 15.12 that the motion of a body about a fixed point—or about its mass center—may be represented by the motion of a body cone rolling on a space cone. In the case of steady precession, the two cones are circular, since the angles γ and $\theta - \gamma$ that the angular velocity $\boldsymbol{\omega}$ forms, respectively, with the axis of symmetry of the body and with the precession axis are constant. Two cases should be distinguished:

1. $I < I'$. This is the case of an elongated body, such as the space vehicle of Fig. 18.23. By (18.49) we have $\gamma < \theta$; the vector $\boldsymbol{\omega}$ lies inside the angle ZGz; the space cone and the body cone are tangent externally; the spin and the precession are both observed as counterclockwise from the positive z axis. The precession is said to be *direct*.

2. $I > I'$. This is the case of a flattened body, such as the satellite of Fig. 18.24. By (18.49) we have $\gamma > \theta$; since the vector $\boldsymbol{\omega}$ must lie outside the angle ZGz, the vector $\dot{\psi}\mathbf{k}$ has a sense opposite to that of the z axis; the space cone is inside the body cone; the precession and the spin have opposite senses; the precession is said to be *retrograde*.

SAMPLE PROBLEM 18.6

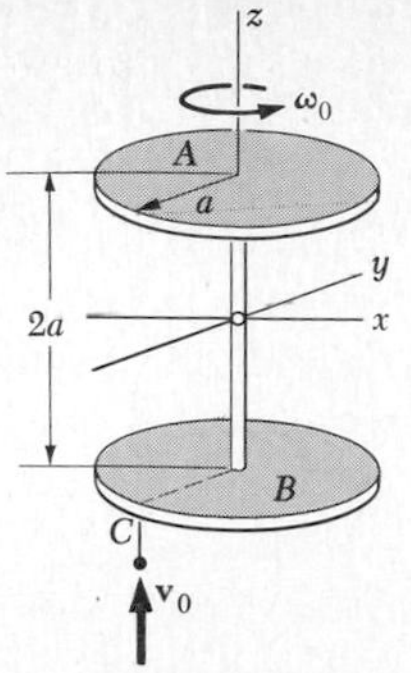

A space satellite of mass m is known to be dynamically equivalent to two thin disks of equal mass. The disks are of radius $a = 800$ mm and are rigidly connected by a light rod of length $2a$. Initially the satellite is spinning freely about its axis of symmetry at the rate $\omega_0 = 60$ rpm. A meteorite, of mass $m_0 = m/1000$ and traveling with a velocity $\mathbf{v}_0$ of 2000 m/s relative to the satellite, strikes the satellite and becomes embedded at C. Determine (*a*) the angular velocity of the satellite immediately after impact, (*b*) the precession axis of the ensuing motion, (*c*) the rates of precession and spin of the ensuing motion.

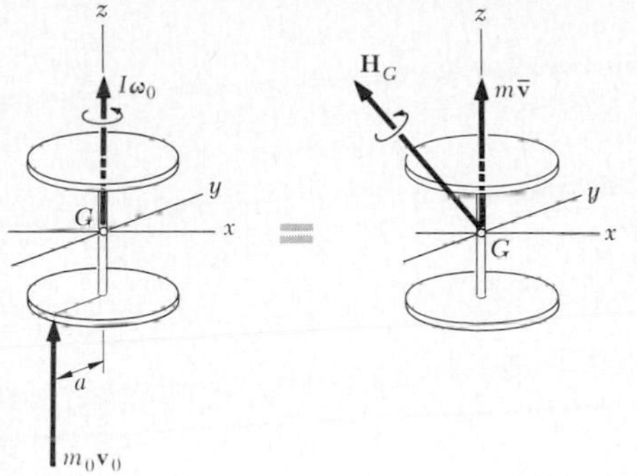

Solution. ***Moments of Inertia.*** We note that the axes shown are principal axes of inertia for the satellite and write

$$I = I_z = \tfrac{1}{2}ma^2 \qquad I' = I_x = I_y = 2[\tfrac{1}{4}(\tfrac{1}{2}m)a^2 + (\tfrac{1}{2}m)a^2] = \tfrac{5}{4}ma^2$$

Principle of Impulse and Momentum. We consider the satellite and the meteorite as a single system. Since no external force acts on this system, the momenta before and after impact are equipollent. Taking moments about G we write

$$-a\mathbf{j} \times m_0 v_0 \mathbf{k} + I\omega_0 \mathbf{k} = \mathbf{H}_G$$

$$\mathbf{H}_G = -m_0 v_0 a \mathbf{i} + I\omega_0 \mathbf{k} \qquad (1)$$

Angular Velocity after Impact. Substituting the values obtained for the components of $\mathbf{H}_G$ and for the moments of inertia into

$$H_x = I_x\omega_x \qquad H_y = I_y\omega_y \qquad H_z = I_z\omega_z$$

we write

$$-m_0 v_0 a = I'\omega_x = \tfrac{5}{4}ma^2\omega_x \qquad 0 = I'\omega_y \qquad I\omega_0 = I\omega_z$$

$$\omega_x = -\frac{4}{5}\frac{m_0 v_0}{ma} \qquad \omega_y = 0 \qquad \omega_z = \omega_0 \qquad (2)$$

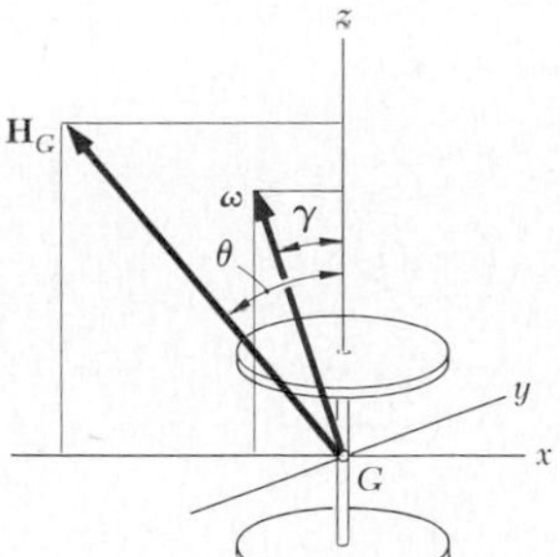

For the satellite considered we have $\omega_0 = 60 \text{ rpm} = 6.283$ rad/s, $m_0/m = \frac{1}{1000}$, $a = 0.800$ m, and $v_0 = 2000$ m/s; we find

$$\omega_x = -2 \text{ rad/s} \qquad \omega_y = 0 \qquad \omega_z = 6.283 \text{ rad/s}$$

$$\omega = \sqrt{\omega_x^2 + \omega_z^2} = 6.594 \text{ rad/s} \qquad \tan\gamma = \frac{-\omega_x}{\omega_z} = +0.3183$$

$$\omega = 63.0 \text{ rpm} \qquad \gamma = 17.7° \blacktriangleleft$$

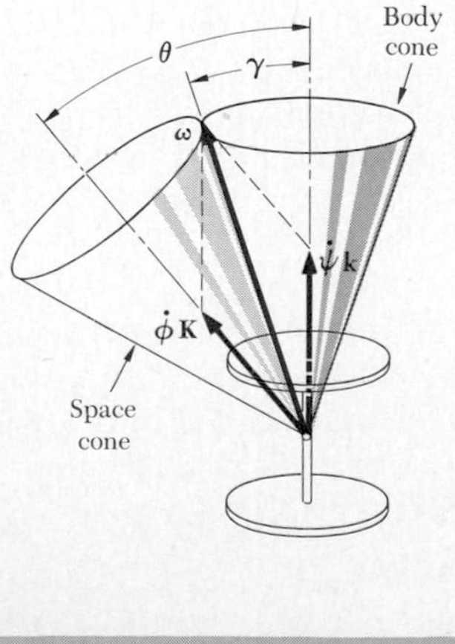

Precession Axis. Since in free motion the direction of the angular momentum $\mathbf{H}_G$ is fixed in space, the satellite will precess about this direction. The angle θ formed by the precession axis and the z axis is

$$\tan\theta = \frac{-H_x}{H_z} = \frac{m_0 v_0 a}{I\omega_0} = \frac{2m_0 v_0}{ma\omega_0} = 0.796 \qquad \theta = 38.5° \blacktriangleleft$$

Rates of Precession and Spin. We sketch the space and body cones for the free motion of the satellite. Using the law of sines, we compute the rates of precession and spin.

$$\frac{\omega}{\sin\theta} = \frac{\dot{\phi}}{\sin\gamma} = \frac{\dot{\psi}}{\sin(\theta - \gamma)}$$

$$\dot{\phi} = 30.8 \text{ rpm} \qquad \dot{\psi} = 35.9 \text{ rpm} \blacktriangleleft$$

Problems

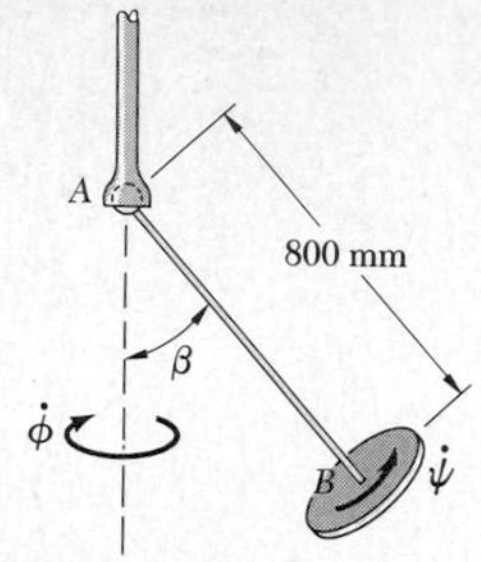

Fig. P18.82

18.82 A 2-kg disk of 200-mm diameter is attached to the end of a rod AB of negligible mass which is supported by a ball-and-socket joint at A. If the disk is observed to precess about the vertical in the sense indicated and at the constant rate of 24 rpm, determine the rate of spin $\dot{\psi}$ of the disk about AB when $\beta = 60°$.

18.83 Solve Prob. 18.82, assuming the same rate of precession and $\beta = 30°$.

18.84 The top shown weighs 3 oz and is supported at the fixed point O. The radii of gyration of the top with respect to its axis of symmetry and with respect to a transverse axis through O are 0.84 in. and 1.80 in., respectively. It is known that $c = 1.50$ in. and that the rate of spin of the top about its axis of symmetry is 1800 rpm. Determine the two possible rates of precession corresponding to $\theta = 30°$.

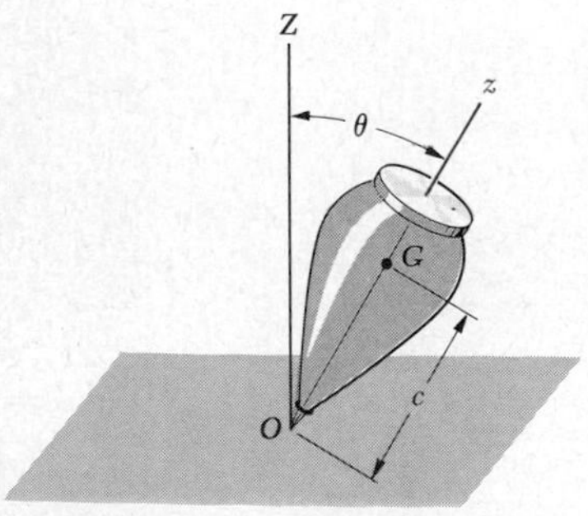

Fig. P18.84 and P18.85

18.85 The top shown is supported at the fixed point O and its moments of inertia about its axis of symmetry and about a transverse axis through O are denoted, respectively, by I and I'. (*a*) Show that the condition for steady precession of the top is

$$(I\omega_z - I'\dot{\phi}\cos\theta)\dot{\phi} = Wc$$

where $\dot{\phi}$ is the rate of precession and ω_z is the component of the angular velocity along the axis of symmetry of the top. (*b*) Show that if the rate of spin $\dot{\psi}$ of the top is very large compared with its rate of precession $\dot{\phi}$, the condition for steady precession is $I\dot{\psi}\dot{\phi} \approx Wc$. (*c*) Determine the percent error introduced when the slower of the two rates of precession obtained for the top of Prob. 18.84 is approximated by this last relation.

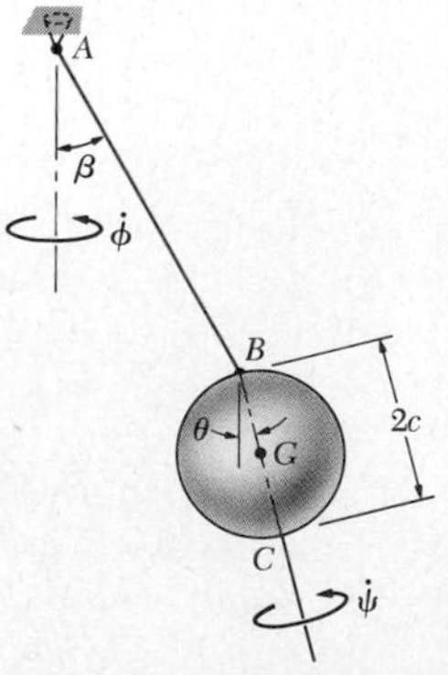

Fig. P18.86 and P18.87

18.86 A homogeneous sphere of radius $c = 30$ mm is attached as shown to a cord AB. The cord forms an angle $\beta = 30°$ with the vertical and is observed to precess at the constant rate $\dot{\phi} = 6$ rad/s about the vertical through A. Determine the angle θ that the diameter BC forms with the vertical, knowing that the sphere (*a*) has no spin, (*b*) spins about its diameter BC at the rate $\dot{\psi} = 30$ rad/s, (*c*) spins about BC at the rate $\dot{\psi} = -30$ rad/s.

18.87 A homogeneous sphere of radius $c = 30$ mm is attached as shown to a cord AB. The cord forms an angle β with the vertical and precesses at the constant rate $\dot{\phi}$, while the sphere spins at the constant rate $\dot{\psi}$ about its diameter BC. Determine the angle θ that BC forms with the vertical.

18.88 A homogeneous cone of height h and with a base of diameter $d < h$ is attached as shown to a cord AB. The cone spins about its axis BC at the constant rate $\dot{\psi}$ and precesses about the vertical through A at the constant rate $\dot{\phi}$. Determine the angle β for which the axis BC of the cone is aligned with the cord AB $(\theta = \beta)$.

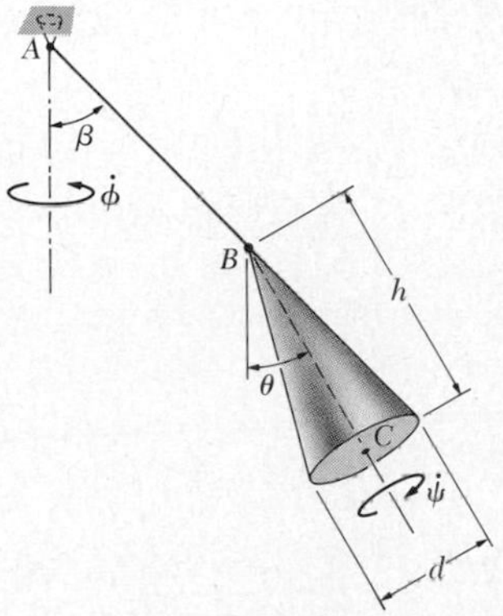

Fig. P18.88 and P18.89

18.89 A homogeneous cone of height $h = 12$ in. and base diameter $d = 6$ in. is attached as shown to a cord AB. Knowing that the angles that the cord AB and the axis BC of the cone form with the vertical are, respectively, $\beta = 45°$ and $\theta = 30°$, and that the cone precesses at the constant rate $\dot{\phi} = 8$ rad/s in the sense indicated, determine (*a*) the rate of spin $\dot{\psi}$ of the cone about its axis BC, (*b*) the length of cord AB.

18.90 If the earth were a sphere, the gravitational attraction of the sun, moon, and planets would at all times be equivalent to a single force **R** acting at the mass center of the earth. However, the earth is actually an oblate spheroid and the gravitational system acting on the earth is equivalent to a force **R** and a couple **M**. Knowing that the effect of the couple **M** is to cause the axis of the earth to precess about the axis GA at the rate of one revolution in 25,800 years, determine the average magnitude of the couple **M** applied to the earth. Assume that the average density of the earth is 5.51, that the average radius of the earth is 3960 mi, and that $\bar{I} = \frac{2}{5}mR^2$. (*Note.* This forced precession is known as the precession of the equinoxes and is not to be confused with the free precession discussed in Prob. 18.98.)

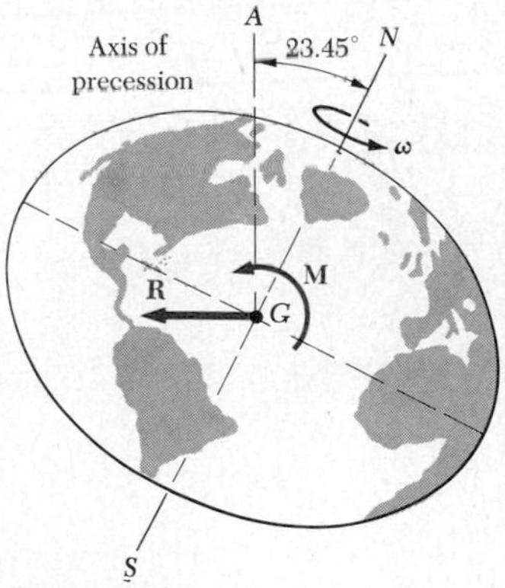

Fig. P18.90

18.91 A high-speed photographic record shows that a certain projectile was fired with a horizontal velocity $\bar{\mathbf{v}}$ of 600 m/s and with its axis of symmetry forming an angle $\beta = 3°$ with the horizontal. The rate of spin $\dot{\psi}$ of the projectile was 6000 rpm, and the atmospheric drag was equivalent to a force **D** of 120 N acting at the center of pressure C_P located at a distance $c = 150$ mm from G. (*a*) Knowing that the projectile has a mass of 20 kg and a radius of gyration of 50 mm with respect to its axis of symmetry, determine its approximate rate of steady precession. (*b*) If it is further known that the radius of gyration of the projectile with respect to a transverse axis through G is 200 mm, determine the exact values of the two possible rates of precession.

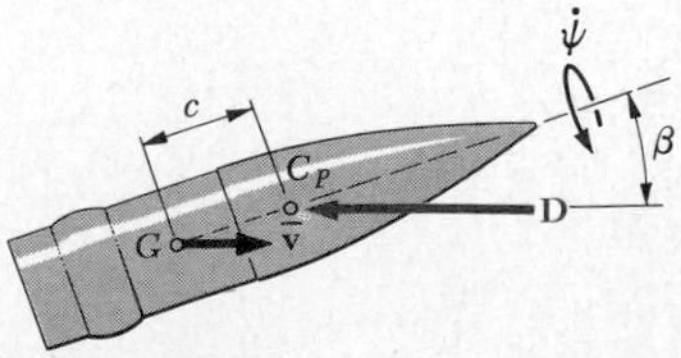

Fig. P18.91

18.92 The essential features of the gyrocompass are shown. The rotor spins at the rate $\dot{\psi}$ about an axis mounted in a single gimbal, which may rotate freely about the vertical axis AB. The angle formed by the axis of the rotor and the plane of the meridian is denoted by θ and the latitude of the position on the earth is denoted by λ. We note that the line OC is parallel to the axis of the earth and we denote by $\boldsymbol{\omega}_e$ the angular velocity of the earth about its axis.

(*a*) Show that the equations of motion of the gyrocompass are

$$I'\ddot{\theta} + I\omega_z\omega_e \cos\lambda \sin\theta - I'\omega_e^2 \cos^2\lambda \sin\theta \cos\theta = 0$$

$$I\dot{\omega}_z = 0$$

where ω_z is the component of the total angular velocity along the axis of the rotor, and I and I' are the moments of inertia of the rotor with respect to its axis of symmetry and a transverse axis through O, respectively.

(*b*) Neglecting the term containing ω_e^2, show that for small values of θ, we have

$$\ddot{\theta} + \frac{I\omega_z\omega_e \cos\lambda}{I'}\theta = 0$$

and that the axis of the gyrocompass oscillates about the north-south direction.

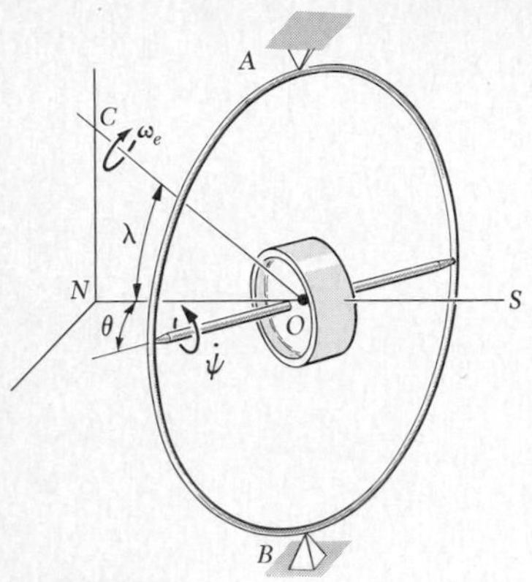

Fig. P18.92

18.93 Show that for an axisymmetrical body under no force, the rates of precession and spin may be expressed, respectively, as

$$\dot{\phi} = \frac{H_G}{I'}$$

and

$$\dot{\psi} = \frac{H_G \cos\theta\,(I' - I)}{II'}$$

where H_G is the constant value of the angular momentum of the body.

18.94 (*a*) Show that for an axisymmetrical body under no force, the rate of precession may be expressed as

$$\dot{\phi} = \frac{I\omega_z}{I' \cos\theta}$$

where ω_z is the component of $\boldsymbol{\omega}$ along the axis of symmetry of the body. (*b*) Use this result to check that the condition (18.44) for steady precession is satisfied by an axisymmetrical body under no force.

18.95 Show that the angular velocity vector $\boldsymbol{\omega}$ of an axisymmetrical body under no force is observed from the body itself to rotate about the axis of symmetry at the constant rate

$$n = \frac{I' - I}{I'}\omega_z$$

where ω_z is the component of $\boldsymbol{\omega}$ along the axis of symmetry of the body.

18.96 For an axisymmetrical body under no force, prove (*a*) that the rate of retrograde precession can never be less than twice the rate of spin of the body about its axis of symmetry, (*b*) that in Fig. 18.24 the axis of symmetry of the body can never lie within the space cone.

18.97 A coin is tossed into the air. During the free motion the angle β between the plane of the coin and the horizontal is observed to be constant. (*a*) Derive an expression for the angle formed by the angular velocity of the coin and the vertical. (*b*) Denoting by $\dot{\psi}$ the rate of spin of the coin about its axis of symmetry, derive an expression for the rate of precession. (*c*) Solve parts *a* and *b* for the case $\beta = 15°$.

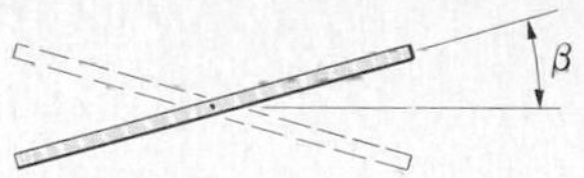

Fig. P18.97

18.98 Using the relation given in Prob. 18.95, determine the period of precession of the north pole of the earth about the axis of symmetry of the earth. The earth may be approximated by an oblate spheroid of axial moment of inertia I and of transverse moment of inertia $I' = 0.9967I$. (*Note.* Actual observations show a period of precession of the north pole of about 432.5 mean solar days; the difference between the observed and computed periods is due to the fact that the earth is not a perfectly rigid body. The free precession considered here should not be confused with the much slower precession of the equinoxes, which is a forced precession. See Prob. 18.90.)

18.99 Determine the precession axis and the rates of precession and spin of the satellite of Prob. 18.30 after the impact, knowing that before the impact the satellite was spinning about its axis of symmetry with the angular velocity $\boldsymbol{\omega}_0 = -(1.25 \text{ rad/s})\mathbf{i}$.

18.100 Determine the precession axis and the rates of precession and spin of the satellite of Prob. 18.30 after the impact.

18.101 An 800-lb geostationary satellite is spinning with the angular velocity $\boldsymbol{\omega}_0 = (1.5 \text{ rad/s})\mathbf{j}$ when it is hit at B by a 6-oz meteorite traveling with the velocity $\mathbf{v}_0 = -(1600 \text{ ft/s})\mathbf{i} + (1300 \text{ ft/s})\mathbf{j} + (4000 \text{ ft/s})\mathbf{k}$ relative to the satellite. Knowing that $b = 20$ in. and that the radii of gyration of the satellite are $\bar{k}_x = \bar{k}_z = 28.8$ in. and $\bar{k}_y = 32.4$ in., determine the precession axis and the rates of precession and spin of the satellite after the impact.

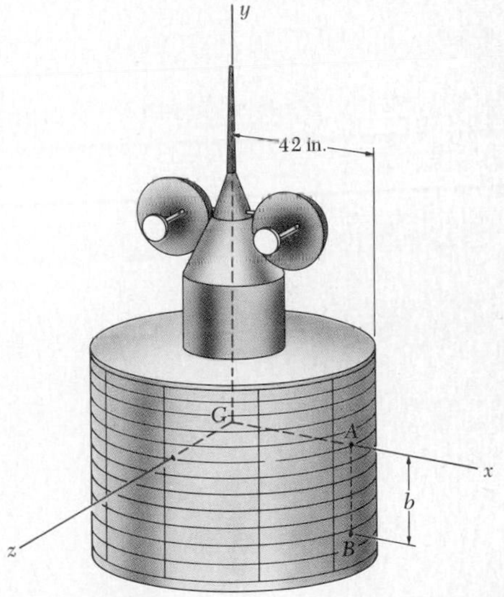

Fig. P18.101

18.102 Solve Prob. 18.101, assuming that the meteorite hits the satellite at A instead of B.

18.103 Solve Sample Prob. 18.6, assuming that the meteorite strikes the satellite at C with a velocity $\mathbf{v}_0 = +(2000 \text{ m/s})\mathbf{i}$.

18.104 After the motion determined in Sample Prob. 18.6 has been established, the rod connecting disks A and B of the satellite breaks, and disk A moves freely as a separate body. Knowing that the rod and the z axis coincide when the rod breaks, determine the precession axis, the rate of precession, and the rate of spin for the ensuing motion of disk A.

18.105 The angular velocity vector of a football which has just been kicked is horizontal, and its axis of symmetry OC is oriented as shown. Knowing that the magnitude of the angular velocity is 200 rpm and that the ratio of the axial and transverse moments of inertia is $I/I' = \frac{1}{3}$, determine (*a*) the orientation of the axis of precession OA, (*b*) the rates of precession and spin.

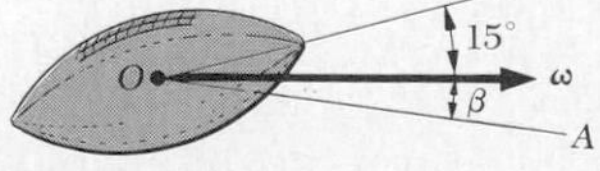

Fig. P18.105

18.106 The slender homogeneous rod AB of mass m and length L is free to rotate about a horizontal axle through its mass center G. The axle is supported by a frame of negligible mass which is free to rotate about the vertical CD. Knowing that, initially, $\theta = \theta_0$, $\dot{\theta} = 0$, and $\dot{\phi} = \dot{\phi}_0$, show that the rod will oscillate about the horizontal axle and determine (*a*) the range of values of the angle θ during this motion, (*b*) the maximum value of $\dot{\theta}$, (*c*) the minimum value of $\dot{\phi}$.

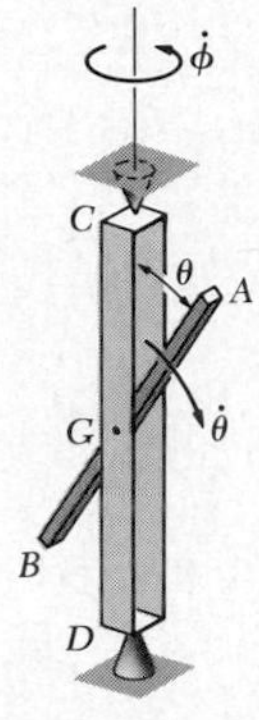

Fig. P18.106

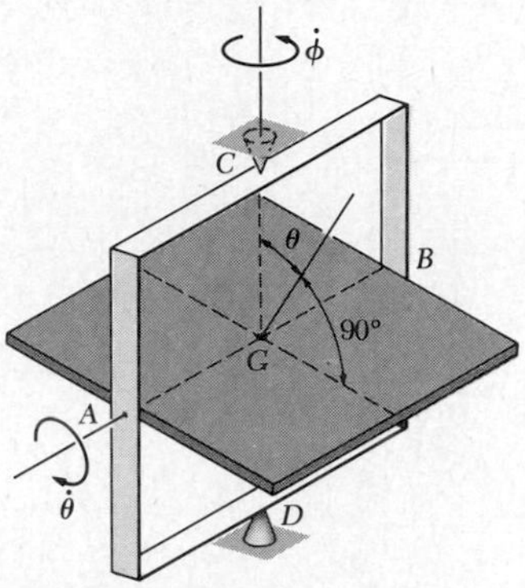

Fig. P18.107

18.107 Solve Prob. 18.106, assuming that the rod is replaced by a thin homogeneous square plate of mass m and side a.

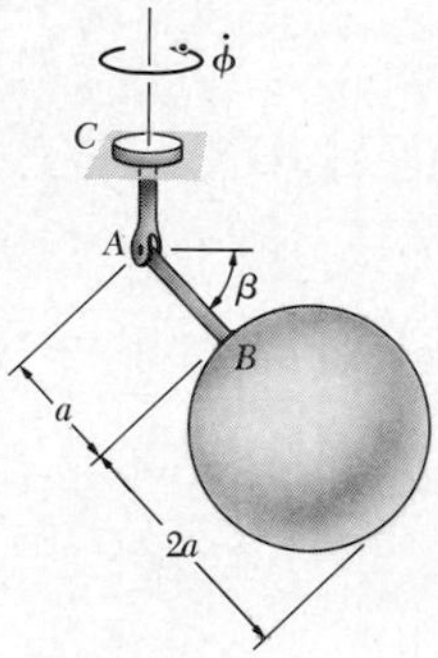

Fig. P18.108

18.108 A homogeneous sphere of mass m and radius a is welded to a rod AB of negligible mass, which is connected by a clevis to the vertical shaft AC. The rod and sphere can rotate freely about a horizontal axis at A, and the shaft AC can rotate freely about a vertical axis. The system is released in the position $\beta = 0$ with an angular velocity $\dot{\phi}_0$ about the vertical axis and no angular velocity about the horizontal axis. Knowing that the largest value of β in the ensuing motion is 30°, determine (*a*) the initial angular velocity $\dot{\phi}_0$, (*b*) the value of $\dot{\phi}$ when $\beta = 30°$.

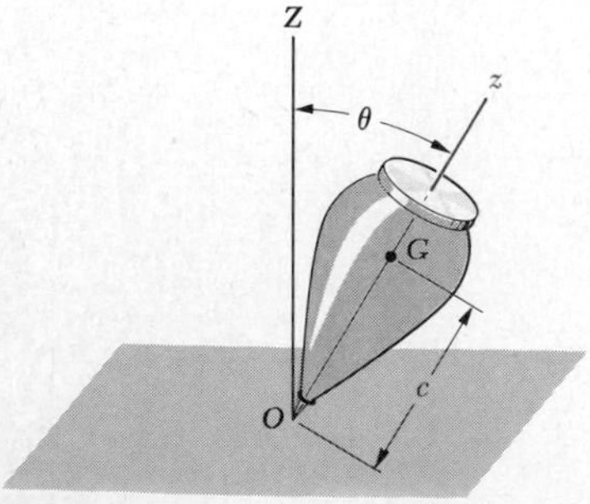

Fig. P18.109

*__18.109__ The top shown is supported at the fixed point O. We denote by ϕ, θ, and ψ the Eulerian angles defining the position of the top with respect to a fixed frame of reference. We shall consider the general motion of the top in which all Eulerian angles vary.

(*a*) Observing that $\Sigma M_Z = 0$ and $\Sigma M_z = 0$, and denoting by I and I', respectively, the moments of inertia of the top about its axis of symmetry and about a transverse axis through O, derive the two first-order differential equations of motion

$$I'\dot{\phi}\sin^2\theta + I(\dot{\psi} + \dot{\phi}\cos\theta)\cos\theta = \alpha \quad (1)$$

$$I(\dot{\psi} + \dot{\phi}\cos\theta) = \beta \quad (2)$$

where α and β are constants depending upon the initial conditions. These equations express that the angular momentum of the top is conserved about both the Z and z axes, i.e., that the rectangular component of $\mathbf{H}_O$ along each of these axes is constant.

(*b*) Use Eqs. (1) and (2) to show that the component ω_z of the angular velocity of the top is constant and that the rate of precession $\dot{\phi}$ depends upon the value of the angle of nutation θ.

***18.110** (*a*) Applying the principle of conservation of energy, derive a third differential equation for the general motion of the top of Prob. 18.109. (*b*) Eliminating the derivatives $\dot{\phi}$ and $\dot{\psi}$ from the equation obtained and from the two equations of Prob. 18.109, express the rate of nutation $\dot{\theta}$ as a function of the angle θ.

***18.111** A thin homogeneous disk is mounted on a light axle OA which is held by a ball-and-socket support at O. The disk is released in the position $\beta = 0$ with a rate of spin $\dot{\psi}_0$, clockwise as viewed from O, and with no precession or nutation. Knowing that the largest value of β in the ensuing motion is $\beta = 30°$, determine (*a*) the rate of spin $\dot{\psi}_0$ of the disk in its initial position, (*b*) the rates of precession and spin as the disk passes through its lowest position. (*Hint.* Use the principle of conservation of energy and the fact that the angular momentum of the body is conserved about both the Z and z axes; see Prob. 18.109, part *a*.)

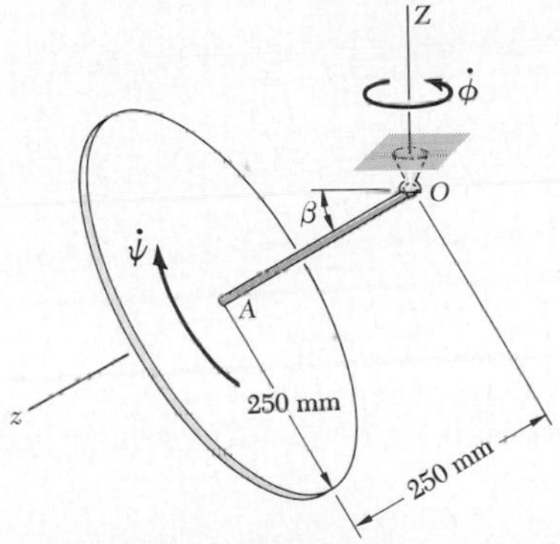

Fig. P18.111 and P18.112

***18.112** A thin homogeneous disk is mounted on a light axle OA which is held by a ball-and-socket support at O. The disk is released in the position $\beta = 0$ with a rate of spin $\dot{\psi}_0 = 15$ rad/s, clockwise as viewed from O, and with no precession or nutation. Determine (*a*) the largest value of β in the ensuing motion, (*b*) the rates of precession and spin as the disk passes through its lowest position. (See hint of Prob. 18.111.)

***18.113** A homogeneous sphere of mass m and radius a is welded to a rod AB of negligible mass, which is held by a ball-and-socket support at A. The sphere is released in the position $\beta = 0$ with a rate of precession $\dot{\phi}_0 = \sqrt{17g/11a}$ and with no spin or nutation. Determine the largest value of β in the ensuing motion. (See hint of Prob. 18.111.)

***18.114** A homogeneous sphere of mass m and radius a is welded to a rod AB of length a and negligible mass, which is held by a ball-and-socket support at A. The sphere is released in the position $\beta = 0$ with a rate of precession $\dot{\phi} = \dot{\phi}_0$ and with no spin or nutation. Knowing that the largest value of β in the ensuing motion is 30°, determine (*a*) the rate of precession $\dot{\phi}_0$ of the sphere in its initial position, (*b*) the rates of precession and spin when $\beta = 30°$. (See hint of Prob. 18.111.)

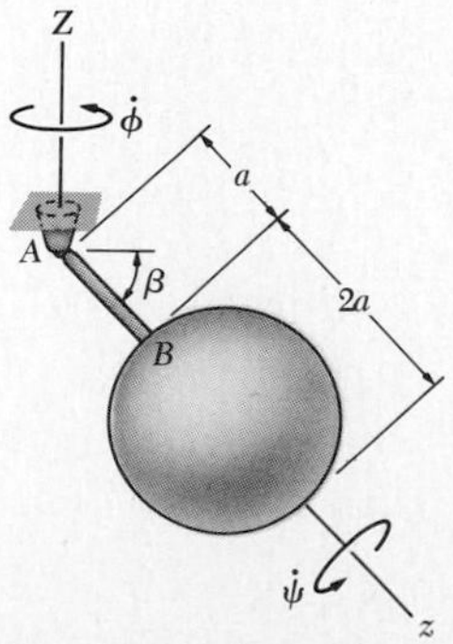

Fig. P18.113 and P18.114

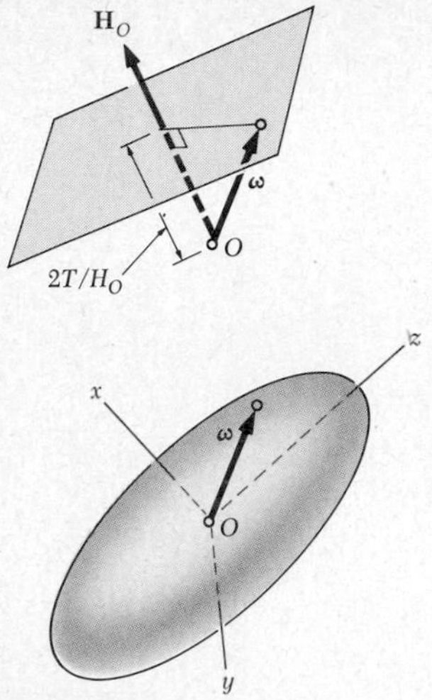

Fig. P18.115

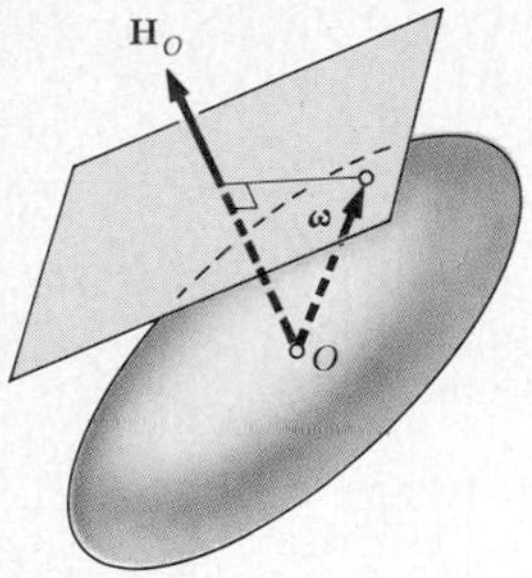

Fig. P18.116

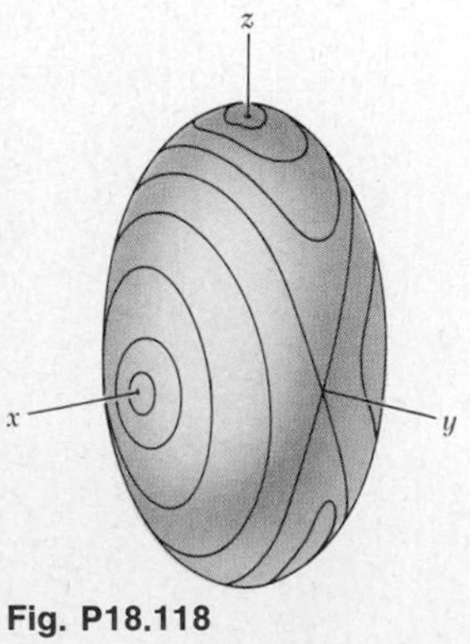

Fig. P18.118

*** 18.115** Consider a rigid body of arbitrary shape which is attached at its mass center O and subjected to no force other than its weight and the reaction of the support at O. (*a*) Prove that the angular momentum $\mathbf{H}_O$ of the body about the fixed point O is constant in magnitude and direction, that the kinetic energy T of the body is constant, and that the projection along $\mathbf{H}_O$ of the angular velocity $\boldsymbol{\omega}$ of the body is constant. (*b*) Show that the tip of the vector $\boldsymbol{\omega}$ describes a curve on a fixed plane in space (called the *invariable plane*), which is perpendicular to $\mathbf{H}_O$ and at a distance $2T/H_O$ from O. (*c*) Show that with respect to a frame of reference attached to the body and coinciding with its principal axes of inertia, the tip of the vector $\boldsymbol{\omega}$ appears to describe a curve on an ellipsoid of equation

$$I_x\omega_x^2 + I_y\omega_y^2 + I_z\omega_z^2 = 2T = \text{constant}$$

This ellipsoid (called the *Poinsot ellipsoid*) is rigidly attached to the body and is of the same shape as the ellipsoid of inertia, but of a different size.

*** 18.116** Referring to Prob. 18.115, (*a*) prove that the Poinsot ellipsoid is tangent to the invariable plane, (*b*) show that the motion of the rigid body must be such that the Poinsot ellipsoid appears to roll on the invariable plane. [*Hint.* In part *a*, show that the normal to the Poinsot ellipsoid at the tip of $\boldsymbol{\omega}$ is parallel to $\mathbf{H}_O$. It is recalled that the direction of the normal to a surface of equation $F(x, y, z) =$ constant at a point P is the same as that of **grad** F at point P.]

*** 18.117** Using the results obtained in Probs. 18.115 and 18.116, show that for an axisymmetrical body attached at its mass center O and under no force other than its weight and the reaction at O, the Poinsot ellipsoid is an ellipsoid of revolution and the space and body cones are both circular and are tangent to each other. Further show that (*a*) the two cones are tangent externally, and the precession is direct, when $I < I'$, where I and I' denote, respectively, the axial and transverse moment of inertia of the body, (*b*) the space cone is inside the body cone, and the precession is retrograde, when $I > I'$.

*** 18.118** Referring to Probs. 18.115 and 18.116, (*a*) show that the curve (called *polhode*) described by the tip of the vector $\boldsymbol{\omega}$ with respect to a frame of reference coinciding with the principal axes of inertia of the rigid body is defined by the equations

$$I_x\omega_x^2 + I_y\omega_y^2 + I_z\omega_z^2 = 2T = \text{constant} \tag{1}$$

$$I_x^2\omega_x^2 + I_y^2\omega_y^2 + I_z^2\omega_z^2 = H_O^2 = \text{constant} \tag{2}$$

and that this curve may, therefore, be obtained by intersecting the Poinsot ellipsoid with the ellipsoid defined by Eq. (2). (*b*) Further show, assuming $I_x > I_y > I_z$, that the polhodes obtained for various values of H_O have the shapes indicated in the figure. (*c*) Using the result obtained in part *b*, show that a rigid body under no force can rotate about a fixed centroidal axis if, and only if, that axis coincides with one of the principal axes of inertia of the body, and that the motion will be stable if the axis of rotation coincides with the major or minor axis of the Poinsot ellipsoid (z or x axis in the figure) and unstable if it coincides with the intermediate axis (y axis).

Review Problems

18.119 Two L-shaped arms, each weighing 6 lb, are welded at the third points of the 3-ft shaft AB. Knowing that shaft AB rotates at a constant rate $\omega = 300$ rpm, determine (*a*) the angular momentum of the body about A, (*b*) the angle formed by the angular momentum and the shaft AB.

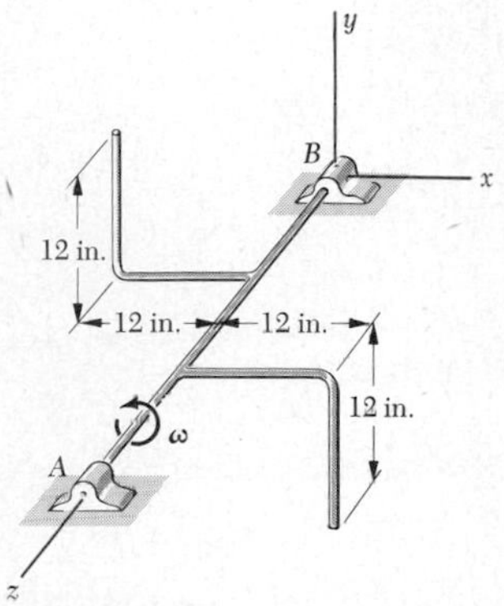

Fig. P18.119

18.120 A thin homogeneous disk of mass 600 g and radius 100 mm rotates at a constant rate $\omega_2 = 30$ rad/s with respect to the arm ABC, which itself rotates at a constant rate $\omega_1 = 15$ rad/s about the x axis. Determine the angular momentum of the disk about point C.

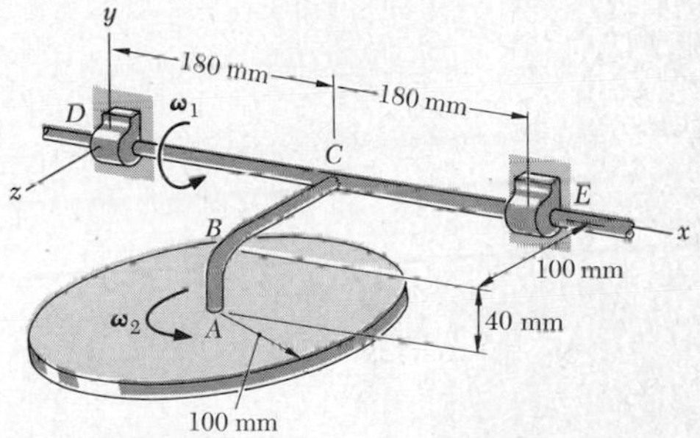

Fig. P18.120

18.121 Three slender homogeneous rods, each of mass m and length d, are welded together to form the assembly shown, which hangs from a wire at G. The assembly is hit at A in a vertical downward direction. Denoting the corresponding impulse by $\mathbf{F}\,\Delta t$, determine immediately after impact (*a*) the velocity of the mass center G, (*b*) the angular velocity of the assembly.

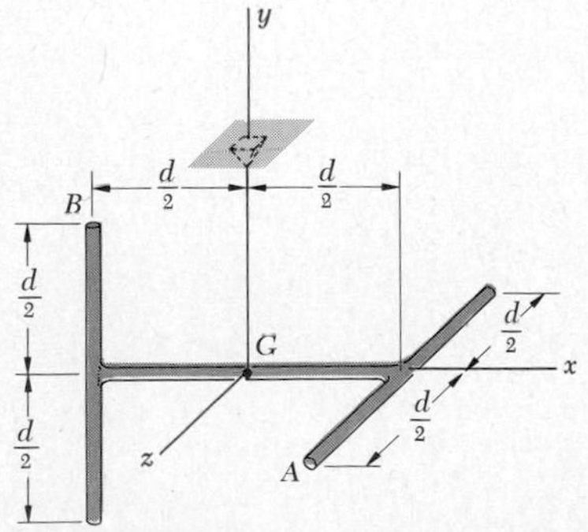

Fig. P18.121

18.122 Solve Prob. 18.121, assuming that the assembly is hit at B in a direction opposite to that of the z axis.

18.123 Gear A rolls on the fixed gear B and rotates about the axle AD of length $L = 500$ mm which is rigidly attached at D to the vertical shaft DE. The shaft DE is made to rotate with a constant angular velocity $\boldsymbol{\omega}_1$ of magnitude 4 rad/s. Assuming that gear A can be approximated by a thin disk of mass 2 kg and radius $a = 100$ mm, and that $\beta = 30°$, determine (*a*) the angular momentum of gear A about point D, (*b*) the kinetic energy of gear A.

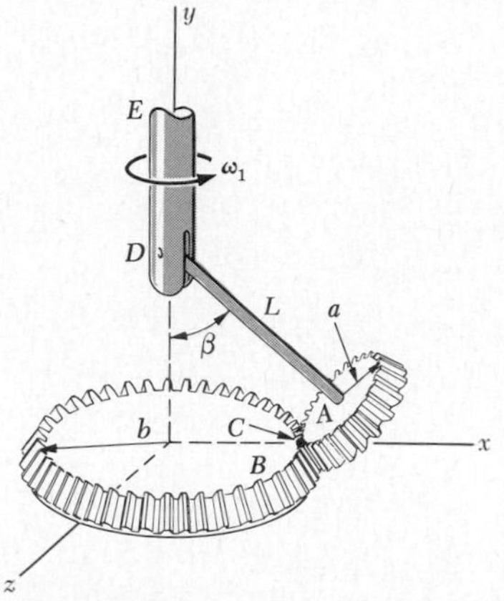

Fig. P18.123

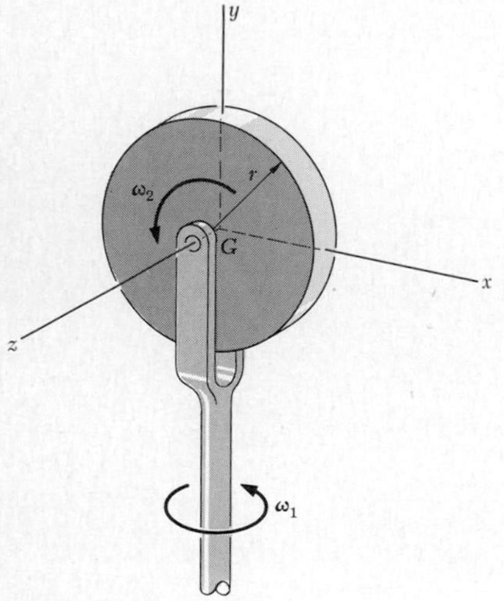

Fig. P18.124

18.124 A thin homogeneous disk of mass m and radius r spins at the constant rate ω_2 about a horizontal axle held by a fork-ended vertical rod which rotates at the constant rate ω_1. Determine the rate of change $\dot{\mathbf{H}}_G$ of the angular momentum of the disk about its mass center.

18.125 Each element of the crankshaft shown is a homogeneous rod of weight w per unit length. Knowing that the crankshaft rotates with a constant angular velocity $\boldsymbol{\omega}$, determine the dynamic reactions at A and B.

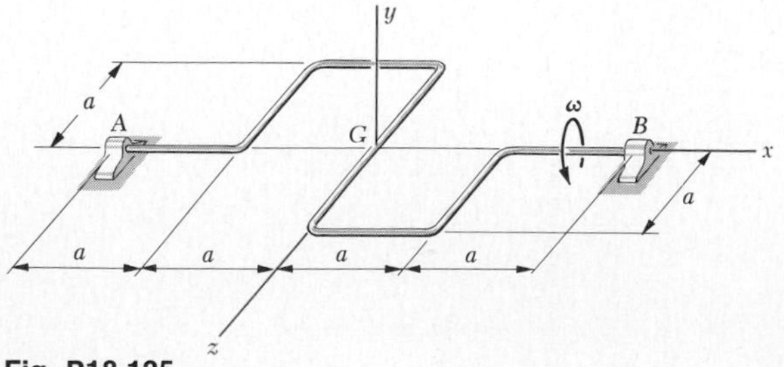

Fig. P18.125

18.126 The shaft of Prob. 18.125 is initially at rest ($\omega = 0$) and is accelerated at the rate $\alpha = \dot{\omega} = 80$ rad/s^2. Knowing that $w = 5$ lb/ft and $a = 4$ in., determine (*a*) the couple $\mathbf{M}$ required to cause the acceleration, (*b*) the corresponding dynamic reactions at A and B.

18.127 A homogeneous sphere of radius a and mass m is attached to a light rod of length $4a$. The rod forms an angle of 30° with the vertical and rotates about AC at the constant rate $\Omega = \sqrt{g/a}$. (a) Assuming that the sphere does not spin about the rod ($\dot{\psi} = 0$), determine the tension in the cord BC and the kinetic energy of the sphere. (b) Determine the spin $\dot{\psi}$ (magnitude and sense) which should be given to the sphere if the tension in the cord BC is to be zero. What is the corresponding kinetic energy of the sphere?

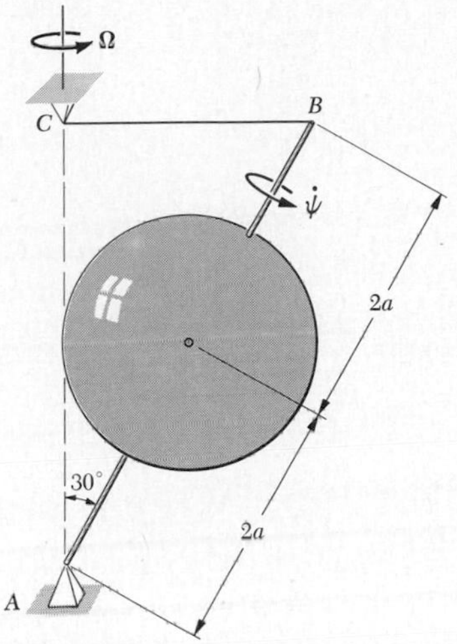

Fig. P18.127

18.128 The space station shown is known to precess about the fixed direction OC at the rate of one revolution per hour. Assuming that the station is dynamically equivalent to a homogeneous cylinder of length 30 m and radius 3 m, determine the rate of spin of the station about its axis of symmetry.

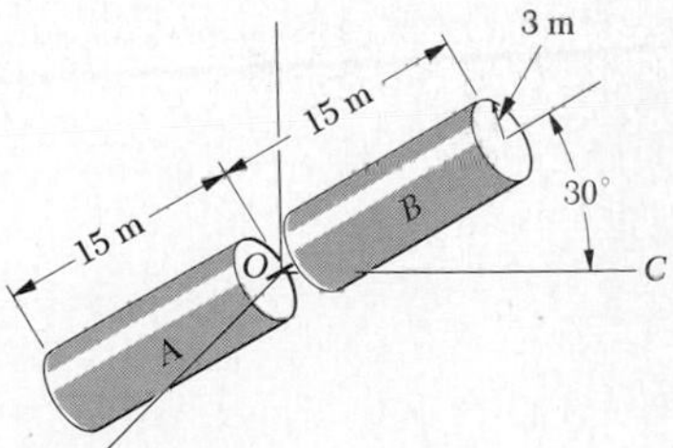

Fig. P18.128

18.129 The link connecting portions A and B of the space station of Prob. 18.128 may be severed to allow each portion to move freely. Each portion of the station is dynamically equivalent to a cylinder of length 15 m and radius 3 m. Knowing that when the link is severed, the station is oriented as shown, determine for portion B the axis of precession, the rate of precession, and the rate of spin about the axis of symmetry.

18.130 A slender homogeneous rod OA of mass m and length L is supported by a ball-and-socket joint at O and may swing freely under its own weight. If the rod is held in a horizontal position ($\theta = 90°$) and given an initial angular velocity $\dot{\phi}_0 = \sqrt{8g/L}$ about the vertical OB, determine (a) the smallest value of θ in the ensuing motion, (b) the corresponding value of the angular velocity $\dot{\phi}$ of the rod about OB. (*Hint.* Apply the principle of conservation of energy and the principle of impulse and momentum, observing that since $\Sigma M_{OB} = 0$, the component of $\mathbf{H}_O$ along OB must be constant.)

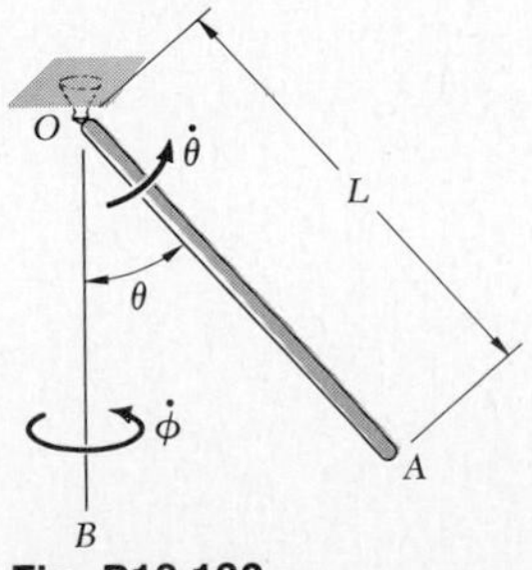

Fig. P18.130

CHAPTER NINETEEN

Mechanical Vibrations

19.1. Introduction. A *mechanical vibration* is the motion of a particle or a body which oscillates about a position of equilibrium. Most vibrations in machines and structures are undesirable because of the increased stresses and energy losses which accompany them. They should therefore be eliminated or reduced as much as possible by appropriate design. The analysis of vibrations has become increasingly important in recent years owing to the current trend toward higher-speed machines and lighter structures. There is every reason to expect that this trend will continue and that an even greater need for vibration analysis will develop in the future.

The analysis of vibrations is a very extensive subject to which entire texts have been devoted. We shall therefore limit our present study to the simpler types of vibrations, namely, the vibrations of a body or a system of bodies with one degree of freedom.

A mechanical vibration generally results when a system is displaced from a position of stable equilibrium. The system tends to return to this position under the action of restoring forces (either elastic forces, as in the case of a mass attached to a spring, or gravitational forces, as in the case of a pendulum). But the system generally reaches its original position with a certain acquired velocity which carries it beyond that position. Since the process can be repeated indefinitely, the system keeps moving back and forth across its position of equilibrium. The time interval required for the system to complete a full cycle of motion is called the *period* of the vibration. The number of cycles per unit time defines the *frequency*, and the maximum displacement of the system from its position of equilibrium is called the *amplitude* of the vibration.

When the motion is maintained by the restoring forces only, the vibration is said to be a *free vibration* (Secs. 19.2 to 19.6). When a periodic force is applied to the system, the resulting motion is described as a *forced vibration* (Sec. 19.7). When the effects of friction may be neglected, the vibrations are said to be *undamped*. However, all vibrations are actually *damped* to some degree. If a free vibration is only slightly damped, its amplitude slowly decreases until, after a certain time, the motion comes to a stop. But damping may be large enough to prevent any true vibration; the system then slowly regains its original position (Sec. 19.8). A damped forced vibration is maintained as long as the periodic force which produces the vibration is applied. The amplitude of the vibration, however, is affected by the magnitude of the damping forces (Sec. 19.9).

VIBRATIONS WITHOUT DAMPING

19.2. Free Vibrations of Particles. Simple Harmonic Motion. Consider a body of mass m attached to a spring of constant k (Fig. 19.1*a*). Since at the present time we are concerned only with the motion of its mass center, we shall refer to this body as a particle. When the particle is in static equilibrium, the forces acting on it are its weight **W** and the force **T** exerted by the spring, of magnitude $T = k\delta_{st}$, where δ_{st} denotes the elongation of the spring. We have, therefore,

$$W = k\delta_{st} \tag{19.1}$$

Suppose now that the particle is displaced through a distance x_m from its equilibrium position and released with no initial velocity. If x_m has been chosen smaller than δ_{st}, the particle will move back and forth through its equilibrium position; a vibration of amplitude x_m has been generated. Note that the vibration may also be produced by imparting a certain initial velocity to the particle when it is in its equilibrium position $x = 0$ or, more generally, by starting the particle from any given position $x = x_0$ with a given initial velocity $\mathbf{v}_0$.

To analyze the vibration, we shall consider the particle in a position P at some arbitrary time t (Fig. 19.1*b*). Denoting by x the displacement OP measured from the equilibrium position O (positive downward), we note that the forces acting on the particle are its weight **W** and the force **T** exerted by the spring which, in this position, has a magnitude $T = k(\delta_{st} + x)$. Recalling (19.1), we find that the magnitude of the resultant **F** of the two forces (positive downward) is

$$F = W - k(\delta_{st} + x) = -kx \tag{19.2}$$

Thus the *resultant* of the forces exerted on the particle is proportional to the displacement *OP measured from the equilibrium position.* Recalling the sign convention, we note that **F** is always directed *toward* the equilibrium position O. Substituting for F into the fundamental equation $F = ma$ and recalling that a is the second derivative $\ddot{x}$ of x with respect to t, we write

$$m\ddot{x} + kx = 0 \tag{19.3}$$

Note that the same sign convention should be used for the acceleration $\ddot{x}$ and for the displacement x, namely, positive downward.

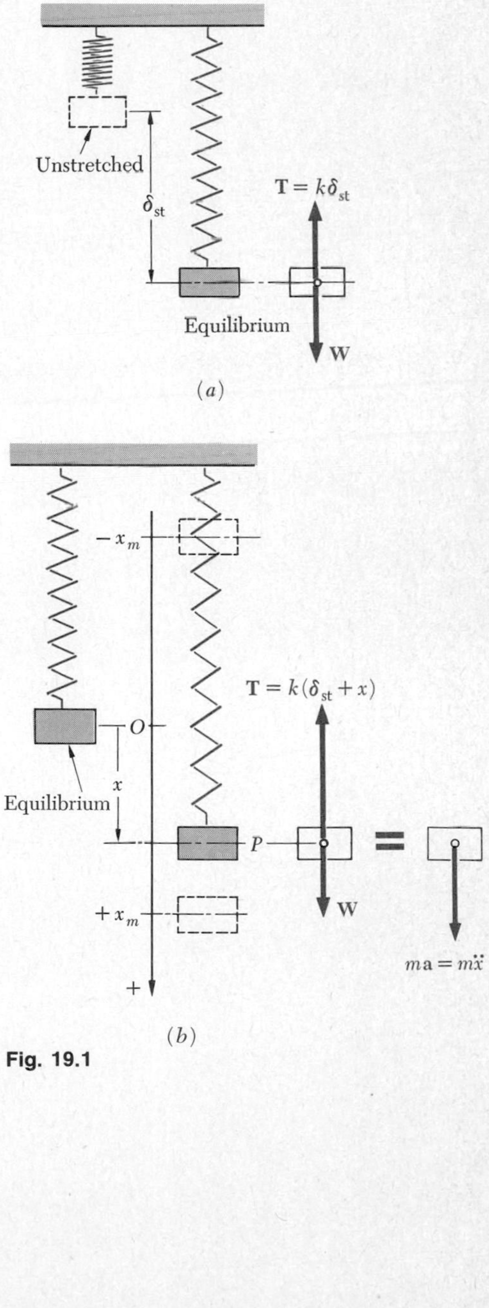

Fig. 19.1

Equation (19.3) is a linear differential equation of the second order. Setting

$$p^2 = \frac{k}{m} \tag{19.4}$$

we write (19.3) in the form

$$\ddot{x} + p^2 x = 0 \tag{19.5}$$

The motion defined by Eq. (19.5) is called *simple harmonic motion*. It is characterized by the fact that *the acceleration is proportional to the displacement and of opposite direction*. We note that each of the functions $x_1 = \sin pt$ and $x_2 = \cos pt$ satisfies (19.5). These functions, therefore, constitute two *particular solutions* of the differential equation (19.5). As we shall see presently, the *general solution* of (19.5) may be obtained by multiplying the two particular solutions by arbitrary constants A and B and adding. We write

$$x = Ax_1 + Bx_2 = A \sin pt + B \cos pt \tag{19.6}$$

Differentiating, we obtain successively the velocity and acceleration at time t:

$$v = \dot{x} = Ap \cos pt - Bp \sin pt \tag{19.7}$$

$$a = \ddot{x} = -Ap^2 \sin pt - Bp^2 \cos pt \tag{19.8}$$

Substituting from (19.6) and (19.8) into (19.5), we verify that the expression (19.6) provides a solution of the differential equation (19.5). Since this expression contains two arbitrary constants A and B, the solution obtained is the general solution of the differential equation. The values of the constants A and B depend upon the *initial conditions* of the motion. For example, we have $A = 0$ if the particle is displaced from its equilibrium position and released at $t = 0$ with no initial velocity, and we have $B = 0$ if P is started from O at $t = 0$ with a certain initial velocity. In general, substituting $t = 0$ and the initial values x_0 and v_0 of the displacement and velocity into (19.6) and (19.7), we find $A = v_0/p$ and $B = x_0$.

The expressions obtained for the displacement, velocity, and acceleration of a particle may be written in a more compact form if we observe that (19.6) expresses that the displacement $x = OP$ is the sum of the x components of two vectors **A** and **B**, respectively, of magnitude A and B, directed as shown in Fig. 19.2*a*. As t varies, both vectors rotate clockwise; we also note that the magnitude of their resultant $\overrightarrow{OQ}$ is equal to the maximum displacement x_m. The simple harmonic motion of P along the x axis may thus be obtained by projecting on this axis the motion of a point Q describing an *auxiliary circle* of radius x_m *with a constant angular velocity* p.

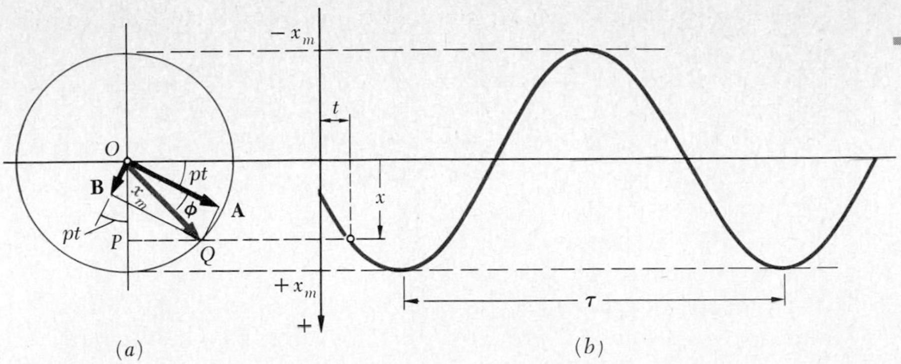

Fig. 19.2

Denoting by ϕ the angle formed by the vectors $\overrightarrow{OQ}$ and **A**, we write

$$OP = OQ \sin (pt + \phi) \tag{19.9}$$

which leads to new expressions for the displacement, velocity, and acceleration of P:

$$x = x_m \sin (pt + \phi) \tag{19.10}$$

$$v = \dot{x} = x_m p \cos (pt + \phi) \tag{19.11}$$

$$a = \ddot{x} = -x_m p^2 \sin (pt + \phi) \tag{19.12}$$

The displacement-time curve is represented by a sine curve (Fig. 19.2*b*), and the maximum value x_m of the displacement is called the *amplitude* of the vibration. The angular velocity p of the point Q which describes the auxiliary circle is known as the *circular frequency* of the vibration and is measured in rad/s, while the angle ϕ which defines the initial position of Q on the circle is called the *phase angle*. We note from Fig. 19.2 that a full *cycle* has been described after the angle pt has increased by 2π rad. The corresponding value of t, denoted by τ, is called the *period* of the vibration and is measured in seconds. We have

$$\text{Period} = \tau = \frac{2\pi}{p} \tag{19.13}$$

The number of cycles described per unit of time is denoted by f and is known as the *frequency* of the vibration. We write

$$\text{Frequency} = f = \frac{1}{\tau} = \frac{p}{2\pi} \tag{19.14}$$

The unit of frequency is a frequency of 1 cycle per second, corresponding to a period of 1 s. In terms of base units the unit of frequency is thus 1/s or s^{-1}. It is called a *hertz* (Hz) in the SI system of units. It also follows from Eq. (19.14) that a frequency of 1 s^{-1} or 1 Hz corresponds to a circular frequency of 2π rad/s. In problems involving angular velocities expressed in revolutions per minute (rpm), we have 1 rpm $= \frac{1}{60}\, s^{-1} = \frac{1}{60}$ Hz, or 1 rpm $= (2\pi/60)$ rad/s.

Recalling that p was defined in (19.4) in terms of the constant k of the spring and the mass m of the particle, we observe that the period and the frequency are independent of the initial conditions and of the amplitude of the vibration. Note that τ and f depend on the *mass* rather than on the *weight* of the particle and thus are independent of the value of g.

The velocity-time and acceleration-time curves may be represented by sine curves of the same period as the displacement-time curve, but with different phase angles. From (19.11) and (19.12), we note that the maximum values of the magnitudes of the velocity and acceleration are

$$v_m = x_m p \qquad a_m = x_m p^2 \tag{19.15}$$

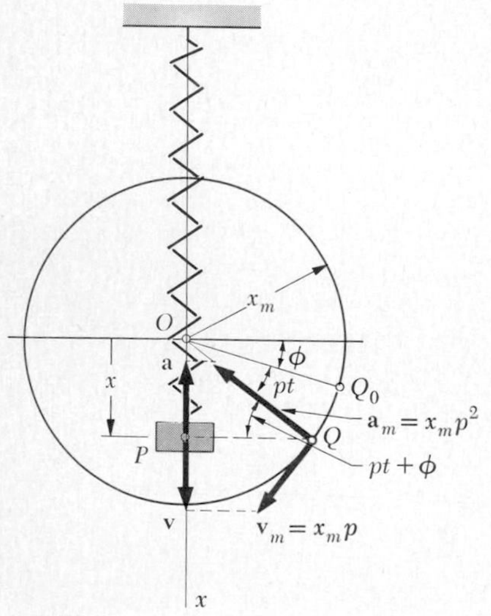

Fig. 19.3

Since the point Q describes the auxiliary circle, of radius x_m, at the constant angular velocity p, its velocity and acceleration are equal, respectively, to the expressions (19.15). Recalling Eqs. (19.11) and (19.12), we find, therefore, that the velocity and acceleration of P may be obtained at any instant by projecting on the x axis vectors of magnitudes $v_m = x_m p$ and $a_m = x_m p^2$ representing, respectively, the velocity and acceleration of Q at the same instant (Fig. 19.3).

The results obtained are not limited to the solution of the problem of a mass attached to a spring. They may be used to analyze the rectilinear motion of a particle *whenever the resultant* **F** *of the forces acting on the particle is proportional to the displacement x and directed toward O.* The fundamental equation of motion $F = ma$ may then be written in the form (19.5), which characterizes simple harmonic motion. Observing that the coefficient of x in (19.5) represents the square of the circular frequency p of the vibration, we easily obtain p and, after substitution into (19.13) and (19.14), the period τ and the frequency f of the vibration.

19.3. Simple Pendulum (Approximate Solution). Most of the vibrations encountered in engineering applications may be represented by a simple harmonic motion. Many others, although of a different type, may be *approximated* by a simple harmonic motion, provided that their amplitude remains small. Consider, for example, a *simple pendulum*, consisting of a bob of mass m attached to a cord of length l, which may oscillate in a vertical plane (Fig. 19.4*a*). At a given time t, the cord forms an angle θ with the vertical. The forces acting on the bob are its weight **W** and the force **T** exerted by the cord (Fig. 19.4*b*). Resolving the vector $m\mathbf{a}$ into tangential and normal components, with $m\mathbf{a}_t$ directed to the right, i.e., in

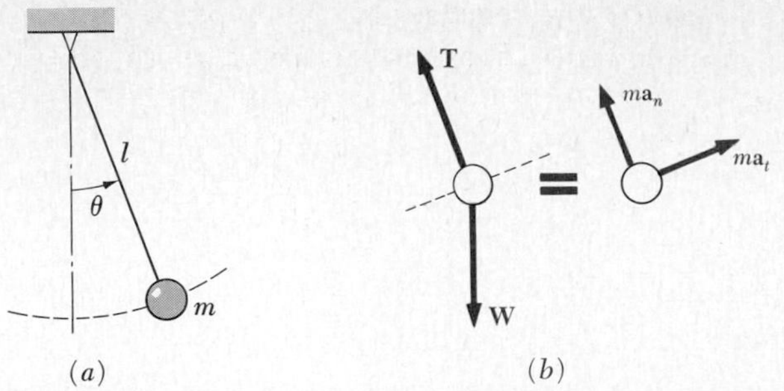

Fig. 19.4

the direction corresponding to increasing values of θ, and observing that $a_t = l\alpha = l\ddot{\theta}$, we write

$$\Sigma F_t = ma_t: \qquad -W \sin\theta = ml\ddot{\theta}$$

Noting that $W = mg$ and dividing through by ml, we obtain

$$\ddot{\theta} + \frac{g}{l}\sin\theta = 0 \tag{19.16}$$

For oscillations of small amplitude, we may replace $\sin\theta$ by θ, expressed in radians, and write

$$\ddot{\theta} + \frac{g}{l}\theta = 0 \tag{19.17}$$

Comparison with (19.5) shows that the equation obtained is that of a simple harmonic motion and that the circular frequency p of the oscillations is equal to $(g/l)^{1/2}$. Substitution into (19.13) yields the period of the small oscillations of a pendulum of length l:

$$\tau = \frac{2\pi}{p} = 2\pi\sqrt{\frac{l}{g}} \tag{19.18}$$

*** 19.4. Simple Pendulum (Exact Solution).** Formula (19.18) is only approximate. To obtain an exact expression for the period of the oscillations of a simple pendulum, we must return to (19.16). Multiplying both terms by $2\dot{\theta}$ and integrating from an initial position corresponding to the maximum deflection, that is, $\theta = \theta_m$ and $\dot{\theta} = 0$, we write

$$\dot{\theta}^2 = \frac{2g}{l}(\cos\theta - \cos\theta_m)$$

or

$$\left(\frac{d\theta}{dt}\right)^2 = \frac{2g}{l}(\cos\theta - \cos\theta_m)$$

Replacing $\cos\theta$ by $1 - 2\sin^2(\theta/2)$ and $\cos\theta_m$ by a similar expression, solving for dt, and integrating over a quarter period from $t = 0$, $\theta = 0$ to $t = \tau/4$, $\theta = \theta_m$, we have

$$\tau = 2\sqrt{\frac{l}{g}}\int_0^{\theta_m}\frac{d\theta}{\sqrt{\sin^2(\theta_m/2) - \sin^2(\theta/2)}}$$

The integral in the right-hand member is known as an *elliptic integral;* it cannot be expressed in terms of the usual algebraic or trigonometric functions. However, setting

$$\sin(\theta/2) = \sin(\theta_m/2)\sin\phi$$

we may write

$$\tau = 4\sqrt{\frac{l}{g}}\int_0^{\pi/2}\frac{d\phi}{\sqrt{1 - \sin^2(\theta_m/2)\sin^2\phi}} \tag{19.19}$$

where the integral obtained, commonly denoted by K, may be found in *tables of elliptic integrals* for various values of $\theta_m/2$.† In order to compare the result just obtained with that of the preceding section, we write (19.19) in the form

$$\tau = \frac{2K}{\pi}\left(2\pi\sqrt{\frac{l}{g}}\right) \tag{19.20}$$

Formula (19.20) shows that the actual value of the period of a simple pendulum may be obtained by multiplying the approximate value (19.18) by the correction factor $2K/\pi$. Values of the correction factor are given in

Table 19.1 Correction Factor for the Period of a Simple Pendulum

θ_m	0°	10°	20°	30°	60°	90°	120°	150°	180°
K	1.571	1.574	1.583	1.598	1.686	1.854	2.157	2.768	∞
$2K/\pi$	1.000	1.002	1.008	1.017	1.073	1.180	1.373	1.762	∞

Table 19.1 for various values of the amplitude θ_m. We note that for ordinary engineering computations the correction factor may be omitted as long as the amplitude does not exceed 10°.

† See, for example, "Standard Mathematical Tables," Chemical Rubber Publishing Company, Cleveland, Ohio.

SAMPLE PROBLEM 19.1

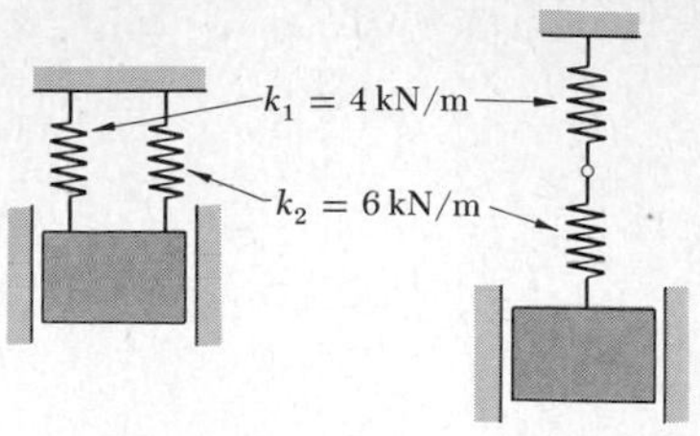

(a) (b)

A 50-kg block moves between vertical guides as shown. The block is pulled 40 mm down from its equilibrium position and released. For each spring arrangement, determine the period of the vibration, the maximum velocity of the block, and the maximum acceleration of the block.

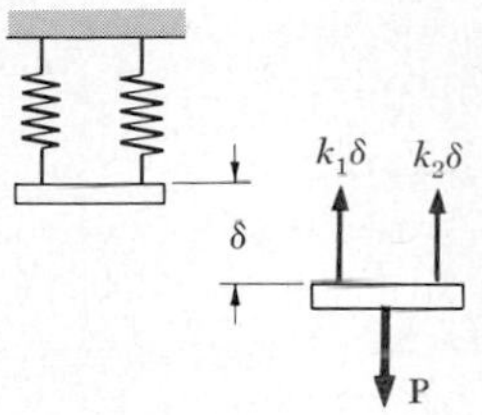

***a.* Springs Attached in Parallel.** We first determine the constant k of a single spring equivalent to the two springs *by finding the magnitude of the force* **P** required to cause a given deflection δ. Since for a deflection δ the magnitudes of the forces exerted by the springs are, respectively, $k_1\delta$ and $k_2\delta$, we have

$$P = k_1\delta + k_2\delta = (k_1 + k_2)\delta$$

The constant k of the single equivalent spring is

$$k = \frac{P}{\delta} = k_1 + k_2 = 4 \text{ kN/m} + 6 \text{ kN/m} = 10 \text{ kN/m} = 10^4 \text{ N/m}$$

Period of Vibration: Since $m = 50$ kg, Eq. (19.4) yields

$$p^2 = \frac{k}{m} = \frac{10^4 \text{ N/m}}{50 \text{ kg}} \qquad p = 14.14 \text{ rad/s}$$

$$\tau = 2\pi/p \qquad \tau = 0.444 \text{ s} \blacktriangleleft$$

Maximum Velocity: $v_m = x_m p = (0.040 \text{ m})(14.14 \text{ rad/s})$

$$v_m = 0.566 \text{ m/s} \qquad \mathbf{v}_m = 0.566 \text{ m/s} \updownarrow \blacktriangleleft$$

Maximum Acceleration: $a_m = x_m p^2 = (0.040 \text{ m})(14.14 \text{ rad/s})^2$

$$a_m = 8.00 \text{ m/s}^2 \qquad \mathbf{a}_m = 8.00 \text{ m/s}^2 \updownarrow \blacktriangleleft$$

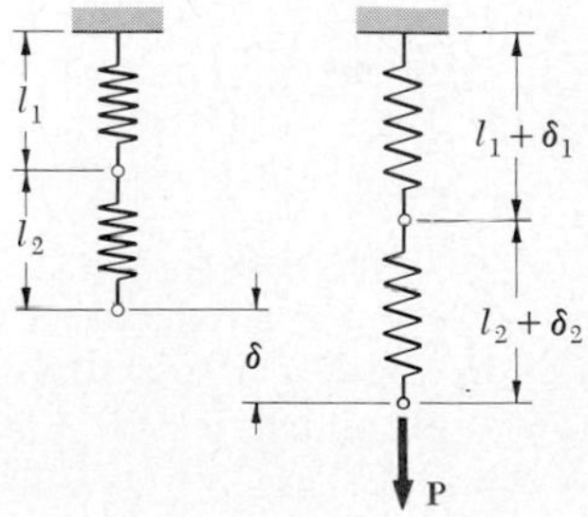

***b.* Springs Attached in Series.** We first determine the constant k of a single spring equivalent to the two springs *by finding the total elongation* δ of the springs under a given static load **P**. To facilitate the computation, a static load of magnitude $P = 12$ kN is used.

$$\delta = \delta_1 + \delta_2 = \frac{P}{k_1} + \frac{P}{k_2} = \frac{12 \text{ kN}}{4 \text{ kN/m}} + \frac{12 \text{ kN}}{6 \text{ kN/m}} = 5 \text{ m}$$

$$k = \frac{P}{\delta} = \frac{12 \text{ kN}}{5 \text{ m}} = 2.4 \text{ kN/m} = 2400 \text{ N/m}$$

Period of Vibration: $p^2 = \dfrac{k}{m} = \dfrac{2400 \text{ N/m}}{50 \text{ kg}} \qquad p = 6.93 \text{ rad/s}$

$$\tau = \frac{2\pi}{p} \qquad \tau = 0.907 \text{ s} \blacktriangleleft$$

Maximum Velocity: $v_m = x_m p = (0.040 \text{ m})(6.93 \text{ rad/s})$

$$v_m = 0.277 \text{ m/s} \qquad \mathbf{v}_m = 0.277 \text{ m/s} \updownarrow \blacktriangleleft$$

Maximum Acceleration: $a_m = x_m p^2 = (0.040 \text{ m})(6.93 \text{ rad/s})^2$

$$a_m = 1.920 \text{ m/s}^2 \qquad \mathbf{a}_m = 1.920 \text{ m/s}^2 \updownarrow \blacktriangleleft$$

Problems

19.1 A particle is known to move with a simple harmonic motion. The maximum acceleration is 3 m/s^2, and the maximum velocity is 150 mm/s. Determine the amplitude and the frequency of the motion.

19.2 Determine the maximum velocity and maximum acceleration of a particle which moves in simple harmonic motion with an amplitude of 150 mm and a period of 0.9 s.

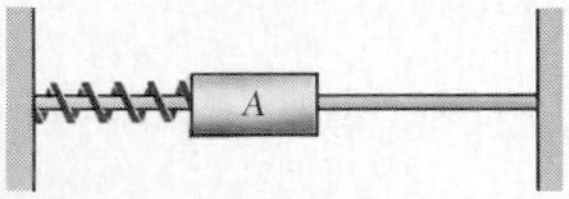

Fig. P19.3 and P19.4

19.3 A 5-lb collar is attached to a spring of constant 3 lb/in. and may slide without friction on a horizontal rod. If the collar is moved 4 in. from its equilibrium position and released, determine the maximum velocity and the maximum acceleration of the collar during the resulting motion.

19.4 A 3-lb collar is attached to a spring of constant 4 lb/in. and may slide without friction on a horizontal rod. The collar is at rest when it is struck with a mallet and given an initial velocity of 50 in./s. Determine the amplitude and the maximum acceleration of the collar during the resulting motion.

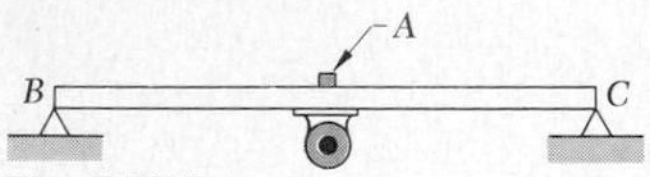

Fig. P19.5

19.5 A variable-speed motor is rigidly attached to the beam BC. The rotor is slightly unbalanced and causes the beam to vibrate with a frequency equal to the motor speed. When the speed of the motor is less than 450 rpm or more than 900 rpm, a small object placed at A is observed to remain in contact with the beam. For speeds between 450 and 900 rpm the object is observed to "dance" and actually to lose contact with the beam. Determine the amplitude of the motion of A when the speed of the motor is (*a*) 450 rpm, (*b*) 900 rpm. Give answers in both SI and U.S. customary units.

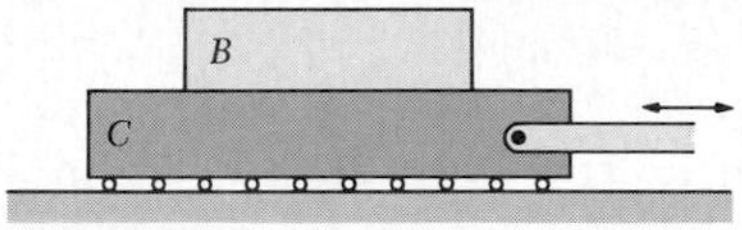

Fig. P19.6

19.6 An instrument package B is placed on the shaking table C as shown. The table is made to move horizontally in simple harmonic motion with a frequency of 3 Hz. Knowing that the coefficient of static friction is $\mu_s = 0.40$ between the package and the table, determine the largest allowable amplitude of the motion if the package is not to slip on the table. Give answers in both SI and U.S. customary units.

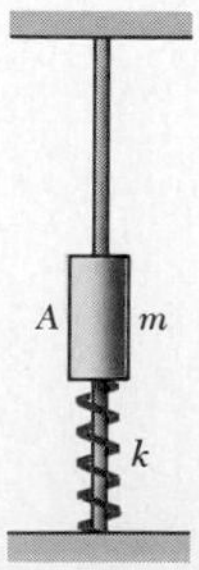

Fig. P19.7

19.7 A 3-kg collar rests on, but is not attached to, the spring shown. The collar is depressed 50 mm and released. If the ensuing motion is to be simple harmonic, determine (*a*) the largest permissible value of the spring constant k, (*b*) the corresponding position, velocity and acceleration of the collar 0.15 s after it has been released.

19.8 A 4-kg collar is attached to a spring of constant $k = 800$ N/m as shown. If the collar is given a displacement of 40 mm downward from its equilibrium position and released, determine (*a*) the time required for the collar to move 60 mm upward, (*b*) the corresponding velocity and acceleration of the collar.

19.9 A 3-lb collar is attached to a spring of constant $k = 5$ lb/in. as shown. If the collar is given a displacement of 2.5 in. downward from its equilibrium position and released, determine (*a*) the time required for the collar to move 2 in. upward, (*b*) the corresponding velocity and acceleration of the collar.

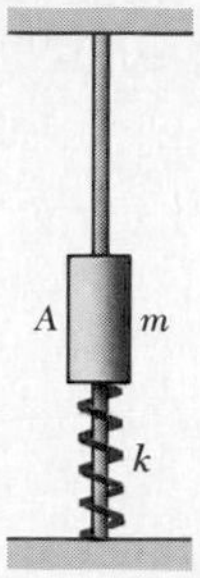

Fig. P19.8 and P19.9

19.10 In Prob. 19.9, determine the position, velocity, and acceleration of the collar 0.20 s after it has been released.

19.11 and 19.12 A 40-lb block is supported by the spring arrangement shown. If the block is moved vertically downward from its equilibrium position and released, determine (*a*) the period and frequency of the resulting motion, (*b*) the maximum velocity and acceleration of the block if the amplitude of the motion is 1.25 in.

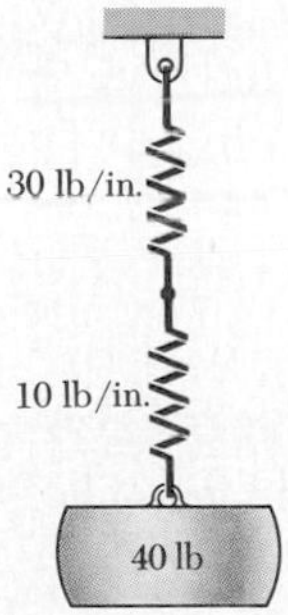

Fig. P19.11

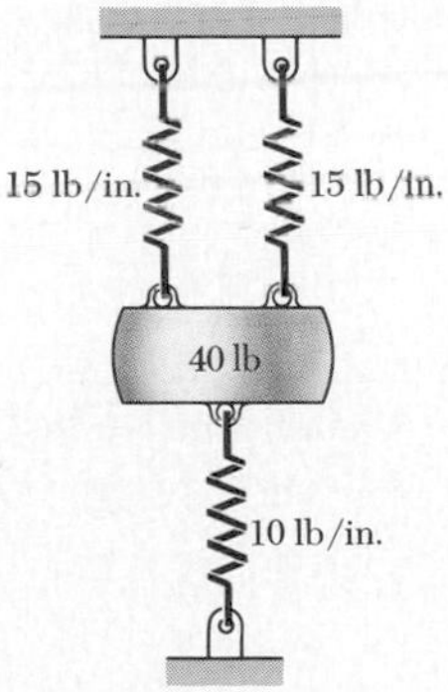

Fig. P19.12

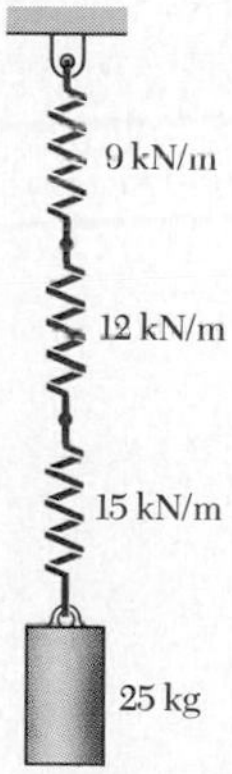

Fig. P19.13

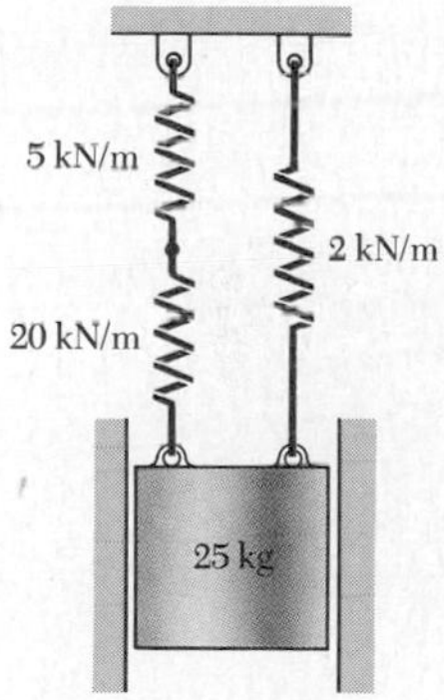

Fig. P19.14

19.13 and 19.14 A 25-kg block is supported by the spring arrangement shown. If the block is moved vertically downward from its equilibrium position and released, determine (*a*) the period and frequency of the resulting motion, (*b*) the maximum velocity and acceleration of the block if the amplitude of the motion is 25 mm.

19.15 The period of vibration of the system shown is observed to be 0.4 s. After cylinder *B* has been removed, the period is observed to be 0.3 s. Determine (*a*) the weight of cylinder *A*, (*b*) the constant of the spring.

19.16 The period of vibration of the system shown is observed to be 1.5 s. After cylinder *B* has been removed and replaced with a 4-lb cylinder, the period is observed to be 1.6 s. Determine (*a*) the weight of cylinder *A*, (*b*) the constant of the spring.

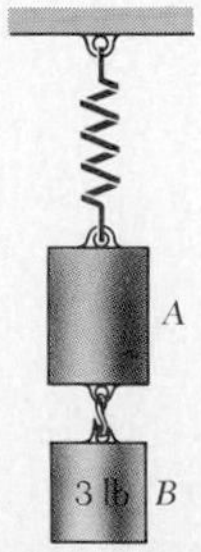

Fig. P19.15 and P19.16

19.17 A tray of mass m is attached to three springs as shown. The period of vibration of the empty tray is 0.5 s. After a 1.5-kg block has been placed in the center of the tray, the period is observed to be 0.6 s. Knowing that the amplitude of the vibration is small, determine the mass m of the tray.

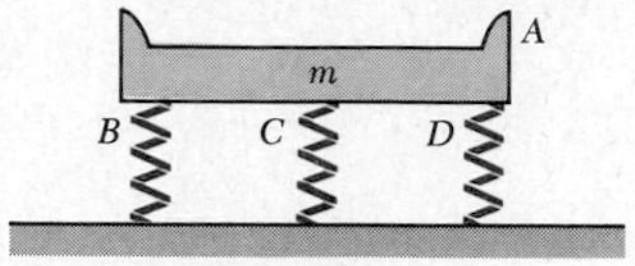

Fig. P19.17 and P19.18

19.18 A tray of mass m is attached to three springs as shown. The period of vibration of the empty tray is 0.75 s. After the center spring C has been removed, the period is observed to be 0.90 s. Knowing that the constant of spring C is 100 N/m, determine the mass m of the tray.

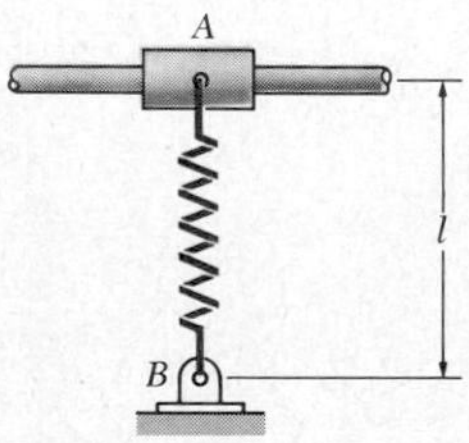

Fig. P19.19

19.19 A collar of mass m slides without friction on a horizontal rod and is attached to a spring AB of constant k. (*a*) If the unstretched length of the spring is just equal to l, show that the collar does not execute simple harmonic motion even when the amplitude of the oscillations is small. (*b*) If the unstretched length of the spring is less than l, show that the motion may be approximated by a simple harmonic motion for small oscillations.

19.20 The rod AB is attached to a hinge at A and to two springs each of constant k. If $h = 600$ mm, $d = 250$ mm, and $m = 25$ kg, determine the value of k for which the period of small oscillations is (*a*) 1 s, (*b*) infinite. Neglect the mass of the rod and assume that each spring can act in either tension or compression.

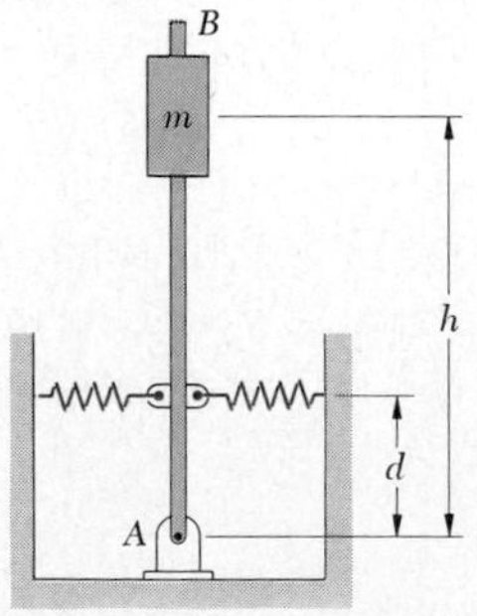

Fig. P19.20 and P19.21

19.21 If $h = 600$ mm and $d = 400$ mm and each spring has a constant $k = 700$ N/m, determine the mass m for which the period of small oscillations is (*a*) 0.50 s, (*b*) infinite. Neglect the mass of the rod and assume that each spring can act in either tension or compression.

19.22 Denoting by δ_{st} the static deflection of a beam under a given load, show that the frequency of vibration of the load is

$$f = \frac{1}{2\pi}\sqrt{\frac{g}{\delta_{st}}}$$

Neglect the mass of the beam, and assume that the load remains in contact with the beam.

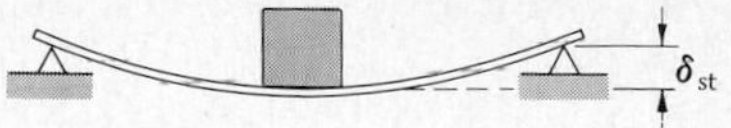

Fig. P19.22

* **19.23** Expanding the integrand in (19.19) into a series of even powers of $\sin\phi$ and integrating, show that the period of a simple pendulum of length l may be approximated by the formula

$$\tau = 2\pi\sqrt{\frac{l}{g}}\left(1 + \tfrac{1}{4}\sin^2\frac{\theta_m}{2}\right)$$

where θ_m is the amplitude of the oscillations.

* **19.24** Using the formula given in Prob. 19.23, determine the amplitude θ_m for which the period of a simple pendulum is 1 percent longer than the period of the same pendulum for small oscillations.

* **19.25** Using the data of Table 19.1, determine the period of a simple pendulum of length $l = 750$ mm (*a*) for small oscillations, (*b*) for oscillations of amplitude $\theta_m = 60°$, (*c*) for oscillations of amplitude $\theta_m = 90°$.

* **19.26** Using a table of elliptic integrals, determine the period of a simple pendulum of length $l = 750$ mm if the amplitude of the oscillations is $\theta_m = 50°$.

19.5. Free Vibrations of Rigid Bodies. The analysis of the vibrations of a rigid body or of a system of rigid bodies possessing a single degree of freedom is similar to the analysis of the vibrations of a particle. An appropriate variable, such as a distance x or an angle θ, is chosen to define the position of the body or system of bodies, and an equation relating this variable and its second derivative with respect to t is written. If the equation obtained is of the same form as (19.5), i.e., if we have

$$\ddot{x} + p^2x = 0 \qquad \text{or} \qquad \ddot{\theta} + p^2\theta = 0 \tag{19.21}$$

the vibration considered is a simple harmonic motion. The period and frequency of the vibration may then be obtained by identifying p and substituting into (19.13) and (19.14).

In general, a simple way to obtain one of Eqs. (19.21) is to express that the system of the external forces is equivalent to the system of the effective forces by drawing a diagram of the body for an arbitrary value of the variable and writing the appropriate equation of motion. We recall that our goal should be *the determination of the coefficient* of the variable x or

θ, *not* the determination of the variable itself or of the derivative $\ddot{x}$ or $\ddot{\theta}$. Setting this coefficient equal to p^2, we obtain the circular frequency p, from which τ and f may be determined.

The method we have outlined may be used to analyze vibrations which are truly represented by a simple harmonic motion, or vibrations of small amplitude which can be *approximated* by a simple harmonic motion. As an example, we shall determine the period of the small oscillations of a square plate of side $2b$ which is suspended from the midpoint O of one side (Fig. 19.5*a*). We consider the plate in an arbitrary position defined by the angle θ that the line OG forms with the vertical and draw a diagram to express that the weight $\mathbf{W}$ of the plate and the components $\mathbf{R}_x$ and $\mathbf{R}_y$ of the reaction at O are equivalent to the vectors $m\mathbf{a}_t$ and $m\mathbf{a}_n$ and to the couple $\bar{I}\boldsymbol{\alpha}$ (Fig. 19.5*b*). Since the angular velocity and angular acceleration of the plate are equal, respectively, to $\dot{\theta}$ and $\ddot{\theta}$, the magnitudes of the two vectors are, respectively, $mb\ddot{\theta}$ and $mb\dot{\theta}^2$, while the moment of the couple is $\bar{I}\ddot{\theta}$. In previous applications of this method (Chap. 16), we tried whenever possible to assume the correct sense for the acceleration. Here, however, we must assume the same positive sense for θ and $\ddot{\theta}$ in order to obtain an equation of the form (19.21). Consequently, the angular acceleration $\ddot{\theta}$ will be assumed positive counterclockwise, even though this assumption is obviously unrealistic. Equating moments about O, we write

$$+\circlearrowleft \qquad -W(b \sin\theta) = (mb\ddot{\theta})b + \bar{I}\ddot{\theta}$$

Noting that $\bar{I} = \frac{1}{12}m[(2b)^2 + (2b)^2] = \frac{2}{3}mb^2$ and $W = mg$, we obtain

$$\ddot{\theta} + \frac{3}{5}\frac{g}{b}\sin\theta = 0 \qquad (19.22)$$

For oscillations of small amplitude, we may replace $\sin\theta$ by θ, expressed in radians, and write

$$\ddot{\theta} + \frac{3}{5}\frac{g}{b}\theta = 0 \qquad (19.23)$$

Comparison with (19.21) shows that the equation obtained is that of a simple harmonic motion and that the circular frequency p of the oscillations is equal to $(3g/5b)^{1/2}$. Substituting into (19.13), we find that the period of the oscillations is

$$\tau = \frac{2\pi}{p} = 2\pi\sqrt{\frac{5b}{3g}} \qquad (19.24)$$

The result obtained is valid only for oscillations of small amplitude. A more accurate description of the motion of the plate is obtained by comparing Eqs. (19.16) and (19.22). We note that the two equations are identical if we choose l equal to $5b/3$. This means that the plate will oscillate as a simple pendulum of length $l = 5b/3$, and the results of Sec. 19.4 may be used to correct the value of the period given in (19.24). The point A of the plate located on line OG at a distance $l = 5b/3$ from O is defined as the *center of oscillation* corresponding to O (Fig. 19.5*a*).

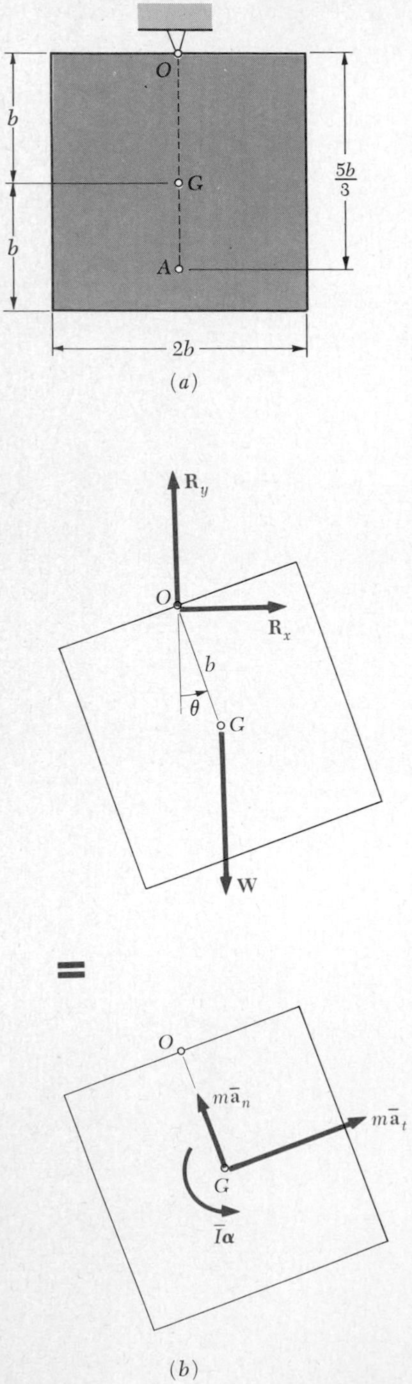

Fig. 19.5

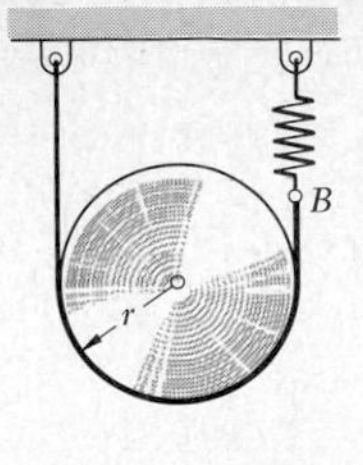

SAMPLE PROBLEM 19.2

A cylinder of weight W and radius r is suspended from a looped cord as shown. One end of the cord is attached directly to a rigid support, while the other end is attached to a spring of constant k. Determine the period and frequency of vibration of the cylinder.

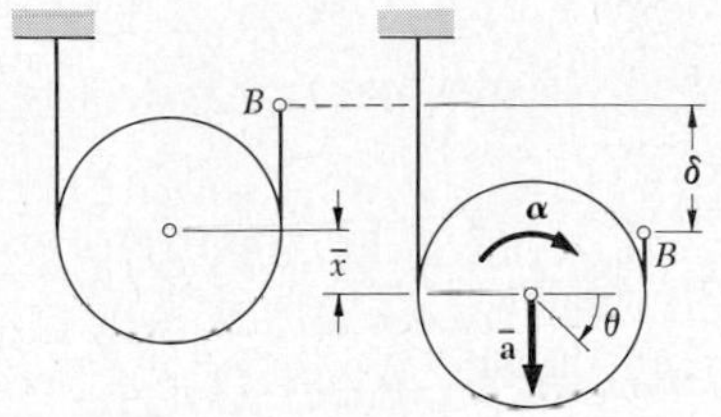

Kinematics of Motion. We express the linear displacement and the acceleration of the cylinder in terms of the angular displacement θ. Choosing the positive sense clockwise and measuring the displacements from the equilibrium position, we write

$$\bar{x} = r\theta \qquad \delta = 2\bar{x} = 2r\theta$$

$$\boldsymbol{\alpha} = \ddot{\theta} \downarrow \qquad \bar{a} = r\alpha = r\ddot{\theta} \qquad \bar{\mathbf{a}} = r\ddot{\theta} \downarrow \tag{1}$$

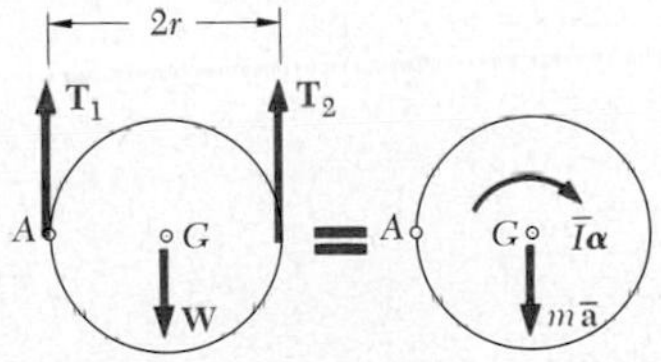

Equations of Motion. The system of external forces acting on the cylinder consists of the weight $\mathbf{W}$ and of the forces $\mathbf{T}_1$ and $\mathbf{T}_2$ exerted by the cord. We express that this system is equivalent to the system of effective forces represented by the vector $m\bar{\mathbf{a}}$ attached at G and the couple $\bar{I}\boldsymbol{\alpha}$.

$$+\downarrow \Sigma M_A = \Sigma (M_A)_{\text{eff}}: \qquad Wr - T_2(2r) = m\bar{a}r + \bar{I}\alpha \tag{2}$$

When the cylinder is in its position of equilibrium, the tension in the cord is $T_0 = \frac{1}{2}W$. We note that for an angular displacement θ, the magnitude of $\mathbf{T}_2$ is

$$T_2 = T_0 + k\delta = \tfrac{1}{2}W + k\delta = \tfrac{1}{2}W + k(2r\theta) \tag{3}$$

Substituting from (1) and (3) into (2), and recalling that $\bar{I} = \frac{1}{2}mr^2$, we write

$$Wr - (\tfrac{1}{2}W + 2kr\theta)(2r) = m(r\ddot{\theta})r + \tfrac{1}{2}mr^2\ddot{\theta}$$

$$\ddot{\theta} + \frac{8}{3}\frac{k}{m}\theta = 0$$

The motion is seen to be simple harmonic, and we have

$$p^2 = \frac{8}{3}\frac{k}{m} \qquad p = \sqrt{\frac{8}{3}\frac{k}{m}}$$

$$\tau = \frac{2\pi}{p} \qquad \tau = 2\pi\sqrt{\frac{3}{8}\frac{m}{k}} \quad ◄$$

$$f = \frac{p}{2\pi} \qquad f = \frac{1}{2\pi}\sqrt{\frac{8}{3}\frac{k}{m}} \quad ◄$$

SAMPLE PROBLEM 19.3

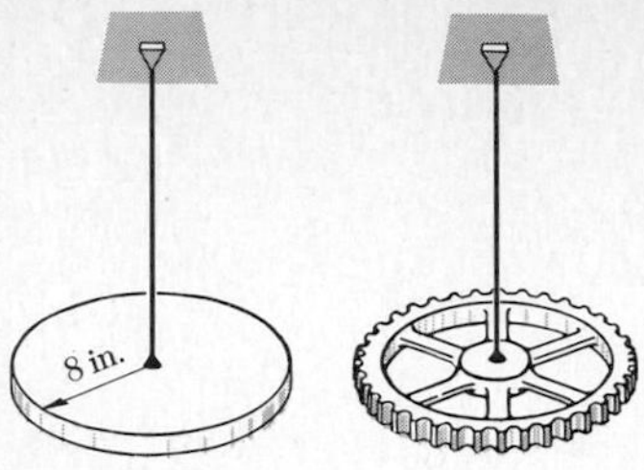

A circular disk, weighing 20 lb and of radius 8 in., is suspended from a wire as shown. The disk is rotated (thus twisting the wire) and then released; the period of the torsional vibration is observed to be 1.13 s. A gear is then suspended from the same wire, and the period of torsional vibration for the gear is observed to be 1.93 s. Assuming that the moment of the couple exerted by the wire is proportional to the angle of twist, determine (*a*) the torsional spring constant of the wire, (*b*) the centroidal moment of inertia of the gear, (*c*) the maximum angular velocity reached by the gear if it is rotated through 90° and released.

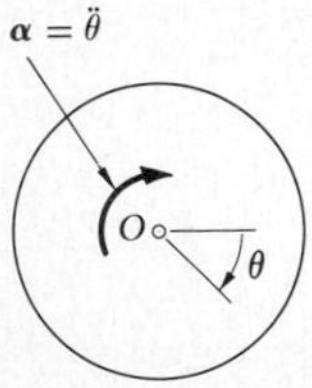

***a.* Vibration of Disk.** Denoting by θ the angular displacement of the disk, we express that the magnitude of the couple exerted by the wire is $M = K\theta$, where K is the torsional spring constant of the wire. Since this couple must be equivalent to the couple $\bar{I}\boldsymbol{\alpha}$ representing the effective forces of the disk, we write

$$+\circlearrowleft \Sigma M_O = \Sigma (M_O)_{\text{eff}}: \qquad +K\theta = -\bar{I}\ddot{\theta}$$

$$\ddot{\theta} + \frac{K}{\bar{I}}\theta = 0$$

The motion is seen to be simple harmonic, and we have

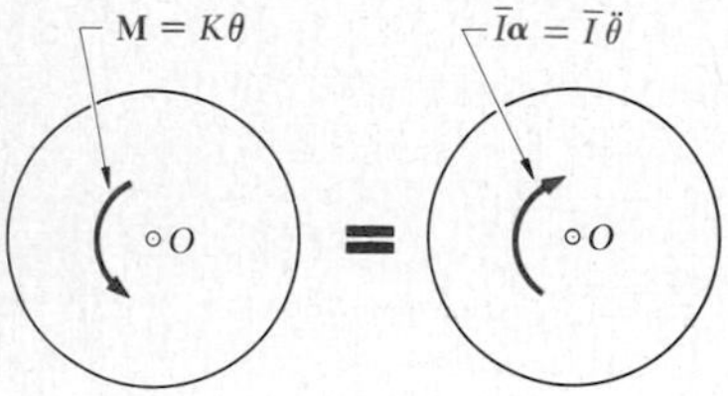

$$p^2 = \frac{K}{\bar{I}} \qquad \tau = \frac{2\pi}{p} \qquad \tau = 2\pi\sqrt{\frac{\bar{I}}{K}} \tag{1}$$

For the disk, we have

$$\tau = 1.13 \text{ s} \qquad \bar{I} = \tfrac{1}{2}mr^2 = \frac{1}{2}\left(\frac{20 \text{ lb}}{32.2 \text{ ft/s}^2}\right)\left(\frac{8}{12}\text{ ft}\right)^2 = 0.138 \text{ lb}\cdot\text{ft}\cdot\text{s}^2$$

Substituting into (1), we obtain

$$1.13 = 2\pi\sqrt{\frac{0.138}{K}} \qquad K = 4.27 \text{ lb}\cdot\text{ft/rad} \blacktriangleleft$$

***b.* Vibration of Gear.** Since the period of vibration of the gear is 1.93 s and $K = 4.27$ lb·ft/rad, Eq. (1) yields

$$1.93 = 2\pi\sqrt{\frac{\bar{I}}{4.27}} \qquad \bar{I}_{\text{gear}} = 0.403 \text{ lb}\cdot\text{ft}\cdot\text{s}^2 \blacktriangleleft$$

***c.* Maximum Angular Velocity of the Gear.** Since the motion is simple harmonic, we have

$$\theta = \theta_m \sin pt \qquad \omega = \theta_m p \cos pt \qquad \omega_m = \theta_m p$$

Recalling that $\theta_m = 90° = 1.571$ rad and $\tau = 1.93$ s, we write

$$\omega_m = \theta_m p = \theta_m\left(\frac{2\pi}{\tau}\right) = (1.571 \text{ rad})\left(\frac{2\pi}{1.93 \text{ s}}\right)$$

$$\omega_m = 5.11 \text{ rad/s} \blacktriangleleft$$

Problems

19.27 The 3-kg uniform rod shown is attached to a spring of constant $k = 900$ N/m. If end A of the rod is depressed 25 mm and released, determine (*a*) the period of vibration, (*b*) the maximum velocity of end A.

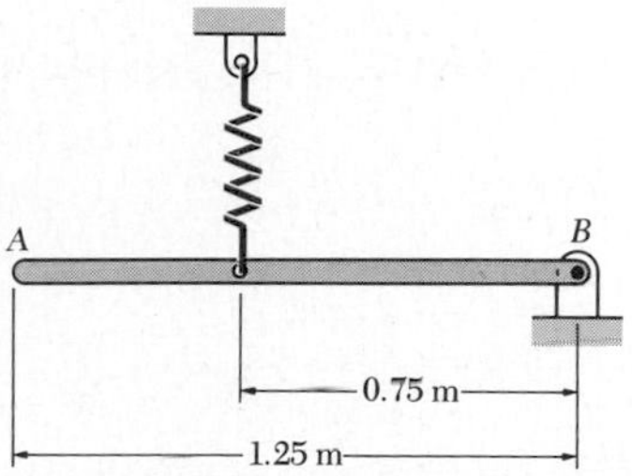

Fig. P19.27

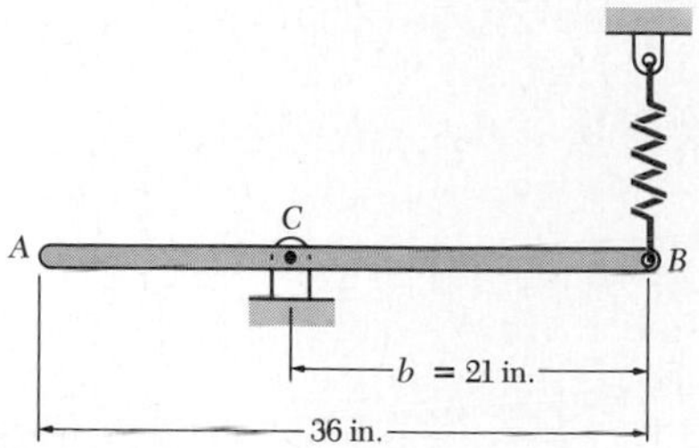

Fig. P19.28

19.28 The uniform rod shown weighs 12 lb and is attached to a spring of constant $k = 3$ lb/in. If end B of the rod is depressed 0.5 in. and released, determine (*a*) the period of vibration, (*b*) the maximum velocity of end B.

19.29 A 12-lb slender rod AB is riveted to a 10-lb uniform disk as shown. A belt is attached to the rim of the disk and to a spring which holds the rod at rest in the position shown. If end A of the rod is moved 1.5 in. down and released, determine (*a*) the period of vibration, (*b*) the maximum velocity of end A.

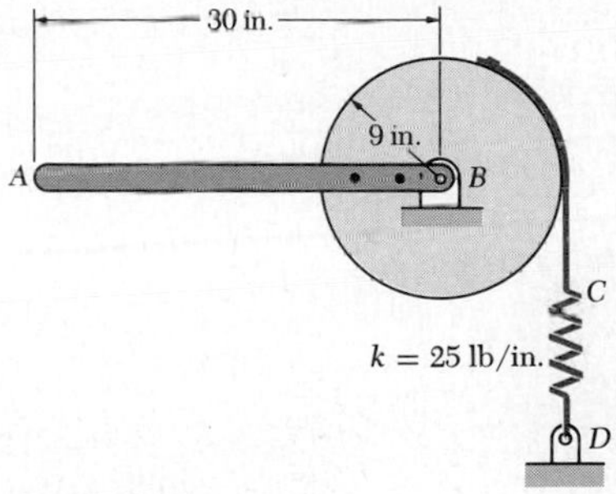

Fig. P19.29

19.30 A 5-kg uniform cylinder rolls without sliding on an incline and is attached to a spring AB as shown. If the center of the cylinder is moved 10 mm down the incline from its equilibrium position and released, determine (*a*) the period of vibration, (*b*) the maximum velocity of the center of the cylinder.

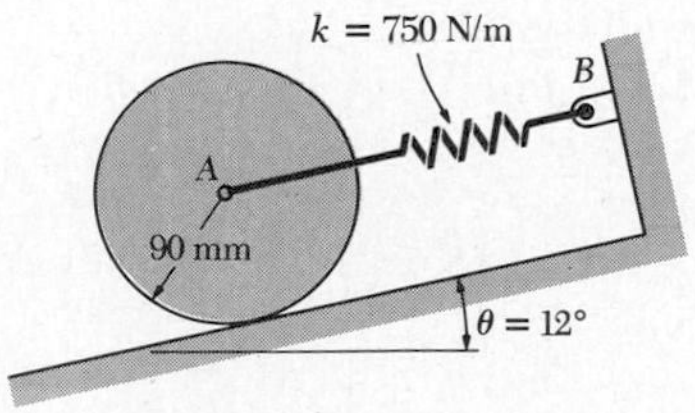

Fig. P19.30

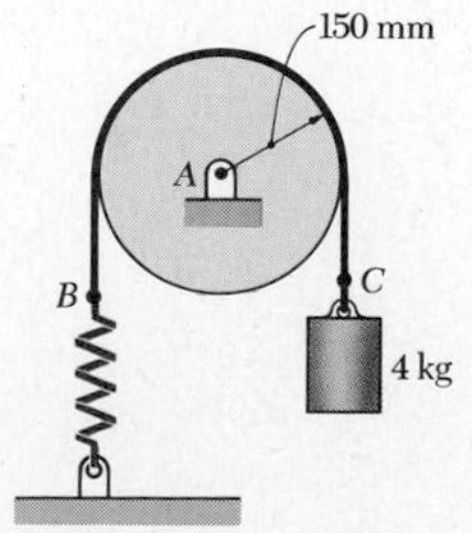

Fig. P19.31

19.31 A belt is placed over the rim of a 12-kg disk as shown and then attached to a 4-kg cylinder and to a spring of constant $k = 500$ N/m. If the cylinder is moved 75 mm down from its equilibrium position and released, determine (*a*) the period of vibration, (*b*) the maximum velocity of the cylinder. Assume friction is sufficient to prevent the belt from slipping on the rim.

19.32 In Prob. 19.31, determine (*a*) the frequency of vibration, (*b*) the maximum tension which occurs in the belt at *B* and at *C*.

19.33 In Prob. 19.28, determine (*a*) the value of *b* for which the smallest period of vibration occurs, (*b*) the corresponding period of vibration.

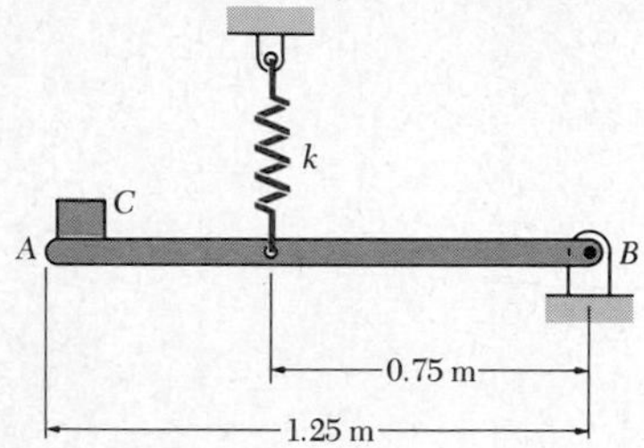

Fig. P19.34

19.34 The 3-kg uniform rod *AB* is attached as shown to a spring of constant $k = 900$ N/m. A small 0.5-kg block *C* is placed on the rod at *A*. (*a*) If end *A* of the rod is then moved down through a small distance δ_0 and released, determine the period of the vibration. (*b*) Determine the largest allowable value of δ_0 if block *C* is to remain at all times in contact with the rod.

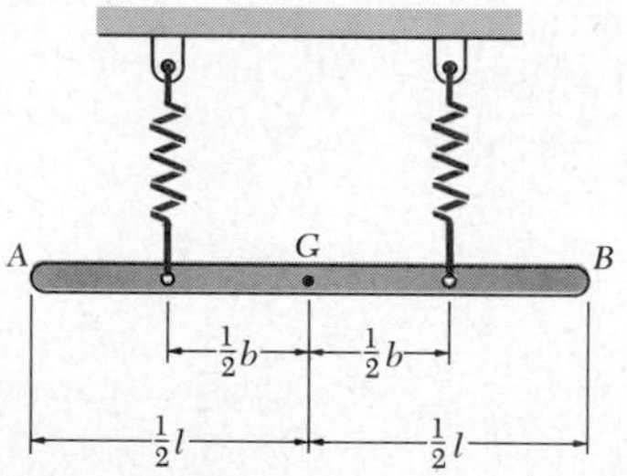

Fig. P19.35

19.35 A slender rod of mass *m* and length *l* is held by two springs, each of constant *k*. Determine the frequency of the resulting vibration if the rod is (*a*) given a small vertical displacement and released, (*b*) rotated through a small angle about a horizontal axis through *G* and released. (*c*) Determine the ratio *b*/*l* for which the frequencies found in parts *a* and *b* are equal.

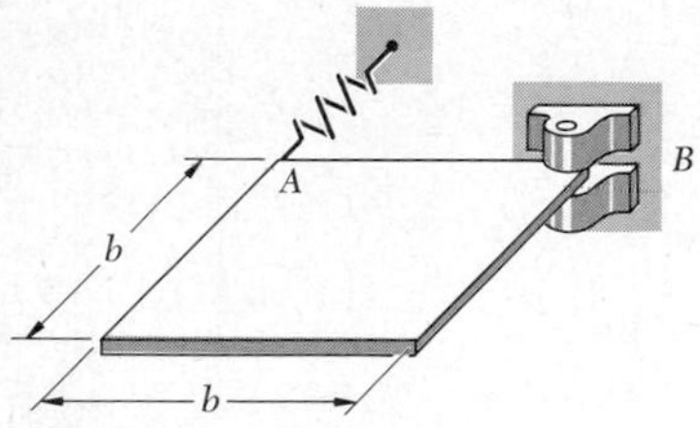

Fig. P19.36

19.36 A uniform square plate of mass *m* is supported in a horizontal plane by a vertical pin at *B* and is attached at *A* to a spring of constant *k*. If corner *A* is given a small displacement and released, determine the period of the resulting motion.

19.37 A *compound pendulum* is defined as a rigid slab which oscillates about a fixed point *O*, called the center of suspension. Show that the period of oscillation of a compound pendulum is equal to the period of a simple pendulum of length *OA*, where the distance from *A* to the mass center *G* is $GA = \bar{k}^2/\bar{r}$. Point *A* is defined as the center of oscillation and coincides with the center of percussion defined in Prob. 17.66.

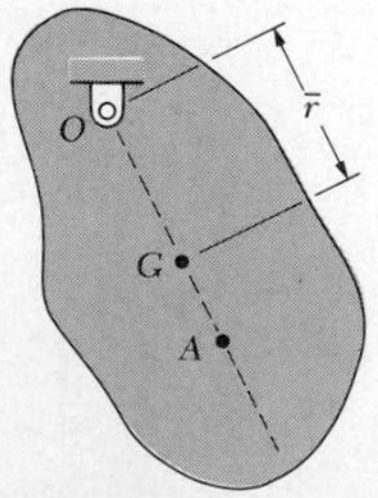

Fig. P19.37 and P19.39

19.38 Show that if the compound pendulum of Prob. 19.37 is suspended from *A* instead of *O*, the period of oscillation is the same as before and the new center of oscillation is located at *O*.

19.39 A rigid slab oscillates about a fixed point *O*. Show that the smallest period of oscillation occurs when the distance $\bar{r}$ from point *O* to the mass center *G* is equal to $\bar{k}$.

19.40 If either a simple or a compound pendulum is used to determine experimentally the acceleration of gravity g, difficulties are encountered. In the simple pendulum, the string is not truly weightless, while in the compound pendulum, the exact location of the mass center is difficult to establish. In the case of a compound pendulum, the difficulty may be eliminated by using a reversible, or Kater, pendulum. Two knife-edges A and B are placed so that they are obviously not at the same distance from the mass center G, and the distance l is measured with great precision. The position of a counterweight D is then adjusted so that the period of oscillation τ is the same when either knife-edge is used. Show that the period τ obtained is equal to that of a true simple pendulum of length l and that $g = 4\pi^2 l/\tau^2$.

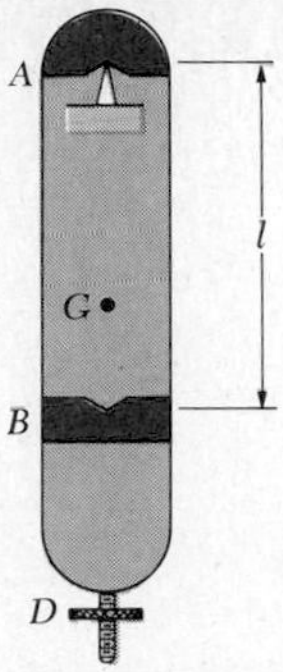

Fig. P19.40

19.41 A uniform semicircular disk of radius r is suspended from a hinge. Determine the period of small oscillations if the disk (*a*) is suspended from A as shown, (*b*) is suspended from point B.

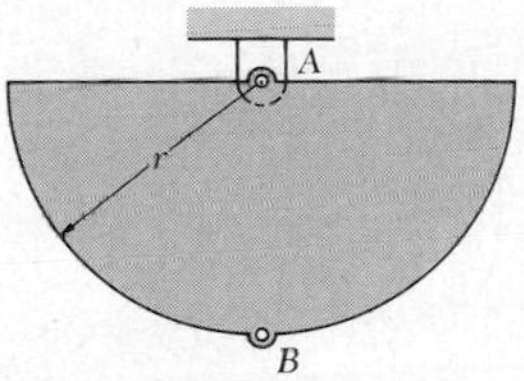

Fig. P19.41

19.42 A uniform bar of length l may oscillate about a hinge A located at a distance c from the mass center G of the bar. (*a*) Determine the frequency of small oscillations if $c = \frac{1}{2}l$. (*b*) Determine a second value of c for which the frequency of small oscillations is the same as that found in part *a*.

19.43 A uniform bar of length l may oscillate about a hinge A located at a distance c from the mass center G of the bar. Determine (*a*) the distance c for which the period of oscillation is minimum, (*b*) the corresponding value of the period.

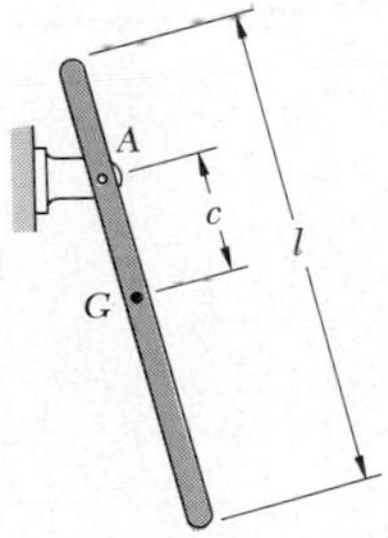

Fig. P19.42 and P19.43

19.44 A thin homogeneous wire is bent into the shape of an equilateral triangle of side $l = 250$ mm. Determine the period of small oscillations if the wire (*a*) is suspended as shown, (*b*) is suspended from a pin located at the midpoint of one side.

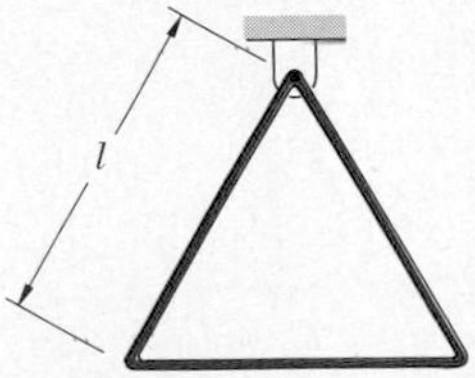

Fig. P19.44

19.45 Two uniform rods, each of mass m and length l, are welded together to form the T-shaped assembly shown. Determine the frequency of small oscillations of the assembly.

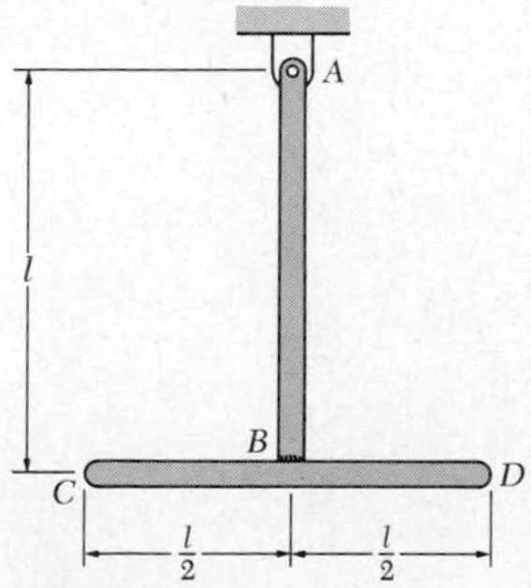

Fig. P19.45

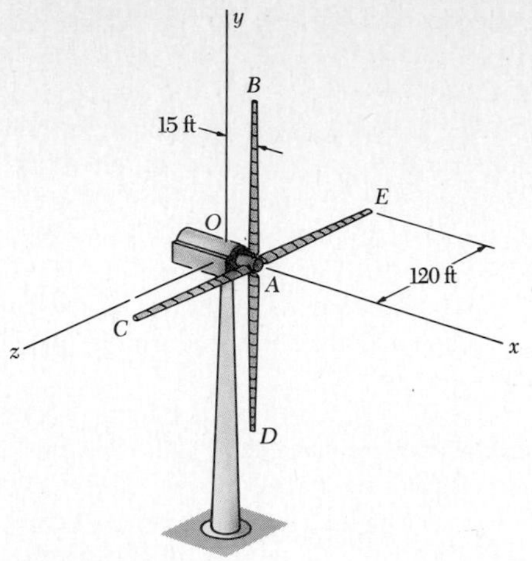

Fig. P19.46

19.46 Blade AB of the wind-turbine generator shown is to be temporarily removed. Motion of the turbine generator about the y axis is prevented, but the remaining three blades may oscillate as a unit about the x axis. Assuming that each blade is equivalent to a 120-ft slender rod, determine the period of small oscillations of the blades in the absence of wind.

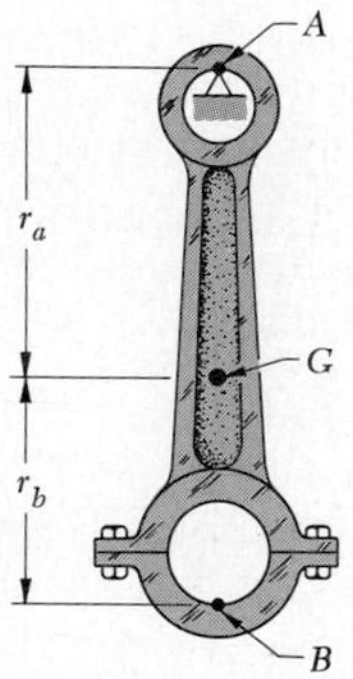

Fig. P19.47 and P19.48

19.47 The period of small oscillations about A of a connecting rod is observed to be 1.06 s. Knowing that the distance r_a is 170 mm, determine the centroidal radius of gyration of the connecting rod.

19.48 A connecting rod is supported by a knife-edge at point A; the period of small oscillations is observed to be 0.895 s. The rod is then inverted and supported by a knife-edge at point B and the period of small oscillations is observed to be 0.805 s. Knowing that $r_a + r_b = 270$ mm, determine (*a*) the location of the mass center G, (*b*) the centroidal radius of gyration $\bar{k}$.

19.49 and 19.50 A thin disk of radius r may oscillate about the axis AB located as shown at a distance b from the mass center G. (*a*) Determine the period of small oscillations if $b = r$. (*b*) Determine a second value of b for which the period of oscillation is the same as that found in part *a*.

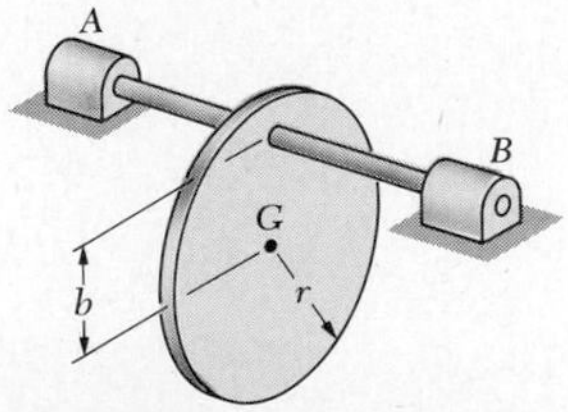

Fig. P19.49

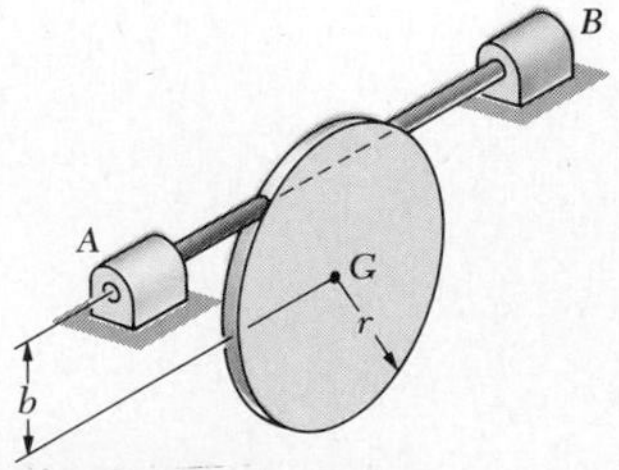

Fig. P19.50

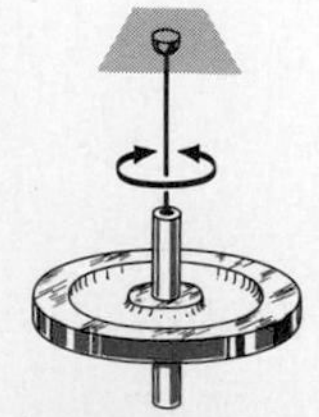

Fig. P19.51

19.51 A period of 3.60 s is observed for the angular oscillations of a 750-g gyroscope rotor suspended from a wire as shown. Knowing that a period of 5.10 s is obtained when a 60-mm-diameter steel sphere is suspended in the same fashion, determine the centroidal radius of gyration of the rotor. (Density of steel = 7850 kg/m^3.)

19.52 A 6-kg slender rod is suspended from a steel wire which is known to have a torsional spring constant $K = 1.75$ N·m/rad. If the rod is rotated through 180° about the vertical and then released, determine (*a*) the period of oscillation, (*b*) the maximum velocity of end *A* of the rod.

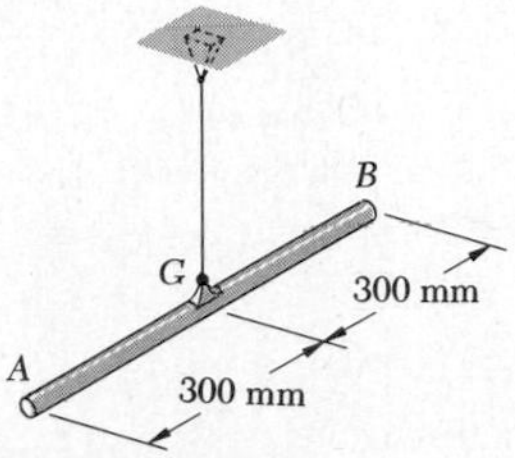

Fig. P19.52

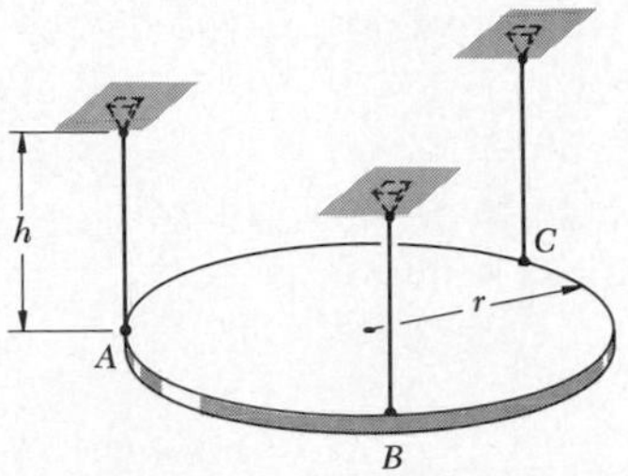

Fig. P19.53

19.53 A thin circular plate of radius r is suspended from three vertical wires of length h equally spaced around the perimeter of the plate. Determine the period of oscillation when (*a*) the plate is rotated through a small angle about a vertical axis passing through its mass center and released, (*b*) the plate is given a small horizontal translation and released.

19.54 Solve Prob. 19.53, assuming that $r = 750$ mm and $h = 600$ mm.

19.55 A uniform disk of radius 10 in. and weighing 18 lb is attached to a vertical shaft which is rigidly held at *B*. It is known that the disk rotates through 3° when a 40-lb·in. static couple is applied to the disk. If the disk is rotated through 8° and then released, determine (*a*) the period of the resulting vibration, (*b*) the maximum velocity of a point on the rim of the disk.

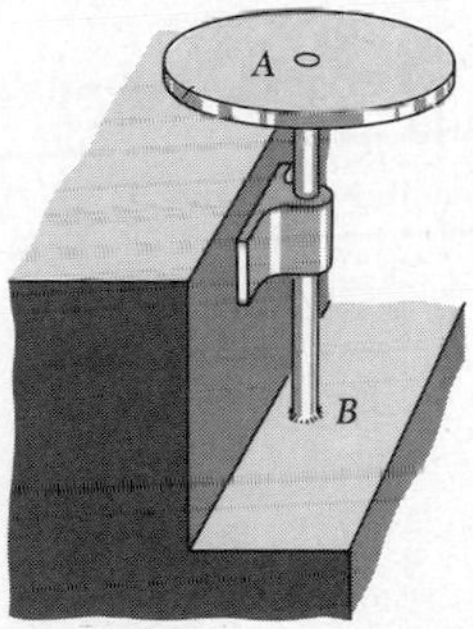

Fig. P19.55

19.56 A steel casting is rigidly bolted to the disk of Prob. 19.55. Knowing that the period of torsional vibration of the disk and casting is 0.90 s, determine the moment of inertia of the casting with respect to the shaft *AB*.

19.6. Application of the Principle of Conservation of Energy. We saw in Sec. 19.2 that when a particle of mass m is in simple harmonic motion, the resultant **F** of the forces exerted on the particle has a magnitude proportional to the displacement x measured from the position of equilibrium O and is directed toward O; we write $F = -kx$. Referring to Sec. 13.6, we note that **F** is a *conservative force* and that the corresponding potential energy is $V = \frac{1}{2}kx^2$, where V is assumed equal to zero in the equilibrium position $x = 0$. Since the velocity of the particle is equal to $\dot{x}$, its kinetic energy is $T = \frac{1}{2}m\dot{x}^2$ and we may express that the total energy of the particle is conserved by writing

$$T + V = \text{constant} \qquad \tfrac{1}{2}m\dot{x}^2 + \tfrac{1}{2}kx^2 = \text{constant}$$

Setting $p^2 = k/m$ as in (19.4), where p is the circular frequency of the vibration, we have

$$\dot{x}^2 + p^2x^2 = \text{constant} \tag{19.25}$$

Equation (19.25) is characteristic of simple harmonic motion; it may be obtained directly from (19.5) by multiplying both terms by $2\dot{x}$ and integrating.

The principle of conservation of energy provides a convenient way for determining the period of vibration of a rigid body or of a system of rigid bodies possessing a single degree of freedom, once it has been established that the motion of the system is a simple harmonic motion, or that it may be approximated by a simple harmonic motion. Choosing an appropriate variable, such as a distance x or an angle θ, we consider two particular positions of the system:

1. *The displacement of the system is maximum;* we have $T_1 = 0$, and V_1 may be expressed in terms of the amplitude x_m or θ_m (choosing $V = 0$ in the equilibrium position).
2. *The system passes through its equilibrium position;* we have $V_2 = 0$, and T_2 may be expressed in terms of the maximum velocity $\dot{x}_m$ or $\dot{\theta}_m$.

We then express that the total energy of the system is conserved and write $T_1 + V_1 = T_2 + V_2$. Recalling from (19.15) that for simple harmonic motion the maximum velocity is equal to the product of the amplitude and of the circular frequency p, we find that the equation obtained may be solved for p.

As an example, we shall consider again the square plate of Sec. 19.5. In the position of maximum displacement (Fig. 19.6*a*), we have

$$T_1 = 0 \qquad V_1 = W(b - b\cos\theta_m) = Wb(1 - \cos\theta_m)$$

or, since $1 - \cos\theta_m = 2\sin^2(\theta_m/2) \approx 2(\theta_m/2)^2 = \theta_m^2/2$ for oscillations of small amplitude,

$$T_1 = 0 \qquad V_1 = \tfrac{1}{2}Wb\theta_m^2 \tag{19.26}$$

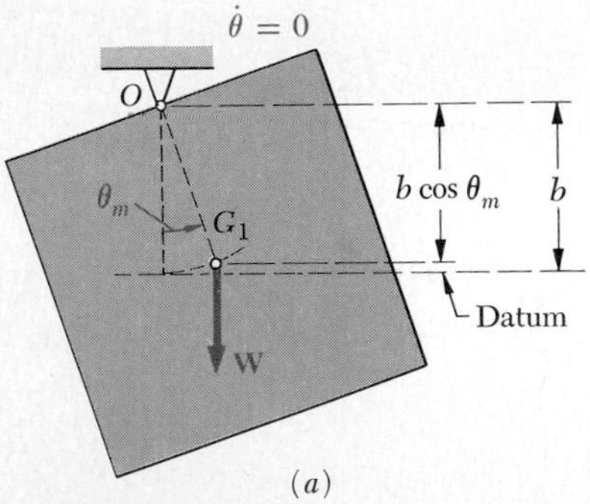

(*a*)

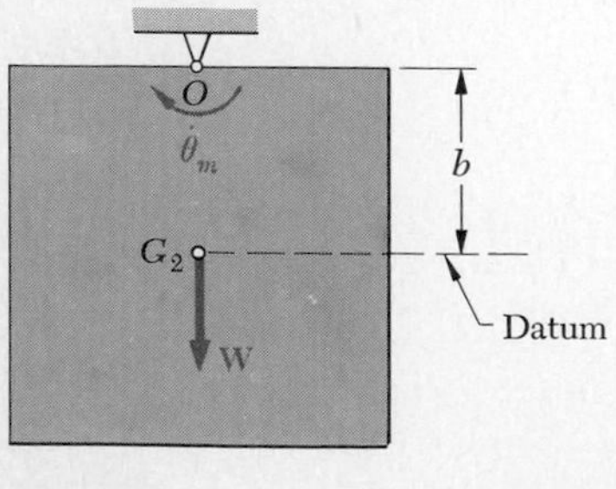

(*b*)

Fig. 19.6

As the plate passes through its position of equilibrium (Fig. 19.6*b*), its velocity is maximum and we have

$$T_2 = \tfrac{1}{2}m\bar{v}_m^2 + \tfrac{1}{2}\bar{I}\omega_m^2 = \tfrac{1}{2}mb^2\dot{\theta}_m^2 + \tfrac{1}{2}\bar{I}\dot{\theta}_m^2 \qquad V_2 = 0$$

or, recalling from Sec. 19.5 that $\bar{I} = \tfrac{2}{3}mb^2$,

$$T_2 = \tfrac{1}{2}(\tfrac{5}{3}mb^2)\dot{\theta}_m^2 \qquad V_2 = 0 \tag{19.27}$$

Substituting from (19.26) and (19.27) into $T_1 + V_1 = T_2 + V_2$, and noting that the maximum velocity $\dot{\theta}_m$ is equal to the product $\theta_m p$, we write

$$\tfrac{1}{2}Wb\theta_m^2 = \tfrac{1}{2}(\tfrac{5}{3}mb^2)\theta_m^2 p^2 \tag{19.28}$$

which yields $p^2 = 3g/5b$ and

$$\tau = \frac{2\pi}{p} = 2\pi\sqrt{\frac{5b}{3g}} \tag{19.29}$$

as previously obtained.

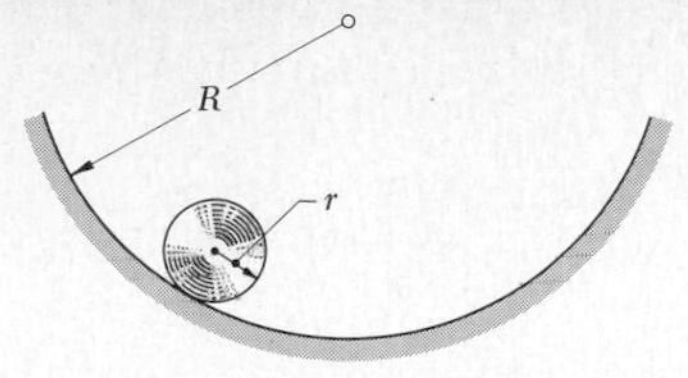

SAMPLE PROBLEM 19.4

Determine the period of small oscillations of a cylinder of radius r which rolls without slipping inside a curved surface of radius R.

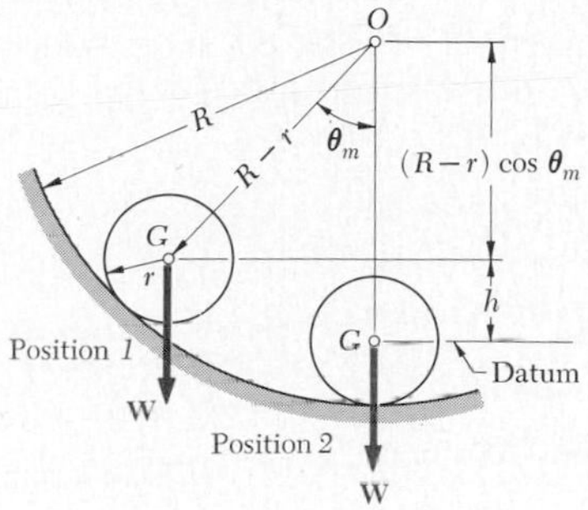

Solution. We denote by θ the angle which line OG forms with the vertical. Since the cylinder rolls without slipping, we may apply the principle of conservation of energy between position *1*, where $\theta = \theta_m$, and position *2*, where $\theta = 0$.

Position *1*. *Kinetic Energy.* Since the velocity of the cylinder is zero, we have $T_1 = 0$.

Potential Energy. Choosing a datum as shown and denoting by W the weight of the cylinder, we have

$$V_1 = Wh = W(R - r)(1 - \cos\theta)$$

Noting that for small oscillations $(1 - \cos\theta) = 2\sin^2(\theta/2) \approx \theta^2/2$, we have

$$V_1 = W(R - r)\frac{\theta_m^2}{2}$$

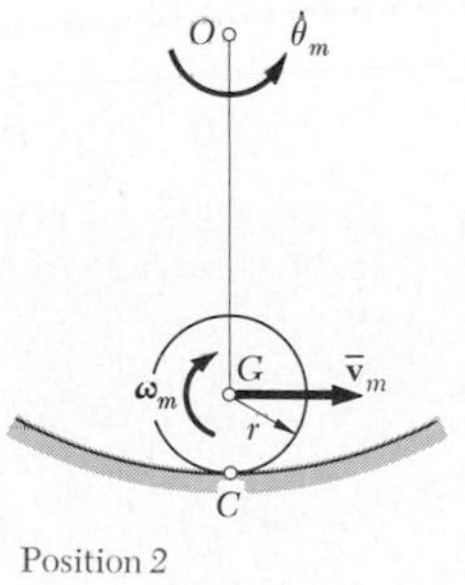

Position *2*. Denoting by $\dot{\theta}_m$ the angular velocity of line OG as the cylinder passes through position *2*, and observing that point C is the instantaneous center of rotation of the cylinder, we write

$$\bar{v}_m = (R - r)\dot{\theta}_m \qquad \omega_m = \frac{\bar{v}_m}{r} = \frac{R - r}{r}\dot{\theta}_m$$

Kinetic Energy

$$\begin{aligned} T_2 &= \tfrac{1}{2}m\bar{v}_m^2 + \tfrac{1}{2}\bar{I}\omega_m^2 \\ &= \tfrac{1}{2}m(R - r)^2\dot{\theta}_m^2 + \tfrac{1}{2}(\tfrac{1}{2}mr^2)\left(\frac{R - r}{r}\right)^2\dot{\theta}_m^2 \\ &= \tfrac{3}{4}m(R - r)^2\dot{\theta}_m^2 \end{aligned}$$

Potential Energy $\qquad V_2 = 0$

Conservation of Energy

$$T_1 + V_1 = T_2 + V_2$$

$$0 + W(R - r)\frac{\theta_m^2}{2} = \tfrac{3}{4}m(R - r)^2\dot{\theta}_m^2 + 0$$

Since $\dot{\theta}_m = p\theta_m$ and $W = mg$, we write

$$mg(R - r)\frac{\theta_m^2}{2} = \tfrac{3}{4}m(R - r)^2(p\theta_m)^2 \qquad p^2 = \frac{2}{3}\frac{g}{R - r}$$

$$\tau = \frac{2\pi}{p} \qquad \tau = 2\pi\sqrt{\frac{3}{2}\frac{R - r}{g}} \quad ◄$$

Problems

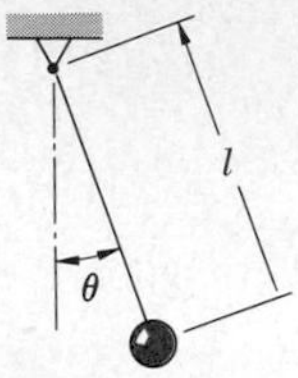

Fig. P19.57

19.57 Using the method of Sec. 19.6, determine the period of a simple pendulum of length l.

19.58 The springs of an automobile are observed to expand 200 mm to an undeformed position as the body is lifted by several jacks. Assuming that each spring carries an equal portion of the weight of the automobile, determine the frequency of the free vertical vibrations of the body.

19.59 Using the method of Sec. 19.6, solve Prob. 19.3.

19.60 Using the method of Sec. 19.6, solve Prob. 19.4.

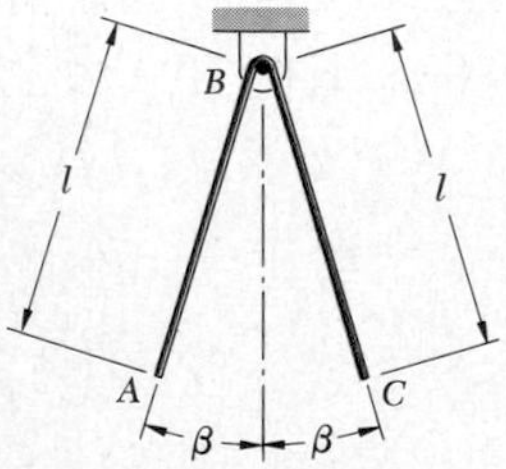

Fig. P19.61 and P19.62

19.61 A homogeneous wire of length $2l$ is bent as shown and allowed to oscillate about a frictionless pin at B. Denoting by τ_0 the period of small oscillations when $\beta = 0$, determine the angle β for which the period of small oscillations is $2\tau_0$.

19.62 Knowing that $l = 750$ mm and $\beta = 40°$, determine the period of small oscillations of the bent homogeneous wire shown.

19.63 A homogeneous wire is bent to form a square of side l which is supported by a ball-and-socket joint at A. Determine the period of small oscillations of the square (a) in the plane of the square, (b) in a direction perpendicular to the square.

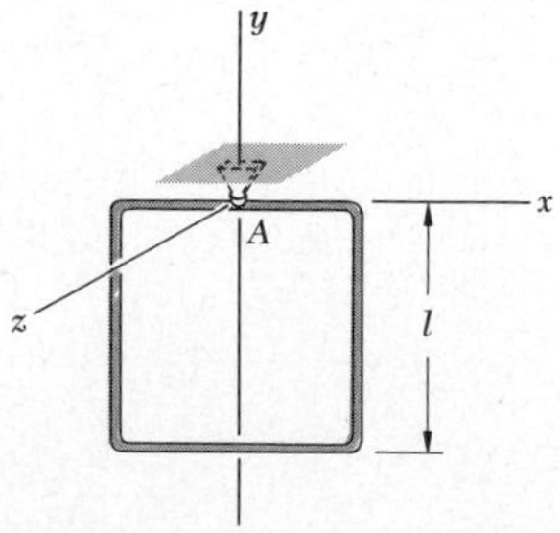

Fig. P19.63

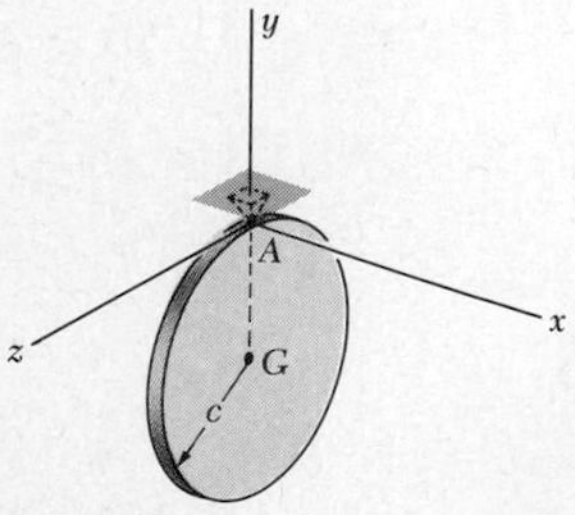

Fig. P19.65

19.64 Solve Prob. 19.63, assuming that the square is supported by a ball-and-socket joint located at one corner of the square.

19.65 A uniform disk of radius c is supported by a ball-and-socket joint at A. Determine the frequency of small oscillations of the disk (a) in the plane of the disk, (b) in a direction perpendicular to the disk.

19.66 It is observed that when an 8-lb weight is attached to the rim of a 6-ft-diameter flywheel, the period of small oscillations of the flywheel is 22 s. Neglecting axle friction, determine the centroidal moment of inertia of the flywheel.

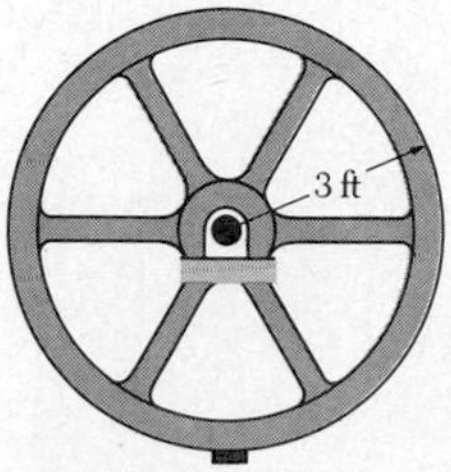

Fig. P19.66

19.67 Using the method of Sec. 19.6, solve Prob. 19.45.

19.68 The uniform rod *ABC* weighs 5 lb and is attached to two springs as shown. If end *C* is given a small displacement and released, determine the frequency of vibration of the rod.

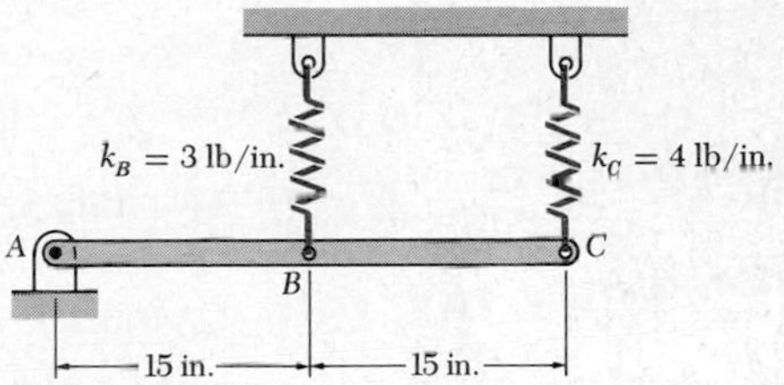

Fig. P19.68

19.69 For the rod of Prob. 19.68, determine the frequency of vibration of the rod if the springs are interchanged so that $k_B = 4$ lb/in. and $k_C = 3$ lb/in.

19.70 The 5-kg slender rod *AB* is welded to the 8-kg uniform disk. A spring of constant 450 N/m is attached to the disk and holds the rod at rest in the position shown. If end *B* of the rod is given a small displacement and released, determine the period of vibration of the rod.

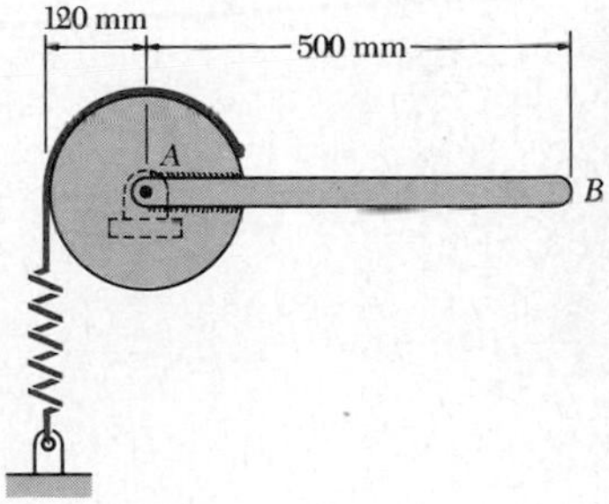

Fig. P19.70

19.71 For the rod and disk of Prob. 19.70, determine the constant of the spring for which the period of vibration of the rod is 1.5 s.

19.72 The slender rod *AB* of mass *m* is attached to two collars of negligible mass. Knowing that the system lies in a horizontal plane and is in equilibrium in the position shown, determine the period of vibration if collar *A* is given a small displacement and released.

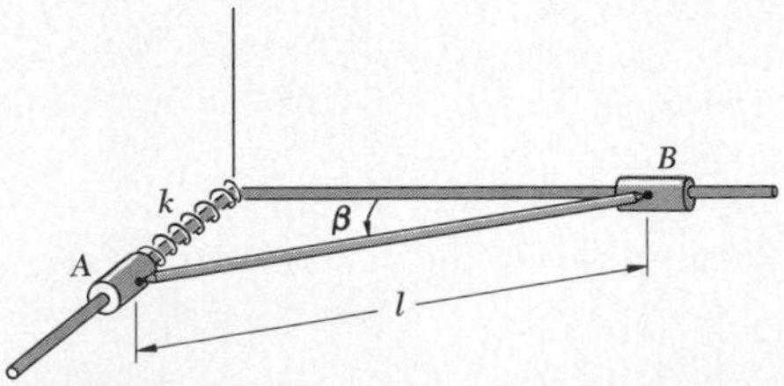

Fig. P19.72

19.73 Disks *A* and *B* are of mass 3 kg and 8 kg, respectively, and a small block *C* of mass 750 g is attached to the rim of disk *B*. Assuming that no slipping occurs between the disks, determine the period of small oscillations of the system.

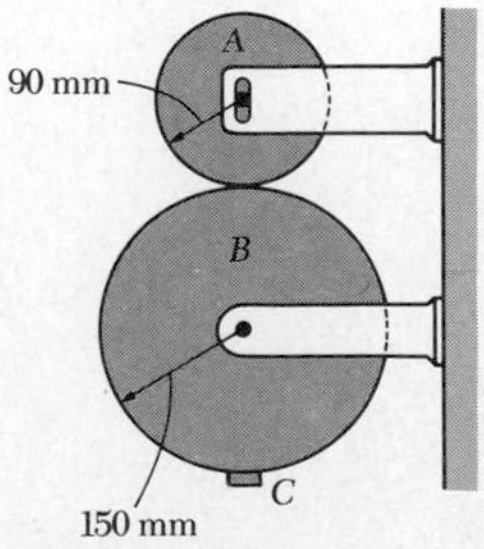

Fig. P19.73

19.74 Two 5-kg uniform disks are attached to the 8-kg rod AB as shown. Knowing that the constant of the spring is 4 kN/m and that the disks roll without sliding, determine the frequency of vibration of the system.

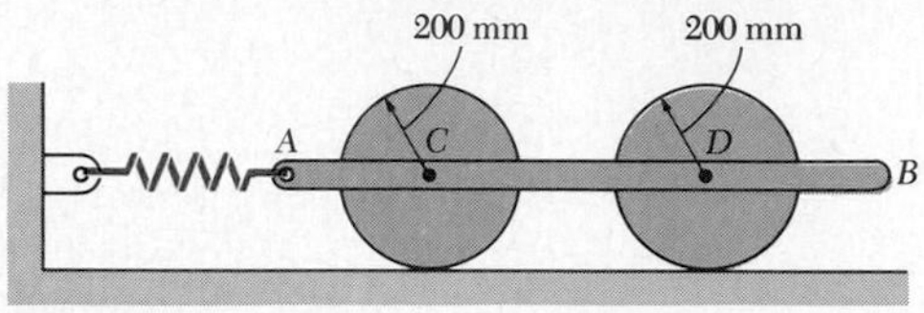

Fig. P19.74

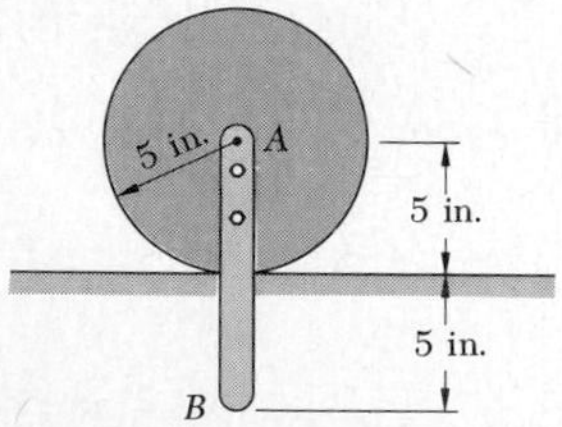

Fig. P19.75 and P19.76

19.75 The 4-lb rod AB is bolted to the 6-lb disk. Knowing that the disk rolls without sliding, determine the period of small oscillations of the system.

19.76 The 4-lb rod AB is bolted to the 5-in.-radius disk as shown. Knowing that the disk rolls without sliding, determine the weight of the disk for which the period of small oscillations of the system is 1.5 s.

19.77 Three identical rods are connected as shown. If $b = \frac{3}{4}l$, determine the frequency of small oscillations of the system.

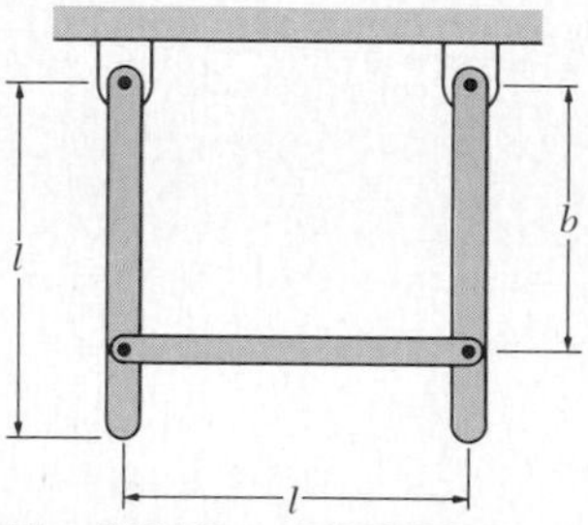

Fig. P19.77 and P19.78

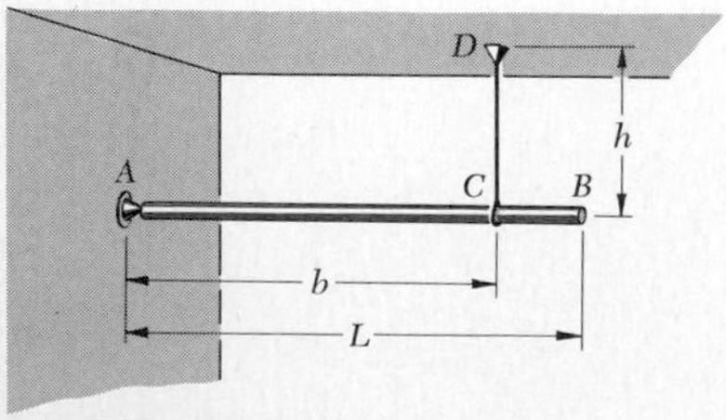

Fig. P19.79

19.78 Three identical rods are connected as shown. Determine (*a*) the distance b for which the frequency of oscillation is maximum, (*b*) the corresponding maximum frequency.

19.79 A uniform rod of length L is supported by a ball-and-socket joint at A and by the vertical wire CD. Derive an expression for the period of oscillation of the rod if end B is given a small horizontal displacement and then released.

19.80 Solve Prob. 19.79, assuming that $L = 3$ m, $b = 2.5$ m, and $h = 2$ m.

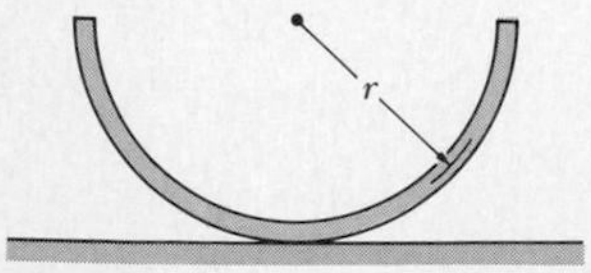

Fig. P19.81

19.81 A half section of pipe is placed on a horizontal surface, rotated through a small angle, and then released. Assuming that the pipe section rolls without sliding, determine the period of oscillation.

19.82 A slender rod of length l is suspended from two vertical wires of length h, each located a distance $\frac{1}{2}b$ from the mass center G. Determine the period of oscillation when (a) the rod is rotated through a small angle about a vertical axis passing through G and released, (b) the rod is given a small horizontal translation along AB and released.

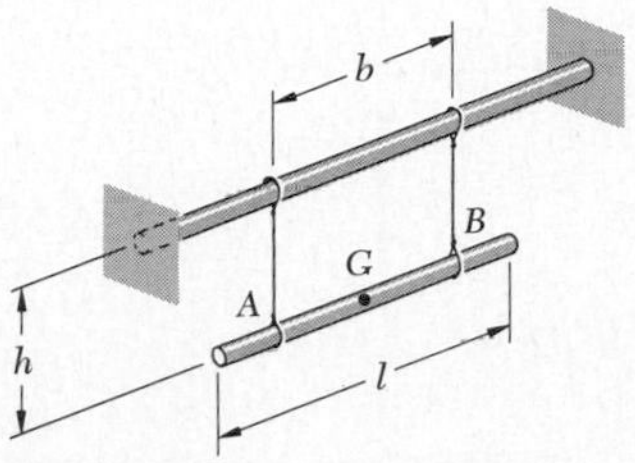

Fig. P19.82

Fig. P19.83

* **19.83** As a submerged body moves through a fluid, the particles of the fluid flow around the body and thus acquire kinetic energy. In the case of a sphere moving in an ideal fluid, the total kinetic energy acquired by the fluid is $\frac{1}{4}\rho V v^2$, where ρ is the density of the fluid, V the volume of the sphere, and v the velocity of the sphere. Consider a 500-g hollow spherical shell of radius 75 mm which is held submerged in a tank of water by a spring of constant 600 N/m. (a) Neglecting fluid friction, determine the period of vibration of the shell when it is displaced vertically and then released. (b) Solve part a, assuming that the tank is accelerated upward at the constant rate of 3 m/s^2.

* **19.84** A thin plate of length l rests on a half cylinder of radius r. Derive an expression for the period of small oscillations of the plate.

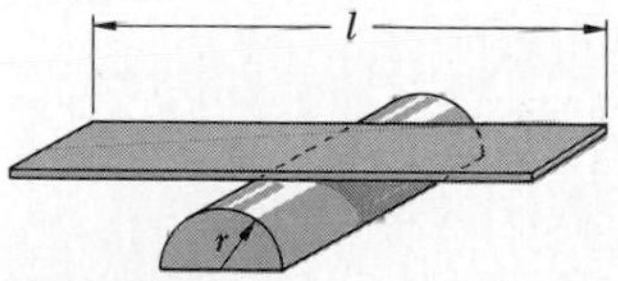

Fig. P19.84

19.7. Forced Vibrations. The most important vibrations from the point of view of engineering applications are the *forced vibrations* of a system. These vibrations occur when a system is subjected to a periodic force or when it is elastically connected to a support which has an alternating motion.

Consider first the case of a body of mass m suspended from a spring and subjected to a periodic force **P** of magnitude $P = P_m \sin \omega t$ (Fig. 19.7). This force may be an actual external force applied to the body, or it may be a centrifugal force produced by the rotation of some unbalanced part of the body (see Sample Prob. 19.5). Denoting by x the displacement of the body measured from its equilibrium position, we write the equation of motion,

$+\downarrow\Sigma F = ma$: $\qquad P_m \sin \omega t + W - k(\delta_{st} + x) = m\ddot{x}$

Recalling that $W = k\delta_{st}$, we have

$$m\ddot{x} + kx = P_m \sin \omega t \tag{19.30}$$

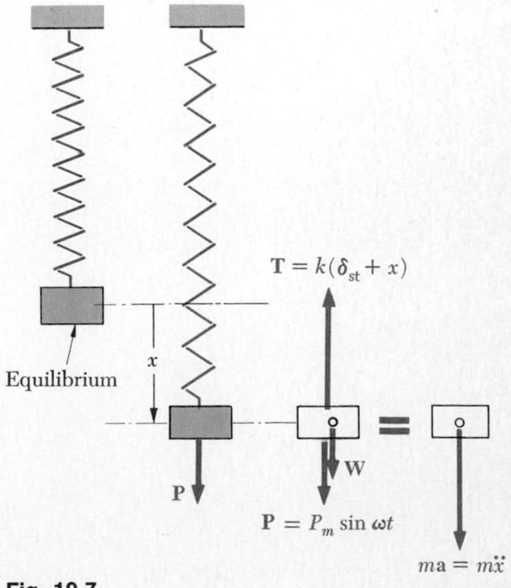

Fig. 19.7

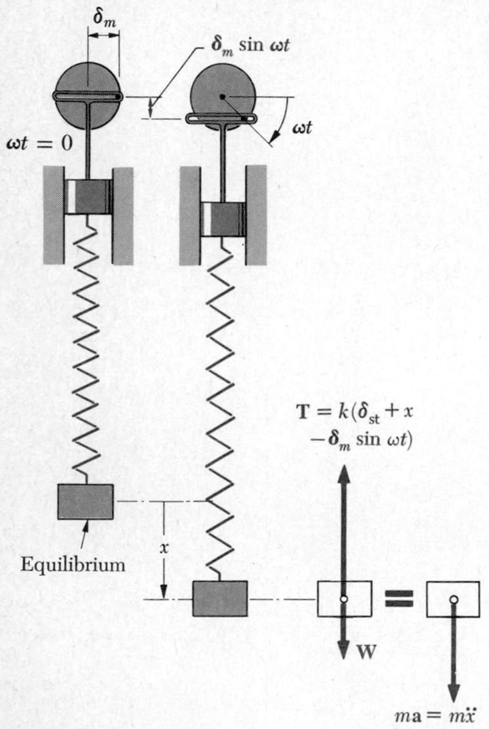

Fig. 19.8

Next we consider the case of a body of mass m suspended from a spring attached to a moving support whose displacement δ is equal to $\delta_m \sin \omega t$ (Fig. 19.8). Measuring the displacement x of the body from the position of static equilibrium corresponding to $\omega t = 0$, we find that the total elongation of the spring at time t is $\delta_{st} + x - \delta_m \sin \omega t$. The equation of motion is thus

$$+\downarrow \Sigma F = ma: \qquad W - k(\delta_{st} + x - \delta_m \sin \omega t) = m\ddot{x}$$

Recalling that $W = k\delta_{st}$, we have

$$m\ddot{x} + kx = k\delta_m \sin \omega t \tag{19.31}$$

We note that Eqs. (19.30) and (19.31) are of the same form and that a solution of the first equation will satisfy the second if we set $P_m = k\delta_m$.

A differential equation such as (19.30) or (19.31), possessing a right-hand member different from zero, is said to be *nonhomogeneous*. Its general solution is obtained by adding a particular solution of the given equation to the general solution of the corresponding *homogeneous* equation (with right-hand member equal to zero). A *particular solution* of (19.30) or (19.31) may be obtained by trying a solution of the form

$$x_{\text{part}} = x_m \sin \omega t \tag{19.32}$$

Substituting x_{part} for x into (19.30), we find

$$-m\omega^2 x_m \sin \omega t + kx_m \sin \omega t = P_m \sin \omega t$$

which may be solved for the amplitude,

$$x_m = \frac{P_m}{k - m\omega^2}$$

Recalling from (19.4) that $k/m = p^2$, where p is the circular frequency of the free vibration of the body, we write

$$x_m = \frac{P_m/k}{1 - (\omega/p)^2} \tag{19.33}$$

Substituting from (19.32) into (19.31), we obtain in a similar way

$$x_m = \frac{\delta_m}{1 - (\omega/p)^2} \tag{19.33'}$$

The homogeneous equation corresponding to (19.30) or (19.31) is Eq. (19.3), defining the free vibration of the body. Its general solution, called the *complementary function*, was found in Sec. 19.2,

$$x_{\text{comp}} = A \sin pt + B \cos pt \tag{19.34}$$

Adding the particular solution (19.32) and the complementary function (19.34), we obtain the *general solution* of Eqs. (19.30) and (19.31):

$$x = A \sin pt + B \cos pt + x_m \sin \omega t \tag{19.35}$$

We note that the vibration obtained consists of two superposed vibrations. The first two terms in (19.35) represent a free vibration of the system. The frequency of this vibration, called the *natural frequency* of the system, depends only upon the constant k of the spring and the mass m of the body, and the constants A and B may be determined from the initial conditions. This free vibration is also called a *transient* vibration, since in actual practice, it will soon be damped out by friction forces (Sec. 19.9).

The last term in (19.35) represents the *steady-state* vibration produced and maintained by the impressed force or impressed support movement. Its frequency is the *forced frequency* imposed by this force or movement, and its amplitude x_m, defined by (19.33) or (19.33′), depends upon the *frequency ratio* ω/p. The ratio of the amplitude x_m of the steady-state vibration to the static deflection P_m/k caused by a force P_m, or to the amplitude δ_m of the support movement, is called the *magnification factor.* From (19.33) and (19.33′), we obtain

$$\text{Magnification factor} = \frac{x_m}{P_m/k} = \frac{x_m}{\delta_m} = \frac{1}{1 - (\omega/p)^2} \tag{19.36}$$

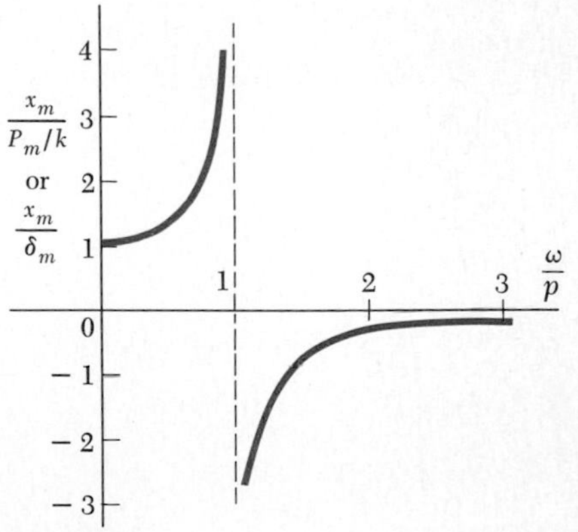

Fig. 19.9

The magnification factor has been plotted in Fig. 19.9 against the frequency ratio ω/p. We note that when $\omega = p$, the amplitude of the forced vibration becomes infinite. The impressed force or impressed support movement is said to be in *resonance* with the given system. Actually, the amplitude of the vibration remains finite because of damping forces (Sec. 19.9); nevertheless, such a situation should be avoided, and the forced frequency should not be chosen too close to the natural frequency of the system. We also note that for $\omega < p$ the coefficient of $\sin \omega t$ in (19.35) is positive, while for $\omega > p$ this coefficient is negative. In the first case the forced vibration is *in phase* with the impressed force or impressed support movement, while in the second case it is 180° *out of phase.*

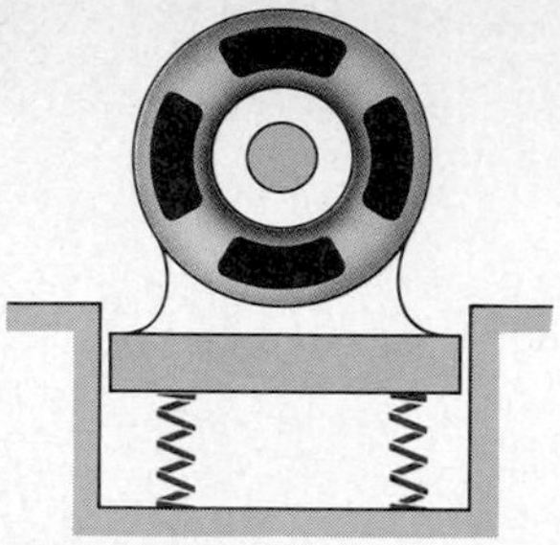

SAMPLE PROBLEM 19.5

A motor weighing 350 lb is supported by four springs, each having a constant of 750 lb/in. The unbalance of the rotor is equivalent to a weight of 1 oz located 6 in. from the axis of rotation. Knowing that the motor is constrained to move vertically, determine (*a*) the speed in rpm at which resonance will occur, (*b*) the amplitude of the vibration of the motor at a speed of 1200 rpm.

***a.* Resonance Speed.** The resonance speed is equal to the circular frequency (in rpm) of the free vibration of the motor. The mass of the motor and the equivalent constant of the supporting springs are

$$m = \frac{350 \text{ lb}}{32.2 \text{ ft/s}^2} = 10.87 \text{ lb}\cdot\text{s}^2/\text{ft}$$

$$k = 4(750 \text{ lb/in.}) = 3000 \text{ lb/in.} = 36{,}000 \text{ lb/ft}$$

$$p = \sqrt{\frac{k}{m}} = \sqrt{\frac{36{,}000}{10.87}} = 57.5 \text{ rad/s} = 549 \text{ rpm}$$

Resonance speed = 549 rpm ◄

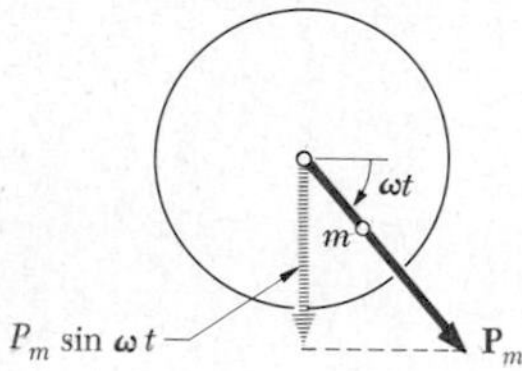

***b.* Amplitude of Vibration at 1200 rpm.** The angular velocity of the motor and the mass of the equivalent 1-oz weight are

$$\omega = 1200 \text{ rpm} = 125.7 \text{ rad/s}$$

$$m = (1 \text{ oz})\frac{1 \text{ lb}}{16 \text{ oz}}\frac{1}{32.2 \text{ ft/s}^2} = 0.00194 \text{ lb}\cdot\text{s}^2/\text{ft}$$

The magnitude of the centrifugal force due to the unbalance of the rotor is

$$P_m = ma_n = mr\omega^2 = (0.00194 \text{ lb}\cdot\text{s}^2/\text{ft})(\tfrac{6}{12} \text{ ft})(125.7 \text{ rad/s})^2 = 15.3 \text{ lb}$$

The static deflection that would be caused by a constant load P_m is

$$\frac{P_m}{k} = \frac{15.3 \text{ lb}}{3000 \text{ lb/in.}} = 0.00510 \text{ in.}$$

Substituting the value of P_m/k together with the known values of ω and p into Eq. (19.33), we obtain

$$x_m = \frac{P_m/k}{1-(\omega/p)^2} = \frac{0.00510 \text{ in.}}{1-(125.7/57.5)^2} = -0.00135 \text{ in.}$$

$x_m = 0.00135$ in. (out of phase) ◄

Note. Since $\omega > p$, the vibration is 180° out of phase with the centrifugal force due to the unbalance of the rotor. For example, when the unbalanced mass is directly below the axis of rotation, the position of the motor is $x_m =$ 0.00135 in. above the position of equilibrium.

Problems

19.85 A 5-kg cylinder is suspended from a spring of constant 320 N/m and is acted upon by a vertical periodic force of magnitude $P = P_m \sin \omega t$, where $P_m = 14$ N. Determine the amplitude of the motion of the cylinder if (*a*) $\omega = 6$ rad/s, (*b*) $\omega = 12$ rad/s.

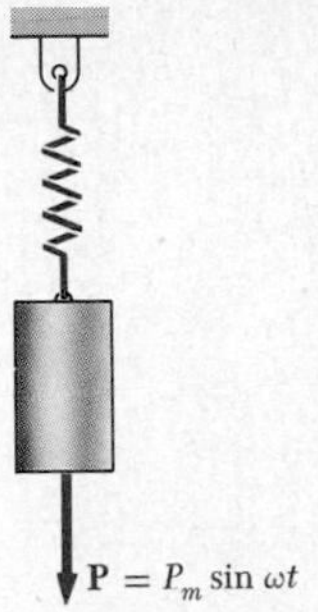

Fig. P19.85 and P19.86

19.86 A cylinder of mass m is suspended from a spring of constant k and is acted upon by a vertical periodic force of magnitude $P = P_m \sin \omega t$. Determine the range of values of ω for which the amplitude of the vibration exceeds twice the static deflection caused by a constant force of magnitude P_m.

19.87 In Prob. 19.86, determine the range of values of ω for which the amplitude of the vibration is less than the static deflection caused by a constant force of magnitude P_m.

19.88 A simple pendulum of length l is suspended from a collar C which is forced to move horizontally according to the relation $x_C = \delta_m \sin \omega t$. Determine the range of values of ω for which the amplitude of the motion of the bob is less than δ_m. (Assume that δ_m is small compared with the length l of the pendulum.)

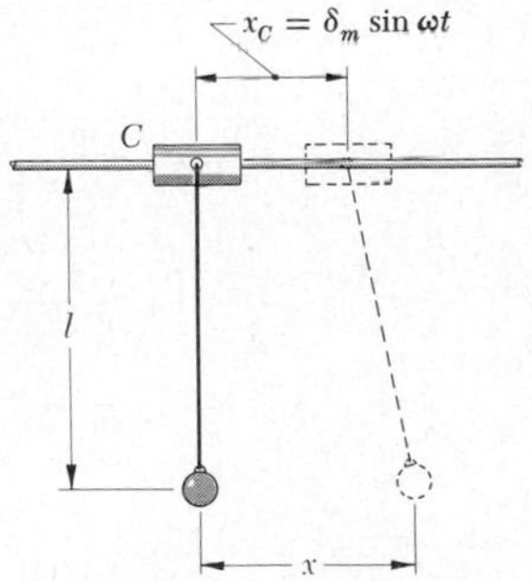

Fig. P19.88

19.89 In Prob. 19.88, determine the range of values of ω for which the amplitude of the motion of the bob exceeds $3\delta_m$.

19.90 A 125-kg motor is supported by a light horizontal beam. The unbalance of the rotor is equivalent to a mass of 25 g located 200 mm from the axis of rotation. Knowing that the static deflection of the beam due to the mass of the motor is 6.9 mm, determine (*a*) the speed (in rpm) at which resonance will occur, (*b*) the amplitude of the steady-state vibration of the motor at a speed of 720 rpm.

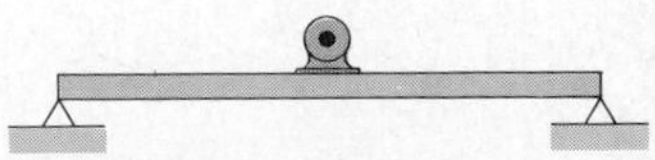
Fig. P19.90

19.91 Solve Prob. 19.90, assuming that the 125-kg motor is supported by a nest of springs having a total constant of 50 kN/m.

19.92 As the speed of a spring-supported motor is slowly increased from 300 to 400 rpm, the amplitude of the vibration due to the unbalance of the rotor is observed to decrease continuously from 0.075 to 0.040 in. Determine the speed at which resonance will occur.

19.93 For the spring-supported motor of Prob. 19.92, determine the speed of the motor for which the amplitude of the vibration is 0.100 in.

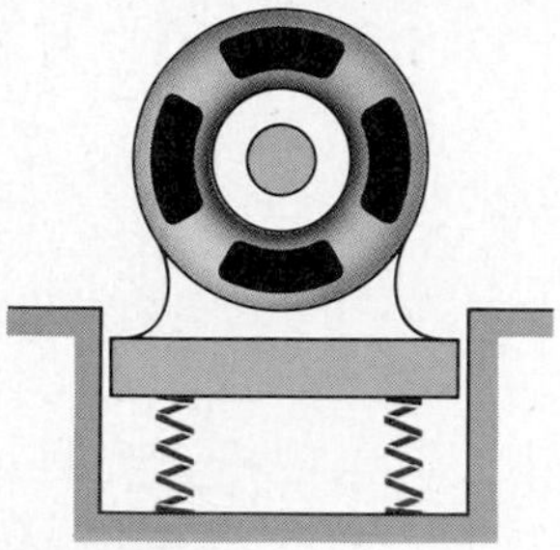
Fig. P19.94

19.94 A motor of mass 9 kg is supported by four springs, each of constant 20 kN/m. The motor is constrained to move vertically, and the amplitude of its motion is observed to be 1.2 mm at a speed of 1200 rpm. Knowing that the mass of the rotor is 2.5 kg, determine the distance between the mass center of the rotor and the axis of the shaft.

19.95 In Prob. 19.94, determine the amplitude of the vertical motion of the motor at a speed of (*a*) 450 rpm, (*b*) 1600 rpm, (*c*) 900 rpm.

19.96 Rod *AB* is rigidly attached to the frame of a motor running at a constant speed. When a collar of mass *m* is placed on the spring, it is observed to vibrate with an amplitude of 10 mm. When two collars, each of mass *m*, are placed on the spring, the amplitude is observed to be 12 mm. What amplitude of vibration should be expected when three collars, each of mass *m*, are placed on the spring? (Obtain two answers.)

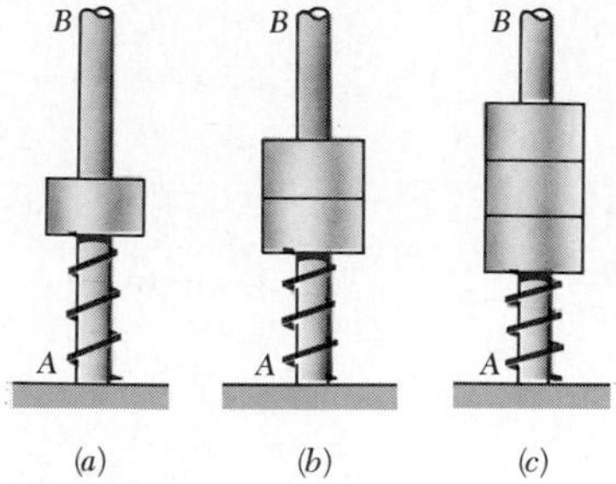

Fig. P19.96

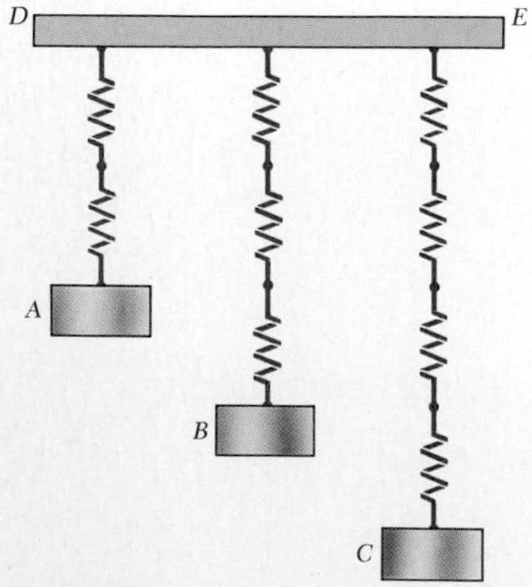

Fig. P19.98

19.97 Solve Prob. 19.96, assuming that the speed of the motor is changed and that one collar has an amplitude of 12 mm and two collars have an amplitude of 4 mm.

19.98 Three identical cylinders *A*, *B*, and *C* are suspended from a bar *DE* by two or more identical springs as shown. Bar *DE* is known to move vertically according to the relation $y = \delta_m \sin \omega t$. Knowing that the amplitudes of vibration of cylinders *A* and *B* are, respectively, 1.5 in. and 0.75 in., determine the expected amplitude of vibration of cylinder *C*. (Obtain two answers.)

19.99 Solve Prob. 19.98, assuming that the amplitudes of vibration of cylinders *A* and *B* are, respectively, 0.8 in. and 1.2 in.

19.100 A variable-speed motor is rigidly attached to the beam BC. When the speed of the motor is less than 750 rpm or more than 1500 rpm, a small object placed at A is observed to remain in contact with the beam. For speeds between 750 and 1500 rpm the object is observed to "dance" and actually to lose contact with the beam. Determine the speed at which resonance will occur.

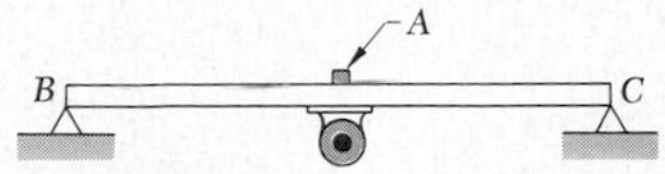

Fig. P19.100

19.101 A disk of mass m is attached to the midpoint of the vertical shaft AB, which revolves at a constant angular velocity ω. Denoting by k the spring constant of the system for a horizontal movement of the disk and by e the eccentricity of the disk with respect to the shaft, show that the deflection of the center of the shaft may be written in the form

$$r = \frac{e(\omega/p)^2}{1 - (\omega/p)^2}$$

where $p = \sqrt{k/m}$.

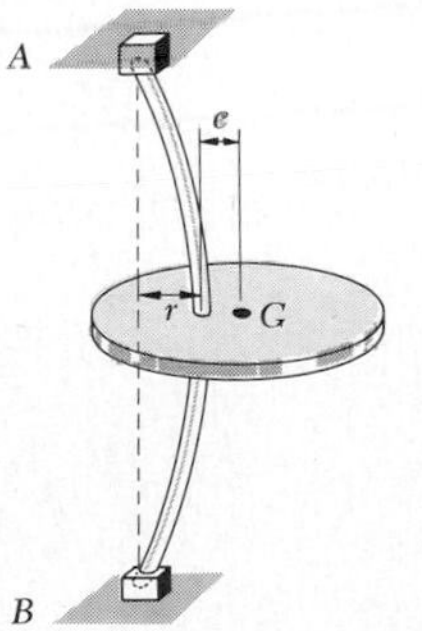

Fig. P19.101 and P19.102

19.102 A disk of mass 20 kg is attached with an eccentricity $e = 0.25$ mm to the midpoint of the vertical shaft AB, which revolves at a constant angular velocity ω. Knowing that a static force of 150 N will deflect the shaft 0.4 mm, determine (*a*) the angular velocity ω at which resonance occurs, (*b*) the deflection r of the shaft when $\omega = 1200$ rpm.

19.103 The amplitude of the motion of the pendulum bob shown is observed to be 60 mm, when the amplitude of the motion of collar C is 15 mm. Knowing that the length of the pendulum is $l = 0.90$ m, determine the two possible values of the frequency of the horizontal motion of collar C.

19.104 A pendulum is suspended from collar C as shown. Knowing that the amplitude of the motion of the bob is 12 mm for $l = 750$ mm and 17 mm for $l = 500$ mm, determine the frequency and amplitude of the horizontal motion of collar C.

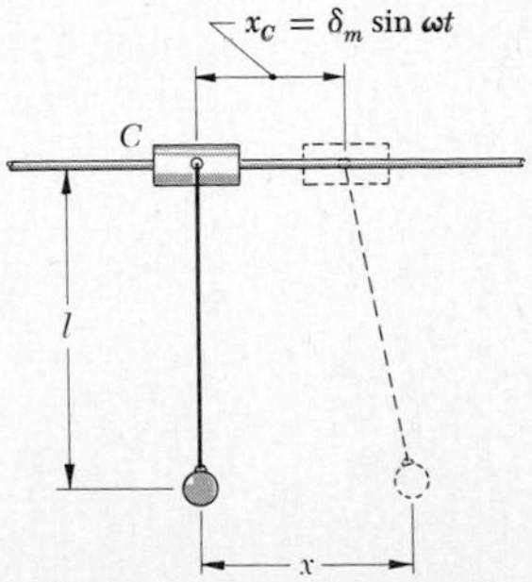

Fig. P19.103 and P19.104

19.105 A small trailer weighing 600 lb with its load is supported by two springs, each of constant 100 lb/in. The trailer is pulled over a road, the surface of which may be approximated by a sine curve of amplitude $1\frac{1}{2}$ in. and of period 16 ft (i.e., the distance between two successive crests is 16 ft, and the vertical distance from a crest to a trough is 3 in.). Determine (*a*) the speed at which resonance will occur, (*b*) the amplitude of the vibration of the trailer at a speed of 40 mi/h.

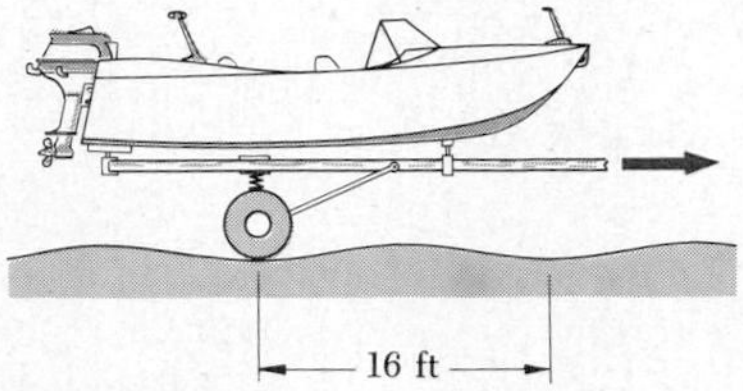

Fig. P19.105

19.106 Knowing that the amplitude of the vibration of the trailer of Prob. 19.105 is not to exceed $\frac{3}{4}$ in., determine the smallest speed at which the trailer can be pulled over the road.

DAMPED VIBRATIONS

* 19.8. Damped Free Vibrations.

The vibrating systems considered in the first part of this chapter were assumed free of damping. Actually all vibrations are damped to some degree by friction forces. These forces may be caused by *dry friction*, or *Coulomb friction*, between rigid bodies, by *fluid friction* when a rigid body moves in a fluid, or by *internal friction* between the molecules of a seemingly elastic body.

A type of damping of special interest is the *viscous damping* caused by fluid friction at low and moderate speeds. Viscous damping is characterized by the fact that the friction force is *directly proportional to the speed* of the moving body. As an example, we shall consider again a body of mass m suspended from a spring of constant k, and we shall assume that the body is attached to the plunger of a dashpot (Fig. 19.10). The magnitude of the friction force exerted on the plunger by the surrounding fluid is equal to $c\dot{x}$, where the constant c, expressed in N·s/m or lb·s/ft and known as the *coefficient of viscous damping*, depends upon the physical properties of the fluid and the construction of the dashpot. The equation of motion is

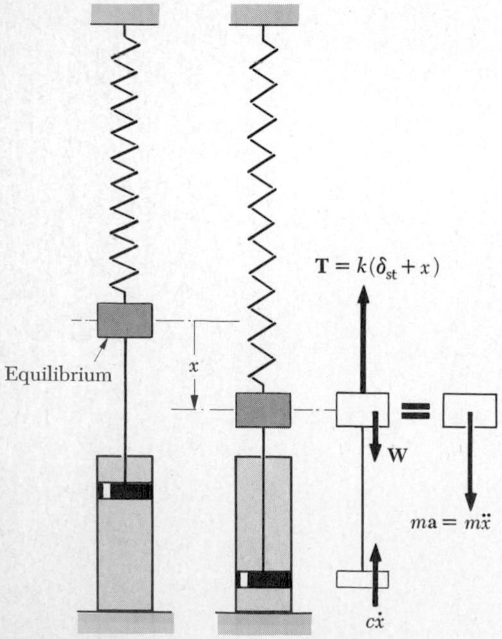

Fig. 19.10

$$+\downarrow\Sigma F = ma: \qquad W - k(\delta_{st} + x) - c\dot{x} = m\ddot{x}$$

Recalling that $W = k\delta_{st}$, we write

$$m\ddot{x} + c\dot{x} + kx = 0 \tag{19.37}$$

Substituting $x = e^{\lambda t}$ into (19.37) and dividing through by $e^{\lambda t}$, we write the *characteristic equation*

$$m\lambda^2 + c\lambda + k = 0 \tag{19.38}$$

and obtain the roots

$$\lambda = -\frac{c}{2m} \pm \sqrt{\left(\frac{c}{2m}\right)^2 - \frac{k}{m}} \tag{19.39}$$

Defining the *critical damping coefficient* c_c as the value of c which makes the radical in (19.39) equal to zero, we write

$$\left(\frac{c_c}{2m}\right)^2 - \frac{k}{m} = 0 \qquad c_c = 2m\sqrt{\frac{k}{m}} = 2mp \tag{19.40}$$

where p is the circular frequency of the system in the absence of damping. We may distinguish three different cases of damping, depending upon the value of the coefficient c.

1. *Heavy damping:* $c > c_c$. The roots λ_1 and λ_2 of the characteristic equation (19.38) are real and distinct, and the general solution of the differential equation (19.37) is

$$x = Ae^{\lambda_1 t} + Be^{\lambda_2 t} \tag{19.41}$$

This solution corresponds to a nonvibratory motion. Since λ_1 and λ_2 are both negative, x approaches zero as t increases indefinitely. However, the system actually regains its equilibrium position after a finite time.

2. *Critical damping:* $c = c_c$. The characteristic equation has a double root $\lambda = -c_c/2m = -p$, and the general solution of (19.37) is

$$x = (A + Bt)e^{-pt} \tag{19.42}$$

The motion obtained is again nonvibratory. Critically damped systems are of special interest in engineering applications since they regain their equilibrium position in the shortest possible time without oscillation.

3. *Light damping:* $c < c_c$. The roots of (19.38) are complex and conjugate, and the general solution of (19.37) is of the form

$$x = e^{-(c/2m)t}(A \sin qt + B \cos qt) \tag{19.43}$$

where q is defined by the relation

$$q^2 = \frac{k}{m} - \left(\frac{c}{2m}\right)^2$$

Substituting $k/m = p^2$ and recalling (19.40), we write

$$q = p\sqrt{1 - \left(\frac{c}{c_c}\right)^2} \tag{19.44}$$

where the constant c/c_c is known as the *damping factor.* A substitution similar to the one used in Sec. 19.2 enables us to write the general solution of (19.37) in the form

$$x = x_m e^{-(c/2m)t} \sin(qt + \phi) \tag{19.45}$$

The motion defined by (19.45) is vibratory with diminishing amplitude (Fig. 19.11). Although this motion does not actually repeat itself, the time interval $\tau = 2\pi/q$, corresponding to two successive points where the curve (19.45) touches one of the limiting curves shown in Fig. 19.11, is commonly referred to as the period of the damped vibration. Recalling (19.44), we observe that τ is larger than the period of vibration of the corresponding undamped system.

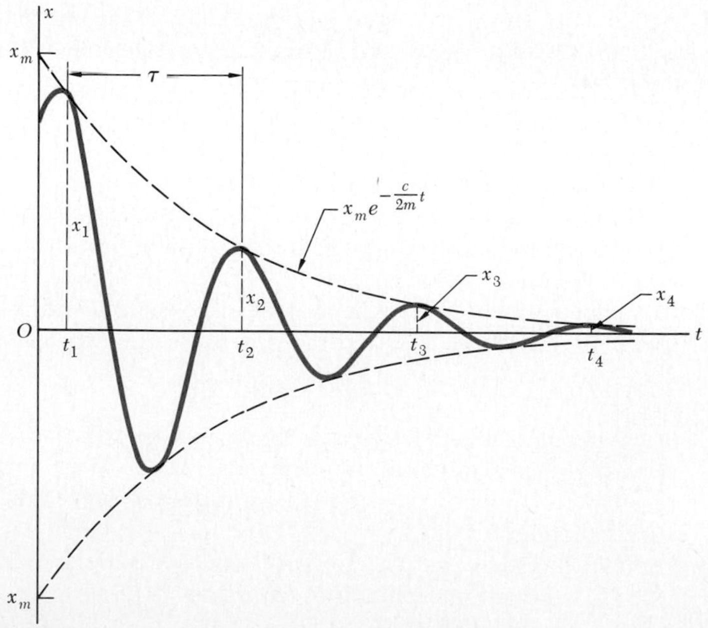

Fig. 19.11

***19.9. Damped Forced Vibrations.** If the system considered in the preceding section is subjected to a periodic force **P** of magnitude $P = P_m \sin \omega t$, the equation of motion becomes

$$m\ddot{x} + c\dot{x} + kx = P_m \sin \omega t \tag{19.46}$$

The general solution of (19.46) is obtained by adding a particular solution of (19.46) to the complementary function or general solution of the homogeneous equation (19.37). The complementary function is given by (19.41), (19.42), or (19.43), depending upon the type of damping considered. It represents a *transient* motion which is eventually damped out.

Our interest in this section is centered on the steady-state vibration represented by a particular solution of (19.46) of the form

$$x_{\text{part}} = x_m \sin (\omega t - \varphi) \tag{19.47}$$

Substituting x_{part} for x into (19.46), we obtain

$$-m\omega^2 x_m \sin (\omega t - \varphi) + c\omega x_m \cos (\omega t - \varphi) + kx_m \sin (\omega t - \varphi) = P_m \sin \omega t$$

Making $\omega t - \varphi$ successively equal to 0 and to $\pi/2$, we write

$$c\omega x_m = P_m \sin \varphi \tag{19.48}$$

$$(k - m\omega^2)x_m = P_m \cos \varphi \tag{19.49}$$

Squaring both members of (19.48) and (19.49) and adding, we have

$$[(k - m\omega^2)^2 + (c\omega)^2]x_m^2 = P_m^2 \tag{19.50}$$

Solving (19.50) for x_m and dividing (19.48) and (19.49) member by member, we obtain, respectively,

$$x_m = \frac{P_m}{\sqrt{(k - m\omega^2)^2 + (c\omega)^2}} \qquad \tan \varphi = \frac{c\omega}{k - m\omega^2} \tag{19.51}$$

Recalling from (19.4) that $k/m = p^2$, where p is the circular frequency of the undamped free vibration, and from (19.40) that $2mp = c_c$, where c_c is the critical damping coefficient of the system, we write

$$\frac{x_m}{P_m/k} = \frac{x_m}{\delta_m} = \frac{1}{\sqrt{[1 - (\omega/p)^2]^2 + [2(c/c_c)(\omega/p)]^2}} \tag{19.52}$$

$$\tan \varphi = \frac{2(c/c_c)(\omega/p)}{1 - (\omega/p)^2} \tag{19.53}$$

Formula (19.52) expresses the magnification factor in terms of the frequency ratio ω/p and damping factor c/c_c. It may be used to determine the amplitude of the steady-state vibration produced by an impressed force of magnitude $P = P_m \sin \omega t$ or by an impressed support movement $\delta = \delta_m \sin \omega t$. Formula (19.53) defines in terms of the same parameters the *phase difference* φ between the impressed force or impressed support movement and the resulting steady-state vibration of the damped system. The magnification factor has been plotted against the frequency ratio in Fig. 19.12 for various values of the damping factor. We observe that the amplitude of a forced vibration may be kept small by choosing a large coefficient of viscous damping c or by keeping the natural and forced frequencies far apart.

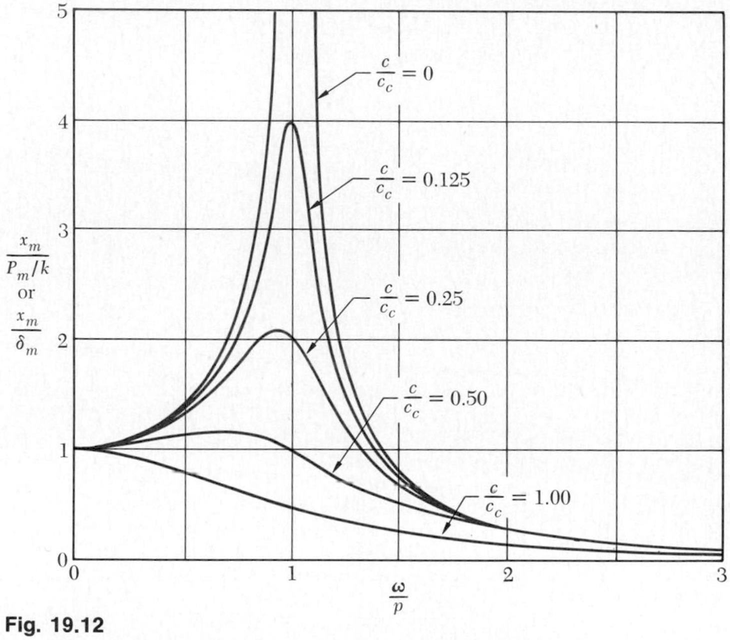

Fig. 19.12

*** 19.10. Electrical Analogues.** Oscillating electrical circuits are characterized by differential equations of the same type as those obtained in the preceding sections. Their analysis is therefore similar to that of a mechanical system, and the results obtained for a given vibrating system may be readily extended to the equivalent circuit. Conversely, any result obtained for an electrical circuit will also apply to the corresponding mechanical system.

Consider an electrical circuit consisting of an inductor of inductance L, a resistor of resistance R, and a capacitor of capacitance C, connected in series with a source of alternating voltage $E = E_m \sin \omega t$ (Fig. 19.13). It is recalled from elementary circuit theory† that if i denotes the current in the circuit and q the electric charge on the capacitor, the drop in potential is $L(di/dt)$ across the inductor, Ri across the resistor, and q/C across the capacitor. Expressing that the algebraic sum of the applied voltage and of the drops in potential around the circuit loop is zero, we write

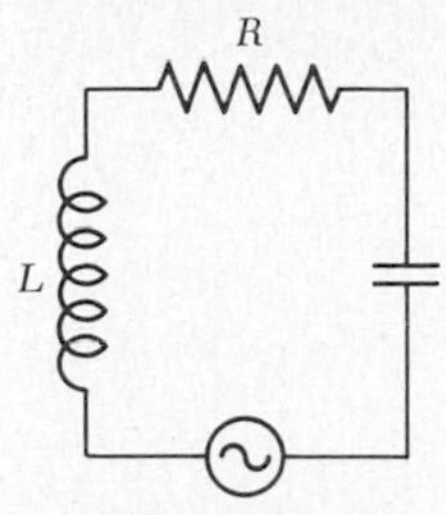

Fig. 19.13

$$E_m \sin \omega t - L\frac{di}{dt} - Ri - \frac{q}{C} = 0 \tag{19.54}$$

Rearranging the terms and recalling that at any instant the current i is equal to the rate of change $\dot{q}$ of the charge q, we have

$$L\ddot{q} + R\dot{q} + \frac{1}{C}q = E_m \sin \omega t \tag{19.55}$$

We verify that Eq. (19.55), which defines the oscillations of the electrical circuit of Fig. 19.13, is of the same type as Eq. (19.46), which characterizes the damped forced vibrations of the mechanical system of Fig. 19.10. By comparing the two equations, we may construct a table of the analogous mechanical and electrical expressions.

Table 19.2 may be used to extend to their electrical analogues the results obtained in the preceding sections for various mechanical systems. For instance, the amplitude i_m of the current in the circuit of Fig. 19.13 may be obtained by noting that it corresponds to the maximum value v_m of the velocity in the analogous mechanical system. Recalling that $v_m = \omega x_m$,

Table 19.2 Characteristics of a Mechanical System and of Its Electrical Analogue

Mechanical System		Electrical Circuit	
m	Mass	L	Inductance
c	Coefficient of viscous damping	R	Resistance
k	Spring constant	$1/C$	Reciprocal of capacitance
x	Displacement	q	Charge
v	Velocity	i	Current
P	Applied force	E	Applied voltage

†See A. E. Fitzgerald, D. E. Higginbotham, and A. Grabel, *Basic Electrical Engineering*, 5th ed., McGraw-Hill, New York, 1981; or J. W. Nilsson, *Electric Circuits*, Addison-Wesley, Reading, Mass., 1983.

substituting for x_m from Eq. (19.51), and replacing the constants of the mechanical system by the corresponding electrical expressions, we have

$$i_m = \frac{\omega E_m}{\sqrt{\left(\frac{1}{C} - L\omega^2\right)^2 + (R\omega)^2}}$$

$$i_m = \frac{E_m}{\sqrt{R^2 + \left(L\omega - \frac{1}{C\omega}\right)^2}} \tag{19.56}$$

The radical in the expression obtained is known as the *impedance* of the electrical circuit.

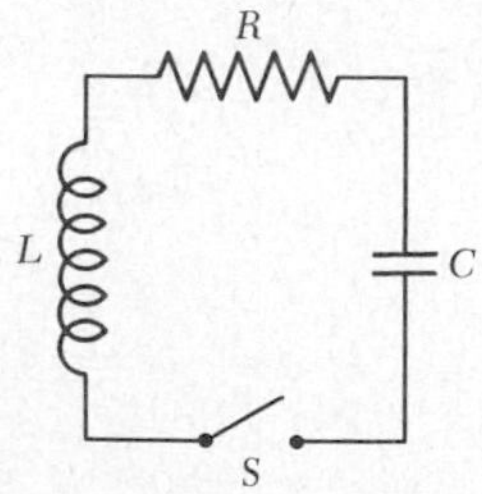

Fig. 19.14

The analogy between mechanical systems and electrical circuits holds for transient as well as steady-state oscillations. The oscillations of the circuit shown in Fig. 19.14, for instance, are analogous to the damped free vibrations of the system of Fig. 19.10. As far as the initial conditions are concerned, we may note that closing the switch S when the charge on the capacitor is $q = q_0$ is equivalent to releasing the mass of the mechanical system with no initial velocity from the position $x = x_0$. We should also observe that if a battery of constant voltage E is introduced in the electrical circuit of Fig. 19.14, closing the switch S will be equivalent to suddenly applying a force of constant magnitude P to the mass of the mechanical system of Fig. 19.10.

The discussion above would be of questionable value if its only result were to make it possible for mechanics students to analyze electrical circuits without learning the elements of circuit theory. It is hoped that this discussion will instead encourage students to apply to the solution of problems in mechanical vibrations the mathematical techniques they may learn in later courses in circuit theory. The chief value of the concept of electrical analogue, however, resides in its application to *experimental methods* for the determination of the characteristics of a given mechanical system. Indeed, an electrical circuit is much more easily constructed than is a mechanical model, and the fact that its characteristics may be modified by varying the inductance, resistance, or capacitance of its various components makes the use of the electrical analogue particularly convenient.

To determine the electrical analogue of a given mechanical system, we shall focus our attention on each moving mass in the system and observe which springs, dashpots, or external forces are applied directly to it. An equivalent electrical loop may then be constructed to match each of the mechanical units thus defined; the various loops obtained in that way will form together the desired circuit. Consider, for instance, the mechanical system of Fig. 19.15. We observe that the mass m_1 is acted upon by two springs of constants k_1 and k_2 and by two dashpots characterized by the coefficients of viscous damping c_1 and c_2. The electrical circuit should therefore include a loop consisting of an inductor of inductance L_1 proportional to m_1, of two capacitors of capacitance C_1 and C_2 inversely proportional to k_1 and k_2, respectively, and of two resistors of resistance R_1 and R_2, proportional to c_1 and c_2, respectively. Since the mass m_2 is acted upon by the spring k_2 and the dashpot c_2, as well as by the force $P = P_m \sin \omega t$, the circuit should also include a loop containing the capacitor C_2, the resistor R_2, the new inductor L_2, and the voltage source $E = E_m \sin \omega t$ (Fig. 19.16).

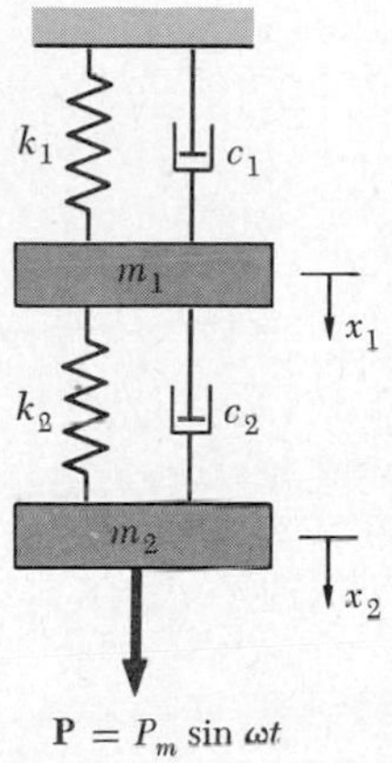

Fig. 19.15

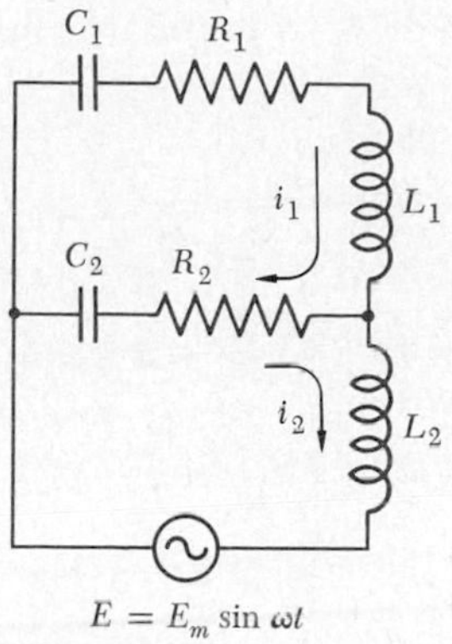

Fig. 19.16

To check that the mechanical system of Fig. 19.15 and the electrical circuit of Fig. 19.16 actually satisfy the same differential equations, we shall first derive the equations of motion for m_1 and m_2. Denoting, respectively, by x_1 and x_2 the displacements of m_1 and m_2 from their equilibrium positions, we observe that the elongation of the spring k_1 (measured from the equilibrium position) is equal to x_1, while the elongation of the spring k_2 is equal to the relative displacement $x_2 - x_1$ of m_2 with respect to m_1. The equations of motion for m_1 and m_2 are therefore

$$m_1\ddot{x}_1 + c_1\dot{x}_1 + c_2(\dot{x}_1 - \dot{x}_2) + k_1x_1 + k_2(x_1 - x_2) = 0 \tag{19.57}$$

$$m_2\ddot{x}_2 + c_2(\dot{x}_2 - \dot{x}_1) + k_2(x_2 - x_1) = P_m \sin \omega t \tag{19.58}$$

Consider now the electrical circuit of Fig. 19.16; we denote, respectively, by i_1 and i_2 the current in the first and second loops, and by q_1 and q_2 the integrals $\int i_1\,dt$ and $\int i_2\,dt$. Noting that the charge on the capacitor C_1 is q_1, while the charge on C_2 is $q_1 - q_2$, we express that the sum of the potential differences in each loop is zero:

$$L_1\ddot{q}_1 + R_1\dot{q}_1 + R_2(\dot{q}_1 - \dot{q}_2) + \frac{q_1}{C_1} + \frac{q_1 - q_2}{C_2} = 0 \tag{19.59}$$

$$L_2\ddot{q}_2 + R_2(\dot{q}_2 - \dot{q}_1) + \frac{q_2 - q_1}{C_2} = E_m \sin \omega t \tag{19.60}$$

We easily check that Eqs. (19.59) and (19.60) reduce to (19.57) and (19.58), respectively, when the substitutions indicated in Table 19.2 are performed.

Problems

19.107 Show that in the case of heavy damping ($c > c_c$), a body never passes through its position of equilibrium O (*a*) if it is released with no initial velocity from an arbitrary position or (*b*) if it is started from O with an arbitrary initial velocity.

19.108 Show that in the case of heavy damping ($c > c_c$), a body released from an arbitrary position with an arbitrary initial velocity cannot pass more than once through its equilibrium position.

19.109 In the case of light damping, the displacements x_1, x_2, x_3, and so forth, shown in Fig. 19.11 may be assumed equal to the maximum displacements. Show that the ratio of any two successive maximum displacements x_n and x_{n+1} is a constant and that the natural logarithm of this ratio, called the *logarithmic decrement*, is

$$\ln \frac{x_n}{x_{n+1}} = \frac{2\pi(c/c_c)}{\sqrt{1-(c/c_c)^2}}$$

19.110 In practice, it is often difficult to determine the logarithmic decrement of a system with light damping defined in Prob. 19.109 by measuring two successive maximum displacements. Show that the logarithmic decrement may also be expressed as $(1/k) \ln (x_n/x_{n+k})$, where k is the number of cycles between readings of the maximum displacement.

19.111 In a system with light damping ($c < c_c$), the period of vibration is commonly defined as the time interval $\tau = 2\pi/q$ corresponding to two successive points where the displacement-time curve touches one of the limiting curves shown in Fig. 19.11. Show that the interval of time (*a*) between a maximum positive displacement and the following maximum negative displacement is $\frac{1}{2}\tau$, (*b*) between two successive zero displacements is $\frac{1}{2}\tau$, (*c*) between a maximum positive displacement and the following zero displacement is greater than $\frac{1}{4}\tau$.

19.112 Successive maximum displacements of a spring-mass-dashpot system similar to that shown in Fig. 19.10 are 50, 40, 32, and 25.6 mm. Knowing that m = 12 kg and k = 1500 N/m, determine (*a*) the damping factor c/c_c, (*b*) the value of the coefficient of viscous damping c. (*Hint.* See Probs. 19.109 and 19.110.)

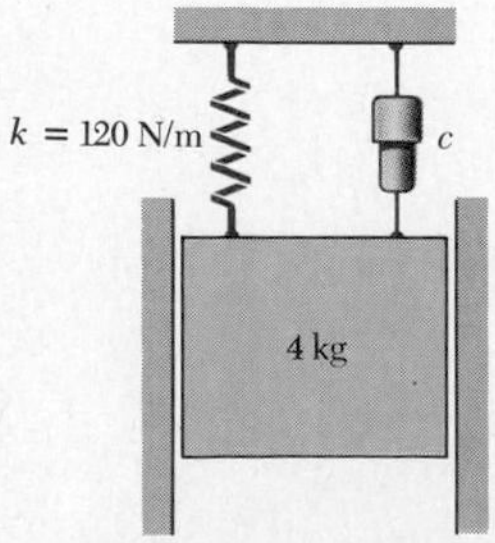

Fig. P19.113

19.113 The block shown is depressed 20 mm from its equilibrium position and released. Knowing that after eight cycles the maximum displacement of the block is 12 mm, determine (*a*) the damping factor c/c_c, (*b*) the value of the coefficient of viscous damping. (*Hint.* See Probs. 19.109 and 19.110.)

19.114 The barrel of a field gun weighs 1400 lb and is returned into firing position after recoil by a recuperator of constant $k = 10{,}000$ lb/ft. Determine (*a*) the value of the coefficient of damping of the recuperator which causes the barrel to return into firing position in the shortest possible time without oscillation, (*b*) the time needed for the barrel to move from its maximum-recoil position halfway back to its firing position.

19.115 Assuming that the barrel of the gun of Prob. 19.114 is modified, with a resulting increase in weight of 400 lb, determine (*a*) the constant k which should be used for the recuperator if the barrel is to remain critically damped, (*b*) the time needed for the modified barrel to move from its maximum-recoil position halfway back to its firing position.

19.116 In the case of the forced vibration of a system with a given damping factor c/c_c, determine the frequency ratio ω/p for which the amplitude of the vibration is maximum.

19.117 Show that for a small value of the damping factor c/c_c, (*a*) the maximum amplitude of a forced vibration occurs when $\omega \approx p$, (*b*) the corresponding value of the magnification factor is approximately $\frac{1}{2}(c_c/c)$..

19.118 A 30-lb motor is directly supported by a light horizontal beam which has a static deflection of 0.05 in. due to the weight of the motor. Knowing that the unbalance of the rotor is equivalent to a weight of 1 oz located 7.5 in. from the axis of rotation, determine the amplitude of vibration of the motor at a speed of 900 rpm, assuming (*a*) that no damping is present, (*b*) that the damping factor c/c_c is equal to 0.075.

19.119 A motor weighing 50 lb is supported by four springs, each having a constant of 1000 lb/in. The unbalance of the rotor is equivalent to a weight of 1 oz located 5 in. from the axis of rotation. Knowing that the motor is constrained to move vertically, determine the amplitude of the steady-state vibration of the motor at a speed of 1800 rpm, assuming (*a*) that no damping is present, (*b*) that the damping factor c/c_c is equal to 0.125.

19.120 Solve Prob. 19.94, assuming that a dashpot having a coefficient of damping $c = 200$ N·s/m has been connected to the motor and to the ground.

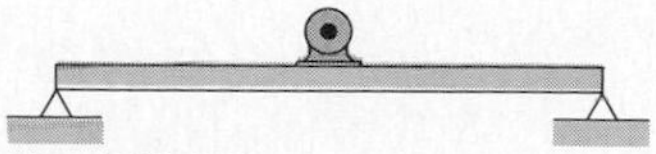

Fig. P19.121

19.121 A 50-kg motor is directly supported by a light horizontal beam which has a static deflection of 6 mm due to the weight of the motor. The unbalance of the rotor is equivalent to a mass of 100 g located 75 mm from the axis of rotation. Knowing that the amplitude of the vibration of the motor is 0.9 mm at a speed of 400 rpm, determine (*a*) the damping factor c/c_c, (*b*) the coefficient of damping c.

19.122 A machine element having a mass of 400 kg is supported by two springs, each having a constant of 38 kN/m. A periodic force of maximum magnitude equal to 135 N is applied to the element with a frequency of 2.5 Hz. Knowing that the coefficient of damping is 1400 N·s/m, determine the amplitude of the steady-state vibration of the element.

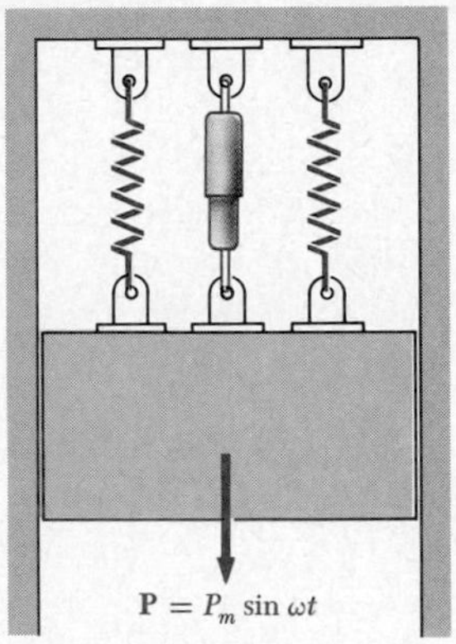

Fig. P19.122

19.123 In Prob. 19.122, determine the required value of the coefficient of damping if the amplitude of the steady-state vibration of the element is to be 3.5 mm.

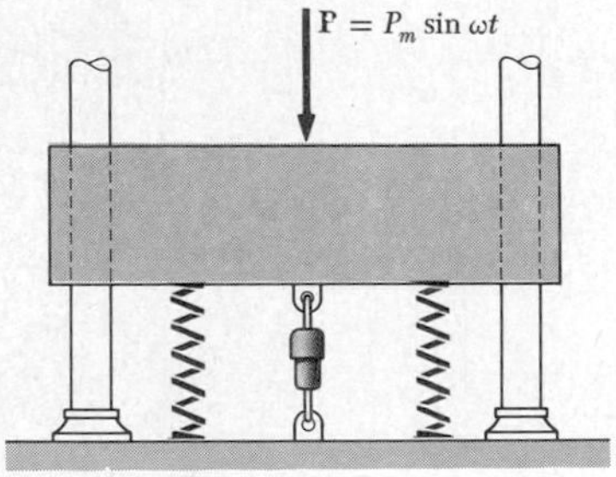

Fig. P19.124

19.124 A platform of weight 200 lb, supported by two springs each of constant $k = 250$ lb/in., is subjected to a periodic force of maximum magnitude equal to 125 lb. Knowing that the coefficient of damping is 10 lb·s/in., determine (*a*) the natural frequency in rpm of the platform if there were no damping, (*b*) the frequency in rpm of the periodic force corresponding to the maximum value of the magnification factor, assuming damping, (*c*) the amplitude of the actual motion of the platform for each of the frequencies found in parts *a* and *b*.

19.125 Solve Prob. 19.124, assuming that the coefficient of damping is increased to 15 lb·s/in.

***19.126** The suspension of an automobile may be approximated by the simplified spring-and-dashpot system shown. (*a*) Write the differential equation defining the absolute motion of the mass m when the system moves at a speed v over a road of sinusoidal cross section as shown. (*b*) Derive an expression for the amplitude of the absolute motion of m.

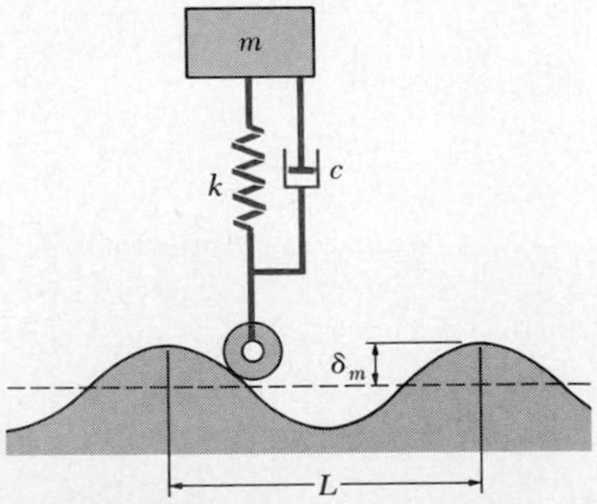

Fig. P19.126

***19.127** Two blocks A and B, each of mass m, are suspended as shown by means of five springs of the same constant k and are connected by a dashpot of coefficient of damping c. Block B is subjected to a force of magnitude $P = P_m \sin \omega t$. Write the differential equations defining the displacements x_A and x_B of the two blocks from their equilibrium positions.

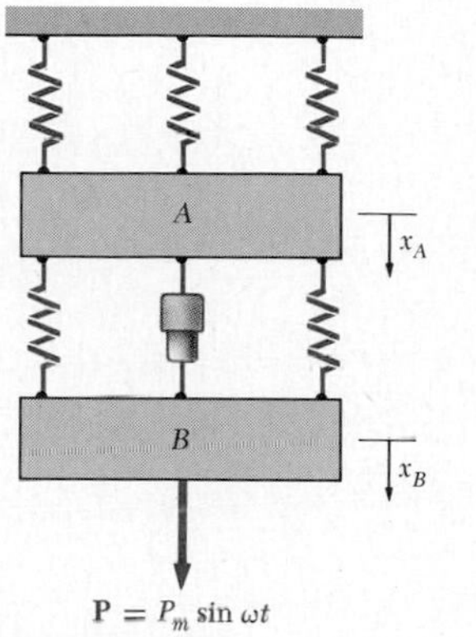

Fig. P19.127

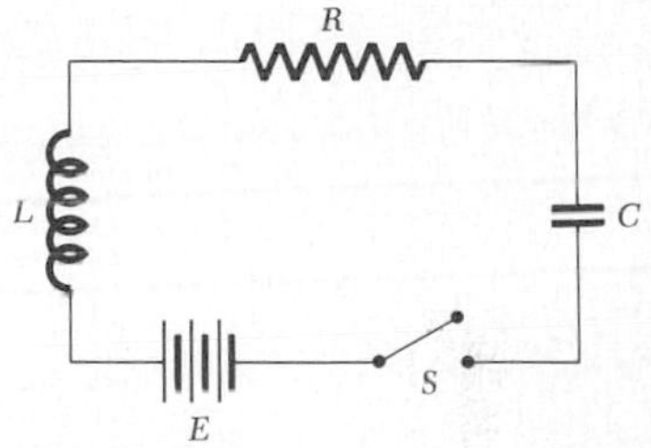

Fig. P19.128

19.128 Determine the range of values of the resistance R for which oscillations will take place in the circuit shown when the switch S is closed.

19.129 Consider the circuit of Prob. 19.128 when the capacitor C is removed. If the switch S is closed at time $t = 0$, determine (*a*) the final value of the current in the circuit, (*b*) the time t at which the current will have reached $(1 - 1/e)$ times its final value. (The desired value of t is known as the *time constant* of the circuit.)

19.130 and 19.131 Draw the electrical analogue of the mechanical system shown. (*Hint.* Draw the loops corresponding to the free bodies m and A.)

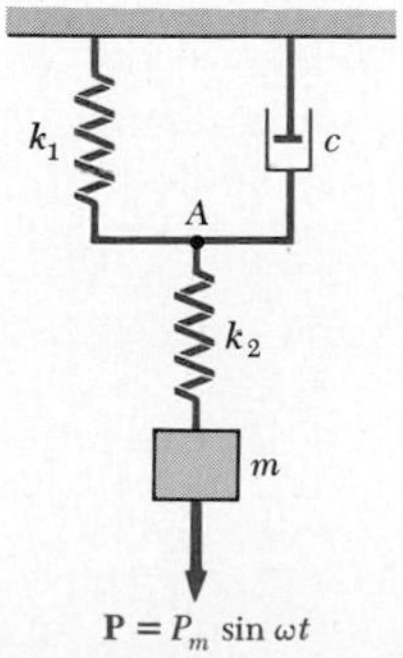

Fig. P19.130 and P19.132

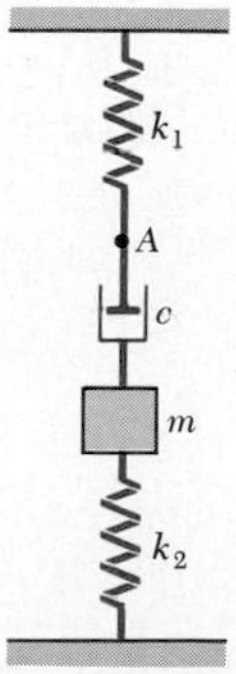

Fig. P19.131 and P19.133

19.132 and 19.133 Write the differential equations defining (*a*) the displacements of mass m and point A, (*b*) the currents in the corresponding loops of the electrical analogue.

19.134 and 19.135 Draw the electrical analogue of the mechanical system shown.

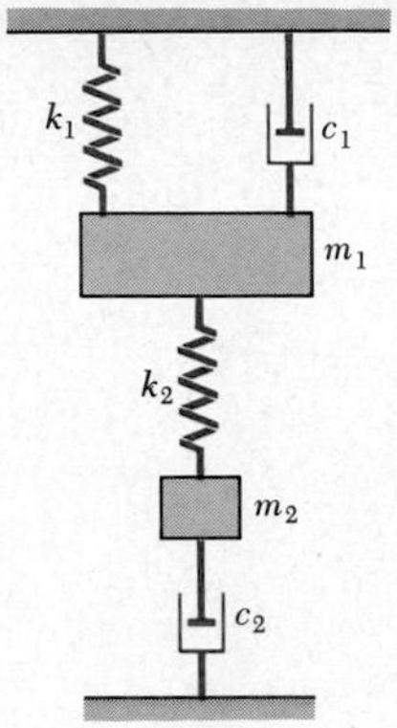

Fig. P19.134 and P19.136

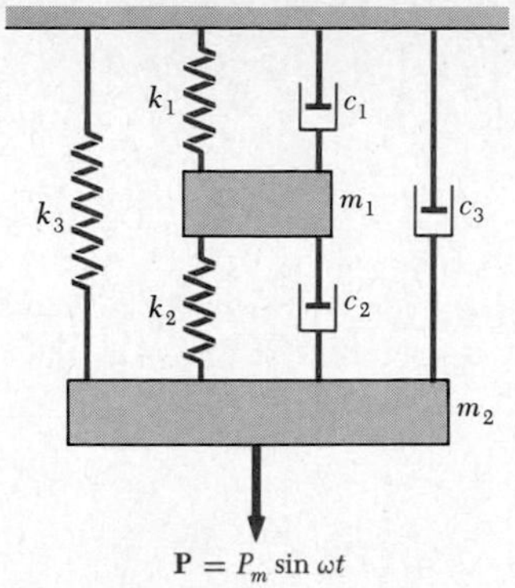

Fig. P19.135 and P19.137

19.136 and 19.137 Write the differential equations defining (*a*) the displacements of the masses m_1 and m_2, (*b*) the currents in the corresponding loops of the electrical analogue.

Review Problems

19.138 A 4-lb block is spring-mounted on the frame of a motor rotating at 1250 rpm. The rotor is unbalanced and the amplitude of the resulting motion of the block is measured and found to be 0.40 in. Knowing that the constant of the spring mounting is $k = 150$ lb/in., determine the amplitude of the motion of the motor.

19.139 A slender bar of length l is attached by a smooth pin A to a collar of negligible mass. Determine the period of small oscillations of the bar, assuming that the coefficient of friction between the collar and the horizontal rod (*a*) is sufficient to prevent any movement of the collar, (*b*) is zero.

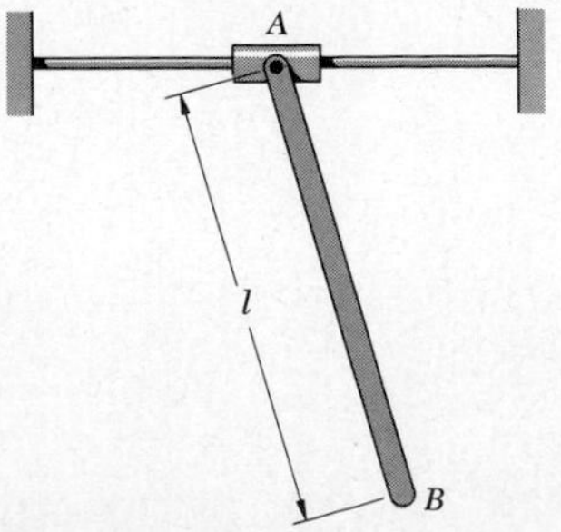

Fig. P19.139

19.140 The 10-kg rod AB is attached to 4-kg disks as shown. Knowing that the disks roll without sliding, determine the frequency of small oscillations of the system.

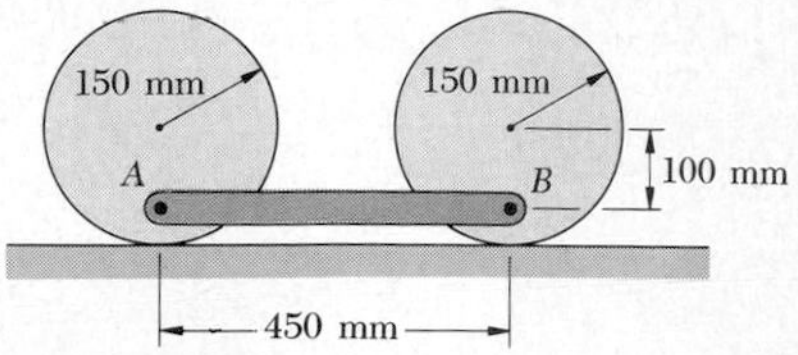

Fig. P19.140

19.141 A particle is placed with no initial velocity on a frictionless *plane* tangent to the surface of the earth. (*a*) Show that the particle will theoretically execute simple harmonic motion with a period of oscillation equal to that of a simple pendulum of length equal to the radius of the earth. (*b*) Compute the theoretical period of oscillation and show that it is equal to the periodic time of an earth satellite describing a low-altitude circular orbit. [*Hint.* See Eq. (12.44).]

19.142 A 1.5-kg collar is attached to a spring of constant k = 750 N/m as shown. If the collar is depressed 20 mm and released, determine (*a*) the maximum velocity of the resulting motion, (*b*) the position and velocity of the collar 0.08 s after release.

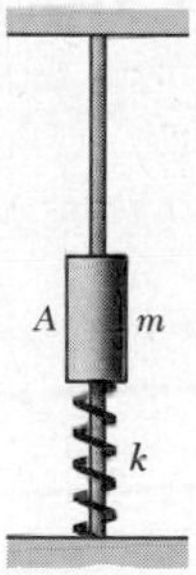

Fig. P19.142

19.143 A rod of mass m and length L rests on the two pulleys A and B which rotate in the directions shown. Denoting by μ_k the coefficient of kinetic friction between the rod and the pulleys, determine the frequency of vibration if the rod is given a small displacement to the right and released.

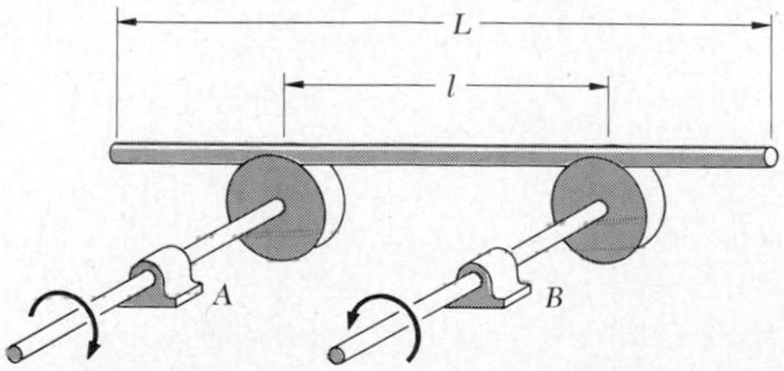

Fig. P19.143

19.144 A torsion pendulum may be used to determine experimentally the moment of inertia of a given object. The horizontal platform P is held by several rigid bars connected to a vertical wire. The period of oscillation of the platform is found equal to τ_0 when the platform is empty and to τ_A when an object of known moment of inertia I_A is placed on the platform so that its mass center is directly above the center of the plate. (*a*) Show that the moment of inertia I_0 of the platform and its supports may be expressed as $I_0 = I_A\tau_0^2/(\tau_A^2 - \tau_0^2)$. (*b*) If a period of oscillation τ_B is observed when an object B of unknown moment of inertia I_B is placed on the platform, show that $I_B = I_A(\tau_B^2 - \tau_0^2)/(\tau_A^2 - \tau_0^2)$.

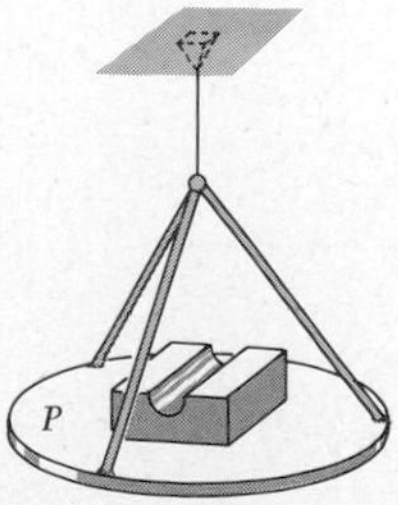

Fig. P19.144

19.145 A 15-kg beam is supported as shown by two uniform disks, each of mass 10 kg and radius 100 mm. Knowing that the disks roll without sliding, determine the period of vibration of the system if the beam is given a small displacement to the right and released.

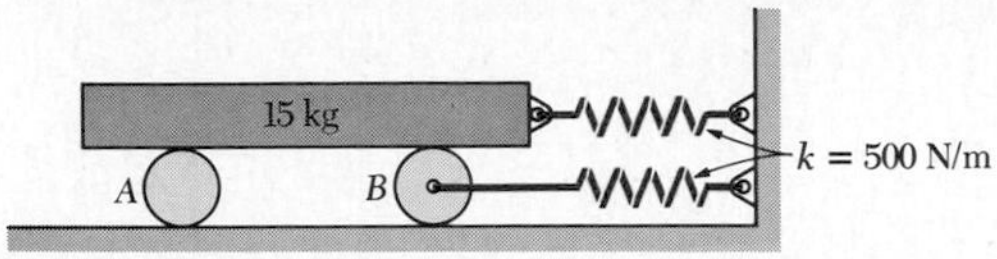

Fig. P19.145

19.146 Solve Prob. 19.145, assuming that the spring attached to the beam is removed.

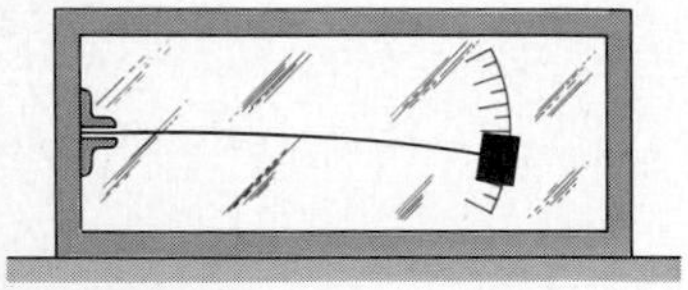

Fig. P19.147

19.147 A certain vibrometer used to measure vibration amplitudes consists essentially of a box containing a slender rod to which a mass m is attached; the natural frequency of the mass-rod system is known to be 8 Hz. When the box is rigidly attached to the casing of a motor rotating at 960 rpm, the mass is observed to vibrate with an amplitude of 0.08 in. relative to the box. Determine the amplitude of the vertical motion of the motor.

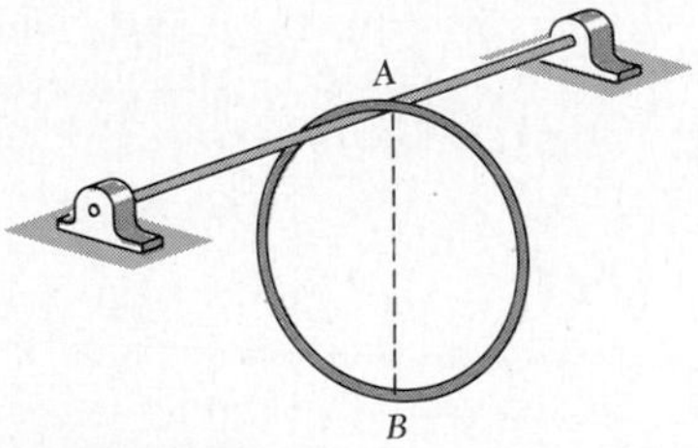

Fig. P19.148

19.148 A thin hoop of radius r and mass m is suspended from a rough rod as shown. Determine the frequency of small oscillations of the hoop (*a*) in the plane of the hoop, (*b*) in a direction perpendicular to the plane of the hoop. Assume that friction is sufficiently large to prevent slipping at A.

19.149 A 400-lb flywheel has a diameter of 32 in. and a radius of gyration of 14 in. A belt is placed around the rim and attached to two springs, each of constant $k = 60$ lb/in. The initial tension in the belt is sufficient to prevent slipping. If the end C of the belt is pulled 1.25 in. down and released, determine (*a*) the period of vibration, (*b*) the maximum angular velocity of the flywheel.

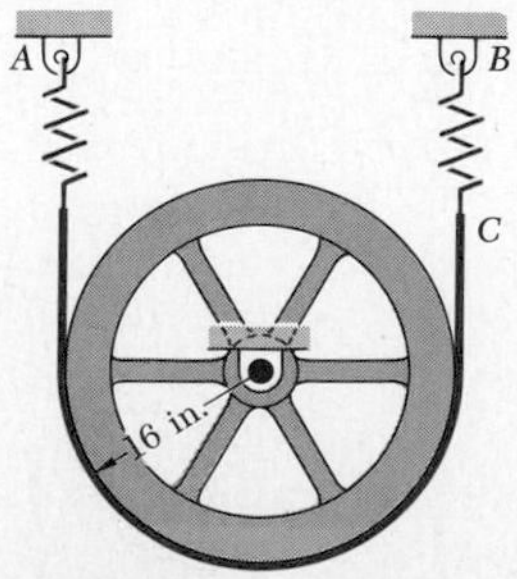

Fig. P19.149

Index

Index

Answers to Even-Numbered Problems

Answers to Even-Numbered Problems

CHAPTER 2

2.2 707 lb ∠ 50.8°.
2.4 17.00 kN ∠ 84.8°.
2.6 18.95°.
2.8 (*a*) 585 lb. (*b*) 896 lb.
2.10 (*a*) 78.9 N. (*b*) 169.2 N.
2.12 $P = 72.1$ N; $\alpha = 44.7°$.
2.14 414 lb ∠ 72.0°.
2.16 (45 lb) +25.8 lb, +36.9 lb;
(60 lb) +49.1 lb, +34.4 lb;
(75 lb) +48.2 lb, −57.5 lb.
2.18 (530 N) +450 N, +280 N;
(510 N) −240 N, +450 N.
2.20 (*a*) 261 lb. (*b*) −167.8 lb.
2.22 −120 lb, +350 lb.
2.26 1449 N ∠ 78.4°.
2.28 760 N ∠ 74.0°.
2.30 25.6 kN.
2.32 (*a*) 36.9°. (*b*) 80 lb.
2.34 $T_{AC} = 352$ lb; $T_{BC} = 261$ lb.
2.36 $T_{AC} = 265$ N; $T_{BC} = 175.0$ N.
2.38 $T_{AC} = 326$ N; $T_{BC} = 369$ N.
2.40 $T_A = 231$ lb; $T_B = 577$ lb.
2.42 $T_{AB} = 600$ N; $T_{AC} = 344$ N.
2.44 (*a*) $\alpha = 35°$; $T_{AC} = 410$ N; $T_{BC} = 287$ N.
(*b*) $\alpha = 55°$; $T_{AC} = T_{BC} = 305$ N.
2.46 30 in.
2.48 (*a*) 1472 N. (*b*) 1472 N. (*c*) 981 N.
(*d*) 981 N. (*e*) 736 N.
2.50 913 N ∠ 82.5°.
2.52 (*a*) 18 lb. (*b*) 64 lb.
2.54 (*a*) +278 N, +383 N, +160.7 N.
(*b*) 56.2°, 40.0°, 71.3°.
2.56 (*a*) −175.8 N, −257 N, +251 N.
(*b*) 116.1°, 130.0°, 51.1°.
2.58 (*a*) −1861 lb, +3360 lb, +677 lb.
(*b*) 118.5°, 30.5°, 80.0°.
2.60 1050 N; 51.8°, 107.7°, 43.6°.
2.62 (*a*) $F_x = +107.7$ lb, $F_z = -267$ lb;
$F = 416$ lb. (*b*) 43.9°.
2.64 −6.30 kN, +6.06 kN, +4.85 kN.
2.66 −225 lb, +150 lb, +90 lb.
2.68 $R = 623$ lb; $\theta_x = 37.4°$, $\theta_y = 122.0°$,
$\theta_z = 72.6°$.
2.70 $R = 498$ N; $\theta_x = 68.9°$, $\theta_y = 26.3°$,
$\theta_z = 75.1°$.
2.72 510 N.
2.74 2025 lb.
2.76 9.32 kN.
2.78 $T_{AB} = 500$ N; $T_{AC} = 459$ N; $T_{AD} = 516$ N.
2.80 $T_{AB} = 2.88$ kN; $T_{AC} = 5.76$ kN;
$T_{AD} = 3.60$ kN.
2.82 $T_{AB} = T_{AC} = 3.35$ lb; $T_{AD} = 5.80$ lb.
2.84 $T_{AB} = 31.7$ lb; $T_{AC} = 64.3$ lb.
2.86 $P = 138$ N; $T_{AB} = 270$ N; $T_{AC} = 196$ N.
2.88 $P = 131.2$ N; $Q = 29.6$ N.
2.90 $T_{BAC} = 57.6$ lb; $T_{AD} = 20.1$ lb; $T_{AE} = 36.9$ lb.
2.92 605 mm.
2.94 $x = 134.2$ mm, $z = 67.1$ mm.
2.96 (*a*) 991 N. (*b*) 973 N.
2.98 62.8 lb $\leq P \leq$ 258 lb.
2.100 141.5°, 124.4°, 74.9°.
2.102 179.3 N $< P <$ 669 N.
2.104 (*a*) 337 lb ∠ 10.7°. (*b*) 296 lb ∠ 3.1°.
2.106 $T_{AD} = 40$ lb; $T_{BD} = T_{CD} = 34$ lb.

CHAPTER 3

3.2 16.4°.
3.4 116.0 N·m ↻.

3.6 (*a*) 27.4 N·m ↺. (*b*) 228 N ∠ 42.0°.
3.8 18.48 N·m ↺.
3.10 156.0 lb.
3.12 $d = (xF_y - yF_x)/(F_x^2 + F_y^2)^{\frac{1}{2}}$.
3.14 (*a*) +3**i** − 26**j** − 18**k**. (*b*) 0. (*c*) −15**i** − 20**j**.
3.16 (7.5 N·m)**i** − (6 N·m)**j** − (10.39 N·m)**k**.
3.18 (196.8 lb·ft)**i** + (57.6 lb·ft)**j** − (148.8 lb·ft)**k**.
3.20 100.8 mm.
3.22 9.21 ft.
3.24 207 mm.
3.26 **P**·**Q** = +1; **P**·**S** = −11; **Q**·**S** = +10.
3.28 77.9°.
3.30 26.8°.
3.32 (*a*) 59.1°. (*b*) 648 N.
3.34 +46; +46; −46.
3.36 $M_x = 0$, $M_y = -162$ N·m, $M_z = +270$ N·m.
3.38 44.1 lb.
3.40 $P = 125$ N; $\phi = 73.7°$; $\theta = 53.1°$.
3.42 −90.0 N·m.
3.44 +207 lb·ft.
3.46 $+Pa/\sqrt{2}$.
3.48 (*b*) $d = a/\sqrt{2}$.
3.50 0.268 m.
3.52 12.69 in.
3.54 (*a*) 16.67 N. (*b*) 31.25 N. (*c*) 14.71 N.
3.56 1.25 in.
3.58 (*a*) $M = 13.63$ N·m; $\theta_x = 27.8°$, $\theta_y = 62.2°$, $\theta_z = 90°$. (*b*) 18.17 N ∠ 62.2° at *B* and 18.17 N ∠ 62.2° at *C*.
3.60 $M = 3610$ lb·ft; $\theta_x = 159.4°$, $\theta_y = 90.8°$, $\theta_z = 69.4°$.
3.62 **F** = 260 lb ∠ 67.4°, **M** = 200 lb·in. ↻.
3.64 (*a*) **F** = 250 N ∠ 25°, **M** = 57.5 N·m ↻.
(*b*) **A** = 375 N ∠ 25°, **B** = 625 N ∠ 25°.
3.66 **F** = 750 N ↓; $x = 40$ mm.
3.68 (*a*) = 30°. (*b*) 65.7°.
3.70 **F** = −(300 lb)**i** + (120 lb)**j** − (200 lb)**k**,
M = (3000 lb·ft)**j** + (1800 lb·ft)**k**.
3.72 **F** = −(122.9 N)**j** − (86.0 N)**k**,
M = (22.6 N·m)**i** + (15.49 N·m)**j** − (22.1 N·m)**k**.
3.74 Force-couple systems at *A* and *C*.
3.76 *c* and *g*.
3.78 (*a*) 500 N ↓; 2.2 m. (*b*) 500 N ↑; 1 m.
(*c*) 500 N ↓; 0.4 m.
3.80 (*a*) 1.50 ft. (*b*) 3.15 ft. (*c*) 5.55 ft.
3.82 (*a*) 391 lb ∠ 39.8°.
(*b*) 12 in. to the right of *C* and 10 in. above *C*.
3.84 (*a*) 136.0 N ∠ 28.0°. (*b*) 18.40 mm to the right of *A* and 50.0 mm above *C*.
3.86 (*a*) **R** = 3.79 kN →, $\mathbf{M}_D^R$ = 22.8 kN·m ↺.
(*b*) **R** = 3.79 kN →, 6 m below *DE*.
3.88 (*a*) **B** = −(80 N)**k**; **C** = −(30 N)**i** + (40 N)**k**.
(*b*) $R_y = 0$, $R_z = -40$ N. (*c*) Vertical.
3.90 (*a*) 60°. (*b*) **R** = (15.00 lb)**i** − (26.0 lb)**j**,
$\mathbf{M}_C^R$ = (312 lb·in.)**i**.
3.92 **R** = (15.00 lb)**i** − (26.0 lb)**j**,
$\mathbf{M}_E^R$ = (312 lb·in.)**i** + (89.7 lb·in.)**k**.
(*a*) Tighten. (*b*) Loosen.
3.94 75 kips at $x = +4.80$ ft and $z = -1.386$ ft.
3.96 140 kN; 0.714 m from *A* on edge *AD*.
3.98 (*a*) **R** = −(60 lb)**j**. (*b*) −0.6 in.
(*c*) $x = 0$, $z = 9$ in.
3.100 (*a*) −(40 N)**j**. (*b*) 2.5 mm.
(*c*) $x = -5$ mm, $z = 0$.
3.102 (*a*) $-\frac{1}{2}P\mathbf{k}$. (*b*) $R = \frac{3}{2}P$; $\theta_x = \theta_y = 48.2°$, $\theta_z = 109.5°$. (*c*) $y = a$, $z = \frac{1}{2}a$.
3.104 (*a*) $R = 491$ N; $\theta_x = 148.7°$, $\theta_y = 95.8°$, $\theta_z = 120.6°$. (*b*) 16.40 mm.
(*c*) $x = +358$ mm, $z = +137.8$ mm.
3.106 (*a*) **R** = (40 lb)**i** + (30 lb)**j**. (*b*) +3.08 in.
(*c*) $x = 0$, $z = +1.44$ in.
3.112 4.06 kN.
3.114 910 lb.
3.116 1165 lb·in. ↻.
3.118 (*a*) 500 N ∠ 60°, 68.9 N·m ↻.
(*b*) **A** = 689 N ↑; **B** = 1150 N ∠ 77.4°.
3.120 60.1 kN ↓, at $x = 4.99$ m, $z = 3.33$ m.
3.122 (*a*) **A** = 25 lb ↓; **C** = 25 lb ↑.
(*b*) **B** = 36 lb ↘; **D** = 36 lb ↖.
(*c*) **A** = 23.1 lb ↘; **D** = 23.1 lb ↖.

CHAPTER 4

4.2 (*a*) 23.4 lb ↑. (*b*) 78.3 lb ↑.
4.4 (*a*) 64.6 kN. (*b*) 106.6 kN ↑.
4.6 62.0°.
4.8 0.5 kN ≤ Q ≤ 11 kN.
4.10 1 kN ≤ Q ≤ 3.25 kN.
4.12 **A** = 489 N ↓; **C** = 669 N ∠ 81.4°.
4.14 230 lb.
4.16 (*a*) **A** = 4.27 kN ∠ 20.6°; **B** = 4.50 kN ↑.
(*b*) **A** = 1.50 kN ↑; **B** = 6.02 kN ∠ 48.4°.
(*c*) **A** = 2.05 kN ∠ 47.0°;
B = 5.20 kN ∠ 60.0°.
4.18 (*a*) **A** = 25 lb ↓; **B** = 25 lb ↑.

(*b*) $\mathbf{A} = 37.5$ lb $\leftarrow$; $\mathbf{B} = 37.5$ lb $\rightarrow$.
(*c*) $\mathbf{A} = 20.8$ lb $\angle$ 60°; $\mathbf{B} = 20.8$ lb $\angle$ 60°.

4.20 $T_{BE} = 196.2$ N; $\mathbf{A} = 73.6$ N $\rightarrow$; $\mathbf{D} = 73.6$ N $\leftarrow$.

4.22 $\mathbf{A} = 69.3$ lb $\rightarrow$; $\mathbf{B} = 34.6$ lb $\angle$ 60°; $\mathbf{C} = 173.2$ lb $\angle$ 60°.

4.24 (*a*) 1842 N $\rightarrow$. (*b*) 1250 N $\angle$ 40°.

4.26 (*a*) 128 lb. (*b*) $\mathbf{A} = 80$ lb $\uparrow$; $\mathbf{B} = 64$ lb $\rightarrow$.

4.28 (*a*) $\mathbf{M} = 4.82$ N·m $\circlearrowleft$.
(*b*) $\mathbf{A}_x = 40.2$ N $\rightarrow$; $\mathbf{A}_y = 150$ N $\uparrow$.

4.30 $51.6 \text{ lb} \leq T_2 \leq 158.0$ lb.

4.32 $\theta = \tan^{-1}(Q/3P)$.

4.34 (*a*) $\theta = 2\sin^{-1}(W/2P)$. (*b*) $\theta = 29.0°$.

4.36 (*a*) $\theta = 2\sin^{-1}[\frac{1}{2}kl/(kl - P)]$. (*b*) 83.6°.

4.38 (*a*) $y\left(1 - \dfrac{a}{\sqrt{a^2 + y^2}}\right) = \dfrac{W}{k}$. (*b*) 1000 N/m.

4.40 (1) Completely constrained; determinate; $\mathbf{A} = 12.02$ kips $\angle$ 56.3°; $\mathbf{B} = 6.67$ kips $\leftarrow$. (2) Improperly constrained; indeterminate; no equil. (3) Completely constrained; determinate; $\mathbf{A} = \mathbf{C} = 5$ kips $\uparrow$. (4) Completely constrained; indeterminate; $\mathbf{A}_x = 6.67$ kips $\rightarrow$; $\mathbf{B}_x = 6.67$ kips $\leftarrow$; $(\mathbf{A}_y + \mathbf{B}_y = 10$ kips $\uparrow)$. (5) Improperly constrained; indeterminate; no equil. (6) Partially constrained; determinate; equil.; $\mathbf{A} = \mathbf{C} = 5$ kips $\uparrow$. (7) Completely constrained; determinate; $\mathbf{A} = 5$ kips $\uparrow$; $\mathbf{B} = 8.33$ kips $\angle$ 36.9°; $\mathbf{C} = 6.67$ kips $\rightarrow$. (8) Completely constrained; indeterminate; $\mathbf{A}_y = 5$ kips $\uparrow$.

4.42 $\mathbf{A} = 2230$ N $\angle$ 7.73°; $\mathbf{B} = 2210$ N $\rightarrow$.

4.44 $\mathbf{A} = 62.5$ lb $\angle$ 53.1°; $\mathbf{B} = 37.5$ lb $\rightarrow$.

4.46 (*a*) $\mathbf{A} = P\sqrt{2}$ $\angle$ 45°; $\mathbf{B} = P \uparrow$.
(*b*) $\mathbf{A} = P/\sqrt{2} \leftarrow$; $\mathbf{B} = P/\sqrt{2} \uparrow$.

4.48 $\mathbf{A} = 82.5$ lb $\angle$ 14.04°; $T = 100$ lb.

4.52 $\mathbf{A} = 170$ N $\angle$ 33.9°; $\mathbf{C} = 160$ N $\angle$ 28.1°.

4.54 (*a*) 120 lb. (*b*) $\mathbf{C} = 150$ lb $\angle$ 36.9°.

4.56 $\theta + \alpha = 90°$.

4.58 (*a*) 225 mm. (*b*) 23.1 N. (*c*) 12.21 N $\rightarrow$.

4.60 34.3°.

4.62 $\mathbf{A} = (30.0 \text{ lb})\mathbf{j} - (2.89 \text{ lb})\mathbf{k}$; $\mathbf{B} = (20.0 \text{ lb})\mathbf{j} - (11.55 \text{ lb})\mathbf{k}$; $\mathbf{C} = (14.43 \text{ lb})\mathbf{k}$.

4.64 $\mathbf{A} = (104 \text{ N})\mathbf{j} + (128 \text{ N})\mathbf{k}$; $\mathbf{B} = (52 \text{ N})\mathbf{j} + (160 \text{ N})\mathbf{k}$.

4.66 (*a*) 750 lb. (*b*) $\mathbf{C} = (250 \text{ N})\mathbf{i} + (750 \text{ N})\mathbf{j}$; $\mathbf{D} = -(1000 \text{ N})\mathbf{i} - (300 \text{ N})\mathbf{j}$.

4.68 $T_A = 24.5$ N; $T_B = 73.6$ N; $T_C = 98.1$ N.

4.70 $W = 90$ lb at $x = 0.889$ ft, $z = 2.44$ ft.

4.72 (*a*) $T_{BD} = T_{BE} = 16.60$ kN.
(*b*) $\mathbf{A} = (17.00 \text{ kN})\mathbf{i} + (45.1 \text{ kN})\mathbf{j}$.

4.74 (*a*) $T_{BD} = 700$ lb; $T_{BC} = 975$ lb.
(*b*) $\mathbf{A} = (1500 \text{ lb})\mathbf{i} + (425 \text{ lb})\mathbf{j}$.

4.76 (*a*) $T_{DE} = T_{DF} = 1.124$ kN.
(*b*) $\mathbf{A} = -(2.95 \text{ kN})\mathbf{i} + (5.77 \text{ kN})\mathbf{j}$.

4.78 (*a*) 157.1 lb.
(*b*) $\mathbf{A} = -(98.2 \text{ lb})\mathbf{j} - (64.7 \text{ lb})\mathbf{k}$; $\mathbf{B} = (45.0 \text{ lb})\mathbf{j}$.

4.80 (*a*) 3.00 lb.
(*b*) $\mathbf{A} = -(2.94 \text{ lb})\mathbf{i} + (6.90 \text{ lb})\mathbf{j}$; $\mathbf{B} = (7.50 \text{ lb})\mathbf{j}$.

4.82 (*a*) 689 N. (*b*) $\mathbf{A} = (503 \text{ N})\mathbf{i} + (424 \text{ N})\mathbf{j}$; $\mathbf{B} = (150.9 \text{ N})\mathbf{i} + (557 \text{ N})\mathbf{j} - (229 \text{ N})\mathbf{k}$.

4.84 (*a*) 1548 N. (*b*) $\mathbf{A} = (503 \text{ N})\mathbf{i} + (1094 \text{ N})\mathbf{j}$; $\mathbf{B} = (150.9 \text{ N})\mathbf{i} - (113.2 \text{ N})\mathbf{j} - (1003 \text{ N})\mathbf{k}$.

4.86 (*a*) 3.00 lb. (*b*) $\mathbf{B} = -(2.94 \text{ lb})\mathbf{i} + (14.40 \text{ lb})\mathbf{j}$; $\mathbf{M}_B = -(20.7 \text{ lb·ft})\mathbf{i} - (8.82 \text{ lb·ft})\mathbf{j}$.

4.88 (*a*) 689 N.
(*b*) $\mathbf{B} = (654 \text{ N})\mathbf{i} + (981 \text{ N})\mathbf{j} - (229 \text{ N})\mathbf{k}$; $\mathbf{M}_B = -(883 \text{ N·m})\mathbf{i} + (1046 \text{ N·m})\mathbf{j}$.

4.90 (*a*) 12.50 lb.
(*b*) $\mathbf{A} = (6.00 \text{ lb})\mathbf{j} + (4.50 \text{ lb})\mathbf{k}$; $\mathbf{B} = (10 \text{ lb})\mathbf{i}$.

4.92 (*a*) $D = 225$ N. (*b*) $S = 300$ N; $d = 0.75$ m.

4.94 $\mathbf{B} = (30.5 \text{ lb})\mathbf{i} - (45.0 \text{ lb})\mathbf{k}$; $\mathbf{C} = -(0.5 \text{ lb})\mathbf{j} + (45.0 \text{ lb})\mathbf{k}$; $\mathbf{D} = -(30.5 \text{ lb})\mathbf{i} + (6 \text{ lb})\mathbf{j}$.

4.96 $T_{BD} = 2.90$ kN; $T_{BE} = 5.25$ kN; $T_{CD} = 2.00$ kN.

4.98 $(18 \text{ lb})\,\mathbf{j}$.

4.100 $(45 \text{ N})\mathbf{k}$.

4.102 652 N.

4.104 (*a*) 450 N. (*b*) $\mathbf{A} = 3$ kN $\leftarrow$; $\mathbf{B} = 3$ kN $\rightarrow$.

4.106 62.5 lb.

4.108 $\mathbf{B} = 571$ N $\angle$ 57.7°; $\mathbf{C} = 356$ N $\angle$ 31.0°.

4.110 $\mathbf{A} = (375 \text{ N})\mathbf{j} - (500 \text{ N})\mathbf{k}$; $\mathbf{B} = (625 \text{ N})\mathbf{i} + (500 \text{ N})\mathbf{k}$; $\mathbf{C} = -(625 \text{ N})\mathbf{i} + (375 \text{ N})\mathbf{j}$.

4.112 19.4°.

4.114 $\mathbf{A} = 58.5$ kN $\angle$ 60°; $\mathbf{B} = 39.6$ kN $\angle$ 83.8°.

CHAPTER 5

5.2 $\bar{X} = 3.27$ in., $\bar{Y} = 2.82$ in.

5.4 $\bar{X} = 70$ mm, $\bar{Y} = 60$ mm.

5.6 $\bar{X} = 16.21$ mm, $\bar{Y} = 31.9$ mm.

5.8 $\bar{X} = \bar{Y} = 5.06$ in.

5.10 $\bar{X} = \bar{Y} = 13.40$ mm.

5.12 $\bar{X} = \bar{Y} = 9.00$ in.

5.14 $\bar{X} = 321$ mm, $\bar{Y} = 53.1$ mm.
5.16 $\bar{X} = \frac{1}{3}a(h_1 + 2h_2)/(h_1 + h_2)$.
5.18 0.520.
5.20 0.494.
5.22 42.25×10^3 mm^3 and -42.25×10^3 mm^3.
5.24 (a) $Q_x = \frac{1}{2}b(c^2 - y^2)$.
(b) $y = 0$; $(Q_x)_{max} = \frac{1}{2}bc^2$.
5.26 $\bar{X} = 3.38$ in., $\bar{Y} = 2.93$ in.
5.28 $\bar{X} = 17.77$ mm, $\bar{Y} = 29.8$ mm.
5.30 120 mm.
5.32 (a) $\sqrt{3}\,r$. (b) $\sqrt{5}\,r$.
5.36 15.3 mm.
5.38 $\bar{x} = 0.7424a$; -1.01 percent.
5.40 $\bar{x} = 2a/5$, $\bar{y} = 3b/7$.
5.42 $\bar{x} = 3a/5$, $\bar{y} = 12h/35$.
5.50 $\bar{y} = 0.48\,h$.
5.52 $\bar{x} = L/\pi$, $\bar{y} = \pi a/8$.
5.54 $\bar{x} = h$, $\bar{y} = \frac{1}{4}h$.
5.56 (a) 584 in^3. (b) 679 in^3.
5.58 $A = 4\pi^2 rR$; $V = 2\pi^2 r^2 R$.
5.60 494 in^3; 11.44 lb.
5.62 6.03 m^2.
5.64 21.3×10^3 mm^3, 153.6 g.
5.66 14.52 in^2.
5.68 (a) 158 ft^3. (b) 146 ft^2.
5.70 (a) $\pi R^2 h$. (b) $\frac{2}{3}\pi R^2 h$. (c) $\frac{1}{2}\pi R^2 h$. (d) $\frac{1}{3}\pi R^2 h$.
(e) $\frac{1}{6}\pi R^2 h$.
5.72 $\mathbf{R} = 4.20$ kN $\downarrow$, 1.143 m to the right of A;
$\mathbf{A} = 2.20$ kN $\uparrow$, $\mathbf{B} = 2.00$ kN $\uparrow$.
5.74 $\mathbf{A} = 630$ lb $\uparrow$; $\mathbf{M}_A = 4680$ lb·in. $\circlearrowleft$.
5.76 $\mathbf{A} = 2.04$ kN $\uparrow$; $\mathbf{B} = 8.16$ kN $\uparrow$.
5.78 $\mathbf{A} = 6.86$ kN $\uparrow$; $\mathbf{B} = 1.776$ kN $\uparrow$.
5.80 $\mathbf{B} = 3770$ lb $\uparrow$; $\mathbf{C} = 429$ lb $\uparrow$.
5.82 $w_A = 10$ kN/m; $w_B = 50$ kN/m.
5.84 (a) $\mathbf{H} = 99.3$ kN $\rightarrow$, $\mathbf{V} = 324$ kN $\uparrow$; 2.03 m to the right of A. (b) $\mathbf{R} = 104.7$ kN $\measuredangle$ 18.4°.
5.86 (a) 36.4 kN, 0.864 m below A. (b) 21.0 kN $\leftarrow$.
5.88 12 in.
5.90 $\mathbf{A} = 129.8$ kN $\measuredangle$ 64.2°; $\mathbf{B} = 127.9$ kN $\uparrow$.
5.92 277 lb $\swarrow$.
5.94 491 N.
5.96 4.87 ft.
5.98 $\mathbf{A} = 153.0$ N $\uparrow$; $\mathbf{B} = 789$ N $\uparrow$.
5.100 $\frac{1}{8}b$ to the right of base of cone.
5.102 27.8 mm above base.
5.104 $\bar{X} = \bar{Z} = 0$, $\bar{Y} = -0.608h$.
5.106 $\bar{Y} = -19.02$ mm.
5.108 $\bar{Z} = 3.47$ in.
5.110 $\bar{Z} = 51.5$ mm.
5.112 $\bar{X} = 125$ mm, $\bar{Y} = 167.0$ mm, $\bar{Z} = 33.5$ mm.
5.114 $\bar{X} = \bar{Z} = 3.62$ in., $\bar{Y} = 4.83$ in.
5.116 $\bar{X} = 0$, $\bar{Y} = 5.02$ in., $\bar{Z} = 2.57$ in.
5.118 33.5 mm above base.
5.124 $\bar{x} = 5h/8$.
5.126 $\bar{x} = \frac{1}{2}a(1 - 4/\pi^2)$.
5.128 $\bar{x} = 3a/8$, $\bar{y} = h/4$, $\bar{z} = 3b/8$.
5.130 42.5 mm.
5.132 $\bar{x} = 0$, $\bar{y} = 5h/16$, $\bar{z} = -a/4$.
5.134 $\bar{X} = 58.3$ mm, $\bar{Y} = 83.5$ mm.
5.136 (a) $\bar{X} = 12$ in., $\bar{Y} = -4.67$ in., $\bar{Z} = 0$.
(b) 1.807.
5.138 $\bar{x} = 90$ mm, $\bar{y} = 15$ mm, $\bar{z} = 60$ mm.
5.140 (a) $+155.5$ m^3. (b) $+83.3$ m^3.
5.142 $\bar{X} = 2.53$ in., $\bar{Y} = 0.615$ in., $\bar{Z} = 2$ in.
5.144 $\cos\alpha = \frac{1}{2}\csc\theta$.

CHAPTER 6

6.2 $F_{AB} = 148$ kN T; $F_{AC} = 72$ kN C;
$F_{AD} = 74$ kN T; $F_{BC} = F_{CD} = 70$ kN C.
6.4 $F_{AB} = 900$ lb T; $F_{AC} = 780$ lb C;
$F_{BC} = 720$ lb T.
6.6 $F_{AB} = 4$ kN T; $F_{AD} = 15$ kN T; $F_{BD} = 9$ kN C;
$F_{BE} = 5$ kN T; $F_{CD} = 16$ kN C; $F_{DE} = 4$ kN C.
6.8 $F_{AB} = 750$ lb T; $F_{AC} = 750$ lb T;
$F_{AD} = 1200$ lb C; $F_{BD} = 210$ lb T;
$F_{CD} = 720$ lb T.
6.10 $F_{AB} = 420$ lb C; $F_{AC} = F_{CE} = 400$ lb T;
$F_{AD} = 260$ lb C; $F_{BC} = F_{CD} = 125$ lb T;
$F_{BE} = 832$ lb C.
6.12 $F_{AB} = 10$ kN C;
$F_{AC} = F_{BC} = F_{CD} = F_{CE} = 11.55$ kN T;
$F_{AD} = F_{BE} = 5.77$ kN C.
6.14 BD, DG, FG, GH, HI, IJ.
6.16 AC, CE, EF, FG.
6.18 (a) $\mathbf{A} = -(112.5\text{ lb})\mathbf{i} - (60\text{ lb})\mathbf{k}$; $\mathbf{B} = (60\text{ lb})\mathbf{k}$;
$\mathbf{C} = (112.5\text{ lb})\mathbf{i} + (170\text{ lb})\mathbf{k}$.
(b) $F_{AB} = F_{AD} = F_{BC} = 0$; $F_{AC} = 170$ lb C;
$F_{AE} = 212.5$ lb T; $F_{BD} = 127.5$ lb T;
$F_{BE} = F_{CD} = 112.5$ lb C; $F_{DE} = 60$ lb C.
6.20 (a) $\mathbf{B} = -(2700\text{ N})\mathbf{k}$;
$\mathbf{C} = -(1800\text{ N})\mathbf{i} - (3375\text{ N})\mathbf{j}$;
$\mathbf{D} = (1800\text{ N})\mathbf{i} + (3375\text{ N})\mathbf{j}$.
(b) $F_{AB} = F_{CD} = 0$; $F_{AC} = 4275$ N T;

$F_{AD} = 4275$ N C; $F_{BC} = 4270$ N C;
$F_{BD} = 4270$ N T.

6.22 $F_{FG} = 5$ kN T; $F_{FH} = 20$ kN T.

6.24 $F_{CE} = 36$ kN T; $F_{EF} = 15$ kN C.

6.26 $F_{FG} = 4.53$ kips T; $F_{FH} = 10.67$ kips T.

6.28 $F_{EI} = 54$ kN T; $F_{EG} = 60$ kN C.

6.30 $F_{DG} = 102$ kips C; $F_{EG} = 48$ kips C.

6.32 $F_{BD} = 20$ kN C; $F_{BE} = 25$ kN C; $F_{CE} = 0$.

6.34 $F_{GJ} = 30$ kN C; $F_{GK} = 50$ kN C;
$F_{IK} = 15$ kN T.

6.36 $F_{DG} = 15$ kips C; $F_{FG} = 11.9$ kips T;
$F_{FH} = 15.3$ kips T.

6.38 $F_{AC} = 11.25$ kN T; $F_{BE} = 11.25$ kN C.

6.40 $F_{AE} = 142.5$ kN C; $F_{DE} = 170$ kN T.

6.42 $F_{DE} = 0$; $F_{BE} = 10$ kips T; $F_{EF} = 5$ kips T.

6.44 $F_{DE} = 1.5$ kips C; $F_{BE} = 2.5$ kips T;
$F_{DG} = 2.5$ kips T.

6.46 (*a*) Completely constrained; determinate.
(*b*) Completely constrained; indeterminate.
(*c*) Improperly constrained.

6.48 (*a*) Completely constrained; determinate.
(*b*) Improperly constrained.
(*c*) Completely constrained; indeterminate.

6.50 $F_{BD} = 300$ N C; $\mathbf{C}_x = 164$ N $\leftarrow$, $\mathbf{C}_y = 288$ N $\downarrow$.

6.52 $\mathbf{A}_x = 900$ lb $\leftarrow$, $\mathbf{A}_y = 75$ lb $\uparrow$; $\mathbf{B} = 825$ lb $\downarrow$;
$\mathbf{D}_x = 900$ lb $\rightarrow$, $\mathbf{D}_y = 750$ lb $\uparrow$.

6.54 $\mathbf{A} = 150$ N $\rightarrow$; $\mathbf{B} = 0$; $\mathbf{C}_x = 150$ N $\leftarrow$,
$\mathbf{C}_y = 100$ N $\uparrow$; $\mathbf{D} = 100$ N $\downarrow$.

6.56 (*a*) 4.46 kN $\measuredangle$ 8.4°. (*b*) 3.53 kN $\measuredangle$ 51.3°.

6.58 $\mathbf{B} = 82$ lb $\downarrow$; $\mathbf{C}_x = 75$ lb $\leftarrow$, $\mathbf{C}_y = 115$ lb $\uparrow$;
$\mathbf{D}_x = 75$ lb $\rightarrow$, $\mathbf{D}_y = 12$ lb $\uparrow$.

6.60 (*a*) 1524 kN T. (*b*) 1133 kN C.
(*c*) 1593 kN $\measuredangle$ 62.1°.

6.62 (*a*) $\mathbf{A} = 65$ lb $\measuredangle$ 22.6°; $\mathbf{C} = 120$ lb $\rightarrow$;
$\mathbf{G} = 60$ lb $\leftarrow$; $\mathbf{I} = 25$ lb $\uparrow$.
(*b*) $\mathbf{A} = 65$ lb $\measuredangle$ 22.6°; $\mathbf{C} = 100$ lb $\rightarrow$;
$\mathbf{G} = 80$ lb $\leftarrow$; $\mathbf{I} = 25$ lb $\uparrow$.

6.64 (*a*) $\mathbf{D}_x = 750$ N $\rightarrow$, $\mathbf{D}_y = 250$ N $\downarrow$;
$\mathbf{E}_x = 750$ N $\leftarrow$, $\mathbf{E}_y = 250$ N $\uparrow$.
(*b*) $\mathbf{D}_x = 375$ N $\rightarrow$, $\mathbf{D}_y = 250$ N $\downarrow$;
$\mathbf{E}_x = 375$ N $\leftarrow$, $\mathbf{E}_y = 250$ N $\uparrow$.

6.66 (*a*) A' and B'. (*b*) 268 N T.

6.68 (*a*) 176.9 lb T. (*b*) $\mathbf{C} = 262$ lb $\measuredangle$ 85.0°.

6.70 $\mathbf{D}_x = 13.60$ kN $\rightarrow$, $\mathbf{D}_y = 7.50$ kN $\uparrow$;
$\mathbf{E}_x = 13.60$ kN $\leftarrow$, $\mathbf{E}_y = 2.70$ kN $\downarrow$.

6.72 (*a*) $\mathbf{E}_x = 3$ kips $\leftarrow$, $\mathbf{E}_y = 1.5$ kips $\uparrow$.
(*b*) $\mathbf{C}_x = 3$ kips $\leftarrow$, $\mathbf{C}_y = 6.5$ kips $\uparrow$.

6.74 (*a*) *At each wheel:* $\mathbf{A} = 2.5$ kips $\uparrow$;
$\mathbf{B} = 37.5$ kips $\uparrow$. (*b*) $\mathbf{C} = 6$ kips $\leftarrow$;
$\mathbf{D}_x = 6$ kips $\rightarrow$, $\mathbf{D}_y = 15$ kips $\downarrow$.

6.76 (*a*) 2125 N. (*b*) *At each wheel:* $\mathbf{A} = 4740$ N $\uparrow$;
$\mathbf{B} = 2730$ N $\uparrow$; $\mathbf{C} = 3570$ N $\uparrow$.

6.78 $\mathbf{A}_x = 1110$ lb $\leftarrow$, $\mathbf{A}_y = 600$ lb $\uparrow$:
$\mathbf{B}_x = 1110$ lb $\leftarrow$, $\mathbf{B}_y = 800$ lb $\downarrow$;
$\mathbf{D}_x = 2220$ lb $\rightarrow$, $\mathbf{D}_y = 200$ lb $\uparrow$.

6.80 $\mathbf{A}_x = 4$ kN $\leftarrow$, $\mathbf{A}_y = 2$ kN $\uparrow$; $\mathbf{B}_x = 6$ kN $\leftarrow$,
$\mathbf{B}_y = 1$ kN $\uparrow$; $\mathbf{D}_x = 10$ kN $\rightarrow$, $\mathbf{D}_y = 3$ kN $\downarrow$.

6.82 $F_{AF} = \frac{1}{2}P$ comp.; $F_{BG} = P/\sqrt{2}$ ten.;
$F_{GD} = \sqrt{2}\,P$ ten.; $F_{EH} = \frac{1}{2}P$ comp.

6.84 $F_{CF} = 7.88$ kN C; $F_{DG} = 5.25$ kN T.

6.86 $F_{BG} = 5.25$ kN T; $F_{CH} = 2.63$ kN C.

6.88 $\mathbf{A} = 40$ lb $\uparrow$; $\mathbf{C}_x = 360$ lb $\rightarrow$, $\mathbf{C}_y = 40$ lb $\downarrow$;
$\mathbf{D} = 360$ lb $\leftarrow$.

6.90 (*a*) $\mathbf{C}_x = 90$ kips $\leftarrow$, $\mathbf{C}_y = 84$ kips $\uparrow$.
(*b*) $\mathbf{B}_x = 90$ kips $\leftarrow$, $\mathbf{B}_y = 6$ kips $\downarrow$.

6.92 $\mathbf{A} = P/15 \uparrow$; $\mathbf{D} = 2P/15 \uparrow$; $\mathbf{E} = 8P/15 \uparrow$;
$\mathbf{H} = 4P/15 \uparrow$.

6.94 (*a*) Rigid; $\mathbf{A} = 2.24P$ $\measuredangle$ 26.6°; $\mathbf{B} = 2P \rightarrow$.
(*b*) and (*c*) Not rigid.

6.96 $a \geq 0.6$ m.

6.98 (*a*) 96 lb $\downarrow$. (*b*) 100 lb $\measuredangle$ 73.7°.

6.100 (*a*) 2020 N $\downarrow$. (*b*) 1875 N $\measuredangle$ 66.2°.

6.102 $\mathbf{D} = 9360$ N $\leftarrow$; $\mathbf{F} = 6770$ N $\measuredangle$ 15.1°.

6.104 $\mathbf{C} = 17.36$ kips $\rightarrow$; $\mathbf{E} = 10.18$ kips $\measuredangle$ 23.1°.

6.106 $\mathbf{A}_x = 180$ N $\leftarrow$, $\mathbf{A}_y = 2250$ N $\downarrow$;
$\mathbf{B} = 2550$ N $\measuredangle$ 61.9°; $\mathbf{C} = 1020$ N $\leftarrow$.

6.108 (*a*) 2.14 kN $\downarrow$. (*b*) 5.00 kN $\downarrow$.

6.110 (*a*) $\mathbf{M}_C = 920$ lb·in. $\circlearrowleft$.
(*b*) $\mathbf{C}_x = 30$ lb $\leftarrow$, $\mathbf{C}_y = 40$ lb $\uparrow$.

6.112 (*a*) 3 kN. (*b*) 3.35 kN $\measuredangle$ 26.6°.

6.114 60 N.

6.116 29.4 lb.

6.118 200 N.

6.120 35.7 lb $\downarrow$.

6.122 (*a*) Zero.
(*b*) $\mathbf{C}_x = 5000$ lb $\leftarrow$, $\mathbf{C}_y = 1200$ lb $\downarrow$.

6.124 $F_{AB} = 1.484$ kN C; $F_{DE} = 31.3$ kN T;
$F_{FI} = 4.55$ kN C.

6.126 (*a*) (420 lb·in.)$\mathbf{i}$. (*b*) $\mathbf{A} = 0$,
$\mathbf{M}_A = -(288$ lb·in.$)\mathbf{i}$; $\mathbf{B} = 0$,
$\mathbf{M}_B = -(432$ lb·in.$)\mathbf{i}$.

6.128 (*a*) 0.8. (*b*) 2.6 M_A $\circlearrowright$.

6.130 (*a*) $M_G = 26.6$ N·m. (*b*) $\mathbf{B} = +(113.7$ N$)\mathbf{i}$;
$\mathbf{C} = -(113.7$ N$)\mathbf{i}$; $\mathbf{E} = 0$.

6.132 663 lb *C*.
6.134 (*a*) 126.0 N·m ↺. (*b*) 116.9 N·m ↺.
6.136 (*a*) **A** = 1.8 kips ↑, $\mathbf{M}_A$ = 14.4 kip·ft ↺;
D = 1.2 kips ↑.
(*b*) **A** = 0, $\mathbf{M}_A$ = 18 kip·ft ↻; **D** = 3 kips ↑.
6.138 (*a*) 1300 N. (*b*) 13.5 N·m.
6.140 $\mathbf{A}_x$ = 90 N ←, $\mathbf{A}_y$ = 60 N ↓; $\mathbf{B}_x$ = 90 N →,
$\mathbf{B}_y$ = 540 N ↑.
6.142 F_{CE} = 6 kips *T*; F_{DE} = 20 kips *T*;
F_{DF} = 18 kips *C*.
6.144 **A** = 666 N ∠ 63.2°; **F** = 325 N ∠ 22.6°.

CHAPTER 7

7.2 (On *JD*) **F** = 1106 lb ∠ 20.6°;
V = 386 lb ∠ 69.4°; **M** = 1650 lb·ft ↻.
7.4 (On *JD*) **F** = 0; **V** = 100 N ↑; **M** = 15 N·m ↺.
7.6 (On *KCD*) **F** = 750 N ↓; **V** = 400 N →;
M = 170 N·m ↻.
7.8 (On *CD*) **F** = 270 N →; **V** = 90 N ↑;
M = 43.2 N·m ↻.
7.10 (On *JCD*) **F** = 750 lb ←, **V** = 0;
M = 6 kip·in. ↻.
7.12 (*a*) (On *AC*) **F** = 520 N ←; **V** = 540 N ↑;
M = 0. (*b*) (On *AJ*) **F** = 520 N ←;
V = 540 N ↑; **M** = 54 N·m ↺.
7.14 (On *AJ*) **F** = 1040 lb ∠ 21.8°;
V = 92.8 lb ∠ 68.2°; **M** = 2000 lb·in. ↻.
7.16 (On *JC*) **M** = 4.28 N·m ↻.
7.18 (On *JC*) **F** = $(\frac{1}{2}W - W\theta/\pi)\cos\theta$ ↘;
V = $(\frac{1}{2}W - W\theta/\pi)\sin\theta$ ↙;
M = $\frac{1}{2}Wr(1 - \cos\theta) -$
$(Wr/\pi)(\sin\theta - \theta\cos\theta)$ ↻.
7.20 0.1786 *W*; $\theta = 49.3°$.
7.22 $M = +\frac{1}{8}wL^2$ at center of the beam.
7.24 $V = 0$, everywhere;
$M = +T$ between *B* and *C*; $M = 0$, elsewhere.
7.26 $V_C = -2P$; $M_C = -3Pa$.
7.28 M_D = +41.6 kip·ft.
7.30 M_C = +180 N·m.
7.32 0.4 m.
7.34 M = −20 kN·m at center.
7.36 M_C = −6 kip·ft.
7.38 *Just to the right of C:* V = +120 N;
M = +24 N·m.
7.40 M = +400 lb·in. to the left of *F*.
7.42 M = +4.8 N·m in portion *CD*.
7.44 (*a*) $\frac{1}{3}P$. (*b*) $\frac{1}{4}P$.
7.58 M_D = +1230 lb·in.
7.60 M_C = +14 kN·m.
7.62 M = +5.76 kN·m, 2.4 m from *A*.
7.64 M = +3125 lb·ft, 5 ft from *A*.
7.66 M = +44.7 kN·m, 3.28 m from *A*.
7.68 $V = w_0(L/\pi)\cos(\pi x/L)$;
$M = w_0(L/\pi)^2 \sin(\pi x/L)$;
$M_{max} = w_0(L/\pi)^2$, at $x = \frac{1}{2}L$.
7.70 $V = \frac{1}{2}w_0L[\frac{1}{3} - (x/L)^2]$;
$M = \frac{1}{6}w_0L^2[(x/L) - (x/L)^3]$;
$M_{max} = +0.0642w_0L^2$, at $x = 0.577L$.
7.72 **P** = 23 kips ↓, **Q** = 79 kips ↓; V_D = +24 kips,
M_F = +252 kip·ft.
7.74 **P** = 250 N ↓, **Q** = 350 N ↓; V_D = +1.8 kN,
M_C = −2.88 kN·m.
7.76 $L/6$.
7.78 2 m.
7.80 (*a*) $\mathbf{E}_x$ = 25 kips →, $\mathbf{E}_y$ = 12 kips ↑.
(*b*) 27.7 kips.
7.82 (*a*) 9 m. (*b*) $\mathbf{E}_x$ = 5 kN →, $\mathbf{E}_y$ = 3 kN ↑.
7.84 a = 2.55 m; b = 5.95 m.
7.86 (*a*) $\mathbf{D}_x$ = 280 lb →, $\mathbf{D}_y$ = 210 lb ↑.
(*b*) 1170 lb.
(*c*) T_{AB} = 1200 lb, T_{BC} = 750 lb, T_{CD} = 350 lb.
7.88 (*a*) 66,900 kips. (*b*) 4353 ft.
7.90 (*a*) 4.74 m. (*b*) 1760 N.
7.92 (*a*) 125 mm. (*b*) −2.9°.
7.94 (*a*) 750 mm from *A*. (*b*) 8.57 N.
7.96 15.07 kips; 17.79 kips.
7.102 49.86 m.
7.104 (*a*) 9.89 ft. (*b*) 15.38 lb.
7.106 (*a*) 15.02 m. (*b*) 4.42 kg/m.
7.108 80.4 ft; 92.5 lb.
7.110 1.822 in.
7.112 24.3 lb.
7.114 (*a*) $L = 1.326\, T_m/w$. (*b*) 8.04 mi.
7.116 0.1408.
7.118 (*a*) 138.1 m. (*b*) 649 N.
7.120 M_C = +1800 lb·ft; M_E = −2400 lb·ft.
7.122 (*a*) 2.6 m. (*b*) 6.71 kN.
7.124 (On *FB*) (*a*) **V** = 0; **M** = 0.8 kN·m ↻.
(*b*) **V** = 0; **M** = 0.4 kN·m ↻.
(*c*) **V** = 2 kN ↓; **M** = 0.6 kN·m ↻.
(*d*) **V** = 2 kN ↑; **M** = 1.4 kN·m ↻.
7.126 (*c*) $M_D = M_E$ = +0.8 kN·m.
(*d*) M_D = +0.8 kN·m; M_E = +1.2 kN·m.
7.128 M_C = +41 lb·in.

CHAPTER 8

8.2 Block moves up; $\mathbf{F} = 151.7$ N ∠ 20°.
8.4 Equilibrium; $\mathbf{F} = 23.5$ lb ∠ 30°.
8.6 (*a*) 462 N. (*b*) 447 N. (*c*) 276 N.
8.8 (*a*) 35.9 lb ∠ 36.7°. (*b*) 3.45 lb ∠ 3.30°.
8.10 51.8°.
8.12 (*a*) Does not move; 800 N ←.
(*b*) Moves; 491 N ←.
8.14 All packages move; $\mathbf{F}_A = 7.58$ N ↗;
$\mathbf{F}_B = 3.03$ N ↗; $\mathbf{F}_C = 7.58$ N ↗.
8.16 31.0°.
8.18 (*a*) 176.7 N. (*b*) 147.3 N. (*c*) 63.1 N.
8.20 (*a*) 66.4°. (*b*) 104.3 lb.
8.22 (*a*) 468 lb →. (*b*) 612 lb →.
8.24 $M = Wr\mu_s(1 + \mu_s)/(1 + \mu_s^2)$.
8.26 (*a*) 30.2°. (*b*) $0.232W$.
8.28 (*a*) 71.9 mm. (*b*) 39.9 mm.
8.30 (*a*) 38.6°. (*b*) $0.330W$.
8.32 $4.84a \leq L \leq 29.9a$.
8.34 $5.94a \leq L \leq 12.24a$.
8.36 Rod not in equilibrium.
8.38 Rod in equilibrium; $\mathbf{F}_B = 18.39$ N ↑.
8.40 2200 N ↓.
8.42 18.75 in.
8.44 0.0533.
8.46 (*a*) $11.6° \leq \theta \leq 78.4°$. (*b*) $0 \leq \theta \leq 63.4°$.
8.48 27.4 N $\leq Q \leq$ 51.8 N.
8.50 20.6 lb $\leq Q \leq$ 46.0 lb.
8.52 200 N.
8.54 190.9 lb $\leq P \leq$ 315 lb.
8.56 (*a*) 0.592. (*b*) At *B*.
8.58 (*a*) 0.479 m. (*b*) Plank does not move.
8.60 (*a*) 283 N ←.
(*b*) $\mathbf{B}_x = 413$ N ←, $\mathbf{B}_y = 480$ N ↓.
8.62 33.4°; 55.0 lb.
8.64 1430 N ↓.
8.66 (*a*) 12,040 lb →. (*b*) 4500 lb ←.
8.68 (*a*) 246 N. (*b*) 246 N.
8.70 3.05 N.
8.72 (*a*) 197.0 lb →. (*b*) Machine will not move.
8.74 14.97 lb·ft.
8.76 (*a*) 11.71 kN·m. (*b*) 2.58 kN·m.
8.78 19.64 N.
8.80 (*a*) Screw *A*. (*b*) 9.22 lb·in.
8.82 14.47 N·m.
8.84 21.3 kN.
8.86 (*a*) 72.8 lb. (*b*) 62.8 lb.
8.88 3.55°.
8.90 0.367.
8.92 (*a*) 920 kN. (*b*) 690 kN.
8.94 0.185.
8.96 208 lb · in.
8.102 0.060 in.
8.104 (*a*) 1288 kN. (*b*) 1058 kN.
8.106 (*a*) 0.344. (*b*) 2.55 turns.
8.108 0.1936.
8.110 (*a*) 22.8 kg. (*b*) 291 N.
8.112 $T_A = 176.3$ lb; $T_B = 76.3$ lb.
8.114 107.3 N·m.
8.116 112.4 lb·ft.
8.118 $T_A = 57.6$ N; $T_B = 92.4$ N.
8.120 0.441.
8.122 (*a*) 85.5 lb. (*b*) 48.0 lb. (*c*) 85.5 lb.
8.124 0.253.
8.128 $T_A = 103.24$ lb; $T_B = 3.24$ lb.
8.130 (*a*) 952 mm. (*b*) 3.96 m.
8.132 10 lb $< P <$ 36.7 lb.
8.134 (*a*) 2.94 N. (*b*) 4.41 N.
8.136 0.304.
8.138 39 lb.
8.140 Rod in equilibrium; $\mathbf{F} = 0.518P$ ←.

CHAPTER 9

9.2 $3a^3b/10$.
9.4 $2a^3b/15$.
9.6 $ab^3/6$.
9.8 $2ab^3/7$.
9.10 $2ab^3/51$; $b\sqrt{7/51}$.
9.12 $2a^3b/11$; $a\sqrt{7/11}$.
9.14 $a^3b/21$; $a\sqrt{1/7}$.
9.16 $43a^4/48$; $a\sqrt{43/72}$.
9.18 (*a*) $\frac{1}{2}\pi(R_2^4 - R_1^4)$. (*b*) $\frac{1}{4}\pi(R_2^4 - R_1^4)$.
9.22 403×10^6 mm^4; 92.8 mm.
9.24 6.99 in^4; 1.127 in.
9.26 $\bar{I}_x = \bar{I}_y = 2.005 \times 10^6$ mm^4.
9.28 46.2×10^6 mm^4.
9.30 9120 in^4.
9.32 $\bar{I}_x = 13.89 \times 10^6$ mm^4;
$\bar{I}_y = 20.88 \times 10^6$ mm^4.
9.34 (*a*) 408 in^4. (*b*) 81.6 in^4.
9.36 $\bar{I}_x = 44.5$ in^4, $\bar{I}_y = 27.7$ in^4; $\bar{k}_x = 2.16$ in.,
$\bar{k}_y = 1.707$ in.
9.38 170.7 mm.

9.40 1.595×10^6 mm^4; 17.84×10^6 mm^4.
9.42 96.6 in^4; 26.6 in^4.
9.44 $3\pi r/16$.
9.46 $B = 13.15$ kN; $C = D = 17.09$ kN.
9.48 (*a*) $\frac{1}{2}\gamma\pi r^2 h$. (*b*) $\frac{1}{4}\gamma\pi r^4$.
9.50 30 mm.
9.54 $\frac{1}{4}a^2b^2/(n+1)$.
9.56 $\frac{1}{8}a^4$.
9.58 $+12.15 \times 10^6$ mm^4.
9.60 $+10.31$ in^4.
9.62 $+159.6 \times 10^3$ mm^4.
9.64 13.79 in^4, 21.16 in^4, $+14.24$ in^4.
9.66 374×10^3 mm^4, 243×10^3 mm^4, $+206 \times 10^3$ mm^4.
9.68 $+25.1°$; 42.1×10^6 mm^4, 10.51×10^6 mm^4.
9.70 $-22.3°$; 32.2 in^4, 2.77 in^4.
9.72 $+23.8°$; 524×10^3 mm^4, 92.5×10^3 mm^4.
9.74 13.79 in^4, 21.16 in^4, $+14.24$ in^4.
9.76 374×10^3 mm^4, 243×10^3 mm^4, $+206 \times 10^3$ mm^4.
9.78 $+25.1°$; 42.1×10^6 mm^4, 10.51×10^6 mm^4.
9.80 $-22.3°$; 32.2 in^4, 2.77 in^4.
9.82 $+23.8°$; 524×10^3 mm^4, 92.5×10^3 mm^4.
9.84 (*a*) -1.171×10^6 mm^4. (*b*) $+19.7°$. (*c*) 4.35×10^6 mm^4.
9.86 $I_{xy} = \sqrt{I_x I_y - I_{max} I_{min}}$; 6.07 in^4.
9.88 (*a*) $\frac{1}{4}m(r_1^2 + r_2^2)$. (*b*) $\frac{1}{2}m(r_1^2 + r_2^2)$.
9.90 $I_{DD'} = 3ma^2/8$; $I_{EE'} = 5ma^2/12$.
9.92 (*a*) $7ma^2/6$. (*b*) $\frac{1}{2}ma^2$.
9.94 $m(3a^2 + L^2)/12$.
9.96 $m(a^2 + 3h^2)/6$; $\sqrt{(a^2 + 3h^2)/6}$.
9.98 $m(a^2 + b^2)/5$.
9.100 (*a*) $m(b^2 + 4a^2 \sin^2\theta)/12$. (*b*) $m(b^2 + 4a^2\cos^2\theta)/12$.
9.102 1.743 lb·ft·s^2, 6.54 in.
9.104 $27ma^2/50$; $0.735a$.
9.106 80.2 g·m^2.
9.108 0.1310 lb·ft·s^2; 3.87 in.
9.110 (*a*) 7.11 g·m^2. (*b*) 16.96 g·m^2. (*c*) 15.27 g·m^2.
9.112 $I_x = 5.35 \times 10^{-3}$ lb·ft·s^2; $I_y = I_z = 21.7 \times 10^{-3}$ lb·ft·s^2.
9.114 30.5×10^{-3} lb·ft·s^2.
9.116 171.0 g·m^2.
9.118 2.06 kg·m^2.
9.120 $I_{xy} = +9.17$ g·m^2; $I_{yz} = I_{zx} = 0$.
9.122 $I_{xy} = -3.57 \times 10^{-3}$ lb·ft·s^2; $I_{yz} = 0$; $I_{zx} = +3.57 \times 10^{-3}$ lb·ft·s^2.
9.124 $I_{xy} = -0.1125$ kg·m^2; $I_{yz} = I_{zx} = 0$.
9.126 (*a*) $I_{zx} = mca/20$. (*b*) $I_{xy} = mab/20$. (*c*) $I_{yz} = mbc/20$.
9.128 $\frac{1}{6}m(2a^2b^2 + 2b^2c^2 + c^2a^2)/(a^2 + c^2)$.
9.130 7.83×10^{-3} kg·m^2.
9.132 $5ma^2/12$.
9.134 $10ma^2/27$.
9.136 (*a*) $1/\sqrt{3}$. (*b*) $2/\sqrt{3}$.
9.138 (*a*) $ma^2/6$. (*b*) $ma^2/6$, $11ma^2/12$.
9.140 (*a*) $0.227mr^2$. (*b*) $0.727mr^2$.
9.142 74.8 in^4.
9.144 $\bar{I}_x = 1.274 \times 10^6$ mm^4, $\bar{I}_y = 3.93 \times 10^6$ mm^4; $\bar{k}_x = 22.8$ mm, $\bar{k}_y = 40.1$ mm.
9.146 1.409 g·m^2; 44.7 mm.
9.148 $8ml^2/9$; $2\sqrt{2}l/3$.
9.150 $5ma^2/12$; $a\sqrt{5/12}$.

CHAPTER 10

10.2 65 N.
10.4 1200 lb·in. ⤴.
10.6 750 N ↑.
10.8 $Q = \frac{3}{2}P\tan\theta$.
10.10 $Q = 3P\tan\theta$.
10.12 $Q = (Pl/a)\cos\theta\sin^2\theta$.
10.14 $M = \frac{1}{2}Pl\cot\theta$.
10.16 $M = 4Pl\sin 2\theta$.
10.18 $M = -4Pl\cos 2\theta$.
10.20 (*a*) 14.70 lb. (*b*) 56.7 lb.
10.22 178.5 N·m.
10.24 64.9°.
10.26 14.9°.
10.28 52.2°.
10.30 13.2 in.
10.32 19.4°.
10.34 79.4°; 317.7°; 386.1°.
10.36 (*a*) $12(BD)$. (*b*) 52.8°.
10.38 6.77 kips ↘.
10.42 180.3 N·m; 57.0 N·m.
10.44 1.361 kN ↑; 550 N·m ⤴.
10.46 (*a*) $T = \frac{1}{2}W/(1 - \tan\theta)$. (*b*) 36.9°.
10.48 0.469 in. →.
10.56 33.7°, stable; −146.3°, unstable.
10.58 14.5°, stable; 165.5°, unstable.
10.60 79.4° and 386.1°, stable; 317.7°, unstable.
10.62 72.7°.
10.64 12.5° and 90°, stable; 30.8°, unstable.

10.66 (a) $\cos\frac{1}{2}\theta - \sin\frac{1}{2}\theta = (1 - Wl/ka^2)\cos\theta$.
(b) 34.2° and 145.8°, stable; 90°, unstable.
10.68 52.2°.
10.70 180°, stable.
10.72 9.6°, unstable; 40.3°, stable.
10.74 $d > 309$ mm.
10.76 $m_1 = m_2(c^2/ab)$.
10.78 $P < \frac{1}{2}kl$.
10.80 $P < 430$ lb.
10.82 60 N.
10.84 60°.
10.86 11.3°, stable; 78.7°, unstable.
10.88 14.2°.
10.90 (a) $P\cos\theta = W\cos\frac{1}{2}\theta$. (b) 65.1°.
10.92 $\dfrac{\cos\theta}{\cos\frac{1}{2}\theta} = \dfrac{m_B}{m_A}\sin\beta$.

CHAPTER 11

11.2 $x = 3$ in., $v = 28$ in./s, $a = 54$ in./s^2.
11.4 $t = 1$ s, $x = 25$ m, $a = -12$ m/s^2;
$t = 5$ s, $x = -7$ m, $a = +12$ m/s^2.
11.6 (a) 1 s, 4 s. (b) +1.5 m, 24.5 m.
11.8 $a = 2t$; $v = t^2 - 9$; $x = \frac{1}{3}t^3 - 9t + 30$.
11.10 (a) 3 ft/s^2. (b) $a = 3t^2$; $v = t^3 - 24$;
$x = \frac{1}{4}t^4 - 24t + 50$.
11.12 (a) ±2.21 m/s. (b) 1.219 m.
11.14 (a) ±20 in./s. (b) +6.71 in. (c) 3.87 in.
11.16 (a) 7.05 m/s. (b) 10.95 m/s. (c) 25.3 m/s.
11.18 (a) 241 ft. (b) Infinite.
11.20 412 m.
11.22 (b) 4 m/s; $\frac{1}{6}$ s.
11.24 (a) -2.43×10^6 ft/s^2. (b) 1.366×10^{-3} s.
11.26 $\sqrt{2gR}$.
11.30 (a) 0.486 m/s^2. (b) 32.7 km/h.
11.32 (a) 70.6 ft. (b) 28.3 ft/s.
11.34 (a) 5.61 s. (b) 60.2 km/h.
11.36 (b) 1.529 s; 11.47 m.
11.38 (a) $t = 16$ s; $x = 960$ ft
(b) $v_A = 51.8$ mi/h; $v_B = 28.6$ mi/h
11.40 (a) 1.25 m/s^2 ↑. (b) 2.5 m/s^2 ↓.
(c) 7.5 m/s ↑.
11.42 (a) $\mathbf{a}_A = 2$ in./s^2 ←; $\mathbf{a}_B = 3$ in./s^2 ←.
(b) 10 in./s ←; 25 in. ←.
11.44 (a) 300 mm/s →. (b) 600 mm/s ←.
(c) 450 mm/s ←.
11.46 (a) 2.5 mm/s^2 ↑; 20 mm/s ↑. (b) 900 mm ↓.
11.48 (a) 2 s. (b) 1.500 in. ↓.
11.50 $\mathbf{v}_A = 125$ mm/s ↑; $\mathbf{v}_B = 75$ mm/s ↓;
$\mathbf{v}_C = 175$ mm/s ↓.
11.52 +25 ft/s; +91.5 ft; 143.5 ft.
11.54 (a) 44 m. (b) 4 s, 18 s.
11.56 Accelerate for 11.84 s; 76.7 km/h.
11.58 15 s.
11.60 (a) -1.464 m/s^2; 0.75 m/s^2. (b) 32.7 km/h.
11.62 8.04 s.
11.64 (a) and (b) 23.3 ft/s^2.
11.66 (a) 3 s. (b) 1000 m.
11.68 (a) 24 s; 300 ft. (b) 100 s; 1800 ft.
11.70 (a) 8 s. (b) 24 m. (c) 3 m/s.
11.72 (a) 272 m/s. (b) 679 m.
11.74 (a) -75.6 in./s^2. (b) -88 in./s^2.
11.76 (a) 2.7 s. (b) 12.15 m.
11.78 51.5 ft.
11.80 (a) 5.66 ft/s ∠45°; 12.65 ft/s^2 ∠71.6°.
(b) 6.12 ft/s ∠11.2°; 4.65 ft/s^2 ∠30.7°.
11.82 (a) 80 mm/s ↑; 307 mm/s^2 ∠15.1°.
(b) 247 mm/s ∠40.3°; 80 mm/s^2 ↑.
11.84 (a) $\mathbf{v} = -(4\pi$ in./s$)\mathbf{i}$; $\mathbf{a} = (4\pi^2$ in./s$^2)\mathbf{j}$.
(b) Path is $y = \frac{1}{8}x^2 - 1$.
11.86 (a) 1 m/s; 2 m/s^2.
(b) 12.69 m/s; 12.17 m/s^2.
11.88 (a) 46.3 ft/s. (b) 44.0 ft/s $\le v_0 \le$ 56.7 ft/s.
11.90 4.70 m/s $\le v_0 \le$ 7.23 m/s.
11.92 5.15 ft $\le h \le$ 10.92 ft.
11.94 4.12 m from B.
11.96 (a) 214 m/s. (b) 231 m/s.
11.98 14.91 ft/s $\le v_0 \le$ 21.5 ft/s.
11.100 (a) 213 ft/s. (b) 29.5°.
11.102 51.6° or 69.3°.
11.104 4.34 m.
11.106 54.0 km/h ∠31.2°.
11.108 (a) 91.0 ft/s ∠63.0°.
(b) 273 ft ∠63.0°. (c) 434 ft.
11.110 (a) 80.7 mm/s^2 ∠68.3°.
(b) 323 mm ∠68.3°.
11.112 (a) 9.91 in./s ∠66.2°.
(b) 14.87 in./s^2 ∠66.2°.
11.114 (a) 63.9°. (b) 1195 ft/s ↑; 32.2 ft/s^2 ↓.
11.116 2.45 m/s ∠35.3°.
11.118 23.7 mi/h.
11.120 44.1 km/h.
11.122 (a) 5 ft/s. 240 ft/s^2.
11.124 93.8 m; 17.68 s.

11.126 43.8 s.
11.128 (*a*) 60.1 km/h ⦣8.6°.
(*b*) 3.21 m/s² ⦤5.9°.
11.130 22,800 ft; 58,100 ft.
11.132 (*a*) 14.01 m/s. (*b*) 12.80 m.
11.134 (*a*) 0.5 m. (*b*) 76.1 m.
11.136 17,390 mi/h.
11.138 3670 km/h.
11.140 (*a*) $-(0.2 \text{ m/s})\mathbf{e}_r - (1.100 \text{ m/s})\mathbf{e}_\theta$.
(*b*) $-(6.35 \text{ m/s}^2)\mathbf{e}_r - (2.04 \text{ m/s}^2)\mathbf{e}_\theta$.
(*c*) $-(6.35 \text{ m/s}^2)\mathbf{e}_r$.
11.142 (*a*) $\mathbf{v} = (10 \text{ in./s})\mathbf{e}_r$; $\mathbf{a} = (125.7 \text{ in./s}^2)\mathbf{e}_\theta$.
(*b*) $\mathbf{v} = (10 \text{ in./s})\mathbf{e}_r + (15.71 \text{ in./s})\mathbf{e}_\theta$;
$\mathbf{a} = -(98.7 \text{ in./s}^2)\mathbf{e}_r + (125.7 \text{ in./s}^2)\mathbf{e}_\theta$.
11.144 (*a*) $\mathbf{v} = (62.8 \text{ mm/s})\mathbf{e}_\theta$;
$\mathbf{a} = -(112.8 \text{ mm/s}^2)\mathbf{e}_r$.
(*b*) $\mathbf{v} = (82.5 \text{ mm/s})\mathbf{e}_r + (110.0 \text{ mm/s})\mathbf{e}_\theta$;
$\mathbf{a} = (43.2 \text{ mm/s}^2)\mathbf{e}_r + (518 \text{ mm/s}^2)\mathbf{e}_\theta$.
11.146 $(v_0/h)\sin^2\theta$.
11.148 $(2v_0^2/h^2)\sin^3\theta\cos\theta$.
11.150 437 mi/h ⦤5.9°.
11.152 $v = e^{b\theta}\sqrt{1 + b^2}\,\dot{\theta}$.
11.154 $a = (1 + b^2)\omega^2 e^{b\theta}$.
11.156 (*b*) Toward A.
11.158 (*b*) $(A^2 + B^2)/A$; $A^2/\sqrt{A^2 + B^2}$.
11.160 (*a*) $v = \sqrt{A^2 + B^2}$;
$a = \sqrt{(1 + 16\pi^2)A^2 + B^2}$.
(*b*) $v = 2\pi A$; $a = 4\pi^2 A$.
11.162 (*a*) Helix, vertical axis at $\theta = 0$ and $R = k$.
(*b*) $v = \sqrt{4k^2 + p^2}$; $a = 4k$.
(*c*) $\rho = k + p^2/4k$.
11.164 (*a*) $\theta_x = \theta_z = 90°$, $\theta_y = 180°$.
(*b*) $\theta_x = 153.1°$, $\theta_y = 94.3°$, $\theta_z = 63.5°$.
11.166 (*a*) 63.7 mm/s. (*b*) 127.3 mm.
(*c*) 31.8 mm/s.
11.168 88.4 ft.
11.170 13.59 m.
11.172 (*a*) 86.4 m; 4.20 s. (*b*) 166.6 m; 5.83 s.
11.174 (*a*) 11.09 s. (*b*) 688 ft.
11.176 80 mm/s ↓.

CHAPTER 12

12.2 (*a*) 4.987 lb at 0°; 5.000 lb at 45°; 5.013 lb at 90°. (*b*) 5.000 lb at all latitudes.
(*c*) 0.1554 lb·s²/ft at all latitudes.
12.4 2.84×10^6 kg·m/s.
12.6 (*a*) 78.1 km/h. (*b*) 63.8 km/h.
12.8 1.610 ft/s².
12.10 (*a*) $\mathbf{a}_A = 8.40 \text{ ft/s}^2$ ↗; $\mathbf{a}_B = 4.20 \text{ ft/s}^2$ ↓.
(*b*) 152.2 lb.
12.12 (*a*) 3.60 m/s² ←. (*b*) 180 N →.
12.14 (*a*) 5.21 m/s² ↙. (*b*) 215 N.
12.16 (*a*) 1.2 m/s →. (*b*) 0.6 m/s ←.
12.18 (*a*) 8.05 ft/s² ↓; 12.88 ft/s² ↓; 1.342 ft/s² ↓.
(*b*) 16.10 ft/s ↓; 25.8 ft/s ↓; 2.68 ft/s ↓.
(*c*) 11.35 ft/s ↓; 14.36 ft/s ↓; 4.63 ft/s ↓.
12.20 (*a*) Load will shift. (*b*) 10.64 ft/s ←.
12.22 (*a*) 3.37 m/s. (*b*) 1.372 s.
12.24 1.981 km.
12.26 $x = Pt/k - (Pm/k^2)(1 - e^{-kt/m})$.
12.28 (*a*) 1.656 lb. (*b*) 20.83 lb.
12.30 (*a*) 22.4 N. (*b*) 1.121 m/s² →.
12.32 (*a*) 6.17 m/s² ⦤30°. (*b*) 144.0 N.
12.34 (*a*) 8.94 ft/s² ←; 18.06 lb.
(*b*) and (*c*) 12.38 ft/s² ←; 15.38 lb.
12.36 (*a*) 24.2°. (*b*) 32.3 N.
12.38 27.1 ft/s.
12.40 $2.82 \text{ m/s} \le v \le 3.15 \text{ m/s}$.
12.42 $2.61 \text{ m/s} \le v \le 3.36 \text{ m/s}$.
12.44 (*a*) 0.395*W*. (*b*) 0.766*W*.
12.46 (*a*) $a_t = g\cos\theta$; $v = \sqrt{2gl\sin\theta}$. (*b*) 41.8°.
12.48 (*a*) 0.303 lb ↑. (*b*) 9.01°.
12.50 $-58.2° \le \theta \le 238.2°$.
12.52 0.45.
12.54 (*a*) 43.7°. (*b*) 0.68. (*c*) 56.1 km/h.
12.56 (*a*) 0.455 N ⦣70°; does not slide.
(*b*) 0.556 N ⦦30°; slides.
12.58 (*a*) 1.010 s. (*b*) 6.06 ft/s.
12.60 (*a*) 1.207 m/s. (*b*) 21.8° and 158.2°.
12.62 $\delta = eVlL/mv_0^2 d$.
12.64 Length l should be doubled.
12.66 $F_r = +0.400$ N, $F_\theta = +3.20$ N.
12.68 (*a*) $F_r = -(11.65 \text{ lb})\cos\theta$,
$F_\theta = -(11.65 \text{ lb})\sin\theta$.
(*b*) $\mathbf{P} = 0$; $\mathbf{Q} = 11.65$ lb ⦦2θ.
12.70 (*a*) $a_r = -72 \text{ m/s}^2$, $a_\theta = +43.2 \text{ m/s}^2$.
(*b*) $\ddot{r} = -14.40 \text{ m/s}^2$. (*c*) $F_\theta = +8.64$ N.
12.72 (*a*) $v_r = -3.00$ m/s, $v_\theta = +3.60$ m/s.
(*b*) $a_r = -62.1 \text{ m/s}^2$, $a_\theta = -72.0 \text{ m/s}^2$.
(*c*) $F_\theta = -14.40$ N.
12.74 $v_r = 2v_0\sin 2\theta$, $v_\theta = v_0\cos 2\theta$.
12.76 (*b*) $3mv_0^2/r_0$.
12.78 383×10^3 km, 238,000 mi.
12.80 (*a*) 1.90×10^{27} kg. (*b*) 422×10^3 km.

12.82 (*a*) 448 m/s. (*b*) 2290 m/s.
12.84 4800 ft/s.
12.86 (*a*) 5.74 in. (*b*) 15.68 ft/s.
12.88 (*a*) 2.45 m/s. (*b*) -12.25 rad/s^2.
12.94 18.82 km/s.
12.96 1.150.
12.98 (*a*) 26.0×10^3 ft/s. (*b*) 173.3 ft/s.
12.100 $\beta = \sqrt{2/(2 + \alpha)}$.
12.102 (*a*) 125×10^3 km. (*b*) 128.3 m/s; 2370 m/s.
12.104 19 h 10 min 7 s.
12.106 (*a*) 1 h 29 min 14 s. (*b*) 43 min 43 s.
12.108 3.29×10^9 mi.
12.110 123.0 m/s.
12.112 28.8 mi.
12.114 (*a*) $\varepsilon = (r_1 - r_0)/(r_1 + r_0)$.
(*b*) $\varepsilon_1 = 0.778$; $\varepsilon_2 = 0.852$.
12.118 3.50 m/s.
12.120 For $P \leq 18.75$ lb: $a = 0.429P$;
For $P \geq 18.75$ lb: $a = 1.073(P - 9 \text{ lb})$.
12.122 $m_B = 2.59m_A$; $m_C = 1.337m_A$.
12.124 190.9 s.
12.126 (*a*) 18,180 mi/h. (*b*) 16,900 mi.
12.128 (*a*) $\mathbf{a}_A = 3.46$ m/s^2 ←; $\mathbf{a}_B = 2.31$ m/s^2 ←.
(*b*) 19.23 N.

CHAPTER 13

13.2 (*a*) 224 ft·lb; 55.9 ft.
(*b*) 224 ft·lb; 339 ft.
13.4 100.9 km/h.
13.6 3.39 m/s.
13.8 (*a*) 14.81 ft. (*b*) 15.17 ft/s. (*c*) 107.4 ft·lb.
13.10 (*a*) 322 ft.
(*b*) $F_{AB} = 5400$ lb C; $F_{BC} = 12{,}600$ lb C.
13.12 (*a*) 156.25 m. (*b*) 2100 N →.
13.16 (*a*) 5.26 m/s. (*b*) 27.2 m.
13.18 (*a*) 2.21 m/s ↓.
(*b*) $\mathbf{F}_A = 91.5$ N ↑; $\mathbf{F}_B = 98.2$ N ↑.
13.20 (*a*) $\mathbf{F}_A = 149.8$ lb ↑; $\mathbf{F}_D = 240.4$ lb ↑.
(*b*) 0.15. (*c*) 453 ft·lb.
13.22 (*a*) 6.95 ft/s ↓. (*b*) 5.54 ft/s ↓.
13.24 $v_1 = 3.43$ m/s; $v_2 = 4.72$ m/s.
13.26 1.180 m/s ↓.
13.28 (*a*) 3.2 in. (*b*) 28.7 in./s.
13.30 134.7 mm.
13.32 A: 5.66 in.; B: 7.5 in.
13.34 (*a*) 32.7 mm; 98.1 N. (*b*) 30.4 mm; 104.9 N.
13.36 5.38 m/s.
13.38 (*a*) 77.2 km. (*b*) 80.7 km.
13.40 867 mi.
13.42 (*a*) 3.13 m/s. (*b*) 3.50 m/s.
13.44 (*a*) 1120 lb ↑; 160 lb ↓. (*b*) 80 ft.
13.46 (*a*) $0.75r$. (*b*) $0.1676r$.
13.48 2.40 MW.
13.50 (*a*) and (*b*) 63.3 kW.
13.52 For $t \leq 50$ s: $P = (132.5 \text{ kW/s})t$.
At $t = 50$ s: $P = 6.63$ MW.
For $t > 50$ s: $P = 375$ kW.
13.54 (*a*) 36.4 hp. (*b*) 60.1 hp.
13.56 (*a*) 20 s; 207 m. (*b*) 33.3 s; 563 m.
13.58 (*a*) 440 mm; 4.80 m/s.
(*b*) 440 mm; 2.40 m/s.
13.60 12.04 ft/s.
13.62 44.0 ft/s.
13.64 (*a*) 4.58 m/s.
(*b*) $x = 160$ mm, $y = 300$ mm, $z = 120$ mm.
13.66 (*a*) 37.1 ft/s. (*b*) 21.3 ft.
13.68 1.546 m.
13.70 (*a*) 2.48 m/s. (*b*) 1.732 m/s.
13.72 51.4 N ↑.
13.74 2.33 lb/in.
13.76 $3l/5$.
13.80 (*a*) Nonconservative; $-ka^2$.
(*b*) Conservative; 0.
13.82 (*a*) 62.5 MJ. (*b*) 11.18 km/s.
13.84 (*b*) $V_g = Wy[1 - (y/R)]$.
13.86 (*a*) 67.3×10^9 ft·lb. (*b*) 154.6×10^9 ft·lb.
13.88 (*a*) 0.919 m/s. (*b*) 8.27 m/s.
13.90 7.35 m/s.
13.92 $v_r = 74.6$ in./s. $v_\theta = 9.00$ in./s.
13.98 (*a*) 1704 m/s. (*b*) 29.8 m/s.
13.100 1555 m/s; 79.3°.
13.102 25.2×10^3 ft/s; 86.4°.
13.104 5440 km.
13.110 $(1 \pm \sin \alpha)r_0$.
13.112 (*a*) $h = r_{min}v_{max}$; $E/m = \frac{1}{2}v_{max}^2 - GM/r_{min}$.
13.114 (*a*) 3.40 s. (*b*) 25.5 s.
13.116 (*a*) 5.49 s. (*b*) $-(92.6 \text{ ft/s})\mathbf{j} - (41.0 \text{ ft/s})\mathbf{k}$.
13.120 (*a*) 22.8 s.
(*b*) $F_{AB} = 12{,}600$ lb C; $F_{BC} = 5400$ lb C.
13.124 (*a*) 15.19 m/s. (*b*) 16.80 m/s.
13.126 (*a*) 38.6 ft/s →. (*b*) 12.08 ft/s →.
13.128 24 kN.
13.130 33.0 kips ⦪79.6°.
13.132 (*a*) 6.17 m/s. (*b*) 0.01235.

13.134 (*a*) 1.333 m/s ←. (*b*) 1.467 m/s ←. (*c*) 1.456 m/s ←.
13.136 (*a*) Car *B*. (*b*) 52.9 mi/h.
13.138 (*a*) 4.80 N·s; 14.40 J. (*b*) 4.00 N·s; 12.00 J.
13.140 (*a*) 0.6 m/s →. (*b*) 0.75.
13.142 (*a*) $\mathbf{v}'_A$ = 1 ft/s ←; $\mathbf{v}'_B$ = 15 ft/s →. (*b*) 3.35 ft·lb.
13.144 $\mathbf{v}'_A$ = 0.209 m/s →; $\mathbf{v}'_B$ = 0.235 m/s →; $\mathbf{v}'_C$ = 1.156 m/s →.
13.146 $\mathbf{v}'_A$ = 5.27 ft/s $\angle$ 27.1°; $\mathbf{v}'_B$ = 7.17 ft/s $\angle$ 50°.
13.148 (*a*) 45°. (*b*) $(1 + e)v_0/4$.
13.152 0.425h.
13.154 (*a*) $h_1 = d_1$ = 6.40 in. (*b*) h_2 = 4.10 in; d_2 = 5.12 in.
13.156 (*a*) 0.80. (*b*) 283 mm.
13.158 $\mathbf{v}'_A = 0.4v_0$ $\angle$ 60°; $\mathbf{v}'_B = 0.7v_0$ ←.
13.160 $\mathbf{v}'_A$ = 2.28 m/s $\angle$ 18.9°; $\mathbf{v}'_B$ = 0.719 m/s ←.
13.162 (*a*) 37.8°. (*b*) 40.9°.
13.164 (*a*) 0.50. (*b*) 29.0°.
13.166 (*a*) 2.90 m. (*b*) 75.6%.
13.168 439 m/s.
13.170 79.0 kN ↑.
13.172 (*a*) 0.80. (*b*) 12 oz.
13.174 6.37 in.
13.178 5.08 m/s.
13.180 103.1 mi/h.
13.182 $h = d/(1 - e^2)$.
13.184 (*a*) 1.048 kN·s; 2.75 kJ. (*b*) 0.749 kN·s; 1.962 kJ.
13.186 0.861.
13.188 (*a*) 2.53 in. (*b*) 3.00 ft·lb.

CHAPTER 14

14.2 (*a*) 3.97 m/s. (*b*) 362 m/s.
14.4 (*a*) $\mathbf{v}_A = \mathbf{v}_B = \mathbf{v}_C$ = 0.50 km/h →. (*b*) $\mathbf{v}_A = \mathbf{v}_B$ = 2.75 km/h ←; $\mathbf{v}_C$ = 7.00 km/h →.
14.6 (*a*) $\mathbf{v}_A = \mathbf{v}_B$ = 3 mi/h →; $\mathbf{v}_C$ = 0. (*b*) $\mathbf{v}_A = \mathbf{v}_B$ = 1 mi/h →; $\mathbf{v}_C$ = 4 mi/h →.
14.8 (*a*) v_x = +6.33 m/s, v_z = +3.75 m/s. (*b*) −44.7 kg·m²/s.
14.10 (*a*) (6 ft)**i** + (3 ft)**j** + (3 ft)**k**. (*b*) (2.11 lb·s)**i** + (2.36 lb·s)**j** − (0.621 lb·s)**k**. (*c*) (0.497 ft·lb·s)**i** + (6.09 ft·lb·s)**j** + (6.34 ft·lb·s)**k**.
14.12 x = 32,100 ft, y = 650 ft.
14.14 x = 253 m, y = 6.61 m, z = 4.67 m.
14.16 $\mathbf{v}_C$ = (1650 ft/s)**i** − (150 ft/s)**j** + (270 ft/s)**k**.
14.18 −(677 m/s)**i**.
14.20 (*a*) 2.21 kg. (*b*) 9.35 m/s.
14.26 (*a*) 18,040 ft·lb. (*b*) 12,020 ft·lb.
14.28 491 kJ.
14.32 $\mathbf{v}_A$ = 0.990 m/s ←; $\mathbf{v}_B$ = 1.485 m/s →.
14.34 (*a*) and (*c*) $\mathbf{v}_A = \mathbf{v}_B = \frac{1}{2}v_0$ →. (*b*) $\mathbf{v}_A = v_0$ →; $\mathbf{v}_B$ = 0.
14.36 (*a*) 11.09 ft/s $\angle$ 30°. (*b*) 2.74 ft/s →.
14.38 (*a*) $\mathbf{v}_A = 0.433v_0$ $\angle$ 30°; $\mathbf{v}_B = 0.661v_0$ $\angle$ 19.1°, $\mathbf{v}_C$ = 0. (*b*) $\frac{3}{8}$.
14.40 $\mathbf{v}_A$ = (160 ft/s)**i** + (160 ft/s)**j** + (1040 ft/s)**k**.
14.42 (*a*) $\mathbf{v}_A$ = 0.520 m/s ↑; $\mathbf{v}_B$ = 0.520 m/s ↓; $\mathbf{v}_C$ = 0.900 m/s →. (*b*) 150 mm.
14.44 (*a*) $\mathbf{v}_A$ = 5.92 ft/s ↑; $\mathbf{v}_B$ = 7.75 ft/s $\angle$ 49.8°. (*b*) a = 5.49 ft.
14.46 4.50 kN $\angle$ 30°.
14.48 (*a*) 4100 lb. (*b*) 6450 lb.
14.50 540 N.
14.52 (*a*) 36.9°. (*b*) 82.9 lb ↓.
14.54 **C** = 31.0 lb ←; $\mathbf{D}_x$ = 58.7 lb ←, $\mathbf{D}_y$ = 51.8 lb ↓.
14.58 $\mathbf{C}_x$ = 90 N →, $\mathbf{C}_y$ = 2360 N ↑; **D** = 2900 N ↑.
14.60 75 kg/s.
14.62 37.2 ft/s.
14.64 5500 kg.
14.66 (*a*) 847 km/h. (*b*) 638 km/h.
14.68 (*a*) 47.4 kJ/s. (*b*) 0.316.
14.70 (*a*) $\frac{1}{2}v_A$ →. (*b*) $\frac{1}{4}A\rho(1 - \cos\theta)v_A^3$. (*c*) $2(V/v_A)[1 - (V/v_A)](1 - \cos\theta)$.
14.74 9.21 m³/s.
14.76 (*a*) $\frac{1}{2}mg + mv^2/l$. (*b*) $\frac{1}{2}mg$.
14.78 $h = v^2/g$.
14.80 (*a*) 1.192 m/s. (*b*) 1.320 m.
14.82 (*a*) m_0e^{qL/m_0v_0}. (*b*) v_0e^{-qL/m_0v_0}.
14.84 51.6 ft/s².
14.86 894 kips.
14.88 (*a*) 31.9 m/s² ↑. (*b*) 240 m/s² ↑.
14.90 1890 kg.
14.92 27,200 ft/s.
14.94 709,000 ft.
14.96 136.9 km.
14.100 6.31 lb.
14.102 (*a*) v_B = 1.371 m/s; v_C = 1.733 m/s. (*b*) 9.58%.

14.104 (a) $\mathbf{v}_A = \mathbf{v}_B = 0$; $\mathbf{v}_C = v_0 \rightarrow$.
(b) $\mathbf{v}_A = 0$; $\mathbf{v}_B = \mathbf{v}_C = \frac{1}{2}v_0 \rightarrow$.
(c) $\mathbf{v}_A = \frac{1}{3}v_0 \leftarrow$; $\mathbf{v}_B = \mathbf{v}_C = \frac{2}{3}v_0 \rightarrow$.
14.106 (a) 30.1×10^3 km/h. (b) 141.4 km.
14.108 (a) $P = qv$.
14.110 (a) 8.67 ft/s →. (b) 4.81 ft/s ∠56.3°.
(c) $b = 5.83$ ft.

CHAPTER 15

15.2 (a) 0; 0.2 rad/s; -0.05 rad/s^2.
(b) 0.506 rad; 0.0736 rad/s; -0.01839 rad/s^2.
(c) 0.8 rad; 0; 0
15.4 (a) 75 rev. (b) 1350 rev.
15.6 $\mathbf{v}_B = -(9 \text{ m/s})\mathbf{j} + (15 \text{ m/s})\mathbf{k}$.
$\mathbf{a}_B = -(204 \text{ m/s}^2)\mathbf{i} - (189 \text{ m/s}^2)\mathbf{j} - (195 \text{ m/s}^2)\mathbf{k}$.
15.8 $\mathbf{v}_F = -(0.4 \text{ m/s})\mathbf{i} - (1.4 \text{ m/s})\mathbf{j} - (0.7 \text{ m/s})\mathbf{k}$.
$\mathbf{a}_F = (8.4 \text{ m/s}^2)\mathbf{i} + (3.3 \text{ m/s}^2)\mathbf{j} - (11.4 \text{ m/s}^2)\mathbf{k}$.
15.10 $\mathbf{v}_D = -(24 \text{ in./s})\mathbf{i} - (21 \text{ in./s})\mathbf{j} - (28 \text{ in./s})\mathbf{k}$.
$\mathbf{a}_D = -(271 \text{ in./s}^2)\mathbf{i} - (12 \text{ in./s}^2)\mathbf{j} - (16 \text{ in./s}^2)\mathbf{k}$.
15.12 (a) 1525 ft/s; 0.1112 ft/s^2.
(b) 1168 ft/s; 0.0852 ft/s^2. (c) 0; 0.
15.14 ± 6.52 rad/s^2.
15.16 (a) 100 mm/s^2. (b) 111.8 mm/s^2.
(c) 224 mm/s^2.
15.18 (a) 67.7 ft/s ←; 24,500 ft/s^2 ↓.
(b) 6.77 ft/s ←; 245 ft/s^2 ∠86.8°.
15.20 (a) 9.6 rad/s^2 ↻. (b) 16.04 rev ↻.
15.22 (a) 0.509 rad/s^2. (b) 20.6 s.
15.24 (a) 15.28 rev. (b) 10.14 s.
15.26 (a) 9.75 s.
(b) $\boldsymbol{\omega}_A = 279$ rpm ↺; $\boldsymbol{\omega}_B = 419$ rpm ↻.
15.28 (a) $\boldsymbol{\alpha}_A = 4.19$ rad/s^2 ↻; $\boldsymbol{\alpha}_B = 2.36$ rad/s^2 ↻.
(b) 6.00 s
15.30 $\mathbf{v}_B = 132$ ft/s →, $\mathbf{v}_C = 0$,
$\mathbf{v}_D = 127.5$ ft/s ∠15°,
$\mathbf{v}_E = 93.3$ ft/s ∠45°.
15.32 (a) 1.570 rad/s ↺. (b) 1.703 m/s ∠25°.
15.34 (a) 3 rad/s ↻.
(b) $\mathbf{v}_A = (300 \text{ mm/s})\mathbf{i} + (270 \text{ mm/s})\mathbf{j}$.
15.36 (a) $\mathbf{v}_A = (24 \text{ in./s})\mathbf{i} + (12 \text{ in./s})\mathbf{j}$.
(b) $x = 3$ in., $y = 0$.
15.38 (a) $\frac{1}{2}\omega_A$ ↺. (b) $\frac{1}{4}\omega_A$ ↻.
15.40 (a) 240 rpm ↺. (b) Zero.
15.42 $\boldsymbol{\omega}_B = 144$ rpm ↻; $\boldsymbol{\omega}_C = 80$ rpm ↺.
15.44 (a) $\boldsymbol{\omega}_{BD} = 0$; $\mathbf{v}_D = 1.885$ m/s ←.
(b) $\boldsymbol{\omega}_{BD} = 7.26$ rad/s ↺; $\mathbf{v}_D = 1.088$ m/s ←.
(c) $\boldsymbol{\omega}_{BD} = 0$; $\mathbf{v}_D = 1.885$ m/s →.
15.46 $\mathbf{v}_P = 12.40$ ft/s →; $\boldsymbol{\omega}_{BD} = 20.6$ rad/s ↺.
15.48 $\boldsymbol{\omega}_{BD} = 0$; $\boldsymbol{\omega}_{DE} = 1.6$ rad/s ↻.
15.50 $\boldsymbol{\omega}_{BD} = 1.5$ rad/s ↻; $\boldsymbol{\omega}_{DE} = 3.75$ rad/s ↻.
15.52 (a) $\mathbf{v}_A = 3$ m/s →; $\boldsymbol{\omega} = 0$.
(b) $\mathbf{v}_A = 1.113$ m/s →; $\boldsymbol{\omega} = 1.291$ rad/s ↻.
15.54 $\omega_C = (r_A/r_C)\omega_A + [1 - (r_A/r_C)]\omega_{ABC}$.
15.56 (a) 3 in. below D. (b) 16 in./s →.
15.58 (a) 10 mm to the right of A. (b) 40 mm/s ↓.
(c) Outer pulley: unwrapped, 240 mm/s; inner pulley: unwrapped, 120 mm/s.
15.60 (a) 6.93 rad/s ↺. (b) 1.039 m/s ←.
(c) 2.11 m/s ∠34.7°.
15.62 (a) 1.5 rad/s ↺. (b) 18 in./s ↑.
(c) 11.25 in./s ∠53.1°.
15.64 (a) 1.867 rad/s ↺. (b) 1.029 m/s ∠76.5°.
15.66 (a) $\boldsymbol{\omega}_{AB} = 1.4$ rad/s ↺; $\boldsymbol{\omega}_{BD} = 4$ rad/s ↺.
(b) 0.7 m/s ∠36.9°.
15.68 (a) 24.9 in./s ∠60°. (b) 20.9 in./s →.
15.70 (a) 64.8°. (b) 60.6 in./s ←.
15.72 (a) $\boldsymbol{\omega}_{AB} = 1.2$ rad/s ↺; $\boldsymbol{\omega}_{DE} = 0.80$ rad/s ↺.
(b) 140 mm/s ↑.
15.74 Space centrode: horizontal surface. Body centrode: circumference of drum.
15.80 $\mathbf{a}_A = 2.6$ m/s^2 ↑; $\mathbf{a}_B = 0.2$ m/s^2 ↑.
15.82 (a) 2.5 in. from B on AB. (b) 2 in. from A on AB.
15.84 14.04 m/s^2 ∠36.7°.
15.86 6.63 ft/s.
15.88 2.30 m/s^2 ∠43.2°.
15.90 $\mathbf{a}_D = 7400$ ft/s^2 ∠45°;
$\mathbf{a}_E = 1529$ ft/s^2 ∠45°.
15.92 32.0 m/s^2 ↘.
15.94 $\boldsymbol{\omega}_{DE} = \frac{1}{2}\omega_0$ ↺; $\boldsymbol{\alpha}_{DE} = 0$;
$\boldsymbol{\omega}_{EF} = 0$; $\boldsymbol{\alpha}_{EF} = \frac{3}{4}\omega_0^2$ ↻.
15.96 (a) 6.90 rad/s^2 ↺. (b) 5.52 rad/s^2 ↺.
15.98 (a) 1.925 rad/s^2 ↻. (b) 3.86 m/s^2 ∠40.3°.
15.100 (a) 1.974 rad/s^2 ↺. (b) 10.03 ft/s^2 ∠70.3°.
15.102 $\omega = (v_B \sin\beta)/(l\cos\theta)$.
15.104 (a) $\mathbf{v}_B = v_0 \cos\frac{1}{2}\theta$ ↖.
(b) $\mathbf{a}_B = \frac{1}{2}(v_0^2/b)\sin\frac{1}{2}\theta$ ↘.
15.106 (a) $v_A = b\omega\cos\theta$.
(b) $a_A = b\alpha\cos\theta - b\omega^2\sin\theta$.
15.108 $\boldsymbol{\omega} = -v_0/(2l\sin\theta)$ ↻;
$\boldsymbol{\alpha} = -(v_0/2l)^2\cos\theta/\sin^3\theta$ ↻.

15.110 (*a*) $\boldsymbol{\omega} = (v_0/b)\sin^2\theta$ ↻.
(*b*) $(\mathbf{v}_A)_x = (v_0 l/b)\sin^2\theta\cos\theta \rightarrow$.
$(\mathbf{v}_A)_y = (v_0 l/b)\sin^3\theta \uparrow$.
15.112 $v_x = v[1 - \cos(vt/r)]$, $v_y = v\sin(vt/r)$.
15.114 (*a*) 1.815 rad/s ↻. (*b*) 16.42 in./s ∠20°.
15.116 $\boldsymbol{\omega}_A = 4.21$ rad/s ↺; $\boldsymbol{\omega}_B = 1.274$ rad/s ↺.
15.118 706 mm/s ∠40.9°.
15.120 $\mathbf{a}_1 = r\omega^2\mathbf{i} - 2u\omega\mathbf{j}$; $\mathbf{a}_2 = 2u\omega\mathbf{i} - r\omega^2\mathbf{j}$;
$\mathbf{a}_3 = -(r\omega^2 + u^2/r + 2u\omega)\mathbf{i}$;
$\mathbf{a}_4 = (r\omega^2 + 2u\omega)\mathbf{j}$.
15.122 (*a*) 25.3 in./s ∠78.4°.
(*b*) 2.31 in./s^2 ∠3.7°.
15.124 (*a*) 240 m/s^2 ∠51.9°.
(*b*) 351 m/s^2 ∠32.5°.
15.126 (*a*) 742 mm/s ∠76.0°.
(*b*) 876 mm/s^2 ∠9.5°.
15.128 (*a*) 24.7 in./s ∠76.0°.
(*b*) 74.2 in./s^2 ∠14.0°.
15.130 $\boldsymbol{\omega}_S = 4.77$ rad/s ↻; $\boldsymbol{\alpha}_S = 127.2$ rad/s^2 ↻.
15.132 3.61 rad/s^2 ↺.
15.134 317 ft/s^2 ∠41.8°.
15.136 (*a*) 0.291 rad/s ↺. (*b*) 0.1206 rad/s^2 ↺.
15.138 5.45 m/s^2 ∠51.6°.
15.140 (*a*) 3.61 rad/s ↺. (*b*) 86.6 in./s ∠30°.
(*c*) 563 in./s^2 ∠46.1°.
15.142 (*a*) (2 rad/s)**i** − (1 rad/s)**j** − (3 rad/s)**k**.
(*b*) (22 in./s)**i** − (16 in./s)**j** + (20 in./s)**k**.
15.144 −300 mm/s.
15.146 (230 rad/s^2)**i** − (2.5 rad/s^2)**k**.
15.148 86.4 rad/s^2 toward the east.
15.150 (*a*) $-(R\omega_1/r)\mathbf{i} + \omega_1\mathbf{j}$. (*b*) $(R\omega_1^2/r)\mathbf{k}$.
15.152 (*a*) −(525 mm/s)**i** + (909 mm/s)**j** +
(438 mm/s)**k**.
(*b*) −(3.43 m/s^2)**i** − (1.456 m/s^2)**j** +
(4.55 m/s^2)**k**.
15.154 (*a*) $\omega_1\omega_2\mathbf{i}$. (*b*) $-r(\omega_1^2 + \omega_2^2)\mathbf{i}$.
(*c*) $-r\omega_2^2\mathbf{j} + 2r\omega_1\omega_2\mathbf{k}$.
15.156 $\mathbf{v}_C$ = (3.17 in./s)**i** − (6.80 in./s)**j** +
(4.08 in./s)**k**.
$\mathbf{a}_C$ = (2.31 in./s^2)**i** + (0.792 in./s^2)**j** −
(0.951 in./s^2)**k**.
15.158 (*a*) (1.5 rad/s)**i** + (3 rad/s)**j**.
(*b*) (300 mm/s)**i** − (150 mm/s)**j**.
(*c*) (60 mm/s)**i** − (30 mm/s)**j** −
(90 mm/s)**k**.
15.160 (*a*) −(0.75 rad/s^2)**i** − (1.5 rad/s^2)**j** +
(4.5 rad/s^2)**k**.
(*b*) −(480 mm/s^2)**i** + (375 mm/s^2)**j** −
(180 mm/s^2)**k**.
15.162 (*a*) −(113.3 rad/s^2)**i**.
(*b*) (109.9 in./s^2)**i** − (5166 in./s^2)**j**.
15.164 (*a*) $\omega_1/\sin\beta$. (*b*) $-(\omega_1/\tan\beta)\mathbf{i}$.
(*c*) $(\omega_1^2/\tan\beta)\mathbf{k}$.
15.166 (12 in./s)**k**.
15.168 −(23 mm/s)**i**.
15.170 −(300 mm/s)**i**.
15.172 $\omega_2 = \omega_1\cos 20°$.
15.174 (*a*) −(5.85 rad/s)**i** − (1.829 rad/s)**j** −
(1.463 rad/s)**k**.
(*b*) −(300 mm/s)**i**.
15.176 −(340 in./s^2)**k**.
15.178 −(27.8 m/s^2)**i**.
15.180 (*a*) (22 in./s)**i** − (60 in./s)**j** + (25 in./s)**k**.
(*b*) (200 in./s^2)**i** − (88 in./s^2)**k**.
15.182 (*a*) (0.6 m/s)**i** + (1.039 m/s)**j** − (1.247 m/s)**k**.
(*b*) −(16.90 m/s^2)**i** + (4 m/s^2)**j** −
(9.6 m/s^2)**k**.
15.184 (*a*) (0.6 m/s)**i** + (1.039 m/s)**j** − (1.247 m/s)**k**.
(*b*) (18.41 m/s^2)**i** + (1.402 m/s^2)**j** −
(5.70 m/s^2)**k**.
15.190 (*a*) −(0.270 rad/s^2)**i**.
(*b*) (6.24 in./s)**i** + (3.6 in./s)**j** − (7.2 in./s)**k**.
(*c*) −(2.70 in./s^2)**i** − (2.81 in./s^2)**j** −
(7.48 in./s^2)**k**.
15.192 $\mathbf{v}_D$ = (0.6 m/s)**j** − (0.48 m/s)**k**.
$\mathbf{a}_D$ = −(3 m/s^2)**i** + (1.92 m/s^2)**j** +
(4.8 m/s^2)**k**.
15.194 (0.1104 m/s^2)**i** − (0.205 m/s^2)**j** +
(0.246 m/s^2)**k**.
15.196 (*a*) (0.6 m/s)**i** − (0.72 m/s)**j** + (0.57 m/s)**k**.
(*b*) (1.260 m/s^2)**i** − (0.486 m/s^2)**j** −
(2.196 m/s^2)**k**.
15.198 (*a*) −(0.262 rad/s^2)**k**.
(*b*) $\mathbf{v}_B$ = (313 ft/s)**k**;
$\mathbf{a}_B$ = (62.7 ft/s^2)**i** − (822 ft/s^2)**j**.
15.200 (*a*) (0.81 m/s)**i** − (0.72 m/s)**j** + (0.76 m/s)**k**.
(*b*) (0.64 m/s^2)**i** − (1.392 m/s^2)**j** −
(1.920 m/s^2)**k**.
15.202 (*a*) (0.24 m/s^2)**j** − (1.856 m/s^2)**k**.
(*b*) −(2.81 m/s^2)**i** + (0.935 m/s^2)**k**.
(*c*) −(0.24 m/s^2)**j** + (3.73 m/s^2)**k**.
15.204 11.78 rad/s^2 with angular-acceleration vector pointing south.
15.206 (*a*) 2 rad/s ↻. (*b*) 6 rad/s ↻.
15.208 −(0.468 m/s^2)**i** − (0.1299 m/s^2)**k**.
15.210 1.513 m/s^2 ∠4.55°.
15.212 −(30 in./s)**j**.
15.214 (*a*) 15 in./s^2 ←. (*b*) 6 in. from *B*.

CHAPTER 16

16.2 (*a*) 9.24 lb. (*b*) 18.59 ft/s^2 ←.
16.4 (*a*) 3.33 m/s^2 →.
(*b*) **A** = 7.55 N ↑; **B** = 7.16 N ↑.
16.6 (*a*) 7.85 m/s^2 ←. (*b*) 3.89 m/s^2 ←.
(*c*) 3.95 m/s^2 ←.
16.8 35.6 N.
16.10 (*a*) 0.882*g*. (*b*) 4.
16.12 At each wheel:
A = 15.63 kN ↑; **B** = 2.52 kN ↑.
16.14 (*a*) 16.1 ft/s^2 →. (*b*) 10 in. $\leq h \leq$ 46 in.
16.16 $h \leq$ 841 mm.
16.18 F_{AD} = 41.5 N *T*; F_{BE} = 4.41 N *T*.
16.20 (*a*) 27.9 ft/s^2 ∡60°. (*b*) 1.708 lb ↑.
16.22 F_{AB} = 36.1 N *T*; F_{DE} = 71.2 N *T*.
16.24 10.03 lb.
16.26 (*a*) 16.1 ft/s^2 ∡30°. (*b*) $\mathbf{a}_B$ = 20.1 ft/s^2 ↓;
$\mathbf{a}_P$ = 40.2 ft/s^2 ∡30°.
16.28 V_D = +13.24 N; M_D = −2.65 N·m.
16.32 3660 rev.
16.34 (*a*) 69.8 rad/s^2 ↻. (*b*) 62.8 rad/s^2 ↻.
16.36 382 mm.
16.38 82.5 lb.
16.40 26.6 s.
16.42 (*a*) 6.37 rad/s^2 ↻. (*b*) 1.912 m/s ↑.
16.44 63.4 lb·ft·s^2.
16.46 (*a*) 2.07 m/s^2 ↓. (*b*) 3.52 m/s ↓.
16.48 $I_R = \left(n + \dfrac{1}{n}\right)^2 I_0 + n^4 I_C$.
16.50 (*a*) $\boldsymbol{\alpha}_A$ = 19.32 rad/s^2 ↺; $\boldsymbol{\alpha}_B$ = 7.25 rad/s^2 ↺.
(*b*) $\boldsymbol{\omega}_A$ = 300 rpm ↻; $\boldsymbol{\omega}_B$ = 225 rpm ↺.
16.52 (*a*) $\boldsymbol{\alpha}_A$ = 11.77 rad/s^2 ↺; $\boldsymbol{\alpha}_B$ = 49.1 rad/s^2 ↺.
(*b*) **C** = 41.2 N ↑; $\mathbf{M}_C$ = 1.130 N·m ↺.
16.54 $\boldsymbol{\alpha} = \dfrac{2\,\mu_k g \sin\phi}{r(\sin 2\phi - \mu_k \cos 2\phi)}$ ↺.
16.58 (*a*) 19.32 ft/s^2 →. (*b*) 6.44 ft/s^2 ←.
16.60 (*a*) 300 mm from *A*.
(*b*) $\boldsymbol{\alpha}$ = 5.33 rad/s^2 ↺; $\bar{\mathbf{a}}$ = 2.4 m/s^2 →.
16.62 $\mathbf{a}_A$ = 0.430 m/s^2 ↑; $\mathbf{a}_B$ = 1.966 m/s^2 ↑.
16.64 T_A = 593 lb; T_B = 505 lb.
16.66 (*a*) *W*. (*b*) $rg/\bar{k}^2$ ↺.
16.68 (*a*) 0.575 ft/s^2 ↑. (*b*) 0.431 rad/s^2 ↺.
16.70 (*a*) 3*g*/*L* ↻. (*b*) *g* ↑. (*c*) 2*g* ↓.
16.72 (*a*) *g*/*L* ↻. (*b*) 0. (*c*) *g* ↓.
16.74 (*a*) $t_1 = 2\omega_0 r/7\,\mu_k g$.
(*b*) $\bar{\mathbf{v}}_1 = 2r\omega_0/7$ →; $\boldsymbol{\omega}_1 = 2\omega_0/t$ ↻.
16.76 (*a*) $t_1 = 2v_1/7\,\mu_k g$.
(*b*) $\bar{\mathbf{v}}_1 = 2v_1/7$ →; $\boldsymbol{\omega}_1 = 5v_1/7r$ ↺.
16.80 (*a*) 18.40 rad/s^2 ↻.
(*b*) $\mathbf{C}_x$ = 0.714 lb ←; $\mathbf{C}_y$ = 7.5 lb ↑.
16.82 (*a*) 34.3 rad/s^2 ↻.
(*b*) $\mathbf{A}_x$ = 4.71 N ←; $\mathbf{A}_y$ = 58.9 N ↑.
16.84 (*a*) 600 mm. (*b*) 13.33 rad/s^2 ↻.
16.86 (*a*) 244 kN. (*b*) 187.0 kN.
16.88 $\mathbf{A}_x$ = (6.64 lb)**i**; $\mathbf{A}_z$ = −(11.63 lb)**k**.
16.90 3280 rpm.
16.92 (*a*) $\frac{3}{2}g$ ↓. (*b*) $\frac{1}{4}W$ ↑.
16.94 (*a*) $\frac{3}{4}g/b$ ↻. (*b*) 1.061*g* ∡45°. (*c*) $\frac{5}{8}W$ ↑.
16.96 (*a*) 5.01 lb ↑. (*b*) $\mathbf{A}_x$ = 77.6 lb ←; $\mathbf{A}_y$ = 5.47 lb ↑.
16.98 (*a*) 29.4 rad/s^2 ↻.
(*b*) $\mathbf{A}_x$ = 96.0 N ←; $\mathbf{A}_y$ = 38.0 N ↑.
16.100 (*a*) 43.3 lb·ft ↺. (*b*) 3.31 lb ∡30°.
16.102 (*a*) 8.20 m/s^2 ↓. (*b*) 9.81 m/s^2 ↓.
16.104 6.0° $< \beta <$ 67.3°.
16.108 600 mm.
16.110 5.02 ft or 1.530 m.
16.112 (*a*) 7.73 rad/s^2 ↻; 5.15 ft/s^2 →. (*b*) 0.340.
16.114 (*a*) 7.73 rad/s^2 ↺; 5.15 ft/s^2 ←. (*b*) 0.320.
16.116 (*a*) Slides. (*b*) 3.65 rad/s^2 ↺; 2.13 m/s^2 →.
16.118 (*a*) Does not slide.
(*b*) 7.2 rad/s^2 ↺; 1.152 m/s^2 ←.
16.120 2.22 m/s^2 ←.
16.122 (*a*) 75.5 rad/s^2 ↺. (*b*) 7.55 m/s^2 ↓.
16.124 $\mathbf{a}_A$ = 4.19 ft/s^2 ←; $\mathbf{a}_B$ = 3.26 ft/s^2 ←.
16.126 (*a*) $\bar{\mathbf{a}}_A = \bar{\mathbf{a}}_B$ = 5.57 ft/s^2 ←. (*b*) 0.779 lb ←.
16.128 1.592 lb ←.
16.130 23.7 rad/s^2 ↺.
16.132 (*a*) 13.07 rad/s^2 ↻. (*b*) 9.06 N ↑.
16.134 (*a*) 7.81 lb ←. (*b*) 5.71 lb.
16.136 (*a*) 2.98 rad/s^2 ↻. (*b*) 1.396 lb ∡50°.
16.138 49.6 N ↑.
16.140 3.50 lb →.
16.142 52.1 N ↓.
16.144 (*a*) 33.5 m/s^2 →. (*b*) 16.14 N ↓.
16.146 **A** = 426 N ←; **B** = 805 N →.
16.148 $\mathbf{D}_x = \frac{7}{2}m'r^2\omega_0^2$ →, $\mathbf{D}_y = \mathbf{E}_y = m'rg$ ↑,
$\mathbf{E}_x = m'r^2\omega_0^2$ ←.
16.150 (*a*) 5.19 m/s^2 ∡20°. (*b*) 32.5 rad/s^2 ↺.
16.152 (*a*) 4.91 ft/s^2 ←. (*b*) 9.03 rad/s^2 ↺.
16.154 (*a*) 12.14 rad/s^2 ↻. (*b*) 11.21 m/s^2 ∡30°.
(*c*) 14.56 N ∡60°.
16.156 (*a*) $\frac{6}{7}g$ ↓. (*b*) $\frac{9}{7}g$ ↓.
16.158 $M_{max} = WL/27$ at $\frac{1}{3}L$ from *A*.
16.160 (*a*) 12*g*/17 ↓. (*b*) $\frac{1}{2}g$ ↓.
16.162 20.6 ft.
16.164 (*a*) $\frac{1}{2}a$ →. (*b*) 2*d* →.

16.166 (*a*) 1.5 m/s^2 $\rightarrow$. (*b*) 9.5 m/s^2 $\rightarrow$.
16.168 $T_{AC} = 4.25$ N; $T_{BC} = 16.99$ N.
16.170 (*a*) 66.8 ft/s; 21.5 ft/s. (*b*) 63.4°; 116.6°.

CHAPTER 17

17.2 5230 rev.
17.4 342 mm.
17.6 (*a*) 19.20 lb·ft·s^2. (*b*) 11.46 rev.
17.8 40.6 rev.
17.10 100.8 lb.
17.12 (*a*) 22.8 rev. (*b*) $-(0.702\text{ lb})\mathbf{j}$.
17.14 770 mm/s $\downarrow$.
17.16 2.19 m/s $\downarrow$.
17.18 (*a*) $\sqrt{24g/7L}$ $\circlearrowright$; $13W/7$ $\uparrow$.
(*b*) 6.07 rad/s $\circlearrowright$; 9.29 lb $\uparrow$.
17.20 (*a*) 6.53 rad/s $\circlearrowleft$. (*b*) 3.00 rad/s $\circlearrowleft$.
17.22 (*a*) 55.2°. (*b*) $0.756\sqrt{gr}$ $\measuredangle$ 55.2°.
17.24 5.87 rad/s $\circlearrowright$.
17.26 (*a*) 10.00 rad/s $\circlearrowright$. (*b*) 10.98 rad/s.
17.28 689 mm.
17.30 (*a*) $\sqrt{\frac{10}{7}g(R-r)(1-\cos\beta)}$.
(*b*) $\frac{1}{7}W(17 - 10\cos\beta)$.
17.32 (*a*) $1.324\sqrt{g/r}$ $\circlearrowleft$. (*b*) $2.12mg$ $\uparrow$.
17.34 1.738 m/s $\leftarrow$.
17.36 (*a*) 7.43 ft/s $\downarrow$. (*b*) 4.00 lb.
17.38 (*a*) 421 mm/s $\leftarrow$. (*b*) 141.5 N $\uparrow$.
17.40 14.32 ft/s $\downarrow$.
17.42 $\boldsymbol{\omega} = 3.21$ rad/s $\circlearrowleft$; $\mathbf{v}_B = 1.923$ m/s $\uparrow$.
17.44 (*a*) $\boldsymbol{\omega} = 1.225\sqrt{g/R}$ $\circlearrowright$;
$\overline{\mathbf{v}} = 0.612\sqrt{gR}$ $\measuredangle$ 60°.
(*b*) $\boldsymbol{\omega} = 0$; $\overline{\mathbf{v}} = \sqrt{gR}$ $\rightarrow$.
17.46 3.60 m/s $\rightarrow$.
17.48 3.23 ft/s $\leftarrow$.
17.50 (*a*) 39.8 N·m. (*b*) 99.5 N·m.
(*c*) 249 N·m.
17.52 (*a*) 7.62 hp. (*b*) 38.1 hp.
17.54 169.1 N·m.
17.56 4.11 rad/s $\circlearrowright$.
17.58 4.59 s.
17.60 (*a*) 3.11 m/s. (*b*) 24.3 N.
17.62 (*a*) $\boldsymbol{\omega}_A = 150$ rpm $\circlearrowleft$; $\boldsymbol{\omega}_B = 225$ rpm $\circlearrowright$.
(*b*) 4.71 N·s $\rightarrow$.
17.64 (*a*) 3.21 lb. (*b*) 1.474 lb.
17.70 (*a*) 3 m/s $\rightarrow$. (*b*) 0.505.
17.72 16.51 ft/s $\uparrow$.
17.74 0.385 lb·ft $\circlearrowright$.
17.76 (*a*) 27.6 ft/s $\downarrow$. (*b*) 4.00 lb.
17.78 1.389 m/s $\leftarrow$.
17.80 (*a*) 58.1 rpm. (*b*) 4.58 J.
17.82 $3\omega_1/8$.
17.84 $7\omega_0/11$.
17.86 (*a*) $\boldsymbol{\omega}_A = 0$; $\boldsymbol{\omega}_B = 168.9$ rpm $\circlearrowright$;
$\boldsymbol{\omega}_P = 11.10$ rpm $\circlearrowleft$.
(*b*) $\boldsymbol{\omega}_A = \boldsymbol{\omega}_P = 10.45$ rpm $\circlearrowleft$;
$\boldsymbol{\omega}_B = 169.5$ rpm $\circlearrowright$.
17.88 6.02 ft/s; 27.1 lb.
17.90 1.136 m/s.
17.92 $0.645L\omega_1$.
17.94 (*a*) 4.98 ft/s $\rightarrow$. (*b*) 1.246 ft/s $\leftarrow$.
17.96 (*a*) 3.6 m/s $\leftarrow$. (*b*) 3600 N $\leftarrow$.
17.98 (*a*) 12 in. (*b*) 6 ft/s $\rightarrow$.
17.100 242 mm/s $\rightarrow$.
17.102 $6v_1/7L$ $\circlearrowleft$; $3\sqrt{2}v_1/7$ $\measuredangle$ 45°.
17.104 $3v_1/4L$ $\circlearrowright$.
17.106 $\omega_2 = \frac{1}{2}\omega_1$.
17.108 $L/\sqrt{3}$.
17.110 (*a*) $3\overline{v}_1/L$ $\circlearrowright$; $\frac{1}{2}\overline{v}_1$ $\downarrow$. (*b*) $3\overline{v}_1/L$ $\circlearrowleft$; $\frac{1}{2}\overline{v}_1$ $\uparrow$.
(*c*) 0; $\overline{v}_1$ $\uparrow$.
17.112 $\sqrt{2}$.
17.114 83.2°.
17.116 $h\left(\dfrac{3m}{6m + m_P}\right)^2$
17.118 (*a*) $0.7\overline{v}_1$ $\rightarrow$. (*b*) $1.8\overline{v}_1/L$ $\circlearrowleft$. (*c*) $0.3\overline{v}_1$ $\rightarrow$.
17.120 (*a*) $v_1/5$ $\downarrow$. (*b*) $12v_1/5L$ $\circlearrowleft$.
17.122 3.01 ft/s $\measuredangle$ 30°.
17.124 (*a*) $\overline{\mathbf{v}}_A = v_0 \sin\theta\,\mathbf{j}$,
$\boldsymbol{\omega}_A = (v_0/r)(-\sin\theta\,\mathbf{i} + \cos\theta\,\mathbf{j})$; $\overline{\mathbf{v}}_B =$
$v_0\cos\theta\,\mathbf{i}$, $\boldsymbol{\omega}_B = 0$. (*b*) $\overline{\mathbf{v}}_B = \frac{5}{7}v_0\cos\theta\,\mathbf{i}$.
17.126 (*a*) $\overline{\mathbf{v}}_A = \frac{2}{7}v_0\cos\theta\,\mathbf{i} + v_0\sin\theta\,\mathbf{j}$.
(*b*) $\tan\phi = \frac{7}{2}\tan\theta$.
17.128 $5v_A/4r$.
17.130 (*a*) 5.5 m/s $\rightarrow$. (*b*) 2.99 m/s $\rightarrow$.
17.132 $9\omega_1/14$ $\circlearrowright$; $9\overline{v}_1/14$ $\measuredangle$ 60°.
17.134 (*a*) $\frac{3}{2}W$. (*b*) $\frac{3}{4}W$.
17.136 (*a*) 1.022 m/s. (*b*) 1.095 m.
17.138 $\dfrac{2n}{n^2+1}\sqrt{\dfrac{\pi M_0}{I_0}}$.
17.140 (*a*) and (*b*) 4.00 rad/s.
17.142 36.4°.

CHAPTER 18

18.2 $-(53.8\text{ g}\cdot\text{m}^2/\text{s})\mathbf{i} + (184.3\text{ g}\cdot\text{m}^2/\text{s})\mathbf{j}$.
18.4 11.9°.
18.6 $(0.0431\text{ ft}\cdot\text{lb}\cdot\text{s})\mathbf{j} + (0.259\text{ ft}\cdot\text{lb}\cdot\text{s})\mathbf{k}$.

18.8 (*a*) 3.49 rad/s. (*b*) 0.0661 rad/s.
18.10 (3.98 kg·m²/s)**j** + (1.250 kg·m²/s)**k**.
18.14 (*a*) $m'r^3\omega(\pi\mathbf{i} - 4\mathbf{k})$. (*b*) 51.9°.
18.16 (2.4 kg·m²/s)**i** + (4.8 kg·m²/s)**j**.
18.18 (*a*) −(1.041 ft·lb·s)**i** − (1.041 ft·lb·s)**j** + (2.31 ft·lb·s)**k**. (*b*) 32.5°.
18.20 (*a*) $-(F\Delta t/m)\mathbf{k}$.
(*b*) $(12F\Delta t/7ma)(-3\mathbf{i} - 2\mathbf{j})$.
18.22 (*a*) $-(F\Delta t/m)\mathbf{k}$.
(*b*) $(F\Delta t/8ma)(30\mathbf{i} - 15\mathbf{j} + 10\mathbf{k})$.
18.24 (*a*) $(3F\Delta t/2ma)(\mathbf{i} - 8\mathbf{j})$. (*b*) Axis through A, in *xy* plane, forming ∢82.9° with *x* axis.
18.26 (*a*) $\frac{1}{8}\omega_0(-\mathbf{i} + \mathbf{j})$. (*b*) $(\omega_0 a/8\sqrt{2})\mathbf{k}$.
18.28 (*a*) $-\frac{3}{7}\bar{v}_0\left(\frac{1}{c}\mathbf{i} + \frac{1}{a}\mathbf{k}\right)$. (*b*) $-\frac{6}{7}\bar{v}_0\mathbf{j}$.
18.30 (0.625 rad/s)**i** − (0.281 rad/s)**j** + (0.0844 rad/s)**k**.
18.32 −(1600 ft/s)**i** + (1300 ft/s)**j** + (4000 ft/s)**k**.
18.36 0.461 J.
18.38 $0.228mr^2\omega^2$.
18.40 5.17 ft·lb.
18.42 $(\frac{7}{192})ma^2\omega_0^2$.
18.44 3.92 J.
18.46 (0.269 N·m)**k**.
18.48 $(\frac{1}{8}mr^2\omega^2 \sin 2\beta)\mathbf{k}$.
18.50 (1.035 lb·ft)**i**.
18.52 $\frac{1}{8}mr^2[\alpha(3 + \cos 2\beta)\mathbf{i} + \alpha \sin 2\beta\,\mathbf{j} + \omega^2 \sin 2\beta\,\mathbf{k}]$.
18.54 $-(\frac{1}{6}mb^2\omega^2 \sin 2\beta)\mathbf{k}$.
18.56 **A** = −(12.00 N)**i**; **B** = −(4.00 N)**i**.
18.58 (*a*) (18.11 rad/s²)**k**.
(*b*) **A** = −(0.375 lb)**i** + (0.375 lb)**j**;
B = −(0.375 lb)**i** − (0.375 lb)**j**.
18.60 (*a*) (10.56 mN·m)**k**.
(*b*) **A** = −**B** = (4 mN)**i** + (8 mN)**j**.
18.62 (*a*) (6 rad/s)**k**.
(*b*) **A** = −**B** = −(20 mN)**i** + (20 mN)**j**.
18.64 **A** = (1.178 N)**j**; **B** = −(1.178 N)**j**.
18.66 5.18 lb ↑.
18.68 (*a*) $\cos^{-1}(3g/2l\omega^2)$. (*b*) $0 \le \omega \le \sqrt{3g/2l}$.
18.70 $\sqrt{8g/11a}$.
18.72 (*a*) **C** = **D** = 0.
(*b*) **C** = (122.6 lb)**j**; **D** = −(122.6 lb)**j**.
18.74 **A** = −(18 N)**i** − (13.5 N)**k**,
$\mathbf{M}_A$ = (3.75 N·m)**i**.
18.76 (*a*) (7.97 N·m)**j**.
(*b*) **A** = −(9.00 N)**i** − (25.5 N)**k**,
$\mathbf{M}_A$ = (3.75 N·m)**i**.
18.78 (*a*) $\frac{1}{4}ma^2\alpha_1\mathbf{i}$.
(*b*) $\mathbf{A} = -\frac{1}{4}ma^2\omega_1\omega_2\mathbf{k}$; $\mathbf{B} = \frac{1}{4}ma^2\omega_1\omega_2\mathbf{k}$.
18.80 (*a*) **A** = (1.072 kN)**i** + (143.7 kN)**j**;
B = −(1.072 kN)**i** + (150.6 kN)**j**.
(*b*) −(22.9 kN·m)**k**.
18.82 4430 rpm.
18.84 45.1 rpm and 534 rpm.
18.86 (*a*) 30°. (*b*) 25.3°. (*c*) 36.5°.
18.88 $\cos\beta = \frac{2d^2}{h^2 - d^2}\frac{\dot\psi}{\dot\phi}$.
18.90 16.01×10^{21} lb·ft.
18.98 302 days.
18.100 Precession axis: $\theta_x = 39.9°$, $\theta_y = 127.9°$, $\theta_z = 79.4°$; precession, 0.458 rad/s; spin, 0.273 rad/s.
18.102 Precession axis: $\theta_x = 90°$, $\theta_y = 26.0°$, $\theta_z = 64.0°$; precession, 0.845 rad/s (retrograde); spin, 0.1591 rad/s.
18.104 Precession axis: $\theta_x = 99.05°$, $\theta_y = 90°$, $\theta_z = 9.05°$; precession, 121.8 rpm (retrograde); spin, 60.2 rpm.
18.106 (*a*) $\theta_0 \le \theta \le 180° - \theta_0$.
(*b*) $\dot\phi_0 \sin\theta_0 \cos\theta_0$. (*c*) $\dot\phi_0 \sin^2\theta_0$.
18.108 (*a*) $\sqrt{17g/11a}$. (*b*) $\sqrt{44g/17a}$.
18.110 (*a*) $\frac{1}{2}[I'(\dot\phi^2 \sin^2\theta + \dot\theta^2) + I(\dot\psi + \dot\phi\cos\theta)^2] + mgc\cos\theta = E$.
(*b*) $\dot\theta^2 = \frac{1}{I'}\left(2E - \frac{\beta^2}{I} - 2mgc\cos\theta\right) - \left(\frac{\alpha - \beta\cos\theta}{I'\sin\theta}\right)^2$.
18.112 (*a*) 48.9°. (*b*) $\dot\phi = 4$ rad/s; $\dot\psi = 17$ rad/s.
18.114 (*a*) $\sqrt{15g/11a}$.
(*b*) $\dot\phi = 2\sqrt{20g/33a}$; $\dot\psi = \sqrt{20g/33a}$.
18.120 (126.9 g·m²/s)**i** + (90 g·m²/s)**j**.
18.122 (*a*) $-(F\Delta t/3m)\mathbf{k}$. (*b*) $(3F\Delta t/4md)(-4\mathbf{i} - \mathbf{j})$.
18.124 $\frac{1}{2}mr^2\omega_1\omega_2\mathbf{i}$.
18.126 (*a*) (1.534 lb·ft)**i**.
(*b*) **A** = (0.690 lb)**j**; **B** = −(0.690 lb)**j**.
18.128 14.01 rev/h.
18.130 (*a*) 70.5°. (*b*) $9\sqrt{g/8L}$.

CHAPTER 19

19.2 1.047 m/s; 7.31 m/s².
19.4 2.20 in.; 94.6 ft/s².
19.6 11.04 mm or 0.435 in.
19.8 (*a*) 0.1481 s. (*b*) 0.490 m/s ↑; 4 m/s² ↓.

19.10 0.888 in. below equilibrium position; 4.94 ft/s ↓; 47.6 ft/s^2 ↑.
19.12 (a) 0.320 s; 3.13 Hz. (b) 2.05 ft/s; 4.03 ft/s^2.
19.14 (a) 0.406 s; 2.47 Hz. (b) 0.387 m/s; 6 m/s^2.
19.16 (a) 4.26 lb. (b) 3.96 lb/ft.
19.18 4.66 kg.
19.20 (a) 4.02 kN/m. (b) 1.777 kN/m.
19.24 23.1°.
19.26 1.824 s.
19.28 (a) 0.329 s. (b) 0.795 ft/s.
19.30 (a) 0.628 s. (b) 100.0 mm/s.
19.32 (a) 1.125 Hz. (b) $T_B = 76.7$ N; $T_C = 54.2$ N.
19.34 (a) 0.428 s. (b) 45.4 mm.
19.36 (a) $\tau = 2\pi\sqrt{2m/3k}$.
19.42 (a) $f = (1/2\pi)\sqrt{3g/2l}$. (b) $\frac{1}{6}l$.
19.44 (a) and (b) 0.933 s.
19.46 17.15 s.
19.48 (a) $r_a = 163.5$ mm. (b) 76.2 mm.
19.50 (a) $\tau = 2\pi\sqrt{5r/4g}$. (b) $\frac{1}{4}r$.
19.52 (a) 2.015 s. (b) 2.94 m/s.
19.54 (a) 1.099 s. (b) 1.554 s.
19.56 1.112 lb·ft·s^2.
19.58 1.115 Hz.
19.62 1.621 s.
19.64 (a) $6.82\sqrt{l/g}$. (b) $6.10\sqrt{l/g}$.
19.66 292 lb·ft·s^2.
19.68 5.28 Hz.
19.70 1.700 s.
19.72 $\tau = 2\pi\sqrt{m/3k\cos^2\beta}$.
19.74 2.10 Hz.
19.76 10.86 lb.
19.78 (a) 0.291 l. (b) $0.209\sqrt{g/l}$.
19.80 2.54 s.
19.82 (a) $\tau = (2\pi l/b)\sqrt{h/3g}$. (b) $\tau = 2\pi\sqrt{h/g}$.
19.84 $\tau = \pi l/\sqrt{3gr}$.
19.86 $\sqrt{k/2m} < \omega < \sqrt{3k/2m}$.
19.88 $\omega > \sqrt{2g/l}$.
19.90 (a) 360 rpm. (b) 53.1 μm.
19.92 245 rpm.
19.94 1.888 mm.
19.96 15 mm or 3.75 mm.
19.98 0.5 in. or 0.3 in.
19.100 813 rpm.
19.102 (a) 1308 rpm. (b) 1.334 mm.
19.104 0.642 Hz; 2.91 mm.
19.106 34.1 mi/h.
19.112 (a) 0.0355. (b) 9.53 N·s/m.
19.114 (a) 1319 lb·s/ft. (b) 0.1107 s.
19.116 $\sqrt{1 - 2(c/c_c)^2}$.
19.118 (a) 0.1202 in. (b) 0.0818 in.
19.120 2.04 mm.
19.122 4.28 mm.
19.124 (a) 297 rpm. (b) 267 rpm. (c) 0.402 in.; 0.423 in.
19.126 (a) $m\dfrac{d^2x}{dt^2} + c\dfrac{dx}{dt} + kx = (k\sin\omega t + c\omega\cos\omega t)\delta_m$.
(b) $x_m = \dfrac{\sqrt{k^2 + c^2\omega^2}\,\delta_m}{\sqrt{(k - m\omega^2)^2 + (c\omega)^2}}$, with $\omega = 2\pi v/L$.
19.128 $R < 2\sqrt{L/C}$.
19.132 (a) $m\dfrac{d^2x_m}{dt^2} + k_2(x_m - x_A) = P_m\sin\omega t$,
$c\dfrac{dx_A}{dt} + k_1x_A + k_2(x_A - x_m) = 0$.
(b) $L\dfrac{d^2q_m}{dt^2} + \dfrac{(q_m - q_A)}{C_2} = E_m\sin\omega t$,
$R\dfrac{dq_A}{dt} + \dfrac{q_A}{C_1} + \dfrac{(q_A - q_m)}{C_2} = 0$.
19.136 (a) $m_1\dfrac{d^2x_1}{dt^2} + c_1\dfrac{dx_1}{dt} + k_1x_1 + k_2(x_1 - x_2) = 0$,
$m_2\dfrac{d^2x_2}{dt^2} + c_2\dfrac{dx_2}{dt} + k_2(x_2 - x_1) = 0$.
(b) $L_1\dfrac{d^2q_1}{dt^2} + R_1\dfrac{dq_1}{dt} + \dfrac{q_1}{C_1} + \dfrac{(q_1 - q_2)}{C_2} = 0$,
$L_2\dfrac{d^2q_2}{dt^2} + R_2\dfrac{dq_2}{dt} + \dfrac{(q_2 - q_1)}{C_2} = 0$.
19.138 0.0730 in.
19.140 0.918 Hz.
19.142 (a) 447 mm/s. (b) 4.33 mm above the equilibrium position; 437 mm/s ↑.
19.146 2.67 s.
19.148 (a) $(1/2\pi)\sqrt{g/2r}$. (b) $(1/2\pi)\sqrt{2g/3r}$.

Centroids of Common Shapes of Areas and Lines

Shape		$\bar{x}$	$\bar{y}$	Area
Triangular area	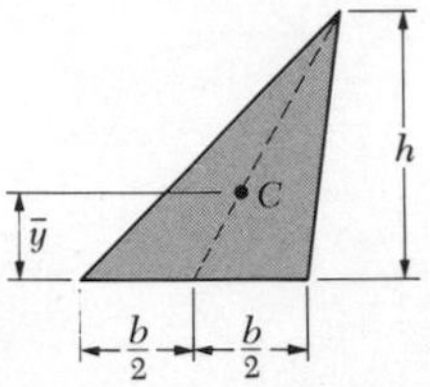		$\frac{h}{3}$	$\frac{bh}{2}$
Quarter-circular area	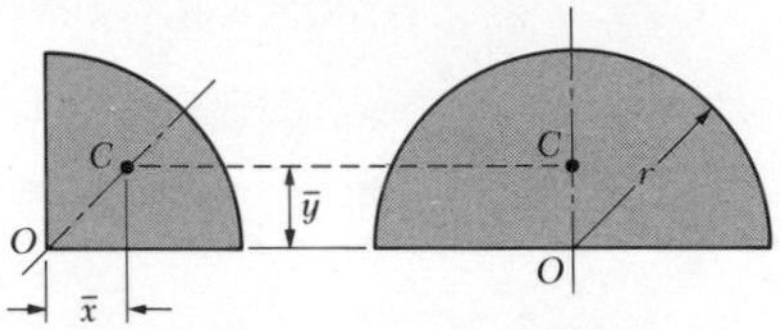	$\frac{4r}{3\pi}$	$\frac{4r}{3\pi}$	$\frac{\pi r^2}{4}$
Semicircular area		0	$\frac{4r}{3\pi}$	$\frac{\pi r^2}{2}$
Semiparabolic area	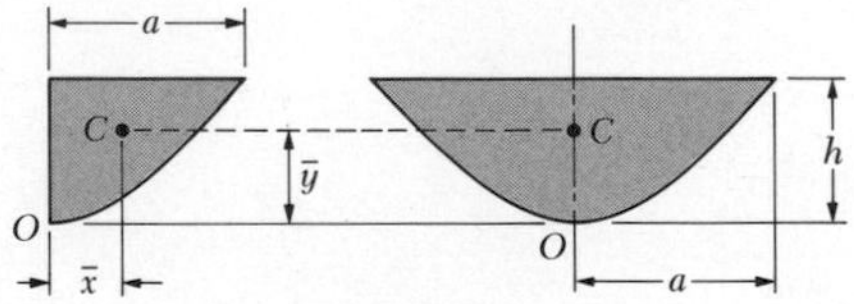	$\frac{3a}{8}$	$\frac{3h}{5}$	$\frac{2ah}{3}$
Parabolic area		0	$\frac{3h}{5}$	$\frac{4ah}{3}$
Parabolic spandrel	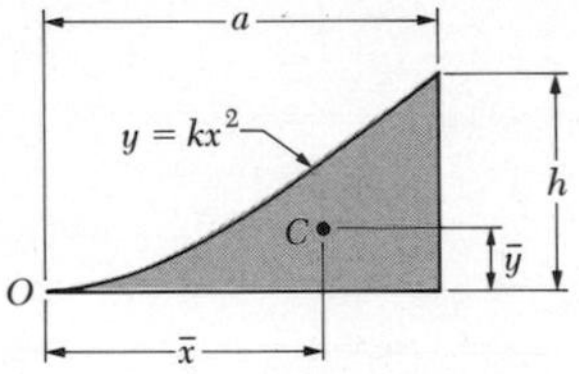	$\frac{3a}{4}$	$\frac{3h}{10}$	$\frac{ah}{3}$
Circular sector	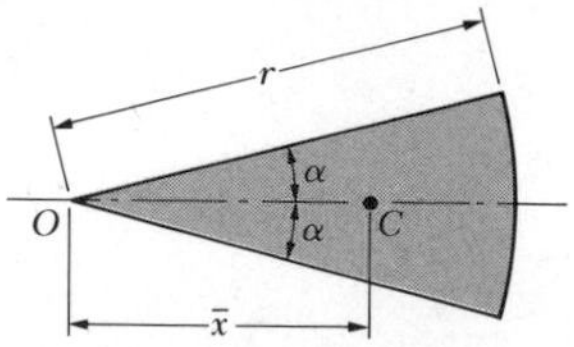	$\frac{2r \sin \alpha}{3\alpha}$	0	αr^2
Quarter-circular arc	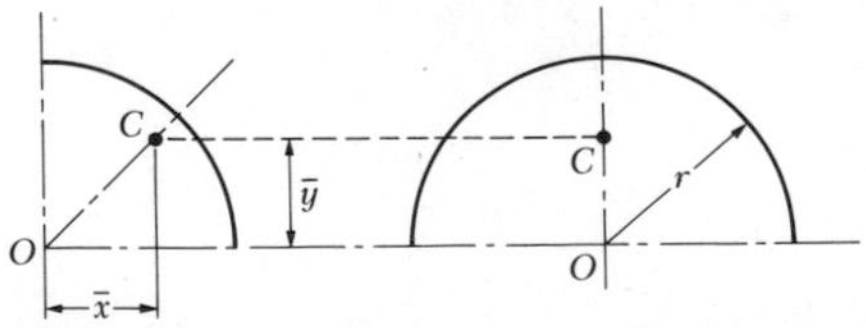	$\frac{2r}{\pi}$	$\frac{2r}{\pi}$	$\frac{\pi r}{2}$
Semicircular arc		0	$\frac{2r}{\pi}$	πr
Arc of circle	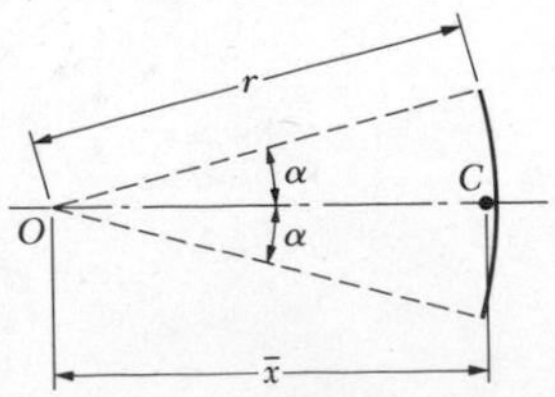	$\frac{r \sin \alpha}{\alpha}$	0	$2\alpha r$